CD-ROM with all content from 2008 *HVAC Systems and Equipmer*

2006 REFRIGERATIO

2005 FUNDAMENTALS

2008 ASHRAE® HANDBOOK

Heating, Ventilating, and Air-Conditioning SYSTEMS AND EQUIPMENT

Inch-Pound Edition

American Society of Heating, Refrigerating and Air-Conditioning Engineers, Inc.

1791 Tullie Circle, N.E., Atlanta, GA 30329

(404) 636-8400 http://www.ashrae.org

DEDICATED TO THE ADVANCEMENT OF

THE PROFESSION AND ITS ALLIED INDUSTRIES

Volunteer members of ASHRAE Technical Committees and others compiled the information in this handbook, and it is generally reviewed and updated every four years. Comments, criticisms, and suggestions regarding the subject matter are invited. Any errors or omissions in the data should be brought to the attention of the Editor. Additions and corrections to Handbook volumes in print will be published in the Handbook published the year following their verification and, as soon as verified, on the ASHRAE Internet Web site.

DISCLAIMER

ASHRAE has compiled this publication with care, but ASHRAE has not investigated, and ASHRAE expressly disclaims any duty to investigate, any product, service, process, procedure, design, or the like that may be described herein. The appearance of any technical data or editorial material in this publication does not constitute endorsement, warranty, or guaranty by ASHRAE of any product, service, process, procedure, design, or the like. ASHRAE does not warrant that the information in this publication is free of errors. The entire risk of the use of any information in this publication is assumed by the user.

ISBN 978-1-933742-33-5
ISSN 1078-6066

The paper for this book is acid free and was manufactured with post-consumer pulp from sources using sustainable forestry practices.

CONTENTS

HEATING EQUIPMENT AND COMPONENTS

COOLING EQUIPMENT AND COMPONENTS

GENERAL COMPONENTS

PACKAGED, UNITARY, AND SPLIT-SYSTEM EQUIPMENT

GENERAL

Additions and Corrections

Index

Composite index to the 2005 Fundamentals, 2006 Refrigeration, 2007 HVAC Applications, and 2008 HVAC Systems and Equipment volumes

Comment Pages

CONTRIBUTORS

In addition to the Technical Committees, the following individuals contributed significantly to this volume. The appropriate chapter numbers follow each contributor's name.

Howard McKew (1, 4)
RDK Engineers, Inc.

Amanda E. McKew (2)
RDK Engineers, Inc.

John Vucci (3)
University of Maryland, College Park

Stephen W. Duda (4, 37, 42)
Ross & Baruzzini, Inc.

Sarah E. Maston (5)
RDK Engineers, Inc.

Ainul Abedin (7, 17)
Ainul Abedin Consulting Engineers

Gearoid Foley (7)
Integrated CHP Systems Corp.

Joseph Orlando (7)
Platinum Energy, Inc.

Ian Spanswick (7)
JCI-York International

Richard Sweetser (7)
Exergy Partners

Timothy C. Wagner (7)
United Technologies Research Center

Abdi Zaltash (7)
Oak Ridge National Laboratory

Doug Cane (8)
Caneta Research, Inc.

Xiaobing Liu (8)
Climatemaster, Ltd.

Frank Pucciano (8)
Jacob Energy Systems

Alan C. Shedd (8)
Jackson EMC

John A. Shonder (8)
Oak Ridge National Laboratory

Vernon Meyer (11)
Heat Distribution Solutions

Victor Penar (11)
Perma-Pipe

Gary Phetteplace (11)
GWA Research LLC

Steve Tredinnick (11)
Syska Hennessy Group, Inc.

Mark C. Hegberg (12, 31, 47)
ITT Bell & Gossett

Chuck Dunn (16)
Lumalier Corporation

Stuart Engel (16)
Sanuvox Technologies, Inc.

Forrest B. Fencl (16)
UV Resources

Karin K. Foarde (16)
Research Triangle Institute

James D. Freihaut (16)
The Pennsylvania State University

Jaak Geboers (16)
Philips Lighting

Stephen B. Martin, Jr. (16)
CDC NIOSH

Richard Vincent (16)
Saint Vincent Hospital

Derald Welles (16)
Steril-Aire, Inc.

David L. Witham (16)
Ultraviolet Devices, Inc.

John Andrepont (17)
The Cool Solutions Company

Patricia Graef (17)
Munters Corporation

John Kraft (17)
Caldwell Energy Company

Dharam V. Punwani (17)
Avalon Consulting, Inc.

Bass Abushakra (18)
Milwaukee School of Engineering

Herman Behls (18)
Behls and Associates

Vernon Peppers (18)
Peppers Engineering

Gus Faris (19)
Nailor Industries, Inc.

David John (19)
METALAIRE

Fred Lorch (19)
Phoenix Controls Corporation

Ken Loudermilk (19)
TROX USA, Inc.

Reinhold Kittler (24)
Hudson Industrial Consulting, LLC

Harry Milliken (24)
Desert Aire

John Murphy (24)
Trane Company

Wayne M. Lawton (27, 29)
X-nth

Hall Virgil (30, 34)

Thomas A. Butcher (30, 34)
Brookhaven National Laboratory

William F. Raleigh (30)

Charles Gaston (32)
The Pennsylvania State University–York

Paul Haydock (32)
Carrier Corporation

Stephen Kowalski (32)
Trane Company

William J. Roy (34)
Lochinvar

Charles Cromer (36)
Florida Solar Energy Center

Mark Hertel (36)
SunEarth, Inc.

Janice Means (36)
Lawrence Technological University

Rick Heiden (37, 42)
Trane Company

Phillip A. Johnson (37, 42)
McQuay International

Thomas E. Watson (37, 42)
McQuay International

Forrest Yount (37, 42)
Carrier Corporation

Michael H. Zamalis (37, 42)
Johnson Controls

Richard L. Hall (37)
Battelle

Matthew T. Irons (37)
Emerson Climate Technologies

Alexander Leyderman (37)
Fairchild Controls

Dan M. Manole (37)
Ingersoll Rand

Joe Huber (38, 41)
Ketema LP

Harry Li (38, 41)
Carrier Corporation

Ramachandran Narayanamurthy (38)
Ice Energy

Gursaran D. Mathur (40)
Calsonic Kansei North America

Mike Scofield (40)
Energy Conservation Products

Johnny Douglass (44)
Washington State University Extension Energy Program

Tom Lowery (44)
Direct Air Systems

Riyaz Papar (44)
Hudson Technologies Company

John Tolbert (44)
Bristol Compressors, Inc.

Russell Tavolacci (48)
Daikin AC (Americas), Inc.

Michael W. Woodford (48)
ARI

Ramez Afify (48, 49)
Clifford Dias Consulting Engineering

Dutch Uselton (49)
Lennox Industries, Inc.

ASHRAE TECHNICAL COMMITTEES, TASK GROUPS, AND TECHNICAL RESOURCE GROUPS

SECTION 1.0—FUNDAMENTALS AND GENERAL

1.1 Thermodynamics and Psychrometrics
1.2 Instruments and Measurements
1.3 Heat Transfer and Fluid Flow
1.4 Control Theory and Application
1.5 Computer Applications
1.6 Terminology
1.7 Business, Management, and General Legal Education
1.8 Mechanical Systems Insulation
1.9 Electrical Systems
1.10 Cogeneration Systems
1.11 Electric Motors and Motor Control
1.12 Moisture Management in Buildings
TG1 Exergy Analysis for Sustainable Buildings (EXER)

SECTION 2.0—ENVIRONMENTAL QUALITY

2.1 Physiology and Human Environment
2.2 Plant and Animal Environment
2.3 Gaseous Air Contaminants and Gas Contaminant Removal Equipment
2.4 Particulate Air Contaminants and Particulate Contaminant Removal Equipment
2.5 Global Climate Change
2.6 Sound and Vibration Control
2.7 Seismic and Wind Restraint Design
2.8 Building Environmental Impacts and Sustainability
2.9 Ultraviolet Air and Surface Treatment
TG2 Blast, Chemical, and Biological Remediation

SECTION 3.0—MATERIALS AND PROCESSES

3.1 Refrigerants and Secondary Coolants
3.2 Refrigerant System Chemistry
3.3 Refrigerant Contaminant Control
3.4 Lubrication
3.6 Water Treatment
3.8 Refrigerant Containment
TG3 HVAC&R Contractors and Design-Build Firms (CDBF)

SECTION 4.0—LOAD CALCULATIONS AND ENERGY REQUIREMENTS

4.1 Load Calculation Data and Procedures
4.2 Climatic Information
4.3 Ventilation Requirements and Infiltration
4.4 Building Materials and Building Envelope Performance
4.5 Fenestration
4.7 Energy Calculations
4.10 Indoor Environmental Modeling
TRG4 Sustainable Building Guidance and Metrics (SBGM)

SECTION 5.0—VENTILATION AND AIR DISTRIBUTION

5.1 Fans
5.2 Duct Design
5.3 Room Air Distribution
5.4 Industrial Process Air Cleaning (Air Pollution Control)
5.5 Air-to-Air Energy Recovery
5.6 Control of Fire and Smoke
5.7 Evaporative Cooling
5.8 Industrial Ventilation Systems
5.9 Enclosed Vehicular Facilities
5.10 Kitchen Ventilation
5.11 Humidifying Equipment

SECTION 6.0—HEATING EQUIPMENT, HEATING AND COOLING SYSTEMS AND APPLICATIONS

6.1 Hydronic and Steam Equipment and Systems
6.2 District Energy
6.3 Central Forced-Air Heating and Cooling Systems
6.5 Radiant and In-Space Convective Heating and Cooling
6.6 Service Water Heating
6.7 Solar Energy Utilization
6.8 Geothermal Energy Utilization
6.9 Thermal Storage
6.10 Fuels and Combustion

SECTION 7.0—BUILDING PERFORMANCE

7.1 Integrated Building Design
7.3 Operation and Maintenance Management
7.4 Building Operation Dynamics
7.5 Smart Building Systems
7.6 Systems Energy Utilization
7.7 Testing and Balancing
7.8 Owning and Operating Costs
7.9 Building Commissioning
TRG7 Tools for Sustainable Building Operations, Maintenance, and Cost Analysis (SBOMC)
TRG7 Underfloor Air Distribution (UFAD)

SECTION 8.0—AIR-CONDITIONING AND REFRIGERATION SYSTEM COMPONENTS

8.1 Positive Displacement Compressors
8.2 Centrifugal Machines
8.3 Absorption and Heat-Operated Machines
8.4 Air-to-Refrigerant Heat Transfer Equipment
8.5 Liquid-to-Refrigerant Heat Exchangers
8.6 Cooling Towers and Evaporative Condensers
8.8 Refrigerant System Controls and Accessories
8.9 Residential Refrigerators and Food Freezers
8.10 Mechanical Dehumidification Equipment and Heat Pipes
8.11 Unitary and Room Air Conditioners and Heat Pumps
8.12 Desiccant Dehumidification Equipment and Components
TG8 Variable Refrigerant Flow (VRF)

SECTION 9.0—BUILDING APPLICATIONS

9.1 Large-Building Air-Conditioning Systems
9.2 Industrial Air Conditioning
9.3 Transportation Air Conditioning
9.4 Applied Heat Pump/Heat Recovery Systems
9.5 Residential and Small-Building Applications
9.6 Healthcare Facilities
9.7 Educational Facilities
9.8 Large-Building Air-Conditioning Applications
9.9 Mission-Critical Facilities, Technology Spaces and Electronic Equipment
9.10 Laboratory Systems
9.11 Clean Spaces
9.12 Tall Buildings
TG9 Justice Facilities

SECTION 10.0—REFRIGERATION SYSTEMS

10.1 Custom-Engineered Refrigeration Systems
10.2 Automatic Icemaking Plants and Skating Rinks
10.3 Refrigerant Piping
10.4 Ultralow-Temperature Systems and Cryogenics
10.5 Refrigerated Distribution and Storage Facilities
10.6 Transport Refrigeration
10.7 Commercial Food and Beverage Cooling Display and Storage
10.8 Refrigeration Load Calculations
10.9 Refrigeration Application for Foods and Beverages
10.10 Management of Lubricant in Circulation

ASHRAE Research: Improving the Quality of Life

The American Society of Heating, Refrigerating and Air-Conditioning Engineers is the world's foremost technical society in the fields of heating, ventilation, air conditioning, and refrigeration. Its members worldwide are individuals who share ideas, identify needs, support research, and write the industry's standards for testing and practice. The result is that engineers are better able to keep indoor environments safe and productive while protecting and preserving the outdoors for generations to come.

One of the ways that ASHRAE supports its members' and industry's need for information is through ASHRAE Research. Thousands of individuals and companies support ASHRAE Research annually, enabling ASHRAE to report new data about material properties and building physics and to promote the application of innovative technologies.

Chapters in the ASHRAE Handbook are updated through the experience of members of ASHRAE Technical Committees and through results of ASHRAE Research reported at ASHRAE meetings and published in ASHRAE special publications and in *ASHRAE Transactions*.

For information about ASHRAE Research or to become a member, contact ASHRAE, 1791 Tullie Circle, Atlanta, GA 30329; telephone: 404-636-8400; www.ashrae.org.

Preface

The 2008 *ASHRAE Handbook—HVAC Systems and Equipment* discusses various systems and the equipment (components or assemblies) that comprise them, and describes features and differences. This information helps system designers and operators in selecting and using equipment. An accompanying CD-ROM contains all the volume's chapters in both I-P and SI units.

This edition includes *two* new chapters, described as follows:

- Chapter 16, Ultraviolet Lamp Systems, includes a review of the fundamentals of UVC germicidal energy's impact on microorganisms; how UVC lamps generate germicidal radiant energy; common approaches to the application of UVGI systems for upper-air room, in-duct, and surface cleansing; and a review of human safety and maintenance issues.
- Chapter 17, Combustion Turbine Inlet Cooling (CTIC), provides a detailed discussion of how CTIC is used to help improve combustion turbine performance.

Some of the revisions and additions to the remainder of the volume are as follows:

- Chapters 1 to 5 have each been revised to include new system and process flow diagrams, plus new discussion content on commissioning, building automation, maintenance management, sustainability/green design, security, and various systems (e.g., underfloor air distribution, chilled beams).
- Chapter 7, Combined Heat and Power Systems, formerly entitled Cogeneration Systems and Engine and Turbine Drives, was reorganized, as well as updated for new technology.
- Chapter 11, District Heating and Cooling, has new guidance on construction cost considerations, central plants, and distribution systems.
- Chapter 12, Hydronic Heating and Cooling, has revised text and figures on all aspects of system design, including design procedure, water temperatures, heat transfer, distribution losses, constant- and variable-speed pumping, sizing control valves, and terminal units.
- Chapter 18, Duct Construction, has new guidance for installation of flexible ducts.
- Chapter 19, Room Air Distribution Equipment, was reorganized to coordinate with its companion chapter in *HVAC Applications*, with added content on equipment for stratified and partially stratified systems.
- Chapter 24, Mechanical Dehumidifiers and Related Components, has new content on installation and service, indoor pool dehumidifiers, and application considerations for various equipment types.
- Chapter 26, Air-Heating Coils, has new text on installation guidelines.
- Chapter 27, Unit Ventilators, Unit Heaters, and Makeup Air Units, has updated content on makeup air units.
- Chapter 30, Automatic Fuel-Burning Systems, extensively reorganized and revised, contains updated information on new technology and code requirements.
- Chapter 31, Boilers, has new material on condensing boilers, burner types, and operating and safety controls.
- Chapter 32, Furnaces, has been thoroughly revised to reflect new technology and code requirements.
- Chapter 34, Chimney, Vent, and Fireplace Systems, has been reorganized for clarity and has new content on designing fireplaces and their chimneys.
- Chapter 36, Solar Energy Equipment, has been reorganized and has new content on photovoltaic systems and testing/rating.
- Chapter 37, Compressors, has been reorganized and has updates on bearings and variable-speed drive technology.
- Chapter 38, Condensers, contains revised content on air-cooled condensers, particularly on type descriptions, heat transfer, pressure drop, testing/rating, and installation and maintenance.
- Chapter 40, Evaporative Air Cooling Equipment, has a rewritten section on indirect coolers.
- Chapter 42, Liquid-Chilling Systems, has new discussion on both refrigerant selection and variable-flow chilled-water systems, as well as new and improved figures.
- Chapter 44, Motors, Motor Controls, and Variable-Speed Drives, has updates for new technology and codes.
- Chapter 48, Unitary Air Conditioners and Heat Pumps, has new content on multisplit units, variable-refrigerant-flow (VRF) equipment, certification, and sustainability.

This volume is published, both as a bound print volume and in electronic format on a CD-ROM, in two editions: one using inch-pound (I-P) units of measurement, the other using the International System of Units (SI).

Corrections to the 2005, 2006, and 2007 Handbook volumes can be found on the ASHRAE Web site at http://www.ashrae.org and in the Additions and Corrections section of this volume. Corrections for this volume will be listed in subsequent volumes and on the ASHRAE Web site.

Reader comments are enthusiastically invited. To suggest improvements for a chapter, **please comment using the form on the ASHRAE Web site** or, using one of the cutout comment pages at the end of this volume's index, write to Handbook Editor, ASHRAE, 1791 Tullie Circle, Atlanta, GA 30329, or fax 678-539-2187, or e-mail mowen @ashrae.org.

Mark S. Owen
Editor

CHAPTER 1

HVAC SYSTEM ANALYSIS AND SELECTION

AN HVAC system maintains desired environmental conditions in a space. In almost every application, many options are available to the design engineer to satisfy a client's building program and design intent. In the analysis, selection, and combination of these options, the design engineer should consider the criteria defined here, as well as project-specific parameters to achieve the functional requirements associated with the project design intent. The design engineer should consider sustainability as it pertains to responsible energy and environmental design, as well as constructability of the design.

HVAC systems are categorized by the method used to produce, deliver, and control heating, ventilating, and air conditioning in the conditioned area. This chapter addresses procedures for selecting the appropriate system for a given application while taking into account pertinent issues associated with designing, building, commissioning, operating, and maintaining the system. It also describes and defines the design concepts and characteristics of basic HVAC systems. Chapters 2 to 5 describe specific systems and their attributes, based on their heating and cooling medium and commonly used variations, constructability, commissioning, operation, and maintenance.

This chapter is intended as a guide for the design engineer, builder, facility manager, and student needing to know or reference the analysis and selection process that leads to recommending the optimum system for the job. The approach applies to HVAC conversion, building system upgrades, system retrofits, building renovations and expansion, and new construction for any building: small, medium, large, below grade, at grade, low-rise, and high-rise. This system analysis and selection process (Figure 1) helps determine the optimum system(s) for any building. Regardless of facility type, analysis examines objective, subjective, short-term, and long-term goals.

SELECTING A SYSTEM

The design engineer is responsible for considering various systems and recommending one or two systems that will meet the project goals and perform as desired. It is imperative that the design engineer and owner collaborate to identify and prioritize criteria associated with the design goal. In addition, if the project has preconstruction services, the designer and operator should consult with the construction manager to take advantage of the constructability and consider value-engineered options. Occupant comfort, process heating, and cooling or ventilation criteria may be considered, including the following:

- Temperature
- Humidity
- Air motion
- Air purity or quality

DATA COLLECTION

DATA ANALYSIS

NO

YES

SOLUTION PLANNING

SOLUTIONS/ RECOMMENDATIONS REPORT

PROJECT DELIVERY

PERFORMANCE CONTRACTING

DESIGN-BID-BUILD

DESIGN-BUILD

CONSTRUCTION MANAGEMENT

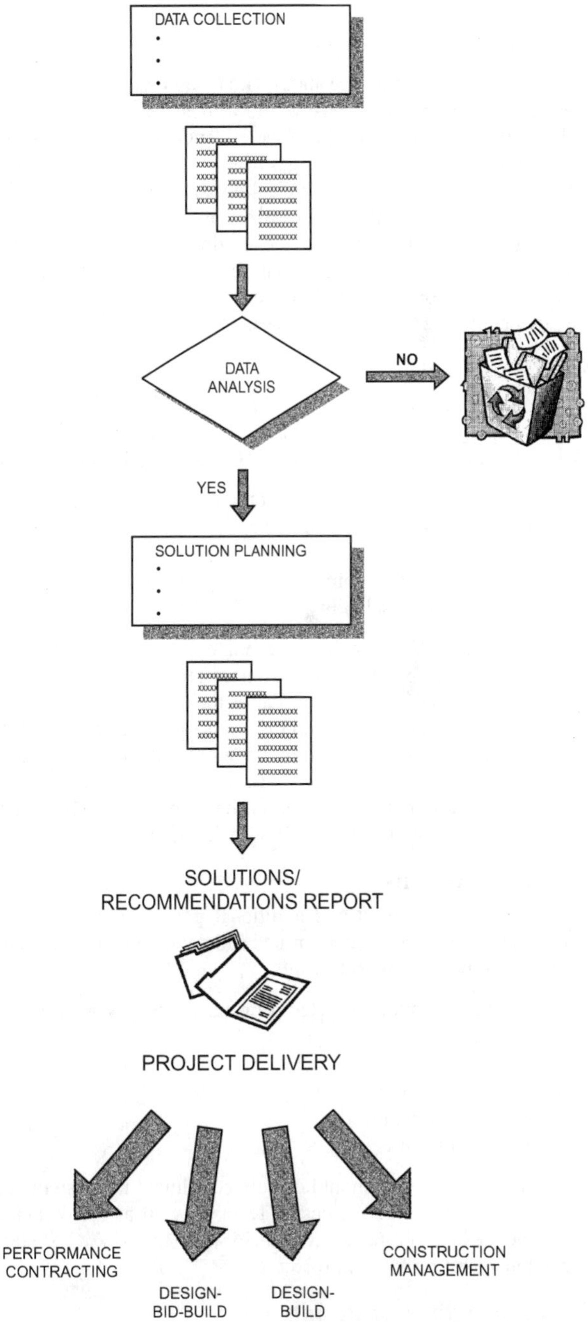

Fig. 1 Process Flow Diagram
(Courtesy RDK Engineers)

The preparation of this chapter is assigned to TC 9.1, Large Building Air-Conditioning Systems.

- Air changes per hour
- Air and/or water velocity requirements
- Local climate
- Space pressure requirements
- Capacity requirements, from a load calculation analysis
- Redundancy
- Spatial requirements
- Security concerns
- First cost
- Operating cost, including energy and power costs
- Maintenance cost
- Reliability
- Flexibility
- Life-cycle analysis
- Sustainability of design
- Acoustics and vibration
- Mold and mildew prevention

Because these factors are interrelated, the owner, design engineer, and operator must consider how these criteria affect each other. The relative importance of factors such as these varies with different owners, and often changes from one project to another for the same owner. For example, typical owner concerns include first cost compared to operating cost, extent and frequency of maintenance and whether that maintenance requires entering the occupied space, expected frequency of system failure, effect of failure, and time required to correct the failure. Each concern has a different priority, depending on the owner's goals.

Additional Goals

In addition to the primary goal of providing the desired environment, the design engineer should be aware of and account for other goals the owner may require. These goals may include the following:

- Supporting a process, such as operation of computer equipment
- Promoting a germ-free environment
- Increasing sales
- Increasing net rental income
- Increasing property salability

The owner can only make appropriate value judgments if the design engineer provides complete information on the advantages and disadvantages of each option. Just as the owner does not usually know the relative advantages and disadvantages of different HVAC systems, the design engineer rarely knows all the owner's financial and functional goals. Hence, the owner must be involved in system selection in the conceptual phase of the job. The same can be said for operator participation so that the final design is sustainable.

System Constraints

Once the goal criteria and additional goal options are listed, many constraints must be determined and documented. These constraints may include the following:

- Performance limitations (e.g., temperature, humidity, space pressure)
- Available capacity
- Available space
- Available utility source
- Available infrastructure
- Building architecture

The design engineer should closely coordinate the system constraints with the rest of the design team, as well as the owner, to overcome design obstacles associated with the HVAC systems under consideration for the project.

Constructability Constraints

The design engineer should take into account HVAC system issues before the project reaches the construction document phase. Some of these constraints may significantly affect the success of the design and cannot be overlooked in the design phase. Some issues and concerns associated with constructability are

- Existing conditions
- Maintaining existing building occupancy and operation
- Construction budget
- Construction schedule
- Ability to phase HVAC system installation
- Equipment availability (i.e., delivery lead times)

Few projects allow detailed quantitative evaluation of all alternatives. Common sense, historical data, and subjective experience can be used to narrow choices to one or two potential systems.

Heating and air-conditioning loads often contribute to constraints, narrowing the choice to systems that fit in available space and are compatible with building architecture. Chapters 29 and 30 of the 2005 *ASHRAE Handbook—Fundamentals* describe methods to determine the size and characteristics of heating and air-conditioning loads. By establishing the capacity requirement, equipment size can be determined, and the choice may be narrowed to those systems that work well on projects within a size range.

Loads vary over time based on occupied and unoccupied periods, and changes in weather, type of occupancy, activities, internal loads, and solar exposure. Each space with a different use and/or exposure may require a different control zone to maintain space comfort. Some areas with special requirements (e.g., ventilation requirements) may need individual systems. The extent of zoning, degree of control required in each zone, and space required for individual zones also narrow system choices.

No matter how efficiently a particular system operates or how economical it is to install, it can only be considered if it (1) maintains the desired building space environment within an acceptable tolerance under all conditions and occupant activities and (2) physically fits into, on, or adjacent to the building without being objectionable.

Cooling and humidity control are often the basis of sizing HVAC components and subsystems, but the system may also be determined based on **ventilation** criteria. For example, if large quantities of outside air are required for ventilation or to replace air exhausted from the building, only systems that transport large air volumes need to be considered.

Effective heat delivery to an area may be equally important in selection. A distribution system that offers high efficiency and comfort for cooling may be a poor choice for heating. The cooling, humidity, and/or heat delivery performance compromises may be small for one application in one climate, but may be unacceptable in another that has more stringent requirements.

HVAC systems and associated distribution systems often occupy a significant amount of **space**. Major components may also require special support from the structure. The size and appearance of terminal devices (e.g., grilles, registers, diffusers, fan-coil units, radiant panels) affect architectural design because they are visible in the occupied space.

Construction budget constraints can also influence the choice of HVAC systems. Based on historical data, some systems may not be economically feasible for an owner's building program. In addition, annual maintenance and operating budget (utilities, labor, and materials) should be an integral part of any system analysis and selection process. This is particularly important for building owners who will retain the building for a substantial number of years. Value-engineered solutions can offer (1) cost-driven performance, which may provide for a better solution for lower first cost; (2) a more sustainable solution over the life of the equipment; or (3) best value based on a reasonable return on investment.

Sustainable energy consumption can be compromised and long-term project success can be lost if building operators are not trained to efficiently and effectively operate and maintain the building systems. For projects in which the design engineer used some form of energy software simulation, these data should be passed on to the building owner so that goals and expectations can be measured and benchmarked against actual system performance. HVAC design is not complete without continuous system performance years after the system selection and analysis has been completed and the systems installed and turned over to the building owner.

Narrowing the Choices

The following chapters in this volume present information to help the design engineer narrow the choices of HVAC systems:

- Chapter 2 focuses on a distributed approach to HVAC.
- Chapter 3 provides guidance for large equipment centrally located in or adjacent to a building.
- Chapter 4 addresses all-air systems.
- Chapter 5 covers building piping distribution, including in-room terminal systems.

Each chapter summarizes positive and negative features of various systems. Comparing the criteria, other factors and constraints, and their relative importance usually identifies one or two systems that best satisfy project goals. In making choices, notes should be kept on all systems considered and the reasons for eliminating those that are unacceptable.

Each selection may require combining a primary system with a secondary (or distribution) system. The primary system converts energy from fuel or electricity into a heating and/or cooling medium. The secondary system delivers heating, ventilation, and/or cooling to the occupied space. The systems are independent to a great extent, so several secondary systems may work with a particular primary system. In some cases, however, only one secondary system may be suitable for a particular primary system.

Once subjective analysis has identified one or two HVAC systems (sometimes only one choice remains), detailed quantitative evaluations must be made. All systems considered should provide satisfactory performance to meet the owner's essential goals. The design engineer should provide the owner with specific data on each system to make an informed choice. Consult the following chapters to help narrow the choices:

- Chapter 9 of the 2005 *ASHRAE Handbook—Fundamentals* covers physiological principles, comfort, and health.
- Chapter 32 of the 2005 *ASHRAE Handbook—Fundamentals* covers methods for estimating annual energy costs.
- Chapter 35 of the 2007 *ASHRAE Handbook—HVAC Applications* covers methods for energy management.
- Chapter 36 of the 2007 *ASHRAE Handbook—HVAC Applications* covers owning and operating costs.
- Chapter 38 of the 2007 *ASHRAE Handbook—HVAC Applications* covers mechanical maintenance.
- Chapter 47 of the 2007 *ASHRAE Handbook—HVAC Applications* covers sound and vibration control.

Other guidelines to consult are ASHRAE standards; local, state, and federal guidelines; and special agency requirements [e.g., U.S. General Services Administration (GSA), Food and Drug Administration (FDA), Joint Commission on Accreditation of Healthcare Organizations (JCAHO), Leadership in Energy and Environmental Design (LEED™)].

Selection Report

As the last step, the design engineer should prepare a summary report that addresses the following:

- The goal
- Criteria for selection
- Important factors, including advantages and disadvantages
- Other goals
- Security concerns
- Basis of design
- HVAC system analysis and selection matrix
- System narratives
- Budget costs
- Recommendation

A brief outline of each of the final selections should be provided. In addition, HVAC systems deemed inappropriate should be noted as having been considered but not found applicable to meet the owner's primary HVAC goal.

The report should include an HVAC system selection matrix that identifies the one or two suggested HVAC system selections (primary and secondary, when applicable), system constraints, and other constraints. In completing this matrix assessment, the design engineer should have the owner's input to the analysis. This input can also be applied as weighted multipliers, because not all criteria carry the same weighted value.

Many grading methods are available to complete an analytical matrix analysis. Probably the simplest is to rate each item excellent, very good, good, fair, or poor. A numerical rating system such as 0 to 10, with 10 equal to excellent and 0 equal to poor or not applicable, can provide a quantitative result. The HVAC system with the highest numerical value then becomes the recommended HVAC system to accomplish the goal.

The system selection report should include a summary followed by a more detailed account of the HVAC system analysis and system selection. This summary should highlight key points and findings that led to the recommendation(s). The analysis should refer to the system selection matrix (such as in Table 1) and the reasons for scoring.

With each HVAC system considered, the design engineer should note the criteria associated with each selection. Issues such as close temperature and humidity control may eliminate some HVAC systems from consideration. System constraints, noted with each analysis, should continue to eliminate potential HVAC systems. Advantages and disadvantages of each system should be noted with the scoring from the HVAC system selection matrix. This process should reduce HVAC selection to one or two optimum choices to present to the owner. Examples of similar installations for other owners should be included with this report to support the final recommendation. Identifying a third party for an endorsement allows the owner to inquire about the success of other HVAC installations.

HVAC SYSTEMS AND EQUIPMENT

The majority of buildings built, expanded, and /or renovated may be ideally suited for decentralized HVAC systems, with equipment located in, throughout, adjacent to, or on top of the building. The alternative is primary equipment located in a central plant (either inside or outside the building) and distributing air and/or water for HVAC needs from this plant.

Decentralized System Characteristics

Temperature, Humidity, and Space Pressure Requirements. A decentralized system may be able to fulfill any or all of these design parameters.

Capacity Requirements. A decentralized system usually requires each piece of equipment to be sized for zone peak capacity, unless the systems are variable-volume. Depending on equipment type and location, decentralized systems do not benefit as much from equipment sizing diversity as centralized systems do.

Redundancy. A decentralized system may not have the benefit of back-up or standby equipment. This limitation may need review.

Facility Management. A decentralized system can allow the building manager to maximize performance using good business/facility management techniques in operating and maintaining the HVAC equipment and systems.

Spatial Requirements. A decentralized system may or may not require equipment rooms. Because of space restrictions imposed on

Table 1 Sample HVAC System Analysis and Selection Matrix (0 to 10 Score)

Goal: Furnish and install an HVAC system that provides moderate space temperature control with minimum humidity control at an operating budget of 70,000 Btu/h per square foot per year				
Categories	**System #1**	**System #2**	**System #3**	**Remarks**
1. Criteria for Selection: • 78°F space temperature with ±3°F control during occupied cycle, with 40% rh and ±5% rh control during cooling. • 68°F space temperature with ±2°F, with 20% rh and ±5% rh control during heating season. • First cost • Equipment life cycle				
2. Important Factors: • First-class office space stature • Individual tenant utility metering				
3. Other Goals: • Engineered smoke control system • ASHRAE *Standard* 62.1 ventilation rates • Direct digital control building automation				
4. System Constraints: • No equipment on first floor • No equipment on ground adjacent to building				
5. Other Constraints: • No perimeter finned-tube radiation or other type of in-room equipment				
TOTAL SCORE				

Source: RDK Engineers.

the design engineer or architect, equipment may be located on the roof and/or the ground adjacent to the building. Depending on system components, additional space may be required in the building for chillers and boilers. Likewise, a decentralized system may or may not require duct and pipe shafts throughout the building.

First Cost. A decentralized system probably has the best first-cost benefit. This feature can be enhanced by phasing in the purchase of decentralized equipment as needed (i.e., buying equipment as the building is being leased/occupied).

Operating Cost. A decentralized system can save operating cost by strategically starting and stopping multiple pieces of equipment. When comparing energy consumption based on peak energy draw, decentralized equipment may not be as attractive as larger, more energy-efficient centralized equipment.

Maintenance Cost. A decentralized system can save maintenance cost when equipment is conveniently located and equipment size and associated components (e.g., filters) are standardized. When equipment is located outdoors, maintenance may be difficult during bad weather.

Reliability. A decentralized system usually has reliable equipment, although the estimated equipment service life may be less than that of centralized equipment. Decentralized system equipment may require maintenance in the occupied space, however.

Flexibility. A decentralized system may be very flexible because it may be placed in numerous locations.

Level of Control. Decentralized systems often use direct refrigerant expansion (DX) for cooling, and on/off or staged heat. This step control results in greater variation in space temperature and humidity, where close control is not desired or necessary. As a caution, oversizing DX or stepped cooling can allow high indoor humidity levels and mold or mildew problems.

Sound and Vibration. Decentralized systems often locate noisy machinery close to building occupants, although equipment noise may be less than that produced by large central systems.

Constructability. Decentralized systems frequently consist of multiple and similar-in-size equipment that makes standardization a construction feature, as well as purchasing units in large quantities.

Centralized System Characteristics

Temperature, Humidity, and Space Pressure Requirements. A central system may be able to fulfill any or all of these design parameters.

Capacity Requirements. A central system usually allows the design engineer to consider HVAC diversity factors that reduce installed equipment capacity. As a result, this offers some attractive first-cost and operating-cost benefits.

Redundancy. A central system can accommodate standby equipment that decentralized configurations may have trouble accommodating.

Facility Management. A central system usually allows the building manager to maximize performance using good business/facility management techniques in operating and maintaining the HVAC equipment and systems.

Spatial Requirements. The equipment room for a central system is normally located outside the conditioned area: in a basement, penthouse, service area, or adjacent to or remote from the building. A disadvantage with this approach may be the additional cost to furnish and install secondary equipment for the air and/or water distribution. A second consideration is the access and physical constraints throughout the building to furnish and install this secondary distribution network of ducts and/or pipes and for equipment replacement.

First Cost. Even with HVAC diversity, a central system may not be less costly than decentralized HVAC systems. Historically, central system equipment has a longer equipment service life to compensate for this shortcoming. Thus, a life-cycle cost analysis is very important when evaluating central versus decentralized systems.

Operating Cost. A central system usually has the advantage of larger, more energy-efficient primary equipment compared to decentralized system equipment. In addition, with multiple pieces of HVAC equipment, a central system allows strategic planning and management of the HVAC systems through staging equipment based on HVAC demands.

Maintenance Cost. The equipment room for a central system provides the benefit of maintaining HVAC equipment away from occupants in an appropriate service work environment. Access to occupant workspace is not required, thus eliminating disruption to the space environment, product, or process. Because of the typically larger capacity of central equipment, there are usually fewer pieces of HVAC equipment to service.

Reliability. Centralized system equipment generally has a longer service life.

Flexibility. Flexibility can be a benefit when selecting equipment that provides an alternative or back-up source of HVAC.

Level of Control. Centralized systems generally use chilled water for cooling, and steam or hydronic heat. This usually allows for close control of space temperature and humidity where desired or necessary.

Sound and Vibration. Centralized systems often locate noisy machinery sufficiently remote from building occupants or noise-sensitive processes.

Constructability. Centralized systems usually require more coordinated installation than decentralized systems. However, consolidation of the primary equipment in a central location also has benefits.

Among the largest centralized systems are HVAC plants serving groups of large buildings. These plants improve diversity and generally operate more efficiently, with lower maintenance costs, than individual central plants. Economic considerations of larger centralized systems require extensive analysis. The utility analysis may consider multiple fuels and may also include gas and steam turbine-driven equipment. Multiple types of primary equipment using multiple fuels and types of HVAC-generating equipment (e.g., centrifugal and absorption chillers) may be combined in one plant. Chapters 12 to 14 provide design details for central plants.

Primary Equipment

The type of decentralized and centralized equipment selected for buildings depends on a well-organized HVAC analysis and selection report. The choice of primary equipment and components depends on factors presented in the selection report (see the section on Selecting a System). Primary HVAC equipment includes refrigeration equipment; heating equipment; and air, water, and steam delivery equipment.

Many HVAC designs recover internal heat from lights, people, and equipment to reduce the size of the heating plant. In buildings with core areas that require cooling while perimeter areas require heating, one of several heat reclaim systems can heat the perimeter to save energy. Sustainable design is also important when considering recovery and reuse of materials and energy. Chapter 8 describes heat pumps and some heat recovery arrangements, Chapter 36 describes solar energy equipment, and Chapter 25 introduces air-to-air energy recovery. In the 2007 *ASHRAE Handbook—HVAC Applications*, Chapter 35 covers energy management and Chapter 40 covers building energy monitoring. Chapter 17 of the 2005 *ASHRAE Handbook—Fundamentals* provides information on sustainable design.

The search for energy savings has extended to **cogeneration** or **total energy [combined heat and power (CHP)]** systems, in which on-site power generation is added to the HVAC project. The economic viability of this function is determined by the difference between gas and electric rates and by the ratio of electric to heating demands for the project. In these systems, waste heat from generators can be transferred to the HVAC equipment (e.g., to drive turbines of centrifugal compressors, serve an absorption chiller, etc.). Chapter 7 covers cogeneration or total energy systems. Alternative fuel sources, such as waste heat boilers, are now being included in fuel evaluation and selection for HVAC applications.

Thermal storage is another cost-saving concept, which provides the possibility of off-peak generation of chilled water or ice. Thermal storage can also be used for storing hot water for heating. Many electric utilities impose severe charges for peak summer power use or offer incentives for off-peak use. Storage capacity installed to level the summer load may also be available for use in winter, thus making heat reclaim a viable option. Chapter 34 of the 2007 *ASHRAE Handbook—HVAC Applications* has more information on thermal storage.

With ice storage, colder supply air can be provided than that available from a conventional chilled-water system. This colder air allows use of smaller fans and ducts, which reduces first cost and (in some locations) operating cost. These life-cycle savings can offset the first cost for storage provisions and the energy cost required to make ice. Similarly, thermal storage of hot water can be used for heating.

Refrigeration Equipment

Chapters 2 and 3 of this volume summarize the primary types of refrigeration equipment for HVAC systems.

When chilled water is supplied from a central plant, as on university campuses and in downtown areas of large cities, the utility service provider should be contacted during system analysis and selection to determine availability, cost, and the specific requirements of the service.

Heating Equipment

Steam boilers and heating-water boilers are the primary means of heating a space using a centralized system, as well as some decentralized systems. These boilers may be (1) used both for comfort and process heating; (2) manufactured to produce high or low pressure; and (3) fired with coal, oil, electricity, gas, and sometimes waste material. Low-pressure boilers are rated for a working pressure of either 15 or 30 psig for steam, and 160 psig for water, with a temperature limit of 250°F. Packaged boilers, with all components and controls assembled at the factory as a unit, are available. Electrode or resistance electric boilers that generate either steam or hot water are also available. Chapter 31 has further information on boilers, and Chapter 26 details air-heating coils.

Where steam or hot water is supplied from a central plant, as on university campuses and in downtown areas of large cities, the utility service entering the building must conform to the utility's standards. The utility provider should be contacted during project system analysis and selection to determine availability, cost, and specific requirements of the service.

When primary heating equipment is selected, the fuels considered must ensure maximum efficiency. Chapter 30 discusses design, selection, and operation of the burners for different types of primary heating equipment. Chapter 18 of the 2005 *ASHRAE Handbook—Fundamentals* describes types of fuel, fuel properties, and proper combustion factors.

Air Delivery Equipment

Primary air delivery equipment for HVAC systems is classified as packaged, manufactured and custom-manufactured, or field-fabricated (built-up). Most air delivery equipment for large systems uses centrifugal or axial fans; however, plug or plenum fans are often used. Centrifugal fans are frequently used in packaged and manufactured HVAC equipment. Axial fans are more often part of a custom unit or a field-fabricated unit. Both types of fans can be used as industrial process and high-pressure blowers. Chapter 20 describes fans, and Chapters 18 and 19 provide information about air delivery components.

SPACE REQUIREMENTS

In the initial phase of building design, the design engineer seldom has sufficient information to render the optimum HVAC design for the project, and its space requirements are often based on percentage of total area or other experiential rule of thumb. The final design is usually a compromise between what the engineer recommends and what the architect can accommodate. At other times, the building owner, who may prefer a centralized or decentralized system, may dictate final design and space requirements. This section discusses some of these requirements.

Equipment Rooms

Total mechanical and electrical space requirements range between 4 and 9% of gross building area, with most buildings in the 6 to 9% range. These ranges include space for HVAC, electrical, plumbing, and fire protection equipment and may also include vertical shaft space for mechanical and electrical distribution through the building.

Most equipment rooms should be centrally located to (1) minimize long duct, pipe, and conduit runs and sizes; (2) simplify shaft layouts; and (3) centralize maintenance and operation. With shorter duct and pipe runs, a central location could also reduce pump and fan motor power requirements, which reduces building operating costs. But, for many reasons, not all mechanical and electrical equipment rooms can always be centrally located in the building. In any case, equipment should be kept together whenever possible to minimize space requirements, centralize maintenance and operation, and simplify the electrical system.

Equipment rooms generally require clear ceiling height ranging from 10 to 18 ft, depending on equipment sizes and the complexity of air and/or water distribution.

The main electrical transformer and switchgear rooms should be located as close to the incoming electrical service as practical. If there is an emergency generator, it should be located considering (1) proximity to emergency electrical loads and sources of combustion and cooling air and fuel, (2) ease of properly venting exhaust gases to the outdoors, and (3) provisions for noise control.

Primary Equipment Rooms. The heating equipment room houses the boiler(s) and possibly a boiler feed unit (for steam boilers), chemical treatment equipment, pumps, heat exchangers, pressure-reducing equipment, air compressors, and miscellaneous equipment. The refrigeration equipment room houses the chiller(s) and possibly chilled-water and condenser water pumps, heat exchangers, air-conditioning equipment, air compressors, and miscellaneous equipment. Design of these rooms needs to consider (1) equipment size and weight, (2) installation and replacement, (3) applicable regulations relative to combustion air and ventilation air, and (4) noise and vibration transmission to adjacent spaces. ASHRAE *Standard* 15 should be consulted for refrigeration equipment room safety requirements.

Some air-conditioned buildings require a cooling tower or other type of heat rejection equipment. If the cooling tower or water-cooled condenser is located at ground level, it should be at least 100 ft away from the building to (1) reduce tower noise in the building, (2) keep discharge air and moisture carryover from fogging the building's windows and discoloring the building facade, and (3) keep discharge air and moisture carryover from contaminating outside air being introduced into the building. Cooling towers should be kept a similar distance from parking lots to avoid staining car finishes with atomized water treatment chemicals. Chapters 38 and 39 have further information on this equipment.

It is often economical to locate the heating and/or refrigeration plant in the building, on an intermediate floor, in a roof penthouse, or on the roof. Electrical service and structural costs are higher, but these may be offset by reduced costs for piping, pumps and pumping energy, and chimney requirements for fuel-fired boilers. Also, initial cost of equipment in a tall building may be less for equipment located on a higher floor because some operating pressures may be lower with boilers located in a roof penthouse.

Regulations applicable to both gas and fuel oil systems must be followed. Gas fuel may be more desirable than fuel oil because of the physical constraints on the required fuel oil storage tank, as well as specific environmental and safety concerns related to oil leaks. In addition, the cost of an oil leak detection and prevention system may be substantial. Oil pumping presents added design and operating problems, depending on location of the oil tank relative to the oil burner.

Energy recovery systems can reduce the size of the heating and/or refrigeration plant. Well-insulated buildings and electric and gas utility rate structures may encourage the design engineer to consider energy conservation concepts such as limiting demand, ambient cooling, and thermal storage.

Fan Rooms

Fan rooms house HVAC air delivery equipment and may include other miscellaneous equipment. The room must have space for removing the fan(s), shaft(s), coils, and filters. Installation, replacement, and maintenance of this equipment should be considered when locating and arranging the room.

Fan rooms in a basement that has an airway for intake of outside air present a potential problem. Low air intakes are a security concern, because harmful substances could easily be introduced (see the section on Security). Placement of the air intake louver(s) is also a concern because debris and snow may fill the area, resulting in safety, health, and fan performance concerns. Parking areas close to the building's outside air intake may compromise ventilation air quality.

Fan rooms on the second floor and above have easier access for outside and exhaust air. Depending on the fan room location, equipment replacement may be easier. The number of fan rooms required depends largely on the total floor area and whether the HVAC system is centralized or decentralized. Buildings with large floor areas may have multiple decentralized fan rooms on each or alternate floors. High-rise buildings may opt for decentralized fan rooms for each floor, or for more centralized service with one fan room serving the lower 10 to 20 floors, one serving the middle floors of the building, and one at the roof serving the top floors.

Life safety is a very important factor in HVAC fan room location. Chapter 52 of the 2007 *ASHRAE Handbook—HVAC Applications* discusses fire and smoke management. State and local codes have additional fire and smoke detection and damper criteria.

Horizontal Distribution

Many decentralized systems and central systems rely on horizontal distribution. To accommodate this need, the design engineer needs to take into account the duct and/or pipe distribution criteria for installation in a ceiling space or below a raised floor space. Water systems usually require the least amount of ceiling or raised floor depth, whereas air distribution systems have the largest demand for horizontal distribution height. Steam systems need to accommodate pitch of steam pipe, end of main drip, and condensate return pipe pitch. Another consideration in the horizontal space cavity is accommodating the structural members, light fixtures, rain leaders, cable trays, etc., that can fill up this space.

Vertical Shafts

Buildings over three stories high usually require vertical shafts to consolidate mechanical, electrical, and telecommunication distribution through the facility.

Vertical shafts in the building provide space for air distribution ducts and for pipes. Air distribution includes HVAC supply air, return air, and exhaust air ductwork. If a shaft is used as a return air plenum, close coordination with the architect is necessary to ensure that the shaft is airtight. If the shaft is used to convey outside air to decentralized systems, close coordination with the architect is also necessary to ensure that the shaft is constructed to meet mechanical code requirements and to accommodate the anticipated internal pressure. Pipe distribution includes heating water, chilled water, condenser water, and steam supply and condensate return. Other distribution systems found in vertical shafts or located vertically in the building include electric conduits/closets, telephone cabling/closets, uninterruptible power supply (UPS), plumbing, fire protection piping, pneumatic tubes, and conveyers.

Vertical shafts should not be adjacent to stairs, electrical closets, and elevators unless at least two sides are available to allow access to ducts, pipes, and conduits that enter and exit the shaft while allowing maximum headroom at the ceiling. In general, duct shafts with an aspect ratio of 2:1 to 4:1 are easier to develop than large square shafts. The rectangular shape also facilitates transition from the equipment in the fan rooms to the shafts.

In multistory buildings, a central vertical distribution system with minimal horizontal branch ducts is desirable. This arrangement (1) is usually less costly; (2) is easier to balance; (3) creates less conflict with pipes, beams, and lights; and (4) enables the architect to design lower floor-to-floor heights. These advantages also apply to vertical water and steam pipe distribution systems.

The number of shafts is a function of building size and shape. In larger buildings, it is usually more economical in cost and space to have several small shafts rather than one large shaft. Separate HVAC supply, return, and exhaust air duct shafts may be desired to reduce the number of duct crossovers. The same applies for steam supply and condensate return pipe shafts because the pipe must be pitched in the direction of flow.

When future expansion is a consideration, a pre-agreed percentage of additional shaft space should be included. The need for access doors into shafts and gratings at various locations throughout the height of the shaft should also be considered.

Rooftop Equipment

For buildings three stories or less, system analysis and selection frequently locates HVAC equipment on the roof or another outside location, where the equipment is exposed to the weather. Decentralized equipment and systems are more advantageous than centralized HVAC for smaller buildings, particularly those with multiple tenants with different HVAC needs. Selection of rooftop equipment is usually driven by first cost versus operating cost and/or maximum service life of the equipment.

Equipment Access

Properly designed mechanical and electrical equipment rooms must allow for moving large, heavy equipment in, out, and through the building. Equipment replacement and maintenance can be very costly if access is not planned properly. Access to rooftop equipment should be by means of a ship's ladder and not by a vertical ladder. Use caution when accessing equipment on sloped roofs.

Because systems vary greatly, it is difficult to estimate space requirements for refrigeration and boiler rooms without making block layouts of the system selected. Block layouts allow the design engineer to develop the most efficient arrangement of the equipment with adequate access and serviceability. Block layouts can also be used in preliminary discussions with the owner and architect. Only then can the design engineer verify the estimates and provide a workable and economical design.

AIR DISTRIBUTION

Ductwork should deliver conditioned air to an area as directly, quietly, and economically as possible. Structural features of the building generally require some compromise and often limit the depth of space available for ducts. Chapter 9 discusses air distribution design for small heating and cooling systems. Chapter 33 of the 2005 *ASHRAE Handbook—Fundamentals* discusses space air distribution and duct design.

The design engineer must coordinate duct layout with the structure as well as other mechanical, electrical, and communication systems. In commercial projects, the design engineer is usually encouraged to reduce floor-to-floor dimensions. The resultant decrease in available interstitial space for ducts is a major design challenge. In institutional and information technology buildings, higher floor-to-floor heights are required because of the sophisticated, complex mechanical, electrical, and communication distribution systems.

Exhaust systems, especially those serving fumes exhaust, dust and/or particle collection, and other process exhaust, require special design considerations. Capture velocity, duct material, and pertinent duct fittings and fabrication are a few of the design parameters necessary for this type of distribution system to function properly, efficiently, and per applicable codes. Refer to Chapters 29 and 30 of the 2007 *ASHRAE Handbook—HVAC Applications* for additional information.

Air Terminal Units

In some instances, such as in low-velocity, all-air systems, air may enter from the supply air ductwork directly into the conditioned space through a grille, register, or diffuser. In medium- and high-velocity air systems, an intermediate device normally controls air volume, reduces duct pressure, or both. Various types of air terminal units are available, including (1) a fan-powered terminal unit, which uses an integral fan to mix ceiling plenum air and primary air from the central or decentralized fan system rather than depending on induction (mixed air is delivered to low-pressure ductwork and then to the space); (2) a variable-air-volume (VAV) terminal unit, which varies the amount of air delivered to the space (this air may be delivered to low-pressure ductwork and then to the space, or the terminal may contain an integral air diffuser); or (3) other in-room terminal type (see Chapter 5). Chapter 19 has more information about air terminal units.

Duct Insulation

In new construction and renovation upgrade projects, HVAC supply air ducts should be insulated in accordance with energy code requirements. ASHRAE *Standard* 90.1 and Chapter 26 of the 2005 *ASHRAE Handbook—Fundamentals* have more information about insulation and calculation methods.

Ceiling and Floor Plenums

Frequently, the space between the suspended ceiling and the floor slab above it is used as a return air plenum to reduce distribution ductwork. Check regulations before using this approach in new construction or a renovation because most codes prohibit combustible material in a ceiling return air plenum. Ceiling plenums and raised floors can also be used for supply air displacement systems to minimize horizontal distribution, along with other features discussed in Chapter 4.

Some ceiling plenum applications with lay-in panels do not work well where the stack effect of a high-rise building or high-rise elevators creates a negative pressure. If the plenum leaks to the low-pressure area, tiles may lift and drop out when the outside door is opened and closed.

Return air temperature in a return air plenum directly below a roof deck is usually higher by 3 to 5°F during the air-conditioning season than in a ducted return. This can be an advantage to the occupied space below because heat gain to the space is reduced. Conversely, return air plenums directly below a roof deck have lower return air temperatures during the heating season than a ducted return and may require supplemental heat in the plenum.

Raised floors using an air distribution system are popular for computer rooms and cleanrooms, and are now being used in other HVAC applications. Underfloor air displacement (UFAD) systems in office buildings use the raised floor as a supply air plenum, which could reduce overall first cost of construction and ongoing improvement costs for occupants. This UFAD system improves air circulation to the occupied area of the space. See Chapter 17 of the 2007 *ASHRAE Handbook—HVAC Applications* and Chapter 33 of the 2005 *ASHRAE Handbook—Fundamentals* for more information on displacement ventilation and underfloor air distribution.

PIPE DISTRIBUTION

Piping should deliver refrigerant, heating water, chilled water, condenser water, fuel oil, gas, steam, and condensate drainage and return to and from HVAC equipment as directly, quietly, and economically as possible. Structural features of the building generally require mechanical and electrical coordination to accommodate P-traps, pipe pitch-draining of low points in the system, and venting of high points. When assessing application of pipe distribution to air distribution, the floor-to-floor height requirement can influence the pipe system: it requires less ceiling space to install pipe. An alternative to horizontal piping is vertical pipe distribution, which may further reduce floor-to-floor height criteria. Chapter 36 of the 2005 *ASHRAE Handbook—Fundamentals* addresses pipe distribution and design.

Pipe Systems

HVAC piping systems can be divided into two parts: (1) piping in the central plant equipment room and (2) piping required to deliver refrigerant, heating water, chilled water, condenser water, fuel oil, gas, steam, and condensate drainage and return to and from decentralized HVAC and process equipment throughout the building. Chapters 10 to 14 discuss piping for various heating and cooling systems. Chapters 1 to 4 and 33 of the 2006 *ASHRAE Handbook—Refrigeration* discuss refrigerant piping practices.

Pipe Insulation

In new construction and renovation projects, certain HVAC piping may or may not be insulated depending on code requirements. ASHRAE *Standard* 90.1 and Chapter 26 of the 2005 *ASHRAE Handbook—Fundamentals* have information on insulation and calculation methods.

SECURITY

Since September 11, 2001, much attention has been given to protecting buildings' HVAC systems against terrorist attacks. The first consideration should be risk assessment of the particular facility, which may be based on usage, size, population, and/or significance. Risk assessment is a subjective judgment by the building owner of whether the building is at low, medium, or high risk. An example of low-risk buildings may be suburban office buildings or shopping malls. Medium-risk buildings may be hospitals, educational institutions, or major office buildings. High-risk buildings may include major government buildings. The level of protection designed into these buildings may include enhanced particulate filtration, gaseous-phase filtration, and various controlled schemes to allow purging of the facility.

Enhanced particulate filtration for air-handling systems to the level of MERV 14 to 16 filters not only tends to reduce circulation of dangerous substances (e.g., anthrax), but also provides better indoor air quality (IAQ). Gaseous-phase filtration can remove harmful substances such as sarin and other gaseous threats. Low-risk buildings may only include proper location of outdoor air intakes and separate systems for mailrooms and other vulnerable spaces. Medium-risk buildings should consider adding enhanced particulate filtration, and high-risk buildings might also add gaseous filtration. The extent to which the HVAC system designer should use these measures depends on the perceived level of risk.

In any building, consideration should be given to protecting outside air intakes against insertion of dangerous materials by locating the intakes on the roof or substantially above grade level. Separate systems for mailrooms, loading docks, and other similar spaces should be considered so that any dangerous material received cannot be spread throughout the building from these vulnerable spaces. Emergency ventilation systems for these types of spaces should be designed so that upon detection of suspicious material, these spaces can be quickly purged.

A more extensive discussion of this topic can be found in ASHRAE's *Guideline* 29.

AUTOMATIC CONTROLS AND BUILDING MANAGEMENT SYSTEM

Basic HVAC system controls are available in electric, pneumatic, or electronic versions. Depending on the application, the design engineer may recommend a simple and basic system strategy as a cost-effective solution to an owner's heating, ventilation, and cooling needs. Chapter 46 of the 2007 *ASHRAE Handbook—HVAC Applications* and Chapter 15 of the 2005 *ASHRAE Handbook—Fundamentals* discuss automatic control in more detail.

The next level of HVAC system management is direct digital control (DDC), with either pneumatic or electric control damper and valve actuators. This automatic control enhancement may include energy monitoring and energy management software. Controls may also be accessible by the building manager using a modem to a remote computer at an off-site location. Building size has little to no effect on modern computerized controls: programmable controls can be furnished on the smallest HVAC equipment for the smallest projects. Chapter 41 of the 2007 *ASHRAE Handbook—HVAC Applications* covers building operating dynamics.

Automatic controls can be prepackaged and prewired on the HVAC equipment. In system analysis and selection, the design engineer needs to include the merits of purchasing prepackaged versus traditional building automation systems. Current HVAC controls and their capabilities need to be compatible with other new and existing automatic controls. Chapter 39 of the 2007 *ASHRAE Handbook—HVAC Applications* discusses computer applications, and ASHRAE *Standard* 135 discusses interfacing building automation systems.

Using computers and proper software, the design engineer and building manager can provide complete facility management. A comprehensive building management system may include HVAC system control, energy management, operation and maintenance management, medical gas system monitoring, fire alarm system, security system, lighting control, and other reporting and trending software. This system may also be integrated and accessible from the owner's computer network and the Internet.

The building management system is an important factor in choosing the optimum HVAC system. It can be as simple as a time clock to start and stop equipment, or as sophisticated as a computerized building automation system serving a decentralized HVAC system, multiple building systems, central plant system, and/or a large campus. With a focus on energy management, the building management system can be an important business tool in achieving sustainable facility management that begins with using the system selection matrix.

Security should be an integral part of system design and building management. Hazardous materials and contaminated air can be introduced into the building through ventilation systems. When recommending the optimum HVAC system for the project, security should not be overlooked, no matter what the application.

Planning in the design phase the early compilation of record documents (e.g., computer-aided drawing and electronic word files, checklists, digital photos taken during construction) is also integral to successful building management and maintenance.

MAINTENANCE MANAGEMENT SYSTEM

Whereas building management systems focus on operation of HVAC, electrical, plumbing, and other systems, maintenance management systems focus on maintaining assets, which include mechanical and electrical systems along with the building structure. A rule of thumb is that 20% of the cost of the building is in the first cost, with the other 80% being operation, maintenance, and rejuvenation of the building and building systems over the life cycle.

When considering the optimum HVAC selection and recommendation at the start of a project, a maintenance management system should be considered for HVAC systems with an estimated long useful service life.

Another maintenance management business tool is a **computerized maintenance management software (CMMS)** system. The CMMS system can include an equipment database, parts and material inventory, project management software, labor records, etc., pertinent to sustainable management of the building over its life. CMMS also can integrate computer-aided drawing (CAD), digital photography and audio/video systems, equipment run-time monitoring and trending, and other proactive facility management systems.

In scoring the HVAC system selection matrix selection, consideration should also be given to the potential for interface of the building management system with the maintenance management system.

BUILDING SYSTEM COMMISSIONING

When compiling data to complete the HVAC system selection matrix to analytically determine the optimum HVAC system for the project, a design engineer should begin to produce the design intent document/basis of design that identifies the project goals. This process is the beginning of building system commissioning and should be an integral part of the job documentation. As design progresses and the contract documents take shape, the commissioning process will continue to be built into what will eventually be the final commissioning report approximately one year after the construction phase has been completed and the warranty phase comes to an end. For more information, see Chapter 42 in the 2007 *ASHRAE Handbook—HVAC Applications* or ASHRAE *Guideline* 1.

Building commissioning contributes to successful sustainable HVAC design by incorporating the system training requirements necessary for building management staff to efficiently take ownership and operate and maintain the HVAC systems over the installation's useful service life.

In addition to building system commissioning, air and water balancing is required to achieve peak building system performance. Review in the design phase of a project should consider both, and both commissioning and balancing should continue through the construction and warranty phases. Based on the systems selected, commissioning and balancing can cost from $0.50 to $2.00 or more per square foot.

With building certification programs (e.g., LEED™), commissioning is a prerequisite because of the importance of ensuring that high-performance energy and environmental designs are long-term successes.

REFERENCES

ASHRAE. 1996. The HVAC commissioning process. *Guideline* 1-1996.

ASHRAE. 2007. Risk management of public health and safety in buildings. Draft *Guideline* 29.

ASHRAE. 2004. Safety standard for refrigeration systems. ANSI/ASHRAE *Standard* 15-2004.

ASHRAE. 2007. Ventilation for acceptable indoor air quality. ANSI/ASHRAE *Standard* 62.1-2007.

ASHRAE. 2004. Energy standard for buildings except low-rise residential buildings. ANSI/ASHRAE/IESNA *Standard* 90.1-2004.

ASHRAE. 2004. BACnet®—A data communication protocol for building automation and control networks. ASHRAE *Standard* 135-2004.

CHAPTER 2

DECENTRALIZED COOLING AND HEATING

FOR MOST small to mid-size installations, decentralized cooling and heating is usually preferable to a centralized system (see Chapter 3). Frequently classified as packaged unit systems (although many are far from being a single packaged unit), decentralized systems can be found in almost all classes of buildings. They are especially suitable for smaller projects with no central plant, where low initial cost and simplified installation are important. These systems are installed in office buildings, shopping centers, manufacturing plants, schools, health care facilities, hotels, motels, apartments, nursing homes, and other multiple-occupancy dwellings. They are also suited to air conditioning existing buildings with limited life or income potential. Applications also include facilities requiring specialized high performance levels, such as computer rooms and research laboratories.

Although some of the equipment addressed here can be applied as a single unit, this chapter covers applying multiple units to form a complete air-conditioning system for a building and the distribution associated with some of these systems. For guidance on HVAC system selection, see Chapter 1.

SYSTEM CHARACTERISTICS

Decentralized systems can be one or more individual HVAC units, each with an integral refrigeration cycle, heating source, and direct or indirect outside air ventilation. Components are factory-designed and assembled into a package that includes fans, filters, heating source, cooling coil, refrigerant compressor(s), controls, and condenser. Equipment is manufactured in various configurations to meet a wide range of applications. Examples of decentralized HVAC equipment include the following:

- Window air conditioners
- Through-the-wall room HVAC units
- Air-cooled heat pump systems
- Water-cooled heat pump systems
- Multiple-unit systems
- Residential and light commercial split systems
- Self-contained (floor-by-floor) systems
- Outside package systems
- Packaged, special-procedure units (e.g., for computer rooms)

For details on window air conditioners and through-the-wall units, see Chapter 49; the other examples listed here are discussed further in Chapter 48. (Multiple-unit systems are also covered in Chapter 4.)

Commercial-grade unitary equipment packages are available only in preestablished increments of capacity with set performance parameters, such as the sensible heat ratio at a given room condition or the airflow per ton of refrigeration capacity. Components are matched and assembled to achieve specific performance objectives. These limitations make manufacture of low-cost, quality-controlled, factory-tested products practical. For a particular kind and capacity of unit, performance characteristics vary among manufacturers. All characteristics should be carefully assessed to ensure that the equipment performs as needed for the application. Several trade associations have developed standards by which manufacturers may test and rate their equipment. See Chapters 48 and 49 for more specific information on pertinent industry standards and on decentralized cooling and heating equipment used in multiple-packaged unitary systems.

Large commercial/industrial-grade equipment can be custom-designed by the factory to meet specific design conditions and job requirements. This equipment carries a higher first cost and is not readily available in smaller sizes.

Self-contained units can use multiple compressors to control refrigeration capacity. For variable-air-volume (VAV) systems, compressors are turned on or off or unloaded to maintain discharge air temperature. As zone demand decreases, the temperature of air leaving the unit can often be reset upward so that a minimum ventilation rate is maintained.

Multiple packaged-unit systems for perimeter spaces are frequently combined with a central all-air or floor-by-floor system. These combinations can provide better humidity control, air purity, and ventilation than packaged units alone. Air-handling systems may also serve interior building spaces that cannot be conditioned by wall or window-mounted units.

For supplementary data on air-side design of decentralized systems, see Chapter 4.

Advantages

- Heating and cooling can be provided at all times, independent of the mode of operation of other building spaces.
- Manufacturer-matched components have certified ratings and performance data.
- Assembly by a manufacturer helps ensure better quality control and reliability.
- Manufacturer instructions and multiple-unit arrangements simplify installation through repetition of tasks.
- Only one zone of temperature control is affected if equipment malfunctions.
- The system is readily available.
- One manufacturer is responsible for the final equipment package.
- For improved energy control, equipment serving vacant spaces can be turned off locally or from a central point, without affecting occupied spaces.
- System operation is simple. Trained operators are not usually required.
- Less mechanical and electrical room space is required than with central systems.
- Initial cost is usually low.
- Equipment can be installed to condition one space at a time as a building is completed, remodeled, or as individual areas are occupied, with favorable initial investment.

The preparation of this chapter is assigned to TC 9.1, Large Building Air-Conditioning Systems.

- Energy can be metered directly to each tenant.
- Air- or water-side economizers may be applicable, depending on type of decentralized system used.

Disadvantages

- Performance options may be limited because airflow, cooling coil size, and condenser size are fixed.
- Larger total building installed cooling capacity is usually required because diversity factors used for moving cooling needs do not apply to dedicated packages.
- Temperature and humidity control may be less stable, especially with mechanical cooling at very low loads.
- Standard commercial units are not generally suited for large percentages of outside air or for close humidity control. Custom or special-purpose equipment, such as packaged units for computer rooms, or large custom units, may be required.
- Energy use is usually greater than for central systems if efficiency of the unitary equipment is less than that of the combined central system components.
- Low-cost cooling by outside air economizers is not always available or practical.
- Air distribution control may be limited.
- Operating sound levels can be high, and noise-producing machinery is often closer to building occupants than with central systems.
- Ventilation capabilities are fixed by equipment design.
- Equipment's effect on building appearance can be unappealing.
- Air filtration options may be limited.
- Discharge temperature varies because of on/off or step control.
- Condensate drain is required with each air-conditioning unit.
- Maintenance may be difficult or costly because of multiple pieces of equipment and their location.

DESIGN CONSIDERATIONS

Rating classifications and typical sizes for equipment addressed in this chapter can be found in Chapters 48 and 49, which also address available components, equipment selection, distribution piping, and ductwork.

Selection of a decentralized system should follow guidance provided in Chapter 1. The design engineer can use the HVAC system analysis selection matrix to analytically assess and select the optimum decentralized system for the project. Combined with the design criteria in Chapters 48 and 49, the basis of design can be documented.

Unlike centralized cooling and heating equipment, capacity diversity is limited with decentralized equipment, because each piece of equipment must be sized for peak capacity.

Noise from this type of equipment may be objectionable and should be checked to ensure it meets sound level requirements. Chapter 47 of the 2007 *ASHRAE Handbook—HVAC Applications* has more information on HVAC-related sound and vibration concerns.

Air-Side Economizer

With some decentralized systems, an air-side economizer is an option, if not an energy code requirement (check state code for criteria). The air-side economizer uses cool outside air to either assist mechanical cooling or, if the outside air is cool enough, provide total cooling. It requires a mixing box designed to allow 100% of the supply air to be drawn from outside. It can be a field-installed accessory that includes an outside air damper, relief damper, return air damper, filter, actuator, and linkage. Controls are usually a factory-installed option.

Self-contained units usually do not include return air fans. A barometric relief, fan-powered relief fan, or return/exhaust fan may be provided as an air-side economizer. The relief fan is off and discharge/exhaust dampers are closed when the air-side economizer is inactive.

Advantages

- Substantially reduces compressor, cooling tower, and condenser water pump energy requirements, generally saving more energy than a water-side economizer.
- Has a lower air-side pressure drop than a water-side economizer.
- Reduces tower makeup water and related water treatment.
- May improve indoor air quality by providing large amounts of outside air during mild weather.

Disadvantages

- In systems with larger return air static pressure requirements, return or exhaust fans are needed to properly exhaust building air and take in outside air.
- If the unit's leaving air temperature is also reset up during the air-side economizer cycle, humidity control problems may occur and the fan may use more energy.
- Humidification may be required during winter.
- More and/or larger air intake louvers, ducts, or shafts may be required.

Water-Side Economizer

The water-side economizer is another option for reducing energy use. ASHRAE *Standard* 90.1 addresses its application, as do some state energy codes. The water-side economizer consists of a water coil in a self-contained unit upstream of the direct-expansion cooling coil. All economizer control valves, piping between economizer coil and condenser, and economizer control wiring can be factory installed.

The water-side economizer uses the low cooling tower or evaporative condenser water temperature to either (1) precool entering air, (2) assist mechanical cooling, or (3) provide total system cooling if the cooling water is cold enough. If the economizer is unable to maintain the air-handling unit's supply air or zone set point, factory-mounted controls integrate economizer and compressor operation to meet cooling requirements. For constant condenser water flow control using a economizer energy recovery coil and the unit condenser, two control valves are factory-wired for complementary control, with one valve driven open while the other is driven closed. This keeps water flow through the condenser relatively constant. In variable-flow control, condenser water flow varies during unit operation. The valve in bypass/energy recovery loop is an on/off valve and is closed when the economizer is enabled. Water flow through the economizer coil is modulated by its automatic control valve, allowing variable cooling water flow as cooling load increases (valve opens) and reduced flow on a decrease in cooling demand. If the economizer is unable to satisfy the cooling requirements, factory-mounted controls integrate economizer and compressor operation. In this operating mode, the economizer valve is fully open. When the self-contained unit is not in cooling mode, both valves are closed. Reducing or eliminating cooling water flow reduces pumping energy.

Advantages

- Compressor energy is reduced by precooling entering air. Often, building load can be completely satisfied with an entering condenser water temperature of less than 55°F. Because the wet-bulb temperature is always less than or equal to the dry-bulb temperature, a lower discharge air temperature is often available.
- Building humidification does not affect indoor humidity by introducing outside air.
- No external wall penetration is required for exhaust or outside air ducts.
- Controls are less complex than for air-side economizers, because they are often inside the packaged unit.
- The coil can be mechanically cleaned.
- More net usable floor area is available because large outside and relief air ducts are unnecessary.

Disadvantages

- Cooling tower water treatment cost is greater.
- Air-side pressure drop may increase with the constant added resistance of a economizer coil in the air stream.
- Condenser water pump pressure may increase slightly.
- The cooling tower must be designed for winter operation.
- The increased operation (including in winter) required of the cooling tower may reduce its life.

WINDOW-MOUNTED AND THROUGH-THE-WALL ROOM HVAC UNITS AND AIR-COOLED HEAT PUMPS

Window air conditioners (air-cooled room conditioners) and through-the-wall room air conditioners with supplemental heating are designed to cool or heat individual room spaces. Window units are used where low initial cost, quick installation, and other operating or performance criteria outweigh the advantages of more sophisticated systems. Room units are also available in through-the-wall sleeve mountings. Sleeve-installed units are popular in low-cost apartments, motels, and homes. Ventilation can be through operable windows or limited outside air ventilation introduced through the self-contained room HVAC unit. These units are described in more detail in Chapter 49.

Window units may be used as auxiliaries to a central heating or cooling system or to condition selected spaces when the central system is shut down. These units usually serve only part of the spaces conditioned by the basic system. Both the basic system and window units should be sized to cool the space adequately without the other operating.

A through-the-wall air-cooled room air conditioner is designed to cool or heat individual room spaces. Design and manufacturing parameters vary widely. Specifications range from appliance grade through heavy-duty commercial grade, the latter known as packaged terminal air conditioners (PTACs) or packaged terminal heat pumps (PTHPs) (ARI *Standard* 310/380). With proper maintenance, manufacturers project an equipment life of 10 to 15 years for these units.

Air-cooled heat pumps located on roofs or adjacent to buildings are another type of package equipment with most of the features noted here, with the additional benefit of supply air distribution and equipment outside the occupied space. This improved ductwork arrangement makes equipment accessible for servicing out the occupied space, unlike in-room units. See Chapter 48 for additional design, operating, and constructability discussion.

Advantages

- Installation of in-room unit is simple. It usually only requires an opening in the wall or displacement of a window to mount the unit, and connection to electrical power.
- Installation of outside heat pumps is simple with rigging onto concrete pad at grade level or on the roof.
- Generally, the system is well-suited to spaces requiring many zones of individual temperature control.
- Designers can specify electric, hydronic, or steam heat or use an air-to-air heat pump design.
- Service of in-room equipment can be quickly restored by replacing a defective chassis.

Disadvantages

- Equipment life may be less than for large central equipment, typically 10 to 15 years, and units are built to appliance standards, rather than building equipment standards.
- Energy use may be relatively high.
- Direct access to outside air is needed for condenser heat rejection; thus, these units cannot be used for interior rooms.
- The louver and wall box must stop wind-driven rain from collecting in the wall box and leaking into the building.
- The wall box should drain to the outside, which may cause dripping on walls, balconies, or sidewalks.
- Temperature control is usually two-position, which causes swings in room temperature.
- Ventilation and economy cycle capabilities are fixed by equipment design.
- Humidification, when required, must be provided by separate equipment.
- Noise and vibration levels vary considerably and are not generally suitable for sound-critical applications.
- Routine maintenance is required to maintain capacity. Condenser and cooling coils must be cleaned, and filters must be changed regularly.

Design Considerations

A through-the-wall or window-mounted air-conditioning unit incorporates a complete air-cooled refrigeration and air-handling system in an individual package. Each room is an individual occupant-controlled zone. Cooled or warmed air is discharged in response to thermostatic control to meet room requirements (see the discussion on controls following in this section).

Each PTAC or PTHP has a self-contained, air-cooled direct-expansion or heat pump cooling system; a heating system (electric, hot water, steam, and/or a heat pump cycle); and controls. See Figure 3 in Chapter 49 for unit configuration.

A through-the-wall air conditioner or heat pump system is installed in buildings requiring many temperature control zones such as office buildings, motels and hotels, apartments and dormitories, schools and other education buildings, and areas of nursing homes or hospitals where air recirculation is allowed.

These units can be used for renovation of existing buildings, because existing heating systems can still be used. The equipment can be used in both low- and high-rise buildings. In buildings where a stack effect is present, use should be limited to areas that have dependable ventilation and a tight wall of separation between the interior and exterior.

Room air conditioners are often used in parts of buildings primarily conditioned by other systems, especially where spaces to be conditioned are (1) physically isolated from the rest of the building and (2) occupied on a different time schedule (e.g., clergy offices in a church, ticket offices in a theater).

Ventilation air through each terminal may be inadequate in many situations, particularly in high-rise structures because of the stack effect. Chapter 27 of the 2005 *ASHRAE Handbook—Fundamentals* explains combined wind and stack effects. Electrically operated outside air dampers, which close automatically when the equipment is stopped, can reduce heat losses in winter.

Refrigeration Equipment. Room air conditioners are generally supplied with hermetic reciprocating or scroll compressors. Capillary tubes are used in place of expansion valves in most units.

Some room air conditioners have only one motor to drive both the evaporator and condenser fans. The unit circulates air through the condenser coil whenever the evaporator fan is running, even during the heating season. Annual energy consumption of a unit with a single motor is generally higher than one with a separate motor, even when the energy efficiency ratio (EER) or the coefficient of performance (COP) is the same for both. Year-round, continuous flow of air through the condenser increases dirt accumulation on the coil and other components, which increases maintenance costs and reduces equipment life.

Because through-the-wall conditioners are seldom installed with drains, they require a positive and reliable means of condensate disposal. Conditioners are available that spray condensate in a fine mist over the condenser coil. These units dispose of more condensate than can be developed without any drip, splash, or spray. In heat pumps, provision must be made for disposal of condensate generated from the outside coil during defrost.

Many air-cooled room conditioners experience evaporator icing and become ineffective when outside temperatures fall below about 65°F. Units that ice at a lower outside temperature may be required to handle the added load created by high lighting levels and high solar radiation found in contemporary buildings.

Heating Equipment. The air-to-air heat pump cycle described in Chapter 48 is available in through-the-wall room air conditioners. Application considerations are similar to conventional units without the heat pump cycle, which is used for space heating when the outside temperature is above 35 to 40°F. Electric resistance elements supply heating below this level and during defrost cycles.

The prime advantage of the heat pump cycle is that it reduces annual energy consumption for heating. Savings in heat energy over conventional electric heating ranges from 10 to 60%, depending on the climate.

Controls. All controls for through-the-wall air conditioners are included as a part of the conditioner. The following control configurations are available:

- **Thermostat control** is either unit-mounted or remote wall-mounted.
- **Motel and hotel guest room controls** allow starting and stopping units from a central point.
- **Occupied/unoccupied controls** (for occupancies of less than 24 h) start and stop the equipment at preset times with a time clock. Conditioners operate normally with the unit thermostat until the preset cutoff time. After this point, each conditioner has its own reset control, which allows the occupant of the conditioned space to reset the conditioner for either cooling or heating, as required.
- **Master/slave control** is used when multiple conditioners are operated by the same thermostat.
- **Emergency standby control** allows a conditioner to operate during an emergency, such as a power failure, so that the room-side blowers can operate to provide heating. Units must be specially wired to allow operation on emergency generator circuits.

When several units are used in a single space, controls should be interlocked to prevent simultaneous heating and cooling. In commercial applications (e.g., motels), centrally operated switches can de-energize units in unoccupied rooms.

WATER-SOURCE HEAT PUMP SYSTEMS

Water-source heat pump systems use multiple cooling/heating units distributed throughout the building. For more in-depth discussion of water-source heat pumps, see Chapter 8. Outside air ventilation requires either direct or indirect supply air from an additional air-handling system.

Designed to cool and heat individual rooms or multiple spaces grouped together by zone, water-source heat pumps may be installed along the perimeter with a combination of horizontal and vertical condenser water piping distribution, or stacked vertically with condenser water piping also stacked vertically to minimize equipment space effects on the rooms they serve. They can also be ceiling mounted or concealed above the ceiling with duct distribution to the area served.

Water-source heat pump systems also require further decentralized equipment that includes a source of heat rejection such as a cooling tower or ground water installation. A supplemental heating source such as a boiler may be required, depending on the installation's location (e.g., in colder winter climates). Data on condenser water systems and the necessary heat rejection equipment can be found in Chapters 13 and 39.

Advantages

- Unit installation is simple for both vertical and horizontal installation and connection to electrical power.
- Outdoor heat pumps can be installed with a lift to rig horizontal units at or above a ceiling.
- Generally, the system is well suited to spaces requiring many zones of individual temperature control.
- Designers can specify energy recovery.

Disadvantages

- Energy use may be relatively high compared with other types of systems.
- Introducing outside air to the building can be a problem.
- Condensate drain piping installation and routine maintenance can be a problem, particularly for units installed above ceilings.
- No air-side economy cycle capabilities.
- Humidification, when required, must be provided by separate equipment.
- Noise and vibration levels vary considerably and are not generally suitable for sound-critical applications.
- Routine maintenance within occupied space is required to maintain capacity.

Design Considerations

Units are usually furnished with individual electric controls. However, control strategy considerations are similar to those for air-cooled heat pumps.

These units can be used for renovation of existing buildings where limited ceiling space would prevent other types of HVAC systems (e.g., all-air systems) from being installed. The equipment can be used in both low- and high-rise buildings; both applications require some form of outdoor ventilation to serve the occupants.

See Chapter 8 for data on refrigeration cycle, heating cycle, automatic controls, and other information on design and operation and maintenance.

MULTIPLE-UNIT SYSTEMS

Multiple-unit systems generally use single-zone unitary HVAC with a unit for each zone (Figure 1). Zoning is determined by (1) cooling and heating loads, (2) occupancy, (3) flexibility requirements, (4) appearance, and (5) equipment and duct space availability. Multiple-unit systems are popular for office buildings, manufacturing plants, shopping centers, department stores, and apartment buildings. Unitary self-contained units are excellent for renovation.

The system configuration can be horizontal distribution of equipment and associated ductwork and piping, or vertical distribution of

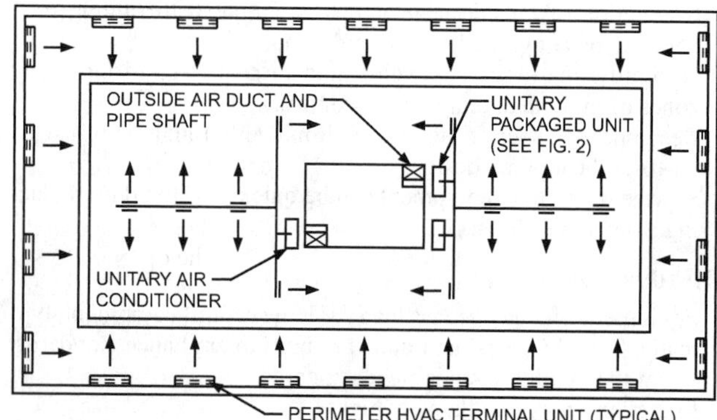

Fig. 1 Multiple-Unit Systems Using Single-Zone Unitary HVAC Equipment
(Courtesy RDK Engineers)

equipment and piping with horizontal distribution of ductwork. Outside air ventilation requires either direct or indirect supply air from an additional air-handling system.

Heating media may be steam or hot water piped to the individual units or electric heat at the unit and/or at individual air terminals that provide multiple-space temperature-controlled zones. Cooling media may be chilled water, refrigerant/direct expansion (DX), or heat pump condenser water piped to individual units. A typical system features zone-by-zone equipment with a central plant chiller and boiler, although electric heat and DX refrigerant cooling may be used.

Supply air may be constant or variable volume; outside ventilation is probably a fixed minimum, possibly with limited variable volume based on carbon dioxide ventilation control. Usually, multiple units do not come with a return air fan; the supply fan must overcome return air and supply air duct static resistance. A supplemental exhaust fan or multiple exhaust fans may be required to complete the design.

Multiple-unit systems require a localized equipment room where one or more units can be installed. This arrangement takes up floor space, but allows equipment maintenance to occur out of the building's occupied areas, minimizing interruptions to occupants.

Advantages

- Installation is simple. Equipment is readily available in sizes that allow easy handling.
- Equipment and components can be standardized.
- Relocation of units to other spaces or buildings is practical, if necessary.
- Energy efficiency can be quite good, particularly where climate or building use results in a balance of heating and cooling zones.
- Units are available with complete, self-contained control systems that include variable-volume control, night setback, and morning warm-up.
- Equipment is out of the occupied space, making the system quieter.
- Easy access to equipment facilitates routine maintenance.
- System is repetitive and simple, facilitating operator training.

Disadvantages

- Fans may have limited static pressure ratings.
- Air filtration options are limited.
- Humidification can be impractical on a unit-by-unit basis and may need to be provided by a separate system.
- Integral air-cooled condensing units for some direct-expansion cooling installations should be located outdoors within a limited distance.
- Multiple units and equipment closets or rooms may occupy rentable floor space.
- Multiple pieces of equipment may increase maintenance requirements.
- Redundant equipment or easy replacement may not be practical.

Design Considerations

Unitary systems can be used throughout a building or to supplement perimeter packaged terminal units (Figures 1 and 2). Because core areas frequently have little or no heat loss, unitary equipment with air- or water-cooled condensers can be applied. The equipment can be used in both low- and high-rise buildings; both applications require some form of outside ventilation to serve the occupants.

Typical application may be a interior work area, computer room, or other space requiring continual cooling. Special-purpose unitary equipment is frequently used to cool, dehumidify, humidify, and reheat to maintain close control of space temperature and humidity in computer areas (Figure 2). For more information, see Chapters 16 and 17 of the 2007 *ASHRAE Handbook—HVAC Applications*, as well as Chapters 48 and 49 of this volume.

In the multiple-unit system shown in Figure 3, one unit may be used to precondition outside air for a group of units. This all-outside-air unit prevents hot, humid air from entering the conditioned space during periods of light load. The outside unit should have sufficient capacity to cool the required ventilation air from outside design conditions to interior design dew point. Zone units are then sized to handle only the internal load for their particular area.

Units under 20 tons of cooling are typically constant-volume units. VAV distribution may be accomplished on these units with a bypass damper that allows excess supply air to bypass to the return air duct. The bypass damper ensures constant airflow across the direct-expansion cooling coil to avoid coil freeze-up caused by low airflow. The damper is usually controlled by supply duct pressure. Variable-frequency drives can also be used.

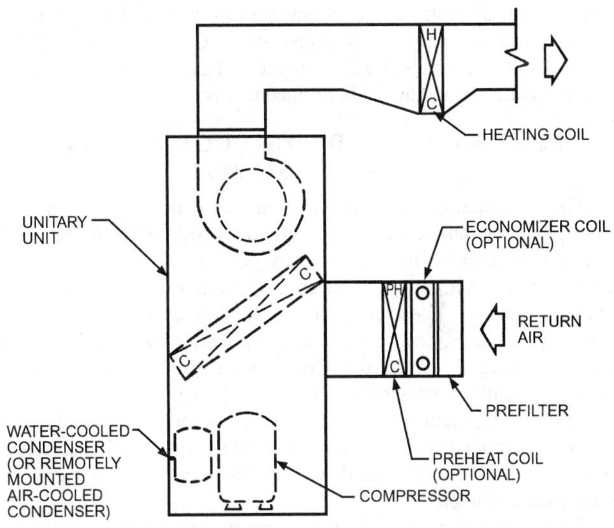

Fig. 2 Vertical Self-Contained Unit
(Courtesy RDK Engineers)

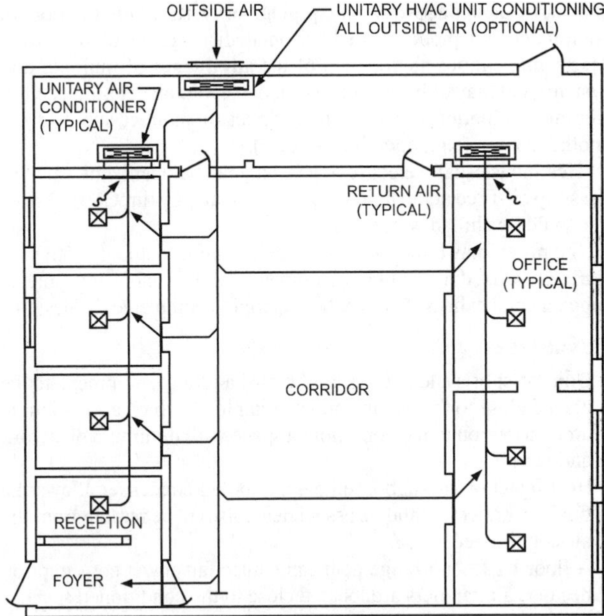

Fig. 3 Multiroom, Multistory Office Building with Unitary Core and Through-the-Wall Perimeter Air Conditioners (Combination Similar to Figure 1)
(Courtesy RDK Engineers)

Controls. Units are usually furnished with individual electric controls, but can be enhanced to a more comprehensive building management system.

Economizer Cycle. When outside temperature allows, energy use can be reduced by cooling with outside air in lieu of mechanical refrigeration. Units must be located close to an outside wall or outside air duct shafts. Where this is not possible, it may be practical to add a water-side economizer cooling coil, with cold water obtained by sending the condenser water through a winterized cooling tower. Chapter 39 has further details.

Acoustics and Vibration. Because these units are typically located near occupied space, they can affect acoustics. The designer must study both the airflow breakout path and the unit's radiated sound power when coordinating selection of wall and ceiling construction surrounding the unit. Locating units over noncritical work spaces such as restrooms or storage areas around the equipment room helps reduce noise in occupied space. Chapter 47 of the 2007 *ASHRAE Handbook—HVAC Applications* has more information on HVAC-related sound and vibration concerns.

RESIDENTIAL AND LIGHT COMMERCIAL SPLIT SYSTEMS

These systems distribute cooling and heating equipment throughout the building. A split system consists of an indoor unit with air distribution and temperature control with either a water-cooled condenser, integral air-cooled condenser, or remote air-cooled condenser. These units are commonly used in single-story or low-rise buildings, and in residential applications where condenser water is not readily available. Commercial split systems are well-suited to small projects with variable occupancy schedules.

Indoor equipment is generally installed in service areas adjacent to the conditioned space. When a single unit is required, the indoor unit and its related ductwork constitute a central air system, as described in Chapter 4.

Typical components of a split-system air conditioner include an indoor unit with evaporator coils, economizer coils, heating coils, filters, valves, and a condensing unit with the compressors and condenser coils.

The configuration can be horizontal distribution of equipment and associated ductwork and piping, or vertical distribution of equipment and piping with horizontal distribution of ductwork. These applications share some of the advantages of multiple-unit systems, but may only have one system installation per project. Outside air ventilation requires either direct or indirect supply from another source (e.g., operable window).

Heating is usually electric or gas, but may be steam, hot water, or possibly oil-fired at the unit. Cooling is usually by direct expansion, but could be chilled water.

Supply air may be constant or variable volume; outdoor ventilation is usually a fixed minimum or barometric relief economizer cycle. A supplemental exhaust fan may be required to completed the design.

Advantages

- Unitary split-system units allow air-handling equipment to be placed close to the heating and cooling load, which allows ample air distribution to the conditioned space with minimum ductwork and fan power.
- Heat rejection through a remote air-cooled condenser allows the final heat rejector (and its associated noise) to be remote from the air-conditioned space.
- A floor-by-floor arrangement can reduce fan power consumption because air handlers are located close to the conditioned space.
- Large vertical duct shafts and fire dampers may be reduced or eliminated.
- Equipment is generally located in the building interior near elevators and other service areas and does not interfere with the building perimeter.

Disadvantages

- The proximity of the air handler to the conditioned space requires special attention to unit inlet and outlet airflow and to building acoustics around the unit.
- Ducting ventilation air to the unit and removing condensate from the cooling coil should be considered.
- A unit that uses an air-side economizer must be located near an outside wall or outside air shaft. Split-system units do not generally include return air fans.
- A separate method of handling and controlling relief air may be required.
- Filter options and special features may be limited.
- Discharge temperature varies because of on/off or step control.

Design Considerations

Characteristics that favor split systems are their low first cost, simplicity of installation, and simplicity of training required for operation. Servicing is also relatively inexpensive.

The modest space requirements of split-system equipment make it excellent for renovations or for spot cooling a single zone. Control is usually one- or two-step cooling and one- or two-step or modulating heat. VAV operation is possible with a supply air bypass. Some commercial units can modulate airflow, with additional cooling modulation using hot-gas bypass.

Commercial split-system units are available as constant-volume equipment for use in atriums, public areas, and industrial applications. Basic temperature controls include a room-mounted or return-air-mounted thermostat that cycles the compressor(s) as needed. Upgrades include fan modulation and VAV control. When applied with VAV terminals, commercial split systems provide excellent comfort and individual zone control.

COMMERCIAL SELF-CONTAINED (FLOOR-BY-FLOOR) SYSTEMS

Commercial self-contained (floor-by-floor) systems are a type of multiple-unit, decentralized cooling and heating system. Equipment is usually configured vertically, but may be horizontal. Supply air distribution may be a discharge air plenum, raised-floor supply air plenum (air displacement), or a limited amount of horizontal duct distribution, installed on a floor-by-floor basis. Outside air ventilation requires either direct or indirect supply air from an additional air-handling system.

Typical components include compressors, water-cooled condensers, evaporator coils, economizer coils, heating coils, filters, valves, and controls (Figure 4). To complete the system, a building needs cooling towers and condenser water pumps. See Chapter 39 for more information on cooling towers.

Advantages

- This equipment integrates refrigeration, heating, air handling, and controls into a factory package, thus eliminating many field integration problems.
- Units are well suited for office environments with variable occupancy schedules.
- Floor-by-floor arrangement can reduce fan power consumption.
- Large vertical duct shafts and fire dampers are eliminated.
- Electrical wiring, condenser water piping, and condensate removal are centrally located.
- Equipment is generally located in the building interior near elevators and other service areas, and does not interfere with the building perimeter.
- Integral water-side economizer coils and controls are available, which allow interior equipment location and eliminate large outside air and exhaust ducts and relief fans.
- An acoustical discharge plenum is available, which allows lower fan power and lower sound power levels.

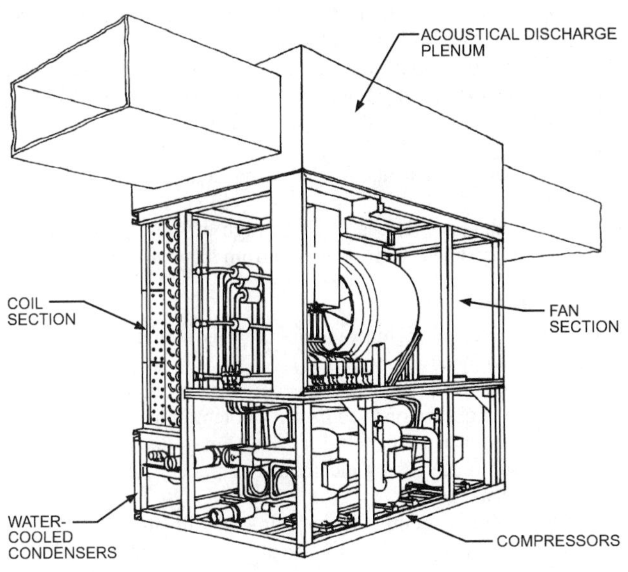

Fig. 4 Commercial Self-Contained Unit with Discharge Plenum

Disadvantages

- Units must be located near an outside wall or outside air shaft to incorporate an air-side economizer.
- A separate relief air system and controls must be incorporated if an air-side economizer is used.
- Close proximity to building occupants requires careful analysis of space acoustics.
- Filter options may be limited.
- Discharge temperature varies because of on/off or step control.

Design Considerations

Commercial self-contained units have criteria similar to those for multiple and light commercial units, and can serve either VAV or constant-volume systems. These units contain one or two fans inside the cabinet. The fans are commonly configured in a draw-through arrangement.

The size and diversity of the zones served often dictate which system is optimal. For comfort applications, self-contained VAV units coupled with terminal boxes or fan-powered terminal boxes are popular for their energy savings, individual zone control, and acoustic benefits. Constant-volume self-contained units have low installation cost and are often used in noncomfort or industrial air-conditioning applications or in single-zone comfort applications.

Unit airflow is reduced in response to terminal boxes closing. Several common methods used to modulate airflow delivered by the fan to match system requirements include inlet guide vanes, fan speed control, inlet/discharge dampers, and multiple-speed fan motors.

Appropriate outside air and exhaust fans and dampers work in conjunction with the self-contained unit. Their operation must be coordinated with unit operation to maintain design air exchange and building pressurization.

Refrigeration Equipment. Commercial self-contained units usually feature reciprocating or scroll compressors, although screw compressors are available in some equipment. Thermostatic or electronic expansion valves are used. Condensers are water-cooled and usually reject heat to a common condenser-water system serving multiple units. A separate cooling tower or other final heat rejection device is required.

Self-contained units may control capacity with multiple compressors. For VAV systems, compressors are turned on or off or unloaded to maintain discharge air temperature. Hot-gas bypass is often incorporated to provide additional capacity control. As system airflow decreases, the temperature of air leaving the unit is often reset upward so that a minimum ventilation rate can be maintained. Resetting the discharge air temperature limits the unit's demand, thus saving energy. However, increased air temperature and volume increase fan energy.

Heating Equipment. In many applications, heating is done by perimeter radiation, with heating installed in the terminal boxes or other such systems when floor-by-floor units are used. If heating is incorporated in these units (e.g., preheat or morning warm-up), it is usually provided by hot-water coils or electric resistance heat, but could be by a gas- or oil-fired heat exchanger.

Controls. Self-contained units typically have built-in capacity controls for refrigeration, economizers, and fans. Although units under 15 tons of cooling tend to have basic on/off/automatic controls, many larger systems have sophisticated microprocessor controls that monitor and take action based on local or remote programming. These controls provide for stand-alone operation, or they can be tied to a building automation system (BAS).

A BAS allows more sophisticated unit control by time-of-day scheduling, optimal start/stop, duty cycling, demand limiting, custom programming, etc. This control can keep units operating at peak efficiency by alerting the operator to conditions that could cause substandard performance.

The unit's control panel can sequence the modulating valves and dampers of an economizer. A water-side economizer is located upstream of the evaporator coil, and when condenser water temperature is lower than entering air temperature to the unit, water flow is directed through the economizer coil to either partially or fully meet building load. If the coil alone cannot meet design requirements, but the entering condenser water temperature remains cool enough to provide some useful precooling, the control panel can keep the economizer coil active as stages of compressors are activated. When entering condenser water exceeds entering air temperature to the unit, the coil is valved off, and water is circulated through the unit's condensers only.

Typically, in an air-side economizer an enthalpy or dry-bulb temperature switch energizes the unit to bring in outside air as the first stage of cooling. An outside air damper modulates flow to meet design temperature, and when outside air can no longer provide sufficient cooling, compressors are energized.

A temperature input to the control panel, either from a discharge air sensor or a zone sensor, provides information for integrated economizer and compressor control. Supply air temperature reset is commonly applied to VAV systems.

In addition to capacity controls, units have safety features for the refrigerant-side, air-side, and electrical systems. Refrigeration protection controls typically consist of high and low refrigerant pressure sensors and temperature sensors wired into a common control panel. The controller then cycles compressors on and off or activates hot-gas bypass to meet system requirements.

Constant-volume units typically have high-pressure cut-out controls, which protect the unit and ductwork from high static pressure. VAV units typically have some type of static pressure probe inserted in the discharge duct downstream of the unit. As terminal boxes close, the control modulates airflow to meet the set point, which is determined by calculating the static pressure required to deliver design airflow to the zone farthest from the unit.

Acoustics and Vibration. Because self-contained units are typically located near occupied space, their performance can significantly affect occupant comfort. Units of less than 15 tons of cooling are often placed inside a closet, with a discharge grille penetrating the common wall to the occupied space. Larger units have their own equipment room and duct system. Common sound paths to consider include the following:

- Fan inlet and compressor sound radiates through the unit casing to enter the space through the separating wall.
- Fan discharge sound is airborne through the supply duct and enters the space through duct breakout and diffusers.
- Airborne fan inlet sound enters the space through the return air ducts, or ceiling plenum if unducted.

Discharge air transition from the self-contained unit is often accomplished with a plenum located on top of the unit. This plenum facilitates multiple duct discharges that reduce the amount of airflow over a single occupied space adjacent to the equipment room (see Figure 4). Reducing airflow in one direction reduces the sound that breaks out from the discharge duct. Several feet of internally lined round duct immediately off the discharge plenum significantly reduces noise levels in adjacent areas.

In addition to the airflow breakout path, the system designer must study unit-radiated sound power when determining equipment room wall and door construction. A unit's air-side inlet typically has the highest radiated sound. The inlet space and return air ducts should be located away from the critical area to reduce the effect of this sound path.

Selecting a fan that operates near its peak efficiency point helps design quiet systems. Fans are typically dominant in the first three octave bands, and selections at high static pressures or near the fan's surge region should be avoided.

Units may be isolated from the structure with neoprene pads or spring isolators. Manufacturers often isolate the fan and compressors internally, which generally reduces external isolation requirements.

COMMERCIAL OUTDOOR PACKAGED SYSTEMS

Commercial outdoor packaged systems are similar to air-cooled heat pump and commercial self-contained (floor-to-floor) systems, and are usually of horizontal configuration.

Heating is usually electric or gas but may be steam, hot water, or possibly oil-fired at the unit. Cooling is usually by direct expansion, but could be chilled water.

Supply air distribution usually has a limited amount of horizontal duct distribution and is installed on a floor-by-floor basis. Outdoor air ventilation can be provided by barometric relief, fan-powered relief, or return air/exhaust air fan.

Equipment is generally mounted on the roof [rooftop units (RTUs)], but can also be mounted at grade level. RTUs are designed as central-station equipment for single-zone, multizone, and VAV applications.

Systems are available in several levels of design sophistication, from simple factory-standard light commercial packaged equipment, to double-wall commercial packaged equipment with upgraded features, up to fully customized industrial-quality packages. Often, factory-standard commercial rooftop unit(s) are satisfactory for small and medium-sized office buildings. On large projects and highly demanding systems, the additional cost of a custom packaged unit can be justified by life-cycle cost analyses. Custom systems offer great flexibility and can be configured to satisfy almost any requirement. Special features such as heat recovery, service vestibules, boilers, chillers, and space for other mechanical equipment can be designed into the unit.

For additional information, see Chapter 48.

Advantages

- Equipment location allows easy service access without maintenance staff entering or disturbing occupied space.
- Construction costs are offset toward the end of the project because the unit can be one of the last items installed.
- Installation is simplified and field labor costs are reduced because most components are assembled and tested in a controlled factory environment.
- A single source has responsibility for design and operation of all major mechanical systems in the building.
- Valuable building space for mechanical equipment is conserved.
- It is suitable for floor-by-floor control in low-rise office buildings.
- Outside air is readily available for ventilation and economy cycle use.
- Combustion air intake and flue gas exhaust are facilitated if natural gas heat is used.
- Upgraded design features, such as high-efficiency filtration or heat recovery devices, are available from some manufacturers.

Disadvantages

- Maintaining or servicing outdoor units is sometimes difficult, especially in inclement weather.
- With all rooftop equipment, safe access to the equipment is a concern. Even slightly sloped roofs are a potential hazard.
- Frequent removal of panels for access may destroy the unit's weatherproofing, causing electrical component failure, rusting, and water leakage.
- Rooftop unit design must be coordinated with structural design because it may represent a significant building structural load.
- In cold climates, provision must be made to keep snow from blocking air intakes and access doors, and the potential for freezing of hydronic heating or steam humidification components must be considered.
- Casing corrosion is a potential problem. Many manufacturers prevent rusting with galvanized or vinyl coatings and other protective measures.
- Outdoor installation can reduce equipment life.
- Depending on building construction, sound levels and transmitted vibration may be excessive.
- Architectural considerations may limit allowable locations or require special screening to minimize visual effect.

Design Considerations

Centering the rooftop unit over the conditioned space reduces fan power, ducting, and wiring. Avoid installation directly above spaces where noise level is critical.

All outdoor ductwork should be insulated, if not already required by associated energy codes. In addition, ductwork should be (1) sealed to prevent condensation in insulation during the heating season and (2) weatherproofed to keep it from getting wet.

Use multiple single-zone, not multizone, units where feasible to simplify installation and improve energy consumption. For large areas such as manufacturing plants, warehouses, gymnasiums, etc., single-zone units are less expensive and provide protection against total system failure.

Use units with return air fans whenever return air static pressure loss exceeds 0.5 in. of water or the unit introduces a large percentage of outside air via an economizer.

Units are also available with relief fans for use with an economizer in lieu of continuously running a return fan. Relief fans can be initiated by static pressure control.

In a rooftop application, the air handler is outside and needs to be weatherproofed against rain, snow, and, in some areas, sand. In coastal environments, enclosure materials' resistance (e.g., to salt spray) must also be considered. In cold climates, fuel oil does not atomize and must be warmed to burn properly. Hot-water or steam heating coils and piping must be protected against freezing. In some areas, enclosures are needed to maintain units effectively during inclement weather. A permanent safe access to the roof, as well as a roof walkway to protect against roof damage, are essential.

Rooftop units are generally mounted using (1) integral frames or (2) lightweight steel structures. Integral support frames are designed by the manufacturer to connect to the base of the unit. Separate openings for supply and return ducts are not required. The completed installation must adequately drain condensed water. Lightweight steel structures allow the unit to be installed above the roof using separate, flashed duct openings. Any condensed water can be drained through the roof drains.

Accessories such as economizers, special filters, and humidifiers are available. Factory-installed and wired economizer packages are also available. Other options offered are return and exhaust fans, variable-volume controls with hot-gas bypass or other form of coil frost protection, smoke and fire detectors, portable external service enclosures, special filters, and microprocessor-based controls with various control options.

For projects with custom-designed equipment, it may be desirable to require additional witnessed factory testing to ensure performance and quality of the final product.

Refrigeration Equipment. Large systems incorporate reciprocating, screw, or scroll compressors. Chapter 37 has information about compressors and Chapters 42 and 48 discuss refrigeration equipment, including the general size ranges of available equipment. Air-cooled or evaporative condensers are built integral to the equipment.

Air-cooled condensers pass outside air over a dry coil to condense the refrigerant. This results in a higher condensing temperature and, thus, a larger power input at peak conditions. However, this peak time may be relatively short over 24 h. The air-cooled condenser is popular in small reciprocating systems because of its low maintenance requirements.

Evaporative condensers pass air over coils sprayed with water, using adiabatic saturation to lower the condensing temperature. As with the cooling tower, freeze prevention and close control of water treatment are required for successful operation. The lower power consumption of the refrigeration system and the much smaller footprint from using an evaporative versus air-cooled condenser are gained at the expense of the cost of water used and increased maintenance costs.

Heating Equipment. Natural-gas, propane, oil, electricity, hot-water, steam, and refrigerant gas heating options are available. These are normally incorporated directly into the air-handling sections. Custom equipment can also be designed with a separate prepiped boiler and circulating system.

Controls. Multiple outdoor units are usually single-zone, constant-volume, or VAV units. Zoning for temperature control determines the number of units; each zone has a unit. Zones are determined by the cooling and heating loads for the space served, occupancy, allowable roof loads, flexibility requirements, appearance, duct size limitations, and equipment size availability. These units can also serve core areas of buildings, with perimeter spaces served by PTACs.

Most operating and safety controls are provided by the equipment manufacturer. Although remote monitoring panels are optional, they are recommended to allow operating personnel to monitor performance.

Acoustics and Vibration. Most unitary equipment is available with limited separate vibration isolation of rotating equipment. Custom equipment is available with several (optional) degrees of internal vibration isolation. Isolation of the entire unit casing is rarely required; however, care should be taken when mounting on lightweight structures. If external isolation is required, it should be coordinated with the unit manufacturer to ensure proper separation of internal versus external isolation deflection.

Outdoor noise from unitary equipment should be reduced to a minimum. Sound power levels at all property lines must be evaluated. Indoor-radiated noise from the unit's fans, compressors, and condensers travels directly through the roof into occupied space below. Mitigation usually involves adding mass, such as two layers of gypsum board, inside the roof curb beneath the unit. Airborne duct discharge noise, primarily from the fans themselves, can be attenuated by silencers in the supply and return air ducts or by acoustically lined ductwork.

AUTOMATIC CONTROLS AND BUILDING MANAGEMENT SYSTEMS

A building management system can be an important tool in achieving sustainable facility energy management. Basic HVAC system controls are electric or electronic, and usually are prepackaged and prewired with equipment at the factory. Controls may also be accessible by the building manager using a remote off-site computer. The next level of HVAC system management is to integrate the manufacturer's control package with the building management system. If the project is an addition or major renovation, prepackaged controls and their capabilities need to be compatible with existing automated controls. Chapter 39 of the 2007 *ASHRAE Handbook—HVAC Applications* discusses computer applications, and ANSI/ASHRAE *Standard* 135 discusses interfacing building automation systems.

MAINTENANCE MANAGEMENT

Because they are simpler and more standardized than centralized systems, decentralized systems can often be maintained by less technically trained personnel. Maintenance management for many packaged equipment systems can be specified with a service contract from a local service contracting firm. Frequently, small to midlevel construction projects do not have qualified maintenance technicians on site once the job is turned over to a building owner, and service contracts can be a viable option. These simpler decentralized systems allow competitive solicitation of bids for annual maintenance to local companies.

BUILDING SYSTEM COMMISSIONING

Commissioning a building system that has an independent control system to be integrated with individual packaged control systems requires that both control contractors participate in the process. Before the commissioning functional performance demonstrations to the client, it is important to obtain the control contractors' individual point checkout sheets, program logic, and list of points that require confirmation with another trade (e.g., fire alarm system installer).

Frequently, decentralized systems are installed in phases, requiring multiple commissioning efforts based on the construction schedule and owner occupancy. This applies to new construction and expansion of existing installations. During the warranty phase, decentralized system performance should be measured, benchmarked, and course-corrected to ensure the design intent can be achieved. If an energy analysis study is performed as part of the comparison between decentralized and centralized concepts, or life-cycle comparison of the study is part of a Leadership in Energy and Environmental Design (LEED™) project, the resulting month-to-month energy data should be a good electronic benchmark for actual energy consumption using the measurement and verification plan implementation.

Ongoing commissioning or periodic recommissioning further ensures that design intent is met, and that cooling and heating are reliably delivered. Retro- or recommissioning should be considered whenever the facility is expanded or an additional connection made to the existing systems, to ensure the original design intent is met.

The initial testing, adjusting, and balancing (TAB) also contributes to sustainable operation and maintenance. The TAB process should be repeated periodically to ensure levels are maintained.

When completing TAB and commissioning, consider posting laminated system flow diagrams at or adjacent to cooling and heating equipment indicating operating instructions, TAB performance, commissioning functional performance tests, and emergency shutoff procedures. These documents also should be filed electronically in the building manager's computer server for quick reference.

Original basis of design and design criteria should be posted as a constant reminder of design intent, and to be readily available in case troubleshooting, expansion, or modernization is needed.

As with all HVAC applications, to be a sustainable design success, building commissioning should include the system training requirements necessary for building management staff to efficiently take ownership and operate and maintain the HVAC systems over the useful service life of the installation.

Commissioning should continue up through the final commissioning report, approximately one year after the construction phase has been completed and the warranty phase comes to an end. For further details on commissioning, see Chapter 42 of the 2007 *ASHRAE Handbook—HVAC Applications*.

BIBLIOGRAPHY

AHAM. 2003. Room air conditioners. ANSI/AHAM *Standard* RAC-1. Association of Home Appliance Manufacturers, Chicago.

ARI. 2004. Packaged terminal air conditioners and heat pumps. *Standard* 310/380-2004. Air-Conditioning and Refrigeration Institute, Arlington, VA.

ASHRAE. 2004. Energy standard for buildings except low-rise residential buildings. ANSI/ASHRAE *Standard* 90.1-2004.

ASHRAE. 2004. BACnet®—A data communication protocol for building automation and control networks. ANSI/ASHRAE *Standard* 135-2004.

CHAPTER 3

CENTRAL COOLING AND HEATING

CENTRAL cooling and/or heating plants generate cooling and/or heating in one location for distribution to multiple locations in one building or an entire campus or neighborhood, and represent approximately 25% of HVAC system applications. Central cooling and heating systems are used in almost all classes of buildings, but particularly in very large buildings and complexes or where there is a high density of energy use. They are especially suited to applications where maximizing equipment service life and using energy and operational workforce efficiently are important.

The following facility types are good candidates for central cooling and/or heating systems:

- Campus environments with distribution to several buildings
- High-rise facilities
- Large office buildings
- Large public assembly facilities, entertainment complexes, stadiums, arenas, and convention and exhibition centers
- Urban centers (e.g., city centers/districts)
- Shopping malls
- Large condominiums, hotels, and apartment complexes
- Educational facilities
- Hospitals and other health care facilities
- Industrial facilities (e.g., pharmaceutical, manufacturing)
- Large museums and similar institutions
- Locations where waste heat is readily available (result of power generation or industrial processes)

This chapter addresses design alternatives that should be considered when centralizing a facility's cooling and heating sources. Distribution system options and equipment are discussed when they relate to the central equipment, but more information on distribution systems can be found in Chapters 10 to 14.

SYSTEM CHARACTERISTICS

Central systems are characterized by large chilling and/or heating equipment located in one facility or multiple smaller installations interconnected to operate as one. Equipment configuration and ancillary equipment vary significantly, depending on the facility's use. See Chapter 1 for information on selecting a central cooling or heating plant.

Equipment can be located adjacent to the facility, or in remote stand-alone plants. Also, different combinations of centralized and decentralized systems (e.g., a central cooling plant and decentralized heating and ventilating systems) can be used.

Primary equipment (i.e., chillers and boilers) is available in different sizes, capacities, and configurations to serve a variety of building applications. Operating a few pieces of primary equipment (often with back-up equipment) gives central plants different benefits from decentralized systems (see Chapter 2).

Multiple types of equipment and fuel sources may be combined in one plant. The heating and cooling energy may be a combination of electricity, natural gas, oil, coal, solar, geothermal, etc. This energy is converted into chilled water, hot water, or steam that is distributed through the facility for air conditioning, heating, and processes. The operating, maintenance, and first costs of all these options should be discussed with the owner before final selection. When combining heating generation systems, it is important to note the presence of direct-firing combustion systems or chilled-water production systems using HFC or HCFC refrigerants, because the safety requirements in ASHRAE *Standard* 15 must be met.

A central plant can be customized without sacrificing the standardization, flexibility, and performance required to support the primary cooling and heating equipment through careful selection of ancillary equipment, automatic control, and facility management. Plant design varies widely based on building use, life-cycle costs, operating economies, and the need to maintain reliable building HVAC, process, and electrical systems. These systems can require more extensive engineering, equipment, and financial analysis than decentralized systems do.

In large buildings with interior areas that require cooling while perimeter areas require heating, one of several types of centralized heat reclaim units can meet both these requirements efficiently. Chapter 8 describes these combinations, and Chapters 12 to 14 give design details for central plants.

Central plants can be designed to accommodate both occupied/unoccupied and constant, year-round operation. Maintenance can be performed with traditional one-shift operating crews, but usually requires 24 h coverage. Higher-pressure steam boiler plants (usually greater than 15 psig) or combined cogeneration and steam heating plants require multiple-operator, 24 h shift coverage.

Advantages

- Primary cooling and heating can be provided at all times, independent of the operation mode of equipment and systems outside the central plant.
- Using larger but fewer pieces of equipment generally reduces the facility's overall operation and maintenance cost. It also allows wider operating ranges and more flexible operating sequences.
- A centralized location minimizes restrictions on servicing accessibility.
- Energy-efficient design strategies, energy recovery, thermal storage, and energy management can be simpler and more cost-effective to implement.
- Multiple energy sources can be applied to the central plant, providing flexibility and leverage when purchasing fuel.
- Standardizing equipment can be beneficial for redundancy and stocking replacement parts. However, strategically selecting different-sized equipment for a central plant can provide better part-load capability and efficiency.
- Standby capabilities (for firm capacity/redundancy) and back-up fuel sources can easily be added to equipment and plant when planned in advance.

The preparation of this chapter is assigned to TC 9.1, Large Building Air-Conditioning Systems.

- Equipment operation can be staged to match load profile and taken offline for maintenance.
- District cooling and heating can be provided.
- A central plant and its distribution can be economically expanded to accommodate future growth (e.g., adding new buildings to the service group).
- Load diversity can substantially reduce the total equipment capacity requirement.
- Submetering secondary distribution can allow individual billing of cooling and heating users outside the central plant.
- Major vibration and noise-producing equipment can be grouped away from occupied spaces, making acoustic and vibration controls simpler. Acoustical treatment can be applied in a single location instead of many separate locations.
- Issues such as cooling tower plume and plant emissions are centralized, allowing a more economic solution.

Disadvantages

- Equipment may not be readily available, resulting in long lead-time for production and delivery.
- Equipment may be more complicated than decentralized equipment, and thus require a more knowledgeable equipment operator.
- A central location within or adjacent to the building is needed.
- Additional equipment room height may be needed.
- Depending on the fuel source, large underground or surface storage tanks may be required on site. If coal is used, space for storage bunker(s) will be needed.
- Access may be needed for large deliveries of fuel (oil or coal).
- Heating plants require a chimney and possibly emission permits, monitoring, and treatments.
- Multiple equipment manufacturers are required when combining primary and ancillary equipment.
- System control logic may be complex.
- First costs can be higher.
- Special permitting may be required.
- Safety requirements are increased.
- A large pipe distribution system may be necessary (which may actually be an advantage for some applications).

DESIGN CONSIDERATIONS

Cooling and Heating Loads

Design cooling and heating loads are determined by considering individual and simultaneous loads. The simultaneous peak or instantaneous load for the entire portion or block of the building served by the HVAC and/or process load is less than the sum of the individual cooling and heating loads (e.g., buildings do not receive peak solar load on the east and west exposures at the same time). This difference between design load and peak load, called the **central equipment diversity factor**, can be as little as 5% less than the sum of individual loads (e.g., 95% diversity factor) or represent a more significant portion of the load (e.g., 45% diversity factor), as is common in academic campus applications. The peak central plant load can be based on this diversity factor, reducing the total installed equipment capacity needed to serve larger building cooling and heating loads. It is important for the design engineer to evaluate the full point-of-use load requirements of each facility served by the central system. Opportunities for improving energy efficiency include

- **Staging** multiple chillers or boilers for part-load operation. Using correctly sized equipment is imperative to accurately provide the most flexible and economical sequencing of equipment.
- For central chiller plants, consider incorporating **variable-frequency drives (VFDs)** onto at least one base-loaded primary chiller. Multiple VFD installations on chillers allow more flexibility in energy control of chiller plant operation.

Discrete loads (e.g., server rooms) are best served independently; a small, independent system designed for the discrete load may be the most cost-effective approach. Central plants sized for minimum part-load operation may not be able to reliably serve a discrete load. For example, a 2000 ton chiller plant serving multiple facilities could be selected with four 500 ton chillers. In a remote facility connected to the central distribution system, an independent computer room server has a year-round 5 ton load. Operating one 500 ton chiller at less than 100 tons reliably may be extremely inefficient and possibly detrimental to the chiller plant operation. Serving the constant remote load independently allows the designer the flexibility to evaluate the chiller plant as a whole and individual subsystems independently, so as to not adversely affect both systems.

Peak cooling load time is affected by outside ventilation, outside dry- and wet-bulb temperatures, period of occupancy, interior equipment heat gain, and relative amounts of north, east, south, and west exposures. For buildings with a balanced distribution of solar exposures, the peak usually occurs on a midsummer afternoon when the west solar load and outside wet-bulb temperature are at or near concurrent maximums. However, for buildings with much more solar exposure on one side than on another, the peak cooling period can change significantly with time of day and month.

The diversity of building occupancies served can significantly affect the peak cooling load diversity factor. For example, in a system serving an entire college campus, the peak cooling period for a classroom is different from that for a dormitory or an administration building. Special consideration for planning load profiles at academic facilities should be identified. Unlike office and residential applications, universities and colleges typically have peak cooling loads during late summer and fall.

Peak heating load has less opportunity to accommodate a diversity factor, so equipment is most likely to be selected on the sum of individual heating loads. This load may occur when the building must be warmed back up to a higher occupied space temperature after an unoccupied weekend setback period. Peak demand may also occur during unoccupied periods when the ambient environment is harshest and there is little internal heat gain to assist the heating system, or during occupied times if significant outside air must be preconditioned or some other process (e.g., process heating) requires significant heat. To accommodate part-load conditions and energy efficiency, variable-flow may be the best economical choice. It is important for the designer to evaluate plant operation and system use.

System Flow Design

The configuration of a central system is based on use and application. Two types of energy-efficient designs used today are primary variable flow and primary/secondary variable flow.

Primary variable flow uses variable flow through the production energy equipment (chiller or heating-water generator) and directly pumps the medium, usually water, to the point of use. Variable-flow can be achieved using two-way automatic control valves at terminal equipment and either variable-frequency drive (VFD) pumping (Figure 1) or distribution pressure control with bypass valve (Figure 2). Both concepts function based on maintaining system pressure, usually at the farthest point (last control valve and terminal unit) in the water system.

Primary/secondary variable flow hydraulically decouples the primary production system (chilled- or heating-water source), which is commonly constant flow. A variable-flow secondary piping system distributes the chilled or heating medium to the point of use (Figures 3 and 4).

Another design is a straight **constant-volume primary** system. Hydronic pumps distribute water through the energy equipment

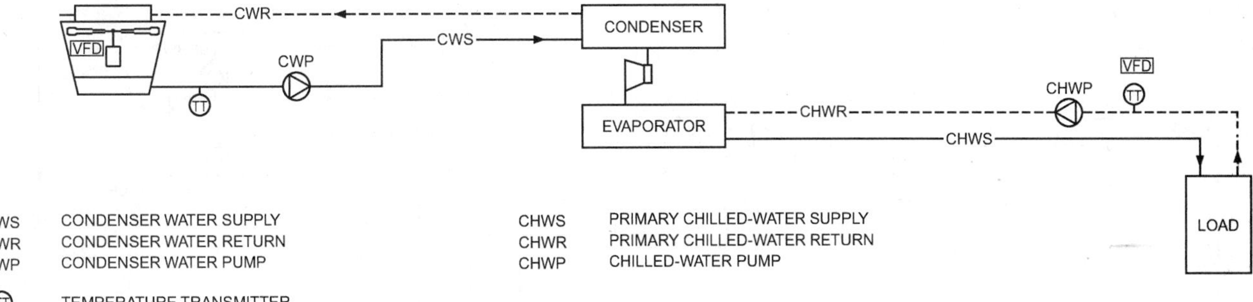

Fig. 1 Primary Variable-Flow System
(Courtesy RDK Engineers)

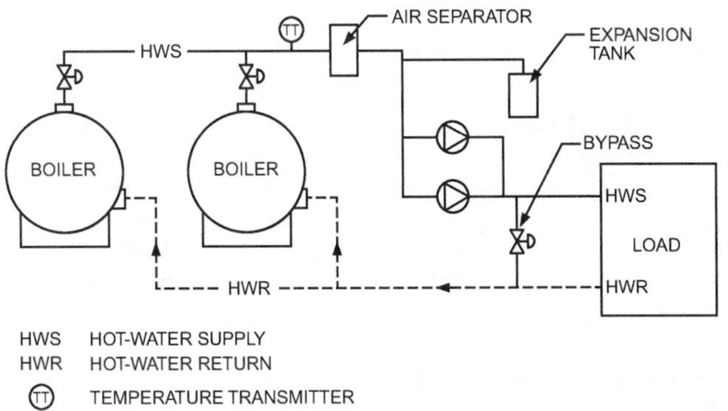

Fig. 2 Primary (Limited) Variable-Flow System Using Distribution Pressure Control
(Courtesy RDK Engineers)

straight to the point of use. Pumping energy is constant, as is the distribution flow, which generally requires a means to maintain the design flow rate through the system while in operation. Thus, these systems are generally more expensive to operate and less attractive in central system designs.

When using either primary/secondary or primary variable-flow designs, the engineer should understand the design differences between two- and three-way modulating valves. Variable-flow designs vary the flow of chilled or heating water through the distribution loop. As terminal units satisfy demand, the valve modulates toward the closed position. In the distribution system, the pump, if at design operating conditions, increases system pressure above the design point. To compensate, a VFD is typically installed on the distribution pump. This VFD is usually controlled by a pressure differential sensor, which may be located approximately two-thirds of the way down the piping loop, at the critical, farthest point in the system, to maintain the minimum pressure needed to provide design flow through the last terminal device and associated control valve. As the pressure differential increases, the sensor sends a signal to the VFD to reduce speed. The affinity laws then allow flow to reduce in direct proportion to the change in flow. As system demand requires increased flow, the control valve modulates open, reducing system differential pressure. The reduction in pressure difference measured causes a corresponding increase in pump speed (and therefore flow) to meet the plant demand. With a primary/secondary design, minimum system flow for the chiller or heating plant is accomplished by using a decoupler bypass across the primary and secondary systems. Because the primary system is constant volume (i.e., the system maintains design flow during operation as designed when production flow exceeds distribution flow), water recirculates within the plant to maintain the minimum design flow required by the production plant. Care should be taken during design to ensure minimum flow is achieved under part-load operation throughout the system. When selecting the distribution system pump, ensure distribution flow does not exceed production flow, which could cause a temperature drop if flow exceeds demand (**low ΔT syndrome**). The engineer should evaluate operational conditions at part load to minimize the potential for this to occur.

With primary variable-flow designs, the primary chilled- or heating-water pumps are of a variable-flow design, again using a VFD. Primary variable-flow systems use modulating two-way control valves to reduce water flow across each heat transfer device (as in the primary/secondary system). At minimum flow conditions, as required by the equipment and plant design, a way is needed to prevent flow from becoming laminar; typically, this is done with a minimum flow bypass around the plant.

With a straight constant-volume system, flow is constant at the design requirement. These systems commonly use three-way control valves, and throttle closed (as in a two-way valve) as terminal unit demand is satisfied. However, the three-way valve also has a bypass port to allow up to design flow to flow around the terminal heat transfer device and return to the central plant. With constant-volume designs, temperature swings typically occur across the plant and terminal heat exchange devices, because design flow does not change.

Energy Recovery and Thermal Storage

Depending on the operations schedule of the building(s) served, energy recovery and thermal storage strategies can be applied to a central cooling and heating plant. Water-to-water energy recovery systems are common and readily available. Thermal water or ice storage also can be very adaptable to central plants. See Chapter 25 in this volume and to Chapters 32 to 35 and 40 in the 2007 *ASHRAE Handbook—HVAC Applications* for more information on energy-related opportunities. Thermal energy storage (TES) systems may offer multiple strategies for both part-load and peak-load efficiency control. Examples such as base-loading plant operation for a more flat-line energy consumption profile may help in negotiating energy costs with a utility company supplier. Thermal energy storage of chilled water, ice, or heating water offers a medium for redundant capacity, with potential reductions in both heating and cooling infrastructure equipment sizing.

EQUIPMENT

Primary Refrigeration Equipment

Central cooling plant refrigeration equipment falls into two major categories: (1) vapor-compression refrigeration cycle (compressorized) chillers, and (2) absorption-cycle chillers. In some cases, the chiller plant may be a combination of these machines, all installed in the same plant. Cooling towers, air-cooled condensers, evaporative condensers, or some combination are also needed to reject heat from this equipment. The energy for the prime driver of

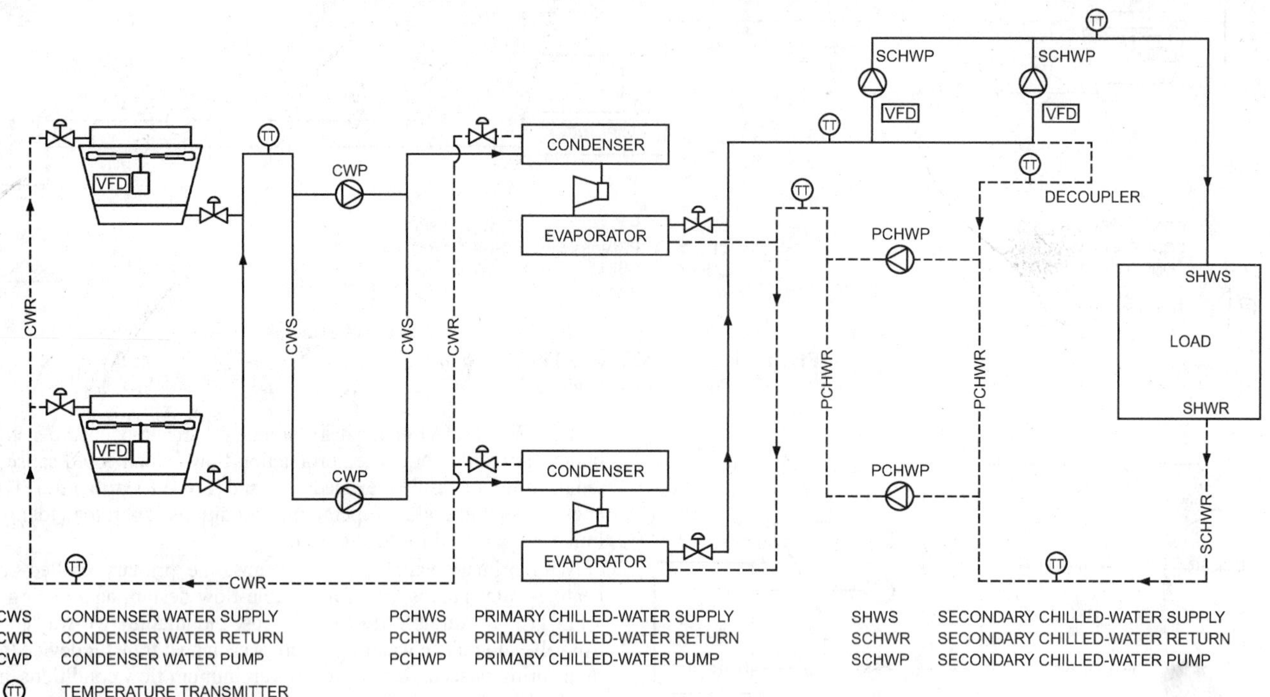

Fig. 3 Primary/Secondary Pumping Chilled-Water System
(Courtesy RDK Engineers)

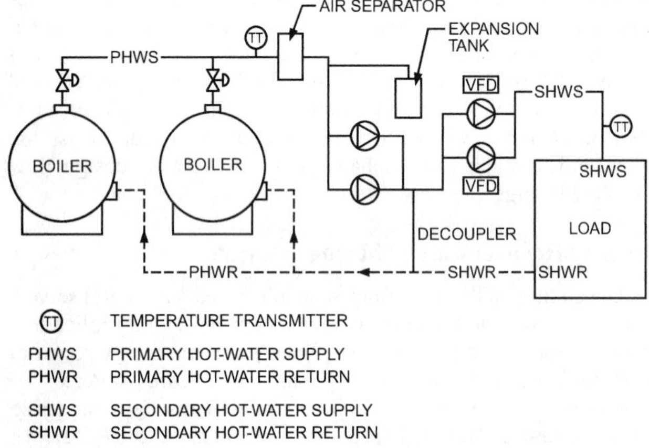

Fig. 4 Primary/Secondary Pumping Hot-Water System
(Courtesy RDK Engineers)

cooling equipment may also come from waste heat from a combined heat and power (CHP) system. Typically, electricity production efficiency is improved when waste heat can be reclaimed and transferred to a heat source. A heat recovery generator supplying either steam or hot water to a driver (e.g., a steam turbine or absorption-cycle chiller) allows a facility to track a thermal load from the heat rejection of electric generation, which can be an attractive economic model.

Chapter 42 of this volume and Chapters 41 and 43 of the 2006 *ASHRAE Handbook—Refrigeration* discuss refrigeration equipment, including the size ranges of typical equipment.

Compressorized chillers feature reciprocating, helical rotary (screw), and centrifugal compressors, which may be driven by electric motors; natural gas-, diesel-, or oil-fired internal combustion engines; combustion turbines; or steam turbines.

Compressors may be purchased as part of a refrigeration chiller that also includes the drive, evaporator, condenser, and necessary safety and operating controls. Reciprocating and helical rotary compressor units can be field-assembled and include air- or water-cooled (evaporative) condensers arranged for remote installation. Centrifugal compressors are usually included in packaged chillers, although they can be very large and sometimes require field erection. These types of chillers also require remote air- or water-cooled ancillary equipment. Chapter 37 has information about compressors.

Absorption chillers may be single- or double-effect, fired by steam or direct-fired by gas, oil, or waste heat. Like centrifugal chillers, absorption chillers are built to perform with remote water-cooled ancillary heat rejection equipment (e.g., cooling towers). These absorption chillers use a lithium bromide/water cycle in which water is the refrigerant, and are generally available in the following configurations: (1) natural gas direct-fired, (2) indirect-generated by low-pressure steam or high-temperature water, (3) indirect-generated by high-pressure steam or hot water, and (4) indirect-generated by hot exhaust gas. Chapter 41 of the 2006 *ASHRAE Handbook—Refrigeration* discusses absorption air-conditioning and refrigeration equipment in more detail.

Ancillary Refrigeration Equipment

Ancillary equipment for central cooling plants consists primarily of heat-rejection equipment (air-cooled condensers, evaporative condensers, and cooling towers), pumps (primary, secondary, and tertiary), and heat exchangers (water-to-water), as well as water pumps and possibly heat exchanger(s). For more detailed information on this additional equipment, see Chapters 25, 38, 39, and 41 to 46.

Air-cooled condensers pass outside air over a dry coil to condense the refrigerant. This results in a higher condensing temperature and thus a larger power input at peak condition (although peak time may be relatively short over 24 h). Air-cooled condensers are popular in small reciprocating and helical rotary compressor units because of their low maintenance requirements.

Evaporative condensers pass outside air over coils sprayed with water, thus taking advantage of adiabatic saturation to lower the condensing temperature. As with cooling towers, freeze prevention and close control of water treatment are required for successful operation. The lower power consumption of the refrigeration system and much smaller footprint of the evaporative condenser are gained at the expense of the cost of water and water treatment used and increased maintenance cost.

Cooling towers provide the same means of heat rejection as evaporative condensers but pass outside air through an open condenser return water spray to achieve similar adiabatic cooling performance. Either natural or mechanical-draft cooling towers or spray ponds can be used; the mechanical-draft tower (forced-draft, induced-draft, or ejector) can be most easily designed for most conditions because it does not depend on wind. Cooling tower types and sizes range from packaged units to field-erected towers with multiple cells in unlimited sizes. Location of heat rejection equipment should consider issues such as reingestion or short-circuiting of discharge heat rejection air because of location in a confined area, tower plume, and the effect of drift on adjacent roadways, buildings, and parking lots.

Makeup water subtraction meters should be evaluated for both evaporative condensers and cooling towers by the design engineer or owner. In many areas, sewage costs are part of the overall water bill, and most domestic water supplied to a typical facility goes into the sewage system. However, evaporated condenser water is not drained to sewer, and if a utility grade meter is used, potential savings in plant operating costs can be made available to the owner. Metering should be evaluated with chilled-water plants, especially those in operation year-round. Consult with the water supplier to identify compliance requirements. Consider piping cooling tower water blowdown and drainage (e.g., to remove dissolved solids or allow maintenance when installing subtraction meters) to the storm drainage system, but note that the environmental effects of the water chemistry must be evaluated. Where chemical treatment conditions do not meet environmental outfall requirements, drainage to the sanitary system may still be required unless prefiltration can be incorporated.

Water pumps move both chilled and condenser water to and from the refrigeration equipment and associated ancillary equipment. See Chapter 43 for additional information on centrifugal pumps, and Chapters 11 to 13 for system design.

Heat exchangers provide operational and energy recovery opportunities for central cooling plants. Operational opportunities generally involve heat transfer between building systems that must be kept separated because of different pressures, media, or other characteristics. For example, heat exchangers can thermally link a low-pressure, low-rise building system with a high-pressure, high-rise building system. Heat exchangers can also transfer heat between chemically treated or open, contaminated water systems and closed, clean water systems (e.g., between a central cooling plant chilled-water system and a high-purified, process cooling water system, or between potentially dirty pond water and a closed-loop condenser water system).

To conserve energy, water-to-water heat exchangers can provide water-side economizer opportunities. When outside conditions allow, condenser water can cool chilled water through a heat exchanger, using the cooling tower and pumps rather than a compressor. This approach should be considered when year-round chilled water is needed to satisfy a process load, or when an air-side economizer is not applicable to the air distribution systems of connected facilities in a central system distribution loop.

Plant Controls. Direct digital control (DDC) systems should be considered for control accuracy and reliability. Temperature, flow, and energy use are best measured and controlled with modern DDC technology. Pneumatic control should be considered where torque or rapid response of power actuation is required; medium pressure (typically 30 to 60 psig) is best. Programmable logic controllers (PLCs) should be considered for larger central plants or where future expansion/growth is possible.

Codes and Standards. Specific code requirements and standards apply when designing central cooling plants. For cooling equipment, refer to ASHRAE *Standard* 15; Chapter 51 of this volume provides a comprehensive list of codes and standards associated with cooling plant design, installation, and operation. Manufacturers' recommendations and federal, state, and local codes and standards should also be followed.

Primary Heating Equipment

The major heating equipment used in central heating plants, **boilers** vary in type and application, and include combined heat and power (CHP) and waste heat boilers. Chapter 31 discusses boilers in detail, including the size ranges of typical equipment.

A boiler adds heat to the **working medium**, which is then distributed throughout the building(s) and/or campus. The working medium may be either water or steam, which can further be classified by its temperature and pressure range. Steam, often used to transport energy long distances, is converted to low-temperature hot water in a heat exchanger near the point of use. Although steam is an acceptable medium for heat transfer, low-temperature hot water is the most common and more uniform medium for providing heating and process heat (e.g., heating water to 200°F for domestic hot water). Elevated steam pressures and high-temperature hot-water boilers are also used. Both hot-water and steam boilers have the same type of construction criteria, based on temperature and pressure.

A boiler may be purchased as a package that includes the burner, fire chamber, heat exchanger section, flue gas passage, fuel train, and necessary safety and operating controls. Cast-iron and water-side boilers can be field-assembled, but fire-tube, scotch marine, and waste-heat boilers are usually packaged units.

Energy Sources. The energy used by a boiler may be electricity, natural gas, oil, coal, or combustible waste material, though natural gas and fossil-fuel oil (No. 2, 4, or 6 grade) are most common, either alone or in combination. Selecting a fuel source requires detailed analysis of energy prices and availability (e.g., natural gas and primary electrical power). The availability of fuel oil or coal delivery affects road access and storage for deliveries. Security around a central plant must be considered in the design and include standby electrical power generation and production of central cooling and heating for critical applications in times of outages or crisis.

Energy for heating may also come from waste heat from a CHP system. Typically, electricity production efficiency is improved when waste heat can be reclaimed and transferred to a heat source. A heat recovery generator converting the heat byproduct of electric generation to steam or hot water to meet a facility's heating needs when the thermal load meets or exceeds the heat rejection of the electric generation can be a cost-effective plant operation.

Additionally, attention to the design of an efficient and maintainable **condensate return** system for a central steam distribution plant is important. A well-designed condensate return system increases overall efficiencies and reduce both chemical and makeup water use.

Codes and Standards. Specific code requirements and standards apply when designing central heating boiler plants. It is important to note that some applications (e.g., high-pressure boilers) require continuous attendance by licensed operators. Operating cost considerations should be included in determining such applications. Numerous codes, standards, and manufacturers' recommendations need to be followed. Refer to Chapter 51 for a comprehensive list of codes and standards associated with this equipment, plant design, installation, and operation.

Ancillary Heating Equipment

Steam Plants. Ancillary equipment associated with central steam heating plants consists primarily of the boiler feed unit, deaerator unit (both with receiver and pumps), chemical feed, and possibly a surge tank and/or heat exchanger(s).

Boiler feed equipment, including the receiver(s) and associated pump(s), serve as a reservoir for condensate and makeup water waiting to be used by the steam boiler. The boiler feed pump provides system condensate and water makeup back to the boiler on an as-needed basis.

Deaerators help eliminate oxygen and carbon dioxide from the feed water (see Chapter 48 of the 2007 *ASHRAE Handbook—HVAC Applications* for more information).

Chemicals can be fed using several methods or a combination of methods, depending on the chemical(s) used (e.g., chelants, amines, oxygen scavengers). Continuous-feed pumps are the preferred and most reliable method for high-pressure steam systems.

Surge tanks are also applicable to steam boiler plants to accommodate large quantities of condenser water return and are primarily used where there is a rapid demand for steam (e.g., morning start-up of a central heating plant).

See Chapter 10 for more information on steam systems.

Hot-Water Plants. Ancillary equipment associated with central hot-water heating plants consists primarily of pumps and possibly heat exchanger(s). Water pumps move boiler feed water and hot-water supply and return to and from the boiler equipment and associated ancillary equipment. See Chapter 43 for additional information on centrifugal pumps, as well as Chapters 10 to 12 and Chapter 14 for system design.

Heat Exchangers. Heat exchangers offer operational and energy recovery opportunities for central heating plants. In addition to the heat exchange and system separation opportunities described in the section on Refrigeration Equipment, the operational opportunities for heat exchangers involve combining steam heating capabilities with hot-water heating capabilities.

Air-to-water and water-to-water heat exchangers provide opportunities for economizing and heat recovery in a central heating plant (e.g., flue gas exhaust heat recovery and boiler blowdown heat recovery).

For more detailed information on ancillary equipment, see Chapters 30 to 32 and Chapter 34.

DISTRIBUTION SYSTEMS

The major piping in a central cooling plant may include, but is not limited to, chilled-water, condenser water, city water, natural gas, fuel oil, refrigerant, vent, and drainpipe systems. For a central steam heating plant, the major piping includes steam supply, condensate return, pumped condensate, boiler feed, city water, natural gas, fuel oil, vent, and drainpipe systems. For a central hot-water heating plant, it includes hot-water supply and return, city water, natural gas, fuel oil, vent, and drainpipe systems. In the 2005 *ASHRAE Handbook—Fundamentals*, see Chapter 35 for information on sizing pipes, and Chapter 37 for identification, color-coding, abbreviations, and symbols for piping systems.

Design selection of cooling and heating temperature set points (supply and return water) can affect first and operating costs. For water systems with a large temperature difference between supply and return water, the resulting flow can allow smaller pipe sizing and smaller valves, fittings, and insulation, which can lower installation cost. However, these savings maybe offset by the larger coils and heat exchangers at the point of use needed to accomplish the same heat transfer. A similar design strategy can be achieved with steam pressure differential.

Determining the optimum cooling and heating water supply and return temperatures requires design consideration of equipment performance, particularly the energy required to produce the supply water temperature. Although end users set water temperatures, the colder the cooling water, the more energy is needed by the chiller. Similar issues affect hot-water supply temperature and steam operating pressure. When designing for energy management, supply water reset may be considered when peak capacity is not needed, potentially reducing energy consumption. This reduces chiller power input, but those savings may be offset by increased pump power input because of a higher water flow rate with higher supply temperature. Energy implications on the whole system must be considered. The design engineer should consider using higher temperature differences between supply and return to reduce the pump energy required by the distribution system, for both heating and chilled-water systems. Additionally, central plant production systems (e.g., boilers, steam-to-heating-water converters, chillers) operate more efficiently with higher return water temperatures. During conceptual planning and design, or as early in design as possible, the engineer should evaluate the type(s) of existing facilities to which the central system will be connected. When using variable-flow distribution with a constant design temperature split between the supply and return medium, it is critical to avoid low ΔT syndrome (see the section on System Characteristics). The engineer should attempt to ensure the condition is minimized or avoided, or at minimum perform due diligence to make the owner aware of the potential shortfalls where it is allowed to occur. With existing constant-flow/variable-temperature systems, measures should be taken to avoid a loss of the conceptual design strategy of a variable-flow/ constant-temperature split. Examples of connection strategies less costly than full renovation of a connected facility are (1) return recirculation control, to maintain a design temperature across a facility connection, or (2) separation of primary distribution supply from the secondary facility connection by a plate-and-frame heat exchanger with a control valve on the primary return controlled by the secondary supply to the facility. If the temperature split is allowed to fall, increased flow of the medium through the distribution system will be required to accomplish the same capacity. This increase in flow increases pump horsepower and chiller plant energy, compromising the available capacity and operation of the system.

Hydraulically modeling the cooling and heating media (chilled water, heating water, steam, domestic water, natural gas, etc.) should be considered. With an emphasis on centralizing the source of cooling and heating, a performance template can be created for large plants by computerized profiling of cooling and heating delivery. Hydraulic modeling provides the economic benefits of predesigning the built-out complete system before installing the initial phase of the project. The model also helps troubleshoot existing systems, select pumps, project energy usage, and develop operation strategy.

Energy conservation and management can best be achieved with computerized design and facility management resources to simulate delivery and then monitor and measure the actual distribution performance.

ACOUSTIC, VIBRATION, AND SEISMIC CONSIDERATIONS

Sound and Vibration

Proper space planning is a key to sound and vibration control in central plant design. For example, central plants are frequently located at or below grade. This provides a very stable platform for vibration isolation and greatly reduces the likelihood of vibration being transmitted into the occupied structure, where it can be regenerated as noise. Also, locating the central ventilation louvers or other openings well distant from noise- and vibration-sensitive areas greatly reduces the potential for problems in those areas. Louvers should be installed to prevent unwanted ambient air or water from being entrained into a facility. Louvers should be installed high enough to prevent security breaches (above ground level, where possible). As a guide, maintain free area around louvers as specified in ASHRAE *Standard* 15.

Vibration and noise transmitted both into the space served by the plant and to neighboring buildings and areas should be considered in determining how much acoustical treatment is appropriate for the design, especially if the plant is located near sensitive spaces such as conference rooms or sleeping quarters. See the section on Space Considerations for further discussion of space planning.

Acoustical considerations must also be considered for equipment outside the central plant. For example, roof-mounted cooling tower fans sometimes transmit significant vibration to the building structure and generate ambient noise. Many communities limit machinery sound pressure levels at the property line, which affects the design and placement of equipment. See Chapter 47 of the 2007 *ASHRAE Handbook—HVAC Applications* for more detailed information on sound and vibration.

Seismic Issues

Depending on code requirements and the facility's location with respect to seismic fault lines, seismic bracing may be required for the central plant equipment and distribution. For instance, a hospital located in a seismically active area must be able to remain open and operational to treat casualties in the aftermath of an earthquake. Most building codes require that measures such as anchors and bracing be applied to the HVAC system. Refer to the local authority responsible for requirements, and to Chapter 54 of the 2007 *ASHRAE Handbook—HVAC Applications* for design guidance.

SPACE CONSIDERATIONS

In the very early phases of building design, architects, owners, and space planners often ask the engineer to estimate how much building space will be needed for mechanical equipment. The type of mechanical system selected, building configuration, and other variables govern the space required, and many experienced engineers have developed rules of thumb to estimate the building space needed. Although few buildings are identical in design and concept, some basic criteria apply to most buildings and help approximate final space allocation requirements. These space requirements are often expressed as a percentage of the total building floor area; the combined mechanical and electrical space requirement of most buildings is 6 to 9%.

A central system reduces mechanical space requirements for each connected building. Where a central plant is remote or standalone, space requirements for equipment and operation are 100% for the plant. In these cases, it may be typical to have space requirements of 2.5 to 3.5 ft^2 per ton of refrigeration.

Space for chillers, pumps, and towers should not only include installation footprints but should also account for adequate clearance to perform routine and major maintenance. Generally, 4 ft service clearance (or the equipment manufacturer's minimum required clearance, whichever is greater) around equipment for operator maintenance and service is sufficient. For chillers, one end of the chiller barrels should be provided with free space the length of the evaporator and condenser barrels, to allow for tube pull clearance. In many cases, designers provide service bay roll-up doors or ventilation louvers (if winter conditions do not cause freeze damage issues) to allow tube access. Overhead service height is also required, especially where chillers are installed. Provision for 20 ft ceilings in a central plant is not uncommon, to accommodate piping and service clearance dimensions.

Plant designs incorporating steam supply from a separate central boiler plant may use steam-to-heating-water converters and the heating distribution equipment (e.g., pumps) along with the chilled-water production equipment and distribution infrastructure, increasing the general space to 4 to 6 ft^2 per ton. Where a boiler installation as well as heating distribution equipment and appurtenances are required, the plant's physical size increases to account for the type of boiler and required exhaust emissions treatment. Generally, for central heating plants, steam or heating water and chilled-water production systems are separated during design, which may further increase the overall footprint of the plant.

The arrangement and strategic location of the mechanical spaces during planning affects the percentage of space required. For example, the relationship between outside air intakes and loading docks, exhaust, and other contaminating sources should be considered during architectural planning. The final mechanical room size, orientation, and location are established after discussion with the architect and owner. The design engineer should keep the architect, owner, and facility engineer informed, whenever possible, about the HVAC analysis and system selection. Space criteria should satisfy both the architect and the owner or owner's representative, though this often requires some compromise. The design engineer should strive to understand the owner's needs and desires and the architect's vision for the building, while fully explaining the advantages, disadvantages, risks, and rewards of various options for mechanical and electrical room size, orientation, and location. All systems should be coordinated during the space-planning stage to safely and effectively operate and maintain the central cooling and heating plant.

In addition, the mechanical engineer sometimes must represent other engineering disciplines in central plant space planning. If so, it is important for the engineer to understand the basics of electrical and plumbing central plant equipment. The main electrical transformer and switchgear rooms should be located as close to the incoming electrical service as practical. The main electric transformers and switchgear for the plant and the mechanical equipment switchgear panels should be in separate rooms that only authorized electricians can enter. If there is an emergency generator, it should be located considering (1) proximity to emergency electrical loads, (2) sources of combustion and cooling air, (3) fuel sources, (4) ease of properly venting exhaust gases outside, and (5) provisions for noise control.

The main plumbing equipment usually contains gas and domestic water meters, the domestic hot-water system, the fire protection system, and elements such as compressed air, special gases, and vacuum, ejector, and sump pumps. Some water and gas utilities require a remote outside meter location.

The heating and air-conditioning equipment room houses the (1) boiler, pressure-reducing station, or both; (2) refrigeration machines, including chilled-water and condensing-water pumps; (3) converters for furnishing hot or cold water for air conditioning; (4) control air compressors, if any; (5) vacuum and condensate pumps; and (6) miscellaneous equipment. For both chillers and boilers, especially in a centralized application, full access is needed on all sides (including overhead) for extended operation, maintenance, annual inspections of tubes, and tube replacement and repair. Where appropriate, the designer should include overhead structural elements to allow safe rigging of equipment. Ideally, large central facilities should include overhead cranes or gantries. It is critical that local codes and ASHRAE *Standard* 15 are consulted for special equipment room requirements. Many jurisdictions require monitoring, alarms, evacuation procedures, separating refrigeration and fuel-fired equipment, and high rates of purge ventilation.

A proper operating environment for equipment and those maintaining it must also be provided. This may involve heating or cooling the space for freeze protection and to prevent overheating motors and controls. HVAC equipment serving the central plant itself may be housed in its own equipment room and serve the chiller, boiler room, and adjacent rooms (e.g., switchgear room, office space, generator room, pump room, machine shop). Where ventilation and maximum temperature limits are not defined in code requirements, a ventilation rate of 0.5 cfm/ft^2 and a maximum temperature rise of 122°F are recommended (ASHRAE 2007). Where refrigeration equipment is installed, follow design requirements in ASHRAE *Standard* 15.

Location of Central Plant and Equipment

Although large central plants are most often located at or below grade, it is often economical to locate the refrigeration plant at the top of the building, on the roof, or on intermediate floors. For central plants, on-grade should be the first choice, followed by below grade. The designer must always ensure access for maintenance and replacement. Locating major equipment in roof penthouses may pose significant access problems for repair and replacement. For single high-rise central plants, intermediate-floor locations that are closer to the load may allow pumping equipment to operate at a lower pressure. A life-cycle cost analysis (LCCA) should include differences in plant location to identify the most attractive plant sustainability options during the planning stage. The LCCA should consider equipment maintenance access, repair, and replacement during the life of the plant, as determined in the owner's criteria. If not identified, an engineer should suggest to the owner or client that maintenance criteria be included as a component in developing the LCCA (see the section on Operations and Maintenance Considerations). Electrical service and structural costs are greater when intermediate-floor instead of ground-level locations are used, but may be offset by reduced energy consumption and condenser and chilled-water piping costs. The boiler plant may also be placed on the roof, eliminating the need for a chimney through the building.

Benefits of locating the air-cooled or evaporative condenser and/or cooling tower on the ground versus the roof should be evaluated. Personnel safety, security, ambient noise, and contamination from hazardous water vapors are some of the considerations that help determine final equipment location. Also, structural requirements (e.g., steel to support roof-mounted equipment, or a concrete pad and structural steel needed to locate equipment at or near grade) require evaluation. When locating a cooling tower at or near grade, the net positive suction head and overflow of condenser water out of the cooling tower sump should be studied if the tower is below, at, or slightly above the level of the chiller.

Numerous variables should be considered when determining the optimum location of a central cooling or heating plant. When locating the plant, consider the following:

- Operating weight of the equipment and its effect on structural costs
- Vibration from primary and ancillary equipment and its effect on adjacent spaces in any direction
- Noise level from primary and ancillary equipment and its effect on adjacent spaces in any direction
- Location of electrical utilities for the central plant room, including primary electric service and associated switchgear and motor control center, as well as electrical transformer location and its entrance into the building
- Location of city water and fire pump room (it may be desirable to consolidate these systems near the central plant room)
- Accessibility into the area and clearances around equipment for employee access, equipment and material delivery, and major equipment replacement, repair, scheduled teardown, and rigging
- Location of cooling, refrigerant relief piping, heating, vents, and boiler flue and stack distribution out of the central plant and into the building, along with the flow path of possible vented hazardous chemical, steam, or combustion exhaust products
- Need for shafts to provide vertical distribution of cooling and heating services in the building
- Future expansion plans of the central plant (e.g., oversizing the central plant now for adding more primary equipment later, based on master planning of the facility)
- Architectural effect on the site
- Location of boiler chimney
- Loading dock for materials and supplies
- Roadway and parking considerations
- Storage of fuel oil, propane gas, and/or coal
- Electrical transformer location
- Underground and/or overhead utility and central cooling and heating system distribution around the central plant and to the building(s)
- Wind effect on cooling tower plume or other volatile discharges such as refrigerant or boiler emissions.

Central Plant Security

Restricted access and proper location of exposed intakes and vents must be designed into the central plant layout to protect the facility from attack and protect people from injury. Care must be taken in locating exposed equipment, vents, and intakes, especially at ground level. Above ground and at least 10 to 15 ft from access to intake face is preferred. When this is not possible, fencing around exposed equipment, such as cooling towers and central plant intakes, should be kept locked at all times to prevent unauthorized access. Ensure that fencing is open to airflow so it does not adversely affect equipment performance. Air intakes should be located above street level if possible, and vents should be directed so they cannot discharge directly on passing pedestrians or into an air intake of the same or an adjacent facility.

AUTOMATIC CONTROLS AND BUILDING MANAGEMENT SYSTEMS

One advantage of central cooling and heating plants is easier implementation of building automation because the major and ancillary equipment is consolidated in one location. Computerized automatic controls can significantly affect system performance. A facility management system to monitor system points and overall system performance should be considered for any large, complex air-conditioning system. This allows a single operator to monitor performance at many points in a building and make adjustments to increase occupant comfort and to free maintenance staff for other duties. Chapter 46 of the 2007 *ASHRAE Handbook—HVAC Applications* describes design and application of controls.

Software to consider when designing, managing, and improving central plant performance should include the following:

- Automatic controls that can interface with other control software (e.g., equipment manufacturers' unit-mounted controls)
- Energy management system (EMS) control
- Hydraulic modeling, as well as metering and monitoring of distribution systems
- Using VFDs on equipment (e.g., chillers, pumps, cooling tower fans) to improve system control and to control energy to consume only the energy required to meet the design parameter (e.g., temperature, flow, pressure).
- Computer-aided facility management (CAFM) for integrating other software (e.g., record drawings, operation and maintenance manuals, asset database)
- Computerized maintenance management software (CMMS)
- Automation from other trades (e.g., fire alarm, life safety, medical gases, etc.)
- Regulatory functions (e.g., refrigerant management, federal, state and local agencies, etc.)

Automatic controls for central cooling and/or heating plants may include standard equipment manufacturer's control logic along with optional, enhanced energy-efficiency control logic. These specialized control systems can be based on different architectures such as distributed controls, programmable logic controllers, or microprocessor-based systems. Beyond standard control technology, the following control points and strategies may be needed for primary equipment, ancillary equipment, and the overall system:

- Discharge temperature and/or pressure
- Return distribution medium temperature
- Head pressure for refrigerant and/or distribution medium
- Stack temperature

- Carbon monoxide and/or carbon dioxide level
- Differential pressures
- Flow rate of distribution medium
- Peak and hourly refrigeration output
- Peak and hourly heat energy output
- Peak and hourly steam output
- Flow rate of fuel(s)
- Reset control of temperature and/or pressure
- Night setback
- Economizer cycle
- Variable flow through equipment and/or system control
- Variable-frequency drive control
- Thermal storage control
- Heat recovery cycle

See Chapter 41 of the 2007 *ASHRAE Handbook—HVAC Applications* for more information on control strategies and optimization.

The **coefficient of performance (COP)** for the entire chilled-water plant can be monitored and allows the plant operator to determine the overall operating efficiency of a plant. Central plant COP can be expressed in the following terms:

- Annual heating or refrigeration per unit of building area (Btu per hour per square foot per year)
- Energy used per unit of refrigeration (kilowatt-hours per ton)
- Annual power required per unit of refrigeration per unit of building area (kilowatts per ton per square foot per year)

Instrumentation

All instrument operations where cooling or heating output are measured should have instrumentation calibration that is traceable to the National Institute of Standards and Technology (NIST).

The importance of local gages and indicating devices, with or without a facility management system, should not be overlooked. All equipment must have adequate pressure gages, thermometers, flow meters, balancing devices, and dampers for effective performance, monitoring, and commissioning. In addition, capped thermometer wells, gage cocks, capped duct openings, and volume dampers should be installed at strategic points for system balancing. Chapter 37 of the 2007 *ASHRAE Handbook—HVAC Applications* indicates the locations and types of fittings required. Chapter 14 of the 2005 *ASHRAE Handbook—Fundamentals* has more information on measurement and instruments.

MAINTENANCE MANAGEMENT SYSTEMS

A review with the end user (owner) should be done, to understand the owner's requirements for operation (e.g., if an owner has an in-house staff, more frequent access may be required, which may affect the extent to which a designer incorporates access). Reviewing ASHRAE *Guideline* 4 and *Standard* 15 with the owner can provide insight to the development of a maintenance plan. If maintenance is outsourced as required, extensive access may not be as high a priority. In some cases, regulatory and code access may be the only determining factors.

Operations and maintenance considerations include the following:

- Accessibility around equipment, as well as above and below when applicable, with minimum clearances per manufacturers' recommendations and applicable codes
- Clearances for equipment removal
- Minimizing tripping hazards (e.g., drain piping extending along the floor)
- Adequate headroom to avoid injuries
- Trenching in floor, if necessary
- Cable trays, if applicable
- Adequate lighting levels
- Task lighting, when needed
- Eyewash stations for safety
- Exterior access for outside air supply and for exhaust
- Storage of mechanical and electrical parts and materials
- Documentation storage and administrative support rooms
- Proper drainage for system maintenance
- Outlets for service maintenance utilities (e.g., water, electricity) in locations reasonably accessible to equipment operators
- Adequate lines of sight to view thermometers, pressure gages, etc.
- Structural steel elements for major maintenance rigging of equipment

Typical operator maintenance functions include cleaning of condensers, evaporators, and boiler tubes. Motor electrical testing such as annual megohm testing should be considered. Cooling tower catwalk safety railing and ladders should be provided to comply with U.S. Occupational Safety and Health Administration (OSHA) requirements.

BUILDING SYSTEM COMMISSIONING

Because a central plant consumes a major portion of the annual energy operating budget, building system commissioning is imperative for new construction and expansion of existing installations. During the warranty phase, central-plant performance should be measured, benchmarked, and course-corrected to ensure design intent is achieved. If an energy analysis study is performed as part of the comparison between decentralized and centralized concepts, and/or life-cycle comparison of the study is part of a Leadership in Energy and Environmental Design (LEED®) project, the resulting month-to-month energy data should be a good electronic document to benchmark actual energy consumption using the measurement and verification plan implementation.

Ongoing commissioning or periodic recommissioning further ensures that the design intent is met, and that cooling and heating are reliably delivered after the warranty phase. Many central systems are designed and built with the intent to connect to facilities in a phased program, which also requires commissioning. Retro- or re-commissioning should be considered whenever the plant is expanded or an additional connection made to the existing systems, to ensure the original design intent is met.

Initial testing, adjusting, and balancing (TAB) also contribute to sustainable operation and maintenance. The TAB process should be repeated periodically to ensure levels are maintained.

When completing TAB and commissioning, consider posting laminated system flow diagrams at or adjacent to the central cooling and heating equipment, indicating operating instructions, TAB performance, commissioning functional performance tests, and emergency shutoff procedures. These documents also should be filed electronically in the central plant computer server for quick reference.

Original basis of design and design criteria should be posted as a constant reminder of design intent, and to be readily available in case troubleshooting, expansion, or modernization is needed.

As with all HVAC applications, for a maintainable, long-term design success, building commissioning should include the system training requirements necessary for building management staff to efficiently take ownership and operate and maintain the HVAC systems over the useful service life of the installation.

REFERENCES

ASHRAE. 1993. Preparation of operating and maintenance documentation for building systems. *Guideline* 4-1993.

ASHRAE. 2007. Safety standard for refrigeration systems. ANSI/ASHRAE *Standard* 15-2007.

CHAPTER 4

AIR HANDLING AND DISTRIBUTION

VERY early in the design of a new or retrofit building project, the HVAC design engineer must analyze and ultimately select the basic systems, as discussed in Chapter 1, and whether production of primary heating and cooling should be decentralized (see Chapter 2) or central (see Chapter 3). This chapter covers the options, processes, available equipment, and challenges of all-air systems; for all-water, air-and-water, and local terminal systems, see Chapter 5.

Building air systems can be designed to provide complete sensible and latent cooling, preheating, and humidification capacity in air supplied by the system. No additional cooling or humidification is then required at the zone, except for certain industrial systems. Heating may be accomplished by the same airstream, either in the central system or at a particular zone. In some applications, heating is accomplished by a separate heat source. The term *zone* implies the provision of, or the need for, separate thermostatic control, whereas the term *room* implies a partitioned area that may or may not require separate control.

The basic all-air system concept is to supply air to the room at conditions such that the sensible and latent heat gains in the space, when absorbed by supply air flowing through the space, bring the air to the desired room conditions. Because heat gains in the space vary with time, a mechanism to vary the energy removed from the space by the supply air is necessary. There are two such basic mechanisms: (1) vary the amount of supply air delivered to the space by varying the flow rate or supplying air intermittently; or (2) vary the temperature of air delivered to the space, either by modulating the temperature or conditioning the air intermittently.

All-air systems may be adapted to many applications for comfort or process work. They are used in buildings of all sizes that require individual control of multiple zones, such as office buildings, schools and universities, laboratories, hospitals, stores, hotels, and even ships. All-air systems are also used virtually exclusively in special applications for close control of temperature, humidity, space pressure, and/or air quality, including cleanrooms, computer rooms, hospital operating rooms, research and development facilities, and many industrial/manufacturing facilities.

Advantages

- Operation and maintenance of major equipment can be performed in an unoccupied area (e.g., a central mechanical room). It also maximizes choices of filtration equipment, vibration and noise control, humidification options, and the selection of high-quality and durable equipment.
- Piping, electrical equipment, wiring, filters, and vibration- and noise-producing equipment are away from the conditioned area, minimizing (1) disruption for service needs and (2) potential harm to occupants, furnishings, and processes.
- These systems offer the greatest potential for using outside air for economizer cooling instead of mechanical refrigeration.
- Seasonal changeover is simple and adapts readily to automatic control.
- A wide choice of zoning, flexibility, and humidity control under all operating conditions is possible. Simultaneous heating of one zone and cooling of another zone during off-season periods is available.
- Air-to-air and other heat recovery may be readily incorporated.
- Designs are flexible for optimum air distribution, draft control, and adaptability to varying local requirements.
- The systems are well-suited to applications requiring unusual exhaust or makeup air quantities (negative or positive pressurization, etc.).
- All-air systems adapt well to winter humidification.
- All-air systems take advantage of load diversity. In other words, a central air-handling unit serving multiple zones needs to be sized only for the peak coincident load, not the sum of the peak loads of each individual zone. In buildings with significant fenestration loads, diversity can be significant, because the sun cannot shine on all sides of a building simultaneously.
- By increasing the air change rate and using high-quality controls, these systems can maintain the closest operating condition of ±0.25°F dry bulb and ±0.5% rh. Today, some systems can maintain essentially constant space conditions.
- Removal and disposal of cold condensate from cooling coils, and capture and return of steam condensate from heating coils, is generally simpler and more practical in an all-air system.

Disadvantages

- Ducts installed in ceiling plenums require additional duct clearance, sometimes reducing ceiling height and/or increasing building height. In retrofits, these clearances may not be available.
- Larger floor plans may be necessary to allow adequate space for vertical shafts (if required for air distribution). In a retrofit application, shafts may be impractical.
- In commercial buildings, air-handling equipment rooms represent nonrentable or non-revenue-generating spaces.
- Accessibility to terminal devices, duct-balancing dampers, etc., requires close cooperation between architectural, mechanical, and structural designers.
- Air balancing, particularly on large systems, can be cumbersome.
- Permanent heating is not always available sufficiently early to provide temporary heat during construction.
- Mechanical failure of a central air-handling component, such as a fan or a cooling-coil control valve, affects all zones served by that unit.

The preparation of this chapter is assigned to TC 9.1, Large Building Air-Conditioning Systems.

Heating and Cooling Calculations

Basic calculations for airflow, temperatures, relative humidity, loads, and psychrometrics are covered in Chapters 6 and 30 of the 2005 *ASHRAE Handbook—Fundamentals*. System selection should be related to the need, as indicated by the load characteristics. The designer should understand the operation of system components, their relationship to the psychrometric chart, and their interaction under various operating conditions and system configurations. The design engineer must properly determine an air-handling system's required supply air temperature and volume; outside air requirements; desired space pressures; heating and cooling coil capacities; humidification and dehumidification capacities; return, relief, and exhaust air volume requirements; and required pressure capabilities of the fan(s).

The HVAC designer should work closely with the architect to optimize the building envelope design. Close cooperation of all parties during design can result in reduced building loads, which allows the use of smaller mechanical systems.

Zoning

Exterior zones are affected by weather conditions (e.g., wind, temperature, sun) and, depending on the geographic area and season, may require both heating and cooling at different times. The system must respond to these variations. The need for separate perimeter zone heating is determined by the following:

- Severity of heating load (i.e., geographic location)
- Nature and orientation of building envelope
- Effects of downdraft at windows and radiant effect of cold glass surfaces (i.e., type of glass, area, height, U-factor)
- Type of occupancy (i.e., sedentary versus transient).
- Operating costs (i.e., in buildings such as offices and schools that are unoccupied for considerable periods, fan operating cost can be reduced by heating with perimeter radiation during unoccupied periods rather than operating the main or local unit supply fans.)

Separate perimeter heating can operate with any all-air system. However, its greatest application has been in conjunction with VAV systems for cooling-only service. Careful design must minimize simultaneous heating and cooling. See the section on Variable Air Volume for further details.

Interior spaces have relatively constant conditions because they are isolated from external influences. Cooling loads in interior zones may vary with changes in the operation of equipment and appliances in the space and changes in occupancy, but usually interior spaces require cooling throughout the year. A VAV system has limited energy advantages for interior spaces, but it does provide simple temperature control. Interior spaces with a roof exposure, however, may require treatment similar to perimeter spaces that require heat.

Space Heating

Although steam is an acceptable medium for central system preheat or reheat coils, low-temperature hot water provides a simple and more uniform means of perimeter and general space heating. Individual automatic control of each terminal provides the ideal space comfort. A control system that varies water temperature inversely with the change in outside temperature provides water temperatures that produce acceptable results in most applications. For best results, the most satisfactory ratio can be set after installation is completed and actual operating conditions are ascertained.

Multiple perimeter spaces on one exposure served by a central system may be heated by supplying warm air from the central system. Areas with heat gain from lights and occupants and no heat loss require cooling in winter, as well as in summer. In some systems, very little heating of return and outside air is required when the space is occupied. Local codes dictate the amount of outside air required (see ASHRAE *Standard* 62.1 for recommended optimum outside air ventilation). For example, with return air at 75°F and outside air at 0°F, the temperature of a 25% outside/75% return air mixture would be 56°F, which is close to the temperature of air supplied to cool such a space in summer. In this instance, a preheat coil installed in the minimum outside airstream to warm outside air can produce overheating, unless it is sized so that it does not heat the air above 35 to 40°F. Assuming good mixing, a preheat coil in the mixed airstream prevents this problem. The outside air damper should be kept closed until room temperatures are reached during warm-up. A return air thermostat can terminate warm-up.

When a central air-handling unit supplies both perimeter and interior spaces, supply air must be cool to handle interior zones. Additional control is needed to heat perimeter spaces properly. Reheating the air is the simplest solution, but is not acceptable under most energy codes. An acceptable solution is to vary the volume of air to the perimeter and to combine it with a terminal heating coil or a separate perimeter heating system, either baseboard, overhead air heating, or a fan-powered terminal unit with supplemental heat. The perimeter heating should be individually controlled and integrated with the cooling control. Lowering the supply water temperature when less heat is required generally improves temperature control. For further information, refer to Chapter 12 in this volume and Chapter 46 of the 2007 *ASHRAE Handbook—HVAC Applications*.

Air Temperature Versus Air Quantity

Designers have considerable flexibility in selecting supply air temperature and corresponding air quantity within the limitations of the procedures for determining heating and cooling loads. The difference between supply air temperature and desired room temperature is often referred to as the ΔT of the all-air system. The relationship between ΔT and air quantity (by volume) is approximately linear and inverse: doubling the ΔT results in halving of the air quantity. ASHRAE *Standard* 55 addresses the effect of these variables on comfort.

The traditional all-air system is typically designed to deliver approximately 55°F supply air, for a conventional building with a desired indoor temperature of approximately 75°F. That supply air temperature is commonplace because the air is low enough in absolute moisture to result in reasonable space relative humidity in conventional buildings with modest latent heat loads. However, lower supply air temperatures may be required in spaces with high latent loads, such as gymnasiums or laundries, and higher supply air temperatures can be applied selectively with caution. Obviously, not all buildings are conventional or typical, and designers are expected not to rely on these conventions unquestioningly. Commercially available load calculation software programs, when applied correctly, help the designer find the optimum supply air temperature for each application.

In cold air-systems, the supply air temperature is designed significantly lower than 55°F (perhaps as low as 44°F) in an effort to reduce the size of ducts and fans. In establishing supply air temperature, the initial cost of lower airflow and low air temperature (smaller fan and duct systems) must be calculated against potential problems of distribution, condensation, air movement, and decreased removal of odors and gaseous or particulate contaminants. Terminal devices that use low-temperature air can reduce the air distribution cost. These devices mix room and primary air to maintain reasonable air movement in the occupied space. Because the amount of outside air needed is the same for any system, the percentage in low-temperature systems is high, requiring special care in design to avoid freezing preheat or cooling coils.

Advantages of cold-air systems include lower humidity levels in the building, because colder air has a lower maximum absolute moisture content, and reduced fan energy consumption. However, these low-temperature air supply systems might actually increase overall building energy consumption, because the cold-air process strips more moisture from the air (i.e., greater latent heat removal) than is otherwise required in comfort applications. Again, commercially available software can help the designer evaluate the overall energy effects of these decisions.

Space Pressure

Designers faced with the need to provide space pressure control along with temperature, humidity, and/or air filtration control will most likely find that all-air systems are the only systems capable of achieving this pressure control. Many special applications, such as isolation rooms, research labs, and cleanrooms, require constant-volume supply and exhaust air to the space to ensure space pressure control. Some of these applications allow the designer to consider reduced air volume during unoccupied periods while still maintaining space pressure control.

Variable-air-volume (VAV) space pressure control is another way to ensure constant space pressure. In addition to individual rooms with fixed positive or negative pressure, entire areas and/or floors may require space pressure control, with individual rooms set for one condition and adjacent areas and/or corridors with opposite space pressure control, so that the entire area is air-balanced to a recommended, slightly positive pressure. With each of these applications, the testing, adjusting, and balancing firm plays an important role in the commissioning process.

Other Considerations

All-air systems operate by maintaining a temperature differential between the supply air and the space. Any load that affects this differential and the associated airflow must be calculated and considered, including the following:

- All **fans** (supply, return, and supplemental) add heat. All of the fan shaft power eventually converts to heat in the system, either initially as fan losses or downstream as duct friction losses. Motor inefficiencies are an added load if that motor is in the airstream. Whether the fan is upstream of the cooling coil (blow-through) or downstream (draw-though) affects how this load must be accounted for. The effect of these gains can be considerable, particularly in process applications. Heat gain in medium-pressure systems is about 0.5°F per inch of water static pressure.
- The **supply duct** may gain or lose heat from the surroundings. Most energy codes require that the supply duct be insulated, which is usually good practice regardless of code requirements. Uninsulated supply ducts delivering cool air are subject to condensation formation, leading to building water damage and potential mold growth, depending on the dew-point temperature of surrounding air.
- Controlling **humidity** in a space can affect the air quantity and become the controlling factor in selecting supply airflow rate. VAV systems provide only limited humidity control, so if humidity is critical, extra care must be taken in design.

First, Operating, and Maintenance Costs

As with all systems, the initial cost of an air-handling system varies widely (even for identical systems), depending on location, condition of the local economy, and contractor preference. For example, a dual-duct system is more expensive because it requires essentially twice the amount of material for ducts as that of a comparable single-duct system. Systems requiring extensive use of terminal units are also comparatively expensive. The operating cost depends on the system selected, the designer's skill in selecting and correctly sizing components, efficiency of the duct design, and effect of building design and type on the operation. All-air systems can greatly minimize operating cost.

Because an all-air system separates the air-handling equipment from occupied space, maintenance on major components in a central location is more economical. Also, central air-handling equipment requires less maintenance than a similar total capacity of multiple small packaged units. The many terminal units used in an all-air system do, however, require periodic maintenance. Because these units (including reheat coils) are usually installed throughout a facility, maintenance costs for these devices must be considered.

Energy

The engineer's early involvement in the design of any facility can considerably affect the building's energy consumption. Careful design minimizes system energy costs. In practice, however, a system might be selected based on a low first cost or to perform a particular task. In general, single-duct systems may consume less energy than dual-duct systems, and VAV systems are more energy-efficient than constant-air-volume systems. Savings from a VAV system come from the savings in fan power and because the system does not overheat or overcool spaces, and reheat is minimized.

The air distribution system for an all-air system consists of two major subsystems: (1) air-handling units that generate conditioned air under sufficient positive pressure to circulate it to and from the conditioned space, and (2) a distribution system that only carries air from the air-handling unit to the space being conditioned. The air distribution subsystem often includes means to control the amount or temperature of air delivered to each space.

AIR-HANDLING UNITS

The basic air-handling system is an all-air, single-zone HVAC system consisting of an air-handling unit and an air distribution system. The air-handling unit may be designed to supply constant or variable air volume for low-, medium-, or high-velocity air distribution. Normally, the equipment is located outside the conditioned area in a basement, penthouse, or service area. It can, however, be installed in the area if conditions permit. The equipment can be adjacent to the primary heating and refrigeration equipment or at considerable distance, with refrigerant, chilled water, hot water, or steam circulated to it for energy transfer.

Figure 1 shows a typical draw-through central system that supplies conditioned air to a single zone. A blow-through configuration may also be used if space or other conditions dictate. The quantity and quality of supplied air are fixed by space requirements and determined as indicated in Chapters 29 and 30 of the 2005 *ASHRAE Handbook—Fundamentals*. Air gains and loses heat by contacting heat transfer surfaces and by mixing with air of another condition. Some of this mixing is intentional, as at the outside air intake; other mixing results from the physical characteristics of a particular component, as when untreated air passes through a coil without contacting the fins (bypass factor).

All treated and untreated air must be well mixed for maximum performance of heat transfer surfaces and for uniform temperatures

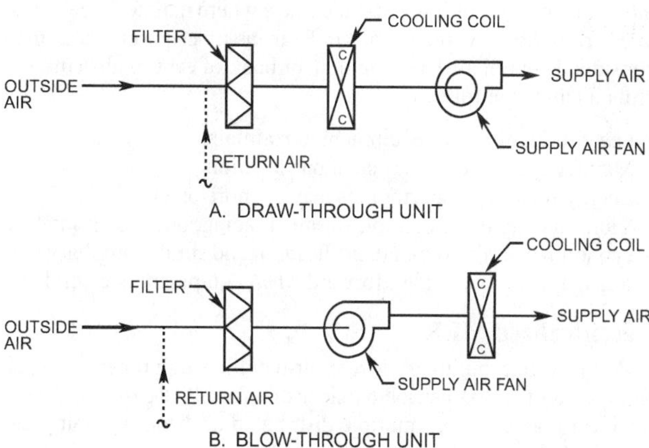

Fig. 1 Typical Air-Handling Unit Configurations
(Courtesy RDK Engineers)

in the airstream. Stratified, parallel paths of treated and untreated air must be avoided, particularly in the vertical plane of systems using double-inlet or multiple-wheel fans. Because these fans may not completely mix the air, different temperatures can occur in branches coming from opposite sides of the supply duct.

Primary Equipment

Cooling. Either central station or localized equipment, depending on the application, can provide cooling. Most large systems with multiple central air-handling units use a central refrigeration plant. Small, individual air-handling equipment can (1) be supplied with chilled water from central chillers, (2) use direct-expansion cooling with a central condensing (cooling tower) system, or (3) be air-cooled and totally self-contained. The decision to provide a central plant or local equipment is based on factors similar to those for air-handling equipment, and is addressed in Chapters 1 to 3.

Heating. The same criteria described for cooling are usually used to determine whether a central heating plant or a local one is desirable. Usually, a central, fuel-fired plant is more desirable for heating large facilities. In small facilities, electric heating is a viable option and is often economical, particularly where care has been taken to design energy-efficient systems and buildings.

Air-Handling Equipment

Packaged air-handling equipment is commercially available in many sizes, capacities, and configurations using any desired method of cooling, heating, humidification, filtration, etc. These systems can be suitable for small and large buildings. In large systems (over 50,000 cfm), air-handling equipment is usually custom-designed and fabricated to suit a particular application. Air handlers may be either centrally or remotely located.

Air-handling units (AHUs) can be one of the more complicated pieces of equipment to specify or order, because a vast array of choices are available, and because there is no single-number identifier (e.g., a "50 ton unit" or a "40,000 cfm unit") that adequately describes the desired product. Regardless of size or type, the designer must properly determine an air-handling unit's required supply air temperature and volume; outside air requirements; desired space pressures; heating and cooling coil capacities; humidification and dehumidification capacities; return, relief, and exhaust air volume requirements; and required pressure capabilities of the fan(s). Typically, these parameters and more must be specified or scheduled by the design engineer before an installer or equipment supplier can provide an AHU.

Central Mechanical Equipment Rooms (MERs)

The type of facility and other factors help determine where the air-handling equipment is located. Central fan rooms are more common in laboratory or industrial facilities, where maintenance is isolated from the conditioned space. Reasons a design engineer may consider a central air-handling unit, or bank of central air-handling units, include the following:

- Fewer total pieces of equipment to maintain
- Maintenance is concentrated at one location
- Energy recovery opportunities may be more practical
- Vibration and noise control, seismic bracing, outside air intakes, economizers, filtration, humidification, and similar auxiliary factors may be more straightforward when equipment is centralized

Decentralized MERs

Many office buildings locate air-handling equipment at each floor, or at other logical subdivisions of a facility. Reasons a design engineer may consider multiple distributed air-handling unit locations include the following:

- Reduced size of ducts reduces space required for distribution ductwork and shafts.
- Reduced equipment size as a result of decentralized systems allows use of less expensive packaged equipment and reduces the necessary sophistication of training for operating and maintenance personnel.
- For facilities with varied occupancy, multiple decentralized air-handling units can be set back or turned off in unoccupied areas.
- Failure of an air-handling unit affects only the part of the building served by that one unit.

Fans

Both packaged and built-up air-handling units can use any type of fan. Centrifugal fans may be forward-curved, backward-inclined, or airfoil, and single-width/single-inlet (SWSI) or double-width/double-inlet (DWDI). Many packaged air-handlers feature a single fan, but it is possible for packaged and custom air handlers to use multiple DWDI centrifugal fans on a common shaft with a single drive motor. SWSI centrifugal plug fans without a scroll are sometimes used on larger packaged air handlers to make them more compact. Vaneaxial fans, both adjustable- and variable-pitch during operation, are often used on very large air-handling units. Fan selection should be based on efficiency and sound power level throughout the anticipated range of operation, as well as on the ability of the fan to provide the required flow at the anticipated static pressure. Chapter 20 further discusses fans and fan selection.

AIR-HANDLING UNIT PSYCHROMETRIC PROCESSES

Cooling

The basic methods used for cooling and dehumidification include the following:

- **Direct expansion** (refrigerant) takes advantage of the latent heat of the refrigerant fluid, and cools as shown in the psychrometric diagram in Figure 2.
- **Chilled-water** (fluid-filled) coils use temperature differences between the fluid and air to exchange energy by the same process as in Figure 2 (see the section on Dehumidification).
- **Direct spray of water in the airstream** (Figure 3), an adiabatic process, uses the latent heat of evaporation of water to reduce dry-bulb temperature while increasing moisture content. Both sensible and latent cooling are also possible by using chilled water. A conventional evaporative cooler uses the adiabatic process by

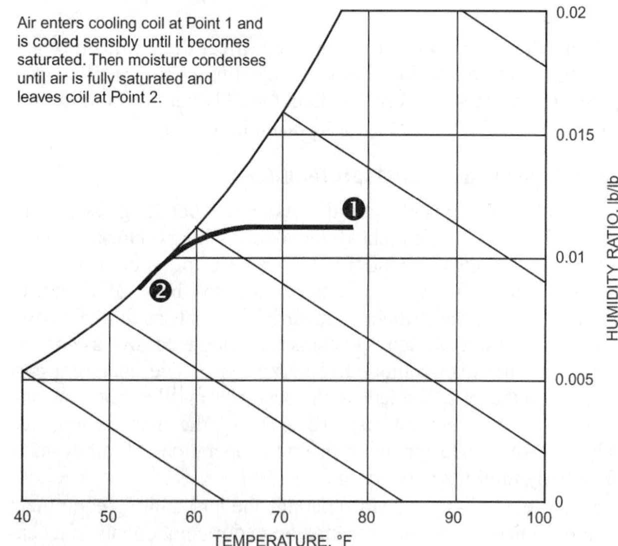
Air enters cooling coil at Point 1 and is cooled sensibly until it becomes saturated. Then moisture condenses until air is fully saturated and leaves coil at Point 2.
1
2
0.02
0.015
0.01
0.005
0
HUMIDITY RATIO, lb/lb
40
50
60
70
80
90
100
TEMPERATURE, °F

Fig. 2 Direct-Expansion or Chilled-Water Cooling and Dehumidification

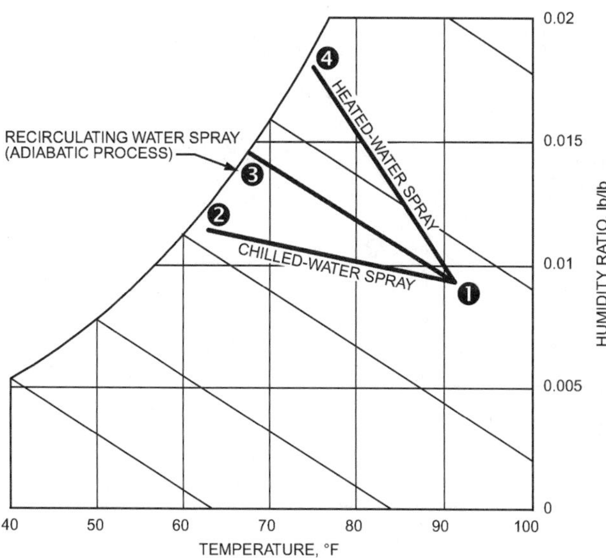

Fig. 3 Direct Spray of Water in Airstream Cooling

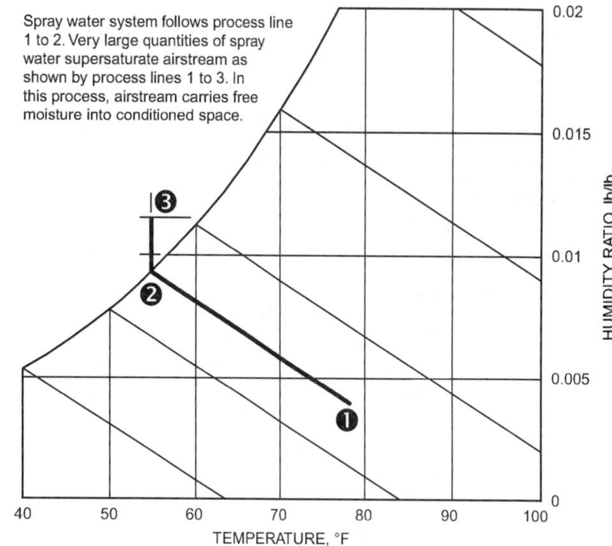

Fig. 4 Supersaturated Evaporative Cooling

spraying or dripping recirculated water onto a filter pad (see the section on Humidification).

- The **wetted duct** or **supersaturated** system is a variation of direct spray. In this system, tiny droplets of free moisture are carried by the air into the conditioned space, where they evaporate and provide additional cooling, reducing the amount of air needed for cooling the space (Figure 4).
- **Indirect evaporation** adiabatically cools outside or exhaust air from the conditioned space by spraying water, then passes that cooled air through one side of a heat exchanger. Air to be supplied to the space is cooled by passing through the other side of the heat exchanger. Chapter 40 has further information on this method of cooling.

Chapter 6 of the 2005 *ASHRAE Handbook—Fundamentals* details the psychrometric process of these methods.

Heating

The basic methods used for heating include the following:

- **Steam** uses the latent heat of the fluid.
- **Hot-water** (fluid-filled) coils use temperature differences between the warm fluid and the cooler air.
- **Electric heat** also uses the temperature difference between the heating coil and the air to exchange energy.
- **Direct or indirect gas- or oil-fired heat exchangers** can also be used to add sensible heat to the airstream.

The effect on the airstream for each of these processes is the same and is shown in Figure 5. For basic equations, refer to Chapter 6 of the 2005 *ASHRAE Handbook—Fundamentals.*

Humidification

Methods used to humidify air include the following:

- **Direct spray of recirculated water** into the airstream (air washer) reduces the dry-bulb temperature while maintaining an almost constant wet bulb, in an adiabatic process (see Figure 3, path 1 to 3). The air may also be cooled and dehumidified, or heated and humidified, by changing the spray water temperature.
  In one variation, the surface area of water exposed to the air is increased by spraying water onto a cooling/heating coil. The coil surface temperature determines leaving air conditions. Another method is to spray or distribute water over a porous medium, such as those in evaporative coolers. This method requires careful monitoring of the water condition to keep biological contaminants from the airstream (Figure 6).

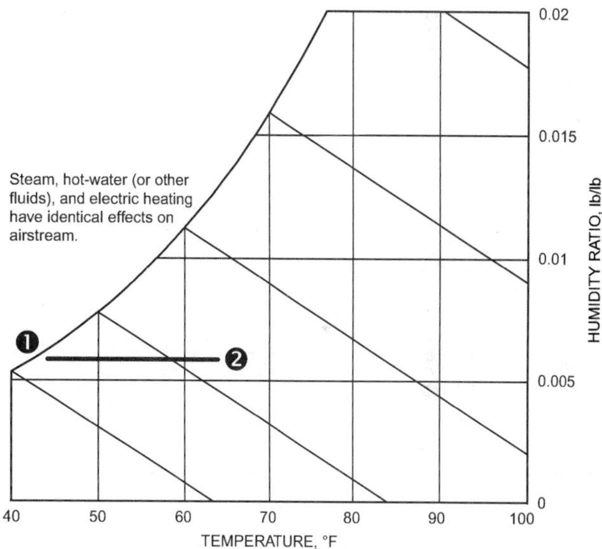

Fig. 5 Steam, Hot-Water, and Electric Heating, and Direct and Indirect Gas- and Oil-Fired Heat Exchangers

- **Compressed air** that forces water through a nozzle into the airstream is essentially a constant wet-bulb (adiabatic) process. The water must be treated to keep particles from entering the airstream and contaminating or coating equipment and furnishings. Many types of nozzles are available.
- **Steam injection** is a constant dry-bulb process (Figure 7). As the steam injected becomes superheated, the leaving dry-bulb temperature increases. If live steam is injected into the airstream, the boiler water treatment chemical must be nontoxic to occupants and nondamaging to building interior and furnishings.

Dehumidification

Moisture condenses on a cooling coil when its surface temperature is below the air's dew point, thus reducing the humidity of the

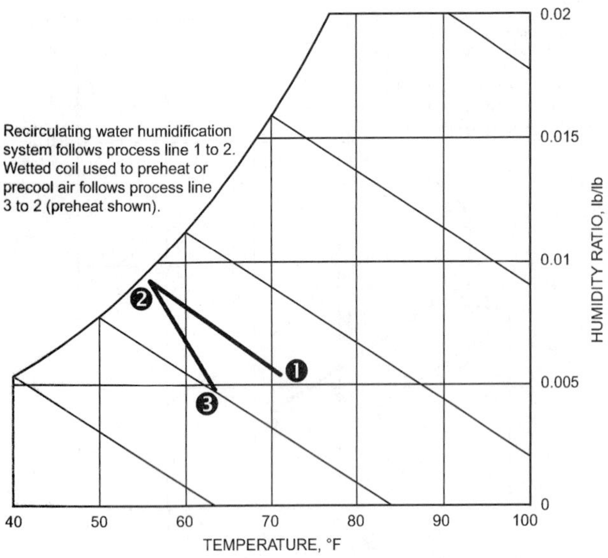

Fig. 6 Direct Spray Humidification

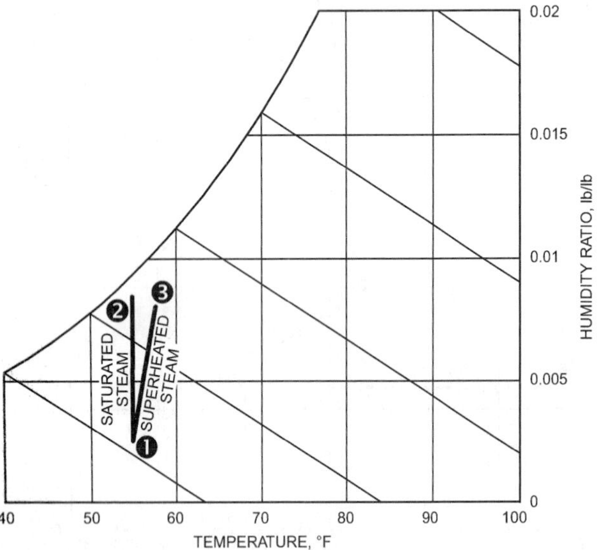

Fig. 7 Steam Injection Humidification

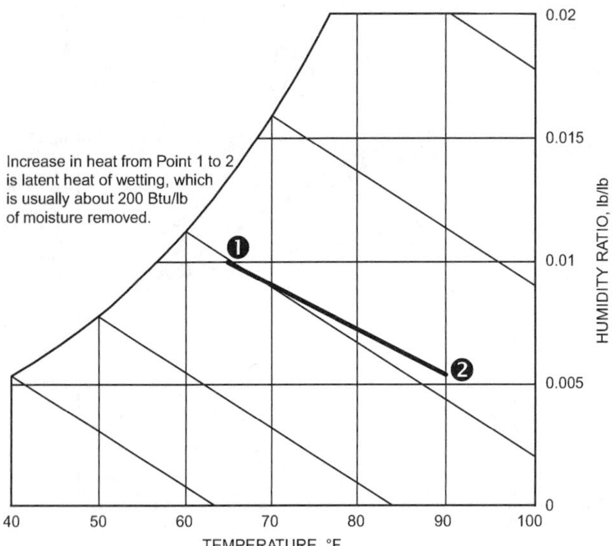

Fig. 8 Chemical Dehumidification

air. Similarly, air will also be dehumidified if a fluid with a temperature below the airstream dew point is sprayed into the airstream (see the section on Air Washers in Chapter 40). The process is identical to that shown in Figure 2, except that the moisture condensed from the airstream condenses on, and dissolves in, the spray droplets instead of on the solid coil surface.

Chemical dehumidification involves either passing air over a solid desiccant or spraying the air with a solution of desiccant and water. Both of these processes add heat, often called the **latent heat of wetting**, to the air being dehumidified. Usually about 200 Btu/lb of moisture is removed (Figure 8). These systems should be reviewed with the user to ensure that the space is not contaminated. Chapter 23 has more information on this topic.

Air Mixing or Blending

Adiabatic mixing of two or more airstreams (e.g., outside and return air) into a common airstream can be shown on a psychrometric chart with reasonable accuracy (see Chapter 6 of the 2005 *ASHRAE Handbook—Fundamentals*).

AIR-HANDLING UNIT COMPONENTS

The following sections describe many commonly available air-handling unit components. Not all of these components will necessarily be used in any one system.

To determine the system's air-handling requirement, the designer must consider the function and physical characteristics of the space to be conditioned, and the air volume and thermal exchange capacities required. Then, the various components may be selected and arranged by considering the fundamental requirements of the central system.

Figure 1 shows one possible general arrangement of air-handling unit components for a single-zone, all-air central system suitable for year-round air conditioning. This arrangement allows close control of temperature and humidity. Although Figure 1 indicates a built-up system, most of these components are available from many manufacturers completely assembled or in subassembled sections that can be bolted together in the field. When selecting central system components, specific design parameters must be evaluated to balance cost, controllability, operating expense, maintenance, noise, and space. The sizing and selection of primary air-handling units substantially affect the results obtained in the conditioned space.

The equipment must be adequate, accessible for easy maintenance, and not overly complex in its arrangement and control to provide the required conditions. Further, the designer should consider economics in component selection. Both initial and operating costs affect design decisions. For example, the designer should not arbitrarily design for a 500 fpm face velocity, which has been common for selecting cooling coils and other components. Filter and coil selection at 300 to 400 fpm, with resultant lower pressure loss, could produce a substantial payback on constant-volume systems. Chapter 36 of the 2007 *ASHRAE Handbook—HVAC Applications* has further information on energy and life-cycle costs.

Return Air Fan

A return air fan is optional on small systems, but is essential for proper operation of air economizer systems for free cooling from outside air if the return path has a significant pressure drop (greater than approximately 0.3 in. of water). It provides a positive return and exhaust from the conditioned area, particularly when mixing dampers allow cooling with outside air in intermediate seasons and winter. The return air fan ensures that the proper volume of air returns from the conditioned space. It prevents excess building pressure when economizer cycles introduce more than the minimum quantity of outside air, and reduces the static pressure against which the supply fan must work. The return air fan should handle a slightly smaller amount of air to account for fixed exhaust systems, such as toilet exhaust, and to ensure a slight positive pressure in the conditioned space. Chapter 46 of the 2007 *ASHRAE Handbook—HVAC Applications* provides design details.

Relief Air Fan

In many situations, a relief (or exhaust) air fan may be used instead of a return fan. A relief air fan relieves ventilation air introduced during air economizer operation and operates only when this control cycle is in effect. When a relief air fan is used, the supply fan must be designed for the total supply and return pressure losses in the system. During economizer mode, the relief fan must be controlled to ensure a slight positive pressure in the conditioned space, as with the return air fan system. The section on Economizers describes the required control for relief air fans.

Automatic Dampers

The section on Mixing Plenums discusses conditions that must be considered when choosing, sizing, and locating automatic dampers for this critical mixing process. These dampers throttle the air with parallel- or opposed-blade rotation. These two forms of dampers have different airflow throttling characteristics (see Chapter 15 of the 2005 *ASHRAE Handbook—Fundamentals*). Pressure relationships between various sections of this mixing process must be considered to ensure that automatic dampers are properly sized for wide-open and modulating pressure drops. See ASHRAE *Guideline* 16 for additional detail.

Relief Openings

Relief openings in large buildings should be constructed similarly to outside air intakes, but may require motorized or self-acting backdraft dampers to prevent high wind pressure or stack action from causing airflow to reverse when the automatic dampers are open. Pressure loss through relief openings should be 0.10 in. of water or less. Low-leakage dampers, such as those for outside intakes, prevent rattling and minimize leakage. The relief air opening should be located so that air does not short-circuit to the outside air intake.

Return Air Dampers

Negative pressure in the outside air intake plenum is a function of the resistance or static pressure loss through the outside air louvers, damper, and duct. Positive pressure in the relief air plenum is likewise a function of the static pressure loss through the relief damper, the relief duct between the plenum and outside, and the relief louver. The pressure drop through the return air damper must accommodate the pressure difference between the positive-pressure relief air plenum and the negative-pressure outside air plenum. Proper sizing of this damper facilitates better control and mixing. An additional manual damper may be required for proper air balancing.

Outside Air Intakes

Resistance through outside air intakes varies widely, depending on construction. Frequently, architectural considerations dictate the type and style of louver. The designer should ensure that the louvers selected offer minimum pressure loss, preferably not more than 0.10 in. of water. High-efficiency, low-pressure louvers that effectively limit carryover of rain are available. Flashing installed at the outside wall and weep holes or a floor drain will carry away rain and melted snow entering the intake. Cold regions may require a snow baffle to direct fine snow to a low-velocity area below the dampers. Outside air dampers should be low-leakage types with special gasketed edges and endseals. A damper section and damper operator are strongly recommended for ensuring minimum ventilation. The maximum outside air damper controls the air needed for economizer cycles.

The location of intake and exhaust louvers should be carefully considered; in some jurisdictions, location is governed by codes. For example, if heat recovery devices are used, intake and exhaust airstreams may need to be run in parallel, such as through air-to-air plate heat exchangers. Louvers must be separated enough to avoid short-circuiting air. Furthermore, intake louvers should not be near a potential source of contaminated air, such as a boiler stack or hood exhaust. Relief air should also not interfere with other systems.

A common complaint in buildings is a lack of outside air. This is especially a concern in VAV systems in which outside air quantities are established for peak loads and are then reduced in proportion to the air supplied during periods of reduced load. A simple control added to the outside-air damper can eliminate this problem and keep the amount of outside air constant, regardless of VAV system operation. However, the need to preheat outside air must be considered if this control is added.

Also, although some codes require as little as 5 cfm per person (about 0.05 cfm/ft^2) of outside air, this amount may be too low for a building using modern construction materials. Higher outside air quantities may be required to reduce odors, volatile organic compounds (VOCs), and potentially dangerous pollutants. ASHRAE *Standard* 62.1 provides information on ventilation for acceptable indoor air quality. Air quality (i.e., control or reduction of contaminants such as VOCs, formaldehyde from furnishings, and dust) must be reviewed by the engineer.

Economizers

An air-side economizer uses outside air to reduce refrigeration requirements. Whereas a logic circuit maintains a fixed minimum of ventilation outside air in all weather, the air-side economizer is an attractive option for reducing energy costs when weather conditions allow. The air-side economizer takes advantage of cool outside air either to assist mechanical cooling or, if outside air is cool enough, to provide total system cooling. When weather permits, temperature controls systems can modulate outside air and return air in the correct proportion to produce desired supply air temperatures without the use of mechanical heating or cooling.

To exhaust the extra outside air brought in by the economizer, a method of variable-volume relief must be provided. The relief volume may be controlled by modulating the relief air dampers in response to building space pressure. Another common approach is opening the relief/exhaust and outside air intake dampers simultaneously, although this alone does not address space pressurization. A powered relief or return/relief fan may also be used. The relief system is off and relief dampers are closed when the air-side economizer is inactive.

For details on advantages and disadvantages of air-side economizers, see Chapter 2.

Mixing Plenums

Mixing plenums provide space for airstreams with different properties to mix as they are introduced into a common section of ductwork or air-handling unit, allowing the system to operate as intended. If the airstreams are not sufficiently mixed, the resulting stratification adversely affects system performance. Some problems associated with stratification are nuisance low-temperature safety cutouts, frozen cooling coils, excess energy use by the preheat coil, inadequate outside air, control hunting, and poor outside air distribution throughout occupied spaces.

A common example of a mixing plenum is the air-handling unit mixing box, in which outside and recirculated airstreams are combined. In air-handling units, mixing boxes typically have one inlet, with control dampers, for each airstream.

There are no performance standards for mixing boxes or mixing plenums. Thus, it is difficult to know whether a particular mixing box design will provide sufficient mixing. In the absence of performance data, many rules of thumb have been developed to increase the mixing provided by mixing boxes. It is important to note that few supporting data exist; the following suggestions are based largely on common-sense solutions and anecdotal evidence:

- The minimum outside air damper should be located as close as possible to the return air damper.
- An outside air damper sized for 1500 fpm gives good control.
- Low-leakage outside air dampers minimize leakage during system shutdown.
- A higher velocity through the return air damper facilitates air balance and may increase mixing.
- Parallel-blade dampers may aid mixing. Positioning the dampers so that the return and outside airstreams are deflected toward each other may increase mixing.
- Mixing dampers should be placed across the full width of the unit, even if the location of the return duct makes it more convenient to return air through the side. Return air entering through the side of an air-handling unit can pass through one side of a double-inlet fan while outside air passes through the other side. This same situation can exist whenever two parallel fans are used in an air-handling unit receiving two different airstreams. Wherever there are two fans and two airstreams, an air mixer should be used.
- Field-built baffles may be used to create additional turbulence and to enhance mixing. Unfortunately, the mixing effectiveness and pressure drop of field-built solutions are unknown.

If stratification is anticipated in a system, then special mixing equipment that has been tested by the manufacturer (see the section on Static Air Mixers) should be specified and used in the air-handling system.

Static Air Mixers

Static air mixers are designed to enhance mixing in the mixing plenum to reduce or eliminate problems associated with stratification. These devices have no moving parts and create turbulence in the airstream, which increases mixing. They are usually mounted between the mixing box and the heating or cooling coil; the space required depends on the amount of mixing that is required. Typical pressure loss for these devices is 0.10 to 0.30 in. of water.

There are no performance standards for air mixers. Thus, manufacturers of air mixers and air-handling units should demonstrate that their devices provide adequate mixing.

Filter Section

A system's overall performance depends heavily on the filter. Unless the filter is regularly maintained, system resistance increases and airflow diminishes. Accessibility for replacement is an important consideration in filter arrangement and location. In smaller air-handling units, filters are often placed in a slide-out rack for side-access replacement. In larger units and built-up systems with internal or front-loading access, there should be at least 3 ft between the upstream face of the filter bank and any obstruction. Other requirements for filters can be found in Chapter 28 and in ASHRAE *Standard* 52.2.

Good mixing of outside and return air is also necessary for good filter performance. A poorly placed outside air duct or a bad duct connection to the mixing plenum can cause uneven filter loading and poor distribution of air through the coil section.

Particulate filters are rated according to ASHRAE *Standard* 52.2's minimum efficiency rating value (MERV) system, a numeric ranking from 1 (least) to 20 (highest). A particulate filter bank of at least MERV 6 should be placed upstream of the first coil, to maintain coil cleanliness. Depending on the spaces served, many applications demand higher-efficiency filters. Some studies suggest filters up to MERV 14 can pay for themselves in reduced coil maintenance and better heat transfer effectiveness. Where higher-MERV filters are used, many designers specify a lower-MERV prefilter as an inexpensive sacrificial filter to capture bulk particulate and extend the life of the more expensive final filter.

The location of the filter bank(s) may be governed by codes. For example, many prevailing health care codes mandate a prefilter upstream of all fans, coils, and humidifiers, plus a final filter bank downstream of all fans, coils, and humidifiers.

Designers are not limited to particulate filters. Electronic air cleaners and gaseous-phase (e.g., activated carbon) filters are available for added protection. For example, ASHRAE *Standard* 62.1 requires use of gaseous-phase filters for certain, usually urban, regions where outside air quality has been measured to exceed threshold values for ozone or other gaseous contaminants. Odor control using activated carbon or potassium permanganate as a filter medium is also available. Chapters 12 and 13 of the 2005 *ASHRAE Handbook—Fundamentals* have more information on odor control.

Preheat Coil

Preheat coils are heating coils placed upstream of a cooling coil. Preheat coils can use steam, hot water, or electric resistance as a medium. Some air-handling units do not require a preheat coil at all, particularly if the percentage of outside air is low and if building heating is provided elsewhere (e.g., perimeter baseboard). Where used, a preheat coil should have wide fin spacing, be accessible for easy cleaning, and be protected by filters. If a preheat coil is located in the minimum outside airstream rather than in the mixed airstream as shown in Figure 1, it should not heat the air to an exit temperature above 35 to 45°F; preferably, it should become inoperative at outside temperatures above 45°F. For use with steam, inner distributing tube or integral face-and-bypass coils are preferable. Hot-water preheat coils should be piped for counterflow so that the coldest air contacts the warmest part of the coil surface first. Consider a constant-flow recirculating pump if the local climate and anticipated percentage of outside air may result in freezing conditions at a hot-water preheat coil. Chapter 26 provides more detailed information on heating coils.

Cooling Coil

Sensible and latent heat are removed from the air by the cooling coils. The cooling medium can be either chilled water or refrigerant, in which case the refrigerant coil serves as the evaporator in a vapor-compression refrigeration cycle. The psychrometrics of cooling and dehumidification were described earlier in this chapter.

In all finned coils, some air passes through without contacting the fins or tubes. The amount of this bypass can vary from 30% for a four-row coil at 700 fpm, to less than 2% for an eight-row coil at 300 fpm. The dew point of the air mixture leaving a four-row coil might satisfy a comfort installation with 25% or less outside air, a small internal latent load, and sensible temperature control only. For close control of room conditions for precision work, a deeper coil may be required. Chapter 22 provides more information on cooling coils and their selection.

Coil freezing can be a serious problem with chilled-water coils. Full-flow circulation of chilled water during freezing weather, or even reduced flow with a small recirculating pump, minimizes coil freezing and eliminates stratification. Further, continuous full-flow circulation can provide a source of off-season chilled water in air-and-water systems. Antifreeze solutions or complete coil draining also prevent coil freezing. However, because it is difficult (if not impossible) to drain most cooling coils completely, caution should be used if this option is considered.

Another design consideration is the drain pan. ASHRAE *Standard* 62.1 calls for drain pans to be sloped to a drain, to avoid holding standing water in the air-handling unit. Because of the constant presence of moisture in the cooling coil drain pan and nearby casing, many designers require stainless steel construction in that portion of the air-handling unit.

Reheat Coil

Reheat coils are heating coils placed downstream of a cooling coil. Reheat systems are strongly discouraged, unless recovered energy is used (see ASHRAE *Standard* 90.1). Reheating is limited to laboratory, health care, or similar applications where temperature and relative humidity must be controlled accurately. Heating coils located in the reheat position, as shown in Figure 1, are frequently used for warm-up, although a coil in the preheat position is preferable. Hot-water coils provide a very controllable source of reheat energy. Inner-distributing-tube coils are preferable for steam applications. Electric coils may also be used. See Chapter 26 for more information.

Humidifiers

Humidifiers may be installed as part of the air-handling unit, or in terminals at the point of use, or both. Where close humidity control of selected spaces is required, the entire supply airstream may be humidified to a low humidity level in the air handler. Terminal humidifiers in the supply ducts serving selected spaces bring humidity up to the required levels. For comfort installations not requiring close control, moisture can be added to the air by mechanical atomizers or point-of-use electric or ultrasonic humidifiers. Proper location of this equipment prevents stratification of moist air in the system.

Steam grid humidifiers with dew-point control usually are used for accurate humidity control. Air to a laboratory or other space that requires close humidity control must be reheated after leaving a cooling coil before moisture can be added. Humidifying equipment capacity should not exceed the expected peak load by more than 10%. If humidity is controlled from the room or return air, a limiting humidistat and fan interlock may be needed in the supply duct. This prevents condensation and mold or mildew growth in the ductwork. Humidifiers add some sensible heat that should be accounted for in the psychrometric evaluation. See Chapter 21 for additional information.

An important question for air-handling unit specifiers is where to place the humidification grid. Moisture cannot be successfully added to cold air, so placement is typically downstream of a preheat coil. For general building humidification, one satisfactory location is between a preheat coil and cooling coil.

Another consideration is absorption distance (i.e., the distance required for the steam to be absorbed into the airstream). This can vary from 18 in. to 5 ft and must be allowed for in the layout and dimensioning of the air-handling unit.

Dehumidifiers

For most routine applications, such as offices, residences, and schools, the air-handling unit's cooling coil provides adequate dehumidification. Where a specialty application requires additional moisture removal, desiccant dehumidifiers are an available accessory. Dust can be a problem with solid desiccants, and lithium contamination is a concern with spray equipment. Chapter 22 discusses dehumidification by cooling coils, and Chapter 23 discusses desiccant dehumidifiers.

Energy Recovery Devices

Energy recovery devices are in greater demand as outside air percentage increases. With some exceptions, ASHRAE *Standard* 90.1 requires energy recovery devices for air-handling units exceeding 5000 cfm and 70% or more outside air. They are used extensively in research and development facilities and in hospitals and laboratories with high outside air requirements. Many types are available, and the type of facility usually determines which is most suitable. Choices include heat pipes, runaround loops, fixed-plate energy exchangers, and rotary wheel energy exchangers. See Chapter 25 for detail.

Most manufacturers of commercial factory-packaged air-handling units now offer optional energy recovery modules for both small and large unit applications, which were formerly the domain of large custom air-handling units.

Many countries with extreme climates provide heat exchangers on outside/relief air, even for private homes. This trend is now appearing in both modest and large commercial buildings worldwide. Under certain circumstances, heat recovery devices can save energy and reduce the required capacity of primary cooling and heating plants by 20% or more.

Sound Control Devices

Where noise control is important, air-handling units can be specified with a noise control section, ranging from a plenum lined with acoustic duct liner to a full bank of duct silencers. This option is available in the smallest to largest units. Sound attenuation can be designed into the discharge (supply) end of the air-handling unit to reduce ductborne fan noise. Remember to consider ductborne noise traveling down the return or outside air paths in a noise-sensitive application, and use a sound attenuation module if necessary at the inlet end of an air-handling unit. See Chapter 47 of the 2007 *ASHRAE Handbook—HVAC Applications* for detail.

Supply Air Fan

Axial-flow, centrifugal, or plenum (plug) fans may be chosen as supply air fans for straight-through flow applications. In factory-fabricated units, more than one centrifugal fan may be tied to the same shaft. If headroom permits, a single-inlet fan should be chosen when air enters at right angles to the flow of air through the equipment. This allows direct airflow from the fan wheel into the supply duct without abrupt change in direction and loss in efficiency. It also allows a more gradual transition from the fan to the duct and increases static regain in the velocity pressure conversion. To minimize inlet losses, the distance between casing walls and fan inlet should be at least the diameter of the fan wheel. With a single-inlet fan, the length of the transition section should be at least half the width or height of the casing, whichever is longer. If fans blow through the equipment, air distribution through the downstream components needs analyzing, and baffles should be used to ensure uniform air distribution. See Chapter 20 for more information.

Two placements of the supply fan section are common. A supply fan placed downstream of the cooling coil is known as a draw-through arrangement, because air is drawn, or induced, across the cooling coil. Similarly, a supply fan placed upstream of the cooling coil is called the blow-through position. Either arrangement is possible in both small and large air-handling units, and in factory-packaged and custom field-erected units.

A **draw-through system** (illustrated in Figure 1) draws air across the coils. A draw-through system usually provides a more even air distribution over all parts of the coil. However, some fan heat is added to the airstream after the air has crossed the cooling coil and must be taken into account when calculating the desired supply air temperature.

A **blow-through system** (illustrated in Figure 1) requires some caution on the part of the designer, because the blast effect of the supply fan outlet can concentrate a high percentage of the total air over a small percentage of the downstream coil surfaces. Air diffusers or diverters may be required. Consequently, blow-through air-handling units may tend to be longer overall than comparable draw-through units. This arrangement offers the advantage of placing the fan before the cooling coil, allowing the cooling coil to remove fan heat from the system. NFPA 54-2006, *National Fuel Gas Code*, requires the blow-through arrangement where natural-gas-fired heat exchangers are used for heating.

Miscellaneous Components

Vibration and sound isolation equipment is required for many central fan installations. Standard mountings of fiberglass, ribbed rubber, neoprene mounts, and springs are available for most fans and prefabricated units. The designer must account for seismic restraint requirements for the seismic zone in which the project is located (see Chapter 54 of the 2007 *ASHRAE Handbook—HVAC Applications*). In some applications, fans may require concrete inertia blocks in addition to spring mountings. Steel springs require sound-absorbing material inserted between the springs and the foundation. Horizontal discharge fans operating at a high static pressure frequently require thrust arrestors. Ductwork connections to fans should be made with fireproof fiber cloth sleeves having considerable slack, but without offset between the fan outlet and rigid duct. Misalignment between the duct and fan outlet can cause turbulence, generate noise, and reduce system efficiency. Electrical and piping connections to vibration-isolated equipment should be made with flexible conduit and flexible connections.

Equipment noise transmitted through ductwork can be reduced by sound-absorbing units, acoustical lining, and other means of attenuation. Sound transmitted through the return and relief ducts should not be overlooked. Acoustical lining sufficient to adequately attenuate objectionable system noise or locally generated noise should be considered. Chapter 47 of the 2007 *ASHRAE Handbook—HVAC Applications*, Chapter 7 of the 2005 *ASHRAE Handbook—Fundamentals*, and ASHRAE *Standard* 68 have further information on sound and vibration control. Noise control, both in occupied spaces and outside near intake or relief louvers, must be considered. Some local ordinances may limit external noise produced by these devices.

AIR DISTRIBUTION

Once air-handling system and air-handling equipment have been selected, air must be distributed to the zone(s) served. Ductwork should deliver conditioned air to each zone as directly, quietly, and economically as possible. Air distribution ductwork and terminal devices selected must be compatible or the system will either fail to operate effectively or incur high first, operating, and maintenance costs.

Ductwork Design

Chapter 35 of the 2005 *ASHRAE Handbook—Fundamentals* describes ductwork design in detail and gives several methods of sizing duct systems, including static regain, equal friction, and T-method. Duct sizing is often performed manually for simple systems, but commercially available duct-sizing software programs are often used for larger and complex systems. It is imperative that the designer coordinate duct design with architectural and structural design. Structural features of the building generally require some compromise and often limit depth. In commercially developed projects, great effort is made to reduce floor-to-floor dimensions. In architecturally significant buildings, high ceilings, barrel-vault ceilings, rotundas and domes, ceiling coves, and other architectural details can place obstacles in the path of ductwork. The resultant decrease in available interstitial space left for ductwork can be a major design challenge. Layout of ductwork in these buildings requires experience, skill, and patience on the part of the designer.

Considerations. Duct systems can be designed for high or low velocity. A high-velocity system has smaller ducts, which save space but require higher pressures and may result in more noise. In some low-velocity systems, medium or high fan pressures may be required for balancing or to overcome high pressure drops from terminal devices. In any variable-flow system, changing operating conditions can cause airflow in the ducts to differ from design flow. Thus, varying airflow in the supply duct must be carefully analyzed to ensure that the system performs efficiently at all loads. This precaution is particularly needed with high-velocity air. Return air ducts are usually sized by the equal friction method.

In many applications, the space between a suspended ceiling and the floor slab or roof above it is used as a return air plenum, so that return air is collected at a central point. Governing codes should be consulted before using this approach in new design, because most codes prohibit combustible material in a ceiling space used as a return air plenum. For example, the *National Electrical Code® Handbook* (NFPA 2005) requires that either conduit or PTFE-insulated wire (often called plenum-rated cable) be installed in a return air plenum. In addition, regulations often require that return air plenums be divided into smaller areas by firewalls and that fire dampers be installed in ducts, which increases first cost.

In research and some industrial facilities, return ducting must be installed to avoid contamination and growth of biological contaminants in the ceiling space. Lobby ceilings with lay-in panels may not work well as return plenums where negative pressure from high-rise elevators or stack effects of high-rise buildings may occur. If the plenum leaks to the low-pressure area, the tiles may lift and drop out when the outside door is opened and closed. Return plenums directly below a roof deck have substantially greater return air temperature increases or decreases than a duct return.

Corridors should not be used for return air because they spread smoke and other contaminants. Although most codes ban returning air through corridors, the method is still used in many older facilities.

All ductwork should be sealed. Energy waste because of leaks in the ductwork and terminal devices can be considerable. Unsealed ductwork in many commercial buildings can have significant leakage.

Air systems are classified in two categories:

- **Single-duct systems** contain the main heating and cooling coils in a series-flow air path. A common duct distribution system at a common air temperature feeds all terminal apparatus. Capacity can be controlled by varying the air temperature or volume.
- **Dual-duct systems** contain the main heating and cooling coils in parallel-flow or series/parallel-flow air paths with either (1) a separate cold- and warm-air duct distribution system that blends air at the terminal apparatus (dual-duct systems), or (2) a separate supply air duct to each zone with the supply air blended at the main unit with mixing dampers (multizone). Dual-duct systems generally vary the supply air temperature by mixing two airstreams of different temperatures, but they can also vary the volume of supply air in some applications.

These categories are further divided and described in the following sections.

SINGLE-DUCT SYSTEMS

Constant Volume

While maintaining constant airflow, single-duct constant volume systems change the supply air temperature in response to the space load (Figure 9).

Single-Zone Systems. The simplest all-air system is a supply unit serving a single zone. The unit can be installed either in or remote from the space it serves, and may operate with or without distribution ductwork. Ideally, this system responds completely to the space needs, and well-designed control systems maintain temperature and humidity closely and efficiently. Single-zone systems often involve short ductwork with low pressure drop and thus low fan energy, and can be shut down when not required without affecting operation of adjacent areas, offering further energy savings. A return or relief fan may be needed, depending on system capacity and whether 100% outside air is used for cooling as part of an economizer cycle. Relief fans can be eliminated if overpressurization can be relieved by other means, such as gravity dampers.

Multiple-Zone Reheat Systems. Multiple-zone reheat is a modification of the single-zone system. It provides (1) zone or space control for areas of unequal loading, (2) simultaneous heating or cooling of perimeter areas with different exposures, and (3) close control for temperature, humidity, and space pressure in process or comfort applications. As the word *reheat* implies, heat is added as a secondary simultaneous process to either preconditioned (cooled, humidified, etc.) primary air or recirculated room air. Relatively small low-pressure systems place reheat coils in the ductwork at each zone. More complex designs include high-pressure primary distribution ducts to reduce their size and cost, and pressure reduction devices to maintain a constant volume for each reheat zone.

The system uses conditioned air from a central unit, generally at a fixed cold-air temperature that is low enough to meet the maximum cooling load. Thus, all supply air is always cooled the maximum amount, regardless of the current load. Heat is added to the airstream in each zone to avoid overcooling that zone, for every zone except the zone experiencing peak cooling demand. The result is very high energy use, and therefore use of this system is restricted by ASHRAE *Standard* 90.1. However, the supply air temperature from the unit can be varied, with proper control, to reduce the amount of reheat required and associated energy consumption. Care must be taken to avoid high internal humidity when the temperature of air leaving the cooling coil is allowed to rise during cooling.

In cold weather, when a reheat system heats a space with an exterior exposure, the reheat coil must not only replace the heat lost from the space, but also must offset the cooling of the supply air (enough cooling to meet the peak load for the space), further increasing energy consumption. If a constant-volume system is oversized, reheat cost becomes excessive.

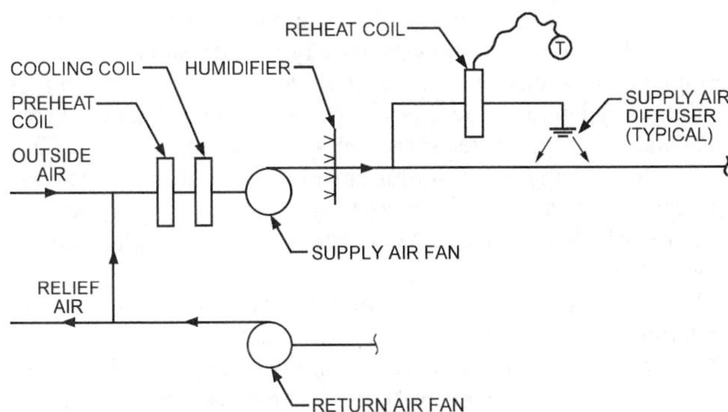

Fig. 9 Constant-Volume System with Reheat

In commercial applications, use of a constant-volume reheat system is generally discouraged in favor of variable-volume or other systems. Constant-volume reheat systems may continue to be applied in hospitals, laboratories, and other critical applications where variable airflow may be detrimental to proper pressure relationships (e.g., for infection control).

Variable Air Volume (VAV)

A VAV system (Figure 10) controls temperature in a space by varying the quantity of supply air rather than varying the supply air temperature. A VAV terminal unit at the zone varies the quantity of supply air to the space. The supply air temperature is held relatively constant. Although supply air temperature can be moderately reset depending on the season, it must always be low enough to meet the cooling load in the most demanding zone and to maintain appropriate humidity. VAV systems can be applied to interior or perimeter zones, with common or separate fans, with common or separate air temperature control, and with or without auxiliary heating devices. The greatest energy saving associated with VAV occurs at the perimeter zones, where variations in solar load and outside temperature allow the supply air quantity to be reduced.

Humidity control is a potential problem with VAV systems. If humidity is critical, as in certain laboratories, process work, etc., constant-volume airflow may be required.

Other measures may also maintain enough air circulation through the room to achieve acceptable humidity levels. The human body is more sensitive to elevated air temperatures when there is little air movement. Minimum air circulation can be maintained during reduced load by (1) raising the supply air temperature of the entire system, which increases space humidity, or supplying reheat on a zone-by-zone basis; (2) providing auxiliary heat in each room independent of the air system; (3) using individual-zone recirculation and blending varying amounts of supply and room air or supply and ceiling plenum air with fan-powered VAV terminal units, or, if design permits, at the air-handling unit; (4) recirculating air with a VAV induction unit; or (5) providing a dedicated recirculation fan to increase airflow.

VAV reheat can ensure close room space pressure control with the supply terminal functioning in sync with associated room exhaust. A typical application might be a fume hood VAV exhaust with constant open sash velocity (e.g., 85 or 100 fpm) or occupied/unoccupied room hood exhaust (e.g., 100 fpm at sash in occupied periods and 60 fpm in unoccupied periods).

Dual-Conduit. This method is an extension of the single-duct VAV system: one supply duct offsets exterior transmission cooling or heating loads by its terminal unit with or without auxiliary heat, and the other supply air path provides cooling throughout the year. The

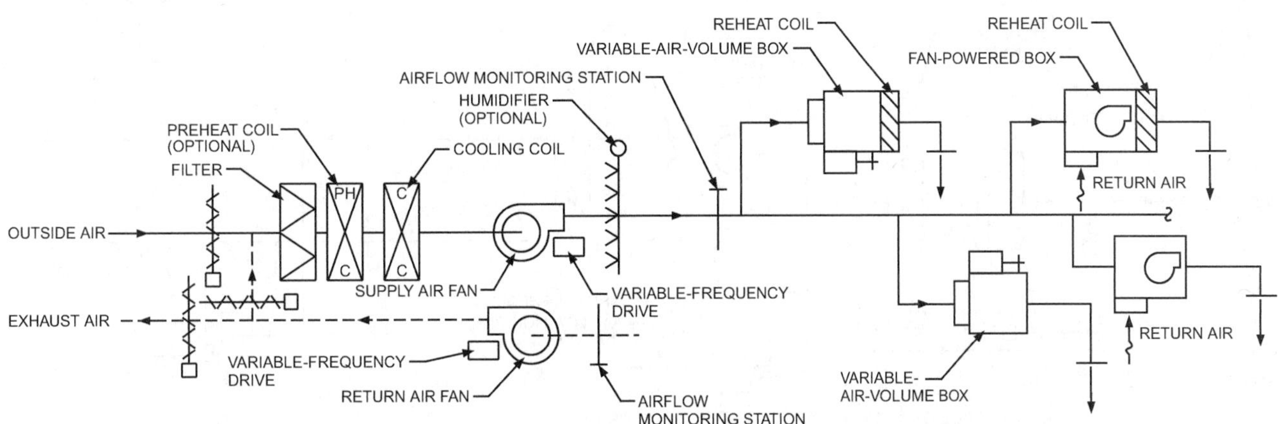

Fig. 10 Variable-Air-Volume System with Reheat and Induction and Fan-Powered Devices
(Courtesy RDK Engineers)

first airstream (primary air) operates as a constant-volume system, and the air temperature is varied to offset transmission only (i.e., it is warm in winter and cool in summer). Often, however, the primary-air fan is limited to operating only during peak heating and cooling periods to further reduce energy use. When calculating this system's heating requirements, the cooling effect of secondary air must be included, even though the secondary system operates at minimum flow. The other airstream, or secondary air, is cool year-round and varies in volume to match the load from solar heating, lights, power, and occupants. It serves both perimeter and interior spaces.

Variable Diffuser. The discharge aperture of this diffuser is reduced to keep discharge velocity relatively constant while reducing conditioned supply airflow. Under these conditions, the induction effect of the diffuser is kept high, cold air mixes in the space, and the room air distribution pattern is more nearly maintained at reduced loads. These devices are of two basic types: one has a flexible bladder that expands to reduce the aperture, and the other has a diffuser plate that moves. Both devices are pressure-dependent, which must be considered in duct-distribution system design. They are either powered by the system or pneumatically or electrically driven.

DUAL-DUCT SYSTEMS

A dual-duct system conditions all the air in a central apparatus and distributes it to conditioned spaces through two ducts, one carrying cold air and the other carrying warm air. In each conditioned zone, air valve terminals mix warm and cold air in proper proportion to satisfy the space temperature and pressure control. Dual-duct systems may be designed as constant volume or variable air volume; a dual-duct, constant-volume system uses more energy than a single-duct VAV system. As with other VAV systems, certain primary-air configurations can cause high relative humidity in the space during the cooling season.

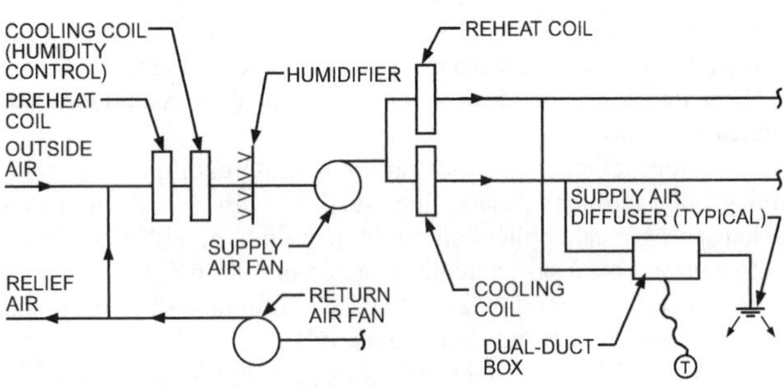

Fig. 11 Single-Fan, Dual-Duct System

Constant Volume

Dual-duct, constant-volume systems using a single supply fan were common through the mid-1980s, and were used frequently as an alternative to constant-volume reheat systems. Today, dual-fan, dual-duct are preferred over the former, based on energy performance. There are two types of dual-duct, single-fan application: with reheat, and without.

Single Fan With Reheat. There are two major differences between this and a conventional terminal reheat system: (1) reheat is applied at a central point in the fan unit hot deck instead of at individual zones (Figure 11), and (2) only part of the supply air is cooled by the cooling coil (except at peak cooling demand); the rest of the supply is heated by the hot-deck coil during most hours of operation. This uses less heating and cooling energy than the terminal reheat system where all the air is cooled to full cooling capacity for more operating hours, and then all of it is reheated as required to match the space load. Fan energy is constant because airflow is constant.

Single Fan Without Reheat. This system has no heating coil in the fan unit hot deck and simply pushes a mixture of outside and recirculated air through the hot deck. A problem occurs during periods of high outside humidity and low internal heat load, causing the space humidity to rise rapidly unless reheat is added. This system has limited use in most modern buildings because most occupants demand more consistent temperature and humidity. A single-fan, no-reheat dual-duct system does not use any extra energy for reheat, but fan energy is constant regardless of space load.

Variable Air Volume

Dual-duct VAV systems blend cold and warm air in various volume combinations. These systems may include single-duct VAV terminal units connected to the cold-air duct distribution system for cooling only interior spaces (see Figures 11 and 12), and the cold duct may serve perimeter spaces in sync with the hot duct. This saves reheat energy for the air for those cooling-only zones because space temperature control is by varying volume, not supply air temperature, which may save some fan energy to the extent that the airflow matches the load.

Newer dual-duct air terminals provide two damper operators per air terminal, which allows the unit to function like a single-duct VAV cooling terminal unit (e.g., 10 in. inlet damper) and a single-duct VAV heating terminal unit (e.g., 6 in. inlet damper) in one physical terminal package. This arrangement allows the designer to

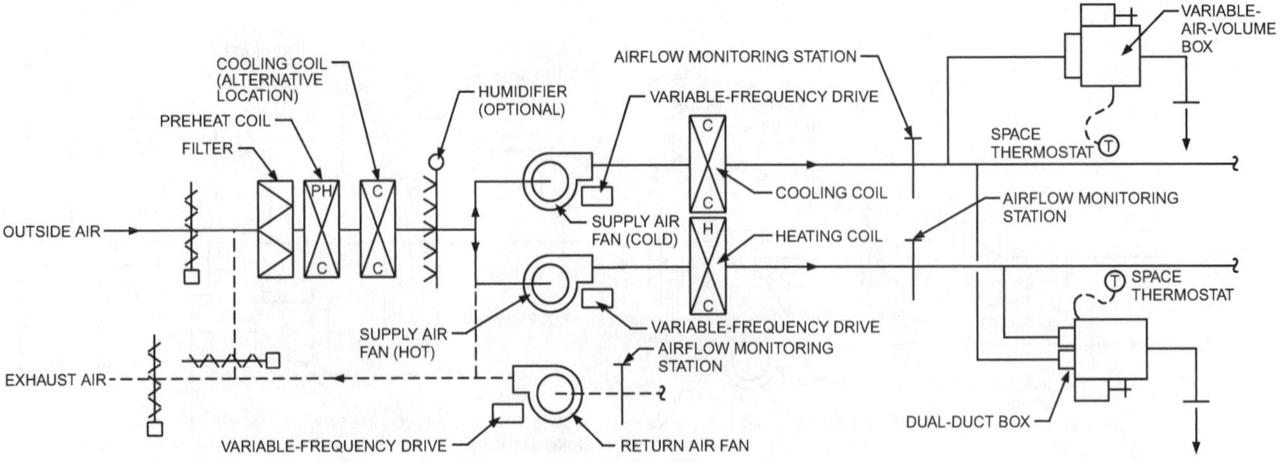

Fig. 12 Dual-Fan, Dual-Duct System
(Courtesy RDK Engineers)

specify the correct cold-air supply damper as part of the dual-duct terminal, which is usually a large supply air quantity in sync with a smaller hot-air supply damper in the same terminal unit. This use of significant minimum airflow levels, providing temperature control by means of dual-duct box mixing at minimum airflow. This variation saves both heating and cooling energy and fan energy because the terminal damper sizes are more appropriate for the design flow versus antiquated dual-duct terminals that provided same-size cold and hot dampers and a single damper operator that did not allow space temperature and airflow controllability.

Single-Fan, Dual-Duct System. This system (Figure 11), frequently used as a retrofit to an antiquated dual-duct, single-fan application during the 1980s and 1990s, uses a single supply fan sized for the peak cooling load or the coincident peak of the hot and cold decks. Fan control is from two static-pressure controllers, one located in the hot deck and the other in the cold deck. The duct requiring the highest pressure governs the airflow by signaling the supply fan VFD speed control. An alternative is to add discharge supply air duct damper control to both the cold and hot ducts to vary flow while the supply fan operates up and down its fan curve. Return air fan tracking of discharge supply air must be assessed with this application.

Usually, the cold deck is maintained at a fixed temperature, although some central systems allow the temperature to rise during warmer weather to save refrigeration. The hot-deck temperature is often raised during periods of low outside temperature and high humidity to increase the flow over the cold deck for dehumidification. Other systems, particularly in laboratories, use a precooling coil to dehumidify the total airstream or just the outside air to limit humidity in the space. Return air quantity can be controlled either by flow-measuring devices in the supply and return duct or by static-pressure controls that maintain space static pressure.

Dual-Fan, Dual-Duct System. Supply air volume of each supply fan is controlled independently by the static pressure in its respective duct (Figure 12). The return fan is controlled based on the sum of the hot and cold fan volumes using flow-measuring stations. Each fan is sized for the anticipated maximum coincident hot or cold volume, not the sum of the instantaneous peaks. The cold deck can be maintained at a constant temperature either with mechanical refrigeration, when minimum outside air is required, or with an air-side economizer, when outside air is below the temperature of the cold-deck set point. This operation does not affect the hot deck, which can recover heat from the return air, and the heating coil need operate only when heating requirements cannot be met using return air. Outside air can provide ventilation air via the hot duct when the outside air is warmer than the return air. However, controls should be used to prohibit introducing excessive amounts of outside air beyond the required minimum when that air is more humid than the return air.

MULTIZONE SYSTEMS

The multizone system (Figure 13) supplies several zones from a single, centrally located air-handling unit. Different zone requirements are met by mixing cold and warm air through zone dampers at the air handler in response to zone thermostats. The mixed, conditioned air is distributed throughout the building by single-zone ducts. The return air is handled conventionally. The multizone system is similar to the dual-duct system and has the same potential problem with high humidity levels. This system can provide a smaller building with the advantages of a dual-duct system, and it uses packaged equipment, which is less expensive.

Packaged equipment is usually limited to about 12 zones, although built-up systems can include as many zones as can be physically incorporated in the layout. A multizone system is somewhat more energy-efficient than a terminal reheat system because not all the air goes through the cooling coil, which reduces the amount of reheat required. But a multizone system uses essentially the same fan energy as terminal reheat because the airflow is constant.

Two common variations on the multizone system are the three-deck multizone and the Texas multizone. A **three-deck multizone** system is similar to the standard multizone system, but bypass zone dampers are installed in the air-handling unit parallel with the hot- and cold-deck dampers. In the **Texas multizone** system, the hot-deck heating coil is removed from the air handler and replaced with an air-resistance plate matching the cooling coil's pressure drop. Individual heating coils are placed in each perimeter zone duct. These heating coils are usually located in the equipment room for ease of maintenance. This system is common in humid climates where the cold deck often produces 48 to 52°F air for humidity control. Using the air-handling units' zone dampers to maintain zone conditions, supply air is then mixed with bypass air rather than heated air. Heat is added only if the zone served cannot be maintained by delivering return air alone. These arrangements can save considerable reheat energy.

SPECIAL SYSTEMS

Primary/Secondary

Some processes use two interconnected all-air systems (Figure 14). In these situations, space gains are very high and/or a large number of air changes are required (e.g., in sterile or cleanrooms or where close-tolerance conditions are needed for process work). The primary system supplies the conditioned outside air requirements for the process to the secondary system. The secondary system provides additional cooling and humidification (and heating, if required) to offset space and fan power gains. Normally, the secondary cooling coil is designed to be dry (i.e., sensible cooling only) to reduce the possibility of bacterial growth, which can create air quality problems. The alternative is to have the primary system supply conditioned outside air [e.g., 20 air changes per hour (ach)] to the ceiling return air plenum, where fan-powered HEPA filter units provide the total supply air (e.g., 120 ach) to the occupied space. Consideration must be given to the total heat gain from the numerous fan-powered motors in the return air plenum.

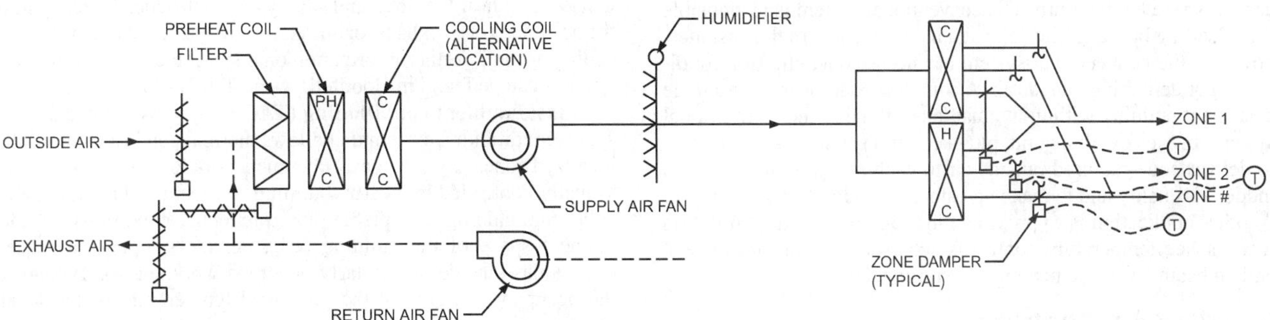

Fig. 13 Multizone System
(Courtesy RDK Engineers)

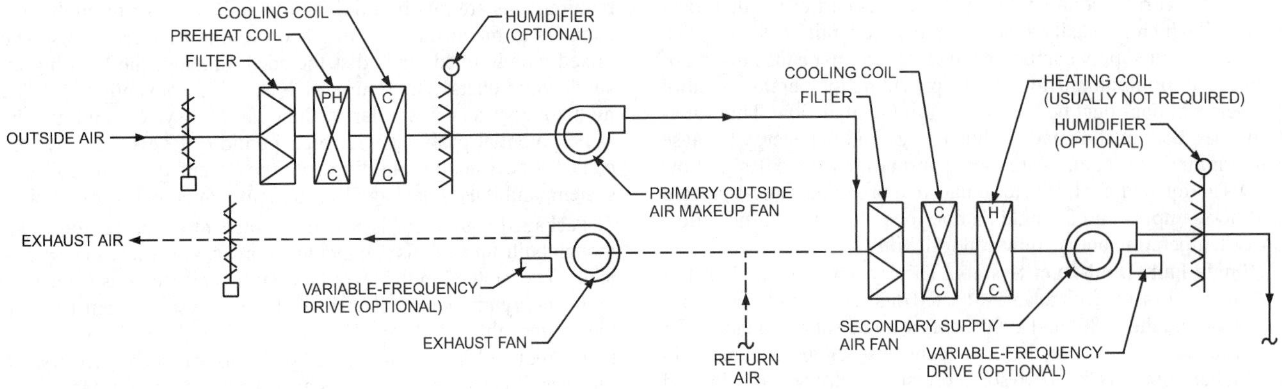

Fig. 14 Primary/Secondary System
(Courtesy RDK Engineers)

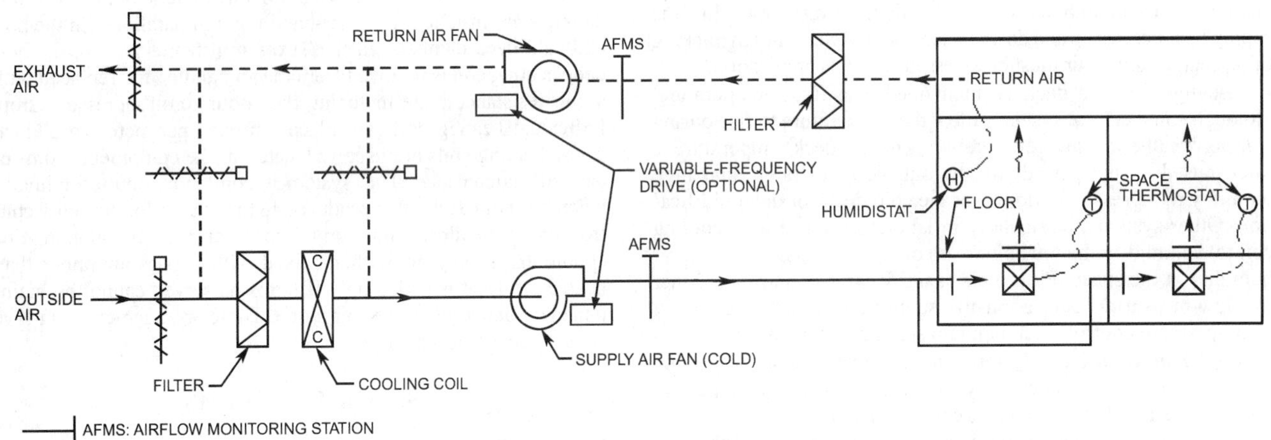

Fig. 15 Underfloor Air Distribution
(Courtesy RDK Engineers)

Dedicated Outdoor Air

Similar in some respects to the primary/secondary system, the dedicated outdoor air system (DOAS) decouples air-conditioning of the outside air from conditioning of the internal loads. Long popular in hotels and multifamily residential buildings, DOAS is now gaining popularity in commercial buildings and many other applications. The DOAS introduces 100% outside air, heats or cools it, may humidify or dehumidify it, and filters it, then supplies this treated air to each of its assigned spaces. Air volume is sized in response to minimum ventilation standards, such as ASHRAE *Standard* 62.1, or to meet makeup air demands. Often, the DOAS serves multiple spaces and is designed not necessarily to control space temperature, but to provide thermally neutral air to those spaces. A second, more conventional system serves those same spaces and is intended to control space temperature. The conventional system is responsible for offsetting building envelope and internal loads. In this instance, however, the conventional system has no responsibility for conditioning or delivering outside air. A common example may be a large apartment building with individual fan-coil units (the conventional system) in each dwelling unit, plus a common building-wide DOAS to deliver code-required outside air to each housing unit for good indoor air quality and to make up bathroom and/or kitchen exhaust. Another application is to provide minimum outside air ventilation while space temperature control is achieved with radiant panels, chilled beams, valance heating and cooling, etc.

Underfloor Air Distribution

An underfloor air distribution (UFAD) system (Figure 15) uses the open space between a structural floor slab and the underside of a raised-floor system to deliver conditioned air to supply outlets at or near floor level. Floor diffusers make up the large majority of installed UFAD supply outlets, which can provide different levels of individual control over the local thermal environment, depending on diffuser design and location. UFAD systems use the same basic types of equipment at the cooling and heating plants and primary air-handling units as conventional overhead systems do. Variations of UFAD include displacement ventilation and task/ambient systems.

Displacement ventilation is a type of UFAD system that uses a large number of low-volume supply air outlets to create laminar flow. Conditioned air slightly cooler than the desired room air temperature in the occupied zone is supplied from air outlets at low air velocities (~100 fpm or less). Because of buoyancy, the cooler air spreads along the floor and floods the room's lower zone. Typically, outlets are located at or near the floor, and supply air is introduced directly into the occupied zone. Exhaust or air returns are located at or close to the ceiling or roof. Further information on displacement air distribution systems can be found in Goodfellow and Tahti (2001).

A **task/ambient conditioning (TAC)** system is defined as any space-conditioning system that allows thermal conditions in small, localized zones (e.g., regularly occupied work locations) to be individually controlled by nearby building occupants while automatically maintaining acceptable environmental conditions in the ambient space of the building (e.g., corridors, open-use space, other areas outside of regularly occupied work space). Typically, the occupant can control the perceived temperature of the local environment by adjusting the speed and direction, and in some cases the temperature, of incoming air supply, much like on the dashboard of a car. TAC systems are distinguished from standard

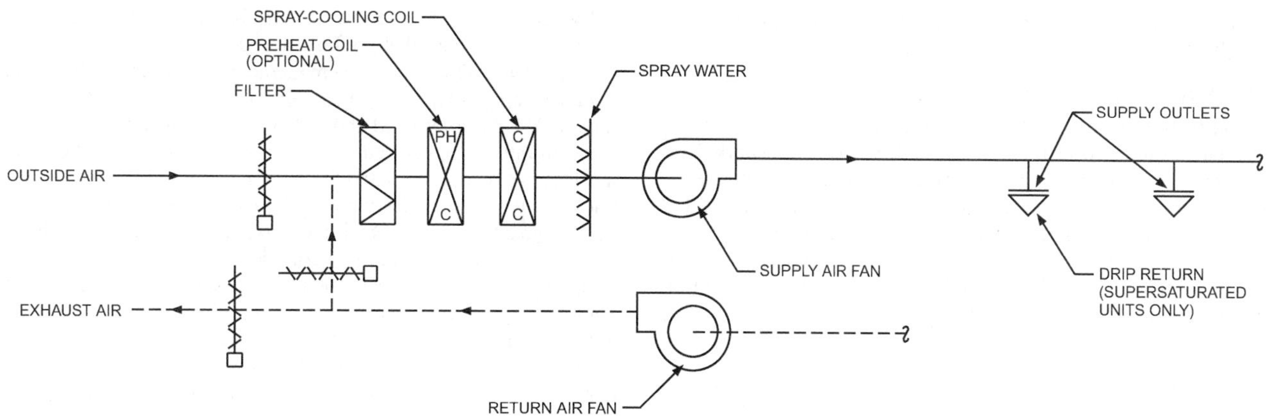

Fig. 16 Supersaturated/Wetted Coil
(Courtesy RDK Engineers)

UFAD systems by the higher degree of personal comfort control provided by the localized supply outlets. TAC supply outlets use direct-velocity cooling to achieve this level of control, and are therefore most commonly configured as fan-driven (active) jet-type diffusers that are part of the furniture or partitions. Active floor diffusers are also possible.

Unlike conventional HVAC design, in which conditioned air is both supplied and exhausted at ceiling level, UFAD removes supply ducts from the ceilings and allows a smaller overhead ceiling cavity. Less stratification and improved ventilation effectiveness may be achieved with UFAD because air flows from a floor outlet to a ceiling inlet, rather than from a ceiling outlet to a ceiling inlet. UFAD has long been popular in computer room and data center applications, and in Europe in conventional office buildings. It is now increasingly applied in North American offices and other commercial buildings.

Because raised floors contain electrical conduits, floor drains, wall partitions, and other items, installed by many trades, underfloor plenums require careful pressure-testing for leakage.

For more information, see Chapter 32 of the 2005 *ASHRAE Handbook—Fundamentals*, Chapters 17 and 29 of the 2007 *ASHRAE Handbook—HVAC Applications*, or Bauman (2003).

Wetted Duct/Supersaturated

Some industries spray water into the airstream at central air-handling units in sufficient quantities to create a controlled carryover (Figure 16). This supercools the supply air, normally equivalent to an oversaturation of about 10 grains per pound of air, and allows less air to be distributed for a given space load. These are used where high humidity is desirable, such as in the textile or tobacco industry, and in climates where adiabatic cooling is sufficient.

Compressed-Air and Water Spray

This is similar to the wetted duct system, except that the water is atomized with compressed air and provides limited cooling while maintaining a relatively humid environment. Nozzles are sometimes placed inside the conditioned space and independent of the cooling air supply, and can be used for large, open manufacturing facilities where humidity is needed to avoid static electricity in the space. Several nozzle types are available, and the designer should understand their advantages and disadvantages. Depending on the type of nozzle, compressed-air and water spray systems can require large and expensive air compressors. The extra first cost may or may not be offset by energy cost savings, depending on the application.

Low-Temperature

Ice storage is often used to reduce peak electrical demand. Low-temperature systems (where air is supplied at as low as 40°F) are sometimes used. Benefits include smaller central air-handling units and associated fan power, and smaller supply air duct distribution. At air terminals, fan-powered units are frequently used to increase supply air to the occupied space. Attention to detail is important because of the low air temperature and the potential for excessive condensate on supply air duct distribution in the return plenum space. These fan-powered terminals mix return air or room air with cold supply air to increase the air circulation rate in the space.

Smoke Management

Air-conditioning systems are often used for smoke control during fires. Controlled airflow can provide smoke-free areas for occupant evacuation and fire fighter access. Space pressurization creates a low-pressure area at the smoke source, surrounding it with high-pressure spaces. The air-conditioning system can also be designed to provide makeup air for smoke exhaust systems in atria and other large spaces (Duda 2004). For more information, see Chapter 52 of the 2007 *ASHRAE Handbook—HVAC Applications*. Klote and Milke (2002) also has detailed information on this topic.

TERMINAL UNITS

Air systems have two types of devices between the primary air distribution system and the conditioned space: (1) passive devices, such as supply outlets (registers or diffusers) and return inlets (grilles), and (2) active devices, which are often called *terminal units*. The register or diffuser should deliver supply air throughout the conditioned space without occupants sensing a draft and without creating excessive noise. The terminal unit controls the quantity and/or temperature of the supply air to maintain desired conditions in the space. Both types of devices are discussed in Chapter 19 of this volume and Chapter 33 of the 2005 *ASHRAE Handbook—Fundamentals*. Terminal units are discussed here briefly in terms of how they fit into the system concept, but Chapter 19 should be consulted for more complete information.

In some instances, such as in low-velocity all-air systems, air may enter from the supply air ductwork directly into the conditioned space through a grille or diffuser. In medium- and high-velocity air systems, air terminal units normally control air volume, reduce duct pressure, or both. Various types are available. A VAV terminal varies the amount of air delivered. Air may be delivered to low-pressure ductwork and then to the space, or the terminal may contain an integral air diffuser. A fan-powered VAV terminal varies the amount of primary air delivered, but it also uses a fan to mix ceiling plenum or return air with primary supply air before it is delivered to the space. An all-air induction terminal controls the volume of primary air, induces a flow of ceiling plenum or space air, and distributes the mixture through low-velocity ductwork to the space. An air-water induction terminal includes a coil or coils in the induced airstream to condition the return air before it mixes with the primary air and enters the space.

Constant-Volume Reheat

Constant-volume reheat terminal boxes are used mainly in terminal reheat systems with medium- to high-velocity ductwork. The unit serves as a pressure-reducing valve and constant-volume regulator to maintain a constant supply air quantity to the space, and is generally fitted with an integral reheat coil that controls space temperature. The constant supply air quantity is selected to provide cooling to match the peak load in the space, and the reheat coil is controlled to raise the supply air temperature as necessary to maintain the desired space temperature at reduced loads.

Variable Air Volume

VAV terminal units are available in many configurations, all of which control the space temperature by varying the volume of cool supply air from the air handler to match the actual cooling load. VAV terminal units are fitted with automatic controls that are either pressure-dependent or pressure-independent. **Pressure-dependent** units control damper position in response to room temperature, and flow may increase and decrease as the main duct pressure varies. **Pressure-independent** units measure actual supply airflow and control flow in response to room temperature. Pressure-independent units may sometimes be fitted with a velocity-limit control that overrides the room temperature signal to limit the measured supply velocity to some selected maximum. Velocity-limit control can be used to prevent excess airflow through units nearest the air handler. Excessive airflow at units close to the air handler can draw so much supply air that units far from the air handler do not get enough air.

Throttling Without Reheat. The throttling (or pinch-off) box without reheat is essentially an air valve or damper that reduces supply airflow to the space in response to falling space temperature. The unit usually includes some means of sound attenuation to reduce air noise created by the throttling action. It is the simplest and least expensive VAV terminal unit, but is suitable for use only where no heat is required and if the unit can go to the completely closed position at reduced cooling loads. If this type of unit is set up with a minimum position, it will constantly provide cooling to the space, whether the space needs it or not, and can overcool the space. This approach offers the lowest fan energy use, because it minimizes airflow to just the amount required by the cooling load.

Throttling With Reheat. This simple VAV system integrates heating at the terminal unit with the same type of air valve. It is applied in interior and exterior zones where full heating and cooling flexibility is required. These terminal units can be set to maintain a predetermined minimum air quantity necessary to (1) offset the heating load, (2) limit maximum humidity, (3) provide reasonable air movement in the space, and (4) provide required ventilation air. The reheat coil is most commonly hot water or electric resistance.

Variable air volume with reheat allows airflow to be reduced as the first step in control; heat is then initiated as the second step. Compared to constant-volume reheat, this procedure reduces operating cost appreciably because the amount of primary air to be cooled and secondary air to be heated is reduced. Many types of controls can provide control sequences with more than one minimum airflow. This type of control allows the box to go to a lower flow rate that just meets ventilation requirements at the lightest cooling loads, then increase to a higher flow rate when the heating coil is energized. A feature can be provided to isolate the availability of reheat during the summer, except in situations where even low airflow would overcool the space and should be avoided or where increased humidity causes discomfort (e.g., in conference rooms when the lights are turned off).

Because the reheat coil requires some minimum airflow to deliver heat to the space, and because the reheat coil must absorb all of the cooling capacity of that minimum airflow before it starts to deliver heat to the space, energy use can be significantly higher than with throttling boxes that go fully closed.

Dual-Duct. Dual-duct systems typically feature throttling dual-duct VAV boxes. These terminal units are very similar to the single-duct VAV boxes discussed above, but as the name implies, two primary air inlets are provided. This allows connection of one primary air inlet to a heating duct and the other to a cooling duct. The dual-duct box then modulates both air dampers in response to instructions from a thermostat. Dual-duct boxes are generally available in a constant-volume output, with cooling and heating dampers operating in tandem but inversely such that the sum total of heating plus cooling is always relatively constant. Dual-duct boxes are also available in a variable-volume output, with only the cooling or heating damper permitted to stroke open at any given time, such that cooling damper must be closed prior to allowing the heating damper to stroke open. Minimum positions are available on these dampers to meet minimum ventilation airflow requirements even when little or no airflow would otherwise be required.

Induction. The VAV induction system uses a terminal unit to reduce cooling capacity by simultaneously reducing primary air and inducing room or ceiling air (replacing the reheat coil) to maintain a relatively constant room supply volume. This operation is the reverse of the bypass box. The primary-air quantity decreases with load, retaining the savings of VAV, and the air supplied to the space is kept relatively constant to avoid the effect of stagnant air or low air movement. VAV induction units require a higher inlet static pressure, which requires more fan energy, to achieve the velocities necessary for induction. Today, induction units have for the most part been displaced by fan-powered terminals, which allow reduction of inlet static pressure and, in turn, reduced central air-handling unit fan power.

Fan-Powered. Fan-powered systems are available in either parallel or series airflow. In **parallel-flow** units, the fan is located outside the primary airstream to allow intermittent fan operation. A backdraft damper on the terminal fan prevents conditioned air from escaping into the return air plenum when the terminal fan is off. In **series** units, the fan is located in the primary airstream and runs continuously when the zone is occupied. These constant-airflow fan boxes in a common return plenum can help maintain indoor air quality by recirculating air from overventilated zones to zones with greater outside air ventilation requirements.

Fan-powered systems, both series and parallel, are often selected because they maintain higher air circulation through a room at low loads but still retain the advantages of VAV systems. As the cold primary-air valve modulates from maximum to minimum (or closed), the unit recirculates more plenum air. In a perimeter zone, a hot-water heating coil, electric heater, baseboard heater, or remote radiant heater can be sequenced with the primary-air valve to offset external heat losses. Between heating and cooling operations, the fan only recirculates ceiling air. This allows heat from lights to be used for space heating, for maximum energy saving. During unoccupied periods, the main supply air-handling unit remains off and individual fan-powered heating zone terminals are cycled to maintain required space temperature, thereby reducing operating cost during unoccupied hours.

Fans for fan-powered air-handling units operated in series are sized and operated to maintain minimum static pressures at the unit inlet connections. This reduces the fan energy for the central air handler, but the small fans in fan-powered units are less efficient than the large air handler fans. As a result, the series fan-powered unit (where small fans operate continuously) may use more fan energy than a throttling unit system. However, the extra fan energy may be more than offset by the reduction in reheat through the recovery of plenum heat and the ability to operate a small fan to deliver heat during unoccupied hours where heat is needed.

Because fan-powered boxes involve an operating fan, they may generate higher sound levels than throttling boxes. Acoustical ceilings generally are not very effective sound barriers, so extra care should be taken in considering the sound level in critical spaces near fan-powered terminal units.

Both parallel and series fan-powered terminal units should be provided with filters. A disadvantage of this type of terminal unit is the need to periodically change these filters, making them unsuitable for installation above inaccessible ceilings. A large building could contain hundreds of fan-powered terminal units, some of which might be located in inconvenient locations above office furniture or executive offices. Select installed locations carefully for maximum accessibility.

The constant (series) fan VAV terminal can accommodate minimum (down to zero) flow at the primary-air inlet while maintaining constant airflow to the space.

Both types of fan-powered units and induction terminal units are usually located in the ceiling plenum to recover heat from lights. This sometimes allows these terminals to be used without reheat coils in internal spaces. Perimeter-zone units are sometimes located above the ceiling of an interior zone where heat from the lights maintains a higher plenum temperature. Provisions must still be made for morning warm-up and night heating. Also, interior spaces with a roof load must have heat supplied either separately in the ceiling or at the terminal.

Terminal Humidifiers

Most projects requiring humidification use steam. This can be centrally generated as part of the heating plant, where potential contamination from water treatment of the steam is more easily handled and therefore of less concern. Where there is a concern, local generators (e.g., electric or gas) that use treated water are used. Compressed-air and water humidifiers are used to some extent, and supersaturated systems are used exclusively for special circumstances, such as industrial processes. Spray-type washers and wetted coils are also more common in industrial facilities. When using water directly, particularly in recirculating systems, the water must be treated to avoid dust accumulation during evaporation and the build-up of bacterial contamination.

Terminal Filters

In addition to air-handling unit filters, terminal filters may used at the supply outlets to protect particular conditioned spaces where an extra-clean environment is desired (e.g., in a hospital's surgery suite). Chapter 28 discusses this topic in detail.

AIR DISTRIBUTION SYSTEM CONTROLS

Controls should be automatic and simple for best operating and maintenance efficiency. Operations should follow a natural sequence. Depending on the space need, one controlling thermostat closes a normally open heating valve, opens the outside air mixing dampers, or opens the cooling valve. In certain applications, an enthalpy controller, which compares the heat content of outside air to that of return air, may override the temperature controller. This control opens the outside air damper when conditions reduce the refrigeration load. On smaller systems, a dry-bulb control saves the cost of the enthalpy control and approaches these savings when an optimum changeover temperature, above the design dew point, is established. Controls are discussed in more detail in Chapter 46 of the 2007 *ASHRAE Handbook—HVAC Applications*.

Air-handling systems, especially variable-air-volume systems, should include means to measure and control the amount of outside air being brought in to ensure adequate ventilation for acceptable indoor air quality. Strategies include the following:

- Separate constant-volume 100% outside air ventilation systems
- Outside air injection fan
- Directly measuring the outside air flow rate
- Modulating the return damper to maintain a constant pressure drop across a fixed outside air orifice
- Airflow-measuring systems that measure both supply and return air volumes and maintain a constant difference between them.
- CO_2- and/or VOC-based demand-controlled ventilation

A minimum outside air damper with separate motor, selected for a velocity of 1500 fpm, is preferred to one large outside air damper with minimum stops. A separate damper simplifies air balancing. Proper selection of outside, relief, and return air dampers is critical for efficient operation. Most dampers are grossly oversized and are, in effect, unable to control. One way to solve this problem is to provide maximum and minimum dampers. A high velocity across a wide-open damper is essential to its providing effective control.

A mixed-air temperature control can reduce operating costs and also reduce temperature swings from load variations in the conditioned space. Chapter 46 of the 2007 *ASHRAE Handbook—HVAC Applications* shows control diagrams for various arrangements of central system equipment. Direct digital control (DDC) is common, and most manufacturers offer either a standard or optional DDC package for equipment, including air-handling units, terminal units, etc. These controls offer considerable flexibility. DDC controls offer the additional advantage of the ability to record actual energy consumption or other operating parameters of various components of the system, which can be useful for optimizing control strategies.

Constant-Volume Reheat. This system typically uses two subsystems for control: one controls the discharge air conditions from the air-handling unit, and the other maintains the space conditions by controlling the reheat coil.

Variable Air Volume. Air volume can be controlled by duct-mounted terminal units serving multiple air outlets in a control zone or by units integral to each supply air outlet.

Pressure-independent volume-regulator units control flow in response to the thermostat's call for heating or cooling. The required flow is maintained regardless of fluctuation of the VAV unit inlet or system pressure. These units can be field- or factory-adjusted for maximum and minimum (or shutoff) air settings. They operate at inlet static pressures as low as 0.2 in. of water.

Pressure-dependent devices control air volume in response to a unit thermostatic (or relative humidity) device, but flow varies with the inlet pressure variation. Generally, airflow oscillates when pressure varies. These units do not regulate flow but position the volume-regulating device in response to the thermostat. They are the least expensive units but should only be used where there is no need for maximum or minimum limit control and when the pressure is stable.

The type of controls available for VAV units varies with the terminal device. Most use either pneumatic or electric controls and may be either self-powered or system-air-actuated. Self-powered controls position the regulator by using liquid- or wax-filled power elements. System-powered devices use air from the air supplied to the space to power the operator. Components for both control and regulation are usually contained in the terminal device.

To conserve power and limit noise, especially in larger systems, fan operating characteristics and system static pressure should be controlled. Many methods are available, including fan speed control, variable-inlet vane control, fan bypass, fan discharge damper, and variable-pitch fan control. The location of pressure-sensing devices depends, to some extent, on the type of VAV terminal unit used. Where pressure-dependent units without controllers are used, the system pressure sensor should be near the static pressure midpoint of the duct run to ensure minimum pressure variation in the system. Where pressure-independent units are installed, pressure controllers may be at the end of the duct run with the highest static pressure loss. This sensing point ensures maximum fan power savings while maintaining the minimum required pressure at the last terminal.

As flow through the various parts of a large system varies, so does static pressure. Some field adjustment is usually required to find the best location for the pressure sensor. In many systems, the initial location is two-thirds to three-fourths of the distance from the supply fan to the end of the main trunk duct. As the pressure at the system control point increases as terminal units close, the pressure controller signals the fan controller to position the fan volume control, which reduces flow and maintains constant pressure. Many systems measure flow rather than pressure and, with the development of economical DDC, each terminal unit (if necessary) can be monitored and the supply and return air fans modulated to exactly match the demand.

Dual Duct. Because dual-duct systems are generally more costly to install than single-duct systems, their use is less widespread. DDC, with its ability to maintain set points and flow accurately, can make dual-duct systems worthwhile for certain applications. They should be seriously considered as alternatives to single-duct systems.

Personnel. The skill levels of personnel operating and maintaining the air conditioning and controls should be considered. In large research and development or industrial complexes, experienced personnel are available for maintenance. On small and sometimes even large commercial installations, however, office managers are often responsible, so designs must be in accordance with their capabilities.

Water System Interface. On large hydronic installations where direct blending is used to maintain (or reset) the secondary-water temperature, the system valves and coils must be accurately sized for proper control. Many designers use variable flow for hydronic as well as air systems, so the design must be compatible with the air system to avoid operating problems.

Relief Fans. In many applications, relief or exhaust fans can be started in response to a signal from the economizer control or to a space pressure controller. The main supply fan must be able to handle the return air pressure drop when the relief fan is not running.

AUTOMATIC CONTROLS AND BUILDING MANAGEMENT SYSTEM

Central air-handling units increasingly come with prepackaged and prewired automatic control systems. Controls may also be accessible by the building manager using a modem to an off-site computer. The next level of HVAC system management is to integrate manufacturers' control packages with the building management system (BMS). If the project is an addition or major renovation of space, prepackaged controls and their capabilities need to be compatible with existing automatic controls. Chapter 39 of the 2007 *ASHRAE Handbook—HVAC Applications* discusses computer applications, and ASHRAE *Standard* 135 discusses interfacing building automation systems.

Automatic temperature controls can be important in establishing a simple or complex control system, more so with all-air systems than with air-and-water and all-water systems. Maintaining these controls can be challenging to building management staff. With a focus on energy management and indoor air quality, the building management system can be an important business tool in achieving sustainable facility management success.

MAINTENANCE MANAGEMENT SYSTEM

Maintenance management for central air-handling units involves many component and devices, with a varied lists of tasks (e.g., check belts, lube fittings, replace filters, adjust dampers), and varied times and frequencies, depending on components and devices (e.g., check damper linkage monthly, change filters based on pressure drop) Small installations may be best served by local service contractors, in lieu of in-house personnel; larger installations may be best served with in-house technicians. See Chapter 38 of the 2007 *ASHRAE Handbook—HVAC Applications* for further discussion.

BUILDING SYSTEM COMMISSIONING

Prepackaged control systems use a different automatic control checkout process than traditional control contractors. When commissioning a building system that integrates an independent control system with individual packaged control systems, the process can be more cumbersome because both control contractors need to participate. Air and water balancing for each all-air system are also important. With the complexity of the air systems and the numerous modes of operation (e.g., economizer cycle, minimum outside air, smoke control mode), it is essential to adjust and balance systems before system commissioning.

Ongoing commissioning or recommissioning should be integral to maintaining each central air system. During the warranty phase, all-air system performance should be measured and benchmarked to ensure continuous system success. Retro- or recommissioning should be considered whenever the facility is expanded or an additional connection made to the existing systems, to ensure the original design intent is met.

When completing TAB and commissioning, consider posting laminated system flow diagrams at or adjacent to the air-handling unit, indicating operating instructions, TAB performance, commissioning functional performance tests, and emergency shutoff procedures. These documents also should be filed electronically in the building manager's computer server for quick reference.

Original basis of design and design criteria should be posted as a constant reminder of design intent, and be readily available in case troubleshooting, expansion, or modernization is needed.

For the HVAC design to succeed, commissioning should include the system training requirements necessary for building management staff to efficiently take ownership and operate and maintain the HVAC systems over the service life of the installation.

Commissioning continues until the final commissioning report, approximately one year after the construction phase has been completed and the warranty phase comes to an end.

REFERENCES

ASHRAE. 2003. Selecting outside, return, and relief dampers for air-side economizer systems. *Standard* 16.

ASHRAE. 2007. Method of testing general ventilation air-cleaning devices for removal efficiency by particle size. ANSI/ASHRAE *Standard* 52.2-2007.

ASHRAE. 2004. Thermal environmental conditions for human occupancy. ANSI/ASHRAE *Standard* 55-2004.

ASHRAE. 2007. Ventilation for acceptable indoor air quality. ANSI/ASHRAE *Standard* 62.1-2007.

ASHRAE. 1997. Laboratory method to determine the sound power in a duct. ANSI/ASHRAE *Standard* 68-1997.

ASHRAE. 2001. Energy standard for buildings except low-rise residential buildings. ANSI/ASHRAE *Standard* 90.1-2001.

ASHRAE. 2004. BACnet®—A data communication protocol for building automation and control networks. ANSI/ASHRAE *Standard* 135-2004.

Duda, S.W. 2004. Atria smoke exhaust: 3 approaches to replacement air delivery. *ASHRAE Journal* 46(6):21-27.

NFPA. 2006. *National fuel gas code.* ANSI Z223.1/NFPA 54-2006. National Fire Protection Association, Quincy, MA.

NFPA. 2005. *National electrical code® handbook*. National Fire Protection Association, Quincy, MA.

BIBLIOGRAPHY

Bauman, F.S. 2003. *Underfloor air distribution design guide*. ASHRAE.

Goodfellow, H. and E. Tahti, eds. 2001. *Industrial ventilation design guidebook.* Academic Press, New York.

Kirkpatrick, A.T. and J.S. Elleson. 1996. *Cold air distribution system design guide.* ASHRAE.

Klote, J.H. and J.A. Milke. 2002. *Principles of smoke management*, 2nd ed. ASHRAE and the Society of Fire Protection Engineers.

Lorsch, H.G., ed. 1993. *Air-conditioning systems design manual.* ASHRAE.

McKew, H. 1978. Double duct design—A better way. *Heating/Piping/Air-Conditioning*, December.

CHAPTER 5

IN-ROOM TERMINAL SYSTEMS

VERY early in the design of a new building project or renovation, the HVAC design engineer must analyze and ultimately select appropriate systems, as discussed in Chapter 1. Next, production of primary heating and cooling is selected as decentralized (see Chapter 2) or centralized (see Chapter 3). Finally, distribution of heating and cooling to the end-use space can be done by an all-air system (see Chapter 4), or a variety of all-water or air-water systems and local terminals, as discussed in this chapter.

SYSTEM CHARACTERISTICS

Terminal-unit systems can be designed to provide complete sensible and latent cooling and heating to an end-use space; however, most terminal systems are best used with a central ventilation system providing tempered air to the space. Heat can be provided by hot water, steam, or an electric heating coil. Cooling can be provided by either a chilled-water coil or a direct-expansion (DX) coil. Heat pumps (discussed in Chapter 2) can also be used, either with a piped water loop (water-source) or air cooled. In-room terminals usually condition a single space, but some (e.g., a large fan-coil unit) may serve several spaces. Separate thermostats provide individual zone temperature control.

A terminal unit system **with central ventilation** provides the cooling or heating necessary to equalize only the sensible and latent heat gain or loss caused by the building envelope design and occupancy. Ventilation, or primary, air is provided by a separate ducted system, either to the terminal unit or ducted directly to the space.

Terminal units **without central ventilation** require additional coil capacity to heat or cool the ventilation air required for the end space. This is often difficult because of limited row capacity of coils in terminal units. In addition, care must be taken to minimize the risk of frozen coils in the winter, and to have enough cooling capacity to dehumidify ventilation air in the summer. Ventilation air is ducted to the unit through an opening in the building skin. In most climates, these systems are used in perimeter spaces of buildings with high sensible loads and where close control of humidity is not required. Simultaneous heating and cooling in different parts of the building during intermediate seasons is also possible.

In-room terminal systems are used in almost all classes of buildings. They are especially suitable for smaller projects with little space for a central mechanical room, or projects in which low initial cost and simplified installation are important. These systems are installed in

- Office buildings
- Shopping centers
- Manufacturing plants
- Schools
- Health care facilities
- Hotels and motels
- Apartments
- Nursing homes
- Other multiple-occupancy dwellings

They are also suited to air conditioning existing buildings with limited life or income potential. Although the equipment can be applied as a single unit, this chapter covers applying multiple units to form a complete air-conditioning system for a building.

Advantages

- Individual room temperature control allows each thermostat to be adjusted for a different temperature at relatively low cost.
- Separate heating and cooling sources for ventilation air and secondary heating or cooling give occupants a choice of heating or cooling.
- Less space is required for the air distribution system when the only air required is that for ventilation. Space heating and cooling needs are met by piped hot or chilled water.
- The central air-handling apparatus is smaller than that of an all-air system because less air must be conditioned at that location.
- Dehumidification, filtration, and humidification are performed in a central location remote from conditioned spaces.
- Ventilation air is positively supplied and can accommodate constant recommended outside air quantities.
- Space can be heated without operating the air system, using the secondary-water system. Nighttime fan operation is avoided in an unoccupied building. Emergency power for heating, if required, is much lower than for most all-air systems.

Disadvantages

- For many buildings, in-room terminals are limited to perimeter space; separate systems are required for other areas.
- More controls are needed than for many all-air systems.
- Primary-air supply usually is constant with no provision for shutoff. This is a disadvantage in residential applications, where tenants or hotel room guests may prefer to turn off the air conditioning, or where management may desire to do so to reduce operating expense.
- Low primary chilled-water temperature and/or deep chilled-water coils are needed to control space humidity adequately.
- The system is not appropriate for spaces with high exhaust requirements (e.g., research laboratories) unless supplementary ventilation air is provided.

The preparation of this chapter is assigned to TC 9.1, Large Building Air-Conditioning Systems.

- Central dehumidification eliminates condensation on the secondary-water heat transfer surface under maximum design latent load, but abnormal moisture sources (e.g., open windows, cooking, or people congregating) can cause annoying or damaging condensation. Therefore, a condensate pan should be provided as for other systems.
- Low primary-air temperatures require heavily insulated ducts.

Heating and Cooling Calculations

Basic calculations for airflow, temperatures, relative humidity, loads, and psychrometrics are covered in Chapters 6, 29, and 30 of the 2005 *ASHRAE Handbook—Fundamentals*. The designer should understand the interaction of the terminal units with the central ventilation system (if applicable). Central ventilation systems should be designed to provide air cold enough to satisfy all latent cooling requirements in the end space to maintain dry coils in the terminal units. Some terminal units, such as radiant cooling panels and chilled-beams systems, do not provide latent cooling; building finishes can be damaged by condensation if latent cooling requirements are not addressed.

The HVAC designer should work closely with the architect to optimize building envelope design. Close cooperation of all parties during design can reduce building loads, which allows use of smaller mechanical systems.

In-room systems are used primarily in perimeter spaces of buildings with high sensible loads and where close control of humidity is not required. Variation in air-conditioning load for perimeter building spaces causes significant variations in space cooling and heating requirements, even in rooms that have the same exposure. Accordingly, accurate environmental control in perimeter spaces requires individual control.

Terminal units are often sized for the difference in the internal and external cooling and heating loads. **Internal** loads that should be accounted for include heat gain from lights, occupants, computers and other heat-generating equipment.

External loads vary considerably because of solar heat gain, which varies throughout the day. The magnitude and rate of change of this load depend on building orientation, glass area, capacity to store heat, and cloud cover. Constantly changing shade patterns from adjacent buildings, trees, or exterior columns and nonuniform overhangs can cause significant variations in solar load between adjacent offices on the same solar exposure.

Transmission load can be either a heat loss or a heat gain, depending on outside temperature.

Moderate, uniformly positive pressurization of the building with ventilation air is normally sufficient to offset summer infiltration. In winter, however, infiltration can cause significant heat loss, particularly on lower floors of high-rise buildings. The magnitude of this component varies with wind and stack effect, as well as with the temperature difference across the outside wall.

Space Heating

Some in-room terminal units provide only heating to the end space. Equipment such as cabinet or unit heaters, radiant panels, radiant floors, and finned-tube radiators are designed for heating only. Extreme care must be used with these systems if they are incorporated into a two-pipe changeover piping distribution system, or any other system in which secondary water being piped is not consistently over 100°F. The heating coils in these units are not designed to handle condensation, and there is no drain pipe in the unit. If cold water is provided to these units, dripping condensation from units, valves, or piping may damage building finishes or saturate the insulation, leading to mold growth. Ball valves tied into the automatic temperature control (ATC) system and/or aquastats should be provided to prevent cold water from reaching heating-only terminal units.

Central Ventilation Systems

Generally, the supply air volume from the central apparatus is constant and is called primary or ventilation air to distinguish it from recirculated room air or secondary air. The quantity of primary air supplied to each space is determined by (1) the amount of outside air required for ventilation and (2) the required sensible cooling capacity at maximum room cooling load (if used for sensible cooling). In this approach, during the cooling season, air is dehumidified sufficiently in the central conditioning unit to maintain comfortable humidity conditions and to prevent condensation on the room cooling coil from the normal room latent load. In winter, moisture can be added centrally to limit dryness. As the primary air is dehumidified, it is also cooled to offset part of the room sensible loads. The air may be from outdoors, or may be mixed outside and return air. A heating coil may be required in the central air handler, as well as a preheater in areas with freezing weather.

In the ideal in-room terminal unit design, the secondary cooling coil is always dry; this greatly extends terminal unit life and eliminates odors and the possibility of bacterial growth in the unit in the occupied space. In this case, in-room terminals may be replaced by radiant panels (see Chapter 6) or chilled beams and panels. The primary air normally controls the space humidity. Therefore, the moisture content of supply air must be low enough to offset the room's latent heat gain and to maintain a room dew point low enough to preclude condensation on the secondary cooling surface. Even though some systems operate successfully with little or no condensate, a condensate drain is recommended. In systems that shut down during off hours, start-up load may include considerable dehumidification, producing moisture to be drained away.

Piping Distribution

The chilled or hot water piped to terminal units is called primary or secondary water. The primary air and water are cooled or heated remotely in central equipment rooms.

The water side, in its basic form, consists of a pump and piping to convey water to the heat transfer surface in the unit in each conditioned space. In-room terminals are categorized as two- or four-pipe, and the water may provide heating, cooling, or both, depending on the type of in-room terminal system. They are similar in function and include both cooling and heating capabilities for year-round air conditioning. These piping arrangements are discussed in greater detail in the section on Fan-Coil Units and in Chapter 12.

Other Considerations

In-room terminal systems can provide heating, cooling, and ventilation air to the end space. The amount of ventilation air required depends on the number of occupants in the space and on other factors (see ASHRAE *Standard* 62.1). The rate of airflow per person or per unit area is also usually dictated by state codes, based on activity in the space and contaminant loads. If the amount of ventilation air required is considerable (i.e., 10% or more of a building's entire air volume), the designer needs to provide exhaust or other means to avoid overpressurizing the space.

First, Operating, and Maintenance Costs

As with all systems, the initial cost of an in-room terminal system varies widely, depending on location, local economy, and contractor preference (even for identical systems). For example, a unit ventilator system is less expensive than fan-coil units with a central ventilation system, because it does not require extensive ductwork distribution. The operating cost depends on the system selected, the designer's skill in selecting and correctly sizing components, and efficiency of the duct and piping design. A terminal unit design without a central ventilation system is often one of the less expensive systems to install.

Because in-room terminal equipment is in occupied spaces, maintenance may be more time consuming, depending on the size of the facility. The equipment is less complex, though, and often units are simply replaced, minimizing the time spent in the occupied space.

Energy

The engineer's early involvement in design of any facility can considerably reduce the building's energy consumption. Careful design minimizes system energy costs. In practice, however, a system might be selected based on a low first cost or to perform a particular task. In general, terminal units can save energy if the building automation system (BAS) controls operation of the units and can deenergize them if the space is unoccupied. This adds significant cost to the controls system, but saves on operation. If a central ventilation system is used, energy recovery in the air handler could be another option.

SYSTEM COMPONENTS AND CONFIGURATIONS

Components

Terminal units have many common components, mainly a fan, coil(s), filter, dampers and controls (although some units only have coils and controls).

Automatic Damper. If a terminal unit is providing ventilation air through the envelope of the building, an automatic damper is needed to stop airflow when the room is not occupied. Because in-room terminal units often have a ducted primary or central ventilation air system, a damper on this system allows airflow to be balanced.

Filtration. In-room terminals typically come with a 2 in. throwaway filter. The components that comprise a terminal unit are usually assembled into a cabinet with little room to spare.

Heating and Cooling Coils. Coils in terminal units are usually available in one-, two-, three-, and sometimes four-row coils for cooling and one- or two-row coils for heating.

Fan. Terminal units typically are not complex. Most units have a three-position (low/medium/high) fan that requires manual adjustment. Some larger units serving multiple spaces may have a variable-frequency drive (VSD) on the fan. Fans are also available in direct-drive and belt-drive models.

Piping. Depending on the type of in-room terminal application, terminal units may be furnished and installed with refrigeration, chilled-water, hot-water, and/or condenser water pipe distribution.

Duct Distribution. Terminal units work best without extensive ductwork. With ducts, static pressure on the fan (instead of the coil capacity) may be the determining factor for sizing the terminal units.

Automatic Controls. Most terminal units are controlled with a standard electronic thermostat, either provided by the manufacturer or packaged by the automatic temperature controls (ATC) contractor. The thermostat should be capable of seven-day programming and night setback. Terminal units can be programmed into a BAS, but the cost to do so may be prohibitive, depending on the number of terminal units in the building.

Capacity Control. Terminal unit capacity is usually controlled by coil water flow, fan speed, or both. Water flow can be thermostatically controlled by return air temperature or a wall thermostat and two- or three-way valve. Unit controls may be a self-contained direct digital microprocessor, line voltage or low-voltage electric, or pneumatic. Fan speed control may be automatic or manual; automatic control is usually on/off, with manual speed selection. Units are available with variable-speed motors for modulated speed control. Room thermostats are preferred where automatic fan speed control is used. Return air thermostats do not give a reliable index of room temperature when the fan is off. Residential fan-coil units have manual three-speed fan control, with water temperature (both heating and cooling) scheduled based on outside temperature. On/off speed control is poor because (1) alternating shifts in fan noise level are more obvious than the sound of a constantly running fan, and (2) air circulation patterns in the room are noticeably affected.

Summer room humidity levels tend to be relatively high, particularly if modulating chilled-water control valves are used for room temperature control. Alternatives are two-position control with variable-speed fans (chilled water is either on or off, and airflow is varied to maintain room temperature) and the bypass unit variable chilled-water temperature control (chilled-water flow is constant, and face and bypass dampers are modulated to control room temperature).

Configurations

Terminal units are available in many different configurations; however, not all of the configurations discussed here are available for all types of terminal units. Figure 1 shows several configurations that are particularly applicable for fan-coils.

Low-profile vertical units are available for use under windows with low sills; however, in some cases, the low silhouette is achieved by compromising features such as filter area, motor serviceability, and cabinet style.

Floor-to-ceiling, **chase-enclosed units** are available in which the water and condensate drain risers are part of the factory-furnished unit. Stacking units with integral prefabricated risers directly one above the other can substantially reduce field labor for installation, an important cost factor. These units are used extensively in hotels and other residential buildings. For units serving multiple rooms, the supply and return air paths must be isolated from each other to prevent air and sound interchange between rooms.

Vertical and chase-enclosed models at the perimeter give better results in climates or buildings with high heating requirements. Heating is enhanced by under-window or exterior wall locations. Vertical units can operate as convectors with the fans turned off during night setback, and overheating can become an issue.

Horizontal overhead units may be fitted with ductwork on the discharge to supply several outlets. A single unit may serve several rooms (e.g., in an apartment house where individual room control is not essential and a common air return is feasible). Units must have larger fan motors designed to handle the higher static pressure resistance of the connected ductwork.

Horizontal models conserve floor space and usually cost less, but when located in furred ceilings, they can create problems such as condensate collection and disposal, mixing return air from other rooms, leaky pans damaging ceilings, and difficult access for filter and component removal. In addition, possible condensate leakage may present air quality concerns.

PIPING ARRANGEMENTS

For terminal units requiring chilled and/or hot water, the piping arrangement determines the performance quality, ease of operation, and initial cost of the system. Each piping arrangement is briefly discussed here; for further details, see the sections on Two-Pipe Systems with Central Ventilation, and Four-Pipe Systems.

Four-Pipe Distribution

Four-pipe distribution of secondary water has dedicated supply and return pipes for chilled and hot water. The four-pipe system generally has a high initial cost compared to a two-pipe system, but it has the best system performance. It provides (1) all-season availability of heating and cooling at each unit, (2) no summer/winter changeover requirement, (3) simpler operation, and (4) hot-water heating that uses any heating fuel, heat recovery, or solar heat. In addition, it can be controlled to maintain a dead band between heating and cooling so simultaneous heating and cooling cannot occur.

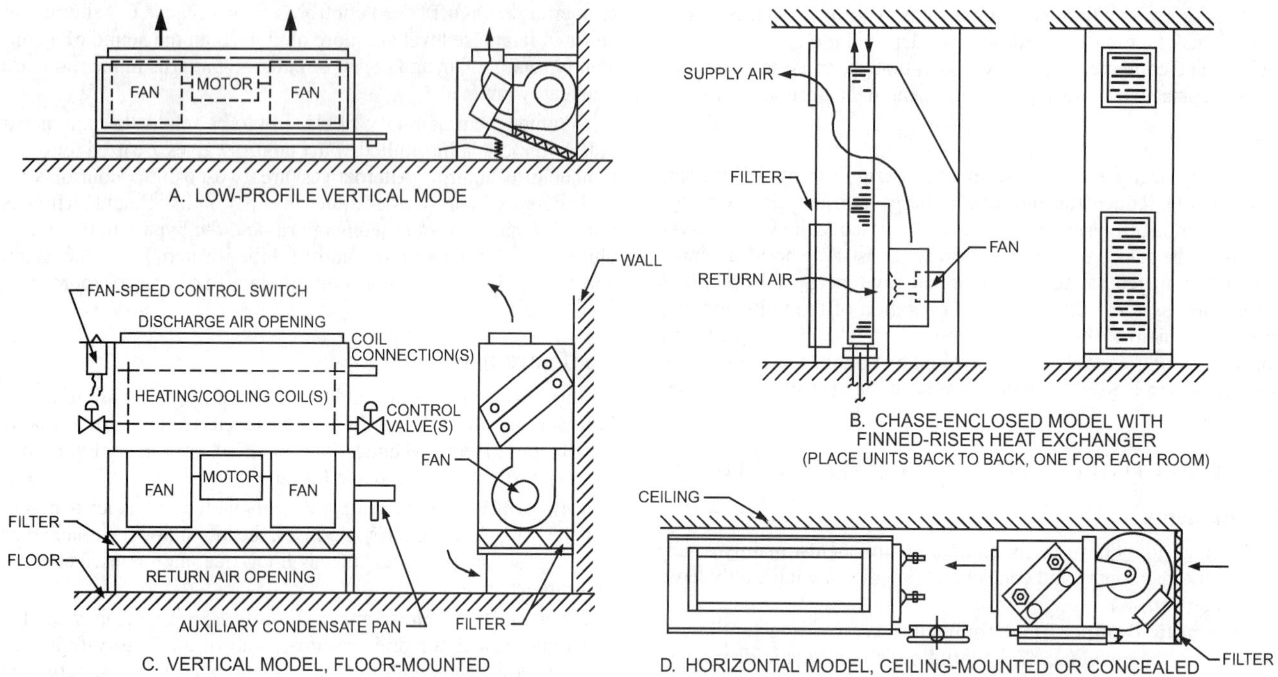

Fig. 1 Typical Fan-Coil Arrangements
(Courtesy RDK Engineers)

Two-Pipe Distribution

Two-Pipe Changeover Without Central Ventilation. In this system, either hot or cold water is supplied through the same piping. The terminal unit has a single coil. The simplest system with the lowest initial cost is the two-pipe changeover with (1) outside air introduced through building apertures, (2) manual three-speed fan control, and (3) hot- and cold-water temperatures scheduled by outside temperatures. This system is generally used in residential buildings or hotels with operable windows and relies on the occupant to control fan speed and open or close windows. The changeover temperature is set at some predetermined set point. If a thermostat is used to control water flow, it must reverse its action depending on whether hot or cold water is available.

The two-pipe system cannot simultaneously heat and cool, which is required for most projects during intermediate seasons when some rooms need cooling and others need heat. This problem can be especially troublesome if a single piping zone supplies the entire building. This deficiency may be partly overcome by dividing the piping into zones based on solar exposure. Then each zone may be operated to heat or cool, independent of the others. However, one room may still require cooling while another room on the same solar exposure requires heating, particularly if the building is partially shaded by an adjacent building or tree.

Another deficiency is the need for frequent changeover from heating to cooling, which complicates operation and increases energy consumption to the extent that it may become impractical. For example, two-pipe changeover system hydraulics must consider the water expansion (and relief) that occurs during cycling from cooling to heating.

Caution must be used with this system when outside air is directly introduced into spaces with widely varying internal loads. Continuous introduction of outside air with reduced load can introduce unconditioned outside air, which can cause very high space humidity levels that may not be able to be handled without reheat capability. The outside air damper in the unit must be motor-operated so it can be closed during unoccupied periods when minimal cooling is required.

For these reasons, the designer should consider the disadvantages of the two-pipe system carefully; many installations of this type waste energy, and have been unsatisfactory in climates where frequent changeover is required and where interior loads require cooling simultaneously as exterior spaces require heat.

Two-Pipe Changeover with Partial Electric Strip Heat. This arrangement provides simultaneous heating and cooling in intermediate seasons by using a small electric strip heater in the terminal unit. The unit can handle heating requirements in mild weather, typically down to 40°F, while continuing to circulate chilled water to handle any cooling requirements. When the outside temperature drops sufficiently to require heating beyond the electric strip heater capacity, the water system must be changed over to hot water.

Two-Pipe Nonchangeover with Full Electric Strip Heat. This system may not be recommended for energy conservation, but it may be practical in areas with a small heating requirement.

Three-Pipe Distribution

Three-pipe distribution uses separate hot- and cold-water supply pipes. A common return pipe carries both hot and cold water back to the central plant. The terminal unit control introduces hot or cold water to the common unit coil based on the need for heating or cooling. This type of distribution is not recommended because of its energy inefficiency from constantly reheating and recooling water, and it does not comply with most recognized energy codes.

FAN-COIL UNIT SYSTEMS

Fan-coil units (1) can provide cooling as well as heating, (2) normally move air by forced convection through the conditioned space, (3) filter circulating air, and (4) may introduce outside ventilation air. These units are available in various configurations to fit under windowsills, above furred ceilings, in vertical pilasters built into walls, etc. Individual room thermostats are usually tied to the piping valve or fan controller in each unit to maintain room temperature. Ventilation air may be provided by a louver in the outside wall and then ducted to the terminal unit, or by a central ventilation system. Fan-coils are also often used in residential applications where ventilation requirements are met by using operable windows.

Basic elements of fan-coil units are a finned-tube heating/cooling coil, filter, and fan section (Figure 2). The fan recirculates air continuously from the space through the coil, which contains either hot or chilled water. The unit may contain an additional electric resistance, steam, or hot-water heating coil. The electric heater is often sized for fall and spring to avoid changeover problems in two-pipe systems; it may also provide reheat for humidity control. A cleanable or replaceable moderate-efficiency filter upstream of the fan prevents clogging of the coil with dirt or lint entrained in recirculated air. It also protects the motor and fan, and reduces the level of airborne contaminants in the conditioned space. The fan and motor assembly is arranged for quick removal for servicing. The fan-coil unit is also equipped with an insulated drain pan.

Most manufacturers furnish units with cooling performance certified as meeting Air-Conditioning and Refrigeration Institute (ARI) standards. The unit prototypes have been tested and labeled by Underwriters Laboratories (UL) or Engineering Testing Laboratories (ETL), as required by some codes. Requirements for testing and standard rating of room fan-coils with air-delivery capacities of 2000 cfm or below are described in ARI *Standard* 440 and ANSI/ASHRAE *Standard* 79.

Fan-coil units for the U.S. market are generally available in nominal sizes of 200, 300, 400, 600, 800, and 1200 cfm, often with multispeed, high-efficiency fan motors. A major advantage of fan-coil unit systems is that the delivery system (piping versus duct systems) requires less building space [a smaller central fan room (or none) and little duct space]. The system has all the benefits of a central water chilling and heating plant, but allows local terminals to be shut off in unused areas. It gives individual room control with little cross-contamination of recirculated air. Extra capacity for quick pulldown response may be provided. Because this system can heat with low-temperature water, it is particularly suitable for use with solar or heat recovery equipment. For existing building retrofit, it is often easier to install piping and wiring for a fan-coil unit system than the large ductwork required for an all-air system.

Fan-coil systems are best applied where individual space temperature control or cross-contamination prevention is needed. Suitable applications are hotels, motels, apartment buildings, and office buildings. Fan-coil systems are used in many hospitals, but they are less desirable because of the low-efficiency filtration and difficulty in maintaining adequate cleanliness in the unit and enclosure. In addition, limits set by *Guidelines for Design and Construction of Hospital and Health Care Facilities* (AIA 2001) do not allow air recirculation in certain types of spaces.

Types and Location

Several types of fan-coils are discussed in the section on System Components and Configurations, and illustrated in Figure 1.

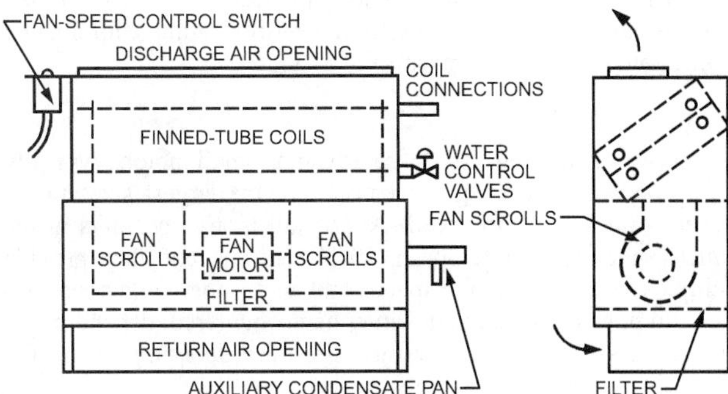

Fig. 2 Typical Fan-Coil Unit

Ventilation Air Requirements

Fan-coil units receive ventilation air from a penetration in the outside wall or from a central air handler. Units that have outside air ducted to them from an aperture in the building envelope are not suitable for commercial buildings because wind pressure allows no control over the amount of outside air admitted. Ventilation rates can be affected by stack effect and wind direction and speed. Also, freeze protection may be required in cold climates. Fan-coils are, however, often used in residential construction because of their simple operation and low first cost, and because residential rooms are often ventilated by opening windows or by outside wall apertures, if not handled by a central system. Operable windows can cause imbalance in a ducted ventilation air system.

When outside air is introduced from a central ventilation system, it may be connected to the inlet plenum of the fan-coil or introduced directly into the space. If introduced directly, ensure that this air is pretreated, dehumidified, and held at a temperature equal to the room temperature so as not to cause occupant discomfort when the space unit is off. One way to prevent air leakage is to provide a spring-loaded motorized damper that closes off ventilation air when the unit's fan is off.

Selection

Some designers size fan-coil units for nominal cooling at medium speed when a three-speed control switch is provided. This method ensures quieter operation in the space and adds a safety factor (capacity can be increased by operating at high speed). Sound power ratings are available from many manufacturers.

If using a horizontal overhead fan coil with ducted supply and return, fan capacity may be the factor that decides the unit's size, not the coil's capacity. Static pressure as little as 0.3 in. of water can significantly affect fan capacity.

Only the internal space heating and cooling loads need to be handled by terminal fan-coil units when outside air is pretreated by a central system to a neutral air temperature of about 70°F. This pretreatment should reduce the size and cost of the terminal units. All loads must be considered in unit selection when outside air is introduced directly through building apertures into the terminal unit.

Wiring

Fan-coil blower fans are driven by small motors, generally shaded pole or capacitor start with inherent overload protection. Operating wattage of even the largest sizes rarely exceeds 300 W at high speed. Running current rarely exceeds 2.5 A. Almost all motors on units in the United States are 120 V, single-phase, 60 Hz current, and they provide multiple (usually three) fan speeds and an off position. Other voltages and power characteristics may be encountered, depending on location, and should be investigated before determining the fan motor characteristics.

In planning the wiring circuit, required codes must be followed. The preferred wiring method generally provides separate electrical circuits for fan-coil units and does not connect them into the lighting circuit.

Separate electrical circuits connected to a central panel allow an energy management system or the building operator to turn off unit fans from a central point during unoccupied hours. Although this panel costs more initially, it can lower operating costs in buildings that do not have 24 h occupancy. In hot, humid climates, care must be taken to avoid excess humidity when units are off, to avoid mildew formation. Using separate electrical circuits allows a single remote thermostat to be mounted in a well-exposed perimeter space to operate unit fans. Another method is to operate the fan-coil continuously on low speed during unoccupied periods.

Condensate

Even when outside air is pretreated, a condensate removal system should be installed for fan-coil units. This precaution ensures that moisture condensed from air from an unexpected source that bypasses the ventilation system (e.g., an open window) is removed. Drain pans are integral for all units. Condensate drain lines should be oversized to avoid clogging with dirt and other materials, and drains should be cleaned periodically. Condensation may occur on the outside of drain piping, which requires that these pipes be insulated. Many building codes have outlawed systems without condensate drain piping because of the potential damage and possibility of mold growth in stagnant water accumulated in the drain pan.

Capacity Control

Fan-coil unit capacity is usually controlled by coil water flow, fan speed, or a combination of these, as discussed in the section on System Components and Configurations.

Maintenance

Fan-coil unit systems require much more maintenance than central all-air systems, and this work must be done in occupied areas. Units operating at low dew points require condensate pans and a drain system that must be cleaned and flushed periodically to prevent overflow and microbial build-up. Drain pans should be trapped to prevent any gaseous back-up. Condensate disposal can be difficult and costly. It is also difficult to clean the coil.

Room fan-coil units are equipped with filters that should be cleaned or replaced when dirty. Filters are small and low-efficiency, and require frequent changing to maintain air volume. Good filter maintenance improves sanitation and provides full airflow, ensuring full-capacity delivery. Cleaning frequency varies with the application. Units in apartments, hotels, and hospitals usually require more frequent filter service because of lint. Fan-coil unit motors may require periodic lubrication. Motor failures are not common, but when they occur, the entire fan can be quickly replaced with minimal interruption in the conditioned space.

UNIT VENTILATOR SYSTEMS

Unit ventilators are similar to fan-coil units, except that unit ventilators are designed to provide up to 100% outside air to the space. They (1) can provide cooling as well as heating, (2) normally move air by forced convection through the conditioned space, (3) filter circulating air, and (4) introduce outside ventilation air as required to meet the needs of the space. These units are available in two main configurations: floor-mounted below a window, or horizontal overhead with ducted supply and return. Most frequently used in classrooms, which need a high percentage of outside air for proper ventilation, unit ventilators are often located under a window along the perimeter wall. Originally, unit ventilators provided only heating, often using steam as the heating source. Now they are available in a two-pipe configuration with changeover, two-pipe with electric heating, four-pipe, or with heating and DX coils for spaces, such as computer rooms, that may require year-round cooling. Limited ductwork is required, allowing for higher and often exposed ceiling systems. The drawback to these units is that they are often used as shelving in the classroom; books and paperwork may be stacked on top of them, impeding airflow to the space. Also, ventilation air intake louvers can become choked by vegetation if they are not properly maintained, which can lead to indoor air quality issues. In addition, because the fans are sized to accommodate 100% ventilation air, they are typically noisier than fan-coils; however, recent research has led to new units that are much quieter.

Basic elements of a unit ventilator are a finned-tube coil, filter, and fan section, and often also a face-and-bypass damper. The fan recirculates air continuously from the space through the coil, which contains either hot or chilled water. The unit may contain an additional electric resistance, steam, or hot-water heating coil. The electric heater is often sized for fall and spring to avoid changeover problems in two-pipe systems, and may also be used to provide reheat for humidity control. A cleanable or replaceable moderate-efficiency filter upstream of the fan prevents clogging of the coil with dirt or lint entrained in recirculated air, protects the motor and fan, and reduces the level of airborne contaminants in the conditioned space. The fan and motor assembly is arranged for quick removal for servicing. The fan-coil unit is also equipped with an insulated drain pan.

Unit ventilators for the U.S. market are generally available in nominal sizes of 750, 1000, 1500, and 2000 cfm, often with multispeed, high-efficiency fan motors. A major advantage of terminal unit systems is that the delivery system (piping versus duct systems) requires less building space [a smaller central fan room (or none) and little duct space]. The system has all the benefits of a central water chilling and heating plant, but allows local terminals to be shut off in unused areas. It gives individual room control with little cross-contamination of recirculated air. Because this system can heat with low-temperature water, it is particularly suitable for use with solar or heat recovery equipment. For existing building retrofit, it is easiest to replace unit ventilators in kind. If a building did not originally use unit ventilators, installing multiple ventilation air intake louvers to accommodate the unit ventilators may be cost-prohibitive. Likewise, attempting to install a different type of system in a building originally fitted with unit ventilators would require bricking up intake louvers and installing exposed ductwork (if there is no ceiling plenum) or creating a ceiling plenum in which to run ductwork. Unit ventilators are best applied where individual space temperature control with large amounts of ventilation air is needed.

Types and Location

Unit ventilators are available in two main configurations. **Floor-mounted units** have different ventilation air ductwork connections, including from the back or ducted collar on the top of the cabinet. **Ceiling-mounted units** can be mounted completely exposed, partially exposed in a soffit, fully recessed, or concealed. Ventilation air connections can be made in the back or top of the unit.

The heating/cooling coils in the unit ventilators differ considerably from fan-coils. Coils in unit ventilators are much deeper, because the unit ventilator needs to be able to heat, cool, and dehumidify up to 100% ventilation air. Coil selection must be based on the temperature of the entering mixture of primary and recirculated air, and air leaving the coil must satisfy the room's sensible and latent cooling and heating requirements.

Ventilation Air Requirements

Unlike fan-coils, unit ventilators are capable of providing the entire volume of ventilation air that is required. Unit ventilators can supply 100% outside air when necessary, because coils in a unit ventilator are deeper than those in fan-coil units.

Selection

Some designers size unit ventilators to provide ventilation air at 50% or less of the total unit airflow. In addition, the coils in the unit ventilator must be selected for the entire heating, cooling, and ventilation loads for the space.

Wiring

Unit ventilator blower fans are driven by small motors, typically 1/2 hp or less. Operating power of even the largest sizes rarely exceeds 400 W at high speed. Almost all motors on units in the United States are single-phase, 60 Hz current, and they provide multiple (usually three) fan speeds and an off position. Other voltages and power characteristics may be encountered, depending on location, and should be investigated before determining the fan motor characteristics.

In planning the wiring circuit, required codes must be followed. The preferred wiring method generally provides separate electrical circuits for unit ventilators and does not connect them into the lighting circuit.

Separate electrical circuits connected to a central panel allow an energy management system or the building operator to turn off unit fans from a central point during unoccupied hours. Although this panel costs more initially, it can lower operating costs in buildings that do not have 24 h occupancy. In hot, humid climates, care must be taken to avoid excess humidity when units are off, to avoid mildew formation. Using separate electrical circuits allows a single remote thermostat to be mounted in a well-exposed perimeter space to operate unit fans. Another method is to operate the unit ventilator continuously on low speed during unoccupied periods.

Condensate

Because unit ventilators can intake up to 100% outside air, condensate piping is required. For floor-mounted units along the perimeter of the building, condensate piping can run from the drain pan to the exterior grade. For a ceiling-mounted unit, the piping could be pitched to the outside wall if there is space available to accommodate the pitch, or a condensate pump could be used. Drain pans are integral for all units. Condensate drain lines should be oversized to avoid clogging with dirt and other materials, and condensate drains should be cleaned periodically. Because condensation may occur on the outside of the drain piping, these pipes must be insulated.

Capacity Control

Unit capacity is usually controlled by coil water flow, fan speed, damper configurations, or a combination of these. In addition, unit ventilators often come equipped with a face-and-bypass damper, which allows for another form of capacity control. For additional information, see the discussion on capacity control in the section on System Components and Configurations.

Maintenance

Maintenance on unit ventilators, such as filter replacement and coil cleaning, must be done in the occupied space. Condensate pans and the drain system must be cleaned and flushed periodically to prevent overflow and microbial build-up. Drain pans should be trapped to prevent any gaseous back-up. Good filter maintenance improves sanitation and provides full airflow, ensuring full-capacity delivery. Cleaning frequency varies with the application. Motors may require periodic lubrication.

CHILLED-BEAM SYSTEMS

Chilled beams are an advancement of chilled ceiling panels. As interior loads have increased with the use of computers and other high-load electrical equipment, so has the need for higher-capacity cooling equipment. Passive chilled beams are the first step. They consist of a chilled-water coil mounted inside a cabinet. Chilled water is piped to the convective coil at between 58 and 60°F. Passive beams use convection currents to cool the space. As air that has been cooled by the beam's chilled-water coil falls into the space, warmer air is displaced, rises into the coil, and is cooled. Passive beams are best used with finned-tube radiation along the space perimeter to provide heat and a separate ventilation system to provide tempered, dehumidified air to the space.

Active chilled beams can provide up to approximately 800 Btu/ft. They operate with induction nozzles that entrain room air and mix it with the primary or ventilation air that is ducted to the beam. Chilled water is piped to the coil at between 55 and 60°F. Primary air should be ducted to the beam at around 55°F and dehumidified. The primary air is then mixed with induced room air at a ratio of 2:1. For example, 50 cfm of primary air at 55°F may be mixed with 100 cfm of recirculated room air, and the active beam would distribute 150 cfm at around 65°F. Active beams can have either a two- or four-pipe distribution system. The two-pipe system may be cooling only or two-pipe changeover. Active beams can be designed to heat and cool the occupied space, but finned-tube radiation is still commonly used to provide heating in a space that is cooled with active beams.

Both active and passive beams are designed to operate dry, without condensate. In some models of active beams, a drain pan may be available if the coil is in a vertical configuration. Horizontal coils in passive beams cannot have drain pans, because the area directly below the coil is needed to allow the air in the convection current to circulate.

Chilled beams can be used in various application; however, they are best used in applications with high sensible loads, such as laboratory spaces with high internal heat gains. See manufacturers' information for beam cooling capacities at various water temperatures and flow rates.

Types and Location

Passive beams are available in sections up to 10 ft long and 18 to 24 in. wide. They can be located above the ceiling with perforated panels below it, mounted into the frame of an acoustical tile ceiling, or mounted in the conditioned space. The perforated panels must have a minimum 50% free area and extend beyond either side of the beam for usually half of the unit's width, so the convection current is not hindered. Also, care must be taken to not locate passive beams too close to window treatments, which can also hinder air movement around the beam.

Active beams are available in sections up to 10 ft long and 12 to 24 in. wide. They can be mounted into the frame of an acoustical tile ceiling or in the conditioned space.

Ventilation Air Requirements

Passive beams require a separate ventilation system to provide tempered and dehumidified air to the space. The ventilation air should be ducted to low-wall diffusers or in an underfloor distribution system so that the ventilation air does not disturb the convection currents in the conditioned space. Ventilation air can be ducted directly into the active beams. If more ventilation air is needed to meet the space requirements, the volume of air can be split by the active beams and high-diffusion diffusers.

Selection

Chilled beams are selected based on the calculated heat gain for the space.

Wiring

Chilled beams only require controls wiring. There is no fan or other electrical equipment to be wired.

Condensate

Chilled beams are designed to operate dry, with few exceptions. In some active beams with vertical coils, a drain pan may be installed. However, as a rule, a separate ventilation system should be sized to handle the latent cooling load in the space, and the relative humidity should be closely monitored.

Capacity Control

Capacity of the chilled beam is controlled by a two-way valve on the chilled-water pipe, which is wired to a room thermostat. There is typically one valve per zone (e.g., office, lecture hall). Beams should be piped directly in a reverse/return piping design. Beams are not typically piped in series.

Maintenance

Maintenance on chilled beams is virtually nonexistent. Most manufacturers require blowing off the coils on a regular basis, but some installations may not need this step.

RADIANT-PANEL HEATING SYSTEMS

Radiant heating panels can use either hot water or electricity. The panels are manufactured in standard 24 by 24 or 24 by 48 in. panels that can be mounted into the frame of an acoustical tile ceiling or directly to an exposed ceiling. Radiant panels are designed for all types of applications. They are very energy efficient, providing a comfortable heat without drying out the room air the way a forced hot-air system may. Occupants in a space heated by radiant heat are comfortable at lower room temperatures, which frequently reduces operational costs. See Chapter 6 for more information on these systems.

Types and Location

Radiant panels are typically mounted on the ceiling in a metal frame. Unlike finned-tube radiation, they do not limit furniture placement. Electric radiant panels are available from 250 to 750 W in standard single-phase voltages.

Ventilation Air Requirements

Radiant panels provide space heating. Ventilation air must be supplied by a central ventilation unit that can provide tempered, dehumidified air to the space.

Selection

Radiant panels are selected based on the calculated heat loss for the space.

Wiring

Electric radiant panels are available in standard single-phase voltages. Panels are often prewired, including the ground wire, with lead wires housed in flexible metal conduit and connector for junction box mounting.

Capacity Control

Panel capacity is usually controlled by coil water flow, or in the case of electric heat, capacity steps. Most radiant panels are controlled with a wall-mounted thermostat located in the space.

Maintenance

Because they have no moving parts, radiant panels require little maintenance.

Other Radiant Panel Options

Radiant panels can also provide space cooling; see the section on Chilled-Beam Systems for design and performance guidelines. Another heating and cooling in-room terminal, the **valance unit** has similar design guidelines.

RADIANT-FLOOR HEAT SYSTEMS

Radiant-floor heat is best applied under a finished floor that is typically cold to the touch. Radiant-floor heat systems in the past used flexible copper pipe heating loops encased in concrete. Unfortunately, the soldered joints could fail or the concrete could corrode the pipes, causing them to leak. However, new technologies include flexible plastic tubing (often referred to as PEX, or cross-linked polyethylene) to replace the old flexible copper tubing. PEX tubing is also available with an oxygen diffusion barrier, because oxygen entrained in the radiant heat tubing can cause corrosion on the ferrous connectors between the tubing and the manifold system. PEX tubing is also available in longer lengths than the flexible copper, which eliminates buried joints. The tubing is run back to a manifold system, which includes valves to balance and shut down the system and a small circulator pump. Multiple zones can be terminated at the same manifold.

Radiant-floor heat is commonly designed for residential applications, where ventilation requirements are often met by operable windows, and cooling is not mandatory. Other applications include large open buildings, such as airplane hangars, where providing heat at the floor is much more cost-effective than heating the entire volume of air in the space. Water in the radiant-floor loop is often around 90°F, depending on the floor finish. This is a lower temperature than forced hot-air systems, and reduces the energy required to heat the building. Buildings that have high ceilings, large windows, or high infiltration rates or that require high air change rates can save additional energy by using radiant-floor heat.

Radiant-floor systems are commonly zoned by room. Each room may require multiple pipe circuits, depending on the room's area and the manifold's location. Maximum tubing lengths are determined based on tubing diameter and desired heat output. If tubing is installed in slab on or below grade, rigid insulation should be incorporated to minimize heat losses to the ground.

See Chapter 6 for more information on these systems.

Types and Location

Radiant-floor heating is located in the flooring or just below it with a heat-reflecting wrap. If the final floor finish is hardwood flooring, the radiant-floor heat can be installed in plywood tracks with a heat reflector, below the finished floor. Care must be taken to ensure that, as the final flooring is nailed down, the flexible tubing is not punctured. If the final floor is a ceramic tile or other surface requiring a poured concrete, the radiant floor can be laid out in the concrete. Radiant-floor heating systems can also be mounted below the floor joists, with a heat reflector below the piping. This method is often used in renovations where removing existing flooring is not feasible.

Ventilation Air Requirements

Ventilation air must be supplied by a central ventilation unit that can provide tempered, dehumidified air to the space.

Selection

Spacing between rows of tubing that make up the radiant floor varies, depending on the heat loss of the space. Usually, the entire heat loss of the space is calculated and the tubing spaced accordingly. Another method is to place tubing closer together near the room's perimeter and increase the spacing in the interior. This method is more time consuming, and the difference is only noticeable in large spaces. Supply water temperature in the tubing is determined based on the flooring materials' resistance to heat flow; the temperature is higher for carpeting and a pad than for ceramic tile. The tubing is also available in different nominal diameters, the most common being 3/8 or 1/2 in.

Wiring

Circulator pumps at the manifolds require power.

Capacity Control

Because radiant floors heat the mass of the floor, these systems are typically slow to respond to environmental changes. The circulator pumps start on a call for heat from a thermostat; however, rapid solar gains to a space with many windows could cause the space to overheat.

Maintenance

The circulator pumps, valves, and manifolds are the only components requiring maintenance. In almost all cases, it is easier to replace a failed component than to repair it. Once the tubing is laid out, it should be pressure-tested for leaks; once covered, it is extremely difficult and/or expensive to access.

INDUCTION-UNIT SYSTEMS

Centrally conditioned primary air is supplied to an induction unit's plenum at medium to high pressure. The acoustically treated plenum attenuates part of the noise generated in the unit and duct. High-velocity induction unit nozzles typically generate significant high-frequency noise. A balancing damper adjusts the primary-air quantity within limits.

Medium- to high-velocity air flows through the induction nozzles and induces secondary air from the room through the secondary coil. Thus, the primary air provides the energy required to circulate the secondary air over the coil in the terminal unit. This secondary air is either heated or cooled at the coil, depending on the season, room requirement, or both. Ordinarily, the room coil does no latent cooling, but a drain pan without a piped drain collects condensed moisture from temporary latent loads such as at start-up. This condensed moisture then re-evaporates when the temporary latent loads are no longer present. Primary and secondary (induced) air is mixed and discharged to the room.

Secondary airflow can cause induction-unit coils to become dirty enough to affect performance. Lint screens used to protect these terminals require frequent in-room maintenance and reduce unit thermal performance.

Induction units are installed in custom enclosures, or in standard cabinets provided by the manufacturer. These enclosures must allow proper flow of secondary air and discharge of mixed air without imposing excessive pressure loss. They must also allow easy servicing. Although induction units are usually installed under a window at a perimeter wall, units designed for overhead installation are also available. During the heating season, the floor-mounted induction unit can function as a convector during off hours, with hot water to the coil and without a primary-air supply. Numerous induction unit configurations are available, including units with low overall height or with larger secondary-coil face areas to suit particular space or load needs.

Induction units may be noisier than fan-coil units, especially in frequencies that interfere with speech. On the other hand, white noise from the induction unit enhances acoustical privacy by masking speech from adjacent spaces.

In-room terminals operate dry, with an anticipated life of 15 to 25 years. The piping and ductwork longevity should equal that of the building. Individual induction units do not contain fans, motors, or compressors. Routine service is generally limited to temperature controls, cleaning lint screens, and infrequently cleaning the induction nozzles.

In existing induction systems, conserving energy by raising the chilled-water temperature on central air-handling cooling coils can damage the terminal cooling coil, causing it to be used constantly as a dehumidifier. Unlike fan-coil units, the induction unit is not designed or constructed to handle condensation. Therefore, it is critical that an induction terminal operates dry.

Induction units are rarely used in new construction. They consume more energy because of the increased power needed to deliver primary air against the pressure drop in the terminal units, and they generate high-frequency noise from the induction nozzles. In addition, the initial cost for a four-pipe induction system is greater than for most all-air systems. However, induction units are still used for direct replacement renovation: because the architecture was originally designed to accommodate the induction unit, other systems may not be easily installed.

SUPPLEMENTAL HEATING UNITS

In-room supplemental heating units come in all sizes. Units can have either electric or hot water heat, and sometimes steam; they can be surface-mounted, semirecessed, or recessed in the walls on the floor or horizontally along the ceiling. Baseboard radiation is usually located at the source of the heat loss, such as under a window or along a perimeter wall, and is usually rated for between 400 and 600 Btu/ft at 170°F. Other supplemental heating units include unit heaters, wall heaters, and cabinet heaters.

All supplemental heating units can be supplied with an integral or separate wall-mounted thermostat. If the heater is located low in the space, an integral thermostat is sufficient most of the time; however, if the unit is mounted horizontally near the ceiling, the thermostat should be wired so that it is located in the space, for accurate space temperature readings. In addition, units may have a summer fan option, which allows the fan to turn on for ventilation. Water flow to space supplemental heaters should be cut off anytime the water temperature to the coil is below 80°F, to avoid condensation and consequent damage or mold growth.

CENTRAL PLANT EQUIPMENT

Central equipment size is based on the block load of the entire building at the time of the building peak load, not on the sum of individual in-room terminal-unit peak loads. Cooling load should include appropriate diversity factors for lighting and occupant loads. Heating load is based on maintaining the unoccupied building at design temperature, plus an additional allowance for pickup capacity if the building temperature is set back at night. For additional information, see Chapter 3.

If water supply temperatures or quantities are to be reset at times other than at peak load, the adjusted settings must be adequate for the most heavily loaded space in the building. Analysis of individual room load variations is required.

If the side of the building exposed to the sun or interior zone loads requires chilled water in cold weather, consider using condenser water with a water-to-water heat exchanger. Varying refrigeration loads require the water chiller to operate satisfactorily under all conditions.

VENTILATION

Central fan equipment is primarily used for an in-room terminal unit system to provide the correct amount of ventilation or makeup air to the various spaces served by terminal units.

Ventilation air is generally the most difficult factor to control and represents a major load component. The designer must select the method that meets all applicable codes, performance requirements, cost constraints, and health requirements.

A central, outside air pretreatment system, which maintains neutral air at about 70°F, best controls ventilation air with the greatest freedom from problems related to the building's stack effect and infiltration. Ventilation air may then be introduced to the room through the terminal unit, or directly into the room as shown in Figure 3. Any type of terminal unit in any location may be used if the outside air ventilation system has separate air outlets.

Ventilation air contributes significantly to the room latent cooling load, so a dehumidifying coil should be installed in the central ventilation system to reduce room humidity during periods of high outside moisture. Centrally supplied air can be supplied at a low

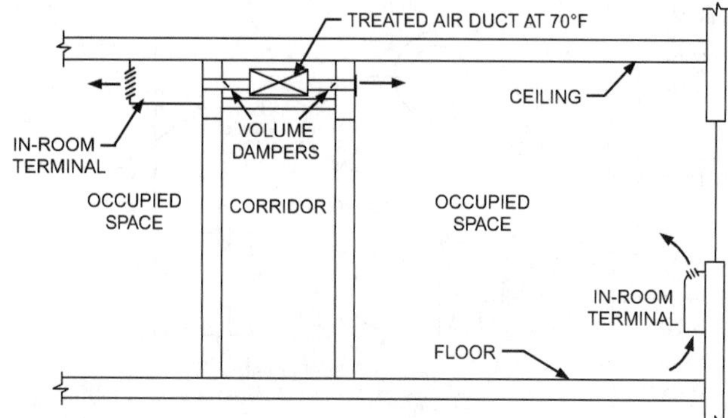

Fig. 3 Ventilation from Separate Duct System

enough dew point to absorb moisture generated in the space, but as a minimum should be supplied at a neutral condition so that the room terminal unit has to remove only the space-generated latent load.

An additional advantage of central ventilation is that, if its supply air dew point is selected to handle the internal latent load, the terminal cooling coil remains dry, extending the unit's life. However, a piped condensate drain is still recommended. This neutral temperature removes the outside air load from the terminal unit, so it can switch from heating to cooling and back without additional internal or external heat loads.

In buildings where terminal units only serve exterior zones and a separate all-air system serves interior zones, exterior zone ventilation air can be provided through the interior zone system. This arrangement can provide desirable room humidity control, as well as temperature control of the ventilation air. In addition, ventilation air held at the neutral condition of 70°F at 50% rh can be introduced into any terminal unit without affecting comfort conditions.

PRIMARY-AIR SYSTEMS

Figure 4 illustrates a primary-air system for in-room terminal systems. The components are described in Chapter 4. Some primary-air systems operate with 100% outside air at all times. Systems using return air should have provision for operating with 100% outside air (economizer cycle) to reduce operating cost during some seasons. In some systems, when the quantity of primary air supplied exceeds the ventilation or exhaust required, excess air is recirculated by a return system common with the interior system. A good-quality filter (55% efficiency or higher) is desirable in the central air treatment apparatus. If it is necessary to maintain a given humidity level in cold weather, a humidifier can usually be installed. Steam humidifiers have been used successfully. Water-spray humidifiers must be operated in conjunction with (1) the preheat coil elevating the temperature of the incoming air or (2) heaters in the spray water circuit. Water-spray humidifiers should be used with caution, however, because of the possible growth of undesirable organisms in untreated water.

The cooling coil is usually selected to provide primary air at a dew point low enough to dehumidify the total system. Supply air must leave the cooling coil at about 50°F or less, and be almost completely saturated.

The supply fan should be selected at a point near maximum efficiency to reduce power consumption, supply air heating, and noise. Sound absorbers may be required at the fan discharge to attenuate fan noise.

Reheat coils are required in a two-pipe system. Reheat is not required for primary-air supply of four-pipe systems. Formerly, many primary-air distribution systems for induction units were designed with 8 to 10 in. of water static pressure. With energy use restrictions, this is no longer economical. Good duct design and

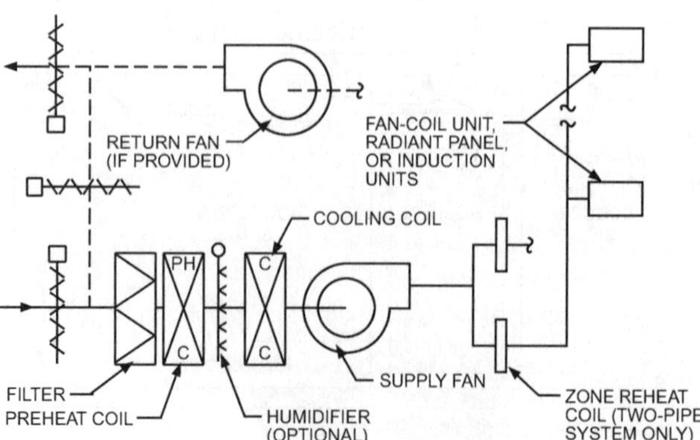

Fig. 4 Primary-Air System

elimination of unnecessary restrictions (e.g., sound traps) can result in primary systems that operate at 4.5 to 6.0 in. of water. Primary-air distribution systems serving fan-coil systems can operate at pressures 1.0 to 1.5 in. of water lower. Careful selection of the primary-air cooling coil and induction units for reasonably low air-pressure drops is necessary to achieve a medium-pressure primary-air system. Distribution for fan-coil systems may be low-velocity or a combination of low- and medium-velocity systems. See Chapter 35 in the 2005 *ASHRAE Handbook—Fundamentals* for a discussion of duct design. Variations in pressure between the first and last terminals should be minimized to limit the pressure drop required across balancing dampers.

Room sound characteristics vary depending on unit selection, air system design, and equipment manufacturer. Units should be selected by considering the unit manufacturer's sound power ratings, desired maximum room noise level, and the room's acoustical characteristics. Limits of sound power level can then be specified to obtain acceptable acoustical performance.

PERFORMANCE UNDER VARYING LOAD

Under peak load conditions, the psychrometrics of induction-unit and fan-coil unit systems are essentially identical for two- and four-pipe systems. Primary air mixes with secondary air conditioned by the room coil in an induction unit before delivery to a room. Mixing also occurs in a fan-coil unit with a direct-connected primary-air supply. If primary air is supplied to the space separately, as in fan-coil systems with independent primary-air supplies, the same effect occurs in the space.

During cooling, the primary-air system provides part of the sensible capacity and all of the dehumidification. The rest of the sensible capacity is accomplished by the cooling effect of secondary water circulating through the unit cooling coils. In winter, primary air is provided at a low temperature, and if humidity control is provided, the air is humidified. All room heating is supplied by the secondary-water system. All factors that contribute to the cooling load of perimeter space in the summer, except transmission, add heat in the winter. The transmission factor becomes negative when the outside temperature falls below room temperature. Its magnitude is directly proportional to the difference between the room and outside temperatures.

For in-room terminal unit systems, it is important to note that in applications where primary air enters at the terminal unit, primary air is provided at summer design temperature during winter. If the economizer cycle is used, heating and cooling energy is not duplicated by reheating primary air that has already been mechanically cooled. For systems where primary air does not enter at the terminal unit, the primary air should be reset to room temperature in winter. A limited amount of cooling can be accomplished by the primary air operating without supplementary cooling from the secondary coil. As long as internal heat gains are not high, this amount of cooling is usually adequate to satisfy east and west exposures during the fall, winter, and spring, because solar heat gain is typically reduced during these seasons. In the northern hemisphere, the north exposure is not a significant factor because solar gain is very low; for south, southeast, and southwest exposures, the peak solar heat gain occurs in winter, coincident with a lower outside temperature (Figure 5). This cooling opportunity may not be available where primary air is supplied directly to the space, because this air could overcool spaces where solar heat gain or internal heat gain is low.

In buildings with large areas of glass, heat transmitted from indoors to the outside, coupled with the normal supply of cool primary air, does not balance internal heat and solar gains until an outside temperature well below freezing is reached. Double-glazed windows with clear or heat-absorbing glass aggravate this condition because this type of glass allows constant inflow of solar radiation during the winter. However, the insulating effect of the

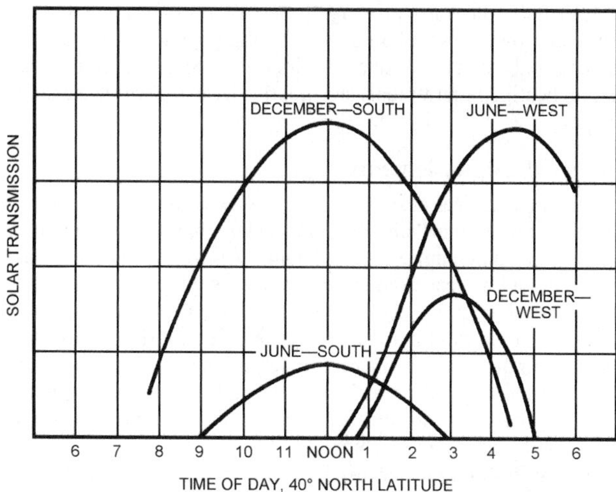

Fig. 5 Solar Radiation Variations with Seasons

double glass reduces reverse transmission; therefore, cooling must be available at lower outside temperatures. In buildings with very high internal heat gains from lighting or equipment, the need for cooling from the room coil, as well as from the primary air, can extend well into winter. In any case, the changeover temperature at which the cooling capacity of the secondary-water system is no longer required for a given space is an important calculation. All these factors should be considered when determining the proper changeover temperature.

CHANGEOVER TEMPERATURE

For all systems using a primary-air system for outside air, there is an outside temperature (a **balance temperature**) at which secondary cooling is no longer required. The system can cool by using outside air at lower temperatures, and heating rather than cooling is needed. For all-air systems operating with up to 100% outside air, mechanical cooling is seldom required at outside temperatures below 55°F. An important characteristic of in-room terminal unit systems, however, is that secondary-water cooling may still be needed, even when the outside temperature is considerably less than 50°F. This cooling may be provided by the mechanical refrigeration unit or by a thermal economizer cycle. Full-flow circulation of primary air through the cooling coil below 50°F often provides all the necessary cooling while preventing coil freeze-up and reducing the preheat requirement. Alternatively, secondary-water-to-condenser-water heat exchangers function well. Some systems circulate condenser water directly through the secondary chilled-water system. This system should be used with caution, recognizing that the vast secondary-water system is being operated as an open recirculating system with the potential hazards that may accompany improper water treatment.

The outside temperature at which the heat gain to every space can be satisfied by the combination of cold primary air and the transmission loss is called the **changeover temperature**. Below this temperature, cooling is not required.

The following empirical equation approximates the changeover temperature at sea level. It should be fine-tuned after system installation (Carrier 1965):

$$t_{co} = t_r - \frac{q_{is} + q_{es} - 1.1Q_p(t_r - t_p)}{\Delta q_{td}} \tag{1}$$

where

t_{co} = temperature of changeover point, °F

t_r = room temperature at time of changeover, normally 72°F

t_p = primary-air temperature at unit after system is changed over, normally 56°F

Q_p = primary-air quantity, cfm

q_{is} = internal sensible heat gain, Btu/h

q_{es} = external sensible heat gain, Btu/h

Δq_{td} = heat transmission per degree of temperature difference between room and outside air

In two-pipe changeover systems, the entire system is usually changed from winter to summer operation at the same time, so the room with the lowest changeover point should be identified. In northern latitudes, this room usually has a south, southeast, or southwest exposure because the solar heat gains on these exposures reach their maximum during winter.

If the calculated changeover temperature is below approximately 48°F, an economizer cycle should operate to allow the refrigeration plant to shut down.

Although factors controlling the changeover temperature of induction unit systems are understood by the design engineer, the basic principles may not be readily apparent to system operators. Therefore, it is important that the concept and calculated changeover temperature are clearly explained in operating instructions given before operating the system. Some increase from the calculated changeover temperature is normal in actual operation. Also, a range or band of changeover temperatures, rather than a single value, is necessary to preclude frequent change in seasonal cycles and to grant some flexibility in operation. The difficulties associated with operator understanding and the need to perform changeover several times a day in many areas have severely limited the acceptability of the two-pipe changeover system.

REFRIGERATION LOAD

The design refrigeration load is determined by considering the entire portion or block of the building served by the air-and-water system at the same time. Because the load on the secondary-water system depends on the simultaneous demand of all spaces, the sum of the individual room or zone peaks is not considered.

The peak load time is influenced by the outside wet-bulb temperature, period of building occupancy, and relative amounts of east, south (in the northern hemisphere), and west exposures. Where the solar load's magnitude is about equal for each exposure, the building peak usually occurs in midsummer afternoon when the west solar load and outside wet-bulb temperature are at or near concurrent maximums.

At sea level, the refrigeration load equals the primary-air cooling coil load plus the secondary system heat pickup:

$$q_{re} = q_s + 4.5Q_p(h_{ea} - h_{la}) - 1.1Q_p(t_r - t_s) \tag{2}$$

where

q_{re} = refrigeration load, Btu/h

q_s = block room sensible heat for all spaces at time of peak, Btu/h

h_{ea} = enthalpy of primary air upstream of cooling coil at time of peak, Btu/lb

h_{la} = enthalpy of primary air leaving cooling coil, Btu/lb

Q_p = primary-air quantity, cfm

t_r = average room temperature for all exposures at peak time, °F

t_s = average primary-air temperature at point of delivery to rooms, °F

Because the latent load is absorbed by the primary air, the resultant room relative humidity can be determined by calculating the block room latent load of all spaces at the time of the peak load. Then, recalling that there are 7000 grains in a pound, the rise in moisture content of the primary air at sea level is

$$W = \frac{7000 v_a q_L}{60 h_{fg} Q_p} = 1.48 \frac{q_L}{Q_p} \quad (3)$$

where

W = moisture content rise per lb dry air, grains
v_a = specific volume of air = 13.3 ft³/lb at sea level
q_L = block room latent load of all spaces at time of peak load, Btu/h
h_{fg} = latent heat of vaporization = 1050 Btu/lb

Then a psychrometric analysis can be performed.

The secondary-water cooling load may be determined by subtracting the primary-air cooling coil load from the total refrigeration load.

TWO-PIPE SYSTEMS WITH CENTRAL VENTILATION

Two-pipe systems for induction and fan-coil systems derive their name from the water-distribution circuit, which consists of one supply and one return pipe. Each unit or conditioned space is supplied with secondary water from this distribution system and with conditioned primary air from a central apparatus. The system design and control of primary-air and secondary-water temperatures must be such that all rooms on the same system (or zone, if applicable) can be satisfied during both heating and cooling seasons. The heating or cooling capacity of any unit at a particular time is the sum of its primary-air output plus its secondary-water output.

The primary-air quantity is fixed, and the primary-air temperature is varied in inverse proportion to the outside temperature to provide the necessary amount of heating during summer and intermediate seasons. During winter, primary air is preheated and supplied at approximately 50°F to provide cooling. All room terminals in a given primary-air preheated zone must be selected to operate satisfactorily with the common primary-air temperature.

The secondary-water coil (cooling-heating) in each space is controlled by a space thermostat and can vary from 0 to 100% of coil capacity, as required to maintain space temperature. The secondary water is cold in summer and intermediate seasons and warm in winter. All rooms on the same secondary-water zone must operate satisfactorily with the same water temperature.

Figure 6 shows the capacity ranges available from a typical two-pipe system. On a hot summer day, loads from about 25 to 100% of the design space cooling capacity can be satisfied. On a 50°F intermediate-season day, the unit can satisfy a heating requirement by closing off the secondary-water coil and using only the output of warm primary air. A lesser heating or net cooling requirement is satisfied by the cold secondary-water coil output, which offsets the warm primary air to obtain cooling. In winter, the unit can provide a small amount of cooling by closing the secondary coil and using only the cold primary air. Smaller cooling loads and all heating requirements are satisfied by using warm secondary water.

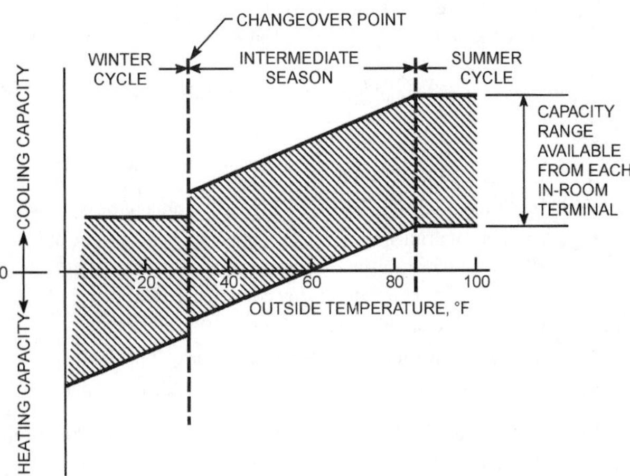

Fig. 6 Capacity Ranges of In-Room Terminal Operating on Two-Pipe System

Critical Design Elements

The most critical design elements of a two-pipe system are the calculation of primary-air quantities and the final adjustment of the primary-air temperature reset schedule. All rooms require a minimum amount of heat from the primary-air supply during the intermediate season. Using the ratio of primary air to transmission per degree (A/T ratio) to maintain a constant relationship between the primary-air quantity and the heating requirements of each space fulfills this need. The A/T ratio determines the primary-air temperature and changeover point, and is fundamental to proper design and operation of a two-pipe system.

Transmission per Degree. The relative heating requirement of every space is determined by calculating the transmission heat flow per degree temperature difference between the space temperature and the outside temperature (assuming steady-state heat transfer). This is the sum of the (1) glass heat transfer coefficient times the glass areas, (2) wall heat transfer coefficient times the wall area, and (3) roof heat transfer coefficient times the roof area.

Air-to-Transmission (A/T) Ratio. The A/T ratio is the ratio of the primary airflow to a given space divided by the transmission per degree of that space: A/T ratio = Primary air/Transmission per degree.

Spaces on a common primary-air zone must have approximately the same A/T ratios. The design base A/T ratio establishes the primary-air reheat schedule during intermediate seasons. Spaces with A/T ratios higher than the design base A/T ratio tend to be overcooled during light cooling loads at an outside temperature in the 70 to 90°F range, whereas spaces with an A/T ratio lower than design lack sufficient heat during the 40 to 60°F outside temperature range when primary air is warm for heating and secondary water is cold for cooling.

The minimum primary-air quantity that satisfies the requirements for ventilation, dehumidification, and both summer and winter cooling is used to calculate the minimum A/T ratio for each space. If the system operates with primary-air heating during cold weather, the heating capacity can also be the primary-air quantity determinant for two-pipe systems.

The design base A/T ratio is the highest A/T ratio obtained, and the primary airflow to each space is increased as required to obtain a uniform A/T ratio in all spaces. An alternative approach is to locate the space with the highest A/T ratio requirement by inspection, establish the design base A/T ratio, and obtain the primary airflow for all other spaces by multiplying this A/T ratio by the transmission per degree of all other spaces.

For each A/T ratio, there is a specific relationship between outside air temperature and temperature of the primary air that maintains the room at 72°F or more during conditions of minimum room cooling load. Figure 7 illustrates this variation based on an assumed minimum room load equivalent to 10°F times the transmission per degree. A primary-air temperature over 122°F at the unit is seldom used. The reheat schedule should be adjusted for hospital rooms or other applications where a higher minimum room temperature is desired, or where a space has no minimum cooling load.

Deviation from the A/T ratio is sometimes permissible. A minimum A/T ratio equal to 0.7 of the maximum A/T is suitable, if the building is of massive construction with considerable heat storage effect (Carrier 1965). The heating performance when using warm primary air becomes less satisfactory than that for systems with a uniform A/T ratio. Therefore, systems designed for A/T ratio deviation should be suitable for changeover to warm secondary water for heating whenever the outside temperature falls below 40°F. A/T ratios should be more closely maintained on buildings with large

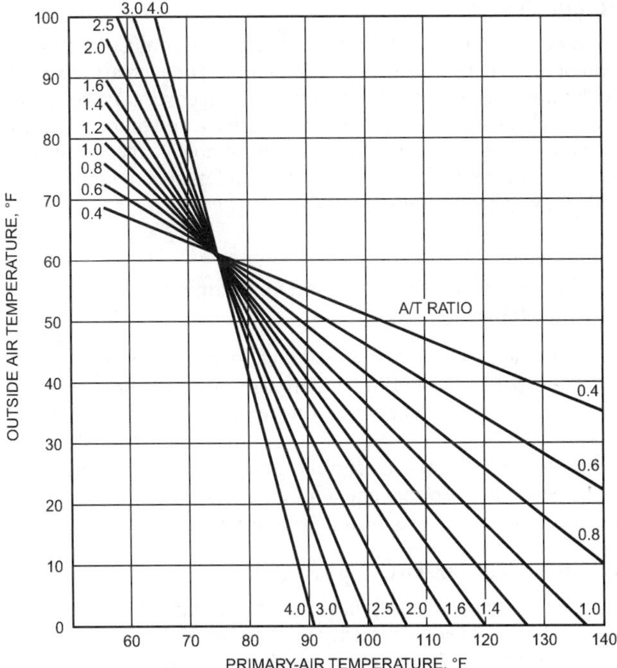
OUTSIDE AIR TEMPERATURE, °F
PRIMARY-AIR TEMPERATURE, °F
A/T RATIO
Note: These temperatures are required at the units, and thermostat settings must be adjusted to allow for duct heat gains or losses. Temperatures are based on
1. Minimum average load in the space, equivalent to 10°F multiplied by the transmission per degree.
2. Preventing the room temperature from dropping below 72°F. These values compensate for the radiation and convection effect of the cold outside wall.

Fig. 7 Primary-Air Temperature Versus Outside Air Temperature

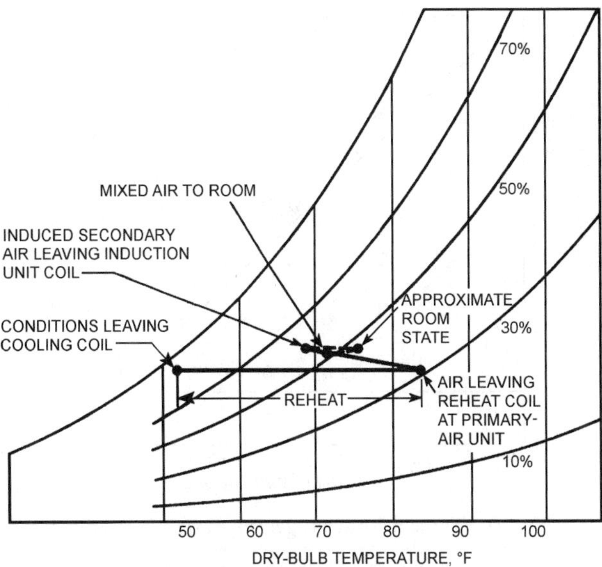
MIXED AIR TO ROOM
INDUCED SECONDARY AIR LEAVING INDUCTION UNIT COIL
CONDITIONS LEAVING COOLING COIL
APPROXIMATE ROOM STATE
REHEAT
AIR LEAVING REHEAT COIL AT PRIMARY-AIR UNIT
70%
50%
30%
10%
DRY-BULB TEMPERATURE, °F
Note: Based on A/T ratio = 1.2, 52°F outside, reheat schedule on.

Fig. 8 Psychrometric Chart, Two-Pipe System, Off-Season Cooling

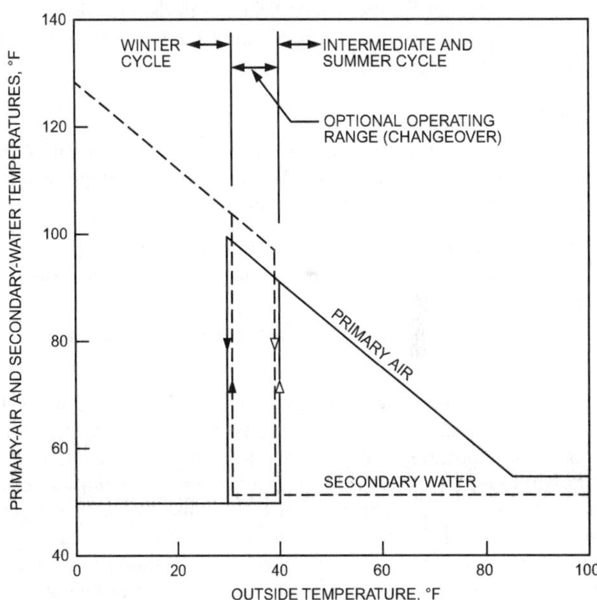
WINTER CYCLE
INTERMEDIATE AND SUMMER CYCLE
OPTIONAL OPERATING RANGE (CHANGEOVER)
PRIMARY AIR
SECONDARY WATER
PRIMARY-AIR AND SECONDARY-WATER TEMPERATURES, °F
OUTSIDE TEMPERATURE, °F

Fig. 9 Typical Changeover System Temperature Variation

glass areas or with curtain wall construction, or on systems with low changeover temperature.

Changeover Temperature Considerations

Transition from summer operations to intermediate-season operation is done by gradually raising the primary-air temperature as the outside temperature falls, to keep rooms with small cooling loads from becoming too cold. The secondary water remains cold during both summer and intermediate seasons. Figure 8 illustrates the psychrometrics of summer operation near the changeover temperature. As the outside temperature drops further, the changeover temperature is reached. The secondary-water system can then be changed over to provide hot water for heating.

If the primary airflow is increased to some spaces to elevate the changeover temperature, the A/T ratio for the reheat zone is affected. Adjustments in primary-air quantities to other spaces on that zone will probably be necessary to establish a reasonably uniform ratio.

System changeover can take several hours and usually temporarily upsets room temperatures. Good design, therefore, includes provision for operating the system with either hot or cold secondary water over a range of 15 to 20°F below the changeover point. This range makes it possible to operate with warm air and cold secondary water when the outside temperature rises above the daytime changeover temperature. Changeover to hot water is limited to times of extreme or protracted cold weather.

Optional hot- or cold-water operation below the changeover point is provided by increasing the primary-air reheater capacity to provide adequate heat at a colder outside temperature. Figure 9 shows temperature variation for a system operating with changeover, indicating the relative temperature of the primary air and secondary water throughout the year and the changeover temperature range. The solid arrows show the temperature variation when changing over from the summer to the winter cycle. The open arrows show the variation when going from the winter to the summer cycle.

Nonchangeover Design

Nonchangeover systems should be considered to simplify operation for buildings with mild winter climates, or for south exposure zones of buildings with a large winter solar load. A nonchangeover system operates on an intermediate-season cycle throughout the heating season, with cold secondary water to the terminal unit coils and with warm primary air satisfying all the heating requirements. Typical temperature variation is shown in Figure 10.

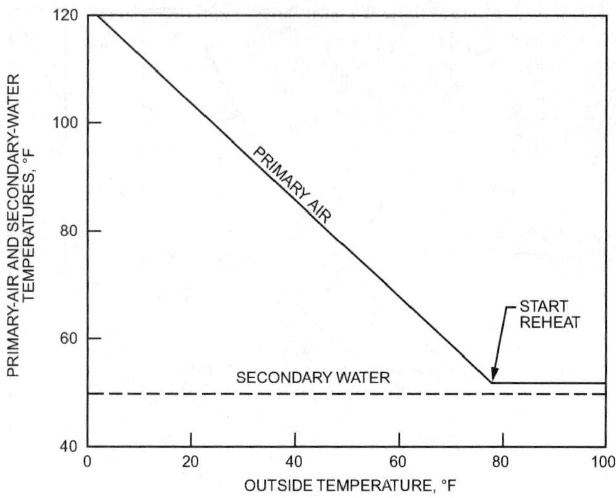

Fig. 10 Typical Nonchangeover System Variations

Spaces may be heated during unoccupied hours by operating the primary-air system with 100% return air. This feature is necessary because nonchangeover design does not usually include the ability to heat the secondary water. In addition, cold secondary water must be available throughout the winter. Primary-air duct insulation and observance of close A/T ratios for all units are essential for proper heating during cold weather.

Zoning

A two-pipe system can provide good temperature control most of the time, on all exposures during the heating and cooling seasons. Operating cost can be improved by zoning

- Primary air to allow different A/T ratios on different exposures
- Primary air to allow solar compensation of primary-air temperature
- Both air and water to allow a different changeover temperature for different exposures

All spaces on the same primary-air zone must have the same A/T ratio. The minimum A/T ratios often are different for spaces on different solar exposures, thus requiring the primary-air quantities on some exposures to be increased if they are placed on a common zone with other exposures. The primary-air quantity to units serving spaces with less solar exposure can usually be reduced by using separate primary-air zones with different A/T ratios and reheat schedules. Primary-air quantity should never be reduced below minimum ventilation requirements.

The peak cooling load for the south exposure occurs during fall or winter when outside temperatures are lower. If shading patterns from adjacent buildings or obstructions are not present, primary-air zoning by solar exposure can reduce air quantities and unit coil sizes on the south. Units can be selected for peak capacity with cold primary air instead of reheated primary air. Primary-air zoning and solar compensators save operating cost on all solar exposures by reducing primary-air reheat and secondary-water refrigeration penalty.

Separate air and water zoning may save operating cost by allowing spaces with less solar exposure to operate on the winter cycle with warm secondary water at outside temperatures as high as 60°F during the heating season. Systems with a common secondary-water zone must operate with cold secondary water to cool heavier solar exposures. Primary airflow can be lower because of separate A/T ratios, resulting in reheat and refrigeration cost savings.

Room Control

When room temperature rises, the thermostat must increase the output of the cold secondary coil (in summer) or decrease the output of the warm secondary coil (in winter). Changeover from cold to hot water in the unit coils requires changing the action of the room temperature control system. Room control for nonchangeover systems does not require the changeover action, unless it is required to provide gravity heating during shutdown.

Evaluation

Characteristics of two-pipe in-room terminal unit systems include the following:

- Usually less expensive to install than four-pipe systems
- Less capable of handling widely varying loads or providing a widely varying choice of room temperatures than four-pipe systems
- Present operational and control changeover problems, increasing the need for competent operating personnel
- More costly to operate than four-pipe systems

Electric Heat for Two-Pipe Systems

Electric heat can be supplied with a two-pipe in-room terminal unit system by a central electric boiler and hot-water terminal coils, or by individual electric-resistance heating coils in the terminal units. One method uses small electric-resistance terminal heaters for intermediate-season heating and a two-pipe changeover chilled-water/hot-water system. The electric terminal heater heats when outside temperatures are above 40°F, so cooling can be kept available with chilled water in the chilled-water/hot-water system. System or zone reheating of primary air is greatly reduced or eliminated entirely. When the outside temperature falls below this point, the chilled-water/hot-water system is switched to hot water, providing greater heating capacity. Changeover is limited to a few times per season, and simultaneous heating/cooling capacity is available, except in extremely cold weather, when little, if any, cooling is needed. If electric-resistance terminal heaters are used, they should be prevented from operating whenever the secondary-water system is operated with hot water.

Another method is to size electric resistance terminal heaters for the peak winter heating load and operate the chilled-water system as a nonchangeover cooling-only system. This avoids the operating problem of chilled-water/hot-water system changeover. In fact, this method functions like a four-pipe system, and, in areas where the electric utility establishes a summer demand charge and has a low unit energy cost for high winter consumption, it may have a lower life-cycle cost than hydronic heating with fossil fuel. A variation, especially appropriate for well-insulated office buildings with induction units where cooling is needed in perimeter offices for almost all occupied hours because of internal heat gain, is to use electric heaters in the terminal unit during occupied hours and to provide heating during unoccupied hours by raising primary-air temperature on an outside reset schedule.

FOUR-PIPE SYSTEMS

Four-pipe systems have a chilled-water supply, chilled-water return, hot-water supply, and hot-water return. The terminal unit usually has two independent secondary-water coils: one served by hot water, the other by cold water. The primary air is cold and remains at the same temperature year-round. During peak cooling and heating, the four-pipe system performs in a manner similar to the two-pipe system, with essentially the same operating characteristics. Between seasons, any unit can be operated at any level from maximum cooling to maximum heating, if both cold and warm water are being circulated, or between these extremes without regard to other units' operation.

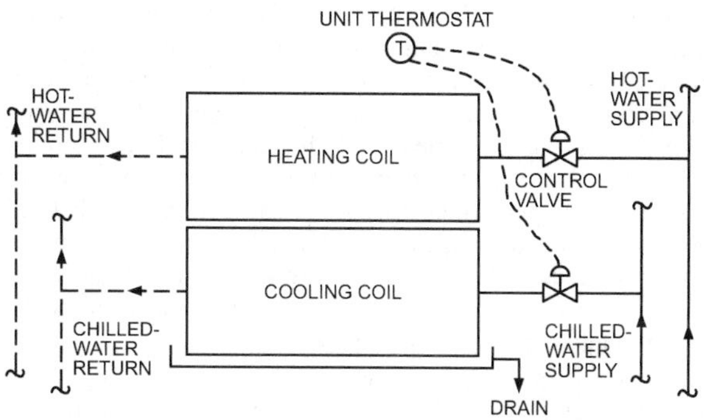

A. FOUR-PIPE SYSTEM

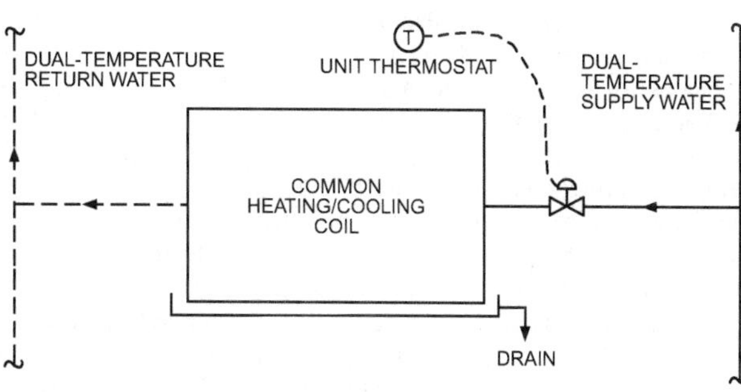

B. TWO-PIPE SYSTEM

Fig. 11 Fan-Coil Unit Control

In-room terminal units are selected by their peak capacity. The A/T ratio does not apply to four-pipe systems. There is no need to increase primary-air quantities on units with low solar exposure beyond the amount needed for ventilation and to satisfy cooling loads. The available net cooling is not reduced by heating the primary air. The changeover point is still important, though, because cooling spaces on the sunny side of the building may still require secondary-water cooling to supplement the primary air at low outside temperatures.

Because primary air is supplied at a constant cool temperature at all times, it is sometimes feasible for fan-coil unit systems to extend the interior system supply to the perimeter spaces, eliminating the need for a separate primary-air system. The type of terminal unit and characteristics of the interior system are determining factors.

Zoning

Zoning primary-air or secondary-water systems is not required with four-pipe systems. All terminal units can heat or cool at all times, as long as both hot and cold secondary pumps are operated and sources of heating and cooling are available.

Room Control

The four-pipe terminal usually has two completely separated secondary-water coils: one receiving hot water and the other receiving cold water. The coils are operated in sequence by the same thermostat; they are never operated simultaneously. The unit receives either hot or cold water in varying amounts, or else no flow is present, as shown in Figure 11A. Adjustable, dead-band thermostats further reduce operating cost.

Figure 11B illustrates another unit and control configuration. A single secondary-water coil at the unit and three-way valves located at the inlet and outlet admit water from either the hot- or cold-water supply, as required, and divert it to the appropriate return pipe. This arrangement requires a special three-way modulating valve, originally developed for one form of the three-pipe system. It controls the hot or cold water selectively and proportionally, but does not mix the streams. The valve at the coil outlet is a two-position valve open to either the hot or cold water return, as required.

Overall, the two-coil arrangement provides a superior four-pipe system. Operation of the induction and fan-coil unit controls is the same year-round.

Evaluation

Compared to the two-pipe system, the four-pipe air-and-water system has the following characteristics:

- More flexible and adaptable to widely differing loads, responding quickly to load changes
- Simpler to operate
- Operates without the summer-winter changeover and primary-air reheat schedule
- Efficiency is greater and operating cost is lower, though initial cost is generally higher
- Can be designed with no interconnection of hot- and cold-water secondary circuits, and the secondary system can be completely independent of the primary-water piping

SECONDARY-WATER DISTRIBUTION

Secondary-water system design applies to induction and fan-coil systems. The secondary-water system includes the part of the water distribution system that circulates water to room terminal units when the water has been cooled or heated either by extraction from or heat exchange with another source in the primary circuit. In the primary circuit, water is cooled by flow through a chiller or is heated by a heat input source. Primary water is limited to the cooling cycle and is the source of the secondary-water cooling. Water flow through the unit coil performs secondary cooling when the room air (secondary air) gives up heat to the water. Secondary-water system design differs for two- and four-pipe systems. Secondary-water systems are discussed in Chapter 12.

AUTOMATIC CONTROLS AND BUILDING MANAGEMENT SYSTEMS

Basic HVAC system controls are available in electric, pneumatic, or electronic control systems. Depending on the application, a simple, basic system strategy may be a cost-effective solution to an owner's heating, ventilation, and cooling needs. Chapter 46 of the 2007 *ASHRAE Handbook—HVAC Applications* and Chapter 15 of the 2005 *ASHRAE Handbook—Fundamentals* discuss automatic control in more detail.

The next level of HVAC system management is **direct digital control (DDC),** with either pneumatic or electric control damper and valve actuators. This automatic control enhancement may include energy monitoring and energy management software, and may also be accessible off-site by modem. Building size has little to no effect on modern computerized controls: programmable controls can be furnished on the smallest HVAC equipment for the smallest projects. Chapter 41 of the 2007 *ASHRAE Handbook—HVAC Applications* covers building operating dynamics.

Automatic controls can be prepackaged and prewired on the HVAC equipment. In system analysis and selection, the design engineer should compare purchasing a prepackaged versus a traditional building automation system (BAS). HVAC controls must be compatible with other new and existing automatic controls. Chapter 39 of the 2007 *ASHRAE Handbook—HVAC Applications* discusses computer applications, and ASHRAE *Standard* 135 discusses interfacing building automation systems.

Using computers and proper software, the design engineer and building manager can provide complete facility management. A comprehensive building management system may include HVAC system control, energy management, operation and maintenance management, medical gas system monitoring, fire alarm, security, lighting control, and other reporting and trending software. This system may also be integrated and accessible from the owner's computer network and the Internet.

The building management system is an important factor in choosing the optimum HVAC system. It can be as simple as a time clock to start and stop equipment, or as sophisticated as a computerized building automation system serving a decentralized HVAC system, multiple building systems, central plant system, and/or a large campus. With a focus on energy management, the building management system can be an important business tool in achieving sustainable facility management success that begins with the System Selection Matrix selection, recommendation, and implementation.

MAINTENANCE MANAGEMENT SYSTEMS

Whereas building management systems focus on HVAC system operation, as well as electrical, plumbing, and other system operation, maintenance management systems focus on maintaining the assets, which include the mechanical and electrical systems along with the building structure, custodial, etc. A rule of thumb is that 20% of the cost of the building is in first cost, with the other 80% in operation, maintenance, and rejuvenation of the building and systems over the life of the building. During initial selection of an HVAC system, maintenance management systems should be considered for HVAC systems with an estimated long useful service life.

A **computerized maintenance management software (CMMS)** system can include an equipment database, parts and material inventory, project management software, labor records, and other relevant information for management of the building. The CMMS system also can integrate computer-aided drawing (CAD), digital photography and audio/video systems, and other proactive facility management systems that an owner may want or need for efficient and effective building management. Interfacing the building management system with the maintenance management system by specifying automatic control monitoring and trending helps the CMMS provide predictive, preventive, and real-time maintenance guidance.

BUILDING SYSTEM COMMISSIONING

With most in-room terminal units, prepackaged automatic controls may use a different automatic control checkout process than traditional control contractors. When commissioning a building system that integrates an independent control system with individual packaged control systems, commissioning can be more cumbersome because both control contractors need to participate. It is also important to obtain the control contractors' individual point checkout sheets, program logic, and list of points that require confirmation with another trade (e.g., fire alarm system installer) before the commissioning performance demonstration.

Frequently in new construction and expansion of existing installations, in-room equipment is installed in phases, requiring multiple commissioning efforts based on the construction schedule and owner occupancy. During the warranty phase, system performance should be measured, benchmarked, and course-corrected to ensure achievement of design intent. If an energy analysis study is performed as part of the comparison between decentralized and centralized concepts and/or life cycle comparison of the study is part of a LEED™ project, the resulting month-to-month energy data should be a good electronic document to benchmark actual energy consumption.

Ongoing commissioning or periodic recommissioning further ensures the original design intent is met, as well as reliable delivery of cooling and heating production. Recommissioning should be considered whenever the facility is expanded or an additional connection made to the existing systems.

Testing, adjusting, and balancing (TAB), performed during commissioning and periodically over the building's life, also contributes to successful operation and maintenance. When completing the TAB and commissioning, consider posting laminated system flow diagrams at or adjacent to the cooling and heating equipment, indicating operating instructions, TAB performance, commissioning functional performance test procedures, and emergency shutoff procedures. These documents should also be filed electronically in the building manager's computer server for quick reference.

Original basis of design and design criteria should be posted as a reminder of design intent, and be readily available in case troubleshooting, expansion, or modernization is needed.

As with all HVAC applications, for design success, building commissioning should include the system training requirements necessary for building management staff to efficiently take ownership and operate and maintain the HVAC systems over the useful service life of the installation.

REFERENCES

AIA. 2001. *Guidelines for design and construction of hospital and health care facilities.* American Institute of Architects, Washington, D.C.

ARI. 2005. Standard for room fan-coils. ARI *Standard* 440-2005.

ASHRAE. 2002. Method of testing for rating fan-coil conditioners. ANSI/ASHRAE *Standard* 79-2002 (RA 2006).

ASHRAE. 2004. BACnet®: A data communication protocol for building automation and control networks. ANSI/ASHRAE *Standard* 135-2004.

Carrier Air Conditioning Company. 1965. *Handbook of air conditioning system design.* McGraw-Hill, New York.

BIBLIOGRAPHY

ASHRAE. 2007. Ventilation for acceptable indoor air quality. ANSI/ASHRAE *Standard* 62.1-2007.

Menacker, R. 1977. Electric induction air-conditioning system. *ASHRAE Transactions* 83(1):664.

McFarlan, A.I. 1967. Three-pipe systems: Concepts and controls. *ASHRAE Journal* 9(8):37.

CHAPTER 6

PANEL HEATING AND COOLING

PANEL heating and cooling systems use temperature-controlled indoor surfaces on the floor, walls, or ceiling; temperature is maintained by circulating water, air, or electric current through a circuit embedded in or attached to the panel. A temperature-controlled surface is called a **radiant panel** if 50% or more of the design heat transfer on the temperature-controlled surface takes place by thermal radiation. Panel systems are characterized by controlled surface temperatures below 300°F. Panel systems may be combined either with a central forced-air system of one-zone, constant-temperature, constant-volume design, or with dual-duct, reheat, multizone or variable-volume systems, decentralized convective systems, or in-space fan-coil units. These combined systems are called **hybrid (load-sharing) HVAC systems**.

This chapter covers temperature-controlled surfaces that are the primary source of sensible heating and cooling in the conditioned space. For snow-melting and freeze-protection applications, see Chapter 50 of the 2007 *ASHRAE Handbook—HVAC Applications*. Chapter 15 covers high-temperature panels over 300°F, which may be energized by gas, electricity, or high-temperature water.

PRINCIPLES OF THERMAL RADIATION

Thermal radiation (1) is transmitted at the speed of light, (2) travels in straight lines and can be reflected, (3) elevates the temperature of solid objects by absorption but does not noticeably heat the air through which it travels, and (4) is exchanged continuously between all bodies in a building environment. The rate at which thermal radiation occurs depends on the following factors:

- Temperature of the emitting surface and receiver
- Emittance of the radiating surface
- Reflectance, absorptance, and transmittance of the receiver
- View factor between the emitting and receiver surfaces (viewing angle of the occupant to the thermal radiation source)

ASHRAE research project RP-876 (Lindstrom et al. 1998) concluded that surface roughness and texture have insignificant effects on thermal convection and thermal radiation, respectively. Surface emittance (the ratio of the radiant heat flux emitted by a body to that emitted by a blackbody under the same conditions) for typical indoor surfaces, such as carpets, vinyl texture paint, and plastic, remained between 0.9 and 1.0 for panel surface temperatures of 86 to 131°F.

The structure of the radiation surface is critical. In general, rough surfaces have low reflectance and high emittance/absorptance characteristics. Conversely, smooth or polished metal surfaces have high reflectance and low absorptance/emittance.

One example of heating by thermal radiation is the feeling of warmth when standing in the sun's rays on a cool, sunny day. Some of the rays come directly from the sun and include the entire electromagnetic spectrum. Other rays are absorbed by or reflected from surrounding objects. This generates secondary rays that are a combination of the wavelength produced by the temperature of the objects and the wavelength of the reflected rays. If a cloud passes in front of the sun, there is an instant sensation of cold. This sensation is caused by the decrease in the amount of heat received from solar radiation, although there is little, if any, change in the ambient air temperature.

Thermal comfort, as defined in ASHRAE *Standard* 55, is "that condition of mind which expresses satisfaction with the thermal environment." No system is completely satisfactory unless the three main factors controlling heat transfer from the human body (radiation, convection, and evaporation) result in thermal neutrality. Maintaining correct conditions for human thermal comfort by thermal radiation is possible for even the most severe climatic conditions (Buckley 1989). Chapter 8 of the 2005 *ASHRAE Handbook—Fundamentals* has more information on thermal comfort.

Panel heating and cooling systems provide an acceptable thermal environment by controlling surface temperatures as well as indoor air temperature in an occupied space. With a properly designed system, occupants should not be aware that the environment is being heated or cooled. The **mean radiant temperature (MRT)** has a strong influence on human thermal comfort. When the temperature of surfaces comprising the building (particularly outdoor exposed walls with extensive fenestration) deviates excessively from the ambient temperature, convective systems sometimes have difficulty counteracting the discomfort caused by cold or hot surfaces. Heating and cooling panels neutralize these deficiencies and minimize radiation losses or gains by the human body.

Most building materials have relatively high surface emittance and, therefore, absorb and reradiate heat from active panels. Warm ceiling panels are effective because heat is absorbed and reflected by the irradiated surfaces and not transmitted through the construction. Glass is opaque to the wavelengths emitted by active panels and, therefore, transmits little long-wave thermal radiation outside. This is significant because all surfaces in the conditioned space tend to assume temperatures that result in an acceptable thermal comfort condition.

GENERAL EVALUATION

Principal **advantages** of panel systems are the following:

- Because not only indoor air temperature but also mean radiant temperature can be controlled, total human thermal comfort may be better satisfied.
- Because the operative temperature for required human thermal comfort may be maintained by primarily controlling the mean radiant temperature of the conditioned indoor space, dry-bulb air temperature may be lower (in heating) or higher (in cooling),

---

The preparation of this chapter is assigned to TC 6.5, Radiant and In-Space Convective Heating and Cooling.

which reduces sensible heating or cooling loads (see Chapter 15 for the definition and calculation of operative and mean radiant temperatures).

- Hydronic panel systems may be connected in series, following other hydronic heating or cooling systems (i.e., their return water may be used), increasing exergetic efficiency.
- Comfort levels can be better than those of other space-conditioning systems because thermal loads are satisfied directly and air motion in the space corresponds to required ventilation only.
- Waste and low-enthalpy energy sources and heat pumps may be directly coupled to panel systems without penalty on equipment sizing and operation. Being able to select from a wide range of moderate operation temperatures ensures optimum design for minimum cost and maximum thermal and exergetic efficiency.
- Seasonal thermal distribution efficiency in buildings may be higher than in other hydronic systems.
- In terms of simple payback period, ceiling cooling panels and chilled beams have the highest technical energy savings potential (DOE 2002).
- Part or all of the building structure may be thermally activated (Meierhans and Olesen 2002).
- Space-conditioning equipment is not required at outdoor exposed walls, simplifying wall, floor, and structural systems.
- Almost all mechanical equipment may be centrally located, simplifying maintenance and operation.
- No space within the conditioned space is required for mechanical equipment. This feature is especially valuable in hospital patient rooms, offices, and other applications where space is at a premium, if maximum cleanliness is essential or legally required.
- Draperies and curtains can be installed at outdoor exposed walls without interfering with the space-conditioning system.
- When four-pipe systems are used, cooling and heating can be simultaneous, without central zoning or seasonal changeover.
- Supply air requirements usually do not exceed those required for ventilation and humidity control.
- Reduced airflow requirements help mitigate bioterrorism risk, especially in large buildings.
- Modular panels provide flexibility to meet changes in partitioning.
- A 100% outdoor air system may be installed with smaller penalties in refrigeration load because of reduced air quantities.
- A common central air system can serve both the interior and perimeter zones.
- Wet-surface cooling coils are eliminated from the occupied space, reducing the potential for septic contamination.
- Modular panel systems can use the automatic sprinkler system piping (see NFPA *Standard* 13, Section 3.6). The maximum water temperature must not fuse the heads.
- Panel heating and cooling with minimum supply air quantities provide a draft-free environment.
- Noise associated with fan-coil or induction units is eliminated.
- Peak loads are reduced as a result of thermal energy storage in the panel structure, as well as walls and partitions directly exposed to panels.
- Panels can be combined with other space-conditioning systems to decouple several indoor requirements (e.g., humidity control, indoor air quality, air velocity) and optimally satisfy them without compromises.
- In-floor heating creates inhospitable living conditions for house dust mites compared to other heating systems (Sugawara et al. 1996).

Disadvantages are similar to those listed in Chapter 3 of the 2005 *ASHRAE Handbook—Fundamentals*. In addition,

- Response time can be slow if controls and/or heating elements are not selected or installed correctly.
- Improper selection of panel heating or cooling tube or electrical heating element spacing and/or incorrect sizing of heating/cooling source can cause nonuniform surface temperatures or insufficient sensible heating or cooling capacity.
- Panels can satisfy only sensible heating and cooling loads unless hybrid panels are used [see the section on Hybrid (Load-Sharing) HVAC Systems]. In a stand-alone panel cooling system, dehumidification and panel surface condensation may be of prime concern. Unitary dehumidifiers should be used, or a latent air-handling system should be introduced to the indoor space.

HEAT TRANSFER BY PANEL SURFACES

Sensible heating or cooling panels transfer heat through temperature-controlled (active) surface(s) to or from an indoor space and its enclosure surfaces by thermal radiation and natural convection.

Heat Transfer by Thermal Radiation

The basic equation for a multisurface enclosure with gray, diffuse isothermal surfaces is derived by radiosity formulation methods (see Chapter 3 of the 2005 *ASHRAE Handbook—Fundamentals*). This equation may be written as

$$q_r = J_p - \sum_{j=1}^{n} F_{pj} J_j \qquad (1)$$

where

q_r = net heat flux because of thermal radiation on active (heated or cooled) panel surface, Btu/h·ft²
J_p = total radiosity leaving or reaching panel surface, Btu/h·ft²
J_j = radiosity from or to another surface in room, Btu/h·ft²
F_{pj} = radiation angle factor between panel surface and another surface in room (dimensionless)
n = number of surfaces in room other than panel(s)

Equation (1) can be applied to simple and complex enclosures with varying surface temperatures and emittances. The net heat flux by thermal radiation at the panel surfaces can be determined by solving the unknown J_j if the number of surfaces is small. More complex enclosures require computer calculations.

Radiation angle factors can be evaluated using Figure 6 in Chapter 3 of the 2005 *ASHRAE Handbook—Fundamentals*. Fanger (1972) shows room-related angle factors; they may also be developed from algorithms in ASHRAE's *Energy Calculations* I (1976).

Several methods have been developed to simplify Equation (1) by reducing a multisurface enclosure to a two-surface approximation. In the **MRT method**, the thermal radiation interchange in an indoor space is modeled by assuming that the surfaces radiate to a fictitious, finite surface that has an emittance and surface temperature that gives about the same heat flux as the real multisurface case (Walton 1980). In addition, angle factors do not need to be determined in evaluating a two-surface enclosure. The MRT equation may be written as

$$q_r = \sigma F_r [T_p^4 - T_r^4] \qquad (2)$$

where

σ = Stefan-Boltzmann constant = 0.1712×10^{-8} Btu/h·ft²·°R⁴
F_r = radiation exchange factor (dimensionless)
T_p = effective temperature of heating (cooling) panel surface, °R
T_r = temperature of fictitious surface (unheated or uncooled), °R

The temperature of the fictitious surface is given by an area emittance weighted average of all surfaces other than the panel(s):

$$T_r = \frac{\sum_{j=p}^{n} A_j \varepsilon_j T_j}{\sum_{j=p}^{n} A_j \varepsilon_j} \qquad (3)$$

where

A_j = area of surfaces other than panels, ft²
ε_j = thermal emittance of surfaces other than panel(s) (dimensionless)

When the surface emittances of an enclosure are nearly equal, and surfaces directly exposed to the panel are marginally unheated (uncooled), then Equation (3) becomes the area-weighted average unheated (uncooled) temperature (AUST) of such surfaces exposed to the panels. Therefore, any unheated (uncooled) surface in the same plane with the panel is not accounted for by AUST. For example, if only part of the floor is heated, the remainder of the floor is not included in the calculation of AUST, unless it is observed by other panels in the ceiling or wall.

The radiation interchange factor for two-surface radiation heat exchange is given by the Hottel equation:

$$F_r = \frac{1}{\frac{1}{F_{p-r}} + \left(\frac{1}{\varepsilon_p} - 1\right) + \frac{A_p}{A_r}\left(\frac{1}{\varepsilon_r} - 1\right)} \qquad (4)$$

where

F_{p-r} = radiation angle factor from panel to fictitious surface (1.0 for flat panel)

A_p, A_r = area of panel surface and fictitious surface, respectively

ε_p, ε_r = thermal emittance of panel surface and fictitious surface, respectively (dimensionless)

In practice, the thermal emittance ε_p of nonmetallic or painted metal nonreflecting surfaces is about 0.9. When this emittance is used in Equation (4), the radiation exchange factor F_r is about 0.87 for most indoor spaces. Substituting this value in Equation (2), σF_r becomes 0.15×10^{-8}. Min et al. (1956) showed that this constant was 0.152×10^{-8} in their test room. Then the equation for heat flux from thermal radiation for panel heating and cooling becomes approximately

$$q_r = 0.15 \times 10^{-8}[(t_p + 459.67)^4 - (\text{AUST} + 459.67)^4] \qquad (5)$$

where

t_p = effective panel surface temperature, °F

AUST = area-weighted temperature of all indoor surfaces of walls, ceiling, floor, windows, doors, etc. (excluding active panel surfaces), °F

Equation (5) establishes the general sign convention for this chapter, which states that heating by the panel is positive and cooling by the panel is negative.

Radiation exchange calculated from Equation (5) is given in Figure 1. The values apply to ceiling, floor, or wall panel output. Radiation removed by a cooling panel for a range of normally encountered temperatures is given in Figure 2.

In many specific instances where normal multistory commercial construction and fluorescent lighting are used, the indoor air temperature at the 5 ft level closely approaches the AUST. In structures where the main heat gain is through the walls or where incandescent lighting is used, wall surface temperatures tend to rise considerably above the indoor air temperature.

Heat Transfer by Natural Convection

Heat flux from natural convection q_c occurs between the indoor air and the temperature-controlled panel surface. Thermal convection coefficients are not easily established. In natural convection, warming or cooling the boundary layer of air at the panel surface generates air motion. In practice, however, many factors, such as the indoor space configuration, interfere with or affect natural convection. Infiltration/exfiltration, occupants' movement, and mechanical ventilating systems can introduce some forced convection that disturbs the natural convection.

Parmelee and Huebscher (1947) included the effect of forced convection on heat transfer occurring on heating or cooling panel surfaces as an increment to be added to the natural-convection coefficient. However, increased heat transfer from forced convection should not be factored into calculations because the increments are unpredictable in pattern and performance, and forced convection does not significantly increase the total heat flux on the active panel surface.

Natural-convection heat flux in a panel system is a function of the effective panel surface temperature and the temperature of the air layer directly contacting the panel. The most consistent measurements are obtained when the dry-bulb air layer temperature is measured close to the region where the fully developed boundary layer begins, usually 2 to 2.5 in. from the panel surfaces.

Min et al. (1956) determined natural-convection coefficients 5 ft above the floor in the center of a 12 by 24.5 ft room (D_e = 16.1 ft). Equations (6) through (11), derived from this research, can be used to calculate heat flux from panels by natural convection.

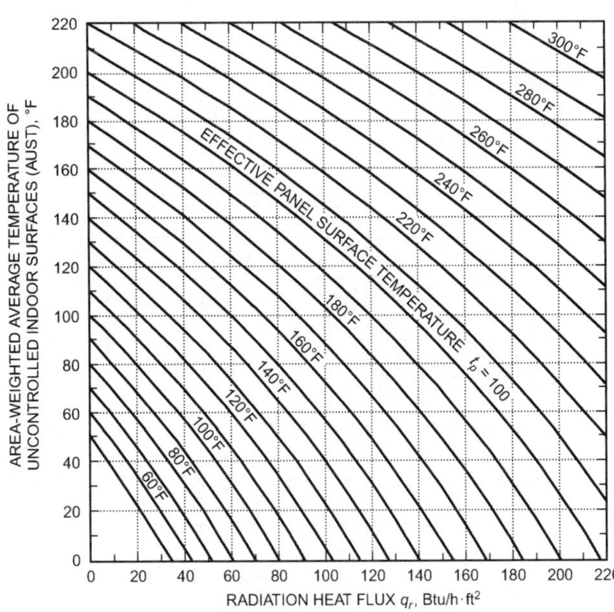

Fig. 1 Radiation Heat Flux at Heated Ceiling, Floor, or Wall Panel Surfaces

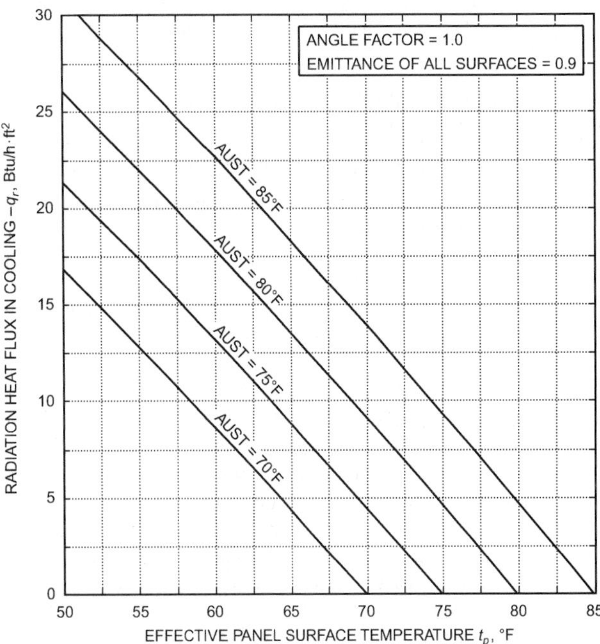

Fig. 2 Heat Removed by Radiation at Cooled Ceiling or Wall Panel Surface

Natural-convection heat flux between **an all-heated ceiling surface and indoor air**

$$q_c = 0.041 \frac{(t_p - t_a)^{1.25}}{D_e^{0.25}} \quad (6)$$

Natural-convection heat flux between **a heated floor or cooled ceiling surface and indoor air**

$$q_c = 0.39 \frac{|t_p - t_a|^{0.31}(t_p - t_a)}{D_e^{0.08}} \quad (7)$$

Natural-convection heat flux between **a heated or cooled wall panel surface and indoor air**

$$q_c = 0.29 \frac{|t_p - t_a|^{0.32}(t_p - t_a)}{H^{0.05}} \quad (8)$$

where

q_c = heat flux from natural convection, Btu/h·ft²
t_p = effective temperature of temperature-controlled panel surface, °F
t_a = indoor space dry-bulb air temperature, °F
D_e = equivalent diameter of panel (4 × area/perimeter), ft
H = height of wall panel, ft

Schutrum and Humphreys (1954) measured panel performance in furnished test rooms that did not have uniform panel surface temperatures and found no variation in performance large enough to be significant in heating practice. Schutrum and Vouris (1954) established that the effect of room size was usually insignificant *except for very large spaces like hangars and warehouses*, for which Equations (6) and (7) should be used. Otherwise, Equations (6), (7), and (8) can be simplified to the following by D_e = 16.1 ft and H = 8.85 ft:

Natural-convection heat flux between **an all-heated ceiling surface and indoor air**

$$q_c = 0.20(t_p - t_a)^{0.25}(t_p - t_a) \quad (9a)$$

Natural-convection heat flux from a heated ceiling may be augmented by leaving cold strips (unheated ceiling sections), which help initiate natural convection. In this case, Equation (9a) may be replaced by Equation (9b) (Kollmar and Liese 1957):

$$q_c = 0.13(t_p - t_a)^{0.25}(t_p - t_a) \quad (9b)$$

For large spaces such as aircraft hangars, if panels are adjoined, Equation (9b) should be adjusted with the multiplier $(16.1/D_e)^{0.25}$.

Natural-convection heat flux between **a heated floor or cooled ceiling surface and indoor air**

$$q_c = 0.31 \quad (10)$$

Natural-convection heat flux between **a heated or cooled wall panel surface and indoor air**

$$q_c = 0.26|t_p - t_a|^{0.32}(t_p - t_a) \quad (11)$$

There are no confirmed data for floor cooling, but Equation (9b) may be used for approximate calculations. Under normal conditions, t_a is the dry-bulb indoor air temperature. In floor-heated or ceiling-cooled spaces with large proportions of outdoor exposed fenestration, t_a may be taken to equal AUST.

In cooling, t_p is less than t_a, so q_c is negative. Figure 3 shows heat flux by natural convection at floor, wall, and ceiling heating panels as calculated from Equations (10), (11), (9a), and (9b), respectively.

Figure 4 compares heat removal by natural convection at cooled ceiling panel surfaces, as calculated by Equation (10), with data from Wilkes and Peterson (1938) for specific panel sizes. An additional curve illustrates the effect of forced convection on the latter data.

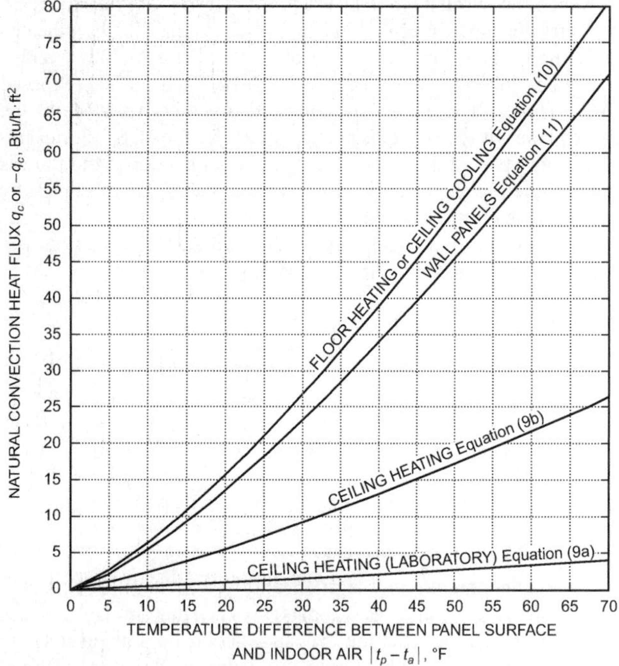

Fig. 3 Natural-Convection Heat Transfer at Floor, Ceiling, and Wall Panel Surfaces

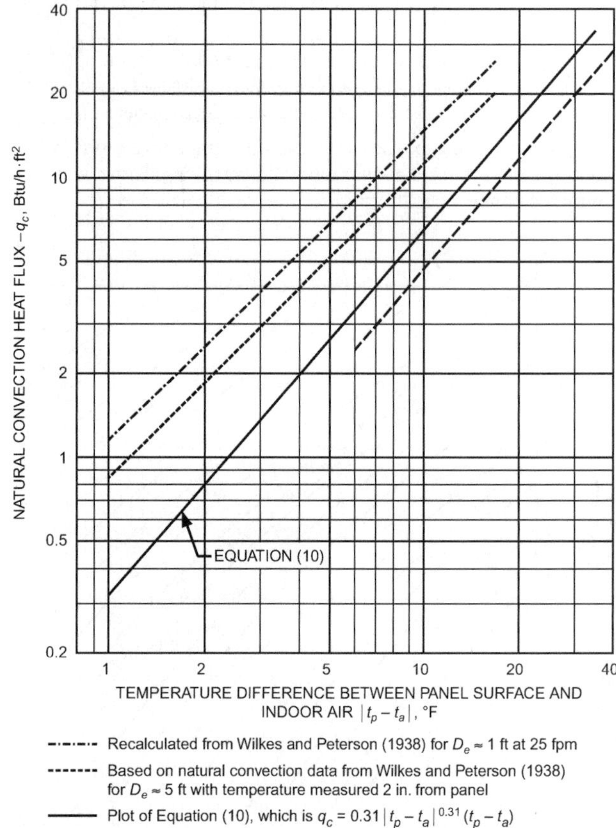

Fig. 4 Empirical Data for Heat Removal by Ceiling Cooling Panels from Natural Convection

Similar adjustment of the data from Min et al. (1956) is inappropriate, but the effects would be much the same.

Combined Heat Flux (Thermal Radiation and Natural Convection)

The combined heat flux on the active panel surface can be determined by adding the thermal-radiation heat flux q_r as calculated by Equation (5) (or from Figures 1 and 2) to the natural-convection heat flux q_c as calculated from Equations (9a), (9b), (10), or (11) or from Figure 3 or Figure 4, as appropriate.

Equation (5) requires the AUST for the indoor space. In calculating the AUST, the surface temperature of interior walls may be assumed to be the same as the dry-bulb indoor air temperature. The inside surface temperature t_w of outdoor exposed walls and outdoor exposed floors or ceilings can be calculated from the following relationship:

$$h(t_a - t_u) = U(t_a - t_o) \tag{12}$$

or

$$t_u = t_a - \frac{U}{h}(t_a - t_o) \tag{13}$$

where

h = natural-convection coefficient of the inside surface of an outdoor exposed wall or ceiling
U = overall heat transfer coefficient of wall, ceiling, or floor, Btu/h·ft²·°F
t_a = dry-bulb indoor space design air temperature, °F
t_u = inside surface temperature of outdoor exposed wall, °F
t_o = dry-bulb outdoor design air temperature, °F

From Table 1 in Chapter 25 of the 2005 *ASHRAE Handbook—Fundamentals*,

h = 1.63 Btu/h·ft²·°F for a horizontal surface with heat flow up
h = 1.46 Btu/h·ft²·°F for a vertical surface (wall)
h = 1.08 Btu/h·ft²·°F for a horizontal surface with heat flow down

Figure 5 is a plot of Equation (13) for a vertical outdoor wall with 70°F dry-bulb indoor air temperature and h = 1.46 Btu/h·ft²·°F. For rooms with dry-bulb air temperatures above or below 70°F, the values in Figure 5 can be corrected by the factors plotted in Figure 6.

Tests by Schutrum et al. (1953a, 1953b) and simulations by Kalisperis (1985) based on a program developed by Kalisperis and Summers (1985) show that the AUST and indoor air temperature are almost equal, if there is little or no outdoor exposure. Steinman et al. (1989) noted that this may not apply to enclosures with large fenestration or a high percentage of outdoor exposed wall and/or ceiling surface area. These surfaces may have a lower (in heating) or higher (in cooling) AUST, which increases the heat flux from thermal radiation.

Figure 7 shows the combined heat flux from thermal radiation and natural convection for cooling, as given in Figures 2 and 4. The data in Figure 7 do not include solar, lighting, occupant, or equipment heat gains.

In suspended-ceiling panels, heat is transferred from the back of the ceiling panel to the floor slab above (in heating) or vice versa (in cooling). The ceiling panel surface temperature is affected because of heat transfer to or from the panel and the slab by thermal radiation and, to a much smaller extent, by natural convection. The thermal-radiation component of the combined heat flux can be approximated using Equation (5) or Figure 1. The natural-convection component can be estimated from Equation (9b) or (10) or from Figure 3 or 4. In this case, the temperature difference is that between the top of the ceiling panel and the midspace of the ceiling. The temperature of the ceiling space should be determined by testing, because it varies with

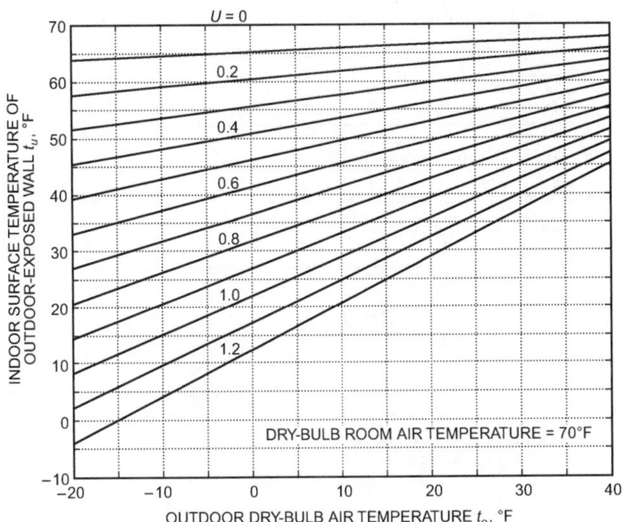

Fig. 5 Relation of Inside Surface Temperature to Overall Heat Transfer Coefficient

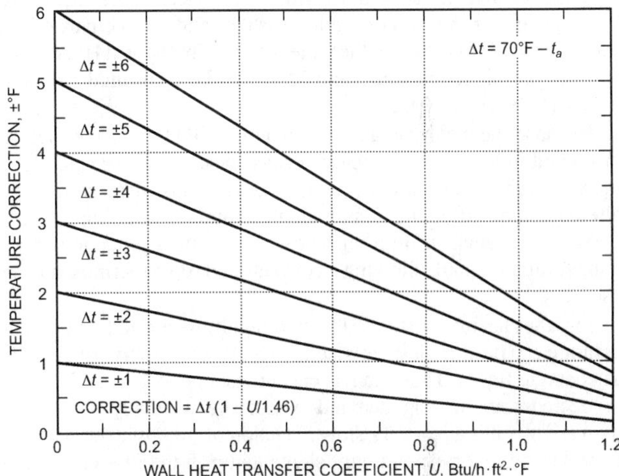

Fig. 6 Inside Surface Temperature Correction for Exposed Wall at Dry-Bulb Air Temperatures Other Than 70°F

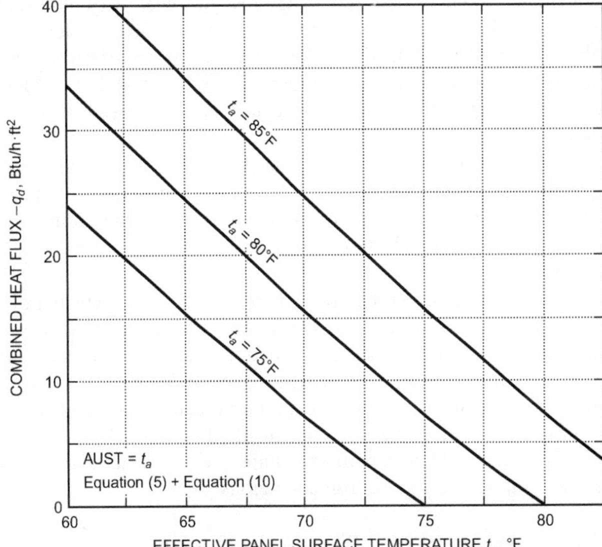

Fig. 7 Cooled Ceiling Panel Performance in Uniform Environment with No Infiltration and No Internal Heat Sources

different panel systems. However, much of this heat transfer is nullified when insulation is placed over the ceiling panel, which, for perforated metal panels, also provides acoustical control.

If artificial lighting fixtures are recessed into the suspended ceiling space, radiation from the top of the fixtures raises the overhead slab temperature and heat is transferred to the indoor air by natural convection. This heat is absorbed at the top of the cooled ceiling panels both by thermal radiation, in accordance with Equation (5) or Figure 2, and by thermal convection, generally in accordance with Equation (9b). The amount the top of the panel absorbs depends on the system. Similarly, panels installed under a roof absorb additional heat, depending on configuration and insulation.

GENERAL DESIGN CONSIDERATIONS

Panel systems and hybrid HVAC systems (typically a combination of panels and forced-convection systems) are similar to other air-water systems in the arrangement of system components. With panel systems, indoor thermal conditions are maintained primarily by thermal radiation, rather than by natural or forced-convection heat transfer. Sensible indoor space heating and cooling loads are calculated conventionally. In hybrid HVAC systems, the latent load is assigned to a forced-convection system, and a large part of the sensible load is assigned to the panel system. In a hybrid HVAC system, indoor air temperature and MRT can be controlled independently (Kilkis et al. 1995).

Because the mean radiant temperature (MRT) in a panel-heated or cooled indoor space increases or decreases as the sensible load increases, the indoor air temperature during this increase may be altered without affecting human thermal comfort. In ordinary structures with normal infiltration loads, the required reduction in air temperature is small, allowing a conventional room thermostat to be used.

In panel heating systems, lowered nighttime air and panel temperature can produce less satisfactory results with heavy panels such as concrete floors. These panels cannot respond to a quick increase or decrease in heating demand within the relatively short time required, resulting in a very slow reduction of the space temperature at night and a correspondingly slow pickup in the morning. Light panels, such as plaster or metal ceilings and walls, may respond to changes in demand quickly enough for satisfactory results from lowered nighttime air and panel temperatures. Berglund et al. (1982) demonstrated the speed of response on a metal ceiling panel to be comparable to that of convection systems. However, very little fuel savings can be expected even with light panels unless the lowered temperature is maintained for long periods. If temperatures are lowered when the area is unoccupied, a way to provide a higher-than-normal rate of heat input for rapid warm-up (e.g., fast-acting ceiling panels) is necessary.

Metal heating panels, hydronic and electric, are applied to building perimeter spaces for heating in much the same way as finned-tube convectors. Metal panels are usually installed in the ceiling and are integrated into the design. They provide a fast-response system (Watson et al. 1998).

Partitions may be erected to the face of hydronic panels but not to the active heating portion of electric panels because of possible element overheating and burnout. Electric panels are often sized to fit the building module with a small removable filler or dummy panel at the window mullion to accommodate future partitions. Hydronic panels also may be cut and fitted in the field; however, modification should be kept to a minimum to keep installation costs down. Hydronic panels can run continuously.

Panel Thermal Resistance

Any thermal resistance between the indoor space and the active panel surface, as well as between the active panel surface and the hydronic tubing or electric circuitry in the panel, reduces system performance. Thermal resistance to heat transfer may vary considerably among different panels, depending on the type of bond between the tubing (electric cabling) and the panel material. Factors such as corrosion or adhesion defects between lightly touching surfaces and the method of maintaining contact may change the bond with time. The actual thermal resistance of any proposed system should be verified by testing. Specific resistance and performance data, when available, should be obtained from the manufacturer. Panel thermal resistances include

r_t = thermal resistance of tube wall per unit tube spacing in a hydronic system, ft²·h·°F/Btu·ft

r_s = thermal resistance between tube (electric cable) and panel body per unit spacing between adjacent tubes (electric cables), ft²·h·°F/Btu·ft

r_p = thermal resistance of panel body, ft²·h·°F/Btu

r_c = thermal resistance of active panel surface covers, ft²·h·°F/Btu

r_u = characteristic (combined) panel thermal resistance, ft²·h·°F/Btu

For a given adjacent tube (electric cable) spacing M,

$$r_u = r_t M + r_s M + r_p + r_c \tag{14}$$

When the tubes (electric cables) are embedded in the slab, r_s may be neglected. However, if they are externally attached to the body of the panel, r_s may be significant, depending on the quality of bonding. Table 1 gives typical r_s values for various ceiling panels.

The value of r_p may be calculated if the characteristic panel thickness x_p and the thermal conductivity k_p of the panel material are known.

If the tubes (electric cables) are embedded in the panel,

$$r_p = \frac{x_p - D_o/2}{k_p} \tag{15a}$$

where D_o = outside diameter of the tube (electric cable). Hydronic floor heating by a heated slab and gypsum-plaster ceiling heating are typical examples.

Table 1 Thermal Resistance of Ceiling Panels

Type of Panel	Thermal Resistance r_p, ft²·h·°F/Btu	r_s, ft²·h·°F/Btu·ft
STEEL PIPE; PAN EDGE HELD AGAINST PIPE BY SPRING CLIP; D_o; x_p; ALUMINUM PAN	$\frac{x_p}{k_p}$	0.55
M; D_o; x_p; COPPER TUBE SECURED TO ALUMINUM SHEET	$\frac{x_p}{k_p}$	0.22
M; D_o; x_p; COPPER TUBE SECURED TO ALUMINUM EXTRUSION	$\frac{x_p}{k_p}$	0.17
x_p; METAL OR GYPSUM LATH; D_o; TUBES	$\frac{x_p - D_o/2}{k_p}$	≈ 0
x_p; M; TUBES OR PIPES; D_o; METAL LATH	$\frac{x_p - D_o/2}{k_p}$	≤0.20

If the tubes (electric cable) are attached to the panel,

$$r_p = \frac{x_p}{k_p} \quad (15b)$$

Metal ceiling panels (see Table 1) and tubes under subfloor (see Figure 23) are typical examples.

Thermal resistance per unit on-center spacing (M = unity) of circular tubes with inside diameter D_i and thermal conductivity k_t is

$$r_t = \frac{\ln(D_o/D_i)}{2\pi k_t} \quad (16a)$$

For an elliptical tube with semimajor and semiminor axes of a and b, respectively, measured at both the outside and inside of the tube,

$$k_t = \ln\frac{(a_o + b_o)/(a_i + b_i)}{2\pi k_t} \quad (16b)$$

In an electric cable, $r_t = 0$.

In metal pipes, r_t is virtually the fluid-side thermal resistance:

$$r_t = \frac{1}{hD_i} \quad (16c)$$

If the tube has multiple layers, Equations (16a) or (16b) should be applied to each individual layer and then summed to calculate the tube's total thermal resistance. Thermal resistance of capillary tube mats can also be calculated from either Equation (16a) or (16b). Typically, capillary tubes are circular, 1/12 in. in internal diameter, and 1/2 in. apart.

Capillary tube mats can be easily applied to existing ceilings in a sand plaster cover layer. Capillary tubes operate under negative pressure, so they do not leak.

If the tube material is nonmetallic, oxygen ingress may be a problem, especially in panel heating. To avoid oxygen corrosion in the heating system, either (1) tubing with an oxygen barrier layer, (2) corrosion-inhibiting additives in the hydronic system, or (3) a heat exchanger separating the panel circuit from the rest of the system should be used.

Table 9 in Chapter 3 of the 2005 *ASHRAE Handbook—Fundamentals* may be used to calculate the forced-convection heat transfer coefficient h. Table 2 gives values of k_t for different tube and pipe materials.

Effect of Floor Coverings

Active panel surface coverings like carpets and pads on the floor can have a pronounced effect on a panel system's performance. The added thermal resistance r_c reduces the panel surface heat flux. To reestablish the required performance, the water temperature must be increased (decreased in cooling). Thermal resistance of a panel covering is

$$r_c = \frac{x_c}{k_c} \quad (17)$$

where

x_c = thickness of each panel covering, ft
k_c = thermal conductivity of each panel cover, Btu/h·ft·°F

If the active panel surface has more than one cover, individual r_c values should be added. Table 3 gives typical r_c values for floor coverings.

If there are covered and bare floor panels in the same hydronic system, it may be possible to maintain a sufficiently high water temperature to satisfy the covered panels and balance the system by throttling the flow to the bare slabs. In some instances, however, the increased water temperature required when carpeting is applied over floor panels makes it impossible to balance floor panel systems in which only some rooms have carpeting unless the piping is arranged to permit zoning using more than one water temperature.

Table 2 Thermal Conductivity of Typical Tube Material

Material	Thermal Conductivity k_t, Btu/h·ft·°F
Carbon steel (AISI 1020)	30
Aluminum	137
Copper (drawn)	225
Red brass (85 Cu-15 Zn)	92
Stainless steel (AISI 202)	10
Low-density polyethylene (LDPE)	0.18
High-density polyethylene (HDPE)	0.24
Cross-linked polyethylene (VPE or PEX)	0.22
Textile-reinforced rubber hose (HTRH)	0.17
Polypropylene block copolymer (PP-C)	0.13
Polypropylene random copolymer (PP-RC)	0.14

Table 3 Thermal Resistance of Floor Coverings

Description	Thermal Resistance r_c, ft²·h·°F/Btu
Bare concrete, no covering	0
Asphalt tile	0.05
Rubber tile	0.05
Light carpet	0.60
Light carpet with rubber pad	1.00
Light carpet with light pad	1.40
Light carpet with heavy pad	1.70
Heavy carpet	0.80
Heavy carpet with rubber pad	1.20
Heavy carpet with light pad	1.60
Heavy carpet with heavy pad	1.90
3/8 in.hardwood	0.54
5/8 in.wood floor (oak)	0.57
1/2 in.oak parquet and pad	0.68
Linoleum	0.12
Marble floor and mudset	0.18
Rubber pad	0.62
Prime urethane underlayment, 3/8 in.	1.61
48 oz. waffled sponge rubber	0.78
Bonded urethane, 1/2 in.	2.09

Notes:
1. Carpet pad thickness should not be more than 1/4 in.
2. Total thermal resistance of carpet is more a function of thickness than of fiber type.
3. Generally, thermal resistance (R-value) is approximately 2.6 times the total carpet thickness in inches.
4. Before carpet is installed, verify that the backing is resistant to long periods of continuous heat up to 120°F

Panel Heat Losses or Gains

Heat transferred from the upper surface of ceiling panels, the back surface of wall panels, the underside of floor panels, or the exposed perimeter of any panel is considered a panel heat loss (gain in cooling). Panel heat losses (gains) are part of the building heat loss (gain) if the heat transfer is between the panel and the outside of the building. If heat transfer is between the panel and another conditioned space, the panel heat loss (gain) is a positive conditioning contribution for that space instead. In either case, the magnitude of panel loss (gain) should be calculated.

Panel heat loss (gain) to (from) space outside the conditioned space should be kept to a reasonable amount by insulation. For example, a floor panel may overheat the basement below, and a ceiling panel may cause the temperature of a floor surface above it to be too high for comfort unless it is properly insulated.

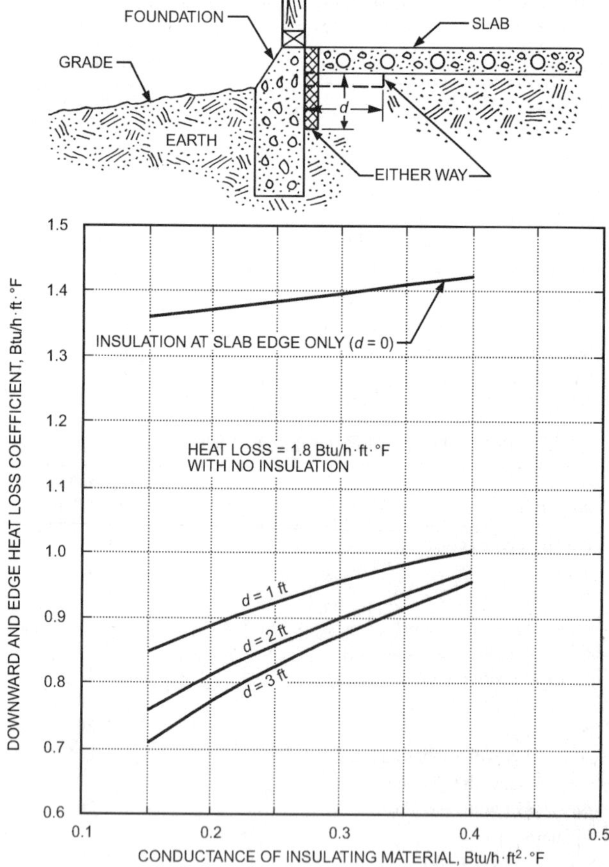

Fig. 8 Downward and Edgewise Heat Loss Coefficient for Concrete Floor Slabs on Grade

The heat loss from most panels can be calculated by using the coefficients given in Table 4 in Chapter 25 of the 2005 *ASHRAE Handbook—Fundamentals*. These coefficients should not be used to determine downward heat loss from panels built on grade because heat flow from them is not uniform (ASHAE 1956, 1957; Sartain and Harris 1956). The heat loss from panels built on grade can be estimated from Figure 8 or from Equation (6) in Chapter 27 of the 1997 *ASHRAE Handbook—Fundamentals*.

Panel Performance

As with other electric or hydronic terminal units, panel performance can be described by the following equation:

$$q = C\Delta T|\Delta T|^{m-1} \tag{18}$$

where

q = combined heat flux on panel surface, Btu/h·ft²
C = characteristic performance coefficient, Btu/h·ft²·°F^m
ΔT = temperature difference, either $t_p - t_o$ in electric heating or $t_w - t_o$ in hydronic heating, °F
t_o = operative temperature, °F
m = $2 + r_c/2r_p$

C and m for a particular panel may be either experimentally determined or calculated from the design material given in this chapter. In either case, sufficient data or calculation points must be gathered to cover the entire operational design range (ASHRAE Draft *Standard* 138P).

PANEL DESIGN

Either hydronic or electric circuits control the active panel surface temperature. The required effective surface temperature t_p necessary to maintain a combined heat flux q (where $q = q_r + q_c$) at steady-state design conditions can be calculated by using applicable heat flux equations for q_r and q_c, depending on the position of the panel. At a given t_a, AUST must be predicted first. Figures 9 and 10 can also be used to find t_p when q and AUST are known. The next step is to determine the required average water (brine) temperature t_w in a hydronic system. It depends primarily on t_p, tube spacing M, and the characteristic panel thermal resistance r_u. Figure 9 provides design information for heating and cooling panels, positioned either at the ceiling or on the floor.

The combined heat flux for ceiling and floor panels can be read directly from Figure 9. Here q_u is the combined heat flux on the floor panel and q_d is the combined heat flux on the ceiling panel. For an electric heating system, t_w scales correspond to the skin temperature of the cable. The following algorithm (TSI 1994) may also be used to design and analyze panels under steady-state conditions:

$$t_d \approx t_a + \frac{(t_p - t_a)M}{2W\eta + D_o} + q(r_p + r_c + r_s M) \tag{19}$$

where

t_d = average skin temperature of tubing (electric cable), °F
q = combined heat flux on panel surface, Btu/h·ft²
t_a = dry-bulb indoor air design temperature, °F. In floor-heated or ceiling-cooled indoor spaces that have large fenestration, t_a may be replaced with AUST.
D_o = outside diameter of embedded tube or characteristic contact width of attached heating or cooling element with panel (see Table 1), ft
M = on-center spacing of adjacent tubes (electric cables), ft
$2W$ = net spacing between tubing (electric cables), $M - D_o$, ft
η = fin efficiency, dimensionless

The first two terms in Equation (19) give the maximum (minimum in cooling) value of the panel surface temperature profile; consequently, if tube spacing M is too large, hot strips along the panel surface may occur, or local condensation occur on these strips in sensible cooling mode.

$$\eta = \frac{\tanh(fW)}{fW} \tag{20a}$$

$$\eta \approx 1/fW \qquad \text{for } fW > 2 \tag{20b}$$

The following equation, which includes transverse heat diffusion in the panel and surface covers, may be used to calculate the fin coefficient f:

$$f \approx \left[\frac{q}{m(t_p - t_a)\sum_{i=1}^{n} k_i x_i}\right]^{1/2} \qquad \text{for } t_p \neq t_a \tag{21}$$

where

f = fin coefficient
m = $2 + r_c/2r_p$
n = total number of different material layers, including panel and surface covers
x_i = characteristic thickness of each material layer i, ft
k_i = thermal conductivity of each layer i, Btu/h·ft·°F

For a hydronic system, the required average water (brine) temperature is

$$t_w = (q + q_b)Mr_t + t_d \tag{22}$$

where q_b is the flux of back and perimeter heat losses (positive) in a heated panel or gains (negative) in a cooling panel.

This algorithm may be applied to outdoor slab heating systems, provided that the combined heat flux q is calculated according to Chapter 50 in the 2007 *ASHRAE Handbook—HVAC Applications*, with the following conditions: no snow, no evaporation, q (in panel heating) = q_h (radiation and convection heat flux in snow-melting calculations). With a careful approach, outdoor slab cooling systems may be analyzed by incorporating the solar radiation gain, thermal radiation, and forced convection from the sky and ambient air.

HEATING AND COOLING PANEL SYSTEMS

The following are the most common forms of panels applied in panel heating systems:

- Metal ceiling panels
- Embedded tubing in ceilings, walls, or floors
- Electric ceiling panels
- Electrically heated ceilings or floors
- Air-heated floors

Residential heating applications usually consist of tubes or electric elements embedded in masonry floors or plaster ceilings. This construction is suitable where loads are stable and building design minimizes solar effects. However, in buildings where fenestration is large and thermal loads change abruptly, the slow response, thermal lag, and override effect of masonry panels are unsatisfactory. Metal ceiling panels with low thermal mass respond quickly to load changes (Berglund et al. 1982).

Panels are preferably located in the ceiling because it is exposed to all other indoor surfaces, occupants, and objects in the conditioned indoor space. It is not likely to be covered, as floors are, and higher surface temperatures can be maintained. Figure 10 gives design data for ceiling and wall panels with effective surface temperatures up to 300°F.

Example 1. An in-slab, on-grade panel (see Figure 20) will be used for both heating and cooling. M = 1 ft (12 in.), r_u = 0.5 ft²·h·°F/Btu, and r_c/r_p is less than 4. t_a is 68°F in winter and 76°F in summer.

AUST is expected to be 2°F less than t_a in winter heating and 1°F higher than t_a in summer cooling.

What is the average water temperature and effective floor temperature (1) for winter heating when q_u = 40 Btu/h·ft², and (2) for summer cooling when $-q_u$ = 15 Btu/h·ft²?

Solution:

Winter heating

To obtain the average water temperature using Figure 9, start on the left axis where q_u = 40 Btu/h·ft². Proceed right to the intersect r_u = 0.5 and then down to the M = 12 in. line. The reading is AUST + 56, which is the solid line value because r_c/r_p < 4. As stated in the initial problem, AUST = t_a − 2 or AUST = 68 − 2 = 66°F. Therefore, the average water temperature would be t_w = 66 + 56 = 122°F.

To find the effective floor temperature, start at q_u = 40 Btu/h·ft² in Figure 9 and proceed right to AUST = t_a − 2°F. The solid line establishes 21°F as the temperature difference between the panel and the

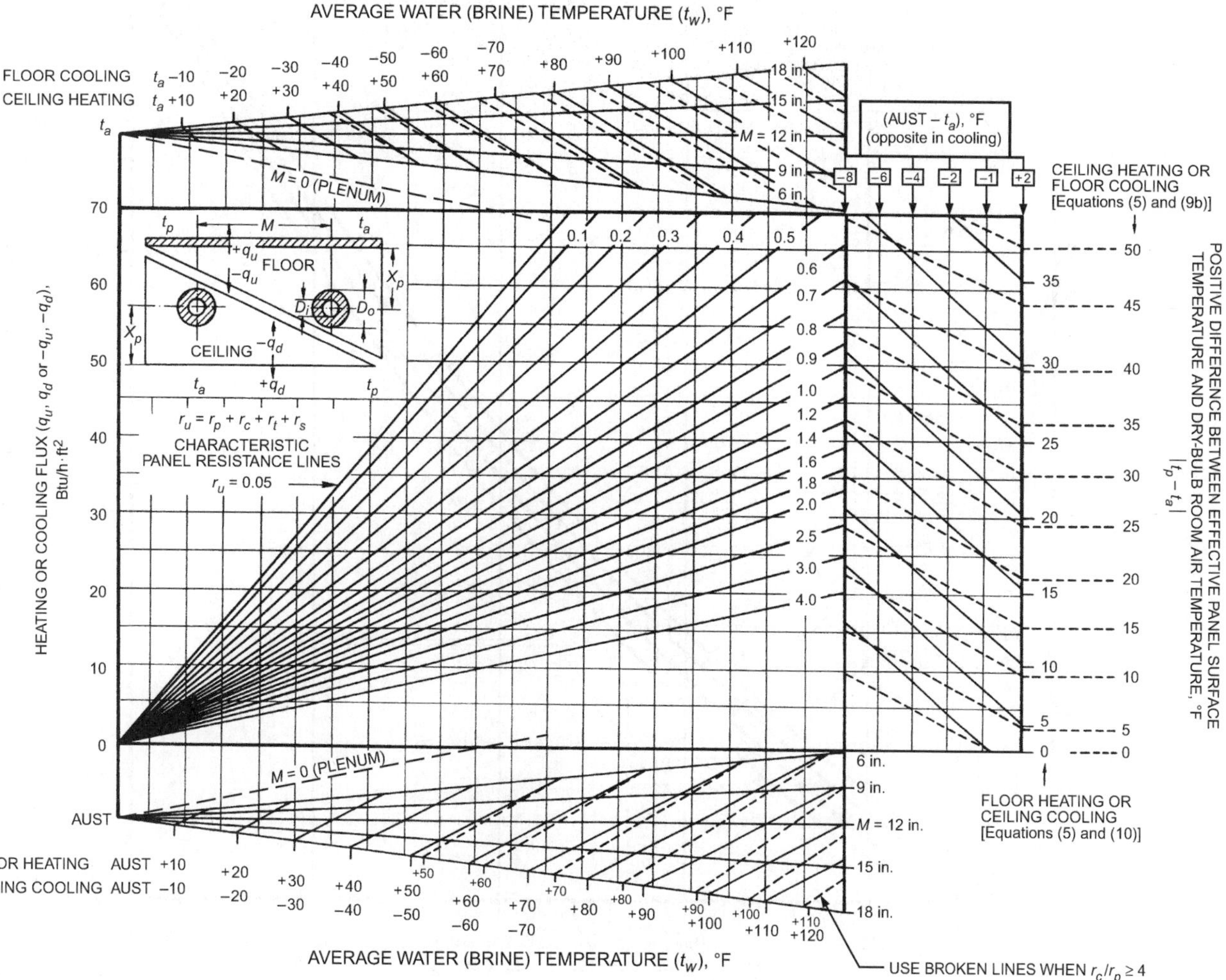

Fig. 9 Design Graph for Sensible Heating and Cooling with Floor and Ceiling Panels

indoor air. Therefore, the effective floor temperature $t_p = t_a + 21$ or $t_p = 68 + 21 = 89$°F.

Summer cooling

Using Figure 9, start at the left axis at $-q_u = 15$ Btu/h·ft². Proceed to $r_u = 0.5$, and then up (for cooling) to $M = 12$ in., which reads $t_a - 23$ or 76 − 23 = 53°F average water temperature for cooling.

To obtain the effective floor temperature at $-q_u = 15$ Btu/h·ft², proceed to AUST − t_a = +1°F, which reads −11°F. Therefore, the effective floor temperature is 76 − 11 = 65°F.

Example 2. An aluminum extrusion panel, which is 0.05 in. thick with heat element spacing of $M = 0.5$ ft (6 in.), is used in the ceiling for heating. If a ceiling heat flux q_d of 400 Btu/h·ft² is required to maintain room temperature t_a at 70°F, what is the required heating element skin temperature t_d and effective panel surface temperature t_p?

Solution:

Using Figure 10 enter the left axis heat flux q_d at 400 Btu/h·ft². Proceed to the line corresponding to $t_a = 70$°F and then move up to the $M = 6$ in. line. The ceiling heating element temperature t_d at the intersection point is 320°F. From the bottom axis of Figure 10, the effective panel surface temperature t_p is 265°F.

Special Cases

Figure 9 may also be used for panels with tubing not embedded in the panel:

- $x_p = 0$ if tubes are externally attached.
- In spring-clipped external tubing, $D_i = 0$ and D_o is the clip thickness.

Warm air and electric heating elements are two design concepts influenced by local factors. The warm-air system has a special cavity construction where air is supplied to a cavity behind or under the panel surface. The air leaves the cavity through a normal diffuser arrangement and is supplied to the indoor space. Generally, these

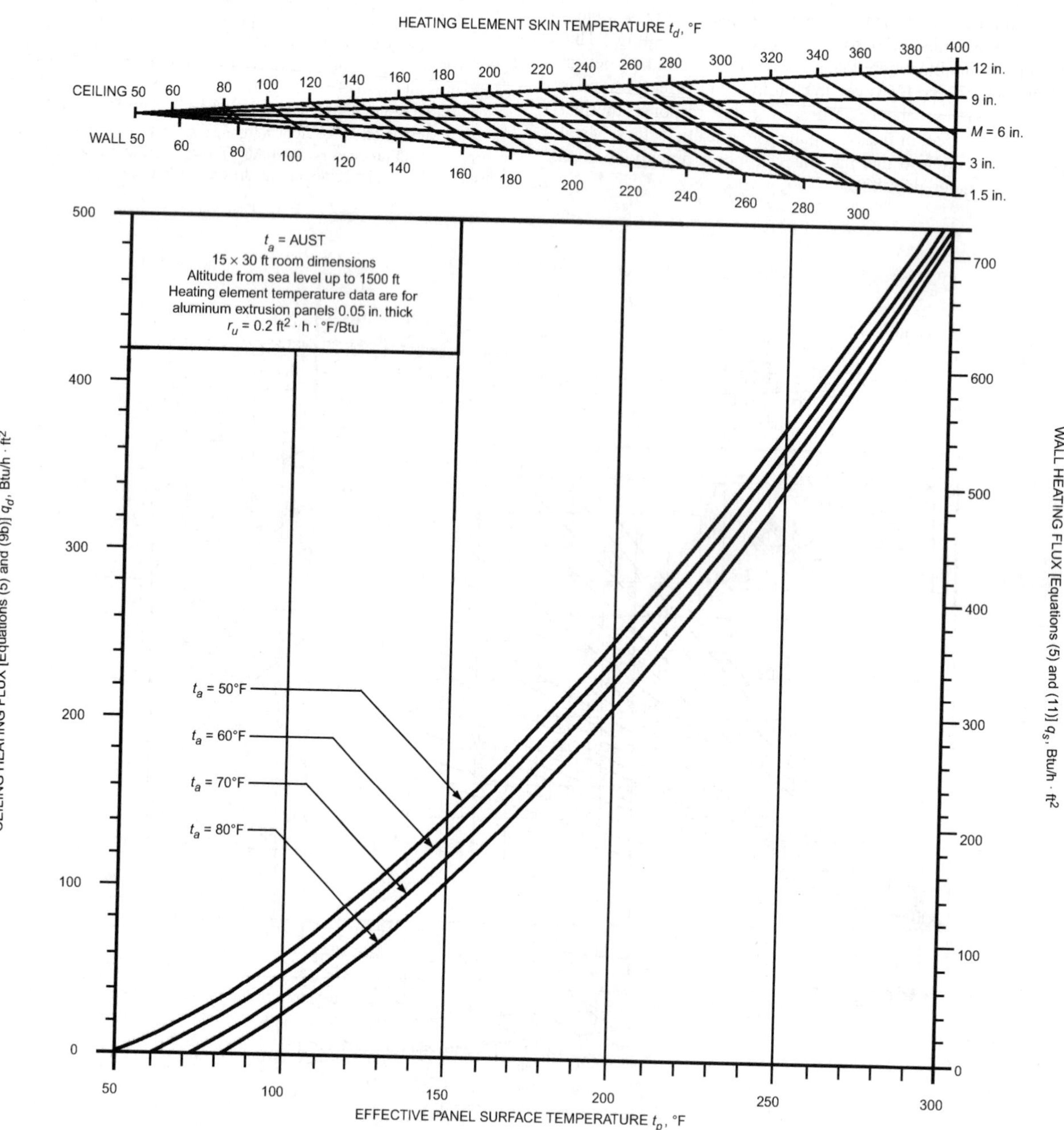

Fig. 10 Design Graph for Heating with Aluminum Ceiling and Wall Panels

systems are used as floor panels in schools and in floors subject to extreme cold, such as in an overhang. Cold outdoor temperatures and heating medium temperatures must be analyzed with regard to potential damage to the building construction. Electric heating elements embedded in the floor or ceiling construction and unitized electric ceiling panels are used in various applications to provide both full heating and spot heating of the space.

HYDRONIC PANEL SYSTEMS

Design Considerations

Hydronic panels can be used with two- and four-pipe distribution systems. Figure 11 shows a typical system arrangement. It is common to design for a 20°F temperature drop for heating across a given grid and a 5°F rise for cooling, but larger temperature differentials may be used, if applicable.

Panel design requires determining panel area, panel type, supply water temperature, water flow rate, and panel arrangement. Panel performance is directly related to indoor space conditions. Air-side design also must be established. Heating and cooling loads may be calculated by procedures covered in Chapters 29 and 30 of the 2005 *ASHRAE Handbook—Fundamentals*. The procedure is as follows:

Sensible Cooling

1. Determine indoor design dry-bulb temperature, relative humidity, and dew point.
2. Calculate sensible and latent heat gains.
3. Establish minimum supply air quantity for ventilation.
4. Calculate latent cooling available from supply air.
5. Calculate sensible cooling available from supply air.
6. Determine remaining sensible cooling load to be satisfied by panel system.
7. Determine minimum permissible effective cooling panel surface temperature that will not lead to surface condensation at design conditions.
8. Determine AUST.
9. Determine necessary panel area for remaining sensible cooling.

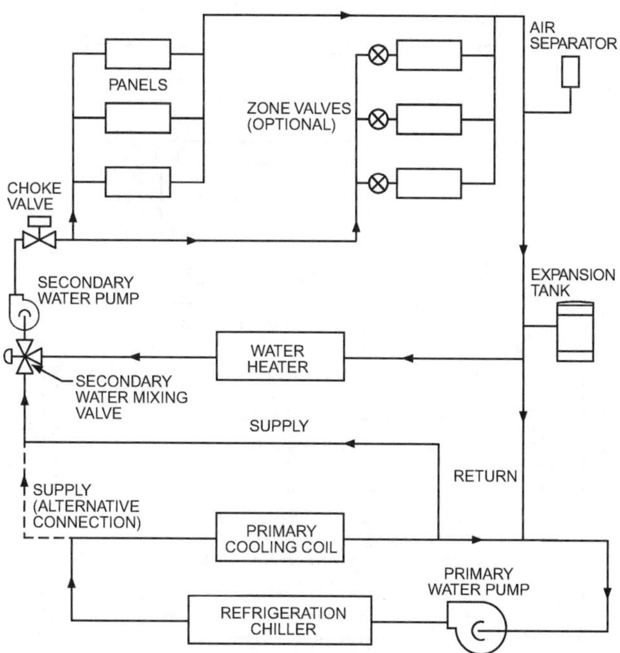

Fig. 11 Primary/Secondary Water Distribution System with Mixing Control

10. Determine average panel cooling water (brine) temperature for given tube spacing, or determine necessary tube spacing if average panel cooling water (brine) temperature is known.

Sensible Heating

1. Designate indoor design dry-bulb temperature for panel heating.
2. Calculate room heat loss.
3. Determine AUST. Use Equation (13) to find surface temperatures of exterior walls and exposed floors and ceilings. Interior walls are assumed to have surface temperatures equal to indoor air temperature.
4. Calculate required effective surface temperature of panel. Refer to Figures 9 and 10 if AUST does not greatly differ from indoor air temperature. Otherwise, use Equations (5), (9a), (9b), (10), and (11) or refer to Figures 1 and 3.
5. Determine panel area. Refer to Figures 9 and 10 if AUST does not vary greatly from indoor air temperature.
6. Refer to manufacturers' data for panel surface temperatures higher than those given in Figures 9 and 10. For panels with several covers, average temperature of each cover and effective panel surface temperature must be calculated and compared to temperature-withstanding capacity for continuous operation of every cover material. For this purpose, Equation (18) may be used: add thermal resistances of all cover layers between panel surface and cover layer in question, then multiply by q and add to first two terms in Equation (18). This is t_{ha}, approximated temperature of the particular layer at design.
7. Determine tube spacing for a given average water temperature or select electric cable properties or electric mat size.
8. In a hydronic panel system, if tube spacing is known, determine required average water temperature.
9. Design panel arrangement.

Other Steps Common for Sensible Heating and Cooling

1. Check thermal comfort requirements in the following steps [see Chapter 8 of the 2005 *ASHRAE Handbook—Fundamentals* and NRB (1981)].
   (a) Determine occupant's clothing insulation value and metabolic rate (see Tables 4, 7, and 8 in Chapter 8 of the 2005 *ASHRAE Handbook—Fundamentals*).
   (b) Determine optimum operative temperature at coldest point in conditioned space (see the Comfort Equations for Radiant Heating section in Chapter 8 of the 2005 *ASHRAE Handbook—Fundamentals*. Note that the same equations may be adopted for panel cooling).
   (c) Determine MRT at the coldest point in the conditioned space [see Chapter 15 and Fanger (1972)].
   Note: If indoor air velocity is less than 1.3 fps and MRT is less than 122°F, operative temperature may be approximated as the average of MRT and t_a.
   (d) From the definition of operative temperature, establish optimum indoor design air temperature at coldest point in the room. If optimum indoor design air temperature varies greatly from designated design temperature, designate a new temperature.
   (e) Determine MRT at hottest point in conditioned space.
   (f) Calculate operative temperature at hottest point in conditioned space.
   (g) Compare operative temperatures at hottest and coldest points. For light activity and normal clothing, the acceptable operative temperature range is 68 to 75°F [see NRB (1981) and ANSI/ASHRAE *Standard* 55-1992R]. If the range is not acceptable, the panel system must be modified.
   (h) Calculate radiant temperature asymmetry (NRB 1981). Acceptable ranges are less than 9°F for warm ceilings, 27°F for cool ceilings, 18°F for cool walls, and 49°F for warm walls at

10% local discomfort dissatisfaction (ANSI/ASHRAE *Standard* 55-1992).

2. Determine water flow rate and pressure drop. Refer to manufacturers' guides for specific products, or use the guidelines in Chapter 36 of the 2005 *ASHRAE Handbook—Fundamentals*. Chapter 12 of this volume also has information on hydronic heating and cooling systems.
3. The supply and return manifolds need to be carefully designed. If there are circuits of unequal coil lengths, the following equations may be used (Hansen 1985; Kilkis 1998) for a circuit i connected to a manifold with n circuits:

$$Q_i = (L_{eq}/L_i)^{1/r} Q_{tot} \tag{23}$$

where

$L_{eq} = \left[\sum_{i=1}^{n} L_i^{-1/r} \right]^{-r}$, ft

Q_i = flow rate in circuit i, gpm

Q_{tot} = total flow rate in supply manifold, gpm

L_i = coil length of hydronic circuit i, ft

r = 1.75 for hydronic panels (Siegenthaler 1995)

Application, design, and installation of panel systems have certain requirements and techniques:

- As with any hydronic system, look closely at the piping system design. Piping should be designed to ensure that water of the proper temperature and in sufficient quantity is available to every grid or coil at all times. Proper piping and system design should minimize the detrimental effects of oxygen on the system. Reverse-return systems should be considered to minimize balancing problems.
- Individual panels can be connected for parallel flow using headers, or for sinuous or serpentine flow. To avoid flow irregularities in a header-type grid, the water channel or lateral length should be greater than the header length. If the laterals in a header grid are forced to run in a short direction, using a combination series-parallel arrangement can solve this problem. Serpentine flow ensures a more even panel surface temperature throughout the heating or cooling zone.
- Noise from entrained air, high-velocity or high-pressure-drop devices, or pump and pipe vibrations must be avoided. Water velocities should be high enough to prevent separated air from accumulating and causing air binding. Where possible, avoid automatic air venting devices over ceilings of occupied spaces.
- Design piping systems to accept thermal expansion adequately. Do not allow forces from piping expansion to be transmitted to panels. Thermal expansion of ceiling panels must be considered.
- In hydronic systems, thermoplastic, rubber tubes, steel, or copper pipes are used widely in ceiling, wall, or floor panel construction. Where coils are embedded in concrete or plaster, no threaded joints should be used for either pipe coils or mains. Steel pipe should be the all-welded type. Copper tubing should be a soft-drawn coil. Fittings and connections should be minimized. Bending should be used to change direction. Solder-joint fittings for copper tube should be used with a medium-temperature solder of 95% tin, 5% antimony, or capillary brazing alloys. All piping should be subjected to a hydrostatic test of at least three times the working pressure. Maintain adequate pressure in embedded piping while pouring concrete.
- Placing the thermostat on a wall where it can observe both the outdoor exposed wall and the warm panel should be considered. The normal thermostat cover reacts to the warm panel, and thermal radiation from the panel to the cover tends to alter the control point so that the thermostat controls 2 to 3°F lower when the outdoor temperature is a minimum and the panel temperature is a maximum. Experience indicates that panel-heated spaces are more comfortable under these conditions than when the thermostat is located on a back wall.
- If throttling valve control is used, either the end of the main should have a fixed bypass, or the last one or two rooms on the mains should have a bypass valve to maintain water flow in the main. Thus, when a throttling valve modulates, there will be a rapid response.
- When selecting heating design temperatures for a ceiling panel surface, the design parameters are as follows:
  - Excessively high temperatures over the occupied zone cause the occupant to experience a "hot head effect."
  - Temperatures that are too low can result in an oversized, uneconomical panel and a feeling of coolness at the outside wall.
  - Locate ceiling panels adjacent to perimeter walls and/or areas of maximum load.
  - With normal ceiling heights of 8 to 9 ft, panels less than 3 ft wide at the outside wall can be designed for 235°F surface temperature. If panels extend beyond 3 ft into the indoor space, the panel surface temperature should be limited to the values given in Figure 16. The surface temperature of concrete or plaster panels is limited by construction.
- Floor panels are limited to surface temperatures of less than 84°F in occupied spaces for comfort reasons. Subfloor temperature may be limited to the maximum exposure temperature specified by the floor cover manufacturer.
- When the panel chilled-water system is started, the circulating water temperature should be maintained at indoor air temperature until the air system is completely balanced, the dehumidification equipment is operating properly, and building relative humidity is at design value.
- When the panel area for cooling is greater than the area required for heating, a two-panel arrangement (Figure 12) can be used. Panel HC (heating and cooling) is supplied with hot or chilled water year-round. When chilled water is used, the controls activate panel CO (cooling only) mode, and both panels are used for cooling.
- To prevent condensation on the cooling panels, the panel water supply temperature should be maintained at least 1°F above the

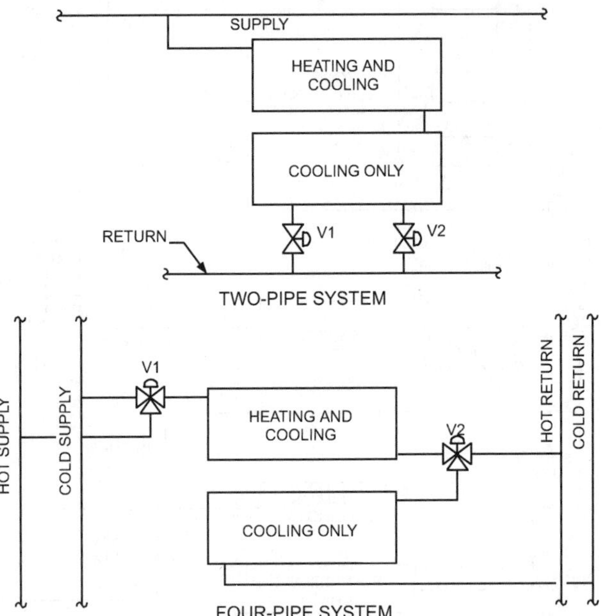

Fig. 12 Split Panel Piping Arrangement for Two-Pipe and Four-Pipe Systems

indoor design dew-point temperature. This minimum difference is recommended to allow for the normal drift of temperature controls for water and air systems, and also to provide a factor of safety for temporary increase in indoor relative humidity.

- Selection of summer design indoor dew point below 50°F generally is not economical.
- The most frequently applied method of dehumidification uses cooling coils. If the main cooling coil is six rows or more, the dew point of leaving air will approach the leaving water temperature. The cooling water leaving the dehumidifier can then be used for the panel water circuit.
- Several chemical dehumidification methods are available to control latent and sensible loads separately. In one application, cooling tower water is used to remove heat from the chemical drying process, and additional sensible cooling is necessary to cool the dehumidified air to the required system supply air temperature.
- When chemical dehumidification is used, hygroscopic chemical dew-point controllers are required at the central apparatus and at various zones to monitor dehumidification.
- When cooled ceiling panels are used with a variable air volume (VAV) system, the air supply rate should be near maximum volume to ensure adequate dehumidification before the cooling ceiling panels are activated.

Other factors to consider when using panel systems are

- Evaluate the panel system to take full advantage in optimizing the physical building design.
- Select recessed lighting fixtures, air diffusers, hung ceilings, and other ceiling devices to provide the maximum ceiling area possible for use as panels.
- The air-side design must be able to maintain relative humidity at or below design conditions at all times to eliminate any possibility of condensation on the panels. This becomes more critical if indoor space dry- and wet-bulb temperatures are allowed to drift for energy conservation, or if duty cycling of the fans is used.
- Do not place cooling panels in or adjacent to high-humidity areas.
- Anticipate thermal expansion of the ceiling and other devices in or adjacent to the ceiling.
- The design of operable windows should discourage unauthorized opening.

HYDRONIC METAL CEILING PANELS

Metal ceiling panels can be integrated into a system that heats and cools. In such a system, a source of dehumidified ventilation air is required in summer, so the system is classed as an air-and-water system. Also, various amounts of forced air are supplied year-round. When metal panels are applied for heating only, a ventilation system may be required, depending on local codes.

Ceiling panel systems are an outgrowth of perforated metal, suspended acoustical ceilings. These ceiling panel systems are usually designed into buildings where the suspended acoustical ceiling can be combined with panel heating and cooling. The panels can be designed as small units to fit the building module, which provides extensive flexibility for zoning and control, or the panels can be arranged as large continuous areas for maximum economy. Some ceiling installations require active panels to cover only part of the indoor space and compatible matching acoustical panels for the remaining ceiling area.

Three types of metal ceiling systems are available. The first consists of light aluminum panels, usually 12 by 24 in., attached in the field to 0.5 in. galvanized pipe coils. Figure 13 illustrates a metal ceiling panel system that uses 0.5 in. pipe laterals on 6, 12, or 24 in. centers, hydraulically connected in a sinuous or parallel-flow welded system. Aluminum ceiling panels are clipped to these pipe laterals and act as a heating panel when warm water is flowing or as a cooling panel when chilled water is flowing.

The second type of panel consists of a copper coil secured to the aluminum face sheet to form a modular panel. Modular panels are available in sizes up to about 36 by 60 in. and are held in position by various types of ceiling suspension systems, most typically a standard suspended T-bar 24 by 48 in. exposed grid system. Figure 14 illustrates metal panels using a copper tube pressed into an aluminum extrusion, although other methods of securing the copper tube have proven equally effective.

Metal ceiling panels can be perforated so that the ceiling becomes sound absorbent when acoustical material is installed on the back of the panels. The acoustical blanket is also required for thermal reasons, so that reverse loss or upward flow of heat from the metal ceiling panels is minimized.

The third type of panel is an aluminum extrusion face sheet with a copper tube mechanically fastened into a channel housing on the back. Extruded panels can be manufactured in almost any shape and size. Extruded aluminum panels are often used as long, narrow panels at the outside wall and are independent of the ceiling system. Panels 15 or 20 in. wide usually satisfy a typical office building's heating requirements. Lengths up to 20 ft are available. Figure 15 illustrates metal panels using a copper tube pressed into an aluminum extrusion.

Performance data for extruded aluminum panels vary with the copper tube/aluminum contact and test procedures used. Hydronic ceiling panels have a low thermal resistance and respond quickly to changes in space conditions. Table 1 shows thermal resistance values for various ceiling constructions.

Metal ceiling panels can be used with any of the all-air cooling systems described in Chapter 2. Chapters 29 and 30 of the 2005 *ASHRAE Handbook—Fundamentals* describe how to calculate

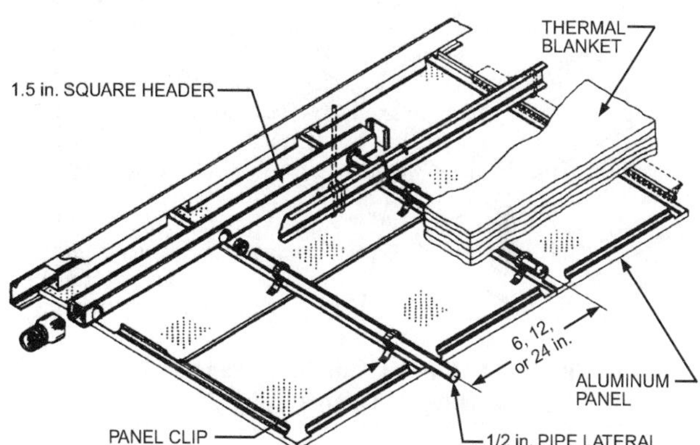

Fig. 13 Metal Ceiling Panels Attached to Pipe Laterals

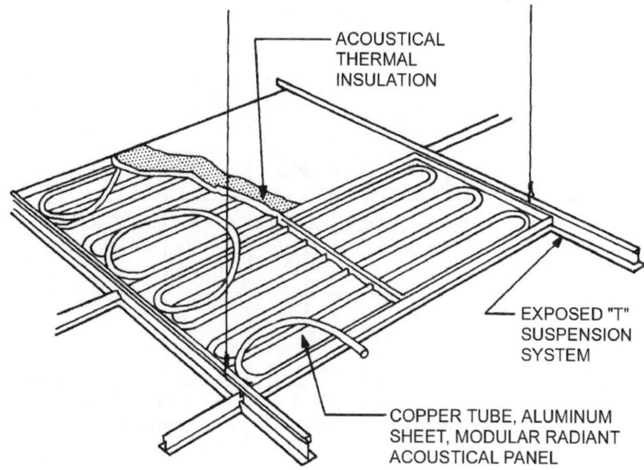

Fig. 14 Metal Ceiling Panels Bonded to Copper Tubing

heating loads. Double-glazing and heavy insulation in outside walls reduce transmission heat losses. As a result, infiltration and reheat have become of greater concern. Additional design considerations include the following:

- Perimeter radiant heating panels extending not more than 3 ft into the indoor space may operate at higher temperatures, as described in the section on Hydronic Panel Systems.
- Hydronic panels operate efficiently at low temperature and are suitable for condenser water heat reclaim systems.
- Locate ceiling panels adjacent to the outside wall and as close as possible to the areas of maximum load. The panel area within 3 ft of the outside wall should have a heating capacity equal to or greater than 50% of the wall transmission load.
- Ceiling system designs based on passing return air through perforated modular panels into the plenum space above the ceiling are not recommended because much of this heat is lost to the return air system in heating mode.

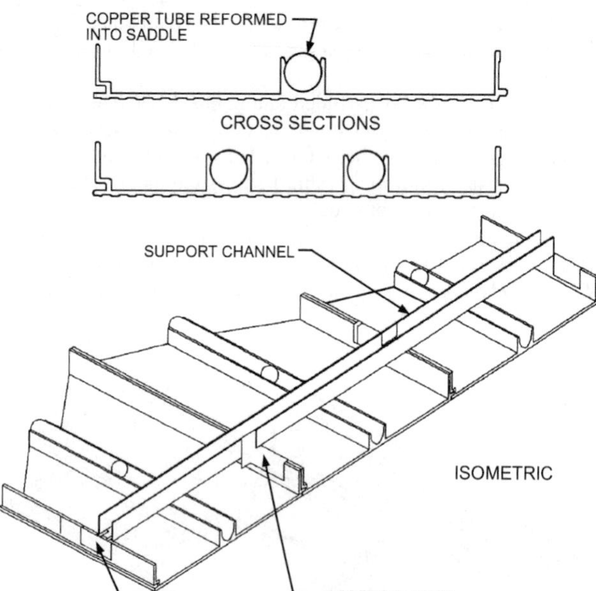

Fig. 15 Extruded Aluminum Panels with Integral Copper Tube

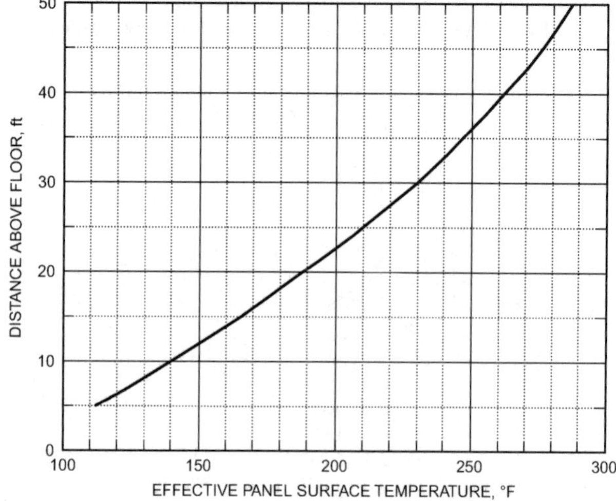

Fig. 16 Permitted Design Ceiling Surface Temperatures at Various Ceiling Heights

- When selecting heating design temperatures for a ceiling panel surface or average water temperature, the design parameters are as follows:
  - Excessively high temperatures over the occupied zone will cause the occupant to experience a "hot head effect."
  - Temperatures that are too low can result in an oversized, uneconomical panel and a feeling of coolness at the outside wall.
  - Give ceiling panel location priority.
  - With normal ceiling heights of 8 to 9 ft, panels less than 3 ft wide at the outside wall can be designed for 235°F surface temperature. If panels extend beyond 3 ft into the indoor space, the panel surface temperature should be limited to the values as given in Figure 16.
- Allow sufficient space above the ceiling for installation and connection of the piping that forms the panel ceiling.

Metal acoustic panels provide heating, cooling, sound absorption, insulation, and unrestricted access to the plenum space. They are easily maintained, can be repainted to look new, and have a life expectancy in excess of 30 years. The system is quiet, comfortable, draft-free, easy to control, and responsive. The system is a basic air-and-water system. First costs are competitive with other systems, and life-cycle cost analysis often shows that the long life of the equipment makes it the least expensive in the long run. The system has been used in hospitals, schools, office buildings, colleges, airports, and exposition facilities.

Metal panels can also be integrated into the ceiling design to provide a narrow band of panel heating around the perimeter of the building. The panel system offers advantages over baseboard or overhead air in appearance, comfort, operating efficiency and cost, maintenance, and product life.

ASHRAE/ANSI *Standard* 138P, Method of Testing for Rating Hydronic Ceiling Panels, discusses testing and rating of ceiling panels.

DISTRIBUTION AND LAYOUT

Chapters 3 and 12 apply to panels. Layout and design of metal ceiling panels for heating and cooling begin early in the job. The type of ceiling chosen influences the panel design and, conversely, thermal considerations may dictate what ceiling type to use. Heating panels should be located adjacent to the outside wall. Cooling panels may be positioned to suit other elements in the ceiling. In applications with normal ceiling heights, heating panels that exceed 160°F should not be located over the occupied area. In hospital applications, valves should be located in the corridor outside patient rooms.

One of the following types of construction is generally used:

- Pipe or tube is embedded in the lower portion of a concrete slab, generally within 1 in. of its lower surface (Figure 17). If plaster is to be applied to the concrete, the piping may be placed directly on the wood forms. If the slab is to be used without plaster finish, the piping should be installed not less than 0.75 in. above the undersurface of the slab. The minimum coverage must comply with local building code requirements.

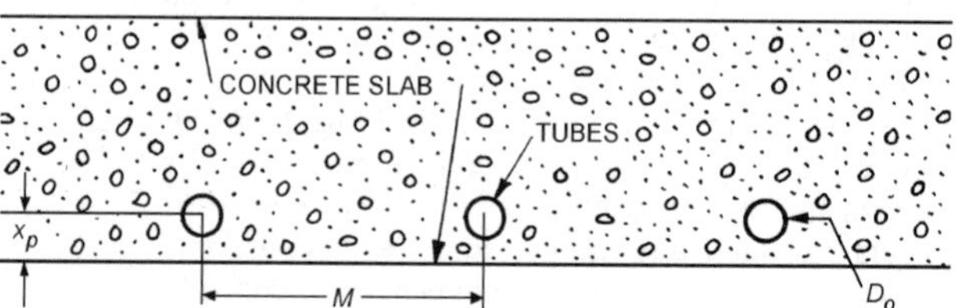

Fig. 17 Coils in Structural Concrete Slab

- Pipe or tube is embedded in a metal lath and plaster ceiling. If the lath is suspended to form a hung ceiling, the lath and heating coils are securely wired to the supporting members so that the lath is below, but in good contact with, the coils. Plaster is then applied to the metal lath, carefully embedding the coil as shown in Figure 18.
- Smaller-diameter copper or thermoplastic tube is attached to the underside of wire or gypsum lath. Plaster is then applied to the lath to embed the tube, as shown in Figure 19.
- Other forms of ceiling construction are composition board, wood paneling, etc., with warm-water piping, tube, or channels built into the panel sections.

Coils are usually laid in a sinusoidal pattern, although some header or grid-type coils have been used in ceilings. Coils may be thermoplastic, ferrous, or nonferrous pipe or tube, with coil pipes spaced from 4.5 to 9 in. on centers, depending on the required heat flux, pipe or tube size, and other factors.

Where plastering is applied to pipe coils, a standard three-coat gypsum plastering specification is followed, with a minimum of 3/8 in. of cover below the tubes when they are installed below the lath. Generally, the surface temperature of plaster panels should not exceed 120°F, which may be satisfied by limiting the average water temperature to a maximum temperature of 140°F.

Insulation should be placed above the coils to reduce back loss, which is the difference between heat supplied to the coil and net useful heat output to the heated indoor space.

To protect the plaster installation and to ensure proper air drying, heat must not be applied to the panels for two weeks after all plastering work has been completed. When the system is started for the first time, water supplied to the panels should not be more than 20°F above the prevailing indoor air temperature and not in excess of 90°F.

Water should be circulated at this temperature for about two days, and then increased at a rate of about 5°F per day to 140°F.

During the air-drying and preliminary warm-up periods, there should be adequate ventilation to carry moisture from the panels. No paint or paper should be applied to the panels before these periods have been completed or while the panels are being operated. After paint and paper have been applied, an additional shorter warm-up period, similar to first-time starting, is also recommended.

Hydronic Wall Panels

Although piping embedded in walls is not as widely used as floor and ceiling panels, it can be constructed by any of the methods outlined for ceilings or floors. Its design is similar to other hydronic panels [see Equations (18) to (21)]. Equations (5) and (11) give the heat flux at the surface of wall panels.

Hydronic Floor Panels

Interest has increased in floor heating with the introduction of nonmetallic tubing and new design, application, and control techniques. Whichever method is used for optimum floor output and comfort, it is important that heat be evenly distributed over the floor. Spacing is generally 4 to 12 in on centers for the coils. Wide spacing under tile or bare floors can cause uneven surface temperatures.

Embedded Tubes or Pipes in Concrete Slab. Thermoplastic, rubber, ferrous, and nonferrous pipes, or composite tubes (e.g., thermoplastic tubes with aluminum sleeves) are used in floor slabs that rest on grade. Hydronic coils are constructed as sinusoidal-continuous coils or arranged as header coils with a spacing of 6 to 18 in. on centers. The coils are generally installed with 1.5 to 4 in. of cover above them. Insulation is recommended to reduce perimeter and back losses. Figure 20 shows application of hydronic coils in slabs resting on grade. Coils should be embedded completely and should not rest on an interface. Any supports used for positioning heating coils should be nonabsorbent and inorganic. Reinforcing steel, angle iron, pieces of pipe or stone, or concrete mounds can be used. No wood, brick, concrete block, or similar materials should support coils. A waterproofing layer is desirable to protect insulation and piping.

Where coils are embedded in structural load-supporting slabs above grade, construction codes may affect their position. Otherwise, the coil piping is installed as described for slabs resting on grade.

The warm-up and start-up period for concrete panels are similar to those outlined for plaster panels.

Embedded systems may fail sometime during their life. Adequate valves and properly labeled drawings help isolate the point of failure.

Suspended Floor Tubing or Piping. Piping may be applied on or under suspended wood floors using several construction methods. Piping may be attached to the surface of the floor and embedded in

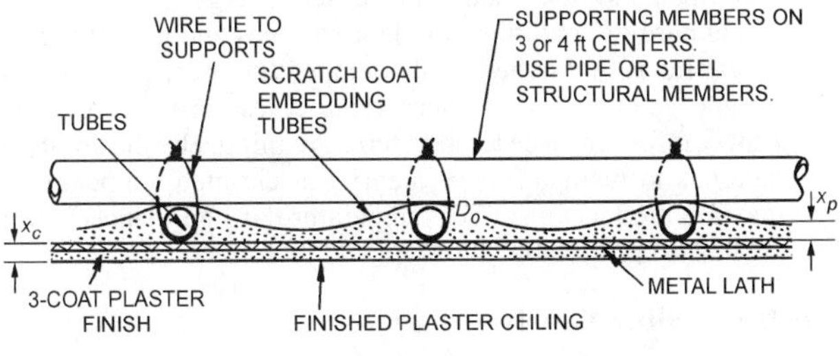

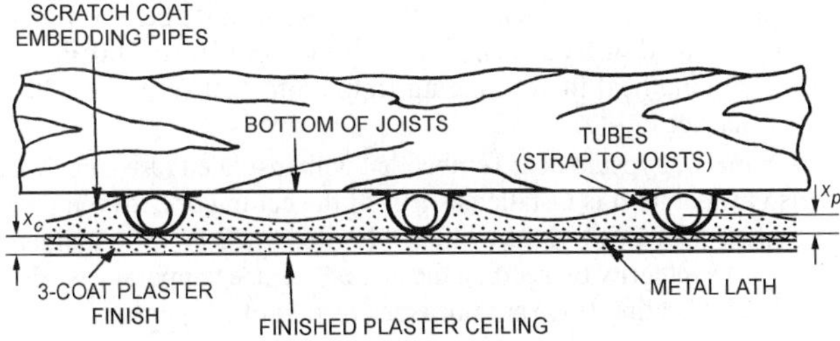

Fig. 18 Coils in Plaster Above Lath

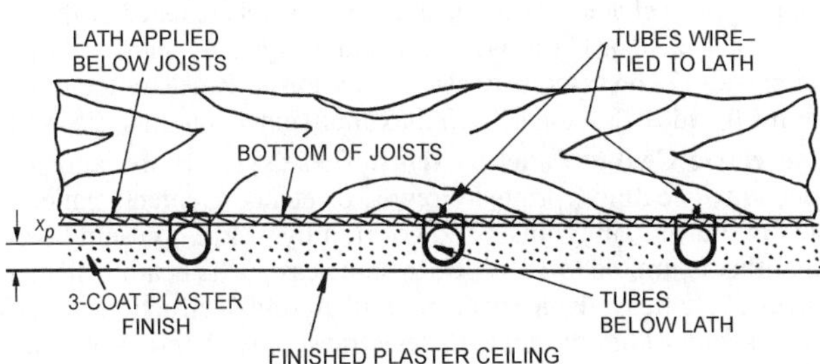

Fig. 19 Coils in Plaster Below Lath

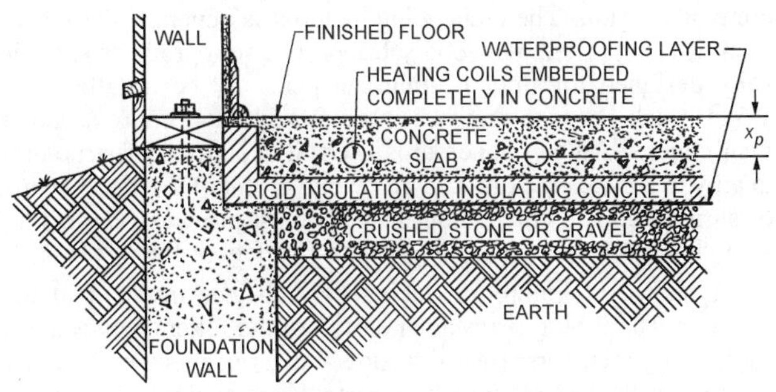

Fig. 20 Coils in Floor Slab on Grade

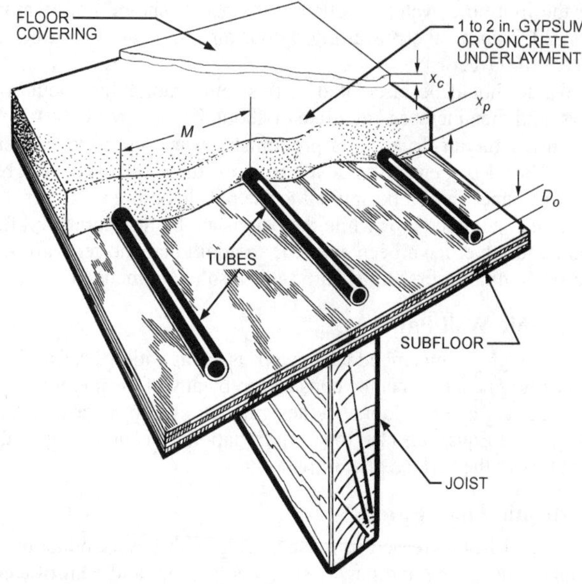

Fig. 21 Embedded Tube in Thin Slab

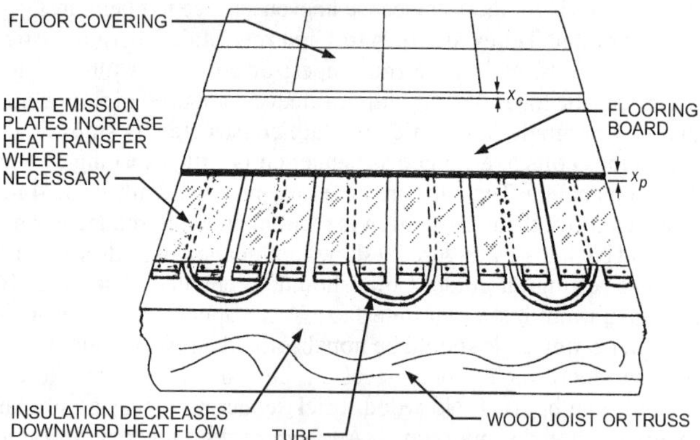

Fig. 22 Tube in Subfloor

a layer of concrete or gypsum, mounted in or below the subfloor, or attached directly to the underside of the subfloor using metal panels to improve heat transfer from the piping. An alternative method is to install insulation with a reflective surface and leave an air gap of 2 to 4 in. to the subfloor. Whichever method is used for optimum floor output and comfort, it is important that heat be evenly distributed throughout the floor. Tubing generally has a spacing of 4 to 12 in. on centers. Wide spacing under tile or bare floors can cause uneven surface temperatures.

Figure 21 illustrates construction with piping embedded in concrete or gypsum. The embedding material is generally 1 to 2 in. thick when applied to a wood subfloor. Gypsum products specifically designed for floor heating can generally be installed 1 to 1.5 in. thick because they are more flexible and crack-resistant than concrete. When concrete is used, it should be of structural quality to reduce cracking caused by movement of the wood frame or shrinkage. The embedding material must provide a hard, flat, smooth surface that can accommodate a variety of floor covers.

As shown in Figure 22, tubing may also be installed in the subfloor. The tubing is installed on top of the rafters between the subflooring members. Heat diffusion and surface temperature can be improved uniformly by adding metal heat transfer plates, which spread heat beneath the finished flooring.

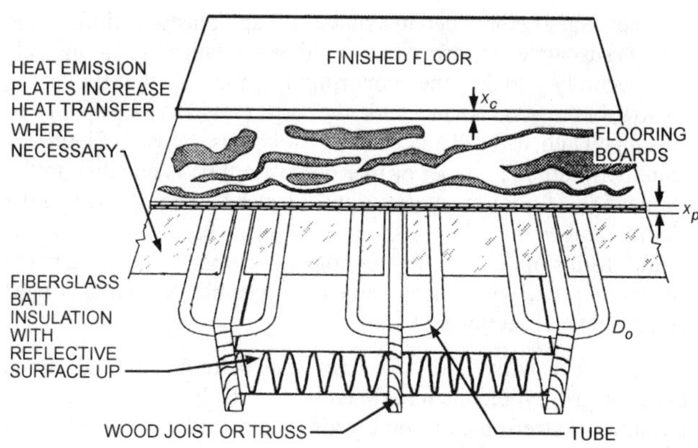

Fig. 23 Tube Under Subfloor

A third construction option is to attach the tube to the underside of the subfloor with or without metal heat transfer plates. The construction is illustrated in Figure 23.

Transfer from the hot-water tube to the surface of the floor is the important consideration in all cases. The floor surface temperature affects the actual heat transfer to the space. Any hindrance between the heated water tube and the floor surface reduces system effectiveness. The method that transfers and spreads heat evenly through the subfloor with the least resistance produces the best results.

ELECTRICALLY HEATED PANEL SYSTEMS

Several panel systems convert electrical energy to heat, raising the temperature of conditioned indoor surfaces and the indoor air. These systems are classified by the temperature of the heated system. Higher-temperature surfaces require less area to maintain occupant comfort. Surface temperatures are limited by the ability of the materials to maintain their integrity at elevated temperatures. The maximum effective surface temperature of floor panels is limited to what is comfortable to occupants' feet.

Electric Ceiling Panels

Prefabricated Electric Ceiling Panels. These panels are available in sizes 1 to 6 ft wide by 2 to 12 ft long by 0.5 to 2 in. thick. They are constructed with metal, glass, or semirigid fiberglass board or vinyl. Heated surface temperatures range from 100 to 300°F, with corresponding heat fluxes ranging from 25 to 100 W/ft^2 for 120 to 480 V services.

A panel of gypsum board embedded with insulated resistance wire is also available. It is installed as part of the ceiling or between joists in contact with a ceiling. Heat flux is limited to 22 W/ft^2 to maintain the board's integrity by keeping the heated surface temperature below 100°F. Nonheating leads are furnished as part of the panel.

Some panels can be cut to fit; others must be installed as received. Panels may be either flush or surface-mounted, and in some cases, they are finished as part of the ceiling. Rigid 2 by 4 ft panels for lay-in ceilings (Figure 24) are about 1 in. thick and weigh from 6 to 25 lb. Typical characteristics of an electric panel are listed in Table 4. Panels may also be (1) surface-mounted on gypsum board and wood ceilings or (2) recessed between ceiling joists. Panels range in size from 4 ft wide to 8 ft long. Their maximum power output is 95 W/ft^2.

Electric Ceiling Panel Systems. These systems are laminated conductive coatings, printed circuits, or etched elements nailed to the bottom of ceiling joists and covered by 1/2 in. gypsum board. Heat flux is limited to 18 W/ft^2. In some cases, the heating element can be cut to fit available space. Manufacturers' instructions specify how to connect the system to the electric supply. Appropriate codes should be followed when placing partitions, lights, and air grilles adjacent to or near electric panels.

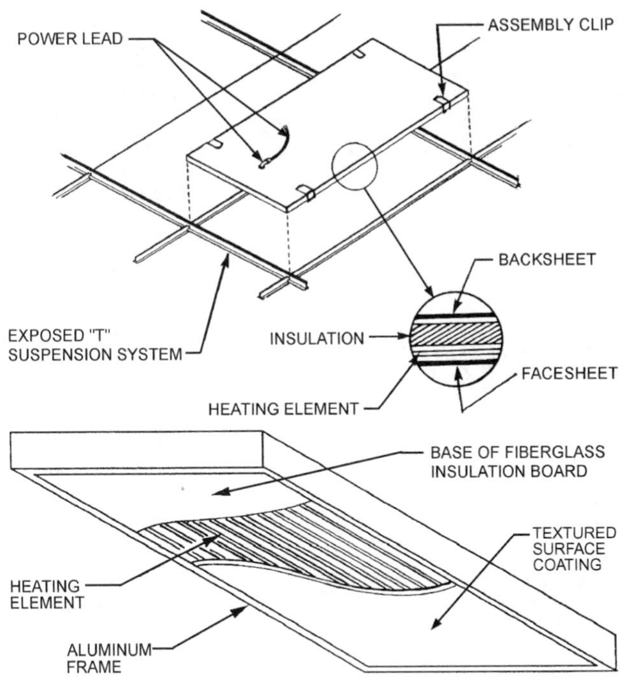

Fig. 24 Electric Heating Panels

Table 4 Characteristics of Typical Electric Panels

Resistor material	Graphite or nichrome wire
Relative heat intensity	Low, 50 to 125 W/ft^2
Resistor temperature	180 to 350°F
Envelope temperature (in use)	160 to 300°F
Thermal-radiation-generating ratio[a]	0.7 to 0.8
Response time (heat-up)	240 to 600 s
Luminosity (visible light)	None
Thermal shock resistance	Excellent
Vibration resistance	Excellent
Impact resistance	Excellent
Resistance to drafts or wind[b]	Poor
Mounting position	Any
Envelope material	Steel alloy or aluminum
Color blindness	Very good
Flexibility	Good—wide range of heat flux, length, and voltage practical
Life expectancy	Over 10,000 h

[a]Ratio of radiation heat flux to input power density (elements only).
[b]May be shielded from wind effects by louvers, deep-drawn fixtures, or both.

Electric Cables Embedded in Ceilings. Electric heating cables for embedded or laminated ceiling panels are factory-assembled units furnished in standard lengths of 75 to 1800 ft. These cable lengths cannot be altered in the field. The cable assemblies are normally rated at 2.75 W per linear foot and are supplied in capacities from 200 to 5000 W in roughly 200 W increments. Standard cable assemblies are available for 120, 208, and 240 V. Each cable unit is supplied with 7 ft nonheating leads for connection at the thermostat or junction box.

Electric cables for panel heating have electrically insulated sleeves resistant to medium temperature, water absorption, aging effects, and chemical action with plaster, cement, or ceiling lath material. This insulation is normally made of polyvinyl chloride (PVC), which may have a nylon jacket. The thickness of the insulation layer is usually about 0.12 in.

For plastered ceiling panels, the heating cable may be stapled to gypsum board, plaster lath, or similar fire-resistant materials with rust-resistant staples (Figure 25). With metal lath or other conducting surfaces, a coat of plaster (brown or scratch coat) is applied to completely cover the metal lath or conducting surface before the cable is attached. After the lath is fastened on and the first plaster coat is applied, each cable is tested for continuity of circuit and for insulation resistance of at least 100 kΩ measured to ground.

The entire ceiling surface is finished with a cover layer of thermally noninsulating sand plaster about 0.50 to 0.75 in. thick, or other approved noninsulating material applied according to manufacturer's specifications. The plaster is applied parallel to the heating cable rather than across the runs. While new plaster is drying, the system should not be energized, and the range and rate of temperature change should be kept low by other heat sources or by ventilation until the plaster is thoroughly cured. Vermiculite or other insulating plaster causes cables to overheat and is contrary to code provisions.

For laminated drywall ceiling panels, the heating cable is placed between two layers of gypsum board, plasterboard, or other thermally noninsulating fire-resistant ceiling lath. The cable is stapled directly to the first (or upper) lath, and the two layers are held apart by the thickness of the heating cable. It is essential that the space between the two layers of lath be completely filled with a noninsulating plaster or similar material. This fill holds the cable firmly in place and improves heat transfer between the cable and the finished ceiling. Failure to fill the space between the two layers of plasterboard completely may allow the cable to overheat in the resulting voids and may cause cable failure. The plaster fill should be applied according to manufacturer's specifications.

Electric heating cables are ordinarily installed with a 6 in. nonheating border around the periphery of the ceiling. An 8 in. clearance must be provided between heating cables and the edges of the outlet or junction boxes used for surface-mounted lighting fixtures. A 2 in. clearance must be provided from recessed lighting fixtures, trim, and ventilating or other openings in the ceiling.

Heating cables or panels must be installed only in ceiling areas that are not covered by partitions, cabinets, or other obstructions. However, it is permissible for a single run of isolated embedded cable to pass over a partition.

The *National Electrical Code*® (NFPA *Standard* 70) requires that all general electrical power and lighting wiring be run above the thermal insulation or at least 2 in. above the heated ceiling surface, or that the wiring be derated.

In drywall ceilings, the heating cable is always installed with the cable runs parallel to the joist. A 2.5 in. clearance between adjacent cable runs must be left centered under each joist for nailing. Cable runs that cross over the joist must be kept to a minimum. Where possible, these crossings should be in a straight line at one end of the indoor space

For cable having a heat flux of 2.75 W/ft, the minimum permissible spacing is 1.5 in. between adjacent runs. Some manufacturers recommend a minimum spacing of 2 in. for drywall construction.

The spacing between adjacent runs of heating cable can be determined using the following equation:

$$M = 12A_n/C \qquad (24)$$

where

M = cable spacing, in.
A_n = net panel heated area, ft^2
C = length of cable, ft

Net panel area A_n in Equation (24) is the net ceiling area available after deducting the area covered by the nonheating border, lighting fixtures, cabinets, and other ceiling obstructions. For simplicity, Equation (24) contains a slight safety factor, and small lighting fixtures are usually ignored in determining net ceiling area.

Electrical resistance of the electric cable must be adjusted according to its temperature at design conditions (Ritter and Kilkis 1998):

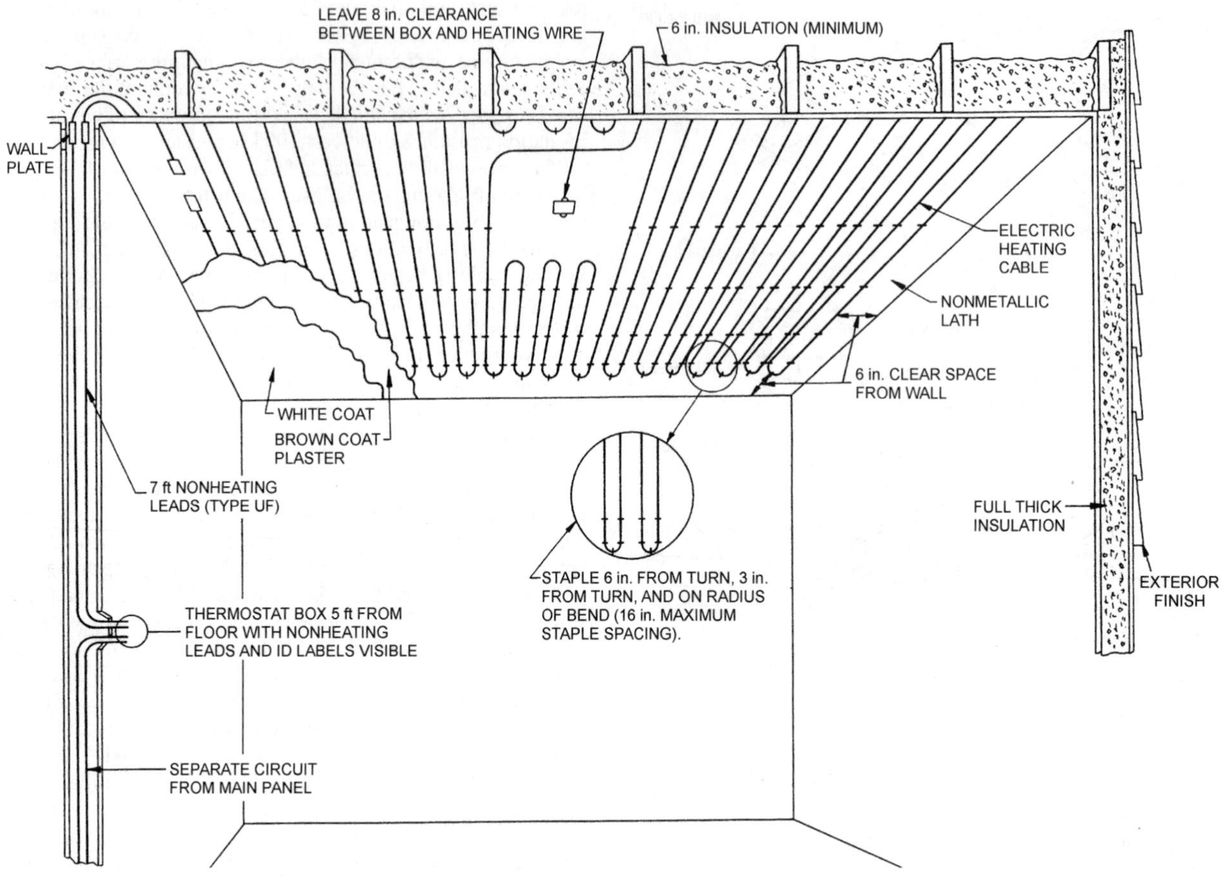

Fig. 25 Electric Heating Panel for Wet Plaster Ceiling

$$R' = R\frac{[1 + \alpha_e(t_d - 68)]}{[1 + \alpha_o(t_d - 68)]} \quad (25)$$

where

R = electrical resistance of electric cable at standard temperature (68°F), Ω/ft

α_e = thermal coefficient for material resistivity, °F^{-1}

α_o = thermal expansion coefficient, °F^{-1}

t_d = surface temperature of electric cable at operating conditions [see Equation (18)], °F

The 2.5 in. clearance required under each joist for nailing in drywall applications occupies one-fourth of the ceiling area if the joists are 16 in. on centers. Therefore, for drywall construction, the net area A_n must be multiplied by 0.75. Many installations have a spacing of 1.5 in. for the first 2 ft from the cold wall. Remaining cable is then spread over the balance of the ceiling.

Electric Wall Panels

Cable embedded in walls similar to ceiling construction is used in Europe. Because of possible damage from nails driven for hanging pictures or from building alterations, most U.S. codes prohibit such panels. Some of the prefabricated panels described in the preceding section are also used for wall panel heating.

Electric Floor Panels

Electric heating cable assemblies such as those used for ceiling panels are sometimes used for concrete floor heating systems. Because the possibility of cable damage during installation is greater for concrete floor slabs than for ceiling panels, these assemblies must be carefully installed. After the cable has been placed, all unnecessary traffic should be eliminated until the concrete layer has been placed and hardened.

Preformed mats are sometimes used for electric floor slab heating systems. These mats usually consist of PVC-insulated heating cable woven into or attached to metallic or glass fiber mesh. Such mats are available as prefabricated assemblies in many sizes from 2 to 100 ft^2 and with heat fluxes from 15 to 25 W/ft^2. When mats are used with a thermally treated cavity beneath the floor, a heat storage system is provided, which may be controlled for off-peak heating.

Mineral-insulated (MI) heating cable is another effective method of slab heating. MI cable is a small-diameter, highly durable, flexible heating cable composed of solid electric-resistance heating wire or wires surrounded by tightly compressed magnesium oxide electrical insulation and enclosed by a metal sheath. MI cable is available in stock assemblies in a variety of standard voltages, heat fluxes (power densities), and lengths. A cable assembly consists of the specified length of heating cable, waterproof hot-cold junctions, 7 ft cold sections, UL-approved end fittings, and connection leads. Several standard MI cable constructions are available, such as single conductor, twin conductor, and double cable. Custom-designed MI heating cable assemblies can be ordered for specific installations.

Other outer-sleeve materials that are sometimes specified for electric floor heating cable include (1) silicone rubber, (2) lead, and (3) tetrafluoroethylene (Teflon®).

For a given floor heating cable assembly, the required cable spacing is determined from Equation (24). In general, cable heat flux and spacing should be such that floor panel heat flux is not greater than 15 W/ft^2. Check the latest edition of the *National Electrical Code*® (NFPA *Standard* 70) and other applicable codes to

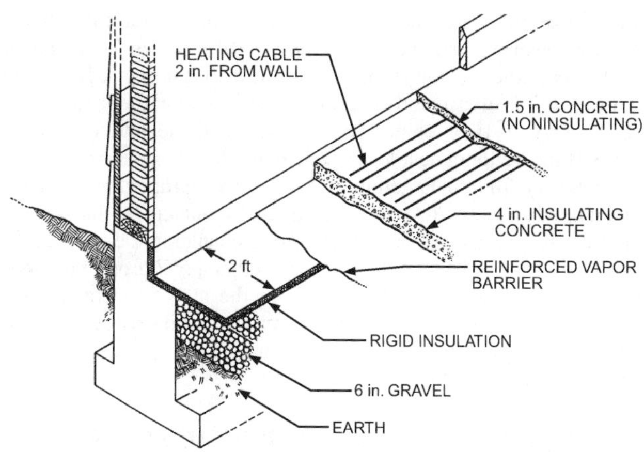

Fig. 26 Electric Heating Cable in Concrete Slab

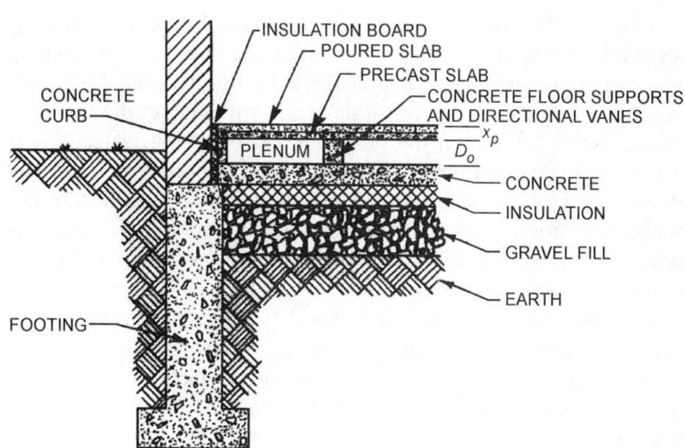

Fig. 27 Warm Air Floor Panel Construction

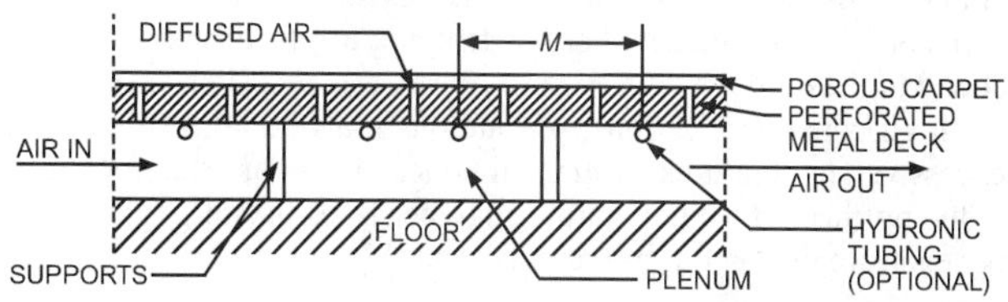

Fig. 28 Typical Hybrid Panel Construction

obtain information on maximum panel heat flux and other required criteria and parameters.

Floor Heating Cable Installation. When PVC-jacketed electric heating cable is used for floor heating, the concrete slab is laid in two pours. The first pour should be at least 3 in. thick and, where practical, should be insulating concrete to reduce downward heat loss. For a proper bond between the layers, the finish slab should be placed within 24 h of the first pour, with a bonding grout applied. The finish layer should be between 1.5 and 2 in. thick. This top layer must not be insulating concrete. At least 1 in. of perimeter insulation should be installed as shown in Figure 26.

The cable is installed on top of the first pour of concrete no closer than 2 in. from adjoining walls and partitions. Methods of fastening the cable to the concrete include the following:

- Staple the cable to wood nailing strips fixed in the surface of the rough slab. Daubs of cement, plaster of paris, or tape maintain the predetermined cable spacing.
- In light or uncured concrete, staple the cable directly to the slab using hand-operated or powered stapling machines.
- Nail special anchor devices to the first slab to hold the cable in position while the top layer is being poured.

Preformed mats can be embedded in the concrete in a continuous pour. The mats are positioned in the area between expansion and/or construction joints and electrically connected to a junction box. The slab is poured to within 1.5 to 2 in. of the finished level. The surface is rough-screeded and the mats placed in position. The final cap is applied immediately. Because the first pour has not set, there is no adhesion problem between the first and second pours, and a monolithic slab results. A variety of contours can be developed by using heater wire attached to glass fiber mats. Allow for circumvention of obstructions in the slab.

MI electric heating cable can be installed in concrete slab using either one or two pours. For single-pour applications, the cable is fastened to the top of the reinforcing steel before the pour is started. For two-layer applications, the cable is laid on top of the bottom structural slab and embedded in the finish layer. Proper spacing between adjacent cable runs is maintained by using prepunched copper spacer strips nailed to the lower slab.

AIR-HEATED OR AIR-COOLED PANELS

Several methods have been devised to warm interior surfaces by circulating heated air through passages in the floor. In some cases, the heated air is recirculated in a closed system. In others, all or part of the air is passed through the indoor space on its way back to the furnace to provide supplementary heating and ventilation. Figure 27 indicates one common type of construction. Compliance with applicable building codes is important.

In principle, the heat transfer equations for the panel surface and the design algorithm explained in the section on Panel Design apply. In these systems, however, the fluid (air) moving in the duct has a virtually continuous contact with the panel. Therefore, $\eta \approx 1$, $D_o = 0$, and $M = 1$. Equation (19) gives the required surface temperature t_d of the plenum. The design of the air side of the system can be carried out by following the principles given in Chapters 27 and 35 of the 2005 *ASHRAE Handbook—Fundamentals*.

If the floor surface is porous, air-heated or air-cooled panels may also be used to satisfy at least part of the latent loads. In this case, the heating or cooling air is conditioned in a central plant. Part of this air diffuses into the indoor conditioned space, as shown in Figure 28. Additional hydronic tubing or electric heating elements may also be attached to the back side of the floor panel to enable independent control of sensible and latent load handling (Kilkis 2002).

CONTROLS

Automatic controls for panel heating may differ from those for convective heating because of the thermal inertial characteristics of the panel and the increase in mean radiant temperature in the space under increasing loads. However, low-mass systems using thin metal panels or thin underlay with low thermal heat capacity may be successfully controlled with conventional control technology using indoor sensors. Many of the control principles for hot-water heating systems described in Chapters 12 and 14 also apply to panel heating. Because panels do not depend on air-side equipment to distribute energy, many control methods have been used successfully; however, a control interface between heating and cooling should be installed to prevent simultaneous heating and cooling.

High-thermal-mass panels such as concrete slabs require a control approach different from that for low-mass panels. Because of thermal inertia, significant time is required to bring such massive panels from one operating point to another, say from vacation setback to standard operating conditions. This will result in long periods of discomfort during the delay, then possibly periods of uncomfortable and wasteful overshoot. Careful economic analysis may reveal that nighttime setback is not warranted.

Once a slab is at operating conditions, the control strategy should be to supply the slab with heat at the rate that heat is being lost from the space (MacCluer et al. 1989). For hydronic slabs with constant circulator flow rates, this means modulating the temperature difference between the outgoing and returning water; this is accomplished with mixing valves, fuel modulation, or, for constant thermal power sources, pulse-width modulation (on-off control). Slabs with embedded electric cables can be controlled by pulse-width modulators such as the common round thermostat with anticipator or its solid-state equivalent.

Outdoor reset control, another widely accepted approach, measures the outdoor air temperature, calculates the supply water temperature required for steady operation, and operates a mixing valve or boiler to achieve that supply water temperature. If the heating load of the controlled space is primarily a function of outdoor air temperature, or indoor temperature measurement of the controlled space is impractical, then outdoor reset control alone is an acceptable control strategy. When other factors such as solar or internal gains are also significant, indoor temperature feedback should be added to the outdoor reset.

In all panel applications, precautions must be taken to prevent excessive temperatures. A manual boiler bypass or other means of reducing the water temperature may be necessary to prevent new panels from drying out too rapidly.

Sensible Cooling Panel Controls

The average water (brine) temperature in the hydronic circuit of panels can be controlled either by mixing, by heat exchange, or by using the water leaving the dehumidifier. Other considerations are listed in the section on General Design Considerations. It is imperative to dry out the building space before starting the panel water system, particularly after extended down periods such as weekends. Such delayed starting action can be controlled manually or by a device.

Panel cooling systems require the following basic areas of temperature control: (1) exterior zones; (2) areas under exposed roofs, to compensate for transmission and solar loads; and (3) each typical interior zone, to compensate for internal loads. For optimum results, each exterior corner zone and similarly loaded face zone should be treated as a separate subzone. Panel cooling systems may also be zoned to control temperature in individual exterior offices, particularly in applications where there is a high artificial lighting load or for indoor spaces at a corner with large fenestration on both walls.

Temperature control of the indoor air and panel water supply should not be a function of the outdoor weather. Normal thermostat drift is usually adequate compensation for the slightly lower temperatures desirable during winter weather. This drift should result in an indoor air temperature change of no more than 1.5°F. Control of interior zones is best accomplished by devices that reflect the actual presence of the internal load elements. Frequently, time clocks and current-sensing devices are used on lighting feeders.

Because air quantities are generally small, constant-volume supply air systems should be used. With the apparatus arranged to supply air at an appropriate dew point at all times, comfortable indoor conditions can be maintained year-round with a panel cooling system. As with all systems, to prevent condensation on window surfaces, the supply air dew point should be reduced during extremely cold weather according to the type of glazing installed.

Heating Slab Controls

In comfort heating, the effective surface temperature of a heated floor slab is held to a maximum of 80 to 84°F. As a result, when the heated slab is the primary heating system, thermostatic controls sensing air temperature should not be used to control the heated slab temperature; instead, the heating system should be wired in series with a slab-sensing thermostat. The remote sensing thermostat in the slab acts as a limit switch to control maximum surface temperatures allowed on the slab. The ambient sensing thermostat controls the comfort level. For supplementary slab heating, as in kindergarten floors, a remote sensing thermostat in the slab is commonly used to tune in the desired comfort level. Indoor-outdoor thermostats are used to vary the floor temperature inversely with the outdoor temperature. If the building heat loss is calculated for an outdoor temperature between 70 to 0°F, and the effective floor temperature range is maintained between 70 to 84°F with a remote sensing thermostat, the ratio of change in outdoor temperature to change in the heated slab temperature is 70:14, or 5:1. This means that a 5°F drop in outdoor temperature requires a 1°F increase in the slab temperature. An ambient sensing thermostat is used to vary the ratio between outdoor and slab temperatures. A time clock is used to control each heating zone if off-peak slab heating is desirable.

HYBRID (LOAD-SHARING) HVAC SYSTEMS

In general, any HVAC system relies on a single heat transfer mode as its major heat transfer mechanism. For example, central air conditioning is a forced-convection system, and a high-intensity radiant system operates almost purely with thermal radiation. Additionally, every system has a typical range for satisfactory and economical operation, with its own advantages, limitations, and disadvantages. A single system may not be sufficient to encompass all requirements of a given building in the most efficient and economical way. Under these circumstances, it may be more desirable to decouple several components of indoor space conditioning, and satisfy them by several dedicated systems. A hybrid heating and cooling system is an optimum partnership of multiple, collocated, simultaneously operating heating and cooling systems, each of which is based on one of the primary heat transfer modes (i.e., radiation and convection). In its most practical form, hybrid heating and cooling may consist of a panel and a forced-air system. The forced-air component may be a central HVAC system or a hydronic system such as fan-coils, as shown in Figure 29. Here, a panel system is added downstream of the condensing fan-coils, and ventilation is provided by a separate system with substantially reduced duct size. Space heating and cooling is achieved by a ground-source heat pump. A hybrid HVAC system is more responsive to control and maintains required operative temperatures for human thermal comfort, because both the operative air and mean radiant temperature can be independently controlled and zoned. Dehumidification and ventilation problems that may be associated with standalone panel cooling systems may be eliminated by hybrid HVAC systems.

ASHRAE research project RP-1140 (Scheatzle 2003) successfully demonstrated the use of panel systems for both heating and cooling, in conjunction with a forced-convection system, to economically achieve year-round thermal comfort in a residence using both active and passive performance of the building and a ground-source heat pump. Skylighting, an energy recovery ventilator, and packaged dehumidifiers were also used. Twenty-four-month-long tests indicated that a radiant/convective system can offer substantial cost savings, given proper design and control.

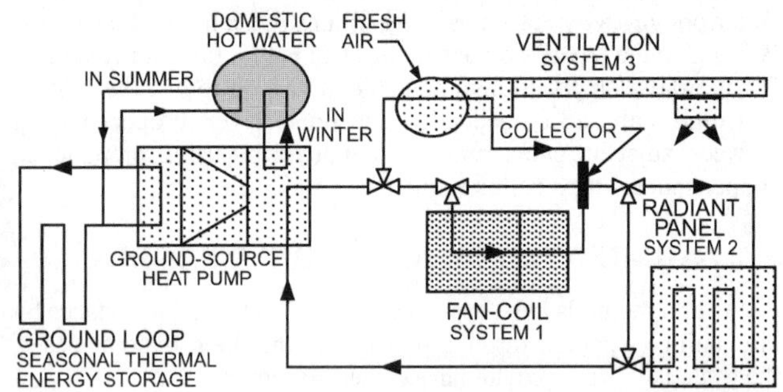

Fig. 29 Typical Residential Hybrid HVAC System

REFERENCES

ASHAE. 1956. Thermal design of warm water ceiling panels. *ASHAE Transactions* 62:71.

ASHAE. 1957. Thermal design of warm water concrete floor panels. *ASHAE Transactions* 63:239.

ASHRAE. 1976. Energy Calculations I—Procedures for determining heating and cooling loads for computerizing energy calculations.

ASHRAE. 1992. Thermal environmental conditions for human occupancy. ANSI/ASHRAE *Standard* 55-1992R.

Berglund, L., R. Rascati, and M.L. Markel. 1982. Radiant heating and control for comfort during transient conditions. *ASHRAE Transactions* (88):765-775.

BSR/ASHRAE. 2001. Method of testing for rating hydronic ceiling panels. Draft *Standard* 138P, 2nd public review.

Buckley, N.A. 1989. Application of radiant heating saves energy. *ASHRAE Journal* 31(9):17-26.

DOE. 2002. *Energy consumption characteristics of commercial building HVAC systems, vol. III: Energy savings potential.* TIAX ref. no. 68370-00, for Building Technologies Program, U.S. Department of Energy. Contract no. DE-AC01-96CE23798. Washington, D.C.

Fanger, P.O. 1972. *Thermal comfort analysis and application in environmental engineering*. McGraw-Hill, New York.

Hansen, E.G. 1985. *Hydronic system design and operation.* McGraw-Hill, New York.

Kalisperis, L.N. 1985. *Design patterns for mean radiant temperature prediction.* Department of Architectural Engineers, Pennsylvania State University, University Park.

Kalisperis, L.N. and L.H. Summers. 1985. MRT33GRAPH—A CAD program for the design evaluation of thermal comfort conditions. Tenth National Passive Solar Conference, Raleigh, NC.

Kilkis, B.I. 1998. Equipment oversizing issues with hydronic heating systems. *ASHRAE Journal* 40(1):25-31.

Kilkis, B.I. 2002. Modeling of a hybrid HVAC panel for library buildings. *ASHRAE Transactions* 108(2):693-698.

Kilkis, B.I., A.S.R. Suntur, and M. Sapci. 1995. Hybrid HVAC systems. *ASHRAE Journal* 37(12):23-28.

Kollmar, A. and W. Liese. 1957. *Die Strahlungsheizung*, 4th ed. R. Oldenburg, Munich.

Lindstrom, P.C., D. Fisher, and C. Pedersen. 1998. Impact of surface characteristics on radiant panel output. ASHRAE Research Project RP-876.

MacCluer, C.R., M. Miklavcic, and Y. Chait. 1989. The temperature stability of a radiant slab-on-grade. *ASHRAE Transactions* 95(1):1001-1009.

Meierhans R. and B.W. Olesen. 2002. Art museum in Bregenz—Soft HVAC for a strong architecture. *ASHRAE Transactions* 108(2).

Min, T.C., L.F. Schutrum, G.V. Parmelee, and J.D. Vouris. 1956. Natural convection and radiation in a panel heated room. *ASHAE Transactions* 62:337.

NFPA. 1999. Installation of sprinkler systems. *Standard* 13-99. National Fire Protection Association, Quincy, MA.

NFPA. 1999. *National electrical code®*. *Standard* 70-99. National Fire Protection Association, Quincy, MA.

NRB. 1981. Indoor climate. *Technical Report* no. 41. The Nordic Committee on Building Regulations, Stockholm.

Parmelee, G.V. and R.G. Huebscher. 1947. Forced convection, heat transfer from flat surfaces. *ASHVE Transactions* 53:245.

Ritter, T.L. and B.I. Kilkis. 1998. An analytical model for the design of in-slab electric heating panels. *ASHRAE Transactions* 104(1B):1112-1115.

Sartain, E.L. and W.S. Harris. 1956. Performance of covered hot water floor panels, Part I—Thermal characteristics. *ASHAE Transactions* 62:55.

Scheatzle, D.G. 2003. Establishing a baseline data set for the evaluation of hybrid (radiant/convective) HVAC systems. ASHRAE Research Project RP-1140, *Final Report.*

Schutrum, L.F. and C.M. Humphreys. 1954. Effects of non-uniformity and furnishings on panel heating performance. *ASHVE Transactions* 60:121.

Schutrum, L.F. and J.D. Vouris. 1954. Effects of room size and non-uniformity of panel temperature on panel performance. *ASHVE Transactions* 60:455.

Schutrum, L.F., G.V. Parmelee, and C.M. Humphreys. 1953a. Heat exchangers in a ceiling panel heated room. *ASHVE Transactions* 59:197.

Schutrum, L.F., G.V. Parmelee, and C.M. Humphreys. 1953b. Heat exchangers in a floor panel heated room. *ASHVE Transactions* 59:495.

Siegenthaler, J. 1995. *Modern hydronic heating.* Delmar Publishers, Boston.

Steinman, M., L.N. Kalisperis, and L.H. Summers. 1989. The MRT-correction method—An improved method for radiant heat exchange. *ASHRAE Transactions* 95(1):1015-1027.

Sugawara, F., M. Nobushisa, and H. Miyazawa. 1996. Comparison of mite-allergen and fungal colonies in floor dust in Seoul (Korea) and Koriyama (Japan) dwellings. *Journal of Architecture, Planning and Environmental Engineering, Architectural Institute of Japan* 48:35-42.

TSI. 1994. Fundamentals of design for floor heating systems (in Turkish). Turkish *Standard* 11261. Turkish Standards Institute, Ankara.

Walton, G.N. 1980. A new algorithm for radiant interchange in room loads calculations. *ASHRAE Transactions* 86(2):190-208.

Watson, R.D., K.S. Chapman, and J. DeGreef. 1998. Case study: Seven-system analysis of thermal comfort and energy use for a fast-acting radiant heating system. *ASHRAE Transactions* 104(1B):1106-1111.

Wilkes, G.B. and C.M.F. Peterson. 1938. Radiation and convection from surfaces in various positions. *ASHVE Transactions* 44:513.

BIBLIOGRAPHY

ALI. 2003. Fundamentals of panel heating and cooling, short course. ASHRAE Learning Institute.

BSR/ASHRAE. 2003. Method of test for determining the design and seasonal efficiencies of residential thermal distribution systems. Draft *Standard* 152P, 2nd Public Review.

Buckley, N.A. and T.P. Seel. 1987. Engineering principles support an adjustment factor when sizing gas-fired low-intensity infrared equipment. *ASHRAE Transactions* 93(1):1179-1191.

Chapman, K.S. and P. Zhang. 1995. Radiant heat exchange calculations in radiantly heated and cooled enclosures. *ASHRAE Transactions* 101(2): 1236-1247.

Chapman, K.S., J. Ruler, and R.D. Watson. 2000. Impact of heating systems and wall surface temperatures on room operative temperature fields. *ASHRAE Transactions* 106(1).

Hanibuchi, H. and S. Hokoi. 2000. Simplified method of estimating efficiency of radiant and convective heating systems, *ASHRAE Transactions* 106(1).

Hogan, R.E., Jr. and B. Blackwell. 1986. Comparison of numerical model with ASHRAE designed procedure for warm-water concrete floor-heating panels. *ASHRAE Transactions* 92(1B):589-601.

Jones, B.W. and K.S. Chapman. 1994. Simplified method to factor mean radiant temperature (MRT) into building and HVAC system design. ASHRAE *Research Project* 657, *Final Report*

Kilkis, B.I. 1993. Computer-aided design and analysis of radiant floor heating systems. Paper no. 80. *Proceedings of Clima 2000*, London (Nov. 1-3).

Kilkis, B.I. 1993. Radiant ceiling cooling with solar energy: Fundamentals, modeling, and a case design. *ASHRAE Transactions* 99(2):521-533.

Kilkis, B.I., S.S. Sager, and M. Uludag. 1994. A simplified model for radiant heating and cooling panels. *Simulation Practice and Theory Journal* 2:61-76.

Ramadan, H.B. 1994. Analysis of an underground electric heating system with short-term energy storage. *ASHRAE Transactions* 100(2):3-13.

Sprecher, P., B. Gasser, O. Böck, and P. Kofoed. 1995. Control strategy for cooled ceiling panels. *ASHRAE Transactions* 101(2).

Watson, R.D. and K.S. Chapman. 2002. *Radiant heating and cooling handbook.* McGraw-Hill, New York.

For additional literature on high-temperature radiant heating, see the Bibliography in Chapter 15.

CHAPTER 7

COMBINED HEAT AND POWER SYSTEMS

COMBINED heat and power (CHP) is the simultaneous production of electrical or mechanical energy (power) and useful thermal energy from a single energy source. By capturing and using the recovered heat energy from an effluent stream that would otherwise be rejected to the environment, CHP (or *cogeneration*) systems can operate at utilization efficiencies greater than those achieved when heat and power are produced in separate processes, thus contributing to sustainable building solutions.

Recovered thermal energy from fuel used in reciprocating engines, steam or combustion turbines (including microturbines, which are typically less than 500 kW in capacity), Stirling engines, or fuel cells can be used in the following applications:

- Direct heating: exhaust gases or coolant fluids used directly for drying processes, to drive an exhaust-fired absorption chiller, to regenerate desiccant materials in a dehumidifier, or to operate a bottoming cycle
- Indirect heating: exhaust gases or coolant fluids used to heat a secondary fluid (e.g., steam or hot water) for devices, to generate power, or to power various thermally activated technologies
- Latent heat: extracting the latent heat of condensation from a recovered flow of steam when the load served allows condensation (e.g., a steam-to-water exchanger) instead of rejecting the latent heat to a cooling tower (e.g., a full condensing turbine with a cooling tower)

There are many potential applications, including base-load power, peaking power where on-site power generation (distributed generation) is used to reduce the demand or high on-peak energy charges imposed by the electric energy supplier, back-up power, remote power, power quality, and CHP, providing both electricity and thermal needs to the site. Usually, customers own the small-scale, on-site power generators, but third parties may own and operate the equipment. Table 1 provides an overview of typical applications, technologies and uses of **distributed generation (DG)** and CHP systems.

On-site CHP systems are small compared to typical central station power plants. DG systems are inherently modular, which makes distributed power highly flexible and able to provide power where and when it is needed. DG and CHP systems can offer significant benefits, depending on location, rate structures, and application. Typical advantages of an on-site CHP plant include improved power reliability and quality, reduced energy costs, increased predictability of energy costs, lowered financial risk, use of renewable energy sources, reduced emissions, and faster response to new power demands because capacity additions can be made more quickly.

CHP system efficiency is not as simple as adding outputs and dividing by fuel inputs. Nevertheless, using what is normally waste exhaust heat yields overall efficiencies (η_O) of 50 to 70% or more (for a definition of overall efficiency, see the section on Performance Parameters).

CHP can operate on a topping, bottoming, or combined cycle. Figure 1 shows an example of topping and bottoming configurations. In a **topping cycle**, energy from the fuel generates shaft or electric power first, and thermal energy from the exiting stream is recovered for other applications such as process heat for cooling or heating systems. In a **bottoming cycle**, shaft or electric power is generated last from thermal energy left over after higher-level thermal energy has been used to satisfy thermal loads. *A typical topping cycle recovers heat from operation of a prime mover and uses this thermal energy for the process (cooling and/or heating). A bottoming cycle recovers heat from the process to generate power.* A **combined cycle** uses thermal output from a prime mover to generate additional shaft power (e.g., combustion turbine exhaust generates steam for a steam turbine generator).

Grid-isolated CHP systems, in which electrical output is used on site to satisfy all site power and thermal requirements, are referred to as **total energy systems**. Grid-parallel CHP systems, which are actively tied to the utility grid, can, on a contractual or tariff basis, exchange power with or reduce load on (thus reducing capacity demand) the public utility. This may eliminate or lessen the need for redundant on-site back-up generating capacity and allows operation at maximum thermal efficiency when satisfying the facility's thermal load; this may produce more electric power than the facility needs.

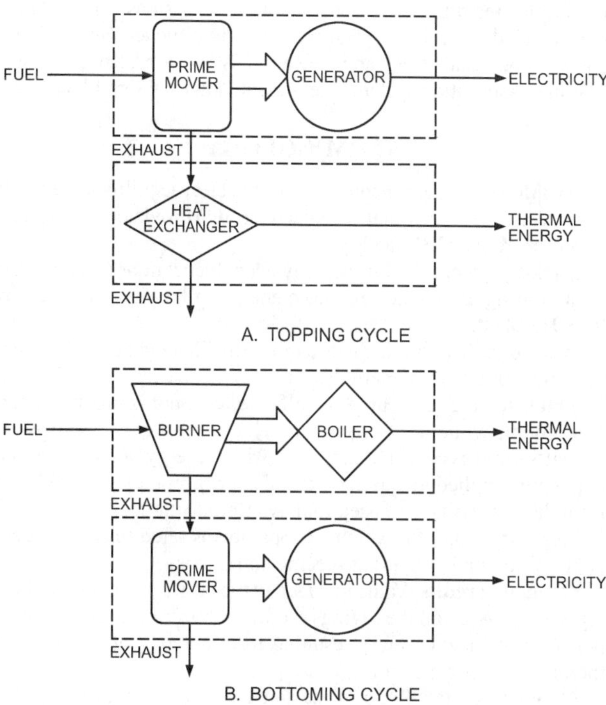

Fig. 1 CHP Cycles

The preparation of this chapter is assigned to TC 1.10, Cogeneration Systems.

Table 1 Applications and Markets for DG/CHP Systems

DG Technologies	Standby Power	Base-Load Power Only	Demand Response Peaking	Customer Peak Shaving	Premium Power	Utility Grid Support	CHP	Applicable Market Sectors
Reciprocating engines: 50 kW to 16 MW	X	X	X	X	X	X	X	Commercial buildings, institutional, industrial, utility grid (larger units), waste fuels
Gas turbines: 500 kW to 50 MW		X		X	X	X	X	Large commercial, institutional, industrial, utility grid, waste fuels
Steam turbines: 500 kW to 100 MW		X			X		X	Institutional buildings/campuses, industrial, waste fuels
Microturbines: 30 to 500 kW	X	X	X	X	X	X	X	Commercial buildings, light industrial, waste fuels
Fuel cells: 5 kW to 2 MW		X			X	X	X	Residential, commercial, light industrial

Source: Adapted from NREL (2003).

CHP feasibility and design depend on the magnitude, duration, and coincidence of electrical and thermal loads, as well as on the selection of the prime mover, waste heat recovery system, and thermally activated technologies. Integrating design of the project's electrical and thermal requirements with the CHP plant is required for optimum economic performance. Matching the CHP plant's thermal/electric ratio with that of the building load is required for optimum economic benefit. The basic components of the CHP plant are the (1) prime mover and its fuel supply system, (2) generator and accessories, including interconnection and protection systems, (3) waste heat recovery system, (4) thermally activated technologies, (5) control system, (6) electrical and thermal transmission and distribution systems, and (7) connections to mechanical and electrical services.

This chapter describes the increasing role of CHP in sustainable design strategies, presents typical system designs, provides means and methods to understand system performance, and describes prime movers, such as reciprocating and Stirling engines, combustion and steam turbines, and fuel cells, and their characteristics for various uses. It also describes thermally activated technologies (TAT) such as heat recovery, absorption chillers, steam turbine-driven chillers, and desiccant dehumidifiers, as well as organic Rankine cycle (ORC) machines for waste heat recovery. Related issues, such as fuels, lubricants, instruments, noise, vibration, emissions, and maintenance, are discussed for each type of prime mover. Siting, interconnection, installation, and operation issues are also discussed. Thermal distribution systems are presented in Chapters 11 and 12.

TERMINOLOGY

Avoided cost. Incremental cost for the electric utility to generate or purchase electricity that is avoided through provision or purchase of power from a CHP facility.

Back-up power. Electric energy available from or to an electric utility during an outage to replace energy ordinarily generated by the CHP plant.

Base load. Minimum electric or thermal load generated or supplied over one or more periods.

Black start. A start-up of an off-line, idle, non-spinning generation source without the electric utility.

Bottoming cycle. CHP facility in which energy put into the system is first applied to another thermal energy process; rejected heat from the process is then used for power production.

Capacity. Load for which an apparatus is rated (electrical generator or thermal system) at specific conditions.

Capacity credits. Value included in the utility's rate for purchasing energy, based on the savings accrued through reduction or postponement of new capacity resulting from purchasing electrical or thermal from cogenerators.

Capacity factor. Ratio of the actual annual output to the rated output over a specified time period.

Coefficient of performance (COP). Refrigeration or refrigeration plus thermal output energy divided by the energy input to the absorption device, refrigeration compressor, or steam turbine. See also **Combined heat and power (CHP)**.

Combined heat and power (CHP). Simultaneous production of electrical or mechanical energy and useful thermal energy from a single energy stream.

Coproduction. Conversion of energy from a fuel (possibly including solid or other wastes) into shaft power (which may be used to generate electricity) and a second or additional useful form of energy. The process generally entails a series of topping or bottoming cycles to generate shaft power and/or useful thermal output. CHP is a form of coproduction.

Demand. Rate at which electric energy is delivered at a given instant or averaged over any designated time, generally over a period of less than 1 h.

Annual demand. Greatest of all demands that occur during a prescribed demand interval billing cycle in a calendar year.

Billing demand. Demand on which customer billing is based, as specified in a rate schedule or contract. It can be based on the contract year, a contract minimum, or a previous maximum, and is not necessarily based on the actual measured demand of the billing period.

Coincident demand. Sum of two or more demands occurring in the same demand interval.

Peak demand. Demand at the instant of greatest load.

Demand charge. Specified charge for electrical capacity on the basis of billing demand.

Demand factor. Average demand over specific period divided by peak demand over the same period (e.g., monthly demand factor, annual demand factor).

Demand-side management (DSM). Process of managing the consumption of energy, generally to optimize available and planned generation resources.

Economic coefficient of performance (ECOP). Output energy in terms of economic costs divided by input fuel in terms of energy purchased or produced, expressed in consistent units (e.g., Btu/h) for each energy stream.

Efficiency. See the section on Performance Parameters.

Energy charge. Portion of the billed charge for electric service based on electric energy (kilowatt-hours) supplied, as contrasted with the demand charge.

Energy Information Administration (EIA). Independent agency in the U.S. Department of Energy that develops surveys, collects energy data, and analyzes and models energy issues.

Electric tariff. Statement of electrical rate and terms and conditions governing its application.

Grid. System of interconnected distribution and transmission lines, substations, and generating plants of one or more utilities.

Grid interconnection. System that manages the flow of power and serves as communication, control, and safety gateway between a CHP plant and an electric utility's distribution network.

Harmonics. Wave forms with frequencies that are multiples of the fundamental (60 or 50 Hz) wave. The combination of harmonics and the fundamental wave causes a nonsinusoidal, periodic wave. Harmonics in power systems result from nonlinear effects. Typically associated with rectifiers and inverters, arc furnaces, arc welders, and transformer magnetizing current. Both voltage and current harmonics occur.

Heat rate. Measure of generating station thermal efficiency, generally expressed in Btu per net kilowatt-hour or lb steam/kWh.

Heating value. Energy content in a fuel that is available as useful heat. The **higher heating value (HHV)** includes the energy needed to vaporize water formed during combustion, whereas the **lower heating value (LHV)** deducts this energy because it does not contribute to useful work.

Intermediate load. Range from base electric or thermal load to a point between base load and peak.

Interruptible power. Electric energy supplied by an electric utility subject to interruption by the electric utility under specified conditions.

Isolated plant. A CHP plant not connected to the electric utility grid.

Load factor. Load served by a system over a designated period divided by system capacity over the same period.

Off-peak. Periods when power demands are below average; for electric utilities, generally nights and weekends; for gas utilities, summer months.

Peak load. Maximum electric or thermal load in a stated period of time.

Peak shaving. Reduction of peak power demand using on-site power generation, thermally activated technologies, or other load-shifting device.

Point of common coupling. Point where a CHP system is connected to the local grid.

Power factor. Ratio of real power (kW) to apparent power (kVA) for any load and time; generally expressed as a decimal.

Premium power. High reliability power supply and/or high voltage/current power quality.

Reactive power. Reactive power exists in all alternating current (AC) power systems as current leads or lags voltages; inadequate reactive power reserves can contribute to voltage sags or even voltage collapse. CHP and/or on-site power systems can provide reactive power support where they are connected to the grid.

Selective energy systems. Form of CHP in which part, but not all, of the site's electrical needs are met solely with on-site generation, with additional electricity purchased from a utility as needed.

Shaft efficiency. Prime mover's shaft energy output divided by its energy input, in equivalent, consistent units. For a steam turbine, input can be the thermal value of the steam or the fuel value required to produce the steam. For a fuel-fired prime mover, it is the fuel input to the prime mover.

Standby power. Electric energy supplied by a cogenerator or other on-site generator during a grid outage.

Supplemental thermal. Heat required when recovered engine heat is insufficient to meet thermal demands.

Supplemental firing. Injection and combustion of additional fuel into an exhaust gas stream to raise its energy content (heat).

Transmission and distribution (T&D). System of wires, transformers, and switches that connects the power-generating plant and end user.

Topping cycle. CHP facility in which energy input to the facility is first used to produce useful power, and rejected heat from production is used for other purposes.

Total energy system. Form of CHP in which all electrical and thermal energy needs are met by on-site systems. A total energy system can be completely isolated or switched over to a normally disconnected electrical utility system for back-up.

Transmission. Network of high-voltage lines, transformers, and switches used to move electric power from generators to the distribution system.

Voltage flicker. Significant fluctuation of voltage (see Figure 28 in Chapter 55 of the 2007 *ASHRAE Handbook—HVAC Applications*).

Wheeling. Using one system's transmission facilities to transmit gas or power for another system.

See Chapter 55 of the 2007 *ASHRAE Handbook—HVAC Applications* for a more detailed discussion of electricity terminology.

CHP SYSTEM CONCEPTS

CUSTOM-ENGINEERED SYSTEMS

Historically, CHP systems were one-of-a-kind, custom-engineered systems. Because of the high cost of engineering, and technical, economic, environmental, and regulatory complexities, extraordinary care and skill are needed to successfully design and build custom-engineered CHP systems. Custom-engineered systems continued to be the norm among systems greater than 5 MW and where process requirements require significant customization.

PACKAGED AND MODULAR SYSTEMS

Packaged/modular CHP systems are available from 5 to over 5000 kW. Packaged systems are defined as the integration of one or more power component (engines, microturbines, combustion turbines, and fuel cells), powered component (generators, compressors, pumps, etc.), heat recovery device [heat exchangers, heat recovery steam generators (HRSGs)], and/or thermally activated technology (absorption/adsorption chillers, steam turbines, ORCs, desiccant dehumidifiers), prefabricated on a single skid. A modular system is two or more packaged systems designed to be easily interconnected in the field. Modular systems are used largely because of shipping or installation size limitations. Packaged and modular systems often can be functionally tested in the factory before shipment, increasing the ease of commissioning. The simplest packaged and modular CHP systems are found in tightly integrated systems in general categories such as the following:

Reciprocating engine systems:

- Small engine generators (under 500 kW) recover jacket and exhaust heat in the form of hot water. Packaged systems include electronic safety and interconnection equipment.
- Small packaged and split-system engine heat pumps, integrating engines with complete vapor compression heat pumps and engine jacket and exhaust heat recovery, available under 50 rated tons (RT).
- Engine packaged systems (typically 100 to 2000 kW) can drive generators, compressors, or pumps. Engine/generators recover at least jacket heat, and several modular systems integrate jacket water and exhaust systems to directly power single- and two-stage absorption chillers, providing power, heating, and cooling.

Microturbine systems:

- Microturbines integrate exhaust heat recovery (hot water) with electronic safety and interconnection equipment in a single compact package.
- Microturbines are easily integrated electrically and thermally (exhaust), providing ideal CHP systems delivering multiple power offerings, typically up to 500 kW, with hot-water systems or using integrated single- and two-stage absorption chillers.

Combustion turbine systems:

- Modular systems have been developed in the 1 to 6 MW range, combining turbine generators, inlet cooling, exhaust control, heat recovery steam generators and/or absorption chillers.

LOAD PROFILING AND PRIME MOVER SELECTION

Selection of a prime mover is determined by the facility's thermal or electrical load profile. The choice depends on (1) the ability of the prime mover's thermal/electric ratio to match the facility loads, (2) the decision whether to parallel with the public utility or be totally independent, (3) the decision whether to sell excess power to the utility, and (4) the desire to size to the thermal baseload. The form and quality of the required thermal energy is very important. If high-pressure steam is required, the reciprocating engine is less attractive as a thermal source than a combustion turbine.

Regardless of how the prime mover is chosen, the degree of use of the available heat determines the overall system efficiency; this is the critical factor in economic feasibility. Therefore, the prime mover's thermal/electric ratio and load must be analyzed as a first step towards making the best choice. Maximizing efficiency is generally not as important as thermal and electric use.

CHP paralleled with the utility grid can operate at peak efficiency if (1) the electric generator can be sized to meet the valley of the thermal load profile, operate at a base electrical load (100% full load) at all times, and purchase the balance of the site's electric needs from the utility; or (2) the electric generators are sized for 100% of the site's electrical demand and recovered heat can be fully used at that condition, with additional thermal demands met by supplementary means and excess power sold to a utility or other electric energy supplier or broker.

Heat output to the primary process is determined by the engine type and load. It must be balanced with actual requirements by supplementation or by rejecting excess heat through peripheral devices such as cooling towers. Similarly, if more than one level of heat is required, controls are needed to (1) reduce heat from a higher level, (2) supplement heat if it is not available, or (3) reject heat when availability exceeds requirements.

In plants with more than one prime mover, controls must be added to balance the power output of the prime movers and to balance reactive power flow between the generators. Generally, an isolated system requires that the prime movers supply the needed electrical output, with heat availability controlled by the electrical output requirements. Any imbalance in heat requirements results in burning supplemental fuels or wasting surplus recoverable heat through the heat rejection system.

Supplemental firing and heat loss can be minimized during parallel operation of the generators and the electric utility system grid by **thermal load following** (adjusting the prime mover throttle for the required amount of heat). The amount of electrical energy generated then depends on heat requirements; imbalances between the thermal and electrical loads are carried by the electric utility either through absorption of excess generation or by delivering supplemental electrical energy to the electrical system.

Similarly, electrical load tracking controls the electric output of the generator(s) to follow the site's electrical load, while using, selling, storing, or discarding (or any combination of these methods) the thermal energy output. To minimize waste of thermal energy, the plant can be sized to track the electrical load profile up to the generator capacity, which is selected for a thermal output that matches the valley of the thermal profile. Careful selection of the prime mover type and model is critical in providing the correct thermal/ electric ratio to minimize electric and thermal waste. Supplemental electric power is purchased and/or thermal energy generated by other means when the thermal load exceeds the generator's maximum capacity.

Analysis of these tracking scenarios requires either a fairly accurate set of coincident electric and thermal profiles typical for a variety of repetitious operating modes, or a set of daily, weekend, and holiday accumulated electrical and thermal consumption requirements. Where thermal load profiles are not necessarily coincident with electric load profiles, thermal energy storage may be used to maximize load factor.

PEAK SHAVING

High on-peak energy charges and/or electrical demand charges in many areas, ratchet charges (minimum demand charge for 11 months = x% of the highest annual peak, leading to an actual payment that in many months is more than the demand charge based on the actual measured usage), and utility capacity shortages have led to **demand-side management (DSM)** by utilities and their consumers. In these cases, a strategy of peak shaving or generating power only during peak cost or peak demand situations may be used.

CONTINUOUS-DUTY STANDBY

An engine that drives a refrigeration compressor can be switched over automatically to drive an electrical generator in the event of a power failure (Figure 2), if loss of compressor service can be tolerated in an emergency.

If engine capacity of a dual-service system equals 150 or 200% of the compressor load, power available from the generator can be delivered to the utility grid during normal operation. Although induction generators may be used for this application, a synchronous generator is required for emergency operation if there is a grid outage.

Dual-service arrangements have the following advantages:

- In comparison to two engines in single service, required capital investment, space, and maintenance are lower, even after allowing for the additional controls needed with dual-service installations.
- Because they operate continuously or on a regular basis, dual-service engines are more reliable than emergency (reserved) single-service engines.
- Engines that are in service and running can be switched over to emergency power generation with minimal loss of continuity.

POWER PLANT INCREMENTAL HEAT RATE

Typically, CHP power plants are rated against incremental heat rates (and thus incremental thermal efficiency) by comparing the incremental fuel requirements with the base case energy needs of a

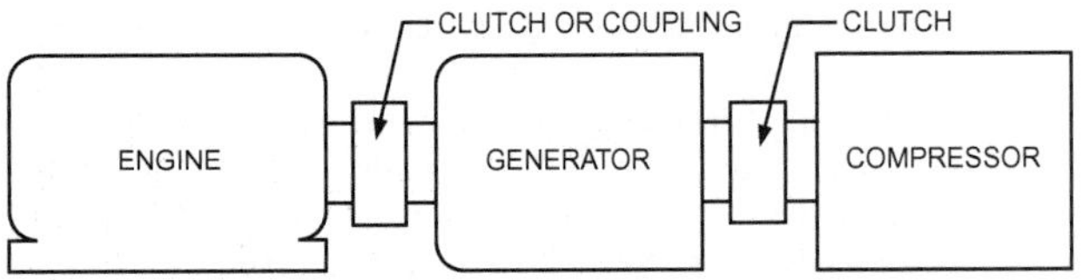

GENERATOR ROTOR ATTACHED TO ENGINE CRANKSHAFT
(Generator rotor requires little energy while generator is under load. In essence, the rotor is a flywheel during this period.)

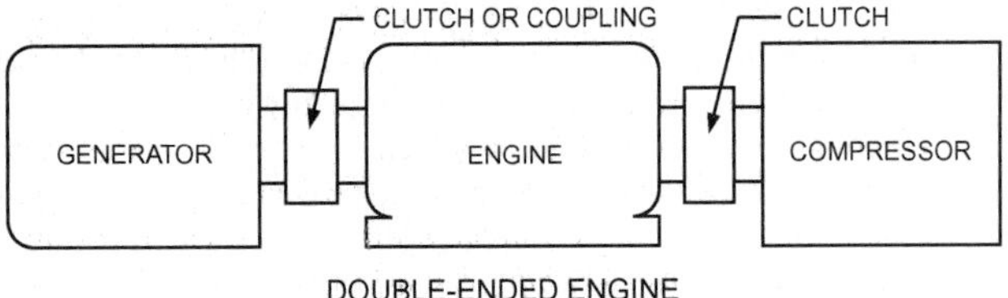

DOUBLE-ENDED ENGINE

Fig. 2 Dual-Service Applications

particular site. For example, if a gas engine generator with a design heat rate of 10,000 Btu/kWh (34% efficiency) provided steam or hot water through waste heat recovery to a particular system that would save 4000 Btu/h energy input, the incremental heat rate of the CHP power plant would be only 6000 Btu/kWh, which translates to an efficiency of 57%. If the same system is applied to another site where only 2000 Btu/h of the recovered heat can be used (against the availability of 4000 Btu/h), the incremental heat rate for the same power plant rises to 8000 Btu/kWh, with efficiency dropping to 43%. Thus, CHP power plant performance really depends on the required thermal/electric ratio for a particular application, and it is only according to this ratio that the type of CHP configuration should be chosen.

A system that requires 1000 kWh electrical energy and 7000 lb low-pressure steam at 30 psig (thermal/electric ratio of about 2.4, or 7 lb steam per kilowatt-hour) can be used to further illustrate measuring CHP system performance. In this example, a 1000 kW gas engine-generator CHP system provides a maximum of 1500 lb/h steam, with the balance of 5500 lb/h to be met by a conventional boiler with 75% thermal efficiency. Thus, the total power requirement is about 10×10^6 Btu/h (34% efficient) for the gas engine and 7.4×10^6 Btu/h for the boiler input, for a total of 17.4×10^6 Btu/h.

If a gas turbine CHP system with the same power and heat requirements is used instead, with a heat rate of 13,650 Btu/h per kilowatt (efficiency of 25%), it would supply both 1000 kW of power and 7000 lb/h steam with only 13.65×10^6 Btu/h fuel input. Thus, the gas engine-generator, although having a very high overall efficiency, is not suitable for the combination system because it would use 3.75×10^6 Btu/h (nearly 28%) more power than the gas turbine for the same total output.

PERFORMANCE PARAMETERS

HEATING VALUE

Natural gas is often selected as the fuel for CHP systems, although the same considerations discussed here apply to biofuels and fossil fuels (Peltier 2001). There are two common ways to define the energy content of fuel: higher heating value and lower heating value.

Turbine, microturbine, engine, and fuel cell manufacturers typically rate their equipment using **lower heating value (LHV)**, which accurately measures combustion efficiency; however, LHV neglects the energy in water vapor formed by combustion of hydrogen in the fuel. This water vapor typically represents about 10% of the energy content. LHVs for natural gas are typically 900 to 950 Btu/ft^3.

Higher heating value (HHV) for a fuel includes the full energy content as defined by bringing all products of combustion to 77°F. Natural gas typically is delivered by the local distribution company with values of 1000 to 1050 Btu/ft^3 on this HHV basis. Because the actual value may vary from month to month, some gas companies convert to therms (1 therm = 100,000 Btu). These measures all represent higher heating values.

Consumers purchase natural gas in terms of its HHV; therefore, performances of CHP systems as well as the electric grid for comparison are calculated in HHV.

The **net electric efficiency** η_E of a generator can be defined by the first law of thermodynamics as net electrical output W_E divided by fuel consumed Q_{fuel} in terms of kilowatt-hours of thermal energy content.

$$\eta_E = \frac{W_E}{Q_{fuel}}$$

A CHP system, by definition, produces useful thermal energy (heat) as well as electricity. If the first law is applied, adding the useful thermal energy Q_{TH} to the net electrical output and dividing by the fuel consumed (which is how virtually all CHP system efficiencies are reported), the resulting overall efficiency η_O does not account for the relative value of the two different energy streams:

$$\eta_O = \frac{W_E + \sum Q_{TH}}{Q_{fuel}}$$

According to the second law of thermodynamics, the two different energy streams have different relative values; heat and electricity are not interchangeable. The first law describes the quantity of the two energy streams, whereas the second law describes their quality or value (exergy). Electrical energy is generally of higher value because it can do many types of work, and, in theory, 100% of it can be converted into thermal energy. Thermal energy is more limited in use and is converted to work at rates usually much lower than 100% conversion. The theoretical maximum efficiency at which thermal energy can be converted to work is the Carnot efficiency, which is a function of the quality, or temperature, of the thermal energy and is defined as $(T_{high} - T_{low})/T_{high}$.

CHP ELECTRIC EFFECTIVENESS

The current methodology of using net electric efficiency η_E and overall efficiency η_O either separately or in combination does not adequately describe CHP performance because

- η_E gives no value to thermal output
- η_O is an accurate measure of fuel use but does not differentiate the relative values of the energy outputs, and is not directly comparable to any performance metric representing separate power and thermal generation

CHP electric effectiveness ε_{EE} is a new, single metric that recognizes and adequately values the multiple outputs of CHP systems and allows direct comparison of system performance to the conventional electric grid and competing technologies. This more closely balances the output values of CHP systems and allows CHP system development to be evaluated over time.

CHP electric effectiveness views the CHP system as primarily providing thermal energy, with electricity as a by-product. It is then defined as net electrical output divided by incremental fuel consumption of the CHP system above the fuel that would have been required to produce the system's useful thermal output by conventional means. This approach credits the system's fuel consumption to account for the value of the thermal energy output, and measures how effective the CHP system is at generating power (or mechanical energy) once the thermal needs of a site have been met. This metric is most effective when used on a consistent and standardized basis, meaning

- The metric measures a single point of performance (design point)
- The design point for power generation is measured at ISO conditions (for combustion turbines, microturbines, and fuel cells, 59°F, 60% rh, sea level, per ISO *Standard* 3977-2; for reciprocating engines, 77°F, 30% rh, and 14.5 psia per ISO *Standard* 3046-1)
- The performance evaluates fuel input and CHP outputs at design point only
- HHV is used because it measures the true values of performance in relation to fuel use and fuel cost (HHV is more commonly used to compare energy systems, is the basis of fuel purchases, and is the basis of emissions regulation)

Power and Heating Systems

For CHP systems delivering power and heating (steam and/or hot water, or direct heating), the CHP electric effectiveness is defined as

$$\varepsilon_{EE} = \frac{W_E}{Q_{fuel} - \sum(Q_{TH}/\alpha)}$$

where α is the efficiency of the conventional technology that otherwise would be used to provide the useful thermal energy output of the system (for steam or hot water, a conventional boiler); see Table 2.

Examples 1 to 5 demonstrate how to apply this metric. The basis for comparison is a 25% HHV efficient electric power source. Performance values for larger combustion turbines, reciprocating engines, and fuel cells vary significantly.

Example 1. Separate Power and Conventional Thermal Generation. A facility supplies its power and thermal requirements by two separate systems: a conventional boiler for its thermal needs and a power-only generator for electricity.

Conventional Boiler: 100 units of fuel are converted into 80 units of heat and 20 units of exhaust energy as shown in Figure 3.

Power-Only Generator: A 25% HHV efficient electric generator consumes 160 units of fuel and produces 40 units of electricity and 120 units of exhaust energy (Figure 4).

The performance metrics for this separate approach to energy supply are as follows:

$$\eta_E = \frac{W_E}{Q_{fuel}} = \frac{40}{160} = 0.25$$

$$\eta_O = \frac{W_E + \sum Q_{TH}}{Q_{fuel}} = \frac{40 + 80}{160 + 100} = 0.46$$

$$\varepsilon_{EE} = \frac{W_E}{Q_{fuel} - \sum(Q_{TH}/\alpha)} = \frac{40}{260 - (80/0.80)} = 0.25$$

Example 2. Combined Power and Thermal Generation (Hot Water/Steam). A CHP system is used to meet the same power and thermal requirements as in Example 1, with a 25% HHV efficient generator and a 67% efficient heat recovery heat exchanger (e.g., a 600°F airstream reduced to 240°F exhaust and yielding 200°F hot water). The performance parameters for this combined system are shown in Figure 5.

$$\eta_E = \frac{W_E}{Q_{fuel}} = \frac{40}{160} = 0.25$$

$$\eta_O = \frac{W_E + \sum Q_{TH}}{Q_{fuel}} = \frac{40 + 80}{160} = 0.75$$

$$\varepsilon_{EE} = \frac{W_E}{Q_{fuel} - \sum(Q_{TH}/\alpha)} = \frac{40}{160 - (80/0.80)} = 0.67$$

Note that η_E for both systems (Example 1's separate generation and Example 2's CHP) is the same, but the CHP system uses less fuel to produce the required outputs, as shown by the differences in overall efficiency (η_O = 75% for CHP versus η_O = 46% for separate systems); see Figure 6. However, this metric does not adequately account for the relative values of the thermal and electric outputs. The electric effectiveness metric, on the other hand, nets out the thermal energy, leaving an ε_{EE} of 67% for the CHP system.

Example 3. Combined Power and Thermal Generation (Direct Exhaust Heat). In some cases, exhaust gases are clean enough to be used for heating directly (e.g., greenhouses and drying where microturbine and gas turbine exhaust is used). For these cases, the thermal recovery efficiency is the difference between exhaust gas temperature and ambient temperature, where delivered exhaust gas is divided by the difference between exhaust gas temperature and outdoor ambient temperature. Direct exhaust gas delivery is a direct-contact form of heating; thus, heat transfer losses are minimal. For a 25% efficient electric generator

Table 2 Values of α for Conventional Thermal Generation Technologies

Fuel	α
Natural gas boiler	0.80
Biomass boiler	0.65
Direct exhaust*	1.0

*Direct drying using exhaust gas is analogous to using flue stack gas for the same purpose; therefore, a direct one-to-one equivalence is best for comparison.

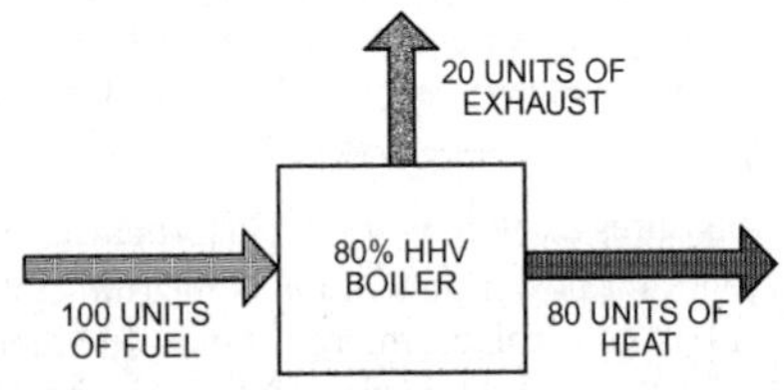

Fig. 3 Conventional Boiler for Example 1

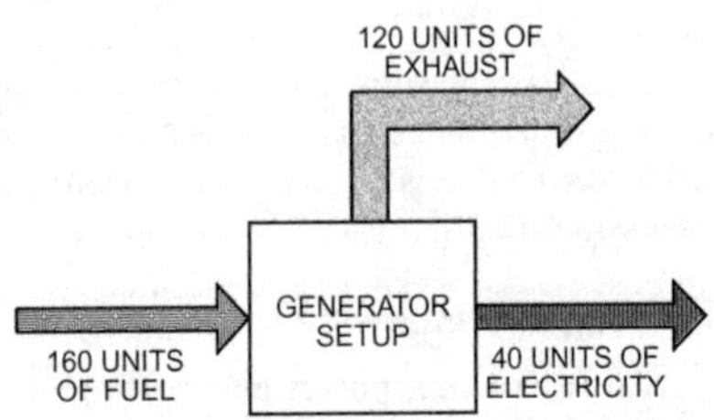

Fig. 4 Power-Only Generator for Example 1

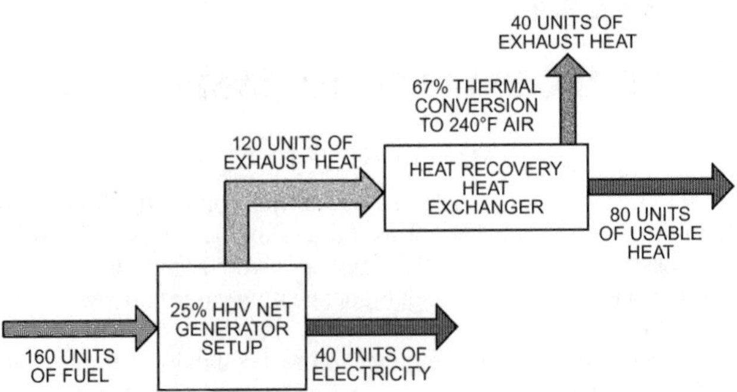

Fig. 5 Performance Parameters for Combined System for Example 2

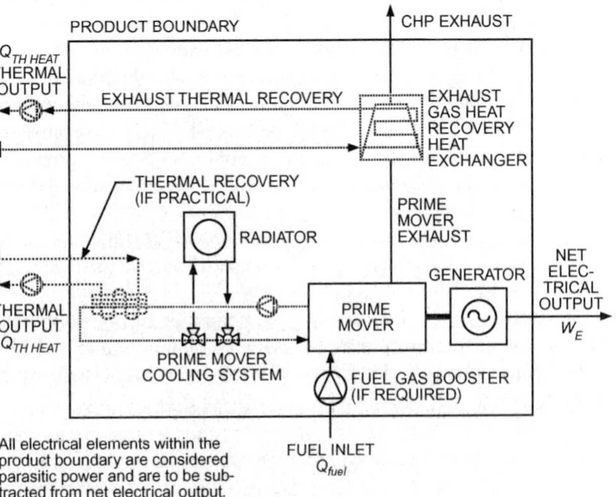

Fig. 6 CHP Power and Heating Energy Boundary Diagram for Example 2

exhausting into a greenhouse with an internal temperature of 100°F, thermal recovery efficiency = (600°F – 100°F)/(600°F – 59°F) = 92% (note that 59°F is the ISO rating condition for microturbines per ISO *Standard* 3977-2).

Performance parameters for this combined system are shown in Figure 7, and system boundaries are shown in Figure 8.

$$\eta_E = \frac{W_E}{Q_{fuel}} = \frac{40}{160} = 0.25$$

$$\eta_O = \frac{W_E + \sum Q_{TH}}{Q_{fuel}} = \frac{40 + 110}{160} = 0.94$$

$$\varepsilon_{EE} = \frac{W_E}{Q_{fuel} - \sum(Q_{TH}/\alpha)} = \frac{40}{160 - (110/1.00)} = 0.80$$

Example 4. Combined Power and Thermal Generation (Combustion Turbine [CT] Without Cofired Duct Burner). In this example, exhaust gas from the 25% efficient electrical combustion turbine generator setup is assessed first without any exhaust enhancement, and then the same system is assessed with temperature and energy content enhancement using cofiring of additional fuel in a duct burner placed in the exhaust and using a heat recovery steam generator (HRSG). The basis for this example is using fuel input and steam output data from a simple cycle 12,000 Btu/h gas turbine, as shown in Figures 9 and 10.

$$\eta_E = \frac{W_E}{Q_{fuel}} = \frac{40}{160} = 0.25$$

$$\eta_O = \frac{W_E + \sum Q_{TH}}{Q_{fuel}} = \frac{40 + 69}{160} = 0.68$$

$$\varepsilon_{EE} = \frac{W_E}{Q_{fuel} - \sum(Q_{TH}/\alpha)} = \frac{40}{160 - (69/0.80)} = 0.54$$

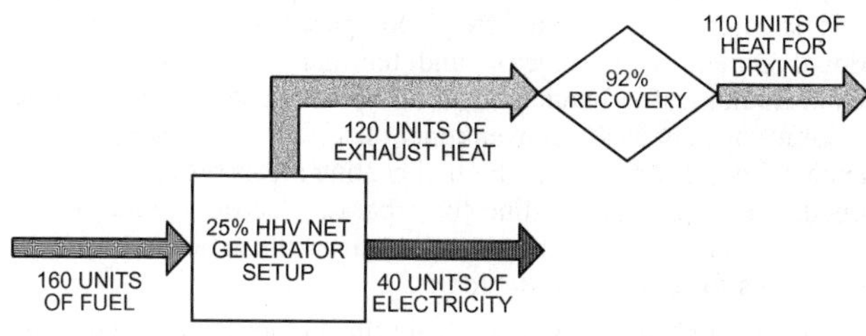

Fig. 7 Performance Parameters for Example 3

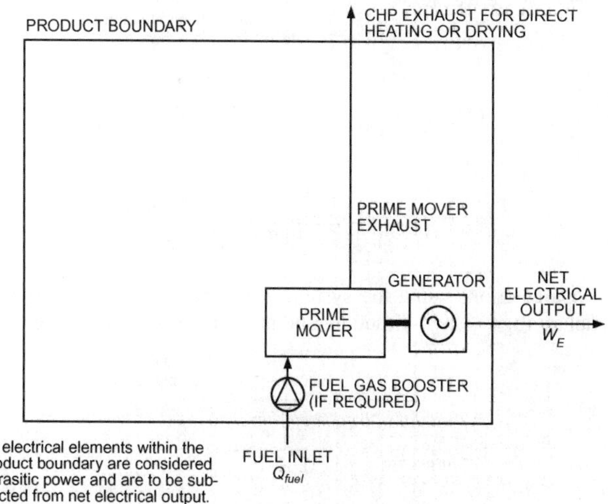

Fig. 8 CHP Power and Direct Heating Energy Boundary Diagram for Example 3

Note that, based on this system approach, cofiring has no effect on η_E because no fuel flows to the duct burner in power-only mode. However, ε_{EE} increases from 0.54 to 0.71 (Figure 11) as shown in Example 5.

Example 5. Combined Power and Thermal Generation (Combustion Turbine [CT] with Cofired Duct Burner) (Figure 12).

$$\eta_E = \frac{W_E}{Q_{fuel}} = \frac{40}{160} = 0.25$$

$$\eta_O = \frac{W_E + \sum Q_{TH}}{Q_{fuel}} = \frac{40 + 182}{284} = 0.78$$

$$\varepsilon_{EE} = \frac{W_E}{Q_{fuel} - \sum(Q_{TH}/\alpha)} = \frac{40}{284 - (182/0.80)} = 0.71$$

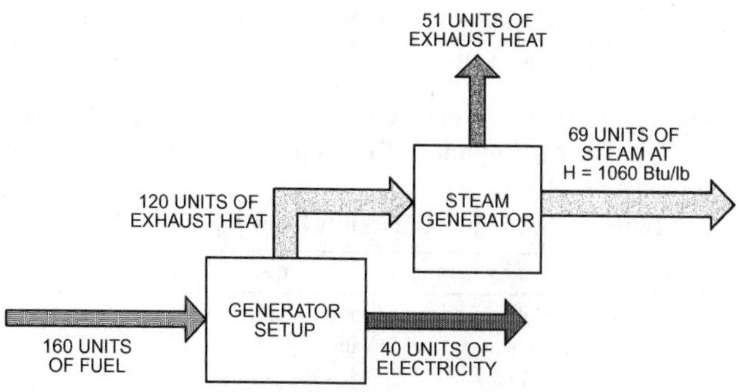

Fig. 9 Performance Parameters for Example 4

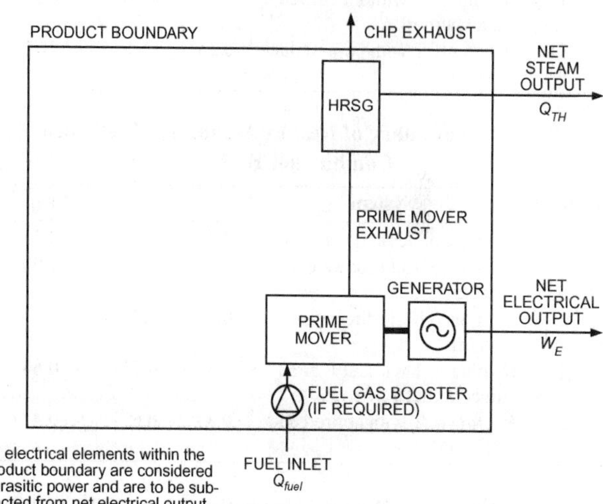

Fig. 10 CHP Power and HRSG Heating Without Duct Burner Energy Boundary Diagram for Example 4

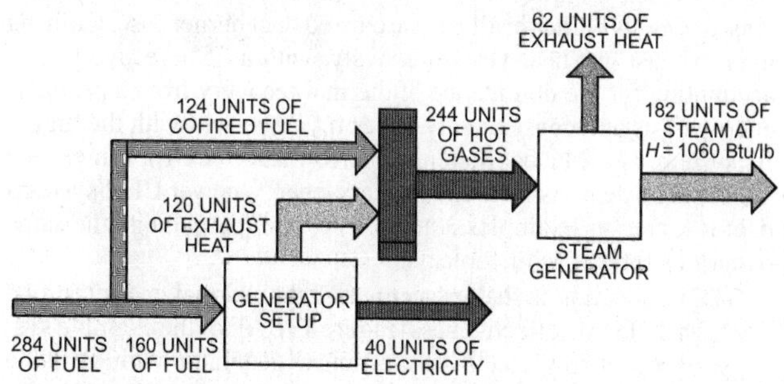

Fig. 11 Cofiring Performance Parameters for Example 4

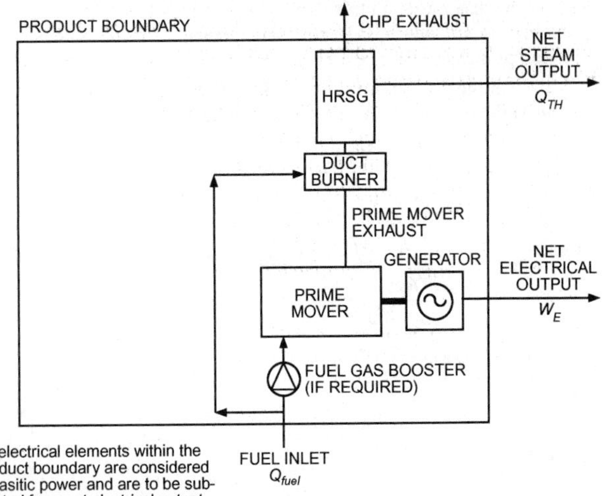

Fig. 12 CHP Power and HRSG Heating with Duct Burner Energy Boundary Diagram for Example 5

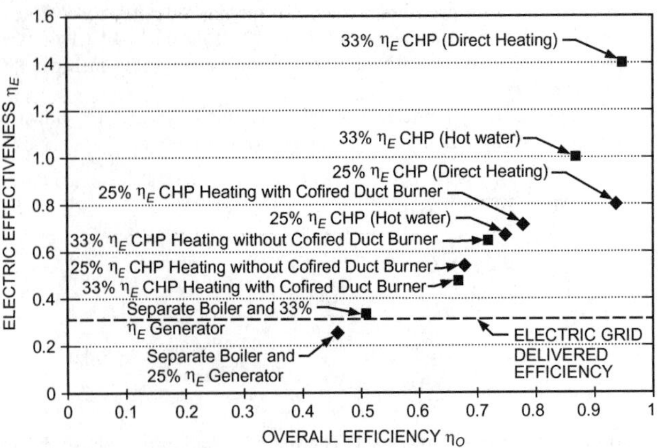

Fig. 13 Electric Effectiveness η_E Versus Overall Efficiency η_O

Table 3 Summary of Results from Examples 1 to 5

Example	System	η_E	η_O	ε_{EE}
1	Separate boiler and generator	0.25	0.46	0.25
2	Combined heat (hot water) and power	0.25	0.75	0.67
3	Combined heat (direct) and power	0.25	0.94	0.80
4	Heating (CT without cofired duct burner)	0.25	0.68	0.54
5	Heating (CT with cofired duct burner)	0.25	0.78	0.71

Table 4 Summary of Results Assuming 33% Efficient Combustion Turbine

Example	System	η_E	η_O	ε_{EE}
1	Separate boiler and generator	0.33	0.51	0.33
2	Combined heat (hot water) and power	0.33	0.87	1.00
3	Combined heat (direct) and power	0.33	0.95	1.40
4	Heating (CT without cofired duct burner)	0.33	0.72	0.64
5	Heating (CT with cofired duct burner)	0.33	0.67	0.47

Table 3 shows a summary of the performance metric results of Examples 1 to 5.

Demonstrating the effect of increasing efficiency, Table 4 presents the results from using a combustion turbine (CT) that delivers 33% efficiency. Notice that η_E is, as expected, higher in all examples, and η_O is higher in all but the cofired duct burner case. Cofiring uses exhaust gas heat very effectively, with a 75% recovery rate, accounting for the dominance of thermal recovery from a primary-energy basis and consequently lower η_O associated with the higher η_E system. ε_{EE}, like coefficient of performance (COP), can exceed 1.00, which demonstrates the primary-energy power CHP systems demonstrated in Examples 3 and 4. ε_{EE}, like η_O, exhibits the same pattern in Table 4 as in Table 3.

The implication is that greater use of the thermal energy results in a higher electric effectiveness (Figure 13). All of the example systems, except a low electrically efficient generator with separate boiler, are superior in electrical effectiveness to the delivered efficiency of the electric grid.

Table 5 Typical ψ Values

Electric Generation Source	Generator η LHV	Generator η HHV	T&D Losses*	ψ
EIA average national grid	—	—	—	0.32*
High-efficiency combined cycle combustion turbine (CT)	0.600	0.540	0.050	0.49
Simple-cycle CT	0.380	0.340	0.068	0.27

*Calculated from DOE/EIA (2005).

FUEL ENERGY SAVINGS

Electrical effectiveness provides a reasonable metric for CHP system comparison; however, it alone cannot provide a measure of fuel savings or emissions effects compared to separate, conventional generation of electric and thermal energy requirements. Understanding fuel savings or emissions effects requires further assumptions about the conventional, or reference, systems. Reference baselines for separate thermal heating are as previously developed. The reference baseline for separate electric generation is a conventional power plant's **electric generation efficiency** ψ. Typical values for ψ are listed in Table 5.

Fuel energy savings $\theta_{savings}$ then reflects fuel savings associated with generating the CHP system power and thermal output through CHP Q_{fuel} compared to using separate heating and electric power sources $\text{FUEL}_{reference}$:

$$\text{FUEL}_{reference} = \underbrace{\left\{\frac{W_E}{\psi}\right\}}_{\text{Separate electricity}} + \underbrace{\left\{\frac{\sum Q_{TH}}{\text{HEAT}_\alpha}\right\}}_{\text{Separate thermal heating}}$$

$$\theta_{savings} = \frac{\text{FUEL}_{reference} - Q_{fuel}}{\text{FUEL}_{reference}}$$

Calculations using the system in Example 2 show a projected fuel savings of 29%, based on operation at the system design point:

$$\text{FUEL}_{reference} = \underbrace{\left\{\frac{40}{0.32}\right\}}_{\text{Electricity}} + \underbrace{\left\{\frac{80}{0.80}\right\}}_{\text{Thermal heating}} = 225$$

$$\theta_{savings} = \frac{225 - 160}{225} = 0.289$$

Note the rated design point, to show the utility of the approach in comparing equipment performance on a consistent basis (i.e., for program management and performance metrics). The same methodology could be applied to different design points (e.g., part load, different ambient temperatures) as long as system outputs and fuel inputs are all determined on a consistent basis (e.g., power output, fuel input, thermal output, and recovery all estimated based on 100°F ambient temperature and 1000 ft altitude). Reference system performance should also be considered on the same basis (e.g., it would not be fair to compare CHP electric effectiveness or fuel savings as calculated on a 100°F day to combined cycle efficiencies calculated at ISO conditions). Similarly, the methodology could be applied to actual application performance if system outputs and utilization are considered on a consistent basis (e.g., evaluating actual system power output, fuel input, and thermal energy used over some specified time period).

Fuel savings from the same direct heating and power CHP system as in Example 3.

$$\text{FUEL}_{reference} = \underbrace{\left\{\frac{40}{0.32}\right\}}_{\text{Electricity}} + \underbrace{\left\{\frac{110}{1.0}\right\}}_{\text{Thermal heating}} = 235$$

$$\theta_{savings} = \frac{235 - 160}{235} = 0.32$$

Applying the same process for the 25% power generator CHP systems (Examples 1 to 5) and using each of the three referenced electric comparisons gives the results presented in Table 6. Table 7 presents results obtained if a 33% generator is assumed.

Table 6 Summary of Fuel Energy Savings for 25% Power Generator in Examples 1 to 5

Example	CHP System	Simple Cycle Peaker	Grid Average	Adv. Combined Cycle
1	Separate boiler and generator	0.00	0.00	0.00
2	Combined heat (hot water) and power	0.35	0.29	0.12
3	Combined heat (direct) and power	0.38	0.32	0.17
4	Heating (CT without cofired duct burner)	0.32	0.24	0.05
5	Heating (CT with cofired duct burner)	0.24	0.19	0.08

Table 7 Summary of Fuel Energy Savings for 33% Power Generator in Examples 1 to 5

Example	CHP System	Simple Cycle Peaker	Grid Average	Adv. Combined Cycle
1	Separate boiler and generator	0.00	0.00	0.00
2	Combined heat (hot water) and power	0.47	0.41	0.26
3	Combined heat (direct) and power	0.50	0.44	0.31
4	Heating (CT without cofired duct burner)	0.41	0.33	0.13
5	Heating (CT with cofired duct burner)	0.23	0.16	(0.01)

FUEL-TO-POWER COMPONENTS

This section describes devices that convert fuel energy to useful power. These energy-conversion technologies provide an important by-product in useful thermal energy that, when harnessed, enables thermal-to-power and thermal-to-thermal devices to operate. These devices are combined with thermal-to-power and/or thermal-to-thermal devices to form CHP systems.

RECIPROCATING ENGINES

Types

Two primary reciprocating engine designs are relevant to stationary power generation applications: the spark ignition (SI) Otto-cycle engine and the compression ignition diesel-cycle engine. The essential mechanical components of Otto-cycle and diesel-cycle engines are the same. Both have cylindrical combustion chambers, in which closely fitting pistons travel the length of the cylinders. The pistons are connected to a crankshaft by connecting rods that transform the linear motion of the pistons into the rotary motion of the crankshaft. Most engines have multiple cylinders that power a single crankshaft.

The primary difference between the Otto and diesel cycles is the method of igniting the fuel. Otto-cycle [or spark-ignition (SI)] engines use a spark plug to ignite the premixed air-fuel mixture after it is introduced into the cylinder. Diesel-cycle engines compress the air introduced into the cylinder, raising its temperature above the auto-ignition temperature of the fuel, which is then injected into the cylinder at high pressure.

Reciprocating engines are further categorized by crankshaft speed (rpm), operating cycle (2- or 4-stroke), and whether turbocharging is used. These engines also are categorized by their original design purpose: automotive, truck, industrial, locomotive, or marine. Engines intended for industrial use are four-stroke Otto-cycle engines, and are designed for durability and for a wide range of mechanical drive and electric power applications. Stationary engine sizes range from 27 to more than 20,000 hp, including industrialized truck engines in the 270 to 800 hp range and industrially applied marine and locomotive engines to more than 20,000 hp. Marine engines (two-stroke diesels) are available in capacities over 100,000 hp.

Both the spark-ignition and the diesel four-stroke engines, most prevalent in stationary power generation applications, complete a power cycle involving the following four piston strokes for each power stroke:

Intake stroke—Piston travels from top dead center (the highest position in the cylinder) to bottom dead center (the lowest position in the cylinder) and draws fresh air into the cylinder during the stroke. The intake valve is kept open during this stroke.

Compression stroke—Piston travels from the cylinder bottom to the top with all valves closed. As the air is compressed, its temperature increases. Shortly before the end of the stroke, a measured quantity of diesel fuel is injected into the cylinder. Fuel combustion begins just before the piston reaches top dead center.

Power stroke—Burning gases exert pressure on the piston, pushing it to bottom dead center. All valves are closed until shortly before the end of the stroke, when the exhaust valves are opened.

Exhaust stroke—Piston returns to top dead center, venting products of combustion from the cylinder through the exhaust valves.

Table 8 Reciprocating Engine Types by Speed (Available Ratings)

Speed Classification	Engine Speed, rpm	Stoichiometric/Rich Burn, Spark Ignition (Natural Gas)	Lean Burn, Spark Ignition (Natural Gas)	Dual Fuel	Diesel
High	1000 to 3600	13 to 2011 hp	201 to 4021 hp	1340 to 4692 hp	13 to 4692 hp
Medium	275 to 1000	None	1340 to 21,448 hp	1340 to 33,512 hp	670 to 46,917 hp
Low	60 to 275	None	None	2681 to 87,131 hp	2681 to 107,239 hp

Source: Adapted from EPA (2002).

The range of medium- and high-speed industrial engines and low-speed marine diesel reciprocating engines are listed in Table 8 by type, fuel, and speed.

Performance Characteristics

Important performance characteristics of an engine include its power rating, fuel consumption, and thermal output. Manufacturers base their engine ratings on the engine duty: prime power, standby operations, and peak shaving. Because a CHP system is most cost-effective when operating at its base load, the rating at prime power (i.e., when the engine is the primary source of power) is usually of greatest interest. This rating is based on providing extended operating life with minimum maintenance. When used for standby, the engine produces continuously (24 h/day) for the length of the primary source outage. Peak power implies an operation level for only a few hours per day to meet peak demand in excess of the prime power capability.

Many manufacturers rate engine capacities according to ISO *Standard* 3046-1, which specifies that continuous net brake power under standard reference conditions (total barometric pressure 14.5 psi, corresponding to approximately 330 ft above sea level, air temperature 77°F, and relative humidity 30%) can be exceeded by 10% for 1 h, with or without interruptions, within a period of 12 h of operation. ISO *Standard* 3046-1 defines prime power as power available for continuous operation under varying load factors and 10% overload as previously described. The standard defines standby power as power available for operation under normal varying load factors, not overloadable (for applications normally designed to require a maximum of 300 h of service per year).

However, the basis of the manufacturer's ratings (ambient temperature, altitude, and atmospheric pressure of the test conditions) must be known to determine the engine rating at site conditions. Various derating factors are used. Naturally aspirated engine output typically decreases 3% for each 1000 ft increase in altitude, whereas turbocharged engines lose 2% per 1000 ft. Output decreases 1% per 10°F increase in ambient temperature, so it is important to avoid using heated air for combustion. In addition, an engine must be derated for fuels with a heating value significantly greater than the base specified by the manufacturer. For CHP applications, natural gas is the baseline fuel, although use of propane, landfill gas, digester gas, and biomass is increasing.

Natural gas spark-ignition engines are typically less efficient than diesel engines because of their lower compression ratios. However, large, high-performance lean-burn engine efficiencies approach those of diesel engines of the same size. Natural gas engine efficiencies range from about 25% HHV (28% LHV) for engines smaller than 50 kW, to 37% HHV (41% LHV) for larger, high-performance, lean-burn engines (DOE 2003).

Power rating is determined by a number of engine design characteristics, the most important of which is displacement; other factors include rotational speed, method of ignition, compression ratio, aspiration, cooling system, jacket water temperature, and intercooler temperature. Most engine designs are offered in a range of displacements achieved by different bore and stroke, but with the same number of cylinders in each case. Many larger engine designs retain the same basic configuration, and displacement increases are achieved by simply lengthening the block and adding more cylinders.

Figure 14 illustrates the efficiency of typical SI natural gas engine/shaft power operating at the prime power rating (HHV).

Fuel consumption is the greatest contributor to operating cost and should be carefully considered during planning and design of a CHP system. It is influenced by combustion cycle, speed, compression ratio, and type of aspiration. It is often expressed in terms of power (Btu/h) for natural gas engines, but for purposes of comparison, it may be expressed as a ratio such as Btu/h per brake horsepower or Btu/kWh. The latter is known as the **heat rate** and equals 3412/

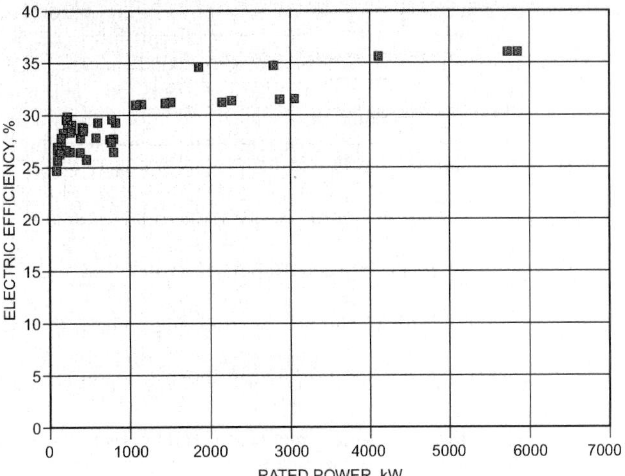

Fig. 14 Efficiency (HHV) of Spark Ignition Engines

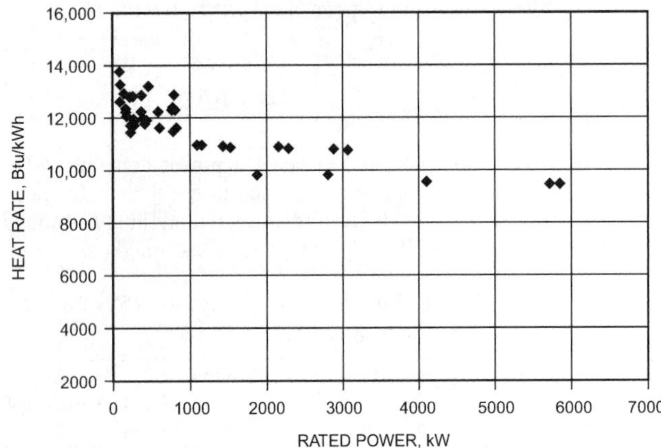

Fig. 15 Heat Rate (HHV) of Spark Ignition Engines

efficiency. The heat rate is the heat input per unit of power output based on either the low or high heating value of the fuel.

Heat rates of several SI engines are shown in Figure 15. The heat rate for an engine of a given size is affected by design and operating factors other than displacement. The most efficient (lowest heat rate) of these engines is naturally aspirated and achieves its increased performance because of the slightly higher compression ratio.

The **thermal-to-electric ratio** is a measure of the useful thermal output for the electrical power being generated. For most reciprocating engines, the recoverable thermal energy is that of the exhaust and jacket. Figure 16 shows the thermal-to-electric ratio of the SI engines.

Ideally, a CHP plant should operate at full output to achieve maximum cost effectiveness. In plants that must operate at part load some of the time, part-load fuel consumption and thermal output are important factors that must be considered in the overall economics of the plant. Figures 17 and 18 show the part-load heat rate and thermal-to-electric ratio as a function of load for 1430, 425, and 85 kW engines.

Fuels and Fuel Systems

Fuel Selection. Fuel specifications, grade, and characteristics have a marked effect on engine performance. Fuel standards for internal combustion engines are designated by the American Society for Testing and Materials (ASTM).

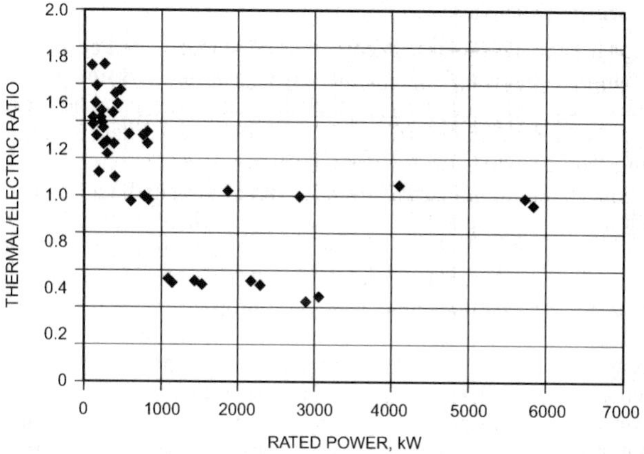

Fig. 16 Thermal-to-Electric Ratio of Spark Ignition Engines (Jacket and Exhaust Energy)

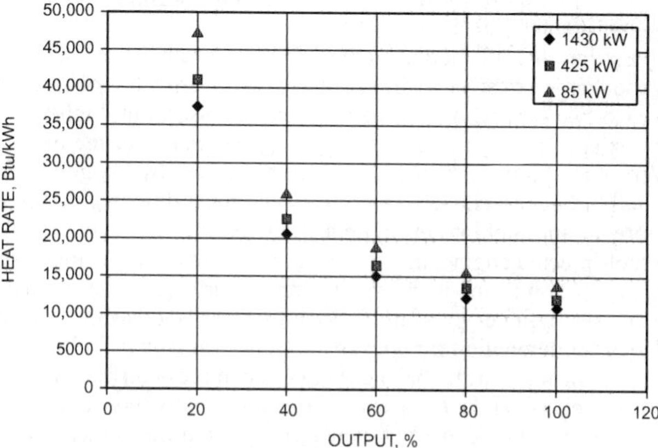

Fig. 17 Part-Load Heat Rate (HHV) of 1430, 425, and 85 kW Gas Engines

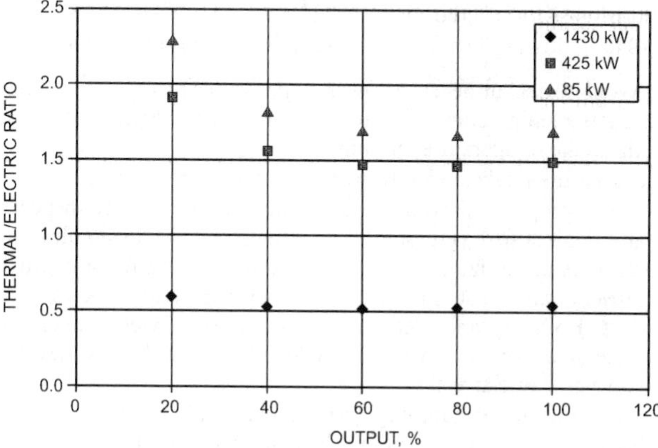

Fig. 18 Part-Load Thermal-to-Electric Ratio of 1430, 425, and 85 kW Gas Engines

Engines may be fueled with a wide variety of fuels and fuel blends, including natural gas, propane, landfill gas, digester gas, bio-oils, biodisel, diesel, or heavier oils. A small amount of diesel oil is used as the compression ignition agent when dual-fuel engines are operated in gaseous-fuel mode.

Gasoline engines are generally not used because of fuel storage hazards, fuel cost, and the higher maintenance required because of deposits of combustion products on internal parts.

Methane-rich gas from wastewater treatment or anaerobic digesters can be used as a fuel for both engines and other heating services. The fuel must be dried and cleaned before injection into engines. Because methane-rich gas has a lower heat content (approximately 600 to 700 Btu/ft^3), it is sometimes mixed with natural gas. The large amount of hydrogen sulfide in the fuel requires using special materials, such as aluminum in the bearings and bushings and low-friction plastic in the O rings and gaskets.

The final choice of fuel should be based on fuel availability, cost, storage requirements, emissions requirements, and fuel rate. Except for gasoline engines, maintenance costs tend to be similar for all engines.

Fuel Heating Value. Fuel consumption data may be reported in terms of either high heating value (HHV) or low heating value (LHV). HHV is used by the gas utility industry and is the basis for evaluating most gaseous fuel. The cost of natural gas is based on HHV. Most natural gases have an LHV/HHV factor of 0.9 to 0.95, ranging from 0.87 for hydrogen to 1.0 for carbon monoxide. For fuel oils, LHV/HHV ranges from 0.96 (heavy oils) to 0.93 (light oils). The HHV is customarily used for oil (including pilot oil in dual-fuel engines).

Fuel Oil Systems. Storage, handling, and cleaning of liquid fuels are covered in Chapter 30. Most oil-fueled CHP is with No. 2 diesel fuel, which is much simpler to handle than the heavier grades. Residual oil is used only for large industrial projects with low-speed engines.

The fuel injection system is the heart of the diesel cycle. Performance functions are as follows:

- Meter a constant quantity to each cylinder at any load during each combustion cycle.
- Inject fuel (1) with a precise and rapid beginning and ending at the correct timing point in each cycle and (2) at a rate needed for controlled combustion and pressure rise.
- Atomize the fuel and distribute it evenly through the air in the combustion chamber.

Atomization can be by high-pressure air injection and mechanical injection. Older systems used air injection until satisfactory mechanical injection systems were developed that avoid the high initial cost and parasitic operating cost of the air compressor.

Earlier mechanical injection pressurized a header to provide a common fuel pressure near 5000 psi. A camshaft opened a spray nozzle in each cylinder, with the length of spray time proportioned to the load through a governor or throttle control. With this system, a leaky valve allowed a steady drip into the cylinder throughout the cycle, which caused poor fuel economy and smoking.

Currently, three injection designs are used:

- Individual plunger pumps for each cylinder, with controlled bypass, controlled suction, variable-suction orifice, variable stroke, or port-and-helix metering
- Common high-pressure metering pump with a separate distribution line to each cylinder that delivers fuel to each cylinder in firing-order sequence
- Common low-pressure metering pump and distributor with a mechanically operated, high-pressure pump and nozzle at each cylinder (excess fuel is recirculated back to the tank through a cooler to remove heat)

Spark Ignition Gas Systems. Fuels vary widely in composition and cleanliness, from pipeline natural gas requiring only a meter, a pressure regulating valve, and safety devices to those from wastewater or biomass, which may also require scrubbers and holding tanks. Gas system accessories include the following.

Line-Type Gas Pressure Regulators. Turbocharged (and aftercooled) engines, as well as many naturally aspirated units, are equipped with line regulators designed to control gas pressure to the engine regulator, as shown in Table 9. The same regulators (both line

Table 9 Line Regulator Pressures

Line Regulator	Turbocharged Engine[a]	Naturally Aspirated Engine[a]
Inlet	14 to 20 psig	2 to 30 psig
Outlet[b]	12 to 15 psig[c]	7 to 10 in. of water

[a]Overall ranges, not variation for individual installations.
[b]Also inlet to engine regulator.
[c]Turbocharger boost plus 7 to 10 in. of water.

and engine) used on naturally aspirated gas engines may be used on turbocharged equipment.

Line-type gas pressure regulators are commonly called service regulators (and field regulators). They are usually located just upstream of the engine regulator to ensure that the required pressure range exists at the inlet to the engine regulator. A remote location is sometimes specified; authorities having jurisdiction should be consulted. Although this intermediate regulation does not constitute a safety device, it does allow initial regulation (by the gas utility at the meter inlet) at a higher outlet pressure, thus allowing an extra cushion of gas between the line regulator and meter for both full gas flow at engine start-ups and delivery to any future branches from the same supply line. The engine manufacturer specifies the size, type, orifice size, and other regulator characteristics based on the anticipated gas pressure range.

Engine-Type Gas Pressure Regulators. This engine-mounted pressure regulator, also called a *carburetor regulator* (and sometimes a *secondary* or *B regulator*), controls fuel pressure to the carburetor. Regulator construction may vary with fuel used. The unit is similar to a zero governor.

Air/Fuel Control. The flow of air/fuel mixtures must be controlled in definite ratios under all load and speed conditions required of engines.

Air/Fuel Ratios. High-rated, naturally aspirated, spark ignition engines require closely controlled air/fuel ratios. Excessively lean mixtures cause excessive lubricating oil consumption and engine overheating. Engines using pilot oil ignition can run at rates above 0.18 ppm/hp without misfiring. Air rates may vary with changes in compression ratio, valve timing, and ambient conditions.

Carburetors. In these venturi devices, the airflow mixture is controlled by a governor-actuated butterfly valve. This air/fuel control has no moving parts other than the butterfly valve. The motivating force in naturally aspirated engines is the vacuum created by the intake strokes of the pistons. Turbocharged engines, on the other hand, supply the additional energy as pressurized air and pressurized fuel.

Fuel Injectors. These electromechanical devices, which are essentially solenoids through which fuel is metered, spray atomized fuel directly into the combustion chamber. Electric current applied to the injector coil creates a magnetic field, which causes the armature to move upward and unseats a spring-loaded ball or pintle valve. Pressurized fuel can then flow out of the injector nozzle in a cone-shaped pattern (caused by the pintle valve's shape). When the injector is deenergized, the ball or valve reseats itself, stopping fuel flow.

Ignition. An electrical system or pilot oil ignition may be used. Electrical systems are either low-tension (make-and-break) or high-tension (jump spark). Systems with breakerless ignition distribution are also used.

Dual-Fuel. Engines using gas with pilot oil for ignition (but that can also operate on 100% diesel) are commonly classified as *dual-fuel engines*. Dual-fuel engines operate either on full oil or on gas and pilot oil, with automatic online switchover when appropriate. In sewage gas systems, a blend with natural gas may be used to maintain a minimum LHV or to satisfy fuel demand when sewage gas production is short.

Combustion Air

All internal combustion engines require clean, cool air for optimum performance. High humidity does not hurt performance, and may even help by slowing combustion and reducing cylinder pressure and temperature. Provisions must be made to silence air noise and provide adequate air for combustion.

Smaller engines generally use engine-mounted impingement filters, often designed for some silencing, whereas larger engines commonly use a cyclone filter or various oil-bath filters. Selection considerations are (1) efficiency or dirt removal capacity; (2) airflow resistance (high intake pressure drop affects performance); (3) ease, frequency, and cost of cleaning or replacement; and (4) first cost. Many of the filter types and media common in HVAC systems are used, but they are designed specifically for engine use. Air piping is designed for low pressure drop (more important for naturally aspirated than for supercharged engines) to maintain high engine performance. For engine intakes, conventional velocities range from 3000 to 7200 fpm, governed by the engine manufacturer's recommended pressure drop of approximately 5.5 in. of water. Evaporative coolers are sometimes used to cool the air before it enters the engine intake.

Essentially, all modern industrial engines above 300 kW are turbocharged to achieve higher power densities. A turbocharger is basically a turbine-driven intake air compressor; hot, high-velocity exhaust gases leaving the engine cylinders power the turbine. Very large engines typically are equipped with two large or four small turbochargers. On a carbureted engine, turbocharging forces more air and fuel into the cylinders, increasing engine output. On a fuel-injected engine, the mass of fuel injected must be increased in proportion to the increased air input. Turbocharging normally increases cylinder pressure and temperature, increasing the tendency for detonation for both spark ignition and dual-fuel engines and requiring a careful balance between compression ratio and turbocharger boost level. Turbochargers normally boost inlet air pressure by a factor of 3 to 4. A wide range of turbocharger designs and models is used. Aftercoolers or intercoolers are often used to cool combustion air exiting the turbocharger compressor, to keep the temperature of air to the engine under a specified limit and to increase the air density. A recirculated water coolant recovery system is used for aftercooling.

The following factors apply to combustion air requirements:

- Avoid heated air because power output varies by $(T_r/T_a)^{0.5}$, where T_r is the temperature at which the engine is rated and T_a is engine air intake temperature, both in °R.
- Locate the intake away from contaminated air sources.
- Install properly sized air cleaners that can be readily inspected and maintained (pressure drop indicators are available). Air cleaners minimize cylinder wear and piston ring fouling. About 90% of valve, piston ring, and cylinder wall wear is the result of dust. Both dry and wet cleaners are used. If wet cleaners are undersized, oil carryover may reduce filter life. Filters may also serve as flame arresters.
- Engine room air-handling systems may include supply and exhaust fans, louvers, shutters, bird screens, and air filters. The maximum total static pressure opposing the fan should be 0.35 in. of water. Table 10 gives sample ventilation air requirements.
- On large engines, intake air is taken from outside the building, enabling a reduction in the building HVAC system. An intake silencer is typically used to eliminate engine noise.

Lubricating Systems

All engines use the lubricating system to remove some heat from the machine. Some configurations cool only the piston skirt with oil; other designs remove more engine heat with the lubricating system. The engine's operating temperature may be significant

Table 10 Ventilation Air for Engine Equipment Rooms

Room Air Temperature Rise,[a] °F	Airflow, cfm/hp		
	Muffler and Exhaust Pipe[b]	Muffler and Exhaust Pipe[c]	Air- or Radiator-Cooled Engine[d]
10	140	280	550
20	70	140	280
30	50	90	180

[a]Exhaust minus inlet.
[b]Insulated or enclosed in ventilated duct.
[c]Not insulated.
[d]Heat discharged in engine room.

in determining the proportion of engine heat removed by the lubricant. Between 5 and 10% of the total fuel input is converted to heat that must be extracted from the lubricating oil; this may warrant using oil coolant at temperatures high enough to allow economic use in a process such as domestic water heating.

Radiator-cooled units generally use the same fluid to cool the engine water jacket and the lubricant; thus, the temperature difference between oil and jacket coolant is not significant. If oil temperature rises in one area (such as around the piston skirts), heat may be transferred to other engine oil passages and then removed by the jacket coolant. When engine jacket temperatures are much higher than lubricant temperatures, the reverse process occurs, and the oil removes heat from the engine oil passages.

Determining the lubricant cooling effect is necessary in the design of heat exchangers and coolant systems. Heat is dissipated to the lubricant in a four-cycle engine with a high-temperature (225 to 250°F) jacket water coolant at a rate of about 7 or 8 Btu/min·bhp; oil heat is rejected in the same engine at 3 to 4 Btu/min·bhp. However, this engine uses more moderate (180°F) coolant temperatures for both lubricating oil and engine jacket.

The characteristics of each lubricant, engine, and application are different, and only periodic laboratory analysis of oil samples can establish optimum lubricant service periods. Consider the following factors in selecting an engine:

- High-quality lubricating oils are generally required for operation between 160 and 200°F, with longer oil life expected at lower temperatures. Moisture may condense in the crankcase if the oil is too cool, which reduces the useful life of the oil.
- Contact with copper can cause oil breakdown, so copper piping should be avoided in oil-side surfaces in oil coolers and heat exchangers.
- A full-flow filter provides better security against oil contamination than one that filters only a portion of circulated lubricating oil and bypasses the rest.

Starting Systems

Larger engines are frequently started with compressed air, either by direct cylinder injection or by air-driven motors. In large plants, one of the smallest multiple compressors is usually engine-driven for a "black start." The same procedure is used for fuel oil systems when the main storage tank cannot gravity-feed the day tank. However, storage tanks must have the capacity for several starting procedures on any one engine, in case of repeated failure to start.

Another start-up concept eliminates all auxiliary engine drives and powers the motor-driven auxiliaries directly through a segregated circuit served by a separate, smaller engine-driven emergency generator. This circuit can be sized for the black-start power and control requirements as well as for emergency lighting and receptacles for power tools and welding devices. For a black start, or after any major damage causing a plant failure, this circuit can be used for repairs at the plant and at other buildings in the complex.

Cooling Systems

Jacket Water Systems. Circulating water and oil systems must be kept clean because the internal coolant passages of the engine are not readily accessible for service. Installation of piping, heat exchangers, valves, and accessories must include provisions for internally cleaning these circuits before they are placed in service and, when possible, for maintenance access afterwards.

Coolant fluids must be noncorrosive and free from salts, minerals, or chemical additives that can deposit on hot engine surfaces or form sludge in relatively inactive fluid passages. Generally, engines cannot be drained and flushed effectively without major disassembly, making any chemical treatment of the coolant fluid that can produce sediment or sludge undesirable.

An initial step toward maintaining clean coolant surfaces is to limit fresh water makeup. The coolant system should be tight and leak-free. Softened or mineral-free water is effective for initial fill and makeup. Forced-circulation hot-water systems may require only minor corrosion-inhibiting additives to ensure long, trouble-free service. This feature is one of the major benefits of hot-water heat recovery systems.

Water-Cooled Engines. Heat in the engine coolant should be removed by heat exchange to a separate water system. Recirculated water can then be cooled in open-circuit cooling towers, where water is added to make up for evaporation. Closed-circuit coolant of all types (e.g., for closed-circuit evaporative coolers, radiators, or engine-side circuits of shell-and-tube heat exchangers) should be treated with a rust inhibitor and/or antifreeze to protect the engine jacket. Because engine coolant is best kept in a protected closed loop, it is usually circulated on the shell side of an exchanger. A minimum fouling factor of 0.002 should be assigned to the tube side.

Jacket water outlet and inlet temperature ranges of 175 to 190°F and 165 to 175°F, respectively, are generally recommended, except when the engines are used with a heat recovery system. These temperatures are maintained by one or more thermostats that bypass water as required. A 10 to 15°F temperature rise is usually accompanied by a circulating water rate of about 0.5 to 0.7 gpm per engine horsepower.

Size water piping according to the engine manufacturer's recommendations, avoiding restrictions in the water pump inlet line. Piping must not be connected rigidly to the engine. Provide shutoffs to facilitate maintenance.

If the cost of pumping water at conventional jacket water temperature can be absorbed in the external system, a hot-water system is preferred. However, the cooling tower must be sized for the full jacket heat rejection if there are periods when no load is available to absorb it (see Chapter 39 for information on cooling tower design). Most engines are designed for forced circulation of the coolant.

Where several engines are used in one process, independent coolant systems for each machine can be used to avoid complete plant shutdown from a common coolant system component failure. This independence can be a disadvantage because unused engines are not maintained at operating temperature, as they are when all units are in a common circulating system. If idle machine temperature drops below combustion products' dew point, corrosive condensate may form in the exhaust gas passages each time the idle machine is started.

When substantial water volume and machinery mass must be heated to operating temperature, the condensate volume is quite significant and must be drained. Some contaminants will get into the lubricant and reduce the service life. If the machinery gets very cold, it may be difficult to start. Units that are started and stopped frequently require an off-cycle heating sequence to lessen exposure to corrosion.

Still another concept for avoiding a total plant shutdown, and minimizing the risk of any single-engine shutdown, requires the following arrangement:

- A common, interconnected jacket water piping system for all the engines.
- An extra standby device for all auxiliaries (e.g., pumps, heat exchangers). Valves must be installed to isolate any auxiliary that fails or is out for preventive maintenance, to allow continued operation.
- Header isolation valves to allow continued plant operation while any section of the common piping is serviced or repaired.

This common piping arrangement permits continuous full-load plant operation if any one auxiliary suffers an outage or needs maintenance; no more than one engine in the battery can have a forced outage if the headers suffer a problem. On the other hand, independent, dedicated auxiliaries for each engine can force an engine outage whenever an auxiliary is down, unless each such auxiliary is provided with a standby, which is not a practical option. Furthermore, using common headers avoids the possibility of a second engine outage when any of the second engine's support components fails while the first is out for major repair. It also allows a warm start of any engine by circulation of a moderate flow of the hot jacket water through any idle engine.

Exhaust Systems

Engine exhaust must be safely conveyed from the engine through piping and any auxiliary equipment to the atmosphere at an allowable pressure drop and noise level. Allowable back pressures, which vary with engine design, range from 2 to 25 in. of water. For low-speed engines, this limit is typically 6 in. of water; for high-speed engines, it is typically 12 in. Adverse effects of excessive pressure drops include power loss, poor fuel economy, and excessive valve temperatures, all of which result in shortened service life and jacket water overheating.

General installation recommendations include the following:

- Install a high-temperature, flexible connection between the engine and exhaust piping. Exhaust gas temperature does not normally exceed 1200°F, but may reach 1400°F for short periods. An appropriate stainless steel connector may be used.
- Adequately support the exhaust system downstream from the connector. At maximum operating temperature, no weight should be exerted on the engine or its exhaust outlet.
- Minimize the distance between the silencer and engine.
- Use a 30 to 45° tailpipe angle to reduce turbulence.
- Specify tailpipe length (in the absence of other criteria) in odd multiples of $12.5(T_e^{0.5}/P)$, where T_e is the temperature of the exhaust gas (°R), and P is exhaust frequency (pulses per second). The value of P is calculated as follows:

  P = rpm/120 = (rev/s)/2 for four-stroke engines

  P = rpm/60 = rev/s for two-stroke engines

  Note that for V-engines with two exhaust manifolds, rpm or rev/s equals engine speed.
- A second, but less desirable, exhaust arrangement is a Y-connection with branches entering the single pipe at about a 60° angle; never use a T-connection, because pulses of one branch will interfere with pulses from the other.
- Use an engine-to-silencer pipe length that is 25% of the tailpipe length.
- Install a separate exhaust for each engine to reduce the possibility of condensation in an engine that is not running.
- Install individual silencers to reduce condensation resulting from an idle engine.
- Limit heat radiation from exhaust piping with a ventilated sleeve around the pipe or with high-temperature insulation.
- Use large enough fittings to minimize pressure drop.
- Allow for thermal expansion in exhaust piping, which is about 0.09 in. per foot of length.
- Specify muffler pressure drops to be within the back-pressure limits of the engine.
- Do not connect the engine exhaust pipe to a chimney that serves natural-draft gas appliances.
- Slope exhaust away from the engine to prevent condensate backflow. Drain plugs in silencers and drip legs in long, vertical exhaust runs may also be required. Raincaps may prevent entrance of moisture but might add back pressure and prevent adequate upward ejection velocity.

Proper effluent discharge and weather protection can be maintained in continuously operated systems by maintaining sufficient discharge velocity (over 2500 fpm) through a straight stack; in intermittently operated systems, protection can be maintained by installing drain-type stacks.

Drain-Type Stack. Drain-type stacks effectively eliminate rainfall entry into a vertical stack terminal without destroying the upward ejection velocity as a rain cap does. This design places a stack head, rather than a stack cap, over the discharge stack. The height of the upper section is important for adequate rain protection, just as the height of the stack is important for adequate dispersal of effluent. Stack height should be great enough to discharge above the building eddy zone (see Chapter 44 of the 2007 *ASHRAE Handbook—HVAC Applications* for more information on exhaust stack design). Bolts for inner stack fastening should be soldered, welded, or brazed, depending on the tack material.

Powerhouse Stack. In this design, a fan discharge intersects the stack at a 45° angle. A drain lip and drain are added in the fume discharge version.

Offset Design. This design is recommended for round ductwork and can be used with sheet metal or glass-fiber-reinforced polyester ductwork.

The exhaust pipe may be routed between an interior engine installation and a roof-mounted muffler through (1) an existing unused flue or one serving power-vented gas appliances only (this should not be used if exhaust gases may be returned to the interior); (2) an exterior fireproof wall with provision for condensate drip to the vertical run; or (3) the roof, provided that a galvanized thimble with flanges and an annular clearance of 4 to 5 in. is used. Sufficient clearance is required between the flue terminal and rain cap on the pipe to allow flue venting. A clearance of 30 in. between the muffler and roof is common. Vent passages and chimneys should be checked for resonance.

When interior mufflers must be used, minimize the distance between the muffler and engine, and insulate inside the muffler portion of the flue. Flue runs more than 25 ft may require power venting, but vertical flues help to overcome the pressure drop (natural draft).

The following design and installation features should be used for flexible connections:

- Material: Convoluted steel (Grade 321 stainless steel) is favored for interior installation.
- Location: Principal imposed motion (vibration) should be at right angles to the connector axis.
- Assembly: The connector (not an expansion joint) should not be stretched or compressed; it should be secured without bends, offsets, or twisting (using float flanges is recommended).
- Anchor: The exhaust pipe should be rigidly secured immediately downstream of the connector in line with the downstream pipe.
- Exhaust piping: Some alloys and standard steel alloy or steel pipe may be joined by fittings of malleable cast iron. Table 11 shows exhaust pipe sizes. The exhaust pipe should be at least as large as the engine exhaust connection. Stainless steel double-wall liners may be used.

Table 11 Exhaust Pipe Diameter*

Output Power, hp	Minimum Pipe Diameter, in. Equivalent Length of Exhaust Pipe			
	25 ft	50 ft	75 ft	100 ft
25	3	4	4	4
50	4	4	5	5
75	4	5	5	5
100	5	5	6	6
200	6	6	7	7
400	7	8	9	9
600	9	9	10	11
800	11	11	11	12
1000	12	12	12	13
1500	13	13	15	15
2000	15	16	17	17

*Minimum exhaust pipe diameter to limit engine exhaust back pressure to 8 in. of water.

Emissions

Exhaust emissions are the major environmental concern with reciprocating engines. The main pollutants are oxides of nitrogen (NO_x), carbon monoxide (CO), and volatile organic compounds (VOCs; unburned or partially burned nonmethane hydrocarbons). Other pollutants, such as oxides of sulfur (SO_x) and particulate matter (PM), depend on the fuel used. Emissions of sulfur compounds (particularly SO_2) are directly related to the fuel's sulfur content. Engines operating on natural gas or distillate oil, which has been desulfurized in the refinery, emit insignificant levels of SO_x. In general, SO_x emissions are an issue only in larger, lower-speed diesel engines firing heavy oils.

Particulate matter can be important for liquid-fueled engines. Ash and metallic additives in fuel and lubricating oil contribute to PM concentrations in exhaust.

NO_x emissions, usually the major concern with natural gas engines, are mostly a mixture of NO and NO_2. Measurements of NO_x are reported as parts per million by volume, in which both species count equally (e.g., ppmv at 15% O_2, dry). Other common units for reporting NO_x in reciprocating engines are specific output-based emission factors, such as g/hp·h and g/kW·h, or as total output rates, such as lb/h. Among the engine options without exhaust aftertreatment, lean-burn natural gas engines produce the lowest NO_x emissions; diesel-fueled engines produce the highest. In many localities, emissions pollutant reduction is mandatory. Three-way catalytic reduction is the general method for stoichiometric and rich-burn engines, and selective catalytic reduction (SCR) for lean-burn engines.

Instruments and Controls

Starting Systems. Start/stop control may include manual or automatic activation of the engine fuel supply, engine cranking cycle, and establishment of the engine heat removal circuits. Stop circuits always shut off the fuel supply, and, for spark ignition engines, the ignition system is generally grounded as a precaution against incomplete fuel valve closing.

Alarm and Shutdown Controls. The prime mover is protected from malfunction by alarms that warn of unusual conditions and by safety shutdown under unsafe conditions. The control system must protect against failure of (1) speed control (underspeed or overspeed), (2) lubrication (low oil pressure, high oil temperature), (3) heat removal (high coolant temperature or lack of coolant flow), (4) combustion process (fuel, ignition), (5) lubricating oil level, and (6) water level.

Controls for alarms preceding shutdown are provided as needed. Monitored alarms without shutdown include lubricating oil and fuel filter, lubricating oil temperature, manifold temperature, jacket water temperature, etc. Automatic start-up of the standby engine when an alarm/shutdown sequence is triggered is often provided.

Both a low-lubrication pressure switch and a high jacket water temperature cutout are standard for most gas engines. Other safety controls used include (1) an engine speed governor, (2) ignition current failure shutdown (battery-type ignition only), and (3) the safety devices associated with a driven machine. These devices shut down the engine to protect it against mechanical damage. They do not necessarily shut off the gas fuel supply unless they are specifically set to do so.

Governors. A governor senses speed (and sometimes load), either directly or indirectly, and acts by means of linkages to control the flow of gas and air through engine carburetors or other fuel-metering devices to maintain a desired speed. Speed control with electronic, hydraulic, or pneumatic governors extends engine life by minimizing forces on engine parts, allows automatic throttle response without operator attention, and prevents destructive overspeeding. A separate overspeed device, sometimes called an overspeed trip, prevents runaway in the event of a failure that disables the governor. Both constant and variable engine speed controls are available. For constant speed, the governor is set at a fixed position, which can be reset manually.

Gas Leakage Prevention. The first method of avoiding gas leakage caused by engine regulator failure is to install a solenoid shutdown valve with a positive cutoff either upstream or downstream of the engine regulator. The second method is a sealed combustion system that carries any leakage gas directly to the outdoors (i.e., all combustion air is ducted to the engine directly from the outdoors).

Noise and Vibration

Engine-driven machines installed indoors, even where the background noise level is high, usually require noise attenuation and isolation from adjoining areas. Air-cooled radiators, noise radiated from surroundings, and exhaust heat recovery boilers may also require silencing. Boilers that operate dry do not require separate silencers. Installations in more sensitive areas may be isolated, receive sound treatment, or both.

Because engine exhaust must be muffled to reduce ambient noise levels, most recovery units also act as silencers. Figure 19 illustrates a typical exhaust noise curve. Figure 20 shows typical attenuation curves for various silencers. Table 42 in Chapter 47 of the 2007 *ASHRAE Handbook—HVAC Applications* lists acceptable noise level criteria for various applications.

Basic attenuation includes (1) turning air intake and exhaust openings away (usually up) from the potential listener; (2) limiting blade-tip speed (if forced-draft air cooling is used) to 12,000 fpm for industrial applications, 10,000 fpm for commercial applications, and 8000 fpm for critical locations; (3) acoustically treating the fan shroud and plenum between blades and coils; (4) isolating (or covering) moving parts, including the unit, from their shelter (where used); (5) properly selecting the gas meter and regulator(s) to prevent singing; and (6) adding sound traps or silencers on ventilation air intake, exhaust, or both.

Further attenuation means include (1) lining intake and exhaust manifolds with sound-absorbing materials; (2) mounting the unit, particularly a smaller engine, on vibration isolators, thereby reducing foundation vibration; (3) installing a barrier (often a concrete block enclosure) between the prime mover and the listener; (4) enclosing the unit with a cover of absorbing material; and (5) locating the unit in a building constructed of massive materials, paying particular attention to the acoustics of the ventilating system and doors.

Noise levels must meet legal requirements (see Chapter 47 of the 2007 *ASHRAE Handbook—HVAC Applications* for details).

Foundations. Multicylinder, medium-speed engines may not require massive concrete foundations, although concrete offers advantages in cost and in maintaining alignment for some driven

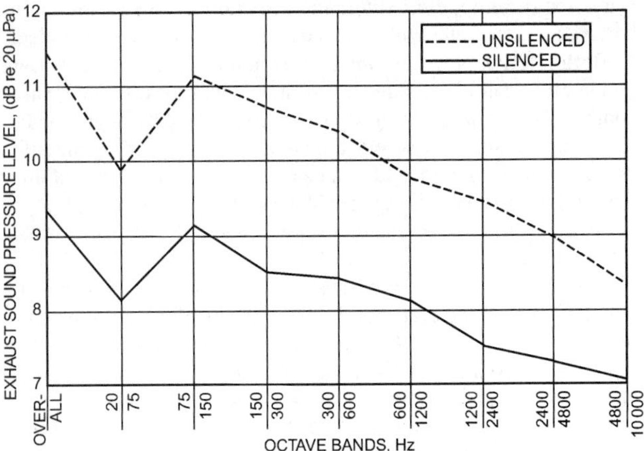

Fig. 19 Typical Reciprocating Engine Exhaust Noise Curves

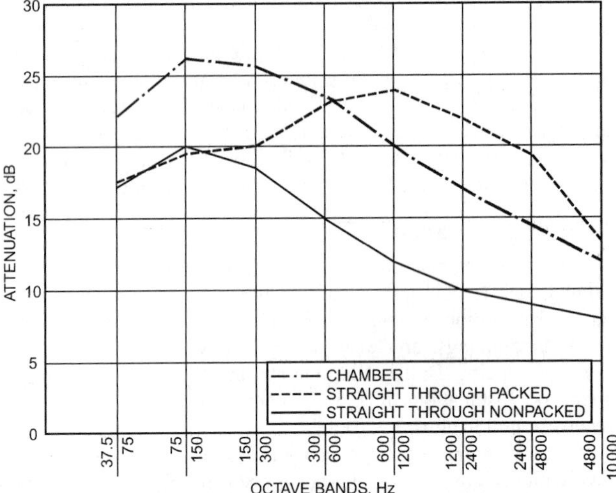

Fig. 20 Typical Attenuation Curves for Engine Silencers

equipment. Fabricated steel bases are satisfactory for direct-coupled, self-contained units, such as electric sets. Steel bases mounted on steel spring or equal-vibration isolators are adequate and need no special foundation other than a floor designed to accommodate the weight. Concrete bases are also satisfactory for such units, if the bases are equally well isolated from the supporting floor or subfloor.

Glass-fiber blocks are effective as isolation material for concrete bases, which should be thick enough to prevent deflection. Excessively thick bases only increase subfloor or soil loading, and still should be supported by a concrete subfloor. In addition, some acceptable isolation material should be placed between the base and the floor. To avoid vibration transmission, an engine base or foundation should never rest directly on natural rock formations. Under some conditions, such as shifting soil on which an outlying CHP plant might be built, a single, very thick concrete pad for all equipment and auxiliaries may be required to avoid a catastrophic shift between one device and another (see Chapters 47 and 54 of the 2007 *ASHRAE Handbook—HVAC Applications* for more information on vibration isolation and seismic design, respectively).

Alignment and Couplings. Proper alignment of the prime mover to the driven device is necessary to prevent undue stresses to the shaft, coupling, and seals of the assembly. Installation instructions usually suggest that alignment of the assembly be performed and measured at maximum load condition and maximum heat input to the turbine or engine.

Torsional vibrations can be a major problem when matching these components, particularly when matching a reciprocating engine to any higher speed centrifugal device. The stiffness of a coupling and its dynamic response to small vibrations from an angular misalignment at critical speed(s) affect the natural frequencies in the assembly. Encotech (1992) describes how changing the coupling's mass, stiffness, or damping can alter the natural frequency during start-up, synchronization, or load change or when a natural frequency exists within the assembly's operating envelope. Proper grouting is needed to preserve the alignment.

Flexible Connections. Greater care is required in the design of piping connections to turbines and engines than for other HVAC equipment because the larger temperature spread causes greater expansion. (See Chapter 45 for further information on piping.)

Installation Ventilation Requirements

In addition to dissipating heat from the jacket water system, exhaust system, lubrication and piston cooling oil, turbocharger, and air intercooler, radiation and convection losses from the surfaces of the engine components and accessories and piping must be dissipated by ventilation. If the radiated heat is more than 8 to 10% of the fuel input, an air cooler may be required. In some cases, rejected heat can be productively applied as tempered makeup air in an adjacent space, with consideration given to life/fire safety requirements, but in most instances it is simply vented.

This heat must be removed to maintain acceptable working conditions and to avoid overloading electrical systems with high ambient conditions. Heat can be removed by outside air ventilation systems that include dampers and fans and thermostatic controls regulated to prevent overheating or excessively low temperatures in extreme weather. The manner and amount of heat rejection vary with the type, size, and make of engine and the extent of engine loading.

An **air-cooled engine** installation includes the following:

- An outside air entrance at least as large as the radiator face and 25 to 50% larger if protective louvers impede airflow.
- Auxiliary means (e.g., a hydraulic, pneumatic, or electric actuator) to open louvers blocking the heated air exit, rather than a gravity-operated actuator.
- Control of jacket water temperature by radiator louvers in lieu of a bypass for freeze protection.
- Thermostatically controlled shutters that regulate airflow to maintain the desired temperature range. In cold climates, louvers should be closed when the engine is shut down to help maintain engine ambient temperature at a safe level. A crankcase heater can be installed on back-up systems located in unheated spaces.
- Positioning the engine so that the radiator face is in a direct line with an air exit leeward of the prevailing wind.
- An easily removable shroud so that exhaust air cannot reenter the radiator.
- Separation of units in a multiple-unit installation to avoid air short-circuiting among them.
- Low-temperature protection against snow and ice formation.
- Propeller fans cannot be attached to long ducts because they can only achieve low static pressure.
- Radiator cooling air directed over the engine promotes good circulation around the engine; thus, the engine runs cooler than for airflow in the opposite direction.
- Adequate sizing to dissipate the other areas of heat emissions to the engine room.

Sufficient ventilation must also be provided to protect against minor fuel supply leaks (not rupture of the supply line). Table 12's minimum ventilation air requirements may be used. Ventilation may be provided by a fan that induces the draft through a sleeve

Table 12 Ventilation Air for Engine Equipment Rooms

Room Air Temperature Rise,[a] °F	Airflow, cfm/hp		
	Muffler and Exhaust Pipe[b]	Muffler and Exhaust Pipe[c]	Air- or Radiator-Cooled Engine[d]
10	140	280	550
20	70	140	280
30	50	90	180

[a]Exhaust minus inlet.
[b]Insulated or enclosed in ventilated duct.
[c]Not insulated.
[d]Heat discharged in engine room.

surrounding the exhaust pipe. Slightly positive pressure should be maintained in the engine room.

Ventilation efficiency for operator comfort and equipment reliability is improved by (1) taking advantage of a full wiping effect across sensitive components (e.g., electrical controls and switchgear) with the coolest air; (2) taking cool air in as low as possible and forcing it to travel at occupancy level; (3) letting cooling air pass subsequently over the hottest components; (4) exhausting from the upper, hotter strata; (5) avoiding short-circuiting of cool air directly to the exhaust while bypassing equipment; and (6) arranging equipment locations, when possible, to allow the desired an airflow path.

Larger engines with off-engine filters should accomplish the following:

- Temper cold outside combustion air when its temperature is low enough to delay ignition timing and inhibit good combustion, which leads to a smoky exhaust.
- Allow the silencer and/or recovery device's hot surfaces to warm cold air to an automatically controlled temperature. As this air enters the machine room, it also provides some cooling to a hot machine room or heating to a cold room.
- Manipulate dampers with a thermostat, which is reset by room temperature, at the inlet to the machine room.
- Cool hot combustion air with a cooling device downstream of the air filter to increase engine performance. This is particularly helpful with large, slow-speed, naturally aspirated engines. The Diesel Engine Manufacturers Association (DEMA 1972) recommended that engines rated at 90°F and 1500 ft above sea level be derated in accordance with the particular manufacturer's ratings. In a naturally aspirated engine, a rating of 100 hp at 1500 ft drops to 50 hp at 16,000 ft. Also, a rating can drop from 100 hp at 90°F to 88 hp at 138°F.

Operation and Maintenance

Preventive Maintenance. One of the most important provisions for healthy and continuous plant operation is implementing a comprehensive preventive maintenance program. This should include written schedules of daily wipedown and observation of equipment, weekly and periodic inspection for replacement of degradable components, engine oil analysis, and maintenance of proper water treatment. Immediate access to repair services may be furnished by subcontract or by in-house plant personnel. Keep an inventory of critical parts on site. (See Chapters 36 to 43 of the 2007 *ASHRAE Handbook—HVAC Applications* for further information on building operation and maintenance.)

Predictive Maintenance. Given the tremendous advancement and availability of both fixed and portable instrumentation for monitoring sound, vibration, temperatures, pressures, flow, and other online characteristics, many key aspects of equipment and system performance can be logged manually or by computer to observe trends. Factors such as fuel rate, heat exchanger approach, and cylinder operating condition can be compared against new and/or optimized baseline conditions to indicate when maintenance may be required. This monitoring allow periods between procedures to be longer, catches incipient problems before they create outages or major repairs, and avoids unnecessary maintenance.

Table 13 Recommended Engine Maintenance

Procedure	Hours Between Procedures	
	Diesel Fueled Engine	Gaseous Fueled Engine
1. Take lubricating oil sample	Once per month plus once at each oil change	Once per month plus once at each oil change
2. Change lubricating oil filters	350 to 750	500 to 1000
3. Clean air cleaners, fuel	350 to 750	350 to 750
4. Clean fuel filters	500 to 750	n.a.
5. Change lubricating oil	500 to 1000	1000 to 2000
6. Clean crankcase breather	350 to 700	350 to 750
7. Adjust valves	1000 to 2000	1000 to 2000
8. Lubricate tachometer, fuel priming pump, and auxiliary drive bearings	1000 to 2000	1000 to 2000 (fuel pump n.a.)
9. Service ignition system; adjust breaker gap, timing, spark plug gap, and magneto	n.a.	1000 to 2000
10. Check transistorized magneto	n.a.	6000 to 8000
11. Flush lubrication oil piping system	3000 to 5000	3000 to 5000
12. Change air cleaner	2000 to 3000	2000 to 3000
13. Replace turbocharger seals and bearings	4000 to 8000	4000 to 8000
14. Replace piston rings, cylinder liners (if applicable), connecting rod bearings, and cylinder heads; recondition or replace turbochargers; replace gaskets and seals	8000 to 12,000	8000 to 12,000
15. Same as item 14, plus recondition or replace crankshaft; replace all bearings	24,000 to 36,000	24,000 to 36,000

Equipment Rotation. When more than one engine, pump, or other component is serving a given distribution system, it is undesirable to operate the units with the equal life approach mode, which puts each unit in the same state of wear and component deterioration. If only one standby unit (or none) is available to a given battery of equipment, and one unit suffers a major failure or shutdown, all units now needed to carry the full load would be prone to additional failure while the first failed unit undergoes repair.

The preferred operating procedure is to keep one unit in continuous reserve, with the shortest possible running hours between overhauls or major repair, and to schedule operation of all others for unequal running hours. Thus any two units would have a minimum statistical chance of a simultaneous failure. All units, however, should be used for several hours in any week.

Engines require periodic servicing and replacement of parts, depending on usage and the type of engine. Transmission drives require periodic gearbox oil changes and the operation and care of external lubricating pumps. Log records should be kept of all servicing; checklists should be used for this purpose.

Table 13 shows ranges of typical maintenance routines for both diesel and gas-fired engines, based on the number of hours run. The actual intervals vary according to the cleanliness of the combustion air, cleanliness of the engine room, engine manufacturer's recommendations, number of engine starts and stops, and lubricating conditions indicated by oil analysis. With some engines and some operating conditions, the intervals between procedures listed in Table 13 may be extended.

A preventive maintenance program should include inspections for

- Leaks (a visual inspection, which is facilitated by a clean engine)

- Abnormal sounds and odors
- Unaccountable speed changes
- Condition of fuel and lubricating oil filters

Daily logs should be kept on all pertinent operating parameters, such as

- Water and lubricating oil temperatures
- Individual cylinder compression pressures, which are useful in indicating blowby
- Changes in valve tappet clearance, which indicate the extent of wear in the valve system

Lubricating oil analysis is a low-cost method of determining the engine's physical condition and a guide to maintenance procedures. Commercial laboratories providing this service are widely available. The analysis should measure the concentration of various elements found in the lubricating oil, such as bearing metals, silicates, and calcium. It should also measure the dilution of the oil, suspended and nonsuspended solids, water, and oil viscosity. The laboratory can often help interpret readings and alert the user to impending problems.

Lubricating oil manufacturers' recommendations should be followed. Both the crankcase oil and oil filter elements should be changed at least once every six months.

COMBUSTION TURBINES

Types

Combustion gas turbines, although originally used for aircraft propulsion, have been developed for stationary use as prime movers. Turbines are available in sizes from 38 to 644,000 hp and can burn a wide range of liquid and gaseous fuels. Turbines, and some dual-fuel engines, can shift from one fuel to another without loss of service. **Microturbine** systems typically have capacities below 670 hp, although this is a subjective cutoff between micro- and **miniturbine** (671 hp to about 1340 hp).

Combustion turbines consist of an air compressor section to boost combustion air pressure, a combination fuel/air mixing and combustion chamber (combustor), and an expansion power turbine section that extracts energy from the combustion gases. Simple-cycle combustion gas turbines have thermal efficiency levels of 25 to 32% HHV (28 to 36% LHV). Recuperative combustion gas turbines have thermal efficiency levels of 35% HHV (39% LHV).

In addition to these components, some turbines use regenerators and recuperators as heat exchangers to preheat combustion air with heat from the turbine discharge gas, thereby increasing machine efficiency. Most microturbine units are currently designed for continuous-duty operation and are recuperated to obtain higher electric efficiencies levels of 23 to 30% HHV (26 to 33% LHV) for sizes 335 hp and below. Unrecuperated engines or simple cycles have lower electric efficiencies but higher exhaust temperatures, which make them better suited for some CHP applications. Most turbines are the single-shaft type, (i.e., air compressor, turbine, and generator on a common shaft). However, dual-shaft machines that use one turbine stage on the same shaft as the compressor and a separate power turbine driving the output shaft are available.

Turbines rotate at speeds varying from 3600 to 100,000 rpm and often need speed-reduction gearboxes to obtain shaft speeds suitable for generators or other machinery. Many microturbine designs directly couple a high-speed turbine to a wide-frequency alternator whose high-frequency ac power is converted to dc by a rectifier and then to 60 Hz ac by an inverter to utility voltage and frequency specifications. Turbine motion is completely rotary and relatively vibration-free. This feature, coupled with low mass and high power output, provides an advantage over reciprocating engines in space, foundation requirements, and ease of start-up.

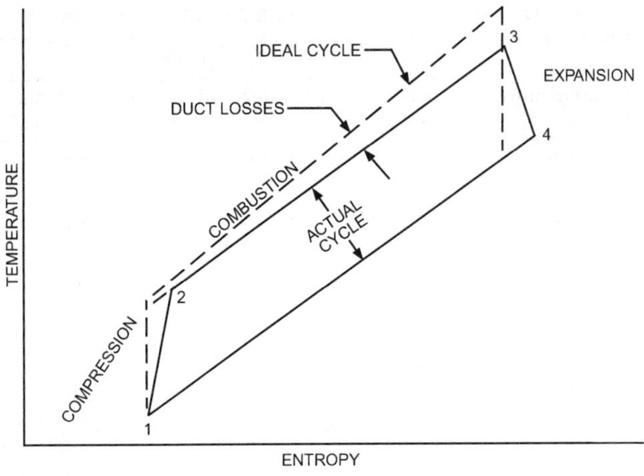

Fig. 21 Temperature-Entropy Diagram for Brayton Cycle

Combustion turbines have the following advantages and disadvantages compared to other internal combustion engine drivers:

Advantages

- Small size, high power-to-weight ratio
- Ability to burn a variety of fuels, though more limited than reciprocating engines
- Ability to meet stringent pollution standards
- High reliability
- Available in self-contained packages
- No cooling water required
- Vibration-free operation
- Easy maintenance
- Low installation cost
- Clean, dry exhaust
- Lubricating oil not contaminated by combustion oil

Disadvantages

- Tend to be more complicated devices
- Lower electrical output efficiency compared to internal combustion engines
- Higher capital (first) costs

Gas Turbine Cycle

The basic gas turbine cycle (Figure 21) is the **Brayton cycle (open cycle)**, which consists of adiabatic compression, constant-pressure heating, and adiabatic expansion. Figure 21 shows that the thermal efficiency of a gas turbine falls below the ideal value because of inefficiencies in the compressor and turbine and because of duct losses. Entropy increases during the compression and expansion processes, and the area enclosed by points 1, 2, 3, and 4 is reduced. This loss of area is a direct measure of the loss in efficiency of the cycle.

Nearly all turbine manufacturers present gas turbine engine performance in terms of power and specific fuel consumption. A comparison of fuel consumption in specific terms is the quickest way to compare overall thermal efficiencies of gas turbines (ASME 2005).

Components

Figure 22 shows the major components of the gas turbine unit, which include the air compressor, combustor, and power turbine. Atmospheric air is compressed by the air compressor. Fuel is then injected into the airstream and ignited in the combustor, with leaving gases reaching temperatures between 1600 and 2500°F. These high-pressure hot gases are then expanded through a turbine, which provides not only the power required by the air compressor, but also power to drive the load.

Gas turbines are available in two major classifications—single-shaft (Figure 22) and dual-shaft (Figure 23). The **single-shaft turbine** has the air compressor, gas-producer turbine, and power turbine on the same shaft. The **dual-shaft** or **split-shaft turbine** has the section required for air compression on one shaft and the section producing output power on a separate shaft. For a dual-shaft turbine, the portion that includes the compressor, combustion chamber, and first turbine section is the hot-gas generator or gas producer. The second turbine section is the power turbine.

The turbine used depends on job requirements. Single-shaft engines are usually selected when a constant-speed drive is required, as in generator drives, and when starting torque requirements are low. A single-shaft engine can be used to drive centrifugal compressors, but the starting system and the compressor match point must be considered. Dual-shaft engines allow for variable speed at full load and can easily be started with a high torque load connected to the power output shaft, and the power turbine can be more optimally configured to match load requirements.

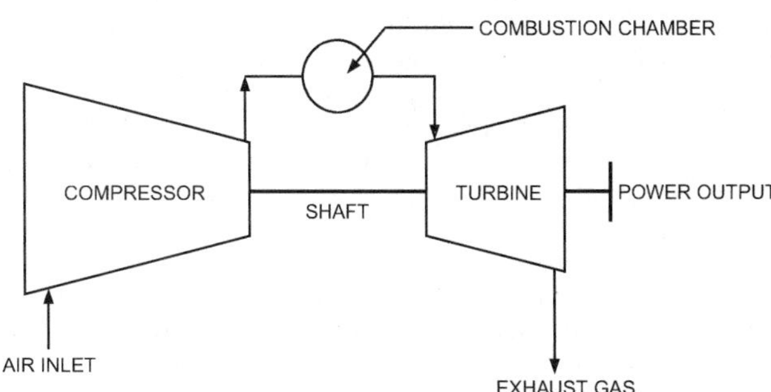

Fig. 22 Simple-Cycle Single-Shaft Turbine

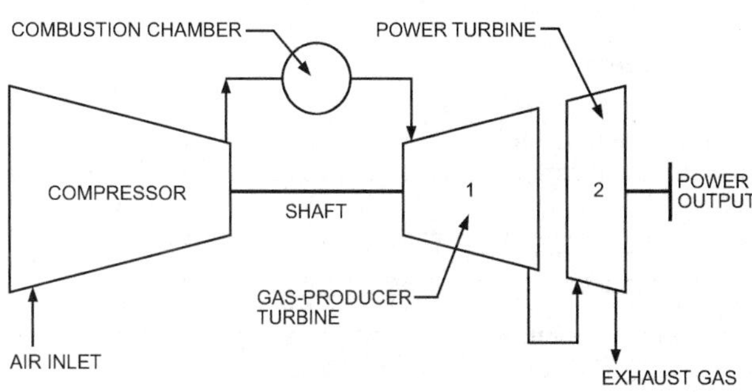

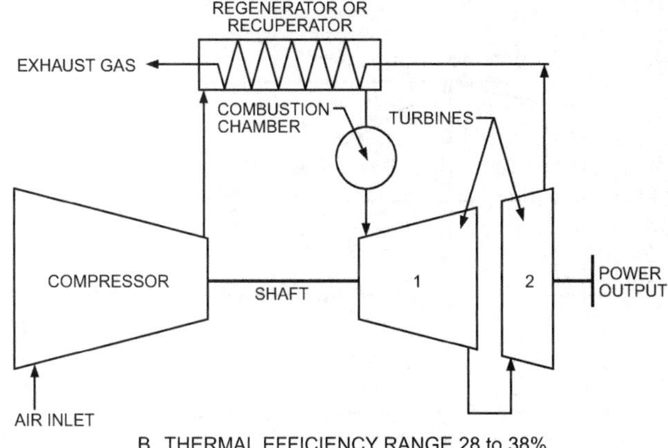

Fig. 23 Simple-Cycle Dual-Shaft Turbines

PERFORMANCE CHARACTERISTICS

A gas turbine's rating is greatly affected by altitude, ambient temperature, inlet pressure to the air compressor, and exhaust pressure from the turbine. In most applications, filters and silencers must be installed in the air inlet. Silencers, waste heat recovery units, or both are used on the exhaust. The pressure drop of these accessories and piping losses must be considered when determining the power output of the unit.

Gas turbine ratings are usually given at standard conditions defined by the International Organization for Standardization (ISO): 59°F, 60% rh, and sea-level pressure at the air compressor's inlet flange and turbine's exhaust flange. Corrections for other conditions must be obtained from the manufacturer, because they vary with each model, depending primarily on gas turbine efficiency. Inlet air cooling has been used to increase capacity. The following approximations may be used for design considerations:

- Each 18°F rise in inlet temperature typically decreases power output by 9%.
- An increase of 1000 ft in altitude decreases power output by approximately 3.5%.
- Inlet pressure loss in filter, silencer, and ducting decreases power output by approximately 0.5% for each inch of water pressure loss.
- Discharge pressure loss in waste heat recovery units, silencer, and ducting decreases power output by approximately 0.3% for each inch of water pressure loss.

Gas turbines operate with a wide range of fuels.

Figure 24 shows a typical performance curve for a 10,000 hp turbine engine. For example, at an air inlet temperature of 86°F, the engine develops its maximum power at about 82% of maximum speed. The shaft thermal efficiency of the prime mover is 18 to 36% with exhaust gases from the turbine ranging from 806 to 986°F. If the exhaust heat can be used, overall thermal utilization efficiency can increase.

Figure 23B shows a regenerator that uses exhaust gas heat to heat air from the compressor before combustion. Overall shaft efficiency can be increased from 25 to 36% HHV (28 to 39% LHV) by using a regenerator or recuperator.

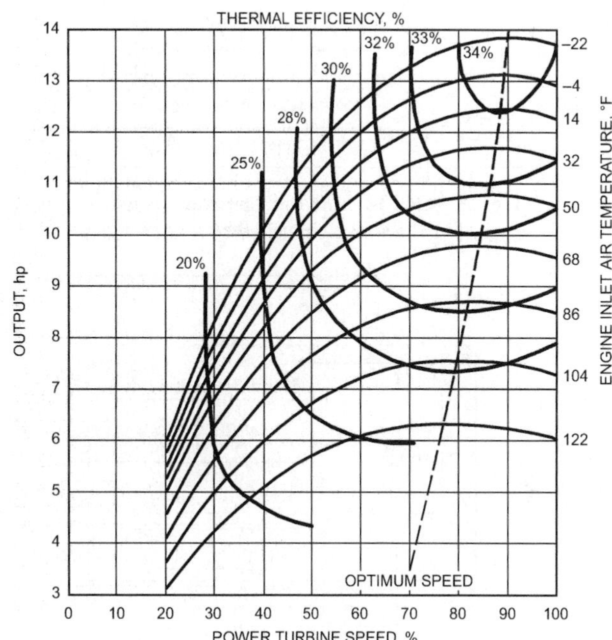

Fig. 24 Turbine Engine Performance Characteristics

If process heat is required, turbine exhaust can satisfy some of that heat, and the combined system is a CHP system. The exhaust can be used (1) directly as a source of hot air, to drive an exhaust-fired absorption chiller, regenerate desiccant materials in a direct-fired dehumidifier, activate a thermal enthalpy wheel, or operate a bottoming cycle (e.g., Rankine) to generate electric or shaft power; (2) in a large boiler or furnace as a source of preheated combustion air or cofiring an absorption chiller (exhaust typically contains 16% or higher oxygen levels); or (3) to heat a process or working fluid such as the steam system shown in Figure 25. Overall thermal efficiency around the CHP system boundary (Figure 26) is [(net electrical energy output + useful thermal energy output) × 100]/(energy input from fuel). Thermal efficiencies of these systems vary from 50% to greater than 70%. The exhaust of a gas turbine has about 4000 to 8000 Btu/h of available heat energy per horsepower output.

Additionally, because of the high oxygen content, the exhaust stream can typically support the combustion of an additional 30,000 Btu/h of fuel per horsepower output. This additional heat can then be used for processes in general manufacturing operations.

Fuels and Fuel Systems

The ability to burn almost any combustible fluid is a key advantage of the gas turbine. Natural gas is preferred over other gaseous fuels because it is readily available, has good combustion characteristics, and is relatively easy to handle. A typical fuel gas control system is a two-stage system that uses pressure control in combination with flow control to achieve a turndown ratio of about 100:1. Other fuel gases include liquefied petroleum gases, which are considered "wet" gases because they can form condensables at normal gas turbine operating conditions, and a wide range of refinery waste and coal-derived gases, which have a relatively high fraction of hydrogen. Both of these features lead to problems in fuel handling and preparation, as well as in gas turbine operation. Heat tracing to heat the piping and jacketing of valves is required to prevent condensation at start-up. Piping runs should be designed to eliminate pockets where condensate might drop out and collect. Low-heat-value fuels from waste gas sites such as landfills or wastewater treatment plants are beginning to be used by gas turbines, although significant fuel conditioning is often needed to remove moisture and contaminants such as siloxane.

Distillate oil is the most common liquid fuel and, except for a few installations where natural gas is not available, is primarily used as a back-up and alternative start-up fuel. Crude oils are common primary fuels in many oil-producing countries because of their abundance. Both crude and residual oils require treatment for sodium salts and vanadium contamination. The most common multiple-fuel combination is natural gas and distillate. The combustion turbine may be started on either fuel, and can transfer from one fuel to the other any time after completing the starting sequence without interrupting operation.

Steam and demineralized water injection are sometimes used in gas turbines for NO_x abatement in quantities up to 2% of compressor inlet airflow. An additional 3% steam may be injected independently at the compressor discharge for power augmentation. The required steam conditions are 300 to 350 psig and a temperature no more than 150°F above compressor discharge temperature, but not less than 50°F of superheat. Steam contaminants should be guarded against, and the steam supply system should be designed to supply dry steam under all operating conditions.

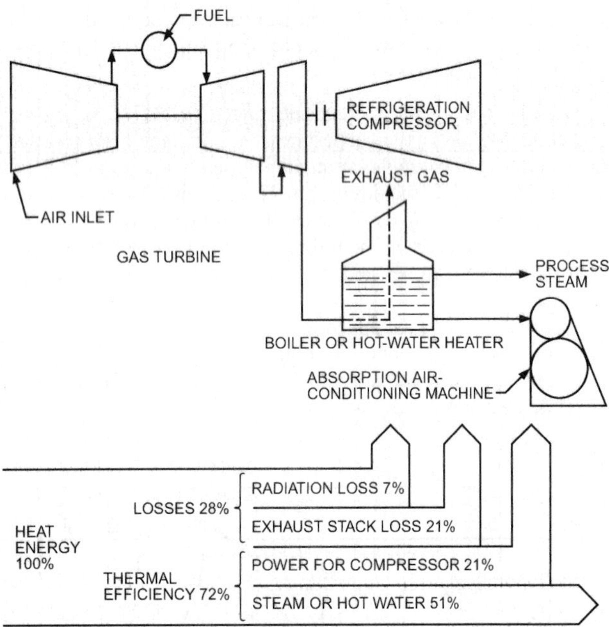

Fig. 25 Gas Turbine Refrigeration System Using Exhaust Heat

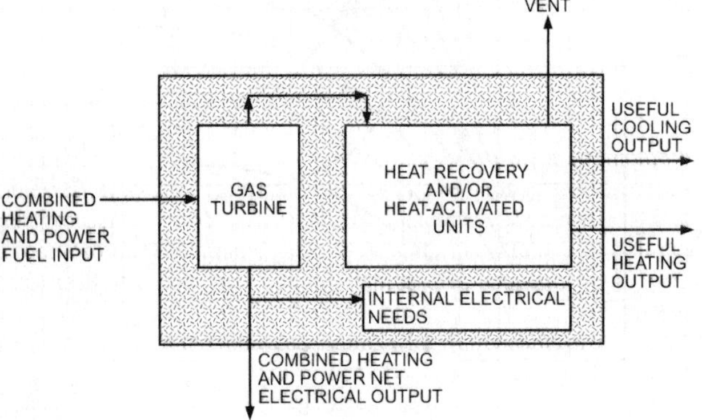

Fig. 26 CHP System Boundary

Combustion Air

Combustion turbine inlet cooling (CTIC) systems increase the capacity of turbine-generators by increasing combustion air density. Because volumetric flow to most turbines is constant, increasing air density increases mass flow rate. As inlet air temperature increases, (e.g., on hot summer days), capacity decreases (MacCracken 1994).

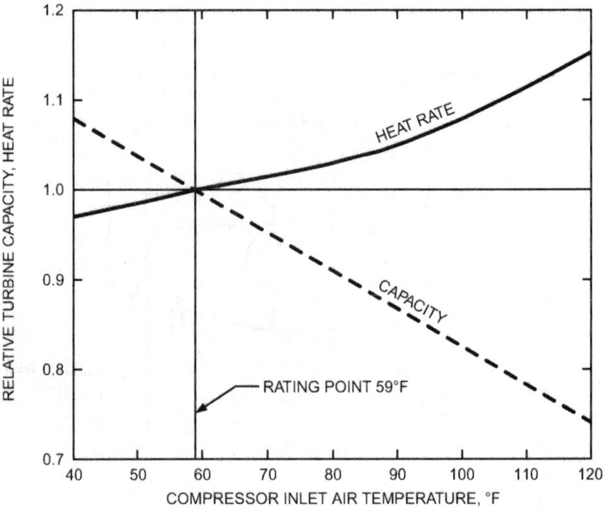

Fig. 27 Relative Turbine Power Output and Heat Rate Versus Inlet Air Temperature

Cooling inlet air increases the power and typically decreases the heat rate, after all parasitic cooling power usage is considered (Figure 27). Factors that affect CTIC installation and operation include turbine type, climate, hours of operation, ratio of airflow rate to power generated, ratio of generation increase with increased airflow, and monetary value of power generated. (More information is available in Chapter 17 and from sources such as the Turbine Inlet Cooling Association.)

Cooling Methods. Some CTIC designs are for turbines that operate only a few hours per year, to demonstrate power reserve or provide peak demand power. **Peaking** combustion turbines operate when utilities experience the greatest demand. Both inlet evaporative cooling systems (wetted media and fogging) and thermal energy storage (TES) systems allow CTIC during turbine operation with no coincident parasitic power usage except for pumps. TES systems allow the use of small-capacity refrigeration systems, operated only during off-peak hours. Utilities and independent power producers may operate turbines at **base load** and need continuous cooling for a significant number of hours per year. For turbines operating continuously or for several hours per day, fuel cost and availability are important factors that favor on-line cooling systems such as direct refrigeration without thermal storage (Brown and Somasundaram 1997).

The most prevalent CTIC system is **evaporative cooling** using wetted media, because of low installation and operating costs. Ideal evaporative cooling occurs at a constant wet-bulb temperature, cooling the air to near 100% rh. The typical evaporative cooling system allows the air/water vapor mixture to reach 85 to 95% of the difference between the dry-bulb air temperature and the wet-bulb temperature. Evaporative cooling can be used before or after indirect (secondary fluid) cooling. If a combination of sensible (cooling coils) and evaporative cooling is used, sensible cooling should be used before evaporative cooling, to reach the minimum temperature without latent cooling by the cooling coils.

Chilled water or direct refrigeration can also be used. These processes decrease the enthalpy (and temperature) of the air/water vapor mixture. The water vapor content (humidity ratio) remains constant as the mixture cools to near the dew-point temperature. Continued cooling follows the cooling coil performance curve, lowering the humidity ratio by forcing part of the water vapor to condense out from the mixture, while holding the mixture's relative humidity near 100%.

Chilled-water systems can be used in conjunction with either ice, chilled-water, or low-temperature fluid TES (Andrepont 2001; Ebeling 1994). From a cost standpoint, the TES system is usually justified for turbines that operate relatively few hours per day or to increase reserve power. In systems designed for long periods of cooling per day, the TES system typically has a higher capital cost and uses more energy than a direct refrigeration cooling system because of the secondary fluid loop, required pumping, and increased size of cooling coils. The chilled-water system, however, requires less refrigerant piping and inventory and is therefore less susceptible to refrigerant leakage. Thermal storage can reduce refrigeration equipment size and on-peak parasitic energy usage, which can decrease overall system capital costs. Parasitic loads for a TES system that operates only a few hours per day usually do not severely affect the economic value of a CTIC system.

A **direct** refrigerant cooling system consists of either a vapor compression system or an absorption system where the liquid refrigerant is used directly in air-cooling coils. The cooling process is identical to that of a chilled-water system. A direct system can provide cooling during all hours of turbine operation but must be sized to meet peak cooling; therefore, it is larger than a TES system (ASHRAE 1997).

For base-load combined-cycle power plants under continuous, varying loads, a direct online chilled-water cooling system using low-pressure steam from back-pressure steam turbines for absorption chillers is an economical option during peak load operation. The incremental cost of low-pressure steam is very low because the original high-pressure steam has done major work already and only a small part of the steam enthalpy remains for use in absorption chillers.

Because all base-load plants operate under part-load conditions for many hours during the day, the CTIC cooling coil increases heat rate because of pressure drop, raising specific fuel consumption for the whole plant. This has discouraged use of CTIC for combined-cycle power plants, especially plants operating under low part-load conditions for extended periods.

One solution is to heat combustion inlet air (a patented process), using the same coil, piping, and pumps, to raise combustion turbine exhaust air temperature and reduce mass flow, for better waste heat recovery and thus higher CHP efficiency. This technique also either avoids the need for variable air inlet guide vanes, thus reducing pressure drop and heat rate, or complements them under very low part-loads. This combustion turbine inlet air conditioning can provide cooling during peak-load operation and heating during part-load operation, save considerable fuel costs, and reduce pollution because of higher CHP efficiencies (Abedin 2003).

Advantages of CTIC.

Capacity Enhancement. Using CTIC for newer turbines, with lower airflow rates per unit of power generated, is even more economical than for older turbines. The lower flow rates require less cooling capacity to lower inlet air temperatures, and therefore smaller evaporative coolers or refrigeration equipment, including TES systems.

Heat Rate Improvement. Fuel mass flow rates increase with inlet airflow and turbine output, but typically at a lower rate. CTIC systems may be used primarily for decreased heat rate and corresponding fuel cost savings.

Turbine Life Extension. Turbines operating at lower inlet air temperatures have extended life and reduced maintenance. Lower and constant turbine inlet air temperatures reduce wear on turbines and turbine components.

Increased Combined-Cycle Efficiency. Lower inlet air temperatures result in lower exhaust gas temperatures, potentially decreasing the capacity of the heat recovery steam generator to provide heat to steam turbines and absorption equipment. However, the greater airflow rate of a CTIC system usually produces an overall increase in capacity because the effect of increased exhaust mass flow rate exceeds the effect of decreased temperature.

Delayed Capacity Addition. The increased generation capacity provided by a CTIC system can delay addition of actual or reserve generation capacity.

Baseload Efficiency Improvements. An ice or chilled-water TES system can help level the baseload of a power generation facility by storing energy using electric chiller equipment during off-peak periods; this tends to increase the efficiency of power production. Electric chillers operated at cooler nighttime temperatures are more efficient and operate at reduced condenser temperatures, which can also use less source energy.

When maximum power is desired every hour of the year, a continuous CTIC system is justified in warm climates to maximize turbine output and minimize heat rate.

Other Benefits. Other advantages include the following:

- Evaporative media filter the inlet air.
- CTIC systems that reduce air temperature below saturation can produce a significant amount of condensed water, a potentially valuable resource that can also provide makeup water for cooling towers or evaporative condensers.
- CTIC systems are simple, energized only when required.
- Emissions can decrease because of increased overall efficiency.
- A CTIC system can match inlet air temperature to required turbine generating capacity, allowing 100% open inlet guide vanes, which eliminate inlet guide vane pressure loss penalties.

Disadvantages.

- CTIC systems require additional space, and increase capital costs and maintenance.
- Evaporative media or cooling coils pose a constant inlet air pressure loss.

Lubricating Systems

Lubricating systems provide filtered and cooled oil to the gas turbine, driven equipment, and gear reducer, and typically include a motor-driven, start-up/coast-down oil pump; a primary oil pump mounted on and driven by the gear reducer; filters; oil reservoir; oil cooler; and automatic controls. Along with lubricating the gas turbine bearings, gear reducer, and driven equipment bearings, the lubrication system sometimes provides hydraulic oil to the gas turbine control system and can be used to drive a hydraulic starter.

The start-up/coast-down oil pump circulates oil until the gas turbine reaches a speed at which the primary pump can take over. If emergency alternating current (ac) power is not available in case of lost outside power, an emergency direct current (dc) motor-driven pump may be required to provide lubricant during start-up and coast-down.

Oil filters serve the pumps' full flow. Two filters are sometimes provided so that one can be changed while the other remains in operation.

The oil reservoir is mounted in the base of the gas turbine's supporting structure. Heater systems in the reservoir maintain oil temperature above a minimum level. A combination mechanical/coalescer filter in the reservoir's vent removes oil from the vent. The oil cooler can be either water- or air-cooled, depending on cooling water availability.

Starting Systems

Starting systems can use pneumatic, hydraulic, or electric motor starters. A pneumatic starting system uses either compressed air or fuel gas to power a pneumatic starter motor or a starting subsystem integrated directly with the rotating components. A hydraulic system uses a hydraulic motor for starting. Hydraulic fluid is provided to the hydraulic motor by either an ac-motor-driven pump or a diesel-engine-driven pump. An electric motor system couples an electric motor directly to the gas engine for starting. All direct-coupled systems use a one-way clutch to couple the starter motor to the gas engine so that, as the engine accelerates above the start speed, the starter can shut down. To perform black starts, include some form of external energy storage or drive system in the starting system.

Exhaust Systems

Exhaust systems of gas turbines used in CHP systems consist of gas ducts, expansion joints, an exhaust silencer, a dump (or bypass) stack, and a diverter valve (or damper). The exhaust silencer, if needed, is installed in a dump stack. The diverter valve modulates the flow of exhaust gas into the heat recovery equipment or diverts 100% of the exhaust gas to the dump stack when heat recovery is not required.

Emissions

Gas turbine power plants emit relatively low levels of CO_x and NO_x compared to other internal combustion engines; however, for each application, the gas turbine manufacturer should be consulted to ensure that applicable codes are met (ASME 1994). Special care should be taken if high-sulfur fuel is being used, because gas turbine exhaust stacks are typically not high, and dilution is not possible.

Instruments and Controls

Control systems are typically microprocessor-based. The control system sequences all systems during starting and normal operation, monitors performance, and protects the equipment. The operator interface is a monitor and keyboard, with analog gages for redundancy.

When operating the gas turbine engine at its maximum rating is desirable, the load is controlled based on the temperature of combustion gases in the turbine section and on the ambient air temperature. When the engine combustion gas temperature reaches a set value, the control system begins to control the engine so that the load (and therefore the temperature) does not increase further. With changes in ambient air temperature, the control system adjusts the load to maintain the set temperature value in the gas engine's turbine section. When maintaining a constant load level is desirable, the control system allows the operator to dial in any load, and the system controls the engine accordingly.

Noise and Vibration

Noise. Gas turbine manufacturers have developed sound-attenuated enclosures that cover the turbine and gear package. Turbine drivers, when properly installed with a sound-attenuated enclosure, inlet silencer, and exhaust silencer, meet the strictest noise standards. The turbine manufacturer should be consulted for detailed noise level data and recommendations on the least expensive method of attenuation for a particular installation.

Operation and Maintenance

Industrial gas turbines are designed to operate for 12,000 to 40,000 h between overhauls, with normal maintenance, which includes checking filters and oil level, inspecting for leaks, etc., all of which can be done by the operator with ordinary mechanics' tools. However, factory-trained service personnel are required to inspect engine components such as combustors and nozzles. These inspections, depending on the manufacturer's recommendations, are required as frequently as every 4000 h of operation.

Most gas turbines are maintained by condition monitoring and inspection (predictive maintenance) rather than by specific overhaul intervals. Gas turbines specifically designed for industrial applications may have an indefinite life for the major housings, shafts, and low-temperature components. Hot-section repair intervals for combustor and turbine components can vary from 10,000 to 100,000 h. The total cost of maintaining a gas turbine includes (1) cost of operator time, (2) normal parts replacement, (3) lubricating oil, (4) filter changes (combustion inlet air, fuel, and lubricating oil), (5) overhauls, and (6) factory service time (to conduct engine inspections). The cost of all these items can be estimated by the manufacturer and must be taken into account to determine the total operating cost.

Chapter 38 of the 2007 *ASHRAE Handbook—HVAC Applications* has more information on operation and maintenance management.

FUEL CELLS

Types

Fuel cells convert chemical energy of a hydrogen-based fuel directly into electricity without combustion. In the cell, a hydrogen-rich fuel passes over the anode, while an oxygen-rich gas (air) passes over the cathode. Catalysts help split the hydrogen into hydrogen ions and electrons. The hydrogen ions move through an external circuit, thus providing a direct current at a fixed voltage potential. A typical packaged fuel cell power plant consists of a fuel reformer (processor), which generates hydrogen-rich gas from fuel; a power section (stack) where the electrochemical process occurs; and a power conditioner (inverter), which converts the dc power generated in the fuel cell into ac power. Most fuel cell applications involve interconnectivity with the electric grid; thus, the power conditioner must synchronize the fuel cell's electrical output with the grid (ASHRAE 2001; ASME *Standard* PTC 50). A growing number of fuel cell applications are grid independent to reliably power remote or critical systems.

Table 14 Overview of Fuel Cell Characteristics

	Phosphoric Acid (PAFC)	Solid Oxide (SOFC)	Molten Carbonate (MCFC)	Proton Exchange Membrane (PEMFC)
Commercially available	Yes	No	Yes	Yes
Size range	100 to 200 kW	1 kW to 10 MW	250 kW to 10 MW	500 W to 250 kW
Efficiency (LHV)	40%	45 to 60%	45 to 55%	30 to 40%
Efficiency (HHV)	36%	40 to 54%	40 to 50%	27 to 36%
Average operating temperature	400°F	1800°F	1200°F	200°F
Heat recovery characteristics	Hot water	Hot water/steam	Hot water, steam	140°F water

Source: Adapted from Foley and Sweetser (2002).

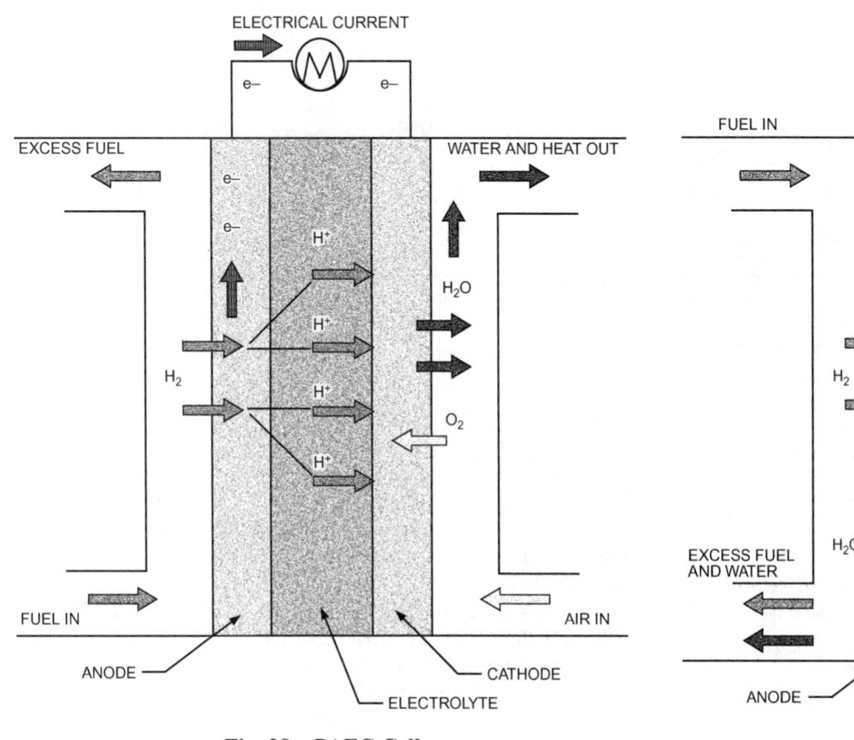

Fig. 28 PAFC Cell
(DOE 2007)

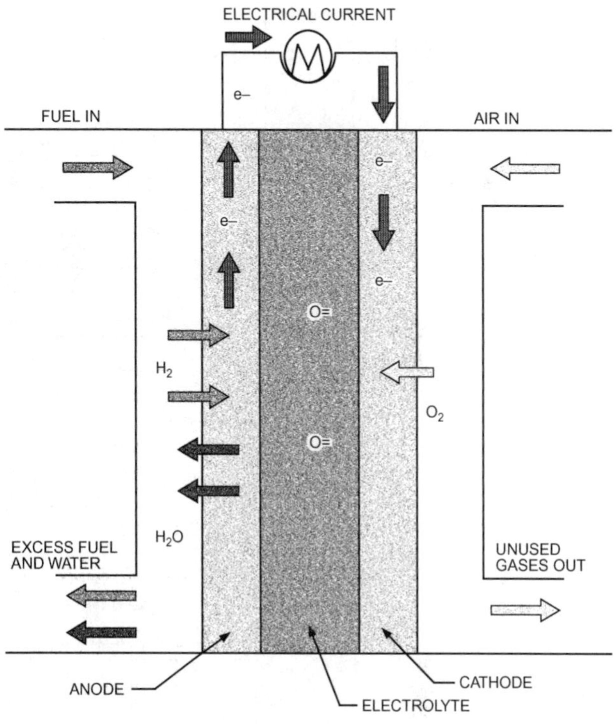

Fig. 29 SOFC Cell
(DOE 2007)

Most fuel cells have similar designs, but differ in the type of electrolyte used. The main types of fuel cells, classified by their electrodes, are (1) phosphoric acid (PAFC), (2) molten carbonate (MCFC), (3) solid oxide (SOFC), (4) proton exchange membrane (PEMFC), and (5) alkaline (AFC). PAFCs, MCFCs, and PEMFCs are commercially available, AFCs are used for space power, and SOFCs are in development and testing. The most significant research and development activities focus on PEMFC for automotive and home use and SOFC for stationary applications. Efficiencies of several types of fuel cells are shown in Table 14 (Foley and Sweetser 2002). Emissions from fuel cells are very low; NO_x emissions are less than 20 ppm. Large phosphoric acid fuel cells are commercially available.

Phosphoric Acid Fuel Cells (PAFCs). PAFCs are generally considered first-generation technology (Figure 28). They operate at about 400°F and achieve 40% LHV efficiencies. PAFCs use liquid phosphoric acid as an electrolyte. Platinum-catalyzed, porous-carbon electrodes are used for both the cathode and anode. For each type of fuel cell, the reformer supplies hydrogen gas to the anode through a process in which hydrocarbons, water, and oxygen react to produce hydrogen, carbon dioxide, and carbon monoxide. At the anode, hydrogen is split into two hydrogen ions (H^+) and two electrons. The ions pass through the electrolyte to the cathode, and the electrons pass through the external circuit to the cathode. At the cathode, the hydrogen, electrons, and oxygen combine to form water.

Solid-Oxide Fuel Cells (SOFCs). SOFCs operate at temperatures up to 1800°F, offering enhanced heat recovery performance (Figure 29). A solid-oxide system typically uses a hard ceramic material instead of a liquid electrolyte. The solid-state ceramic construction is a more stable and reliable design, enabling high temperatures and more flexibility in fuel choice. SOFCs can reach 54% HHV (60% LHV) efficiencies. Combined-cycle applications could reach system efficiencies up to 77% HHV (85% LHV).

SOFCs can use carbon monoxide as well as hydrogen as direct fuel. Hydrogen and carbon monoxide in the fuel stream react with oxide ions from the electrolyte, producing water and carbon dioxide, and releasing electrons into the anode. The electrons pass outside the fuel cell, through the load, and back to the cathode. At the cathode, oxygen molecules from the air receive the electrons and the molecules are converted into oxide ions. These ions are injected back into the electrolyte.

Molten-Carbonate Fuel Cells (MCFCs). MCFCs can reach 50% HHV (55% LHV) efficiencies (Figure 30). Combined-cycle applications could reach system thermal efficiencies of 77% HHV (85% LHV). MCFCs operate on hydrogen, carbon monoxide, natural gas, propane, landfill gas, marine diesel, and simulated coal gasification products. Operating temperatures are around 1200°F.

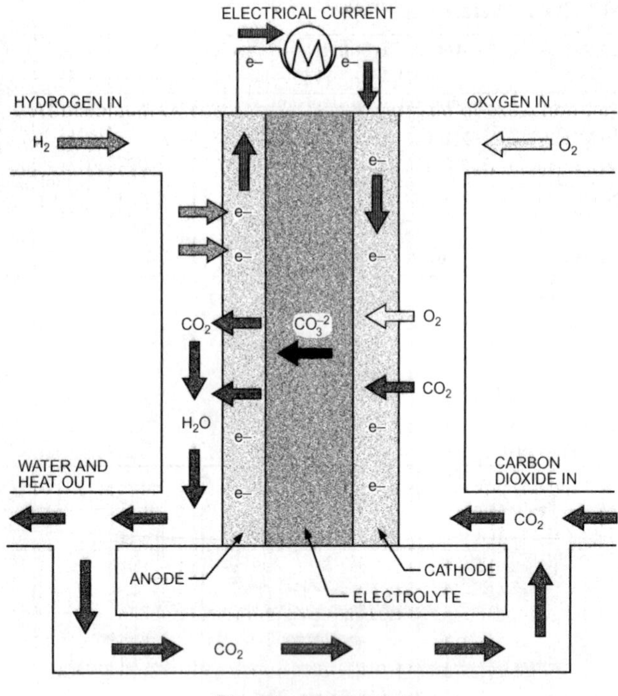

Fig. 30 MCFC Cell
(DOE 2007)

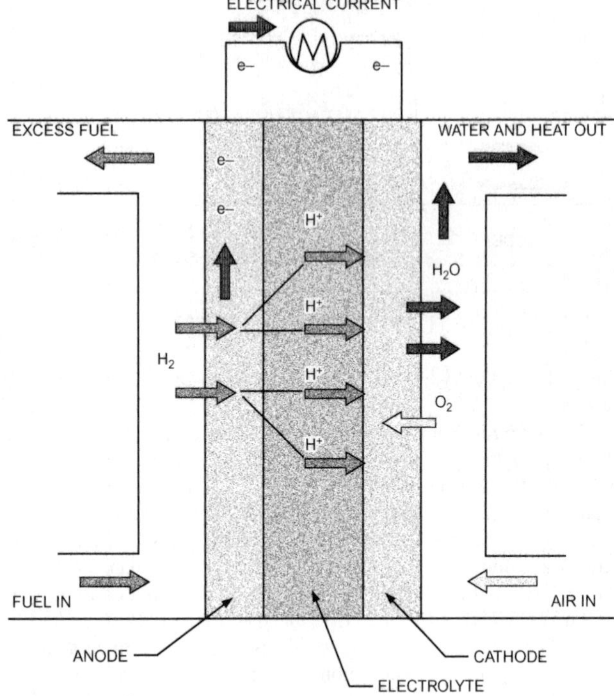

Fig. 31 PEMFC Cell
(DOE 2007)

This high operating temperature makes direct operation on gaseous hydrocarbon fuels (e.g., natural gas) possible. Natural gas can be reformed internally in MCFCs to produce hydrogen.

MCFCs use a molten carbonate salt mixture as an electrolyte; the electrolyte composition varies, but usually consists of lithium carbonate and potassium carbonate. The salt mixture is liquid and a good ionic conductor at the MCFC's high operating temperature. An electrochemical reaction occurs at the anode between the hydrogen

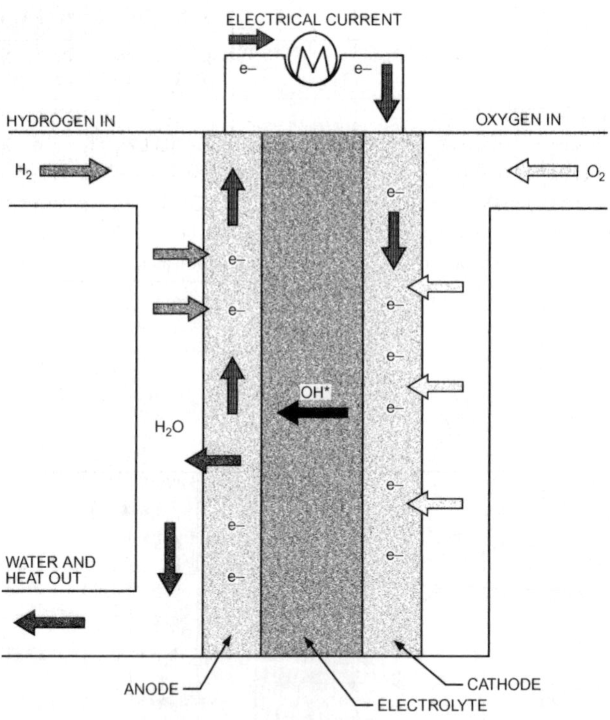

Fig. 32 AFC Cell
(DOE 2007)

fuel and carbonate ions from the electrolyte. This reaction produces water and carbon dioxide, and releases electrons to the anode. At the cathode, oxygen and carbon dioxide from the oxidant stream combine with electrons from the anode to produce carbonate ions, which enter the electrolyte.

Proton Exchange Membrane Fuel Cells (PEMFCs). PEMFCs can reach 36% HHV (40% LHV) efficiencies. PEMFCs contain a thin plastic polymer membrane through which hydrogen ions can pass. The membrane is coated on both sides with highly dispersed metal alloy particles (mostly platinum) that are active catalysts (Figure 31). Because the electrolyte is a solid polymer, electrolyte loss does not affect stack life. Using a solid electrolyte eliminates the safety concerns and corrosive effects associated with liquid electrolytes. PEMFCs operate at relatively low temperatures (approximately 200°F).

Electrode reactions in the PEMFC are analogous to those in the PAFC. Hydrogen ions and electrons are produced from the fuel gas at the anode. At the cathode, oxygen combines with electrons from the anode and hydrogen ions from the electrolyte to produce water. The solid electrolyte does not absorb the water, which is rejected from the back of the cathode into the oxidant gas stream (Hodge and Hardy 2002).

Hydrogen is delivered to the anode side of the membrane-electrode assembly (MEA), where it is catalytically split into protons and electrons. The protons move through the polymer electrolyte membrane to the cathode side, and the electrons travel to the cathode side along an external load circuit, creating the fuel cell's current output.

Meanwhile, oxygen is delivered to the cathode side of the MEA. The oxygen molecules react with protons and electrons coming from the anode side to form water molecules.

Alkaline Fuel Cells (AFCs). Alkaline fuel cells (AFCs) use a solution of potassium hydroxide in water as the electrolyte and various nonprecious metals as a catalyst at the anode and cathode (Figure 32). High-temperature AFCs operate between 212 and 482°F, although, newer designs operate at roughly 74 to 158°F.

AFCs' high performance derives from the rate at which chemical reactions take place in the cell. They have demonstrated efficiencies near 60% in space applications. AFC stacks maintain stable operation for 8000 operating hours, which limits their commercial viability.

THERMAL-TO-POWER COMPONENTS

These devices convert thermal energy to useful power. They do not convert the fuel source directly to power; this separation of energy conversion enables these devices to operate with an enormous variety of fuels and waste heat.

STEAM TURBINES

Steam turbines, which are among the oldest prime mover technologies, convert thermal energy from pressurized steam into useful power. They require a source of high-pressure steam produced in a boiler or HRSG. Most electricity in the United States is generated by conventional steam turbine power plants. Steam turbines are made in a variety of sizes, ranging from small 1 hp units to several hundred megawatts. In CHP applications, low-pressure steam from the turbine can be used directly in a process, or for district heating, or can be converted to other useful thermal energy.

Types

Axial Flow Turbines. Conventional axial flow steam turbines direct steam axially through the peripheral blades of one or more staged turbine wheels (much like a pinwheel) one after another on the same shaft. Figure 33 shows basic types of axial turbines. NEMA *Standard* SM 24 defines these and further subdivisions of their basic families as follows:

Noncondensing (Back-Pressure) Turbine. A steam turbine designed to operate with an exhaust steam pressure at any level that may be required by a downstream process, where all condensing takes place.

Condensing Turbine. A steam turbine with an exhaust steam pressure below atmospheric pressure, such that steam is directly and completely condensed.

Automatic Extraction Turbine. A steam turbine that has opening(s) in the turbine casing for extracting steam and means for directly regulating the extraction steam pressure.

Nonautomatic Extraction Turbine. A steam turbine that has opening(s) in the turbine casing for extracting steam without a means for controlling its pressure.

Induction (Mixed-Pressure) Turbine. A steam turbine with separate inlets for steam at two pressures, with an automatic device for controlling the pressure of the secondary steam induced into the turbine and means for directly regulating the flow of steam to the turbine stages below the induction opening.

Induction-Extraction Turbine. A steam turbine that can either exhaust or admit a supplemental flow of steam through an intermediate port in the casing, thereby maintaining a process heat balance. Extraction and induction-extraction turbines may have several casing openings, each passing steam at a different pressure.

The necessary rotative force for shaft power in a turbine may be imposed through the steam's velocity, pressure energy, or both. If velocity energy is used, the movable wheels are usually fitted with crescent-shaped blades. A row of fixed nozzles in the steam chest increases steam velocity into the blades with little or no steam pressure drop across them, and causes wheel rotation. These combinations of nozzles and velocity-powered wheels are characteristic of an **impulse turbine**.

Reaction Turbine. A **reaction turbine** uses alternate rows of fixed and moving blades, generally of an airfoil shape. Steam velocity increases in the fixed nozzles and drops in the movable ones, and steam pressure drops through both.

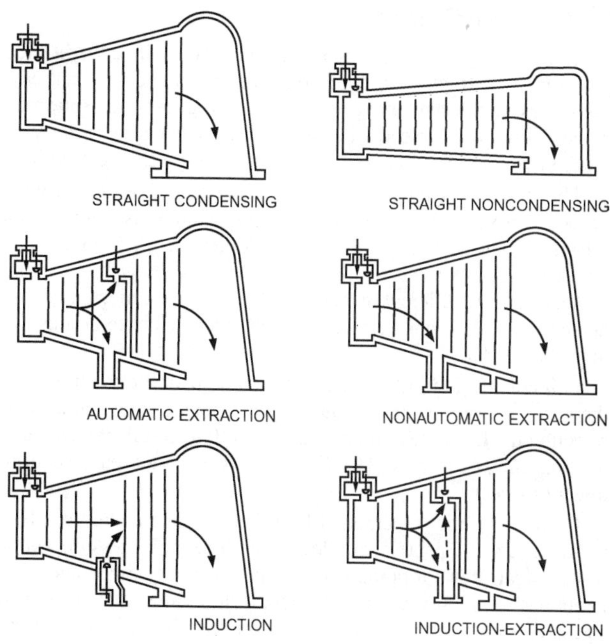

Fig. 33 Basic Types of Axial Flow Turbines

The power capability of a reaction turbine is maximum when the moving blades travel at about the velocity of the steam passing through them; in the impulse turbine, maximum power is produced with a blade velocity of about 50% of steam velocity. Steam velocity is related directly to pressure drop. To achieve the desired relationship between steam velocity and blade velocity without resorting to large wheel diameters or high rotative speeds, most turbines include a series of impulse or reaction stages or both, thus dividing the total steam pressure drop into manageable increments. A typical commercial turbine may have two initial rows of rotating impulse blading with an intervening stationary row (called a *Curtis stage*), followed by several alternating rows of fixed and movable impulse- or reaction-type blading. Most multistage turbines use some degree of reaction.

Construction. Turbine manufacturers' standards prescribe casing materials for various limits of steam pressure and temperature. The choice between built-up or solid rotors depends on turbine speed or inlet steam temperature. Water must drain from pockets in the turbine casing to prevent damage from condensate accumulation. Carbon rings or closely fitted labyrinths prevent steam leakage between pressure stages of the turbine, outward steam leakage, and inward air leakage at the turbine glands. The erosive and corrosive effect of moisture entering with the supply steam must be considered. Heat loss is controlled by installation (often at the manufacturer's plant) of thermal insulation and protective metal jacketing on hotter portions of the turbine casing.

Radial Inflow Turbines. Radial inflow turbines have a radically different configuration from axial flow machines. Steam enters through the center or eye of the impeller and exits from the periphery, much like the path of fluid through a compressor or pump, but in this case the steam actuates the wheel, instead of the wheel actuating the air or water.

Radial, multistage arrangements comprise separate, single-stage wheels connected with integral reduction gearing in a factory-assembled package. Induction, extraction, and moisture elimination are accomplished in the piping between stages, giving the radial turbine a greater tolerance of condensate.

Performance Characteristics

The topping CHP-cycle steam turbine is typically either a back-pressure or extraction condensing type that makes downstream, low-pressure thermal energy available for process use. Bottoming cycles commonly use condensing turbines because these yield more power, having lower-grade throttle energy to begin with.

The highest steam plant efficiency is obtainable with a back-pressure turbine when 100% of its exhaust steam is used for thermal processes. The only inefficiencies are the gear drive, alternator, and inherent steam-generating losses. A large steam-system topping cycle using an efficient water-tube boiler, economizer, and pre-heater can easily achieve an overall efficiency (fuel to end use) of more than 90%.

Full condensing turbine heat rates (Btu/hp·h) are the highest in the various steam cycles because the turbine's exhaust condenses, rejecting the latent heat of condensation (1036 Btu/lb at a condensing pressure of 1 psia and 101°F) to a waste heat sink (e.g., a cooling tower or river).

Conversely, the incremental heat that must be added to a low-pressure (e.g., 30 psia) steam flow to produce high-pressure, superheated steam for a topping cycle is only a small percentage of its latent heat of vaporization. For example, to produce 250°F, 30 psia saturated steam for a single-stage absorption chiller in a low-pressure boiler requires 1164 Btu/lb, of which 945 Btu/lb is the heat of vaporization. To boost this to 600°F, 320 psia requires an additional 146 Btu/lb, which is only 15% of the latent heat at 30 psia, for an enthalpy of 1310 Btu/lb.

A low-pressure boiler generating 30 psia steam directly to the absorber has a 75% fuel-to-steam efficiency, which is 15% lower than the 90% efficiency of a high-pressure boiler used in the CHP cycle. Therefore, from the standpoint of fuel cost, the power generated by the back-pressure turbine is virtually free when its 30 psia exhaust is discharged into the absorption chiller.

The potential power-generating capacity and size of the required turbine are determined by its efficiency and steam rate (or water rate). This capacity is, in turn, the system's maximum steam load, if the turbine is sized to satisfy this demand. Efficiencies range from 55 to 80% and are the ratio of actual to theoretical steam rate, or actual to theoretical enthalpy drop from throttle to exhaust conditions.

NEMA *Standard* SM 24 defines the theoretical steam rate as the quantity of steam per unit of power required by an ideal Rankine cycle, which is an isentropic or reversible adiabatic process of expansion. This can best be seen graphically on an enthalpy-entropy (Mollier) chart. Expressed algebraically, the steam rate is

$$w_t = \frac{2546}{h_i - h_e} \tag{1}$$

where

w_t = theoretical steam rate, lb/hp·h
h_e = enthalpy of steam at exhaust pressure and inlet entropy, Btu/lb
h_i = enthalpy of steam at throttle inlet pressure and temperature, Btu/lb
2546 = Btu/hp·h

This isentropic expansion through the turbine represents 100% conversion efficiency of heat energy to power. An example is shown on the Mollier chart in Figure 34 as the vertical line from 320 psia, 600°F, 1310 Btu/lb to 30 psia, 250°F, 93% quality, 1100 Btu/lb.

On the other hand, zero efficiency is a throttling, adiabatic, non-reversible horizontal line terminating at 30 psia, 552°F, 1310 Btu/lb. An actual turbine process would lie between 0 and 100% efficiency, such as the one shown at actual exhaust condition of saturated steam h_a at 30 psia, 250°F, 1164 Btu/lb; the actual turbine efficiency is

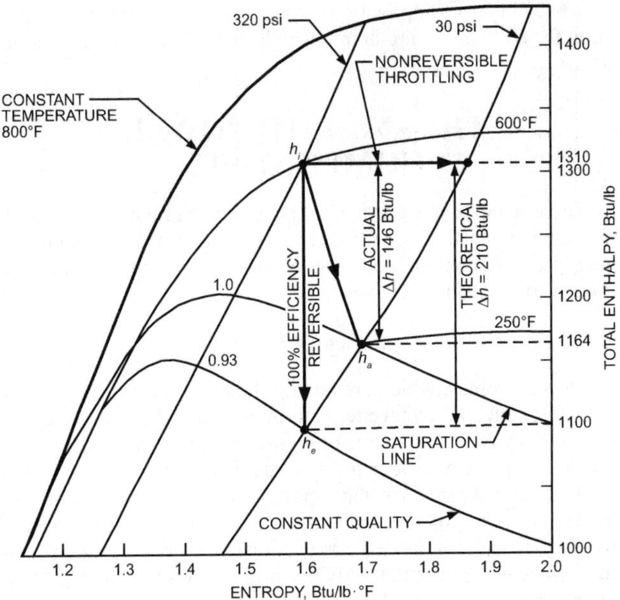

Fig. 34 Isentropic Versus Actual Turbine Process

$$E_a = \frac{h_i - h_a}{h_i - h_e} \tag{2}$$

and the actual steam rate is

$$w_a = \frac{3412}{h_i - h_a} \tag{3}$$

where

w_a = actual steam rate, lb/kWh
h_a = enthalpy of steam at actual exhaust conditions, Btu/lb
3412 = Btu/kWh

For the case described,

$$E_a = \frac{1310 - 1164}{1310 - 1100} = 0.70 \text{ or } 70\%$$

As a CHP cycle, if the previously described absorption chiller has a capacity of 2500 tons, which requires 45,000 lb/h of 30 psia saturated steam, it can be provided by the 69% efficient turbine at an actual steam rate of 3412/146 = 23.4 lb/kWh; the potential power generation is

$$\frac{45{,}000}{23.4} = 1923 \text{ kWh}$$

The incremental turbine heat rate to generate this power is only

$$\frac{146 \times 45{,}000}{1923} = 3417 \text{ Btu/kWh}$$

instead of a typical 9000 Btu/kWh (thermal efficiency of 38%) for an efficient steam power plant with full condensing turbines and cooling towers.

Turbine performance tests should be conducted in accordance with the appropriate American Society of Mechanical Engineers (ASME) *Performance Test Code:* PTC 6, PTC 6S Report, or PTC 6A. The steam rate of a turbine is reduced with higher turbine speeds, a greater number of stages, larger turbine size, and a higher difference in heat content between entering and leaving steam conditions. Often, one or more of these factors can be improved with only a nominal increase in initial capital cost. CHP applications range, with equal flow turbines, from approximately 100 to 10,000 hp and from

3000 to 10,000 rpm, with high speeds generally associated with lower power outputs, and low speeds with higher power outputs. (Some typical characteristics of turbines driving centrifugal water chillers are shown in Figures 35 to 39.)

Initial steam pressures for small turbines commonly fall in the 100 to 250 psig range, but wide variations are possible. Turbines in the range of 2000 hp and above commonly have throttle pressures of 400 psig or greater.

Back pressure associated with noncondensing turbines generally ranges from 50 psig to atmospheric, depending on the use for the exhaust steam. Raising the initial steam temperature by superheating improves steam rates.

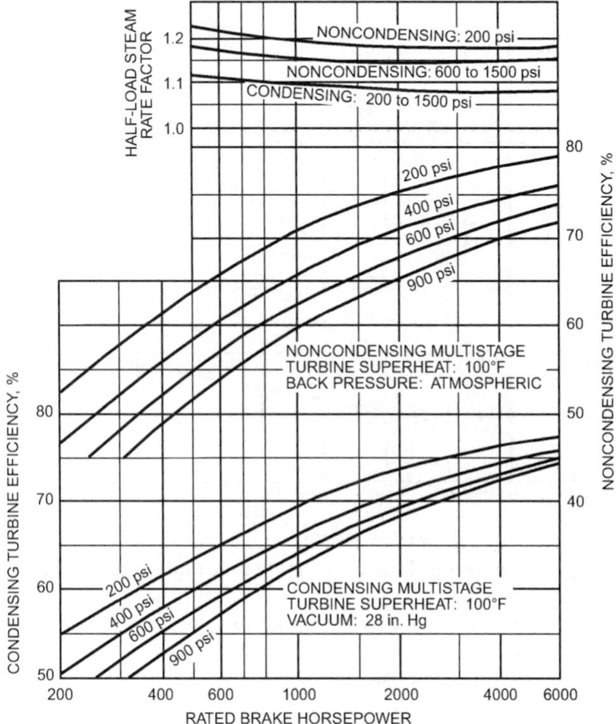

Fig. 35 Efficiency of Typical Multistage Turbines

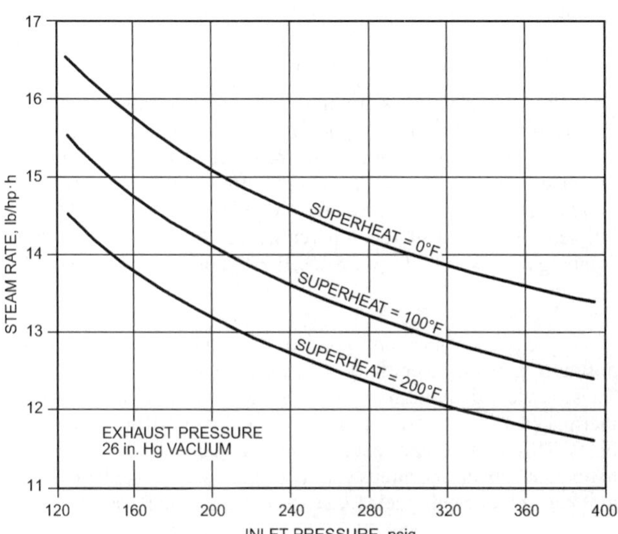

Fig. 36 Effect of Inlet Pressure and Superheat on Condensing Turbine

NEMA *Standards* SM 23 and SM 24 govern allowable deviations from design steam pressures and temperatures. Because of possible unpredictable variations in steam conditions and load requirements, turbines are selected for a power capability of 105 to 110% of design shaft output and speed capabilities of 105% of design rpm.

Because no rigid standards prevail for the turbine inlet steam pressure and temperature, fixed design conditions proposed by ASME/IEEE should be used to size the steam system initially. These values are 400 psig at 750°F, 600 psig at 825°F, 850 psig at 900°F, and 1250 psig at 950 or 1000°F.

Table 15 lists theoretical steam rates for steam turbines at common conditions. If project conditions dictate different throttle/exhaust conditions from the steam tables, theoretical steam rate tables or graphical Mollier chart analysis may be used.

Steam rates for multistage turbines depend on many variables and require extensive computation. Manufacturers provide simple tables and graphs to estimate performance, and these data are good guides for preliminary sizing of turbines and associated auxiliaries for the complete system.

Using the entire exhaust steam flow from a base-loaded back-pressure turbine achieves the maximum efficiency of a steam turbine CHP cycle. However, if the facility's thermal/electrical load

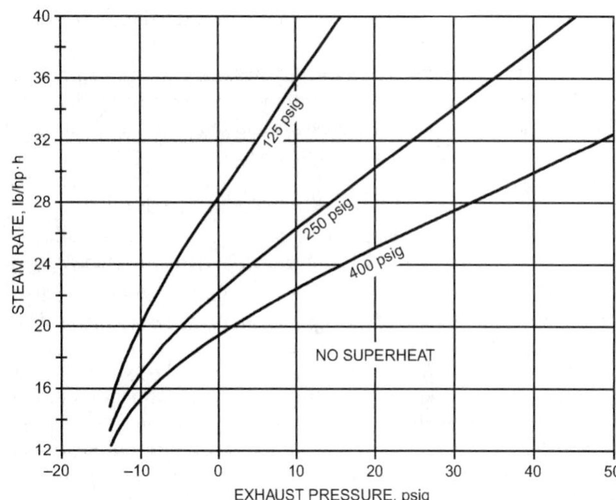

Fig. 37 Effect of Exhaust Pressure on Noncondensing Turbine

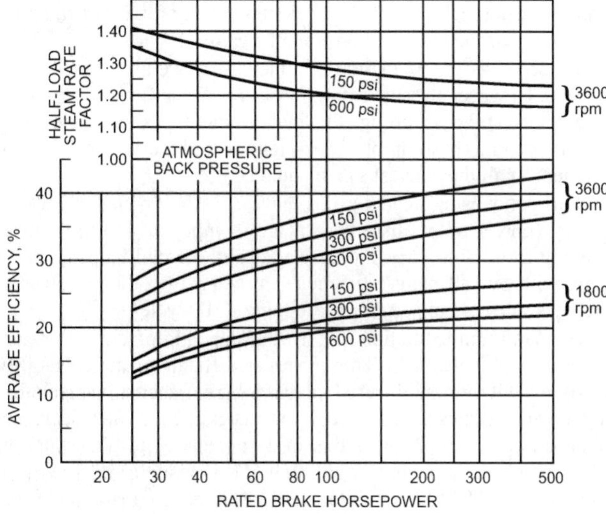

Fig. 38 Single-Stage Noncondensing Turbine Efficiency

Table 15 Theoretical Steam Rates for Steam Turbines at Common Conditions, lb/kWh

	Throttle Steam Conditions							
Exhaust Pressure	150 psig, 366°F, Saturated	200 psig, 388°F, Saturated	250 psig, 500°F, 94°F Superheat	400 psig, 750°F, 302°F Superheat	600 psig, 750°F, 261°F Superheat	600 psig, 825°F, 336°F Superheat	850 psig, 825°F, 298°F Superheat	850 psig, 900°F, 373°F Superheat
2 in. Hg (absolute)	10.52	10.01	9.07	7.37	7.09	6.77	6.58	6.28
4 in. Hg (absolute)	11.76	11.12	10.00	7.99	7.65	7.28	7.06	6.73
0 psig	19.37	17.51	15.16	11.20	10.40	9.82	9.31	8.81
10 psig	23.96	21.09	17.90	12.72	11.64	10.96	10.29	9.71
30 psig	33.60	28.05	22.94	15.23	13.62	12.75	11.80	11.07
50 psig	46.00	36.00	28.20	17.57	15.36	14.31	13.07	12.21
60 psig	53.90	40.40	31.10	18.75	16.19	15.05	13.66	12.74
70 psig	63.50	45.60	34.10	19.96	17.00	15.79	14.22	13.25
75 psig	69.30	48.50	35.80	20.59	17.40	16.17	14.50	13.51

cannot absorb the fully loaded output of the turbine, whichever profile is lower can be tracked, and the power output or steam flow is reduced unless the output remaining is exported. Annual efficiency can still be high if the machine operates at significant combined loads for substantial periods. Straight steam condensing turbines offer no opportunity for topping cycles but are not unusual in bottoming cycles because waste steam from the process can be most efficiently used by full-condensing turbines when there is no other use for low-pressure steam. Either back-pressure or extraction condensing turbines may be used as extraction turbines.

The steam in an extraction turbine expands part of the way through the turbine until the pressure and temperature required by the external thermal load are attained. The remaining steam continues through the low-pressure turbine stages; however, it is easier to adjust for noncoincident electrical and thermal loads.

Because steam cycles operate at pressures above those allowed by ASME and local codes for unattended operation, their use in CHP plants is limited to large systems where attendants are required for other reasons or the labor burden of operating personnel does not seriously affect overall economics.

Figure 39 shows the performance of a 1500 kW extraction condensing turbine, indicating the effect of various extraction rates on total steam requirements as follows: at zero extraction and 1500 kW, 17,500 lb/h or a water rate of 11.67 lb/kWh is required. When 45,000 lb/h is extracted at 100 psig, only 4000 lb/h more (49,000 – 45,000) is required at the throttle condition of 400 psig to develop the same 1500 kW, chargeable to the generation of electric power. The portion of input energy chargeable to the power is represented by the sum of the enthalpy of this 4000 lb/h at throttle conditions and the difference in enthalpy between the throttle and extraction conditions of the extracted portion of steam.

In effect, as the extraction rate increases, overall efficiency increases. However at "full" extraction, a significant flow of "cooling" steam must still pass through the final turbine stages for condensing. At this condition, a simple back-pressure turbine would be more efficient, if all the exhaust steam could be used.

Full-condensing steam turbines have a maximum plant shaft efficiency (power output as a percentage of input fuel to the boiler) ranging from 20 to 36%, but have no useful thermal output. Therefore, with overall plant efficiencies no better than their shaft efficiencies, they are unsuitable for topping CHP cycles.

At maximum extraction, the heat/power ratio of extraction turbines is relatively high. This makes it difficult to match facility loads, except those with very high base thermal loads, if reasonable annual efficiencies are to be achieved. As extraction rates decrease, plant efficiency approaches that of a condensing turbine, but can never reach it. Thus, the 17,500 lb/h (11.67 lb/kWh) illustrated in Figure 39 at 1500 kW and zero extraction represents a steam-to-electric efficiency of 28.6%, but a fuel-to-electric efficiency of 25%, with a boiler plant efficiency of 85%, developed as follows:

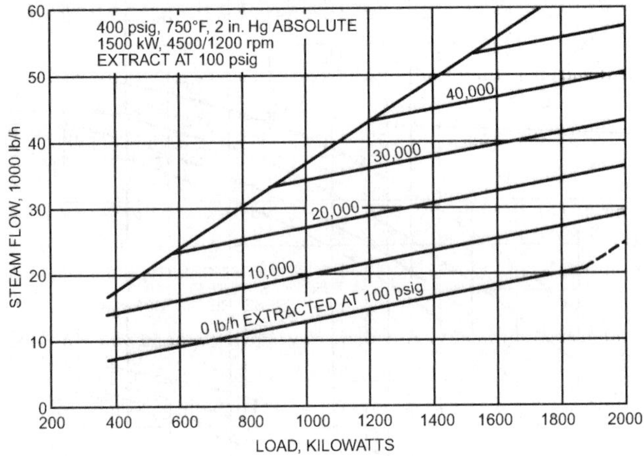

Fig. 39 Effect of Extraction Rate on Condensing Turbine

Isentropic Δh from 400 psig steam to 2 in. Hg (absolute) condensing pressure = 1022 Btu/lb output.

$$\text{Actual } \Delta h = \frac{3412 \text{ Btu/kWh}}{\text{Actual steam rate (lb/kWh)}}$$

Actual Δh = 3412/11.67 = 292 Btu/lb

Plant shaft efficiency = (100 × 292 × 0.85)/1022
= 24.3% (electric output to fuel energy input)

Radial inflow turbines are more efficient than single-stage axial flow turbines of the same output. They are available up to 15,000 hp from several manufacturers, with throttle steam up to 2100 psig, wheel speeds up to 60,000 rpm, and output shaft speeds as low as 3600 or 1800 rpm. It is these high wheel speeds that yield turbine efficiencies of 70 to 80%, compared with single-stage axial turbines spinning at only 10,000 rpm with efficiencies of up to 40%.

Fuel Systems

Unlike gas turbines and reciprocating engines, steam turbines normally generate electricity as a by-product of heat generation. These devices do not directly convert fuel to useful power; instead, thermal energy is transferred from the boiler or HRSG to these turbines. This separation of energy conversion enables steam turbines to operate with an enormous variety of fuels such as coal, oil, or natural gas or renewable fuels like wood or municipal waste.

Lubricating Oil Systems

Small turbines often have only simple oil rings to handle bearing lubrication, but most turbines for CHP service have a complete pressure lubrication system. Basic components include a shaft-driven oil pump, oil filter, oil cooler, means of regulating oil pressure, reservoir, and interconnecting piping. Turbines with a hydraulic governor may use oil from the lubrication circuit or, with some types of governors, use a self-contained governor oil system. To ensure an adequate supply of oil to bearings during acceleration and deceleration, many turbines include an auxiliary motor or turbine-driven oil pump. Oil pressure-sensing devices act in two ways: (1) to stop the auxiliary pump once the shaft-driven pump has attained proper flow and pressure or (2) to start the auxiliary pump if the shaft-driven pump fails or loses pressure when decelerating. In some industrial applications, the lubrication systems of the turbine and driven compressor are integrated. Proper oil pressure, temperature, and compatibility of lubricant qualities must be maintained.

Power Systems

Steam turbines are used for power only or in CHP applications. In CHP applications, low-pressure steam leaving the turbine is used directly in a process or for district heating, or converted to other forms of thermal energy. In power-only applications, condensing turbines are generally used to maximize electricity production by using fuels that otherwise would go to waste. Condensing turbines are also used in bottoming cycles for gas turbines to produce additional electricity in a combined cycle (gas turbine exhaust is used in a HRSG to generate high-pressure steam to drive the steam turbine) and improve the cycle's electric efficiency.

Exhaust Systems

Steam turbines can be driven with a boiler firing with various fuels, or they can be used in a combined cycle with a gas turbine and HRSG. Boiler emissions depend on the fuel used to generate the high-pressure steam to drive the steam turbine.

Instruments and Controls

Starting Systems. Unlike reciprocating engines and combustion turbines, steam turbines do not require auxiliary starting systems. Steam turbines are started through controlled opening of the main steam valve, which is in turn controlled by the turbine governing system. Larger turbines with multiple stages and/or dual shafting arrangements are started gradually to allow for controlled expansion and thermal stressing. Many of these turbines are provided with electrically powered turning gears that slowly rotate the shaft(s) during the initial stages of start-up.

Governing Systems. The wide variety of available governing systems allows a governor to be ideally matched to the characteristics of the driven machine and load profiles. The principal and most common function of a fixed-speed steam turbine governing system is to maintain constant turbine speed despite load fluctuations or minor variations in supply steam pressure. This arrangement assumes that close control of the output of the driven component, such as a generator in a power plant, is primary to plant operation, and that the generator can adjust its capacity to varying loads.

Often it is desirable to vary turbine speed in response to an external signal. In centrifugal water-chilling systems, for example, reduced speed generally reduces steam rate at partial load. An electric, electronic, or pneumatic device responds to the system load or temperature of fluid leaving the water-chilling heat exchanger (evaporator). To avoid compressor surge and optimize the steam rate, the speed is controlled initially down to some part load, then controlled in conjunction with the compressor's built-in capacity control (e.g., inlet vanes).

Process applications frequently require placing an external signal on the turbine governing system to reset the speed control point. External signals may be needed to maintain a fixed compressor discharge pressure, regardless of load or condenser water temperature variations. Plants relying on a closely maintained heat balance may control turbine speed to maintain an optimum pressure level of steam entering, being extracted from, or exhausting from the turbine. One example is the combination turbine absorption plant, where control of pressure of the steam exhausting from the turbine (and feeding the absorption unit) is an integral part of the plant control system.

Components. The steam turbine governing system consists of (1) a speed governor (mechanical, hydraulic, electrical, or electronic), (2) a speed control mechanism (relays, servomotors, pressure- or power-amplifying devices, levers, and linkages), (3) governor-controlled valve(s), (4) a speed changer, and (5) external control devices, as required.

The **speed governor** responds directly to turbine speed and initiates action of the other parts of the governing system. The simplest speed governor is the direct-acting flyball, which depends on changes in centrifugal force for proper action. Capable of adjusting speeds through an approximate 20% range, it is widely used on single-stage, mechanical-drive steam turbines with speeds of up to 5000 rpm and steam pressure of up to 600 psig.

The most common speed governor for centrifugal water-chilling system turbines is the oil pump type. In its direct-acting form, oil pressure, produced by a pump either directly mounted on the turbine shaft or in some form responsive to turbine speed, actuates the inlet steam valve.

The oil relay hydraulic governor (Figure 40), has greater sensitivity and effective force. Here, the speed-induced oil pressure changes are amplified in a **servomotor** or **pilot-valve relay** to produce the motive effort required to reposition the steam inlet valve or valves.

The least expensive turbine has a single governor-controlled steam admission **throttle valve**, perhaps augmented by one or more small auxiliary valves (usually manually operated), which close off nozzles supplying the turbine steam chest for better part-load efficiency. Figure 41 shows the effect of auxiliary valves on part-load turbine performance.

For more precise speed governing and maximum efficiency without manual valve adjustment, multiple automatic nozzle control is used (Figure 42). Its principal application is in larger turbines where a single governor-controlled steam admission valve would be too large to allow sensitive control. The greater power required to actuate the multiple-valve mechanism dictates using hydraulic servomotors. **Speed changers** adjust the setting of the governing systems while the turbine is in operation. Usually, they comprise either a means of changing spring tension or a means of regulating

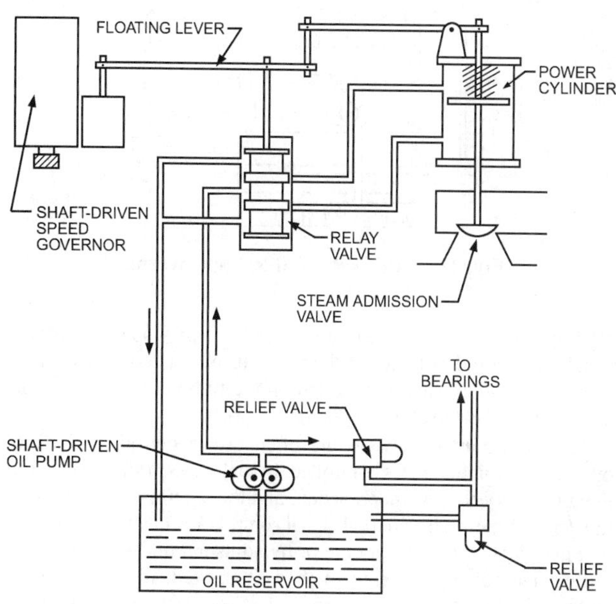

Fig. 40 Oil Relay Governor

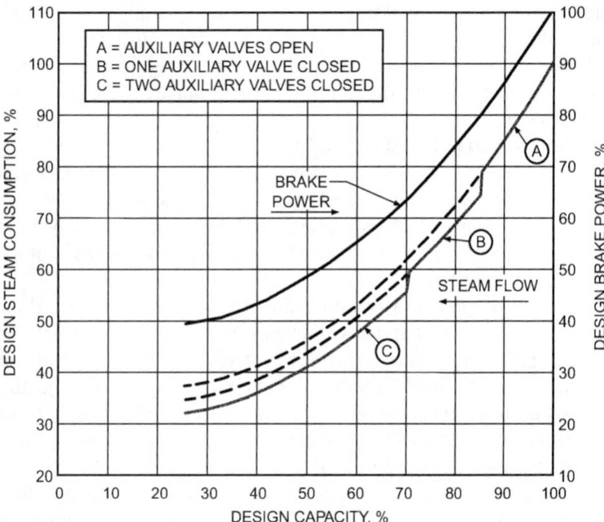

Fig. 41 Part-Load Turbine Performance Showing Effect of Auxiliary Valves

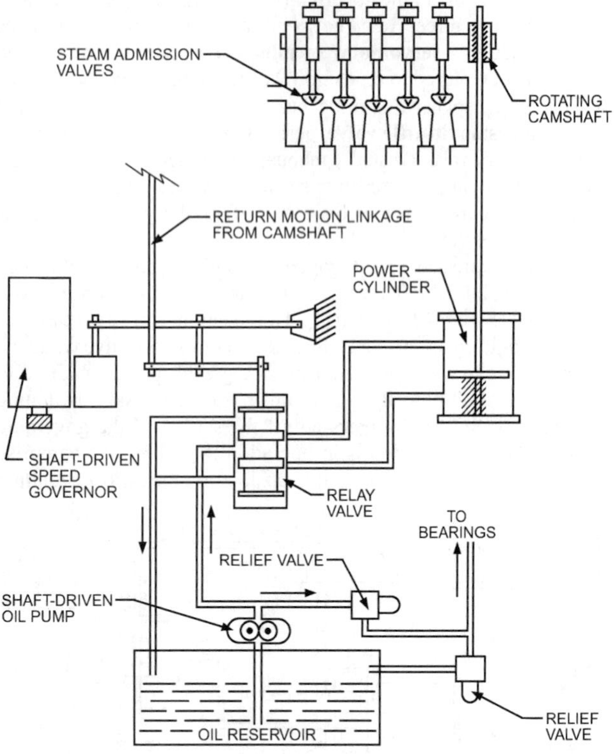

Fig. 42 Multivalve Oil Relay Governor

oil flow by a needle valve. The upper limit of a speed changer's capability should not exceed the rated turbine speed. These speed changers, though usually mounted on the turbine, may sometimes be remotely located at a central control point.

External control devices are often used when some function other than turbine speed is controlled. In such cases, a signal overrides the turbine speed governor's action, and the latter assumes a speed-limiting function. The external signal controls steam admission either by direct inlet valve positioning or by adjusting the speed governor setting. The valve-positioning method either exerts mechanical force on the valve-positioning mechanism or, if power has to be amplified, regulates the pilot valve in a hydraulic servomotor system.

Table 16 NEMA Classification of Speed Governors

Class of Governor	Range of Speed Changer Adjustment, %	Maximum Steady-State Speed Regulation, %	Maximum Speed Variation, % Plus or Minus	Maximum Speed Rise, %	Trip Speed, % Above Rated Speed
A	10 to 65	10	0.75	13	15
B	10 to 80	6	0.50	7	10
C	10 to 80	4	0.25	7	10
D	10 to 90	0.50	0.25	7	10

Source: NEMA *Standard* SM 24.

Where more precise control is required, the speed governor adjusting method is preferred. Although the external signal continually resets the governor as required, the speed governor always provides ideal turbine speed control. Thus, it maintains the particular set speed, regardless of load or steam pressure variations.

Classification. The National Electrical Manufacturers Association (NEMA) *Standard* SM 24 classifies steam turbine governors as shown in Table 16. **Range of speed changer adjustment**, expressed as a percentage of rated speed, is the range through which the turbine speed may be adjusted downward from rated speed by the speed changer, with the turbine operating under control of the speed governor and passing a steam flow equal to the flow at rated power, output, and speed. The range of the speed changer adjustment, expressed as a percentage of rated speed, is derived from the following equation:

$$\text{Range (\%)} = \frac{(\text{Rated speed}) - (\text{Minimum speed setting})}{\text{Rated speed}} \times 100$$

Steady-state speed regulation, expressed as a percentage of rated speed, is the change in sustained speed when the power output of the turbine is gradually changed from rated power output to zero power output under the following conditions:

- Steam conditions (initial pressure, initial temperature, and exhaust pressure) are set at rated values and held constant.
- Speed changer is adjusted to give rated speed with rated power output.
- Any external control device is rendered inoperative and blocked open to allow free flow of steam to the governor-controlled valve(s).

The steady-state speed regulation is derived from the following equation:

$$\text{Regulation (\%)} = \frac{\begin{pmatrix}\text{Speed at zero}\\ \text{power output}\end{pmatrix} - \begin{pmatrix}\text{Speed at rated}\\ \text{power output}\end{pmatrix}}{\text{Speed at rated power output}} \times 100$$

Steady-state speed regulation of automatic extraction or mixed pressure turbines is derived with zero extraction or induction flow and with the pressure-regulating system(s) inoperative and blocked in the position corresponding to rated extraction or induction pressure(s) at rated power output.

Speed variation, expressed as a percentage of rated speed, is the total magnitude of speed change or fluctuations from the speed setting. It is defined as the difference in speed variation between the governing system in operation and the governing system blocked to be inoperative, with all other conditions constant. This characteristic includes dead band and sustained oscillations. Expressed as a percentage of rated speed, the speed variation is derived from the following equation:

$$\text{Speed Variation (\%)} = \frac{\begin{pmatrix}\text{Speed change}\\ \text{above set speed}\end{pmatrix} - \begin{pmatrix}\text{Speed change}\\ \text{below set speed}\end{pmatrix}}{\text{Rated speed}} \times 100$$

Dead band, also called **wander**, is a characteristic of the speed-governing system. It is the insensitivity of the speed-governing system and the total speed change during which the governing valve(s) do not change position to compensate for the speed change.

Stability is a measure of the speed-governing system's ability to position the governor-controlled valve(s); thus, sustained oscillations of speed are not produced during a sustained load demand or following a change to a new load demand. Speed oscillations, also called hunt, are characteristics of the speed-governing system. A governing system's ability to minimize sustained oscillations is measured by its stability.

Maximum speed rise, expressed as a percentage of rated speed, is the maximum momentary increase in speed obtained when the turbine is developing rated power output at rated speed and the load is suddenly and completely reduced to zero. The maximum speed rise, expressed as a percentage of rated speed, is derived from the following equation:

$$\text{Speed rise (\%)} = \frac{\begin{pmatrix}\text{Maximum speed at}\\ \text{zero power output}\end{pmatrix} - \begin{pmatrix}\text{Rated}\\ \text{speed}\end{pmatrix}}{\text{Rated speed}} \times 100$$

Protective Devices. In addition to speed-governing controls, certain safety devices are required on steam turbines. These include an overspeed mechanism, which acts through a quick-tripping valve independent of the main governor valve to shut off the steam supply to the turbine, and a pressure relief valve in the turbine casing. Overspeed trip devices may act directly, through linkages to close the steam valve, or hydraulically, by relieving oil pressure to allow the valve to close. Also, the turbine must shut down if other safety devices, such as oil pressure failure controls or any of the driven system's protective controls, so dictate. These devices usually act through an electrical interconnection to close the turbine trip valve mechanically or hydraulically. To shorten the coast-down time of a tripped condensing turbine, a vacuum breaker in the turbine exhaust opens to admit air on receiving the trip signal.

Operation and Maintenance

Maintenance requirements for steam turbines vary greatly with complexity of design, throttle pressure rating, duty cycle, and steam quality (both physical and chemical). Typically, several common factors can be attributed to operational problems with steam turbines. These include erosion of high-pressure turbine nozzle and blades by solid particles transported from steam lines, superheaters, or reheaters, especially during cycling operation, that weakens rotating blades; fouling of high-pressure turbine with copper deposits (in case of mixed metallurgy of feedwater train); and stress corrosion cracking and corrosion fatigue of low-pressure turbine. The latter failures occur during both steady-state and cycling operation and are attributed to formation of early condensate in the so-called phase transition region where steam changes from superheated to saturated condition. The concentration of impurities in the early condensate is higher than in the steam that enters the turbine. Early condensate droplets may precipitate on blade surface and form liquid films, which can evaporate during flow and form the highly concentrated solutions on the surfaces inside turbines. The level of corrosive impurities in these films may be much higher than in the early condensate. In addition, steam moisture may result in water droplet erosion and, in combination with other parameters, flow-accelerated corrosion.

The best way to minimize both corrosion and erosion in steam turbines is to maintain proper feedwater boiler and steam chemistry. The requirements to cycle chemistry are more stringent with steam pressure. To protect steam turbines from unnecessary damage, the level of impurities in the steam from the boiler or HRSG must be kept within the limits specified by corresponding cycle chemistry guidelines.

Turbines subject to cyclical operation should be examined carefully every 18 to 36 months. Usually, nondestructive testing is used to establish material loss trends and predictable maintenance requirements for sustained planned outages.

Turbine seals, glands, and bearings are also common areas of deterioration and maintenance. Bearings require frequent examination, especially in cyclical duty systems. Oil samples from the lubrication system should be taken regularly to determine concentration of solid particle contamination and changes in viscous properties. Filters and oil should be recycled according to manufacturers' recommendations and the operational history of the turbine system.

Large multistage steam turbines usually contain instrumentation that monitors vibration within the casing. As deposits build on blades, blade material erodes or corrodes; as mechanical tolerance of bearing surfaces increases, nonuniform rotation increases turbine vibration. Vibration instrumentation, consequently, is used to determine maintenance intervals for turbines, especially those subject to extensive base-load operations where visual examination is not feasible.

ORGANIC RANKINE CYCLES

Organic Rankine cycle (ORC) technology is based on the Rankine process, except that instead of water/steam, an organic working fluid is used. ORCs can convert high- or low-quality thermal energy into useful mechanical or electrical energy and thus improve the efficiency of integrated energy systems by using waste heat from DG equipment. One problem with ORCs is that their inherent thermal conversion (Carnot) efficiency is very low for lower-quality thermal sources. However, this disadvantage is offset by the fact that the fuel is energy that would otherwise be wasted.

ORCs began to be studied and applied as energy conversion engines over 100 years ago to use low-grade heat from geothermal waters, solar energy from liquid and vapor collectors, industrial waste heat, biomass combustion, etc.

ORCs are used for low-temperature waste heat recovery, efficiency improvement in power stations, and recovery of geothermal and solar heat. Small-scale ORCs have been used commercially or as pilot plants. About 30 commercial ORC plants were built before 1984 with an output of 100 kW. ORCs allow efficient exploitation of low-temperature heat sources to produce electricity in a wide range of power outputs, from a few kilowatts to 3 MW. Several organic compounds have been used in ORCs (e.g., CFCs, halocarbons, isopentane, ammonia) to match the temperature of available waste heat sources. Efficiency is estimated to be between 10 and 30%, depending on operating temperatures.

EXPANSION ENGINES/TURBINES

Expansion turbines use compressed gases available from gas-producing processes to produce shaft power. One application expands natural gas from high-pressure pipelines in much the same way that a steam turbine replaces a pressure-reducing valve. These engines are used mainly, however, for cryogenic applications to about −320°F (e.g., oxygen for steel mills, low-temperature chemical processes, the space program). Relatively high-pressure air or gas expanded in an engine drives a piston and is cooled in the process. At the shaft, about 42 Btu/hp is removed. Available units, developing as much as 600 hp, handle flows ranging from 100 to 10,000 cfm. Throughput at a given pressure is controlled by varying the cutoff point, engine speed, or both. The conversion efficiency of heat energy to shaft work ranges from 65 to 85%. A 5-to-1 pressure ratio and an inlet pressure of 3000 psig are recommended. Outlet temperatures as low as −450°F have been handled satisfactorily.

This process consumes no fuel, but it does provide shaft energy, as well as expansion refrigeration, which fits the definition of CHP. Similarly, a back-pressure steam turbine, used instead of a pressure-reducing valve, generates productive shaft power and allows use of the residual thermal energy.

Natural gas from a high-pressure pipeline can be run through a turbine to produce shaft energy and then burned downstream. See the section on Combustion Turbines for more information.

STIRLING ENGINES

Stirling-cycle engines were patented in 1816 and were commonly used before World War I. They were popular because they had a better safety record than steam engines and used air as the working fluid. As steam engines improved and the competing compact Otto-cycle engine was invented, Stirling engines lost favor. Recent attention to distributed energy resources (DER), used by the space and marine industries, has revived interest in Stirling engines and increased research and development efforts.

Stirling engines, classified as external combustion engines, are sealed systems with an inert, reusable working fluid (either helium or hydrogen). Rather than burning fuel inside the cylinder, the engine uses external heat to expand the gas contained inside the cylinder and push against its pistons. The engine then recycles the same captive working fluid by cooling and compressing it, then reheating it to expand and drive the pistons, which in turn drive a generator. These engines are generally found in small sizes (1 to 55 kW). The maximum Carnot efficiency based on the second law of thermodynamics $[1 - (T_{cold}/T_{hot})]$ for a heat source of 1600°F (2060°R) and a cold sink of 77°F (537°R) is approximately 74%. Overall, electrical efficiency of these engines are usually in the range of 12 to 30% (difference is due to friction losses and heat transfer ineffectiveness); overall efficiency in a combined heat and power mode is around 80%.

Types

Stirling engines can be divided into two major types: kinematic and free-piston engines (NREL 2003).

Kinematic Stirling Engines. Pistons of these engines are connected by rods and crankshaft or with a wobble plate (also known as a swash plate). The wobble plate allows for variable output while maintaining thermal input, allowing faster response to changes in demand (Figure 43).

Kinematic Stirling engines can be designed as either single- or double-acting engines. Single-acting kinematic Stirling engines can have one or more cylinders with varying designs for mechanical linkage. Double-acting kinematic engines require several cylinders. The working fluid in a double-acting system operates on opposite sides of the piston.

Free-Piston Stirling Engines. These engines do not have any mechanical coupling between the piston and the power output. The pistons are mounted in flexures and oscillate freely (Figure 44). Flexures are springs that are flexible in the axial direction, but very stiff in the radial direction. A fluid or a linear alternator receives the reciprocating output via pneumatic transfer. Free-piston Stirling engines are single-acting, with the working fluid only operating on one side of a piston. Free-piston configurations are limited to sizes less than 12.5 kW.

Performance Characteristics

Most manufacturers promoting Stirling engines for distributed generation do not yet offer a standard product line. Although Stirling engines work on simple principles, designers are faced with challenges in determining design tradeoffs between complexity, cost, durability/reliability, and efficiency. The main difficulties with recent Stirling engines are their long-term durability/reliability and high cost. Durability concerns for Stirling engines include the following:

- Shaft seals separating the high-pressure hydrogen space from lubrication in the mechanical drive train
- Low-leakage piston rings and bearings for operation in the unlubricated working engine space
- Minimizing material stress and corrosion in the high-temperature, high-pressure heater head, which must operate at internal conditions of >2000 psi and 1300°F.
- Blockage of fine-meshed heat matrices used in regenerator assemblies by particles generated by rubbing piston rings

Although these problems have delayed adoption of Stirling engine technology, run times are beginning to approach acceptable lengths for some distributed power applications. The electrical efficiency of current engines is approximately 27% HHV (30% LHV).

Fuel Systems

Stirling engines absorb heat from a wide range of fuel sources (e.g., natural gas, propane, oil, biomass, waste fuels) and convert it to electricity with minimal emissions and low maintenance requirements. Typical heat sources include standard gaseous and liquid fuels, with options to accept low-heating-value landfill and digester gas, petroleum flare gas, and other low-grade gaseous, liquid, and solid waste fuels. Raw heat from solar concentrators or flue gas stacks can also be converted to electricity, with no fuel costs and no incremental emissions.

Stirling engines usually require a heat source ranging between 1500 and 1800°F.

Contaminants are a concern with some waste fuels, particularly acid gas components such as hydrogen sulfide (H_2S content should be less than 7%).

Power Systems

Fuel flexibility of these engines makes them candidates for use as heat recovery units in industrial processes. These engines generally

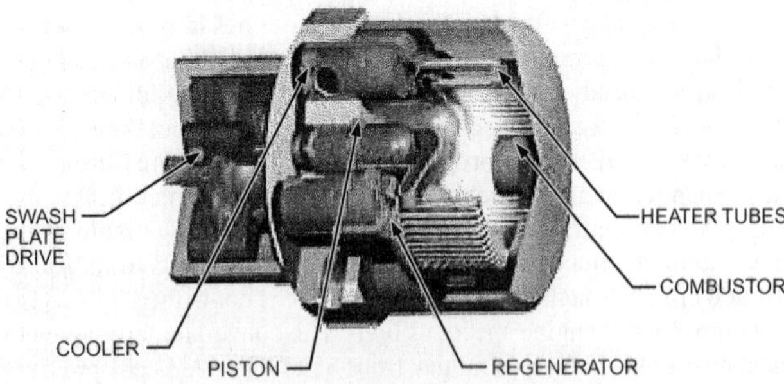

Fig. 43 Cutaway Core of a Kinematic Stirling Engine

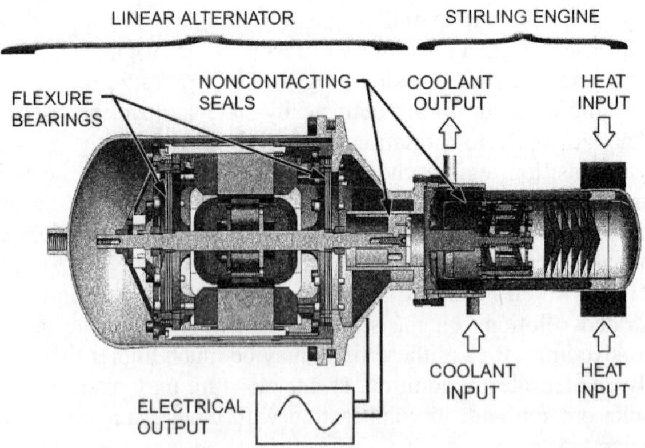

Fig. 44 Cutaway Core of a Free-Piston Stirling Engine
(Courtesy Infinia Corporation)

require a heat source between 1500 and 1800°F, noncorrosive, and relatively free of particulates. Their electric efficiencies are approximately 30% (based on LHV), leaving approximately 70% of the heat input to be rejected in the cooling system and exhaust gases. As with any combustion system, there performance drops as elevation increases, because of density changes in the fuel and combustion air.

Part-load operation is a design trade-off issue. Devices designed for distributed generation applications are usually capable of a 3:1 turn-down.

Exhaust Systems

The combustor of a Stirling engine can be customized to achieve low emissions. With a hot-end operating temperature of 1600°F, the combustor needs to operate only a few hundred degrees hotter, which puts it in the lower end of the thermal NO_x formation range. NO_x-reducing techniques are available, such as lean burn, staged combustion, or flue gas recirculation. NO_x emissions improve at part load because of lower combustor temperatures. When firing with natural gas, SO_x should not be an issue.

Coolant Systems

Stirling engines usually convert approximately 30% of heat input to electric power, with almost 50% going to the coolant and the rest dissipated as exhaust or other losses. Because Stirling engines are liquid-cooled, it is relatively easy to capture the low-temperature waste heat for CHP applications (thermal output of approximately 140°F) such as water heating or other low-temperature heating applications. The addition of a heat exchanger (with related controls) is the only difference between a power-only unit and CHP unit. Power-only units are usually equipped with a radiator and fan to dissipate coolant heat into the environment.

Operation and Maintenance

Maintenance procedures and associated costs have not yet been developed because of the lack of long-term operating experience. The engine is usually designed for a 40,000 h life. The engine would be replaced at the end of this period. Periodic maintenance after 8000 h usually includes oil and oil filter and air filter changes. Another important maintenance item is working fluid makeup: the engine operates at a relatively high working pressure, and the small molecular size of the working fluid means it can leak through the engine seals. Occasional working fluid injection may be required. Economic life of these systems is estimated to be around 10 years.

THERMAL-TO-THERMAL COMPONENTS

THERMAL OUTPUT CHARACTERISTICS

By definition, CHP systems use the fuel energy that the prime mover does not convert into shaft energy. If site heat energy requirements can be met effectively by the thermal by-product at the level it is available from the prime mover, this salvaged heat reduces the normal fuel requirements of the site and increase overall plant efficiency. The ability to use prime mover waste heat determines overall system efficiency and is one of the critical factors in economic feasibility.

This section describes devices that convert prime mover "waste" thermal energy into energy streams suited to meet typical thermal loads. The section begins with a description of the thermal output characteristics and heat recovery opportunities of key prime movers.

Reciprocating Engines

In all reciprocating internal combustion engines except small air-cooled units, heat can be reclaimed from the jacket cooling system, lubricating system, turbochargers, exhaust, and aftercoolers. These engines require extensive cooling to remove excess heat not conducted into the power train during combustion and the heat resulting from friction. Coolant fluids and lubricating oil are circulated to remove this engine heat.

Waste heat in the form of hot water or low-pressure steam is recovered from the engine jacket manifolds and exhaust, and additional heat can be recovered from the lubrication system (see Figures 47 to 51).

Provisions similar to those used with gas turbines are necessary if supplemental heat is required, except an engine exhaust is rarely fired with a booster because it contains insufficient oxygen. If electrical supplemental heat is used, the additional electrical load is reflected back to the prime mover, which reacts accordingly by producing additional waste heat. This action creates a feedback effect, which can stabilize system operation under certain conditions. The approximate distribution of input fuel energy under selective control of the thermal demand for an engine operating at rated load is as follows:

Shaft power	33%
Convection and radiation	7%
Rejected in jacket water	30%
Rejected in exhaust	30%

These amounts vary with engine load and design. Four-cycle engine heat balances for naturally aspirated (Figure 45) and turbocharged gas engines (Figure 46) show typical heat distribution. The exhaust gas temperature for these engines is about 1200°F at full load and 1000°F at 60% load.

Two-cycle lower-speed (900 rpm and below) engines operate at lower exhaust gas temperatures, particularly at light loads, because the scavenger air volumes remain high through the entire range of capacity. High-volume, lower-temperature exhaust gas offers less efficient exhaust gas heat recovery possibilities. The exhaust gas temperature is approximately 700°F at full load and drops below 500°F at low loads.

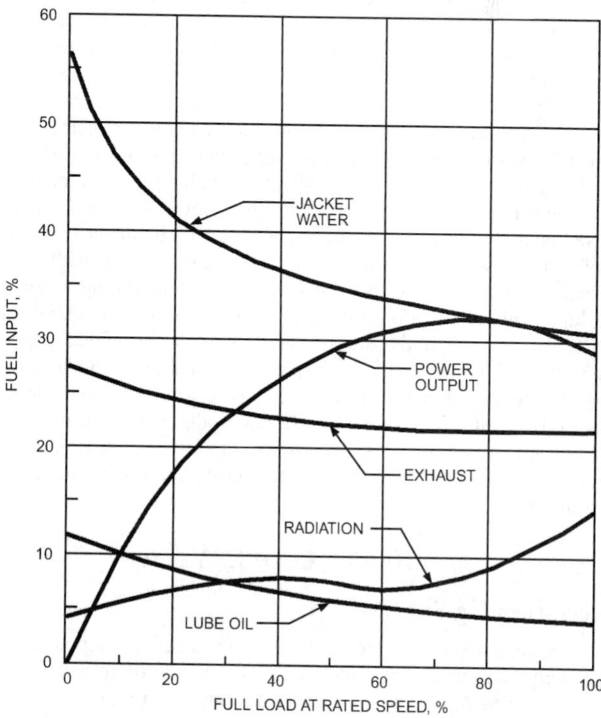

Fig. 45 Heat Balance for Naturally Aspirated Engine

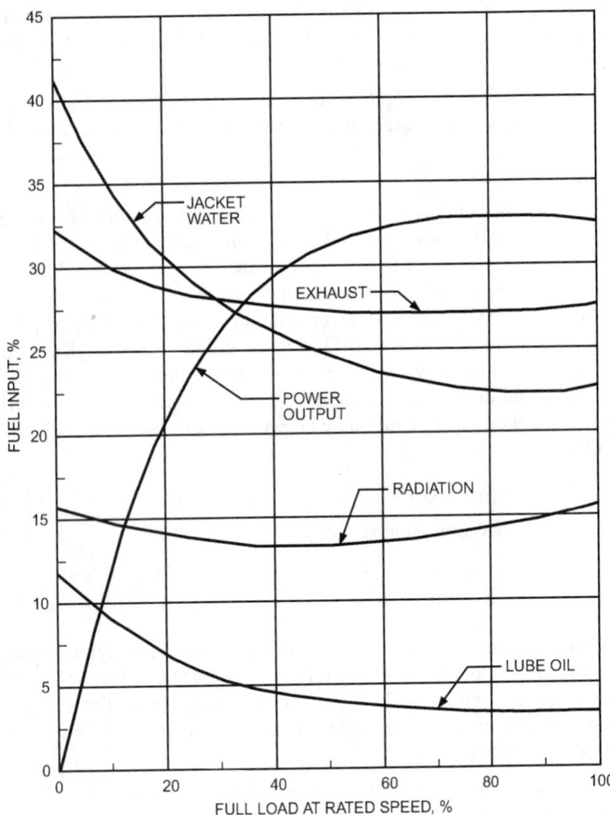

Fig. 46 Heat Balance for Turbocharged Engine

Combustion Turbines

In the gas turbine cycle, the average fuel-to-electrical-shaft efficiency ranges from approximately 12% to above 35%, with the rest of the fuel energy discharged in the exhaust and through radiation or internal coolants in large turbines. A minimum stack exhaust temperature of about 300°F is required to prevent condensation. Because the heat rate efficiency is lower, the quantity of heat recoverable per unit of power is greater for a gas turbine than for a reciprocating engine. This heat is generally available at the higher temperature of the exhaust gas. The net result is an overall thermal efficiency of 50% and higher. Because gas turbine exhaust contains a large percentage of excess air (high O_2 content), afterburners or boost burners may be installed in the exhaust to create a supplementary boiler system or to cofire an absorption chiller. This system can provide additional steam, or level the steam production or cooling during reduced turbine loads. Absorption chillers able to operate directly off turbine exhausts (exhaust-fired absorption chillers) with coefficients of performance (COPs) of 1.30 or more are also available. The COP is the cooling energy output divided by energy input. The conventional method of controlling steam or hot-water production in a heat recovery system at part load is to bypass some exhaust gases around the boiler tubes and out the exhaust stack through a gas bypass valve assembly.

HEAT RECOVERY

Reciprocating Engines

Engine Jacket Heat Recovery. Engine jacket cooling passages for reciprocating engines, including the water-cooling circuits in the block, heads, and exhaust manifolds, must remove about 30% of the heat input to the engine. If the machine operates above 180°F coolant temperature, condensation of combustion products should produce no ill effects. Some engines have modified gaskets and seals to enable satisfactory operation up to 250°F at 30 psia. Keeping temperature rise through the jacket to less than 15°F helps avoid thermal stress. Keep flow rates within engine manufacturers' design limits to avoid erosion from excessive flow or inadequate distribution in the engine from low flow rates.

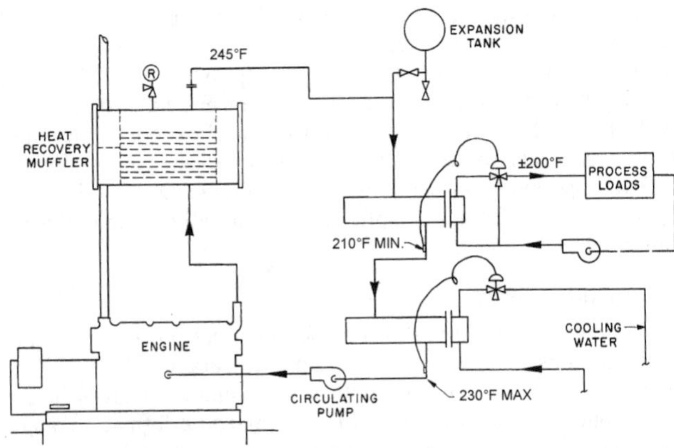

Fig. 47 Hot-Water Heat Recovery

Engine-mounted water-circulating pumps driven from an auxiliary shaft can be modified with proper shaft seals and bearings to give good service life at elevated temperatures. Configurations that have a circulating pump for each engine increase reliability because the remaining engines can operate if one engine pump assembly fails. An alternative design uses an electric-drive pump battery to circulate water to several engines and has a standby pump assembly in reserve, interconnected so that any engine or pump can be cut off without disabling the jacket water system.

Forced-circulation hot water at 250°F and 30 psia, can be used for many loads, including water-heating systems for comfort and process loads, absorption refrigeration chillers, desiccant dehumidifiers, and service water heating. Engine jacket coolant distribution must be limited to reduce the risk of leaks, contamination, or other failures in downstream equipment that could prevent engine cooling.

One solution is to confine each engine circuit to its individual engine, using a heat exchanger to transfer salvaged heat to another circuit that serves several engines through an extensive distribution system. An additional heat exchanger is needed in each engine circuit to remove heat whenever the waste heat recovery circuit does not extract all the heat produced (Figure 47). This approach is highly reliable, but because the salvage circuit must be at a lower temperature than the engine operating level, it requires either larger heat exchangers, piping, and pumps or a sacrifice of system efficiency that might result from these lower temperatures.

Low-temperature limit controls prevent excessive system heat loads from overloading the system and seriously reducing the engines' operating temperature, which could cause the engine castings to crack. A heat storage tank is an excellent buffer, because it can provide a high rate of heat for short periods to protect machinery serving the heat loads. The heat level can be controlled with supplementary input, such as an auxiliary boiler.

A limited distribution system that distributes steam through nearby heat exchangers can salvage heat without contaminating the main engine cooling system (Figures 48 and 49). The salvage heat temperature is kept high enough for most low-pressure steam loads. Returning condensate must be treated to prevent engine oxidation. The flash tank can accumulate the sediment from treatment chemicals.

Lubricant Heat Recovery. Lubricant heat exchangers should keep oil temperatures at 190°F, with the highest coolant temperature consistent with economical use of the salvaged heat. Engine manufacturers usually size oil cooler heat transfer surfaces based on 130°F

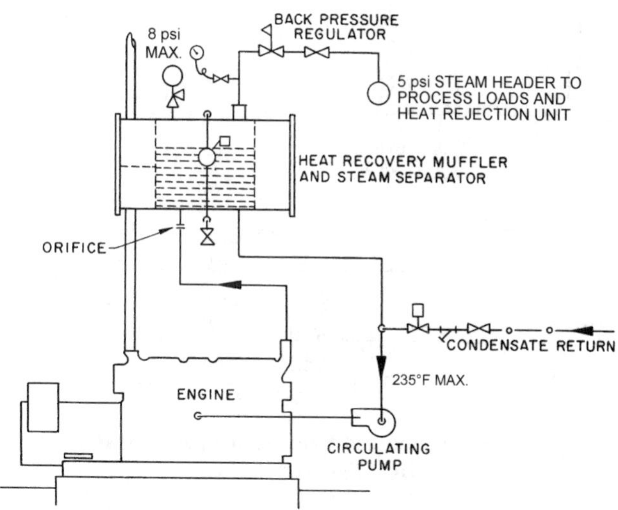

Fig. 48 Hot-Water Engine Cooling with Steam Heat Recovery (Forced Recirculation)

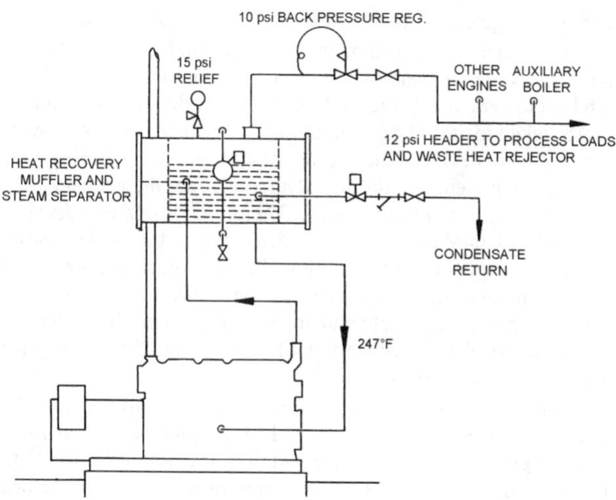

Fig. 49 Engine Cooling with Gravity Circulation and Steam Heat Recovery

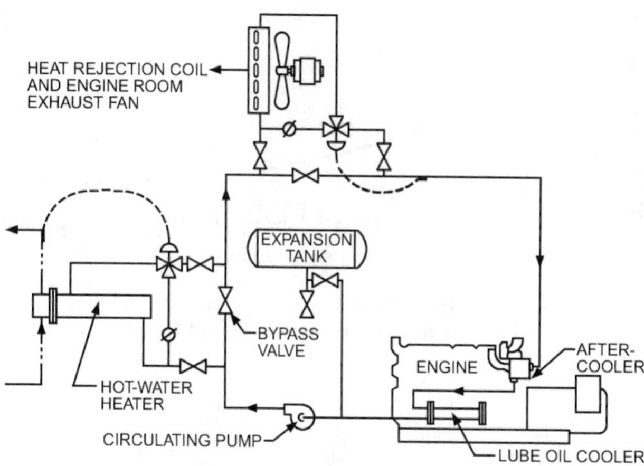

Fig. 50 Lubricant and Aftercooler System

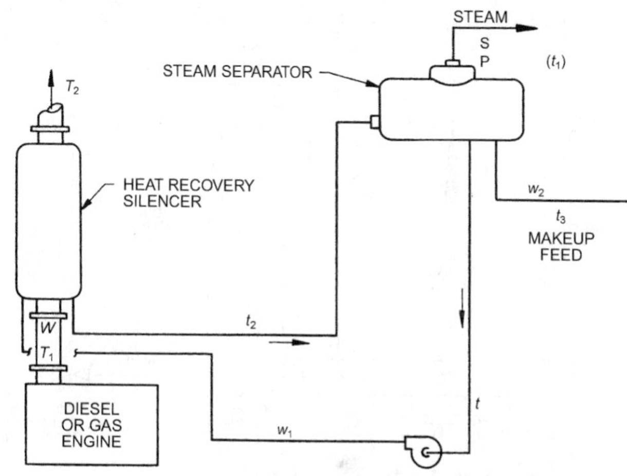

Fig. 51 Exhaust Heat Recovery with Steam Separator

entering coolant water, without provision for additional lubricant heat gains from high engine operating temperatures. Compare the cost of a reliable supply of lower-temperature cooling water with that of increasing the oil-cooling heat exchanger size and using cooling water at 165°F. If engine jacket coolant temperatures are above 220°F and heat at 155 to 165°F can be used, heat from the lubricant can be recovered profitably.

The coolant should not foul the oil cooling heat exchanger. Untreated water should not be used unless it is free of silt, calcium carbonates, sand, and other contaminants. A good solution is a closed-circuit, treated water system using an air-cooled heat transfer coil with freeze-protected coolant. A domestic water heater can be installed on the closed circuit to act as a reserve heat exchanger and to salvage some useful heat when needed. Inlet air temperatures are not as critical with diesel engines as with turbocharged natural gas engines, and aftercooler water on diesel engines can be run in series with the oil cooler (Figure 50). A double-tube heat exchanger can also be used to prevent contamination from a leaking tube.

Turbocharger Heat Recovery. Natural gas engine turbochargers need medium fuel gas pressure (12 to 20 psi) and low aftercooler water temperatures (90°F or less) for high compression ratios and best fuel economy. Aftercooler water at 90°F is a premium coolant in many applications, because usual sources are raw domestic water and evaporative cooling systems such as cooling towers. Aftercooler water as warm as 135°F can be used, although this derates the engine. Using domestic water may be expensive because (1) coolant is continuously needed while the engine is running, and (2) available heat exchanger designs require a lot of water even though the load is less than 200 Btu/hp·h. A cooling tower can be used, but it can increase initial costs; it also requires freeze protection and water quality control. If a cooling tower is used, the lubricant cooling load must be included in the tower design load for periods when there is no use for salvage lubricant heat.

Exhaust Gas Heat Recovery. Almost all heat transferred to the engine jacket cooling system can be reclaimed in a standard jacket cooling process or in combination with exhaust gas heat recovery. However, only part of the exhaust heat can be salvaged because of the limitations of heat transfer equipment and the need to prevent flue gas condensation (Figure 51).

Energy balances are often based on standard air at 60°F; however, exhaust temperature cannot easily be reduced to this level.

A minimum exhaust temperature of 250°F was established by the Diesel Engine Manufacturers Association (DEMA). Many heat recovery boiler designs are based on a minimum exhaust temperature of 300°F to avoid condensation and acid formation in the exhaust piping. Final exhaust temperature at part load is important for generator sets that frequently operate at part load. Depending on the initial exhaust temperature, about 50 to 60% of the available exhaust heat can be recovered.

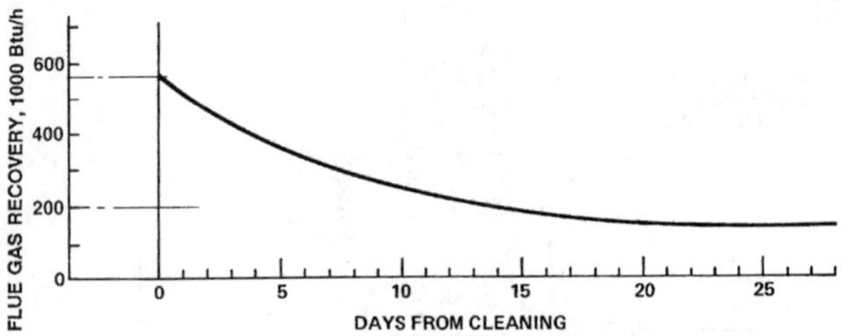

Fig. 52 Effect of Soot on Energy Recovery from Flue Gas Recovery Unit on Diesel Engine

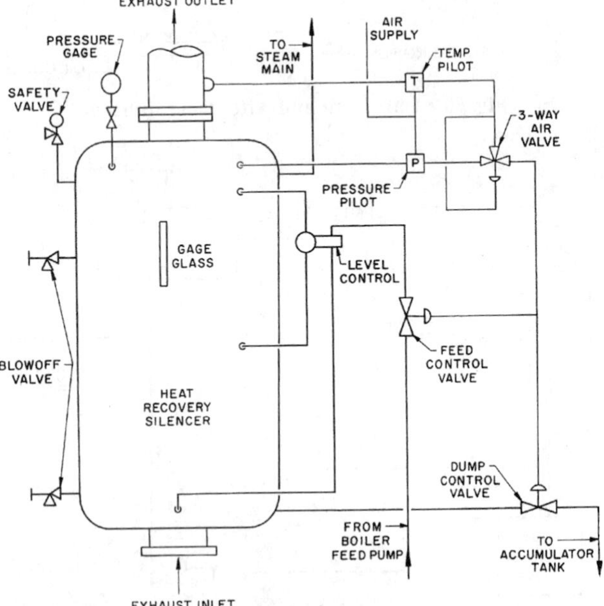

Fig. 53 Automatic Boiler System with Overriding Exhaust Temperature Control

A complete heat recovery system, which includes jacket water, lubricant, turbocharger, and exhaust, can increase the overall thermal efficiency from 30% for the engine generator alone to approximately 75%. Exhaust heat recovery equipment is available in the same categories as standard fire-tube and water-tube boilers.

Other heat recovery silencers and boilers include coil-type water heaters with integral silencers, water-tube boilers with steam separators for gas turbine, and engine exhausts and steam separators for high-temperature cooling of engine jackets. Recovery boiler design should facilitate inspection and cleaning of the exhaust gas and water sides of the heat transfer surface. Diesel engine units should have a means of soot removal, because soot deposits can quickly reduce heat exchanger effectiveness (Figure 52). These recovery boilers can also serve other requirements of the heat recovery system, such as surge tanks, steam separators, and fluid level regulators.

In many applications for heat recovery equipment, the demand for heat requires some method of **automatic control**. In vertical recovery boilers, control can be achieved by varying the water level in the boiler. Figure 53 shows a control system using an air-operated pressure controller with diaphragm or bellows control valves. When steam production begins to exceed demand, the feed control valve begins to close, throttling the feedwater supply. Concurrently, the dump valve begins to open, and the valves reach an equilibrium position that maintains a level in the boiler to match the steam demand.

This system can be fitted with an overriding exhaust temperature controller that regulates boiler output to maintain a preset minimum

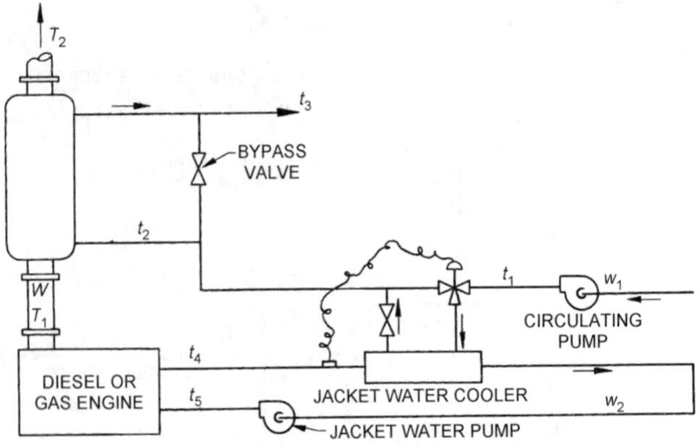

Fig. 54 Combined Exhaust and Jacket Water Heat Recovery System

exhaust temperature at the outlet. This type of automatic control is limited to vertical boilers because the ASME (1998) *Boiler and Pressure Vessel Code* does not allow horizontal boilers to be controlled by varying the water level. Instead, a control condenser, radiator, or thermal storage can be used to absorb excess steam production.

In hot-water units, a temperature-controlled bypass valve can divert the water or exhaust gas to achieve automatic modulation with heat load demand (Figure 54). If water is diverted, precautions must be taken to limit the temperature rise of the lower flow of water in the recovery device, which could otherwise cause steaming. Heat recovery equipment should not adversely affect the primary function of the engine to produce work. Therefore, design of waste heat recovery boilers should begin by determining the back pressure imposed on engine exhaust gas flow. Limiting back pressures vary widely with the make of engine, but the typical value is 6 in. of water gage. The next step is to calculate the heat transfer area that gives the most economical heat recovery without reducing the final exhaust temperature below 300°F.

Heat recovery silencers are designed to adapt to all engines; efficient heat recovery depends on the **initial exhaust temperature**. Most designs can be modified by adding or deleting heat transfer surface to suit the initial exhaust conditions and to maximize heat recovery down to a minimum temperature of 300°F.

Figure 55 illustrates the effect of lowering exhaust temperature below 300°F. This curve is based on a specific heat recovery silencer design with an initial exhaust temperature of 1000°F. Lowering the final temperature from 300°F to 200°F increases heat recovery 14% but requires a 28% surface increase. Similarly, a reduction from 300°F to 100°F increases heat recovery 29% but requires a 120% surface increase. Therefore, the cost of heat transfer surface must be considered when determining the final temperature.

Another factor to consider is water vapor condensation and acid formation if exhaust gas temperature falls below the dew point. This point varies with fuel and atmospheric conditions and is usually in the range of 125 to 150°F. This gives an adequate margin of safety for the 250°F minimum temperature recommended by DEMA. Also, it allows for other conditions that could cause condensation, such as an uninsulated boiler shell or other cold surface in the exhaust system, or part loads on an engine.

The quantity of water vapor varies with the type of fuel and the intake air humidity. Methane fuel, under ideal conditions and with only the correct amount of air for complete combustion, produces 2.25 lb of water vapor for every pound of methane burned. Similarly, diesel fuel produces 1.38 lb of water vapor per pound of fuel. In the gas turbine cycle, these relationships would not hold true because of the large quantities of excess air. Condensates formed at low exhaust temperatures can be highly acidic. Sulfuric acid from diesel fuels and carbonic acid from natural gas fuels can cause

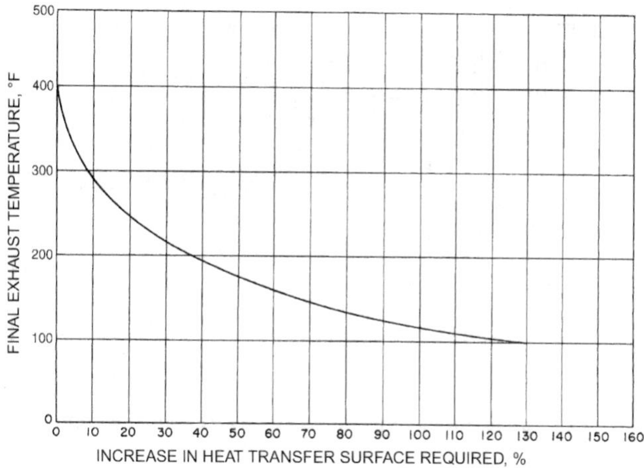

Fig. 55 Effect of Lowering Exhaust Temperature below 300°F

severe corrosion in the exhaust stack as well as in colder sections of the recovery device.

If engine exhaust flow and temperature data are available, and maximum recovery to 300°F final exhaust temperature is desired, the basic **exhaust recovery equation** is

$$q = \dot{m}_e (c_p)_e (t_e - t_f) \tag{4}$$

where

q = heat recovered, Btu/h
$\dot{m}_e$ = mass flow rate of exhaust, lb/h
t_e = exhaust temperature, °F
t_f = final exhaust temperature, °F
$(c_p)_e$ = specific heat of exhaust gas = 0.25 Btu/lb·°F

Exhaust recovery equation applies to both steam and hot water units. To estimate the quantity of steam obtainable, the total heat recovered q is divided by the latent heat of steam at the desired pressure. The latent heat value should include an allowance for the temperature of the feedwater return to the boiler. The basic equation is

$$q = \dot{m}_s (h_s - h_f) \tag{5}$$

where

$\dot{m}_s$ = mass flow rate of steam, lb/h
h_s = enthalpy of steam, Btu/lb
h_f = enthalpy of feedwater, Btu/lb

Similarly, the quantity of hot water can be determined by

$$q = \dot{m}_w (c_p)_w (t_o - t_i) \tag{6}$$

where

$\dot{m}_w$ = mass flow rate of water, lb/h
$(c_p)_w$ = specific heat of water = 1.0 Btu/lb·°F
t_o = temperature of water out, °F
t_i = temperature of input water, °F

If shaft power is known but engine data are not, heat available from the exhaust is about 1000 Btu/h per horsepower output or 1 lb/h steam per horsepower output. The exhaust recovery equations also apply to gas turbines, although the flow rate is much greater. The estimate for gas turbine boilers is 8 to 10 lb/h of steam per horsepower output. These values are reasonably accurate for steam pressures in the range of 15 to 150 psig.

Normal procedure is to design and fabricate heat recovery boilers to the ASME (1998) *Boiler and Pressure Vessel Code* (Section VIII) for the working pressure required. Because temperatures in most exhaust systems are not excessive, it is common to use flange or firebox-quality steels for pressure parts and low-carbon steels for the nonpressure components. Wrought iron or copper can be used for extended-fin surfaces to improve heat transfer capacities.

In special applications such as sewage gas engines, where exhaust products are highly corrosive, wrought iron or special steels are used to improve corrosion resistance. Exhaust heat may be used to make steam, or it may be used directly for drying or other processes. The steam provides space heating, hot water, and absorption refrigeration, which may supply air conditioning and process refrigeration. Heat recovery systems generally involve equipment specifically tailored for the job, although conventional fire-tube boilers are sometimes used. Exhaust heat may be recovered from reciprocating engines by a muffler-type exhaust heat recovery unit. Table 17 gives the temperatures normally required for various heat recovery applications.

In some engines, exhaust heat rejection exceeds jacket water rejection. Generally, gas engine exhaust temperatures run from 700 to 1200°F, as shown in Table 18. About 50 to 75% of the sensible heat in the exhaust may be considered recoverable. The economics of exhaust heat boiler design often limits the temperature differential between exhaust gas and generated steam to a minimum of 100°F. Therefore, in low-pressure steam boilers, gas temperature can be reduced to 300 to 350°F; the corresponding final exhaust temperature range in high-pressure steam boilers is 400 to 500°F.

Because they require higher airflows to purge their cylinders, two-cycle engines have lower temperatures than four-cycle engines, and thus are less desirable for heat recovery. Gas turbines have even larger flow rates, but at high enough temperatures to make heat recovery worthwhile when the recovered heat can be efficiently used.

Table 17 Temperatures Normally Required for Various Heating Applications

Application	Temperature, °F
Absorption refrigeration machines	190 to 245
Space heating	120 to 250
Water heating (domestic)	120 to 200
Process heating	150 to 250
Evaporation (water)	190 to 250
Residual fuel heating	212 to 330
Auxiliary power producers, with steam turbines or binary expanders	190 to 350

Table 18 Full-Load Exhaust Mass Flows and Temperatures for Various Engines

Type of Four-Cycle Engine	Mass Flow, lb/bhp·h	Temperature, °F
Naturally aspirated gas	9	1200
Turbocharged gas	10	1200
Naturally aspirated diesel	12	750
Turbocharged diesel	13	850
Gas turbine, nonregenerative	18 to 48*	800 to 1050*

*Lower mass flows correspond to more efficient gas turbines.

Combustion Turbines

The information in the section on Exhaust Gas Heat Recovery applies to combustion turbines as well.

Steam Turbines

Noncondensing Turbines. The back-pressure or noncondensing turbine is the simplest turbine. It consists of a turbine in which steam is exhausted at atmospheric pressure or higher. They are generally used when there is a process need for high-pressure steam, and all steam condensation takes place downstream of the turbine cycle and in the process. Figure 56 illustrates the steam path for a noncondensing turbine. The back-pressure steam turbine operates on the enthalpy difference between steam inlet and exhaust conditions.

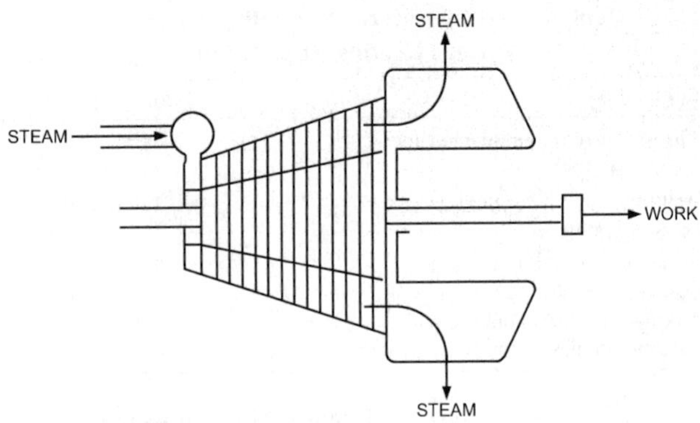

Fig. 56 Back-Pressure Turbine

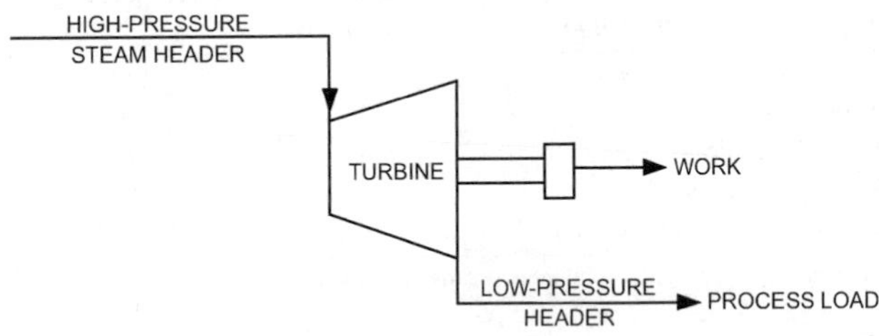

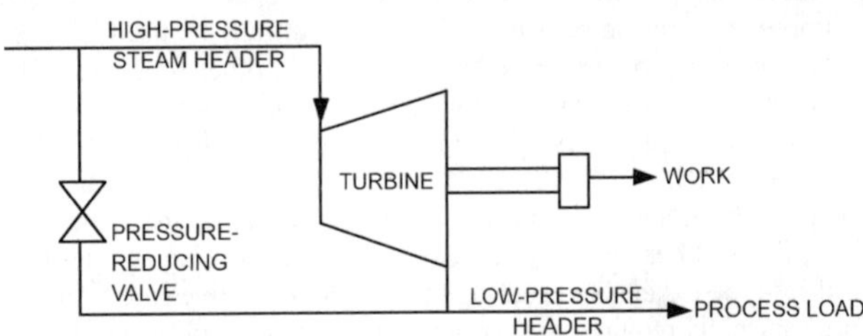

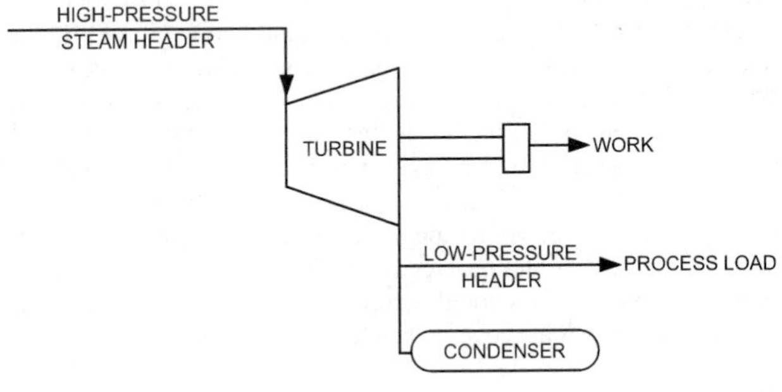

Fig. 57 Integration of Back-Pressure Turbine with Facility

The noncondensing turbine's Carnot cycle efficiency tends to be lower than is possible with other turbines because the difference between turbine inlet and exhaust temperatures tends to be lower. Because much of the steam's heat, including the latent heat of vaporization, is exhausted and then used in a process, the back-pressure CHP system process or total energy efficiency can be very high. One application for back-pressure turbines is as a substitute for pressure-reducing valves; they provide the same function (pressure regulation), but also produce a useful product (power).

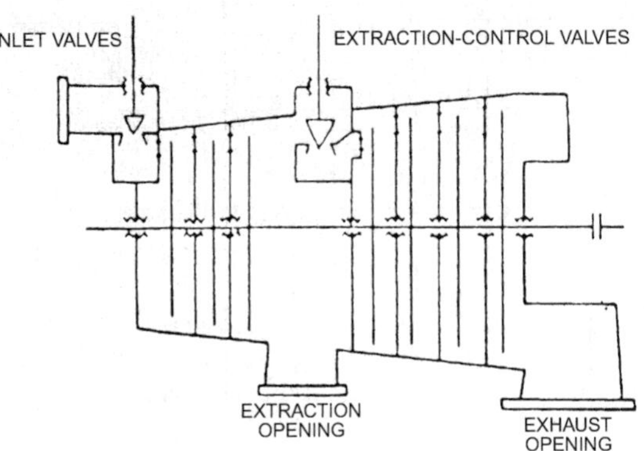

Fig. 58 Condensing Automatic Extraction Turbine

The back-pressure turbine has one major disadvantage in CHP: because the process load is the heat sink for the steam, the amount of steam passed through the turbine depends on the heat load at the site. Thus, the back-pressure turbine provides little flexibility in directly matching electrical output to electrical requirements; electrical output is controlled by the thermal load. Direct linkage of site steam requirements and electrical output can result in electric utility charges for standby service or increased supplemental service unless some measures are taken to increase system flexibility.

Figure 57 illustrates several ways to achieve flexible performance when electrical and thermal loads do not match the back-pressure turbine's capability. Figure 57A shows the basic arrangement of a noncondensing steam turbine and its relationship to the facility. Figure 57B illustrates the addition of a pressure-reducing valve (PRV) to bypass some or all of the steam around the turbine. Thus, if the process steam demand exceeds the turbine's capability, the additional steam can be provided through the PRV. Figure 57C illustrates use of a load condenser to allow electricity generation, even when there is no process steam demand. These techniques to match thermal and electrical loads are very inefficient, and operating time at these off-design conditions must be minimized by careful analysis of the coincident, time-varying process steam and electrical demands.

Process heat recovered from the noncondensing turbine can be easily estimated using steam tables combined with knowledge of the steam flow, steam inlet conditions, steam exit pressure, and turbine isentropic efficiency.

Extraction Turbines. Figure 58 illustrates the internal arrangement of an automatic extraction turbine that exhausts steam at one or more stations along the steam flow path. Conceptually, the extraction turbine is a hybrid of condensing and noncondensing turbines. Its advantage is that it allows extraction of the quantity of steam required at each temperature or pressure needed by the industrial process. Multiple extraction ports allow great flexibility in matching the CHP cycle to thermal requirements at the site. Extracted steam can also be used in the power cycle for feedwater heating or powerhouse pumps. Depending on cycle constraints and process requirements, the extraction turbine's final exit conditions can be either back-pressure or condensing.

A diaphragm in the automatic extraction turbine separates the high- and low-pressure sections. All the steam passes through the high-pressure section just as it does in a single back-pressure turbine. A throttle, controlled by the pressure in the process steam line, controls steam flow into the low-pressure section. If pressure drops, the throttle closes, allowing more steam to the process. If pressure rises, the throttle opens to allow steam to flow through the low-pressure section, where additional power is generated.

An automatic extraction turbine (Figure 59) is uniquely designed to meet the specific power and heat capability of a given site; therefore, no simple relationship generally applies. For preliminary design analyses, the procedures presented by Newman

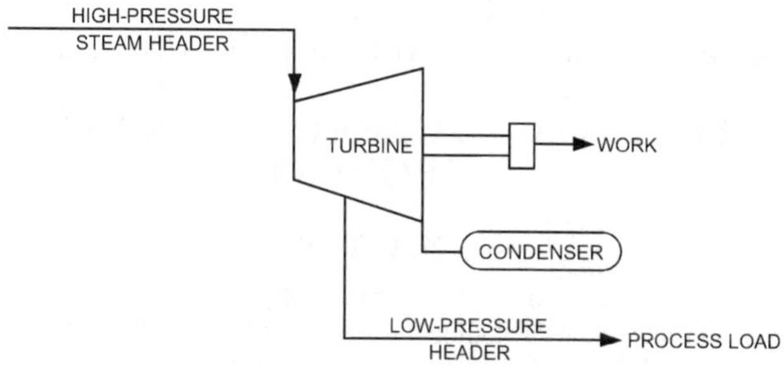

Fig. 59 Automatic Extraction Turbine CHP System

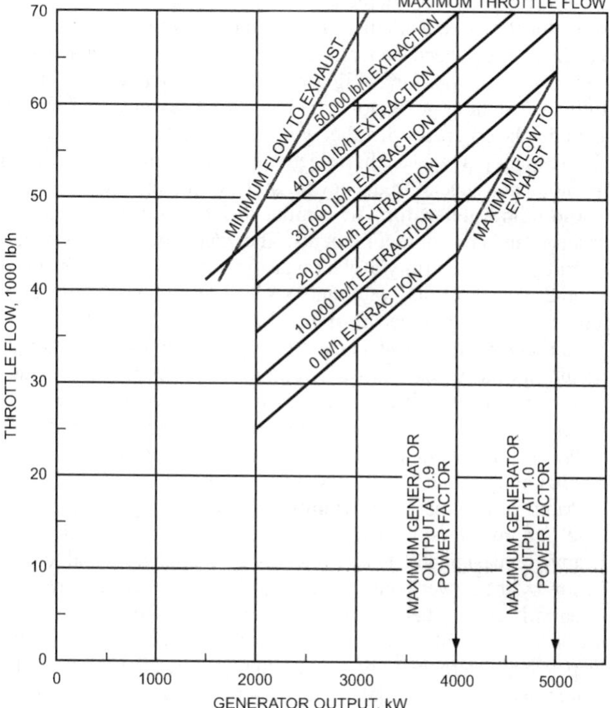

Fig. 60 Performance Map of Automatic Extraction Turbine

(1945) can be used to estimate performance. The product of such an analysis is a performance map similar to that shown in Figure 60 for a 5000 kW generator. The performance map provides steam flow to the turbine as a function of generator output with extraction flow as a parameter.

THERMALLY ACTIVATED TECHNOLOGIES

Heat-Activated Chillers

Waste heat may be converted and used to produce chilled water by several methods. The conventional method is to use hot water (>200°F) or low-pressure steam (<15 psig) in single-stage absorption chillers. These single-stage absorbers have a COP of 0.7 or less; 18,000 Btu/h of recovered heat can produce about 1 ton of cooling (12,000 Btu/h). Note that the following equations are based on full-load operation. Part-load output can be calculated by substituting the part-load COPs provided by the manufacturer for part-load COPs given here.

Hot water at 190 to 220°F, recovered from the cooling jacket loop of the prime mover, is used to produce chilled water in an indirect-fired, single-stage absorption machine. The equation to estimate the cooling produced from the heat recovered from the water is

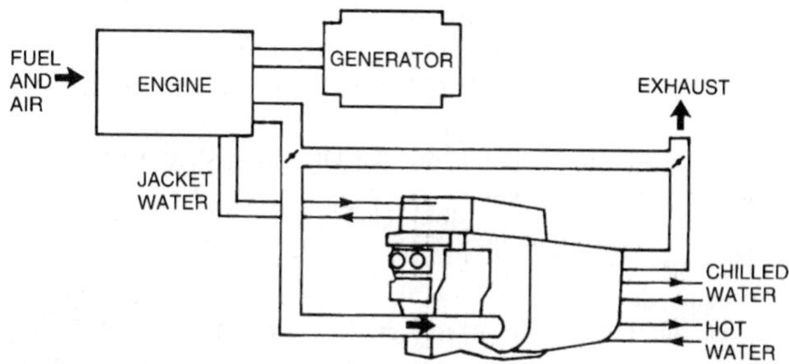

Fig. 61 Hybrid Heat Recovery Absorption Chiller-Heater

$$q = \frac{\text{COP} \times \dot{m}_w (c_p)_w (t_1 - t_2)}{12{,}000}$$

where

$\dot{m}_w$ = mass flow of water, lb/h
$(c_p)_w$ = specific heat of water, 1.0 Btu/lb·°F
t_1 = water temperature out of engine, °F
t_2 = water temperature returned to engine, °F [typically, $(t_1 - t_2) = 15$°F]
COP = coefficient of performance (typically 0.7)
12,000 = Btu/h·ton

For engines under 2000 kW, jacket water heat and exhaust energy are usually combined to maximize cooling output from the system. In this case, jacket water passes through the engine and picks up available waste heat, and is then directed to an exhaust-to-water heat exchanger where its temperature is elevated by the exhaust energy. Flow rate is determined by the engine manufacturer and should be set to maximize the temperature differential as much as possible; the temperature differential across the engine loop is typically 15°F, and the exhaust energy can add a further 7 to 10°F, for a total differential of 25°F.

Heat may be recovered from engines and gas turbines as high-pressure steam, depending on exhaust temperature. Steam pressures from 15 to 200 psig are common. When steam is produced at pressures over 43 psig, two-stage steam absorption chillers or steam-turbine-driven chillers can also be considered. The full-load COP at standard conditions of a steam-fired two-stage absorption chiller is typically 1.20. The steam input required is 10 lb/ton·h. This compares to 18 lb/ton·h for the single-stage absorption machine using 15 psig steam.

Steam turbine drive chillers can accept steam pressures exceeding 200 psi. The turbine drive centrifugal chiller typically incorporates a variable expansion valve which allows full-load operation at low cooling tower temperatures down to 50°F.

Providing cooling with engine exhaust heat is described in the section on Exhaust Gas Heat Recovery, and some absorption machines have been designed specifically for heat recovery in CHP applications. These units use both the jacket water and exhaust gas directly (see Figure 61).

Another type of absorption chiller uses gas engine or turbine exhaust directly (exhaust-fired absorption chiller). If a direct-exhaust, two-stage absorption chiller is used, the equation to estimate cooling produced from recoverable heat is

$$q = \frac{\dot{m}_e c_p (t_1 - t_2)(\text{COP} \times 0.97)}{12{,}000}$$

where

q = cooling produced, tons
$\dot{m}_e$ = mass flow of exhaust gas, lb/h
c_p = specific heat of gas = 0.268 Btu/lb·°F
t_1 = exhaust temperature in, °F

t_2 = exhaust temperature out = 375°F
COP = coefficient of performance (typically 1.3)
0.97 = connecting duct system efficiency
12,000 = Btu/h·ton

If available, the manufacturer's COP rating should be used to replace the assumed value.

Desiccant Dehumidification

Desiccant dehumidification is another essential thermally activated technology to convert low-grade waste heat into useful dehumidification that helps maximize energy savings and economic return. Both solid and liquid desiccant units have been successfully applied for many years for humidity control. Desiccants are particularly well suited to CHP because their operation requires mainly thermal energy, which can be provided from most prime movers. The energy required to remove humidity can be relatively low compared to the heat available from a prime mover, so desiccants can work in conjunction with other thermally activated technologies. Their temperature requirements are also relatively low (160 to 250°F), so they can be configured for a bottoming cycle. For more information, see Chapter 22.

Hot Water and Steam Heat Recovery

CHP heat recovery equipment encompasses all forms of heat exchangers that capture or recover waste heat or exhaust gas of a prime mover, such as a combustion turbines or natural gas or diesel engines, to generate steam or hot water. Heat recovery steam generators (HRSGs) are used to recover energy from the hot exhaust gases in power generation. Water is pumped and circulated through tubes heated by the exhaust gases and can be held under high pressure and temperature, and boiled to produce steam. HRSGs are found in may combined-cycle power plants. They were originally designed to produce steam at one pressure level; modern HRSGs may have up to three pressure levels with superheat and reheat, and may be once-through or recirculating.

Thermal Energy Storage Technologies

Thermal energy storage (TES) can decouple power generation from the production of process heat, allowing production of dispatchable power while fully using the thermal energy available from the prime mover. Thermal energy from the prime mover exhaust can be stored as sensible or latent heat and used during peak demand periods to produce electric power or process steam/hot water. However, the additional material and equipment necessary for a TES system add to the capital costs (though there can be added value from the resulting increase in peaking capacity). As a result, evaluation of economic benefits of adding TES to a conventional CHP system must consider the increased cost of the combined system and the value of peaking capacity.

Selection of a specific storage system depends on the quality and quantity of recoverable thermal energy and on the nature of the thermal load to be supplied from the storage system. Chapter 34 of the 2007 *ASHRAE Handbook—HVAC Applications* has more information on thermal storage.

TES systems and technologies for power generation applications can be categorized by storage temperature. High-temperature storage can be used to store thermal energy from sources (e.g., gas turbine exhaust) at high temperatures (**heat storage**). Storage options such as oil/rock, molten nitrate salt, and combined molten salt and oil/rock are well developed and commercially available (Somasundaram et al. 1996). Medium-temperature storage with hot water can also be used.

Low-temperature TES technologies store thermal energy below ambient temperature (**cool storage**) and can be used for cooling air entering gas turbines, or for storing cooling captured from waste heat by absorption and/or steam-turbine-driven chillers. Examples include commercially available options such as diurnal ice, chilled-water, and low-temperature-fluid storage, as well as more advanced schemes represented by complex, compound chemisorption TES systems.

ELECTRICAL GENERATORS AND COMPONENTS

GENERATORS

Criteria for selecting alternating current (ac) generators for CHP systems are (1) machine efficiency in converting mechanical input into electrical output at various loads; (2) electrical load requirements, including frequency, power factor, voltage, and harmonic distortion; (3) phase balance capabilities; (4) equipment cost; and (5) motor-starting current requirements.

For prime movers coupled to a generator, generator speed is a direct function of the number of poles and the output frequency. For 60 Hz output, speed can range from 3600 rpm for a two-pole machine to 900 rpm for an eight-pole machine. There is a wide latitude in matching generator speed to prime mover speed without reducing the efficiency of either unit. This range in speed and resultant frequency suggests that electrical equipment with improved operation at a special frequency might be accommodated.

Induction generators are similar to induction motors in construction and control requirements. The generator draws excitation current from the utility's electrical system and produces power when driven above its synchronous speed. In the typical induction generator, full output occurs at 5% above synchronous speed.

To prevent large transient overvoltage in the induction generator circuit, special precautions are required to prevent the generator from being isolated from the electrical system while connected to power factor correction capacitors. Also, an induction generator cannot operate without excitation current from the utility; only the **synchronous generator** has its own excitation.

Prime movers that use **alternators** (to convert high-frequency ac to dc) and **inverters** (dc to ac) to produce 60/50 Hz ac power rely on the inverter design both to meet electrical specifications and for protection by the power conversion controls. The generator's rotating speed is irrelevant because the inverter creates the appropriate ac power to either follow the utility electric system voltage and frequency characteristics when operating grid-parallel, or meet the requirements of local loads when operating grid-isolated during system outages. Note that in these situations, the unit is clearly separated from the utility system by breakers or switches.

The combined efficiency of the generator, alternator, and inverter is a nonlinear function of the load and is usually maximized at or near the rated load (Figure 62).

The rated load estimate should include a safety factor to cover transient conditions such as short-term peaking and equipment start-up. Because combustion turbine engines can produce more power under cold ambient conditions than their ISO rating, generators are typically sized to match the engine's expected maximum power output. For combustion turbines driving alternator and inverter systems, the maximum power capability of the inverter typically limits system output. The inverter design may not support power output above the engine's ISO ratings because of the expense of the power electronics (UL 1999). Industrial generators are designed to handle a steady-state overload of 20 to 25% for several hours of continuous operation. If sustained overloads are possible, the generator ventilation system must be able to relieve the temperature rise of the windings, and the prime mover must be able to accommodate the overload.

Proper phase balance is extremely important. Driving three-phase motors and other loads from the three-phase generator presents the best phase balance, assuming that power factor requirements are met. Motor start-up can be a problem for inverter-based microturbines because they are current-limited to protect the inverter. Driving single-phase motors and building lighting or distribution systems may cause an unbalanced distribution of the single-phase loads that leads to harmonic distortion, overheating, and electrical imbalance of the generator. In practice, maximum phase imbalance can be held within 5 to 10% by proper distribution system loading.

In grid-isolated operation, voltage for synchronous generators is regulated by using static converters or rotating dc generators to excite the generator. Voltage regulation should be within 0.5% of nominal voltage and frequency regulation within 0.3 Hz from full load to no load during steady-state conditions. Good electronic three-phase voltage sensing is necessary to control the system response to load changes and excitation of paralleled alternators to ensure reactive load division.

The system power factor is reflected to the generator and should be no less than 0.8 for reasonable generator efficiency. To fall within this limit, the planned electrical load may have to be adjusted so the combined leading power factor substantially offsets the combined lagging power factor. Although more expensive, individual power factor correction at each load with properly sized capacitors is preferred to total power factor correction on the bus. On-site generators can correct some power factor problems, and for CHP systems interconnected to the grid, they improve the power factor seen by the grid.

Generators operating in **parallel** with the utility system grid have different control requirements than those that operate **isolated** from the utility grid. A system that operates in parallel and provides emergency standby power if a utility system source is lost must also be able to operate in the same control mode as the system that normally operates isolated from the electric utility grid.

U.S. *National Electrical Code*® (NFPA *Standard* 70), Article 250, discusses grounding neutral connection of on-site generators. Section 250-5 covers emergency generators in electrical systems, with a four-pole transfer switch that prevents a solid neutral connection from service equipment to generator. For "separately derived systems," grounding requirements of Section 250-5(d) apply only where the generator has no direct electrical connection, including a solidly grounded circuit conductor to the normal service. The rule of grounding applies to a generator that feeds its load without any tie-in through a transfer switch to any other system, but does not apply to one with a solidly connected neutral from it to the service through a three-pole, solid neutral transfer switch.

Section 250-27(b) requires that a neutral that might function as an equipment grounding conductor have cross-sectional area of at least 12.5% of the cross-sectional area of the largest phase conductor of the generator circuit to the transfer switch.

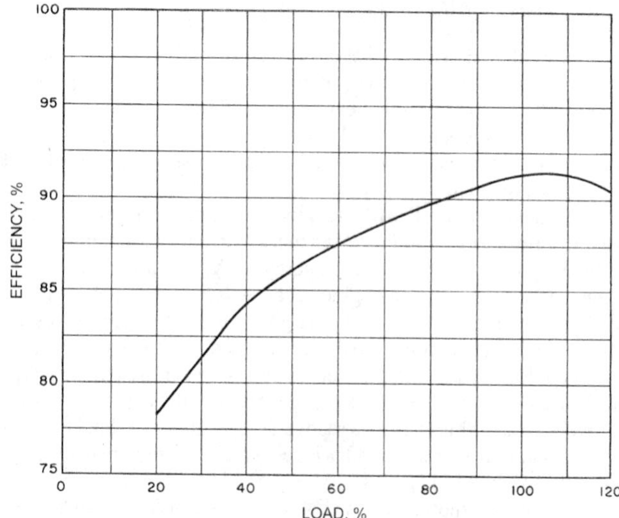

Fig. 62 Typical Generator Efficiency

Control requirements for systems that provide electricity and heat for equipment and electronic processors differ, depending on the number of energy sources and type of operation relative to the electric utility grid. Isolated systems generally use more than one prime mover during normal operation to allow for load following and redundancy.

Table 19 shows the control functions required for systems isolated from and in parallel with the utility grid, with single or multiple prime movers. Frequency and voltage are directly controlled in a single-engine isolated system. Power is determined by the load characteristics and is met by automatic adjustment of the throttle. Reactive power is also determined by the load and is automatically met by the exciter in conjunction with voltage control.

In parallel operation, both frequency and voltage are determined by the utility service. Power output is determined by the throttle setting, which responds to system heat requirements if thermal tracking governs, or to system power load if that governs. Only the reactive power flow is independently controlled by the generator controls.

When additional generators are added to the system, there must be a way to control the power division between multiple prime movers and for continuing to divide and control the reactive power flow. All units require synchronizing equipment.

The generator system must be protected from overload, overheating, short-circuit faults, and reverse power. The minimum protection is a properly sized circuit breaker with a shunt-trip coil for immediate automatic disconnect in the event of low voltage, overload, or reverse power. The voltage regulation control must prevent overvoltage. Circuit breakers for low voltage (below 600 V) are typically air-type, and circuit breakers for medium voltage (up to 12,000 V) should be vacuum-type. The Institute of Electrical and Electronics Engineers' (IEEE) *Standard* 1547 defines interconnection requirements for a paralleling local generator and the utility's electrical power system, and describes specific design, operation, and testing requirements for interconnecting generators below 10 MW and typical radial or spot network primary or secondary electrical distribution systems.

In grid-isolated operation, voltage must be held to close tolerances by the voltage regulator from no load to full load. A tolerance of 0.5% is realistic for steady-state conditions from no load to full load. The voltage regulator must allow the system to respond to load changes with minimum transient voltage variations. During parallel operation, the reactive load must be divided through the voltage regulator to maintain equal excitation of the alternators connected to the bus. True reactive load sensing is of prime importance to good reactive load division. An electronic voltage control responds rapidly and, if all three phases are sensed, better voltage regulation is obtained even if the loads are unbalanced on the phases. The construction of a well-designed voltage regulator dictates the transient voltage variation.

Engine sizing can be influenced by the control system's accuracy in dividing real load. If one engine lags another in carrying its share of the load, the capacity that it lags is never used. Therefore, if the load-sharing tolerance is small, the engines can be sized more closely to the power requirements. A load-sharing tolerance of less than 5% of unit rating is necessary to use the engine capacity to good advantage.

A load-sharing tolerance of 5% is also true for reactive load sharing and alternator sizing. If reactive load sharing is not close, a circulating current results between the alternators. The circulating current uses up alternator capacity, which is determined by the heat generated by the alternator current. The heat generated, and thus the alternator capacity, is proportional to the square of the current. Therefore, a precise control system should be installed; the added cost is justified by the possible installation of smaller engines and alternators.

Table 19 Generator Control Functions

	Isolated		Parallel	
Control Functions	**One Engine**	**Two or More Engines**	**One Engine**	**Two or More Engines**
Frequency	Yes	Yes	No	No
Voltage	Yes	Yes	No	No
Power	No (Load following)	Yes (Division of load)	No	Yes (Division of load)
Reactive kVAR	No (Load following)	Yes (Division of kVAR)	Yes	Yes
Heat t_1	Supplement only	Supplement only	Load following	Load following
Heat $t_2 - t_x$	Reduce from t_1 or supplement	Reduce from t_1 or supplement	Reduce from t_1 and load following	Reduce from t_1 and load following
Cooling	Remove excess heat (tower, fan, etc.)	Remove excess heat (tower, fan, etc.)	Normally no (Emergency yes)	Normally no (Emergency yes)
Synchronizing	No	Yes	Yes	Yes
Black start	Yes	Yes (one engine)	Emergency use	Emergency use (one engine)

SYSTEM DESIGN

CHP ELECTRICITY-GENERATING SYSTEMS

Good CHP planning responds to the end user's requirements and strives to maximize use of the equipment and the energy it produces. For CHP to be economically feasible, the energy recovered must match the site requirements well and avoid as much waste as possible. Depending on the design and operating decisions, users may tie into the electric utility grid for some or all of their electric energy needs.

Thermal Loads

For maximum heat recovery, the thermal load must remove sufficient energy from the heat recovery medium to lower its temperature to that required to cool the prime mover effectively. A supplementary means for rejecting heat from the prime mover may be required if the thermal load does not provide adequate cooling during all modes of operation or as a back-up to thermal load loss.

Internal combustion reciprocating engines have the lowest heat/power ratio, yielding most of the heat at a maximum temperature of 200 to 250°F. This jacket water heat can be used by applications requiring low-temperature heat.

Gas turbines can provide a larger quantity and better quality of heat per unit of power, whereas extraction steam turbines can provide even greater flexibility in both quantity and quality (temperature and pressure) of heat delivered.

If a gas turbine plant is designed to serve a variety of loads (e.g., direct drying, steam generation for thermal heating or cooling, and shaft power), it is even more flexible than one that serves only one or two such loads. Of course, such diverse equipment service must be economically justified.

Prime Mover Selection

Selection of the prime mover depends on the thermal and power profiles required by the end user and on the contemporaneous relationship of these profiles. Ideally, the recoverable heat is fully used as the prime mover follows the power load, but this ideal condition seldom occurs over extended periods.

For maximum equipment use and least energy waste, use the following methods produce only the power and thermal energy that is required on site:

- Match the prime mover's heat/power ratio to that of the user's hourly load profiles.
- Store excess power as chilled water or ice when thermal demand exceeds coincident power demand.

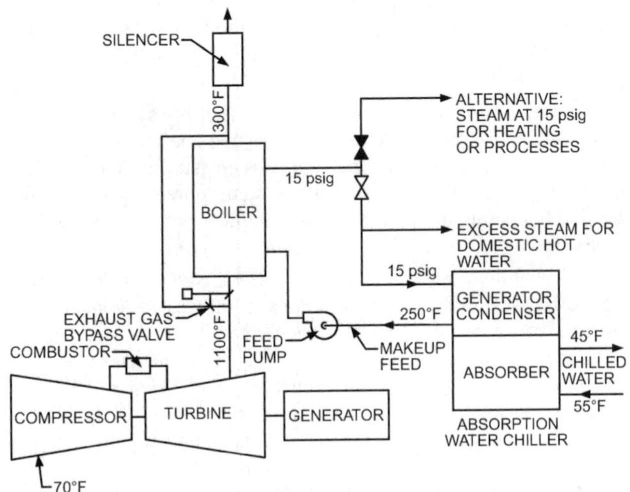

Fig. 63 Typical Heat Recovery Cycle for Gas Turbine

- Store excess thermal production as heat when power demand exceeds heat demand. Either cool or heat storage must be able to productively discharge most of its energy before it is dissipated to the environment.
- Sell excess power or heat on a contract basis to a user outside of the host facility. Usually the buyer is the local utility, but sometimes it is a nearby facility.

The quality of recovered energy is the second major determinant in selecting the prime mover. If the quantity of high-temperature (above 260°F) recoverable heat available from an engine's exhaust is significantly less than that demanded by end users, a combustion or steam turbine may be preferred. Figure 63 shows a typical heat recovery cycle.

Low heat/power ratios of 1 to 3 lb/h steam per horsepower output of the prime mover indicate the need for one with a high shaft efficiency of 30 to 45% (shaft energy/LHV fuel input). This efficiency is a good fit for an engine because its heat output is available as 15 psig steam or 250°F water. Higher temperatures/pressures are available, but only from an exhaust gas recovery system, separated from the low-temperature jacket water system. However, for a typical case in which 30% of the fuel energy is in the exhaust, approximately 50% of this energy is recoverable (with 300°F final exhaust gas temperature); less than 50% is recoverable if higher steam pressures are required.

Medium heat/power ratios of 4 to 11 lb steam/hp·h can be provided by combustion turbines, which are inherently low in shaft efficiency. Smaller turbines, for example, are only 27 to 34% efficient, with 66 to 73% of their fuel energy released into the exhaust. For a typical case with 70% of the fuel energy in the exhaust, assuming that approximately 55% of this energy is recoverable (with 300°F final exhaust gas temperature), net result is an overall thermal efficiency of 69% (excluding power required to drive auxiliary equipment such as fans and pumps to recover and use the exhaust heat).

High heat/power ratios of 8 to 40 lb steam/hp·h, provided by various steam turbine configurations, make this prime mover highly flexible for higher thermal demands. The designer can vary throttle, exhaust and/or extraction conditions, and turbine efficiency to attain the most desirable ratio for varying heat/power loads in many applications, thus furnishing a wide variety of thermal energy quality levels.

Air Systems

Large, central air handlers with deep and/or suitably circuited coils that operate with a large cooling and heating Δt (e.g., 24°F) are available to reduce distribution piping and pumping costs. These units serve multiple control zones or large single-zone spaces with air distribution ductwork. Smaller units such as perimeter fan-coil units directly condition spaces with small lengths of duct or no ductwork. Thus, they are totally decentralized from the air side of the system. The maximum Δt through the coil is only 12 to 14°F.

Both central and decentralized air handlers can be coupled with CHP in mildly cold climates in a two-pipe changeover configuration with a small, intermediate-season electric heating coil. This arrangement can heat or cool different zones simultaneously during intermediate seasons. Chilled water is available to cool any zone. Zones needing heat can cut off the coil and turn on electric heat. A four-pipe system is unnecessary in these conditions.

When the building's balance point is reached (i.e., when all zones need heat), the pumping system is changed over to hot water. The concept applies best (and mostly) to perimeter zone layouts and to large air handlers with economizer cycles that do not need chilled water below the ambient changeover point (i.e., when economizer cooling can satisfy their loads). However, the application must have a high enough thermal demand for process or other non-space-heating loads to absorb the extra thermal energy produced by the engine generator for this additional electric heating load.

If the predominant thermal load during this period is for space heating and cooling (both at a low demand level), it makes no sense to exacerbate the already low heat/power ratio by designing for more electrical load with no use for the heat generated. Hospitals and apartment houses with high process heating demands are examples of suitable applications, but single-function office buildings are not.

The significance of this CHP design is that the prime mover's electrical output can be swung from a motor-driven refrigeration load, which is less during intermediate seasons, to the electric heating function as long as the additional thermal energy can be absorbed. This can work well with engine-generators and electric refrigeration.

An absorption chiller might be a better match where the site's heat/power ratio is low, such as for an office building, but a mechanical chiller without CHP may offer an even better return than an absorption system.

Hydronic Systems

Hydronics, particularly in buildings with no need for process or high-pressure steam, is a much more widely used transport medium than steam. See Chapter 12 for more information on hydronic system design; information on various types of terminals and systems may be found in Chapter 5. Loosely defined, hydronics covers all liquid transport systems, including (1) chilled water, hot water, and thermal fluids that convey energy to locations where space and process heating or cooling occur; (2) domestic or service hot water; (3) coolant for refrigeration or a process; (4) fresh or raw water for potable or process purposes; and (5) wastewater.

From a CHP design standpoint, all these applications are relevant, but some are not HVAC applications. Each application may offer an opportunity to improve the CHP system. For example, a four-pipe, two-coil system and a two-pipe, common-coil system offer similar options in plant design, with the four-pipe system offering superior flexibility for individual control of space conditions. All-electric, packaged terminal air conditioners offer little opportunity for a sizable plant, unless a substantial thermal demand (e.g., for process heat) exists in addition to normal comfort space-conditioning needs. Without the thermal demand, the only option is to install a plant that generates a fraction of the total electrical demand while satisfying service water heating requirements, for example, and to buy the bulk of the electricity required. If the site's heat/power ratio is so low that it cannot support the lowest-ratio prime mover for a large portion of the electrical demand, then a smaller plant that can operate close to a base-loaded electric generating condition might be considered.

The temperature required by the site loads also influences the feasibility of CHP. If a temperature of 110 to 120°F can satisfy most of the site's heating requirements (with air handlers, fan-coil units, or multitiered finned radiators), a central motor-driven heat pump might offer a more cost-effective alternative to a prime mover in a CHP plant that produces more heat than required.

Even if refrigeration from the heat pump is not used, the heat pump takes only 42% of the source fuel from the electric utility's boiler to produce the same heat energy as an 80% efficient on-site boiler.

The heat pump is even more effective if there is simultaneous demand for both refrigeration and heating. To the extent that refrigeration is in excess of demand, it can be used instead of air handler economizer cooling or for fan-coil units not equipped with economizers. If a CHP plant has a heat pump, it may produce too much heat for the site to absorb, thereby reducing the heat pump utilization factor. More information on heat pumps may be found in Chapter 8.

Service Water Heating

Service hot-water systems can be a major and preferred user of thermal energy from prime movers, and often constitute a fairly level year-round load, when averaged over a 24 to 48 h period. Service hot-water use in hospitals, domiciliary facilities, etc., is usually variable in a 24 h weekday or 48 h weekend profile; heat storage allows expanded use of the thermal output and justification for larger prime movers.

The service hot-water demand often provides a strong case for consuming the entire thermal output with packages sized for the 24 h demand, instead of for space cooling or heating.

District Heating and Cooling

The high cost of a central plant and distribution system generally mandates that significant economic returns develop soon after the plant and distribution systems are complete. Because of the high risk, the developer must have satisfactory assurance that there are enough buyers for the product. Furthermore, the developer must know what distribution media and quality are best for connection to existing buyers and must install a system that is flexible enough for future buyers.

Generally, the load on a district system tends to level out because of the great diversity factor of the many loads and noncoincident peaking. This variety also makes plant sizing and consumption estimate aspects difficult. Statistical data from case studies and broad assumptions may be the only source of information. Chapter 11 has further information on district systems.

Utility Interfacing

All electric utility interfacing requires safety on the electric grid and the ability to meet the operating problems of the electric grid and its generating system. Additional control functions depend on the desired operating method during loss of interconnection. For example, the throttle setting on a single generator operating grid-parallel with the utility is determined by either heat recovery requirements or power requirements, whichever govern, and its exciter current, which is set by the reactive power flow through the interconnection. When interconnection with the utility is lost, the generator control system must detect that loss, assume voltage and frequency control, and immediately disconnect the intertie to prevent an unsynchronized reconnection. The Institute of Electrical and Electronics Engineers (IEEE) *Standard* 1547 establishes requirements for operating the distributed generation (DG) in parallel with the grid both for normal and fault conditions. Depending on the design, the system may be able to continue operating while isolated from the grid, providing power to some or all of a facility's electrical loads as an intentional island. A key concern of this standard is to prevent **islanding** (a situation in which generator closely matches the local and/or neighboring load). Anti-islanding precautions are both for safety and protection.

With throttle control now determined solely by the electrical load, the heat produced may not match the requirements for supplemental or discharge heat from the system. When the utility source is reestablished, the system must be manually or automatically synchronized, and the control functions restored to normal operation.

Loss of the utility source may be sensed through the following factors: overfrequency, underfrequency, overcurrent, overvoltage, undervoltage, reverse power, or any combination of these. The most severe condition occurs when the generator is delivering all electrical requirements of the system up to the point of disconnection, whether it is on the electric utility system or at the plant switchgear. Under such conditions, the generator tends to operate until the load changes, at which point, it either speeds up or slows down. This allows the over- or underfrequency device to sense loss of source and reprogram the generator controls to isolated system operation. The interconnection is normally disconnected during such a change and automatically prevented from reclosing to the electric system until the electric source is reestablished and stabilized and the generator is brought back to synchronous speed. Additional utility interfacing aspects are covered in following sections.

Power Quality

Electrical energy can be delivered to the utility grid, directly to the user, or to both. Generators for on-site power plants can deliver electrical energy equal in quality to that provided by the electric utility in terms of voltage regulation, frequency control, harmonic content, reliability, and phase balance. They can be more capable than the utility of satisfying stringent requirements imposed by user computer applications, medical equipment, high-frequency equipment, and emergency power supplies because other end users on the utility grid can create quality problems. The generator's electrical interface should be designed according to user or utility electrical characteristics.

Output Energy Streams

Interconnects may have to be made to electrical systems of one or more voltages; to low-, medium-, or high-pressure steam systems; to chilled-water or secondary coolant systems; to low-, medium-, or high-temperature hot-water or thermal fluid systems; or to thermal energy storage systems. Each variation should be addressed in the planning stages.

Electrical. Electrical energy can produce work, heating, or cooling; it is the most transmittable form of energy. As a CHP output, it can be used to refrigerate or to supplement the prime mover's thermal output during periods of high thermal/low cooling demand. Mechanical aspects of a CHP system must be coordinated with electrical system designers who are familiar with power plant switchgear and utility and building interface requirements. See the section on Utility Interfacing for more information.

Steam. Engines and combustion and steam turbines can provide a range of pressure/temperature characteristics encountered in almost all steam systems. Their selection is basically a matter of choosing the prime mover and heat recovery steam generator combination that best suits the economic and physical goals.

Steam can also provide work, heating, or cooling, but with somewhat less range and flexibility than electric power. Distribution to remote users is more expensive than for electricity and is less adaptable for remote production of work.

Economics limit the pressure and/or temperature (P/T) of steam available from gas turbine exhaust because the incremental cost/benefit ratio of increasing the heat recovery generator surface to yield a higher P/T is limited by a relatively fixed exhaust gas temperature, unless the turbine is equipped with supplemental firing with an auxiliary duct burner. However, steam turbines are not similarly limited, except by throttle conditions, because extraction can be accomplished from any point in the P/T reduction process of the turbine. Chapters 10 and 11 have further information on steam systems.

Chilled Water. The entire output of any prime mover can be converted to refrigeration and then chilled water, serving the wide variety of terminal units in conventional systems. In widely spread service distribution systems, choices must be made whether to serve outlying facilities with electric, steam, or hot water. All three can be used directly, for building or process heating and/or cooling, or indirectly, through heat exchangers and mechanical or thermally-activated chillers or desiccant dehumidifiers located at remote facilities.

Central chilled water production and distribution to existing individual or multibuilding complexes is most practical if a chilled water network already exists and all that is required is an interconnect at or near the CHP plant. If the buildings already have one or more types of chillers in good condition, CHP and chilled water distribution may have diminished economic prospects unless applied on a small scale to individual buildings.

If chilled-water distribution is feasible, central CHP is easier to justify, and several techniques can be used to improve the viability of a cogenerated chilled-water system by significantly reducing the owning and operating costs of piping and pumping systems and their associated components (e.g., valves, insulation, etc.). Such systems have been widely discussed, successfully developed, modified, and specified by many firms (Avery et al. 1990; Becker 1975; Mannion 1988).

Cost-reduction concepts for variable-flow water systems include the following:

- Let the main pump(s) and primary distribution system flow rate match the instantaneous sum of the demand flows from all cooling coils served by the primary loop. Chilled water should not be pumped off the primary loop in such a way as to circulate more chilled water through the secondary pump of the outlying buildings than the flow that it draws from the primary loop.
- Use two-way throttling control valves on all coils. Avoid three-way control valves for coil control or for bypassing chilled-water supply into the chilled-water return (e.g., end-of-line bypass to maintain a constant pump flow or system pressure differential). Valves must have suitable control characteristics for the system and full shutoff capability at the maximum pressure differential encountered.
- Select and circuit cooling coils for a large chilled-water temperature difference (Δt as much as 24°F) while maintaining coil tube velocities of 5 to 10 fps and the required supply air conditions off the coil. Such coils may require 8 to 10 rows, but the additional pressure drop and cost are offset by the lower cost of the pump and distribution piping of long distribution systems. A system

with Δt of 24°F requires only one-third the flow rate required by one with a Δt of 8°F.

- Care is necessary for successful implementation of these concepts. The *Air-Conditioning Systems Design Manual* (Lorsch 1993) has further design information.

Hot Water. CHP thermal output is well adapted to low-, medium-, or high-temperature water (LTW, MTW, HTW) distribution systems. The major difference between chilled- and hot-water systems is that even LTW systems (up to 250°F) can be designed with a Δt as high as 100°F with low flow by using different series-parallel terminal circuiting, as described in Chapter 12. This way, even equipment that is limited to Δt = 20°F (e.g., radiators, convectors) can be adapted to large system temperature differences. For example, beyond those circuits given in Chapter 12, unit heaters can be piped in series and parallel on a single hot-water building loop without a conventional supply and return line. Parallel runs of five heaters each can drop in 20°F increments. The first group drops the temperature from 250 to 230°F, the last from 170 to 150°F, and all are sized at the 170 to 150°F range. Fan cycling off each local thermostat maintains control despite the different temperatures.

Similarly, larger heaters with conventional small-row coils (not metal cores) can be fitted with three-way modulating bypass valves, sized for a 10 to 40°F drop, with the through-flow and bypass flow from the first flowing to the second, and so forth, using only one primary loop.

Medium- (250 to 350°F) and high- (350°F and higher) temperature water systems are designed with an even higher Δt, but are not customarily connected directly to the primary loop. These systems can be connected to steam generators in outlying buildings that have steam distribution and steam terminal devices.

When a choice can be made, a prime mover's thermal output should be used according to the following priorities. Apply the output first for useful work, second for an efficient form of thermal conversion, and third for productive thermal use. For example, if a combustion turbine's exhaust can cost-effectively produce shaft power or, if not, heat some process, it should be considered for these functions. Case-by-case analysis of applying heat in other thermally activated technologies is helpful, because usefulness depends on specific economic conditions.

For both hot- and chilled-water distribution systems, a common approach is to lower the hot-water supply temperature as ambient temperature rises and to raise the chilled-water supply temperature as loads are reduced. Both techniques reduce pipe transmission and fuel or electrical costs for heating or cooling and stabilize valve control, but the effect of increased pumping costs is often overlooked.

Below some part-load condition in both hot- and chilled-water systems, the cost of reducing flow by varying pump speed and air volume may be more than the energy savings. This is especially true for chilled-water systems, when the cascading effect of VAV fan power reduction from lower supply air temperatures, together with pumping savings, becomes more significant than the low-load chiller savings from raising the chilled-water system temperature. Also, humidity control for the space limits how much the chilled-water temperature can rise.

These factors need to be examined in determining the part-load condition at which hot- and chilled-water system scheduling might be advantageously modified.

CHP SHAFT-DRIVEN HVAC AND REFRIGERATION SYSTEMS

Engine-Driven Systems

HVAC Equipment. Engine-driven chillers are rated according to ARI *Standard* 550/590 conditions. Manufacturers offer performance curves for other conditions. As with any chiller, performance

Table 20 Coefficient of Performance (COP) of Engine-Driven Chillers

Heat Recovery Option	COP at Full Load
No heat recovery	1.2 to 2.0
Jacket water heat recovery	1.5 to 2.25
Jacket water and exhaust heat recovery	1.7 to 2.4

is largely a function of design conditions for the condenser and chilled-water supply temperatures. The engine size required for a given chiller capacity is typically 1 hp per ton of cooling.

Table 20 provides a typical range of engine-driven chiller coefficient of performance (COPs) with and without heat recovery. The COP is cooling energy output divided by fuel energy input. Engine fuel input is based on HHV of the fuel.

Heat recovered from the jacket coolant and exhaust gas is added to the cooling load produced by the chiller, increasing useful thermal output and the COP. Because no standards exist for calculating the COP of an engine-driven chiller when considering heat recovery, most manufacturers present COPs with and without heat recovery.

Reciprocating Compressors. Engine-driven reciprocating compressor water-chiller units may be packaged or field-assembled from commercially available equipment for comfort service, low-temperature refrigeration, and heat pump applications. Both direct-expansion and flooded chillers are used. Some models achieve low operating cost and high flexibility by combining speed variation with cylinder unloading. Part-load capacity is controlled by modulating engine speed down to about 30 to 50% of rated speed, which improves fuel economy. Some reciprocating engine chillers also use cylinder unloading to reduce capacity further. Engine speed should not be reduced below the minimum specified by the manufacturer for adequate lubrication or good fuel economy.

Most engine-driven reciprocating compressors are equipped with a cylinder loading mechanism for idle (unloaded) starting. This arrangement may be required because the starter may not have sufficient torque to crank both the engine and the loaded compressor. With some compressors, not all the cylinders (e.g., 4 out of 12) unload; in this case, a bypass valve must be installed for a fully unloaded start. The engine first reaches one-half or two-thirds of full speed. Then, a gradual cylinder load is added, and engine speed increases over a period of 2 to 3 min. In some applications, such as an engine-driven heat pump, low-speed starting may cause oil accumulation and sludge. As a result, a high-speed start is required.

These systems operate at specific fuel consumptions (SFCs) of approximately 8 to 13 ft^3/h of pipeline-quality natural gas (HHV = 1000 Btu/ft^3) per horsepower in sizes down to 25 tons. Comparable heat rates for diesel engines run from 7000 to 9000 Btu/hp·h. Smaller units are also available. Coolant pumps can also be driven by the engine. These direct-connected pumps never circulate tower water through the engine jacket. Figure 64 illustrates the fuel economy effected by varying prime mover speed with reciprocating compressor load until the machine is operating at about half its capacity. Below this level, the load is reduced at essentially constant engine speed by unloading the compressor cylinders.

Frequent operation at low engine idling speed may require an auxiliary oil pump for the compressor. To reduce wear and assist in starts, a tank-type lubricant heater or a crankcase heater and a motor-driven auxiliary oil pump should be installed to lubricate the engine with warm oil when it is not running. Refrigerant piping practices for engine-driven units are the same as for motor-driven units.

Centrifugal Compressors. Packaged, engine-driven centrifugal chillers that do not require field assembly are available in capacities up to 2100 tons. Automotive-derivative engines modified for use on natural gas are typical of these packages because of their compact size and mass. These units may be equipped with manual or automatic start/stop systems and engine speed controls.

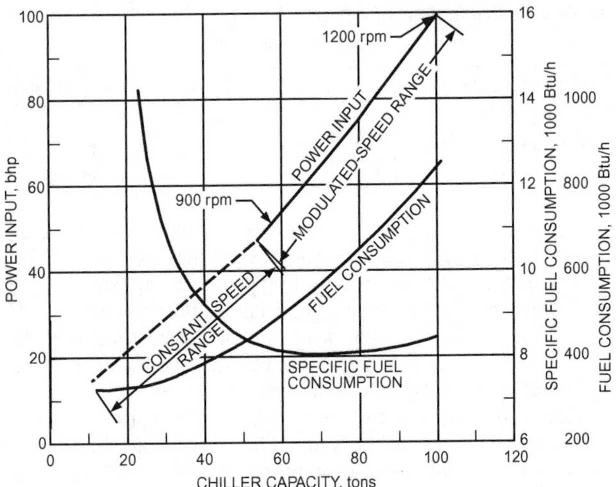

Fig. 64 Performance Curve for Typical 100 Ton, Gas-Engine-Driven, Reciprocating Chiller

Larger open-drive centrifugal chillers are usually field-assembled and include a compressor mounted on an individual base and coupled by flanged pipes to an evaporator and a condenser. The centrifugal compressor is driven through a speed increaser. Many of these compressors operate at about six times the speed of the engines; compressor speeds of up to 14,000 rpm have been used.

To affect the best compromise between the initial cost of the equipment (engine, couplings, and transmission) and maintenance cost, engine speeds between 900 and 1200 rpm are generally used. Engine output can be modulated by reducing engine speed. If operation at 100% of rated speed produces 100% of rated output, approximately 60% of rated output is available at 75% of rated speed. Capacity control of the centrifugal compressor can be achieved by variable inlet guide vane control with constant compressor speed, or a combination of variable-speed control and inlet guide vane control, the latter providing the greatest operating economy.

Heat Pumps. An additional economic gain can result from operating an engine-driven refrigeration cycle as a heat pump, if the facility has a thermal load profile that can adequately absorb its 100 to 120°F low-quality heat. Using the same equipment for both heating and cooling reduces capital investment. A gas engine drive for heat pump operation also makes it possible to operate in a CHP mode, which requires a somewhat larger thermal load. Unless a major portion of this larger thermal recovery can be absorbed, the cycle may not be economical.

The classic definition of reverse-cycle performance is (heat out)/(work in). The definition does not recognize the fuel input to the engine, just as it ignores the fuel input to generate the electricity for the motor of a motor-driven compressor. No coefficient of performance is really defined for the cycle that captures jacket and exhaust heat.

Screw Compressors. Chiller packages with these compressors are available for refrigeration applications. Manufacturers offer water chillers that use screw compressors driven directly by natural gas engines. Capacity control is achieved by varying engine speed and adjusting the slide valve on the compressor. Units have an ECOP near 1.45 at rated cooling load. Chapter 37 has more information on compressors.

Because refrigeration equipment operates at low evaporator temperatures (20 to –70°F), refrigerants such as ammonia and other cycles that improve efficiency over single-stage cycles are used. Besides the standard, single-stage vapor compression cycle, a multistage or cascade refrigeration cycle may be chosen. The multistage cycle is the most common cycle used to efficiently provide refrigeration from –10 to –70°F. The section on Compression Refrigeration

Table 21 Typical Efficiency of Engine-Driven Refrigeration Equipment (Ammonia Screw Compressor)

	Saturation Suction Temperature/ Saturation Discharge Temperature		
	–40/95°F	–12/95°F	20/95°F
Electric COP	1.32	2.66	4.62
Engine-driven COP without heat recovery	0.44	0.78	1.32
Engine-driven COP with jacket water heat recovery	0.74	1.08	1.62
Engine-driven COP with jacket water and exhaust heat recovery	0.89	1.23	1.77

Source: AGCC (1999).

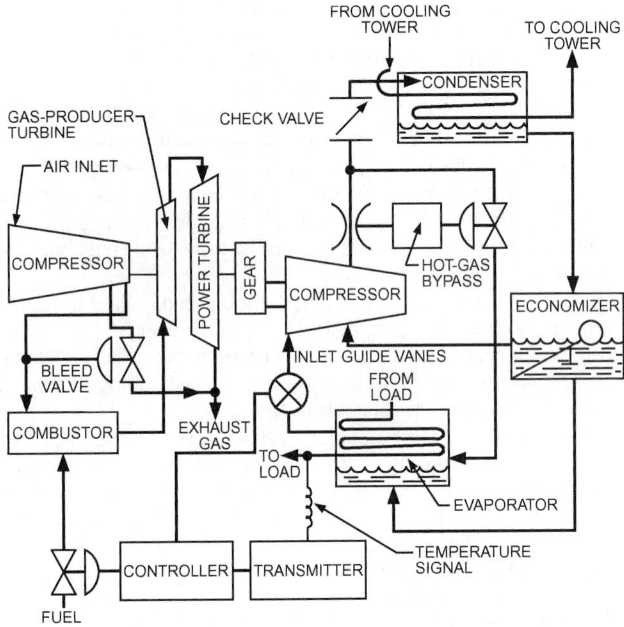

Fig. 65 Typical Gas Turbine Refrigeration Cycle

Cycles in Chapter 1 of the 2005 *ASHRAE Handbook—Fundamentals* describes this cycle.

As with engine-driven chillers, heat recovery from the jacket coolant and exhaust gas boosts overall energy use and efficiency. Table 21 lists the coefficient of performance and effect of heat recovery for a range of conditions.

Combustion-Turbine-Driven Systems

The gas turbine has become increasingly important as a prime mover for electric power generation up to more than 240 MW and for shaft power drives up to more than 108,000 hp. Figure 65 shows a typical gas turbine refrigeration cycle, with optional combustion air precooling. A gas turbine must be brought up to speed by an auxiliary starter. With a single-shaft turbine, the air compressor, turbine, speed reducer gear, and refrigeration compressor must all be started and accelerated by this starter. The refrigeration compressor must also be unloaded to ease the starting requirement. Sometimes, this may be done by making sure the capacity control vanes close tightly. At other times, it may be necessary to depressurize the refrigeration system to get started.

With a split-shaft design, only the air compressor and the gas producer turbine must be started and accelerated. The rest of the unit starts rotating when enough energy has been supplied to the blades of the power turbine. At this time, the gas producer turbine is up to speed, and the fuel supply is ignited. Electric starters are usually available as standard equipment. Reciprocating engines, steam turbines, and hydraulic or pneumatic motors may also be used. The output shaft of the gas turbine must rotate in the direction required by the refrigeration compressor; in many cases, manufacturers of split-shaft engines can provide the power turbine with either direction of rotation.

At low loads, both the gas turbine unit and centrifugal refrigeration machine are affected by surge, a characteristic of all centrifugal and axial flow compressors. At a certain pressure ratio, a minimum flow through the compressor is necessary to maintain stable operation. In the unstable area, a momentary backward flow of gas occurs through the compressor. Stable operation can be maintained, however, by using a hot-gas bypass valve.

The turbine manufacturer normally includes automatic surge protection, either as a bleed valve that bypasses some of the air directly from the axial compressor into the exhaust duct or by providing for a change in the position of the axial compressor stator vanes. Both methods are used in some cases.

The assembly should be prevented from rotating backward, which may occur if the unit is suddenly stopped by one of the safety controls. The difference in pressure between the refrigeration condenser and cooler can make the compressor suddenly become a turbine and cause it to rotate in the opposite direction. This rotation can force hot turbine gases back through the air compressor, causing considerable damage. Reverse flow through the refrigeration compressor may be prevented in a variety of ways, depending on the system's components.

When there is no refrigerant receiver, quick-closing inlet guide vanes are usually satisfactory because there is very little high-pressure refrigerant to cause reverse rotation. However, when there is a receiver, a substantial amount of energy is available to cause reverse rotation. This can be reduced by opening the hot-gas bypass valve on shutdown and installing a discharge check valve on the compressor.

The following safety controls are usually supplied with a gas turbine:

- Overspeed
- Compressor surge
- Overtemperature during operation under load
- Low oil pressure
- Failure to light off during start cycle
- Underspeed during operation under load

A fuel supply regulator can maintain a single-shaft gas turbine at a constant speed. With the split-shaft design, the turbine's output shaft runs at the speed required by the refrigeration compressor. The temperature of chilled water or brine leaving the cooler of the refrigeration machine controls the fuel. See also the section on Fuels and Fuel Systems.

Steam-Turbine-Driven Systems

Steam turbines in air conditioning and refrigeration are mainly used to drive centrifugal compressors, which are usually part of a water or secondary-coolant chilling system using one of the newer or halogenated hydrocarbon refrigerants. In addition, many industrial processes use turbine-driven centrifugal compressors with various other refrigerants such as ammonia, propane, butane, or other process gases.

Related applications of steam turbines include driving chilled-water and condenser-water circulating pumps and serving as prime movers for electrical generators in CHP systems. In industrial applications, the steam turbine may be advantageous, serving either as a work-producing steam pressure reducer or as a scavenger using otherwise wasted low-pressure steam.

Many steam turbines are used in urban areas where commercial buildings are served with steam from a central public utility or municipal source. Institutions where large central plants serve a multitude of buildings with heating and cooling also use steam-turbine-driven equipment.

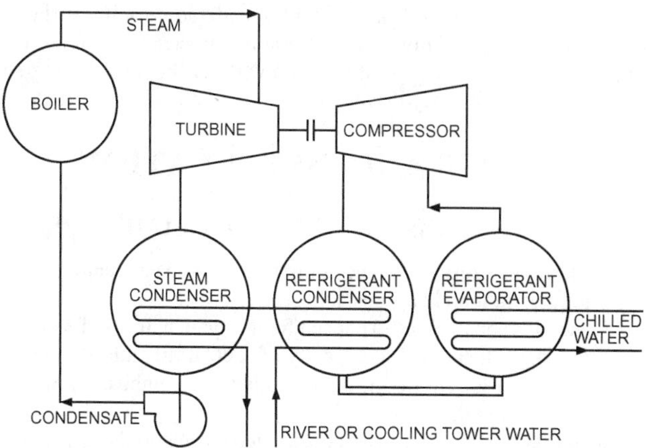

Fig. 66 Condensing Turbine-Driven Centrifugal Compressor

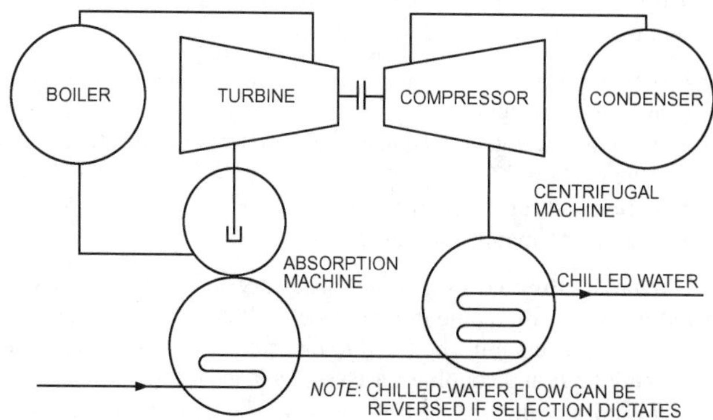

Fig. 67 Combination Centrifugal-Absorption System

Most steam turbines driving centrifugal compressors for air conditioning are multistage condensing turbines (Figure 66), which provide good steam economy at reasonable initial cost. Usually, steam is available at 50 psig or higher, and there is no demand for exhaust steam. However, turbines may work equally well where an abundance of low-pressure steam is available. As an example of the wide range of application of this turbine, one industrial firm drives a sizable capacity of water-chilling centrifugal compressors with an initial steam pressure of less than 4 psig, thus balancing summer cooling against winter heating with steam from generator-turbine exhausts.

Aside from wide industrial use, the noncondensing (back-pressure) turbine is most often used in water-chilling plants to drive a centrifugal compressor that shares the cooling load with one or more absorption units (Figure 67). Exhaust steam from the turbine, commonly at about 15 psig, serves as the heat source for the absorption unit's generator (concentrator). This dual use of heat energy in the steam generally results in a lower energy input per unit of refrigeration output than is attained by either machine operating alone. An important aspect in design of combined systems is balancing turbine exhaust steam flow with absorption input steam requirements over the full range of load.

Extraction and mixed-pressure turbines are used mainly in industry or in large central plants. Extracted steam is often used for boiler feedwater heating or other processes where steam with lower heat content is needed. Most motor-driven centrifugal refrigeration compressors are driven at constant speed (some with variable-frequency drives). However, governors on steam turbines can maintain a constant or variable speed without the need for expensive variable-frequency drives.

CODES AND INSTALLATION

GENERAL INSTALLATION PARAMETERS

Structural support for the equipment must be adequate for the operating weight and any inertial forces.

Depending on the fuel(s) chosen for the system, it may be important to address location and capacity of **fuel supply** and storage.

Ventilation is required to supply adequate combustion air and prevent plant overheating.

Exhaust systems must be constructed of materials resistant to corrosion by exhaust gases, and condensate drains are usually required. Locate exhaust outlets carefully to avoid noise and pollution problems.

Monitoring systems should allow operators to monitor CHP performance for maintenance and economic purposes. As a minimum, the following should be monitored:

- CHP run time
- Fuel consumption
- Electrical power generated in each period (minus parasitic losses)
- Usable heat produced
- Duration and cause of any plant failure
- Cost of fuels, maintenance, etc.

Acoustic attenuation is often necessary for CHP units. Packaged plants usually come in a purpose-built acoustic enclosure. In addition, silencers must be fitted to the exhaust system. Antivibration mountings and couplings are usually standard.

Commissioning and **testing** are key. The plant must be tested under various loads, according to the manufacturer's instructions. Pumps, electrical switchgear, and controls need careful commissioning to ensure they operate as designed. Test that electrical output can be synchronized, paralleled, and disconnected safely. Where installed, heat rejection systems should be tested across a range of loads to establish that sufficient heat can be rejected and overheating avoided. Also, ensure that the CHP unit interacts correctly with the existing heating system.

At project completion, proof of commissioning and operating instruction documentation must be provided. In particular, the CHP system must be included in the building log book required by building regulations. Permits and approvals also should be handed over, and appropriate staff trained to monitor the plant. Most of this is usually done by the CHP supplier, but is ultimately the client's responsibility.

It is important to monitor the immediate and ongoing system performance. As soon as operation begins, detailed monitoring is essential to confirm efficiency and economic viability. Continuous monitoring is required to maintain long-term performance, particularly as fuel costs and electricity prices change.

UTILITY INTERCONNECTION

Requirements for interconnection with public utilities' grids vary by cogenerator and by the individual electric utility, depending on generation equipment, size, and host utility systems. Interconnection equipment requirements increase with generator size and voltage level. Generally, complexity of the utility interface depends on the mode of transition between paralleling and stand-alone operation. The plant connection to the electric grid must have an automatic utility tie-breaker and associated protective relays. When utility power is lost, this tie-breaker opens and isolates the generator and its loads from the utility. Protective relays at the entrance and generator must be coordinated so that the utility tie-breaker opens before the generator's breaker. An automatic load control system is needed to shed noncritical loads to match generator's capacity. When utility power returns, the generator must be synchronized with the utility across the utility tie-breaker, whereas under normal start-up conditions the generator is synchronized across the generator circuit breaker. The synchronizing equipment must accommodate both situations.

When a CHP system is integrated into the utility system, the following issues must be understood and managed:

- Control and monitoring
- Metering
- Protection
- Stability
- Voltage, frequency, synchronization, and reactive compensation for power factors
- Safety
- Power system imbalance
- Voltage flicker
- Harmonics
- Grounding

Many jurisdictions have developed standard interconnections requirements, most of which are based on IEEE *Standard* 1547 and UL *Standard* 1741. IEEE *Standard* 1547 addresses performance, operation, testing, and safety of interconnection products and services, such as hardware and software for distributed power control and communication. It also discusses product quality, interoperability, design, engineering, installation, and certification, and provides a platform for standardized interconnection throughout the United States.

AIR PERMITS

Basic air permitting and emission control requirements are as follows:

- De minimis exemptions
- State minor source permitting
- Major source permitting
- Emergency generators

Most states have a **de minimis exemption**, a threshold below which units are either small enough or have emissions low enough that they do not have to apply for a permit of any kind. Requirements and conditions vary by state. Even when the CHP system does not require a construction permit, it may be advisable to file an information copy with the permitting authority.

Sources that are not exempted must obtain a permit. An important factor in determining how a source is permitted is its **potential to emit**, which is the measure of its maximum possible emissions if operated at full capacity for 8760 h per year. If a source's potential emissions exceed certain thresholds, it is considered a major source and, in the United States, is subject to federal new source review permitting. The trigger threshold depends on local air quality.

Intermediate-level sources are generally subject to state minor source permitting.

Both minor and major source permits are likely to require some kind of emission limitations or controls, which could be anything from raising a unit's stack height to installing the most stringent control technologies available. The permitting process also can range from a simple application to a complex cost-based technology evaluation. Requirements vary by state and the type of unit proposed.

In addition, most states have special treatment for emergency back-up generators. The EPA recommends calculating emergency units' the potential to emit based on 500 h of operation per year.

Note that air permits are required before any construction can begin.

BUILDING, ZONING, AND FIRE CODES

U.S. building design and construction codes (except for federal buildings and facilities) are typically developed at the state and local level. Because power generation in the United States has typically been regulated from within the utility industry, few provisions exist in standards, codes, and building construction regulations for CHP systems; in fact, many codes explicitly exempt facilities used exclusively for generating electricity. As a result, there is very little regulatory guidance for power generation technologies and installations, and most code officials are not experienced in design, construction, installation, or operation of power plants, and particularly generators serving commercial buildings.

Important considerations include the following.

Zoning

Zoning regulations are locally implemented and enforced, but typically are based on national standards. Topics addressed range from noise regulations to visual impact. Blanket prohibitions on electricity generation, limits on operation of back-up generators, and height restrictions on towers for wind generators, are not uncommon. In these cases, the developer must obtain an exception, which increases the expense and time needed for approval.

Building Code/Structural Design

The site must structurally support the system load and vibration. If a unit is installed within a building, the whole building may need to be reclassified for occupancy. The unit cannot block egress from the building.

Mechanical/Plumbing Code

Some localities do not allow installation of gas-fired units not listed as gas appliances by an approved agency. In these cases, a registered engineer must certify that the installation meets all applicable standards and is in safe operating condition, and must seek an exemption. Codes also regulate use of gas meters, piping and emergency shutoff valves.

Fire Code

Fire prevention and firefighters' access to equipment during an emergency are strongly regulated by code rules. The type of fuel (natural gas, oil, propane, etc.) used must be determined, and its flammability and combustibility analyzed. The fire department must know a unit's location and how to disconnect it in case of a fire. A fire suppression system may be necessary for buildings with interior CHP systems.

Applicable codes and standards of the National Fire Protection Association (NFPA) include the following:

- *Standard* 850, Electric Generating Plants
- *Standard* 37, Installation and Use of Stationary Combustion Engines and Gas Turbines, which also covers temporary, portable supplementary engines
- *Standard* 54, National Fuel Gas Code
- *Standard* 70, National Electrical Code®

Electrical Connection

The most important electrical decision is the point of connection. Connection on the line side creates the fewest design problems. When connecting on the load side, the service equipment's ampacity, ground fault protection, and system neutral bonding must be considered.

ECONOMIC FEASIBILITY

ECONOMIC ASSESSMENT

The economics of CHP systems is evaluated based on the costs of equipment, installation, operation (fuel and maintenance), and capital compared to the cost of grid electricity and fuel for meeting thermal loads. With this information, CHP can be compared to a conventional system on the basis of energy cost. Once a utilization factor has been established, the CHP system can be evaluated on the basis of simple payback. For systems with dual-mode capability (grid parallel and grid islanding), the value of avoided power outages should also be included.

Some CHP analysis can be performed with as little as the last 12 months' utility bills, but the more site information that can be obtained, the more accurate the analysis.

Understanding the facility and its needs is essential. For a new building, this includes the building design and energy model. For an existing building, a site visit is necessary to gather information on energy usage, utility costs, operating schedules, electric service, and existing heating and cooling equipment size, configuration, and location. During a site walk-through, the following questions may be relevant:

- Can important infrastructure issues (e.g., lack of a centralized, facility-wide distribution system; lack of space for CHP equipment near the central plant; presence of many electricity meters) be handled at reasonable expense?
- Are there other justifications for a CHP system beyond cost savings by reducing energy bills? Examples include the need to add back-up power, and an unreliable existing power supply.
- Are there any specific financial opportunities that make CHP more attractive? Examples include a 12-month heating load that is large compared to the heating output of the potential generating package, or a planned retirement or upgrade of the existing heating and cooling equipment.

Two project assessment levels are discussed here: simple CHP screening, and feasibility analysis.

Simple CHP Screening Analysis. This level may be adequate to provide a quick understanding of whether a site is suitable for CHP, but investment decisions are not normally made as a result.

Energy consumption and utility rate information gathered during the walk-through are used in a rule-of-thumb evaluation, including the type and approximate size of the CHP system, a rough first cost of the CHP system, and the estimated range of annual energy savings from CHP. Equipment costs are estimates, and their accuracy often depends on the analyst's level of experience.

After obtaining estimates of the energy savings and first cost, an estimated simple payback for the CHP system can be calculated. A packaged screening analysis tool, such as a computer program or a spreadsheet using an established methodology, is recommended. This analysis should be inexpensive. Averaged costs for electricity, particularly if there are demand charges, ratchet and/or time of day rates, should not be used in economic evaluations.

CHP Feasibility Analysis. This more detailed level of analysis includes three steps: energy analysis, conceptual development, and financial pro-forma analysis.

The **energy analysis** should be based on engine or turbine net useful heat output. Generally, most CHP systems cannot recover more than 60 to 70% of the engine's net useful output for most commercial applications (although hospitals and some industrial recovery may be up to about 85%). Note that operating CHP systems when off-peak rates apply (i.e., night, weekends, and designated holidays) may not be beneficial. Usually, operation should focus on on-peak utility rate hours.

Consider the size of the CHP system compared to the facility's peak load. Find the shortest CHP paybacks by running repeated financial analyses across practical system size ranges. In general, for commercial-class buildings applications, the best payback occurs when the CHP system is 40 to 60% of the facility's peak electric load. A 100% (island) system should only be considered if needed by the owner.

The analysis should be run both with and without using heat rejected from the CHP system for water and space heating and for absorption cooling, unless the application has a specific difficulty with any of these technologies. Desiccant dehumidification should also be considered.

Normalize the load (electric and thermal) characterization to the local average weather year. This is particularly important for commercial buildings with loads dominated by space heating and cooling, and if the load was developed using only the previous year's utility consumption and bills, which may not reflect usage in an average year and could introduce error in extended (five- to seven-year) forecasts.

Provide a detailed list of all assumed first costs, as well as costs for electricity and natural gas (both currently, and reasonable projections over the next five to seven years). Standby charges from the local electric utility should also be addressed, and are usually based on the CHP system's total capacity or that of the largest single generator in the CHP system, unless another solution for engine outages is used.

The **conceptual development analysis** should provide one-line and block diagrams of electrical and mechanical layouts, as well as the major planned components and issues noticed during the walk-through. Specifics, such as type and size of equipment, should be included.

The **financial pro forma analysis** should show the following, on a year-by-year basis:

- Initial and additional out-year investments, if any.
- Engine generator maintenance allocations, with charges based on projected equipment operating hours.
- Energy cost savings.
- Capital repayment and carrying charges.
- Depreciation.
- Tax effects (can be zeroed out for nontaxable owners)
- Internal rate of return (IRR) on the investment, without leverage; this includes all annual energy cost savings, maintenance costs, and tax savings, as well as the initial cost of the CHP system (as a single lump sum outflow in year zero). It should not include principal or interest payments on financing.

Do not use fuel or electricity price escalators or leveraged rates of return in the financial analysis. Projected rates need to be considered, particularly when their rates diverge. Most financial decision makers require this analysis.

PRELIMINARY FEASIBILITY BIN ANALYSIS EXAMPLES

Planning a CHP system is considerably more involved than planning an HVAC system. HVAC systems must be sized to meet peak loads; CHP systems need not. Also, HVAC systems do not have to be coordinated and integrated with other energy systems as extensively as do CHP systems.

First Estimates

Becker (1988) suggested a quick way to determine whether a study should be undertaken: if the cost of electricity expressed in U.S. \$/kWh is more than 0.013 times the cost of fuel expressed in U.S. \$/10^6 Btu, a study should be considered. If it is 0.026 times or more, the chances are excellent for simple payback in three years or less.

The economic coefficient of performance (ECOP) is a methodology that enables each energy stream to be valued on a comparable economic basis and at prevailing rates, with 1 kWh of electrical energy taken as 3412 Btu and fuel input as the high heating value (HHV) in Btu. Then, a direct comparison can be made. Rates with step charges based on the load factor must be carefully evaluated to be sure the appropriate incremental cost is used.

For example, with energy from the utility at \$0.08/kWh and natural gas supply at \$5.50 per million Btu (HHV of 1000 Btu/ft^3), 1000 Btu of electrical energy costs \$0.023 (0.08/3.412), and 1000 Btu of natural gas costs \$0.0055. Thus, the ECOP can be defined as all output energy in desired output forms, converted in terms of economic costs, divided by all energy input (fuel input based on HHV), again converted in terms of economic costs of each energy stream. For this example, the electrical energy costs (0.023/0.0055) = 4.18 times more than the equivalent energy from natural gas. This ratio is used to calculate the ECOP in the following examples.

Example 6. Calculate the ECOP of a low-pressure steam absorption chiller with motor auxiliaries totaling 25 hp per 1000 tons. The on-site boiler generates 19 lb/ton·h steam at 15 psig (1164 Btu/lb enthalpy) at 80% efficiency from feedwater at 0 psig, 212°F (180 Btu/lb enthalpy).

Solution:

$$\text{On-site fuel input} = 19(1164 - 180)/0.8 = 23{,}400 \text{ Btu/ton·h}$$

The electrical input generated off site supplies 25 hp/1000 tons at a motor efficiency of 0.87.

$$\text{Off-site electrical input} = (0.746 \text{ kW/hp} \times 0.025 \text{ hp/ton})/0.87 = 0.0214 \text{ kW input/ton} = 3412 \times 0.0214 = 73 \text{ Btu/ton·h}$$

The equivalent total input per unit of output (cooling only) is

$$23{,}400 + (73 \times 4.18) = 23{,}705 \text{ Btu/ton·h(equivalent energy)}$$

Thus, the ECOP for 12,000 Btu/ton·h output and 23,705 Btu/ton·h equivalent input (at the preceding power costs) is

$$12{,}000/23{,}705 = 0.506$$

Example 7. Calculate and compare the ECOP for the same cooling output, using an engine-driven, vapor-compression chiller and piggyback absorption chiller. The engine has an 8600 Btu/hp·h heat rate (30% shaft thermal efficiency) and 3470 Btu/hp·h of saturated steam at 15 psig heat recovery (40% heat recovery rate). Heat rate in this example is based on HHV.

Solution: The total cooling output is

From engine-chiller at 1 hp/ton cooling	12,000 Btu
From the absorption chiller at 19(1164 – 180) = 18,700 Btu/ton · h and for 3470 Btu/ton · h (cooling), 12,000 × 3470/18,700 =	2,227 Btu
Total cooling	14,227 Btu

Off-site electrical input for absorption chiller auxiliaries at 25 hp per 1000 tons, as detailed in Example 6, is 73 Btu/ton·h. The equivalent total input energy is

$$8600 + (73 \times 4.18) = 8905 \text{ Btu}$$

Thus, the ECOP for above is 14,227/8905 = 1.598, which is more than three times that of the conventional system covered in Example 6.

A similar approach can produce ECOPs for different configurations and with different electrical and fuel (gas or oil) costs. But an ECOP should be considered an indicator only and should be followed with a life-cycle cost analysis to make a final decision.

Load Duration Curve Analysis

A much more comprehensive energy analysis, combined with an economic analysis, must be used to select a CHP system that maximizes efficiency and economic return on investment. For better identification and screening of potential candidates, a simplified but accurate performance analysis must be conducted that considers the dynamics of the facility's electrical and thermal loads, as well as the size and fuel consumption of the prime mover.

Accurate analysis is especially important for commercial and institutional CHP applications because of the large time-dependent changes in magnitude of load and the noncoincident nature of the power and thermal loads. A facility containing a generator sized and operated to meet the thermal demand may occasionally have to purchase supplemental power and sometimes may produce power in excess of facility demand.

Even in the early planning stages, a reasonably accurate estimate of the following must be determined:

- Fuel consumed (if it is a topping cycle)
- Amount of supplemental electricity that must be purchased
- Amount of supplemental boiler fuel (if any) that must be purchased
- Amount of excess power available for sale
- Electrical capacity required from the utility for supplemental and standby power.
- Electrical capacity represented by any excess power if the utility offers capacity credits

Obtaining estimates of the these performance values for multiple time-varying loads is difficult, and is further complicated by utility rate structures that may be based on time-of-day or time-of-year purchase and sale of power. Data must be collected at intervals short enough to give the desired levels of accuracy, yet taken over a long period and/or a well-selected group of sampling periods.

A basic method for analyzing HVAC system performance is the bin method. The basic tool for sizing and evaluating power system performance is the load duration curve, which contains the same information as bins, but arranges the load data in a slightly different manner. The load duration curve is a plot of hourly averaged instantaneous load data over a period; the plot is rearranged to indicate the frequency, or hours per period, that the load is at or below the stated value. The load duration curve is constructed by sorting the hourly averaged load values of the facility into descending order. Large volumes of load data can be easily sorted with spreadsheets or databases. The load duration curve produces a visually intuitive tool for sizing CHP systems and for accurately estimating system performance.

Figure 68 shows a hypothetical steam load profile for a plant operating with two shifts each weekday and one shift each weekend day; no steam is consumed during nonworking hours for this example. The data provide little information for thermally sizing a generator, except to indicate that the peak demand for steam is about 46,000 lb/h and the minimum demand is about 13,000 lb/h.

Figure 69 is a load duration curve (a descending-order sort) of the steam load data in Figure 70. Mathematically, the load duration curve shows the frequency with which load equals or exceeds a given value; the curve is one minus the integral of the frequency distribution function for a random variable. Because the frequency distribution is a continuous representation of a histogram, the load duration curve is simply another arrangement of bin data.

In the frequency domain, or load duration curve form, the base load and peak load can be readily identified. Note that the practical base load at the "knee" of the curve is about 21,000 lb/h rather than the 13,000 lb/h absolute minimum identified on the load profile curve.

Sizing a cogenerator at baseload achieves the greatest efficiency and best use of capital. However, it may not offer the shortest payback because of the high value of electrical power. An appropriately sized CHP plant might be sized somewhat larger than base load to minimize payback time through increased electrical savings. A combination of load analysis and economic analysis must be performed to determine the most economical plant. For this example, the maximum economical plant size is arbitrarily assumed to be 28,000 lb/h. CHP systems sized this way produce high equipment utilization and depend on the utility to serve peak loads above the plant's 28,000 lb/h capacity.

The load duration curve allows the designer to estimate the total amount of steam generated within the interval. Because the total amount of steam is the area under the curve, the calculation may be performed by using either equations for rectangles and triangles or an appropriate curve analysis program. For a thermally tracked plant sized at the 21,000 lb/h baseload, the hours of operation are about 93 h per week. Therefore, based on the area under the rectangle, the total steam produced by the CHP plant is

$$\text{Cogenerated steam} = (93 \text{ h per week})(21 \times 10^3 \text{ lb/h}) = 1.953 \times 10^6 \text{ lb per week}$$

If the plant is shut down during nonworking hours, no steam is wasted. In addition, if the electrical load profile is always above that needed to produce the 21,000 lb/h steam, the plant can run at full capacity for power and steam, while an electric utility provides peak power and a supplementary boiler provides peak steam. However, if

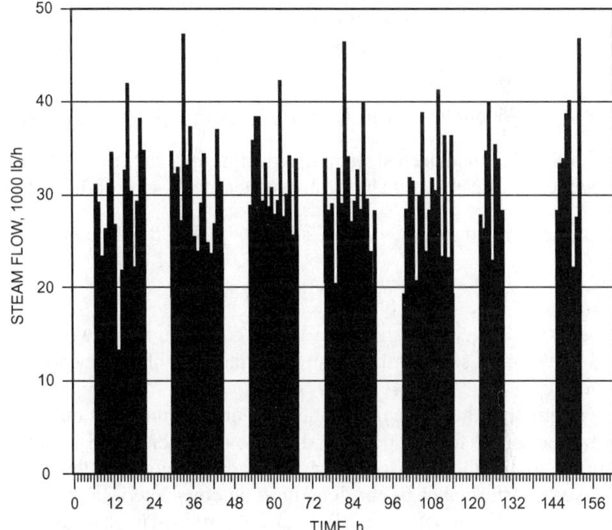

Fig. 68 Hypothetical Steam Load Profile

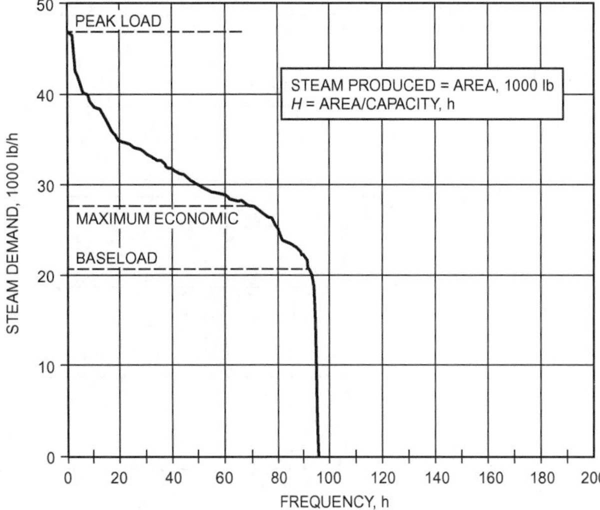

Fig. 69 Load Duration Curve

the facility's steam load profile is unable to absorb the steam produced at continuous full power, electrical output can be reduced accordingly to avoid steam waste.

Where it is more cost-effective to generate excess power than to suffer the parasitic cost of condensing the steam, the plant can still be operated at full power. Therefore, it is important to examine whether the electrical load profile matches or exceeds the electrical output. If it does, then the load duration curve reveals the quality of boiler-generated steam required. This value, which can be estimated by calculating the size of the triangular area above the baseload, is

$$\begin{aligned}\text{Boiler steam required} &= 93(47{,}000 - 21{,}000)/2 \\ &= 1.21 \times 10^6 \text{ lb per week}\end{aligned}$$

$$\text{Cogenerated kW} = (21 \times 10^3 \text{ lb/h})/(\text{Steam-to-Electric Ratio})$$

where the ratio is that of the prime mover, in lb of steam/kWh. The total weekly electrical production is

$$\text{Cogenerated kWh} = 1.953 \times 10^6/\text{Ratio}$$

Fuel consumption equals cogenerated kilowatt-hours times the full-load heat in Btu, to which is added the boiler fuel consumption.

The **equivalent full-load hours (EFLH)** of both steam and electric CHP production in this base-load sizing and operating mode is 93 h per week, so that

$$\text{Electric production} = \text{EFLH} \times \text{Rated capacity (kW) at full load}$$

If the plant had been sized at the maximum economic return, the cogenerated steam would be the area under the 28×10^3 lb/h level, or

$$\begin{aligned}\text{Cogenerated steam} &= 1.953 \times 10^6 + 70(28{,}000 - 21{,}000) \\ &\quad + (93 - 70)(28{,}000 - 21{,}000)/2 \\ &= 2.52 \times 10^6 \text{ lb per week}\end{aligned}$$

For this larger size, there is a higher electrical and lower boiler production, but some of the steam is condensed without any productive use.

Estimating the fuel consumption and electrical energy output of a system sized for the peak load or above the baseload is not as simple as for the base-load design because changes in the prime mover's performance at part load must be considered. In this case, average value estimates of performance at part load must be used for preliminary studies.

In some cases, an installation has only one prime mover; however, several smaller units operating in parallel provide increased reliability and performance during part-load operation. Figure 70 illustrates the use of three prime movers rated at 10,000 lb/h each. For this example, generators #1 and #2 operate fully loaded, and #3 operates between full-load capacity and 50% capacity while tracking facility thermal demand. Because operation at less than 50% load is inefficient, further reduction in total output must be achieved by part-load operation of generators #1 and #2.

The advantages of multiple units are offset by their higher specific investment and maintenance costs, the control complexity, and the usually lower efficiency of smaller units.

Facilities such as hospitals often seek to reduce their utility costs by using existing standby generators to share the electrical peak (peak shaving). Such an operation is not strictly CHP because heat recovery is rarely justified. Figure 71 illustrates a hypothetical electrical load duration curve with frequency as a percent of total hours in the year (8760). A generator rated at 1000 kW, for example, reduces the peak demand by 500 kW if it is operated between 500 and 1000 kW to avoid extended hours of low-efficiency operation.

Some electrical energy saving as well as a reduction in demand are obtained by operating the generator. However, this saving is idealistic because it can only be obtained if the operators or control system can anticipate when facility demand will exceed 1400 kW in time to bring the generator up to operating condition. The peak shaver should not be started too early, because it would waste fuel.

Also, many utilities include ratchet clauses in their rate schedules. As a result, if the peak-shaving generator is inoperative for any reason when the facility's monthly peak occurs, the ratchet is set for a year hence, and the demand savings potential of the peaking generator will not be realized until a year later. Even though an existing standby generator may seem to offer "free" peak shaving capacity, careful planning and operation are required to secure its full potential.

Conversely, continuous-duty/standby systems offer the benefits of heat recovery while satisfying the facility's standby requirements. During emergencies, the generator load is switched from its normal nonessential load to the emergency load.

Two-Dimensional Load Duration Curve

A two-dimensional load duration curve becomes necessary when the designer must consider simultaneous steam and electrical load variations, such as when facility electrical demand drops below the output of a steam-tracking generator, and the excess power capacity cannot be exported. During these periods, the generator is throttled to curtail electrical output to that of facility demand (i.e., it operates in electrical tracking mode).

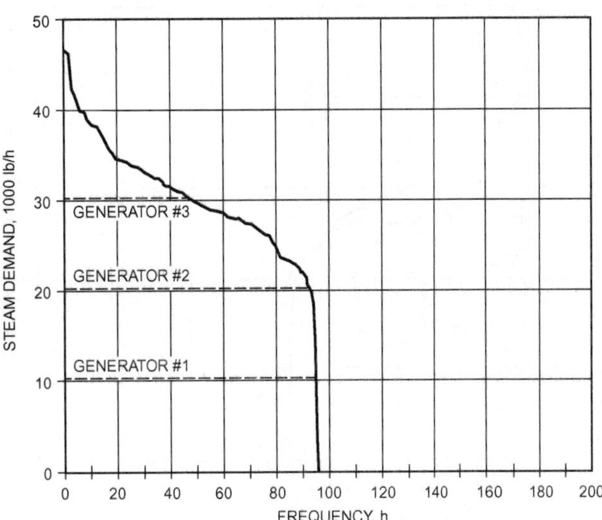

Fig. 70 Load Duration Curve with Multiple Generators

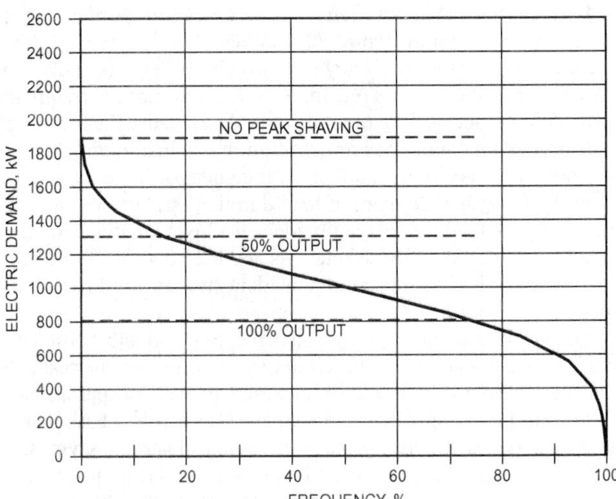

Fig. 71 Hypothetical Peaking Generator

To develop the two-dimensional load duration curve method for simultaneous loads, either the electrical or steam demand must be broken into discrete periods defined by the number of hours per year when the demand is within a certain range of values, or bins.

Duration curves for the remaining load are then created for each period as before, using coincident values. Figure 72 illustrates such a representation for three bins. The electrical load values indicated in the figure represent the center value of each bin. In general, a large number of periods gives a more accurate representation of facility loads. Also, the greater the load fluctuation, the greater the number of periods required for accurate representation. Note that the total number of hours for all periods adds up to 8760, the number of hours in a year.

Another situation that requires a two-dimensional load duration curve is when the facility buys or sells power, the price of which depends on the time of day or time of the year. Many electric rate structures contain explicit time periods for the purchase or sale of power. In some cases, there is only a summer/winter distinction. Other rate schedules may have several periods to reflect time-of-use or time-of-sale rates. The two-dimensional analysis is required to consider these rate schedule periods when defining the load bins because the operating schedule with the greatest annual savings is influenced by energy prices as well as energy demands.

Using Figure 72 as an illustration of a summer-peaking utility, the first two bins might coincide with the winter/fall/spring off-peak rate and the third bin with the on-peak rate. Because analysis becomes burdensome when there are large load swings and several rate periods, the calculations are run by computers.

Analysis by Simulations

The load duration curve is a convenient, intuitive graphical tool for preliminary sizing and analysis of a CHP system; it lacks the capability, however, for detailed analysis. A commercial or institutional facility, for example, can have as many as four different loads that must be considered simultaneously: cooling, noncooling electrical, steam or high-temperature hot water for space heating, and low-temperature service hot water. These loads are never in balance at any instant, which complicates sizing equipment, establishing operating modes, and determining the quantity of heat rejected from the cogenerator that is usefully applied to the facility loads.

The fact that prime movers rarely operate at full rated load further complicates evaluation of commercial systems; therefore, part-load operating characteristics such as fuel consumption, exhaust mass flow, exhaust temperature, and heat rejected from the jacket and intercooler of internal combustion engines must be considered.

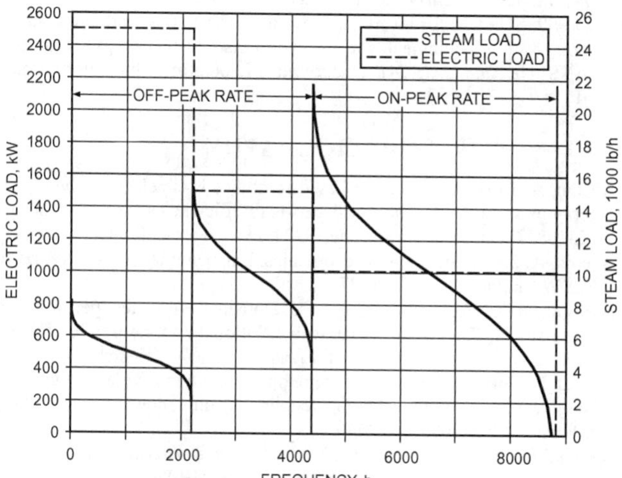

Fig. 72 Example of Two-Dimensional Load Duration Curve

If the prime mover is a combustion gas turbine, then the effects of ambient temperature on full-load capacity and part-load fuel consumption and exhaust characteristics (or on cooling capacity if gas turbine inlet air cooling is used) should be considered.

In addition to the prime movers, a commercial or institutional CHP system may include absorption chillers that use jacket and exhaust heat to produce chilled water at two different COPs. Some commercial systems use thermal energy storage to store hot water, chilled water, or ice to reduce the recovered thermal energy that must be dumped during periods of low demand. Internal combustion engines coupled with steam compressors and steam-injected gas turbines have been produced to allow some variability of the output heat/power ratio from a single package.

Computer screening programs that analyze CHP systems are readily available. Many of these programs emphasize the financial aspects of CHP with elaborate rate structures, energy price forecasts, and economic models. Other programs emphasize equipment part-load performance, load schedules, and other technical characteristics. Although these programs can be useful, they do not remove the need for detailed analysis before undertaking a project.

The primary consideration in selecting a computer program for analyzing commercial CHP technical feasibility is the ability to handle multiple, time-varying loads. The four methods of modeling the thermal and electrical loads are as follows:

- Hourly average values for a complete year
- Monthly average values
- Truncated year consisting of hourly averaged values for one or more typical (usually working and nonworking) days of each month
- Bin methods

CHP simulation using hourly averaged values for load representation provides the greatest accuracy; however, the 8760 values for each type of load can make managing load data a formidable task. Guinn (1987) describes a public-domain simulation that can consider a full year of multiple-load data.

The data needed to perform a monthly averaged load representation can often be obtained from utility billings. This model should only be used as an initial analysis; accuracy is greatest when thermal and electric profiles are relatively consistent.

The truncated-year model is a compromise between the accuracy offered by the full hourly model and the minimal data-handling requirements of the monthly average model. It involves developing hourly values for each load over a typical average day of each month. Usually, two typical days are considered: working and nonworking. Thus, instead of 8760 values to represent a load over a year, only 576 values are required. This type of load model is often used in CHP computer programs. Pedreyra (1988) and Somasundaram (1986) describe programs that use this method of load modeling.

Bin methods are based on the frequency distribution, or histogram, of load values. The method determines the number of hours per year the load was in different ranges, or bins. This method of representing weather data is widely used to perform building energy analysis. It is a convenient way to condense a large database into a smaller set of values, but it is no more accurate than the time resolution of the original set. Furthermore, bin methods become unacceptably cumbersome for CHP analysis if more than two loads must be considered.

REFERENCES

Abedin, A. 2003. Cogeneration systems: Balancing the heat-power ratio. *ASHRAE Journal* 45(8):24-27.

AGCC. 1999. *Application engineering manual for engine-driven commercial and industrial refrigeration systems.* American Gas Cooling Center, Washington, D.C.

Andrepont, J.S. 2001. Combustion turbine inlet air cooling (CTIAC): Benefits and technology options in district energy applications. *ASHRAE Transactions* 107(1):892-899.

ARI. 2003. Standard for performance of water-chilling packages using the vapor compression cycle. *Standard* 550/590-2003. Air-Conditioning and Refrigeration Institute, Arlington, VA.

ASHRAE. 1997. *Design guide for combustion turbine inlet air cooling systems*.

ASHRAE. 2001. *Fuel cells for building applications*.

ASME. 2001. Test code for steam turbines—Appendix to PTC 6. ANSI/ASME *Standard* PTC 6A-2001. American Society of Mechanical Engineers, New York.

ASME. 1985. Guidance for evaluation of measurement uncertainty in performance tests of steam turbines. ANSI/ASME *Standard* PTC 6 Report-85 (R 1997). American Society of Mechanical Engineers, New York.

ASME. 1988. Procedures for routine performance tests of steam turbines. ANSI/ASME *Standard* PTC 6S Report-88 (R 1995). American Society of Mechanical Engineers, New York.

ASME. 1994. Measurement of exhaust emissions from stationary gas turbine engines. *Standard* B133.9. American Society of Mechanical Engineers, New York.

ASME. 2004. Steam turbines. *Standard* PTC 6-2004. American Society of Mechanical Engineers, New York.

ASME. 2005. Performance test code on gas turbines. *Standard* PTC 22-2005. American Society of Mechanical Engineers, New York.

ASME. 2004. Rules for the construction of pressure vessels. *Boiler and Pressure Vessel Code*, Section VIII-2004, Division 1. American Society of Mechanical Engineers, New York.

ASME. 2002. Fuel cell power systems performance. *Standard* PTC 50. American Society of Mechanical Engineers, New York.

Avery, G., W.C. Stethem, W.J. Coad, R.A. Hegberg, F.L. Brown, and R. Petitjean. 1990. The pros and cons of balancing a variable flow water system. *ASHRAE Journal* 32(10):32-59.

Becker, H.P. 1975. Energy conservation analysis of pumping systems. *ASHRAE Journal* 17(4):43-51.

Becker, H.P. 1988. Is cogeneration right for you? *Hotel and Resort Industry* (Sept.).

Brown, D.R. and S. Somasundaram. 1997. Recuperators, regenerators, and storage: Thermal energy storage applications in gas-fired power plants. Chapter 13B in *CRC handbook of energy efficiency*. Frank Kreith and Ronald E. West, eds. CRC Press, Boca Raton, FL.

DEMA. 1972. *Standard practices for low and medium stationary diesel and gas engines*, 6th ed. Diesel Engine Manufacturers Association.

DOE. 2003. *Gas-fired distributed energy resource technology characterizations: Reciprocating engines*. U.S. Department of Energy, Washington, D.C.

DOE. 2007. *Hydrogen, fuel cells and infrastructure technologies program—Fuel cells*. U.S. Department of Energy, Washington, D.C. http://www1.eere.energy.gov/hydrogenandfuelcells/fuelcells/fc_types.html (24 Mar. 2008).

Ebeling, J.A. 1994. Combustion turbine inlet air cooling alternatives and case histories. *Proceedings of the 27th Annual Frontiers of Power Conference*, Stillwater, OK (Oct. 24-25).

EIA. 2005. Annual Energy Review 2005. DOE/EIA-0384(2005). July 2006. Washington, D.C. Energy Information Administration, U.S. Department of Energy, Washington, D.C.

Encotech. 1992. Coupling effects on the torsional vibration of turbine generator power trains. *Encotech Technical Topics* 7(1). Schenectady, NY.

EPA. 2002. *Technology characterization: Reciprocating engines*. U.S. Environmental Protection Agency, Washington, D.C. http://www.epa.gov/chp/documents/internal_combustion.pdf.

Foley, G. and R. Sweetser. 2002. Emerging role for absorption chillers in integrated energy systems in America. *Proceedings of the ASME International Mechanical Engineering Congress & Exposition*, New Orleans, LA. pp. 1-11.

Guinn, G.R. 1987. Analysis of cogeneration systems using a public domain simulation. *ASHRAE Transactions* 93(2):333-366.

Hodge, B.K. and J.D. Hardy. 2002. Cooling, heating, and power for buildings (CHP-B) instructional module. Department of Mechanical Engineering, Mississippi State University.

Honeywell Laboratories. 2006. Modular integrated energy systems. *Final Report*. Oak Ridge National Laboratory, Oak Ridge, TN. http://www.ornl.gov/sci/engineering_science_technology/cooling_heating_power/pdf/2006_June_4,00001 1476_FinalReport.pdf.

IEEE. 2003. Standard for interconnecting distributed resources with electric power systems. IEEE *Standard* 1547. Institute of Electrical and Electronics Engineers, New York.

ISO. 2002. Reciprocating internal combustion engines—Performance—Part 1: Declarations of power, fuel, and lubricating oil consumptions, and test methods—Additional requirements for engines for general use. *Standard* 3046-1:2002. International Organization for Standardization, Geneva, Switzerland.

ISO. 1997. Gas turbines—Procurement—Part 2: Standard reference conditions and ratings. *Standard* 3977-1:1997. International Organization for Standardization, Geneva, Switzerland.

Lorsch, H. 1993. *Air-conditioning system design manual*. ASHRAE.

MacCracken, C.D. 1994. Overview of the progress and the potential of thermal storage in off-peak turbine inlet cooling. *ASHRAE Transactions* 100(1):569-571.

Mannion, G.F. 1988. High temperature rise piping design for variable volume pumping systems: Key to chiller energy management. *ASHRAE Transactions* 94(2):1427-1443.

NEMA. 1991. Steam turbines for mechanical drive service. ANSI/NEMA *Standard* SM 23-91 (R 2002). National Electrical Manufacturers Association, Rosslyn, VA.

NEMA. 1991. Land based steam turbine generator sets 0 to 33,000 kW. *Standard* SM 24-1991 (R1997, R2002). National Electrical Manufacturers Association, Rosslyn, VA.

NFPA. 2006. Installation and use of stationary combustion engines and gas turbines. *Standard* 37. National Fire Prevention Association, Quincy, MA.

NFPA. 2006. National fuel gas code. *Standard* 54. National Fire Prevention Association, Quincy, MA.

NFPA. 2008. National electrical code®. *Standard* 70. National Fire Prevention Association, Quincy, MA.

NFPA. 2005. Recommended practice for fire protection for electric generating plants and high voltage direct current converter stations. *Standard* 850. National Fire Prevention Association, Quincy, MA.

Newman, L.E. 1945. Modern extraction turbines, Part II—Estimating the performance of a single automatic extraction condensing steam turbine. *Power Plant Engineering* (Feb.).

NREL. 2003. Gas-fired distributed energy resource technology characterizations. *Report*. National Renewable Energy Laboratory (NREL), Golden, CO, U.S. Department of Energy, Office of Energy Efficiency and Renewable Energy (EERE), Washington, D.C., and Gas Research Institute (GRI), Des Plaines, IL.

Pedreyra, D.C. 1988. A microcomputer version of a large mainframe program for use in cogeneration analysis. *ASHRAE Transactions* 94(1):1617-1625.

Peltier, R.V. 2001. Cogeneration: How efficient is "efficiency"? *Power Magazine* (March/April).

Somasundaram, S. 1986. An analysis of a cogeneration system at T.W.U. *Cogeneration Journal* 1(4):16-26.

Somasundaram, S., M.K. Drost, D.R. Brown, and Z.I. Antoniak. 1996. Thermal energy storage in utility-scale applications. *ASME Journal of Engineering for Gas Turbines and Power* 118(1):32-37.

UL. 1999. Standard for inverters, converters, controllers, and interconnection system equipment for use with distributed energy resources. ANSI/UL *Standard* 1741. Underwriters Laboratories Inc., Northbrook, IL.

BIBLIOGRAPHY

Abedin, A. 1997. A gas turbine-based combined cycle electric power generation system with increased part-load efficiencies. U.K. Patent GB 2318832, date of filing 10 June 1997 (www.cogen-unlimited.com).

Abedin, A. 2003. Cogeneration economics: Balancing the heat-power ratio. *ASHRAE Journal* 45(8):24-27.

Burns & McDonnell Engineering. 2005. Domain CHP system performance assessment. *Final Report*. Oak Ridge National Laboratory, Oak Ridge, TN, and U.S. Department of Energy, Office of Energy Efficiency and Renewable Energy, Washington, D.C. http://www.ornl.gov/sci/engineering_science_technology/cooling_heating_power/pdf/05-1208_Domain_May_2005_Final.pdf.

National Engine Use Council. 1967. *NEUC engine criteria*. Chicago.

Orlando, J. 1996. *Cogeneration design guide*. ASHRAE.

UL. 1998. Standard for stationary engine generator assemblies. *Standard* 2200. Underwriters Laboratories Inc., Northbrook, IL.

CHAPTER 8

APPLIED HEAT PUMP AND HEAT RECOVERY SYSTEMS

TERMINOLOGY

BALANCED heat recovery. Occurs when internal heat gain equals recovered heat and no external heat is introduced to the conditioned space. Maintaining balance may require raising the temperature of recovered heat.

Break-even temperature. The outdoor temperature at which total heat losses from conditioned spaces equal internally generated heat gains.

Changeover temperature. The outdoor temperature the designer selects as the point of changeover from cooling to heating by the HVAC system.

Coefficient of performance (COP). The ratio of heat transferred at the condenser of a heat pump to the energy used to power the heat pump.

External heat. Heat generated from sources outside the conditioned area. This heat from gas, oil, steam, electricity, or solar sources supplements internal heat and internal process heat sources. Recovered internal heat can reduce the demand for external heat.

Internal heat. Total passive heat generated within the conditioned space. It includes heat generated by lighting, computers, business machines, occupants, and mechanical and electrical equipment such as fans, pumps, compressors, and transformers.

Internal process heat. Heat from industrial activities and sources such as wastewater, boiler flue gas, coolants, exhaust air, and some waste materials. This heat is normally wasted unless equipment is included to extract it for further use.

Pinch technology. An energy analysis tool that uses vector analysis to evaluate all heating and cooling utilities in a process. Composite curves created by adding the vectors allow identification of a "pinch" point, which is the best thermal location for a heat pump.

Recovered (or reclaimed) heat. Comes from internal heat sources. It is used for space heating, domestic or service water heating, air reheat in air conditioning, process heating in industrial applications, or other similar purposes. Recovered heat may be stored for later use.

Stored heat. Heat from external or recovered heat sources that is held in reserve for later use.

System coefficient of performance. Ratio of heat recovery system output to entire system energy input, including compressor, pumps, etc.

Usable temperature. Temperature or range of temperatures at which heat energy can be absorbed, rejected, or stored for use within the system.

Waste heat. Heat rejected from the building (or process) because its temperature is too low for economical recovery or direct use.

The preparation of this chapter is assigned to TC 9.4, Applied Heat Pump/Heat Recovery Systems.

APPLIED HEAT PUMP SYSTEMS

A heat pump extracts heat from a source and transfers it to a sink at a higher temperature. According to this definition, all pieces of refrigeration equipment, including air conditioners and chillers with refrigeration cycles, are heat pumps. In engineering, however, the term **heat pump** is generally reserved for equipment that heats for beneficial purposes, rather than that which removes heat for cooling only. Dual-mode heat pumps alternately provide heating or cooling. Heat reclaim heat pumps provide heating only, or simultaneous heating and cooling. An applied heat pump requires competent field engineering for the specific application, in contrast to the use of a manufacturer-designed unitary product. Applied heat pumps include built-up heat pumps (field- or custom-assembled from components) and industrial process heat pumps. Most modern heat pumps use a vapor compression (modified Rankine) cycle or absorption cycle. Any of the other refrigeration cycles discussed in Chapter 1 of the 2005 *ASHRAE Handbook—Fundamentals* are also suitable. Although most heat pump compressors are powered by electric motors, limited use is also made of engine and turbine drives. Applied heat pumps are most commonly used for heating and cooling buildings, but they are gaining popularity for efficient domestic and service water heating, pool heating, and industrial process heating.

Applied heat pumps with capacities from 24,000 to 150,000,000 Btu/h operate in many facilities. Some machines are capable of output water temperatures up to 220°F and steam pressures up to 60 psig.

Compressors in large systems vary from one or more reciprocating or screw types to staged centrifugal types. A single or central system is often used, but in some instances, multiple heat pump systems are used to facilitate zoning. Heat sources include the ground, well water, surface water, gray water, solar energy, the air, and internal building heat. Compression can be single-stage or multistage. Frequently, heating and cooling are supplied simultaneously to separate zones.

Decentralized systems with water loop heat pumps are common, using multiple water-source heat pumps connected to a common circulating water loop. They can also include ground coupling, heat rejectors (cooling towers and dry coolers), supplementary heaters (boilers and steam heat exchangers), loop reclaim heat pumps, solar collection devices, and thermal storage. The initial cost is relatively low, and building reconfiguration and individual space temperature control are easy.

Community and district heating and cooling systems can be based on both centralized and distributed heat pump systems.

HEAT PUMP CYCLES

Several types of applied heat pumps (both open- and closed-cycle) are available; some reverse their cycles to deliver both

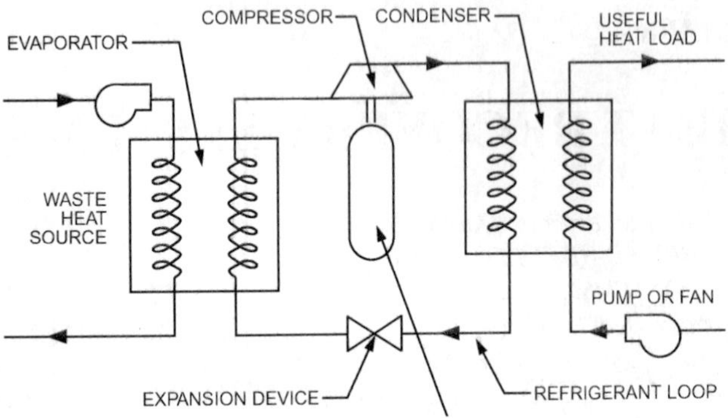

Fig. 1 Closed Vapor Compression Cycle

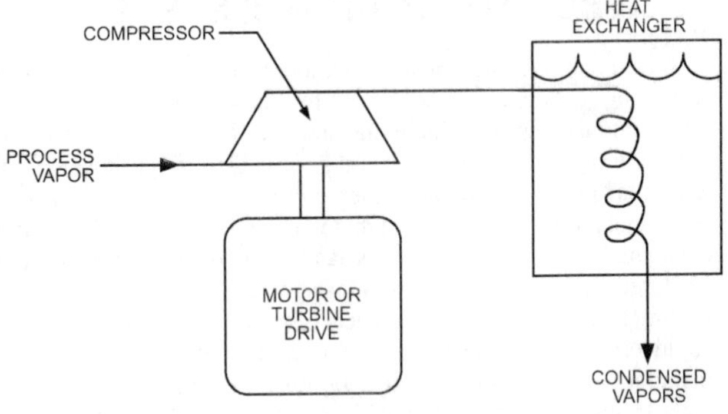

Fig. 2 Mechanical Vapor Recompression Cycle with Heat Exchanger

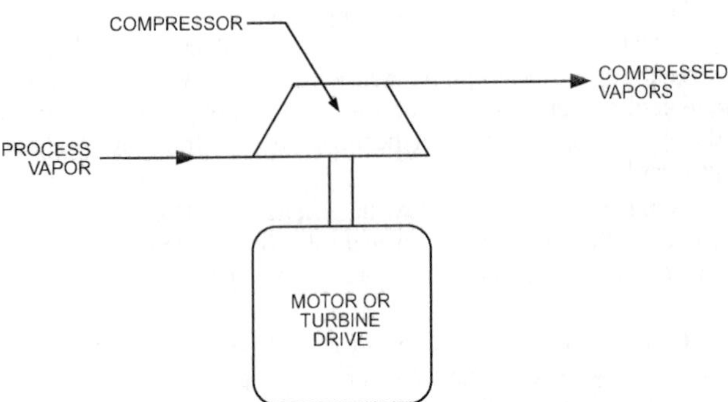

Fig. 3 Open Vapor Recompression Cycle

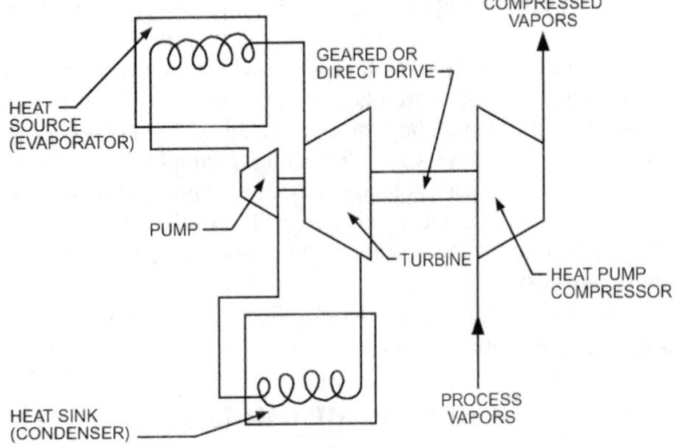

Fig. 4 Heat-Driven Rankine Cycle

heating and cooling in HVAC systems, and others are for heating only in HVAC and industrial process applications. The following are the four basic types of heat pump cycles:

- **Closed vapor compression cycle** (Figure 1). This is the most common type in both HVAC and industrial processes. It uses a conventional, separate refrigeration cycle that may be single-stage, compound, multistage, or cascade.
- **Mechanical vapor recompression (MVR) cycle with heat exchanger** (Figure 2). Process vapor is compressed to a temperature and pressure sufficient for reuse directly in a process. Energy consumption is minimal, because temperatures are optimum for the process. Typical applications include evaporators (concentrators) and distillation columns.
- **Open vapor recompression cycle** (Figure 3). A typical application is in an industrial plant with a series of steam pressure levels and an excess of steam at a lower-than-desired pressure. Heat is pumped to a higher pressure by compressing the lower-pressure steam.
- **Heat-driven Rankine cycle** (Figure 4). This cycle is useful where large quantities of heat are wasted and energy costs are high. The heat pump portion of the cycle may be either open or closed, but the Rankine cycle is usually closed.

HEAT SOURCES AND SINKS

Table 1 shows the principal media used as heat sources and sinks. Selecting a heat source and sink for an application is primarily influenced by geographic location, climate, initial cost, availability, and type of structure. Table 1 presents various factors to be considered for each medium.

Air

Outdoor air is a universal heat source and sink medium for heat pumps and is widely used in residential and light commercial systems. Extended-surface, forced-convection heat transfer coils transfer heat between the air and refrigerant. Typically, the surface area of outdoor coils is 50 to 100% larger than that of indoor coils. The volume of outdoor air handled is also greater than the volume of indoor air handled by about the same percentage. During heating, the temperature of the evaporating refrigerant is generally 10 to 20°F less than the outdoor air temperature. Air heating and cooling coil performance is discussed in more detail in Chapters 22 and 26.

When selecting or designing an air-source heat pump, two factors in particular must be considered: (1) the local outdoor air temperature and (2) frost formation.

As the outdoor temperature decreases, the heating capacity of an air-source heat pump decreases. This makes equipment selection for a given outdoor heating design temperature more critical for an air-source heat pump than for a fuel-fired system. Equipment must be sized for as low a balance point as is practical for heating without having excessive and unnecessary cooling capacity during the summer. A procedure for finding this balance point, which is defined as the outdoor temperature at which heat pump capacity matches heating requirements, is given in Chapter 48.

When the surface temperature of an outdoor air coil is 32°F or less, with a corresponding outside air dry-bulb temperature 4 to 10°F higher, frost may form on the coil surface. If allowed to accumulate, frost inhibits heat transfer; therefore, the outdoor coil must be defrosted periodically. The number of defrosting operations is influenced by the climate, air-coil design, and the hours of operation. Experience shows that, generally, little defrosting is required when outdoor air conditions are below 17°F and 60% rh. This can be confirmed by psychrometric analysis using the principles given in Chapter 22. However, under very humid conditions, when small suspended water droplets are present in the air, the rate of frost deposit may be about three times as great as predicted from psychrometric theory and the heat pump may require defrosting after as little

Table 1 Heat Pump Sources and Sinks

Medium	Examples	Suitability: Heat Source	Suitability: Heat Sink	Availability: Location Relative to Need	Availability: Coincidence with Need	Cost: Installed	Cost: Operation and Maintenance	Temperature: Level	Temperature: Variation	Common Practice: Use	Common Practice: Limitations
AIR											
Outdoor	Ambient air	Good, but efficiency and capacity in heating mode decrease with decreasing outdoor air temperature	Good, but efficiency and capacity in cooling mode decrease with increasing outdoor air temperature	Universal	Continuous	Low	Moderate	Variable	Generally extreme	Most common, many standard products	Defrosting and supplemental heat usually required
Exhaust	Building ventilation	Excellent	Fair	Excellent if planned for in building design	Excellent	Low to moderate	Low unless exhaust is laden with dirt or grease	Excellent	Very low	Excellent as energy-conservation measure	Insufficient for typical loads
WATER											
Well*	Ground-water well may also provide potable water source	Excellent	Excellent	Poor to excellent; practical depth varies by location	Continuous	Low if existing well used or shallow wells suitable; can be high otherwise	Low, but periodic maintenance required	Generally excellent; varies by location	Extremely stable	Common	Water disposal and required permits may limit; may require double-wall exchangers; may foul or scale
Surface	Lakes, rivers, oceans	Excellent for large water bodies or high flow rates	Excellent for large water bodies or high flow rates	Limited; depends on proximity	Usually continuous	Depends on proximity and water quality	Depends on proximity and water quality	Usually satisfactory	Depends on source	Available, particularly for fresh water	Often regulated or prohibited; may clog, foul, or scale
Tap (city)	Municipal water supply	Excellent	Excellent	Excellent	Continuous	Low	Low energy cost, but water use and disposal may be costly	Excellent	Usually very low	Use is decreasing because of regulations	Use or disposal may be regulated or prohibited; may corrode or scale
Condensing	Cooling towers, refrigeration systems	Excellent	Poor to good	Varies	Varies with cooling loads	Usually low	Moderate	Favorable as heat source	Depends on source	Available	Suitable only if heating need is coincident with heat rejection
Closed loops	Building water-loop heat pump systems	Good; loop may need supplemental heat	Favorable; may need loop heat rejection	Excellent if designed as such	As needed	Low	Low to moderate	As designed	As designed	Very common	Most suitable for medium or large buildings
Waste	Raw or treated sewage, gray water	Fair to excellent	Fair; varies with source	Varies	Varies; may be adequate	Depends on proximity; high for raw sewage	Varies; may be high for raw sewage	Excellent	Usually low	Uncommon; practical only in large systems	Usually regulated; may clog, foul, scale, or corrode
GROUND*											
Ground-coupled	Buried or submerged fluid loops	Good if ground is moist; otherwise poor	Fair to good if ground is moist; otherwise poor	Depends on soil suitability	Continuous	High to moderate	Low	Usually good	Low, particularly for vertical systems	Rapidly increasing	High initial costs for ground loop
Direct-expansion	Refrigerant circulated in ground coil	Varies with soil conditions	Varies with soil conditions	Varies with soil conditions	Continuous	High	High	Varies by design	Generally low	Extremely limited	Leak repair very expensive; requires large refrigerant quantities
SOLAR ENERGY											
Direct or heated water	Solar collectors and panels	Fair	Poor; usually unacceptable	Universal	Highly intermittent; night use requires storage	Extremely high	Moderate to high	Varies	Extreme	Very limited	Supplemental source or storage required
INDUSTRIAL PROCESS											
Process heat or exhaust	Distillation, molding, refining, washing, drying	Fair to excellent	Varies; often impractical	Varies	Varies	Varies	Generally low	Varies	Varies	Varies	May be costly unless heat need is near rejected source

*Groundwater-source heat pumps are also considered ground-source heat pump systems.

as 20 min of operation. The loss of available heating capacity caused by frosting should be considered when sizing an air-source heat pump.

Following commercial refrigeration practice, early designs of air-source heat pumps had relatively wide fin spacing of 4 to 5 fins/in., based on the theory that this would minimize defrosting frequency. However, experience has shown that effective hot-gas defrosting allows much closer fin spacing and reduces the system's size and bulk. In current practice, fin spacings of 10 to 20 fins/in. are widely used.

In many institutional and commercial buildings, some air must be continuously exhausted year-round. This exhaust air can be used as a heat source, although supplemental heat is generally necessary.

High humidity caused by indoor swimming pools causes condensation on ceiling structural members, walls, windows, and floors and causes discomfort to spectators. Traditionally, outside air and dehumidification coils with reheat from a boiler that also heats the pool water are used. This is ideal for air-to-air and air-to-water heat pumps because energy costs can be reduced. Suitable materials must be chosen so that heat pump components are resistant to corrosion from chlorine and high humidity.

Water

Water can be a satisfactory heat source, subject to the considerations listed in Table 1. City water is seldom used because of cost and municipal restrictions. Groundwater (well water) is particularly attractive as a heat source because of its relatively high and nearly constant temperature. Water temperature depends on source depth and climate, but, in the United States, generally ranges from 40°F in northern areas to 70°F in southern areas. Frequently, sufficient water is available from wells (water can be reinjected into the aquifer). This use is nonconsumptive and, with proper design, only the water temperature changes. Water quality should be analyzed, and the possibility of scale formation and corrosion should be considered. In some instances, it may be necessary to separate the well fluid from the equipment with an additional heat exchanger. Special consideration must also be given to filtering and settling ponds for specific fluids. Other considerations are the costs of drilling, piping, pumping, and a means for disposal of used water. Information on well water availability, temperature, and chemical and physical analysis is available from U.S. Geological Survey offices in many major cities.

Heat exchangers may also be submerged in open ponds, lakes, or streams. When surface or stream water is used as a source, the temperature drop across the evaporator in winter may need to be limited to prevent freeze-up.

In industrial applications, waste process water (e.g., spent warm water in laundries, plant effluent, warm condenser water) may be a heat source for heat pump operation.

Sewage, which often has temperatures higher than that of surface or groundwater, may be an acceptable heat source. Secondary effluent (treated sewage) is usually preferred, but untreated sewage may be used successfully with proper heat exchanger design.

Use of water during cooling follows the conventional practice for water-cooled condensers.

Water-to-refrigerant heat exchangers are generally direct-expansion or flooded water coolers, usually shell-and-coil or shell-and-tube. Brazed-plate heat exchangers may also be used. In large applied heat pumps, the water is usually reversed instead of the refrigerant.

Ground

The ground is used extensively as a heat source and sink, with heat transfer through buried coils. Soil composition, which varies widely from wet clay to sandy soil, has a predominant effect on thermal properties and expected overall performance. The heat transfer process in soil depends on transient heat flow. Thermal diffusivity is a dominant factor and is difficult to determine without local soil data. Thermal diffusivity is the ratio of thermal conductivity to the product of density and specific heat. The soil's moisture content influences its thermal conductivity.

There are three primary types of ground-source heat pumps: (1) groundwater, which is discussed in the previous section; (2) direct-expansion, in which the ground-to-refrigerant heat exchanger is buried underground; and (3) ground-coupled (also called closed-loop ground-source), in which a secondary loop with a brine connects the ground-to-water and water-to-refrigerant heat exchangers (see Figure 5).

Ground loops can be placed either horizontally or vertically. A horizontal system consists of single or multiple serpentine heat exchanger pipes buried 3 to 6 ft apart in a horizontal plane at a depth 3 to 6 ft below grade. Pipes may be buried deeper, but excavation costs and temperature must be considered. Horizontal systems can also use coiled loops referred to as **slinky coils**. A vertical system uses a concentric tube or U-tube heat exchanger. The design of ground-coupled heat exchangers is covered in Chapter 32 of the 2007 *ASHRAE Handbook—HVAC Applications*.

Solar Energy

Solar energy may be used either as the primary heat source or in combination with other sources. Air, surface water, shallow groundwater, and shallow ground-source systems all use solar energy indirectly. The principal advantage of using solar energy directly is that, when available, it provides heat at a higher temperature than the indirect sources, increasing the heating coefficient of performance. Compared to solar heating without a heat pump, the collector efficiency and capacity are increased because a lower collector temperature is required.

Research and development of solar-source heat pumps has been concerned with two basic types of systems: direct and indirect. The **direct** system places refrigerant evaporator tubes in a solar collector, usually a flat-plate type. Research shows that a collector without glass cover plates can also extract heat from the outdoor air. The same surface may then serve as a condenser using outdoor air as a heat sink for cooling.

An **indirect** system circulates either water or air through the solar collector. When air is used, the collector may be controlled in such a way that (1) the collector can serve as an outdoor air preheater, (2) the outdoor air loop can be closed so that all source heat is derived from the sun, or (3) the collector can be disconnected from the outdoor air serving as the source or sink.

TYPES OF HEAT PUMPS

Heat pumps are classified by (1) heat source and sink, (2) heating and cooling distribution fluid, (3) thermodynamic cycle, (4) building structure, (5) size and configuration, and (6) limitation of the source and sink. Figure 5 shows the more common types of closed vapor-compression cycle heat pumps for heating and cooling service.

Air-to-Air Heat Pumps. This type of heat pump is the most common and is particularly suitable for factory-built unitary heat pumps. It is widely used in residential and commercial applications (see Chapter 48). The first diagram in Figure 5 is a typical refrigeration circuit.

In air-to-air heat pump systems, air circuits can be interchanged by motor-driven or manually operated dampers to obtain either heated or cooled air for the conditioned space. In this system, one heat exchanger coil is always the evaporator, and the other is always the condenser. Conditioned air passes over the evaporator during the cooling cycle, and outdoor air passes over the condenser. Damper positioning causes the change from cooling to heating.

Water-to-Air Heat Pumps. These heat pumps rely on water as the heat source and sink, and use air to transmit heat to or from the conditioned space. (See the second diagram in Figure 5.) They include the following:

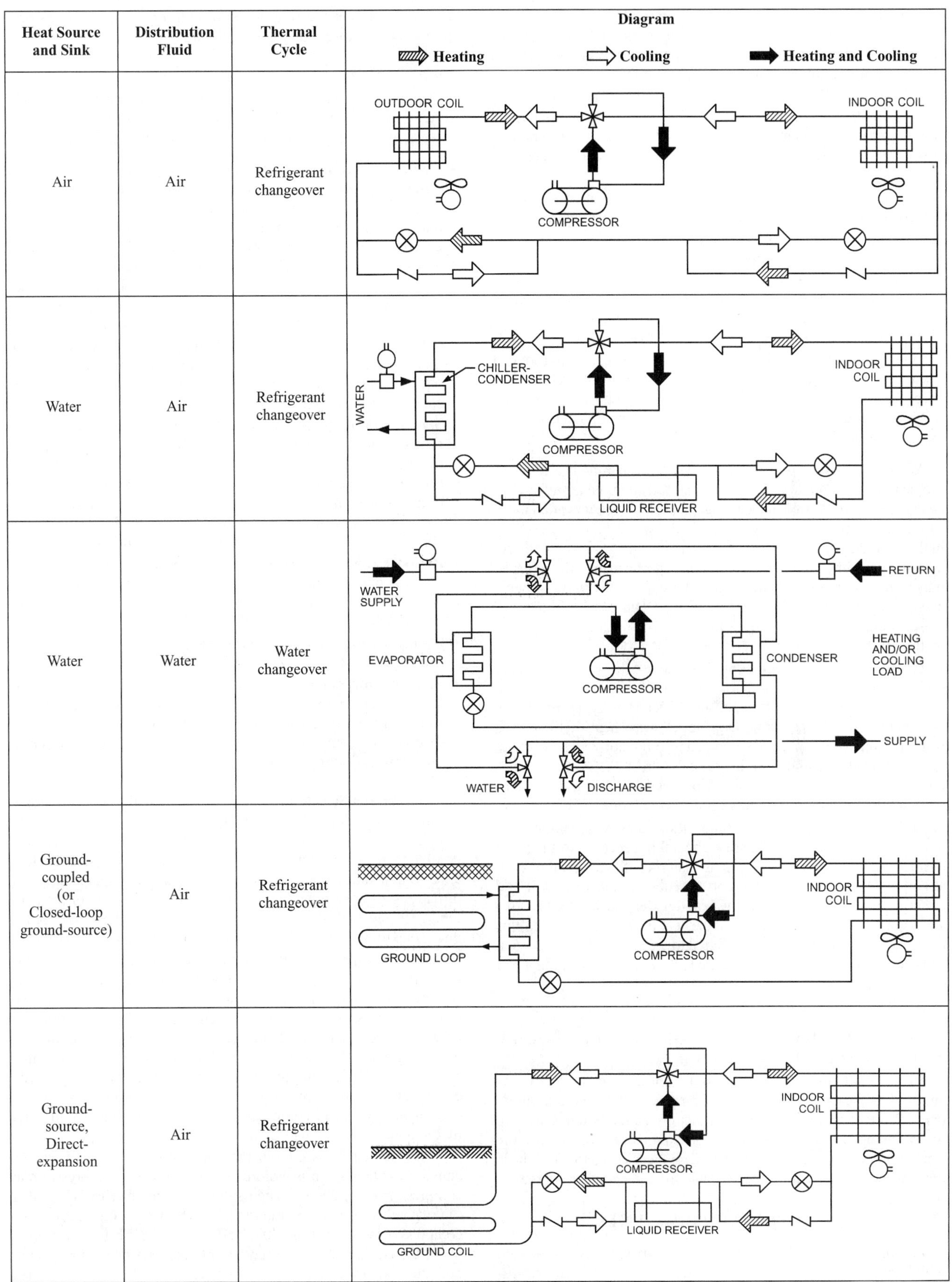

Fig. 5 Heat Pump Types

- ***Groundwater heat pumps***, which use groundwater from wells as a heat source and/or sink. They can either circulate source water directly to the heat pump or use an intermediate fluid in a closed loop, similar to the ground-coupled heat pump.
- ***Surface water heat pumps***, which use surface water from a lake, pond, or stream as a heat source or sink. As with ground-coupled and groundwater heat pumps, these systems can either circulate source water directly to the heat pump or use an intermediate fluid in a closed loop.
- ***Internal-source heat pumps***, which use the high internal cooling load generated in modern buildings either directly or with storage. These include water-loop heat pumps.
- ***Solar-assisted heat pumps***, which rely on low-temperature solar energy as the heat source. Solar heat pumps may resemble water-to-air, or other types, depending on the form of solar heat collector and the type of heating and cooling distribution system.
- ***Wastewater-source heat pumps***, which use sanitary waste heat or laundry waste heat as a heat source. Waste fluid can be introduced directly into the heat pump evaporator after waste filtration, or it can be taken from a storage tank, depending on the application. An intermediate loop may also be used for heat transfer between the evaporator and the waste heat source.

Water-to-Water Heat Pumps. These heat pumps use water as the heat source and sink for cooling and heating. Heating/cooling changeover can be done in the refrigerant circuit, but it is often more convenient to perform the switching in the water circuits, as shown in the third diagram of Figure 5. Although the diagram shows direct admittance of the water source to the evaporator, in some cases, it may be necessary to apply the water source indirectly through a heat exchanger (or double-wall evaporator) to avoid contaminating the closed chilled-water system, which is normally treated. Another method uses a closed-circuit condenser water system.

Ground-Coupled Heat Pumps. These use the ground as a heat source and sink. A heat pump may have a refrigerant-to-water heat exchanger or may be direct-expansion (DX). Both types are shown in Figure 5. In systems with refrigerant-to-water heat exchangers, a water or antifreeze solution is pumped through horizontal, vertical, or coiled pipes embedded in the ground. Direct-expansion ground-coupled heat pumps use refrigerant in direct-expansion, flooded, or recirculation evaporator circuits for the ground pipe coils.

Soil type, moisture content, composition, density, and uniformity close to the surrounding field areas affect the success of this method of heat exchange. With some piping materials, the material of construction for the pipe and the corrosiveness of the local soil and underground water may affect the heat transfer and service life. In a variation of this cycle, all or part of the heat from the evaporator plus the heat of compression are transferred to a water-cooled condenser. This condenser heat is then available for uses such as heating air or domestic hot water.

Additional heat pump types include the following:

Air-to-Water Heat Pumps Without Changeover. These are commonly called *heat pump water heaters*.

Refrigerant-to-Water Heat Pumps. These condense a refrigerant by the cascade principle. Cascading pumps the heat to a higher temperature, where it is rejected to water or another liquid. This type of heat pump can also serve as a condensing unit to cool almost any fluid or process. More than one heat source can be used to offset those times when insufficient heat is available from the primary source.

HEAT PUMP COMPONENTS

For the most part, the components and practices associated with heat pumps evolved from work with low-temperature refrigeration. This section outlines the major components and discusses characteristics or special considerations that apply to heat pumps in combined room heating and air-conditioning applications or in higher temperature industrial applications.

Compressors

The principal types of compressors used in applied heat pump systems are briefly described in this section. For more details on these compressor types and others, refer to Chapter 37.

- **Centrifugal compressors.** Most centrifugal applications in heat pumps have been limited to large water-to-water or refrigerant-to-water heat pump systems, heat transfer systems, storage systems, and hydronically cascaded systems. With these applications, the centrifugal compressor allows heat pump use in industrial plants, as well as in large multistory buildings; many installations have double-bundle condensers. The transfer cycles allow low pressure ratios, and many single- and two-stage units with various refrigerants are operational with high coefficients of performance (COPs). Centrifugal compressor characteristics do not usually meet the needs of air-source heat pumps. High pressure ratios, or high lifts, associated with low gas volume at low load conditions cause the centrifugal compressor to surge.
- **Screw compressors.** Screw compressors offer high pressure ratios at low to high capacities. Capacity control is usually provided by variable porting or sliding vanes. Generally, large oil separators are required, because many compressors use oil injection. Screw compressors are less susceptible to damage from liquid spillover and have fewer parts than do reciprocating compressors. They also simplify capacity modulation.
- **Rotary vane compressors.** These compressors can be used for the low stage of a multistage plant; they have a high capacity but are generally limited to lower pressure ratios. They also have limited means for capacity reduction.
- **Reciprocating compressors.** These compressors are the most common for 0.5 to 100 ton systems.
- **Scroll compressors.** These are a type of orbital motion positive-displacement compressor. Their use in heat pump systems is increasing. Their capacity can be controlled by varying the drive speed or the compressor displacement. Scroll compressors have low noise and vibration levels.

Compressor Selection. A compressor used for comfort cooling usually has a medium clearance volume ratio (the ratio of gas volume remaining in the cylinder after compression to the total swept volume). For an air-source heat pump, a compressor with a smaller clearance volume ratio is more suitable for low-temperature operation and provides greater refrigerating capacity at lower evaporator temperatures. However, this compressor requires somewhat more power under maximum cooling load than one with medium clearance.

More total heat capacity can be obtained at low outdoor temperatures by deliberately oversizing the compressor. This allows some capacity reduction for operation at higher outdoor temperatures by multispeed or variable-speed drives, cylinder cutouts, or other methods. The disadvantage is that the greater number of operating hours that occur at the higher suction temperatures must be served with the compressor unloaded, which generally lowers efficiency and raises annual operating cost. The additional initial cost of the oversized compressor must be economically justified by the gain in heating capacity.

One method proposed for increasing heating output at low temperatures uses **staged compression**. For example, one compressor may compress from −30°F saturated suction temperature (SST) to 40°F saturated discharge temperature (SDT), and a second compressor compresses the vapor from 40°F SST to 120°F SDT. In this arrangement, the two compressors may be interconnected in parallel, with both pumping from about 45°F SST to 120°F SDT for cooling. Then, at some predetermined outdoor temperature on heating, they are reconnected in series and compress in two successive stages.

Figure 6 shows the performance of a pair of compressors for units of both medium and low clearance volume. At low suction temperatures, the reconnection in series adds some capacity. However, the motor for this case must be selected for the maximum loading conditions for summer operation, even though the low-stage compressor has a greatly reduced power requirement under the heating condition.

Compressor Floodback Protection. A suction line separator, similar to that shown in Figure 7, combined with a liquid-gas heat exchanger can be used to minimize migration and harmful liquid floodback to the compressor. The solenoid valve should be controlled to open when the compressor is operating, and the hand valve should then be adjusted to provide an acceptable bleed rate into the suction line.

Heat Transfer Components

Refrigerant-to-air and refrigerant-to-water heat exchangers are similar to heat exchangers used in air-conditioning refrigeration systems.

Defrosting Air-Source Coils. Frost accumulates rather heavily on outdoor-air-source coils when the outdoor air temperature is less than approximately 40°F, but lessens somewhat with the simultaneous decrease in outdoor temperature and moisture content. Most systems defrost by reversing the cycle.

Another method of defrosting is spraying heated water over an outdoor coil. The water can be heated by the refrigerant or by auxiliaries.

Draining Heat Source and Heat Sink Coils. Direct-expansion indoor and outdoor heat transfer surfaces serve as condensers and evaporators. Both surfaces, therefore, must have proper refrigerant distribution headers to serve as evaporators and have suitable liquid refrigerant drainage while serving as condensers.

Flooded systems use the normal float control to maintain the required liquid refrigerant level.

Liquid Subcooling Coils. A refrigerant subcooling coil can be added to the heat pump cycle (Figure 8) to preheat ventilation air during heating and, at the same time, to lower the liquid refrigerant temperature. Depending on the refrigerant circulation rate and quantity and temperature of ventilation air, heating capacity can be increased as much as 15 to 20%, as indicated in Figure 9.

Refrigeration Components

Refrigerant piping, receivers, expansion devices, and refrigeration accessories in heat pumps are usually the same as components found in other types of refrigeration and air-conditioning systems.

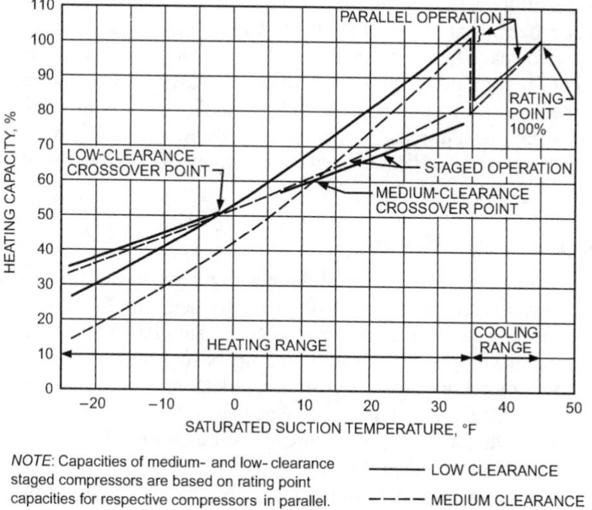

Fig. 6 Comparison of Parallel and Staged Operation for Air-Source Heat Pumps

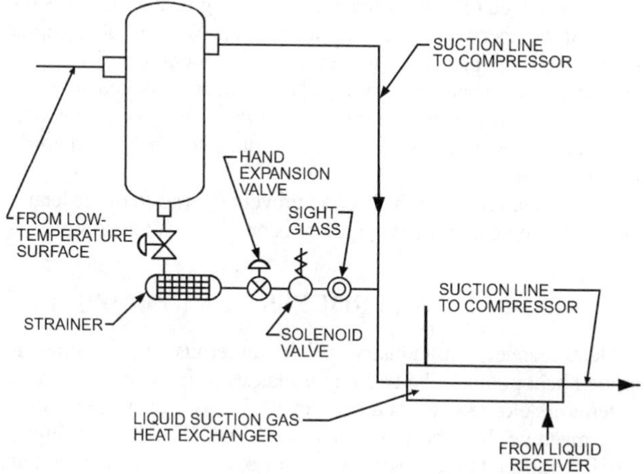

Fig. 7 Suction Line Separator for Protection Against Liquid Floodback

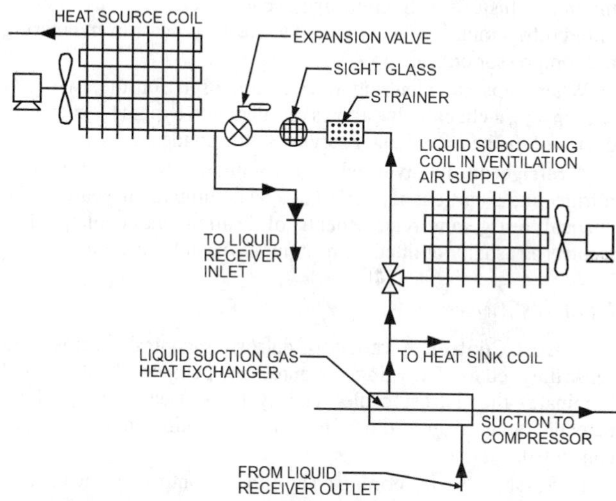

Fig. 8 Liquid Subcooling Coil in Ventilation Air Supply to Increase Heating Effect and Heating COP

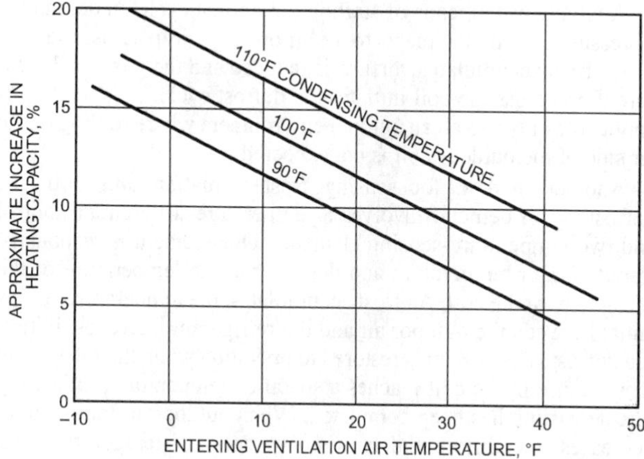

Fig. 9 Typical Increase in Heating Capacity Resulting from Using Liquid Subcooling Coil

A **reversing valve** changes the system from cooling to heating mode. This changeover requires using a valve(s) in the refrigerant circuit, except where the change occurs in fluid circuits external to the refrigerant circuit. Reversing valves are usually pilot-operated by solenoid valves, which admit compressor head and suction pressures to move the operating elements.

The **expansion device** for controlling the refrigerant flow is normally a **thermostatic expansion valve**, which is described in Chapter 44 of the 2006 *ASHRAE Handbook—Refrigeration.* The control bulb must be located carefully. If the circuit is arranged so that the refrigerant line on which the control bulb is placed can become the compressor discharge line, the resulting pressure developed in the valve power element may be excessive, requiring a special control charge or pressure-limiting element. When a thermostatic expansion valve is applied to an outdoor air coil, a special cross-charge to limit the superheat at low temperatures allows better use of the coil. In its temperature-sensing element, a cross-charged thermostatic expansion valve uses a fluid, or mixture of fluids, that is different from the refrigerant. An **electronic expansion valve** improves control of refrigerant flow.

A **capillary tube** (used as a metering device for an air-source heat pump that operates over a wide range of evaporating temperatures) may pass refrigerant at an excessive rate at low back pressures, causing liquid floodback to the compressor. Suction line accumulators or charge-control devices are sometimes added to minimize this effect. Suction line accumulators may also prevent liquid refrigerant that has migrated to the evaporator from entering the compressor on start-up.

When separate metering devices are used for the two heat exchangers, a **check valve** allows refrigerant to bypass the metering device of the heat exchanger serving as the condenser.

A **refrigerant receiver**, which is commonly used to store liquid refrigerant, is particularly useful in a heat pump to take care of the unequal refrigerant requirements of heating and cooling. The receiver is usually omitted on heat pumps used for heating only.

Controls

Defrost Control. A variety of defrosting control schemes can sense the need for defrosting air-source heat pumps and initiate (and terminate) the defrost cycle. The cycle is usually initiated by demand rather than a timer, though termination may be timer-controlled.

Defrost can also be terminated by either a control sensing the coil pressure or a thermostat that measures the temperature of liquid refrigerant in the outdoor coil. Completion of defrosting is ensured when the temperature (or corresponding saturation pressure) of the liquid leaving the outdoor coil rises to about 70°F.

A widely used means of starting the defrost cycle on demand is a pressure control that reacts to the air pressure drop across the coil. When frost accumulates, airflow is reduced and the increased pressure drop across the coil initiates the defrost cycle. This method is applicable only in a clean outdoor environment where fouling of the air side of the outdoor coil is not expected.

Another method for sensing frost formation and initiating defrosting on demand involves a temperature differential control and two temperature-sensing elements. One element is responsive to outdoor air temperature and the other to the temperature of the refrigerant in the coil. As frost accumulates, the temperature differential between the outdoor air and the refrigerant increases, initiating defrost. The system is restored to operation when the refrigerant temperature in the coil reaches a specified temperature, indicating that defrosting has been completed. When outdoor air temperature decreases, the differential between outdoor air temperature and refrigerant temperature decreases because of reduced heat pump capacity. Thus, unless compensation is provided, the defrost cycle is not initiated until a greater amount of frost has built up.

The following controls may be used to change from heating to cooling:

- **Conditioned space thermostats** on residences and small commercial applications.
- **Outdoor air thermostats** (with provision for manual overriding for variable solar and internal load conditions) on larger installations, where it may be difficult to find a location in the conditioned space that is representative of the total building.
- **Manual changeover** using a heat-off-cool position on the indoor thermostat. A single thermostat is used for each heat pump unit.
- **Sensing devices**, which respond to the greater load requirement, heating or cooling, are generally used on simultaneous heating and cooling systems.
- **Dedicated microcomputers** to automate changeover and perform all the other control functions needed, and to simultaneously monitor the performance of the system. This may be a stand-alone device, or it may be incorporated as part of a larger building automation system.

On the heat pump system, it is important that space thermostats are interlocked with ventilation dampers so that both operate on the same cycle. During the heating cycle, the outside air damper should be positioned for the minimum required ventilation air, with the space thermostat calling for increased ventilation air only if the conditioned space becomes too warm. Fan and/or pump interlocks are generally provided to prevent the heat pump system from operating if the accessory equipment is not available. On commercial and industrial installations, some form of head pressure control is required on the condenser when cooling at outdoor air temperatures below 60°F.

Supplemental Heating

Heating needs may exceed the heating capacity available from equipment selected for the cooling load, particularly if outdoor air is used as the heat source. When this occurs, supplemental heating or additional compressor capacity should be considered. The additional compressor capacity or the supplemental heat is generally used only in the severest winter weather and, consequently, has a low usage factor. Both possibilities must be evaluated to determine the most economical selection.

When supplemental heaters are used, the elements should always be located in the air or water circuit downstream from the heat pump condenser. This allows the system to operate at a lower condensing temperature, increasing heating capacity and improving the COP. Controls should sequence the heaters so that they are energized after all heat pump compressors are fully loaded. An outdoor thermostat is recommended to limit or prevent energizing heater elements during mild weather when they are not needed. Where 100% supplemental heat is provided for emergency operation, it may be desirable to keep one or more stages of the heaters locked out whenever the compressor is running. In this way, the cost of electrical service to the building is reduced by limiting the maximum coincidental demand.

A flow switch should be used to prevent operation of the heating elements and heat pump when there is no air or water flow.

INDUSTRIAL PROCESS HEAT PUMPS

Heat recovery in industry offers numerous opportunities for applied heat pumps. The two major classes of industrial heat pump systems are closed-cycle and open-cycle. Factory-packaged, closed-cycle machines have been built to heat fluids to 120°F and as high as 220°F. Skid-mounted open- and semi-open-cycle machines have been used to produce low-level saturated and superheated steam.

Industrial heat pumps are generally used for process heating rather than space heating. Each heat pump system must be designed for the particular application. Rather than being dictated by weather or design standards, the selection of size and output temperature for a system is often affected by economic restraints, environmental standards, or desired levels of product quality or output. This gives the designer more flexibility in equipment selection because the systems are frequently applied in conjunction with a more traditional process heating system such as steam.

ASHRAE research project RP-656 (Cane et al. 1994) gathered information on energy performance, economics of operation, operating difficulties, operator and management reactions, and design details for various **heat recovery heat pump** (HRHP) systems. The most common reason given for installing HRHP systems was reducing energy cost. Other reasons cited were

- Need to eliminate bacterial growth in storage tanks
- Need to increase ammonia refrigeration system capacity
- Reduction of makeup water use
- Need for flexibility in processing
- Year-round processing (drying) possible
- Superior drying quality compared with conventional forced-air kilns
- Process emissions eliminated without the need for costly pollution control equipment
- Recovery and reuse of product from process
- Reduced effluent into the environment

Economics associated with energy reductions were calculated for most of the test sites. Economic justifications for the other reasons for installation were difficult to estimate. Only half the survey sites reported actual payback periods. Half of these had simple paybacks of less than 5 years.

The most frequently cited problem was widely fluctuating heat source or sink flow rates or temperatures. Considerable differences between design parameters and actual conditions were also mentioned frequently. In some cases, the difference resulted in oversizing, which caused poor response to load variation, nuisance shutdowns, and equipment failure. Other problems were significant process changes after installation and poor placement of the HRHP.

The number of projects that had overstated savings based on overstated run hours demonstrated the need for accurate prediction method. Although the low hours of use in some cases resulted from first-year start-up and balancing issues, in most cases it was due to plant capacity reductions, process modifications, or other factors that were not understood during the design phase.

Closed-Cycle Systems

Closed-cycle systems use a suitable working fluid, usually a refrigerant in a sealed system. They can use either absorption or vapor compression. Traditionally, vapor compression systems have used CFC-11, CFC-12, CFC-113, and CFC-114 to obtain the desired temperature; however, production of these refrigerants has been phased out. ASHRAE research project RP-1308 is investigating potential refrigerant replacements for high-temperature applications. Heat is transferred to and from the system through heat exchangers similar to refrigeration system heat exchangers. Closed-cycle heat pump systems are often classed with industrial refrigeration systems except that they operate at higher temperatures.

Heat exchangers used must comply with federal and local codes. For example, some jurisdictions require a double separation between potable water and refrigerant. Heat exchangers must be resistant to corrosion and fouling conditions of the source and sink fluids. The refrigerant and oil must be (1) compatible with component materials and (2) mutually compatible at the expected operating temperatures. In addition, the viscosity and foaming characteristics of the oil and refrigerant mixtures must be consistent with the lubrication requirements at the specific mechanical load imposed on the equipment. Proper oil return and heat transfer at the evaporators and condensers must also be considered.

The specific application of a closed-cycle heat pump frequently dictates the selection considerations. In this section, the different types of closed-cycle systems are reviewed, and factors important to the selection process are addressed.

Air-to-air heat pumps or **dehumidification heat pumps** (Figure 10) are most frequently used in industrial operations to dry or cure products. For example, dehumidification kilns are used to dry lumber to improve its value. Compared to conventional steam kilns, the heat pump provides two major benefits: improved product quality and reduced percent degrade. With dehumidification, lumber can be dried at a lower temperature, which reduces warping, cracking, checking, and discoloration. The system must be selected according to the type of wood (i.e., hard or soft) and the required dry time. Dehumidification heat pumps can also be used to dry agricultural products; poultry, fish, and meat; textiles; and other products.

Air-to-water heat pumps, also called **heat pump water heaters**, are a special application of closed-cycle systems that are usually unitary. See Chapter 48 of this volume and Chapter 49 of the 2007 *ASHRAE Handbook—HVAC Applications* for more information.

Water-to-water heat pumps may have the most widespread application in industry. They can use cooling tower water, effluent streams, and even chilled-water makeup streams as heat sources. The output hot water can be used for product rinse tanks, equipment cleanup water systems, and product preheaters. The water-to-water heat pump system may be simple, such as that shown in Figure 11,

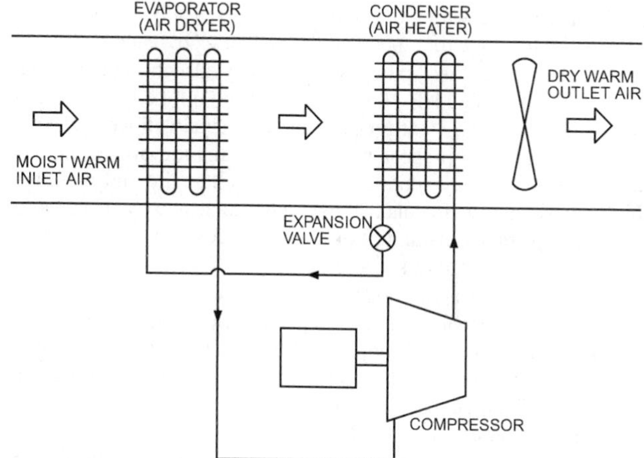

Fig. 10 Dehumidification Heat Pump

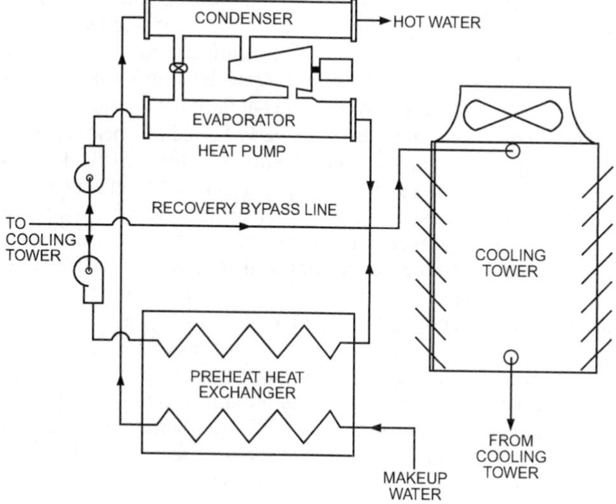

Fig. 11 Cooling Tower Heat Recovery Heat Pump

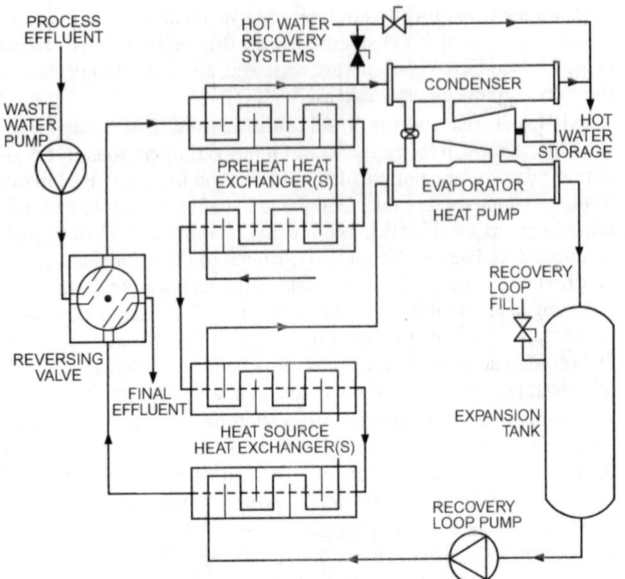

Fig. 12 Effluent Heat Recovery Heat Pump

which recovers heat from a process cooling tower to heat water for another process. Figure 11 also shows the integration of a heat exchanger for preheating the process water. Typically, the heat pump COP is in the range of 4 to 6, and the system COP reaches 8 to 15.

Water-to-water heat pump systems may also be complicated, such as the cascaded HRHP system (Figure 12) for a textile dyeing and finishing plant. Heat recovered from various process effluent streams is used to preheat makeup water for the processes. The effluent streams may contain materials, such as lint and yarn, and highly corrosive chemicals that may foul the heat exchanger; therefore, special materials and antifouling devices may be needed for the heat exchanger for a successful design.

Most food-processing plants, which use more water than a desuperheater or a combination desuperheater/condenser can provide, can use a water-to-water heat pump (Figure 13). A water-cooled condenser recovers both sensible and latent heat from the high-pressure refrigerant. The water heated in the condenser is split into two streams: one as a heat source for the water-to-water heat pump, the other to preheat the makeup process and cleanup water. Preheated water is blended with hot water from the storage tank to limit the temperature difference across the heat pump condenser to about 20°F, which is standard for many chiller applications. The heated water is then piped back to the storage tank, which is typically sized for 1.5 to 5 h of holding capacity. Because the refrigeration load may be insufficient to provide all the water heating, existing water heaters (usually steam) provide additional heat and control for process water at the point of use.

Three major tasks must be addressed when adding heat recovery to existing plant refrigeration systems: (1) forcing the hot-gas refrigerant to flow to the desired heat exchanger, (2) scheduling the refrigeration processes to provide an adequate heat source over time while still meeting process requirements, and (3) integrating water-cooled, shell-and-tube condensers with evaporative condensers. The refrigerant direction can be controlled either by series piping of the two condenser systems or by three pressure-regulating valves (PRVs): one to the hot-gas defrost (lowest-pressure setting), one to the recovery system (medium-pressure setting), and one to the evaporative condensers (highest-pressure setting). The PRVs offer good control but can be mechanically complicated. Series piping is simple but can cost more because of the pipe size required for all the hot gas to pass through the water-cooled condenser.

Each refrigeration load should be reviewed for its required output and production requirements. For example, ice production can

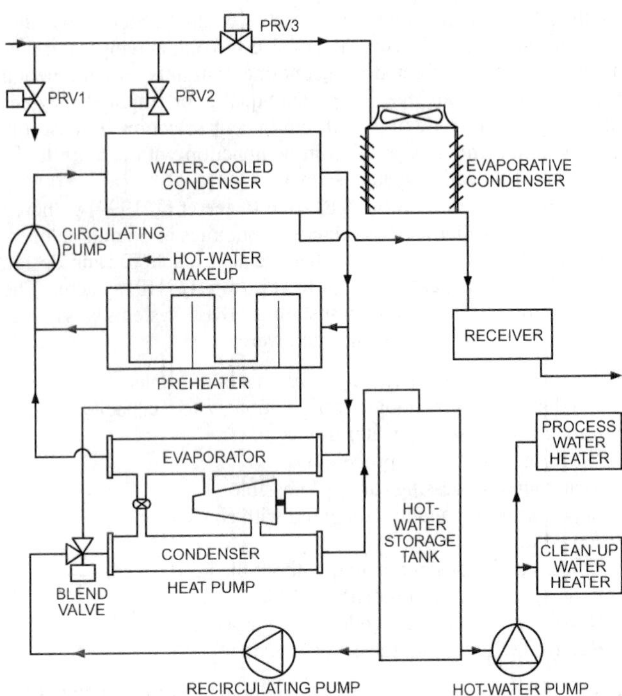

Fig. 13 Refrigeration Heat Recovery Heat Pump

frequently be scheduled during cleanup periods, and blast freezing for the end of shifts.

In multisystem integration, as in expanded system integration, equalization lines, liquid lines, receivers, and so forth must be designed according to standard refrigeration practices.

Process-fluid-to-process-fluid heat pumps can be applied to evaporation, concentration, crystallization, and distillation process fluids that contain chemicals that would destroy a steam compressor. A closed-cycle vapor compression system (Figure 14) is used for the separation of a solid and a liquid in an evaporator system and for the separation of two liquids in a distillation system. Both systems frequently have COPs of 8 to 10 and have the added benefit of cooling tower elimination. These systems can frequently be specified and supplied by the column or evaporator manufacturer as a value-added system.

The benefits of water-to-water, refrigerant-to-water, and other fluid-to-fluid systems may include the following:

- Lower energy costs because of the switch from fuel-based water heating to heat recovery water heating
- Reduced costs for water-treatment chemicals for the boiler because of the switch from a steam-based system
- Reduced emissions of NO_x, SO_x, CO, CO_2, and other harmful chemicals because of reduced boiler loading
- Decreased effluent temperature, which improves the effectiveness of the water treatment process that breaks down solids
- Increased production because of the increased water temperature available at the start of process cycles (for processes requiring a cooler start temperature, preheated water may be blended with ambient water)
- Higher product quality from rinsing with water at a higher temperature (blending preheated water with ambient water may be necessary if a cooler temperature is required)
- Reduced water and chemical consumption at cooling towers and evaporative condensers
- More efficient process cooling if heat recovery can be used to reduce refrigeration pressure or cooling tower return temperature

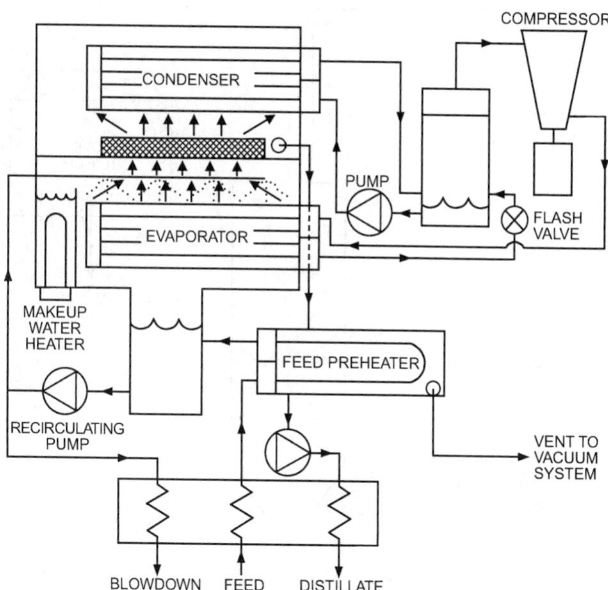

Fig. 14 Closed-Cycle Vapor Compression System

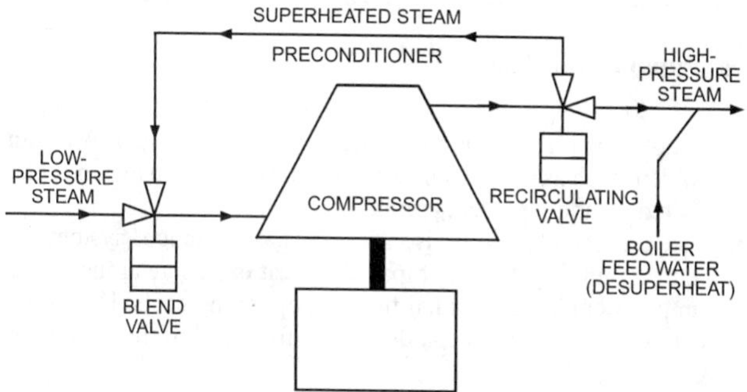

Fig. 15 Recompression of Boiler-Generated Process Steam

- Reduced scaling of heat exchangers because of the lower surface temperatures in heat pump systems compared to steam coils, forced-air gas systems, and resistance heaters

Open-Cycle and Semi-Open-Cycle Heat Pump Systems

Open- and **semi-open-cycle heat pump systems** use process fluid to raise the temperature of available heat energy by vapor compression, thus eliminating the need for chemical refrigerants. The most important class of applications is steam recompression. Compression can be provided with a mechanical compressor or by a thermocompression ejector driven by the required quantity of high-pressure steam. The three main controlling factors for this class of systems are vapor quality, boiling point elevation, and chemical makeup.

Recompression of boiler-generated process steam (Figure 15) has two major applications: (1) large facilities with substantial steam pressure drop due to line losses and (2) facilities with a considerable imbalance between steam requirements at low, medium, and high levels. Boiler-generated process steam usually conforms to cleanliness standards that ensure corrosion-free operation of the compression equipment. Evaluation of energy costs and steam value can be complicated with these applications, dictating the use of analytical tools such as pinch technology.

Application of open-cycle heat pumps to evaporation processes is exceptionally important. Single-effect **evaporators** (Figure 16)

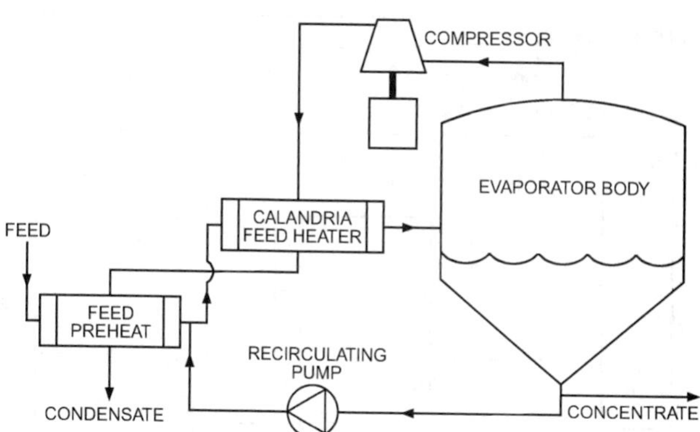

Fig. 16 Single-Effect Heat Pump Evaporator

are common when relatively small volumes of water and solid need to be separated. The most frequent applications are in the food and dairy industries. The overhead vapors of the evaporator are compressed, and thus heated, and piped to the system heater (i.e., calandria). The heat is transferred to the dilute solution, which is then piped to the evaporator body. Flashing occurs upon entry to the evaporator body, sending concentrate to the bottom and vapors to the top. System COPs reach 10 to 20, and long-term performance of the evaporator improves as less of the product (and thus less buildup) occurs in the calandria because of lower operating temperatures. Multiple-effect evaporators (Figure 17), usually applied in the paper and chemical industries, use the same principles, only on a much larger scale. The multiple evaporator bodies can be piped in series or in parallel. System COPs can exceed 30.

Using open-cycle heat pumps in distillation processes (Figure 18) is similar to evaporation. However, compression of flammable gaseous compounds can be dangerous, so great care must be taken. The overhead vapors are compressed to a higher pressure and temperature, and then condensed in the reboiler. This eliminates the need for boiler steam in the reboiler and reduces overall energy consumption. As the pressure of the condensed vapor is reduced through the expansion valve, some of the liquid at a lower temperature is returned to the column as reflux, and the balance forms the overhead product, both as liquid and as vapor. Vapor is recycled as necessary. System COPs can reach 30 or 40.

Process emission-to-steam heat pump systems can be used for cooking, curing, and drying systems that operate with low-pressure saturated or superheated steam. The vapors from a cooking process, such as at a rendering plant, can be recovered to generate the steam required for cooking (Figure 19). The vapors are compressed with a screw compressor because noncondensable materials removed from the process by the vapor could erode or damage reciprocating or centrifugal compressors. Compressed steam is supplied as the heat source to the cooker, and the steam condenses. The condensate is then supplied to a heat exchanger to heat process water. Noncondensables are usually scrubbed or incinerated, and the water is treated or discharged to a sewer.

Contaminants in some processes require the use of a semi-open-cycle heat pump (Figure 20). This system uses a heat exchanger, frequently called a *reboiler*, to recover heat from the stack gases. The reboiler produces low-pressure steam, which is compressed to the desired pressure and temperature. A clean-in-place (CIP) system may be used if the volume of contaminants is substantial.

Variable-speed drives can be specified with all these systems for closer, more efficient capacity control. Additional capacity for emergencies can be made available by temporarily overspeeding the drive, while sizing for nominal operating conditions to retain optimal efficiency. Proper integration of heat pump controls and system controls is essential.

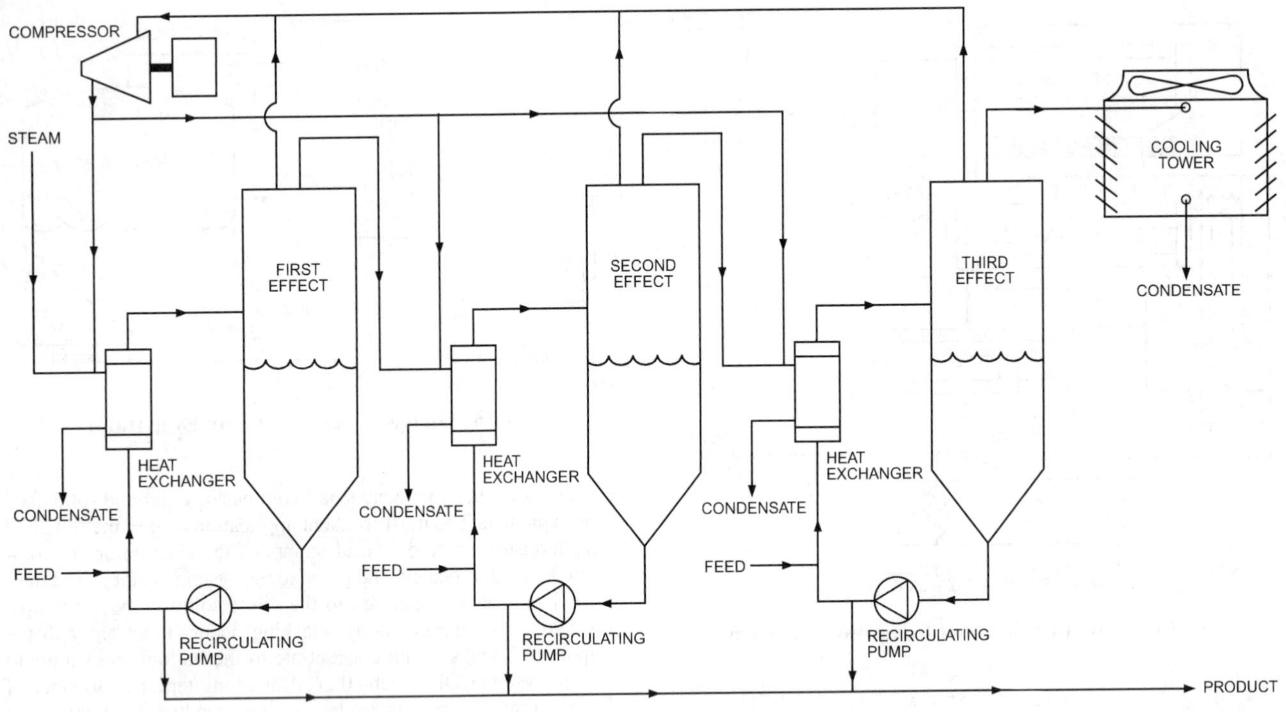

Fig. 17 Multiple-Effect Heat Pump Evaporator

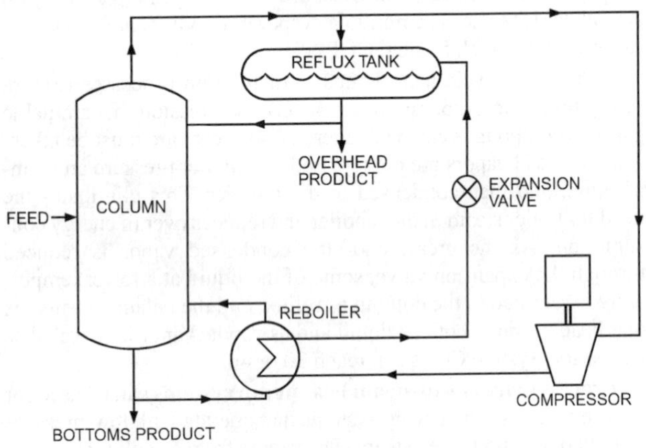

Fig. 18 Distillation Heat Pump System

Heat Recovery Design Principles

The following basic principles should be applied when designing heat recovery systems:

- Use second-law, pinch technology, or other thermodynamic analysis methods, especially for complex processes, before detailed design to ensure proper thermodynamic placement of the HRHP.
- Design for base-load conditions. Heat recovery systems are designed for reduced operating costs. Process scheduling and thermal storage (usually hot water) can be used for better system balancing. Existing water-heating systems can be used for peak load periods and for better temperature control.
- Exchange heat first, then pump the heat. If a heat exchanger is to provide the thermal work in a system, it should be used by itself. If additional cooling of the heat source stream and/or additional heating of the heat sink stream are needed, then a heat pump should be added.
- Do not expect a heat pump to solve a design problem. A design problem such as an unbalanced refrigeration system may be exacerbated by adding a heat pump.
- Make a complete comparison of the heat exchanger system and the heat pump system. Compared to heat exchange alone, a heat pump system has additional first and operating costs. If feasible, heat exchange should be added to the front end of the heat pump system.
- Evaluate the cost of heat displaced by the heat pump system. If the boiler operation is not changed, or if it makes the boiler less efficient, the heat pump may not be economical.
- Investigate standard fuel-handling and heat exchanger systems already used in an industry before designing the heat pump.
- Measure the flow and temperature profile of both the heat source and heat sink over an extended period of time. The data may help prevent overestimating the requirements and economics of the system.
- Investigate future process changes that may affect the thermal requirements and/or availability of the system. Determine whether the plant is changing to a cold-water cleanup system or to a process to produce less effluent at lower temperatures.
- Design the system to give plant operators the same or better control. Thoroughly review manual versus automatic controls.
- Determine whether special material specifications are needed for handling any process flows. Obtain a chemical analysis for any flow of unknown makeup.
- Inform equipment suppliers of the full range of ambient conditions to which the equipment will be exposed and the expected loading requirements.

ASHRAE research project RP-807 (Caneta Research 1998) produced guidelines for evaluating environmental benefits (e.g., energy, water, and plant emission reductions) of heat recovery heat pumps. Calculation procedures for energy and water savings and plant emission reductions were outlined. Other, less easily quantified benefits were also documented. Evaluation guidelines were provided for a heat pump water heater in a hospital kitchen; a heat recovery heat pump at a resort with a spa, pool, and laundry service building; heat recovery from refrigeration compressor superheat for makeup water

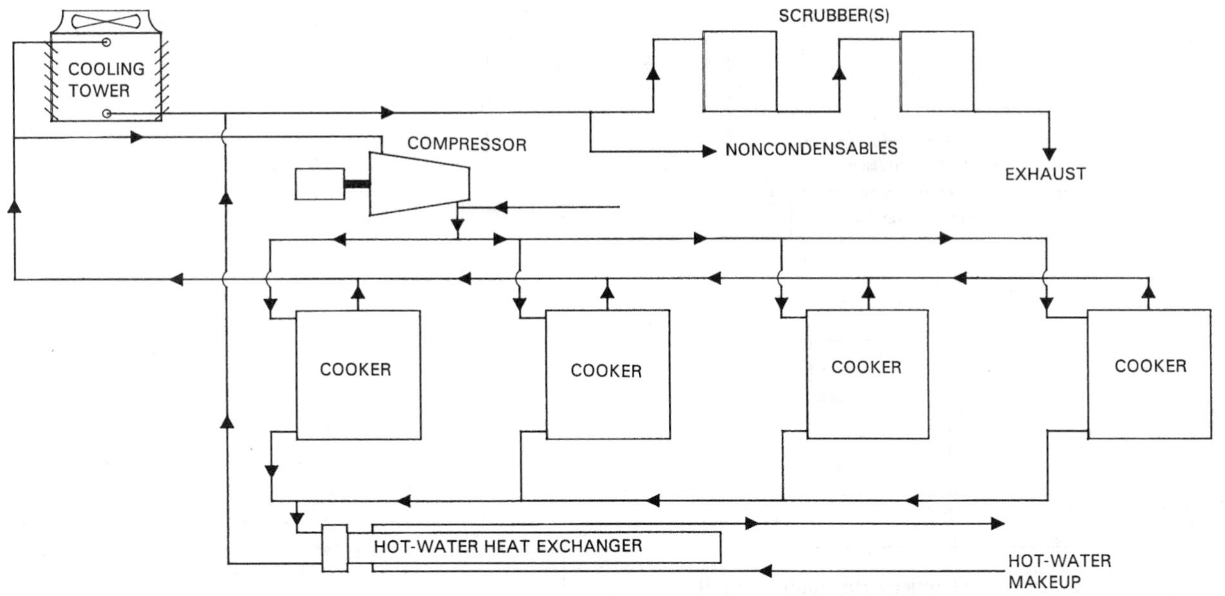

Fig. 19 Heat Recovery Heat Pump System in a Rendering Plant

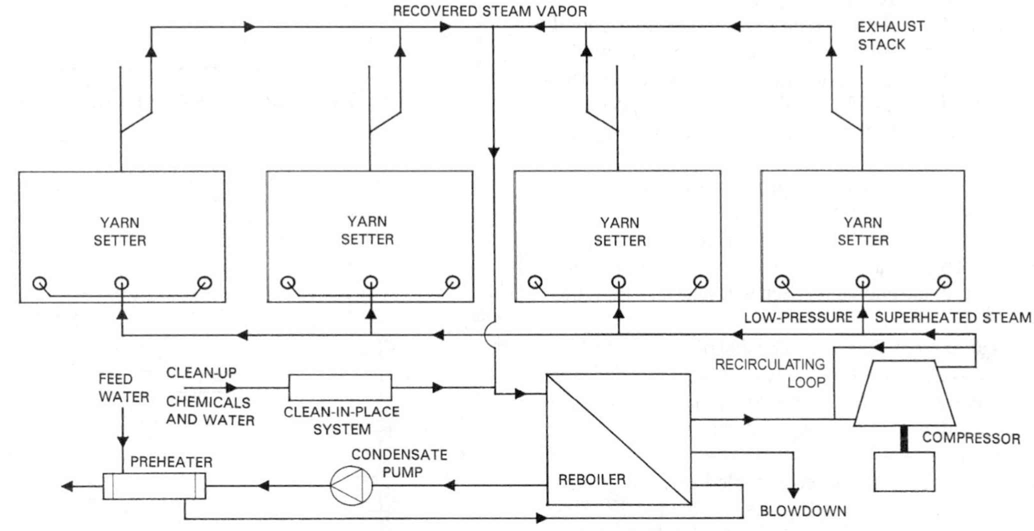

Fig. 20 Semi-Open-Cycle Heat Pump in a Textile Plant

heating; and various drying processes using heat pumps and concentration processes. These guidelines can be used to quantify energy, water, and emission reductions and help justify using heat recovery systems in commercial, institutional, and industrial buildings.

APPLIED HEAT RECOVERY SYSTEMS

WASTE HEAT RECOVERY

In many large buildings, internal heat gains require year-round chiller operation. The chiller condenser water heat is often wasted through a cooling tower. Figure 21 illustrates an HRHP installed in the water line from the chiller's condenser before rejection at the cooling tower. This arrangement uses the otherwise wasted heat to provide heat at the higher temperatures required for space heating, reheat, and domestic water heating.

Prudent design may dictate cascade systems with chillers in parallel or series. Custom components are available to meet a

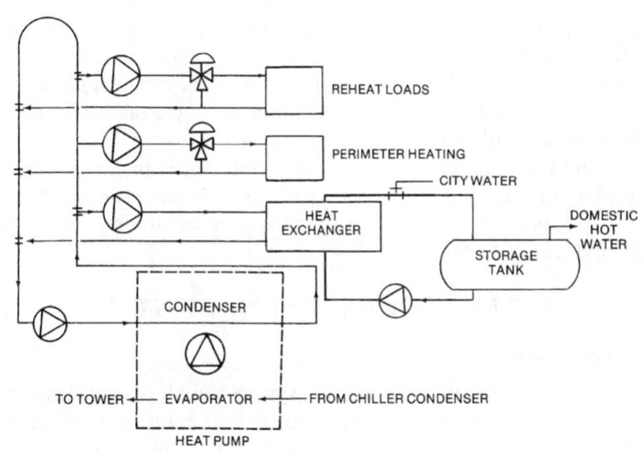

Fig. 21 Heat Recovery Heat Pump

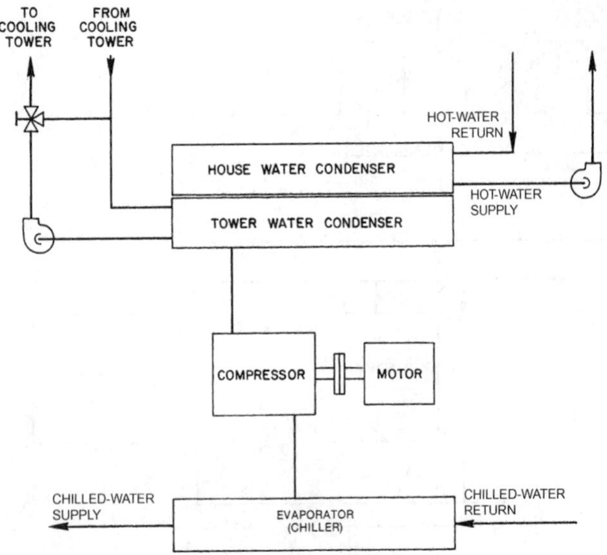

Fig. 22 Heat Recovery Chiller with Double-Bundle Condenser

wide range of load and temperature requirements. The double-bundle condenser working with a reciprocating or centrifugal compressor is most often used in this application. Figure 22 shows the basic configuration of this system, which makes heat available in the range of 100 to 130°F. Warm water is supplied as a secondary function of the heat pump and represents recovered heat.

Figure 23 shows a similar cycle, except that a storage tank has been added, enabling the system to store heat during occupied hours by raising the water temperature in the tank. During unoccupied hours, water from the tank is gradually fed to the evaporator providing load for the compressor and condenser that heat the building during off hours.

Figure 24 shows a heat transfer system capable of generating 130 to 140°F or warmer water whenever there is a cooling load, by cascading two refrigeration systems. In this configuration, Machine No. 1 can be considered as a chiller only and Machine No. 2 as a heating-only heat pump.

An indication of the magnitude of recoverable heat in a modern multistory office building is shown in Figures 25, 26, and 27. Figures 25 and 26 show the gross heat loss and the internal heat gain of the exterior and interior zones during occupied periods. Figure 27 shows the total amount of heat available for recovery. Heat recovered from internal zones can be used to provide all or some of the external zones' heating requirements. Excess recovered heat can be diverted to thermal storage for later use during unoccupied periods. During the occupied periods, no outside heat source or supplemental heat is needed at outdoor temperatures of 23°F or above for this hypothetical building.

It must be understood that relative performances are indicated in these figures to illustrate general cases. Of course, each building will have parameters that differ from those of the building used for the figures.

WATER LOOP HEAT PUMP SYSTEMS

Description

A water loop heat pump (WLHP) system combines load transfer characteristics with multiple water-to-air heat pump units (Figure 28). Each zone, or space, has one or more water-to-air heat pumps. The units in both the building core and perimeter areas are connected hydronically with a common two-pipe system. Each unit

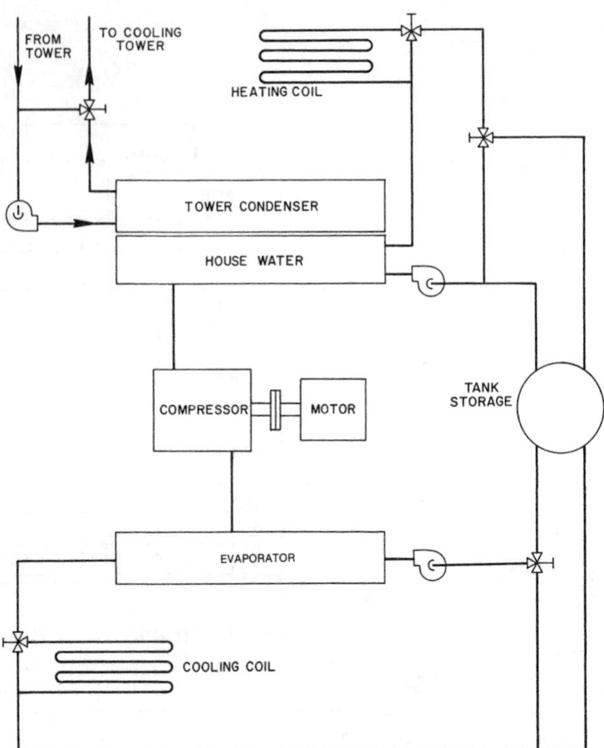

Fig. 23 Heat Recovery Chiller with Storage Tank

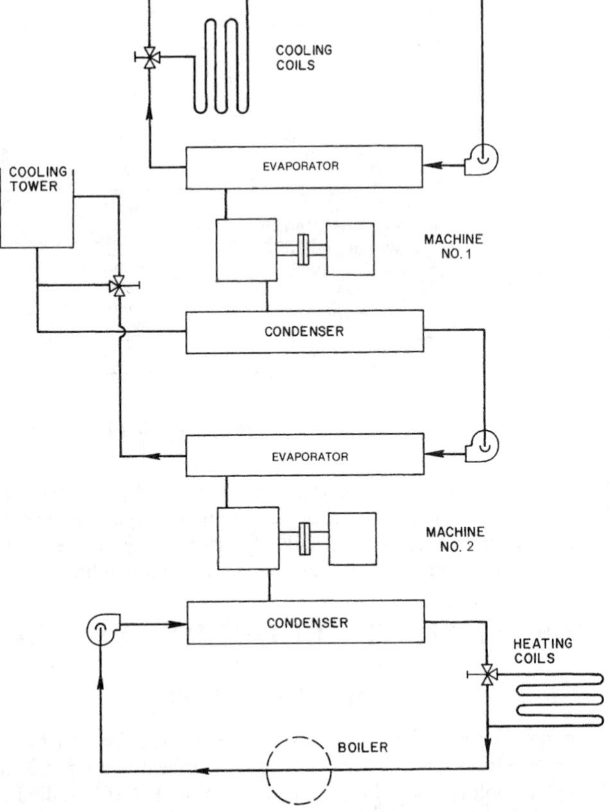

Fig. 24 Multistage (Cascade) Heat Transfer System

cools conventionally, supplying air to the individual zone and rejecting the heat removed to the two-pipe system through its integral condenser. Excess heat gathered by the two-pipe system is expelled

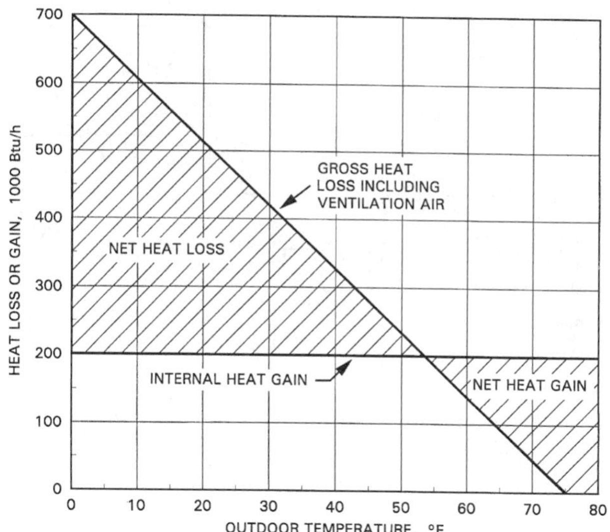

Fig. 25 Heat Loss and Heat Gain for Exterior Zones During Occupied Periods

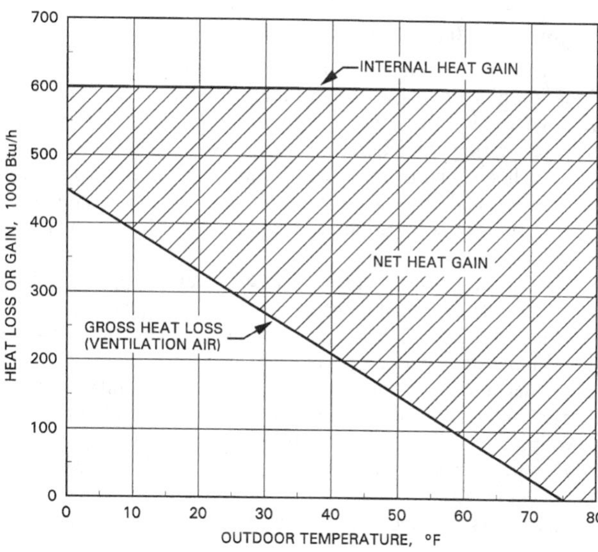

Fig. 26 Heat Loss and Heat Gain for Interior Zones During Occupied Periods

through a common heat rejection device, which often includes a closed-circuit evaporative cooling tower with an integral spray pump. If and when some of the zones, particularly on the northern side of the building, require heat, the individual units switch (by means of reversing refrigerant valves) into the heating cycle. These units then extract heat from the two-pipe water loop, a relatively high-temperature source that is totally or partially maintained by heat rejected from the condensers of the units that provide cooling to other zones. When only heating is required, all units are in the heating cycle and, consequently, an external heat input to the loop is needed to maintain the loop temperature. The water loop temperature has traditionally been maintained in the range of 60 to 90°F and, therefore, seldom requires piping insulation. However, extended-range water loop heat pumps are becoming more common because they provide high efficiency over a wider operating range. The expanded operating range (45 to 110°F) reduces boiler and cooling tower operating costs. The lower operating temperatures may require insulating the main supply and return lines. These units

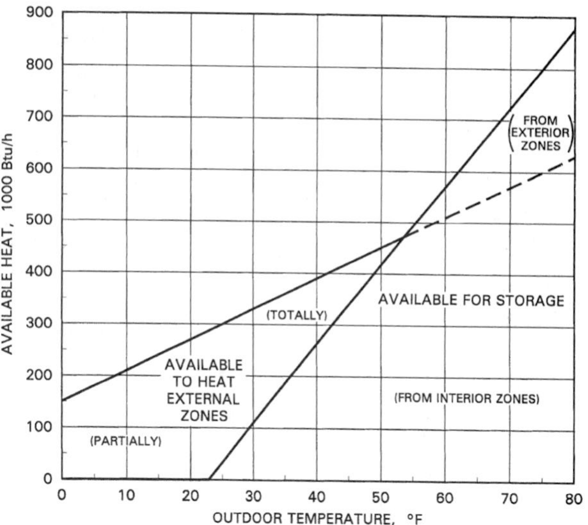

Fig. 27 Internal Heat Available for Recovery During Occupied Periods

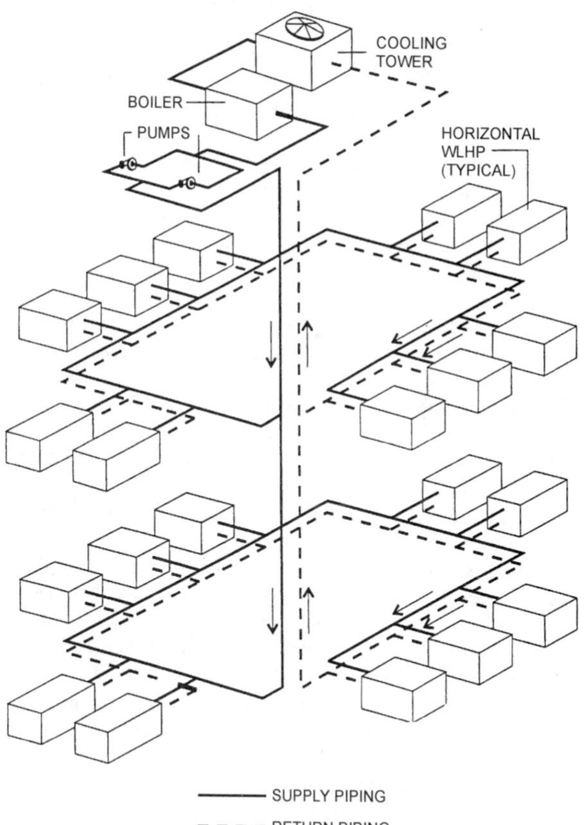

Fig. 28 Heat Recovery System Using Water-to-Air Heat Pumps in a Closed Loop

typically come equipped with insulated water-to-refrigerant heat exchangers, loss-of-flow protection, and thermoexpansion valves instead of the traditional capillary tube expansion device.

Any number of water-to-air heat pumps may be installed in such a system. Water circulates through each unit via the closed loop.

The water circuit usually includes two circulating pumps (one pump is 100% standby) and a means for adding and rejecting heat to and from the loop. Each heat pump can either heat or cool to maintain the comfort level in each zone.

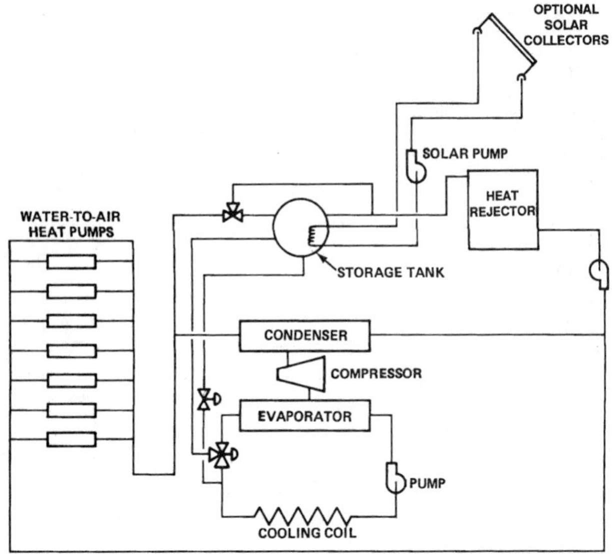

Fig. 29 Closed-Loop Heat Pump System with Thermal Storage and Optional Solar-Assist Collectors

Units in heating mode extract heat from the circulated water, whereas those in cooling mode reject heat to the water. Thus, the system recovers and redistributes heat, where needed. Unlike air-source heat pumps, heating output for this system does not depend on outdoor temperature. The water loop conveys rejected heat, but a secondary heat source, typically a boiler, is usually provided.

Another WLHP version uses a coil buried in the ground as a heat source and sink. This ground-coupled system does not normally need the boiler and cooling tower incorporated in conventional WLHP systems to keep circulating water within acceptable temperature limits. However, ground-coupled heat pumps may operate at lower entering water (or antifreeze solution) temperatures. Some applications may require a heat pump tolerant of entering water temperatures ranging from 25 to 110°F. In climates where air conditioning dominates, a cooling tower is sometimes combined with ground coupling to reduce overall installed costs.

Figure 29 illustrates a system with a storage tank and solar collectors. The storage tank in the condenser circuit can store excess heat during occupied hours and provide heat to the loop during unoccupied hours. During this process, solar heat may also be added within the limitations of the temperatures of the condenser system. The solar collectors are more effective and efficient under these circumstances because of the temperature ranges involved.

For further heat reclaim, a water-to-water heat pump can be added in the closed water loop before the heat rejection device. This heat pump provides domestic hot water or elevates loop water temperatures in a storage tank so the water can be bled back into the loop when needed during the heating cycle. Figure 30 illustrates such a system.

Many facilities require large cooling loads (e.g., for interior zones, lights, people, business machines, computers, switchgear, and production machinery) that result in net loop heat rejection during all or most of the year, particularly during occupied hours. This waste heat rejection often occurs while other heating loads in the facility (e.g., ventilation air, reheat, domestic hot water) are using external purchased energy to supply heat. By including a secondary water-to-water heat pump (as in Figure 30), additional balanced heat recovery can be economically achieved. The secondary heat pump can effectively reclaim this otherwise rejected heat, raise its temperature, and use it to serve other heating loads, thus minimizing use of purchased energy.

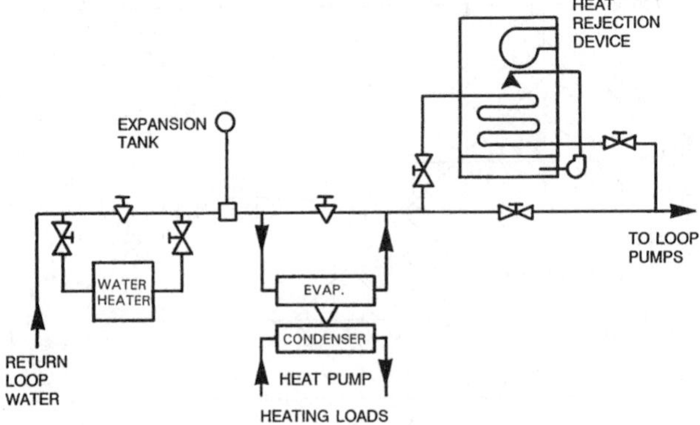

Fig. 30 Secondary Heat Recovery from WLHP System
(Adapted from *Marketing the Industrial Heat Pump*, Edison Electric Institute 1989)

In another WLHP system variation, the building sprinkler system is used as part of the loop water distribution system.

Some aspects of systems described in this section may be proprietary and should not be used without appropriate investigation.

Design Considerations

Water loop heat pump (WLHP) systems are used in many types of multiroom buildings. A popular application is in office buildings, where heat gains from interior zones can be redistributed to the perimeter during winter. Other applications include hotels and motels, schools, apartment buildings, nursing homes, manufacturing facilities, and hospitals. Operating costs for these systems are most favorable in applications with simultaneous heating and cooling requirements.

Accurate design tools are needed to model and predict WLHP energy use and electrical demand. ASHRAE research project RP-620 (Cane et al. 1993) evaluated computer models for water-loop heat pump systems and compared the accuracy of the test models with actual monitored data from buildings with similar systems. Although the computer programs predicted whole-building energy use within 1 to 15% of measured total energy use over a 12-month period, much larger variations were found at the HVAC system and component levels.

Cane et al. (1993) concluded that the models could be improved by

- Modeling each heat pump rather than lumping performance characteristics of all heat pumps in one zone
- Assuming water-to-air heat pumps cycle on/off to satisfy loads, which increases energy consumption, rather than assuming they unload as a chiller would
- Being able to model thermal storage added to the water loop and service water preheat from the loop

Unit Types. Chapter 48 describes various types and styles of units and the control options available.

Zoning. The WLHP system offers excellent zoning capability. Because equipment can be placed within the zones, future relocation of partitions can be accommodated with minimum duct changes. Some systems use heat pumps for perimeter zones and the top floor, with cooling-only units serving interior zones; all units are connected into the same loop water circuit.

Heat Recovery and Heat Storage. WLHPs work well for heat storage. Installations, such as schools, that cool most of the day in winter and heat at night make excellent use of heat storage. Water may be stored in a large tank in the closed-loop circuit ahead of the boiler. In this application, the loop temperature is allowed to build up to 90°F during the day. Stored water at 90°F can be used during unoccupied hours to maintain heat in the building, with the loop temperature allowed to drop to 60°F. The water heater (or boiler) would not be used until the loop temperature had dropped the entire 30°F. The storage tank operates as a flywheel to prolong the period of operation where neither heat makeup nor heat rejection is required.

Concealed Units. Equipment installed in ceiling spaces must have access for maintenance and servicing filters, control panels, compressors, and so forth. Adequate condensate drainage must be provided.

Ventilation. Outdoor air for ventilation may be (1) ducted from a ventilation supply system to the units or (2) drawn in directly through a damper into the individual units. To operate satisfactorily, air entering the water-source heat pumps should be above 60°F. In cold climates, ventilation air must be preheated. The quantity of ventilation air entering directly through individual units can vary greatly because of stack effect, wind, and balancing difficulties.

Secondary Heat Source. The secondary heat source for heat makeup may be electric, gas, oil, ground, solar energy, and/or waste heat. Normally, a water heater or boiler is used; however, electric resistance heat in the individual heat pumps, with suitable controls, may also be used. The control may be an aquastat set to switch from heat pump to resistance heaters when the loop water approaches the minimum 60°F.

An electric boiler is readily controllable to provide 60°F outlet water and can be used directly in the loop. With a gas- or oil-fired combustion boiler, a heat exchanger may be used to transfer heat to the loop or, depending on the type of boiler used, a modulating valve may blend hot water from the boiler into the loop.

Solar or ground energy can supply part or all of the secondary heat. Water or antifreeze solution circulated through collectors can add heat to the system directly or indirectly via a secondary heat exchanger.

A building with night setback may need a supplementary heater boiler sized for the installed capacity, not the building heat loss, because the morning warm-up cycle may require every heat pump to operate at full heating capacity until the building is up to temperature. In this case, based on a typical heating COP of 4.0 for a water-source heat pump, the boiler would be sized to provide about 75% of the total heating capacity of all water-source heat pumps installed in the building.

Heat Rejector Selection. A closed-loop circuit requires a heat rejector that is either a heat exchanger (loop water to cooling tower water), a closed-circuit evaporative cooler, or a ground coil. The heat rejector is selected in accordance with manufacturer's selection curves, using the following parameters:

- **Water flow rates.** Manufacturers' recommendations on water flow rates vary between 2 and 3 gpm per ton of installed cooling capacity. Lower flow rates are generally preferred in regions with a relatively low summer outdoor design wet-bulb temperature. In more humid climates, a higher flow rate allows a higher water temperature to be supplied from the heat rejector to the heat pumps, without a corresponding increase in temperature leaving the heat pumps. Thus, cooling tower or evaporative cooler size and cost are minimized without penalizing the performance of the heat pumps.
- **Water temperature range.** Range (the difference between leaving and entering water temperatures at the heat rejector) is affected by heat pump cooling efficiency, water flow rate, and diversity. It is typically between 10 and 15°F.
- **Approach.** Approach is the difference between the water temperature leaving the cooler and the wet-bulb temperature of the outside air. The maximum water temperature expected in the loop supply is a function of the design wet-bulb temperature.
- **Diversity.** Diversity is the maximum instantaneous cooling load of the building divided by the installed cooling capacity:

$$D = Q_m/Q_i \tag{1}$$

where

D = diversity
Q_m = maximum instantaneous cooling load
Q_i = total installed cooling capacity

Diversity times the average range of the heat pumps is the applied range of the total system (the rise through all units and the drop through the heat rejector). For systems with a constant pumping rate regardless of load,

$$R_s = DR_p \tag{2}$$

where

R_s = range of system
R_p = average range of heat pumps

For systems with a variable pumping rate and with each pump equipped with a solenoid valve to start and stop water flow through the unit with compressor operation,

$$R_s \approx R_p \tag{3}$$

The average leaving water temperature of the heat pumps is the entering water temperature of the heat rejector. The leaving water temperature of the heat rejector is the entering water temperature of the heat pumps.

- **Winterization.** For buildings with some potential year-round cooling (e.g., office buildings), loop water may be continuously pumped through the heat rejector. This control procedure reduces the danger of freezing. However, it is important to winterize the heat rejector to minimize the heat loss.

  In northern climates, the most important winterization step for evaporative coolers is installing a discharge air plenum with positive-closure, motorized, ice-proof dampers. The entire casing that houses the tube bundle and discharge plenum may be insulated. The sump, if outside the heated space, should be equipped with electric heaters. The heat pump equipment manufacturer's instructions will help in the selection and control of the heat rejector.

  If sections of the water circuit will be exposed to freezing temperatures, consider adding an antifreeze solution. In a serpentine pipe circuit with no automatic valves that could totally isolate individual components, an antifreeze solution prevents bursting pipes, with minimal effect on system performance.

  An open cooling tower with a separate heat exchanger is a practical alternative to the closed water cooler (Figure 31). An additional pump is required to circulate the tower water through the heat exchanger. In such an installation, where no tubes are exposed to the atmosphere, it may not be necessary to provide freeze protection on the tower. The sump may be indoors or, if outdoors, may be heated to keep the water from freezing. This arrangement allows the use of small, remotely located towers. Temperature control necessary for tower operation is maintained by a sensor in the water loop system controlling operation of the tower fan(s).

  The combination of an open tower, heat exchanger, and tower pump frequently has a lower first cost than an evaporative cooler. In addition, operating costs are lower because no heat is lost from the loop in winter and, frequently, less power is required for the cooling tower fans.

Ductwork Layout. Often, a WLHP system has ceiling-concealed units, and the ceiling area is used as the return plenum. Troffered light fixtures are a popular means of returning air to the ceiling plenum.

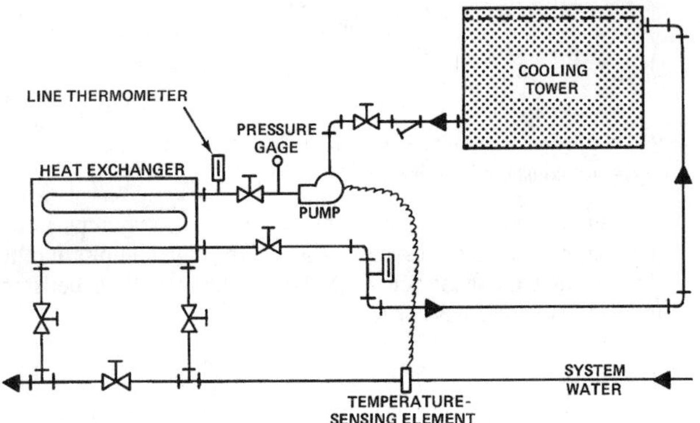

Fig. 31 Cooling Tower with Heat Exchanger

Air supply from the heat pumps should be designed for quiet operation. Heat pumps connected to ductwork must be capable of overcoming external static pressure. Heat pump manufacturers' recommendations should be consulted for the maximum and minimum external static pressure allowable with each piece of equipment.

Piping Layout. Reverse-return piping should be used wherever possible with the WLHP system, particularly when all units have essentially the same capacity. Balancing is then minimized except for each of the system branches. If direct-return piping is used, balancing water flow is required at each individual heat pump. The entire system flow may circulate through the boiler and heat rejector in series. Water makeup should be at the constant-pressure point of the entire water loop. Piping system design is similar to the secondary water distribution of air-and-water systems.

Pumping costs for a WLHP system can be significant. Because system loads vary considerably, variable-speed pumping should be considered. This requires an automatic valve at each heat pump that allows water flow through the heat pump coil only during compressor operation.

Clean piping is vital to successful performance of the water-source heat pump system. The pipe should be clean when installed, kept clean during construction, and thoroughly cleaned and flushed upon completion of construction. Start-up water filters in the system bypass (pump discharge to suction) should be included on large, extensive systems.

Controls

The WLHP system has simpler controls than totally central systems. Each heat pump is controlled by a thermostat in the zone. There are only two centralized temperature control points: one to add heat when the water temperature approaches a prescribed lower temperature (45 to 60°F), and the other to reject heat when the water temperature approaches a prescribed upper temperature (typically about 90 to 110°F).

The boiler controls should be checked to be sure that outlet water is controlled at 60°F, because controls normally supplied with boilers are in a much higher range.

An evaporative cooler should be controlled by increasing or decreasing heat rejection capacity in response to the loop water temperature leaving the cooler. A reset schedule that operates the system at a lower water temperature (to take advantage of lower outdoor wet-bulb temperatures) can save energy when heat from the loop storage is not likely to be used.

Abnormal condition alarms typically operate as follows:

- On a fall in loop temperature to 50°F, initiate an alert. Open heat pump control circuits at 45°F.
- On a rise in loop temperature to 105°F, initiate an alert. Open heat pump control circuits at 110°F.
- On sensing insufficient system water flow, a flow switch initiates an alert and opens the heat pump control circuits.

Outside ambient control should be provided to prevent operation of the cooling tower sump pump at freezing temperatures.

Optional system control arrangements include the following:

- Night setback control
- Automatic unit start/stop, with after-hour restart as a tenant option
- Warm-up cycle
- Pump alternator control

Advantages of a WLHP System

- Affords opportunity for energy conservation by recovering heat from interior zones and/or waste heat and by storing excess heat from daytime cooling for nighttime or other heating uses.
- Allows recovery of solar energy at a relatively low fluid temperature where solar collector efficiency is likely to be greater.
- The building does not require wall penetrations to provide for the rejection of heat from air-cooled condensers.
- Provides environmental control in scattered occupied zones during nights or weekends without the need to start a large central refrigeration machine.
- Units are not exposed to outdoor weather, which allows installation in coastal and other corrosive atmospheres.
- Units have a longer service life than air-cooled heat pumps.
- Noise levels can be lower than those of air-cooled equipment because condenser fans are eliminated and the compression ratio is lower.
- Two-pipe boiler/chiller systems are potentially convertible to this system.
- The entire system is not shut down when a unit fails. However, loss of pumping capability, heat rejection, or secondary heating could affect the entire system.
- Energy usage by the heat pumps can be metered for each tenant. However, this metering would not include energy consumed by the central pump, heat rejector, or boiler.
- Total life-cycle cost of this system frequently compares favorably to that of central systems when considering installed cost, operating costs, and system life.
- Units can be installed as space is leased or occupied.

Limitations of a WLHP System

- Space is required for the boiler, heat exchangers, pumps, and heat rejector.
- Initial cost may be higher than for systems that use multiple unitary HVAC equipment.
- Reduced airflow can cause the heat pump to overheat and cut out. Therefore, periodic filter maintenance is imperative.
- The piping loop must be kept clean.

BALANCED HEAT RECOVERY SYSTEMS

Definition

In an ideal heat recovery system, all components work year-round to recover all the internal heat before adding external heat. Any excess heat is either stored or rejected. Such an idealized goal is identified as a balanced heat recovery system.

When the outdoor temperature drops significantly, or when the building is shut down (e.g., on nights and weekends), internal heat gain may be insufficient to meet the space heating requirements. Then, a balanced system provides heat from storage or an external source. When internal heat is again generated, the external heat is automatically reduced to maintain proper temperature in the space. There is a time delay before equilibrium is reached. The size of the equipment and the external heat source can be reduced in a balanced system that includes storage. Regardless of the system, a heat balance analysis establishes the merits of balanced heat recovery at various outdoor temperatures.

Outdoor air less than 55 to 65°F may be used to cool building spaces with an air economizer cycle. When considering this method of cooling, the space required by ducts, air shafts, and fans, as well as the increased filtering requirements to remove contaminants and the hazard of possible freeze-up of dampers and coils must be weighed against alternatives such as using deep row coils with antifreeze fluids and efficient heat exchange. Innovative use of heat pump principles may give considerable energy savings and more satisfactory human comfort than an air economizer. In any case, hot and cold air should not be mixed (if avoidable) to control zone temperatures because it wastes energy.

Heat Redistribution

Many buildings, especially those with computers or large interior areas, generate more heat than can be used for most of the year. Operating cost is minimized when the system changes over from net heating to net cooling at the break-even outdoor temperature at which the building heat loss equals the internal heat load. If heat is unnecessarily rejected or added to the space, the changeover temperature varies from the natural break-even temperature, and operating costs increase. Heating costs can be reduced or eliminated if excess heat is stored for later distribution.

Heat Balance Concept

The concept of ideal heat balance in an overall building project or a single space requires that one of the following takes place on demand:

- Heat must be removed.
- Heat must be added.
- Heat recovered must exactly balance the heat required, in which case heat should be neither added nor removed.

In small air-conditioning projects serving only one space, either cooling or heating satisfies the thermostat demand. If humidity control is not required, operation is simple. Assuming both heating and cooling are available, automatic controls will respond to the thermostat to supply either. A system should not heat and cool the same space simultaneously.

Multiroom buildings commonly require heating in some rooms and cooling in others. Optimum design considers the building as a whole and transfers excess internal heat from one area to another, as required, without introducing external heat that would require waste heat disposal at the same time. The heat balance concept is violated when this occurs.

Humidity control must also be considered. Any system should add or remove only enough heat to maintain the desired temperature and control the humidity. Large percentages of outdoor air with high wet-bulb temperatures, as well as certain types of humidity control, may require reheat, which could upset the desirable balance. Usually, humidity control can be obtained without upsetting the balance. When reheat is unavoidable, internally transferred heat from heat recovery should always be used to the extent it is available, before using an external heat source such as a boiler. However, the effect of the added reheat must be analyzed, because it affects the heat balance and may have to be treated as a variable internal load.

When a building requires heat and the refrigeration plant is not in use, dehumidification is not usually required and the outdoor air is dry enough to compensate for any internal moisture gains. This should be carefully reviewed for each design.

Heat Balance Studies

The following examples illustrate situations that can occur in nonrecovery and unbalanced heat recovery situations. Figure 32 shows the major components of a building that comprise the total air-conditioning load. Values above the zero line are cooling loads, and values below the zero line are heating loads. On an individual basis, the ventilation and conduction loads cross the zero line, which indicates that these loads can be a heating or a cooling load, depending on outdoor temperature. Solar and internal loads are always a cooling load and are, therefore, above the zero line.

Figure 33 combines all the loads shown in Figure 32. The graph is obtained by plotting the conduction load of a building at various outdoor temperatures, and then adding or subtracting the other loads at each temperature. The project load lines, with and without solar effect, cross the zero line at 16 and 30°F, respectively. These are the outdoor temperatures for the plotted conditions when the naturally created internal load exactly balances the loss.

As plotted, this heat balance diagram includes only the building loads with no allowance for additional external heat from a boiler or other source. If external heat is necessary because of system design, the diagram should include the additional heat.

Figure 34 illustrates what happens when heat recovery is not used. It is assumed that at a temperature of 70°F, heat from an external source is added to balance conduction through the building's skin in increasing amounts down to the minimum outdoor temperature winter design condition. Figure 34 also adds the heat required for the outdoor air intake. The outdoor air comprising part or all of the supply air must be heated from outdoor temperature to room temperature. Only the temperature range above the room temperature is effective for heating to balance the perimeter conduction loss.

These loads are plotted at the minimum outdoor winter design temperature, resulting in a new line passing through points A, D, and E. This line crosses the zero line at –35°F, which becomes the artificially created break-even temperature rather than 30°F, when not allowing for solar effect. When the sun shines, the added solar heat at the minimum design temperature would further drop the –35°F break-even temperature. Such a design adds more heat than the overall project requires and does not use balanced heat

CONDUCTION LOAD

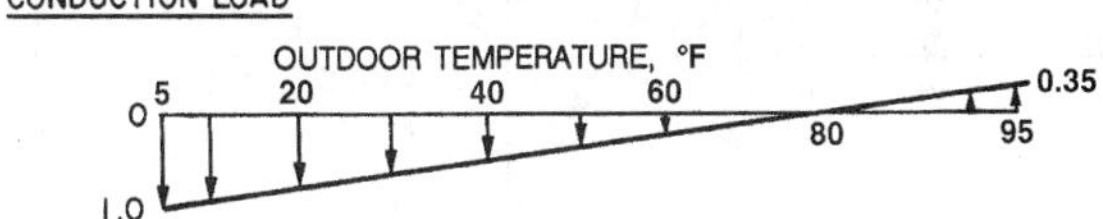

VENTILATION LOAD

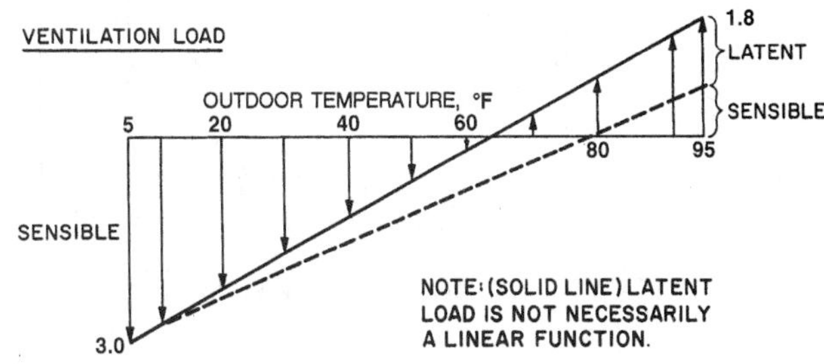

SOLAR, EQUIPMENT, LIGHTS AND PEOPLE

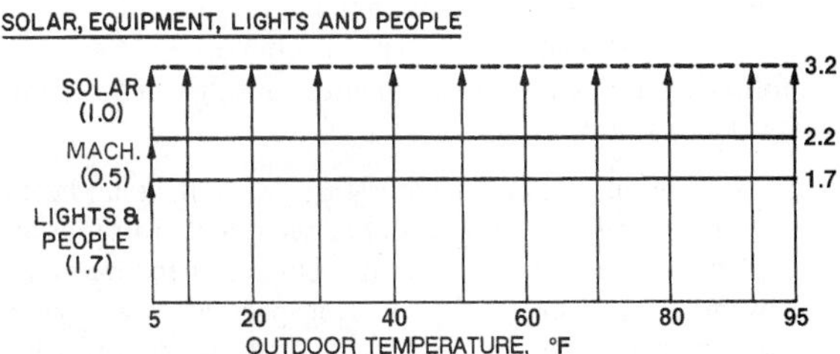

NOTE: ALL LOADS ARE IN 10^6 Btu/h

Fig. 32 Major Load Components

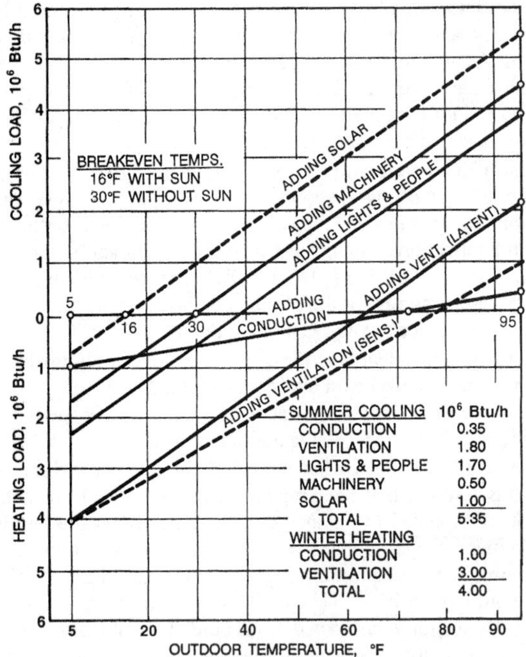

Fig. 33 Composite Plot of Loads in Figure 32 (Adjust for Internal Motor Heat)

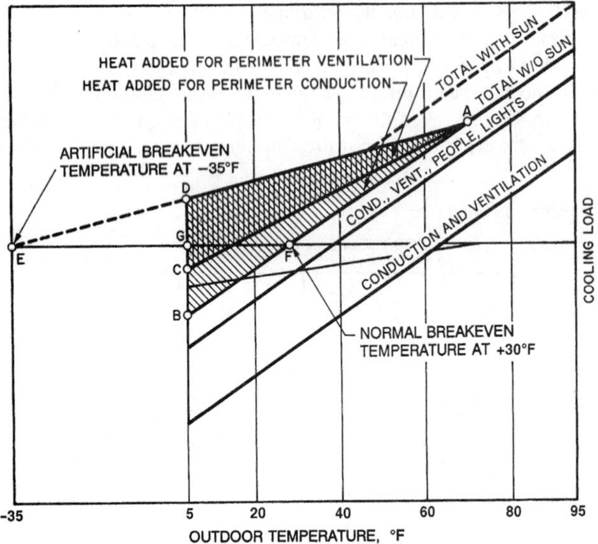

Fig. 34 Non-Heat-Recovery System

recovery to use the available internal heat. This problem is most evident during mild weather on systems not designed to take full advantage of internally generated heat year-round.

The following are two examples of situations that can be shown in a heat balance study:

1. As the outdoor air wet-bulb temperature drops, the total heat of the air falls. If a mixture of outdoor and recirculated air is cooled to 55°F in summer and the same dry-bulb temperature is supplied by an economizer cycle for interior space cooling in winter, there will be an entirely different result. As the outdoor wet-bulb temperature drops below 55°F, each unit volume of air introduced does more cooling. To make matters more difficult, this increased cooling is latent cooling, which requires adding latent heat to prevent too low a relative humidity, yet this air is intended to cool. The extent of this added external heat for free cooling is shown to be very large when plotted on a heat balance analysis at 0°F outside temperature.

   Figure 34 is typical for many current non-heat-recovery systems. There may be a need for cooling, even at the minimum design temperature, but the need to add external heat for humidification can be eliminated by using available internal heat. When this asset is thrown away and external heat is added, operation is inefficient.

   Some systems recover heat from exhaust air to heat the incoming air. When a system operates below its natural break-even temperature t_{be} such as 30 or 16°F (shown in Figure 33), the heat recovered from exhaust air is useful and beneficial. This assumes that only the available internal heat is used and that no supplementary heat is added at or above t_{be}. Above t_{be}, the internal heat is sufficient and any recovered heat would become excessive heat to be removed by more outdoor air or refrigeration.

   If heat is added to a central system to create an artificial t_{be} of −35°F as in Figure 34, any recovered heat above −35°F requires an equivalent amount of heat removal elsewhere. If the project were in an area with a minimum design temperature of 0°F, heat recovery from exhaust air could be a liability at all times for the conditions stipulated in Figure 33. This does not mean that the value of heat recovered from exhaust air should be forgotten. The emphasis should be on recovering heat from exhaust air rather than on adding external heat.
2. A heat balance shows that insulation, double glazing, and so forth can be extremely valuable on some projects. However, these practices may be undesirable in some regions during the heating season, when excess heat must usually be removed from large buildings. For instance, for minimum winter design temperatures of approximately 35 to 40°F, it is improbable that the interior core of a large office building will ever reach its break-even temperature. The temperature lag for shutdown periods, such as nights and weekends, at minimum design conditions could never economically justify the added cost of double-pane windows. Therefore, double-pane windows merely require the amount of heat saved to be removed elsewhere.

General Applications

A properly applied heat reclaim system automatically responds to make a balanced heat recovery. An example is a reciprocating water chiller with a hot-gas diverting valve and both a water-cooled and an air-cooled condenser. Hot gas from the compressor is rejected to the water-cooled condenser. This hot water provides internal heat as long as it is needed. At a predetermined temperature, the hot gas is diverted to the air-cooled condenser, rejecting excess heat from the total building system. Larger projects with centrifugal compressors use double-condenser chiller units, which are available from many manufacturers. For typical buildings, chillers normally provide hot water for space heating at 105 to 110°F.

Many buildings that run chillers all or most of the year reclaim some of the condenser heat to provide domestic hot water.

Designers should include a source of external heat for back-up. The control system should ensure that back-up heat is not injected unless all internal heat has been used. For example, if electric back-up coils are in series with hot-water coils fed from a hot-water storage tank, they may automatically start when the system restarts after the building temperature has dropped to a night low-limit setting. An adjustable time delay in the control circuit gives the stored hot water time to warm the building before energizing the electric heat.

This type of heat reclaim system is readily adaptable to smaller projects using a reciprocating chiller with numerous air terminal units or a common multizone air handler. The multizone air handler should have individual zone duct heating coils and controls arranged to prevent simultaneous heating and cooling in the same zone.

Properly applied heat reclaim systems not only meet all space heating needs, but also provide hot water required for showers, food service facilities, and reheat in conjunction with dehumidification cycles.

Heat reclaim chillers or heat pumps should not be used with air-handling systems that have modulating damper economizer control. This free cooling may result in a higher annual operating cost than a minimum fresh air system with a heat reclaim chiller. Careful study shows whether the economizer cycle violates the heat balance concept.

Heat reclaim chillers or heat pumps are available in many sizes and configurations. Combinations include (1) centrifugal, reciprocating, and screw compressors; (2) single- and double-bundle condensers; (3) cascade design for higher temperatures (up to 220°F); and (4) air- or water-cooled, or both.

The designer can make the best selection after both a heating and cooling load calculation and a preliminary economic analysis, and with an understanding of the building, processes, operating patterns, and available energy sources.

ASHRAE research project RP-620 (Cane et al. 1993) evaluated computer models for heat recovery chillers. Whole-building energy use predictions were 5 to 15% of actual monitored data for one building used in the evaluation. Heat recovery chiller estimates were within 10 to 15% of measured values in the same building.

Applications of heat reclaim chillers or heat pumps range from simple systems with few control modes to complex systems having many control modes and incorporating two- or four-pipe circulating systems. Some systems using double- and single-bundle condensers coupled with exterior closed-circuit coolers have been patented. Potential patent infringements should be checked early in the planning stage.

Successful heat recovery design depends on the performance of the total system, not just the chiller or heat pump. Careful, thorough analysis is often time-consuming and requires more design time than a nonrecovery system. The balanced heat recovery concept should guide all phases of planning and design, and the effects of economic compromise should be studied. There may be little difference between the initial (installed) cost of a heat recovery system and a nonrecovery system, especially in larger projects. Also, in view of energy costs, life-cycle analysis usually shows dramatic savings when using balanced heat recovery.

Multiple Buildings

A multiple-building complex is particularly suited to heat recovery. Variations in occupancy and functions provide an abundance of heat sources and uses. Applying the balanced heat concept to a large multibuilding complex can save substantial energy. Each building captures its own total heat by interchange. Heat rejected from one building could possibly heat adjacent buildings.

REFERENCES

Cane, R.L.D., S.B. Clemes, and D.A. Forgas. 1993. Validation of water-loop heat pump system modeling (RP-620). *ASHRAE Transactions* 99(2):3-12.

Cane, R.L.D., S.B. Clernes, and D.A. Forgas. 1994. Heat recovery heat pump operating experiences. *ASHRAE Transactions* 100(2):165-172.

Caneta Research. 1998. Guidelines for the evaluation of resource and environmental benefits of heat recovery heat pumps (RP-807). ASHRAE Research Project, *Final Report*.

Edison Electric Institute. 1989. *Marketing the industrial heat pump*. Washington, D.C.

BIBLIOGRAPHY

ASHRAE. 1989. Heat recovery. *Technical Data Bulletin* 5(6).

Ashton, G. J., H.R. Cripps, and H.D. Spriggs. 1987. Application of "pinch" technology to the pulp and paper industry. *TAPPI Journal* (August):81-85.

Becker, F.E. and A.I. Zakak. 1985. Recovering energy by mechanical vapor recompression. *Chemical Engineering Progress* (July):45-49.

Bose, J.E., J.D. Parks, and F.C. McQuiston. 1985. *Design/data manual for closed-loop ground-coupled heat pump systems*. ASHRAE.

Cane, R.L.D., J.M. Garnett, and C.J. Ireland. 2000. Development of guidelines for assessing the environmental benefits of heat recovery heat pumps. *ASHRAE Transactions* 106(2).

Ekroth, I.A. 1979. Thermodynamic evaluation of heat pumps working with high temperatures. *Proceedings of the Inter-Society Energy Conversion Engineering Conference* 2:1713-19.

Eley Associates. 1992. *Water-loop heat pump systems*, vol. 1: *Engineering design guide*. Electric Power Research Institute, Palo Alto, CA.

EPRI. 1984. Heat pumps in distillation processes. *Report* EM-3656. Electric Power Research Institute, Palo Alto, CA.

EPRI. 1986. Heat pumps in evaporation processes. *Report* EM-4693. Electric Power Research Institute, Palo Alto, CA.

EPRI. 1988. Industrial heat pump manual. *Report* EM-6057. Electric Power Research Institute, Palo Alto, CA.

EPRI. 1989. Heat pumps in complex heat and power systems. *Report* EM-4694. Electric Power Research Institute, Palo Alto, CA.

International Energy Agency Heat Pump Centre. 1992. *Heat Pump Centre Newsletter* 10(1).

International Energy Agency Heat Pump Centre. 1993. *Heat Pump Centre Newsletter* 11(1).

Karp, A. 1988. Alternatives to industrial cogeneration: A pinch technology perspective. Presented at the International Energy Technology Conference (IETC), Houston, TX.

Linnhoff, B. and G.T. Polley. 1988. General process improvement through pinch technology. *Chemical Engineering Progress* (June):51-58.

Mashimo, K. 1992. Heat pumps for industrial process and district heating. Presented at the Annual Meeting of International Users Club of Absorption Systems, Thisted, Denmark.

Oil and Gas Journal. 1988. Refiners exchange conservation experiences. *Oil and Gas Journal* (May 23).

Phetteplace, G.E. and H.T. Ueda. 1989. Primary effluent as a heat source for heat pumps. *ASHRAE Transactions* 95(1):141-146.

Pucciano, F.J. 1995. Heat recovery in a textile plant. *IEA Heat Pump Centre Newsletter* 13(2):34-36.

Tjoe, T.N. and B. Linnhoff. 1986. Using pinch technology for process retrofit. *Chemical Engineering* (April 28):47-60.

Zimmerman, K.H., ed. 1987. Heat pumps—Prospects in heat pump technology and marketing. *Proceedings of the 1987 International Energy Agency Heat Pump Conference, Orlando, FL.*

CHAPTER 9

DESIGN OF SMALL FORCED-AIR HEATING AND COOLING SYSTEMS

THIS chapter describes the basics of design and component selection of small forced-air heating and cooling systems, explains their importance, and describes the system's parametric effects on energy consumption. It also gives an overview of test methods for thermal distribution system efficiency, and considers the interaction between the building thermal/pressure envelope and the forced-air heating and cooling system, which is critical to the energy efficiency and cost-effectiveness of the overall system. This chapter pertains to residential and certain small commercial systems; large commercial systems are beyond the scope of this chapter.

COMPONENTS

Forced-air systems are heating and/or cooling systems that use motor-driven blowers to distribute heated, cooled, and otherwise treated air for the comfort of individuals in confined spaces. A typical residential or small commercial system includes (1) a heating and/or cooling unit, (2) accessory equipment, (3) supply and return ductwork, (4) supply and return registers and grilles, and (5) controls. These components are described briefly in the following sections and are illustrated in Figure 1.

Heating and Cooling Units

Three types of forced-air heating and cooling devices are (1) furnaces, (2) air conditioners, and (3) heat pumps.

Furnaces are the basic component of most forced-air heating systems. They are augmented with an air-conditioning coil when cooling is included, and are manufactured to use specific fuels such as oil, natural gas, or liquefied petroleum gas. The fuel used dictates installation requirements and safety considerations (see Chapter 28).

Common **air-conditioning** systems use a split configuration with an air-handling unit, such as a furnace. The air-conditioning evaporator coil (indoor unit) is installed on the discharge air side of the air handler. The compressor and condensing coil (outdoor unit) are located outside the structure, and refrigerant lines connect the outdoor and indoor units.

Self-contained air conditioners contain all necessary air-conditioning components, including circulating air blowers, and may or may not include fuel-fired heat exchangers or electric heating elements.

The **heat pump** cools and heats using the refrigeration cycle. It is available in split and packaged (self-contained) configurations. Generally, the air-source heat pump requires supplemental heating; therefore, electric heating elements are usually included with the heat pump as part of the forced-air system. Careful consideration should be given to sizing both the heat pump and supplemental heat capacity to minimize operation of electric heat elements, especially if night setback is used. This issue is discussed in Bouchelle et al. (2000), Bullock (1978), Ellison (1977), and FSEC (2001).

Heat pumps are also combined with fossil-fuel furnaces to take advantage of their high efficiency at mild temperatures to minimize heating cost. Heat pump supplemental heating may also be provided by thermostat-controlled, AFUE-rated gas heating appliances (e.g., fireplaces, free-standing stoves).

Ground-source heat pumps (GSHPs) are becoming more common in residential housing, especially in colder climates. GSHPs typically do not use supplemental heating except in emergency mode.

Accessory Equipment

Forced-air systems may be equipped to humidify and dehumidify the indoor environment, remove contaminants from recirculated air, and provide circulation of outside air in economizer operation. The following accessories affect airflow and pressure requirements. Losses must be taken into account when selecting heating and cooling equipment and sizing ductwork.

Humidifiers. Several types of humidifiers are available, including self-contained steam, atomizing, evaporative, and heated pan. The Air-Conditioning and Refrigeration Institute (ARI) provides a method of testing and rating various humidifiers in ARI *Standard* 610.

Humidifiers must match the heating unit. Discharge air temperatures on heating systems vary, and some humidifiers do not provide their own heat source for humidification. These humidifiers should be applied with caution to heat pumps and other heating units with low air temperature rise.

Structures with complete vapor retarders (walls, ceilings, and floors) normally require no supplemental moisture during the heating season if the internally generated moisture maintains an acceptable relative humidity of 20 to 60%.

Chapter 20 contains more information on humidifiers.

Dehumidifiers. Dehumidifiers may be used when air conditioning in hot, humid climates or outside air heating in colder, humid climates is insufficient to control indoor humidity alone. Also see the section on Dehumidifiers in Chapter 1 of 2007 *ASHRAE Handbook—HVAC Applications*.

Electronic air cleaners. These units attract oppositely charged particles, fine dust, smoke, and other particles to collecting plates in the air cleaner. Electronic air cleaners usually have a washable prefilter to trap lint and larger particles as they enter the unit. The remaining particles take on an electric charge in a charging section, then travel to the collector section where they are drawn to and trapped by the oppositely charged collecting plates. A nearly constant pressure drop can be expected unless the cleaner collecting plates and/or filters become severely loaded with dust. See Chapter 28 for more information on air cleaners.

The preparation of this chapter is assigned to TC 6.3, Central Forced Air Heating and Cooling Systems.

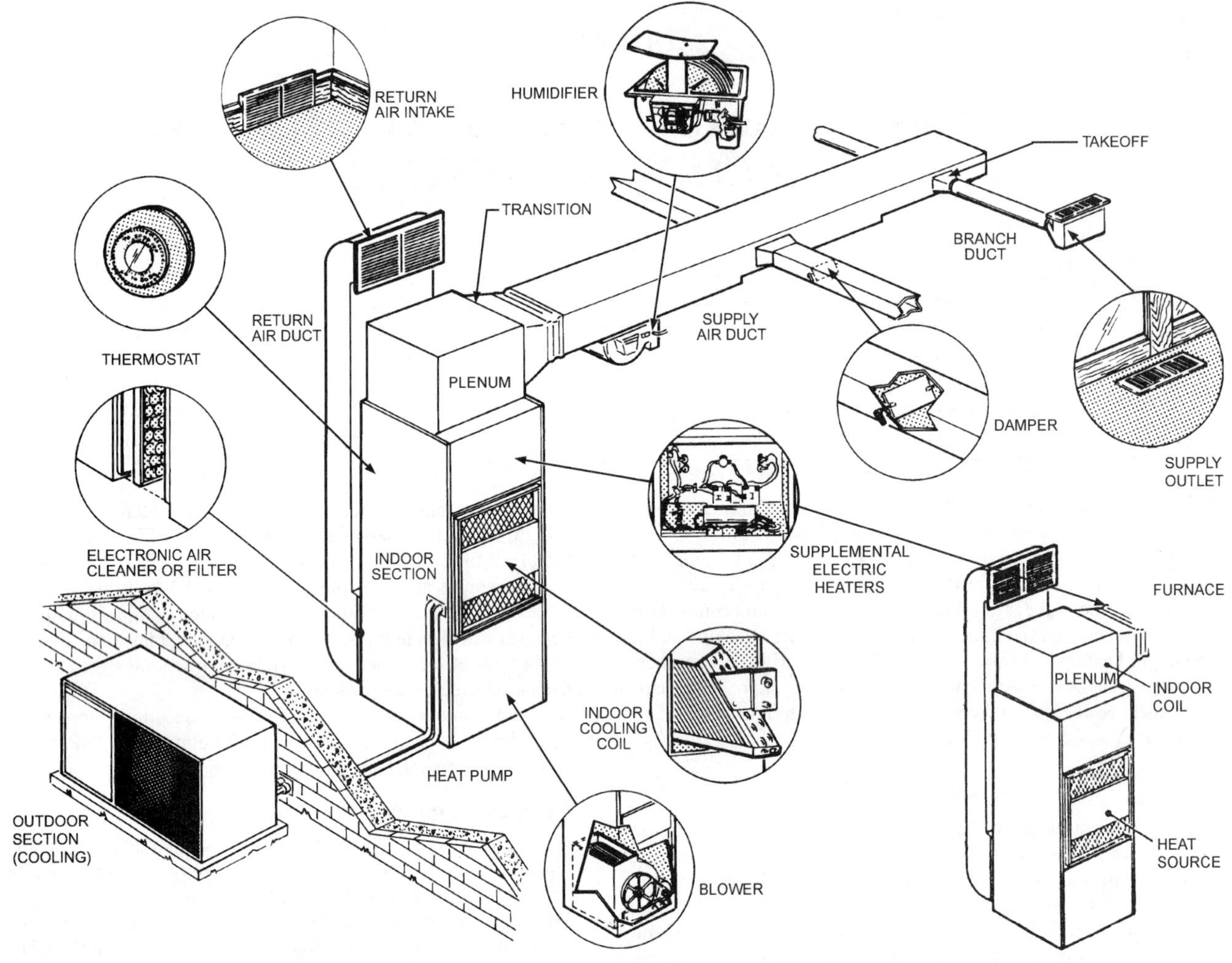

Fig. 1 Heating and Cooling Components

Custom accessories. Solar, off-peak storage, and other custom systems are not covered in this chapter. However, their components may be classified as duct system accessories.

Energy/heat recovery ventilators. These devices provide ventilation air to the conditioned space and recover energy/heat from the air being exhausted outdoors. They can be operated as standalone devices or installed with forced-air distribution.

Economizer control. This device monitors outdoor temperature and humidity and automatically shuts down the air-conditioning unit when a preset outdoor condition is met. Damper motors open outdoor return air dampers, letting outside air enter the system to provide comfort cooling. When outdoor air conditions are no longer acceptable, the outdoor air dampers close and the air-conditioning unit comes back on.

Ducts

In small commercial and residential applications, ductwork design depends on the air-moving characteristics of the blower included with the selected equipment. It is important to recognize this difference between small commercial or residential systems and large commercial and industrial systems. The designer of smaller systems must determine resistances to air movement and adjust duct sizes to limit the static pressure against which the blower operates. Manufacturers publish static pressure versus flow rate information so the designer can determine the maximum static pressure against which the blower will operate while delivering the proper volume of air (see Chapter 18). Ductwork and HVAC equipment location, size, length, surface area, insulation level, and air leakage rates all affect systems' ability to maintain thermal comfort and optimize energy efficiency (Modera 1989).

Materials. Duct materials affect both thermal and mechanical performance. A typical installation may use combinations of sheet metal, fiberglass duct board, and flexible ducting. Chapter 18 briefly discusses some of the materials used for ducts. Information manuals and design guides are available from the Air Conditioning Contractors of America (ACCA), Air Diffusion Council (ADC), National Association of Home Builders (NAHB), North American Insulation Manufacturers Association (NAIMA), and Sheet Metal and Air Conditioning Contractors National Association (SMACNA). Installation codes should also be reviewed for accepted duct materials and installation practices.

Noise and Vibration. Performance aspects of sound and vibration must be considered when designing forced-air heating and cooling systems. Sound and vibration are generated by mechanical equipment (fan, etc.) and propagate throughout the building through the air, building structure, and duct systems. The duct system may also generate noise at fittings and outlets, and may transmit noise from one room to another. Ebbing and Blazier (1998) discuss noise characteristics of forced-air equipment and duct systems, and also present general design guidelines. Chapter 47 of the 2007 *ASHRAE Handbook—HVAC Applications* includes information on sound and vibration design criteria as well as a discussion of design and analysis. Sound performance of equipment and components should be specified with reference to appropriate industry test standards. Forced-air space conditioning equipment such as furnaces should be tested, rated, and specified for sound performance in accordance with ARI *Standard* 260. Ebbing and Blazier (1998) provide references to other relevant test standards.

Duct Insulation. Duct insulation can improve comfort and lower utility bills and equipment cost. The need for insulation can be reduced if ducts are located within the conditioned space. In this location, any conductive losses and gains are minimal because ducts are exposed to indoor air temperatures. Some insulation is still required to ensure that conditioned air is delivered at the desired temperature. Insulation is especially needed in hot, humid areas to prevent condensation from forming on cold duct surfaces. The amount of insulation required is less than for ducts in unconditioned spaces.

Insulation R-values should be selected based on climate and duct location. Ductwork located outdoors, in attics, in crawlspaces, and in basements must be insulated as outlined in Chapter 24 of the 2005 *ASHRAE Handbook—Fundamentals*, to minimum levels required by energy code jurisdictions such as the *International Residential Code® for One- and Two-Family Dwellings* (IRC 2000). A minimum level of R-8 duct insulation is recommended for all externally exposed ducts. The Environmental Protection Agency Residential ENERGY STAR Program recommends R-11 duct insulation in colder climates (EPA 2000).

Poorly applied and/or uninsulated ducts in unconditioned spaces such as attics, crawlspaces, garages, or unfinished basements may lose a significant percentage of heating or cooling energy through conduction through duct surfaces (Andrews and Modera 1991; Modera 1989). Uninsulated and/or poorly insulated ducts can also cause occupant discomfort, especially during winter. When conditioned air moves through uninsulated ducts, it loses heat through conduction, which can cause occupants in rooms served by long duct runs to experience "cold blow" between cycles.

Duct Sealing

Benefits of duct sealing include improved comfort and indoor air quality, better humidity control, and lower utility bills and equipment cost. If properly sealed, the duct system in a house can significantly improve heating and cooling system efficiency and performance (Modera 1989).

All duct sections should be properly connected and all connections and seams properly sealed to minimize duct leakage. Residential ducts typically leak 15 to 20% of the air they convey (Modera 1989). Conditioned air that leaks out of supply ducts is lost in the surrounding spaces. Typically, heating and cooling equipment is designed to condition return air that is at or near room temperature. Leaky return ducts can draw air out of unconditioned spaces that is hotter or colder than the return air, thus increasing loads on heating and cooling systems. This problem is most pronounced in attics where, during summer, air temperatures can be 150°F or higher. Even when furnaces or air conditioners are not operating, leaky ducts waste energy by contributing to the overall air leakage of a house. In new, tightly constructed houses, ducts can account for a significant portion of the total air leakage. Leaky ducts in unconditioned spaces can also introduce airborne pollutants, moisture, and unpleasant odors into homes, thus reducing indoor air quality.

Duct leakage often results from improper installation and poor materials. Duct tape, which is commonly used, does not adequately seal joints between ducts over the system's life (Sherman et al. 2000). More stable and permanent sealing materials are needed, such as mastic, duct-manufacturer-approved foil tape, fiberglass tape, and aerosol-applied duct-sealing polymers (Modera et al. 1996).

Section M1601.3.1 of the *International Residential Code®* provides more information on making ducts substantially airtight using tapes, mastics, gaskets, or other approved closure systems. Even when ducts are in conditioned spaces, sealing is still required to ensure proper air distribution (IRC 2000).

Supply and Return Registers and Grilles

Supply air should be directed to the sources of greatest heat loss and/or gain to offset their effects. However, thermal efficiency must also be considered when determining duct run and supply and return registers/grille locations. Registers and grilles for supply should accommodate all aspects of air distribution patterns such as throw, spread, and drop (see Chapter 19 for more information). Noise generated at registers and grilles must be considered in system noise and vibration analysis, and recommended limits on face velocities should be maintained to keep noise levels in rooms to acceptable levels. General guidelines are provided in this chapter; for additional discussion, see Chapter 47 of the 2007 *ASHRAE Handbook—HVAC Applications* and Sections 10 and 11 of ACCA *Manual* T (1992).

Controls

Forced-air heating and/or cooling systems may be controlled in several ways. Simple on/off cycling of central equipment is frequently adequate to maintain comfort. Spaces with large load variations may require zone control or multiple units with separate ductwork. Systems with minimal load variations in the space may function adequately with one central wall thermostat or a return air thermostat. Residential conditioning systems of 60,000 Btu/h capacity or less are typically operated with one central thermostat.

Forced-air control may require several devices, depending on system complexity and the accessories used (see Chapter 46 of the 2007 *ASHRAE Handbook—HVAC Applications*). Energy conservation has increased the importance of control, so methods that were once considered too expensive for small systems may now be cost-effective.

Temperature control, the primary consideration in forced-air systems, may be accomplished by a single-stage thermostat. When properly located, and in some cases with correctly adjusted heat anticipators, this device accurately controls temperature. Multistage thermostats are required on many systems (e.g., a heat pump with auxiliary heating) and may improve temperature regulation. Outdoor thermostats, in series with indoor control, can stage heating increments adequately. Energy use can increase if outdoor thermostats are set significantly higher than manufacturer recommendations.

Indoor thermostats may incorporate many control capabilities in one device, including continuous or automatic fan control and automatic or manual changeover between heating and cooling. Where more than one system conditions a common space, manual control is preferred to prevent simultaneous heating and cooling.

Thermostats with programmable temperature control allow occupants to vary the temperature set point for different periods. These devices can save substantial energy by applying automatic night and/or daytime temperature setback for all systems when used by occupants and appropriately matched to the HVAC system. With heat pumps, however, night setback of thermostats with morning setup can significantly increase supplemental heating, as noted in Bouchelle et al. (2000), Bullock (1978), Ellison (1977), and FSEC (2001).

Two-speed fan control may be desirable for fossil-fueled equipment, though caution should be used in applying it with ducts that are outside the conditioned space (Andrews 2003). Such control should not be applied to heat pumps unless recommended by the manufacturer. Humidistats should be specified for humidifier control or thermostats incorporating humidity sensors. When applying an unusual control, manufacturer recommendations should be followed to prevent equipment damage or misuse.

DESIGN

The size and performance characteristics of components are interrelated, and the overall design should proceed in the organized manner described. For example, furnace selection depends on heat gain and loss and is also affected by duct location (attic, basement, etc.), duct materials, night setback, and humidifier use. Here is a recommended procedure:

1. Estimate heating and cooling loads, including target values for duct losses.
2. Determine preliminary ductwork location and materials of ductwork and outlets.
3. Determine heating and cooling unit location.
4. Select accessory equipment. Accessory equipment is not generally provided with initial construction; however, the system may be designed for later addition of these components.
5. Select control components.
6. Select heating/cooling equipment.
7. Determine maximum airflow (cooling or heating) for each supply and return location.
8. Determine airflow at reduced heating and cooling loads (two-speed and variable-speed fans).
9. Select heating/cooling equipment.
10. Select control system.
11. Finalize duct design and size.
12. Select supply and return grilles.
13. When the duct system is in place, measure duct leakage and compare results with target values used in step 1.

This procedure requires certain preliminary information such as location, weather conditions, and architectural considerations. The following sections cover the preliminary considerations and discuss how to follow this recommended procedure.

Estimating Heating and Cooling Loads

Design heating and cooling loads can be calculated by following the procedures outlined in Chapters 29 and 30 of the 2005 *ASHRAE Handbook—Fundamentals* When calculating design loads, heat losses or gains from the air distribution system must be included in the total load for each room. In residential applications, local codes often require outdoor air ventilation, which is added to the building load. Target values for duct losses may be set by codes, voluntary programs, or other recommendations. If ducts are located in the conditioned space, losses can be reduced essentially to zero. If this is not possible, losses should be limited to 10% of the heating or cooling load.

Locating Outlets, Returns, Ducts, and Equipment

The characteristics of a residence determine the appropriate type of forced-air system and where it can be installed. The presence or absence of particular areas in a residence directly influences equipment and duct location. The structure's size, room or area use, and air-distribution system determine how many central systems will be needed to maintain comfort temperatures in all areas.

For maximum energy efficiency, ductwork and equipment should be installed in the conditioned space. ASHRAE *Standard* 90.2 gives a credit for installation in this location. The next best location is in a full basement. If a residence has an insulated, unvented, and sealed crawlspace, the ductwork and equipment can be located there (with appropriate provision for combustion air, if applicable), or the equipment can be placed in a closet or utility room. Vented attics and vented crawlspaces are the least preferred location for ductwork and HVAC equipment. The equipment's enclosure must meet all fire and safety code requirements; adequate service clearance must also be provided. In a home built on a concrete slab, equipment could be located in the conditioned space (for systems that do not require combustion air), in an unconditioned closet, in an attached garage, in the attic space, or outdoors. Ductwork normally is located in a furred space, in the slab, or in the attic. Cummings et al. (2003) tested air leakage in 30 air handler cabinets and at connections to supply and return ductwork and found leakage rates averaged 6.3% of overall system airflow.

Duct construction must conform to local code requirements, which often reference NFPA *Standard* 90B or the *Residential Comfort System Installation Standards Manual* (SMACNA 1998).

Weather should be considered when locating equipment and ductwork. Packaged outdoor units for houses in severely cold climates must be installed according to manufacturer recommendations. Most houses in cold climates have basements, making them well-suited for indoor furnaces and split-system air conditioners or heat pumps. In mild and moderate climates, ductwork is frequently in the attic or crawlspace.

Locating and Selecting Outlets and Returns. Although the principles of air distribution discussed in Chapter 19 of this volume and Chapter 33 of the 2005 *ASHRAE Handbook—Fundamentals* apply in forced-air system design, simplified methods of selecting outlet size and location are generally used.

Supply outlets fall into four general groups, defined by their air discharge patterns: (1) horizontal high, (2) vertical nonspreading, (3) vertical spreading, and (4) horizontal low. Straub and Chen (1957) and Wright et al. (1963) describe these types and their performance characteristics under controlled laboratory and actual residence conditions. Table 1 lists the general characteristics of supply outlets. It includes the performance of various outlet types for cooling as well as heating, because one of the advantages of forced-air systems is that they may be used for both heating and cooling. However, as indicated in Table 1, no single outlet type is best for both heating and cooling.

The best outlets for heating are located near the floor at outside walls and provide a vertical spreading air jet, preferably under windows, to blanket cold areas and counteract cold drafts. Called **perimeter heating**, this arrangement mixes warm supply air with both cool air from the area of high heat loss and cold air from infiltration, preventing drafts.

The best outlet types for cooling are located in the ceiling and have a horizontal air discharge pattern. For year-round systems, supply outlets are located to satisfy the more critical load.

Figure 2 illustrates preferred return locations for different supply outlet positions and system functions and typical temperature profiles. These return locations are based on the presence of the stagnant layer in a room, which is beyond the influence of the supply outlet and thus experiences little air motion (e.g., smoke "hanging"

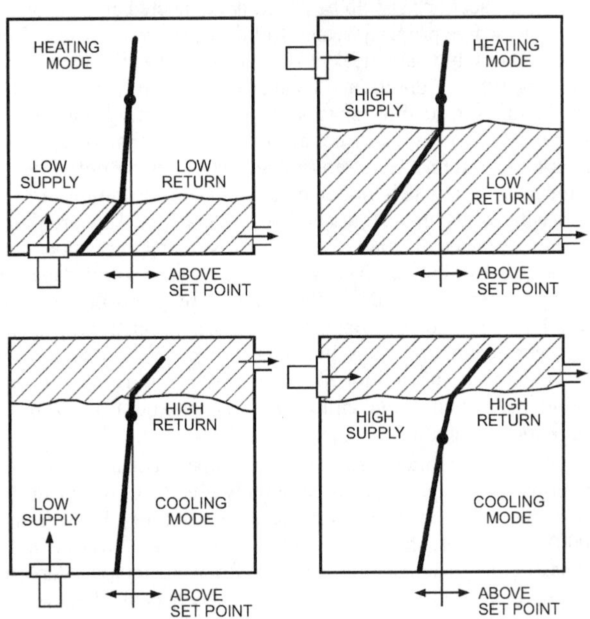

Fig. 2 Preferred Return Locations for Various Supply Outlet Positions

Table 1 General Characteristics of Supply Outlets

Group	Outlet Type	Outlet Flow Pattern	Conditioning Mode	Most Effective Application	Selection Criteria (see Figure 2)
1	Ceiling and high sidewall	Horizontal	Cooling	*Ceiling outlets*	
				Full-circle or widespread type	Select for throw equal to distance from outlet to nearest wall at design flow rate and pressure limitations.
				Narrow spread type	Select for throw equal to 0.75 to 1.2 times distance from outlet to nearest wall at design flow rate and pressure limitations.
				Two adjacent ceiling outlets	Select each so that throw is about 0.5 times distance between them at design flow rate and pressure limits.
				High sidewall outlets	Select for throw equal to 0.75 to 1.2 times distance to nearest wall at design flow rate and pressure limits. If pressure drop is excessive, use several smaller outlets rather than one large one to reduce pressure drop.
2	Floor diffusers, baseboard, and low sidewall	Vertical, nonspreading	Cooling and heating		Select for 6 to 8 ft throw at design flow rate and pressure limitations.
3	Floor diffusers, baseboard, and low sidewall	Vertical, spreading	Heating and cooling		Select for 4 to 6 ft throw at design flow rate and pressure limitations.
4	Baseboard, and low sidewall	Horizontal	Heating only		Limit face velocity to 300 fpm.

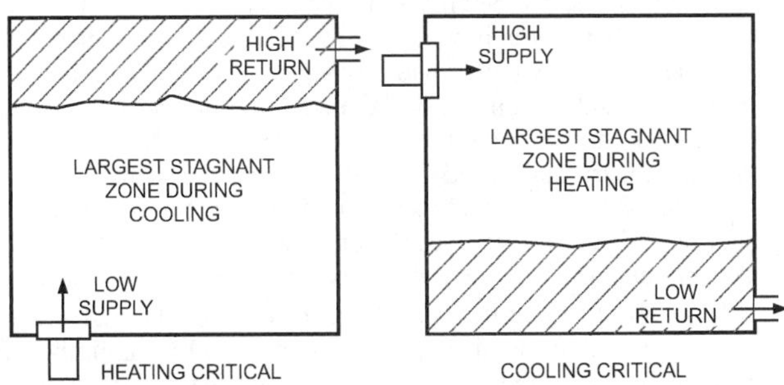

Fig. 3 Best Compromise Return Locations for Year-Round Heating and Cooling

in a spot in a room is evidence of a stagnant region). The stagnant layer degrades room comfort.

The stagnant layer develops near the floor during heating and near the ceiling during cooling. Returns help remove air from this region if the return face is placed in the stagnant zone. Thus, for heating, returns should be placed low; for cooling, returns should be placed high. However, in a year-round heating and cooling application, a compromise must be made by placing returns where the largest stagnant zone develops (Figure 3). With low supply outlets, the largest stagnant zone develops during cooling, so returns should be placed high or opposite the supply locations. Conversely, high supply outlets do not perform as well during heating; therefore, returns should be placed low to be of maximum benefit.

If central return is used, the airflow between supply registers and the return should not be impeded even when interior doors are closed.

Determining Heating and Cooling Loads

Design heating and cooling loads can be calculated by following the procedures outlined in Chapters 29 and 30 of the 2005 *ASHRAE Handbook—Fundamentals*. When calculating design loads, heat losses or gains from the air distribution system must be included in the total load for each room. In residential applications, local codes often require outside air ventilation, which is added to the building load.

Selecting Equipment

Furnace heating output should match or slightly exceed the estimated design load. A 40% limit on oversizing has been recommended by the Air Conditioning Contractors of America (ACCA) for fossil-fuel furnaces. This limit minimizes venting problems associated with oversized equipment and improves part-load performance. Note that the calculated load must include duct loss, humidification load, and night setback recovery load, as well as building conduction and infiltration heat losses. Chapter 28 has detailed information on how to size and select a furnace.

To help conserve energy, manufacturers have added features to improve furnace efficiency. Electric ignition has replaced the standing pilot; vent dampers and more efficient motors are also available. Furnaces with fan-assisted combustion systems (FACSs) and condensing furnaces also improve efficiency. Two-stage heating and cooling, variable-speed heat pumps, and two-speed and variable-speed blowers are also available.

Research on the effect of blower performance on residential forced-air heating system performance suggested reductions of 180 to 250 kWh/yr for automatic furnace fan operation and 2600 kWh/yr for continuous fan operation by changing from permanent split capacitor (PSC) blower motors to brushless permanent electronically commutated magnet motors (ECMs) (Phillips 1998).

A system designed to both heat and cool and that cycles cooling equipment on and off by sensing dry-bulb temperature alone should be sized to match the design heat gain as closely as possible. Oversizing under this control strategy could lead to higher-than-desired indoor humidity levels. Chapter 29 of the 2005 *ASHRAE Handbook—Fundamentals* recommends that cooling units not be oversized. Other sources suggest limiting oversizing to 15% of the sensible load. A heat pump should be sized for the cooling load with supplemental heat provided to meet heating requirements. Air-source heat pumps should be sized in accordance with the equipment manufacturer recommendations. ACCA *Manual* S can also be used to assist in the selection and sizing of equipment.

Determining Airflow Requirements

After the equipment is selected and before duct design, the following decisions must be made:

1. Determine the air quantities required for each room or space during heating and cooling based on each room's heat loss or heat gain. The air quantity selected should be the greater of the heating or cooling requirement.
2. Determine the number of supply outlets needed for each space to supply the selected air quantity, considering discharge velocity, spread, throw, terminal velocity, occupancy patterns, location of heat gain and loss sources, and register or diffuser design.

3. Determine the type of return (multiple or central), availability of space for grilles, filtering, maximum velocity limits for sound, efficient filtration velocity, and space use limitations.

Detailing the Duct Configuration

The next major decision is to select a generic duct system. In order of decreasing efficiency, the three main types are

1. Ducts in conditioned space
2. Minimum-area ductwork
3. Traditional designs

Ductwork costs and system energy use can be reduced when the home designer/architect, builder, subcontractors, and HVAC installer collaborate to place ducts in conditioned spaces and minimize duct runs. Residential duct systems in unconditioned spaces can lose a significant percent of the energy in the air they distribute. These losses can be almost entirely eliminated by simply locating ducts in the conditioned space (insulated building envelope), which is a cost-effective way to increase heating and cooling equipment efficiency and lower utility bills (Modera 1989). Benefits include improved comfort, improved indoor air quality, and lower utility bills and equipment cost.

Any losses (air or conductive) from ducts in conditioned space still provide space conditioning. Ducts in conditioned space are also subjected to much less severe conditions, reducing conductive losses and the effect of return air leaks. There are a number of approaches that can be used to accomplish this:

- Trunks and branches can be located between floors of a two-story residence or along the wall-ceiling intersections in a single-story dwelling. Care must be taken to seal the rim joist between floors, and/or the wall-to-ceiling intersection. Figures 4A and 4B illustrate the planning required for locating ductwork between floors in a two-story residence and townhouse.
- In some houses, the ceiling in a central hallway can be lowered. The air barrier is still provided at the higher level, bringing the space between the ceiling and air barrier into conditioned space. Ducts are installed in this space, with supply registers located on the walls of adjacent spaces. The ceiling can be dropped in closets, bathrooms, or, if necessary, a soffit to get ducts to rooms that are not adjacent to the central hallway.
- A slab-on-grade foundation is common in mild or moderate climates. With this type of foundation, supply air ducts are typically located in the attic. During the winter, attic air temperatures tend to match outdoor air temperatures. During the summer, solar heat gains can raise attic air temperatures over 150°F. These temperature extremes increase heat losses and gains from conduction and radiation and decrease duct efficiency. In addition, any conditioned air that leaks out of the duct is lost into the attic.
- Figures 5A, 5B, and 5C show that constructing a ceiling plenum in the hallway allows ducts to be located in the conditioned space. Air temperatures in this location are typically between 55 and 85°F, which minimizes conduction and radiation losses. Air that leaks out of the ducts goes into the conditioned space.
- Attics can be included in the conditioned space by relocating the thermal barrier to the roof and eliminating ridge and soffit vents to provide an air barrier at the roof line. Insulation can be installed at the roofline by, for example, installing netting material between trusses, and installing blown-in cellulose insulation. Ductwork can then be installed in the attic in conditioned space. In cold climates, care must be taken to avoid condensation on the inside of the roof deck; in hot climates, the lack of roof venting may argue against using asphalt-shingle roofing.
- A plenum space can be created in the attic by using roof trusses that do not have a traditional flat bottom chord. A modified scissors truss design, which provides space between the bottom chord of the truss and the top chord of the wall framing, provides a duct space that can be brought into conditioned space. The bottom chord of the trusses is used to install an air barrier, with insulation blown in on top. Ductwork is installed in the plenum space, with supply registers located near interior walls (because the space may not extend all the way to the exterior walls).

It is important that the ducts be located inside thermal and air barriers, and that the air barrier be well sealed to minimize air communication with the outdoors. The duct space is rarely completely in conditioned space (other than in exposed ductwork systems). When there is an air barrier between the ducts and the occupied space, some fraction of air and thermal losses from the duct system goes to the outdoors rather than to the occupied space. High-quality air sealing on the exterior air barrier minimizes these losses to the outdoors.

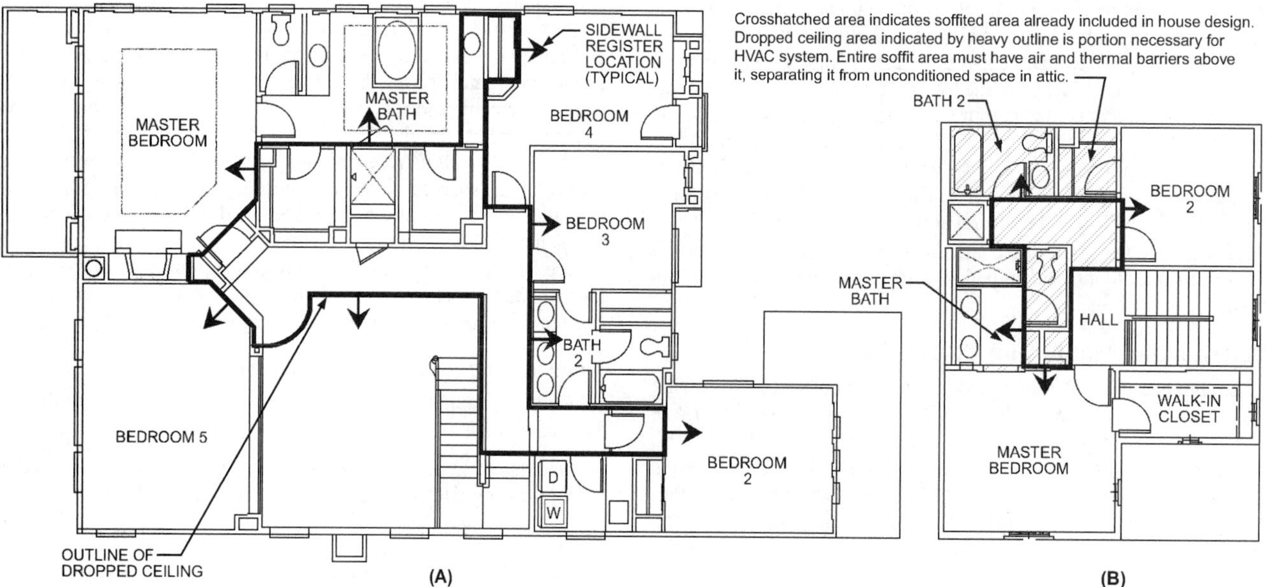

Fig. 4 Sample Floor Plans for Locating Ductwork in Second Floor of (A) Two-Story House and (B) Townhouse
(Hedrick 2002)

Many new buildings have well-insulated envelopes or sufficient thermal integrity so that supply registers do not have to be located next to exterior walls. Placing registers in interior walls can reduce duct surface area by 50% or more, with similar reductions in leakage and conductive losses. This option also offers significant first-cost savings. Minimum-area ductwork systems are used in most houses built with ducts in conditioned space, including those using a dropped ceiling.

Figures 4 and 5 are improved duct designs for new energy-efficient residential construction. These residences are designed with tighter envelopes/ducts, increased insulation, and high-performance windows, resulting in wall, window, floor, and ceiling temperatures that are warmer in winter and cooler in summer, and are more comfortable and less drafty.

In traditional designs for standard residential construction, supply ducts are typically run in unconditioned spaces, with supplies located near the perimeter of a house to offset drafts from cold exterior surfaces, especially windows (Figures 6A and 6B). Because this is the least efficient option overall, particular care should be taken to seal and insulate the ductwork. Any air leaks on the supply side of the system allow conditioned supply air to escape to the outdoors. Return-side leaks draw air at extreme temperatures into the system instead of tempered room air. Return leaks can also have indoor air quality effects if the return ducts are located in garages or other spaces where contaminants may be present. In humid climates, return leaks bringing in humid outdoor air can raise the humidity in the space, increasing the risk of mold and mildew.

Detailing the Distribution Design

The major goal in duct design is to provide proper air distribution throughout a residence. To achieve this in an energy-efficient manner, ducts must be sized and laid out to facilitate airflow and minimize friction, turbulence, and heat loss and gain. The optimal air distribution system has "right-sized" ducts, minimal runs, the smoothest interior surfaces possible, and the fewest possible direction and size changes. Figure 5C provides an example of right-sized ducts design.

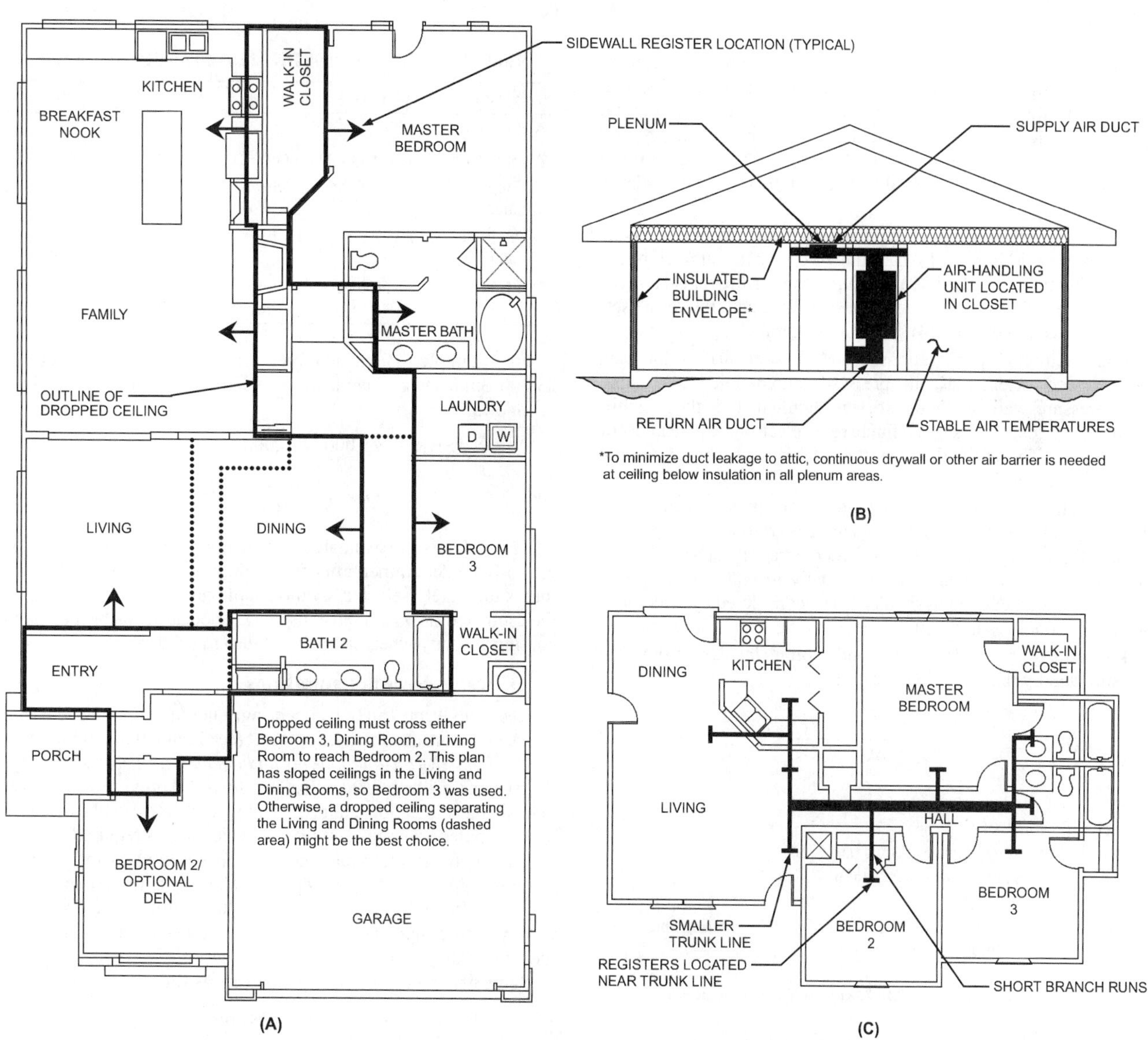

Fig. 5 Sample Floor Plans for One-Story House with (A) Dropped Ceilings, (B) Ducts in Conditioned Spaces, and (C) Right-Sized Air Distribution in Conditioned Spaces
(EPA 2000)

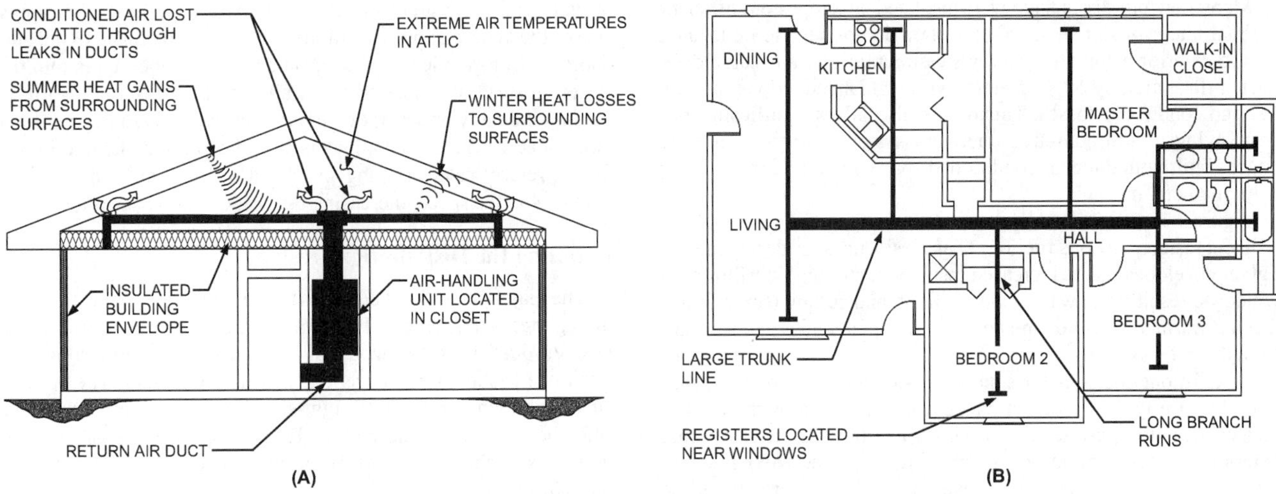

Fig. 6 (A) Ducts in Unconditioned Spaces and (B) Standard Air Distribution System in Unconditioned Spaces
(EPA 2000)

The required airflow and blower's static pressure limitation are the parameters around which the duct system is designed. The heat loss or gain for each space determines the proportion of the total airflow supplied to each space. Static pressure drop in supply registers should be limited to about 0.03 in. of water. The required pressure drop must be deducted from the static pressure available for duct design.

The flow delivered by a single supply outlet should be determined by considering the (1) space limitations on the number of registers that can be installed, (2) pressure drop for the register at the flow rate selected, (3) adequacy of air delivery patterns for offsetting heat loss or gain, and (4) space use pattern.

Manufacturers' specifications include blower airflow for each blower speed and external static pressure combination. Determining static pressure available for duct design should include the possibility of adding accessories in the future (e.g., electronic air cleaners or humidifiers). Therefore, the highest available fan speed should not be used for design.

For systems that heat only, the blower rate may be determined from the manufacturer's data. The temperature rise of air passing through the heat exchanger of a fossil-fuel furnace must be within the manufacturer's recommended range (usually 40 to 80°F). The possible later addition of cooling should also be considered by selecting a blower that operates in the midrange of the fan speed and settings.

For cooling only, or for heating and cooling, the design flow can be estimated by the following equation:

$$Q = \frac{q_s}{60\rho c_p \Delta t} = \frac{q_s}{1.1\Delta t} \tag{1}$$

where

Q = flow rate, cfm
ρ = air density assumed to equal 0.075 lb/ft^3
c_p = specific heat of air = 0.24 Btu/lb·°F
q_s = sensible load, Btu/h
Δt = dry-bulb temperature difference between air entering and leaving equipment, °F

For preliminary design, an approximate Δt is as follows:

Sensible Heat Ratio (SHR)	Δt, °F
0.75 to 0.79	21
0.80 to 0.85	19
0.85 to 0.90	17

SHR = Calculated sensible load/Calculated total load

Table 2 Recommended Division of Duct Pressure Loss

	System Characteristics	Supply, %	Return, %
A	Single return at blower	90	10
B	Single return at or near equipment	80	20
C	Single return with appreciable return duct run	70	30
D	Multiple return with moderate return duct system	60	40
E	Multiple return with extensive return duct system	50	50

For example, if calculation indicates the sensible load is 23,000 Btu/h and the latent load is 4900 Btu/h, the SHR is calculated as follows:

$$\text{SHR} = 23{,}000/(23{,}000 + 4900) = 0.82$$

and

$$Q = \frac{23{,}000}{1.1 \times 19} = 1100 \text{ cfm}$$

This value is the estimated design flow. The exact design flow can only be determined after the cooling unit is selected. The unit that is ultimately selected should supply an airflow in the range of the estimated flow, and must also have adequate sensible and latent cooling capacity when operating at design conditions.

Duct Design Recommendations

Residential construction duct design should be approached using duct calculators and the friction chart (see Figure 9 in Chapter 35 of the 2005 *ASHRAE Handbook—Fundamentals*). Chapters 8 to 11 of the ACCA Residential Duct System *Manual* D provide step-by-step duct sizing calculation examples and worksheets. Hand calculators and computer programs simplify the calculations required.

The ductwork distributes air to spaces according to the space heating and/or cooling requirements. The return air system may be single, multiple, or any combination that returns air to the equipment within design static pressure and with satisfactory air movement patterns (Table 2).

Some general rules in duct design are as follows:

- Keep main ducts as straight as possible.
- Streamline transitions.
- Design elbows with an inside radius of at least one-third the duct width. If this inside radius is not possible, include turning vanes.
- Seal ducts to limit air leakage.

- Insulate and/or line ducts, where necessary, to conserve energy and limit noise.
- Locate branch duct takeoffs at least 4 ft downstream from a fan or transition, if possible.
- Isolate air-moving equipment from the duct using flexible connectors to isolate noise.

Large air distribution systems are designed to meet specific noise criteria (NC) levels. Small systems should also be designed to meet appropriate NC levels; however, acceptable duct noise levels can often be achieved by limiting air velocities in mains and branches to the following:

Main ducts	700 to 900 fpm
Branch ducts	600 fpm
Branch risers	500 fpm

Considerable difference may exist between the cooling and heating flow requirements. Because many systems cannot be rebalanced seasonally, a compromise must be made in the duct design to accommodate the most critical need. For example, a kitchen may require 165 cfm for cooling but only 65 cfm for heating. Because the kitchen may be used heavily during design cooling periods, the cooling flow rate should be used. Normally, the maximum design flow should be used, as register dampers do allow some optional reduction in airflows.

Zone Control for Small Systems

In residential applications, some complaints about rooms that are too cold or too hot are related to the system's limitations. No matter how carefully a single-zone system is designed, problems will occur if the control is unable to accommodate the various load conditions that occur simultaneously throughout the house at any time of day and/or during any season.

Single-zone control works as long as the various rooms are open to each other. In this case, room-to-room temperature differences are minimized by convection currents between the rooms. For small rooms, an open door is adequate. For large rooms, openings in partitions should be large enough to ensure adequate air interchange for single-zone control.

When rooms are isolated from each other, temperature differences cannot be moderated by convection currents, and conditions in the room with the thermostat may not be representative of conditions in the other rooms. In this situation, comfort can be improved by continuous blower operation, but this strategy may not completely solve the problem.

Zone control is required when conditions at the thermostat are not representative of all the rooms. This situation will almost certainly occur if any of the following conditions exists:

- House has more than one level
- One or more rooms are used for entertaining large groups
- One or more rooms have large glass areas
- House has an indoor swimming pool and/or hot tub
- House has a solarium or atrium

In addition, zoning may be required when several rooms are isolated from each other and from the thermostat. This situation is likely to occur when

- House spreads out in many directions (wings)
- Some rooms are distinctly isolated from rest of house
- Envelope only has one or two exposures
- House has a room or rooms in a finished basement or attic
- House has one or more rooms with slab or exposed floor

Zone control can be achieved by installing

- Discrete heating/cooling ducts for each zone requiring control
- Automatic zone damper in a single heating/cooling duct system

The rate of airflow delivered to each room must be able to offset the peak room load during cooling. The peak room load can be determined using Chapter 29 of the 2005 *ASHRAE Handbook—Fundamentals*. The same supply air temperature difference used to size equipment can be substituted into Equation (1) to find airflow. The design flow rate for any zone is equal to the sum of the peak room flow rates assigned to a zone.

Duct Sizing for Zone Damper Systems

The following guidelines are proposed in ACCA *Manual* D to size various duct runs.

1. Use the design blower airflow rate to size a plenum or a main trunk that feeds the zone trunks. Size plenum and main trunk ducts at 800 fpm.
2. Use zone airflow rates (those based on the sum of the peak room loads) to size the zone trunk ducts. Size all zone trunks at 800 fpm.
3. Use the peak room airflow rate (those based on the peak room loads) to size the branch ducts or runouts. Size all branch runouts at a friction rate of 0.10 in. of water per 100 ft.
4. Size return ducts for 600 fpm air velocity.

Box Plenum Systems Using Flexible Duct

In some climates, an overhead duct with a box plenum feeding a series of individual, flexible-duct, branch runouts is popular. The pressure drop through a flexible duct is higher than through a rigid sheet metal duct, however. Recognizing this larger loss is important when designing a box plenum/flexible duct system.

The design of the box plenum is critical to avoid excessive pressure loss and to minimize unstable air rotation in the plenum, which can change direction between blower cycles. This in turn may change the air delivery through individual branch takeoffs. Unstable rotation can be avoided by having the air enter the box plenum from the side and by using a special splitter entrance fitting.

Gilman et al. (1951) proposed box plenum dimensions and entrance fitting designs to minimize unstable conditions as summarized in Figures 7 and 8. For residential systems with less than 2250 cfm capacity, pressure loss through the box plenum is approximately 0.05 in. of water. This loss should be deducted from the available static pressure to determine the static pressure available for duct branches. In terms of equivalent length, add approximately 50 ft to the measured branch runs.

Embedded Loop Ducts

In cold climates, floor slab construction requires that the floor and slab perimeter be heated to provide comfort and prevent condensation. The temperature drop (or rise) in the supply air is significant, and special design tables must be used to account for the different supply air temperatures at distant registers. Because duct heat losses may cause a large temperature drop, feed ducts need to be placed at critical points in the loop.

A second aspect of a loop system is installation. The building site must be well drained and the surrounding grade sloped away from the structure. A vapor retarder must be installed under the slab. The bottom of the embedded duct must not be lower than the finished grade. Because a concrete slab loses heat from its edges outward through the foundation walls and downward through the earth, the edge must be properly insulated.

A typical loop duct is buried in the slab 2 to 18 in. from the outside edge and about 2.5 in. beneath the slab surface. If galvanized sheet metal is used for the duct, it must be coated on the outside to comply with Federal Specification SS-A-701. Other special materials used for ducts must be installed according to the manufacturer's instructions. In addition, care must be taken when the slab is poured not to puncture the vapor retarder or to crush or dislodge the ducts.

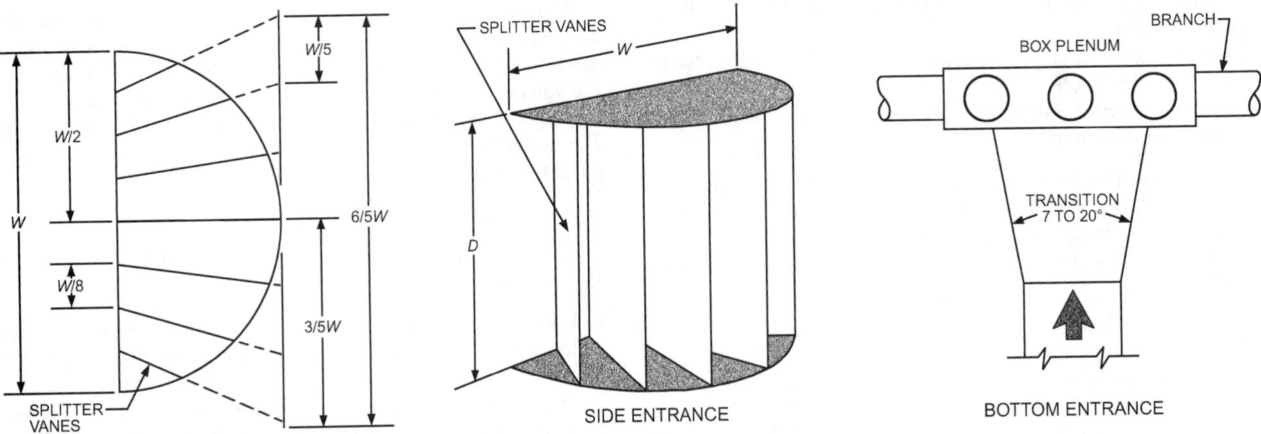

Fig. 7 Entrance Fittings to Eliminate Unstable Airflow in Box Plenum

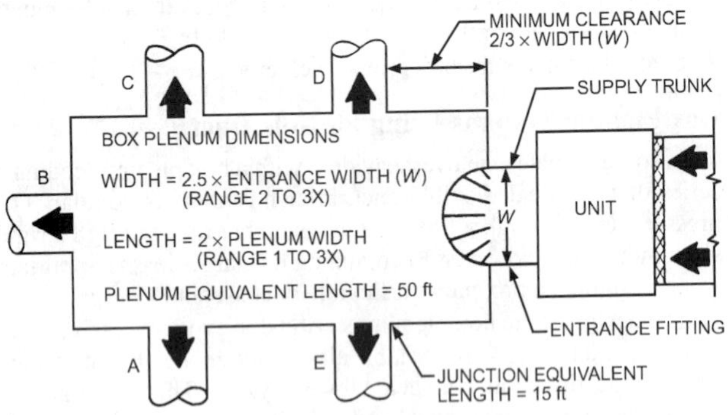

Fig. 8 Dimensions for Efficient Box Plenum

SELECTING SUPPLY AND RETURN GRILLES AND REGISTERS

Grilles and registers are selected from a manufacturer's catalog with appropriate engineering data after the duct design is completed. Rule-of-thumb selection should be avoided. Carefully determine the suitability of the register or grille selected for each location according to its performance specification for the quantity of air to be delivered and the discharge velocity from the duct.

Generally, in small commercial and residential applications, the selection and application of registers and grilles is particularly important because system size and air-handling capacity are small in energy-efficient structures. Proper selection ensures satisfactory delivery of heating and/or cooling. Table 1 summarizes selection criteria for common types of supply outlets. Pressure loss is usually limited to 0.03 in. of water or less.

Return grilles are usually sized to provide a face velocity of 400 to 600 fpm, or 2.7 to 4.1 cfm per square inch of free area. Some central return grilles are designed to hold an air filter. This design allows the air to be filtered close to the occupied area and also allows easy access for filter maintenance. Easy access is important when the furnace is in a remote area such as a crawlspace or attic. The air velocity through a filter grille should not exceed 300 fpm, which means that the volume of air should not exceed 2.1 cfm per square inch of free filter area.

COMMERCIAL SYSTEMS

The duct design procedure described in this chapter can be applied to small commercial systems using residential equipment, provided that the application does not include moisture sources that create a large latent load.

In commercial applications that do not require low noise, air duct velocities may be increased to reduce duct size. Long throws from supply outlets are also required for large areas, and higher velocities may be required for that reason.

Commercial systems with significant variation in airflow for cooling, heating, and large internal loads (e.g., kitchens, theaters) should be designed in accordance with Chapters 33 and 35 of the 2005 *ASHRAE Handbook—Fundamentals*, Chapters 4 and 31 in the 2007 *ASHRAE Handbook—HVAC Applications,* and Chapters 18, 19, and 20 of this volume.

Air Distribution in Small Commercial Buildings

According to Andrews et al. (2002), forced-air thermal distribution systems in small commercial buildings tend to be similar in many ways to those in residential buildings. As in most residential systems, there is often a single air handler that transfers heat or cooling from the equipment to an airstream that is then circulated through the building by means of ducts. Two major differences, however, may affect the performance of small commercial buildings: (1) significant (and often multiple) connections with outside air, and (2) the ceiling-space configuration.

Outside-Air Connections. In small commercial buildings, many forced-air systems have an outside-air duct leading from the outside to the return side of the ductwork, used to provide ventilation. Ventilation may also be provided by a separate exhaust-air system consisting of a duct and fan blowing air out of the building. Finally, there may also be a makeup air system blowing air into the building, used to balance out all the other airflows. A malfunction in any of these components can compromise the energy efficiency and thermal comfort performance of the entire system.

Ceiling-Space Configuration. Knowing the layout of the ceiling space is key to understanding uncontrolled airflows. The overhead portions of the air and thermal barriers can either be together, at the ceiling or at the roof, or separate, with the thermal barrier at the ceiling and the air barrier at the roof.

One configuration, typical of residential buildings but uncommon in commercial buildings, has a tight gypsum-board ceiling with insulation directly above and a vented attic. Ductwork is often placed in the vented attic space, though it may be elsewhere (e.g., in the conditioned space, under the building, or outside). The ducts are outside both the air and thermal barriers.

A more common configuration is similar except that it has a suspended T-bar ceiling instead of gypsum board. Air leakage through this type of ceiling tends to be quite high because the (usually leaky) suspended ceiling and attic vent provide an easy airflow path between the building and the outside. Efficiency is also compromised by duct placement in a very hot and humid location. Because of these two factors, uncontrolled airflows can strongly affect energy use, ventilation rates, and indoor humidity. This configuration should be avoided.

A third configuration is also similar, except that the ceiling space is not vented. This puts the ducts inside the air barrier (desirable) but outside the thermal barrier (undesirable). During the cooling season, the ceiling space tends to be very hot and dry. Uncontrolled airflows increase energy use but not ventilation rates or humidity levels.

The best configuration has insulation at the roof plane, leaving the space below the roof unvented, with or without a dropped ceiling. This design is very forgiving of uncontrolled airflow as long as the ductwork is inside the building. Duct leakage and unbalanced return air have little effect on energy use, ventilation rates, and indoor humidity, because conditions in the space below the roof deck are not greatly different from those in the rooms.

Controlling Airflow in New Buildings

Airflow control should be a key objective in designing small commercial buildings. Designers and builders can plan for proper airflows at the outset. The following design goals are recommended:

- *Design the building envelope to minimize effects of uncontrolled airflows.* Place the air and thermal barriers together in the roof, with ducts inside the conditioned zone. There are several good options for placing insulation at the roof level, including sprayed polymer foam, rigid insulation board on the roof deck beneath a rubber membrane, and insulating batts attached to the underside of the roof.
- *Minimize duct leakage to outside.* Make sure as much ductwork is in the conditioned envelope as possible. Do not vent the ceiling space. If possible, dispense with the dropped ceiling altogether and use exposed ductwork. Avoid using building cavities as part of the air distribution system.
- *Minimize unbalanced return air.* The best way is to provide a ducted return for each zone, and then balance these with the supply ducts serving the respective zones. Where that is not possible, transfer ducts or grilles may be provided to link a zone without a return duct to another zone with a return duct, provided that these have a minimum of 70 in^2 of net free area per 100 cfm of return airflow. This approach should only be used if the thermal and air barriers are in the same plane (roof or ceiling).
- *Minimize unbalanced airflows across the building envelope.* Design the exhaust system with the smallest airflow rates necessary to capture and remove targeted air contamination sources and meet applicable standards. Ensure that the sum of makeup and outdoor airflow rates exceeds the exhaust airflow rate, not only for the building as a whole but also for any zones that can be isolated. Unconditioned makeup air should equal 75 to 85% of exhaust airflow, where possible. In buildings where continuous ventilation is required and the climate is especially humid, special design options may be needed (e.g., a dedicated makeup air unit could be provided with its own desiccant dehumidifier).
- *Ensure proper operation of outside air dampers.* Outside air dampers on air handlers or rooftop air-conditioning units are frequently stuck or rusted shut, even on recently installed equipment. Proper performance helps ensure proper air quality and thermal comfort. Inspect space conditioning equipment annually to ensure proper operation of these dampers.

Further information can be found in Andrews et al. (2002).

TESTING FOR DUCT EFFICIENCY

ASHRAE and other research organizations have conducted significant research and published numerous articles about methods for testing, and measured performance of the design and seasonal efficiencies of residential duct systems in the heating and cooling modes. A compilation of this research is provided in the Bibliography.

Although duct leakage is a major cause of duct inefficiency, other factors, such as heat conduction through duct walls, influence of fans on pressure in the house, and partial regain of lost heat, must also be taken into account. The following summary of information needed to evaluate the efficiency of a duct system also describes the results that a test method should provide.

Data Inputs

The variables that are known or can be measured to provide the basis for calculating duct system efficiency include the following:

Local climate data. Three outside design temperatures are needed to describe an area's climate: one dry-bulb and one wet-bulb temperature for cooling, and one dry-bulb for heating.

Dimensions of living space. The volume of the conditioned space must be known to estimate the impact of the duct system on air infiltration. Typical, average values have been developed and if default options are used, the floor area of the conditioned space must be known.

Surface areas of ducts and R-values of insulation. The total surface area of supply and return ducts and the insulation R-value of each are needed for calculating conductive heat losses through the duct walls. Also needed is the fraction of supply and return ducts in each type of buffer zone (e.g., an attic, basement, or crawlspace).

Fan flow rate. The airflow through the fan must be measured to determine duct leakage (a major factor of efficiency) as a percentage of fan flow. An adjustable, calibrated fan flowmeter is the most accurate device for measuring airflow. Other measurement methods include a pitot tube traverse, flow grids inserted at the filter housing (Palmiter and Francisco 2000), calculations based on the temperature rise caused by a known heat input, and measurement of the concentration of a tracer gas.

Duct leakage to the outdoors. Air leakage from supply ducts to the outdoors and from the outdoors and buffer spaces into return ducts is another major factor that affects efficiency. Typically, 17% or more of the total airflow is leakage.

Data Output

Distribution efficiency is the main output of a test method. This figure of merit is the ratio of the input energy that would be needed to heat or cool the house if the duct system had no losses to the actual energy input required. Distribution efficiency also accounts for the effect the duct system has on equipment efficiency and the space conditioning load. Thus, distribution efficiency differs from delivery effectiveness, which is the simple output-to-input ratio for a duct system.

Four types of distribution efficiencies are typically considered. They relate to efficiency during either heating or cooling and for either design conditions or seasonal averages.

- Design distribution efficiency, heating
- Seasonal distribution efficiency, heating
- Design distribution efficiency, cooling
- Seasonal distribution efficiency, cooling

Design values of distribution efficiency are peak-load values that should be used when sizing equipment. Seasonal values should be used for determining annual energy use and subsequent costs.

SYSTEM PERFORMANCE

Both furnace performance and the interaction of the furnace and the building's distribution system determine how much fuel energy input to the furnace beneficially heats the conditioned space. Performance depends on the definition of the space in which it applies. In conditioned space, temperature is actively controlled by a thermostat. A building can contain other space (attic, basement, or crawlspace) that may influence the thermal performance of the conditioned space, but it is not defined as part of the conditioned space.

For houses with basements, it is important to decide whether the basement is part of the conditioned space because it typically receives some fraction of the HVAC output. In this analysis the basement is part of the conditioned space only if it is under active thermostat control and warm-air registers are provided to maintain comfort. Otherwise, the basement is not part of the conditioned space even if some heat is provided with fixed open registers, for example. The following performance examples show designs for improving efficiency, along with their effect on temperature in the unconditioned basement.

"HOUSE" Dynamic Simulation Model

The dynamic response and interactions between components of central forced warm-air systems are sufficiently complex that the effects of system options on annual fuel use are not easily evaluated. ASHRAE special project SP-43 assessed the effects of system component and control mode options. The resulting simulation model accounts for the dynamic and thermal interactions of equipment and loads in response to varying weather patterns. Both single-zone (HOUSE-I) and multizone (HOUSE-II) versions of the model have been developed.

Fischer et al. (1984) described the HOUSE simulation model. Herold et al. (1987), Jakob et al. (1986a), and Jakob et al. (1987) described the validation model for the heating mode through field experiments in two houses. Herold et al. (1986) summarized both the project and the model.

Jakob et al. (1986b) and Locklin et al. (1987) presented the model's predictions of overall performance of the forced warm-air heater. These variables include furnace and venting types, furnace installation location and combustion air source, furnace sizing, night setback, thermostat cycling rate, blower operating strategy, basement insulation, duct sealing and insulation, house and foundation type, and climate.

HOUSE models assume that duct leakage must be specified as an input. Subsequent research has developed methods for calculating losses from leakage areas and operating pressures. Moreover, in all the SP-43 runs, supply and return leakage values were set equal to each other, a situation that often does not hold in practice. Nevertheless, the generic results generally confirm the usefulness of the annual fuel utilization efficiency (AFUE) as a measure of furnace efficiency and point the way toward improvements in efficiency. Also, the parameters used to characterize energy flows and efficiency ratios are the basis for a proposed standard test method for thermal distribution efficiency. ASHRAE research project RP-852 (Gu et al. 1998) provided insights on modeling the performance of residential duct systems.

SYSTEM PERFORMANCE FACTORS

A series of system performance factors, consisting of both efficiency factors and dimensionless energy factors, describe dynamic performance of the individual components and the overall system over any period of interest. Jakob et al. (1986b) and Locklin et al. (1987) described the factors in detail.

Table 3 identifies the performance factors and their mathematical definitions in four categories: (1) equipment-component efficiency factors, (2) equipment-system performance factors, (3) equipment-load interaction factors, and (4) energy cost factors. The detailed results of the analysis may be found in the references mentioned previously. The following discussion focuses on insights gained from ASHRAE SP-43.

Equipment-Component Efficiency Factors

Furnace Efficiency E_F. This factor is the ratio of the energy delivered to the plenum during cyclic operation of the furnace to the total input energy on an annual basis. E_F includes summertime pilot losses and blower energy.

This factor is similar in concept to the AFUE in ASHRAE *Standard* 103, which provides an estimate of annual energy, taking into account assumed system dynamics. However, E_F differs importantly from AFUE in that

- The AFUE for a given furnace is defined by a single predetermined cyclic condition with standard dynamics; E_F is based on the integrated cyclic performance over a year.
- The AFUE does not include auxiliary electric input, and gives credit for jacket losses, except when the furnace is an outdoor unit; E_F and the other efficiency factors defined here include auxiliary electric input.
- Several other effects regarding combustion-induced infiltration and dynamics were investigated, but they had little effect on performance.

Duct Efficiency E_D. This is the ratio of the energy intentionally delivered to the conditioned space through the supply registers to the energy delivered to the furnace plenum, on an annual basis.

Equipment-System Performance Factors

Heat Delivery Efficiency E_{HD}. This is the product of furnace efficiency E_F and duct efficiency E_D. It is the ratio of the energy intentionally delivered to the conditioned space to the total input energy. It is a measure of how effectively the HVAC delivers heat directly to the conditioned space on an annual basis.

Miscellaneous Gain Factor F_{MG}. This is the total heat delivered to the conditioned space divided by the energy intentionally delivered to the conditioned space through the duct registers.

System Efficiency E_S. This is the product of E_{HD} and F_{MG}. It is the total energy delivered to the conditioned space divided by the total energy input to the furnace. Thus, E_S includes intentional and unintentional energy gains.

Equipment-Load Interaction Factors

Load Modification Factor F_{LM}. This is the ratio of the total heat delivered for a base case to the total heat delivered for a case of interest. It adjusts E_S to account for the effect of operation on the heating load. It accounts for the effect of combustion-induced infiltration and off-period infiltration because of draft hood flow, as well as effects of temperature changes of unconditioned spaces adjacent to the conditioned space.

System Index I_S. This index is the product of system efficiency E_S and the load modification factor F_{LM}. I_S is an energy-based figure of merit that adjusts E_S for any credits or debits from system-induced loads relative to a base case load. I_S is a powerful tool for comparing alternative systems. However, high values of I_S are sometimes associated with a low basement temperature because more of the furnace output is delivered directly to the conditioned space. The ratio of the system indexes for two systems being compared is the inverse of the ratio of their annual energy use (AEU).

Energy Cost Factors

Table 3 also defines cost factors, which are discussed in the cited literature and not reviewed here.

Implications

The following implications apply to the definitions for the various performance factors.

- The defined conditioned space is important to the comparisons of system index I_S. Because F_{LM} and I_S are based on the same reference equipment and house configuration, performance of various furnaces installed in basements or in the conditioned space (i.e., closet installations) may be compared. However, performance of a furnace installed in a basement or crawlspace cannot

Table 3 Definitions of System Performance Factors

Definition	Comments
Equipment-Component efficiency factors	
E_F = Furnace Efficiency = $100 \frac{\text{Furnace Output}}{\text{Total Energy Input}} = 100 \frac{\text{Duct Input}}{\text{Total Energy Input}}$	• Integrated energy over all operating cycles
E_D = Duct Efficiency = $100 \frac{\text{Duct Output}}{\text{Duct Input}}$	
Equipment-System performance factors	
E_{HD} = Heat Delivery Efficiency = $\frac{E_F \times E_D}{100} = \frac{\text{Duct Output}}{\text{Total Energy Input}}$	• Efficiency of the furnace/duct subsystem
F_{MG} = Miscellaneous Gain Factor = $\frac{\text{Total Heat Delivered}^a}{\text{Duct Output}}$	• Accounts for heating by fugitive gains
E_S = System Efficiency = $E_{HD} \times F_{MG} = 100 \frac{\text{Total Heat Delivered}}{\text{Total Energy Input}}$	• Efficiency of the combined HVAC system
Equipment-Load interaction factors	
F_{IL} = Induced Load Factor[b] = $\frac{\text{System Induced Load}^c}{\text{Total Heat Delivered}}$	• Accounts for added loads from equipment operation
F_{LM} = Load Modification Factor = $1.0 - F_{IL} = \frac{\text{Total Heat Delivered} - \text{System Induced Load}}{\text{Total Heat Delivered}}$	
I_S = System Index[d] = $\frac{E_S \times F_{LM}}{100} = \frac{\text{Total Heat Delivered} - \text{System Induced Load}}{\text{Total Energy Input}}$	• Common index for ranking system. **Not** an efficiency.
Energy cost factors	
R_{AE} = Auxiliary Energy Ratio = $\frac{\text{Auxiliary Energy Input}}{\text{Primary Energy Input}} = \frac{\text{Electrical Energy Input}}{\text{Fuel Energy Input}}$	• System energy characteristics
R_{CL} = Local Energy Cost Ratio = $\frac{\text{Electrical Cost per Energy Unit}}{\text{Reference Fuel Cost per Energy Unit}}$ (in common units)	
F_{CR} = Cost Ratio Factor = $\frac{\text{Fuel Energy Input} + \text{Electric Energy Input}}{\text{Fuel Energy Input} + R_{CL}(\text{Electric Energy Input})} = \frac{1.0 + R_{AE}}{1.0 + R_{CL} \times R_{AE}}$ Special Case (Fuel = 0): $F_{CR} = 1/R_{CL}$	• Economics
I_{SCM} = Cost-Modified System Index[d] = $I_S \times F_{CR} = \frac{\text{Total Heat Delivered} - \text{System Induced Load}}{\text{Primary Energy Input} + R_{CL}(\text{Auxiliary Energy Input})}$	• Common economic index for ranking systems
Annual energy use	
AEU = Annual Energy Use (fuel and electricity) predicted by the HOUSE model, in common energy units	
Annual Fuel Used = AEU / $(1.0 + R_{AE})$	
Annual Electricity Used = AEU / $(1.0 + 1/R_{AE})$	
Percent savings	
% Energy Saving = $100\,[I_S - (I_S)_{BC}]/I_S$, where $(I_S)_{BC} = I_S$ for base case	• Saving relative to base case
% Cost Saving = $100\,[I_{SCM} - (I_{SCM})_{BC}]/I_{SCM}$, where $(I_{SCM})_{BC}$ = Cost-modified I_S for base case	
Other factors for dynamic performance	
AFUE = Annual Fuel Utilization Efficiency by ANSI/ASHRAE *Standard* 103 efficiency rating, applicable to specific furnaces. Values in this chapter are for generic furnaces.	
SSE = Steady-State Efficiency value for a given furnace by ANSI Z21.47/ CSA 2.3 test procedure.	

Note: Energy inputs and outputs are integrated over an annual period. Efficiencies (E) are expressed as percents. Indexes (I), factors (F), and ratios (R) are expressed as fractions.

[a]The *Total Heat Delivered* is the integration over time of all the energy supplied to the conditioned space by the HVAC equipment. By definition, it is exactly equal to the space-heating load.

[b]The *Induced Load Factor* may be positive or negative, depending on the value of the load relative to the selected base case.

[c]The *System Induced Load* is the difference between the space-heating load for a particular case and the space-heating load for the base case. For the base case, the System Induced Load is, by definition, zero.

[d]Indexes are referenced as "base cases" from which improvements are measured.

be compared with that of heating systems installed only in the conditioned space.

Because I_S depends on a reference equipment and house configuration, it may be used only as a ranking index from which the relative benefits of different features can be derived. That is, it can be used to compare the costs and savings of various features in specific applications to those of a base case.

- The miscellaneous gain factor F_{MG} includes only those heating losses that go *directly* to the conditioned space.
- The equipment-system performance factors relate strictly to the subject equipment, whereas the equipment-load interaction factors draw comparisons between the subject equipment and an explicitly defined base case. This base case is a specific load and equipment configuration to which all alternatives are compared.

Systems with the best total energy economy have the highest I_S. Those with leaky and uninsulated ducts could have a higher efficiency, even though fuel use would be higher, if basement duct losses that become gains to the conditioned space were included in the miscellaneous gain factor F_{MG}. The foregoing definitions prevent this possibility.

System Performance Examples

The ASHRAE SP-43 study was limited to certain house configurations and climates and to gas-fired equipment. Electric heat pump and zoned baseboard systems were not studied. For this reason, these data should not be used to compare systems or select a heating fuel. Several factors addressed are not references to performance; instead, they are figures of merit, which represent the effect of various components on a system. As such, these factors should not be applied outside the scope of these examples.

The following examples of overall thermal performance illustrate how the furnace, vent, duct system, and building can interact. Table 4 summarizes HOUSE-I simulation model predictions of the annual system performance for a base case (a conventional, natural-draft gas furnace with an intermittent ignition device) and an example case (a noncondensing, fan-assisted combustion furnace). Each is installed in a typical three-bedroom, ranch-style house of frame construction, located in Pittsburgh, Pennsylvania, constructed according to HUD minimum property standards circa 1980. Table 5 shows the base case operating assumption for the simulation predictions.

Base Case. Referring to Table 4, the annual furnace efficiency E_F of the conventional, natural-draft furnace is predicted to be 75.5%. Air leakage and heat loss from the uninsulated duct result in a duct efficiency E_D of 60.9%. The heat delivery efficiency E_{HD}, which is the ratio of energy intentionally delivered to the conditioned space through supply registers to total input energy, is 46.0%. The miscellaneous gain factor F_{MG}, which is 1.004, accounts for the small heat gain to the conditioned space from the heated masonry chimney passing through the conditioned space. The system efficiency E_S (the ratio of total heat delivered or space-heating load to total energy input) is 46.1%. Because this case is designated as the base case, the load modification factor F_{LM} is 1.0. Thus, the system index I_S is also 1.000 (by definition).

Duct loss and jacket loss are accounted for in the energy balance on the basement air and in the energy flow between the basement and conditioned space. The increase in infiltration caused by the need for combustion air and vent dilution air is also accounted for in energy balances on the living space air and basement air. In the base case, the temperature in the unconditioned basement is nearly the same (68°F) as in the first floor where the thermostat is located. This condition is caused by heat loss of exposed ducts in the basement and by the low outdoor infiltration into the basement achieved by sealing construction cracks.

Alternative Case. Again referring to Table 4, the furnace efficiency E_F for the noncondensing fan-assisted combustion furnace being compared is 85.5%. The duct efficiency E_D is 59.3%, slightly

Table 4 System Performance Examples

Performance Factor	Base Case: Typical Conventional, Natural-Draft Furnace with IID	Alternative Case: Typical Noncondensing Fan-Assisted Furnace
ASHRAE 103-93 AFUE (indoor) per DOE rules, %	69	81.5
Furnace efficiency E_F, %	75.5	85.5
Duct efficiency E_D, %	60.9	59.3
Heat delivery efficiency E_{HD}, %	46.0	50.7
Miscellaneous gain factor F_{MG}	1.004	0.983
System efficiency E_S, %	46.1	49.8
Load modification factor F_{LM}	1.000 (Base case)	1.099
System index I_S (Base case = 1.00)	1.000	1.189
Annual energy use AEU, 10^6 Btu	73.0	61.5
Auxiliary energy ratio R_{AE}	0.027	0.028
Energy saving from base case, %	—	15.9
Cost saving from base case, % (with R_{CL} = 4)	—	15.6

Note: The values presented here do not represent only this class of equipment; electric furnaces and heat pumps in a similar installation and under similar conditions would incur similar losses. The system index for any central air system can be improved, in comparison to the examples, by insulating the ducts, locating the ductwork inside the conditioned space, or both.

Table 5 Base Case Assumptions for Simulation Predictions

	Base Case
Furnace, Adjustments, and Controls	
Furnace size	1.4 × DHL
Circulating air temperature rise	60°F
Thermostat set point	68°F
Thermostat cycling rate at 50% on-time	6 cycles/h
Night setback, 8 h	None
Blower control	
On	80 s
Off	90°F
Duct-Related Factors	
Insulation	None
Leakage, relative to duct flow	10%
Location	Basement
Load-Related Factors	
Nominal infiltration*	
Conditioned space	0.75 ach
Basement	0.25 ach
Occupancy, persons	3 during evening and night, 1 during day
Internal loads	Typical appliances, day and evening only (20 kWh/day)
Shading by adjacent trees or houses	None

*Model runs used variable infiltration, as driven by indoor-outdoor temperature differences, wind, and burner operation. Values shown above are nominal.
ach = Air changes per hour; DHL = Design heat loss

lower than that for the base case. Therefore, the heat delivery efficiency E_{HD} is 50.7%, which reflects the higher furnace efficiency.

The miscellaneous gain factor F_{MG} is 0.983, reflecting the small heat loss from the conditioned space to the colder masonry chimney (due to reduced off-cycle vent flow). The system efficiency E_S (i.e., $E_{HD} \times F_{MG}$) is 49.8%.

For this furnace system, compared to the base case system of the conventional, natural-draft furnace, the load modification factor F_{LM} is 1.099. Therefore, the space-heating load for the house with the noncondensing fan-assisted combustion furnace is 1/1.099, or 91% of the space-heating load for the house with the conventional,

Table 6 Effect of Furnace Type on Annual Heating Performance

	Furnace Characterization Typical Values,[a] AFUE/SSE	Predicted by HOUSE-II Model: Annual Performance Factors E_F	E_D	F_{MG}	F_{LM}	I_S	I_{SCM} (R_{CL} = 4)	AEU, 10^6 Btu	Auxiliary Energy Ratio	Average Basement, °F
Conventional, natural-draft										
Pilot	64.5/77	72.9	60.9	1.006	1.000	0.970	0.971	75.5	0.026	67.9
Intermittent ignition device (Base case)[b]	69/77	75.5	60.9	1.004	1.000	1.000	1.000	73.0	0.027	67.8
IID + Thermal vent damper	78/77	75.4	61.0	1.002	1.086	1.087	1.085	67.3	0.027	68.2
IID + Electric vent damper	78/77	75.4	61.2	0.988	1.105	1.093	1.091	66.9	0.027	68.3
Fan-assisted types										
Noncondensing	81.5/82.5	85.5	59.3	0.983	1.099	1.189	1.185	61.5	0.028	67.6
Condensing[c]	92.5/93.1	95.5	62.0	1.000	1.050	1.349	1.322	54.2	0.034	66.9
Electric furnace	n.a.	99.5	60.6	1.000	1.079	1.412	0.380	51.8	0.020[d]	67.1

Note: Values in table are figures of merit to be considered within confines of SP-43 project, and should not be applied outside scope of examples.
[a]AFUE=Annual Fuel Utilization Efficiency by ANSI/ASHRAE *Standard* 103.
SSE=Steady-state efficiency by ANSI Z21.47/CSA 2.3 test procedure.
[b]Ranch-style house with basement in Pittsburgh, PA, climate and base conditions of60°F circulating air temperature rise, 6 cycles/h, no setback, 10% duct air leakage.
[c]Direct vent uses outdoor air for combustion (includes preheat).
[d]Blower energy is treated as auxiliary energy.

Table 7 Effect of Climate and Night Setback on Annual Heating Performance

Furnace Type and Location	Setback,[a] °F	Average % On-Time	Average Basement, °F	Furnace Efficiency E_F, %	Duct Efficiency E_D, %	System Index I_S	AEU,[b] 10^6 Btu	% Energy Saved by Setback
Conventional, Natural-Draft (Base Case)								
Nashville	0	12.7	64.6	73.9	56.2	1.000	55.6	
	10	10.7	63.1	74.8	58.9	1.192	46.7	16.0
Pittsburgh (base city)	0	18.0	67.8	75.5	60.9	1.000	73.0	
	10	15.6	65.7	75.9	62.4	1.154	63.4	13.2
Minneapolis	0	20.9	68.0	76.8	63.2	1.000	99.1	
	10	18.7	65.7	77.0	62.4	1.121	88.4	10.7
Direct, Condensing								
Nashville	0	11.1	63.9	94.0	57.0	1.000	41.0	
	10	9.5	62.6	93.7	59.5	1.174	35.0	14.8
Pittsburgh	0	16.2	67.8	93.3	61.7	1.000	55.1	
	10	14.2	65.0	93.0	63.2	1.144	48.1	12.6
Minneapolis	0	18.2	66.8	95.1	64.5	1.000	73.7	
	10	16.4	64.7	94.8	65.6	1.115	66.2	10.2

Note:
1. Ranch-style house, basement, and base conditions: 60°F circulating air temperature rise, 6 cycles/h, 10% duct air leakage. Thermal envelope typical of each city; for example, no basement insulation in Nashville, TN.
2. Values in table are figures of merit to be considered within confines of SP-43 project, and should not be applied outside scope of examples.
[a]The base case is 0°F setback in each city.
[b]AEU = Annual energy use.

natural-draft furnace. This reduction in heating load is mainly due to the reduction in off-cycle vent flow.

The system index I_S for the example case is 1.189. Note that I_S is also the inverse of the ratio of the annual energy use AEU ($61.5 \times 10^6 / 73 \times 10^6 = 0.842$).

Effect of Furnace Type

Table 6 summarizes the energy effects of several furnaces. Note that the system indexes I_S for both thermal and electric vent dampers are similar, although thermal vent dampers are slower reacting and less effective at blocking the vent. Also, the ratio of a furnace's AFUE to the base case AFUE closely matches the I_S values for the furnaces. The exceptions are the vent damper cases, where the improvement in I_S suggests a smaller AFUE credit. In general, the study found that the furnace AFUE is a good indication of relative annual performance of furnaces in typical systems.

The results reported in Table 6 are for homes that do not include the basement in the conditioned space (i.e., energy lost to the basement contributes only indirectly to the useful heating effect). If the basement is considered part of the conditioned space, the miscellaneous gain factor, load modification factor, and subsequent calculated efficiencies and indexes are adjusted to account for the beneficial effects of equipment (furnace jacket and duct system) heat losses that contribute to heating the basement. The system performance factors increase by 60% when the basement is considered part of the conditioned space. The indexes, however, retain their same relative ranking of systems.

Effect of Climate and Night Setback

Table 7 covers the effects of climate (insulation levels change by location) and night setback on performance for two furnaces. The improvements in I_S with higher percent on-time (colder climates) follow improvements in duct efficiency. Furnace efficiency E_F appears to be relatively uniform in houses representative of typical construction practices in each city and where the furnace is sized at 1.4 times the design heat loss. Also, the saving from night setback increases in magnitude with a warmer climate. The percent energy saved in the three climates varies with the magnitude of energy use (from 10 to 16% for the natural-draft cases).

Effect of Furnace Sizing

Furnace sizing affects I_S depending on how the venting and ducting are designed. As Table 8 indicates, if the ducts and vent are sized according to the furnace size (referred to as the new case), I_S drops about 10% as the furnace capacity is varied between 1.0 and 2.5 times the design heat loss for a given application. In the retrofit case, where the vent and duct are sized at a furnace capacity of 1.4 times

Table 8 Effect of Sizing, Setback, and Design Parameters on Annual Heating Performance—Conventional, Natural-Draft Furnace

Furnace Multiplier[a]	Duct Design	Setback, °F	Annual Performance Factors E_F	E_D	F_{MG}	F_{LM}	I_S[e]	Temperature Swing, °F	Average Room, °F	Recovery Time, h[b]	Average Basement, °F
1.00	New[c]	0	76.1	59.2	1.016	1.104	1.095	3.4	67.7	n.a.[d]	67.9
1.15	New	0	75.6	59.0	1.009	1.059	1.032	4.0	67.8	n.a.	67.9
	Retrofit	0	75.5	59.0	1.002	1.032	0.998	4.0	67.8	n.a.	67.9
	Retrofit	10	76.1	60.7	0.999	1.155	1.155	4.1	65.8	2.02	65.7
1.40	New	0	75.5	60.8	1.010	1.029	1.034	4.8	68.1	n.a.	67.9
	Retrofit	0	75.5	60.9	1.004	1.000	1.000[e]	4.9	68.0	n.a.	67.8
	Retrofit	10	75.9	62.4	1.001	1.120	1.152	5.2	66.0	1.02	65.7
1.70	New	0	74.8	61.3	1.013	1.017	1.023	5.9	68.2	n.a.	68.1
	Retrofit	0	75.0	61.7	1.006	0.983	0.992	5.9	68.2	n.a.	67.9
	Retrofit	10	75.6	63.5	1.003	1.095	1.143	6.3	66.2	0.54	65.8
2.50	New	0	74.9	63.4	1.008	0.943	0.978	8.2	68.6	n.a.	68.1
	Retrofit	0	75.2	64.9	1.009	0.937	1.000	8.6	68.6	n.a.	67.8
	Retrofit	10	75.7	66.5	1.006	1.042	1.144	9.3	66.6	0.24	65.8

Notes:
1. Ranch-style house with basement in Pittsburgh, PA, climate with base conditions of 60°F circulating air temperature rise, 6 cycles/h, 10% duct air leakage.
2. Values in table are figures of merit to be considered in confines of SP-43 project, and should not be applied outside scope of examples.

[a]Furnace output rating or heating capacity (Furnace multiplier) × (Design heat loss).
[b]Longest recovery time during winter (lowest outdoor temperature = 5°F).
[c]Retrofit case was not run for furnace multiplier 1.00.
[d]n.a. indicates not applicable.
[e]Base case = 1.0.

Table 9 Effect of Furnace Sizing on Annual Heating Performance—Condensing Furnace with Preheat

Furnace Multiplier[a]	Annual Performance Factors E_F	E_D	F_{MG}	F_{LM}	I_S[b]	Temp. Swing, °F	Avg. Room, °F	Avg. Bsmt., °F
1.15	95.1	60.9	1.000	1.076	1.351	3.5	67.9	66.7
1.40	95.5	62.0	1.000	1.050	1.347	4.3	68.1	66.7
2.50	95.4	64.8	1.000	0.990	1.325	7.3	68.7	66.7

Notes:
1. Ranch-style house with basement in Pittsburgh climate with base conditions of 60°F circulating air temperature rise, 6 cycles/h, 10% duct air leakage.
2. Values in table are figures of merit to be considered in confines of SP-43 project, and should not be applied outside scope of examples.

[a]Furnace output rating or heating capacity = (Furnace multiplier) × (Design heat loss).
[b]Base case = 1.0.

the design heat loss, I_S changes little with increased furnace capacity, indicating little energy savings. In a new case, where the duct is designed for cooling and the vent size does not change between furnace capacities, the SP-43 study indicates that there is essentially no effect on I_S.

Table 9 shows similar results in condensing furnaces for the new case of ducts and vents sized according to the furnace capacity. In this case, the decrease in I_S is smaller, about 2% over the range of 1.15 to 2.5 times the design heat loss.

Finally, both Table 8 and Table 9 show that duct efficiency E_D increases with furnace capacity because higher-capacity furnaces are on less than lower-capacity furnaces.

Effects of Furnace Sizing and Night Setback

Table 8 also shows the relationship between furnace sizing and night setback for the retrofit case. The energy saving from night setback, 8 h per day at 10°F, is nearly constant at 15% and independent of furnace size. Table 7 covers the effect of climate variation on energy saving from night setback.

Duct Treatment. Table 10 shows the effect of duct treatment on furnace performance. Duct treatment includes sealants to reduce leaks and interior or exterior insulation to reduce heat loss from conduction. Sealing and insulation improve system performance, as indicated by I_S. For cases with no duct insulation, reducing duct leakage from 10% to zero increases I_S by 2.6%. R-5 insulation on the exterior of the ducts increases I_S by 4.4%, and 5 insulation on the interior of the duct increases I_S by 8.5%.

Basement Configuration. Table 11 covers the effect of basement configuration and duct treatment on system performance.

Table 10 Effect of Duct Treatment on System Performance

Case	Duct Configuration 1	2[a]	3	4	5	6	7
Condition							
Duct insulation	None	None	None	R-5	R-5	R-5	R-5[b]
Duct leakage, %	0	10	20	0	10	20	10
Basement insulation							
Ceiling	None	None	None	None	None	None	None
Wall	R-8	R-8	R-8	R-8	R-8	R-8	R-8
Performance							
Burner on-time, %	17.5	18.0	18.6	16.8	17.2	17.8	16.6
Blower on-time, %	23.8	24.3	24.9	23.0	23.4	24.1	22.9
Average basement temperature, °F	66.8	67.8	68.8	65.2	66.3	67.4	65.1
Furnace efficiency E_F, %	75.4	75.5	75.7	75.0	75.2	75.4	75.0
Duct efficiency E_D, %	66.8	60.9	54.8	77.4	70.4	63.2	79.6
Load modification factor	0.94	1.00	1.07	0.85	0.91	0.97	0.84
I_S (base case = 1.0)	1.026	1.000	0.970	1.070	1.044	1.010	1.085

[a]Case 2 is the base case.
[b]Case 7 is interior insulation (liner); Cases 1 through 6 are exterior insulation (wrap).

Table 11 Effect of Duct Treatment and Basement Configuration on System Performance

Case	Duct Configuration 1	2	3	4	5
Condition					
Duct insulation	None	None	None	None	R-5
Duct leakage, %	20	10	10	10	0
Basement insulation					
Ceiling	None	None	R-11	R-11	R-11
Wall	None	None	None	R-8	R-8
Performance					
Burner on-time, %	23.6	22.7	21.8	18.0	16.3
Blower on-time, %	29.8	29.2	28.0	24.2	22.2
Average basement temperature, °F	64.4	63.3	62.4	67.9	64.3
Furnace efficiency E_F, %	75.3	75.2	75.2	75.7	75.0
Duct efficiency E_D, %	50.9	56.9	56.7	61.7	77.0
Load modification factor	0.91	0.85	0.89	0.99	0.88
I_S (base case = 1.0)	0.765	0.794	0.825	1.001	1.103

Note: See Table 10 for base case.

Insulating and sealing ducts reduces basement temperature. More heat is then required in the conditioned space to make up for losses to the colder basement. Where ducts pass through the attic or ventilated crawlspace, insulation and sealing improve duct performance, although the total system performance is poorer. On the other hand, installing ducts in the conditioned space significantly improves F_{MG} because duct losses are added directly to the conditioned space. In this case, I_S would also improve.

REFERENCES

ACCA. 1999. Residential duct systems. *Manual* D. Air Conditioning Contractors of America, Washington, D.C.

ACCA. 1995. Residential load calculations. *Manual* J. Air Conditioning Contractors of America, Washington, D.C.

ACCA. 1995. Residential equipment selection. *Manual* S. Air Conditioning Contractors of America, Washington, D.C.

ACCA. 1992. Air distribution basics for residential and small commercial buildings. *Manual* T. Air Conditioning Contractors of America, Washington, D.C.

Andrews, J.W. 2003. Effect of airflow and heat input rates on thermal distribution efficiency. *ASHRAE Transactions* 109(2).

Andrews, J.W., J.B. Cummings, and M.P. Modera. 2002. *Controlling airflow: Better air distribution for small commercial buildings*. U.S. Department of Energy, Building Technologies Program.

Andrews, J.W. and M.P. Modera. 1991. Energy savings potential for advanced thermal distribution technology in residential and small commercial buildings. LBL-31042.

ARI. 2001. Sound rating of ducted air moving and conditioning equipment. ARI *Standard* 260. Air-Conditioning and Refrigeration Institute, Arlington, VA.

ARI. 1996. Central system humidifiers for residential applications. *Standard* 610-96. Air-Conditioning and Refrigeration Institute, Arlington, VA.

ASHRAE. 1993. Energy-efficient design of new low-rise residential buildings. *Standard* 90.2-1993.

ASHRAE. 1993. Methods of testing for annual fuel utilization efficiency of residential central furnaces and boilers. *Standard* 103-1993.

ASHRAE. 2001. Method of test for determining the design and seasonal efficiencies of residential thermal distribution systems. *Standard* 152P Advanced Working Draft.

Bouchelle, M., D.S. Parker, M.T. Anello, and K.M. Richardson. 2000. Factors influencing space heat and heat pump efficiency from a large-scale residential monitoring study. *Proceedings of 2000 Summer Study on Energy Efficiency in Buildings*, American Council for an Energy-Efficient Economy, Washington, D.C.

Bullock, C. 1978. Energy savings through thermostat setback with residential heat pumps. *ASHRAE Transactions* 84(2):352-363.

CSA. 1998. Gas-fired central furnaces. ANSI Z21.47-1998/CSA 2.3-M98. Canadian Standards Association International, Cleveland, OH.

Cummings, J.B., C. Withers. J. McIlvaine, J. Sonne, and M. Lombardi. 2003. Air handler leakage: Field testing results in residences. CH-03-7-3. Presented at the 2003 ASHRAE Winter Meeting, Chicago.

Ebbing, C. and W. Blazier. 1998. *Application of manufacturers' sound data.* ASHRAE.

Ellison, R.D. 1977. The effects of reduced indoor temperature and night setback on energy consumption of residential heat pumps. *ASHRAE Journal.*

EPA. 2000. *Duct insulation*. EPA 430-F-97-028. U.S. Environmental Protection Agency, Washington, D.C.

Federal Specification. 1974. Asphalt, petroleum (primer, roofing, and waterproofing). SS-A-70.

FSEC. 2001. Factors influencing space heat and heat pump efficiency from a large-scale residential monitoring study. FSEC-PF362-01. Florida Solar Energy Center, Cocoa.

Fischer, R.D., F.E. Jakob, L.J. Flanigan, D.W. Locklin, and R.A. Cudnik. 1984. Dynamic performance of residential warm-air heating systems—Status of ASHRAE Project SP43. *ASHRAE Transactions* 90(2B):573-590.

Gilman, S.F., R.J. Martin, and S. Konzo. 1951. Investigation of the pressure characteristics and air distribution in box-type plenums for air conditioning duct systems. University of Illinois Engineering Experiment Station *Bulletin* no. 393 (July), Urbana-Champaign.

Gu, L., J.E. Cummings, P.W. Fairey, and M.V. Swami. 1998. Comparison of duct computer models that could provide input to the thermal distribution standard method of test (ASHRAE Research Project RP-852). *ASHRAE Transactions* 104(1):1349-1359.

Hedrick, R.L. 2002. *Alternative design details for building houses with ducts in conditioned space.* Prepared by Gard Analytics for New Buildings Institute on behalf of California Energy Commission, Public Interest Energy Research (PIER) Program.

Hedrick, R.L. 2003. Building homes with ducts in conditioned space. *Draft Report* to the California Energy Commission, Public Interest Energy Research (PIER) Program.

Herold, K.E., F.E. Jakob, and R.D. Fischer. 1986. The SP43 simulation model: Residential energy use. *Proceedings of ASME Conference*, Anaheim, pp. 81-87.

Herold, K.E., R.A. Cudnik, L.J. Flanigan, and R.D. Fischer. 1987. Update on experimental validation of the SP43 simulation model for forced-air heating systems. *ASHRAE Transactions* 93(2):1919-1933.

IRC. 2000. *International residential code® for one- and two-family dwellings*. International Code Council, Falls Church, VA.

Jakob, F.E., R.D. Fischer, L.J. Flanigan, D.W. Locklin, K.E. Harold, and R.A. Cudnik. 1986a. Validation of the ASHRAE SP43 dynamic simulation model for residential forced-warm air systems. *ASHRAE Transactions* 92(2B):623-643.

Jakob, F.E., D.W. Locklin, R.D. Fischer, L.J. Flanigan, and R.A. Cudnik. 1986b. SP43 evaluation of system options for residential forced-air heating. *ASHRAE Transactions* 92(2B):644-673.

Jakob, F.E., R.D. Fischer, and L.J. Flanigan. 1987. Experimental validation of the duct submodel for the SP43 simulation model. *ASHRAE Transactions* 93(1):1499-1515.

Locklin, D.W., K.E. Herold, R.D. Fischer, F.E. Jakob, and R.A. Cudnik. 1987. Supplemental information from SP43 evaluation of system options for residential forced-air heating. *ASHRAE Transactions* 93(2):1934-1958.

Modera, M.P. 1989. Residential duct system leakage: Magnitude, impacts, and potential for reduction. *ASHRAE Transactions* 95(2).

Modera, M.P., D. Dickerhoff, O. Nilssen, H. Duquette, and J. Geyselaers. 1996. Residential field testing of an aerosol-based technology for sealing ductwork. *Proceedings of the 1996 ACEEE Summer Study on Energy Efficiency in Buildings*, pp. 1.169-1.175.

NFPA. 1993. Installation of warm air heating and air-conditioning systems. *Standard* 90B-93. National Fire Protection Association, Quincy, MA.

NFPA. 2002. Standard on manufactured housing. *Standard* 501-MEC. National Fire Protection Association, Quincy, MA.

Palmiter, L. and P.W. Francisco. 2000. A new device for field measurement of air handler flows. *Proceedings of the ACEEE 2000 Summer Study on Energy Efficiency in Buildings*, Washington, D.C.

Phillips, B.G. 1998. Impact of blower performance on residential forced-air heating system performance. *ASHRAE Transactions* 104(1B):1817-1825.

Sherman, M., I. Walker, and D. Dickerhoff. 2000. Stopping duct quacks: Longevity of residential duct sealants. *Proceedings of the 2000 ACEEE Summer Study on Energy Efficiency in Buildings*, vol. 1, pp. 273-284.

SMACNA. 1998. *Residential comfort system installation standards manual*, 7th ed. Sheet Metal and Air Conditioning Contractors' National Association, Chantilly, VA.

Straub, H.E. and M. Chen. 1957. Distribution of air within a room for year-round air conditioning—Part II. University of Illinois Engineering Experimental Station *Bulletin* no. 442 (March), Urbana-Champaign.

Wright, J.R., D.R. Bahnfleth, and E.G. Brown. 1963. Comparative performance of year-round systems used in air conditioning research no. 2. University of Illinois Engineering Experimental Station *Bulletin* no. 465 (January), Urbana-Champaign.

BIBLIOGRAPHY

Abushkara, B., I.S. Walker, and M.H. Sherman 2002. A study of pressure losses in residential air distribution systems. *Proceedings of the 2002 ACEEE Summer Study on Energy Efficiency in Buildings*, pp. 1.1-1.14.

Andrews, J.W. 1992. Optimal forced-air distribution in new housing. *Proceedings of the 1992 ACEEE Summer Study on Energy Efficiency in Buildings*, pp. 2.1-2.4.

Andrews, J.W. 1996. Field comparison of design and diagnostic pathways for duct efficiency evaluation. *Proceedings of the ACEEE 1996 Summer Study on Energy Efficiency in Buildings*, pp. 1.21-1.30. Washington, D.C.

Andrews, J.W. 2000. Reducing measurement uncertainties in duct leakage testing. *Proceedings of the 2000 ACEEE Summer Study on Energy Efficiency in Buildings*, vol. 1, pp. 13-28.

Andrews, J.W. 2002. Laboratory evaluation of the delta Q test for duct leakage. *Proceedings of the 2002 ACEEE Summer Study on Energy Efficiency in Buildings*, pp. 1.15-1.28.

Andrews, J.W., R. Hedrick, M. Lubliner, B. Reid, B. Pierce, and D. Saum. 1998. Reproducibility of ASHRAE *Standard* 152P: Results of a round-robin test. *ASHRAE Transactions* 104(1):1376-1388.

Andrews, J.W., R.F. Krajewski, and J.J. Strasser. 1996. Electric co-heating in the ASHRAE standard method of test for thermal distribution efficiency: Test results in two New York state homes. *ASHRAE Transactions* 102(1).

ARI. Updated annually. *Directory of certified unitary products; unitary air conditioners; unitary air-source heat pumps; and sound rated outdoor unitary equipment*. Air-Conditioning and Refrigeration Institute, Arlington, VA.

Asiedu, Y., R.W. Besant, and P. Gu. 2000. A simplified procedure for HVAC duct sizing. *ASHRAE Transactions* 106(1):124-142.

ASTM. 1994. Standard test methods for determining external air leakage of air distribution systems by fan pressurization. ASTM *Standard* E 1554-94. American Society for Testing and Materials, Philadelphia.

Axley, J. 1999. Passive ventilation for residential air quality control. *ASHRAE Transactions* 105(2):864-876.

Caffey, G. 1979. Residential air infiltration. *ASHRAE Transactions* 85(1).

Crisafulli, J.C., R.A. Cudnik, and L.R. Brand. 1989. Investigation of regional differences in residential HVAC system performance using the SP43 simulation model. *ASHRAE Transactions* 95(1):915-929.

CEE. 2000. *Specification of energy efficient installation practices for residential HVAC systems*. Consortium for Energy Efficiency, Boston.

Colto, F. and G. Syphers. 1998. Are your ducts all in a row? Duct efficiency testing and analysis for 150 new homes in northern California. *Proceedings of the 1998 ACEEE Summer Study on Energy Efficiency in Buildings*, pp. 1.33-1.41.

Conlin, F. 1996. An analysis of air distribution system losses in contemporary HUD-code manufactured homes. *Proceedings of the 1996 ACEEE Summer Study on Energy Efficiency in Buildings*, pp. 1.53-1.59.

Cummings, J.B. and J.J. Tooley, Jr. 1989. Infiltration and pressure differences induced by forced air systems in Florida residences. *ASHRAE Transactions* 95(2).

Cummings, J.B., J.J. Tooley, Jr., and R. Dunsmore. 1990. Impacts of duct leakage on infiltration rates, space conditioning energy use, and peak electrical demand in Florida homes. *Proceedings of the ACEEE 1990 Summer Study on Energy Efficiency in Buildings*, Washington, D.C.

Cummings, J. and C. Withers, Jr. 1998. Building cavities used as ducts: Air leakage characteristics and impacts in light commercial buildings. *ASHRAE Transactions* 104(2):743-752.

Cummings, J.B., C.R. Withers, Jr., and N. Moyer. 2000. Evaluation of the duct leakage estimation procedures of ASHRAE *Standard* 152P. *ASHRAE Transactions* 106(2).

Cummings, J.B., C.R. Withers Jr., N.A. Moyer, P.W. Fairey, and B.B. McKendry 1996. Field measurement of uncontrolled airflow and depressurization in restaurants. *ASHRAE Transactions* 102(1):859-869.

Davis, B., J. Siegel, and L. Palmiter. 1996. Field measurements of heating system efficiency and air leakage in energy-efficient manufactured homes. *Proceedings of the ACEEE 1996 Summer Study on Energy Efficiency in Buildings,* Washington, D.C.

Delp, W.W., N.E. Matson, and M.P. Modera 1998. Exterior exposed ductwork: Delivery effectiveness and efficiency. *ASHRAE Transactions* 104(2):709-721.

Delp, W.W., N.E. Matson, E. Tschudy, M.P. Modera, and R.C. Diamond. 1998. Field investigation of duct system performance in California light commercial buildings. *ASHRAE Transactions* 104(2):722-731.

Downey, T. and J. Proctor. 2002. What can 13,000 air conditioners tell us? *Proceedings of the 2002 ACEEE Summer Study on Energy Efficiency in Buildings*, pp. 1.53-1.67.

Fischer, R.D. and R.A. Cudnik. 1993. The HOUSE-II computer model for dynamic and seasonal performance simulation of central forced-air systems in multi-zone residences. *ASHRAE Transactions* 99(1):614-626.

Francisco, P.W. and L. Palmiter. 1998. Measured and modeled duct efficiency in manufactured homes: Insights for *Standard* 152P. *ASHRAE Transactions* 104(1B):1389-1401.

Francisco, P.W. and L. Palmiter. 1999. Field validation of ASHRAE *Standard* 152. *Final Report*, ASHRAE Research Project RP-1056.

Francisco, P.W. and L. Palmiter. 2000. Field validation of *Standard* 152P. *ASHRAE Transactions* 106(2).

Francisco, P.W. and L. Palmiter. 2000. Performance of duct leakage measurement techniques in estimating duct efficiency: Comparison to measured results. *Proceedings of the ACEEE 2000 Summer Study on Energy Efficiency in Buildings,* Washington, D.C.

Francisco, P.W. and L.S. Palmiter. 2001. The nulling test: A new measurement technique of estimating duct leakage in residential homes. *ASHRAE Transactions* 107(1):297-303.

Francisco, P.W., L. Palmiter, and B. Davis. 1998. Modeled vs. measured duct distribution efficiency in six forced-air gas-heated homes. *Proceedings of the 1998 ACEEE Summer Study on Energy Efficiency in Buildings*, pp. 1.103-1.114.

Francisco, P.W., L. Palmiter, and B. Davis. 2002. Field performance of two new residential duct leakage measurement techniques. *Proceedings of the 2002 ACEEE Summer Study on Energy Efficiency in Buildings*, pp. 1.69-1.80.

GAMA. Updated annually. *Consumer's directory of certified efficiency ratings for residential heating and water heating equipment*. Gas Appliance Manufacturers Association, Arlington, VA.

Gammage, R.B., A.R. Hawthorne, and D.A. White. 1984. Parameters affecting air infiltration and air tightness in thirty-one east Tennessee homes. *Proceedings of the ASTM Symposium on Measured Air Leakage Performance of Buildings*, Philadelphia.

Griffiths, D., R. Aldrich, W. Zoeller, and M. Zuluaga. 2002. An innovative approach to reducing duct heat gains for a production builder in a hot and humid climate—How we got there. *Proceedings of the 2002 ACEEE Summer Study on Energy Efficiency in Buildings*, pp. 1.81-1.90.

Guyton, M.L. 1993. Measured performance of relocated air distribution systems in an existing residential building. *ASHRAE Transactions* 99(2).

Hammarlund, J., J. Proctor, G. Kast, and T. Ward. 1992. Enhancing the performance of HVAC and distribution systems in residential new construction. *Proceedings of the 1992 ACEEE Summer Study on Energy Efficiency in Buildings*, pp. 2.85-2.88.

Hedrick, R.L., M.J. Witte, N.P. Leslie, and W.W. Bassett. 1992. Furnace sizing criteria for energy-efficient setback strategies. *ASHRAE Transactions* 98(1):1239-1246.

Herold, K.E., R.D. Fischer, and L.J. Flanigan. 1987. Measured cooling performance of central forced-air systems and validation of the SP43 simulation model. *ASHRAE Transactions* 93(1):1443-1457.

Hise, E.C. and A.S. Holman. 1977. Heat balance and efficiency measurements of central, forced-air, residential gas furnaces. *ASHRAE Transactions* 83(1):865-880.

Jump, D.A., I.S. Walker, and M.P. Modera. 1996. Field measurements of efficiency and duct retrofit effectiveness in residential forced-air distribution systems. *Proceedings of the ACEEE 1996 Summer Study on Energy Efficiency in Buildings,* Washington, D.C.

Lubliner, M., R. Kunkle, J. Devine, and A. Gordon. 2002. Washington state residential ventilation and indoor air quality code (VIAQ): Whole house ventilation systems field research report. *Proceedings of the 2002 ACEEE Summer Study on Energy Efficiency in Buildings*, pp. 1.115-1.130.

Lubliner, M., D.T. Stevens, and B. Davis. 1997. Mechanical ventilation in HUD-code manufactured housing in the Pacific northwest. *ASHRAE Transactions* 103(1):693-705.

Matthews, T.G., C.V. Thompson, K.P. Monar, and C.S. Dudney. 1990. Impact of heating and air-conditioning system operation and leakage on ventilation and intercompartmental transport: Studies in occupied and unoccupied Tennessee Valley homes. *Journal of the Air Waste Management Association* 40:194-198.

Modera, M.P., J. Andrews, and E. Kweller. 1992. A comprehensive yardstick for residential thermal distribution efficiency. *Proceedings of the 1992 ACEEE Summer Study on Energy Efficiency in Buildings*, Washington, D.C.

Oppenheim, P. 1992. Energy-saving potential of a zoned forced-air heating and cooling system. *ASHRAE Transactions* 98(1):1247-1257.

Palmiter, L. and T. Bond. 1992. Impact of mechanical systems on ventilation and infiltration in homes. *Proceedings of the ACEEE 1992 Summer Study on Energy Efficiency in Buildings,* Washington, D.C.

Palmiter, L. and P.W. Francisco 1994. Measured efficiency of forced-air distribution systems in 24 homes. *Proceedings of the ACEEE 1994 Summer Study on Energy Efficiency in Buildings,* Washington, D.C.

Palmiter, L. and P.W. Francisco. 1996. A practical method for estimating the thermal efficiency of residential forced-air distribution systems. *Proceedings of the ACEEE 1996 Summer Study on Energy Efficiency in Buildings,* Washington, D.C.

Palmiter, L. and P.W. Francisco. 1998. Modeled duct distribution efficiency: Insights for heat pumps. *Proceedings of the 1998 ACEEE Summer Study on Energy Efficiency in Buildings*, pp. 1.223-1.234.

Palmiter, L. and P.W. Francisco. 2002. Measuring duct leakage with a blower door: Field results. *Proceedings of the 2002 ACEEE Summer Study on Energy Efficiency in Buildings*, pp. 1.207-1.218.

Palmiter, L.S., I.A. Brown, and T.C. Bond 1991. Measured infiltration and ventilation in 472 all-electric homes. *ASHRAE Transactions* 97(2):979-987.

Parker, D.S. 1989. Evidence of increase levels of space heat consumption and air leakage associated with forced air heating systems in houses in the Pacific northwest. *ASHRAE Transactions* 95(2):527-533.

Proctor, J. 1998. Monitored in-situ performance of residential air-conditioning systems. *ASHRAE Transactions* 104(1B):1833-1840.

Proctor, J.P. 1998. Verification test of ASHRAE *Standard* 152P. *ASHRAE Transactions* 104(1B):1402-1412.

Proctor, J.P., and R.K. Pernick. 1992. Getting it right the second time: Measured savings and peak reduction from duct and appliance repairs. *Proceedings of the ACEEE 1992 Summer Study on Energy Efficiency in Buildings,* Washington, D.C.

Robison, D.H. and P.E. Lambert. 1989. Field investigation of residential infiltration and heating duct leakage. *ASHRAE Transactions* 95(2):542-550.

Rutkowski, H. and J.H. Healy. 1990. Selecting residential air-cooled cooling equipment based on sensible and latent performance. *ASHRAE Transactions* 96(2):851-856.

Rudd, A.F. and J.W. Lstiburek 2000. Measurement of ventilation and interzonal distribution in single-family houses. *ASHRAE Transactions* 106(2):709-718.

Sachs, H.M., T. Kubo, S. Smith, and K. Scott. 2002. Residential HVAC fans and motors are bigger than refrigerators. *Proceedings of the 2002 ACEEE Summer Study on Energy Efficiency in Buildings*, pp. 1.261-1.272.

Saunders, D.H., T.M. Kenney, and W.W. Bassett. 1992. Factors influencing thermal stratification and thermal comfort in four heated residential buildings. *ASHRAE Transactions* 98(1):1258-1265.

Saunders, D.H., T.M. Kenney, and W.W. Bassett. 1993. Evaluation of the forced-air distribution effectiveness in two research houses. *ASHRAE Transactions* 99(1):1407-1419.

Strunk, P.R. 2000. Validation of ASHRAE *Standard* 152P in basement warm-air distribution systems. *ASHRAE Transactions* 106(2).

Treidler, B. and M.P. Modera. 1996. Thermal performance of residential duct systems in basements. *ASHRAE Transactions* 102(1):847-858.

Treidler, B., M.P. Modera, R.G. Lucas, and J.D. Miller 1996. Impact of residential duct insulation on HVAC energy use and life-cycle costs to consumers. *ASHRAE Transactions* 102(1):881-894.

Walker, I.S. 1998. Technical background for default values used for forced-air systems in proposed ASHRAE *Standard* 152P. *ASHRAE Transactions* 104(1B):1360-1375.

Walker, I.S. 1999. Distribution system leakage impacts on apartment building ventilation rates. *ASHRAE Transactions* 105(1):943-950.

Walker, I.S. and M.P. Modera. 1996. Energy effectiveness of duct sealing and insulation in two multifamily buildings. *Proceedings of the 1996 ACEEE Summer Study on Energy Efficiency in Buildings*, pp. 1.247-1.254.

Walker, I.S. and M.P. Modera. 1998. Field measurements of interactions between furnaces and forced-air distribution systems. *ASHRAE Transactions* 104(1B):1805-1816.

Walker, I.S., D.J. Dickerhoff, and M.H. Sherman 2002. The delta *Q* method of testing the air leakage of ducts. *Proceedings of the 2002 ACEEE Summer Study on Energy Efficiency in Buildings*, pp. 1.327-1.338.

Wray, C., I. Walker, and M. Sherman. 2002. Accuracy of flow hoods in residential applications. *Proceedings of the 2002 ACEEE Summer Study on Energy Efficiency in Buildings*, pp. 1.339-1.350.

Withers, C.W., Jr. and J.B. Cummings. 1998. Ventilation, humidity, and energy impacts of uncontrolled airflow in a light commercial building. *ASHRAE Transactions* 104(2):733-742.

CHAPTER 10

STEAM SYSTEMS

STEAM systems use the vapor phase of water to supply heat or kinetic energy through a piping system. As a source of heat, steam can heat a conditioned space with suitable terminal heat transfer equipment such as fan-coil units, unit heaters, radiators, and convectors (finned tube or cast iron), or steam can heat through a heat exchanger that supplies hot water or some other heat transfer medium to the terminal units. In addition, steam is commonly used in heat exchangers (shell-and-tube, plate, or coil types) to heat domestic hot water and supply heat for industrial and commercial processes such as in laundries and kitchens. Steam is also used as a heat source for certain cooling processes such as single-stage and two-stage absorption refrigeration machines.

ADVANTAGES

Steam offers the following advantages:

- Steam flows through the system unaided by external energy sources such as pumps.
- Because of its low density, steam can be used in tall buildings where water systems create excessive pressure.
- Terminal units can be added or removed without making basic changes to the design.
- Steam components can be repaired or replaced by closing the steam supply, without the difficulties associated with draining and refilling a water system.
- Steam is pressure-temperature dependent; therefore, the system temperature can be controlled by varying either steam pressure or temperature.
- Steam can be distributed throughout a heating system with little change in temperature.

In view of these advantages, steam is applicable to the following facilities:

- Where heat is required for process and comfort heating, such as in industrial plants, hospitals, restaurants, dry-cleaning plants, laundries, and commercial buildings
- Where the heating medium must travel great distances, such as in facilities with scattered building locations, or where the building height would result in excessive pressure in a water system
- Where intermittent changes in heat load occur

FUNDAMENTALS

Steam is the vapor phase of water and is generated by adding more heat than required to maintain its liquid phase at a given pressure, causing the liquid to change to vapor without any further increase in temperature. Table 1 illustrates the pressure-temperature relationship and various other properties of steam.

Temperature is the thermal state of both liquid and vapor at any given pressure. The values shown in Table 1 are for dry saturated steam. The vapor temperature can be raised by adding more heat, resulting in superheated steam, which is used (1) where higher temperatures are required, (2) in large distribution systems to compensate for heat losses and to ensure that steam is delivered at the desired saturated pressure and temperature, and (3) to ensure that the steam is dry and contains no entrained liquid that could damage some turbine-driven equipment.

Enthalpy of the liquid h_f (sensible heat) is the amount of heat in Btu required to raise the temperature of a pound of water from 32°F to the boiling point at the pressure indicated.

Enthalpy of evaporation h_{fg} (latent heat of vaporization) is the amount of heat required to change a pound of boiling water at a given pressure to a pound of steam at the same pressure. This same amount of heat is released when the vapor is condensed back to a liquid.

Enthalpy of the steam h_g (total heat) is the combined enthalpy of liquid and vapor and represents the total heat above 32°F in the steam.

Specific volume, the reciprocal of density, is the volume of unit mass and indicates the volumetric space that 1 lb of steam or water occupies.

An understanding of the above helps explain some of the following unique properties and advantages of steam:

- Most of the heat content of steam is stored as latent heat, which permits large quantities of heat to be transmitted efficiently with little change in temperature. Because the temperature of saturated steam is pressure-dependent, a negligible temperature reduction occurs from the reduction in pressure caused by pipe friction losses as steam flows through the system. This occurs regardless of the insulation efficiency, as long as the boiler maintains the initial pressure and the steam traps remove the condensate. Conversely, in a hydronic system, inadequate insulation can significantly reduce fluid temperature.
- Steam, as all fluids, flows from areas of high pressure to areas of low pressure and is able to move throughout a system without an external energy source. Heat dissipation causes the vapor to condense, which creates a reduction in pressure caused by the dramatic change in specific volume (1600:1 at atmospheric pressure).
- As steam gives up its latent heat at the terminal equipment, the condensate that forms is initially at the same pressure and temperature as the steam. When this condensate is discharged to a lower

The preparation of this chapter is assigned to TC 6.1, Hydronic and Steam Equipment and Systems.

Table 1 Properties of Saturated Steam

Pressure, psi		Saturation Temperature, °F	Specific Volume, ft^3/lb		Enthalpy, Btu/lb		
Gage	Absolute		Liquid v_f	Steam v_g	Liquid h_f	Evaporation h_{fg}	Steam h_g
25 in. Hg vac.	2.47	134	0.0163	142.2	101	1018	1119
9.6 in. Hg vac.	10.0	193	0.0166	38.4	161	982	1143
0	14.7	212	0.0167	26.8	180	970	1150
2	16.7	218	0.0168	23.8	187	966	1153
5	19.7	227	0.0168	20.4	195	961	1156
15	29.7	250	0.0170	13.9	218	946	1164
50	64.7	298	0.0174	6.7	267	912	1179
100	114.7	338	0.0179	3.9	309	881	1190
150	164.7	366	0.0182	2.8	339	857	1196
200	214.7	388	0.0185	2.1	362	837	1200

Note: Values are rounded off or approximated to illustrate various properties discussed in text. For calculation and design, use values of thermodynamic properties of water shown in Chapter 6 of the 2005 *ASHRAE Handbook—Fundamentals* or a similar table.

pressure (as when a steam trap passes condensate to the return system), the condensate contains more heat than necessary to maintain the liquid phase at the lower pressure; this excess heat causes some of the liquid to vaporize or "flash" to steam at the lower pressure. The amount of liquid that flashes to steam can be calculated as follows:

$$\% \text{ Flash Steam} = \frac{100(h_{f1} - h_{f2})}{h_{fg2}} \tag{1}$$

where

h_{f1} = enthalpy of liquid at pressure p_1
h_{f2} = enthalpy of liquid at pressure p_2
h_{fg2} = latent heat of vaporization at pressure p_2

Flash steam contains significant and useful heat energy that can be recovered and used (see the section on Heat Recovery). This reevaporation of condensate can be controlled (minimized) by subcooling the condensate within the terminal equipment before it discharges into the return piping. The volume of condensate that is subcooling should not be so large as to cause a significant loss of heat transfer (condensing) surface.

EFFECTS OF WATER, AIR, AND GASES

The enthalpies shown in Table 1 are for dry saturated steam. Most systems operate near these theoretically available values, but the presence of water and gases can affect enthalpy, as well as have other adverse operating effects.

Dry saturated steam is pure vapor without entrained water droplets. However, some amount of water usually carries over as condensate forms because of heat losses in the distribution system. **Steam quality** describes the amount of water present and can be determined by calorimeter tests.

While steam quality might not have a significant effect on the heat transfer capabilities of the terminal equipment, the backing up or presence of condensate can be significant because the enthalpy of condensate h_f is negligible compared with the enthalpy of evaporation h_{fg}. If condensate does not drain properly from pipes and coils, the rapidly flowing steam can push a slug of condensate through the system. This can cause water hammer and result in objectionable noise and damage to piping and system components.

The presence of air also reduces steam temperature. Air reduces heat transfer because it migrates to and insulates heat transfer surfaces. Further, oxygen in the system causes pitting of iron and steel surfaces. Carbon dioxide (CO_2) traveling with steam dissolves in condensate, forming **carbonic acid**, which is extremely corrosive to steam heating pipes and heat transfer equipment.

The combined adverse effects of water, air, and CO_2 necessitate their prompt and efficient removal.

HEAT TRANSFER

The quantity of steam that must be supplied to a heat exchanger to transfer a specific amount of heat is a function of (1) the steam temperature and quality, (2) the character and entering and leaving temperatures of the medium to be heated, and (3) the heat exchanger design. For a more detailed discussion of heat transfer, see Chapter 3 of the 2005 *ASHRAE Handbook—Fundamentals*.

BASIC STEAM SYSTEM DESIGN

Because of the various codes and regulations governing the design and operation of boilers, pressure vessels, and systems, steam systems are classified according to operating pressure. **Low-pressure systems** operate up to 15 psig, and **high-pressure systems** operate over 15 psig. There are many subclassifications within these broad classifications, especially for heating systems such as one- and two-pipe, gravity, vacuum, or variable vacuum return systems. However, these subclassifications relate to the distribution system or temperature-control method. Regardless of classification, all steam systems include a source of steam, a distribution system, and terminal equipment, where steam is used as the source of power or heat.

STEAM SOURCE

Steam can be generated directly by boilers using oil, gas, coal, wood, or waste as a fuel source, or by solar, nuclear, or electrical energy as a heat source. Steam can be generated indirectly by recovering heat from processes or equipment such as gas turbines and diesel or gas engines. The cogeneration of electricity and steam should always be considered for facilities that have year-round steam requirements. Where steam is used as a power source (such as in turbine-driven equipment), the exhaust steam may be used in heat transfer equipment for process and space heating.

Steam can be provided by a facility's own boiler or cogeneration plant or can be purchased from a central utility serving a city or specific geographic area. This distinction can be very important. A facility with its own boiler plant usually has a closed-loop system and requires the condensate to be as hot as possible when it returns to the boiler. Conversely, condensate return pumps require a few degrees of subcooling to prevent cavitation or flashing of condensate to vapor at the suction eye of pump impellers. The degree of subcooling varies, depending on the hydraulic design or characteristics of the pump in use. [See the section on Pump Suction Characteristics (NPSH) in Chapter 43.]

Central utilities often do not take back condensate, so it is discharged by the using facility and results in an open-loop system. If a utility does take back condensate, it rarely gives credit for its heat content. If condensate is returned at 180°F, and a heat recovery system reduces this temperature to 80°F, the heat remaining in the condensate represents 10 to 15% of the heat purchased from the utility. Using this heat effectively can reduce steam and heating costs by 10% or more (see the section on Heat Recovery).

Boilers

Fired and waste heat boilers are usually constructed and labeled according to the ASME *Boiler and Pressure Vessel Code* because pressures normally exceed 15 psig. Details on design, construction, and application of boilers can be found in Chapter 31. Boiler selection is based on the combined loads, including heating processes and equipment that use steam, hot water generation, piping losses, and pickup allowance.

The Hydronics Institute standards (HYDI 1989) are used to test and rate most low-pressure heating boilers that have net and gross ratings. In smaller systems, selection is based on a net rating. Larger system selection is made on a gross load basis. The occurrence and nature of the load components, with respect to the total load, determine the number of boilers used in an installation.

Heat Recovery and Waste Heat Boilers

Steam can be generated by waste heat, such as exhaust from fuel-fired engines and turbines. Figure 1 schematically shows a typical exhaust boiler and heat recovery system used for diesel engines. A portion of the water used to cool the engine block is diverted as preheated makeup water to the exhaust heat boiler to obtain maximum heat and energy efficiency. Where the quantity of steam generated by the waste heat boiler is not steady or ample enough to satisfy the facility's steam requirements, a conventional boiler must generate supplemental steam.

Heat Exchangers

Heat exchangers are used in most steam systems. Steam-to-water heat exchangers (sometimes called converters or storage tanks with steam heating elements) are used to heat domestic hot water and to supply the terminal equipment. These heat exchangers are the plate type or the shell-and-tube type, where the steam is admitted to the shell and the water is heated as it circulates through the tubes. Condensate coolers (water-to-water) are sometimes used to subcool the condensate while reclaiming the heat energy.

Water-to-steam heat exchangers (steam generators) are used in high-temperature water (HTW) systems to provide process steam. Such heat exchangers generally consist of a U-tube bundle, through which the HTW circulates, installed in a tank or pressure vessel.

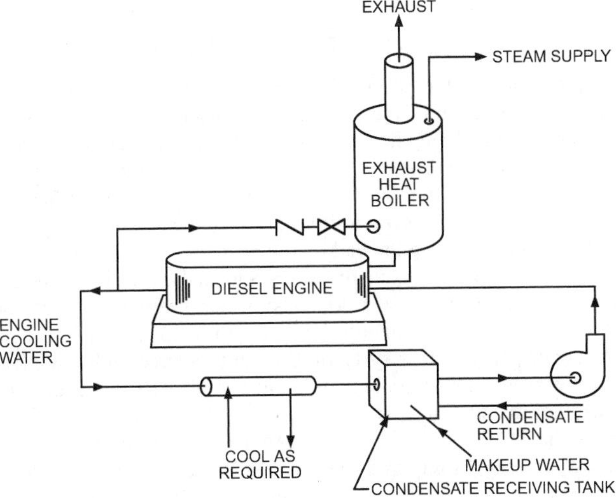

Fig. 1 Exhaust Heat Boiler

All heat exchangers should be constructed and labeled according to the applicable ASME *Boiler and Pressure Vessel Code*. Many jurisdictions require double-wall construction in shell-and-tube heat exchangers between the steam and potable water. Chapter 43 discusses heat exchangers in detail.

BOILER CONNECTIONS

Figure 2 shows recommended boiler connections for pumped and gravity return systems; local codes should be checked for specific legal requirements.

Supply Piping

Small boilers usually have one steam outlet connection sized to reduce steam velocity to minimize carryover of water into supply lines. Large boilers can have several outlets that minimize boiler water entrainment. The boiler manufacturer's recommendations concerning near-boiler piping should be followed because this piping may act as a steam/liquid separator for the boiler.

Figure 2 shows piping connections to the steam header. Although some engineers prefer to use an enlarged steam header for additional storage space, if there is no sudden demand for steam except during the warm-up period, an oversized header may be a disadvantage. The boiler header can be the same size as the boiler connection or the pipe used on the steam main. The horizontal runouts from the boiler(s) to the header should be sized by calculating the heaviest load that will be placed on the boiler(s). The runouts should be sized on the same basis as the building mains. Any change in size after the vertical uptakes should be made by reducing elbows.

Return Piping

Cast-iron boilers have return tappings on both sides, and steel boilers may have one or two return tappings. Where two tappings are provided, both should be used to effect proper circulation through the boiler. Condensate in boilers can be returned by a pump or a gravity return system. Return connections shown in Figure 2 for a multiple-boiler gravity return installation may not always maintain the correct water level in all boilers. Extra controls or accessories may be required.

Recommended return piping connections for systems using gravity return are detailed in Figure 3. Dimension A must be at least 28 in. for each 1 psig maintained at the boiler to provide the pressure required to return the condensate to the boiler. To provide a reasonable safety factor, make dimension A at least 14 in. for small systems with piping sized for a pressure drop of 1/8 psi, and at least 28 in. for larger systems with piping sized for a pressure drop of 0.5 psi. The **Hartford loop** protects against a low water condition, which can occur if a leak develops in the wet return portion of the piping. The Hartford loop takes the place of a check valve on the wet return; however, certain local codes require check valves. Because of hydraulic pressure limitations, gravity return systems are only suitable for systems operating at a boiler pressure between 0.5 and 1 psig. However, because these systems have minimum mechanical equipment and low initial installed cost, they are appropriate for many small systems. Kremers (1982) and Stamper and Koral (1979) provide additional design information on piping for gravity return systems.

Recommended piping connections for steam boilers with pump-returned condensate are shown in Figure 2. Common practice provides an individual condensate or boiler feedwater pump for each boiler. Pump operation is controlled by the boiler water level control on each boiler. However, one pump may be connected to supply the water to each boiler from a single manifold by using feedwater control valves regulated by the individual boiler water level controllers. When such systems are used, the condensate return pump runs continuously to pressurize the return header.

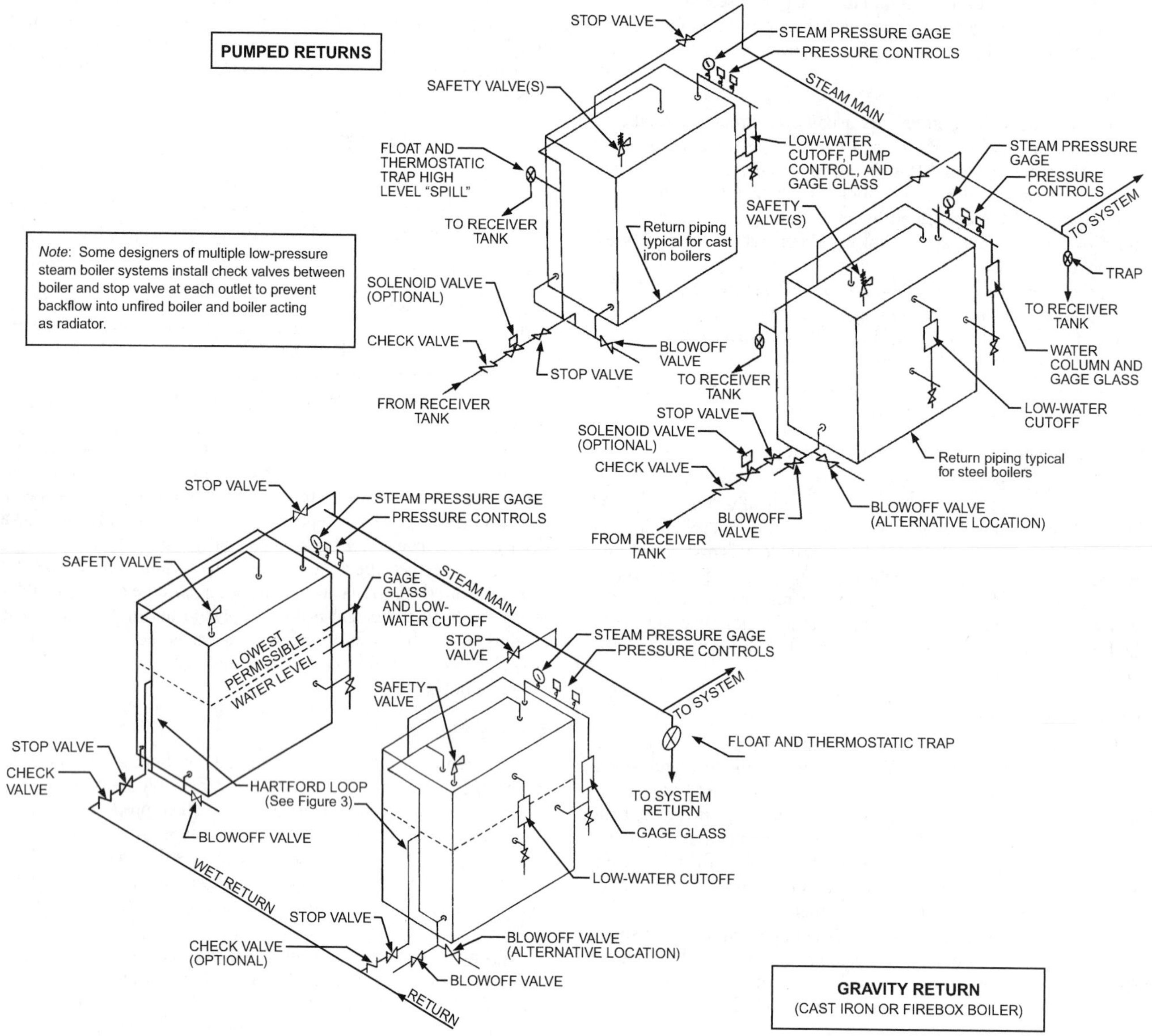

PUMPED RETURNS

Note: Some designers of multiple low-pressure steam boiler systems install check valves between boiler and stop valve at each outlet to prevent backflow into unfired boiler and boiler acting as radiator.

GRAVITY RETURN
(CAST IRON OR FIREBOX BOILER)

Fig. 2 Typical Boiler Connections

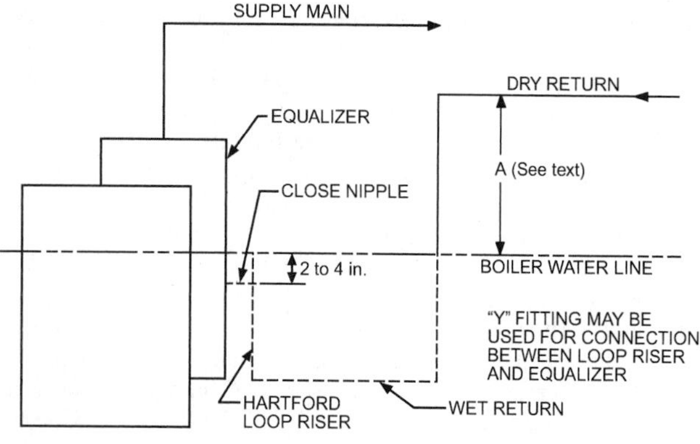

Fig. 3 Boiler with Gravity Return

Return piping should be sized based on total load. The line between the pump and boiler should be sized for a very small pressure drop and the maximum pump discharge flow rate.

DESIGN STEAM PRESSURE

One of the most important decisions in the design of a steam system is the selection of the generating, distribution, and utilization pressures. Considering investment cost, energy efficiency, and control stability, the pressure should be held to the minimum values above atmospheric pressure that accomplish the required heating task, unless detailed economic analysis indicates advantages in higher pressure generation and distribution.

The first step in selecting pressures is to analyze the load requirements. Space heating and domestic water heating can best be served, directly or indirectly, with low-pressure steam less than 15 psig or 250°F. Other systems that can be served with low-pressure steam include single-stage absorption units (10 psig), cooking, warming, dishwashing, and snow melting heat exchangers. Thus, from the standpoint of load requirements, high-pressure steam (above 15 psig) is required only for loads such as dryers, presses, molding dies, power drives, and other processing, manufacturing, and power requirements. The load requirement establishes the pressure requirement.

When the source is close to the load(s), the generation pressure should be high enough to provide the (1) load design pressure, (2) friction losses between the generator and the load, and (3) control range. Losses are caused by flow through the piping, fittings, control valves, and strainers. If the generator(s) is located remote from the loads, there could be some economic advantage in distributing the steam at a higher pressure to reduce pipe size. When this is considered, the economic analysis should include the additional investment and operating costs associated with a higher pressure generation system. When an increase in the generating pressure requires a change from below to above 15 psig, the generating system equipment changes from low-pressure class to high-pressure class and there are significant increases in both investment and operating cost.

Where steam is provided from a nonfired device or prime mover such as a diesel engine cooling jacket, the source device can have an inherent pressure limitation.

PIPING

The piping system distributes the steam, returns the condensate, and removes air and noncondensable gases. In steam heating systems, it is important that the piping distribute steam, not only at full design load, but at partial loads and excess loads that can occur on system warm-up. The usual average winter steam demand is less than half the demand at the lowest outdoor design temperature. However, when the system is warming up, the load on the steam mains and returns can exceed the maximum operating load for the coldest design day, even in moderate weather. This load comes from raising the temperature of the piping to the steam temperature and that of the building to the indoor design temperature. Supply and return piping should be sized according to Chapter 36 of the 2005 *ASHRAE Handbook—Fundamentals*.

Supply Piping Design Considerations

1. Size pipe according to Chapter 36 of the 2005 *ASHRAE Handbook—Fundamentals*, taking into consideration pressure drop and steam velocity.
2. Pitch piping uniformly down in the direction of steam flow at 0.25 in. per 10 ft. If piping cannot be pitched down in the direction of the steam flow, refer to Chapter 36 of the 2005 *ASHRAE Handbook—Fundamentals* for rules on pipe sizing and pitch.
3. Insulate piping well to avoid unnecessary heat loss (see Chapters 23 and 24 of the 2005 *ASHRAE Handbook—Fundamentals*).
4. Condensate from unavoidable heat loss in the distribution system must be removed promptly to eliminate water hammer and degradation of steam quality and heat transfer capability. Install drip legs at all low points and natural drainage points in the system, such as at the ends of mains and the bottoms of risers, and ahead of pressure regulators, control valves, isolation valves, pipe bends, and expansion joints. On straight horizontal runs with no natural drainage points, space drip legs at intervals not exceeding 300 ft when the pipe is pitched down in the direction of the steam flow and at a maximum of 150 ft when the pipe is pitched up, so that condensate flow is opposite of steam flow. These distances apply to systems where valves are opened manually to remove air and excess condensate that forms during warm-up conditions. Reduce these distances by about half in systems that are warmed up automatically.
5. Where horizontal piping must be reduced in size, use eccentric reducers that permit the continuance of uniform pitch along the bottom of piping (in downward pitched systems). Avoid concentric reducers on horizontal piping, because they can cause water hammer.
6. Take off all branch lines from the top of the steam mains, preferably at a 45° angle, although vertical 90° connections are acceptable.
7. Where the length of a branch takeoff is less than 10 ft, the branch line can be pitched back 0.5 in. per 10 ft, providing drip legs as described previously in (4).
8. Size drip legs properly to separate and collect the condensate. Drip legs at vertical risers should be full-size and extend beyond the riser, as shown in Figure 4. Drip legs at other locations should be the same diameter as the main. In steam mains 6 in. and over, this can be reduced to half the diameter of the main, but to no less than 4 in. Where warm-up is supervised, the length of the collecting leg is not critical. However, the recommended length is 1.5 times the pipe diameter and not less than 8 in. For automatic warm-up, collecting legs should always be the same size as the main and should be at least 28 in. long to provide the hydraulic pressure differential necessary for the trap to discharge before a positive pressure is built up in the steam main.
9. Condensate should flow by gravity from the trap to the return piping system. Where the steam trap is located below the return line, the condensate must be lifted. In systems operating above 40 psig, the trap discharge can usually be piped directly to the return system (Figure 5). However, back pressure at the trap discharge (return line pressure plus hydraulic pressure created by height of lift) must not exceed steam main pressure, and the trap must be sized after considering back pressure. A collecting leg must be used and the trap discharge must flow by gravity to a vented condensate receiver, from which it is pumped to the overhead return in systems (1) operating under 40 psig, (2) where the temperature is regulated by modulating the steam control valves,

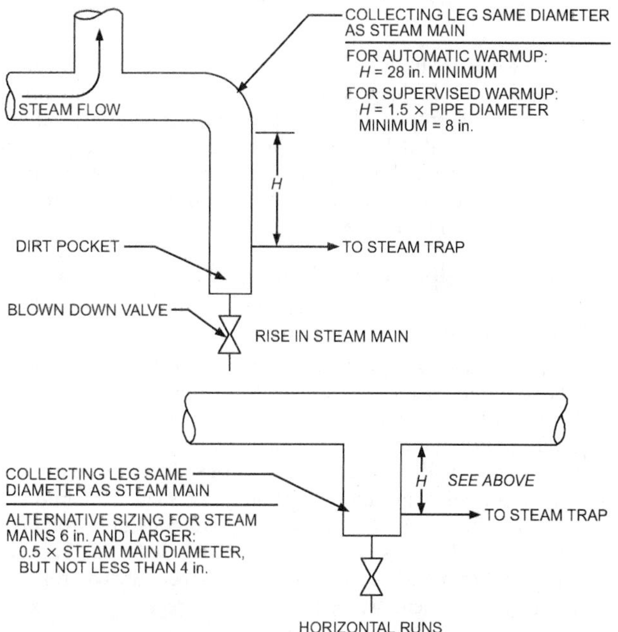

Fig. 4 Method of Dripping Steam Mains

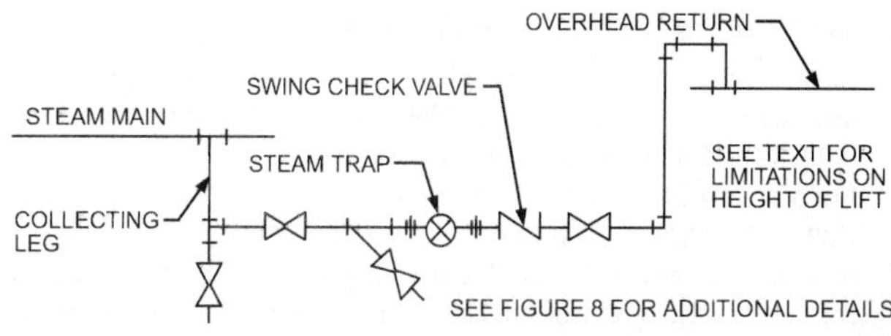

Fig. 5 Trap Discharging to Overhead Return

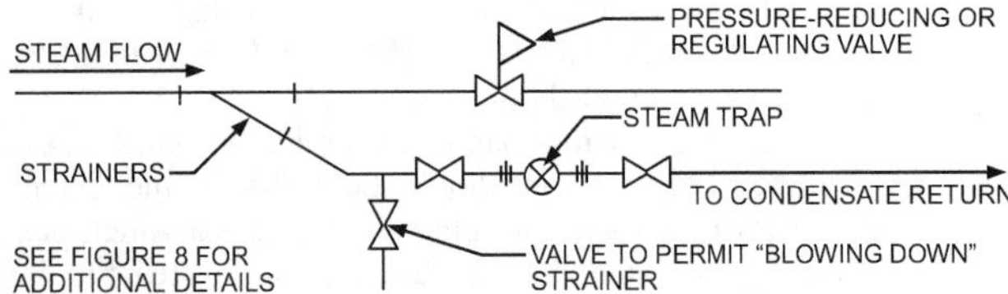

Fig. 6 Trapping Strainers

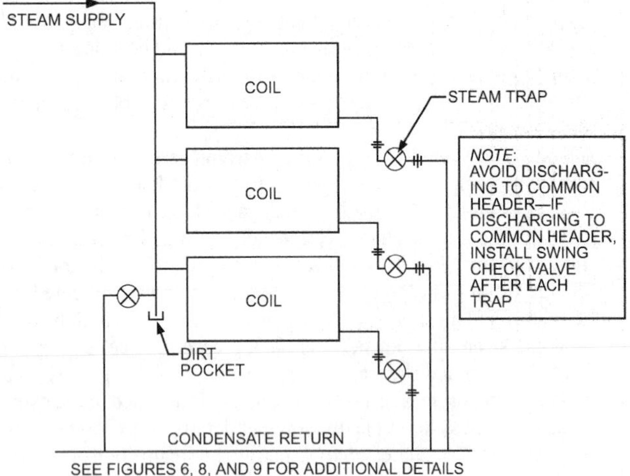

Fig. 7 Trapping Multiple Coils

or (3) where the back pressure at the trap is close to system pressure. The trap discharge should flow by gravity to a vented condensate receiver from which it is pumped to the overhead return.

10. Strainers installed before the pressure-reducing and control valves are a natural water collection point. Since water carryover can erode the valve seat, install a trap at the strainer blowdown connection (Figure 6).

Terminal Equipment Piping Design Considerations

1. Size piping the same as the supply and return connections of the terminal equipment.
2. Keep equipment and piping accessible for inspection and maintenance of the steam traps and control valves.
3. Minimize strain caused by expansion and contraction with pipe bends, loops, or three elbow swings to take advantage of piping flexibility, or with expansion joints or flexible pipe connectors.
4. In multiple-coil applications, separately trap each coil for proper drainage (Figure 7). Piping two or more coils to a common header served by a single trap can cause condensate backup, improper heat transfer, and inadequate temperature control.
5. Terminal equipment, where temperature is regulated by a modulating steam control valve, requires special consideration. Refer to the section on Condensate Removal from Temperature-Regulated Equipment.

Return Piping Design Considerations

1. Flow in the return line is two-phase, consisting of steam and condensate. See Chapter 36 of the 2005 *ASHRAE Handbook—Fundamentals* for sizing considerations.
2. Pitch return lines downward in the direction of the condensate flow at 0.5 in. per 10 ft to ensure prompt condensate removal.
3. Insulate the return line well, especially where the condensate is returned to the boiler or the condensate enthalpy is recovered. In vacuum systems, the return lines are not insulated since condensate subcooling is required.

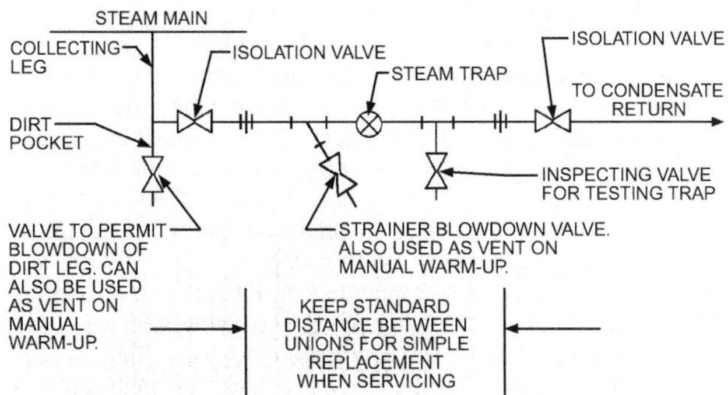

Fig. 8 Recommended Steam Trap Piping

4. Where possible and practical, use heat recovery systems to recover the condensate enthalpy. See the section on Heat Recovery. In vacuum systems, the return lines are not insulated since condensate subcooling is required.
5. Equip dirt pockets of the drip legs and strainer blowdowns with valves to remove dirt and scale.
6. Install steam traps close to drip legs and make them accessible for inspection and repair. Servicing is simplified by making the pipe sizes and configuration identical for a given type and size of trap. The piping arrangement in Figure 8 facilitates inspection and maintenance of steam traps.
7. When elevating condensate to an overhead return, consider the pressure at the trap inlet and the fact that it requires approximately 1 psi to elevate condensate 2 ft. See (9) in the section on Supply Piping Design Considerations for a complete discussion.

CONDENSATE REMOVAL FROM TEMPERATURE-REGULATED EQUIPMENT

When air, water, or another product is heated, the temperature or heat transfer rate can be regulated by a modulating steam pressure control valve. Because pressure and temperature do not vary at the same rate as load, the steam trap capacity, which is determined by the pressure differential between the trap inlet and outlet, may be adequate at full load, but not at some lesser load.

Analysis shows that steam pressure must be reduced dramatically to achieve a slight lowering of temperature. In most applications, this can result in subatmospheric pressure in the coil, while as much as 75% of the full condensate load has to be handled by the steam trap. This is especially important for coils exposed to outside air, because subatmospheric conditions can occur in the coil at outside temperatures below 32°F, and the coil will freeze if the condensate is not removed.

Armers (1985) provides detailed methods for determining condensate load under various operating conditions. However, in most cases, this load need not be calculated if the coils are piped as shown in Figure 9 and this procedure is followed:

1. Place the steam trap 1 to 3 ft below the bottom of the steam coil to provide a pressure of approximately 0.5 to 1.5 psig. Locating the trap at less than 12 in. usually results in improper drainage and operating difficulties.
2. Install vacuum breakers between the coil and trap inlet to ensure that the pressure can drain the coil when it is atmospheric or subatmospheric. The vacuum breaker should respond to a differential pressure of no greater than 3 in. of water. For atmospheric returns, the vacuum breaker should be opened to the atmosphere, and the return system must be designed to ensure no pressurization of the return line. In vacuum return systems, the vacuum breaker should be piped to the return line.
3. Discharge from the trap must flow by gravity, without any lifts in the piping, to the return system, which must be vented properly

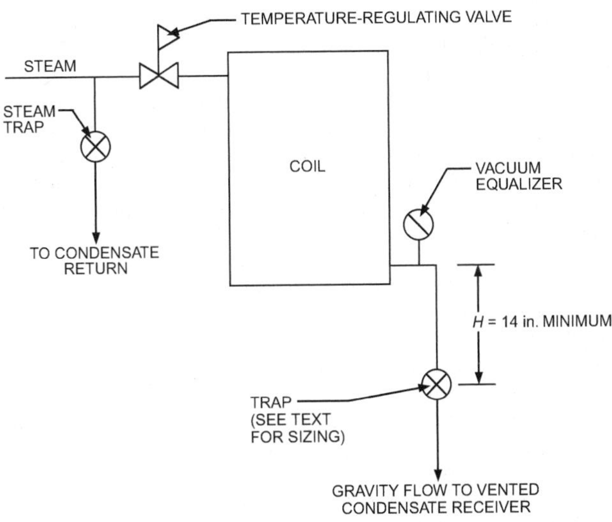

Fig. 9 Trapping Temperature-Regulated Coils

to the atmosphere to eliminate any back pressure that could prevent the trap from draining the coil. Where the return main is overhead, the trap discharge should flow by gravity to a vented receiver, from which it is then pumped to the overhead return.

4. Design traps to operate at maximum pressure at the control valve inlet and size them to handle the full condensate load at a pressure differential equal to the hydraulic pressure between the trap and coil. The actual condensate load can vary from the theoretical design load because of the safety factors used in coil selection and the fact that condensate does not always form at a uniform steady rate; therefore, size steam traps according to the following:

   For an actual steam pressure p_1 in the coil at full condensate load w, the proportion X of full load needing atmospheric pressure at the coil is

$$X = \frac{212 - t_c}{t_s - t_c} \tag{2}$$

   where

   t_c = control temperature, °F
   t_s = steam temperature at p_1, °F

   Then, the steam trap must be sized both to pass the full load w at a differential pressure equal to p_1 and to pass $X \cdot w$ (the proportion of full load) at 0.5 psi.

5. To reduce the possibility of a steam coil freezing, the temperature-regulating valve is often left wide open, and the leaving air temperature is controlled by a face-and-bypass damper on the steam coil.
6. For air temperatures below freezing, traps selected for draining steam coils should fail open (e.g., bucket traps) or two traps in parallel should be used.

STEAM TRAPS

Steam traps are an essential part of all steam systems, except one-pipe steam heating systems. Traps discharge condensate, which forms as steam gives up some of its heat, and direct the air and noncondensable gases to a point of removal. Condensate forms in steam mains and distribution piping because of unavoidable heat losses through less-than-perfect insulation, as well as in terminal equipment such as radiators, convectors, fan-coil units, and heat exchangers, where steam gives up heat during normal operation. Condensate must always be removed from the system as soon as it accumulates for the following reasons:

- Although condensate contains some valuable heat, using this heat by holding the condensate in the terminal equipment reduces the heat transfer surface. It also causes other operating problems because it retains air, which further reduces heat transfer, and noncondensable gases such as CO_2, which cause corrosion. As discussed in the section on Steam Source, recovering condensate heat is usually only desirable when the condensate is not returned to the boiler. Methods for this are discussed in the section on Heat Recovery.
- Steam moves rapidly in mains and supply piping, so when condensate accumulates to the point where the steam can push a slug of it, serious damage can occur from the resulting water hammer.

Ideally, the steam trap should remove all condensate promptly, along with air and noncondensable gases that might be in the system, with little or no loss of live steam. A steam trap is an automatic valve that can distinguish between steam and condensate or other fluids. Traps are classified as follows:

- **Thermostatic traps** react to the difference in the temperatures of steam and condensate.
- **Mechanical traps** depend on the difference in the densities of steam and condensate.
- **Kinetic traps** rely on the difference in the flow characteristics of steam and condensate.

The following points apply to all steam traps:

- No single type of steam trap is best suited to all applications, and most systems require more than one type of trap.
- Steam traps, regardless of type, should be carefully sized for the application and condensate load to be handled, because both undersizing and oversizing can cause serious problems. Undersizing can result in undesirable condensate backup and excessive cycling, which can lead to premature failure. Oversizing might appear to solve this problem and make selection much easier because fewer different sizes are required, but if the trap fails, excessive steam can be lost.

Steam traps should be installed between two unions to facilitate maintenance and/or replacement.

Thermostatic Traps

In thermostatic traps, a bellows or bimetallic element operates a valve that opens in the presence of subcooled condensate and closes in the presence of steam (Figure 10). Because condensate is initially at the same temperature as the steam from which it was condensed, the thermostatic element must be designed and calibrated to open at a temperature below the steam temperature; otherwise, the trap would blow live steam continuously. Therefore, the condensate is subcooled by allowing it to back up in the trap and in a portion of the upstream drip leg piping, both of which are left uninsulated. Some thermostatic traps operate with a continuous water leg behind the trap so there is no steam loss; however, this prohibits the discharge of air and noncondensable gases, and can cause excessive condensate to back up into the mains or terminal equipment, thereby resulting in operating problems. Devices that operate without significant backup can lose steam before the trap closes.

Although both bellows and bimetallic traps are temperature-sensitive, their operations are significantly different. The **bellows thermostatic trap** has a fluid with a lower boiling point than water. When the trap is cold, the element is contracted and the discharge port is open. As hot condensate enters the trap, it causes the contents of the bellows to boil and vaporize before the condensate temperature rises to steam temperature. Because the contents of the bellows boil at a lower temperature than water, the vapor pressure inside the bellows element is greater than the steam pressure outside, causing the element to expand and close the discharge port.

Assuming the contained liquid has a pressure-temperature relationship similar to that of water, the balance of forces acting on the bellows element remains relatively constant, no matter how the steam pressure varies. Therefore, this is a balanced pressure device that can be used at any pressure within the operating range of the device. However, this device should not be used where superheated steam is present, because the temperature is no longer in step with the pressure and damage or rupture of the bellows element can occur.

Bellows thermostatic traps are best suited for steady light loads on low-pressure service. They are most widely used in radiators and convectors in HVAC applications.

The **bimetallic thermostatic trap** has an element made from metals with different expansion coefficients. Heat causes the element to change shape, permitting the valve port to open or close. Because a bimetallic element responds only to temperature, most traps have the valve on the outlet so that steam pressure is trying to open the valve. Therefore, by properly designing the bimetallic element, the trap can operate on a pressure-temperature curve approaching the steam saturation curve, although not as closely as a balanced pressure bellows element.

Unlike the bellows thermostatic trap, bimetallic thermostatic traps are not adversely affected by superheated steam or subject to damage by water hammer, so they can be readily used for high-pressure applications. They are best suited for steam tracers, jacketed piping, and heat transfer equipment, where some condensate backup is tolerable. If they are used on steam main drip legs, the element should not back up condensate.

Mechanical Traps

Mechanical traps are buoyancy operated, depending on the difference in density between steam and condensate. The **float and thermostatic trap** (Figure 10) is commonly called the F&T trap and is actually a combination of two types of traps in a single trap body: (1) a bellows thermostatic element operating on temperature difference, which provides automatic venting, and (2) a float portion, which is buoyancy operated. Float traps without automatic venting should not be used in steam systems.

On start-up, the float valve is closed and the thermostatic element is open for rapid air venting, permitting the system or equipment to rapidly fill with steam. When steam enters the trap body, the thermostatic element closes and, as condensate enters, the float rises and the condensate discharges. The float regulates the valve opening so it continuously discharges the condensate at the rate at which it reaches the trap.

The F&T trap has large venting capabilities, continuously discharges condensate without backup, handles intermittent loads very well, and can operate at extremely low pressure differentials. Float and thermostatic traps are suited for use with temperature-regulated steam coils. They also are well suited for steam main and riser drip legs on low-pressure steam heating systems. Although F&T traps are available for pressures to 250 psig or higher, they are susceptible to water hammer, so other traps are usually a better choice for high-pressure applications.

Bucket traps operate on buoyancy, but they use a bucket that is either open at the top or inverted instead of a closed float. Initially, the bucket in an open bucket trap is empty and rests on the bottom of the trap body with the discharge vented, and, as condensate enters the trap, the bucket floats up and closes the discharge port. Additional condensate overflows into the bucket, causing it to sink and open the discharge port, which allows steam pressure to force the condensate out of the bucket. At the same time, it seals the bottom of the discharge tube, prohibiting air passage. Therefore, to prevent air binding, this device has an automatic air vent, as does the F&T trap.

Inverted bucket traps eliminate the size and venting problems associated with open bucket traps. Steam or air entering the submerged inverted bucket causes it to float and close the discharge port. As more condensate enters the trap, it forces air and steam out of the vent on top of the inverted bucket into the trap body where the steam condenses by cooling. When the mass of the bucket exceeds the buoyancy effect, the bucket drops, opening the discharge port, and steam pressure forces the condensate out, and the cycle repeats.

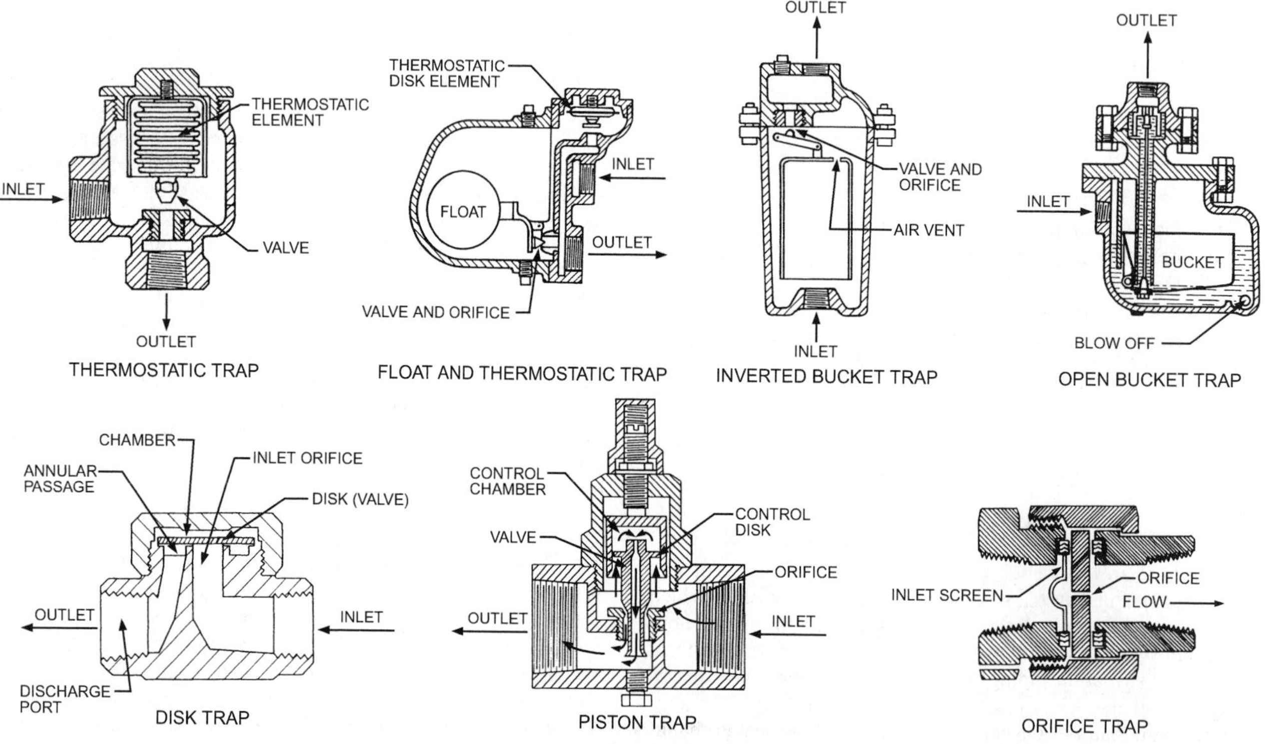

Fig. 10 Thermostatic Traps

Unlike most cycling-type traps, the inverted bucket trap continuously vents air and noncondensable gases. Although it discharges condensate intermittently, there is no condensate backup in a properly sized trap. Inverted bucket traps are made for all pressure ranges and are well suited for steam main drip legs and most HVAC applications. Although inverted bucket traps can be used for temperature-regulated steam coils, the F&T trap is usually better because it has the high venting capability desirable for such applications.

Kinetic Traps

Numerous devices operate on the difference between the flow characteristics of steam and condensate and on the fact that condensate discharging to a lower pressure contains more heat than necessary to maintain the liquid phase. This excess heat causes some of the condensate to flash to steam at the lower pressure.

Thermodynamic traps or **disk traps** are simple devices with only one moving part. When air or condensate enters the trap on start-up, it lifts the disk off its seat and is discharged. When steam or hot condensate (some of which "flashes" to steam upon exposure to a lower pressure) enters the trap, the increased velocity of this vapor flow decreases the pressure on the underside of the disk and increases the pressure above the disk, causing it to snap shut. Pressure is then equalized above and below the disk, but because the area exposed to pressure is greater above than below it, the disk remains shut until the pressure above is reduced by condensing or bleeding, thus permitting the disk to snap open and repeat the cycle. This device does not cycle open and shut as a function of condensate load, it is a time-cycle device that opens and shuts at fixed intervals as a function of how fast the steam above the disk condenses. Because disk traps require a significant pressure differential to operate properly, they are not well suited for low-pressure systems or for systems with significant back pressure. They are best suited for high-pressure systems and are widely applied to steam main drip legs.

Impulse traps or **piston traps** continuously pass a small amount of steam or condensate through a "bleed" orifice, changing the pressure positions within the piston. When live steam or very hot condensate that flashes to steam enters the control chamber, the increased pressure closes the piston valve port. When cooler condensate enters, the pressure decreases, permitting the valve port to open. Most impulse traps cycle open and shut intermittently, but some modulate to a position that passes condensate continuously.

Impulse traps can be used for the same applications as disk traps; however, because they have a small "bleed" orifice and close piston tolerances, they can stick or clog if dirt is present in the system.

Orifice traps have no moving parts. All other traps have discharge ports or orifices, but in the traps described previously, this opening is oversized, and some type of closing mechanism controls the flow of condensate and prevents the loss of live steam.

The orifice trap has no such closing mechanism, and the flow of steam and condensate is controlled by two-phase flow through an orifice. A simple explanation of this theory is that an orifice of any size has a much greater capacity for condensate than it does for steam because of the significant differences in their densities and because "flashing" condensate tends to choke the orifice. An orifice is selected larger than required for the actual condensate load; therefore, it continuously passes all condensate along with the air and noncondensable gases, plus a small controlled amount of steam. The steam loss is usually comparable to that of most cycling-type traps.

Orifice traps must be sized more carefully than cycling-type traps. On light condensate loads, the orifice size is small and, like impulse traps, tends to clog. Orifice traps are suitable for all system pressures and can operate against any back pressure. They are best suited for steady pressure and load conditions such as steam main drip legs.

PRESSURE-REDUCING VALVES

Where steam is supplied at pressures higher than required, one or more pressure-reducing valves (pressure regulators) are required. The pressure-reducing valve reduces pressure to a safe point and regulates pressure to that required by the equipment. The district heating industry refers to valves according to their functional use. There are two classes of service: (1) the steam must be shut off completely (dead-end valves) to prevent buildup of pressure on the low-pressure side during no load (single-seated valves should be used) and (2) the low-pressure lines condense enough steam to prevent buildup of pressure from valve leakage (double-seated valves can be used). Valves available for either service are direct-operated, spring-loaded, weight-loaded, air-loaded, or pilot-controlled, using either the flowing steam or auxiliary air or water as the operating medium. The direct-operated, double-seated valve is less affected by varying inlet steam pressure than the direct-operated, single-seated valve. Pilot-controlled valves, either single- or double-seated, tend to eliminate the effect of variable inlet pressures.

Installation

Pressure-reducing valves should be readily accessible for inspection and repair. There should be a bypass around each reducing valve equal to the area of the reducing-valve seat ring. The globe valve in a bypass line should have plug disk construction and must have an absolutely tight shutoff. Steam pressure gages, graduated up to the initial pressure, should be installed on the low-pressure side and on the high-pressure side. The low-pressure gage should be ahead of the shutoff valve because the reducing valve can be adjusted with the shutoff valve closed. A similar gage should be installed downstream from the shutoff valve for use during manual operation. Typical service connections are shown in Figure 11 for low-pressure service and Figure 12 for high-pressure service. In the smaller sizes, the standby pressure-regulating valve can be removed, a filler installed, and the inlet stop valves used for manual pressure regulation until repairs are made.

Strainers should be installed on the inlet of the primary pressure-reducing valve and before the second-stage reduction if there is considerable piping between the two stages. If a two-stage reduction is made, it is advisable to install a pressure gage immediately before the reducing valve of the second-stage reduction to set and check the operation of the first valve. A drip trap should be installed before the two reducing valves.

Where pressure-reducing valves are used, one or more relief devices or safety valves must be provided, and the equipment on the low-pressure side must meet the requirements for the full initial pressure. The relief or safety devices are adjoining or as close as possible to the reducing valve. The combined relieving capacity must be adequate to avoid exceeding the design pressure of the low-pressure system if the reducing valve does not open. In most

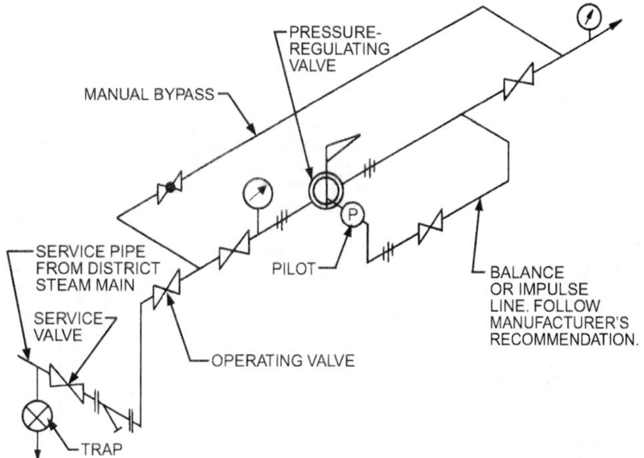

Fig. 11 Pressure-Reducing Valve Connections—Low Pressure

areas, local codes dictate the safety relief valve installation requirements.

Safety valves should be set at least 5 psi higher than the reduced pressure if the reduced pressure is under 35 psig and at least 10 psi higher than the reduced pressure if the reduced pressure is above 35 psig or the first-stage reduction of a double reduction. The outlet from relief valves should not be piped to a location where the discharge can jeopardize persons or property or is a violation of local codes.

Figure 13 shows a typical service installation, with a separate line to various heating zones and process equipment. If the initial pressure is below 50 psig, the first-stage pressure-reducing valve can be omitted. In Assembly A (Figure 13), the single-stage pressure-reducing valve is also the pressure-regulating valve.

When making a two-stage reduction (e.g., 150 to 50 psig and then 50 to 2 psig), allow for the expansion of steam on the low-pressure side of each reducing valve by increasing the pipe area to about double the area of a pipe the size of the reducing valve. This also allows steam to flow at a more uniform velocity. It is recommended that the valves be separated by a distance of at least 20 ft to reduce excessive hunting action of the first valve, although this should be checked with the valve manufacturer.

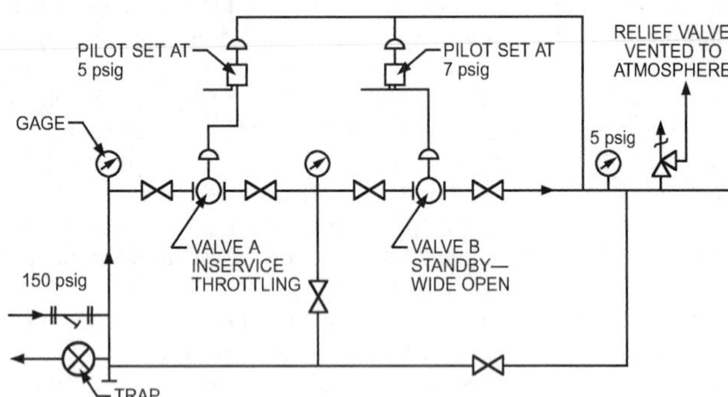

Fig. 12 Pressure-Reducing Valve Connections—High Pressure

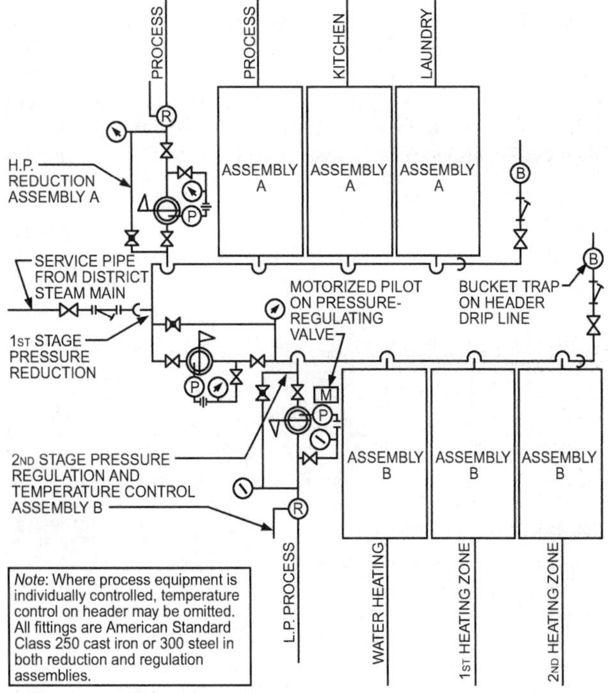

Fig. 13 Steam Supply

Figure 14 shows a typical double-reduction installation where the pressure in the district steam main is higher than can safely be applied to building heating systems. The first or pressure-reducing valve effects the initial pressure reduction. The pressure-regulating valve regulates the steam to the desired final pressure.

Pilot-controlled or air-loaded direct-operated reducing valves can be used without limitation for all reduced-pressure settings for all heating, process, laundry, and hot water services. Spring-loaded direct-operated valves can be used for reduced pressures up to 50 psig, providing they can pass the required steam flow without excessive deviation in reduced pressure.

Weight-loaded valves may be used for reduced pressures below 15 psig and for moderate steam flows.

Pressure-equalizing or impulse lines must be connected to serve the type of valve selected. With direct-operated diaphragm valves having rubber-like diaphragms, the impulse line should be connected to the reduced-pressure steam line to allow for maximum condensate on the diaphragm and in the impulse or equalizing line. If it is connected to the top of the steam line, a condensate accumulator should be used to reduce variations in the pressure of condensate on the diaphragm. The impulse line should not be connected to the bottom of the reduced pressure line since it could become clogged. Equalizing or impulse lines for pilot-controlled and direct-operated reducing valves using metal diaphragms should be connected into the expanded outlet piping approximately 2 to 4 ft from the reducing valve and pitched away from the reducing valve to prevent condensate accumulation. Pressure impulse lines for externally pilot-controlled reducing valves using compressed air or fresh water should be installed according to manufacturer's recommendations.

Valve Size Selection

Pressure-regulating valves should be sized to supply the maximum steam requirements of the heating system or equipment. Consideration should be given to rangeability, speed of load changes, and regulation accuracy required to meet system needs, especially with temperature control systems using intermittent steam flow to heat the building.

The reducing valve should be selected carefully. The manufacturer should be consulted. Piping to and from the reducing valve should be adequate to pass the desired amount of steam at the desired maximum velocity. A common error is to make the size of the reducing valve the same as the service or outlet pipe size; this makes the reducing valve oversized and causes wiredrawing or erosion of the valve and seat because of the high-velocity flow caused by the small lift of the valve.

On installations where the steam requirements are large and variable, wiredrawing and cycling control can occur during mild

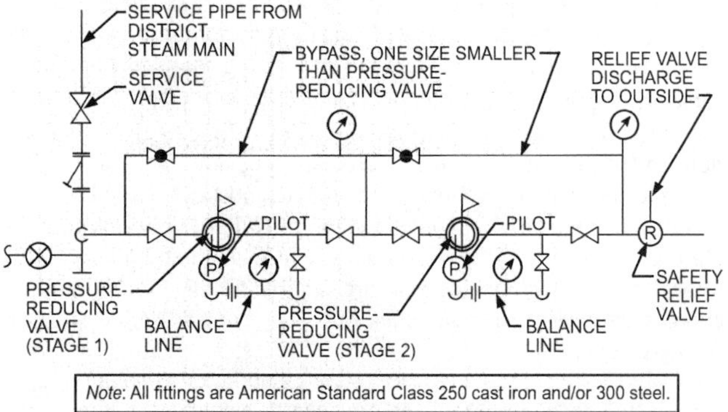

Fig. 14 Two-Stage Pressure-Regulating Valve
(Used where high-pressure steam is supplied for low-pressure requirements)

weather or during reduced demand periods. To overcome this condition, two reducing valves are installed in parallel, with the sizes selected on a 70 and 30% proportion of maximum flow. For example, if 10,000 lb of steam per hour are required, the size of one valve is based on 7000 lb of steam per hour, and the other is based on 3000 lb of steam per hour. During mild weather (spring and fall), the larger valve is set for slightly lower reduced pressure than the smaller one and remains closed as long as the smaller one can meet the demand. During the remainder of the heating season, the valve settings are reversed to keep the smaller one closed, except when the larger one is unable to meet the demand.

TERMINAL EQUIPMENT

A variety of terminal units are used in steam heating systems. All are suited for use on low-pressure systems. Terminal units used on high-pressure systems have heavier construction, but are otherwise similar to those on low-pressure systems. Terminal units are usually classified as follows:

1. **Natural convection units** transfer most heat by convection and some heat by radiation. The equipment includes cast-iron radiators, finned-tube convectors, and cabinet and baseboard units with convection-type elements (see Chapter 35).
2. **Forced-convection units** employ a forced air movement to increase heat transfer and distribute the heated air within the space. The devices include unit heaters, unit ventilators, induction units, fan-coil units, the heating coils of central air-conditioning units, and many process heat exchangers. When such units are used for both heating and cooling, there is a steam coil for heating and a separate chilled water or refrigerant coil for cooling. See Chapters 1, 3, 4, 5, 22, 26, and 27.
3. **Radiant panel systems** transfer some heat by convection. Because of the low-temperature and high-vacuum requirements, this type of unit is rarely used on steam systems.

Selection

The primary consideration in selecting terminal equipment is comfortable heat distribution. The following briefly describes suitable applications for specific types of steam terminal units.

Natural Convection Units

Radiators, convectors, convection-type cabinet units, and baseboard convectors are commonly used for (1) facilities that require heating only, rather than both heating and cooling, and (2) in conjunction with central air conditioning as a source of perimeter heating or for localized heating in spaces such as corridors, entrances, halls, toilets, kitchens, and storage areas.

Forced-Convection Units

Forced-convection units can be used for the same types of applications as natural convection units but are primarily used for facilities that require both heating and cooling, as well as spaces that require localized heating.

Unit heaters are often used as the primary source of heat in factories, warehouses, and garages and as supplemental freeze protection for loading ramps, entrances, equipment rooms, and fresh air plenums.

Unit ventilators are forced-convection units with dampers that introduce controlled amounts of outside air. They are used in spaces with ventilation requirements not met by other system components.

Cabinet heaters are often used in entranceways and vestibules that can have high intermittent heat loads.

Induction units are similar to fan-coil units, but the air is supplied by a central air system rather than individual fans in each unit. Induction units are most commonly used as perimeter heating for facilities with central systems.

Fan-coil units are designed for heating and cooling. On water systems, a single coil can be supplied with hot water for heating and chilled water for cooling. On steam systems, single- or dual-coil units can be used. Single-coil units require steam-to-water heat exchangers to provide hot water to meet heating requirements. Dual-coil units have a steam coil for heating and a separate coil for cooling. The cooling media for dual-coil units can be chilled water from a central chiller or refrigerant provided by a self-contained compressor. Single-coil units require the entire system to be either in a heating or cooling mode and do not permit simultaneous heating and cooling to satisfy individual space requirements. Dual-coil units can eliminate this problem.

Central air-handling units are used for most larger facilities. These units have a fan, a heating and cooling coil, and a compressor if chilled water is not available for cooling. Multizone units are arranged so that each air outlet has separate controls for individual space heating or cooling requirements. Large central systems distribute either warm or cold conditioned air through duct systems and employ separate terminal equipment such as mixing boxes, reheat coils, or variable volume controls to control the temperature to satisfy each space. Central air-handling systems can be factory assembled, field erected, or built at the job site from individual components.

CONVECTION STEAM HEATING

Any system that uses steam as the heat transfer medium can be considered a steam heating system; however, the term "steam heating system" is most commonly applied to convection systems using radiators or convectors as terminal equipment. Other types of steam heating systems use forced convection, in which a fan or air-handling system is used with a convector or steam coil such as unit heaters, fan-coil units and central air-conditioning and heating systems.

Convection-type steam heating systems are used in facilities that have a heating requirement only, such as factories, warehouses, and apartment buildings. They are often used in conjunction with central air-conditioning systems to heat the perimeter of the building. Also, steam is commonly used with incremental units that are designed for cooling and heating and have a self-contained air-conditioning compressor.

Steam heating systems are classified as one-pipe or two-pipe systems, according to the piping arrangement that supplies steam to and returns condensate from the terminal equipment. These systems can be further subdivided by (1) the method of condensate return (gravity flow or mechanical flow by means of condensate pump or vacuum pump) and (2) by the piping arrangement (upfeed or downfeed and parallel or counterflow for one-pipe systems).

One-Pipe Steam Heating Systems

The one-pipe system has a single pipe through which steam flows to and condensate is returned from the terminal equipment (Figure 15). These systems are designed as gravity return, although a condensate pump can be used where there is insufficient height above the boiler water level to develop enough pressure to return condensate directly to the boiler.

A one-pipe system with gravity return does not have steam traps; instead it has air vents at each terminal unit and at the ends of all supply mains to vent the air so the system can fill with steam. In a system with a condensate pump, there must be an air vent at each terminal unit and steam traps at the ends of each supply main.

The one-pipe system with gravity return has low initial cost and is simple to install, because it requires a minimum of mechanical equipment and piping. One-pipe systems are most commonly used in small facilities such as small apartment buildings and office buildings. In larger facilities, the larger pipe sizes required for two-phase flow, problems of distributing steam quickly and evenly throughout the system, the inability to zone the system, and

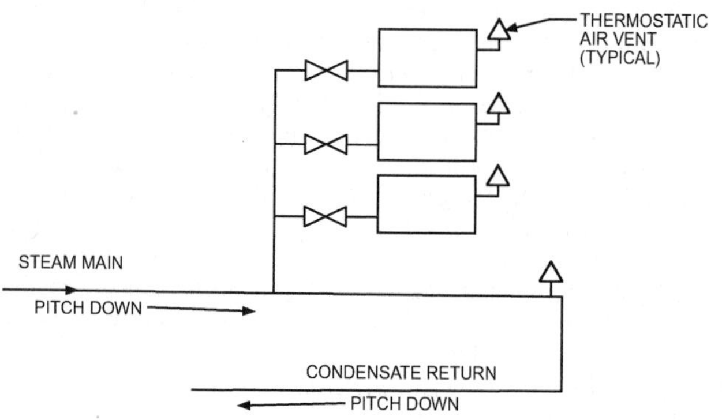

Fig. 15 One-Pipe System

difficulty in controlling the temperature make the one-pipe system less desirable than the two-pipe system.

The heat input to the system is controlled by cycling the steam on and off. In the past, temperature control in individual spaces has been a problem. Many systems have adjustable vents at each terminal unit to help balance the system, but these are seldom effective. A practical approach is to use a self-contained thermostatic valve in series with the air vent (as explained in the section on Temperature Control) that provides limited individual thermostatic control for each space.

Many designers do not favor one-pipe systems because of their distribution and control problems. However, when a self-contained thermostatic valve is used to eliminate the problems, one-pipe systems can be considered for small facilities, where initial cost and simple installation and operation are prime factors.

Most one-pipe gravity return systems are in facilities that have their own boiler, and, because returning condensate must overcome boiler pressure, these systems usually operate from a fraction of a psi to a maximum of 5 psig. The boiler hookup is critical and the Hartford loop (described in the section on Boiler Connections) is used to avoid problems that can occur with boiler low-water condition. Stamper and Koral (1979) and Hoffman give piping design information for one-pipe systems.

Two-Pipe Steam Heating Systems

The two-pipe system uses separate pipes to deliver the steam and return the condensate from each terminal unit (Figure 16). Thermostatic traps are installed at the outlet of each terminal unit to keep the steam in the unit until it gives up its latent heat, at which time the trap cycles open to pass the condensate and permits more steam to enter the radiator. If orifices are installed at the inlet to each terminal unit, as discussed in the sections on Steam Distribution and Temperature Control, and if the system pressure is precisely regulated so only the amount of steam is delivered to each unit that it is capable of condensing, the steam traps can be omitted. However, omitting steam traps is generally not recommended for initial design.

Two-pipe systems can have either gravity or mechanical returns; however, gravity returns are restricted to use in small systems and are generally outmoded. In larger systems that require higher steam pressures to distribute steam, some mechanical means, such as a condensate pump or vacuum pump, must return condensate to the boiler. A **vacuum return system** is used on larger systems and has the following advantages:

- The system fills quickly with steam. The steam in a gravity return system must push the air out of the system, resulting in delayed heat-up and condensate return that can cause low-water problems. A vacuum return system can eliminate these problems.
- The steam supply pressure can be lower, resulting in more efficient operation.

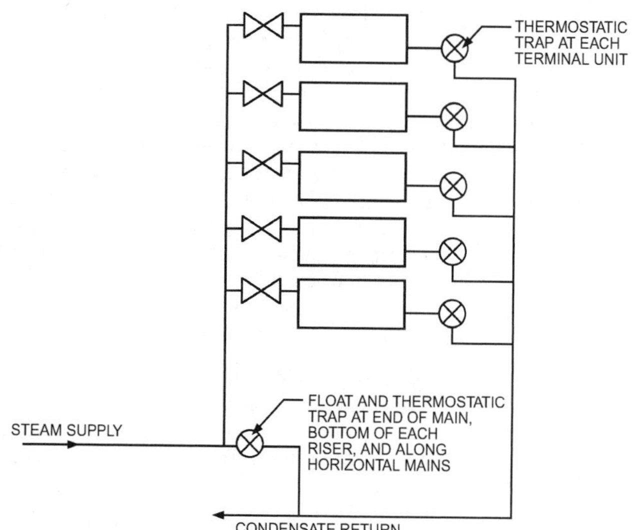

Fig. 16 Two-Pipe System

Variable vacuum or subatmospheric systems are a variation of the vacuum return system in which a controllable vacuum is maintained in both the supply and return sides. This permits using the lowest possible system temperature and prompt steam distribution to the terminal units. The primary purpose of variable vacuum systems is to control temperature as discussed in the section on Temperature Control.

Unlike one-pipe systems, two-pipe systems can be simply zoned where piping is arranged to supply heat to individual sections of the building that have similar heating requirements. Heat is supplied to meet the requirements of each section without overheating other sections. The heat also can be varied according to factors such as the hours of use, type of occupancy, and sun load.

STEAM DISTRIBUTION

Steam supply piping should be sized so that the pressure drops in all branches of the same supply main are nearly uniform. Return piping should be sized for the same pressure drop as supply piping for quick and even steam distribution. Because it is impossible to size piping so that pressure drops are exactly the same, the steam flows first to those units that can be reached with the least resistance, resulting in uneven heating. Units farthest from the source of steam will heat last, while other spaces are overheated. This problem is most evident when the system is filling with steam. It can be severe on systems in which temperature is controlled by cycling the steam on and off. The problem can be alleviated or eliminated by balancing valves or inlet orifices.

Balancing valves are installed at the unit inlet and contain an adjustable valve port to control the amount of steam delivered. The main problem with such devices is that they are seldom calibrated accurately, so that variations in orifice size as small as 0.003 in^2 can make a significant difference.

Inlet orifices are thin brass or copper plates installed in the unit inlet valve or pipe unions (Figure 17). Inlet orifices can solve distribution problems, because they can be drilled for appropriate size and changed easily to compensate for unusual conditions. Properly sized inlet orifices can compensate for oversized heating units, reduce energy waste and system control problems caused by excessive steam loss from defective steam traps, and provide a means of temperature control and balancing within individual zones. Manufacturers of orifice plates provide data for calculating the required sizes for most systems. Also, Sanford and Sprenger (1931) developed data for sizing orifices for low-pressure, gravity return systems

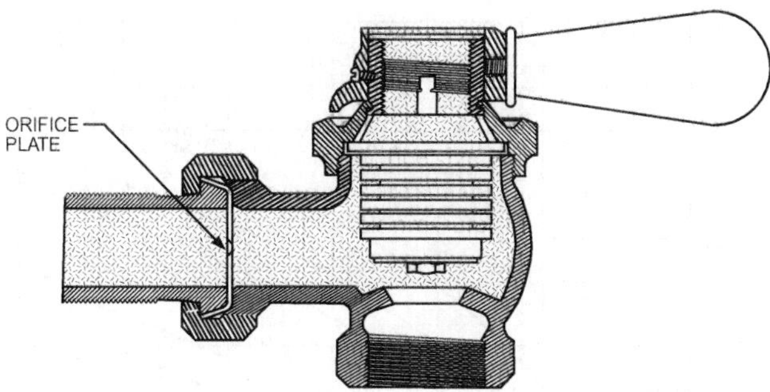

Fig. 17 Inlet Orifice

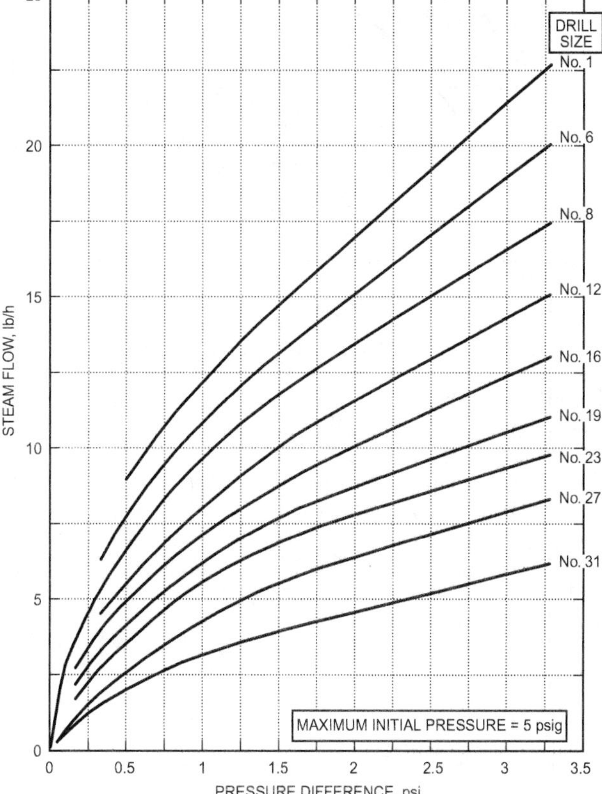

Fig. 18 Orifice Capacities for Different Pressure Differentials
(Schroeder 1950)

See Figure 18 for capacities of orifices at various pressure differentials.

If orifices are installed in valves or pipe unions that are conveniently accessible, minor rebalancing among individual zones can be accomplished through replacement of individual orifices.

TEMPERATURE CONTROL

All heating systems require some means of temperature control to achieve the desired comfort conditions and operating efficiencies. In convection-type steam heating systems, the temperature and resulting heat output of the terminal units must be increased or decreased. This can be done by (1) permitting the steam to enter the heating unit intermittently, (2) varying the steam temperature delivered to each unit, and (3) varying the amount of steam delivered to each unit.

There are two types of controls: those that control the temperature or heat input to the entire system and those that control the temperature of individual spaces or zones. Often, both types are used together for maximum control and operating efficiency.

Intermittent flow controls, commonly called **heat timers**, control the temperature of the system by cycling the steam on and off during a certain portion (or fraction) of each hour as a function of the outdoor and/or indoor temperature. Most of these devices provide for night setback, and computerized or electronic models optimize morning start-up as a function of outdoor and indoor temperature and make anticipatory adjustments. Used independently, these devices do not permit varying heat input to different parts of the building or spaces, so they should be used with zone control valves or individual thermostatic valves for maximum energy efficiency.

Zone control valves control the temperature of spaces with similar heating requirements as a function of outdoor and/or indoor temperature, as well as permit duty cycling. These intermittent flow control devices operate full open or full closed. They are controlled by an indoor thermostat and are used in conjunction with a heat timer, variable vacuum, or pressure differential control that controls heat input to the system.

Individual thermostatic valves installed at each terminal unit can provide the proper amount of heat to satisfy the individual requirements of each space and eliminate overheating. Valves can be actuated electrically, pneumatically, or by a self-contained mechanism that has a wax- or liquid-filled sensing element requiring no external power source.

Individual thermostatic valves can be and are often used as the only means of temperature control. However, these systems are always "on," resulting in inefficiency and no central control of heat input to the system or to the individual zones. Electronic, electric, and pneumatic operators allow centralized control to be built into the system. However, it is desirable to have a supplemental system control with self-contained thermostatic valves in the form of zone control valves, or a system that controls the heat input to the entire system, relying on self-contained valves as a local high-temperature shutoff only.

Self-contained thermostatic valves can also be used to control temperature on one-pipe systems that cycle on and off when installed in series with the unit air vent. On initial start-up, the air vent is open, and inherent system distribution problems still exist. However, on subsequent on-cycles where the room temperature satisfies the thermostatic element, the vent valve remains closed, preventing the unit from filling with steam.

Variable vacuum or subatmospheric systems control the temperature of the system by varying the steam temperature through pressure control. These systems differ from the regular vacuum return system in that the vacuum is maintained in both supply and return lines. Such systems usually operate at a pressure range from 2 psig to 25 in. Hg vacuum. Inlet orifices are installed at the terminal equipment for proper steam distribution during mild weather.

Design-day conditions are seldom encountered, so the system usually operates at a substantial vacuum. Because the specific volume of steam increases as the pressure decreases, it takes less steam to fill the system and operating efficiency results. The variable vacuum system can be used in conjunction with zone control valves or individual self-contained valves for increased operating efficiency.

Pressure differential control (two-pipe orifice) systems provide centralized temperature control with any system that has properly sized inlet orifices at each heating unit. This method operates on the principle that flow through an orifice is a function of the pressure differential across the orifice plate. The pressure range is selected to fill each heating unit on the coldest design day; on warmer days, the supply pressure is lowered so that heating units are only partially filled with steam, thereby reducing their heat output. An advantage of this method is that it virtually eliminates all heating unit steam trap losses because the heating units are only partially filled with steam on all but the coldest design days.

Table 2 Pressure Differential Temperature Control

Outdoor Temperature, °F	Required Pressure Differential*, in. Hg.
0	6.0
10	4.5
20	3.2
30	2.1
40	1.2
50	0.5
60	0.1

*To maintain 70°F indoors for 0°F outdoor design.

The required system pressure differential can be achieved manually with throttling valves or with an automatic pressure differential controller. Table 2 shows the required pressure differentials to maintain 70°F indoors for 0°F outdoor design, with orifices according to Figure 18. Required pressure differential curves can be established for any combination of supply and return line pressures by sizing orifices to deliver the proper amount of steam for the coldest design day, calculating the amount of steam required for warmer days using Equation (3), and then selecting the pressure differential that will provide this flow rate.

$$Q_r = Q_d \frac{(t_i)_{design} - t_o}{(t_i)_{design} - (t_o)_{design}} \tag{3}$$

where

Q_r = required flow rate
Q_d = design day flow rate
$(t_i)_{design}$ = design indoor temperature
$(t_o)_{design}$ = design outdoor temperature
t_o = outside temperature

Pressure differential systems can be used with zone control valves or individual self-contained thermostatic valves to increase operating efficiency.

HEAT RECOVERY

Two methods are generally employed to recover heat from condensate: (1) the enthalpy of the liquid condensate (sensible heat) can be used to vaporize or "flash" some of the liquid to steam at a lower pressure, or (2) it can be used directly in a heat exchanger to heat air, fluid, or a process.

The particular methods used vary with the type of system. As explained in the section on Basic Steam System Design, facilities that purchase steam from a utility generally do not have to return condensate and, therefore, can recover heat to the maximum extent possible. On the other hand, facilities with their own boiler generally want the condensate to return to the boiler as hot as possible, limiting heat recovery because any heat removed from condensate has to be returned to the boiler to generate steam again.

Flash Steam

Flash steam is an effective use for the enthalpy of the liquid condensate. It can be used in any facility that has a requirement for steam at different pressures, regardless of whether steam is purchased or generated by a facility's own boiler. Flash steam can be used in any heat exchange device to heat air, water, or other liquids or directly in processes with lower pressure steam requirements. Equation (1) may be used to calculate the amount of flash steam generated, and Figure 19 provides a graph for calculating the amount of flash steam as a function of system pressures.

Although flash steam can be generated directly by discharging high-pressure condensate to a lower pressure system, most designers prefer a **flash tank** to control flashing. Flash tanks can be mounted either vertically or horizontally, but the vertical arrangement shown in Figure 21 is preferred because it provides better separation of steam and water, resulting in the highest possible steam quality.

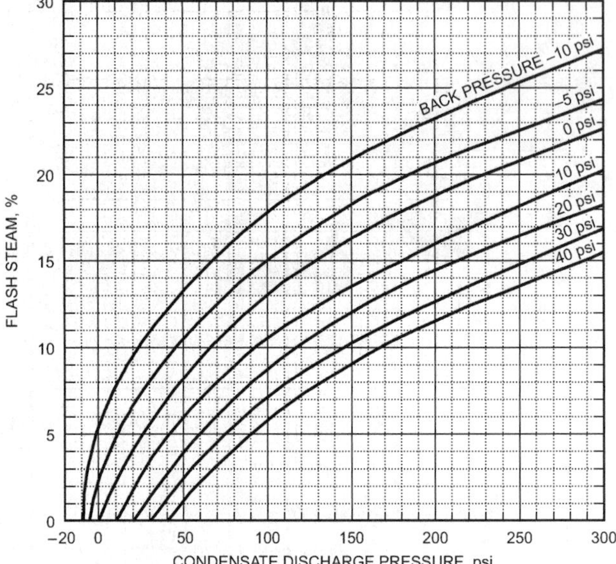

Fig. 19 Flash Steam

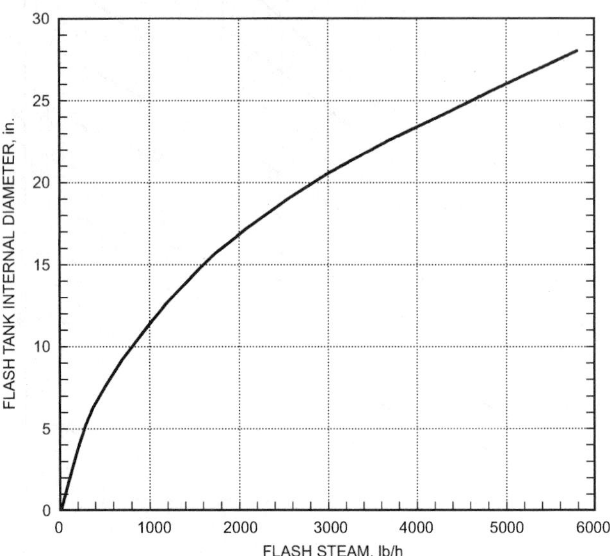

Fig. 20 Flash Tank Diameters

The most important dimension in the design of vertical flash tanks is the internal diameter, which must be large enough to ensure a low upward velocity of flash to minimize water carryover. If this velocity is low enough, the height of the tank is not important, but it is good practice to use a height of at least 2 to 3 ft. The graph in Figure 20 can be used to determine the internal diameter and is based on a steam velocity of 10 ft/s, which is the maximum velocity in most systems.

Installation is important for proper flash tank operation. Condensate lines should pitch towards the flash tank. If more than one condensate line discharges to the tank, each line should be equipped with a swing check valve to prevent backflow. Condensate lines and the flash tank should be well insulated to prevent any unnecessary heat loss. A thermostatic air vent should be installed at the top of the tank to vent any air that accumulates. The tank should be trapped at

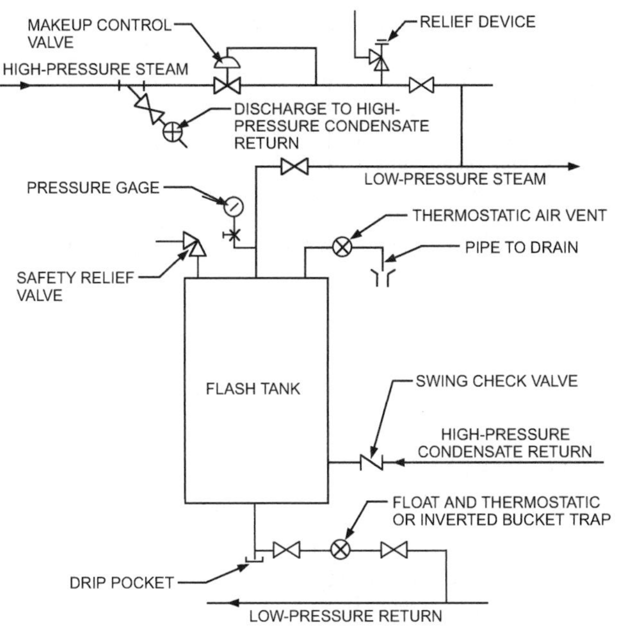

Fig. 21 Vertical Flash Tank

the bottom with an inverted bucket or float and thermostatic trap sized to triple the condensate load.

The demand load must always be greater than the amount of flash steam available to prevent the low-pressure system from becoming overpressurized. A safety relief valve should always be installed at the top of the flash tank to preclude such a condition.

Because the flash steam available is generally less than the demand for low-pressure steam, a makeup valve ensures that the low-pressure system maintains design pressure.

Flash tanks are considered pressure vessels and must be constructed in accordance with ASME and local codes.

Direct Heat Recovery

Direct heat recovery that uses the enthalpy of the liquid in some type of heat exchange device is appropriate when condensate is not returned to a facility's own boiler; any lowering of condensate temperature below saturation temperature requires reheating at the boiler to regenerate steam.

The enthalpy of the condensate can be used in fan-coil units, unit heaters, or convectors to heat spaces where temperature control is not critical such as garages, ramps, loading docks, and entrance halls or in a shell-and-tube heat exchanger to heat water or other fluids.

In most HVAC applications, the enthalpy of the liquid condensate may be used most effectively and efficiently to heat domestic hot water with a shell-and-tube or plate-type heat exchanger, commonly called an economizer. Many existing economizers do not use the enthalpy of the condensate effectively because they are designed only to preheat makeup water flowing directly through the heat exchanger at the time of water usage. Hot water use seldom coincides with condensate load, and most of the enthalpy is wasted in these preheat economizer systems.

A storage-type water heater can be effectively used for heat recovery. This can be a shell-and-tube heat exchanger with a condensate coil for heat recovery and a steam or electric coil when the condensate enthalpy cannot satisfy the demand. Another option is a storage-type heat exchanger incorporating only a coil for condensate with a supplemental heat exchanger to satisfy peak loads. Note that many areas require a double-wall heat exchanger between steam and the potable water.

Chapter 49 of the 2007 *ASHRAE Handbook—HVAC Applications* provides useful information on determining hot water loads for various facilities. In general, however, the following provide optimum heat recovery:

- Install the greatest storage capacity in the available space. Although all systems must have supplemental heaters for peak load conditions, with ample storage capacity these heaters may seldom function if all the necessary heat is provided by the enthalpy of the condensate.
- For maximum recovery, permit stored water to heat to 180°F or higher, using a mixing valve to temper the water to the proper delivery temperature.

COMBINED STEAM AND WATER SYSTEMS

Combined steam and water systems are often used to take advantage of the unique properties of steam, which are described in the sections on Advantages and Fundamentals.

Combined systems are used where a facility must generate steam to satisfy the heating requirements of certain processes or equipment. They are usually used where steam is available from a utility and economic considerations of local codes preclude the facility from operating its own boiler plant. There are two types of combined steam and water systems: (1) steam is used directly as a heating medium; the terminal equipment must have two separate coils, one for heating with steam and one for cooling with chilled water, and (2) steam is used indirectly and is piped to heat exchangers that generate the hot water for use at terminal equipment; the exchanger for terminal equipment may use either one coil or separate coils for heating and cooling.

Combined steam and water systems may be two-, three-, or four-pipe systems. Chapters 3 and 5 have further descriptions.

COMMISSIONING

After design and construction of a system, care should be taken to ensure correct performance of the system and that building personnel understand the operating and maintenance procedure required to maintain operating efficiency.

REFERENCES

Armers, A. 1985. Sizing procedure for steam traps with modulating steam pressure control. Spirax Sarco Inc., Concord, Ontario.

ASME. 2001. *Boiler and pressure vessel code*. American Society of Mechanical Engineers, New York.

ASME. 1998. Power piping. ANSI/ASME *Standard* B31.1-98. American Society of Mechanical Engineers, New York.

Hoffman steam heating systems design manual. *Bulletin* no. TES-181, Hoffman Specialty ITT Fluid Handling Division, Indianapolis, IN.

HYDI. 1989. *Testing and rating heating boilers*. Hydronics Institute, Berkeley Heights, NJ.

Kremers, J.A. 1982. Modulating steam pressure in coils compound steam trap selection procedures. *Armstrong Trap Magazine* 50(1).

Sanford, S.S. and C.B. Sprenger. 1931. Flow of steam through orifices. *ASHVE Transactions* 37:371-394.

Schroeder, D.E. 1950. Balancing a steam heating system by the use of orifices. *ASHVE Transactions* 56:325-340.

Stamper, E. and R.L. Koral. 1979. *Handbook of air conditioning, heating and ventilating*, 3rd ed. Industrial Press, New York.

CHAPTER 11

DISTRICT HEATING AND COOLING

DISTRICT heating and cooling (DHC) distributes thermal energy from a central source to residential, commercial, and/or industrial consumers for use in space heating, cooling, water heating, and/or process heating. The energy is distributed by steam or hot- or chilled-water lines. Thus, thermal energy comes from a distribution medium rather than being generated on site at each facility.

Whether the system is a public utility or user owned, such as a multibuilding campus, it has economic and environmental benefits depending somewhat on the particular application. Political feasibility must be considered, particularly if a municipality or governmental body is considering a DHC installation. Historically, successful DHC systems have had the political backing and support of the community.

Applicability

District heating and cooling systems are best used in markets where (1) the thermal load density is high and (2) the annual load factor is high. A high load density is needed to cover the capital investment for the transmission and distribution system, which usually constitutes most of the capital cost for the overall system, often ranging from 50 to 75% of the total cost for district heating systems (normally lower for district cooling applications).

The annual load factor is important because the total system is capital intensive. These factors make district heating and cooling systems most attractive in serving (1) industrial complexes, (2) densely populated urban areas, and (3) high-density building clusters with high thermal loads. Low-density residential areas have usually not been attractive markets for district heating, although there have been some successful applications. District heating is best suited to areas with a high building and population density in relatively cold climates. District cooling applies in most areas that have appreciable concentrations of cooling loads, usually associated with tall buildings.

Components

District heating and cooling systems consist of three primary components: the central plant, the distribution network, and the consumer systems (Figure 1).

The **central source** or **production plant** may be any type of boiler, a refuse incinerator, a geothermal source, solar energy, or thermal energy developed as a by-product of electrical generation. The last approach, called combined heat and power (CHP), has a high energy utilization efficiency; see Chapter 7 for information on CHP.

Chilled water can be produced by

- Absorption refrigeration machines

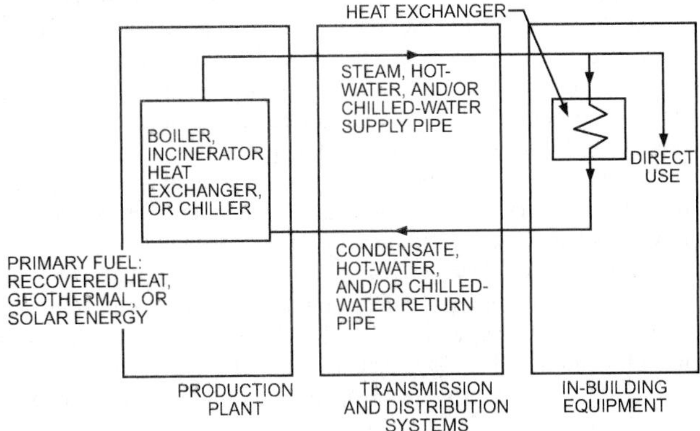

Fig. 1 Major Components of District Heating System

- Electric-driven compression equipment (reciprocating, rotary screw or centrifugal chillers)
- Gas/steam turbine- or engine-driven compression equipment
- Combination of mechanically driven systems and thermal-energy-driven absorption systems

The second component is the **distribution** or **piping network** that conveys the energy. The piping is often the most expensive portion of a district heating or cooling system. The piping usually consists of a combination of preinsulated and field-insulated pipe in both concrete tunnel and direct burial applications. These networks require substantial permitting and coordinating with nonusers of the system for right-of-way if not on the owner's property. Because the initial cost is high, it is important to optimize use.

The third component is the **consumer system**, which includes in-building equipment. When steam is supplied, it may be (1) used directly for heating; (2) directed through a pressure-reducing station for use in low-pressure (0 to 15 psig) steam space heating, service water heating, and absorption cooling; or (3) passed through a steam-to-water heat exchanger. When hot or chilled water is supplied, it may be used directly by the building systems or isolated by a heat exchanger (see the section on Consumer Interconnections).

BENEFITS

Environmental Benefits

Emissions from central plants are easier to control than those from individual plants and, in aggregate, are lower because of higher quality of equipment, seasonal efficiencies and level of maintenance, and lower system heat loss. A central plant that burns high-sulfur coal can economically remove noxious sulfur emissions,

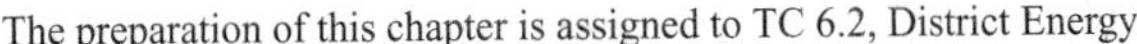

The preparation of this chapter is assigned to TC 6.2, District Energy.

where individual combustors could not. Similarly, the thermal energy from municipal wastes can provide an environmentally sound system. Cogeneration of heat and electric power allows for combined efficiencies of energy use that greatly reduce emissions and also allow for fuel flexibility. In addition, refrigerants and other CFCs can be monitored and controlled more readily in a central plant. Where site conditions allow, remote location of the plant reduces many of the concerns with use of ammonia systems for cooling.

Consumer Economic Benefits

A district heating and cooling system offers the following economic benefits. Even though the basic costs are still borne by the central plant owner/operator, because the central plant is large the customer can realize benefits of economies of scale.

Operating Personnel. One of the primary advantages to a building owner is that operating personnel for the HVAC system can be reduced or eliminated. Most municipal codes require operating engineers to be on site when high-pressure boilers are in operation. Some older systems require trained operating personnel to be in the boiler/mechanical room at all times. When thermal energy is brought into the building as a utility, depending on the sophistication of the building HVAC controls, there may be opportunity to reduce or eliminate operating personnel.

Insurance. Both property and liability insurance costs are significantly reduced with the elimination of a boiler in the mechanical room, because risk of a fire or accident is reduced.

Usable Space. Usable space in the building increases when a boiler and/or chiller and related equipment are no longer necessary. The noise associated with such in-building equipment is also eliminated. Although this space usually cannot be converted into prime office space, it does provide the opportunity for increased storage or other use.

Equipment Maintenance. With less mechanical equipment, there is proportionately less equipment maintenance, resulting in less expense and a reduced maintenance staff.

Higher Thermal Efficiency. A larger central plant can achieve higher thermal and emission efficiencies than can several smaller units. When strict regulations must be met, additional pollution control equipment is also more economical for larger plants. Cogeneration of heat and electric power results in much higher overall efficiencies than are possible from separate heat and power plants.

Partial load performance of central plants may be more efficient than that of many isolated small systems because the larger plant can operate one or more capacity modules as the combined load requires and can modulate output. Central plants generally have efficient base-load units and less costly peaking equipment for use in extreme loads or emergencies.

PRODUCER ECONOMICS

Available Fuels. Smaller heating plants are usually designed for one type of fuel, which is generally gas or oil. Central DHC plants can operate on less expensive coal or refuse. Larger facilities can often be designed for more than one fuel (e.g., coal and oil), and combined with power generation (see Chapter 7 for information on combined heat and power systems).

Energy Source Economics. If an existing facility is the energy source, the available temperature and pressure of the thermal fluid is predetermined. If exhaust steam from an existing electrical generating turbine is used to provide thermal energy, the conditions of the bypass determine the maximum operating pressure and temperature of the DHC system. A tradeoff analysis must be conducted to determine what percentage of the energy will be diverted for thermal generation and what percentage will be used for electrical generation. Based on the marginal value of energy, it is critical to determine the operating conditions in the economic analysis.

If a new central plant is being considered, a decision of whether to cogenerate electrical and thermal energy or to generate thermal energy only must be made. An example of cogeneration is a diesel or natural gas engine-driven generator with heat recovery equipment. The engine drives a generator to produce electricity, and heat is recovered from the exhaust, cooling, and lubrication systems. Other systems may use one of several available steam turbine designs for cogeneration. These turbine systems combine the thermal and electrical output to obtain the maximum amount of available energy. Chapter 7 has further information on cogeneration.

The selection of temperature and pressure is crucial because it can dramatically affect the economic feasibility of a DHC system design. If the temperature and/or pressure level chosen is too low, a potential customer base might be eliminated. On the other hand, if there is no demand for absorption chillers or high-temperature industrial processes, a low-temperature system usually provides the lowest delivered energy cost.

The availability and location of fuel sources must also be considered in optimizing the economic design of a DHC system. For example, a natural gas boiler might not be feasible where abundant sources of natural gas are not available.

Initial Capital Investment

The initial capital investment for a DHC system is usually the major economic driving force in determining whether there is acceptable payback for implementation. Normally, the initial capital investment includes the four components of (1) concept planning, (2) design, (3) construction, and (4) consumer interconnections.

Concept Planning. In concept planning, three areas are generally reviewed. First, the **technical feasibility** of a DHC system must be considered. Conversion of an existing heat source, for example, usually requires the services of an experienced power plant or DHC engineering firm.

Financial feasibility is the second consideration. For example, a municipal or governmental body must consider availability of bond financing. **Alternative energy choices** for potential customers must be reviewed because consumers are often asked to sign long-term contracts in order to justify a DHC system.

Design. The distribution system accounts for a significant portion of the initial investment. Distribution design depends on the heat transfer medium chosen, its operating temperature and pressure, and the routing. Failure to consider these key variables results in higher-than-planned installation costs. An analysis must be done to optimize insulating properties. The section on Economical Thickness for Pipe Insulation discusses determining insulation values.

Construction. The construction costs of the central plant and distribution system depend on the quality of the concept planning and design. Although the construction cost usually accounts for most of the initial capital investment, neglect in any of the other three areas could mean the difference between economic success and failure. Field changes usually increase the final cost and delay start-up. Even a small delay in start-up can adversely affect both economics and consumer confidence. It is extremely important that the contractors have experience commensurate with the project. DHC project costs vary greatly and depend on local construction environment and site conditions such as

- Labor rates
- Construction environment (i.e., slow or busy period)
- Distance to ship equipment
- Permits and fees (e.g., franchise fees)
- Local authorities (e.g., traffic control, times of construction in city streets)
- Soil conditions (e.g., clay, bedrock)
- Quality of equipment and controls (e.g., commercial or industrial)
- Availability of materials

- Size of distribution piping system
- Type of insulation or cathodic protection for piping system
- Type of distribution system installation (e.g., direct buried, tunnel)
- Depth of bury and restoration of existing conditions (e.g., city streets, green areas)
- Below-grade conflict resolutions
- Economies of scale

Sample construction cost unit pricing is as follows, but the designer is cautioned that cost can vary widely:

- Cooling plant (building, chillers, cooling towers, pumps, piping, controls) = $1500 to $2600 per ton
- Boiler plant (building, boilers, stacks, pumps, piping, controls) = $1500 to $2300 per boiler horsepower
- Distribution systems (includes excavation, backfill, surface restoration, piping, etc.):
  - Direct-buried chilled-water systems = $500 to $1250 per foot of trench
  - Direct-buried preinsulated heating piping (steam/condensate and hot water) = $750 to $1500 per foot of trench
  - Inaccessible tunnels = $500 to $1000 per foot of trench
  - Walkable tunnels = $3500 to $15,000 per foot of trench

Lead time needed to obtain equipment generally determines the time required to build a DHC system. In some cases, lead time on major components in the central plant can be over a year.

Installation time of the distribution system depends in part on the routing interference with existing utilities. A distribution system in a new industrial park is simpler and requires less time to install than a system being installed in an established business district.

Consumer Interconnection. These costs are usually borne by the consumer. High interconnection costs may favor an in-building plant instead of a DHC system. For example, if an existing building is equipped for steam service, interconnection to a hot-water DHC system may be too costly, even though the cost of energy is lower.

CENTRAL PLANT

The central plant may include equipment to provide heat only, cooling only, both heat and cooling, or any of these three options in conjunction with electric power generation. In addition to the central plant, small so-called satellite plants are sometimes used in situations where a customer's building is located in an area where distribution piping is not yet installed.

HEATING AND COOLING PRODUCTION

Heating Medium

In plants serving hospitals, industrial customers, or those also generating electricity, steam is the usual choice for production in the plant and, often, for distribution to customers. For systems serving largely commercial buildings, hot water is an attractive medium. From the standpoint of distribution, hot water can accommodate a greater geographical area than steam because of the ease with which booster pump stations can be installed. The common attributes and relative merits of hot water and steam as heat-conveying media are described as follows.

Heat Capacity. Steam relies primarily on the latent heat capacity of water rather than on sensible heat. The net heat content for saturated steam at 100 psig (338°F) condensed and cooled to 180°F is approximately 1040 Btu/lb. Hot water cooled from 350 to 250°F has a net heat effect of 103 Btu/lb, or only about 10% as much as that of steam. Thus, a hot-water system must circulate about 10 times more mass than a steam system of similar heat capacity.

Pipe Sizes. Despite the fact that less steam is required for a given heat load, and flow velocities are greater, steam usually requires a larger pipe size for the supply line because of its lower density (Aamot and Phetteplace 1978). This is compensated for by a much smaller condensate return pipe. Therefore, piping costs for steam and condensate are often comparable with those for hot-water supply and return.

Return System. Condensate return systems require more maintenance than hot-water return systems because hot-water systems function as a closed loop with very low makeup water requirements. For condensate return systems, corrosion of piping and other components, particularly in areas where feedwater is high in bicarbonates, is a problem. Nonmetallic piping has been used successfully in some applications, such as systems with pumped returns, where it has been possible to isolate the nonmetallic piping from live steam.

Similar concerns are associated with condensate drainage systems (steam traps, condensate pumps, and receiver tanks) for steam supply lines. Condensate collection and return should be carefully considered when designing a steam system. Although similar problems with water treatment occur in hot-water systems, they present less of a concern because makeup rates are much lower.

Pressure and Temperature Requirements. Flowing steam and hot water both incur pressure losses. Hot-water systems may use intermediate booster pumps to increase the pressure at points between the plant and the consumer. Because of the higher density of water, pressure variations caused by elevation differences in a hot-water system are much greater than for steam systems. This can adversely affect the economics of a hot-water system by requiring the use of a higher pressure class of piping and/or booster pumps.

Regardless of the medium used, the temperature and pressure used for heating should be no higher than needed to satisfy consumer requirements; this cannot be overemphasized. Higher temperatures and pressures require additional engineering and planning to avoid higher heat losses. Safety and comfort levels for operators and maintenance personnel also benefit from lower temperature and pressure. Higher temperatures may require higher pressure ratings for piping and fittings and may preclude the use of materials such as polyurethane foam insulation and nonmetallic conduits.

Hot-water systems are divided into three temperature classes:

- High-temperature systems supply temperatures over 350°F.
- Medium-temperature systems supply temperatures in the range of 250 to 350°F.
- Low-temperature systems supply temperatures of 250°F or lower.

The temperature drop at the consumer end should be as high as possible, preferably 40°F or greater. A large temperature drop allows the fluid flow rate through the system, pumping power, return temperatures, return line heat loss, and condensing temperatures in cogeneration power plants to be reduced. A large customer temperature drop can often be achieved by cascading loads operating at different temperatures.

In many instances, existing equipment and processes require the use of steam. See the section on Consumer Interconnections for further information.

Heat Production

Fire-tube and water-tube boilers are available for gas/oil firing. If coal is used, either package-type coal-fired boilers in small sizes (less than 20,000 to 25,000 lb/h) or field-erected boilers in larger sizes are available. Coal-firing underfeed stokers are available up to a 30,000 to 35,000 lb/h capacity; traveling grate and spreader stokers are available up to 160,000 lb/h capacity in single-boiler installations. Fluidized-bed boilers can be installed for capacities over 300,000 lb/h. Larger coal-fired boilers are typically multiple installations of the three types of stokers or larger, pulverized fired or fluidized-bed boilers. Generally, the complexity of fluidized bed or pulverized firing does not lend itself to small central heating plant operation.

Cooling Supply

Chilled water may be produced by an absorption refrigeration machine or by vapor-compression equipment driven by electric, turbine (steam or combustion), or internal combustion engines. The chilled-water supply temperature for a conventional system ranges from 40 to 44°F. A 12°F temperature difference (Δt) results in a flow rate of 2 gpm/ton of refrigeration. Because of the cost of the distribution system piping, large chilled-water systems are sometimes operated at lower supply water temperatures to allow a larger Δt to be achieved, thereby reducing chilled-water flow per ton of capacity. For systems involving stratified chilled-water storage, a practical lower limit is 39°F because of water density considerations; however, chemical additives can suppress this temperature below 28°F. For ice storage systems, temperatures as low as 34°F have been used.

Multiple air-conditioning loads interconnected with a central chilled-water system provide some economic advantages and energy conservation opportunities. In addition, central plants afford the opportunity to consider the use of refrigerants such as ammonia that may be impractical for use in individual buildings. The size of air-conditioning loads served, as well as the diversity among the loads and their distance from the chilling plant, are principal factors in determining the feasibility of large central plants. The distribution system pipe capacity is directly proportional to the operating temperature difference between the supply and return lines, and it benefits additionally from increased diversity in the connected loads.

For extremely large district systems, several plants are required to meet the loads economically and provide redundancy. In some areas, plants over 20,000 tons are common for systems exceeding 50,000 tons. Another reason for multiple plants is that single plants over 30,000 tons can require large piping headers (over 48 in. in diameter) within the plant as well as large distribution headers in streets already congested with other utilities, making piping layout problematic.

An economic evaluation of piping and pumping costs versus chiller power requirements can establish the most suitable supply water temperature. When sizing piping and calculating pumping cost the heat load on the chiller generated by the frictional heating of the flowing fluid should be considered because most of the pumping power adds to the system heat load. For high chiller efficiency, it is often more efficient to use isolated auxiliary equipment for special process requirements and to allow the central plant supply water temperature to float up at times of lower load. As with heating plants, optimum chilled-water control may require a combination of temperature modulation and flow modulation. However, the designer must investigate the effects of higher chilled-water supply temperatures on chilled-water secondary system distribution flows and air-side system performance (humidity control) before applying this to individual central water plants.

Thermal Storage

Both hot- and chilled-water thermal storage can be implemented for district systems. In North America, the current economic situation primarily results in chilled-water storage applications. Depending on the plant design and loading, thermal storage can reduce chiller equipment requirements and lower operating costs. By shifting a part of the chilling load, chillers can be sized closer to the average load than the peak load. Shifting the entire refrigeration load to off peak requires the same (or slightly larger) chiller machine capacity, but removes all of the electric load from the peak period. Because many utilities offer lower rates during off-peak periods, operating costs for electric-driven chillers can be substantially reduced.

Both ice and chilled-water storage have been applied to district-sized chiller plants. In general, the largest systems (>20,000 ton-hour capacity) use chilled-water storage and small- to moderate-sized systems use ice storage. Storage capacities in the 10,000 to 30,000 ton-hour range are now common and systems have been installed up to 125,000 ton-hour for district cooling systems.

In Europe, several cooling systems use naturally occurring underground aquifers (caverns) for storage of chilled water. Selection of the storage configuration (chilled-water steel tank above grade, chilled-water concrete tank below grade, ice direct, ice indirect) is often influenced by space limitations. Chilled-water storage requires four to six times the volume of ice storage for the same capacity. For chilled-water storage, the footprint of steel tanks (depending on height) can be less than concrete tanks for the same volume (Andrepont 1995); furthermore, the cost of above-grade tanks is usually less than below-grade tanks. Chapter 50 has information on thermal storage.

Auxiliaries

Numerous pieces of auxiliary support equipment related to the boiler and chiller operations are not unique to the production plant of a DHC system and are found in similar installations. Some components of a DHC system deserve special consideration because of their critical nature and potential effect on operations.

Although instrumentation can be either electronic or pneumatic, electronic instrumentation systems offer the flexibility of combining control systems with data acquisition systems. This combination brings improved efficiency, better energy management, and reduced operating staff for the central heating and/or cooling plant. For systems involving multiple fuels and/or thermal storage, computer-based controls are indispensable for accurate decisions about boiler and chiller operation.

Boiler feedwater treatment has a direct bearing on equipment life. Condensate receivers, filters, polishers, and chemical feed equipment must be accessible for proper management, maintenance, and operation. Depending on the temperature, pressure, and quality of the heating medium, water treatment may require softeners, alkalizers, and/or demineralizers for systems operating at high temperatures and pressures.

Equipment and layout of a central heating and cooling plant should reflect what is required for proper plant operation and maintenance. The plant should have an adequate service area for equipment and a sufficient number of electrical power outlets and floor drains. Equipment should be placed on housekeeping pads. Figure 2 presents a layout for a large hot-water/chilled-water plant.

Expansion Tanks and Water Makeup. The expansion tank is usually located in the central plant building. To control pressure, either air or nitrogen is introduced to the air space in the expansion tank. To function properly, the expansion tank must be the single point of the system where no pressure change occurs. Multiple, air-filled tanks may cause erratic and possibly harmful movement of air though the piping. Although diaphragm expansion tanks eliminate air movement, the possibility of hydraulic surge should be considered. On large chilled-water systems, a makeup water pump generally is used to makeup water loss. The pump is typically controlled from level switches on the expansion tank or from a desired pump suction pressure.

A conventional water meter on the makeup line can show water loss in a closed system. This meter also provides necessary data for water treatment. The fill valve should be controlled to open or close and not modulate to a very low flow, so that the water meter can detect all makeup.

Emission Control. Environmental equipment, including electrostatic precipitators, baghouses, and scrubbers, is required to meet emission standards for coal-fired or solid-waste-fired operations. Proper control is critical to equipment operation, and it should be designed and located for easy access by maintenance personnel.

A baghouse gas filter provides good service if gas flow and temperature are properly maintained. Because baghouses are designed

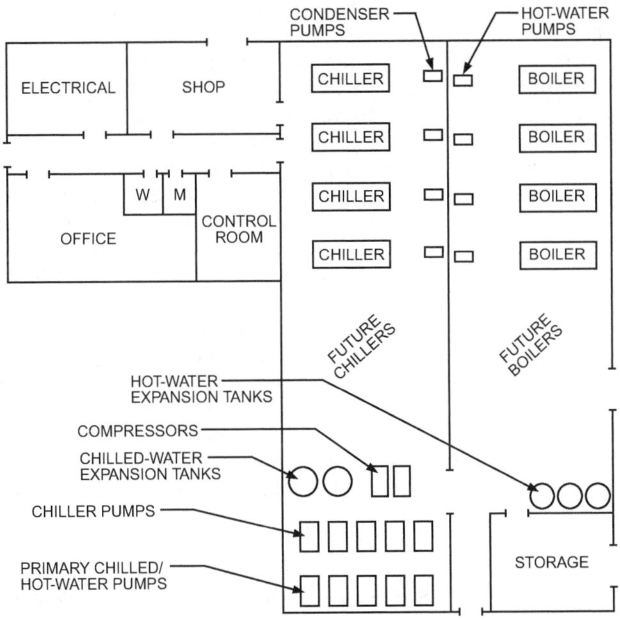

Fig. 2 Layout for Hot-Water/Chilled-Water Plant

for continuous online use, they are less suited for cyclic operation. Heating and cooling significantly reduces the useful life of the bags through acidic condensation. Using an economizer to preheat boiler feedwater and help control flue gas temperature may enhance baghouse operation. Contaminants generated by plant operation and maintenance, such as washdown of floors and equipment, may need to be contained.

Local codes and regulations may also require low-NO_x burners on gas- or oil-fired boilers or engine generators. Chapter 18 of the 2005 *ASHRAE Handbook—Fundamentals* and Chapter 29 have information on air pollution and its control.

DISTRIBUTION DESIGN CONSIDERATIONS

Water distribution systems are designed for either constant flow (variable return temperature) or variable flow (constant return temperature). The design decision between constant or variable volume flow affects the (1) selection and arrangement of the chiller(s), (2) design of the distribution system, and (3) design of the customer connection to the distribution system. Unless very unusual circumstances exist, most systems large enough to be considered in the district category are likely to benefit from variable flow design.

Constant Flow

Constant flow is generally applied only to smaller systems where simplicity of design and operation are important and where distribution pumping costs are low. Chillers may be arranged in series to handle higher temperature differentials. Flow volume through a full-load distribution system depends on the type of constant-flow system used. One technique connects the building and its terminals across the distribution system. The central plant pump circulates chilled water through three-way valve controlled air-side terminal units (constant-volume direct primary pumping). Balancing problems may occur in this design when many separate flow circuits are interconnected (Figure 3).

Constant-flow distribution is also applied to in-building circuits with separate pumps. This arrangement isolates the flow balance problem between buildings. In this case, flow through the distribution system can be significantly lower than the sum of the flows needed by the terminal if the in-building system supply temperature is higher than the distribution system supply temperature (Figure 3).

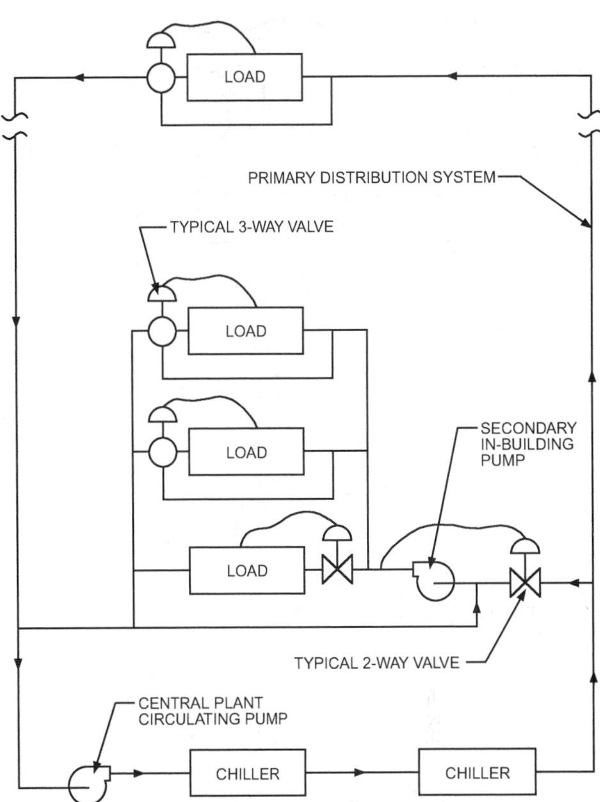

Fig. 3 Constant-Flow Primary Distribution with Secondary Pumping

The water temperature rise in the distribution system is determined by the connected in-building systems and their controls.

In constant-flow design, chillers arranged in parallel have decreased entering water temperatures at part load; thus, several machines may need to run simultaneously, each at a reduced load, to produce the required chilled-water flow. In this case, chillers in series are better because constant flow can be maintained though the chilled-water plant at all times, with only the chillers required for producing chilled water energized. Constant-flow systems should be analyzed thoroughly when considering multiple chillers in a parallel arrangement, because the auxiliary electric loads of condenser water pumps, tower fans, and central plant circulating pumps are a significant part of the total energy input. Modern control systems mitigate the reason for constant flow through the chillers, and variable flow should be investigated for larger systems.

Variable Flow

Variable-flow design can improve energy use and expand the capacity of the distribution system piping by using diversity. To maintain a high temperature differential at part load, the distribution system flow rate must track the load imposed on the central plant. Multiple parallel pumps or, more commonly, variable-speed pumps can reduce flow and pressure, and lower pumping energy at part load. Terminal device controls should be selected to ensure that variable flow objectives are met. Flow-throttling (two-way) valves provide the continuous high return temperature needed to correlate the system load change to a system flow change.

Systems in each building are usually two-pipe, with individual in-building pumping. In some cases, the pressure of the distribution system may cause flow through the in-building system without in-building pumping. Distribution system pumps can provide total building system pumping if (1) the distribution system pressure drops are minimal, and (2) the distribution system is relatively

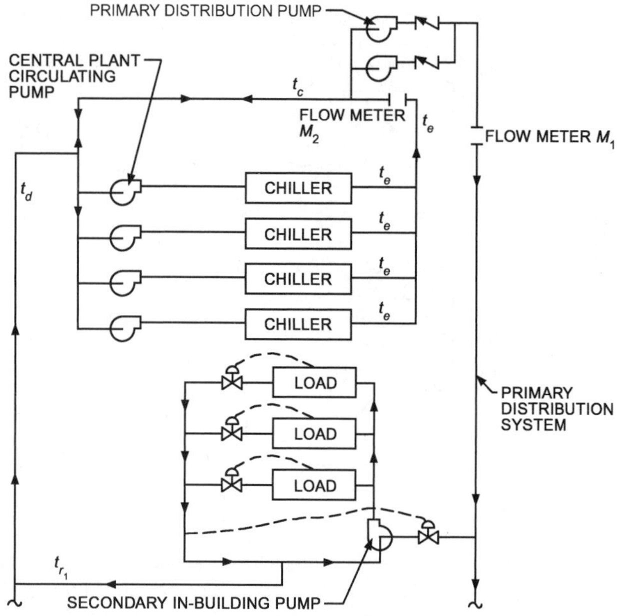

Fig. 4 Variable-Flow Primary/Secondary Systems

short-coupled (3000 ft or less). To implement this pumping method, the total flow must be pumped at a pressure sufficient to meet the requirements of the building with the largest pressure requirement. Consequently, all buildings on the system should have their pressure differentials monitored and transmitted to the central plant, where pump speeds are adjusted to provide adequate pressure to the building with the lowest margin of pressure differential. If the designer has control over the design of each in-building system, this pumping method can be achieved in a reasonable manner. In retrofit situations where existing buildings under different ownership are connected to a new central plant, coordination is difficult and individual building pumps are more practical.

When buildings have separate circulating pumps, hydraulic isolating piping and pumping design should be used to ensure that two-way control valves are subjected only to the differential pressure established by the in-building pump. Figure 4 shows a connection using in-building pumping with hydraulic isolation from the primary loop.

When in-building pumps are used, all series interconnections between the distribution system pump and the in-building pumps must be removed. Without adequate instrumentation and controls, a series connection can cause the distribution system return to operate at a higher pressure than the distribution system supply and disrupt flow to adjacent buildings. Series operation often occurs during improper use of three-way mixing valves in the distribution-to-building connection.

In very large systems, a design known as **distributed pumping** may be used. Under this approach, the distribution pumps in the central plant are eliminated. Instead, the distribution system pumping load is borne by the pumps in the user buildings. In cases where the distribution network piping constitutes a significant pressure loss (systems covering a large area), this design allows the distributed pumps in the buildings to be sized for just the pressure loss imposed at that particular location. Ottmer and Rishel (1993) found that this approach reduces total chilled-water pump power by 20 to 25% in very large systems. It is best applied in new construction where the central plant and distributed building systems can be planned fully and coordinated initially. Note that this system is not the best approach for a system that expects dynamic growth, such as a college and university campus, because the addition of a large load anywhere in the system affects the pressure drop of each distribution pump.

Usually, a positive pressure must be maintained at the highest point of the system at all times. This height determines the static pressure at which the expansion tank operates. Excessively tall buildings that do not isolate the in-building systems from the distribution system can impose unacceptable static pressure on the distribution system. To prevent excessive operating pressure in distribution systems, heat exchangers have been used to isolate the in-building system from the distribution system. To ensure reasonable temperature differentials between supply and return temperatures, flow must be controlled on the distribution system side of the heat exchanger.

In high-rise buildings, all piping, valves, coils, and other equipment may be required to withstand high pressure. Where system static pressure exceeds safe or economical operating pressure, either the heat exchanger method or pressure sustaining valves in the return line with check valves in the supply line may be used to minimize pressure. However, the pressure sustaining/check valve arrangement may overpressurize the entire distribution system if a malfunction of either valve occurs.

Design Guidelines

Guidelines for plant design and operation include the following:

- Variable-speed pumping saves energy and should be considered for distribution system pumping.
- Design chilled-water systems for a minimum temperature differential of 12 to 16°F. A 12 to 20°F maximum temperature differential with a 10 to 12°F minimum temperature differential can be achieved with this design.
- Limit the use of constant-flow systems to relatively small central chilled-water plants. Investigate chillers arranged in series.
- Larger central chilled-water plants can benefit from primary/secondary or primary/secondary/tertiary pumping with constant flow in central plant and variable flow in the distribution system. Size the distribution system for a low overall total pressure loss. Short-coupled distribution systems (3000 ft or less) can be used for a total pressure loss of 20 to 40 ft of water. With this maximum differential between any points in the system, size the distribution pumps to provide the necessary pressure to circulate chilled water through the in-building systems, eliminating the need for in-building pumping systems. This decreases the complexity of operating central chilled-water systems. Newer controls on chillers enable all-variable-flow systems. Minimum flows on chiller evaporators should be investigated with the manufacturer to achieve stable operation over all load ranges.
- All two-way valves must have proper close-off ratings and a design pressure drop of at least 20% of the maximum design pressure drop for controllability. Commercial quality automatic temperature control valves generally have low shutoff ratings; but industrial valves can achieve higher ratings. See Chapter 42 for more information on valves.
- The lower practical limit for chilled water supply temperatures is 39°F. Temperatures below that should be carefully analyzed, although systems with thermal energy storage may operate advantageously at lower temperatures.

DISTRIBUTION SYSTEM

HYDRAULIC CONSIDERATIONS

Objectives of Hydraulic Design

Although the distribution of a thermal utility such as hot water encompasses many of the aspects of domestic hot-water distribution, many dissimilarities also exist; thus the design should not be approached in the same manner. Thermal utilities must supply sufficient energy at the appropriate temperature and pressure to meet consumer needs. Within the constraints imposed by the consumer's end use and equipment, the required thermal energy can be delivered with various combinations of temperature and pressure. Computer-aided design methods are available for thermal piping networks (Bloomquist et al. 1999; COWIconsult 1985; Rasmussen and Lund 1987; Reisman 1985). Using these methods allows rapid evaluation of many alternative designs.

General steam system design can be found in Chapter 10, as well as in IDHA (1983). For water systems, consult Chapter 12 and IDHA (1983).

Water Hammer

The term water hammer is used to describe several phenomena that occur in fluid flow. Although these phenomena differ in nature, they all result in stresses in the piping that are higher than normally encountered. Water hammer can have a disastrous effect on a thermal utility by bursting pipes and fittings and threatening life and property.

In steam systems (IDHA 1983), water hammer is caused primarily by condensate collecting in the bottom of the steam piping. Steam flowing at velocities 10 times greater than normal water flow picks up a slug of condensate and accelerates it to a high velocity. The slug of condensate subsequently collides with the pipe wall at a point where flow changes direction. To prevent this type of water hammer, condensate must be prevented from collecting in steam pipes by using proper steam pipe pitch and adequate condensate collection and return facilities.

Water hammer also occurs in steam systems because of rapid condensation of steam during system warm-up. Rapid condensation decreases the specific volume and pressure of steam, which precipitates pressure shock waves. This form of water hammer is prevented by controlled warm-up of the piping. Valves should be opened slowly and in stages during warm-up. Large steam valves should be provided with smaller bypass valves to slow the warm-up.

Water hammer in hot- and chilled-water distribution systems is caused by sudden changes in flow velocity, which causes pressure shock waves. The two primary causes are pump failure and sudden valve closures. A simplified method to determine maximum resultant pressure may be found in Chapter 36 of the 2005 *ASHRAE Handbook—Fundamentals*. More elaborate methods of analysis can be found in Fox (1977), Stephenson (1981), and Streeter and Wylie (1979). Preventive measures include operational procedures and special piping fixtures such as surge columns.

Pressure Losses

Friction pressure losses occur at the interface between the inner wall of a pipe and a flowing fluid due to shear stresses. In steam systems, these pressure losses are compensated for with increased steam pressure at the point of steam generation. In water systems, pumps are used to increase pressure at either the plant or intermediate points in the distribution system. The calculation of pressure loss is discussed in Chapters 2 and 36 of the 2005 *ASHRAE Handbook—Fundamentals*.

Pipe Sizing

Ideally, the appropriate pipe size should be determined from an economic study of the life-cycle cost for construction and operation. In practice, however, this study is seldom done because of the effort involved. Instead, criteria that have evolved from practice are frequently used for design. These criteria normally take the form of constraints on the maximum flow velocity or pressure drop. Chapter 36 of the 2005 *ASHRAE Handbook—Fundamentals* provides velocity and pressure drop constraints. Noise generated by excessive flow velocities is usually not a concern for thermal utility distribution systems outside of buildings. For steam systems, maximum flow velocities of 200 to 250 fps are recommended (IDHA 1983). For water systems, Europeans use the criterion that pressure losses should be limited to 0.44 psi per 100 ft of pipe (Bøhm 1988). Other studies indicate that higher levels of pressure loss may be acceptable (Stewart and Dona 1987) and warranted from an economic standpoint (Bøhm 1986; Koskelainen 1980; Phetteplace 1989).

When establishing design flows for thermal distribution systems, the diversity of consumer demands should be considered (i.e., the various consumers' maximum demands do not occur at the same time). Thus, the heat supply and main distribution piping may be sized for a maximum load that is somewhat less than the sum of the individual consumers' maximum demands. For steam systems, Geiringer (1963) suggests diversity factors of 0.80 for space heating and 0.65 for domestic hot-water heating and process loads. Geiringer also suggests that these factors may be reduced by approximately 10% for high-temperature water systems. Werner (1984) conducted a study of the heat load on six operating low-temperature hot-water systems in Sweden and found diversity factors ranging from 0.57 to 0.79, with the average being 0.685.

Network Calculations

Calculating flow rates and pressures in a piping network with branches, loops, pumps, and heat exchangers can be difficult without the aid of a computer. Methods have been developed primarily for domestic water distribution systems (Jeppson 1977; Stephenson 1981). These may apply to thermal distribution systems with appropriate modifications. Computer-aided design methods usually incorporate methods for hydraulic analysis as well as for calculating heat losses and delivered water temperature at each consumer. Calculations are usually carried out in an iterative fashion, starting with constant supply and return temperatures throughout the network. After initial estimates of the design flow rates and heat losses are determined, refined estimates of the actual supply temperature at each consumer are computed. Flow rates at each consumer are then adjusted to ensure that the load is met with the reduced supply temperature, and the calculations are repeated.

Condensate Drainage and Return

Condensate forms in operating steam lines as a result of heat loss. When a steam system's operating temperature is increased, condensate also forms as steam warms the piping. At system start-up, these loads usually exceed any operating heat loss loads; thus, special provisions should be made.

To drain the condensate, steam piping should slope toward a collection point called a **drip station**. Drip stations are located in access areas or buildings where they are accessible for maintenance. Steam piping should slope toward the drip station at least 1 in. in 40 ft. If possible, the steam pipe should slope in the same direction as steam flow. If it is not possible to slope the steam pipe in the direction of steam flow, increase the pipe size to at least one size greater than would normally be used. This reduces the flow velocity of the steam and provides better condensate drainage against the steam flow. Drip stations should be spaced no further than 500 ft apart in the absence of other requirements.

Drip stations consist of a short piece of pipe (called a **drip leg**) positioned vertically on the bottom of the steam pipe, as well as a steam trap and appurtenant piping. The drip leg should be the same diameter as the steam pipe. The length of the drip leg should provide a volume equal to 50% of the condensate load from system start-up for steam pipes of 4 in. diameter and larger and 25% of the start-up condensate load for smaller steam pipes (IDHA 1983). Steam traps should be sized to meet the normal load from operational heat losses only. Start-up loads should be accommodated by manual operation of the bypass valve.

Steam traps are used to separate the condensate and noncondensable gases from the steam. For drip stations on steam distribution piping, use inverted bucket or bimetallic thermostatic traps. Some steam traps have integral strainers; others require separate strainers. Ensure that drip leg capacity is adequate when thermostatic traps are used because they will always accumulate some condensate.

If it is to be returned, condensate leaving the steam trap flows into the condensate return system. If steam pressure is sufficiently high, it may be used to force the condensate through the condensate return system. With low-pressure steam or on systems where a large pressure exists between drip stations and the ultimate destination of the condensate, condensate receivers and pumps must be provided.

Schedule 80 steel piping is recommended for condensate lines because of the extra allowance for corrosion its provides. Steam traps have the potential of failing in an open position, thus nonmetallic piping must be protected from live steam where its temperature/pressure would exceed the limitations of the piping. Nonmetallic piping should not be located so close to steam pipes that heat losses from the steam pipes could overheat it. Additional information on condensate removal may be found in Chapter 10. Information on sizing condensate return piping may be found in Chapter 36 of the 2005 *ASHRAE Handbook—Fundamentals*.

THERMAL CONSIDERATIONS

Thermal Design Conditions

Three thermal design conditions must be met to ensure satisfactory system performance:

1. The "normal" condition used for the life-cycle cost analysis determines appropriate insulation thickness. Average values for the temperatures, burial depth, and thermal properties of the materials are used for design. If the thermal properties of the insulating material are expected to degrade over the useful life of the system, appropriate allowances should be made in the cost analysis.
2. Maximum heat transfer rate determines the load on the central plant due to the distribution system. It also determines the temperature drop (or increase, in the case of chilled-water distribution), which determines the delivered temperature to the consumer. For this calculation, the thermal conductivity of each component must be taken at its maximum value, and the temperatures must be assumed to take on their extreme values, which would result in the greatest temperature difference between the carrier medium and the soil or air. The burial depth will normally be at its lowest value for this calculation.
3. During operation, none of the thermal capabilities of the materials (or any other materials in the area influenced thermally by the system) must exceed design conditions. To satisfy this objective, each component and the surrounding environment must be examined to determine whether thermal damage is possible. A heat transfer analysis may be necessary in some cases.

   The conditions of these analyses must be chosen to represent the worst-case scenario from the perspective of the component being examined. For example, in assessing the suitability of a coating material for a metallic conduit, the thermal insulation is assumed to be saturated, the soil moisture is at its lowest probable level, and the burial depth is maximum. These conditions, combined with the highest anticipated pipe and soil temperatures, give the highest conduit surface temperature to which the coating could be exposed.

Heat transfer in buried systems is influenced by the thermal conductivity of the soil and by the depth of burial, particularly when the insulation has low thermal resistance. Soil thermal conductivity changes significantly with moisture content; for example, Bottorf (1951) indicated that soil thermal conductivity ranges from 0.083 Btu/h·ft·°F during dry soil conditions to 1.25 Btu/h·ft·°F during wet soil conditions.

Thermal Properties of Pipe Insulation and Soil

Uncertainty in heat transfer calculations for thermal distribution systems results from uncertainty in the thermal properties of materials involved. Generally, the designer must rely on manufacturers' data to obtain approximate values. The data in this chapter should only be used as guidance in preliminary calculations until specific products have been identified; then specific data should be obtained from the manufacturer of the product in question.

Insulation. Insulation provides the primary thermal resistance against heat loss or gain in thermal distribution systems. Thermal properties and other characteristics of insulations normally used in thermal distribution systems are listed in Table 1. Material properties such as thermal conductivity, density, compressive strength, moisture absorption, dimensional stability, and combustibility are typically reported in ASTM standard for the respective material.

Table 1 Comparison of Commonly Used Insulations in Underground Piping Systems

	Calcium Silicate Type I/II ASTM C533	Urethane Foam	Cellular Glass ASTM C552	Mineral Fiber/ Preformed Glass Fiber Type 1 ASTM C547
Thermal conductivity[a] (Values in parenthesis are maximum permissible by ASTM standard listed), Btu/h·ft·°F				
Mean temp. = 100°F	0.028	0.013	0.033 (0.030)	0.022 (0.021)
200°F	0.031 (0.038/0.045)	0.014	0.039 (0.037)	0.025 (0.026)
300°F	0.034 (0.042/0.048)		0.046 (0.045)	0.028 (0.033)
400°F	0.038 (0.046/0.051)		0.053 (0.054)	(0.043)
Density (max.), lb/ft^3	15/22		6.7 to 9.2	8 to 11
Maximum temperature, °F	1200	250	800	850
Compressive strength (min),[b] kPa	700 at 5% deformation		450	N/A
Dimensional stability, linear shrinkage at maximum use temperature	2%		N/A	2%
Flame spread	0		5	25
Smoke index	0		0	50
Water absorption	As shipped moisture content, 20% max. (by weight)		0.5	Water vapor sorption, 5% max. (by weight)

[a]Thermal conductivity values in this table are from previous editions of this chapter and have been retained as they were used in examples. Thermal conductivity of insulation may vary with temperature, temperature gradient, moisture content, density, thickness, and shape. ASTM maximum values given are comparative for establishing quality control compliance, and are suggested for preliminary calculations where available. They may not represent installed performance of insulation under actual conditions that differ substantially from test conditions. The manufacturer should have design values.

[b]Compressive strength for cellular glass shown is for flat material, capped as per ASTM C240.

Table 2 Effect of Moisture on Underground Piping System Insulations

Characteristics	Polyurethane[a]	Cellular Glass	Mineral Wool[b]	Fibrous Glass
Heating Test	Pipe temp. 35°F to 260°F Water bath 46°F to 100°F	Pipe temp. 35°F to 420°F Water bath 46°F to 100°F	Pipe temp.35°F to 450°F Water bath 46°F to 100°F	Pipe temp. 35°F to 450°F Water bath 46°F to 100°F
Length of submersion time to reach steady-state *k*-value	70 days	See Note C	10 days	2 h
Effective *k*-value increase from dry conditions after steady state achieved in submersion	14 to 19 times at steady state. Estimated water content of insulation 70% (by volume).	Avg. 10 times, process unsteady (Note C). Insulation showed evidence of moisture zone on inner diameter.	Up to 50 times at steady state. Insulation completely saturated	52 to 185 times. Insulation completely saturated at steady state.
Primary heat transfer mechanism	Conduction	See Note C	Conduction and convection	Conduction and convection
Length of time for specimen to return to dry steady-state *k*-value after submersion	Pipe at 260°F, after 16 days moisture content 10% (by volume) remaining	Pipe at 420°F, 8 h	Pipe at 450°F, 9 days	Pipe at 380°F, 6 days
Cooling Test	Pipe temp. 37°F Water bath at 52°F	Pipe temp. 36°F Water bath 46°F to 58°F	Pipe temp. 35°F to 45°F Water bath 55°F	Insulation 35°F to 450°F Water bath 46°F to 100°F
Length of submersion time to reach steady-state conditions for *k*-value	16 days	Data recorded at 4 days constant at 12 days	6 days	1/2 h
Effective *k*-value increase from dry conditions after steady state achieved	2 to 4 times. Water absorption minimal, ceased after 7 days.	None. No water penetration.	14 times. Insulation completely saturated at steady state.	20 times. Insulation completely saturated at steady state.
Primary heat transfer mechanism	Conduction	Conduction	Conduction and convection	Conduction and convection
Length of time for specimen to return to dry steady-state *k*-value after submersion	Pipe at 37°F, data curve extrapolated to 10+ days	Pipe at 33°F, no change	Pipe at 35°F, data curve extrapolated to 25 days	Pipe at 35°F, 15 days

Source: Chyu et al. (1997a, 1997b; 1998a, 1998b).
[a]Polyurethane material tested had a density of 2.9 lb/ft^3.
[b]Mineral wool tested was a preformed molded basalt designed for pipe systems operating up to 1200°F. It was specially formulated to withstand the Federal Agency Committee 96 h boiling water test.
[c]Cracks formed on heating for all samples of cellular glass insulation tested. Flooded heat loss mechanism involved dynamic two-phase flow of water through cracks; the period of dynamic process was about 20 min. Cracks had negligible effect on the thermal conductivity of dry cellular glass insulation before and after submersion. No cracks formed during cooling test.

Table 3 Soil Thermal Conductivities

Soil Moisture Content (by mass)	Thermal Conductivity, Btu/h·ft·°F		
	Sand	Silt	Clay
Low, <4%	0.17	0.08	0.08
Medium, 4 to 20%	1.08	0.75	0.58
High, >20%	1.25	1.25	1.25

Some properties have more than one associated standard. For example, thermal conductivity for insulation material in block form may be reported using ASTM C177, C518, or C1114. Thermal conductivity for insulation material fabricated or molded for use on piping is reported using ASTM C335.

Chyu et al. (1997a, 1997b, 1998a, 1998b) studied the effect of moisture on the thermal conductivity of insulating materials commonly used in underground district energy systems (ASHRAE Research Project RP-721). The results are summarized in Table 2. The insulated pipe was immersed in water maintained at 46 to 100°F to simulate possible conduit water temperatures during a failure. The fluid temperature in the insulated pipe ranged from 35 to 450°F. All insulation materials were tested unfaced and/or unjacketed to simulate installation in a conduit.

Painting chilled-water piping before insulating is recommended in areas of high humidity. Insulations used today for chilled water include polyurethane and polyisocyanurate cellular plastics, phenolics, and fiberglass. With the exception of fiberglass, the rest can form acidic solutions (pH 2 to 3) once they hydrolyze in the presence of water. The acids emanate from the chlorides, sulfates, and halogens added during manufacturing to increase fire retardancy or expand the foam. Phenolics can be more than six times more corrosive than polyurethane because of acids used in their manufacture and can develop environments to pH 1.8. The easiest way to mitigate corrosion is to paint the pipe exterior with a strong rust-preventative coating (two-coat epoxy) before insulating. This is good engineering practice and most insulation manufacturers suggest this, but it may not be in their literature. In addition, a good vapor barrier is required.

Soil. If an analysis of the soil is available or can be done, the thermal conductivity of the soil can be estimated from published data (e.g., Farouki 1981; Lunardini 1981). The thermal conductivity factors in Table 3 may be used as an estimate when detailed information on the soil is not known. Because dry soil is rare in most areas, a low moisture content should be assumed only for system material design, or where it can be validated for calculation of heat losses in the normal operational condition. Values of 0.8 to 1 Btu/h·ft·°F are commonly used where soil moisture content is unknown. Because moisture will migrate toward a chilled pipe, use a thermal conductivity value of 1.25 Btu/h·ft·°F for chilled-water systems in the absence of any site-specific soil data. For steady-state analyses, only the thermal conductivity of the soil is required. If a transient analysis is required, the specific heat and density are also required.

METHODS OF HEAT TRANSFER ANALYSIS

Because heat transfer in piping is not related to the load factor, it can be a large part of the total load. The most important factors affecting heat transfer are the difference between earth and fluid temperatures and the thermal insulation. For example, the extremes might be a 6 in., insulated, 400°F water line in 40°F earth with 100 to 200 Btu/h·ft loss; and a 6 in., uninsulated, 55°F chilled-water return in 60°F earth with 10 Btu/h·ft gain. The former requires analysis to determine the required insulation and its effect on the total heating system; the latter suggests analysis and insulation needs might be minimal. Other factors that affect heat transfer are (1) depth of burial, related to the earth temperature and soil thermal resistance; (2) soil thermal conductivity, related to soil moisture content and density; and (3) distance between adjacent pipes.

To compute transient heat gains or losses in underground piping systems, numerical methods that approximate any physical problem and include such factors as the effect of temperature on thermal properties must be used. For most designs, numerical analyses may not be warranted, except where the potential exists to thermally damage something adjacent to the distribution system. Also, complex geometries may require numerical analysis. Albert and Phetteplace (1983), Minkowycz et al. (1988), and Rao (1982) have further information on numerical methods.

Steady-state calculations are appropriate for determining the annual heat loss/gain from a buried system if average annual earth temperatures are used. Steady-state calculations may also be appropriate for worst-case analyses of thermal effects on materials. Steady-state calculations for a one-pipe system may be done without a computer, but it becomes increasingly difficult for a two-, three-, or four-pipe system.

The following steady-state methods of analysis use resistance formulations developed by Phetteplace and Meyer (1990) that simplify the calculations needed to determine temperatures within the system. Each type of resistance is given a unique subscript and is defined only when introduced. In each case, the resistances are on a unit length basis so that heat flows per unit length result directly when the temperature difference is divided by the resistance. *Note*: For consistency and simplicity, all thermal conductivities are given in Btu/h·ft·°F with all dimensions in feet (not the more traditional Btu·in/h·ft²·°F).

Calculation of Undisturbed Soil Temperatures

Before any heat loss/gain calculations may be conducted, the undisturbed soil temperature at the site must be determined. The choice of soil temperature is guided primarily by the type of calculation being conducted; see the section on Thermal Design Considerations. For example, if the purpose of the calculation is to determine if a material will exceed its temperature limit, the maximum expected undisturbed ground temperature is used. The appropriate choice of undisturbed soil temperature also depends on the location of the site, time of year, depth of burial, and thermal properties of the soil. Some methods for determining undisturbed soil temperatures and suggestions on appropriate circumstances to use them are as follows:

1. Use the average annual air temperature to approximate the average annual soil temperature. This estimate is appropriate when the objective of the calculation is to yield the average heat loss over the yearly weather cycle. Mean annual air temperatures may be obtained from various sources of climatic data (e.g., CRREL 1999).
2. Use the maximum/minimum air temperature as an estimate of the maximum/minimum undisturbed soil temperature for pipes buried at a shallow depth. This approximation is an appropriate conservative assumption when checking the temperatures to determine if the temperature limits of any of materials proposed for use will be exceeded. Maximum and minimum expected air temperatures may be calculated from the information found in Chapter 28 of the 2005 *ASHRAE Handbook—Fundamentals*.
3. For systems that are buried at other than shallow depths, maximum/minimum undisturbed soil temperatures may be estimated as a function of depth, soil thermal properties, and prevailing climate. This estimate is appropriate when checking the temperatures in a system to determine if the temperature limits of any of the materials proposed for use will be exceeded. The following equations may be used to estimate the minimum and maximum expected undisturbed soil temperatures:

$$\text{Maximum temperature} = t_{s,z} = t_{ms} + A_s e^{-z\sqrt{\pi/\alpha\tau}} \quad (1)$$

$$\text{Minimum temperature} = t_{s,z} = t_{ms} - A_s e^{-z\sqrt{\pi/\alpha\tau}} \quad (2)$$

where

$t_{s,z}$ = temperature, °F
z = depth, ft
τ = annual period, 365 days
α = thermal diffusivity of the ground, ft²/day
t_{ms} = mean annual surface temperature, °F
A_s = surface temperature amplitude, °F

CRREL (1999) lists values for the climatic constants t_{ms} and A_s for various regions of the United States.

Thermal diffusivity for soil may be calculated as follows:

$$\alpha = \frac{24k_s}{\rho_s[c_s + c_w(w/100)]} = \frac{24k_s}{\rho_s[c_s + (w/100)]} \quad (3)$$

where

ρ_s = soil density, lb/ft³
c_s = dry soil specific heat, Btu/lb·°F
c_w = specific heat of water = 1.0 Btu/lb·°F
w = moisture content of soil, % (dry basis)
k_s = soil thermal conductivity, Btu/h·ft·°F

Because the specific heat of dry soil is nearly constant for all types of soil, c_s may be taken as 0.175 Btu/lb·°F.

4. For buried systems, the undisturbed soil temperatures may be estimated for any time of the year as a function of depth, soil thermal properties, and prevailing climate. This temperature may be used in lieu of the soil surface temperature normally called for by the steady-state heat transfer equations [i.e., Equations (6) and (7)] when estimates of heat loss/gain as a function of time of year are desired. The substitution of the undisturbed soil temperatures at the pipe depth allows the steady-state equations to be used as a first approximation to the solution to the actual transient heat transfer problem with its annual temperature variations at the surface. The following equation may be used to estimate the undisturbed soil temperature at any depth at any point during the yearly weather cycle (ASCE 1996). (*Note*: The argument for the sine function is in radians.)

$$t_{s,z} = t_{ms} + A_s e^{-z\sqrt{\pi/\alpha\tau}} \sin\left[\frac{2\pi(\theta - \theta_{lag})}{\tau} - z\sqrt{\frac{\pi}{\alpha\tau}}\right] \quad (4)$$

where

θ = Julian date, days
θ_{lag} = phase lag of soil surface temperature, days

CRREL (1999) lists values for the climatic constants t_{ms}, A_s, and θ_{lag} for various regions of the United States. Equation (3) may be used to calculate soil thermal diffusivity.

Equation (4) does not account for latent heat effects from freezing, thawing, or evaporation. However, for soil adjacent to a buried heat distribution system, the equation provides a good estimate, because heat losses from the system tend to prevent the adjacent ground from freezing. For buried chilled-water systems, freezing may be a consideration; therefore, systems that are not used or drained during the winter months should be buried below the seasonal frost depth. For simplicity, the ground surface temperature is assumed to equal the air temperature, which is an acceptable assumption for most design calculations. If a more accurate calculation is desired, the following method may be used to compensate for the convective thermal resistance to heat transfer at the ground surface.

Convective Heat Transfer at Ground Surface

Heat transfer between the ground surface and the ambient air occurs by convection. In addition, heat transfer with the soil occurs due to precipitation and radiation. The heat balance at the ground surface is too complex to warrant detailed treatment in the design of buried district heating and cooling systems. However, McCabe et al. (1995) observed significant temperature variations caused by the type of surfaces over district heating and cooling systems.

As a first approximation, an **effectiveness thickness** of a fictitious soil layer may be added to the burial depth to account for the effect of the convective heat transfer resistance at the ground surface. The effective thickness is calculated as follows:

$$\delta = k_s/h \tag{5}$$

where

δ = effectiveness thickness of fictitious soil layer, ft
k_s = thermal conductivity of soil, Btu/h·ft·°F
h = convective heat transfer coefficient at ground surface, Btu/h·ft²·°F

The effective thickness calculated with Equation (5) is simply added to the actual burial depth of the pipes in calculating the soil thermal resistance using Equations (6), (7), (20), (21), and (27).

Single Uninsulated Buried Pipe

For this case (Figure 5), an estimate for soil thermal resistance may be used. This estimate is sufficiently accurate (within 1%) for the ratios of burial depth to pipe radius indicated next to Equations (6) and (7). Both the actual resistance and the approximate resistance are presented, along with the depth/radius criteria for each.

$$R_s = \frac{\ln\{(d/r_o) + [(d/r_o)^2 - 1]^{1/2}\}}{2\pi k_s} \quad \text{for } d/r_o > 2 \tag{6}$$

$$R_s = \frac{\ln(2d/r_o)}{2\pi k_s} \quad \text{for } d/r_o > 4 \tag{7}$$

where

R_s = thermal resistance of soil, h·ft·°F/Btu
k_s = thermal conductivity of soil, Btu/h·ft·°F
d = burial depth to centerline of pipe, ft
r_o = outer radius of pipe or conduit, ft

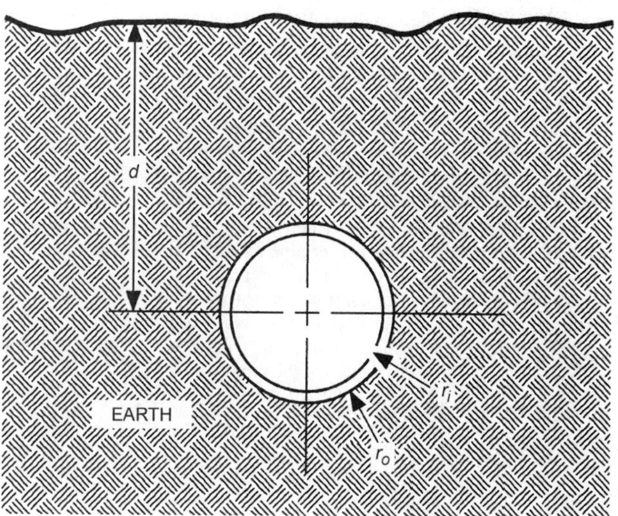

Fig. 5 Single Uninsulated Buried Pipe

The thermal resistance of the pipe is included if it is significant when compared to the soil resistance. The thermal resistance of a pipe or any concentric circular region is given by

$$R_p = \frac{\ln(r_o/r_i)}{2\pi k_p} \tag{8}$$

where

R_p = thermal resistance of pipe wall, h·ft·°F/Btu
k_p = thermal conductivity of pipe, Btu/h·ft·°F
r_i = inner radius of pipe, ft

Example 1. Consider an uninsulated, 3 in. Schedule 40 PVC chilled-water supply line carrying 45°F water. Assume the pipe is buried 3 ft deep in soil with a thermal conductivity of 1 Btu/h·ft·°F, and no other pipes or thermal anomalies are within close proximity. Assume the average annual soil temperature is 60°F.

Solution: Calculate the thermal resistance of the pipe using Equation (8):

$$R_p = 0.21 \text{ h} \cdot \text{ft} \cdot °\text{F/Btu}$$

Calculate the thermal resistance of the soil using Equation (7). [*Note*: $d/r_o = 21$ is greater than 4; thus Equation (7) may be used in lieu of Equation (6).]

$$R_s = 0.59 \text{ h} \cdot \text{ft} \cdot °\text{F/Btu}$$

Calculate the rate of heat transfer by dividing the overall temperature difference by the total thermal resistance:

$$q = \frac{t_f - t_s}{R_t} = \frac{(45 - 60)}{0.80 \text{ h} \cdot \text{ft} \cdot °\text{F/Btu}} = -19 \text{ Btu/h} \cdot \text{ft}$$

where

R_t = total thermal resistance, (i.e., $R_s + R_p$ in this case of pure series heat flow), h·ft·°F/Btu
t_f = fluid temperature, °F
t_s = average annual soil temperature, °F
q = heat loss or gain per unit length of system, Btu/h·ft

The negative result indicates a heat gain rather than a loss. Note that the thermal resistance of the fluid/pipe interface has been neglected, which is a reasonable assumption because such resistances tend to be very small for flowing fluids. Also note that, in this case, the thermal resistance of the pipe comprises a significant portion of the total thermal resistance. This results from the relatively low thermal conductivity of PVC compared to other piping materials and the fact that no other major thermal resistances exist in the system to overshadow it. If any significant amount of insulation were included in the system, its thermal resistance would dominate, and it might be possible to neglect that of the piping material.

Single Buried Insulated Pipe

Equation (8) can be used to calculate the thermal resistance of any concentric circular region of material, including an insulation layer (Figure 6). When making calculations using insulation thickness, actual thickness rather than nominal thickness should be used to obtain the most accurate results.

Example 2. Consider the effect of adding 1 in. of urethane foam insulation and a 1/8 in. thick PVC jacket to the chilled-water line in Example 1. Calculate the thermal resistance of the insulation layer from Equation (8) as follows:

$$R_i = \frac{\ln(0.229/0.146)}{2\pi \times 0.0125} = 5.75 \text{ h} \cdot \text{ft} \cdot °\text{F/Btu}$$

For the PVC jacket material, use Equation (8) again:

$$R_j = \frac{\ln(0.240/0.229)}{2\pi \times 0.10} = 0.07 \text{ h} \cdot \text{ft} \cdot {}^\circ\text{F/Btu}$$

The thermal resistance of the soil as calculated by Equation (7) decreases slightly to R_s = 0.51 h·ft·°F/Btu because of the increase in the outer radius of the piping system. The total thermal resistance is now

$$R_t = R_p + R_i + R_j + R_s = 0.21 + 5.75 + 0.07 + 0.51$$
$$= 6.54 \text{ h} \cdot \text{ft} \cdot {}^\circ\text{F/Btu}$$

The heat gain by the chilled-water pipe is reduced to about 2 Btu/h·ft. In this case, the thermal resistance of the piping material and the jacket material could be neglected with a resultant error of <5%. Considering that the uncertainties in the material properties are likely greater than 5%, it is usually appropriate to neglect minor resistances such as those of piping and jacket materials if insulation is present.

Single Buried Pipe in Conduit with Air Space

Systems with air spaces may be treated by adding an appropriate resistance for the air space. For simplicity, assume a heat transfer coefficient of 3 Btu/h·ft²·°F (based on the outer surface area of the insulation), which applies in most cases. The resistance caused by this heat transfer coefficient is then

$$R_a = 1/(3 \times 2\pi \times r_{oi}) = 0.053/r_{oi} \tag{9}$$

where

r_{oi} = outer radius of insulation, ft
R_a = resistance of air space, h·ft·°F/Btu

A more precise value for the resistance of an air space can be developed with empirical relations available for convection in enclosures such as those given by Grober et al. (1961). The effect of radiation in the annulus should also be considered when high temperatures are expected within the air space. For the treatment of radiation, refer to Siegel and Howell (1981).

Example 3. Consider a 6 in. nominal diameter (6.625 in. outer diameter) high-temperature water line operating at 375°F. Assume the pipe is insulated with 2.5 in. of mineral wool with a thermal conductivity k_i = 0.026 Btu/h·ft·°F at 200°F and k_i = 0.030 Btu/h·ft·°F at 300°F.

The pipe will be encased in a steel conduit with a concentric air gap of 1 in. The steel conduit will be 0.125 in. thick and will have a corrosion resistant coating approximately 0.125 in. thick. The pipe will be buried 4 ft deep to pipe centerline in soil with an average annual temperature of 60°F. The soil thermal conductivity is assumed to be 1 Btu/h·ft·°F. The thermal resistances of the pipe, conduit, and conduit coating will be neglected.

Solution: Calculate the thermal resistance of the pipe insulation. To do so, assume a mean temperature of the insulation of 250°F to establish its thermal conductivity, which is equivalent to assuming the insulation outer surface temperature is 125°F. Interpolating the data listed previously, the insulation thermal conductivity k_i = 0.028 Btu/h·ft·°F. Then calculate insulation thermal resistance using Equation (8):

$$R_i = \frac{\ln(0.484/0.276)}{2\pi \times 0.028} = 3.19 \text{ h} \cdot \text{ft} \cdot {}^\circ\text{F/Btu}$$

Calculate the thermal resistance of the air space using Equation (9):

$$R_a = 0.053/0.484 = 0.11 \text{ h} \cdot \text{ft} \cdot {}^\circ\text{F/Btu}$$

Calculate the thermal resistance of the soil from Equation (7):

$$R_s = \frac{\ln(8.0/0.589)}{2\pi \times 1} = 0.42 \text{ h} \cdot \text{ft} \cdot {}^\circ\text{F/Btu}$$

The total thermal resistance is

$$R_t = R_i + R_a + R_s = 3.19 + 0.11 + 0.42 = 3.72 \text{ h} \cdot \text{ft} \cdot {}^\circ\text{F/Btu}$$

The first estimate of the heat loss is then

$$q = (375 - 60)/3.72 = 84.7 \text{ Btu/h} \cdot \text{ft}$$

Now repeat the calculation with an improved estimate of the mean insulation temperature obtained, and calculate the outer surface temperature as

$$t_{io} = t_{po} - (qR_i) = 375 - (84.7 \times 3.19) = 105^\circ\text{F}$$

where t_{po} = outer surface temperature of pipe, °F.

The new estimate of the mean insulation temperature is 240°F, which is close to the original estimate. Thus, the insulation thermal conductivity changes only slightly, and the resulting thermal resistance is R_i = 3.24 h·ft·°F/Btu. The other thermal resistances in the system remain unchanged, and the heat loss becomes q = 83.8 Btu/h·ft. The insulation surface temperature is now approximately t_{io} = 104°F, and no further calculations are needed.

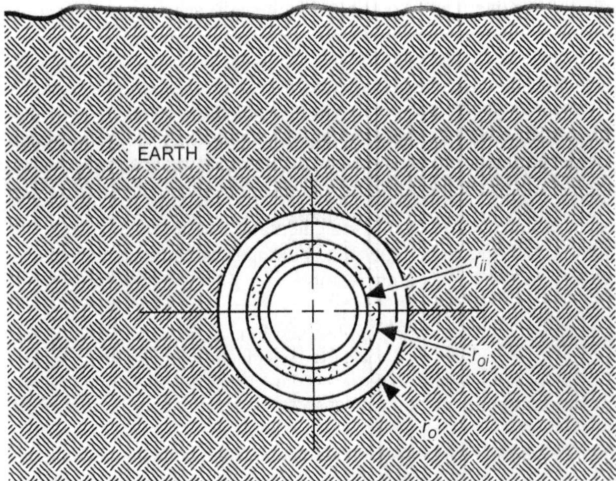

Fig. 6 Single Insulated Buried Pipe

Single Buried Pipe with Composite Insulation

Many systems are available that use more than one insulating material. The motivation for doing so is usually to use an insulation that has desirable thermal properties or lower cost, but it is unable to withstand the service temperature of the carrier pipe (e.g., polyurethane foam). Another insulation with acceptable service temperature limits (e.g., calcium silicate or mineral wool) is normally placed adjacent to the carrier pipe in sufficient thickness to reduce the temperature at its outer surface to below the limit of the insulation that is desirable to use (e.g., urethane foam). The calculation of the heat loss and/or temperature in a composite system is straightforward using the equations presented previously.

Example 4. A high-temperature water line operating at 400°F is to be installed in southern Texas. It consists of a 6 in. nominal diameter (6.625 in. outer diameter) carrier pipe insulated with 1.5 in. of calcium silicate. In addition, 1 in. polyurethane foam insulation will be placed around the calcium silicate insulation. The polyurethane insulation will be encased in a 0.5 in. thick fiberglass-reinforced plastic (FRP) jacket. The piping system will be buried 10 ft deep to the pipe centerline. Neglect the thermal resistances of the pipe and FRP jacket.

Calcium silicate thermal conductivity
0.038 Btu/h·ft·°F at 200°F
0.042 Btu/h·ft·°F at 300°F
0.046 Btu/h·ft·°F at 400°F
Polyurethane foam thermal conductivity
0.013 Btu/h·ft·°F at 100°F
0.014 Btu/h·ft·°F at 200°F

0.015 Btu/h·ft·°F at 275°F

Assumed soil properties

Thermal conductivity = 0.2 Btu/h·ft·°F

Density (dry soil) = 105 lb/ft³

Moisture content = 5% (dry basis)

What is the maximum operating temperature of the polyurethane foam insulation?

Solution: Because the maximum operating temperature of the materials is sought, the maximum expected soil temperature rather than the average annual soil temperature must be found. Also, the lowest anticipated soil thermal conductivity and the deepest burial depth are assumed. These conditions produce the maximum internal temperatures of the components.

To solve the problem, first calculate the maximum soil temperature expected at the installation depth using Equation (1). CRREL (1999) shows the values for climatic constants for this region as t_{ms} = 71.8°F and A_s = 15.0°F. Soil thermal diffusivity may be estimated using Equation (3).

$$\alpha = \frac{24 \times 0.2}{105[0.175 + 1.0(5/100)]} = 0.20 \text{ ft}^2/\text{day}$$

Then Equation (1) is used to calculate the maximum soil temperature at the installation depth as follows:

$$t_{s,z} = 71.8 + 15.0 \exp\left[-10\sqrt{\frac{\pi}{0.20 \times 365}}\right] = 73.7°\text{F}$$

Now calculate the first estimates of the thermal resistances of the pipe insulations. For the calcium silicate, assume a mean temperature of the insulation of 300°F to establish its thermal conductivity. From the data listed previously, the calcium silicate thermal conductivity k_i = 0.042 Btu/h·ft·°F at this temperature. For the polyurethane foam, assume the mean temperature of the insulation is 200°F. From the data listed previously, the polyurethane foam thermal conductivity k_i = 0.014 Btu/h·ft·°F at this temperature. Now the insulation thermal resistances are calculated using Equation (8):

$$\text{Calcium silicate} \quad R_{i,1} = \frac{\ln(0.401/0.276)}{2\pi(0.042)} = 1.42 \text{ h·ft·°F/Btu}$$

$$\text{Polyurethane foam} \quad R_{i,2} = \frac{\ln(0.484/0.401)}{2\pi(0.014)} = 2.14 \text{ h·ft·°F/Btu}$$

Calculate the thermal resistance of the soil from Equation (7):

$$R_s = \frac{\ln(2 \times 10/0.526)}{2\pi(0.2)} = 2.89 \text{ h·ft·°F/Btu}$$

The total thermal resistance is

$$R_t = 1.42 + 2.14 + 2.89 = 6.45 \text{ h·ft·°F/Btu}$$

The first estimate of the heat loss is then

$$q = (400 - 73.7)/6.45 = 50.6 \text{ Btu/h·ft}$$

Next, calculate the estimated insulation surface temperature with this first estimate of the heat loss. Find the temperature at the interface between the calcium silicate and polyurethane foam insulations.

$$t_{io,1} = t_{po} - qR_{i,1} = 400 - (50.6 \times 1.42) = 328°\text{F}$$

where t_{po} is the outer surface temperature of the pipe and $t_{io,1}$ is the outer surface temperature of first insulation (calcium silicate).

$$t_{io,2} = t_{po} - q(R_{i,1} + R_{i,2}) = 400 - 50.6(1.42 + 2.14) = 220°\text{F}$$

where $t_{io,2}$ is the outer surface temperature of second insulation (polyurethane foam).

The new estimate of the mean insulation temperature is (400 + 328)/2 = 364°F for the calcium silicate and (328 + 220)/2 = 274°F for the polyurethane foam. Thus, the insulation thermal conductivity for the calcium silicate would be interpolated to be 0.045 Btu/h·ft·°F, and the resulting thermal resistance is $R_{i,1}$ = 1.32 h·ft·°F/Btu.

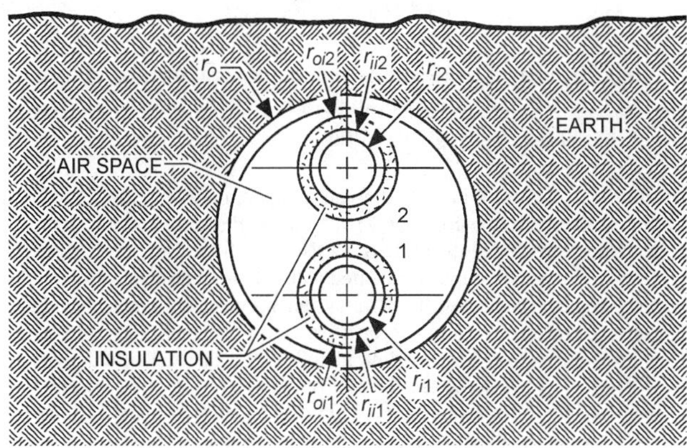

Fig. 7 Two Pipes Buried in Common Conduit with Air Space

For the polyurethane foam insulation, the thermal conductivity is interpolated to be 0.015 Btu/h·ft·°F, and the resulting thermal resistance is $R_{i,2}$ = 2.00 h·ft·°F/Btu. The soil thermal resistance remains unchanged, and the heat loss is recalculated as q = 52.5 Btu/h·ft.

The calcium silicate insulation outer surface temperature is now approximately $t_{io,1}$ = 331°F, and the outer surface temperature of the polyurethane foam is calculated to be $t_{io,2}$ = 226°F. Because these temperatures are within a few degrees of those calculated previously, no further calculations are needed.

The maximum temperature of the polyurethane insulation of 331°F occurs at its inner surface (i.e., the interface with the calcium silicate insulation). This temperature clearly exceeds the maximum accepted 30-year service temperature of polyurethane foam of 250°F (EuHP 1991). Thus, the amount of calcium silicate insulation needs to be increased significantly to achieve a maximum temperature for the polyurethane foam insulation within its long term service temperature limit. Under the conditions of this example, it would take about 5 in. of calcium silicate insulation to reduce the insulation interface temperature to less than 250°F.

Two Pipes Buried in Common Conduit with Air Space

For this case (Figure 7), make the same assumption as made in the previous section, Single Buried Pipe in Conduit with Air Space. For convenience, add some of the thermal resistances as follows:

$$R_1 = R_{p1} + R_{i1} + R_{a1} \tag{10}$$

$$R_2 = R_{p2} + R_{i2} + R_{a2} \tag{11}$$

Subscripts 1 and 2 differentiate between the two pipes within the conduit. The combined heat loss is then given by

$$q = \frac{[(t_{f1} - t_s)/R_1] + [(t_{f2} - t_s)/R_2]}{1 + (R_{cs}/R_1) + (R_{cs}/R_2)} \tag{12}$$

where R_{cs} is the total thermal resistance of conduit shell and soil.

Once the combined heat flow is determined, calculate the bulk temperature in the air space:

$$t_a = t_s + qR_{cs} \tag{13}$$

Then calculate the insulation outer surface temperature:

$$t_{i1} = t_a + (t_{f1} - t_a)(R_{a1}/R_1) \tag{14}$$

$$t_{i2} = t_a + (t_{f2} - t_a)(R_{a2}/R_2) \tag{15}$$

The heat flows from each pipe are given by

$$q_1 = (t_{f1} - t_a)/R_1 \tag{16}$$

$$q_2 = (t_{f2} - t_a)/R_2 \tag{17}$$

When the insulation thermal conductivity is a function of its temperature, as is usually the case, an iterative calculation is required, as illustrated in the following example.

Example 5. A pair of 4 in. NPS medium-temperature hot-water supply and return lines run in a common 21 in. outside diameter conduit. Assume that the supply temperature is 325°F and the return temperature is 225°F. The supply pipe is insulated with 2.5 in. of mineral wool insulation, and the return pipe has 2 in. of mineral wool insulation. This insulation has the same thermal properties as those given in Example 3. The pipe is buried 4 ft to centerline in soil with a thermal conductivity of 1 Btu/h·ft·°F. Assume the thermal resistance of the pipe, the conduit, and the conduit coating are negligible. As a first estimate, assume the bulk air temperature within the conduit is 100°F. In addition, use this temperature as a first estimate of the insulation surface temperatures to obtain the mean insulation temperatures and subsequent insulation thermal conductivities.

Solution: By interpolation, estimate the insulation thermal conductivities from the data given in Example 3 at 0.0265 Btu/h·ft·°F for the supply pipe and 0.0245 Btu/h·ft·°F for the return pipe. Calculate the thermal resistances using Equations (10) and (11):

$$R_1 = R_{i1} + R_{a1} = \ln(0.396/0.188)/(2\pi \times 0.0265) + 0.053/0.396$$
$$= 4.48 + 0.13 = 4.61 \text{ h} \cdot \text{ft} \cdot °\text{F/Btu}$$

$$R_2 = R_{i2} + R_{a2} = \ln(0.354/0.188)/(2\pi \times 0.0245) + 0.053/0.354$$
$$= 4.11 + 0.15 = 4.26 \text{ h} \cdot \text{ft} \cdot °\text{F/Btu}$$

$$R_{cs} = R_s = \ln(8/0.875)/(2\pi \times 1) = 0.352 \text{ h} \cdot \text{ft} \cdot °\text{F/Btu}$$

Calculate first estimate of combined heat flow from Equation (12):

$$q = \frac{(325 - 60)/4.61 + (225 - 60)/4.26}{1 + (0.352/4.61) + (0.352/4.26)} = 83.0 \text{ Btu/h} \cdot \text{ft}$$

Estimate bulk air temperature in the conduit with Equation (13):

$$t_a = 60 + (83.0 \times 0.352) = 89.2°\text{F}$$

Then revise estimates of the insulation surface temperatures with Equations (14) and (15):

$$t_{i1} = 89.2 + (325 - 89.2)(0.13/4.61) = 95.8°\text{F}$$
$$t_{i2} = 89.2 + (225 - 89.2)(0.15/4.26) = 94.0°\text{F}$$

These insulation surface temperatures are close enough to the original estimate of 100°F that further iterations are not warranted. If the individual supply and return heat losses are desired, calculate them using Equations (16) and (17).

Two Buried Pipes or Conduits

This case (Figure 8) may be formulated in terms of the thermal resistances used for a single buried pipe or conduit and some correction factors. The correction factors needed are

$$\theta_1 = (t_{p2} - t_s)/(t_{p1} - t_s) \tag{18}$$

$$\theta_2 = 1/\theta_1 = (t_{p1} - t_s)/(t_{p2} - t_s) \tag{19}$$

$$P_1 = \frac{1}{2\pi k_s} \ln\left[\frac{(d_1 + d_2)^2 + a^2}{(d_1 - d_2)^2 + a^2}\right]^{0.5} \tag{20}$$

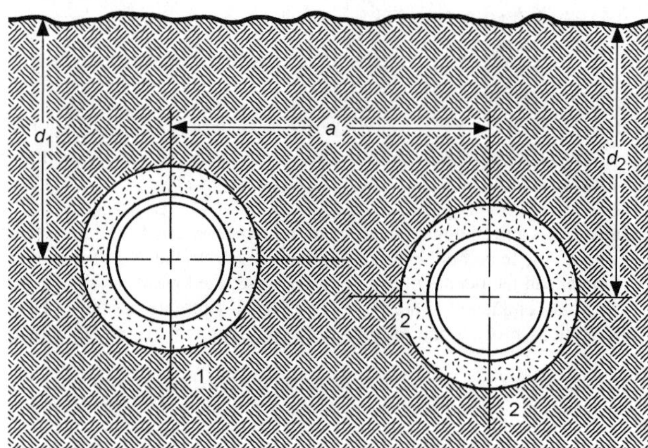

Fig. 8 Two Buried Pipes or Conduits

$$P_2 = \frac{1}{2\pi k_s} \ln\left[\frac{(d_2 + d_1)^2 + a^2}{(d_2 - d_1)^2 + a^2}\right]^{0.5} \tag{21}$$

where a = horizontal separation distance between centerline of two pipes, ft.

The thermal resistance for each pipe or conduit is given by

$$R_{e1} = \frac{R_{t1} - (P_1^2/R_{t2})}{1 - (P_1\theta_1/R_{t2})} \tag{22}$$

$$R_{e2} = \frac{R_{t2} - (P_2^2/R_{t1})}{1 - (P_2\theta_2/R_{t1})} \tag{23}$$

where

θ = temperature correction factor, dimensionless
P = geometric/material correction factor, h·ft·°F/Btu
R_e = effective thermal resistance of one pipe/conduit in two-pipe system, h·ft·°F/Btu
R_t = total thermal resistance of one pipe/conduit if buried separately, h·ft·°F/Btu

Heat flow from each pipe is then calculated from

$$q_1 = (t_{p1} - t_s)/R_{e1} \tag{24}$$

$$q_2 = (t_{p2} - t_s)/R_{e2} \tag{25}$$

Example 6. Consider buried supply and return lines for a low-temperature hot-water system. The carrier pipes are 4 in. NPS (4.5 in. outer diameter) with 1.5 in. of urethane foam insulation. The insulation is protected by a 0.25 in. thick PVC jacket. The thermal conductivity of the insulation is 0.013 Btu/h·ft·°F and is assumed constant with respect to temperature. The pipes are buried 4 ft deep to the centerline in soil with a thermal conductivity of 1 Btu/h·ft·°F and a mean annual temperature of 60°F. The horizontal distance between the pipe centerlines is 2 ft. The supply water is at 250°F, and the return water is at 150°F.

Solution: Neglect the thermal resistances of the carrier pipes and the PVC jacket. First, calculate the resistances from Equations (7) and (8) as if the pipes were independent of each other:

$$R_{s1} = R_{s2} = \frac{\ln(8.0/0.333)}{2\pi \times 1.0} = 0.51 \text{ h} \cdot \text{ft} \cdot °\text{F/Btu}$$

$$R_{i1} = R_{i2} = \frac{\ln(0.313/0.188)}{2\pi \times 0.013} = 6.25 \text{ h} \cdot \text{ft} \cdot °\text{F/Btu}$$

$$R_{t1} = R_{t2} = 0.51 + 6.25 = 6.76 \text{ h} \cdot \text{ft} \cdot °\text{F/Btu}$$

From Equations (20) and (21), the correction factors are

$$P_1 = P_2 = \frac{1}{2\pi \times 1} \ln\left[\frac{(4+4)^2 + 2^2}{(4-4)^2 + 2^2}\right]^{0.5} = 0.225 \text{ h} \cdot \text{ft} \cdot {}^\circ\text{F/Btu}$$

$$\theta_1 = (150-60)/(250-60) = 0.474$$

$$\theta_2 = 1/\theta_1 = 2.11$$

Calculate the effective total thermal resistances as

$$R_{e1} = \frac{6.76 - (0.225^2/6.76)}{1 - (0.225 \times 0.474/6.76)} = 6.87 \text{ h} \cdot \text{ft} \cdot {}^\circ\text{F/Btu}$$

$$R_{e2} = \frac{6.76 - (0.225^2/6.76)}{1 - (0.225 \times 2.11/6.76)} = 7.32 \text{ h} \cdot \text{ft} \cdot {}^\circ\text{F/Btu}$$

The heat flows are then

$$q_1 = (250-60)/6.87 = 27.7 \text{ Btu/h} \cdot \text{ft}$$

$$q_2 = (150-60)/7.32 = 12.3 \text{ Btu/h} \cdot \text{ft}$$

$$q_t = 27.7 + 12.3 = 40.0 \text{ Btu/h} \cdot \text{ft}$$

Note that when the resistances and geometry for the two pipes are identical, the total heat flow from the two pipes is the same if the temperature corrections are used or if they are set to unity. The individual losses will vary somewhat, however. These equations may also be used with air space systems. When the thermal conductivity of the pipe insulation is a function of temperature, iterative calculations must be done.

Pipes in Buried Trenches or Tunnels

Buried rectangular trenches or tunnels (Figure 9) require several assumptions to obtain approximate solutions for the heat transfer. First, assume that the air within the tunnel or trench is uniform in temperature and that the same is true for the inside surface of the trench/tunnel walls. Field measurements by Phetteplace et al. (1991) on operating shallow trenches indicate maximum spatial air temperature variations of about 10°F. Air temperature variations of this magnitude within a tunnel or trench do not cause significant errors for normal operating temperatures when using the following calculation methods.

Unless numerical methods are used, an approximation must be made for the resistance of a rectangular region such as the walls of a trench or tunnel. One procedure is to assume linear heat flow through the trench or tunnel walls, which yields the following resistance for the walls (Phetteplace et al. 1981):

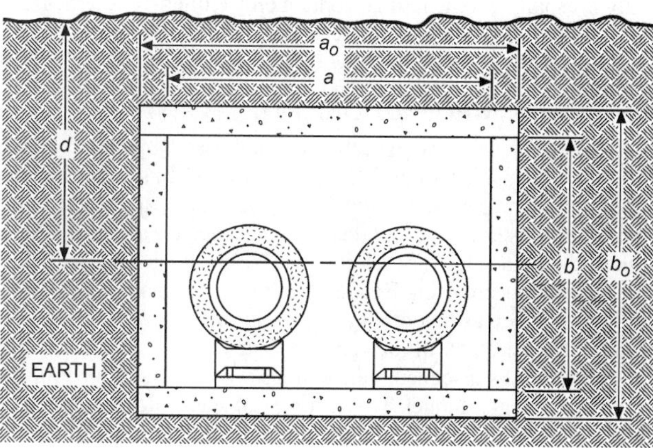

Fig. 9 Pipes in Buried Trenches or Tunnels

$$R_w = \frac{x_w}{2k_w(a+b)} \tag{26}$$

where

R_w = thermal resistance of trench/tunnel walls, h·ft·°F/Btu
x_w = thickness of trench/tunnel walls, ft
a = width of trench/tunnel inside, ft
b = height of trench/tunnel inside, ft
k_w = thermal conductivity of trench/tunnel wall material, Btu/h·ft·°F

As an alternative to Equation (26), the thickness of the trench/tunnel walls may be included in the soil burial depth. This approximation is only acceptable when the thermal conductivity of the trench/tunnel wall material is similar to that of the soil.

The thermal resistance of the soil surrounding the buried trench/tunnel is calculated using the following equation (Rohsenow 1998):

$$R_{ts} = \frac{\ln[3.5d/(b_o^{0.75} a_o^{0.25})]}{k_s[(a_o/2b_o) + 5.7]} \qquad a_o > b_o \tag{27}$$

where

R_{ts} = thermal resistance of soil surrounding trench/tunnel, h·ft·°F/Btu
a_o = width of trench/tunnel outside, ft
b_o = height of trench/tunnel outside, ft
d = burial depth of trench to centerline, ft

Equations (26) and (27) can be combined with the equations already presented to calculate heat flow and temperature for trenches/tunnels. As with the conduits described in earlier examples, the heat transfer processes in the air space inside the trench/tunnel are too complex to warrant a complete treatment for design purposes. The thermal resistance of this air space may be approximated by several methods. For example, Equation (9) may be used to calculate an approximate resistance for the air space.

Thermal resistances at the pipe insulation/air interface can also be calculated from heat transfer coefficients as done in the section on Pipes in Air. If the thermal resistance of the air/trench wall interface is also included, use Equation (28):

$$R_{aw} = 1/[2h_t(a+b)] \tag{28}$$

where

R_{aw} = thermal resistance of air/trench wall interface, h·ft·°F/Btu
h_t = total heat transfer coefficient at air/trench wall interface, Btu/h·ft²·°F

The total heat loss from the trench/tunnel is calculated from the following relationship:

$$q = \frac{(t_{p1} - t_s)/R_1 + (t_{p2} - t_s)/R_2}{1 + (R_{ss}/R_1) + (R_{ss}/R_2)} \tag{29}$$

where

R_1, R_2 = thermal resistances of two-pipe/insulation systems within trench/tunnel, h·ft·°F/Btu
R_{ss} = total thermal resistance on soil side of air within trench/tunnel, h·ft·°F/Btu

Once the total heat loss has been found, the air temperature within the trench/tunnel may be found as

$$t_{ta} = t_s + qR_{ss} \tag{30}$$

where t_{ta} is the air temperature within trench/tunnel.

The individual heat flows for the two pipes within the trench/tunnel are then

$$q_1 = (t_{p1} - t_{ta})/R_1 \tag{31}$$

$$q_2 = (t_{p2} - t_{ta})/R_2 \quad (32)$$

If the thermal conductivity of the pipe insulation is a function of temperature, assume an air temperature for the air space before starting calculations. Iterate the calculations if the air temperature calculated with Equation (30) differs significantly from the initial assumption.

Example 7. The walls of a buried trench are 6 in. thick, and the trench is 3 ft wide and 2 ft tall. The trench is constructed of concrete, with a thermal conductivity of k_w = 1 Btu/h·ft·°F. The soil surrounding the trench also has a thermal conductivity of k_s = 1 Btu/h·ft·°F. The centerline of the trench is 4 ft below grade, and the soil temperature is assumed be 60°F. The trench contains supply and return lines for a medium-temperature water system with the physical and operating parameters identical to those in Example 5.

Solution: Assuming the air temperature within the trench is 100°F, the thermal resistances for the pipe/insulation systems are identical to those in Example 5, or

$$R_1 = 4.61 \text{ h} \cdot \text{ft} \cdot °\text{F/Btu}$$

$$R_2 = 4.26 \text{ h} \cdot \text{ft} \cdot °\text{F/Btu}$$

The thermal resistance of the soil surrounding the trench is given by Equation (27):

$$R_{ts} = \frac{\ln[(3.5 \times 4)/(3^{0.75} \times 4^{0.25})]}{1[(4/6) + 5.7]} = 0.231 \text{ h} \cdot \text{ft} \cdot °\text{F/Btu}$$

The trench wall thermal resistance is calculated with Equation (26):

$$R_w = 0.5/[2(3+2)] = 0.050 \text{ h} \cdot \text{ft} \cdot °\text{F/Btu}$$

If the thermal resistance of the air/trench wall is neglected, the total thermal resistance on the soil side of the air space is

$$R_{ss} = R_w + R_{ts} = 0.050 + 0.231 = 0.281 \text{ h} \cdot \text{ft} \cdot °\text{F/Btu}$$

Find a first estimate of the total heat loss using Equation (29):

$$q = \frac{(325-60)/4.61 + (225-60)/4.26}{1 + (0.281/4.61) + (0.281/4.26)} = 85.4 \text{ Btu/h} \cdot \text{ft}$$

The first estimate of the air temperature in the trench is given by Equation (30):

$$t_{ta} = 60 + (85.4 \times 0.281) = 84.0°\text{F}$$

Refined estimates of the pipe insulation surface temperatures are then calculated using Equations (14) and (15):

$$t_{i1} = 84.0 + [(325 - 84.0)(0.13/4.61)] = 90.8°\text{F}$$

$$t_{i2} = 84.0 + [(225 - 84.0)(0.15/4.26)] = 89.0°\text{F}$$

From these estimates, calculate the revised mean insulation temperatures to find the resultant resistance values. Repeat the calculation procedure until satisfactory agreement between successive estimates of the trench air temperature is obtained. Calculate the individual heat flows from the pipes with Equations (31) and (32).

If the thermal resistance of the trench walls is added to the soil thermal resistance, the thermal resistance on the soil side of the air space is

$$R_{ss} = \frac{\ln[14/(2^{0.75} \times 3^{0.25})]}{1[(3/4) + 5.7]} = 0.286 \text{ h} \cdot \text{ft} \cdot °\text{F/Btu}$$

The result is less than 2% higher than the resistance previously calculated by treating the trench walls and soil separately. In the event that the soil and trench wall material have significantly different thermal conductivities, this simpler calculation will not yield as favorable results and should not be used.

Pipes in Shallow Trenches

The cover of a shallow trench is exposed to the environment. Thermal calculations for such trenches require the following assumptions: (1) the interior air temperature is uniform as discussed in the section on Pipes in Buried Trenches or Tunnels, and (2) the soil and the trench wall material have the same thermal conductivity. This assumption yields reasonable results if the thermal conductivity of the trench material is used, because most of the heat flows directly through the cover. The thermal resistance of the trench walls and surrounding soil is usually a small portion of the total thermal resistance, and thus the heat losses are not usually highly dependent on this thermal resistance. Using these assumptions, Equations (27) and (29) to (32) may be used for shallow trench systems.

Example 8. Consider a shallow trench having the same physical parameters and operating conditions as the buried trench in Example 7, except that the top of the trench is at grade level. Calculate the thermal resistance of the shallow trench using Equation (27):

$$R_{ts} = R_{ss} = \frac{\ln[(3.5 \times 1.5)/(2^{0.75} \times 3^{0.25})]}{1[(3/4) + 5.7]} = 0.134 \text{ h} \cdot \text{ft} \cdot °\text{F/Btu}$$

Use thermal resistances for the pipe/insulation systems from Example 7, and use Equation (24) to calculate q:

$$q = \frac{(325-60)/4.61 + (225-60)/4.26}{1 + (0.134/4.61) + (0.134/4.26)} = 90.7 \text{ Btu/h} \cdot \text{ft}$$

From this, calculate the first estimate of the air temperature, using Equation (30):

$$t_{ta} = 60 + (90.7 \times 0.134) = 72.2°\text{F}$$

Then refine estimates of the pipe insulation surface temperatures using Equations (14) and (15):

$$t_{i1} = 72.2 + [(325 - 72.2)(0.13/4.61)] = 79.3°\text{F}$$

$$t_{i2} = 72.2 + [(225 - 72.2)(0.15/4.26)] = 77.6°\text{F}$$

From these, calculate the revised mean insulation temperatures to find resultant resistance values. Repeat the calculation procedure until satisfactory agreement between successive estimates of the trench air temperatures are obtained. If the individual heat flows from the pipes are desired, calculate them using Equations (31) and (32).

Another method for calculating the heat losses in a shallow trench assumes an interior air temperature and treats the pipes as pipes in air (see the section on Pipes in Air). Interior air temperatures in the range of 70 to 120°F have been observed in a temperate climate (Phetteplace et al. 1991).

It may be necessary to drain the system or provide heat tracing in areas with significant subfreezing air temperatures in winter, especially for shallow trenches that contain only chilled-water lines (and thus have no source of heat) if the system is not in operation, or is in operation at very low flow rates.

Buried Pipes with Other Geometries

Other geometries not specifically addressed by the previous cases have been used for buried thermal utilities. In some instances, the equations presented previously may be used to approximate the system. For instance, the soil thermal resistance for a buried system with a half-round clay tile on a concrete base could be approximated as a circular system using Equation (6) or (7). In this case, the outer radius r_o is taken as that of a cylinder with the same circumference as the outer perimeter of the clay tile system. The remainder of the resistances and subsequent calculations would be similar to those for a buried trench/tunnel. The accuracy of such calculations varies inversely with the proportion of the total thermal resistance that the thermal resistance in question comprises. In most instances, the thermal resistance of the pipe insulation overshadows other resistances, and the errors induced by approximations in the other resistances are acceptable for design calculations.

Pipes in Air

Pipes surrounded by gases may transfer heat via conduction, convection, and/or thermal radiation. Heat transfer modes depend mainly on the surface temperatures and geometry of the system being considered. For air, conduction is usually dominated by the other modes and thus may be neglected. The thermal resistances for cylindrical systems can be found from heat transfer coefficients derived using the equations that follow. Generally, the piping systems used for thermal distribution have sufficiently low surface temperatures to preclude any significant heat transfer by thermal radiation.

Example 9. Consider a 6 in. nominal (6.625 in. outside diameter), high-temperature hot-water pipe that operates in air at 375°F with 2.5 in. of mineral wool insulation. The surrounding air annually averages 60°F (520°R). The average annual wind speed is 4 mph. The insulation is covered with an aluminum jacket with an emittance ε = 0.26. The thickness and thermal resistance of the jacket material are negligible. Because the heat transfer coefficient at the outer surface of the insulation is a function of the temperature there, initial estimates of this temperature must be made. This temperature estimate is also required to estimate mean insulation temperature.

Solution: Assuming that the insulation surface is at 100°F (560°R) as a first estimate, the mean insulation temperature is calculated as 238°F. Using the properties of mineral wool given in Example 3, the insulation thermal conductivity is k_i = 0.0275 Btu/h·ft·°F. Using Equation (8), the thermal resistance of the insulation is R_i = 3.25 h·ft·°F/Btu. The forced convective heat transfer coefficient at the surface of the insulation can be found using the following equation (ASTM 1995):

$$h_{cv} = 1.016\left(\frac{1}{d}\right)^{0.2}\left(\frac{2}{t_{oi}+t_a}\right)^{0.181}(t_{oi}-t_a)^{0.266}(1+1.277V)^{0.5}$$

$$= 1.016\left(\frac{1}{11.625}\right)^{0.2}\left(\frac{2}{160}\right)^{0.181}(40)^{0.266}(6.11)^{0.5}$$

$$= 1.86 \text{ Btu/h} \cdot \text{ft}^2 \cdot {}^\circ\text{F}$$

where

d = outer diameter of surface, in.
t_a = ambient air temperature, °F
V = wind speed, mph

The radiative heat transfer coefficient must be added to this convective heat transfer coefficient. Determine the radiative heat transfer coefficient as follows (ASTM 1995):

$$h_{rad} = \varepsilon\sigma\frac{(T_a^4 - T_s^4)}{T_a - T_s}$$

$$h_{rad} = 0.26 \times 1.713 \times 10^{-9}\frac{(560^4 - 520^4)}{560-520} = 0.28 \text{ Btu/h·ft}^2\text{·°F}$$

Add the convective and radiative coefficients to obtain a total surface heat transfer coefficient h_t of 2.14 Btu/h·ft²·°F. The equivalent thermal resistance of this heat transfer coefficient is calculated from the following equation:

$$R_{surf} = \frac{1}{2\pi r_{oi} h_t} = \frac{1}{2\pi \times 0.484 \times 2.14} = 0.15 \text{ h·ft·°F/Btu}$$

With this, the total thermal resistance of the system becomes R_t = 3.40 h·ft·°F/Btu, and the first estimate of the heat loss is q = 92.6 Btu/h·ft.

An improved estimate of the insulation surface temperature is t_{oi} = 375 – (92.6 × 3.25) = 74°F. From this, a new mean insulation temperature, insulation thermal resistance, and surface resistance can be calculated. The heat loss is then 90.0 Btu/h·ft, and the insulation surface temperature is calculated as 77°F. These results are close enough to the previous results that further iterations are not warranted.

Note that the contribution of thermal radiation to the heat transfer could have been omitted with negligible effect on the results. In fact, the entire surface resistance could have been neglected and the resulting heat loss would have increased by only about 4%.

In Example 9, the convective heat transfer was forced. In cases with no wind, where the convection is free rather than forced, the radiative heat transfer is more significant, as is the total thermal resistance of the surface. However, in instances where the piping is well insulated, the thermal resistance of the insulation dominates, and minor resistances can often be neglected with little resultant error. By neglecting resistances, a conservative result is obtained (i.e., the heat transfer is overpredicted).

Economical Thickness for Pipe Insulation

A life-cycle cost analysis may be run to determine the economical thickness of pipe insulation. Because the insulation thickness affects other parameters in some systems, each insulation thickness must be considered as a separate system. For example, a conduit system or one with a jacket around the insulation requires a larger conduit or jacket for greater insulation thicknesses. The cost of the extra conduit or jacket material may exceed that of the additional insulation and is therefore usually included in the analysis. It is usually not necessary to include excavation, installation, and backfill costs in the analysis.

The life-cycle cost of a system is the sum of the initial capital cost and the present worth of the subsequent cost of heat lost or gained over the life of the system. The initial capital cost needs only to include those costs that are affected by insulation thickness. The following equation can be used to calculate the life-cycle cost:

$$\text{LCC} = \text{CC} + (q t_u C_h \text{PWF}) \tag{33}$$

where

LCC = present worth of life-cycle costs associated with pipe insulation thickness, $/ft
CC = capital costs associated with pipe insulation thickness, $/ft
q = annual average rate of heat loss, Btu/h·ft
t_u = utilization time for system each year, h
C_h = cost of heat lost from system, $/Btu
PWF = present worth factor for future annual heat loss costs, dimensionless

The present worth factor is the reciprocal of the capital recovery factor, which is found from the following equation:

$$\text{CRF} = \frac{i(1+i)^n}{(1+i)^n - 1} \tag{34}$$

where

CRF = cost recovery factor, dimensionless
i = interest rate
n = useful lifetime of system, years

If heat costs are expected to escalate, the present worth factor may be multiplied by an appropriate escalation factor and the result substituted in place of PWF in Equation (33).

Example 10. Consider a steel conduit system with an air space. The insulation is mineral wool with thermal conductivity as given in Examples 3 and 4. The carrier pipe is 4 in. NPS and operates at 350°F for the entire year (8760 h). The conduit is buried 4 ft to the centerline in soil with a thermal conductivity of 1 Btu/h·ft·°F and an annual average temperature of 60°F. Neglect the thermal resistance of the conduit and carrier pipe. The useful lifetime of the system is assumed to be 20 years and the interest rate is taken as 10%.

Solution: Find CRF from Equation (34):

$$\mathrm{CRF} = \frac{0.10(1+0.10)^{20}}{(1+0.10)^{20}-1} = 0.11746$$

The value C_h of heat lost from the system is assumed to be $10 per million Btu. The following table summarizes the heat loss and cost data for several available insulation thicknesses.

Insulation thickness, in.	1.5	2.0	2.5	3.0	3.5	4.0
Insulation outer radius, ft	0.313	0.354	0.396	0.438	0.479	0.521
Insulation k, Btu/h·ft·°F	0.027	0.027	0.027	0.027	0.027	0.027
R_i, Eq. (8), h·ft·°F/Btu	3.00	3.73	4.39	4.99	5.52	6.00
R_a, Eq. (9), h·ft·°F/Btu	0.17	0.15	0.13	0.12	0.11	0.10
Conduit outer radius, ft	0.448	0.448	0.531	0.531	0.583	0.667
R_s, Eq. (7), h·ft·°F/Btu	0.46	0.46	0.43	0.43	0.42	0.40
R_t, h·ft·°F/Btu	3.63	4.34	4.95	5.54	6.04	6.50
q, heat loss rate, Btu/h·ft	79.9	66.8	58.6	52.3	48.0	44.6
Conduit system cost, $/ft	23.00	24.50	28.25	30.00	33.00	40.00
LCC, Eq. (33), $/ft	82.59	74.32	71.95	69.00	68.80	73.27

The table indicates that 3.5 in. of insulation yields the lowest life-cycle cost for the example. Because the results depend highly on the economic parameters used, they must be accurately determined.

EXPANSION PROVISIONS

All piping moves because of temperature changes, whether it contains chilled water, steam, or hot water. The piping's length increases or decreases with its temperature. Field conditions and the type of system govern the method used to absorb the movement. Turns where the pipe changes direction must be used to provide flexibility. When the distance between changes in direction becomes too large for the turns to compensate for movement, expansion loops are positioned at appropriate locations. If loops are required, additional right-of-way may be required. If field conditions allow, the flexibility of the piping should be used to allow expansion. Where space constraints do not allow expansion loops and/or changes in direction, mechanical methods, such as expansion joints or ball joints, must be used. However, because ball joints change the direction of a pipe, a third joint may be required to reduce the length between changes in direction.

Chapter 45 covers the design of the pipe bends, loops, and the use of expansion joints. However, the chapter uses conservative stress values. Computer-aided programs that calculate stress from pressure, thermal expansion, and weight simultaneously allow the designer to meet the requirements of ASME *Standard* B31.1. When larger pipe diameters are required, a computer program should be used when the pipe will provide the required flexibility. For example, Table 11 in Chapter 45 indicates that a 12 in. standard weight pipe with 12 in. of movement requires a 15.5 ft wide by 31 ft high expansion loop. One computer-aided program recommends a 13 ft wide by 26 ft high loop; and if equal height and width is specified, the loop is 23.5 ft in each direction.

Although the inherent flexibility of the piping should be used to handle expansion as much as possible, expansion joints must be used where space is too small to allow a loop to be constructed to handle the required movement. For example, expansion joints are often used in walk-through tunnels because there is seldom space to construct pipe loops. Either pipe loops or expansion joints can be used for aboveground, concrete shallow trench, direct-buried, and poured-envelope systems. The manufacturer of the conduit or envelope material should design loops and offsets in conduit and poured envelope systems because clearance and design features are critical to the performance of both the loop and the pipe.

All expansion joints require maintenance, and should therefore always be accessible for service. Joints in direct-buried and poured-envelope systems and trenches without removable covers should be located in access ports.

Cold springing is normally used when thermal expansion compensation is used. In DHC systems with natural flexibility, cold springing minimizes the clearance required for pipe movement only. The pipes are sprung 50% of the total amount of movement, in the direction toward the anchor. However, ASME *Standard* B31.1 does not allow cold springing in calculating the stresses in the piping. When expansion joints are used, they are installed in an extended position to achieve maximum movement. However, the manufacturer of the expansion joint should be contacted for the proper amount of extension.

In extremely hot climates, anchors may also be required, to compensate for pipe contraction when pipes are installed in high ambient temperatures and then filled with cold water. This can affect buried tees in the piping, especially at branch service line runouts to buildings. Crushable insulation may be used in the trench as part of the backfill, to compensate for the contractions. Anchors should be sized using computer-aided design software.

Pipe Supports, Guides, and Anchors

For conduit and poured-envelope systems, the system manufacturer usually designs the pipe supports, guides, and anchors in consultation with the expansion joint manufacturer, if such devices are used. For example, the main anchor force of an in-line axial expansion joint is the sum of the pressure thrust (system pressure times the cross-sectional area of the expansion joint and the joint friction or spring force) and the pipe friction forces. The manufacturer of the expansion device should be consulted when determining anchor forces. Anchor forces are normally less when expansion is absorbed through the system instead of with expansion joints.

Pipe guides used with expansion joints should be spaced according to the manufacturer's recommendations. They must allow longitudinal or axial motion and restrict motion perpendicular to the axis of the pipe. Guides with graphite or low-friction fluorocarbon slide surfaces are often desirable for long pipe runs. In addition, these surface finishes do not corrode or increase sliding resistance in aboveground installations. Guides should be selected to handle twice the expected movement, so they may be installed in a neutral position without the need for cold-springing the pipe.

DISTRIBUTION SYSTEM CONSTRUCTION

The combination of aesthetics, first cost, safety, and life-cycle cost naturally divide distribution systems into two distinct categories: aboveground and underground distribution systems. The materials needed to ensure long life and low heat loss further classify DHC systems into low-temperature, medium-temperature, and high-temperature systems. The temperature range for medium-temperature systems is usually too high for the materials that are used in low-temperature systems; however, the same materials that are used in high-temperature systems are typically used for medium-temperature systems. Because low-temperature systems have a lower temperature differential between the working fluid temperatures and the environment, heat loss is inherently less. In addition, the selection of efficient insulation materials and inexpensive pipe materials that resist corrosion is much greater for low-temperature systems.

The aboveground system has the lowest first cost and the lowest life-cycle cost because it can be maintained easily and constructed with materials that are readily available. Generally, aboveground systems are acceptable where they are hidden from view or can be hidden by landscaping. Poor aesthetics and the risk of vehicle damage to the aboveground system remove it from contention for many projects.

Although the aboveground system is sometimes partially factory prefabricated, more typically it is entirely field fabricated of components such as pipes, insulation, pipe supports, and insulation jackets or protective enclosures that are commercially available. Other common systems that are completely field fabricated include the walk-through tunnel (Figure 10), the concrete surface trench (Figure 11), the deep-burial small tunnel (Figure 12), and underground systems that use poured insulation (Figure 13) or cellular glass (Figure 14) to form an envelope around the carrier pipes.

Field-assembled systems must be designed in detail, and all materials must be specified by the project design engineer. Evaluation of the project site conditions indicates which type of system should be considered for the site. For instance, the shallow trench system is best where utilities that are buried deeper than the trench bottom need to be avoided and where the covers can serve as sidewalks. Direct-buried conduit, with a thicker steel casing, may be the only system that can be used in flooded sites. The conduit system is used where aesthetics is important. It is often used for short distances between buildings and the main distribution system, and where the owner is willing to accept higher life-cycle costs.

Direct-buried conduit, concrete surface trench, and other underground systems must be routed to avoid existing utilities, which requires a detailed site survey and considerable design effort. In the absence of a detailed soil temperature distribution study, direct-buried heating systems should be spaced more than 15 ft from other utilities constructed of plastic pipes because the temperature of the soil during dry conditions can be high enough to reduce the strength of plastic pipe to an unacceptable level. Rigid, extruded polystyrene insulation may be used to insulate adjacent utilities from the impact of a buried heat distribution pipe; however, the temperature limit of the extruded polystyrene insulation must not be exceeded. A numerical analysis of the thermal problem may be required to ensure that the desired effect is achieved.

Tunnels that provide walk-through or crawl-through access can be buried in nearly any location without causing future problems because utilities are typically placed in the tunnel. Regardless of the type of construction, it is usually cost effective to route distribution piping through the basements of buildings, but only after liability issues are addressed. In laying out the main supply and return piping, redundancy of supply and return should be considered. If a looped system is used to provide redundancy, flow rates under all possible failure modes must be addressed when sizing and laying out the piping.

Access ports for underground systems should be at critical points, such as where there are

- High or low points on the system profile that vent trapped air or where the system can be drained
- Elevation changes in the distribution system that are needed to maintain the required constant slope
- Major branches with isolation valves
- Steam traps and condensate drainage points on steam lines
- Mechanical expansion devices

To facilitate leak location and repair and to limit damage caused by leaks, access points generally should be spaced no farther than 500 ft apart. Special attention must be given to the safety of personnel who come in contact with distributions systems or who must enter spaces occupied by underground systems. The regulatory authority's definition of a confined space and the possibility of exposure to high-temperature or high-pressure piping can have a significant impact on the access design, which must be addressed by the designer. Gravity venting of tunnels is good practice, and access ports and tunnels should have lighting and convenience outlets to aid in inspection and maintenance of anchors, expansion joints, and piping.

Piping Materials and Standards

Supply Pipes for Steam and Hot Water. Adequate temperature and pressure ratings for the intended service should be specified for any piping. All piping, fittings, and accessories should be in accordance with ASME *Standard* B31.1 or with local requirements if more stringent. For steam and hot water, all joints should be welded and pipe should conform to either ASTM A53 seamless or ERW, Grade B; or ASTM A106 seamless, Grade B. Care should be taken to exclude ASTM A53, Type F, because of its lower allowable stress and because of the method by which its seams are manufactured. Mechanical joints of any type are not recommended for steam or hot water. Pipe wall thickness is determined by the maximum operating temperature and pressure. In the United States, most piping for steam and hot water is Schedule 40 for 10 in. NPS and below and standard weight for 12 in. NPS and above.

Many European low-temperature water systems have piping with a wall thickness similar to Schedule 10. The reduced piping material in these systems means they are not only less expensive, but they also develop reduced expansion forces and thus require simpler methods of expansion compensation, a point to be considered when choosing between a high-temperature and a low-temperature system. Welding pipes with thinner walls requires extra care and may require additional inspection. Also, extra care must be taken to avoid internal and external corrosion because the thinner wall provides a much lower corrosion allowance.

Condensate Return Pipes. Condensate pipes require special consideration because condensate is much more corrosive than steam. This corrosive nature is caused by the oxygen that the condensate accumulates. The usual method used to compensate for condensate corrosiveness is to select steel pipe that is thicker than the steam pipe. For highly corrosive condensate, stainless steel and/or other corrosion resistant materials should be considered. Materials that are corrosion resistant in air may not be corrosion resistant when exposed to condensate; therefore, a material with good experience handling condensate should be selected. Fiberglass-reinforced plastic (FRP) or glass-reinforced plastic (GRP) pipe used for condensate return has not performed well. Failed steam traps, pipe resin solubility, deterioration at elevated temperatures, and thermal expansion are thought to be the cause of premature failure.

Chilled-Water Distribution. For chilled-water systems, a variety of pipe materials, such as steel, ductile iron, HDPE, PVC, and FRP/GRP, have been used successfully. If ductile iron or steel is used, the designer must resolve the internal and external corrosion issue, which may be significant unless cement-lined ductile iron is used. The soil temperature is usually highest when the chilled-water loads are highest; therefore, it is usually life-cycle cost-effective to insulate the chilled-water pipe. A life-cycle cost analysis often favors a factory prefabricated product that is insulated with a plastic foam with a waterproof casing or a field-fabricated system that is insulated with ASTM C552 cellular glass insulation. If a plastic product is selected, care must be taken to maintain an adequate distance from any high-temperature underground system that may be near. Damage to the chilled-water system would most likely occur from elevated soil temperatures when the chilled-water circulation stopped. The heating distribution system should be at least 15 ft from a chilled-water system containing plastic unless a detailed study of the soil temperature distribution indicates otherwise. Rigid extruded polystyrene insulation may be used to insulate adjacent chilled-water lines from the impacts of a buried heat distribution pipe; however, care must be taken not to exceed the temperature limits of the extruded polystyrene insulation; numerical analysis of the thermal problem may be required. Finally, when transitioning from ductile or plastic piping to steel at the buildings, flanged connections are usually best, but should be located inside the building and not buried. Proper gasket selection and bolt torque are also critical.

Aboveground Systems

An aboveground system consists of a distribution pipe, insulation that surrounds the pipe, and a protective jacket that surrounds the insulation. The jacket may have an integral vapor retarder. When the distribution system carries chilled water or other cold media, a vapor retarder is required for all types of insulation except cellular glass. In an ASHRAE test by Chyu et al. (1998a), cellular glass absorbed essentially zero water in a chilled-water application. In heating applications, the vapor retarder is not needed nor recommended; however, a reasonably watertight jacket is required to keep storm water out of the insulation. The jacket material can be aluminum, stainless steel, galvanized steel, or plastic sheet; a multilayered fabric and organic cement composite; or a combination of these. Plastics and organic cements exposed to sunshine must be ultraviolet-light resistant.

Structural columns and supports are typically made of wood, steel, or concrete. A crossbar is often placed across the top of the column when more than one distribution pipe is supported from one column. Sidewalk and road crossings require an elaborate support structure to elevate the distribution piping above traffic. Pipe expansion and contraction is taken up in loops, elbows, and bends. Manufactured expansion joints may be used, but they are usually not recommended because of a shorter life or a higher frequency of required maintenance than the rest of the system. Supports that attach the distribution pipes to the support columns are commercially available as described in MSS *Standards* SP-58 and SP-69. The distribution pipes should have welded joints.

An aboveground system has the lowest first cost and is the easiest to inspect and maintain; therefore, it has the lowest life-cycle cost. It is the standard against which all other systems are compared. Its major drawbacks are its poor aesthetics, its safety hazard if struck by vehicles and equipment, and its susceptibility to freezing in cold climates if circulation is stopped or if heat is not added to the working medium. These drawbacks often remove this system from contention as a viable alternative.

Underground Systems

An underground system solves the problems of aesthetics and exposure to vehicles of the aboveground system; however, burying a system causes other problems with materials, design, construction, and maintenance that have historically been difficult to solve. An underground heat distribution system is not a typical utility like gas, domestic water, and sanitary systems. It requires an order of magnitude more design effort and construction inspection accuracy when compared to gas, water, and sanitary distribution projects. The thermal effects and difficulty of keeping the insulation dry make it much more difficult to design and construct when compared to systems operating near the ambient temperature. Underground systems cost almost 10 times as much to build, and require much more to operate and maintain. Heat distribution systems must be designed for zero leakage and must account for thermal expansion, degradation of material as a function of temperature, high pressure and transient shock waves, heat loss restrictions, and accelerated corrosion. In the past, resolving one problem in underground systems often created a new, more serious problem that was not recognized until premature failure occurred. Segan and Chen (1984) describe the types of premature failures that may occur if this guidance is ignored.

Common types of underground systems are the walk-through tunnel, concrete surface trench, deep-buried small tunnel, poured insulation envelope, cellular glass, and conduit system.

Walk-Through Tunnel. This system (Figure 10) consists of a field-erected tunnel large enough for a someone to walk through after the distribution pipes are in place. It is essentially an aboveground system enclosed with a tunnel. The tunnel is buried deep enough to cover the top with earth, and is large enough for routine maintenance and inspection to be done easily without excavation. The preferred construction material for the tunnel walls and top cover is reinforced concrete. Masonry units and metal preformed sections have been used to construct the tunnel and top with less success, because of groundwater leakage and metal corrosion. The distribution pipes are supported from the tunnel wall or floor with pipe supports that are commonly used on aboveground systems or in buildings. Some groundwater will penetrate the top and walls of the tunnel; therefore, a water drainage system must be provided. Usually, electric lights and electric service outlets are provided for ease of inspection and maintenance. This system has the highest first cost of all underground systems; however, it can have the lowest life-cycle cost because of its ease of maintenance, the ability to correct construction errors easily, and an extremely long life.

Shallow Concrete Surface Trench. This system (Figure 11) is a partially buried system. The floor is usually about 3 ft below the surface grade. It is only wide enough for the carrier pipes and the pipe insulation plus some additional width to allow for pipe movement and possibly enough room for a person to stand on the floor. The trench usually is about as wide as it is deep. The top is constructed of reinforced concrete covers that protrude slightly above the surface and may also serve as a sidewalk. The floor and walls are usually cast-in-place reinforced concrete and the top is either precast or cast-in-place concrete. Precast concrete floor and wall sections have not been successful because of the large number of oblique joints and nonstandard sections required to follow the surface topography and to slope the floor to drain. This system is designed to handle the storm water and groundwater that enters the system, so the floor is always sloped toward a drainage point. A drainage system is required at all floor low points.

Cross beams that attach to the side walls are preferred to support the carrier pipes. This keeps the floor free of obstacles that would interfere with drainage and allows the distribution pipes to be assembled before lowering them on the pipe supports. Also, floor-mounted pipe supports tend to corrode.

The carrier pipes, pipe supports, expansion loops and bends, and insulation jacket are similar to aboveground systems, with the exception of pipe insulation. Experience with these systems indicates that flooding will occur several times during their design life; therefore, the insulation must be able to survive flooding and boiling and then return to near its original thermal efficiency. The pipe insulation is covered with a metal or plastic jacket to protect the insulation from abuse and from storm water that enters at the top cover butt joints. Small inspection ports of about 12 in. diameter

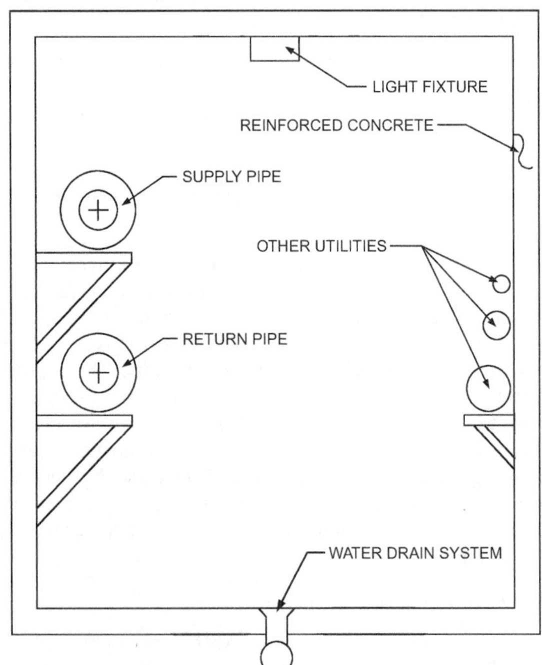

Fig. 10 Walk-Through Tunnel

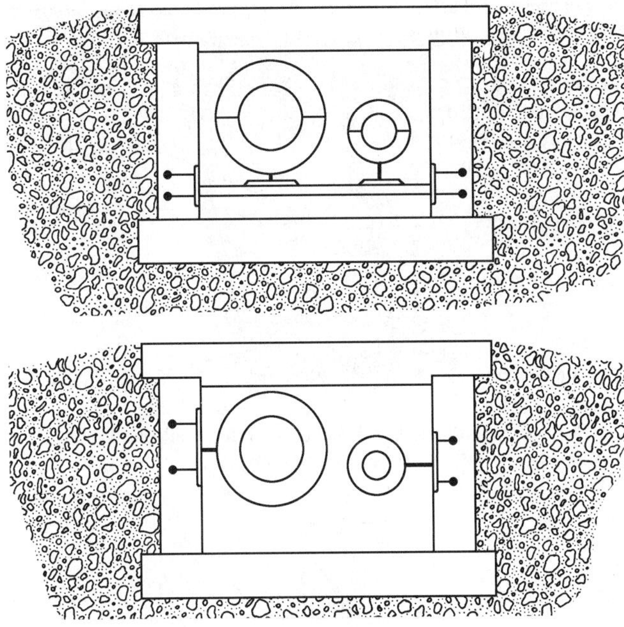

Fig. 11 Concrete Surface Trench

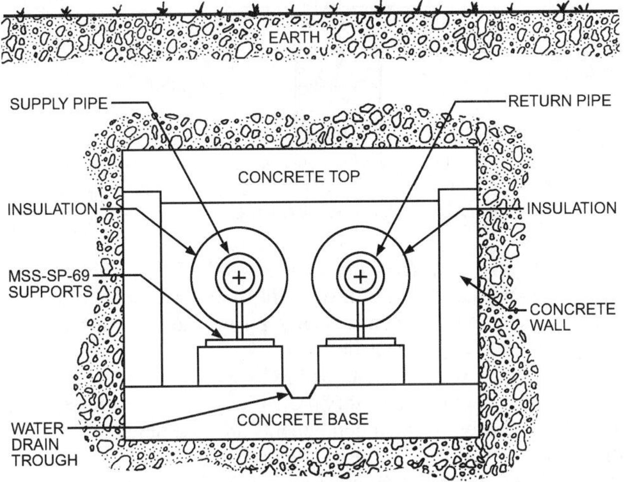

Fig. 12 Deep-Bury Small Tunnel

may be cast into the top covers at key locations so that the system can be inspected without removing the top covers. All replaceable elements, such as valves, condensate pumps, steam traps, strainers, sump pumps, and meters, are located in valve vaults. The first cost of this system is among the lowest for underground systems because it uses typical construction techniques and materials. The life-cycle cost is often the lowest because it is easy to maintain, correct construction deficiencies, and repair leaks.

Deep-Bury Tunnel. The tunnel in this system (Figure 12) is only large enough to contain the distribution piping, pipe insulation, and pipe supports. One type of deep-bury tunnel is the shallow concrete surface trench covered with earth and sloped independent of the topography. Because the system is covered with earth, it is essentially not maintainable between valve vaults without major excavation. All details of this system must be designed and all materials must be specified by the project design engineer.

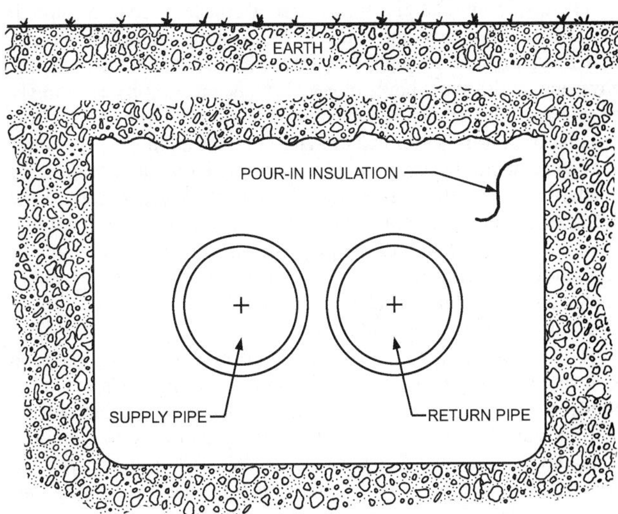

Fig. 13 Poured Insulation System

Because this system is not maintainable between valve vaults, great care must be taken to select materials that will last for the intended life and to ensure that the groundwater drainage system will function reliably. This system is intended to be used on sites where the groundwater elevation is typically lower than the bottom of the tunnel. The system can tolerate some groundwater saturation, depending on the watertightness of the construction and the capacity and reliability of the internal drainage. But even in desert areas, storms occur that expose underground systems to flooding; therefore, as with other types of underground systems, it must be designed to handle groundwater or storm water that enters the system. The distribution pipe insulation must be of the type that can withstand flooding and boiling and still retain its thermal efficiency.

Construction of this system is typically started in an excavated trench by pouring a cast-in-place concrete base that is sloped so intruding groundwater can drain to the valve vaults. The slope selected must also be compatible with the pipe slope requirements of the distribution system. The concrete base may have provisions for the supports for the distribution pipes, the groundwater drainage system, and the mating surface for the side walls. The side walls may have provisions for the pipe supports if the pipes are not bottom supported. If the upper portion is to have cast-in-place concrete walls, the bottom may have reinforcing steel for the walls protruding upward. The pipe supports, the distribution pipes, and the pipe insulation are all installed before the top cover is installed.

The groundwater drainage system may be a trough formed into the concrete bottom, a sanitary drainage pipe cast into the concrete bottom, or a sanitary pipe that is located slightly below the concrete base. The cover for the system is typically either of cast-in-place concrete or preformed sections such as precast concrete sections or half-round clay tile sections. The top covers must mate to the bottom and each other as tightly as possible to limit the entry of groundwater. After the covers are installed, the system is covered with earth to match the existing topography.

Poured Insulation. This system (Figure 13) is buried with the distribution system pipes encased in an envelope of insulating material and the insulation envelope covered with a thick layer of earth as required to match existing topography. This system is used on sites where the groundwater is typically below the system. Like other underground systems, experience indicates that it will be flooded because the soil will become saturated with water several times during the design life; therefore, the design must accommodate flooded condition.

The insulation material serves several functions. It may support the distribution pipes, and it must support earth loads. The insulation must prevent groundwater from entering the interior of the envelope, and it must have long-term resistance to physical breakdown caused by heat and water. The insulation envelope must allow the distribution pipes to expand and contract axially as the pipes change temperature. In elbows, expansion loops, and bends, the insulation must allow formed cavities for lateral movement of the pipes, or be able to migrate around the pipe without significant distortion of the insulation envelope while still retaining the required structural load carrying capacity. Special attention must be given to corrosion of metal parts and water penetration at anchors and structural supports that penetrate the insulation envelope.

Hot distribution pipes tend to drive moisture out of the insulation as steam; however, pipes used to distribute a cooling medium tend to condense water in the insulation, which reduces the insulation thermal resistance. A groundwater drainage system may be required, depending on the insulation material selected and the severity of the groundwater; however, if such a drainage is needed, it is a strong indicator that this is not the proper system for the site conditions.

This system is constructed by excavating a trench with a bottom slope that matches the desired slope of the distribution piping. The width of the bottom of the trench is usually the same as the width of the insulation envelope because it serves as a form. The distribution piping is then assembled in the trench and supported at the anchors and by blocks that are removed as the insulation is poured in place. The form for the insulation can be the trench bottom and sides, wooden forms, or sheets of plastic, depending on the type of insulation used and the site conditions. The insulation envelope is covered with earth to complete the installation.

The project design engineer is responsible for finding an insulation material that fulfills all of the previously mentioned requirements. At present, no standards have been developed for insulation used in this type of application. **Hydrophobic powders**, which are a special type of pulverized rock that is treated to be water repellent, have been used successfully. The hydrophobic characteristic of this powder prevents water from dampening the powder and has some capability as a barrier for preventing water from entering the insulation envelope. This insulating powder typically has a much higher thermal conductivity than mineral wool or fiberglass pipe insulation; therefore, the thickness of the poured envelope must be significantly greater.

Field-Installed, Direct-Buried Cellular Glass. In this system (Figure 14), cellular glass insulation is covered with an asphaltic jacket. The insulation supports the pipe. Oversized loops with internal support elements provide for expansion. The project design engineer must make provisions for movement of the pipes in the expansion loops. As shown in Figure 14, a drain should be installed to drain groundwater away. A waterproof jacket is recommended for all buried applications.

When used for heating applications, the dry soil condition must be investigated to determine if the temperature of the jacket exceeds the material allowable temperature (see the section on Methods of Heat Transfer Analysis). This is one of the controlling conditions to determine how high the carrier pipe fluid temperature will be allowed to be without exceeding the temperature limits of the jacket material. The thickness of insulation depends on the thermal operating parameters of the carrier pipe. For maximum system integrity under extreme operating conditions, such as groundwater flooding, the jacketing may be applied to the insulation segments in the fabrication shop with hot asphalt. As with other underground systems, experience indicates that this system may be flooded several times during its projected life.

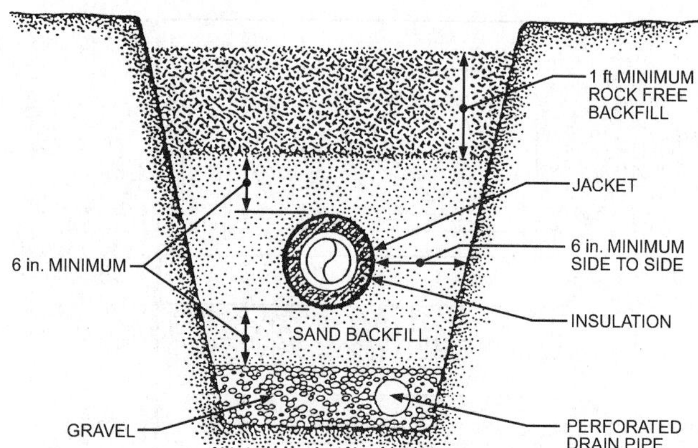

Fig. 14 Field-Installed, Direct-Buried Cellular Glass Insulated System

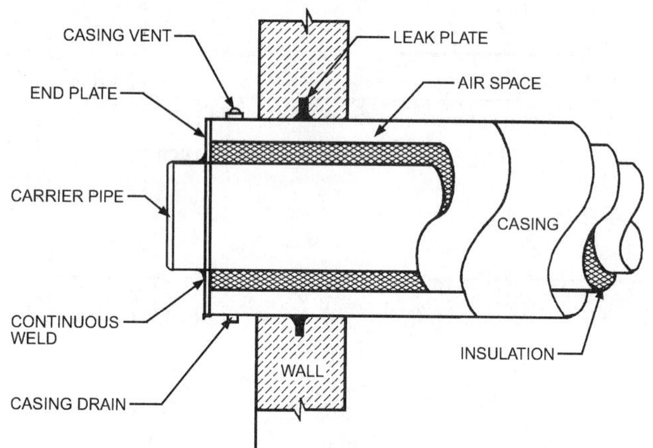

Fig. 15 Conduit System Components

Conduits

The term conduit denotes an entire assembly, which consists of a carrier pipe, the pipe insulation, the casing, and the exterior casing coating (Figure 15). The conduit is assembled in a factory and shipped as unit called a **conduit section**. The pipe that carries the working medium is called a **carrier pipe** and the outermost perimeter enclosure is called a **casing**.

Each conduit section is shipped in lengths up to 40 ft. Elbows, tees, loops, and bends are factory prefabricated to match the straight sections. The prefabricated components are assembled at the construction site; therefore, a construction contract is typically required for trenching, backfilling, connecting to buildings, connecting to distribution systems, constructing valve vaults, and performing some electrical work associated with sump pumps, power receptacles, and lights.

Much of the design work is done by the factory that manufactures the prefabricated sections; however, the field work must be designed and specified by the project design engineer or architect. Prefabricated components create a serious problem with accountability. For comparison, when systems are entirely field assembled, the design responsibility clearly belongs to the project design engineer, and system assembly is clearly the responsibility of the construction contractor. When a condition arises where a conduit system cannot be built without modifying prefabricated components, or if the construction contractor does not follow the instructions from the prefabricator, a serious conflict of responsibility arises. For these reasons, it is imperative that the project design engineer or architect clearly delineate the responsibilities of the factory prefabricator.

Crushing loads have been used (erroneously) to size the casing thickness, assuming that corrosion was not a factor. However, corrosion rate is usually the controlling factor because the casing temperature can range from less than 100°F to more than 300°F, a range

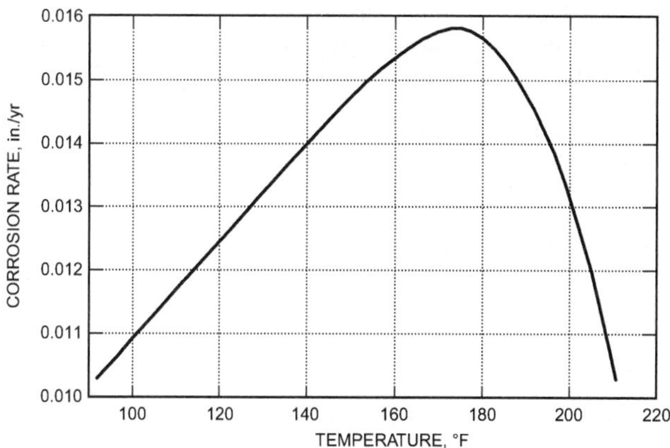

Fig. 16 Corrosion Rate in Aggressive Environment Similar to Mild Steel Casings in Soil

that encompasses the maximum corrosion rate of steel (Figure 16). As shown in the figure, the steel casing of a district heating pipe experiences corrosion rates several times that of domestic water pipes. The temperature of the casing varies with burial depth, soil conditions, carrier pipe temperature, and pipe insulation thickness. The casing must be strong and thick enough to withstand expansion and contraction forces and corrosion degradation.

All insulation must be kept dry for it to maintain its thermal insulating properties; the exception is cellular glass in cold applications. Because underground systems may be flooded several times during their design life, even on sites that are thought to be dry, a reliable water intrusion removal system is necessary in the valve vaults. Two designs are used to ensure that the insulation performs satisfactorily for the life of the system. In the **air space system**, an annular air space between the pipe insulation and the casing allows the insulation to be dried out if water enters. In the **water spread limiting (WSL) system**, which has no air space, the conduit is designed to keep water from entering the insulation. In the event that water enters one section, a WSL system prevents its spread to adjacent sections of piping.

The air space conduit system (Figures 17 and 18) should have an insulation that can survive short-term flooding without damage. The conduit manufacturer usually runs a boiling test with the insulation installed in the typical factory casing. No U.S. standard has been approved for this boiling test; however, the U.S. government has been using a Federal Agency Committee 96 h boiling test for conduit insulation. The insulation must have demonstrated that it can be dried with air flowing through an annular air space, and it must retain nearly new thermal insulating properties when dried.

Insulation fails because the bonding agents, called binders, that hold the principal insulation material in the desired shape degrade. The annular air space around the insulation, typically more than 1 in. wide, allows air to flow outside the insulation to dry it. Unfortunately, the air space has a serious detrimental side effect: it allows unwanted water to flow freely to other parts of the system.

The WSL system (Figure 19) encloses the insulation in an envelope that will not allow water to contact the insulation. The typical insulation is polyurethane foam, which will be ruined if excess water infiltration occurs. Polyurethane foam is limited to a temperature of about 250°F for a service life of 30 years or more. Europe has the most successful of the WSL systems, which are typically used in low-temperature water applications. These systems, available in the United States as well, meet European *Standard* EN 253 with regard to all major construction and design details. Standards also have been established for fittings (EN 448), preinsulated valves (EN 488), and the field joint assemblies (EN 489). With this system, the carrier pipe, insulation, and casing are bonded together to form a single unit. Forces caused by thermal expansion are

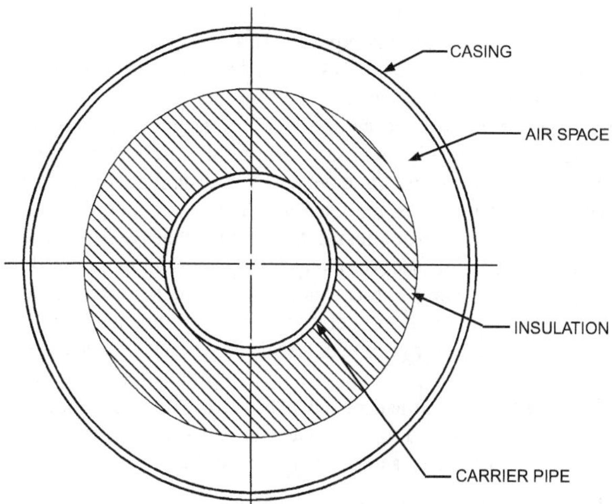

Fig. 17 Conduit System with Annular Air Space and Single Carrier Pipe

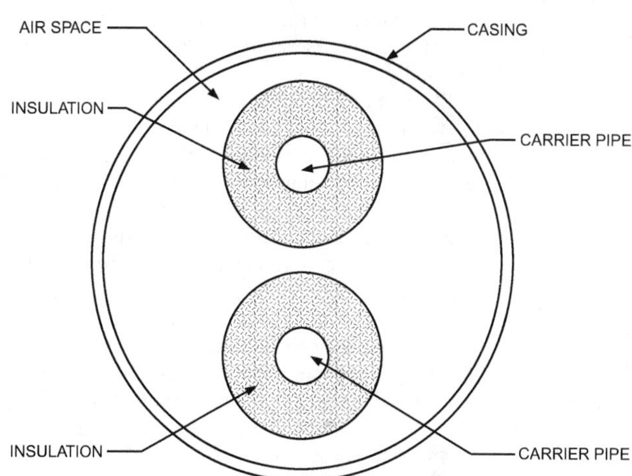

Fig. 18 Conduit System with Two Carrier Pipes and Annular Air Space

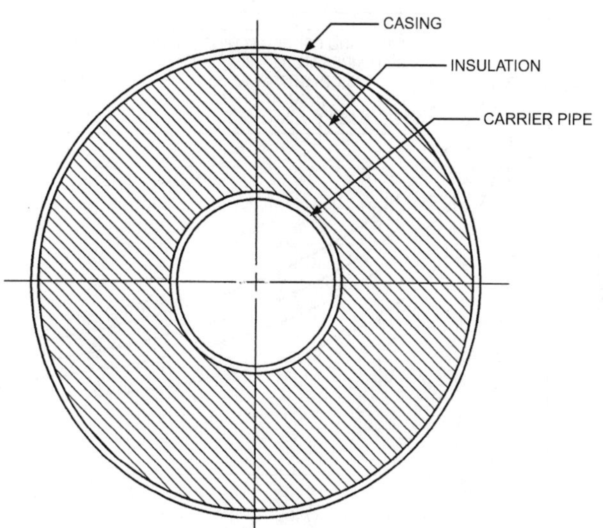

Fig. 19 Conduit System with Single Carrier Pipe and No Air Space (WSL)

passed as shear forces to the mating component and ultimately to the soil; thus, no additional expansion provisions are required.

This type of WSL system is feasible because of the small temperature differential and because the thinner carrier pipe wall creates smaller expansion forces because of its low cross-sectional area. Some systems rely on a watertight field joint between joining casing sections to extend the envelope to a distant envelope termination point where the casing is sealed to the carrier pipe. Other systems form the waterproof insulation envelope in each individual prefabricated conduit section using the casing and carrier pipe to form part of the envelope and a waterproof bulkhead to seal the casing to the carrier pipe. In another type of construction, a second pipe fits tightly over the carrier pipe and seals the insulation between the second pipe and casing to achieve a watertight insulation envelope.

Conduit Design Conditions. The following three design conditions must be addressed to have reasonable assurance that the system selected will have a satisfactory service life:

- **Maximum heat loss** occurs when the soil is wettest and the conduit is shallowly buried (minimum burial depth), usually with about 2 ft of earth cover. This condition represents the highest gross heat transfer and is used to size the distribution piping and equipment in the central heating plant. For heating piping, because the casing is coldest during start-up, the relative movement of the carrier pipe with respect to the casing may be maximum during this condition.
- A **dry-soil condition** may occur when the conduit is buried deep. The soil plays a more significant role in the heat transfer than the pipe insulation because of the soil's thickness (and thus, insulating value). The highest temperature of the insulation, casing, and casing coating occur during this condition. Paradoxically, the minimum heat loss occurs during this condition because the soil is acting as a good insulator. This condition is used to select temperature-sensitive materials and to design for casing expansion. The relative movement of the carrier pipe with respect to the casing may be minimum during this condition; however, if the casing is not restrained, its movement with respect to the soil will be maximum. If restrained, the casing axial stresses and axial forces will be highest and the casing allowable stresses will be lowest because of the high casing temperature.

  Figure 20A shows the effect of burial depth on casing temperature as a function of soil thermal conductivity for a typical system. Figure 20B shows the effect of insulation thickness on casing temperature, again as a function of soil thermal conductivity. Analysis of Figure 20A and 20B suggests some design solutions that could lower the effects of the dry-soil condition. Possible solutions are to lower the carrier pipe temperature, use thicker carrier pipe insulation, provide a device to keep the soil wet, or minimize the burial depth. However, if these solutions are not feasible or cost effective, a different type of material or an alternative system should be considered.

  Although it is possible that the soil will never dry out, given the variability of climate in most areas, it is likely that a drought will occur during the life of the system. Only one very dry condition can cause permanent damage to the insulation if the soil thermal conductivity drops below the assumed design value. On-site measurement of the driest soil condition likely to cause insulation damage is not feasible. As a result, the designer is left with the conservative choice of using the lowest thermal conductivity from Table 3 to calculate the highest temperature to be used to select materials.
- The **nominal** or **average condition** occurs when the soil is at its average water content and the conduit is buried at the average depth. This condition is used to compute the yearly energy consumption from heat loss to the soil (see Table 3 for soil thermal conductivity and the previous section on Thermal Properties).

Cathodic Protection of Direct-Buried Conduits

Corrosion is an electrochemical process that occurs when a corrosion cell is formed. A corrosion cell consists of an anode, a cathode, a connecting path between them, and an electrolyte (soil or water). The structure of this cell is the same as a dry-cell battery, and, like a battery, it produces a direct electrical current. The anode and cathode in the cell may be dissimilar metals, and because of differences in their natural electrical potentials, a current flows from anode to cathode. When current leaves an anode, it destroys the anode material at that point. The anode and cathode may also be the same material. Differences in composition, environment, temperature, stress, or shape makes one section anodic and an adjacent section cathodic. With a connection path and the presence of an electrolyte, this combination also generates a direct electrical current and causes corrosion at the anodic area.

Cathodic protection is a standard method used by the underground pipeline industry to further protect coated steel against

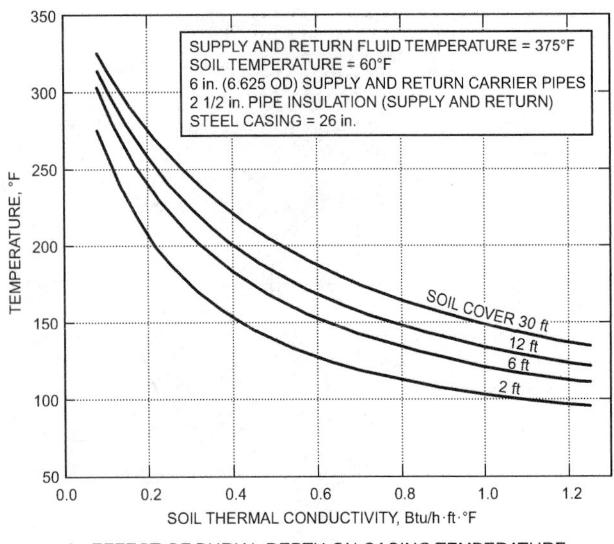

A. EFFECT OF BURIAL DEPTH ON CASING TEMPERATURE

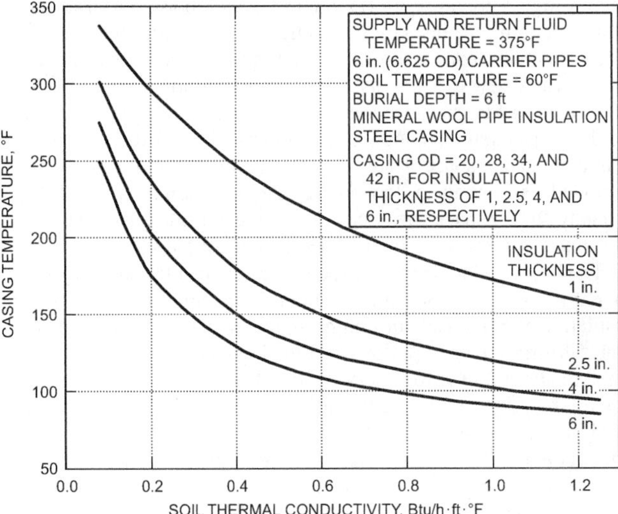

B. EFFECT OF INSULATION THICKNESS ON CASING TEMPERATURE

Fig. 20 Conduit Casing Temperature Versus Soil Thermal Conductivity

corrosion. Cathodic protection systems are routinely designed for a minimum life of 20 years. Cathodic protection may be achieved by the sacrificial anode method or the impressed current method.

Sacrificial anode systems are normally used with well-coated structures. A direct current is applied to the outer surface of the steel structure with a potential driving force that prevents the current from leaving the steel structure. This potential is created by connecting the steel structure to another metal, such as magnesium, aluminum, or zinc, which becomes the anode and forces the steel structure to be the cathode. The moist soil acts as the electrolyte. These deliberately connected materials become the sacrificial anode and corrode. If they generate sufficient current, they adequately protect the coated structure, and their low current output is not apt to corrode other metallic structures in the vicinity.

Impressed current systems use a rectifier to convert an alternating current power source to usable direct current. The current is distributed to the metallic structure to be protected through relatively inert anodes such as graphite or high-silicon cast iron. The rectifier allows the current to be adjusted over the life of the system. Impressed current systems, also called rectified systems, are used on long pipelines in existing systems with insufficient coatings, on marine facilities, and on any structure where current requirements are high. They are installed selectively in congested pipe areas to ensure that other buried metallic structures are not damaged.

The design of effective cathodic protection requires information on the diameter of both the carrier pipe and conduit casing, length of run, number of conduits in a common trench, and number of system terminations in access areas, buildings, etc. Soil from the construction area should be analyzed to determine the soil resistivity, or the ease at which current flows through the soil. Areas of low soil resistivity require fewer anodes to generate the required cathodic protection current, but the life of the system depends on the weight of anode material used. The design life expectancy of the cathodic protection must also be defined. All anode material is theoretically used up at the end of the cathodic protection system life. At this point, the corrosion cell reverts to the unprotected system and corrosion occurs at points along the conduit system or buried metallic structure. Anodes may be replaced or added periodically to continue the cathodic protection and increase the conduit life.

A cathodically protected system must be electrically isolated at all points where the pipe is connected to building or access port piping and where a new system is connected to an existing system. Conduits are generally tied to another building or access piping with flanged connections. Flange isolation kits, including dielectric gaskets, washers, and bolt sleeves, electrically isolate the cathodically protected structure. If an isolation flange is not used, any connecting piping or metallic structure will be in the protection system, *but protection may not be adequate*.

The effectiveness of cathodic protection can only be determined by an installation survey after the system has been energized. Cathodically protected structures should be tested at regular intervals to determine the continued effectiveness and life expectancy of the system. Sacrificial anode cathodic protection is monitored by measuring the potential (voltage) between the underground metallic structure and the soil versus a stable reference. This potential is measured with a high resistance voltmeter and a reference cell. The most commonly used reference cell material is copper/copper sulfate. One criterion for protection of buried steel structures is a negative voltage of at least 0.85 V as measured between the structure surface and a saturated copper/copper sulfate reference electrode in contact with the electrolyte. Impressed current systems require more frequent and detailed monitoring than sacrificial anode systems. The rectified current and potential output and operation must be verified and recorded at monthly intervals.

NACE *Standard* RPO169 has further information on control of external corrosion on buried metallic structures.

Leak Detection

The conduit may require excavation to repair construction errors after burial. Various techniques are available for detecting leaks in district heating and cooling piping. They range from performing periodic pressure tests on the piping system to installing a sensor cable along the entire length of the piping to continuously detect and locate leaks. Pressure testing should be performed on all piping to verify integrity during installation and the life of the piping. Chilled-water systems should be pressure tested during the winter, and hot-water and steam systems tested during the summer.

A leak is difficult to locate without the aid of a cable type leak detector. Finding a leak typically involves excavating major sections between valve vaults. Infrared detectors and acoustic detectors can help narrow down the location of a leak, but they do not work equally well for all underground systems. Also, they are not as accurate with underground systems as with an aboveground system.

Chilled- and Hot-Water Systems. Chilled-water piping systems are usually insulated with urethane foam with a vaporproof jacket (HDPE, urethane, PVC, CPVC, etc.). Copper wires can be installed during fabrication to aid in detecting and locating liquid leaks. The wires may be insulated or uninsulated, depending on the manufacturer. Some systems monitor the entire wire length, whereas others only monitor at the joints of the piping system. The detectors either look for a short in the circuit using Ohm's law or monitor for impedance change using time domain reflectometry (TDR).

Steam, High-Temperature Hot-Water, and Other Conduit Systems. Air gap designs, which have a gap between the inner wall of the outer casing and the insulation, can have probes installed at the low points of drains or at various points to detect leaks. Leaks can also be detected with a continuous cable that monitors liquid leakage. The cable is installed at the bottom of the conduit with a minimum air gap required, typically 1 in. Pull points or access ports are installed every 400 to 500 ft on straight runs, with changes in direction reducing the length between pull points. Systems monitor either by looking for a short on the cable using Ohm's law or by sensing the impedance on a coaxial cable using resistance temperature devices (RTD). During installation, care must be taken to keep the system clean and dry to keep any contamination from the leak detection system that might cause it to fail. The system must be sealed airtight to prevent condensation from accumulating in the piping at the low points.

Valve Vaults and Entry Pits

Valve vaults allow a user to isolate problems to one area rather than analyze the entire line. This feature is important if the underground distribution system cannot be maintained between valve vaults without excavation. The optimum number of valve vaults is that which affords the lowest life-cycle cost and still meets all design requirements, usually no more than 500 ft apart. Valve vaults provide a space in which to put valves, steam traps, carrier pipe drains, carrier pipe vents, casing vents and drains, condensate pumping units, condensate cooling devices, flash tanks, expansion joints, groundwater drains, electrical leak detection equipment and wiring, electrical isolation couplings, branch line isolation valves, carrier pipe isolation valves, and flowmeters. Valve vaults allow for elevation changes in the distribution system piping while maintaining an acceptable slope on the system; they also allow the designer to better match the topography and avoid unreasonable and expensive burial depths.

Ponding Water. The most significant problem with valve vaults is that water ponds in them. Ponding water may be from either carrier pipe leaks or intrusion of surface or groundwater into the valve vault. When the hot- and chilled-water distribution systems share the same valve vault, plastic chilled-water lines often fail because ponded water heats the plastic to failure. Water gathers in the valve vault irrespective of climate; therefore, design strives to eliminate the water for the entire life of the underground distribution system. The most successful water removal systems are those that drain to sanitary or storm drainage systems; this technique is successful because the system is affected very little by corrosion and has no moving parts to fail. Backwater valves are recommended in case the drainage system backs up.

Duplex sump pumps with lead-lag controllers and a failure annunciation system are used when storm drains and sanitary drains are not accessible. Because pumps have a history of frequent failures, duplex pumps help eliminate short cycling and provide standby pumping capacity. Steam ejector pumps can be used only if the distribution system is never shut down because the carrier pipe insulation can be severely damaged during even short outages. A labeled, lockable, dedicated electrical service should be used for electric pumps. The circuit label should indicate what the circuit is used for; it should also warn of the damage that will occur if the circuit is deenergized.

Electrical components have experienced accelerated corrosion in the high heat and humidity of closed, unventilated vaults. A pump that works well at 50°F often performs poorly at 200°F and 100% rh. To resolve this problem, one approach specifies components that have demonstrated high reliability at 200°F and 100% rh with a damp-proof electrical service. The pump should have a corrosion-resistant shaft (when immersed in water) and impeller and have demonstrated 200,000 cycles of successful operation, including the electrical switching components, at the referenced temperature and humidity. The pump must also pass foreign matter; therefore, the requirement to pass a 3/8 in. ball should be specified.

Another method drains the valve vaults into a separate sump adjacent to the vault. Then the pumps are placed in this sump, which is cool and more nearly a sump pump environment. Redundant methods may be necessary if maximum reliability is needed or future maintenance is questionable. The pump can discharge to the sanitary or storm drain or to a splash block near the valve vault. Water pumped to a splash block has a tendency to enter the vault, but this is not a significant problem if the vault construction joints have been sealed properly. Extreme caution must be exercised if the bottom of the valve vault has French drains. These drains allow groundwater to enter the vault and flood the insulation on the distribution system during high-groundwater conditions. Adequate ventilation of the valve vault is also important.

Crowding of Components. The valve vault must be laid out in three dimensions, considering standing room for the worker, wrench swings, the size of valve operators, variation between manufacturers in the size of appurtenances, and all other variations that the specifications allow with respect to any item placed in the vault. To achieve desired results, the vault layout must be shown to scale on the contract drawings.

High Humidity. High humidity develops in a valve vault when it has no positive ventilation. Gravity ventilation is often provided in which cool air enters the valve vault and sinks to the bottom. At the bottom of the vault, the air warms, becomes lighter, and rises to the top of the vault, where it exits. In the past, some designers used a closed-top valve vault with an exterior ventilation pipe with an elbow that directs the exiting air down. However, the elbowed-down vent hood tends to trap the exiting air and prevent gravity ventilation from working. Open structural grate tops are the most successful covers for ventilation purposes. Open grates allow rain to enter the vault; however, the techniques mentioned previously in the section on Ponding Water are sufficient to handle the rainwater. Open grates with sump basins have worked well in extremely cold climates and in warm climates. Some vaults have a closed top and screened, elevated sides to allow free ventilation. In this design, the solid vault sides extend slightly above grade; then, a screened window is placed in the wall on at least two sides. The overall above-grade height may be only 18 in.

High Temperatures. The temperature in the valve vault rises when no systematic way is provided to remove heat losses from the distribution system. The gravity ventilation rate is usually not sufficient to transport heat from the closed vaults. Part of this heat transfers to the earth; however, an equilibrium temperature is reached that may be higher than desired. Ventilation techniques discussed in the section on High Humidity can resolve the problem of high temperature if the heat loss from the distribution system is near normal. Typical problems that greatly increase the amount of heat released include

- Leaks from a carrier pipe, gaskets, packings, or appurtenances
- Insulation that has deteriorated because of flooding or abuse
- Standing water in a vault that touches the distribution pipe
- Steam vented to the vault from partial flooding between valve vaults
- Vents from flash tanks
- Insulation removed during routine maintenance and not replaced

To prevent heat release in a new system, a workable ventilation system must be designed. On existing valve vaults, the valve vault must be ventilated properly, all leaks corrected, and all insulation that was damaged or left off replaced. Commercially available insulation jackets that can easily be removed and reinstalled from fittings and valves should be installed. If flooding occurs between valve vaults, portions of the distribution system may have to be excavated and repaired or replaced. Vents from vault appurtenances that exhaust steam into the vault may have to be routed aboveground if the ventilation technique is insufficient to handle the quantity of steam exhausted.

Deep Burial. When a valve vault is buried too deeply, (1) the structure is exposed to groundwater pressures, (2) entry and exit often become a safety problem, (3) construction becomes more difficult, and (4) the cost of the vault is greatly increased. Valve vault spacing should be no more than 500 ft (NAS 1975). If greater spacing is desired, use an accurate life-cycle analysis to determine spacing. The most common way to limit burial depth is to place the valve vaults closer together. Steps in the distribution system slope are made in the valve vault (i.e., the carrier pipes come into the valve vault at one elevation and leave at a different elevation). If the slope of the distribution system is changed to more nearly match the earth topography, the valve vaults will be shallower; however, the allowable range of slope of the carrier pipes restricts this method. In most systems, the slope of the distribution system can be reversed in a valve vault, but not out in the system between valve vaults. The minimum slope for the carrier pipes is 1 in. in 20 ft. Lower slopes are outside the range of normal construction tolerance. If the entire distribution system is buried too deeply, the designer must determine the maximum allowable burial depth of the system and survey the topography of the distribution system to determine where the maximum and minimum depth of burial will occur. All elevations must be adjusted to limit the minimum and maximum allowable burial depths.

Freezing Conditions. Failure of distribution systems caused by water freezing in components is common. The designer must consider the coldest temperature that may occur at a site and not the 99% or 99.6% condition used in building design (as discussed in Chapter 28 of the 2005 *ASHRAE Handbook—Fundamentals*). Drain legs or vent legs that allow water to stagnate are usually the cause of failure. Insulation should be on all items that can freeze, and it must be kept in good condition. Electrical heat tape and pipe-type heat tracing can be used under insulation. If part of a chilled-water system is in a ventilated valve vault, the chilled water may have to be circulated or be drained if not used in winter.

Safety and Access. Some of the working fluids used in underground distribution systems can cause severe injury and death if accidentally released in a confined space such as a valve vault. The shallow valve vault with large openings is desirable because it allows personnel to escape quickly in an emergency. The layout of the pipes and appurtenances must allow easy access for maintenance without requiring maintenance personnel to crawl underneath or between other pipes. The goal of the designer is to keep clear work spaces for maintenance personnel so that they can work efficiently and, if necessary, exit quickly. Engineering drawings must show pipe insulation thickness; otherwise they will give a false impression of the available space.

The location and type of ladder is important for safety and ease of egress. It is best to lay out the ladder and access openings when laying out the valve vault pipes and appurtenances as a method of exercising control over safety and ease of access. Ladder steps, when cast in the concrete vault walls, may corrode if not constructed of the correct material. Corrosion is most common in steel rungs. Either cast-iron or prefabricated, OSHA-approved, galvanized steel ladders that sit on the valve vault floor and are anchored near the top to hold it into position are best. If the design uses lockable access doors, the locks must be operable from inside or have some keyed-open device that allows workers to keep the key while working in the valve vault.

Vault Construction. The most successful valve vaults are those constructed of cast-in-place reinforced concrete. These vaults conform to the earth excavation profile and show little movement when backfilled properly. Leakproof connections can be made with mating tunnels and conduit casings even though they may enter or leave at oblique angles. In contrast, prefabricated valve vaults may settle and move after construction is complete. Penetrations for prefabricated vaults, as well as the angles of entry and exit, are difficult to locate exactly. As a result, much of the work associated with penetrations is not detailed and must be done by construction workers in the field, which greatly lowers the quality and greatly increases the chances of a groundwater leak.

Construction deficiencies that go unnoticed in the buildings can destroy a heating and cooling distribution system; therefore, the designer must clearly convey to the contractor that a valve vault does not behave like a sanitary access port. A design that is sufficient for a sanitary access port will prematurely fail if used for a heating and cooling distribution system.

CONSUMER INTERCONNECTIONS

The thermal energy produced at the plant is transported via the distribution network and is finally transferred to the consumer. When thermal energy (hot water, steam, or chilled water) is supplied, it may be used directly by the building HVAC system or process loads, or indirectly via a heat exchanger that transfers energy from one media to another. When energy is used directly, it may be reduced in pressure that is commensurate to the buildings' systems. The design engineer must perform an analysis to determine which connection type is best.

For commercially operated systems, a contract boundary or point of delivery divides responsibilities between the energy provider and the customer. This point can be at a piece of equipment, as in a heat exchanger with an indirect connection, or flanges as in a direct connection. A chemical treatment analysis must be performed (regardless of the type of connection) to determine the compatibility of each side of the system (district and consumer) before energizing.

Direct Connection

Because a direct connection offers no barrier between the district water and the building's own system (e.g., air-handling unit cooling and heating coils, fan-coils, radiators, unit heaters, and process loads), water circulated at the district plant has the same quality as the customer's water. Direct connections, therefore, are at a greater risk of incurring damage or contamination based on the poor water quality of either party. Typically, district systems have contracts with water treatment vendors and monitor water quality continuously. This may not be the case with all consumers. A direct connection is often more economical than an indirect connection because the consumer is not burdened by the installation of heat exchangers, additional circulation pumps, or water treatment systems; therefore, investment costs are reduced and return temperatures identical to design values are possible.

Figure 21 shows the simplest form of direct connection, which includes a pressure differential regulator, a thermostatic control valve on each terminal unit, a pressure relief valve, and a check valve. Most commercial systems have a flowmeter installed as well as temperature sensors and transmitters to calculate the energy used. The location of each device may vary from system to system, but all of the major components are indicated. The control valve is the capacity regulating device that restricts flow to maintain either a chilled-water supply or return temperature on the consumer's side.

Particular attention must be paid to connecting high-rise buildings because they induce a static head. Pressure control devices should be investigated carefully. It is not unusual to have a water-based district heating or cooling system with a mixture of direct and indirect connections in which heat exchangers isolate the systems hydraulically.

In a direct system, the pressure in the main distribution system must meet local building codes to protect the customer's installation and the reliability of the district system. To minimize noise, cavitation, and control problems, constant-pressure differential control valves should be installed in the buildings. Special attention should be given to potential noise problems at the control valves. These valves must correspond to the design pressure differential in a system that has constantly varying distribution pressures because of load shifts. Multiple valves may be required, to serve the load under all flow and pressure ranges. Industrial-quality valves and actuators should be used for this application.

If the temperature in the main distribution system is lower than that required in the consumer cooling systems, a larger temperature differential between supply and return occurs, thus reducing the required pipe size. The consumer's desired supply temperature can be attained by mixing the return water with the district cooling supply water. Depending on the size and design of the main system, elevation differences, and types of customers and building systems, additional safety equipment, such as automatic shutoff valves on both supply and return lines, may be required.

When buildings have separate circulation pumps, primary/secondary piping, and pumping, isolating techniques are used (cross-connection between return and supply piping, decouplers, and bypass lines). This ensures that two-way control valves are subjected only to the differential pressure established by the customer's

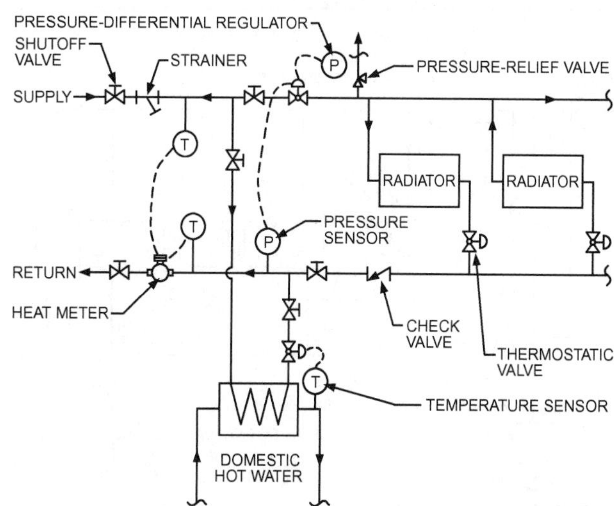

Fig. 21 Direct Connection of Building System to District Hot Water

building (tertiary) pump. Figure 4 shows a primary/secondary connection using an in-building pumping scheme.

When tertiary pumps are used, all series connections between the district system pumps must be removed. A series connection can cause the district system return to operate at a higher pressure than the distribution system supply and disrupt normal flow patterns. Series operation usually occurs during improper use of three-way mixing valves in the primary to secondary connection.

Indirect Connection

Many of the components are similar to those used in the direct connection applications, with the exception that a heat exchanger performs one or more of the following functions: heat transfer, pressure interception, and buffer between potentially different quality water treatment.

Identical to the direct connection, the rate of energy extraction in the heat exchanger is governed by a control valve that reacts to the building load demand. Once again, the control valve usually modulates to maintain a temperature set point on either side of the heat exchanger, depending on the contractual agreement between the consumer and the producer.

The three major advantages of using heat exchangers are (1) the static head influences of a high-rise building are eliminated, (2) the two water streams are separated, and (3) consumers must make up all of their own lost water. The disadvantages of using an indirect connection are the (1) additional cost of the heat exchanger and (2) temperature loss and increased pumping pressure because of the addition of another heat transfer surface.

COMPONENTS

Heat Exchangers

Heat exchangers act as the line of demarcation between ownership responsibility of the different components of an indirect system. They transfer thermal energy and act as pressure interceptors for the water pressure in high-rise buildings. They also keep fluids from each side (which may have different chemical treatments) from mixing. Figure 22 shows a basic building schematic including heat exchangers and secondary systems.

Reliability of the installation is increased if multiple heat exchangers are installed. The number selected depends on the types of loads present and how they are distributed throughout the year. When selecting all equipment for the building interconnection, but specifically heat exchangers, the designer should

- Size the unit's capacity to match the given load and estimated load turndown as close as possible (oversized units may not perform as desired at maximum turndown; therefore, several smaller units will optimize the installation).
- Assess the critical nature of the load/operation/process to address reliability and redundancy. For example, if a building has 24 h process loads (i.e., computer room cooling, water-cooled equipment, etc.), consider adding a separate heat exchanger for this load. Also, consider operation and maintenance of the units. If the customer is a hotel, hospital, casino, or data center, select a minimum of two units at 50% load each to allow one unit to be cleaned without interrupting building service. Separate heat exchangers should be capable of automatic isolation during low-load conditions to increase part-load performance.
- Determine customer's temperature and pressure design conditions. Some gasket materials for plate heat exchangers have limits for low pressure and temperature.
- Evaluate customer's water quality (i.e., use appropriate fouling factor).
- Determine available space and structural factors of the mechanical room.
- Quantify design temperatures. The heat exchanger may require rerating at a higher inlet temperature during off-peak hours.
- Calculate the allowable pressure drop on both sides of heat exchangers. The customer's side is usually the most critical for pressure drop. The higher the pressure drop, the smaller and less expensive the heat exchanger. However, the pressure drop must be kept in reasonable limits (15 psig or below) if the existing pumps are to be reused. Investigate the existing chiller evaporator pressure drop in order to assist in this evaluation.

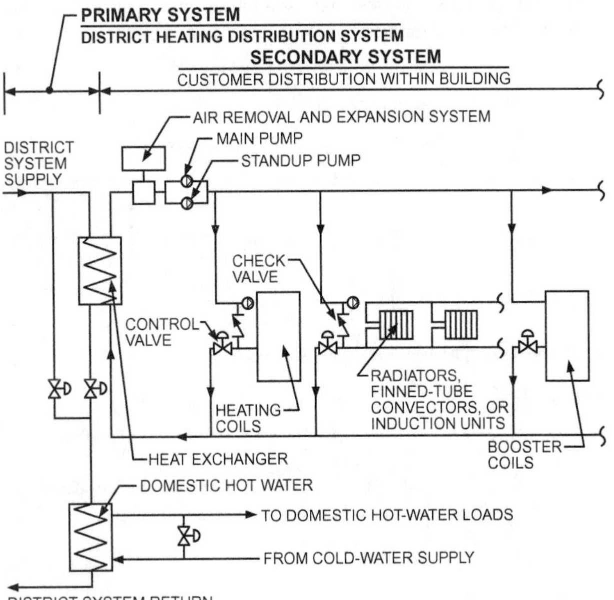

Fig. 22 Basic Heating-System Schematic

All heat exchangers should be sized with future expansion in mind. When selecting heat exchangers, be cognizant that closer approach temperatures or low pressure drop require more heat transfer area and hence cost more and take up more space. Strainers should be installed in front of any heat exchanger and control valve to keep debris from fouling surfaces.

Plate, shell-and-coil, and shell-and-tube heat exchangers are all used for indirect connection. Whatever heat transfer device is selected must meet the appropriate temperature and pressure duty, and be stamped/certified accordingly as pressure vessels.

Plate Heat Exchangers (PHEs). These exchangers, which are used for either steam, hot-water, or chilled-water applications, are available as gasketed units and in two gasket-free designs (brazed and all or semiwelded construction). All PHEs consist of metal plates compressed between two end frames and sealed along the edges. Alternate plates are inverted and the gaps between the plates form the liquid flow channels. Fluids never mix as hot fluid flows on one side of the plate and cool fluid flows countercurrent on the other side. Ports at each corner of the end plates act as headers for the fluid. One fluid travels in the odd-numbered plates and the other in the even-numbered plates.

Because PHEs require turbulent flow for good heat transfer, pressure drops may be higher than those for comparable shell-and-tube models. High efficiency leads to a smaller package. The designer should consider specifying that the frame be sized to hold 20% additional plates. PHEs require very little maintenance because the high velocity of the fluid in the channels tends to keep the surfaces clean. PHEs generally have a cost advantage and require one-third to one-half the surface required by shell-and-tube units for the same operating conditions. PHEs are also capable of closer approach temperatures.

Gasketed PHEs (also called plate-and-frame heat exchangers) consist of a number of gasketed embossed metal plates bolted together between two end frames. Gaskets are placed between the plates to contain the two media in the plates and to act as a boundary. Gasket failure will not cause the two media to mix; instead, the media leak to the atmosphere. Gaskets can be either glued or clip-on. Gasketed PHEs are suitable for steam-to-liquid and liquid-to-liquid applications. Designers should select the appropriate gasket material for the design temperatures and pressures expected. Plates are typically stainless steel; however, plate material can be varied based on the chemical makeup of the heat transfer fluids.

PHEs are typically used for district heating and cooling with water and for cooling tower water heat recovery (free cooling). Double-wall plates are also available for potable-water heating, chemical processes, and oil quenching. PHEs have three to five times greater heat transfer coefficients than shell-and-tube units and are capable of achieving 1°F approach. This type of PHE can be disassembled in the field to clean the plates and replace the gaskets. Typical applications go up to 365°F and 400 psig.

Brazed PHEs are suitable for steam, vapor, or water solutions. They feature a close approach temperature (within 2°F), large temperature drop, compact size, and a high heat transfer coefficient. Construction materials are stainless steel plates and frames brazed together with copper or nickel. Tightening bolts are not required as in the gasketed design. These units cannot be disassembled and cleaned; therefore, adequate strainers must be installed ahead of an exchanger and it must be periodically flushed clean in a normal maintenance program. Brazed PHEs usually peak at a capacity of under 200,000 Btu/h (about 200 plates and 600 gpm) and suitable for 435 psig and 435°F. Typical applications are district heating using hot water and refrigeration process loads. Double-wall plates are also available. Applications where the PHE may be exposed to large, sudden, or frequent changes in temperature and load must be avoided because of risk of thermal fatigue.

Welded PHEs can be used in any application for which shell-and-tube units are used that are outside the accepted range of gasketed PHE units in liquid-to-liquid, steam-to-liquid, gas-to-liquid, gas-to-gas, and refrigerant applications. Construction is very similar to gasketed units except gaskets are replaced with laser welds. Materials are typically stainless steel, but titanium, monel, nickel, and a variety of alloys are available. Models offered have design ratings that range from 500°F at 150 psig to 1000°F at 975 psig; however, they are available only in small sizes. Normally, these units are used in ammonia refrigeration and aggressive process fluids. They are more suitable to pressure pulsation or thermal cycling because they are thermal fatigue resistant. A semiwelded PHE is a hybrid of the gasketed and the all-welded units in which the plates are alternatively sealed with gaskets and welds.

Shell-and-Coil Heat Exchangers. These European-designed heat exchangers are suitable for steam-to-water and water-to-water applications and feature an all-welded-and-brazed construction. This counter/crossflow heat exchanger consists of a hermetically sealed (no gaskets), carbon-steel pressure vessel with hemispherical heads. Copper or stainless helical tubes within are installed in a vertical configuration. This type of heat exchanger offers a high temperature drop and close approach temperature. It requires less floor space than other designs and has better heat transfer characteristics than shell-and-tube units.

Shell-and-Tube Heat Exchangers. These exchangers are usually a multiple-pass design. The shell is usually constructed from steel and the tubes are often of U-bend construction, usually 3/4 in. (nominal) OD copper, but other materials are available. These units are ASME U-1 stamped for pressure vessels.

Flow Control Devices

In commercial systems, after the flowmeter, control valves are the most important element in the interface with the district energy system because proper valve adjustment and calibration save energy. High-quality, industrial-grade control valves provide more precise control, longer service life, and minimum maintenance.

All control valve actuators should take longer than 60 s to close from full open to mitigate pressure transients or water hammer, which occurs when valves slam closed. Actuators should also be sized to close against the anticipated system pressure so the valve seats are not forced open, thus forcing water to bypass and degrading temperature differential.

The wide range of flows and pressures expected makes selection of control valves difficult. Typically, only one control valve is required; however, for optimal response to load fluctuations and to prevent cavitation, two valves in parallel are often needed. The two valves operate in sequence and for a portion of the load (i.e., one valve is sized for two-thirds of peak flow and the other sized for one-third of peak flow). The designer should review the occurrence of these loads to size the proportions correctly. The possibility of overstating customer loads complicates the selection process, so accurate load information is important. It is also important that the valve selected operates under the extreme pressure and flow ranges foreseen. Because most commercial-grade valves will not perform well for this installation, industrial-quality valves are specified.

Electronic control valves should remain in a fixed position when a power failure occurs and should be manually operable. Pneumatic control valves should close upon loss of air pressure. A manual override on the control valves allows the operator to control flow. All chilled-water control valves must fail in the closed position. Then, when any secondary in-building systems are deenergized, the valves close and will not bypass chilled water to the return system. All steam pressure-reducing valves should close as well.

Oversizing reduces valve life and causes valve hunting. Select control valves having a wide range of control; low leakage; and proportional-plus-integral control for close adjustment, balancing, temperature accuracy, and response time. Control valves should have actuators with enough force to open and close under the maximum pressure differential in the system. The control valve should have a pressure drop through the valve equal to at least 10 to 30% of the static pressure drop of the distribution system. This pressure drop gives the control valve the "authority" it requires to properly control flow. The relationship between valve travel and capacity output should be linear, with an equal percentage characteristic.

In hot-water systems, control valves are normally installed in the return line because the lower temperature in the line reduces the risk of cavitation and increases valve life. In chilled-water systems, control valves can be installed in either location; typically, however, they are also installed the return line.

Instrumentation

In many systems, where energy to the consumer is measured for billing purposes, temperature sensors assist in calculating the energy consumed as well as in diagnosing performance. Sensors and their transmitters should have an accuracy range commensurate to the accuracy of the flowmeter. In addition, pressure sensors are required for variable-speed pump control (water systems) or valve control for pressure-reducing stations (steam and water).

Temperature sensors need to be located by the exchangers being controlled rather than in the common pipe. Improperly located sensors will cause one control valve to open and others to close, resulting in unequal loads in the exchangers.

Controller

The controller performs several functions, including recording demand and the amount of energy used for billing purposes, monitoring the differential pressure for plant pump control, energy calculations, alarming for parameters outside normal, and monitoring and control of all components.

Typical control strategies include regulating district flow to maintain the customer's supply temperature (which results in a fluctuating customer return temperature) or maintaining the customer's return temperature (which results in a fluctuating customer supply temperature). When controlling return flow for cooling, the effect on the customer's ability to dehumidify properly with an elevated entering coil temperature should be investigated carefully.

Pressure Control Devices

If the steam or water pressure delivered to the customer is too high for direct use, it must be reduced. Similarly, pressure-reducing or pressure-sustaining valves may be required if building height creates a high static pressure and influences the district system's return water pressure. Water pressure can also be reduced by control valves or regenerative turbine pumps. The risk of using pressure-regulating devices to lower pressure on the return line is that if they fail, the entire distribution system is exposed to their pressure, and overpressurization will occur.

In high-rise buildings, all piping, valves, coils, and other equipment may be required to withstand higher design pressures. Where system static pressure exceeds safe or economical operating pressure, either the heat exchanger method or pressure-sustaining valves in the return line may be used to minimize the impact of the pressure. Vacuum vents should be provided at the top of the building's water risers to introduce air into the piping in case the vertical water column collapses.

HEATING CONNECTIONS

Steam Connections

Although higher pressures and temperatures are sometimes used, most district heating systems supply saturated steam at pressures between 5 and 250 psig to customers' facilities. The steam is pretreated to maintain a neutral pH, and the condensate is both cooled and discharged to the building sewage system or returned back to the central plant for recycling. Many consumers run the condensate through a heat exchanger to heat the domestic hot-water supply of the building before returning it to the central plant or to the building drains. This process extracts the maximum amount of energy out of the delivered steam.

Interconnection between the district and the building is simple when the building uses the steam directly in heating coils or radiators or for process loads (humidification, kitchen, laundry, laboratory, steam absorption chillers, or turbine-driven devices). Other buildings extract the energy from the district steam via a steam-to-water heat exchanger to generate hot water and circulate it to the air-side terminal units. Typical installations are shown in Figures 23 and 24. Chapter 10 has additional information on building distribution piping, valving, traps, and other system requirements. The type of steam chemical treatment should be considered in applications for the food industry and for humidification.

Other components of the steam connection may include condensate pumps, flowmeters (steam and/or condensate), and condensate conductivity probes, which may dump condensate if they are contaminated by unacceptable debris. Many times, energy meters are installed on both the steam and condensate pipes to allow the district energy supplier to determine how much energy is used directly and how much energy (condensate) is not returned back to the plant. The use of customer energy meters for both steam and condensate is desirable for the following reasons:

- Offers redundant metering (if the condensate meter fails, the steam meter can detect flow or vice versa)

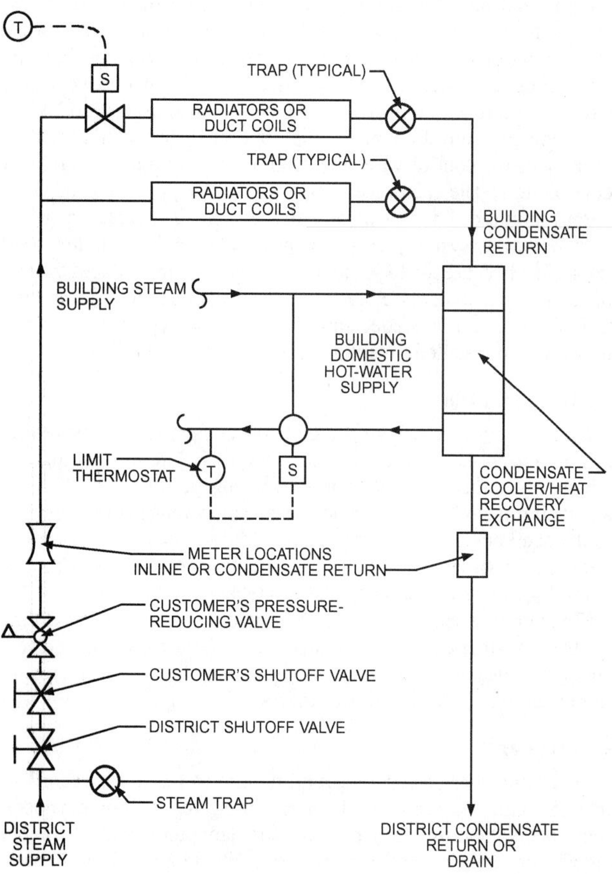

Fig. 23 District/Building Interconnection with Heat Recovery Steam System

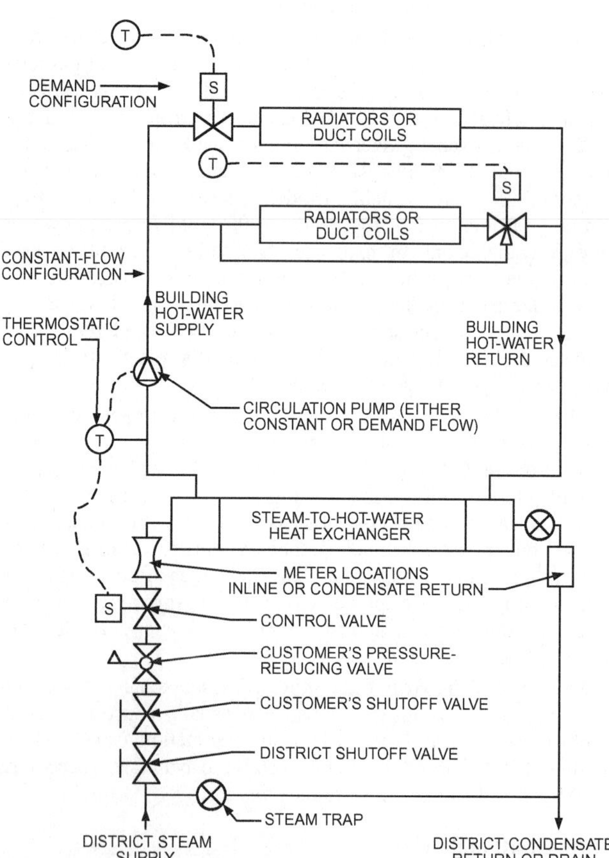

Fig. 24 District/Building Interconnection with Heat Exchange Steam System

- Bills customer accordingly for makeup water and chemical treatment on all condensate that is not returned or is contaminated
- Meter is in place if customer requires direct use of steam in the future
- Assists in identifying steam and condensate leaks
- Improves customer relations (may ease customer's fears of overbilling because of a faulty meter)
- Provides a more accurate reading for peak demand measurements and charges

Each level of steam pressure reduction should also be monitored as well as the temperature of the condensate. Where conductivity probes are used to monitor the quality of the water returned to the steam plant, adequate drainage and cold-water quenching equipment may be required to satisfy local plumbing code requirements (temperature of fluid discharging into a sewer). The probe status should also be monitored at the control panel, to communicate high conductivity alarms to the plant and, when condensate is being "dumped," to notify the plant a conductivity problem exists at a customer.

Hot-Water Connections

Figure 25 illustrates a typical indirect connection using a heat exchanger between the district hot-water system and the customer's system. It shows the radiator configurations typically used in both constant- and variable-flow systems. Figure 26 shows a typical direct connection between the district hot-water system and the building. It includes the typical configurations for both demand flow and constant-flow systems and the additional check valve and piping required for the constant-flow system. Figure 27 shows an indirect connection for both space and hot-water heating.

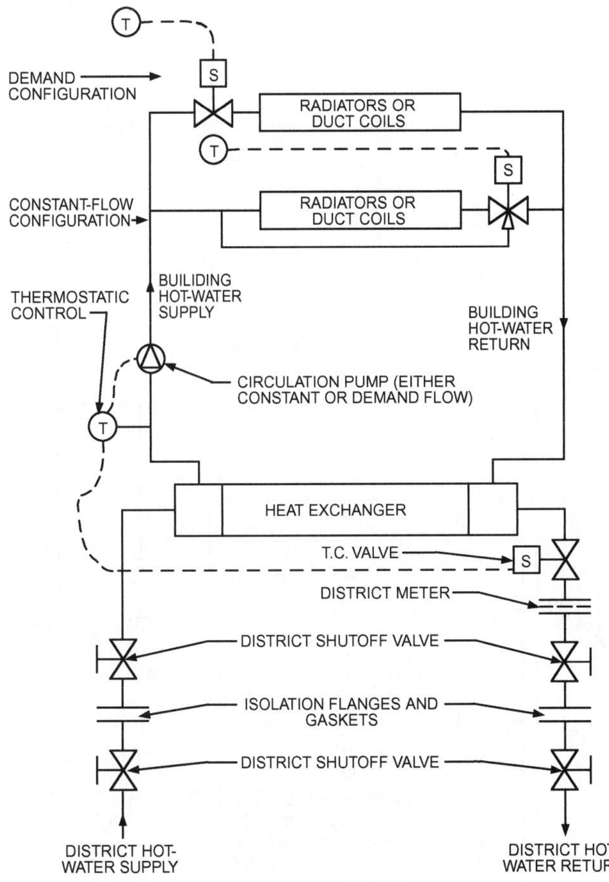

Fig. 25 District/Building Indirect Interconnection Hot-Water System

Lines on these systems must be sized using the same design used for the main feed lines of an in-building power plant. In general, demand flow systems permit better energy transfer efficiency and smaller line size for a given energy transfer requirement. Line sizing should account for any future loads on the building, etc. To keep return temperature low, water flow through heating equipment should be controlled according to the heating demand in the space.

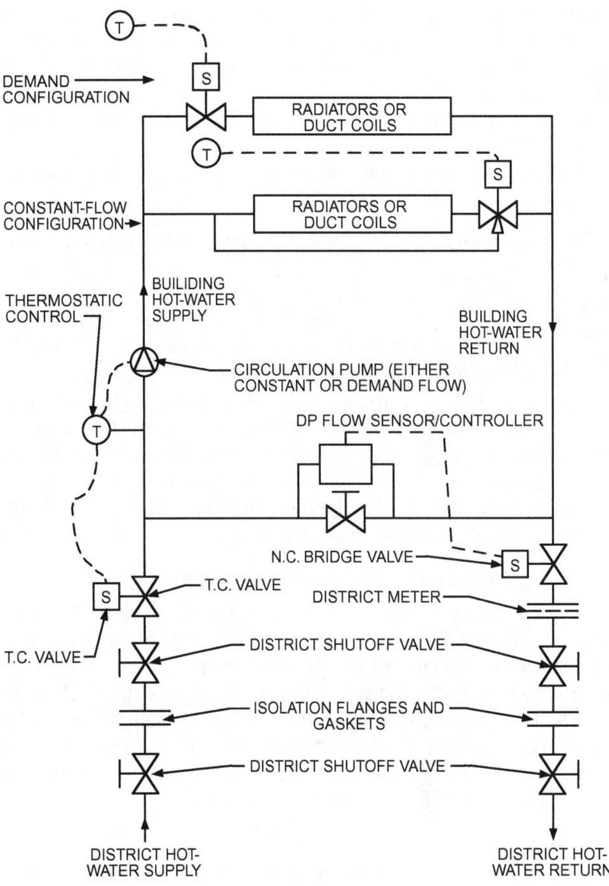

Fig. 26 District/Building Direct Interconnection Hot-Water System

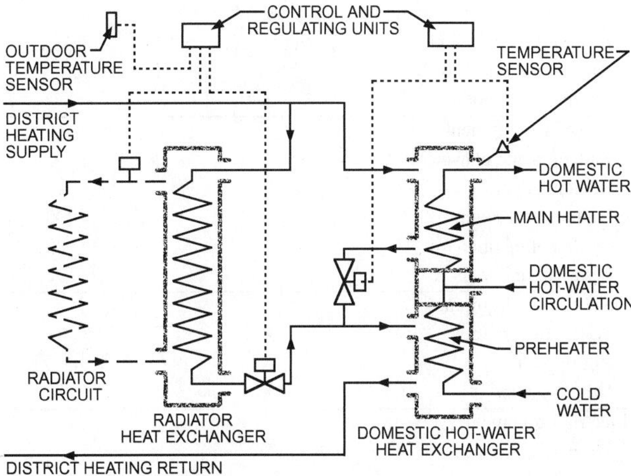

Fig. 27 Building Indirect Connection for Both Heating and Domestic Hot Water

The secondary supply temperature must be controlled in a manner similar to the primary supply temperature. To ensure a low primary return temperature and large temperature difference, the secondary system should have a low design return temperature. This design helps reduce costs through smaller pipes, pumps, amounts of insulation, and pump motors. A combined return temperature on the secondary side of 130°F or lower is reasonable, although this temperature may be difficult to achieve in smaller buildings with only baseboard or radiator space heating.

A hot-water district heating distribution system has many advantages over a steam system. Because of the efficiency of the relatively low temperature, energy can be saved over an equivalent steam system. Low- and medium-temperature hot-water district heating allows a greater flexibility in the heat source, lower-cost piping materials, and more cost-effective ways to compensate for expansion and new customer connections.

Major advantages of a hot-water distribution system include the following:

- Changing both temperature and flow can vary the delivered thermal power.
- Hot water can be pumped over a greater distance with minimal energy loss. This is a major disadvantage of a steam system.
- Operates at lower supply temperatures over the year. This results in lower line losses. Steam systems operate at constant pressure and temperature year round, which increases energy production and system losses while decreasing annual system efficiency.
- Has much lower operating and maintenance costs than steam systems (leaks at steam traps, chemical treatment, spares, etc.).
- Uses prefabricated insulated piping, which has a lower initial capital cost as compared to steam systems.

Building Conversion to District Heating

Table 4 (Sleiman et al. 1990) summarizes the suitability or success rate of converting various heating systems to be served by a district hot-water system. As can be seen, the probability is high for water-based systems, lower for steam, and lowest for fuel oil or electric based systems. Systems that are low on suitability usually require the high expense of replacing the entire heating terminal and generating units with suitable water-based equipment, including piping, pumps, controls, and heat transfer media.

Table 4 Conversion Suitability of Heating System by Type

Type of System	Low	Medium	High
Steam Equipment			
One-pipe cast iron radiation	X		
Two-pipe cast iron radiation		X	
Finned-tube radiation		X	
Air-handling unit coils		X	
Terminal unit coils	X		
Hot-water Equipment			
Radiators and convectors			X
Radiant panels			X
Unitary heat pumps			X
Air-handling unit coils			X
Terminal unit coils			X
Gas/Oil-fired Equipment			
Warm-air furnaces	X		
Rooftop units	X		
Other systems	X		
Electric Equipment			
Warm-air furnaces	X		
Rooftop units	X		
Air-to-air heat pumps	X		
Other systems	X		

CHILLED-WATER CONNECTIONS

Similar to district hot-water systems, chilled-water systems can operate with either constant or variable flow. Variable-flow systems can interconnect with either building demand or constant-flow systems. Variable flow is best if dehumidification is required with comfort cooling because the supply temperature remains relatively constant. Figure 28 illustrates a typical configuration for a variable-flow building interconnection using return water control or temperature differential control. However, the district return water side may be satisfied at the expense of an increase in the building supply water temperature. Typically, this connection also monitors the building-side supply water temperature to determine if it increases too much to control building humidity. The best method of connection is the simplest, with no control valve, but with high return water temperature at varying flows. However, this method requires the building design engineer and controls contractor to implement a design that operates per the design intent.

Typical constant-flow systems are found in older buildings and may be converted to simulate a variable-flow system by blocking off the bypass line around the air handler heat exchanger coil three-way control valve. At low operating pressures, this potentially may convert a three-way bypass-type valve to a two-way modulating shutoff valve. Careful analysis of the valve actuator must be undertaken, because shut-off requirements and control characteristics are totally different for a two-way valve compared to a three-way valve.

Peak demand requirements must be determined for the building at maximum design conditions (above the ASHRAE 0.4% design values). These conditions usually include direct bright sunlight on the building, 95 to 100°F dry-bulb temperature, and 73 to 78°F wet-bulb temperature occurring at peak conditions.

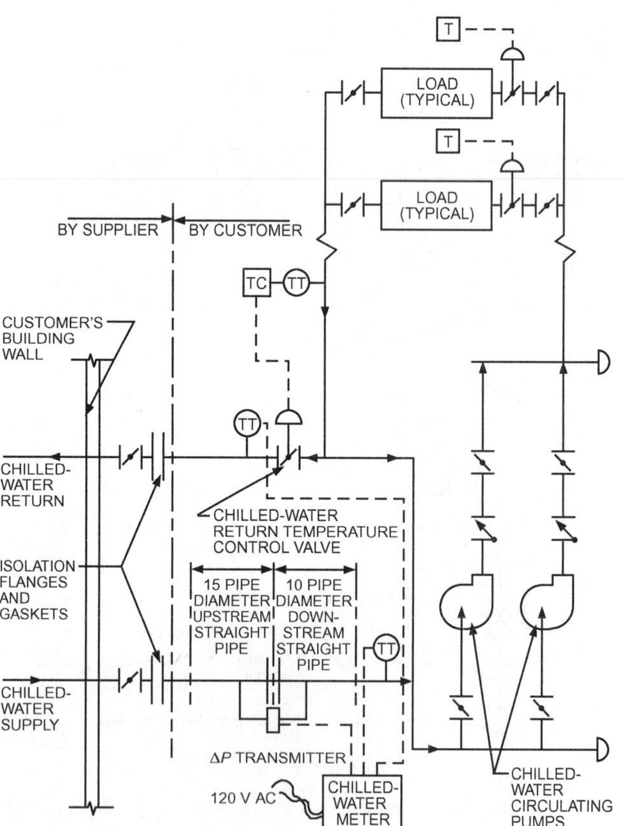

Fig. 28 Typical Chilled-Water Piping and Metering Diagram

The designer must consider the effect of return water temperature control. This is the single most important factor in obtaining a high temperature difference and providing an efficient plant. In theory, a partially loaded cooling coil should have higher return water temperature than at full load because the coil is oversized for the duty and hence has closer approach temperatures. In many real systems, as the load increases, the return water temperature tends to rise and, with a low-load condition, the supply water temperature rises. Consequently, process or critical humidity control systems may suffer when connected to a system where return water temperature control is used to achieve high temperature differentials. Other techniques, such as separately pumping each chilled water coil, may be used where constant supply water temperatures are necessary year-round.

TEMPERATURE DIFFERENTIAL CONTROL

The success of the district heating and cooling system efficiency is usually measured in terms of the temperature differential for water systems. Proper control of the district heating and cooling temperature differential is not dictated at the plant but at the consumer. If the consumer's system is not compatible with the temperature parameters of the DHC system, operating efficiency will suffer unless components in the consumer's system are modified.

Generally, maintaining a high temperature differential (Δt) between supply and return lines is most cost effective because it allows smaller pipes to be used in the primary distribution system. These savings must be weighed against any higher building conversion cost that may result from the need for a low primary return temperature. Furthermore, optimization of the Δt is critical to the successful operation of the district energy system. That is the reason the customer's Δt must be monitored and controlled.

To optimize the Δt and meet the customer's chilled-water demand, the flow from the plant should vary. Varying the flow also saves pump energy. Chilled-water flow in the customer's side must be varied as well. Terminal units in the building connected to the chilled-water loop (i.e., air-handling units, fan-coils, etc.) may require modifications (change three-way valves to two-way, etc.) to operate with variable water flow to ensure a maximum return water temperature.

For cooling coils, six-row 12 to 14 fins-per-inch coils are the minimum size coil applied to central station air-handling units to provide adequate performance. With this type of coil, the return water temperature rise should range from 12 to 16°F at full load. Coil performance at reduced loads should be considered as well; therefore, fluid velocity in the tube should remain high to stay in the turbulent flow range. To maintain a reasonable temperature differential at design conditions, fan-coil units are sized for an entering water temperature several degrees above the main chilled-water plant supply temperature. This requires that temperature-actuated diversity control valves be applied to the primary distribution cross connection between the supply and return piping.

METERING

All thermal energy or power delivered to customers or end users for billing or revenue by a commercially operated district energy system must be metered. The type of meter selected depends on the fluid to be measured, accuracy required, and expected turndown of flow to meet the low-flow and maximum-flow conditions. It is important that the meter be sized accurately for the anticipated loads and not oversized, because this will lead to inaccuracies. Historical, metered, or benchmarked data should be used when available if the actual load is not accurately known.

Steam may be used for direct comfort heating or to power absorption chillers for cooling. It is typically measured by using the differential pressure across calibrated orifices, nozzles, or venturi tubes; or by pitot tubes, vortex-shedding meters, or condensate meters. In the United States, the customary commercial unit is pounds of steam per hour with a heat equivalent typically assumed to be 1000 Btu/lb, but for more precise thermal metering, the meters are coupled with steam quality (temperature and pressure) differential heat content measuring devices. Care must be taken to deliver dry steam (superheated or saturated without free water) to the customer. Dry steam is delivered by installing an adequate trap just ahead of the customer's meter (or ahead of the customer's process when condensate meters are used).

When condensate meters are used, care must be taken to ensure that all condensate from the customer's process, but *only* such condensate, goes to the condensate meter. This may not be possible if the steam is used directly for humidification purposes and hence no condensate is returned. In this case, steam meters are preferable. IDHA (1969) and Stultz and Kitto (1992) have more information on steam metering. For steam, as with hot- and chilled-water system metering, electronic and computer technology provide direct, integrating, and remote input to central control/measurement energy management systems.

Hot- and chilled-water systems are metered by measuring the temperature differential between the supply and return lines and the flow rate of the energy transfer medium. Thermal (Btu or kWh) meters compensate for the actual volume and heat content characteristics of the energy transfer medium. Thermal transducers, resistance thermometer elements, or liquid expansion capillaries are usually used to measure the differential temperature of the energy transfer medium in supply and return lines.

Water flow can be measured with a variety of meters, usually pressure differential, turbine or propeller, or displacement meters. Chapter 14 of the 2005 *ASHRAE Handbook—Fundamentals*, the *District Heating Handbook* (IDHA 1983), and Pomroy (1994) have more information on measurement. Ultrasonic meters are sometimes used to check performance of installed meters. Various flowmeters are available for district energy billing purposes. Critical characteristics for proper installation include clearances and spatial limitations as well as the attributes presented in Table 5. The data in the table only provide general guidance, and the manufacturers of meters should be contacted for data specific to their products.

The meter should be located upstream of the heat exchanger and the control valve(s) should be downstream from the heat exchanger. This orientation minimizes the possible formation of bubbles in the flow stream and provide a more accurate flow indication. The transmitter should be calibrated for zero and span as recommended by the manufacturer.

Wherever possible, the type and size of meters selected should be standardized to reduce the number of stored spare parts, technician training, etc.

Displacement meters are more accurate than propeller meters, but they are also larger. They can handle flow ranges from less than 2% up to 100% of the maximum rated flow with claimed ±1% accuracy. Turbine-type meters require the smallest physical space for a given maximum flow. However, like many meters, they require at least 10 diameters of straight pipe upstream and downstream of the meter to achieve their claimed accuracy.

The United States has no performance standards for thermal meters. ASHRAE *Standard* 125 describes a test method for rating liquid thermal meters. Several European countries have developed performance standards and/or test methods for thermal meters, and CEN *Standard* EN 1434, developed by the European Community, is a performance and testing standard for heat meters.

District energy plant meters intended for billing or revenue require means for verifying performance periodically. Major meter manufacturers, some laboratories, and some district energy companies maintain facilities for this purpose. In the absence of a single performance standard, meters are typically tested in accordance with their respective manufacturers' recommendations. Primary

Table 5 Flowmeter Characteristics

Meter Type	Accuracy	Range of Control	Pressure Loss	Straight Piping Requirements (Length in Pipe Diameters)
Orifice plate	±1% to 5% full scale	3:1 to 5:1	High (>5 psi)	10 D to 40 D upstream; 2 D to 6 D downstream
Electromagnetic	±0.15% to 1% rate	30:1 to 100:1	Low (<3 psi)	5 D to 10 D upstream; 3 D downstream
Vortex	±0.5% to 1.25% rate	10:1 to 25:1	Medium (3 to 5 psi)	10 D to 40 D upstream; 2 D to 6 D downstream
Turbine	±0.15% to 0.5% rate	10:1 to 50:1	Medium (3 to 5 psi)	10 D to 40 D upstream; 2 D to D downstream
Ultrasonic	±1% to 5% rate	>10:1 to 100:1	Low (<3 psi)	10 D to 40 D upstream; 2 D to 6 D downstream

measurement elements used in these laboratories frequently obtain calibration traceability to the National Institute of Standards and Technology (NIST).

For district energy cogeneration systems that send out and/or accept electric power to or from a utility grid, the demand and usage meters must meet the existing utility requirements. For district energy systems that send out electric power directly to customers, the electric demand and usage meters must comply with local and state regulations. American National Standards Institute (ANSI) standards are established for all customary electric meters.

OPERATION AND MAINTENANCE

As with any major capital equipment, care must be exercised in operating and maintaining district heating and cooling systems. Both central plant equipment and terminal equipment located in the consumer's building must be operated within intended parameters and maintained on a schedule as recommended by the manufacturer. Thermal distribution systems, especially buried systems, which are out of sight, may suffer from inadequate maintenance. To maintain the thermal efficiency of the distribution system as well as its reliability, integrity, and service life, periodic preventative maintenance is strongly recommended. In the case of steam and hot-water distribution systems, it may also be a matter of due diligence on the part of the owner/operator to ensure system integrity, to avoid thermal damage to adjacent property or harm to individuals coming in contact with the system or its thermal effects. Establishing an adequate preventive maintenance program for the distribution system requires consulting the recommendations of the system manufacturer (where applicable) as well as considering the type, age, and condition of the system and its operating environment.

REFERENCES

Aamot, H. and G. Phetteplace. 1978. Heat transmission with steam and hot water. *Publication* H00128. American Society of Mechanical Engineers, New York.

Albert, M.R. and G.E. Phetteplace. 1983. Computer models for two-dimensional steady-state heat conduction. CRREL *Report* 83-10. U.S. Army Cold Regions Research and Engineering Laboratory, Hanover, NH.

Andrepont, J.S. 1995. Chilled water storage: A suite of benefits for district cooling. *District Energy* 81(1). International District Energy Association, Washington, D.C.

ASCE. 1996. *Cold regions utilities monograph.* American Society of Civil Engineers, Reston, VA.

ASHRAE. 1992. Method of testing thermal energy meters for liquid streams in HVAC systems. ANSI/ASHRAE *Standard* 125-1992 (RA 2000).

ASME. 1998. Power piping. *Standard* B31.1. American Society of Mechanical Engineers, New York.

ASTM. 1995. Standard practice for determination of heat gain or loss and the surface temperatures of insulated pipe and equipment systems by the use of a computer program. *Standard* C680-98 (R 1995). American Society for Testing and Materials, West Conshohocken, PA.

Bloomquist, R.G., R. O'Brien, and M. Spurr. 1999. *Geothermal district energy at co-located sites.* WSU-EEP 99007, Washington State University Energy Office.

Bøhm, B. 1986. On the optimal temperature level in new district heating networks. *Fernwärme International* 15(5):301-306.

Bøhm, B. 1988. *Energy-economy of Danish district heating systems: A technical and economic analysis.* Laboratory of Heating and Air Conditioning, Technical University of Denmark, Lyngby.

Bottorf, J.D. 1951. *Summary of thermal conductivity as a function of moisture content.* Thesis, Purdue University, West Lafayette, IN.

CEN. 1997. Heat meters. *Standard* EN 1434. Comité Européen de Normalisation.

Chyu, M.-C., X. Zeng, and L. Ye. 1997a. Performance of fibrous glass insulation subjected to underground water attack. *ASHRAE Transactions* 103(1):303-308.

Chyu, M.-C., X. Zeng, and L. Ye. 1997b. The effect of moisture content on the performance of polyurethane insulation on a district heating and cooling pipe. *ASHRAE Transactions* 103(1):309-317.

Chyu, M.-C., X. Zeng, and L. Ye. 1998a. Behavior of cellular glass insulation on a DHC pipe subjected to underground water attack. *ASHRAE Transactions* 104(2):161-167.

Chyu, M.-C., X. Zeng, and L. Ye. 1998b. Effect of underground water attack on the performance of mineral wool pipe insulation. *ASHRAE Transactions* 104(2):168-175.

COWIconsult. 1985. *Computerized planning and design of district heating networks.* COWIconsult Consulting Engineers and Planners AS, Virum, Denmark.

CRREL. 1999. *Regional climatic constants for Equation 6 of the Corps of Engineers Guide Spec 02695.* (Best fit to mean monthly temperatures averaged for the period 1895-1996). U.S. Army Cold Regions Research and Engineering Laboratory, Hanover, NH.

EuHP. 1991. *Preinsulated bonded pipe systems.* EN253, European District Heating Pipe Manufacturers Association, Brussels.

Farouki, O.T. 1981. Thermal properties of soils. CRREL *Monograph* 81-1. U.S. Army Cold Regions Research and Engineering Laboratory, Hanover, NH.

Fox, J.A. 1977. *Hydraulic analysis of unsteady flow in pipe networks.* John Wiley & Sons, New York.

Geiringer, P.L. 1963. *High temperature water heating: Its theory and practice for district and space heating applications.* John Wiley & Sons, New York.

Grober, H., S. Erk, and U. Grigull. 1961. *Fundamentals of heat transfer.* McGraw-Hill, New York.

IDHA. 1969. *Code for steam metering.* International District Energy Association, Washington, D.C.

IDHA. 1983. *District heating handbook*, 4th ed. International District Energy Association, Washington, D.C.

Jeppson, R.W. 1977. *Analysis of flow in pipe networks.* Ann Arbor Science, MI.

Koskelainen, L. 1980. Optimal dimensioning of district heating networks. *Fernwärme International* 9(4):84-90.

Lunardini, V.J. 1981. *Heat transfer in cold climates.* Van Nostrand Reinhold, New York.

McCabe R.E., J.J. Bender, and K.R. Potter. 1995. Subsurface ground temperature—Implications for a district cooling system. *ASHRAE Journal* 37(12):40-45.

Minkowycz, W.J., E.M. Sparrow, G.E. Schneider, and R.H. Pletcher. 1988. *Handbook of numerical heat transfer.* John Wiley & Sons, New York.

MSS. 1993. Pipe hangers and supports—Materials, design and manufacture. *Standard* MS-58-1993. Manufacturers Standardization Society of the Valve and Fittings Industry, Vienna, VA.

MSS. 1996. Pipe hangers and supports—Selection and application. *Standard* MS-69-1996. Manufacturers Standardization Society of the Valve and Fittings Industry, Vienna, VA.

NACE. 1996. Control of external corrosion on underground or submerged metallic piping systems. *Standard* RPO169-96. National Association of Corrosion Engineers, Houston, TX.

NAS. 1975. *Technical Report No.* 66. National Academy of Sciences National Research Council Building Research Advisory Board (BRAB), Federal Construction Council.

Ottmer, J.H. and J.B. Rishel. 1993. Airport pumping system horsepower requirements take a nose dive. *Heating, Piping and Air Conditioning* 65(10).

Phetteplace, G.E. 1989. Simulation of district heating systems for piping design. International Symposium on District Heat Simulation, Reykjavik, Iceland.

Phetteplace, G.E., D. Carbee, and M. Kryska. 1991. *Field measurement of heat losses from three types of heat distribution systems.* CRREL SR 631. U.S. Army Cold Regions Research and Engineering Laboratory, Hanover, NH.

Phetteplace, G. and V. Meyer. 1990. Piping for thermal distribution systems. CRREL *Internal Report* 1059. U.S. Army Cold Regions Research and Engineering Laboratory, Hanover, NH.

Phetteplace, G.E., W. Willey, and M.A. Novick. 1981. *Losses from the Fort Wainwright heat distribution system.* CRREL SR 81-14. U.S. Army Cold Regions Research and Engineering Laboratory, Hanover, NH.

Pomroy, J. 1994. Selecting flowmeters. *Instrumentation & control systems.* Chilton Publications.

Rao, S.S. 1982. *The finite element method in engineering.* Pergamon, New York.

Rasmussen, C.H. and J.E. Lund. 1987. *Computer aided design of district heating systems.* District Heating Research and Technological Development in Denmark, Danish Ministry of Energy, Copenhagen.

Reisman, A.W. 1985. *District heating handbook*, vol. 2: *A handbook of district heating and cooling models.* International District Energy Association, Washington, D.C.

Rohsenow, W.M. 1998. *Handbook of heat transfer.* McGraw-Hill, New York.

Segan, E.G. and C.-P. Chen. 1984. Investigation of tri-service heat distribution systems. CERL *Technical Report* M-347. U.S. Army Construction Engineering Research Laboratory (CERL), Champaign, IL.

Siegel, R. and J.R. Howell. 1981. *Thermal radiation heat transfer.* Hemisphere, New York.

Sleiman, A.H. et al. 1990. *Guidelines for converting building heating systems for hot water district heating.* NOVEM, Sittard, The Netherlands.

Stephenson, D. 1981. *Pipeline design for water engineers.* Elsevier Scientific, New York.

Stewart, W.E. and C.L. Dona. 1987. Water flow rate limitations. *ASHRAE Transactions* 93(2):811-825.

Streeter, V.L. and E.B. Wylie. 1979. *Fluid mechanics.* McGraw-Hill, New York.

Stultz, S.C. and J.B. Kitto, eds. 1992. *Steam: Its generation and use*, 40th ed. Chapter 40-13, Measurement of steam quality and purity. Chapter 40-19, Measurement of steam flow. Babcock & Wilcox, Barberton, OH.

Werner, S.E. 1984. *The heat load in district heating systems.* Chalmers University of Technology, Göteborg, Sweden.

BIBLIOGRAPHY

Bowling, T. 1990. *A technology assessment of potential telemetry technologies for district heating.* International Energy Agency and NOVEM, Sittard, The Netherlands.

Holtse, C. and P. Randlov, eds. 1989. *District heating and cooling R&D project review.* International Energy Agency and NOVEM, Sittard, The Netherlands.

Kusuda, T. and P.R. Achenbach. 1965. Earth temperature and thermal diffusivity at selected stations in the United States. *ASHRAE Transactions* 71(1):61-75.

Mørck, O. and T. Pedersen, eds. 1989. *Advanced district heating production technologies.* International Energy Agency and NOVEM, Sittard, The Netherlands.

Ulseth, R., ed. 1990. *Heat meters: Report of research activities.* International Energy Agency and NOVEM, Sittard, The Netherlands.

U.S. Air Force, Army, and Navy. 1978. Engineering weather data. Dept. of the Air Force *Manual* AFM 88-29, Dept. of the Army *Manual* TM 5-785, and Dept. of the Navy *Manual* NAVFAC P-89.

CHAPTER 12

HYDRONIC HEATING AND COOLING SYSTEM DESIGN

WATER systems that convey heat to or from a conditioned space or process with hot or chilled water are frequently called *hydronic systems*. Water flows through piping that connects a boiler, water heater, or chiller to suitable terminal heat transfer units located at the space or process.

Water systems can be classified by (1) operating temperature, (2) flow generation, (3) pressurization, (4) piping arrangement, and (5) pumping arrangement.

Classified by flow generation, hydronic heating systems may be (1) **gravity systems**, which use the difference in density between the supply and return water columns of a circuit or system to circulate water; or (2) **forced systems**, in which a pump, usually driven by an electric motor, maintains flow. Gravity systems are seldom used today and are therefore not discussed in this chapter. See the *ASHVE Heating Ventilating Air Conditioning Guide* issued before 1957 for information on gravity systems.

Water systems can be either once-through or recirculating systems. This chapter describes forced recirculating systems.

Successful water system design depends on awareness of the many complex interrelationships between various elements. In a practical sense, no component can be selected without considering its effect on the other elements. For example, design water temperature and flow rates are interrelated, as are the system layout and pump selection. The type and control of heat exchangers used affect the flow rate and pump selection, and the pump selection and distribution affect the controllability. The designer must thus work back and forth between tentative points and their effects until a satisfactory integrated design has been reached. Because of these relationships, rules of thumb usually do not lead to a satisfactory design.

Principles

Effective and economical water system design is affected by complex relationships between the various system components. The design water temperature, flow rate, piping layout, pump selection, terminal unit selection, and control method are all interrelated. System size and complexity determine the importance of these relationships to the total system operating success. In the United States, present hydronic heating system design practice originated in residential heating applications, where a temperature drop Δt of 11 K was used to determine flow rate. However, almost universal use of hydronic systems for both heating and cooling of large buildings and building complexes has rendered this simplified approach obsolete.

The preparation of this chapter is assigned to TC 6.1, Hydronic and Steam Equipment and Systems.

TEMPERATURE CLASSIFICATIONS

Water systems can be classified by operating temperature as follows.pump

Low-temperature water (LTW) systems operate within the pressure and temperature limits of the ASME *Boiler and Pressure Vessel Code* for low-pressure boilers. The maximum allowable working pressure for low-pressure boilers is 1100 kPa (gage), with a maximum temperature of 120°C. The usual maximum working pressure for boilers for LTW systems is 200 kPa, although boilers specifically designed, tested, and stamped for higher pressures are frequently used. Steam-to-water or water-to-water heat exchangers are also used for heating low-temperature water. Low-temperature water systems are used in buildings ranging from small, single dwellings to very large and complex structures.

Medium-temperature water (MTW) systems operate between 120 and 175°C, with pressures not exceeding 1100 kPa. The usual design supply temperature is approximately 120 to 160°C, with a usual pressure rating of 1 MPa for boilers and equipment.

High-temperature water (HTW) systems operate at temperatures over 175°C and usual pressures of about 2 MPa. The maximum design supply water temperature is usually about 200°C, with a pressure rating for boilers and equipment of about 2 MPa. The pressure-temperature rating of each component must be checked against the system's design characteristics.

Chilled-water (CW) systems for cooling normally operate with a design supply water temperature of 4 to 13°C (usually 7°C), and at a pressure of up to 830 kPa. Antifreeze or brine solutions may be used for applications (usually process applications) that require temperatures below 4°C or for coil freeze protection. Well-water systems can use supply temperatures of 15°C or higher.

Dual-temperature water (DTW) systems combine heating and cooling, and circulate hot and/or chilled water through common piping and terminal heat transfer apparatus. These systems operate within the pressure and temperature limits of LTW systems, with usual winter design supply water temperatures of about 38 to 65°C and summer supply water temperatures of 4 to 7°C.

Terminal heat transfer units include convectors, cast-iron radiators, baseboard and commercial finned-tube units, fan-coil units, unit heaters, unit ventilators, central station air-handling units, radiant panels, and snow-melting panels. A large storage tank may be included in the system to store energy to use when heat input devices such as the boiler or a solar energy collector are not supplying energy.

This chapter covers the principles and procedures for designing and selecting piping and components for low-temperature water, chilled water, and dual-temperature water systems. See Chapter 14 for information on medium- and high-temperature water systems.

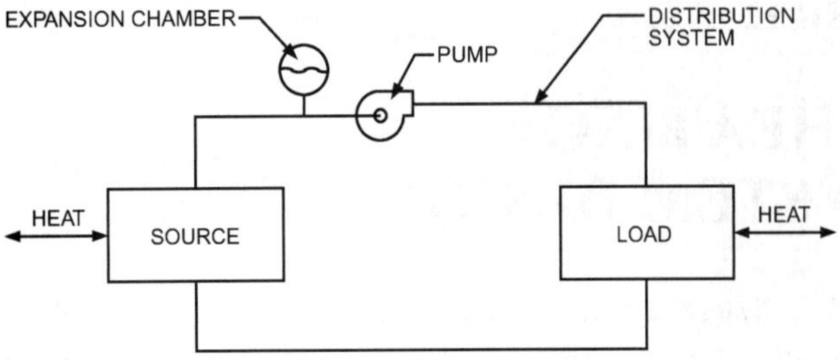

Fig. 1 Fundamental Components of Hydronic System

CLOSED WATER SYSTEMS

Because most hot- and chilled-water systems are closed, this chapter addresses only closed systems. The fundamental difference between a closed and an open water system is the interface of the water with a compressible gas (such as air) or an elastic surface (such as a diaphragm). A **closed water system** is defined as one with no more than one point of interface with a compressible gas or surface, and that will not create system flow by changes in elevation. This definition is fundamental to understanding the hydraulic dynamics of these systems. Earlier literature referred to a system with an open or vented expansion tank as an "open" system, but this is actually a closed system; the atmospheric interface of the tank simply establishes the system pressure.

An **open system**, on the other hand, has more than one such interface. For example, a cooling tower system has at least two points of interface: the tower basin and the discharge pipe or nozzles entering the tower. One major difference in hydraulics between open and closed systems is that some hydraulic characteristics of open systems cannot occur in closed systems. For example, in contrast to the hydraulics of an open system, in a closed system (1) flow cannot be motivated by static head differences, (2) pumps do not provide static lift, and (3) the entire piping system is always filled with water.

Figure 1 shows the fundamental components of a closed hydronic system. Actual systems generally have additional components such as valves, vents, regulators, etc., but these are not essential to the basic principles underlying the system.

These fundamental components are

- Loads
- Source
- Expansion chamber
- Pump
- Distribution system

Theoretically, a hydronic system could operate with only these five components.

The components are subdivided into two groups: thermal and hydraulic. Thermal components consist of the load, source, and expansion chamber. Hydraulic components consist of the distribution system, pump, and expansion chamber. The expansion chamber is the only component that serves both a thermal and a hydraulic function.

METHOD OF DESIGN

This section outlines general steps a designer may follow to complete system design. The methodology is not a rigid framework, but rather a flexible outline that should be adapted by the designer to suit current needs. The general order as shown is approximately chronological, but it is important to note that succeeding steps often affect preceding steps, so a fundamental reading of this entire chapter is required to fully understand the design process.

1. **Determine system and zone loads.** Loads are covered in Chapters 27 to 32 of the 2005 *ASHRAE Handbook—Fundamentals*. Several load calculation procedures have been developed, with varying degrees of calculation accuracy. The load determines the flow of the hydronic system, which ultimately affects the system's heat transfer ability and energy performance. Designers should apply the latest computerized calculation methods for optimal system design. Load calculation should also detail the facility's loading profile facility to enhance the hydronic system control strategy.
2. **Select comfort heat transfer devices.** This often means a coil- or water-to-air heat exchanger (terminal). Coil selection and operation has the single largest influence on hydronic system design. Coils implement the design criteria of flow, temperature drop, and control ability. Coil head loss and location affects pipe design and sizing, control devices, and pump selection. For details on coils, see Chapters 22 and 26.
3. **Select system distribution style(s).** Based on the load and its location, different piping styles may be appropriate for a given design. Styles may be comingled in a successful hydronic system design to optimize building performance. Schematically lay out the system to establish a preliminary design.
4. **Size branch piping system.** Based on the selection of the coil, its controlling devices, style of installation, and location, branch piping is sized to provide required flow, and head loss is calculated.
5. **Calculate distribution piping head loss.** Although the criteria for pipe selection in branch and distribution system piping may be similar, understanding the relationship and effect of distribution system head loss is important in establishing that all terminals get the required flow for the required heat transfer.
6. **Lay out piping system and size pipes.** After preliminary calculations of target friction loss for the pipes, sketch the system. After the piping system is laid out and the calculations of actual design head loss are complete, note the losses on the drawings for the commissioning process.
7. **Select pump specialties.** Any devices required for operation or measurement are identified, so their head loss can be determined and accounted for in pump selection.
8. **Select air management methodology.** All hydronic systems entrain air in the circulated fluid. Managing the collection of that air as it leaves the working fluid is essential to management of system pressure and the safe operation of system components.
9. **Select pump (hydraulic components).** Unless a system is very small (e.g., a residential hot-water heating system), the pump is selected to fit the system. A significant portion of energy use in a hydronic system is transporting the fluid through the distribution system. Proper pump selection limits this energy use, whereas improper selection leads to energy inefficiency and poor distribution and heat transfer.
10. **Determine installation details, iterate design.** Tuning the design to increase performance and cost effectiveness is an important last step. Documenting installation details is also important, because this communication is necessary for well-built designs and properly operated systems.

THERMAL COMPONENTS

Loads

The load is the device that causes heat to flow out of or into the system to or from the space or process; it is the independent variable to which the remainder of the system must respond. Outward heat flow characterizes a heating system, and inward heat flow characterizes a cooling system. The quantity of heating or cooling is calculated by one of the following means.

Sensible Heating or Cooling. The rate of heat entering or leaving an airstream is expressed as follows:

$$q = 60Q_a \rho_a c_p \Delta t \quad (1)$$

where

q = heat transfer rate to or from air, Btu/h
Q_a = airflow rate, cfm
ρ_a = density of air, lb/ft^3
c_p = specific heat of air, Btu/lb·°F
Δt = temperature increase or decrease of air, °F

For standard air with a density of 0.075 lb/ft^3 and a specific heat of 0.24 Btu/lb·°F, Equation (1) becomes

$$q = 1.1Q_a \Delta t \quad (2)$$

The heat exchanger or coil must then transfer this heat from or to the water. The rate of sensible heat transfer to or from the heated or cooled medium in a specific heat exchanger is a function of the heat transfer surface area; the mean temperature difference between the water and the medium; and the overall heat transfer coefficient, which itself is a function of the fluid velocities, properties of the medium, geometry of the heat transfer surfaces, and other factors. The rate of heat transfer may be expressed by

$$q = UA(\text{LMTD}) \quad (3)$$

where

q = heat transfer rate through heat exchanger, Btu/h
U = overall coefficient of heat transfer, Btu/h·ft^2·°F
A = heat transfer surface area, ft^2
LMTD = logarithmic mean temperature difference, heated or cooled medium to water, °F

Cooling and Dehumidification. The rate of heat removal from the cooled medium when both sensible cooling and dehumidification are present is expressed by

$$q_t = w\Delta h \quad (4)$$

where

q_t = total heat transfer rate from cooled medium, Btu/h
w = mass flow rate of cooled medium, lb/h
Δh = enthalpy difference between entering and leaving conditions of cooled medium, Btu/lb

Expressed for an air-cooling coil, this equation becomes

$$q_t = 60Q_a \rho_a \Delta h \quad (5)$$

which, for standard air with a density of 0.075 lb/ft^3, reduces to

$$q_t = 4.5Q_a \Delta h \quad (6)$$

Heat Transferred to or from Water. The rate of heat transfer to or from the water is a function of the flow rate, specific heat, and temperature rise or drop of the water as it passes through the heat exchanger. The heat transferred to or from the water is expressed by

$$q_w = \dot{m} c_p \Delta t \quad (7)$$

where

q_w = heat transfer rate to or from water, Btu/h
$\dot{m}$ = mass flow rate of water, lb/h
c_p = specific heat of water, Btu/lb·°F
Δt = water temperature increase or decrease across unit, °F

With water systems, it is common to express the flow rate as volumetric flow, in which case Equation (7) becomes

$$q_w = 8.02\rho_w c_p Q_w \Delta t \quad (8)$$

where

Q_w = water flow rate, gpm
ρ_w = density of water, lb/ft^3

For standard conditions in which the density is 62.4 lb/ft^3 and the specific heat is 1 Btu/lb·°F, Equation (8) becomes

$$q_w = 500Q_w \Delta t \quad (9)$$

Equation (8) or (9) can be used to express the heat transfer across a single load or source device, or any quantity of such devices connected across a piping system. In the design or diagnosis of a system, the load side may be balanced with the source side using these equations.

Heat-Carrying Capacity of Piping. Equations (8) and (9) are also used to express the heat-carrying capacity of the piping or distribution system or any portion thereof. The existing temperature differential Δt, sometimes called the temperature range, is identified; for any flow rate Q_w through the piping, q_w is called the **heat-carrying capacity**.

Terminal Heating and Cooling Units

Many types of terminal units are used in central water systems, and may be classified in several different ways:

- **Natural convection units** include cabinet convectors, baseboard, and finned-tube radiation. Older systems may have cast-iron radiators, which are sometimes sought out for architectural restoration.
- **Forced-convection units** include unit heaters, unit ventilators, fan-coil units, air-handling units, heating and cooling coils in central station units, and most process heat exchangers. Fan-coil units, unit ventilators, and central station units can be used for heating, ventilating, and cooling.
- **Radiation units** include panel systems, unit radiant panels, in-floor or wall piping systems, and some older styles of radiators. All transfer some convective heat. These units are generally used in low-temperature water systems, with lower design temperatures. Similarly, chilled panels are also used for sensible cooling and in conjunction with central station air-handling units isolating outdoor air conditioning.

Terminal units must be selected for sufficient capacity to match the calculated heating and cooling loads. Manufacturers' ratings should be used with reference to actual operating conditions. Ratings are either computer selected, or cataloged by water temperature, temperature drop or rise, entering air temperatures, water velocity, and airflow. Ratings are usually given for standard test conditions with correction factors or curves, and rating tables are given covering a range of operating conditions. Because the choice of terminal units for any particular building or type of system is so wide, the designer must carefully consider the advantages and disadvantages of the various alternatives so the end result is maximum comfort and economy.

Most **load devices** (in which heat is conveyed to or from the water for heating or cooling the space or process) are a water-to-air finned-coil heat exchanger or a water-to-water exchanger. The specific configuration is usually used to describe the load device. The most common configurations include the following:

Heating load devices
Preheat coils in central units
Heating coils in central units
Zone or central reheat coils
Finned-tube radiators
Baseboard radiators

Convectors
Unit heaters
Fan-coil units
Water-to-water heat exchangers
Radiant heating panels
Snow-melting panels

Cooling load devices
Coils in central units
Fan-coil units
Induction unit coils
Radiant cooling panels
Water-to-water heat exchangers

Source

The source is the point where heat is added to (heating) or removed from (cooling) the system. Ideally, the amount of energy entering or leaving the source equals the amount entering or leaving through the load. Under steady-state conditions, the load energy and source energy are equal and opposite. Also, when properly measured or calculated, temperature differentials and flow rates across the source and loads are all equal. Equations (8) and (9) express the source capacities as well as the load capacities.

Any device that can be used to heat or cool water under controlled conditions can be used as a source device. The most common source devices for heating and cooling systems are the following:

Heating source devices
Hot-water generator or boiler
Steam-to-water heat exchanger
Water-to-water heat exchanger
Solar heating panels
Heat recovery or salvage heat device
(e.g., water jacket of an internal combustion engine)
Exhaust gas heat exchanger
Incinerator heat exchanger
Heat pump condenser
Air-to-water heat exchanger

Cooling source devices
Electric compression chiller
Thermal absorption chiller
Heat pump evaporator
Air-to-water heat exchanger
Water-to-water heat exchanger

The two primary considerations in selecting a source device are the design capacity and the part-load capability, sometimes called the **turndown ratio**. The turndown ratio, expressed in percent of design capacity, is

$$\text{Turndown ratio} = 100\,\frac{\text{Minimum capacity}}{\text{Design capacity}} \tag{10}$$

The reciprocal of the turndown ratio is sometimes used (for example, a turndown ratio of 25% may also be expressed as a turndown ratio of 4).

The turndown ratio has a significant effect on system performance; lack of consideration of the source system's part-load capability has been responsible for many systems that either do not function properly or do so at the expense of excess energy consumption. The turndown ratio has a significant effect on the ultimate equipment and/or system design selection.

Note that the turndown ratio for a source is different from that specified for a control valve. Turndown for a valve is comparable to valve rangeability. Whereas **rangeability** is the relationship of the maximum controllable flow to the minimum controllable flow based on testing, **turndown** is the relationship of the valve's normal maximum flow to minimum controllable flow.

System Temperatures. Design temperatures and temperature ranges are selected by consideration of the performance requirements and the economics of the components. For a cooling system that must maintain 50% rh at 75°F, the dew-point temperature is 55°F, which sets the maximum return water temperature at something near 55°F (60°F maximum); on the other hand, the lowest practical temperature for refrigeration, considering the freezing point and economics, is about 40°F. This temperature spread then sets constraints for a chilled-water system. Pedersen et al. (1998) describe a classic method for calculating the required temperature of chilled water from the psychrometric chart.

The entering water temperature follows the relationship

$$t_w = 2t_{ad} - t_{wb} \tag{11}$$

where

t_w = Coil entering water temperature
t_{ad} = Apparatus dew point
t_{wb} = Coil leaving air wet-bulb temperature

The designer should note that there are also constraints imposed on the temperature and differential temperature selection by the chiller selection (e.g., refrigerant choice), and on the coil's flow tolerance with respect to heat transfer. Consult with the chiller manufacturer so that performance requirements of the chiller are taken into account.

For a heating system, the maximum hot-water temperature is normally established by the ASME *Boiler and Pressure Vessel Code* as 250°F, and with space temperature requirements of little above 75°F, the actual operating supply temperatures and temperature ranges are set by the design of the load devices. Most economic considerations relating to distribution and pumping systems favor using the maximum possible temperature range Δt.

Expansion Chamber

The expansion chamber (also called an **expansion** or **compression tank**) serves both a thermal and a hydraulic function. In its thermal function, the tank provides a space into which the noncompressible liquid can expand or from which it can contract as the liquid undergoes volumetric changes with changes in temperature. To allow for this expansion or contraction, the expansion tank provides an interface point between the system fluid and a compressible gas. By definition, a closed system can have only one such interface; thus, a system designed to function as a closed system can have only one expansion chamber.

Expansion tanks are of three basic configurations: (1) a closed tank, which contains a captured volume of compressed air and water, with an air/water interface (sometimes called a **plain steel tank**); (2) an open tank (i.e., a tank open to the atmosphere); and (3) a diaphragm tank, in which a flexible membrane is inserted between the air and the water (a modified version is the **bladder tank**).

Properly installed, a closed or diaphragm tank serves the purpose of system pressurization control with a minimum of exposure to air in the system. Open tanks, commonly used in older systems, tend to introduce air into the system, which can enhance piping corrosion. Open tanks are generally not recommended for application in current designs. Older-style steel compression tanks tend to be larger than diaphragm expansion tanks. In some cases, there may be economic considerations that make one tank preferable over another. These economics usually are relatively straightforward (e.g., initial cost), but there can be significant size differences, which affect placement and required building space and structural support, and these effects should also be considered.

Sizing the tank is the primary thermal consideration in incorporating a tank into a system. However, before sizing the tank, air control or elimination must be considered. The amount of air that will be absorbed and can be held in solution with the water is expressed by Henry's equation (Pompei 1981):

$$x = p/H \qquad (12)$$

where

x = solubility of air in water (% by volume)
p = absolute pressure
H = Henry's constant

Henry's constant, however, is constant only for a given temperature (Figure 2). Combining the data of Figure 2 (Himmelblau 1960) with Equation (12) results in the solubility diagram of Figure 3.

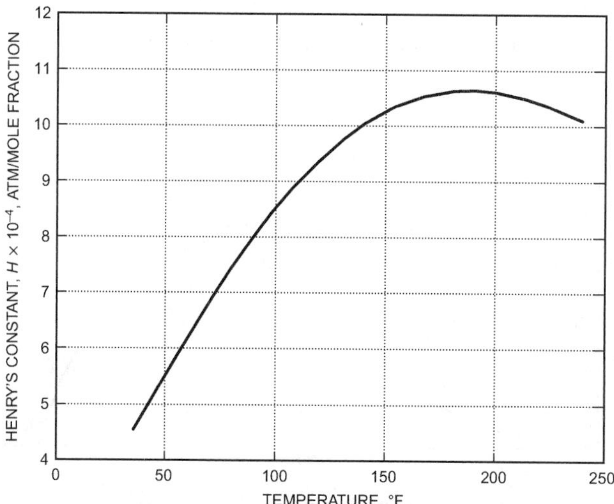

Fig. 2 Henry's Constant Versus Temperature for Air and Water
(Coad 1980a)

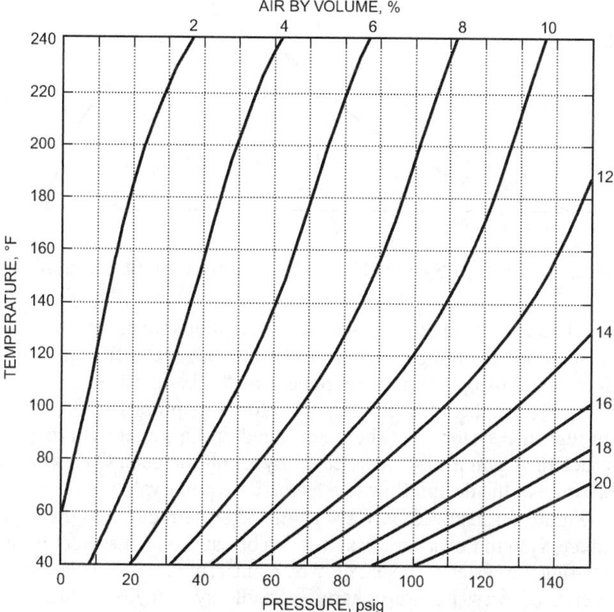

Fig. 3 Solubility Versus Temperature and Pressure for Air/Water Solutions
(Coad 1980a)

With that diagram, the solubility can be determined if the temperature and pressure are known.

If the water is not saturated with air, it will absorb air at the air/water interface until the point of saturation has been reached. Once absorbed, the air will move throughout the body of water either by mass migration or by molecular diffusion until the water is uniformly saturated. If the air/water solution changes to a state that reduces solubility, excess air will be released as a gas. For example, if the air/water interface is at high pressure, the water will absorb air to its limit of solubility at that point; if at another point in the system the pressure is reduced, some of the dissolved air will be released.

In the design of systems with open or plain steel expansion tanks, it is common practice to use the tank as the major air control or release point in the system.

Equations for sizing the three common configurations of expansion tanks follow (Coad 1980b):

For closed tanks with air/water interface

$$V_t = V_s \frac{[(v_2/v_1) - 1] - 3\alpha\,\Delta t}{(P_a/P_1) - (P_a/P_2)} \qquad (13)$$

For open tanks with air/water interface

$$V_t = 2V_s\left[\left(\frac{v_2}{v_1} - 1\right) - 3\alpha\,\Delta t\right] \qquad (14)$$

For diaphragm tanks

$$V_t = V_s \frac{[(v_2/v_1) - 1] - 3\alpha\Delta t}{1 - (P_1/P_2)} \qquad (15)$$

where

V_t = volume of expansion tank, gal
V_s = volume of water in system, gal
t_1 = lower temperature, °F
t_2 = higher temperature, °F
P_a = atmospheric pressure, psia
P_1 = pressure at lower temperature, psia
P_2 = pressure at higher temperature, psia
v_1 = specific volume of water at lower temperature, ft^3/lb
v_2 = specific volume of water at higher temperature, ft^3/lb
α = linear coefficient of thermal expansion, in/in·°F
= 6.5×10^{-6} in/in·°F for steel
= 9.5×10^{-6} in/in·°F for copper
$\Delta t = (t_2 - t_1)$, °F

As an example, the lower temperature for a heating system is usually normal ambient temperature at fill conditions (e.g., 50°F) and the higher temperature is the operating supply water temperature for the system. For a chilled-water system, the lower temperature is the design chilled-water supply temperature, and the higher temperature is ambient temperature (e.g., 95°F). For a dual-temperature hot/chilled system, the lower temperature is the chilled-water design supply temperature, and the higher temperature is the heating water design supply temperature.

For specific volume and saturation pressure of water at various temperatures, see Table 3 in Chapter 6 of the 2005 *ASHRAE Handbook—Fundamentals*.

At the tank connection point, the pressure in closed-tank systems increases as the water temperature increases. Pressures at the expansion tank are generally set by the following parameters:

- The lower pressure is usually selected to hold a positive pressure at the highest point in the system (usually about 10 psig).
- The higher pressure is normally set by the maximum pressure allowable at the location of the safety relief valve(s) without opening them.

Other considerations are to ensure that (1) the pressure at no point in the system will ever drop below the saturation pressure at the operating system temperature and (2) all pumps have sufficient net positive suction head (NPSH) available to prevent cavitation.

Example 1. Size an expansion tank for a heating water system that will operate at a design temperature range of 180 to 220°F. The minimum pressure at the tank is 10 psig (24.7 psia) and the maximum pressure is 25 psig (39.7 psia). (Atmospheric pressure is 14.7 psia.) The volume of water is 3000 gal. The piping is steel.

1. Calculate the required size for a closed tank with an air/water interface.

Solution: For lower temperature t_1, use 40°F.

From Table 3 in Chapter 6 of the 2005 *ASHRAE Handbook—Fundamentals*,

$$v_1(\text{at } 40°\text{F}) = 0.01602 \text{ ft}^3/\text{lb}$$

$$v_2(\text{at } 220°\text{F}) = 0.01677 \text{ ft}^3/\text{lb}$$

Using Equation (13),

$$V_t = 3000 \times \frac{[(0.01677/0.01602) - 1] - 3(6.5 \times 10^{-6})(220 - 40)}{(14.7/24.7) - (14.7/39.7)}$$

$$V_t = 578 \text{ gal}$$

2. If a diaphragm tank were used in lieu of the plain steel tank, what tank size would be required?

Solution: Using Equation (15),

$$V_t = 3000 \times \frac{[(0.01677/0.01602) - 1] - 3(6.5 \times 10^{-6})(220 - 40)}{1 - (24.7/39.7)}$$

$$V_t = 344 \text{ gal}$$

HYDRAULIC COMPONENTS

Pump or Pumping System

Centrifugal pumps are the most common type in hydronic systems (see Chapter 43). Circulating pumps used in water systems can vary in size from small in-line circulators delivering 5 gpm at 6 or 7 ft head to base-mounted or vertical pumps handling hundreds or thousands of gallons per minute, with pressures limited only by the system characteristics. Pump operating characteristics must be carefully matched to system operating requirements.

Pump Curves and Water Temperature for Constant-Speed Systems. Performance characteristics of centrifugal pumps are described by pump curves, which plot flow versus head or pressure, as well as by efficiency and power information, as shown in Figure 4. Large pumps tend to have a series of curves, designated with a numerical size in inches (10 to 13.5 in. in diameter), to represent performance of the pump impeller and outline the envelope of pump operation. Intersecting elliptical lines designate the pump's efficiency. The **net positive suction head (NPSH)** required line represents the required entering operating pressure for the pump to operate satisfactorily. Diagonal lines represent the required power of the pump motor. The point at which a pump operates is the point at which the pump curve intersects the system curve (Figure 5).

In Figure 4, note that each performance curve has a defined end point. Small circulating pumps, which may exhibit a pump curve as shown in Figure 6, may actually extend to the abscissa showing a run-out flow, and may also not show multiple impellers, efficiency, or NPSH. Large pumps do not exhibit the run-out flow characteristic. In a large pump, the area to the right of the curve is an area of unsatisfactory performance, and may represent pump operation in a state of cavitation. It is important that the system curve always intersect the pump curve in operation, and the design must ensure that system operation stays on the pump curve.

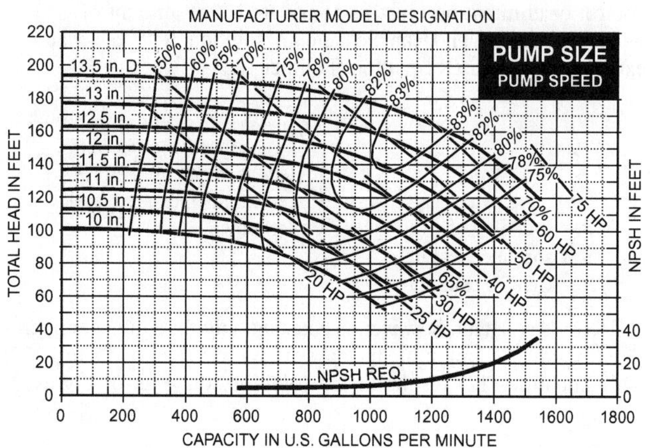

Fig. 4 Example of Manufacturer's Published Pump Curve

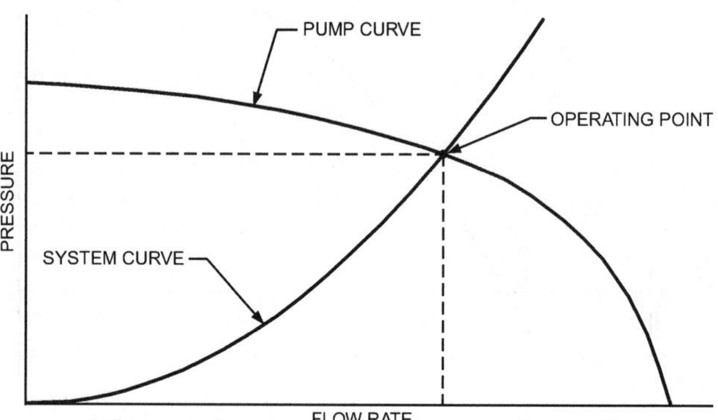

Fig. 5 Pump Curve and System Curve

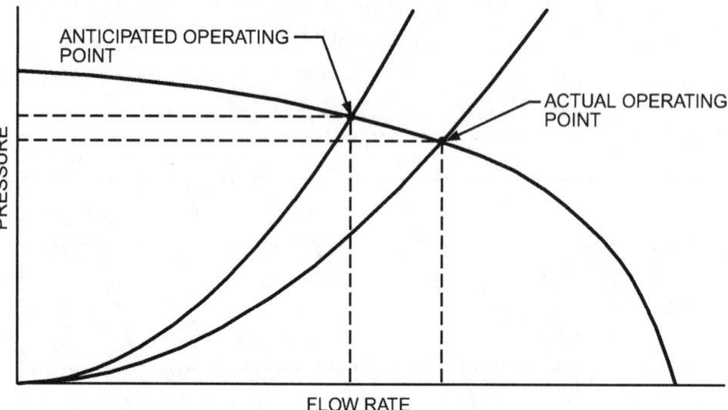

Fig. 6 Shift of System Curve Caused by Circuit Unbalance

Chapter 43 also discusses system and pump curves.

A complete piping system follows the same water flow/pressure drop relationships as any component of the system [see Equation (17)]. Thus, the pressure required for any proposed flow rate through the system may be determined and a system curve constructed. A pump may be selected by using the calculated system pressure at the design flow rate as the base point value.

Figure 6 illustrates how a shift of the system curve to the right affects system flow rate. This shift can be caused by incorrectly calculating the system pressure drop by using arbitrary safety factors or overstated pressure drop charts. Variable system flow caused by control valve operation, a larger than required control valve, or improperly balanced systems (subcircuits having substantially lower pressure drops than the longest circuit) can also cause a shift to the right.

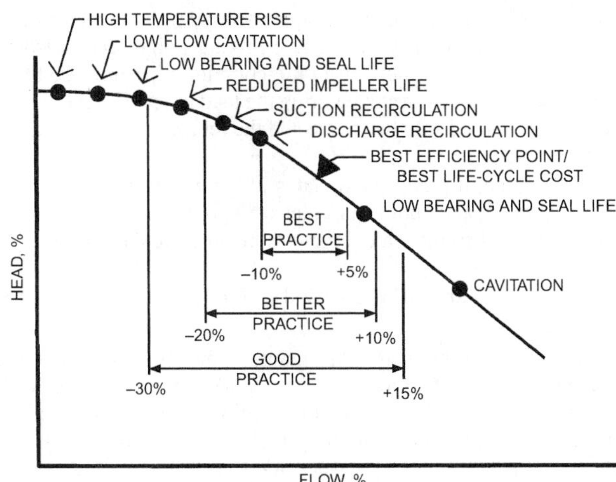

Fig. 7 General Pump Operating Condition Effects
(Hydraulic Institute)

As described in Chapter 43, pumps for closed-loop piping systems should have a flat pressure characteristic and should operate slightly to the left of the peak efficiency point on their curves. This allows the system curve to shift to the right without causing undesirable pump operation, overloading, or reduction in available pressure across circuits with large pressure drops.

Many dual-temperature systems are designed so that the chillers are bypassed during winter. The chiller pressure drop, which may be quite high, is thus eliminated from the system pressure drop, and the pump shift to the right may be quite large. For such systems, system curve analysis should be used to check winter operating points.

Operating points may be highly variable, depending on (1) load conditions, (2) the types of control valves used, and (3) the piping circuitry and heat transfer elements. In general, the best selection in smaller systems is

- For design flow rates calculated using pressure drop charts that illustrate actual closed-loop hydronic system piping pressure drops
- To the left of the maximum efficiency point of the pump curve to allow shifts to the right caused by system circuit unbalance, direct-return circuitry applications, and modulating three-way valve applications
- A pump with a flat curve to compensate for unbalanced circuitry and to provide a minimum pressure differential increase across two-way control valves

As system sizes and corresponding pump sizes increase in size, more care is needed in analysis of the pump selection. The Hydraulic Institute (HI 2000) offers a detailed discussion of pump operation and selection to optimize life cycle costs. The HI guide covers all types of pumping systems, including those that are much more sophisticated than a basic HVAC closed-loop circulating system, and use much more power. There are direct parallels, though. HI's discussion of pump reliability sensitivity includes a chart similar to that shown in Figure 7.

HVAC pumps tend to be low-energy devices compared to industrial or process pumps, which might be responsible for a wide variety of different fluids and operating conditions. Select a pump as close as possible to the best efficiency point, to optimize life-cycle costs and maximize operating life with a minimum of maintenance. HI's recommendations are also appropriate for HVAC pump selection.

Parallel Pumping. When pumps are applied in parallel, each pump operates at the same head, and provides its share of the system flow at that pressure (Figure 8). Generally, pumps of equal size are used, and the parallel-pump curve is established by doubling the flow of the single-pump curve (with identical pumps).

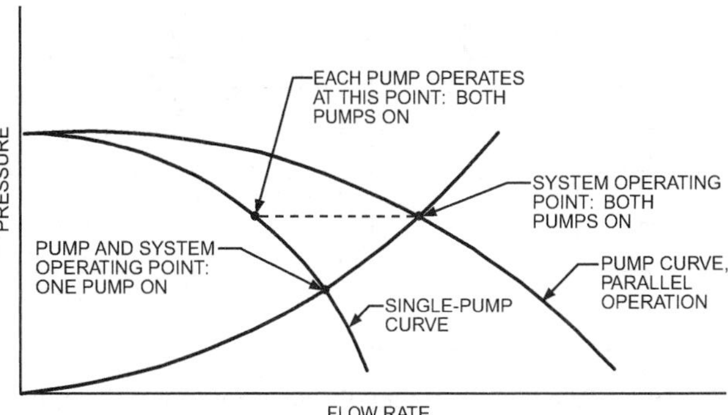

Fig. 8 Operating Conditions for Parallel-Pump Installation

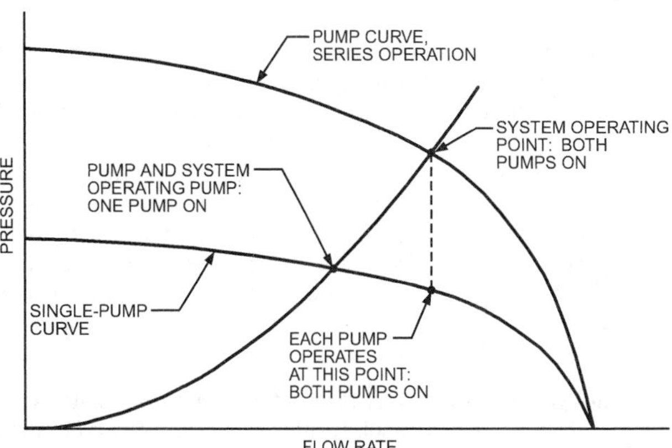

Fig. 9 Operating Conditions for Series-Pump Installation

Plotting a system curve across the parallel-pump curve shows the operating points for both single- and parallel-pump operation (Figure 8). Note that single-pump operation does not yield 50% flow. The system curve crosses the single-pump curve considerably to the right of its operating point when both pumps are running. This leads to two important concerns: (1) the pumps must be powered to prevent overloading during single-pump operation, and (2) a single pump can provide standby service of up to 80% of design flow; the actual amount depends on the specific pump curve and system curve. As pumps become larger, or more than two pumps are placed in parallel operation, it is still very important to ensure in the design that the operating system intersects the operating pump curve, and that there are safeties in place to ensure that, should a pump be turned off, the remaining pumps and system curve still intersect one another.

Series Pumping. When pumps are operated in series, each pump operates at the same flow rate and provides its share of the total pressure at that flow. A system curve plotted across the series-pump curve shows the operating points for both single- and series-pump operation (Figure 9). Note that the single pump can provide up to 80% flow for standby and at a lower power requirement.

Series-pump installations are often used in heating and cooling systems so that both pumps operate during the cooling season to provide maximum flow and head, whereas only a single pump operates during the heating season. Note that both parallel- and series-pump applications require that the actual pump operating points be used to accurately determine the pumping point. Adding artificial safety factor head, using improper pressure drop charts, or incorrectly calculating pressure drops may lead to an unwise selection.

Multiple-Pump Systems. Care must be taken in designing systems with multiple pumps to ensure that, if pumps ever operate in either parallel or series, such operation is fully understood and considered by the designer. Pumps performing unexpectedly in

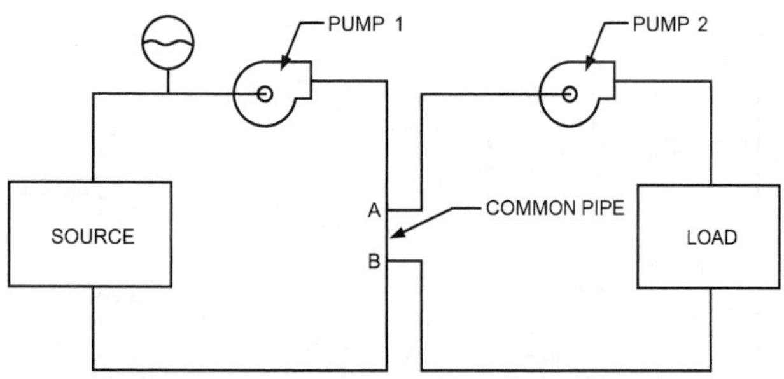

Fig. 10 Compound Pumping (Primary-Secondary Pumping)

series or parallel have caused performance problems in hydronic systems, such as the following:

Parallel. With pumps of unequal pressures, one pump may create a pressure across the other pump in excess of its cutoff pressure, causing flow through the second pump to diminish significantly or to cease. This can cause flow problems or pump damage.

Series. With pumps of different flow capacities, the pump of greater capacity may overflow the pump of lesser capacity, which could cause damaging cavitation in the smaller pump and could actually cause a pressure drop rather than a pressure rise across that pump. In other circumstances, unexpected series operation can cause excessively high or low pressures that can damage system components.

Standby Pump Provision. If total flow standby capacity is required, a properly valved standby pump of equal capacity is installed to operate when the normal pump is inoperable. A single standby may be provided for several similarly sized pumps. Parallel- or series-pump installation can provide up to 80% standby, which is often sufficient.

Compound Pumping. In larger systems, compound pumping, also known as **primary-secondary pumping**, is often used to provide system advantages that would not be available with a single pumping system. Compound pumping is illustrated in Figure 10.

In Figure 10, pump 1 can be referred to as the **source** or **primary pump** and pump 2 as the **load** or **secondary pump**. The short section of pipe between A and B is called the **common pipe** (also called the **decoupling line** or **neutral bridge**) because it is common to both the source and load circuits. In the design of compound systems, the common pipe should be kept as short and as large in diameter as practical to minimize pressure loss between those two points. Care must be taken, however, to ensure adequate length in the common pipe to prevent recirculation from entry or exit turbulence. There should never be a valve or check valve in the common pipe. If these conditions are met and the pressure loss in the common pipe can be assumed to be zero, then neither pump will affect the other. Then, except for the system static pressure at any given point, the circuits can be designed and analyzed and will function dynamically independently of one another.

In Figure 10, if pump 1 has the same flow capacity in its circuit as pump 2 has in its circuit, all of the flow entering point A from pump 1 will leave in the branch supplying pump 2, and no water will flow in the common pipe. Under this condition, the water entering the load will be at the same temperature as that leaving the source.

If the flow capacity of pump 1 exceeds that of pump 2, some water will flow downward in the common pipe. Under this condition, Tee A is a diverting tee, and Tee B becomes a mixing tee. Again, the temperature of the fluid entering the load is the same as that leaving the source. However, because of the mixing taking place at point B, the temperature of water returning to the source is between the source supply temperature and the load return temperature.

On the other hand, if the flow capacity of pump 1 is less than that of pump 2, then point A becomes a mixing point because some water must recirculate upward in the common pipe from point B. The temperature of the water entering the load is between the supply water temperature from the source and the return water temperature from the load.

For example, if pump 1 circulates 25 gpm of water leaving the source at 200°F, and pump 2 circulates 50 gpm of water leaving the load at 100°F, then the water temperature entering the load is

$$t_{load} = 200 - (25/50)(200 - 100) = 150°\text{F} \qquad (16)$$

Mixing is a primary reason against application of compound pumping systems, particularly in chilled-water systems with constant-speed pumping applied to the primary pump circuit of the source, and variable-speed pumping applied to the secondary system connected to the loads. The issues associated with this are generally source- and control-related. At one time, chiller manufacturers restricted water flow variation through a chiller, despite the fact that designers and system operators wanted this ability, to increase energy operating economy and reduce operating costs by reducing flow. Mixing was an inevitable by-product, reducing the chiller's operating efficiency. This situation was exacerbated when variable-speed drives were applied to pumps, while chiller pumps stayed constant-speed. Eventually, changes to chiller operating controls allowed flow rate through the chiller evaporator to be varied. Some system designs have shifted away from compound pumping to variable-speed, variable-flow primary pumping to enhance system efficiency. Depending on locale, system size, and selection criteria for the chiller, compound pumping may still be beneficial to system design and operating stability. Chiller selections often are limited to maximum velocities of 7 fps, and minimum velocities of 3 fps. Equal-percentage valve characteristics, though, should operate most often at valve strokes less than 80%, which is a flow rate of less than 40%, and less than the 42% that might represent a low-flow limit for the chiller in a 3 fps operation criterion. Variable-speed pumping in both primary and secondary circuits can be applied to reduce mixing that might occur as chillers are sequenced on and off to compensate for the loads.

The following are some advantages of compound circuits:

- They allow different water temperatures and temperature ranges in different elements of the system. Compound pumping can be used to vary coil capacity by controlling coil temperature, which leads to better latent energy control.
- They decouple the circuits hydraulically, thereby making the control, operation, and analysis of large systems much less complex. Hydraulic decoupling also prevents unwanted series or parallel operation. Large water circuits often experience pressure imbalances, but compound pumping allows a design to be hydraulically organized into separate smaller subsystems, making troubleshooting and operation easier.

Variable-Speed Pumping Application

Centrifugal pumps may also be operated with variable-frequency drives, which adjust the speed of the electric motor, changing the pump curve. In this application, a pump controller and typically one frequency drive per pump are applied to control the required system variable.

The most typical application is to control differential pressure across one or more branches of the piping network. In a common application, the differential pressure is sensed across a control valve or, alternatively, the valve, coil, and branch (Figure 11). The controller is given a set point equal to the pressure loss of the sensed components at design flow. For example, a 5 psi differential pressure was used to size the control valve; the sensor is connected to sense differential pressure across the valve, so a 5 psi set point is given to

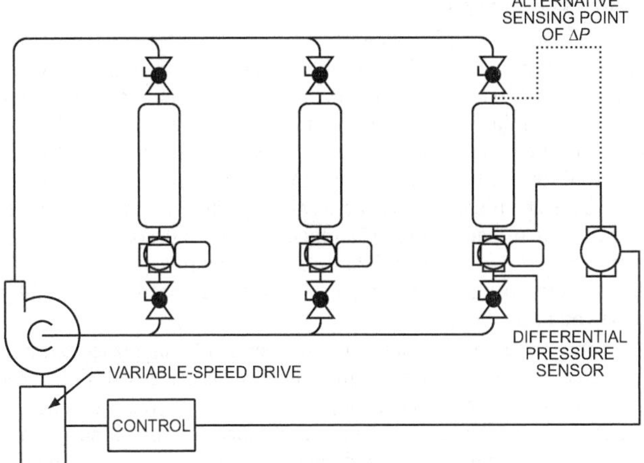

Fig. 11 Example of Variable-Speed Pump System Schematic

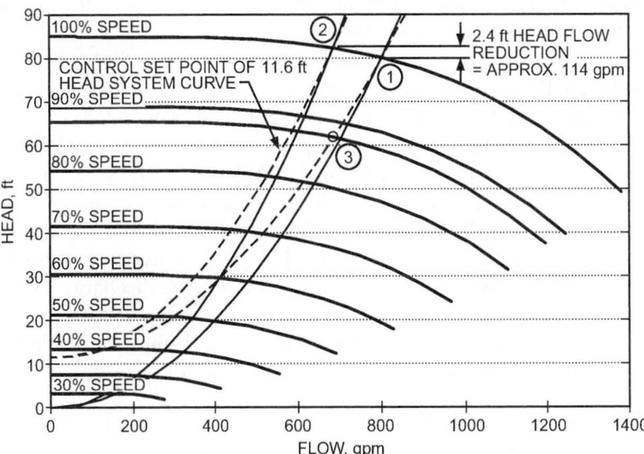

Fig. 12 Example of Variable-Speed Pump and System Curves

the controller. As the control valve closes in reaction to a control signal (Figure 12) from 1 to 2, the differential pressure across the branch rises as the system curve shifts counterclockwise up the pump curve. The pump controller, sensing an increase in differential, decreases the speed of the pump to about point 3, roughly 88% speed. Note that each system curve is shown with two representations. The system curve is the simple relationship of the flow ratio squared to the head ratio. Under control of a pump controller with a single sensor across the control valve, as shown in Figure 11, the pump decreases in speed until the theoretical zero-flow point, at which the pump goes just fast enough to maintain a differential pressure under no flow. In this case, the speed is about 88%, and the power is about 68%. In operation, expect there to be a difference in performance. Figure 12 shows the change as a step function, opposite of the way in which the pump controller functions. Based on the control method, the controller adjustment settings (gain, integral time, and derivative time), system hydraulics, valve time constant, sensor sensitivity, etc., it is far more likely that a small incremental rise in differential pressure will have a corresponding small decrease in pump speed, as the valve repositions itself in reaction to its control signal. This appears as a small sawtoothed series of system curve and pump curve intersection points, and is not of real importance. What is important is that many factors influence operation, and theoretical variable-speed pumping is different from reality.

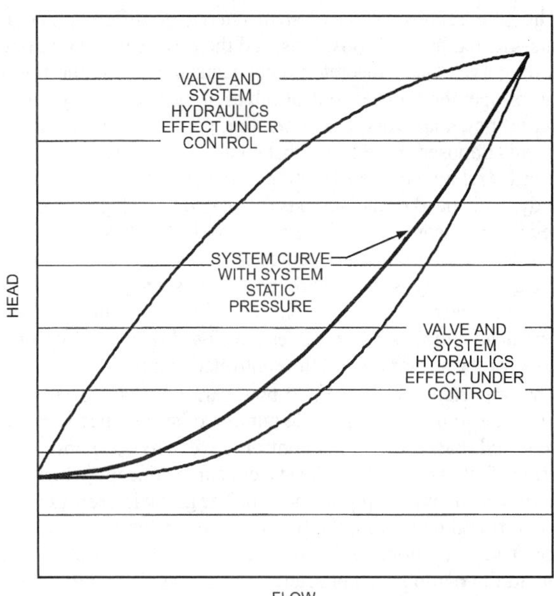

Fig. 13 System Curve with System Static Pressure (Control Area)

Decreasing pump speed is analogous to using a smaller impeller size in the volute of the pump, and the result is that motor power is reduced by the cube of the speed reduction, closely following the affinity laws. (Variable-speed pump curves are shown in Chapter 43.) Design flow conditions should be minimal hours of operation per year; as such, the energy savings potential for a variable-speed pump is great. In the general control valve application, a reasonably selected equal percentage valve with a 50% valve authority should reduce flow by about 30% when the valve is positioned from 100% open (design flow, minimal hours per year) to 90% open. The 10% change in stroke should reduce pump operating power about 70%. Hydronic systems should operate most of the time at flow rates well below design. The coil characteristic suggests that, in sensible applications, a small percentage of flow yields an exceptionally high degree of coil design heat transfer, adequate for most of the year's operation.

Depending on system design, direct digital control may also allow more advanced control strategies. There are various reset control strategies (e.g., cascade control) to optimize pump speed and flow performance. Many of these monitor valve position and drive the pump to a level that keeps one valve open while maintaining set point for comfort conditions. Physical operational requirements of the components must be taken into account. There are some concerns over operating pumps and their motor drives at speeds less than 30% of design, particularly about maintaining proper lubrication of the pump mechanical seals and motor bearings. From a practical perspective, 30% speed is a scant 3% of design power, so it may be unnecessary to reduce speed any further. Consult manufacturers for information on device limits, to maintain the system in good operating condition.

Exceptional energy reduction potential and advances in variable-frequency drive technology that have reduced drive costs have made the application of variable-speed drives common on closed hydronic distribution systems. Successful application of a variable-speed drive to a pump is not a given, however, and depends on the designer's skill and understanding of system operation.

For instance, the designer should understand the system curve with system static pressure, as shown in Figure 13. This control phenomenon of variable-speed, variable-flow pumping systems is often called the **control area**; Hegberg (2003) describes the analysis used to create these graphs.

The plot represents the system curves of different operation points, created by valve positions, and their intersection points with a pump curve. These discrete points occur under specific imposed points of operation on the control valves of the hydronic system, and represent potential worst-case operation points in the system. Typically, this imposed sequence is that control valves are closed in specific order relative to their location to the controlled pump, and the total dynamic head of the pump is calculated at each position. When there is one sensor of control, the boundaries as shown are created. The upper system boundary represents system head as valves are closed sequentially, starting with those nearest the pump and ending with those closer to the sensor. The lower boundary curve represents system head and flow when valves are closed sequentially from the location of the sensor toward the controlled pump.

The importance of these two boundaries is that, for any given system flow, there are an infinite number of system heads that may be required, based on which control valves are open and to what position. This is opposite of the system curve concept, which is one flow, one head loss in a piping system. The graph is a representation of a controlled operation; the piping system follows the principles of the Darcy equation, but the intervening feedback control adjustments to the pump speed produce an installed characteristic in the system different from what might otherwise be expected. The implication of any boundary above the generic system curve relationship to design flow is that, as flow decreases in the system, it does not follow the affinity law relationship in reducing operating power. Conversely, the boundary below the system curve operates with a greater reduction in power than the generic system curve.

Although operating below the system curve may seem to be advantageous, saving energy and pumping costs, the designer is cautioned to remember that these represent a system flow less than that required for comfort control. This operating series of conditions represents control valve interaction. One valve flow affects flow to all other circuits between it and the pump. This interaction can be quite large. Depending on distribution system head loss and system balancing techniques, one valve may reduce system flow by two to three times the individual valve's rated flow. This can negatively affect the control and sequencing of the source, and possibly also the control of the affected terminal units. Controlling this interaction is done by engineering the hydraulic losses and distribution pipe friction head losses, using pressure-independent valves (both balancing and control), using sensors at critical points of the hydronic system, and combinations of these strategies.

These issues are part of what has caused debate over system balancing, pump method of control, etc. Simply put, system design dictates the requirements of adjustment and control. Balance of a system is more than having a balancing valve; it involves pipe selection and location combined with valves and coils and, in some cases, pumps, as well as measurement and adjustment devices. Similarly, the effects of one method of pump control over another (e.g., differential pressure control, valve position of pump speed reset) must also be considered. Traditional system design and feedback control techniques use a series of single-loop controllers. These techniques also work on variable-speed pumping systems, but the interaction of all of the individual controlled systems requires analysis by the designer and calculated design choices based on a thorough understanding of the loads, both theoretical and practical, and should also take into account that operators may run systems in a manner not intended by the designer.

Address these issues during design, because they are difficult to understand in the field, and the required instrumentation to monitor the effects is rarely installed. Despite these challenges, application of variable-speed, variable-flow pumping can be very satisfactory. However, variable-speed pumping designs require that the designer allow adequate time in calculation and design iteration to mitigate potential operation issues, and ensure the design criteria of a comfort providing system with operating energy and cost efficiency.

Pump Connection

Pump suction piping should be at least as large as the nozzle serving the pump, and there should be minimal fittings or devices in the suction to obstruct flow. Typically, pump manufacturers prefer five to eight pipe diameters of unobstructed (i.e., no fittings) straight pipe entering the pump. Fittings such as tees and elbows, especially when there is a change in planar direction, cause water to swirl in the pipe; this can be detrimental to pump performance, and may also lead to pump damage. Manufacturers may recommend special fittings in applications where piping space is unavailable to overcome geometry factors. These fittings need to be carefully reviewed in application.

Piping to the pump should be independently supported, adding no load to the pump flanges, which, unless specifically designed for the purpose, are incapable of supporting the system piping. Supporting the pipe weight on the flange can cause serious damage (e.g., breaking the flange), or may induce stresses that misalign the pump. Similar results can also occur when the pipe and pump are improperly aligned to each other, or pipe expansion and contraction are unaccounted for. Flexible couplings are one way to overcome some of these issues, when both the pump and the pipe are supported independently of each other, and the flexible coupling is not arbitrarily connected to the pump and pipe.

When pumps are piped in parallel, these requirements are extended. Pump entering and discharge pressures should be equal in operation. In addition to maintaining the recommended straight inlet pipe to the pump, manifold pipe serving the inlets should also have a minimum of two manifold pipe diameters between pump suction center lines. Pump discharge manifolds should be constructed to keep discharge velocities less than 10 to 15 fps, and lower when a check valve is applied, because it is necessary to prevent hydraulic shock (water hammer). Soft-seating discharge check valves are required on the pumps to prevent reverse flow from one pump to another. Various specialty valves and fittings are available for serving one or more of these functions.

Distribution System

The distribution system is the piping connecting the various other components of the system. The primary considerations in designing this system are (1) sizing the piping to handle the heating or cooling capacity required and (2) arranging the piping to ensure flow in the quantities required at design conditions and at all other loads.

The flow requirement of the pipe is determined by Equation (8) or (9). After Δt is established based on the thermal requirements, either of these equations (as applicable) can be used to determine the flow rate. First-cost economics and energy consumption make it advisable to design for the greatest practical Δt because the flow rate is inversely proportional to Δt; that is, if Δt doubles, the flow rate is reduced by half.

The three related variables in sizing the pipe are flow rate, pipe size, and pressure drop. The primary consideration in selecting a design pressure drop is the relationship between the economics of first cost and energy costs.

Once the distribution system is designed, the pressure loss at design flow is calculated by the methods discussed in Chapter 36 of the 2005 *ASHRAE Handbook—Fundamentals*. The relationship between flow rate and pressure loss can be expressed by

$$Q = C_v \sqrt{\Delta p} \tag{17}$$

where

Q = system flow rate, gpm
Δp = pressure drop in system, psi
C_v = system constant (sometimes called valve coefficient, discussed in Chapter 42)

Equation (17) may be modified as follows:

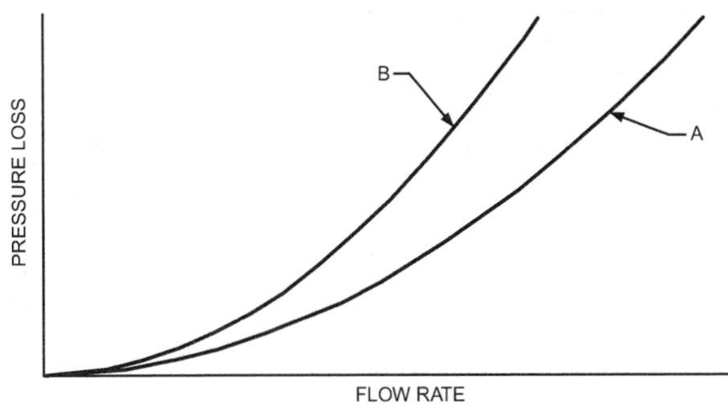

Fig. 14 Typical System Curves for Closed System

$$Q = C_s\sqrt{Dh} \tag{18}$$

where

Δh = system head loss, ft of fluid [$\Delta h = \Delta p/\rho$]
C_s = system constant [$C_s = 0.67C_v$ for water with density $\rho = 62.4$ lb/ft^3]

Equations (17) and (18) are the system constant form of the Darcy-Weisbach equation. If the flow rate and head loss are known for a system, Equation (18) may be used to calculate the system constant C_v. From this calculation, the pressure loss can be determined at any other flow rate. Equation (18) can be graphed as a system curve (Figure 14).

The system curve changes if anything occurs to change the flow/pressure drop characteristics. Examples include a strainer that starts to block or a control valve closing, either of which increases the head loss at any given flow rate, thus changing the system curve in a direction from curve A to curve B in Figure 14.

This type of evaluation can help determine the effects of control and component selection in system operation. The designer cannot assume that the control system will compensate for poor selections and unnoticed installation mistakes. Anecdotal data suggest that numerous adjustments are required when selection and operation techniques are poor. There is also a need for system balance when the pump is correctly sized: no excess head is added to the pump; the balance device adds head only to the circuits for coils 1 and 2 to compensate for the difference in friction loss as water goes from one terminal takeoff to another. System flow is reduced to design, and no extra energy input is required for the pump.

Expansion Chamber

As a hydraulic device, the expansion tank serves as the reference pressure point in the system, analogous to a ground in an electrical system (Lockhart and Carlson 1953). Where the tank connects to the piping, the pressure equals the pressure of the air in the tank plus or minus any fluid pressure caused by the elevation difference between the tank liquid surface and the pipe (Figure 15).

A closed system should have only one expansion chamber. Having more than one chamber or excessive amounts of undissolved air in a piping system can cause the closed system to behave in unintended (but understandable) ways, causing extensive damage from shock waves or water hammer.

With a single chamber on a system, assuming isothermal conditions for the air, the air pressure can change only as a result of displacement by the water. The only thing that can cause the water to move into or out of the tank (assuming no water is being added to or removed from the system) is expansion or shrinkage of the water in the system. Thus, in sizing the tank, thermal expansion is related to the pressure extremes of the air in the tank [Equations (13), (14), and (15)].

The point of connection of the tank should be based on the pressure requirements of the system, remembering that the pressure at

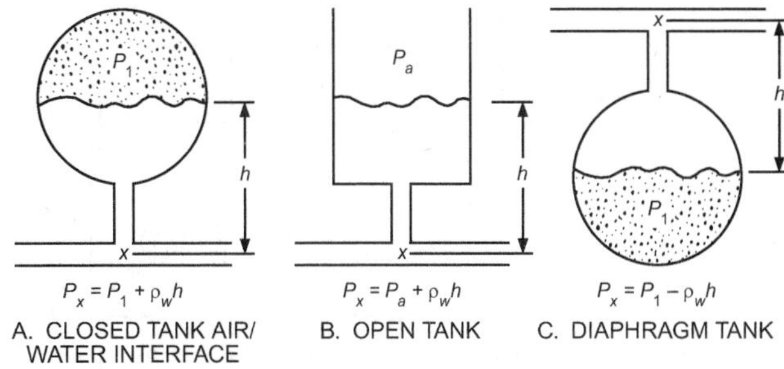

Fig. 15 Tank Pressure Related to System Pressure

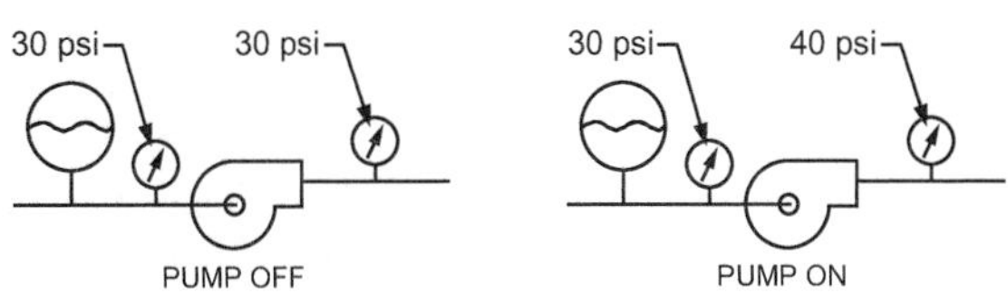

A. TANK ON PUMP SUCTION SIDE

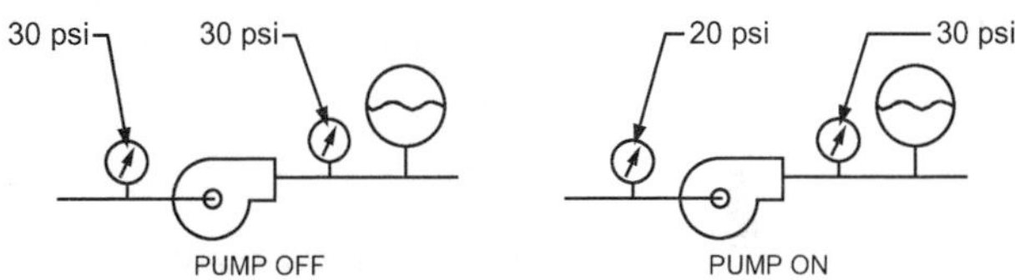

B. TANK ON PUMP DISCHARGE SIDE

Fig. 16 Effect of Expansion Tank Location with Respect to Pump Pressure

the tank connection will not change as the pump is turned on or off. For example, consider a system containing an expansion tank at 30 psig and a pump with a pump head of 23.1 ft (10 psig). Figure 16 shows alternative locations for connecting the expansion tank; in either case, with the pump off, the pressure will be 30 psig on both the pump suction and discharge. With the tank on the pump suction side, when the pump is turned on, the pressure increases on the discharge side by an amount equal to the pump pressure (Figure 16A). With the tank on the discharge side of the pump, the pressure decreases on the suction side by the same amount (Figure 16B).

Other tank connection considerations include the following:

- A tank open to the atmosphere must be located above the highest point in the system, or be equipped with pressure-sustaining valves (as used in thermal storage applications).
- A tank with an air/water interface is generally used with an air control system that continually revents air into the tank. For this reason, it should be connected at a point where air can best be released.
- Within reason, the lower the pressure in a tank, the smaller the tank is [see Equations (13) and (15)]. Thus, in a vertical system, the higher the tank is placed, the smaller it can be.

PIPING CIRCUITS

Hydronic systems are designed with many different configurations of piping circuits. In addition to simple preference by the design engineer, the method of arranging circuiting can be dictated by such factors as the shape or configuration of the building, the economics of installation, energy economics, the nature of the load, part-load capabilities or requirements, and others.

Each piping system is a network; the more extensive the network, the more complex it is to understand, analyze, or control. Thus, a major design objective is to maximize simplicity.

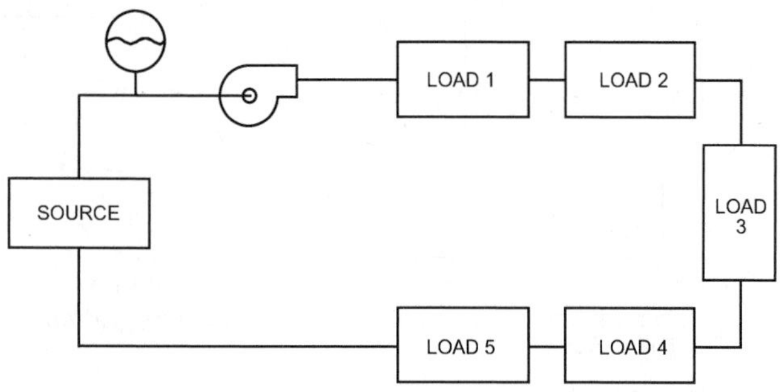

Fig. 17 Flow Diagram of Simple Series Circuit

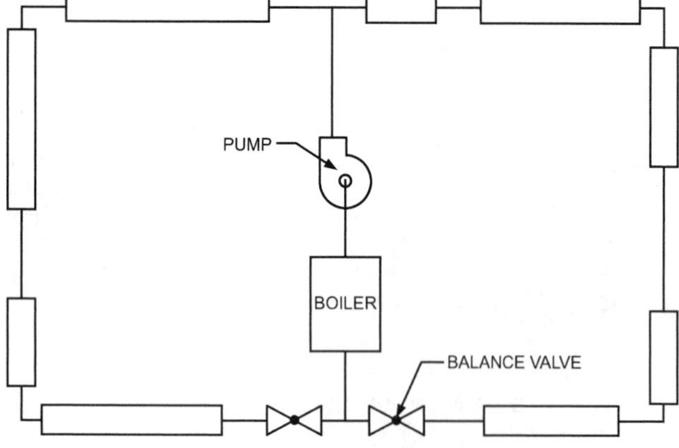

Fig. 18 Series Loop System

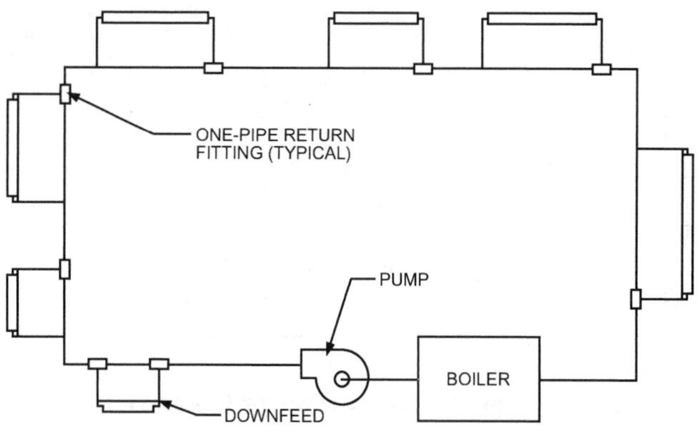

Fig. 19 One-Pipe Diverting Tee System

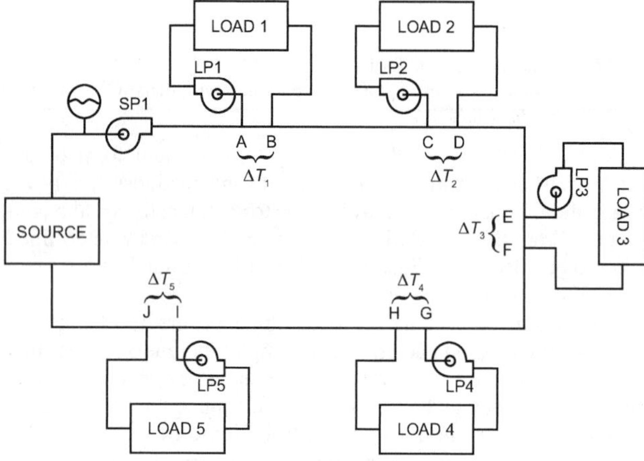

Fig. 20 Series Circuit with Load Pumps

Load distribution circuits are of four general types:

- Full series
- Diverting series
- Parallel direct return
- Parallel reverse return

A simple **series** circuit is shown in Figure 17. Series loads generally have the advantages of lower piping costs and higher temperature drops, resulting in smaller pipe size and lower energy consumption. A disadvantage is that the different circuits cannot be controlled separately. Simple series circuits are generally limited to residential and small commercial standing radiation systems. Figure 18 shows a typical layout of such a system with two zones for residential or small commercial heating.

The simplest **diverting series** circuit diverts some flow from the main piping circuit through a special diverting tee to a load device (usually standing radiation) that has a low pressure drop. This system is generally limited to heating systems in residential or small commercial applications.

Figure 19 illustrates a typical one-pipe diverting tee circuit. For each terminal unit, a supply and a return tee are installed on the main. One of the two tees is a special diverting tee that creates pressure drop in the main flow to divert part of the flow to the unit. One (return) diverting tee is usually sufficient for upfeed (units above the main) systems. Two special fittings (supply and return tees) are usually required to overcome thermal pressure in downfeed units. Special tees are proprietary; consult the manufacturer's literature for flow rates and pressure drop data on these devices. Unit selection can be only approximate without these data.

One-pipe diverting series circuits allow manual or automatic control of flow to individual heating units. On/off rather than flow modulation control is preferred because of the relatively low pressure drop allowable through the control valve in the diverted flow circuit. This system is likely to cost more than the series loop because extra branch pipe and fittings, including special tees, are required. Each unit usually requires a manual air vent because of the low water velocity through the unit. The length and load imposed on a one-pipe circuit are usually small because of these limitations.

Because only a fraction of the main flow is diverted in a one-pipe circuit, the flow rate and pressure drop are less variable as water flow to the load is controlled than in some other circuits. When two or more one-pipe circuits are connected to the same two-pipe mains, the circuit flow may need to be mechanically balanced. After balancing, sufficient flow must be maintained in each one-pipe circuit to ensure adequate flow diversion to the loads.

When coupled with compound pumping systems, series circuits can be applied to multiple control zones on larger commercial or institutional systems (Figure 20). Note that in the series circuit with compound pumping, the load pumps need not be equal in capacity to the system pump. If, for example, load pump LP1 circulates less flow (Q_{LP1}) than system pump SP1 (Q_{SP1}), the temperature difference across Load 1 would be greater than the circuit temperature difference between A and B (i.e., water would flow in the common pipe from A to B). If, on the other hand, the load pump LP2 is equal in flow capacity to the system pump SP1, the temperature differentials across Load 2 and across the system from C to D would be equal and no water would flow in the common pipe. If Q_{LP3} exceeds Q_{SP1}, mixing occurs at point E and, in a heating system, the temperature entering pump LP3 would be lower than that available from the system leaving load connection D.

Thus, a series circuit using compound or load pumps offers many design options. Each of the loads shown in Figure 20 could also be a complete piping circuit or network.

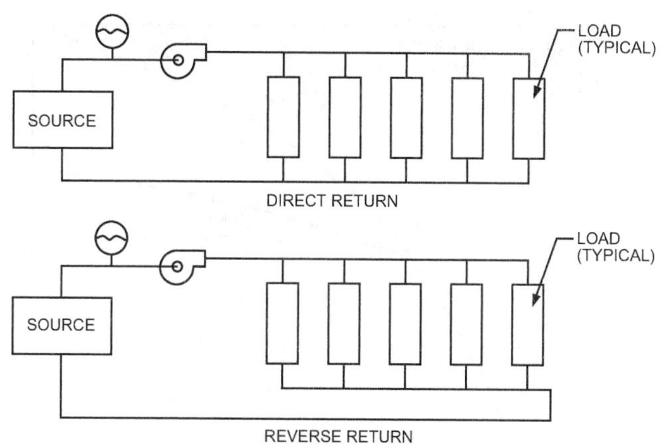

Fig. 21 Direct- and Reverse-Return Two-Pipe Systems

Parallel piping networks are the most commonly used in hydronic systems because they allow the same temperature water to be available to all loads. The two types of parallel networks are direct-return and reverse-return (Figure 21).

In the **direct-return** system, the length of supply and return piping through the subcircuits is unequal, which may cause unbalanced flow rates and require careful balancing to provide each subcircuit with design flow. Ideally, the **reverse-return** system provides nearly equal total lengths for all terminal circuits.

Direct-return piping has been successfully applied where the designer has guarded against major flow unbalance by

- Providing for pressure drops in the subcircuits or terminals that are significant percentages of the total, usually establishing pressure drops for close subcircuits at higher values than those for the far subcircuits
- Minimizing distribution piping pressure drop (in the limit, if the distribution piping loss is zero and the loads are of equal flow resistance, the system is inherently balanced)
- Including balancing devices and some means of measuring flow at each terminal or branch circuit
- Using control valves with a high head loss at the terminals

Carlson (1968) described the effects of distribution piping friction loss on total system flow to terminals in constant speed pumping systems through a **branch-to-riser pressure drop ratio (BRPDR)**. This concept helps minimize distribution losses. In constant-speed pumping systems, a BRPDR of 4:1 yielded a flow of 95% of design in all terminals at part-load conditions, 90% at 2:1, and 80% at 1:1. Application of this ratio helps alleviate system balancing problems without adjusting a controlling device. When used with variable-speed, variable-flow pumping systems, the control area can be minimized and control interaction can be alleviated. It also helps minimize overall friction losses, which reduces required pump horsepower and energy use.

CAPACITY CONTROL OF LOAD SYSTEM

The two alternatives for controlling the capacity of hydronic systems are on-/off control and variable-capacity or modulating control. The on/off option is generally limited to smaller systems (e.g., residential or small commercial) and individual components of larger systems. In smaller systems where the entire building is a single zone, control is accomplished by cycling the source device (the boiler or chiller) on and off. Usually a space thermostat allows the chiller or boiler to run, then a water temperature thermostat (aquastat) controls the capacity of the chiller(s) or boiler(s) as a function of supply or return water temperature. The pump can be either cycled with the load device (usually the case in a residential heating system) or left running (usually done in commercial hot- or chilled-water systems).

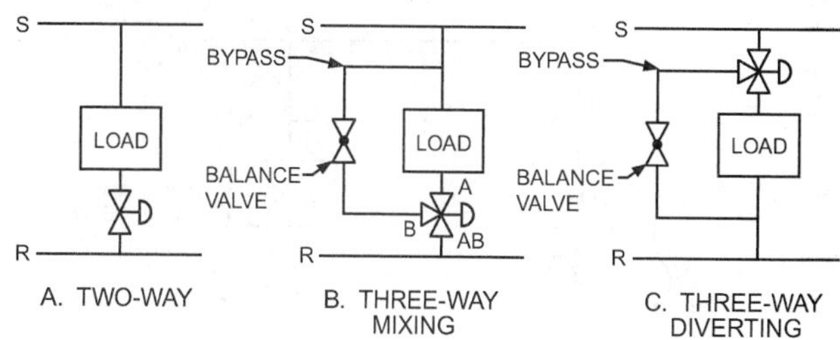

Fig. 22 Load Control Valves

In these single-zone applications, the piping design requires no special consideration for control. Where multiple zones of control are required, the various load devices are controlled first; then the source system capacity is controlled to follow the capacity requirement of the loads.

Control valves are commonly used to control loads. These valves control the capacity of each load by varying the amount of water flow through the load device when load pumps are not used. Control valves for hydronic systems are straight-through (two-way) valves and three-way valves (Figure 22). The effect of either valve is to vary the amount of water flowing through the load device.

With a two-way valve (Figure 22A), as the valve strokes from full-open to full-closed, the quantity of water flowing through the load gradually decreases from design flow to no flow. With a three-way mixing valve (Figure 22B) in one position, the valve is open from Port A to AB, with Port B closed off. In that position, all the flow is through the load. As the valve moves from the A-AB position to the B-AB position, some of the water bypasses the load by flowing through the bypass line, thus decreasing flow through the load. At the end of the stroke, Port A is closed, and all of the fluid flows from B to AB with no flow through the load. Thus, the three-way mixing valve has the same effect on the load as the two-way valve—as the load reduces, the quantity of water flowing through the load decreases.

The effect on load control with the three-way diverting valve (Figure 22C) is the same as with the mixing valve in a closed system: the flow is either directed through the load or through the bypass in proportion to the load. Because of the dynamics of valve operation, diverting valves are more complex in design and are thus more expensive than mixing valves; because they accomplish the same function as the simpler mixing valve, they are seldom used in closed hydronic systems.

In terms of load control, a two-way valve and a three-way valve perform identical functions: varying flow through the load as the load changes. The fundamental difference is that as the source or distribution system sees the load, the two-way valve provides a variable-flow load response and the three-way valve provides a constant-flow load response.

According to Equation (9), load q is proportional to the product of Q and Δt. Ideally, as the load changes, Q changes, while Δt remains fixed. However, as the system sees it, as the load changes with the two-way valve, Q varies and Δt is fixed, whereas with a three-way valve, Δt varies and Q is fixed. This principle is illustrated in Figure 23. Understanding this concept is fundamental to design or analysis of hydronic systems.

The flow characteristics of two- and three-way valve ports are described in Chapter 15 of the 2005 *ASHRAE Handbook—Fundamentals* and in Chapter 46 of this volume, and must be understood. The equal percentage characteristic is recommended for proportional control of load flow for two- and three-way valves; the bypass flow port of three-way valves should have the linear characteristic to maintain a uniform flow during part-load operation.

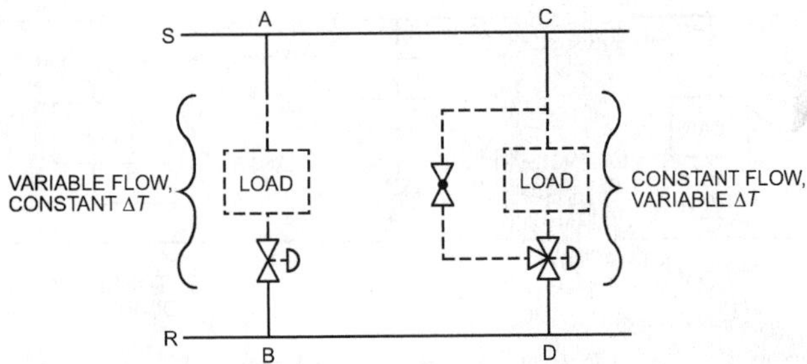

Fig. 23 System Flow with Two-Way and Three-Way Valves

Sizing Control Valves

For stable control, the pressure drop in the control valve at the full-open position should be no less than one-third of the pump head, or controlled branch differential pressure in a variable speed pumping system. This a simple rule of thumb recommendation, and caution should be used in application. Systems designed around closed-loop, constant-speed pumping systems with low-differential-temperature coils are considered forgiving in operation; as system complexity increases, however, the advanced strategies used require diligence and examination to evaluate valve performance and selection.

General pressure drops are commonly applied to valve sizing. Values chosen roughly correspond to the pressure loss on a coil, although there is no literature to explain why these values are used.

Using the concept of valve authority, the relationship is

$$\beta = \frac{\Delta P_{valve}}{\Delta P_{system}} \times 100 \qquad (19)$$

where β is authority.

Common practice has been to assume that, if the coil and the valve have the same pressure drop, they have equal authority of 50%, and that stable control is established. This belief is a rule of thumb, and depends on the type of pumped system and intervening control actions. Using flow coefficient analysis, however, results in a slightly modified definition for authority, comparing the flow coefficient of the valve (C_v) to the coefficient of the remaining system components (C_s). The authority of 50%, then, is representative of the two components having equal flow coefficients, which implies that the valve pressure drop is much greater than the coil pressure drop.

Selection of the proper valve and its corresponding pressure drop requires analysis of the method of control to be used based on the desired response of the controlled system. Depending on the applied control theory and required level of performance, some systems may be run with an on/off control mode, in which case a two-position valve is used. The pressure drop applied on the valve may be minimal, as long as flow through the corresponding circuit is limited in some way to design flow and heat transfer flow tolerance.

More complex systems with fast time constants or that attempt to match seasonal production to load use modulating throttling valves. In these applications, heat transfer characteristic of the coil and the flow control characteristic of the installed control valve (under the effects of authority) and the desired controller gain are analyzed (Figures 24 and 25). The graphic solution of the two characteristics (coil and valve authority) has traditionally yielded a linear function so that the simple proportional controller could have a constant gain of one over the operation of the modulating control valve (Figures 26 to 28). This should not imply that this is the only method for sizing a valve, or that linearity is the only allowable control characteristic. Controllers represent devices implementing a mathematical argument. Knowledge of control theory or programming of different

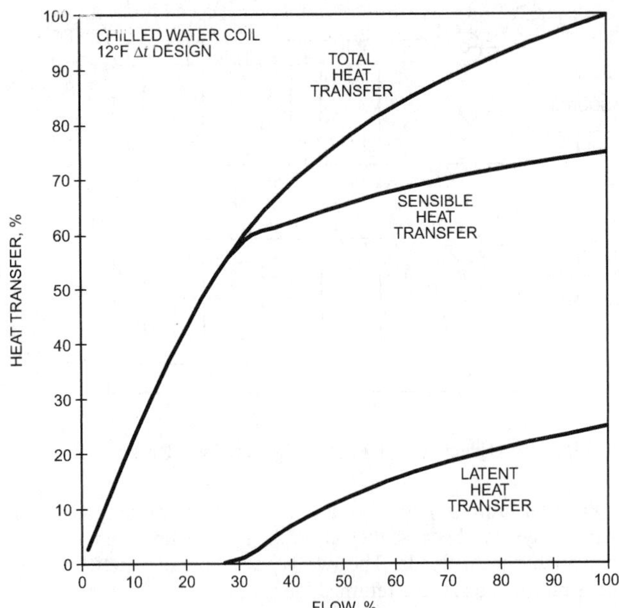

Fig. 24 Chilled-Water Coil Heat Transfer Characteristic

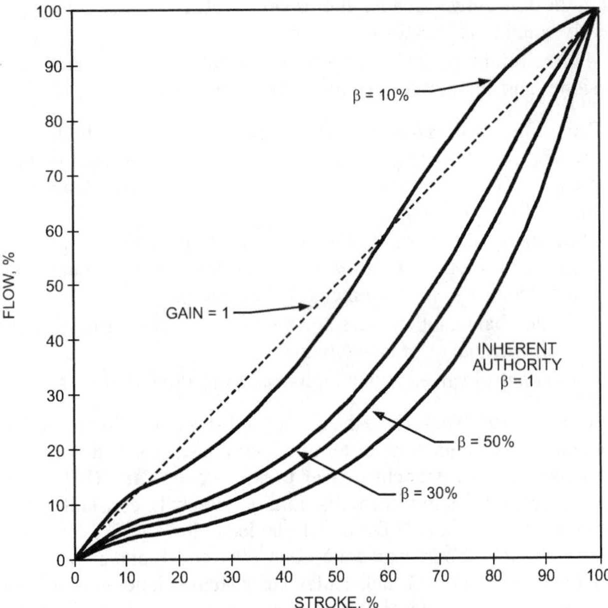

Fig. 25 Equal-Percentage Valve Characteristic with Authority

mathematical relationships is acceptable as long as all of the effects of such action are determined by the designer. However, linear control algorithms such as proportional with integral (PI) and proportional with integral and derivative (PID) control are the most common in HVAC systems, and should be considered in the application of the devices applied to the hydronic system.

Experience in of applied control systems shows the following general guidance.

Proportional control is adequate for slowly changing, single-variable systems such as space temperature control. Many of these applications may have slow response time, with time constants for temperature change on the order of 10 to 50 min or more, so a properly applied proportional controller with a properly applied valve and coil characteristic can perform adequately, and may approach

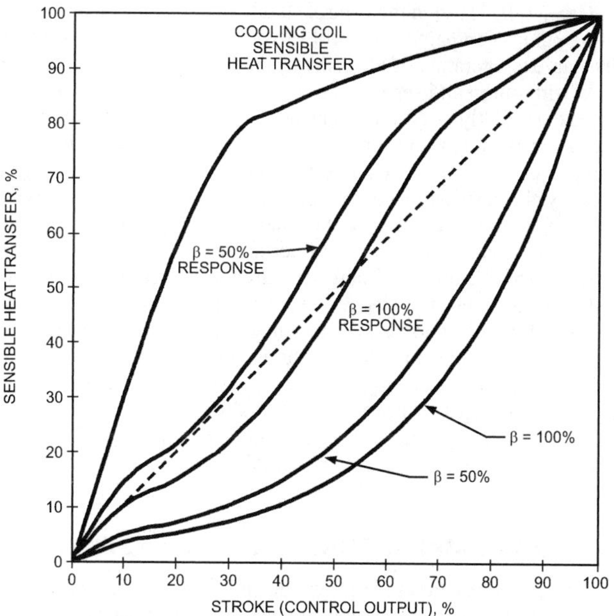

Fig. 26 Control Valve and Coil Response, Inherent and 50% Authority

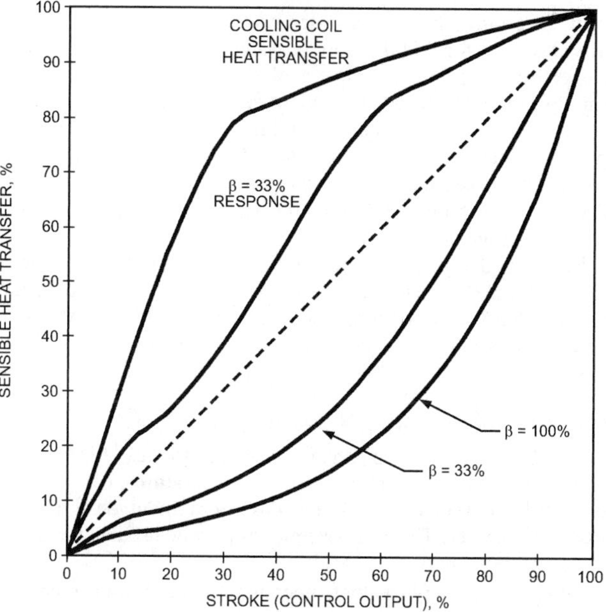

Fig. 27 Control Valve and Coil Response, 33% Authority

simple on/off or valve-open/valve-closed implementation with minimal noticeable temperature hunting in the space. There is a reasonable history of the application of PI control in this application, also. In systems with extra effects, such as variable-speed pumping systems, PI control may be necessary to deal with the control errors introduced when variables outside the direct control loop modify the valve. Note that applying variable-speed drives to pumps and blowers significantly affects the ability of a controlled device to achieve its design differential pressure, because there is an extra variable in the system, with corresponding downstream effects. The effects are the result of the changes that occur in piping system pressure distribution because of changes in flow and corresponding pressures, as affected by the control set point and measurement

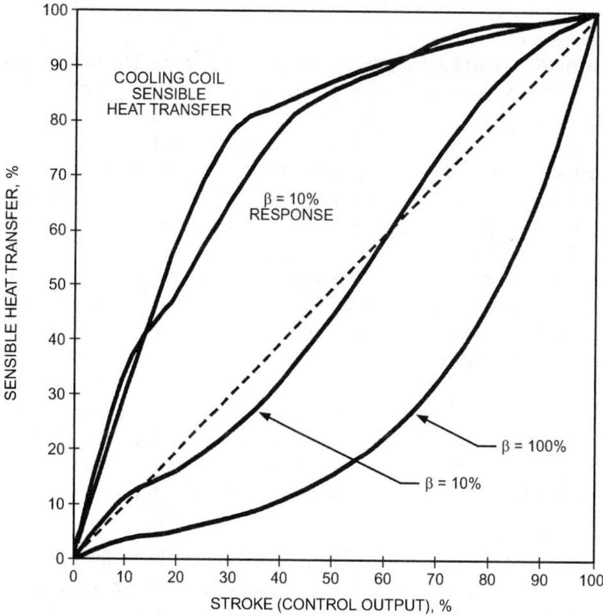

Fig. 28 Coil Valve and Coil Response, 10% Authority

responses in the pump controller. These effects can be significant or negligible, depending on how the piping system friction loss has been designed and distributed. Understanding variable-speed pump system control area and distribution of predicted dynamic flow in the building through energy analysis requires significant calculation, but may be necessary in certain applications.

Although **proportional-integral controllers** can be applied to simple systems, they are more often required for primary space and fluid conditioning systems such as discharge temperature control of an air-handling unit, or pressure/flow control of a pumping system. In practice, devices that exhibit fast response time and have low time constants are candidates for PI controller application, and possibly also for additional derivative control action. Generally, these control modes are applied through application of direct digital controls (DDC). Thorough understanding of the applied controller's implementation of the PID algorithm is required to understand the complete effects of each control action. In some cases, the cyclic nature of the generic DDC device or a specific manufacturer's feature may limit unwanted responses, or add stabilization to the output signal very different from the generic mathematical PID controller algorithm.

For example, in Figure 23, the pressure drop at full-open position for the two-way valve should be great enough from A to B that, when the flow coefficient analysis is performed and combined with the coil characteristic (Figure 24) a straight line is developed (Figures 25 and 26). Typically, authorities in the range of 30 to 50% (for piping systems with 2:1 branch-to-riser pressure drop ratios) are adequate for HVAC coil applications, with higher Δt coils requiring less valve authority. Authorities less than 30% should be avoided because they lead to a flow characteristic shift that makes the equal-percentage valve appear to be more linear in nature (Figures 27 and 28). For the three-way valve shown in Figure 23, the full-open pressure drop should be from C to D. The pressure drop in the bypass balancing valve in the three-way valve circuit should be set to equal that in the coil (load), Chapter 46 discusses valve authority in three way modulating applications.

Control valves should be sized on the basis of the required valve coefficient C_v for the required pressure drop and flow. For more information, see the section on Control Valve Sizing under Automatic Valves in Chapter 46. Briefly, in variable-speed pumping systems using differential pressure control of the branch, the pressure drop of concern is the controlled differential, and is generally placed to measure the valve and coil pressure drops. The valve pressure drop for control, then, is that required to provide complementary characteristic. In the previous coils, a 12°F Δt was used, and valve authority of 50 to 100% provides good results, implying a pressure drop on the valve equal to that of the coil. Control valves that follow are thus dependent on the controlled hydraulic performance. In these valves, attention must also be paid to valve construction (especially the maximum pressure allowed on the valve body, and maximum allowed pressure drop across the valve).

If a system is to be designed with multiple zones of control such that load response is to be by constant flow through the load and variable Δt, control cannot be achieved by valve control alone; a load pump is required.

Several control arrangements of load pump and control valve configurations are shown in Figure 29. Note that, in all three configurations, the common pipe has no restriction or check valve. In all configurations, there is no difference in control as seen by the load. However, the basic differences in control are

- With the two-way valve configuration (Figure 29A), the distribution system sees variable flow and constant Δt, whereas with both three-way configurations, the distribution system sees constant flow and variable Δt.
- Configuration B differs from C in that the pressure required through the three-way valve in Figure 29B is provided by the load pump, whereas in Figure 29C it is provided by the distribution pump(s).

Alternatives to Control Valves

Use of variable-frequency drives has greatly increased during the past few years because advances in technology and decreases in cost have made these drives an attractive alternative to using control valves for heat transfer control (Figures 30).

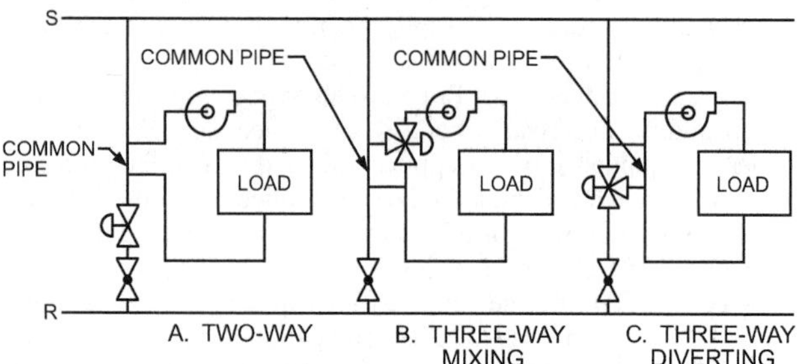

Fig. 29 Load Pumps with Valve Control

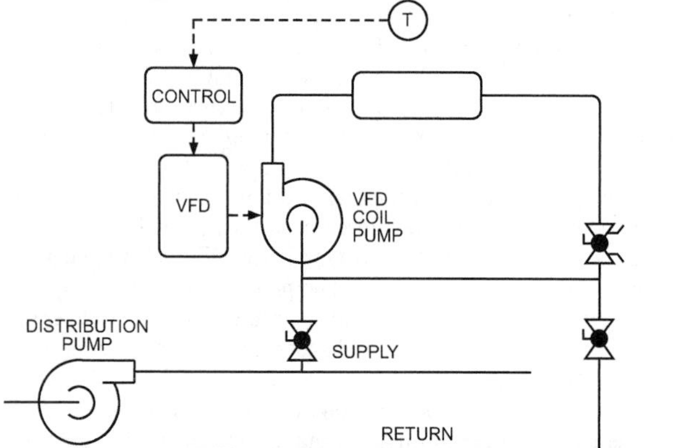

Fig. 30 Schematic of Variable-Speed Pump Coil Control

Green (1994) tested the control stability of variable-speed circulating pumps compared to control valves. Stability of the controlled discharge temperature of the coil was on the order of ±0.5°F. Large coils with pump and separate drive may also be economical compared to similarly sized valves and actuators. The stability of on/off pump control was also found to be reasonable. The attraction of a pump over a control valve is the reduction of head and thus energy used by the distribution pumping system. For the valve to work properly, pump head must be throttled to control flow. Any throttling process is inherently energy-inefficient. Direct load control by a pump provides only the energy required to overcome the friction loss of the load heat transfer device and piping. Properly sequenced, the distribution pumping system can reduce flow, which causes higher coil entering water temperatures, which causes the circulating pump to operate at a higher flow. This is helpful in allowing for higher percentages of heat transfer capability of the sensible and latent loads through the coil. Applications of this control method must also take into account the potential of gravity flow in the piping scheme.

LOW-TEMPERATURE HEATING SYSTEMS

These systems are used for heating spaces or processes directly, as with standing radiation and process heat exchangers, or indirectly, through air-handling unit coils for preheating, for reheating, or in hot-water unit heaters. They are generally designed with supply water temperatures from 180 to 240°F and temperature drops from 20 to 100°F.

In the United States, hot-water heating systems were historically designed for a 200°F supply water temperature and a 20°F temperature drop. This practice evolved from earlier gravity system designs and provides convenient design relationships for heat transfer coefficients related to coil tubing and finned-tube radiation and for calculations (1 gpm conveys 10,000 Btu/h at 20°F Δt). Because many terminal devices still require these flow rates, it is important to recognize this relationship in selecting devices and designing systems.

However, the greater the temperature range (and related lower flow rate) that can be applied, the less costly the system is to install and operate. A lower flow rate requires smaller and less expensive piping, less secondary building space, and smaller pumps. Also, smaller pumps require less energy, so operating costs are lower.

Nonresidential Heating Systems

Possible approaches to enhancing the economics of large heating systems include (1) higher supply temperatures, (2) primary-secondary pumping, and (3) terminal equipment designed for smaller flow rates. The three techniques may be used either singly or in combination.

Using higher supply water temperatures achieves higher temperature drops and smaller flow rates. Terminal units with a reduced heating surface can be used. These smaller terminals are not necessarily less expensive, however, because their required operating temperatures and pressures may increase manufacturing costs and the problems of pressurization, corrosion, expansion, and control. System components may not increase in cost uniformly with temperature, but rather in steps conforming to the three major temperature classifications. Within each classification, the most economical design uses the highest temperature in that classification.

Primary/secondary or compound pumping reduces the size and cost of the distribution system and also may use larger flows and lower temperatures in the terminal or secondary circuits. A primary pump circulates water in the primary distribution system while one or more secondary pumps circulate the terminal circuits. The connection between primary and secondary circuits provides complete hydraulic isolation of both circuits and allows a controlled interchange of water between the two. Thus, a high supply water temperature can be used in the primary circuit at a low flow rate and high temperature drop, while a lower temperature and conventional temperature drop can be used in the secondary circuit(s).

For example, a system could be designed with primary-secondary pumping in which the supply temperature from the boiler was 240°F, the supply temperature in the secondary was 200°F, and the return temperature was 180°F. This design results in a conventional 20°F Δt in the secondary zones, but allows the primary circuit to be sized on the basis of a 60°F drop. This primary-secondary pumping arrangement is most advantageous with terminal units such as convectors and finned radiation, which are generally unsuited for small flow rate design.

Many types of terminal heat transfer units are being designed to use smaller flow rates with temperature drops up to 100°F in low-temperature systems and up to 150°F in medium-temperature systems. Fan apparatus, the heat transfer surface used for air heating in fan systems, and water-to-water heat exchangers are most adaptable to such design.

A fourth technique is to put certain loads in series using a combination of control valves and compound pumping (Figure 31). In the system illustrated, the capacity of the boiler or heat exchanger is 2×10^6 Btu/h, and each of the four loads is 0.5×10^6 Btu/h. Under design conditions, the system is designed for an 80°F water temperature drop, and the loads each provide 20°F of the total Δt. The loads in these systems, as well as the smaller or simpler systems in residential or commercial applications, can be connected in a direct-return or a reverse-return piping system. The different features of each load are as follows:

- The domestic hot-water heat exchanger has a two-way valve and is thus arranged for variable flow (the main distribution circuit provides constant flow for the boiler circuit).
- The finned-tube radiation circuit is a 20°F Δt circuit with the design entering water temperature reduced to and controlled at 200°F.
- The reheat coil circuit takes a 100°F temperature drop for a very low flow rate.
- The preheat coil circuit provides constant flow through the coil to keep it from freezing.

When loads such as water-to-air heating coils in LTW systems are valve-controlled (flow varies), they have a heating characteristic of flow versus capacity as shown in Figure 32 for 20°F and 60°F temperature drops. For a 20°F Δt coil, 50% flow provides approximately 90% capacity; valve control will tend to be unstable. For this reason, proportional temperature control is required, and equal percentage characteristic two-way valves should be selected such that 10% flow is achieved with 50% valve lift. This combination of the valve characteristic and the heat transfer characteristic of the coil makes control linear with respect to the control signal. This type of control can be obtained only with equal percentage two-way valves and can be further enhanced if piped with a secondary pump arrangement as shown in Figure 29A. See Chapter 46 of the 2007 *ASHRAE Handbook—HVAC Applications* for further information on automatic controls.

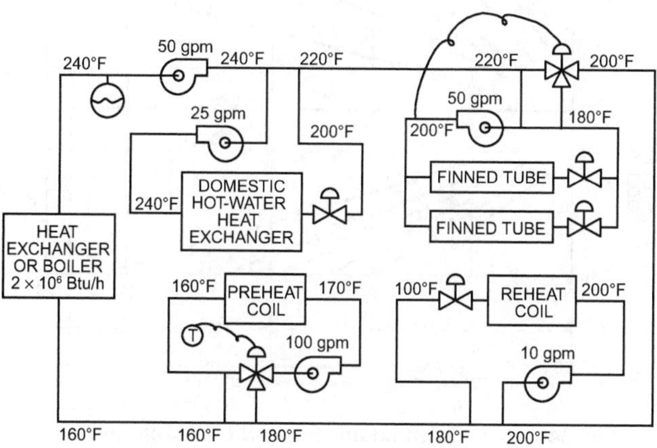

Fig. 31 Example of Series-Connected Loading

CHILLED-WATER SYSTEMS

Designers have less latitude in selecting supply water temperatures for cooling applications because there is only a narrow range of water temperatures low enough to provide adequate dehumidification and high enough to avoid chiller freeze-up. Circulated water quantities can be reduced by selecting proper air quantities and heat transfer surface at the terminals. Terminals suited for a 16 to 20°F rise rather than a 10 to 12°F rise reduce circulated water quantity and pump power by one-third and increase chiller efficiency.

A proposed system should be evaluated for the desired balance between installation cost and operating cost. Table 1 shows the effect of coil circuiting and chilled-water temperature on water flow and temperature rise. This yields a coil characteristic not unlike the heating coil shown in Figure 33. The characteristic, though, should be shown as total, sensible, and latent heat transfer. The coil rows, fin

Table 1 Chilled-Water Coil Performance

Coil Circuiting	Chilled-Water Inlet Temp., °F	Coil Pressure Drop, psi	Chilled-Water Flow, gpm/ton	Chilled-Water Temp. Rise, °F
Full[a]	45	1.0	2.2	10.9
Half[b]	45	5.5	1.7	14.9
Full[a]	40	0.5	1.4	17.1
Half[b]	40	2.5	1.1	21.8

Note: Table is based on cooling air from 81°F db, 67°F wb to 58°F db, 56°F wb.
[a] Full circuiting (also called single circuit). Water at the inlet temperature flows simultaneously through all tubes in a plane transverse to airflow; it then flows simultaneously through all tubes, in unison, in successive planes (i.e., rows) of the coil.
[b] Half circuiting. Tube connections are arranged so there are half as many circuits as there are tubes in each plane (row), thereby using higher water velocities through the tubes. This circuiting is used with small water quantities.

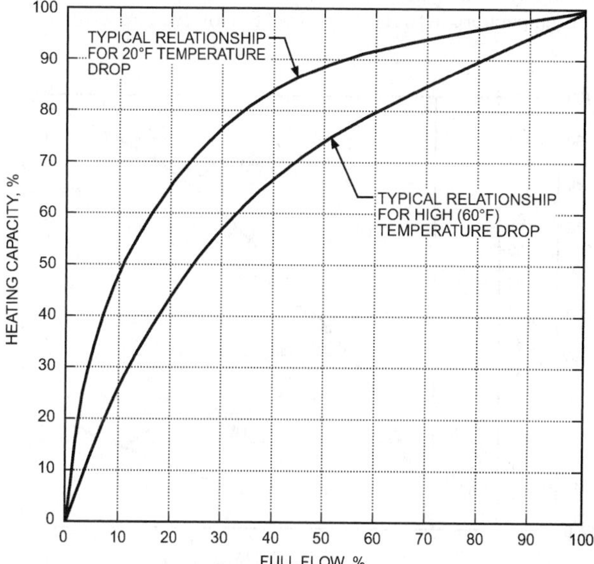

Fig. 32 Heat Emission Versus Flow Characteristic of Typical Hot Water Heating Coil

spacing, air-side performance, and cost are identical for all selections. Morabito (1960) showed how such changes in coil circuiting affect the overall system. Considering the investment cost of piping and insulation versus the operating cost of refrigeration and pumping motors, higher temperature rises, (i.e., 16 to 24°F temperature rise at about 1.0 to 1.5 gpm per ton of cooling) can be applied on chilled-water systems with long distribution piping runs; larger flow rates should be used only where reasonable in close-coupled systems.

For the most economical design, the minimum flow rate to each terminal heat exchanger is calculated. For example, if one terminal can be designed for an 18°F rise, another for 14°F, and others for 12°F, the highest rise to each terminal should be used, rather than designing the system for an overall temperature rise based on the smallest capability.

The control system selected also influences the design water flow. The expense of operating chilled-water systems, combined with the complexity of interrelated system variables, demands direct digital control. Evaluation suggests that, in many cases, a minimum of PI control is required, if not PID. To control flow, the designer should apply two-way modulating control valves, matching the required characteristic to the coil's operational characteristic. The control system programmer may need some method of linearizing through programming the controller output to the combined characteristic of the control valve and the selected coil.

For systems with multiple terminal units, use care in applying diversity factors, which can negatively affect not only chiller selection, but also pump selection. Designers should apply variable-speed pumping techniques to the system pumps. System balancing is best accomplished with automatic flow-limiting valves, with the pumps capable of either providing the coil block flow load, or limiting the stroke of the control valves in conditions where single-loop controllers would tend to open all valves at once (e.g., at start-up). This is necessary to prevent a no-flow condition from occurring in coils farther from the pump.

A primary consideration with chilled-water system design is the control of the source systems at reduced loads. The constraints on the temperature parameters are (1) a water freezing temperature of 32°F, (2) economics of the refrigeration system in generating chilled water, and (3) the dew-point temperature of the air at nominal indoor comfort conditions (55°F dew point at 75°F and 50% rh).

These parameters have led to the common practice of designing for a supply chilled-water temperature of 42 to 45°F and a return water temperature between 55 and 66°F. However, Pederson et al. (1998) suggest an alternative classic design methodology for selecting entering water temperature to the chilled-water coil, based on the coil's dew-point temperature.

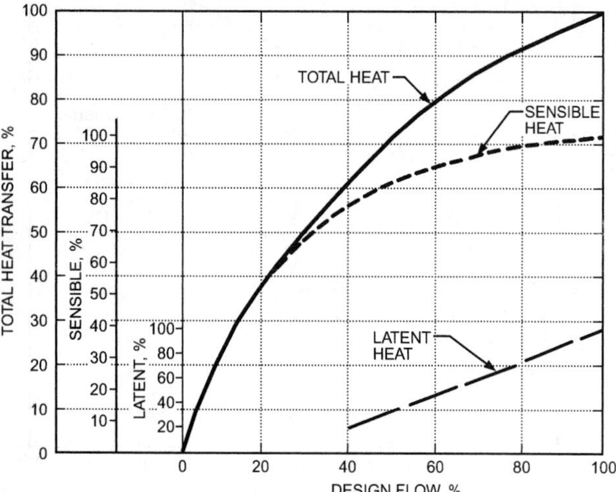

Fig. 33 Generic Chilled-Water Coil Heat Transfer Characteristic

Final selection of the design criteria should be judged against the required flow tolerance of the coil to the required heat transfer characteristics of the coil. Carlson (1981) notes relationships for entering and differential temperatures of the coil as it compares to providing 97% of design heat transfer, and compares the required flow as a percentage of design for the purpose of specifying hydronic system balancing, and selection of the distribution system friction loss.

With increasing entering water temperatures and selection differential temperatures, the coil characteristic becomes very sensitive to changes in flow at the design condition. Carlson (1981) showed that flow tolerance to the coil could be as high as ±20% flow for low Δt designs, but the tolerance became much closer (±5% flow or less) for coils with higher differential-temperature designs (Figure 34). The concern was providing adequate flow for adequate coil heat transfer at a design condition, without grossly overflowing the coil. Carlson (1981) suggested a maximum allowable flow tolerance of ±10% design flow at design flow conditions to enhance operating system energy efficiency. Many devices are available to help achieve this. However, because coil heat transfer sensitivity to flow increases as water-side differential temperature increases, designers must weigh coil design criteria, pipe sizing decisions, and balancing interactions. From a practical perspective, many balancing devices can achieve ±10% flow adjustment or better, and in some cases meet or exceed ±5% flow adjustment. However, as Figure 34 indicates, changes in water-side design Δt and entering air conditions to the coil offer the potential for losing some high-flow heat transfer if design flows cannot be accomplished. This requires that flow be balanced as closely to design as possible so the required heat transfer for design condition can be met. If the designer has concerns

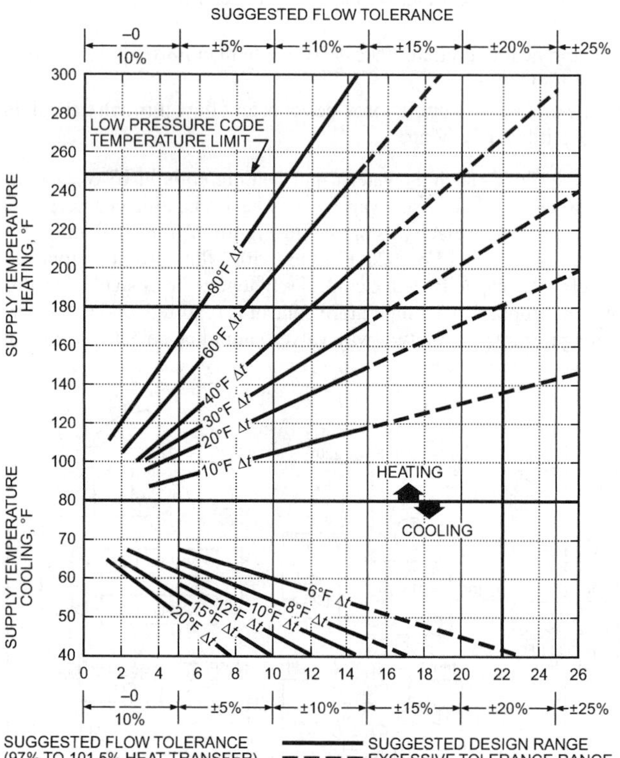

Fig. 34 Recommendations for Coil Flow Tolerance to Maintain 97% Design Heat Transfer

(Carlson 1981)

with temperature and humidity control in the space, the maximum potential operating energy efficiency of the system (characterized by higher coil design differential temperatures and lower flows) must be balanced with the required comfort conditions of the space and the ability to control them based on the operating interactions of all system components (pipe, valve, fitting friction loss, coil operation, and control system sensitivity to adjustment). Care must be taken by the designer to achieve the desired operation. Combined with energy-sensible concepts such as variable-speed, variable-flow pumping and the potential flow interaction of the control valve, the designer must carefully analyze the system loads and consider all system effects. Figure 34 shows one approach for of this type of analysis. However, it shows that heat transfer, and potentially comfort, is affected if flow to a coil varies too much. System designers should consider similar types of analysis based on their selected design parameters.

Figure 35 shows a typical configuration of small chilled-water systems, using two parallel chillers and loads with three-way valves. Note that flow should be essentially constant, although the valve characteristic and authority could cause more or less than design flow at part-load (valve stroke) conditions. A simple energy balance [Equation (9)] dictates that, with a constant flow rate, at one-half of design load, the water temperature differential drops to one-half of design. At this load, if one of the chillers is turned off, the return water circulating through the *off* chiller mixes with the supply water. This mixing raises the temperature of the supply chilled water and can cause a loss of control if the designer does not consider this operating mode. A better approach is to use a modulated system instead, to enhance operating efficiency and meet the requirements of building energy codes as presented in ASHRAE *Standard* 90.1.

A typical configuration of a large chilled-water system with multiple chillers and loads and compound piping is shown in Figure 36. This system provides variable flow, essentially constant-supply-temperature chilled water, multiple chillers, more stable two-way control valves, and the advantage of adding chilled-water storage with little additional complexity. As mentioned, mixing in the transition of chillers can be of concern, but using variable-speed drives can help eliminate the issue while keeping the benefit of hydraulic organization.

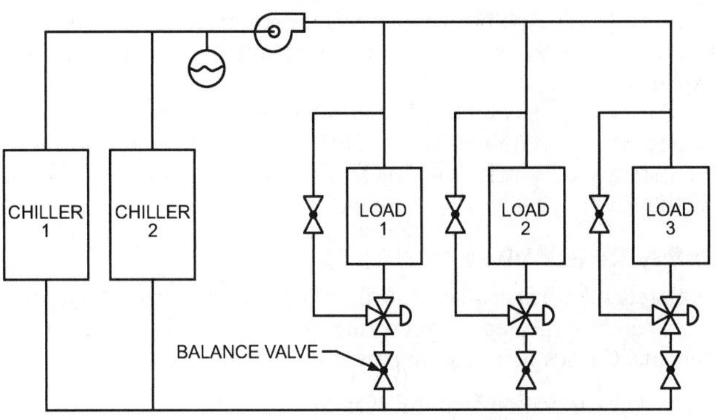

Fig. 35 Constant-Flow Chilled-Water System

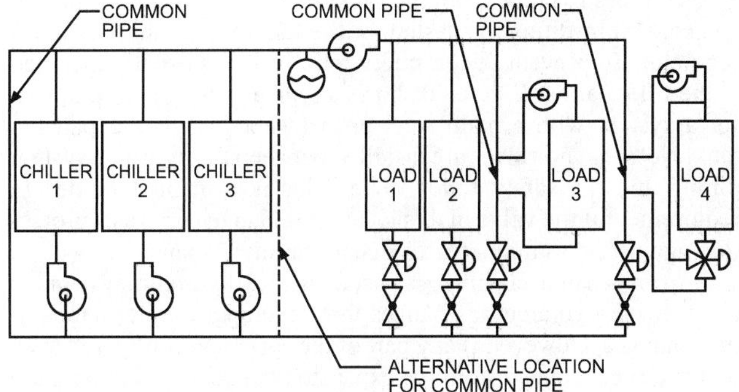

Fig. 36 Variable-Flow Chilled-Water System

One design issue illustrated in Figure 36 is the placement of the common pipe for the chillers. With the common pipe as shown, the chillers unload from left to right. With the common pipe in the alternative location shown, the chillers unload equally in proportion to their capacity (i.e., equal percentage).

The **one-pipe chilled-water system**, also called the **integrated decentralized chilled-water system**, is another system that has seen considerable use in campus-type chilled-water systems with multiple chillers and multiple buildings (Coad 1976). A single pumped main circulates water in a closed loop through all the connected buildings. Each load and/or chiller is connected to the loop, with the chillers usually downstream from a load connection. The loop capacity is limited only by the fact that the flow capacity for any single load or chiller connection cannot exceed the flow rate of the loop. Because the loads are in series, the cooling coils must be sized for higher entering water temperatures than are normally used.

DUAL-TEMPERATURE SYSTEMS

Dual-temperature systems are used when the same load devices and distribution systems are used for both heating and cooling (e.g., fan-coil units and central station air-handling unit coils). In dual-temperature system design, the cooling cycle design usually dictates the requirements of the load heat exchangers and distribution systems. Dual-temperature systems are basically of three different configurations, each requiring different design techniques:

1. Two-pipe systems
2. Four-pipe common load systems
3. Four-pipe independent load systems

Two-Pipe Systems

In a dual-temperature two-pipe system, the load devices and the distribution system circulate chilled water when cooling is required and hot water when heating is required (Figure 37). Design considerations for these systems include the following:

- Loads must all require cooling or heating coincidentally; that is, if cooling is required for some loads and heating for other loads at a given time, this type of system should not be used.
- When designing the system, the flow and temperature requirements for both cooling and heating media must be calculated first. The load and distribution system should be designed for the more stringent, and the water temperatures and temperature differential should be calculated for the other mode.
- Changeover should be designed such that the chiller evaporator is not exposed to damaging high water temperatures and the boiler is not subjected to damaging low water temperatures. To accom-

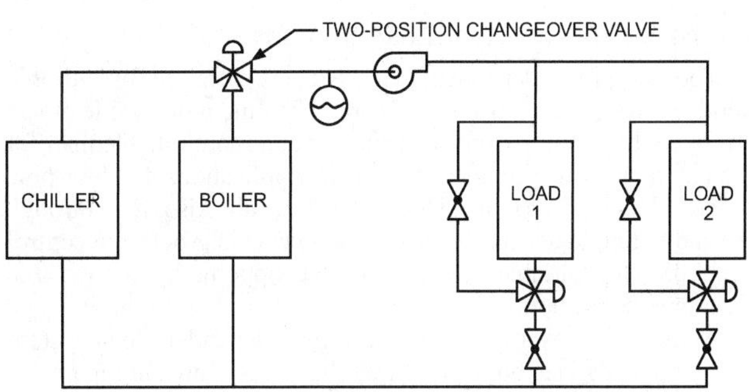

Fig. 37 Simplified Diagram of Two-Pipe System

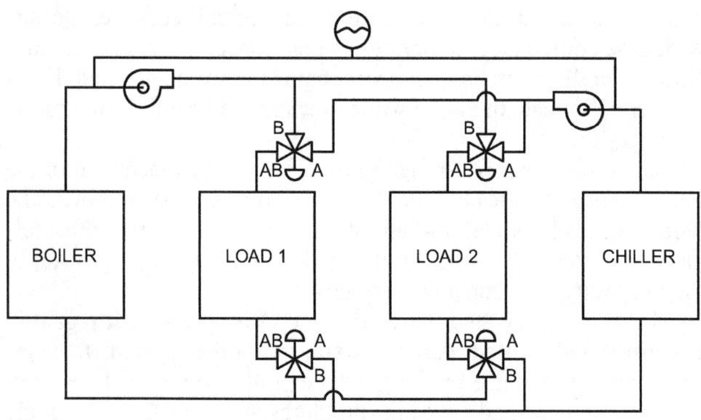

Fig. 38 Four-Pipe Common Load System

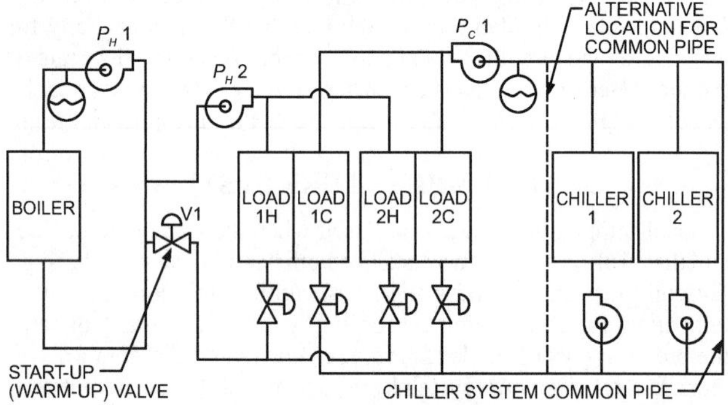

Fig. 39 Four-Pipe Independent Load System

modate these limiting requirements, changeover of a system from one mode to the other requires considerable time. If rapid load swings are anticipated, a two-pipe system should not be selected, although it is the least costly of the three options.

Four-Pipe Common Load Systems

In the four-pipe common load system, load devices are used for both heating and cooling as in the two-pipe system. The four-pipe common load system differs from the two-pipe system in that both heating and cooling are available to each load device, and changeover from one mode to the other takes place at each individual load device, or grouping of load devices, rather than at the source. Thus, some load systems can be in cooling mode while others are in heating mode. Figure 38 is a flow diagram of a four-pipe common load system, with multiple loads and a single boiler and chiller.

Many of these systems were installed and did not perform successfully because of problems in implementing the design concepts, and leakage through the control valves. These systems are not allowed for new application in energy standards such as ASHRAE 90.1, and are mentioned here as historical reference.

Four-Pipe Independent Load Systems

The four-pipe independent load system is preferred for hydronic applications in which some loads are in heating mode while others are in cooling mode. Control is simpler and more reliable than for the common load systems and, in many applications, the four-pipe independent load system is less costly to install. Also, flow through the individual loads can be modulated, providing both the control capability for variable capacity and the opportunity for variable flow in either or both circuits.

A simplified example of a four-pipe independent load system with two loads, one boiler, and two chillers is shown in Figure 39. Note that both hydronic circuits are essentially independent, so that each can be designed with disregard for the other system.

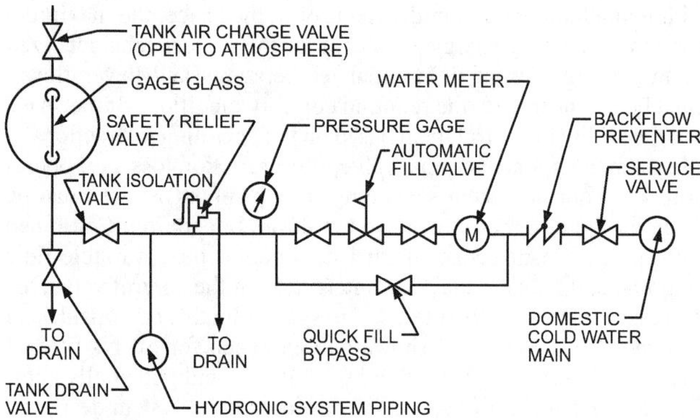

Fig. 40 Typical Makeup Water and Expansion Tank Piping Configuration for Plain Steel Expansion Tank

Although both circuits in the figure are shown as variable-flow distribution systems, they could be constant-flow (three-way valves) or one variable and one constant. Generally, the control modulates the two load valves in sequence with a dead band at the control midpoint.

This type of system offers additional flexibility when some selective loads are arranged for heating only or cooling only, such as unit heaters or preheat coils. Then, central station systems can be designed for humidity control with reheat through configuration at the coil locations and with proper control sequences.

OTHER DESIGN CONSIDERATIONS

Makeup and Fill Water Systems

Generally, a hydronic system is filled with water through a valved connection to a domestic water source, with a service valve, a backflow preventer, and a pressure gage. (The domestic water source pressure must exceed the system fill pressure.)

Because the expansion chamber is the reference pressure point in the system, the water makeup point is usually located at or near the expansion chamber.

Many designers prefer to install automatic makeup valves, which consist of a pressure-regulating valve in the makeup line. However, the quantity of water being made up must be monitored to avoid scaling and oxygen corrosion in the system.

Safety Relief Valves

Safety relief valves should be installed at any point at which pressures can be expected to exceed the safe limits of the system components. Causes of excessive pressures include

- Overpressurization from fill system
- Pressure increases caused by thermal expansion
- Surges caused by momentum changes (shock or water hammer)

Overpressurization from the fill system could occur because of an accident in filling the system or the failure of an automatic fill regulator. To prevent this, a safety relief valve is usually installed at the fill location. Figure 40 shows a typical piping configuration for a system with a plain steel or air/water interface expansion tank. Note that no valves are installed between the hydronic system piping and the safety relief valve. This is a mandatory design requirement if the valve in this location is also to serve as a protection against pressure increases due to thermal expansion.

An expansion chamber is installed in a hydronic system, to allow for the volumetric changes that accompany water temperature changes. However, if any part of the system is configured such that it can be isolated from the expansion tank and its temperature can increase while it is isolated, then overpressure relief should be provided.

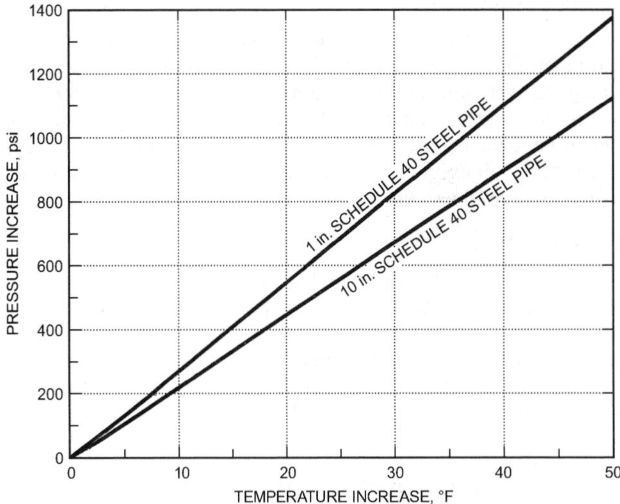

Fig. 41 Pressure Increase Resulting from Thermal Expansion as Function of Temperature Increase

The relationship between pressure change caused by temperature change and the temperature change in a piping system is expressed by the following equation:

$$\Delta p = \frac{(\beta - 3\alpha)\Delta t}{(5/4)(D/E\,\Delta r) + \gamma} \tag{20}$$

where

Δp = pressure increase, psi
β = volumetric coefficient of thermal expansion of water, 1/°F
α = linear coefficient of thermal expansion for piping material, 1/°F
Δt = water temperature increase, °F
D = pipe diameter, in.
E = modulus of elasticity of piping material, psi
γ = volumetric compressibility of water, in^2/lb
Δr = thickness of pipe wall, in.

Figure 41 shows a solution to Equation (23) demonstrating the pressure increase caused by any given temperature increase for 1 in. and 10 in. steel piping. If the temperature in a chilled water system with piping spanning sizes between 1 and 10 in. were to increase by 15°F, the pressure would increase between 340 and 420 psi, depending on the average pipe size in the system.

Safety relief should be provided to protect boilers, heat exchangers, cooling coils, chillers, and the entire system when the expansion tank is isolated for air charging or other service. As a minimum, the ASME *Boiler and Pressure Vessel Code* requires that a dedicated safety relief valve be installed on each boiler and that isolating or service valves be provided on the supply and return connections to each boiler.

Potential forces caused by shock waves or water hammer should also be considered in design. Chapter 36 of the 2005 *ASHRAE Handbook—Fundamentals* discusses the causes of shock forces and the methodology for calculating the magnitude of these forces.

Air Elimination

If air and other gases are not eliminated from the flow circuit, they may slow or stop the flow through the terminal heat transfer elements and cause corrosion, noise, reduced pumping capacity, and loss of hydraulic stability (see the section on Principles at the beginning of the chapter). A closed tank without a diaphragm can be installed at the point of the lowest solubility of air in water. With a diaphragm tank, air in the system can be removed by an air separator and air elimination valve installed at the point of lowest solubility. This type of system is called an air elimination system. Manual vents should be installed at high points to remove all air trapped during initial operation. Shutoff valves should be installed on any automatic air removal device to allow servicing without draining the system.

Air elimination devices are most effective at low velocities. Thus, the pipe leading up to the air elimination device often is smaller than the device piping, and this size difference should be accounted for in the design. Alternatively, with variable-speed pumping systems, the air elimination device could be commissioned at a flow rate less than full design flow, allowing for a lower entering velocity to the device. After commissioning, when air has been purged from the system, the device can be operated at the higher flow. This is less effective for air removal, but air is unlikely to reenter the closed-loop pumping system.

Standard steel vessels used as system compression tanks are air management systems. An air separator installed at the point of lowest solubility collects air recovered from the system and transfers it to the compression tank. The tank must have a monolithic tank fitting, allowing water and air to enter and leave the tank, and preventing gravity flow between the system and the tank. Automatic air removal devices should never be used in these systems, because collected system air provides the gas cushion for water system expansion and contraction.

Drain and Shutoff

All low points should have drains. Separate shutoff and draining of individual equipment and circuits should be possible so that the entire system does not have to be drained to service a particular item. Whenever a device or section of the system is isolated and water in that section or device could increase in temperature following isolation, overpressure safety relief protection must be provided.

Balance Fittings

Balance fittings or valves and a means of measuring flow quantity should be applied as needed to allow balancing of individual terminals and subcircuits. Balance, however, cannot be achieved through fittings or devices alone. In a balanced system, at design flow conditions, all terminals receive as a minimum enough flow to create the required design heat transfer. Carlson (1968, 1981) suggested that 97% heat transfer was a reasonable value for required heat transfer accuracy, implying a flow tolerance to the coil based on heat transfer. On a 200°F entering water temperature (EWT) hot-water coil with a 20°F Δt, flow could be ±25% to the coil, and the required heat transfer would be achieved. However, as Δt and coil EWT increase, this flow tolerance decreases. Carlson suggested that, for system efficiency, circuits should be balanced to ±10% flow, which is commonly accepted. However, Carlson also noted that, for some systems, flow tolerance is tighter (±5% or, in some cases, +10% and 0%), particularly in cases where the log mean temperature difference (LMTD) of the coil is limited (e.g., in chilled-water systems) and flow tolerance becomes an issue in proper system operation. One method Carlson suggested to overcome these issues in constant-speed pumping systems was to use a branch-to-riser pressure drop ratio as a criterion for selecting distribution system friction loss. A 4:1 ratio allowed 95% of design flow to always reach the terminal, a 2:1 ratio yielded 90% flow, and 1:1 yielded 80% flow. In variable-speed pumping systems, this concept is helpful to analyze, but yields different results. Regardless, system distribution piping losses should be analyzed with respect to their effects on system flow, balance, and control valve operation, which leads to control stability. Application of flow measurement fittings requires stabilized pipe flow, and is affected by the location and planar geometry of the device to fittings both up- and downstream. In general, allow reasonable straight pipe lengths (anything from a few pipe diameters to 20 diameters) before flow-measurement devices.

Pitch

Piping need not pitch but can run level, providing that flow velocities exceeding 1.5 fps are maintained or a diaphragm tank is used.

Strainers

Strainers should be used where necessary to protect system elements, and sparingly to enhance energy efficiency. Strainers in the pump suction must be checked carefully to avoid pump cavitation. Designers should consider strainer designs that can trap particles during the commissioning (flush and clean) phase, and allow more particle trap size modification after commissioning, to reduce system pressure loss. Large separating chambers can serve as main air venting points and dirt strainers ahead of pumps. Automatic control valves or other devices operating with small clearances require protection from pipe scale, gravel, and welding slag, which may readily pass through the pump and its protective separator. Individual fine mesh strainers may therefore be required ahead of each control valve. An alternative is to use two manual three-way valves entering a coil with connected bypasses. During commissioning, all coils are isolated from the system, with the three-way valve bypasses connected, allowing flushing water to serve main distribution piping to the coil, but not carrying particulates through the coil. After piping is flushed and cleaned, the valves are positioned to allow flow to and from the coil. This method removes most system particulates, thus protecting the coil, and, if individual coil strainers are used, greatly reduces labor in flushing each individual strainer.

Thermometers

Thermometers or thermometer wells should be installed to assist the system operator in routine operation and troubleshooting. Permanent thermometers, with the correct scale range and separate sockets, should be used at all points where temperature readings are regularly needed. Thermometer wells should be installed where readings will be needed only during start-up and infrequent troubleshooting. If a central monitoring system is provided, a calibration well should be installed adjacent to each sensing point in insulated piping systems.

Flexible Connectors and Pipe Expansion Compensation

Flexible connectors are sometimes installed at pumps and machinery to reduce pipe stress. See Chapter 47 of the 2007 *ASHRAE Handbook—HVAC Applications* for vibration isolation information. Expansion, flexibility, and hanger and support information is in Chapter 45 of this volume. Piping systems should be supported independently of flexible connections.

Gage Cocks

Gage cocks or quick-disconnect test ports should be installed at points requiring pressure or temperature readings. Gages permanently installed in the system will deteriorate because of vibration and pulsation and may become unreliable. It is good practice to install gage cocks and provide the operator with several high-quality gages for diagnostic purposes. Avoid overuse of gage cocks and test ports, because they represent points of potential system leakage. In general, one port entering a coil and leaving the coil is adequate, and can be combined with other functions to produce data required to verify system operation.

Insulation

Insulation should be applied to minimize pipe thermal loss and to prevent condensation during chilled-water operation (see Chapter 24 of the 2005 *ASHRAE Handbook—Fundamentals*). On chilled-water systems, special rigid metal sleeves or shields should be installed at all hanger and support points, and all valves should be provided with extended bonnets to allow for the full insulation thickness without interference with the valve operators.

Condensate Drains

Condensate drains from dehumidifying coils should be trapped and piped to an open-sight plumbing drain. Traps should be deep enough to overcome the air pressure differential between drain inlet and room, which ordinarily will not exceed 2 in. of water. Pipe should be noncorrosive and insulated to prevent moisture condensation. Depending on the quantity and temperature of condensate, plumbing drain lines may require insulation to prevent sweating.

Common Pipe

In compound (primary-secondary) pumping systems, the common pipe is used to dynamically decouple the two pumping circuits. Ideally, there is no pressure drop in this section of piping; however, in actual systems, it is recommended that this section of piping be a minimum of 10 diameters in length to reduce the likelihood of unwanted mixing resulting from velocity (kinetic) energy or turbulence.

OTHER DESIGN PROCEDURES

Preliminary Equipment Layout

Flows in Mains and Laterals. Regardless of the method used to determine the flow through each item of terminal equipment, the desired result should be listed on the preliminary plans or in a schedule of flow rates for the piping system.

In an equipment schedule or on the plans, starting from the most remote terminal and working toward the pump, progressively list the cumulative flow and head loss in each of the mains and branch circuits in the distribution system. It is helpful in system commissioning to have a piping system schematic noting the friction loss calculations of each flow path. This should be included with the design drawings of the project.

The designer is responsible for selecting control valves and coordinating them with heat transfer devices. A schedule of devices and detailed connection schematic should be given.

Preliminary Pipe Sizing. For each portion of the piping circuit, select a tentative pipe size from the unified flow chart (Figure 1 in Chapter 36 of the 2005 *ASHRAE Handbook—Fundamentals*), using a value of pipe friction loss ranging from 0.75 to 4 ft per 100 ft of straight pipe. Velocity through the pipe should also be examined, and should not exceed 10 fps in general application. Air management techniques suggest that velocity should also be kept above 2 fps, so entrained air is carried through the system, although piping layout considerations may make this less of an issue. The copper piping trade association suggests establishing maximum velocities of 5 fps for hot-water piping and 8 fps for cold-water piping (CDA 2006). Others suggest that no greater than 4 fps be used on pipes less than 1.5 in. in diameter. Other suggestions are in Chapter 36 of the 2005 *ASHRAE Handbook—Fundamentals*.

Residential piping size is often based on pump preselection using pipe sizing tables, which are available from the Hydronics Institute or from manufacturers. Allow adequate space for water entrance to pump volute, to reduce pump suction losses in manifolds.

Preliminary Pressure Drop. Using the preliminary pipe sizing, determine the pressure drop through each portion of piping. The total pressure drop in the longest circuits determines the maximum pressure drop through the piping, including the terminals and control valves, that must be available in the form of pump pressure.

Preliminary Pump Selection. The preliminary selection should be based on the pump's ability to fulfill the determined capacity requirements. It should be selected as close as possible to best efficiency on the pump curve and should not overload the motor. Because pressure drop in a flow system varies as the square of the flow rate, the flow variation between the nearest size of stock pump and an exact point selection will be relatively minor. Note that although efficiency is important, it is secondary to the pipe sizing criteria. Proper pipe sizing reduces overall head losses, allowing a lower-power pump to be used for the same flow rate, and should be considered first.

Final Pipe Sizing and Pressure Drop Determination

Final Piping Layout. Examine the overall piping layout to determine whether pipe sizes in some areas need to be readjusted. Several principal circuits should have approximately equal pressure drops so that excessive pressures are not needed to serve a small portion of the building.

Consider both the initial cost of the pump and piping system and the pump's operating cost when determining final system friction loss. Generally, lower heads and larger piping are more economical when longer amortization periods are considered, especially in larger systems. However, in small systems such as in residences, it may be most economical to select the pump first and design the piping system to meet the available pressure. In all cases, adjust the piping system design and pump selection until the optimum design is found.

Final Pressure Drop. When the final piping layout has been established, determine the friction loss for each section of the piping system from the pressure drop charts (Chapter 36 of the 2005 *ASHRAE Handbook—Fundamentals*) for the mass flow rate in each portion of the piping system.

After calculating friction loss at design flow for all sections of the piping system and all fittings, terminal units, and control valves, sum them for several of the longest piping circuits to determine the pressure against which the pump must operate at design flow.

Final Pump Selection. After completing the final pressure drop calculations, select the pump by plotting a system curve and pump curve and selecting the pump or pump assembly that operates closest to the calculated design point.

Freeze Prevention

All circulating water systems require precautions to prevent freezing, particularly in makeup air applications in temperate climates where (1) coils are exposed to outside air at freezing temperatures, (2) undrained chilled-water coils are in the winter airstream, or (3) piping passes through unheated spaces. Freezing will not occur as long as flow is maintained and the water is at least warm. Unfortunately, during extremely cold weather or in the event of a power failure, water flow and temperature cannot be guaranteed. Additionally, continuous pumping can be energy-intensive and cause system wear. The following are precautions to avoid flow stoppage or damage from freezing:

- Select all load devices (such as preheat coils) subjected to outside air temperatures for constant-flow, variable Δt control.
- Position the coil valves of all cooling coils with valve control that are dormant in winter to the full-open position at those times.
- If intermittent pump operation is used as an economy measure, use an automatic override to operate both chilled-water and heating-water pumps in below-freezing weather.
- Select pump starters that automatically restart after power failure (i.e., maintain-contact control).
- Select nonoverloading pumps.
- Instruct operating personnel never to shut down pumps in subfreezing weather.
- Do not use aquastats, which can stop a pump, in boiler circuits.
- Avoid sluggish circulation, which may cause air binding or dirt deposit. Properly balance and clean systems. Provide proper air control or means to eliminate air.
- Install low-temperature-detection thermostats that have phase change capillaries wound in a serpentine pattern across the leaving face of the upstream coil.

In fan equipment handling outside air, take precautions to avoid stratification of air entering the coil. The best methods for proper mixing of indoor and outdoor air are the following:

- Select dampers for pressure drops adequate to provide stable control of mixing, preferably with dampers installed several equivalent diameters upstream of the air-handling unit.
- Design intake and approach duct systems to promote natural mixing.
- Select coils with circuiting to allow parallel flow of air and water.

Freeze-up may still occur with any of these precautions. If an antifreeze solution is not used, water should circulate at all times. Valve-controlled elements should have low-limit thermostats, and sensing elements should be located to ensure accurate air temperature readings. Primary/secondary pumping of coils with three-way valve injection (as in Figures 29B and 29C) is advantageous. Use outdoor reset of water temperature wherever possible.

ANTIFREEZE SOLUTIONS

In systems in danger of freeze-up, water solutions of ethylene glycol and propylene glycol are commonly used. Freeze protection may be needed (1) in snow-melting applications (see Chapter 50 of the 2007 *ASHRAE Handbook—HVAC Applications*); (2) in systems subjected to 100% outside air, where the methods outlined above may not provide absolute antifreeze protection; (3) in isolated parts or zones of a heating system where intermittent operation or long runs of exposed piping increase the danger of freezing; and (4) in process cooling applications requiring temperatures below 40°F. Although using ethylene glycol or propylene glycol is comparatively expensive and tends to create corrosion problems unless suitable inhibitors are used, it may be the only practical solution in many cases.

Solutions of triethylene glycol, as well as certain other heat transfer fluids, may also be used. However, ethylene glycol and propylene glycol are the most common substances used in hydronic systems because they are less costly and provide the most effective heat transfer.

Effect on Heat Transfer and Flow

Tables 6 to 13 and Figures 9 to 16 in Chapter 21 of the 2005 *ASHRAE Handbook—Fundamentals* show density, specific heat, thermal conductivity, and viscosity of various aqueous solutions of ethylene glycol and propylene glycol. Tables 4 and 5 of that chapter indicate the freezing points for the two solutions.

System heat transfer rate is affected by relative density and specific heat according to the following equation:

$$q_w = 500Q(\rho/\rho_w)c_p\Delta t \qquad (21)$$

where

q_w = total heat transfer rate, Btu/h
Q = flow rate, gpm
ρ = fluid density, lb/ft^3
ρ_w = density of water at 60°F, lb/ft^3
c_p = specific heat of fluid, Btu/lb·°F
Δt = temperature increase or decrease, °F

Effect on Heat Source or Chiller

Generally, ethylene glycol solutions should not be used directly in a boiler because of the danger of chemical corrosion caused by glycol breakdown on direct heating surfaces. However, properly inhibited glycol solutions can be used in low-temperature water systems directly in the heating boiler if proper operation can be ensured. Automobile antifreeze solutions are not recommended because the silicate inhibitor can cause fouling, pump seal wear, fluid gelation, and reduced heat transfer. The area or zone requiring antifreeze protection can be isolated with a separate heat exchanger or converter. Glycol solutions are used directly in water chillers in many cases.

Glycol solutions affect the output of a heat exchanger by changing the film coefficient of the surface contacting the solution. This change in film coefficient is caused primarily by viscosity changes. Figure 42 illustrates typical changes in output for two types of heat exchangers, a steam-to-liquid converter and a refrigerant-to-liquid chiller. The curves are plotted for one set of operating conditions only and reflect the change in ethylene glycol concentration as the only variable. Propylene glycol has a similar effect on heat exchanger output.

Because many other variables (e.g., liquid velocity, steam or refrigerant loading, temperature difference, and unit construction) affect the overall coefficient of a heat exchanger, designers should consult manufacturers' ratings when selecting such equipment. The curves indicate only the magnitude of these output changes.

Effect on Terminal Units

Because the effect of glycol on the capacity of terminal units may vary widely with temperature, the manufacturer's rating data should be consulted when selecting heating or cooling units in glycol systems.

Effect on Pump Performance

Centrifugal pump characteristics are affected to some degree by glycol solutions because of viscosity changes. Figure 43 shows these effects on pump capacity, head, and efficiency. Figures 12 and 16 in Chapter 21 of the 2005 *ASHRAE Handbook—Fundamentals* plot the viscosity of aqueous ethylene glycol and propylene glycol. Centrifugal pump performance is normally cataloged for water at 60 to 80°F. Hence, absolute viscosity effects below 1.1 centipoise can safely be ignored as far as pump performance is concerned. In intermittently operated systems, such as snow-melting applications, viscosity effects at start-up may decrease flow enough to slow pickup.

Effect on Piping Pressure Loss

Friction loss in piping also varies with viscosity changes. Figure 44 gives correction factors for various ethylene glycol and propylene glycol solutions. These factors are applied to the calculated pressure loss for water [Equation (23)]. No correction is needed for ethylene glycol and propylene glycol solutions above 160°F.

Installation and Maintenance

Because glycol solutions are comparatively expensive, the smallest possible concentrations to produce the desired antifreeze properties should be used. The system's total water content should be calculated carefully to determine the required amount of glycol (Craig et al. 1993). The solution can be mixed outside the system in drums or barrels and then pumped in. Air vents should be watched during filling to prevent loss of solution. The system and cold-water supply should not be permanently connected, so automatic fill valves are usually not used.

Ethylene glycol and propylene glycol normally include an inhibitor to help prevent corrosion. Solutions should be checked each year using a suitable refractometer to determine glycol concentration. Certain precautions regarding the use of inhibited ethylene glycol solutions should be taken to extend their service life and to preserve equipment:

- Before injecting the glycol solution, thoroughly clean and flush the system.
- Use waters that are soft and low in chloride and sulfate ions to prepare the solution whenever possible.

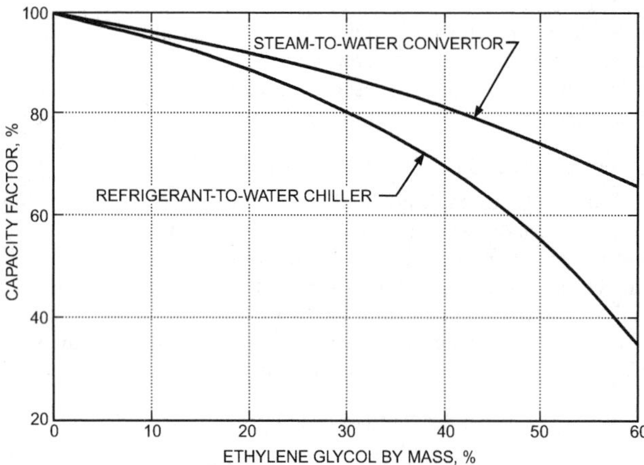

Fig. 42 Example of Effect of Aqueous Ethylene Glycol Solutions on Heat Exchanger Output

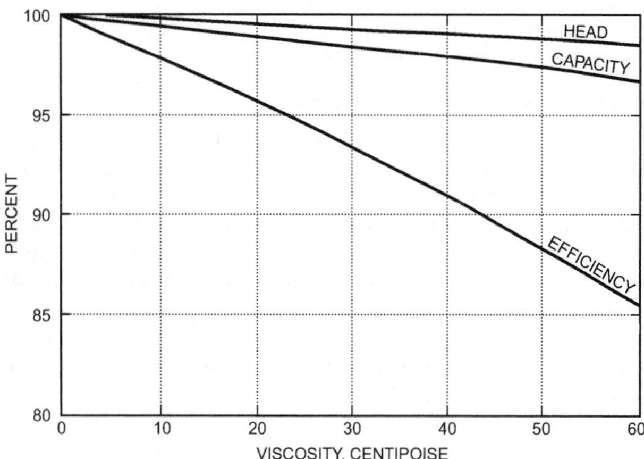

Fig. 43 Effect of Viscosity on Pump Characteristics

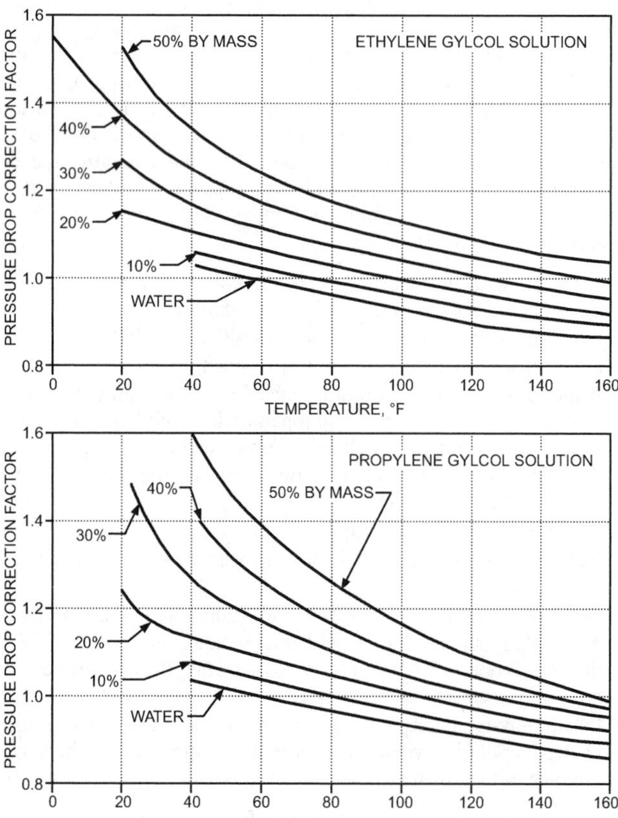

Fig. 44 Pressure Drop Correction for Glycol Solutions

- Limit the maximum operating temperature to 250°F in a closed hydronic system. In a heat exchanger, limit glycol film temperatures to 300 to 350°F (steam pressures 120 psi or less) to prevent deterioration of the solution.
- Check the concentration of inhibitor periodically, following procedures recommended by the glycol manufacturer.

REFERENCES

ASME. 1998. *Boiler and pressure vessel code.* American Society of Mechanical Engineers, New York.

Bahnfleth, W. and E. Peyer. 2004. *Variable-primary flow chilled water systems: Potential benefits and application issues*. Air-Conditioning and Refrigeration Technology Institute, Arlington, VA.

Carlson, G.F. 1968. Hydronic systems: Analysis and evaluation—Part II. *ASHRAE Journal* 10(11):45-51.

Carlson, G.F. 1981. Pump energy conservation and flow balance analysis. *ASHRAE Transactions* 87(1):985-999.

CDA. 2006. *The copper tube handbook*. Copper Development Association, New York.

Coad, W.J. 1976. Integrated decentralized chilled water systems. *ASHRAE Transactions* 82(1):566-574.

Coad, W.J. 1980a. Air in hydronic systems. *Heating/Piping/Air Conditioning Engineering* (July).

Coad, W.J. 1980b. Expansion tanks. *Heating/Piping/Air Conditioning* (May).

Coad, W.J. 1985. Variable flow in hydronic systems for improved stability, simplicity, and energy economics. *ASHRAE Transactions* 91(1B):224-237.

Craig, N.C., B.W. Jones, and D.L. Fenton. 1993. Glycol concentration requirements for freeze burst protection. *ASHRAE Transactions* 99(2): 200-209.

Green, R.H. 1994. An air conditioning control system using variable speed water pumps. *ASHRAE Transactions* 100(1):463-470.

Hegberg, M.C. 2003. What is the control area? *ASHRAE Transactions* 109(1):361-372.

HI. 2000. *Guide to LLC analysis for pumping systems*. Hydraulic Institute, Parsippany, NJ.

Himmelblau, D.M. 1960. Solubilities of inert gases in water. *Journal of Chemical and Engineering Data* 5(1).

Hull, R.F. 1981. Effect of air on hydraulic performance of the HVAC system. *ASHRAE Transactions* 87(1):1301-1325.

Lockhart, H.A. and G.F. Carlson. 1953. Compression tank selection for hot water heating systems. *ASHVE Journal* 25(4):132-139. Also in *ASHVE Transactions* 59:55-76.

Morabito, B.P. 1960. How higher cooling coil differentials affect system economics. *ASHRAE Journal* 2(8):60.

Pedersen, C.O., D.E. Fisher, J.D. Spitler, and R.J. Liesen. 1998. *Cooling and heating load calculation principles (RP-875)*. ASHRAE.

Pierce, J.D. 1963. Application of fin tube radiation to modern hot water heating systems. *ASHRAE Journal* 5(2):72.

Pompei, F. 1981. Air in hydronic systems: How Henry's law tells us what happens. *ASHRAE Transactions* 87(1):1326-1342.

Stewart, W.E. and C.L. Dona. 1987. Water flow rate limitations. *ASHRAE Transactions* 93(2):811-825.

Williams, G.J. 2005. Specifying chilled water cooling coils. *Heating, Piping & Air Conditioning* (November).

BIBLIOGRAPHY

Carlson, G.F. 1968–1969. Hydronic systems: Analysis and evaluation—Parts I–VI. *ASHRAE Journal* (October to March).

Carlson, G.F. 1972. Central plant chilled water systems—Pumping and flow balance, Parts I–III. *ASHRAE Journal* (February to April).

Carlson, G.F. 1981. The design influence of air on hydronic systems. *ASHRAE Transactions* 87(1):1293-1300.

Hegberg, M.C. 2000. Control valve selection for hydronic systems. *ASHRAE Journal* (November):33-39.

Hegberg, R.A. 2000. *Fundamentals of water system design.* ASHRAE.

CHAPTER 13

CONDENSER WATER SYSTEMS

CONDENSER water systems for refrigeration processes are classified as (1) once-through systems, such as city water systems; or (2) recirculating or cooling tower systems.

ONCE-THROUGH CITY WATER SYSTEMS

Figure 1 shows a water-cooled condenser using city water. The return is run higher than the condenser so that the condenser is always full of water. Water flow through the condenser is modulated by a control valve in the supply or discharge line, usually actuated from condenser head pressure to (1) maintain a constant condensing temperature with load variations and (2) close when the refrigeration compressor turns off. City water systems should always include approved backflow prevention devices and open (air gap) drains. When more than one condenser is used on the same circuit, individual control valves are used.

Once-through city water systems are discouraged in most localities because of the waste of city water and the burden on the sewage or wastewater system. Some localities allow their use as a standby or emergency condenser water system for critical refrigeration needs such as for computer rooms, research laboratories, or critical operating room or life support machinery.

Piping materials for these systems are generally nonferrous, usually copper but sometimes high-pressure plastic because corrosion-protective chemicals cannot be used. Scaling can be a problem with higher-temperature condensing surfaces when the water has a relatively high calcium content. In these applications, mechanically cleanable straight tubes should be used.

Piping should be sized according to the principles outlined in Chapter 36 of the 2005 *ASHRAE Handbook—Fundamentals*, with velocities of 5 to 10 fps for design flow rates. A pump is not required where city water is used. **Well water** can be used in lieu of city water, connected on the service side of the pumping/pressure control system. Because most well water has high calcium content, scaling on the condenser surfaces can be a problem.

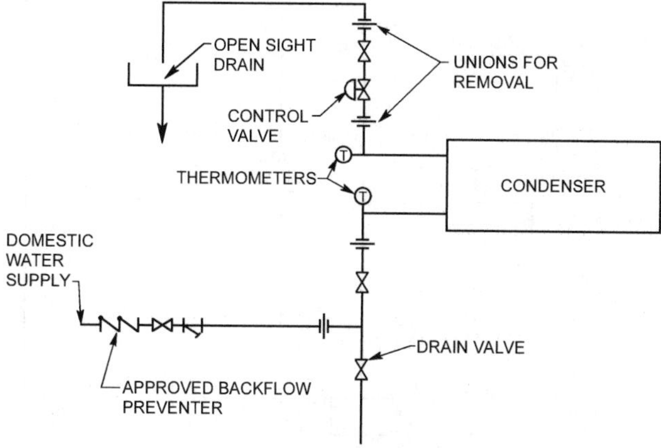

Fig. 1 Condenser Connections for Once-Through City Water System

The preparation of this chapter is assigned to TC 6.1, Hydronic and Steam Equipment and Systems.

OPEN COOLING TOWER SYSTEMS

Open systems have at least two points of interface between the system water and the atmosphere; they require a different approach to hydraulic design, pump selection, and sizing than do closed hot-water and chilled-water systems. Some heat conservation systems rely on a split condenser heating system that includes a two-section condenser. One section of the condenser supplies heat for closed-circuit heating or reheat systems; the other section serves as a heat rejection circuit, which is an open system connected to a cooling tower.

In selecting a pump for a cooling tower/condenser water system, consideration must be given to the static head and the system friction loss. The pump inlet must have an adequate net positive suction head (see Chapter 39). In addition, continuous contact with air introduces oxygen into the water and concentrates minerals that can cause scale and corrosion on a continuing basis. Fouling factors and an increased pressure drop caused by aging of the piping must be taken into account in the condenser piping system design (see Chapter 36 of the 2005 *ASHRAE Handbook—Fundamentals*).

The required water flow rate depends on the refrigeration unit used and on the temperature of the available condenser water. Cooling tower water is available for return to the condenser at a temperature several degrees above the design wet-bulb temperature, depending on tower performance. An approach of 7°F to the design wet-bulb temperature is frequently considered an economically sound design. In city, lake, river, or well water systems, the maximum water temperature that occurs during the operating season must be used for equipment selection and design flow rates and temperature ranges.

The required flow rate through a condenser may be determined with manufacturers' performance data for various condensing temperatures and capacities. With air-conditioning refrigeration applications, a return or leaving condenser water temperature of 95°F is considered standard practice. If economic feasibility analyses can justify it, higher leaving water temperatures may be used.

Figure 2 shows a typical cooling tower system for a refrigerant condenser. Water flows to the pump from the tower basin or sump and is discharged under pressure to the condenser and then back to the tower. When it is desirable to control condenser water temperature or maintain it above a predetermined minimum, water is diverted through a control valve directly back to the tower basin.

Piping from the tower sump to the pump requires some precautions. The sump level should be above the top of the pump casing for positive prime, and piping pressure drop should be such that there is always adequate net positive suction head on the pump. All piping must pitch up to the tower basin, if possible, to eliminate air pockets.

If used, suction strainers should be equipped with inlet and outlet gages to indicate when cleaning is required. In-line pipe strainers are not recommended for cooling tower systems because they tend to become blocked and turn into a reliability problem in themselves. Many designers depend on large mesh screens in the tower sump and condenser heads designed with settling volumes to remove particulate matter. If a strainer is deemed necessary, two-large capacity basket strainers, installed in parallel such that they can be alternately put into service and valved out for cleaning, are recommended.

Air and Vapor Precautions

Both vapor and air can create serious problems in open cooling tower systems. Water vaporizes in the pump impeller if adequate net positive suction head is not available. When this occurs, the pump loses capacity, and serious damage to the impeller can result. Equally damaging **vaporization** can occur in other portions of the system where pressure in the pipe can drop below the vapor pressure at the water temperature. On shutdown, these very low pressures can result from a combination of static pressure and momentum. Vaporization is often followed by an implosion, which causes destructive water hammer. To avoid this problem, all sections of the piping system except the return line to the upper tower basin should be kept below the basin level. When this cannot be achieved, a thorough dynamic analysis of the piping system must be performed for all operating conditions, and a soft start and stop control such as a variable-frequency drive on the pump motor is recommended as an additional precaution.

Air release is another characteristic of open condenser water systems that must be addressed. Because the water/air solution in the tower basin is saturated at atmospheric pressure and cold-water basin temperature, the system should be designed to maintain the pressure at all points in the system sufficiently above atmospheric that no air will be released in the condenser or in the piping system (see Figures 2 and 3 in Chapter 12).

Another cause of air in the piping system is **vortexing** at the tower basin outlet. This can be avoided by ensuring that the maximum flow does not exceed that recommended by the tower manufacturer. Release of air in condenser water systems is the major cause of corrosion, and it causes decreased pump flow (similar to cavitation), water flow restrictions in some piping sections, and possible water hammer.

Piping Practice

The elements of required pump head are illustrated in Figure 3. Because there is an equal head of water between the level in the tower sump or interior reservoir and the pump on both the suction and discharge sides, these static heads cancel each other and can be disregarded.

The elements of pump head are (1) static head from tower sump or interior reservoir level to the tower header, (2) friction loss in suction and discharge piping, (3) pressure loss in the condenser, (4) pressure loss in the control valves, (5) pressure loss in the strainer, and (6) pressure loss in the tower nozzles, if used. Added together, these elements determine the required pump total dynamic head.

Normally, piping is sized for water velocities between 5 and 12 fps. Refer to Chapter 36 of the 2005 *ASHRAE Handbook—Fundamentals*, for piping system pressure losses. Friction factors for 15-year-old pipe are commonly used. Manufacturers' data contain pressure drops for the condenser, cooling tower, control valves, and strainers.

If multiple cooling towers are to be connected, the piping should be designed so that the pressure loss from the tower to the pump suction is *exactly* equal for each tower. Additionally, large equalizing lines or a common reservoir can be used to ensure the same water level in each tower. However, for reliability and ease of maintenance, multiple basins are often preferred.

Evaporation in a cooling tower concentrates the dissolved solids in the circulating water. This concentration can be limited by discharging a portion of the water as overflow or blowdown.

Makeup water is required to replace water lost by evaporation, blowdown, and drift. Automatic float valves or level controllers are usually installed to maintain a constant water level.

Water Treatment

Water treatment is necessary to prevent scaling, corrosion, and biological fouling of the condenser and circulating system. The extent and nature of the treatment depends on the chemistry of the available water and on the system design characteristics. On large systems, fixed continuous-feeding chemical treatment systems are frequently installed in which chemicals, including acids for pH control, must be diluted and blended and then pumped into the condenser water system. Corrosion-resistant materials may be required for surfaces that come in contact with these chemicals. In piping

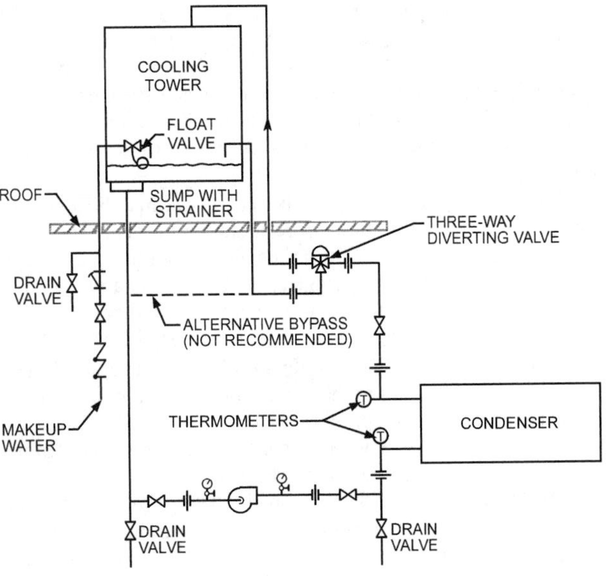

Fig. 2 Cooling Tower Piping System

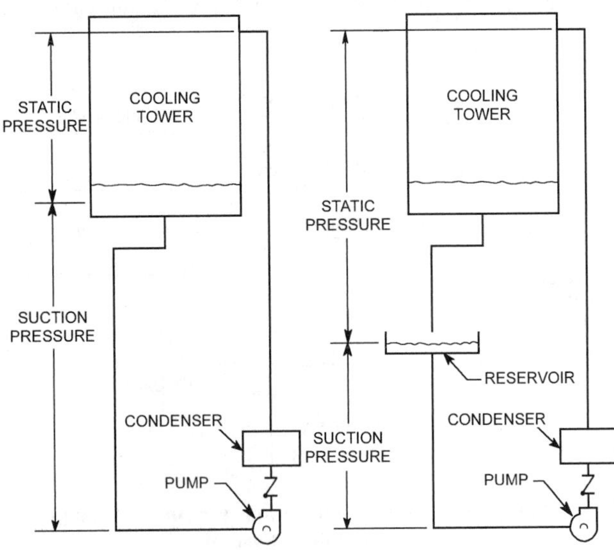

Fig. 3 Schematic Piping Layout Showing Static and Suction Head

system design, provisions for feeding the chemicals, blowdowns, drains, and testing must be included. For further information on water treatment, refer to Chapter 39 of this volume and Chapter 48 of the 2007 *ASHRAE Handbook—HVAC Applications*.

Freeze Protection

Outdoor piping must be protected or drained when a tower operates intermittently during cold weather. The most satisfactory arrangement is to provide an indoor receiving tank into which the cold-water basin drains by gravity, as shown in Figure 4A. The makeup, overflow, and pump suction lines are then connected to the indoor reservoir tank rather than to the tower basin.

A control sequence for the piping arrangement of Figure 4A with rising water temperature would be as follows:

- A temperature sensor measures the temperature of the water leaving the indoor sump.
- As the temperature starts to rise, the diverting valve begins directing some of the water over the tower.
- After the water is in full flow over the tower, and the temperature continues to rise above set point, the fan is started on low speed, with the speed increasing until the water temperature reaches set point.
- With falling water temperature, the opposite sequence occurs.

Tower basin heaters or heat exchangers connected across the supply and return line to the tower can also be selected. Steam or electric basin heaters are most commonly used. An arrangement incorporating an indoor heater is shown in Figure 4B.

LOW-TEMPERATURE (WATER ECONOMIZER) SYSTEMS

When open cooling tower systems are used for generating chilled water directly, through an indirect heat exchanger such as a plate frame heat exchanger, or with a chiller thermocycle circuit, the bulk water temperature is such that icing can occur on or within the cooling tower, with destructive effects. The piping circuit precautions are similar to those described in the section on Open Cooling Tower Systems, but they are much more critical. For precautions in protecting cooling towers from icing damage, see Chapter 39.

Water from open cooling tower systems should not be piped directly through cooling coils, unitary heat pump condensers, or plate frame heat exchangers unless the water is first filtered through a high-efficiency filtering system and the water treatment system is managed carefully to minimize the dissolved solids. Even with these precautions, cooling coils should have straight tubes arranged for visual inspection and mechanical cleaning; unitary heat pumps should be installed for easy removal and replacement; and plate frame heat exchangers should be installed for ready accessibility for disassembly and cleaning.

CLOSED-CIRCUIT EVAPORATIVE COOLERS

Because of the potential for damaging freezing of cooling towers, some designers prefer to use closed-circuit evaporative/dry coolers for water economizer chilled-water systems. One of the many different configurations of these systems is shown in Figure 5. The open water is simply recirculated from the basin to the sprays of the evaporative cooler, and the cooling water system is a closed hydronic system usually using a glycol-water mixture for freeze protection. The open water system is then drained in freezing weather and the cooling heat exchanger unit is operated as a dry heat exchanger. This type of system in some configurations can be used to generate chilled water through the plate frame heat exchanger shown or to remove heat from a closed heat pump circuit. The glycol circuit is generally not used directly either for building cooling or for the heat pump circuit because of the economic penalty of the extensive glycol system. The closed circuit is designed in accordance with the principles and procedures described in Chapter 12.

OVERPRESSURE CAUSED BY THERMAL FLUID EXPANSION

When open condenser water systems are used at low temperatures for winter cooling, special precautions should be taken to prevent damaging overpressurization due to thermal expansion. This phenomenon has been known to cause severe damage when a section of piping containing water at a lower temperature than the surrounding space is isolated while cold. This isolation could be intentional, such as isolation by two service valves, or it could be as subtle as a section of piping isolated between a check valve and a control valve. If such an isolation occurs when water at 45°F is in a 1 in. pipe passing through a 75°F space, Figure 41 in Chapter 12 reveals that the pressure would increase by 815 psi, which would be destructive to many components of most condenser water systems. Refer to the section on Other Design Considerations in Chapter 12 for a discussion of this phenomenon.

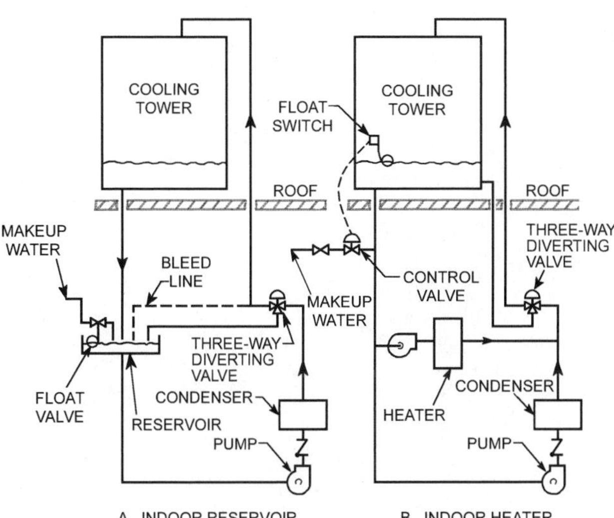

Fig. 4 Cooling Tower Piping to Avoid Freeze-Up

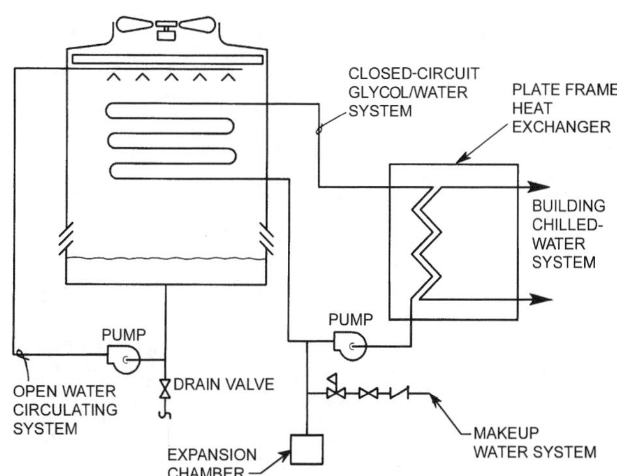

Fig. 5 Closed-Circuit Cooler System

CHAPTER 14

MEDIUM- AND HIGH-TEMPERATURE WATER HEATING SYSTEMS

MEDIUM-TEMPERATURE water systems have operating temperatures below 350°F and permit design to a pressure rating of 125 to 150 psig. High-temperature water systems are classified as those operating with supply water temperatures above 350°F and designed to a pressure rating of 300 psig. The usual practical temperature limit is about 450°F because of pressure limitations on pipe fittings, equipment, and accessories. The rapid pressure rise that occurs as the temperature rises above 450°F increases cost because components rated for higher pressures are required (see Figure 1). The design principles for both medium- and high-temperature systems are basically the same. In this chapter, "high-temperature water" (HTW) refers to both systems.

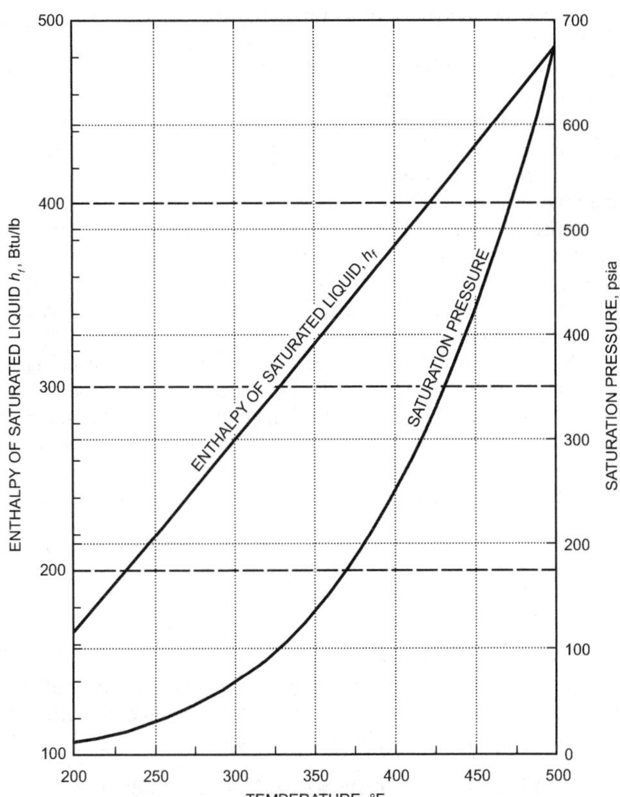

Fig. 1 Relation of Saturation Pressure and Enthalpy to Water Temperature

The preparation of this chapter is assigned to TC 6.1, Hydronic and Steam Equipment and Systems.

This chapter presents the general principles and practices that apply to HTW and distinguishes them from low-temperature water systems operating below 250°F. See Chapter 12 for basic design considerations applicable to all hot-water systems.

SYSTEM CHARACTERISTICS

The following characteristics distinguish HTW systems from steam distribution or low-temperature water systems:

- The system is a completely closed circuit with supply and return mains maintained under pressure. There are no losses from flashing, and heat that is not used in the terminal heat transfer equipment is returned to the HTW generator. Tight systems have minimal corrosion.
- Mechanical equipment that does not control performance of individual terminal units is concentrated at the central station.
- Piping can slope up or down or run at a variety of elevations to suit the terrain and the architectural and structural requirements without provision for trapping at each low point. This may reduce the amount of excavation required and eliminate drip points and return pumps required with steam.
- Greater temperature drops are used and less water is circulated than in low-temperature water systems.
- The pressure in any part of the system must always be well above the pressure corresponding to the temperature at saturation in the system to prevent the water flashing into steam.
- Terminal units requiring different water temperatures can be served at their required temperatures by regulating the flow of water, modulating water supply temperature, placing some units in series, and using heat exchangers or other methods.
- The high heat content of the water in the HTW circuit acts as a thermal flywheel, evening out fluctuations in the load. The heat storage capacity can be further increased by adding heat storage tanks or by increasing the temperature in the return mains during periods of light load.
- The high heat content of the heat carrier makes high-temperature water unsuitable for two-pipe dual-temperature (hot and chilled water) applications and for intermittent operation if rapid start-up and shutdown are desired, unless the system is designed for minimum water volume and is operated with rapid response controls.
- Higher engineering skills are required to design a HTW system that is simple, yet safer and more convenient to operate than are required to design a comparable steam or low-temperature water system.
- HTW system design requires careful attention to basic laws of chemistry and physics as these systems are less forgiving than standard hydronic systems.

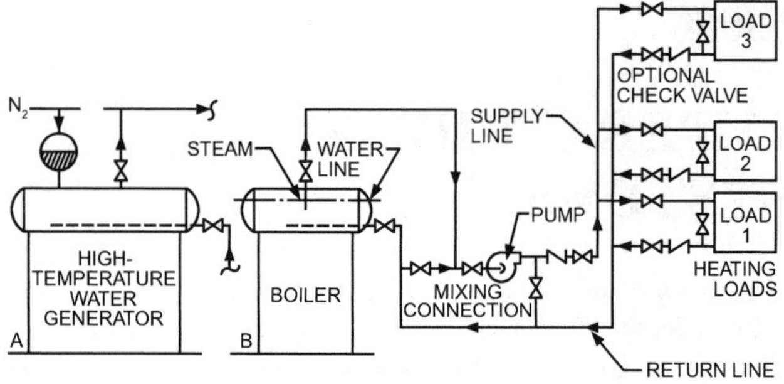

Fig. 2 Elements of High-Temperature Water System

BASIC SYSTEM

HTW systems are similar to conventional forced hot-water heating systems. They require a heat source (which can be a direct-fired HTW generator, a steam boiler, or an open or closed heat exchanger) to heat the water. The expansion of the heated water is usually taken up in an expansion vessel, which simultaneously pressurizes the system. Heat transport depends on circulating pumps. The distribution system is closed, comprising supply and return pipes under the same basic pressure. Heat emission at the terminal unit is indirect by heat transfer through heat transfer surfaces. The basic system is shown in Figure 2.

The main differences of HTW systems from low-temperature water systems are the higher pressure, heavier equipment, generally smaller pipe sizes, and manner in which water pressure is maintained.

Most systems are either (1) a saturated steam cushion system, in which the high-temperature water develops its own pressure, or (2) a gas- or pump-pressurized system, in which the pressure is imposed externally.

HTW generators and all auxiliaries (such as water makeup and feed equipment, pressure tanks, and circulating pumps) are usually located in a central station. Cascade HTW generators sometimes use an existing steam distribution system and are installed remote from the central plant.

DESIGN CONSIDERATIONS

Selection of the system pressure, supply temperature, temperature drop, type of HTW generator, and pressurization method are the most important initial design considerations. The following are some of the determining factors:

- Type of load (space heating and/or process); load fluctuations during a 24 h period and a 1 year period. Process loads might require water at a given minimum supply temperature continuously, whereas space heating can permit temperature modulation as a function of outdoor temperature or other climatic influences.
- Terminal unit temperature requirements.
- Distance between heating plant and space or process requiring heat.
- Quantity and pressure of steam used for power equipment in the central plant.
- Elevation variations within the system and the effect of basic pressure distribution.

Usually, distribution piping is the major investment in an HTW system. A distribution system with the widest temperature spread (Δt) between supply and return will have the lowest initial and operating costs. Economical designs have a Δt of 150°F or higher.

The requirements of terminal equipment or user systems determine the system selected. For example, if the users are 10 psig steam generators, the return temperatures would be 250°F. A 300 psig rated system operated at 400°F would be selected to serve the load. In another example, where the primary system serves predominantly 140 to 180°F hot-water heating systems, an HTW system that operates at 325°F could be selected. The supply temperature is reduced by blending with 140°F return water to the desired 180°F hot water supply temperature in a direct-connected hot-water secondary system. This highly economical design has a 140°F return temperature in the primary water system and a Δt of 185°F.

Because the danger of water hammer is always present when the pressure drops to the point at which pressurized hot water flashes to steam, the primary HTW system should be designed with steel valves and fittings of 150 psi. The secondary water, which operates below 212°F and is not subject to flashing and water hammer, can be designed for 125 psi and standard HVAC equipment.

Theoretically, water temperatures up to about 350°F can be provided using equipment suitable for 125 psi. But in practice, unless push-pull pumping is used, maximum water temperatures are limited by the system design, pump pressures, and elevation characteristics to values between 300 and 325°F.

Many systems designed for self-generated steam pressurization have a steam drum through which the entire flow is taken, and which also serves as an expansion vessel. A circulating pump in the supply line takes water from the tank. The temperature of the water from the steam drum cannot exceed the steam temperature in the drum that corresponds to its pressure at saturation. The point of maximum pressure is at the discharge of the circulating pump. If, for example, this pressure is to be maintained below 125 psig, the pressure in the drum that corresponds to the water temperature cannot exceed 125 psig minus the sum of the pump pressure and the pressure that is caused by the difference in elevation between the drum and the circulating pump.

Most systems are designed for inert gas pressurization. In most of these systems, the pressurizing tank is connected to the system by a single balance line on the suction side of the circulating pump. The circulating pump is located at the inlet side of the HTW generator. There is no flow through the pressurizing tank, and a reduced temperature will normally establish itself inside. A special characteristic of a gas-pressurized system is the apparatus that creates and maintains gas pressure inside the tank.

In designing and operating an HTW system, it is important to maintain a pressure that always exceeds the vapor pressure of the water, even if the system is not operating. This may require limiting the water temperature and thereby the vapor pressure, or increasing the imposed pressure.

Elevation and the pressures required to prevent water from flashing into steam in the supply system can also limit the maximum water temperature that may be used and must therefore be studied in evaluating the temperature-pressure relationships and method of pressurizing the system.

The properties of water that govern design are as follows:

- Temperature versus pressure at saturation (Figure 1)
- Density or specific volume versus temperature
- Enthalpy or sensible heat versus temperature
- Viscosity versus temperature
- Type and amount of pressurization

The relationships among temperature, pressure, specific volume, and enthalpy are all available in steam tables. Some properties of water are summarized in Table 1 and Figure 3.

Direct-Fired High-Temperature Water Generators

In direct-fired HTW generators, the central stations are comparable to steam boiler plants operating within the same pressure range. The generators should be selected for size and type in keeping with the load and design pressures, as well as the circulation requirements peculiar to high-temperature water. In some systems, both steam for power or processing and high-temperature water are

Table 1 Properties of Water, 212 to 400°F

Temperature, °F	Absolute Pressure, psia*	Density, lb/ft³	Specific Heat, Btu/lb·°F	Total Heat above 32°F		Dynamic Viscosity, Centipoise
				Btu/lb*	Btu/ft³	
212	14.70	59.81	1.007	180.07	10,770	0.2838
220	17.19	59.63	1.009	188.13	11,216	0.2712
230	20.78	59.38	1.010	198.23	11,770	0.2567
240	24.97	59.10	1.012	208.34	12,313	0.2436
250	29.83	58.82	1.015	218.48	12,851	0.2317
260	35.43	58.51	1.017	228.64	13,378	0.2207
270	41.86	58.24	1.020	238.84	13,910	0.2107
280	49.20	57.94	1.022	249.06	14,430	0.2015
290	57.56	57.64	1.025	259.31	14,947	0.1930
300	67.01	57.31	1.032	269.59	15,450	0.1852
310	77.68	56.98	1.035	279.92	15,950	0.1779
320	89.66	56.66	1.040	290.28	16,437	0.1712
330	103.06	56.31	1.042	300.68	16,931	0.1649
340	118.01	55.96	1.047	311.13	17,409	0.1591
350	134.63	55.59	1.052	321.63	17,879	0.1536
360	153.04	55.22	1.057	332.18	18,343	0.1484
370	173.37	54.85	1.062	342.79	18,802	0.1436
380	195.77	54.47	1.070	353.45	19,252	0.1391
390	220.37	54.05	1.077	364.17	19,681	0.1349
400	247.31	53.65	1.085	374.97	20,117	0.1308

**Source*: *Thermodynamic Properties of Steam*, J.H. Keenan and F.G. Keyes, John Wiley & Sons, 1936 edition.

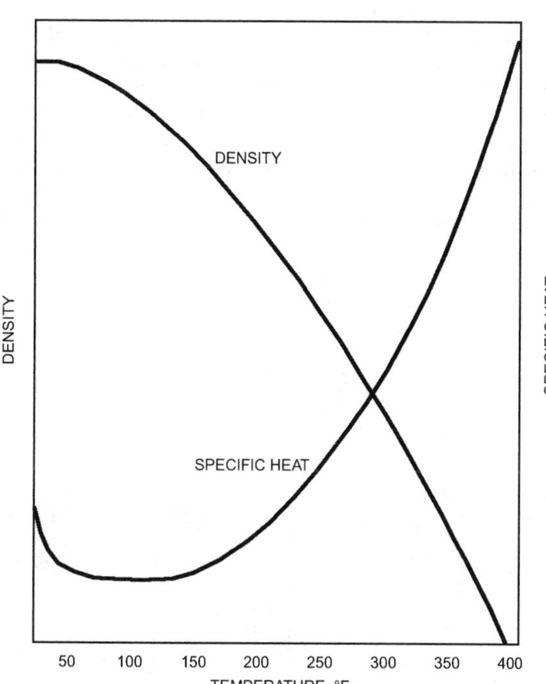

Fig. 3 Density and Specific Heat of Water

supplied from the same boiler; in others, steam is produced in the boilers and used for generating high-temperature water; and in many others, the burning fuel directly heats the water.

HTW generators can be the water-tube or fire-tube type, and can be equipped with any conventional fuel-firing apparatus. Water-tube generators can have either forced circulation, gravity circulation, or a combination of both. The recirculating pumps of forced-circulation generators must operate continuously while the generator is being fired. Steam boilers relying on natural circulation may require internal baffling when used for HTW generation. In scotch marine boilers, thermal shock may occur, caused by a sudden drop in the temperature of the return water or when the Δt exceeds

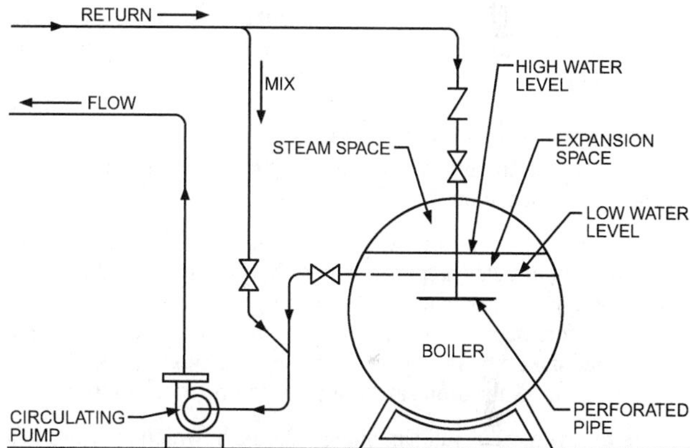

Fig. 4 Arrangement of Boiler Piping

40°F. Forced-circulation HTW generators are usually the once-through type and rely solely on pumps to achieve circulation. Depending on the design, internal orifices in the various circuits may be required to regulate the water flow rates in proportion to the heat absorption rates. Circulation must be maintained at all times while the generator is being fired, and the flow rate must never drop below the minimum indicated by the manufacturer.

Where gravity circulation steam boilers are used for HTW generation, the steam drum usually serves as an expansion vessel. In steam-pressurized forced-circulation HTW generators, a separate vessel is commonly used for expansion and for maintaining the steam pressure cushion. A separate vessel is always used when the system is cushioned by an inert gas or auxiliary steam. Proper internal circulation is essential in all types of boilers to prevent tube failures caused by overheating or unequal expansion.

In HTW systems, the generator is a steam boiler with an integral steam drum used for pressurization and for expansion of the water level. A dip pipe removes water below the water line (Figure 4) (Applegate 1958). This dip pipe should be installed so that it picks up water at or near the saturation temperature, without too many steam bubbles. If a pipe breaks somewhere in the system, the boiler must not empty to a point where heating surfaces are bared and a boiler explosion occurs. The same precautions must be taken with the return pipe. If the return pipe is connected in the lower part of the boiler, a check valve should be placed in the connecting line to the boiler to preclude the danger of emptying the boiler.

When two or more such boilers supply a common system, the same steam pressure and water level must be maintained in each. Water and steam balance pipes are usually installed between the drums (Figure 5). These should be liberally sized. The following table shows recommended sizes:.

Boiler Rating, million Btu/h	Balance Pipe Diameter, in.
2.5	3
5	3.5
10	4
15	5
20	6
30	8

A difference of only 0.25 psi in the system pressure between two boilers operated at 100 psig would cause a difference of 9 in. in the water level. The situation is further aggravated because an upset is not self-balancing. Rather, when too high a heat release in one of the boilers has caused the pressure to rise and the water level to fall in this boiler, the decrease in the flow of colder return water into it causes a further pressure rise, while the opposite happens in the other boiler. It is therefore important that the firing rates match the flow through each boiler at all times. Modern practice is to use either flooded HTW heaters with a single external pressurized expansion drum common to all the generators, or the combination of steam boilers and a direct-contact (cascade) heater.

Expansion and Pressurization

In addition to the information in Chapter 12, the following factors should be considered:

- The connection point of the expansion tank used for pressurization greatly affects the pressure distribution throughout the system and the avoidance of HTW flashing.
- Proper safety devices for high and low water levels and excessive pressures should be incorporated in the expansion tank and interlocked with combustion safety and water flow rate controls.

The following four fundamental methods, in which pressure in a given hydraulic system can be kept at a desired level, amplify the discussion in Chapter 12 (Blossom and Ziel 1959, National Academy of Sciences 1959).

1. **Elevating the storage tank** is a simple pressurization method, but because of the great heights required for the pressure encountered, it is generally impractical.

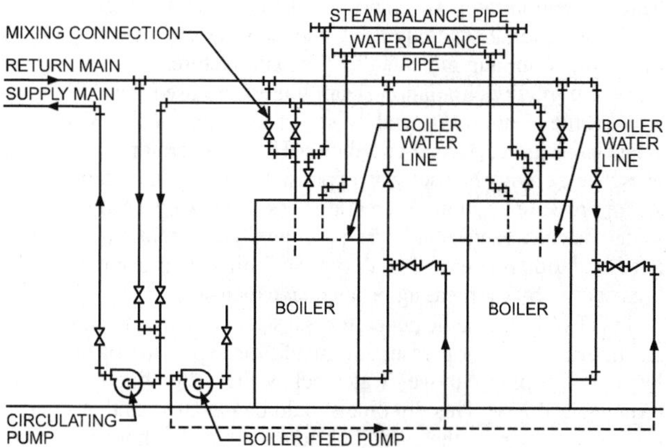

Fig. 5 Piping Connections for Two or More Boilers in HTW System Pressurized by Steam

2. **Steam pressurization** requires the use of an expansion vessel separate from the HTW generator. Because firing and flow rates can never be perfectly matched, some steam is always carried. Therefore, the vessel must be above the HTW generators and connected in the supply water line from the generator. This steam, supplemented by flashing of the water in the expansion vessel, provides the steam cushion that pressurizes the system.

The expansion vessel must be equipped with steam safety valves capable of relieving the steam generated by all the generators. The generators themselves are usually designed for a substantially higher working pressure than the expansion drum, and their safety relief valves are set for the higher pressure to minimize their lift requirement.

The basic HTW pumping arrangements can be either single-pump, in which one pump handles both the generator and system loads, or two-pump, in which one pump circulates high-temperature water through the generator and a second pump circulates high-temperature water through the system (see Figures 6 and 7). The circulating pump moves the water from the expansion vessel to the system and back to the generator. The vessel must be elevated to increase the net positive suction pressure to prevent cavitation or flashing in the pump suction. This arrangement is critical. A bypass from the HTW system return line to the pump suction helps prevent flashing. Cooler return water is then mixed with hotter water from the expansion vessel to give a resulting temperature below the corresponding saturation point in the vessel.

In the two-pump system, the boiler recirculation should always exceed the system circulation, because excessive cooling of the water in the drum by the cooler return water entering the drum, in case of overcirculation, can cause pressure loss and flashing in the distribution system. Backflow into the drum can be prevented by installing a check valve in the balance line from the drum to the boiler recirculating pumps. Higher cushion pressure may be maintained by auxiliary steam from a separate generator.

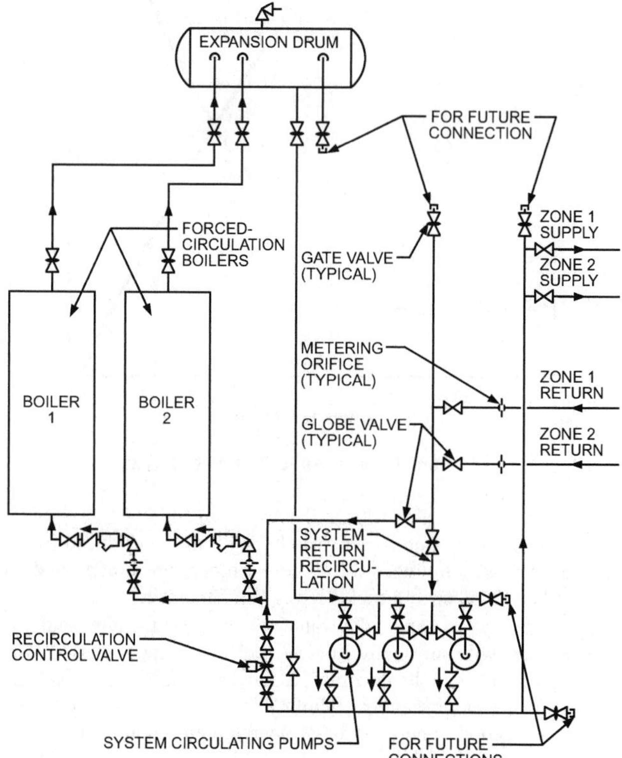

Fig. 6 HTW Piping for Combined (One-Pump) System (Steam Pressurized)

Sizing. Steam-pressurized vessels should be sized for a total volume V_T, which is the sum of the volume V_1 required for the steam space, the volume V_2 required for water expansion, and the volume V_3 required for sludge and reserve. An allowance of 20% of the sum of V_2 and V_3 is a reasonable estimate of the volume V_1 required for the steam space.

The volume V_2 required for water expansion is determined from the change in water volume from the minimum to the maximum operating temperatures of the complete cycle. It is not necessary to allow for expansion of the total water volume in the system from a cold initial start. It is necessary during a start-up period to bleed off the volume of water caused by expansion from the initial starting temperature to the lowest average operating temperature.

The volume V_3 for sludge and reserve varies greatly depending on the size and design of system and generator capacity. An allowance of 40% of the volume V_2 required for water expansion is a reasonable estimate of the volume required for V_3.

3. **Nitrogen**, the most commonly used inert gas, is used for gas pressurization. Air is not recommended because the oxygen in air contributes to corrosion in the system.

The expansion vessel is connected as close as possible to the suction side of the HTW pump by a balance line. The inert gas used for pressurization is fed into the top of the cylinder, preferably through a manual fill connection using a reducing station connected to an inert gas cylinder. Locating the relief valve below the minimum water line is advantageous, because it is easier to keep it tightly sealed with water on the pressure side. If the valve is located above the water line, it is exposed to the inert gas of the system.

To reduce the area of contact between gas and water and the resulting absorption of gas into the water, the tank should be installed vertically. It should be located in the most suitable place in the central station. Similar to the steam-pressurized system, the pumping arrangements can be either one- or two-pump (see Figures 8 and 9).

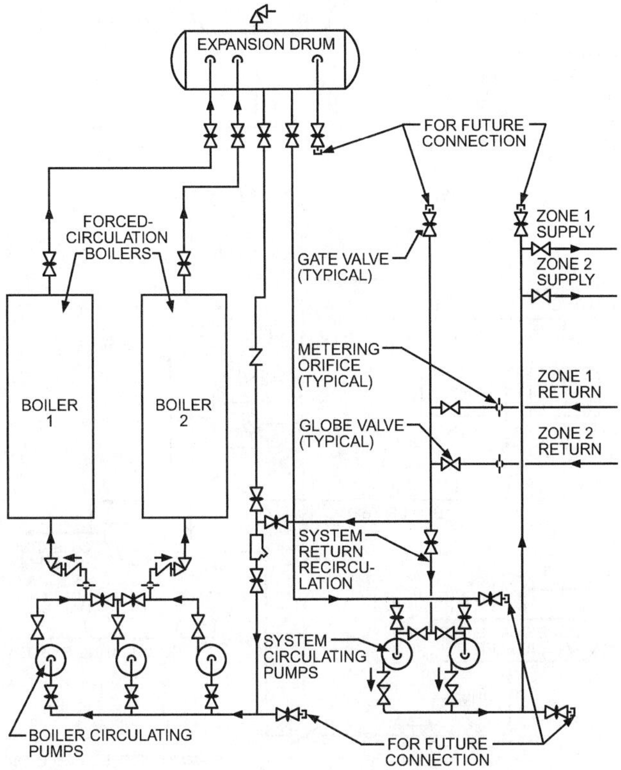

Fig. 7 HTW Piping for Separate (Two-Pump) System (Steam Pressurized)

The ratings of fittings, valves, piping, and equipment are considered in determining the maximum system pressure. A minimum pressure of about 25 to 50 psi over the maximum saturation pressure can be used. The imposed additional pressure above the vapor pressure must be large enough to prevent steaming in the HTW generators at all times, even under conditions when flow and firing rates in generators operated in parallel, or flow and heat absorption in parallel circuits within a generator, are not evenly matched. This is critical, because gas-pressurized systems do not have steam separating means and safety valves to evacuate the steam generated.

The simplest type of gas-pressurization system uses a variable gas quantity with or without gas recovery (Figure 10) (National Academy of Sciences 1959). When the water rises, the inert gas is relieved from the expansion vessel and is wasted or recovered in a low-pressure gas receiver from which the gas compressor pumps it into a high-pressure receiver for storage. When the water level drops in the expansion vessel, the control cycle adds inert gas from bottles or from the high-pressure receiver to the expansion vessel to maintain the required pressure.

Gas wastage can significantly affect the operating cost. The gas recovery system should be analyzed based on the economics of each application. Gas recovery is generally more applicable to larger systems.

Sizing. The vessel should be sized for a total volume V_T, which is the sum of the volume V_1 required for pressurization, the volume V_2 required for water expansion, and the volume V_3 required for sludge and reserve.

Calculations made on the basis of pressure-volume variations following Boyle's Law are reasonably accurate, assuming that the tank

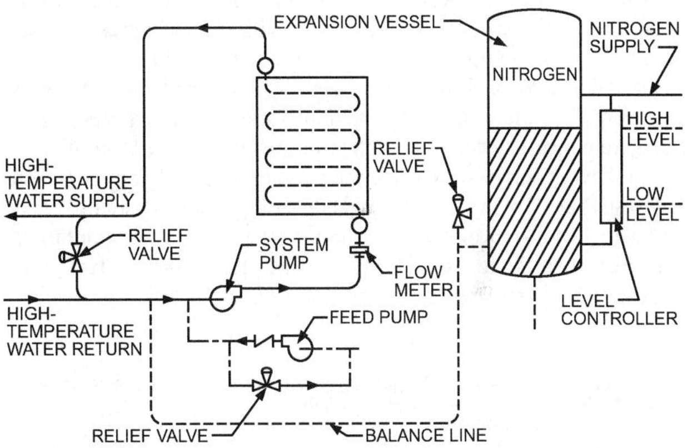

Fig. 8 Inert Gas Pressurization for One-Pump System

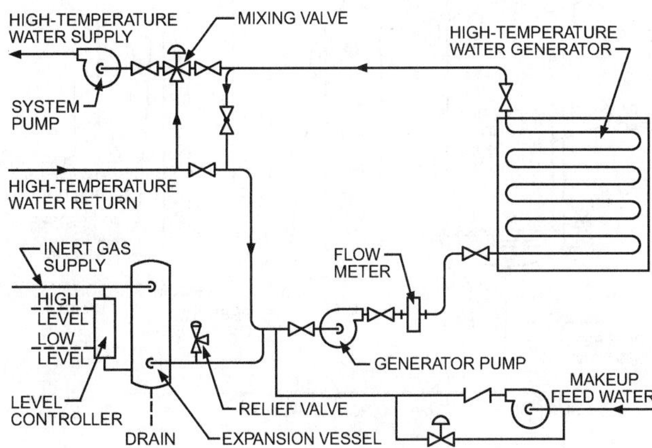

Fig. 9 Inert Gas Pressurization for Two-Pump System

operates at a relatively constant temperature. The minimum gas volume can be determined from the expansion volume V_2 and from the control range between the minimum tank pressure P_1 and the maximum tank pressure P_2. The gas volume varies from the minimum V_1 to a maximum, which includes the water expansion volume V_2.

The minimum gas volume V_1 can be obtained from

$$V_1 = P_1V_2/(P_2 - P_1)$$

where P_1 and P_2 are units of absolute pressure.

An allowance of 10% of the sum of V_1 and V_2 is a reasonable estimate of the sludge and reserve capacity V_3. The volume V_2 required for water expansion should be limited to the actual expansion that occurs during operation through its minimum to maximum operating temperatures. It is necessary to bleed off water during a start-up cycle from a cold start. It is practicable on small systems (e.g., under 1,000,000 to 10,000,000 Btu/h) to size the expansion vessel for the total water expansion from the initial fill temperature.

4. **Pump pressurization** in its simplest form consists of a feed pump and a regulator valve. The pump operates continuously, introducing water from the makeup tank into the system. The pressure regulator valve bleeds continuously back into the makeup tank. This method is usually restricted to small process heating systems. However, it can be used to temporarily pressurize a larger system to avoid shutdown during inspection of the expansion tank.

In larger central HTW systems, pump pressurization is combined with a fixed-quantity gas compression tank that acts as a buffer. When the pressure rises above a preset value in the buffer tank, a control valve opens to relieve water from the balance line into the makeup storage tank. When the pressure falls below a preset second value, the feed pump is started automatically to pump water from the makeup tank back into the system. The buffer tank is designed to absorb only the limited expansion volume that is required for the pressure control system to function properly; it is usually small.

To prevent corrosion-causing elements, principally oxygen, from entering the HTW system, the makeup storage tank is usually closed and a low-pressure nitrogen cushion of 1 to 5 psig is maintained. The gas cushion is usually the variable gas quantity type with release to the atmosphere.

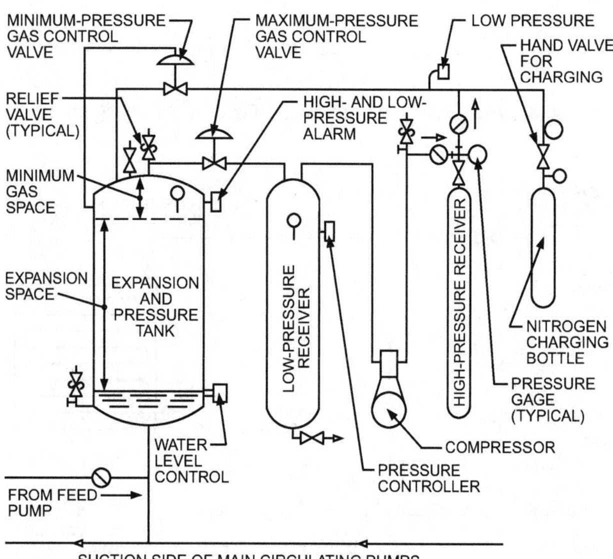

Fig. 10 Inert Gas Pressurization Using Variable Gas Quantity with Gas Recovery

Direct-Contact Heaters (Cascades)

High-temperature water can be obtained from direct-contact heaters in which steam from turbine exhaust, extraction, or steam boilers is mixed with return water from the system. The mixing takes place in the upper part of the heater where the water cascading from horizontal baffles comes in direct contact with steam (Hansen 1966). The basic systems are shown in Figures 11 and 12.

The steam space in the upper part of the heater serves as the steam cushion for pressurizing the system. The lower part of the heater serves as the system's expansion tank. Where the water heater and the boiler operate under the same pressure, the surplus water is usually returned directly into the boiler through a pipe connecting the outlet of the high-temperature water-circulating pump to the boiler.

The cascade system is also applicable where both steam and HTW services are required (Hansen and Liddy 1958). Where heat and power production are combined, the direct-contact heater becomes the mixing condenser (Hansen and Perrsall 1960).

System Circulating Pumps

Forced-circulation boiler systems can be either one-pump or two-pump. These terms do not refer to the number of pumps but to the number of groups of pumps installed. In the one-pump system (see Figure 6), a single group of pumps assures both generator and distribution system circulation. In this system, both the distribution system and the generators are in series (Carter and Sturdevant 1958).

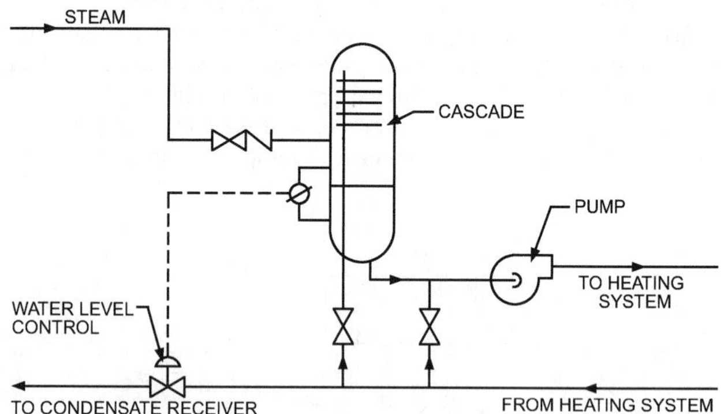

Fig. 11 Cascade HTW System

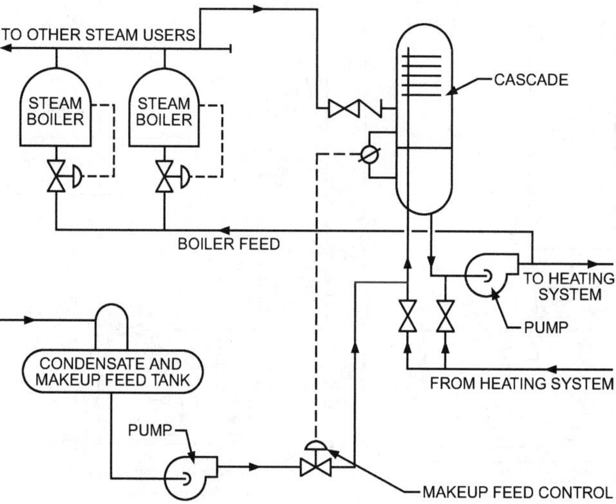

Fig. 12 Cascade HTW System Combined with Boiler Feedwater Preheating

However, to ensure the minimum flow through the boiler at all times, a bypass around the distribution system must be provided. The one-pump method usually applies only to systems in which the total friction pressure is relatively low, because the energy loss of available circulating pressure from throttling in the bypass at times of reduced flow requirements in the district can substantially increase the operating cost.

In the two-pump system (see Figure 7), an additional group of recirculating pumps is installed solely to provide circulation for the generators (Carter and Sturdevant 1958). One pump is often used for each generator to draw water from either the expansion drum or the system return and to pump it through the generator into the expansion drum. The system circulating pumps draw water from the expansion drum and circulate it through the distribution system only. The supply temperature to the distribution system can be varied by mixing water from the return into the supply on the pump suction side. Where zoning is required, several groups of pumps can be used with a different pressure and different temperature in each zone. The flow rate can also be varied without affecting the generator circulation and without using a system bypass.

In steam-pressurized systems, the circulating pump is installed in the supply line to maintain all parts of the distributing system at pressures exceeding boiler pressure. This minimizes the danger of flashing into steam.

It is common practice to install a mixing connection from the return to the pump suction that bypasses the HTW generator. This connection is used for start-up and for modulating the supply temperature; it should not be relied on for increasing the pressure at the pump inlet. Where it is impossible to provide the required submergence by proper design, a separate small-bore premixing line should be provided.

Hansen (1966) describes push-pull pumping, which divides the circulating pressure equally between two pumps in series. One is placed in the supply and is sized to overcome frictional resistance in the supply line of the heat distribution system. The second pump is placed in the return and is sized to overcome frictional resistance in the return. The expansion tank pressure is impressed on the system between the pumps. The HTW generator is either between the pumps or in the supply line from the pumps to the distribution system (see Figure 13).

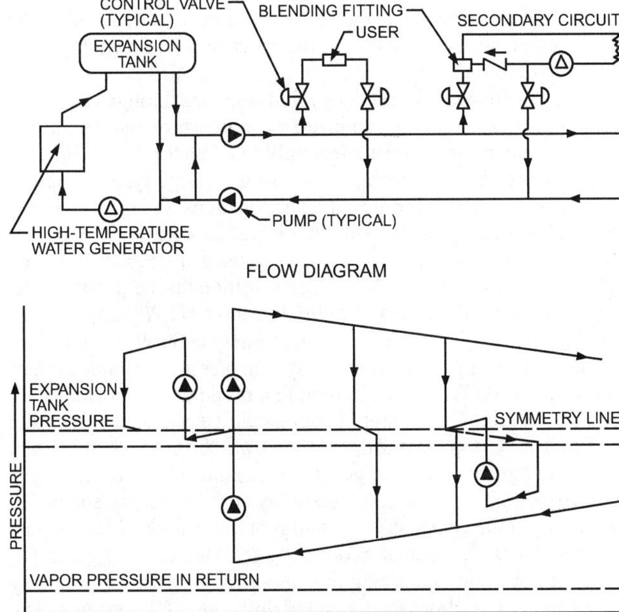

Fig. 13 Typical HTW System with Push-Pull Pumping

In the push-pull system, the pressures in the supply and the return mains are symmetrical in relation to a line representing the pressure imposed on the system by the pressurizing source (expansion tank). This pressure becomes the system pressure when the pumps are stopped. The heat supply to equipment or secondary circuits is controlled by two equal regulating valves, one on the inlet and the other on the outlet side, instead of the customary single valve on the leaving side. Both valves are operated in unison from a common controller; there are equal frictional resistances on both sides. Therefore, the pressure in the user circuits or equipment is maintained at all times halfway between the pressures in the supply and return mains. Because the halfway point is located on the symmetry line, the pressure in the user equipment or circuits is always equal to that of the pressurizing source (expansion tank) plus or minus static pressure caused by elevation differences. In other words, no system distribution pressure is reflected against the user circuit or equipment.

While the pressure in the supply system is higher than that of the expansion tank, the pressure in the return system, being symmetrical to the former, is lower. Therefore, the push-pull method is applicable only where the temperature in the return is always significantly lower than that in the supply. Otherwise, flashing could occur. This is critical and requires careful investigation of the temperature-pressure relationship at all points. The push-pull method is not applicable in reverse-return systems.

Push-pull pumping permits use of standard 125 psi fittings and equipment in many medium-temperature water (MTW) systems. Such systems, combined with secondary pumping, can be connected directly to low-temperature terminal equipment in the building heating system. Temperature drops normally obtainable only in HTW systems can be achieved with MTW. For example, 330°F water can be generated at 90 psig and distributed at less than 125 psig. Its temperature can be reduced to 200°F by secondary pumping. The pressure in the terminal equipment then is 90 psig and the MTW is returned to the primary system at 180°F. The temperature difference between supply and return in the primary MTW system is 150°F, which is comparable to that of an HTW system. In addition, conventional heat exchangers, expansion tanks, and water makeup equipment are eliminated from the secondary systems.

DISTRIBUTION PIPING DESIGN

Data for pipe friction are presented in Chapter 36 of the 2005 *ASHRAE Handbook—Fundamentals*. These pipe friction and fitting loss tables are for a 60°F water temperature. When applied to HTW systems, the values obtained are excessively high. The data should be used for preliminary pipe sizing only. Final pressure drop calculations should be made using the fundamental Darcy-Weisbach equation [Equation (1) in Chapter 36 of the 2005 *ASHRAE Handbook—Fundamentals*] in conjunction with friction factors, pipe roughness, and fitting loss coefficients presented in the section on Flow Analysis in Chapter 2 of the 2005 *ASHRAE Handbook—Fundamentals*.

The conventional conduit or tunnel distribution systems are used with similar techniques for installation (see Chapter 11). A small valved bypass connection between the supply and the return pipe should be installed at the end of long runs to maintain a slight circulation in the mains during periods of minimum or no demand.

All pipe, valves, and fittings used in HTW systems should comply with the requirements of ASME *Standard* B-31.1, Power Piping, and the *National Fuel Gas Code* (NFPA 54/ANSI Z223.1). These codes state that hot-water systems should be designed for the highest pressure and temperature actually existing in the piping under normal operation. This pressure equals cushion pressure plus pump pressure plus static pressure. Schedule 40 steel pipe is applicable to most HTW systems with welded steel fittings and steel valves. A minimum number of joints should be used. In many installations, all valves in the piping system are welded or brazed. Flange connections used at major equipment can be serrated, raised flange facing, or ring joint. It is desirable to have back-seating valves with special packing suitable for this service.

The ratings of valves, pipe, and fittings must be checked to determine the specific rating point for the given application. The pressure rating for a standard 300 psi steel valve operating at 400°F is 665 psi. Therefore, it is generally not necessary to use steel valves and fittings over 300 psi ratings in HTW systems.

Because high-temperature water is more penetrating than low-temperature water, leakage caused partly by capillary action should not be ignored because even a small amount of leakage vaporizes immediately. This slight leakage becomes noticeable only on the outside of the gland and stem of the valve where thin deposits of salt are left after evaporation. Avoid screwed joints and fittings in HTW systems. Pipe unions should not be used in place of flange connections, even for small-bore piping and equipment.

Individual heating equipment units should be installed with separate valves for shutoff. These should be readily accessible. If the unit is to be isolated for service, valves are needed in both the supply and return piping to the unit. Valve trim should be stainless steel or a similar alloy. Do not use brass and bronze.

High points in piping should have air vents for collecting and removing air, and low points should have provision for drainage. Loop-type expansion joints, in which the expansion is absorbed by deflection of the pipe loop, are preferable to the mechanical type. Mechanical expansion joints must be properly guided and anchored.

HEAT EXCHANGERS

Heat exchangers or converters commonly use steel shells with stainless steel, admiralty metal, or cupronickel tubes. Copper should not be used in HTW systems above 250°F. Material must be chosen carefully, considering the pressure-temperature characteristics of the particular system. All connections should be flanged or welded. On larger exchangers, water box-type construction is desirable to remove the tube bundle without breaking piping connections. Normally, HTW is circulated through the tubes, and because the heated water contains dissolved air, the baffles in the shell should be constructed of the same material as the tubes to control corrosion.

AIR HEATING COILS

In HTW systems over 400°F, coils should be cupronickel or all-steel construction. Below this point, other materials (e.g., red brass) can be used after determining their suitability for the temperatures involved. Coils in outdoor air connections need freeze protection by damper closure or fan shutdown controlled by a thermostat. It is also possible to set the control valve on the preheat coil to a minimum position. This protects against freezing, as long as there is no unbalance in the tube circuits where parallel paths of HTW flow exist. A better method is to provide constant flow through the coil and to control heat output with face and bypass dampers or by modulating the water temperature with a mixing pump.

SPACE HEATING EQUIPMENT

In industrial areas, space heating equipment can be operated with the available high-temperature water. Convectors and radiators may require water temperatures in the low- and medium-temperature range 120 to 180°F or 200 to 250°F, depending on their design pressure and proximity to the occupants. The water velocity through the heating equipment affects its capacity. This must be considered in selecting the equipment because, if a large water temperature drop is used, the circulation rate is reduced and, consequently, the flow velocity may be reduced enough to appreciably lower the heat transfer rate.

Convectors, specially designed to provide low surface temperatures, are now available to operate with water temperatures from 300 to 400°F.

These high temperatures are suitable for direct use in radiant panel surfaces. Because radiant output is a fourth-power function of the surface temperature, the surface area requirements are reduced over low-temperature water systems. The surfaces can be flat panels consisting of a steel tube, usually 0.38 or 0.5 in., welded to sheet steel turned up at the edges for stiffening. Several variations are available. Steel pipe can also be used with an aluminum or similar reflector to reflect the heat downward and to prevent smudging the surfaces above the pipe.

INSTRUMENTATION AND CONTROLS

Pressure gages should be installed in the pump discharge and suction and at locations where pressure readings will assist operation and maintenance. Thermometers (preferably dial-type) or thermometer wells should be installed in the flow and return pipes, the pump discharge, and at any other points of major temperature change or where temperatures are important in operating the system. It is desirable to have thermometers and gages in the piping at the entrance to each building converter.

On steam-pressurized cycles, the temperature of the water leaving the generator should control the firing rate to the generator. A master pressure control operating from the steam pressure in the expansion vessel should be incorporated as a high-limit override. Inert gas-pressurized systems should be controlled from the generator discharge temperature. Combustion controls are discussed in detail in Chapter 30.

In the water-tube generators most commonly used for HTW applications, the flow of water passes through the generator in seconds. The temperature controller must have a rapid response to maintain a reasonably uniform leaving water temperature. In steam-pressurized units, the temperature variation must not exceed the antiflash pressure margin. At 300°F, a 5°F temperature variation corresponds to a 5 psi variation in the vapor pressure. At 350°F, the same temperature variation results in a vapor pressure variation of 10 psi. At 400°F, the variation increases to 15 psi and at 450°F, to 22 psi. The permissible temperature swing must be reduced as the HTW temperature increases, or the pressure margin must be increased to avoid flashing.

Keep the controls simple. The rapid response through the generator makes it necessary to modulate the combustion rate on all systems with a capacity of over a few million Btu/h. In the smaller size range, this can be done by high/low firing. In large systems, particularly those used for central heating applications, full modulation of the combustion rate is desirable through at least 20% of full capacity. On/off burner control is generally not used in steam-pressurized cycles because the system loses pressurization during the off cycle, which can cause flashing and cavitation at the HTW pumps.

All generators should have separate safety controls to shut down the combustion apparatus when the system pressure or water temperature is high. HTW generators require a minimum water flow at all times to prevent tube failure. Means should be provided to measure the flow and to stop combustion if the flow falls below the minimum value recommended by the generator manufacturer. For inert gas-pressurized cycles, a low-pressure safety control should be included to shut down the combustion system if pressurization is lost. Figure 14 shows the basic schematic control diagram for an HTW generator.

Valve selection and sizing are important because of relatively high temperature drops and smaller flows in HTW systems. The valve must be sized so that it is effective over its full range of travel. The valve and equipment must be sized to absorb, in the control valve at full flow, not less than half the available pressure difference

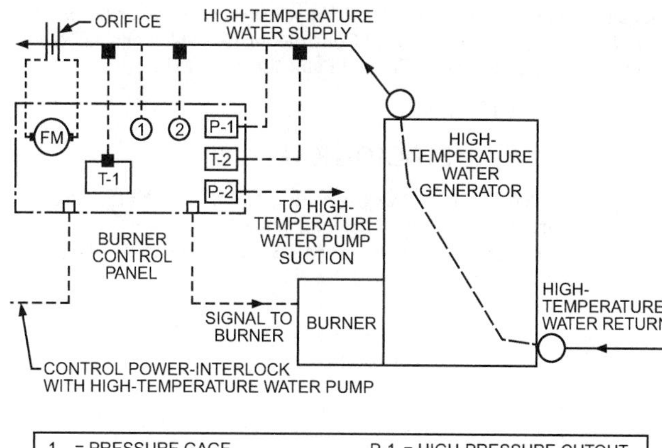

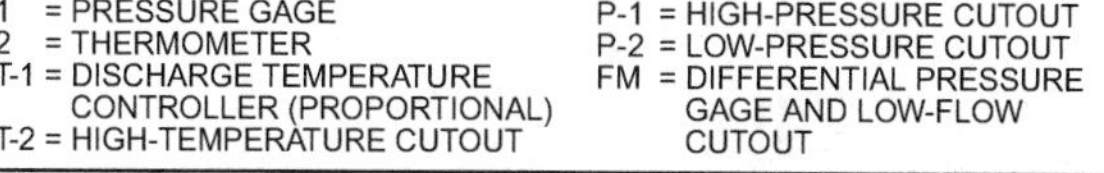

Fig. 14 Control Diagram for HTW Generator

between supply and return mains where the equipment is served. A valve with equal percentage flow characteristics is needed. Sometimes two small valves provide better control than one large valve. Stainless steel trim is recommended, and all valve body materials and packing should be suitable for high temperatures and pressures. The valve should have a close-off rating at least equal to the maximum pressure produced by the circulating pump. Generally, two-way valves are more desirable than three-way valves because of equal percentage flow characteristics and the smaller capacities available in two-way valves. Single-seated valves are preferable to double-seated valves because the latter do not close tightly.

Control valves are commonly located in the return lines from heat transfer units to reduce the valve operating temperature and to prevent plug erosion caused by high-temperature water flashing to steam at lower discharge pressure. A typical application is for controlling the temperature of water being heated in a heat exchanger where the heating medium is high-temperature water. The temperature-measuring element of the controller is installed on the secondary side and should be located where it can best detect changes to prevent overheating of outlet water. When the measuring element is located in the outlet pipe, there must be a continuous flow through the exchanger and past the element. The controller regulates the HTW supply to the primary side by means of the control valve in the HTW return. If the water leaving the exchanger is used for space heating, the set point of the thermostat in this water can be readjusted according to outdoor temperature.

Another typical application is to control a low- or medium-pressure steam generator, usually less than 50 psig, using high-temperature water as the heat source. In this application, a proportional pressure controller measures the steam pressure on the secondary side and positions the HTW control valve on the primary side to maintain the desired steam pressure. For general information on automatic controls, refer to Chapter 46 of the 2007 *ASHRAE Handbook—HVAC Applications*.

Where submergence is sufficient to prevent flashing in the **vena contracta**, control valves can be in the HTW supply instead of in the return to water heaters and steam generators (Figure 15). When used in conjunction with a check valve in the return, this arrangement shuts off the high-temperature water supply to the heat exchangers if a tube bundle leaks or ruptures.

WATER TREATMENT

Water treatment for HTW systems should be referred to a specialist. Oxygen introduced in makeup water immediately oxidizes steel at these temperatures, and over a period of time the corrosion

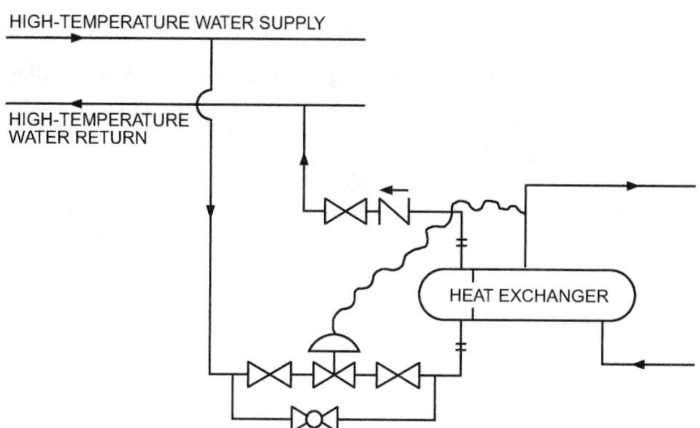

Fig. 15 Heat Exchanger Connections

can be substantial. Other impurities can also harm boiler tubes. Solids in impure water left by invisible vapor escaping at packings increase maintenance requirements. The condition of the water and the steel surfaces should be checked periodically in systems operating at these temperatures.

For further information, see Chapter 48 of the 2007 *ASHRAE Handbook—HVAC Applications,* especially the section on Water Heating Systems under Selection of Water Treatment.

HEAT STORAGE

The high heat storage capacity of water produces a flywheel effect in most HTW systems that evens out load fluctuations. Systems with normal peaks can obtain as much as 15% added capacity through such heat storage. Excessive peak and low loads of a cyclic nature can be eliminated by an HTW accumulator, based on the principle of stratification. Heat storage in an extensive system can sometimes be increased by bypassing water from the supply into the return at the end of the mains, or by raising the temperature of the returns during periods of low load.

SAFETY CONSIDERATIONS

A properly engineered and operated HTW system is safe and dependable. Careful selection and arrangement of components and materials are important. Piping must be designed and installed to prevent undue stress. When high-temperature water is released to atmospheric pressure, flashing takes place, which absorbs a large portion of the energy. Turbulent mixing of the liquid and vapor with room air reduces the temperature well below 212°F. With low mass flow rates, the temperature of the escaping mixture can fall to 125 to 140°F within a short distance, compared with the temperature of the discharge of a low-temperature water system, which remains essentially the same as the temperature of the working fluid (Hansen 1959, Armstrong and Harris 1966).

If large mass flow rates of HTW are released to atmospheric pressure in a confined space (e.g., rupture of a large pipe or vessel) a hazardous condition could exist, similar to that occurring with the rupture of a large steam main. Failures of this nature are rare if good engineering practice is followed in system design.

REFERENCES

Applegate, G. 1958. British and European design and construction methods. *ASHRAE Journal* section, *Heating, Piping & Air Conditioning* 30(3):169.

Armstrong, C.P. and W.S. Harris. 1966. Temperature distributions in steam and hot water jets simulating leaks. *ASHRAE Transactions* 72(1):147.

Blossom, J.S. and P.H. Ziel. 1959. Pressurizing high-temperature water systems. *ASHRAE Journal* 1(11):47.

Carter, C.A. and B.L. Sturdevant. 1958. Design of high temperature water systems for military installations. *ASHRAE Journal* section, *Heating, Piping & Air Conditioning* 29(2):109.

Hansen, E.G. 1959. Safety of high temperature hot water. *Actual Specifying Engineer* (July).

Hansen, E.G. 1966. Push-pull pumping permits use of MTW in building radiation. *Heating, Piping & Air Conditioning* 38(5):97.

Hansen, E.G. and W. Liddy. 1958. A flexible high pressure hot water and steam boiler plant. *Power* (May):109.

Hansen, E.G. and N.E. Perrsall. 1960. Turbo-generators supply steam for high-temperature water heating. *Air Conditioning, Heating and Ventilating* 32(6):90.

Keenan, J.H. and F.G. Keyes. 1936. *Thermodynamic properties of steam.* John Wiley & Sons, New York.

National Academy of Sciences. 1959. *High temperature water for heating and light process loads.* Federal Construction Council Technical *Report* no. 37. National Research Council Publication no. 753.

BIBLIOGRAPHY

ANSI/NFPA. 2006. *National fuel gas code.* ANSI Z223.1/NFPA 54-2006. National Fire Protection Association, Quincy, MA.

ASME. 2001. Power piping. ANSI/ASME *Standard* B31.1-01. American Society of Mechanical Engineers, New York.

CHAPTER 15

INFRARED RADIANT HEATING

INFRARED radiant heating principles discussed in this chapter apply to equipment with thermal radiation source temperatures ranging from 300 to 5000°F. (Equipment with source temperatures starting from below the indoor air temperature to 300°F is classified as panel heating and cooling equipment, discussed in Chapter 6.) Infrared radiant heaters with source temperatures in this range are categorized into three groups as follows:

- **Low-intensity** source temperatures range from 300 to 1200°F. A typical low-intensity heater is mounted on the ceiling and may be constructed of a 4 in. steel tube 10 to 80 ft long. A gas burner inserted into the end of the tube raises the tube surface temperature, and because most units are equipped with a reflector, thermal radiation is directed down to the heated space.
- **Medium-intensity** source temperatures range from 1200 to 1800°F. Typical equipment types include porous matrix gas-fired infrared heaters or metal-sheathed electric heaters.
- **High-intensity** source temperatures range from 1800 to 5000°F. A typical high-intensity heater is an electrical reflector lamp with a resistor temperature of 4050°F.

Low-, medium-, and high-intensity infrared heaters are frequently applied in aircraft hangars, factories, warehouses, foundries, greenhouses, and gymnasiums. They are applied to open areas such as loading docks, racetrack stands, under marquees, vestibules, outdoor restaurants, carwashes, and around swimming pools. Infrared heaters are also used for snow and ice melting (see Chapter 50 of the 2007 *ASHRAE Handbook—HVAC Applications*), condensation control, and industrial process heating. Reflectors are frequently used to control the distribution of heat flux from thermal radiation in specific patterns.

When infrared radiant heating is used, the environment is characterized by

- A directional thermal radiation field created by the infrared heaters
- A thermal radiation field consisting of reradiation and reflection from the walls and/or other enclosing surfaces
- Ambient air temperatures often lower than those found with convective systems

The combined action of these factors determines occupant thermal comfort and the thermal acceptability of the environment.

ENERGY CONSERVATION

Infrared heaters are effective for spot heating. However, because of their efficient performance, they are also used for total heating of large areas and entire buildings (Buckley 1989). Radiant heaters transfer heat directly to solid objects. Little heat is lost during transmission because air is a poor absorber of radiant heat. Because an intermediate transfer medium such as air or water is not required, fans or pumps are not needed.

As thermal radiation warms floors, walls, and objects, they in turn release heat to the air by convection. Reradiation to surrounding objects also contributes to comfort in the area. An energy-saving advantage is that infrared heaters can be turned off when not needed; when turned on again, they are effective in minutes. Even when the infrared heater is off, the heated surrounding objects at occupant level continue to contribute to comfort by reradiating heat and releasing heat by convection.

Human thermal comfort is primarily governed by the operative temperature of the heated space (ASHRAE *Standard* 55). Operative temperature may be approximated by the arithmetic average of the mean radiant temperature (MRT) of the heated space and dry-bulb air temperature, if air velocity is less than 1.3 fps and MRT is less than 120°F. See Chapter 53 of the 2007 *ASHRAE Handbook—HVAC Applications* for further details. In radiant heating, the dry-bulb air temperature may be kept lower for a given comfort level than with other forms of heating because of increased MRT. As a result, heat lost to ventilating air and via conduction through the shell of the structure is proportionally smaller, as is energy consumption. Infiltration loss, which is a function of dry-bulb air temperature, is also reduced.

Because of the unique split of radiant and convective components in radiant heating, air movement and stratification in the heated space is minimal. This further reduces infiltration and transmission heat losses.

Buckley and Seel (1987) compared energy savings of infrared heating with those of other types of heating systems. Recognizing the reduced fuel requirement for these applications, Buckley and Seel (1988) noted that it is desirable for manufacturers of radiant heaters to recommend installation of equipment with a rated output that is 80 to 85% of the heat loss calculated by methods described in Chapters 29 and 30 of the 2005 *ASHRAE Handbook—Fundamentals*. BSR/ASHRAE *Standard* 138P describes a rated output system for ceiling radiant heaters.

Chapman and Zhang (1995) developed a three-dimensional mathematical model to compute radiant heat exchange between surfaces. A building comfort analysis program (BCAP) was developed as part of ASHRAE research project RP-657 (Jones and Chapman 1994). The BCAP program was later enhanced to analyze the effect of radiant heaters over 300°F on thermal comfort calculations, and to analyze the thermal comfort effect of obstacles in the heated space in ASHRAE research project RP-1037 (Chapman 2002).

INFRARED ENERGY SOURCES

Gas Infrared

Modern gas-fired infrared heaters burn gas to heat a specific radiating surface. The surface is heated by direct flame contact or with combustion gases. Studies by the Gas Research Board of London (1944), Haslam et al. (1925), and Plyler (1948) reveal that only 10 to 20% of the energy produced by open combustion of a gaseous fuel is infrared radiant energy. The wavelength span over which radiation from a heated surface is distributed can be controlled by design. The specific radiating surface of a properly designed unit directs

The preparation of this chapter is assigned to TC 6.5, Radiant and In-Space Convective Heating and Cooling.

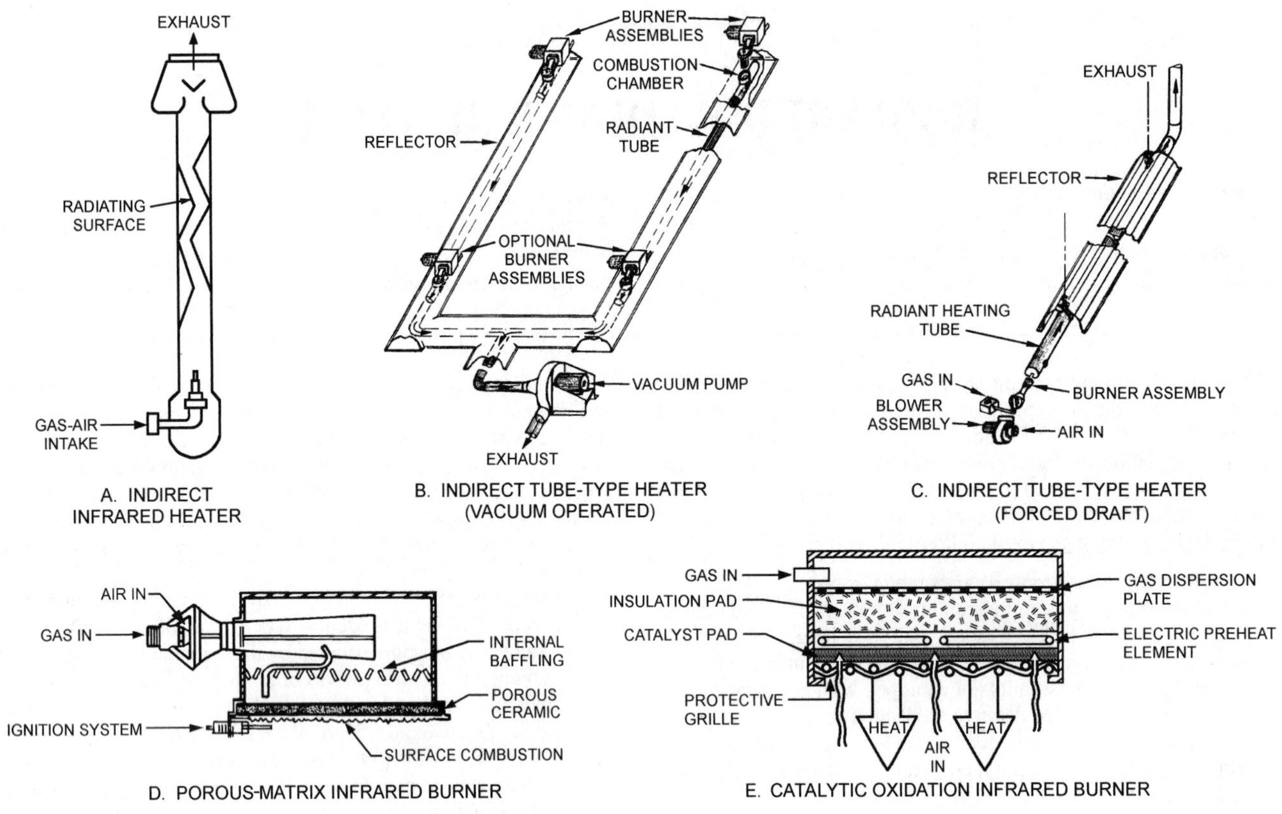

Fig. 1 Types of Gas-Fired Infrared Heaters

Table 1 Characteristics of Typical Gas-Fired Infrared Heaters

Characteristics	Indirect	Porous Matrix	Catalytic Oxidation
Operating source temperature	Up to 1200°F	1600 to 1800°F	650 to 700°F
Relative heat flux,[a] Btu/h·ft²	Low, up to 7500	Medium, 17,000 to 32,000	Low, 800 to 3000
Response time (heat-up)	180 s	60 s	300 s
Thermal radiation-energy input ratio[b]	0.35 to 0.55	0.35 to 0.60	No data
Thermal shock resistance	Excellent	Excellent	Excellent
Vibration resistance	Excellent	Excellent	Excellent
Color blindness[c]	Excellent	Very good	Excellent
Luminosity (visible light)	To dull red	Yellow red	None
Mounting height	9 to 50 ft	12 to 50 ft	To 10 ft
Wind or draft resistance	Good	Fair	Very good
Venting	Optional	Nonvented	Nonvented
Flexibility	Good	Excellent—wide range of heat fluxes and mounting possibilities available	Limited to low-heat-flux applications

[a]Heat flux emitted at burner surface.
[b]Ratio of thermal radiation to energy output to input.
[c]Color blindness refers to absorptance by various loads of energy emitted by different sources.

radiation toward the load. Gas-fired infrared radiation heaters are available in the following types (see Table 1 for characteristics).

Indirect infrared radiation heaters (Figures 1A, 1B, and 1C) are internally fired and have the radiating surface between the hot gases and the load. Combustion takes place within the radiating elements, which operate with surface temperatures up to 1200°F. The elements may be tubes or panels with metal or ceramic components. Indirect infrared radiation units are usually vented and may require eductors.

Porous-matrix infrared radiation heaters (Figure 1D) have a refractory material that may be porous ceramic, drilled port ceramic, stainless steel, or a metallic screen. The units are enclosed, except for the major surface facing the load. A combustible gas-air mixture enters the enclosure, flows through the refractory material to the exposed face, and is distributed evenly by the porous character of the refractory. Combustion occurs evenly on the exposed surface. The flame recedes into the matrix, which adds radiant energy to the flame. If the refractory porosity is suitable, an atmospheric burner can be used, resulting in a surface temperature approaching 1650°F. Power burner operation may be required if refractory density is high. However, the resulting surface temperature may also be higher (1800°F).

Catalytic oxidation infrared radiation heaters (Figure 1E) are similar to porous-matrix units in construction, appearance, and operation, but the refractory material is usually glass wool, and the radiating surface is a catalyst that causes oxidation to proceed without visible flames.

Electric Infrared

Electric infrared heaters use heat produced by electric current flowing in a high-resistance wire, graphite ribbon, or film element. The following are the most commonly used types (see Table 2 for characteristics).

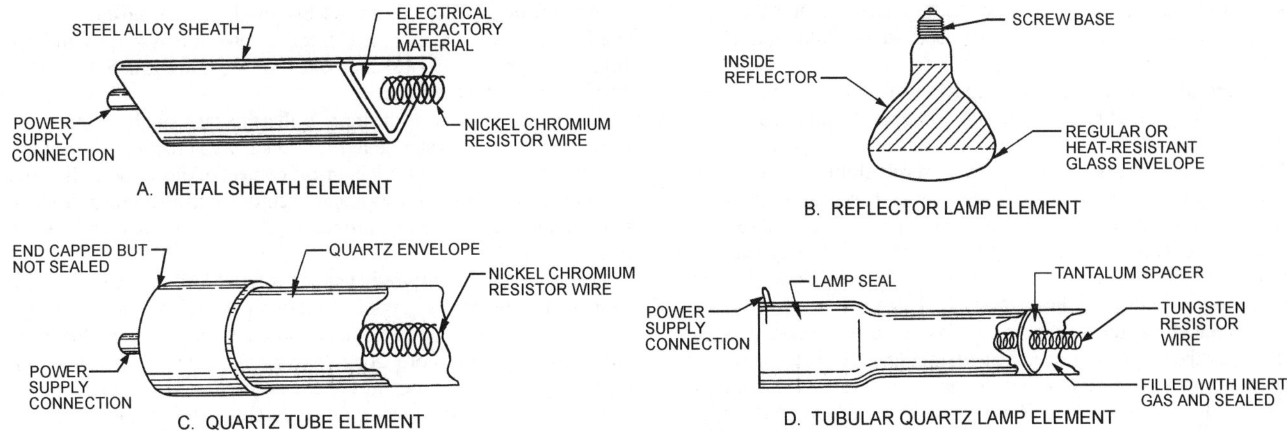

Fig. 2 Common Electric Infrared Heaters

Table 2 Characteristics of Four Electric Infrared Elements

Characteristic	Metal Sheath	Reflector Lamp	Quartz Tube	Quartz Lamp
Resistor material	Nickel-chromium alloy	Tungsten wire	Nickel-chromium alloy	Tungsten wire
Relative linear heat flux	Medium, 60 W/in., 0.5 in. diameter	High, 125 to 375 W/spot	Medium to high, 75 W/in., 0.5 in. diameter	High, 100 W/in., 3/8 in. diameter
Resistor temperature	1750°F	4050°F	1700°F	4050°F
Envelope temperature (in use)	1550°F	525 to 575°F	1200°F	1100°F
Thermal radiation-energy input ratio[a]	0.58	0.86	0.81	0.86
Response time (heat-up)	180 s	A few seconds	60 s	A few seconds
Luminosity (visible light)	Very low (dull red)	High (8 lm/W)	Low (orange)	High (7.5 lm/W)
Thermal shock resistance	Excellent	Poor to excellent (heat-resistant glass)	Excellent	Excellent
Vibration resistance	Excellent	Medium	Medium	Medium
Impact resistance	Excellent	Medium	Poor	Poor
Wind or draft resistance[b]	Medium	Excellent	Medium	Excellent
Mounting position	Any	Any	Horizontal[c]	Horizontal
Envelope material	Steel alloy	Regular or heat-resistant glass	Translucent quartz	Clear, translucent, or frost quartz and integral red filter glass
Color blindness	Very good	Fair	Very good	Fair
Flexibility	Good—wide range of power density, length, and voltage practical	Limited to 125-250 and 375 W at 120 V	Excellent—wide range of power density, diameter, length, and voltage practical	Limited—1 to 3 W for each V; 1 length for each capacity
Life expectancy	Over 5000 h	5000 h	5000 h	5000 h

[a]Ratio of thermal radiation output to energy input (elements only).
[b]May be shielded from wind effects by louvers, deep-drawn fixtures, or both.
[c]May be provided with special internal supports for other than horizontal use.

Metal sheath infrared radiation elements (Figure 2A) are composed of a nickel-chromium heating wire embedded in an electrical insulating refractory, which is encased by a metal tube. These elements have excellent resistance to thermal shock, vibration, and impact, and can be mounted in any position. At full voltage, the elements attain a sheath surface temperature of 1200 to 1800°F. Higher temperatures are attained by configurations such as a hairpin shape. These units generally contain a reflector, which directs radiation to the load. Higher radiosity is obtained if the elements are shielded from wind because the surface-cooling effect of the wind is reduced.

Reflector lamp infrared radiation heaters (Figure 2B) have a coiled tungsten filament, which approximates a point source radiator. The filament is enclosed in a clear, frosted, or red heat-resistant glass envelope, which is partially silvered inside to form an efficient reflector. Units that may be screwed into a light socket are common.

Quartz tube infrared radiation heaters (Figure 2C) have a coiled nickel-chromium wire lying unsupported within an unevacuated, fused quartz tube, which is capped (not sealed) by porcelain or metal terminal blocks. These units are easily damaged by impact and vibration but stand up well to thermal shock and splashing. They must be mounted horizontally to minimize coil sag, and they are usually mounted in a fixture that contains a reflector. Normal operating temperatures are from 1300 to 1800°F for the coil and about 1200°F for the tube.

Tubular quartz lamp units (Figure 2D) consist of a 0.38 in. diameter fused quartz tube containing an inert gas and a coiled tungsten filament held in a straight line and away from the tube by tantalum spacers. Filament ends are embedded in sealing material at the ends of the envelope. Lamps must be mounted horizontally, or nearly so, to minimize filament sag and overheating of the sealed ends. At normal design voltages, quartz lamp filaments operate at about 4050°F, while the envelope operates at about 1100°F.

Oil Infrared

Oil-fired infrared radiant heaters are similar to gas-fired indirect infrared radiant heaters (Figures 1A, 1B, and 1C). Oil-fired units are vented.

SYSTEM EFFICIENCY

Because many factors contribute to the performance of a specific infrared radiant heating system, a single criterion should not be used to evaluate comparable systems. Therefore, at least two of the following indicators should be used when evaluating system performance.

Thermal radiation-energy input ratio is the thermal energy transferred by radiation in the infrared wavelength spectrum divided by the total energy input.

Fixture efficiency is an index of a fixture's ability to radiate thermal energy from the source; it is usually based on total energy input. The housing, reflector, and other parts of a fixture absorb some infrared energy and convert it to heat, which is lost through convection. A fixture that controls direction and distribution of energy effectively may have higher fixture efficiency.

Pattern efficiency is an index of a fixture's effectiveness in directing infrared energy into a specific pattern. This effectiveness, plus effective application of the pattern to the thermal load, influences the system's total effectiveness (Boyd 1963). Typical thermal radiation-energy input ratios of gas infrared heaters are shown in Table 1. Limited test data indicate that the amount of thermal radiation from gas infrared units ranges from 35 to 60% of the amount of convective heat. The Stefan-Boltzmann law can be used to estimate thermal radiation if reasonably accurate values of true surface temperature, emitting area, and surface emittance are available (DeWerth 1960). DeWerth (1962) also addresses the spectral distribution of energy curves for several gas sources.

Table 2 lists typical thermal radiation-energy input ratios of electric infrared heaters. Fixture efficiencies are typically 80 to 95% of the thermal radiation-energy input ratios.

Infrared heaters should be operated at rated input. A small reduction in input causes a larger decrease in radiant output because of the fourth-power dependence of radiant output on source temperature. Because a variety of infrared units with a variety of reflectors and shields are available, the manufacturers' information should be consulted.

REFLECTORS

Radiation from most infrared heating devices is directed by the emitting surface and can be concentrated by reflectors. Mounting height and whether spot heating or total heating is used usually determine which type of reflector will achieve the desired heat flux pattern at floor level. Four types of reflectors can be used: (1) parabolic, which produce essentially parallel beams of energy; (2) elliptical, which direct all energy that is received or generated at the first focal point through a second focal point; (3) spherical, which are a special class of elliptical reflectors with coincident foci; and (4) flat, which redirect the emitted energy without concentrating or collimating the rays.

Energy data furnished by the manufacturer should be consulted to apply a heater properly.

CONTROLS

Normally, all controls (except the thermostat) are built into gas-fired infrared heaters, whereas electric infrared fixtures usually have no built-in controls. Because of the effects of direct radiation, as well as higher MRT of the heated space and decreased air temperature compared to convective systems, infrared heating requires careful selection and location of the thermostat or sensor. A thermostat sensing the operative temperature rather than the air temperature is desirable; placing the thermostat or sensor in the radiation pattern increases accuracy. The nature of the system, type of infrared heaters, and nature of the thermostat or sensor dictate the appropriate approach. Furthermore, no single location appears to be equally effective after a cold start and after a substantial period of operation. To reduce high and low temperature swings, a long rather than a short thermostat cycling time is preferred. A properly sized system and modulating or dual-stage operation can improve comfort conditions.

An infrared heater controlled by low-limit thermostats can be used for freeze protection. For freeze protection systems and heat flux requirement, refer to Chapter 50 of the 2007 *ASHRAE Handbook—HVAC Applications*.

On gas-fired infrared heaters, a thermostat usually controls an electronic ignition system that monitors the gas flame and operates the automatic valve to provide on-off control of gas flow to all burners. For all gas-fired infrared heaters, conform to all applicable standards and codes covering proper venting, safety, and indoor air quality.

Gas and electric infrared systems for full-building heating may have a zone thermostatic control system in which a thermostat representative of one outside exposure operates heaters along that outside wall. Two or more zone thermostats may be required for extremely long wall exposures. Heaters for an internal zone may be grouped around a thermostat representative of that zone. Manual switches or thermostats are usually used for spot or area heating, but input controllers may also be used.

Input controllers effectively control electric infrared heaters having metal sheath or quartz tube elements. An input controller is a motor-driven cycling device in which *on* time per cycle can be set. A 30 s cycle is normal. When a circuit's capacity exceeds an input controller's rating, the controller can be used to cycle a pilot circuit of contactors adequate for the load.

Input controllers work well with metal sheath heaters because the sheath mass smooths the pulses into even radiation. The control method decreases the efficiency of infrared generation slightly. When controlled with these devices, quartz tube elements, which have a warm-up time of several seconds, have perceptible but not normally disturbing pulses of infrared, with only moderate reduction in generation efficiency.

Input controllers should not be used with quartz lamps because the cycling luminosity would be distracting. Instead, quartz lamp output can be controlled by changing the voltage applied to the lamp element, using modulating transformers or by switching the power supply from hot-to-hot to hot-to-ground potential.

Electrical power consumed by the tungsten filament of the quartz lamp varies approximately as the 1.5 power of the applied voltage, whereas that of metal sheath or quartz tube elements (using nickel-chromium wire) varies as the square of the applied voltage. Multiple circuits for electric infrared systems can be manually or automatically switched to provide multiple stages of heat. Three circuits or control stages are usually adequate. For areas with fairly uniform radiation, one circuit should be controlled with input control or voltage variation control on electric units, while the other two are full on or off control. This arrangement gives flexible, staged control with maximum efficiency of infrared generation. The variable circuit alone provides zero to one-third capacity. Adding another circuit at full on provides one-third to two-thirds capacity, and adding the third circuit provides two-thirds to full capacity.

PRECAUTIONS

Precautions for applying infrared heaters include the following:

- All infrared heaters covered in this chapter have high surface temperatures when operating and should, therefore, not be used when the atmosphere contains ignitable dust, gases, or vapors in hazardous concentrations.
- Manufacturers' recommendations for clearance between a fixture and combustible material should be followed. If combustible material is being stored, warning notices defining proper clearances should be posted near the fixture.
- Manufacturers' recommendations for clearance between a fixture and personnel areas should be followed to prevent personnel stress from local overheating.
- Infrared fixtures should not be used if the atmosphere contains gases, vapors, or dust that decompose to hazardous or toxic

materials in the presence of high temperature and air. For example, infrared units should not be used in an area with a degreasing operation that uses trichloroethylene unless the area has a suitable exhaust system that isolates the contaminant. Trichloroethylene, when heated, forms phosgene (a toxic compound) and hydrogen chloride (a corrosive compound).

- Humidity must be controlled in areas with unvented gas-fired infrared units because water formed by combustion increases humidity. Sufficient ventilation [NFPA 54 (ANSI Z223.1), *National Fuel Gas Code*], direct venting, or insulation on cold surfaces helps control moisture problems.
- Lamp holders and electrical grounding for infrared heating lamps should comply with Section 422-15 of the *National Electrical Code®* (NFPA *Standard* 70).
- Sufficient makeup air (NFPA 54, *National Fuel Gas Code*) must be provided to replace the air used by combustion heaters, regardless of whether units are direct vented.
- If unvented combustion infrared heaters are used, the area must have adequate ventilation to ensure that combustion products in the air are held to an acceptable level (Prince 1962). See Chapter 45 of the 2007 *ASHRAE Handbook—HVAC Applications* for information on IAQ concerns.
- For comfort in areas such as hangars and docks, conditioned space should be protected from substantial wind or drafts. Suitable wind shields seem to be more effective than increased radiation heat flux (Boyd 1960).

In the United States, refer to Occupational Safety and Health Administration (OSHA) guidelines for additional information. For nonvented infrared radiant heaters, IAQ and occupant comfort are important. For IAQ concerns, see Chapter 45 of the 2007 *ASHRAE Handbook—HVAC Applications*.

MAINTENANCE

Gas- and oil-fired infrared heaters require periodic cleaning to remove dust, dirt, and soot. Reflecting surfaces must be kept clean to remain efficient. Annual cleaning of heat exchangers, radiating surfaces, burners, and reflectors with compressed air is usually sufficient. Chemical cleaners must not leave a film on reflector surfaces.

Air ports of gas-fired units should be kept free of lint and dust. The nozzle, draft tube, and nose cone of oil-fired unit burners are designed to operate in a particular combustion chamber, so they must be replaced carefully when they are removed.

Electric infrared heaters require little care beyond cleaning the reflectors. Quartz and glass elements must be handled carefully because they are fragile, and fingerprints must be removed (preferably with alcohol) to prevent etching at operating temperature, which causes early failure.

DESIGN CONSIDERATIONS FOR BEAM RADIANT HEATERS

Chapter 53 of the 2007 *ASHRAE Handbook—HVAC Applications* introduces design principles for spot beam radiant heating. The effective radiant flux (ERF) represents the radiant energy absorbed by an occupant from all temperature sources different from the ambient. ERF is defined as

$$\text{ERF} = h_r(\bar{t}_r - t_a) \tag{1}$$

where

ERF = effective radiant flux, Btu/h·ft²
h_r = linear radiation heat transfer coefficient, Btu/h·ft²·°F
$\bar{t}_r$ = mean radiant temperature affecting occupant, °F
t_a = ambient dry-bulb air temperature near occupant, °F

ERF may be measured as the heat absorbed at the skin-clothing surface from a beam heater treated as a point source:

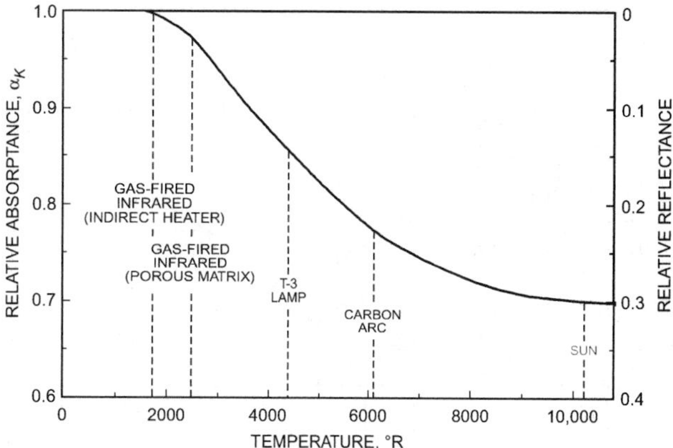

Fig. 3 Relative Absorptance and Reflectance of Skin and Typical Clothing Surfaces

$$\text{ERF} = \frac{\alpha_K I_K(A_p/d^2)}{A_D} \tag{2}$$

where

α_K = absorptance of skin-clothing surface at emitter temperature (Figure 3), dimensionless
I_K = irradiance from beam heater, Btu/h·sr
A_p = projected area of occupant on plane normal to direction of heater beam, ft²
d = distance from beam heater to center of occupant, ft
A_D = body surface area of occupant, ft²

A_p/d^2 is the solid angle subtended by the projected area of the occupant from the beam heater. See Figure 5 in Chapter 53 of the 2007 *ASHRAE Handbook—HVAC Applications* for a representation of these variables. The value of the DuBois area A_D has been defined as follows:

$$A_D = 0.108W^{0.425}H^{0.725}$$

where

W = weight of occupant, lb
H = height of occupant, in.

Two radiation area factors are defined as

$$f_{eff} = A_{eff}/A_D \tag{3}$$

$$f_p = A_p/A_{eff} \tag{4}$$

where A_{eff} is the effective radiating area of the total body surface. Equation (2) becomes

$$\text{ERF} = \alpha_K f_{eff} f_p I_K/d^2 \tag{5}$$

Fanger (1972) developed precise optical methods to evaluate the angle factors f_{eff} and f_p for both sitting and standing positions and for males and females. An average value for f_{eff} of 0.71 for both sitting and standing is accurate within ±2%. Variations in angle factor f_p over various azimuths and elevations for seated and standing positions are illustrated in Figures 4 and 5, and according to Fanger, apply equally to both males and females.

Manufacturers of infrared heaters usually supply performance specifications for their equipment (Gagge et al. 1967). Design information on sizing infrared heating units is also available (Howell and Suryanarayana 1990), as is the relation between color temperature of heaters and the applied electric potential (voltage) or electrical power. Gas-fired radiators usually operate at constant source temperatures of 1340 to 1700°F (1800 to 2160°R). Figure 3 relates the

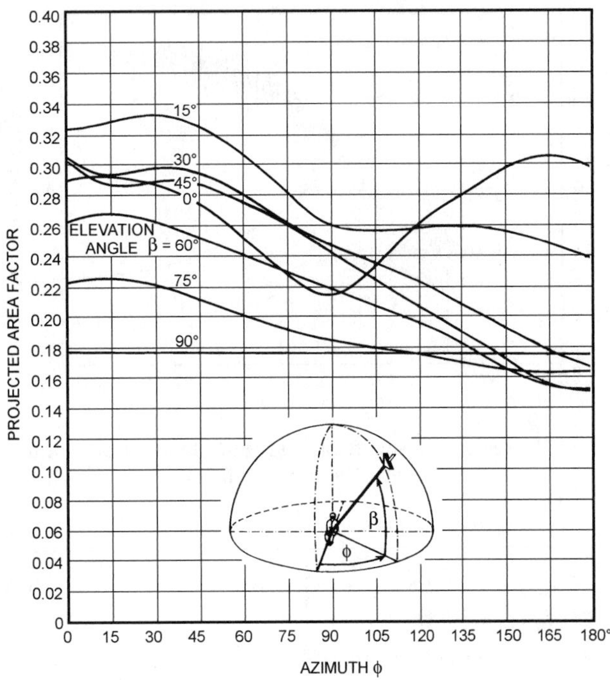

Fig. 4 Projected Area Factor for Seated Persons, Nude and Clothed
(Fanger 1972)

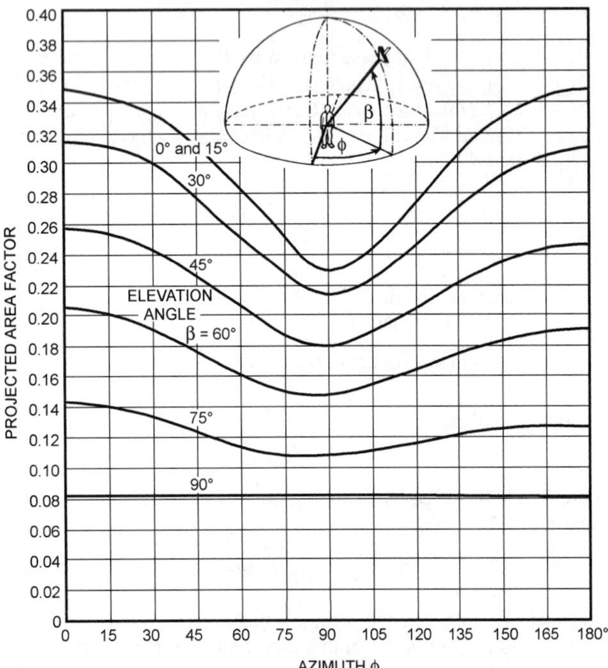

Fig. 5 Projected Area Factor for Standing Persons, Nude and Clothed
(Fanger 1972)

absorptance α_K to the radiating temperature of the radiant source. Manufacturers also supply the heat flux distribution of beam heaters with and without reflectors. Figures 6 to 9 illustrate the heat flux distribution for four typical electric and gas-fired radiant heaters. Generally, in electrical beam heaters, 70 to 80% of the total heat transfer occurs by thermal radiation, in contrast to 40% for gas-fired

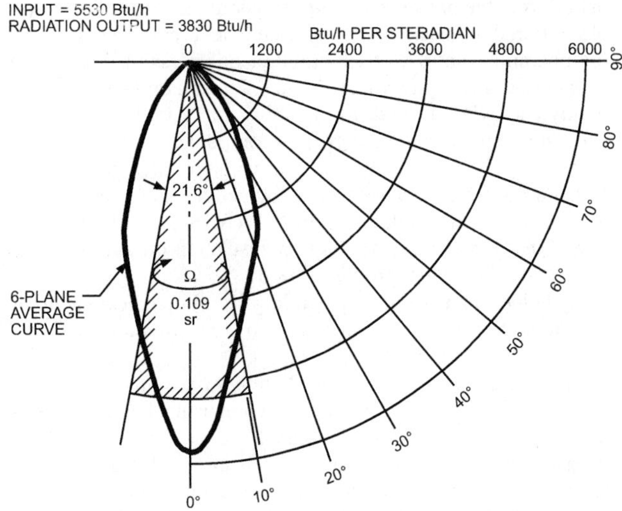

Fig. 6 Radiant Heat Flux Distribution Curve of Typical Narrow-Beam High-Intensity Electric Infrared Heaters

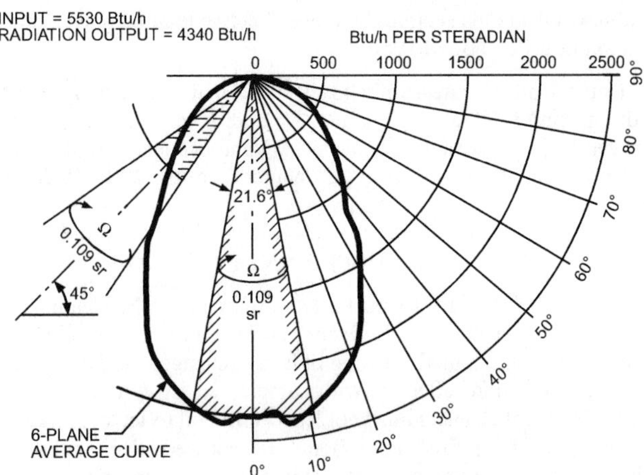

Fig. 7 Radiant Heat Flux Distribution Curve of Typical Broad-Beam High-Intensity Electric Infrared Heaters

types. In practice, the designer should choose a beam heater that will illuminate the subject with acceptable uniformity. Even with complete illumination by a beam 0.109 sr (21.6°) wide, Figure 6 shows that only 8% (100 × 1225 × 0.109/1620) of the initial electrical power input to the heater is usable for specifying the necessary I_K in Equation (5). The corresponding percentages for Figures 7, 8, and 9 are 5%, 4%, and 2%, respectively. The last two are for gas-fired beams.

The input energy to the beam heater not used for directly irradiating the occupant ultimately increases the ambient air temperature and mean radiant temperature of the heated space This increase will reduce the original ERF required for comfort and acceptability. The continuing reradiation and convective heating of surrounding walls and the presence of air movement make precise calculations of radiant heat exchange difficult.

The basic principles of beam heating are illustrated by the following examples.

Example 1. Determine the irradiance required for comfort from a quartz lamp (Figure 6) when the worker is sedentary, lightly clothed (0.5 clo), and seated. The dry-bulb indoor air temperature t_a near the

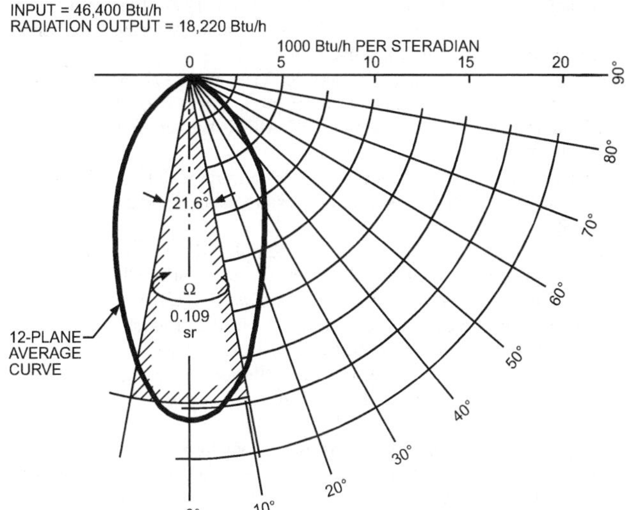

Fig. 8 Radiant Heat Flux Distribution Curve of Typical Narrow-Beam High-Intensity Atmospheric Gas-Fired Infrared Heaters

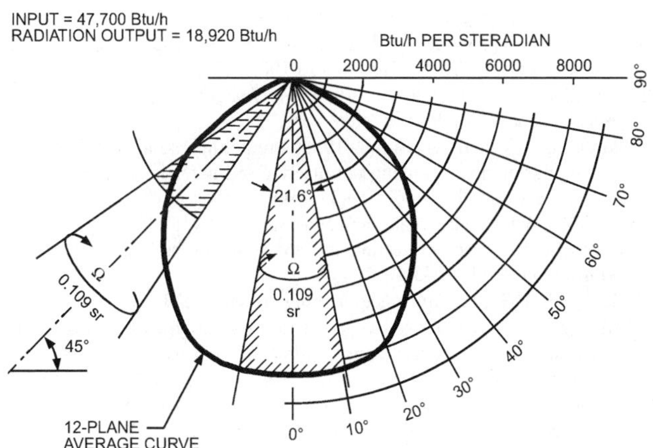

Fig. 9 Radiant Heat Flux Distribution Curve of Typical Broad-Beam High-Intensity Atmospheric Gas-Fired Infrared Heaters

occupant is 59°F, with air movement at 30 fpm. The lamp is mounted on the 8 ft high ceiling and is directed at the back of the seated person so that the elevation angle β is 45° and the azimuth angle ϕ is 180°. Assume the ambient and mean radiant temperatures of the unheated space are equal. The lamp operates at 240 V and has a source temperature of 4500°R.

Solution: The ERF for comfort can be calculated as 19.3 Btu/h·ft² by the procedures outlined in the section on Design Criteria for Acceptable Radiant Heating in Chapter 53 of the 2007 *ASHRAE Handbook—HVAC Applications*. $\alpha_K = 0.85$ at 4500°R (from Figure 3); $f_{eff} = 0.71$; $f_p = 0.17$ (Figure 3); $d = 8 - 2 = 6$ ft, where 2 ft is sitting height of occupant.

From Equation (5), the irradiance I_K from the beam heater necessary for comfort is

$$I_K = \text{ERF}(d^2/\alpha_K f_{eff} f_p)$$

$$= 19.3(6)^2/(0.85 \times 0.71 \times 0.17) = 6772 \text{ Btu/h} \cdot \text{sr}$$

Example 2. For the same occupant in Example 1, when two beams located on the ceiling and operating at half of rated voltage are directed downward at the subject at 45° and at azimuth angle 90° on each side, what would be the I_K required from each heater?

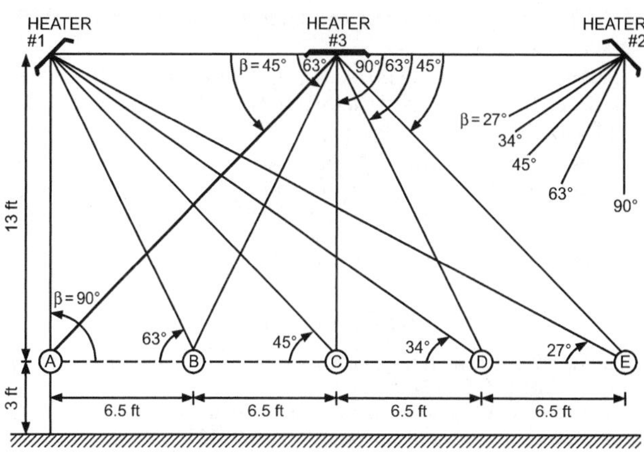

Fig. 10 Calculation of Total ERF from Three Gas-Fired Heaters on Worker Standing at Positions A Through E

Solution: The ERF for comfort from each beam is 19.3/2 or 9.65 Btu/h·ft². The value of f_p is 0.25 (from Figure 3). At half power, $V = 170$, $R \approx 3600$, and $\alpha_K \approx 0.9$. Hence, the required irradiation from each beam is

$$I_K = 9.65(6)^2/(0.9 \times 0.71 \times 0.25) = 2174 \text{ Btu/h} \cdot \text{sr}$$

This estimate indicates that two beams similar to Figure 7, each operating at half of rated power, can produce the necessary ERF for comfort. A comparison between the I_K requirements in Examples 1 and 2 shows that irradiating a sitting person from the back is much less efficient than irradiating from the side.

Example 3. A broad-beam gas-fired infrared heater is mounted 16 ft above the floor. The heater is directed 45° downward toward a standing subject 13 ft away (see Heater #1 in Figure 10, position C).

Question (1): What is the resulting ERF from the beam acting on the subject?

Solution: From Equation (5),

$$\text{ERF} = \alpha_K f_{eff} f_p I_K / d^2 = 4.2 \text{ Btu/h} \cdot \text{ft}^2$$

where

$\alpha_K = 0.97$ (Figure 3)
$f_{eff} = 0.71$
$f_p = 0.26$ (Figure 5 at $\beta = 45°$ and $\phi = 0°$)
$I_K = 8000$ Btu/h·sr (Figure 9)
$d^2 = 13^2 + 13^2 = 338$ ft² (center of standing subject is 3 ft above floor)

If the heater were 10 ft above the center of the standing subject (13 ft above floor), the ERF would be 5.3 Btu/h·ft².

Question (2): How does the ERF vary along the 0° azimuth, every 6.5 ft beginning at a point directly under Heater #1 (Figure 10) and for elevations β = 90° at A, 63.4° at B, 45° at C, 33.6° at D, and 26.6° at E? From Figure 9, the values for f_p for the five positions are (A) 0.08, (B) 0.19, (C) 0.26, (D) 0.30, and (E) 0.33.

Solution: Because the beam is directed 45° downward, the respective deviations from the beam center for a person standing at the five positions A through E are 45°, 18.4°, 0°, 11.4°, and 18.4°; the corresponding I_K values from Figure 9 are 5000, 8000, 8000, 8000, and 8000 Btu/h·sr. The respective d^2 are 169, 211, 338, 549, and 845 ft². The ERFs for a person standing in the five positions are (A) 1.6, (B) 4.9, (C) 4.2, (D) 3.0, and (E) 2.1 Btu/h·ft².

Question (3): How will the total ERF at each of the five locations A through E vary if two additional heaters (#2 and #3 in Figure 10) are added 16 ft above the floor over positions C and E? The center heater is directed downward, the outer one directed as above, 45° towards the center of the heated space.

Solution: At each of five locations in the heated space (A, B, C, D, E), add the ERF from each of the three radiators to determine the total ERF affecting the standing person.

A	1.6 + 2.5 + 2.1	or	6.2
B	4.9 + 4.2 + 3.0	or	12.1
C	4.2 + 2.5 + 4.2	or	10.9
D	3.0 + 4.2 + 4.9	or	12.1
E	2.1 + 2.5 + 1.6	or	6.2

REFERENCES

ASHRAE. 1992. Thermal environmental conditions for human occupancy. ANSI/ASHRAE *Standard* 55-1992.

Boyd, R.L. 1960. What do we know about infrared comfort heating? *Heating, Piping and Air Conditioning* (November):133.

Boyd, R.L. 1963. Control of electric infrared energy distribution. *Electrical Engineering* (February):103.

BSR/ASHRAE. 2002. Method of testing for rating ceiling panels for sensible heating and cooling. BSR/ASHRAE *Standard* 138P. American National Standards Institute Board of Standards Review, New York, and ASHRAE.

Buckley, N.A. 1989. Applications of radiant heating saves energy. *ASHRAE Journal* 31(9):17.

Buckley, N.A. and T.P. Seel. 1987. Engineering principles support an adjustment factor when sizing gas-fired low-intensity infrared equipment. *ASHRAE Transactions* 93(1):1179-1191.

Buckley, N.A. and T.P. Seel. 1988. Case studies support adjusting heat loss calculations when sizing gas-fired, low-intensity, infrared equipment. *ASHRAE Transactions* 94(1):1848-1858.

Chapman, K.S. 2002. Development of a simplified methodology to incorporate radiant heaters over 300°F into thermal comfort calculations (RP-1037). *Final Report*. ASHRAE.

Chapman, K.S. and P. Zhang. 1995. Radiant heat exchange calculations in radiantly heated and cooled enclosures. *ASHRAE Transactions* 101(1): 1236-1247.

DeWerth, D.W. 1960. Literature review of infra-red energy produced with gas burners. *Research Bulletin* 83. American Gas Association, Cleveland.

DeWerth, D.W. 1962. A study of infra-red energy generated by radiant gas burners. *Research Bulletin* 92. American Gas Association, Cleveland.

Fanger, P.O. 1972. *Thermal comfort*. McGraw-Hill, New York.

Gagge, A.P., G.M. Rapp, and J.D. Hardy. 1967. The effective radiant field and operative temperature necessary for comfort with radiant heating. *ASHRAE Transactions* 73(1) and *ASHRAE Journal* 9:63-66.

Gas Research Board of London. 1944. The use of infra-red radiation in industry. *Information Circular* 1.

Haslam, W.G. et al. 1925. Radiation from non-luminous flames. *Industrial and Engineering Chemistry* (March).

Howell, R. and S. Suryanarayana. 1990. Sizing of radiant heating systems: Part II—Heated floors and infrared units. *ASHRAE Transactions* 96(1): 666-675.

Jones, B.W. and K.S. Chapman. 1994. Simplified method to factor mean radiant temperature (MRT) into building and HVAC design (RP-657). *Final Report*. ASHRAE.

NFPA. 1996. National fuel gas code. ANSI/NFPA 54-99. National Fire Protection Association, Quincy, MA. ANSI Z223.1-99. American Gas Association, Cleveland.

NFPA. 1998. National electrical code®. ANSI/NFPA *Standard* 70-98. National Fire Protection Association, Quincy, MA.

Plyler, E.K. 1948. Infrared radiation from bunsen flames. National Bureau of Standards *Journal of Research* 40(February):113.

Prince, F.J. 1962. Selection and application of overhead gas-fired infrared heating devices. *ASHRAE Journal* (October):62.

BIBLIOGRAPHY

Carlson, B.G. and K.D. Lathrop. 1963. *Transport theory—The method of discrete-ordinates in computing methods in reactor physics*. Grenspan, Keller, and Okrent, eds. Gordon and Breach, New York.

Chapman, K.S., J.M. DeGreef, and R.D. Watson. 1997. Thermal comfort analysis using BCAP for retrofitting a radiantly heated residence. *ASHRAE Transactions* 103(1):959-965.

Chapman, K.S., S. Ramadhyani, and R. Viskanta. 1992. Modeling and parametric studies of heat transfer in a direct-fired furnace with impinging jets. Presented at the 1992 ASME Winter Annual Meeting, Anaheim.

Chapman, K.S. and P. Zhang. 1996. Energy transfer simulation for radiantly heated and cooled enclosures. *ASHRAE Transactions* 102(1): 76-85.

DeGreef, J.M. and K.S. Chapman. 1998. Simplified thermal comfort evaluation of MRT gradients and power consumption predicted with the BCAP methodology. *ASHRAE Transactions* 104(1B):1090-1097.

Fanger, P. 1967. Calculation of thermal comfort: Introduction of a basic comfort equation. *ASHRAE Transactions* 73(2):III.4.1-20.

Fiveland, W.A. 1984. Discrete-ordinates solutions of the radiative transport equation for rectangular enclosures. *Journal of Heat Transfer* 106:699-706.

Fiveland, W.A. 1987. Discrete-ordinates methods for radiative heat transfer in isotropically and anisotropically scattering media. *Journal of Heat Transfer* 109:809-812.

Fiveland, W.A. 1988. Three dimensional radiative heat-transfer solutions by the discrete-ordinates method. *Journal of Thermophysics and Heat Transfer* 2(4):309-316.

Fiveland, W.A. and A.S. Jamaluddin. 1989. Three-dimensional spectral radiative heat transfer solutions by the discrete-ordinates method. *ASME Heat Transfer Conference Proceedings, Heat Transfer Phenomena in Radiation, Combustion, and Fires.* HTD-106:43-48. American Society of Mechanical Engineers, New York.

Gan, G. and D.J. Croome. 1994. Thermal comfort models based on field measurements. *ASHRAE Transactions* 100(1):782-794.

Incropera, F.P. and D.P. DeWitt. 1990. *Fundamentals of heat and mass transfer.* John Wiley & Sons, New York.

Jamaluddin, A.S. and P.J. Smith. 1988. Predicting radiative transfer in rectangular enclosures using the discrete ordinates method. *Combustion Science and Technology* 59:321-340.

Jones, B.W., W.F. Niedringhaus, and M.R. Imel. 1989. Field comparison of radiant and convective heating in vehicle repair buildings. *ASHRAE Transactions* 95(1):1045-1051.

Modest, M.F. 1993. *Radiative heat transfer.* McGraw-Hill, New York.

NAHB. 1994. Enerjoy case study: A comparative analysis of thermal comfort conditions and energy consumption for Enerjoy PeopleHeaters™ and a conventional heating system. *Project Report* 4159. National Association of Home Builders Research Center, Upper Marlboro, MD.

Özisik, M.N. 1977. *Basic heat transfer.* McGraw-Hill, New York.

Patankar, S.V. 1980. *Numerical heat transfer and fluid flow.* McGraw-Hill, New York.

Sanchez, A. and T.F. Smith. 1992. Surface radiation exchange for two-dimensional rectangular enclosures using the discrete-ordinates method. *Journal of Heat Transfer* 114:465-472.

Siegel, R. and J.R. Howell. 1981. *Thermal radiation heat transfer.* McGraw-Hill, New York.

Truelove, J.S. 1987. Discrete-ordinates solutions of the radiative transport equation. *Journal of Heat Transfer* 109:1048-1051.

Truelove, J.S. 1988. Three-dimensional radiation in absorbing-emitting media. *Journal of Quantitative Spectroscopy & Radiative Transfer* 39(1):27-31.

Viskanta, R. and M.P. Mengüc. 1987. Radiation heat transfer in combustion systems. *Progress in Energy and Combustion Science* 13.

Viskanta, R. and S. Ramadhyani. 1988. *Radiation heat transfer in directly-fired natural gas furnaces: A review of literature.* GRI Report GRI-88/0154. Gas Research Institute, Chicago.

Watson, R.D., K.S. Chapman, and J. DeGreef. 1998. Case study: Seven-system analysis of thermal comfort and energy use for a fast-acting radiant heating system. *ASHRAE Transactions* 104(1B):1106-1111.

Yücel, A. 1989. Radiative transfer in partitioned enclosures. *ASME Heat Transfer Conference Proceedings, Heat Transfer Phenomena in Radiation, Combustion, and Fires.* HTD-106:35-41. American Society of Mechanical Engineers, New York.

CHAPTER 16

ULTRAVIOLET LAMP SYSTEMS

USE of ultraviolet (UV) lamps and lamp systems to disinfect room air and air streams dates to about 1900; see Riley (1988) and Schechmeister (1991) for extensive reviews of UV disinfection. Early work established that the most effective UV wavelength range for inactivation of microorganisms was between 220 to 300 nm, with peak effectiveness near 265 nm.

UV energy is electromagnetic radiation with a wavelength shorter than that of visible light, but longer than soft x-rays. All UV ranges and bands are invisible to the human eye. The UV spectrum can be subdivided into following bands:

- UVA (long-wave; 400 to 315 nm): the most abundant in sunlight, responsible for skin tanning and wrinkles
- UVB (medium-wave; 315 to 280 nm): primarily responsible for skin reddening and skin cancer
- UVC (short-wave; 280 to 200 nm): the most effective wavelengths for germicidal control
- Far or vacuum UV (200 to 30 nm)

UVC energy disrupts the DNA of a wide range of microorganisms, rendering them harmless (Brickner 2003; CIE 2003). Figure 1 shows the DNA response to UV energy. Most, if not all commercial germicidal lamps are low-pressure mercury lamps that emit UV energy at 253.7 nm, very close to the optimal wavelength.

Ultraviolet germicidal irradiation (UVGI) has been used in air ducts for some time, and its use is becoming increasingly frequent as concern about indoor air quality increases. UVGI is being used as an engineering control to interrupt the transmission of pathogenic organisms, such as *Mycobacterium tuberculosis* (TB), influenza viruses, mold, and possible bioterrorism agents (Brickner 2003; CDC 2002, 2005; General Services Administration 2003).

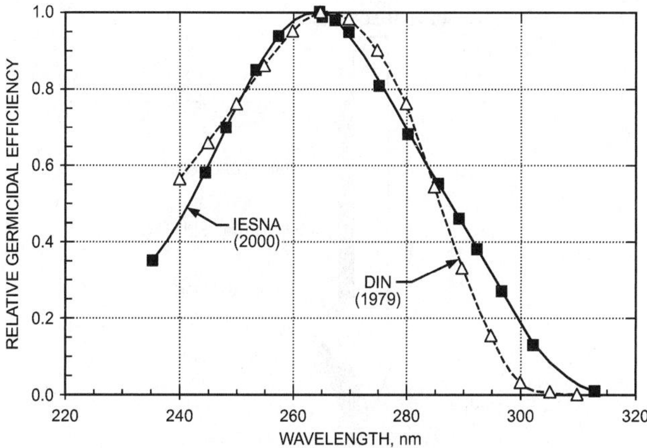

Fig. 1 Relative Germicidal Efficiency

This chapter includes a review of the fundamentals of UVC germicidal energy's impact on microorganisms; how UVC lamps generate germicidal radiant energy; common approaches to the application of UVGI systems for upper-air room, in-duct, and surface cleansing; and a review of human safety and maintenance issues.

TERMINOLOGY

Burn-in time. Period of time that UV lamps are powered on before being put into service, typically 100 h.

Droplet nuclei. Microscopic particles produced when a person coughs, sneezes, shouts, or sings. The particles can remain suspended for prolonged periods and can be carried on normal air currents in a room and beyond to adjacent spaces or areas receiving exhaust air.

Erythema (actinic). Reddening of the skin, with or without inflammation, caused by the actinic effect of solar radiation or artificial optical radiation. See CIE (1987) for details. (Nonactinic erythema can be caused by various chemical or physical agents.)

Exposure. Being subjected to something (e.g., infectious agents, irradiation, particulates, chemicals) that could have harmful effects. For example, a person exposed to *M. tuberculosis* does not necessarily become infected.

Exposure dose. Radiant exposure (J/m^2, unweighted) incident on biologically relevant surface.

Fluence. Radiant flux passing from all directions through a unit area in J/m^2 or J/cm^2; includes backscatter.

Germicidal radiation. Optical radiation able to kill pathogenic microorganisms.

Irradiance. Power of electromagnetic radiation incident on a surface per unit surface area, typically reported in microwatts per square centimeter ($\mu W/cm^2$). See CIE (1987) for details.

Mycobacterium tuberculosis. The namesake member of *M. tuberculosis* complex of microorganisms, and the most common cause of tuberculosis (TB) in humans. In some instances, the species name refers to the entire *M. tuberculosis* complex, which includes *M. bovis*, *M. africanum*, *M. microti*, *M. canettii*, *M. caprae*, and *M. pinnipedii*.

Optical radiation. Electromagnetic radiation at wavelengths between x-rays ($\lambda \approx 1$ nm) and radio waves ($\lambda \approx 1$ mm). See CIE (1987) for details.

Permissible exposure time (PET). Calculated time period that humans, with unprotected eyes and skin, can be exposed to a given level of UV irradiance without exceeding the NIOSH recommended exposure limit (REL) or ACGIH Threshold Limit Value® (TLV®) for UV radiation.

Personal protective equipment (PPE). Protective clothing, helmets, goggles, respirators, or other gear designed to protect the

The preparation of this chapter is assigned to TC 2.9, Ultraviolet Air and Surface Treatment.

wearer from injury from a given hazard, typically used for occupational safety and health purposes.

Photokeratitis. Defined by CIE (1993) as corneal inflammation after overexposure to ultraviolet radiation.

Photoconjunctivitis. Defined by CIE (1993) as a painful conjunctival inflammation that may occur after exposure of the eye to ultraviolet radiation.

Photokeratoconjunctivitis. Inflammation of cornea and conjunctiva after exposure to UV radiation. Wavelengths shorter than 320 nm are most effective in causing this condition. The peak of the action spectrum is approximately at 270 nm. See CIE (1993) for details. *Note:* Different action spectra have been published for photokeratitis and photoconjuctivitis (CIE 1993); however, the latest studies support the use of a single action spectrum for both ocular effects.

Threshold Limit Value® (TLV®). An exposure level under which most people can work consistently for 8 h a day, day after day, without adverse effects. Used by the ACGIH to designate degree of exposure to contaminants. TLVs can be expressed as approximate milligrams of particulate per cubic meter of air (mg/m^3). TLVs are listed either for 8 h as a time-weighted average (TWA) or for 15 min as a short-term exposure limit (STEL).

Ultraviolet radiation. Optical radiation with a wavelength shorter than that of visible radiation. [See CIE (1987) for details.] The range between 100 and 400 nm is commonly subdivided into

UVA 315 to 400 nm
UVB 280 to 315 nm
UVC 100 to 280 nm

Ultraviolet germicidal irradiation (UVGI). Use of ultraviolet radiation to kill or inactivate microorganisms. UVGI is generated by germicidal lamps that kill or inactivate microorganisms by emitting ultraviolet germicidal radiation, predominantly at a wavelength of 253.7 nm.

UV dose. Product of UV irradiance and exposure time on a given microorganism or surface, typically reported in millijoules per square centimeter (mJ/cm^2).

Wavelength. Distance between repeating units of a wave pattern, commonly designated by the Greek letter lambda (λ).

UVGI FUNDAMENTALS

Microbial Dose Response

Lamp manufacturers have published design guidance documents for in-duct use (Philips Lighting 1992; Sylvania 1982; Westinghouse 1982). Bahnfleth and Kowalski (2004) and Scheir and Fencl (1996) summarized the literature and discussed in-duct applications. These and other recent papers were based on case studies and previously published performance data. The Air-Conditioning and Refrigeration Technology Institute (ARTI) funded a research project to evaluate UV lamps' availability to inactivate microbial aerosols in ventilation equipment, using established bioaerosol control device performance measures (VanOsdell and Foarde 2002). The data indicated that UVC systems can be used to inactivate a substantial fraction of environmental bioaerosols in a single pass.

For constant and uniform irradiance, the disinfection effect of UVGI on a single microorganism population can be expressed as follows (Phillips Lighting 1992):

$$N_t/N_0 = \exp(-kE_{ff}\Delta t) = \exp(-k \times \text{Dose}) \quad (1)$$

where

N_0 = initial number of microorganisms
N_t = number of microorganisms after any time Δt
N_t/N_0 = fraction of microorganisms surviving
k = microorganism-dependent rate constant, cm^2/(μW·s)
E_{ff} = effective (germicidal) irradiance received by microorganism, μW/cm^2
Dose = $E_{ff} \times \Delta t$, (μW·s)/cm^2

The units shown are common, but others are used as well, including irradiance in W/m^2 and dose in J/m^2.

Equation (1) describes an exponential decay in the number of living organisms as a constant level of UVGI exposure continues. The same type of equation is used to describe the effect of disinfectants on a population of microorganisms, with the dose in that case being a concentration-time product. The fractional kill after time t is $(1 - N_t/N_0)$. In an air duct, the use of Equation (1) is complicated by the movement of the target microorganisms in the airstream and the fact that the UVGI irradiance is not constant within the duct. In addition, the physical parameters of the duct, duct airflow, and UV installation have the potential to affect both the irradiance and the microorganisms' response to it. As is the case with upper-room UV installation design, the design parameters for UVGI in in-duct applications are not simple because of some uncertainty in the data available to analyze them, and because of secondary effects.

A key difference between surface decontamination and airborne inactivation of organisms is exposure time. Residence time in in-duct devices is on the order of seconds or fractions of seconds. In a moving airstream, exposure time is limited by the effective distance in which the average irradiance was calculated; for instance, at 500 fpm, 1 ft of distance takes 0.12 s. Therefore, neutralization methods against an airborne threat must be effective in seconds or fractions of a second, depending on the device's characteristics, and high UV intensity is generally required. Conversely, when irradiating surfaces in an HVAC system, exposure time is often continuous, so much lower levels of UV intensity may be required.

Susceptibility of Microorganisms to UV Energy

Organisms differ in their susceptibility to UV inactivation; Figure 2 shows the general ranking of susceptibility by organism groups. Viruses are a separate case and are not included in Figure 2, because, as a group, their susceptibility to inactivation is even broader than bacteria or fungi. A few examples of familiar pathogenic organisms are included in each group for information (see Table 1). Note that it is impossible to list all of the organisms of interest in each group. Depending on the application, a public health or medical professional, microbiologist, or other individual with knowledge of the threat or organisms of concern should be consulted.

As shown in Figure 2, the vegetative bacteria are the most susceptible, followed by the *Mycobacteria*, bacterial spores, and, finally, fungal spores, which are the most resistant. Within each group, an individual species may be significantly more resistant or susceptible than others, so care should be taken, using this ranking only as a guideline. Note that spore-forming bacteria and fungi also

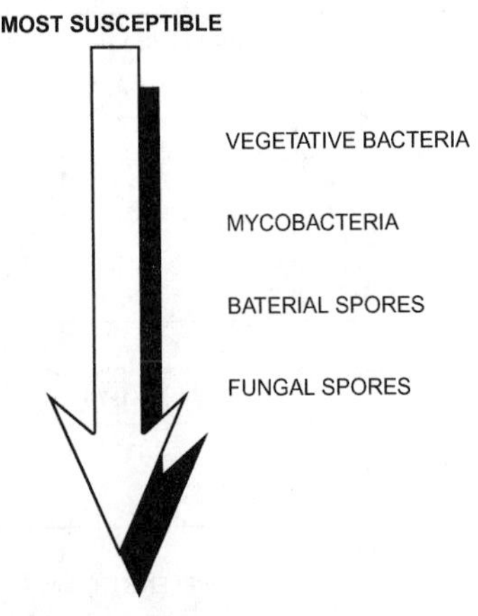

Fig. 2 General Ranking of Susceptibility to UVC Inactivation of Microorganisms by Group

Table 1 Representative Members of Organism Groups

Organism Group	Member of Group
Vegetative Bacteria	*Staphylococcus aureus*
	Streptococcus pyogenes
	Escherichia coli
	Pseudomonas aeruginosa
	Serratia marcescens
Mycobacteria	*Mycobacterium tuberculosis*
	Mycobacterium bovis
	Mycobacterium leprae
Bacterial Spores	*Bacillus anthracis*
	Bacillus cereus
	Bacillus subtilis
Fungal Spores	*Aspergillus versicolor*
	Penicillium chrysogenum
	Stachybotrys chartarum

have vegetative forms, which are markedly more susceptible to inactivation than the spore forms. Using Equation (1), it is clear that larger values of k represent more susceptible microorganisms and smaller values represent less susceptible ones. Units of k are the inverse of the units used for dose.

Using k values to design HVAC duct systems can be challenging. Values of k vary over several orders of magnitude, depending on organism susceptibility, and values reported in the literature for the same microorganism sometimes differ greatly. For example, Luckiesh (1946) reported a k for *Staphylococcus aureus* of 0.9602 m^2/J [0.009602 $cm^2/(\mu W \cdot s)$] and 0.00344 m^2/J for *Aspergillus amstelodami* spores. However, k values for *S. aureus* as small as 0.419 m^2/J were reported by Abshire and Dunton (1981). The wide variation for a single species is the result of a number of factors, the most important of which is differences in the conditions under which measurements were conducted (in air, in water, on plates). Especially for many of the vegetative organisms, the amount of protection offered by organic matter, humidity, and components of ambient air can significantly affect their susceptibility to UVGI (VanOsdell and Foarde (2002). Kowalski (2002) has an extensive compilation of published k values, and research to obtain more reliable design values is ongoing. Take care when using published values, and obtain the original papers to evaluate the relevance of the k value of any particular organism to a specific application.

LAMPS AND BALLASTS

Types of Germicidal Lamps

UV lamps are based on a low-pressure mercury discharge. These lamps contain mercury, which vaporizes when the lamp is lighted. The mercury atoms accelerate because of the electrical field in the discharge colliding with the noble gas, and reach an excited stage. The excited mercury atoms emit almost 85% of their energy at 253.7 nm wavelength. The remaining energy is emitted at various wavelengths in the UV region (mainly 185 nm); very little is emitted in the visible region.

UV lamps exist in different shapes, which are mostly based on general lighting fluorescent lamps:

- **Cylindrical** lamps may be any length or diameter. Like fluorescent lamps, most UV lamps have electrical connectors at both ends, but single-ended versions also exist. Typical diameters are 1.5 in. T12, 1.1 in. T8, 0.79 in. T6, and 0.63 in. T5.
- **Biaxial** lamps are essentially two cylindrical lamps that are interconnected at the outer end. These lamps have an electrical connector at only one end.

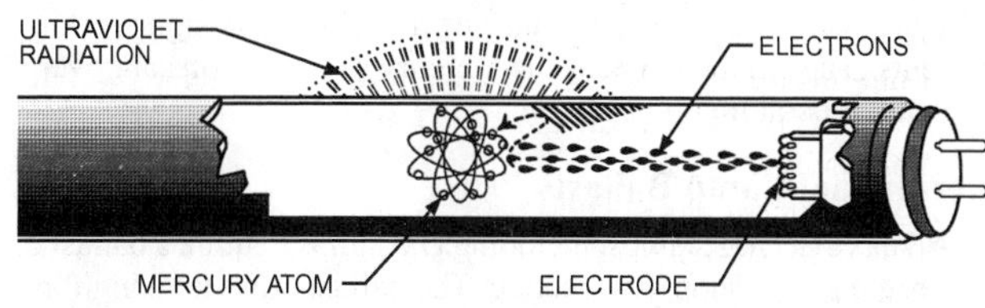

Fig. 3 Typical UVGI Lamp

- **U-tube** lamps are similar to biaxial lamps having the electrical connector at one end. They have a continuously curved bend at the outer end.

UV lamps can be grouped into the following three output types:

- **Standard-output** lamps operate typically at 425 mA.
- **High-output** lamps have hot cathode filaments sized to operate from 800 up to 1200 mA. Gas mixture and pressure are optimized to deliver a much higher UVC output while maintaining long lamp life, in the same lamp dimensions as standard-output lamps.
- **Amalgam** lamps have hot cathode filaments sized to operate at 1200 mA or higher. The gas mixture, pressure, and sometimes lamp diameter have been optimized for delivering an even higher UV without deteriorating lamp life.

As shown in Figure 3, UV lamps use electrodes between which the electrical discharge runs and are filled with a noble gas such as argon, neon, or a mix thereof. The outer envelope is made out of a UV transmitting material such as quartz or a special soda barium glass (sometimes called soft glass). A small amount of mercury is present in the envelope.

The electrodes are very important for the lamp behavior. There are two major types:

- A **cold-cathode** lamp usually contains a pair of cathodes parallel to one another. The cathodes are not heated in order to excite the electrons. A high voltage potential is needed to ionize the gas in the tube and to cause current flow in an ambient temperature. Cold-cathode lamps offer instant starting, and life is not affected by on/off cycles. Cold-cathode UV lamps provide less UVGI output than hot-cathode UV lamps, but consume less energy and last several thousand hours longer, thus requiring less costly maintenance.
- A **hot cathode** emits electrons through thermo-ionic emission. The electrode consists of an electrical filament coated with a special material (emittor) that lowers the emission potential. The electrodes are heated by current before starting the discharge and, once started, by the discharge current itself. Hot-cathode lamps typically allow much higher power densities than cold-cathode lamps, and thus generate much more UVGI intensity.

The outer envelope of UV lamps is made of UV-transmitting soft glass or quartz. Special wires are vacuum sealed into this envelope to allow transmission of electrical energy to the electrodes.

Soft glass can be used to produce UV lamps that emit 253.7 nm, but it is not suitable for producing the ozone wavelength of 185 nm. Fused quartz silica can be used to produce UV lamps with either only 253.7 nm output or with 185 nm/253.7 nm output by changing its transmission properties.

To maintain UV output over time, the inside of the glass/quartz tube can be coated with a special layer to slow down the decrease of UV transmission over time.

Mercury can be present in UV lamps as a pure metal or as an amalgam. The amount of mercury is always (slightly) overdosed because some mercury will be chemically bound during the life of the lamp. Depending on the application, the amount of mercury in the lamp can be less than 5 mg. An amalgam is used in lamps having a higher wall temperature because of their higher design working currents. The amalgam keeps the mercury pressure constant over a certain temperature range.

Germicidal Lamp Ballasts

All gas discharge lamps, including UV lamps, require a ballast or electronic power supply to operate. The ballast provides a high initial voltage to initiate the discharge, and then rapidly limits the lamp current to safely sustain the discharge. Most lamp manufacturers recommend one or more ballasts to operate their lamps, and the American National Standards Institute (ANSI) publishes recommended lamp input specifications for all ANSI type lamps. This information, together with operating conditions such as line voltage, number of switches, etc., allows users to select proper ballast. Ballasts are designed to optimally operate a unique lamp type; however, modern electronic ballasts often adequately operate more than one type of lamp.

It is strongly advised to use the recommended ballast for each lamp type because less than optimum conditions will affect the lamp's starting characteristics, light output, and operating life.

Circuit Type and Operating Mode. Ballasts for low-pressure mercury lamps are designed according to the following three primary lamp operation modes:

- In **preheat**, lamp electrodes are heated before beginning discharge. No auxiliary power is applied across the electrodes during operation.
- In **rapid start**, lamp electrodes are heated before and during operation. The ballast transformers have two special secondary windings to provide the proper low voltage to the electrodes. The advantages include smooth starting, long life, and dimming capabilities.
- **Instant-start** ballasts do not heat the electrodes before operation. Ballasts for instant-start lamps are designed to provide a relatively high starting voltage (compared to preheat and rapid-start lamps) to initiate discharge across the unheated electrodes. They are not recommended if frequent switching is needed.

Preheat mode is more efficient than rapid start, because separate power is not required to continuously heat the electrodes. If operated with glow starters, lamps tend to flicker during starting. Electronic ballasts with preheat offer smooth starting, long life, and good switching behavior.

Instant-start operation is more efficient than rapid start, but lamp life is shorter, especially when lamps are frequently switched on and off.

Energy Efficiency. UV lamps are very efficient at converting input power to UV output; nevertheless, much of the power supplied into a UV lamp-ballast system produces waste heat energy. There are three primary ways to improve efficiency of a UV lamp-ballast system:

- Reduce ballast losses
- Operate lamp(s) at a high frequency
- Reduce losses attributable to lamp electrodes

Newer, more energy-efficient ballasts, both magnetic and electronic, use one or more of these techniques to improve lamp-ballast system efficacy, measured in lumens per watt.

Losses in magnetic ballasts have been reduced by using higher-grade magnetic components. Some rapid-start magnetic ballasts improve efficacy by removing power to the lamp electrodes after starting. Using a single ballast to drive three or four lamps, instead of only one or two, may also reduce ballast losses.

Electronic ballasts operate lamps at high frequency (typically more than 20 kHz), allowing the lamps to convert power to UV more efficiently than if operated by electromagnetic ballasts. For example, lamps operated on electronic ballasts can produce over 10% more UV than if operated on electromagnetic ballasts at the same power levels.

Ballast Factor. The ballast factor is a measure of the actual output for a specific lamp/ballast system relative to the rated output measured with reference ballast under ANSI test conditions (open air at 77°F). Ballast factor is not a measure of energy efficiency. Although a lower ballast lumen factor reduces lamp lumen output, it also consumes proportionally less input power. A low ballast lumen factor may drastically reduce lamp life, because the lamp electrodes run too cold.

For new equipment, high ballast factors are generally the best choice because fewer lamps and ballasts are needed to reach the system's required UV output.

Audible Noise. Iron-cored electromagnetic ballasts operating at 60 Hz generate audible noise, because of electromagnetic action in the core and coil assembly of the ballast. Noise can increase at high temperatures, and it can be amplified by some luminaries' designs. The best ballasts use high-quality materials and construction to reduce noise. Noise is rated A, B, C, or D, in decreasing order of preference: an A-rated ballast hums softly; a D-rated ballast makes a loud buzz. Virtually all energy-efficient magnetic ballasts for G36T5L and G30T8 lamps are A-rated, with a few exceptions (e.g., low-temperature ballasts).

Because electronic high-frequency ballasts have smaller magnetic components, they typically have a lower sound rating and should not emit perceptible hum. All electronic ballasts are A-rated for sound.

EMI/RFI. Because they operate at high frequency, electronic ballasts may produce electromagnetic interference (EMI), which can affect any operating frequency, or radiofrequency interference (RFI), which applies only to radio and television frequencies. This interference could affect the operation of sensitive electrical equipment, such as system controls, televisions, or medical equipment. Good-quality electronic ballasts should incorporate features necessary to maximize protection for the operating environment and to operate well within regulatory limits.

Inrush Current. All electrical devices, including ballasts, have an initial current surge that is greater than their steady-state operating current. National Electrical Manufacturers Association (NEMA) *Standard* 410 covers worst-case ballast inrush currents. All circuit breakers and light switches are designed for inrush currents. The electrical system should be designed with this issue in mind.

Total Harmonic Distortion (THD). Harmonic distortion occurs when the wave-shape of current or voltage varies from a pure sine wave. Except for a simple resistor, all electronic devices, including electromagnetic and electronic ballasts, contribute to power line distortion. For ballasts, THD is generally considered the percent of harmonic current the ballast adds to the power distribution system. The ANSI standard for electronic ballasts specifies a maximum THD of 32%. However, most electric utilities now require that the THD of electronic ballasts be 20% or less.

Dimming. Unlike incandescent lamps, UV lamps cannot be properly dimmed with a simple wall box device. For a UV lamp to be dimmed over a full range without reducing lamp life, its electrode temperature must be maintained while the lamp arc current is reduced.

Dimming ballasts are available in both magnetic and electronic versions. **Magnetic dimming ballasts** require control gear containing expensive, high-power switching devices that condition the input power delivered to the ballasts. This is economically viable only when controlling large numbers of ballasts on the same branch circuit.

Electronic dimming ballasts alter the output power to the lamps within the ballast itself, driven by a low-voltage signal into the driver circuit. This allows control of one or more ballasts independent of the electrical distribution system. With dimming electronic ballast systems, a low-voltage control network can be used to group ballasts into arbitrarily sized control zones. Dimming range differs greatly; most electronic dimming ballasts can vary output levels between 100% and about 10% of full output, but ballasts are also available that operate lamps down to 1% of full output.

Germicidal Lamp Cooling and Heating Effects

Output of UV lamps is critically dependent on mercury vapor pressure within the lamp envelope. The mercury vapor pressure is controlled by the temperature of the cold spot (the coldest portion of the UV lamp during lamp operation), as shown in Figure 4. If the mercury vapor pressure is too low, UV output is low because there are not enough mercury atoms to generate UV radiation. Too high a mercury vapor pressure also decreases UV output, because the excess evaporated mercury absorbs ultraviolet rays generated in the UVGI lamp.

In low-pressure mercury lamps, mercury vapor pressure reaches its optimal level in still air at 77°F. Depending on lamp type, the cold-spot temperature must be between 103 and 122°F to reach maximum UV output. In moving air, the cold-spot temperature of standard lamps is too low to reach the necessary UV output (Figure 5). Special windchill-corrected lamps are designed to make the lamps function optimally in moving air.

By introducing mercury into the lamp in the form of amalgams, the cold-spot temperature can be increased to between 158 and 248°F, making it possible to reach optimum UV output at higher temperatures. Amalgams reduce mercury vapor pressure relative to that of pure mercury at any given temperature, and also provide a broadened peak in UV output versus temperature (see curve), so that near-optimum light output is obtained over an extended range of ambient temperatures.

Germicidal Lamp Aging

Output of UV lamps decreases over time. UV lamps are rated in effective hours of UV emission, and not in end of electrical life hours. Many UV lamps are designed to emit intensity levels at the end of their useful life that are 50 to 85% or more of that measured at initial operation (after 100 h), although current models continue to emit blue light long after they have passed their useful life. Lamp manufacturers' specification data can verify depreciation over useful life. UVGI systems should be designed for the output at the end of effective life.

Cold-cathode UV lamps can have useful life rating of approximately 20,000 h. Hot-cathode UV lamps have a wide range of useful life hours, depending on the type of glass envelope used, any protective internal glass wall coatings, filament current load design, gas pressure, and gas mixture. Consequently, hot-cathode UV lamps can have useful lives ranging from 6000 to 13,000 h.

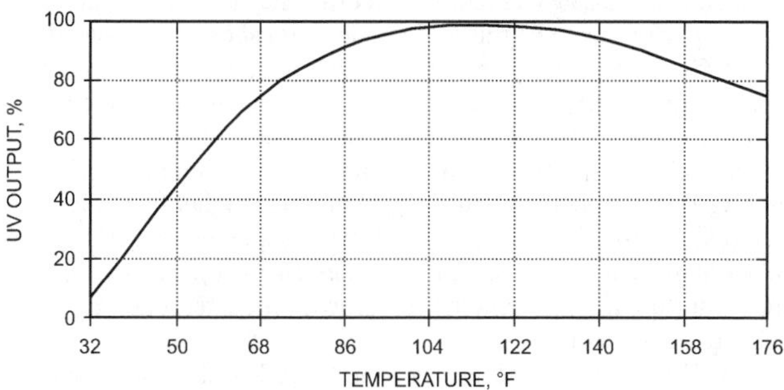

Fig. 4 Example of Lamp Efficiency as Function of Cold-Spot Temperature

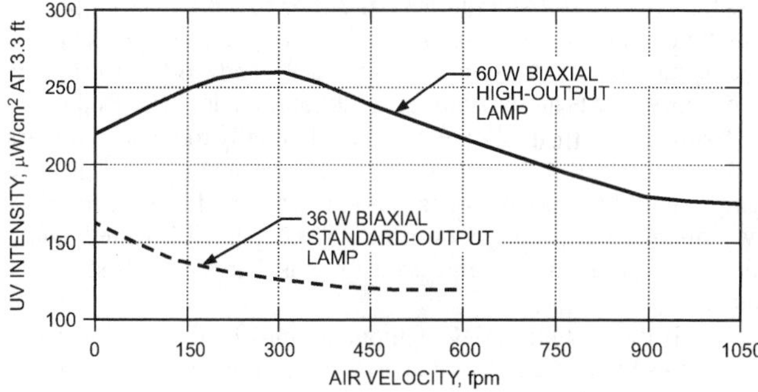

Fig. 5 Windchill Effect on UVC Lamp Efficiency

UVGI Lamp Irradiance

UVGI lamp intensity is measured in μW/cm². Often, manufacturers obtain their lamp intensity measurements by taking a UV intensity reading 3.3 ft from the center of a UVGI lamp, in an open-air ambient of approximately 75°F and with approximately zero air velocity. Test standard protocols for measuring UVC lamps' ability to inactivate airborne and surface microorganisms are currently being developed by ASHRAE Standard Project Committee SPC 185.

The irradiance E on a small surface in point P on a distance a from a linear UV lamp length AB = L can be calculated with the following equation if the UV output of the lamp is represented by ϕ (Figure 6):

$$E = \frac{\phi}{2\pi^2 La}(2\alpha + \sin 2\alpha) \tag{2}$$

At shorter distances ($a < 0.5l$), the irradiance is inversely proportional to the distance of the measurement point from the lamp, as can be seen from the following simplified equation.

$$E = \frac{\phi}{2\pi La} \tag{3}$$

APPLICATION

Ultraviolet Fixture Configurations

Upper Air Fixtures. Wall-mounted fixtures with louvers are designed to keep ultraviolet rays above eye and head level. Specially designed louvers keep the rays from bouncing off the ceiling and in a vertical path above 7 ft. These fixtures reduce airborne microorganisms as normal air convection moves them into the path of the ultraviolet rays, where they are inactivated. Some fixtures also use small fans to help circulate the air by the UV fixture.

Ceiling-suspended fixtures are designed in the same way as wall mounted fixtures, incorporating louvers and/or UV traps to keep the rays out of eye and head level.

In-Duct Units. In-duct UVGI systems are designed to mount inside an air-conditioning duct; the UV energy is confined to the inside of the duct. They are normally installed with a safety interlock,

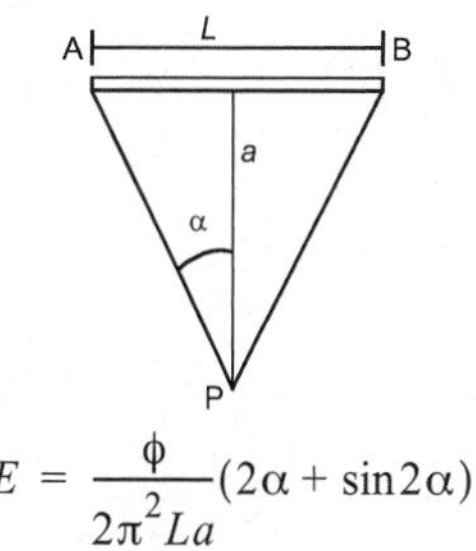

$$E = \frac{\phi}{2\pi^2 La}(2\alpha + \sin 2\alpha)$$

Fig. 6 Diagram of Irradiance Calculation

so if the fixture is removed from the duct or an access door is opened, the lamps turn off to avoid accidental human exposure to UV energy.

In-duct UVGI systems can be used to irradiate cooling coils, drain pans, and other HVAC components, or to disinfect moving air.

In-Duct Airstream Disinfection

In-duct UVGI systems disinfect an airstream in a building or room ventilation system. These systems are designed to treat the airflow and use available space within the duct. In-duct systems are generally engineered to achieve a required level of air disinfection and are often unique to each installation.

Numerous variables must be factored in to properly size and apply a UVGI system for airstream disinfection, including the following:

- Duct height and width
- Duct length where airstream is exposed to UV
- Air velocity
- Air temperature
- Lamp cooling effect of temperature and air velocity
- Lamp fouling (decreases the UV delivered)
- Biocontaminants and their k values (sensitivity to UV)
- Disinfection performance required
- Lamp age
- Type of power supply driving the UV lamp
- Reflectivity of duct material or duct lining
- Location of lamps with respect to duct
- Humidity

Outside makeup air cannot be treated if it is brought in downstream from where the UV fixtures are mounted, unless the makeup air is treated separately. Mounting the UVGI fixtures in the supply plenum ensures that both return and makeup air are treated.

UV lamps may be installed in different orientations, including the following:

- Perpendicular to airflow
- Parallel to airflow, with lamp(s) radiating outwardly
- Parallel to airflow, with lamp(s) radiating inwardly

In any UV installation, performance can be improved by increasing the UVC reflectivity of the duct walls. Note that air ducts may have different reflective properties, but typical galvanized duct material has a UVC reflectivity of about 57%. Aluminum and other more UVC-reflective materials may be used as lines to improve reflectivity. Most manufacturers have this information, and it should be included in the dosage calculations. Table 2 provides typical approximations, but actual material should be tested.

Table 2 Material Reflectivity

Material	Reflectance, %
Aluminum, etched	88
Aluminum foil	73
Aluminum paint	40 to 75
Chromium	45
Galvanized	57
Glass	4
Magnesium oxide	75
Nickel	38
Silver	22
Stainless steel	20 to 30
Tin-plated steel	28
Typical duct liner	0 to 1
White cotton	30
White oil paints	5 to 10
White paper	25
White porcelain enamel	5
White wall plaster	40 to 60
White water paint	10 to 35

The number and location of UVGI fixtures are dictated by the average percentage of reduction desired of the targeted biocontaminants in the duct, taking into account all of the variables mentioned previously.

Maintenance procedures should be as specified by the UVGI equipment manufacturer.

Air Handler Component Surface Disinfection

Applications of UV include uses in hospitals, schools, office buildings, food, pharmaceutical and commercial buildings, and even homes (Bernstein et al. 2006; Blatt et al. 2006). Since 2000, the General Services Administration (GSA) has required UV to be applied to the air-conditioning coils and drain pans of all HVAC systems in every GSA-funded new construction project (Department of General Services; GSA 2003).

As HVAC equipment ages, its performance can degrade, and so may the quality of air it delivers to occupied spaces (Kowalski 2006). Cooling coils can act as filters to collect and retain a substantial amount of particulates, including microbes (Siegel et al. 2002). These materials are quite small, so this occurs even in a system with reasonable or good filtration. Between 30 and 100% rh, damp coil and drain pan conditions are excellent forums for the growth of bacteria and mold (Levetin et al. 2001). Coil fouling also increases coil pressure drop and reduces airflow, reducing heat transfer from coil fins to lessen the amount of work a system can perform (Montgomery and Baker 2006) and reducing indoor environmental quality (IEQ). It can contribute to sick building syndrome and building-related illnesses ranging from mild irritations to the spread of infectious agents (Menzies et al. 2003). The decaying accumulation is often a source of odor, as well (Kowalski 2006).

Periodically cleaning the coils and drain pan is recommended by system manufacturers, though it is not always practical. Traditional coil cleaning, such as steam cleaning and chemical pressure washing, can restore system performance, improve air quality, and reduce energy consumption (Montgomery and Baker 2006). However, chemical and mechanical cleaning can be dangerous, costly, and difficult. Chemicals can contribute to air quality problems, and mechanical cleaning can reduce coil efficiency and life. In addition, system performance begins to degrade again shortly after cleaning.

UVGI can be readily applied to HVAC systems to help maintain system cleanliness (Blatt et al. 2006). It is used to complement system maintenance by keeping coils, drain pans, and other surfaces clean and free of microbial contamination. Stationary surfaces receive UVC doses many orders of magnitude higher than microbes in moving air do, making it relatively easy, using lower levels of UV, to maintain heat exchange efficiency, design airflow, and to improve indoor air quality by reducing the growth of bacteria and mold on system components.

UVGI reduces microbial levels on HVAC surfaces and often in the air (RLW Analytics 2006). Coil pressure drop is reduced and, therefore, airflow is restored (Witham 2005). Because heat transfer also is restored, this combination can result in energy savings (Levetin et al. 2001), which can be significant, with payback of possibly less than two years (Montgomery and Baker 2006). In addition, the associated improvements in air quality may reduce respiratory distress symptoms and thus improve attendance and work performance in occupied spaces (Bernstein et al. 2006; Menzies et al. 2003).

Installation. Coils should be cleaned initially to reduce biomass and to accelerate systemwide cleaning and energy savings. UV lamps should be mounted near cooling coils and spaced to allow even distribution of energy over the surface to be disinfected. Qualified UV equipment manufacturers or consultants can assist in system design.

UVGI fixtures for HVAC equipment must be designed to withstand moisture and condensate (from the coil or caused by reduced operating temperatures) and to operate properly over the full range of system operating temperatures. Care must be taken at the installation

Table 3 Advantages and Disadvantages of UVC Fixture Location Relative to Coil

Location	Advantages	Disadvantages
Downstream	More space to install fixtures. Allows fixtures to better irradiate surface where condensation is highest. Allows fixtures to irradiate generally most contaminated part of coil and drain pan.	Lamp and fixture must be rated for damp location. Lamp cooling effects may reduce UV output, or require windchill correction or more lamps and fixtures for a given result.
Upstream	Lamp and fixture may be subjected to less moisture. May be the only location to apply fixtures. Fewer lamps and fixtures may be needed than on downstream side.	May not allow enough space to install fixtures. May initially take longer to clean coil and may not disinfect drain pan.

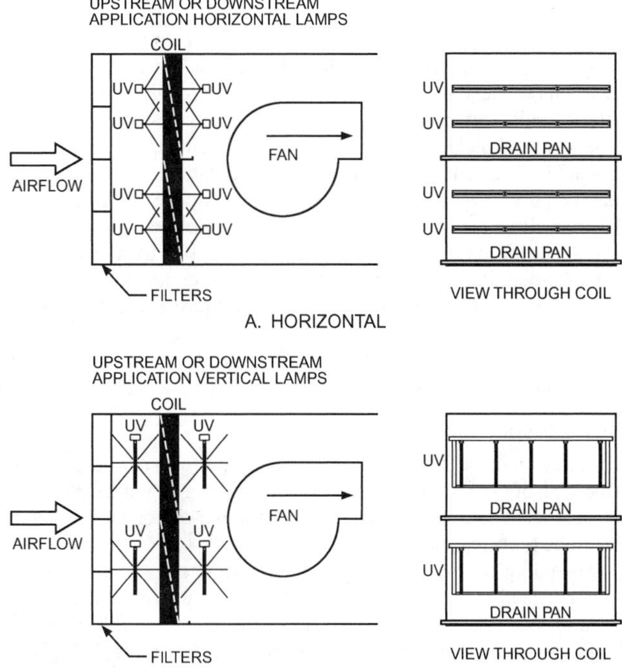

Fig. 7 UV Lamps Upstream or Downstream of Coil and Drain Pan

site to ensure that electrical interlocks are included to deenergize the UV system when it is accessed. UV systems should operate continuously to maximize UV's benefits and to improve lamp life, and to counteract mold and bacteria growth that occurs when an HVAC system is not operating.

UVGI systems can be installed upstream or downstream of the cooling coil (Figure 7). Both locations have advantages and disadvantages, as shown in Table 3. Figure 8 shows an actual installation at a coil.

Upper-Air UVGI Systems

Upper-air irradiation systems are designed to irradiate only air in the upper part of the room. Their narrow, focused beam is placed parallel to the plane of the ceiling and prevents stray ultraviolet rays from impinging on occupants below. Upper-air systems rely on air convection and mixing to move air from the lower to the upper portion of the room, where it can be irradiated and airborne microorganisms inactivated (Kethley and Branch 1972). Many fixtures

Fig. 8 Horizontal Lamp Placement for Coil Surface Disinfection

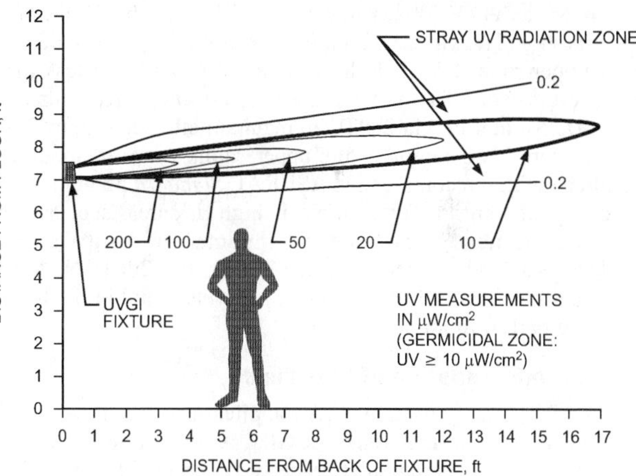

Fig. 9 Typical Elevation View

incorporate a safety switch that breaks the circuit when fixtures are opened for servicing, and should contain baffles or louvers appropriately positioned to direct UV irradiation to the upper air space. Baffles and louvers must never be bent or deformed.

Upper-room UVGI fixtures typically use low-pressure UVC lamps in tubular and compact shapes, and require a variety of electrical wattages. Beyond lamp size, shape, and ballast, fixtures are designed to be open or restricted in distribution, depending on the physical space to be treated.

Ceiling heights above 10 ft allow more for more open fixtures, which are more efficient. For occupied spaces with lower ceilings (less than 10 ft), various louvered upper-room UVGI fixtures (wall, pendant, and corner) are available to be mounted in combinations at least 7 ft from the floor to the bottom of the fixture. Figure 9 shows some typical elevations and corresponding UV levels, and Figure 10 illustrates distribution in a room.

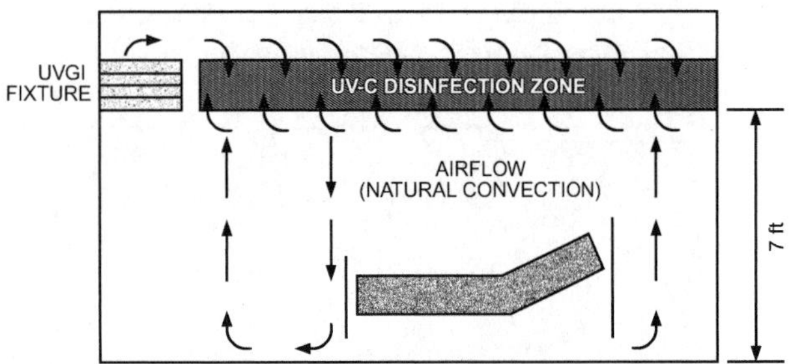

Fig. 10 Room Distribution

Application guidance with placement criteria for UV equipment is provided by Boyce (2003), CDC (2005), CIE (2003), Coker (2001), First (1999), and IESNA (2000). Additionally, manufacturer-specific advice on product operations should be followed.

A basic criterion for upper-room installations has been one 30 W (electrical input) fixture for every 18.6 m^2 (200 ft^2) (Riley et al. 1976). A UVGI installation that produces a maintained, uniform distribution of UV averaging between 30 and 50 μW/cm^2 should be effective in inactivating most airborne droplet nuclei containing mycobacteria, and is presumably effective against viruses, as well (First et al. 2007a, 2007b; Miller et al. 2002; Xu and Peccia 2003). Beyond UVC emission strength, effectiveness of upper-air UVGI is related to air mixing, relative humidity, and the inherent characteristics of the pathogenic organisms being addressed (Ka et al. 2004; Ko et al. 2000; Rudnick 2007). Effectiveness improves greatly with well-mixed air (First et al. 2007a, 2007b; Miller et al. 2002; Riley et al. 1971a, 1971b). Ventilation systems should maximize air mixing to receive the greatest benefit from upper-room UVGI. Relative humidity should be less than 60%; levels over 80% rh may reduce effectiveness (Kujundzic et al. 2007; Xu and Peccia 2003). To maintain efficient output from low-pressure UVC lamps in the upper room, room temperature should be within recommended ASHRAE *Standard* 55 guidelines for occupant comfort. For example, in high-risk areas such as corridors of infectious disease wards, a minimum of 0.4 μW/cm^2 at eye level is a good engineering guide (Coker et al. 2001). No long-term health effects of UVC exposure at levels found in the lower occupied part of rooms are known.

UV Photodegradation of Materials

The UVC energy used in HVAC applications can be very detrimental to organic materials (ACGIH 1999; Bolton 2001). As such, if the UV is not applied properly and vulnerable materials are not shielded or substituted, substantial degradation can occur (NEHC 1992). Easily degraded materials include synthetic filter media, gaskets, rubber, motor windings, electrical insulation, internal insulation, and plastic piping; many of these can degrade in days under sufficient irradiation. This degradation can result in decreased filtration efficiency, defective seals, and damaged system components, causing a loss in system performance and/or potential safety concerns. As a simple, practical approach, it is wise to shield all organic material components within about 5 ft of the UV lamp.

If materials cannot be shielded, then the UVC dose that will cause failure of the part must be determined. To ensure that materials can withstand UVC exposure over the life of the product, measure the level of UVC incident upon the part. If, over its lifetime, the part will receive a UVC dosage greater than the failure level, a more resistant material must be substituted. Materials that are located on the opposite side of the coil from the UV lamps are generally safer, because the level of UV penetrating a cooling coil is low.

Inorganic components such as metal and glass are not affected by UV. Many components that contain a lot of black coloring (lampblack), such as drive belts, are quite resistant to UV and are generally acceptable to use if UV levels are reasonable.

MAINTENANCE

Lamp Replacement

UV lamps should be replaced at the end of their useful life, based on recommendations of the equipment manufacturer. It may be prudent to simply change lamps annually (8760 h when lamps are run continuously) to ensure that adequate UV energy is supplied. Lamps can operate long after their useful life, but at greatly reduced performance. The typical rated life of UV lamps is in the range of 6000 to 10,000 h of operation. Switching lamps on and off too often may lead to early lamp failure, depending on the ballast type used. Consult the lamp manufacturer for specific information on expected lamp life and effects of switching.

Lamp Disposal

UV lamps should be treated the same as other mercury-containing devices, such as fluorescent bulbs. Most lamps must be treated as hazardous waste and cannot be discarded with regular waste. Low-mercury bulbs often can be discarded as regular waste; however, some state and local jurisdictions classify these lamps as hazardous waste. The U.S. EPA's universal waste regulations allow users to treat mercury lamps as regular waste for the purpose of transporting to a recycling facility (EPA 2008). This simplified process was developed to promote recycling. The National Electrical Manufacturers Association (NEMA) maintains a list of companies claiming to recycle or handle used mercury lamps at http://www.lamprecycle.org.

The most stringent of local, state, or federal regulations for disposal should be followed.

Visual Inspection

Maintenance personnel should routinely perform periodic visual inspection of the UV lamp assembly. Typically, a viewing port or an access door window is sufficient. Any burned-out or failing lamp should be replaced immediately.

Depending on the application and environment, a maintenance plan may need to include direct physical inspection of the fixture. If the lamp has become dirty because of inadequate prefiltration, it should be cleaned with a lint-free cloth and commercial glass cleaner or alcohol.

Future UVGI systems may include a feedback component to alert maintenance personnel to UVC lamp output decline.

SAFETY

Hazards of Ultraviolet Radiation to Humans

UVC is a low-penetrating form of UV compared to UVA or UVB. Measurements of human tissue show that 4 to 7% of UVC (along with a wide range of wavelengths, 250 to 400 nm) is reflected (Diffey 1983) and absorbed in the first 2 μm of the stratum cornea (outer dead layer of human skin), thus minimizing the amount of UVC transmitted through the epidermis (Bruls 1984).

Although UV is more energetic than the visible portion of the electromagnetic spectrum, UV is invisible to humans. Therefore, exposure to ultraviolet energy may result in ocular damage, which may initially go unnoticed.

Ocular damage generally begins with **photokeratitis** (inflammation of the cornea), but can also result in **keratoconjunctivitis** [inflammation of the conjunctiva (ocular lining)]. Symptoms, which may not be evident until several hours after exposure, may include an abrupt sensation of sand in the eyes, tearing, and eye pain, possibly severe. These symptoms usually appear within 6 to 12 h after UV exposure, and resolve within 24 to 48 h.

Cutaneous damage consists of erythema, a reddening of the skin. It is like sunburn with no tanning. The maximum effect of erythema occurs at a wavelength of 296.7 nm in the UVB band. UVC radiation at a wavelength of 253.7 nm is less effective, but is still a skin hazard.

Acute **overexposure** to UVC band radiation is incapacitating, but generally regresses after several days, leaving no permanent damage.

Sources of UV Exposure

UVC energy does not normally penetrate through solid substance, and is attenuated by most materials. Quartz glass and TFPE plastic have high transmissions for UVC radiation.

UVC energy can reflect from polished metals and several types of painted and nonpainted surfaces; however, a surface's ability to reflect visible light cannot be used to indicate its UV reflectance. The fact that a blue glow can be observed on the metal surface from an operating low-pressure UV fixture lamp could indicate the presence of UV, and a measurement should be performed to ensure there is no exposure risk. The lack of reflected blue light clearly indicates the absence of UV energy.

Well-designed and commissioned UVGI installations, education of maintenance personnel, signage, and safety switches can avoid overexposure. During commissioning and before operation of the UVGI installation, hand-held radiometers with sensors tuned to the read the specific 254 nm wavelength should be used to measure stray UVC energy (primarily in upper-air systems).

Exposure Limits

In 1972, the Centers for Disease Control and Prevention (CDC) and National Institute for Occupational Safety and Health (NIOSH) published a **recommended exposure limit (REL)** for occupational exposure to UV radiation. REL is intended to protect workers from the acute effects of UV exposure, although photosensitive persons and those exposed concomitantly to photoactive chemicals might not be protected by the recommended standard.

Table 4 lists some permissible exposure times for different levels of UVC irradiance. Exposures exceeding CDC/NIOSH REL levels require use of personal protective equipment (PPE), which consists of eyewear and clothing known to be nontransparent to UVC penetration and which covers exposed eyes and skin.

UV inspection, maintenance, and repair workers typically do not remain in one location during the course of their workday, and therefore are not exposed to UV irradiance levels for 8 h. Threshold Limit Value® (TLV®) consideration should be based on occupancy use of spaces treated by UVGI (ACGIH 2007).

Some plants do not tolerate prolonged UVC exposure and should not be hung in the upper room.

At 253.7 nm, the CDC/NIOSH REL is 6 mJ/cm^2 (6000 μJ/cm^2) for a daily 8 h work shift. ACGIH's (2007) TLV for UV radiation is identical to the REL for this spectral region. Permissible exposure times (PET) can be calculated for various irradiance levels using the following equation:

$$\text{PET, s} = \frac{\text{REL of 6000 } \mu\text{J/cm}^2 \text{ at 254 nm}}{\text{Measured irradiance level at 254 nm in } \mu\text{W/cm}^2} \qquad (4)$$

UV Radiation Measurements

UV levels can be measured with a UV radiometer directly facing the device at eye height at various locations in a room, and must be taken in the same location each time. If the readings indicate a dosage exceeding 6 mJ/cm^2, the UV systems must be deactivated until adjustments can be made or the manufacturer can be contacted. UV radiation measurements should be taken

- At initial installation
- Whenever new tubes are installed (newer tube designs may have increased irradiance)
- Whenever modifications are made to the UVGI system or room (e.g., adjustment of fixture height, location or position of louvers, addition of UV-absorbing or -reflecting materials, room dimension changes, modular partition height changes)

Table 4 Permissible Exposure Times for Given Effective Irradiance Levels of UVC Energy at 253.7 nm

Permissible Exposure Time*	Effective Irradiance, μW/cm^2
24 h	0.07
18 h	0.09
12 h	0.14
10 h	0.17
8 h	0.2
4 h	0.4
2 h	0.8
1 h	1.7
30 min	3.3
15 min	6.7
10 min	10
5 min	20
1 min	100
30 s	200
15 s	400
5 s	1200
1 s	6000

Source: ACGIH (2007).

Safety Design Guidance

In-duct systems should be fully enclosed to prevent leakage of UV radiation to unprotected persons or materials outside of the HVAC equipment.

All access panels or doors to the lamp chamber and panels or doors to adjacent chambers where UV radiation may penetrate or be reflected should have warning labels in appropriate languages. Labels should be on the outside of each panel or door, in a prominent location visible to people accessing the system.

Lamp chambers should have electrical disconnect devices. Positive disconnection devices are preferred over switches. Disconnection devices must be able to be locked or tagged out, and should be located outside the lamp chamber, next to the chamber's primary access panel or door. Switches should be wired in series so that opening any access deenergizes the system. On/off switches for UV lamps must not be located in the same location as general room lighting; instead, they must be in a location that only authorized persons can access, and should be locked to ensure that they are not accidentally turned on or off.

The lamp chamber should have one or more viewports of UVC-absorbing materials. Viewports should be sized and located to allow an operating UV system to be viewed from outside of the HVAC equipment.

Upper-air systems should have on/off switches and an electrical disconnect device on the louvers. If UV radiation measurements at the time of initial installation exceed the recommended exposure limit, all highly UV-reflecting materials should be removed, replaced, or covered. UV-absorbing paints containing titanium oxide can be used on ceilings and walls to minimize reflectance in the occupied space.

Warning labels must be posted on all upper-air UV fixtures to alert personnel of potential eye and skin hazards. Damaged or illegible labels must be replaced as a high priority. Warning labels must contain the following information:

- Wall sign for upper-air UVGI
  Caution: Ultraviolet energy. Switch off lamps before entering upper room.

- General warning posted near UVGI lamps
  Caution: Ultraviolet energy. Protect eyes and skin.
- Warning posted on the door of air handlers where UVGI is present in ductwork
  Caution: Ultraviolet energy in duct. Do not switch off safety button or activate lamps with door open.

Personnel Safety Training

Workers should be provided with as much training as necessary, including health and safety training, and some degree of training in handling lamps and materials. Workers should be made aware of hazards in the work area and trained in precautions to protect themselves. Training topics include

- UV exposure hazards
- Electrical safety
- Lock-out/tag-out
- Health hazards of mercury
- Rotating machinery
- Slippery condensate pans
- Sharp unfinished edges
- Confined-space entry (if applicable)
- Emergency procedures

Workers expected to clean up broken lamps should be trained in proper protection, cleanup, and disposal.

No personnel should be subject to direct UV exposure, but if exposure is unavoidable, personnel should wear protective clothing (no exposed skin), protective eyewear, and gloves. Most eyewear, including prescription glasses, are sufficient to protect eyes from UV, but not all offer complete coverage; standard-issue protective goggles may be the best alternative.

If individual lamp operating condition must be observed, this should preferably be done using the viewing window(s).

Access to lamps should only be allowed when lamps are deenergized. The lamps should be turned off before air-handling unit (AHU) or fan shutdown to allow the lamps to cool and to purge any ozone in the lamp chamber (if ozone-producing lamps are used). If AHUs or fans are deenergized first, the lamp chamber should be opened and allowed to ventilate for several minutes. Workers should always wear protective eyewear and puncture-resistant gloves for protection in case a lamp breaks.

Access to the lamp chamber should follow a site-specific lock-out/tag-out procedure. Do not rely on panel and door safety switches as the sole method to ensure lamp deenergizing. Doors may be inadvertently closed or switches may be inadvertently contacted, resulting in unexpected lamp activation.

If workers will enter the condensate area of equipment, the condensate pan should be drained and any residual water removed.

In general, avoid performing readings with the fan running and workers inside an AHU (e.g., to test for output reduction caused by air cooling). Tests of this nature should be instrumented and monitored from outside the equipment.

During maintenance, renovation, or repair work in rooms where upper-air UV systems are present, all UVGI systems must be deactivated before personnel enter the upper part of the room.

Lamp Breakage

If a lamp breaks, all workers must exit the HVAC equipment. Panels or doors should be left open and any additional lamp chamber access points should also be opened. Do not turn air-handling unit fans back on. After a period of 15 minutes, workers may reenter the HVAC equipment to begin bulb clean-up.

If a lamp breaks in a worker's hand, the worker should not exit the HVAC equipment with the broken bulb. Carefully set the broken bulb down, then exit the equipment. When possible, try not to set the broken lamp in any standing condensate water. Follow standard ventilation and reentry procedures.

Cleanup requires special care because of mercury drop proliferation, and should be performed by trained workers. As a minimum, workers should wear cut-resistant gloves, as well as safety glasses to protect eyes from glass fragments. Large bulb pieces should be carefully picked up and placed in an impervious bag. HEPA-vacuum the remaining particles, or use other means to avoid dust generation.

UNIT CONVERSIONS

Just as it is customary to express the size of aerosols in micrometres and electrical equipment's power consumption in watts, regardless of the prevailing unit system, it is also customary to express total lamp UV output, UV fluence, and UV dose using SI units.

Multiply I-P	By	To Obtain SI
Btu/ft^2 (International Table)	1135.65	mJ/cm^2
Btu/h·ft^2	315.46	μW/cm^2
To Obtain I-P	**By**	**Divide SI**

REFERENCES

Abshire, R.L. and H. Dunton. 1981. Resistance of selected strains of *Pseudomonas aeruginosa* to low-intensity ultraviolet radiation. *Applied Environmental Microbiology* 41(6):1419-1423.

ACGIH. 2007. *TLVs® and BEIs®*. American Conference of Governmental Industrial Hygienists, Cincinnati.

ASHRAE. 2004. Thermal environmental conditions for human occupancy. ANSI/ASHRAE *Standard* 55-2004.

Bahnfleth, W.P. and W.J. Kowalski. 2004. Clearing the air on UVGI systems. *RSES Journal*, pp. 22-24.

Bernstein J.A., R.C. Bobbitt, L. Levin, R. Floyd, M.S. Crandall, R.A. Shalwitz, A. Seth, and M. Glazman. 2006. Health effects of ultraviolet irradiation in asthmatic children's homes. *Journal of Asthma* 43(4):255-262.

Blatt, M.S., T. Okura, and B. Meister. 2006. Ultraviolet light for coil cleaning in schools. *Engineered Systems* (March):50-61.

Bolton, J.R. 2001. *Ultraviolet applications handbook*. Photosciences, Ontario.

Boyce, P. 2003. *Controlling tuberculosis transmission with ultraviolet irradiation*. Rensselaer Polytechnic Institute, Troy, NY.

Brickner, P.W., R.L. Vincent, M. First, E. Nardell, M. Murray, and W. Kaufman. 2003. The application of ultraviolet germicidal irradiation to control transmission of airborne disease: Bioterrorism countermeasure. *Public Health Report* 118(2):99-114.

Bruls, W. 1984. Transmission of human epidermis and stratum corneum as a function of thickness in the ultraviolet and visible wavelengths. *Journal of Photochemistry and Photobiology* 40:485-494.

CDC. 2002. *Comprehensive procedures for collecting environmental samples for culturing* Bacillus anthracis. Centers for Disease Control and Prevention, Atlanta, GA. http://www.bt.cdc.gov/agent/anthrax/environmental-sampling-apr2002.asp.

CDC. 2005. Guidelines for preventing the transmission of *Mycobacterium tuberculosis* in health-care settings. *Morbidity and Mortality Weekly Report (MMWR)* 37-38, 70-75.

CIE. 1987. *International lighting vocabulary*, 4th ed. Commission Internationale de L'Eclairage, Vienna.

CIE. CIE collection in photobiology and photochemistry. *Publications* 106/1 (Determining ultraviolet action spectra), 106/2 (Photokeratitis), and 106/3 (Photoconjuctivitis). Commission Internationale de L'Eclairage, Vienna.

CIE. 2003. *Ultraviolet air disinfection*. Commission Internationale de L'Eclairage, Vienna.

Coker, A., E. Nardell, P. Fourie, W. Brickner, S. Parsons, N. Bhagwandin, and P. Onyebujoh. 2001. *Guidelines for the utilization of ultraviolet germicidal irradiation (UVGI) technology in controlling the transmission of tuberculosis in health care facilities in South Africa*. South African Centre for Essential Community Services and National Tuberculosis Research Programme, Medical Research Council, Pretoria.

Department of General Services. 2001. *Working with ultraviolet germicidal irradiation (UVGI) lighting systems: Code of safe practice*. County of Sacramento, CA.

Diffey, B.L. 1983. A mathematical model for ultraviolet optics in skin. *Physics in Medicine and Biology* 28:657-747.

DIN. 1979. Optical radiation physics and illumination engineering. *Standard* 5031. German Institute for Standardization, Berlin.

EPA. 2008. *Universal waste*. Available at http://www.epa.gov/epaoswer/hazwaste/id/univwaste/.

First, M.W., E.A. Nardell, W.T. Chaisson, and R.L. Riley. 1999. Guidelines for the application of upper-room ultraviolet irradiation for preventing transmission of airborne contagion—Part 1: Basic principles. *ASHRAE Transactions* 105(1):869-876.

First, M.W, F.M. Rudnick, K. Banahan, R.L. Vincent, and P.W. Brisker. 2007a. Fundamental factors affecting upper-room ultraviolet germicidal irradiation—Part 1: Experimental. *Journal of Environmental Health* 4:1-11.

First, M.W., K. Banahan, and T.S. Dumyahn. 2007b. Performance of ultraviolet light germicidal irradiation lamps and luminaires in long-term service. *Leukos* 3:181-188.

GSA. 2003. *The facilities standards for the Public Buildings Service*. Public Buildings Service of the General Services Administration, Washington, D.C.

IESNA. 2000. *IESNA lighting handbook*, 9th ed. M. Rea, ed. Illuminating Engineering Society of North America, New York.

Ka, M., H.A.B. Lai, and M.W. First. 2004. Size and UV germicidal irradiation susceptibility of *Serratia marcescens* when aerosolized from different suspending media. *Applied and Environmental Microbiology* (Apr.): 2021-2027.

Kethley, T.W. and K. Branc. 1972. Ultraviolet lamps for room air disinfection: Effect of sampling location and particle size of bacterial aerosol. *Archives of Environmental Health* 25(3):205-214.

Ko, G., M.W. First, and H.A. Burge. 2000. Influence of relative humidity on particle size and UV sensitivity of *Serratia marcescens* and *Mycobacterium bovis* BCG aerosols. *Tubercle and Lung Disease* 80(4-5):217-228.

Kowalski, W.J. 2002. *Immune building systems technology*. McGraw-Hill, New York.

Kowalski, W.J. 2006. *Aerobiological engineering handbook*. McGraw-Hill, New York.

Kujundzic, E., M. Hernandez, and S.L. Miller. 2007. Ultraviolet germicidal irradiation inactivation of airborne fungal spores and bacteria in upper-room air and HVAC in-duct configurations. *Journal of Environmental Engineering Science* 6:1-9.

Levetin, E., R. Shaughnessy, C. Rogers, and R. Scheir. 2001. Effectiveness of germicidal UV radiation for reducing fungal contamination within air-handling units. *Applied and Environmental Microbiology* 67(8):3712-3715.

Luckiesh, M. 1946. *Applications of germicidal, erythemal and infrared energy*. D. Van Nostrand, New York.

Menzies, D., J. Popa, J. Hanley, T. Rand, and D. Milton. 2003. Effect of ultraviolet germicidal lights installed in office ventilation systems on workers' health and well being: Double-blind multiple cross over trial. *Lancet* 363:1785-1792.

Miller, S.L., M. Fennelly, M. Kernandez, K. Fennelly, J. Martyny, J. Mache, E. Kujundzic, P. Xu, P. Fabian, J. Peccia, and C. Howard. 2002. Efficacy of ultraviolet irradiation in controlling the spread of tuberculosis. *Final Report*, Centers for Disease Control, Atlanta, GA, and National Institute for Occupational Safety and Health, Washington, D.C.

Montgomery, R. and R. Baker. 2006. Study verifies coil cleaning saves energy. *ASHRAE Journal* 48(11):34-36.

NEHC. 1992. *Ultraviolet radiation guide*. Navy Environmental Health Center, Bureau of Medicine and Surgery, Norfolk, VA.

NEMA. 2004. Performance testing for lighting controls and switching devices with electronic fluorescent ballasts. *Standard* 410-2004. National Electrical Manufacturers Association, Rosslyn, VA.

NIOSH. 1972. Criteria for a recommended standard: Occupational exposure to ultraviolet radiation. *Publication* 73-11009. National Institute for Occupational Safety and Health, Washington, D.C.

Philips Lighting. 1992. *Disinfection by UV-radiation*. Eindhoven, the Netherlands.

Riley, R.L. 1988. Ultraviolet air disinfection for control of respiratory contagion. In *Architectural design and indoor microbial pollution*, pp. 179-197. Oxford University Press, New York.

Riley, R.L. and S. Permutt. 1971. Room air disinfection by ultraviolet irradiation of upper air: Air mixing and germicidal effectiveness. *Archives of Environmental Health* 22(2):208-219.

Riley, R.L., M. Knight, and G. Middlebrook. 1976. Ultraviolet susceptibility of BCG and virulent tubercle bacilli. *American Review of Respiratory Disease* 113:413-418.

Riley, R.L., S. Permutt, and J.E. Kaufman. 1971. Convection, air mixing, and ultraviolet air disinfection in rooms. *Archives of Environmental Health* 22(2):200-207.

Rudnick, F. 2007. Fundamental factors affecting upper-room germicidal irradiation—Part 2: Predicting effectiveness. *Journal of Occupational and Environmental Hygiene* 4(5):352-362.

RLW Analytics. 2006. Improving indoor environment quality and energy performance of California K-12 schools, project 3: Effectiveness of UVC light for improving school performance. *Final Report*, California Energy Commission Contract 59903-300.

Schechmeister, I.L. 1991. Sterilization by ultraviolet radiation. In *Disinfection, sterilization and preservation*, pp. 535-565. Lea and Febiger, Philadelphia.

Scheir, R. and F.B. Fencl. 1996. Using UVGI technology to enhance IAQ. *Heating, Piping and Air Conditioning* 68:109-14, 117-8, 123-4.

Siegel, J., I. Walker, and M. Sherman. 2002. Dirty air conditioners: Energy implications of coil fouling. *Proceedings of the ACEEE Summer Study on Energy Efficiency in Buildings*. pp. 287-299.

Sylvania. 1982. Germicidal and short-wave ultraviolet radiation. Sylvania *Engineering Bulletin* 0-342.

VanOsdell, D. and K. Foarde. 2002. Defining the effectiveness of UV lamps installed in circulating air ductwork. *Final Report*, Air-Conditioning and Refrigeration Technology Institute 21-CR Project 610-40030.

Westinghouse. 1982. Westinghouse sterilamp germicidal ultraviolet tubes. Westinghouse *Engineering Notes* A-8968.

Witham, D. 2005. Ultraviolet—A superior tool for HVAC maintenance. *IUVA Congress, Tokyo*.

Xu, P., J. Peccia, P. Fabian, J.W. Martyny, K.P. Fennelly, M. Hernandez, and S.L. Miller. 2003. Efficacy of ultraviolet germicidal irradiation of upper-room air in inactivating airborne bacterial spores and mycobacteria in full-scale studies. *Atmospheric Environment* 37(3):405-419.

CHAPTER 17

COMBUSTION TURBINE INLET COOLING

POWER OUTPUT capacity of all combustion turbines (CTs) varies with ambient air temperature and site elevation. The rated capacities of all CTs are based on standard ambient air at 59°F, 60% rh, 14.7 psia at sea level, and zero inlet and exhaust pressure drops, as selected by the International Organization for Standardization (ISO). For all CTs, increased ambient air temperature or site elevation decreases power output; increased ambient air temperature also reduces fuel efficiency (i.e., increases the heat rate, defined as fuel energy required per unit of electric energy produced). However, the extent of the effect of these changes on output and efficiency varies with CT design. This chapter provides a detailed discussion on combustion turbine inlet cooling (CTIC). Additional information on applying CTIC to combined heat and power systems (cogeneration) is provided in Chapter 7.

There are two types of CTs: aeroderivative and industrial/frame. Figures 1 and 2 show typical effects of ambient air temperature on power output and heat rate, respectively, for these types of turbines. The actual performance of a specific CT at different inlet air temperatures depends on its design. Figures 1 and 2 show that aeroderivative CTs are more sensitive to ambient air temperature than are industrial/frame CTs. Figure 1 (Punwani and Hurlbert 2005) shows that, for a typical aeroderivative CT, an increase in inlet air temperature from 59 to 100°F on a hot summer day decreases power output to about 81% of its rated capacity: a loss of 19% of the rated capacity. Figure 2 (Punwani 2003) shows that, for the same change in ambient air temperature, the heat rate of a typical aeroderivative CT increases (i.e., fuel efficiency decreases) by about 4% of the rated heat rate at ISO conditions. Increasingly, industrial/frame CTs are using aeroderivative technology to improve performance; thus, their performance curves are moving toward those of the classic aeroderivative CT.

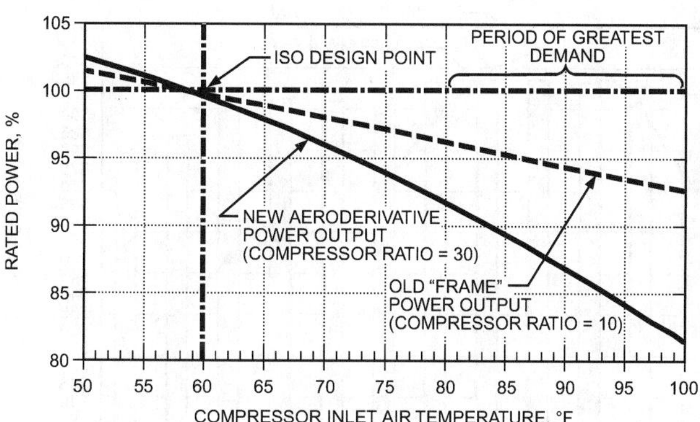

Fig. 1 Effect of Ambient Temperature on CT Output
(Punwani and Hurlbert 2005)

The preparation of this chapter is assigned to TC 1.10, Cogeneration Systems.

In cogeneration and combined-cycle systems that use thermal energy in CT exhaust gases for steam generation, heating, cooling, or more power generation, increases in ambient air temperature also reduce the total thermal energy available for these applications, as shown in Figure 3 (Orlando 1996).

CTs, in simple- and combined-cycle systems are particularly qualified to meet peak electricity demand because of their ability to start and stop more quickly than steam-turbine based thermal power generation systems using coal, oil or gas, and nuclear plants. For fossil fuel power generation, combined-cycle systems are the most fuel-efficient (lowest heat rate of typically 7000 Btu/kWh) and

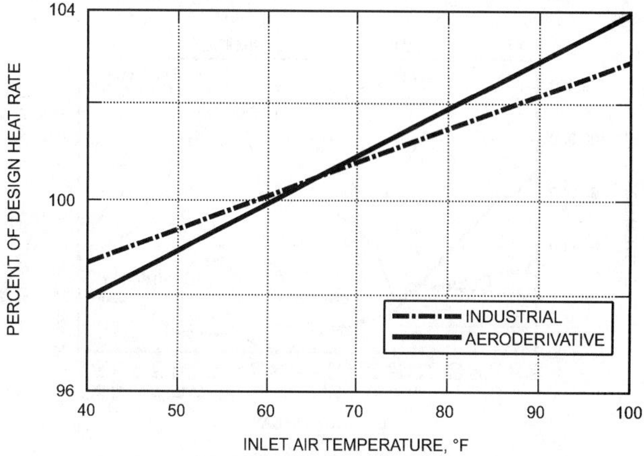

Fig. 2 Effect of Ambient Temperature on CT Heat Rate
(Punwani 2003)

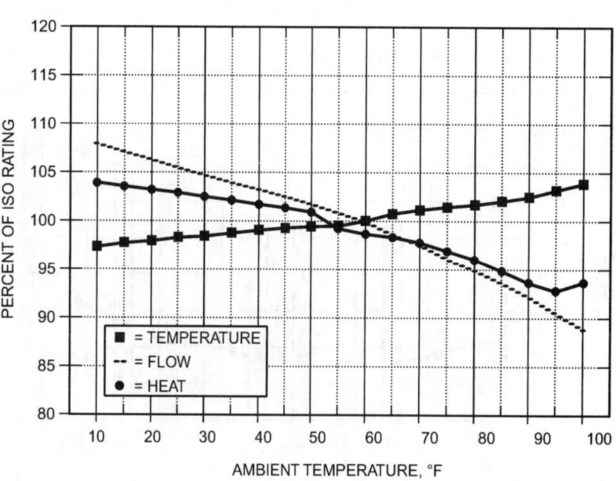

Fig. 3 Effects of Ambient Temperature on Thermal Energy, Mass Flow Rate and Temperature of CT Exhaust Gases
(Orlando 1996)

steam-turbine-based systems are the least efficient (highest heat rate range between 12,000 to 20,000 Btu/kWh, depending on turbine age). A typical heat rate for a simple-cycle system is about 10,000 Btu/kWh. Therefore, to minimize the fuel cost for power generation, the preferred order of dispatching power to meet the market demand is to operate combined-cycle systems first, simple-cycle systems next, and steam turbines as the last resort.

Electric power demand is generally high when ambient temperatures are high. An example of an hourly profile of ambient temperature, system load, and CT output is shown in Figure 4 (Punwani and Hurlbert 2005).

When high ambient temperatures drive up power demand, the use of less efficient (high-fuel-cost) generation plants is required and that drives up the market price of electric energy. Figure 5 (Hilberg 2006) shows the hourly load profile in one of the US reliability regions for a single day in the summer. Although the peak electricity demand increases by 80%, the peak power price increases by over 400%. Figures 4 and 5 show that power output capacity decreases just when it is most needed, and when power is also most valuable.

The trends shown in Figures 4 and 5 are not unique to the United States. The Middle East is seeing much higher growth rates in power demand, and that demand is also directly linked to hot weather power usage. In some countries in the Middle East, over 40% of power usage is linked to air conditioning.

CTIC is used by thousands of CT-based power plants to overcome the ill effects of increased ambient temperature on CT performance.

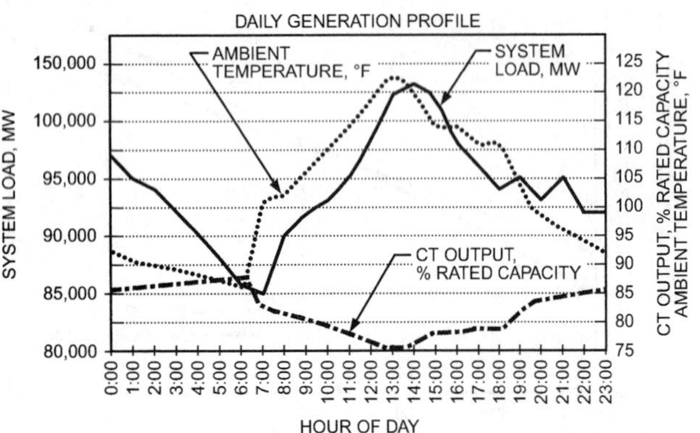

Fig. 4 Typical Hourly Power Demand Profile
(Punwani and Hurlbert 2005)

It can provide economic and environmental benefits for plant owners, ratepayers, and the general public.

ADVANTAGES

CTIC offers economic as well as environmental benefits.

Economic Benefits

- Maximizes power output when most needed and most valuable
- Reduces capital cost ($/kW) for the incremental capacity
- Increases CT fuel efficiency (lowers heat rate)
- Minimizes use of less efficient steam-turbine-based systems, thus helping to minimize increase in rates to electricity users

Environmental Benefits

- Allows minimum use of inefficient and polluting power plants by allowing maximum use of efficient and cleaner CT plants
  - Conserves natural fuel resources
  - Reduces emissions of pollutants (SO_x, NO_x, particulates, and hydrocarbons)
  - Reduces emissions of global warming/climate change gas (CO_2)
- Minimizes/eliminates new power plant siting issues

Emissions reductions from CTIC result from its displacement of the very-high-heat-rate steam turbine peaker power plants (consuming as much as 20,000 Btu/kWh operating on boilers and steam turbines), as shown in an example in Table 1.

DISADVANTAGES

- Permanently higher CT inlet pressure drop that results in a small drop in the CT output capacity even when CTIC is not being used (magnitude of pressure drop varies with CTIC technology)
- Additional maintenance cost for CTIC equipment

DEFINITION AND THEORY

A schematic flow diagram of a combustion turbine (CT) system is shown in Figure 6.

Power output of a CT is directly proportional to and limited by the mass flow rate of the compressed air available to it from the air compressor that provides high-pressure air to the combustion chamber of the CT system. An air compressor has a fixed capacity for handling a volumetric flow rate of air for a given rotational speed of

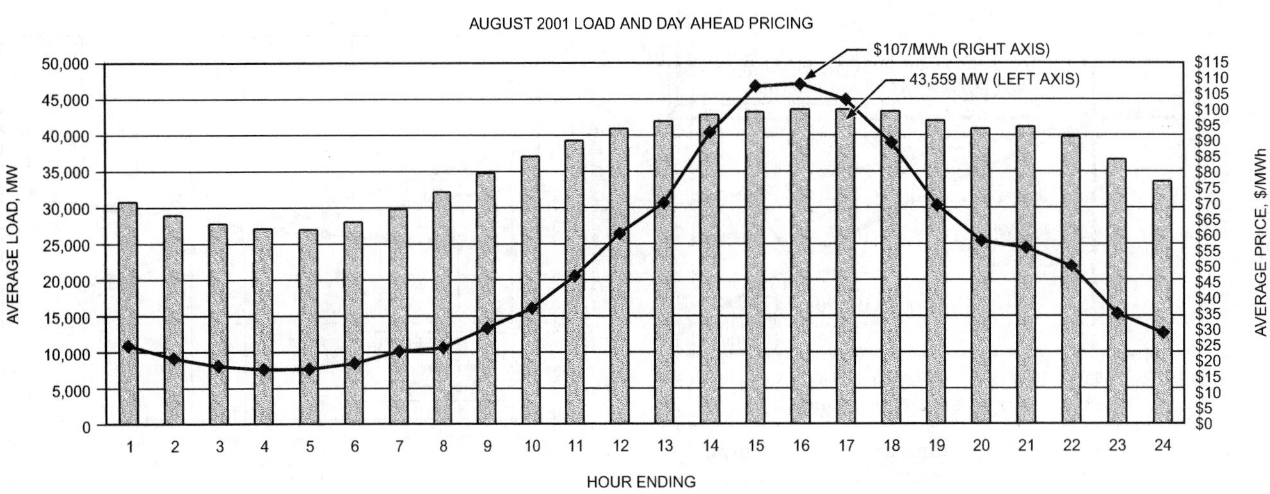

Fig. 5 Example of Daily System Load and Electric Energy Pricing Profiles
(Hilberg 2006)

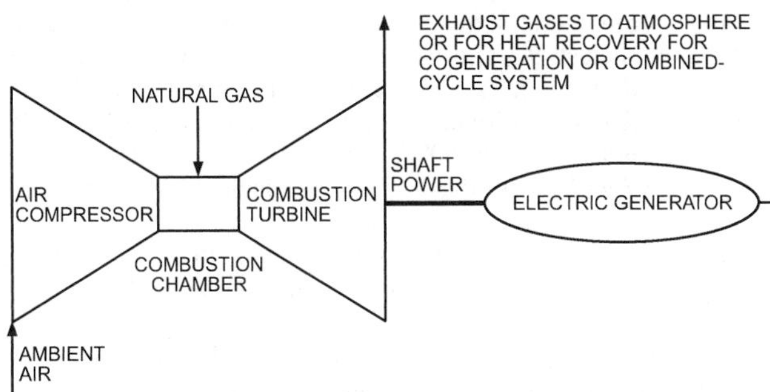

Fig. 6 Schematic Flow Diagram of Typical Combustion Turbine System

Table 1 Examples of Emissions from Typical Combined-Cycle, Simple-Cycle, and Steam Turbine Systems

Unit Type	TIC Candidates: Combined-Cycle CT	TIC Candidates: Simple-Cycle CT	Existing Older Plants: Boiler + Steam Turbine
Prime mover	Frame CT-STG	Frame CT	Condensing STG
Fuel	Gas	Gas	No. 6 oil
Fuel sulfur, % weight	0	0	0.01
Plant age, yr	<5	<5	>30
Heat rate, Btu/kWh	7000	10,750	13,000
NO_x control	DLN-SCR	DLN	LNB with FRG
NO_x target, ppm	3	9	N/A
NO_x, lb/10^6 Btu	0.0114	0.0341	0.3
SO_2, lb/10^6 Btu	0	0	1.02
Incremental capacity, MW	100	100	100
Hours of operation	400	400	400
Fuel, 10^6 Btu	280,000	430,000	520,000
CO_2 emissions, tons	16,275	24,993	44,730
NO_x emissions, tons	1.595	7.3315	78
SO_x emissions, tons	0	0	265.2

LNB = low NO_x burners DLN = dry low NO_x
FGR = fuel gas recirculation SCR = selective catalytic NO_x reduction

the compressor. Even though the volumetric capacity of a compressor remains constant, the mass flow rate of air it delivers to the CT changes with ambient air temperature. The mass flow rate of air decreases as the air density decreases with increasing ambient temperature. CTIC reduces the inlet air temperature, resulting in increases in air density mass flow rate and power output. For more details, consult Stewart (1999).

SYSTEM TYPES

Many technologies are commercially available for CTIC, but the overall approaches can be divided into three major groups:

- Evaporative systems
- Chiller systems
- Liquefied natural gas (LNG) vaporization systems

Each approach has advantages and disadvantages; for more information, see Stewart (1999).

Evaporative Systems

Evaporative cooling systems rely on cooling produced by evaporation of water added into the inlet air. An ideal evaporative cooling process occurs at a constant wet-bulb temperature and cools the air to a higher relative humidity (i.e., water vapor content increases). When the warm ambient inlet air comes in contact with the added water, it transfers some of its heat to the liquid water and evaporates some of the water. The process of heat transfer from inlet air to water cools the inlet air. Water added in the evaporative systems also acts as an air washer by cleaning the inlet air stream of airborne particulates and soluble gases, which could be a significant benefit to downstream filter elements. Studies show that evaporative cooling also reduces NO_x emissions because of the increase in moisture added to the air. The psychrometry of CTIC using evaporative systems has been described in detail by Stewart (1999).

Evaporative systems can cool the inlet air up to 98% of the difference between the ambient dry-bulb and wet-bulb temperatures. Therefore, the most cooling can be achieved during hot and/or dry weather. Design and hourly wet-bulb temperatures for many locations can be found in Chapter 28 of the 2005 *ASHRAE Handbook—Fundamentals* and the *Weather Year for Energy Calculations 2 (WYEC2) Data and Toolkit CD* (ASHRAE 1997). This information is useful in evaluating the cooling potential for locations in many climates.

Evaporative systems have the lowest capital costs among CTIC systems, and are the most common type in use. Their primary disadvantage is that the extent of cooling produced is limited by the wet-bulb temperature (and thus, cooling is weather dependent). In arid climates, these systems consume large quantities of water (e.g., about 5 gal/h to cool 10,000 cfm of air by 20°F), which comprises the major component of the system operating cost.

Two primary system types are commercially used for evaporative cooling: wetted media and fogging.

Wetted Media. As the first approach adopted for CTIC, these systems have a long history of success in a wide range of operating climates. Inlet air is exposed to a thin film of water on the extended surface of honeycomb-like wetted media. Water used for wetting may or may not require treatment, depending on its quality and the medium manufacturer's specifications.

Typical pressure drop across the wetted media is about 0.3 in. of water. The higher the degree of saturation required, the more media required and higher the pressure drop. A wetted-media system requires proper control of the chemistry of recirculating water (e.g. contaminant absorption) and monitoring of media degradation (Graef 2004). The media may need to be replaced every 5 to 10 years, depending on water quality, air quality, and hours of operation.

Fogging. This approach adds moisture to the inlet air by spraying very fine droplets of water. High-pressure fogging systems can be designed to produce droplets of variable sizes, depending on the desired evaporation time and ambient conditions. The water droplet size is generally less than 40 μm, and averages about 20 μm. Fogging systems typically require higher-quality water than wetted-media systems do. Generally, reverse-osmosis or demineralized water is used to ensure cleanliness throughout the system. Typical pressure drop across the fogging section is about 0.1 in. of water. Fogging nozzles may need to be replaced every 5 to 10 years because of erosion. The high-pressure pumps require servicing at least annually.

A fogging system that sprays more water than could evaporate before the inlet air enters the compressor is known as a **wet compression**, **overspray**, **high-fogging**, or **over-fogging system** (Jolly 2005; Kraft 2004; Schwieger 2004). Water in excess of that required by fogging does not further reduce inlet air temperature. It is ingested into the compressor section of the CT, and provides a threefold effect for the system:

- It ensures inlet air achieves maximum evaporative cooling.
- The additional water evaporates in the compression stages of the compressor, allowing the compressor to be cooler and require less work for air compression.
- The excess water increases the total mass flow rate of gases entering the CT and helps increase power output beyond that possible with fogging alone.

There is debate in the industry whether wet compression is truly a CTIC technology; accordingly, this chapter addresses only the inlet fogging contribution of wet compression technology.

Chiller Systems

Chiller systems cool inlet air by exchange of heat through indirect or direct contact between warm ambient inlet air and a cold fluid produced by chillers. In **indirect heat exchange systems**, chilled fluid flows inside a coil while the inlet air flows across the coil face. Typical inlet-air-side pressure drop across the heat exchange coil is about 1 in. of water. The water vapor content (humidity ratio) of the inlet air remains constant as its dry-bulb temperature decreases. Ammonia is most commonly used chilled fluid in direct refrigerant applications; for indirect-contact heat exchange systems, the preferred chilled fluid is often water, though water/glycol, HFCs, HCFCs, and other aqueous fluids are also used.

Recently, a **direct heat exchange system** (also known as a **bulk air cooler**) has been introduced. It uses wetted media in combination with chillers, and allows the flexibility of using wetted media either with ambient temperature water or with chilled water, depending on the desired power output capacity. A direct-contact heat exchanger operating without chilling the added water performs just as a wetted-media evaporative system does, except that the pressure drop is about 0.8 in. of water. Stewart (1999) describes the psychrometry of CTIC for chilled-fluid systems in detail.

The primary advantage of chiller systems is that they can cool inlet air to much lower temperatures (thus enhancing power output capacity more than evaporative cooling systems can), and can maintain any desired inlet air temperature down to 42°F, independent of ambient dry-bulb and wet-bulb temperature. Additional inlet air cooling of up to 8 to 10°F may occur in the bell-mouth section of the air compressor inlet housing, depending on housing design; during hot and humid weather, this additional temperature drop may be enough to cause the temperature of moisture-saturated air to fall below 32°F and form ice crystals, which could damage compressor blades. If, however, the ambient air is very dry, inlet air could be cooled below 42°F (Stewart 2001).

The primary disadvantages of chiller systems are that their installed costs and inlet air pressure drops (which lead to power loss even when CTIC is not being used) are much higher than those for evaporative systems. Chiller systems could be air or water cooled; water-cooled systems are the most common. Most water-cooled chiller systems require cooling towers, although some directly use seawater. Systems that use cooling towers require makeup water, which may require treatment, although usually less than is required by evaporative systems. Makeup water required for cooling towers is about 20 to 30 gpm per 1000 ton of chiller capacity. During hot and humid weather conditions, most chiller systems produce saturated air and require mist eliminators downstream of the cooling to minimize the potential for water carryover from the cooling coil. It is also common to collect the condensate that forms on the inlet air coils; this can offset a substantial portion of the cooling tower makeup water requirements.

Chiller systems can use mechanical and/or absorption chillers, with or without thermal energy storage (TES).

Mechanical chillers used in CTIC systems could be centrifugal, screw, or reciprocating, and could driven by electric motors, steam turbines, or engines. These chillers can use CFCs, HCFCs, or ammonia as a refrigerant and can cool the inlet air to 42°F (or below, if desired) without the risk of forming ice crystals. Installed cost of motor-driven mechanical chillers is lower than that for any other chiller, but their parasitic loads are the highest: 0.6 to 0.8 kW per ton of refrigeration. Mechanical chillers driven by steam turbines or engines have lower parasitic electrical loads (0.03 kW/ton), but require fuel or heat inputs and have a higher installed cost. Mechanical chillers are more commonly used for CTIC than are absorption chillers.

Absorption chillers require thermal energy from steam, hot water, hot exhaust gases, or natural gas as the primary source of energy, and need much less electric energy than mechanical chillers. If water is used as the refrigerant, as is most common for CTIC, inlet air can be cooled to about 50°F; if ammonia is the refrigerant, air temperatures will be similar to those available using mechanical chillers.

Absorption chillers can be single or double effect. Single-effect absorption chillers use hot water or 15 psig steam (18 lb/ h·ton), whereas double-effect chillers require less steam (10 lb/ h·ton) but need it at higher pressure (typically 115 psig). The advantage of an absorption chiller system is that it has much less parasitic electric load (typically, 0.03 kW/ton), but its capital cost is higher than that for mechanical chiller systems. Their main successful application is in power plants where excess waste thermal energy is available, and where use of this energy saves higher-value electric energy.

Thermal energy storage (TES) is used in chilled-fluid systems to increase the available net power output capacity during on-peak periods. It is typically used when only a small number of hours per day require inlet air cooling, or when power enhancement is highly valued. TES systems incorporate tanks that store chilled fluid or ice, which is produced by chillers or refrigeration systems during off-peak periods, when the market value of electric energy is low. The stored chilled water or ice is used to cool the inlet air during on-peak periods, when the value of electric energy is high. The primary advantage of a TES system is that it can reduce total system capital cost by reducing chiller capacity requirements from that required to match the instantaneous on-peak demand for cooling. It also increases on-peak capacity and revenues for the power plant, because little or no electric energy is required to operate the chillers during the high-demand, high-value, on-peak periods. The disadvantage of a TES system is that it requires a larger site footprint for the TES tank.

There are two types of TES systems: full and partial shift. In **full-shift systems**, chillers are not operated during on-peak periods; all the required cooling is achieved by using the chilled fluid or ice from the TES tank. In **partial-shift systems**, the down-sized chiller system also runs during on-peak periods to complement the cooling capacity available from the stored chilled fluid or ice.

Several publications compare the performance and economics of TES options for CTIC (Andrepont 1994) or provide methodology for their analysis (Andrepont and Pasteris 2003) and case studies (Liebendorfer and Andrepont 2005). A database of TES-CTIC applications has been developed and analyzed, illustrating the trends in TES-CTIC technology, capacity, and geographic locales, as well as demonstrating the performance and economic advantages of combining TES with TIC in many applications (Andrepont 2005; Dec and Andrepont 2006).

LNG Vaporization Systems

These systems are useful for power plants located near a liquefied natural gas (LNG) import facility. In supplying natural gas for pipelines, power plants, or other applications, LNG must be vaporized by some heat source. For applications in CTIC, the inlet air is used as such a heat source. Cho et al. (2003) and Punwani and Pasteris (2004) give examples of LNG vaporization systems.

CALCULATION OF POWER CAPACITY ENHANCEMENT AND ECONOMICS

A CTIC system's power capacity enhancement and economics depends on several parameters, including the following:

- CT design and characteristics
  - CT heat rate versus compressor inlet air temperature
  - CT power output versus compressor inlet air temperature
  - CT airflow (not an independent variable; dependent on compressor inlet air temperature)

- CTIC design characteristics
  - Parasitic load
  - Pressure drop across the component inserted upstream of the compressor (insertion loss)
  - Water usage
- Hourly weather data (dry-bulb and coincident wet-bulb temperatures) for the geographic location of the CT
- Selected ambient design conditions
- Selected cooled air temperature upstream of compressor
- Cost of fuel
- Cost of water
- Power demand profile
- Hourly market value of electric energy
- Hourly market value of plant capacity

Preliminary estimates of net capacity enhancement by CTIC and the associated costs can be made by the following calculation procedure, which is based on several rules of thumb. More accurate calculations can require sophisticated combustion turbine models and site-specific cost analyses.

Cooling. Cooling achieved is easily calculated for all of the systems.

Media and fogger evaporative coolers cool a percentage of the difference between wet-bulb (WB) and dry-bulb (DB) temperatures:

$$\text{Degrees of cooling} = \text{Cooling efficiency} \times (\text{DB} - \text{WB})$$

Chillers are sized to cool to the selected design temperature:

$$\text{CCL} = AF_m(H_a - H_c)/12{,}000$$

where

CCL = Chiller cooling load, tons
AF_m = Mass flow rate of cooled air, lb/h
H_a = Enthalpy of ambient air, Btu/lb
H_c = Enthalpy of cooled air, Btu/lb
12,000 = Ton of cooling, Btu/h

Parasitic Loss. Parasitic loss is the power required to run the CTIC system.

Media evaporative coolers consume the same amount of power any time the pump is energized, regardless of cooling produced. The total power required varies depending on pumping head, pressure drop of valves, and other restrictions, but a reasonable estimate is 0.02 W/cfm.

Fogging systems power consumption is a function of the amount of water injected into the air, and water pressure. For a water pressure of 3000 psi, the parasitic loss is approximately 2 kW/gpm).

Chillers consume power based on the cooling load and condenser temperature.

$$\text{Chiller parasitic load, kW} = 0.7 \times \text{CCL}$$

The cooling load of the cooling tower (CTL), in tons, is 125% of the chiller load. Estimate the water flow rate capacity (CTC) of the cooling tower at a rate of 3 gpm × CTL. The power of the pump and fan is estimated by the following equation:

$$\text{Tower parasitic load, kW} = 6 \times \text{CTC}/(38 \times 1.341)$$

For air-cooled chillers, power usage is greater than for water-cooled chillers, and is a function of the dry-bulb (rather than the wet-bulb) ambient air temperature.

For chillers integrated with TES, most or all of the chiller parasitic load (excluding some water pumps) can be eliminated during peak periods of CTIC operation, with the chiller operating (and the associated parasitic load) during off-peak times of lower power demand and lower power value.

Insertion Loss. All CTs have an insertion loss, which reduces the turbine's power output by approximately 0.25% per in. of water of pressure differential across the inserted component. The pressure differential for the evaporative cooler when no water is being injected (cooler not in operation) is about half as much as when it is running. Insertion loss on a cooling coil also decreases when it is not running or there is no condensate on the coil.

Net and Gross Power Increase. Figure 1 shows typical power degradation caused by increased ambient temperature. For a new aeroderivative CT, power can be lost or restored at an estimated rate of 0.44% per °F of temperature change. Using these numbers, the power output of the CT can be estimated with and without CTIC. The gross power increase is the difference between these two values. The net power increase is the gross increase less all parasitic loads.

Fuel Usage. Figure 2 shows the change in heat rate with increase in inlet air temperature. The curve for the industrial turbine can be estimated to be 0.08% per °F. Fuel usage of the CT at ISO standard conditions ranges between 8 and 12,000 Btu/kWh.

Fuel usage must be calculated for the turbine with and without cooling. Fuel consumption is greater (but more efficient) when the air has been cooled, and the cost of this additional fuel should be deducted from the gross increase in revenue.

Water Usage. Water is often a precious commodity at industrial sites. Usage can be estimated as follows.

For media and fogging evaporative coolers, water usage is a function of the amount of cooling and the airflow:

$$\text{Water usage, gph} = \frac{1.2 \times \Delta t \times AF_v}{10{,}000}$$

where

Δt = Degrees of cooling, °F
AF_v = Volumetric airflow rate, cfm

Water evaporation can also be calculated by multiplying the difference in pounds of moisture of the entering (v_1) and leaving (v_2) air times the weight per minute of airflow:

$$\text{Evaporation, gpm} = \frac{(v_2 - v_1) \times AF_m}{8.337}$$

Bleedoff from media evaporative coolers should be added to the evaporated water when calculating the cost of water. Typical bleed rates range between 10 and 50% of the evaporation rate. Foggers use high-purity water produced by reverse osmosis or demineralization; this water produces a waste stream of 10 to 30% that may need to be included in the water usage calculation.

Water-cooled chillers use water at the cooling tower. Usage is a function of the amount of cooling, and is about 0.02 to 0.03 gpm per ton of chiller capacity.

Bleedoff and drift from the cooling tower should be added to the evaporated water when refining the calculation for water cost. The bleed depends on water treatment and quality, and drift is a function of the tower's water recirculation rate. For more information, refer to Chapter 39.

In applications using cooling coils or bulk air coolers to cool the turbine inlet air stream, water may be produced by condensing (if ambient air is cooled below its dew point) and capturing moisture contained in the inlet air. This water can then be used in other locations of the power plant.

CTIC Cost. There are many variables in the cost, including turbine capacity, location, local labor rates, applicable local codes, and customer specifications.

Total installed cost of the CTIC system increases (although the cost per unit of capacity decreases) with increases in CT output and airflow rate capacities, because an increase in CT capacity increases the face area of the media for an evaporative cooler, and increases capacity and the heat exchange coil for the chiller system. For inlet fogging systems, increased airflow means an increase in pumping capacity and in the number of nozzles, pumps and stages. For the same reason, fogging system cost also increases proportionately with increasing differences between the wet- and dry-bulb temperatures. This can be equated to the enthalpy difference between air entering and leaving the coil. When sizing the chiller system, calculations should be based on the actual airflow of the cooled air.

The cost of chiller systems depends heavily on the cooling load for the design day, and on whether (and in what configuration) thermal energy storage (TES) is integrated with the CTIC chiller system.

Operation and Maintenance (O&M) Cost. O&M cost must also be considered, and varies considerably with the type of CTIC technology used. For an overview of O&M costs in general, see Chapter 36 of the 2007 *ASHRAE Handbook—HVAC Applications*.

REFERENCES

Andrepont, J.S. 1994. Performance and economics of combustion turbine (CT) inlet air cooling using chilled water storage. *ASHRAE Transactions* 100(1):587-594.

Andrepont, J.S. 2005. Combustion turbine plant power augmentation using turbine inlet cooling with thermal energy storage. *Proceedings of Power-Gen International.*

Andrepont, J.S. and R.M. Pasteris. 2003. Methodology and case studies of enhancing the performance and economics of gas turbine power plants. *Proceedings of Power-Gen International.*

ASHRAE. 1997. *Weather year for energy calculations 2 (WYEC).*

Cho, J., et al. 2003. Economic benefits of integrated LNG-to-power. Presented at AIChE 2003 Spring National Meeting, 3rd Topical Conference on Natural Gas Utilization.

Graef, P.T. 2004. Operation and maintenance of wetted-media evaporative coolers. *Energy-Tech* (December):19-20.

Hilberg, G.R. 2006. Case study for retrofit of turbine inlet cooling in Batam, Indonesia. *POWER-GEN Asia 2006 Proceedings.*

Jolly, S., S. Cloyd, and J. Hinrichs. 2005. Wet compression adds power, flexibility to aeroderivative GTs. *Power* 149(4).

Kraft, J.E. 2004. Evaporative cooling and wet compression technologies. *Energy-Tech* (February).

Liebendorfer, K.M. and J.S. Andrepont. 2005. Cooling the hot desert wind: Turbine inlet cooling with thermal energy storage (TES) increases net power plant output by 30%. *ASHRAE Transactions* 111(2):545-550.

Orlando, J. 1996. *ASHRAE cogeneration design guide.*

Punwani, D.V. and C.M. Hurlbert. 2005. Unearthing hidden treasure. *Power Engineering.*

Punwani, D.V. and R.M. Pasteris. 2004. Hybrid systems & LNG for turbine inlet cooling. *Energy-Tech* (October):19-20.

Schwieger, R.G. 2004. Recent experience indicates wet compression meets expectations when done correctly. *Combined Cycle Journal* (Spring).

Stewart, W.E. 2001. Condensation and icing in gas turbine systems: Inlet air temperature and humidity limits. *ASHRAE Transactions* 107(1):887-891.

Stewart, W.E. 1999. *Design guide: Combustion turbine inlet air cooling systems.* ASHRAE.

BIBLIOGRAPHY

Andrepont, J.S. 2000. Combustion turbine inlet air cooling (CTIAC): Benefits, technology options, and applications for district energy. *Proceedings of the 91st Annual IDEA Conference.*

Andrepont, J.S. 2001. Combustion turbine inlet cooling: Benefits and options for district energy. *IDEA District Energy* 87(3):16-19.

Andrepont, J.S. 2002. Demand-side and supply-side load management: Optimizing with thermal energy storage (TES) for the restructuring energy marketplace. *Proceedings of 24th Industrial Energy Technology Conference.*

Andrepont, J.S. 2004. Thermal energy storage: Benefits and examples in high-tech industrial applications. *Proceedings of the AEE World Energy Engineering Congress (WEEC).*

Andrepont, J.S. 2004. Thermal energy storage—Large new applications capture dramatic savings in both operating and capital costs. *Climatização* 4(44):40-46. (In Portuguese.)

Andrepont, J.S. 2004. Thermal energy storage technologies for turbine inlet cooling, *Energy-Tech* (August):18-19.

Andrepont, J.S. 2004. Turbine inlet cooling success stories, by technology. *Energy-Tech* (October):1-16

Andrepont, J.S. 2005. Developments in thermal energy storage: Large applications, low temps, high efficiency, and capital savings. *Proceedings of the AEE World Energy Engineering Congress (WEEC).*

Andrepont, J.S. 2006. Maximizing power augmentation while lowering capital cost per MW via turbine inlet cooling (TIC) with thermal energy storage (TES). *Proceedings of Electric Power.*

Andrepont, J.S. and S.L. Steinmann. 1994. Summer peaking capacity via chilled water storage cooling of combustion turbine inlet air., *Proceedings of the American Power Conference*, Chicago.

Clark, K.M., et al. 1998. The application of thermal energy storage for district cooling and combustion turbine inlet air cooling. *Proceedings of the IDEA 89th Annual Conference*, San Antonio.

Cross, J.K., W.A. Beckman, J.W. Mitchell, D.T. Reindl, and D.E. Knebel. 1995. Modeling of hybrid combustion turbine inlet air cooling systems. *ASHRAE Transactions* 101(2):1335-1341.

Evans, K. 2007. Gas turbine technology: Reaching the peak. *Middle East Energy* 4(1).

Farmer, R. 2003. Evap cooling and wet compression boost steam injected Fr6B output. *Gas Turbine World* (Summer).

Kraft, J.E. 2004. Combustion turbine inlet air cooling—Check your design point. *Power Engineering* (May).

Kraft, J.E. 2006. Turbine inlet cooling system comparisons. *Energy-Tech* (August).

Kraft, J.E. 2006. Beating the heat with inlet cooling. *Power* (July/August).

Liebendorfer, K.M. and J.S. Andrepont. 2005. Cooling the hot desert wind: Turbine inlet cooling with thermal energy storage (TES) increases net power plant output by 30%. *ASHRAE Transactions* 111(2):545-550.

Mercer, M. 2004. Wet compression technologies for combustion turbines. *Diesel and Gas Turbine Worldwide* (May).

Punwani, D.V. 2003. GT inlet air cooling boosts output on warm days to increase revenues. *Combined-Cycle Journal* (October).

Punwani, D.V. 2005. Turbine inlet cooling for power augmentation in combined heat & power (CHP) systems. *POWER-GEN International Proceedings.*

Punwani, D.V. and C.M. Hurlbert. 2006. Cool or not to cool. *Power Engineering,* pp. 18-23.

Punwani, D.V., T. Pierson, J. Bagley, and W.A. Ryan. 2001. A hybrid system for combustion turbine inlet cooling for a cogeneration plant in Pasadena, TX. *ASHRAE Transactions* 107(1):875-881.

Stewart, W.E. 2000. Air temperature depression and potential icing at the inlet of stationary combustion turbines. *ASHRAE Transactions* 106(2): 318-327.

Tillman, T.C., et al. 2003. Comparisons of power enhancement options for greenfield combined cycle power plants. *Power-Gen International Proceedings.*

Zheng, Q., Y. Sun, S. Li, and Y. Wang. 2002. Thermodynamic analysis of wet compression process in the compressor of gas turbine. *Proceedings of ASME EXPO 2002*, Amsterdam.

Zwillenberg, M.L., A. Cohn, W. Major, I. Oliker, and D. Smith. 1991. Assessment of refrigeration-type cooling of inlet air for Essex unit no. 9. ASME *Paper* 91-JPGC-GT-4. Presented at International Power Generation Conference, San Diego.

CHAPTER 18

DUCT CONSTRUCTION

THIS chapter covers construction of HVAC and exhaust duct systems for residential, commercial, and industrial applications. Technological advances in duct construction should be judged relative to the construction requirements described here and to appropriate codes and standards. Although the construction materials and details shown in this chapter may coincide, in part, with industry standards, they are not in an ASHRAE standard.

BUILDING CODE REQUIREMENTS

In the U.S. private sector, each new construction or renovation project is normally governed by state laws or local ordinances that require compliance with specific health, safety, property protection, and energy conservation regulations. Figure 1 illustrates relationships between laws, ordinances, codes, and standards that can affect design and construction of HVAC duct systems (note that it may not list all applicable regulations and standards for a specific locality). Specifications for U.S. federal government construction are promulgated by agencies such as the Federal Construction Council, the General Services Administration, the Department of the Navy, and the Veterans Administration.

Because safety codes, energy codes, and standards are developed independently, the most recent edition of a code or standard may not have been adopted by a local jurisdiction. HVAC designers must know which code compliance obligations affect their designs. If a provision conflicts with the design intent, the designer should resolve the issue with local building officials. New or different construction methods can be accommodated by the provisions for equivalency incorporated into codes. Staff engineers from the model code agencies are available to assist in resolving conflicts, ambiguities, and equivalencies.

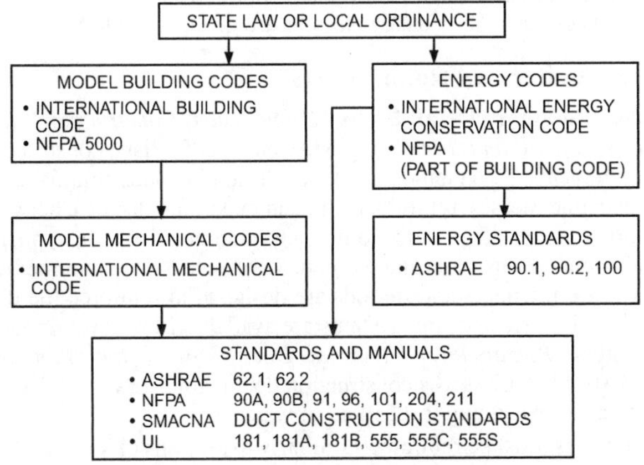

Fig. 1 Hierarchy of Building Codes and Standards

Smoke management is covered in Chapter 52 of the 2007 *ASHRAE Handbook—HVAC Applications*. The designer should consider flame spread, smoke development, combustibility, and toxic gas production from ducts and duct insulation materials. Code documents for ducts in certain locations in buildings rely on a criterion of limited combustibility (see NFPA *Standard* 90A), which is independent of the generally accepted criteria of 25 flame spread and 50 smoke development; however, certain duct construction protected by extinguishing systems may be accepted with higher levels of combustibility by code officials.

Combustibility and toxicity ratings are normally based on tests of new materials; little research is reported on ratings of aged duct materials or of dirty, poorly maintained systems.

CLASSIFICATIONS

Duct construction is classified by application and pressure. HVAC systems in public assembly, business, educational, general factory, and mercantile buildings are usually designed as *commercial*. Air pollution control systems, industrial exhaust systems, and systems outside the pressure range of commercial system standards are classified as *industrial*.

Classifications are as follows:

Residences	±0.5 in. of water
	±1 in. of water
Commercial Systems	±0.5 in. of water
	±1 in. of water
	±2 in. of water
	±3 in. of water
	±4 in. of water
	±6 in. of water
	±10 in. of water
Industrial Systems	Any pressure

Air conveyed by a duct adds both static pressure and velocity pressure loads on the duct's structure. The load from mean static pressure differential across the duct wall normally dominates and is generally used for duct classification. Turbulent airflow adds relatively low but rapidly pulsating loading on the duct wall.

Static pressure at specific points in an air distribution system is not necessarily the static pressure rating of the fan; the actual static pressure in each duct section must be obtained by computation. Therefore, the designer should specify the pressure classification of the various duct sections in the system. All modes of operation must be considered, especially in systems used for smoke management and those with fire dampers that must close when the system is running.

The preparation of this chapter is assigned to TC 5.2, Duct Design.

Table 1 Recommended Duct Seal Levels*

Duct Location	Duct Type: Supply ≤2 in. water	Duct Type: Supply >2 in. water	Exhaust	Return
Outdoors	A	A	A	A
Unconditioned spaces	A	A	B	B
Conditioned spaces	A	A	B	B

*See Table 2 for definition of seal level.

Table 2 Duct Seal Levels*

Seal Level	Sealing Requirements
A	All transverse joints, longitudinal seams, and duct wall penetrations
B	All transverse joints and longitudinal seams

*Transverse joints are connections of two ducts oriented perpendicular to flow. Longitudinal seams are joints oriented in the direction of airflow. Duct wall penetrations are openings made by screws, non-self-sealing fasteners, pipe, tubing, rods, and wire. Round and flat oval spiral lock seams need not be sealed. All other connections are considered transverse joints, including but not limited to spin-ins, taps and other branch connections, access door frames, and duct connections to equipment.

DUCT CLEANING

Ducts may collect dirt and moisture, which can harbor or transport microbial contaminants. Design, construct, and maintain ducts to minimize the opportunity for growth and dissemination of microorganisms. Recommended control measures include using access for cleaning, using proper filtration, and preventing moisture and dirt accumulation. NADCA (2006) and NAIMA (2002a) have specific information and procedures for cleaning ducts. Owners should routinely conduct inspections for cleanliness.

LEAKAGE

Predicted leakage rates for unsealed and sealed ducts are reviewed in Chapter 35 of the 2005 *ASHRAE Handbook—Fundamentals*. Project specifications should define allowable duct leakage, specify the need for leak testing, and require the duct installer to perform a leak test after installing an initial portion of the duct. Ducts should be sealed in compliance with Table 1; duct seal levels are defined in Table 2. Exposed supply ducts in conditioned spaces should be seal level A to prevent dirt smudges. Leakage classifications for ducts are given in Table 6 in Chapter 35 of the 2005 *ASHRAE Handbook—Fundamentals*. Procedures in the *HVAC Air Duct Leakage Test Manual* (SMACNA 1985) should be followed for leak testing. If a test indicates excess leakage, corrective measures should be taken to ensure quality.

ASHRAE Research Project RP-1132 (Sahu and Idem 2003) established baseline data on the effects of sealing ducts to diffusers and supply and return grilles. Test setups were designed to simulate typical field conditions. For flow rates from 250 to 1300 cfm, the leakage flow rate generally varied from 1 to 8% of the total approach flow rate, depending on sealing conditions. Moderate sealing of the connection between the air terminal collar and the duct significantly reduced leakage. In general, for rigid ducts, mounting the terminal collar to the duct with sheet metal screws (as prescribed by the terminal manufacturer) reduced leakage by more than 50%. Likewise, using draw bands to mount air terminals to flexible ducts can reduce leakage at the collar to virtually zero. It is recommended that all rigid duct/terminal connections be sealed.

Responsibility for proper assembly and sealing belongs to the installing contractor. The most cost-effective way to control leakage is to follow proper installation procedures. However, the incremental cost of achieving 1% or less leakage becomes prohibitively high, particularly for large duct systems. Because access for repairs is usually limited, poorly installed duct systems that must later be resealed can cost more than a proper installation.

Table 3 Residential Metal Duct Construction[1]

Shape of Duct and Exposure	Galvanized Steel Minimum Thickness, in. (gage)	Aluminum[2] Minimum Thickness, in.
Round and enclosed rectangular ducts		
14 in. or less	0.013 (30)	0.014
Over 14 in.	0.016 (28)	0.0175
Exposed rectangular ducts		
14 in. or less	0.016 (28)	0.0175
Over 14 in.	0.019 (26)	0.0215

[1]NFPA *Standard* 90B.
[2]ASTM B-209; Alloy 3003-H14

RESIDENTIAL DUCT CONSTRUCTION

NFPA *Standard* 90B, ICC's (2006) *International Residential Code for One- and Two-Family Dwellings*, or a local code is used for duct systems in single-family dwellings. Generally, local authorities use NFPA *Standard* 90A for multifamily homes.

Supply ducts may be steel, aluminum, or a material with a UL *Standard* 181 listing. Sheet metal ducts should be of the minimum thickness shown in Table 3 and installed in accordance with *HVAC Duct Construction Standards—Metal and Flexible* (SMACNA 2005). Fibrous glass ducts should be installed in accordance with the *Fibrous Glass Duct Construction Standards* (NAIMA 2002b; SMACNA 2003). For return ducts, alternative materials, and other exceptions, consult NFPA *Standard* 90B.

COMMERCIAL DUCT CONSTRUCTION

Materials

Many building code agencies use NFPA *Standard* 90A as a guide. NFPA *Standard* 90A invokes UL *Standard* 181, which classifies ducts as follows:

Class 0: Zero flame spread, zero smoke developed
Class 1: 25 flame spread, 50 smoke developed

NFPA *Standard* 90A states that ducts must be iron, steel, aluminum, concrete, masonry, or clay tile. However, ducts may be UL *Standard* 181 Class 1 materials when they are not used as vertical risers of more than two stories or in systems with air temperatures higher than 250°F. Many manufactured flexible and fibrous glass ducts are UL listed as Class 1. For galvanized ducts, a G60 coating is recommended (see ASTM *Standard* A653). The minimum thickness and weight of sheet metal sheets are given in Tables 4A, 4B, and 4C.

External duct-reinforcing members are formed from sheet metal or made from hot-rolled or extruded structural shapes. The size and weights of commonly used members are given in Table 5.

Rectangular and Round Ducts

Rectangular Metal Ducts. *HVAC Duct Construction Standards—Metal and Flexible* (SMACNA 2005) lists construction requirements for rectangular steel ducts and includes combinations of duct thicknesses, reinforcement, and maximum distance between reinforcements. Transverse joints (e.g., standing drive slips, pocket locks, and companion angles) and, when necessary, intermediate structural members and tie rods are designed to reinforce the duct system. Proprietary joint systems are available from several manufacturers. *Rectangular Industrial Duct Construction Standards* (SMACNA 2007) gives construction details for ducts up to 144 in. wide at a pressure up to ±150 in. of water.

Fittings must be reinforced similarly to sections of straight duct. On size change fittings, the greater fitting dimension determines material thickness. Where fitting curvature or internal member attachments provide equivalent rigidity, such features may be credited as reinforcement.

Table 4A Galvanized Sheet Thickness

Galvanized Sheet Gage	Thickness, in. Nominal	Minimum*	Nominal Weight, lb/ft²
30	0.0157	0.0127	0.656
28	0.0187	0.0157	0.781
26	0.0217	0.0187	0.906
24	0.0276	0.0236	1.156
22	0.0336	0.0296	1.406
20	0.0396	0.0356	1.656
18	0.0516	0.0466	2.156
16	0.0635	0.0575	2.656
14	0.0785	0.0705	3.281
13	0.0934	0.0854	3.906
12	0.1084	0.0994	4.531
11	0.1233	0.1143	5.156
10	0.1382	0.1292	5.781

*Minimum thickness is based on thickness tolerances of hot-dip galvanized sheets in cut lengths and coils (per ASTM *Standard* A924). Tolerance is valid for 48 and 60 in. wide sheets.

Table 4B Uncoated Steel Sheet Thickness

Manufacturers' Standard Gage	Thickness, in. Nominal	Minimum* Hot-Rolled	Minimum* Cold-Rolled	Nominal Weight, lb/ft²
28	0.0149		0.0129	0.625
26	0.0179		0.0159	0.750
24	0.0239		0.0209	1.000
22	0.0299		0.0269	1.250
20	0.0359		0.0329	1.500
18	0.0478	0.0428	0.0438	2.000
16	0.0598	0.0538	0.0548	2.500
14	0.0747	0.0677	0.0697	3.125
13	0.0897	0.0827	0.0847	3.750
12	0.1046	0.0966	0.0986	4.375
11	0.1196	0.1116	0.1136	5.000
10	0.1345	0.1265	0.1285	5.625

Note: Table is based on 48 in. width coil and sheet stock; 60 in. coil has same tolerance, except that 16 gage is ±0.007 in. in hot-rolled coils and sheets.
*Minimum thickness is based on thickness tolerances of hot- and cold-rolled sheets in cut lengths and coils (per ASTM *Standards* A568, A1008, and A1011).

Table 4C Stainless Steel Sheet Thickness

Gage	Thickness, in. Nominal	Minimum*	Nominal Weight, lb/ft² Stainless Steel 300 Series	400 Series
28	0.0151	0.0131	0.634	0.622
26	0.0178	0.0148	0.748	0.733
24	0.0235	0.0205	0.987	0.968
22	0.0293	0.0253	1.231	1.207
20	0.0355	0.0315	1.491	1.463
18	0.0480	0.0430	2.016	1.978
16	0.0595	0.0535	2.499	2.451
14	0.0751	0.0681	3.154	3.094
13	0.0900	0.0820	3.780	3.708
12	0.1054	0.0964	4.427	4.342
11	0.1200	0.1100	5.040	4.944
10	0.1350	0.1230	5.670	5.562

*Minimum thickness is based on thickness tolerances of hot-rolled sheets in cut lengths and cold-rolled sheets in cut lengths and coils (per ASTM *Standard* A480).

Round Metal Ducts. Round ducts are inherently strong and rigid, and are generally the most efficient and economical ducts for air systems. The dominant factor in round duct construction is the material's ability to withstand the physical abuse of installation and negative pressure requirements. SMACNA (2005) lists construction requirements as a function of static pressure, type of seam (spiral or longitudinal), and diameter. Proprietary joint systems are available from several manufacturers.

Table 5 Steel Angle Weight per Unit Length (Approximate)

Angle Size, in.	Weight, lb/ft
3/4 × 3/4 × 1/8	0.59
1 × 1 × 0.0466 (minimum)	0.36
1 × 1 × 0.0575 (minimum)	0.44
1 × 1 × 1/8	0.80
1 1/4 × 1 1/4 × 0.0466 (minimum)	0.45
1 1/4 × 1 1/4 × 0.0575 (minimum)	0.55
1 1/4 × 1 1/4 × 0.0854 (minimum)	0.65
1 1/4 × 1 1/4 × 1/8	1.01
1 1/2 × 1 1/2 × 0.0575 (minimum)	0.66
1 1/2 × 1 1/2 × 1/8	1.23
1 1/2 × 1 1/2 × 3/16	1.80
1 1/2 × 1 1/2 × 1/4	2.34
2 × 2 × 0.0575 (minimum)	0.89
2 × 2 × 1/8	1.65
2 × 2 × 3/16	2.44
2 × 2 × 1/4	3.19
2 1/2 × 2 1/2 × 3/16	3.07
2 1/2 × 2 1/2 × 1/4	4.10

Nonferrous Ducts. SMACNA (2005) lists construction requirements for rectangular (±3 in. of water) and round (±2 in. of water) aluminum ducts. *Round Industrial Duct Construction Standards* (SMACNA 1999) gives construction requirements for round aluminum duct systems for pressures up to ±30 in. of water.

Flat Oval Ducts

SMACNA (2005) also lists flat oval duct construction requirements. Seams and transverse joints are generally the same as those allowed for round ducts. However, proprietary joint systems are available from several manufacturers. Flat oval duct is for positive-pressure applications only, unless special designs are used. Hanger designs and installation details for rectangular ducts generally also apply to flat oval ducts.

Fibrous Glass Ducts

Fibrous glass ducts are a composite of rigid fiberglass and a factory-applied facing (typically aluminum or reinforced aluminum), which serves as a finish and vapor retarder. This material is available in molded round sections or in board form for fabrication into rectangular or polygonal shapes. Duct systems of round and rectangular fibrous glass are generally limited to 2400 fpm and ±2 in. of water. Molded round ducts are available in higher pressure ratings. *Fibrous Glass Duct Construction Standards* (NAIMA 2002b; SMACNA 2003) and manufacturers' installation instructions give details on fibrous glass duct construction. SMACNA (2003) also covers duct and fitting fabrication, closure, reinforcement, and installation, including installation of duct-mounted HVAC appurtenances (e.g., volume dampers, turning vanes, register and grille connections, diffuser connections, access doors, fire damper connections, electric heaters). AIA (2006) includes guidelines for using fibrous glass duct in hospital and health care facilities.

Flexible Ducts

Flexible ducts connect mixing boxes, light troffers, diffusers, and other terminals to the air distribution system. SMACNA (2005) has an installation standard and a specification for joining, attaching, and supporting flexible duct. ADC (2003) has another installation standard.

The routing, the number and sharpness of bends, and the amount of sag allowed between support joints significantly affect system performance because of the increased resistance each introduces. Use the minimum length of flexible duct needed to make connections. Excess length of flexible ducts should not be installed to allow for possible future relocations of air terminal devices. Constructability-related flow restrictions should be avoided (e.g., duct-hanging wires should not reduce the effective duct diameter). Avoid bending ducts across sharp corners or incidental contact with metal fixtures, pipes, or conduits. The turn radius at duct centerline should not be less than one duct diameter.

At terminal units, splices, and collars, pull back the jacket and insulation from the core and connect to the collar in accordance with ADC (2003) or SMACNA (2005) installation standards. After the flexible duct is connected to a collar or splice by appropriate duct tape and/or clamps, pull the jacket and insulation back over the core. For a collar, tape the jacket with at least two wraps of duct tape. A clamp may be used in place of or in combination with the duct tape. For a splice, tape jackets together with at least two wraps of duct tape.

UL *Standard* 181 covers testing materials used to fabricate flexible ducts that are separately categorized as air ducts or connectors. NFPA *Standard* 90A defines acceptable use of these products. The flexible duct connector has less resistance to flame penetration, has lower puncture and impact resistance, and is subject to many restrictions listed in NFPA *Standard* 90A. Only flexible ducts that are air duct rated should be specified. Tested products are listed in the UL *Online Certifications Directory.*

Plenums and Apparatus Casings

SMACNA (2005) shows details on field-fabricated plenum and apparatus casings. Sheet metal thickness and reinforcement for plenum and casing pressure outside the range of –3 to 10 in. of water can be based on *Rectangular Industrial Duct Construction Standards* (SMACNA 2007).

Carefully analyze plenums and apparatus casings on the discharge side of a fan for maximum operating pressure in relation to the construction detail being specified. On the fan's suction side, plenums and apparatus casings are normally constructed to withstand negative air pressure at least equal to the total upstream static pressure loss. Accidentally stopping intake airflow can apply a negative pressure as great as the fan shutoff pressure. Conditions such as malfunctioning dampers or clogged louvers, filters, or coils can collapse a normally adequate casing. To protect large casing walls or roofs from damage, it is more economical to provide fan safety interlocks, such as damper end switches or pressure limit switches, than to use heavier sheet metal construction.

Apparatus casings can perform two acoustical functions. If the fan is completely enclosed within the casing, fan noise transmission through the fan room to adjacent areas is reduced substantially. An acoustically lined casing also reduces airborne noise in connecting ductwork. Acoustical treatment may consist of a single metal wall with a field-applied acoustical liner or thermal insulation, or a double-walled panel with an acoustical liner and a perforated metal inner liner. Double-walled casings are marketed by many manufacturers, who publish data on structural, acoustical, and thermal performance and also prepare custom designs.

Acoustical Treatment

Metal ducts are frequently lined with acoustically absorbent materials to reduce noise. Although many materials are acoustically absorbent, duct liners must also be resistant to erosion and fire and have properties compatible with the ductwork fabrication and erection process. For higher-velocity ducts, double-walled construction using a perforated metal inner liner is frequently specified. Chapter 47 of the 2007 *ASHRAE Handbook—HVAC Applications* addresses design considerations, including external lagging. ASTM *Standard* C423 covers laboratory testing of duct liner materials to determine their sound absorption coefficients, and ASTM *Standard* E477 covers acoustical insertion loss of duct liner materials. Designers should review all of the tests in ASTM *Standard* C1071. A wide range of performance attributes (e.g., vapor adsorption and resistance to erosion, temperature, bacteria, and fungi) is covered. Health and safety precautions are addressed, and manufacturers' certifications of compliance are also covered. AIA (2006) includes guidelines for using duct liner in hospital and health care facilities.

Rectangular duct liners should be secured by mechanical fasteners and installed in accordance with *HVAC Duct Construction Standards—Metal and Flexible* (SMACNA 2005). Adhesives should be Type I, in conformance to ASTM *Standard* C916, and should be applied to the duct, with at least 90% coverage of mating surfaces. Good workmanship prevents delamination of the liner and possible blockage of coils, dampers, flow sensors, or terminal devices. Avoid uneven edge alignment at butted joints to minimize unnecessary resistance to airflow (Swim 1978).

Rectangular metal ducts are susceptible to rumble from flexure in the duct walls during start-up and shutdown. For a system that must switch on and off frequently (for energy conservation) while buildings are occupied, duct construction that reduces objectionable noise should be specified.

Hangers

SMACNA (2005) covers commercial HVAC system hangers for rectangular, round, and flat oval ducts. When special analysis is required for larger ducts or loads or for other hanger configurations than are given, AISC and AISI design manuals should be consulted. To hang or support fibrous glass ducts, the methods detailed by NAIMA (2002b) and SMACNA (2003) are recommended. UL *Standard* 181 discusses maximum support intervals for UL listed ducts.

INDUSTRIAL DUCT CONSTRUCTION

NFPA *Standard* 91 is widely used for duct systems conveying particulates and removing flammable vapors (including paint-spraying residue), and corrosive fumes. Particulate-conveying duct systems are generally classified as follows:

- **Class 1** covers nonparticulate applications, including makeup air, general ventilation, and gaseous emission control.
- **Class 2** is imposed on moderately abrasive particulate in light concentration, such as that produced by buffing and polishing, woodworking, and grain handling.
- **Class 3** consists of highly abrasive material in low concentration, such as that produced from abrasive cleaning, dryers and kilns, boiler breeching, and sand handling.
- **Class 4** is composed of highly abrasive particulates in high concentration, including materials conveying high concentrations of particulates listed under Class 3.
- **Class 5** covers corrosive applications such as acid fumes.

For contaminant abrasiveness ratings, see *Round Industrial Duct Construction Standards* (SMACNA 1999). Consult Chapters 12 to 31 of the 2007 *ASHRAE Handbook—HVAC Applications* for specific processes and uses.

Materials

Galvanized steel, uncoated carbon steel, or aluminum are most frequently used for industrial air handling. Aluminum ducts are not used for conveying abrasive materials; when temperatures exceed 400°F, galvanized steel is not recommended. Duct material for handling corrosive gases, vapors, or mists must be selected carefully. For the application of metals and use of protective coatings in corrosive environments, consult *Accepted Industry Practice for Industrial Duct Construction* (SMACNA 1975), the *Pollution Engineering Practice Handbook* (Cheremisinoff and Young 1975), and publications of the National Association of Corrosion Engineers (NACE) and ASM International (asminternational.org).

Round Ducts

SMACNA (1999) gives information on selecting material thickness and reinforcement members for spiral and nonspiral industrial ducts. (Spiral-seam ducts are only for Class 1 and 2 applications.) The tables in this manual are presented as follows:

Class. *Steel*: Classes 1, 2, 3, 4, and 5. *Aluminum*: Class 1 only. *Stainless steel*: Classes 1 and 5.

Pressure classes for steels and aluminum. ±2 to ±30 in. of water, in increments of 2 in. of water.

Duct diameter for steels and aluminum. 4 to 96 in., in increments of 2 in. Equations are available for calculating construction requirements for diameters over 96 in.

Software is also available from SMACNA for design with steel, stainless steel, and aluminum. For other spiral duct applications, consult manufacturers' construction schedules, such as those listed in the *Industrial Duct Engineering Data and Recommended Design Standards* (United McGill Corporation 1985).

Rectangular Ducts

Rectangular Industrial Duct Construction Standards (SMACNA 2007) is available for selecting material thickness and reinforcement members for industrial ducts. The data in this manual give the duct construction for any pressure class and panel width. Each side of a rectangular duct is considered a panel, each of which is usually built of material with the same thickness. Ducts (usually those with heavy particulate accumulation) are sometimes built with the bottom plate thicker than the other three sides to save material.

The designer selects a combination of panel thickness, reinforcement, and reinforcement member spacing to limit the deflection of the duct panel to a design maximum. Any shape of transverse joint or intermediate reinforcement member that meets the minimum requirement of both section modulus and moment of inertia may be selected. The SMACNA data, which may also be used for designing apparatus casings, limit the combined stress in either the panel or structural member to 24,000 psi and the maximum allowable deflection of the reinforcement members to 1/360 of the duct width.

Construction Details

Recommended manuals for other construction details are *Industrial Ventilation: A Manual of Recommended Practice* (ACGIH 2007), NFPA *Standard* 91, and *Accepted Industry Practice for Industrial Ventilation* (SMACNA 1975). For industrial duct Classes 2, 3, and 4, transverse reinforcing of ducts subject to negative pressure below –3 in. of water should be welded to the duct wall rather than relying on mechanical fasteners to transfer the static load.

Hangers

The *Steel Construction Manual* (AISC 2006) and the *Cold-Formed Steel Design Manual* (AISI 2002) give design information for industrial duct hangers and supports. SMACNA standards for round and rectangular industrial ducts (SMACNA 1999, 2007) and manufacturers' schedules include duct design information for supporting ducts at intervals of up to 35 ft.

ANTIMICROBIAL-TREATED DUCTS

Antimicrobial-treated ducts are made of steel that is coated (as a precoating, or after fabrication) with a substance that inhibits the growth of bacteria, mold, and fungi (including mildew). Either galvanized or stainless steel ducts can be used, if service temperatures of the antimicrobial compound are not exceeded. Prefabricated coatings allow the metal to be pressed, drawn, bent, and rollformed without coating loss. Some coatings are still effective even with small scratches. Large imperfections, such as spot welds and welded joints, can be repaired with a touch-up paint of the antimicrobial compound.

All antimicrobial coatings or touch-up paint should be an EPA-registered antimicrobial compound, should be tested under ASTM *Standard* E84, survive minimum and maximum service temperature limits, and comply with NFPA *Standards* 90A and 90B. Coatings should have flame spread/smoke developed ratings not exceeding 25/50, and meet local building code requirements.

DUCT CONSTRUCTION FOR GREASE- AND MOISTURE-LADEN VAPORS

Installation and construction of ducts for removing smoke or grease-laden vapors from cooking equipment should be in accordance with NFPA *Standard* 96 and SMACNA's round and rectangular industrial duct construction standards (SMACNA 1999, 2007). Kitchen exhaust ducts that conform to NFPA *Standard* 96 must (1) be constructed from carbon steel with a minimum thickness of 0.054 in. (16 gage) or stainless steel sheet with a minimum thickness of 0.043 in. (18 gage); (2) have all longitudinal seams and transverse joints continuously welded; and (3) be installed without dips or traps that may collect residues, except where traps with continuous or automatic removal of residue are provided. Because fires may occur in these systems (producing temperatures in excess of 2000°F), provisions are necessary for expansion in accordance with the following table. Ducts that must have a fire resistance rating are usually encased in materials with appropriate thermal and durability ratings.

Kitchen Exhaust Duct Material	Duct Expansion at 2000°F, in/ft
Carbon steel	0.19
Type 304 stainless steel	0.23
Type 430 stainless steel	0.13

Ducts that convey moisture-laden air must have construction specifications that properly account for corrosion resistance, drainage, and waterproofing of joints and seams. No nationally recognized standards exist for applications in areas such as kitchens, swimming pools, shower rooms, and steam cleaning or washdown chambers. Galvanized steel, stainless steel, aluminum, and plastic materials have been used. Wet and dry cycles increase metal corrosion. Chemical concentrations affect corrosion rate significantly. Chapter 48 of the 2007 *ASHRAE Handbook—HVAC Applications* addresses material selection for corrosive environments. Conventional duct construction standards are frequently modified to require welded or soldered joints, which are generally more reliable and durable than sealant-filled, mechanically locked joints. The number of transverse joints should be minimized, and longitudinal seams should not be located on the bottom of the duct. Risers should drain and horizontal ducts should pitch in the direction most favorable for moisture control. ACGIH (2007) covers hood design.

RIGID PLASTIC DUCTS

The *Thermoplastic Duct (PVC) Construction Manual* (SMACNA 1995) covers thermoplastic (polyvinyl chloride, polyethylene, polypropylene, acrylonitrile butadiene styrene) ducts used in commercial and industrial installations. SMACNA's manual provides comprehensive polyvinyl chloride duct construction details for positive or negative 2, 4, 6, and 10 in. of water. NFPA *Standard* 91 provides construction details and application limitations for plastic ducts. Model code agencies publish evaluation reports indicating terms of acceptance of manufactured ducts and other ducts not otherwise covered by industry standards and codes.

Physical properties, manufacture, construction, installation, and methods of testing for fiberglass-reinforced thermosetting plastic (FRP) ducts are described in the *Thermoset FRP Duct Construction Manual* (SMACNA 1997). These ducts are intended for air conveyance in corrosive environments as manufactured by hand lay-up, spray-up, and filament winding fabrication techniques. The term *FRP* also refers to fiber-reinforced plastic (fibers other than glass). Other terms for FRP are reinforced thermoset plastic (RTP) and glass-reinforced plastic (GRP), which is commonly used in Europe and Australia. SMACNA (1997) has construction standards for pressures up to ±30 in. of water and duct sizes from 4 to 72 in. round and 12 to 96 in. rectangular.

FABRIC DUCTS

Fabric ducts can distribute air through built-in diffusers or (with a permeable fabric) can diffuse air directly. Use is limited to areas where the fabric duct is exposed within an environment.

Consult the manufacturer for design criteria for selecting air dispersion type (vents, orifices, or porous fabric), fabric (porous or nonporous, color, weight, and construction), suspension options (cable or track options), and installation instructions. In design, consider velocities of 1000 to 1800 fpm at static pressures of 0.3 to 1.0 in. of water to ensure proper airflow performance. Excessively turbulent airflow (from metal fittings or fans) or higher inlet velocities can cause fabric fluttering, excessive noise, premature material failure, and poor air dispersion. Fabric airflow restriction devices are available to help balance static regain, reduce turbulence, reduce abrupt inflation, and balance airflow into branch ducts.

UNDERGROUND DUCTS

No comprehensive standards exist for underground air duct construction. Coated steel, asbestos cement, plastic, tile, concrete, reinforced fiberglass, and other materials have been used. Underground duct and fittings should always be round and have a minimum thickness as listed in SMACNA (2005), although greater thickness may be needed for individual applications. Specifications for construction and installation of underground ducts should account for the following: water tables, ground surface flooding, the need for drainage piping beneath ductwork, temporary or permanent anchorage to resist flotation, frost heave, backfill loading, vehicular traffic load, corrosion, cathodic protection, heat loss or gain, building entry, bacterial organisms, degree of water- and airtightness, inspection or testing before backfill, and code compliance. Chapter 11 has information on cathodic protection of buried metallic conduits. *Installation Techniques for Perimeter Heating and Cooling* (ACCA 1990) covers residential systems and gives five classifications of duct material related to particular performance characteristics. Residential installations may also be subject to the requirements in NFPA *Standard* 90B. Commercial systems also normally require compliance with NFPA *Standard* 90A.

DUCTS OUTSIDE BUILDINGS

Location and construction of ducts exposed to outdoor atmospheric conditions are generally regulated by building codes. Exposed ducts and their sealant/joining systems must be evaluated for the following:

- Waterproofing
- Resistance to external loads (wind, snow, and ice)
- Degradation from corrosion, ultraviolet radiation, or thermal cycles
- Heat transfer, solar reflectance, and thermal emittance
- Susceptibility to physical damage
- Hazards at air inlets and discharges
- Maintenance needs

In addition, supports must be custom-designed for rooftop, wall-mounted, and bridge or ground-based applications. Specific requirements must also be met for insulated and uninsulated ducts.

SEISMIC QUALIFICATION

Seismic analysis of duct systems may be required by building codes or federal regulations. Provisions for seismic analysis are given by the Federal Emergency Management Agency (FEMA 2000a, 2000b). Ducts, duct hangers, fans, fan supports, and other duct-mounted equipment are generally evaluated independently. Chapter 54 of the 2007 *ASHRAE Handbook—HVAC Applications* gives design details. SMACNA (1998) provides guidelines for seismic restraints of mechanical systems and gives bracing details for ducts, pipes, and conduits that apply to the model building codes and ASCE *Standard* 7. FEMA (2002, 2004a, 2004b) has three fully illustrated guides that show equipment installers how to attach mechanical and electrical equipment or ducts and pipes to a building to minimize earthquake damage.

SHEET METAL WELDING

AWS (2006) covers sheet metal arc welding and braze welding procedures. It also addresses the qualification of welders and welding operators, workmanship, and the inspection of production welds.

THERMAL INSULATION

Insulation materials for ducts, plenums, and apparatus casings are covered in Chapter 24 of the 2005 *ASHRAE Handbook—Fundamentals*. Codes generally limit factory-insulated ducts to UL *Standard* 181, Class 0 or 1. *Commercial and Industrial Insulation Standards* (MICA 2006) gives insulation details. ASTM *Standard* C1290 gives specifications for fibrous glass blanket external insulation for ducts.

MASTER SPECIFICATIONS

Master specifications for duct construction and most other elements in building construction are produced and regularly updated by several organizations. Two examples are *MASTERSPEC* by the American Institute of Architects (AIA) and *SPECTEXT* by the Construction Sciences Research Foundation (CSRF). *MasterFormat*™ (CSI 2004) is the organization standard for specifications. These documents are model project specifications that require little editing to customize each application for a project.

Nationally recognized model specifications

- Focus industry practice on a uniform set of requirements in a widely known format
- Reduce the need to prepare new specifications for each project
- Remain relatively current and automatically incorporate new and revised editions of construction, test, and performance standards published by other organizations
- Are adaptable to small or large projects
- Are performance- or prescription-oriented as the designer desires
- Provide lists of products and equipment by name and number or descriptions that are deemed equal
- Are divided into subsections that are coordinated with other sections of related work
- Are increasingly being used by government agencies to replace separate and often different agency specifications

REFERENCES

ACCA. 1990. Installation techniques for perimeter heating and cooling. *Manual* 4. Air Conditioning Contractors of America, Arlington, VA.

ACGIH. 2007. *Industrial ventilation—A manual of recommended practice for design*, 26th ed. American Conference of Governmental Industrial Hygienists, Cincinnati.

ADC. 2003. *Flexible duct performance and installation standards*, 4th ed. Air Diffusion Council, Schaumburg, IL.

AIA. Updated quarterly. *MASTERSPEC*. American Institute of Architects, Washington, D.C.

AIA. 2006. *Guidelines for design and construction of hospital and health care facilities*. American Institute of Architects, Washington, D.C.

AISC. 2006. *Steel construction manual*, 13th ed. American Institute of Steel Construction, Chicago.

AISI. 2002. *Cold-formed steel design manual*. American Iron and Steel Institute, Washington, D.C.

ASCE. 2005. Minimum design loads for buildings and other structures. ANSI/ASCE *Standard* 7-05. American Society of Civil Engineers, Reston, VA.

ASTM. 2006. Standard specification for general requirements for flat-rolled stainless and heat-resisting steel plate, sheet, and strip. *Standard* A480-06b. American Society for Testing and Materials, West Conshohocken, PA.

ASTM. 2007. Standard specification for steel, sheet, carbon, structural, and high-strength, low-alloy, hot-rolled and cold-rolled, general requirements for. *Standard* A568/A568M-07a. American Society for Testing and Materials, West Conshohocken, PA.

ASTM. 2007. Standard specification for sheet steel, zinc-coated (galvanized) or zinc-iron alloy-coated (galvannealed) by the hot-dip process. *Standard* A653/A653M-07. American Society for Testing and Materials, West Conshohocken, PA.

ASTM. 2007. Standard specification for general requirements for steel sheet, metallic-coated by the hot-dip process. *Standard* A924-07. American Society for Testing and Materials, West Conshohocken, PA.

ASTM. 2007. Standard specification for steel, sheet, cold-rolled, carbon, structural, high-strength low-alloy and high-strength low-alloy with improved formability, solution hardened, and bake hardenable. *Standard* A1008/A1008M-07a. American Society for Testing and Materials, West Conshohocken, PA.

ASTM. 2007. Standard specification for steel, sheet and strip, hot-rolled, carbon, structural, high-strength low-alloy and high-strength low-alloy with improved formability, and ultra-high strength. *Standard* A1011/A1011M-07. American Society for Testing and Materials, West Conshohocken, PA.

ASTM. 2007. Standard specification for aluminum and aluminum-alloy sheet and plate. *Standard* B209-07. American Society for Testing and Materials, West Conshohocken, PA.

ASTM. 2007. Standard test method for sound absorption and sound absorption coefficients by the reverberation room method. *Standard* C423-07a. American Society for Testing and Materials, West Conshohocken, PA.

ASTM. 2007. Standard specification for adhesives for duct thermal insulation. *Standard* C916-85(2007). American Society for Testing and Materials, West Conshohocken, PA.

ASTM. 2005. Standard specification for fibrous glass duct lining insulation (thermal and sound absorbing material). *Standard* C1071-05. American Society for Testing and Materials, West Conshohocken, PA.

ASTM. 2006. Standard specification for flexible fibrous glass blanket insulation used to externally insulate HVAC ducts. *Standard* C1290-06. American Society for Testing and Materials, West Conshohocken, PA.

ASTM. 2007. Standard test method for surface burning characteristics of building materials. *Standard* E84-07b. American Society for Testing and Materials, West Conshohocken, PA.

ASTM. 2006. Standard test method for measuring acoustical and airflow performance of duct liner materials and prefabricated silencers. *Standard* E477-06a. American Society for Testing and Materials, West Conshohocken, PA.

AWS. 2006. Sheet metal welding code. *Standard* D9.1M-2006. American Welding Society, Miami.

Cheremisinoff, P.N. and R.A. Young. 1975. *Pollution engineering practice handbook*. Ann Arbor Science Publishers, Inc., Ann Arbor, MI.

CSI. 2004. *MasterFormat*™. Construction Specifications Institute, Alexandria, VA.

CSRF. *SPECTEXT.* Construction Sciences Research Foundation, Baltimore.

FEMA. 2000a. NEHRP (National Earthquake Hazards Reduction Program) recommended provisions for seismic regulations for new buildings and other structures—Part 1: Provisions. *Publication* 368. Federal Emergency Management Agency, Washington, D.C.

FEMA. 2000b. NEHRP (National Earthquake Hazards Reduction Program) recommended provisions for seismic regulations for new buildings and other structures—Part 2: Commentary. *Publication* 369. Federal Emergency Management Agency, Washington, D.C.

FEMA. 2002. Installing seismic restraints for mechanical equipment. *Publication* 412. Federal Emergency Management Agency, Washington, D.C.

FEMA. 2004a. Installing seismic restraints for electrical equipment. *Publication* 413. Federal Emergency Management Agency, Washington, D.C.

FEMA. 2004b. Installing seismic restraints for duct and pipe. *Publication* 414. Federal Emergency Management Agency, Washington, D.C.

ICC. 2006. *International residential code for one- and two-family dwellings*. International Code Council, Washington, D.C.

MICA. 2006. *National commercial and industrial insulation standards*, 6th ed. Midwest Insulation Contractors Association. Omaha, NE.

NADCA. 2006. Assessment, cleaning, and restoration of HVAC systems. *Standard* ACR. National Air Duct Cleaners Association, Washington, D.C.

NAIMA. 2002a. *Cleaning fibrous glass insulated air duct systems—Recommended practices*. North American Insulation Contractors Association, Alexandria, VA.

NAIMA. 2002b. *Fibrous glass duct construction standards*, 5th ed. North American Insulation Contractors Association, Alexandria, VA.

NFPA. 2002. Installation of air conditioning and ventilating systems. ANSI/NFPA *Standard* 90A. National Fire Protection Association, Quincy, MA.

NFPA. 2006. Installation of warm air heating and air-conditioning systems. ANSI/NFPA *Standard* 90B. National Fire Protection Association, Quincy, MA.

NFPA. 2004. Exhaust systems for air conveying of vapors, gases, mists, and noncombustible particulate solids. NFPA *Standard* 91. National Fire Protection Association, Quincy, MA.

NFPA. 2008. Ventilation control and fire protection of commercial cooking operations. ANSI/NFPA *Standard* 96. National Fire Protection Association, Quincy, MA.

Sahu, S. and S.A. Idem. 2003. Leakage of ducted air terminal connections: Part 1—Experimental procedure and data reduction; Part 2—Experimental results (RP-1132). *ASHRAE Transactions* 109(2):185-206.

SMACNA. 1975. *Accepted industry practice for industrial duct construction*, 1st ed. Sheet Metal and Air Conditioning Contractors' National Association, Chantilly, VA.

SMACNA. 1985. *HVAC air duct leakage test manual*, 1st ed. Sheet Metal and Air Conditioning Contractors' National Association, Chantilly, VA.

SMACNA. 1995. *Thermoplastic duct (PVC) construction manual*, 1st ed. Sheet Metal and Air Conditioning Contractors' National Association, Chantilly, VA.

SMACNA. 1997. *Thermoset FRP duct construction manual*, 1st ed. Sheet Metal and Air Conditioning Contractors' National Association, Chantilly, VA.

SMACNA. 1998. *Seismic restraint manual: Guidelines for mechanical systems*, 2nd ed. ANSI/SMACNA *Standard*. Sheet Metal and Air Conditioning Contractors' National Association, Chantilly, VA.

SMACNA. 1999. *Round industrial duct construction standards*, 2nd ed. ANSI/SMACNA *Standard*. Sheet Metal and Air Conditioning Contractors' National Association, Chantilly, VA.

SMACNA. 2003. *Fibrous glass duct construction standards*, 7th ed. Sheet Metal and Air Conditioning Contractors' National Association, Chantilly, VA.

SMACNA. 2005. *HVAC duct construction standards—Metal and flexible*, 3rd ed. ANSI/SMACNA *Standard*. Sheet Metal and Air Conditioning Contractors' National Association, Chantilly, VA.

SMACNA. 2007. *Rectangular industrial duct construction standards*, 2nd ed. Sheet Metal and Air Conditioning Contractors' National Association, Chantilly, VA.

Swim, W.B. 1978. Flow losses in rectangular ducts lined with fiberglass. ASHRAE *Transactions* 84(2).

UL. 2005. Factory-made air ducts and air connectors, 10th ed. ANSI/UL *Standard* 181. Underwriters Laboratories, Northbrook, IL.

UL. (Ongoing.) *UL online certifications directory*. Underwriters Laboratories, Northbrook, IL. http://database.ul.com/cgi.bin/XYV/template/LISEXT/1FRAME/index.htm.

United McGill Corporation. 1985. *Industrial duct engineering data and recommended design standards*. Form No. SMP-IDP (February):24-28. Westerville, OH.

BIBLIOGRAPHY

ASHRAE. 2007. Ventilation for acceptable indoor air quality. ANSI/ ASHRAE *Standard* 62.1-2007.

ASHRAE. 2007. Ventilation for acceptable indoor air quality in low-rise residential buildings. ANSI/ASHRAE *Standard* 62.2-2007.

ASHRAE. 2007. Energy standard for buildings except low-rise residential buildings. ANSI/ASHRAE/IESNA *Standard* 90.1.-2007

ASHRAE. 2004. Energy-efficient design of low-rise residential buildings. ANSI/ASHRAE *Standard* 90.2.

ASHRAE. 2006. Energy conservation in existing buildings. ANSI/ ASHRAE/IESNA *Standard* 100.

ICC. 2006. *International building code.* International Code Council, Washington, D.C.

ICC. 2006. *International energy conservation code.* International Code Council, Washington, D.C.

ICC. 2006. *International mechanical code*. International Code Council. Washington, D.C.

NFPA. 2006. Life safety code. ANSI/NFPA *Standard* 101. National Fire Protection Association, Quincy, MA.

NFPA. 2007. Smoke and heat venting. ANSI/NFPA *Standard* 204. National Fire Protection Association, Quincy, MA.

NFPA. 2006. Chimneys, fireplaces, vents, and solid fuel-burning appliances. ANSI/NFPA *Standard* 211. National Fire Protection Association, Quincy, MA.

NFPA. 2006. Building construction and safety code. ANSI/NFPA *Standard* 5000. National Fire Protection Association, Quincy, MA.

UL. 2005. Closure systems for use with rigid air ducts, 3rd ed. ANSI/UL *Standard* 181A. Underwriters Laboratories, Northbrook, IL.

UL. 2005. Closure systems for use with flexible air ducts and air connectors, 2nd ed. ANSI/UL *Standard* 181B. Underwriters Laboratories, Northbrook, IL.

UL. 2006. Fire dampers, 7th ed. ANSI/UL *Standard* 555. Underwriters Laboratories, Northbrook, IL.

UL. 2006. Ceiling dampers, 3rd ed. ANSI/UL *Standard* 555C. Underwriters Laboratories, Northbrook, IL.

UL. 1999. Smoke dampers, 4th ed. ANSI/UL *Standard* 555S. Underwriters Laboratories, Northbrook, IL.

CHAPTER 19

ROOM AIR DISTRIBUTION EQUIPMENT

SUPPLY air outlets and diffusing equipment introduce air into a conditioned space to obtain a desired indoor atmospheric environment. Return and exhaust air are removed from a space through return and exhaust inlets (*inlet* and *outlet* are defined relative to the duct system and not the room, as shown in Figure 1). Various types of air outlets and inlets are available as standard manufactured products. This chapter describes this equipment, details its proper use, and is intended to help HVAC designers select room air distribution equipment applicable to the air distribution methods outlined in Chapter 56 of the 2007 *ASHRAE Handbook—HVAC Applications.*

Room air distribution systems can be classified according to their primary objective and the method used to accomplish that objective. The objective of any air distribution system is to condition and/or ventilate the space for occupants' thermal comfort, or to support processes within the space, or both.

Methods used to condition a space can be classified as one of the following:

- **Mixed** systems have little or no thermal stratification of air within the occupied and/or process space. Overhead air distribution is an example of this type of system.
- **Full thermal stratification** systems have little or no mixing of air within the occupied and/or process space. Thermal displacement ventilation is an example of this type of system.
- **Partially mixed** systems provide limited mixing of air within the occupied and/or process space. Most underfloor air distribution designs are examples of this type of system.
- **Task/ambient** air distribution focuses on conditioning only a portion of the space for thermal comfort and/or process control. Examples of task/ambient systems are personally controlled desk outlets and spot-conditioning systems. Because task/ambient distribution requires a high level of individual control, it is not covered in this chapter, but is discussed in Chapter 33 of the 2005 *ASHRAE Handbook—Fundamentals*. Additional design guidance is also provided in Bauman (2003).

Figure 2 illustrates the spectrum between the two extremes (full mixing and full stratification) of room air distribution strategies.

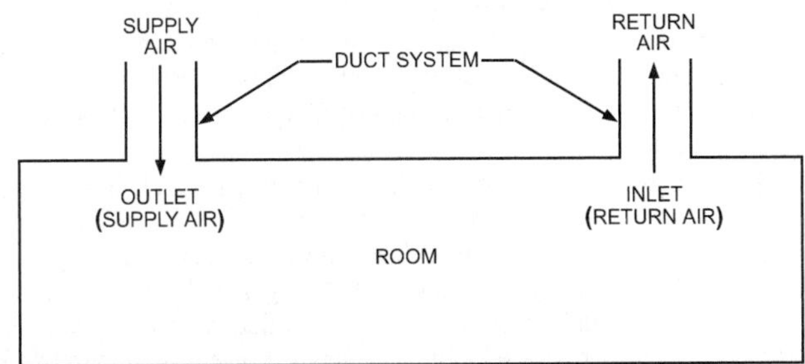

Fig. 1 Designations for Inlet and Outlet

The preparation of this chapter is assigned to TC 5.3, Room Air Distribution.

The following publications should be reviewed when selecting systems and equipment for room air distribution:

- ANSI/ASHRAE *Standard* 55-2004 establishes indoor thermal environmental and personal factors for the occupied space.
- ANSI/ASHRAE *Standard* 62.1-2007 specifies ventilation requirements for acceptable indoor environmental quality. This standard is adopted as part of many building codes.
- ANSI/ASHRAE/IESNA *Standard* 90.1-2004 provides minimum energy efficiency requirements that affect supply air characteristics.
- ANSI/ASHRAE *Standard* 113-2005 defines a method for testing the steady-state air diffusion performance of various room air distribution systems.
- Chapter 47 of the 2007 *ASHRAE Handbook—HVAC Applications* recommends ranges for HVAC-related background noise in various spaces.

Local codes should also be checked for applicability to each of these subjects.

Other useful references on selecting air distribution equipment include Chapter 33 of the 2005 *ASHRAE Handbook—Fundamentals*, Chapter 56 of the 2007 *ASHRAE Handbook—HVAC Applications*, as well as Bauman (2003), Chen and Glicksman (2003), Rock and Zhu (2002), and Skistad et al. (2002).

SUPPLY OUTLETS

FULLY MIXED SYSTEMS

In fully mixed systems, supply air outlets, properly sized and located, control the air pattern to obtain proper air mixing and temperature equalization in the space.

Accessories used with an outlet regulate the volume of supply air and control its flow pattern. For example, an outlet cannot discharge air properly and uniformly unless the air enters it in a straight and uniform manner. Accessories may also be necessary for proper air distribution in a space, so they must be selected and used according to the manufacturers' recommendations.

Primary airflow from an outlet entrains room air into the jet. This entrained air increases the total air in the jet stream. Because the momentum of the jet remains constant, velocity decreases as the mass increases. As the two air masses mix, the temperature of the jet approaches the room air temperature (Rock and Zhu 2002). Outlets should be sized to project air so that its velocity and temperature reach acceptable levels before entering the occupied zone.

Outlet locations and patterns also affect a jet's throw, entrainment, and temperature equalization capabilities. Some general characteristics include the following:

- When outlets are located close to a surface, entrainment may be restricted, which can result in a longer throw.
- When the air pattern is spread horizontally, throw is reduced.
- Outlets with horizontally radial airflow patterns typically have shorter throws than outlets with directional patterns.

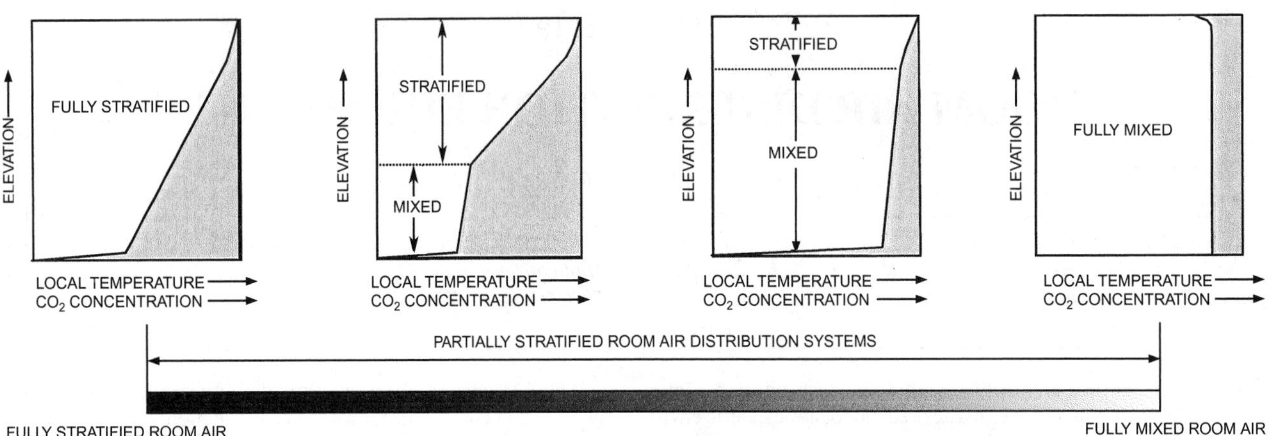

Fig. 2 Classification of Air Distribution Strategies

Ceiling or sidewall outlets in cooling applications are most commonly selected with supply air temperatures at or above 52°F. Special high-induction outlets are available for use with low-temperature air distribution systems (i.e., those with supply air temperature below 52°F). These outlets include special features that rapidly mix cold supply air with room air at the outlet and effectively reduce the temperature differential between the supply and room air. For further information, designers can consult ASHRAE's *Cold Air Distribution System Design Guide* (ASHRAE 1996).

Outlet Selection Procedure

The following procedure is generally used in selecting and locating an outlet in a fully mixed system. More details and examples are available in Rock and Zhu (2002).

1. Determine the amount of air to be supplied to each room. (See Chapters 29 and 30 of the 2005 *ASHRAE Handbook—Fundamentals* to determine air quantities for heating and cooling.)
2. Select the type and quantity of outlets for each room, considering factors such as air quantity required, distance available for throw or radius of diffusion, structural characteristics, and architectural concepts. Table 1, which is based on experience and typical ratings of various outlets, may be used as a guide for using outlets in rooms with various heating and cooling loads. Special conditions, such as ceiling height less than 8 or greater than 12 ft, exposed duct mounting, product modifications, and unusual conditions of room occupancy, should be considered. Manufacturers' performance data should be consulted to determine the suitability of the outlets used.
3. Outlets may be sized and located to distribute air in the space to achieve acceptable temperature and velocity in the occupied zone.
4. Select the proper size outlet from the manufacturers' performance data according to air quantity, neck and discharge velocity, throw, distribution pattern, and sound level. Note manufacturers' recommendations with regard to use. In an open space, the interaction of airstreams from multiple air outlets may alter a single outlet's throw, air temperature, or air velocity. As a result, manufacturers' data may be insufficient to predict air motion in a particular space. Also, obstructions to the primary air distribution pattern require special study.

Factors that Influence Selection

Coanda (Surface or Ceiling) Effect. An airstream moving adjacent to or in contact with a wall or ceiling creates a low-pressure area immediately adjacent to that surface, causing the air to remain in contact with the surface substantially throughout the length of throw. This **Coanda effect**, also referred to as the **surface** or **ceiling effect**, counteracts the drop of a horizontally projected cool airstream.

Round and four-way horizontal-throw ceiling outlets exhibit a high Coanda effect because the discharge air pattern blankets the ceiling area surrounding each outlet. This effect diminishes with a directional discharge that does not blanket the full ceiling surface surrounding the outlet. Sidewall grilles exhibit varying degrees of Coanda effect, depending on the spread of the particular air pattern and the proximity and angle of airstream approach to the surface.

When outlets are mounted on an exposed duct discharging into a free space, the airstream entrains air around the entire perimeter of the jet. As a result, a higher rate of entrainment is obtained and the isothermal throw is shortened by approximately 30%. When outlets are installed on exposed ducts for cooling applications, the supply air tends to drop. Outlets can be selected to counteract this effect.

Multiple parallel jets in close proximity tend to combine into a single jet, increasing the throw distance of the combined jet. More information on this subject can be found in Chapter 33 of the 2005 *ASHRAE Handbook—Fundamentals*.

Temperature Differential. The greater the temperature differential between the supply air projected into a space and the air in the space, the greater the buoyancy effect on the path of the supply airstream. Because heated, horizontally projected air rises and cooled air falls, consideration should be given to this effect during outlet selection. See Chapter 33 of the 2005 *ASHRAE Handbook—Fundamentals* for further discussion of the buoyancy effect.

Low-temperature supply air or cold building start-up in a humid environment may cause condensation. Consideration should be given to the effect of condensation on outlet and space surfaces during outlet selection.

Sound Level. The sound level from an outlet is largely a function of its discharge velocity and transmission of system noise. For a given air capacity, a larger outlet has a lower discharge velocity and corresponding lower generated sound. A larger outlet also allows a higher level of sound to pass through the outlet, which may appear as outlet-generated noise. High-frequency noise can result from excessive outlet velocity but may also be generated in the duct by the moving airstream. Low-frequency noise is generally mechanical equipment sound and/or terminal box or balancing damper sound transmitted through the duct and outlet to the room.

The cause of the noise can usually be pinpointed as outlet or system sounds by removing the outlet core during operation. If the noise remains essentially unchanged, the system is the source. If the noise is significantly reduced, the outlet is the source. The noise may be caused by a highly irregular velocity profile at the entrance to the outlet. The velocity profile should be measured. If the velocity varies less than 10% in the air outlet entrance neck, the outlet is causing the noise. If the velocity profile at the entrance indicates peak velocities significantly higher than average, check the manufacturer's data for sound at the peak velocity. If this value approximates the observed noise, the velocity profile in the duct must be corrected to achieve design performance.

Smudging. Smudging is the deposition of particles on the air outlet or a surface near the outlet. Particles are entrained into the primary discharge jet and impinged into the device or ceiling surface in areas of lower pressure. Smudging tends to be heavier in high-traffic areas near building entrances, where particulates are brought into a space on the bottom of occupants' shoes. In well-maintained systems, filtered supply air contributes little to ceiling smudging. Smudging is typically more prevalent with ceiling-mounted outlets and linear outlets that discharge parallel to the mounting surface than with outlets that discharge perpendicular to the surface.

Variable Air Volume. Outlet(s) should be selected based on the total range of airflow for the space served. Outlet performance characteristics should be evaluated at both the minimum and maximum flow. More information regarding selection of outlets can be found in Chapter 56 of the 2007 *ASHRAE Handbook—HVAC Applications*.

FULLY STRATIFIED SYSTEMS

Stratified room air distribution systems generally rely on supply outlets with very low discharge velocities (70 to 80 fpm based on total face area) to produce minimal room air entrainment so that much of the temperature difference between supply and ambient air is preserved. Thus, cool supply air accumulates in the lower levels of the space. Horizontal movement of air in the space occurs at minimal velocities that are insufficient to produce mixing with room air; thus, the supply airstream maintains its thermal integrity. Heat sources in the space create convection plumes that originate around the boundaries of the heat source and rise naturally because of their buoyancy. If these sources are near the supply airstream, supply air is entrained to fill the void of the rising convection plume.

Although the supply airstream is several degrees cooler than the room air when it enters the occupied zone, the temperature differential between supply and room air is generally less than that commonly used for fully mixed systems. Stratified systems used in transient spaces such as transportation terminals, lobbies, and industrial spaces may, however, use air temperature differentials similar to those in fully mixed systems.

Outlet Selection Procedure

Supply outlets used in stratified air systems tend to be mounted in low sidewall or floor locations. To produce adequately low discharge velocities, the outlets also tend to be quite large. Because discharge velocities are very low, the supply airstream produces little momentum, and obstacles (other than heat sources) in its path have little or no effect on its travel. Selection and application of air outlets for these systems is based primarily on the following considerations:

- Maintaining **vertical temperature gradients** within the occupied space that conform to ASHRAE *Standard* 55-2004. Further guidance on designing for conformance to this standard is presented in Chapter 56 of the 2007 *ASHRAE Handbook—HVAC Applications*.
- Maintaining a **near zone** adjacent to the outlets that is acceptable to the use and occupancy of the space.
- Providing **acoustical performance** that conforms to the requirements of the space.

Supply outlets used in fully stratified systems are typically selected for a maximum face velocity of 70 to 80 fpm. Limitation of the face velocity is determined by noise requirements and proximity of occupants to the outlet. Where space noise requirements are not so stringent and stationary occupants are far away from diffusers, higher face velocities can be used. It is also important that the supply air outlet be designed to distribute airflow evenly across its entire discharge area to avoid excessive velocity deviations.

The area adjacent to the supply outlet where local velocities may exceed 40 fpm is defined as the near zone, where local velocity/temperatures may combine to create drafty conditions. Manufacturers of outlets specifically intended for fully stratified systems publish predicted near-zone values that depend on the outlet supply airflow rate and the initial temperature difference between the supply and room air. Stationary occupants should not be located in the near zone.

For applications requiring very low noise criteria, such as broadcast or recording studios or performing arts venues, acoustical performance can be an important consideration. Because of their low velocity discharge, these supply outlets can generally be selected to meet acoustical criteria for these applications.

Factors that Influence Selection

Space Considerations. To maximize system efficiencies, the preferred locations for supply outlets are in the low sidewall or floor. These supply outlets typically take up considerably more space than outlets used in fully mixed systems. To increase the face area, these outlets are often configured as quarter-round, semicircular, or cylindrical outlets. The latter configuration is generally mounted in open space, whereas the other configurations mentioned are mounted in corners or adjacent to the sidewall, respectively.

Space Heating Considerations. Skistad et al. (2002) reported that displacement ventilation can be combined successfully with radiators and convectors at exterior walls to offset space heat losses. Radiant heating panels and heated floors also can be used with displacement ventilation. When a secondary heating system is used, displacement outlets can supply air with a supply-to-room cooling differential as low as 4°F and still maintain a displacement airflow to the space.

When warm air is supplied through displacement outlets, its performance is similar to a fully mixed system in a heating application.

System and Terminal Considerations. Low-velocity supply air outlets used in fully stratified systems can function properly with either constant- or variable-air-volume (VAV) supply air systems. For VAV applications, the supply air volume should be determined by a thermostat located in the space at a height of four to five feet. Because stratification results in cooler temperatures below the thermostat, its set point can be maintained 2.5 to 3°F warmer than is typical in fully mixed systems.

When fully stratified systems are applied in humid climates, the use of series fan-powered terminal units or other mixing zone devices may allow supply air to be delivered from the central HVAC system at conventional temperatures and then blended with return/plenum air in the terminal to bring the air to an appropriate discharge temperature. However, this may compromise the space contaminant removal benefits of the displacement system.

PARTIALLY MIXED SYSTEMS

Partially mixed room air distribution systems are those whose design intent is to create mixed conditions in a portion of the room while maintaining thermal stratification in the remainder of the space. Supply outlets for these systems are usually designed and selected to discharge cool supply air from low sidewall or floor locations. These outlets produce high room air entrainment such that velocity and temperature differentials between supply and room air can be quickly dissipated. This results in relatively well-mixed room air conditions in some or all of the occupied space, while stratified conditions are maintained throughout the remainder. Although most underfloor air distribution systems should be classified as partially mixed systems, underfloor air delivery can also produce fully mixed or fully stratified room air distribution.

Because supply air is introduced in the occupied zone, the supply air temperature is generally 62 to 65°F for cooling.

Outlet Selection Procedures

Supply outlets used in partially mixed air systems tend to be mounted in the low sidewall or floor. They may also be mounted in the floor or risers beneath seats in public assembly facilities. Because of their high degree of mixing supply and room air, these outlets can be selected for much higher discharge velocities than those used in fully stratified systems, resulting in significantly smaller outlet discharge areas. Selection and application of air outlets for partially mixed systems is based primarily on the following considerations:

- Maintaining **vertical temperature gradients** in the occupied space that conform to ASHRAE *Standard* 55-2004. Further guidance on designing for conformance to this standard is presented in Chapter 56 of the 2007 *ASHRAE Handbook—HVAC Applications*.
- Maintaining a **near zone** adjacent to the outlets that is acceptable to the use and occupancy of the space.
- Providing **acoustical performance** that conforms to the requirements of the space.

Supply outlets should be selected such that their vertical projection achieves comfort and ventilation objectives. Limiting the vertical projection (to a terminal velocity of 50 fpm) of the supply air to below the respiration level allows convective heat plume formation around occupants to convey respiratory contaminants out of the occupied zone. This creates breathing-level CO_2 concentrations similar to those associated with fully stratified room air distribution systems. Projections that exceed this level discourage formation of such plumes and result in space ventilation similar to that of fully mixed systems.

The area adjacent to the supply outlet where local velocities may exceed 40 fpm is defined as the near zone. This is the area where local velocity/temperatures may combine to create drafty conditions. Manufacturers of outlets specifically intended for partially mixed systems publish predicted near-zone values that depend on the outlet supply airflow rate and the initial temperature difference between supply and room air. Stationary occupants should not generally be located in the near zone.

For applications requiring very low noise criteria, such as broadcast or recording studios or performing arts venues, acoustical performance can be an important consideration. These supply outlets may be suitable to meet acoustical criteria for these applications.

Factors that Influence Selection

Space Considerations. Partially mixed air distribution systems often rely on a pressurized plenum to deliver conditioned air to the supply air outlets; therefore, most of these outlets are not individually ducted. For example, underfloor air distribution systems commonly use the cavity beneath a raised access floor as a pressurized plenum. The supply outlets are mounted in the access floor tiles and can easily be relocated in response to space changes and workstation relocation.

Most supply outlets used in pressurized floor plenum applications can be easily adjusted for airflow by the space occupant. In such applications, it is usually effective to provide an adjustable outlet in every office or workstation to afford occupants control of their own environment. Many of these outlets can also be fitted with a thermostatically controlled damper that provides variable air volume to the space.

System and Terminal Considerations. Supply air outlets designed specifically for partially mixed room air distribution systems can function properly with either constant- or variable-air-volume (VAV) systems. For VAV applications, the supply air volume should be determined by a thermostat located in the space at a height of four to five feet. Because stratification typically results in cooler temperatures below the thermostat, its set point can usually be maintained slightly warmer than is typical in fully mixed systems.

TYPES OF SUPPLY AIR OUTLETS

Table 1 introduces the types of supply air outlets and provides guidance for best use practices. This table is for guidance only; designers should consult manufacturers' literature for additional application information.

Grilles

A supply air grille usually consists of a frame enclosing a set of either vertical or horizontal vanes (for a single-deflection grille) or both (for a double-deflection grille). These are typically used in sidewall, ceiling, sill, and floor applications.

Types.

Adjustable-Blade Grille. This is the most common type of grille used as a supply outlet. A single-deflection grille includes a set of either vertical or horizontal vanes or blades. Vertical vanes deflect the airstream in the horizontal plane; horizontal vanes deflect the airstream in the vertical plane. A double-deflection grille has a second set of vanes typically installed behind and at right angles to the face vanes, and controls the airstream in both the horizontal and vertical planes.

Fixed-Blade Grille. This grille is similar to the adjustable-blade grille except that the vanes or blades are not adjustable and may be straight or angled. The angle(s) at which air is discharged depends on the deflection of the vanes.

Linear Bar Grille. This outlet has fixed bars at its face. The bars normally run parallel to the length of the outlet and may be straight or angled. These devices supply air in a constant direction and are usually attached to a separate supply air plenum that has its own inlet. Linear bar grilles can be installed in multiple sections to achieve long, continuous lengths or installed as a discrete length. Typically designed for supply applications, they are also commonly used as return inlets to provide a consistent architectural appearance. Also commonly used in underfloor air distribution systems, some linear bar grilles allow the discharge pattern to be changed using removable cores. Many also incorporate a damper/actuator for automatic control of the supply air volume by a space thermostat.

Accessories. Various accessories, designed to modify the performance of grille outlets, are available:

- **Dampers** should be attached to the backs of grilles or installed as separate units in the duct to regulate airflow. (The combination of a supply air grille and a damper is called a **register**.) **Opposed-blade** damper vanes rotate in opposite directions (Figure 3A) **Parallel-blade** damper vanes rotate in the same direction (Figure 3B). Dampers deflect the airstream, and when located near the grille, they may cause nonuniform airflow and increase pressure drop and sound.
- **Extractors** are installed in collar connections to the outlet and are used to improve the flow distribution into the grille or register. The device shown in Figure 3C has vanes that pivot such that the supply airflow to the grille or register remains perpendicular to the face of the outlet. The device shown in Figure 3D has fixed vanes. Both devices restrict the area of the duct in which they are installed and should be used only when the duct is large enough to

Table 1 Typical Applications for Supply Air Outlets

	Fully Mixed			Fully Stratified		Partially Mixed		
Outlet Types	**Ceiling Mounted**	**Wall Mounted**	**Floor/Sill**	**Wall Mounted**	**Floor/Sill**	**Ceiling Mounted**	**Wall Mounted**	**Floor/Sill**
Grilles								
Adjustable blade	◒	●	⊗	⊗	⊗	⊗	⊙	⊙
Fixed blade	⊙	◒	⊗	◒	⊙	⊗	⊙	⊗
Linear bar	⊗	●	●	⊙	⊙	⊗	⊙	●
Nozzle	◒	●	⊗	⊗	⊗	⊗	⊗	⊗
Diffusers								
Round	●	⊗	⊗	⊗	⊗	⊗	⊗	⊗
Square	●	⊗	⊗	⊗	⊗	⊗	⊗	⊗
Perforated face	●	⊗	⊗	⊗	⊗	⊗	⊗	⊗
Louvered face	●	⊗	⊗	⊗	⊗	⊗	⊗	⊗
Plaque face	●	⊗	⊗	⊗	⊗	⊗	⊗	⊗
Hemispherical	⊗	⊗	⊗	⊗	⊗	●	⊗	⊗
Laminar flow	⊗	⊗	⊗	⊗	⊗	●	⊗	⊗
Variable geometry	●	⊗	⊗	⊗	⊗	⊗	⊗	⊗
Linear slot	●	●	⊗	⊗	⊗	⊗	⊗	⊗
T-bar slot	●	⊗	⊗	⊗	⊗	⊗	⊗	⊗
Light troffer	●	⊗	⊗	⊗	⊗	⊗	⊗	⊗
Swirl	●	⊗	⊗	⊗	◒	⊗	●	●
Displacement	⊗	⊗	⊗	●	●	⊙	⊗	⊗
Active chilled beam	●	⊗	⊗	⊗	⊗	⊗	⊗	⊗
Air dispersion duct	◒	⊗	⊗	⊗	⊗	●	⊗	⊗

● = often used ◒ = sometimes used ⊙ = seldom used ⊗ = not recommended

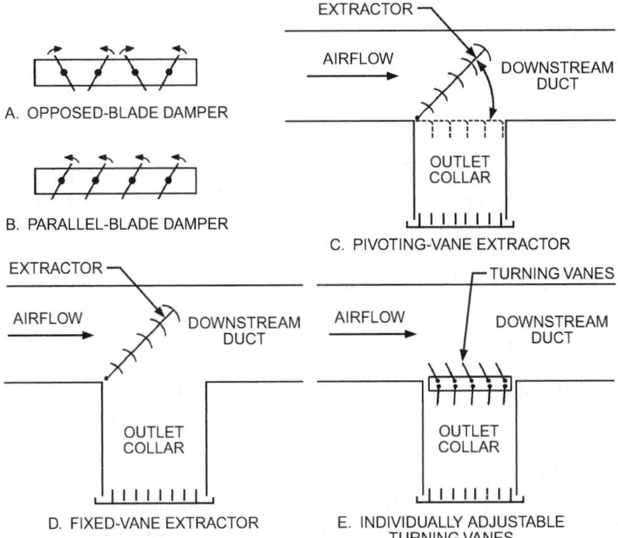

Fig. 3 Accessory Controls for Supply Air Grilles

allow the device to open to its maximum position without unduly restricting airflow to the downstream duct. These devices may increase the system pressure requirement, thereby limiting downstream airflow and increasing the sound level.

- The device shown in Figure 3E has individually **adjustable vanes.** Typically, two sets of vanes are used. One set equalizes flow across the collar, and the other set turns the air. The vanes should not be adjusted to act as a damper, because balancing requires removing the grille to gain access and then reinstalling it to measure airflow.
- Other miscellaneous accessories, such as **remote control devices** to operate the dampers, and **travel limit stops**, are also available.

Applications. Typically, supply air grilles are used in high-sidewall, ceiling, sill, or floor applications.

High Sidewall. An adjustable double-deflection grille usually provides the most satisfactory solution. The vertical louvers of the grille can be set for approximately 50° maximum deflection to either side, which can usually cover the conditioned space. Horizontal louvers can be set to control the elevation of the discharge pattern. Upward deflection minimizes thermal drop in cooling applications.

Ceiling. Such installation is generally limited to grilles with curved vanes that discharge parallel to the mounting surface. For high mounting locations (greater than 10 ft above the occupied zone), vertical discharge may be the preferred application. Grilles installed in 8 to 10 ft high ceilings, discharging the airstream into occupied zone, usually result in unacceptable comfort conditions. Satisfactory performance can be obtained if special allowances are made for terminal velocities and temperature differentials in the occupied space. Grille or register selections for heating and cooling applications from the same device should be carefully examined for use in both applications.

Sill. Linear bar grilles are commonly used in sill applications. The grille should be installed with the supply air jet directed vertically away from the occupied space. When the device is mounted 12 in. or less from the wall, a device with 0° deflection is suitable. If the device is installed more than 12 in. from the vertical surface, a linear bar grille with a fixed 15 to 30° deflection is recommended. This device should be installed with the jet directed toward the wall. These grilles are typically available with doors or other means of access to mechanical equipment that may be installed in the sill enclosure. The presence of window draperies or blinds and the effect of an impinging airstream must be considered in the selection.

Floor. Linear bar grilles are typically used for floor applications. The designer should determine the traffic and floor loading on the grille and consult the manufacturer's load limit for the grille. The grille should be placed in low-traffic areas. A floor-mounted grille is usually selected to discharge supply air along a wall or exterior surface. A floor grille is appropriate along exterior surfaces for heating. The grille should be installed with the supply air jet directed vertically away from the occupied space. When the device is mounted 12 in. or less from the wall, a device with 0° deflection is suitable.

If the device is installed more than 12 in. from the vertical surface, a linear bar grille with a fixed 15 to 30° deflection is recommended. This device should be installed with the jet directed toward the wall. The presence of window draperies or blinds and the effect of an impinging airstream must be considered in the selection.

Nozzles

Ball, drum, and other simple, usually adjustable nozzles allow air jets to be directed. These devices typically have no or few vanes in their airflow paths. Low pressure losses, moderate sound generation, and long throws are commonly produced by nozzles. They are often installed in buildings as horizontally discharging high-sidewall outlets, or in fur-downs (interior duct soffits). Nozzles are typically used in large spaces. Aircraft and automobile ventilation systems also use these types of outlets (Rock and Zhu 2002).

The equations in Chapter 33 of the 2005 *ASHRAE Handbook—Fundamentals* can be used to estimate throw from many such simple duct terminations or orifices. For complex situations, physical experiments or computational fluid dynamics (CFD) modeling may be helpful in selection and design.

Diffusers

Diffusers usually generate a radial or directional discharge pattern. For ceiling applications, this pattern is typically parallel to the mounting surface. Diffusers may also include adjustable deflectors that allow discharge to be directed perpendicular to the mounting surface. A diffuser typically consists of an outer shell, which contains a duct collar, and internal deflector(s), which define the diffuser's performance, including discharge pattern and direction.

Types.

Round Diffuser. This diffuser is a series of concentric conical rings, typically installed either in gypsum-board ceilings or on exposed ducts. Round ceiling diffusers are available in a broad range of sizes and capacities, with adjustable inner cones that allow the diffuser to discharge air either parallel or perpendicular to the ceiling or mounting surface.

Square Diffusers. This diffuser is consists of concentric square, drawn louvers that radiate from the center of the diffuser. Available with faces that are flush with the ceiling plane or with dropped inner cones, these diffusers have a fixed horizontal radial discharge pattern or an adjustable discharge pattern that allows the direction to be either horizontal or vertical. Special borders can be selected to accommodate various mounting applications.

Perforated Diffuser. This diffuser consists of a duct collar and a single perforated plate that forms the diffuser's face, with typical free area of about 50%. The perforated face tends to create a slightly higher pressure drop and sound than other square ceiling diffusers. They are available with deflection devices mounted at the neck or on the face plate. The deflectors may be adjustable, to provide horizontal air discharge in one, two, three, or four directions. Special borders can be selected to accommodate various mounting applications.

Louvered-Face Diffuser. This diffuser consists of an outer border, which includes an integral duct collar, and a series of louvers. Louvered-face diffusers typically provide a horizontal discharge perpendicular to the louver length. The louvers may be arranged to provide four-way, three-way, two-way opposite, two-way corner, or one-way discharge. Special borders can be selected to accommodate various mounting applications. Some louvered-face diffusers are available with adjustable louvers that can change the discharge direction from horizontal to vertical.

Plaque-Face Diffuser. This diffuser is constructed with a duct collar and a single plaque that forms the diffuser's face. This air outlet typically has a horizontal, radial discharge pattern. Typically, performance is similar to that of a square-face, round-neck diffuser. Special borders can be selected to accommodate various mounting applications.

Hemispherical Flow Diffuser. This diffuser provides a vertically radial air discharge pattern. The discharge penetrates the conditioned space perpendicular to the mounting surface. These diffusers are typically used for applications needing high air change rates, and/or low local velocities. Some outlet models are flush to the mounting surface; others intrude into the space below the ceiling. Most function similarly with or without an adjacent ceiling surface.

Laminar-Flow Diffuser. This diffuser provides a unidirectional discharge perpendicular to the mounting surface. The free area of the perforated face is typically less than 35%. Most outlets include a means to develop a uniform velocity profile over the full face. This minimizes mixing with surrounding ambient air and reduces entrainment of any surrounding contaminants. These diffusers are typically used in hospital operating rooms, cleanrooms, or laboratories.

Variable-Geometry Diffuser. This diffuser assembly can vary its discharge area in response to changes in space temperature or supply air temperature. As the terminal's airflow rate changes, the diffuser's discharge area can change to minimize changes in the diffuser's throw.

Linear Slot Diffuser. Long and narrow, linear slot diffusers may be installed in multiple sections to achieve long, continuous lengths or installed as a discrete length. They can consist of a single slot or multiple slots, and are available in configurations that provide vertical to horizontal airflow. Typically, a supply air plenum is provided separately and attached during installation. Other applications include field mounting the linear slot diffuser directly into a supply duct.

T-Bar Slot Diffuser. This diffuser is manufactured with an integral plenum and normally is installed in modular T-bar ceilings. Available with either fixed-deflection or adjustable-pattern controllers, these devices can discharge air from fully vertical to fully horizontal.

Light Troffer Diffuser. A light troffer diffuser serves as the combined plenum, inlet, and attachment device to an air-handling light fixture, which has a slot to receive the diffuser at or near the face of the lighting device, to discharge supply air into the space. Normally, only the air-handling slot is visible from the occupied space.

Swirl Diffuser. These diffusers feature a series of linear openings arranged in a radial pattern around the center of the diffuser face. This promotes a high degree of entrainment of room air, resulting in very high induction ratios, which maximize mixing in the area adjacent to the diffuser face. Swirl diffusers may be mounted in ceiling, sidewall, or floor locations. When mounted in floor locations, care should be taken to ensure that the diffuser can meet the floor loading requirements.

Displacement Diffuser. Typically located in floor or sidewall locations, these diffusers are designed to limit discharge velocities to 70 to 80 fpm, to minimize mixing between supply and room air. These outlets tend be large, to generate the low velocities required for thermal displacement ventilation. The low-velocity discharge allows the cooler supply air to fall to the floor and remain there because of its reduced buoyancy with respect to the ambient room air above it. These outlets are available in various shapes and configurations that facilitate flush or adjacent mounting to the sidewalls or floor of the space.

Air Dispersion Duct. This outlet system is designed to both convey and disperse air within the space being conditioned. Diffusion options include outlets selected for a full range of entrainment. Typically, these systems are made of fabric, sheet metal, or plastic film.

Active Chilled Beams. These devices typically use integral slot diffusers as their supply air outlets in a fully mixed room air distribution system.

Accessories. Various performance-modifying devices are available for use with diffusers:

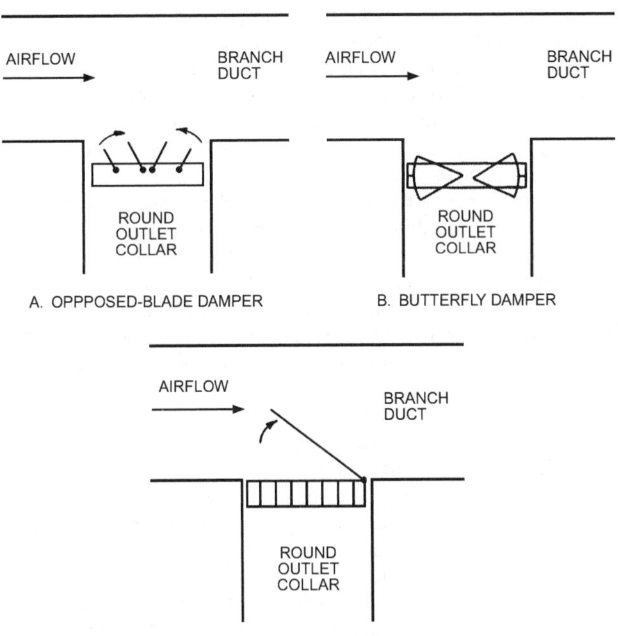

Fig. 4 Accessory Controls for Ceiling Diffusers

- **Dampers** can be attached to the inlets of diffusers or installed as separate units in the duct, to regulate the volume of air being discharged. **Opposed-blade** damper vanes rotate in opposite directions (Figure 3A) and are available for round, square, or rectangular necks. **Parallel-blade** dampers rotate in the same direction (Figure 3B). **Adjustable-vane** dampers have individually adjustable vanes (Figure 4A). **Butterfly** dampers are constructed with opposing damper plates that move in opposite directions and are adjusted from a center point (Figure 4B). A **splitter** damper is a single-blade device, hinged at one edge and usually located at the branch connection of a duct or outlet (Figure 4C). The device is designed to allow adjustment at the branch connection of a duct or outlet to adjust flow. **Radial** dampers are made up of multiple overlapping flat blades that rotate in the horizontal plane to deflect the airstream. When installed on the diffuser inlet, these dampers are operated through the face of the diffuser. When these dampers are located near the grille, they may cause nonuniform airflow and increase pressure drop and sound. Refer to Chapter 47 of the 2007 *ASHRAE Handbook—HVAC Applications* for the effects of damper location on sound level.
- **Equalizing** or **flow-straightening devices** or **grids** allow adjustment of the airstream to obtain more uniform flow to the diffuser.
- **Other balancing devices** are available. Consult manufacturers' literature as a source of information for other air-balancing devices.

Applications. For the following applications, the manufacturer's catalog data should be checked to select the air outlet that meets throw, pressure, and sound requirements.

Perimeter Zone Ceiling. In perimeter ceiling applications, the air outlet must handle the exterior surface load as well as the interior zone load generated along the perimeter. Refer to Chapter 56 of the 2007 *ASHRAE Handbook—Applications* for more details.

Interior Zone Ceiling. Typical interior ceilings require an air outlet that produces a horizontal pattern along the ceiling. Many ceiling diffusers are well suited for this application. Selection can be made based on performance, appearance, and cost. The air outlet selected should be sized to keep the supply air jet away from the occupied zone.

Vertical Projection. Downward projection of air from the outlet is often required in a high-ceiling application or in a high-load area. Vertical projection may be selected to meet an individual's comfort requirements. When selecting an outlet for vertical projection for both heating and cooling, its performance under both conditions must be taken into consideration. This is especially necessary when air outlet discharge characteristics are not automatically changed from heating to cooling.

Spot Heating or Cooling. Spot heating or cooling typically requires using an outlet that projects the air jet to a specific location. When selecting an outlet for both heating and cooling, its performance under both conditions must be taken into consideration. This is especially necessary when air outlet discharge characteristics are not automatically changed when switching, from heating to cooling.

Exposed Duct. In exposed-duct (no ceiling) applications, most ceiling diffusers can be used; however, the throw or radius of diffusion should be derated to allow for the influence of obstructions. Other approaches for exposed-duct applications include using perforated and/or fabric ducts.

RETURN AND EXHAUST AIR INLETS

Return air inlets may either be connected to a duct or simply cover openings that transfer air from one area to another. Exhaust air inlets remove air directly from a building and, therefore, are most always connected to a duct. Velocity, sound, and pressure requirements for inlets are determined by size and configuration. In general, the same type of equipment, grilles, slot diffusers, and ceiling diffusers used for supplying air may also be used for air return and exhaust. However, inlets do not require the deflection, flow equalizing, and turning devices necessary for supply outlets.

TYPES OF INLETS

For discussion of **adjustable-blade**, **fixed-blade**, and **linear bar grilles**, see the section on types of supply air outlets.

V-Bar Grille

Made with bars in the shape of inverted Vs stacked within the grille frame, this grille has the advantage of being sightproof. Door grilles are usually v-bar grilles. The airflow capacity of the grille decreases with increased sight-tightness.

Lightproof Grille

This grille is used to transfer air to or from darkrooms. The bars of this type of grille form a labyrinth and are painted black. The bars may be several sets of v-bars or be an interlocking louver design to provide the required labyrinth.

Stamped Grilles

Stamped grilles are also frequently used as return and exhaust inlets, particularly in restrooms and utility areas.

Eggcrate and Perforated-Face Grilles

These return grilles typically have large free areas. Perforated grilles often have free areas around 50%; eggcrate grilles can exceed 90% free area.

APPLICATIONS

Return and exhaust inlets may be mounted in practically any location (e.g., ceilings, high or low sidewalls, floors, and doors). To fully obtain their inherent contaminant removal benefits, return and exhaust opening serving spaces conditioned by fully stratified and partially mixed systems should always be located above the occupied zone. Chapter 33 of the 2005 *ASHRAE Handbook—Fundamentals* and Rock and Zhu (2002) discuss factors and the effect of inlet location on the system.

Dampers like the one shown in Figure 3A are sometimes used in conjunction with grille return and exhaust inlets to aid system balancing.

TERMINAL UNITS

In air distribution systems, special control and acoustical equipment is frequently required to introduce conditioned air into a space properly. Airflow controls for these systems consist principally of terminal units (historically called "boxes"), with which the airflow can be varied by pressure-modulating valves, fan controls or both. Terminal units may be classified as single or dual duct, with or without cooling, fans, heat or reheat, and having either constant or variable primary airflow rate. The constant primary air volume may provide for a variable discharge air volume to the conditioned space. Terminal units may employ plenum or room air induction to affect space temperature control while maintaining a constant or variable discharge airflow rate. Terminal units often include sound attenuators, or heaters, reheaters, or cooling coils.

This section discusses control equipment for single- and dual-duct air conditioning systems. Chapter 18 covers duct construction details. Chapter 47 of the 2007 *ASHRAE Handbook—HVAC Applications* includes information on sound control in air-conditioning systems and sound rating for air outlets.

General

Terminal units are factory-made assemblies for air distribution. A terminal unit manually or automatically performs one or more of the following functions: (1) controls air velocity, pressure, or temperature; (2) controls airflow rate; (3) mixes airstreams of different temperatures or humidities; (4) mixes, within the assembly, primary air from the duct system with air from the treated space or from a secondary duct system; and (5) heats or cools the air. To achieve these functions, terminal unit assemblies are made from an appropriate selection of the following components: casing, mixing section, manual or automatic damper, heat exchanger, induction section (with or without fan), coils, and flow controller.

A terminal unit may include a sound attenuation chamber to reduce sound. At the same time, the terminal unit can reduce velocity and pressure at the inlet to a lower velocity and pressure. The sound attenuation chamber is typically lined with thermal and sound insulating material and may be equipped with baffles.

Additional sound absorption material may be required in low-pressure distribution ducts connected to the discharge of larger units. Smaller units may not require additional sound absorption; however, manufacturers' catalogs should be consulted for specific performance information.

Terminal units are typically classified by the function of their flow controllers, which are generally constant- or variable-flow devices. They are further categorized as either pressure-dependent, where airflow through the assembly varies in response to changes in system pressure, or pressure-independent (pressure-compensating), where airflow through the device does not vary in response to changes in system pressure.

Variable-air-volume (VAV) reset controllers can control the VAV damper to regulate airflow to a constant fixed amount or to a variable modulating value that is calculated by the room demand. These controllers can be electric (pressure dependent), electronic (pressure independent) or pneumatic (pressure dependent or pressure independent). Pressure independent controllers require a pressure or velocity signal input to reset the VAV damper, which controls airflow. Temperature inputs are also required for calculating room demand for comfort conditioning. Variable flow may also be obtained by decreasing the flow through a constant-volume regulator with a modulating damper ahead of the regulator. This arrangement typically allows for variations in flow between the high- and low-capacity limits or between a high limit and shutoff. These units are pressure-dependent and volume-limiting in function.

Terminal units are also categorized as (1) system-powered, in which the assembly derives all the energy necessary for operation from supply air within the distribution system; or (2) externally powered, in which the assembly derives part or all of the energy from a pneumatic or electric outside source. In addition, assemblies may be self-contained (furnished with all necessary controls for their operation, including actuators, regulators, motors, and thermostats or space temperature sensors) or non-self-contained assemblies (part or all of the necessary controls for operation may be furnished by someone other than the assembly manufacturer). In the latter case, the controls may be mounted on the assembly by the assembly manufacturer or mounted by others after delivery.

The damper or flow controller in the unit can be adjusted manually, automatically, or by a pneumatic or electric motor. The unit is actuated by a signal from a thermostat or flow regulator, depending on the desired function of the box.

Air from the unit may be discharged through a single opening suitable for connection to a low-pressure rigid branch duct, or through a plenum to several round outlets suitable for connecting to flexible ducts. A single supply air outlet connected directly to the discharge end or bottom of the unit is an optional arrangement; however, the acoustic performance of this close-coupled arrangement must be carefully considered.

Single-Duct Terminal Units

Single-duct terminal units can be cooling only, cooling/heating if the primary air unit provides both or reheat if a heater is present. Reheat terminal units add sensible heat to the supply air. Water or steam coils or electric resistance heaters are placed in or attached directly to the air discharge of the unit. These are single-duct or bypass units that can operate as either constant- or variable-volume units. However, if they are variable-volume, they must maintain some minimum airflow for reheat. Some have a dual minimum flow, with one minimum being either zero during the no-occupancy cooling cycle or the airflow required for minimum ventilation during the cooling cycle, and the second minimum being the capacity required for reheat capacity during the reheat cycle. This type of equipment can provide local individual reheat without a central equipment station or zone change.

Dual-Duct Terminal Units

Dual-duct terminal units are typically controlled by a room thermostat. They receive warm or neutral and cold air from separate air supply ducts in accordance with room requirements to obtain room control without zoning. Volume-regulated units have individual modulating dampers and operators to regulate the amount of warm and cool air. When a single modulating damper operator regulates the amount of both warm and cold air, a separate pressure-reducing damper or volume controller is suggested in the unit to reduce pressure and limit airflow. Specially designed baffles may be required inside the unit or at its discharge to mix varying amounts of warm and cold air and/or to provide uniform flow and temperature equalization downstream. Dual-duct units can be equipped with constant- or variable-flow devices. These are usually pressure-independent, to provide a number of volume and temperature control functions. Dual-duct terminals may also be used as outside air terminals in which the neutral air inlet is used to control and maintain the required volumetric flow of ventilation air into the space. Dual-duct units with cooling and neutral air may need a local heating device.

Air-to-Air Induction Terminal Units

Induction terminals supply primary air or a mixture of primary and recirculated air to the conditioned space. They achieve this function with a primary air orifice that induces air from the ceiling plenum or individual rooms (via a return duct). Primary air cool enough to satisfy zone design cooling loads is ducted to the terminal The induction unit contains dampers that are actuated in response to a thermostat, to modulate the mixture of cool primary air and warm induced air. As less cooling is required, the primary airflow is gradually reduced, as the induced air rate generally increases. To meet interior load requirements, reheat coils can be installed in the primary supply air duct.

Chilled Beams

Chilled beams may be classified as active or passive. **Active chilled beams** are ceiling-mounted induction terminals that consist of a primary air duct connection, a series of induction nozzles, a hydronic heat transfer coil, a supply outlet, and a room air inlet section. Primary air discharged through the induction nozzles entrains room air through the inlet section and across a chilled- and/or hot-water coil where it is reconditioned before being mixed into the primary airstream. The free area of the room air inlet should be as high as possible, and at a minimum 50% of its total face area. Chilled-water supply temperatures are maintained at or above the room air dew point to prevent condensation from forming on the heat transfer coil. The mixture of primary and reconditioned room air is then discharged to the room through linear slots. These terminals typically discharge a constant-volume airflow with its temperature modulated (by the room air reconditioning) in accordance with the space thermostat demand. Active chilled beams generally produce a fully mixed room air distribution.

Passive chilled beams are sheet metal enclosures that contain finned-tube (chilled-water) convection coils. They rely on thermal stratification in the space to deliver warm air to its upper boundaries, some of which enters the passive beam casing. As this air passes over the back of the passive beam, it is drawn through the chilled-water coil and much of its heat is rejected. The reduced buoyancy that results causes the cooled air to drop back into the space. Because passive beams are not connected to a supply air duct, outside supply air must be provided to the space by a separate system. Supply air outlets used with passive beams should be selected to provide a fully stratified or partially mixed room air distribution, because turbulent airflow near passive beams disturbs the natural convection currents that deliver warm air to the beams.

Fan-Powered Terminal Units

Fan-powered terminal units are used in primary-secondary HVAC systems as secondary-level air handlers, and are typically installed in return air plenums. They are also frequently used as small, stand-alone air handlers. They differ from air-to-air induction units in that they include a blower, driven by a small motor, which draws air from the space, ceiling plenum, or floor plenum, that may be mixed with the cool air from the main air handler. The advantages of fan-powered units over straight VAV units are (1) during the heating mode, the primary air is mixed with warmer plenum air to blend the air temperature entering the heater to a level at or above room temperature, thus eliminating reheat; (2) downstream air pressure can be boosted to deliver air to areas that otherwise would be short of airflow; (3) room airflow volumes can be varied to improve occupant comfort; (4) perimeter zones can be heated without operating the main air handler fan during unoccupied periods; (5) depending on construction of the building envelope, the air in the plenum may be warm enough for low to medium heating loads; and (6) main air handler unit operating pressure can be reduced with series units, reducing the air distribution system's energy consumption.

In thermal storage and other systems with supply air temperatures below 52°F, fan-powered terminal units are used to mix cold supply air with induced return or plenum air, to moderate the supply air temperature. Some units are equipped with special insulation and a vapor barrier to prevent condensation with these low supply temperatures. Manufacturers' catalogs provide further information on these special features.

Fan-powered terminal units can be divided into two categories: (1) series, with all primary and induced air passing through a blower operating continuously during the occupied mode; and (2) parallel, in which the blower operates only on demand when induced air or heat is required.

A **series** unit typically has two inlets, one for cool primary air from the central fan system and one for secondary or plenum air. All air delivered to the space passes through the blower. The blower operates continuously whenever the primary air fan is on and can be cycled to deliver heat, as required, when the primary fan is off. As cooling load decreases, a damper throttles the amount of primary air delivered to the blower. The blower makes up for this reduced amount of primary air by drawing air in from the space or ceiling plenum through the return or secondary air opening. Sometimes a series unit has two ducted inlets, like a dual-duct terminal unit, in addition to the induction air inlet. The second duct is typically used for dedicated outdoor air systems. Fan air as well as primary air can be varied when the units are in part-load condition, but fan air should never be less than the total amount of air supplied by the ducted inlets.

Parallel fan-powered terminal units bypass the cool primary air around the blower portion of the unit so that the primary air flows directly to the space. The blower section draws in plenum air and is mounted in parallel with the primary air damper. A backdraft damper keeps primary air from flowing through the blower section when the blower is not energized. The blower in these units is generally energized after the primary air damper is partially or completely throttled closed. Some electronically controlled units gradually increase fan speed as the primary air damper is throttled, to maintain constant airflow while allowing the fan to shut off when it approaches full-cooling mode. Parallel units are typically limited to one ducted supply inlet.

Fan-powered terminals in supply air plenums (for underfloor air distribution). Fan terminal units are commonly used for perimeter area heating and/or cooling in underfloor air distribution applications. Although the general operation of these terminals is very similar to that for other fan terminals, location of these terminals in the supply air plenum results in certain design and operational considerations that differentiate these fan powered terminals. Decisions regarding the employment of fan terminals and their associated ductwork in these systems should also consider their potential effect on the future relocation power, voice, and data services to the space, because these services are also housed in the supply air plenum.

Series fan-powered terminals mounted in the (underfloor) supply air plenum typically have two dampered inlets located upstream of the integral blower unit. One inlet provides conditioned supply air from either the underfloor air plenum, or ducted from the central air handling unit. The second inlet allows the integral fan to induce ducted room or return plenum air. The blower operates continuously in the occupied mode while the dampers cycle to deliver heated, cooled, or recirculated air to the space. The terminal unit fan may be modulated based on the space load.

Alternatively, perimeter areas in underfloor air distribution applications can be served by booster fan terminals that are continuously or intermittently operated. Variable-speed fan terminals can be used to vary the delivery of cool air from the supply air plenum is accordance with the cooling demand indicated by the space thermostat or building automation system. Reheat operation (using air from the supply plenum) can be accomplished with either constant or variable airflow rate. Intermittent (cycled) fan operation can be used to satisfy perimeter area heating or heating/cooling demands. Intermittent fan terminals used only for perimeter heating may be combined with variable-volume cooling terminals that deliver cool air directly from the supply air plenum. The fan (and its reheat coil) remains off during cooling operation. Intermittent fan terminals used for perimeter heating and cooling are cycled (in accordance to the space thermostat or building automation system demand) during periods of heating and cooling, delivering conditioned air from the supply air plenum, reheated as required.

Locating series terminals in the supply air plenum may significantly alter their benefits. Using return air for heat requires that a duct be provided from floor-based return air inlets to the fan terminal inlet plenum. This ductwork may create a path for excessive fan noise to be transmitted to the space. In addition, room air induced at the floor level is only slightly warmer than the alternative air source, the underfloor supply air plenum. Careful analysis of the advantage (minimal energy savings) and disadvantages (inlet ductwork potential effects on noise transmission and relocation of other services housed in the supply air plenum) of using floor level room air recirculation in this application should be performed.

Series fan-powered terminal units may also be used to mix cold supply air from a central air-handling unit with induced return air to moderate the supply air temperature for injection into the supply air plenum. This enables transport of supply air from the air handler to supply plenum inlet locations at lower temperatures and volume flow rates, which may result in smaller supply and return air ducts.

Fan-powered terminal units installed under the raised floor must be selected to fit in the space between the structural slab and the raised access floor tiles and not interfere with the floor support structure.

Bypass Terminal Units

A bypass terminal unit handles a constant supply of primary air through its inlet; with a diverting damper, it bypasses the primary air to the ceiling plenum so that the amount of cool air delivered to the conditioned space meets the thermal requirement. Bypass air is diverted into the ceiling plenum and returned to the central air handler. The pressure requirement through the supply air path to the conditioned space and through the bypass path is equalized so that the fan handles a constant flow. This method provides a low first cost with minimum fan controls, but it is energy-inefficient compared to a VAV fan system. Its most frequent application is on small systems.

REFERENCES

Rock, B.A. and D. Zhu. 2002. *Designer's guide to ceiling-based air diffusion* (RP-1065). ASHRAE.

Skistad, H., E. Mundt, P. Nielsen, K. Hagström, and J. Railio. 2002. Displacement ventilation in non-industrial premises. REHVA *Guidebook* 1. Federation of European Heating and Air-Conditioning Associations, Brussels.

BIBLIOGRAPHY

ASHRAE. 1996. *Cold air distribution system design guide* (RP-849).

ASHRAE. 2004. Thermal environmental conditions for human occupancy. ANSI/ASHRAE *Standard* 55-2004.

ASHRAE. 2007. Ventilation for acceptable indoor air quality. ANSI/ASHRAE *Standard* 62.1-2007.

ASHRAE. 2007. Energy standard for buildings except low-rise residential buildings. ANSI/ASHRAE/IESNA *Standard* 90.1-2007.

ASHRAE. 2005. Method of testing for room air diffusion. ANSI/ASHRAE *Standard* 113-2005.

Bauman, F. 2003. *Underfloor air distribution design guide* (RP-1064). ASHRAE.

Chen, Q.Y. and L. Glicksman. 2003. *System performance evaluation and design guidelines for displacement ventilation*. ASHRAE.

CHAPTER 20

FANS

A FAN is an air pump that creates a pressure difference and causes airflow. The impeller does work on the air, imparting to it both static and kinetic energy, which vary in proportion, depending on the fan type.

Fan efficiency ratings are based on ideal conditions; some fans are rated at more than 90% total efficiency. However, actual connections often make it impossible to achieve ideal efficiencies in the field.

TYPES OF FANS

Fans are generally classified as centrifugal or axial flow according to the direction of airflow through the impeller. Figure 1 shows the general configuration of a centrifugal fan. The components of an axial-flow fan are shown in Figure 2. Table 1 compares typical characteristics of some of the most common fan types.

Two modified versions of the centrifugal fan are used but are not listed in Table 1 as separate fan types. Unhoused centrifugal fan impellers are used as circulators in some industrial applications (e.g., heat-treating ovens) and are identified as plug fans. In this case, there is no duct connection to the fan because it simply circulates the air within the oven. In some HVAC installations, the unhoused fan impeller is located in a plenum chamber with the fan inlet connected to an inlet duct from the system. Outlet ducts are connected to the plenum chamber. This fan arrangement is identified as a plenum fan.

PRINCIPLES OF OPERATION

All fans produce pressure by altering the airflow's velocity vector. A fan produces pressure and/or airflow because the rotating blades of the impeller impart kinetic energy to the air by changing its velocity. Velocity change is in the tangential and radial velocity components for centrifugal fans, and in the axial and tangential velocity components for axial-flow fans.

Centrifugal fan impellers produce pressure from the (1) centrifugal force created by rotating the air column contained between the blades and (2) kinetic energy imparted to the air by its velocity leaving the impeller. This velocity is a combination of rotational velocity of the impeller and airspeed relative to the impeller. When the blades are inclined forward, these two velocities are cumulative; when backward, oppositional. Backward-curved blade fans are generally more efficient than forward-curved blade fans.

Axial-flow fan impellers produce pressure principally by the change in air velocity as it passes through the impeller blades, with none being produced by centrifugal force. These fans are divided into three types: propeller, tubeaxial, and vaneaxial. Propeller fans, customarily used at or near free air delivery, usually have a small hub-to-tip-ratio impeller mounted in an orifice plate or inlet ring. Tubeaxial fans usually have reduced tip clearance and operate at higher tip speeds, giving them a higher total pressure capability than the propeller fan. Vaneaxial fans are essentially tubeaxial fans with guide vanes and reduced running blade tip clearance, which give improved pressure, efficiency, and noise characteristics.

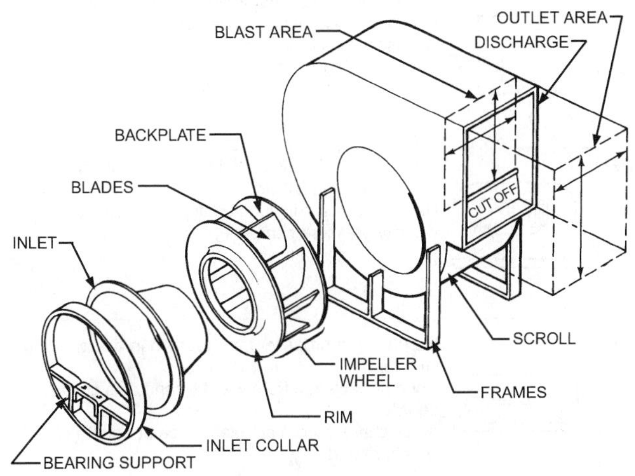

Fig. 1 Centrifugal Fan Components

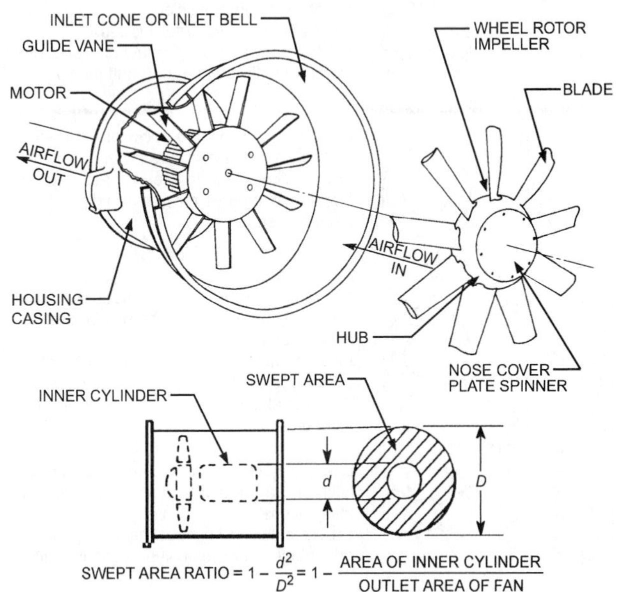

Fig. 2 Axial Fan Components

The preparation of this chapter is assigned to TC 5.1, Fans.

Table 1 Types of Fans

	TYPE	IMPELLER DESIGN	HOUSING DESIGN
CENTRIFUGAL FANS	AIRFOIL	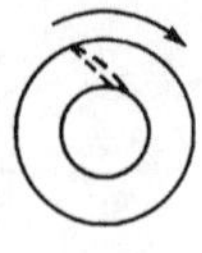 Highest efficiency of all centrifugal fan designs. Ten to 16 blades of airfoil contour curved away from direction of rotation. Deep blades allow efficient expansion within blade passages. Air leaves impeller at velocity less than tip speed. For given duty, has highest speed of centrifugal fan designs.	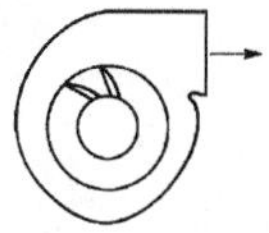 Scroll design for efficient conversion of velocity pressure to static pressure. Maximum efficiency requires close clearance and alignment between wheel and inlet.
CENTRIFUGAL FANS	BACKWARD-INCLINED BACKWARD-CURVED	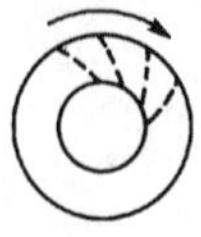 Efficiency only slightly less than airfoil fan. Ten to 16 single-thickness blades curved or inclined away from direction of rotation. Efficient for same reasons as airfoil fan.	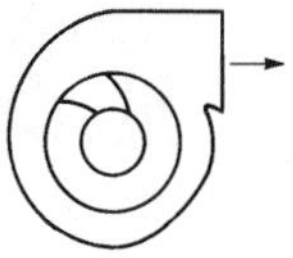 Uses same housing configuration as airfoil design.
CENTRIFUGAL FANS	RADIAL	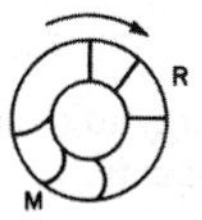 Higher pressure characteristics than airfoil, backward-curved, and backward-inclined fans. Curve may have a break to left of peak pressure and fan should not be operated in this area. Power rises continually to free delivery.	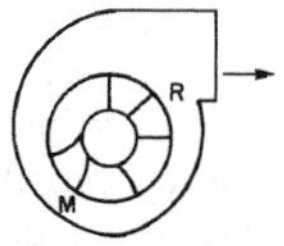 Scroll. Usually narrowest of all centrifugal designs. Because wheel design is less efficient, housing dimensions are not as critical as for airfoil and backward-inclined fans.
CENTRIFUGAL FANS	FORWARD-CURVED	 Flatter pressure curve and lower efficiency than the airfoil, backward-curved, and backward-inclined. Do not rate fan in the pressure curve dip to the left of peak pressure. Power rises continually toward free delivery. Motor selection must take this into account.	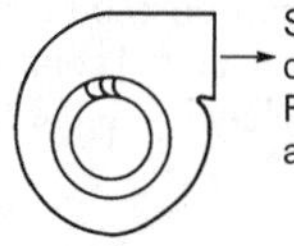 Scroll similar to and often identical to other centrifugal fan designs. Fit between wheel and inlet not as critical as for airfoil and backward-inclined fans.
AXIAL FANS	PROPELLER	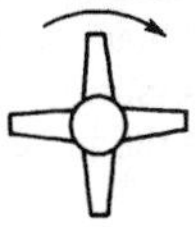 Low efficiency. Limited to low-pressure applications. Usually low-cost impellers have two or more blades of single thickness attached to relatively small hub. Primary energy transfer by velocity pressure.	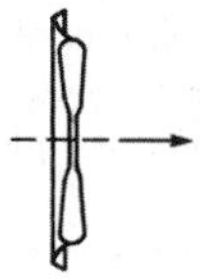 Simple circular ring, orifice plate, or venturi. Optimum design is close to blade tips and forms smooth airfoil into wheel.
AXIAL FANS	TUBEAXIAL	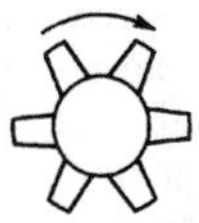 Somewhat more efficient and capable of developing more useful static pressure than propeller fan. Usually has 4 to 8 blades with airfoil or single-thickness cross section. Hub is usually less than half the fan tip diameter.	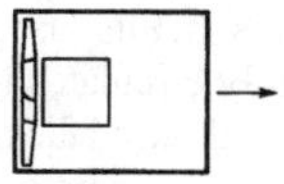 Cylindrical tube with close clearance to blade tips.
AXIAL FANS	VANEAXIAL	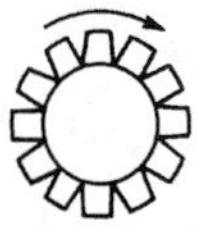 Good blade design gives medium- to high-pressure capability at good efficiency. Most efficient have airfoil blades. Blades may have fixed, adjustable, or controllable pitch. Hub is usually greater than half fan tip diameter.	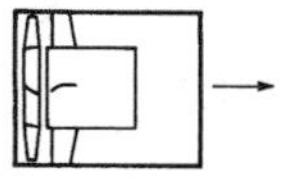 Cylindrical tube with close clearance to blade tips. Guide vanes upstream or downstream from impeller increase pressure capability and efficiency.
SPECIAL DESIGNS	TUBULAR CENTRIFUGAL	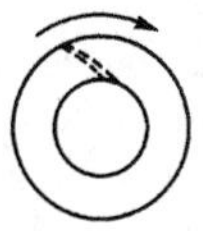 Performance similar to backward-curved fan except capacity and pressure are lower. Lower efficiency than backward-curved fan. Performance curve may have a dip to the left of peak pressure.	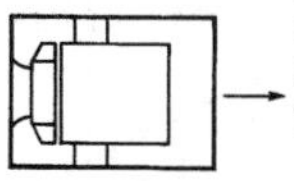 Cylindrical tube similar to vaneaxial fan, except clearance to wheel is not as close. Air discharges radially from wheel and turns 90° to flow through guide vanes.
SPECIAL DESIGNS	POWER ROOF VENTILATORS — CENTRIFUGAL	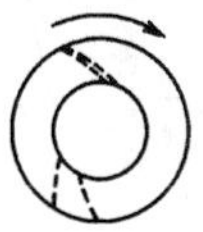 Low-pressure exhaust systems such as general factory, kitchen, warehouse, and some commercial installations. Provides positive exhaust ventilation, which is an advantage over gravity-type exhaust units. Centrifugal units are slightly quieter than axial units.	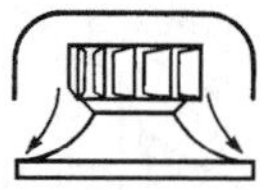 Normal housing not used, because air discharges from impeller in full circle. Usually does not include configuration to recover velocity pressure component.
SPECIAL DESIGNS	POWER ROOF VENTILATORS — AXIAL	 Low-pressure exhaust systems such as general factory, kitchen, warehouse, and some commercial installations. Provides positive exhaust ventilation, which is an advantage over gravity-type exhaust units.	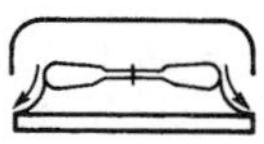 Essentially a propeller fan mounted in a supporting structure. Hood protects fan from weather and acts as safety guard. Air discharges from annular space at bottom of weather hood.

Table 1 Types of Fans (*Concluded*)

PERFORMANCE CURVES*	PERFORMANCE CHARACTERISTICS	APPLICATIONS
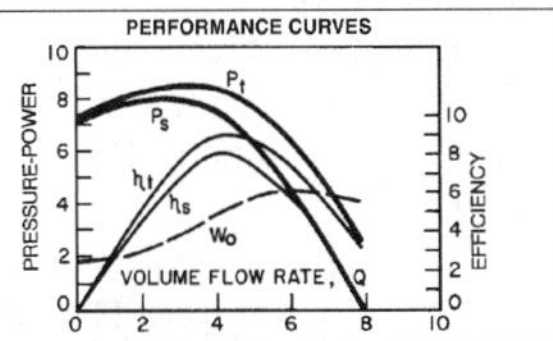 PERFORMANCE CURVES; PRESSURE-POWER; EFFICIENCY; VOLUME FLOW RATE, Q	Highest efficiencies occur at 50 to 60% of wide-open volume. This volume also has good pressure characteristics. Power reaches maximum near peak efficiency and becomes lower, or self-limiting, toward free delivery.	General heating, ventilating, and air-conditioning applications. Usually only applied to large systems, which may be low-, medium-, or high-pressure applications. Applied to large, clean-air industrial operations for significant energy savings.
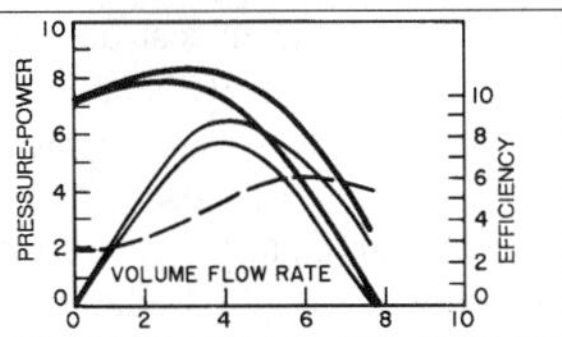 PRESSURE-POWER; EFFICIENCY; VOLUME FLOW RATE	Similar to airfoil fan, except peak efficiency slightly lower.	Same heating, ventilating, and air-conditioning applications as airfoil fan. Used in some industrial applications where environment may corrode or erode airfoil blade.
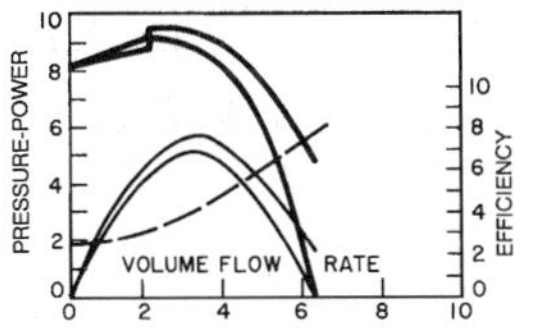 PRESSURE-POWER; EFFICIENCY; VOLUME FLOW RATE	Higher pressure characteristics than airfoil and backward-curved fans. Pressure may drop suddenly at left of peak pressure, but this usually causes no problems. Power rises continually to free delivery.	Primarily for materials handling in industrial plants. Also for some high-pressure industrial requirements. Rugged wheel is simple to repair in the field. Wheel sometimes coated with special material. Not common for HVAC applications.
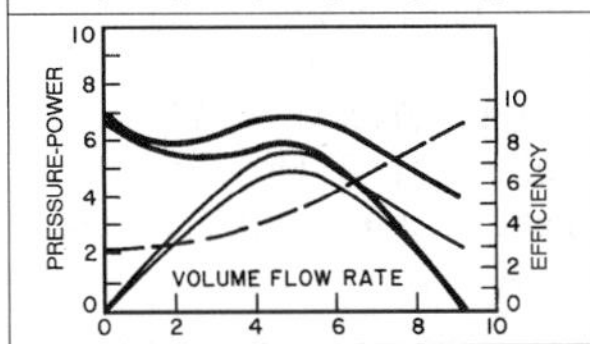 PRESSURE-POWER; EFFICIENCY; VOLUME FLOW RATE	Pressure curve less steep than that of backward-curved fans. Curve dips to left of peak pressure. Highest efficiency to right of peak pressure at 40 to 50% of wide-open volume. Rate fan to right of peak pressure. Account for power curve, which rises continually toward free delivery, when selecting motor.	Primarily for low-pressure HVAC applications, such as residential furnaces, central station units, and packaged air conditioners.
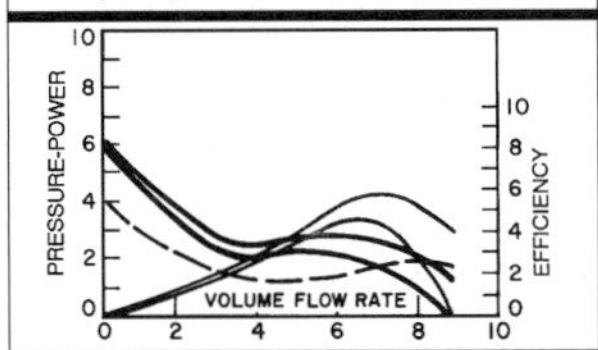 PRESSURE-POWER; EFFICIENCY; VOLUME FLOW RATE	High flow rate, but very low-pressure capabilities. Maximum efficiency reached near free delivery. Discharge pattern circular and airstream swirls.	For low-pressure, high-volume air moving applications, such as air circulation in a space or ventilation through a wall without ductwork. Used for makeup air applications.
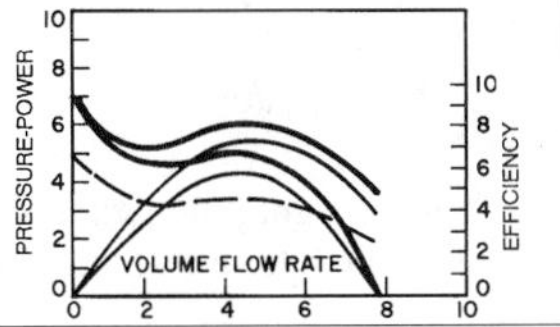 PRESSURE-POWER; EFFICIENCY; VOLUME FLOW RATE	High flow rate, medium-pressure capabilities. Performance curve dips to left of peak pressure. Avoid operating fan in this region. Discharge pattern circular and airstream rotates or swirls.	Low- and medium-pressure ducted HVAC applications where air distribution downstream is not critical. Used in some industrial applications, such as drying ovens, paint spray booths, and fume exhausts.
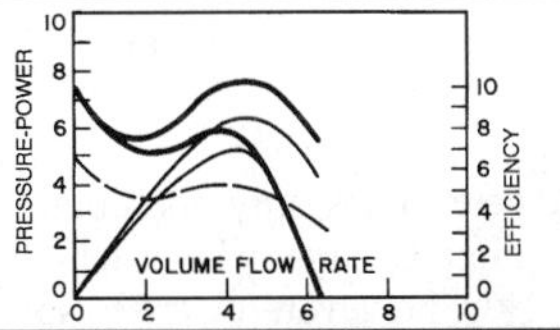 PRESSURE-POWER; EFFICIENCY; VOLUME FLOW RATE	High-pressure characteristics with medium-volume flow capabilities. Performance curve dips to left of peak pressure because of aerodynamic stall. Avoid operating fan in this region. Guide vanes correct circular motion imparted by wheel and improve pressure characteristics and efficiency of fan.	General HVAC systems in low-, medium-, and high-pressure applications where straight-through flow and compact installation are required. Has good downstream air distribution. Used in industrial applications in place of tubeaxial fans. More compact than centrifugal fans for same duty.
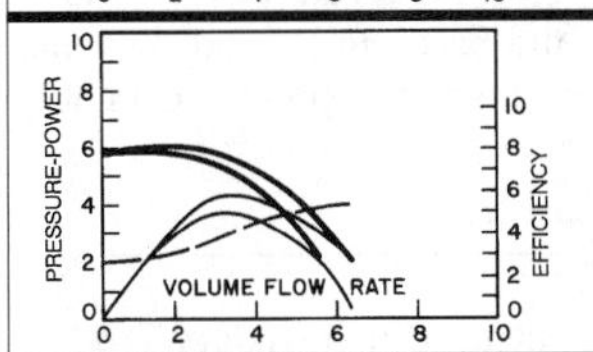 PRESSURE-POWER; EFFICIENCY; VOLUME FLOW RATE	Performance similar to backward-curved fan, except capacity and pressure is lower. Lower efficiency than backward-curved fan because air turns 90°. Performance curve of some designs is similar to axial flow fan and dips to left of peak pressure.	Primarily for low-pressure, return air systems in HVAC applications. Has straight-through flow.
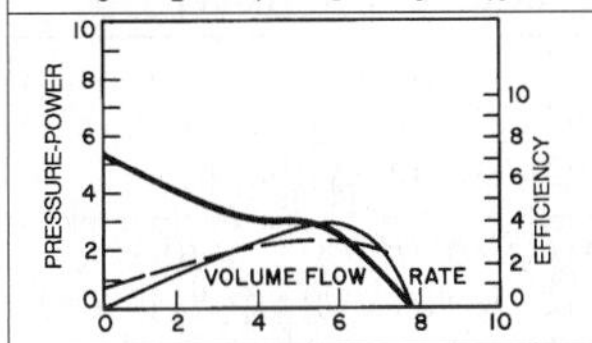 PRESSURE-POWER; EFFICIENCY; VOLUME FLOW RATE	Usually operated without ductwork; therefore, operates at very low pressure and high volume. Only static pressure and static efficiency are shown for this fan.	Low-pressure exhaust systems, such as general factory, kitchen, warehouse, and some commercial installations. Low first cost and low operating cost give an advantage over gravity flow exhaust systems. Centrifugal units are somewhat quieter than axial flow units.
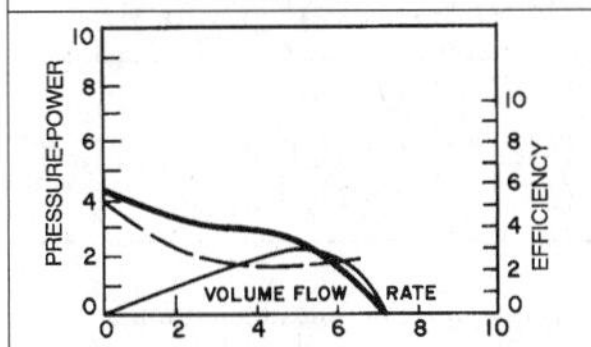 PRESSURE-POWER; EFFICIENCY; VOLUME FLOW RATE	Usually operated without ductwork; therefore, operates at very low pressure and high volume. Only static pressure and static efficiency are shown for this fan.	Low-pressure exhaust systems, such as general factory, kitchen, warehouse, and some commercial installations. Low first cost and low operating cost give an advantage over gravity-flow exhaust systems.

*These performance curves reflect general characteristics of various fans as commonly applied. They are not intended to provide complete selection criteria, because other parameters, such as diameter and speed, are not defined.

Table 1 includes typical performance curves for various types of fans. These performance curves show the general characteristics of various fans as they are normally used; they do not reflect fan characteristics reduced to common denominators such as constant speed or constant propeller diameter, because fans are not selected on the basis of these constants. The efficiencies and power characteristics shown are general indications for each type of fan. A specific fan (size, speed) must be selected by evaluating actual characteristics.

TESTING AND RATING

ANSI/ASHRAE *Standard* 51 (ANSI/AMCA *Standard* 210) specifies the procedures and test setups to be used in testing fans and other air-moving devices. Figure 3 diagrams one of the most common procedures for developing characteristics of a fan tested from **shutoff** conditions to nearly **free delivery** conditions. At shutoff, the duct is completely blocked off; at free delivery, the outlet resistance is reduced to zero. Between these two conditions, various airflow restrictions are placed on the end of the duct to simulate various operating conditions on the fan. Sufficient points are obtained to define the curve between shutoff and free air delivery conditions. Pitot-static tube traverses of the test duct are performed with the fan operating at constant rotational speed. The point of rating may be any point on the fan performance curve. For each case, the specific point on the curve must be defined by referring to the airflow rate and corresponding total or static pressure. Other test setups described in ANSI/ASHRAE *Standard* 51 should produce the same performance curve.

Fans designed for use with duct systems are tested with a length of duct between the fan and measuring station. This length of duct smooths the air discharged from the fan and provides stable, uniform airflow conditions at the plane of measurement. Measured pressures are corrected back to fan outlet conditions. Fans designed for use without ducts, including almost all propeller fans and power roof ventilators, are tested without ductwork.

Not all sizes are tested for rating. Test information may be used to calculate performance of larger fans that are geometrically similar, but such information should not be extrapolated to smaller fans. For performance of one fan to be determined from the known performance of another, the two fans must be dynamically similar. Strict dynamic similarity requires that the important nondimensional parameters (those that affect aerodynamic characteristics, such as Mach number, Reynolds number, surface roughness, and gap size) vary in only insignificant ways. (For more specific information, consult the manufacturer's application manual or engineering data.)

FAN LAWS

The fan laws (see Table 2) relate performance variables for any dynamically similar series of fans. The variables are fan size D, rotational speed N, gas density ρ, volume airflow rate Q, pressure P_{tf} or P_{sf}, power W, and mechanical efficiency η_t. **Fan Law 1** shows the effect of changing size, speed, or density on volume airflow rate, pressure, and power level. **Fan Law 2** shows the effect of changing size, pressure, or density on volume airflow rate, speed, and power. **Fan Law 3** shows the effect of changing size, volume airflow rate, or density on speed, pressure, and power.

The fan laws apply only to a series of aerodynamically similar fans at the same point of rating on the performance curve. They can be used to predict the performance of any fan when test data are available for any fan of the same series. Fan laws may also be used with a particular fan to determine the effect of speed change. However, caution should be exercised in these cases, because the laws apply only when all flow conditions are similar. Changing the speed of a given fan changes parameters that may invalidate the fan laws.

Unless otherwise identified, fan performance data are based on dry air at standard conditions: 14.696 psi and 70°F (0.075 lb/ft^3). In actual applications, the fan may be required to handle air or gas at some other density. The change in density may be caused by temperature, composition of the gas, or altitude. As indicated by the fan laws, fan performance is affected by gas density. With constant size and speed, power and pressure vary in accordance with the ratio of gas density to standard air density.

Figure 4 illustrates the application of the fan laws for a change in fan speed N for a specific-sized fan (i.e., $D_1 = D_2$). The computed P_t curve is derived from the base curve. For example, point E ($N_1 = 650$) is computed from point D ($N_2 = 600$) as follows:

At point D,

$$Q_2 = 6000 \text{ cfm and } P_{tf_2} = 1.13 \text{ in. of water}$$

Using Fan Law 1a at point E,

$$Q_1 = 6000(650/600) = 6500 \text{ cfm}$$

Using Fan Law 1b ($\rho_1 = \rho_2$),

$$P_{tf_1} = 1.13(650/600)^2 = 1.33 \text{ in. of water}$$

The total pressure curve P_{tf_1} at $N = 650$ rpm may be generated by computing additional points from data on the base curve, such as point G from point F.

If equivalent points of rating are joined, as shown by the dashed lines in Figure 4, they form parabolas, which are defined by the relationship expressed in Equation (2).

Each point on the base P_{tf} curve determines only one point on the computed curve. For example, point H cannot be calculated from either point D or point F. Point H is, however, related to some point between these two points on the base curve, and only that point can be used to locate point H. Furthermore, point D cannot be

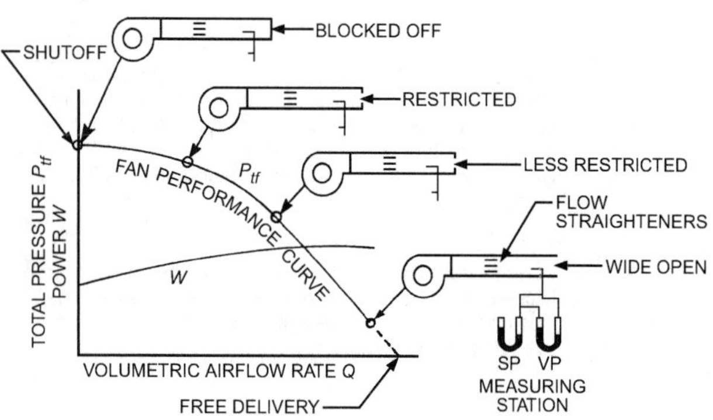

Fig. 3 Method of Obtaining Fan Performance Curves

Table 2 Fan Laws

Law No.	Dependent Variables				Independent Variables
1a	Q_1	=	Q_2	×	$(D_1/D_2)^3\,(N_1/N_2)$
1b	P_1	=	P_2	×	$(D_1/D_2)^2\,(N_1/N_2)^2\,\rho_1/\rho_2$
1c	W_1	=	W_2	×	$(D_1/D_2)^5\,(N_1/N_2)^3\,\rho_1/\rho_2$
2a	Q_1	=	Q_2	×	$(D_1/D_2)^2\,(P_1/P_2)^{1/2}\,(\rho_2/\rho_1)^{1/2}$
2b	N_1	=	N_2	×	$(D_2/D_1)\,(P_1/P_2)^{1/2}\,(\rho_2/\rho_1)^{1/2}$
2c	W_1	=	W_2	×	$(D_1/D_2)^2\,(P_1/P_2)^{3/2}\,(\rho_2/\rho_1)^{1/2}$
3a	N_1	=	N_2	×	$(D_2/D_1)^3\,(Q_1/Q_2)$
3b	P_1	=	P_2	×	$(D_2/D_1)^4\,(Q_1/Q_2)^2\,\rho_1/\rho_2$
3c	W_1	=	W_2	×	$(D_2/D_1)^4\,(Q_1/Q_2)^3\,\rho_1/\rho_2$

Notes:
1. Subscript 1 denotes fan under consideration. Subscript 2 denotes tested fan.
2. For all fans laws $(\eta_t)_1 = (\eta_t)_2$ and (Point of rating)$_1$ = (Point of rating)$_2$.
3. P equals either P_{tf} or P_{sf}.

used to calculate point F on the base curve. The entire base curve must be defined by test.

FAN AND SYSTEM PRESSURE RELATIONSHIPS

As previously stated, a fan impeller imparts static and kinetic energy to the air. This energy is represented in the increase in total pressure and can be converted to static or velocity pressure. These two quantities are interdependent: fan performance cannot be evaluated by considering one alone. Energy conversion, indicated by changes in velocity pressure to static pressure and vice versa, depends on the efficiency of conversion. Energy conversion occurs in the discharge duct connected to a fan being tested in accordance with ANSI/ASHRAE *Standard* 51, and the efficiency is reflected in the rating.

Fan total pressure rise P_{tf} is a true indication of the energy imparted to the airstream by the fan. System pressure loss (ΔP) is the sum of all individual total pressure losses imposed by the air distribution system duct elements on both the inlet and outlet sides of the fan. An energy loss in a duct system can be defined only as a total pressure loss. The measured static pressure loss in a duct element equals the total pressure loss only in the special case where air velocities are the same at both the entrance and exit of the duct element. By using total pressure for both fan selection and air distribution system design, the design engineer ensures proper design. These fundamental principles apply to both high- and low-velocity systems. (Chapter 35 of the 2005 *ASHRAE Handbook—Fundamentals* has further information.)

Fan static pressure rise P_{sf} is often used in low-velocity ventilating systems where the fan outlet area essentially equals the fan outlet duct area, and little energy conversion occurs. When fan performance data are given in terms of P_{sf}, the value of P_{tf} may be calculated from catalog data.

To specify the pressure performance of a fan, the relationship of P_{tf}, P_{sf}, and P_{vf} must be understood, especially when negative pressures are involved. Most importantly, P_{sf} is defined in ANSI/ASHRAE *Standard* 51 as $P_{sf} = P_{tf} - P_{vf}$. Except in special cases, P_{sf} is not necessarily the measured difference between static pressure on the inlet side and static pressure on the outlet side.

Figures 5 to 8 depict the relationships among these various pressures. Note that, as defined, $P_{tf} = P_{t2} - P_{t1}$. Figure 5 illustrates a fan with an outlet system but no connected inlet system. In this case, fan static pressure P_{sf} equals the static pressure rise across the fan. Figure 6 shows a fan with an inlet but no outlet system. Figure 7 shows a fan with both an inlet and an outlet system. In both cases,

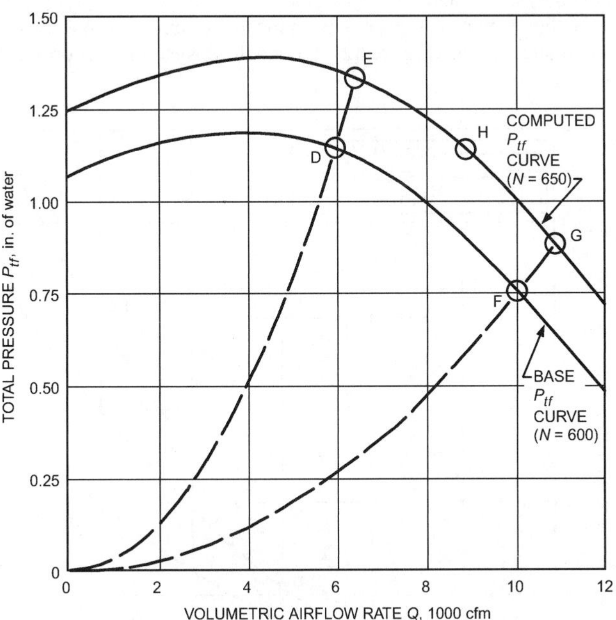

Fig. 4 Example Application of Fan Laws

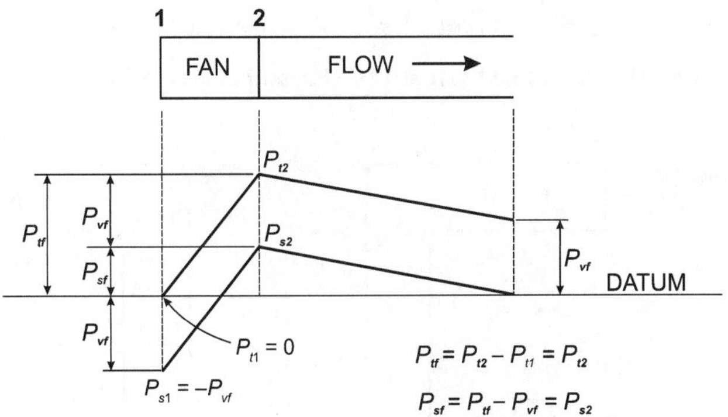

Fig. 5 Pressure Relationships of Fan with Outlet System Only

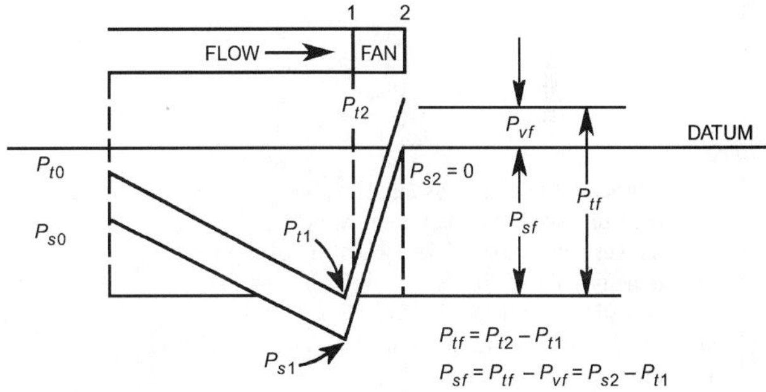

Fig. 6 Pressure Relationships of Fan with Inlet System Only

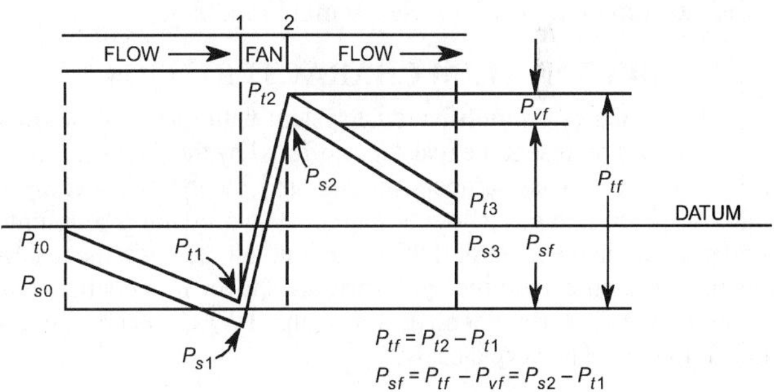

Fig. 7 Pressure Relationships of Fan with Equal-Sized Inlet and Outlet Systems

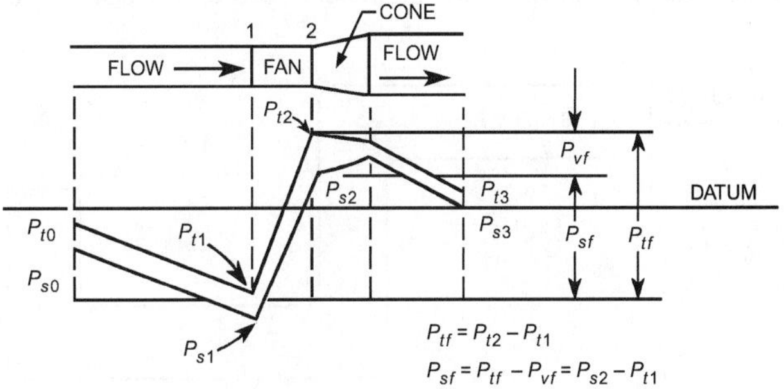

Fig. 8 Pressure Relationships of Fan with Diverging Cone Outlet

the measured difference in static pressure across the fan ($P_{s2} - P_{s1}$) is not equal to the fan static pressure (P_{sf}).

All the systems illustrated in Figures 5 to 7 have inlet or outlet ducts that match the fan connections in size. Usually the duct size is not identical to the fan outlet or inlet, so that a further complication is introduced. To illustrate the pressure relationships in this case, Figure 8 shows a diverging outlet cone, which is a common type of fan connection. In this case, the pressure relationships at the fan outlet do not match the pressure relationships in the airflow section. Furthermore, static pressure in the cone actually increases in the direction of airflow. The static pressure changes throughout the system, depending on velocity. The total pressure, which, as noted in the figure, decreases in the direction of airflow, more truly represents the loss introduced by the cone or by flow in the duct. Only the fan changes this trend. Total pressure, therefore, is a better indication of fan and duct system performance. In this normal fan situation, the static pressure across the fan ($P_{s2} - P_{s1}$) does not equal the fan static pressure P_{sf}.

TEMPERATURE RISE ACROSS FANS

In certain applications, it may be desirable to calculate the temperature rise across the fan. For low pressure rises (<10 in. of water), the temperature rise may be found by the following:

$$\Delta T = \frac{\Delta P C_p}{\rho c_p J \eta} \tag{1}$$

where

ΔT = temperature rise across fan, °F
ΔP = pressure rise across fan, in. of water
C_p = conversion factor = 5.193 $lb_f/ft^2 \cdot$in. of water
ρ = density = 0.075 lb_m/ft^3
c_p = specific heat = 0.24 $Btu/lb_m \cdot °F$
J = mechanical equivalent of heat = 778.2 $ft \cdot lb_f/Btu$
η = efficiency, decimal

If the motor is not in the airstream, the efficiency is the fan total efficiency. If the motor is in the airstream, the efficiency is the set efficiency (combined efficiencies of motor and fan).

DUCT SYSTEM CHARACTERISTICS

Figure 9 shows a simplified duct system with three 90° elbows. These elbows represent the resistance offered by the ductwork, heat exchangers, cabinets, dampers, grilles, and other system components. A given rate of airflow through a system requires a definite total pressure in the system. If the rate of airflow changes, the resulting total pressure required will vary, as shown in Equation (2), which is true for turbulent airflow systems. HVAC systems generally follow this law very closely.

$$(\Delta P_2/\Delta P_1) = (Q_2/Q_1)^2 \tag{2}$$

This chapter covers only turbulent flow (the flow regime in which most fans operate). In some systems, particularly constant- or variable-volume air conditioning, the air-handling devices and associated controls may produce effective system resistance curves that deviate widely from Equation (2), even though each element of the system may be described by this equation.

Equation (2) permits plotting a turbulent flow system's pressure loss (ΔP) curve from one known operating condition (see Figure 4). The fixed system must operate at some point on this system curve as the volume flow rate changes. As an example, in Figure 10, at point A of curve A, when the flow rate through a duct system such as that shown in Figure 9 is 10,000 cfm, the total pressure drop is 3 in. of water. If these values are substituted in Equation (2) for ΔP_1 and Q_1, other points of the system's ΔP curve (Figure 10) can be determined.

For 6000 cfm (Point D on Figure 10):

$$\Delta P_2 = 3(6000/10{,}000)^2 = 1.08 \text{ in. of water}$$

If a change is made within the system so that the total pressure at design flow rate is increased, the system will no longer operate on the previous ΔP curve, and a new curve will be defined.

For example, in Figure 11, an elbow added to the duct system shown in Figure 9 increases the total pressure of the system. If the total pressure at 10,000 cfm is increased by 1.00 in. of water, the system total pressure drop at this point is now 4.00 in. of water, as shown by point B in Figure 10.

If the system in Figure 9 is changed by removing one of the schematic elbows (Figure 12), the resulting system total pressure

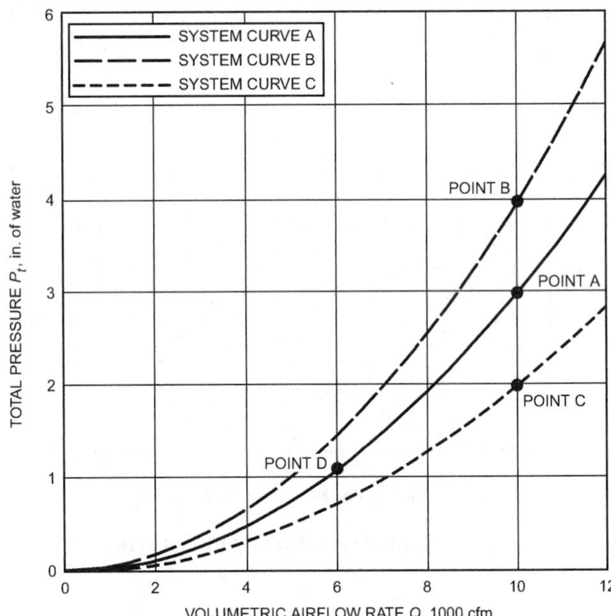

Fig. 10 Example System Total Pressure Loss (ΔP) Curves

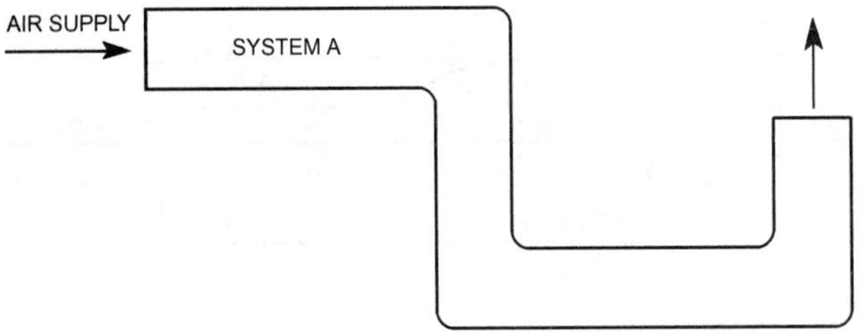

Fig. 9 Simple Duct System with Resistance to Flow Represented by Three 90° Elbows

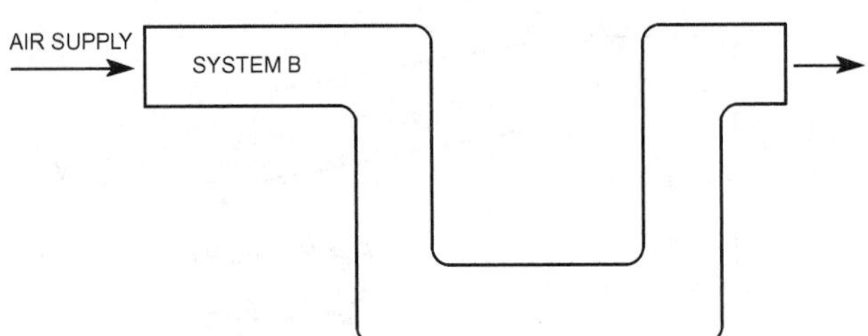

Fig. 11 Resistance Added to Duct System of Figure 9

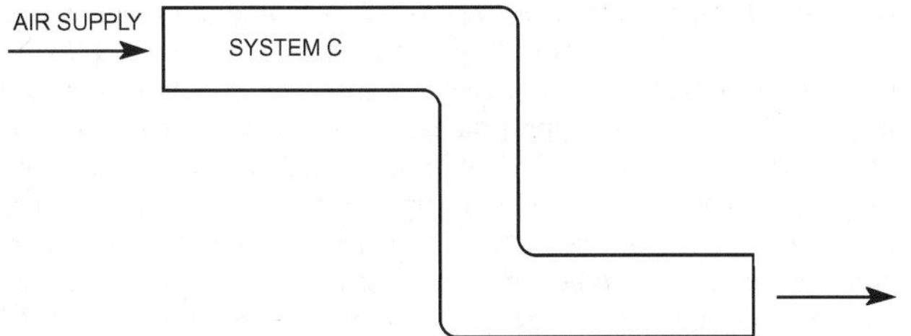

Fig. 12 Resistance Removed from Duct System of Figure 9

drops below the total pressure resistance, and the new ΔP curve is curve C of Figure 10. For curve C, a total pressure reduction of 1.00 in. of water has been assumed when 10,000 cfm flows through the system; thus, the point of operation is at 2.00 in. of water, as shown by point C.

These three ΔP curves all follow the relationship expressed in Equation (2). These curves result from changes in the system itself and do not change fan performance. During design, such system total pressure changes may occur because of alternative duct routing, differences in duct sizes, allowance for future duct extensions, or the design safety factor being applied to the system.

In an actual operating system, these three ΔP curves can represent three system characteristic lines caused by three different positions of a throttling control damper. Curve C is the most open position, and curve B is the most closed. A control damper forms a continuous series of these ΔP curves as it moves from wide open to completely closed and covers a much wider range of operation than is illustrated here. Such curves can also represent the clogging of turbulent flow filters in a system.

SYSTEM EFFECTS

Normally, a fan is tested with open inlets, and a section of straight duct is attached to the outlet. This setup results in uniform flow into the fan and efficient static pressure recovery on the fan outlet. If good inlet and outlet conditions are not provided in the actual installation, fan performance suffers. To select and apply the fan properly, these effects must be considered and the pressure requirements of the fan, as calculated by standard duct design procedures, must be increased.

These calculated system effect factors are only an approximation, however. Fans of different types, and even fans of the same type but supplied by different manufacturers, do not necessarily react to a system in the same way. Therefore, judgment based on experience must be applied to any design. Chapter 35 of the 2005 *ASHRAE Handbook—Fundamentals* gives information on calculating the system effect factors and lists loss coefficients for a variety of fittings. Clarke et al. (1978) and AMCA *Publication* 201 provide further information.

SELECTION

After the system pressure loss curve of the air distribution system has been defined, a fan can be selected to meet the system requirements (Graham 1966, 1972). Fan manufacturers present performance data in either graphic (curve) (Figure 13) or tabular form (multirating tables). Multirating tables usually provide only performance data within the recommended operating range. The optimum selection range or peak efficiency point is identified in various ways by different manufacturers.

Performance data as tabulated in the usual fan tables are based on arbitrary increments of flow rate and pressure. In these tables, adjacent data, either horizontally or vertically, represent different points of operation (i.e., different points of rating) on the fan performance curve. These points of rating depend solely on the fan's

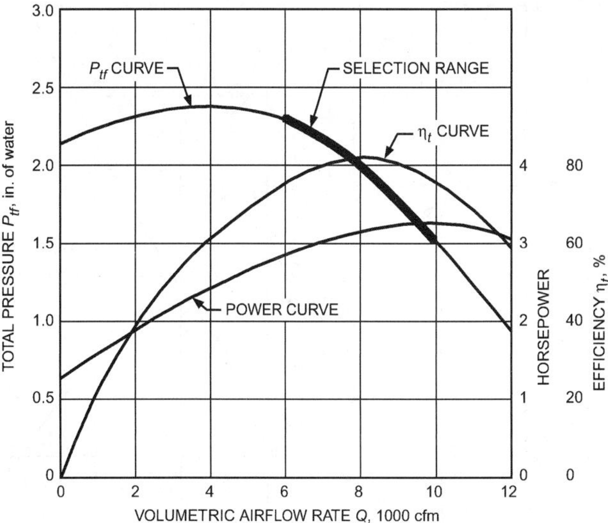

Fig. 13 Conventional Fan Performance Curve Used by Most Manufacturers

characteristics; they cannot be obtained from each other by the fan laws. However, points of operation listed in multirating tables are usually close together, so intermediate points may be interpolated arithmetically with adequate accuracy for fan selection.

Selecting a fan for a particular air distribution system requires that the fan pressure characteristics fit the system pressure characteristics. Thus, the total system must be evaluated and airflow requirements, resistances, and system effect factors at the fan inlet and outlet must be known (see Chapter 35 of the 2005 *ASHRAE Handbook—Fundamentals*). Fan speed and power requirements are then calculated, using multirating tables or single or multispeed performance curves or graphs.

In using curves, it is necessary that the point of operation selected (Figure 14) represent a desirable point on the fan curve, so that maximum efficiency and resistance to stall and pulsation can be attained. In systems where more than one point of operation is encountered during operation, it is necessary to look at the range of performance and evaluate how the selected fan reacts within this complete range. This analysis is particularly necessary for variable-volume systems, where not only the fan undergoes a change in performance, but the entire system deviates from the relationships defined in Equation (2). In these cases, it is necessary to look at actual losses in the system at performance extremes.

PARALLEL FAN OPERATION

The combined performance curve for two fans operating in parallel may be plotted by using the appropriate pressure for the ordinates and the sum of the volumes for the abscissas. When two fans having a pressure reduction to the left of the peak pressure point are operated in parallel, a fluctuating load condition may result if one of the fans operates to the left of the peak static point on its performance curve.

The P_t curves of a single fan and of two identical fans operating in parallel are shown in Figure 15. Curve A-A shows the pressure characteristics of a single fan. Curve C-C is the combined performance of the two fans. The unique figure-8 shape is a plot of all possible combinations of volume airflow at each pressure value for the individual fans. All points to the right of CD are the result of each fan operating at the right of its peak point of rating. Stable performance results for all systems with less obstruction to airflow than is shown on the ΔP curve D-D. At points of operation to the left of CD,

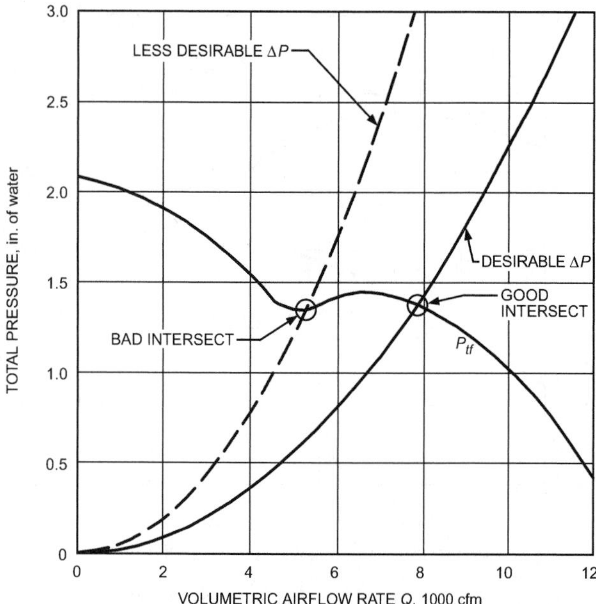

Fig. 14 Desirable Combination of P_{tf} and ΔP Curves

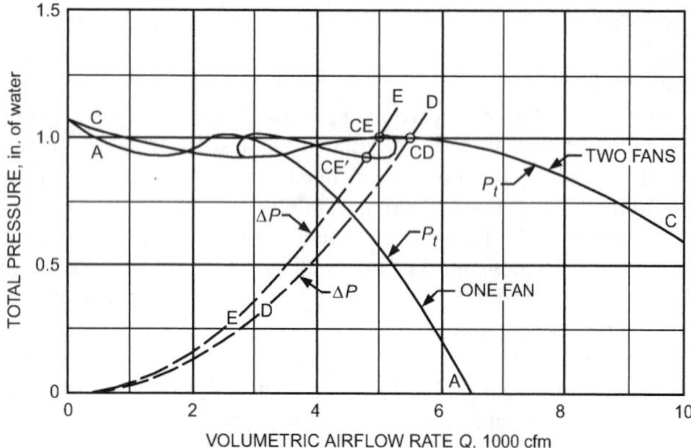

Fig. 15 Two Forward-Curved Centrifugal Fans in Parallel Operation

system requirements can be satisfied with each fan operating at a different point of rating. For example, consider ΔP curve E-E, which requires a pressure of 1.00 in. of water and a volume of 5000 cfm. The requirements of this system can be satisfied with each fan delivering 2500 cfm at 1.00 in. of water pressure, Point CE. The system can also be satisfied at Point CE′ with one fan operating at 1400 cfm at 0.9 in. of water, while the second fan delivers 3400 cfm at the same 0.9 in. of water.

Note that system curve E-E passes through the combined performance curve at two points. Under such conditions, unstable operation can result. Under conditions of CE′, one fan is underloaded and operating at poor efficiency. The other fan delivers most of the system requirements and uses substantially more power than the underloaded fan. This imbalance may reverse and shift the load from one fan to the other.

NOISE

Fan noise is a function of the fan design, volume airflow rate Q, total pressure P_t, and efficiency η_t. After a decision has been made regarding the proper type of fan for a given application (keeping in mind the system effects), the best size selection of that fan must be based on efficiency, because the most efficient operating range for a specific line of fans is normally the quietest. Low outlet velocity does not necessarily ensure quiet operation, so selections made on this basis alone are not appropriate. Also, noise comparisons of different types of fans, or fans offered by different manufacturers, made on the basis of rotational or tip speed are not valid. The only valid basis for comparison are the actual sound power levels generated by the different types of fans when they are all producing the required volume airflow rate and total pressure. Sound power level data should be obtained from the fan manufacturer for the specific fan being considered.

The data are reported by fan manufacturers as sound power levels in eight octave bands. These levels are determined by using a reverberant room for the test facility and comparing the sound generated by the fan to the sound generated by a reference source of known sound power. The measuring technique is described in AMCA *Standard* 300, Reverberant Room Method for Sound Testing of Fans. ANSI/ASHRAE *Standard* 68 (AMCA *Standard* 330), Laboratory Method of Testing to Determine the Sound in a Duct, describes an alternative test to determine the sound power a duct fan radiates into a supply and/or return duct terminated by an anechoic chamber. These standards do not fully evaluate the pure tones generated by some fans; these tones can be quite objectionable when they are radiated into occupied spaces. On critical installations, special allowance should be made by providing extra sound attenuation in the octave band containing the tone.

Discussions of sound and sound control may be found in Chapter 7 of the 2005 *ASHRAE Handbook—Fundamentals* and Chapter 47 of the 2007 *ASHRAE Handbook—HVAC Applications*.

VIBRATION

All rotating elastic structures, including fans, have certain operating speeds (known as **critical speeds**) at which resonances tend to cause objectionable vibrations. These critical speeds correspond to the various natural frequencies of the fan structure. Vibrations may be induced in a fan by unbalance, motor torque pulsations, and aerodynamic forces. Balancing fans with a frequency corresponding to the operating speed is desirable, but it is impossible to eliminate these forces completely.

Critical speeds are not solely the property of the shaft. Bearings, supports, foundations, and soil conditions all contribute to the elastic properties of any given system. Characteristics of the supporting system should be specified when a critical speed is calculated. Computer programs are available for calculating critical speeds, and manufacturers should be consulted. In axial-flow fans, the dynamic properties of the blades are of particular concern (see ANSI/ASHRAE *Standard* 87.2- 2002, In-Situ Method of Testing Propeller Fans for Reliability).

Vibration Isolation

During fan operation, some net force is transmitted to the supporting structure, making the supporting structure part of the vibrating system. Although many fans can be installed without vibration isolation, the system must be carefully designed and good balance maintained. Vibration isolation is required whenever vibrations in the supporting structure are annoying or destructive. More information on vibration and applying vibration isolation may be found in Chapter 7 of the 2005 *ASHRAE Handbook—Fundamentals* and Chapter 47 of the 2007 *ASHRAE Handbook—HVAC Applications*.

ARRANGEMENT AND INSTALLATION

Direction of rotation is determined from the drive side of the fan. On single-inlet centrifugal fans, the drive side is usually considered the side opposite the fan inlet. AMCA has published standard nomenclature to define positions.

Fan Isolation

In air-conditioning systems, ducts should be connected to fan outlets and inlets with unpainted canvas or other flexible material. Access should be provided in the connections for periodic removal of any accumulations tending to unbalance the rotor. When operating against high resistance or when low noise levels are required, it is preferable to locate the fan in a room removed from occupied areas or in a room acoustically treated to prevent sound transmission. The lighter building construction common today makes it desirable to mount fans and driving motors on resilient bases designed to prevent vibration transmission through floors to the building structure. Conduits, pipes, and other rigid members should not be attached to fans. Noise that results from obstructions, abrupt turns, grilles, and other items not connected with the fan may be present. Treatments for such problems, as well as the design of sound and vibration absorbers, are discussed in Chapter 47 of the 2007 *ASHRAE Handbook—HVAC Applications*.

CONTROL

In many heating and ventilating systems, the volume of air handled by the fan varies. The proper method for varying airflow for any particular case is influenced by two basic considerations: (1) the frequency with which changes must be made and (2) balancing reduced power consumption against increases in first cost.

To control airflow, the characteristic of either the system or the fan must be changed. The system characteristic curve may be altered by installing dampers or orifice plates. This technique reduces airflow by increasing the system pressure required and, therefore, increases power consumption. Figure 10 shows three different system curves, A, B, and C, such as would be obtained by changing the damper setting or orifice diameter. Dampers are usually the lowest-first-cost method of achieving airflow control; they can be used even in cases where essentially continuous control is needed.

Changing the fan characteristic (P_t curve) for control can reduce power consumption. From this standpoint, the most desirable control method is to vary the fan's rotational speed to produce the desired performance. If change is infrequent, belt-driven units may be adjusted by changing the pulley on the drive motor of the fan. Variable-speed motors or variable-speed drives, whether electrical or hydraulic, may be used when frequent or essentially continuous variations are desired. When speed control is used, the revised P_t curve can be calculated with the fan laws.

Inlet vane control is frequently used. Figure 16 illustrates the change in fan performance with inlet vane control. Curves A, B, C, D, and E are the pressure and power curves for various vane settings between wide open (A) and nearly closed (E).

Tubeaxial and vaneaxial fans are made with adjustable-pitch blades to permit balancing the fan against the system or to make infrequent adjustments. Vaneaxial fans are also produced with controllable-pitch blades (i.e., pitch that can be varied while the fan is in operation) for frequent or continuous adjustment. Varying pitch angle retains high efficiencies over a wide range of conditions. The performance shown in Figure 17 is from a typical vaneaxial fan with variable-pitch blades. From the standpoint of noise, variable speed is somewhat better than variable blade pitch; however, both of these control methods give high operating efficiency control and generate appreciably less noise than inlet vane or damper control.

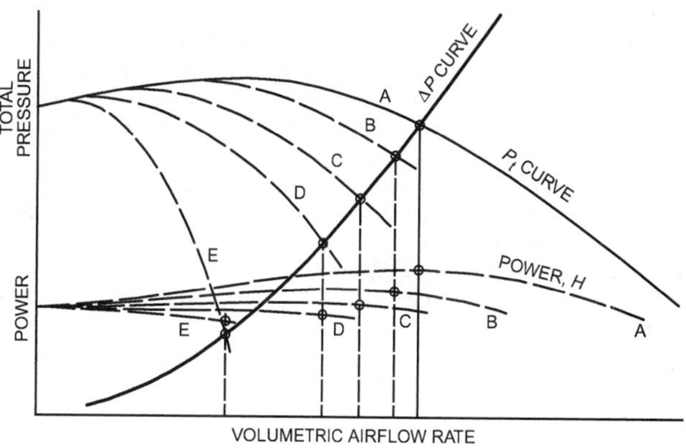

Fig. 16 Effect of Inlet Vane Control on Backward-Curved Centrifugal Fan Performance

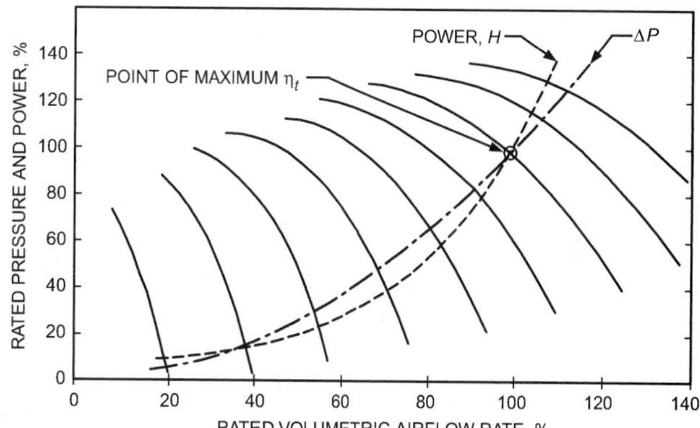

Fig. 17 Effect of Blade Pitch on Controllable-Pitch Vaneaxial Fan Performance

SYMBOLS

A = fan outlet area, ft^2
C_p = constant in Equation (1)
c_p = specific heat in Equation (1), Btu/lb$_m$·°F
D = fan size or impeller diameter
J = mechanical equivalent of heat, ft·lb$_f$/Btu
N = rotational speed, revolutions per minute
Q = volume airflow rate moved by fan at fan inlet conditions, cfm
P_{tf} = fan total pressure: fan total pressure at outlet minus fan total pressure at inlet, in. of water
P_{vf} = fan velocity pressure: pressure corresponding to average velocity determined from volume airflow rate and fan outlet area, in. of water
P_{sf} = fan static pressure: fan total pressure diminished by fan velocity pressure, in. of water. Fan inlet velocity head is assumed equal to zero for fan rating purposes.
P_{sx} = static pressure at given point, in. of water
P_{vx} = velocity pressure at given point, in. of water
P_{tx} = total pressure at given point, in. of water
ΔP = pressure change, in. of water
ΔT = temperature change, °F
V = fan inlet or outlet velocity, fpm
W_o = power output of fan: based on fan volume flow rate and fan total pressure, horsepower
W_i = power input to fan: measured by power delivered to fan shaft, horsepower
η_t = mechanical efficiency of fan (or fan total efficiency): ratio of power output to power input ($\eta_t = W_o/W_i$)
η_s = static efficiency of fan: mechanical efficiency multiplied by ratio of static pressure to fan total pressure, $\eta_s = (P_s/P_t)\eta_t$
ρ = gas (air) density, lb/ft^3

REFERENCES

AMCA. 1997. Laboratory method of testing to determine the sound power in a duct. ANSI/AMCA *Standard* 330-97 (ANSI/ASHRAE *Standard* 68-1997). Air Movement and Control Association, Arlington Heights, IL.

AMCA. 2002. Fans and systems. *Publication* 201-02. Air Movement and Control Association, Arlington Heights, IL.

AMCA. 1996. Reverberant room method for sound testing of fans. *Standard* 300-96. Air Movement and Control Association, Arlington Heights, IL.

AMCA. 1999. Laboratory methods of testing fans for aerodynamic performance rating. ANSI/AMCA *Standard* 210-99 (ANSI/ASHRAE *Standard* 51-1999). Air Movement and Control Association, Arlington Heights, IL.

ASHRAE. 1997. Laboratory method of testing in-duct sound power measurement procedure forms. ANSI/ASHRAE *Standard* 68-1997 (ANSI/AMCA *Standard* 330-86).

ASHRAE. 1999. Laboratory methods of testing fans for aerodynamic performance rating. ANSI/ASHRAE *Standard* 51-1999 (ANSI/AMCA *Standard* 210-99).

ASHRAE. 2002. In-situ method of testing propeller fans for reliability. ANSI/ASHRAE *Standard* 87.2-2002.

ASHRAE. 2001. Methods of testing propeller fan vibration—Diagnostic test methods. ANSI/ASHRAE *Standard* 87.3-2001.

Clarke, M.S., J.T. Barnhart, F.J. Bubsey, and E. Neitzel. 1978. The effects of system connections on fan performance. *ASHRAE Transactions* 84(2):227.

Graham, J.B. 1966. Fan selection by computer techniques. *Heating, Piping and Air Conditioning* (April):168.

Graham, J.B. 1972. Methods of selecting and rating fans. ASHRAE Symposium *Bulletin* SF-70-8, Fan Application—Testing and Selection.

BIBLIOGRAPHY

AMCA. 2003. Drive arrangements for centrifugal fans. *Standard* 99-2404-03. Air Movement and Control Association, Arlington Heights, IL.

AMCA. 2003. Designation for rotation and discharge of centrifugal fans. *Standard* 99-2406-03. Air Movement and Control Association, Arlington Heights, IL.

AMCA. 2003. Motor positions for belt or chain drive centrifugal fans. *Standard* 99-2407-03. Air Movement and Control Association, Arlington Heights, IL.

AMCA. 2003. Drive arrangements for tubular centrifugal fans. *Standard* 99-2410-86-03. Air Movement and Control Association, Arlington Heights, IL.

Howden Buffalo. 1999. *Fan engineering*, 9th ed. R. Jorgensen, ed. Howden Buffalo, Camden, NJ.

CHAPTER 21

HUMIDIFIERS

IN the selection and application of humidifiers, the designer considers (1) the environmental conditions of the occupancy or process and (2) the characteristics of the building enclosure. Because these may not always be compatible, compromise is sometimes necessary, particularly in the case of existing buildings.

ENVIRONMENTAL CONDITIONS

A particular occupancy or process may dictate a specific relative humidity, a required range of relative humidity, or certain limiting maximum or minimum values. The following classifications explain the effects of relative humidity and provide guidance on the requirements for most applications.

Human Comfort

The complete effect of relative humidity on all aspects of human comfort has not yet been established. For thermal comfort, higher temperature is generally considered necessary to offset decreased relative humidity (see ASHRAE *Standard* 55).

Low relative humidity increases evaporation from the membranes of the nose and throat, drying the mucous membranes in the respiratory system; it also dries the skin and hair. The increased incidence of respiratory complaints during winter is often linked to low relative humidity. Epidemiological studies have found lower rates of respiratory illness reported among occupants of buildings with mid-range relative humidity than among occupants of buildings with low humidity.

Extremes of humidity are the most detrimental to human comfort, productivity, and health. Figure 1 shows that the range between 30 and 60% rh (at normal room temperatures) provides the best conditions for human occupancy (Sterling et al. 1985). In this range, both the growth of bacteria and biological organisms and the speed at which chemical interactions occur are minimized.

Prevention and Treatment of Disease

Relative humidity has a significant effect on the control of airborne infection. At 50% rh, the mortality rate of certain organisms is highest, and the influenza virus loses much of its virulence. The mortality rate decreases both above and below this value. High humidity can support the growth of pathogenic or allergenic organisms. Relative humidity in habitable spaces should be maintained between 30 and 60%.

Potential Bacterial Growth

Certain microorganisms are occasionally present in poorly maintained humidifiers. To deter the propagation and spread of these detrimental microorganisms, periodic cleaning of the humidifier and draining of the reservoir (particularly at the end of the humidification season) are required. Cold-water reservoir atomizing room humidifiers have been banned in some hospitals because of germ propagation. Research by Unz et al. (1993) on several types of plenum-mounted residential humidifiers showed no evidence of organism transmission originating from the humidifier. Ruud et al. (1993) also determined that humidifiers did not add particles to the heated airstream.

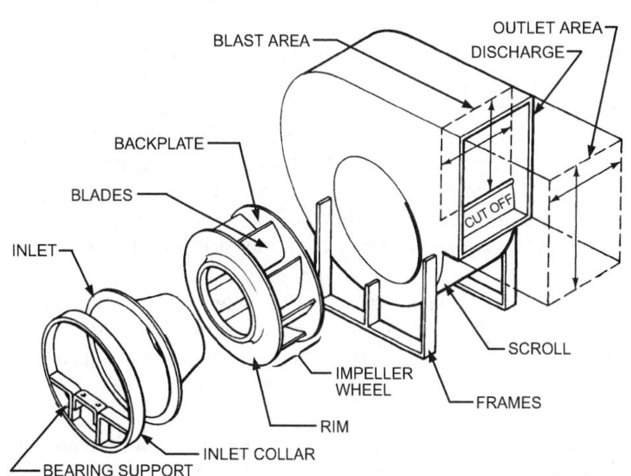

Fig. 1 Optimum Humidity Range for Human Comfort and Health
(Adapted from Sterling et al. 1985)

Electronic Equipment

Electronic data processing equipment requires controlled relative humidity. High relative humidity may cause condensation in the equipment, whereas low relative humidity may promote static electricity. Also, rapid changes in relative humidity should be avoided because of their effect on bar code readers, magnetic tapes, disks, and data processing equipment. Generally, computer systems have a recommended design and operating range of 35 to 55% rh. However, the manufacturer's recommendations should be adhered to for specific equipment operation.

Process Control and Materials Storage

The relative humidity required by a process is usually specific and related to one or more of several factors:

- Control of moisture content or regain
- Rate of chemical or biochemical reactions
- Rate of crystallization
- Product accuracy or uniformity
- Corrosion
- Static electricity

Typical conditions of temperature and relative humidity for the storage of certain commodities and the manufacturing and processing of others may be found in Chapter 12 of the 2007 *ASHRAE Handbook—HVAC Applications*.

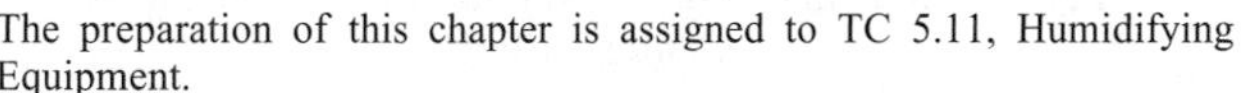

The preparation of this chapter is assigned to TC 5.11, Humidifying Equipment.

Low humidity in winter may cause drying and shrinking of furniture, wood floors, and interior trim. Winter humidification should be considered to maintain relative humidity closer to that experienced during manufacture or installation.

For storing hygroscopic materials, maintaining constant humidity is often as important as the humidity level itself. The design of the structure should always be considered. Temperature control is important because of the danger of condensation on products through a transient lowering of temperature.

Static Electricity

Electrostatic charges are generated when materials of high electrical resistance move against each other. The accumulation of such charges may have a variety of results: (1) unpleasant sparks caused by friction between two materials (e.g., stocking feet and carpet fibers); (2) difficulty in handling sheets of paper, fibers, and fabric; (3) objectionable dust clinging to oppositely charged objects (e.g., negatively charged metal nails or screws securing gypsum board to wooden studding in the exterior walls of a building that attract positively charged dust particles); (4) destruction of data stored on magnetic disks and tapes that require specifically controlled environments; and (5) hazardous situations if explosive gases are present, as in hospitals, research laboratories, or industrial clean rooms.

Increasing the relative humidity of the environment reduces the accumulation of electrostatic charges, but the optimum level of humidity depends to some extent on the materials involved. Relative humidity of 45% reduces or eliminates electrostatic effects in many materials, but wool and some synthetic materials may require a higher relative humidity.

Hospital operating rooms, where explosive mixtures of anesthetics are used, constitute a special and critical case. A relative humidity of at least 50% is usually required, with special grounding arrangements and restrictions on the types of clothing worn by occupants. Conditions of 72°F and 55% rh are usually recommended for comfort and safety.

Sound Wave Transmission

The air absorption of sound waves, which results in the loss of sound strength, is worst at 15 to 20% rh, and the loss increases as the frequency rises (Harris 1963). There is a marked reduction in sound absorption at 40% rh; above 50%, the effect of air absorption is negligible. Air absorption of sound does not significantly affect speech but may merit consideration in large halls or auditoriums where optimum acoustic conditions are required for musical performances.

Miscellaneous

Laboratories and test chambers, in which precise control of relative humidity over a wide range is desired, require special attention. Because of the interrelation between temperature and relative humidity, precise humidity control requires equally precise temperature control.

ENCLOSURE CHARACTERISTICS

Vapor Retarders

The maximum relative humidity level to which a building may be humidified in winter depends on the ability of its walls, roof, and other elements to prevent or tolerate condensation. Condensed moisture or frost on surfaces exposed to the building interior (visible condensation) can deteriorate the surface finish, cause mold growth and subsequent indirect moisture damage and nuisance, and reduce visibility through windows. If the walls and roof have not been specifically designed and properly protected with vapor retarders on the warm side to prevent the entry of moist air or vapor from the inside, concealed condensation within these constructions is likely to occur, even at fairly low interior humidity, and cause serious deterioration.

Visible Condensation

Condensation forms on an interior surface when its temperature is below the dew-point temperature of the air in contact with it. The maximum relative humidity that may be maintained without condensation is thus influenced by the thermal properties of the enclosure and the interior and exterior environment.

Average surface temperatures may be calculated by the methods outlined in Chapter 23 of the 2005 *ASHRAE Handbook—Fundamentals* for most insulated constructions. However, local cold spots result from high-conductivity paths such as through-the-wall framing, projected floor slabs, and metal window frames that have no thermal breaks. The vertical temperature gradient in the air space and surface convection along windows and sections with a high thermal conductivity result in lower air and surface temperatures at the sill or floor. Drapes and blinds closed over windows lower surface temperature further, while heating units under windows raise the temperature significantly.

In most buildings, windows present the lowest surface temperature and the best guide to permissible humidity levels for no condensation. While calculations based on overall thermal coefficients provide reasonably accurate temperature predictions at mid-height, actual minimum surface temperatures are best determined by test. Wilson and Brown (1964) related the characteristics of windows with a **temperature index**, defined as $(t - t_o)/(t_i - t_o)$, where t is the inside window surface temperature, t_i is the indoor air temperature, and t_o is the outdoor air temperature.

The results of limited tests on actual windows indicate that the temperature index at the bottom of a double, residential-type window with a full thermal break is between 0.55 and 0.57, with natural convection on the warm side. Sealed, double-glazed units exhibit an index from 0.33 to 0.48 at the junction of glass and sash, depending on sash design. The index is likely to rise to 0.53 or greater only 1 in. above the junction.

With continuous under-window heating, the minimum index for a double window with a full break may be as high as 0.60 to 0.70. Under similar conditions, the index of a window with a poor thermal break may be increased by a similar increment.

Figure 2 shows the relationship between temperature index and the relative humidity and temperature conditions at which condensation

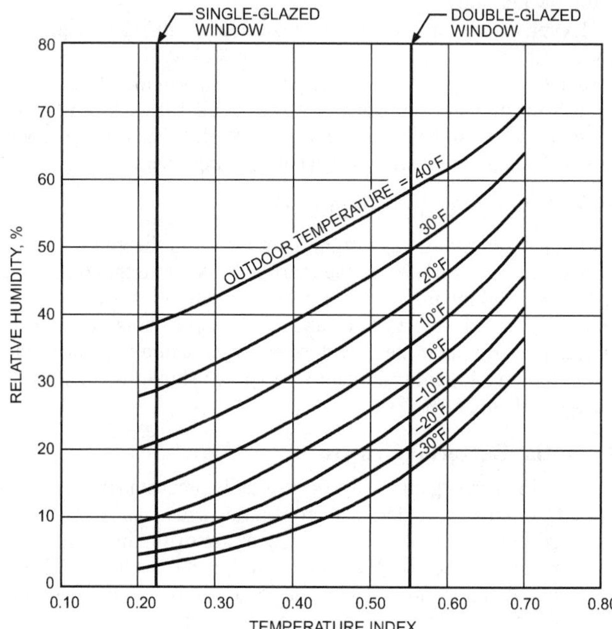

Fig. 2 Limiting Relative Humidity for No Window Condensation

Table 1 Maximum Relative Humidity In a Space for No Condensation on Windows

	Limiting Relative Humidity, %	
Outdoor Temperature, °F	Single Glazing	Double Glazing
40	39	59
30	29	50
20	21	43
10	15	36
0	10	30
−10	7	26
−20	5	21
−30	3	17

Note: Natural convection, indoor air at 74°F.

occurs. The limiting relative humidities for various outdoor temperatures intersect vertical lines representing particular temperature indexes. A temperature index of 0.55 has been selected to represent an average for double-glazed, residential windows; 0.22 represents an average for single-glazed windows. Table 1 shows the limiting relative humidities for both types of windows at various outdoor air temperatures.

Concealed Condensation

Vapor retarders are imperative in certain applications because the humidity level a building is able to maintain without serious concealed condensation may be much lower than that indicated by visible condensation. The migration of water vapor through the inner envelope by diffusion or air leakage brings the vapor into contact with surfaces at temperatures below its dew point. During the design of a building, the desired interior humidity may be determined by the ability of the building enclosure to handle internal moisture. This is particularly important when planning for building humidification in colder climates.

ENERGY CONSIDERATIONS

When calculating the energy requirement for a humidification system, the effect of the dry air on any material supplying it with moisture should be considered. The release of liquid in a hygroscopic material to a vapor state is an evaporative process that requires energy. The source of energy is the heat contained in the air. The heat lost from the air to evaporate moisture equals the heat necessary to produce an equal amount of moisture vapor with an efficient humidifier. If proper humidity levels are not maintained, moisture migration from hygroscopic materials can have destructive effects.

The true energy required for a humidification system must be calculated from the actual humidity level in the building, not from the theoretical level.

A study of residential heating and cooling systems showed a correlation between infiltration and inside relative humidity, indicating a significant energy saving from increasing the inside relative humidity, which reduced infiltration of outside air by up to 50% during the heating season (Luck and Nelson 1977). This reduction is apparently due to the sealing of window cracks by the formation of frost.

To assess accurately the total energy required to provide a desired level of humidity, all elements relating to the generation of humidity and the maintenance of the final air condition must be considered. This is particularly true when comparing different humidifiers. For example, the cost of boiler steam should include generation and distribution losses; costs for an evaporative humidifier include electrical energy for motors or compressors, water conditioning, and the addition of reheat (when the evaporative cooling effect is not required).

Load Calculations

The humidification load depends primarily on the rate of natural infiltration of the space to be humidified or the amount of outside air introduced by mechanical means. Other sources of moisture gain or loss should also be considered. The **humidification load** H can be calculated by the following equations:

For ventilation systems having natural infiltration,

$$H = \rho VR(W_i - W_o) - S + L \qquad (1)$$

For mechanical ventilation systems having a fixed quantity of outside air,

$$H = 60\rho Q_o(W_i - W_o) - S + L \qquad (2)$$

For mechanical systems having a variable quantity of outside air,

$$H = 60\rho Q_t(W_i - W_o)\left(\frac{t_i - t_m}{t_i - t_o}\right) - S + L \qquad (3)$$

where

H = humidification load, lb of water/h
V = volume of space to be humidified, ft^3
R = infiltration rate, air changes per hour
Q_o = volumetric flow rate of outside air, cfm
Q_t = total volumetric flow rate of air (outside air plus return air), cfm
t_i = design indoor air temperature, °F
t_m = design mixed air temperature, °F
t_o = design outside air temperature, °F
W_i = humidity ratio at indoor design conditions, lb of water/lb of dry air
W_o = humidity ratio at outdoor design conditions, lb of water/lb of dry air
S = contribution of internal moisture sources, lb of water/h
L = other moisture losses, lb of water/h
ρ = density of air at sea level, 0.074 lb/ft^3

Design Conditions

Interior design conditions are dictated by the occupancy or the process, as discussed in the preceding sections on Enclosure Characteristics and on Environmental Conditions. Outdoor relative humidity can be assumed to be 70 to 80% at temperatures below 32°F or 50% at temperatures above 32°F for winter conditions in most areas. Additional data on outdoor design data may be obtained from Chapter 28 of the 2005 *ASHRAE Handbook—Fundamentals*. Absolute humidity values can be obtained either from Chapter 6 of the 2005 *ASHRAE Handbook—Fundamentals* or from an ASHRAE psychrometric chart.

For systems handling fixed outside air quantities, load calculations are based on outdoor design conditions. Equation (1) should be used for natural infiltration, and Equation (2) for mechanical ventilation.

For economizers that achieve a fixed mixed air temperature by varying outside air, special considerations are needed to determine the maximum humidification load. This load occurs at an outside air temperature other than the lowest design temperature because it is a function of the amount of outside air introduced and the existing moisture content of the air. Equation (3) should be solved for various outside air temperatures to determine the maximum humidification load. It is also important to analyze the energy use of the humidifier (especially for electric humidifiers) when calculating the economizer setting in order to ensure that the energy saved by "free cooling" is greater than the energy consumed by the humidifier.

In residential load calculations, the actual outdoor design conditions of the locale are usually taken as 20°F and 70% rh, while indoor conditions are taken as 70°F and 35% rh. These values yield an absolute humidity difference ($W_i - W_o$) of 0.0040 lb per pound of dry air for use in Equation (1). However, the relative humidity may need to be less than 35% to avoid condensation at low outdoor temperatures (see Table 1).

Ventilation Rate

Ventilation of the humidified space may be due to either natural infiltration alone or natural infiltration in combination with intentional mechanical ventilation. Natural infiltration varies according to the indoor-outdoor temperature difference, wind velocity, and tightness of construction, as discussed in Chapter 27 of the 2005 *ASHRAE Handbook—Fundamentals*. The rate of mechanical ventilation may be determined from building design specifications or estimated from fan performance data (see ASHRAE *Standard* 62.1).

In load calculations, the water vapor removed from the air during cooling by air-conditioning or refrigeration equipment must be considered. This moisture may have to be replaced by humidification equipment to maintain the desired relative humidity in certain industrial projects where the moisture generated by the process may be greater than that required for ventilation and heating.

Estimates of infiltration rate are made in calculating heating and cooling loads for buildings; these values also apply to humidification load calculations. For residences where such data are not available, it may be assumed that a tight house has an infiltration rate of 0.5 air changes per hour (ach); an average house, 1 ach; and a loose house, as many as 1.5 ach. A tight house is assumed to be well insulated and to have vapor retarders, tight storm doors, windows with weather stripping, and a dampered fireplace. An average house is insulated and has vapor retarders, loose storm doors and windows, and a dampered fireplace. A loose house is generally one constructed before 1930 with little or no insulation, no storm doors, no insulated windows, no weather stripping, no vapor retarders, and often a fireplace without an effective damper. For building construction, refer to local codes and building specifications.

Additional Moisture Losses

Hygroscopic materials, which have a lower moisture content than materials in the humidified space, absorb moisture and place an additional load on the humidification system. An estimate of this load depends on the absorption rate of the particular material selected. Table 2 in Chapter 12 of the 2007 *ASHRAE Handbook—HVAC Applications* lists the equilibrium moisture content of hygroscopic materials at various relative humidities.

In cases where a certain humidity must be maintained regardless of condensation on exterior windows and walls, the dehumidifying effect of these surfaces constitutes a load that may need to be considered, if only on a transient basis. The loss of water vapor by diffusion through enclosing walls to the outside or to areas at a lower vapor pressure may also be involved in some applications. The properties of materials and flow equations given in Chapter 23 of the 2005 *ASHRAE Handbook—Fundamentals* can be applied in such cases. Normally, this diffusion constitutes a small load, unless openings exist between the humidified space and adjacent rooms at lower humidities.

Internal Moisture Gains

The introduction of a hygroscopic material can cause moisture gains to the space if the moisture content of the material is above that of the space. Similarly, moisture may diffuse through walls separating the space from areas of higher vapor pressure or move by convection through openings in these walls (Brown et al. 1963).

Moisture contributed by human occupancy depends on the number of occupants and their degree of physical activity. As a guide for residential applications, the average rate of moisture production for a family of four has been taken as 0.7 lb/h. Unvented heating devices produce about 1 lb of vapor for each pound of fuel burned. These values may no longer apply because of changes in equipment as well as in living habits.

Industrial processes constitute additional moisture sources. Single-color offset printing presses, for example, give off 0.45 lb of water per hour. Information on process contributions can best be obtained from the manufacturer of the specific equipment.

Supply Water for Humidifiers

There are three major categories of supply water: potable (untreated) water, softened potable water, and demineralized [deionized (DI) or reverse osmosis (RO)] water. Either the application or the humidifier may require a certain water type; the humidifier manufacturer's literature should be consulted.

In areas with water having a high mineral content, precipitated solids may be a problem. They clog nozzles, tubes, evaporative elements, and controls. In addition, solids allowed to enter the airstream via mist leave a fine layer of white dust over furniture, floors, and carpets. Some wetted-media humidifiers bleed off and replace some or all of the water passing through the element to reduce the concentration of salts in the recirculating water.

Dust, scaling, biological organisms, and corrosion are all potential problems associated with water in humidifiers. Stagnant water can provide a fertile breeding ground for algae and bacteria, which have been linked to odor and respiratory ailments. Bacterial slime reacts with sulfates in the water to produce hydrogen sulfide and its characteristic bad odor. Regular maintenance and periodic disinfecting with approved microbicides may be required (Puckorius et al. 1995). This has not been a problem with residential equipment; however, regular maintenance is good practice because biocides are generally used only with atomizing humidifiers.

Scaling

Industrial pan humidifiers, when supplied with water that is naturally low in hardness, require little maintenance, provided a surface skimmer bleedoff is used.

Water softening is an effective means of eliminating mineral precipitation in a pan-type humidifier. However, the concentration of sodium left in a pan as a result of water evaporation must be held below the point of precipitation by flushing and diluting the tank with new softened water. The frequency and duration of dilution depend on the water hardness and the rate of evaporation. Dilution is usually accomplished automatically by a timer-operated drain valve and a water makeup valve.

Demineralized or reverse osmosis (RO) water may also be used. The construction materials of the humidifier and the piping must withstand the corrosive effects of this water. Commercial demineralizers or RO equipment removes hardness and other total dissolved solids completely from the humidifier makeup water. They are more expensive than water softeners, but no humidifier purging is required. Sizing is based on the maximum required water flow to the humidifier and the amount of total dissolved solids in the makeup water.

EQUIPMENT

Humidifiers can generally be classified as either residential or industrial, although residential humidifiers can be used for small industrial applications, and small industrial units can be used in large homes. Equipment designed for use in central air systems also differs from that for space humidification, although some units are adaptable to both.

Air washers and direct evaporative coolers may be used as humidifiers; they are sometimes selected for additional functions such as air cooling or air cleaning, as discussed in Chapter 19.

The capacities of residential humidifiers are generally based on gallons per day of operation; capacities of industrial and commercial humidifiers are based on pounds per hour of operation. Published evaporation rates established by equipment manufacturers through test criteria may be inconsistent. Rates and test methods should be evaluated when selecting equipment. The Air-Conditioning and Refrigeration Institute (ARI) developed *Standard* 610 for residential central system humidifiers and *Standard* 640 for commercial and industrial humidifiers. Association of Home Appliance Manufacturers (AHAM) *Standard* HU-1 addresses self-contained residential units.

Residential Humidifiers for Central Air Systems

Residential humidifiers designed for central air systems depend on airflow in the heating system for evaporation and distribution. General principles and description of equipment are as follows:

Pan Humidifiers. Capacity varies with temperature, humidity, and airflow.

- *Basic pan.* A shallow pan is installed within the furnace plenum. Household water is supplied to the pan through a control device.
- *Electrically heated pan.* Similar to the basic unit, this type adds an electric heater to increase water temperature and evaporation rate.
- *Pan with wicking plates.* Similar to the basic unit, this type includes fitted water-absorbent plates. The increased area of the plates provides greater surface area for evaporation to take place (Figure 3A).

Wetted Element Humidifiers. Capacity varies with temperature, humidity, and airflow. Air circulates over or through an open-textured, wetted medium. The evaporating surface may be a fixed pad wetted by either sprays or water flowing by gravity, or a paddle-wheel, drum, or belt rotating through a water reservoir. The various types are differentiated by the way air flows through them:

- *Fan type.* A small fan or blower draws air from the furnace plenum, through the wetted pad, and back to the plenum. A fixed pad (Figure 3B) or a rotating drum-type pad (Figure 3C) may be used.
- *Bypass type.* These units do not have their own fan, but rather are mounted on the supply or return plenum of the furnace with an air connection to the return plenum (Figure 3D). The difference in static pressure created by the furnace blower circulates air through the unit.
- *Duct-mounted type.* These units are designed for installation within the furnace plenum or ductwork with a drum element rotated by either the air movement in the duct or a small electric motor.

Atomizing Humidifiers. The capacity of an atomizing humidifier does not depend on the air conditions. However, it is important not to oversaturate the air and allow liquid water to form in the duct. The ability of the air to absorb moisture depends on the temperature, flow rate, and moisture content of the air moving through the system. Small particles of water are formed and introduced into the airstream in one of the following ways:

- A spinning disk or cone throws a water stream centrifugally to the rim of the disk and onto deflector plates or a comb, where it is turned into a fine fog (Figure 3E).
- Spray nozzles rely on water pressure to produce a fine spray.
- Spray nozzles use compressed air to create a fine mist.
- Ultrasonic vibrations are used as the atomizing force.

Residential Humidifiers for Nonducted Applications

Many portable or room humidifiers are used in residences heated by nonducted hydronic or electric systems, or where the occupant is prevented from making a permanent installation. These humidifiers may be equipped with humidity controllers.

Portable units evaporate water by any of the previously described means, such as heated pan, fixed or moving wetted element, or atomizing spinning disk. They may be tabletop-sized or a larger, furniture-style appliance (Figure 3F). A multispeed motor on the fan or blower may be used to adjust output. Portable humidifiers usually require periodic filling from a bucket or filling hose.

Some portable units are offered with an auxiliary package for semipermanent water supply. This package includes a manual shut-off valve, a float valve, copper or other tubing with fittings, and so forth. Lack of drainage provision for water overflow may result in water damage.

Some units may be recessed into the wall between studs, mounted on wall surfaces, or installed below floor level. These units are permanently installed in the structure and use forced-air circulation. They may have an electric element for reheat when desired. Other types for use with hydronic systems involve a simple pan or pan plate, either installed within a hot-water convector or using the steam from a steam radiator.

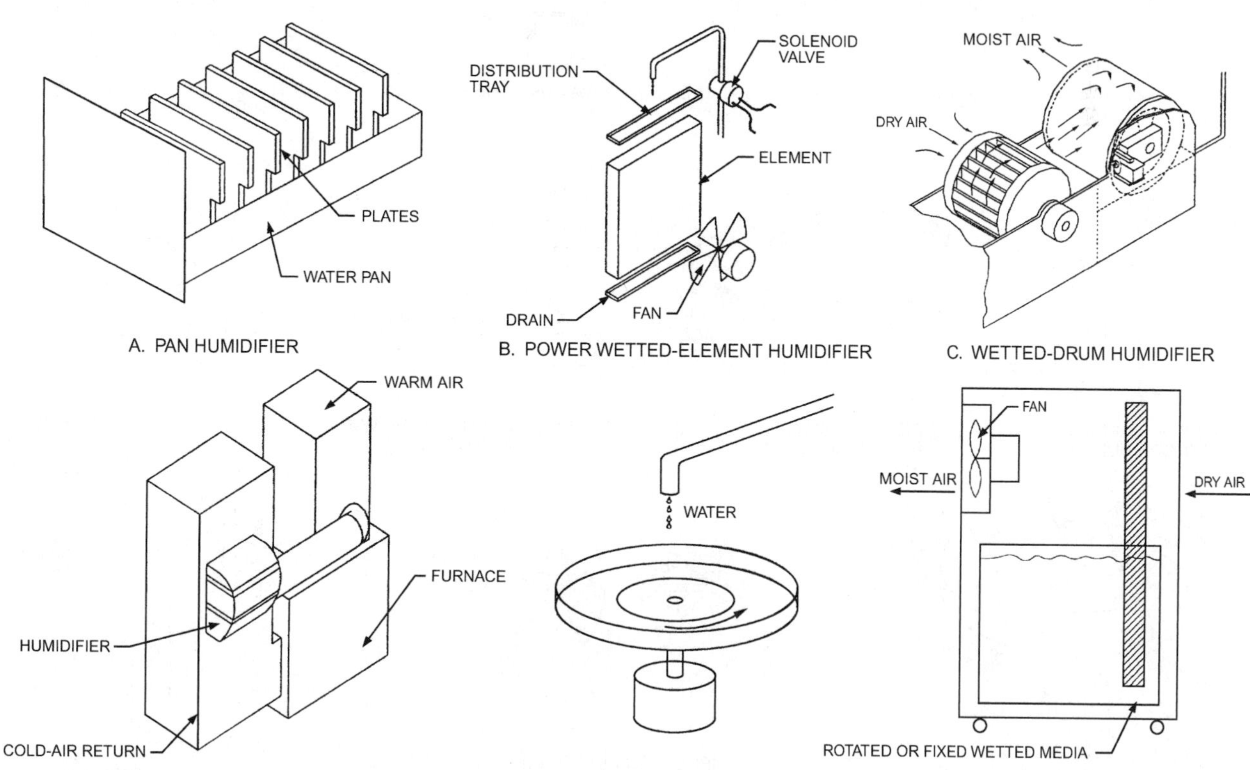

Fig. 3 Residential Humidifiers

Industrial and Commercial Humidifiers for Central Air Systems

Humidifiers must be installed where the air can absorb the vapor; the temperature of the air being humidified must exceed the dew point of the space being humidified. When fresh or mixed air is humidified, the air may need to be preheated to allow absorption to take place.

Heated Pan Humidifiers. These units offer a broad range of capacities and may be heated by a heat exchanger supplied with either steam or hot water (see Figure 4A). They may be installed directly under the duct, or they may be installed remotely and feed vapor through a hose. In either case, a distribution manifold should be used.

Steam heat exchangers are commonly used in heated-pan humidifiers, with steam pressures ranging from 5 to 15 psig. Hot-water heat exchangers are also used in pan humidifiers; a water temperature below 240°F is not practical.

All pan-type humidifiers should have water regulation and some form of drain or flush system. When raw water is used, periodic cleaning is required to remove the buildup of minerals. (Use of softened or demineralized water can greatly extend time between cleanings.) Care should also be taken to ensure that all water is drained off when the system is not in use to avoid the possibility of bacterial growth in the stagnant water.

Direct Steam Injection Humidifiers. These units cover a wide range of designs and capacities. Steam is water vapor under pressure and at high temperature, so the process of humidification can be simplified by adding steam directly into the air. This method is an isothermal process because the temperature of the air remains almost constant as the moisture is added. For this type of humidification system, the steam source is usually a central steam boiler at low pressure. When steam is supplied from a source at a constant supply pressure, humidification responds quickly to system demand. A control valve may be modulating or two-position in response to a humidity sensor/controller. Steam can be introduced into the airstream through one of the following devices:

- *Single or multiple steam-jacketed manifolds* (Figure 4B), depending on the size of the duct or plenum. The steam jacket is designed to reevaporate any condensate droplets before they are discharged from the manifold.
- *Nonjacketed manifold or panel-type distribution systems* (Figure 4C), with or without injection nozzles for distributing steam across the face of the duct or plenum.

Units must be installed where the air can absorb the discharged vapor before it comes into contact with components in the airstream, such as coils, dampers, or turning vanes. Otherwise, condensation can occur in the duct. Absorption distance varies according to the design of the humidifier distribution device and the air conditions

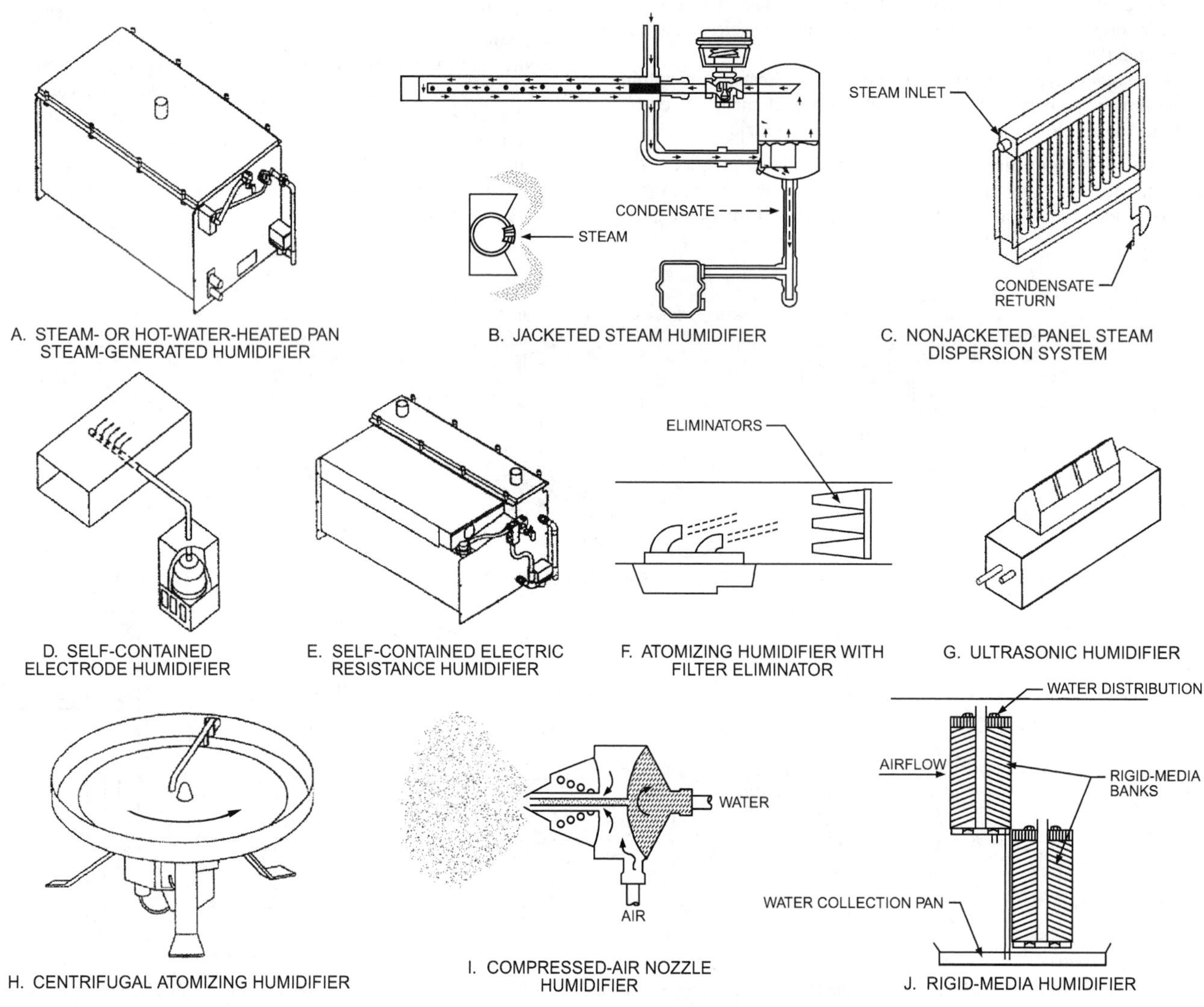

Fig. 4 Industrial Humidifiers

within the duct. For proper psychrometric calculations, refer to Chapter 6 of the 2005 *ASHRAE Handbook—Fundamentals*. Because these humidifiers inject steam from a central boiler source directly into the space or distribution duct, boiler treatment chemicals discharged into the air system may compromise indoor air quality. Chemicals should be checked for safety, and care should be taken to avoid contamination from the water or steam supplies.

Electrically Heated, Self-Contained Steam Humidifiers. These units convert ordinary city tap water to steam by electrical energy using either electrodes or resistance heater elements. The steam is generated at atmospheric pressure and discharged into the duct system through dispersion manifolds; if the humidifier is a freestanding unit, the steam is discharged directly into the air space through a fan unit. Some units allow the use of softened or demineralized water, which greatly extends the time between cleanings.

- *Electrode-type humidifiers* (Figure 4D) operate by passing an electric current directly into ordinary tap water, thereby creating heat energy to boil the water. The humidifier usually contains a polypropylene plastic bottle, either throwaway or cleanable, that is supplied with water through a solenoid valve. Water is drained off periodically to maintain a desirable solids concentration and the correct electrical flow. Manufacturers offer humidifiers with several different features, so their data should be consulted.
- *Resistance-type humidifiers* (Figure 4E) use one or more electrical elements that heat the water directly to produce steam. The water can be contained in a stainless steel or coated steel shell. The element and shell should be accessible for cleaning out mineral deposits. The high and low water levels should be controlled with either probes or float devices, and a blowdown drain system should be incorporated, particularly for off-operation periods.

Atomizing Humidifiers. Water treatment should be considered if mineral fallout from hard water is a problem. Optional filters may be required to remove the mineral dust from the humidified air (Figure 4F). Depending on the application and the water condition, atomizing humidifiers may require a reverse osmosis (RO) or a deionized (DI) water treatment system to remove the minerals. It is also important to note that wetted parts should be able to resist the corrosive effects of DI and RO water.

There are three main categories of atomizing humidifiers:

- *Ultrasonic humidifiers* (Figure 4G) use a piezoelectric transducer submerged in demineralized water. The transducer converts a high-frequency mechanical electric signal into a high-frequency oscillation. A momentary vacuum is created during the negative oscillation, causing the water to cavitate into vapor at low pressure. The positive oscillation produces a high-compression wave that drives the water particle from the surface to be quickly absorbed into the airstream. Because these types use demineralized water, no filter medium is required downstream. The ultrasonic humidifier is also manufactured as a freestanding unit.
- *Centrifugal humidifiers* (Figure 4H) use a high-speed disk, which slings the water to its rim, where it is thrown onto plates or a comb to produce a fine mist. The mist is introduced to the airstream, where it is evaporated.
- *Compressed-air nozzle humidifiers* can operate in two ways:
  1. Compressed air and water are combined inside the nozzle and discharged onto a resonator to create a fine fog at the nozzle tip (Figure 4I).
  2. Compressed air is passed through an annular orifice at the nozzle tip, and water is passed through a center orifice. The air creates a slight vortex at the tip, where the water breaks up into a fine fog on contact with the high-velocity compressed air.
  3. Compressed air is passed through an annular orifice at the nozzle tip, and water is passed through a center orifice. The air creates a slight vortex at the tip, where the water breaks up into a fine fog on contact with the high-velocity compressed air.

Wetted-Media Humidifiers. *Rigid-media humidifiers* (Figure 4J) use a porous core. Water is circulated over the media while air is blown through the openings. These humidifiers are adiabatic, cooling the air as it is humidified. Rigid-media cores are often used for the dual purpose of winter humidification and summer cooling. They depend on airflow for evaporation: the rate of evaporation varies with air temperature, humidity, and velocity.

The rigid media should be located downstream of any heating or cooling coils. For close humidity control, the element can be broken down into several (usually two to four) banks having separate water supplies. Solenoids controlling water flow to each bank are activated as humidification is required.

Rigid-media humidifiers have inherent filtration and scrubbing properties because of the water-washing effect in the filter-like channels. Only pure water is evaporated; therefore, contaminants collected from the air and water must be flushed from the system. A continuous bleed or regular pan flushing is recommended to minimize accumulation of contaminants in the pan and on the media.

Evaporative Cooling. Atomizing and wetted media humidifiers discharge water at ambient temperature. The water absorbs heat from the surrounding air to evaporate the fog, mist, or spray at a rate of 1075 Btu per pound of water. This evaporative cooling effect (see Chapter 19) should be considered in the design of the system and if reheat is required to achieve the final air temperature. The ability of the surrounding air to efficiently absorb the fog, mist, or spray will also depend on its temperature, air velocity, and moisture content.

CONTROLS

Many humidity-sensitive materials are available. Some are organic, such as nylon, human hair, wood, and animal membranes that change length with humidity changes. Other sensors change electrical properties (resistance or capacitance) with humidity.

Mechanical Controls

Mechanical sensors depend on a change in the length or size of the sensor as a function of relative humidity. The most commonly used sensors are synthetic polymers or human hair. They can be attached to a mechanical linkage to control the mechanical, electrical, or pneumatic switching element of a valve or motor. This design is suitable for most human comfort applications, but it may lack the necessary accuracy for industrial applications.

A humidity controller is normally designed to control at a set point selected by the user. Some controllers have a setback feature that lowers the relative humidity set point as outdoor temperature drops to reduce condensation within the structure.

Electronic Controllers

Electrical sensors change electrical resistance as the humidity changes. They typically consist of two conductive materials separated by a humidity-sensitive, hygroscopic insulating material (polyvinyl acetate, polyvinyl alcohol, or a solution of certain salts). Small changes are detected as air passes over the sensing surface. Capacitive sensors use a dielectric material that changes its dielectric constant with relative humidity. The dielectric material is sandwiched between special conducting material that allows a fast response to changes in relative humidity.

Electronic control is common in laboratory or process applications requiring precise humidity control. It is also used to vary fan speed on portable humidifiers to regulate humidity in the space more closely and to reduce noise and draft to a minimum.

Electronic controls are now widely used for residential applications because of low-cost, accurate, and stable sensors that can be used with inexpensive microprocessors. They may incorporate methods of determining outside temperature so that relative humidity can be automatically reset to some predetermined algorithm intended to maximize human comfort and minimize any condensation problems (Pasch et al. 1996).

Along with a main humidity controller, the system may require other sensing devices:

- **High-limit sensors** may be required to ensure that duct humidity levels remain below the saturation or dew-point level. Sometimes cooler air is required to offset sensible heat gains. In these cases, the air temperature may drop below the dew point. Operating the humidifier under these conditions causes condensation in the duct or fogging in the room. High-limit sensors may be combined with a temperature sensor in certain designs.
- **Airflow sensors** should be used in place of a fan interlock. They sense airflow and disable the humidifier when insufficient airflow is present in the duct.
- **Steam sensors** are used to keep the control valve on direct-injection humidifiers closed when steam is not present at the humidifier. A pneumatic or electric temperature-sensing switch is fitted between the separator and the steam trap to sense the temperature of the condensate and steam. When the switch senses steam temperature, it allows the control valve to function normally.

Humidity Control in Variable Air Volume (VAV) Systems

Control in VAV systems is much more demanding than in constant volume systems. VAV systems, common in large, central station applications, control space temperature by varying the volume rather than the temperature of the supply air. Continual airflow variations to follow load changes within the building can create wide and rapid swings in space humidity. Because of the fast-changing nature and cooler supply air temperatures (55°F or lower) of most VAV systems, special modulating humidity controls should be applied.

Best results are obtained by using both space and duct modulating-type humidity sensors in conjunction with an integrating device, which in turn modulates the output of the humidifier. This allows the duct sensor to respond quickly to a rapid rise in duct humidity caused by reduced airflow to the space as temperature conditions are satisfied. The duct sensor at times overrides the space humidistat

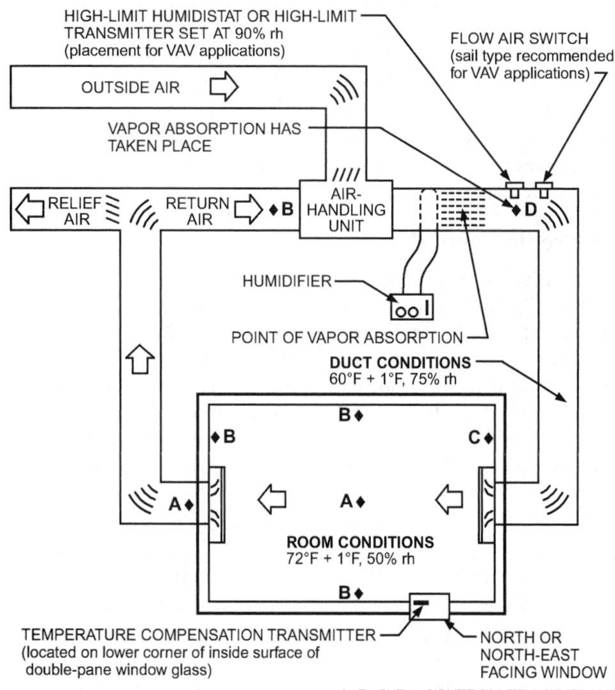

Fig. 5 Recommended Humidity Controller Location

by reducing the humidifier output to correspond to decreasing air volumes. This type of system, commonly referred to as **anticipating control**, allows the humidifier to track the dynamics of the system and provide uniform control. Due to the operating duct static pressures of a VAV system, use of an **airflow proving device** is recommended to detect air movement.

Further information on the evaluation of humidity sensors can be found in ASHRAE (1992a).

Control Location

In centrally humidified structures, the humidity controller is most commonly mounted in a controlled space. Another method is to mount the controller in the return air duct of an air-handling system to sense average relative humidity. Figure 5 shows general recommended locations for the humidistat for a centrally air-conditioned room.

The manufacturer's instructions regarding the use of the controller on counterflow furnaces should be followed because reverse airflow when the fan is off can substantially shift the humidity control point in a home. The sensor should be located where it will not be affected by (1) air that exits the bypass duct of a bypass humidifier or (2) drafts or local heat or moisture sources.

REFERENCES

ASHRAE. 1992a. Control of humidity in buildings. *Technical Data Bulletin* 8(3).

ASHRAE. 1992b. Thermal environmental conditions for human occupancy. ANSI/ASHRAE *Standard* 55-1992.

Brown, W.G., K.R. Solvason, and A.G. Wilson. 1963. Heat and moisture flow through openings by convection. *ASHRAE Journal* 5(9):49.

Harris, C.M. 1963. Absorption of sound in air in the audio-frequency range. *Journal of the Acoustical Society of America* 35(January).

Luck, J.R. and L.W. Nelson. 1977. The variation of infiltration rate with relative humidity in a frame building. *ASHRAE Transactions* 83(1):718-729.

Pasch, R.M., M. Comins, and J.S. Hobbins. 1996. Field experiences in residential humidification control with temperature-compensated automatic humidistats. *ASHRAE Transactions* 102(2):628-632.

Puckorius, P.R., P.T. Thomas, and R.L. Augspurger. 1995. Why evaporative coolers have not caused Legionnaires' disease. *ASHRAE Journal* 37(1): 29-33.

Ruud, C.O., J.W. Davis, and R.F. Unz. 1993. Analysis of furnace-mount humidifier for microbiological and particle emissions—Part II: Particle sampling and results. *ASHRAE Transactions* 99(1):1387-1395.

Sterling, E.M., A. Arundel, and T.D. Sterling. 1985. Criteria for human exposure to humidity in occupied buildings. *ASHRAE Transactions* 91(1B):611-622.

Unz, R.F., J.W. Davis, and C.O. Ruud. 1993. Analysis of furnace-mount humidifiers for microbiological and particle emissions—Part III: Microbiological sampling and results. *ASHRAE Transactions* 99(1):1396-1404.

Wilson, A.G. and W.P. Brown. 1964. Thermal characteristics of double windows. *Canadian Building Digest* no. 58. Division of Building Research, National Research Council, Ottawa, ON.

BIBLIOGRAPHY

AHAM. 2003. Appliance humidifiers. *Standard* HU-1. Association of Home Appliance Manufacturers, Chicago.

ARI. 2001. Central system humidifiers for residential applications. ANSI/ARI *Standard* 610-01. Air-Conditioning and Refrigeration Institute, Arlington, VA.

ARI. 2001. Commercial and industrial humidifiers. ANSI/ARI *Standard* 640-01. Air-Conditioning and Refrigeration Institute, Arlington, VA.

ASHRAE. 2004. Ventilation for acceptable indoor air quality. ANSI/ASHRAE *Standard* 62.1-2004.

Berglund, L.G. 1998. Comfort and humidity. *ASHRAE Journal* 40(8):35-41.

Davis, J.W., C.O. Ruud, and R.F. Unz. 1993. Analysis of furnace-mount humidifier for microbiological and particle emissions—Part I: Test system development. *ASHRAE Transactions* 99(1):1377-1386.

CHAPTER 22

AIR-COOLING AND DEHUMIDIFYING COILS

MOST equipment used today for cooling and dehumidifying an airstream under forced convection incorporates a coil section that contains one or more cooling coils assembled in a coil bank arrangement. Such coil sections are used extensively as components in room terminal units; larger factory-assembled, self-contained air conditioners; central station air handlers; and field built-up systems. Applications of each coil type are limited to the field within which the coil is rated. Other limitations are imposed by code requirements, proper choice of materials for the fluids used, the configuration of the air handler, and economic analysis of the possible alternatives for each installation.

USES FOR COILS

Coils are used for air cooling with or without accompanying dehumidification. Examples of cooling applications without dehumidification are (1) precooling coils that use well water or other relatively high-temperature water to reduce load on the refrigerating equipment and (2) chilled-water coils that remove sensible heat from chemical moisture-absorption apparatus. The heat pipe coil is also used as a supplementary heat exchanger for preconditioning in air-side sensible cooling (see Chapter 25). Most coil sections provide air sensible cooling and dehumidification simultaneously.

The assembly usually includes a means of cleaning air to protect the coil from dirt accumulation and to keep dust and foreign matter out of the conditioned space. Although cooling and dehumidification are their principal functions, cooling coils can also be wetted with water or a hygroscopic liquid to aid in air cleaning, odor absorption, or frost prevention. Coils are also evaporatively cooled with a water spray to improve efficiency or capacity. Chapter 40 has more information on indirect evaporative cooling. For general comfort conditioning, cooling, and dehumidifying, the **extended-surface (finned) cooling coil** design is the most popular and practical.

COIL CONSTRUCTION AND ARRANGEMENT

In finned coils, the external surface of the tubes is primary, and the fin surface is secondary. The primary surface generally consists of rows of round tubes or pipes that may be staggered or placed in line with respect to the airflow. Flattened tubes or tubes with other nonround internal passageways are sometimes used. The inside surface of the tubes is usually smooth and plain, but some coil designs have various forms of internal fins or turbulence promoters (either fabricated or extruded) to enhance performance. The individual tube passes in a coil are usually interconnected by return bends (or hairpin bend tubes) to form the serpentine arrangement of multipass tube circuits. Coils are usually available with different circuit arrangements and combinations offering varying numbers of parallel water flow passes within the tube core (Figure 1).

Cooling coils for water, aqueous glycol, brine, or halocarbon refrigerants usually have aluminum fins on copper tubes, although

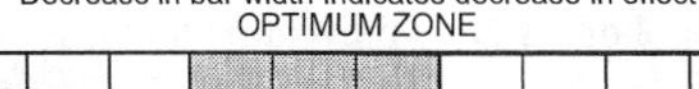

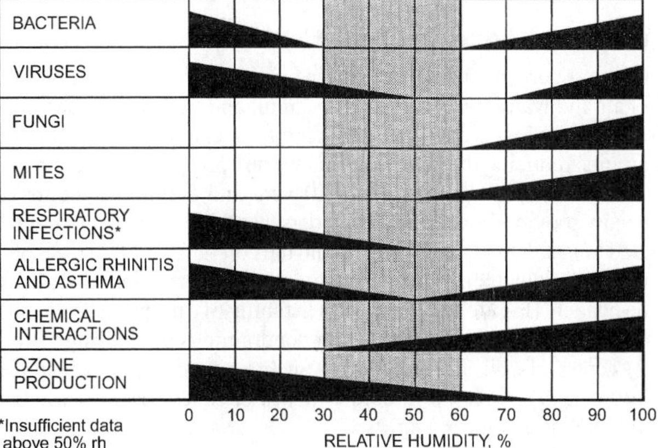

Fig. 1 Typical Water Circuit Arrangements

copper fins on copper tubes and aluminum fins on aluminum tubes (excluding water) are also used. Adhesives are sometimes used to bond header connections, return bends, and fin-tube joints, particularly for aluminum-to-aluminum joints. Certain special-application coils feature an all-aluminum extruded tube-and-fin surface.

Common core tube outside diameters are 5/16, 3/8, 1/2, 5/8, 3/4, and 1 in., with fins spaced 4 to 18 per inch. Tube spacing ranges from 0.6 to 3.0 in. on equilateral (staggered) or rectangular (in-line) centers, depending on the width of individual fins and on other performance considerations. Fins should be spaced according to the job to be performed, with special attention given to air friction; possibility of lint accumulation; and frost accumulation, especially at lower temperatures.

Tube wall thickness and the required use of alloys other than copper are determined mainly by the coil's working pressure and safety factor for hydrostatic burst (pressure). Maximum allowable working pressure (MAWP) for a coil is derived according to ASME's *Boiler and Pressure Vessel Code*, Section VIII, Division 1 and Section II (ASTM material properties and stress tables). Pressure vessel safety standards compliance and certifications of coil construction may be required by regional and local codes before field installation. Fin type and header construction also play a large part in determining wall thickness and material. Local job site codes and applicable nationally recognized safety standards should be consulted in coil design and application.

This type of air-cooling coil normally has a shiny aluminum air-side surface. For special applications, the fin surface may be copper or have a brown or blue-green dip-process coating. These coatings protect the fin from oxidation that occurs when common airborne corrosive contaminants are diluted on a wet (dehumidifying) surface. Corrosion protection is increasingly important as indoor air quality (IAQ) guidelines call for higher percentages of outside air. Baked-on or anodized coating improves the expected service life

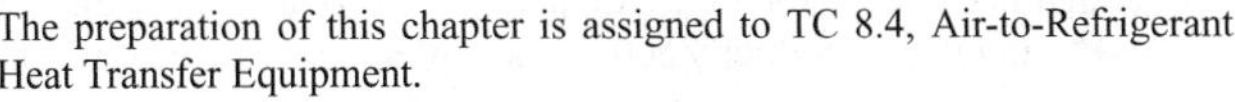

The preparation of this chapter is assigned to TC 8.4, Air-to-Refrigerant Heat Transfer Equipment.

compared to plain aluminum fins under similar conditions. Uncoated fins on non-dehumidifying, dry cooling coils are generally not affected by normal ambient airborne chemicals, except, to some extent, in a saline atmosphere. Once the coil is installed, little can be done to improve air-side protection.

Incoming airstream stratification across the coil face reduces coil performance. Proper air distribution is defined as having a measured airflow anywhere on the coil face that does not vary more than 20%. Moisture carryover at the coil's air leaving side or uneven air filter loading are indications of uneven airflow through the coil. Normal corrective procedure is to install inlet air straighteners, or an air blender if several airstreams converge at the coil inlet face. Additionally, condensate water should never be allowed to saturate the duct liner or stand in the drain pan (trough). The coil frame (particularly its bottom sheet metal member) should not be allowed to sit in a pool of water, to prevent rusting.

Water and Aqueous Glycol Coils

Good performance of water-type coils requires both eliminating all air and water traps in the water circuit and the proper distribution of water. Unless properly vented, air may accumulate in the coil tube circuits, reducing thermal performance and possibly causing noise or vibration in the piping system. Air vent and drain connections are usually provided on coil water headers, but this does not eliminate the need to install, operate, and maintain the coil tube core in a level position. Individual coil vents and drain plugs are often incorporated on the headers (Figure 1). Water traps in tubing of a properly leveled coil are usually caused by (1) improper nondraining circuit design and/or (2) center-of-coil downward sag. Such a situation may cause tube failure (e.g., freeze-up in cold climates or tube erosion because of untreated mineralized water).

Depending on performance requirements, fluid velocity inside the tube usually ranges from approximately 1 to 8 fps for water and 0.5 to 6 fps for glycol. When turbulators or grooved tubes are used, in-tube velocities should not exceed 4 fps. The design fluid pressure pressure drop across the coils varies from about 5 to 50 ft of water head. For nuclear HVAC applications, ASME *Standard* AG-1, Code on Nuclear Air and Gas Treatment, requires a minimum tube velocity of 2 fps. ARI *Standard* 410 requires a minimum of 1 fps or a Reynolds number of 3100 or greater. This yields more predictable performance.

In certain cases, the water may contain considerable sand and other foreign matter (e.g., precooling coils using well water, or where minerals in the cooling water deposit on and foul the tube surface). It is best to filter out such sediment. Some coil manufacturers offer removable water header plates or a removable plug for each tube that allows the tube to be cleaned, ensuring continuing rated performance while the cooling units are in service. Where build-up of scale deposits or fouling of the water-side surface is expected, a scale factor is sometimes included when calculating thermal performance of the coils. Cupronickel, red brass, bronze, and other tube alloys help protect against corrosion and erosion deterioration caused primarily by internal fluid flow abrasive sediment. The core tubes of properly designed and installed coils should feature circuits that (1) have equally developed line length, (2) are self-draining by gravity during the coil's off cycle, (3) have the minimum pressure drop to aid water distribution from the supply header without requiring excessive pumping head, and (4) have equal feed and return by the supply and return header. Design for proper in-tube water velocity determines the circuitry style required. Multirow coils are usually circuited to the cross-counterflow arrangement and oriented for top-outlet/bottom-feed connection.

Direct-Expansion Coils

Coils for halocarbon refrigerants present more complex cooling fluid distribution problems than do water or brine coils. The coil should cool effectively and uniformly throughout, with even refrigerant distribution. Halocarbon coils are used on two types of refrigerated systems: flooded and direct-expansion.

A flooded system is used mainly when a small temperature difference between the air and refrigerant is desired. Chapter 3 of the 2006 *ASHRAE Handbook—Refrigeration* describes flooded systems in more detail.

For direct-expansion systems, two of the most commonly used refrigerant liquid metering arrangements are the capillary tube assembly (or restrictor orifice) and the thermostatic expansion valve (TXV) device. The **capillary tube** is applied in factory-assembled, self-contained air conditioners up to approximately 10 ton capacity, but is most widely used on smaller-capacity models such as window or room units. In this system, the bore and length of a capillary tube are sized so that at full load, under design conditions, just enough liquid refrigerant to be evaporated completely is metered from the condenser to the evaporator coil. Although this type of metering arrangement does not operate over a wide range of conditions as efficiently as a TXV system, its performance is targeted for a specific design condition.

A **thermostatic expansion valve** system is commonly used for all direct-expansion coil applications described in this chapter, particularly field-assembled coil sections and those used in central air-handling units and larger, factory-assembled hermetic air conditioners. This system depends on the TXV to automatically regulate the rate of refrigerant liquid flow to the coil in direct proportion to the evaporation rate of refrigerant liquid in the coil, thereby maintaining optimum performance over a wide range of conditions. Superheat at the coil suction outlet is continually maintained within the usual predetermined limits of 6 to 10°F. Because the TXV responds to the superheat at the coil outlet, superheat within the coil is produced with the least possible sacrifice of active evaporating surface.

The length of each coil's refrigerant circuits, from the TXV's distributor feed tubes through the suction header, should be equal. The length of each circuit should be optimized to provide good heat transfer, good oil return, and a complementary pressure drop across the circuit. The coil should be installed level, and coil circuitry should be designed to self-drain by gravity toward the suction header connection. This is especially important on systems with unloaders or variable-speed-drive compressor(s). When non-self-drain circuitry is used, the circuit and suction connection should be designed for a minimum tube velocity sufficient to avoid compressor lube oil trapping in the coil.

To ensure reasonably uniform refrigerant distribution in multicircuit coils, a distributor is placed between the TXV and coil inlets to divide refrigerant equally among the coil circuits. The refrigerant distributor must be effective in distributing both liquid and vapor because refrigerant entering the coil is usually a mixture of the two, although mainly liquid by weight. Distributors can be placed either vertically or horizontally; however, the vertical down position usually distributes refrigerant between coil circuits better than the horizontal for varying load conditions.

Individual coil circuit connections from the refrigerant distributor to the coil inlet are made of small-diameter tubing; the connections are all the same length and diameter so that the same flow occurs between each refrigerant distributor tube and each coil circuit. To approximate uniform refrigerant distribution, refrigerant should flow to each refrigerant distributor circuit in proportion to the load on that coil. The heat load must be distributed equally to each refrigerant circuit for optimum coil performance. If the coil load cannot be distributed uniformly, the coil should be recircuited and connected with more than one TXV to feed the circuits (individual suction may also help). In this way, refrigerant distribution is reduced in proportion to the number of distributors to have less effect on overall coil performance when design must accommodate some unequal circuit loading. Unequal circuit loading may also be caused by uneven air velocity across the coil's face, uneven entering

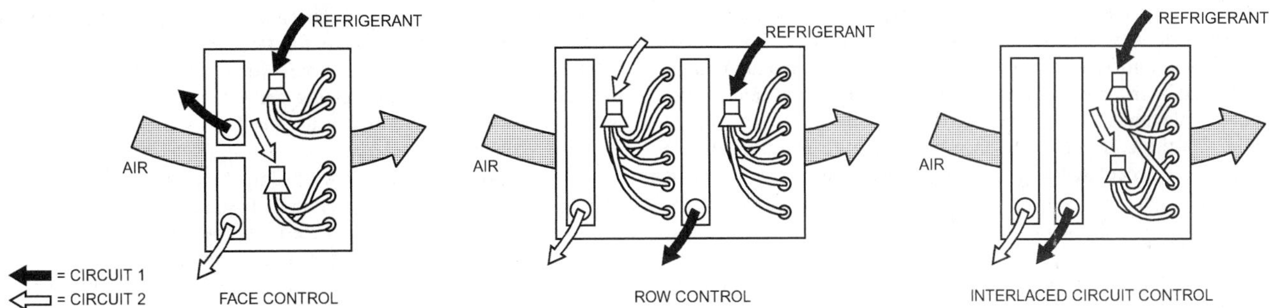

Fig. 2 Arrangements for Coils with Multiple Thermostatic Expansion Valves

air temperature, improper coil circuiting, oversized orifice in distributor, or the TXV's not being directly connected (close-coupled) to the distributor.

Control of Coils

Cooling capacity of water coils is controlled by varying either water flow or airflow. Water flow can be controlled by a three-way mixing, modulating, and/or throttling valve. For airflow control, face and bypass dampers are used. When cooling demand decreases, the coil face damper starts to close, and the bypass damper opens. In some cases, airflow is varied by controlling fan capacity with speed controls, inlet vanes, or discharge dampers.

Chapter 46 of the 2007 *ASHRAE Handbook—HVAC Applications* addresses air-cooling coil control to meet system or space requirements and factors to consider when sizing automatic valves for water coils. Selection and application of refrigerant flow control devices (e.g., thermostatic expansion valves, capillary tube types, constant-pressure expansion valves, evaporator pressure regulators, suction-pressure regulators, solenoid valves) as used with direct-expansion coils are discussed in Chapter 44 of the 2006 *ASHRAE Handbook—Refrigeration*.

For factory-assembled, self-contained packaged systems or field-assembled systems using direct-expansion coils equipped with TXVs, a single valve is sometimes used for each coil; in other cases, two or more valves are used. The thermostatic expansion valve controls the refrigerant flow rate through the coil circuits so refrigerant vapor at the coil outlet is superheated properly. Superheat is obtained with suitable coil design and proper valve selection. Unlike water flow control valves, standard pressure/temperature thermostatic expansion valves alone do not control the refrigeration system's capacity or the temperature of the leaving air, nor do they maintain ambient conditions in specific spaces. However, some electronically controlled TXVs have these attributes.

To match refrigeration load requirements for the conditioned space to the cooling capacity of the coil(s), a thermostat located in the conditioned space(s) or in the return air temporarily interrupts refrigerant flow to the direct-expansion cooling coils by stopping the compressor(s) and/or closing the solenoid liquid-line valve(s). Other solenoids unload compressors by suction control. For jobs with only a single zone of conditioned space, the compressor's on-off control is frequently used to modulate coil capacity. Selection and application of evaporator pressure regulators and similar regulators that are temperature-operated and respond to the temperature of conditioned air are covered in Chapter 44 of the 2006 *ASHRAE Handbook—Refrigeration*.

Applications with multiple zones of conditioned space often use solenoid liquid-line valves to vary coil capacity. These valves should be used where thermostatic expansion valves feed certain types (or sections) of evaporator coils that may, according to load variations, require a temporary but positive interruption of refrigerant flow. This applies particularly to multiple evaporator coils in a unit where one or more must be shut off temporarily to regulate its zone capacity. In such cases, a solenoid valve should be installed directly upstream of the thermostatic expansion valve(s). If more than one expansion valve feeds a particular zone coil, they may all be controlled by a single solenoid valve.

For a coil controlled by multiple refrigerant expansion valves, there are three arrangements: (1) face control, in which the coil is divided across its face; (2) row control; and (3) interlaced circuitry (Figure 2).

Face control, which is the most widely used because of its simplicity, equally loads all refrigerant circuits in the coil. Face control has the disadvantage of permitting condensate reevaporation on the coil portion not in operation and bypassing air into the conditioned space during partial-load conditions, when some of the TXVs are off. However, while the bottom portion of the coil is cooling, some of the advantages of single-zone humidity control can be achieved with air bypasses through the inactive top portion.

Row control, seldom available as standard equipment, eliminates air bypassing during partial-load operation and minimizes condensate reevaporation. Close attention is required for accurate calculation of row-depth capacity, circuit design, and TXV sizing.

Interlaced circuit control uses the whole face area and depth of coil when some expansion valves are shut off. Without a corresponding drop in airflow, modulating refrigerant flow to an interlaced coil increases coil surface temperature, thereby necessitating compressor protection (e.g., suction pressure regulators or compressor multiplexing).

Flow Arrangement

In air conditioning, the relation of the fluid flow arrangement in the coil tubes to coil depth greatly influences performance of the heat transfer surface. Generally, air-cooling and dehumidifying coils are multirow and circuited for **counterflow** arrangement. Inlet air is applied at right angles to the coil's tube face (coil height), which is also at the coil's outlet header location. Air exits at the opposite face (side) of the coil where the corresponding inlet header is located. Counterflow can produce the highest possible heat exchange in the shortest possible (coil row) depth because it has the closest temperature relationships between tube fluid and air at each (air) side of the coil; the temperature of the entering air more closely approaches the temperature of the leaving fluid than the temperature of the leaving air approaches the temperature of the entry fluid. The potential of realizing the highest possible mean temperature difference is thus arranged for optimum performance.

Most direct-expansion coils also follow this general scheme of thermal counterflow, but proper superheat control may require a hybrid combination of parallel flow and counterflow. (Air flows in the same direction as the refrigerant in parallel-flow operation.) Often, the optimum design for large coils is parallel flow in the coil's initial (entry) boiling region followed by counterflow in the superheat (exit) region. Such a hybrid arrangement is commonly used for process applications that require a low temperature difference (low TD).

Coil hand refers to either the right hand (RH) or left hand (LH) for counterflow arrangement of a multirow counterflow coil. There is no convention for what constitutes LH or RH, so manufacturers usually establish a convention for their own coils. Most manufacturers designate the location of the inlet water header or refrigerant distributor as the coil hand reference point. Figure 3 illustrates the more widely accepted coil hand designation for multirow water or refrigerant coils.

Applications

Figure 4 shows a typical arrangement of coils in a field built-up central station system. All air should be filtered to prevent dirt, insects, and foreign matter from accumulating on the coils. The cooling coil (and humidifier, when used) should include a drain pan under each coil to catch condensate formed during cooling (and excess water from the humidifier). The drain connection should be downstream of the coils, be of ample size, have accessible cleanouts, and discharge to an indirect waste or storm sewer. The drain also requires a deep-seal trap so that no sewer gas can enter the system. Precautions must be taken if there is a possibility that the drain might freeze. The drain pan, unit casing, and water piping should be insulated to prevent sweating.

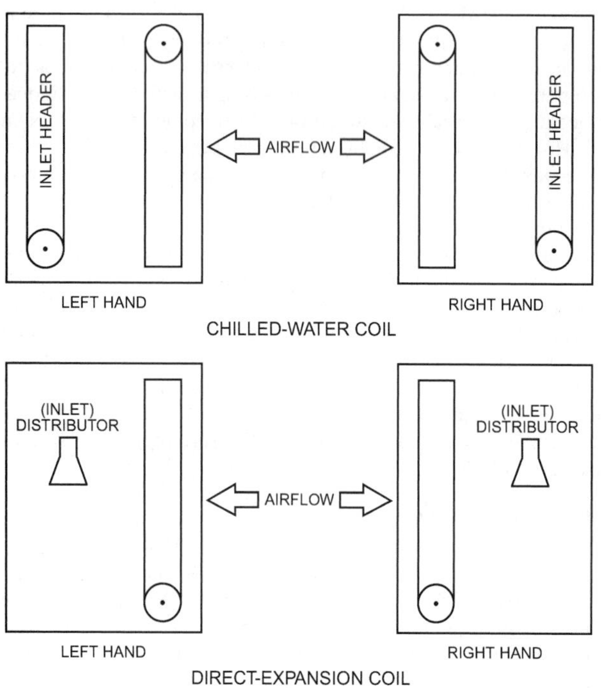

Fig. 3 Typical Coil Hand Designation

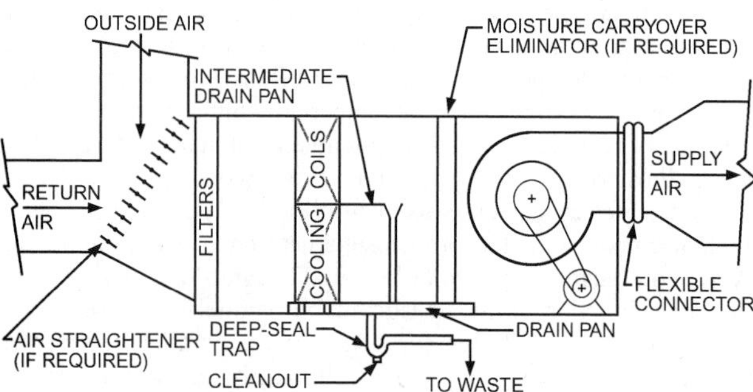

Fig. 4 Typical Arrangement of Cooling Coil Assembly in Built-Up or Packaged Central Station Air Handler

Factory-assembled central station air handlers incorporate most of the design features outlined for field built-up systems. These packaged units can generally accommodate various sizes, types, and row depths of cooling and heating coils to meet most job requirements. This usually eliminates the need for field built-up central systems, except on very large jobs.

The coil's design features (fin spacing, tube spacing, face height, type of fins), together with the amount of moisture on the coil and the degree of surface cleanliness, determine the air velocity at which condensed moisture blows off the coil. Generally, condensate water begins to be blown off a plate fin coil face at air velocities above 600 fpm. Water blowoff from coils into air ductwork external to the air-conditioning unit should be prevented. However, water blowoff is not usually a problem if coil fin heights are limited to 45 in. and the unit is set up to catch and dispose of condensate. When a number of coils are stacked one above another, condensate is carried into the airstream as it drips from one coil to the next. A downstream eliminator section could prevent this, but an intermediate drain pan and/or condensate trough (Figure 5) to collect the condensate and conduct it directly to the main drain pan is preferred. Extending downstream of the coil, each drain pan length should be at least one-half the coil height, and somewhat greater when coil airflow face velocities and/or humidity levels are higher.

When water is likely to carry over from the air-conditioning unit into external air ductwork, and no other means of prevention is provided, eliminator plates should be installed downstream of the coils. Usually, eliminator plates are not included in packaged units because other means of preventing carryover, such as space made available within the unit design for longer drain pan(s), are included in the design.

However, on sprayed-coil units, eliminators are usually included in the design. Such cooling and dehumidifying coils are sometimes sprayed with water to increase the rate of heat transfer, provide outlet air approaching saturation, and continually wash the surface of the coil. Coil sprays require a collecting tank, eliminators, and a recirculating pump (see Figure 6). Figure 6 also shows an air bypass, which helps a thermostat control maintain the humidity ratio by diverting a portion of the return air from the coil.

In field-assembled systems or factory-assembled central station air-handling units, fans are usually positioned downstream from the coil(s) in a draw-through arrangement. This arrangement provides acceptable airflow uniformity across the coil face more often than does the blow-through arrangement. In a blow-through arrangement, fan location upstream from the coils may require air baffles or

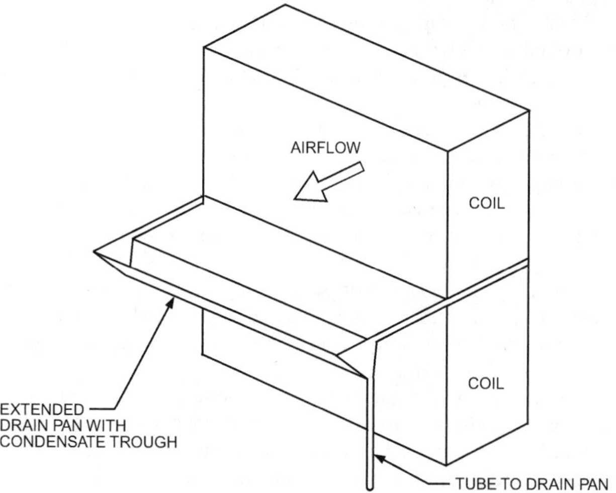

Fig. 5 Coil Bank Arrangement with Intermediate Condensate Pan

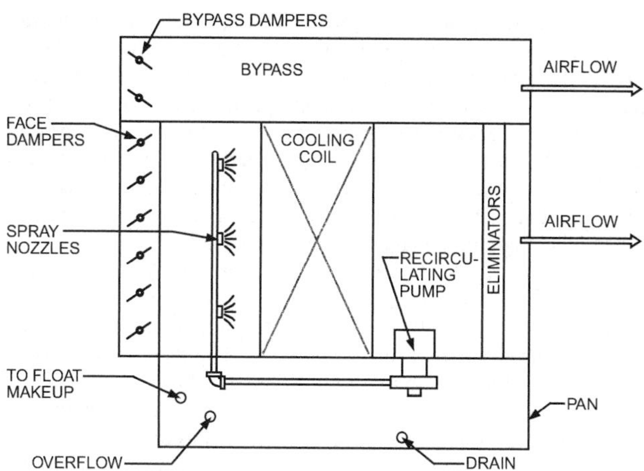

Fig. 6 Sprayed-Coil System with Air Bypass

diffuser plates between the fan discharge and the cooling coil to obtain uniform airflow. This is often the case in packaged multizone unit design. Airflow is considered to be uniform when measured flow across the entire coil face varies no more than 20%.

Air-cooling and dehumidifying coil frames, as well as all drain pans and troughs, should be of an acceptable corrosion-resistant material suitable for the system and its expected useful service life. The air handler's coil section enclosure should be corrosion-resistant; be properly double-wall insulated; and have adequate access doors for changing air filters, cleaning coils, adjusting flow control valves, and maintaining motors.

Where suction line risers are used for air-cooling coils in direct-expansion refrigeration systems, the suction line must be sized properly to ensure oil return from coil to compressor at minimum load conditions. Oil return is normally intrinsic with factory-assembled, self-contained air conditioners but must be considered for factory-assembled central station units or field-installed cooling coil banks where suction line risers are required and are assembled at the job site. Sizing, design, and arrangement of suction lines and their risers are described in Chapter 3 of the 2006 *ASHRAE Handbook—Refrigeration.*

COIL SELECTION

When selecting a coil, the following factors should be considered:

- Job requirements—cooling, dehumidifying, and the capacity required to properly balance with other system components (e.g., compressor equipment in the case of direct-expansion coils)
- Entering air dry-bulb and wet-bulb temperatures
- Available cooling media and operating temperatures
- Space and dimensional limitations
- Air and cooling fluid quantities, including distribution and limitations
- Allowable frictional resistances in air circuit (including coils)
- Allowable frictional resistances in cooling media piping system (including coils)
- Characteristics of individual coil designs and circuitry possibilities
- Individual installation requirements such as type of automatic control to be used; presence of corrosive atmosphere; design pressures; and durability of tube, fins, and frame material

Chapters 29 and 30 of the 2005 *ASHRAE Handbook—Fundamentals* contain information on load calculation.

Air quantity is affected by factors such as design parameters, codes, space, and equipment. Resistance through the air circuit influences fan power and speed. This resistance may be limited to allow the use of a given size fan motor, to keep operating expense low, or because of sound-level requirements. Air friction loss across the cooling coil (in summation with other series air-pressure drops for elements such as air filters, water sprays, heating coils, air grilles, and ductwork) determines the static pressure requirement for the complete airway system. The static pressure requirement is used in selecting fans and drives to obtain the design air quantity under operating conditions. See Chapter 18 for a description of fan selection.

The conditioned-air face velocity is determined by economic evaluation of initial and operating costs for the complete installation as influenced by (1) heat transfer performance of the specific coil surface type for various combinations of face areas and row depths as a function of air velocity; (2) air-side frictional resistance for the complete air circuit (including coils), which affects fan size, power, and sound-level requirements; and (3) condensate water carryover considerations. Allowable friction through the water or brine coil circuitry may be dictated by the head available from a given size pump and pump motor, as well as the same economic factors governing the air side made applicable to the water side. Additionally, the adverse effect of high cooling-water velocities on erosion-corrosion of tube walls is a major factor in sizing and circuitry to keep tube velocity below the recommended maximums. On larger coils, water pressure drop limits of 15 to 20 ft usually keep such velocities within acceptable limits of 2 to 4 fps, depending on circuit design.

Coil ratings are based on a uniform velocity. Design interference with uniform airflow through the coil makes predicting coil performance difficult as well as inaccurate. Such airflow interference may be caused by air entering at odd angles or by inadvertent blocking of a portion of the coil face. To obtain rated performance, the volumetric airflow quantity must be adjusted on the job to that at which the coil was rated and must be kept at that value. At start-up for air balance, the most common causes of incorrect airflow are the lack of altitude correction to standard air (where applicable) and ductwork problems. At commissioning, the most common causes of an air quantity deficiency are filter fouling and dirt or frost collection on the coils. These difficulties can be avoided through proper design, start-up checkout, and regular servicing.

The required total heat capacity of the cooling coil should be in balance with the capacity of other refrigerant system components such as the compressor, water chiller, condenser, and refrigerant liquid metering device. Chapter 43 of the 2006 *ASHRAE Handbook—Refrigeration* describes methods of estimating balanced system capacity under various operating conditions when using direct-expansion coils for both factory- and field-assembled systems.

For dehumidifying coils, it is important that the proper amount of surface area be installed to obtain the ratio of air-side sensible-to-total heat required to maintain air dry-bulb and wet-bulb temperatures in the conditioned space. This is an important consideration when preconditioning is done by reheat arrangement. The method for calculating the sensible and total heat loads and leaving air conditions at the coil to satisfy the sensible-to-total heat ratio required for the conditioned space is covered in Appendix D of *Cooling and Heating Load Calculation Principles* (Pedersen et al. 1998).

The same room air conditions can be maintained with different air quantities (including outside and return air) through a coil. However, for a given total air quantity with fixed percentages of outside and return air, there is only one set of air conditions leaving the coil that will precisely maintain room design air conditions. Once air quantity and leaving air conditions at the coil have been selected, there is usually only one combination of face area, row depth, and air face velocity for a given coil surface that will precisely maintain the required room ambient conditions. Therefore, in making final coil selections it is necessary to recheck the initial selection to ensure that the leaving air conditions, as calculated by a coil selection computer program or other procedure, will match those determined from the cooling load estimate.

Coil ratings and selections can be obtained from manufacturers' catalogs. Most catalogs contain extensive tables giving the performance of coils at various air and water velocities and entering humidity and temperatures. Most manufacturers provide computerized coil selection programs to potential customers. The final choice can then be made based on system performance and economic requirements.

Performance and Ratings

The long-term performance of an extended-surface air-cooling and dehumidifying coil depends on its correct design to specified conditions and material specifications, proper matching to other system components, proper installation, and proper maintenance as required.

In accordance with ARI *Standard* 410, Forced-Circulation Air-Cooling and Air-Heating Coils, dry-surface (sensible cooling) coils and dehumidifying coils (which both cool and dehumidify), particularly those for field-assembled coil banks or factory-assembled packaged units using different combinations of coils, are usually rated within the following parameters:

Entering air dry-bulb temperature: 65 to 100°F
Entering air wet-bulb temperature: 60 to 85°F (if air is not dehumidified in the application, select coils based on sensible heat transfer)
Air face velocity: 200 to 800 fpm
Evaporator refrigerant saturation temperature: 30 to 55°F at coil suction outlet (refrigerant vapor superheat at coil suction outlet is 6°F or higher)
Entering chilled-water temperature: 35 to 65°F
Water velocity: 1 to 8 fps
For ethylene glycol solution: 1 to 6 fps, 0 to 90°F entering dry-bulb temperature, 60 to 80°F entering wet-bulb temperature, 10 to 60% aqueous glycol concentration by weight

The air-side ratio of sensible to total heat removed by dehumidifying coils varies in practice from about 0.6 to 1.0 (i.e., sensible heat is from 60 to 100% of the total, depending on the application). For information on calculating a dehumidifying coil's sensible heat ratio, see the section on Performance of Dehumidifying Coils, or Appendix D of *Cooling and Heating Load Calculation Principles* (Pedersen et al. 1998). For a given coil surface design and arrangement, the required sensible heat ratio may be satisfied by wide variations in and combinations of air face velocity, in-tube temperature, flow rate, entering air temperature, coil depth, and so forth, although the variations may be self-limiting. The maximum coil air face velocity should be limited to a value that prevents water carryover into the air ductwork. Dehumidifying coils for comfort applications are frequently selected in the range of 400 to 500 fpm air face velocity.

Operating ratings of dehumidifying coils for factory-assembled, self-contained air conditioners are generally determined in conjunction with laboratory testing for the system capacity of the complete unit assembly. For example, a standard rating point has been 33.4 cfm per 1000 Btu/h (or 400 cfm per ton of refrigeration effect), not to exceed 37.5 cfm per 100 Btu/h for unitary equipment. Refrigerant (e.g., R-22) duty would be 6 to 10°F superheat for an appropriate balance at 45°F saturated suction. For water coils, circuitry would operate at 4 fps, 42°F inlet water, 12°F rise (or 2 gpm per ton of refrigeration effect). The standard ratings at 80°F db and 67°F wb are representative of the entering air conditions encountered in many comfort operations. Although indoor conditions are usually lower than 67°F wb, it is usually assumed that introduction of outside air brings the air mixture to the cooling coil up to about 80°F db/67°F wb entering air design conditions.

Dehumidifying coils for field-assembled projects and central station air-handling units were formerly selected according to coil rating tables but are now selected by computerized selection programs. Either way, selecting coils from the load division indicated by the load calculation works satisfactorily for the usual human comfort applications. Additional design precautions and refinements are necessary for more exacting industrial applications and for all types of air conditioning in humid areas. One such refinement, the dual-path air process, uses a separate cooling coil to cool and dehumidify ventilation air before mixing it with recirculated air. This process dehumidifies what is usually the main source of moisture: makeup outside air. Condenser heat reclaim (when available) is another refinement required for some industrial applications and is finding greater use in commercial and comfort applications.

Airflow ratings are based on standard air of 0.075 lb/ft^3 at 70°F and a barometric pressure of 29.92 in. Hg. In some mountainous areas with a sufficiently large market, coil ratings and altitude-corrected psychrometrics are available for their particular altitudes.

When checking the operation of dehumidifying coils, climatic conditions must be considered. Most problems are encountered at light-load conditions, when the cooling requirement is considerably less than at design conditions. In hot, dry climates, where the outside dew point is consistently low, dehumidifying is not generally a problem, and the light-load design point condition does not pose any special problems. In hot, humid climates, the light-load condition has a higher proportion of moisture and a correspondingly lower proportion of sensible heat. The result is higher dew points in the conditioned spaces during light-load conditions unless a special means for controlling inside dew points (e.g., reheat or dual path) is used.

Fin surface freezing at light loads should be avoided. Freezing occurs when a dehumidification coil's surface temperature falls below 32°F. Freezing does not occur with standard coils for comfort installations unless the refrigerant evaporating temperature at the coil outlet is below 25 to 28°F saturated; the exact value depends on the design of the coil, its operating dew point, and the amount of loading. With coil and condensing units to balance at low temperatures at peak loads (not a customary design choice), freezing may occur when load suddenly decreases. The possibility of this type of surface freezing is greater if a bypass is used because it causes less air to be passed through the coil at light loads.

AIRFLOW RESISTANCE

A cooling coil's airflow resistance (air friction) depends on the tube pattern and fin geometry (tube size and spacing, fin configuration, and number of in-line or staggered rows), coil face velocity, and amount of moisture on the coil. The coil air friction may also be affected by the degree of aerodynamic cleanliness of the coil core; burrs on fin edges may increase coil friction and increase the tendency to pocket dirt or lint on the faces. A completely dry coil, removing only sensible heat, offers approximately one-third less resistance to airflow than a dehumidifying coil removing both sensible and latent heat.

For a given surface and airflow, increasing the number of rows or fins increases airflow resistance. Therefore, final selection involves economic balancing of the initial cost of the coil against the operating costs of the coil geometry combinations available to adequately meet the performance requirements.

The aluminum fin surfaces of new dehumidifying coils tend to inhibit condensate sheeting action until they have aged for a year. Hydrophilic aluminum fin surface coatings reduce water droplet surface tension, producing a more evenly dispersed wetted surface action at initial start-up. Manufacturers have tried different methods of applying such coatings, including dipping the coil into a tank, coating the fin stock material, or subjecting the material to a chemical etching process. Tests have shown as much as a 30% reduction in air pressure drop across a hydrophilic coil as opposed a new untreated coil.

HEAT TRANSFER

The heat transmission rate of air passing over a clean tube (with or without extended surface) to a fluid flowing within it is impeded principally by three thermal resistances: (1) surface air-side film thermal resistance from the air to the surface of the exterior fin and tube assembly; (2) metal thermal resistance to heat conductance through the exterior fin and tube assembly; and (3) in-tube fluid-side film thermal resistance, which impedes heat flow between the internal surface of the metal and the fluid flowing within the tube. For some applications, an additional thermal resistance is factored in to account for external and/or internal surface fouling. Usually, the combination of metal and tube-side film resistance is considerably lower than the air-side surface resistance.

For a reduction in thermal resistance, the fin surface is fabricated with die-formed corrugations instead of the traditional flat design. At low airflows or wide fin spacing, the air-side transfer coefficient is virtually the same for flat and corrugated fins. Under normal comfort conditioning operation, the corrugated fin surface is designed to reduce the boundary air film thickness by undulating the passing airstream within the coil; this produces a marked improvement in heat transfer without much airflow penalty. Further fin enhancements, including louvered and lanced fin designs, have been driven by the desire to duplicate throughout the coil depth the thin boundary air film characteristic of the fin's leading edge. Louvered fin design maximizes the number of fin surface leading edges throughout the entire secondary surface area and increases the external secondary surface area A_s through the multiplicity of edges.

Where an application allows economical use of coil construction materials, the mass and size of the coil can be reduced when boundary air and water films are lessened. For example, the exterior surface resistance can be reduced to nearly the same as the fluid-side resistance by using lanced and/or louvered fins. External as well as internal tube fins (or internal turbulators) can economically decrease overall heat transfer surface resistances. Also, water sprays applied to a flat fin coil surface may increase overall heat transfer slightly, although they may better serve other purposes such as air and coil cleaning.

Heat transfer between the cooling medium and the airstream across a coil is influenced by the following variables:

- Temperature difference between fluids
- Design and surface arrangement of the coil
- Velocity and character of the airstream
- Velocity and character of the in-tube coolant

With water coils, only the water temperature rises. With coils of volatile refrigerants, an appreciable pressure drop and a corresponding change in evaporating temperature through the refrigerant circuit often occur. Alternative refrigerants to R-22, such as R-407C, which has a temperature glide, will have an evaporation temperature rise of 7 to 12°F through the evaporator. This must be considered in design and performance calculation of the coil. A compensating pressure drop in the coil may partially, or even totally, compensate for the low-side temperature glide of a zeotropic refrigerant blend. Rating direct-expansion coils is further complicated by the refrigerant evaporating in part of the circuit and superheating in the remainder. Thus, for halocarbon refrigerants, a cooling coil is tested and rated with a specific distributing and liquid-metering device, and the capacities are stated with the superheat condition of the leaving vapor.

At a given air mass velocity, performance depends on the turbulence of airflow into the coil and the uniformity of air distribution over the coil face. The latter is necessary to obtain reliable test ratings and realize rated performance in actual installations. Air resistance through the coils helps distribute air properly, but the effect is frequently inadequate where inlet duct connections are brought in at sharp angles to the coil face. Reverse air currents may pass through a portion of the coils. These currents reduce capacity but can be avoided with proper inlet air vanes or baffles. Air blades may also be required. Remember that coil performance ratings (ARI *Standard* 410) represent optimum conditions resulting from adequate and reliable laboratory tests (ASHRAE *Standard* 33).

For cases when available data must be extended, for arriving at general design criteria for a single, unique installation, or for understanding the calculation progression, the following material and illustrative examples for calculating cooling coil performance are useful guides.

PERFORMANCE OF SENSIBLE COOLING COILS

The performance of sensible cooling coils depends on the following factors. See the section on Symbols for an explanation of the variables.

- The overall coefficient U_o of sensible heat transfer between airstream and coolant fluid
- The mean temperature difference Δt_m between airstream and coolant fluid
- The physical dimensions of and data for the coil (such as coil face area A_a and total external surface area A_o) with characteristics of the heat transfer surface

The sensible heat cooling capacity q_{td} of a given coil is expressed by the following equation:

$$q_{td} = U_o F_s A_a N_r \Delta t_m \tag{1a}$$

with

$$F_s = A_o / A_a N_r \tag{1b}$$

Assuming no extraneous heat losses, the same amount of sensible heat is lost from the airstream:

$$q_{td} = w_a c_p (t_{a1} - t_{a2}) \tag{2a}$$

with

$$w_a = \rho_a A_a V_a \tag{2b}$$

The same amount of sensible heat is absorbed by the coolant; for a nonvolatile type, it is

$$q_{td} = w_r c_r (t_{r2} - t_{r1}) \tag{3}$$

For a nonvolatile coolant in thermal counterflow with the air, the mean temperature difference in Equation (1a) is expressed as

$$\Delta t_m = \frac{(t_{a1} - t_{r2}) - (t_{a2} - t_{r1})}{\ln[(t_{a1} - t_{r2})/(t_{a2} - t_{r1})]} \tag{4}$$

Proper temperature differences for various crossflow situations are given in many texts, including Mueller (1973). These calculations are based on various assumptions, among them that U for the total external surface is constant. Although this assumption is generally not valid for multirow coils, using crossflow temperature differences from Mueller (1973) or other texts should be preferable to Equation (4), which applies only to counterflow. However, using the log mean temperature difference is widespread.

The overall heat transfer coefficient U_o for a given coil design, whether bare-pipe or finned-type, with clean, nonfouled surfaces, consists of the combined effect of three individual heat transfer coefficients:

- The **film coefficient** f_a of sensible heat transfer between air and the external surface of the coil
- The **unit conductance** $1/R_{md}$ of the coil material (i.e., tube wall, fins, tube-to-fin thermal resistance)
- The **film coefficient** f_r of heat transfer between the internal coil surface and the coolant fluid within the coil

These three individual coefficients acting in series form an overall coefficient of heat transfer in accordance with the material given in Chapters 3 and 23 of the 2005 *ASHRAE Handbook—Fundamentals*.

For a bare-pipe coil, the overall coefficient of heat transfer for sensible cooling (without dehumidification) can be expressed by a simplified basic equation:

$$U_o = \frac{1}{(1/f_a) + (D_o - D_i)/24k + (B/f_r)} \quad (5a)$$

When pipe or tube walls are thin and of high-conductivity material (as in typical heating and cooling coils), the term $(D_o - D_i)/24k$ in Equation (5a) frequently becomes negligible and is generally disregarded. (This effect in typical bare-pipe cooling coils seldom exceeds 1 to 2% of the overall coefficient.) Thus, the overall coefficient for bare pipe in its simplest form is

$$U_o = \frac{1}{(1/f_a) + (B/f_r)} \quad (5b)$$

For finned coils, the equation for the overall coefficient of heat transfer can be written

$$U_o = \frac{1}{(1/\eta f_a) + (B/f_r)} \quad (5c)$$

where the **fin effectiveness** η allows for the resistance to heat flow encountered in the fins. It is defined as

$$\eta = (EA_s + A_p)/A_o \quad (6)$$

For typical cooling surface designs, the surface ratio B ranges from about 1.03 to 1.15 for bare-pipe coils and from 10 to 30 for finned coils. Chapter 3 of the 2005 *ASHRAE Handbook—Fundamentals* describes how to estimate fin efficiency and calculate the tube-side heat transfer coefficient f_r for nonvolatile fluids. Table 2 in ARI *Standard* 410 lists thermal conductivity k of standard coil materials.

Estimating the air-side heat transfer coefficient f_a is more difficult because well-verified general predictive techniques are not available. Hence, direct use of experimental data is usually necessary. For plate fin coils, some correlations that satisfy several data sets are available (Kusuda 1970; McQuiston 1981). Webb (1980) reviewed air-side heat transfer and pressure drop correlations for various geometries. Mueller (1973) and Chapter 3 of the 2005 *ASHRAE Handbook—Fundamentals* provide guidance on this subject.

For analyzing a given heat exchanger, the concept of **effectiveness** is useful. Expressions for effectiveness have been derived for various flow configurations and can be found in Kusuda (1970) and Mueller (1973). The cooling coils covered in this chapter actually involve various forms of crossflow. However, the case of counterflow is addressed here to illustrate the value of this concept. The air-side effectiveness E_a for counterflow heat exchangers is given by the following equations:

$$q_{td} = w_a c_p (t_{a1} - t_{r1}) E_a \quad (7a)$$

with

$$E_a = \frac{t_{a1} - t_{a2}}{t_{a1} - t_{r1}} \quad (7b)$$

or

$$E_a = \frac{1 - e^{-c_o(1-M)}}{1 - Me^{-c_o(1-M)}} \quad (7c)$$

with

$$c_o = \frac{A_o U_o}{w_a c_p} = \frac{F_s N_r U_o}{60 \rho_a V_a c_p} \quad (7d)$$

and

$$M = \frac{w_a c_p}{w_r c_r} = \frac{60 \rho_a A_a V_a c_p}{w_r c_r} \quad (7e)$$

Note the following two special conditions:

If $M = 0$, then $E_a = 1 - e^{-c_o}$

If $M \geq 1$, then

$$E_a = \frac{1}{(1/c_o) + 1}$$

With a given design and arrangement of heat transfer surface used as cooling coil core material for which basic physical and heat transfer data are available to determine U_o from Equations (5a), (5b), and (5c), the selection, sizing, and performance calculation of sensible cooling coils for a particular application generally fall into either of two categories:

1. Heat transfer surface area A_o or coil row depth N_r for a specific coil size is required and initially unknown. Sensible cooling capacity q_{td}, flow rates for both air and coolant, entrance and exit temperatures of both fluids, and mean temperature difference between fluids are initially known or can be assumed or determined from Equations (2a), (3), and (4). A_o or N_r can then be calculated directly from Equation (1a).
2. Sensible cooling capacity q_{td} for a specific coil is required and initially unknown. Face area and heat transfer surface area are known or can be readily determined. Flow rates and entering temperatures of air and coolant are also known. Mean temperature difference Δt_m is unknown, but its determination is unnecessary to calculate q_{td}, which can be found directly by solving Equation (7a). Equation (7a) also provides a basic means of determining q_{td} for a given coil or related family of coils over the complete rating ranges of air and coolant flow rates and operating temperatures.

The two categories of application problems are illustrated in Examples 1 and 2, respectively:

Example 1. Standard air flowing at a mass rate equivalent to 9000 cfm is to be cooled from 85 to 75°F, using 330 lb/min chilled water supplied at 50°F in thermal counterflow arrangement. Assuming an air face velocity of V_a = 600 fpm and no air dehumidification, calculate coil face area A_a, sensible cooling capacity q_{td}, required heat transfer surface area A_o, coil row depth N_r, and coil air-side pressure drop Δp_{st} for a clean, nonfouled, thin-walled bare copper tube surface design for which the following physical and performance data have been predetermined:

B = surface ratio = 1.07
c_p = 0.24 Btu/lb·°F
c_r = 1.0 Btu/lb·°F
F_s = (external surface area)/(face area)(rows deep) = 1.34
f_a = 15 Btu/h·ft²·°F
f_r = 800 Btu/h·ft²·°F
$\Delta p_{st}/N_r$ = 0.027 in. of water/number of coil rows
ρ_a = 0.075 lb/ft³

Solution: Calculate the coil face area required.

$$A_a = 9000/600 = 15 \text{ ft}^2$$

Neglecting the effect of tube wall, from Equation (5b),

$$U_o = \frac{1}{(1/15) + (1.07/800)} = 14.7 \text{ Btu/h} \cdot \text{ft}^2 \cdot {}^\circ\text{F}$$

From Equations (2a) and (2b), the sensible cooling capacity is

$$q_{td} = 60 \times 0.075 \times 15 \times 600 \times 0.24(85 - 75) = 97{,}200 \text{ Btu/h}$$

From Equation (3),

$$t_{r2} = 50 + 97{,}200/(330 \times 60 \times 1)(1000 \times 2.5 \times 4.18) = 54.9°\text{F}$$

From Equation (4),

$$\Delta t_m = \frac{(85 - 54.9) - (75 - 50)}{\ln[(85 - 54.9)/(75 - 50)]} = 27.5°F$$

From Equations (1a) and (1b), the surface area required is

$A_o = 97{,}200/(14.7 \times 27.5) = 240$ ft² external surface

From Equation (1b), the required row depth is

$N_r = 240/(1.34 \times 15) = 11.9$ rows deep

The installed 15 ft² coil face, 12 rows deep, slightly exceeds the required capacity. The air-side pressure drop for the installed row depth is then

$\Delta p_{st} = (\Delta p_{st}/N_r)N_r = 0.027 \times 12 = 0.32$ in. of water at 70°F

In this example, for some applications where such items as V_a, w_r, t_{r1}, and f_r may be arbitrarily varied with a fixed design and arrangement of heat transfer surface, a trade-off between coil face area A_a and coil row depth N_r is sometimes made to obtain alternative coil selections that produce the same sensible cooling capacity q_{td}. For example, an eight-row coil could be selected, but it would require a larger face area A_a with lower air face velocity V_a and a lower air-side pressure drop Δp_{st}.

Example 2. An air-cooling coil using a finned tube-type heat transfer surface has physical data as follows:

A_a = 10 ft²
A_o = 800 ft² external
B = surface ratio = 20
F_s = (external surface area)/(face area)(rows deep) = 27
N_r = 3 rows deep

Air at a face velocity of V_a = 800 fpm and 95°F entering air temperature is to be cooled by 15 gpm of well water supplied at 55°F. Calculate the sensible cooling capacity q_{td}, leaving air temperature t_{a2}, leaving water temperature t_{r2}, and air-side pressure drop Δp_{st}. Assume clean and nonfouled surfaces, thermal counterflow between air and water, no air dehumidification, standard barometric air pressure, and that the following data are available or can be predetermined:

c_p = 0.24 Btu/lb·°F
c_r = 1.0 Btu/lb·°F
f_a = 17 Btu/h·ft²·°F
f_r = 500 Btu/h·ft²·°F
η = fin effectiveness = 0.9
$\Delta p_{st}/N_r$ = 0.22 in. of water/number of coil rows
ρ_a = 0.075 lb/ft³
ρ_w = 62.4 lb/ft³ (8.34 lb/gal)

Solution: From Equation (5c),

$$U_o = \frac{1}{1/(0.9 \times 17) + (20/500)} = 9.5 \text{ Btu/h·ft}^2\text{·°F}$$

From Equations (7d) and (2b),

$$c_o = \frac{800 \times 9.5}{60 \times 0.075 \times 10 \times 800 \times 0.24} = 0.88$$

From Equation (7e),

$$M = \frac{60 \times 0.075 \times 10 \times 800 \times 0.24}{15 \times 60 \times 8.34 \times 1} = 1.15$$

Substituting in,

$$-c_o(1 - M) = -0.88(1 - 1.15) = 0.132$$

From Equation (7c),

$$E_a = \frac{1 - e^{0.132}}{1 - 1.15e^{0.132}} = 0.452$$

From Equation (7a), the sensible cooling capacity is

$q_{td} = 60 \times 0.075 \times 10 \times 800 \times 0.24(95 - 55) \times 0.452 = 156{,}000$ Btu/h

From Equation (2a), the leaving air temperature is

$$t_{a2} = 95 - \frac{156{,}000}{60 \times 0.075 \times 10 \times 800 \times 0.24} = 76.9°F$$

From Equation (3), the leaving water temperature is

$$t_{r2} = 55 + \frac{156{,}000}{15 \times 60 \times 8.34 \times 1} = 75.8°F$$

The air-side pressure drop is

$\Delta p_{st} = 0.22 \times 3 = 0.66$ in. of water

The preceding equations and examples demonstrate the method for calculating thermal performance of sensible cooling coils that operate with a dry surface. However, when cooling coils operate wet or act as dehumidifying coils, performance cannot be predicted without including the effect of air-side moisture (latent heat) removal.

PERFORMANCE OF DEHUMIDIFYING COILS

A dehumidifying coil normally removes both moisture and sensible heat from entering air. In most air-conditioning processes, the air to be cooled is a mixture of water vapor and dry air gases. Both lose sensible heat when in contact with a surface cooler than the air. Latent heat is removed through condensation only on the parts of the coil where the surface temperature is lower than the dew point of the air passing over it. Figure 2 in Chapter 4 shows the assumed psychrometric conditions of this process. As the leaving dry-bulb temperature drops below the entering dew-point temperature, the difference between leaving dry-bulb temperature and leaving dew point for a given coil, airflow, and entering air condition is lessened.

When the coil starts to remove moisture, the cooling surfaces carry both the sensible and latent heat load. As the air approaches saturation, each degree of sensible cooling is nearly matched by a corresponding degree of dew-point decrease. The latent heat removal per degree of dew-point change is considerably greater. The following table compares the amount of moisture removed from air at standard barometric pressure that is cooled from 60 to 59°F at both wet and dry conditions.

Dew Point	h_s, Btu/lb	Dry Bulb	h_a, Btu/lb
60°F	26.467	60°F	14.415
59°F	25.792	59°F	14.174
Difference	0.675	Difference	0.241

Note: These numerical values conform to Table 2 in Chapter 6 of the 2005 *ASHRAE Handbook—Fundamentals*.

For volatile refrigerant coils, the refrigerant distributor assembly must be tested at the higher and lower capacities of its rated range. Testing at lower capacities checks whether the refrigerant distributor provides equal distribution and whether the control is able to modulate without hunting. Testing at higher capacities checks the maximum feeding capacity of the flow control device at the greater pressure drop that occurs in the coil system.

Most manufacturers develop and produce their own performance rating tables using data obtained from suitable tests. ASHRAE *Standard* 33 specifies the acceptable method of lab-testing coils. ARI *Standard* 410 gives a method for rating thermal performance of dehumidifying coils by extending data from laboratory tests on prototypes to other operating conditions, coil sizes, and row depths. To account for simultaneous transfer of both sensible and latent heat from the airstream to the surface, ARI *Standard* 410 uses essentially the same method for arriving at cooling and dehumidifying coil thermal performance as determined by McElgin and Wiley (1940) and described in the context of *Standard* 410 by Anderson (1970). In systems operating at partial flow such as in thermal storage, accurate performance predictions for 2300 < Re < 4000 flows have been obtained by using the Gnielinski correlation. This work was presented in detail comparable to *Standard* 410 by Mirth et al. (1993).

The potential or driving force for transferring total heat q_t from the airstream to the tube-side coolant is composed of two components in series heat flow: (1) an air-to-surface air enthalpy difference $(h_a - h_s)$ and (2) a surface-to-coolant temperature difference $(t_s - t_r)$.

Figure 7 is a typical thermal diagram for a coil in which the air and a nonvolatile coolant are arranged in counterflow. The top and bottom lines in the diagram indicate, respectively, changes across the coil in the airstream enthalpy h_a and the coolant temperature t_r. To illustrate continuity, the single middle line in Figure 7 represents both surface temperature t_s and the corresponding saturated air enthalpy h_s, although the temperature and air enthalpy scales do not actually coincide as shown. The differential surface area dA_w represents any specific location within the coil thermal diagram where operating conditions are such that the air-surface interface temperature t_s is lower than the local air dew-point temperature. Under these conditions, both sensible and latent heat are removed from the airstream, and the cooler surface actively condenses water vapor.

Neglecting the enthalpy of condensed water vapor leaving the surface and any radiation and convection losses, the total heat lost from the airstream in flowing over dA_w is

$$dq_t = -w_a(dh_a) \tag{8}$$

This same total heat is transferred from the airstream to the surface interface. According to McElgin and Wiley (1940),

$$dq_t = \frac{(h_a - h_s)dA_w}{c_p R_{aw}} \tag{9}$$

The total heat transferred from the air-surface interface across the surface elements and into the coolant is equal to that given in Equations (8) and (9):

$$dq_t = \frac{(t_s - t_r)dA_w}{R_{mw} + R_r} \tag{10}$$

The same quantity of total heat is also gained by the nonvolatile coolant in passing across dA_w:

$$dq_t = -w_r c_r(dt_r) \tag{11}$$

If Equations (9) and (10) are equated and the terms rearranged, an expression for the coil characteristic C is obtained:

$$C = \frac{R_{mw} + R_r}{c_p R_{aw}} = \frac{t_s - t_r}{h_a - h_s} \tag{12}$$

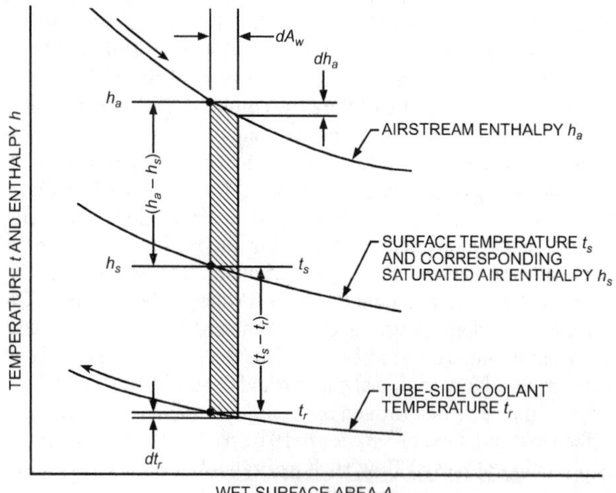

Fig. 7 Two-Component Driving Force Between Dehumidifying Air and Coolant

Equation (12) shows the basic relationship of the two components of the driving force between air and coolant in terms of three principal thermal resistances. For a given coil, these three resistances of air, metal, and in-tube fluid (R_{aw}, R_{mw}, and R_r) are usually known or can be determined for the particular application, which gives a fixed value for C. Equation (12) can then be used to determine point conditions for the interrelated values of airstream enthalpy h_a, coolant temperature t_r, surface temperature t_s, and enthalpy h_s of saturated air corresponding to the surface temperature. When both t_s and h_s are unknown, a trial-and-error solution is necessary; however, this can be solved graphically by a surface temperature chart such as Figure 8.

Figure 9 shows a typical thermal diagram for a portion of the coil surface when it is operating dry. The illustration is for counterflow with a halocarbon refrigerant. The diagram at the top of the figure illustrates a typical coil installation in an air duct with tube passes circuited countercurrent to airflow. Locations of the entering and leaving boundary conditions for both air and coolant are shown.

The thermal diagram in Figure 9 is the same type as in Figure 7, showing three lines to illustrate local conditions for the air, surface, and coolant throughout a coil. The dry-wet boundary conditions are located where the coil surface temperature t_{sb} equals the entering air dew-point temperature t''_{a1}. Thus, the surface area A_d to the left of this boundary is dry, with the remainder A_w of the coil surface area operating wet.

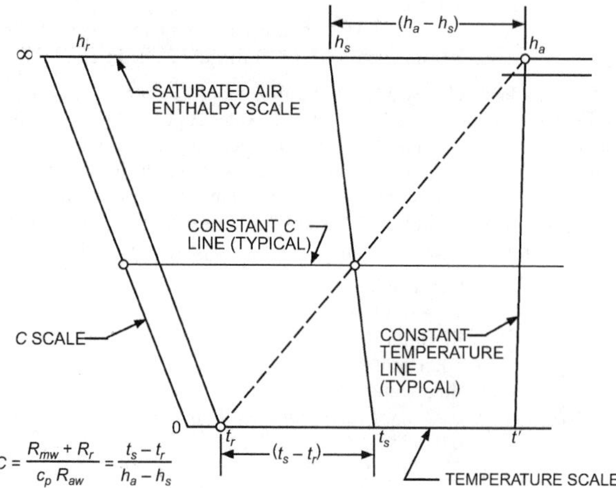

Fig. 8 Surface Temperature Chart

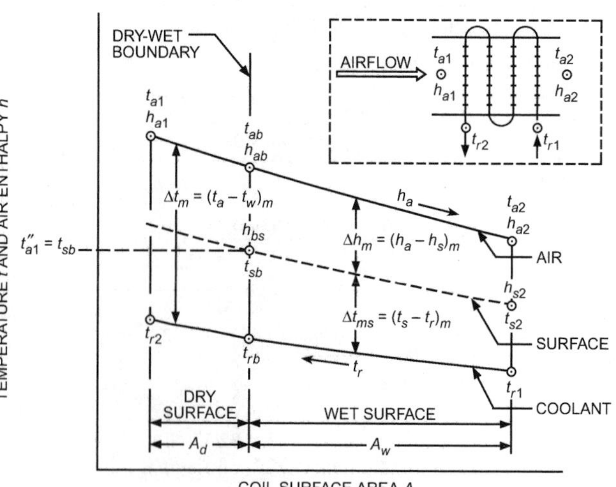

Fig. 9 Thermal Diagram for General Case When Coil Surface Operates Partially Dry

When using fluids or halocarbon refrigerants in a thermal counterflow arrangement as illustrated in Figure 9, the dry-wet boundary conditions can be determined from the following relationships:

$$y = \frac{t_{r2} - t_{r1}}{h_{a1} - h_{a2}} = \frac{w_a}{w_r c_r} \tag{13}$$

$$h_{ab} = \frac{t''_{a1} - t_{r2} + y h_{a1} + C h''_{a1}}{C + y} \tag{14}$$

The value of h_{ab} from Equation (14) serves as an index of whether the coil surface is operating fully wetted, partially dry, or completely dry, according to the following three limits:

1. If $h_{ab} \geq h_{a1}$, the surface is fully wetted.
2. If $h_{a1} > h_{ab} > h_{a2}$, the surface is partially dry.
3. If $h_{ab} \leq h_{a2}$, the surface is completely dry.

Other dry-wet boundary properties are then determined:

$$t_{sb} = t''_{a1} \tag{15}$$

$$t_{ab} = t_{a1} - (h_{a1} - h_{ab})/c_p \tag{16}$$

$$t_{rb} = t_{r2} - y c_p (t_{a1} - t_{ab}) \tag{17}$$

The dry surface area A_d required and capacity q_{td} are calculated by conventional sensible heat transfer relationships, as follows.

The overall thermal resistance R_o comprises three basic elements:

$$R_o = R_{ad} + R_{md} + R_r \tag{18}$$

with

$$R_r = B/f_r \tag{19}$$

The mean difference between air dry-bulb temperature and coolant temperature, using symbols from Figure 9, is

$$\Delta t_m = \frac{(t_{a1} - t_{r2}) - (t_{ab} - t_{rb})}{\ln[(t_{a1} - t_{r2})/(t_{ab} - t_{rb})]} \tag{20}$$

The dry surface area required is

$$A_d = \frac{q_{td} R_o}{\Delta t_m} \tag{21}$$

The air-side total heat capacity is

$$q_{td} = w_a c_p (t_{a1} - t_{ab}) \tag{22a}$$

From the coolant side,

$$q_{td} = w_r c_r (t_{r2} - t_{rb}) \tag{22b}$$

The wet surface area A_w and capacity q_{tw} are determined by the following relationships, using terminology in Figure 9.

For a given coil size, design, and arrangement, the fixed value of the coil characteristic C can be determined from the ratio of the three prime thermal resistances for the job conditions:

$$C = \frac{R_{mw} + R_r}{c_p R_{aw}} \tag{23}$$

Knowing coil characteristic C for point conditions, the interrelations between airstream enthalpy h_a, coolant temperature t_r, and surface temperature t_s and its corresponding enthalpy of saturated air h_s can be determined by using a surface temperature chart (Figure 8) or by a trial-and-error procedure using Equation (24):

$$C = \frac{t_{sb} - t_{rb}}{h_{ab} - h_{sb}} = \frac{t_{s2} - t_{r1}}{h_{a2} - h_{s2}} \tag{24}$$

The mean effective difference in air enthalpy between airstream and surface from Figure 9 is

$$\Delta h_m = \frac{(h_{ab} - h_{sb}) - (h_{a2} - h_{s2})}{\ln[(h_{ab} - h_{sb})/(h_{a2} - h_{s2})]} \tag{25}$$

Similarly, the mean temperature difference between surface and coolant is

$$\Delta t_{ms} = \frac{(t_{sb} - t_{rb}) - (t_{s2} - t_{r1})}{\ln[(t_{sb} - t_{rb})/(t_{s2} - t_{r1})]} \tag{26}$$

The wet surface area required, calculated from air-side enthalpy difference, is

$$A_w = \frac{q_{tw} R_{aw} c_p}{\Delta h_m} \tag{27a}$$

Calculated from the coolant-side temperature difference,

$$A_w = \frac{q_{tw}(R_{mw} + R_r)}{\Delta t_{ms}} \tag{27b}$$

The air-side total heat capacity is

$$q_{tw} = w_a[h_{a1} - (h_{a2} + h_{fw})] \tag{28a}$$

The enthalpy h_{fw} of condensate removed is

$$h_{fw} = (W_1 - W_2) c_{pw} (t'_{a2} - 32) \tag{28b}$$

where c_{pw} = specific heat of water = 1.0 Btu/lb$_w$·°F.

Note that h_{fw} for normal air-conditioning applications is about 0.5% of the airstream enthalpy difference $(h_{a1} - h_{a2})$ and is usually neglected.

The coolant-side heat capacity is

$$q_{tw} = w_r c_r (t_{rb} - t_{r1}) \tag{28c}$$

The total surface area requirement of the coil is

$$A_o = A_d + A_w \tag{29}$$

The total heat capacity for the coil is

$$q_t = q_{td} + q_{tw} \tag{30}$$

The leaving air dry-bulb temperature is found by the method illustrated in Figure 10, which represents part of a psychrometric chart showing the air saturation curve and lines of constant air enthalpy closely corresponding to constant wet-bulb temperature lines.

For a given coil and air quantity, a straight line projected through the entering and leaving air conditions intersects the air saturation curve at a point denoted as the effective coil surface temperature $t_{\bar{s}}$. Thus, for fixed entering air conditions t_{a1} and h_{a1} and a given effective surface temperature t_s, leaving air dry bulb t_{a2} increases but is still located on this straight line if air quantity is increased or coil depth is reduced. Conversely, a decrease in air quantity or an increase in coil depth produces a lower t_{a2} that is still located on the same straight-line segment.

An index of the air-side effectiveness is the heat transfer exponent c, defined as

$$c = \frac{A_o}{w_a c_p R_{ad}} \tag{31}$$

This exponent c, sometimes called the number of air-side transfer units NTU$_a$, is also defined as

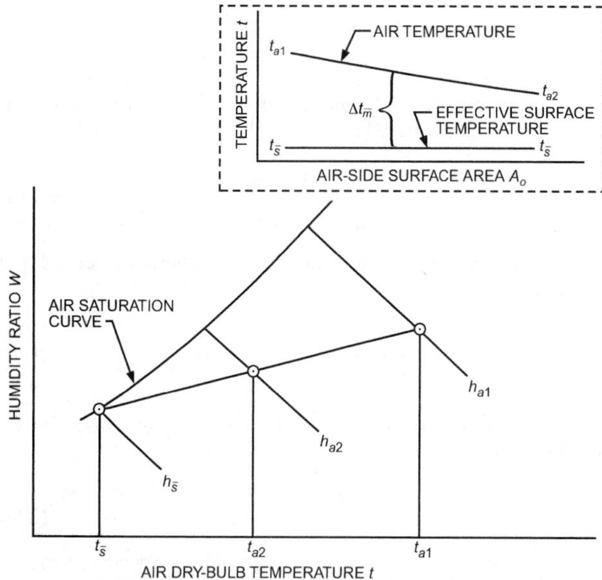

Fig. 10 Leaving Air Dry-Bulb Temperature Determination for Air-Cooling and Dehumidifying Coils

$$c = \frac{t_{a1} - t_{a2}}{\Delta t_{\bar{m}}} \tag{32}$$

The temperature drop $(t_{a1} - t_{a2})$ of the airstream and mean temperature difference $\Delta t_{\bar{m}}$ between air and effective surface in Equation (32) are illustrated at the top of Figure 10.

Knowing the exponent c and entering and leaving enthalpies h_{a1} and h_{a2} for the airstream, the enthalpy of saturated air $h_{\bar{s}}$ corresponding to effective surface temperature $t_{\bar{s}}$ is calculated as follows:

$$h_{\bar{s}} = h_{a1} - \frac{h_{a1} - h_{a2}}{1 - e^{-c}} \tag{33}$$

After finding the value of $t_{\bar{s}}$ that corresponds to $h_{\bar{s}}$ from the saturated air enthalpy tables, the leaving air dry-bulb temperature can be determined:

$$t_{a2} = t_{\bar{s}} + e^{-c}(t_{a1} - t_{\bar{s}}) \tag{34}$$

The air-side sensible heat ratio SHR can then be calculated:

$$\text{SHR} = \frac{c_p(t_{a1} - t_{a2})}{h_{a1} - h_{a2}} \tag{35}$$

For thermal performance of a coil to be determined from the foregoing relationships, values of the following three principal resistances to heat flow between air and coolant must be known:

- Total metal thermal resistances across the fin R_f and tube assembly R_t for both dry R_{md} and wet R_{mw} surface operation
- Air-film thermal resistances R_{ad} and R_{aw} for dry and wet surfaces, respectively
- Tube-side coolant film thermal resistance R_r

In ARI *Standard* 410, metal thermal resistance R_m is calculated based on the physical data, material, and arrangement of the fin and tube elements, together with the fin efficiency E for the specific fin configuration. R_m is variable as a weak function of the effective air-side heat transfer coefficient f_a for a specific coil geometry, as illustrated in Figure 11.

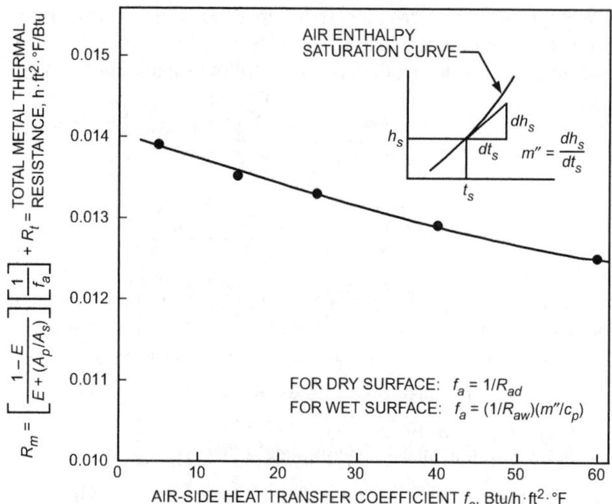

Fig. 11 Typical Total Metal Thermal Resistance of Fin and Tube Assembly

For wetted-surface application, Brown (1954), with certain simplifying assumptions, showed that f_a is directly proportional to the rate of change m'' of saturated air enthalpy h_s with the corresponding surface temperature t_s. This slope m'' of the air enthalpy saturation curve is illustrated in the small inset graph at the top of Figure 11.

The abscissa for f_a in the main graph of Figure 11 is an effective value, which, for a dry surface, is the simple thermal resistance reciprocal $1/R_{ad}$. For a wet surface, f_a is the product of the thermal resistance reciprocal $1/R_{aw}$ and the multiplying factor m''/c_p. ARI *Standard* 410 outlines a method for obtaining a mean value of m''/c_p for a given coil and job condition. The total metal resistance R_m in Figure 11 includes the resistance R_t across the tube wall. For most coil designs, R_t is quite small compared to the resistance R_f through the fin metal.

The air-side thermal resistances R_{ad} and R_{aw} for dry and wet surfaces, together with their respective air-side pressure drops $\Delta p_{st}/N_r$ and $\Delta p_{sw}/N_r$, are determined from tests on a representative coil model over the full range in the rated airflow. Typical plots of experimental data for these four performance variables versus coil air face velocity V_a at 70°F are illustrated in Figure 12.

If water is used as the tube-side coolant, the heat transfer coefficient f_r is calculated from Equation (8) in ARI *Standard* 410. For evaporating refrigerants, many predictive techniques for calculating coefficients of evaporation are listed in Table 2 in Chapter 4 of the 2005 *ASHRAE Handbook—Fundamentals*. The most verified predictive technique is the Shah correlation (Shah 1976, 1982). A series of tests is specified in ARI *Standard* 410 for obtaining heat transfer data for direct-expansion refrigerants inside tubes of a given diameter.

ASHRAE *Standard* 33 specifies laboratory apparatus and instrumentation, including procedure and operating criteria for conducting tests on representative coil prototypes to obtain basic performance data. Procedures are available in ARI *Standard* 410 for reducing these test data to the performance parameters necessary to rate a line or lines of various air coils. This information is available from various coil manufacturers for use in selecting ARI *Standard* 410 certified coils.

The following example illustrates a method for selecting coil size, row depth, and performance data to satisfy specified job requirements. The application is for typical cooling and dehumidifying coil selection under conditions in which a part of the coil surface on the entering air side operates dry, with the remaining surface wet with condensing moisture. Figure 9 shows the thermal diagram, dry-wet boundary conditions, and terminology used in the problem solution.

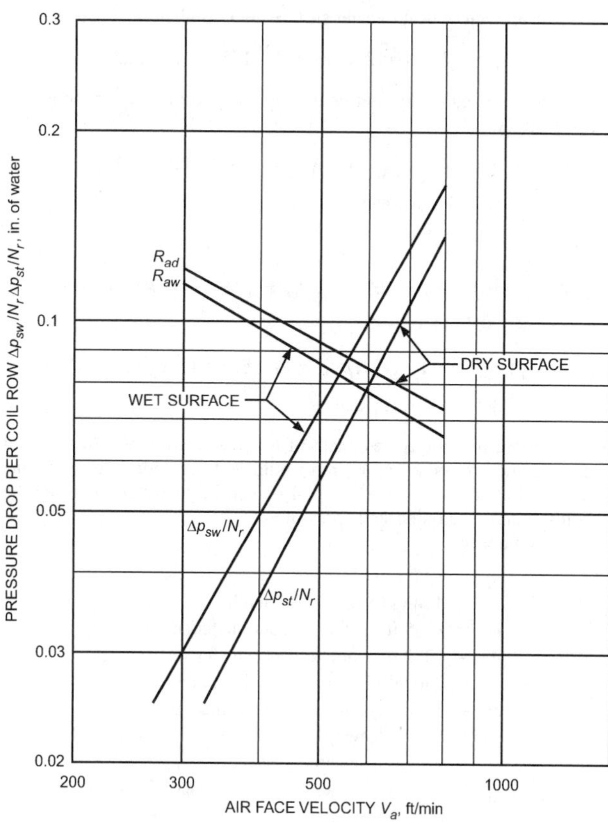

Fig. 12 Typical Air-Side Application Rating Data Determined Experimentally for Cooling and Dehumidifying Water Coils

Example 3. Standard air flowing at a mass rate equivalent of 6700 cfm enters a coil at 80°F db (t_{a1}) and 67°F wb (t'_{a1}). The air is to be cooled to 56°F leaving wet-bulb temperature t'_{a2} using 40 gpm of chilled water supplied to the coil at an entering temperature t_{r1} of 44°F, in thermal counterflow arrangement. Assume a standard coil air face velocity of V_a = 558 fpm and a clean, nonfouled, finned-tube heat transfer surface in the coil core, for which the following physical and performance data (such as illustrated in Figures 11 and 12) can be predetermined:

$$
\begin{aligned}
B &= \text{surface ratio} = 25.9 \\
c_p &= 0.243 \text{ Btu/lb}\cdot°\text{F} \\
F_s &= \text{(external surface area)/(face area)(row deep)} = 32.4 \\
f_r &= 750 \text{ Btu/h}\cdot\text{ft}^2\cdot°\text{F} \\
\Delta p_{sd}/N_r &= 0.165 \text{ in. of water/number of coil rows, dry surface} \\
\Delta p_{sw}/N_r &= 0.27 \text{ in. of water/number of coil rows, wet surface} \\
R_{ad} &= 0.073°\text{F}\cdot\text{ft}^2\cdot\text{h/Btu} \\
R_{aw} &= 0.066°\text{F}\cdot\text{ft}^2\cdot\text{h/Btu} \\
R_{md} &= 0.021°\text{F}\cdot\text{ft}^2\cdot\text{h/Btu} \\
R_{mw} &= 0.0195°\text{F}\cdot\text{ft}^2\cdot\text{h/Btu} \\
\rho_a &= 0.075 \text{ lb/ft}^3 \\
\rho_w &= 8.34 \text{ lb/gal}
\end{aligned}
$$

Referring to Figure 9 for the symbols and typical diagram for applications in which only a part of the coil surface operates wet, determine (1) coil face area A_a, (2) total refrigeration load q_t, (3) leaving coolant temperature t_{r2}, (4) dry-wet boundary conditions, (5) heat transfer surface area required for dry A_d and wet A_w sections of the coil core, (6) leaving air dry-bulb temperature t_{a2}, (7) total number N_{ri} of installed coil rows, and (8) dry Δp_{st} and wet Δp_{sw} coil air friction.

Solution: The psychrometric properties and enthalpies for dry and moist air are based on Figure 1 (ASHRAE Psychrometric Chart No. 1) and Tables 2 and 3 in Chapter 6 of the 2005 *ASHRAE Handbook—Fundamentals* as follows:

$$
\begin{aligned}
h_{a1} &= 31.52 \text{ Btu/lb}_a & h''_{a1} &= 26.67 \text{ Btu/lb}_a \\
W_1 &= 0.0112 \text{ lb}_w/\text{lb}_a & {}^*h_{a2} &= 23.84 \text{ Btu/lb}_a \\
t''_{a1} &= 60.3°\text{F} & {}^*W_2 &= 0.0095 \text{ lb}_w/\text{lb}_a
\end{aligned}
$$

*As an approximation, assume leaving air is saturated (i.e., $t_{a2} = t'_{a2}$).

Calculate coil face area required:

$$A_a = 6700/558 = 12 \text{ ft}^2$$

From Equation (28b), find condensate heat rejection:

$$h_{fw} = (0.0112 - 0.0095)(1)(56 - 32) = 0.04 \text{ Btu/lb}_a$$

Compute the total refrigeration load from the following equation:

$$
\begin{aligned}
q_t &= 60\rho_a w_a[h_{a1} - (h_{a2} + h_{fw})] \\
&= 60 \times 0.075 \times 6700[31.52 - (23.84 + 0.04)] = 230{,}000 \text{ Btu/h}
\end{aligned}
$$

From Equation (3), calculate coolant temperature leaving coil:

$$
\begin{aligned}
t_{r2} &= t_{r1} + q_t/w_r c_r \\
&= 44 + 230{,}000/(40 \times 60 \times 8.34 \times 1.0) = 55.5°\text{F}
\end{aligned}
$$

From Equation (19), determine coolant film thermal resistance:

$$R_r = 25.9/750 = 0.0345°\text{F}\cdot\text{ft}^2\cdot\text{h/Btu}$$

Calculate the wet coil characteristic from Equation (23):

$$C = \frac{0.0195 + 0.0346}{0.243 \times 0.066} = 3.37 \text{ lb}_a\cdot°\text{F/Btu}$$

Calculate from Equation (13):

$$y = \frac{55.5 - 44}{31.52 - 23.84} = 1.50 \text{ lb}_a\cdot°\text{F/Btu}$$

The dry-wet boundary conditions are determined as follows: From Equation (14), the boundary airstream enthalpy is

$$
\begin{aligned}
h_{ab} &= \frac{(60.3 - 55.5) + (1.50 \times 31.52) + (3.37 \times 26.67)}{3.37 + 1.50} \\
&= 29.15 \text{ Btu/lb}_a
\end{aligned}
$$

According to limit (2) under Equation (14), part of the coil surface on the entering air side will be operating dry, because $h_{a1} > 29.15 > h_{a2}$ (see Figure 8).

From Equation (16), the boundary airstream dry-bulb temperature is

$$t_{ab} = 80 - (31.52 - 29.15)/0.243 = 70.25°\text{F}$$

The boundary surface conditions are

$$t_{sb} = t''_{a1} = 60.3°\text{F} \quad \text{and} \quad h_{sb} = h''_{a1} = 26.67 \text{ Btu/lb}_a$$

From Equation (17), the boundary coolant temperature is

$$t_{rb} = 55.5 - 1.50 \times 0.243(80 - 70.25) = 51.9°\text{F}$$

The cooling load for the dry surface part of the coil is now calculated from Equation (22b):

$$q_{td} = 40 \times 60 \times 8.34 \times 1.0 \times (55.5 - 51.9) = 72{,}000 \text{ Btu/h}$$

From Equation (18), the overall thermal resistance for the dry surface section is

$$R_o = 0.073 + 0.021 + 0.0346 = 0.129°\text{F}\cdot\text{ft}^2\cdot\text{h/Btu}$$

From Equation (20), the mean temperature difference between air dry bulb and coolant for the dry surface section is

$$\Delta t_m = \frac{(80 - 55.5) - (70.25 - 51.92)}{\ln[(80 - 55.5)/(70.25 - 51.92)]} = 21.3°\text{F}$$

The dry surface area required is calculated from Equation (21):

$$A_d = 72{,}000 \times 0.129/21.3 = 436 \text{ ft}^2$$

From Equation (30), the cooling load for the wet surface section of the coil is

$$q_{tw} = 230{,}000 - 72{,}000 = 158{,}000 \text{ Btu/h}$$

Knowing C, h_{a2}, and t_{r1}, the surface condition at the leaving air side of the coil is calculated by trial and error using Equation (24):

$$C = 3.37 = (t_{s2} - 44)/(23.84 - h_{s2})$$

The numerical values for t_{s2} and h_{s2} are then determined directly by using a surface temperature chart (as shown in Figure 8 or in Figure 9 of ARI *Standard* 410) and saturated air enthalpies from Table 2 in Chapter 6 of the 2005 *ASHRAE Handbook—Fundamentals*:

$$t_{s2} = 51.03°\text{F} \quad \text{and} \quad h_{s2} = 20.88 \text{ Btu/lb}_a$$

From Equation (25), the mean effective difference in air enthalpy between airstream and surface is

$$\Delta h_m = \frac{(29.15 - 26.67) - (23.84 - 20.88)}{\ln[(29.15 - 26.67)/(23.84 - 20.88)]} = 2.713 \text{ Btu/lb}_a$$

From Equation (27a), the wet surface area required is

$$A_w = 158{,}000 \times 0.066 \times 0.243/2.713 = 934 \text{ ft}^2$$

From Equation (29), the net total surface area requirement for the coil is then

$$A_o = 436 + 934 = 1370 \text{ ft}^2 \text{ external}$$

From Equation (31), the net air-side heat transfer exponent is

$$c = 1370/(60 \times 0.075 \times 6700 \times 0.243 \times 0.073) = 2.56$$

From Equation (33), the enthalpy of saturated air corresponding to the effective surface temperature is

$$h_{\bar{s}} = 31.52 - \frac{31.52 - 23.84}{1 - e^{-2.56}} = 23.20 \text{ Btu/lb}_a$$

The effective surface temperature that corresponds to $h_{\bar{s}}$ is then obtained from Table 2 in Chapter 6 of the 2005 *ASHRAE Handbook—Fundamentals* as $t_{\bar{s}} = 54.95°\text{F}$.

The leaving air dry-bulb temperature is calculated from Equation (34):

$$t_{a2} = 54.95 + e^{-2.56}(80 - 54.95) = 56.9°\text{F}$$

The air-side sensible heat ratio is then found from Equation (35):

$$\text{SHR} = \frac{0.243(80 - 56.9)}{31.52 - 23.84} = 0.731$$

From Equation (1b), the calculated coil row depth N_{rc} to match job requirements is

$$N_{rc} = A_o/A_aF_s = 1370/(12 \times 32.4) = 3.5 \text{ rows deep}$$

In most coil selection problems of this type, the initial calculated row depth to satisfy job requirements is usually a noninteger value. In many cases, there is sufficient flexibility in fluid flow rates and operating temperature levels to recalculate the required row depth of a given coil size to match an available integer row depth more closely. For this example, if the calculated row depth is $N_{rc} = 3.5$, and coils of three or four rows deep are commercially available, the coil face area, operating conditions, and fluid flow rates and/or velocities could possibly be changed to recalculate a coil depth close to either three or four rows. Although core tube circuitry has limited possibilities on odd (e.g., three or five) row coils, alternative coil selections for the same job are often made desirable by trading off coil face size for row depth.

Most coil manufacturers have computer programs to run the iterations needed to predict operating values for specific coil performance requirements. The next highest integral row depth than computed is then selected for a commercially available coil with an even number of circuits, same end connected. For this example, assume that the initial coil selection requiring 3.5 rows deep is sufficiently refined that no recalculation is necessary, and that a 4-row coil with 4-pass coil circuitry is available. Thus, the installed row depth N_{ri} is

$$N_{ri} = 4 \text{ rows deep}$$

The amount of heat transfer surface area installed is

$$A_{oi} = A_aF_sN_{ri} = 12 \times 32.4 \times 4 = 1555 \text{ ft}^2 \text{ external}$$

The completely dry and completely wetted air-side frictions are, respectively,

$$\Delta p_{sd} = (A_d/A_o)(\Delta p_{sd}/N_{ri})N_{ri} = (436/1370) \times 0.165 \times 4 = 0.21 \text{ in. of water}$$

and

$$\Delta p_{sw} = (A_w/A_o)(\Delta p_{sw}/N_{ri})N_{ri} = (934/1370) \times 0.27 \times 4 = 0.74 \text{ in. of water}$$

$$\Delta p_{st} = (A_{oi}/A_o)\Delta p_{sd} + \Delta p_{sw} = (1555/1370) \times (0.21 + 0.74) = 1.08 \text{ in. of water total}$$

A more realistic Δp estimate of a coil operating at >70% wetted surface and a velocity > 400 fpm would be to consider the entire surface as wetted. Therefore, 0.27 × 4 = 1.08 in. of water would be the coil's operating air-side static pressure.

In summary,

A_a = 12 ft² coil face area
N_{ri} = 4 rows installed coil depth
A_{oi} = 1555 ft² installed heat transfer surface area
A_o = 1370 ft² required heat transfer surface area
q_t = 230,000 Btu/h total refrigeration load
t_{r2} = 55.5°F leaving coolant temperature
t_{a2} = 56.9°F leaving air dry-bulb temperature
SHR = 0.731 air sensible heat ratio
Δp_{sw} = 0.74 in. of water wet-coil surface air friction
Δp_{sd} = 0.21 in. of water dry-coil surface air friction
Δp_{st} = 1.08 in. of water total coil surface air friction

DETERMINING REFRIGERATION LOAD

The following calculation of refrigeration load distinguishes between the true sensible and latent heat loss of the air, which is accurate within the data's limitations. This division will not correspond to load determination obtained from approximate factors or constants.

The total refrigeration load q_t of a cooling and dehumidifying coil (or air washer) per unit mass of dry air is indicated in Figure 13 and consists of the following components:

- The sensible heat q_s removed from the dry air and moisture in cooling from entering temperature t_1 to leaving temperature t_2
- The latent heat q_e removed to condense the moisture at the dew-point temperature t_4 of the entering air

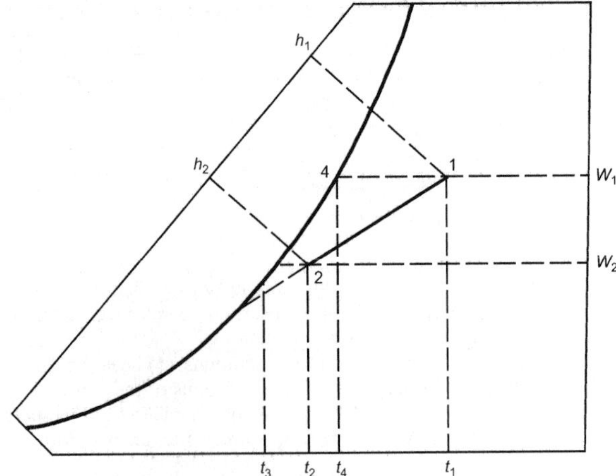

Fig. 13 Psychrometric Performance of Cooling and Dehumidifying Coil

- The heat q_w removed to further cool the condensate from its dew point t_4 to its leaving condensate temperature t_3

The preceding components are related by the following equation:

$$q_t = q_s + q_e + q_w \tag{36}$$

If only the total heat value is desired, it may be computed by

$$q_t = (h_1 - h_2) - (W_1 - W_2)h_{w3} \tag{37}$$

where

h_1 and h_2 = enthalpy of air at points 1 and 2, respectively
W_1 and W_2 = humidity ratio at points 1 and 2, respectively
h_{w3} = enthalpy of saturated liquid at final temperature t_3

If a breakdown into latent and sensible heat components is desired, the following relations may be used.

Latent heat may be found from

$$q_e = (W_1 - W_2)h_{fg4} \tag{38}$$

where

h_{fg4} = enthalpy representing latent heat of water vapor at condensing temperature t_4

Sensible heat may be shown to be

$$q_s + q_w = (h_1 - h_2) - (W_1 - W_2)h_{g4} + (W_1 - W_2)(h_{w4} - h_{w3}) \tag{39a}$$

or

$$q_s + q_w = (h_1 - h_2) - (W_1 - W_2)(h_{fg4} + h_{w3}) \tag{39b}$$

where

$h_{g4} = h_{fg4} + h_{w4}$ = enthalpy of saturated water vapor at condensing temperature t_4
h_{w4} = enthalpy of saturated liquid at condensing temperature t_4

The last term in Equation (39a) is the heat of subcooling the condensate from t_4 to its final temperature t_3. Then,

$$q_w = (W_1 - W_2)(h_{w4} - h_{w3}) \tag{40}$$

The final condensate temperature t_3 leaving the system is subject to substantial variations, depending on the method of coil installation, as affected by coil face orientation, airflow direction, and air duct insulation. In practice, t_3 is frequently the same as the leaving wet-bulb temperature. Within the normal air-conditioning range, precise values of t_3 are not necessary because the heat q_t of condensate removed from the air usually represents about 0.5 to 1.5% of the total refrigeration cooling load.

Example 4. Air enters a coil at 90°F db, 75°F wb; it leaves at 61°F db, 58°F wb; leaving water temperature is assumed to be 54°F, which is between the leaving air dew point and coil surface temperature. Find the total, latent, and sensible cooling loads on the coil with air at standard barometric pressure.

Solution: Using Figure 1 (or the indicated equations) from Chapter 6 of the 2005 *ASHRAE Handbook—Fundamentals*,

h_1 =	38.37 Btu/lb$_a$	(32)
h_2 =	25.06 Btu/lb$_a$	(32)
W_1 =	0.01523 lb$_w$/lb$_a$	(35)
W_2 =	0.00958 lb$_w$/lb$_a$	(35)
t_4 =	69.04°F dew point of entering air	(39)
h_{w4} =	37.04 Btu/lb	(34)
h_{w3} =	22.00 Btu/lb	(34)
h_{g4} =	1091.66 Btu/lb	(31)
h_{fg4} =	1054.61 Btu/lb	(31)

From Equation (37), the total heat is

$$q_t = (38.37 - 25.06) - (0.01523 - 0.00958)22.00 = 13.19 \text{ Btu/lb}_a$$

From Equation (38), the latent heat is

$$q_e = (0.01523 - 0.00958)1054.61 = 5.96 \text{ Btu/lb}_a$$

The sensible heat is therefore

$$q_s + q_w = q_t - q_e = 13.19 - 5.96 = 7.23 \text{ Btu/lb}_a$$

The sensible heat may be computed from Equation (39a) as

$$q_s + q_w = (38.37 - 25.06) - (0.01523 - 0.00958)1091.66 + (0.01523 - 0.00958)(37.04 - 22.00) = 7.22 \text{ Btu/lb}_a$$

The same value is found using Equation (39b). The subcooling of the condensate as a part of the sensible heat is indicated by the last term of the equation, 0.08 Btu/lb$_a$.

MAINTENANCE

If the coil is to deliver its full cooling capacity, both its internal and external surfaces must be clean. The tubes generally stay clean in pressurized water or brine systems. Tube surfaces can be cleaned in a number of ways, but are often washed with low-pressure water spray and mild detergent. Water coils should be completely drained if freezing is possible. When coils use built-up system refrigerant evaporators, oil can accumulate. Check and drain oil occasionally, and check for leaks and refrigerant dryness.

Air Side. The best maintenance for the outside finned area is consistent inspection and service of inlet air filters. Surface cleaning of the coil with pressurized hot water and a mild detergent should be done only when necessary (primarily when a blockage occurs under severe fin-surface-fouling service conditions, or bacterial growth is seen or suspected). Pressurized cleaning is more thorough if done first from the air exit side of the coil and then from the air entry side. Foaming chemical sprays and washes should be used instead of high pressure on fragile fins, or when fin density is too restrictive to allow proper in-depth cleaning with pressurized water spray. In all cases, limit spray water temperature to below 150°F on evaporator coils containing refrigerant. In cases of marked neglect or heavy-duty use (especially in restaurants where grease and dirt have accumulated) coils must sometimes be removed and the accumulation washed off with steam, compressed air and water, or hot water.

The surfaces can also be brushed and vacuumed. Best practice is to inspect and service the filters frequently. Also, condensate drain pan(s) and their drain lines, including open drain areas, should be kept clean and clear at all times.

Water Side. The best service for the inside tube coil surface is keeping the circulated fluid (water or glycol) free of sediment, corrosive products, and biological growth. Maintaining proper circulated water chemistry and velocity and filtering out solids should minimize the water-side fouling factor. If large amounts of scale form when untreated water is used as coolant, chemical or mechanical (rod) cleaning of internal surfaces at frequent intervals is necessary. A properly maintained chilled-water system using a glycol solution as the circulated fluid is not considered to ever have water-side fouling, as such, but glycol solutions must be analyzed seasonally to determine alkalinity, percent concentration, and corrosion inhibitor condition. Consult a glycol expert or the manufacturer's agent for detailed recommendations on proper use and control of glycol solutions.

Refrigerant Side. Moisture content of the refrigerant should be checked yearly, and acidity of the compressor oil as often as monthly. For built-up direct-expansion (DX) systems, normal is ≤50 ppm of moisture for systems with mineral oil, and ≤100 ppm for some refrigerant types in systems with polyol ester (POE) compressor oil. The actual values are set by the compressor manufacturer, and should be checked and verified during operation by moisture-indicating sight-glass viewing, and yearly by laboratory sample analysis reports. Depending on temperature and velocity, excessive moisture coming to the cooling coil through refrigerant might cause internal freeze-up around the coil's expansion valve. Generally, moisture in a system contributes to formation of acids, sludge, copper plating, and corrosion. Oil breakdown (by excessive compressor overheat) can form organic, hydrofluoric, and hydrochloric acids in the lubricant oil, all of which can corrode copper.

The refrigerant must be of high quality and meet ARI *Standard* 700 purity requirements. Application design ratings of the coil or coil bank are based on purity and dryness. This is a primary requirement of refrigerant evaporator coil maintenance.

DX coil replacement often coincides with refrigerant change-out to a chlorine-free refrigerant. Special care should be taken to ensure that the system is retrofitted in accordance with the compressor and refrigerant manufacturers' written procedures.

SYMBOLS

A_a = coil face or frontal area, ft^2
A_d = dry external surface area, ft^2
A_o = total external surface area, ft^2
A_p = exposed external prime surface area, ft^2
A_s = external secondary surface area, ft^2
A_w = wet external surface area, ft^2
B = ratio of external to internal surface area, dimensionless
C = coil characteristic as defined in Equations (12) and (23), $lb_a \cdot °F/Btu$
c = heat transfer exponent, or NTU_a, as defined in Equations (31) and (32), dimensionless
c_o = heat transfer exponent, as defined in Equation (7d), dimensionless
c_p = specific heat of humid air = 0.243 $Btu/lb_a \cdot °F$ for cooling coils
c_{pw} = specific heat of water = 1.0 $Btu/lb_w \cdot °F$
c_r = specific heat of nonvolatile coolant, $Btu/lb_a \cdot °F$
D_i = tube inside diameter, in.
D_o = tube outside diameter, in.
E_a = air-side effectiveness defined in Equation (7b), dimensionless
F_s = coil core surface area parameter = (external surface area)/(face area) (no. of rows deep)
f = convection heat transfer coefficient, $Btu/h \cdot ft^2 \cdot °F$
h = air enthalpy (actual in airstream or saturation value at surface temperature), Btu/lb_a
Δh_m = mean effective difference of air enthalpy, as defined in Equation (25), Btu/lb_a
k = thermal conductivity of tube material, $Btu/h \cdot ft \cdot °F$
M = ratio of nonvolatile coolant-to-air temperature changes for sensible heat cooling coils, as defined in Equation (7e), dimensionless
m'' = rate of change of air enthalpy at saturation with air temperature, $Btu/lb \cdot °F$
N_r = number of coil rows deep in airflow direction, dimensionless
Δp_{sd} = isothermal dry surface air-side pressure drop at standard conditions (70°F, 29.92 in. Hg), in. of water
Δp_{sw} = wet surface air-side pressure drop at standard conditions (70°F, 29.92 in. Hg), in. of water
q = heat transfer capacity, Btu/h
q_e = latent heat removed from entering air to condense moisture, Btu/lb_a
q_s = sensible heat removed from entering air, Btu/lb_a
q_t = total refrigeration load of cooling and dehumidifying coil, Btu/lb_a
q_w = sensible heat removed from condensate to cool it to leaving temperature, Btu/lb_a
R = thermal resistance, referred to external area A_o, $h \cdot °F \cdot ft^2/Btu$
SHR = ratio of air sensible heat to air total heat, dimensionless
t = temperature, °F
Δt_m = mean effective temperature difference, air dry bulb to coolant temperature, °F
Δt_{ms} = mean effective temperature difference, surface-to-coolant, °F
$\Delta t_{\bar{m}}$ = mean effective temperature difference, air dry bulb to effective surface temperature $t_{\bar{s}}$, °F
U_o = overall sensible heat transfer coefficient, $Btu/h \cdot ft^2 \cdot °F$
V_a = coil air face velocity at 70°F, fpm
W = air humidity ratio, pounds of water per pound of air
w = mass flow rate, lb/h
y = ratio of nonvolatile coolant temperature rise to airstream enthalpy drop, as defined in Equation (13), $lb_a \cdot °F/Btu$
η = fin effectiveness, as defined in Equation (6), dimensionless
ρ_a = air density = 0.075 lb/ft^3 at 70°F at sea level

Superscripts

$'$ = wet bulb
$''$ = dew point

Subscripts

1 = condition entering coil
2 = condition leaving coil
a = airstream
ab = air, dry-wet boundary
ad = dry air
aw = wet air
b = dry-wet surface boundary
d = dry surface
e = latent
f = fin (with R); saturated liquid water (with h)
g = saturated water vapor
i = installed, selected (with A_o, N_r)
m = metal (with R) and mean (with other symbols)
md = dry metal
mw = wet metal
o = overall (except for A)
r = coolant
rb = coolant dry-wet boundary
s = surface (with d, t, and w) and saturated (with h)
$\bar{s}$ = effective surface
sb = surface dry-wet boundary
t = tube (with R) and total (with s and q)
td = total heat capacity, dry surface
tw = total heat capacity, wet surface
w = water (with ρ), condensate (with h and subscript number), and wet surface (with other symbols)

REFERENCES

Anderson, S.W. 1970. Air-cooling and dehumidifying coil performance based on ARI *Industrial Standard* 410-64. In *Heat and mass transfer to extended surfaces*, ASHRAE Symposium CH-69-3, pp. 22-28.

ARI. 2001. Forced-circulation air-cooling and air-heating coils. ANSI/ARI *Standard* 410-01. Air-Conditioning and Refrigeration Institute, Arlington, VA.

ARI. 1999. Specification for fluorocarbon refrigerants. *Standard* 700-99. Air-Conditioning and Refrigeration Institute, Arlington, VA.

ASHRAE. 2000. Methods of testing forced circulation air-cooling and air heating coils. *Standard* 33-2000.

ASME. 1997. Code on nuclear air and gas treatment. ANSI/ASME *Standard* AG-1-97. American Society of Mechanical Engineers, New York.

Brown, G. 1954. Theory of moist air heat exchangers. Royal Institute of Technology *Transactions* no. 77, Stockholm, Sweden, 12.

Kusuda, T. 1970. Effectiveness method for predicting the performance of finned tube coils. In *Heat and mass transfer to extended surfaces*, ASHRAE Symposium CH-69-3, pp. 5-14.

McElgin, J. and D.C. Wiley. 1940. Calculation of coil surface areas for air cooling and dehumidification. *Heating, Piping and Air Conditioning* (March):195.

McQuiston, F.C. 1981. Finned tube heat exchangers: State of the art for air side. *ASHRAE Transactions* 87(1):1077-1085.

Mirth, D.R., S. Ramadhyani, and D.C. Hittle. 1993. Thermal performance of chilled water cooling coils operating at low water velocities. *ASHRAE Transactions* 99(1):43-53.

Mueller, A.C. 1998. Heat exchangers. *Handbook of heat transfer,* 3rd ed. Rohsenow, Hartnett, and Cho, eds. McGraw-Hill, New York.

Pedersen, C.O., D.E. Fisher, J.D. Spitler, and R.J. Liesen. 1998. *Cooling and heating load calculation principles*. ASHRAE.

Shah, M.M. 1976. A new correlation for heat transfer during boiling flow through pipes. *ASHRAE Transactions* 82(2):66-75.

Shah, M.M. 1982. CHART correlation for saturated boiling heat transfer: Equations and further study. *ASHRAE Transactions* 88(1):185-196.

Webb, R.L. 1980. Air-side heat transfer in finned tube heat exchangers. *Heat Transfer Engineering* 1(3):33.

BIBLIOGRAPHY

Shah, M.M. 1978. Heat transfer, pressure drop, visual observation, test data for ammonia evaporating inside pipes. *ASHRAE Transactions* 84(2): 38-59.

CHAPTER 23

DESICCANT DEHUMIDIFICATION AND PRESSURE-DRYING EQUIPMENT

DEHUMIDIFICATION is the removal of water vapor from air, gases, or other fluids. There is no pressure limitation in this definition, and sorption dehumidification equipment has been designed and operated successfully for system pressures ranging from subatmospheric to as high as 6000 psi. In common practice, *dehumidification* usually refers to equipment operating at essentially atmospheric pressures and built to standards similar to other types of air-handling equipment. For drying gases under pressure, or liquids, the term **dryer** or **dehydrator** is normally used.

This chapter mainly covers equipment and systems that dehumidify air rather than those that dry other gases or liquids. Both liquid and solid desiccants are used; they either adsorb water on the desiccant's surface (adsorption) or chemically combine with water (absorption).

Nonregenerative equipment uses hygroscopic salts such as calcium chloride, urea, or sodium chloride. **Regenerative** systems usually use a form of silica or alumina gel, activated alumina, molecular sieves, lithium chloride salt or solution, or glycol solution. In regenerative equipment, the water removal mechanism is reversible. The choice of desiccant depends on installation requirements, equipment design, and chemical compatibility with the gas to be treated or impurities in the gas. Chapter 22 of the 2005 *ASHRAE Handbook—Fundamentals* has more information on desiccant materials and how they operate.

Some commercial applications of desiccant dehumidification include

- Keeping buildings and HVAC systems dry to prevent mold growth
- Lowering relative humidity to facilitate manufacturing and handling of hygroscopic materials
- Lowering the dew point to prevent condensation on products manufactured in low-temperature processes
- Providing protective atmospheres for heat treatment of metals
- Controlling humidity in warehouses and caves used for storage
- Preserving ships, aircraft, and industrial equipment that would otherwise deteriorate
- Maintaining a dry atmosphere in a closed space or container, such as the cargo hold of a ship or numerous static applications
- Eliminating condensation and subsequent corrosion
- Drying air to speed drying of heat-sensitive products, such as candy, seeds, and photographic film
- Drying natural gas
- Drying gases that are be liquefied
- Drying instrument and plant air
- Drying process and industrial gases
- Dehydration of liquids
- Frost-free cooling for low-temperature process areas such as brewery fermenting, aging, filtering, and storage cellars; blast freezers; and refrigerated warehouses
- Frost-free dehumidification for processes that require air at a subfreezing dew-point humidity

This chapter covers (1) the types of dehumidification equipment for liquid and solid desiccants, including high-pressure equipment; (2) performance curves; (3) variables of operation; and (4) some typical applications. Using desiccants to dry refrigerants is addressed in Chapter 6 of the 2006 *ASHRAE Handbook—Refrigeration*.

METHODS OF DEHUMIDIFICATION

Air may be dehumidified by (1) cooling it or increasing its pressure, reducing its capacity to hold moisture, or (2) removing moisture by attracting the water vapor with a liquid or solid desiccant. Frequently, systems use a combination of these methods to maximize operating efficiency and minimize installed cost.

Figure 1 illustrates three methods to dehumidify with desiccant materials or equipment. Air in the condition at Point A is dehumidified and cooled to Point B. In a liquid-desiccant unit, air is simultaneously cooled and dehumidified directly from Point A to Point B. In a solid-desiccant unit, this process can be completed by precooling and dehumidifying from Point A to Point C, then desiccating from Point C to Point E, and finally cooling to Point B. It could also be done with solid desiccant equipment by dehumidifying from Point A to Point D and then cooling from Point D to Point B.

Compression

Compressing air reduces its capacity to hold moisture. The resulting condensation reduces the air's moisture content in absolute terms,

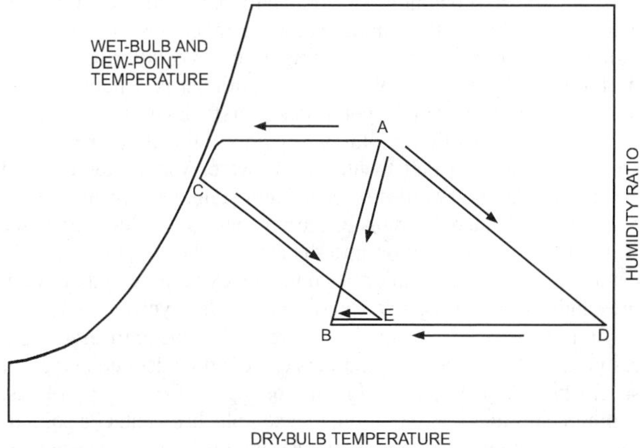

Fig. 1 Methods of Dehumidification

The preparation of this chapter is assigned to TC 8.12, Desiccant Dehumidification Equipment and Components.

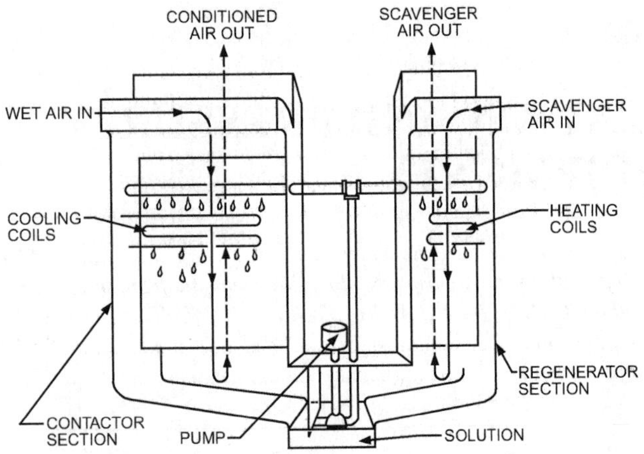

Fig. 2 Flow Diagram for Liquid-Absorbent Dehumidifier

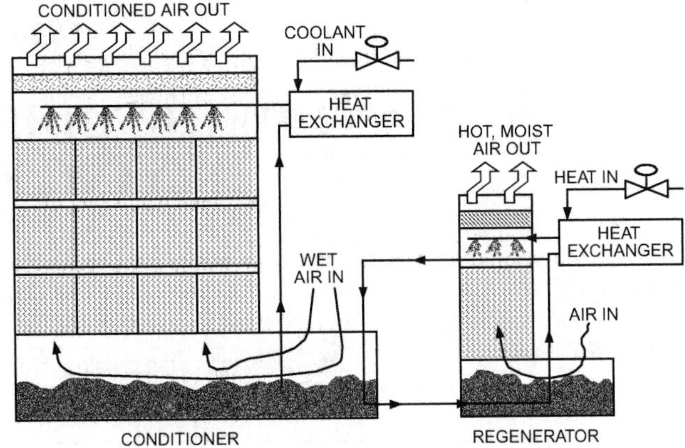

Fig. 3 Flow Diagram for Liquid-Absorbent Unit with Extended Surface Air Contact Medium

but produces a saturated condition: 100% relative humidity at elevated pressure. In atmospheric-pressure applications, this method is too expensive, but is worthwhile in pressure systems such as instrument air. Other dehumidification equipment, such as coolers or desiccant dehumidifiers, often follows the compressor to avoid problems associated with high relative humidity in compressed-air lines.

Cooling

Refrigerating air below its dew point is the most common method of dehumidification. This is advantageous when the gas is comparatively warm, has a high moisture content, and the desired outlet dew point is above 40°F. Frequently, refrigeration is combined with desiccant dehumidification to obtain an extremely low dew point at minimum cost.

Liquid Desiccants

Liquid-desiccant conditioners (absorbers) contact the air with a liquid desiccant, such as lithium chloride or glycol solution (Figures 2 and 3). The water vapor pressure of the solution is a function of its temperature and concentration. Higher concentrations and lower temperatures result in lower water vapor pressures.

A simple way to show this relationship is to graph the humidity ratio of air in equilibrium with a liquid desiccant as a function of its concentration and temperature. Figure 4 presents this relationship for lithium chloride/water solutions in equilibrium with air at 14.7 psi. The graph has the same general shape as a psychrometric chart, with the relative humidity lines replaced by desiccant concentration lines.

Liquid-desiccant conditioners typically have a high contact efficiency, so air leaves the conditioner at a temperature and humidity ratio very close to the entering temperature and equilibrium humidity ratio of the desiccant. When the conditioner is dehumidifying, moisture absorbed from the conditioned airstream dilutes the desiccant solution. The diluted solution is reconcentrated in the regenerator, where it is heated to elevate its water vapor pressure and equilibrium humidity ratio. A second airstream, usually outside air, contacts the heated solution in the regenerator; water evaporates from the desiccant solution into the air, and the solution is reconcentrated. Desiccant solution is continuously recirculated between the conditioner and regenerator to complete the cycle.

Liquid desiccants are typically a very effective antifreeze. As a result, liquid-desiccant conditioners can continuously deliver air at subfreezing temperatures without frosting or freezing problems. Lithium chloride/water solution, for example, has a eutectic point of –90°F; liquid-desiccant conditioners using this solution can cool air to temperatures as low as –65°F.

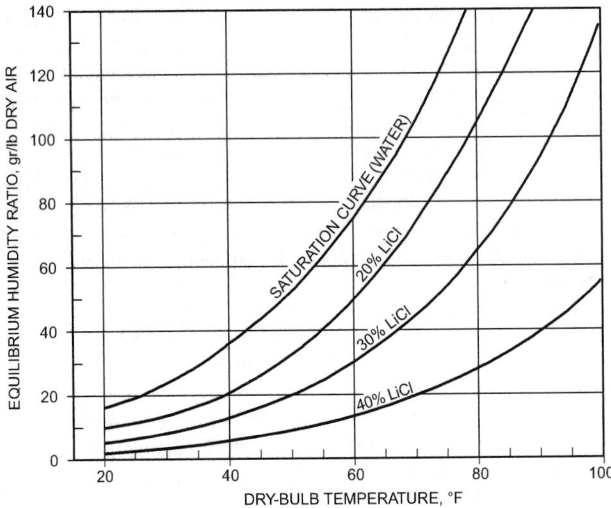

Fig. 4 Lithium Chloride Equilibrium

Solid Sorption

Solid sorption passes air through a bed of granular desiccant or through a structured packing impregnated with desiccant. Humid air passes through the desiccant, which when active has a vapor pressure below that of the humid air. This vapor pressure differential drives water vapor from the air onto the desiccant. After becoming loaded with moisture, the desiccant is reactivated (dried out) by heating, which raises the vapor pressure of the material above that of the surrounding air. With the vapor pressure differential reversed, water vapor moves from the desiccant to a second airstream called the *reactivation air*, which carries moisture away from the equipment.

DESICCANT DEHUMIDIFICATION

Both liquid and solid desiccants may be used in equipment designed for drying air and gases at atmospheric or elevated pressures. Regardless of pressure levels, basic principles remain the same, and only the desiccant towers or chambers require special design consideration.

Desiccant capacity and actual dew-point performance depend on the specific equipment used, characteristics of the various desiccants, initial temperature and moisture content of the gas to be dried, reactivation methods, etc. Factory-assembled units are available up to a capacity of about 80,000 cfm. Greater capacities can be obtained with field-erected units.

LIQUID-DESICCANT EQUIPMENT

Liquid-desiccant dehumidifiers are shown in Figures 2 and 3. In Figure 2, liquid desiccant is distributed onto a **cooling coil**, which acts as both a contact surface and a means of removing heat released when the desiccant absorbs moisture from the air. In Figure 3, liquid desiccant is distributed onto an extended heat and mass transfer surface (a **packing** material similar to that used in cooling towers and chemical reactors). The packing provides a great deal of surface for air to contact the liquid desiccant, and the heat of absorption is removed from the liquid by a heat exchanger outside the airstream. Air can be passed through the contact surface vertically or horizontally to suit the best arrangement of air system equipment.

Depending on the air and desiccant solution inlet conditions, air can be simultaneously cooled and dehumidified, heated and dehumidified, heated and humidified, or cooled and humidified. When the enthalpy of the air is to be increased in the conditioner unit, heat must be added either by preheating the air before it enters the conditioner or by heating the desiccant solution with a second heat exchanger. When the air is to be humidified, makeup water is automatically added to the desiccant solution to keep it at the desired concentration.

Moisture is absorbed from or desorbed into the air because of the difference in water vapor pressure between the air and the desiccant solution. For a given solution temperature, a higher solution concentration results in a lower water vapor pressure. For a given solution concentration, a lower solution temperature results in a lower water vapor pressure. By controlling the temperature and concentration of the desiccant solution, the conditioner unit can deliver air at a precisely controlled temperature and humidity regardless of inlet air conditions. The unit dehumidifies the air during humid weather and humidifies it during dry weather. Thus, liquid-desiccant conditioners can accurately control humidity without face-and-bypass dampers or after-humidifiers.

Heat Removal

When a liquid desiccant absorbs moisture, heat is generated. This heat of absorption consists of the latent heat of condensation of water vapor at the desiccant temperature and the heat of solution (heat of mixing) of the condensed water and the desiccant. The heat of mixing is a function of the equilibrium relative humidity of the desiccant: a lower equilibrium relative humidity produces a greater heat of mixing.

The total heat that must be absorbed by the desiccant solution consists of the (1) heat of absorption, (2) sensible heat associated with reducing the dry-bulb temperature of the air, and (3) residual heat carried to the conditioner by the warm, concentrated desiccant returning from the regenerator unit. This total heat is removed by cooling the desiccant solution in the conditioner heat exchanger (Figure 3). Any coolant can be used, including cooling tower water, ground water, seawater, chilled water or brine, and direct-expansion or flooded refrigerants.

Regenerator residual heat, generally called **regenerator heat dumpback**, can be substantially reduced by using a liquid-to-liquid heat exchanger to precool the warm, concentrated desiccant transferred to the conditioner using the cool, dilute desiccant transferred from the conditioner to the regenerator. This also improves the thermal efficiency of the regenerator, typically reducing heat input by 10 to 15%.

Regeneration

When the conditioner is dehumidifying, water is automatically removed from the liquid desiccant to maintain the desiccant at the proper concentration. Removal takes place in a separate regenerator. A small sidestream of the desiccant solution, typically 8% or less of the flow to the conditioner packing, is transferred to the regenerator unit. In the regenerator, a separate pump continuously circulates the desiccant solution through a heat exchanger and distributes it over the packed bed contactor surface. The heat exchanger heats the desiccant solution with low-pressure steam or hot water so that its water vapor pressure is substantially higher than that of the outside air. Outside air is passed through the packing, and water evaporates into it from the desiccant solution, concentrating the solution. The hot, moist air from the regenerator is discharged to the outdoors. A sidestream of concentrated solution is transferred to the conditioner to replace the sidestream of weak solution transferred from the conditioner and complete the cycle.

The regenerator's water removal capacity is controlled to match the moisture load handled by the conditioner. This is accomplished by regulating heat flow to the regenerator heat exchanger to maintain a constant desiccant solution concentration. This is most commonly done by maintaining a constant solution level in the system with a level controller, but specific-gravity or boiling-point controllers are used under some circumstances. Regenerator heat input is regulated to match the instantaneous water removal requirements, so no heat input is required if there is no moisture load on the conditioner. When the conditioner is used to humidify the air, the regenerator fan and desiccant solution pump are typically stopped to save energy.

Because the conditioner and regenerator are separate units, they can be in different locations and connected by piping. This can substantially lower ductwork cost and required mechanical space. Commonly, a single regenerator services several conditioner units (Figure 5). In the simplest control arrangement, concentrated desiccant solution is metered to each conditioner at a fixed rate. The return flow of weak solution from each conditioner is regulated to maintain a constant operating level in the conditioner. A level controller on the regenerator regulates heat flow to the regenerator solution heater to maintain a constant volume of desiccant solution, and hence a practically constant solution concentration.

The regenerator can be sized to match the dehumidification load of the conditioner unit or units. Regenerator capacity is affected by regenerator heat source temperatures (higher source temperatures increase capacity) and by desiccant concentration (higher concentrations reduce capacity). The relative humidity of air leaving the conditioner is practically constant for a given desiccant concentration, so regenerator capacity can be shown as a function of delivered air relative humidity and regenerator heat source temperature. Figure 6 is a normalized graph showing this relationship. For a given moisture load, a variety of regenerator heat sources may be used if the regenerator is sized for the heat source selected. In many cases, the greater capital cost of a larger regenerator is paid back very quickly by reduced operating cost when a lower-cost or waste-heat source (e.g., process or turbine tailsteam, jacket heat from an engine-driven generator or compressor, or refrigeration condenser heat) is used.

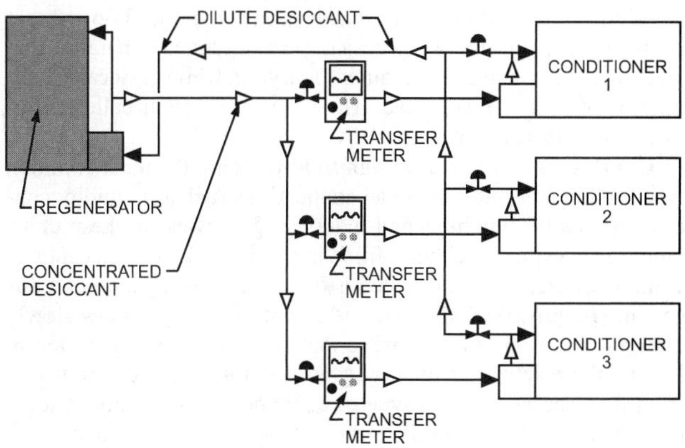

Fig. 5 Liquid Desiccant System with Multiple Conditioners

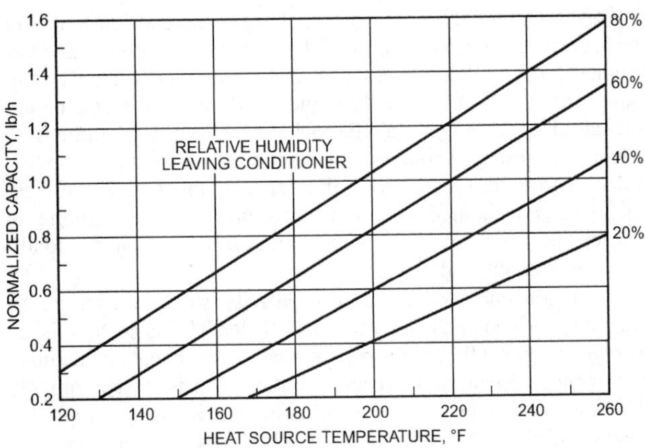

Fig. 6 Liquid Desiccant Regenerator Capacity

SOLID-SORPTION EQUIPMENT

Solid desiccants, such as silica gel, zeolites (molecular sieves), activated alumina, or hygroscopic salts, are generally used to dehumidify large volumes of moist air, and are continuously reactivated. Dry desiccants can also be used in (1) nonreactivated, disposable packages and (2) periodically reactivated desiccant cartridges.

Disposable packages of solid desiccant are often sealed into packaging for consumer electronics, pharmaceutical tablets, and military supplies. Disposable desiccant packages rely entirely on vapor diffusion to dehumidify, because air is not forced through the desiccant. This method is used only in applications where there is no anticipated moisture load at all (such as hermetically sealed packages) because the moisture absorption capacity of any nonreactivated desiccant is rapidly exceeded if a continuous moisture load enters the dehumidified space. Disposable packages generally serve as a form of insurance against unexpected, short-term leaks in small, sealed packages.

Periodically reactivated cartridges of solid desiccant are used where the expected moisture load is continuous, but very small. A common example is the breather, a tank of desiccant through which air can pass, compensating for changes in liquid volume in petroleum storage tanks or drums of hygroscopic chemicals. Air dries as it passes through the desiccant, so moisture will not contaminate the stored product. When the desiccant is saturated, the cartridge is removed and heated in an oven to restore its moisture sorption capacity. Desiccant cartridges are used where there is no requirement for a constant humidity control level and where the moisture load is likely to exceed the capacity of a small, disposable package of desiccant.

Desiccant dehumidifiers for drying liquids and gases other than air often use a variation of this reactivation technique. Two or more pressurized containers of dry desiccant are arranged in parallel, and air is forced through one container for drying, while desiccant in the other container is reactivated. These units are often called **dual-tower** or **twin-tower dehumidifiers**.

Continuous reactivation dehumidifiers are the most common type used in high-moisture-load applications such as humidity control systems for buildings and industrial processes. In these units, humid **process air** is dehumidified in one part of the desiccant bed while a different part of the bed is dried for reuse by a second airstream (**reactivation air**). The desiccant generally rotates slowly between these two airstreams, so that dry, high-capacity desiccant leaving the reactivation air is always available to remove moisture from the process air. This type of equipment is generally called a **rotary desiccant dehumidifier**. It is most commonly used in building air-handling systems, and the section on Rotary Solid-Desiccant Dehumidifiers describes its function in greater detail.

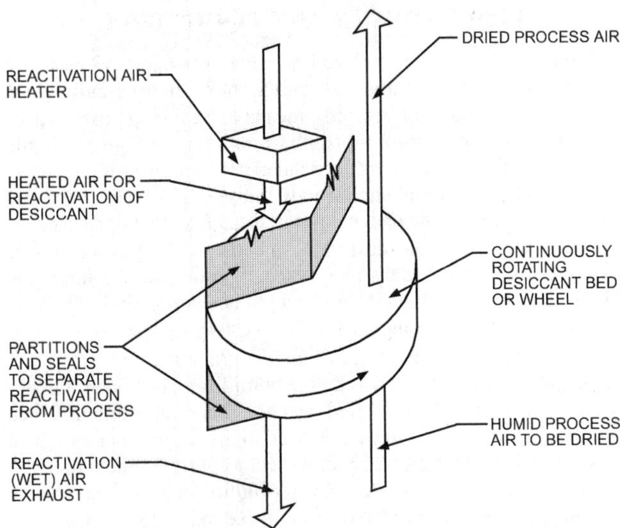

Fig. 7 Typical Rotary Dehumidification Wheel

ROTARY SOLID-DESICCANT DEHUMIDIFIERS

Operation

Figure 7 illustrates the principle of operation and arrangement of major components of a typical rotary solid-desiccant dehumidifier. The desiccant can be beads of granular material packed into a bed, or it can be finely divided and impregnated throughout a structured medium. The structured medium resembles corrugated cardboard rolled into a drum, so that air can pass freely through flutes aligned lengthwise through the drum.

In both granular and structured-medium units, the desiccant itself can be either a single material, such as silica gel, or a combination, such as dry lithium chloride mixed with zeolites. The wide range of dehumidification applications requires flexibility in selecting dessicants to minimize operating and installed costs.

In rotary desiccant dehumidifiers, more than 20 variables can affect performance. In general, equipment manufacturers fix most of these to provide predictable performance in common applications for desiccant systems. Primary variables left to the system designer to define include the following for both process and reactivation air:

- Inlet air temperature
- Moisture content
- Velocity at face of the desiccant bed

In any system, these variables change because of weather, variations in moisture load, and fluctuations in reactivation energy levels. It is useful for the system designer to understand the effect of these normal variations on dehumidifier performance.

Figures 8 to 12 show changes in process air temperature and moisture leaving a generic rotary desiccant dehumidifier as modeled by a finite difference analysis program (Worek and Zheng 1991). Commercial unit performance differs from this model because such units are generally optimized for very deep drying. However, for illustration purposes, the model accurately reflects the relationships between the key variables.

To further illustrate relationships between these variables, ASHRAE Technical Committee 8.12 developed an interactive wheel performance estimator (available on the committee's section of the ASHRAE Web site at www.ashrae.org; see Figure 13). Entering different values for moisture and temperature of process and reactivation airstreams gives the performance of a typical desiccant dehumidification wheel. This program is generic and not based on

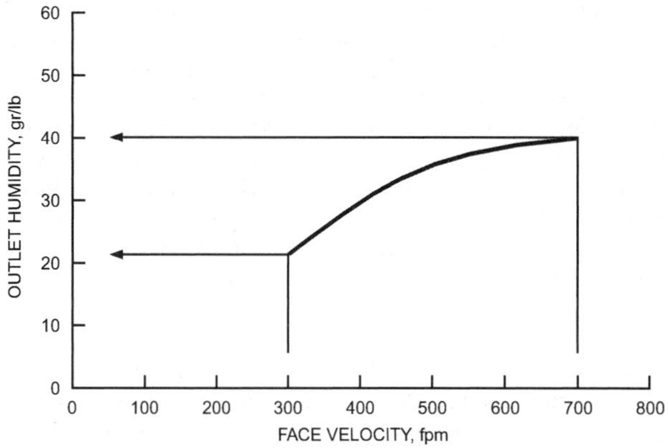

Fig. 8 Effect of Changes in Process Air Velocity on Dehumidifier Outlet Moisture

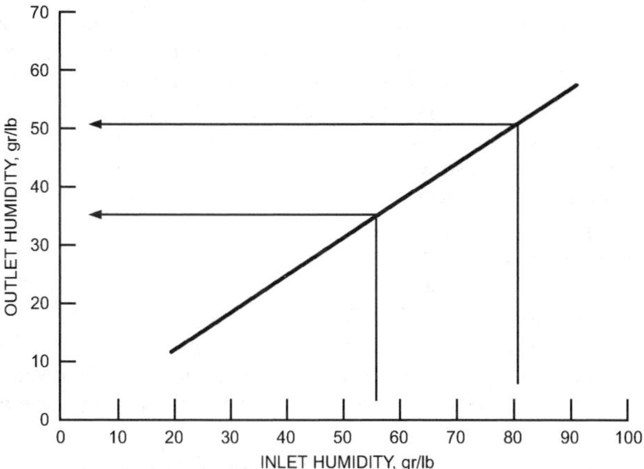

Fig. 9 Effect of Changes in Process Air Inlet Moisture on Dehumidifier Outlet Moisture

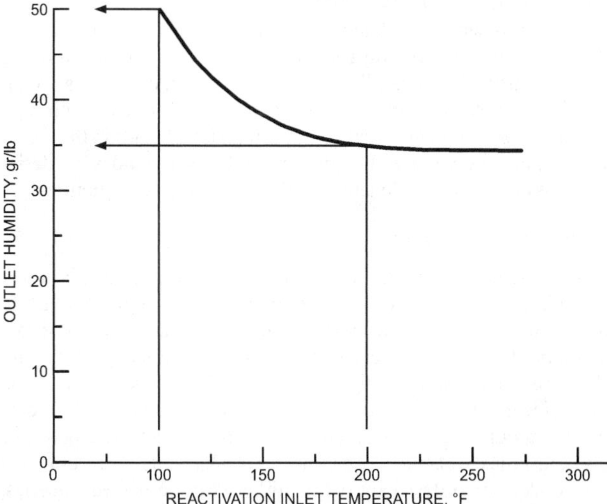

Fig. 10 Effect of Changes in Reactivation Air Inlet Temperature on Dehumidifier Outlet Moisture

any specific equipment. Performance of units made by different manufacturers varies widely. Consequently, this estimator is useful only for education, not for design.

For example, Figure 13 shows that when the velocity through the desiccant bed increases, the process air is not as dry as it leaves the

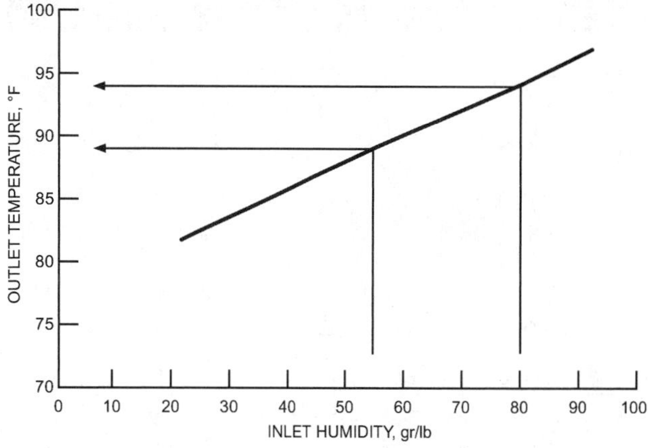

Fig. 11 Effect of Changes in Process Air Inlet Moisture on Dehumidifier Outlet Temperature

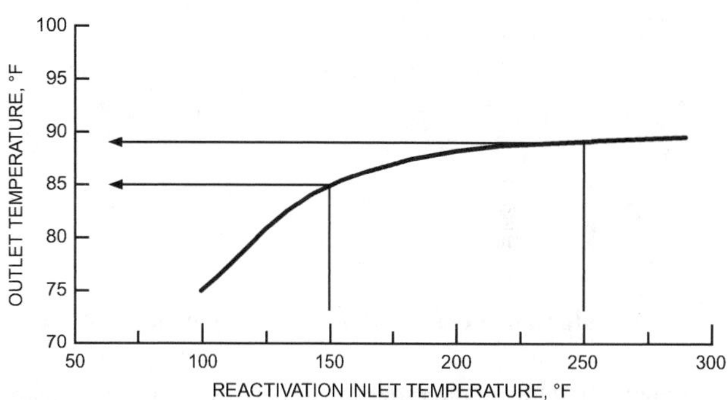

Fig. 12 Effect of Changes in Reactivation Air Inlet Temperature on Dehumidifier Outlet Temperature

unit. But, at the same time, as that greater mass of air flows through the wheel (higher air velocity), the dehumidifier actually removes more total pounds of water per hour. This relationship explains why industrial applications, which often require low dew points, generally use lower velocities through the dehumidifier. Commercial applications, which commonly have much larger moisture loads and less need for very low dew points, usually use higher velocities through desiccant equipment.

The desiccant used for the model is silica gel; the bed is a structured, fluted medium; the bed depth is 16 in. in the direction of airflow; and the ratio of process air to reactivation air is approximately 3:1. Process air enters at normal comfort conditions of 70°F, 50% rh (56 grains per pound of dry air).

Process air velocity through the desiccant bed strongly affects leaving moisture. As shown in Figure 8, if the entering moisture is 56 gr/lb and all other variables are held constant, the outlet moisture varies from 22 gr/lb at 300 fpm to 40 gr/lb at 700 fpm. Thus, air that passes through the bed more slowly is dried more deeply. Therefore, if air must be dried very deeply, a large unit (slower air velocities) must be used.

Process air inlet moisture content affects outlet moisture: if air is more humid entering the dehumidifier, it will be more humid leaving the unit. For example, Figure 9 indicates that for an inlet humidity of 56 gr/lb, the outlet humidity will be 35 gr/lb. If inlet moisture content rises to 80 gr/lb, the outlet humidity rises to 50 gr/lb. Therefore, if constant outlet humidity is necessary, the dehumidifier needs capacity control unless the process inlet airstream does not vary in temperature or moisture throughout the year (a rare circumstance).

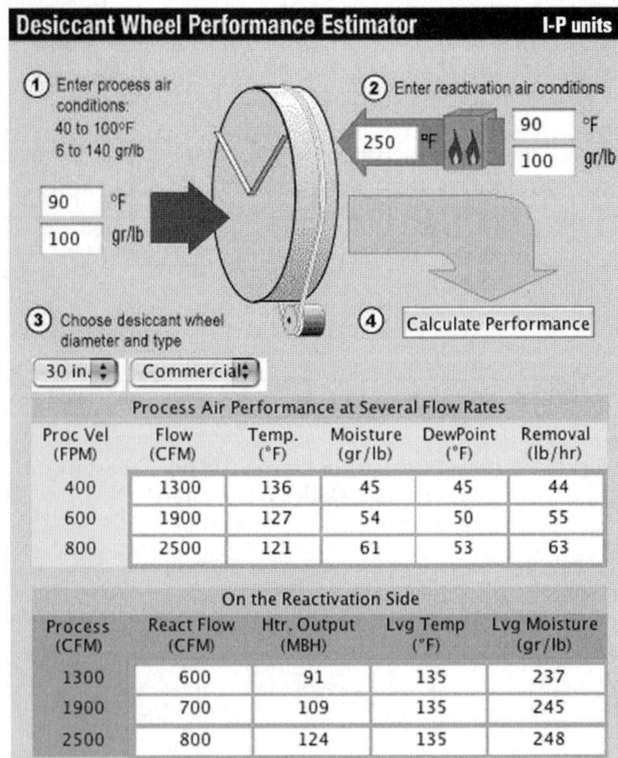

Process Air Performance at Several Flow Rates

Proc Vel (FPM)	Flow (CFM)	Temp. (°F)	Moisture (gr/lb)	DewPoint (°F)	Removal (lb/hr)
400	1300	136	45	45	44
600	1900	127	54	50	55
800	2500	121	61	53	63

On the Reactivation Side

Process (CFM)	React Flow (CFM)	Htr. Output (MBH)	Lvg Temp (°F)	Lvg Moisture (gr/lb)
1300	600	91	135	237
1900	700	109	135	245
2500	800	124	135	248

Fig. 13 Interactive Desiccant Wheel Performance Estimator

Reactivation air inlet temperature changes the outlet moisture content of the process air. From 100 to 250°F, as more heat is added to the reactivation air, the desiccant dries more completely, which means that it can attract more moisture from the process air (Figure 10). If reactivation air is only heated to 100°F, process outlet moisture is 50 gr/lb, or only 6 gr/lb lower than the entering humidity. In contrast, if reactivation air is heated to 200°F, the outlet moisture is 35 gr/lb, so that almost 40% of the original moisture is removed.

This relationship has two important consequences. If the design needs dry air, it is generally more economical to use high reactivation air temperatures. Conversely, if leaving humidity from the dehumidifier need not be especially low, inexpensive, low-grade heat sources (e.g., waste heat, cogeneration heat, or rejected heat from refrigeration condensers) can be used to reactivate the desiccant.

Process air outlet temperature is higher than the inlet air temperature primarily because the heat of sorption of moisture removed from the air is converted to sensible heat. The heat of sorption includes the latent heat of condensation of the removed moisture, plus additional chemical heat, which varies depending on the desiccant type and process air outlet humidity. Also, some heat is carried over to the process air from the reactivation sector because the desiccant is warm as it enters the relatively cooler process air. Generally, 80 to 90% of the temperature rise of process air is from the heat of sorption, and the balance is from heat carried over from reactivation.

Process outlet temperature versus inlet humidity is illustrated in Figure 11. Note that as more moisture is removed (higher inlet humidity), outlet temperature rises. Air entering at room comfort conditions of 70°F, 56 gr/lb leaves the dehumidifier at 89°F. If the dehumidifier removes more moisture, such as when the inlet humidity is 80 gr/lb, outlet temperature rises to 94°F. The increase in temperature rise is roughly proportional to the increase in moisture removal.

Process outlet temperature versus reactivation air temperature is illustrated in Figure 12, which shows the effect of increasing reactivation temperature when the moisture content of the process inlet air stays constant. If the reactivation sector is heated to elevated temperatures, more moisture is removed on the process side, so the temperature rise from latent-to-sensible heat conversion is slightly greater. In this constant-moisture inlet situation, if the reactivation sector is very hot, more heat is carried from reactivation to process as the desiccant mass rotates from reactivation to process. Figure 12 shows that if reactivation air is heated to 150°F, the process air leaves the dehumidifier at 85°F. If reactivation air is heated to 250°F, the process air outlet temperature rises to 89°F. The 4°F increase in process air temperature is primarily caused by the increase in heat carried over from reactivation.

One consequence of this relationship is that desiccant equipment manufacturers constantly seek to minimize the "waste mass" in a desiccant dehumidifier, to avoid heating and cooling extra, nonfunctional material such as heavy desiccant support structures or extra desiccant that air cannot reach. Theoretically, the most efficient desiccant dehumidifier has an infinitely large effective desiccant surface combined with an infinitely low mass.

Use of Cooling

In process drying applications, desiccant dehumidifiers are sometimes used without additional cooling because the temperature increase from dehumidification helps the drying process. In semi-process applications such as controlling frost formation in supermarkets, excess sensible cooling capacity may be present in the system as a whole, so warm air from a desiccant unit is not a major consideration. However, in most other applications for desiccant dehumidifiers, provision must be made to remove excess sensible heat from process air after dehumidification.

In a liquid-desiccant system, heat is removed by cooling the liquid desiccant itself, so process air emerges from the desiccant medium at the appropriate temperature. In a solid-desiccant system, cooling is accomplished downstream of the desiccant bed with cooling coils. The source of this cooling can affect the system's operating economics.

In some systems, postcooling is accomplished in two stages, with cooling tower water as the primary source followed by compression or absorption cooling. Alternatively, various combinations of indirect and direct evaporative cooling equipment are used to cool the dry air leaving the desiccant unit.

In systems where the latent and sensible loads peak at different times, the sensible cooling capacity of the basic air-conditioning system is sufficient to handle the process air temperature rise without additional equipment. Systems in moderate climates with high ventilation requirements often combine high latent loads in the morning, evening, and night with high sensible loads at midday, so desiccant subsystems to handle latent loads are especially economical.

Using Units in Series

Dry-desiccant dehumidifiers are often used to provide air at low dew points. Applications requiring large volumes of air at moisture contents of 5 gr/lb (0°F dew point) are quite common and can be easily achieved by rotary desiccant units in a single pass beginning with inlet moisture contents as high as 45 gr/lb (45°F dew point). Some dry-desiccant units commonly deliver air at 2 gr/lb (–18°F dew point) without special design considerations. Where extremely low dew points must be achieved, or where air leakage inside the unit may be a concern, two desiccant dehumidifiers can be placed in series, with dry air from one unit feeding both process and reactivation air to a second unit. The second unit delivers very dry air, because there is reduced risk of moisture being carried over from reactivation to process air when dry air is used to reactivate the second unit.

Industrial Rotary Desiccant Dehumidifier Performance

Figures 8 through 12 are based on the generalized model of a desiccant dehumidifier described by Worek and Zheng (1991). The model, however, differs somewhat from commercial products.

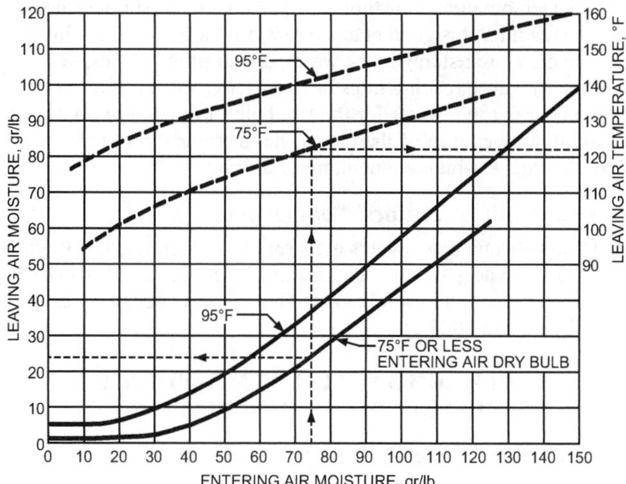

Fig. 14 Typical Performance Data for Rotary Solid Desiccant Dehumidifier

Figure 14 shows typical performance of an industrial desiccant dehumidifier.

EQUIPMENT OPERATING RECOMMENDATIONS

Desiccant equipment tends to be very durable if maintained properly, often operating at high efficiency 30 years after it was originally installed. Required maintenance is specific to the type of desiccant equipment, the application, and the installation. Each system requires a somewhat different maintenance and operational routine. The information in this section does not substitute for or supersede any recommendations of equipment manufacturers, and it is not a substitute for owners' experience with specific applications.

Process Air Filters

Clean filters are the most important item in a maintenance routine. If a solid desiccant is clogged with particulates, or if a liquid desiccant's sorption characteristics are changed by entrained particulates, the material may have to be replaced prematurely. Filters are much less expensive and much easier to change than the desiccant. Although each application is different, the desiccant usually must be replaced, replenished, or reconditioned after 5 to 10 years of operation. Without attention to filters, desiccant life can be reduced to 1 or 2 years of operation or less. Filters should be checked at least four times per year, and more frequently when airstreams are heavily laden with particulates.

The importance of filter maintenance requires that filter racks and doors on desiccant systems be freely accessible and that enough space be allowed to inspect, remove, and replace filters. Optimal design ensures that filter locations, as well as the current condition of each filter, are clearly visible to maintenance personnel.

Reactivation/Regeneration Filters

Air is filtered before entering the heater of a desiccant unit. If filters are clogged and airflow is reduced, unit performance may be reduced because there is not enough air to carry all the moisture away from the desiccant. If electrical elements or gas burners are used to heat the air, reducing airflow may damage the heaters. Thus, the previous suggestions for maintaining process air filters also apply to reactivation/regeneration filters.

Reactivation/Regeneration Ductwork

Air leaving the reactivation/regeneration section is hot and moist. When units first start up in high-moisture-load applications, the reactivation air may be nearly saturated and even contain water droplets. Thus, ductwork that carries air away from the unit should be corrosion-resistant, because condensation can occur inside the ducts, particularly if the ducts pass through unheated areas in cool weather. If heavy condensation seems probable, the ductwork should be designed with drains at low points or arranged to let condensation flow out of the duct where the air is vented to the weather. The high temperature and moisture of the leaving air may make it necessary to use dedicated ductwork, rather than combining the air with other exhaust airflows, unless the other flows have similar characteristics.

Leakage

All desiccant units produce dry air in part of the system. If humid air leaks into either the dry air ductwork or the unit itself, system efficiency is reduced. Energy is also wasted if dry air leaks out of the distribution duct connections. Therefore, duct connections for desiccant systems should be sealed tightly. In applications requiring very low dew points (below 10°F), the ductwork and desiccant system are almost always tested for leaks at air pressures above those expected during normal operation. In applications at higher dew points, similar leak testing is considered good practice and is recommended by many equipment manufacturers.

Because desiccant equipment tends to be durably constructed, workers often drill holes in the dehumidifier unit casing to provide support for piping, ductwork, or instruments. Such holes eventually leak air, desiccant, or both. Designers should provide other means of support for external components so contractors do not puncture the system unnecessarily.

Contractors installing desiccant systems should be aware that any holes made in the system must be sealed tightly using both mechanical means and sealant compounds. Sealants must be selected for long life at the working temperatures of the application and of the casing walls that have been punctured. For example, reactivation/regeneration sections often operate in a range from a cold winter ambient of –40°F to a heated temperature as high as 300°F. Process sections may operate in a range of –40°F at the inlet to 150°F at the outlet.

Airflow Indication and Control

As explained in the section on Rotary Solid-Desiccant Dehumidifiers, performance depends on how quickly air passes through the desiccant; changes in air velocity affect performance. Thus, it is important to quantify the airflow rate through both the process and reactivation/regeneration parts of the unit. Unless both airflows are known, it is impossible to determine whether the unit is operating properly. In addition, if velocity exceeds the maximum design value, the air may carry desiccant particles or droplets out of the unit and into the supply air ductwork. Thus, manufacturers often provide airflow gages on larger equipment so the owner can be certain the unit is operating within the intended design parameters.

Smaller equipment is not always provided with airflow indicators because precise performance may be less critical in applications such as small storage rooms. However, in any system using large equipment, or if performance is critical in smaller systems, unit airflow should be quantified and clearly indicated, so operating personnel can compare current flow rates through the system with design values.

Many desiccant units are equipped with manual or automatic flow control dampers to control the airflow rate. If these are not provided with the unit, they should be installed elsewhere in the system. Airflows for process and reactivation/regeneration must be correctly set after all ductwork and external components are attached, but before the system is put into use.

Commissioning

Heat and moisture on the dry-air side of desiccant equipment is balanced equally by the heat and moisture on the regeneration/

reactivation side. To confirm that a solid-desiccant system is operating as designed, the commissioning technician must measure airflow, temperature, and moisture on each side to calculate a mass balance. In liquid systems, these six measurements are taken on the process-air side. On the regenerator side, the liquid temperature is read in the sump and at the spray head to confirm the regenerator's heat transfer rate at peak-load conditions.

If the dehumidification unit does not provide the means, the system should be designed to facilitate taking the readings that are essential to commissioning and troubleshooting. Provisions must be made to measure flow rates, temperatures, and moisture levels of airstreams as they enter and leave the desiccant. For liquid systems, provisions must be made for measuring the solution temperature and concentration at different points in the system. Four precautions for taking these readings at different points in a desiccant system follow.

Airflow. Airflow instruments measure the actual volumetric flow rate, which must be converted to standard flow rate to calculate mass flow. Because temperatures in a desiccant system are often well above or below standard temperature, these corrections are essential.

Air temperature. Most airstreams in a desiccant system have temperatures between 0 and 300°F, but temperature can be widely varied and stratified as air leaves the desiccant in solid-desiccant systems. Air temperature readings must be averaged across the duct for accurate calculations. Readings taken after a fan tend to be more uniform, but corrections must be made for heat added by the fan itself.

Process air moisture leaving dry desiccant. In solid-desiccant equipment, air leaving the desiccant bed or wheel is both warm and dry: usually below 20% rh, often below 10%, and occasionally below 2%. Most low-cost instruments have limited accuracy below 15% rh, and all but the most costly instruments have an error of ±2% rh. Consequently, to measure relative humidities near 2%, technicians use very accurate instruments such as manual dew cups or automated optical dew-point hygrometers. ASHRAE *Standard* 41.6 describes these instruments and procedures for their proper use. When circumstances do not allow the use of dew-point instruments, other methods may be necessary. For example, an air sample may need to be cooled to produce a higher, more easily measured relative humidity.

Low humidity readings can be difficult to take with wet-bulb thermometers because the wet wick dries out very quickly, sometimes before the true wet-bulb reading is reached. Also, when the wet-bulb temperature is below the freezing point of water, readings take much longer, which may allow the wick to dry out, particularly in solid-desiccant systems where there may be considerable heat in the air leaving the desiccant. Therefore, wicks must be monitored for wetness. Many technicians avoid wet-bulb readings in air leaving a solid-desiccant bed, partly for these reasons, and partly because of the difficulty and time required to obtain average readings across the whole bed.

Like air temperature, air moisture level leaving a solid-desiccant bed varies considerably; if taken close to the bed, readings must be averaged to obtain a true value for the whole air mass.

When very low dew points are expected, the commissioning technician should be especially aware of limitations of the air-sampling system and the sensor. Even the most accurate sensors require more time to come to equilibrium at low dew points than at moderate moisture levels. For example, at dew points below –20°F, the sensor and air sample tubing may take many hours rather than a few minutes to equilibrate with the air being measured. Time required to come to equilibrium also depends on how much moisture is on the sensor before it is placed into the dry airstream. For example, taking a reading in the reactivation/regeneration outlet essentially saturates the sensor, so it will take much longer than normal to equilibrate with the low relative humidity of the process leaving air.

Reactivation/regeneration air moisture leaving desiccant. Air leaving the reactivation/regeneration side of the desiccant is warm and close to saturation. If the humidity measurement sensor is at ambient temperature, moisture may condense on its surface, distorting the reading. It is good practice to warm the sensor (e.g., by taking the moisture reading in the warm, dry air of the process-leaving airstream) before reading moisture in reactivation air. If a wet-bulb instrument is used, water for the wet bulb must be at or above the dry-bulb temperature of the air, or the instrument will read lower than the true wet-bulb temperature of the air.

Owners' and Operators' Perspectives

Designers and new owners are strongly advised to consult other equipment owners and the manufacturer's service department early in design to gain the useful perspective of direct operating experience (Harriman 2003).

APPLICATIONS FOR ATMOSPHERIC-PRESSURE DEHUMIDIFICATION

Preservation of Materials in Storage

Special moisture-sensitive materials are sometimes kept in dehumidified warehouses for long-term storage. Tests by the Bureau of Supplies and Accounts of the U.S. Navy concluded that 40% rh is a safe level to control deterioration of materials. Others have indicated that 60% rh is low enough to control microbiological attack. With storage at 40% rh, no undesirable effects on metals or rubber compounds have been noted. Some organic materials such as sisal, hemp, and paper may lose flexibility and strength, but they recover these characteristics when moisture is regained.

Commercial storage relies on similar equipment for applications that include beer fermentation rooms, meat storage, and penicillin processing, as well as storage of machine tools, candy, food products, furs, furniture, seeds, paper stock, and chemicals. For recommended conditions of temperature and humidity, refer to the food refrigeration section (Chapters 17 to 29) of the 2006 *ASHRAE Handbook—Refrigeration*.

Process Dehumidification

Requirements for dehumidification in industrial processes are many and varied. Some of these processes are as follows:

- Metallurgical processes, in conjunction with controlled-atmosphere annealing of metals
- Conveying hygroscopic materials
- Film drying
- Manufacturing candy, chocolate, and chewing gum
- Manufacturing drugs and chemicals
- Manufacturing plastic materials
- Manufacturing laminated glass
- Packaging moisture-sensitive products
- Assembling motors and transformers
- Solid propellant mixing
- Manufacturing electronic components, such as transistors and microwave components

For information about the effect of low-dew-point air on drying, refer to Chapters 18, 20, 23, and 28 of the 2007 *ASHRAE Handbook—HVAC Applications*.

Ventilation Air Dehumidification

Over a full year, ventilation air loads a cooling system with much more moisture than heat. Except in desert and high-altitude regions, ventilation moisture loads in the United States exceed sensible loads by at least 3:1, and often by as much as 5:1 (Harriman et al. 1997). Consequently, desiccant systems are used to dehumidify ventilation air before it enters the main air-conditioning system.

Drying ventilation air has gained importance because building codes mandate larger amounts of ventilation air than in the past, in an effort to improve indoor air quality. Large amounts of humid ventilation air carry enough moisture to upset the operation of high-efficiency cooling equipment, which is generally designed to remove more sensible heat than moisture (Kosar et al. 1998). Removing excess moisture from the ventilation air with a ventilation dehumidification system improves both humidity control and cooling system effectiveness. For example, field tests suggest that when the environment is kept dry, occupants prefer warmer temperatures, which in turn saves cooling operational costs (Fischer and Bayer 2003). Also, cooling equipment is often oversized to remove ventilation-generated moisture. Predrying with a desiccant system may reduce the building's construction cost, if excess cooling capacity is removed from the design (Spears and Judge 1997).

Figures 15, 16, and 17 show the relative importance of moisture load from ventilation, how a commercial building can use a desiccant system to remove that load, and how such a system is applied in the field (Harriman et al. 2001).

Ventilation dehumidification is most cost-effective for buildings with high ventilation airflow rather than high sensible loads from internal heat or from heat transmitted through the building envelope. As a result, this approach is most common in densely occupied buildings such as schools, theaters, elder care facilities, large-scale retail buildings, and restaurants (Harriman 2003).

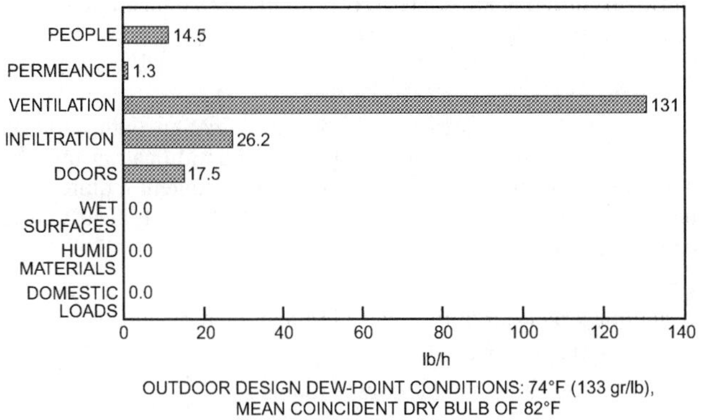

PEOPLE
14.5
PERMEANCE
1.3
VENTILATION
131
INFILTRATION
26.2
DOORS
17.5
WET SURFACES
0.0
HUMID MATERIALS
0.0
DOMESTIC LOADS
0.0
0
20
40
60
80
100
120
140
lb/h
OUTDOOR DESIGN DEW-POINT CONDITIONS: 74°F (133 gr/lb), MEAN COINCIDENT DRY BULB OF 82°F

Fig. 15 Typical Peak Moisture Loads for Medium-Sized Retail Store in Atlanta, Georgia
(Harriman et al. 2001)

VENTILATION AIR DEHUMIDIFIER
RETURN-AIR CONDITIONER
EXHAUST FAN
A
B
C
HUMID-SEASON OPERATION
DEHUMIDIFICATION SET POINT
130
110
90
70
50
30
10
HUMIDITY RATIO, gr/lb
A. OUTSIDE AIR
Hot and humid in summertime, it must be cooled and dehumidified.
B. VENTILATION DEHUMIDIFIER
Dries incoming air to a condition below the desired humidity set point.
C. DRY VENTILATION AIR
Removes moisture loads generated inside building.

Fig. 16 Predrying Ventilation Air to Dehumidify a Commercial Building
(Harriman et al. 2001)

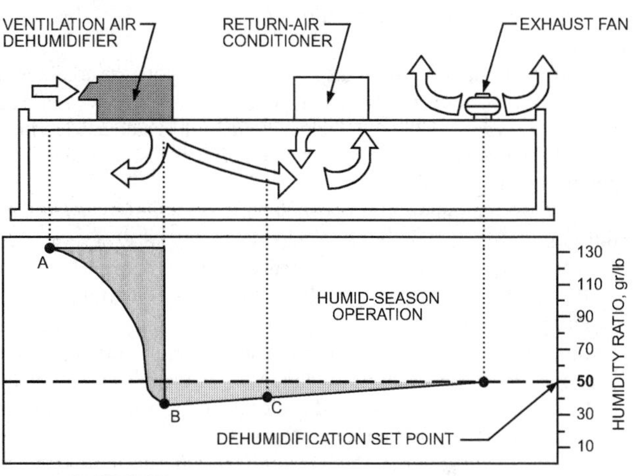

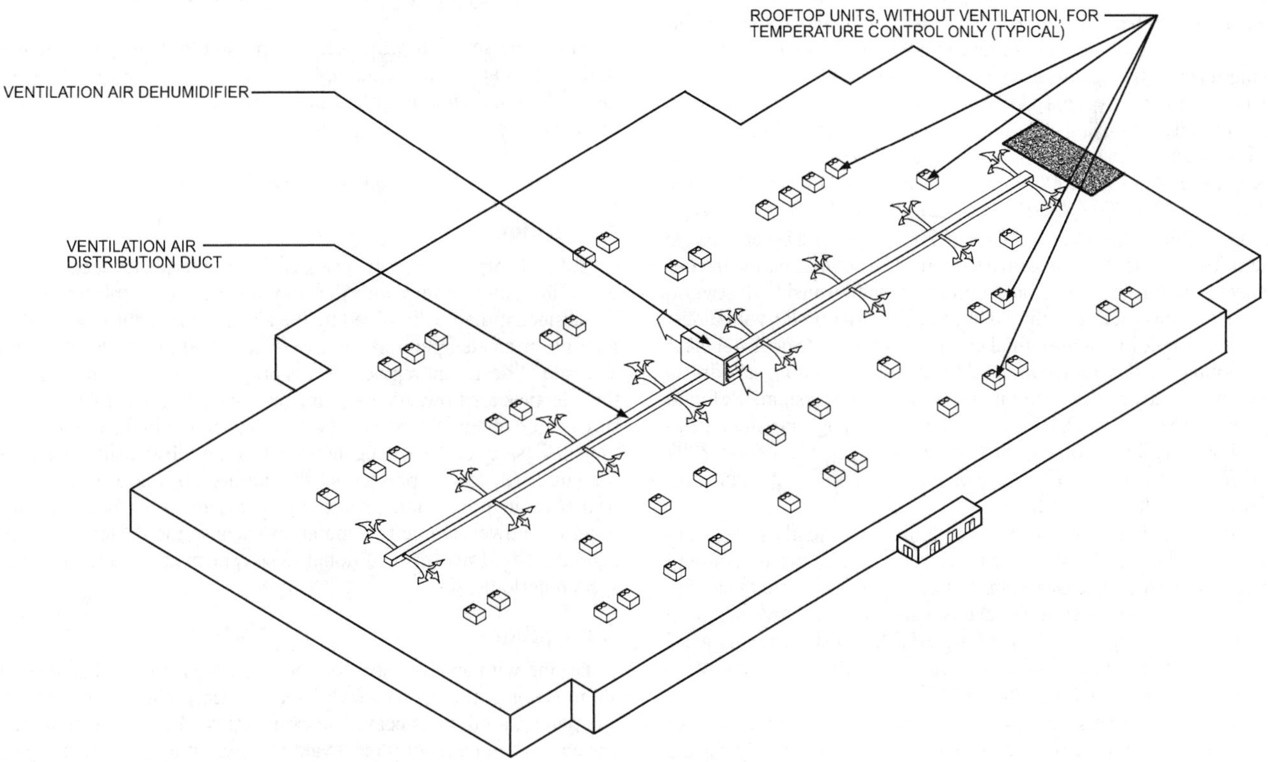

ROOFTOP UNITS, WITHOUT VENTILATION, FOR TEMPERATURE CONTROL ONLY (TYPICAL)
VENTILATION AIR DEHUMIDIFIER
VENTILATION AIR DISTRIBUTION DUCT

Fig. 17 Typical Rooftop Arrangement for Drying Ventilation Air Centrally, Removing Moisture Load from Cooling Units
(Harriman et al. 2001)

Condensation Prevention

Many applications require moisture control to prevent condensation. Airborne moisture condenses on cold cargo in a ship's hold when it reaches a moist climate. Moisture condenses on a ship when the moist air in its cargo hold is cooled by the hull and deck plates as the ship passes from a warm to a cold climate.

A similar problem occurs when aircraft descend from high, cold altitudes into a high dew point at ground level. Desiccant dehumidifiers are used to prevent condensation inside the airframe and avionics that leads to structural corrosion and failure of electronic components.

In pumping stations and sewage lift stations, moisture condenses on piping, especially in the spring when the weather warms and water in the pipes is still cold. Dehumidification is also used to prevent airborne moisture from dripping into oil and gasoline tanks and open fermentation tanks.

Electronic equipment is often cooled by refrigeration, and dehumidifiers are required to prevent internal condensation of moisture. Electronic and instrument compartments in missiles are purged with low-dew-point air before launching to prevent malfunctioning caused by condensation.

Waveguides and radomes are also usually dehumidified, as are telephone exchanges and relay stations. For proper operation of their components, missile and radar sites depend largely on prevention of condensation on interior surfaces.

Dry Air-Conditioning Systems

Cooling-based air-conditioning systems remove moisture from air by condensing it onto cooling coils, producing saturated air at a lower absolute moisture content. In many circumstances, however, there is a benefit to using a desiccant dehumidifier to remove the latent load from the system, avoiding problems caused by condensation, frost, and high relative humidity in air distribution systems.

For example, low-temperature product display cases in supermarkets operate less efficiently when humidity in the store is high because condensate freezes on the cooling coils, increasing operating cost. Desiccant dehumidifiers remove moisture from the air, using rejected heat from refrigeration condensers to reduce the cost of desiccant reactivation. Combining desiccants and conventional cooling can lower installation and operating costs (Calton 1985). For information on the effect of humidity on refrigerated display cases, see Chapter 46 of the 2006 *ASHRAE Handbook—Refrigeration* and Chapter 2 of the 2007 *ASHRAE Handbook—HVAC Applications*.

Air conditioning in hospitals, nursing homes, and other medical facilities is particularly sensitive to biological contamination in condensate drain pans, filters, and porous insulation inside ductwork. These systems often benefit from drying ventilation air with a desiccant dehumidifier before final cooling. Condensate does not form on cooling coils or drain pans, and filters and duct lining stay dry so that mold and mildew cannot grow inside the system. Refer to ASHRAE *Standard* 62.1 for guidance concerning maximum relative humidity in air distribution systems. Chapter 7 of the 2007 *ASHRAE Handbook—HVAC Applications* has information on ventilation of health care facilities.

Hotels and large condominium buildings historically suffer from severe mold and mildew problems caused by excessive moisture in the building structure. Desiccant dehumidifiers are sometimes used to dry ventilation air so it can act as a sponge to remove moisture from walls, ceilings, and furnishings (AHMA 1991). See Chapter 5 of the 2007 *ASHRAE Handbook—HVAC Applications* for more information on ventilating hotels and similar structures.

Like supermarkets, ice rinks have large exposed cold surfaces that condense and freeze moisture in the air, particularly during spring and summer. Desiccant dehumidifiers remove excess humidity from air above the rink surface, preventing fog and improving both the ice surface and operating economics of the refrigeration plant. For recommended temperature and humidity for ice rinks, see Chapter 35 of the 2006 *ASHRAE Handbook—Refrigeration*.

Indoor Air Quality Contaminant Control

Desiccant sorption is not restricted to water vapor. Both liquid and solid desiccants collect both water and large organic molecules at the same time (Hines et al. 1993). As a result, desiccant systems can be used to remove volatile organic compound (VOC) emissions from building air systems.

In addition to preventing growth of mold, mildew, and bacteria by keeping buildings dry, desiccant systems can supplement filters to remove bacteria from the air itself. This is particularly useful for hospitals, medical facilities, and related biomedical manufacturing facilities where airborne microorganisms can cause costly problems. The usefulness of certain liquid and solid desiccants in such systems stems from their ability to either kill microorganisms or avoid sustaining their growth (Battelle 1971; SUNY Buffalo School of Medicine 1988).

Testing

Many test procedures require dehumidification with sorption equipment. Frequently, other means of dehumidification may be used with sorbent units, but the low moisture content required can be obtained only by liquid or solid sorbents. Some typical testing applications are as follows:

- Wind tunnels
- Spectroscopy rooms
- Paper and textile testing
- Bacteriological and plant growth rooms
- Dry boxes
- Environmental rooms and chambers

DESICCANT DRYING AT ELEVATED PRESSURE

The same sorption principles that pertain to atmospheric dehumidification apply to drying high-pressure air and process or other gases. The sorbents described previously can be used with equal effectiveness.

EQUIPMENT

Absorption

Solid absorption systems use a calcium chloride desiccant, generally in a single-tower unit that requires periodic replacement of the desiccant that is dissolved by the absorbed moisture. Normally, inlet air or gas temperature does not exceed 90 to 100°F saturated. The rate of desiccant replacement is proportional to the moisture in the inlet process flow. A dew-point depression of 20 to 40°F at pressure can be obtained when the system is operated in the range of 60 to 100°F saturated entering temperature and 100 psig operating pressure. At lower pressures, the ability to remove moisture decreases in proportion to absolute pressure. Such units do not require a power source for operation because the desiccant is not regenerated. However, additional desiccant must be added to the system periodically.

Adsorption

Drying with an adsorptive desiccant such as silica gel, activated alumina, or a molecular sieve usually incorporates regeneration equipment, so the desiccant can be reactivated and reused. These desiccants can be readily reactivated by heat, purging with dry gas, or both. Depending on the desiccant selected, dew-point performance expected is in the range of –40 to –100°F measured at the operating pressure with inlet conditions of 90 to 100°F saturated and

100 psig. Figure 18 shows typical performance using activated alumina or silica gel desiccant.

Equipment design may vary considerably in detail, but most basic adsorption units use twin-tower construction for continuous operation, with an internal or external heat source, with air or process gas as the reactivation purge for liberating moisture adsorbed previously. A single adsorbent bed may be used for intermittent drying requirements. Adsorption units are generally constructed in the same manner as atmospheric-pressure units, except that the vessels are suitable for the operating pressure. Units have been operated successfully at pressures as high as 6000 psig.

Prior compression or cooling (by water, brine, or refrigeration) to below the dew point of the gas to be dried reduces the total moisture load on the sorbent, permitting the use of smaller drying units. The cost of compression, cooling, or both must be balanced against the cost of a larger adsorption unit.

The many different dryer designs can be grouped into the following basic types:

Heat-reactivated, purge dryers. Normally operating on 4 h (or longer) adsorption periods, these dryers are generally designed with heaters embedded in the desiccant. They use a small portion of dried process gas as a purge to remove the moisture liberated during reactivation heating. (See Figure 19.)

Heatless dryers. These dryers operate on a short adsorption period (usually 60 to 300 s). Depressurizing gas in the desiccant tower lowers the vapor pressure, so adsorbed moisture is liberated from the desiccant and removed by a high purge rate of the dried process gas. Using an ejector reduces the purge gas requirements.

Convection dryers. These dryers usually operate on 4 h (or longer) adsorption periods and are designed with an external heater and cooler as the reactivation system. Some designs circulate reactivation process gas through the system by a blower; others divert some or all of the process gas flow through the reactivation system before adsorption. Both heating and cooling are by convection.

Radiation dryers. Also operating on 4 h (or longer) adsorption periods, radiation dryers are designed with an external heater and blower to force heated atmospheric air through the desiccant tower for reactivation. Desiccant tower cooling is by radiation to atmosphere.

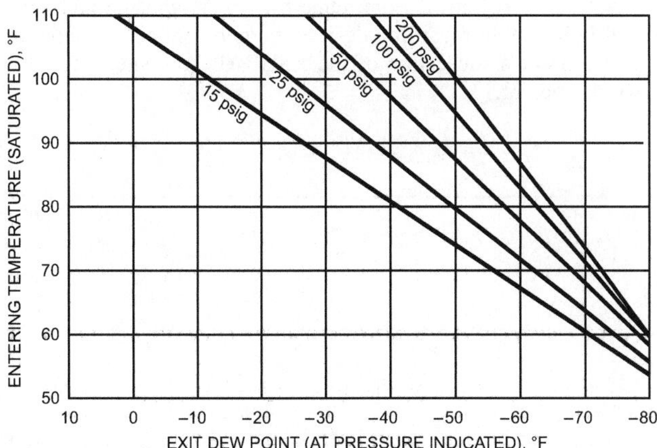

Fig. 18 Typical Performance Data for Solid Desiccant Dryers at Elevated Pressures

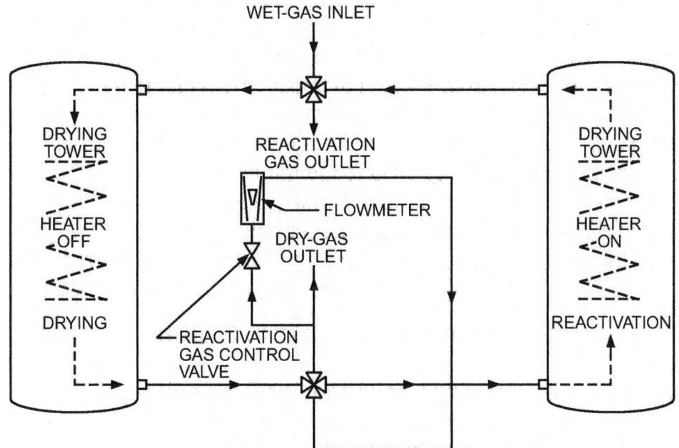

Fig. 19 Typical Adsorption Dryer for Elevated Pressures

APPLICATIONS

Material Preservation

Generally, materials in storage are preserved at atmospheric pressure, but a few materials are stored at elevated pressures, especially when the dried medium is an inert gas. These materials deteriorate when subjected to high relative humidity or oxygen content in the surrounding medium. Drying high-pressure air, subsequently reduced to 3.5 to 10 psig, has been used most effectively in pressurizing coaxial cables to eliminate electrical shorts caused by moisture infiltration. This same principle, at somewhat lower pressures, is also used in waveguides and radomes to prevent moisture film on the envelope.

Process Drying of Air and Other Gases

Drying instrument air to a dew point of –40°F, particularly where air lines are outdoors or exposed to temperatures below the dew point of air leaving the aftercooler, prevents condensation or freeze-up in instrument control lines.

To prevent condensation and freezing, it is necessary to dry plant air used for pneumatically operated valves, tools, and other equipment where piping is exposed to low ambient temperatures. Additionally, dry air prevents rusting of the air lines, which produces abrasive impurities, causing excessive wear on tools.

Industrial gases or fuels such as natural gas are dried. For example, fuels (including natural gas) are cleaned and dried before storage underground to ensure that valves and transmission lines do not freeze from condensed moisture during extraordinarily cold weather, when the gas is most needed. Propane must also be clean and dry to prevent ice accumulation. Other gases, such as bottled oxygen, nitrogen, hydrogen, and acetylene, must have a high degree of dryness. In liquid oxygen and ozone manufacturing, the air supplied to the process must be clean and dry.

Drying air or inert gas for conveying hygroscopic materials in a liquid or solid state ensures continuous, trouble-free plant operation. Normally, gases for this purpose are dried to a –40°F dew point. Purging and blanketing operations in the petrochemical industry depend on using dry inert gas for reducing problems such as explosive hazards and the reaction of chemicals with moisture or oxygen.

Equipment Testing

Dry, high-pressure air is used extensively for testing refrigeration condensing units to ensure tightness of components and to prevent moisture infiltration. Similarly, dry inert gas is used in testing copper tubing and coils to prevent corrosion or oxidation. Manufacture and assembly of solid-state circuits and other electronic components require exclusion of all moisture, and final testing in dry boxes must be carried out in moisture-free atmospheres. Simulation of dry high-altitude atmospheres for testing aircraft and missile components in wind tunnels requires extremely low dew-point conditions.

REFERENCES

AHMA. 1991. *Mold and mildew in hotels and motels*. Executive Engineers Committee Report. American Hotel and Motel Association, Washington, D.C.

Battelle Memorial Institute. 1971. *Project* N-0914-5200-1971. Battelle Memorial Institute, Columbus, OH.

Calton, D.S. 1985. Application of a desiccant cooling system to supermarkets. *ASHRAE Transactions* 91(1B):441-446.

Fischer, J.C. and C.W. Bayer. 2003. Report card on humidity control. *ASHRAE Journal* 45(5):30-39.

Harriman, L.G., III. 2003. 20 years of commercial desiccant systems: Where they've been, where they are now and where they're going. *Heating/Piping/Air Conditioning Engineering* (June & July):43-54.

Harriman, L.G., III, G. Brundrett, and R. Kittler. 2001. *Humidity control design guide for commercial and institutional buildings*. ASHRAE.

Harriman, L.G., III, D. Plager, and D. Kosar. 1997. Dehumidification and cooling loads from ventilation air. *ASHRAE Journal* 39(11):37-45.

Hines, A.L., T.K. Ghosh, S.K. Loyalka, and R.C. Warder, Jr. 1993. *Investigation of co-sorption of gases and vapors as a means to enhance indoor air quality*. Gas Research Institute, Chicago. Available from the National Technical Information Service, Springfield, VA. Order PB95-104675.

Kosar, D.R., M.J. Witte, D.B. Shirey, and R.L. Hedrick. 1998. Dehumidification issues of *Standard* 62-1989. *ASHRAE Journal* 40(5):71-75.

Spears, J.W. and J.J. Judge. 1997. Gas-fired desiccant system for retail superstore. *ASHRAE Journal* 39(10):65-69.

SUNY Buffalo School of Medicine. 1988. Effects of glycol solutions on microbiological growth. Niagara Blower *Report* 03188.

Worek, W. and W. Zheng. 1991. *UIC IMPLICIT rotary desiccant dehumidifier finite difference program*. University of Illinois at Chicago, Department of Mechanical Engineering.

BIBLIOGRAPHY

ASHRAE. 2006. Standard method for measurement of moist air properties. ANSI/ASHRAE *Standard* 41.6-1994 (RA 2006).

ASHRAE. 2007. Ventilation for acceptable indoor air quality. ANSI/ASHRAE *Standard* 62.1-2007.

ASHRAE. 1992. *Desiccant cooling and dehumidification*, L. Harriman, ed.

Bradley, T.J. 1994. Operating an ice rink year-round by using a desiccant dehumidifier to remove humidity. *ASHRAE Transactions* 100(1): 116-131.

Collier, R.K. 1989. Desiccant properties and their effect on cooling system performance. *ASHRAE Transactions* 95(1):823-827.

Harriman, L.G., III. 1990. *The dehumidification handbook*. Munters Cargocaire, Amesbury, MA.

Harriman, L.G., III. 1996. *Applications engineering manual for desiccant systems*. American Gas Cooling Center, Arlington, VA.

Harriman, L.G., III and J. Judge. 2002. Dehumidification equipment advances. *ASHRAE Journal* 44(8):22-29

Jones, B.W., B.T. Beck, and J.P. Steele. 1983. Latent loads in low humidity rooms due to moisture. *ASHRAE Transactions* 89(1A):35-55.

Lowenstein, A.I. and R.S. Gabruk. 1992. The effect of absorber design on the performance of a liquid-desiccant air conditioner. *ASHRAE Transactions* 98(1):712-720.

Lowenstein, A.I. and R.S. Gabruk. 1992. The effect of regenerator performance on a liquid-desiccant air conditioner. *ASHRAE Transactions* 98(1):704-711.

Meckler, M. 1994. Desiccant-assisted air conditioner improves IAQ and comfort. *Heating/Piping/Air Conditioning Engineering* 66(10):75-84.

Pesaran, A. and T. Penney. 1991. Impact of desiccant degradation on cooling system performance. *ASHRAE Transactions* 97(1):595-601.

Vineyard, E.A., J.R. Sand, and D.J. Durfee. 2000. Parametric analysis of variables that affect the performance of a desiccant dehumidification system. *ASHRAE Transactions* 106(1):87-94.

ADDITIONAL INFORMATION

ASHRAE Technical Committee 8.12 posts updated and additional information regarding desiccant equipment and systems on the committee's subsection of the ASHRAE Web site, located at www.ashrae.org.

CHAPTER 24

MECHANICAL DEHUMIDIFIERS AND RELATED COMPONENTS

THE correct moisture level in the air is important for health and comfort. Controlling humidity and condensation is important to prevent moisture damage and mold or mildew development, thus protecting buildings and occupants, and preserving building contents. This chapter covers mechanical dehumidification using a cooling process only, including basic dehumidifier models (with moisture removal capacity of less than 3 lb/h) used for home basements and small storage areas, as well as larger sizes required for commercial applications. Other methods of dehumidification are covered in Chapter 23.

Commercial applications for mechanical dehumidifiers include the following:

- Indoor swimming pools
- Makeup air treatment
- Ice rinks
- Dry storage
- Schools
- Hospitals
- Office buildings
- Museums, libraries, and archives
- Restaurants
- Hotels and motels
- Assisted living facilities
- Supermarkets
- Manufacturing plants and processes

In addition, an air-to-air heat exchanger (such as a heat pipe, coil runaround loop, fixed-plate heat exchanger, or rotary heat exchanger) may be used to enhance moisture removal by a mechanical dehumidifier or air conditioner. The section on Wraparound Heat Exchangers discusses how dehumidification processes can be improved by using such a device. Other uses of air-to-air heat exchangers are covered in Chapter 25.

MECHANICAL DEHUMIDIFIERS

Mechanical dehumidifiers remove moisture by passing air over a surface that has been cooled below the air's dew point. This cold surface may be the exterior of a chilled-water coil or a direct-expansion refrigerant coil. To prevent overcooling the space (and avoid the need to add heat energy from another source), a mechanical dehumidifier also usually has means to reheat the air, normally using recovered and recycled energy (e.g., recovering heat from hot refrigerant vapor in the refrigeration circuit). Using external energy input for reheat is wasteful and is prohibited or limited in many countries (see ASHRAE *Standard* 90.1).

A mechanical dehumidifier differs from a typical off-the-shelf air conditioner in that the dehumidifier usually has a much lower sensible heat ratio (SHR). The dehumidifier starts the compressor on a call for dehumidification, whereas an air conditioner starts the compressor on a call for sensible cooling. Typically, a room dehumidifier has an SHR of 0.6 or less, compared to a standard air-conditioning system of 0.8 SHR. Dehumidifiers must also allow condensation from the cooling coil to drain easily from the coils. They may need air velocities over the cooling coil lower than those for a typical air conditioner, to improve moisture runoff and minimize carryover of condensed moisture.

In addition, the need to introduce code-mandated ventilation air may require that outside air be treated to avoid introducing excessive moisture. Basic strategies include precooling outside air entering the air-conditioning evaporator coil, or providing a separate system to provide properly conditioned outside air. For some low-dew-point (below 45°F) applications, mechanical dehumidification may be used as the first stage, with desiccant dehumidification for the final stage to maximize efficiency and minimize installed cost.

Although the main purpose of a mechanical dehumidifier is to remove moisture from the air, many features can be incorporated for various applications, such as

- Dehumidifying and cooling (no reheat)
- Dehumidifying with partial reheat (leaving dry-bulb temperature is cooler than with a dehumidifier with full reheat)
- Dehumidifying with full reheat
- Dehumidifying with heat recovery to various heat sinks
- Dehumidification capacity modulation
- Reheat capacity modulation
- Ventilation air introduction
- Auxiliary space or water heating

Often, mechanical dehumidifiers can be incorporated in a system to use waste heat from mechanical cooling (e.g., heat rejection to a swimming pool, whirlpool, domestic hot water, heat pump loop, chilled-water loop, or remote air-cooled condenser).

Outdoor dehumidifiers should be protected against internal moisture condensation when winter conditions are severe, because of the higher dew-point temperature of air circulating in the unit.

Psychrometrics of Dehumidification

Air enters the dehumidifying coil at point A (Figures 1 and 2). The dehumidifying coil removes sensible heat (SH) and latent heat (LH) from the airstream. The dehumidified, cooled air leaves the coil at its saturation temperature at point B. The total heat removed (TH) is the net cooling capacity of the system.

In reheating, the hot gas rejects heat from three sources. First, sensible heat absorbed in the air-cooling process is rejected to air leaving the cooling coil. This air is at point C, which is the same dry-bulb temperature as the entering air minus the moisture content. Second, the latent heat removal that causes the moisture to condense also adds heat to the hot refrigerant gas. This heat is also rejected into the airstream, raising the air temperature to point D. Third, nearly all electric power required to drive the refrigeration cycle is converted to heat. This portion of heat rejection raises the air leaving temperature to point E.

The preparation of this chapter is assigned to TC 8.10, Mechanical Dehumidification Equipment and Heat Pipes.

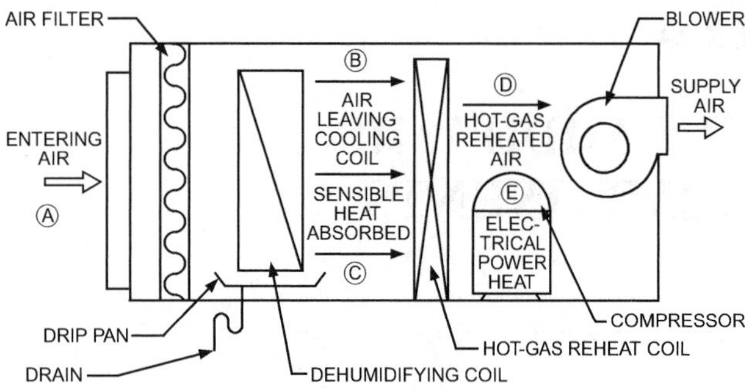

Fig. 1 Dehumidification Process Points

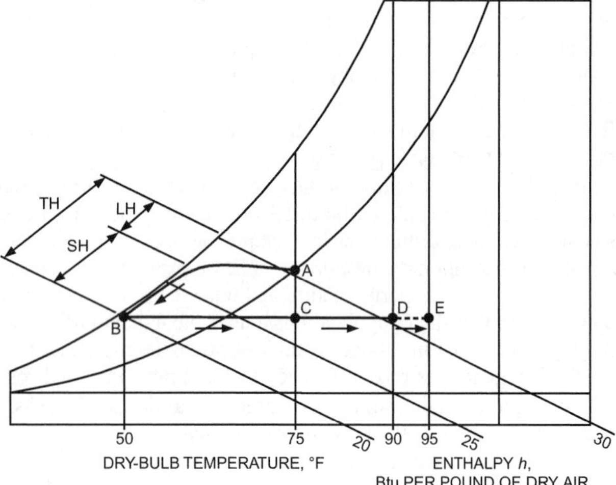

Fig. 2 Psychrometric Diagram of Typical Dehumidification Process

This process assumes that all heat is rejected by the reheat coil. Depending on the refrigerant system complexity, any part of the total heat rejection can be diverted to other heat exchangers (condensers/desuperheaters).

Dehumidifier supply air temperatures can be controlled between 50 and 95°F. However, system design should not rely on a mechanical dehumidifier as a dependable heat source for space heating, because heat is only available when the unit is operating.

Domestic Dehumidifiers

Domestic dehumidifiers are smaller (usually less than 1 ton), simpler versions of commercial dehumidifiers. They are self-contained and easily movable.

As shown in Figure 3, a single fan draws humid room air through the cold coil, removing moisture that either drains into the water receptacle or passes through the cabinet into some other means of disposal. The cooled air passes through the condenser, reheating the air.

Domestic dehumidifiers ordinarily maintain satisfactory humidity levels in an enclosed space when the airflow rate and unit placement move the entire air volume of the space through the dehumidifier once an hour.

Design and Construction. Domestic dehumidifiers use hermetic motor-compressors; the refrigerant condenser is usually conventional finned tube. Refrigerant flow is usually controlled by a capillary tube, although some high-capacity dehumidifiers use an expansion valve. A propeller fan moves air through the unit at typical airflows of 125 to 250 cfm.

The refrigerated surface (evaporator) is usually a bare-tube coil, although finned-tube coils can be used if they are spaced to allow rapid runoff of water droplets. Vertically disposed bare-tube coils

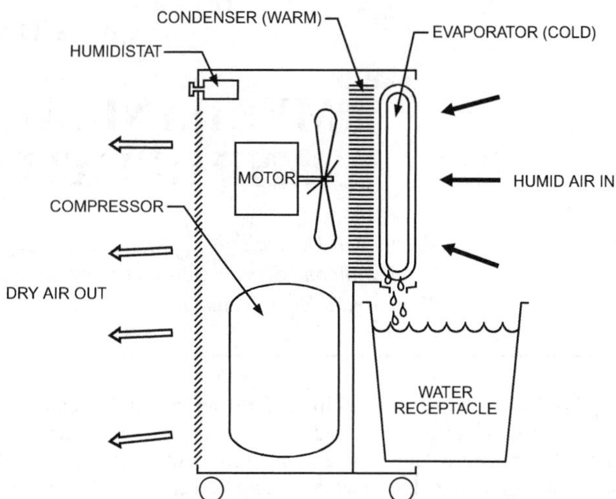

Fig. 3 Typical Domestic Dehumidifier

tend to collect smaller drops of water, promote quicker runoff, and cause less water reevaporation than finned-tube or horizontally arranged bare-tube coils. Continuous bare-tube coils, wound in a flat circular spiral (sometimes with two coil layers) and mounted with the flat dimension of the coil in the vertical plane, are a good design compromise because they have most of the advantages of the vertical bare-tube coil.

Evaporators are protected against corrosion by finishes such as waxing, painting, or anodizing (on aluminum). Waxing reduces the wetting effect that promotes condensate formation; however, tests on waxed versus nonwaxed evaporator surfaces show negligible loss of capacity. Thin paint films do not have an appreciable effect on capacity.

Removable water receptacles, provided with most dehumidifiers, hold 16 to 24 pints and are usually made of plastic to withstand corrosion. Easy removal and handling without spillage are important. Most dehumidifiers also provide either a means of attaching a flexible hose to the water receptacle or a fitting provided specially for that purpose, allowing direct gravity drainage to another means of disposal external to the cabinet.

More expensive dehumidifiers usually have higher output capacities, are more attractively styled, and have various auxiliary features. An adjustable humidistat (30 to 80%) automatically cycles the unit to maintain a preselected relative humidity. The humidistat may also provide a detent setting for continuous operation. Some models also include a sensing and switching device that automatically turns the unit off when the water receptacle is full.

Dehumidifiers are designed to provide optimum performance at standard rating conditions of 80°F db room temperature and 60% rh. When the room is less than 65°F db and 60% rh, the evaporator may freeze. This effect is especially noticeable on units with a capillary tube.

Some dehumidifiers are equipped with defrost controls that cycle the compressor off under frosting conditions. This control is generally a bimetal thermostat attached to the evaporator tubing, allowing dehumidification to continue at a reduced rate when frosting conditions exist. The humidistat can sometimes be adjusted to a higher relative humidity setting, which reduces the number and duration of running cycles and allows satisfactory operation at low-load conditions. Often, especially in the late fall and early spring, supplemental heat must be provided from other sources to maintain above-frosting temperatures in the space.

Capacity and Performance Rating. Domestic dehumidifiers are available with moisture removal capacities of 11 to 60 pints per 24 h, and are operable from ordinary household electrical outlets

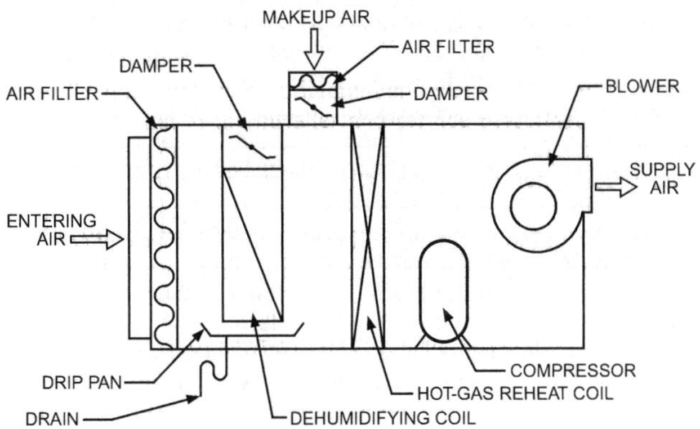

Fig. 4 Typical General-Purpose Dehumidifier

(115 or 230 V, single-phase, 60 Hz). Input varies from 200 to 800 W, depending on the output capacity rating.

AHAM *Standard* DH-1 establishes a uniform procedure for determining the rated capacity of dehumidifiers under specified test conditions and also establishes other recommended performance characteristics. An industry certification program sponsored by AHAM covers the great majority of domestic dehumidifiers and certifies dehumidification capacity.

The U.S. Environmental Protection Agency (EPA) qualifies dehumidifiers to carry its ENERGY STAR® label if they remove the same amount of moisture as similarly sized standard units, but use at least 10% less energy. The EPA's ENERGY STAR web site provides additional information on qualifying products (EPA 2008).

Codes. Domestic dehumidifiers are designed to meet the safety requirements of UL *Standard* 474, *Canadian Electrical Code*, and ASHRAE *Standard* 15. UL-listed and CSA-approved equipment have a label or data plate indicating approval. UL also publishes the *Electrical Appliance and Utilization Equipment Directory*, which covers this type of appliance.

General-Purpose Dehumidifiers

Basic components of general-purpose dehumidifiers are shown in Figure 4. An air filter is required to protect the evaporator. Dehumidifying coils, because of their depth and thoroughly wetted surfaces, are excellent dust collectors and not as easily cleanable as much thinner air-conditioning evaporator coils. However, the large amount of condensate has a self-cleaning effect. The bypass damper above the evaporator coil allows airflow adjustments for the evaporator without decreasing airflow for the reheat coil. Dehumidifying and reheat coils perform best at different airflows.

The compressor may be isolated from the airstream or located in it. Locating the compressor in the airstream may make service more difficult, but this arrangement allows heat lost through the compressor casing to be provided to the conditioned space while reducing the size of the enclosure. During the cooling season, this compressor location reduces the unit's sensible cooling capacity.

Code-required ventilation air may be introduced between the evaporator and reheat coil. The amount of makeup air should be controlled to not adversely affect the refrigeration system's operation. Preheating ventilation air may be required in colder climates.

Computerized controls can sense return air temperature and relative humidity. Remote wall-mounted sensors are also available. More sophisticated controls are desirable to regulate dew-point temperature and maintain the desired relative humidity in the space.

General Considerations. Before considering installation of any type of dehumidification equipment, all latent loads should be identified and quantified. In many cases, this might lead to decisions that reduce the latent load. For example, a storage facility that does not have an adequate vapor retarder in the building envelope should be retrofitted first before attempting to calculate the amount of moisture migration through the structure. The same approach should be taken to reduce the amount of uncontrolled air infiltration.

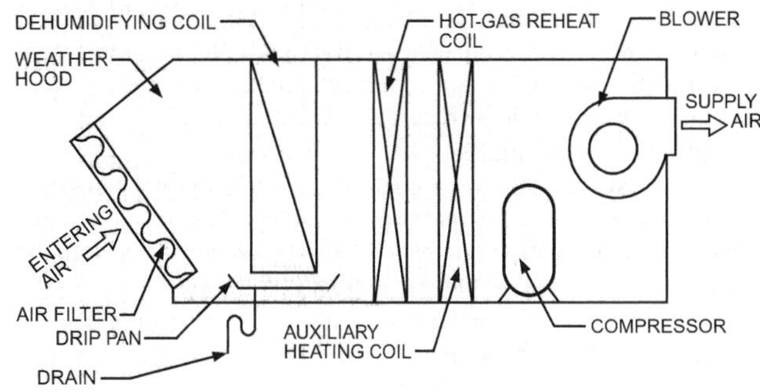

Fig. 5 Typical Makeup Air Dehumidifier

Consider covering large water surfaces, such as vats, and/or providing a local exhaust hood to evacuate concentrated water vapor from where it is generated. Although these corrections seem to add cost to a project, the resulting reduced size of the dehumidifier and its lower operating cost often result in an attractive financial payback.

Other special considerations include the following:

- **High volumes of makeup air.** A project may start out to be suitable for a general-purpose dehumidifier. However, once makeup air requirements are quantified to compensate for exhaust and to pressurize the facility, a general-purpose dehumidifier may no longer be applicable. The maximum acceptable portion of outdoor air for general-purpose dehumidifiers is limited, and depends on climatic conditions and the desired indoor conditions to be maintained. As a general rule, when outdoor air requirements exceed 20% of the dehumidifier's total airflow, the manufacturer should be consulted to determine whether the equipment is suitable for the application. In many cases, a makeup air dehumidifier should be considered instead.
- **Low-temperature applications.** Many storage facilities require humidity control, but do not have specific temperature control requirements. When the space temperature is allowed to drop below 65°F, consult the manufacturer of the dehumidifier to determine whether the equipment is suitable for the application. Some dehumidifiers, such as those used for indoor ice rinks, can operate at ambient conditions as low as 50°F. Lower air temperatures are likely to require a desiccant dehumidifier.

Makeup Air Dehumidifiers

A makeup air dehumidifier [or **dedicated outdoor air system (DOAS)**] is used to separately condition all outdoor air brought into the building for ventilation or to replace air that is being exhausted. (As such, a makeup air dehumidifier should be selected based on its latent dehumidification capacity, not necessarily on its total air-conditioning capacity.) This conditioned outdoor air is then delivered either directly to each occupied space, to small HVAC units located in or near each space, or to central air handlers serving those spaces. Meanwhile, the local or central HVAC equipment is used to maintain space temperature. Treating the outdoor air separately from recirculated return air makes it easy to verify sufficient ventilation airflow and enables enforcement of a maximum humidity limit in the occupied spaces.

Makeup air dehumidifiers may require simultaneous heat rejection to the reheat coil and/or another condenser (air- or water-cooled), because it may not always be possible to reject the total heat of rejection from the dehumidifying coil to the makeup air. A rainproof air intake and cooling capacity modulation (or staging) are important. With constantly changing weather conditions, even throughout the day, compressor capacity must be adjusted to prevent coil freeze-ups. Basic components are shown in Figure 5.

Auxiliary heating may be required for year-round operation in some climates. Water- and steam-heating coils should have freeze protection features. When using indirect-fired gas heaters, the combustion chamber should be resistant to condensation.

Makeup air dehumidifiers may be interfaced with a building automation system (BAS) to control the unit's on/off status and operating mode, because most spaces do not require continuous makeup air. Air exhaust systems must also be synchronized with the makeup air equipment to maintain proper building pressurization.

Combining exhaust air with makeup air systems provides the opportunity to transfer energy between the two airstreams. A typical arrangement is shown in Figure 6.

Types and functions of air-to-air energy recovery devices are covered in Chapter 25. During summer, entering outdoor air is precooled and possibly partially dehumidified to lower the enthalpy of air entering the cooling coil. Conversely, during winter operation, entering outdoor air is preheated and possibly partially humidified. When using air-to-air energy recovery, the required compressor capacity of the makeup air dehumidifier may be significantly reduced. Using a proper damper system, a makeup air dehumidifier may also be able to treat recirculated air, which can allow for humidity control during unoccupied periods.

Efficient use of makeup air dehumidifiers may improve the overall efficiency of a building air-conditioning system.

General Considerations. The sophistication of a makeup air dehumidifier varies greatly with its application. Typically, it requires some type of capacity modulation for dehumidification as well as the heating mode (if so equipped). On/off cycles are normally not acceptable when supplying outdoor air to a conditioned space. Other considerations include the following:

- **Deliver conditioned outdoor air dry.** Regardless of where conditioned outdoor air is delivered, the makeup air dehumidifier should dehumidify the outdoor air so that it is drier than the required space dew point. If the dew-point temperature of the conditioned outdoor air is lower than the dew point in the space, it can also offset some or all of the space latent loads (Morris 2003). This adequately limits indoor humidity at both full and part load without the need for additional dehumidification enhancements in the local HVAC units. The local units only need to offset the space sensible cooling loads. To prevent warm discomfort, Nevins et al. (1975) recommended that, on the warm side of the comfort zone, relative humidity should not exceed 60%. At 77°F, this results in a dew-point temperature of 62°F. Therefore, this type of dehumidifier typically requires only a basic step control to prevent evaporator coil freezing at low loads.

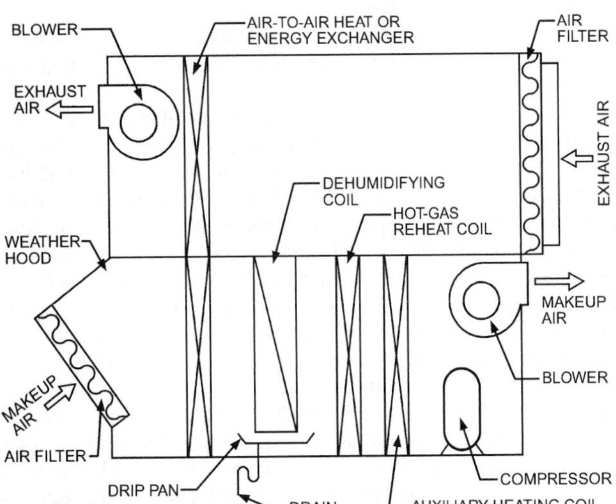

Fig. 6 Typical Makeup Air Dehumidifier with Exhaust Air Heat/Energy Recovery

- **Neutral versus cold leaving air temperature.** Many dedicated outdoor-air systems are designed to dehumidify outdoor air so it is drier than the space, and then reheat it to approximately space temperature (neutral). (Various methods of heat recovery can be used to increase the efficiency of this process by recovering heat for reheat.) This can simplify control of the local HVAC units, eliminate the concern about overcooling the space, and avoid condensation-related problems when conditioned air is delivered to an open ceiling plenum.

  However, when a cooling coil is used to dehumidify outdoor air, the dry-bulb temperature of air leaving the coil is colder than the space. In some applications and under some operating conditions, this air can be delivered at a dry-bulb temperature cooler than the space (not reheated to neutral), thus offsetting part of the sensible cooling load in the space. This means less cooling capacity is required from the local HVAC equipment than if the air is delivered at a neutral temperature. A control sequence can be used to reset the leaving air dry-bulb temperature, and activate reheat when necessary, to avoid overcooling the space (Murphy 2006) while still dehumidifying the outdoor air to the required dew point.
- **Exhaust air heat recovery.** Because all outdoor air is brought in at a central location, consider including an air-to-air heat exchanger to precondition the outdoor air (Figure 6). During summer operation, outdoor air is cooled and possibly dehumidified. During winter, outdoor air is heated and possible humidified. This reduces operating costs and may allow downsizing of the mechanical cooling, dehumidification, heating, and humidification equipment. Types and functions of air-to-air energy recovery devices are covered in Chapter 25.
- **Makeup air for processes.** Commercial processes may require a large amount of makeup air (because a large amount of air is exhausted) and precise dew-point temperature control. As an additional challenge, airflow may be variable. This type of application may require a near-proportional capacity reduction control using several stages of compressor capacity and modulating refrigeration controls.

Indoor Swimming Pool Dehumidifiers

Indoor pools (natatoriums) are an efficient application for mechanical dehumidifiers. Humidity control is required 24 h a day, year-round. Dehumidifiers are available as single- and double-blower units (see Figures 7 to 10).

The latent heat (LH) from dehumidification (see Figure 2) comes nearly exclusively from pool water (excluding humidity from makeup air and latent heat from large spectator areas). Loss of evaporation heat cools the pool water. By returning evaporation heat losses to the pool water, the sensible heat between points C and D of Figure 2 is not rejected into the supply air, which can reduce supply air temperature by approximately 15°F.

Methods for classifying and rating performance of single-blower pool dehumidifiers are published in ARI *Standard* 910. The ARI rating configuration is for a dehumidifier operating in recirculated air mode, with no ventilation air being introduced to the system. Caution is suggested when reviewing the performance of a dehumidifier and at what mode of operation it is being rated. Introducing ventilation air to a dehumidifier and adding exhaust fans for energy recovery between the two airstreams can affect the dehumidifier's performance and ratings. The sensible and latent capacity of the dehumidifier may change from the actual listed performance. All unit performance ratings must be reviewed during each mode of operation against the building load. Where ventilation air is introduced to the system, as well as the location of the exhaust air, may affect dehumidifier performance differently, depending on whether the space is being cooled or heated. The dehumidifier and associated equipment must be selected based on space heating and cooling loads, including ventilation loads, so that both sensible and latent loads are satisfied to the extent possible.

Single-blower pool dehumidifiers (Figure 7) are similar to general-purpose dehumidifiers (see Figure 4), with the following exceptions:

- One or several water heaters may be installed to add recovered heat from the refrigeration circuit into the pool water. Heaters can provide full pool-water heating for several pools maintained at different temperatures.
- All components must be corrosion-resistant to chloramine-laden air.
- The pool water heater must be resistant to chlorinated water.
- Cross-contamination prevention features are not required but should be considered. Pool water must be kept sanitary at all times. Accidental contamination with refrigerant oil should be prevented.

Figure 8 shows a double-blower pool dehumidifier with economizer dampers and a full-sized return fan located upstream of the evaporator coil. This configuration can provide up to 100% makeup air to maintain humidity levels, which can be attractive when the outdoor dew point is below the required indoor dew point during mild weather, or in climates when enough hours of dry- and wet-bulb conditions are below the level to be maintained. Preheating outdoor air may be required to prevent condensation inside the mixing box. Also, this configuration does not recover energy from the warm, moist exhaust air. During dehumidifying coil operation, the amount of makeup and exhaust air is limited by outdoor condition, especially during the heating season. Cold makeup air may lower the mixed-air temperature to the point where the dehumidifying coil cannot extract any moisture.

In some regions, it is economically attractive to remove moisture from the exhaust air to recover its latent heat. In this case, the dehumidifying coil is installed in the return air section. Figure 9 shows a double-blower pool dehumidifier with economizer dampers and return fan located downstream of the evaporator coil. A damper system can also be incorporated to exhaust before the evaporator coil during colder conditions and after the evaporator during warmer conditions. This configuration recovers energy from the warm, moist exhaust air; however, exhausting air from downstream of the evaporator coil also reduces the unit's sensible capacity by the amount of the exhaust air. The ratio of return air to exhaust air must be considered to determine the unit's capacity to remove moisture from the conditioned space.

Figure 10 shows a different unit configuration that addresses concerns related to blower energy use during the various operating modes. This unit can operate with the supply blower only, or with the addition of one or two exhaust blowers.

Some manufacturers also offer air-to-air heat recovery between the exhaust and makeup airstreams (see Chapter 25). During cold weather, this arrangement preheats entering makeup air with heat recovered from the exhaust airstream. Latent heat recovery may not be practical, however, because it transfers moisture to the entering air, thus possibly increasing dehumidification requirements.

Control systems should be compatible with building automation systems; however, the BAS must not disable dehumidifier operation because indoor pools always need some dehumidification, regardless of occupancy, and require specialized control sequences.

General Considerations. The primary function of an indoor swimming pool dehumidifier is to lower the space dew-point temperature to an acceptable level and to provide adequate air circulation to comply with minimum air change rates.

For more information on indoor swimming pool (natatorium) applications, see Chapter 4 of the 2007 *ASHRAE Handbook—HVAC Applications*.

Types of Equipment. Indoor swimming pool dehumidifiers are available in single- and double-blower configurations (see Figures 7 to 10). Heat from the refrigeration circuit can be (1) used to reheat supply air, (2) used to heat pool water, or (3) rejected to the outdoors by an optional air- or water-cooled condenser. Equipment configurations are available to use the heat for any combination of these three purposes.

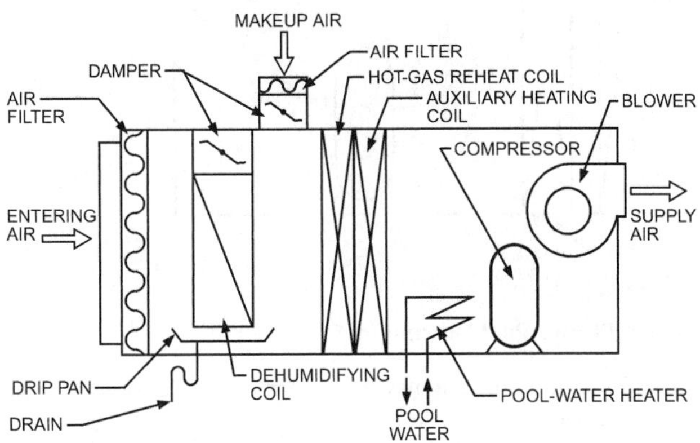

Fig. 7 Typical Single-Blower Pool Dehumidifier

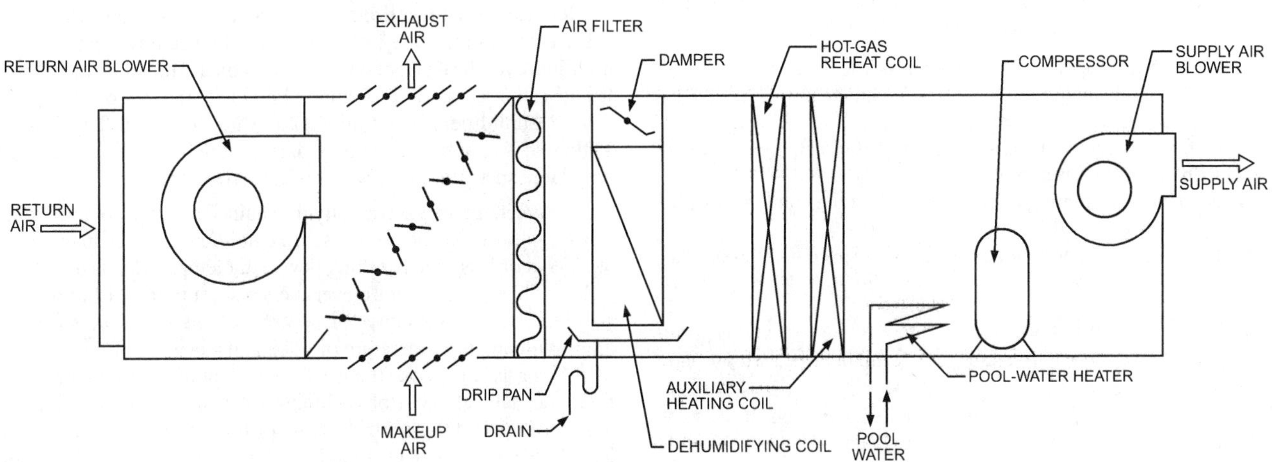

Fig. 8 Typical Double-Blower Pool Dehumidifier with DX Coil in Supply Air Section

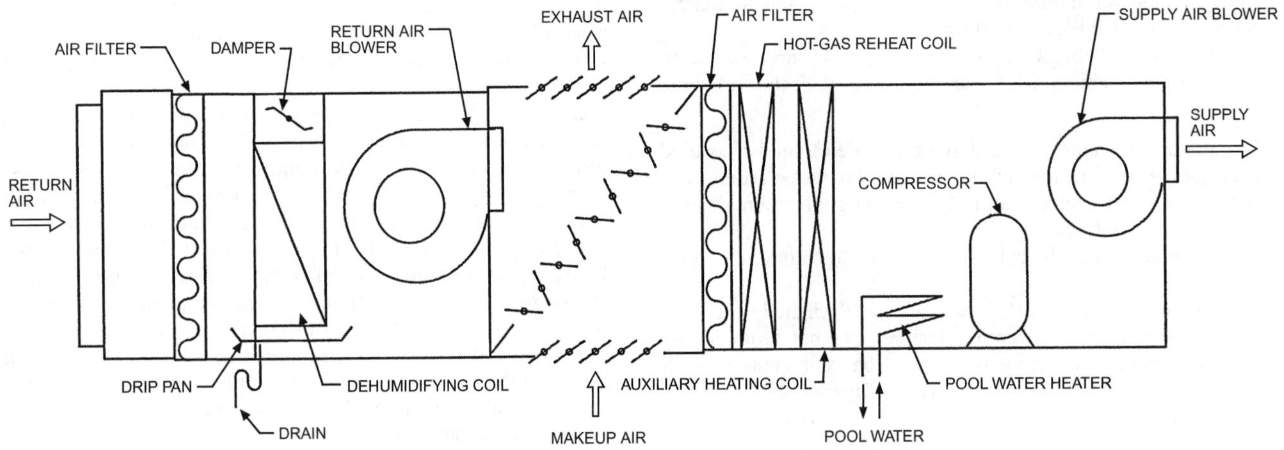

Fig. 9 Typical Double-Blower Pool Dehumidifier with DX Coil in Return Air Section

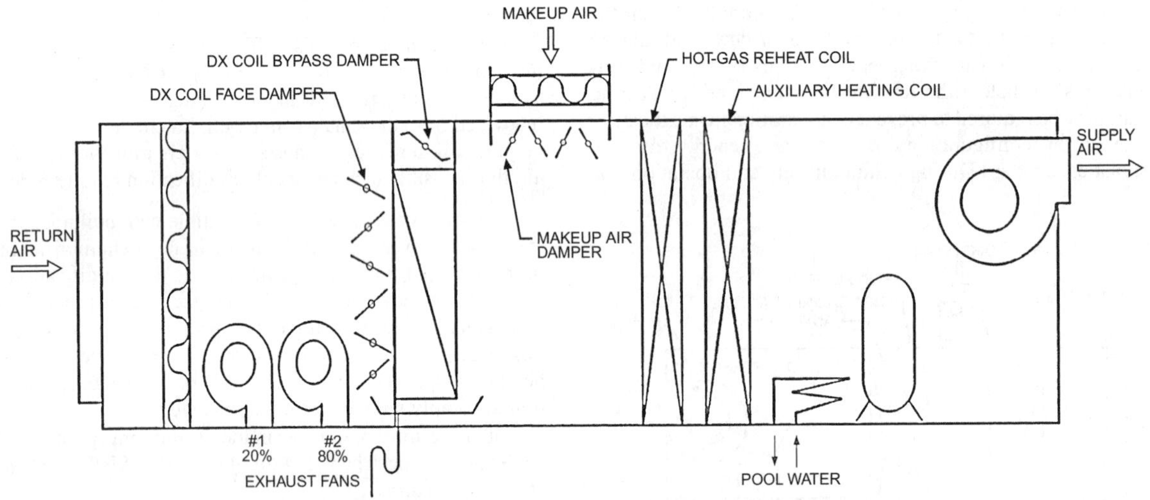

Fig. 10 Supply Blower and Double Exhaust Blower Pool Dehumidifier

The equipment can be located indoors or outdoors, and may be manufactured as a single package or as a split system with a remote condenser. Indoor, air-cooled condensers are typically equipped with a blower-type fan suitable for duct connections. An optional economizer allows for the introduction of up to 100% of outdoor air (turning off the compressors) when conditions are appropriate.

When selecting a dehumidifier for an indoor swimming pool application, several questions need to be addressed:

- In what mode of operation is the dehumidifier rated?
- Does the rating include ventilation air, and what effect, if any, does it have on dehumidifier performance?
- Does the unit include exhaust air, and what effect, if any, does it have on dehumidifier performance?
- Does the cost of running a second fan offset the energy saved by the economizer?
- Will the dehumidifier maintain the desired space conditions during all modes of operation?

Ice Rink Dehumidifiers

Design for ice rink dehumidifiers is similar to that of general-purpose dehumidifiers (see Figure 4). However, because of the lower temperatures, airflow and dehumidifying and reheat coils are selected in accordance with the following conditions:

- The dehumidifying coil may or may not have an air bypass, depending on the location of makeup air intake and/or coil selection.
- The dehumidifying coil may have means to defrost or to prevent frost formation.
- Makeup air treatment is limited.

For large spectator areas, special makeup air dehumidifiers may be required.

General Considerations. For community ice rinks with small spectator areas (or none), it is customary to install two small dehumidifiers over the dasher boards in a diagonal arrangement, 12 to 15 ft above the ice surface (Figure 11). Take care that discharge air from dehumidifiers is not directed toward the ice surface. Forced airflow at any temperature may damage the ice surface. Ice rinks with large spectator areas have different requirements.

The spectator area is typically maintained at 70°F. To limit moisture migration to the ice sheet, space conditions must be maintained at 50% rh or less. The resulting dew-point temperature is then 50°F or less. The air temperature over the ice sheet in the dasher boards, however, is approximately 5°F lower than the air in the spectator area. With an air temperature of 65°F and a dew-point temperature of 50°F or less, fog over the ice sheet cannot develop. As a general rule, mechanical ice rink dehumidifiers are most effective for condensation and fog control when dry-bulb space temperature is at least 15°F above the dew point. For additional fog and condensation prevention methods, see Chapter 35 of the 2006 *ASHRAE Handbook—Refrigeration*.

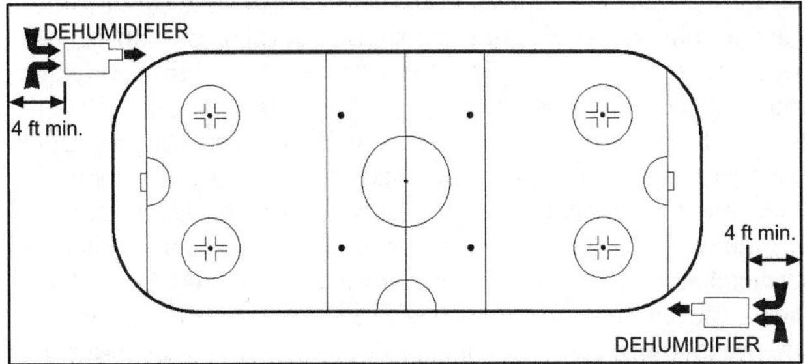

Fig. 11 Typical Installation of Ice Rink Dehumidifiers

INSTALLATION AND SERVICE CONSIDERATIONS

Equipment must be installed properly so that it functions in accordance with the manufacturer's specifications. Interconnecting diagrams for the low-voltage control system should be documented for proper future servicing. Planning is important for installing large, roof-mounted equipment because special rigging is frequently required.

The refrigerant circuit must be clean, dry, and leak-free. An advantage of packaged equipment is that proper installation minimizes the risk of field contamination of the circuit. Take care to properly install split-system interconnecting tubing (e.g., proper cleanliness, brazing, evacuation to remove moisture). Split systems should be charged according to the manufacturer's instructions.

Equipment must be located to avoid noise and vibration problems. Single-package equipment of over 20 tons in capacity should be mounted on concrete pads if vibration control is a concern. Large-capacity equipment should be roof-mounted only after the roof's structural adequacy has been evaluated. Additional installation guidelines include the following:

- In general, install products containing compressors on solid, level surfaces.
- Avoid mounting products containing compressors (e.g., remote units) on or touching the foundation of a building. A separate pad that does not touch the foundation is recommended to reduce noise and vibration transmission through the slab.
- Do not box in outdoor air-cooled units with fences, walls, overhangs, or bushes. Doing so reduces the unit's air-moving ability, reducing efficiency.
- For a split-system remote unit, choose an installation site that is close to the indoor part of the system to minimize refrigerant charge and pressure drop in the connecting refrigerant tubing.
- Contact the equipment manufacturer or consult the installation instructions for further information on installation procedures.

Equipment should be listed or certified by nationally recognized testing laboratories to ensure safe operation and compliance with government and utility regulations. Equipment should also be installed to comply with agency standards' rating and application requirements to ensure that it performs according to industry criteria. Larger and more specialized equipment often does not carry agency labeling. However, power and control wiring practices should comply with the *National Electrical Code*® (NFPA *Standard* 70). Consult local codes before design, and consult local inspectors before installation.

A clear, accurate wiring diagram and well-written service manual are essential to the installer and service personnel. Easy, safe service access must be provided for cleaning, lubrication, and periodic maintenance of filters and belts. In addition, access for replacement of major components must be provided and preserved.

Service personnel must be qualified to repair or replace mechanical and electrical components and to recover and properly recycle or dispose of any refrigerant removed from a system. They must also understand the importance of controlling moisture and other contaminants in the refrigerant circuit; they should know how to clean a hermetic system if it has been opened for service (see Chapter 6 of the 2006 *ASHRAE Handbook—Refrigeration*). Proper service procedures help ensure that the equipment continues operating efficiently for its expected life.

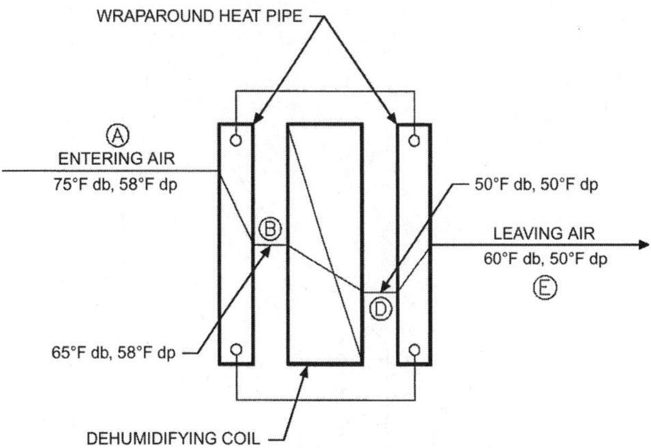

Fig. 12 Dehumidification Enhancement with Wraparound Heat Pipe

WRAPAROUND HEAT EXCHANGERS

An air-to-air heat exchanger (heat pipe, coil runaround loop, fixed-plate heat exchanger, or rotary heat exchanger) in a series (or wraparound) configuration can be used to enhance moisture removal by a mechanical dehumidifier, improving efficiency, and possibly allowing reduced refrigeration capacity in new systems. Other uses of air-to-air heat exchangers are covered in Chapter 25.

Air-to-air heat exchangers are used with a mechanical dehumidification system to passively move heat from one place to another. The most common configuration used for dehumidification is the **runaround** (or **wraparound**) configuration (Figure 12), which removes sensible heat from the entering airstream and transfers it to the leaving airstream. (Points A to E correspond to points labeled in Figure 13.) This improves the cooling coil's latent dehumidification capacity. This method can be applied if design calculations have taken into account the condition of air entering the evaporator coil.

In the runaround or wraparound configuration (Figure 12), one section of the air-to-air heat exchanger is placed upstream of the cooling coil and the other section is placed downstream of the cooling coil. The air is precooled before entering the cooling coil. Heat absorbed by the upstream section of the air-to-air heat exchanger is then transferred to air leaving the cooling coil (or supply airstream) by the downstream section.

Sensible precooling by the air-to-air heat exchanger reduces the sensible load on the cooling coil, allowing an increase in its latent capacity (Figure 13). The combination of these two effects lowers the system SHR, much like the process described in the Mechanical Dehumidifiers section. The addition of the air-to-air heat exchanger brings the condition of air entering the evaporator coil closer to the saturation line on the psychrometric chart (A to B). In new installations, this requires careful evaporator coil design that accounts for the actual range of air conditions after the air-to-air heat exchanger, which may differ significantly from the return air conditions.

In retrofits, the **duct-to-duct** (or **slide-in**) configuration (Figure 14) is sometimes used. One section of the air-to-air heat exchanger is placed in the return airstream, and the other section is placed in the supply airstream. This configuration, however, does not provide

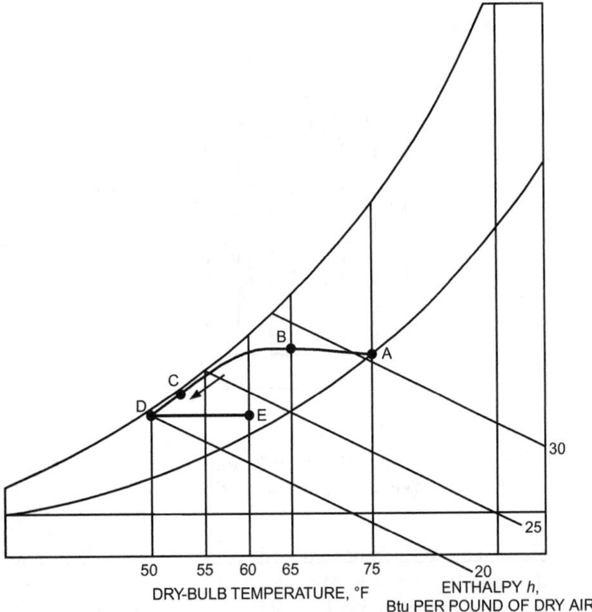

Fig. 13 Enhanced Dehumidification with a Wraparound Heat Pipe

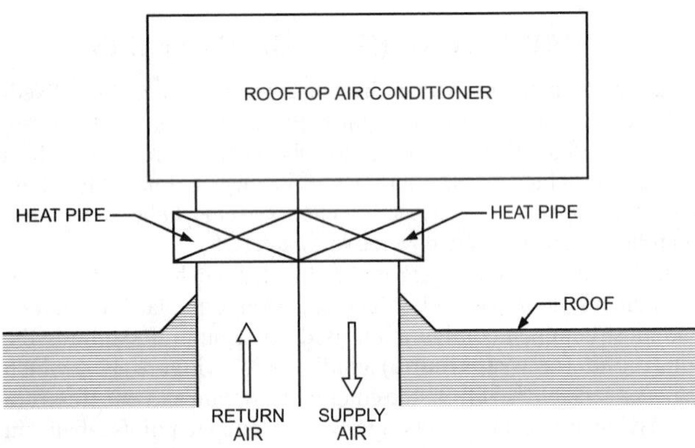

Fig. 14 Slide-in Heat Pipe for Rooftop Air Conditioner Refit
(Kittler 1996)

as much benefit as the wraparound configuration because (1) the upstream side of the heat exchanger is located upstream of where outdoor air enters the system, (2) the higher velocity reduces the effectiveness and increases the air-side pressure drop of the heat exchanger, and (3) it requires an additional filter upstream of the air-to-air heat exchanger.

In retrofits, the lower entering-air temperature at the evaporator coil lowers the temperature of the air leaving the evaporator coil. Evaporator coil capacity is reduced because of the lower entering wet-bulb temperature, changing the operating point of the system. This must be analyzed to ensure that the mechanical refrigeration system still operates correctly. If evaporator coil freeze-up is possible, the system must include some means of deactivating the air-to-air heat exchanger or increasing airflow to prevent evaporator freezing. Some way to modulate the air-to-air heat exchanger's capacity may be incorporated to better meet the load requirement of the mechanical dehumidifier.

Adding an air-to-air heat exchanger typically improves the moisture removal capacity of an existing mechanical dehumidification system by allowing a lower supply air dew point, while providing some reheat without additional energy use. Proper design practices must be followed to ensure that the unit's mechanical refrigeration system will still operate efficiently with the new entering air conditions and additional air-side pressure drop. Also, the added pressure drop of the air-to-air heat exchanger is likely to reduce the airflow delivered, unless fan speed in increased. If increasing fan speed is necessary, verify that the fan motor can handle the added load.

Figure 13 shows the dehumidification process when an air-to-air heat exchanger is added to an existing evaporator coil. Point A to C shows the cooling and dehumidification process of an existing direct-expansion (DX) evaporator coil, without the air-to-air heat exchanger. Point A to B shows precooling by the upstream section of heat exchanger. The process line from B to D (versus B to C, without the heat exchanger) shows how the evaporator coil's dehumidification performance improves (lowering leaving air dew point, from C to D) if an air-to-air heat exchanger is added to the existing system, because the enthalpy of the air entering the evaporator is lowered. Point D to E illustrates that the heat removed from air upstream of the evaporator (A to B) is added back into air leaving the evaporator. The total amount of heat energy (enthalpy) removed in section A-B is equal to the amount of heat added in section D-E.

REFERENCES

AHAM. 2003. Dehumidifiers. *Standard* DH-1-2003. Association of Home Appliance Manufacturers, Chicago, IL.

ARI. 1999. Indoor pool dehumidifiers. *Standard* 910. Air Conditioning and Refrigeration Institute, Arlington, VA.

ASHRAE. 2004. Energy standard for buildings except low-rise residential buildings. ANSI/ASHRAE *Standard* 90.1.

EPA. 2008. *ENERGY STAR®*. http://www.energystar.gov.

Kittler, R. 1996. Mechanical dehumidification control strategies and psychrometrics. *ASHRAE Transactions* 102(2):613-617.

Morris, W. 2003. The ABCs of DOAS. *ASHRAE Journal* (May).

Murphy, J. 2006. Smart dedicated outdoor-air systems. *ASHRAE Journal* (July).

Nevins, R., R.R. Gonzalez, Y. Nishi, and A.P. Gagge. 1975. Effect of changes in ambient temperature and level of humidity on comfort and thermal sensations. *ASHRAE Transactions* 81(2).

BIBLIOGRAPHY

AHAM. Semiannually. *Directory of certified dehumidifiers*. Association of Home Appliance Manufacturers, Chicago, IL.

ASHRAE. 2004. Safety standard for refrigeration systems. ANSI/ASHRAE *Standard* 15-2004.

CSA. 2002. Canadian electrical code. CSA *Standard* C22.1-02. Canadian Standards Association, Toronto.

Harriman, L., G. Brundrett, and R. Kittler. 2001. *Humidity control design guide for commercial and institutional buildings*. ASHRAE.

IEC. 2002. Household and similar electric appliances—Safety—Part 2: Particular requirements for electrical heat pumps, air-conditioners and dehumidifiers. IEC *Standard* 60335-2-40 (2002-12). International Electrotechnical Commission.

Kittler, R. 1983. Indoor natatoriums design and energy recycling. *ASHRAE Transactions* 95(1):521-526.

Kittler, R. 1994. Separate makeup air makes IAQ affordable. *Mechanical Buyer & Specifier* (June).

NFPA. 2008. *National electrical code®*, 2008 edition. NFPA *Standard* 70-2008. National Fire Protection Association, Quincy, MA.

UL. 2002. Leakage current for appliances. *Standard* C 101-02. Underwriters Laboratories, Northbrook, IL.

UL. 2004. Dehumidifiers. *Standard* 474-04. Underwriters Laboratories, Northbrook, IL.

UL. 2007. *Electrical appliances and utilization equipment directory*. Underwriters Laboratories, Northbrook, IL.

Wilson, G., R. Kittler, J. Teskoski, P. Reynolds, and K. Coursin. 1991. Controlling indoor environments for comfort, structure protection. *Aquatics International* (November/December).

CHAPTER 25

AIR-TO-AIR ENERGY RECOVERY

AIR-TO-AIR energy recovery is the process of recovering energy or/and moisture from an airstream at a high temperature or humidity to an airstream at a low temperature or humidity. This process is important in maintaining acceptable indoor air quality (IAQ) while maintaining low energy costs and reducing overall energy consumption. This chapter describes various technologies for air-to-air energy recovery. Thermal and economic performance, maintenance, and related operational issues are presented, with emphasis on energy recovery for ventilation.

Energy can be recovered either in its sensible (temperature only) or latent (moisture) form, or combination of both from multiple sources. Sensible energy can be extracted, for example, from outgoing airstreams in dryers, ovens, furnaces, combustion chambers, and gas turbine exhaust gases to heat supply air. Units used for this purpose are called **sensible heat exchange devices** or **heat recovery ventilators (HRVs)**. Devices that transfer both heat and moisture are known as **energy** or **enthalpy devices** or **energy recovery ventilators (ERVs)**. HRVs and ERVs are available for commercial and industrial applications as well as for residential and small-scale commercial uses.

Air conditioners use much energy to dehumidify moist airstreams. Excessive moisture in the air of a building can result in mold, allergies, and bacterial growth. ERVs can enhance dehumidification with packaged unitary air conditioners. Introducing outside or ventilation air is the primary means of diluting air contaminants to achieve acceptable indoor air quality. ERVs can cost-effectively provide large amounts of outside air to meet minimum ventilation requirements as prescribed in ASHRAE *Standards* 62.1 and 62.2.

Types of ERVs include compact air-to-air cross-flow heat exchangers, rotary wheels, heat pipes, runaround loops, thermosiphons, and twin-tower enthalpy recovery loops. Performance is typically measured by effectiveness, pressure drop or pumping power of fluids, cross-flow, (the amount of air leakage from one stream to the other), and frost control (used to prevent frosting on the heat exchanger). Efficiency, the ratio of output of a device to its input, is also often considered. In energy recovery ventilators, *effectiveness* refers to the ratio of actual energy or moisture recovered to the maximum possible.

Fluid stream pressure drops because of the friction between the fluid and solid surface, and because of the geometrical complexity of the flow passages. Pumping power is the product of the fluid volume flow rate and pressure drop. Economic factors such as cost of energy recovered and capital and maintenance cost (including pumping power cost) play a vital role in determining the economic feasibility of recovery ventilators for a given application.

APPLICATIONS

Air-to-air energy recovery systems may be categorized according to their application as (1) process-to-process, (2) process-to-comfort, or (3) comfort-to-comfort. Typical air-to-air energy recovery applications are listed in Table 1.

In **process-to-process** applications, heat is captured from the process exhaust stream and transferred to the process supply airstream. Equipment is available to handle process exhaust temperatures as high as 1600°F.

Process-to-process recovery devices generally recover only sensible heat and do not transfer latent heat, because moisture transfer is usually detrimental to the process. Process-to-process applications usually recover the maximum amount of energy. In cases involving condensable gases, less recovery may be desired to prevent condensation and possible corrosion.

In **process-to-comfort** applications, waste heat captured from process exhaust heats building makeup air during winter. Typical applications include foundries, strip-coating plants, can plants, plating operations, pulp and paper plants, and other processing areas with heated process exhaust and large makeup air volume requirements.

Although full recovery is usually desired in process-to-process applications, recovery for process-to-comfort applications must be modulated during warm weather to prevent overheating the makeup air. During summer, no recovery is required. Because energy is saved only in the winter and recovery is modulated during moderate weather, process-to-comfort applications save less energy annually than do process-to-process applications.

Process-to-comfort recovery devices generally recover sensible heat only and do not transfer moisture between airstreams.

In **comfort-to-comfort** applications, the heat recovery device lowers the enthalpy of the building supply air during warm weather and raises it during cold weather by transferring energy between the ventilation air supply and exhaust airstreams.

Table 1 Applications for Air-to-Air Energy Recovery

Method	Typical Application
Process-to-process and Process-to-comfort	Dryers Ovens Flue stacks Burners Furnaces Incinerators Paint exhaust Welding exhaust
Comfort-to-comfort	Swimming pools Locker rooms Residential Operating rooms Nursing homes Animal ventilation Plant ventilation General exhaust Smoking exhaust

The preparation of this chapter is assigned to TC 5.5, Air-to-Air Energy Recovery.

Air-to-air energy recovery devices for comfort-to-comfort applications may be sensible heat exchange devices (i.e., transferring sensible energy only) or energy exchange devices (i.e., transferring both sensible energy and moisture). These devices are discussed further in the section on Additional Technical Considerations.

When outside air humidity is low and the building space has an appreciable latent load, an ERV can recover sensible energy while possibly slightly increasing the latent space load because of water vapor transfer within the ERV. It is therefore important to determine whether the given application calls for HRV or ERV.

HRVs are suitable when outside air humidity is low and latent space loads are high for most of the year, and also for use with swimming pools, chemical exhaust, paint booths, and indirect evaporative coolers.

ERVs are suitable for applications in schools, offices, residences and other applications that require year-round economical preheating or/and precooling of outside supply air.

BASIC RELATIONS

The second law of thermodynamics states that heat energy always transfers from a region of high temperature to one of low temperature. This law can be extended to say that mass transfer always occurs from a region of high vapor pressure to one of low vapor pressure. The ERV facilitates this transfer across a separating wall (shown by a thick horizontal line in Figure 1) made of a material that conducts heat and is permeable to water vapor. Moisture is transferred when there is a difference in vapor pressure between the two airstreams.

On a typical summer day, supply air at temperature, humidity, or enthalpy of x_1 and mass flow rate m_s enters the ERV, while exhaust air from the conditioned space enters at conditions x_3 and m_3. Because conditions at x_3 are lower than conditions at x_1, heat and mass transfer from the supply airstream to the exhaust airstream because of differences in temperature and vapor pressures across the separating wall. Consequently, the supply air exit properties decrease, while those of the exhaust air increase. Exit properties of these two streams can be estimated, knowing the flow rates and the effectiveness of the heat exchanger.

ASHRAE *Standard* 84 defines effectiveness as

$$\varepsilon = \frac{\text{Actual transfer of moisture or energy}}{\text{Maximum possible transfer between airstreams}} \tag{1}$$

Heat Recovery Ventilators

From Figure 1, the sensible effectiveness ε_s of a heat recovery ventilator is given as

$$\varepsilon_s = \frac{q_s}{q_{s,max}} = \frac{m_s c_{ps}(t_2 - t_1)}{C_{min}(t_3 - t_1)} = \frac{m_s c_{pe}(t_3 - t_4)}{C_{min}(t_3 - t_1)} \tag{2a}$$

where q_s is the actual sensible heat transfer rate given by

$$q_s = \varepsilon_s\, q_{s,max} \tag{2b}$$

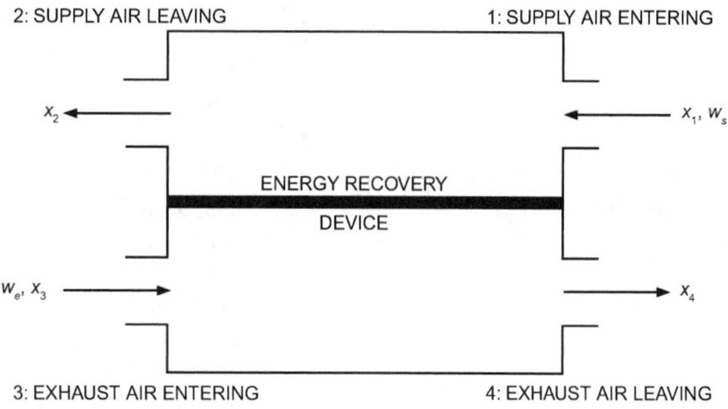

Fig. 1 Airstream Numbering Convention

where $q_{s,max}$ is the maximum sensible heat transfer rate given by

$$q_{s,max} = 60 C_{min}(t_3 - t_1) \tag{2c}$$

where

ε_s = sensible effectiveness
t_1 = dry-bulb temperature at location 1 in Figure 1, °F
m_s = supply dry air mass flow rate, lb/min
m_e = exhaust dry air mass flow rate, lb/min
C_{min} = smaller of $c_{ps}m_s$ and $c_{pe}m_e$
c_{ps} = supply moist air specific heat at constant pressure, Btu/lb·°F
c_{pe} = exhaust moist air specific heat at constant pressure, Btu/lb·°F

Assuming no water vapor condensation in the HRV, the leaving supply air condition is

$$t_2 = t_1 - \varepsilon_s \frac{C_{min}}{m_s c_{ps}}(t_1 - t_3) \tag{3a}$$

and the leaving exhaust air condition is

$$t_4 = t_3 + \varepsilon_s \frac{C_{min}}{m_e c_{pe}}(t_1 - t_3) \tag{3b}$$

Equations (2), (3a), and (3b) assume steady-state operating conditions; no heat or moisture transfer between the heat exchanger and its surroundings; no cross-leakage, and no energy gains from motors, fans, or frost control devices. Furthermore, condensation or frosting does not occur or is negligible. These assumptions are generally nearly true for larger commercial HRV applications. Note that the HRV only allows transfer of sensible heat energy associated with heat transfer because of temperature difference between the airstreams or between an airstream and a solid surface. These equations apply even in winter, if there is no condensation in the HRV.

The sensible heat energy transfer q_s from the heat recovery ventilator can be estimated from

$$q_s = 60 m_s c_{ps}(t_2 - t_1) = 60 Q_s \rho_s c_{ps}(t_2 - t_1) \tag{3c}$$

$$q_s = 60 m_e c_{pe}(t_4 - t_3) = 60 Q_e \rho_e c_{pe}(t_4 - t_3) \tag{3d}$$

$$q_s = 60 \varepsilon_s\, m_{min}\, c_p(t_1 - t_3) \tag{3e}$$

where

Q_s = volume flow rate of supply air, cfm
Q_e = volume flow rate of exhaust air, cfm
ρ_s = density of dry supply air, lb/ft³
ρ_e = density of dry exhaust air, lb/ft³
t_1, t_2, t_3, t_4 = inlet and exit temperatures of supply and exhaust airstreams, respectively
m_{min} = smaller of m_s and m_e

Because c_{ps} and c_{pe} are nearly equal, these terms may be omitted from Equations (1) to (4).

Sensible heat exchangers (HRVs) can be used in virtually all cases, especially for swimming pool, paint booth, and reheat applications. Equations (1) to (3e) apply for both HRVs and ERVs with appropriate selection of x_1, x_2, x_3, and x_4.

Energy Recovery Ventilators

The ERV allows the transfer of both sensible and latent heat, the latter due to the difference in water vapor pressures between the airstreams or between an airstream and a solid surface. ERVs are available as desiccant rotary wheels and also as membrane plate exchangers; although other gases may also pass through the membrane (Sparrow et al. 2001a) of membrane plate energy exchangers, it is assumed in the following equations that only the water vapor is allowed to pass through the membrane.

From Figure 1, assuming no condensation in the ERV, the latent effectiveness ε_L of an energy recovery ventilator is given as

$$\varepsilon_L = \frac{q_L}{q_{L,max}} = \frac{m_s h_{fg}(w_1 - w_2)}{m_{min} h_{fg}(w_1 - w_3)} = \frac{m_e h_{fg}(w_4 - w_3)}{m_{min} h_{fg}(w_1 - w_3)} \quad (4a)$$

where q_L is the actual latent heat transfer rate given by

$$q_L = \varepsilon_L q_{L,max} \quad (4b)$$

where $q_{L,max}$ is the maximum heat transfer rate given by

$$q_{L,max} = 60 m_{min} h_{fg}(w_1 - w_3) \quad (4c)$$

where

ε_L = latent effectiveness
h_{fg} = enthalpy of vaporization, Btu/lb
w = humidity ratios at locations indicated in Figure 1
m_s = supply dry air mass flow rate, lb/min
m_e = exhaust dry air mass flow rate, lb/min
m_{min} = smaller of m_s and m_e

Because the enthalpy of vaporization from Equation (4a) can be dropped out from numerator and denominator, Equation (4a) can be rewritten as

$$\varepsilon_m = \frac{m_w}{m_{w,max}} = \frac{m_s(w_1 - w_2)}{m_{min}(w_1 - w_3)} = \frac{m_e(w_4 - w_3)}{m_{min}(w_1 - w_3)} \quad (4d)$$

where ε_m is moisture effectiveness, numerically equal to latent effectiveness ε_L, and m_w is actual moisture transfer rate given by

$$m_w = \varepsilon_m m_{w,max} \quad (4e)$$

where $m_{s,max}$ is the maximum moisture transfer rate given by

$$m_{s,max} = m_{w,min}(w_1 - w_3) \quad (4f)$$

Assuming no water vapor condensation in the ERV, the leaving humidity ratios can be given as follows. The supply air leaving humidity ratio is

$$w_2 = w_1 - \varepsilon_L \frac{m_{w,min}}{m_s}(w_1 - w_3) \quad (5a)$$

and the leaving exhaust air humidity ratio is

$$w_4 = w_3 + \varepsilon_L \frac{m_{w,min}}{m_s}(w_1 - w_3) \quad (5b)$$

The total effectiveness ε_t of an energy recovery ventilator is given as

$$\varepsilon_t = \frac{q_t}{q_{t,max}} = \frac{m_s(h_2 - h_1)}{m_{min}(h_3 - h_1)} = \frac{m_e(h_3 - h_4)}{m_{min}(h_3 - h_1)} \quad (6a)$$

where q_t is the actual total heat transfer rate given by

$$q_t = \varepsilon_t q_{t,max} \quad (6b)$$

where $q_{t,max}$ is the maximum total heat transfer rate given by

$$q_{t,max} = 60 m_{min}(h_1 - h_3) \quad (6c)$$

where

ε_t = total effectiveness
h = enthalpy at locations indicated in Figure 1, Btu/lb
m_s = supply dry air mass flow rate, lb/min
m_e = exhaust dry air mass flow rate, lb/min
m_{min} = smaller of m_s and m_e

The leaving supply air condition is

$$h_2 = h_1 - \varepsilon_t \frac{m_{min}}{m_s}(h_1 - h_3) \quad (7a)$$

and the leaving exhaust air condition is

$$h_4 = h_3 + \varepsilon_t \frac{m_{min}}{m_e}(h_1 - h_3) \quad (7b)$$

Assuming the stream at state 1 is of higher humidity, the latent heat recovery q_L from the ERV can be estimated from

$$q_L = 60 m_s h_{fg}(w_1 - w_2) = 60 Q_s \rho_s h_{fg}(w_1 - w_2) \quad (8a)$$

$$q_L = 60 m_e h_{fg}(w_4 - w_3) = 60 Q_e \rho_e h_{fg}(w_4 - w_3) \quad (8b)$$

$$q_L = 60 \varepsilon_L m_{min} h_{fg}(w_1 - w_3) \quad (8c)$$

where

h_{fg} = enthalpy of vaporization or heat of vaporization of water vapor, Btu/lb
w_1, w_2, w_3, w_4 = inlet and exit humidity ratios of supply and exhaust airstreams, respectively

The total energy transfer q_t between the streams is given by

$$q_t = q_s + q_L = 60 m_s(h_{1s} - h_{2s}) = 60 Q_s \rho_s(h_{1s} - h_{2s}) = 60[m_s c_{ps}(t_1 - t_2) + m_s h_{fg}(w_1 - w_2)] \quad (9)$$

$$q_t = q_s + q_L = 60 m_e(h_{4e} - h_{3e}) = 60 Q_e \rho_e(h_{4e} - h_{3e}) = 60[m_e c_{pe}(t_4 - t_3) + m_e h_{fg}(w_4 - w_3)] \quad (10a)$$

$$q_t = 60 \varepsilon_t m_{min}(h_{1s} - h_{3e}) \quad (10b)$$

where

h_{1s} = enthalpy of supply air at inlet, Btu/lb
h_{3e} = enthalpy of exhaust air at inlet, Btu/lb
h_{2s} = enthalpy of supply air at outlet, Btu/lb
h_{4e} = enthalpy of exhaust air at outlet, Btu/lb

ERVs can be used where there is an opportunity to transfer heat and mass (water vapor) (e.g., humid areas, schools, offices with large occupancies). Latent energy transfer can be positive or negative depending on the direction of decreasing vapor pressure. An airstream flowing through an ERV may gain heat energy ($+q_s$) from the adjoining stream, but will lose the latent energy ($-q_L$) if it transfers the water vapor to the adjoining stream, because of transfer of moisture. The total net energy gain is the difference between q_s and q_L, as shown in Example 1.

Example 1. Inlet supply air enters an ERV with a flow rate of 9350 cfm at 95°F and 20% rh. Inlet exhaust air enters with a flow rate of 9050 cfm at 75°F and 50% rh. Assume that the energy exchanger was tested under ASHRAE *Standard* 84, which rated the sensible heat transfer effectiveness at 50% and the latent (water vapor) transfer effectiveness at 50%. Assuming the specific heat of air is 0.24 Btu/lb·°F and the latent heat of vaporization to be 1100 Btu/lb, determine the sensible, latent, and net energy gained by the exhaust air.

Solution:

From the psychrometric chart, the properties of air at 95°F and 20% rh are

$V_1 = 14.14$ ft³/lb $\quad h_1 = 30.6$ Btu/lb $\quad w_1 = 0.0071$ lb/lb of dry air

and the properties of air at 75°F and 50% rh are

$V_3 = 13.68$ ft³/lb $\quad h_3 = 28.15$ Btu/lb $\quad w_3 = 0.0093$ lb/lb of dry air

The mass flow rate at state 1 is obtained from

$$m_1 = \frac{Q_1}{V_1} = \frac{9350 \text{ ft}^3/\text{min}}{14.14 \text{ ft}^3/\text{lb}} = 660 \text{ lb/min}$$

Similarly, the mass flow rate at state 3 is obtained from

$$m_3 = \frac{Q_3}{V_3} = \frac{9050 \text{ ft}^3/\text{min}}{13.68 \text{ ft}^3/\text{lb}} = 660 \text{ lb/min}$$

These equal mass flow rates conform with ASHRAE *Standard* 84.

Exit temperatures of the airstreams can be obtained from the Equations (3a) and (3b) as follows:

$$t_2 = 95°\text{F} - 0.5\frac{(660 \text{ lb/min})(0.24 \text{ Btu/lb}\cdot°\text{F})}{(660 \text{ lb/min})(0.24 \text{ Btu/lb}\cdot°\text{F})}(95°\text{F} - 75°\text{F}) = 85°\text{F}$$

$$t_4 = 75°\text{F} + 0.5\frac{(660 \text{ lb/min})(0.24 \text{ Btu/lb}\cdot°\text{F})}{(660 \text{ lb/min})(0.24 \text{ Btu/lb}\cdot°\text{F})}(95°\text{F} - 75°\text{F}) = 85°\text{F}$$

The exit humidity of the airstreams is found from Equations (5a) and (5b) as follows:

$$w_2 = 0.0071 - 0.5\frac{(660 \text{ lb/min})}{(660 \text{ lb/min})}(0.0071 - 0.0093)$$
$$= 0.0082 \text{ lb/lb of dry air}$$

$$w_4 = 0.0093 + 0.5\frac{(660 \text{ lb/min})}{(660 \text{ lb/min})}(0.0071 - 0.0093)$$
$$= 0.0082 \text{ lb/lb of dry air}$$

The sensible heat gained by the exhaust stream is found from Equation (3c) as

$$q_s = (660 \text{ lb/min})(0.24 \text{ Btu/lb}\cdot°\text{F})(85°\text{F} - 75°\text{F}) = 1584 \text{ Btu/min}$$

The latent heat gained by the exhaust stream is found from Equation (8a) as

$$q_L = (660 \text{ lb/min})(1100 \text{ Btu/lb})(0.0082 - 0.0093)$$
$$= -799 \text{ Btu/min}$$

The net heat energy gained by the exhaust airstream is therefore

$$q = q_s + q_L = 1584 - 799 = 785 \text{ Btu/min}$$

If the incoming outdoor air conditions had been at 95°F and 14% rh, then the net energy gained by the exhaust airstream would have been zero. The outlet exhaust airstream enthalpy at 85°F and 0.0082 lb/lb of dry air is given in the psychrometric chart as 29.4 Btu/lb. The net heat gained by the exhaust airstream [found from Equation (10)] is close to 945 Btu/min.

The fan power P_s required by the supply air is estimated from

$$P_s = Q_s\,\Delta p_s/6356\,\eta_f \tag{11}$$

The fan power P_e required by the exhaust air is estimated from

$$P_e = Q_e\,\Delta p_e/6356\,\eta_f \tag{12}$$

where

P_s = fan power for supply air, hp
P_e = fan power for exhaust air, hp
Δp_s = pressure drop of supply air caused by fluid friction, in. of water
Δp_e = pressure drop of exhaust air caused by fluid friction, in. of water
η_f = overall efficiency of fan and motor or product of fan and motor efficiencies

However, the density and viscosity of air vary with the temperature. The variation of viscosity with temperature is given by the Sutherland law as

$$\frac{\mu}{\mu_o} = \left(\frac{T}{T_o}\right)^{3/2}\left(\frac{T_o + S}{T + S}\right) \tag{13}$$

where

T = absolute temperature, °F
T_o = reference temperature, °F
S = constant = 198.7°F

Treating air as an ideal gas, the pressure drop Δp at any temperature T is related to the pressure drop Δp_o at reference temperature T_o and is expressed as

$$\frac{\Delta p}{\Delta p_o} = \left(\frac{m}{m_o}\right)^{1.75}\left(\frac{T}{T_o}\right)^{1.375}\left(\frac{T_o + S}{T + S}\right)^{0.25} \tag{14}$$

Equation (14) is only accurate when the Reynolds number Re_D for airflow through the exchanger is in the range

$$5 \times 10^3 \le Re_D \le 10^5$$

where

$$Re_D = \frac{(\rho V)_{av} D_h}{\mu_{\bar{T}}} \tag{15}$$

where

ρ = air density, lb/ft³
V = average velocity in flow channels, fpm
D_h = hydraulic diameter of flow channels, ft
$\mu_{\bar{T}}$ = dynamic viscosity at average temperature $\bar{T}$, lb/ft·min

For fully developed laminar flow through an energy exchanger (e.g., an energy wheel), the corresponding dimensionless pressure drop relation similar to Equation (14) is given as

$$\frac{\Delta p}{\Delta p_0} = \left(\frac{m}{m_0}\right)\left(\frac{\bar{T}}{\bar{T}_0}\right)^{3/2}\left(\frac{1 + C/\bar{T}_0}{1 + C/\bar{T}}\right) \tag{16}$$

where $Re_D < 2000$.

Equations (14) and (16) cannot be used in the case of (1) flow channels that are not reasonably smooth, (2) flow Reynolds numbers that are out of range, (3) significant exchanger fouling due to condensation, frost, or dust, (4) excessive nonuniform property distributions inside the exchanger, or (5) significant pressure deformation of the flow channels (e.g., some plate cross-flow exchangers).

The total pumping power P of the ERV can be given as

$$P = P_s + P_e \tag{17}$$

Ideal Air-to-Air Energy Exchange

An ideal air-to-air energy exchanger

- Allows temperature-driven heat transfer between participating airstreams
- Allows partial-pressure-driven moisture transfer between the two streams
- Minimizes cross-stream transfer of air, other gases (e.g., pollutants), biological contaminants, and particulates

Heat transfer is an important energy recovery vehicle from airstreams that carry waste heat. The role of moisture transfer as an energy recovery process is less well known and merits explanation.

Consider an air-to-air energy exchanger operating in a hot, humid climate in a comfort air-conditioning application. If the energy exchanger exchanges heat but not moisture, it cools outside ventilation air as it passes through the exchanger to the indoor space. Heat flows from the incoming outside air to the outgoing (and cooler) exhaust air drawn from the indoor conditioned space. This does very little to mitigate the high humidity carried into the indoor space by the outside ventilation air and may even increase the relative humidity in the conditioned space, resulting in increased refrigeration and/or reheat to dehumidify the air and achieve acceptable comfort conditions. On the other hand, if the energy exchanger transfers both heat and moisture, the humid outside supply air transfers moisture to the less-humid inside exhaust air as the streams pass through the energy exchanger. The lower humidity of the entering ventilation air requires less energy input for comfort conditioning.

AIRFLOW ARRANGEMENTS

Heat exchanger effectiveness depends heavily on the airflow direction and pattern of the supply and exhaust airstreams. Parallel-flow exchangers (Figure 2A), in which both airstreams move along heat exchange surfaces in the same direction, have a theoretical maximum effectiveness of 50%. Counterflow exchangers (Figure 2B), in which airstreams move in opposite directions, can have an effectiveness approaching 100%, but typical units have a lower effectiveness. Normal effectiveness for cross-flow heat exchangers is 50 to 70% (Figure 2C) and 60 to 85% for multiple-pass exchangers (Figure 2D).

In practice, construction limitations favor designs that use transverse flow (or cross-flow) over much of the heat exchange surface (Figures 2C and 2D).

Effectiveness

Heat or energy exchange effectiveness as defined in Equation (1) is used to characterize each type of energy transfer in air-to-air exchangers. For a given set of inlet properties and flow rates, knowledge of each effectiveness allows the designer to calculate the sensible, latent, and total energy transfer rates using Equations (3c), (8c), and (10a), respectively. These effectiveness values can be determined either from measured test data or using correlations that have been verified in the peer-reviewed engineering literature. These correlations can also be used to predict energy transfer rates and outlet air properties for operating conditions different from those used for certification purposes. Predicting effectiveness for noncertified operating conditions using certified test data is the most common use of correlations for HVAC designs. Although correlations are not available for all types of air-to-air exchangers under all operating conditions, they are available for the most common types of air-to-air exchanger under operating conditions which have no condensation or frosting.

Rate of Energy Transfer

The rate of energy transfer depends on the operating conditions and the intrinsic characteristics of the energy exchanger, such as the geometry of the exchanger (parallel flow/counterflow/ crossflow, number of passes, fins), thermal conductivity of walls separating the streams, and permeability of walls to various gases. As in a conventional heat exchanger, energy transfer between the airstreams is driven by cross-stream dry-bulb temperature differences. Energy is also transported piggyback-style between the streams by cross-stream mass transfer, which may include air, gases, and water vapor. In another mode of energy transfer, water vapor condenses into liquid in one of the two airstreams of the exchanger. The condensation process liberates the latent heat of condensation, which is transferred to the other stream as sensible heat; this two-step process is also called *latent heat transfer*.

Latent energy transfer between airstreams occurs only when moisture is transferred from one airstream to another without condensation, thereby maintaining the latent heat of condensation. Once moisture has crossed from one airstream to the other, it may either remain in the vapor state or condense in the second stream, depending on the temperature of that stream.

Rotating and permeable-walled flat-plate energy recovery units are used because of their moisture exchange function. Passage of air or other gases (e.g., pollutants) across the exchanger is a negative consequence. As well, some cross-stream mass transfer may occur through leakage even when such transfer is unintended. This may alter exchanger performance from its design value, but for most HVAC applications with exhaust air from occupied spaces, small transfers to the supply air are not important.

Heat transfer differs in principle from mass transfer. Heat transfer only occurs when there is a temperature difference. In the case of air-to-air exchange between the supply and exhaust airflow, heat transfer by conduction and convection only occurs when there is a temperature difference between these airstreams. The following facts about heat/mass exchanger performance must be recognized:

- The effectiveness for moisture transfer may not equal the effectiveness for heat transfer.
- The total energy effectiveness may not equal either the sensible or latent effectiveness.

Net total energy transfer and effectiveness need careful examination when the direction of sensible (temperature-driven) transfer is opposite to that of latent (moisture or water vapor) transfer.

ERV performance is expressed by the magnitudes of pumping power and sensible, latent, or total energy recovered. The energy recovered is estimated from the exit temperatures or humidity ratios, which are directly related to the effectiveness. Effectiveness is a function of two parameters: the number of transfer units (NTU) and thermal flow capacity ratio C_r.

$$\text{NTU} = UA/60C_{min} \tag{18}$$

$$C_r = C_{min}/C_{max} \tag{19}$$

where

U = overall heat transfer coefficient, related to flow rates and dimensions of fluid flow path in heat exchanger, Btu/h·ft²·°F
A = area of heat exchanger, ft²
C_{max} = maximum of $m_s c_{ps}$ and $m_e c_{pe}$

Figure 7 depicts the variation of effectiveness with NTU for a rotary heat wheel.

OUTSIDE AIR
SUPPLY AIR
RETURN AIR
EXHAUST AIR

A. PARALLEL HEAT EXCHANGE

RETURN AIR
OUTSIDE AIR
SUPPLY AIR
EXHAUST AIR

C. CROSS-FLOW HEAT EXCHANGE

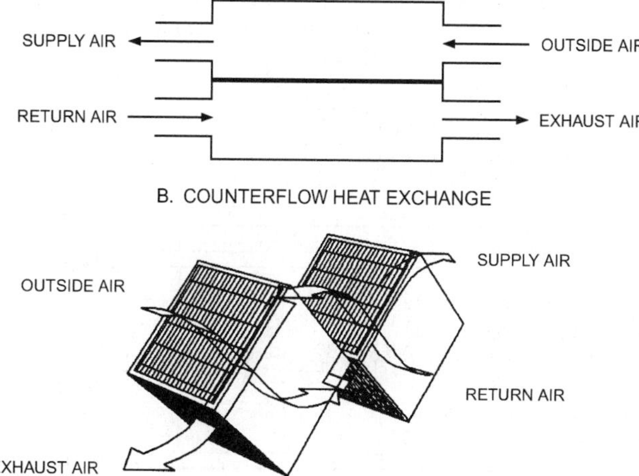

Fig. 2 Heat Exchanger Airflow Configurations

ADDITIONAL TECHNICAL CONSIDERATIONS

The rated effectiveness of energy recovery units is typically obtained under balanced flow conditions (i.e., the flow rates of supply and exhaust airstreams are the same). However, these ideal conditions do not always exist due to design for positive building pressure, the presence of air leakage, fouling, condensation or frosting and several other factors as described below.

Air Leakage

Air leakage refers to any air that enters or leaves the supply or exhaust airstreams. Zero air leakage in either airstream would require m_1 to equal m_2 and m_3 to equal m_4. External air leakage occurs when the ambient air surrounding the energy recovery system leaks into (or exits) either or both airstreams. Cross-flow air leakage results from inadequate sealing construction between ambient and cross-stream seal interfaces. Internal air leakage occurs when holes or passages are open to the other airstream. Internal air leakage occurs when heat or energy exchanger design allows (1) tangential air movement in the wheel's rotational direction and (2) air movement through holes in the barrier between airstreams. Under some pressure differentials, air leaks in and out of each airstream in nearly equal amounts, giving the illusion that there is no air leakage. Heat and water vapor transfer could appear to be greater than it actually is. Air leakage is seldom zero because external and internal air pressures are usually different, causing air to leak from higher-pressure regions to lower-pressure regions.

Cross-flow air leakage is usually caused by pressure differentials between airstreams. Carryover air leakage (specific to wheels) is caused by continuous rotation of trapped exhaust air in cavities in the heat transfer surface, which reverses airflow direction as the wheel rotates and spills this exhaust air into the supply airstream.

Cross-leakage, cross-contamination, or mixing between supply and exhaust airstreams may occur in air-to-air heat exchangers and may be a significant problem if exhaust gases are toxic or odorous. Cross-leakage varies with heat exchanger type and design, airstream static pressure differences, and the physical condition of the heat exchanger (see Table 2).

Air leakage between incoming fresh air and outgoing exhaust air is comprised of two paths called cross-flow and carryover leakage. **Cross-flow leakage** is caused primarily by difference in static pressures between states 2 and 3 and/or between states 1 and 4, as shown in Figure 3. This is a major cause of cross-flow leakage, and underscores the importance of specifying precise locations for fans that circulate the airstreams. Cross-flow can also be caused by factors such as provisions for surging, geometrical irregularities, and local velocity distribution of the airstreams.

Carryover occurs in rotary recovery units because of wheel rotation. The quantitative estimate of the air leakage is expressed by two dimensionless parameters: the exhaust air transfer ratio (EATR) and outside air correction factor (OACF).

$$\text{EATR} = \frac{c_2 - c_1}{c_3 - c_1} \tag{20}$$

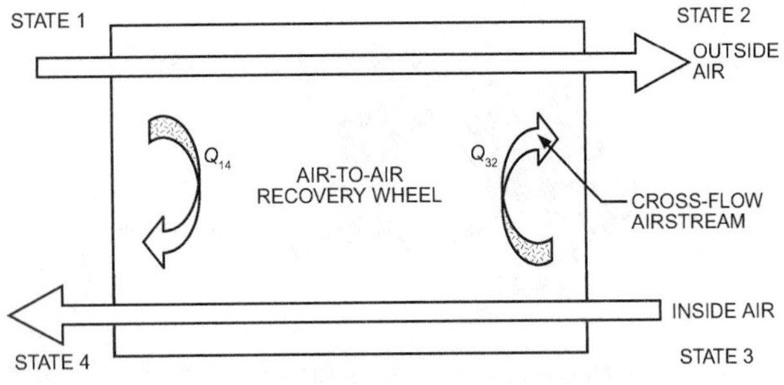

Fig. 3 Air Leakage in Energy Recovery Units

where c_1, c_2, and c_3 are the concentrations of inert gas at states 1, 2 and 3, respectively. Note that EATR represents an exhaust air leakage based on observed relative concentration of inert gas in supply airflow.

$$\text{OACF} = \frac{m_1}{m_2} \tag{21}$$

where m_1 and m_2 are the mass flow rates of the incoming fresh airstream at state 1 and 2, respectively. OACF helps estimate the extra quantity of outside air required at the inlet to compensate for the air that leaks into or out of the exchanger, and to meet the required net supply airflow to the building space. Ideal airflow conditions exist when there is no air leakage between the streams; EATR is close to zero, and OACF approaches 1. Deviations from ideal conditions indicate air leakage between the airstreams, which complicates the determination of accurate values for pressure drop and effectiveness. Methods to estimate actual flow rates at states 1, 2, 3, and 4 when air leakage exists and the values of the parameters EATR and OACF are known are discussed in Friedlander (2003) and Moffitt (2003). EATR and OACF are useful in comparing HRVs and ERVs and in evaluating the actual flow rate of outside air supplied for a given ventilation requirement or in estimating the capacity of ventilator fans, as illustrated by the following.

Air Capacity of Ventilator Fans

For a given ventilation requirement Q_v, the volume flow rate capacity Q_1 of the supply fan is greater if air leakage exists (see Figure 3).

EATR is the percentage of supply air Q_2 that is made up of exhaust air Q_3 that has leaked Q_{32} through the device.

$$Q_{32} = Q_2\left(\frac{\text{EATR}}{100}\right)$$

If the ventilation requirement is Q_v, then the actual volume flow rate of supply air to the space is calculated as

$$Q_2 = Q_v + Q_{32} = Q_v + Q_2\left(\frac{\text{EATR}}{100}\right)$$

This may be simplified as

$$Q_2 = \frac{Q_v}{\left(1 - \frac{\text{EATR}}{100}\right)}$$

which gives the quantity of air entering the space. Assuming steady-state conditions, balanced flow through the energy recovery ventilator, and negligible variation in air densities, the quantity of air leaving the building space (Q_3) should be same as air supplied to the space Q_2.

$$Q_3 = Q_2 = \frac{Q_v}{\left(1 - \frac{\text{EATR}}{100}\right)} \tag{22}$$

Assuming negligible change in air density and after substituting Equation (22) into Equation (21) gives

$$Q_1 = Q_2(\text{OACF}) = \frac{Q_v(\text{OACF})}{\left(1 - \frac{\text{EATR}}{100}\right)} \tag{23}$$

Equation (23) gives the volume flow rate capacity of the intake, which equals the exhaust for balanced airflow. [Fan capacity depends on location. This statement only applies to the supply fan at station 1 (blow-through); its capacity would change at station 2 (draw-through).]

Cross-stream mass transfer of air and water vapor can be driven by two independent types of pressure differences: (1) cross-stream total pressure differences and (2) cross-stream partial pressure differences. Air mass movement is driven primarily by air pressure differences and is minimized by a high bulk-flow resistance of the exchanger wall barrier and adjacent air seals. Moisture mass transfer is driven by a combination of air pressure differences and vapor partial-pressure differences. Cross-stream moisture transfer is maximized by a low bulk-flow resistance, high moisture adsorption/desorption characteristics of desiccant coatings in total energy wheels, high moisture absorption/desorption characteristics of twin-tower coupling desiccant liquid, and high permeability of the exchanger wall barrier in permeable-walled flat-plate energy recovery units. High bulk-flow resistance retards viscous bulk airflow and minimizes the effect of air pressure differences on air mass transfer. Air-pressure- and partial-pressure-driven mass transfer may be additive or subtractive.

Heat, moisture, and air transfer rates are sometimes (but not generally) independent of and separate from one another. Heat is always driven cross-stream from higher to lower temperature. Air is predominantly driven cross-stream from higher to lower air pressure. Water vapor mass is driven cross-stream in an amount and direction influenced by several variables. Design and construction characteristics of the exchanger greatly influence whether the moisture mass is (1) transferred predominantly by riding on the cross-stream air mass or (2) separated from the air mass by a permeable desiccant, selective microporous membrane, or other moisture-separating device. The net effect of vapor pressure differences cross-stream influences the net intensity and direction of moisture exchange.

Pressure Drop

Pressure drop for each airstream through an energy recovery unit depends on many factors, including exchanger design, mass flow rate, temperature, moisture, and inlet and outlet air connections. The pressure drop must be overcome by fans or blowers. Because the power required for circulating airstreams through the recovery unit is directly proportional to the pressure drop, the pressure drop through the energy recovery unit should be known. The pressure drop may be used with the fan efficiency to characterize the energy used by the exchanger and in turn the efficiency (not effectiveness) of an application.

Maintenance

The method used to clean a heat exchanger depends on the transfer medium or mechanism used in the energy recovery unit and on the nature of the material to be removed. Grease build-up from kitchen exhaust, for example, is often removed with an automatic water-wash system. Other kinds of dirt may be removed by vacuuming, blowing compressed air through the passages, steam cleaning, manual spray cleaning, soaking the units in soapy water or solvents, or using soot blowers. The cleaning method should be determined during design so that a compatible heat exchanger can be selected.

Cleaning frequency depends on the quality of the exhaust airstream. Residential and commercial HVAC systems generally require only infrequent cleaning; industrial systems, usually more. Equipment suppliers should be contacted regarding the specific cleaning and maintenance requirements of the systems being considered.

Filtration

Filters should be placed in both the supply and exhaust airstreams to reduce fouling and thus the frequency of cleaning. Exhaust filters are especially important if the contaminants are sticky or greasy or if particulates can plug airflow passages in the exchanger. Supply filters eliminate insects, leaves, and other foreign materials, thus protecting both the heat exchanger and air-conditioning equipment. Snow or frost can block the air supply filter and cause severe problems. Specify steps to ensure a continuous flow of supply air.

Controls

Heat exchanger controls may control frost formation or regulate the amount of energy transferred between airstreams at specified operating conditions. For example, ventilation systems designed to maintain specific indoor conditions at extreme outdoor design conditions may require energy recovery modulation to provide an economizer function, to prevent overheating ventilation supply air during cool to moderate weather or to prevent overhumidification. Modulation methods include tilting heat pipes, changing rotational speeds of (or stopping) heat wheels, or bypassing part of one airstream around the heat exchanger using dampers (i.e., changing the supply-to-exhaust mass airflow ratio).

Fouling

Fouling, an accumulation of dust or condensates on heat exchanger surfaces, reduces heat exchanger performance by increasing resistance to airflow, interfering with mass transfer, and generally decreasing heat transfer coefficients. Increased resistance to airflow increases fan power requirements and may reduce airflow.

Increased pressure drop across the heat exchanger core can indicate fouling and, with experience, may be used to establish cleaning schedules. Reduced mass transfer performance (latent effectiveness) indicates fouling of permeable membranes or desiccant sorption sites. Heat exchanger surfaces must be kept clean to maximize system performance.

Corrosion

Process exhaust frequently contains corrosive substances. If it is not known which construction materials are most corrosion-resistant for an application, the user and/or designer should examine on-site ductwork, review literature, and contact equipment suppliers before selecting materials. A corrosion study of heat exchanger construction materials in the proposed operating environment may be warranted if installation costs are high and the environment is corrosive. Experimental procedures for such studies are described in an ASHRAE symposium (ASHRAE 1982). Often contaminants not directly related to the process are present in the exhaust airstream (e.g., welding fumes or paint carryover from adjacent processes).

Moderate corrosion generally occurs over time, roughening metal surfaces and increasing their heat transfer coefficients. Severe corrosion reduces overall heat transfer and can cause cross-leakage between airstreams because of perforation or mechanical failure.

Condensation and Freeze-Up

Condensation, ice formation, and/or frosting may occur on heat exchange surfaces. If entrance and exit effects are neglected, four distinct air/moisture regimes may occur as the warm airstream cools from its inlet condition to its outlet condition. First, there is a dry region with no condensate. Once the warm airstream cools below its dew point, there is a condensing region, which wets the heat exchange surfaces. If the heat exchange surfaces fall below freezing, the condensation freezes. Finally, if the warm airstream temperature falls below its dewpoint, sublimation causes frost to form. The locations of these regions and rates of condensation and frosting depend on the duration of frosting conditions; airflow rates; inlet air temperature and humidity; heat exchanger core temperature; heat exchanger effectiveness; the geometry, configuration, and orientation; and heat transfer coefficients.

Sensible heat exchangers, which are ideally suited to applications in which heat transfer is desired but humidity transfer is not (e.g., swimming pools, kitchens, drying ovens), can benefit from the latent heat released by the exhaust gas when condensation occurs. One pound of moisture condensed transfers about 1050 Btu to the incoming air at room temperature.

Condensation increases the heat transfer rate and thus the sensible effectiveness; it can also significantly increase pressure drops in heat exchangers with narrow airflow passage spacing. Frosting fouls the heat exchanger surfaces, which initially improves energy transfer but subsequently restricts the exhaust airflow, which in turn reduces the energy transfer rate. In extreme cases, the exhaust airflow (and supply, in the case of heat wheels) can become blocked. Defrosting a fully blocked heat exchanger requires that the unit be shut down for an extended period. As water cools and forms ice, it expands, which may seriously damage the heat exchanger core unless the water is entirely removed.

For frosting or icing to occur, an airstream must be cooled to a dew point below 32°F. Total heat exchangers transfer moisture from the airstream with higher moisture content (usually the warmer airstream) to the less humid one. As a result, frosting or icing occurs at lower supply air temperatures in enthalpy exchangers than in sensible heat exchangers. In enthalpy heat exchangers, which use chemical absorbents, condensation may cause the absorbents to deliquesce, resulting in loss of absorbent.

For these reasons, some form of freeze control must be incorporated into heat exchangers that are expected to operate under freezing conditions. Frosting and icing can be prevented by preheating the supply air, or reducing heat exchanger effectiveness (e.g., reducing heat wheel speed, tilting heat pipes, or bypassing part of the supply air around the heat exchanger). Alternatively, the heat exchanger may be periodically defrosted.

The performance of several freeze control strategies is discussed in ASHRAE research project RP-543 (Phillips et al. 1989a, 1989b). ASHRAE research project RP-544 (Barringer and McGugan 1989a, 1989b) discusses the performance of enthalpy heat exchangers. Many effective defrost strategies have been developed for residential air-to-air heat exchangers. These strategies may also be applied to commercial installations. Phillips et al. (1992) describe frost control strategies and their impact on energy performance in various climates.

For sensible heat exchangers, system design should include drains to collect and dispose of condensation, which occurs in the warm airstream. In comfort-to-comfort applications, condensation may occur in the supply side in summer and in the exhaust side in winter.

Sensible heat exchangers, which are ideally suited to applications in which heat transfer is desired but humidity transfer is not (e.g., swimming pools, kitchens, drying ovens), can benefit from the latent heat released by the exhaust gas when condensation occurs. One pound of moisture condensed transfers about 1050 Btu to the incoming air at room temperature.

Condensation increases the heat transfer rate and thus the sensible effectiveness; it can also significantly increase pressure drops in heat exchangers with narrow airflow passage spacing. Frosting fouls the heat exchanger surfaces, which initially improves energy transfer but subsequently restricts the exhaust airflow, which in turn reduces the energy transfer rate. In extreme cases, the exhaust airflow (and supply, in the case of heat wheels) can become blocked. Defrosting a fully blocked heat exchanger requires that the unit be shut down for an extended period. As water cools and forms ice, it expands, which may seriously damage the heat exchanger core unless it is entirely moved.

For frosting or icing to occur, an airstream must be cooled to a dew point below 32°F. Total heat exchangers transfer moisture from the airstream with higher moisture content (usually the warmer airstream) to the less humid one. As a result, frosting or icing occurs at lower supply air temperatures in enthalpy exchangers than in sensible heat exchangers. In enthalpy heat exchangers, which use chemical absorbents, condensation may cause the absorbents to deliquesce, resulting in loss of absorbent.

For these reasons, some form of freeze control must be incorporated into heat exchangers that are expected to operate under freezing conditions. Frosting and icing can be prevented by preheating the supply air, or reducing heat exchanger effectiveness (e.g., reducing heat wheel speed, tilting heat pipes, or bypassing part of the supply air around the heat exchanger). Alternatively, the heat exchanger may be periodically defrosted.

The performance of several freeze control strategies is discussed in ASHRAE research project RP-543 (Phillips et al. 1989a, 1989b). ASHRAE research project RP-544 (Barringer and McGugan 1989a, 1989b) discusses the performance of enthalpy heat exchangers. Many effective defrost strategies have been developed for residential air-to-air heat exchangers. These strategies may also be applied to commercial installations. Phillips et al. (1989c) describe frost control strategies and their impact on energy performance in various climates.

For sensible heat exchangers, system design should include drains to collect and dispose of condensation, which occurs in the warm airstream. In comfort-to-comfort applications, condensation may occur in the supply side in summer and in the exhaust side in winter.

Frost Blockage and Control in Air-to-Air Exchangers

Air-to-air exchangers are widely used in HVAC building applications where outside air temperatures are often well below freezing in winter. As a result, condensation and freezing may occur in the exhaust airstream or both the exhaust and supply for regenerative heat or energy wheels. Condensation occurs in the exchanger when and where the local exchanger surface temperature is below the dew-point temperature but above the frost point. Freezing occurs when and where the surface temperatures are below the frost point. Low-density frost grows when surface temperatures are well below freezing.

Design should allow for continuous removal of any condensation. Ice and frost must be prevented or cyclically removed. Generally, frost growth is considered to be more serious because it can degrade exchanger performance, sometimes within only minutes. The consequences of frost growth are usually (1) increased pressure drop, (2) decreased mass flow rates of air, (3) decreased exchanger effectiveness, and (4) increased OACF.

Predicting frosting conditions for a given exchanger is usually a function of its operating conditions and performance factors. For example, for counterflow air-to-air exchangers, which essentially includes all types of exchanger except crossflow, an equation for the frost onset threshold of the supply air inlet temperature T_1 is given by:

$$T_{1f}(°\text{F}) = \left(\frac{-6.0\dot{m}_e}{\varepsilon_s \dot{m}_e}\right)(1 + 3.5\varepsilon_L \phi_3)\left[1 + \frac{0.012\dot{m}_e}{\varepsilon_s \dot{m}_e}(T_3 - 20)\right] \quad (24)$$

with no frost control and $T_0 = t_1 < T_{1f}$, where T_{1f} is the threshold temperature when frost will grow without limit. The sensible effectiveness $\varepsilon_s > 0.5$ for energy wheels and $\varepsilon_l = 0.5$ for sensible heat exchangers even though there is no moisture transfer between the airstreams.

This equation suggests several methods for steady-state frost control when outside ambient air temperature T_o is less than T_{1f}: (1) preheat supply air so that $T_1 \geq T_{1f}$; (2) preheat exhaust air so that $T_{1f} \geq T_1 = T_0$; (3) bypass some supply air to increase exhaust to supply air mass flow rate m_e/m_s, which results in a new $T_{1f} \leq T_1 = T_0$; (4) increase exhaust air humidity for energy wheels so that $T_{1f} \leq T_1 = T_0$; (5) decrease the value of the sensible effectiveness (e.g., decreasing the wheel speed of energy wheels, tilt control on heat pipes, changing the liquid flow rate in the loop on runaround systems); or (6) any combination of these five methods.

If cyclic frost methods are used, the defrost period should be long enough to completely remove the frost, ice, and water during each defrost period.

Cross-flow heat exchangers are more frost-tolerant than other plate exchangers because frost blockage only occurs over a fraction of the exchanger exhaust airflow passages. Similar frost control strategies can be used for counterflow exchangers, but Equation

(24) must be adapted: frost forms in the exhaust airflow passages whenever the outside air temperatures bring any part of the exhaust channel surfaces below the frost point temperature of about 23°F.

See the Bibliography for sources of more information on frost growth and control.

PERFORMANCE RATINGS

Standard laboratory rating tests and predictive computer models give exchanger performance values for (1) heat transfer, (2) moisture transfer, (3) cross-stream air transfer, (4) average exhaust mass airflow, and (5) supply mass airflow leaving the exchanger. Effectiveness ratios for heat and mass water vapor transfer must be separately determined by rating tests in a laboratory that is staffed and instrumented to meet requirements of ASHRAE *Standard* 84 and ARI *Standard* 1060. It may be very difficult to adhere to any standard when field tests are made.

ASHRAE *Standard* 84, Method of Testing Air-to-Air Heat Exchangers, (1) establishes a uniform method of testing for obtaining performance data; (2) specifies the data required, calculations to be used, and reporting procedures for testing each of seven independent performance factors and their uncertainty limits; and (3) specifies the types of test equipment. The independent performance factors specified by *Standard* 84 are sensible (ε_s), latent (ε_l), and total (ε_t) effectivenesses; supply (ΔP_s) and exhaust (ΔP_e) air pressure drops; exhaust air transfer ratio (EATR), which characterizes the fraction of exhaust air transferred to the supply air; and outside air correction factor (OACF), which is the ratio of supply inlet to outlet air flow.

ARI *Standard* 1060, Rating Air-to-Air Energy Recovery Ventilation Equipment, is an industry-established standard for rating air-to-air heat/energy exchanger performance for use in energy recovery ventilation equipment. This standard, based on ASHRAE *Standard* 84, establishes definitions, requirements for marking and nameplate data, and conformance conditions intended for the industry, including manufacturers, engineers, installers, contractors, and users. Standard temperature and humidity conditions at which equipment tests are to be conducted are specified for summer and winter conditions. Published ratings must be reported for each of the seven performance factors specified in ASHRAE *Standard* 84. The ARI certification program using *Standard* 1060 is used to verify ratings published by manufacturers.

ARI *Standard* 1060 requires balanced airflow rates (see Figure 3) and the following conditions:

Winter: Outside air at t_1 = 35°F and t_{w1} = 33°F
Inside (room) air at t_3 = 67°F
t_{w3} = 58°F and $p_2 - p_3 = 0$

Summer: Outside air at t_1 = 95°F and t_{w1} = 78°F
Inside (room) air at t_3 = 75°F
t_{w3} = 63°F and $p_2 - p_3 = 0$

Balanced mass airflows, as required for some ASHRAE and ARI standard test methods, are rarely achieved in field operation for air-handling systems. Fans are nearly constant-volume devices usually designed to run at a preset rpm. Significantly more mass airflow will be transported in cold (winter) conditions than in hot (summer) conditions.

For estimating changes in exchanger performance factors at each operating condition, ASHRAE *Standard* 84 specifies knowledge of seven performance factors (i.e., ΔP_s, ΔP_e, ε_s, ε_L, ε_t, EATR, and OACF), but ARI *Standard* 1060 certifies performance at only a few standard operating conditions. At other operating conditions, these performance factors must be extrapolated using accepted correlations.

Variables that can affect these performance factors for total energy transfer or sensible heat transfer devices include (1) water vapor partial-pressure differences; (2) heat transfer area; (3) air velocity or mass flow rates through the heat exchangers; (4) airflow arrangement or geometric configuration, or characteristic dimension of the flow passage through the recovery ventilator; and (6) method of frost control. The effect of frost control method on seasonal performance is discussed in Phillips et al. (1989a), and sensible versus latent heat recovery for residential comfort-to-comfort applications is addressed in Barringer and McGugan (1989b).

Current testing standards do not validate exchanger performance for testing conditions that require freezing or condensing temperatures, unbalanced airflow ratios, high pressure differentials, or air leakage rates based on varying the inputs. An alternative test setup using a multifunction wind tunnel facility is presented by Sparrow et al. (2001) to provide comprehensive performance data for conditions other than existing standards.

DESIGN CONSIDERATIONS OF VARIOUS ERV SYSTEMS

Fixed-Plate Heat Exchangers

Plate exchangers are available in many configurations, materials, sizes, and flow patterns. Many have modules that can be arranged to handle almost any airflow, effectiveness, and pressure drop requirement. Plates are formed with spacers or separators (e.g., ribs, dimples, ovals) constructed into the plates or with external separators (e.g., supports, braces, corrugations). Airstream separations are sealed by folding, multiple folding, gluing, cementing, welding, or any combination of these, depending on the application and manufacturer. Ease of access for examining and cleaning heat transfer surfaces depends on the configuration and installation.

Heat transfer resistance through the plates is small compared to the airstream boundary layer resistance on each side of the plates. Heat transfer efficiency is not substantially affected by the heat transfer coefficient of the plates. Aluminum is the most popular plate construction material because of its nonflammability and durability. Polymer plate exchangers may improve heat transfer by causing some turbulence in the channel flow, and are popular because of their corrosion resistance and cost-effectiveness. Steel alloys are used for temperatures over 400°F and for specialized applications where cost is not a key factor. Plate exchangers normally conduct sensible heat only; however, water-vapor-permeable materials, such as treated paper and microporous polymeric membranes, may be used to transfer moisture, thus providing total (enthalpy) energy exchange.

Most manufacturers offer modular plate exchangers. Modules range in capacity from 25 to 10,000 cfm and can be arranged into configurations exceeding 100,000 cfm. Multiple sizes and configurations allow selections to meet space and performance requirements.

Plate spacing ranges from 0.1 to 0.5 in., depending on the design and application. Heat is transferred directly from the warm airstreams through the separating plates into the cool airstreams. Usually design, construction, and cost restrictions result in the selection of cross-flow exchangers, but additional counterflow patterns can increase heat transfer effectiveness.

Normally, both latent heat of condensation (from moisture condensed as the temperature of the warm exhaust airstream drops below its dew point) and sensible heat are conducted through the separating plates into the cool supply airstream. Thus, energy is transferred but moisture is not. Recovering 80% or more of the available waste exhaust heat is possible.

Fixed-plate heat exchangers can achieve high sensible heat recovery and total energy effectiveness because they have only a primary heat transfer surface area separating the airstreams and are therefore not inhibited by the additional secondary resistance (e.g., pumping liquid, in runaround systems or transporting a heat transfer medium) inherent in some other exchanger types. In a cross-flow arrangement (Figure 4), they usually do not have sensible effectiveness greater than 75% unless two devices are used in series as shown in Figure 2D.

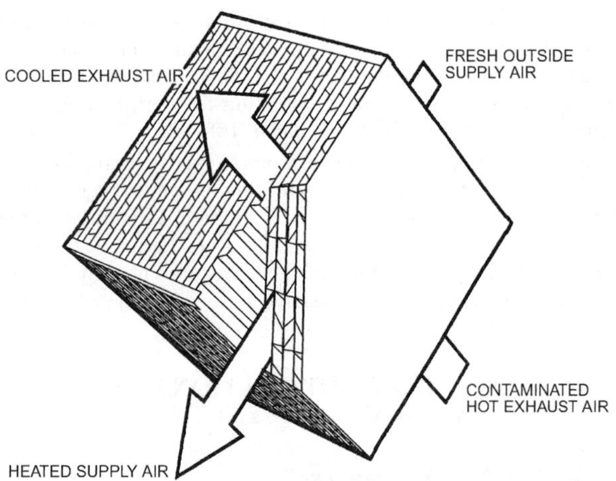

Fig. 4 Fixed-Plate Cross-Flow Heat Exchanger

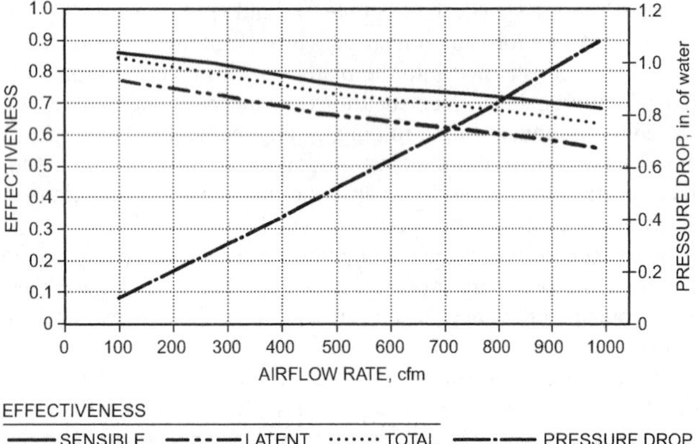

Fig. 5 Variation of Pressure Drop and Effectiveness with Air Flow Rates for a Membrane Plate Exchanger

One advantage of the plate exchanger is that it is a static device with little or no leakage between airstreams. As velocity increases, the pressure difference between the two airstreams increases (Figure 5). High differential pressures may deform the separating plates and, if excessive, can permanently damage the exchanger, significantly reducing the airflow rate on the low-pressure side as well as the effectiveness and possibly causing excessive air leakage. This is not normally a problem because differential pressures in most applications are less than 4 in. of water. In applications requiring high air velocities, high static pressures, or both, plates are not recommended.

Most plate exchangers have condensate drains, which remove condensate and also wastewater in water-wash systems. Heat recovered from a high-humidity exhaust is better returned to a building or process by a sensible heat exchanger rather than an enthalpy exchanger if humidity transfer is not desired.

Frosting can be controlled by preheating incoming supply air, bypassing part of the incoming air, recirculating supply air through the exhaust side of the exchanger, or temporarily interrupting supply air while maintaining exhaust. However, frost on cross-flow heat exchangers is less likely to block the exhaust airflow completely than with other types of exchangers. Generally, frost should be avoided unless a defrost cycle is included.

Fixed-plate heat exchangers can be made from permeable microporous membranes designed to maximize moisture and energy transfer between airstreams while minimizing air transfer. Suitable permeable microporous membranes for this emerging technology include cellulose, polymers, and other synthetic materials such as hydrophilic electrolyte. Hydrophilic electrolytes are made from

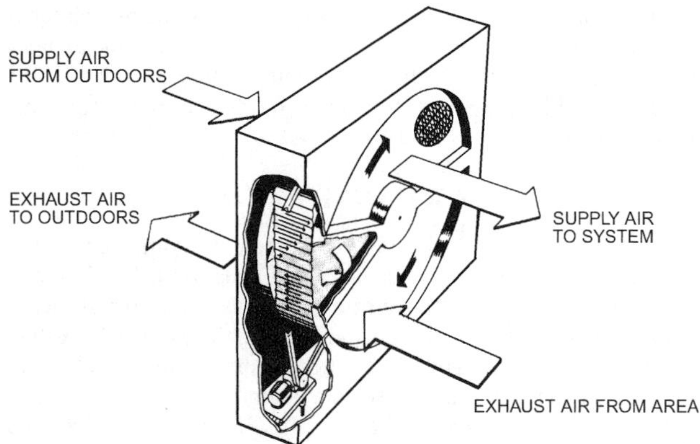

Fig. 6 Rotary Air-to-Air Energy Exchanger

sulphonation chemistry techniques and contain charged ions that attract polar water molecules; adsorption and desorption of water occur in vapor state.

Rotary Air-to-Air Energy Exchangers

A rotary air-to-air energy exchanger, or **rotary enthalpy wheel**, has a revolving cylinder filled with an air-permeable medium having a large internal surface area. Adjacent supply and exhaust airstreams each flow through half the exchanger in a counterflow pattern (Figure 6). Heat transfer media may be selected to recover sensible heat only or total (sensible plus latent) heat.

Sensible heat is transferred as the medium picks up and stores heat from the hot airstream and releases it to the cold one. Latent heat is transferred as the medium adsorbs water vapor from the higher-humidity airstream and desorbs moisture into the lower-humidity airstream, driven in each case by the vapor pressure difference between the airstream and energy exchange medium. Thus, the moist air is dried while the drier air is humidified. In total heat transfer, both sensible and latent heat transfer occur simultaneously. Sensible-only wheels (not coated with desiccant) can also transfer water via a mechanism of condensation and reevaporation driven by dew point and vapor pressure; the effectiveness varies strongly with conditions. Because rotary exchangers have a counterflow configuration and normally use small-diameter flow passages, they are quite compact and can achieve high transfer effectiveness.

Air contaminants, dew point, exhaust air temperature, and supply air properties influence the choice of materials for the casing, rotor structure, and medium of a rotary energy exchanger. Aluminum, steel, and polymers are the usual structural, casing, and rotor materials for normal comfort ventilating systems. Exchanger media are fabricated from metal, mineral, or synthetic materials and provide either random or directionally oriented flow through their structures.

Random-flow media are made by knitting wire into an open woven cloth or corrugated mesh, which is layered to the desired configuration. Aluminum mesh is packed in pie-shaped wheel segments. Stainless steel and monel mesh are used for high-temperature and corrosive applications. These media should only be used with clean, filtered airstreams because they plug easily. Random-flow media also require a significantly larger face area than directionally oriented media for a given airflow and pressure drop.

Directionally oriented media are available in various geometric configurations. The most common consist of small (0.06 to 0.08 in.) air passages parallel to the direction of airflow. Air passages are very similar in performance regardless of their shape (triangular, hexagonal, parallel plate, or other). Aluminum foil, paper, plastic, and synthetic materials are used for low and medium temperatures. Stainless steel and ceramics are used for high temperatures and corrosive atmospheres.

Media surface areas exposed to airflow vary from 100 to 1000 ft^2/ft^3, depending on the type of medium and physical configuration. Media may also be classified by their ability to recover sensible heat only or total heat. Media for sensible heat recovery are made of aluminum, copper, stainless steel, and monel. Media for total heat recovery can be from any of a number of materials and treated with a desiccant (typically zeolites, molecular sieves, silica gels, activated alumina, titanium silicate, synthetic polymers, lithium chloride, or aluminum oxide) to have specific moisture recovery characteristics.

Cross-Leakage. Cross-leakage, cross-contamination, or mixing between supply and exhaust airstreams occurs in all rotary energy exchangers by two mechanisms: carryover and seal leakage. **Carryover** occurs as air is entrained within the rotation medium and is carried into the other airstream. **Leakage** occurs because the differential static pressure across the two airstreams drives air from a higher to a lower static pressure region. Cross-contamination can be reduced by placing the blowers so that they promote leakage of outside air to the exhaust airstream. Carryover occurs each time a portion of the matrix passes the seals dividing the supply and exhaust airstreams. Because carryover from exhaust to supply may be undesirable, a **purge section** can be installed on the heat exchanger to reduce cross-contamination.

In many applications, recirculating some air is not a concern. However, critical applications such as hospital operating rooms, laboratories, and cleanrooms require stringent control of carryover. Carryover can be reduced to less than 0.1% of the exhaust airflow with a purge section but cannot be completely eliminated.

The theoretical carryover of a wheel without a purge section is directly proportional to the speed of the wheel and the void volume of the medium (75 to 95% void, depending on type and configuration). For example, a 10 ft diameter, 8 in. deep wheel with a 90% void volume operating at 14 rpm has a carryover volumetric flow of

$$\pi(10/2)^2(8/12)(0.9)(14) = 660 \text{ cfm}$$

If the wheel is handling a 20,000 cfm balanced flow, the percentage carryover is

$$\frac{660}{20{,}000} \times 100 = 3.3\%$$

The exhaust fan, which is usually located at the exit of the exchanger, should be sized to include leakage and purge air flows.

Control. Two control methods are commonly used to regulate wheel energy recovery. In **supply air bypass** control, the amount of supply air allowed to pass through the wheel establishes the supply air temperature. An air bypass damper, controlled by a wheel supply air discharge temperature sensor, regulates the proportion of supply air permitted to bypass the exchanger.

The second method regulates the energy recovery rate by varying wheel rotational speed. The most frequently used **variable-speed drives** are (1) a silicon controlled rectifier (SCR) with variable-speed dc motor, (2) a constant-speed ac motor with hysteresis coupling, and (3) an ac frequency inverter with an ac induction motor.

Figure 7 shows the effectiveness ε, sensible heat transfer only, with balanced airflow, convection-conduction ratio less than 4, and no leakage or cross-flow, of a regenerative counterflow heat wheel versus number of transfer units (NTU). This simple example of a regenerative wheel also shows that regenerative counterflow rotary effectiveness increases with wheel speed (C_r is proportional to wheel speed), but there is no advantage in going beyond $C_r/C_{min} = 5$ because the carryover of contaminants increases with wheel speed. See Shah (1981) or Kays and Crawford (1993) for details.

Rotary energy or enthalpy wheels are more complex than heat wheels, but recent research has characterized their behavior using laboratory and field data (Johnson et al. 1998).

A dead band control, which stops or limits the exchanger, may be necessary when no recovery is desired (e.g., when outside air temperature is higher than the required supply air temperature but below the exhaust air temperature). When outside air temperature is above the exhaust air temperature, the equipment operates at full capacity to cool incoming air. During very cold weather, it may be necessary to heat the supply air, stop the wheel, or, in small systems, use a defrost cycle for frost control.

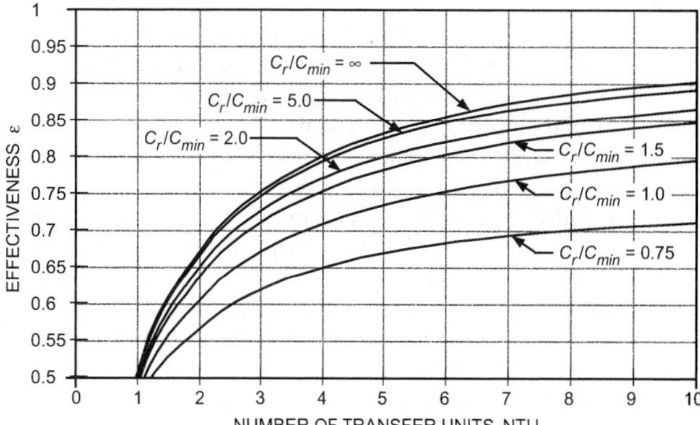

Fig. 7 Effectiveness of Counterflow Regenerator
(Shah 1981)

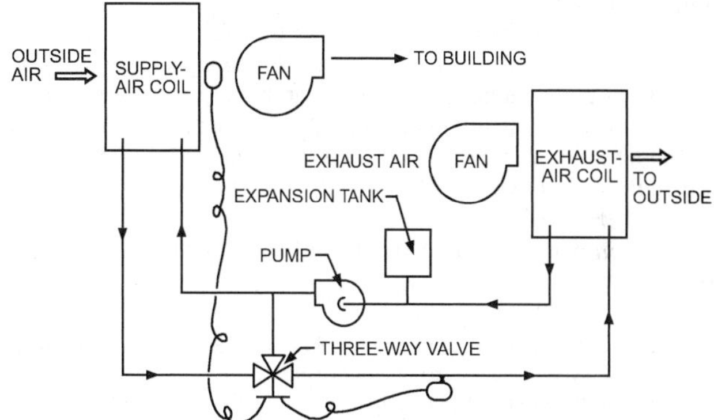

Fig. 8 Coil Energy Recovery Loop

Rotary enthalpy wheels require little maintenance and tend to be self-cleaning because the airflow direction is reversed for each rotation of the wheel. The following maintenance procedures ensure best performance:

- Clean the medium when lint, dust, or other foreign materials build up, following the manufacturer's instructions.
- Maintain drive motor and train according to the manufacturer's recommendations. Speed-control motors that have commutators and brushes require more frequent inspection and maintenance than induction motors. Brushes should be replaced, and the commutator should be periodically turned and undercut.
- Inspect wheels regularly for proper belt or chain tension.
- Refer to the manufacturer's recommendations for spare and replacement parts.

Coil Energy Recovery (Runaround) Loops

A typical coil energy recovery loop (Figure 8) places extended-surface, finned-tube water coils in the supply and exhaust airstreams of a building or process. The coils are connected in a closed loop by counterflow piping through which an intermediate heat transfer fluid (typically water or antifreeze solution) is pumped.

Moisture must not freeze in the exhaust coil air passage. A dual-purpose, three-way temperature control valve prevents the exhaust coil from freezing. The valve is controlled to maintain the temperature of solution entering the exhaust coil at 40°F or above. This condition is maintained by bypassing some of the warmer solution around the supply air coil. The valve can also ensure that a prescribed air temperature from the supply air coil is not exceeded.

Coil energy recovery loops are highly flexible and well suited to renovation and industrial applications. The loop accommodates remote supply and exhaust ducts and allows simultaneous transfer of energy between multiple sources and uses. An expansion tank must be included to allow fluid expansion and contraction. A closed expansion tank minimizes oxidation when ethylene glycol is used.

Standard finned-tube water coils may be used; however, these need to be selected using an accurate simulation model if high effectiveness and low costs are needed (Johnson et al. 1995). Integrating runaround loops in buildings with variable loads to achieve maximum benefits may require combining the runaround simulation with building energy simulation (Dhital et al. 1995). Manufacturers' design curves and performance data should be used when selecting coils, face velocities, and pressure drops, but only when the design data are for the same temperature and operating conditions as in the runaround loop.

The coil energy recovery loop cannot transfer moisture between airstreams; however, indirect evaporative cooling can reduce the exhaust air temperature, which significantly reduces cooling loads. For the most cost-effective operation, with equal airflow rates and no condensation, typical effectiveness values range from 45 to 65%. The highest effectiveness does not necessarily give the greatest net life-cycle cost saving.

The following example illustrates the capacity of a typical system:

Example 2. Runaround Loop Energy Recovery System. A waste heat recovery system heats 48,000 lb/h of air from a 0°F design outside temperature with an exhaust temperature of 75°F db and 60°F wb. Air flows through identical eight-row coils at 400 fpm. A 30% ethylene glycol solution flows through the coils at 3.4 cfm. Assuming the performance characteristic of the runaround loop as shown in Figure 9, determine the sensible effectiveness and exit temperature of the exhaust air (1) when the outside air temperature is at 0°F and (2) when it is at 18°F.

Solution:

Figure 9 shows the effect of outside air temperature on capacity, including the effects of the three-way temperature control valve. For this example, the capacity is constant for outside air temperatures below 18°F. This constant output of 413,000 Btu/h occurs because the valve has to control the temperature of fluid entering the exhaust coil to prevent frosting. As the exhaust coil is the source of heat and has a constant airflow rate, entering air temperature, liquid flow rate, entering fluid temperature (as set by the valve), and fixed coil parameters, energy recovered must be controlled to prevent frosting in the exhaust coil.

(1) Equation (2a) with the numerator equal to 413,000 Btu/h may be used to calculate the sensible heat effectiveness.

$$\varepsilon = \frac{413{,}000 \text{ Btu/h}}{(48{,}000 \text{ lb/h})(0.24 \text{ Btu/lb}\cdot{}^\circ\text{F})(75-0)} = 48\%$$

From Equation (3b), the exhaust airstream exit temperature can be estimated as

$$t_4 = 75 + 0.48\frac{(48{,}000 \text{ lb/h})(0.24 \text{ Btu/lb}\cdot{}^\circ\text{F})}{(48{,}000 \text{ lb/h})(0.24 \text{ Btu/lb}\cdot{}^\circ\text{F})}(0-75) = 39^\circ\text{F}$$

(2) When the three-way control valve operates at outside air temperatures of 18°F or lower, 413,000 Btu/h is recovered. Using the same equations, it can be shown that at 18°F, the sensible heat effectiveness is 64% and the exhaust air leaving dry-bulb temperature is 39°F. Above 60°F outside air temperature, the supply air is cooled with an evaporative cooler located upstream from the exhaust coil.

Typically, the sensible heat effectiveness of a coil energy recovery loop is independent of the outside air temperature. However, when the capacity is controlled, the sensible heat effectiveness decreases.

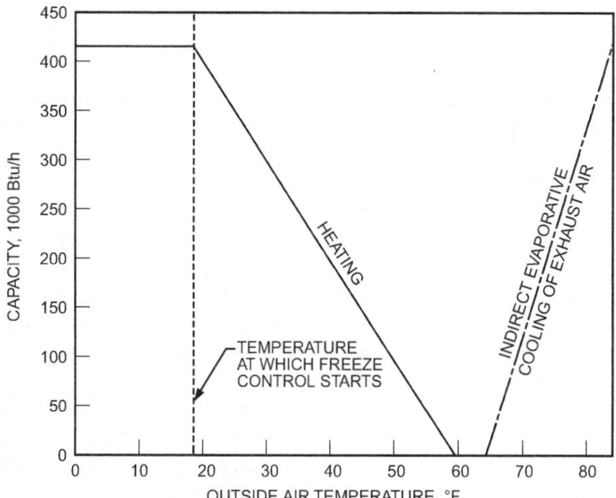

Fig. 9 Energy Recovery Capacity Versus Outside Air Temperature for Typical Loop

Coil energy recovery loops use coils constructed to suit their environment and operating conditions. For typical comfort-to-comfort applications, standard coil construction usually suffices. In process-to-process and process-to-comfort applications, the effect of high temperature, condensable gases, corrosives, and contaminants on the coil(s) must be considered. At temperatures above 400°F, special construction may be required to ensure a permanent fin-to-tube bond. The effects of condensable gases and other adverse factors may require special coil construction and/or coatings. Chapters 21 and 23 discuss the construction and selection of coils in more detail.

Complete separation of the airstreams eliminates cross-contamination between the supply and exhaust air.

Coil energy recovery loops require little maintenance. The only moving parts are the circulation pump and three-way control valve. However, to ensure optimum operation, the air should be filtered, the coil surface cleaned regularly, the pump and valve maintained, and the transfer fluid refilled or replaced periodically. Fluid manufacturers or their representatives should be contacted for specific recommendations.

The thermal transfer fluid selected for a closed-loop exchanger depends on the application and on the temperatures of the two airstreams. An inhibited ethylene glycol solution in water is common when freeze protection is required. These solutions break down to an acidic sludge at temperatures above 275°F. If freeze protection is needed and exhaust air temperatures exceed 275°F, a nonaqueous synthetic heat transfer fluid should be used. Heat transfer fluid manufacturers and chemical suppliers should recommend appropriate fluids.

Heat Pipe Heat Exchangers

Figure 10 shows a typical heat pipe assembly. Hot air flowing over the evaporator end of the heat pipe vaporizes the working fluid. A vapor pressure gradient drives the vapor to the condenser end of the heat pipe tube, where the vapor condenses, releasing the latent energy of vaporization (Figure 11). The condensed fluid is wicked or flows back to the evaporator, where it is revaporized, thus completing the cycle. Thus the heat pipe's working fluid operates in a closed-loop evaporation/condensation cycle that continues as long as there is a temperature difference to drive the process. Using this mechanism, the heat transfer rate along a heat pipe is up to 1000 times greater than through copper (Ruch 1976).

Energy transfer in heat pipes is often considered isothermal. However, there is a small temperature drop through the tube wall, wick, and fluid medium. Heat pipes have a finite heat transfer capacity that is affected by factors such as wick design, tube diameter, working fluid, and tube (heat pipe) orientation relative to horizontal.

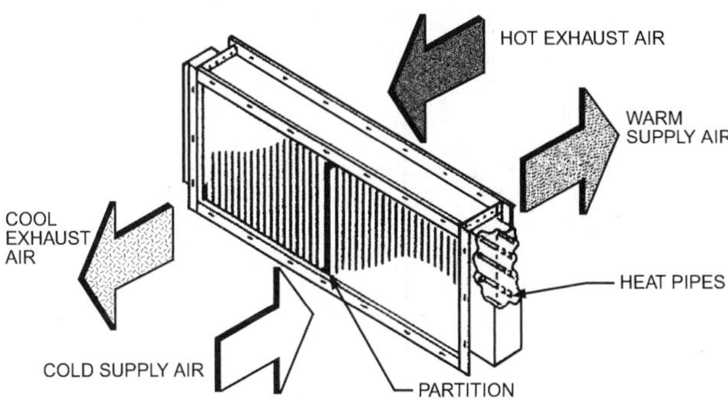

Fig. 10 Heat Pipe Assembly

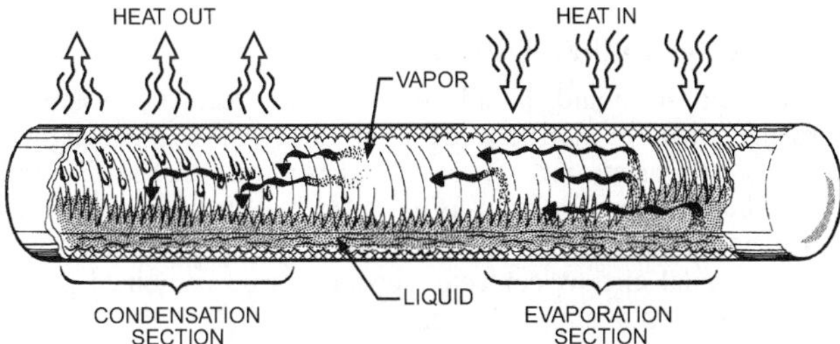

Fig. 11 Heat Pipe Operation

HVAC systems use copper or aluminum heat pipe tubes with aluminum fins. Fin designs include continuous corrugated plate, continuous plain, and spiral. Modifying fin design and tube spacing changes pressure drop at a given face velocity.

For process-to-comfort applications with large temperature changes, tubes and fins are usually constructed of the same material to avoid problems with different thermal expansions of materials. Heat pipe heat exchangers for exhaust temperatures below 425°F are most often constructed with aluminum tubes and fins. Protective coatings allow inexpensive aluminum to replace exotic metals in corrosive atmospheres; these coatings have a minimal effect on thermal performance.

Heat pipe heat exchangers for use above 425°F are generally constructed with steel tubes and fins. The fins are often aluminized to prevent rusting. Composite systems for special applications may be created by assembling units with different materials and/or different working fluids.

Selecting the proper working fluid for a heat pipe is critical to long-term operation. The working fluid should have high latent heat of vaporization, a high surface tension, and a low liquid viscosity over the operating range; it must be thermally stable at operating temperatures. Decomposition of the thermal fluids can form noncondensable gases that deteriorate performance. For low-temperature applications, gases such as helium can be used as working fluid; for moderate temperatures, liquids such as water can be used; and for high temperatures, liquid metals such as sodium or mercury can be used.

Heat pipe heat exchangers typically have no cross-contamination for pressure differentials between airstreams of up to 50 in. of water. A vented double-wall partition between the airstreams can provide additional protection against cross-contamination. If an exhaust duct is attached to the partition space, any leakage is usually withdrawn and exhausted from the space between the two ducts.

Heat pipe heat transfer capacity depends on design and orientation. Figure 12 shows a typical effectiveness curves for various face velocities and rows of tubes. As the number of rows increases, effectiveness increases at a decreasing rate. For example, doubling the number of rows of tubes in a 60% effective heat exchanger increases the effectiveness to 75%. The effectiveness of a counterflow heat pipe heat exchanger depends on the total number of rows such that

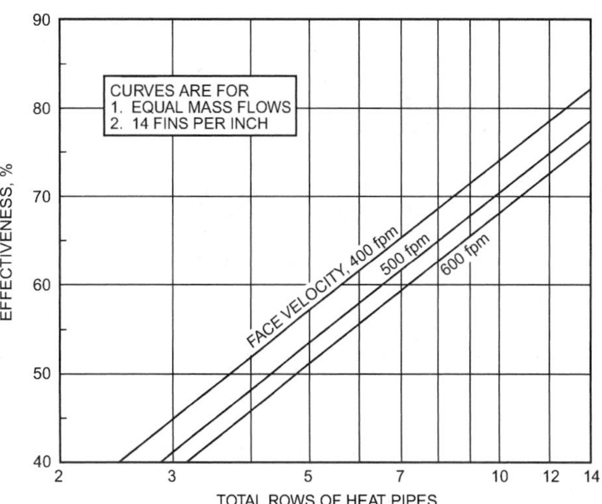

Fig. 12 Heat Pipe Exchanger Effectiveness

two units in series yield the same effectiveness as a single unit of the same total number of rows. Series units are often used to facilitate handling, cleaning, and maintenance. Effectiveness also depends on outside air temperature and the ratio of mass flow rates of the airstreams. Typically, heat capacity in the cooling season increases with a rise in outside air temperature. It has an opposite effect during the heating season. Effectiveness increases with the ratio of mass flow rates of the fluids (flow rate of the fluid with warmer entering temperature over that of cooler entering fluid temperature).

The heat transfer capacity of a heat pipe increases roughly with the square of the inside diameter of the pipe. For example, at a given tilt angle, a 1 in. inside diameter heat pipe will transfer roughly 2.5 times as much energy as a 5/8 in. inside diameter pipe. Consequently, heat pipes with large diameters are used for larger-airflow applications and where level installation is required to accommodate both summer and winter operation.

Heat transfer capacity is virtually independent of heat pipe length, except for very short heat pipes. For example, a 4 ft long heat pipe has approximately the same capacity as an 8 ft pipe. Because the 8 ft heat pipe has twice the external heat transfer surface area of the 4 ft pipe, it will reach its capacity limit sooner. Thus, in a given application, it is more difficult to meet capacity requirements as the heat pipes become longer. A system can be reconfigured to a taller face height and more numerous but shorter heat pipes to yield the same airflow face area while improving system performance.

The selection of fin design and spacing should be based on the dirtiness of the two airstreams and the resulting cleaning and maintenance required. For HVAC applications, 11 to 14 fins per in. (fpi) is common. Wider spacing (8 to 10 fpi) is usually used for industrial applications. Plate-fin heat pipe heat exchangers can easily be constructed with different fin spacing for the exhaust and supply airstreams, allowing wider fin spacing on the dirty exhaust side. This increases design flexibility where pressure drop constraints exist and also prevents deterioration of performance caused by dirt buildup on the exhaust side surface.

Changing the tilt of a heat pipe controls the amount of heat it transfers. Operating the heat pipe on a slope with the hot end below (or above) the horizontal improves (or retards) condensate flow back to the evaporator end of the heat pipe. This feature can be used to regulate the effectiveness of the heat pipe heat exchanger (Guo et al. 1998).

Tilt control is achieved by pivoting the exchanger about the center of its base and attaching a temperature-controlled actuator to one end of the exchanger (Figure 13). Pleated flexible connectors attached to the ductwork allow freedom for the small tilting movement of only a few degrees.

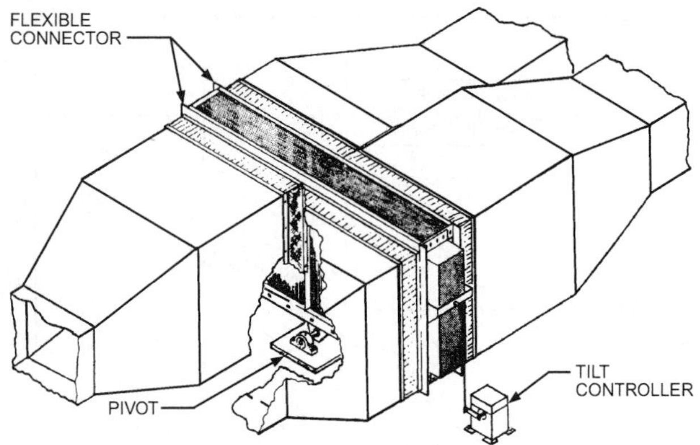

Fig. 13 Heat Pipe Heat Exchanger with Tilt Control

Tilt control may be desired

- To change from supply air heating to supply air cooling (i.e., to reverse the direction of heat flow) during seasonal changeover
- To modulate effectiveness to maintain desired supply air temperature (often required for large buildings to avoid overheating air supplied to the interior zone)
- To decrease effectiveness to prevent frost formation at low outside air temperatures (with reduced effectiveness, exhaust air leaves the unit at a warmer temperature and stays above frost-forming conditions)

Other devices, such as face-and-bypass dampers and preheaters, can also be used to control the rate of heat exchange.

Bidirectional heat transfer in heat pipes is achieved through recent design improvements. Some heat pipe manufacturers have eliminated the need for tilting for capacity control or seasonal changeover. Once installed, the unit is removed only for routine maintenance. Capacity and frost control can also be achieved through bypassing airflow over the heat pipes, as for air coils.

Example 3. Sensible Heat Energy Recovery in a Heat Pipe

Outside air at 50°F enters a six-row heat pipe with a flow rate of 660 lb/min and a face velocity of 500 fpm. Exhaust air enters the heat pipe with the same velocity and flow rate but at 75°F. The pressure drop across the heat pipe is 0.6 in. of water. The supply air density is 0.08 lb/ft^3. The efficiency of the electric motor and the connected fan are 90 and 75%, respectively. Assuming the performance characteristics of the heat pipe are as shown in Figure 12, determine the sensible effectiveness, exit temperature of supply air to the space, energy recovered, and power supplied to the fan motor.

Solution:

From Figure 12, at face velocity of 500 fpm and with six rows, the effectiveness is about 58%. Because the mass flow rate of the airstreams is the same and assuming their specific heat of 0.24 Btu/lb·°F is the same, then the exit temperature of the supply air to the space can be obtained from Equation (3a):

$$t_2 = 50 - 0.58\frac{(660\text{ lb/min})(0.24\text{ Btu/lb}\cdot°\text{F})}{(660\text{ lb/min})(0.24\text{ Btu/lb}\cdot°\text{F})}(50 - 75) = 64.5°\text{F}$$

The sensible energy recovered can be obtained from Equation (3c) as

$$q_s = (60)(660\text{ lb/min})(0.24\text{ Btu/lb}\cdot°\text{F})(64.5 - 50) = 139{,}200\text{ Btu/h}$$

The supply air fan power can be obtained from Equation (11) as

$$Q_s = \frac{660\text{ lb/min}}{0.08\text{ lb/ft}^3} = 8250\text{ ft}^3\text{/min}$$

$$P_s = [(8250\text{ft}^3\text{/min})(0.6)]/[(6356)(0.9)(0.75)] = 1.15\text{ hp}$$

Because there are two airstreams, neglecting the difference in the air densities of the airstreams, the total pumping power of the heat pipe is twice the above value (i.e., 2.3 hp).

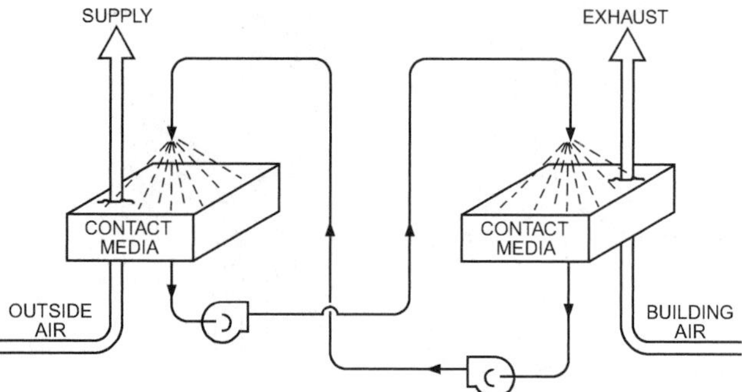

Fig. 14 Twin-Tower Enthalpy Recovery Loop

Twin-Tower Enthalpy Recovery Loops

In this air-to-liquid, liquid-to-air enthalpy recovery system, a sorbent liquid circulated between supply and exhaust contactor towers directly contacts both airstreams, transporting water vapor and energy between the airstreams (Figure 14). Supply air temperatures can be as high as 115°F or as low as –40°F. Any number of vertical and horizontal airflow contactor towers can be combined into a common system of any airflow capacity.

Leaving air passes through demister pads to remove entrained sorbent solution. Airstreams containing lint, animal hair, or other solids should be filtered upstream of the contactor towers. Wetted particles should be filtered from the sorbent solution, which minimizes particulate cross-contamination. Sorbent solutions (typically a halogen salt solution such as lithium chloride and water) usually contain bactericidal and viricidal additives. Testing has shown that contactor towers can effectively remove up to 94% of atmospheric bacteria, a desirable feature in health care applications. Limited gaseous cross-contamination may occur. If either airstream contains gaseous contaminants, their effects on the sorbent solution should be investigated.

In colder climates, moisture losses from the exhaust airstream may overdilute the sorbent solution. Heating the sorbent liquid entering the supply air contactor tower raises the discharge temperature and humidity of the leaving supply air, preventing overdilution. This, coupled with automatic makeup water addition, can maintain sorbent solution concentrations during cold weather, enabling the system to deliver air at a fixed humidity and temperature.

THERMOSIPHON HEAT EXCHANGERS

Two-phase thermosiphon heat exchangers are sealed systems that consist of an evaporator, a condenser, interconnecting piping, and an intermediate working fluid in both liquid and vapor phases. Two types of thermosiphon are used: a sealed tube (Figure 15) and a coil type (Figure 16). In the **sealed-tube thermosiphon**, the evaporator and the condenser are usually at opposite ends of a bundle of straight, individual thermosiphon tubes, and the exhaust and supply ducts are adjacent to each other (this arrangement is similar to that in a heat pipe system). In **coil-type thermosiphons**, evaporator and condenser coils are installed independently in the ducts and are interconnected by the working fluid piping (this configuration is similar to that of a coil energy recovery loop).

A thermosiphon is a sealed system containing a two-phase working fluid. Because part of the system contains vapor and part contains liquid, the pressure in a thermosiphon is governed by the liquid temperature at the liquid/vapor interface. If the surroundings cause a temperature difference between the regions where liquid and vapor interfaces are present, the resulting vapor pressure difference causes vapor to flow from the warmer to the colder region. The flow is sustained by condensation in the cooler region and by evaporation in the warmer region. The condenser and evaporator must be oriented so

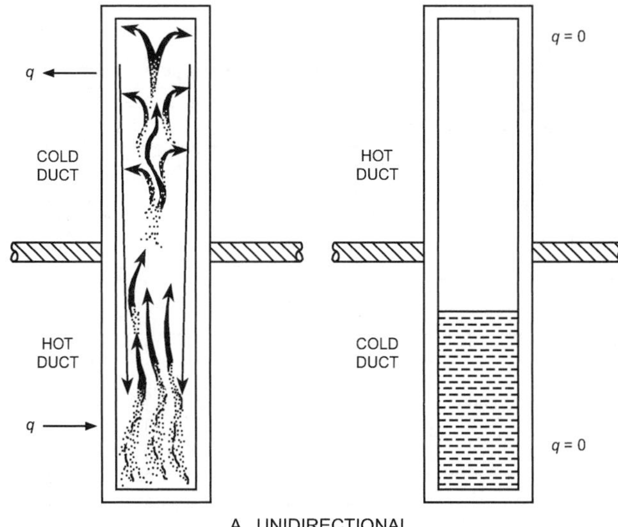

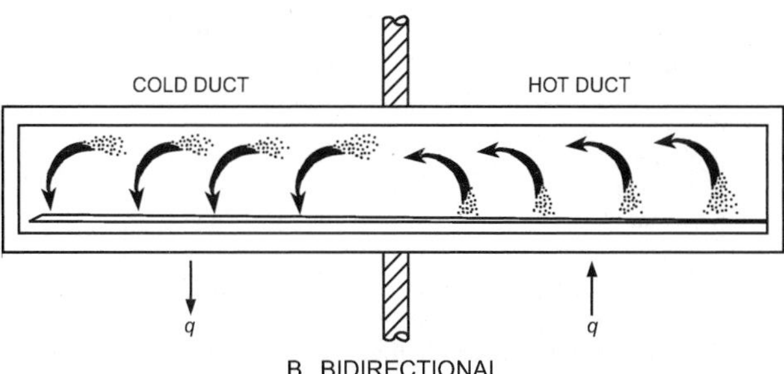

Fig. 15 Sealed-Tube Thermosiphons

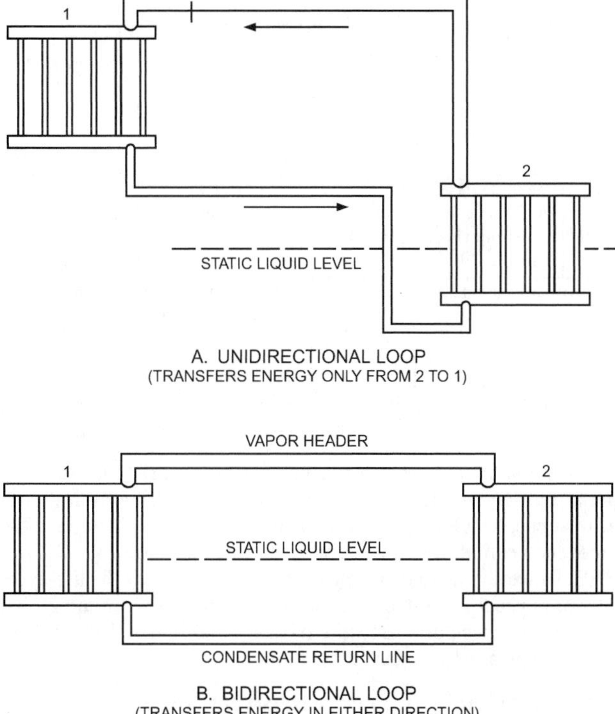

Fig. 16 Coil-Type Thermosiphon Loops

that the condensate can return to the evaporator by gravity (Figures 15 and 16).

In thermosiphon systems, a temperature difference and gravity force are required for the working fluid to circulate between the evaporator and condenser. As a result, thermosiphons may be designed to transfer heat equally in either direction (bidirectional), in one direction only (unidirectional), or in both directions unequally.

Although similar in form and operation to heat pipes, thermosiphon tubes are different in two ways: (1) they have no wicks and hence rely only on gravity to return condensate to the evaporator, whereas heat pipes use capillary forces; and (2) they depend, at least initially, on nucleate boiling, whereas heat pipes vaporize the fluid from a large, ever-present liquid/vapor interface. Thus, thermosiphon heat exchangers may require a significant temperature difference to initiate boiling (Mathur and McDonald, 1987; McDonald and Shivprasad 1989). Thermosiphon tubes require no pump to circulate the working fluid. However, the geometric configuration must be such that liquid working fluid is always present in the evaporator section of the heat exchanger.

Thermosiphon loops differ from other coil energy recovery loop systems in that they require no pumps and hence no external power supply, and the coils must be appropriate for evaporation and condensation. Two-phase thermosiphon loops are used for solar water heating (Mathur 1990a) and for performance enhancement of existing (i.e., retrofit applications) air-conditioning systems (Mathur 1997). Two-phase thermosiphon loops can be used to downsize new air-conditioning systems and thus reduce the overall project costs. Figure 17 shows thermosiphon loop performance (Mathur and McDonald 1986).

COMPARISON OF AIR-TO-AIR ENERGY RECOVERY SYSTEMS

It is difficult to compare different types of air-to-air energy recovery systems based on overall performance. They can be compared based on certified ratings such as sensible, latent, and total effectiveness or on air leakage parameters. To compare them on payback period or maximum energy cost savings, accurate values of their capital cost, life, and maintenance cost, which vary from product to product for the same type of recovery system, must be known. Without such data, and considering the data available in the open literature such as that presented by Besant and Simonson (2003), use Table 2's comparative data for common types of air-to-air energy recovery devices.

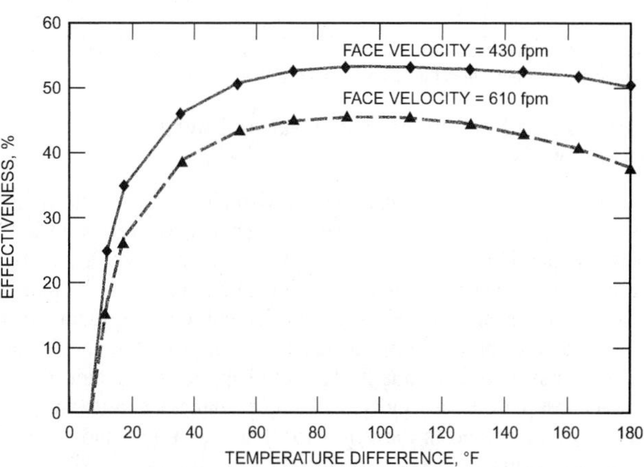

Fig. 17 Typical Performance of Two-Phase Thermosiphon Loop
(Mathur and McDonald 1986)

Table 2 Comparison of Air-to-Air Energy Recovery Devices

	Fixed Plate	Membrane Plate	Energy Wheel	Heat Wheel	Heat Pipe	Runaround Coil Loop	Thermosiphon	Twin Towers
Airflow arrangements	Counterflow Cross-flow	Counterflow Cross-flow	Counterflow Parallel flow	Counterflow	Counterflow Parallel flow	—	Counterflow Parallel flow	—
Equipment size range, cfm	50 and up	50 and up	50 to 74,000 and up	50 to 74,000 and up	100 and up	100 and up	100 and up	—
Typical sensible effectiveness (m_s = m_e), %	50 to 80	50 to 75	50 to 85	50 to 85	45 to 65	55 to 65	40 to 60	40 to 60
Typical latent effectiveness,* %	—	50 to 72	50 to 85	0	—	—	—	—
Total effectiveness,* %	—	50 to 73	50 to 85	—	—	—	—	—
Face velocity, fpm	200 to 1000	200 to 600	500 to 1000	400 to 1000	400 to 800	300 to 600	400 to 800	300 to 450
Pressure drop, in. of water	0.4 to 4	0.4 to 2	0.4 to 1.2	0.4 to 1.2	0.6 to 2	0.6 to 2	0.6 to 2	0.7 to 1.2
EATR, %	0 to 5	0 to 5	0.5 to 10	0.5 to 10	0 to 1	0	0	0
OACF	0.97 to 1.06	0.97 to 1.06	0.99 to 1.1	1 to 1.2	0.99 to 1.01	1.0	1.0	1.0
Temperature range, °F	–75 to 1470	15 to 120	–65 to 1470	–65 to 1470	–40 to 105	–50 to 930	–40 to 105	–40 to 115
Typical mode of purchase	Exchanger only Exchanger in case Exchanger and blowers Complete system	Exchanger only Exchanger in case Exchanger and external blowers Complete system	Exchanger only Exchanger in case Exchanger and blowers Complete system	Exchanger only Exchanger in case Exchanger and blowers Complete system	Exchanger only Exchanger in case Exchanger and blowers Complete system	Coil only Complete system	Exchanger only Exchanger in case	Complete system
Advantages	No moving parts Low pressure drop Easily cleaned	No moving parts Low pressure drop Low air leakage	Moisture or mass transfer Compact large sizes Low pressure drop Available on all ventilation system platforms	Compact large sizes Low pressure drop Easily cleaned	No moving parts except tilt Fan location not critical Allowable pressure differential up to 2 psi	Exhaust airstream can be separated from supply air Fan location not critical	No moving parts Exhaust airstream can be separated from supply air Fan location not critical	Latent transfer from remote airstreams Efficient microbiological cleaning of both supply and exhaust airstreams
Limitations	Large size at higher flow rates	Few suppliers Long-term maintenance and performance unknown	Supply air may require some further cooling or heating Some EATR without purge	Some EATR without purge	Effectiveness limited by pressure drop and cost Few suppliers	Predicting performance requires accurate simulation model	Effectiveness may be limited by pressure drop and cost Few suppliers	Few suppliers Maintenance and performance unknown
Heat rate control (HRC) methods	Bypass dampers and ducting	Bypass dampers and ducting	Bypass dampers and wheel speed control	Bypass dampers and wheel speed control	Tilt angle down to 10% of maximum heat rate	Bypass valve or pump speed control	Control valve over full range	Control valve or pump speed control over full range

*Rated effectiveness values are for balanced flow conditions. Effectiveness values increase slightly if flow rates of either or both airstreams are higher than flow rates at which testing is done.

EATR = Exhaust Air Transfer Ratio
OACF = Outside Air Correction Factor

Long-Term Performance of Heat or Energy Recovery Ventilators

The type of seal used on a rotating wheel affects its performance as well as its capital cost. A measure of energy recovery ventilator performance is the relative magnitude of the actual energy recovered and the power supplied to the fans to circulate the airstreams. The cost of power supplied to the fans depends on the pressure drop of airstreams, volume flow rate, and the combined efficiency of the fan motor systems. The quality of power supplied to the fans is high and its cost per unit energy is much higher than the quality and cost of energy recovered in the ventilator. The magnitude and costs of these two forms of energy vary over the year. Besant and Simonson (2003) suggest that a parameter such as ratio of energy recovered (RER) may be introduced to reflect the long-term performance of the recovery ventilators:

$$\text{RER} = \frac{\int(\text{rate of energy recovered})dt}{\int(\text{rate of power supplied to the fan motors})dt} \tag{25}$$

RER is similar to the energy efficiency ratio (EER) for chillers or unitary air-conditioning equipment. Besant and Simonson (2003) also suggest that the entire system performance, including the recovery ventilator, can be represented by the ratio of COP and RER. However, the true overall system performance is the life-cycle cost or payback period, both of which take into account the capital and maintenance costs. Because of lack of sufficient data on these factors, they are not presented in Table 2.

Selection of Heat or Energy Recovery Ventilators

Heat and energy recovery ventilators are available as heat exchangers only or as a complete system, including the heat exchangers and fan/motor systems, as indicated in Table 2. Energy recovery is also available integrated into unitary air-conditioning equipment or in both standard and custom air-handling systems. Selection of such units primarily is dictated by the quantity of ventilation air. Several manufacturers have developed software or tables to help the user in selection of these units. The user may have to determine the required fan size (see Example 7), if only the heat exchanger is to be purchased.

ENERGY AND/OR MASS RECOVERY CALCULATION PROCEDURE

The rate of energy transfer to or from an airstream depends on the rate and direction of heat transfer and water vapor (moisture) transfer. Under customary design conditions, heat and water vapor transfer will be in the same direction, but the rate of heat transfer will not be the same as the rate of energy transfer by the cross-stream flow of water vapor. This is because the driving potentials for heat and mass transfer are different, as are the respective wall resistances for the two types of transport. Both transfer rates depend on exchanger construction characteristics. Equation (26) is used to determine the rate of energy transfer when sensible (temperature) and latent (moisture) energy transfer occurs; Equation (27) is used for sensible-only energy transfer.

$$q_t = 60Q\rho(h_{in} - h_{out}) \quad (26)$$

$$q_s = 60Q\rho c_p(t_{in} - t_{out}) \quad (27)$$

where

- $q_t = q_s + q_L$ = total energy transfer, Btu/h
- q_s = sensible heat transfer, Btu/h
- Q = airflow rate, cfm
- ρ = air density, lb/ft^3
- c_p = specific heat of air = 0.24 Btu/lb·°F
- t_{in} = dry-bulb temperature of air entering exchanger, °F
- t_{out} = dry-bulb temperature of air leaving exchanger, °F
- h_{in} = enthalpy of air entering heat exchanger, Btu/lb
- h_{out} = enthalpy of air leaving heat exchanger, Btu/lb

The following general procedure may be used to determine the performance and energy recovered in air-to-air energy recovery applications at each operating condition.

Step 1. Determine supply and exhaust air pressure drops Δp_s and Δp_e across exchanger.

Request air pressure drops Δp_s and Δp_e across the heat or energy exchanger from the manufacturer, who may have certified ARI *Standard* 1060 test condition data obtained using ASHRAE *Standard* 84 as a test procedure and analysis guide. These data may be extrapolated to non-ARI conditions by the manufacturer using correlations such as Equations (14) or (16), if their restrictions are satisfied. For other flow conditions, somewhat different correlations may be more accurate to determine the pressure drop.

Step 2. Calculate theoretical maximum moisture and energy transfer rates m_{max} and q_{max}.

The airstream with the lower mass flow m_{min} limits heat and moisture transfer. Some designers specify and prefer working with airflows stated at standard temperature and pressure conditions. To correctly calculate moisture or energy transfer rates, the designer must determine mass flow rates. For this reason, the designer must know whether airflow rates are quoted for the entry conditions specified or at standard temperature and pressure conditions. If necessary, convert flow rates to mass flow rates (e.g., scfm to lb/min) and then determine which airstream has the minimum mass.

The theoretical maximum moisture, sensible heat, latent heat and total energy rates are given by Equations (4f), (2c), (4c), and (6c), respectively.

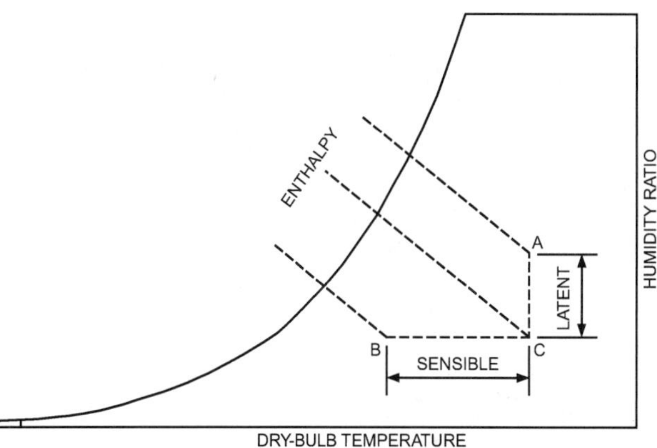

Fig. 18 Maximum Sensible and Latent Heat from Process A-B

The split between latent and sensible energy can be determined by plotting airstream conditions on a psychrometric chart as shown in Figure 18. Maximum sensible heat transfer is represented by a horizontal line drawn between the two dry-bulb temperatures, and maximum latent energy transfer is represented by the vertical line.

Step 3. Establish the moisture, sensible, and total effectiveness ε_s, ε_L, and ε_t.

Each of these ratios is obtained from manufacturers' product data using input conditions and airflows for both airstreams. The effectiveness for airflows depends on (1) exchanger construction, including configuration, heat transfer material, moisture transfer properties, transfer surface area, airflow path, distance between heat transfer surfaces, and overall size; and (2) inlet conditions for both airstreams, including pressures, velocities, temperatures, and humidities. In applications with unequal airflow rates, the enthalpy change will be higher for the airstream with the lesser mass flow.

Step 4. Calculate actual moisture (latent) and energy (sensible, latent or total) transfer rates.

The actual moisture, sensible heat, latent heat, and total energy rates are given by Equations (4e), (2b), (4b), and (6b), respectively. Note that ε_m (mass effectiveness) = ε_L (latent effectiveness), as shown by Equations (4a) and (4d).

Step 5. Calculate leaving air properties for each airstream using Equations (3), (5), and (7).

With an enthalpy or moisture-permeable heat exchanger, moisture (and its inherent latent energy) is transferred between airstreams. With a sensible-only heat exchanger, if the warmer airstream is cooled below its dew point, the resulting condensed moisture transfers additional energy. When condensation occurs, latent heat is released, maintaining that airstream at a higher temperature than if condensation had not occurred. This higher air temperature (potential flux) increases the heat transfer to the other airstream. The sensible and total effectiveness are widely used because the energy flow in the condensate is relatively small in most applications. (Freezing and frosting are unsteady conditions that should be avoided unless a defrost cycle is included.) Equations (5a) and (5b) must be used to calculate the leaving air humidity conditions, and Equations (7a) and (7b) to calculate the enthalpy values for airstreams in which inherent latent energy transfer occurs. Equations (3a) and (3b) may be used for airstreams if only sensible energy transfer is involved.

Step 6. Check the energy transfer balance between airstreams.

Equations (3c) and (3d) can be used to estimate the sensible energy for the two airstreams, and Equations (8a) and (8b) can estimate the total energy for the two airstreams. Total energy transferred from one airstream should equal total heat transferred to the other.

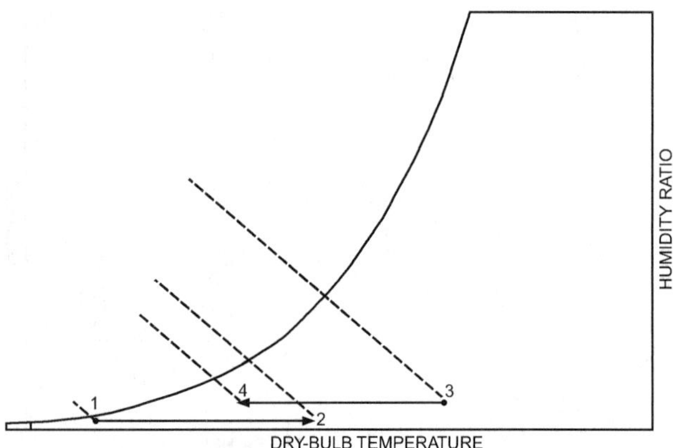

Fig. 19 Sensible Heat Recovery in Winter (Example 4)

Calculate and compare the energy transferred to or from each airstream. Differences between these energy flows are usually because of measurement errors.

Step 7. Plot entering and leaving conditions on psychrometric chart.

Examine the plotted information for each airstream to verify that performance is reasonable and accurate.

(Steps 8 to 10 apply only when EATR ≠ 0 and OACF ≠ 1.)

Step 8. Obtain data on exhaust air transfer ratio (EATR > 0 and typically 0.05 > EATR > 0 for regenerative wheels).

Request the EATR data from the manufacturer, who may have certified ARI *Standard* 1060 test condition data obtained using ASHRAE *Standard* 84 as a test procedure and analysis guide. These data may be extrapolated to non-ARI test conditions using correlations relating EATR to air pressure differences between the supply and exhaust and, for rotary regenerative wheels, carryover due to wheel rotation. Shang et al. (2001a) show that, for regenerative wheels, a correlation may be developed between EATR and carryover ratio, R_c, and OACF, but for other air-to-air exchangers EATR will be very small or negligible.

Step 9. Obtain data on outside air correction factor (OACF ≈ 1 and typically 0.9 < OACF < 1.1 for regenerative wheels).

Request the OACF data from the manufacturer, who may have certified ARI *Standard* 1060 test condition data obtained using ASHRAE *Standard* 84 as a test procedure and analysis guide. These data may be extrapolated to non-ARI test conditions using correlations relating OACF to pressure differences (Shang et al. 2001b), for regenerative wheels; for other exchangers, OACF will be very nearly 1.0.

Step 10. Correct the supply air ventilation rate, moisture transfer rate, and energy transfer rates for EATR ≠ 0 and OACF ≠ 1.0.

Values of EATR significantly larger than zero and OACF significantly different than 1.0 imply the air-to-air exchanger is transferring air between the exhaust and supply airstreams. This transfer may be important, especially for some devices such as regenerative wheels. Shang et al. (2001b) show a method to correct the energy rates when EATR ≠ 0 and OACF ≠ 1. The procedure to correct the supply air ventilation rate is illustrated in Example 7.

Example 4. Sensible Heat Recovery in Winter

Exhaust air at 75°F and 10% rh with a flow rate of 800 lb/min preheats an equal mass flow rate of outdoor air at 0°F and 60% rh ($\rho = 0.087$ lb/ft^3) using an air-to-air heat exchanger with a measured effectiveness of 60%. Airflows are specified in scfm, so an air density of 0.075 lb/ft^3 for both airstreams is appropriate. Assuming EATR = 0 and OACF ≈ 1, determine the leaving supply air temperatures and energy recovered, and check the heat exchange balance.

Solution:

Note: the numbers correspond to the steps in the calculation procedure.

1. Because the data on pressure drop are missing, skip this step.

2. Calculate the theoretical maximum heat transfer.

The two inlet conditions plotted on a psychrometric chart (Figure 19) indicate that, because the exhaust air has low relative humidity, latent energy transfer does not occur. Using Equation (2c), the theoretical maximum sensible heat transfer rate q_s is

$$q_{max} = (60)(800 \text{ lb/min})(0.24 \text{ Btu/lb·°F})(75 - 0) = 864{,}000 \text{ Btu/h}$$

3. Establish the sensible effectiveness.

From manufacturer's literature and certified performance test data, effectiveness is determined to be 60% at the design conditions.

4. Calculate actual heat transfer at given conditions.

Using Equation (2b),

$$q_s = (0.6)(864{,}000 \text{ Btu/h}) = 518{,}400 \text{ Btu/h}$$

5. Calculate leaving air conditions.

Because no moisture or latent energy transfer will occur,

a. Leaving supply air temperature t_2 is given as

$$t_2 = 0°\text{F} + \frac{518{,}400 \text{ Btu/h}}{(60)(800 \text{ lb/min})(0.24 \text{ Btu/lb·°F})} = 45°\text{F}$$

b. Leaving exhaust air temperature t_4 is given as

$$t_4 = 75°\text{F} - \frac{518{,}400 \text{ Btu/h}}{(60)(800 \text{ lb/min})(0.24 \text{ Btu/lb·°F})} = 30°\text{F}$$

6. Using Equations (3c) and (3d), check performance.

$$q_s = (60)(800 \text{ lb/min})(0.24 \text{ Btu/lb·°F})(45 - 0) = 518{,}400 \text{ Btu/h saved}$$

$$q_e = (60)(800 \text{ lb/min})(0.24 \text{ Btu/lb·°F})(75 - 30) = 518{,}400 \text{ Btu/h saved}$$

7. Plot conditions on psychrometric chart to confirm that no moisture exchange occurred (Figure 19).

Because EATR = 0 and OACF ≈ 1, Steps 8 to 10 of the calculation procedure are not presented here.

Example 5. Sensible Heat Recovery in Winter with Water Vapor Condensation

Exhaust air at 75°F and 28% rh ($\rho = 0.075$ lb/ft^3) and flow rate of 10,600 cfm is used to preheat 9500 cfm of outdoor air at 14°F and 50% rh ($\rho = 0.084$ lb/ft^3) using a heat exchanger with a sensible effectiveness of 70%. Assuming EATR = 0 and OACF ≈ 1, determine the leaving supply air conditions and energy recovered, and check the energy exchange balance.

Solution:

The supply airstream has a lower airflow rate than the exhaust airstream, so it may appear that the supply airstream limits heat transfer. However, determination of mass flow rates for the given entry conditions shows that the mass flow rate of the supply airstream (47,900 lb/h) is slightly greater than that of the exhaust airstream (47,600 lb/h), so exhaust is the limiting airstream. Nevertheless, because the mass difference is negligible, it is convenient to use supply air volume as the limiting airstream.

1. Because the data on pressure drop are missing, skip this step.

2. Calculate the theoretical maximum sensible heat transfer.

The limiting airstream, the supply airstream, will be preheated in the heat exchanger, so it is not subject to condensation. Therefore, Equation (2c) is used:

$$q_{max} = (60)(9500 \text{ ft}^3\text{/min})(0.084 \text{ lb/ft}^3)(0.24 \text{ Btu/lb·°F})(75 - 14)$$
$$= 700{,}000 \text{ Btu/h}$$

3. Select sensible effectiveness.

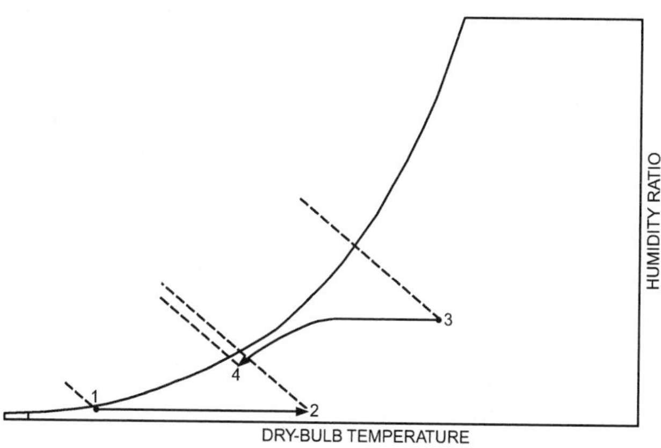

Fig. 20 Sensible Heat Recovery in Winter with Condensate (Example 5)

From manufacturer's literature and performance test data, the sensible effectiveness is determined to be 70% at the design conditions.

4. Calculate actual heat transfer at design conditions using Equation (2b):

$$q_s = (0.7)(700{,}000 \text{ Btu/h}) = 490{,}000 \text{ Btu/h}$$

5. Calculate leaving air conditions.

a. Leaving supply air temperature is calculated by

$$t_2 = 14°\text{F} + \frac{490{,}400 \text{ Btu/h}}{(60)(0.084 \text{ lb/ft}^3)(9500 \text{ ft}^3/\text{min})(0.24 \text{ Btu/lb·°F})} = 56.6°\text{F}$$

b. Because the dew point of exhaust air at inlet is 38.4°F, condensation occurs on the exhaust side, so the leaving exhaust air temperature cannot be determined using Equation (3b). The entering exhaust air enthalpy and humidity ratio are determined for the dry-bulb temperature of 75°F and 28% rh using a psychrometric chart and found to be h_3 = 23.7 Btu/lb and w_3 = 0.0052 lb/lb. However, the leaving exhaust air enthalpy can be determined by

$$h_4 = 23.7 \text{ Btu/lb} - \frac{490{,}000 \text{ Btu/h}}{(60)(0.075 \text{ lb/ft}^3)(10{,}600 \text{ ft}^3/\text{min})} = 13.43 \text{ Btu/lb}$$

Because the air will be saturated at the outlet of exhaust air, the dry-bulb or wet-bulb temperature and humidity ratio corresponding to an enthalpy of 13.43 Btu/lb is found to be t_4 = 35.7°F and w_4 = 0.0044 lb/lb. The rate of moisture condensed m_w is

$$m_w = m_e(w_3 - w_4) = (60)(800 \text{ lb/min})(0.0052 - 0.0044) = 38.4 \text{ lb/h}$$

6. Check performance.

$$q_s = (60)(0.084 \text{ lb/ft}^3)(9500 \text{ft}^3/\text{min})(0.24 \text{ Btu/lb·°F})(56.6 - 14)$$
$$= 489{,}500 \text{ Btu/h}$$

Neglect the enthalpy of the condensed water by adding the energy lost through condensation of vapor to the sensible heat lost of the exhausting air.

$$q_e = (60)(0.075 \text{ lb/ft}^3)(10{,}600 \text{ ft}^3/\text{min})(0.24 \text{ Btu/lb·°F})(75 - 34.5) + (38.4 \text{ lb/h})(1100 \text{ Btu/lb})$$

= 505,900 Btu/h saved, which is very close to 489,500 Btu/lb

7. Plot conditions on psychrometric chart (Figure 20). Note that moisture condenses in the exhaust side of the heat exchanger.

Because EATR = 0 and OACF ≈ 1, Steps 8 to 10 of the calculation procedure are not presented here.

Example 6. Total Heat Recovery in Summer

Exhaust air at 75°F and 63°F wb (ρ = 0.073 lb/ft³) with a flow rate of 10,600 cfm is used to precool 8500 cfm of supply outdoor air at 95°F and 81°F wb (ρ = 0.069 lb/ft³) using a hygroscopic total energy exchanger. The sensible and total effectiveness for this heat exchanger are 70 and 56.7%, respectively. Assuming EATR = 0 and OACF ≈ 1, determine the leaving supply air conditions and energy recovered, and check the energy exchange balance.

Solution:

1. Because the data on pressure drop are missing, skip this step.
2. Calculate the theoretical maximum heat transfer.

The supply airstream is a lesser or limiting airstream for energy and moisture transfer. Determine entering airstream enthalpies and humidity ratio from psychrometric chart.

Supply inlet (95°F db, 81°F wb) h_1 = 44.6 Btu/lb w_1 = 0.0198 lb/lb

Exhaust inlet (75°F db, 63°F wb) h_3 = 28.5 Btu/lb w_3 = 0.0096 lb/lb

The theoretical maximum sensible and total heat transfer rates can be obtained as follows:

$$q_{max}(\text{sensible}) = (60)(0.069 \text{ lb/ft}^3)(8500 \text{ ft}^3/\text{min})(0.24 \text{ Btu/lb·°F}) \times (95 - 75) = 169{,}000 \text{ Btu/h}$$

$$q_{max}(\text{total energy}) = (60)(0.069 \text{ lb/ft}^3)(8500 \text{ ft}^3/\text{min})(44.6 - 28.5) = 566{,}000 \text{ Btu/h}$$

3. Determine supply sensible and total effectiveness.

The manufacturer's selection data for the design conditions provide the following effectiveness ratios:

ε_s = 70% ε_t = 56.7%

4. Calculate energy transfer at design conditions.

$$q_t = (0.567)(566{,}000 \text{ Btu/h}) = 321{,}000 \text{ Btu/h total recovered}$$
$$q_s = -(0.7)(169{,}000 \text{ Btu/h}) = -118{,}000 \text{ Btu/h sensible recovered}$$
$$q_{lat} = 203{,}000 \text{ Btu/h latent recovered}$$

5. Calculate leaving air conditions.

a. Supply air conditions

$$t_2 = 95°\text{F} + \frac{-118{,}000 \text{ Btu/h}}{(60)(0.069 \text{ lb/ft}^3)(8500 \text{ ft}^3/\text{min})(0.24 \text{ Btu/lb·°F})} = 81°\text{F}$$

$$h_2 = 44.6 \text{ Btu/lb} + \frac{-321{,}000 \text{ Btu/h}}{(60)(0.069 \text{ lb/ft}^3)(8500 \text{ ft}^3/\text{min})} = 35.5 \text{ Btu/lb}$$

From the psychrometric chart, the supply air humidity ratio and wet-bulb temperature are found to be w_2 = 0.0145 and t_{w2} = 71.6°F.

b. Exhaust air conditions

$$t_4 = 75°\text{F} + \frac{118{,}000 \text{ Btu/h}}{(60)(0.073 \text{ lb/ft}^3)(10{,}000 \text{ ft}^3/\text{min})(0.24 \text{ Btu/lb·°F})}$$
$$= 86.2°\text{F}$$

$$h_4 = 28.5 \text{ Btu/lb} + \frac{321{,}000 \text{ Btu/h}}{(60)(0.073 \text{ lb/ft}^3)(10{,}600 \text{ ft}^3/\text{min})} = 35.4 \text{ Btu/lb}$$

From the psychrometric chart, the exhaust humidity ratio and wet-bulb temperature are found to be w_4 = 0.0134 and t_{w4} = 71.6°F.

6. Check total performance (Equations 9 and 10).

$$q_t = (60)(0.069 \text{ lb/ft}^3)(8500 \text{ cfm})(44.6 - 35.5) = 320{,}000 \text{ Btu/h saved}$$

$$q_t = (60)(0.073 \text{ lb/ft}^3)(10{,}600 \text{ cfm})[(0.24 \text{ Btu/lb·°F})(86.2 - 75) + (0.0134 - 0.0096)(1100 \text{ Btu/lb})]$$

= 319,000 Btu/h, which is close to 320,000 Btu/h

7. Plot conditions on psychrometric chart (Figure 21).

Because EATR = 0 and OACF ≈ 1, Steps 8 to 10 are not presented here.

Example 7. Total energy recovery with EATR ≠ 0 and OACF ≠ 1.0

An ERV manufacturer claims a product has performance characteristics as shown in Figure 5. A building has a ventilation requirement of 850 cfm and exhaust air at 75°F and 63°F wb (ρ = 0.075 lb/ft³) is used to precool supply outdoor air at 95°F and 81°F wb (ρ = 0.072 lb/ft³).

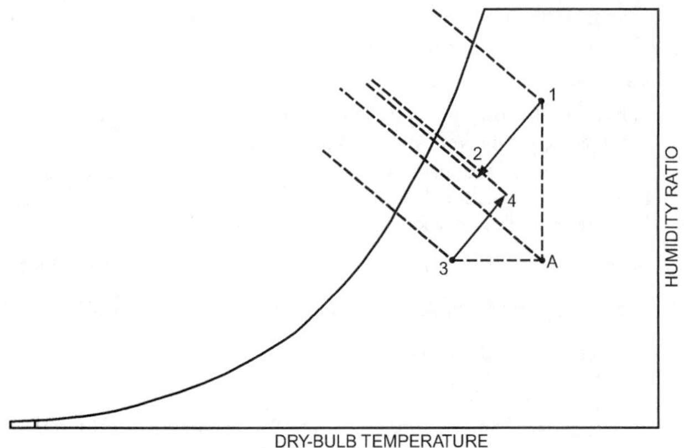

Fig. 21 Total Heat Recovery in Summer (Example 6)

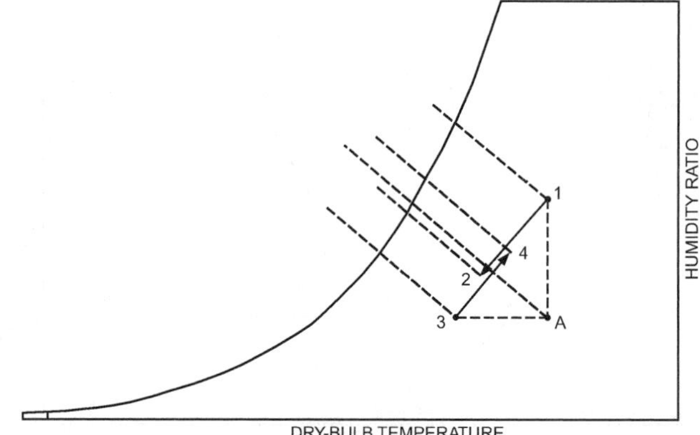

Fig. 22 Total Energy Recovery with EATR ≠ 0 and OACF ≠ 1 (Example 7)

(a) Assuming EATR = 0 and OACF ≈ 1, determine the leaving supply air conditions and energy recovered, and check the energy exchange balance.

(b) If EATR = 5% and OACF = 1.05, determine the actual air flow rates.

Solution:

1. From Figure 5, at a flow rate of 850 cfm, the pressure drop Δp = 0.9 in. of water. Assuming the effective efficiency of the fan motor combination is about 0.6, the power P_s required to circulate the supply air can be obtained from Equation (11) as

$$P_s = [(850\ \text{ft}^3/\text{min})(0.9)]/[(6356)(0.6)] = 0.2\ \text{hp}$$

Assuming the balanced flow the power required to circulate the exhaust air would be same, therefore the total power P required to circulate the airstreams would be twice this amount.

$$P = 0.4\ \text{hp}$$

2. Calculate the theoretical maximum heat transfer.

Determine entering airstream enthalpies and humidity ratio from the psychrometric chart.

Supply inlet (95°F db, 81°F wb) h_1 = 44.6 Btu/lb w_1 = 0.0198 lb/lb

Exhaust inlet (75°F db, 63°F wb) h_3 = 28.5 Btu/lb w_3 = 0.0096 lb/lb

The theoretical maximum heat transfer rates can be obtained as follows:

$$q_{max}(\text{sensible}) = (60)(0.072\ \text{lb/ft}^3)(850\ \text{ft}^3/\text{min})(0.24\ \text{Btu/lb}\cdot°\text{F}) \times (95 - 75) = 17{,}600\ \text{Btu/h}$$

$$q_{max}(\text{latent}) = (60)(0.072\ \text{lb/ft}^3)(850\ \text{ft}^3/\text{min})(11{,}000\ \text{Btu/lb}) \times (0.0194 - 0.0093) = 40{,}800\ \text{Btu/h}$$

$$q_{max}(\text{total}) = (60)(0.072\ \text{lb/ft}^3)(850\ \text{ft}^3/\text{min}) \times (36.6 - 20.6) = 58{,}750\ \text{Btu/h}$$

Note that sum of sensible and latent energy should equal the total energy.

3. Determine supply sensible and total effectiveness.

From Figure 5 at a flow rate of 850 cfm, ε_s = 0.73, ε_L = 0.68, and ε_t = 0.715.

4. Calculate energy transfer at design conditions.

q_s = (0.73)(17,600 Btu/h) = 12,850 Btu/h sensible recovered

q_L = (0.68)(40,800 Btu/h) = 27,750 Btu/h latent recovered

q_t = (0.715)(58,750 Btu/h) = 42,000 Btu/h total recovered

5. Calculate leaving air conditions.

a. Supply air conditions

$$t_2 = 95°\text{F} + \frac{-12{,}850\ \text{Btu/h}}{(60)(0.072\ \text{lb/ft}^3)(850\ \text{ft}^3/\text{min})(0.24\ \text{Btu/lb}\cdot°\text{F})} = 80°\text{F}$$

$$h_2 = 44.6\ \text{Btu/lb} + \frac{-42{,}000\ \text{Btu/h}}{(60)(0.072\ \text{lb/ft}^3)(850\ \text{ft}^3/\text{min})} = 33.16\ \text{Btu/lb}$$

From the psychrometric chart, the supply air humidity ratio and wet-bulb temperature are w_2 = 0.0129 and t_{w2} = 69.2°F.

b. Exhaust air conditions

$$t_4 = 75°\text{F} + \frac{12{,}850\ \text{Btu/h}}{(0.075\ \text{lb/ft}^3)(60)(850\ \text{ft}^3/\text{min})(0.24\ \text{Btu/lb}\cdot°\text{F})} = 89°\text{F}$$

$$h_4 = 28.5\ \text{Btu/lb} + \frac{42{,}000\ \text{Btu/h}}{(0.075\ \text{lb/ft}^3)(60)(850\ \text{ft}^3/\text{min})} = 39.5\ \text{Btu/lb}$$

From the psychrometric chart, the exhaust humidity ratio and wet-bulb temperature are found to be w_4 = 0.0164, t_{w4} = 76.1°F.

6. Check total performance.

q_t = (0.072 lb/ft³)(60)(850 ft³/min)(36.6 – 25.1) = 42,230 Btu/h saved

q_t = (0.075 lb/ft³)(60)(850 ft³/min)[(0.24 Btu/lb·°F)(89 – 75) + (0.0164 – 0.0096)(1100 Btu/lb)] = 41,460 Btu/h, which is close to 42,230 Btu/h

7. Plot conditions on psychrometric chart (Figure 22).

8. Obtain data on EATR. (Given: EATR = 5% or 0.05.)

9. Obtain data on OACF. (Given: OACF = 1.05.)

10. Correct the supply air ventilation rate, the moisture transfer rate, and energy transfer rates EATR ≠ 0 and OACF ≠ 1.0.

The net ventilation rate is 850 cfm and the EATR = 0.05; therefore, the actual flow rate Q_2 to the space can be obtained from Equation (22) as

$$Q_3 = Q_2 = \frac{Q_v}{1 - \text{EATR}} = \frac{850\ \text{ft}^3/\text{min}}{1 - 5/100} = 895\ \text{cfm}$$

Because OACF = 1.05, the actual flow rate Q_1 of fresh air from outside can be calculated from Equation (23) as

$$Q_1 = Q_2(\text{OACF}) = \frac{Q_v(\text{OACF})}{1 - \text{EATR}} = \frac{(850\ \text{ft}^3/\text{min})(1.05)}{1 - 5/100} = 939\ \text{cfm}$$

To balance the flow rates into the ERV, the actual air flow rates at states 3 and 4 are as shown in Figure 23.

Supply and exhaust fan capacity should match the flows required at their locations. Example 7's results are for balanced flow. Assuming flow rates in the ERV are same as the outside air ventilation requirements, then the effectiveness would be same as that for no air leakage. If air leaks at the inlet and outlet of the energy recovery ventilator, then the exit conditions of air temperature and humidity at states 3 and 4 can be calculated as those of the airstream mixture. For instance, the properties at state 2 would be those of an airstream

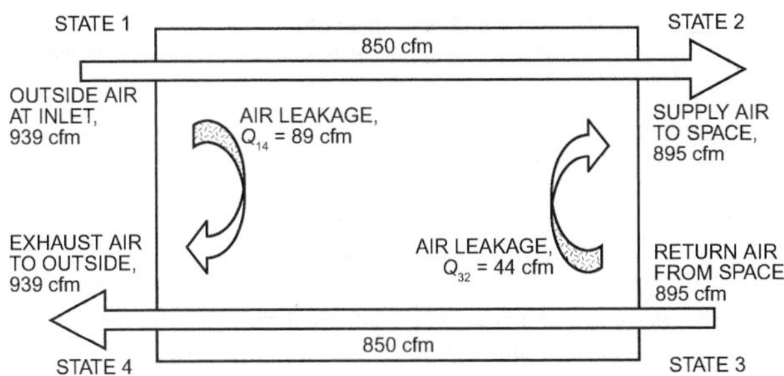

Fig. 23 Actual Airflow Rates at Various State Points (Example 7)

mixture at state 2 for no air leakage and air quantity (Q_{32}) at state 3. The error for these calculations should be less than 5%.

Shang et al. (2001a) show a method to accurately estimate the energy rates when EATR ≠ 0 and OACF ≠ 1.

INDIRECT EVAPORATIVE AIR COOLING

Exhaust air passing through a water spray absorbs water vapor until it becomes nearly saturated. As the water evaporates, it absorbs sensible energy from the air, lowering its temperature. This process follows a constant wet-bulb line on a psychrometric chart. Thus, the airstream enthalpy remains nearly constant, moisture content increases, and dry-bulb temperature decreases. The evaporatively cooled exhaust air can then be used to cool supply air through an air-to-air heat exchanger, which may be used either for year-round energy recovery or exclusively for its evaporative cooling benefits.

Indirect evaporative cooling has been used with heat pipe heat exchangers, two-phase thermosiphon loops, runaround coil loop exchangers, and flat-plate heat exchangers for summer cooling (Dhital et al. 1995; Johnson et al. 1995; Mathur 1990a, 1990b, 1992, 1993; Scofield and Taylor 1986). Exhaust air or a scavenging airstream is cooled by passing it through a water spray, a wet filter, or other wetted media, resulting in a greater overall temperature difference between the supply and exhaust or scavenging airstreams and thus more heat transfer. Energy recovery is further enhanced by improved heat transfer coefficients because of wetted exhaust-side heat transfer surfaces. No moisture is added to the supply airstream, and there are no auxiliary energy inputs other than fan and water pumping power. The COP tends to be high, typically from 9 to 20, depending on available dry-bulb temperature depression. The dry-bulb temperature decrease in the exhaust airstream caused by evaporative cooling tends to be 85 to 95% of the maximum available difference between the exhaust air inlet dry-bulb and wet-bulb temperatures. Therefore, exhaust air evaporative cooling is usually most cost-effective in hot, dry climates where the evaporator can be used frequently to obtain large exhaust air dry-bulb temperature depressions.

Without a bypass scheme for either the evaporator or air-to-air heat exchanger, the net annual energy costs include the extra annual fan power for these devices as well as the benefit of evaporative cooling.

Because less mechanical cooling is required with evaporative cooling, energy consumption and peak demand load are both reduced, yielding lower energy bills. Overall mechanical refrigeration system requirements are reduced, allowing use of smaller mechanical refrigeration systems. In some cases, the mechanical system may be eliminated. Chapter 19 of this volume and Chapter 51 of the 2007 *ASHRAE Handbook—HVAC Applications* have further information on evaporative cooling.

Example 8. Indirect Evaporative Cooling Recovery

Room air at 86°F and 63°F wb (ρ = 0.075 lb/ft^3) with a flow rate of 32,000 cfm is used to precool 32,000 cfm of supply outdoor air at 102°F and 68°F wb (ρ = 0.071 lb/ft^3) using an aluminum fixed-plate heat exchanger and indirect evaporative cooling. The evaporative cooler increases the exhaust air to 90% rh before it enters the heat exchanger. The sensible effectiveness is given by the manufacturer as 78%. Assuming EATR = 0 and OACF ≈ 1, determine the leaving supply air conditions and energy recovered, and check the energy exchange balance.

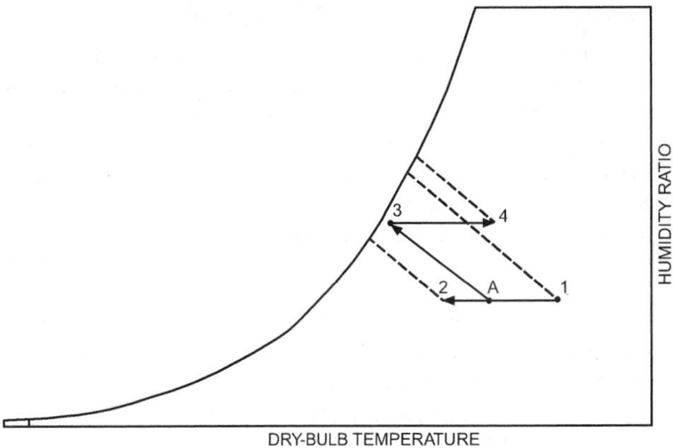

Fig. 24 Indirect Evaporative Cooling Recovery (Example 8)

Solution:

First, determine the exhaust air condition entering the exchanger (i.e., after it is adiabatically cooled). Air at 86°F db, 63°F wb cools to 64°F db, 63°F wb as shown by the process line from point A to point 3 in Figure 24. In this problem the volumetric flows are equal, but the mass flows are not.

1. Because the data on pressure drop are missing, skip this step.

2. Calculate the theoretical maximum heat transfer.

Based on a preliminary assessment, the supply air is not expected to cool below its wet-bulb temperature of 68°F. Thus, use sensible heat Equation (3c).

$$q_{max}\,(\text{sensible}) = (60)(0.071\ \text{lb/ft}^3)(32{,}000\ \text{ft}^3/\text{min})(0.24\ \text{Btu/lb}\cdot°\text{F}) \times (102-64) = 1{,}240{,}000\ \text{Btu/h}$$

3. Establish the sensible effectiveness.

From manufacturer's exchanger selection data for indirect evaporative coolers, an effectiveness of 78% is found to be appropriate.

4. Calculate actual energy transfer at the design conditions.

$$q_{actual} = (0.78)(1{,}240{,}000\ \text{Btu/h}) = 967{,}000\ \text{Btu/h recovered}$$

5. Calculate leaving air conditions.

a. Leaving supply air temperature is

$$t_2 = 102°\text{F} + \frac{-967{,}000\ \text{Btu/h}}{(60)(0.071\ \text{lb/ft}^3)(32{,}000\ \text{ft}^3/\text{min})(0.24\ \text{Btu/lb}\cdot°\text{F})} = 72.4°\text{F}$$

b. Leaving exhaust air temperature is

$$t_4 = 64°\text{F} + \frac{967{,}000\ \text{Btu/h}}{(60)(0.075\ \text{lb/ft}^3)(32{,}000\ \text{ft}^3/\text{min})(0.24\ \text{Btu/lb}\cdot°\text{F})} = 92.0°\text{F}$$

6. Check performance.

$$q_s = (0.071\ \text{lb/ft}^3)(60)(32{,}000\ \text{ft}^3/\text{min})(0.24\ \text{Btu/lb}\cdot°\text{F})(102-72.4) = 968{,}000\ \text{Btu/h recovered}$$

$$q_e = (0.075\ \text{lb/ft}^3)(60)(32{,}000\ \text{ft}^3/\text{min})(0.24\ \text{Btu/lb}\cdot°\text{F})(92-64) = 968{,}000\ \text{Btu/h recovered}$$

7. Plot conditions on psychrometric chart (Figure 24), and confirm that no latent exchange occurred.

Because EATR = 0 and OACF ≈ 1, Steps 8 to 10 are not presented here.

Precooling Air Reheater (Series Application)

In some applications, such as ventilation in hot, humid climates, supply air is cooled below the desired delivery temperature to condense moisture and reduce humidity. Using this overcooled supply air to precool outside air reduces the air-conditioning load, which allows refrigeration equipment to be downsized and eliminates the need to reheat supply air with purchased energy.

In this three-step process, illustrated in Figure 25, outside air passes through an air-to-air heat exchanger, where it is precooled by supply air leaving the cooling coil. It is then further cooled and dehumidified in the cooling coil. After leaving the cooling coil, it passes through the other side of the air-to-air heat exchanger, where the incoming supply air reheats it.

Fixed-plate, heat pipe, and rotary heat wheel exchangers can be used to reheat precooled supply air. At part-load operation, the amount of heat transferred from precooling to reheating may require modulation. This can be done using the heat rate control schemes in Table 2.

Example 9. Precooling Air Reheater Dehumidifier

In this application, 3400 cfm of outdoor supply air at 95°F and 81°F wb ($\rho = 0.069$ lb/ft^3) is used to reheat 3400 cfm of the same air leaving a cooling coil (exhaust) at 52.1°F and 51.8°F wb using a sensible heat exchanger as a precooling air reheater. The reheated air is to be between 75 and 79°F. In this application, the warm airstream is outdoor air and the cold airstream is the same air after it leaves the cooling coil. The unit's manufacturer lists its effectiveness as 58.4%. For EATR = 0 and OACF ≈ 1, determine the leaving precooled and reheated air conditions and energy recovered, and check the energy exchange balance.

Solution:

1. Because the data on pressure drop are missing, skip this step.
2. Calculate the theoretical maximum energy transfer.

The air being reheated will have less mass than the outdoor air entering the pre-cooler because moisture will condense from it as it passes through the precooler and cooling coil. Reheat is sensible heat only, so Equation (2c) is used to determine the theoretical maximum energy transfer.

$$q_s = (60)(0.069 \text{ lb/ft}^3)(3400 \text{ ft}^3\text{/min})(0.24 \text{ Btu/lb}\cdot°\text{F})(95 - 52.1)$$
$$= 145{,}000 \text{ Btu/h saved}$$

3. Establish the sensible effectiveness.

The manufacturer give the effectiveness as 58.4% at the designated operating conditions.

4. Calculate actual energy transfer at design conditions.

$$q_{actual} = (0.584)(145{,}000 \text{ Btu/h}) = 84{,}700 \text{ Btu/h}$$

5. Calculate leaving air conditions.

Because condensation occurs as the outside airstream passes through the precooling side of the heat exchanger, use Equation (26) to determine its leaving enthalpy, which is the inlet condition for the cooling coil. Sensible heat transfer Equation (27) is used to determine the temperature of air leaving the preheat side of the heat exchanger.

a. Precooler leaving air conditions

Entering enthalpy, determined from the psychrometric chart for 95°F db and 81°F wb, is 44.6 Btu/lb.

$$h_s = 44.6 \text{ Btu/lb} + \frac{-84{,}700 \text{ Btu/h}}{(60)(0.069 \text{ lb/ft}^3)(3400 \text{ ft}^3\text{/min})} = 38.6 \text{ Btu/lb}$$

The wet-bulb temperature for saturated air with this enthalpy is 75°F. This is point 2 on the psychrometric chart (Figure 26), which is near saturation. Note that this precooled air is further dehumidified by a cooling coil from point 2 to point 3 in Figure 26.

b. Reheater leaving air conditions

$$t_4 = 52.1°\text{F} + \frac{84{,}700 \text{ Btu/h}}{(60)(0.069 \text{ lb/ft}^3)(3400 \text{ ft}^3\text{/min})(0.24 \text{ Btu/lb}\cdot°\text{F})}$$
$$= 77.2°$$

Entering enthalpy, determined from the psychrometric chart for 52.1°F db and 51.8°F wb, is 21.3 Btu/lb.

$$h_4 = 21.3 \text{ Btu/lb} + \frac{84{,}700 \text{ Btu/h}}{(60)(0.069 \text{ lb/ft}^3)(3400 \text{ ft}^3\text{/min})} = 27.3 \text{ Btu/lb}$$

The wet-bulb temperature for air with this temperature and enthalpy is 61.3°F.

6. Check performance.

$$q_s = (60)(0.069 \text{ lb/ft}^3)(3400 \text{ ft}^3\text{/min})(44.6 - 38.6)$$
$$= 84{,}500 \text{ Btu/h precooling}$$

$$q_s = (60)(0.069 \text{ lb/ft}^3)(3400 \text{ ft}^3\text{/min})(0.24 \text{ Btu/lb}\cdot°\text{F})(77.2 - 52.1)$$
$$= 84{,}800 \text{ Btu/h reheat}$$

7. Plot conditions on psychrometric chart (Figure 26).

Because EATR = 0 and OACF ≈ 1, Steps 8 to 10 are not presented here.

ECONOMIC CONSIDERATIONS

Air-to-air energy recovery systems are used in both new and retrofit applications. These systems should be designed for the maximum cost benefit or least life-cycle cost (LCC) expressed either over the service life or annually and with an acceptable payback period.

The annualized system owning, operating, and maintenance costs are a complex function of the future value of money as well as all the design variables in the energy/heat exchanger. These variables include the mass of each material used, the cost of forming these materials into a highly effective energy/heat exchanger, the cost of auxiliary equipment and controls, and the cost of installation.

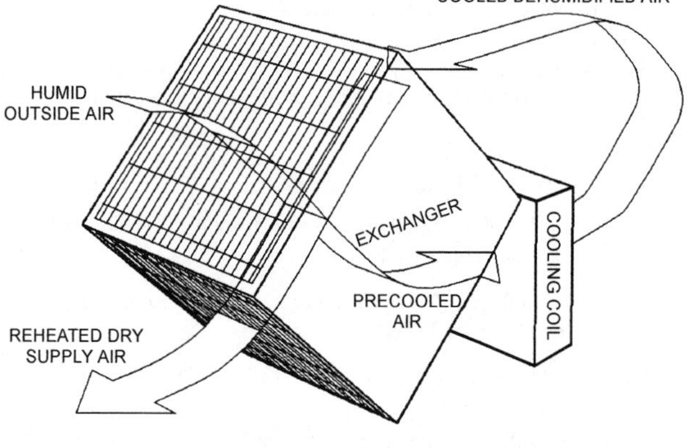

COOLED DEHUMIDIFIED AIR
HUMID OUTSIDE AIR
EXCHANGER
COOLING COIL
PRECOOLED AIR
REHEATED DRY SUPPLY AIR

Fig. 25 Precooling Air Reheater

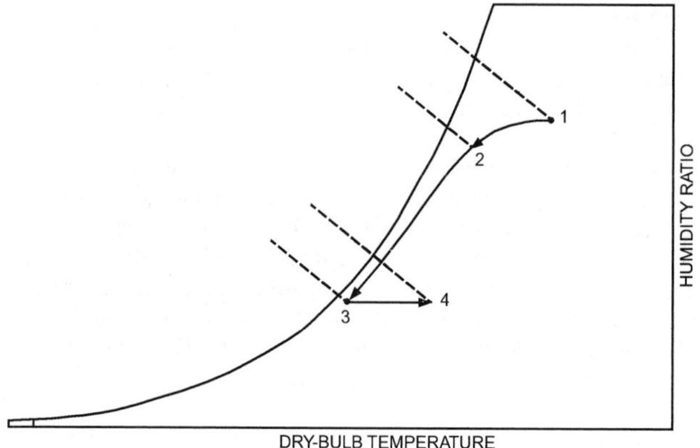

Fig. 26 Precooling Air Reheater Dehumidifier (Example 9)

Owning and operating costs are discussed in more detail in Chapter 36 of the 2007 *ASHRAE Handbook—HVAC Applications*.

The **operating energy cost** for energy recovery systems involves functions integrated over time, including variables such as flow rate, pressure drop, fan efficiency, energy cost, and energy recovery rate. The calculations are complex because air-heating and/or cooling loads are, for a range of supply temperatures, time-dependent in most buildings. Time-of-use schedules for buildings can impose different ventilation rates for each hour of the day. Electrical utility charges often vary with time of day, amount of energy used, and peak power load. For building ventilation air-heating applications, the peak heat recovery rate usually occurs at the outside supply temperature at which frosting control throttling must be imposed. In addition to designing for the winter design temperature, heat recovery systems should be optimized for peak heat recovery rate, taking frost control into account.

Overall exchanger effectiveness ε should be high (see Table 2 for typical values); however, a high ε implies a high capital cost, even when the exchanger is designed to minimize the amount of materials used. Energy costs for fans and pumps are usually very important and accumulate operating cost even when the energy recovery system must be throttled back. For building ventilation, throttling may be required much of the time. Thus, the overall LCC minimization problem for optimal design may involve 10 or more independent design variables as well as several specified constraints and operating conditions [see, e.g., Besant and Johnson (1995)].

In addition, comfort-to-comfort energy recovery systems often operate with much smaller temperature differences than most auxiliary air-heating and cooling heat exchangers. These small temperature differences need more accurate energy transfer models to reach the maximum cost benefit or lowest LCC. Most importantly, recovered energy at design may be used to reduce the required capacity of heating and cooling equipment, which can be significant in both system performance/efficiency and economics.

The **payback period (PP)** is best computed after the annualized costs have been evaluated. It is usually defined as

$$\begin{aligned} PP &= \frac{\text{Capital cost and interest}}{\text{Annual operating energy cost saved}} \\ &= \frac{C_{s,init} - \text{ITC}}{C_e(1 - T_{inc})}\text{CRF}(i'', n) \end{aligned} \tag{28}$$

where

$C_{s,init}$ = initial system cost
ITC = investment tax credit for energy-efficient improvements
C_e = cost of energy to operate the system for one period
T_{inc} = net income tax rate where rates are based on last dollar earned (i.e., marginal rates) = (local + state + federal rate) – (federal rate)(local + state rate)
CRF = capital recovery factor
i'' = effective discount rate adjusted for energy inflation
n = total number of periods under analysis

The inverse of this term is usually called the **return on investment (ROI).** Well-designed energy recovery systems normally have a PP of less than 5 years, and often less than 3 years. Paybacks of less than 1 year are not uncommon in comfort-to-comfort applications in hot, humid climates, primarily because of the reduced size of cooling equipment required.

Other economic factors include the following.

System Installed Cost. Initial installed HVAC system cost is often lower for air-to-air energy recovery devices because mechanical refrigeration and fuel-fired heating equipment can be reduced in size. Thus, a more efficient HVAC system may also have a lower installed total HVAC cost. The installed cost of heat recovery systems becomes lower per unit of flow as the amount of outside air used for ventilation increases.

Life-Cycle Cost. Air-to-air energy recovery cost benefits are best evaluated considering all capital, installation, operating, and energy-saving costs over the equipment life under normal operating conditions in terms of the life-cycle cost. As a rule, neither the most efficient nor the least expensive energy recovery device will be most economical. Optimizing the life-cycle cost for maximum net savings may involve many design variables, requiring careful cost estimates and use of an accurate recovery system model with all its design variables [see, e.g., Besant and Simonson (2000)].

Energy Costs. The absolute cost of energy and relative costs of various energy forms are major economic factors. High energy costs favor high levels of energy recovery. In regions where electrical costs are high relative to fuel prices, heat recovery devices with low pressure drops are preferable.

Amount of Recoverable Energy. Economies of scale favor large installations. Equipment is commercially available for air-to-air energy recovery applications using 50 cfm and above.

Grade of Exhaust Energy. High-grade (i.e., high-temperature) exhaust energy is generally more economical to recover than low-grade energy. Energy recovery is most economical for large temperature differences between the waste energy source and destination.

Coincidence and Duration of Waste Heat Supply and Demand. Energy recovery is most economical when supply coincides with demand and both are relatively constant throughout the year. Thermal storage may be used to store energy if supply and demand are not coincident, but this adds cost and complexity to the system.

Proximity of Supply to Demand. Applications with a large central energy source and a nearby waste energy use are more favorable than applications with several scattered waste energy sources and uses.

Operating Environment. High operating temperatures or the presence of corrosives, condensable gases, and particulates in either airstream results in higher equipment and maintenance costs. Increased equipment costs result from the use of corrosion- or temperature-resistant materials, and maintenance costs are incurred by an increase in the frequency of equipment repair and wash down and additional air filtration requirements.

Effect on Pollution Control Systems. Removing process heat may reduce the cost of pollution control systems by (1) allowing less expensive filter bags to be used, (2) improving the efficiency of electronic precipitators, or (3) condensing out contaminant vapors, thus reducing the load on downstream pollution control systems. In some applications, recovered condensable gases may be returned to the process for reuse.

Effect on Heating and Cooling Equipment. Heat recovery equipment may reduce the size requirements for primary utility equipment such as boilers, chillers, and burners, as well as the size of piping and electrical services to them. Larger fans and fan motors (and hence fan energy) are generally required to overcome increased static pressure loss caused by the energy recovery devices. Auxiliary heaters may be required for frost control.

Effect on Humidifying or Dehumidifying Equipment. Selecting total energy recovery equipment results in the transfer of moisture from the airstream with the greater humidity ratio to the airstream with the lesser humidity ratio. This is desirable in many situations because humidification costs are reduced in cold weather and dehumidification loads are reduced in warm weather.

SYMBOLS

A = area of recovery exchanger, ft^2
c_p = specific heat of moist air, Btu/lb·°F
C = capital cost
C_e = cost of energy
C_r = ratio of C_{min}/C_{max}
$C_{s,int}$ = initial system capital cost
CRF = capital recovery factor
h = enthalpy, Btu/lb

h_{fg} = enthalpy of vaporization, Btu/lb
i = arbitrary state, or discount rate
ITC = income tax credit
m_s = mass flow rate of supply moist air from outside, lb/min
m_e = mass flow rate of exhaust moist air, lb/min
n = number of years in consideration, years
NTU = number of transfer units = UA/C_{min}
p = pressure, in. of water
pp = payback period, years
P = pumping power, Btu/h
q = heat transfer rate, Btu/h
Q = volume flow rate, cfm
S = reference temperature, °F
t = moist air temperature at state i, °F
t_{w3} = wet-bulb temperature of moist air at state 3, °F
T = absolute temperature, °F, or tax
U = overall heat transfer coefficient, Btu/h·ft^2·°F
V = mean velocity, fpm
w = humidity ratio

Greek Letters

ε_s = sensible effectiveness of heat or energy wheel
ε_L = latent effectiveness of energy wheel
ε_t = total effectiveness of ERV
η = efficiency
ρ = density, lb/ft^3
σ = volume fraction
ϕ = relative humidity
ω = rotational speed of the wheel, rpm

Subscripts

a = air
e = exhaust side of heat/energy exchanger, exit or energy
f = fan or fan motor combination
if = threshold temperature of the outside air for freezing to occur
in = indoor conditions of building space
inc = increment
h = hydraulic
L = latent
max = maximum value
min = minimum value
o = reference state or outlet
p = constant pressure
s = supply side or suction side
t = total

REFERENCES

ARI. 2001. Rating air-to-air energy recovery ventilation equipment. ANSI/ARI *Standard* 1060-2001. Air-Conditioning and Refrigeration Institute, Arlington, VA.

ASHRAE. 2001. Ventilation for acceptable indoor air quality. ANSI/ASHRAE *Standard* 62.

ASHRAE. 1991. Method of testing air-to-air heat exchangers. ANSI/ASHRAE *Standard* 84-1991.

Besant, R.W. and C.J. Simonson. 2000. Air-to-air energy recovery. *ASHRAE Journal* 42(5):31-38.

Dhital, P., R. Besant, and G.J. Schoenau. 1995. Integrating run-around heat exchanger systems into the design of large office buildings. *ASHRAE Transactions* 101(2):979-999.

Friedlander, M. 2003. How certified ratings can improve system design. Seminar at ASHRAE Winter Annual Meeting, Chicago.

Guo, P., D.L. Ciepliski, and R.W. Besant. 1998. A testing and HVAC design methodology for air-to-air heat pipe heat exchangers. *International Journal of HVAC&R Research* 4(1):3-26.

Johnson, A.B., R.W. Besant, and G.J. Schoenau. 1995. Design of multi-coil run-around heat exchanger systems for ventilation air heating and cooling. *ASHRAE Transactions* 101(2):967-978.

Johnson, A.B., C.J. Simonson, and R.W. Besant. 1998. Uncertainty analysis in the testing of air-to-air heat/energy exchangers installed in buildings. *ASHRAE Transactions* 104(1B):1639-1650.

Mathur, G.D. 1990a. Long-term performance prediction of refrigerant charged flat plate solar collector of a natural circulation closed loop. ASME *HTD* 157:19-27.

Mathur, G.D. 1990b. Indirect evaporative cooling using heat pipe heat exchangers. ASME Symposium, Thermal Hydraulics of Advanced Heat Exchangers, ASME Winter Annual Meeting, Dallas.

Mathur, G.D. 1990c. Indirect evaporative cooling using two-phase thermosiphon loop heat exchangers. *ASHRAE Transactions* 96(1):1241-1249.

Mathur, G.D. 1992. Indirect evaporative cooling. *Heating/Piping/Air Conditioning* 64(4):60-67.

Mathur, G.D. 1993. Retrofitting heat recovery systems with evaporative coolers. *Heating/Piping/Air Conditioning* 65(9):47-51.

Mathur, G.D. 1997. Performance enhancement of existing air conditioning systems. *Proceedings of Intersociety Energy Conversion Engineering Conference, Honolulu*, American Institute of Chemical Engineers, Paper #97367, pp. 1618-1623.

Mathur, G.D. and T.W. McDonald. 1986. Simulation program for a two-phase thermosiphon-loop heat exchanger. *ASHRAE Transactions* 92(2A): 473-485.

Mathur, G.D, and T.W. McDonald. 1987. Evaporator performance of finned air-to-air two-phase thermosiphon loop heat exchangers. *ASHRAE Transactions* 98(2):247-257.

McDonald, T.W. and D. Shivprasad. 1989. Incipient nucleate boiling and quench study. *Proceedings of CLIMA 2000* 1:347-352. Sarajevo, Yugoslavia.

Moffitt, R. 2003. (Personal communication and reference, *Trane Application Engineering Manual* SYS-APM003-EN.)

Ruch, M.A. 1976. Heat pipe exchangers as energy recovery devices. *ASHRAE Transactions* 82(1):1008-1014.

Scofield, M. and J.R. Taylor. 1986. A heat pipe economy cycle. *ASHRAE Journal* 28(10):35-40.

Shah, R.K. 1981. Thermal design theory for regenerators. In *Heat exchangers: Thermal-hydraulic fundamentals and design*. S. Kakec, A.E. Bergles, and F. Maysinger, eds. Hemisphere Publishing, New York.

Shang, W., M. Wawryk, and R.W. Besant. 2001a. Air crossover in rotary wheels used for air-to-air heat and moisture recovery. *ASHRAE Transactions* 107(2).

Shang, W., H. Chen, R.W. Evitts, and R.W. Besant. 2001b. Frost growth in regerative heat exchangers: Part I—Problem formulation and method of solution; Part II—Simulation and discussion. *Proceedings of ASME International Mechanical Engineering Congress and Expo*, November, New York.

Sparrow, E.M., J.P. Abraham, and J.C.K. Tong. 2001. An experimental investigation on a mass exchanger for transferring water vapor and inhibiting the transfer of other gases. *International Journal of Heat and Mass Transfer* 44(November):4313-4321.

BIBLIOGRAPHY

Andersson, B., K. Andersson, J. Sundell, and P.A. Zingmark. 1992. Mass transfer of contaminants in rotary enthalpy exchangers. *Indoor Air* 93(3): 143-148.

ASHRAE. 1974. Symposium on heat recovery. *ASHRAE Transactions* 80(1):307-332.

ASHRAE. 1982. Symposium on energy recovery from air pollution control. *ASHRAE Transactions* 88(1):1197-1225.

Barringer, C.G. and C.A. McGugan. 1989a. Development of a dynamic model for simulating indoor air temperature and humidity. *ASHRAE Transactions* 95(2):449-460.

Barringer, C.G. and C.A. McGugan. 1989b. Effect of residential air-to-air heat and moisture exchangers on indoor humidity. *ASHRAE Transactions* 95(2):461-474.

Besant, R.W. and A.B. Johnson. 1995. Reducing energy costs using run-around systems. *ASHRAE Journal* 37(2):41-47.

Besant, R.W. and C. Simonson. 2003. Air-to-air exchangers. *ASHRAE Journal* 45(4):42-50.

CSA. 1988. Standard methods of test for rating the performance of heat-recovery ventilators. CAN/CSA-C439-88. Canadian Standards Association, Rexdale, ON.

Ciepliski, D.L., C.J. Simonson, and R.W. Besant. 1998. Some recommendations for improvements to ASHRAE *Standard* 84-1991. *ASHRAE Transactions* 104(1B):1651-1665.

Dehli, F., T. Kuma, and N. Shirahama. 1993. A new development for total heat recovery wheels. *Energy Impact of Ventilation and Air Infiltration, 14th AIVC Conference, Copenhagen, Denmark*, pp. 261-268.

Kays, W.M. and M.E Crawford, 1993. *Convective heat and mass transfer,* 3rd ed. McGraw-Hill, New York.

Ninomura, P.T. and R. Bhargava. 1995. Heat recovery ventilators in multi-family residences in the Arctic. *ASHRAE Transactions* 101(2):961-966.

Phillips, E.G., R.E. Chant, B.C. Bradley, and D.R. Fisher. 1989. A model to compare freezing control strategies for residential air-to-air heat recovery ventilators. *ASHRAE Transactions* 95(2):475-483.

Phillips, E.G., R.E. Chant, D.R. Fisher, and B.C. Bradley. 1989. Comparison of freezing control strategies for residential air-to-air heat recovery ventilators. *ASHRAE Transactions* 95(2):484-490.

Phillips, E.G., D.R. Fisher, R.E. Chant, and B.C. Bradley. 1992. Freeze-control strategy and air-to-air energy recovery performance. *ASHRAE Journal* 34(12):44-49.

Shang, W. and R.W. Besant. 2001. Energy wheel effectiveness evaluation: Part I—Outlet airflow property distributions adjacent to an energy wheel; Part II—Testing and monitoring energy wheels in HVAC applications. *ASHRAE Transactions* 107(2).

Simonson, C.J. and R.W. Besant. 1997. Heat and moisture transfer in desiccant coated rotary energy exchangers: Part I—Numerical model; Part II—Validation and sensitivity studies. *International Journal of HVAC&R Research* 3(4):325-368.

Simonson, C.J. and R.W. Besant. 1998. Heat and moisture transfer in energy wheels during sorption, condensation and frosting. *ASME Journal of Heat Transfer* 120(3):699-708.

Simonson, C.J. and R.W. Besant. 1998. Energy wheel effectiveness: Part I—Development of dimensionless groups; Part II—Correlations, *International Journal of Heat and Mass Transfer*, 42(12):2161-2186.

Simonson, C.J., D.L. Cieplisky, and R.W. Besant. 1999. Determining performance of energy: Part I—Experimental and numerical methods; Part II—Experimental data and validation. *ASHRAE Transactions* 105 (1):177-205.

SMACNA. 1978. *Energy recovery equipment and systems*. Report.

Sparrow, E.M., J.P. Abraham, J.C. Tong, and G.L. Martin. 2001. Air-to-air energy exchanger test facility for mass and energy transfer performance. *ASHRAE Transactions* 107(2):450-456.

Stauder, F.A., Mathur, G.D, and T.W. McDonald. 1985. Experimental and computer simulation study of an air-to-air two-phase thermosiphon-loop heat exchanger. *ASME* 85-WA/HT-15.

Stauder, F.A. and T.W. McDonald. 1986. Experimental study of a two-phase thermosiphon-loop heat exchanger. *ASHRAE Transactions* 92(2A):486-497.

CHAPTER 26

AIR-HEATING COILS

AIR-HEATING coils are used to heat air under forced convection. The total coil surface may consist of a single coil section or several coil sections assembled into a bank. The coils described in this chapter apply primarily to comfort heating and air conditioning using steam, hot water, refrigerant vapor heat reclaim (including heat pumps), and electricity. The choice between the various methods of heating depends greatly on the cost of the various available energy sources. For instance, in areas where electric power is cheaply available and heating requirements are limited, heat pumps are a very viable option. With available power and higher heat requirements, electric heat is used. If electric power is considerably expensive, steam or hot water generated using gas-fired sources is used in larger buildings and district cooling. In smaller buildings, heat is supplied using gas furnaces, which are covered in Chapters 32 and 33. Water and steam heating are also widely used where process waste heat is available.

COIL CONSTRUCTION AND DESIGN

Extended-surface coils consist of a primary and a secondary heat-transfer surface. The primary surface is the external surface of the tubes, generally consisting of rows of round tubes or pipes that may be staggered or parallel (in-line) with respect to the airflow. Flattened tubes or tubes with other nonround internal passageways are sometimes used. The inside of the tube is usually smooth and plain, but some coil designs feature various forms of internal fins or turbulence promoters (either fabricated and then inserted, or extruded) to enhance fluid coil performance. The secondary surface is the fins' external surface, which consists of thin metal plates or a spiral ribbon uniformly spaced or wound along the length of the primary surface. The intimate contact with the primary surface provides good heat transfer. Air-heating fluid and steam coils are generally available with different circuit arrangements and combinations that offer varying numbers of parallel water flow passes in the tube core.

Copper and aluminum are the materials most commonly used for extended-surface coils. Tubing made of steel or various copper alloys is used in applications where corrosive forces might attack the coils from inside or outside. The most common combination for low-pressure applications is aluminum fins on copper tubes. Low-pressure steam coils are usually designed to operate up to 50 psig. Higher-strength tube materials such as red brass, admiralty brass, or cupronickel assembled by brazed construction are usable up to 366°F water or 150 psig saturated steam. Higher operating conditions call for electric welded stainless steel construction, designed to meet Section II and Section VIII requirements of the ASME *Boiler and Pressure Vessel Code.*

Customarily, the coil casing consists of a top and bottom channel (also known as baffles or side sheets), two end supports (also known as end plates or tube sheets), and, on longer coils, intermediate supports (also known as center supports or tube sheets). Designs vary, but most are mounted on ducts or built-up systems. Most often, casing material is spangled zinc-coated (galvanized) steel with a minimum coating designation of G90-U. Some corrosive air conditions may require stainless steel casings or corrosive-resistant coating, such as a baked phenolic applied by the manufacturer to the entire coil surface. Steam coil casings should be designed to accommodate thermal expansion of the tube core during operation (a *floating core* arrangement).

Common core tube diameters vary from 5/16 up to 1 in. outside diameter (OD) and fin spacings from 4 to 18 fins per inch. Fluid heating coils have a tube spacing from 3/4 to 1 3/4 in. and tube diameters from 5/16 to 5/8 in. OD. Steam coils have tube spacing from 1 1/4 to 3 in. and tube diameters from 0.5 to 1 in. OD. The most common arrangements are one- or two-row steam coils and two- to four-row hot-water coils. Fins should be spaced according to the application requirements, with particular attention given to any severe duty conditions, such as inlet temperatures and contaminants in the airstream.

Tube wall thickness and the required use of alloys other than (standard) copper are determined primarily by the coil's specified maximum allowable working pressure (MAWP) requirements. A secondary consideration is expected coil service life. Fin type, header, and connection construction also play a large part in this determination. All applicable local job site codes and national safety standards should be followed in the design and application of heating coils.

Flow direction can strongly affect heat transfer surface performance. In air-heating coils with only one row of tubes, the air flows at right angles to the heating medium. Such a cross-flow arrangement is common in steam heating coils. The steam temperature in the tubes remains uniform, and the mean temperature difference is the same regardless of the direction of flow relative to the air. The steam supply connection is located either in the center or at the top of the inlet header. The steam condensate outlet (return connection) is always at the lowest point in the return header.

When coils have two or more tube rows in the direction of airflow, such as hot-water coils, the heating medium in the tubes may be circuited in various parallel-flow and counterflow arrangements. Counterflow is the arrangement most preferred to obtain the highest possible mean temperature difference, which determines the heat transfer of the coil. The greater this temperature difference, the greater the coil's heat transfer capacity. In multirow coils circuited for counterflow, water enters the tube row on the leaving air side of the coil.

Steam Coils

Steam coils are generally classified, similarly to boilers, by operating pressure: low (≤15 psi) or high (>15 psi). However, various organizations use other pressure classification schemes with differing divisions [e.g., low (≤15 psi), medium (15 to 100 psi), or high (>100 psi)]. Steam coils can also be categorized by operating limits of the tube materials:

Standard steam	≤150 psi	366°F	copper tube
High-pressure steam	≤236 psi	400°F	special material [e.g., cupronickel (CuNi)]

The preparation of this chapter is assigned to TC 8.4, Air-to-Refrigerant Heat Transfer Equipment.

Table 1 Preferred Operating Limits for Continuous-Duty Steam Coil Materials in Commercial and Institutional Applications

Pressure, psi	Material	Tube Wall Thickness, in.
≤5	Copper	0.020
5 to 15		0.025
15 to 30		0.035
30 to 50		0.049
50 to 75	Red brass	0.025
75 to 100		0.035
100 to 150	90/10 CuNi	0.035
150 to 200		0.049

Note: Red brass and CuNi may be interchanged, depending on coil manufacturer's specifications.

Although these operating conditions are allowed by code, long exposures to them will shorten coil tube life. Leaks are less likely when the coil tube core has thicker walls of higher-strength materials. Operational experience suggests preferred limits for continuous-duty steam coils (tube OD ranging from 5/8 to 1 in.) in commercial and institutional applications, as shown in Table 1.

Steam coils also can be categorized by type as basic steam, steam-distributing, or face-and-bypass.

Basic steam coils generally have smooth tubes with fins on the air side. The steam supply connection is at one end and the tubes are pitched toward the condensate return, which is usually at the opposite end. For horizontal airflow, the tubes can be either vertical or horizontal. Horizontal tubes should be pitched within the casing toward the condensate return to facilitate condensate removal. Uniform steam distribution to all tubes is accomplished by careful selection of header size, its connection locations, and positioning of inlet connection distributor plates. Orifices also may be used at the core tube entrances in the supply header.

Steam-distributing coils most often incorporate perforated inner tubes that distribute steam evenly along the entire coil. The perforations perform like small steam ejector jets that, when angled in the inner tube, help remove condensate from the outer tube. An alternative design for short coils is an inner tube with no distribution holes, but with an open end. On all coils, supply and return connections can be at the same end or at opposite ends of the coil. For long, low-pressure coils, supply is usually at both ends and the condensate return on one end only.

Face-and-bypass steam coils have short sections of steam coils separated by air bypass openings. Airflow through the coil or bypass section is controlled by coil face-and-bypass dampers linked together. As a freeze protection measure, large installations use face-and-bypass steam coils with vertical tubes.

For proper performance of all types of steam heating coils, air or other noncondensables in the steam supply must be eliminated. Equally important, condensate from the steam must easily drain from inside the coil. Air vents are located at a high point of the piping and at the coil's inlet steam header. Whether airflow is horizontal or vertical, the coil's finned section is pitched toward the condensate return connection end of the coil. Installers must give particular care in the selection and installation of piping, controls, and insulation necessary to protect the coil from freeze-up caused by incomplete condensate drainage.

When entering air is at or below 32°F, the steam supply to the coil should not be modulated, but controlled as full on or full off. Coils located in series in the airstream, with each coil sized and controlled to be full on or completely off (in a specific sequence, depending on the entering air temperature), are not as likely to freeze. Temperature control with face-and-bypass dampers is also common. During part-load conditions, air is bypassed around the steam coil with full steam flow to the coil. In a face-and-bypass arrangement, high-velocity streams of freezing air must not impinge on the coil when the face dampers are partially closed. The section on Overall Requirements in this chapter and the section on Control of HVAC Elements (Heating Coils) in Chapter 46 of the 2007 *ASHRAE Handbook—HVAC Applications* have more details.

Water/Aqueous Glycol Heating Coils

Normal-temperature hot-water heating coils can be categorized as booster coils or standard heating coils. Booster (duct-mounted or reheat) coils are commonly found in variable-air-volume systems. They are one or two rows deep, have minimal water flow, and provide a small air temperature rise. Casings can be either flanged or slip-and-drive construction. Standard heating coils are used in run-around systems, makeup air units, and heating and ventilating systems. All use standard construction materials of copper tube and aluminum fins.

High-temperature water coils may operate with up to 400°F water, with pressures comparable or somewhat higher than the saturated vapor temperature of the water supply. The temperature drop across the coil may be as high as 150°F. To safely accommodate these fluid temperatures and thermal stresses, the coil requires industrial-grade construction that conforms to applicable boiler and safety codes. These requirements should be listed in detail by the specifying engineer, along with the inspection and certification requirements and a compliance check before coil installation and operation.

Proper water coil performance depends on eliminating air and on good water distribution in the coil and its interconnecting piping. Unless properly vented, air may accumulate in the coil circuits, which reduces heat transfer and possibly causes noise and vibration in the pipes. For this reason, water coils should be constructed with self-venting, drainable circuits. The self-venting design is maintained by field-connecting the water supply connection to the bottom and the water return connection to the top of the coil. Ideally, water is supplied at the bottom, flows upward through the coil, and forces any air out the return connection. Complete fluid draining at the supply connection indicates that coils are self-draining and without air or water traps. Such a design ensures that the coil is always filled with water, and it should completely drain when it is required to be empty. Most manufacturers provide vent and drain fittings on the supply and return headers of each water coil.

When water does get trapped in the coil core, it is usually caused by a sag in the coil core or by a nondraining circuit design. During freezing periods, even a small amount of water in the coil core can rupture a tube. Also, such a static accumulation of either water or glycol can corrode the tube over an extended period. Large multirow, multicircuited coils may not drain rapidly, even with self-draining circuitry; if they are not installed level, complete self draining will not take place. This problem can be prevented by including intermediate drain headers and installing the coil so that it is pitched toward the connections.

To produce desired ratings without excessive water pressure drop, manufacturers use various circuit arrangements. A single-feed serpentine circuit is commonly used on booster coils with low water flows. With this arrangement, a single feed carrying the entire water flow makes a number of passes across the airstream. The more common circuit arrangement is called a **full row feed** or **standard circuit**. With this design, all the core tubes of a row are fed with an equal amount of water from the supply header. Others, such as quarter, half, and double-row feed circuit arrangements, may be available, depending on the total number of tubes and rows of the coil. Uniform flow in each water circuit is obtained by designing each circuit's length as equal to the other as possible.

Generally, higher velocity provides greater capacity and more even discharge air temperature across the coil face, but with diminishing returns. To prevent erosion, 6 fps should not be exceeded for copper coils. At higher velocities, only modest gains in capacity can be achieved at increasingly higher pumping power penalties. Above 8 fps, any gain is negligible.

Velocities with fluid flow Reynolds numbers (Re) between 2000 and 10,000 fall into a transition range where heat transfer capacity predictions are less likely to be accurately computed. Below Re = 2000, flow is laminar, where heat transfer prediction is again reliable, but coil capacity is greatly diminished and tube fouling can become a problem. For further insight on the transition flow effect on capacity, refer to Figure 16 of Air-Conditioning and Refrigeration Institute (ARI) *Standard* 410. Methods of controlling water coils to produce a uniform exit air temperature are discussed in Chapter 46 of the 2007 *ASHRAE Handbook—HVAC Applications*.

In some cases, the hot water circulated may contain a considerable amount of sand and other foreign matter such as minerals. This matter should be filtered from the water circuit. Additionally, some coil manufacturers offer removable water header boxes (some are plates), or a removable plug at the return bends of each tube, allowing the tubes to be rodded clean. In an area where build-up of scale or other deposits is expected, include a fouling factor when computing heating coil performance. Hot-water coil ratings (ARI *Standard* 410) include a 0.00025 h·ft^2·°F/Btu fouling factor. Cupronickel, red brass, admiralty, stainless steel, and other tube alloys are usable to protect against corrosion and erosion, which can be common in hot-water/glycol systems.

Volatile Refrigerant Heat Reclaim Coils

A heat reclaim coil with a volatile refrigerant can function as a condenser either in series or parallel with the primary condenser of a refrigeration system. Heat from condensing or desuperheating vapor warms the airstream. It can be used as a primary source of heat or to assist some other form of heating, such as reheat for humidity control. In the broad sense, a heat reclaim coil functions at half the heat-dissipating capacity of its close-coupled refrigerant system's condenser. Thus, a heat reclaim coil should be (1) piped to be upstream from the condenser and (2) designed with the assumption that some condensate must be removed from the coil. For these reasons, the coil outlet should be located at the lowest point of the coil and trapped if this location is lower than the inlet of the condenser.

Heat reclaim coils are normally circuited for counterflow of air and refrigerant. However, most supermarket heat reclaim coils are two rows deep and use a crossflow design. The section on Air-Cooled Condensers in Chapter 38 has additional information on this topic. Also, because refrigerant heat reclaim involves specialized heating coil design, a refrigerant equipment manufacturer is the best source for information on the topic.

Electric Heating Coils

An electric heating coil consists of a length of resistance wire (commonly nickel/chromium) to which a voltage is applied. The resistance wire may be bare or sheathed in an electrically insulating layer, such as magnesium oxide, and compacted inside a finned steel tube. Sheathed coils are more expensive, have a higher air-side pressure drop, and require more space. A useful comparison for sizing is a heat transfer capacity of 41,000 Btu/h·ft^2 of face area compared to 100,000 Btu/h·ft^2 for bare resistance wire coils. However, the outer surface temperature of sheathed coils is lower, the coils are mechanically stronger, and contact with personnel or housing is not as dangerous. Coils with sheathed heating elements having an extended finned surface are generally preferred (1) for dust-laden atmospheres, (2) where there is a high probability of maintenance personnel contact, or (3) downstream from a dehumidifying coil that might have moisture carryover. Manufacturers can provide further information, including selection recommendations, applications, and maintenance instructions.

COIL SELECTION

The following factors should be considered in coil selection:

- Required duty or capacity considering other components
- Temperature of air entering the coil and air temperature rise
- Available heating media's operating and maximum pressure(s) and temperature(s)
- Space and dimensional limitations
- Air volume, speed, distribution, and limitations
- Heating media volume, flow speed, distribution, and limitations
- Permissible flow resistances for both the air and heating media
- Characteristics of individual designs and circuit possibilities
- Individual installation requirements, such as type of control and material compatibility
- Specified and applicable codes and standards regulating the design and installation.

Load requirements are discussed in Chapters 29 and 30 of the 2005 *ASHRAE Handbook—Fundamentals*. Much is based on the choice of heating medium, as well as operating temperatures and core tube diameter. Also, proper selection depends on whether the installation is new, being modified, or a replacement. Dimensional fit is usually the primary concern of modified and replacement coils; heating capacity is often unknown.

Air quantity is regulated by factors such as design parameters, codes, space, and size of the components. Resistance through the air circuit influences fan power and speed. This resistance may be limited to allow use of a given size fan motor or to keep operating expenses low, or because of sound level requirements. All of these factors affect coil selection. The air friction loss across the heating coil (summed with other series air pressure drops for system component such as air filters, cooling coils, grilles, and ductwork) determines the static pressure requirements of the complete air system. See Chapter 20 for selecting the fan component.

Permissible resistance through the water or glycol coil circuitry may be dictated by the available pressure from a given size pump and motor. This is usually controlled within limits by careful selection of coil header size and the number of tube circuits. Additionally, the adverse effect of high fluid velocity in contributing to erosion/corrosion of the tube wall is a major factor in selecting tube diameter and the circuit. Heating coil performance depends on the correct choice of original equipment and proper application and installation. For steam coils, proper performance relies first on selecting the correct type of steam coil, and then the proper size and type of steam trap. Properly sized connecting refrigerant lines, risers, and traps are critical to heat reclaim coils.

Heating coil thermal performance is relatively simple to derive. It only involves a dry-bulb temperature and sensible heat, without the complications of latent load and wet-bulb temperature for dehumidifying cooling performance. Even simpler, consult coil manufacturers' catalogs for ratings and selection. Most manufacturers provide computerized coil selection and rating programs on request; some are certified accurate within 5% of an application parameter, representative of a normal application range. Many manufacturers participate in the ARI Coil Certification Program, which approves application ratings that conform to all ARI *Standard* 410 requirements, based on qualifying testing to ASHRAE *Standard* 33.

Coil Ratings

Coil ratings are based on uniform face velocity. Nonuniform airflow may be caused by the system, such as air entering at odd angles, or by inadvertent blocking of part of the coil face. To obtain rated performance, the airflow quantity in the field must correspond to the design requirements and the velocity vary no greater than 20% at any point across the coil face.

The industry-accepted method of coil rating is outlined in ARI *Standard* 410. The test requirements for determining standard coil ratings are specified in ASHRAE *Standard* 33. ARI application ratings are derived by extending the ASHRAE standard rating test results for other operating conditions, coil sizes, row depths, and fin count for a particular coil design and arrangement. Steam, water, and glycol heating coils are rated within the following limits (listed in ARI *Standard* 410), which may be exceeded for special applications.

Air face velocity. 200 to 1500 fpm, based on air density of 0.075 lb/ft^3

Entering air temperature. Steam coils: –20 to 100°F
Water coils: 0 to 100°F

Steam pressure. 2 to 250 psig at coil steam supply connection (pressure drop through steam control valve must be considered)

Fluid temperatures. Water: 120 to 250°F
Ethylene glycol: 0 to 200°F

Fluid velocities. Water: 0.5 to 8 fps
Ethylene glycol: 0.5 to 6 fps

Overall Requirements

Individual installations vary widely, but the following values can be used as a guide. The air face velocity is usually between 500 and 1000 fpm. Delivered air temperature varies from about 72°F for ventilation only to about 150°F for complete heating. Steam pressure typically varies from 2 to 15 psig, with 5 psig being the most common. A minimum steam pressure of 5 psig is recommended for systems with entering air temperatures below freezing. Hot-water (or glycol) temperature for comfort heating is commonly between 180 and 200°F, with water velocities between 4 and 6 fps. For high-temperature water, water temperatures can be over 400°F with operating pressures of 15 to 25 psi over saturated water temperature.

Water quantity is usually based on about 20°F temperature drop through the coil. Air resistance is usually limited to 0.4 to 0.6 in. of water for commercial buildings and to about 1 in. for industrial buildings. High-temperature water systems commonly have a water temperature between 300 and 400°F, with up to 150°F drop through the coil.

Steam coils are selected with dry steam velocities not exceeding 6000 fpm and with acceptable condensate loading per coil core tube depending on the type of steam coil. Table 2 shows some typical maximum condensate loads.

Steam coil performance is maximized when the supply is dry, saturated steam, and condensate is adequately removed from the coil and continually returned to the boiler.

Although steam quality may not significantly affect the heat transfer of the coil, the back-up effect of too rapid a condensate rate, augmented by a wet supply stream, can cause a slug of condensate to travel through the coil and condensate return. This situation can result in noise and possible damage.

Complete mixing of return and outdoor air is essential to proper coil operation. The design of the air mixing damper or ductwork connection section is critical to the proper operation of a system and its air temperature delivery. Systems in which the air passes through a fan before flowing through a coil do not ensure proper air mixing. Dampers at the inlet air face of a steam coil should be the opposed-blade type, which are better than in-line blades for controlling air volume and reducing individual blade-directed cold airstreams when modulating in low-heat mode.

Table 2 Typical Maximum Condensate Loads

Tube Outside Diameter	Maximum Allowable Condensate Load, lb/h	
	Basic Coil	Steam Distributing Coil
5/8 in.	68	40
1 in.	168	95

Heat Transfer and Pressure Drops. For air-side heat transfer and pressure drop, the information given in Chapter 22 for sensible cooling coils is applicable. For water (or glycol) coils, the information given in Chapter 22 for water-side heat transfer and pressure drop also applies here. For steam coils, the heat transfer coefficient of condensing steam must be calculated (see Chapter 3 of the 2005 *ASHRAE Handbook—Fundamentals*). For estimating the pressure drop of condensing steam, see Chapter 5 of the 2005 *ASHRAE Handbook—Fundamentals*.

Parametric Effects. The heat transfer performance of a given coil can be changed by varying the airflow rate and/or the temperature of the heating medium, both of which are relatively linear. Understanding the interaction of these parameters is necessary for designing satisfactory coil capacity and control. A review of manufacturers' catalogs and selection programs, many of which are listed in ARI's *Applied Directory of Certified Products* and *Forced-Circulation Air-Cooling and Air-Heating Coils*, shows the effects of varying these parameters.

INSTALLATION GUIDELINES

Steam systems designed to operate at outdoor air temperatures below 32°F should be different from those designed to operate above it. Below 32°F, the steam air-heating system should be designed as a preheat and reheat pair of coils. The preheat coil functions as a nonmodulating basic steam coil, which requires full steam pressure whenever the outdoor temperature is below freezing. The reheat coil, typically a modulated steam-distributing coil, provides the heating required to reach the design air temperature. Above 32°F, the heating coil can be either a basic or steam-distributing type as needed for the duty.

When the leaving air temperature is controlled by modulating steam supply to the coil, steam-distributing tube coils provide the most uniform exit air temperature (see the section on Steam Coils). Correctly designed steam-distributing tube coils can limit the exit air temperature stratification to a maximum of 6°F over the entire length of the coil, even when steam supply is modulated to a fraction of full-load capacity.

Low-pressure steam systems and coils controlled by modulating steam supply should have a vacuum breaker or be drained through a vacuum-return system to ensure proper condensate drainage. It is good practice to install a closed vacuum breaker (where required) connected to the condensate return line through a check valve. This unit breaks the vacuum by equalizing the pressure, yet minimizes the possibility of air bleeding into the system. Steam traps should be located at least 12 in. below the condensate outlet to allow the coil to drain properly. Also, coils supplied with low-pressure steam or controlled by modulating steam supply should not be trapped directly to an overhead return line. Condensate can be lifted to overhead returns only when enough pressure is available to overcome the condensate head and any return line pressure. If overhead returns are necessary, the condensate must be pumped to the higher elevation (see Chapter 10).

Water coils for air heating generally have horizontal tubes to avoid air pockets. Where water or glycol coils may be exposed to a freezing condition, drainability must be considered. If a coil is to be drained and then exposed to below-freezing temperatures, it should first be flushed with a nonfreeze solution.

To minimize the danger of freezing in both steam and water-heating coils, the outside air inlet dampers usually close automatically when the fan is stopped (system shutdown). In steam systems with very cold outside air conditions (e.g., –20°F or below), it is desirable to fully open the steam valve when the system is shut down. If outside air is used for proportioning building makeup air, the outside air damper should be an opposed-blade design.

Heating coils are designed to allow for expansion and contraction resulting from the temperature ranges in which they operate.

Care must be taken to prevent imposing strains from the piping to the coil connections, particularly on high-temperature hot-water applications. Expansion loops, expansion or three-elbow swing joints, or flexible connections usually provide the needed protection (see Chapter 10).

Good practice supports banked coils individually in an angle-iron frame or a similar supporting structure. With this arrangement, the lowest coil is not required to support the weight of the coils stacked above. This design also facilitates removing individual coils in a multiple-coil bank for repair or replacement.

Heat reclaim coils depend on a closely located, readily available source of high-side refrigerant vapor. For example, supermarket rack compressors and the air handler's coil section should be installed close to the store's motor room. Most commonly, the heat reclaim coil section is piped in series with the rack's condenser and sized for 50% heat extraction. Heat reclaim in supermarkets is discussed in Chapter 2 of the 2007 *ASHRAE Handbook—HVAC Applications*; also see Chapter 46 of the 2006 *ASHRAE Handbook—Refrigeration*.

In heat pump applications, the heating coil is usually the indoor coil that is used for cooling in the summer. The heat pump system is normally optimized for cooling load and efficiency. Heating performance is a by-product of cooling performance, and any unsatisfied heating requirement is met using electric heating coils. The refrigerant circuiting is also specialized and involves driving the refrigerant in the reverse direction from the cooling mode. Heat pump coils also have an additional check valve bypass to the expansion device for operating in heat pump mode. For additional details about heat pump coil design and operation, see Chapters 8 and 48.

COIL MAINTENANCE

Both internal and external surfaces must be clean for coils to deliver their full rated capacity. The tubes generally stay clean in glycol systems and in adequately maintained water systems. Should scale be detected in the piping where untreated water is used, chemical or mechanical cleaning of the internal tube surfaces is required. The need for periodic descaling can be minimized by proper boiler water treatment and deaeration.

Internal coil maintenance consists primarily of preventing scale and corrosion in the coil core tubes and piping of potable-water-heating (including steam) systems. In its simplest form, this involves removing dissolved oxygen, maintaining deionized water, and controlling boiler water pH. Boiler water can be deaerated mechanically. Vacuum deaerating simultaneously removes oxygen and carbon dioxide. The last traces of oxygen can be removed chemically by adding sodium sulfide. For steam coils, 100% dry steam contains 0% air. Good boiler water results in the absence of oxygen and a pH maintained at 10.5. If this is not practicable, a pH of 7 to 9 with a corrosion inhibitor is recommended. Because calcium carbonate is less soluble in water at higher pH, the inhibitor most often used for this purpose is sodium nitrate. Usually, the requirements for chemical treatment increase as the temperature of the coil's return flow drops. With few exceptions, boiler water chemical treatment programs use only proportional feeding, which is the recommended way of maintaining a constant concentration at all times. Periodic batch or slug feeding of water treatment chemicals is not an accurate way to treat boiler water, particularly if the system has a high water makeup rate. Only use chemicals known to be compatible with the coil's tube and connection metal(s). Chemical treatment of boiler water is complicated and has environmental effects. For this reason, it is important to consult a water treatment specialist to establish a proper boiler water treatment program. For further information on boilers, see Chapter 31.

The finned surface of heating coils can sometimes be brushed and cleaned with a vacuum cleaner. Coils are commonly surface-cleaned annually using pressurized hot water containing a mild detergent. Reheat coils that contain their refrigerant charge should never be cleaned with a spray above 150°F. In extreme cases of neglect, especially in restaurants where grease and dirt have accumulated, coil(s) may need to be removed to completely clean off the accumulation with steam, compressed air and water, or hot water containing a suitable detergent. Pressurized cleaning is more thorough if first done from the coil's air leaving side, before cleaning from the coil's air entry side. Often, outside makeup air coils have no upstream air filters, so they should be visually checked on a frequent schedule. Overall, coils should be inspected and serviced regularly. Visual observation should not be relied on to judge cleaning requirements for coils greater than three rows deep because airborne dirt tends to pack midway through the depth of the coil.

REFERENCES

ARI. 2001. Forced-circulation air-cooling and air-heating coils. *Standard* 410-2001. Air-Conditioning and Refrigeration Institute, Arlington, VA.

ARI. 2008. *CHC: Certified air-cooling and air-heating coils*. Air-Conditioning and Refrigeration Institute, Arlington, VA. http://www.ahridirectory.org.

ARI. Semiannually. *Applied directory of certified air-conditioning products*. Air-Conditioning and Refrigeration Institute, Arlington, VA. http://www.ahridirectory.org.

ASHRAE. 2000. Methods of testing forced circulation air cooling and air heating coils. *Standard* 33-2000.

ASME. 2001. *Boiler and pressure vessel code*. American Society of Mechanical Engineers, New York.

CHAPTER 27

UNIT VENTILATORS, UNIT HEATERS, AND MAKEUP AIR UNITS

UNIT VENTILATORS

A HEATING **unit ventilator** is an assembly whose principal functions are to heat, ventilate, and cool a space by introducing outside air in quantities up to 100% of its rated capacity. The heating medium may be steam, hot water, gas, or electricity. The essential components of a heating unit ventilator are the fan, motor, heating element, damper, filter, automatic controls, and outlet grille, all of which are encased in a housing.

An **air-conditioning unit ventilator** is similar to a heating unit ventilator; however, in addition to the normal winter function of heating, ventilating, and cooling with outside air, it is also equipped to cool and dehumidify during the summer. It is usually arranged and controlled to introduce a fixed quantity of outside air for ventilation during cooling in mild weather. The air-conditioning unit ventilator may be provided with a variety of combinations of heating and air-conditioning elements. Some of the more common arrangements include

- Combination hot- and chilled-water coil (two-pipe)
- Separate hot- and chilled-water coils (four-pipe)
- Hot-water or steam coil and direct-expansion coil
- Electric heating coil and chilled-water or direct-expansion coil
- Gas-fired furnace with direct-expansion coil

The typical unit ventilator is equipped with controls that allow heating, ventilating, and cooling to be varied while the fans operate continuously. In normal operation, the discharge air temperature from a unit is varied in accordance with the room requirements. The heating unit ventilator can provide **ventilation cooling** by bringing in outside air whenever the room temperature is above the room set point. Air-conditioning unit ventilators can provide refrigeration when the outside air temperature is too high to be used effectively for ventilation cooling.

Unit ventilators are available for floor mounting, ceiling mounting, and recessed applications. They are available with various airflow and capacity ratings, and the fan can be arranged so that air is either blown through or drawn through the unit. With direct-expansion refrigerant cooling, the condensing unit can either be furnished as an integral part of the unit ventilator assembly or be remotely located.

Figure 1A shows a typical heating unit ventilator. The heating coil can be hot water, steam, or electric. Hot-water coils can be provided with face-and-bypass dampers for capacity control, if desired. Valve control of capacity is also available.

Figure 1B shows a typical air-conditioning unit ventilator with a combination hot- and chilled-water coil for use in a two-pipe system. This type of unit is usually provided with face-and-bypass dampers for capacity control.

Figure 1C illustrates a typical air-conditioning unit ventilator with two separate coils, one for heating and the other for cooling with a four-pipe system. The heating coil may be hot water, steam, or electric. The cooling coil can be either a chilled-water coil or a direct-expansion refrigerant coil. Heating and cooling coils are sometimes combined in a single coil by providing separate tube circuits for each function. In such cases, the effect is the same as having two separate coils.

Figure 1D illustrates a typical air-conditioning unit ventilator with a fan section, a gas-fired heating furnace section, and a direct-expansion refrigerant coil section.

APPLICATION

Unit ventilators are used primarily in schools, meeting rooms, offices, and other areas where the density of occupancy requires controlled ventilation to meet local codes.

Floor-model unit ventilators are normally installed on an outside wall near the centerline of the room. Ceiling models are mounted against either the outside wall or one of the inside walls. Ceiling models discharge air horizontally. Best results are obtained if the unit can be placed so that the airflow is not interrupted by ceiling beams or surface-mounted lighting fixtures.

Downdraft can be a problem in classrooms with large window areas in cold climates. Air in contact with the cold glass is cooled and flows down into the occupied space. Floor-standing units often include one of the following provisions to prevent downdraft along the windows (Figure 2):

- **Window sill heating** uses finned radiators of moderate capacity installed along the wall under the window area. Heated air rises upward by convection and counteracts the downdraft by tempering it and diverting it upward.
- **Window sill recirculation** is obtained by installing the return air intake along the window sill. Room or return air to the unit includes the cold downdrafts, takes them from the occupied area, and eliminates the problem.
- **Window sill discharge** directs a portion of the unit ventilator discharge air into a delivery duct along the sill of the window. The discharge air, delivered vertically at the window sill, is distributed throughout the room, and the upwardly directed air combats downdraft.

The preparation of the sections on Unit Ventilators and Unit Heaters is assigned to TC 6.1, Hydronic and Steam Equipment and Systems. The preparation of the section on Makeup Air Units is assigned to TC 5.8, Industrial Ventilation Systems.

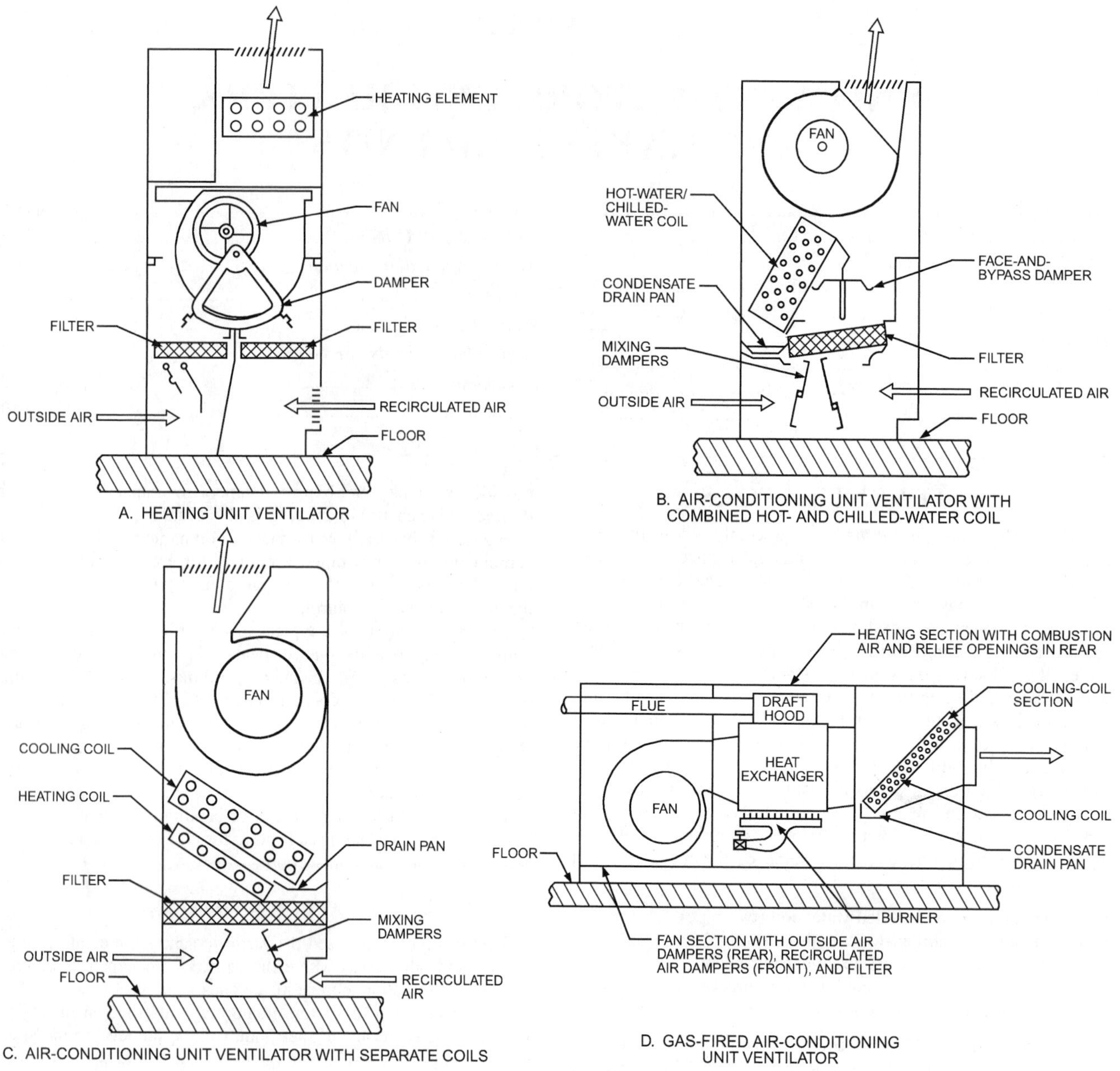

HEATING ELEMENT
FAN
DAMPER
FILTER
FILTER
OUTSIDE AIR
RECIRCULATED AIR
FLOOR
A. HEATING UNIT VENTILATOR
FAN
HOT-WATER/ CHILLED- WATER COIL
FACE-AND- BYPASS DAMPER
CONDENSATE DRAIN PAN
MIXING DAMPERS
FILTER
OUTSIDE AIR
RECIRCULATED AIR
FLOOR
B. AIR-CONDITIONING UNIT VENTILATOR WITH COMBINED HOT- AND CHILLED-WATER COIL
FAN
COOLING COIL
HEATING COIL
DRAIN PAN
FILTER
MIXING DAMPERS
OUTSIDE AIR
FLOOR
RECIRCULATED AIR
C. AIR-CONDITIONING UNIT VENTILATOR WITH SEPARATE COILS
HEATING SECTION WITH COMBUSTION AIR AND RELIEF OPENINGS IN REAR
FLUE
DRAFT HOOD
COOLING-COIL SECTION
HEAT EXCHANGER
FAN
COOLING COIL
FLOOR
CONDENSATE DRAIN PAN
BURNER
FAN SECTION WITH OUTSIDE AIR DAMPERS (REAR), RECIRCULATED AIR DAMPERS (FRONT), AND FILTER
D. GAS-FIRED AIR-CONDITIONING UNIT VENTILATOR

Fig. 1 Typical Unit Ventilators

WINDOW SILL HEATING
WINDOW SILL RECIRCULATING
WINDOW SILL DISCHARGE

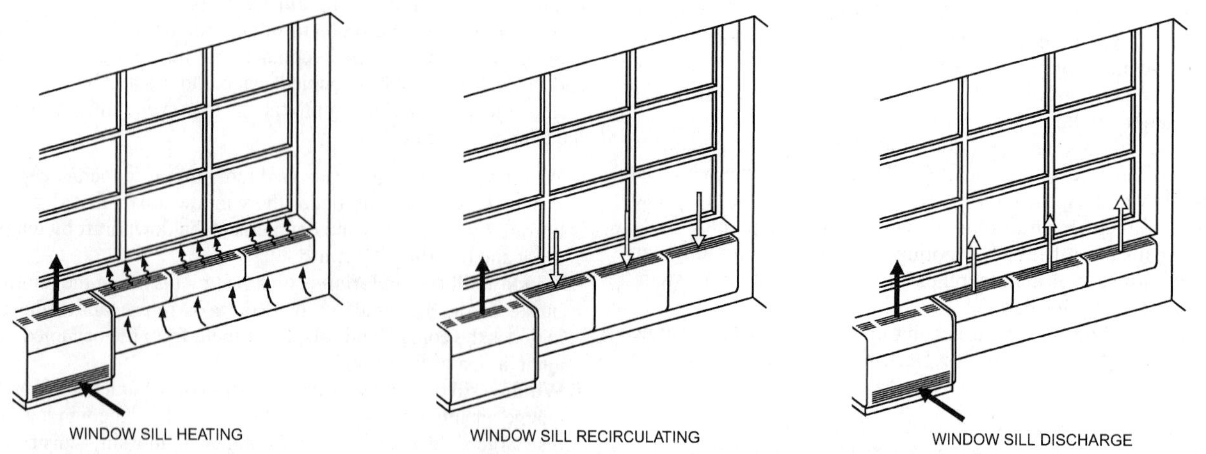

Fig. 2 Methods of Preventing Downdraft along Windows

SELECTION

Items to be considered in the application of unit ventilators are

- Unit air capacity
- Percent minimum outside air
- Heating and cooling capacity
- Cycle of control
- Location of unit

Mild-weather cooling capacity and number of occupants in the space are the primary considerations in selecting the unit's air capacity. Other factors include state and local requirements, volume of the room, density of occupancy, and use of the room. The number of air changes required for a specific application also depends on window area, orientation, and maximum outside temperature at which the unit is expected to prevent overheating.

Rooms oriented to the north (in the northern hemisphere) with small window areas require about 6 air changes per hour (ach). About 9 ach are required in rooms oriented to the south that have large window areas. As many as 12 ach may be required for very large window areas and southern exposures. These airflows are based on preventing overheating at outside temperatures up to about 55°F. For satisfactory cooling at outside air temperatures up to 60°F, airflow should be increased accordingly.

These airflows apply principally to classrooms. Factories and kitchens may require 30 to 60 ach (or more). Office areas may need 10 to 15 ach.

The minimum amount of outside air for ventilation is determined after the total air capacity has been established. It may be governed by local building codes or it may be calculated to meet the ventilating needs of the particular application. For example, ASHRAE *Standard* 62.1 requires 7.5 to 10 cfm of outside air per occupant (0.06 to 0.18 cfm/ft^2) in lecture halls or classrooms, laboratories, and cafeterias, and 5 cfm per occupant in conference rooms.

The heating and cooling capacity of a unit to meet the heating requirement can be determined from the manufacturer's data. Heating capacity should always be determined after selecting the unit air capacity for mild-weather cooling.

Capacity

Manufacturers publish the heating and cooling capacities of unit ventilators. Table 1 lists typical nominal capacities.

Heating Capacity Requirements. Because a unit ventilator has a dual function of introducing outside air for ventilation and maintaining a specified room condition, the required heating capacity is the sum of the heat required to bring outside ventilation air to room temperature and the heat required to offset room losses. The ventilation cooling capacity of a unit ventilator is determined by the air volume delivered by the unit and the temperature difference between the unit discharge and the room temperature.

Example. A room has a heat loss of 24,000 Btu/h at a winter outside design condition of 0°F and an indoor design of 70°F, with 20% outside air. Minimum air discharge temperature from the unit is 60°F. To obtain the specified number of air changes, a 1250 cfm unit ventilator is required. Determine the ventilation heat requirement, the total heating requirement, and the ventilation cooling capacity of this unit with outside air temperature below 60°F.

Solution:

Ventilation heat requirement:

$$q_v = 60\rho c_p Q(t_i - t_o)$$

where

q_v = heat required to heat ventilating air, Btu/h
ρ = density of air at standard conditions = 0.075 lb/ft^3
c_p = air specific heat = 0.24 Btu/lb·°F
Q = ventilating airflow, cfm
t_i = required room air temperature, °F
t_o = outside air temperature, °F

$$q_v = 60 \times 0.075 \times 0.24 \times 1250(20/100)(70 - 0) = 18{,}900 \text{ Btu/h}$$

Total heating requirement:

$$q_t = q_v + q_s$$

where

q_t = total heat requirement, Btu/h
q_s = heat required to make up heat losses, Btu/h

$$q_t = 18{,}900 + 24{,}000 = 42{,}900 \text{ Btu/h}$$

Ventilation cooling capacity:

$$q_c = 60\rho c_p Q(t_i - t_f)$$

where

q_c = ventilation cooling capacity of unit, Btu/h
t_f = unit discharge air temperature, °F

$$q_c = 60 \times 0.075 \times 0.24 \times 1250(70 - 60) = 13{,}500 \text{ Btu/h}$$

Table 1 Typical Unit Ventilator Capacities

Airflow, cfm	Heating Unit Ventilator Total Heating Capacity, Btu/h	A/C Unit Ventilator Total Cooling Capacity, Btu/h
500	38,000	19,000
750	50,000	28,000
1000	72,000	38,000
1250	85,000	47,000
1500	100,000	56,000

CONTROL

Many cycles of control are available. The principal difference in the various cycles is the amount of outside air delivered to the room. Usually, a room thermostat simultaneously controls both a valve, damper, or step controller to regulate the heat supply and an outside and return air damper. A thermostat in the airstream prevents discharge of air below the desired minimum temperature. Unit ventilator controls provide the proper sequence for the following stages:

Warm-Up Stage. All control cycles allow rapid warm-up by having the units generate full heat with the outside damper closed. Thus 100% of the room air is recirculated and heated until the room temperature approaches the desired level.

Heating and Ventilating Stage. As the room temperature rises into the operating range of the thermostat, the outside air damper partially or completely opens to provide ventilation, depending on the cycle used. Auxiliary heating equipment is shut off. As the room temperature continues to rise, the unit ventilator heat supply is throttled.

Cooling and Ventilating Stage. When the room temperature rises above the desired level, the room thermostat throttles the heat supply so that cool air flows into the room. The thermostat gradually shuts off the heat and then opens the outside air damper. The airstream thermostat frequently takes control during this stage to keep the discharge temperature from falling below a set level.

The section on Air Handling, under the Air Systems section, in Chapter 46 of the 2007 *ASHRAE Handbook—HVAC Applications* describes the three cycles of control commonly used for unit ventilators:

Cycle I. 100% outside air is admitted at all times, except during warm-up.

Cycle II. A minimum amount of outside air (normally 20 to 50%) is admitted during the heating and ventilating stage. This percentage is gradually increased to 100%, if needed, during the ventilation cooling stage.

Cycle III. Except during warm-up, a variable amount of outside air is admitted, as needed, to maintain a fixed temperature of air entering the heating element. The amount of air admitted is controlled by the airstream thermostat, which is set low enough (often at 55°F) to provide cooling when needed.

Air-conditioning unit ventilators can include any of the three cycles in addition to the mechanical cooling stage in which a fixed amount of outside air is introduced. The cooling capacity is controlled by the room thermostat.

For maximum heating economy, the building temperature is reduced at night and during weekends and vacations. Several arrangements are used to accomplish this. One arrangement takes advantage of the natural convective capacity of the unit when the fans are off. This capacity is supplemented by cycling the fan with the outside damper closed as required to maintain the desired room temperature.

UNIT HEATERS

A unit heater is an assembly of elements, the principal function of which is to heat a space. The essential elements are a fan and motor, a heating element, and an enclosure. Filters, dampers, directional outlets, duct collars, combustion chambers, and flues may also be included. Some types of unit heaters are shown in Figure 3.

Unit heaters can usually be classified in one or more of the following categories.

- **Heating Medium.** Media include (1) steam, (2) hot water, (3) gas indirect-fired, (4) oil indirect-fired, and (5) electric heating.
- **Type of Fan.** Three types of fans can be considered: (1) propeller, (2) centrifugal, and (3) remote air mover. Propeller fan units may be arranged to blow air horizontally (horizontal blow) or vertically (downblow). Units with centrifugal fans may be small cabinet units or large industrial units. Units with remote air movers are known as duct unit heaters.
- **Arrangement of Elements.** Two types of units can be considered: (1) draw-through, in which the fan draws air through the unit; and (2) blow-through, in which the fan blows air through the heating element. Indirect-fired unit heaters are always blow-through units.

APPLICATION

Unit heaters have the following principal characteristics:

- Relatively large heating capacities in compact casings
- Ability to project heated air in a controlled manner over a considerable distance
- Relatively low installed cost per unit of heat output
- Application where sound level is permissible

They are, therefore, usually placed in applications where the heating capacity requirements, physical volume of the heated space, or both, are too large to be handled adequately or economically by other means. By eliminating extensive duct installations, the space is freed for other use.

Unit heaters are mostly used for heating commercial and industrial structures such as garages, factories, warehouses, showrooms, stores, and laboratories, as well as corridors, lobbies, vestibules, and similar auxiliary spaces in all types of buildings. Unit heaters may often be used to advantage in specialized applications requiring spot or intermittent heating, such as at outside doors in industrial plants or in corridors and vestibules. Cabinet unit heaters may be used where heated air must be filtered.

Unit heaters may be applied to a number of industrial processes, such as drying and curing, in which the use of heated air in rapid circulation with uniform distribution is of particular advantage. They may be used for moisture absorption applications, such as removing fog in dye houses, or to prevent condensation on ceilings or other cold surfaces of buildings in which process moisture is released. When such conditions are severe, unit ventilators or makeup air units may be required.

SELECTION

The following factors should be considered when selecting a unit heater:

- Heating medium to be used
- Type of unit
- Location of unit for proper heat distribution
- Permissible sound level
- Rating of the unit
- Need for filtration

Heating Medium

The proper heating medium is usually determined by economics and requires examining initial cost, operating cost, and conditions of use.

Steam or **hot-water unit heaters** are relatively inexpensive but require a boiler and piping system. The unit cost of such a system generally decreases as the number of units increases. Therefore, steam or hot-water heating is most frequently used (1) in new installations involving a relatively large number of units, and (2) in existing systems that have sufficient capacity to handle the additional load. High-pressure steam or high-temperature hot-water units are normally used only in very large installations or when a high-temperature medium is required for process work. Low-pressure steam and conventional hot-water units are usually selected for smaller installations and for those concerned primarily with comfort heating.

Gas and **oil indirect-fired unit heaters** are frequently preferred in small installations where the number of units does not justify the expense and space requirements of a new boiler system or where individual metering of the fuel supply is required, as in a shopping center. Gas indirect-fired units usually have either horizontal propeller fans or industrial centrifugal fans. Oil indirect-fired units largely have industrial centrifugal fans. Some codes limit the use of indirect-fired unit heaters in some applications.

Electric unit heaters are used when low-cost electric power is available and for isolated locations, intermittent use, supplementary heating, or temporary service. Typical applications are ticket booths, security offices, factory offices, locker rooms, and other isolated rooms scattered over large areas. Electric units are particularly useful in isolated and untended pumping stations or pits, where they may be thermostatically controlled to prevent freezing.

Type of Unit

Propeller fan units are generally used in nonducted applications where the heating capacity and distribution requirements can best be met by units of moderate output and where heated air does not need to be filtered. Horizontal-blow units are usually installed in buildings with low to moderate ceiling heights. Downblow units are used in spaces with high ceilings and where floor and wall space limitations dictate that heating equipment be kept out of the way. Downblow units may have an adjustable diffuser to vary the discharge pattern from a high-velocity vertical jet (to achieve the maximum distance of downward throw) to a horizontal discharge of lower velocity (to prevent excessive air motion in the zone of occupancy). Revolving diffusers are also available.

Cabinet unit heaters are used when a more attractive appearance is desired. They are suitable for free-air delivery or low static pressure duct applications. They may be equipped with filters, and

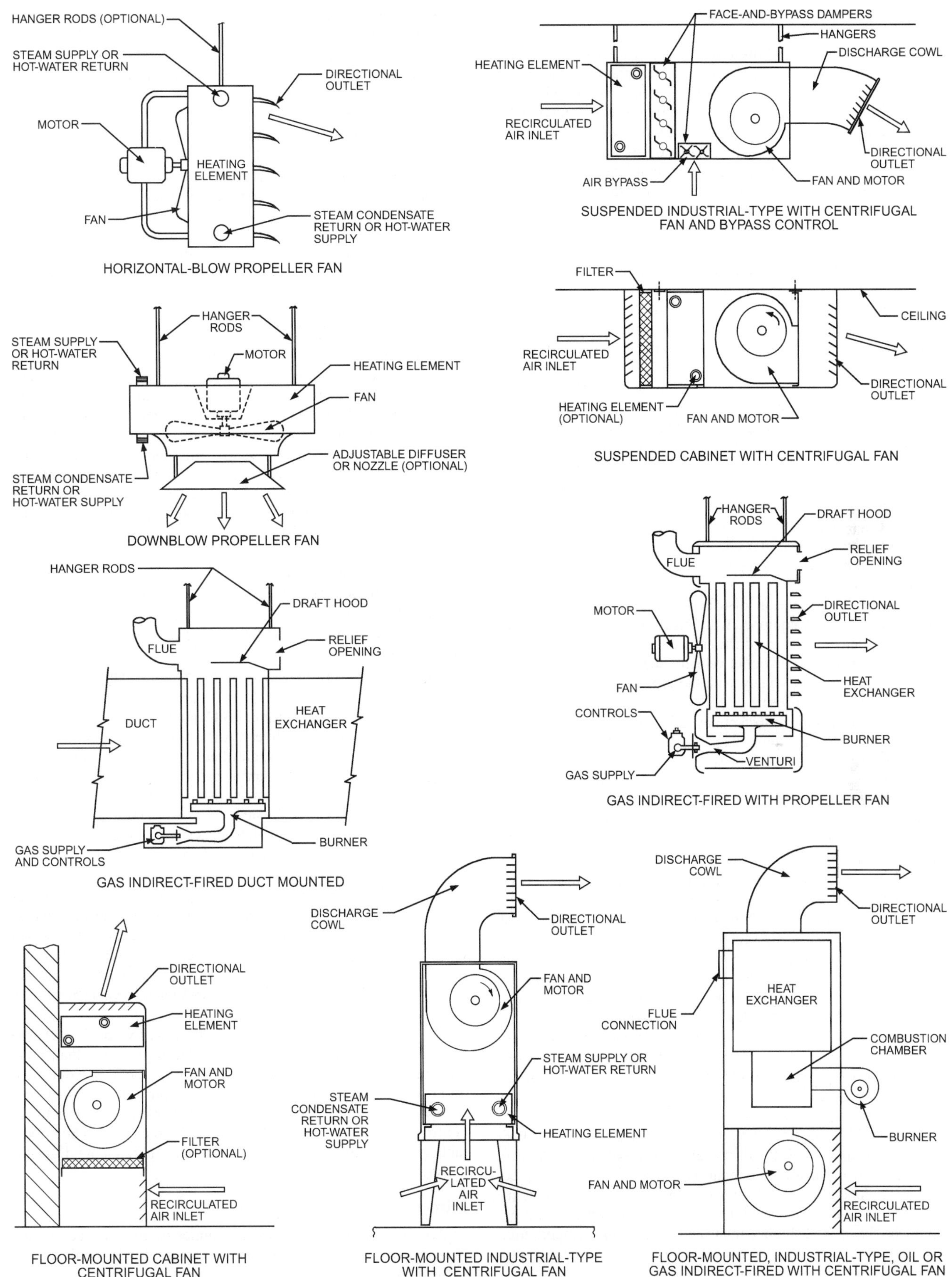

Fig. 3 Typical Unit Heaters

they can be arranged to discharge either horizontally or vertically up or down.

Industrial centrifugal fan units are applied where heating capacities and space volumes are large or where filtration of the heated air or operation against static resistance is required. Downblow or horizontal-blow units may be used, depending on the requirements.

Duct unit heaters are used where the air handler is remote from the heater. These heaters sometimes provide an economical means of adding heating to existing cooling or ventilating systems with ductwork. They require flow and temperature limit controls.

Location for Proper Heat Distribution

Units must be selected, located, and arranged to provide complete heat coverage while maintaining acceptable air motion and temperature at an acceptable sound level in the working or occupied zone. Proper application depends on size, number, and type of units; direction of airflow and type of directional outlet used; mounting height; outlet velocity and temperature; and air volumetric flow. Many of these factors are interrelated.

The mounting height may be governed by space limitations or by the presence of equipment such as display cases or machinery. The higher a downblow heater is mounted, the lower the temperature of air leaving the heater must be to force the heated air into the occupied zone. Also, the distance that air leaving the heater travels depends largely on the air temperature and initial velocity. A high temperature reduces the area of heat coverage.

Unit heaters for high-pressure steam or high-temperature hot water should be designed to produce approximately the same leaving air temperature as would be obtained from a lower temperature heating medium.

To obtain the desired air distribution and heat diffusion, unit heaters are commonly equipped with directional outlets, adjustable louvers, or fixed or revolving diffusers. For a given unit with a given discharge temperature and outlet velocity, the mounting height and heat coverage can vary widely with the type of directional outlet, adjustable louver, or diffuser.

Other factors that may substantially reduce heat coverage include obstructions (such as columns, beams, or partitions) or machinery in either the discharge airstream or approach area to the unit. Strong drafts or other air currents also reduce coverage. Exposures such as large glass areas or outside doors, especially on the windward side of the building, require special attention; units should be arranged so that they blanket the exposures with a curtain of heated air and intercept the cold drafts.

For area heating, horizontal-blow unit heaters in exterior zones should be placed so that they blow either along the exposure or toward it at a slight angle. When possible, multiple units should be arranged so that the discharge airstreams support each other and create a general circulatory motion in the space. Interior zones under exposed roofs or skylights should be completely blanketed. Downblow units should be arranged so that the heated areas from adjacent units overlap slightly to provide complete coverage.

For spot heating of individual spaces in larger unheated areas, single unit heaters may be used, but allowance must be made for the inflow of unheated air from adjacent spaces and the consequent reduction in heat coverage. Such spaces should be isolated by partitions or enclosures, if possible.

Horizontal unit heaters should have discharge outlets located well above head level. Both horizontal and vertical units should be placed so that the heated airstream is delivered to the occupied zone at acceptable temperature and velocity. The outlet air temperature of free-air delivery unit heaters used for comfort heating should be 50 to 60°F higher than the design room temperature. When possible, units should be located so that they discharge into open spaces, such as aisles, and not directly on the occupants. For further information on air distribution, see Chapter 33 of the 2005 *ASHRAE Handbook—Fundamentals*.

Manufacturers' catalogs usually include suggestions for the best arrangements of various unit heaters, recommended mounting heights, heat coverage for various outlet velocities, final temperatures, directional outlets, and sound level ratings.

Sound Level in Occupied Spaces

Sound pressure levels in workplaces should be limited to values listed in Table 42 in Chapter 47 of the 2007 *ASHRAE Handbook—HVAC Applications*. Although the noise level is generated by all equipment within hearing distance, unit heaters may contribute a significant portion of noise level. Both noise and air velocity in the occupied zone generally increase with increased outlet velocities. An analysis of both the diverse sound sources and the locations of personnel stations establishes the limit to which the unit heaters must be held.

Ratings of Unit Heaters

Steam or Hot Water. Heating capacity must be determined at a standard condition. Variations in entering steam or water temperature, entering air temperature, and steam or water flow affect capacity. Typical standard conditions for rating steam unit heaters are dry saturated steam at 2 psig pressure at the heater coil, air at 60°F (29.92 in. Hg barometric pressure) entering the heater, and the heater operating free of external resistance to airflow. Standard conditions for rating hot-water unit heaters are entering water at 200°F, water temperature drop of 20°F, entering air at 60°F (29.92 in. Hg barometric pressure), and the heater operating free of external resistance to airflow.

Gas-Fired. Gas-fired unit heaters are rated in terms of both input and output, in accordance with the approval requirements of the American Gas Association.

Oil-Fired. Ratings of oil-fired unit heaters are based on heat delivered at the heater outlet.

Electric. Electric unit heaters are rated based on the energy input to the heating element.

Effect of Airflow Resistance on Capacity. Unit heaters are customarily rated at free-air delivery. Airflow and heating capacity will decrease if outside air intakes, air filters, or ducts on the inlet or discharge are used. The reduction in capacity caused by this added resistance depends on the characteristics of the heater and on the type, design, and speed of the fans. As a result, no specific capacity reduction can be assigned for all heaters at a given added resistance. The manufacturer should have information on the heat output to be expected at other than free-air delivery.

Effect of Inlet Temperature. Changes in entering air temperature influence the total heating capacity in most unit heaters and the final temperature in all units. Because many unit heaters are located some distance from the occupied zone, possible differences between the temperature of the air actually entering the unit and that of air being maintained in the heated area should be considered, particularly with downblow unit heaters.

Higher-velocity units and units with lower vertical discharge air temperature maintain lower temperature gradients than units with higher discharge temperatures. Valve- or bypass-controlled units with continuous fan operation maintain lower temperature gradients than units with intermittent fan operation. Directional control of the discharged air from a unit heater can also be important in distributing heat satisfactorily and in reducing floor-to-ceiling temperature gradients.

Filters

Air from propeller unit heaters cannot be filtered because the heaters are designed to operate with heater friction loss only. If dust in the building must be filtered, centrifugal fan units or cabinet units should be used. Chapter 28 has further information on air cleaners for particulate contaminants.

CONTROL

The controls for a steam or hot water unit heater can provide either (1) on/off operation of the unit fan, or (2) continuous fan operation with modulation of heat output. For on/off operation, a room thermostat is used to start and stop the fan motor or group of fan motors. A limit thermostat, often strapped to the supply or return pipe, prevents fan operation in the event that heat is not being supplied to the unit. An auxiliary switch that energizes the fan only when power is applied to open the motorized supply valve may also be used to prevent undesirable cool air from being discharged by the unit.

Continuous fan operation eliminates both the intermittent blasts of hot air resulting from on/off operation and the stratification of temperature from floor to ceiling that often occurs during off periods. In this arrangement, a proportional room thermostat controls a valve modulating the heat supply to the coil or a bypass around the heating element. A limit thermostat or auxiliary switch stops the fan when heat is no longer available.

One type of control used with downblow unit heaters is designed to automatically return the warm air, which would normally stratify at the higher level, down to the zone of occupancy. Two thermostats and an auxiliary switch are required. The lower thermostat is placed in the zone of occupancy and is used to control a two-position supply valve to the heater. An auxiliary switch is used to stop the fan when the supply valve is closed. The higher thermostat is placed near the unit heater at the ceiling or roof level where the warm air tends to stratify. The lower thermostat automatically closes the steam valve when its setting is satisfied, but the higher thermostat overrides the auxiliary switch so that the fan continues to run until the temperature at the higher level falls below a point sufficiently high to produce a heating effect.

Indirect-fired and electric units are usually controlled by intermittent operation of the heat source under control of the room thermostat, with a separate fan switch to run the fan when heat is being supplied. For more information on automatic control, refer to Chapter 47 of the 2007 *ASHRAE Handbook—HVAC Applications*.

Unit heaters can be used to circulate air in summer. In such cases, the heat is shut off and the thermostat has a bypass switch, which allows the fan to run independently of the controls.

PIPING CONNECTIONS

Piping connections for steam unit heaters are similar to those for other types of fan blast heaters. Unit heater piping must conform strictly to the system requirements, while allowing the heaters to function as intended. Basic piping principles for steam systems are discussed in Chapter 10.

Steam unit heaters condense steam rapidly, especially during warm-up periods. The return piping must be planned to keep the heating coil free of condensate during periods of maximum heat output, and the steam piping must be able to carry a full supply of steam to the unit to take the place of condensed steam. Adequate pipe size is especially important when a unit heater fan is operated under on/off control because the condensate rate fluctuates rapidly.

Recommended piping connections for unit heaters are shown in Figure 4. In steam systems, the branch from the supply main to the heater must pitch toward the main and be connected to its top in

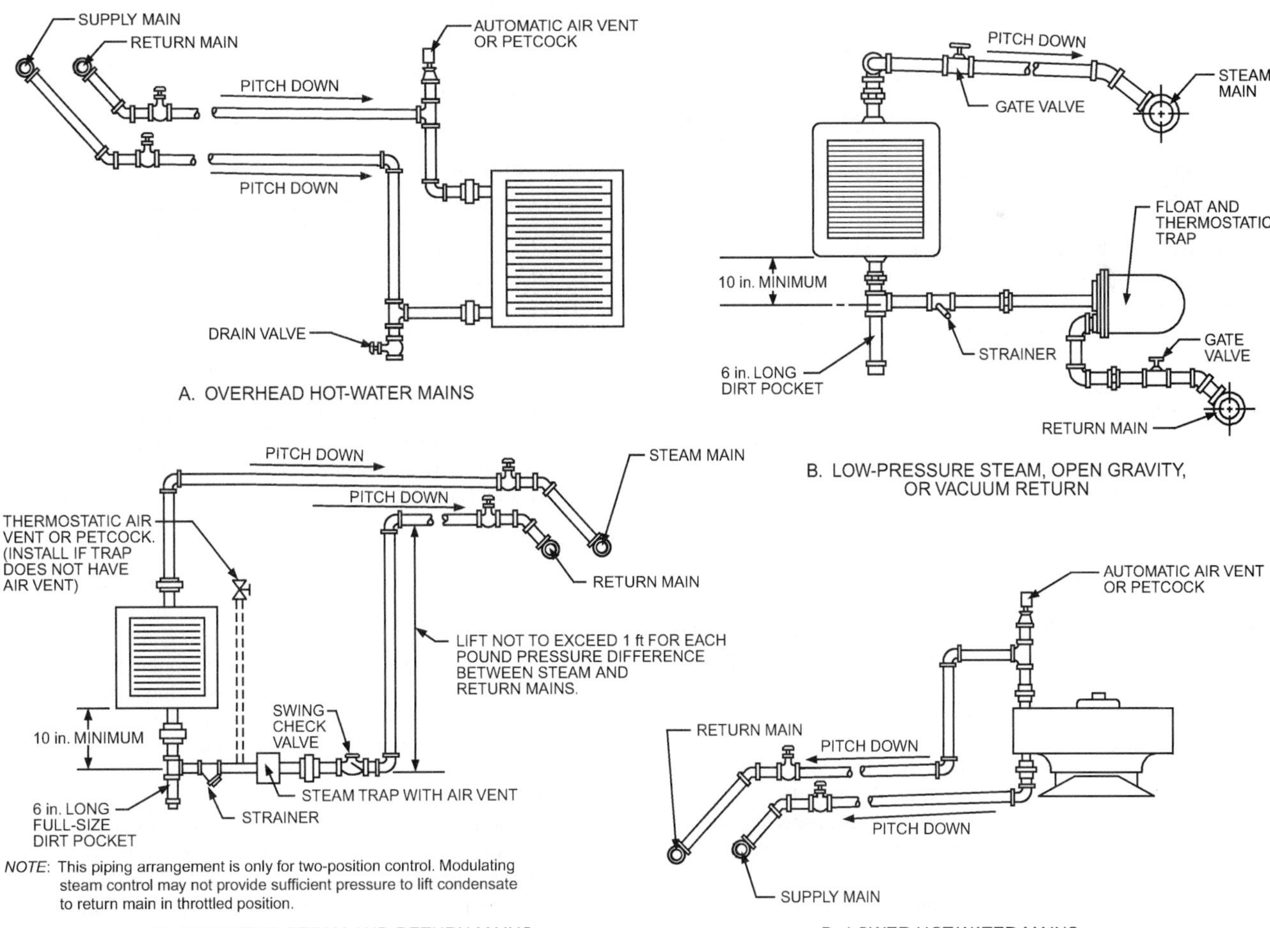

Fig. 4 Hot Water and Steam Connections for Unit Heaters

order to prevent condensate in the main from draining through the heater, where it might reduce capacity and cause noise.

The return piping from steam unit heaters should provide a minimum drop of 10 in. below the heater, so that the pressure of water required to overcome resistances of check valves, traps, and strainers will not cause condensate to remain in the heater.

Dirt pockets at the outlet of unit heaters and strainers with 0.063 in. perforations to prevent rapid plugging are essential to trap dirt and scale that might affect the operation of check valves and traps. Strainers should always be installed in the steam supply line if the heater has steam-distributing coils or is valve controlled.

An adequate air vent is required for low-pressure closed gravity systems. The vertical pipe connection to the air vent should be at least 3/4 in. NPT to allow water to separate from the air passing to the vent. If thermostatic instead of float-and-thermostatic traps are used in vacuum systems, a cooling leg must be installed ahead of the trap.

In high-pressure systems, it is customary to continuously vent the air through a petcock (as indicated in Figure 4C), unless the steam trap has a provision for venting air. Most high-pressure return mains terminate in flash tanks that are vented to the atmosphere. When possible, pressure-reducing valves should be installed to permit operation of the heaters at low pressure. Traps must be suitable for the operating pressure encountered.

When piping is connected to hot-water unit heaters, it must be pitched to permit air to vent to the atmosphere at the high point in the piping. An air vent at the heater is used to facilitate air removal or to vent the top of the heater. The system must be designed for complete drainage, including placing nipple and cap drains on drain cocks when units are located below mains.

MAINTENANCE

Regular inspection, based on a schedule determined by the amount of dirt in the atmosphere, assures maximum operating economy and heating capacity. Heating elements should be cleaned when necessary by brushing or blowing with high-pressure air or by using a steam spray. A portable sheet metal enclosure may be used to partially enclose smaller heaters for cleaning in place with air or steam jets. In certain installations, however, it may be necessary to remove the heating element and wash it with a mild alkaline solution, followed by a thorough rinsing with water. Propeller units do not have filters and are, therefore, more susceptible to dust build-up on the coils.

Dirt on fan blades reduces capacity and may unbalance the blades, which causes noise and bearing damage. Fan blades should be inspected and cleaned when necessary. Vibration and noise may also be caused by improper fan position or loose set screws. A fan guard should be placed on downblow unit heaters that have no diffuser or other device to catch the fan blade if it comes loose and falls from the unit.

The amount of attention required by the various motors used with unit heaters varies greatly. Instructions for lubrication, in particular, must be followed carefully for trouble-free operation: excess lubrication, for example, may damage the motor, and an improper lubricant may cause the bearings to fail. Instructions for care of the motor on any unit heater should be obtained from the manufacturer and kept at the unit.

Fan bearings and drives must be lubricated and maintained according to the instructions specified by the manufacturer. If the unit is direct-connected, the couplings should be inspected periodically for wear and alignment. V-belt drives should have all belts replaced with a matched set if one belt shows wear.

Periodic inspections of traps, inspections of check and air valves, and the replacement of worn fans are other important maintenance functions. Strainers should be cleaned regularly. Filters, if included, must be cleaned or replaced when dirty.

MAKEUP AIR UNITS

DESCRIPTION AND APPLICATIONS

Makeup air units are designed to condition ventilation air introduced into a space or to replace air exhausted from a building. The air exhausted may be from a process or general area exhaust, through either powered exhaust fans or gravity ventilators. The units may be used to prevent negative pressure within buildings or to reduce airborne contaminants in a space.

If temperature and/or humidity in the structure are controlled, the makeup air system must have the capacity to condition the replacement air. In most cases, makeup air units must be used to supply this conditioned makeup air. The units may heat, cool, humidify, dehumidify, and/or filter incoming air. They may be used to replace air in the conditioned space or to supplement or accomplish all or part of the airflow needed to satisfy the heating, ventilating, or cooling airflow requirements.

Makeup air can enter at a fixed flow rate or as a variable volume of outside air. It can be used to accomplish building pressurization or contamination reduction, and may be controlled in a manner that responds directly to exhaust flow. Makeup air units may also be connected to process exhaust with air-to-air heat recovery units.

Buildings under negative pressure because of inadequate makeup air may have the following symptoms:

- Gravity stacks from unit heaters and processes back-vent.
- Exhaust systems do not perform at rated volume.
- The perimeter of the building is cold in winter because of high infiltration.
- Severe drafts occur at exterior doors.
- Exterior doors are hard to open.
- Heating systems cannot maintain comfortable conditions throughout the building because the central core area becomes overheated.

Other Applications

Spot Cooling. High-velocity air jets in the unit may be directed to working positions. During cold weather, supply air must be tempered or reduced in velocity to avoid overcooling workers.

Door Heating. Localized air supply at swinging doors or overhead doors, such as for loading docks, can be provided by makeup air units. Heaters may blanket door openings with tempered air. The temperature may be reset from the outside temperature or with dual-temperature air (low when the door is closed and high when the door is open during cold weather). Heating may be arranged to serve a single door or multiple doors by an air distribution system. Door heating systems may also be arranged to minimize entry of insects during warm weather.

SELECTION

Makeup air systems used for ventilation may be (1) sized to balance air exhaust volumes or (2) sized in excess of the exhaust volume to dilute contaminants. In applications where contaminant levels vary, variable-flow units should be considered so that the supply air varies for contaminant control and the exhaust volume varies to track supply volume. In critical spaces, the exhaust volume may be based on requirements to control pressure in the space.

Location

Makeup air units are defined by their location or the use of a key component. Examples are rooftop makeup air units, truss- or floor-mounted units, and sidewall units. Some manufacturers differentiate their units by heating mode, such as steam or direct gas-fired makeup air units.

Rooftop units are commonly used for large single-story industrial buildings to simplify air distribution. Access (via roof walks) is more convenient than access to equipment mounted in the truss; truss units are only accessible by installing a catwalk adjacent to the air units. Disadvantages of rooftop units are (1) they increase foot traffic on the roof, which reduces its life and increases the likelihood of leaks; (2) inclement weather reduces equipment accessibility; and (3) units are exposed to weather.

Makeup air units can also be placed around the perimeter of a building with air ducted through the sidewall. This approach limits future building expansion, and the effectiveness of ventilating internal spaces decreases as the building gets larger. However, access to the units is good, and minimum support is required because the units are mounted on the ground.

Use caution in selecting the location of the makeup air unit and/or its associated combustion air source, to avoid introducing combustible vapors into the unit. Consult state and local fire codes for specific guidance.

Heating and Cooling Media

Heating. Makeup air units are often identified by the heating or cooling medium they use. Heaters in makeup air systems may be direct gas-fired burners, electric resistance heating coils, indirect gas-fired heaters, steam coils, or hot-water heating coils. (Chapter 26 covers the design and application of heating coils.) Air distribution systems are often required to direct heat to spaces requiring it.

Natural gas can be used to supply an indirect-fired burner, which is the same as a large furnace. (Chapter 32 has more information, including detailed heater descriptions.) In a **nonrecirculating** direct-fired heater, levels of combustion products generated by the heater are very low (CO less than 5.0 ppm, NO_2 less than 0.5 ppm, and CO_2 less than 4000 ppm) and are released directly into the airstream being heated. All air to a nonrecirculating makeup air heater must be ducted directly from outside source. Nonrecirculating direct gas-fired industrial air heaters are typically certified to comply with ANSI *Standard* Z83.4b/CSA3.7b-2006. In a **recirculating** makeup air heater, ventilation air to the heater must be ducted directly from an outside source to limit the concentration of combustion products in the conditioned space to a level below 25 ppm for CO, 3 ppm for NO_2, and 5000 ppm for CO_2. Recirculating direct gas-fired industrial air heaters are typically certified to comply with ANSI *Standard* Z83.18-2000. Installing carbon monoxide detectors to protect building occupants in the event of a heater malfunction is good engineering practice, and may be required by local codes.

Hydronic heating sections in spaces requiring a fully isolated source (100%) of outside air must be protected from freezing in cold climates. Low-temperature protection includes two-position control of steam coils; careful selection of the water coil heating surface and control valves; careful control of water supply temperature; and use of an antifreeze additive.

Cooling. Mechanical refrigeration with direct-expansion or chilled-water cooling coils, direct or indirect evaporative cooling sections, or well water coils may be used. Air distribution systems are often required to direct cooling to specific spaces that experience or create heat gain.

Because industrial facilities often have high sensible heat loads, evaporative cooling can be particularly effective. An evaporative cooler helps clean the air, as well. A portion of the spray water must be bled off to keep the water acceptably clean and to maintain a low solids concentration. Chapter 40 of this volume and Chapter 51 of the 2007 *ASHRAE Handbook—HVAC Applications* cover evaporative cooling in more detail.

Chapter 22 provides information on air-cooling coils. If direct-expansion coils are used in conjunction with direct-fired gas coils, the cooling coils' headers must be isolated from the airstream and directly vented outdoors.

Filters

High-efficiency filters are not normally used in a makeup air unit because of their relatively high cost. Designers should ensure that all filters are easy to change or clean. Appropriate washing equipment should be located near all washable filters. Throwaway filters should be sized for easy removal and disposal. Chapters 28 and 29 have more information on air filters and cleaners.

CONTROL

Controls for a makeup air unit fall into the following categories: (1) local temperature controls, (2) airflow controls, (3) plant-wide controls for proper equipment operation and efficient performance, (4) safety controls for burner gas, and (5) building smoke control systems. For control system information, refer to Chapters 40 and 46 of the 2007 *ASHRAE Handbook—HVAC Applications*.

Safety controls for gas-fired units include components to properly light the burner and to provide a safeguard against flame failure. The heater and all attached inlet ducting must be purged with at least four air changes before initiating an ignition sequence and before reignition after a malfunction. A flame monitor and control system must be used to automatically shut off gas to the burner upon burner ignition or flame failure. Critical malfunctions include flame failure, supply fan failure, combustion air depletion, power failure, control signal failure, excessive or inadequate inlet gas supply pressure, excess air temperature, and gas leaks in motorized valves or inlet gas supply piping.

APPLICABLE CODES AND STANDARDS

A gas-fired makeup air unit must be designed and built in accordance with NFPA *Standard* 54 and the requirements of the owner's insurance underwriter. Local codes must also be observed when using direct-fired gas makeup air units because some jurisdictions prohibit or restrict their use and may also require exhaust fans to be used while the unit is in operation.

The following standards may also apply, depending on the application:

ACCA. 1992. Direct-fired makeup air equipment. *Technical Bulletin* 109. Air-Conditioning Contractors of America, Washington, D.C.

ACGIH. 2007. *Industrial ventilation: A manual of recommended practice*, 26th ed. American Conference of Governmental Industrial Hygienists, Cincinnati, OH.

ANSI. 2006. Non-recirculating direct gas-fired industrial air heaters. *Standard* Z83.4b/CSA3.7b-2006. American National Standards Institute, New York.

ANSI. 2000. Recirculating direct gas-fired industrial heaters. *Standard* Z83.18. American National Standards Institute, New York.

ARI. 1989. Central-station air-handling units. ANSI/ARI *Standard* 430-89. Air-Conditioning and Refrigeration Institute, Arlington, VA.

ASHRAE. 2007. Ventilation for acceptable indoor air quality. ANSI/ASHRAE *Standard* 62.1-2007.

ASHRAE. 2004. Energy standard for buildings except low-rise residential buildings. ANSI/ASHRAE/IESNA *Standard* 90.1-2004.

ASHRAE. 2006. Energy conservation in existing buildings. ASHRAE/IESNA *Standard* 100-2006.

CSA International. 2006. Direct gas-fired make-up air heaters. *Standard* Z83.4b/CSA3.7b-2006.

ICC. 2006. *International mechanical code*. International Code Council, Falls Church, VA.

ICC. 2006. *International fuel gas code*. International Code Council, Falls Church, VA.

COMMISSIONING

Commissioning of makeup air systems is similar to that of other air-handling systems, requiring attention to

- Equipment identification
- Piping system identification
- Belt drive adjustment
- Control system checkout
- Documentation of system installation
- Lubrication
- Electrical system checkout for overload heater size and function
- Cleaning and degreasing of hydronic piping systems
- Pretreatment of hydronic fluids
- Setup of chemical treatment program for hydronic systems and evaporative apparatus
- Start-up of major equipment items by factory-trained technician
- Testing and balancing
- Planning of preventive maintenance program
- Instruction of owner's operating and maintenance personnel

MAINTENANCE

Basic operating and maintenance data required for makeup air systems may be obtained from the ASHRAE Handbook chapters covering the components. Specific operating instructions are required for makeup air heaters that require changeover from winter to summer conditions, including manual fan speed changes, air distribution pattern adjustment, or heating cycle lockout.

Operations handling 100% outside air may require more frequent maintenance, such as changing filters, lubricating bearings, and checking the water supply to evaporative coolers/humidifiers. Filters on systems in locations with dirty air require more frequent changing, so a review may determine whether upgrading filter media would be cost-effective. More frequent cleaning of fans' blades and heat transfer surfaces may be required in such locations to maintain airflow and heat transfer performance.

BIBLIOGRAPHY

Brown, W.K. 1990. Makeup air systems—Energy saving opportunities. *ASHRAE Transactions* 96(2):609-615.

Bridgers, F.H. 1980. Efficiency study—Preheating outdoor air for industrial and institutional applications. *ASHRAE Journal* 22(2):29-31.

Gadsby, K.J. and T.T. Harrje. 1985. Fan pressurization of buildings—Standards, calibration and field experience. *ASHRAE Transactions* 91(2B): 95-104.

Holness, G.V.R. 1989. Building pressurization control: Facts and fallacies. *Heating/Piping/Air Conditioning* (February).

NFPA. 2006. National fuel gas code. *Standard* 54-2006.

Persily, A. 1982. Repeatability and accuracy of pressurization testing. In *Thermal Performance of the Exterior Envelopes of Buildings II, Proceedings of ASHRAE/DOE Conference*. ASHRAE SP 38:380-390.

CHAPTER 28

AIR CLEANERS FOR PARTICULATE CONTAMINANTS

THIS chapter discusses cleaning particulate contaminants from both ventilation and recirculated air for conditioning building interiors. Complete air cleaning may also require removing of airborne particles, microorganisms, and gaseous contaminants, but this chapter only covers removal of airborne particles and briefly discusses bioaerosols.

The total suspended particulate concentration in applications discussed in the chapter seldom exceeds 2 mg/m^3 and is usually less than 0.2 mg/m^3 of air. This contrasts with flue gas or exhaust gas from processes, where dust concentration typically ranges from 200 to 40,000 mg/m^3. Chapter 45 of the 2007 *ASHRAE Handbook—HVAC Applications* covers the removal of gaseous contaminants; Chapter 25 of this volume discusses exhaust-gas control.

Most air cleaners discussed in this chapter are not used in exhaust gas streams, mainly because of the extreme dust concentration and temperature. However, the principles of air cleaning do apply to exhaust streams, and air cleaners discussed in the chapter are used extensively in supplying gases of low particulate concentration to industrial processes.

ATMOSPHERIC DUST

Atmospheric dust is a complex mixture of smokes, mists, fumes, dry granular particles, bioaerosols, and natural and synthetic fibers. When suspended in a gas, this mixture is called an **aerosol**. A sample of atmospheric dust usually contains soot and smoke, silica, clay, decayed animal and vegetable matter, organic materials in the form of lint and plant fibers, and metallic fragments. It may also contain living organisms, such as mold spores, bacteria, and plant pollens, which may cause diseases or allergic responses. (Chapter 12 of the 2005 *ASHRAE Handbook—Fundamentals* contains further information on atmospheric contaminants.) A sample of atmospheric dust gathered at any point generally contains materials common to that locality, together with other components that originated at a distance but were transported by air currents or diffusion. These components and their concentrations vary with the geography of the locality (urban or rural), season of the year, weather, direction and strength of the wind, and proximity of dust sources.

Aerosol sizes range from 0.01 μm and smaller for freshly formed combustion particles and radon progeny; to 0.1 μm for aged cooking and cigarette smokes; to 0.1 to 10 μm for airborne dust, microorganisms, and allergens; and up to 100 μm and larger for airborne soil, pollens, and allergens.

Concentrations of atmospheric aerosols generally peak at submicrometre sizes and decrease rapidly as the particulate size increases above 1 μm. For a given size, the concentration can vary by several orders of magnitude over time and space, particularly near an aerosol source, such as human activities, equipment, furnishings, and pets (McCrone et al. 1967). This wide range of particulate size and concentration makes it impossible to design one cleaner for all applications.

AEROSOL CHARACTERISTICS

The characteristics of aerosols that most affect air filter performance include particle size and shape, mass, concentration, and electrical properties. The most important of these is particle size. Figure 3 in Chapter 12 of the 2005 *ASHRAE Handbook—Fundamentals* gives data on the sizes and characteristics of a wide range of airborne particles that may be encountered.

Particle size in this discussion refers to aerodynamic particle size. Therefore, larger particles with lower densities may be found in the alveolar region of the lungs. Also note that fibers are different from particles in that the fiber's shape, diameter, and density affect where a fiber settles in the body (NIOSH 1973).

Particles less than 2.5 μm in diameter are generally referred to as **fine mode**, and those larger than 2.5 μm as **coarse mode**. Fine- and coarse-mode particles typically originate by separate mechanisms, are transformed separately, have different chemical compositions, and require different control strategies. Fine-mode particles generally originate from condensation or are directly emitted as combustion products. Many microorganisms (bacteria and fungi) either are in this size range or produce components this size. These particles are less likely to be removed by gravitational settling and are just as likely to deposit on vertical surfaces as on horizontal surfaces. Coarse-mode particles are typically produced by mechanical actions such as erosion and friction. Coarse particles are more easily removed by gravitational settling, and thus have a shorter airborne lifetime.

For industrial hygiene, particles ≤5 μm in diameter are considered **respirable particles (RSPs)** because a large percentage of them may reach the alveolar region of the lungs. Willeke and Baron (1993) describe a detailed aerosol sampling technique. They discuss the need for an impactor with a 2.5 μm cutoff for RSPs that can be deposited in the alveolar region. The cutoff for particles affecting respiratory function is considered to be 2.0 or 2.5 μm. See the discussion in the section on Sizes of Airborne Particles in Chapter 12 of the 2005 *ASHRAE Handbook—Fundamentals*. A cutoff of 5.0 μm includes 80 to 90% of the particles that can reach the functional pulmonary region of the lungs (James et al. 1991; Phalen et al. 1991).

Bioaerosols are a diverse class of particulates of biological origin. They are of particular concern in indoor air because of their association with allergies and asthma and their ability to cause disease. Chapters 9 and 12 of the 2005 *ASHRAE Handbook—Fundamentals* contains more detailed descriptions of these contaminants.

Airborne viral and bacterial aerosols are generally transmitted by droplet nuclei, which average about 3 μm in diameter. Fungal spores are generally 2 to 5 μm in diameter (Wheeler 1994). Combinations of proper ventilation and filtration can be used to control indoor bioaerosols. Morey (1994) recommends providing a ventilation rate of

The preparation of this chapter is assigned to TC 2.4, Particulate Air Contaminants and Particulate Contaminant Removal Equipment.

15 to 35 cfm per person to control human-shed bacteria. ACGIH (1989) recommends dilution with a minimum of 15 cfm per person. It also reports 50 to 70% ASHRAE atmospheric dust-spot efficiency filters can remove most microbial agents 1 to 2 μm in diameter. Wheeler (1994) states that 60% ASHRAE atmospheric dust-spot efficiency filters remove 85% or more of particles 2.5 μm in diameter, and 80 to 85% efficiency filters remove 96% of 2.5 μm particles.

AIR-CLEANING APPLICATIONS

Different fields of application require different degrees of air cleaning effectiveness. In industrial ventilation, only removing the larger dust particles from the airstream may be necessary for cleanliness of the structure, protection of mechanical equipment, and employee health. In other applications, surface discoloration must be prevented. Unfortunately, the smaller components of atmospheric dust are the worst offenders in smudging and discoloring building interiors. Electronic air cleaners or medium- to high-efficiency filters are required to remove smaller particles, especially the respirable fraction, which often must be controlled for health reasons. In cleanrooms or when radioactive or other dangerous particles are present, high- or ultrahigh-efficiency filters should be selected. For more information on cleanrooms, see Chapter 16 of the 2007 *ASHRAE Handbook—HVAC Applications*.

Major factors influencing filter design and selection include (1) degree of air cleanliness required, (2) specific particle size range or aerosols that require filtration, (3) aerosol concentration, (4) resistance to airflow through the filter and (5) design face velocity to achieve published performance.

MECHANISMS OF PARTICLE COLLECTION

In particle collection, air cleaners made of fibrous media rely on the following five main principles or mechanisms:

Straining. The coarsest kind of filtration strains particles through an opening smaller than the particle being removed. It is most often observed as the collection of large particles and lint on the filter surface. The mechanism is not adequate to explain the filtration of submicrometre aerosols through fibrous matrices, which occurs through other physical mechanisms, as follows.

Inertial Impingement. When particles are large or dense enough that they cannot follow the airstream around a fiber, they cross over streamlines, hit the fiber, and remain there if the attraction is strong enough. With flat-panel and other minimal-media-area filters having high air velocities (where the effect of inertia is most pronounced), the particle may not adhere to the fiber because drag and bounce forces are so high. In this case, a viscous coating (preferably odorless and nonmigrating) is applied to the fiber to enhance retention of the particles. This adhesive coating is critical to metal mesh impingement filter performance.

Interception. Particles follow the airstream close enough to a fiber that the particle contacts the fiber and remains there mainly because of van der Waals forces (i.e., weak intermolecular attractions between temporary dipoles). The process depends on air velocity through the media being low enough not to dislodge the particles, and is therefore the predominant capture mechanism in extended-media filters such as bag and deep-pleated rigid cartridge types.

Diffusion. The path of very small particles is not smooth but erratic and random within the airstream. This is caused by gas molecules in the air bombarding them (Brownian motion), producing an erratic path that brings the particles close enough to a media fiber to be captured by interception. As more particles are captured, a concentration gradient forms in the region of the fiber, further enhancing filtration by diffusion and interception. The effects of diffusion increase with decreasing particle size and media velocity.

Electrostatic Effects. Particle or media electrostatic charge can produce changes in dust collection affected by the electrical properties of the airstream. Some particles may carry a natural charge. Passive electrostatic (without a power source) filter fibers may be electrostatically charged during their manufacture or (in some materials) by mainly dry air blowing through the media. Charges on the particle and media fibers can produce a strong attracting force if opposite. Efficiency is generally considered to be highest when the media is new and clean, decreasing rapidly as the filter loads.

Some progress has been made in calculating theoretical filter media efficiency from the physical constants of the media by considering the effects of the collection mechanisms (Lee and Liu 1982a, 1982b; Liu and Rubow 1986).

EVALUATING AIR CLEANERS

In addition to criteria affecting the degree of air cleanliness, factors such as cost (initial investment and maintenance), space requirements, and airflow resistance have led to the development of a wide variety of air cleaners. Accurate comparisons of different air cleaners can be made only from data obtained by standardized test methods.

The three distinguishing operating characteristics are efficiency, resistance to airflow, and dust-holding capacity. **Efficiency** measures the ability of the air cleaner to remove particles from an airstream. Minimum efficiency during the life of the filter is the most meaningful characteristic for most filters and applications. **Resistance to airflow** (or simply resistance) is the static pressure drop differential across the filter at a given face velocity. The term *static pressure differential* is interchangeable with pressure drop and resistance if the difference of height in the filtering system is negligible. **Dust-holding capacity** defines the amount of a particular type of dust that an air cleaner can hold when it operates at a specified airflow rate to some maximum resistance value (ASHRAE *Standard* 52.1).

Complete evaluation of air cleaners therefore requires data on efficiency, resistance, dust-holding capacity, and the effect of dust loading on efficiency and resistance. When applied to automatic renewable media devices (roll filters, for example), the evaluation must include the rate at which the media is supplied to maintain constant resistance when standardized test dust is fed at a specified rate.

Air filter testing is complex and no individual test adequately describes all filters. Ideally, performance testing of equipment should simulate operation under actual conditions and evaluate the characteristics important to the equipment user. Wide variations in the amount and type of particles in the air being cleaned make evaluation difficult. Another complication is the difficulty of closely relating measurable performance to the specific requirements of users. Recirculated air tends to have a larger proportion of lint than does outside air. However, performance tests should strive to simulate actual use as closely as possible. In general, five types of tests, together with certain variations, determine air cleaner performance:

Arrestance. A standardized ASHRAE synthetic dust consisting of various particle sizes and types is fed into the test air stream to the air cleaner and the weight fraction of the dust removed is determined. In the ASHRAE *Standard* 52.1 test, summarized in the segment on Air Cleaner Test Methods in this chapter, this measurement is called **synthetic dust weight arrestance** to distinguish it from other efficiency values.

The indicated weight arrestance of air filters, as determined in the arrestance test, depends greatly on the particle size distribution of the test dust, which, in turn, is affected by its state of agglomeration. Therefore, this filter test requires a high degree of standardization of the test dust, the dust dispersion apparatus, and other elements of test equipment and procedures. This test is particularly suited to distinguish between the many types of low- to medium-efficiency air filters that are common in recirculating systems with air handlers and fan-coil units having minimal external static pressure capability. It does not adequately distinguish between higher-efficiency filters.

ASHRAE Atmospheric Dust-Spot Efficiency. Unconditioned atmospheric air is passed into the air cleaner under test and the discoloration level of the cleaned air (downstream of the test filter) on filter paper targets is compared with that of the unfiltered outside air (upstream of the test filter). The dust-spot test measures the ability of a filter to reduce soiling of fabrics and building interior surfaces. Because these effects depend mostly on fine particles, this test is most useful for higher-efficiency filters. The variety and variability of atmospheric dust (Horvath 1967; McCrone et al. 1967; Whitby et al. 1958) may cause the same filter to test at different dust-spot efficiencies at different locations (or even at the same location at different times). Accuracy diminishes on lower-efficiency filters.

Fractional Efficiency or Penetration. Uniform-sized particles are fed into the air cleaner and the percentage removed by the cleaner is determined, typically by a photometer, optical particle counter, or condensation nuclei counter. In fractional efficiency tests, the use of uniform-particle-size aerosols has resulted in accurate measure of the particle size versus efficiency characteristic of filters over a wide atmospheric size spectrum. The method is time-consuming and has been used primarily in research. However, the dioctyl phthalate (DOP) or Emery 3000 test for HEPA filters is widely used for production testing at a narrow particle size range. For more information on the DOP test, see the DOP Penetration Test section.

Efficiency by Particle Size. A polydispersed challenge aerosol such as potassium chloride is metered into the test airstream to the air cleaner. Air samples taken upstream and downstream are drawn through an optical particle counter or similar measurement device to obtain removal efficiency versus particle size at a specific airflow rate.

Dust-Holding Capacity. The true dust-holding capacity of similar air cleaners is a function of environmental conditions as well as the variability of atmospheric dust (size, shape and concentration) and is therefore impossible to duplicate in a laboratory test. For testing purposes, measured amounts of standardized dust are used to artificially load the filters. This procedure shortens the dust-loading cycle from weeks or years to hours. Artificial dusts are not the same as atmospheric dusts, so dust-holding capacity as measured by these accelerated tests is different from that achieved by "life" tests using atmospheric dust. The exact life of a filter in field use is impossible to determine by laboratory testing. However, filter testing under standardized conditions does provide a rough guide to the relative effect of dust loading on the performance of similar units, and is one means used to compare them.

Laboratory filter tests are accurate and reproducible within acceptable tolerances. Differences in reported values generally derive from the variability of test aerosols, measurement devices, and dusts. Because most media are made of random air- or water-laid fibrous materials, the inherent media variations affect filter performance. Awareness of these variations prevents misunderstanding and specification of impossibly close performance tolerances. Caution must be used in interpreting published efficiency data, because two identical air cleaners tested by the same procedure may not give exactly the same results, and the result will not necessarily be exactly duplicated in a later test. Test values from different procedures generally cannot be compared. A performance test value of air cleaner efficiency is only a guide to the rate of soiling of a space or of mechanical equipment.

AIR CLEANER TEST METHODS

Air cleaner test methods have been developed by the heating and air-conditioning industry, the automotive industry, the atomic energy industry, and government and military agencies. Several tests have become standard in general ventilation applications in the United States. In 1968, the test techniques developed by the U.S. National Bureau of Standards [now the National Institute of Standards and Technology (NIST)] and the Air Filter Institute (AFI) were unified (with minor changes) into a single test procedure, ASHRAE *Standard* 52-1968. Dill (1938), Nutting and Logsdon (1953), and Whitby et al. (1956) give details of the original codes. ASHRAE *Standard* 52-1968 was revised and is currently ASHRAE *Standard* 52.1-1992.

In general, the ASHRAE weight arrestance test parallels that of the AFI, making use of a similar test dust. The ASHRAE atmospheric dust-spot efficiency test parallels the AFI and NBS atmospheric dust-spot efficiency tests and specifies a dust-loading technique.

ASHRAE *Standard* 52.1 includes both a weight test and a dust-spot test and requires that values for both be reported. The results may be used for comparisons among air cleaners. ASHRAE *Standard* 52.2 develops **minimum efficiency reporting values (MERVs)** for air cleaner particle size efficiency. Table 3 (from ASHRAE *Standard* 52.2) provides an approximate cross-reference for air cleaners tested under ASHRAE *Standards* 52.1 and 52.2. Currently there is no ASHRAE Standard for testing electronic air cleaners.

Arrestance Test

ASHRAE *Standard* 52.1 defines synthetic test dust as a compounded test dust consisting of (by weight) 72% ISO 12 103-A2 fine test dust, 23% powdered carbon, and 5% No. 7 cotton linters. A known amount of the prepared test dust is fed into the test unit at a known and controlled concentration. The amount of dust in the air leaving the filter is determined by passing the entire airflow through a high-efficiency after-filter and measuring the gain in filter weight. The **arrestance** is calculated using the weights of the dust captured on the final high efficiency filter and the total dust fed.

Atmospheric dust particles range from a small fraction of a micrometre to tens of micrometres in diameter. The artificially generated dust cloud used in the ASHRAE arrestance method is considerably coarser than typical atmospheric dust. It tests the ability of a filter to remove the largest atmospheric dust particles and gives little indication of filter performance in removing the smallest particles. However, where the mass of dust in the air is the primary concern, this is a valid test because most of the mass is contained in the larger, visible particles. Where extremely small particles (such as respirable sizes) are involved, arrestance rating does not differentiate between filters.

Atmospheric Dust-Spot Efficiency Test

One objectionable characteristic of finer airborne dust particles is their capacity to soil walls and other interior surfaces. The discoloring rate of white, filter-paper targets (microfine glass-fiber HEPA filter media) filtering samples of air is an accelerated simulation of this effect. By measuring the change in light transmitted by these targets, the filter efficiency in reducing surface soiling may be computed.

ASHRAE *Standard* 52.1 specifies two equivalent atmospheric dust-spot test procedures, each taking a different approach to correct for the nonlinear relation between the discoloration of target papers and their dust load. In the first procedure, called the **intermittent-flow method**, samples of conditioned atmospheric air are drawn upstream and downstream of the tested filter. These samples are drawn at equal flow rates through identical targets of glass-fiber filter paper. The downstream sample is drawn continuously; the upstream sample is interrupted in a timed cycle so that the average rate of discoloration of the upstream and downstream targets is approximately equal. The percentage of *off*-time approximates the filter efficiency. (See ASHRAE *Standard* 52.1 for details.)

In the alternative procedure, called the **constant-flow method**, conditioned atmospheric air samples are also drawn at equal flow rates through equal-area glass-fiber filter paper targets upstream and downstream, but without interrupting either sample. Discoloration of the upstream target is therefore greater than for the downstream target. Sampling is halted when the upstream target light transmission has dropped by at least 10% but no more than 40%. The **opacities** (percent change in light transmission) are then calculated for the targets as defined for the intermittent flow method. These opacities are next converted into **opacity indices** to correct for nonlinearity.

The advantage of the constant-flow method is that it takes the same length of time to run regardless of the efficiency of the filter, whereas the intermittent-flow method takes longer for higher-efficiency filters. For example, a test on a 90% efficient filter using intermittent flow requires ten times as long as one using constant flow.

The standard allows dust-spot efficiencies to be taken at intervals during an artificial dust-loading procedure. This characterizes the change of dust-spot efficiency as dust builds up on the filter in service.

Dust-Holding Capacity Test

Synthetic test dust is fed to the filter in accordance with ASHRAE *Standard* 52.1 procedures. The pressure drop across the filter (its resistance) rises as dust is fed. The test normally ends when resistance reaches the maximum operating resistance set by the manufacturer. However, not all filters of the same type retain collected dust equally well. The test, therefore, requires that arrestance be measured at least four times during dust loading and that the test be terminated when two consecutive arrestance values of less than 85%, or one value equal to or less than 75% of the maximum arrestance, have been measured. The ASHRAE **dust-holding capacity** is, then, the integrated amount of dust held by the filter up to the time the dust-loading test is terminated. (See ASHRAE *Standard* 52.1 for more detail.)

A typical set of curves for an ASHRAE air filter test report on a fixed cartridge-type filter is shown in Figure 1. Both synthetic dust weight arrestance and atmospheric dust-spot efficiencies are shown. The standard also specifies how self-renewable devices are to be loaded with dust to establish their performance under standard conditions. Figure 2 shows the results of such a test on an automatic roll media filter.

Particle Size Removal Efficiency Test

ASHRAE *Standard* 52.2 prescribes a way to test air-cleaning devices for removal efficiency by particle size while addressing two air cleaner performance characteristics important to users: the ability of the device to remove particles from the airstream and its resistance to airflow. In this method, air cleaner testing is conducted at a specific airflow based on the upper limit of the air cleaner's application range. Airflow must be between 470 and 2990 cfm in the 24 by 24 in. test section (face velocity between 120 and 750 fpm). The test aerosol consists of laboratory-generated potassium chloride particles dispersed in the airstream. An optical particle counter(s) measures and counts the particles in 12 geometric logarithmic-scale, equally distributed particle size ranges both upstream and downstream for efficiency determinations. The size range encompassed by the test is 0.3 to 10 μm polystyrene latex equivalent optical particle size. A method of loading the air cleaner with synthetic dust to simulate field conditions is also specified. The synthetic loading dust is the same as in ASHRAE *Standard* 52.1.

A set of particle size removal efficiency performance curves is developed from the test and, together with an initial clean performance curve, is the basis of a composite curve representing performance in the range of sizes. Points on the composite curve are averaged and these averages are used to determine the MERV of the air cleaner. A complete test report includes (1) a summary section, (2) removal efficiency curves of the clean devices at each of the

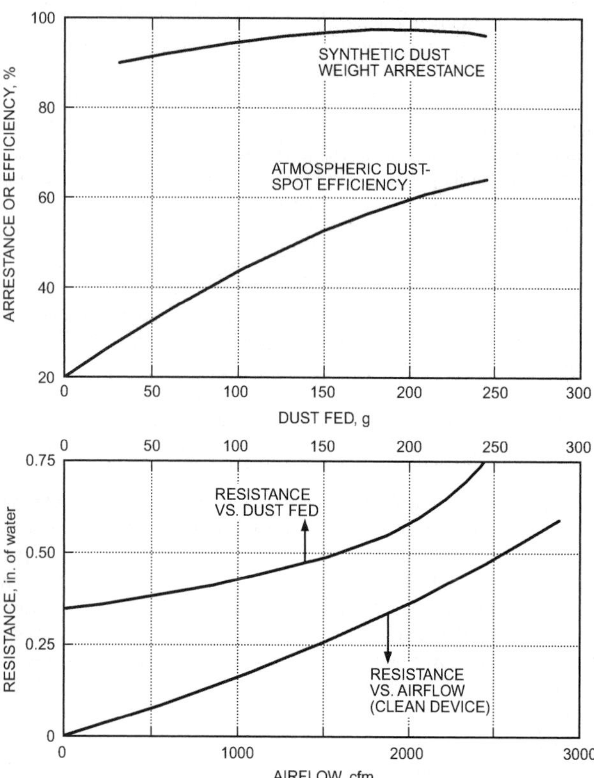

Fig. 1 Typical Performance Curves for Fixed Cartridge-Type Filter According to ASHRAE *Standard* 52.1

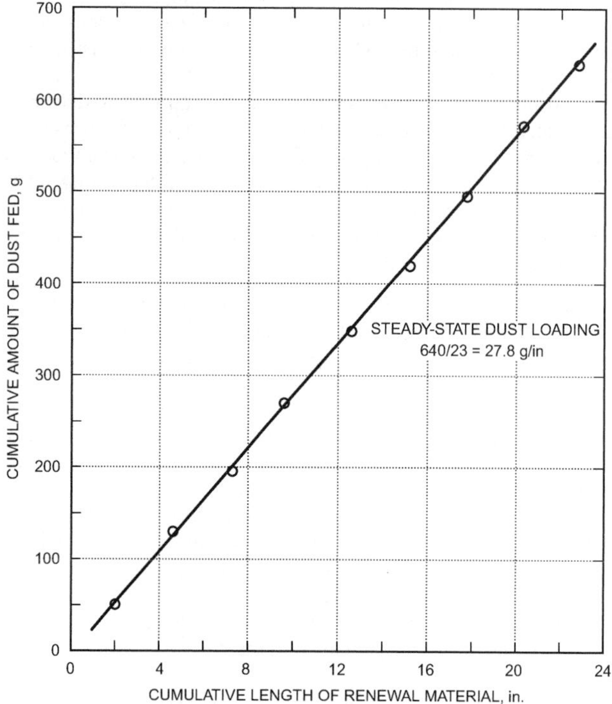

Note: Calculate loading per unit by multiplying steady-state dust loading by 144 and dividing by media width, normally 24 in. For example, 27.8 × 144/24 = 167 g/ft².

Fig. 2 Typical Dust-Loading Graph for Self-Renewable Air Filter

loading steps, and (3) a composite minimum removal efficiency curve.

DOP Penetration Test

For high-efficiency filters of the type used in cleanrooms and nuclear applications (HEPA filters), the normal test in the United States is the thermal DOP method, outlined in U.S. Military Standard MIL-STD-282 (1956) and U.S. Army document 136-300-175A (1965). DOP is dioctyl phthalate or bis-[2-ethylhexyl] phthalate, which is an oily liquid with a high boiling point. In this method, a smoke cloud of DOP droplets condenses from DOP vapor.

The count median diameter for DOP aerosols is about 0.18 μm, and the mass median diameter is about 0.27 μm with a cloud concentration of approximately 80 mg/m^3 under properly controlled conditions. The procedure is sensitive to the mass median diameter, and DOP test results are commonly referred to as efficiency on 0.30 μm particles.

The DOP smoke cloud is fed to the filter, which is held in a special test fixture. Any smoke that penetrates the body of the filter or leaks through gasket cracks passes into the region downstream from the filter, where it is thoroughly mixed. Air leaving the fixture thus contains the average concentration of penetrating smoke. This concentration, as well as the upstream concentration, is measured by a light-scattering photometer. **Filter penetration *P*** (%) is

$$P = 100\left(\frac{\text{Downstream concentration}}{\text{Upstream concentration}}\right) \qquad (1)$$

Penetration, not efficiency, is usually specified in the test procedure because HEPA filters have efficiencies so near 100% (e.g., 99.97% or 99.99% on 0.30 μm particles). The two terms are related by the equation $E = 100 - P$.

U.S. specifications frequently call for testing HEPA filters at both rated flow and 20% of rated flow. This procedure helps detect gasket leaks and pinholes that would otherwise escape notice. Such defects, however, are not located by the DOP penetration test.

The Institute of Environmental Sciences and Technology has published two recommend practices: IEST RP-CC 001.3, HEPA and ULPA Filters, and IEST RP-CC 007.1, Testing ULPA Filters.

Leakage (Scan) Tests

For HEPA filters, leakage tests are sometimes desirable to show that no small "pinhole" leaks exist or to locate any that may exist so they may be patched. Essentially, this is the same technique as used in the DOP penetration test, except that the downstream concentration is measured by scanning the face of the filter and its gasketed perimeter with a moving probe. The exact point of smoke penetration can then be located and repaired. This same test (described in IEST RP-CC 001.3) can be performed after the filter is installed; in this case, a portable but less precise Laskin nozzle aspirator-type DOP generator is used instead of the much larger thermal generator. Smoke produced by a portable generator is not uniform in size, but its average diameter can be approximated as 0.6 μm. Particle diameter is less critical for leak location than for penetration measurement.

Specialized Performance Test

American Home Appliance Manufacturers Association (AHAM) *Standard* AC-1 describes a method for measuring the ability of portable household air cleaners to reduce generated particles suspended in the air in a room-size test chamber. The procedure compares the natural decay of three contaminants (dust, smoke, and pollen) with the reduction in particles by the air cleaner.

Other Performance Tests

The European Standardization Institute (Comité Européen de Normalisation, or CEN) developed EN 779, "Particulate air filters for general ventilation—Requirements, testing, marking" (CEN 1993). Its revision, "Particulate air filters for general ventilation—Determination of the filtration performance" (CEN 2000), was in the formal vote phase as of January, 2001. Eurovent working group 4B (Air Filters) developed Eurovent Document 4/9 (1996), "Method of testing air filters used in general ventilation for determination of fractional efficiency" and is working on the revision of Document 4/10, "In situ determination of fractional efficiency of general ventilation filters." CEN also developed EN 1822 (CEN 1998, 2000), according to which HEPA and ULPA filters must be tested. Also, special test standards have been developed in the United States for respirator air filters (NIOSH/MSHA 1977) and ULPA filters (IEST RP-CC 007.1). *[Editor's note: See issuing organizations for latest editions of publications listed in this section.]*

Environmental Tests

Air cleaners may be subjected to fire, high humidity, a wide range of temperatures, mechanical shock, vibration, and other environmental stress. Several standardized tests exist for evaluating these environmental effects on air cleaners. U.S. Military Standard MIL-STD-282 includes shock tests (shipment rough handling) and filter media water-resistance tests. Several U.S. Atomic Energy Commission agencies (now part of the U.S. Department of Energy) specify humidity and temperature-resistance tests (Peters 1962, 1965).

Underwriters Laboratories has two major standards for air cleaner flammability. The first, for commercial applications, determines flammability and smoke production. UL *Standard* 900 Class 1 filters, when clean, do not contribute fuel when attacked by flame and emit negligible amounts of smoke. UL *Standard* 900 Class 2 filters, when clean, burn moderately when attacked by flame or emit moderate amounts of smoke, or both. In addition, UL *Standard* 586 for flammability of HEPA filters has been established. The UL tests do not evaluate the effect of collected dust on filter flammability; depending on the dust, this effect may be severe. UL *Standard* 867 applies to electronic air cleaners.

ARI Standards

The Air-Conditioning and Refrigeration Institute published ARI *Standards* 680 (residential) and 850 (commercial/industrial) for air filter equipment. These standards establish (1) definitions and classification; (2) requirements for testing and rating (performance test methods are per ASHRAE *Standard* 52.1); (3) specification of standard equipment; (4) performance and safety requirements; (5) proper marking; (6) conformance conditions; and (7) literature and advertising requirements. However, certification of air cleaners is not a part of these standards.

TYPES OF AIR CLEANERS

Common air cleaners are broadly grouped as follows:

In **fibrous media unit filters**, the accumulating dust load causes pressure drop to increase up to some maximum recommended value. During this period, efficiency normally increases. However, at high dust loads, dust may adhere poorly to filter fibers and efficiency drops because of offloading. Filters in this condition should be replaced or reconditioned, as should filters that have reached their final (maximum recommended) pressure drop. This category includes viscous impingement and dry air filters, available in low-efficiency to ultrahigh-efficiency construction.

In **renewable media filters**, fresh media is introduced into the airstream as needed to maintain essentially constant resistance and, consequently, constant average efficiency.

Electronic air cleaners, if maintained properly by regular cleaning, have relatively constant pressure drop and efficiency.

Combination air cleaners combine the other types. For example, an electronic air cleaner may be used as an agglomerator with a fibrous media downstream to catch the agglomerated particles blown off the plates. Electrode assemblies have been installed in air-handling systems, making the filtration system more effective (Frey 1985, 1986). Also, low-efficiency pads, throwaway panels and automatically renewable media roll filters, or low- to medium-efficiency pleated prefilters may be used upstream of a high-efficiency filter to extend the life of the better and more costly final filter. Charged media filters are also available that increase particle deposition on media fibers by an induced electrostatic field. With these filters, pressure loss increases as it does on a non-charged fibrous media filter. The benefits of combining different air cleaning processes vary. ASHRAE *Standard* 52.1 and 52.2 test methods may be used to compare the performance of combination air cleaners.

FILTER TYPES AND PERFORMANCE

Panel Filters

Viscous impingement panel filters are made up of coarse, highly porous fibers. Filter media are generally coated with an odorless, nonmigrating adhesive or other viscous substance, such as oil, which causes particles that impinge on the fibers to stick to them. Design air velocity through the media usually ranges from 200 to 800 fpm. These filters are characterized by low pressure drop, low cost, and good efficiency on lint and larger particles (5 μm and larger), but low efficiency on normal atmospheric dust. They are commonly made 0.5 to 4 in. thick. Unit panels are available in standard and special sizes up to about 24 by 24 in. This type of filter is commonly used in residential furnaces and air conditioning and is often used as a prefilter for higher-efficiency filters.

Filter media materials include metallic wools, expanded metals and foils, crimped screens, random matted wire, coarse (15 to 60 μm diameter) glass fibers, coated animal hair, vegetable or synthetic fibers, and synthetic open-cell foams.

Although viscous impingement filters usually operate at around 300 to 600 fpm, they may be operated at higher velocities. The limiting factor, other than increased flow resistance, is the danger of blowing off agglomerates of collected dust and the viscous coating on the filter.

The loading rate of a filter depends on the type and concentration of dirt in the air being handled and the operating cycle of the system. Manometers, static pressure differential gages, or pressure transducers are often installed to measure pressure drop across the filter bank. This measurement can identify when the filter requires servicing. The final allowable pressure differential may vary from one installation to another; but, in general, viscous impingement filters are serviced when their operating resistance reaches 0.5 in. of water. Life cycle cost (LCC), including energy necessary to overcome the filter resistance, should be calculated to evaluate the overall cost of the filtering system. The decline in filter efficiency caused by dust coating the adhesive, rather than by the increased resistance because of dust load, may be the limiting factor in operating life.

The manner of servicing unit filters depends on their construction and use. Disposable viscous impingement panel filters are constructed of inexpensive materials and are discarded after one period of use. The cell sides of this design are usually a combination of cardboard and metal stiffeners. Permanent unit filters are generally constructed of metal to withstand repeated handling. Various cleaning methods have been recommended for permanent filters; the most widely used involves washing the filter with steam or water (frequently with detergent) and then recoating it with its recommended adhesive by dipping or spraying. Unit viscous filters are also sometimes arranged for in-place washing and recoating.

The adhesive used on a viscous impingement filter requires careful engineering. Filter efficiency and dust-holding capacity depend on the specific type and quantity of adhesive used; this information is an essential part of test data and filter specifications. Desirable adhesive characteristics, in addition to efficiency and dust-holding capacity, are (1) a low percentage of volatiles to prevent excessive evaporation; (2) a viscosity that varies only slightly within the service temperature range; (3) the ability to inhibit growth of bacteria and mold spores; (4) a high capillarity or the ability to wet and retain the dust particles; (5) a high flash point and fire point; and (6) freedom from odorants or irritants.

Typical performance of viscous impingement unit filters operating within typical resistance limits is shown as MERV 1 through 6 in Table 3.

Dry extended-surface filters use media of random fiber mats or blankets of varying thicknesses, fiber sizes, and densities. Bonded glass fiber, cellulose fibers, wool felt, polymers, synthetics, and other materials have been used commercially. Media in these filters are frequently supported by a wire frame in the form of pockets, or V-shaped or radial pleats. In other designs, the media may be self-supporting because of inherent rigidity or because airflow inflates it into extended form (e.g., bag filters). Pleating media provides a high ratio of media area to face area, thus allowing reasonable pressure drop and low media velocities.

In some designs, the filter media is replaceable and is held in position in permanent wire baskets. In most designs, the entire cell is discarded after it has accumulated its maximum dust load.

Efficiency is usually higher than that of panel filters, and the variety of media available makes it possible to furnish almost any degree of cleaning efficiency desired. The dust-holding capacities of modern dry filter media and filter configurations are generally higher than those of panel filters.

Using coarse prefilters upstream of extended-surface filters is sometimes justified economically by the longer life of the main filters. Economic considerations include the prefilter material cost, changeout labor, and increased fan power. Generally, prefilters should be considered only if they can substantially reduce the part of the dust that may plug the protected filter. A prefilter usually has an arrestance of at least 70% (MERV 3) but is commonly rated up to 92% (MERV 6). Temporary prefilters protecting higher-efficiency filters are worthwhile during building construction to capture heavy loads of coarse dust. Filters of 95% DOP efficiency and greater should always be protected by prefilters of 80 to 85% or greater ASHRAE average atmospheric dust-spot efficiency (MERV 13). A single filter gage may be installed when a panel prefilter is placed adjacent to a final filter. Because the prefilter is frequently changed on a schedule, the final filter pressure drop can be read without the prefilter in place every time the prefilter is changed. For maximum accuracy and economy of prefilter use, two gages can be used. Some air filter housings are available with pressure taps between the pre- and final filter tracks to accommodate this arrangement.

Typical performance of some types of filters in this group, when operated within typical rated resistance limits and over the life of the filters, is shown as MERV 7 through 16 in Table 3.

Initial resistance of an extended-surface filter varies with the choice of media and filter geometry. Commercial designs typically have an initial resistance from 0.1 to 1.0 in. of water. It is customary to replace the media when the final resistance of 0.5 in. of water is reached for low-resistance units and 2.0 in. of water for the highest-resistance units. Dry media providing higher orders of cleaning efficiency have a higher average resistance to airflow. The operating resistance of the fully dust-loaded filter must be considered in design, because that is the maximum resistance against which the fan operates. Variable-air-volume and constant-air-volume system controls prevent abnormally high airflows or possible fan motor overloading from occurring when filters are clean.

Flat panel filters with media velocity equal to duct velocity are made only with the lowest-efficiency dry-type media (open-cell foams and textile denier nonwoven media). Initial resistance of this group, at rated airflow, is generally between 0.05 and 0.25 in. of water. They are usually operated to a final resistance of 0.50 to 0.70 in. of water.

In intermediate-efficiency extended-surface filters, the filter media area is much greater than the face area of the filter; hence, velocity through the filter media is substantially lower than the velocity approaching the filter face. Media velocities range from 6 to 90 fpm, although approach velocities run to 750 fpm. Depth in direction of airflow varies from 2 to 36 in.

Intermediate-efficiency filter media include (1) fine glass or synthetic fibers, 0.7 to 10 μm in diameter, in mats up to 0.5 in. thick; (2) wet laid paper or thin nonwoven mats of fine glass fibers, cellulose, or cotton wadding; and (3) nonwoven mats of comparatively large-diameter fibers (more than 30 μm) in greater thicknesses (up to 2 in.).

Electret filters are composed of electrostatically charged fibers. The charges on the fibers augment collection of smaller particles by interception and diffusion (Brownian motion) with Coulomb forces caused by the charges. There are three types of these filters: resin wool, electret, and an electrostatically sprayed polymer. The charge on resin wool fibers is produced by friction during the carding process. During production of the electret, a corona discharge injects positive charges on one side of a thin polypropylene film and negative charges on the other side. These thin sheets are then shredded into fibers of rectangular cross section. The third process spins a liquid polymer into fibers in the presence of a strong electric field, which produces the charge separation. Efficiency of charged-fiber filters is determined by both the normal collection mechanisms of a media filter (related to fiber diameter) and the strong local electrostatic effects (related to the amount of electrostatic charge). The effects induce efficient preliminary loading of the filter to enhance the caking process. However, dust collected on the media can reduce the efficiency of electret filters.

Very high-efficiency dry filters, **HEPA (high-efficiency particulate air) filters**, and **ULPA (ultralow-penetration air) filters** are made in an extended-surface configuration of deep space folds of submicrometre glass fiber paper. These filters operate at duct velocities near 250 fpm, with resistance rising from 0.5 to more than 2.0 in. of water over their service life. These filters are the standard for cleanroom, nuclear, and toxic particulate applications.

Membrane filters are used mainly for air sampling and specialized small-scale applications where their particular characteristics compensate for their fragility, high resistance, and high cost. They are available in many pore diameters and resistances and in flat-sheet and pleated forms.

Renewable-media filters may be one of two types: (1) moving-curtain viscous impingement filters or (2) moving-curtain dry-media roll filter.

In one viscous type, random-fiber (nonwoven) media is furnished in roll form. Fresh media is fed manually or automatically across the face of the filter, while the dirty media is rewound onto a roll at the bottom. When the roll is exhausted, the tail of the media is wound onto the take-up roll, and the entire roll is thrown away. A new roll is then installed, and the cycle repeats.

Moving-curtain filters may have the media automatically advanced by motor drives on command from a pressure switch, timer, or media light-transmission control. A pressure switch control measures the pressure drop across the media and switches on and off at chosen upper and lower set points. This saves media, but only if the static pressure probes are located properly and unaffected by modulating outside and return air dampers. Most pressure drop controls do not work well in practice. Timers and media light-transmission controls help avoid these problems; their duty cycles can usually be adjusted to provide satisfactory operation with acceptable media consumption.

Filters of this replaceable roll design generally have a signal indicating when the roll is nearly exhausted. At the same time, the drive motor is deenergized so that the filter cannot run out of media. Normal service requirements involve inserting a clean roll of media at the top of the filter and disposing of the loaded dirty roll. Automatic filters of this design are not, however, limited to the vertical position; horizontal arrangements are available for makeup air and air-conditioning units. Adhesives must have qualities similar to those for panel viscous impingement filters, and they must withstand media compression and endure long storage.

The second type of automatic viscous impingement filter consists of linked metal mesh media panels installed on a traveling curtain that intermittently passes through an adhesive reservoir. In the reservoir, the panels give up their dust load and, at the same time, take on a new coating of adhesive. The panels thus form a continuous curtain that moves up one face and down the other face. The media curtain, continually cleaned and renewed with fresh adhesive, lasts the life of the filter mechanism. The precipitated captured dirt must be removed periodically from the adhesive reservoir. New installations of this type of filter are rare in North America, but are often found in Europe and Asia.

The resistance of both types of viscous impingement automatically renewable filters remains approximately constant as long as proper operation is maintained. A resistance of 0.4 to 0.5 in. of water at a face velocity of 500 fpm is typical of this class.

Moving-curtain dry-media roll filters use random-fiber (nonwoven) dry media of relatively high porosity for general ventilation service. Operating duct velocities near 200 fpm are generally lower than those of viscous impingement filters.

Special automatic dry filters are also available, designed for removing lint in textile mills, laundries, and dry-cleaning establishments and for collecting lint and ink mist in printing press rooms. The medium used is extremely thin and serves only as a base for the buildup of lint, which then acts as a filter medium. The dirt-laden media is discarded when the supply roll is used up.

Another form of filter designed specifically for dry lint removal consists of a moving curtain of wire screen, which is vacuum cleaned automatically at a position out of the airstream. Recovery of the collected lint is sometimes possible with these devices.

ASHRAE arrestance, efficiency, and dust-holding capacities for typical viscous impingement and dry renewable-media filters are listed in Table 1.

Electronic Air Cleaners

Electronic air cleaners can be highly efficient filters using electrostatic precipitation to remove and collect particulate contaminants such as dust, smoke, and pollen. The term *electronic air cleaner* denotes a precipitator for HVAC air filtration. The filter consists of an ionization section and a collecting plate section.

In the ionization section, small-diameter wires with a positive direct current potential between 6 and 25 kV are suspended equidistant between grounded plates. The high voltage on the wires creates an ionizing field for charging particles. The positive ions created in the field flow across the airstream and strike and adhere to the particles, imparting a charge to them. The charged particles then pass into the collecting plate section.

The collecting plate section consists of a series of parallel plates equally spaced with a positive direct current voltage of 4 to 10 kV applied to alternate plates. Plates that are not charged are at ground potential. As the particles pass into this section, they are attracted to the plates by the electric field on the charges they carry; thus, they are removed from the airstream and collected by the plates. Particle retention is a combination of electrical and intermolecular adhesion forces and may be augmented by special oils or adhesives on the plates. Figure 3 shows a typical electronic air cleaner cell.

In lieu of positive direct current, a negative potential also functions on the same principle, but generates more ozone. With voltages

Table 1 Performance of Renewable Media Filters (Steady-State Values)

Description	Type of Media	ASHRAE Weight Arrestance, %	ASHRAE Atmospheric Dust-Spot Efficiency, %	ASHRAE Dust-Holding Capacity, g/ft^2	Approach Velocity, fpm
20 to 40 μm glass and synthetic fibers, 2 to 2 1/2 in. thick	Viscous impingement	70 to 82	<20	60 to 180	500
Permanent metal media cells or overlapping elements	Viscous impingement	70 to 80	<20	NA (permanent media)	500
Coarse textile denier nonwoven mat, 1/2 to 1 in. thick	Dry	60 to 80	<20	15 to 70	500
Fine textile denier nonwoven mat, 1/2 to 1 in. thick	Dry	80 to 90	<20	10 to 50	200

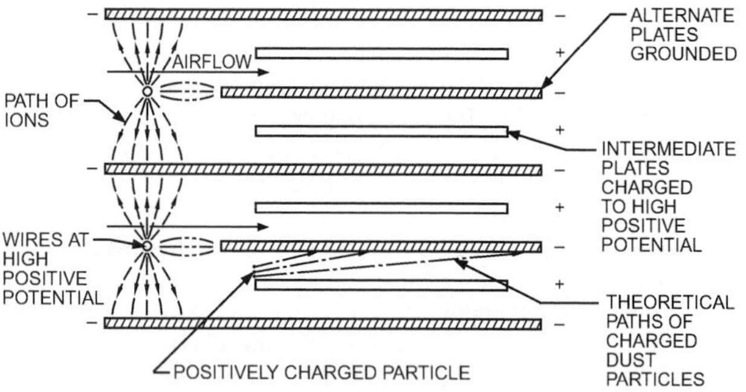

Fig. 3 Cross Section of Ionizing Electronic Air Cleaner

of 4 to 25 kV (dc), safety measures are required. A typical arrangement makes the air cleaner inoperative when the doors are removed for cleaning the cells or servicing the power pack. Electronic air cleaners typically operate from a 120 or 240 V (ac) single-phase electrical service. The high voltage supplied to the air cleaner cells is normally created with solid-state power supplies. The electric power consumption ranges from 20 to 40 W per 1000 cfm of air cleaner capacity.

This type of air filter can remove and collect airborne contaminants with an initial efficiency of up to 98% at low airflow velocities (150 to 350 fpm) when tested according to ASHRAE *Standard* 52.1. Efficiency decreases (1) as collecting plates become loaded with particulates, (2) with higher velocities, or (3) with nonuniform velocity.

As with most air filtration devices, duct approaches to and from the air cleaner housing should be arranged so that airflow is distributed uniformly over the face area. Panel prefilters should also be used to help distribute airflow and to trap large particles that might short out or cause excessive arcing in the high-voltage section of the air cleaner cell. Electronic air cleaner design parameters of air velocity, ionizer field strength, cell plate spacing, depth, and plate voltage must match the application requirements (e.g., contaminant type, particle size, volume of air, required efficiency). Many units are designed for installation into central heating and cooling systems for total air filtration. Other self-contained units are furnished complete with air movers for source control of contaminants in specific applications that need an independent air cleaner.

Electronic air cleaner cells must be cleaned periodically with detergent and hot water. Some designs incorporate automatic wash systems that clean the cells in place; in others, the cells are removed for cleaning. The frequency of cleaning (washing) the cell depends on the contaminant and the concentration. Industrial applications may require cleaning every 8 h, but a residential unit may only require cleaning every one to three months. The timing of the cleaning schedule is important to keep the unit performing at peak efficiency. For some contaminants, special attention must be given to cleaning the ionizing wires.

Optional features are often available for electronic air cleaners. Afterfilters such as roll filters collect particulates that agglomerate and blow off the cell plates. These are used mainly where heavy contaminant loading occurs and extension of the cleaning cycle is desired. Cell collector plates may be coated with special oils, adhesives, or detergents to improve both particle retention and particle removal during cleaning. High-efficiency dry extended-media area filters are also used as afterfilters in special designs. The electronic air cleaner used in this system improves the service life of the dry filter and collects small particles such as smoke.

A **negative ionizer** uses the principle of particle charging but does not use a collecting section. Particles enter the ionizer of the unit, receive an electrical charge, and then migrate to a grounded surface closest to the travel path.

Space Charge. Particulates that pass through an ionizer and are charged, but not removed, carry the electrical charge into the space. If continued on a large scale, a space charge builds up, which tends to drive these charged particles to walls and interior surfaces. Thus, a low-efficiency electronic air cleaner used in areas of high ambient dirt concentrations (or a malfunctioning unit), can blacken walls faster than if no cleaning device were used (Penney and Hewitt 1949; Sutton et al. 1964).

Ozone. All high-voltage devices can produce ozone, which is toxic and damaging not only to human lungs, but to paper, rubber, and other materials. When properly designed and maintained, an electronic air cleaner produces an ozone concentration that only reaches a fraction of the limit acceptable for continuous human exposure and is less than that prevalent in many American cities (EPA 1996). Continuous arcing and brush discharge in an electronic air cleaner may yield an ozone level that is annoying or mildly toxic; this is indicated by a strong ozone odor. Although the nose is sensitive to ozone, only actual measurement of the concentration can determine whether a hazardous condition exists.

ASHRAE *Standard* 62.1 defines acceptable concentrations of oxidants, of which ozone is a major contributor. The U.S. Environmental Protection Agency (EPA) specifies a 1 h average maximum allowable exposure to ozone of 0.12 ppm for outside ambient air. The U.S. Department of Health and Human Services specifies a maximum allowable continuous exposure to ozone of 0.05 ppm for contaminants of indoor origin. Sutton et al. (1976) showed that indoor ozone levels are only 30% of the outdoor level with ionizing air cleaners operating, although Weschler et al. (1989) found that this level increases when outdoor airflow is increased during an outdoor event that creates ozone.

SELECTION AND MAINTENANCE

To evaluate filters and air cleaners properly for a particular application, the following factors should be considered:

1. Types of contaminants present indoors and outdoors
2. Sizes and concentrations of contaminants
3. Air cleanliness levels required in the space
4. Air filter efficiency needed to achieve cleanliness

5. Space available to install and access equipment
6. Life cycle costing, including
   - Operating resistance to airflow (static pressure differential)
   - Disposal or cleaning requirements of spent filters
   - Initial cost of selected system
   - Cost of replacement filters or cleaning
   - Cost of warehousing filter stock and change-out labor

Savings (from reduced housekeeping expenses, protection of valuable property and equipment, dust-free environments for manufacturing processes, improved working conditions, and even health benefits) should be credited against the cost of installing and operating an adequate system. The capacity and physical size of the required unit may emphasize the need for low maintenance cost. Operating cost, predicted life, and efficiency are as important as initial cost because air cleaning is a continuing process.

Panel filters do not have efficiencies as high as can be expected from extended-surface filters, but their initial cost and upkeep are generally low. Compared to moving-curtain filters, panel filters of comparable efficiencies require more attention to maintain the resistance within reasonable limits. However, single-stage, face- or side-access, low- to medium-efficiency filters (25 to 50% dust spot and MERV 6 to 10), from a 2 in. pleat to a 12 in. deep cube, bag, or deep pleated cartridge, require less space with lower initial cost, and have better efficiency. The bag and cartridges generally have a similar service life to that of a roll filter.

If efficiency of 60 to 65% dust spot (MERV 11) or higher is required, extended-surface filters or electronic air cleaners should be considered. The use of very fine glass fiber mats or other materials in extended-surface filters has made these available at the highest efficiency.

Initial cost of an extended-surface filter is lower than for an electronic unit, but higher than for a panel type. Operating and maintenance costs of some extended-surface filters may be higher than for panel types and electronic air cleaners, but efficiencies are always higher than for panel types; the cost/benefit ratio must be considered. Pressure drop of media-type filters is greater than that of electronic-type and slowly increases during their useful life. Advantages include the fact that no mechanical or electrical services are required. Choice should be based on both initial and operating costs (life-cycle costs) as well as on the degree of cleaning efficiency and maintenance requirements.

Although electronic air cleaners have a higher initial cost and maintenance cost, they have high initial efficiencies in cleaning atmospheric air, largely because of their ability to remove fine particulate contaminants. System resistance remains unchanged as particles are collected, and efficiency is reduced until the resulting residue is removed from the collection plates to prepare the equipment for further duty. The manufacturer must supply information on maintenance or cleaning. Also, note that electronic air cleaners may not collect particles greater than 10 μm in diameter.

Table 2 lists some applications of filters classified according to their efficiencies and type, and Table 3 provides an approximate cross-reference between the ASHRAE *Standard* 52.1 and 52.2 reporting methods. A corollary purpose is to provide application guidance for the user and the HVAC designer. HEPA filters are tested by non-ASHRAE standards, but they have been included in the table by arbitrarily assigning minimum efficiency reporting values (MERVs) to Institute of Environmental Sciences and Technology (IEST) test standard ratings. Table 3 combines all the parameters into a single reference covering most types of air cleaners and applications. However, a single performance measurement cannot be applied precisely to all types and styles of air cleaners. Each air cleaner has unique characteristics that change during its useful life.

The typical contaminants listed in Table 3 appear in the general reporting group that removes the smallest known size of that specific contaminant. The order in which they are listed has no significance nor is the list complete. The typical applications and typical air cleaners listed are intended to show where and what type of air cleaner has been traditionally used. Traditional usage may not represent the optimum choice, so using the table as a selection guide is not appropriate when a specific performance requirement is needed. An air cleaner application specialist should then be consulted and manufacturers' performance curves should be reviewed.

Common sense and some knowledge of how air cleaners work help the user achieve satisfactory results. Air cleaner performance varies from the time it is first installed until the end of its service life. Generally, the longer a media-type filter is in service, the higher the efficiency. The accumulation of contaminants begins to close the porous openings, and, therefore, the filter is able to intercept smaller particles. However, there are exceptions that vary with different styles of media-type filters. Electronic air cleaners and charged-fiber media filters start at high efficiency when new (or after proper service, in the case of electronic air cleaners) but their efficiency decreases as contaminants accumulate. Some air cleaners, particularly low-efficiency devices, may begin to shed some collected contaminants after being in service. Testing with standardized synthetic loading dust attempts to predict this, but such testing rarely, if ever, duplicates the air cleaner's performance on atmospheric dust.

Residential Air Cleaners

Filters used for residential applications are often of spun glass and only filter out the largest of particles. These filters may prevent damage to downstream equipment, but they do little to improve air quality in the residence. Offermann (1992) describes a series of tests used to rate residential air cleaners. The tests were run in a test house with environmental tobacco smoke (ETS) as the test particulate (mass mean diameter = 0.5 μm). The typical residential air filter is not effective on these very small respirable-sized particles.

Console-type air cleaners can be used when an air cleaning system cannot be a part of the ducted HVAC equipment. For whole-house applications, in-line units ducted to the central heating and air-conditioning equipment are recommended.

VAV Systems

ASHRAE *Standard* 52.1 tests on numerous different media-type air cleaners under both constant and variable airflow showed no significant performance differences under the different flow conditions, and the air cleaners were not damaged by VAV flow. Low-efficiency air cleaners did show substantial reentrainment for both constant and VAV flow (Rivers and Murphy 2000).

Antimicrobial Treatment of Filter Media

One key factor influencing the ability of a filter to support fungal growth is the filter media itself; using antimicrobials successfully on air cleaners is very complex. Effective assessment of antimicrobial efficacy should include an "as-used" type test, not a test of the antimicrobial alone.

In laboratory and field testing, microbial growth has been seen on both treated and untreated dust-loaded samples for certain media types. Under normal use, fibrous air cleaners are unlikely to become a source of microbial contamination to the space, and antimicrobial treatment does not increase the filtration efficiency for bioaerosols (Foarde et al. 2000).

AIR CLEANER INSTALLATION

Many air cleaners are available in units of convenient size for manual installation, cleaning, and replacement. A typical unit filter may be 20 to 24 in. square, from 1 to 40 in. thick, and of either the dry or viscous impingement type. In large systems, the frames in which these units are installed are bolted or riveted together to form a filter bank. Automatic filters are constructed in sections offering

Table 2 Typical Filter Applications Classified by Filter Efficiency and Type[a]

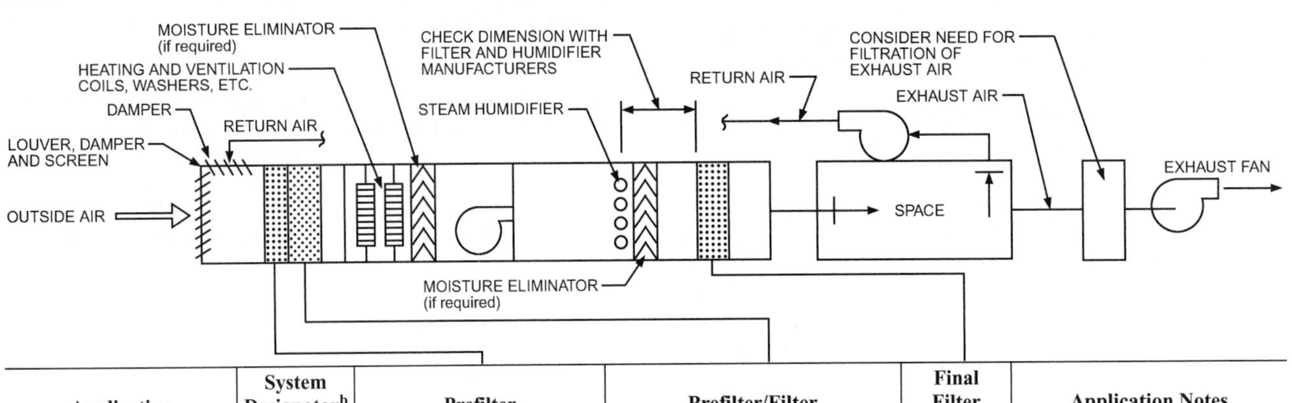

Application	System Designator[b]	Prefilter		Prefilter/Filter		Final Filter	Application Notes
Warehouse, storage, shop and process areas, mechanical equipment rooms, electrical control rooms, protection for heating and cooling coils	A1	None	None	70 to 85% arrestance	Panel-type or automatic roll	None	Reduces larger particle settling. Protects coils from dirt and lint.
	A2	None	None	85 to 92% arrestance, 25 to 50% dust spot	Pleated panel or extended surface	None	
Special process areas, electrical shops, paint shops, average general offices and laboratories	B1	None	None	50 to 65% dust spot	Extended surface, cartridge, bag, or electronic (manually cleaned or replaceable media)	None	Average housecleaning. Reduces lint in airstream. Reduces ragweed pollen >85% at 35%. Removes all pollens at 60%, somewhat effective on particles causing smudge and stain.
Analytical laboratories, electronics shops, drafting areas, conference rooms, above-average general offices	C1	75 to 85% arrestance, 25 to 40% dust spot	Extended surface, cartridge, or bag	80 to 85% dust spot	Bag, cartridge, or electronic (semi-automatic cleaning)	None	Above-average housecleaning. No settling particles of dust. Cartridge and bag types very effective on particles causing smudge and stain, partially effective on tobacco smoke. Electronic types quite effective on smoke.
	C2	None	None	80 to 85% dust spot	Electronic (agglomerator) with bag or cartridge section	None	
Hospitals, pharmaceutical R&D and manufacturing (nonaseptic areas only), some clean ("gray") rooms	D1	75 to 85% arrestance, 25 to 40% dust spot	Extended surface, cartridge, or bag	80 to 85% dust spot	Bag, cartridge, electronic (semi-automatic cleaning)	95% DOP disposable cell	Excellent housecleaning. Very effective on particles causing smudge and stain, smoke and fumes. Highly effective on bacteria.
	D2	None	None	80 to 95% dust spot	Electronic (agglomerator) with bag or cartridge section	None	
Aseptic areas in hospital and pharmaceutical R&D and manufacturing. Cleanrooms in film and electronics manufacturing, radioactive areas, etc.[c]	E1	75 to 85% arrestance, 25 to 40% dust spot	Extended surface, cartridge, or bag	80 to 95% dust spot	Bag, cartridge, electronic (semi-automatic cleaning)	≥99.97% DOP disposable cell	Protects against bacteria, radioactive dusts, toxic dusts, smoke, and fumes.

[a]Adapted from a similar table courtesy of E.I. du Pont de Nemours & Company.
[b]System designators have no significance other than their use in this table.
[c]Electronic agglomerators and air cleaners are not usually recommended for cleanroom applications.

several choices of width up to 70 ft and generally range in height from 40 to 200 in., in 4 to 6 in. increments. Several sections may be bolted together to form a filter bank.

Several manufacturers provide side-loading filter sections for various types of filters. Filters are changed from outside the duct, making service areas in the duct unnecessary, thus saving cost and space.

Of course, in-service efficiency of an air filter is sharply reduced if air leaks through the bypass dampers or poorly designed frames. The higher the filter efficiency, the more attention that must be paid to the rigidity and sealing effectiveness of the frame. In addition, high-efficiency filters must be handled and installed with care. The National Air Filtration Association (NAFA) (1997) suggests some precautions needed for HEPA filters.

Air cleaners may be installed in the outside-air intake ducts of buildings and residences and in the recirculation and bypass air ducts, but prefilters and intermediate filters (in a three-stage system) should be placed upstream of heating or cooling coils and other air-conditioning equipment in the system to protect that equipment from dust. Dust captured in an outside-air intake duct is likely to be mostly greasy particles, whereas lint may predominate in dust from within the building.

Table 3 Cross-Reference and Application Guidelines (Table E-1, ASHRAE *Standard* 52.2)

Std. 52.2 Minimum Efficiency Reporting Value (MERV)	Approx. *Std.* 52.1 Results		Application Guidelines		
	Dust-Spot Efficiency	Arrestance	Typical Controlled Contaminant	Typical Applications and Limitations	Typical Air Filter/Cleaner Type
20	n/a	n/a	**≤0.30 μm Particles** Virus (unattached) Carbon dust Sea salt All combustion smoke Radon progeny	Cleanrooms Radioactive materials Pharmaceutical manufacturing Carcinogenic materials Orthopedic surgery	**HEPA/ULPA Filters** ≥99.999% efficiency on 0.1 to 0.2 μm particles, IEST Type F ≥99.999% efficiency on 0.3 μm particles, IEST Type D ≥99.99% efficiency on 0.3 μm particles, IEST Type C ≥99.97% efficiency on 0.3 μm particles, IEST Type A
19	n/a	n/a			
18	n/a	n/a			
17	n/a	n/a			
16	n/a	n/a	**0.3 to 1.0 μm Particles** All bacteria Most tobacco smoke Droplet nuclei (sneeze) Cooking oil Most smoke Insecticide dust Copier toner Most face powder Most paint pigments	Hospital inpatient care General surgery Smoking lounges Superior commercial buildings	**Bag Filters** Nonsupported (flexible) microfine fiberglass or synthetic media. 12 to 36 in. deep, 6 to 12 pockets. **Box Filters** Rigid style cartridge filters 6 to 12 in. deep may use lofted (air-laid) or paper (wet-laid) media.
15	>95%	n/a			
14	90 to 95%	>98%			
13	80 to 90%	>98%			
12	70 to 75%	>95%	**1.0 to 3.0 μm Particles** Legionella Humidifier dust Lead dust Milled flour Coal dust Auto emissions Nebulizer drops Welding fumes	Superior residential Better commercial buildings Hospital laboratories	**Bag Filters** Nonsupported (flexible) microfine fiberglass or synthetic media. 12 to 36 in. deep, 6 to 12 pockets. **Box Filters** Rigid style cartridge filters 6 to 12 in. deep may use lofted (air-laid) or paper (wet-laid) media.
11	60 to 65%	>95%			
10	50 to 55%	>95%			
9	40 to 45%	>90%			
8	30 to 35%	>90%	**3.0 to 10.0 μm Particles** Mold Spores Hair spray Fabric protector Dusting aids Cement dust Pudding mix Snuff Powdered milk	Commercial buildings Better residential Industrial workplaces Paint booth inlet air	**Pleated Filters** Disposable, extended-surface, 1 to 5 in. thick with cotton/polyester blend media, cardboard frame. **Cartridge Filters** Graded-density viscous-coated cube or pocket filters, synthetic media **Throwaway** Disposable synthetic media panel filters
7	25 to 30%	>90%			
6	<20%	85 to 90%			
5	<20%	80 to 85%			
4	<20%	75 to 80%	**>10.0 μm Particles** Pollen Spanish moss Dust mites Sanding dust Spray paint dust Textile fibers Carpet fibers	Minimum filtration Residential Window air conditioners	**Throwaway** Disposable fiberglass or synthetic panel filters **Washable** Aluminum mesh, latex coated animal hair, or foam rubber panel filters **Electrostatic** Self-charging (passive) woven polycarbonate panel filter
3	<20%	70 to 75%			
2	<20%	65 to 70%			
1	<20%	<65%			

Note: MERV for non-HEPA/ULPA filters also includes test airflow rate, but it is not shown here because it is of no significance for the purposes of this table.

Where high-efficiency filters protect critical areas such as cleanrooms, it is important that the filters be installed as close to the room as possible to prevent pickup of particles between the filters and the outlet. The ultimate is the unidirectional flow room, in which the entire ceiling or one entire wall becomes the final filter bank.

Published performance data for all air filters are based on straight-through unrestricted airflow. Filters should be installed so that the face area is at right angles to the airflow whenever possible. Eddy currents and dead air spaces should be avoided; air should be distributed uniformly over the entire filter surface using baffles, diffusers, or air blenders, if necessary. Filters are sometimes damaged if higher-than-normal air velocities impinge directly on the face of the filter.

Failure of air filter installations to give satisfactory results can, in most cases, be traced to faulty installation, improper maintenance, or both. The most important requirements of a satisfactory and efficiently operating air filter installation are as follows:

- The filter must be of ample capacity for the amount of air and dust load it is expected to handle. An overload of 10 to 15% is regarded as the maximum allowable. When air volume is subject to future increase, a larger filter bank should be installed initially.
- The filter must be suited to the operating conditions, such as degree of air cleanliness required, amount of dust in the entering air, type of duty, allowable pressure drop, operating temperature, and maintenance facilities.

The following recommendations apply to filters installed with central fan systems:

- Duct connections to and from the filter should change size or shape gradually to ensure even air distribution over the entire filter area.
- The filter should be placed far enough from the fan to prevent or reduce reentrainment of particles, especially during start/stop cycles.

- Sufficient space should be provided in front of or behind the filter, or both, depending on its type, to make it accessible for inspection and service. A distance of 20 to 40 in. is required, depending on the filter chosen.
- Access doors of convenient size must be provided to the filter service areas.
- All doors on the clean-air side should be gasketed to prevent infiltration of unclean air. All connections and seams of the sheet-metal ducts on the clean-air side should be airtight. The filter bank must be caulked to prevent bypass of unfiltered air, especially when high-efficiency filters are used.
- Lights should be installed in the plenum in front of and behind the air filter bank.
- Filters installed close to an air inlet should be protected from the weather by suitable louvers or inlet hood. In areas with extreme rainfall or where water can drip over or bounce up in front of the inlet, use drainable track moisture separator sections upstream of the first filter bank. A large-mesh wire bird screen should be placed in front of the louvers or in the hood.
- Filters, other than electronic air cleaners, should have permanent indicators to give notice when the filter reaches its final pressure drop or is exhausted, as with automatic roll media filters.
- Electronic air cleaners should have an indicator or alarm to indicate when high voltage is off or shorted out.

SAFETY CONSIDERATIONS

Safety ordinances should be investigated when the installation of an air cleaner is contemplated. Combustible filtering media may not be permitted by some local regulations. Combustion of dust and lint on filtering media is possible, although the media itself may not burn. This may cause a substantial increase in filter combustibility. Smoke detectors and fire sprinkler systems may be considered for filter bank locations. In some cases, depending on the contaminant, hazardous material procedures must be followed during removal and disposal of the spent filter. Bag-in/bag-out (BI/BO) filter housings should be considered in those cases.

Many air filters are efficient collectors of bioaerosols. When provided moisture and nutrients, the microorganisms can multiply and may become a health hazard for maintenance personnel. Moisture in filters can be minimized by preventing (1) entrance of rain, snow, and fog; (2) carryover of water droplets from coils, drain pans, and humidifiers; and (3) prolonged exposure to elevated humidity. Changing or cleaning filters regularly is important for controlling microbial growth. Good health-safety practices for personnel handling dirty filters include using face masks, thorough washing upon completion of the work, and placing used filters in plastic bags or other containers for disposal.

[Editor's note: Consult the publishers of sources listed in the References and Bibliography for the latest editions.]

REFERENCES

ACGIH. 1989. *Guidelines for the assessment of bioaerosols in the indoor environment.* American Conference of Governmental Industrial Hygienists, Cincinnati, OH.

AHAM. 1988. Performance standard for room air cleaners. *Standard* AC-1-1988. Association of Home Appliance Manufacturers, Chicago, IL.

ARI. 1993. Residential air filter equipment. *Standard* 680-93. Air-Conditioning and Refrigeration Institute, Arlington, VA.

ARI. 1993. Commercial and industrial air filter equipment. *Standard* 850-93. Air-Conditioning and Refrigeration Institute, Arlington, VA.

ASHRAE. 1992. Gravimetric and dust-spot procedures for testing air-cleaning devices used in general ventilation for removing particulate matter. ANSI/ASHRAE *Standard* 52.1-1992.

ASHRAE. 1999. Method of testing general ventilation air-cleaning devices for removal efficiency by particle size. ANSI/ASHRAE *Standard* 52.2-1999.

CEN. 1994. Particulate air filters for general ventilation—Requirements, testing, marking. *Standard* EN 779. Comité Européen de Normalisation, Brussels.

CEN. 1998. High efficiency air filters (HEPA and ULPA)—Part 1: Classification, performance testing, marking; Part 2: Aerosol production, measuring equipment, particle counting statistics; Part 3: Testing flat sheet filter media. *Standard* EN 1822. Comité Européen de Normalisation, Brussels.

CEN. 2000. High efficiency air filters (HEPA and ULPA)—Part 4: Determining leakage of filter element; Part 5: Determining the efficiency of filter element. *Standard* EN 1822. Comité Européen de Normalisation, Brussels.

EPA. 1995. *Air quality criteria for ozone and other photochemical oxidants.* Environmental Protection Agency, Office of Air Noise and Radiation, Office of Air Quality Planning and Standards, Research Triangle Park, NC.

EPA. 1996. *National air quality and emissions trends reports.* 454/R-96.005. U.S. Environmental Protection Agency, Office of Air Quality Planning and Standards, Research Triangle Park, NC.

Foarde, K.K., J.T. Hanley, and A.C. Veeck. 2000. Efficacy of antimicrobial filter treatments. *ASHRAE Journal* 42(12):52.

Frey, A.H. 1985. Modification of aerosol size distribution by complex electric fields. *Bulletin of Environmental Contamination and Toxicology* 34:850-857.

Frey, A.H. 1986. The influence of electrostatics on aerosol deposition. *ASHRAE Transactions* 92(1B):55-64.

Horvath, H. 1967. A comparison of natural and urban aerosol distribution measured with the aerosol spectrometer. *Environmental Science and Technology* (August):651.

McCrone, W.C., R.G. Draftz, and J.G. Delley. 1967. *The particle atlas.* Ann Arbor Science, Ann Arbor, MI.

Morey, P.R. 1994. Suggested guidance on prevention of microbial contamination for the next revision of ASHRAE *Standard* 62. *Indoor Air Quality '94.* ASHRAE.

National Air Pollution Control Administration. 1969. *Air quality criteria for particulate matter.* Available from National Technical Information Service (NTIS), Springfield, VA.

NIOSH. 1973. *The industrial environment—Its evaluation and control.* S/N 017-001-00396-4. U.S. Government Printing Office, Washington, D.C.

Offermann, F.J., S.A. Loiselle, and R.G. Sextro. 1992. Performance of air cleaners in a residential forced air system. *ASHRAE Journal* 34(7):51-57.

Penney, G.W. and G.W. Hewitt. 1949. Electrically charged dust in rooms. *AIEE Transactions* (68):276-282.

Rivers, R.D. and D.J. Murphy. 2000. Air filter performance under variable air volume conditions. *ASHRAE Transactions* 106(2).

Sutton, D.J., H.A. Cloud, P.E. McNall, Jr., K.M. Nodolf, and S.H. McIver. 1964. Performance and application of electronic air cleaners in occupied spaces. *ASHRAE Journal* 6(6):55.

Sutton, D.J., K.M. Nodolf, and H.K. Makino. 1976. Predicting ozone concentrations in residential structures. *ASHRAE Journal* 18(9):21.

UL. 1990. High efficiency, particulate, air filter units. *Standard* 586-90. Underwriters Laboratories, Inc., Northbrook, IL.

UL. 1995. Air filter units. *Standard* 900-95. Underwriters Laboratories, Inc., Northbrook, IL.

U.S. Army. 1965. Instruction manual for the installation, operation, and maintenance of penetrometer, filter testing, DOP, Q107. *Document* 136-300-175A. Edgewood Arsenal, MD.

U.S. Navy. 1956. DOP—Smoke penetration and air resistance of filters. Military *Standard* MIL-STD-282. Department of Navy, Defense Printing Service, Philadelphia, PA.

Weschler, C.J., H.C. Shields, and D.V. Noik. 1989. Indoor ozone exposures. *Journal of the Air Pollution Control Association* 39(12):1562.

Wheeler, A.E. 1994. Better filtration for healthier buildings. *ASHRAE Journal* 36(6):62-69.

Whitby, K.T., A.B. Algren, and R.C. Jordan. 1958. Size distribution and concentration of airborne dust. *ASHRAE Transactions* 64:129.

Willeke, K. and P.A. Baron, eds. 1993. *Aerosol measurement: Principles, techniques and applications.* Van Nostrand Reinhold, New York.

BIBLIOGRAPHY

ASHRAE. 2007. Ventilation for acceptable indoor air quality. ANSI/ASHRAE *Standard* 62.1-2007.

ASHRAE. 2007. Ventilation and acceptable indoor air quality in low-rise residential buildings. ANSI/ASHRAE *Standard* 62.2-2007.

Bauer, E.J., B.T. Reagor, and C.A. Russell. 1973. Use of particle counts for filter evaluation. *ASHRAE Journal* 15(10):53.

Davies, C.N. 1973. *Aerosol filtration.* Academic Press, London.

Dennis, R. 1976. *Handbook on aerosols.* Technical Information Center, Energy Research and Development Administration. Oak Ridge, TN.

Dill, R.S. 1938. A test method for air filters. *ASHRAE Transactions* 44:379.

Dorman, R.G. 1974. *Dust control and air cleaning*. Pergamon, New York.

Duncan, S.F. 1964. Effect of filter media microstructure on dust collection. *ASHRAE Journal* 6(4):37.

Engle, P.M., Jr. and C.J. Bauder. 1964. Characteristics and applications of high performance dry filters. *ASHRAE Journal* 6(5):72.

Eurovent/Cecomaf. 1996. Method of testing air filters used in general ventilation for determination of fractional efficiency. *Document* 4/9. European Committee of Air Handling and Refrigeration Equipment Manufacturers, Brussels (www.eurovent-association.eu).

Foarde, K.K. and D.W. VanOsdell. 1994. Investigate and identify indoor allergens and toxins that can be removed by filtration. *Final Report*, ASHRAE Research Project RP-760.

Gieseke, J.A., E.R. Blosser, and R.B. Rief. 1975. Collection and characterization of airborne particulate matter in buildings. *ASHRAE Transactions* 84(1):572.

Gilbert, H. and J. Palmer. 1965. *High efficiency particulate air filter units.* USAEC, TID-7023. Available from NTIS, Springfield, VA.

Hinds, W.C. 1982. *Aerosol technology: Properties, behavior, and measurement of airborne particles.* John Wiley & Sons, New York.

Howden Buffalo. 1999. *Fan engineering handbook*, 9th ed. Howden Buffalo Company, Buffalo, NY.

Hunt, C.M. 1972. An analysis of roll filter operation based on panel filter measurements. *ASHRAE Transactions* 78(2):227.

IEST. 1992. *Testing ULPA filters.* IES RP-CC 007.1. Institute of Environmental Sciences and Technology, Mount Prospect, IL.

IEST. 1993. *HEPA and ULPA filters.* IES RP-CC 001.3. Institute of Environmental Sciences and Technology, Mount Prospect, IL.

James, A.C., W. Stahlhofen, G. Rudolf, M.J. Egan, W. Nixon, P. Gehr, and J.K. Briant. 1991. The respirator tract deposition model proposed by the ICRP task group. *Radiation Protection Dosimetry* 38:159-165.

Kemp, S.J., T.H. Kuehn, D.Y.H. Pui, D. Vesley, and A.J. Streifel. 1995. Filter collection efficiency and growth of the microorganisms on filters loaded with outdoor air. *ASHRAE Transactions* 101(1):228-238.

Kemp, S.J., T.H. Kuehn, D.Y.H. Pui, D. Vesley, and A.J. Streifel. 1995. Growth of microorganisms on HVAC filters under controlled temperature and humidity conditions. *ASHRAE Transactions* 101(1):305-316.

Kerr, G. and L.C. Nguyen Thi. 2001. Identification of contaminants, exposure effects and control options for construction/renovation activities. *Final Report*, ASHRAE Research Project RP-961.

Kuehn, T.H., D.Y.H. Pui, and D.Vesley. 1991. Matching filtration to health requirements. *ASHRAE Transactions* 97(2):164-169.

Kuehn, T.H., B. Gacek, C-H. Yang, D.T. Grimsrud, K.A. Janni, A.J. Streifel, and M. Pearce. 1996. Identification of contaminants, exposure effects and control options for construction/renovation activities. *ASHRAE Transactions* 102(2):89-101.

Lee, K.W. and B.T. Liu. 1982. Experimental study of aerosol filtration by fibrous filters. *Aerosol Science and Technology* 1(1):35-46.

Lee, K.W. and B.T. Liu. 1982. Theoretical study of aerosol filtration by fibrous filters. *Aerosol Science and Technology* 1(2):147-161.

Liu, B.Y. and K.L. Rubow. 1986. Air filtration by fibrous filter media. In *Fluid-Filtration: Gas*. ASTM STF973. American Society for Testing and Materials, Philadelphia, PA.

Licht, W. 1980. *Air pollution control engineering: Basic calculations for particulate collection.* Marcel Dekker, New York.

Lioy, P.J. and M.J.Y. Lioy, eds. 1983. *Air sampling instruments for evaluation of atmospheric contaminants*, 6th ed. American Conference of Governmental Industrial Hygienists, Cincinnati, OH.

Liu, B.Y.H. 1976. *Fine particles.* Academic Press, New York.

Lundgren, D.A., F.S. Harris, Jr., W.H. Marlow, M. Lippmann, W.E. Clark, and M.D. Durham, eds. 1979. *Aerosol measurement*. University Presses of Florida, Gainesville.

Matthews, R.A. 1963. Selection of glass fiber filter media. *Air Engineering* 5(October):30.

McNall, P.E., Jr. 1986. Indoor air quality—A status report. *ASHRAE Journal* 28(6):39.

NAFA.1993. *NAFA guide to air filtration*. National Air Filtration Association, Washington, D.C.

NAFA. 1997. *Installation, operation and maintenance of air filtration systems*. National Air Filtration Association, Washington, D.C.

NRC. 1981. *Indoor pollutants.* National Research Council, National Academy Press, Washington, D.C.

Nazaroff, W.W. and G.R. Cass. 1989. Mathematical modeling of indoor aerosol dynamics. *Environmental Science and Technology* 23(2):157-166.

NIOSH/MSHA. 1977. *U.S. Federal mine safety and health act of 1977*, Title 30 CFR Parts 11, 70. National Institute for Occupational Safety and Health, Columbus, OH, and Mine Safety and Health Administration, Department of Labor, Washington, D.C.

Nutting, A. and R.F. Logsdon. 1953. New air filter code. *Heating, Piping and Air Conditioning* (June):77.

Ogawa, A. 1984. *Separation of particles from air and gases*, vols. I and II. CRC Press, Boca Raton, FL.

Penney, G.W. and N.G. Ziesse. 1968. Soiling of surfaces by fine particles. *ASHRAE Transactions* 74(1).

Peters, A.H. 1962. *Application of moisture separators and particulate filters in reactor containment.* USAEC-DP812. U.S. Department of Energy, Washington, D.C.

Peters, A.H. 1965. *Minimal specification for the fire-resistant high-efficiency filter unit.* USAEC Health and Safety Information (212). U.S. Department of Energy, Washington, D.C.

Phalen, R.F., R.G. Cuddihy, G.I. Fisher, O.R. Moss, R.B. Schlesinger, D.L. Swift, and H.-C. Yeh. 1991. Main features of the proposed NCRP respiratory tract model. *Radiation Protection Dosimetry* 38:179-184.

Rivers, R.D. 1988. Interpretation and use of air filter particle-size-efficiency data for general-ventilation applications. *ASHRAE Transactions* 88(1).

Rose, H.E. and A.J. Wood. 1966. *An introduction to electrostatic precipitation.* Dover Publishing, New York.

Stern, A.C., ed. 1977. *Air pollution*, 3rd ed., vol. 4, Engineering control of air pollution. Academic Press, New York.

Swanton, J.R., Jr. 1971. Field study of air quality in air-conditioned spaces. *ASHRAE Transactions* 77(1):124.

UL. 2000. Electrostatic air cleaners. ANSI/UL *Standard* 867-2000. Underwriters Laboratories, Inc., Northbrook, IL.

U.S. Government. 1988. Clean room and work station requirements, controlled environments. *Federal Standard* 209D, amended 1991. Available from GSA, Washington, D.C.

USAEC/ERDA/DOE. *Air cleaning conference proceedings.* Available from NTIS, Springfield, VA.

Walsh, P.J., C.S. Dudney, and E.D. Copenhaver, eds. 1984. *Indoor air quality.* CRC Press, Boca Raton, FL.

Whitby, K.T. 1965. Calculation of the clean fractional efficiency of low media density filters. *ASHRAE Journal* 7(9):56.

Whitby, K.T., A.B. Algren, and R.C. Jordan. 1956. The dust spot method of evaluating air cleaners. *Heating, Piping and Air Conditioning* (December):151.

White, P.A.F. and S.E. Smith, eds. 1964. *High efficiency air filtration*. Butterworth, London, UK.

Yocom, J.E. and W.A. Cote. 1971. Indoor/outdoor air pollutant relationships for air-conditioned buildings. *ASHRAE Transactions* 77(1):61.

Ziesse, N.G. and G.W. Penney. 1968. The effects of cigarette smoke on space charge soiling of walls when air is cleaned by a charging type electrostatic precipitator. *ASHRAE Transactions* 74(2).

CHAPTER 29

INDUSTRIAL GAS CLEANING AND AIR POLLUTION CONTROL

INDUSTRIAL gas cleaning performs one or more of the following functions:

- Maintains compliance of an industrial process with the laws or regulations for air pollution
- Reduces nuisance or physical damage from contaminants to individuals, equipment, products, or adjacent properties
- Prepares cleaned gases for processes
- Reclaims usable materials, heat, or energy
- Reduces fire, explosion, or other hazards

Equipment that removes particulate matter from a gas stream may also remove or create some gaseous contaminants; on the other hand, equipment that is primarily intended for removal of gaseous pollutants might also remove or create objectionable particulate matter to some degree. In all cases, gas-cleaning equipment changes the process stream, and it is therefore essential that the engineer evaluate the consequences of those changes to the plant's overall operation.

Equipment Selection

In selecting industrial gas-cleaning equipment, plant operations and the use or disposal of materials captured by the gas-cleaning equipment must be considered. Because the cost of gas-cleaning equipment affects manufacturing costs, alternative processes should be evaluated early to minimize the effect the equipment may have on the total cost of a product. An alternative manufacturing process may reduce the cost of or eliminate the need for gas-cleaning equipment. However, even when gas-cleaning equipment is required, process and system control should minimize the load on the collection device.

An industrial process may be changed from dirty to clean by substituting a process material (e.g., switching to a cleaner-burning fuel or pretreating the existing fuel). Equipment redesign, such as enclosing pneumatic conveyors or recycling noncondensable gases, may also clean the process. Occasionally, additives (e.g., chemical dust suppressants used in quarrying or liquid animal fat applied to dehydrated alfalfa before grinding) reduce the potential for air pollution or concentrate the pollutants so that a smaller, more concentrated process stream may be treated.

Gas streams containing contaminants should not usually be diluted with extraneous air unless the extra air is required for cooling or to condense contaminants to make them collectible. The volume of gas to be cleaned is a major factor in the owning and operating costs of control equipment. Therefore, source capture ventilation, where contaminants are kept concentrated in relatively small volumes of air, is generally preferable to general ventilation, where pollutants are allowed to mix into and be diluted by much of the air in a plant space. Chapters 29 and 30 of the 2007 *ASHRAE Handbook—HVAC Applications* address local and general ventilation of industrial environments. Regulatory authorities generally require the levels of emissions to be corrected to standard conditions taking into account temperature, pressure, moisture content, and factors related to combustion or production rate. However, the air-cleaning equipment must be designed using the actual conditions of the process stream as it will enter the equipment.

In this chapter, each generic type of equipment is discussed on the basis of its primary method for gas or particulate abatement. The development of systems that incorporate several of the devices discussed here for specific industrial processes is left to the engineer.

The preparation of this chapter is assigned to TC 5.4, Industrial Process Air Cleaning (Air Pollution Control).

REGULATIONS AND MONITORING

Gas-Cleaning Regulations

In the United States, industrial gas-cleaning installations that exhaust to the outdoor environment are regulated by the U.S. Environmental Protection Agency (EPA); those that exhaust to the workplace are regulated by the Occupational Safety and Health Administration (OSHA) of the U.S. Department of Labor.

The EPA has established Standards of Performance for New Stationary Sources [New Source Performance Standards (NSPSs), GPO] and more restrictive State Implementation Plans (SIP 1991) and local codes as a regulatory basis to achieve air quality standards. Information on the current status of the NSPSs can be obtained through the Semi-Annual Regulatory Agenda, as published in the Federal Register, and through the regional offices of the EPA. Buonicore and Davis (1992) and Sink (1991) provide additional design information for gas-cleaning equipment.

Where air is not affected by combustion, solvent vapors, and toxic materials, it may be desirable to recirculate the air to the workplace to reduce energy costs or to balance static pressure in a building. High-efficiency fabric or cartridge filters, precipitators, or special-purpose wet scrubbers are typically used in general ventilation systems to reduce particle concentrations to levels acceptable for recirculated air.

The Industrial Ventilation Committee of the American Conference of Governmental Industrial Hygienists (ACGIH) and the National Institute of Occupational Safety and Health (NIOSH) have established criteria for recirculation of cleaned process air to the work area (ACGIH 2006; NIOSH 1978). Fine-particle control by various dust collectors under recirculating airflows has been investigated by Bergin et al. (1989).

Public complaints may occur even when the effluent concentrations discharged to the atmosphere are below the maximum permissible emission rates and opacity limits. Thus, in addition to codes or regulations, the plant location, the contaminants involved, and the meteorological conditions of the area must be evaluated.

In most cases, emission standards require a higher degree of gas cleaning than necessary for economical recovery of process products (if this recovery is desirable). Gas cleanliness is a priority, especially where toxic materials are involved and cleaned gases might be recirculated to the work area.

Measuring Gas Streams and Contaminants

Stack sampling is often required to fulfill requirements of operating and installation permits for gas-cleaning devices, to establish conformance with regulations, and to commission new equipment. Also, it can be used to establish specifications for gas-cleaning equipment and to certify that the equipment is functioning properly. The tests determine the composition and quantity of gases and particulate matter at selected locations along the process stream. The following general principles apply to a stack sampling program:

- The sampling location(s) must be acceptable to all parties who will use the results.
- The sampling location(s) must meet acceptable criteria with respect to temperature, flow distribution and turbulence, and distance from disturbances to the process stream. Exceptions based on physical constraints must be identified and reported.
- Samples that are withdrawn from a duct or stack must represent typical conditions in the process stream. Proper stack traverses must be made and particulate samples withdrawn isokinetically.
- Stack sampling should be performed in accordance with approved methods and established protocols whenever possible.
- Variations in the volumetric flow, temperature, and particulate or gaseous pollutant emissions, along with upset conditions, should be identified.
- A report from stack sampling should include a summary of the process, which should identify any deviations from normal process operations that occurred during testing. The summary should be prepared during the testing phase and certified by the process owner at completion of the tests.
- Disposal methods for waste generated during testing must be identified before testing begins. This is especially critical where pilot plant equipment is being tested because new forms of waste are often produced.
- The regulatory basis for the tests should be established so that the results can be presented in terms of process mass rate, consumption of raw materials, energy use, and so forth.

Analyses of the samples can provide the following types of information about the emissions:

- Physical characteristics of the contaminant: solid dust, liquid mist or "smoke," waxy solids, or a sticky mixture of liquid and solid.
- Distribution of particle sizes: optical, physical, aerodynamic, etc.
- Concentration of particulate matter in the gases, including average and extreme values, and a profile of concentrations in the duct or stack.
- Volumetric flow of gases, including average and extreme values, and a profile of this flow in the duct or stack. The volumetric flow is commonly expressed at actual conditions and at various standard conditions of temperature, pressure, moisture, and process state.
- Chemical composition of gases and particulate matter, including recovery value, toxicity, solubility, acid dew point, etc.
- Particle and bulk densities of particulate matter.
- Handling characteristics of particulate matter, including erosive, corrosive, abrasive, flocculative, or adhesive/cohesive qualities.
- Flammable or explosive limits.
- Electrical resistivity of deposits of particulate matter under stack and laboratory conditions. These data are useful for assessments of electrostatic precipitators and other electrostatically augmented technology.

EPA Reference Methods. The EPA has developed methods to measure the particulate and gaseous components of emissions from many industrial processes and has incorporated these in the NSPS by reference. Appendix A of the New Source Performance Standards lists the reference methods (GPO, annual). These are updated regularly in the Federal Register. Guidance for using these reference methods can be found in the *Quality Assurance Handbook for Air Pollution Measurement Systems* (EPA 1994).

The EPA (2004) has promulgated fine particle standards for ambient air quality. Known as PM-10 and PM-2.5 Standards, these revised standards focus particulate abatement efforts toward the collection and control of airborne particles smaller than 10 μm and 2.5 μm, respectively. Fine particles (with aerodynamic particle size smaller than 2.5 μm) are of concern because they penetrate deeply into the lungs. With the development of these standards, concern has arisen over the efficiency of industrial gas cleaners at various particle sizes.

ASTM Methods. The EPA has also approved ASTM test methods when cited in the NSPSs or the applicable EPA reference methods.

Other Methods. Sometimes, the reference methods must be modified, with the consent of regulatory authorities, to achieve representative sampling under the less-than-ideal conditions of industrial operations. Modifications to test methods should be clearly identified and explained in test reports.

Gas Flow Distribution

The control of gas flow through industrial gas-cleaning equipment is important for good system performance. Because of the large gas flows commonly encountered and the frequent need to retrofit equipment to existing processes, space allocations often preclude gradual expansion and long-radius turns. Instead, elbow splitters, baffles, etc., are used. These components must be designed to limit dust buildup and corrosion to acceptable levels.

Monitors and Controls

Current regulatory trends anticipate or demand continuous monitoring and control of equipment to maintain optimum performance against standards. Under regulatory data requirements, operating logs provide the owner with process control information and others with a baseline for the development or service of equipment.

Larger systems include programmable controllers and computers for control, energy management, data logging, and diagnostics. Increasing numbers of systems have modem connections to support monitoring and service needs from remote locations.

PARTICULATE CONTAMINANT CONTROL

A large range of equipment for the separation of particulate matter from gaseous streams is available. Typical concentration ranges for this equipment are summarized in Table 1.

High-efficiency particulate air (HEPA) and ultralow penetration air (ULPA) filters are often used in cleanrooms to maintain nearly particulate-free environments. In commercial and residential buildings, air cleaners are used to remove nuisance dust. In other instances, air cleaners are selected to control particulate matter that constitutes a health hazard in the workplace (e.g., radioactive particles, beryllium-containing particles, or biological airborne wastes). Recirculation of air in industrial plants could use air cleaners or may require more heavy-duty equipment. Secondary filtration systems

Table 1 Intended Duty of Gas-Cleaning Equipment

	Maximum Concentration
Air cleaners	<0.002 gr/ft^3
Cleanrooms	
Commercial/residential buildings	
Plant air recirculation	
Industrial gas cleaners	<35 gr/ft^3
Product capture in pneumatic conveying	<3500 gr/ft^3

Table 2 Principal Types of Particulate Control Equipment

Gravity and momentum collectors	• Settling chambers • Louvers and baffle chambers
Centrifugal collectors	• Cyclones and multicyclones • Rotating "centrifugal" mist collectors
Electrostatic precipitators	• Tubular or plate-type, wet or dry, high-voltage (single-stage) precipitators • Plate-type, wet or dry, low-voltage (two-stage) mist and smoke precipitators
Fabric filters	• Baghouses; fabric collectors; cartridge filters • Disposable media filters (for dust and/or mist)
Granular-bed filters	• Fixed bed • Moving bed
Particulate scrubbers	• Spray scrubbers • Impingement scrubbers • Centrifugal-type scrubbers • Orifice-type scrubbers • Venturi scrubbers • Packed towers • Mobile bed scrubbers • Electrostatically augmented scrubbers

[typically HEPA or 95% dioctyl phthalate (DOP) filter systems] are sometimes required with recirculation systems.

Air cleaners are discussed in Chapter 28. This chapter is concerned with heavy-duty equipment for the control of emissions from industrial processes. The particulate emissions from these processes generally have a concentration in the range of 0.01 to 35 gr/ft^3. The gas-cleaning equipment is usually installed for air pollution control.

Particulate control technology is selected to satisfy the requirements for specific processes. Available technology differs in basic design, removal efficiency, first cost, energy requirements, maintenance, land use, operating costs, and ability of the collectors to handle various types and sizes of contaminant particles without requiring excessive maintenance. Some of the principal types of particulate control equipment are listed in Table 2 and discussed in the following sections.

Collector Performance

Particulate collectors may be evaluated for their ability to remove particulate matter from a gas stream and for their ability to reduce the emissions of selected particle sizes. The degree to which particulate matter is separated from a gas stream is known as the efficiency of a collector; the fraction of material escaping collection is the penetration. The reduction of particulate matter in a selected particle size range is known as the fractional efficiency of a particulate collector.

The **efficiency** η of a collector is generally expressed as a percent of the mass flow rate of material entering and exiting the collector.

$$\eta = 100(w_i - w_o)/w_i = 100w_c/w_i \quad (1)$$

where

η = efficiency of collector, %
w_i = mass flow rate of contaminant in gases entering collector
w_o = mass flow rate of contaminant in gases exiting collector
w_c = mass flow rate of contaminant captured by collector

Alternatively, the efficiency of a collector can be expressed in terms of the concentrations of particulate matter entering and exiting the equipment. This approach can be unsatisfactory because of changes that occur in the gas stream due to air leakage in the system, condensation, and temperature and pressure changes.

Penetration P is usually measured, and the efficiency for a collector calculated, using the following equations:

$$P = 100w_o/w_i \quad (2)$$

with

$$\eta = 100 - P \quad (3)$$

The **fractional efficiency** of a particulate collector is determined by measuring the mass rate of contaminants entering and exiting the collector in selected particle size ranges. Methods for measuring the fractional efficiency of particulate collectors in industrial applications are only beginning to emerge, largely because of the need to compare the fine-particle performance of collectors used under the PM-10 and PM-2.5 regulations for ambient air quality.

Measures of performance other than efficiency should also be considered in designing industrial gas-cleaning systems. Table 3 compares some of these factors for typical equipment. Note that this table contains only nominal values and is no substitute for experience, trade studies, and an engineering assessment of requirements for specific installations.

Table 4 summarizes the types of collectors that have been used in industrial applications.

MECHANICAL COLLECTORS

Settling Chambers

Particulate matter will fall from suspension in a reasonable time if the particles are larger than about 40 μm. Plenums, dropout boxes, or gravitational settling chambers are thus used for the separation of coarse or abrasive particulate matter from gas streams.

Settling chambers are occasionally used in conjunction with fabric filters or electrostatic precipitators to reduce overall system cost. The settling chamber serves as a precollector to remove coarse particles from the gas stream.

Settling chambers sometimes contain baffles to distribute gas flow and to serve as surfaces for the impingement of coarse particles. Other designs use baffles to change the direction of the gas flow, thereby allowing coarse particles to be thrown from the gas stream by inertial forces.

The fractional efficiency for a settling chamber with uniform gas flow may be estimated by

$$\eta = 100u_t L/HV \quad (4)$$

where

η = efficiency of collector for particles with settling velocity u_t, %
u_t = settling velocity for selected particles, fps
L = length of chamber, ft
H = height of chamber, ft
V = superficial velocity of gases through chamber, fps

The **superficial velocity** of gases through the chamber is determined from measurements of the volumetric flow of gases entering and exiting the chamber and the cross-sectional area of the chamber. This average velocity must be low enough to prevent reentrainment of the deposited dust; a superficial velocity below 60 fpm is satisfactory for many materials.

Table 3 Measures of Performance for Gas-Cleaning Equipment

Type of Particle Collector	Particle Diameter,[a] μm	Max. Loading, gr/ft³	Collection Efficiency, % by mass	Pressure Loss: Gas, in. of water	Pressure Loss: Liquid, psi	Utilities per 1000 cfm (gas)	Comparative Energy Requirement	Superficial Velocity,[b] fpm	Capacity Limits, 1000 cfm	Space Required (Relative)
Dry inertial collectors										
Settling chamber	>40	>5	50	0.1 to 0.5	—	—	1	300 to 600	None	Large
Baffle chamber	>20	>5	50	0.5 to 1.5	—	—	1.5	1000 to 2000	None	Medium
Skimming chamber	>20	>1	70	<1.0	—	—	3.0	2000 to 4000	50	Small
Louver	>10	>1	80	0.3 to 2.0	—	—	1.5 to 6.0	2000 to 4000	30	Medium
Cyclone	>15	>1	85	0.5 to 3.0	—	—	1.5 to 9.0	2000 to 4000	50	Medium
Multicyclone	>5	>1	95	2.0 to 10.0	—	—	6.0 to 20	2000 to 4000	200	Small
Impingement	>10	>1	90	1.0 to 2.0	—	—	3.0 to 6.0	2000 to 4000	None	Small
Dynamic	>10	>1	90	Provides pressure	—	1.0 to 2.0 hp	10 to 20	—	50	—
Electrostatic precipitators										
High-voltage	>0.01	>0.1	99	0.2 to 1.0	—	0.1 to 0.6 kW	0.8 to 20	60 to 400	10 to 2000	Large
Low-voltage	>0.001	0.5	90 to 99	0.2 to 0.5	—	0.03 to 0.06 kW	0.5 to 1.0	200 to 700	0.1 to 100	Medium
Fabric filters										
Baghouses	>0.08	>0.5	99	2.0 to 6.0	—	—	6.0 to 20	1.0 to 20	200	Large
Cartridge filters	>0.05	>0.1	99+	2.0 to 8.0	—			0.5 to 5	40 to 50	Medium
Wet scrubbers										
Gravity spray	>10	>1	70	0.1 to 1.0	20 to 100	0.5 to 2.0 gpm	5.0	100 to 200	100	Medium
Centrifugal	>5	>1	90	2.0 to 8.0	20 to 100	1 to 10 gpm	12 to 26	2000 to 4000	100	Medium
Impingement	>5	>1	95	2.0 to 8.0	20 to 100	1 to 5 gpm	9.0 to 31	3000 to 6000	100	Medium
Packed bed	>5	>0.1	90	0.5 to 10	5 to 30	5 to 15 gpm	4.0 to 34	100 to 300	50	Medium
Dynamic	>2	>1	95	Provides pressure	5 to 30	1 to 5 gpm 3 to 20 hp	30 to 200	3000 to 4000	50	Small
Submerged orifice	>2	>0.1	90	2.0 to 6.0	None	No pumping	9.0 to 21	3000	50	Medium
Jet	>2	>0.1	90	Provides pressure	50 to 100	50 to 100 gpm	15 to 30	2000 to 20,000	100	Small
Venturi	>0.1	>0.1	95 to 99	10 to 60	10 to 30	3 to 10 gpm	30 to 300	12,000 to 42,000	100	Small

Source: IGCI (1964). Information updated by ASHRAE Technical Committee 5.4.
[a]Minimum particle diameter for which the device is effective.
[b]Average speed of gases flowing through the equipment's collection region.

Typical data on settling can be found in Table 5. Because of air inclusions in the particle, the density of a dust particle can be substantially lower than the true density of the material from which it is made.

Inertial Collectors

Louver and Baffle Collectors. Louvers are widely used to control particles larger than about 15 μm in diameter. The louvers cause a sudden change in direction of gas flow. By virtue of their inertia, particles move away from high-velocity gases and are either collected in a hopper or trap or withdrawn in a concentrated sidestream. The sidestream is cleaned using a cyclone or high-efficiency collector, or it is simply discharged to the atmosphere. In general, the pressure drop across inertial collectors with louvers or baffles is greater than that for settling chambers, but this loss is balanced by higher collection efficiency and more compact equipment.

Inertial collectors are occasionally used to control mist. In some applications, the interior of the collector may be irrigated to prevent reentrainment of dry dust and to remove soluble deposits.

Typical louver and baffle collectors are shown in Figure 1.

Cyclones and Multicyclones. A cyclone collector transforms a gas stream into a confined vortex, from which inertia moves suspended particles to the wall of the cyclone's body. The inertial effect of turning the gas stream, as used in the baffle collector, is used continuously in a cyclone to improve collection efficiency. Cyclone collectors are often used as precleaners to reduce the loading of more efficient pollution control devices. Figure 2 shows some typical cyclone collectors.

A low-efficiency cyclone operates with a static pressure drop from 1 to 1.5 in. of water between its inlet and outlet and can remove 50% of the particles from 5 to 10 μm. High-efficiency cyclones operate with static pressure drops from 3 to 8 in. of water between their inlet and outlet and can remove 70% of the particulates of approximately 5 μm.

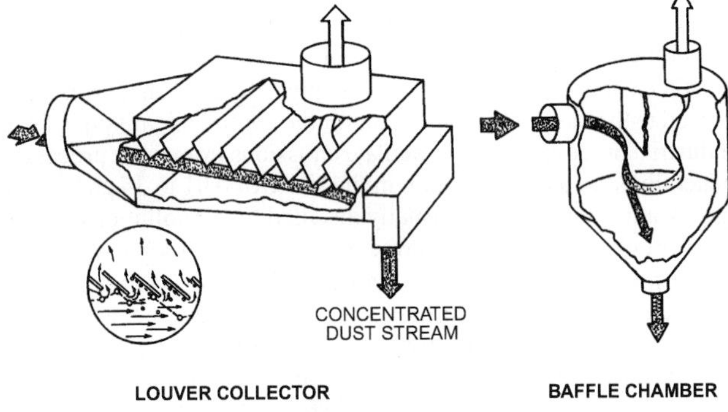

Fig. 1 Typical Louver and Baffle Collectors

The efficiency of a cyclone depends on particle density, shape, and size (aerodynamic size D_p, which is the average of the size range). Cyclone efficiency may be estimated from Figure 3.

The parameter D_{pc}, known as the **cut size**, is defined as the diameter of particles collected with 50% efficiency. The cut size may be estimated using the following equation:

$$D_{pc} = \sqrt{\frac{9\mu b}{2N_e V_i(\rho_p - \rho_g)\pi}} \tag{5}$$

where

D_{pc} = cut size, ft
μ = absolute gas viscosity, centipoise

Table 4 Collectors Used in Industry

Operation	Concentration	Particle Size	Cyclone	High-Efficiency Centrifugal	Rotating Centrifugal Mist	Wet Collectors: Medium-Pressure	Wet Collectors: High-Energy	Self-Cleaning Fabric Filter	Disposable Media Filter	Electrostatic Precipitators: High-Voltage	Electrostatic Precipitators: Low-Voltage	Notes
Ceramics												
a. Raw product handling	Light	Fine	Rare	Seldom	N/A	Frequent	N/U	Frequent	N/A	N/U	N/A	1
b. Fettling	Light	Fine to medium	Rare	Occasional	N/A	Frequent	N/U	Frequent	N/A	N/U	N/A	2
c. Refractory sizing	Heavy	Coarse	Seldom	Occasional	N/A	Frequent	Rare	Frequent	N/A	N/U	N/A	3
d. Glaze and vitreous enamel spray	Moderate	Medium	N/U	N/U	N/A	Usual	N/U	Occasional	N/A	N/A	N/A	—
e. Glass melting	Light	Fine	N/A	N/A	N/A	Occasional	N/U	Occasional	N/A	Usual	N/A	—
f. Frit smelting	Light	Fine	N/A	N/A	N/A	N/U	Often	Often	N/A	Often	N/A	—
g. Fiberglass forming and curing	Light	Fine	N/A	N/A	N/A	Occasional	N/U	N/U	Rare	Usual	N/A	—
Chemicals												
a. Material handling	Light to moderate	Fine to medium	Occasional	Frequent	N/A	Frequent	Frequent	Frequent	N/A	N/U	N/A	4
b. Crushing, grinding	Moderate to heavy	Fine to coarse	Often	Frequent	N/A	Frequent	Occasional	Frequent	N/A	N/U	N/A	5
c. Pneumatic conveying	Very heavy	Fine to coarse	Usual	Occasional	N/A	Rare	Rare	Usual	N/A	N/U	N/A	6
d. Roasters, kilns, coolers	Heavy	Medium to coarse	Occasional	Usual	N/A	Usual	Frequent	Rare	N/A	Often	N/A	7
e. Incineration	Light to medium	Fine	N/U	N/U	N/A	N/U	Frequent	Rare	N/A	Frequent	N/A	8
Coal Mining and Handling												
a. Material handling	Moderate	Medium	Rare	Occasional	N/A	Occasional	N/U	Usual	N/A	N/A	N/A	9
b. Bunker ventilation	Moderate	Fine	Occasional	Frequent	N/A	Occasional	N/U	Usual	N/A	N/A	N/A	10
c. Dedusting, air cleaning	Heavy	Medium to coarse	Occasional	Frequent	N/A	Occasional	N/U	Usual	N/A	N/A	N/A	11
d. Drying	Moderate	Fine	Rare	Occasional	N/A	Frequent	Occasional	N/U	N/A	N/A	N/A	12
Combustion Fly Ash												
a. Coal burning:												
Chain grate	Light	Fine	N/A	Rare	N/A	N/U	N/U	Frequent	N/A	N/U	N/A	13
Spreader stoker	Moderate	Fine to coarse	Rare	Rare	N/A	N/U	N/U	Frequent	N/A	Rare	N/A	14
Pulverized coal	Heavy	Fine	N/A	Frequent	N/A	N/U	N/U	Frequent	N/A	Usual	N/A	14
Fluidized bed	Moderate	Fine	Usual	—	N/A	—	—	Frequent	N/A	Frequent	N/A	—
Coal slurry	Light	—	—	—	N/A	—	—	Often	N/A	Often	N/A	—
b. Wood waste	Varied	Coarse	Usual	Usual	N/A	N/U	N/U	Occasional	N/A	Often	N/A	15
c. Municipal refuse	Light	Fine	N/U	N/U	N/A	Occasional	N/U	Usual	N/A	Frequent	N/A	—
d. Oil	Light	Fine	N/U	N/U	N/A	N/U	N/U	Usual	N/A	Often	N/A	—
e. Biomass	Moderate	Fine to coarse	N/U	N/U	N/A	Occasional	N/U	Usual	N/A	Frequent	N/A	—
Foundry												
a. Shakeout	Light to moderate	Fine	Rare	Rare	N/A	Rare	Seldom	Usual	N/A	N/U	N/A	16
b. Sand handling	Moderate	Fine to medium	Rare	Rare	N/A	Usual	N/U	Rare	N/A	N/U	N/A	17
c. Tumbling mills	Moderate	Medium to coarse	N/A	N/A	N/A	Frequent	N/U	Usual	N/A	N/U	N/A	18
d. Abrasive cleaning	Moderate to heavy	Fine to medium	N/A	Occasional	N/A	Frequent	N/U	Usual	N/A	N/U	N/A	19
Grain Elevator, Flour and Feed Mills												
a. Grain handling	Light	Medium	Usual	Occasional	N/A	Rare	N/U	Frequent	N/A	N/A	N/A	20
b. Grain drying	Light	Coarse	N/A	N/A	N/A	N/U	N/U	See Note 20	N/A	N/A	N/A	21
c. Flour dust	Moderate	Medium	Rare	Often	N/A	Occasional	N/U	Usual	N/A	N/A	N/A	22
d. Feed mill	Moderate	Medium	Often	Often	N/A	Occasional	N/U	Frequent	N/A	N/A	N/A	23
Metal Melting												
a. Steel blast furnace	Heavy	Varied	Frequent	Rare	N/A	Frequent	Frequent	N/U	N/A	Frequent	N/A	24
b. Steel open hearth, basic oxygen furnace	Moderate	Fine to coarse	N/A	N/A	N/A	N/A	Often	Rare	N/A	Frequent	N/A	25
c. Steel electric furnace	Light	Fine	N/A	N/A	N/A	N/A	Occasional	Usual	N/A	Rare	N/A	26
d. Ferrous cupola	Moderate	Varied	N/A	N/A	N/A	Frequent	Often	Frequent	N/A	Occasional	N/A	27
e. Nonferrous reverberatory furnace	Varied	Fine	N/A	N/A	N/A	Rare	Occasional	Usual	N/A	N/U	N/A	28
f. Nonferrous crucible	Light	Fine	N/A	N/A	N/A	Rare	Rare	Occasional	N/A	N/U	N/A	29

Table 4 Collectors Used in Industry (*Continued*)

Operation	Concentration	Particle Size	Cyclone	High-Efficiency Centrifugal	Rotating Centrifugal Mist	Wet Collectors: Medium-Pressure	Wet Collectors: High-Energy	Self-Cleaning Fabric Filter	Disposable Media Filter	Electrostatic Precipitators: High-Voltage	Electrostatic Precipitators: Low-Voltage	Notes
Metal Mining and Rock Products												
a. Material handling	Moderate	Fine to medium	Rare	Occasional	N/A	Usual	N/U	Considerable	N/A	N/U	N/A	30
b. Dryers, kilns	Moderate	Medium to coarse	Frequent	Occasional	N/A	Frequent	Occasional	N/U	N/A	Occasional	N/A	31
c. Cement rock dryer	Moderate	Fine to medium	N/A	Frequent	N/A	Occasional	Rare	N/U	N/A	Occasional	N/A	30
d. Cement kiln	Heavy	Fine to medium	N/A	Frequent	N/A	Rare	N/U	Usual	N/A	Usual	N/A	32
e. Cement grinding	Moderate	Fine	N/A	Rare	N/A	N/U	N/U	Usual	N/A	Rare	N/A	33
f. Cement clinker cooler	Moderate	Coarse	N/A	Occasional	N/A	N/U	N/U	Occasional	N/A	N/U	N/A	34
Metal Working												
a. Production grinding, scratch brushing, abrasive cutoff	Light	Coarse	Occasional	Frequent	N/A	Considerable	N/U	Considerable	N/A	N/U	N/A	35
b. Portable and swing frame	Light	Medium	Rare	Frequent	N/A	Frequent	N/U	Considerable	N/A	N/U	N/A	—
c. Buffing	Light	Varied	Frequent	Rare	N/A	Frequent	N/U	Rare	N/A	N/U	N/A	36
d. Tool room	Light	Fine	Frequent	Frequent	N/A	Frequent	N/U	Frequent	N/A	N/U	N/A	37
e. Cast-iron machining	Moderate	Varied	Rare	Frequent	N/A	Considerable	N/U	Considerable	N/A	N/U	N/A	38
f. Steel, brass, aluminum machining	Light to moderate	Submicron smoke, med. mist to solids	N/A	N/A	Frequent	Occasional	N/U	Occasional	Frequent	N/U	Frequent	39
g. Welding	Light to moderate	Submicron fume to med.	N/A	N/A	N/A	Occasional	N/U	Frequent	Frequent	Rare	Occasional	40
h. Plasma and laser cutting	Moderate		N/A	N/A	N/A	Occasional	N/U	Frequent	Rare	N/A	N/U	41
i. Laser welding	Moderate		N/A	N/A	N/A	Occasional	N/U	Frequent	Rare	N/A	N/U	41
j. Abrasive machining	Moderate to heavy	Fine to submicron	N/A	N/U	Occasional	Occasional	N/U	Rare	Frequent	N/A	Rare	39
k. Milling, turning, cutting tools	Light to moderate	Fine to submicron	N/A	N/U	Frequent	Occasional	N/U	N/A	Frequent	N/A	Frequent	—
l. Annealing, heat treating, induction heating, quenching	Moderate to heavy	Submicron	N/A	N/U	N/A	Rare	Rare	N/A	Rare	N/A	Frequent	—
Pharmaceutical and Food Products												
a. Mixers, grinders, weighting, blending, bagging, packaging	Light	Medium	Rare	Frequent	N/A	Frequent	N/U	Frequent	Occasional	N/U	N/U	42
b. Coating pans	Varied	Fine to medium	Rare	Rare	N/A	Frequent	N/U	Frequent	Rare	N/U	N/U	43
Plastics												
a. Raw material processing	(See comments under Chemicals)											44
b. Plastic finishing	Light to moderate	Varied	Frequent	Frequent	N/A	Frequent	N/U	Frequent	Rare	N/U	N/U	45
c. Molding, extruding, curing	Light to moderate	Submicron smoke	N/A	N/A	N/A	Rare	N/U	N/A	Occasional	N/U	Considerable	46
Pulp and Paper												
a. Recovery boilers:												
Direct contact	Heavy	Medium	N/U	N/U	N/A	N/U	N/U	Occasional	N/A	Usual	N/A	—
Low odor	Heavy	Medium	N/U	N/U	N/A	N/U	N/U	Occasional	N/A	Usual	N/A	—
b. Lime kilns	Heavy	Coarse	N/U	N/U	N/A	N/U	N/U	Often	N/A	Often	N/A	—
c. Wood-chip dryers	Varied	Fine to coarse	N/U	N/U	N/A	N/U	N/U	Occasional	N/A	Often	N/A	—
Rubber Products												
a. Mixers	Moderate	Fine	N/A	N/A	N/A	Frequent	N/U	Usual	Rare	N/U	N/U	47
b. Batch-out rolls	Light	Fine	N/A	N/A	N/A	Usual	N/U	Frequent	N/A	N/U	Rare	48
c. Talc dusting and dedusting	Moderate	Medium	N/A	N/A	N/A	Frequent	N/U	Usual	Rare	N/U	N/U	49
d. Grinding	Moderate	Coarse	Often	Often	N/A	Frequent	N/U	Often	Rare	N/U	N/U	50
e. Molding, extruding, curing	Light to moderate	Submicron smoke	N/A	N/A	N/A	Rare	N/U	N/A	Occasional	N/A	Considerable	46
Wood Particle Board and Hard Board												
a. Particle dryers	Moderate	Fine to coarse	Usual	Occasional	N/A	Frequent	Occasional	Rare	N/A	Occasional	Rare	51
Woodworking												
a. Woodworking machines	Moderate	Varied	Usual	Occasional	N/A	Rare	N/U	Frequent	N/A	N/U	N/A	52
b. Sanding	Moderate	Fine	Frequent	Occasional	N/A	Occasional	N/U	Frequent	Rare	N/U	N/A	53
c. Waste conveying, hogs	Heavy	Varied	Usual	Rare	N/A	Occasional	N/U	Occasional	N/A	N/U	N/A	54

Source: Kane and Alden (1982). Information updated by ASHRAE Technical Committee 5.4.

Notes for Table 4

Definitions
N/A: Not applicable because of inefficiency or process incompatibility.
N/U: Not widely used.
Particle size
Fine: 50% in 0.5 to 7 μm diameter range
Medium: 50% in 7 to 15 μm diameter range
Coarse: 50% over 15 μm diameter range
Concentration of particulate matter entering collector (loading)
Light: <2 gr/ft^3
Moderate: 2 to 5 gr/ft^3
Heavy: >5 gr/ft^3

[1] Dust released from bin filling, conveying, weighing, mixing, pressing, forming. Refractory products, dry pan, and screening operations more severe.
[2] Operations found in vitreous enameling, wall and floor tile, pottery.
[3] Grinding wheel or abrasive cutoff operation. Dust abrasive.
[4] Operations include conveying, elevating, mixing, screening, weighing, packaging. Category covers so many different materials that recommendation will vary widely.
[5] Cyclone and high-efficiency centrifugal collectors often act as primary collectors, followed by fabric filters or wet collectors.
[6] Usual setup uses cyclone as product collector followed by fabric filter for high overall collection efficiency.
[7] Dust concentration determines need for dry centrifugal collector; plant location, product value determine need for final collectors. High temperatures are usual, and corrosive gases not unusual. Liquid smoke emissions may be controlled by condensing precipitator systems using low-voltage, two-stage electrostatic precipitators.
[8] Ionizing wet scrubbers are widely used.
[9] Conveying, screening, crushing, unloading.
[10] Remote from other dust-producing points. Separate collector generally used.
[11] Heavy loading suggests final high-efficiency collector for all except very remote locations.
[12] Loadings and particle sizes vary with different drying methods.
[13] Boiler blowdown discharge is regulated, generally for temperature and, in some places, for pH limits; check local environmental codes on sanitary discharge.
[14] Collection for particulate or sulfur control usually requires a scrubber (dry or wet) and a fabric filter or electrostatic precipitator.
[15] Public nuisance from settled wood char indicates collectors are needed.
[16] Hot gases and steam usually involved.
[17] Steam from hot sand, adhesive clay bond involved.
[18] Concentration very heavy at start of cycle.
[19] Heaviest load from airless blasting because of high cleaning speed. Abrasive shattering greater with sand than with grit or shot. Amounts removed greater with sand castings, less with forging scale removal, least when welding scale is removed.
[20] Operations such as car unloading, conveying, weighing, storing.
[21] Special filters are successful.
[22] In addition to grain handling, cleaning rolls, sifters, purifiers, conveyors, as well as storing, packaging operations are involved.
[23] In addition to grain handling, bins, hammer mills, mixers, feeders, conveyors, bagging operations need control.
[24] Primary dry trap and wet scrubbing usual. Electrostatic precipitators are added where maximum cleaning is required.
[25] Air pollution control is expensive for open hearth, accelerating the use of substitute melting equipment, such as basic oxygen process and electric-arc furnace.
[26] Fabric filters have found extensive application for this air pollution control problem.
[27] Cupola control varies with plant size, location, melt rate, and air pollution emission regulations.
[28] Corrosive gases can be a problem, especially in secondary aluminum.
[29] Zinc oxide plume can be troublesome in certain plant locations.
[30] Crushing screening, conveying, storing involved. Wet ores often introduce water vapor in exhaust airstream.
[31] Dry centrifugal collectors are used as primary collectors, followed by a final cleaner.
[32] Collectors usually permit salvage of material and also reduce nuisance from settled dust in plant area.
[33] Salvage value of collected material is high. Same equipment used on raw grinding before calcining.
[34] Coarse abrasive particles readily removed in primary collector types.
[35] Roof discoloration, deposition on autos can occur with cyclones and, less frequently, with dry centrifugal. Heavy-duty air filter sometimes used as final cleaner.
[36] Linty particles and sticky buffing compounds can cause trouble in high-efficiency centrifugals and fabric filters. Fire hazard is also often present.
[37] Unit collectors extensively used, especially for isolated machine tools.
[38] Dust ranges from chips to fine floats, including graphitic carbon.
[39] Coolant mist and thermal smoke, often with solid swarf particulate entrained.
[40] Submicron smoke. Arc welding creates mostly dry metal oxide particulate, sometimes with liquid oil smoke. Resistance welding usually creates only liquid oil smoke, unless done at extremely high currents that vaporize some of the metal being welded.
[41] Plasma and laser cutting and welding of clean metals usually creates dry submicron smoke, but oily work pieces frequently generate a sticky mix of liquid and solid submicron smoke or fume.
[42] Materials involved vary widely. Collector selection may depend on salvage value, toxicity, sanitation yardsticks.
[43] Controlled temperature and humidity of supply air to coating pans makes recirculation from coating pans desirable.
[44] Manufacture of plastic compounds involves operations allied to many in chemical field and vanes with the basic process used.
[45] Operations are similar to woodworking, and collector selection involves similar considerations.
[46] Submicron liquid smoke is frequently emitted when plastic and rubber products are heated.
[47] Concentration is heavy during feed operation. Carbon black and other fine additions make collection and dust-free disposal difficult.
[48] Often, no collection equipment is used where dispersion from exhaust stack is good and stack location is favorable.
[49] Salvage of collected material often dictates type of high-efficiency collector.
[50] Fire hazard from some operations must be considered.
[51] Granular-bed filters, at times electrostatically augmented, have occasionally been used in this application.
[52] Bulky material. Storage for collected material is considerable; bridging from splinters and chips can be a problem.
[53] Production sanding produces heavy concentrations of particles too fine to be effectively captured by cyclones or dry centrifugal collectors.
[54] Primary collector invariably indicated with concentration and partial size range involved; when used, wet or fabric collectors are used as final collectors.

Table 5 Terminal Settling Velocities of Particles, fps

Particle Density Relative to Water	Particle or Aggregate Diameter, μm									
	1	2	5	10	20	50	100	200	500	1000
0.05	5.9 E–6	2.3 E–5	1.3 E–4	5.2 E–4	2.3 E–3	1.3 E–3	5.2 E–2	0.18	0.75	1.7
0.1	1.2 E–5	4.6 E–5	2.6 E–4	1.0 E–3	4.6 E–3	2.6 E–3	9.8 E–2	0.36	1.3	2.7
0.2	2.4 E–5	9.2 E–5	5.2 E–4	2.1 E–3	9.2 E–3	5.2 E–2	0.19	0.62	2.1	4.3
0.5	5.9 E–5	2.3 E–4	1.3 E–3	5.2 E–3	2.3 E–2	0.13	0.46	1.4	4.0	8.2
1.0	1.2 E–4	4.6 E–4	2.6 E–3	1.0 E–2	3.9 E–2	0.25	0.82	2.3	6.4	12.8
2.0	2.4 E–4	9.2 E–4	5.2 E–3	2.1 E–2	8.2 E–2	0.46	1.5	3.7	10.2	20.5
5.0	5.9 E–4	2.3 E–3	1.3 E–2	4.9 E–2	0.21	1.1	3.2	7.3	18.9	36.1
10.0	1.2 E–3	4.6 E–3	2.6 E–2	0.10	0.39	2.0	5.4	11.5	29.2	56.8

Source: Billings and Wilder (1970). *Note*: E–6 = $\times 10^{-6}$, etc.

b = cyclone inlet width, ft
N_e = effective number of turns within cyclone; approximately 5 for a high-efficiency cyclone and may be from 0.5 to 10 for other cyclones
V_i = inlet gas velocity, fpm
ρ_p = density of particle material, lb/ft^3
ρ_g = density of gas, lb/ft^3

At inlet gas velocity above 4800 fpm, internal turbulence limits improvements in the efficiency of a given cyclone. The pressure drop through a cyclone is proportional to the inlet velocity pressure and hence the square of the volumetric flow.

ELECTROSTATIC PRECIPITATORS

Electrostatic precipitators use the forces acting on charged particles passing through an electric field to separate those particles from the airstream in which they were suspended. In every precipitator, three distinct functions must be accomplished:

1. **Ionization:** Charging contaminant particles
2. **Collection:** Subjecting particles to a precipitating force that moves them toward collecting electrodes
3. **Collector cleaning:** Removal of collected contaminant from precipitator

Units in which ionization and collection are accomplished simultaneously in a single structure are called **single-stage precipitators** (Figure 4). They have widely spaced electrodes (3.15 to 6 in.) and typically operate with high voltages (20 to 60 kV) but relatively low (rarely as high as 11 kV/in.) field gradients.

In **two-stage precipitators**, ionization and collection are performed independently in discrete charging and precipitating structures (Figure 5). Because their ionizing and collecting electrodes are closely spaced (0.7 to 1.5 in.), two-stage precipitators normally operate with high field gradients (usually more than 10 kV/in.) but low voltages (usually 10 kV or less, and never more than 14 kV) (White 1963).

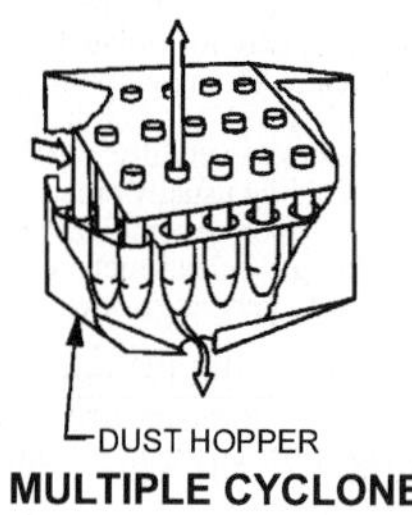

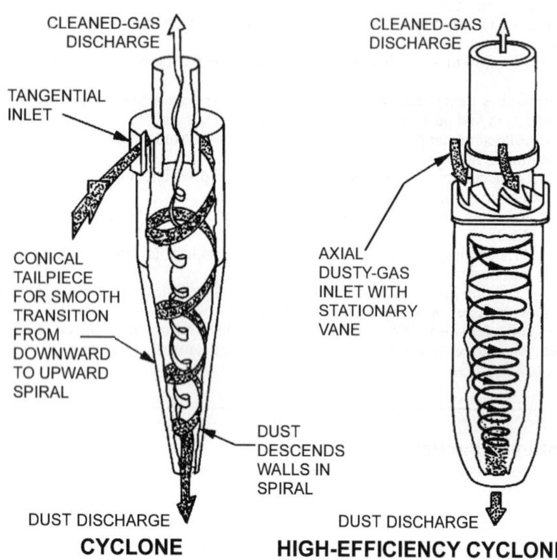

Fig. 2 Typical Cyclone Collectors

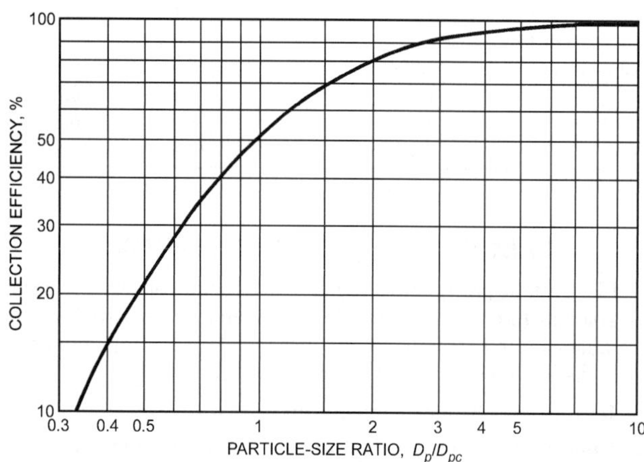

Fig. 3 Cyclone Efficiency
(Lapple 1951)

Because of fundamental differences in their ionization processes and practical differences in the way they are usually constructed, high-voltage (single-stage) and low-voltage (two-stage) precipitators are suited for entirely different air-cleaning requirements.

Single-Stage Designs

Figure 6 shows several types of single-stage precipitators. The charging electrodes are located between parallel collecting plates. The gas flows horizontally through the precipitators. High-voltage precipitators collect larger particles better than small particles (they are less efficient at collecting contaminants smaller than 1 to 2 μm). Their precipitation efficiency depends in part on the relative electrical resistivity of the pollutant being collected; most are less efficient when collecting either conductive or highly dielectric contaminants.

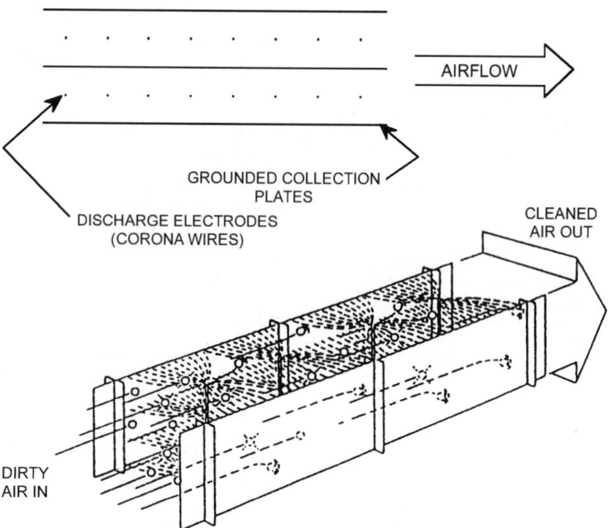

Fig. 4 Typical Single-Stage Electrostatic Precipitator

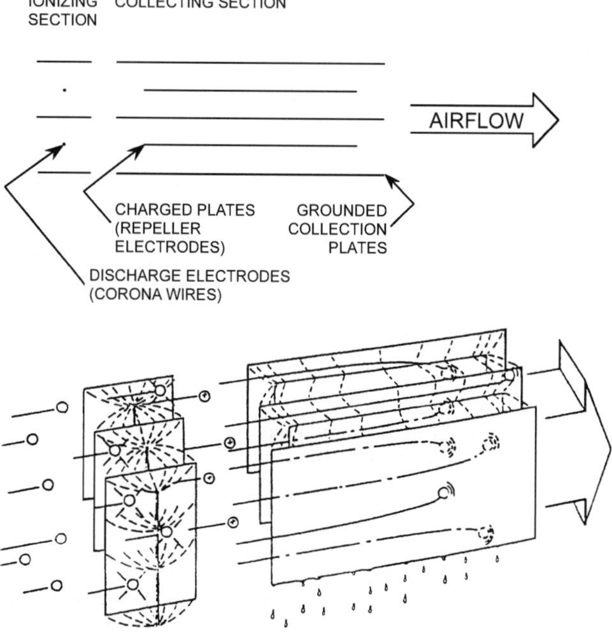

Fig. 5 Typical Two-Stage Electrostatic Precipitators

Single-stage high-voltage precipitators can easily handle heavy loadings of dry dust. Most are configured to operate continuously (using online vibratory or shaker cleaning). They can continuously collect large quantities (hundreds of pounds per hour) of airborne materials such as foundry shakeout, cement, ceramics, chemical dusts, fly ash, blast furnace dust and fumes, and paper mill recovery boiler emissions.

Although they are rarely used to clean exhaust airflows much smaller than 50,000 cfm, single-stage precipitators can be constructed to handle airflows as large as 2,000,000 cfm. Gas velocity through the electrostatic field is ordinarily 60 to 400 fpm, with treatment time in the range of 2 to 10 s. Only special-purpose single-stage "wet" precipitators are configured for collection of liquid contaminants.

Single-stage industrial electrostatic precipitators are distinguished from low-voltage two-stage precipitator designs by several attributes:

1. Separation between the electrodes is larger to secure acceptable collection and electrode cleaning under the high dust loading and hostile conditions of industrial gas streams.
2. Construction is heavy-duty for operation to 850°F and +30 in. of water.
3. They are generally used for exhaust applications where ozone generation is not of concern. Consequently, they may operate with negative ionization, the polarity that gives the maximum electric field strength between the electrodes.
4. They are normally custom-engineered and assembled on location for a particular application.

Single-stage precipitators can be designed to operate at collection efficiencies above 99.9% for closely specified conditions. Properties of the dust, such as particle size distribution and a deposit's electrical resistivity, can affect performance significantly, as can variations in gas composition and flow rate.

Two-Stage Designs

Two-stage precipitators are manufactured in two general forms: (1) electronic air cleaners, which are commonly used for light dust loadings in commercial/residential ventilation and air-conditioning service (see Chapter 28); and (2) heavy-duty industrial electrostatic precipitators designed primarily to handle the heavy loadings of submicron liquid particulates that are emitted from hot industrial processes and machining operations. Two-stage industrial precipitators often include sumps and drainage provisions that encourage continuous gravity runoff of collected liquids (Beck 1975; Shabsin 1985).

Two-stage precipitators are frequently used in industry to collect submicron pollutants that would be difficult (or impossible) to collect in other types of equipment. Two-stage precipitators are frequently sized to handle less than 1000 cfm of air and rarely built to handle more than 50,000 to 75,000 cfm of contaminated air in one unit. Gas velocity through the collecting fields usually ranges from 200 to 700 fpm, with treatment times per pass of 0.015 to 0.05 s in the ionizing fields and 0.06 to 0.25 s in the collecting fields.

Because gravity drainage of precipitated liquid smoke or fog is a dependable collector-cleaning mechanism, low-voltage two-stage precipitators are most often recommended to collect liquefiable pollutants in which few (or no) solids are entrained.

Since the early 1970s, an important application for low-voltage two-stage electrostatic precipitators has been as the gas-cleaning component of **condensing precipitator systems**. Design of these air pollution control systems (and of similar **condensing filtration systems**) is based on the following principle: although the hot gases or fumes emitted by many processes are not easily filtered or precipitated (because they are in the vapor phase), the condensation aerosol fogs or smokes that form as those vaporized pollutants cool can be efficiently collected by filtration or precipitation (Figure 7). Many condensation aerosol smokes consist of submicron liquid droplets, making them a good match for the collecting capability of low-voltage two-stage precipitators (Rossnagel 1973; Sauerland 1976; Thiel 1977).

Low-voltage precipitators can be very effective at collecting the aerosol particles smaller than 1 μm (down to 0.001 to 0.01 μm) that are responsible for most plume opacity and for virtually all blue smoke (blue-tinted) emissions. Condensing precipitator systems may therefore be a good choice for eliminating the residual opacity of blue smoke formed by condensation aerosol plumes from hot processes, dryers, ovens, furnaces, or other exhaust air cleaning devices (Beltran 1972; Beltran and Surati 1976; Thiel 1977).

When condensation aerosol pollutants are odorous in character, precipitation of the submicron droplets of odorant can prevent the long-distance drift of odorous materials, possibly eliminating neighborhood complaints that are associated with submicron particulate

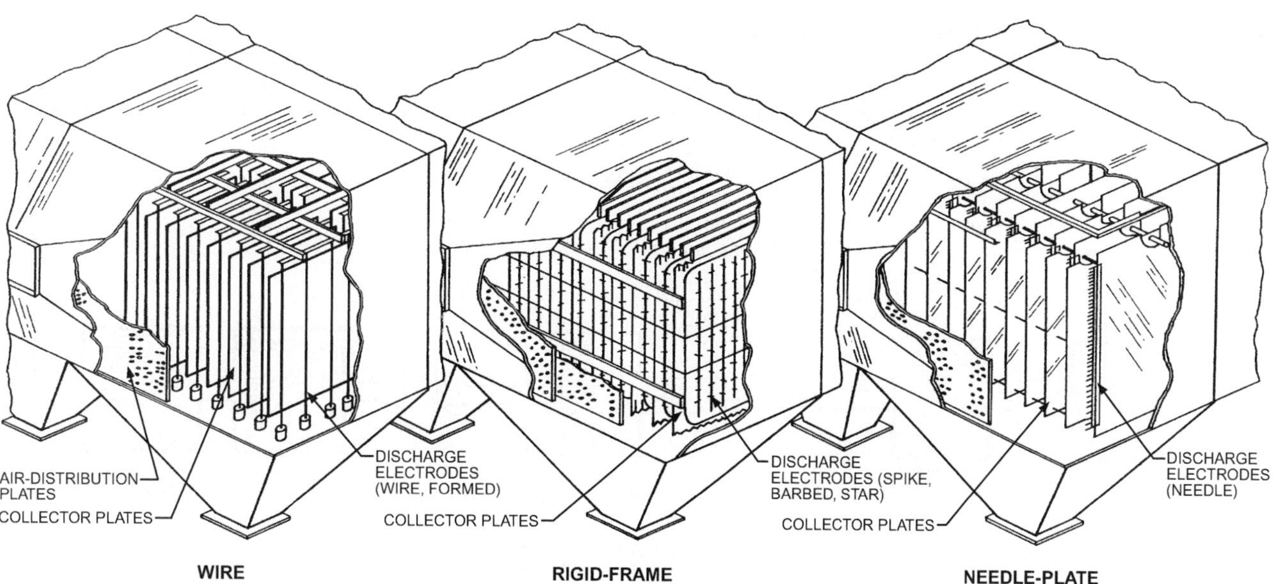

Fig. 6 Typical Single-Stage Precipitators

smokes. Odors that have been successfully controlled by precipitation include asphalt fumes, food frying smokes, meat smokehouse smoke, plasticizer smokes, rubber curing smoke, tar, and textile smokes (Chopyk and Larkin 1982; Thiel 1983).

FABRIC FILTERS

Fabric filters are dry dust collectors that use a stationary medium to capture the suspended dust and remove it from the gas stream. This medium, called fabric, can be composed of a wide variety of materials, including natural and synthetic cloth, glass fibers, and even paper.

Three types of industrial fabric filters are in common use: pulse jet (and, rarely, reverse air or shaker) cleaned baghouses, pulse jet cleaned cartridge filters, and disposable media filters. Although most commercially available fabric filter systems are currently one of these three types, there are a number of design variations among competing products that can greatly influence the suitability of a particular collector for a specific air-cleaning application.

Most **baghouses** (see Figures 9, 10, 12, and 13) use flexible cloth or other fabric-like filter media in the shape of long cylindrical bags. Bags in pulse jet cleaned systems are rarely more than 6.25 in. in diameter but can be up to 20 ft long. Timed pulses of compressed air flex the bags while blowing collected dust off the media surface. Continuous operation, with no downtime for filter cleaning or dust removal, is common (Figure 12).

Cartridge filters, illustrated in Figure 14, use a nearly identical compressed-air media cleaning system. However, their fabric is relatively rigid material, packaged into pleated cylindrical cartridges. Cartridges are self-supported and much easier to handle and replace than are long tubular baghouse bags. Most cartridges are 12.75 in. or less in diameter and rarely longer than 30 in.

With the pleated media construction, a large filter surface area can be packaged in a relatively small housing to reduce both cost and space required. Most cartridge filter units are not nearly as tall as baghouses having comparable capacity. Pleat depth and spacing are critically important variables that determine the suitability and useful lifetime of particular filter cartridges under the conditions of each specific application.

Both baghouse and cartridge filter systems are practical only when airborne contaminants consist almost exclusively of dry dust. The presence of any entrained liquid in the airstream usually creates a severe maintenance problem because the filter self-cleaning systems (i.e., pulse jet, reverse air, or shaker) become less effective, so the filters become plugged or "blinded" by collected material and fail after only brief operation.

Conservatively selected and carefully applied baghouse and cartridge filter systems, on the other hand, can easily provide excellent dust collection performance with a filter service life of more than 1 year. They often require very little maintenance, even when handling heavy dust loads in continuous 24 h, 7 day operation.

Disposable fabric (or **disposable-media**) filter collectors are usually simple and economical units that hold enough fabric or similar media to collect modest quantities of almost any particulate pollutant, including liquid, solid, and sticky or waxy materials, regardless of particle size (at least for particles larger than 0.5 to 1 μm). When each filter has accumulated as much material as it can practically hold, it is discarded and replaced by a new element. Both envelope bag arrays and pleated rectangular cartridge elements are popular media forms for use in disposable filter collectors.

When considering using disposable media filter collectors, serious attention must be given to safe, legal, and ethical disposal of spent and/or contaminated filter elements.

Principle of Operation

When contaminated gases pass through a fabric, particles may attach to the fibers and/or other particles and separate from the gas stream. The particles are normally captured near the inlet side of a fabric to form a deposit known as a **dust cake**. In self-cleaning designs, the dust cake is periodically removed from the fabric to prevent excessive resistance to the gas flow. Finer particles may penetrate more deeply into a fabric and, if not removed during cleaning, may blind it.

The primary mechanisms for particle collection are direct interception, inertial impaction, electrostatic attraction, and diffusion (Billings and Wilder 1970).

Direct interception occurs when the fluid streamline carrying the particle passes within one-half of a particle diameter of a fiber. Regardless of the particle's size, mass, or inertia, it will be collected if the streamline passes sufficiently close.

Inertial impaction occurs when the particle would miss the fiber if it followed the streamline, but its inertia overcomes the resistance offered by the flowing gas, and the particle continues in a direct enough course to strike the fiber.

Electrostatic attraction occurs when the particle, the filter, or both possess sufficient electrical charge to cause the particles to precipitate on the fiber.

Diffusion makes particles more likely to pass near fibers and thus be collected. Once a particle resides on a fiber, it effectively increases the size of the fiber.

Self-cleaning fabric filters have several **advantages** over other high-efficiency dust collectors such as electrostatic precipitators and wet scrubbers:

- They provide high efficiency with lower installed cost.
- The particulate matter is collected in the same state in which it was suspended in the gas stream (a significant factor if product recovery is desired).

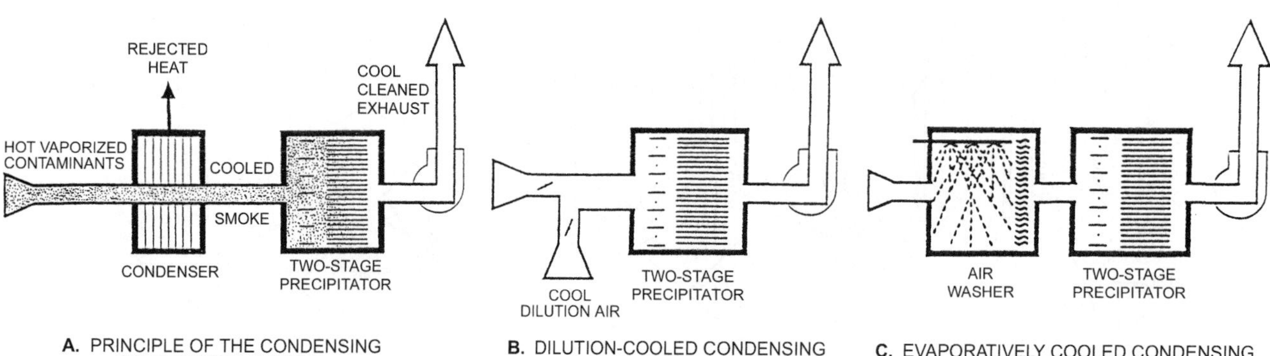

Fig. 7 Condensing Precipitator Systems for Control of Hot Organic Smokes

- Process upsets seldom result in the violation of emission standards. The mass rate of particulate matter escaping collection remains low over the life of the filter media and is insensitive to large changes in the mass-loading of dust entering the collector.

Self-cleaning fabric filter dust collectors also have some **limitations**:

- Liquid aerosols, moist and sticky materials, and condensation blind fabrics and reduce or prevent gas flow. These actions can curtail plant operation.
- Fabric life may be shortened in the presence of acidic or alkaline components in the gas stream.
- Use of fabric filters is generally limited to temperatures below 500°F.
- Should a spark or flame accidently reach the collector, fabrics can contribute to the fire/explosion hazard.
- When a large volume of gas is to be cleaned, the large number of fabric elements (bags, cartridges, or envelopes) required and the maintenance problem of detecting, locating, and replacing a damaged element should be considered. Monitoring equipment can detect leakage in an individual row of filters or an individual bag.

Pressure-Volume Relationships

The size of a fabric filter is based on empirical data on the amount of fabric media required to clean the desired volumetric flow of gas with an acceptable static pressure drop across the media. The appropriate media and conditions for its use are selected by pilot test or from experience with media in similar full-scale installations. These tests provide a recommended range of approach velocities for a specific application.

The **approach velocity** is stated in practical applications as a gas-to-media ratio or filtration velocity (i.e., the velocity at which gas flows through the filter media). The **gas-to-media ratio** is the ratio of the volumetric flow of gases through the fabric filter to the area of fabric that participates in the filtration. It is the average approach velocity of the gas to the surface of the filter media.

It is difficult to estimate the correlation between pressure drop and gas-to-media ratio for a new installation. However, when the flow entering a filter with a dust cake is laminar and uniform, the pressure drop is proportional to the approach velocity:

$$\Delta p = KV_iW = KQW/A \qquad (6)$$

where

Δp = pressure drop, in. of water
K = specific resistance coefficient, pressure drop per unit air velocity and mass of dust per unit area
V_i = approach velocity or gas-to-media ratio, fpm
W = area density of dust cake, oz/ft^2
Q = volumetric flow of gases, cfm
A = area of cloth that intercepts gases, ft^2

Equation (6) suggests that increasing the area of the fabric during initial installation has some advantage. A larger fabric area reduces both the gas-to-media ratio and the thickness of the dust cake, resulting in a decreased pressure drop across the collector and reduced cleaning requirements. A lower gas-to-media ratio generally lowers operational cost for the fabric filter system, extends the useful life of the filter elements, and reduces maintenance frequency and expense. In addition, the lower gas-to-media ratio allows for some expansion of the system and, more importantly, additional surge capacity when upset conditions such as unusually high moisture content occur.

The specific resistance coefficient K is usually higher for fine dusts. The use of a primary collector to remove the coarse fraction seldom causes a significant change in the pressure drop across the collector. In fact, the coarse dust fraction helps reduce operating pressure because it results in a more porous dust cake, which provides better dust cake release.

Figure 8 illustrates the dependence of the pressure drop on time for a single-compartment fabric filter, operating through its cleaning cycle. The volumetric flow, particulate concentration, and distribution of particle sizes are assumed to remain constant. In practice, these assumptions are not usually valid. However, the interval between cleaning events is usually long enough that these variations are insignificant for most systems; between cleaning events, the dependence of pressure drop on time is approximately linear.

The volumetric flow of gases from a process often varies in response to changes in pressure drop across the fabric filter. The degree to which this variation is significant depends on the operating point for the fan and the requirements for the process.

Electrostatic Augmentation

Electrostatic augmentation involves establishing an electric field between the fabric and another electrode, precharging the dust particles, or both. The effect of electrostatic augmentation is that the interstitial openings in the fabric material function as if they were smaller, and hence smaller particles are retained. Its principle advantage has been in the more rapid build-up of the dust layer and somewhat higher efficiency for a given pressure drop. Although tested by many, this technique has not been broadly applied.

Fabrics

Commercially available fabrics, when applied appropriately, will separate 1 μm or larger particles from a gas with an efficiency of 99.9% or better. Particle size is not the major factor influencing efficiency attained from an industrial fabric filter. Most manufacturers of fabric filters will guarantee such efficiencies on applications in which they have prior experience. Lower efficiencies are generally attributed to poor maintenance (torn fabric seams, loose connections, etc.) or the inappropriate selection of lighter/higher-permeability fabrics in an effort to reduce the cost of the collector.

Fabric specifications summarize information on such factors as cost, fiber diameter, type of weave, fabric density, tensile strength, dimensional stability, chemical resistance, finish, permeability, and abrasion resistance. Usually, comparisons are difficult, and the supplier must be relied on to select the appropriate material for the service conditions.

Table 6 summarizes experience with the exposure of fabrics to industrial atmospheres. Although higher temperatures are acceptable for short periods, reduced fabric life can be expected with continued use above the maximum temperature. The filter is often protected from high temperatures by thermostatically controlled air bleed-in or collector bypass dampers.

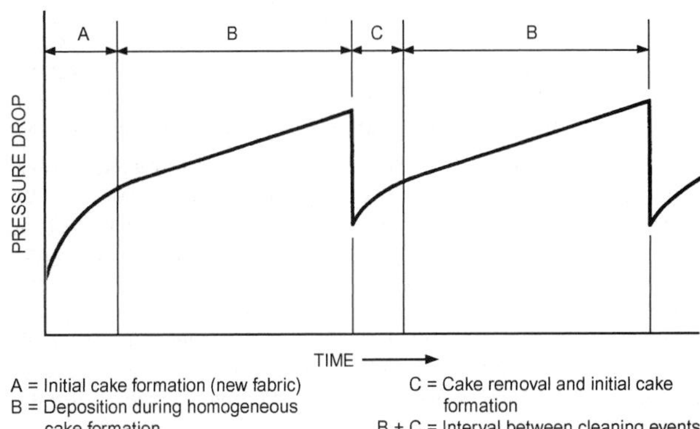

Fig. 8 Time Dependence of Pressure Drop Across Fabric Filter

When the gases are moist or the fabric must collect hygroscopic or sticky materials, **synthetic media** are recommended. They are also recommended for high-temperature gases. Polypropylene is a frequent selection. One limitation of synthetic media is greater penetration of the media during the cleaning cycle.

Woven fabrics are generally porous, and effective filtration depends on prior formation of a dust cake. New cloth collects poorly until particles bridge the openings in the cloth. Once the cake is formed, the initial layers become part of the fabric; they are not destroyed when the bulk of the collected material is dislodged during the cleaning cycle.

Cotton and wool fibers in woven media as well as most felted fabrics accumulate an initial dust cake in a few minutes. Synthetic woven fabrics may require a few hours in the same application because of the smoothness of the monofilament threads. Spun threads in the fill direction, when used, reduce the time required to build up the initial dust cake. Felted fabrics contain no straight-through openings and have a reasonably good efficiency for most particulates, even when clean. After the dust cake builds internally, as well as on the fabric's surface, shaking does not make it porous.

Cartridges manufactured from **pleated paper** and various **synthetic microfibers** (usually spun-bonded, not resin-glued, to form a stiff, nonwoven, microporous media) are also fabric filters because they operate in the same general manner as high-efficiency fabrics. As with cloth filters, the efficiency of the filter is increased by the formation of a dust cake on the medium.

Media used in cartridge type filters is usually manufactured to have many more pores (through which gas flows while being filtered) than do any of the common baghouse fabrics. Initial filtration efficiency of clean new cartridge filters is usually much greater, particularly for submicron-sized dust particles, than that of bare baghouse media (before a significant cake of filtered dust forms) because the pores in cartridge media are much smaller than those in baghouse fabrics. Despite having pores approximately one-tenth the diameter of those in the best baghouse felts, both cellulose (paper) and synthetic (most often polyester) cartridge media have so many pores that their permeability to gas flow is considerably greater than that of commonly used polyester felt baghouse media.

Because cartridge filters pack much more filter media per unit volume into their cartridges than can conventional bag filters, pulse jet cartridges filter collectors usually have less resistance to gas flow and operate with lower pressure drop from inlet to outlet than any of the three baghouse variations.

Pulse jet cleaned cartridge filter dust collectors are usually designed to operate at much lower filtration velocity (typically 1 to 3 fpm) than pulse jet cleaned tubular media baghouses. Submicron dust collection efficiency of cartridge filter media is so high that cartridge-type collectors are often used in applications where cleaned air will be directly recirculated back into the factory to reduce the expense of heating or cooling replacement (makeup) air.

Table 6 Temperature Limits and Characteristics of Fabric Filter Media

	Maximum Continuous Operating Temp., °F	Acid Resistance	Alkali Resistance	Flex Abrasion	Cost Relative to Cotton
Cotton	180	Poor	Very good	Very good	1.00
Wool	200	Good	Poor	Fair to good	2.75
Nylon[a,b]	200	Poor	Excellent	Excellent	2.50
Nomex[a,b]	400	Fair	Very good	Very good	8.00
Acrylic	260	Good	Fair	Good	3.00
Polypropylene	180	Excellent	Excellent	Very good	1.75
Polyethylene	145	Excellent	Excellent	Very good	2.00
Teflon[b]	425	Excellent	Excellent	Good	30.00
Glass fiber	500	Fair to good	Fair to good	Poor	5.00
Polyester[a]	275	Good	Good	Very good	2.50
Cellulose	180	Poor	Good	Good	—

[a]These fibers are subject to hydrolysis when they are exposed to hot, wet atmospheres.
[b]DuPont trademark.

Types of Self-Cleaning Mechanisms for Fabric Dust Collectors

The most common filter cleaning methods are (1) shaking the bags, (2) reversing the flow of gas through the bags, and (3) using an air pulse (pulse jet) to shock the dust cake and break it from the bags. Pulse jet fabric filters are usually cleaned on-line, whereas fabric filters using shakers or reverse air cleaners are usually cleaned off-line. Generally, large installations are compartmented and use off-line cleaning. Other cleaning methods include shake-deflate, a combination of shaker and reverse air cleaning, and acoustically augmented cleaning.

Shaker Collectors. When a single compartment is needed that can be cleaned off-line during shift change or breaks, shaker-type fabric filters are usually the least expensive choice. The fabric medium in a shaker-type fabric filter, whether formed into cylindrical tubes or rectangular envelopes, is mechanically agitated to remove the dust cake. Figures 9 and 10 show typical shaker-type fabric filters using bag and envelope media, respectively.

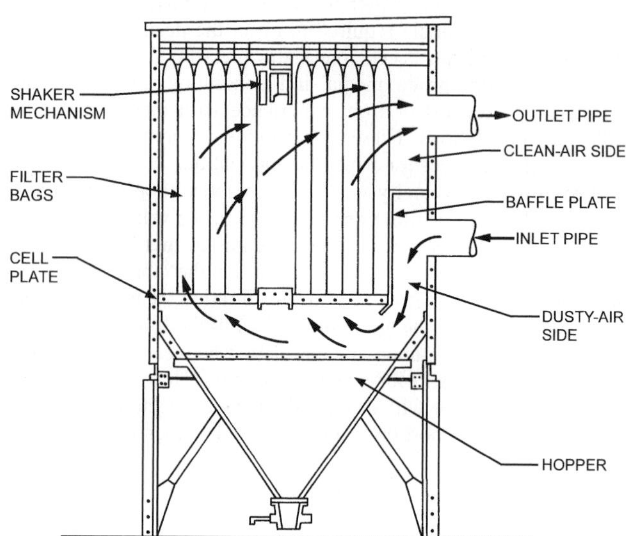

Fig. 9 Bag-Type Shaker Collector

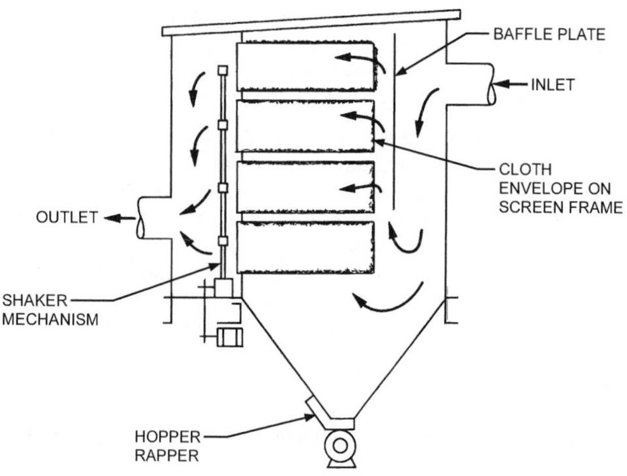

Fig. 10 Envelope-Type Shaker Collector

When the fabric filter cannot be stopped for cleaning, the collector is divided into a number of independent sections that are sequentially taken off-line for cleaning. Because it is usually difficult to maintain a good seal with dampers, relief dampers are often included. The relief dampers introduce a small volume of reverse gas to keep the gas flow at the fabric suitable for cake removal. Use of compartments, with their frequent cleaning cycles, does not permit a substantial increase in flow rates over those of a single-compartment unit cleaned periodically. The best situation for fabric reconditioning is when the system is stopped because even small particles will then fall into the hopper.

Figure 11 shows typical pressure diagrams for four- and six-compartment fabric filters that are continuously cleaned with mechanical shakers. Continuously cleaned units have compartment valves that close for the shaking cycle. This diagram is typical for a multiple-compartment fabric filter where individual compartments are cleaned off-line.

The gas-to-media ratio for a shaker-type dust collector is usually in the range of 2 to 4 fpm; it might be lower where the collector filters particles that are predominantly smaller than 2 μm. The abatement of metallurgical fumes is one example where a shaker collector is used to control particles that are less than 1 μm in size.

The bags are usually 4 to 8 in. in diameter and 10 to 20 ft in length, whereas envelope assemblies may be almost any size or shape. For ambient air applications, a woven cotton or polypropylene fabric is usually selected for shaker-type fabric filters. Synthetics are chosen for their resistance to elevated temperature and to chemicals.

Reverse-Flow Collectors. Reverse-flow cleaning is generally chosen when the volumetric flow of gases is very large. This method of cleaning inherently requires a compartmented design because the reverse flow needed to collapse the bags entrains dust that must be returned to on-line compartments of the fabric filter. Each compartment is equipped with one main shut-off valve and one reverse gas valve (whether the system is blown-through or drawn-through). A secondary blower and duct system is required to reverse the gas flow in the compartment to be cleaned. When a compartment is isolated for cleaning, the reverse gas circuit increases the volumetric flow and dust loading through the collector's active compartments.

The fabric medium is reconditioned by reversing the direction of flow through the bags, which partially collapse. The cleaning action is illustrated in Figure 12. After cleaning, the reintroduction of gas is delayed to allow dislodged dust to fall into the hopper.

Reverse-flow cleaning reduces the number of moving parts in the fabric filter system—a maintenance advantage, especially when large volumetric flows are cleaned. However, the cleaning or reconditioning is less vigorous than other methods, and the residual drag of the reconditioned fabric is higher. Reverse-flow cleaning is particularly suited for fabrics, like glass cloth, that require gentle cleaning.

Reverse-flow bags are usually 8 to 12 in. in diameter and 22 to 33 ft long and are generally operated at flow velocities in the 2 to 4 fpm range. As a consequence, reverse-air dust collectors tend to be substantially larger than pulse-jet-cleaned designs of similar capacity.

For ambient air applications, a woven cotton or polypropylene fabric is the usual selection for reverse-flow cleaning. For higher temperatures, woven polyester, glass fiber, or trademarked fabrics are often selected.

Pulse Jet Collectors. Efforts to decrease fabric filter sizes by increasing the flow rates through the fabric have concentrated on methods of implementing frequent or continuous cleaning cycles without taking major portions of the filter surface out of service. In the pulse jet design shown in Figure 13, a compressed air jet operating for a fraction of a second causes a rapid vibration or ripple in the fabric, which dislodges the accumulated dust cake. Simultaneously, outflow of both compressed cleaning air and entrained air

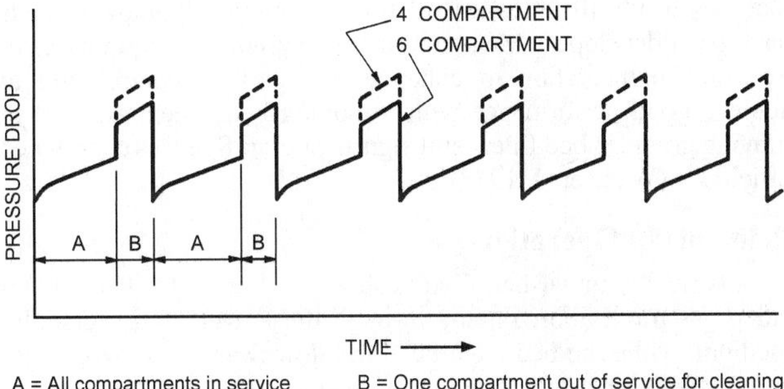

Fig. 11 Pressure Drop Across Shaker Collector Versus Time

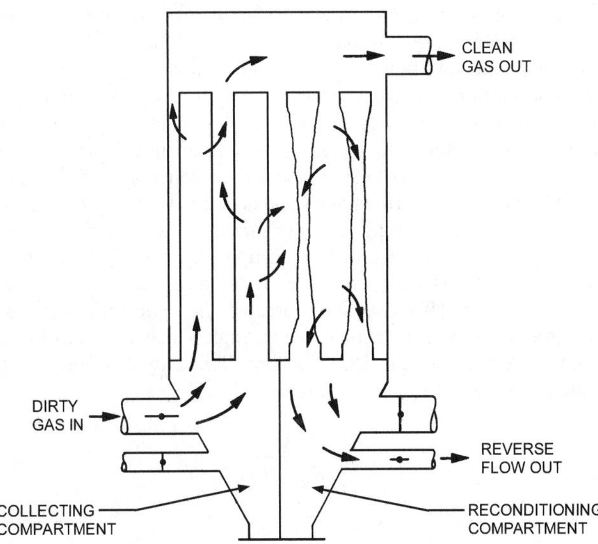

Fig. 12 Draw-Through Reverse-Flow Cleaning of Fabric Filter

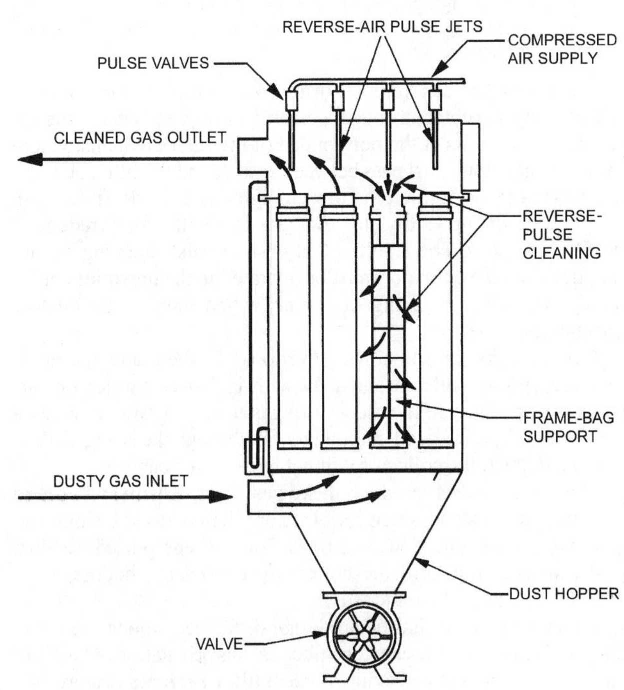

Fig. 13 Typical Pulse Jet Fabric Filter

from the top clean air plenum helps to sweep pulsed-off dust away from the filter surface. The pulse jet design is predominantly used because (1) it is easier to maintain than the reverse-jet mechanism and (2) collectors can be smaller and less costly because the greater useful cleaning energy makes operation with higher filtration velocities practical.

The tubular bags are supported by wire cages during normal operation. The reverse-flow pulse breaks up the dust layer on the outside of the bag, and the dislodged material eventually falls to a hopper. Filtration velocities for reverse jet or pulse jet designs range from 5 to 15 fpm for favorable dusts but require greater fabric area for many materials that produce a low-permeability dust cake. Felted fabrics are generally used for these designs because the jet cleaning opens the pores of woven cloth and produces excessive leakage through the filter.

Most pulse jet designs require high-pressure, dry, compressed (up to 100 psi) air, the cost of which should be considered when air-cleaning systems are designed.

When collecting light or fluffy dust of low bulk density (such as arc welding fumes, plasma or laser cutting fumes, finish sanding of wood, fine plastic dusts, etc.), serious attention must be given to the direction and velocity at which dust-laden air travels as it moves from the collector inlet to approach the filter media surface. Dusty air velocity is called **can velocity** and is defined as the actual velocity of airflow approaching the filter surfaces. Can velocity is computed by subtracting the total area occupied by filter elements (bags or cartridges, measured perpendicularly to the direction of gas flow) from the overall cross-sectional area of the collector's dusty air housing to compute the actual area through which dusty gas flows. The total gas flow being cleaned is then divided by that flow area to yield can velocity:

$$V_c = \frac{Q}{A_h - NA_f} \tag{7}$$

where

V_c = can velocity, fpm
Q = gas flow being cleaned, cfm
A_h = cross-sectional area of collector dusty gas housing, ft^2
N = number of filter bags or cartridges in collector
A_f = cross-sectional area, perpendicular to gas flow, of each filter bag or cartridge, ft^2

The maximum can velocity in **upflow** collectors (i.e., those in which dusty gas enters through a plenum or hopper beneath the filter elements) exists at the bottom end of the filter elements, where the entire gas flow must pass between and around the filters. Unless the maximum can velocity is low enough that pulsed-off dust can fall through the upwardly flowing gas, dust will simply redeposit on filter surfaces. The result is that on-line pulse cleaning cannot function, and the collector must be operated in the downtime pulse mode, with filter cleaning done only when there is no airflow through the collector.

Can velocity is sometimes overlooked when attempting to increase upflow collector capacity with improved fabrics or cartridges. Regardless of the theoretical gas-to-media ratio at which a filter operates, if released dust cannot fall through the rising airflow into the hopper, the collector will not be able to clean itself.

Collector designs in which dusty gas flows **downward** around the filters are much less susceptible to problems caused by high can velocity because the downward gas flow sweeps pulsed-off dust down toward (and into) the bottom dust discharge hopper, from which it can easily be removed.

Perhaps the most significant design difference among the many commercially available cartridge filter units is orientation of the filter cartridges. In collectors having **vertical filter surfaces** (Figure 14), all the pulsed-off dust can fall (by gravity) toward the bottom discharge flange, where it can be removed from the system (presuming that can velocity is low enough in upflow collectors). However, in designs having **horizontal or sloped filter surfaces**, nearly half of all pulsed-off dust (i.e., that from the top surfaces of the cartridges) must fall back onto that top surface. As a result, up to 50% of the total media area (i.e., nearly all media above the horizontal centerline of each filter cartridge) rapidly becomes blinded by a dust cake that is so thick that little or no gas can pass through to be filtered. This means that collectors having horizontal filter surfaces require twice as much installed filter area to achieve the same level of performance and useful filter life as collectors having vertical filter surfaces.

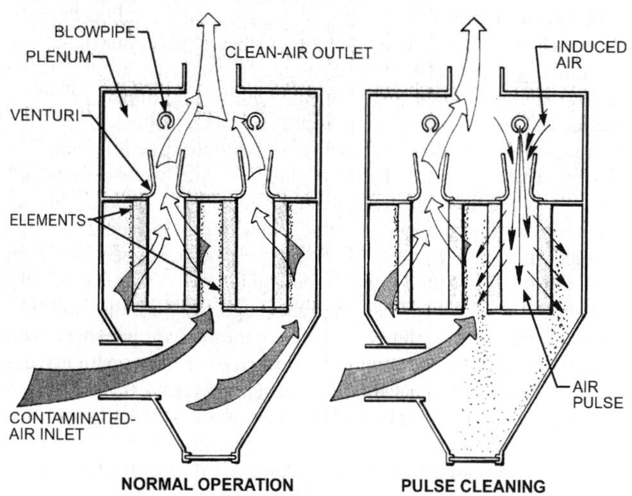

Fig. 14 Pulse Jet Cartridge Filters (Upflow Design with Vertical Filters)

This chapter can not adequately cover all the collector design variables and experience-related factors that must be considered when deciding which baghouse or cartridge self-cleaning dust collector design is best suited for each particular application. Engineers making dust collector selections are encouraged to discuss all aspects of each application in detail with all vendors being considered. It is necessary to judge

- The relative expertise of each prospective vendor
- Which dust collector design is most desirable
- How much media surface is needed in each design for each specified gas flow rate
- Which filter media is best suited to the particular application
- In the case of pulse-jet-cleaned cartridge collectors, what pleat spacing and pleat depth will give optimum or acceptable dust cake removal performance under the particular application conditions

GRANULAR-BED FILTERS

Usually, granular-bed filters use a fixed bed of granular material that is periodically cleaned off-line. Continuously moving beds have been developed. Most commercial systems incorporate electrostatic augmentation to enhance fine particle control and to achieve good performance with a moving bed. Reentrainment in moving granular-bed filters still significantly influences overall bed efficiency (Wade et al. 1978).

Principle of Operation

A typical granular-bed filter is shown in Figure 15. Particulate-laden gas travels horizontally through the louvers and a granular medium, while the bed material flows downward. The gases typically travel with a superficial velocity near 100 to 150 fpm.

The filter medium moves continuously downward by gravity to prevent a filter cake from forming on the face of the filter and to

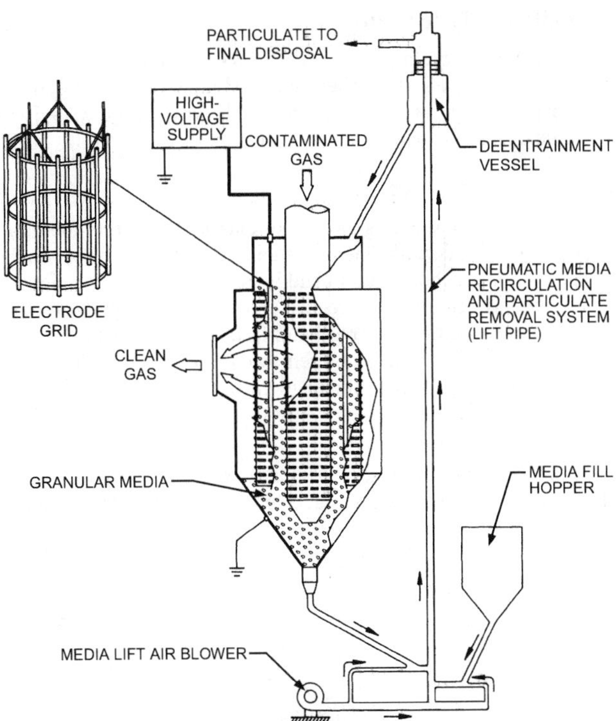

Fig. 15 Typical Granular-Bed Filter

prevent a high pressure drop. To provide complete cleaning of the louver's face, the louvers are designed so that some of the medium falls through each louver opening, thus preventing any bridging or build-up of particulate material.

Electrostatic augmentation gives the granular-bed filter many of the characteristics of a two-stage electrostatic precipitator. The obvious disadvantage of a granular-bed filter is in removal of the collected dust, which requires liquid backwash or circulation and cleaning of the filter material.

PARTICULATE SCRUBBERS (WET COLLECTORS)

Wet-type dust collectors use liquid (usually, but not necessarily, water) to capture and separate particulate matter (dust, mist, and fumes) from a gas stream. Some scrubbers operate by spraying the **scrubbing liquid** into the contaminated air. Others bubble air through the scrubbing liquid. In addition, many hybrid designs exist.

Particle sizes, which can be controlled by a wet scrubber, range from 0.3 to 50 μm or larger. Wet collectors can be classified into three categories: (1) **low-energy** (up to 1 W/cfm, 1 to 6 in. of water); (2) **medium-energy** (1 to 3 W/cfm, 6 to 18 in. of water); and (3) **high-energy** (>3 W/cfm, >18 in. of water). Typical wet-scrubber performance is summarized in Table 3.

Wet collectors may be used for the collection of most particulates from industrial process gas streams where economics allow for collection of the material in a wet state.

Advantages of wet collectors include

- Constant operating pressure
- No secondary dust sources
- Small spare parts requirement
- Ability to collect both gases and particulates
- Ability to handle high-temperature and high-humidity gas streams, as well as to reduce the possibility of fire or explosion
- Reasonably small space requirements for scrubbers
- Ability to continuously collect sticky and hygroscopic solids without becoming fouled

Disadvantages include

- High susceptibility to corrosion (corrosion-resistant construction is expensive)
- High humidity in the discharge gas stream, which may give rise to visible and sometimes objectionable fog plumes, particularly during winter
- Large pressure drops and high power requirement for most designs that can efficiently collect fine (particularly submicron) particles
- Possible difficulty or high cost of disposal of waste water or clarification waste
- Rapidly decreasing fractional efficiency for most scrubbers for particles less than 1 μm in size
- Freeze protection required in many applications in colder environments

Principle of Operation

The more important mechanisms involved in the capture and removal of particulate matter in scrubbers are inertial impaction, Brownian diffusion, and condensation.

Inertial impaction occurs when a dust particle and a liquid droplet collide, resulting in the capture of the particle. The resulting liquid/dust particle is relatively large and may be easily removed from the carrier gas stream by gravitation or impingement on separators.

Brownian diffusion occurs when the dust particles are extremely small and have motion independent of the carrier gas stream. These small particles collide with one another, making larger particles, or collide with a liquid droplet and are captured.

Condensation occurs when the gas or air is cooled below its dew point. When moisture is condensed from the gas stream, fogging occurs, and the dust particles serve as condensation nuclei. The dust particles become larger as a result of the condensed liquid, and the probability of their removal by impaction is increased.

Wet collectors perform two individual operations. The first occurs in the **contact zone**, where the dirty gas comes in contact with the liquid; the second is in the **separation zone**, where the liquid that has captured the particulate matter is removed from the gas stream. All well-designed wet collectors use one or more of the following principles:

- High liquid-to-gas ratio
- Intimate contact between the liquid and dust particles, which may be accomplished by formation of large numbers of small liquid droplets or by breaking up the gas flow into many small bubbles that are driven through a bath of scrubbing liquid, to increase the chances that contaminants will be wetted and collected
- Abrupt transition from dry to wet zones to avoid particle build-up where the dry gas enters the collectors

For a given type of wet collector, the greater the power applied to the system, the higher will be the collection efficiency (Lapple and Kamack 1955). This is the contacting power theory. Figure 16 compares the fractional efficiencies of several wet collectors, and Figure 17 shows the relationship between the pressure drop across a venturi scrubber and the abatement of particulate matter.

Spray Towers and Impingement Scrubbers

Spray towers and impingement scrubbers are available in many different arrangements. The gas stream may be subjected to a single spray or a series of sprays, or the gas may be forced to impinge on a series of irrigated baffles. Except for packed towers, these types of scrubbers are in the low-energy category; thus, they have relatively low particulate removal efficiency. A typical spray tower and an impingement scrubber are illustrated in Figures 18 and 19, respectively.

The efficiency of a spray tower can be improved by adding high-pressure sprays. A spray tower efficiency of 50 to 75% can be improved to 95 to 99% (for dust particles with size near 2 μm) by pressures in the range of 30 to 100 psig.

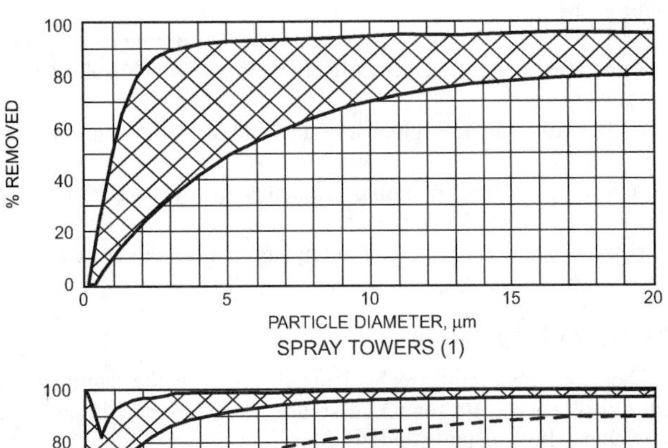

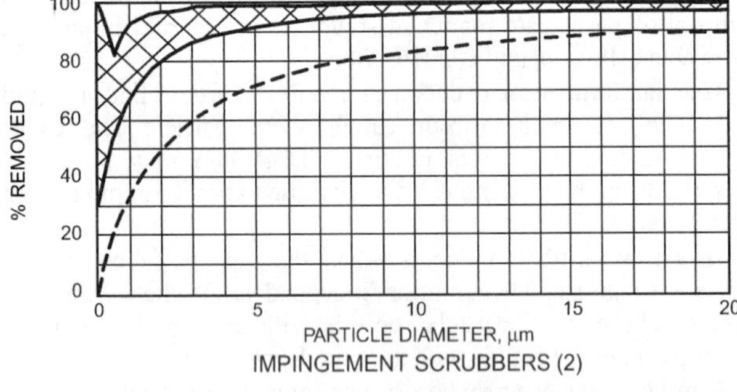

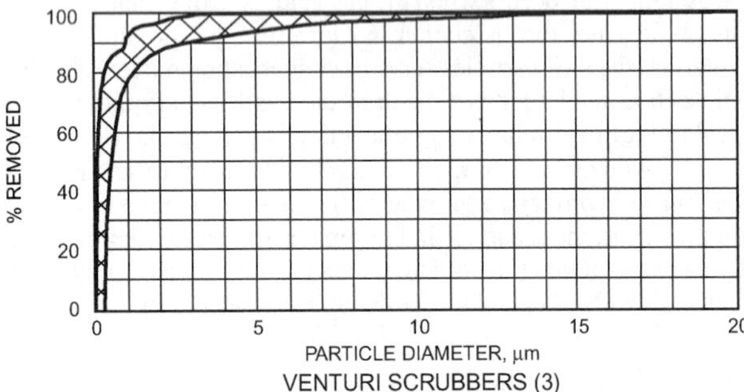

Notes:

1. Efficiency depends on liquid distribution.
2. Upper curve is for packed tower; lower curve is for orifice-type wet collector. Dashed lines indicate performance of low-efficiency, irrigated baffles or rods.
3. Efficiency is directly related to fluid rate and pressure drop.

Fig. 16 Fractional Efficiency of Several Wet Collectors

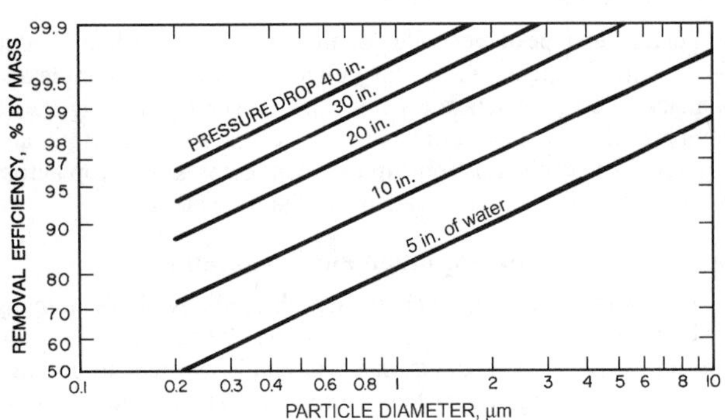

Fig. 17 Efficiency of Venturi Scrubber

Centrifugal-Type Collectors

These collectors are characterized by a tangential entry of the gas stream into the collector. They are classed with medium-energy scrubbers. The impingement scrubber shown in Figure 19 is an example of a centrifugal-type wet collector.

Orifice-Type Collectors

Orifice-type collectors are also classified in the medium-energy category. Usually, the gas stream is made to impinge on the surface of the scrubbing liquid and is forced through constrictions where the gas velocity is increased and where the liquid-gaseous-particulate interaction occurs. Water usage for orifice collectors is limited to evaporation loss and removal of collected pollutants. A typical orifice-type wet collector is illustrated in Figure 20.

Venturi Scrubber

A high-energy venturi scrubber passes the gas through a venturi-shaped orifice where the gas is accelerated to 12,000 fpm or more.

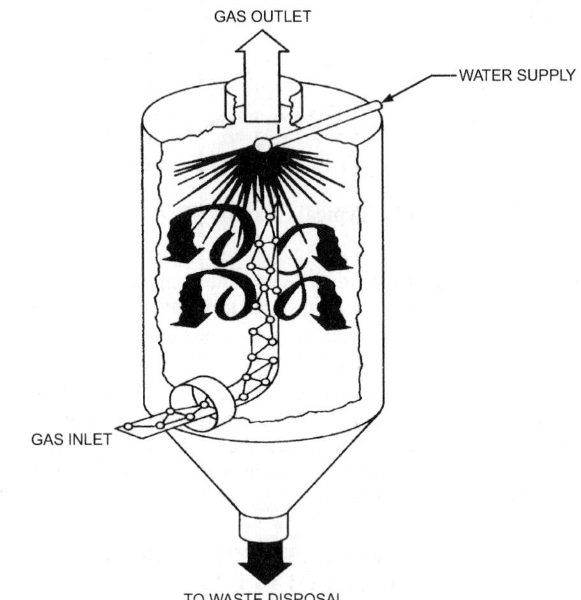

Fig. 18 Typical Spray Tower

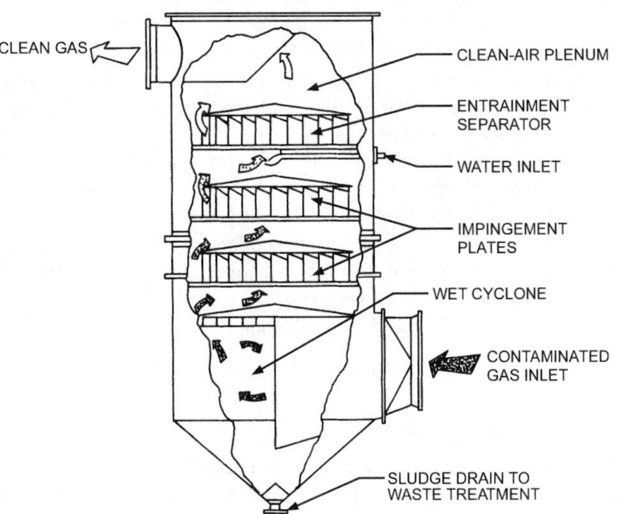

Fig. 19 Typical Impingement Scrubber

Depending on the design, the scrubbing liquid is added at, or ahead of, the throat. The rapid acceleration of the gas shears the liquid into a fine mist, increasing the chance of liquid-particle impaction. Yung has developed a mathematical model for the performance and design of venturi scrubbers (Semrau 1977). Subsequent validation experiments (Rudnick et al. 1986) demonstrated that this model yields a more representative prediction of venturi scrubber performance than other performance models do.

In typical applications, the pressure drop for gases across a venturi is higher than for other types of scrubber. Water circulation is also higher; thus, venturi systems use water reclamation systems. One example of a venturi scrubber is illustrated in Figure 21.

Electrostatically Augmented Scrubbers

Several gas-cleaning devices combine electrical charging of particulate matter with wet scrubbing. Electrostatic augmentation enhances fine particle control by causing an electrical attraction between the particles and the liquid droplets. Compared to venturi scrubbers, electrostatically augmented scrubbers remove particles smaller than 1 μm at a much lower pressure drop.

There are three generic designs for electrostatic augmentation:

1. Unipolar charged aerosols pass through a contact chamber containing randomly oriented packing elements of dielectric material. A typical electrostatically augmented scrubber of this design is shown in Figure 22.
2. Unipolar charged aerosols pass through a low-energy venturi scrubber.
3. Unipolar charged aerosols pass into a spray chamber where they are attracted to oppositely charged liquid droplets.

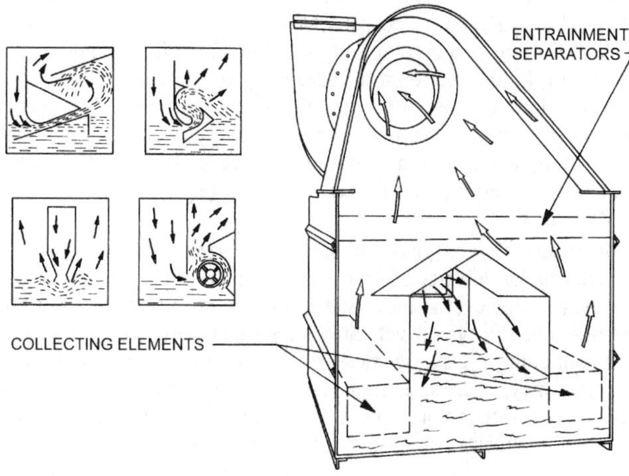

Fig. 20 Typical Orifice-Type Wet Collector

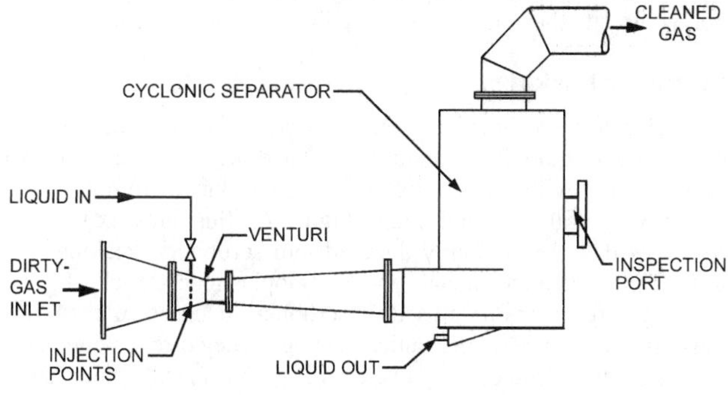

Fig. 21 Typical High-Energy Venturi Scrubber

Collection efficiencies of 50 to 90% can be achieved in a single particle charging and collection stage, depending on the mass loading of fine particles and the superficial velocity of gases in the collector. Higher collection efficiencies can be obtained by using two or more stages. Removal efficiency of gaseous pollutants depends on the mass transfer and absorption design of the scrubber section.

In most applications of electrostatically augmented scrubbers, the dirty gas stream is quenched by adiabatic cooling with liquid sprays; thus, it contains a large amount of water vapor that wets the particulate contaminants. This moisture provides a dominant influence on particle adhesion and the electrical resistivity of deposits within the collector.

Electrical equipment for particle charging is similar to that for electrostatic precipitators. The scrubber section is usually equipped with a liquid recycle pump, recycle piping, and a liquid distribution system.

GASEOUS CONTAMINANT CONTROL

Many industrial processes produce large quantities of gaseous or vaporized contaminants that must be separated from gas streams. Removal of these contaminants is usually achieved through absorption into a liquid or adsorption onto a solid medium. Incineration of the exhaust gas (see the section on Incineration of Gases and Vapors) has also been successfully used to remove organic gases and vapors. Low-vapor-pressure odorous materials that condense to form submicron condensation aerosols after being emitted from hot industrial processes can sometimes be successfully controlled by well-designed condensing filter or condensing precipitator submicron particulate collection systems (see the section on Two-Stage Designs under Electrostatic Precipitators).

SPRAY DRY SCRUBBING

Spray dry scrubbing is used to absorb and neutralize acidic gaseous contaminants in hot industrial gas streams. The system uses an alkali spray to react with the acid gases to form a salt. The process heat evaporates the liquid, resulting in a dry particulate that is removed from the gas stream.

Typical industrial applications of spray dry scrubbing are

- Control of hydrochloric acid (HCl) emissions from biological hazardous-waste incinerators

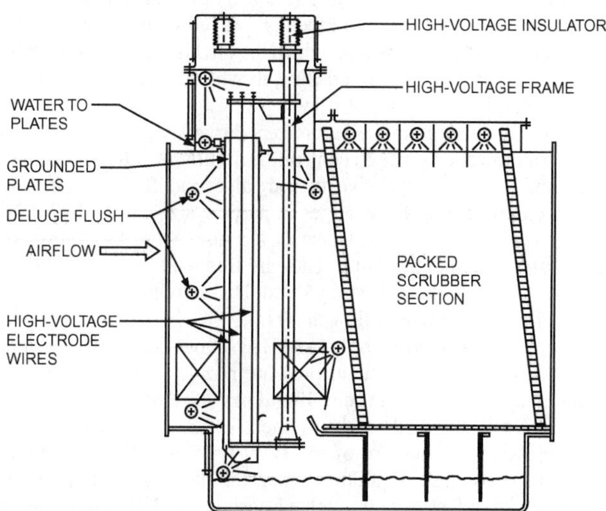

Fig. 22 Typical Electrostatically Augmented Scrubber

- Control of sulfuric acid and sulfur trioxide emissions from burning high-sulfur coal
- Control of sulfur oxides (SO_x), boric acid, and hydrogen fluoride (HF) gases from glass-melting furnaces.

Principle of Operation

Spray drying involves four operations: (1) atomization, (2) gas droplet mixing, (3) drying of liquid droplets, and (4) removal and collection of a dry product. These operations are carried out in a tower or a specially designed vessel.

In any spray dryer design, good mixing and efficient gas droplet contact are desirable. Dryer height is largely determined by the time required to dry the largest droplets produced by the atomizer. Towers used for acid gas control typically have gas residence times of about 10 s, compared to about 3 s for towers designed for evaporative cooling. The longer residence time is needed because drying by itself is not the primary goal for the equipment. Many of the acid/alkali reactions are accelerated in the liquid state. It is, therefore, desirable to cool the gases to as close as possible to the adiabatic saturation temperature (dew point) without risking condensation in downstream particulate collectors. At these low temperatures, droplets survive against evaporation much longer, thus obtaining a better chance of contacting all acidic contaminants while the alkali compounds are in their most reactive state.

Equipment

The **atomizer** must disperse a liquid containing an alkali compound that will react with acidic components of the gas stream. The liquid must be distributed uniformly within the dryer and mixed thoroughly with the hot gases in droplets of a size that will evaporate before striking a dryer surface.

In typical spray dryers used for acid gas control, the droplets have diameters ranging from 50 to 200 μm. The larger droplets are of most concern because these might survive long enough to impinge on equipment surfaces. In general, a tradeoff must be made between the largest amount of liquid that can be sprayed and the largest droplets that can be tolerated by the equipment. The angular distribution or **fan-out** of the spray is also important. In spray drying, the angle is often 60 to 80°, although both lower and higher angles are sometimes required. The fan-out may change with distance from the nozzle, especially at high pressures.

An important aspect of spray dryer design and operation is the production and control of the gas flow patterns within the drying chamber. Because of the importance of the flow patterns, spray dryers are usually classified on the basis of gas flow direction in the chamber relative to the spray. There are three basic designs: (1) **cocurrent**, in which the liquid feed is sprayed with the flow of the hot gas; (2) **countercurrent**, in which the feed is sprayed against the flow of the gas; and (3) **mixed flow**, in which there is a combined cocurrent and countercurrent flow.

There are several types of atomizers. High-speed rotating **disks** achieve atomization through centrifugal motion. Although disks are bulky and relatively expensive, they are also more flexible than nozzles in compensating for changes in particle size caused by variations in feed characteristics. Disks are also used when high-pressure feed systems are not available. They are frequently used when high volumes of liquid must be spray dried. Disks are not well suited to counterflow or horizontal flow dryers.

Nozzles are also commonly used. These may be subdivided into two distinct types: centrifugal pressure nozzles and two-fluid (or pneumatic) nozzles. In the **centrifugal pressure nozzle**, energy for atomization is supplied solely by the pressure of the feed liquid. Most pressure nozzles are of the swirl type, in which tangential inlets or slots spin the liquid in the nozzle. The pressure nozzle satisfactorily atomizes liquids with viscosities of 300 centipoise or higher. It is well suited to counterflow spray dryers and to installations requiring multiple atomizers. Capacities up to 10,000 lb/h through a single nozzle are possible. Pressure nozzles have some disadvantages. For example, pressure, capacity, and orifice size are independent, resulting in a certain degree of inflexibility. Moreover, pressure nozzles (particularly those with small passages) are susceptible to erosion in applications involving abrasive materials. In such instances, tungsten carbide or a similarly tough material is mandatory.

In **two-fluid nozzles**, air (or steam) supplies most of the energy required to atomize the liquid. Liquid, admitted under low pressure, may be mixed with the air either internally or externally. Although energy requirements for this atomizer are generally greater than for spinning disks or pressure nozzles, the two-fluid nozzle can produce very fine atomization, particularly with viscous materials.

The density and viscosity of the feed materials and how these might change at elevated temperature should be considered. Some alkali compounds do not form a solution at the concentrations needed for acid gas control. It is not uncommon that pumps and nozzles must be chosen to handle and meter slurries.

Spray dryer systems include metering valves, pumps and compressors, and controls to assure optimal chemical feed and temperature within the gas-cleaning system.

WET-PACKED SCRUBBERS

Packed scrubbers are used to remove gaseous and particulate contaminants from gas streams. Scrubbing is accomplished by impingement of particulate matter and/or by absorption of soluble gas or vapor molecules on the liquid-wetted surface of the packing. There is no limit to the amount of particulate capture, as long as the properties of the liquid film are unchanged.

Gas or vapor removal is more complex than particulate capture. The contaminant becomes a solute and has a vapor pressure above that of the scrubbing liquid. This vapor pressure typically increases with increasing concentration of the solute in the liquid and/or with increasing liquid temperature. Scrubbing of the contaminant continues as long as the partial pressure of contaminant in the gas is above its vapor pressure with respect to the liquid. The rate of contaminant removal is a function of the difference between the partial pressure and vapor pressure, as well as the rate of diffusion of the contaminant.

In most common gas-scrubbing operations, small increases in the superficial velocity of gases through the collector decrease the removal efficiency; therefore, these devices are operated at the highest possible superficial velocities consistent with acceptable contaminant control. Usually, increasing the liquid rate has little effect on efficiency; therefore, liquid flow is kept near the minimum required for satisfactory operation and removal of particulate matter. As the superficial velocity of gases in the collector increases above a certain value, there is a tendency to strip liquid from the surface of the packing and entrain the liquid from the scrubber. If the pressure drop is not the limiting factor for operation of the equipment, maximum scrubbing capacity will occur at a gas rate just below the rate that causes excessive liquid reentrainment.

Scrubber Packings

Packings are designed to present a large surface area that will wet evenly with liquid. They should also have high void ratio so that pressure drop will be low. High-efficiency packings promote turbulent mixing of the gas and liquid. Figure 23 illustrates six types of packings that are randomly dumped into scrubbers. Packings are available in ceramic, metal, and thermoplastic materials. Plastic packings are extensively used in scrubbers because of their low mass and resistance to mechanical damage. They offer a wide range of chemical resistance to acids, alkalies, and many organic compounds; however, plastic packing can be deformed by excessive temperatures or by solvent attack.

Table 7 Packing Factor *F* for Various Scrubber Packing Materials

Type of Packing	Material	Nominal Size, in. 3/4	1	1.5	2	3 or 3.5
Super Intalox	Plastic		40		21	16
Super Intalox	Ceramic		60		30	
Intalox saddles	Ceramic	145	92	52	40	22
Berl saddles	Ceramic	170	110	65	45	
Raschig rings	Ceramic	255	155	95	65	37
Hy-Pak	Metal		43	26	18	15
Pall rings	Metal		48	33	20	16
Pall rings	Plastic		52	40	24	16
Tellerettes	Plastic		36		18	16
Maspac	Plastic				32	21

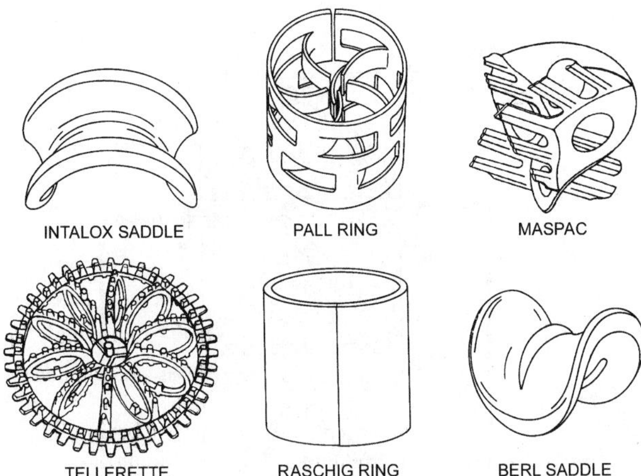

Fig. 23 Typical Packings for Scrubbers

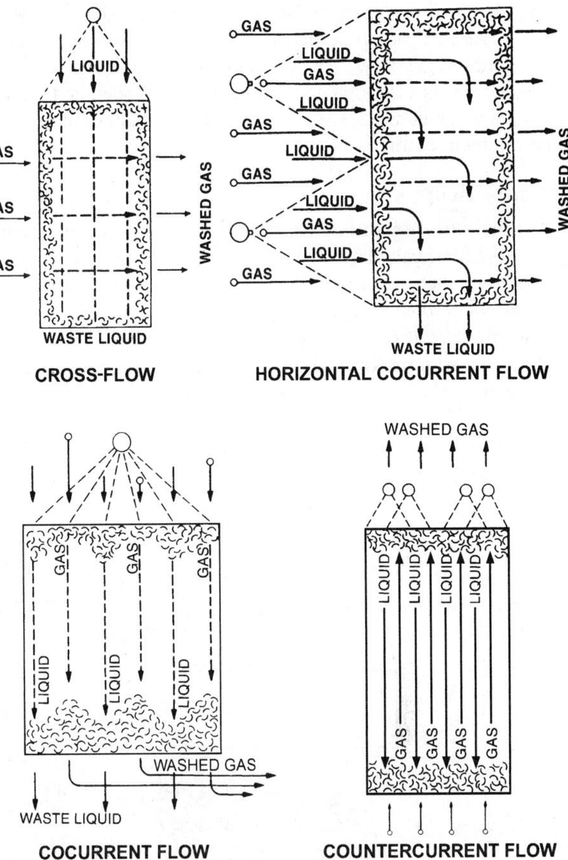

Fig. 24 Flow Arrangements Through Packed Beds

The relative capacity of tower packings at constant pressure drop can be obtained by calculation from the **packing factor *F***. The gas-handling capacity *G* of a packing is inversely proportional to the square root of *F*:

$$G = K/\sqrt{F} \qquad (8)$$

where

G = mass flow rate of gases through scrubber
F = packing factor (surface area of packing per unit volume of gently poured material)

The smaller the packing factor of a given packing, the greater will be its gas-handling capacity. Typical packing factors for scrubber packings are summarized in Table 7.

Arrangements of Packed Scrubbers

The four generic arrangements for wet-packed scrubbers are illustrated in Figure 24:

- Horizontal cocurrent scrubber
- Vertical cocurrent scrubber
- Cross-flow scrubber
- Countercurrent scrubber

Cocurrent flow scrubbers can be operated with either horizontal or vertical gas and liquid flows. A **horizontal cocurrent scrubber** depends on the gas velocity to carry the liquid into the packed bed. It operates as a wetted entrainment separator with limited gas and liquid contact time. The superficial velocity of gases in the collector is limited by liquid reentrainment to about 650 fpm. A **vertical cocurrent scrubber** can be operated at very high pressure drop (1 to 3 in. of water per foot of packing depth) because there is no flooding limit for the superficial velocity. The contact time in a cocurrent scrubber is a function of bed depth. The effectiveness of absorption processes is lower in cocurrent scrubbers than in the other arrangements because the liquid containing contaminant is in contact with the exit gas stream.

Cross-flow scrubbers use downward-flowing liquid and a horizontally moving gas stream. The effectiveness of absorption processes in cross-flow scrubbers lies between those for cocurrent and countercurrent flow scrubbers.

Countercurrent scrubbers use a downward-flowing liquid and an upward-flowing gas. The gas-handling capacity of countercurrent scrubbers is limited by pressure drop or by liquid entrainment. The contact time can be controlled by the depth of packing used. The effectiveness of absorption processes is maximized because the exiting gas is in contact with fresh scrubbing liquid.

The most broadly used arrangement is the countercurrent packed scrubber. This type of scrubber, illustrated in Figure 25, gives the best removal of gaseous contaminants while keeping liquid consumption to a minimum. The effluent liquid has the highest contaminant concentration.

Extended-surface packings have been used successfully for the absorption of highly soluble gases such as HCl because the required contact time is minimal. This type of packing consists of a woven mat of fine fibers of a plastic material that is not affected by chemical exposure. Figure 26 shows an example of a scrubber consisting of three wetted stages of extended surface packing in series with the gas flow. A final dry mat is used as an entrainment eliminator. If solids are present in the inlet gas stream, a wetted impingement stage precedes the wetted mats to prevent plugging of the woven mats.

Figure 27 shows a scrubber with a vertical arrangement of extended surface packing. This design uses three complete stages in series with the gas flow. The horizontal mat at the bottom of each stage operates as a flooded bed scrubber. The flooded bed is used to minimize water consumption. The two inclined upper mats operate as entrainment eliminators.

Pressure Drop

The pressure drop through a particular packing in countercurrent scrubbers can be calculated from the airflow and water flow per unit area. Charts, such as the one shown in Figure 28, are available from manufacturers of each type and size of packing.

The pressure drop for any packing can also be estimated by using the data on packing factors in Table 7 and the modified generalized pressure drop correlation shown in Figure 29. This correlation was developed for a gas stream substantially of air, with water as the scrubbing liquid. It should not be used if the properties of the gas or liquid vary significantly from air or water, respectively. Countercurrent scrubbers are generally designed to operate at pressure drops between 0.25 and 0.65 in. of water per foot of packing depth. Liquid irrigation rates typically vary between 5 and 20 gpm per square foot of bed area.

Absorption Efficiency

The prediction of the absorption efficiency of a packed bed scrubber is much more complex than estimating its capacity because performance estimates involve the mechanics of absorption. Some of the factors affecting efficiency are superficial velocity

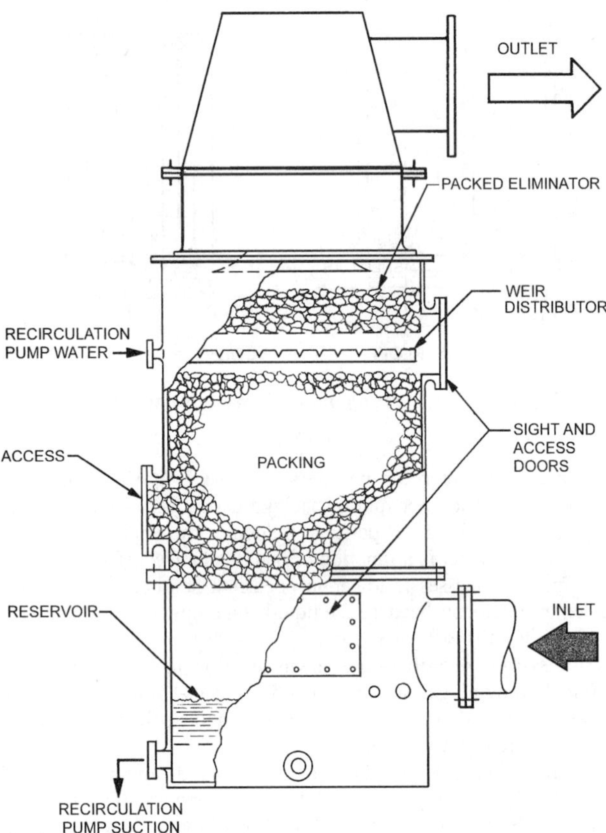

Fig. 25 Typical Countercurrent Packed Scrubber

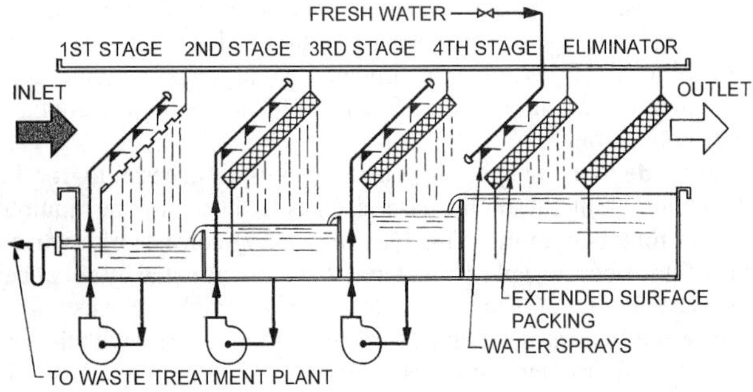

Fig. 26 Horizontal Flow Scrubber with Extended Surface

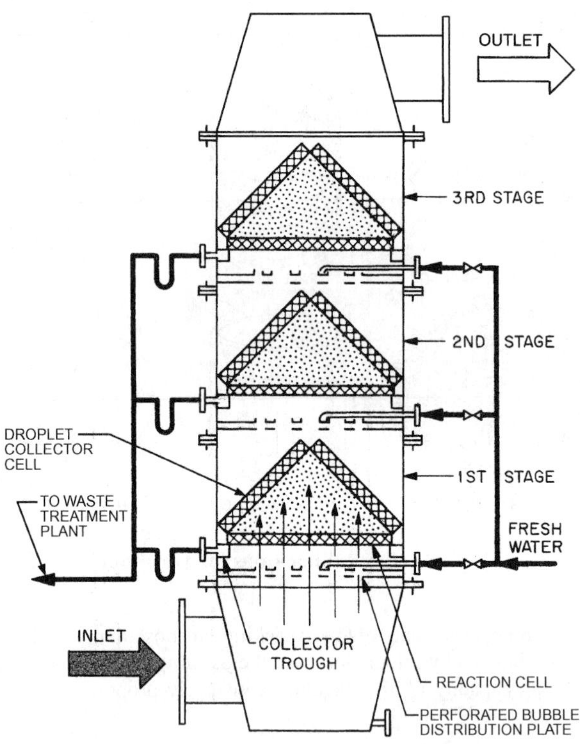

Fig. 27 Vertical Flow Scrubber with Extended Surface

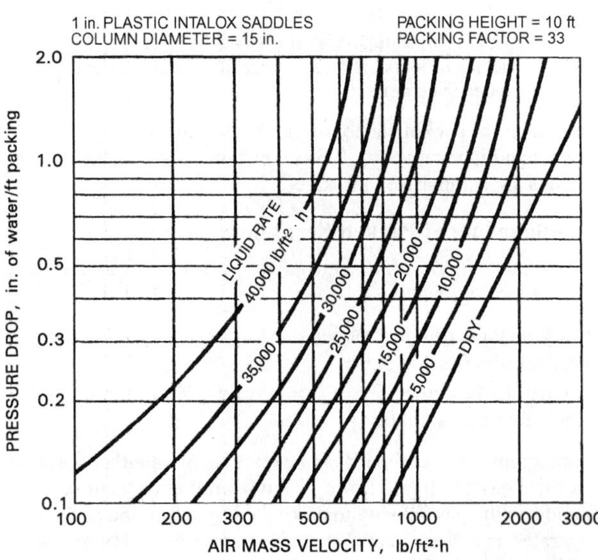

Fig. 28 Pressure Drop Versus Gas Rate for Typical Packing

of gases in the scrubber, liquid injection rate, packing size, type of packing, amount of contaminant to be removed, distribution and amount of absorbent available for reaction, temperature, and reaction rate for absorption.

Practically all commercial packings have been tested for absorption rate (mass transfer coefficient) using standard absorber conditions: carbon dioxide (CO_2) in air and a solution of caustic soda (NaOH) in water. This system was selected because the interaction of the variables is well understood. Further, the mass transfer coefficients for this system are low; thus, they can be determined accurately by experiment. The values of mass transfer coefficients ($K_G a$) for various packings under these standard test conditions are given in Table 8.

The vast majority of wet absorbers are used to control low concentrations (less than 0.005 mole fraction) of contaminants in air. Dilute aqueous solutions of NaOH are usually chosen as the scrubbing fluid. These conditions simplify the design of scrubbers somewhat. Mass transfer from the gas to the liquid is then explained by the **two-film theory**: the gaseous contaminant travels by diffusion from the main gas stream through the gas film, then through the liquid film, and finally into the main liquid stream. The relative influences of the gas and liquid films on the absorption rate depend on the solubility of the contaminant in the liquid. Sparingly soluble gases like hydrogen sulfide (H_2S) and CO_2 are said to be liquid-film-controlled; highly soluble gases such as HCl and ammonia (NH_3) are said to be gas-film-controlled. In liquid-film-controlled systems, the mass transfer coefficient varies with the liquid injection rate but is only slightly affected by the superficial velocity of the gases. In gas-film-controlled systems, the mass transfer coefficient is a function of both the superficial velocity of the gases and the liquid injection rate.

In the absence of leakage, the percentage by volume of the contaminant removed from the air can be found from the inlet and outlet concentrations of contaminant in the airstream:

Table 8 Mass Transfer Coefficients ($K_G a$) for Scrubber Packing Materials

Type of Packing	Material	Nominal Size, in. 1	1.5	2	3 or 3.5
		$K_G a$, lb·mol/h·ft³·atm			
Super Intalox	Plastic	2.19		1.44	0.887
Intalox saddles	Ceramic	1.96	1.71	1.44	0.820
Raschig rings	Ceramic	1.73	1.50	1.21	
Hy-Pak	Metal	2.20	1.87	1.69	1.09
Pall rings	Metal	2.32	1.87	1.62	0.91
Pall rings	Plastic	1.98	1.73	1.46	0.89
Tellerettes	Plastic	2.19		1.98	
Maspac	Plastic			1.44	0.89

System: CO_2 and NaOH; gas rate: 110 cfm/ft²; liquid rate: 4 gpm/ft².

Table 9 Relative $K_G a$ for Various Contaminants in Liquid-Film-Controlled Scrubbers

Gas Contaminant	Scrubbing Liquid	$K_G a$, lb·mol/h·ft³·atm
CO_2	4% (by mass) NaOH	2.0
H_2S	4% (by mass) NaOH	5.92
SO_2	Water	2.96
HCN	Water	5.92
HCHO	Water	5.92
Cl_2	Water	4.55

Note: Data for 2 in. plastic Super Intalox. Temperatures: from 60 to 75°F; liquid rate: 10 gpm/ft²; gas rate: 215 cfm/ft².

$$\% \text{ Removed} = 100(1 - Y_o/Y_i) \quad (9)$$

where

Y_i = mole fraction of contaminants entering scrubber (dry gas basis)
Y_o = mole fraction of contaminants exiting scrubber (dry gas basis)

The driving pressure for absorption (assuming negligible vapor pressure above the liquid) is controlled by the logarithmic mean of inlet and outlet concentrations of the contaminant:

$$\Delta p_{ln} = p\frac{Y_i - Y_o}{\ln(Y_i/Y_o)} \quad (10)$$

where

Δp_{ln} = driving or diffusion pressure acting to absorb contaminants on packing
p = inlet pressure

The rate of absorption of contaminant (mass transfer coefficient) is related to the depth of packing as follows:

$$K_G a = N/HA\Delta p_{ln} \quad (11)$$

where

N = solute absorbed, lb·mol/h
H = depth of packing, ft
A = cross-sectional area of scrubber, ft²

The value of N can be determined from

$$N = G(Y_i - Y_o) \quad (12)$$

where G = mass flow rate of gases through scrubber, lb·mol/h.

The superficial velocity of gases is a function of the unit gas flow rate and the gas density:

$$V = 60TGM_v/C_1A \quad (13)$$

where

V = superficial gas velocity, fpm
M_v = molar volume, ft³/lb·mol
T = exit gas temperature, °R = °F + 460
C_1 = 460°R

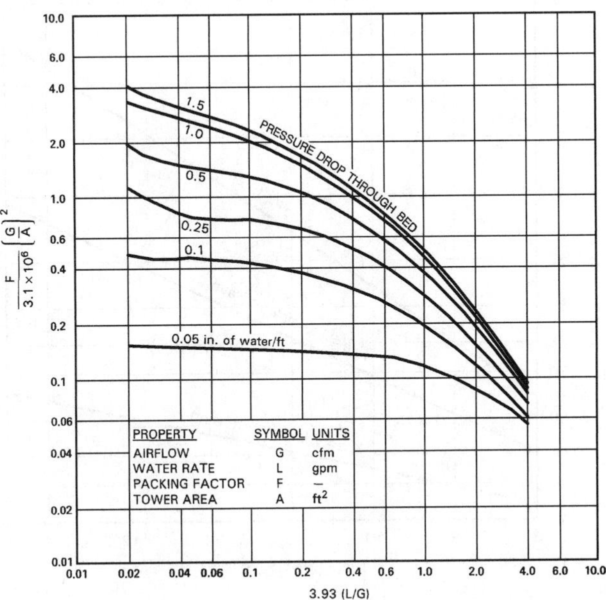

Fig. 29 Generalized Pressure Drop Curves for Packed Beds

By combining these equations and assuming ambient pressure, a graphical solution can be derived for both liquid-film- and gas-film-controlled systems. Figures 30, 31, and 32 show the height of packing required versus percent removal for various mass transfer coefficients at superficial velocities of 120, 240, and 360 fpm, respectively, with liquid-film-controlled systems. Figures 33, 34, and 35 show the height of packing versus percent removal for various mass transfer coefficients at the same three superficial velocities with gas film-controlled systems.

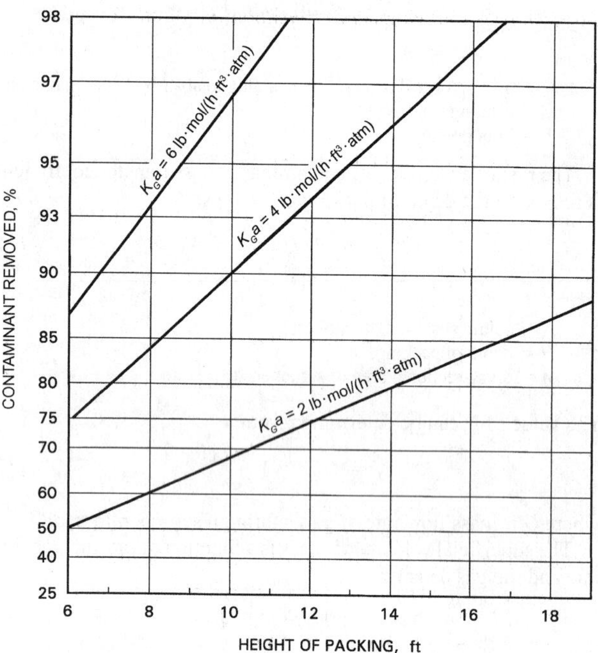

Fig. 30 Contaminant Control at Superficial Velocity = 120 fpm (Liquid-Film-Controlled)

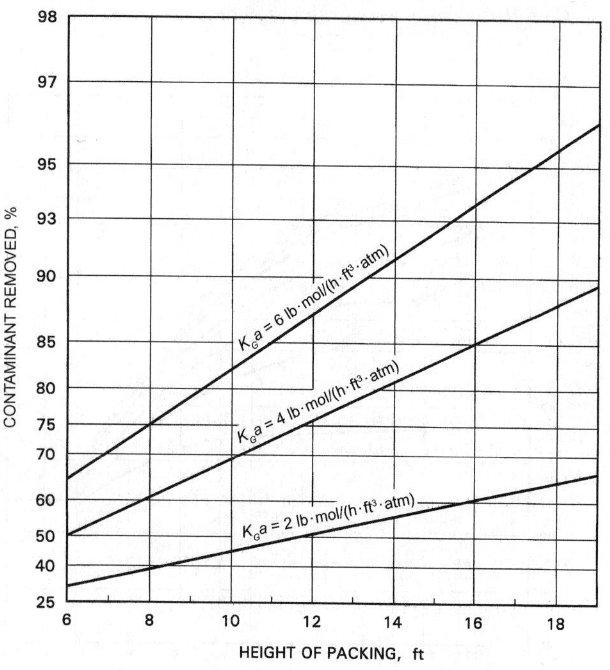

Fig. 31 Contaminant Control at Superficial Velocity = 240 fpm (Liquid-Film-Controlled)

These graphs can be used to determine the height of 2 in. plastic Intalox saddles (see Figure 23) required to give the desired percentage of contaminant removal. The height for any other type or size of packing is inversely proportional to the ratio of standard K_Ga taken

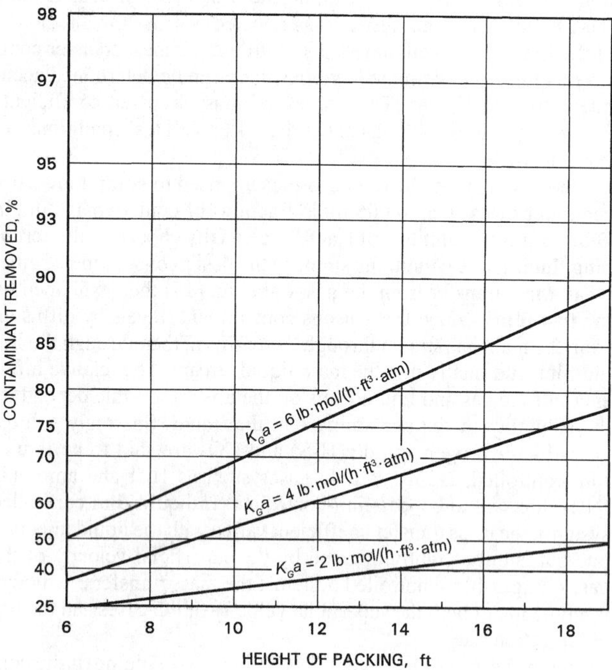

Fig. 32 Contaminant Control at Superficial Velocity = 360 fpm (Liquid-Film-Controlled)

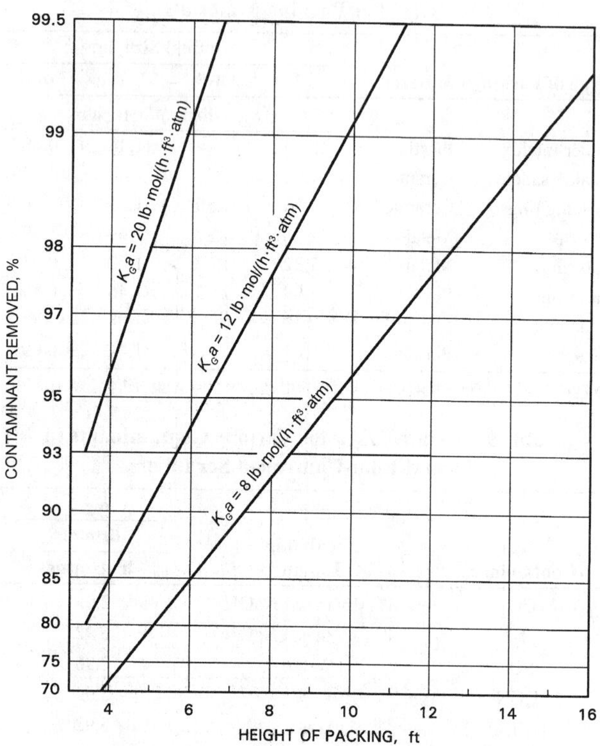

Fig. 33 Contaminant Control at Superficial Velocity = 120 fpm (Gas-Film-Controlled)

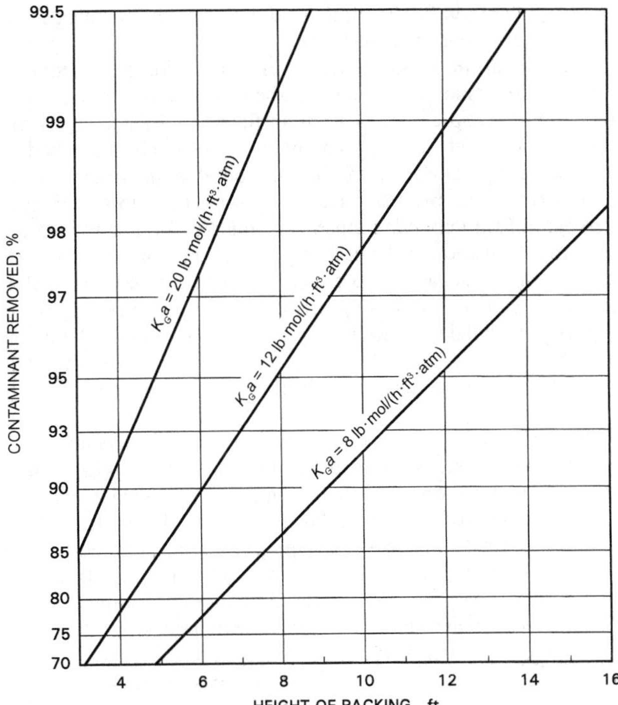

Fig. 34 Contaminant Control at Superficial Velocity = 240 fpm (Gas-Film-Controlled)

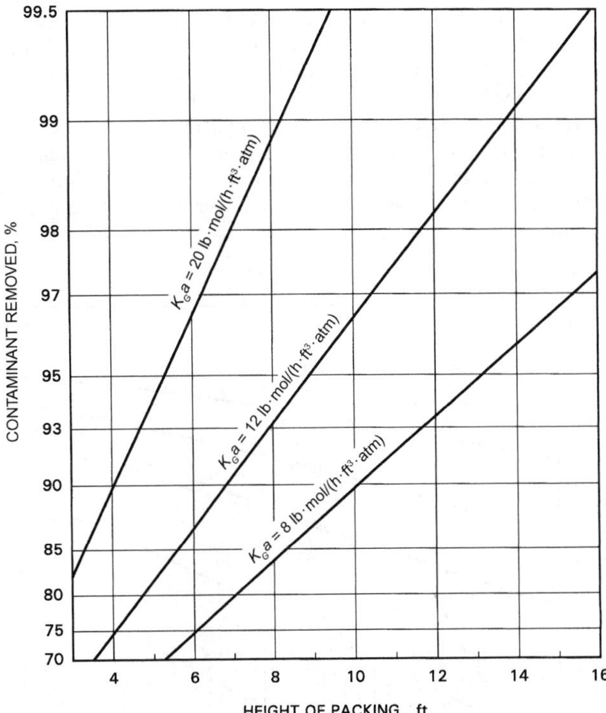

Fig. 35 Contaminant Control at Superficial Velocity = 360 fpm (Gas-Film-Controlled)

from Table 8. Thus, if 13 ft packing depth were required for 95% removal of contaminants, the same efficiency could be obtained with a 9.5 ft depth of 1 in. plastic pall rings (Figure 23), at the same superficial velocity and liquid injection rate. However, the pressure drop would be higher for the smaller-diameter packing.

Table 10 Relative K_Ga for Various Contaminants in Gas-Film-Controlled Scrubbers

Gas Contaminant	Scrubbing Liquid	K_Ga, $\frac{\text{lb·mol}}{\text{h·ft}^3\text{·atm}}$
HCl	Water	18.66
HBr	Water	5.92
HF	Water	7.96
NH_3	Water	17.30
Cl_2	8% (by mass) NaOH	14.33
SO_2	11% (by mass) Na_2CO_3	11.83
Br_2	5% (by mass) NaOH	5.01

Note: Data for 2 in. plastic Super Intalox. Temperatures 60 to 75°F; liquid rate 10 gpm/ft²; gas rate 215 cfm/ft².

Figures 30 to 35 are useful when the value of the mass transfer coefficient for the particular contaminant to be removed is known. Table 9 contains mass transfer coefficients for 2 in. plastic Intalox saddles in typical liquid-film-controlled scrubbers. These values can be compared with the mass transfer coefficients in Table 10 for the same packing used in gas-film-controlled scrubbers. When the scrubbing liquid is not water, the mass transfer coefficients in these tables can only be used if the amount of reagent in the solution exceeds by at least 33% the amount needed to completely absorb the gaseous contaminant.

When HCl is dissolved in water, there is little vapor pressure of HCl above solutions of less than 8% (by mass) concentration. On the other hand, when NH_3 is dissolved in water, there is an appreciable vapor pressure of NH_3 above solutions, even at low concentrations. The height of packing needed for NH_3 removal, obtained from Figures 33 through 35, is based on the use of dilute acid to maintain the pH of the solution below 7.

The following is an example of a typical scrubbing problem.

Example 1. Remove 95% of the HF from air at 90°F. The concentration of HF is 600 ppm on a dry gas basis. The concentration of HF in the exhaust gas should not exceed 30 ppm.

The following design conditions apply:

Total volumetric flow of gas	G	=	4600 cfm
Liquid injection rate		=	3.75 gpm/ft²
Liquid temperature		=	68°F
Packed tower diameter		=	4 ft

Packing material is 2 in. polypropylene Super Intalox

Solution:

Cross-sectional area of absorber

$$A = \pi(4/2)^2 = 12.6 \text{ ft}^2$$

Total liquid flow rate

$$L = 3.75 \times 12.6 = 47.1 \text{ gpm}$$

Packing factor (from Table 7)

$$F = 21$$

Figure 29 may be used to find the pressure drop through the packed tower:

$$x\text{-axis} = 3.93(L/G) = 0.040$$

$$y\text{-axis} = (F/3.1 \times 10^6)(G/A)^2 = 0.90$$

From Figure 29, the pressure drop is about 0.28 in. of water per foot of packing depth.

From Table 10, K_Ga = 0.35 for HF. From Figure 33, the depth of packing required for 95% removal is 13 ft. Thus, the total pressure drop is 13 × 0.28 = 3.6 in. of water.

General Efficiency Comparisons

Figure 35 indicates that, with $K_G a$ = 0.35, 90% removal of HF could be achieved with 10 ft of packing; this is 23% less packing than needed for 96% removal. Furthermore, with the same superficial velocity, both liquid-film- and gas-film-controlled systems require a 43% increase in absorbent depth to raise the removal efficiency from 80 to 90%.

A comparison of Figures 34 and 35 shows that increasing the superficial velocity by 50% in a gas-film-controlled scrubber requires only a 12% increase in bed depth to maintain equal removal efficiencies. In the liquid-film-controlled system (Figures 31 and 32), increasing the superficial velocity by 50% requires an approximately 50% increase in bed depth to maintain equal removal efficiencies.

Thus, in a gas-film-controlled system, the superficial velocity can be increased significantly with only a small increase in bed depth required to maintain the efficiency. In practical terms, gas-film-controlled scrubbers of fixed depth can handle an overload condition with only a minor loss of removal efficiency. The performance of liquid-film-controlled scrubbers degrades significantly under similar overload conditions. This occurs because the mass transfer coefficient is independent of superficial velocity.

Liquid Effects

Some liquids tend to foam when they are contaminated with particulates or soluble salts. In these cases, the pressure drop should be kept in the lower half of the normal range: 0.25 to 0.40 in. of water per foot of packing depth.

In the control of gaseous pollution, most systems do not destroy the pollutant but merely remove it from the air. When water is used as the scrubbing liquid, effluent from the scrubber will contain suspended particulate or dissolved solute. Water treatment is often required to alter the pH and/or remove toxic substances before the solutions can be discharged.

ADSORPTION OF GASEOUS CONTAMINANTS

The surface of freshly broken or heated solids often contains van der Waals (London dispersion) forces that are able to physically or chemically adsorb nearby molecules in a gas or liquid. The captured molecules form a thin layer on the surface of the solid that is typically one to three molecules thick. Commercial adsorbents are solids with an enormous internal surface area. This large surface area enables them to capture and hold large numbers of molecules. For example, each gram of a typical activated carbon adsorbent contains over 10,000 ft^2 of internal surface area. Adsorbents are used for removing organic vapors, water vapor, odors, and hazardous pollutants from gas streams.

The most common adsorbents used in industrial processes include activated carbons, activated alumina, silica gel, and molecular sieves. Activated carbons are derived from coal, wood, or coconut shells. They are primarily selected to remove organic compounds in preference to water. The other three common gas-phase adsorbents have a great affinity for water and will adsorb it to the exclusion of any organic molecules also present in a gas stream. They are used primarily as gas-drying agents. Molecular sieves also find use in several specialized pollution control applications, including removal of mercury vapor, sulfur dioxide (SO_2), or nitrogen oxides (NO_x) from gas streams.The capacity of a particular activated carbon to adsorb any organic vapor from an exhaust gas stream is related to the concentration and molecular weight of the organic compound and the temperature of the gas stream. Compounds with a higher molecular weight are usually more strongly adsorbed than those with lower molecular weight. The capacity of activated carbon to adsorb any given organic compound increases with the concentration of that compound. Reducing the temperature also favors adsorption. Typical adsorption capacities of an activated carbon for toluene (molecular weight 92) and acetone (molecular weight 58) are illustrated in Figure 36 for various temperatures and concentrations.

Regeneration. Adsorption is reversible. An increase in temperature causes some or all of an adsorbed vapor to desorb. The temperature of low-pressure steam is sufficient to drive off most of a low-boiling-point organic compound previously adsorbed at ambient temperature. Higher-boiling-point organic compounds may require high-pressure steam or hot inert gas to secure good desorption. Compounds with a very high molecular weight can require reactivation of the carbon adsorber in a furnace at 1350°F to drive off all the adsorbed material. Regeneration of the carbon adsorbers can also be accomplished, in some instances, by washing with an aqueous solution of a chemical that will react with the adsorbed organic material, making it water soluble. An example is washing carbon containing adsorbed sulfur compounds with NaOH.

The difference between an adsorbent's capacity under adsorbing and desorbing conditions in any application is its **working capacity**. Activated carbon for air pollution control is found in canisters under the hoods of most automobiles. The adsorber in these canisters captures gasoline vapors escaping from the carburetor (when the engine is stopped) and from the fuel tank's breather vent. Desorption of gasoline vapors is accomplished by pulling fresh air through the carbon canister and into the carburetor when the engine is running. Although there is no temperature difference between adsorbing and desorbing conditions in this case, the outside airflow desorbs enough gasoline vapors to give the carbon a substantial working capacity.

For applications where only traces of a pollutant must be removed from exhaust air, the life of a carbon bed is very long. In these cases, it is often more economical to replace the carbon than to invest in regeneration equipment. Larger quantities can be returned to the carbon manufacturer for high-temperature thermal reactivation. Regeneration in place by steam, hot inert gas, or washing with a solution of alkali is sometimes practiced.

Impregnated (chemically reactive) adsorbents are used when physical adsorption alone is too weak to remove a particular gaseous contaminant from an industrial gas stream. Through impregnation, the reactive chemical is spread over the immense internal surface area of an adsorbent.

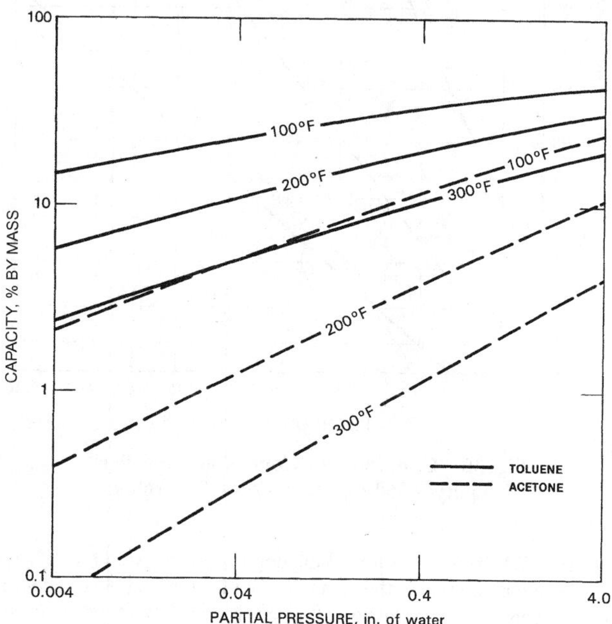

Fig. 36 Adsorption Isotherms on Activated Carbon

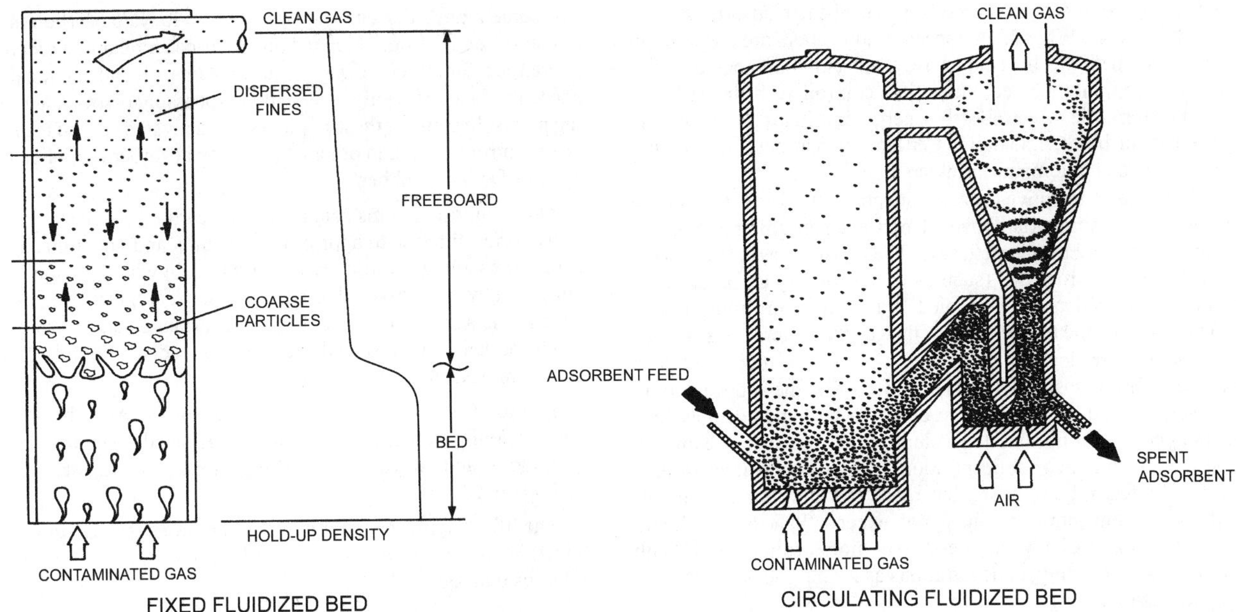

Fig. 37 Fluidized-Bed Adsorption Equipment

Typical applications of impregnated adsorbent include the following:

- Sulfur- or iodine-impregnated carbon removes mercury vapor from air, hydrogen, or other gases by forming mercuric sulfide or iodide.
- Metal oxide-impregnated carbons remove hydrogen sulfide.
- Amine- or iodine-impregnated carbons and silver exchanged zeolites remove radioactive methyl iodide from nuclear power plant work areas and exhaust gases.
- Alkali-impregnated carbons remove acid gases.
- Activated alumina impregnated with potassium permanganate removes acrolein and formaldehyde.

Equipment for Adsorption

Three types of adsorbers are usually found in industrial applications: (1) fixed beds, (2) moving beds, and (3) fluidized beds (Figure 37).

Fixed beds of regenerable or disposable media are most common. Carbon filter elements are a typical example.

Moving beds use granular adsorbers placed on inclined trays or on vertical frames similar to those used in granular bed particulate collectors. Moving beds offer continuous contaminant control and regeneration. Often, moving bed adsorbers and regeneration equipment are integrated components in a process.

Fluidized beds contain a fine granular adsorber, which is continuously mixed with the contaminated gas by suspension in the process gas stream. The bed may be either "fixed" at lower superficial velocities or highly turbulent and conveying (circulating). Illustrations of these types of fluidized beds are shown in Figure 37.

Solvent Recovery

The most common use of adsorption in stationary sources is in recovering solvent vapors from manufacturing and cleaning processes. Typical applications include solvent degreasing, rotogravure printing, dry cleaning, and the manufacture of such products as synthetic fibers, adhesive labels, tapes, coated copying paper, rubber goods, and coated fabrics.

Figure 38 illustrates the components of a typical solvent recovery system using two carbon beds. One bed is used as an adsorber while the other is regenerated with low-pressure steam. Desorbed solvent vapor and steam are recovered in a water-cooled condenser. If the solvent is immiscible with water, an automatic decanter separates the solvent for reuse. A distillation column is used for water-miscible solvents.

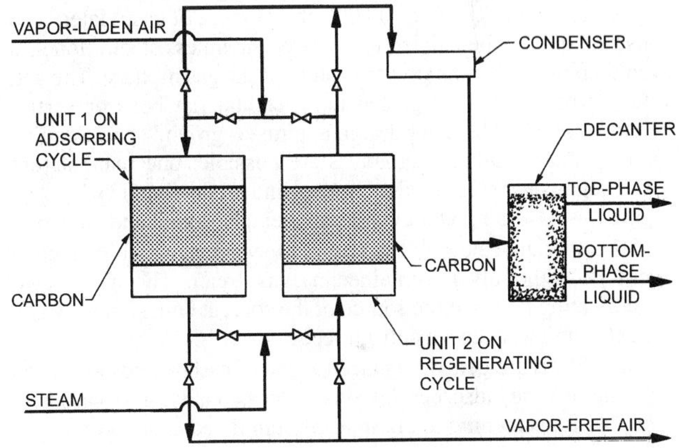

Fig. 38 Schematic of Two-Unit Fixed Bed Adsorber

Adsorption time per cycle typically runs from 30 min to several hours. The adsorbing carbon bed is switched to regeneration by (1) an automatic timer shortly before the solvent vapor breaks through from the bed or (2) an organic vapor-sensing control device in the exhaust gas stream immediately after the solvent breaks through from the bed.

Low-pressure steam consumption for regeneration is generally about 3.5 lb/h per pound of solvent recovered (Boll 1976), but it can range from 2 to over 5 lb/h per pound of solvent recovered, depending on the specific solvent and its concentration in the exhaust gas stream being stripped. Steam with only a slight superheat is normally used, so that it condenses quickly and gives rapid heat transfer.

After steaming, the hot, moist carbon bed is usually cooled and partially dried before being placed back on stream. Heat for drying is supplied by the cooling of the carbon and adsorber, and sometimes by an external air heater. In most cases, it is desirable to leave some moisture in the bed. When solvent vapors are adsorbed, heat is generated. For most common solvents, the heat of adsorption is 40 to 60 Btu/lb·mol. When high-concentration vapors are adsorbed in a dry carbon bed, this heat can cause a substantial temperature rise and can even ignite the bed, unless it is controlled. If the bed contains moisture, the water absorbs energy and helps to prevent an undue rise in bed temperature. Certain applications may require heat sensors and automatic sprinklers.

Because the adsorptive capacity of activated carbon depends on temperature, it is important that solvent-laden air going to a recovery unit be as cool as is practicable. The exhaust gases from many solvent-emitting processes (such as drying ovens) are at elevated temperatures. Water- or air-cooled heat exchangers must be installed to reduce the temperature of the gas that enters the adsorber.

Solvent at very low vapor concentrations can be recovered in an activated carbon system. The size and cost of the recovery unit, however, depend on the volume of air to be handled; it is thus advantageous to minimize the volume of an exhaust stream and keep the solvent vapor concentration as high as possible, consistent with safety requirements. Insurance carriers specify that solvent vapor concentrations must not exceed 25% of the lower explosive limit (LEL) when intermittent monitoring is used. With continuous monitoring, concentrations as high as 50% of the LEL are permissible.

Solvent recovery systems, with gas-handling capacities up to about 11,000 cfm, are available as skid-mounted packages. Several of these packaged units can be used for larger gas flows. Custom systems can be built to handle 200,000 cfm or more of gas. Materials of construction may be painted carbon steel, stainless steel, Monel, or even titanium, depending on the nature of the gas mixture. The activated carbon is usually placed in horizontal flat beds or vertical cylindrical beds. The latter design minimizes ground space required for the system. Other alternatives are possible; one manufacturer uses a segmented horizontal rotating cylinder of carbon in which one segment is adsorbing while others are being steamed and cooled.

Commercial-scale solvent recovery systems typically recover over 99% of the solvent contained in a gas stream. The efficiency of the collecting hoods at the source of the solvent emission is usually the determining factor in solvent recovery.

Dust filters are generally placed ahead of carbon beds to prevent blinding of the adsorber by dust. Occasionally, the carbon is removed for screening to eliminate accumulated dust and fine particles of carbon.

If the solvent mixture contains high-boiling-point components, the working capacity of activated carbon can decrease with time. This occurs when high-boiling-point organic compounds are only partially removed by low-pressure steam. In this situation, two alternatives should be considered:

1. Periodic removal of the carbon and return to its manufacturer for high-temperature furnace reactivation to virgin carbon activity.
2. Use of more rigorous solvent desorbing conditions in the solvent recovery system. High-temperature steam, hot inert gas, or a combination of electrical heating and application of a vacuum may be used. The last method is selected, for example, to recover lithography ink solvents with high boiling points. Note that this method may not remove all of the high-boiling-point compounds.

Odor Control

Incineration and scrubbing are usually the most economical methods of controlling high concentrations of odorous compounds from equipment such as cookers in rendering plants. However, many odors that arise from harmlessly low concentrations of vapors are still offensive. The odor threshold (for 100% response) of acrolein in air, for example, is only 0.21 ppm, whereas that for ethyl mercaptan is 0.001 ppm and that for hydrogen sulfide is 0.0005 ppm (AIHA 1989; MCA 1968). Activated carbon beds effectively overcome many odor emission problems. Activated carbon is used to control odors from chemical and pharmaceutical manufacturing operations, foundries, sewage treating plants, oil and chemical storage tanks, lacquer drying ovens, food processing plants, and rendering plants. In some of these applications, activated carbon is the sole odor control method; in others, the carbon adsorber is applied to the exhaust from a scrubber.

Odor control systems using activated carbon can be as simple as a steel drum fitted with appropriate gas inlet and outlet ducts, or as complex as a large, vertically moving bed, in which carbon is contained between louvered side panels. A typical moving-bed adsorber is shown in Figure 39. In this arrangement, fresh carbon can be added at the top, and spent or dust-laden carbon is periodically removed from the bottom.

Figure 40 shows a fixed-bed odor adsorber. Adsorbers of this general configuration are available as packaged systems, complete with motor and blower. Air-handling capacities range from 500 to 12,000 cfm.

The life of activated carbon in odor control systems ranges from a few weeks to a year or more, depending on the concentration of the odorous emission.

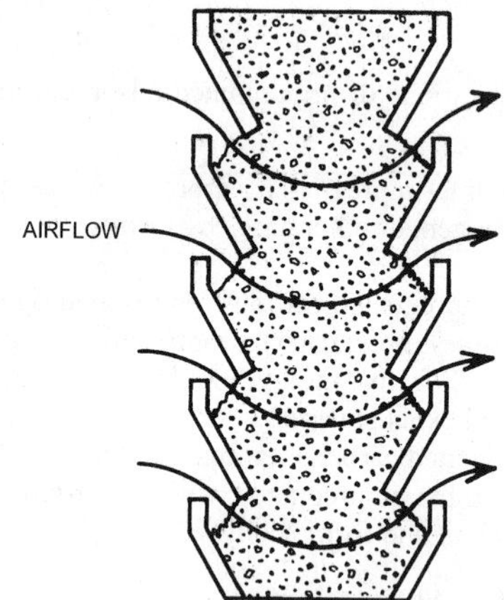

Fig. 39 Moving-Bed Adsorber

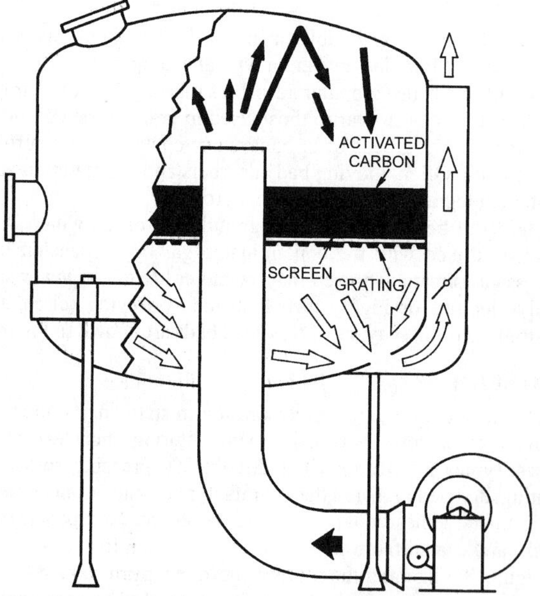

Fig. 40 Typical Odor Adsorber

Applications of Fluidized Bed Adsorbers

The injection of alkali compounds into fluidized bed combustors for control of sulfur-containing compounds is one example of the use of fluidized bed adsorbers. Another example is the control of HF emissions from Søderberg aluminum reduction processes by a fixed or circulating fluidized bed of alumina.

INCINERATION OF GASES AND VAPORS

Incineration is the process by which volatile organic compounds (VOCs), organic aerosols, and most odorous materials in a contaminated gas stream are converted to innocuous carbon dioxide and water vapor using heat energy. Incineration is an effective means for totally eliminating VOCs. The types of incineration commonly used for air pollution control are thermal and catalytic, sometimes with recuperative or regenerative heat recovery.

To differentiate such air-cleaning systems from liquid and solid waste incinerators, the preferred term to describe such gas and aerosol phase air pollution control systems is now **oxidizers**.

Thermal Oxidizers

Thermal oxidizers, also known as **afterburners** or **direct flame incinerators**, consist of an insulated oxidation chamber in which gas and/or oil burners are typically located. The contaminated gas stream enters the chamber and comes into direct contact with the flame, which provides the heat energy necessary to promote oxidation. Under the proper conditions of time, temperature, and turbulence, the gas stream contaminants are oxidized effectively. The contaminated gas stream enters the combustion chamber near the burner, where turbulence-inducing devices are usually installed. The final contaminant conversion efficiency largely depends on good mixing within the contaminated gas stream and on the temperature of the oxidation chamber.

Supplemental fuel is used for start-up, to raise the temperature of the contaminated gas stream enough to initiate contaminant oxidation. Once oxidation begins, the temperature rises further because of the energy released by combustion of the contaminant. The supplementary fuel feed rate is then modulated to maintain the desired oxidizer operating temperature. Most organic gases oxidize to approximately 90% conversion efficiency if a temperature of at least 1200°F and a residence time of 0.3 to 0.5 s are achieved within the oxidation chamber. However, oxidation temperatures are typically maintained in the range of 1400 to 1500°F with residence times of 0.5 to 1 s to ensure conversion efficiencies of 95% or greater.

Although the efficiency of thermal oxidizers can exceed 95% destruction of the contaminant, the reaction may form undesirable products of combustion. For example, oxidation of chlorinated hydrocarbons causes the formation of hydrogen chloride, which can have an adverse effect on equipment. These new contaminants then require additional controls.

Oxidation systems incorporate primary heat recovery to preheat the incoming contaminated gas stream and, in some cases, provide secondary heat recovery for process or building heating. Primary heat recovery is almost always achieved using air-to-air heat exchangers. Use of a regenerable, ceramic medium for heat recovery has increased due to superior heat recovery efficiency. Secondary heat recovery may incorporate an air-to-air heat exchanger or a waste heat boiler (DOE 1979).

Oxidation systems using conventional air-to-air heat exchangers can achieve up to 80% heat recovery efficiency. Regenerative heat exchanger units have claimed as high as 95% heat recovery efficiency and are routinely operated at 85 to 90%. When operated at these high heat recovery efficiencies and with inlet VOC concentrations of 15 to 25% of the LEL, the oxidation process approaches a self-sustaining condition, requiring very little supplementary fuel.

Catalytic Oxidizers

Catalytic oxidizers operate under the same principles as thermal oxidizers, except that they use a catalyst to promote oxidation. The catalyst allows oxidation to occur at lower temperatures than in a thermal oxidizer for the same VOC concentration. Therefore, catalytic oxidizers require less supplemental fuel to preheat the contaminated gas stream and have lower overall operating temperatures.

A catalytic oxidizer generally consists of a preheat chamber followed by the catalyst bed. Residence time and turbulence are not as important as with thermal oxidizers, but it is essential that the contaminated gas stream be heated uniformly to the required catalytic reaction temperature. The required temperature varies, depending on the catalyst material and configuration.

The temperature of the contaminated gas stream is raised in the preheat chamber by a conventional burner. Although the contaminated gas stream contacts the burner flame, the heat input is significantly less than that for a thermal oxidizer, and only a small degree of direct contaminant oxidation occurs. Natural gas is preferred to prevent catalyst contamination, which could occur with sulfur-bearing fuel oils. However, No. 2 fuel oil units have been operated successfully. The most effective catalysts contain noble metals such as platinum or palladium.

Catalysis occurs at the molecular level. Therefore, an available, active catalyst surface area is important for maintaining high conversion efficiencies. If particulate materials contact the catalyst as either discrete or partially oxidized aerosols, they can ash on the catalyst surface and blind it. This problem is usually accompanied by a secondary pollution problem: odorous emissions caused by the partially oxidized organic compounds.

The greatest concern to users of catalytic oxidizers is **catalyst poisoning or deactivation**. Poisoning is caused by specific gas stream contaminants that chemically combine or alloy with the active catalyst material. Poisons frequently cited include phosphorus, bismuth, arsenic, antimony, lead, tin, and zinc. The first five materials are considered fast-acting poisons and must be excluded from the contaminated gas stream. Even trace quantities of the fast-acting poisons can cause rapid catalyst deactivation. The last two materials are slow-acting poisons; catalysts are somewhat tolerant of these materials, particularly at temperatures lower than 1000°F. However, even the slow poisons should be excluded from the contaminated gas stream to ensure continuous, reliable performance. Therefore, galvanized steel, another possible source of the slow poisons, should not be used for the duct leading to the oxidizer.

Sulfur and halogens are also regarded as catalyst poisons. In most cases, their chemical interaction with the active catalyst material is reversible. That is, catalyst activity can be restored by operating the catalyst without the halogen or sulfur-bearing compound in the gas stream. The potential problem of greater concern with respect to the halogen-bearing compounds is the formation of hydrogen chloride or hydrogen fluoride gas, or hydrochloric or hydrofluoric acid emissions.

Some organic compounds, such as polyester amides and imides, are also poisonous.

Catalytic oxidizers generally cost less to operate than thermal oxidizers because of their lower fuel consumption. With the exception of regenerative heat recovery techniques, primary and secondary heat recovery can be incorporated into a catalytic oxidation system to further reduce operating costs. Maintenance costs are usually higher for catalytic units, particularly if frequent catalyst cleaning or replacement is necessary. Concern over catalyst life has been the major factor limiting more widespread application of catalytic oxidizers.

Applications of Oxidizers

Odor Control. All highly odorous pollutant gases are combustible or chemically changed to less odorous pollutants when they are sufficiently heated. Often, the concentration of odorous materials in the waste gas is extremely low, and the only feasible method of control is oxidation. Odors from rendering plants, mercaptans, and organic sulfides from kraft pulping operations are examples of effluents that can be controlled by incineration. Other forms of oxidation can achieve the same ends (see Chapter 44 of the 2007 *ASHRAE Handbook—HVAC Applications*).

Reduction in Emissions of Reactive Hydrocarbons. Some air pollution control agencies regulate the emission of organic gases and vapors because of their involvement in photochemical smog reactions. Flame afterburning is an effective way of destroying these materials.

Reduction in Explosion Hazard. Refineries and chemical plants are among the factories that must dispose of highly combustible or otherwise dangerous organic materials. The safest method of disposal is usually by burning in flares or in specially designed furnaces. However, special precautions and equipment design must be used in the handling of potentially explosive mixtures.

Adsorption and Oxidation

Alternate cycles of adsorption and desorption in an activated carbon bed are used to concentrate solvent or odor vapors before oxidation. This technique greatly reduces the fuel required for burning organic vapor emissions. Fuel savings of 98% compared to direct oxidation are possible. The process is particularly useful in cases where emission levels vary from hour to hour. This technique is common in metal finishing for automotive and office furniture manufacturing.

Contaminated gas is passed through a carbon bed until saturation occurs. The gas stream is then switched to another carbon bed, and the exhausted bed is shut down for desorption. A hot inert gas, usually burner flue gas, is introduced to the adsorber to drive off concentrated organic vapors and to convey them to an oxidizer. The volume of this desorbing gas stream is much smaller than the original contaminated gas volume, so that only a small oxidizer, operating intermittently, is required (Grandjacques 1977).

AUXILIARY EQUIPMENT

DUCTS

Basic duct design is covered in Chapter 35 of the 2005 *ASHRAE Handbook—Fundamentals*. This chapter covers only those duct components or problems that warrant special concern when handling gases that contain particulate or gaseous contaminants.

Duct systems should be designed to allow thermal expansion or contraction as gases move from the process, through gas-cleaning equipment, and on to the ambient environment. Appropriately designed duct expansion joints must be located in proper relation to sliding and fixed duct supports. Besides withstanding maximum possible temperature, duct must also withstand maximum positive or negative pressure, a partial load of accumulated dust, and reasonable amounts of corrosion. Duct supports should be designed to accommodate these overload conditions as well. Bottom entries into duct junctions create a low-velocity area that can allow contaminants to settle out, thus creating a potential fire hazard. Bottom duct entries should be avoided.

Where gases might condense and cause corrosion or sticky deposits, duct should be insulated or fabricated from materials that will survive this environment. A psychrometric analysis of the exhaust gases is useful to determine the dew point. Surface temperatures should then be held above the dew-point temperature by preheating on start-up or by insulating. Slag traps with clean-out doors, inspection doors where direction changes, and dead-end full-sized caps are required for systems having heavy particulate loading. Special attention should be given to high-temperature duct, where the duct might corrode if insulated or become encrusted when molten particulate impacts on cool, uninsulated surfaces. Water-cooled duct or refractory lining is often used where the high operating temperature exceeds the safe limits of low-cost materials.

Gas flow through ducts should be considered as a part of overall system design. Good gas flow distribution is essential for measurements of process conditions and can lead to energy savings and increased system life. The minimum speed of gases in a duct should be sufficiently high to convey the heaviest particulate fraction with a degree of safety.

Slide gates, balance gates, equipment bypass ducts, and clean-out doors should be incorporated in the duct system to allow for maintenance of key gas-cleaning systems. In some cases, emergency bypass circuitry should be included to vent emissions and protect gas-cleaning equipment from process upsets.

Temperature Controls

Control of gas temperature in a gas-cleaning system is often vital to a system's performance and life. In some cases, gases are cooled to concentrate contaminants, condense gases, and recover energy. In other cases, gas-cleaning equipment, such as fabric filters and scrubbers, can only operate at well-controlled temperatures. Cooling exhaust gases through air-to-air heat transfer has been highly successful in many applications. Controlled evaporative cooling is also used, but it increases the dew point and the danger of acid gas condensation and/or the formation of sticky deposits. However, controlled evaporative cooling to within 50°F of dew point has been used with success. Dilution by the injection of ambient air into the duct is expensive because it increases the volumetric flow of gas and, consequently, the size of collector needed to meet gas-cleaning objectives. Water-cooled duct is often used where the gas temperature exceeds the safe limits of the low-cost materials.

For dilution cooling, louver-type dampers are often used to inject ambient air and provide fine temperature control. Controls can be used to provide full modulation of the damper or to provide open or closed operation. Emergency bypass damper systems and bypass duct/stacks are used where limiting excessive temperature is critical.

Fans

Because the static pressure across a gas-cleaning device varies depending on conditions, the fan should operate on the steep portion of the fan pressure-volumetric flow curve. This tends to provide less variation in the volumetric flow. An undersized fan has a steeper characteristic than an oversized fan for the same duty; however, it will be noisier.

In the preferred arrangement, the fan is located on the clean gas side of gas-cleaning equipment. Advantages of placing the fan at this point include the following:

- A fan on the clean gas side handles clean gases and minimizes abrasive exposure from the collected product.
- High-efficiency backward-blade and airfoil designs can be selected because accumulation on the fan wheel is not as great a factor.
- Escape of hazardous materials through leaks is minimized.
- The collector can be installed inside the plant, even near the process, because any leakage in the duct or collector will be into the system and will not increase the potential for exposure. However, the fan itself should be mounted outside, so that the positive-pressure duct is outside the work environment.

For economic reasons, a fan may be located on the contaminant-laden side of the gas-cleaning equipment if the contaminants are relatively nonabrasive, and especially if the equipment can be located outdoors. This arrangement should be avoided because of the potential for leakage of concentrated contaminants to the environment. In some instances, however, the collector housing design, duct design, and energy savings of this arrangement reduce costs.

Most scrubbers are operated on the suction side of the fan. This not only eliminates the leakage of contaminants into the work area, but also allows for servicing the unit while it is in operation.

Additionally, such an arrangement minimizes corrosion of the fan. Stacks on the exhaust streams of scrubbers should be arranged to drain condensate rather than allow it to accumulate and reenter the fan.

Fabric filters require special consideration. When new, clean fabric is installed in a collector, the resistance is low, and the fan motor may be overloaded. This overloading may be prevented during start-up by using a temporary throttling damper in the main duct, for example, on the clean side of the filter in a pull-through system. Overloading may also be prevented by using a backward-curved blade (nonoverloading fan) on the clean side of the collector.

DUST- AND SLURRY-HANDLING EQUIPMENT

Once the particulate matter is collected, new control problems arise from the need to remove, transport, and dispose of the material from the collector. A study of all potential methods for handling the collected material, which might be a hazardous waste, must be an integral part of initial system design.

Hoppers

Dust collector hoppers are intended only to channel collected material to the hopper's outlet, where it is continuously discharged. When hoppers are used to store dust, the plates and charging electrodes of electrostatic precipitators can be shorted electrically, resulting in failure of entire electrical sections. Fabric collectors, particularly those that have the gas inlet in the hopper, are usually designed on the assumption that the hopper is not used for storage and that collected waste will be continuously removed. High dust levels in the hoppers of fabric collectors often result in high dust reentrainment. This reentrainment causes increased operating pressure and the potential for fabric damage. The excessive dust levels not only expose the system to potentially corrosive conditions and fire/explosion hazards, but also place increased structural demands on the system.

Aside from misuse of hoppers for storage, common problems with dust-handling equipment include (1) plugging of hoppers, (2) blockage of dust valves with solid objects, and (3) improper or insufficient maintenance.

Hopper auxiliaries to be considered include (1) insulation, (2) dust level indicators, (3) rapper plates, (4) vibrators, (5) heaters, and (6) "poke" holes.

Hopper Discharge. Dust is often removed continuously from hoppers by means of rotary valves. Alternative equipment includes the double-flap valve, or vacuum system valves. Wet electrostatic precipitators and scrubbers often use sluice valves and drains to ensure that insoluble particulate remains in suspension during discharge.

Dust Conveyors

Larger dust collectors are fitted with one or more conveyors to feed dust to a central discharge location or to return it to a process. Drag, screw, and pneumatic conveyors are commonly used with dust collectors. Sequential start-up of conveyor systems is essential. Motion switches to monitor operation of the conveyor are useful.

Dust Disposal

Several methods are available for disposal of collected dust. It can be emptied into dumpsters in its as-collected dry form or be pelletized and hauled to a landfill. It can also be converted to a slurry and pumped to a settling pond or to clarification equipment. The advantages and disadvantages of each method are beyond the scope of this chapter, but they should be evaluated for each application.

Slurry Treatment

When slurry from wet collectors cannot be returned directly to the process or tailing pond, liquid clarification and treatment systems can be used for recycling the water to prevent stream pollution. Stringent stream pollution regulations make even a small discharge of bleed water a problem. Clarification equipment may include settling tanks, sludge-handling facilities, and, possibly, centrifuges or vacuum filters. Provisions must be provided for handling and disposal of dewatered sludge, so that secondary pollution problems do not develop.

OPERATION AND MAINTENANCE

A planned program for operation and maintenance of equipment is a necessity. Such programs are becoming mandatory because of the need for operators to prove continuous compliance with emission regulations. Good housekeeping and record-keeping will also help prolong the life of the equipment, support a program for positive relations with regulatory and community groups, and aid problem-solving efforts, should nonroutine maintenance or service be necessary.

A typical program includes the following minimum requirements (Stern et al. 1984):

- Central location for filing equipment records, warranties, instruction manuals, etc.
- Lubrication and cleaning schedules
- Planning and scheduling of preventive maintenance (including inspection and major repair)
- Storeroom and inventory system for spare parts and supplies
- Listing of maintenance personnel (including supplier contacts and consultants)
- Costs and budgets for activities associated with operation and maintenance of the equipment
- Storage for special tools and equipment

Corrosion

Because high-temperature gas cleaning often involves corrosive materials, chemical attack on system components must be anticipated. This is especially true if the temperature in a gas-cleaning system falls below the moisture or acid dew point. Housing insulation should be such that the internal metal surface temperature is 20 to 30°F greater than the moisture and/or acid dew point at all times. In applications with fabric filters where alkali materials are injected into the gas stream to react with acid gases, care must be taken to protect the clean side of the housing downstream of the fabric from corrosion.

Fires and Explosions

Industrial gas-cleaning systems often concentrate combustible materials and expose them to environments that are hostile and difficult to control. These environments also make fires difficult to detect and stop. Industrial gas-cleaning systems are, therefore, potential fire or explosion hazards (Billings and Wilder 1970; EEI 1980; Frank 1981).

Fires and explosions in industrial process exhaust streams are not generally limited to gas-cleaning equipment. Ignition may take place in the process itself, in the duct, or in exhaust system components other than the gas-cleaning equipment. Once uncontrolled combustion begins, it may propagate throughout the system. Workers around pollution control equipment should never open access doors to gas-cleaning equipment when a fire is believed to be in process; the fire could easily transform into an explosion.

The following devices help maintain a safe particulate control system.

Explosion Doors. An explosion door or explosion relief valve permits instantaneous pressure relief for equipment when the pressure reaches a predetermined level. Explosion doors are mandatory for certain applications to meet OSHA, insurance, or National Fire Protection Association (NFPA) regulations.

Detectors. Temperature-actuated switches or infrared sensors can be used to detect changes in the inlet-to-outlet temperature difference or a localized, elevated temperature that might signal a fire within the gas-cleaning system or a process upset. These detectors can be used to activate bypass dampers, trigger fire alarm/control systems, and/or shut down fans.

Fire Control Systems. Inert gas and water spray systems can be used to control fires in dust collectors. They are of little value in controlling explosions.

REFERENCES

AIHA. 1989. *Odor thresholds for chemicals with established occupational health standards*. American Industrial Hygiene Association, Akron, OH.

Beck, A.J. 1975. Heat treat engineering. *Heat Treating* (January).

Beltran, M.R. 1972. Smoke abatement for textile finishers. *American Dyestuff Reporter* (August).

Beltran, M.R. and H. Surati. 1976. Heat recovery vs. evaporative cooling on organic electrostatic precipitators. *Proceedings of the American Institute of Plant Engineers/Rossnagel & Associates Sixth Annual Industrial Air Pollution Control Seminar* (April 6, 1976), Cherry Hill, NJ.

Billings, C.E. and J. Wilder. 1970. *Handbook of fabric filter technology*, vol. 1: Fabric filter systems study. NTIS *Publication* PB 200 648, 2-201. National Technical Information Service, Springfield, VA.

Boll, C.H. 1976. Recovering solvents by adsorption. *Plant Engineering* (January).

Chopyk, J. and M.C. Larkin. 1982. Smoke and odor subdued with two-stage precipitator. *Plant Services* (January).

DOE. 1979. The coating industry: Energy savings with volatile organic compound emission control. *Report* TID-28706, U.S. Department of Energy, Washington, D.C.

EEI. 1980. Air preheaters and electrostatic precipitators fire prevention and protection (coal fired boilers). *Report of the Fire Protection Committee* 06-80-07 (September). Edison Electric Institute, Washington, D.C.

EPA. 2004. Air quality criteria for particulate matter. EPA 600/P-99/002aF-bF. U.S. Environmental Protection Agency, Washington, D.C. http://cfpub2.epa.gov/ncea/cfm/recordisplay.cfm?deid=87903. (October 2004)

Frank, T.E. 1981. Fire and explosion control in bag filter dust collection systems. *Proceedings of the Conference on the Hazards of Industrial Explosions from Dusts, New Orleans, LA* (October).

Grandjacques, B. 1977. Carbon adsorption can provide air pollution control with savings. *Pollution Engineering* (August).

IGCI. 1964. Determination of particulate collection efficiency of gas scrubbers. *Publication* 1. Industrial Gas Cleaning Institute, Washington, D.C.

Kane, J.M. and J.L. Alden. 1982. *Design of industrial ventilation systems*. Industrial Press, New York.

Lapple, C.E. 1951. Processes use many collection types. *Chemical Engineering* (May):145-151.

Lapple, C.E. and H.J. Kamack. 1955. Performance of wet scrubbers. *Chemical Engineering Progress* (March).

MCA. 1968. *Odor thresholds for 53 commercial chemicals*. Manufacturing Chemists Association, Washington, D.C. (October).

Rossnagel, W.B. 1973. Condensing/precipitator systems on organic emissions. *Proceedings of the Third Annual Industrial Air Pollution Control Seminar* (May 8, 1973), Paramus, NJ.

Rudnick, S.N., J.L.M. Koehler, K.P. Martin, D. Leith, and D.W. Cooper. 1986. Particle collection efficiency in a venturi scrubber: Comparison of experiments with theory. *Environmental Science & Technology* 20(3): 237-242.

Sauerland, W.A. 1976. Successful application of electrostatic precipitators on asphalt saturator emissions. *Proceedings of the American Institute of Plant Engineers/Rossnagel & Associates Sixth Annual Industrial Air Pollution Control Seminar* (April 6, 1976), Cherry Hill, NJ.

Semrau, K.T. 1977. Practical process design of particulate scrubbers. *Chemical Engineering* (September):87-91.

Shabsin, J. 1985. Clean plant air PLUS energy conservation. *Fastener Technology* (April).

Stern, A.C., R.W. Boubel, D.B. Turner, and D.L. Fox. 1984. *Fundamentals of air pollution control*, 2nd ed. Chapter 25, Control Devices and Systems. Academic Press, San Diego.

Thiel, G.R. 1977. Advances in electrostatic control techniques for organic emissions. *Proceedings of the Seventh Annual Industrial Air Pollution/Contamination Control Seminar* (March 29, 1977), Paramus, NJ.

Thiel, G.R. 1983. Cleaning and recycling plant air . . . Improvement of air cleaner performance and recirculation procedures. *Plant Engineering* (January 6).

Wade, G., J. Wigton, J. Guillory, G. Goldback, and K. Phillips. 1978. Granular bed filter development program. U.S. DOE *Report* FE-2579-19 (April).

White, H.J. 1963. *Industrial electrostatic precipitation.* Addison-Wesley, Reading, MA.

BIBLIOGRAPHY

ACGIH. 2006. *Industrial ventilation: A manual of recommended practice*, 23rd ed. Committee on Industrial Ventilation, American Conference of Governmental Industrial Hygienists, Cincinnati, OH.

Bergin, M.H., D.Y.H. Pui, T.H. Kuehn, and W.T. Fay. 1989. Laboratory and field measurements of fractional efficiency of industrial dust collectors. *ASHRAE Transactions* 95(2):102-112.

Buonicore, A.J. and W.T. Davis, eds. 1992. *Air pollution engineering manual.* Van Nostrand Reinhold, New York.

Crynack, R.B. and J.D. Sherow. 1984. Use of a mobile electrostatic precipitator for pilot studies. *Proceedings of the Fifth Symposium on the Transfer and Utilization of Particulate Control Technology* 2:3-1.

Deutsch, W. 1922. Bewegung und Ladung der Elektrizitätzträger im Zylinderkondensator. *Annalen der Physik* 68:335.

DuBard, J.L. and R.F. Altman. 1984. Analysis of error in precipitator performance estimates. *Proceedings of the Fifth Symposium on the Transfer and Utilization of Particulate Control Technology* 2:2-1.

EPA. 1994. Quality assurance handbook for air pollution measurement systems. *Report* EPA-600R94038A. Environmental Protection Agency, Washington, D.C.

Faulkner, M.G. and J.L. DuBard. 1984. A mathematical model of electrostatic precipitation, 3rd ed. *Publication* EPA-600/7-84-069a. Environmental Protection Agency, Washington, D.C.

GPO. Annual. *Code of federal regulations* 40(60). U.S. Government Printing Office, Washington, D.C. Revised annually and published in July.

Hall, H.J. 1975. Design and application of high voltage power supplies in electrostatic precipitation. *Journal of the Air Pollution Control Association* 25(2).

HEW. 1967. Air pollution engineering manual. *Publication* 999-AP-40. Department of Health and Human Services (formerly Department of Health, Education, and Welfare), Washington, D.C.

NIOSH. 1978. A recommended approach to recirculation of exhaust air. *Publication* 78-124. National Institute of Occupational Safety and Health, Washington, D.C.

Noll, C.G. 1984. Electrostatic precipitation of particulate emissions from the melting of borosilicate and lead glasses. *Glass Technology* (April).

Noll, C.G. 1984. Demonstration of a two-stage electrostatic precipitator for application to industrial processes. *Proceedings of the Second International Conference on Electrostatic Precipitation*, Kyoto, Japan (November), pp. 428-434.

Oglesby, S. and G.B. Nichols. 1970. *A manual of electrostatic precipitator technology.* NTIS PB-196-380. National Technical Information Service, Springfield, VA.

Sink, M.K. 1991. Handbook: Control technologies for hazardous air pollutants. *Report* EPA/625/6-91/014. Environmental Protection Agency, Washington, D.C.

SIP. 1991. State implementation plans and guidance available from the Regional Offices of the U.S. EPA and state environmental authorities.

White, H.J. 1974. Resistivity problems in electrostatic precipitation. *Journal of the Air Pollution Control Association* 24(4).

CHAPTER 30

AUTOMATIC FUEL-BURNING SYSTEMS

FUEL-BURNING systems provide a means to mix fuel and air in the proper ratio, ignite it, control the position of the flame envelope within the combustion chamber, and control a fuel-flow rate for safe combustion-heat energy release for space conditioning, water heating, and other processes. This chapter covers the design and use of automatic fuel-burning systems. The fuel can be gaseous (e.g., natural or liquefied petroleum gas), liquid (primarily the lighter grades of fuel oil or biodiesel), or solid (e.g., coal, or renewable items such as wood or corn). For discussion of some of these fuels, their combustion chemistry, and thermodynamics, see Chapter 18 of the 2005 *ASHRAE Handbook—Fundamentals*.

GENERAL CONSIDERATIONS

TERMINOLOGY

The following terminology for combustion systems, equipment, and fuel-fired appliances is consistent with usage in gas-fired appliance standards of the American National Standards Institute (ANSI) and Canadian Standards Association, the National Fire Protection Association's *National Fuel Gas Code* (ANSI Z223.1/NFPA 54), and the Canadian Standards Association's *Natural Gas and Propane Installation Code* (CSA *Standard* B149.1).

Air, circulating. Air distributed to habitable spaces for heating, cooling, or ventilation.

Air, dilution. Air that enters a draft hood or draft regulator and mixes with flue gas.

Air, excess. Air that passes through the combustion chamber in excess of the amount required for complete (stoichiometric) combustion.

Air, primary. Air introduced into a burner that mixes with fuel gas before the mixture reaches the burner ports.

Air, secondary. Air supplied to the combustion zone downstream of the burner ports.

Appliance. Any device that uses a gas, a liquid, or a solid as a fuel or raw material to produce light, heat, power, refrigeration, or air conditioning.

Draft. Negative static pressure, measured relative to atmospheric pressure; thus, positive draft is negative static pressure. Draft is the force (buoyancy of hot flue gas or other form of energy) that produces flow and causes pressure drop through an appliance combustion system and/or vent system. See Chapter 34 for additional information.

Equipment. Devices other than appliances, such as supply piping, service regulators, sediment traps, and vents in buildings.

Flue. General term for passages and conduit through which flue gases pass from the combustion chamber to the outdoors.

Flue gas. Products of combustion plus excess air in appliance flues, heat exchangers, and vents.

Input rate. Fuel-burning capacity of an appliance in Btu/h as specified by the manufacturer. Appliance input ratings are marked on appliance rating plates.

Vent. Passageway used to convey flue gases from appliances or their vent connectors to the outdoors.

Vent gas. Products of combustion plus excess air and dilution air in vents.

SYSTEM APPLICATION

The following considerations are important in the design, specification, and/or application of systems for combustion of fossil fuels.

Safety. Safety is of prime concern in the design and operation of automatic fuel-burning appliances. For more information, see the sections on Safety and Controls. Appliance standards and installation codes (e.g., ANSI *Standard* Z21.47/CSA *Standard* 2.3 for gas-fired central furnaces and ANSI Z223.1/NFPA 54, *National Fuel Gas Code*, in the United States) provide minimum safety requirements. Appliance manufacturers may include additional safety components to address hazards not covered by appliance standards and installation codes.

Suitability for Application. The system must meet the requirements of the application, not only in heating capacity, but also in its ability to handle the load profile. It must be suitable for its environment and for the substance to be heated.

Combustion System Type. System operation is very much a function of the type of burner(s), means for moving combustion products through the system, proper combustion air supply, and venting of combustion gases to the outdoors.

Efficiency. Efficiency can be specified in various ways, depending on the application. Stack loss and heat output are common measures, but for some applications, transient operation must be considered. In very high-efficiency appliances, heat extraction from combustion products may cool vent gas below its dew point, so condensation of water vapor in the combustion products must be handled, and venting design must consider corrosion by combustion products, as well as their lack of buoyancy.

Operating Control. Heat load or process requirements may occur in batches or may be transient events. The burner control system must accommodate those requirements, and the combustion system must be able to respond to the controls.

The preparation of this chapter is assigned to TC 6.10, Fuels and Combustion.

Emissions. For safety and air quality reasons, combustion products must not contain excessive levels of noxious materials, notably carbon monoxide, oxides of nitrogen, unburned hydrocarbons, and particulate material such as soot.

Fuel Provision. Liquid and solid fuels, liquefied gases, and some gaseous fuels require space for storage. Gaseous and liquid fuels require appropriate piping for the fuel-burning system. Fuel storage and delivery provisions must be of adequate capacity and must be designed to ensure safe operation.

Sustainability. Fuel-burning appliances consume fuel resources and produce combustion products that are emitted to the atmosphere. The effect of these inherent characteristics can be minimized by using highly efficient systems with very low emissions of undesirable substances. (See the section on Efficiency and Emission Ratings.) Appliances using environmentally neutral (nonfossil) fuel, such as biofuels processed from vegetable oils and biomass material (e.g., forestry waste), are available. New technologies are also emerging. Through photosynthesis, biomass chlorophyll captures energy from the sun by converting carbon dioxide from the atmosphere and water from the ground into carbohydrates, complex compounds composed of carbon, hydrogen, and oxygen. When these carbohydrates are burned, they are converted back into carbon dioxide and water, and release the sun's energy. In this way, biomass stores solar energy and is renewable and carbon-neutral (UCS 2007).

Venting, Combustion Air Supply, and Appliance Installation. Combustion product gases must be handled properly to ensure safety and satisfactory system operation. Adequate air supply must be provided for combustion and ventilation. Appliances must be located to provide safe clearance from combustible material and for convenient service.

Standards and Codes. Building codes typically require that fuel-burning appliances be design-certified or listed to comply with nationally recognized standards. Appliance construction, safe operation, installation practices, and emissions requirements are often specified. In some locations, codes require special restraint for seismic or high wind conditions.

Cost. The choice of fuel-burning system is often based on it being the least expensive way to provide the heat needed for a process. The basic cost of energy tends to narrow the choices, but the total cost of purchase and ownership should dictate the final decision. Initial cost is the cost of the appliance(s), associated equipment and controls, and installation labor. Operating cost includes the cost of fuel, other utilities, maintenance, depreciation, and various ongoing charges, taxes, fees, etc. Energy cost analysis may indicate that one fuel is best for some loads and seasons, and another fuel is best for other times. Substantial operating cost is incurred if skilled personnel are required for operation and maintenance. Sometimes these costs can be reduced by appliances and control systems that automate operation and allow remote monitoring of system performance and maintenance requirements. Warranties should be considered. See Chapter 36 of the 2007 *ASHRAE Handbook—HVAC Applications* for a thorough discussion of costs.

SAFETY

All appliance systems must either operate safely or have a way to sense unsafe operation, and safely and promptly shut off the fuel supply before injury or property damage occurs. Safe and unsafe operation sensing and control is generally designed into the combustion control system. Examples of what controls must detect, evaluate, and act on include the following:

- Time to achieve fuel ignition
- Sufficient combustion air and/or flue gas flow rates
- Fuel flow rate (e.g., gas orifice pressure)
- Loss of flame
- Heat exchange operation (e.g., circulating air blower operating speed and timing for furnaces)
- Flame containment (flame rollout)
- Appliance component temperatures
- Loss of control power supply

EFFICIENCY AND EMISSION RATINGS

Heating capacity may be the primary factor in selecting fuel-burning appliances, but efficiency and emission ratings are often of equal importance to building owners and governmental regulators.

Steady-State and Cyclic Efficiency

Efficiency calculations are discussed in Chapter 18 of the 2005 *ASHRAE Handbook—Fundamentals*. Boiler and furnace efficiencies are discussed in Chapters 31 and 32 of this volume.

Stack Efficiency. Stack efficiency is a widely used rating approach based on measurement of the temperature and composition of gases exhausted by fuel-burning appliances. Knowing the oxygen or carbon dioxide concentration of the flue gases and the fuel's hydrocarbon content provides a measure of stack mass flow. In conjunction with flue gas temperature, these data allow determination of energy loss in the flue gas exiting the stack. The difference between stack loss and energy input is assumed to be useful energy, and stack efficiency is the ratio of that useful energy to energy input, expressed as a percentage. Generally, the rating is applied to steady-state combustion processes. Flue gas carbon dioxide and oxygen concentrations are affected mostly by the fuel's hydrocarbon content and by the appliance's combustion system design. Flue gas temperature is mostly affected by the appliance's heat exchanger design.

Heat Output Efficiency. Some rating standards require actual measurement of heat transferred to the substance being heated. Heat output measurement accounts for all heat losses, not just those in the flue gases. Nonstack heat loss, often called **jacket loss**, is difficult to measure and may be quite small, but can be accounted for by measuring heat output. The ratio of heat output to energy input, expressed as a percentage, is the heat output efficiency.

Load Profile Efficiency. U.S. Federal Trade Commission rules require that some types of residential appliances be rated under protocols that consider load profile. Residential and commercial space-heating furnaces and boilers, for example, are rated by their **annual fuel utilization efficiency (AFUE)**, which considers steady-state efficiency, heat-up and cooldown transients, and off-season energy consumption by gas pilot burners. Residential storage-type water heaters are rated under a protocol that requires measurement of energy consumption over a 24 h period, during which prescribed amounts of heated water are drawn. A water heater **energy factor** E_f is calculated from the measurements. The E_f rating accounts for standby losses (i.e., energy loss through the tank and fittings that does not go into the water). Ratings and discussion of AFUE and energy factors can be found in ASHRAE *Standard* 103 and in product directories by the Gas Appliance Manufacturers Association.

Emissions

Regulated Flue Gas Constituents. Appliance safety standards and environmental regulations specify limits for various substances that may be found in combustion flue gases. Substances most often regulated are carbon monoxide, oxides of nitrogen, and soot. Limits for sulfur oxides, unburned hydrocarbons, and other particulate matter may also be specified. Rules vary with location, type of installation, and type of combustion appliance. Regulations that restrict fuel sulfur content are generally intended to reduce sulfur oxide emissions, which may also reduce particulate emissions under certain conditions.

Flue Gas Concentration Limits. Standards and codes often specify maximum concentration levels permitted in flue gases. Because flue gases may be diluted by air, requirements invariably specify that measured concentration levels be adjusted by calculation to a standard condition. In appliance certification standards, that condition is typically the air-free level (i.e., what the concentration would be if there were no excess air). Carbon monoxide, for example, is often limited to 400 parts per million air-free. Air quality regulations sometimes specify the maximum level for a fixed degree of excess air. For oxides of nitrogen, the level is often specified in parts per million at 3% oxygen. The calculation, in effect, adds or subtracts dilution air to reach a condition at which oxygen concentration is 3% of the exhaust gas volume.

Emission per Unit of Useful Heat. In some U.S. jurisdictions, regulations for gas-fired residential furnaces and water heaters require that emission of oxides of nitrogen not exceed a specified level in nanograms per joule of useful heat. Measured flue gas emission levels are compared with the benefit in terms of heat output. Under this rating method, high efficiency is rewarded. Regulations for new installations may differ from those for existing systems and may be more stringent.

Mass Released to Atmosphere. Emissions from large fuel-fired appliances are often regulated at the site in terms of mass released to the atmosphere. Limits may be expressed, for example, in terms of pounds per million Btu burned or tons per year.

GAS-BURNING APPLIANCES

GAS-FIRED COMBUSTION SYSTEMS

Gas-burning combustion systems vary widely, the most significant differences being the type of burner and the means by which combustion products are moved through the system. Gas input rate control also has a substantial effect on combustion system design.

Burners

A primary function of a gas burner is mixing fuel gas and combustion air in the proper ratio before their arrival at the flame. In a **partially-aerated burner (Bunsen burner)**, only part of the necessary combustion air is mixed with the gas ahead of the flame. This primary air is typically about 30 to 50% of the stoichiometric air (i.e., that amount of air necessary for complete combustion of the gas). Combustion occurs at the point where adequate secondary air enters the combustion zone and diffuses into the mixture. In most cases, secondary air entry continues downstream of the burner and heat release is distributed accordingly. The total of primary and secondary air typically ranges from 140 to 180% of the stoichiometric air (i.e., 40 to 80% excess air).

Most often, partially aerated burners are **atmospheric** or **natural-draft burners** (i.e., they operate without power assist of any kind), which have the advantage of quiet operation. Fuel gas is injected from a pressurized gas supply through an injector (orifice) to form a gas jet, which propels discharged gas into the burner throat, entraining primary air by viscous shear. Primary air may also be drawn into the burner throat by venturi action. Fuel gas and air are mixed in a mixing tube before their arrival at the burner ports where burning occurs. A typical partially aerated burner is illustrated in Figure 1.

A **premix burner** is a power burner in which all or nearly all of the combustion air is mixed with the fuel gas before arrival at the flame. Because the necessary air is present at the flame front, combustion and heat release take place in a compact zone and there is no need for secondary aeration. Combustion quality (i.e., emission performance) tends to be better than that of partially aerated burners because of inherent mixing advantages, so premix burners can normally be operated at lower excess air levels (often 15 to 20%). Low excess air increases flame temperature, which enhances heat exchange but imposes greater thermal stress on the combustion chamber and its components. Extremely low excess air may result in higher CO and/or NO_x emissions.

A fan is almost always necessary to force the mixture of gas and air through a premix burner. Airflow is three or four times that through partially aerated burners, and the associated pressure drop is normally too much to be handled by fuel gas entrainment or stack draft. In general, appliances with premix burners are tuned more finely than those with partially aerated burners, to take advantage of their inherent advantages and to ensure reliable operation. A typical premix burner system is illustrated in Figure 2.

Combustion System Flow

In broad terms, flow through the combustion system is by natural draft or is motivated by some sort of fan assist. In a **natural-draft system**, the low density of hot combustion products creates a buoyant flow through the combustion chamber, heat exchanger, and venting system (chimney or stack). Historically, natural-draft systems used atmospheric burners, but emission and efficiency regulations have spawned designs that use fan-assisted burners in conjunction with natural-draft flow through the venting system. Chimneys or venting must be appropriate for natural-draft flow; see the discussion of venting in the Applications section.

Fan-assisted combustion systems have become common. A fan pushes or pulls combustion air and fuel gas through the burner, and combustion products through the combustion chamber and heat exchanger (and, in some cases, the venting). Some fan-assisted systems use atmospheric burners, applying the fan power mainly to force flow through an enhanced heat exchange process. Serpentine heat exchangers, for example, typically require fan assist because they have too much pressure drop to operate in natural-draft mode. In systems with significant burner pressure drop, such as that of a premix burner, fan power is required for the burner as well. Fan-assisted systems can operate with or without pressurizing the vent, depending on the flow rate of the combustion system, flue gas conditions (temperature and buoyancy), resistance of the venting system, and location of an induced-draft fan, if used. For more information, see the discussion of venting in the Applications section.

Push-through or **forced-draft systems** (Figure 3) use a fan to force air into the burner at positive pressure (higher than atmospheric). In some designs, it may also pressurize the combustion chamber and heat exchanger, and, in other designs, also the venting. **Pull-through** or **induced-draft systems** (Figure 4) use a fan in the appliance near the flue collar or vent outlet to pull flue gas through the combustion chamber and heat exchanger. These systems operate with negative pressure at all points before the fan.

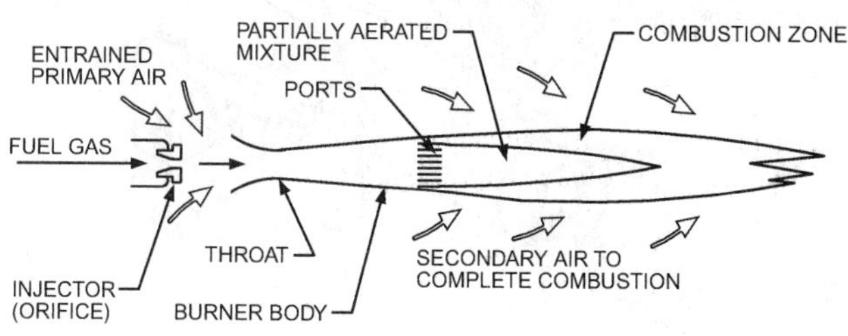

Fig. 1 Partially Aerated (Bunsen) Burner

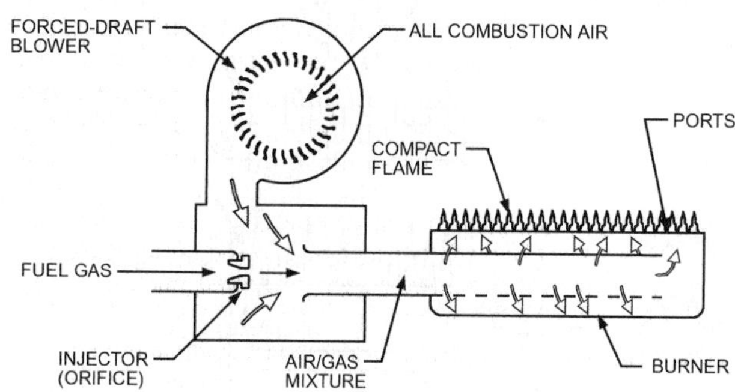

Fig. 2 Premix Burner

Packaged power burners are often used in factory-built heating appliances and in field-assembled installations. These burners include all gas- and air-handling components, along with ignition controls, housing, mechanical means for mounting to the heating appliance, and connection of fuel gas and electric power. They tend to have a gun-type configuration (i.e., gas and air are mixed in an outlet tube with burner head, the latter inserted into the heating appliance's combustion chamber). Packaged power burners may also include special hardware and controls for gas input rate control, and some may include special features to reduce combustion emissions. Flue gas recirculation (i.e., returning some combustion products to the flame) is sometimes used to reduce production of oxides of nitrogen. A typical packaged power burner is shown in Figure 5.

Pulse combustion is a process in which combustion system flow is motivated by low-frequency pressure pulses created in the combustion chamber by cyclic/repetitive self-generating ignition of an air/gas mixture. It provides low emission levels and enhanced heat transfer. The oscillating nature of flow through the system provides a beneficial "scrubbing" effect on heat exchanger surfaces. Additional discussion of pulse combustion is provided in Chapter 18 of the 2005 *ASHRAE Handbook—Fundamentals.*

Ignition

Safety standards and codes specify requirements for ignition and proof of flame presence. Ignition must be immediate, smooth, and complete. Once flame is established, the ongoing presence of the ignition source or the flame itself must be ensured (i.e., flame supervision by the combustion control must detect loss of pilot and/or main flame and immediately shut off fuel gas flow). **Pilot burners** have been used very effectively for decades in appliances such as small residential water heaters and in very large field-assembled installations. The pilot flame may be detected by a temperature sensor (thermocouple) or by various electronic sensing systems. Pilot flames may be continuous or ignited only when there is a demand for heat (**intermittent pilot** operation).

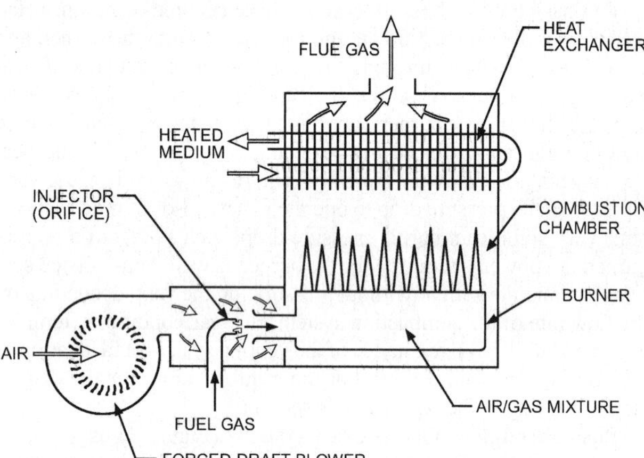

Fig. 3 Forced-Draft Combustion System

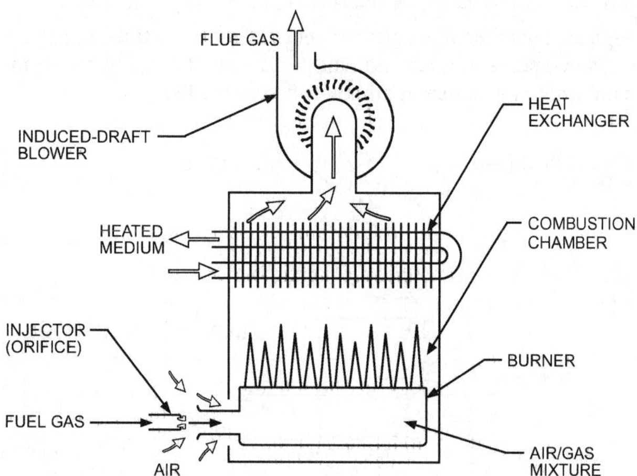

Fig. 4 Induced-Draft Combustion System

In some types of appliances, **direct ignition** is common. A **spark igniter** or a **hot surface igniter** is applied directly at the main burner ports to ignite the gas/air mixture. Direct-ignition systems also include a means, usually electronic, to sense presence of (supervise) the flame.

Ignition system standards and installation codes usually include requirements for other parameters such as flame failure response time, trial for ignition, and combustion chamber purging. For listed or design-certified appliances, applicable requirements have been test-verified by listing or certifying agencies. Other appliances and those installed in some building occupancy classes may be subject to special ignition and flame safety requirements in building and safety codes or by insurance underwriters. (See the section on Controls for more detail.)

Input Rate Control

A wide range of heat input is sometimes required of gas-fired appliances. Space heating, for example, is often done by zones, and must work under a wide range of outdoor conditions. Some water-heating and process applications also require a wide gas input rate range, and transients can be very steep input rate swings. Gas burners with staged or modulating control can be applied to meet these requirements.

Staged systems can be operated at discrete input rate levels, from full rated input to preset lower rates, sometimes with airflow remaining at the full-rate level and sometimes with proportional control of combustion air. Efficiency can be enhanced when combustion air is proportionally controlled, because flame temperature can remain high while the amount of heat exchanger surface per unit of input effectively becomes larger. A common staging approach uses a two-stage gas pressure regulator to change fuel gas pressure at an injector or metering orifice(s). In other designs, staging is accomplished by operating groups of individual burners or combustion chambers, each under control of its own gas valve and, if necessary, having its own ignition control. An extension of this approach is to use multiple individual heating appliances, controlled such that one or more can be called on as needed to meet the demand.

Modulating burner systems vary the input rate continuously, from full rated input to a minimum value. Modulation may be done by a throttling device in the gas burner piping, or with a modulating gas pressure regulator. Modulating systems require special controllers to provide a signal to the gas flow control device that is in some way proportional to the demand. As with staged systems, some designs provide commensurate control of combustion air, whereas

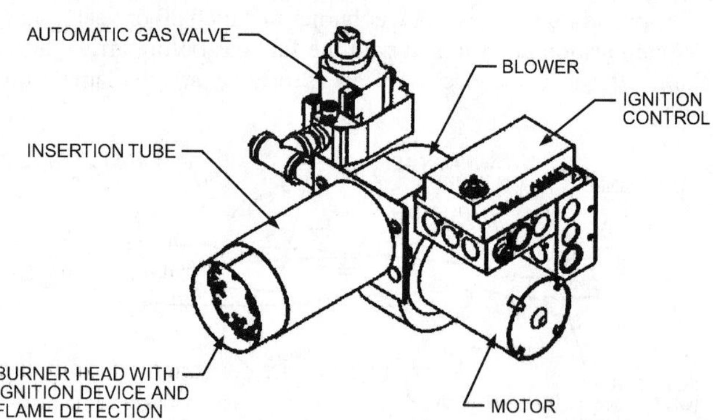

Fig. 5 Packaged Power Burner

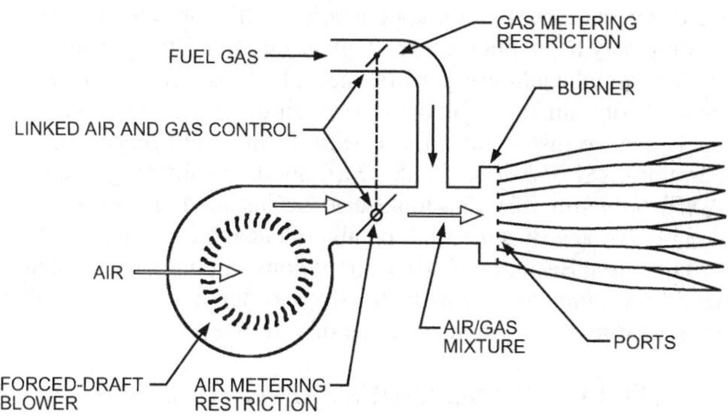

Fig. 6 Combustion System with Linked Air and Gas Flow

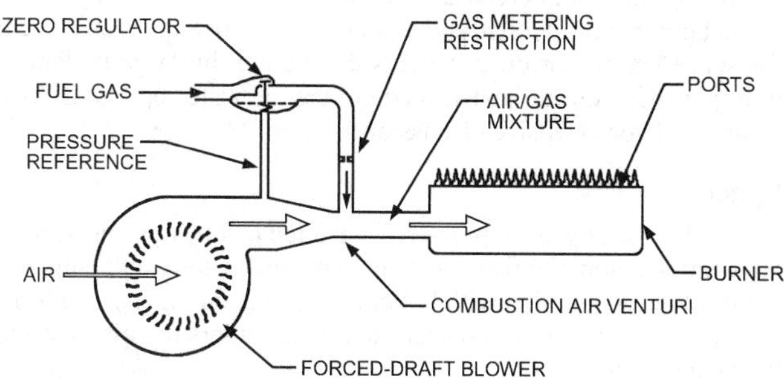

Fig. 7 Tracking Combustion System with Zero Regulator

others control only gas input rate with constant combustion airflow rate. Figure 6 illustrates a system in which combustion air and gas flow rates are throttled and linked.

Tracking burner systems, in which combustion air is controlled and gas follows proportionately, are a variation of modulating systems. These systems are designed to take advantage of the fact that flow of air and gas through orifices, venturis, or similar restrictions follows the Bernoulli equation. Special gas pressure regulators are used in conjunction with equalizer tubes or passages to link gas and air reference pressures. Airflow is controlled by dampers or a variable-speed fan. Changes in airflow bring about Bernoulli-governed changes in air pressure, which in turn change gas pressure proportionately. Usually, the design is such that the air/gas mixture proportions remain essentially constant at all input rates. This maintains high flame temperature and provides higher efficiency as the firing rate decreases. Special gas pressure regulators, referred to as **zero-pressure regulators**, **zero governors**, or **negative-pressure regulators**, are required in these systems. The pressure reference and venturi can be located upstream or downstream of the forced-draft blower; systems are available in either configuration. In either case, the pressure signal at the venturi inlet adjusts the gas pressure as needed to preserve venturi operation. An advantage to placing the venturi at the outlet is that the blower does not handle the combustible gas mixture. Consult manufacturer data for additional information. A tracking burner system is illustrated in Figure 7.

RESIDENTIAL APPLIANCES

Boilers

Residential space heating is often done with gas-fired low-pressure steam or hot-water boilers (i.e., steam boilers operating at 15 psig or less, or hot-water boilers operating at 160 psig or less with 250°F maximum water temperature). Steam or hot water is distributed to convectors, radiators, floor piping, fan coils, or other heat transfer devices in the space to be heated. Space temperature control may be by zone, in which case the boiler and distribution system must be able to accommodate reduced-load operation. Burners and combustion systems can be any of the previously described types, and some designs include input rate control. Rules of the U.S. Federal Trade Commission (FTC) and federal law require residential boilers with input rates less than 300,000 Btu/h to comply with minimum efficiency requirements, following the rating protocol of ASHRAE *Standard* 103. For hot-water boilers, 80% annual fuel utilization efficiency (AFUE) is required; for steam boilers, 75%. For ratings and technical information on rating protocol, see the *Consumers' Directory of Certified Efficiency Ratings* by the Gas Appliance Manufacturers Association (AHRI 2007). Manufacturers' literature also provides technical data and ratings.

Some boilers have low mass and essentially instantaneous response, whereas others have higher water volume and mass, which provides a degree of inherent storage capacity to better handle load change. Both steam and hot-water space-heating boilers are available in models having internal coils for service water heating. These **combination boilers** eliminate the need for a separate water heater, but they must be operated whenever service water may be needed, including times when space heating is not necessary. For a comprehensive discussion of boilers, see Chapter 31.

Forced-Air Furnaces

Central gas-fired, forced-air furnaces are the most common residential space-heating systems in the United States and Canada. Forced-air furnaces are available in configurations for upflow, downflow, and horizontal flow air distribution. Most have induced-draft combustion systems with Bunsen-type burners, and are typically of modular design (i.e., burner and heat exchanger modules are used in multiples to provide appliance models with a range of heating capacities). Some are available with staged or modulated input rate, and some have coordinated control of combustion air and circulating airflow. The FTC and U.S. federal law require furnaces with firing rates less than 225,000 Btu/h to meet minimum efficiency ratings: 75% AFUE for mobile home furnaces and 78% AFUE for other furnaces. The rating protocol for furnaces is different from that for boilers, and is outlined in ASHRAE *Standard* 103 and by AHRI (2007). Manufacturers' literature and AHRI (2007) provide ratings and other technical data. For detailed discussion of furnaces, see Chapter 32.

Water Heaters

In the United States, most residential water heaters are of the storage type (i.e., they have relatively low gas input rates and significant hot-water storage capacity). Typically, a single Bunsen-type burner is applied beneath a flue that rises through the stored water. Flue gas usually flows by natural draft. ANSI and Canadian Standards Association (CSA) appliance standards limit the input rate of these heaters to 75,000 Btu/h and require that water heaters manufactured since mid-2003 be designed to be **flammable vapor ignition resistant (FVIR)** because water heaters are often installed in garages and should not be an ignition source in case gasoline or other volatile or flammable substances may be present.

Instantaneous water heaters, often designed for mounting to a wall, are characterized by low water storage capacity and low mass, and have special burners and control systems designed for immediate response to demand for hot water. Burner input rate is typically linked to water flow rate and temperature, often by both hydraulic and electronic mechanisms. Because there is no storage, input rates tend to be higher than for storage heaters, and must be adequate for instantaneous demand. Standby losses of instantaneous water heaters are typically less than those of storage heaters. However, U.S. water use habits favor storage water heaters, which are used more extensively throughout the country.

In the United States, the efficiency of residential storage-type water heaters is based on a Department of Energy (DOE) protocol that requires measurement of energy consumption over a 24 h period, during which prescribed amounts of heated water are drawn.

An **energy factor** E_f is calculated from the measurements. The rating accounts for standby losses (i.e., heat lost from stored hot water by conduction and convection through tank walls, flue, and pipe fittings to the environment). Hot-water delivery flow also appears in ratings. For storage heaters, the **first-hour rating** (i.e., volume of water that can be drawn in the first hour of use) is provided. For instantaneous heaters, the **maximum flow rate** in gallons per minute is provided. See AHRI (2007) for ratings and details about the energy factor.

Combination Space- and Water-Heating Appliances

Residential appliances that provide both space and water heating are available in a variety of configurations. One configuration consists of a specially designed storage water heater that heats and stores water for washing activities or for use as a heat transfer medium in a space-heating fan-coil unit. Another configuration, common in Europe and Asia, is a wall-hung boiler that provides hot water for use in either mode. Storage or instantaneous heating capacity must be adequate to meet user demand for showers or other peak activity. Control systems normally prioritize hot-water consumption requirements over the need for space heating, which is allowed only when the hot-water demand has been satisfied. Descriptions and technical data for these and other configurations are available from manufacturers. Ratings are provided in a special section of AHRI (2007). The method for testing and rating of combination space and water heating appliances is specified in ASHRAE *Standard* 124.

Pool Heaters

Pool heaters are a special type of water heater designed specifically for handling high flow rates of water at relatively low water temperature. Various burner and combustion system approaches are used in pool heaters. Input rate control is not normally incorporated because swimming pools are of very high mass and do not change temperature rapidly. Consult manufacturer and pool industry technical data for pool heater selection and application factors.

Conversion Burners

Conversion burners are complete burner and control units designed for installation in existing boilers and furnaces. Atmospheric conversion burners may have drilled-port, slotted-port, or single-port burner heads. These burners are either upshot or inshot types. Figure 8 shows a typical atmospheric upshot gas conversion burner.

Several power burners are available in residential sizes. These are of gun-burner design and are desirable for furnaces or boilers with restricted flue passages or with downdraft passages.

Conversion burners for domestic application are available in sizes ranging from 40,000 to 400,000 Btu/h input, the maximum rate being set by ANSI *Standard* Z21.17/CSA 2.7. However, large gas conversion burners for applications such as apartment building heating may have input rates as high as 900,000 Btu/h or more.

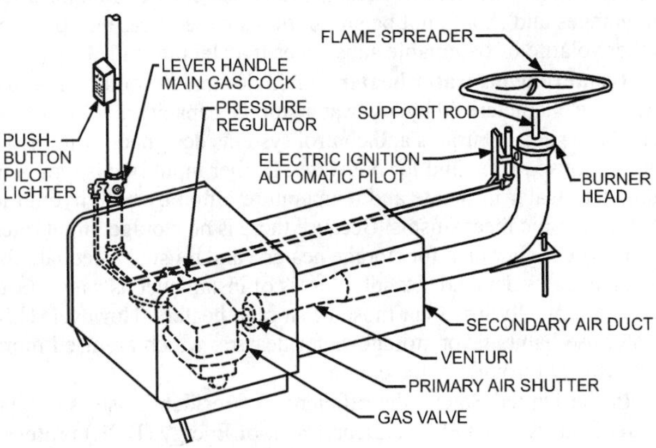

Fig. 8 Typical Single-Port Upshot Gas Conversion Burner

Successful and safe performance of a gas conversion burner depends on numerous factors other than those incorporated in the appliance, so installations must be made in strict accordance with current ANSI *Standard* Z21.8. Draft hoods conforming to current ANSI *Standard* Z21.12 should also be installed (in place of the dampers used with solid fuel) on all boilers and furnaces converted to burn gas. Because of space limitations, a converted appliance with a breeching over 12 in. in diameter is often fitted with a double-swing barometric regulator instead of a draft hood.

COMMERCIAL-INDUSTRIAL APPLIANCES

Boilers

Boilers for commercial and industrial application can be very large, both in physical size and input rate. Virtually any requirement for space heating or other process can be met by large boilers or multiple boilers. The heated medium can be water or steam. (See Chapter 31 for extensive discussion.)

Space Heaters

A wide variety of appliances is available for large air-heating applications. Some of them heat air by means of hot-water coils and are used in conjunction with boilers. Some accomplish space heating by means of a fuel-fired heat exchanger. Others fire directly into the heated space.

Forced-air fuel-burning furnaces for commercial and industrial application are essentially like those for residential use, but have larger heating and air-handling capacity. Input rates for single furnaces certified under ANSI *Standard* Z21.47/CSA *Standard* 2.3 (ANSI 2006) may be as high as 400,000 Btu/h. High capacity can also be provided by parallel (twinned) application of two furnaces. Most manufacturers provide kits to facilitate and address the special safety, mechanical, and control issues posed by twinned application. See manufacturer data and Chapter 32 for additional information about forced-air furnaces and their application.

Duct furnaces are fuel-fired appliances for placement in field-assembled systems with separate air-moving means. Combustion products heat air through heat exchangers mounted in the airstream. The combustion components, heat exchangers, and controls are prepackaged in a cabinet suitable for mounting in a duct system. For proper operation of the duct furnace, the airflow rate must be within the range specified by the manufacturer. See manufacturer data and Chapter 32 for additional information.

Unit heaters are free-standing appliances for heating large spaces without ductwork. They are often placed overhead, positioned to direct heat to specific areas. Typically, they incorporate fuel-fired heat exchangers and a fan or blower to move air through the exchangers and into the space. However, in some designs the fuel-fired exchangers are replaced by air-heating coils using hot water as the heating medium. Chapter 27 provides further information on unit heaters.

Direct gas-fired makeup-air heaters do not have heat exchangers. They heat large spaces by firing combustion products directly into the space, accompanied by a large quantity of dilution airflow. They have special burners capable of operation in high airflow and controls that allow operation over a wide fuel input rate range. As their name implies, they are used in applications requiring heated makeup air in conjunction with building exhaust systems. (See Chapter 27 for additional information.)

Infrared heaters radiate heat directly to surfaces and objects in a space. Air may be heated by convective heat transfer from the objects. They are suitable for overall heating of a building, but their selection is often based on their ability to radiate heat directly to people in limited areas without intentionally heating items or air in the space, such as in work stations in large open spaces that are not otherwise heated. Indirect infrared heaters use a radiating surface, such as a tube, between the combustion products and the space. Direct infrared heaters use the burner surface, typically a glowing ceramic or metal matrix, as the radiator. Chapter 15 provides additional information. See also manufacturers' data.

Water Heaters

Large-load water-heating applications, such as for large residence buildings, schools and hospitals, restaurants, and industrial processes, require water-heating appliances with correspondingly high fuel input rates, usually in conjunction with substantial hot-water storage. Large tank-type heaters with multiple flues are used in many intermediate-sized applications. They may operate as natural-draft systems, but forced- and induced-draft designs are common. Large applications or those with very high short-term water draw are often handled by means of large unfired storage tanks in conjunction with water-tube or other low-volume, high-input heaters. Fan-assisted combustion systems are increasingly common in those heaters. Premix burners may be used mainly to help meet emissions restrictions. Chapter 49 of the 2007 *ASHRAE Handbook—HVAC Applications* extensively discusses water-heating issues and appliances. ASHRAE *Standards* 118.1 and 118.2 provide methods of testing for rating commercial and residential water heaters, respectively.

Pool Heaters

Fuel-fired pool heaters are available in very large sizes, with fuel input rates ranging to several million Btu/h. They are designed to handle low-temperature water at high flow rates, and have sensitive temperature controls to ensure swimmer comfort and energy efficiency. Heating pool water with appliances not designed for that purpose can result in severe problems with combustion product condensation, corrosion, and/or scaling. See manufacturers' data for complete information.

APPLICATIONS

Gas-burning appliances cannot perform as intended unless they are properly installed and set up. Once an appliance of appropriate type, size, and features is selected, the location, fuel supply, air for combustion and ventilation, and venting must be considered and specified correctly. Other factors, notably elevation above sea level, must also be considered and handled.

Location

Listed appliances are provided with rating plate and installation information, with explicit requirements for location. The required clearance to combustible material is particularly important, to eliminate the hazard of fire caused by overheating. Other requirements may be less obvious. Adequate space must be provided for connecting ductwork, piping, and wiring, and for convenient maintenance and service. There must be access to chimneys and vents, and vent terminal locations must comply with specific requirements for safe discharge of combustion products without injury to people or damage to surroundings. Outdoor appliances must be located in consideration of wind effects and similar factors.

Local building codes provide basic rules for unlisted appliances and may impose additional requirements on listed appliances.

Gas Supply and Piping

Natural Gas. Natural gas is usually provided by the local gas utility. Most North American utilities provide substantial and reliable supply pressure, but it is important to verify adequate supply pressure during maximum simultaneous gas consumption by all appliances sharing the supply, and to design adequate supply piping between the utility supply point and the gas-burning appliance. On listed appliances, rating plates specify the minimum supply pressures at which the appliances will operate safely and as intended. This information is also provided in installation instructions, and is available from manufacturers before purchasing appliances.

Gas piping between the utility company meter and the appliance must provide adequate pressure when all concurrent loads operate at their maximum rate. Tables provided in ANSI Z223.1/NFPA 54 (*National Fuel Gas Code*), CSA B149.1 (CSA 2005), local codes, and elsewhere provide procedures for ensuring adequate pressure. For residential and light commercial services, utility companies typically provide gas at 7 in. of water. Building distribution piping is usually designed for a full-load pressure drop of less than 0.3 or 0.5 in. of water. Industrial and large-building applications are often supplied with gas at higher pressures; in that case, the distribution piping can be designed for larger pressure drop, but the end result must supply pressure to an individual appliance within the range required by its manufacturer. A pressure regulator may be required at the appliance to reduce pressure to comply with the rating plate pressure requirement.

Liquefied Petroleum Gas (LPG). LPG can contain a range of gas components. If it is not commercial propane or butane, the actual composition must be ascertained and accommodated. LPG is stored on site as a liquid at moderate pressure; it is vaporized as gas is drawn. A pressure regulator at the tank reduces pressure for distribution to the appliance through piping, subject to the same considerations as for natural gas. In the United States and Canada, normal supply pressure for residential and light commercial applications is 11 in. of water. Appliance rating plates and installation instructions typically require that supply pressure be maintained at or near that level. Piping must be designed to ensure adequate pressure when concurrent connected loads operate at their maximum input rates.

An important but sometimes overlooked issue is the need to provide heat to vaporize liquefied petroleum in the tank to deliver gas. In most residential applications, heat for vaporization is simply taken from outdoor air through the tank walls. This natural heat source may become inadequate, however, as the air temperature falls, draw rate increases, or tank liquid level falls. Commercial propane and butane have boiling point temperatures of –44°F and +32°F, respectively. As the LPG tank approaches the boiling point temperature, tank pressure falls to the point where the gas cannot be supplied at the required rate. In these cases and in high-demand applications, supplemental heat may be necessary to vaporize LPG. Information on selection and application of vaporization equipment is available from LPG dealers and distributors, from the National Propane Gas Association, and in various codes and standards.

Air for Combustion and Ventilation

In application of fuel-burning appliances, inadequate provision of air for combustion and ventilation is a serious mistake. In the worst scenario, shortage of combustion air results in incomplete combustion and production of poisonous carbon monoxide, which can kill. Inadequate ventilation of the space in which an appliance is installed can result in high ambient temperatures that stress the appliance itself or other appliances or materials in the vicinity. For those reasons, building codes and manufacturers' installation instructions include requirements for combustion and ventilation air supply. Requirements vary, with several factors having to do with how easily air can get to the appliance from the outdoors. Infiltration is seldom adequate, and it is usually necessary to provide dedicated means for supply of air for combustion and ventilation. In cold regions, measures should be taken to prevent the freezing of water pipes and other equipment by cold air in the appliance space. For more information, see Ackerman et al. (1995) and Dale et al. (1997).

Draft Control

Natural-draft appliances typically include a **draft hood** or **diverter** to decouple the combustion system from undesirable draft effects, notably the draft or pull of the chimney. These devices provide a path for dilution air to mix with flue gas before entering the connector between the appliance and the chimney, and to accommodate updraft and downdraft variations that occur in the field because of wind. A **barometric draft control** is a similar device that uses a damper to control the flow of dilution air into the vent. The damper of a barometric draft control can be manually adjusted to regulate the draft imposed on the appliance. Often, this is accomplished with special weights, in conjunction with measurement with a draft gage. Draft controls should be supplied by the appliance manufacturer as part of the appliance combustion controls. See Chapter 34 for design considerations for vent and chimney draft control.

Venting

Safety and technical factors must be considered in venting appliances, including some less obvious considerations. Consequences of incorrect venting are very serious, and can include production of lethal carbon monoxide, spilling of heat and combustion moisture indoors, and deterioration of the vent or chimney caused by condensation of water vapor from vent gas. High-efficiency appliances can produce combustion products at temperatures near or below their dew point. Venting those flue gases requires use of special materials and installation practices, and provision for condensate drainage and disposal.

Comprehensive guidance for design of venting systems is provided in Chapter 34. For many North American gas-fired appliances, however, ANSI Z21.47/CSA 2.3 and ANSI 21.13/CSA 4.9 require categorization by the type of vent system necessary for safe and effective operation. Appliances are tested to determine the temperature and pressure of vent gas released into the vent. The categories, which apply only to appliances design-certified as complying with standards having category specifications, are as follows:

Category	Vent Static Pressure	Vent Gas Temperature High Enough to Avoid Excessive Condensate Production in Vent?
I	Nonpositive	Yes
II	Nonpositive	No
III	Positive	Yes
IV	Positive	No

Category I and II appliances with a forced- or induced-draft blower to move combustion air and combustion products through the appliance flue create no pressure at the appliance flue exit (entrance to the venting system), and therefore do not augment draft in the vent.

Most local building codes include vent sizing tables and requirements that must be used for design of venting systems for category I appliances. Those tables are adopted from the *National Fuel Gas Code*, which distinguishes between appliances with draft hoods and appliances having fan-assisted combustion systems without draft hoods. In both cases, vent gases flow into the vent at category I conditions, but there is less dilution air with fan-assisted appliances than with draft-hood-equipped appliances. Vent gas flow and condensation tendencies differ accordingly. The tables specify

- Maximum (NAT Max) input rates for single-appliance vent systems and multiple-appliance vent connectors of given sizes for draft hood-equipped appliances
- Minimum (Fan Min) and maximum (Fan Max) input rates for single-appliance vent systems and multiple-appliance vent connectors of given sizes for fan-assisted appliances
- Maximum (Fan + NAT) input rates for multiple-appliance common vents of given sizes for combinations of draft-hood-equipped and fan-assisted appliance systems
- Maximum (Fan + Fan) input rates for multiple-appliance common vents of given sizes for fan-assisted appliance systems

Category II appliances are rare because it is difficult to vent low-temperature flue gas by its own buoyancy. Category III and IV appliances, with positive vent pressure, are common. Those appliances must be vented in accordance with the manufacturers' installation instructions, and require special venting materials. Category II and IV appliances also require venting designs that provide for collection and disposal of condensate. Condensate tends to be corrosive and may require treatment.

Appliances designed for installation with piping and terminals for both venting of flue gas and intake of combustion air directly to the appliance are called **direct-vent** (and sometimes, erroneously, **sealed combustion**) systems. (Sealed combustion systems take combustion air from outside the space being heated, not necessarily outdoors, and all flue gases are discharged outdoors; this is not a balanced system.) The vent and combustion air intake terminals of direct-vent appliances should be located outdoors, close to each other, so that they form a balanced system that is not adversely affected by winds from various directions. Vent pipe and combustion air intake pipe materials are provided or specified by the appliance manufacturer, and their use is mandatory. A variation in which only vent materials are specified, with combustion air taken directly from inside the conditioned space, is referred to as a **direct-exhaust** system. Most category III and IV systems are direct-vent or direct-exhaust systems.

Unlisted appliances must be vented in accordance with local building codes and the manufacturers' installation instructions. Clearance from vent piping to combustible material, mechanical support of vent piping, and similar requirements are also included in local codes.

Building Depressurization

Appliance operation can be affected by operation of other appliances and equipment in the building that change building pressure with respect to outdoor pressure. Building pressure can be reduced by bathroom and kitchen exhaust fans, cooktop range downdraft exhausters, clothes dryers, fireplaces, other fuel-burning appliances, and other equipment that removes air from the building. If building pressure is significantly lower than outdoor pressure, venting flue gases to the outdoors might be adversely affected and potentially hazardous combustion products may be spilled into the inhabited space, especially from category I and II appliances. Category III and IV appliances are less susceptible to venting and spillage problems, because these appliances produce pressure to force the flue gases through their vents to the outdoors. In addition, direct-vent appliances of all vent categories take their combustion air directly from the outdoors, which makes them even less susceptible to building depressurization. Wind can produce building depressurization, if building infiltration and exfiltration are unfavorably imbalanced.

Gas Input Rate

Gas input rate is the rate of heat energy input to an appliance, measured in Btu per hour.

The unit heating value of the fuel gas is expressed in Btu per cubic foot. **Higher heating value (HHV)** is commonly used to specify the heat available from gas when combusted. HHV includes all of the heat available by burning fuel gas delivered at 60°F and 14.735 psia (30 in. Hg) (i.e., standard conditions for the gas industry in North America) when combustion products are cooled to 60°F and water vapor formed during combustion is condensed at 60°F. 2005 See Chapter 18 of the 2005 *ASHRAE Handbook—Fundamentals* for more information on HHV.

In practical laboratory or field situations, fuel gas is not delivered at standard conditions. Determination of appliance input rate must include compensation for the actual temperature and pressure conditions.

In the laboratory, gas input rate is calculated with the following equation:

$$Q = \text{HHV} \times \text{VFR}\frac{(T_s \times P)}{(T \times P_s)}$$

where

Q = gas input rate, Btu/h
HHV = gas higher heating value at standard temperature and pressure, Btu/ft³
VFR = fuel gas volumetric flow rate at meter temperature and pressure, ft³/h
T_s = standard temperature, 520°R (60°F + 460°R)
P = fuel gas pressure in gas meter, psia
T = absolute temperature of fuel gas in meter, °R (fuel gas temperature in °F + 460°R)
P_s = standard pressure, 14.735 psia

Example 1. Calculate the gas input rate for 1025 Btu/ft³ HHV fuel gas, 75°F fuel gas temperature, 14.175 psia barometer pressure (1000 ft altitude), 100 ft³/h of fuel gas volumetric flow rate clocked at the meter with 7.0 in. of water fuel gas pressure in the gas meter.

$$P = B + P_f$$

where

B = local barometric pressure, psia
P_f = fuel gas pressure in gas meter, in. of water
P = 14.175 psia + (7.0 in. of water × 0.03613 psi/in. of water) = 14.42791 psia
T = 75°F + 460°R = 535°R

Thus,

$$Q = \frac{1025\text{Btu/ft}^3 \times 100\text{ft}^3\text{/h} \times (520°\text{R} \times 14.42791 \text{ psia})}{535°\text{R} \times 14.735 \text{ psia}}$$

$$= 97{,}550 \text{ Btu/h gas input rate}$$

In the field, input can be measured in two ways:

- With a gas meter, usually furnished by the gas supplier at the gas entrance point to a building
- By using the appliance burner gas injector (orifice) size and the manifold pressure [pressure drop through the injector (orifice)]

Gas input rate is calculated with the following equation, when fuel gas volumetric flow rate is measured using the gas supplier's meter.

$$Q = \text{HC} \times \text{VFR}$$

where

Q = gas input rate, Btu/h
HC = local gas heat content, Btu/ft³
VFR = fuel gas volumetric flow rate, ft³/h

Effect of Gas Temperature and Barometric Pressure Changes on Gas Input Rate

In the field, gas temperature is typically unknown, but is sometimes assumed to be a standard temperature such as 60°F; some meters have built-in temperature compensation to that temperature. Some gas suppliers also tabulate data for both local barometric pressure and local metering pressure. Add the meter gage pressure to the barometric pressure to get a correct fuel-gas absolute pressure. This methodology is used in ANSI Z223.1/NFPA 54 (NFPA 2006), the *National Fuel Gas Code*, sections 11.1.1, 11.1.2, and A.11.1.1, and Table A.11.1.1.

Also see the installation codes in the references for methods for using gas meters and for using injector (orifice) size with pressure drop to measure fuel gas flow rate.

Fuel Gas Interchangeability

Gas-burning appliances are normally set up for operation at their rated input with fuel gas of specified properties. Because fuel gases vary greatly, other gases cannot be substituted indiscriminately.

In most gas appliances, input rate is controlled by establishing a specified gas pressure difference (often referred to as **manifold pressure**) across one or more precisely sized orifices. Per the Bernoulli equation, gas velocity through the orifice is proportional to the square root of the pressure difference and inversely proportional to the square root of the density. The input rate also depends on the heat content of the gas and orifice size. Simplified to the basics, it is expressed as follows for particular conditions of temperature and barometric pressure:

$$Q = K \times \text{HHV} \times D_o^2\sqrt{\frac{P_m}{\text{SG}}} \tag{1}$$

where

Q = input rate, Btu/h
K = constant accounting for measurement units, orifice discharge coefficient, and atmospheric conditions
HHV = gas higher heating value, Btu/ft³
D_o = orifice diameter, in.
P_m = manifold pressure, in. of water
SG = gas specific gravity, dimensionless

If a substitute gas is introduced, only the higher heating value and specific gravity change. The orifice diameter, manifold pressure, and factors accounted for in the constant K do not change. Therefore, the new input is directly proportional to the gas higher heating value and inversely proportional to the square root of the specific gravity:

$$\frac{Q_2}{Q_1} = \frac{\left(\frac{\text{HHV}_2}{\sqrt{\text{SG}_2}}\right)}{\left(\frac{\text{HHV}_1}{\sqrt{\text{SG}_1}}\right)} \tag{2}$$

where subscripts 1 and 2 indicate the original and substitute gases, respectively. The ratio of higher heating value to the square root of the specific gravity has been named the **Wobbe index W**. The units of the Wobbe index are the same as those of the heating value because the specific gravity is dimensionless. Substituting for Wobbe index,

$$\frac{\text{HHV}}{\sqrt{\text{SG}}} = W \tag{3}$$

$$\frac{Q_2}{Q_1} = \frac{W_2}{W_1} \tag{4}$$

$$Q_2 = Q_1\frac{W_2}{W_1} \tag{5}$$

In other words, when one gas is substituted for another in an appliance and no other changes are made, the input rate changes in proportion to their Wobbe indices. Preservation of input rate does not necessarily ensure proper operation of an appliance, however. Ignition and burning characteristics can differ significantly for gases that have the same Wobbe index. To ensure safe, efficient operation, appliance manufacturers typically limit the range of gases that may be used.

Converting an appliance for use with a gas substantially different from the gas it was originally set up for requires changing the gas-handling components and/or adjustments. Typical natural gas and commercial propane, for example, have Wobbe indices of about 1335 and 2040 Btu/ft³, respectively. Based on that difference, substituting propane in an unaltered natural gas appliance results in overfiring of the appliance by about 53%, leading to appliance overheating and high production of soot and carbon monoxide. To accommodate the change, it is necessary to change the gas orifice size and, usually, the gas pressure regulator setting to achieve the same gas input rate. Acceptable performance may require additional appliance modifications to avoid other problems (e.g., resonance, poor flame carryover between ganged burners) that manufacturers must identify and resolve. The components necessary for conversion are normally provided by the appliance manufacturer, along with instructions for their installation and checkout of appliance operation.

Natural gas is increasingly being transported across oceans as **liquefied natural gas (LNG)**, shipped at high pressure and low temperature in specially designed ships. It is regasified in facilities at the destination, then distributed by utilities in conventional pipelines and service piping. Depending on the characteristics of a utility's existing gas supply, LNG can differ significantly in its mix of various hydrocarbons, inert gases, etc., and its burning characteristics may be of concern relative to the existing gas. Depending on the extent of the difference, a utility may mix it with other components to improve its compatibility with the existing appliance load. The Wobbe index is useful in evaluating the need for such accommodation.

Altitude

When gas-fired appliances are operated at altitudes substantially above sea level, three notable effects occur:

- Oxygen available for combustion is reduced in proportion to the atmospheric pressure reduction
- With gaseous fuels, the heat of combustion per unit volume of fuel gas (gas heat content) is reduced because of reduced fuel gas density in proportion to the atmospheric pressure reduction
- Reduced air density affects the performance and operating temperature of heat exchangers and appliance cooling mechanisms

In addition to reducing the gas heat content of fuel gas, reduced fuel gas density also causes increased gas velocity through flow metering orifices. The net effect is for gas input rate to decrease naturally with increases in altitude, but at less than the rate at which atmospheric oxygen decreases. This effect is one reason that derating is required when appliances are operated at altitudes significantly above sea level. Early research by American Gas Association Laboratories with draft hood-equipped appliances established that appliance input rates should be reduced at the rate of 4% per 1000 ft above sea level, for altitudes higher than 2000 ft above sea level (Figure 9).

Experience with recently developed appliances having fan-assisted combustion systems demonstrated that the 4% rule may not apply in all cases. It is therefore important to consult the manufacturer's listed appliance installation instructions, which are based on both how the combustion system operates and other factors such as impaired heat transfer. Note also that manufacturers of appliances having tracking-type burner systems may not require derating at altitudes above 2000 ft. In those systems, fuel gas and combustion airflow are affected in the same proportion by density reduction.

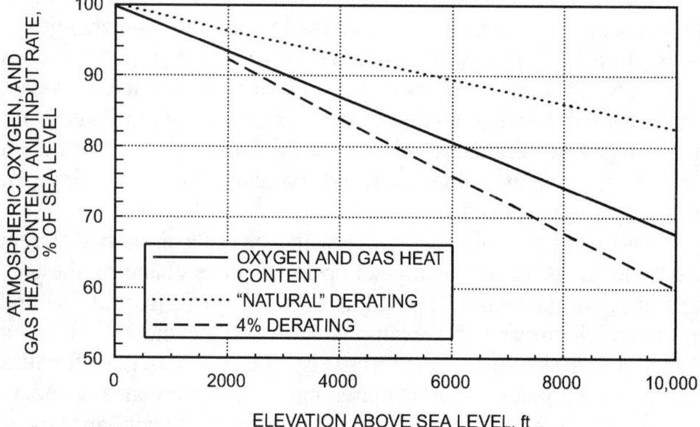

Note: Natural derating applies for fixed injector (orifice) size and pressure.

Fig. 9 Altitude Effects on Gas Combustion Appliances

In terms of end use, it is important for the appliance specifier to be aware that the heating capacity of appliances is substantially reduced at altitudes significantly above sea level. To ensure adequate delivery of heat, derating of heating capacity must also be considered and quantified.

By definition, fuel gas HHV value remains constant for all altitudes because it is based on standard conditions of 14.735 psia (30.00 in. Hg) and 60°F (520°R). Some fuel gas suppliers at high altitudes (e.g., at Denver, Colorado, at 5000 ft) may report fuel gas heat content at local barometric pressure instead of standard pressure. Local gas heat content can be calculated using the following equation:

$$HC = HHV \times \frac{B}{P_s}$$

where

HC = local gas heat content at local barometric pressure and standard temperature conditions, Btu/ft^3
HHV = gas higher heating value at standard temperature and pressure of 520°R and 14.735 psia, respectively, Btu/ft^3
B = local barometric pressure, psia (not corrected to sea level: do not use barometric pressure as reported by weather forecasters, because it is corrected to sea level)
P_s = standard pressure = 14.735 psia

For example, at 5000 ft, the barometric pressure is 12.23 psia. If HHV of a fuel gas sample is 1000 Btu/ft^3 (at standard temperature and pressure), the local gas heat content would be 830 Btu/ft^3 at 12.23 psia barometric pressure 5000 ft above sea level.

$$HC = 1000 \text{ Btu/ft}^3 \times 12.23 \text{ psia}/14.735 \text{ psia} = 830 \text{ Btu/ft}^3$$

Therefore, the local gas heat content of a sample of fuel gas can be expressed as 830 Btu/ft^3 at local barometric pressure of 12.23 psia and standard temperature or as 1000 Btu/ft^3 (HHV). Both gas heat contents are correct, but the application engineer must understand the difference to use each one correctly. As described earlier, the local heat content HC can be used to determine appliance input rate.

When gas heat value (either HHV or HC) is used to determine gas input rate, the gas pressure and temperature in the meter must also be considered. Add the gage pressure of gas in the meter to the local barometric pressure to calculate the heat content of the gas at the pressure in the meter. The gas temperature in the meter also affects the heat content of the gas in the meter. The gas heat value is directly proportional to the gas pressure and inversely proportional to its absolute temperature in accordance with the perfect gas laws, as illustrated in the following example calculations for gas input rate with either the HHV or the local heat content.

Example 2. Calculate the gas input rate for 1000 Btu/ft^3 HHV fuel gas, 100 ft^3/h volumetric flow rate of 75°F fuel gas at 12.23 psia barometer pressure (5000 ft altitude) with 7 in. of water fuel gas gage pressure in the gas meter.

HHV Method:

$$Q = HHV \times VFR_s$$

where

Q = fuel gas input rate, Btu/h
HHV = fuel gas higher heating value at standard temperature and pressure, Btu/ft^3
VFR_s = fuel gas volumetric flow rate adjusted to standard temperature and pressure, ft^3/h = $VFR(T_s \times P)/(T \times P_s)$
VFR = fuel gas volumetric flow rate at local temperature and pressure conditions, ft^3/h
T_s = standard temperature, 520°R (60°F + 460°R)
P = gas meter absolute pressure, psia (local barometer pressure + gas pressure in meter relative to barometric pressure)

= 12.23 psia + (7 in. of water × 0.03613 psi/in. of water)
= 12.48291 psia gas meter absolute pressure

T = absolute temperature of fuel gas, °R (fuel gas temperature in °F + 460°R)

P_s = standard pressure, 14.735 psia

Substituting given values into the equation for VFR_s gives

$$VFR_s = \frac{100 \text{ ft}^3/\text{h} \times 520°\text{F} \times 12.48291 \text{ psia}}{(75°\text{F} + 460°\text{R})14.735 \text{ psia}} = 82.341 \text{ ft}^3/\text{h}$$

Then,

$$Q = 1000 \text{ Btu/ft}^3 \times 82.341 \text{ ft}^3/\text{h} = 82{,}341 \text{ Btu/h}$$

Local Gas Heat Content Method: The local gas heat content is simply the HHV adjusted to local gas meter pressure and temperature conditions. The gas input rate is simply the observed volumetric gas flow rate times the local gas heat content.

$$Q = HC \times VFR$$

where

Q = gas input rate, Btu/h

HC = gas heat content at local gas meter pressure and temperature conditions, Btu/ft^3

VFR = fuel gas volumetric flow rate, referenced to local gas meter pressure and temperature conditions, ft^3/h

$$HC = HHV(T_s \times P)/(T \times P_s)$$

T_s = standard temperature, 520°R (60°F + 460°R)

P = gas meter absolute pressure, psia (local barometer pressure + gas pressure in gas meter relative to barometric pressure)

P_s = standard pressure = 14.735 psia

B = local barometric pressure = 12.230 psia

T = absolute temperature of fuel gas, 535°R (75°F fuel gas temperature + 460°R)

Substituting given values into the equation for HC gives

$$HC = \frac{1000 \times 520[12.23 + (7.0 \times 0.03613)]}{535 \times 14.735} = 823.41 \text{ Btu/ft}^3$$

Then,

$$Q = 823.41 \text{ Btu/ft}^3 \times 100 \text{ ft}^3/\text{h} = 82{,}341 \text{ Btu/h}$$

The gas input rate is exactly the same for both calculation methods.

OIL-BURNING APPLIANCES

An oil burner is a mechanical device for preparing fuel oil to combine with air under controlled conditions for combustion. Fuel oil is atomized at a controlled flow rate. Air for combustion is generally supplied with a forced-draft fan, although natural or mechanically induced draft can be used. Ignition is typically provided by an electric spark, although gas pilot flames, oil pilot flames, and hot-surface igniters may also be used. Oil burners operate from automatic temperature- or pressure-sensing controls.

Oil burners may be classified by application, type of atomizer, or firing rate. They can be divided into two major groups: residential and commercial/industrial. Further distinction is made based on design and operation; different types include pressure atomizing, air or steam atomizing, rotary, vaporizing, and mechanical atomizing. Unvented, portable kerosene heaters are not classified as residential oil burners or as oil heat appliances.

RESIDENTIAL OIL BURNERS

Residential oil burners ordinarily consume fuel at a rate of 0.5 to 3.5 gph, corresponding to input rates of about 70,000 to 500,000 Btu/h. However, burners up to 7 gph (about 1,000,000 Btu/h input) sometimes fall in the residential classification because of basic similarities in controls and standards. (Burners with a capacity of 3.5 gph and above are classified as commercial/industrial.) No. 2 fuel oil is generally used, although burners in the residential size range can also operate on No. 1 fuel oil. Burners in the 0.5 to 2.5 gph range are used not only for boilers and furnaces for space heating, but also for separate tank-type residential water heaters, infrared heaters, space heaters, and other commercial appliances.

Central heating appliances include warm-air furnaces and steam or hot water boilers. Oil-burning furnaces and boilers operate essentially the same way as their gas counterparts. NFPA *Standard* 31 prescribes correct installation practices for oil-burning appliances.

Steam or hot-water boilers are available in cast iron and steel. In addition to supplying space heating, many boilers are designed to provide hot water using tankless integral or external heat exchangers. Residential boilers designed to operate as direct-vent appliances are available.

Over 95% of residential burners manufactured today are high-pressure atomizing gun burners with retention-type heads (Figure 10). This type of burner supplies oil to the atomizing nozzle at pressures that range from 100 to 300 psi. A fan supplies air for combustion, and generally an inlet damper regulates the air supply at the burner. A high-voltage electric spark ignites the fuel by either **constant ignition** (on when the burner motor is on) or **interrupted ignition** (on only to start combustion). Typically, these burners fire into a combustion chamber in which draft is maintained. Increasingly, however, these burners are being fired into applications (including direct-vent and high-efficiency condensing applications) that have a low level of positive combustion chamber pressure.

Modern retention-head oil burners use 3450 rpm motors, and their fans achieve a maximum static pressure of about 3 in. of water. Older residential oil burners, which can still be found in the field, used lower-speed motors (1750 rpm) and achieved a maximum static pressure of about 1 to 1.5 in. of water. With higher fan static pressure, burner airflow is less sensitive to variations in appliance draft conditions. Higher-static-pressure burners operate better in applications with positive combustion chamber pressure. In many cases, these burners can be operated without a flue draft regulator.

Traditionally, oil burners are operated with a fuel pressure of 100 psi, and burners' nozzles are all rated for firing rate and spray angle at this pressure. Increasingly higher pressures (130 to 150 psi) are specified by manufacturers for some burners and applications, to achieve better atomization and combustion performance (Figure 11). Nozzle fuel firing rate increases in proportion to the square root of the fuel pressure.

Oil nozzle line heaters are used in some applications. These are very small electric heaters located adjacent to the nozzle adapter and typically integrated with the burner. These include positive-temperature-coefficient heaters that self-regulate to achieve a fuel temperature in the 120 to 150°F range. Heating fuel in this way improves atomization quality and reduces fuel firing rate.

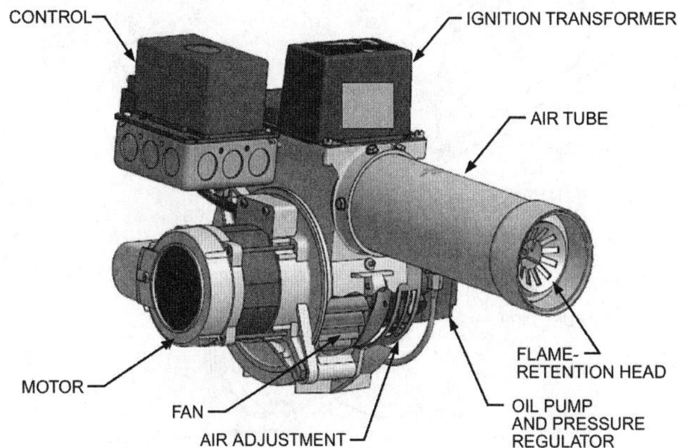

Fig. 10 High-Pressure Atomizing Gun Oil Burner

Electric fuel solenoid valves may be installed in the fuel line between the fuel pump and the nozzle and may be mounted directly on the fuel pump. These valves serve to provide sharp fuel flow starts and stops during cyclic operation and augment the function of the fuel pump and pressure-operated flow control valves. In addition, solenoid valves can be used to achieve pre- and post-purge operation in which the burner motor runs to provide purge airflow through the burner and appliance. The post purge can be useful in reducing nozzle temperatures after shutdown and preventing odors indoors, particularly with non-direct-vent applications.

Pressure-operated valves integrated with oil burner nozzle assemblies are also available. These valves also provide positive fuel cutoff after burner shutdown and avoid after-drip caused by heat transfer from the still-hot combustion chamber to the nozzle assembly and accumulation of fuel vapor pressure in the nozzle assembly and fuel line. When used, these nozzle valves require the fuel pump discharge pressure to be increased, using the integral pump pressure regulator to compensate for the added pressure drop of the valve.

The following designs are still in operation but are not a significant part of the residential market:

- The low-pressure atomizing gun burner differs from the high-pressure type in that it uses air at a low pressure to atomize oil.
- The pressure atomizing induced-draft burner uses the same type of oil pump, nozzle, and ignition system as the high-pressure atomizing gun burner.
- Vaporizing burners are designed for use with No. 1 fuel oil. Fuel is ignited electrically or by manual pilot.
- Rotary burners are usually of the vertical wall flame type.

Beyond the pressure atomizing burner, considerable developmental effort has been put into advanced technologies such as air atomization, fuel prevaporization, and pulsed fuel flow. Some of these are now used commercially in Europe. Reported benefits include reduced emissions, increased efficiency, and the ability to quickly vary the firing rate. See Locklin and Hazard (1980) for a historical review of technology for residential oil burners. More recent developments are described in the *Proceedings of the Oil Heat Technology Conferences and Workshops* (McDonald 1989-1998).

In Europe, there is increasing use of low-NO_x, **blue flame burners** for residential and commercial applications. Peak flame temperatures (and thermal NO_x emissions) are reduced by using high flame zone recirculation rates. In a typical boiler, a conventional yellow flame retention head oil burner has a NO_x emission of 90 to 110 ppm. A blue flame burner emits about 60 ppm. The blue flame burners require flame sensors for the safety control, which responds to the fluctuating light emitted. Blue flame burners are somewhat more expensive than conventional yellow flame burners. See Butcher et al. (1994) for additional discussion of NO_x and small oil burners.

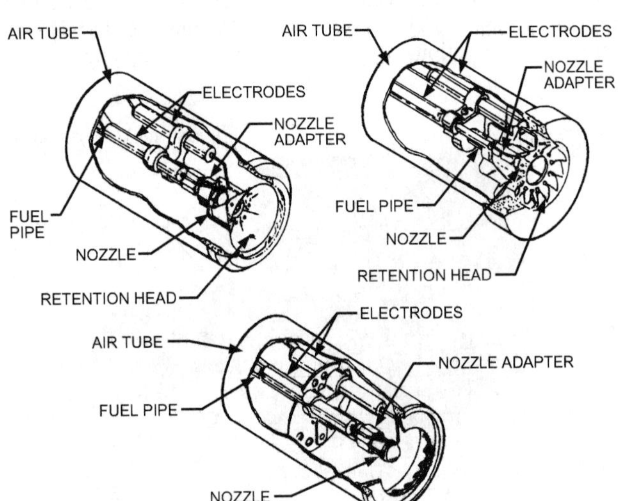

Fig. 11 Details of High-Pressure Atomizing Oil Burner

COMMERCIAL/INDUSTRIAL OIL BURNERS

Commercial and industrial oil burners are designed for use with distillate or residual grades of fuel oil. With slight modifications, burners designed for residual grades can use distillate fuel oils.

The commercial/industrial burners covered here have atomizers, which inject the fuel oil into the combustion space in a fine, conical spray with the apex at the atomizer. The burner also forces combustion air into the oil spray, causing an intimate and turbulent mixing of air and oil. Applied for a predetermined time, an electrical spark, a spark-ignited gas, or an oil igniter ignites the mixture, and sustained combustion takes place. Safety controls are used to shut down the burner upon failure to ignite.

All of these burners are capable of almost complete burning of the fuel oil without visible smoke when they are operated with excess air as low as 20% (approximately 12% CO_2 in the flue gases). Atomizing oil burners are generally classified according to the method used for atomizing the oil, such as pressure atomizing, return-flow pressure atomizing, air atomizing, rotary cup atomizing, steam atomizing, mechanical atomizing, or return-flow mechanical atomizing. Descriptions of these burners are given in the following sections, together with usual capacities and applications. Table 1 lists approximate size range, fuel grade, and usual applications. All burners described are available as gas/oil (dual-fuel) burners.

Pressure-Atomizing Oil Burners

This type of burner is used in most installations where No. 2 fuel oil is burned. The oil is pumped at pressures of 100 to 300 psi through a suitable burner nozzle orifice that breaks it into a fine mist and swirls it into the combustion space as a cone-shaped spray. Combustion air from a fan is forced through the burner air-handling parts surrounding the oil nozzle and is directed into the oil spray.

For smaller-capacity burners, ignition is usually started by an electric spark applied near the discharge of the burner nozzle. For burner capacities above 20 gph, a spark-ignited gas or an oil igniter is used.

Pressure-atomizing burners are designated commercially as forced-draft, natural-draft, or induced-draft burners. The forced-draft burner has a fan and motor with capacity to supply all combustion air to the combustion chamber or furnace at a pressure high enough to force the gases through the heat-exchange equipment without the assistance of an induced-draft fan or a chimney draft. Mixing of the fuel and air is such that a minimum of refractory material is required in the combustion space or furnace to support combustion. The natural-draft burner requires a draft in the combustion space.

Burner range, or variation in burning rate, is changed by simultaneously varying the oil pressure to the burner nozzle and regulating the airflow by a damper. This range is limited to about 1.6 to 1 for any given nozzle orifice. Burner firing mode controls for various capacity burners differ among manufacturers. Usually, larger burners have controls that provide variable heat inputs. If burner capacity is up to 15 gph, a staged control is typically used; if it is up to 25 gph, a **modulation control** is typically used. In both cases, the low burning rate is about 60% of the full-load capacity of the burner.

For pressure-atomizing burners, no preheating is required for burning No. 2 oil. No. 4 oil must be preheated to about 100°F for proper burning. When properly adjusted, these burners operate well with less than 20% excess air (approximately 12% CO_2); no visible smoke (approximately No. 2 smoke spot number, as determined by ASTM *Standard* D2156); and only a trace of carbon monoxide in the flue gas in commercial applications. In these applications, the

Table 1 Classification of Atomizing Oil Burners

Type of Oil Burner	Heat Range, 1000 Btu/h	Flow Volume, gph	Fuel Grade	Usual Application
Pressure-atomizing	70 to 7000	0.5 to 50	No. 2 (less than 25 gph)	Boilers
				Warm-air furnaces
			No. 4 (greater than 25 gph)	Appliances
Return-flow pressure-atomizing or modulating pressure-atomizing	3500 and above	25 and above	No. 2 and heavier	Boilers
				Warm-air furnaces
Air-atomizing	70 to 1000	0.5 to 70	No. 2 and heavier	Boilers
				Warm-air furnaces
Horizontal rotary cup	750 to 37,000	5 to 300	No. 2 for small sizes	Boilers
			No. 4, 5, or 6 for larger sizes	Large warm-air furnaces
Steam-atomizing (register-type)	12,000 and above	80 and above	No. 2 and heavier	Boilers
Mechanical atomizing (register-type)	12,000 and above	80 and above	No. 2 and heavier	Boilers
				Industrial furnaces
Return-flow mechanical atomizing	45,000 to 180,000	300 to 1200	No. 2 and heavier	Boilers

regulation of combustion airflow is typically based on smoke level or flue gas monitoring using analyzers for CO, O_2, or CO_2.

Burners with lower firing rates used to power appliances, residential heating boilers, or warm-air furnaces are usually set up to operate to about 50% excess air (approximately 10% CO_2).

Good operation of these burners calls for (1) a relatively constant draft (either in the furnace or at the breeching connection, depending on the burner selected), (2) clean burner components, and (3) good-quality fuel oil that complies with the appropriate specifications.

Return-Flow Pressure-Atomizing Oil Burners

This burner is a modification of the pressure-atomizing burner; it is also called a modulating-pressure atomizer. It has the advantage of a wide load range for any given atomizer: about 3 to 1 turndown (or variation in load) as compared to 1.6 to 1 for the straight pressure atomizing burner.

This wide range is accomplished by means of a return-flow nozzle, which has an atomizing swirl chamber just ahead of the orifice. Good atomization throughout the load range is attained by maintaining a high rate of oil flow and high pressure drop through the swirl chamber. The excess oil above the load demand is returned from the swirl chamber to the oil storage tank or to the suction of the oil pump.

The burning rate is controlled by varying oil pressure in both the oil inlet and oil return lines. Except for the atomizer, load range, and method of control, the information given for the straight pressure-atomizing burner applies to this burner as well.

Air-Atomizing Oil Burners

Except for the nozzle, this burner is similar in construction to the pressure-atomizing burner. Atomizing air and oil are supplied to individual parts within the nozzle. The nozzle design allows the oil to break up into small droplet form as a result of the shear forces created by the atomizing air. The atomized oil is carried from the nozzle through the outlet orifice by the airflow into the furnace.

The main combustion air from a draft fan is forced through the burner throat and mixes intimately with the oil spray inside the combustion space. The burner igniter is similar to that used on pressure-atomizing burners.

This burner is well suited for heavy fuel oils, including No. 6, and has a wide load range, or turndown, without changing nozzles. Turndown of 3 to 1 for the smaller sizes and about 6 or 8 to 1 for the larger sizes may be expected. Load range is varied by simultaneously varying the oil pressure, the atomizing air pressure, and the combustion air entering the burner. Some designs use relatively low atomizing air pressure (5 psi and lower); other designs use air pressures up to 75 psi. The burner uses from 2.2 to 7.7 ft^3 of compressed air per gallon of fuel oil (on an air-free basis).

Because of its wide load range, this burner operates well on modulating control.

No preheating is required for No. 2 fuel oil. The heavier grades of oil must be preheated to maintain proper viscosity for atomization. When properly adjusted, these burners operate well with less than 15 to 25% excess air (approximately 14 to 12% CO_2, respectively, at full load); no visible smoke (approximately No. 2 smoke spot number); and only a trace of carbon monoxide in the flue gas.

Horizontal Rotary Cup Oil Burners

This burner atomizes the oil by spinning it in a thin film from a horizontal rotating cup and injecting high-velocity primary air into the oil film through an annular nozzle that surrounds the rim of the atomizing cup.

The atomizing cup and frequently the primary air fan are mounted on a horizontal main shaft that is motor-driven and rotates at constant speed (3500 to 6000 rpm) depending on the size and make of the burner. The oil is fed to the atomizing cup at controlled rates from an oil pump that is usually driven from the main shaft through a worm and gear.

A separately mounted fan forces secondary air through the burner windbox. Secondary air should not be introduced by natural draft. The oil is ignited by a spark-ignited gas or an oil-burning igniter (pilot). The load range or turndown for this burner is about 4 to 1, making it well suited for operation with modulating control. Automatic combustion controls are electrically operated.

When properly adjusted, these burners operate well with 20 to 25% excess air (approximately 12.5 to 12% CO_2, respectively, at full load), no visible smoke (approximately No. 2 smoke spot number), and only a trace of carbon monoxide in the flue gas.

This burner is available from several manufacturers as a package comprising burner, primary air fan, secondary air fan with separate motor, fuel oil pump, motor, motor starter, ignition system (including transformer), automatic combustion controls, flame safety equipment, and control panel.

Good operation of these burners requires relatively constant draft in the combustion space. The main assembly of the burner, with motor, main shaft, primary air fan, and oil pump, is arranged for mounting on the boiler front and is hinged so that the assembly can be swung away from the firing position for easy access.

Rotary burners require some refractory in the combustion space to help support combustion. This refractory may be in the form of throat cones or combustion chamber liners.

Steam-Atomizing Oil Burners (Register Type)

Atomization is accomplished in this burner by the impact and expansion of steam. Oil and steam flow in separate channels through the burner gun to the burner nozzle. There, they mix before discharging through an orifice, or series of orifices, into the combustion chamber.

Combustion air, supplied by a forced-draft fan, passes through the directing vanes of the burner register, through the burner throat, and into the combustion space. The vanes give the air a spinning motion, and the burner throat directs it into the cone-shaped oil spray, where intimate mixing of air and oil takes place.

Full-load oil pressure at the burner inlet is generally some 100 to 150 psi, and the steam pressure is usually kept higher than the oil pressure by about 25 psi. Load range is accomplished by varying these pressures. Some designs operate with oil pressure ranging from 150 psi at full load to 10 psi at minimum load, resulting in a turndown of about 8 to 1. This wide load range makes the steam atomizing burner suited to modulating control. Some manufacturers provide dual atomizers within a single register so that one can be cleaned without dropping load.

Depending on the burner design, steam atomizing burners use from 1 to 5 lb of steam to atomize a gallon of oil. This corresponds to 0.5 to 3.0% of the steam generated by the boiler. Where no steam is available for start-up, compressed air from the plant air supply may be used for atomizing. Some designs allow use of a pressure atomizing nozzle tip for start-up when neither steam nor compressed air is available.

This burner is used mainly on water-tube boilers, which generate steam at 150 psi or higher and at capacities above 12,000,000 Btu/h input.

Oils heavier than grade No. 2 must be preheated to the proper viscosity for good atomization. When properly adjusted, these burners operate well with 15% excess air (14% CO_2) at full load, without visible smoke (approximately No. 2 smoke spot number), and with only a trace of carbon monoxide in the flue gas.

Mechanical Atomizing Oil Burners (Register Type)

Mechanical atomizing, as generally used, describes a technique synonymous with pressure atomizing. Both terms designate atomization of the oil by forcing it at high pressure through a suitable stationary atomizer.

The mechanical atomizing burner has a windbox, which is a chamber into which a fan delivers combustion and excess air for distribution to the burner. The windbox has an assembly of adjustable internal air vanes called an air register. Usually, the fan is mounted separately and connected to the windbox by a duct.

Oil pressure of 90 to 900 psi is used, and load range is obtained by varying the pressure between these limits. The operating range or turndown for any given atomizer can be as high as 3 to 1. Because of its limited load range, this type of burner is seldom selected for new installations.

Return-Flow Mechanical Atomizing Oil Burners

This burner is a modification of a mechanical atomizing burner; atomization is accomplished by oil pressure alone. Load ranges up to 6 or 8 to 1 are obtained on a single-burner nozzle by varying the oil pressure between 100 and 1000 psi.

The burner was developed for use in large installations such as on ships and in electric generating stations where wide load range is required and water loss from the system makes use of atomizing steam undesirable. It is also used for firing large hot-water boilers. Compressed air is too expensive for atomizing oil in large burners.

This is a register burner similar to the mechanical atomizing burner. Wide range is possible by using a return-flow nozzle, which has a swirl chamber just ahead of the orifice or sprayer plate. Good atomization is attained by maintaining a high rate of oil flow and a high pressure drop through the swirl chamber. Excess oil above the load demand is returned from the swirl chamber to the oil storage tank or to the oil pump suction. Control of burning rate is accomplished by varying the oil pressure in both the oil inlet and the oil return lines.

DUAL-FUEL GAS/OIL BURNERS

Dual-fuel, combination gas/oil burners are forced-draft burners that incorporate, in a single assembly, the features of the commercial/industrial-grade gas and oil burners described in the preceding sections. These burners have controls to ensure that the burner flame relay or programmer cycles the burner through post- and prepurge before starting again on the other fuel. Burner manufacturers for larger boilers design the special mechanical linkages needed to deliver the correct air/fuel ratios at full fire, low fire, or any intermediate rate. Smaller burners may have straight **on/off** firing. Larger burners may have low-fire starts on both fuels and use a common flame scanner. Smaller dual-fuel burners usually include pressure atomization of the oil. Air atomization systems are included in large oil burners.

Dual-fuel burners often have automatic changeover controls that respond to outdoor temperature. A special temperature control, located outdoors and electrically interlocked with the dual-fuel burner control system, senses outdoor temperature. When the outdoor temperature drops to the outdoor control set point, the control changes fuels automatically after putting the burner through post- and prepurge cycles. A manual fuel selection switch can be retained as a manual override on the automatic feature. These control systems require special design and are generally provided by the burner manufacturer.

The dual-fuel burner is fitted with a gas train and oil piping that is connected to a two-pipe oil system following the principles of the preceding sections. An oil reserve must be maintained at all times for automatic fuel changeover.

Boiler flue chimney connectors are equipped with special double-swing barometric draft regulators or, if required, sequential furnace draft control to operate an automatic flue damper.

Dual-fuel burners and their accessories should be installed by experienced contractors to ensure satisfactory operation.

EQUIPMENT SELECTION

Economic and practical factors (e.g., the degree of operating supervision required by the installation) generally dictate the selection of fuel oil based on the maximum heat input of the oil-burning appliance. For heating loads and where only one oil-burning appliance is operated at any given time, the relationship is as shown in Table 2 (which is only a guide). In many cases, a detailed analysis of operating parameters results in the burning of lighter grades of fuel oil at capacities far above those indicated.

Fuel Oil Storage Systems

All fuel-oil storage tanks should be constructed and installed in accordance with NFPA *Standard* 31 and with local ordinances.

Storage Capacity. Dependable and economical operation of oil-burning appliances requires ample and safe storage of fuel oil at the site. Design responsibility should include analysis of specific storage requirements as follows:

- Rate of oil consumption.
- Dependability of oil deliveries.
- Economical delivery lots. The cost of installing larger storage capacity should be balanced against the savings indicated by accommodating larger delivery lots. Truck lots and railcar lots vary with various suppliers, but the quantities are approximated as follows:

Small truck lots in metropolitan area	500 to 2000 gal
Normal truck lots	3000 to 5000 gal
Transport truck lots	5000 to 9000 gal
Rail tanker lots	8000 to 12000 gal

Tank Size and Location. Standard oil storage tanks range in size from 55 to 50,000 gal and larger. Tanks are usually built of steel; concrete construction may be used only for heavy oil. Unenclosed

Table 2 Guide for Fuel Oil Grades Versus Firing Rate

Maximum Heat Input of Appliance, 1000 Btu/h	Volume Flow Rate, gph	Fuel Grade
Up to 3500	Up to 25	2
3500 to 7000	25 to 50	2, 4, 5
7000 to 15,000	50 to 100	5, 6
Over 15,000	Over 100	6

tanks located in the lowest story, cellar, or basement should not exceed 660 gal capacity each, and the aggregate capacity of such tanks should not exceed 1320 gal unless each 660 gal tank is insulated in an approved fireproof room having a fire resistance rating of at least 2 h.

If storage capacity at a given location exceeds about 1000 gal, storage tanks should be underground and accessible for truck or rail delivery with gravity flow from the delivering carrier into storage. If the oil is to be burned in a central plant such as a boiler house, the storage tanks should be located, if possible, so that the oil burner pump (or pumps) can pump directly from storage to the burners. For year-round operation, except for storage or supply capacities below 2000 gal, at least two tanks should be installed to facilitate tank inspection, cleaning, repairs, and clearing of plugged suction lines.

When the main oil storage tank is not close enough to the oil-burning appliances for the burner pumps to take suction from storage, a supply tank must be installed near the oil-burning appliances and oil must be pumped periodically from storage to the supply tank by a transport pump at the storage location. Supply tanks should be treated the same as storage tanks regarding location within buildings, tank design, etc. On large installations, it is recommended that standby pumps be installed as a protection against heat loss in case of pump failure.

Because all piping connections to underground tanks must be at the top, such tanks should not be more than 10.5 ft from top to bottom to avoid pump suction difficulties. (This dimension may have to be less for installations at high altitudes.) At sea level, the total suction head for the oil pump must not exceed 14 ft.

Connections to Storage Tank. All piping connections for tanks over 275 gal capacity should be through the top of the tank. Figure 12 shows a typical arrangement for a cylindrical storage tank with a heating coil as required for No. 5 or 6 fuel oils. The heating coil and oil suction lines should be located near one end of the tank. The maximum allowable steam pressure in such a heating coil is 15 psi. The heating coil is unnecessary for oils lighter than No. 5 unless a combination of high pour point and low outdoor temperature makes heating necessary.

A watertight access port with internal ladder provides access to the inside of the tank. If the tank is equipped with an internal heating coil, a second access port is required in order to permit withdrawal of the coil.

The fill line should be vertical and should discharge near the end of the tank away from the oil suction line. The inlet of the fill line must be outside the building and accessible to the oil delivery vehicle unless an oil transfer pump is used to fill the tank. When possible, the inlet of the fill line should be at or near grade where filling may be accomplished by gravity. For gravity filling, the fill line should be at least 2 in. in diameter for No. 2 oil and 6 in. in diameter for No. 4, 5, or 6 oils. Where filling is done by pump, the fill line for No. 4, 5, or 6 oils may be 4 in. in diameter.

An oil return line bringing recirculated oil from the burner line to the tank should discharge near the oil suction line inlet. Each storage tank should be equipped with a vent line sized and arranged in accordance with NFPA *Standard* 31.

Each storage tank must have a device for determining oil level. For tanks inside buildings, the gaging device should be designed and installed so that oil or vapor will not discharge into a building from the fuel supply system. No storage tank should be equipped

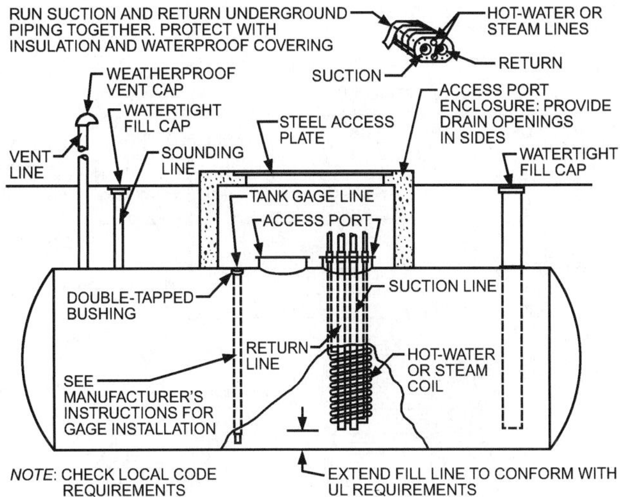

Fig. 12 Typical Oil Storage Tank (No. 6 Oil)

with a glass gage or any gage that, if broken, would allow oil to escape from the tank. Gaging by a measuring stick is permissible for outside or underground tanks.

Fuel-Handling Systems

The fuel-handling system consists of the pumps, valves, and fittings for moving fuel oil from the delivery truck or car into the storage tanks and from the storage tanks to the oil burners. Depending on the type and arrangement of the oil-burning appliances and the grade of fuel oil burned, fuel-handling systems vary from simple to quite complicated arrangements.

The simplest handling system would apply to a single burner and small storage tank for No. 2 fuel oil, similar to a residential heating installation. The storage tank is filled through a hose from the oil delivery truck, and the fuel-handling system consists of a supply pipe between the storage tank and the burner pump. Equipment should be installed on light-oil tanks to indicate visibly or audibly when the tank is full; on heavy-oil tanks, a remote-reading liquid-level gage should be installed.

Figure 13 shows a complex oil supply arrangement for two burners on one oil-burning appliance. For an appliance with a single burner, the change in piping is obvious. For a system with two or more appliances, the oil line downstream of the oil discharge strainer becomes a main supply header, and the branch supply line to each appliance includes a flowmeter, automatic control valve, etc. For light oils requiring no heating, all oil-heating equipment shown in Figure 13 would be omitted. Both a suction and a return line should be used, except for gravity flow in residential installations.

Oil pumps (steam or electrically driven) should deliver oil at the maximum rate required by the burners (this includes the maximum firing rate, the oil required for recirculating, plus a 10% margin).

The calculated suction head at the entrance of any burner pump should not exceed 10 in. Hg for installations at sea level. At higher elevations, the suction head should be reduced in direct proportion to the reduction in barometric pressure.

Oil temperature at the pump inlet should not exceed 120°F. Where oil burners with integral oil pumps (and oil heaters) are used and suction lift from the storage tank is within the capacity of the burner pump, each burner may take oil directly from the storage tank through an individual suction line unless No. 6 oil is used.

Where two or more tanks are used, the piping arrangement into the top of each tank should be the same as for a single tank so that any tank may be used at any time; any tank can be inspected, cleaned, or repaired while the system is in operation.

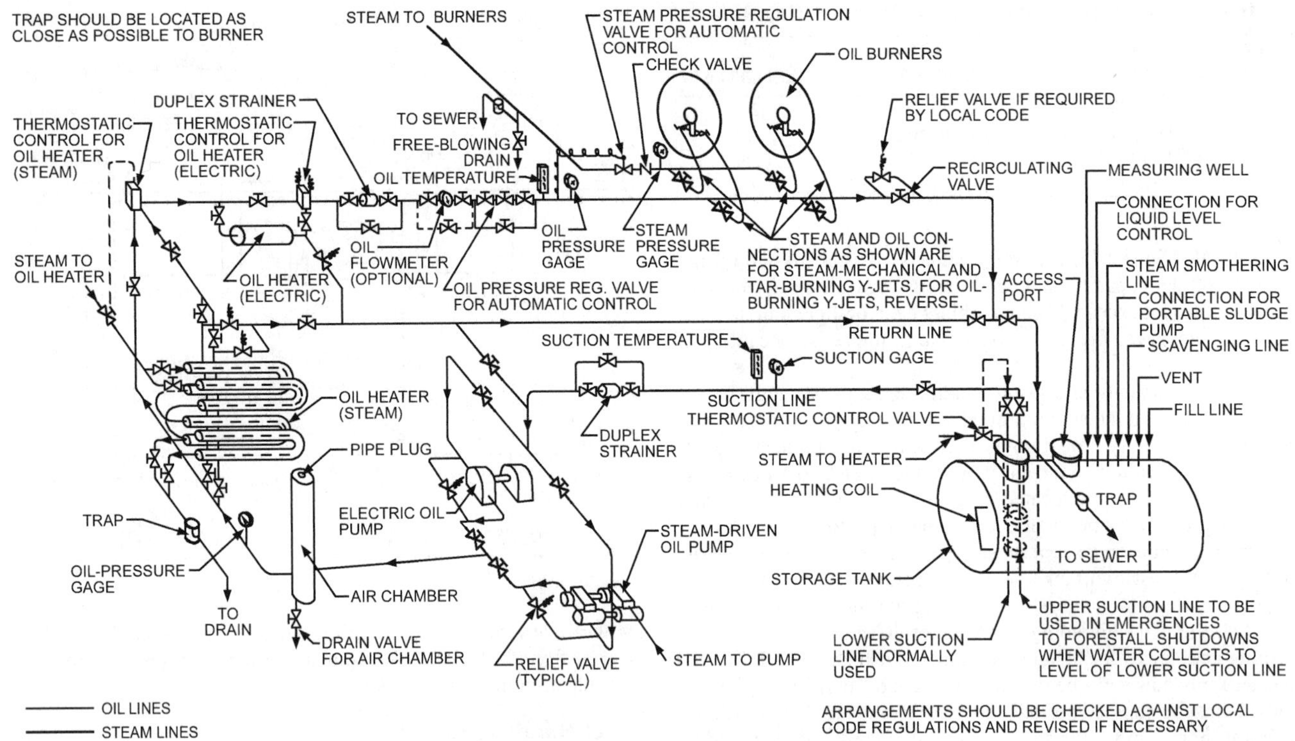

Fig. 13 Industrial Burner Auxiliary Equipment

The length of suction line between storage tank and burner pumps should not exceed 100 ft. If the main storage tank(s) are located more than 100 ft from the pumps, a supply tank should be installed near the pumps, and a transfer pump should be installed at the storage tanks for delivery of oil to the supply tank.

Central oil distribution systems comprising a central storage facility, distribution pumps or provision for gravity delivery, distribution piping, and individual fuel meters are used for residential communities, notably mobile home parks. Provisions of NFPA *Standard* 31, Installation of Oil Burning Equipment, should be followed in installing a central oil distribution system.

Fuel Oil Preparation System

Fuel oil preparation systems consist of oil heater, oil temperature controls, strainers, and associated valves and piping required to maintain fuel oil at the temperatures necessary to control the oil viscosity, facilitate oil flow and burning, and remove suspended matter.

Preparation of fuel oil for handling and burning requires heating the oil if it is No. 5 or 6. This decreases its viscosity so it flows properly through the oil system piping and can be atomized by the oil burner. No. 4 oil occasionally requires heating to facilitate burning. No. 2 oil requires heating only under unusual conditions.

For handling residual oil from the delivering carrier into storage tanks, the viscosity should be about 156 centistokes (cSt). For satisfactory pumping, viscosity of oil surrounding the inlet of the suction pipe must be 444 cSt or lower; for oil with a high pour point, the temperature of the entire oil content of the tank must be above the pour point.

Storage tank heaters are usually made of pipe coils or grids using steam or hot water at or less than 15 psi as the heating medium. Electric heaters are sometimes used. For control of viscosity for pumping, the heated oil surrounds the oil suction line inlet. For heating oils with high pour points, the heater should extend the entire length of the tank. All heaters have suitable thermostatic controls. In some cases, storage tanks may be heated satisfactorily by returning or recirculating some of the oil to the tank after it has passed through heaters located between the oil pump and oil burner.

Heaters to regulate viscosity at the burners are installed between the oil pumps and the burners. When required for small packaged burners, the heaters are either assembled integrally with the individual burners or mounted separately. The heat source may be electricity, steam, or hot water. For larger installations, the heater is mounted separately and is often arranged in combination with central oil pumps, forming a central oil pumping and heating set. The separate or central oil pumping and heating set is recommended for installations that burn heavy oils, have a periodical load demand, and require continuous circulation of hot oil during down periods.

Another system of oil heating to maintain pumping viscosity that is occasionally used for small- or medium-sized installations consists of an electrically heated section of oil piping. Low-voltage current is passed through the pipe section, which is isolated by nonconducting flanges.

The oil-heating capacity for any given installation should be approximately 10% greater than the maximum oil flow. Maximum oil flow is the maximum oil-burning rate plus the rate of oil recirculation.

Controls for oil heaters must be dependable to ensure proper oil atomization and avoid overheating of oil, which results in coke deposits inside the heaters. In steam or electric heating, an interlock should be included with a solenoid valve or switch to shut off the steam or electricity. During periods when the oil pump is not operating, the oil in the heater can become overheated and deposit carbon. Overheating also can be a problem with oil heaters using high-temperature hot water. Provisions must be made to avoid overheating the oil when the oil pump is not operating.

Oil heaters with low- or medium-temperature hot water are not generally subject to coke deposits. Where steam or hot water is used in oil heaters located after the oil pumps, the pressure of the steam or water in the heaters is usually lower than the oil pressure. Consequently, heater leakage between oil and steam causes oil to flow into the water or condensing steam. To prevent oil from entering the boilers, the condensed steam or the water from such heaters should be discarded from the system, or special equipment should be provided for oil removal.

Hot water oil heaters of double-tube-and-shell construction with inert heat transfer oil and a sight glass between the tubes are available. With this type of heater, oil leaks through an oil-side tube appear in the sight glass, and repairs can be made to the oil-side tube before a water-side tube leaks.

This discussion of oil-burning equipment also applies to oil-fired boilers and furnaces (see Chapters 31 and 32).

SOLID-FUEL-BURNING APPLIANCES

A mechanical stoker is a device that feeds a solid fuel into a combustion chamber. It supplies air for burning the fuel under automatic control and, in some cases, incorporates automatic ash and refuse removal.

CAPACITY CLASSIFICATION OF STOKERS

Stokers are classified according to their coal-feeding rates. Although some residential applications still use stokers, their main application is in commercial and industrial areas. The U.S. Department of Commerce, in cooperation with the Stoker Manufacturers Association, use the following classification:

Class 1: Capacity less than 60 lb of coal per hour
Class 2: Capacity 60 to less than 100 lb of coal per hour
Class 3: Capacity 100 to less than 300 lb of coal per hour
Class 4: Capacity 300 to less than 1200 lb of coal per hour
Class 5: Capacity 1200 lb of coal per hour and over

Class 1 stokers are used primarily for residential heating and are designed for quiet, automatic operation. These stokers are usually underfeed types and are similar to those shown in Figure 14, except that they are usually screw-feed. Class 1 stokers feed coal to the furnace intermittently, in accordance with temperature or pressure demands. A special control is needed to ensure stoker operation in order to maintain a fire during periods when no heat is required.

Class 2 and 3 stokers are usually of the screw-feed type, without auxiliary plungers or other means of distributing the coal. They are used extensively for heating plants in apartment buildings, hotels, and industrial plants. They are of the underfeed type and are available in both the hopper and bin-feed type. These stokers are also built in a plunger-feed type with an electric motor, steam, or hydraulic cylinder coal-feed drive.

Class 2 and 3 stokers are available for burning all types of anthracite, bituminous, and lignite coals. The tuyere and retort design varies according to the fuel and load conditions. Stationary grates are used on bituminous models, and clinkers formed from the ash accumulate on the grates surrounding the retort.

Class 2 and 3 anthracite stokers are equipped with moving grates that discharge ash into a pit below the grate. This ash pit may be located on one side or both sides of the grate and, in some installations, is big enough to hold the ash for several weeks of operation.

Class 4 stokers vary in details of design, and several methods of feeding coal are practiced. Underfeed stokers are widely used, although overfeed types are used in the larger sizes. Bin-feed and hopper models are available in underfeed and overfeed types.

Class 5 stokers are spreader, underfeed, chain or traveling grate, and vibrating grate. Various subcategories reflect the type of grate and method of ash discharge.

STOKER TYPES BY FUEL-FEED METHODS

Class 5 stokers are classified according to the method of feeding fuel to the furnace: (1) spreader, (2) underfeed, (3) chain or traveling grate, and (4) vibrating grate. The type of stoker used in a given installation depends on the general system design, capacity required, and type of fuel burned. In general, the spreader stoker is the most widely used in the capacity range of 75,000 to 400,000 lb/h because it responds quickly to load changes and can burn a wide variety of coals. Underfeed stokers are mainly used with small industrial boilers of less than 30,000 lb/h. In the intermediate range, the large underfeed stokers, as well as the chain and traveling grate stokers, are being displaced by spreader and vibrating grate stokers. Table 3 summarizes the major features of the different stokers.

Spreader Stokers

Spreader stokers use a combination of suspension burning and grate burning. As shown in Figure 15, coal is continually projected into the furnace above an ignited fuel bed. The coal fines are partially burned in suspension. Large particles fall to the grate and are burned in a thin, fast-burning fuel bed. Because this firing method provides extreme responsiveness to load fluctuations and because ignition is almost instantaneous on increased firing rate, the spreader stoker is favored over other stokers in many industrial applications.

The spreader stoker is designed to burn about 50% of the fuel in suspension. Thus, it generates much higher particulate loadings than other types of stokers and requires dust collectors to trap particulate material in the flue gas before discharge to the stack. To minimize carbon loss, fly carbon reinjection systems are sometimes used to return particles into the furnace for complete burnout. Because this process increases furnace dust emissions, it can be used only with highly efficient dust collectors.

Grates for spreader stokers may be of several types. All grates are designed with high airflow resistance to avoid formation of blowholes through the thin fuel bed. Early designs were simple stationary grates from which ash was removed manually. Later designs allowed intermittent dumping of the grate either manually or by a power cylinder. Both types of dumping grates are frequently used for small and medium-sized boilers (see Table 3). Also, both types are sectionalized, and there is a separate undergrate air chamber for each grate section and a grate section for each spreader stoker. Consequently, both the air supply and the fuel supply to one section

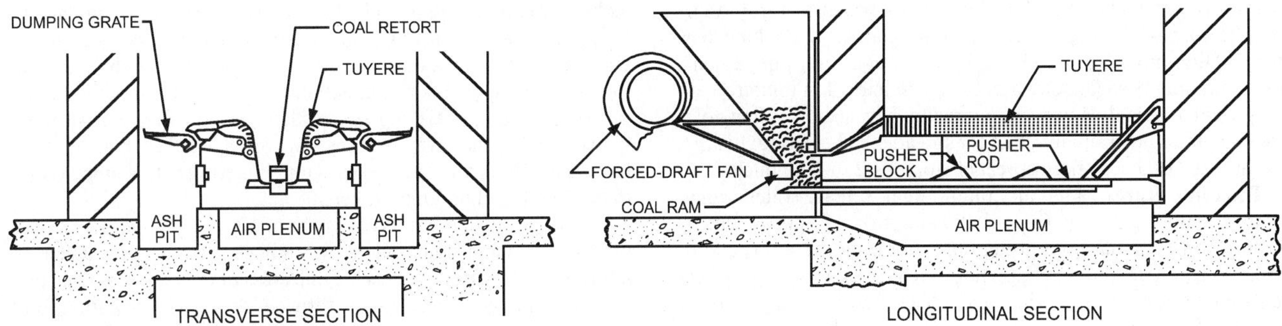

Fig. 14 Horizontal Underfeed Stoker with Single Retort

Table 3 Characteristics of Various Types of Stokers (Class 5)

Stoker Type and Subclass	Typical Capacity Range, lb/h	Maximum Burning Rate, Btu/h·ft²	Characteristics
Spreader			Capable of burning a wide range of coals; best to follow fluctuating loads; high fly ash carryover; low-load smoke
Stationary and dumping grate	20,000 to 80,000	450,000	
Traveling grate	100,000 to 400,000	750,000	
Vibrating grate	20,000 to 100,000	400,000	
Underfeed	20,000 to 30,000	400,000	Capable of burning caking coals and a wide range of coals (including anthracite); high maintenance; low fly ash carryover; suitable for continuous-load operation
Single- or double-retort			
Multiple-retort	30,000 to 500,000	600,000	
Chain grate and *traveling grate*	20,000 to 100,000	500,000	Low maintenance; low fly ash carryover; capable of burning a wide variety of weakly caking coals; smokeless operation over entire range
Vibrating grate	1,400 to 150,000	400,000	Characteristics similar to chain and traveling grate stokers, except that these stokers have no difficulty in burning strongly caking coals

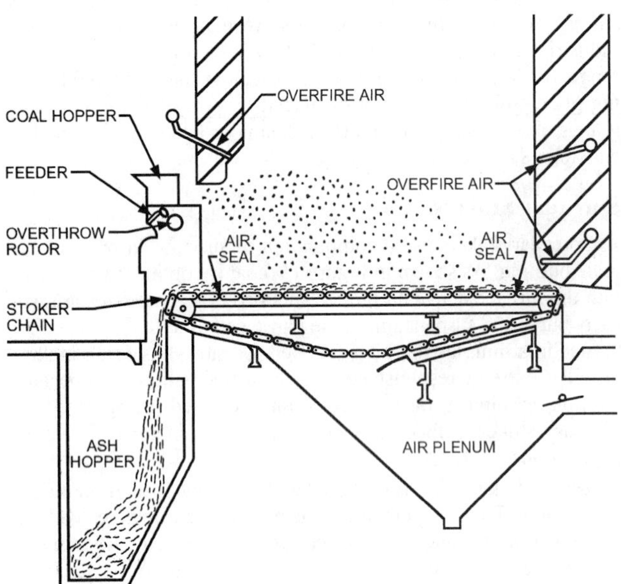

Fig. 15 Spreader Stoker, Traveling Grate Type

can be temporarily discontinued for cleaning and maintenance without affecting operation of other sections of the stoker.

For high-efficiency operation, a continuous ash-discharging grate, such as the traveling grate, is necessary. Introduction of the spreader stoker with the traveling grate increased burning rates by about 70% over the stationary and dumping grate types. Although both reciprocating and vibrating continuous ash discharge grates have been developed, the traveling grate stoker is preferred because of its higher burning rates.

Fuels and Fuel Bed. All spreader stokers (particularly those with traveling grates) can use fuels with a wide range of burning characteristics, including caking tendencies, because the rapid surface heating of the coal in suspension destroys the caking tendency. High-moisture, free-burning bituminous and lignite coals are commonly burned; coke breeze can be burned in mixture with a high-volatile coal. However, anthracite, because of its low volatile content, is not a suitable fuel for spreader stoker firing. Ideally, the fuel bed of a coal-fired spreader stoker is 2 to 4 in. thick.

Burning Rates. The maximum heat release rates range from 400,000 Btu/h·ft² (a coal consumption of approximately 40 lb/h) on stationary, dumping, and vibrating grate designs to 750,000 Btu/h·ft² on traveling grate spreader stokers. Higher heat release rates are practical with some waste fuels in which a greater portion of fuel can be burned in suspension than is possible with coal.

Underfeed Stokers

Underfeed stokers introduce raw coal into a retort beneath the burning fuel bed. They are classified as horizontal or gravity feed. In the horizontal type, coal travels within the furnace in a retort parallel with the floor; in the gravity-feed type, the retort is inclined by 25°. Most horizontal-feed stokers are designed with single or double retorts (and, rarely, triple retorts), whereas gravity-feed stokers are designed with multiple retorts.

In the horizontal stoker (see Figure 14), coal is fed to the retort by a screw (for smaller stokers) or a ram (for larger stokers). Once the retort is filled, the coal is forced upward and spills over the retort to form and feed the fuel bed. Air is supplied through tuyeres at each side of the retort and through air ports in the side grates. Over-fire air provides additional combustion air to the flame zone directly above the bed to prevent smoking, especially at low loads.

Gravity-feed stokers are similar in operating principle. These stokers consist of sloping multiple retorts and have rear ash discharge. Coal is fed into each retort, where it is moved slowly to the rear while simultaneously being forced upward over the retorts.

Fuels and Fuel Bed. Either type of underfeed stoker can burn a wide range of coal, although the horizontal type is better suited for free-burning bituminous coal. These stokers can burn caking coal, if there is not an excess amount of fines. The ash-softening temperature is an important factor in selecting coals because the possibility of excessive clinkering increases at lower ash-softening temperatures. Because combustion occurs in the fuel bed, underfeed stokers respond slowly to load change. Fuel-bed thickness is extremely nonuniform, ranging from 8 to 24 in. The fuel bed often contains large fissures separating masses of coke.

Burning Rates. Single-retort or double-retort horizontal stokers are generally used to service boilers with capacities up to 30,000 lb/h. These stokers are designed for heat release rates of 400,000 Btu/h·ft².

Chain and Traveling Grate Stokers

Figure 16 shows a typical chain or traveling grate stoker. These stokers are often used interchangeably because they are fundamentally the same, except for grate construction. The essential difference is that the links of chain grate stokers are assembled so that they move with a scissors-like action at the return bend of the stoker, whereas in most traveling grates there is no relative movement between adjacent grate sections. Accordingly, the chain grate is more suitable for handling coal with clinkering ash characteristics than is the traveling grate stokers.

Operation of the two types is similar. Coal, fed from a hopper onto the moving grate, enters the furnace after passing under an adjustable gate that regulates the thickness of the fuel bed. The layer of coal on the grate entering the furnace is heated by radiation from the furnace gases or from a hot refractory arch. As volatile matter is driven off by this rapid radiative heating, ignition occurs. The fuel continues to burn as it moves along the fuel bed, and the layer becomes progressively thinner. At the far end of the grate, where combustion of the coal is completed, ash is discharged into the pit as the grates pass downward over a return bend.

Often, furnace arches (front and/or rear) are included with these stokers to improve combustion by reflecting heat to the fuel bed. The front arch also serves as a bluff body, mixing rich streams of volatile gases with air to reduce unburned hydrocarbons. A chain grate stoker with overfire air jets eliminates the need for a front arch for burning volatiles. As shown in Figure 17, the stoker was zoned, or sectionalized, and equipped with individual zone dampers to control the pressure and quantity of air delivered to the various sections.

Fuels and Fuel Bed. The chain grate and traveling grate stokers can burn a variety of fuels (e.g., peat, lignite, subbituminous coal, free-burning bituminous coal, anthracite coal, and coke), as long as the fuel is sized properly. However, strongly caking bituminous coals have a tendency to mat and prevent proper air distribution to the fuel bed. Also, a bed of strongly caking coal may not be responsive to rapidly changing loads. Fuel bed thickness varies with the type and size of the coal burned. For bituminous coal, a 5 to 7 in. bed is common; for small-sized anthracite, the fuel bed is reduced to 3 to 5 in.

Burning Rates. Chain and traveling grate stokers are offered for a maximum continuous burning rate of 350,000 to 500,000 Btu/h·ft^2, depending on the type of fuel and its ash and moisture content.

Vibrating Grate Stokers

The vibrating grate stoker, as shown in Figure 17, is similar to the chain grate stoker in that both are overfeed, mass-burning, continuous ash discharge stokers. However, in the vibrating stoker, the sloping grate is supported on equally spaced vertical plates that oscillate back and forth in a rectilinear direction, causing the fuel to move from the hopper through an adjustable gate into the active combustion zone. Air is supplied to the stoker through laterally exposed areas beneath the stoker formed by the individual flexing of the grate support plates. Ash is automatically discharged into a shallow or basement ash pit. The grates are water cooled and are connected to the boiler circulating system.

The rates of coal feed and fuel bed movement are determined by the frequency and duration of the vibrating cycles and regulated by automatic combustion controls that proportion the air supply to optimize heat release rates. Typically, the grate is vibrated about every 90 s for durations of 2 to 3 s, but this depends on the type of coal and boiler operation. The vibrating grate stoker is increasingly popular because of its simplicity, inherently low fly ash carryover, low maintenance, wide turndown (10 to 1), and adaptability to multiple-fuel firing.

Fuels and Fuel Bed. The water-cooled vibrating grate stoker is suitable for burning a wide range of bituminous and lignite coals. The gentle agitation and compaction of the vibratory actions allow coal having a high free-swelling index to be burned and a uniform fuel bed without blowholes and thin spots to be maintained. The uniformity of air distribution and fuel bed conditions produce both good response to load swings and smokeless operation over the entire load range. Fly ash emission is probably greater than from the traveling grate because of the slight intermittent agitation of the fuel bed. The fuel bed is similar to that of a traveling grate stoker.

Burning Rates. Burning rates of vibrating grate stokers vary with the type of fuel used. In general, however, the maximum heat release rate should not exceed 400,000 Btu/h·ft^2 (a coal use of approximately 40 lb/h) to minimize fly ash carryover.

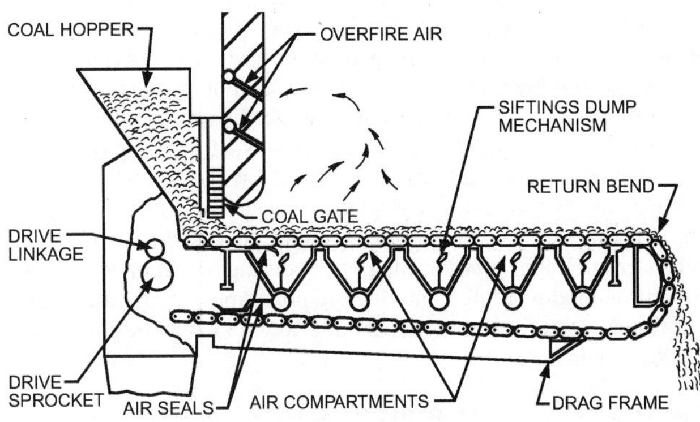

Fig. 16 Chain Grate Stoker

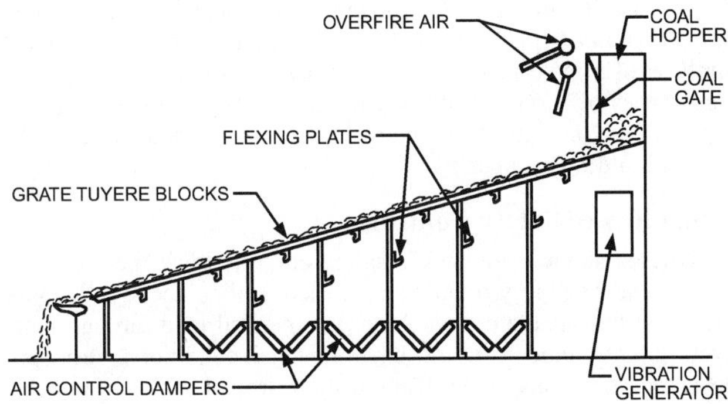

Fig. 17 Vibrating Grate Stoker

CONTROLS

This section covers controls required for automatic fuel-burning systems. Chapter 15 of the 2005 *ASHRAE Handbook—Fundamentals* addresses basic automatic control.

Automatic fuel-burning appliances require control systems that supervise combustion and take proper corrective action in the event of a failure in the appliance or related installation equipment. Requirements are similar for gas, oil, and solid fuel (stoker) burners.

Controls can be classified as safety or operating. **Safety controls** monitor potentially hazardous operating conditions such as ignition, combustion, temperature, and pressure, and function as required to ensure safe operation. **Operating controls** handle appliance operation at the required input rate when heat is required.

Figure 18 illustrates the basic elements of a control system for a fuel-fired appliance. In this diagram, the limit control and the ignition safety control, including its sensor, are safety controls.

SAFETY CONTROLS AND INTERLOCKS

Safety controls protect against hazards related to the combustion process. Personnel and material near the appliance must be

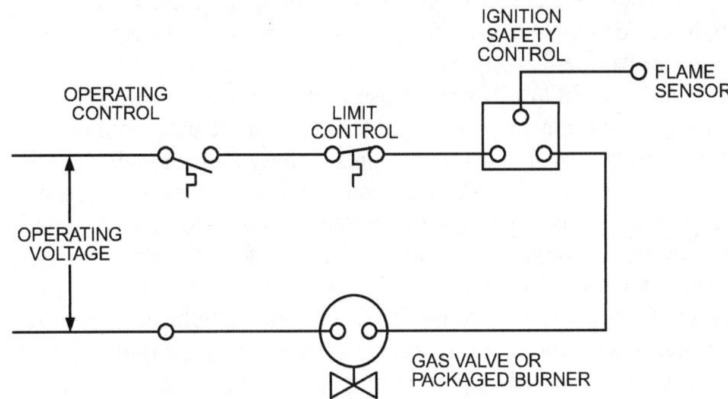

Fig. 18 Basic Control Circuit for Fuel-Burning Appliance

protected against explosion, fire, or excessive temperature. Common safety control functions include the following:

- Ignition and flame monitoring
  - Ignition proving (e.g., gas pilot thermocouple, flame rectification)
  - Proof of flame
- Draft proof (e.g., sail switch, pressure switch)
- Limit controls
  - Excessive temperature of heated medium (e.g., air or water)
  - Excessive pressure of heated medium (e.g., steam or hot water)
  - Low water level in steam or hot-water boilers
  - Low fuel-oil pressure
  - Low or excessive fuel-oil temperature when burning heavy oils
  - Low atomizing air pressure or atomizing steam pressure
  - Low fuel-gas pressure (e.g., empty propane tank)
- Flue gas spillage switch
- Flame rollout detection
- Carbon monoxide sensors
- Similar devices to prevent hazards from abnormal conditions
- Safety-related mechanical malfunction controls (e.g., open furnace blower door, burner out of position or rotary cup burner motor failure)
- Field interlocks having safety function

Ignition and Flame Monitoring

Failure to ignite fuel or failure of an established flame can cause explosion, if unburned fuel is subsequently provided with an ignition source. Ignition and flame safeguard controls, properly applied and used, prevent these occurrences. Some ignition controls monitor existence of an ignition source such as a pilot burner, allowing introduction and continued flow of fuel only when the ignition source is proven. Others use devices that ignite the main flame directly, in conjunction with high-speed flame detection capabilities and rapid shutdown of the fuel supply, if ignition is not immediate or if the flame is lost. Ignition controls often include programming features that govern the entire sequence of operation when a burner is started, operated, and stopped.

In residential gas-fired appliances, burner ignition is generally achieved by a small standing gas pilot flame, by an intermittent pilot system, or by a direct ignition system. Standing pilot systems prove existence of the ignition source by sensing its heat with a thermocouple. In intermittent pilot systems, a spark or hot surface igniter lights the pilot burner when there is a call for heat, thus saving pilot burner fuel when no heat is needed. Flame rectification (the ability of a flame to change an alternating current to direct current) is commonly used to prove the existence of the pilot flame (in intermittent pilot systems) or main burner flame (in direct-ignition systems). In direct-ignition systems, a spark or hot surface device ignites the main burner directly, eliminating the need for a pilot burner. If flame is not immediately established during ignition or if an established flame is lost, the gas supply system is quickly shut off. Flame rollout (flame escaping the appliance combustion chamber) protection detects flame in appliance combustion air inlet areas by sensing high temperatures and shuts off the fuel supply.

Most residential oil-burners use high-energy sparking for ignition. Sparking may be continuous (present for the entire duration of burner operation) or intermittent (present only long enough to establish presence of the flame). Most residential oil burners use an oil primary control that operates the burner motor and ignition spark transformer in conjunction with a flame detector. Flame proof is typically achieved by sensing visible or ultraviolet light emitted by the flame. A cadmium cell is used to detect visible light: its resistance changes when exposed to light from a burner flame, and that change is interpreted and acted upon by the primary control.

Commercial appliances with high fuel input rates normally use some form of a proven ignition source, often a standing or intermittent pilot burner, which itself may be quite large. North American safety standards do not allow use of direct-ignition systems for appliances with input greater than 400,000 Btu/h. Standing pilots or a proved igniter system must be used. The latter may closely resemble the direct-ignition system, except that the existence of the ignition source must be proven before fuel gas is allowed to flow to the main burner. As in the direct-ignition system, main burner gas flow is shut down if flame is not established quickly or if it goes out after ignition. Flame proving in large appliances is typically by flame rectification or ultraviolet light sensing.

Draft Proving

Fuel-burning appliances often include forced- or induced-draft blowers to move combustion air and combustion products through the appliance and venting systems. Sometimes blowers are applied outside the appliance in the venting system. Draft-proving controls supervise proper operation of these components to allow gas flow, ignition, and combustion only if appliance draft is adequate, typically using pressure and/or flow sensors to verify sufficient draft.

Limit Controls

In a warm-air furnace or space heater, excessive air temperature must be prevented. In a hot-water boiler, water temperature and pressure must not exceed the ratings of external piping and terminal equipment. Steam pressure must not exceed ratings of pressure vessels and steam system equipment. Fuel oil must be provided at pressures and temperatures that allow proper operation of oil burners. These requirements and restrictions are the focus of limit controls required by installation codes and appliance standards, and applied by appliance manufacturers and field installers. Design and principle of operation vary greatly, but in all cases limit controls must provide reliable detection of a fault and shut down the fuel-burning appliance.

Other Safety Controls

Requirements for safety devices are a function of the particular appliance and installation circumstances, and must be considered in that context. Appliances listed or design-certified as complying with recognized safety standards generally have appropriate safety devices as required by those standards. Building codes typically require appliances to be listed or design-certified by recognized testing laboratories, and that appliances not listed or design-certified comply with requirements within the code. Field circumstances may impose a need for control interlocks that relate to safety. Mechanical-draft venting systems, for example, must be proven to be in operation before an appliance ignition attempt and during burner operation. An appropriate field-installed interlock to prove mechanical-draft venting system operation before an ignition attempt would be considered a safety control.

Prescriptive Requirements for Safety Controls

Industrial safety codes and insurers often require that fuel-fired appliances be provided with specific components and construction features related to safety. Codes and requirements such as ASME *Standard* CSD-1 for boilers or those of insurance organizations may apply. Prescriptive governmental requirements may also apply to some types of installations. Often, requirements are based on the type of appliance, its capacity or size, or the intended use, such as type of building occupancy.

Reliability of Safety Controls

Devices suitable for use as safety controls must meet special quality and reliability requirements. Electrical contacts, for example, must be capable of switching the intended load through high cycle counts under extreme environmental conditions. Products designed for safety control application are invariably listed, or design-certified as complying with recognized standards that impose a wide range of construction and testing requirements. The degree of reliability of safety controls must be extremely high, and normally far exceeds that required of operating controls to meet appliance safety standards.

OPERATING CONTROLS

Operating controls start and stop burner operation in response to demands of the application load, and often incorporate capabilities for load matching and other application-related features. In contrast to safety controls, their purpose is to satisfy the application's fundamental purpose (e.g., warm air, hot water, steam pressure). Related functions may involve operating ancillary components of the appliance or system, such as blowers or pumps, which must operate at the appropriate speed.

Load matching is the most significant differentiating feature of operating controls. The major categories are on/off, staged, and modulated control. Fuel input rate varies for the latter two modes, but combustion airflow rate may also be controlled.

An **on/off control** starts and stops the flow of fuel to the burner to satisfy heat demand. An appliance with on/off control can operate only at its rated input regardless of the rate at which combustion heat may be needed. In a typical system, demand is sensed by a temperature or pressure sensor. When the temperature or pressure falls to an *on* set point, the control initiates combustion, which proceeds until the temperature or pressure increases to an *off* set point. The difference between the *off* and *on* values is called the **differential**. On/off control is satisfactory for systems with high thermal inertia (i.e., those in which heat input does not rapidly change the controlled temperature or pressure).

A **staged control** operates an appliance at multiple fixed input rates in response to heat demand. The input rates are determined by burner system hardware and settings. A residential gas-fired appliance, for example, may include a gas pressure regulator capable of providing high and low gas pressures to the gas orifice(s). Applying electrical power to a solenoid associated with the pressure regulator enables a second orifice pressure and thus a second gas input rate. In large appliances such as commercial boilers, staging is often accomplished with multiple burner assemblies, operated singly or together as necessary to match the load. As in on/off control, combustion initiation and changes between high and low input rates are in response to demand, as determined by the sensor and control logic.

Staged control logic is sometimes provided within an appliance control to enhance load response of the appliance. For example, a heating appliance control can provide staged burner operation without a staged room thermostat by beginning burner operation on burner stage 1 when a single-stage thermostat calls for heat. If burner stage 1 does not satisfy the thermostat within a specified time period, the appliance control turns on burner stage 2 until the thermostat is satisfied, at which time the burner is completely shut off until the thermostat's next call for heat.

Figure 19 illustrates the characteristics of a three-stage control system that provides heat in response to a temperature sensor and is required to maintain temperature at a set point t_S. If the controlled temperature is above t_S, no heat is provided. When the temperature drops to $t_{1_{On}}$, the first-stage burner is operated; if the temperature then rises back to t_S, the burner is turned off. If the first stage is inadequate, the temperature continues to fall; when it reaches $t_{2_{On}}$, the second stage is operated. A third stage is available if the second stage is not adequate, operating when the temperature falls to its *on* temperature $t_{3_{On}}$. Successful handling of the load results in raising of the controlled temperature and stepped sequential reduction of the heat input as the *off* temperatures $t_{3_{Off}}$, $t_{2_{Off}}$, and $t_{1_{Off}}$ (= t_S) are reached. The individual stages operate subject to a **stage differential**, which is the difference between the temperature at which a stage is shut off and the (lower) temperature at which it is again turned on. Interstage differential and overall differential, as shown in Figure 19, are analogous. The *on* and *off* temperatures may or may not overlap, depending on the relative

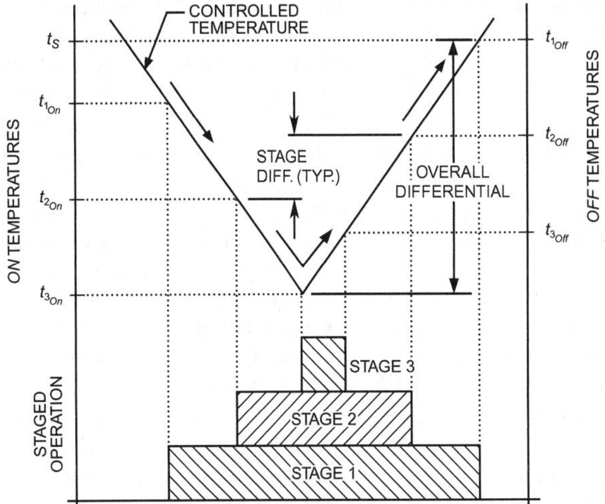

Fig. 19 Control Characteristics of Three-Stage System

size of the stage differentials. In sophisticated electronic controls, the differential settings are adjustable to meet the particular requirements of the application.

Staged control logic often includes features that enhance load response or appliance operation. As shown in Figure 19, load matching might be best if the control initiates the lowest input rate on a call for heat and proceeds to higher inputs only if the demand cannot be met at lower rates. Appliance or application characteristics may dictate other approaches. For example, if appliance ignition is most reliable at high input rate, a staged control might require that ignition begin at a high input rate with an immediate reduction to a lower input rate to meet the actual heat demand.

Combustion airflow may or may not be staged. If not, operation at lower input rates, particularly in appliances with fan-assisted combustion, results in lean operation (a high percentage of excess air). There may be little effect, but in general, heat transfer efficiency suffers if the combustion air-to-fuel ratio (A/F) is not maintained at an optimum level on each stage.

A **modulating control** regulates input rate to follow load demands more closely than on/off or staged controls. Fuel input rate may be varied by throttling (mechanical restriction of flow by a valve) or by pressure regulation to control injector (orifice) pressure. Demand is often indicated by a temperature or pressure sensor, and the input rate increases as the temperature or pressure falls from set point values. In large fuel-oil appliances, a fuel oil control valve and burner damper respond over a range of positions within the operating range of the burner.

Air/fuel proportioning controls are a variation of modulating controls in which the proper ratio of air to fuel is maintained throughout the burner's fuel input range. In response to the need for heat, both fuel input rate and combustion airflow rate are modulated, maintaining an optimal air/fuel ratio to maximize efficiency. The controls may change blower speed or (particularly in large commercial appliances) link air dampers to fuel input rate control.

Tracking controls are proportioning controls that maintain the proper air/fuel ratio by applying basic Bernoulli-equation fluid flow behavior. Gas-fired appliances are especially suitable for this approach. As combustion air or gas flow rate changes, such as through metering orifices or venturis, pressure changes in proportion to the square of the air or gas velocity. The pressure differential created by forcing air through a venturi, for example, can be used to regulate gas flow through an orifice. When the relationship is established, a change of airflow rate results in a pressure differential change that causes a proportional (tracking) change in gas flow rate. In one approach, airflow rate is varied according to heating demand by changing blower speed or adjusting a damper position, and gas flow rate tracks the change. Variations may use intermediate ratio controllers. Other systems control gas input rate, and combustion air rate tracks. In the ideal case, the air-to-fuel ratio remains constant at all input rates, and efficiency increases substantially when firing rate is reduced.

Combustion quality is monitored and controlled in some large installations in order to ensure efficient operation or satisfactory emission characteristics. Air or fuel can be controlled in response to devices that sense concentrations of various combustion product constituents. Butcher (1990) used the measured intensity of light emitted from flames of fixed-input oil burners to judge basic flame quality. Flame steady-state intensity is compared to a predetermined flame intensity set point. Deviation of flame intensity beyond an optimum intensity range indicates poor flame quality.

Operating controls often provide functions not directly related to combustion. These **appliance controls** include things such as operating air-circulating blowers or pumps according to the fuel combustion system's needs. A blower or pump can be operated in response to a temperature or pressure sensor or according to a time protocol imposed with electronic or digital capabilities. Additional algorithms in the appliance control can further enhance appliance operation.

Integrated and Programmed Controls

Integrated Controls. Many appliance controls combine safety and operating functions into electronic microprocessors with various sensors and electromechanical devices to handle most of the control functions necessary to operate fuel-burning appliances. Figure 20 illustrates an integrated control approach typical of residential forced-air furnaces or boilers with fan-assisted combustion and a limit control. Other safety devices, such as a flame roll-out switch, typically are present. For a forced-air furnace, the remote thermostat would be a wall thermostat. For a hot-water boiler, the operating control might be a wall thermostat, possibly with a second sensor to control boiler water temperature. Ignition and flame safety algorithms are provided in the microprocessor-based control. The ignition device and flame sensor are connected to the control. Note that the control operates both the circulating blower or pump and a combustion blower. Both are controlled in accordance with an appropriate sequence of operation. A draft-proving device is connected to the control to confirm operation of the combustion blower prior to igniter operation and opening of the gas valve, and during subsequent burner operation. Typically, light-emitting diodes (LEDs) are provided on the control to indicate operating status and to facilitate diagnosis of operating problems.

Programmed controls for ignition and flame safeguard have been used in large fuel-fired appliances for many years. More recently, programming has been applied in integrated controls for smaller appliances. These controls handle not only the ignition sequence, but also operating functions, and provide status and diagnostic information. Accomplishing those functions with separate components would be difficult and clumsy.

When power is supplied, an integrated control conducts a self-check to ensure that the appliance and control system are in safe working order, and then typically indicates readiness by flashing one or more LEDs. The self-check may include verifying that the flame-proving device does not indicate flame before the ignition sequence has started and the draft-proving device does not indicate draft before the combustion blower has started. Self-checking for safe conditions may continue during burner operation and during standby after the thermostat is satisfied. If unacceptable operation is detected, the control will take corrective action or attempt safe shut-off of the appliance. On a call for heat, an operating sequence typically proceeds as follows:

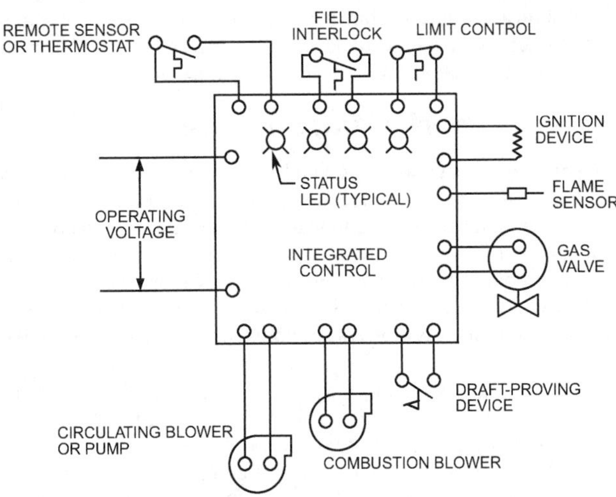

Fig. 20 Integrated Control System for Gas-Fired Appliance

1. The call for heat is indicated by an assigned LED flashing sequence.
2. The combustion blower starts and its operation is verified by the draft-proving device.
3. When the combustion blower has run long enough to purge the combustion chamber of residual fuel and combustion products (i.e., flush it with four air changes), the ignition device energizes.
4. The control allows enough time for the ignition device to become operative. In some applications, operation is verified electronically.
5. The gas valve opens.
6. Flame detection checks for main burner ignition. If flame presence is not proven within a very short predetermined time, or if continuous proof of flame fails at any time while the gas valve is open, the gas valve closes. Programming typically requires rapid detection of flame failure, and gas shutoff in accordance with safety standards.
7. If flame presence is detected, the air circulating blower or hot-water pump is started. Depending on the application, blower or pump operation could start with burner operation or after a time delay.
8. When the call for heat is satisfied, the gas valve closes to extinguish the burner flame.
9. The combustion blower operates for a short period to flush the combustion chamber with air (post purge).
10. The circulating blower or pump stops, usually after an application-specific delay.
11. The status LED(s) return to the standby mode.

This sequence can be varied as necessary to accommodate the requirements of specific appliances and applications. For example, staged or modulated operation may be included.

Status LEDs are turned on and off in particular patterns to display coded diagnostic information for many different normal and abnormal conditions (e.g., failure to prove draft, failure to detect flame, overtemperature). Appliance standards and installation codes require that appliance controls lock out operation of the burner system in response to certain failure conditions, when manual intervention by a qualified service agency is necessary.

REFERENCES

Ackerman, M.Y., J.D. Dale, D.J. Wilson, and N.P. Fleming. 1995. Design guidelines for combustion air systems in cold climates (RP-735). ASHRAE Research Project, *Final Report*.

AHRI. 2007. *Consumers' directory of certified efficiency ratings*. Air Conditioning, Heating and Refrigeration Institute (formerly Gas Appliance Manufacturers Association), Arlington, VA. Available at http://www.gamapower.org/.

ANSI. 2002. Installation of domestic gas conversion burners. *Standard* Z21.8-1994 (R2002). American National Standards Institute, Washington, D.C.

ANSI. 2006. Gas-fired central furnaces. ANSI *Standard* Z21.47-2006/CSA 2.3-2006, ANSI *Standard* Z21.47a-2007/CSA 2.3a-2007. Secretariat: CSA-America, Cleveland, OH. American National Standards Institute, Washington, D.C.

ASHRAE. 1993. Method of testing for annual fuel utilization efficiency of residential central furnaces and boilers. *Standard* 103-1993.

ASHRAE. 2003. Method of testing for rating commercial gas, electric, and oil service water heating equipment. ANSI/ASHRAE *Standard* 118.1-2003.

ASHRAE. 2006. Method of testing for rating residential water heaters. ANSI/ASHRAE *Standard* 118.2-2006.

ASHRAE. 2007. Methods of testing for rating combination space-heating and water-heating appliances. *Standard* 124-2007.

ASME. 2006. Controls and safety devices for automatically fired boilers. *Standard* CSD-1-2006. American Society of Mechanical Engineers, New York.

ASTM. 2003. Test method for smoke density in flue gases from burning distillate fuels. ANSI/ASTM *Standard* D2156-94 (R 2003). American Society for Testing and Materials, West Conshohocken, PA.

Butcher, T. 1990. Performance control strategies for oil-fired residential heating systems. BNL *Report* 52250. Brookhaven National Laboratory, Upton, NY.

Butcher, T., L.A. Fisher, B. Kamath, T. Kirchstetter, and J. Batey. 1994. Nitrogen oxides (NO_x) and oil burners. BNL *Report* 52430. Brookhaven National Laboratory, Upton, NY.

Dale, J.D., D.J. Wilson, M.Y. Ackerman, and N.P. Fleming. 1997. A field study of combustion air systems in cold climates (RP-735). *ASHRAE Transactions* 103(1):910-920.

Locklin, D.W. and H.R. Hazard. 1980. Technology for the development of high-efficiency oil-fired residential heating equipment. BNL *Report* 51325. Brookhaven National Laboratory, Upton, NY.

McDonald, R.J., ed. 1989-2003. Proceedings of the Oil Heat Technology Conferences and Workshops. BNL *Reports* 52217, 52284, 52340, 52392, 52430, 52475, 52506, 52537, 52670, and 71337. Brookhaven National Laboratory, Upton, NY. Available at http://www.osti.gov/bridge.

UCS. 2007. *Clean energy: How biomass energy works*. Union of Concerned Scientists, Cambridge, MA. Available at http://www.ucsusa.org/clean_energy/renewable_energy_basics/offmen-how-biomass-energy-works.html.

BIBLIOGRAPHY

ANSI. 2005. Draft hoods. *Standards* Z21.12-1990 (R 2005), Z21.12a-1993 (R 2005), and Z21.12b-1994 (R 2005). Secretariat: CSA-America, Cleveland, OH.American National Standards Institute, Washington, D.C.

ANSI. 2005. Gas-fired low pressure steam and hot water boilers. *Standard* Z21.13-2004/CSA 4.9-2004, ANSI Standard Z21.13a-2005/CSA 4.9a-2005, and ANSI *Standard* Z21.13b-2007/CSA 4.9b-2007. Secretariat: CSA-America, Cleveland, OH. American National Standards Institute, Washington, D.C.

ANSI. 2002. Installation of domestic gas conversion burners. *Standard* Z21.8-94 (R 2002). Secretariat: CSA-America, Cleveland, OH. American National Standards Institute, Washington, D.C.

ANSI. 2006. Gas unit heaters and gas-fired duct furnaces. *Standard* Z83.8-2006/CSA 2.6-2006. Secretariat: CSA-America, Cleveland, OH. American National Standards Institute, Washington, D.C.

ANSI. 2004. Domestic gas conversion burners. *Standard* Z21.17-98 (R2004)/CSA 2.7-M98. Secretariat: CSA-America, Cleveland, OH. American National Standards Institute, Washington, D.C.

ANSI/NFPA. 2006. National fuel gas code. ANSI Z223.1/NFPA 54-2006. American Gas Association, Washington, D.C., and National Fire Protection Association, Quincy, MA.

CSA. 2005. Natural gas and propane installation code. *National Standard of Canada* CAN/CSA B149.1-05, January 2005; Supplement No. 1, January 2007; and Update No. 1, February 2007. Canadian Standards Association, Mississauga, ON.

NFPA. 2006. Installation of oil-burning equipment. ANSI/NFPA *Standard* 31-2006. National Fire Protection Association, Quincy, MA.

Segeler, C.G., ed. 1965. *Gas engineers handbook*, Section 12, Ch. 2. Industrial Press, New York.

UL. 2006. Commercial-industrial gas-heating equipment. *Standard* 795-06. Underwriters Laboratories, Northbrook, IL.

UL. 2003. Oil burners. ANSI/UL *Standard* 296-10. Underwriters Laboratories, Northbrook, IL.

UL. 1995. Oil-fired boiler assemblies. *Standard* 726-07. Underwriters Laboratories, Northbrook, IL.

UL. 2006. Oil-fired central furnaces. *Standard* 727-09. Underwriters Laboratories, Northbrook, IL.

CHAPTER 31

BOILERS

BOILERS are pressure vessels designed to transfer heat (produced by combustion) to a fluid. The definition has been expanded to include transfer of heat from electrical resistance elements to the fluid or by direct action of electrodes on the fluid. In most boilers, the fluid is usually water in the form of liquid or steam. If the fluid being heated is air, the heat exchange device is called a furnace, not a boiler. The firebox, or combustion chamber, of some boilers is also called a furnace.

Excluding special and unusual fluids, materials, and methods, a boiler is a cast-iron, carbon or stainless steel, aluminum, or copper pressure vessel heat exchanger designed to (1) burn fossil fuels (or use electric current) and (2) transfer the released heat to water (in water boilers) or to water and steam (in steam boilers). Boiler heating surface is the area of fluid-backed surface exposed to the products of combustion, or the fire-side surface. Various manufacturers define allowable heat transfer rates in terms of heating surface based on their specific boiler design and material limitations. Boiler designs provide for connections to a piping system, which delivers heated fluid to the point of use and returns the cooled fluid to the boiler.

Chapters 6, 10, 11, 12, and 14 cover applications of heating boilers. Chapter 7 discusses cogeneration, which may require boilers.

CLASSIFICATIONS

Boilers may be grouped into classes based on working pressure and temperature, fuel used, material of construction, type of draft (natural or mechanical), and whether they are condensing or noncondensing. They may also be classified according to shape and size, application (such as heating or process), and the state of the output medium (steam or water). Boiler classifications are important to the specifying engineer because they affect performance, first cost, and space requirements. Excluding designed-to-order boilers, significant class descriptions are given in boiler catalogs or are available from the boiler manufacturer. The following basic classifications may be helpful.

Working Pressure and Temperature

With few exceptions, boilers are constructed to meet ASME *Boiler and Pressure Vessel Code*, Section IV (SCIV), Rules for Construction of Heating Boilers (low-pressure boilers), or Section I (SCI), Rules for Construction of Power Boilers (high-pressure boilers).

Low-pressure boilers are constructed for maximum working pressures of 15 psig steam and up to 160 psig hot water. Hot water boilers are limited to 250°F operating temperature. Operating and safety controls and relief valves, which limit temperature and pressure, are ancillary devices required to protect the boiler and prevent operation beyond design limits.

High-pressure boilers are designed to operate above 15 psig steam, or above 160 psig and/or 250°F for water boilers. Similarly, operating and safety controls and relief valves are required.

Steam boilers are generally available in standard sizes up to and above 100,000 lb steam/h (60,000 to over 100,000,000 Btu/h), many of which are used for space heating applications in both new and existing systems. On larger installations, they may also provide steam for auxiliary uses, such as hot water heat exchangers, absorption cooling, laundry, and sterilizers. In addition, many steam boilers provide steam at various temperatures and pressures for a wide variety of industrial processes.

Water boilers are generally available in standard sizes from 35,000 to over 100,000,000 Btu/h, many of which are in the low-pressure class and are used primarily for space heating applications in both new and existing systems. Some water boilers may be equipped with either internal or external heat exchangers for domestic water service.

Traditionally, boilers were rated by boiler horsepower, a unit of measurement with one boiler horsepower being equal to 33,475 Btu/h or the evaporation of 34.5 lb of water per hour at standard atmospheric pressure (14.7 psia) and 212°F.

Every steam or water boiler is rated for a maximum working pressure that is determined by the applicable boiler code under which it is constructed and tested. When installed, it also must be equipped at a minimum with operation and safety controls and pressure/temperature-relief devices mandated by such codes.

Fuel Used

Boilers may be designed to burn coal, wood, various grades of fuel oil, waste oil, various types of fuel gas, or to operate as electric boilers. A boiler designed for one specific fuel type may not be convertible to another type of fuel. Some boilers can be adapted to burn coal, oil, or gas. Several designs accommodate firing oil or gas, and other designs permit firing dual-fuel burning equipment. Accommodating various fuel burning equipment is a fundamental concern of boiler manufacturers, who can furnish details to a specifying engineer. The manufacturer is responsible for performance and rating according to the code or standard for the fuel used (see section on Performance Codes and Standards).

Construction Materials

Most noncondensing boilers are made with cast iron sections or steel. Some small boilers are made of copper or copper-clad steel. Condensing boilers are typically made of stainless steel or aluminum because copper, cast iron, and carbon steel will corrode because of acidic condensate.

Cast-iron sectional boilers generally are designed according to ASME SCIV requirements and range in size from 35,000 to 13,975,000 Btu/h gross output. They are constructed of individually cast sections, assembled into blocks (assemblies) of sections. Push or screw nipples, gaskets, and/or an external header join the

The preparation of this chapter is assigned to TC 6.1, Hydronic and Steam Equipment and Systems.

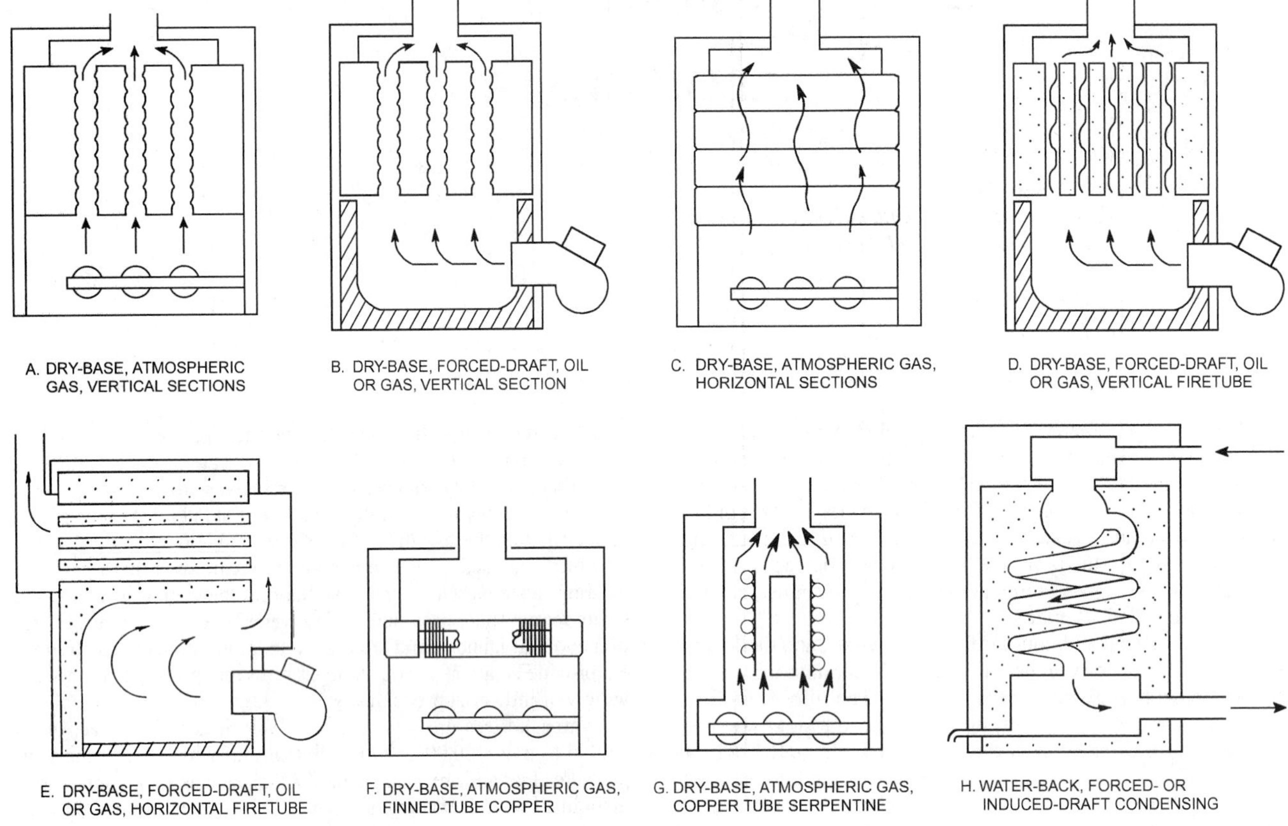

Fig. 1 Residential Boilers

sections pressure-tight and provide passages for the water, steam, and products of combustion. The number of sections assembled determines the boiler size and energy rating. Sections may be vertical or horizontal, the vertical design being more common (Figures 1A and 1C).

The boiler may be **dry-base** (the combustion chamber is beneath the fluid-backed sections), as in Figure 1B; **wet-base** (the combustion chamber is surrounded by fluid-backed sections, except for necessary openings), as in Figure 2A; or **wet-leg** (the combustion chamber top and sides are enclosed by fluid-backed sections), as in Figure 2B.

The three types of boilers can be designed to be equally efficient. Testing and rating standards apply equally to all three types. The wet-base design is easiest to adapt for combustible floor installations. Applicable codes usually demand a floor temperature under the boiler no higher than 90°F above room temperature. A steam boiler at 215°F or a water boiler at 240°F may not meet this requirement without appropriate floor insulation. Large cast-iron boilers are also made as water-tube units with external headers (Figure 2C).

Steel boilers generally range in size from 50,000 Btu/h to the largest boilers made. Designs are constructed to either ASME SCI or SCIV (or other applicable code) requirements. They are fabricated into one assembly of a given size and rating, usually by welding. The heat exchange surface past the combustion chamber is usually an assembly of vertical, horizontal, or slanted tubes. Boilers of the fire-tube design contain flue gases in tubes completely submerged in fluid (Figures 1D and 1E show residential units, and Figures 3A through 3D and Figure 4A show commercial units). Water-tube boilers contain fluid inside tubes with tube pattern arrangement providing for the combustion chamber (Figures 4C and 4D). The internal configuration may accommodate one or more flue gas passes. As with cast-iron sectional boilers, dry-base,

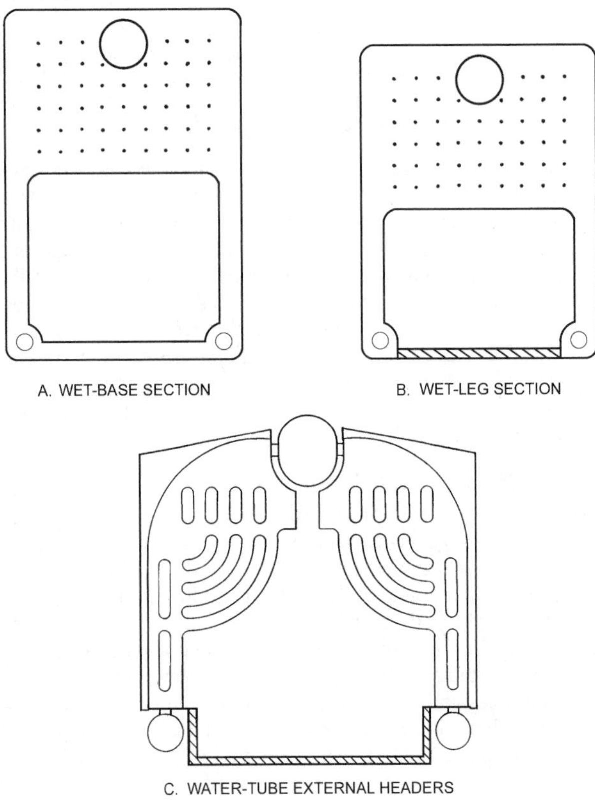

Fig. 2 Cast-Iron Commercial Boilers

wet-leg, or wet-base designs may be used. Most small steel boilers are of the dry-base, vertical fire-tube type (Figure 1D).

Larger boilers usually incorporate horizontal or slanted tubes; both fire-tube and water-tube designs are used. A popular horizontal fire-tube design for medium and large steel boilers is the **scotch marine**, which is characterized by a central fluid-backed cylindrical combustion chamber, surrounded by fire-tubes accommodating two or more flue gas passes, all within an outer shell (Figures 3A through 3D). In another horizontal fire-tube design, the combustion chamber has a similar central fluid-backed combustion chamber surrounded by fire tubes accommodating two or more flue gas passes, all within an outer shell. However, this design uses a dry base and wet-leg (or mud leg) (Figure 4A).

Copper boilers are usually some variation of the water-tube boiler. Parallel finned copper tube coils with headers, and serpentine copper tube units are most common (Figures 1F and 1G). Some are offered as wall-hung residential boilers. The commercial bent water-tube design is shown in Figure 4B. Natural gas is the usual fuel for copper boilers.

Stainless steel boilers usually are designed to operate with condensing flue gases. Most are single-pass, firetube design and are generally resistant to thermal shock. ASME limits operating temperatures to 210°F and 160 psig working pressure.

Aluminum boilers are also usually designed to operate with condensing flue gases. Typical designs incorporate either cast aluminum boiler sections or integrally finned aluminum tubing. ASME limits operating temperatures to 200°F and working pressure to 50 psig.

Type of Draft

Draft is the pressure difference that causes air and/or fuel to flow through a boiler or chimney. A **natural draft boiler** is designed to operate with a negative pressure in the combustion chamber and in the flue connection. The pressure difference is created by the tendency of hot gases to rise up a chimney or by the height of the boiler up to the draft control device. In a **mechanical draft boiler**, a fan or blower or other machinery creates the required pressure difference. These boilers may be either forced draft or induced draft. In a **forced-draft boiler**, air is forced into the combustion chamber to maintain a positive pressure in the combustion chamber and/or the space between the tubing and the jacket (breaching). In an **induced-draft boiler**, air is drawn into the combustion chamber to maintain a negative pressure in the combustion chamber.

Condensing or Noncondensing

Traditionally designed boilers must operate without condensing the flue gas in the boiler. This precaution was necessary to prevent corrosion of cast-iron, steel, or copper parts. Hot-water units were operated at 140°F minimum water temperature to prevent this corrosion and to reduce the likelihood of thermal shock.

Because a higher boiler efficiency can be achieved with a lower water temperature, the condensing boiler allows the flue gas water vapor to condense and drain. Full condensing boilers are now available from a large number of manufacturers. These boilers are specifically designed for operation with the low return water temperatures found in hot-water reset, water-source heat pump, two-pipe fan-coil, and reheat systems. Two types of commercial condensing boilers are shown in Figure 5. Figure 6 shows a typical relationship of overall condensing boiler efficiency to return water temperature. The dew point of 130°F shown in the figure varies with the percentage of hydrogen in the fuel and oxygen-carbon dioxide ratio, or excess air, in the flue gases. A condensing boiler is shown in Figure 1H. Condensing boilers can be of the fire-tube, water-tube, cast-iron, and cast-aluminum sectional design.

Condensing boilers are generally provided with high-turndown modulating burners and are more efficient than noncondensing boilers at any return water temperature (RWT), including noncondensing-temperature applications. Efficiencies of noncondensing boilers must be limited to avoid potential condensing and corrosion. Further efficiencies can be gained by using lower RWT or higher Δt as

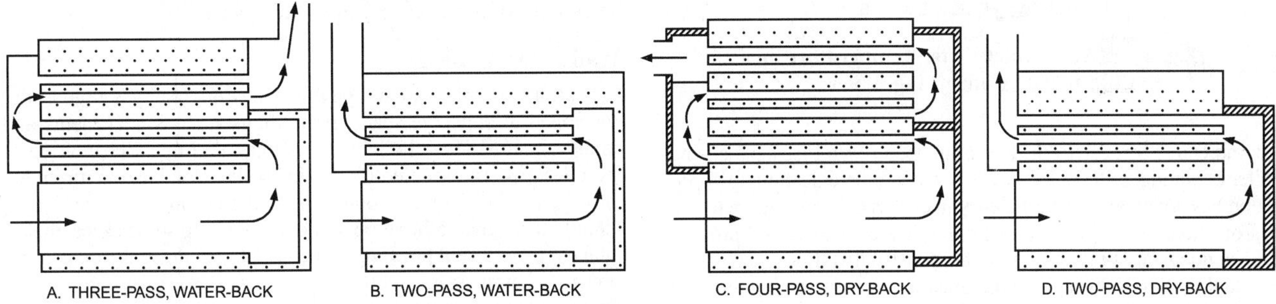

Fig. 3 Scotch Marine Commercial Boilers

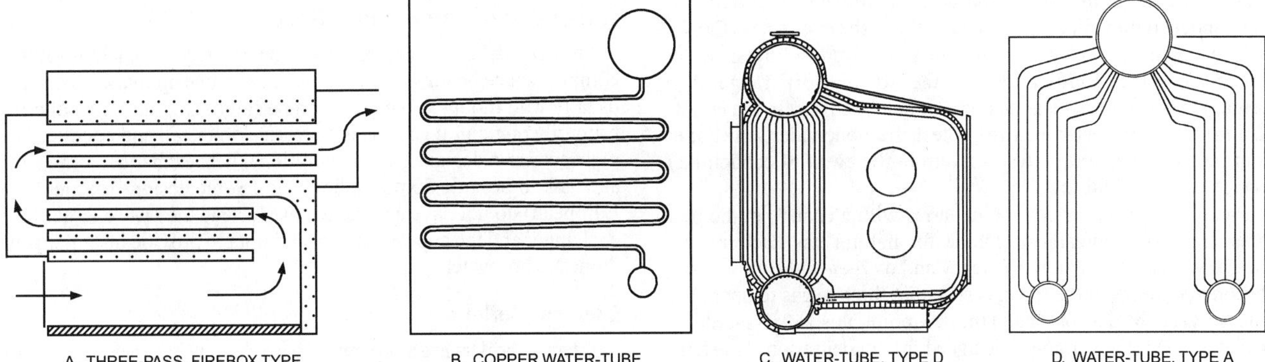

Fig. 4 Commercial Fire-Tube and Water-Tube Boilers

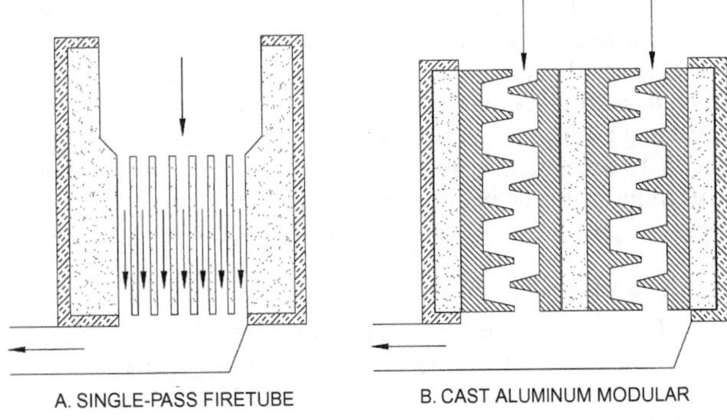

Fig. 5 Commercial Condensing Boilers

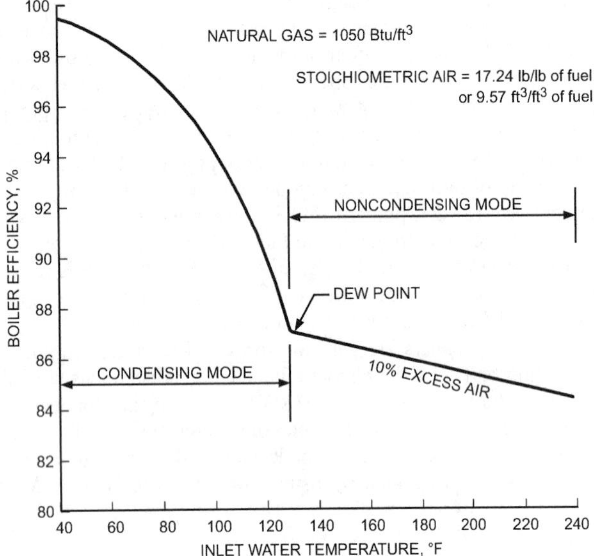

Fig. 6 Effect of Inlet Water Temperature on Efficiency of Condensing Boilers

recommended by ASHRAE. For example, a natural gas condensing boiler operating with 60°F RWT in a water-source heat pump application has potential boiler efficiency in excess of 98% (Figure 6).

For maximum reliability and durability over extended product life, condensing boilers should be constructed from corrosion-resistant materials throughout the fireside combustion chamber and heat exchanger. These materials include certain grades of stainless steel and aluminum.

Noncondensing heat plant efficiency may in some cases be improved with the use of external flue gas-to-water economizers. The condensing medium may include domestic hot-water (DHW) preheat, steam condensate or hot-water return, fresh-water makeup, or other fluid sources in the 70 to 130°F range. The medium can also be used as a source of heat recovery in the HVAC system. Care must be taken to protect the noncondensing boiler from the low-temperature water return in the event of economizer service or control failure.

Figure 7 shows how dew point varies with a change in the percentages of oxygen/carbon dioxide for natural gas. Boilers that operate with a combustion efficiency and oxygen and carbon dioxide concentrations in the flue gas such that the flue gas temperature falls between the dew point and the dew point plus 140°F should be avoided, unless the venting is designed for condensation. This temperature typically occurs with boilers operating between 83 and 87% efficiency and the flue gas has an oxygen concentration of 7 to

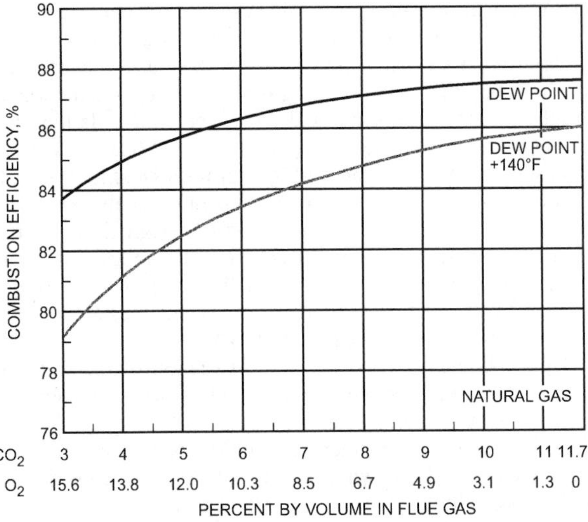

Fig. 7 Relationship of Dew Point, Carbon Dioxide, and Combustion Efficiency for Natural Gas

10% and the carbon dioxide is 6 to 8%. Chapter 30 gives further details on chimneys.

The condensing portion of these boilers requires special material to resist the corrosive effects of the condensing flue gases. Cast iron, carbon steel, and copper are not suitable materials for the condensing section of a boiler. Certain stainless steels and aluminum alloys, however, are suitable. Commercial boiler installations can be adapted to condensing operation by adding a condensing heat exchanger in the flue gas vent.

Heat exchangers in the flue gas venting require a condensing medium such as (1) low pressure steam condensate or hot water return, (2) domestic water service, (3) fresh water makeup, or (4) other fluid sources in the 70 to 130°F range. The medium can also be a source of heat recovery in HVAC systems.

Wall Hung Boilers

Wall hung boilers are a type of small residential gas fired boiler developed to conserve space in buildings such as apartments and condominiums. These boilers are popular in Europe. The most common designs are mounted on outside walls. Combustion air enters through a pipe from the outdoors, and flue products are vented directly through another pipe to the outdoors. In some cases, the air intake pipe and vent pipe are concentric. Other designs mount adjacent to a chimney for venting and use indoor air for combustion. These units may be condensing or noncondensing. As these boilers are typically installed in the living space, provisions for proper venting and combustion air supply are very important.

Integrated (Combination) Boilers

Integrated boilers are relatively small, residential boilers that combine space heating and water heating in one appliance. They are usually wall mounted but may also be floor standing. They operate primarily on natural gas and are practical to install and operate. The most common designs have an additional heat exchanger and a storage tank to provide domestic hot water. Some designs (particularly European) do not have a storage tank. Instead they use a larger heat exchanger and the appropriate burner input to provide instantaneous domestic hot water.

Electric Boilers

Electric boilers are a separate class of boiler. Because no combustion occurs, a boiler heating surface and flue gas venting are unnecessary. The heating surface is the surface of the electric elements or electrodes immersed in the boiler water. The design of electric boilers is largely determined by the shape and heat release rate of the electric elements used. Electric boiler manufacturers' literature describes available size, shapes, voltages, ratings, and methods of control.

SELECTION PARAMETERS

Boiler selection should be based on a competent review of the following parameters:

All Boilers

- Application of terminal unit selection
- Applicable code under which the boiler is constructed and tested
- Gross boiler heat output
- Part- versus full-load efficiency (life-cycle cost)
- Total heat transfer surface area
- Water content weight or volume
- Auxiliary power requirement
- Cleaning and service access provisions for fireside and waterside heat transfer surfaces
- Space requirement and piping arrangement
- Water treatment requirement
- Operating personnel capabilities and maintenance/operation requirements
- Regulatory requirements for emissions, fuel usage/storage

Fuel-Fired Boilers

- Combustion chamber (furnace volume)
- Internal flow pattern of combustion products
- Combustion air and venting requirements
- Fuel availability/capability

Steam Boilers

- Steam quality

The codes and standards outlined in the section on Performance Codes and Standards include requirements for minimum efficiency, maximum temperature, burner operating characteristics, and safety control. Test agency certification and labeling, which are published in boiler manufacturers' catalogs and shown on boiler rating plates, are generally sufficient for determining boiler steady-state operating characteristics. However, for noncondensing commercial and industrial boilers, these ratings typically do not consider part-load or seasonal efficiency, which is less than steady-state efficiency. Condensing boilers, generally provided with modulating burners, provide higher part- and full-load efficiency. Some boilers are not tested and rated by a recognized agency, and, therefore, do not bear the label of an agency. Nonrated boilers (rated and warranted only by the manufacturer) are used when jurisdictional codes or standards do not require a rating agency label. As previously indicated, almost without exception, both rated and nonrated boilers are of ASME Code construction and are marked accordingly.

EFFICIENCY: INPUT AND OUTPUT RATINGS

The efficiency of fuel-burning boilers is defined in the following ways: combustion, overall, and seasonal. However, manufacturers are not required to test or publish efficiencies that coincide with these industry definitions. Further explanation of boiler test requirements is found in the section on Performance Codes and Standards.

Combustion efficiency is input minus stack (flue gas outlet) loss, divided by input, and generally ranges from 75 to 86% for most noncondensing boilers. Condensing boilers generally operate in the range of 88 to 95% combustion efficiency.

Overall (or **thermal**) **efficiency** is gross energy output divided by energy input. Gross output is measured in the steam or water leaving the boiler and depends on the characteristics of the individual installation. Overall efficiency of electric boilers is generally 92 to 96%. Overall efficiency is lower than combustion efficiency by the percentage of heat lost from the outside surface of the boiler (radiation loss or jacket loss) and by off-cycle energy losses (for applications where the boiler cycles on and off). Overall efficiency can be precisely determined only under controlled laboratory test conditions, directly measuring the fuel input and the heat absorbed by the water or steam of the boiler. Precise efficiency measurements are generally not performed under field conditions because of the inability to control the required parameters and the high cost involved in performing such an analysis. An approximate combustion efficiency for noncondensing boilers can be determined under any operating condition by measuring flue gas temperature and percentage of CO_2 or O_2 in the flue gas and by consulting a chart or table for the fuel being used. The approximate combustion efficiency of a condensing boiler must include the energy transferred by condensation in the flue gas.

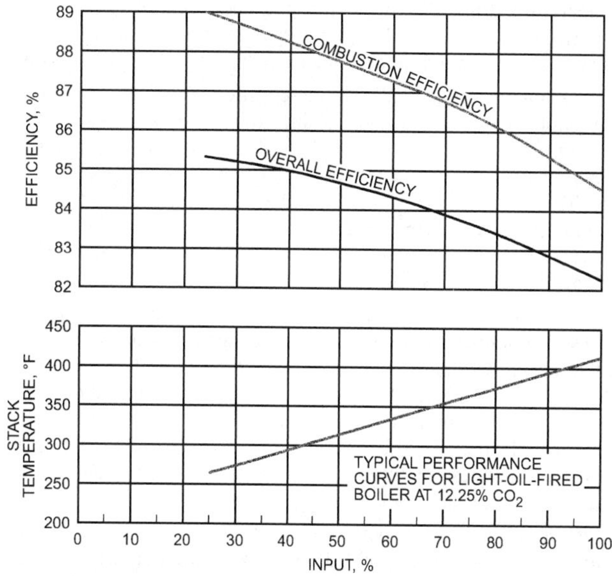

Fig. 8 Boiler Efficiency as Function of Fuel and Air Input

Seasonal efficiency is the actual operating efficiency that the boiler will achieve during the heating season at various loads. Because most heating boilers operate at part load, the part-load efficiency, including heat losses when the boiler is off, has a great effect on the seasonal efficiency. The difference in seasonal efficiency between a boiler with an on/off firing rate and one with modulating firing rate can be appreciable if the airflow through the boiler is modulated along with the fuel input. Figure 8 shows how efficiency increases at part load for a typical boiler equipped with a burner that can fire at reduced inputs while modulating both fuel and air. This increase in efficiency is due to the increase in the ratio of heat exchanger surface area to heat input as the firing rate is reduced.

PERFORMANCE CODES AND STANDARDS

Commercial heating boilers (i.e., boilers with inputs of 300,000 Btu/h and larger) at present are only tested for full-load steady-state efficiency according to standards developed by either (1) the Hydronics Institute Division of the Gas Appliance Manufacturers Association (GAMA) [formerly the Institute of Boiler and Radiator Manufacturers (I-B-R) and the Steel Boiler Institute (SBI)], (2) the American Gas Association (AGA), or (3) Underwriters Laboratories (UL).

The Hydronics Institute (1990) standard for rating cast-iron sectional, steel, and copper boilers bases performance on controlled test conditions for fuel inputs of 300,000 Btu/h and larger. The gross output obtained by the test is limited by such factors as flue gas temperature, draft, CO_2 in the flue gas, and minimum overall efficiency. This standard applies primarily to oil-fired equipment; however, it is also applied to forced draft gas fired or dual-fueled units.

Gas boilers are generally design-certified by an accredited testing laboratory based on tests conducted in accordance with ANSI *Standard* Z21.13 or UL *Standard* 795. Note that the Z21.13 test procedure may be applied to both condensing and noncondensing boilers. This test uses 80°F RWT, 100°F Δt, steady-state, full-load operation, and allows the presence of condensate to be ignored. Efficiencies published under this test procedure are generally not achieved in actual operation.

Instead of the HI-GAMA, AGA, and UL standards, test procedures for commercial-industrial and packaged fire-tube boilers are often performed based on ASME *Performance Test Code* 4.1 (1991). Units are tested for performance under controlled test conditions with minimum required levels of efficiency. Further, the American Boiler Manufacturers Association (ABMA) publishes several guidelines for the care and operation of commercial and industrial boilers and for control parameters.

Residential heating boilers (i.e., all gas- and oil-fired boilers with inputs less than 300,000 Btu/h in the United States) are rated according to standards developed by the U.S. Department of Energy (DOE). The procedure determines both on-cycle and off-cycle losses based on a laboratory test. The test results are applied to a computer program, which simulates an installation and predicts an annual fuel utilization efficiency (AFUE). The steady-state efficiency developed during the test is similar to combustion efficiency and is the basis for determining DOE **heating capacity**, a term similar to gross output. The AFUE represents the part-load efficiency at the average outdoor temperature and load for a typical boiler installed in the United States. Although this value is useful for comparing different boiler models, it is not meant to represent actual efficiency for a specific installation.

SIZING

Boiler sizing is the selection of boiler output capacity to meet connected load. The boiler gross output is the rate of heat delivered by the boiler to the system under continuous firing at rated input. Net rating (I-B-R rating) is gross output minus a fixed percentage (called the piping and pickup factor) to allow for an estimated average piping heat loss, plus an added load for initially heating up the water in a system (sometimes called **pickup**). This I-B-R piping and pickup factor is 1.15 for water boilers and ranges from 1.27 to 1.33 for steam boilers, with the smaller number applying as the boilers get larger. The net rating is calculated by dividing the gross output by the appropriate piping and pickup factor.

Piping loss is variable. If all piping is in the space defined as load, loss is zero. If piping runs through unheated spaces, heat loss from the piping may be much higher than accounted for by the fixed net rating factor. Pickup is also variable. When the actual connected load is less than design load, the pickup factor may be unnecessary.

On the coldest day, extra output (boiler and radiation) is needed to pickup the load from a shutdown or low night setback. If night setback is not used, or if no extended shutdown occurs, no pickup load exists. Standby capacity for pickup, if needed, can be in the form of excess capacity in baseload boilers or in a standby boiler.

If piping and pickup losses are negligible, the boiler gross output can be considered the design load. If piping loss and pickup load are large or variable, those loads should be calculated and equivalent gross boiler capacity added. Boiler capacity must be matched to the terminal unit and system delivery capacity. That is, if the boiler output is greater than the terminal output, the water temperature will rise and the boiler will cycle on the high-limit control, delivering an average input that is much lower than the boiler gross output. Significant oversizing of the boiler may result in a much lower overall boiler efficiency.

BURNER TYPES

Burners for installation on boilers are grouped generally by fuel used and pressure type. Fuel groupings include fuel oil, natural gas, propane, wood, or coal. A **dual-fuel burner** may use two or more fuels (e.g., No. 2 fuel oil and natural gas). The pressure type refers to whether the burner is atmospheric or a fan is used for pressurization. In **atmospheric burners**, firing generally natural gas or propane, the fuel is introduced across a drilled orifice manifold where it contacts combustion air and is ignited. The chimney or flue produces a natural draft to remove the products of combustion. In **power burners**, a fan pushes combustion air into a burner or combustion chamber under positive pressure where it mixes with the fuel and is ignited. The products of combustion are pushed through the combustion chamber and boiler by the fan, then flow through the chimney by natural draft, or induced draft caused by a chimney fan.

Burners may also be classified by method of fuel atomization. In **air atomization**, fuel oil at 80 to 300 psig is pumped through a nozzle orifice to create a fine mist. The fuel-rich mist is mixed with combustion air provided by a fan and is ignited at the burner. In **steam atomization**, generally used on heavy grades of fuel oil, high-pressure steam is mixed with pressurized fuel oil through a nozzle orifice to heat the oil and reduce the oil's viscosity to create a fine mist. The mist is mixed with combustion air provided by a fan and is ignited at the burner.

BOILER CONTROLS

Boiler controls provide automatic regulation of burner and boiler performance to ensure safe and efficient operation. Operating and combustion controls regulate the rate of fuel input in response to a signal representing load change (demand), so that the average boiler output equals the load within some accepted tolerance. Water level and flame safety controls cut off fuel flow when unsafe conditions develop. The National Fire Protection Association (NFPA) Code 85, *Boiler and Combustion Systems Hazard Code* (NFPA 2007), is generally accepted as the governing code for boiler control systems. Other requirements from insurance companies or local governing agencies may also be applicable. Often, the governing agency having jurisdiction may specify specific requirements that the heating system designer or specifying engineer must comply with in the design. It is essential that the designer or engineer determine the applicable codes, and specify the controls and skills needed to complete the control system.

Operating Controls

Steam boilers are operated by boiler-mounted, pressure-actuated controls, which vary fuel input to the boiler. Traditional examples of burner controls were on/off, high/low/off, and modulating. Modulating controls infinitely vary fuel input from 100% down to a selected minimum set point. The ratio of maximum to minimum is the turndown ratio. The minimum input is usually between 5 and 33% (i.e., 20 to 1 down to 3 to 1 ratios); input depends on the size and type of fuel-burning equipment and system. High turndown ratios in noncondensing boilers must be considered carefully to prevent condensation at lower firing rates.

Hot-water boilers are operated by temperature-actuated controls that are usually mounted on the boiler. Traditionally, burner controls were the same as for steam boilers (i.e., on/off, high/low/off, and modulating). Modulating controls typically offer more precise water temperature control and higher efficiency than on/off or high/low controls, if airflow through the boiler is modulated along with fuel input.

Boiler reset controls can enhance the efficiency of hot-water boilers. These controls may operate with any of the burner controls mentioned previously. They automatically change the high-limit set point of the boiler to match the variable building load demands caused by changing outdoor temperatures. By keeping boiler water temperature as low as possible, efficiency is enhanced and standby losses are reduced.

Microprocessor-Based Control Systems. The introduction of microprocessor-based control systems has changed traditional operating controls on boilers. In the past, smaller boilers were equipped with on/off or high/low/off electromagnetic-relay-based burner operating controls with mercury switches, with larger boilers provided with modulating controls. The low cost and greater efficiency of microprocessor-based control systems has resulted in the availability of such controls on small factory-packaged boilers, and nearly all medium and larger boiler installations. The recent introduction of integrated combustion and burner safeguard microprocessor controllers has accelerated this availability.

Traditionally, most burner installations used a single actuator to drive the combustion air damper and fuel ratio valves through common linkages. Such installations were called **single-jackshaft** controls. The fuel ratio valves often used set screw cams to produce efficient combustion throughout the firing range of the burner. Tuning the burner involved positioning the set screws to adjust the cam, which in turn regulated the fuel ratio at various firing rates. Tuning was often cumbersome, and easily lost when set screws loosened. Single-jackshaft control also invariably meant compromises were made in tuning, generally resulting in inefficient combustion with high excess air ratios at low firing rates. Current technology eliminates the single-jackshaft, using "linkless" burners with individual actuators controlling the combustion air damper and each fuel valve. With the high-speed processing ability of the microprocessors, individual actuators can quickly and accurately respond to changes in load, ensuring efficient combustion throughout the full firing range. When oxygen analyzers are installed to measure the oxygen content of the flue gas, microprocessor combustion controllers can modulate the combustion air damper and fuel valve actuators to ensure optimal combustion efficiency.

Water Level Controls

Maintaining proper water level in a boiler is of paramount concern. Should water level drop below a preset limit, damage may occur from overheating of boiler surfaces, resulting in cracking of cast-iron sections, or plastic deformation of steel tubes and tubesheets. Such a condition is known as **dry-firing** of the boiler. The installation of a low-water cutoff switch to stop fuel flow to the burner is necessary to prevent damage from dry firing.

To maintain proper water levels, different methods may be used. Smaller boilers generally use a boiler feedwater controller to cause a feedwater pump to pump water directly to the boiler to maintain proper water level. In larger installations, a feedwater piping loop may serve several boilers in parallel. In such an installation, each boiler has a feedwater valve controlled by a feedwater controller mounted on the boiler. The controller modulates the feedwater valve to maintain proper water level in the boiler. Simple feedwater control systems use water level as the control parameter to modulate the feedwater valve. Such systems are considered **single-element** feedwater systems. In larger boiler installations where steam is generated, the steam flow rate or rate of change of steam pressure may also be monitored with a signal sent to the feedwater controller. If the controller is programmed to modulate the feedwater valve based on both water level and steam flow rate or rate of change of pressure, the feedwater control system is referred to a **two-element** system. A **three-element** system uses water level, steam flow rate, and rate of pressure change to modulate the feedwater valve.

FLAME SAFEGUARD CONTROLS

Flame safeguard controls monitor flame condition and shut fuel flow to the burner in the event of an unsafe condition. The safety circuit of a flame safeguard control system typically includes switch contacts for low-water cutoff, high limits, air proving switches, redundant safety and operating controls, and flame monitors. Flame monitors typically sue either infrared or ultraviolet scanners to monitor flame condition and deactivate the burner in the event of nonignition or other unsafe flame condition. Flame safeguard controllers usually use preprogrammed algorithms to operate a burner and cycle it through stages of operation. The first stage is a purge cycle wherein the boiler's combustion chamber is flushed with combustion air to remove any unspent fuel and products of combustion that remain from the previous cycle. After purging, the pilot flame is ignited and the main fuel introduced into the burner and ignited. In the absence of a pilot fuel, such as with direct electronic spark ignition, the main fuel is introduced and ignited. In either case, the flame monitor determines whether ignition has occurred and whether the resulting flame is proper. If ignition has failed, or flame is not indicated, the flame monitor cuts off fuel flow, causing a postpurge cycle to rid the combustion chamber of unspent fuel and products of combustion. On restart, the burner again starts with a purge cycle and repeats the steps to ignition. When proper flame is established, the flame safeguard control system allows the operating control or combustion control system to control or modulate the burner firing rate.

Traditionally, flame safeguard systems were separate controllers from operating or combustion control systems. With the advent of microprocessor-based control systems, separate microprocessors control the individual functions of flame safeguard and combustion control. Recently introduced burner controllers provide an integrated control with algorithms for both flame safeguard and combustion control.

REFERENCES

ANSI. 2004. Gas-fired low-pressure steam and hot water boilers. *Standard* Z21.13-2004/CSA 4.9-2004. CSA International, Mississauga, Ontario.

ASME. 2007. Rules for construction of power boilers. *Boiler and Pressure Vessel Code*, Section I-2007. American Society of Mechanical Engineers, New York.

ASME. 2007. Rules for construction of heating boilers. *Boiler and Pressure Vessel Code*, Section IV-2007. American Society of Mechanical Engineers, New York.

ASME. 2007. Recommended rules for the care and operation of heating boilers. *Boiler and Pressure Vessel Code*, Section VI-2007. American Society of Mechanical Engineers, New York.

ASME. 2007. Recommended rules for the care of power of boilers. *Boiler and Pressure Vessel Code*, Section VII-2007. American Society of Mechanical Engineers, New York.

Hydronics Institute. 1990. *Testing and rating standard for heating boilers*, 6th ed. Hydronics Institute, Berkeley Heights, NJ.

BIBLIOGRAPHY

ABMA. 1999. *Packaged boiler engineering manual*. American Boiler Manufacturers Association, Arlington, VA.

ASME. 1998. Steam generating units. *Performance Test Code* 4. American Society of Mechanical Engineers, New York.

ASME. 2006. Controls and safety devices for automatically fired boilers, *Standard* CSD-1-2006. American Society of Mechanical Engineers, New York.

NFPA. 2007. Boiler and combustion systems hazard code. NFPA *Code* 85-2007. National Fire Protection Association, Quincy, MA.

Strehlow, R.A. 1984. *Combustion fundamentals*. McGraw-Hill, New York.

Woodruff, E.B., H.B. Lammers, and T.F. Lammers. 1984. *Steam-plant operation*, 5th ed. McGraw-Hill, New York.

UL. 2006. Commercial-industrial gas heating equipment. *Standard* 795. Underwriters Laboratories, Northbrook, IL.

CHAPTER 32

FURNACES

FURNACES are self-enclosed, permanently installed major appliances that provide heated air through ductwork to the space being heated. In addition, a furnace may provide the indoor fan necessary for circulating heated or cooled air from a split or single-package air conditioner or heat pump (see Chapter 8). Furnaces may be used in either residential or commercial applications, and may be grouped according to the following characteristics:

- Heat source: electricity, natural gas/propane (natural draft, fan assisted, or condensing/noncondensing), or oil (forced draft with power atomizing burner)
- Installation location: within conditioned space (indoors), or outside conditioned space [either outdoors, or inside the structure but not within the heated space (isolated combustion systems)]
- Mounting arrangement and airflow: horizontal forced-air, vertical (natural convection, forced-air upflow, forced-air downflow, or forced-air lowboy), or multiposition forced-air

Furnaces that use electricity as a heat source include a resistance-type heating element that heats the circulating air either directly or through a metal sheath that encloses the resistance element. In gas- or oil-fired furnaces, combustion occurs in a combustion chamber. Circulating air passes over the outside surfaces of a heat exchanger so that it does not contact the fuel or the products of combustion, which are passed to the outside atmosphere through a vent.

In North America, natural gas is the most common fuel supplied for residential heating, and the central-system forced-air furnace (Figure 1) is the most common way of heating with natural gas. This type of furnace is equipped with a blower to circulate air through the furnace enclosure, over the heat exchanger, and through the ductwork distribution system. The furnace is categorized as follows:

- Heat source: Gas
- Combustion system: Induced-draft manifold burner
- Installation location: Inside the structure but not within the conditioned space
- Mounting: Vertical
- Airflow: Upflow

COMPONENTS

A typical furnace consists of the following basic components: (1) a cabinet or casing; (2) heat exchangers; (3) combustion systems and other heat sources, including burners and controls; (4) venting components, such as a forced-draft blower, induced-draft blower, or draft hood; (5) a circulating air blower and motor; and (6) an air filter and other accessories such as a humidifier, an electronic air cleaner, an air-conditioning coil, or a combination of these elements.

The preparation of this chapter is assigned to TC 6.3, Central Forced Air Heating and Cooling Systems.

Casing or Cabinet

The furnace casing is most commonly formed from painted cold-rolled steel. Access panels on the furnace allow access to those sections requiring service. The inside of the casing adjacent to the heat exchanger or electric heat elements is lined with a foil-faced blanket insulation and/or a metal radiation shield to reduce heat losses through the casing and to limit the outside surface temperature of the furnace. On some furnaces, the inside of the blower compartment is lined with insulation to acoustically dampen the blower noise. Commercial furnace cabinets may also include the indoor and outdoor air-conditioning or heat pump components.

Heat Exchangers

Furnaces with gas-fired burners have heat exchangers that are made either of mirror-image formed parts that are joined together to form a clam shell or finless tubes bent into a compact form. Standard indoor furnace heat exchangers are generally made of coated or alloy steel. Common corrosion-resistant materials include aluminized steel, ceramic-coated cold-rolled steel, and stainless steel. Furnaces certified for use downstream of a cooling coil must have corrosion-resistant heat exchangers.

Some problems of heat exchanger corrosion and failure have been encountered because of exposure to halogen ions in the flue gas. These problems were caused by combustion air contaminated by substances such as laundry bleach, cleaning solvents, and halogenated hydrocarbon refrigerants.

Research has been done on corrosion-resistant materials for use in condensing (secondary) heat exchangers (Stickford et al. 1985). The presence of chloride compounds in the condensate can cause a

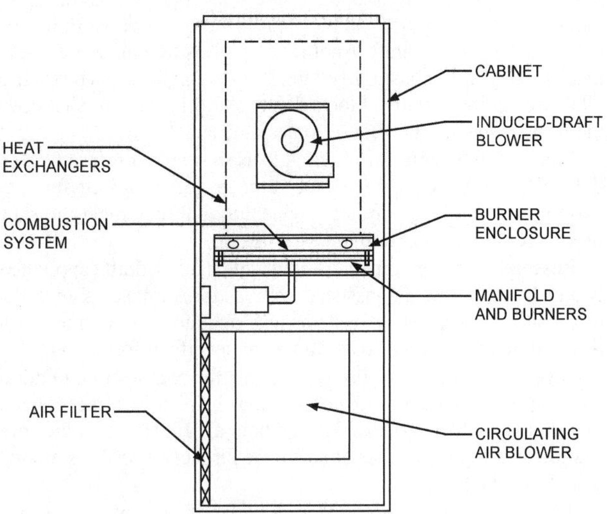

Fig. 1 Induced-Draft Gas Furnace

condensing heat exchanger to fail, unless a corrosion-resistant material is used.

Several manufacturers produce liquid-to-air heat exchangers in which a liquid is heated and is either evaporated or pumped to a condenser section or fan-coil, which heats circulating air.

Heat exchangers of oil-fired furnaces are normally heavy-gage steel formed into a welded assembly. Hot flue products flow through the inside of the heat exchanger into the chimney, and conditioned air flows over the outside of the heat exchanger and into the air supply plenum.

Electric Heat Elements. Elements for electric furnaces are generally either open wire, open ribbon, or wire enclosed in a tube. Current is applied to the element and heats it through resistance of the material.

Burners and Internal Controls. Gas burners are most frequently made of stamped sheet metal, although cast iron is also used. Fabricated sheet metal burners may be made from cold-rolled steel coated with high-temperature paint or from a corrosion-resistant material such as stainless or aluminized steel. Burner material must meet the corrosion protection requirements of the specific application. Gas furnace burners may be of either the monoport or multiport type; the type used with a particular furnace depends on compatibility with the heat exchanger.

Gas furnace controls include an ignition device, gas valve, fan control, limit switch, and other components specified by the manufacturer. These controls allow gas to flow to the burners when heat is required. The most common ignition systems are (1) standing pilot, (2) intermittent pilot, (3) direct spark, and (4) hot-surface ignition (ignites either a pilot or the main burners directly). (Standing-pilot ignition systems are not typically available from manufacturers today because federally mandated efficiency standards preclude their use.) The section on Technical Data has further details on the function and performance of individual control components.

Oil furnaces are generally equipped with pressure-atomizing burners. The pump pressure and size of the injection nozzle orifice regulate the firing rate of the furnace. Electric ignition lights the burners. Other furnace controls, such as the blower switch and the limit switch, are similar to those used on gas furnaces.

Combustion Venting Components

Natural-draft indoor furnaces are equipped with a **draft hood** connecting the heat exchanger flue gas exit to the vent pipe or chimney. The draft hood has a relief air opening large enough to ensure that the exit of the heat exchanger is always at atmospheric pressure. One purpose of the draft hood is to make certain that the natural-draft furnace continues to operate safely without generating carbon monoxide if the chimney is blocked, if there is a downdraft, or if there is excessive updraft. Another purpose is to maintain constant pressure on the combustion system. Residential furnaces built since 1987 are equipped with a blocked-vent shutoff switch to shut down the furnace in case the vent becomes blocked.

Fan-assisted combustion furnaces use a small blower to force or induce flue products through the furnace. Induced-draft furnaces may or may not have a relief air opening, but they meet the same safety requirements regardless.

Research into common venting of natural-draft appliances (water heaters) and fan-assisted combustion furnaces shows that nonpositive vent pressure systems may operate on a common vent. Refer to manufacturers' instructions for specific information.

Direct-vent furnaces use outdoor air for combustion. Outdoor air is supplied to the furnace combustion chamber by direct connections between the furnace and the outdoor air. If the vent or the combustion air supply becomes blocked, the furnace control system will shut down the furnace.

ANSI *Standard* Z21.47/CSA 2.3 classifies venting systems. Central furnaces are categorized by temperature and pressure attained in the vent and by the steady-state efficiency attained by the furnace. Although ANSI *Standard* Z21.47/CSA 2.3 uses 83% as the steady-state efficiency dividing central furnace categories, a general rule of thumb is as follows:

Category I: nonpositive vent pressure and flue loss of 17% or more.

Category II: nonpositive vent pressure and flue loss less than 17%.

Category III: positive vent pressure and flue loss of 17% or more.

Category IV: positive vent pressure and flue loss less than 17%.

Furnaces rated in accordance with ANSI *Standard* Z21.47/CSA 2.3 that are not direct vent are marked to show that they are in one of these four venting categories.

Ducted-system, oil-fired, forced-air furnaces are usually forced draft.

Circulating Blowers and Motors

Centrifugal blowers with forward-curved blades of the double-inlet type are used in most forced-air furnaces. These blowers overcome the resistance of furnace air passageways, filters, and ductwork. They are usually sized to provide the additional air requirement for cooling and the static pressure required for the cooling coil. The blower may be a direct-drive type, with the blower wheel attached directly to the motor shaft, or it may be a belt-drive type, with a pulley and V-belt used to drive the blower wheel.

Electric motors used to drive furnace blowers are usually custom designed for each furnace model or model series. Direct-drive motors may be of the shaded-pole or permanent split-capacitor type. Speed variation may be obtained by taps connected to extra windings in the motor. Belt-drive blower motors are normally split-phase or capacitor-start. The speed of belt-drive blowers is controlled by adjusting a variable-pitch drive pulley.

Electronically controlled, variable-speed motors are also available. This type of motor reduces electrical consumption when operated at low speeds.

Filters and Other Accessories

Air Filters. An air filter in a forced-air furnace removes dust from the air that could reduce the effectiveness of the blower and heat exchanger(s), and may also help provide cleaner air for the indoor environment (see ASHRAE *Standard* 52.2). Filters installed in a forced-air furnace are often disposable. Permanent filters that may be washed or vacuum-cleaned and reinstalled are also used. The filter is always located in the circulating airstream ahead of the blower and heat exchanger. Because the air filter keeps airflow components of the furnace clean, it should be cleaned or replaced regularly to extend the life of the furnace components. See Chapters 9 and 28 for further information on air filters.

Humidifiers. These are not included as a standard part of the furnace package. However, one advantage of a forced-air heating system is that it offers the opportunity to control the relative humidity of the heated space at a comfortable level. Chapter 21 addresses various types of humidifiers used with forced-air furnaces.

Electronic Air Cleaners. These air cleaners may be much more effective than the air filter provided with the furnace, and they filter out much finer particles, including smoke and pollen. Electronic air cleaners create an electric field of high-voltage direct current in which dust particles are given a charge and collected on a plate having the opposite charge. The collected material is then cleaned periodically from the collector plate by the homeowner. Electronic air cleaners are mounted in the airstream entering the furnace. Chapter 28 has detailed information on filters.

Automatic Vent Dampers. This device closes the vent opening on a draft hood-equipped natural-draft furnace when the furnace is

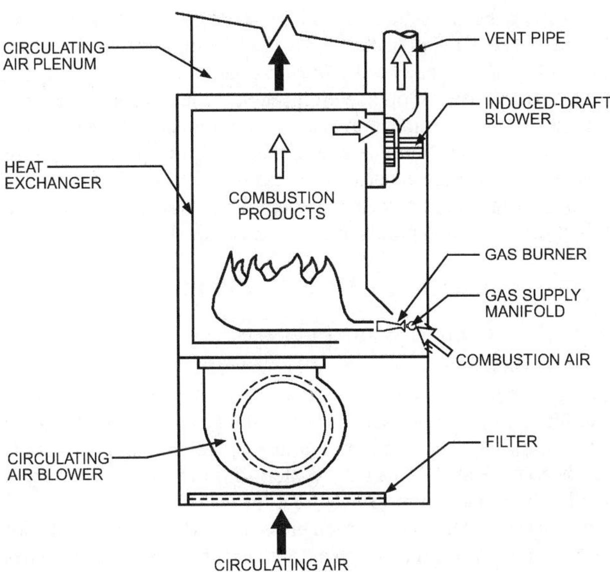

Fig. 2 Upflow Category I Furnace with Induced-Draft Blower

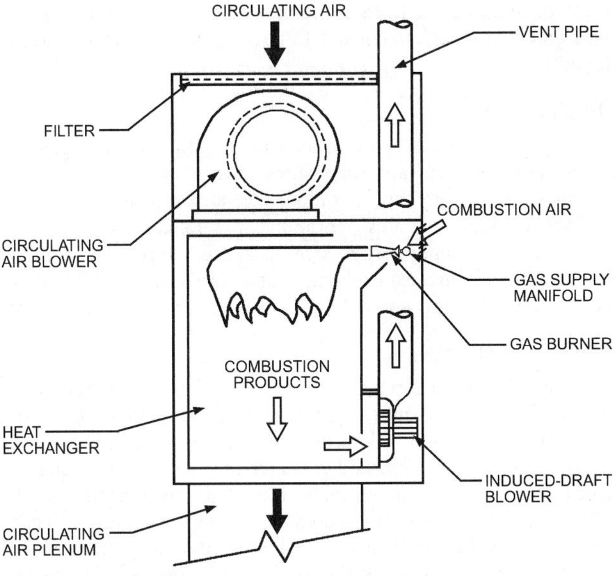

Fig. 3 Downflow (Counterflow) Category I Furnace with Induced-Draft Blower

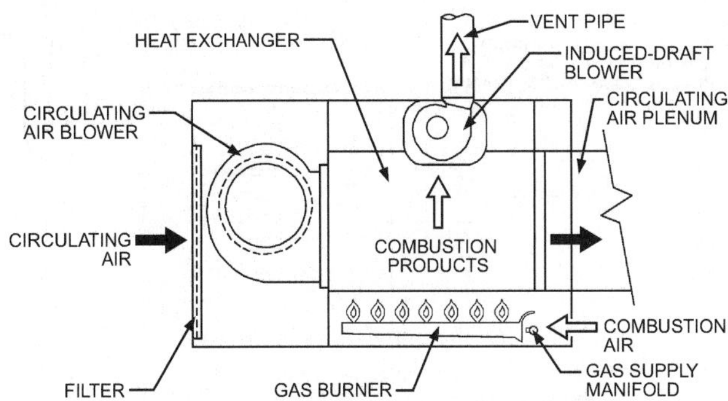

Fig. 4 Horizontal Category I Furnace with Induced-Draft Blower

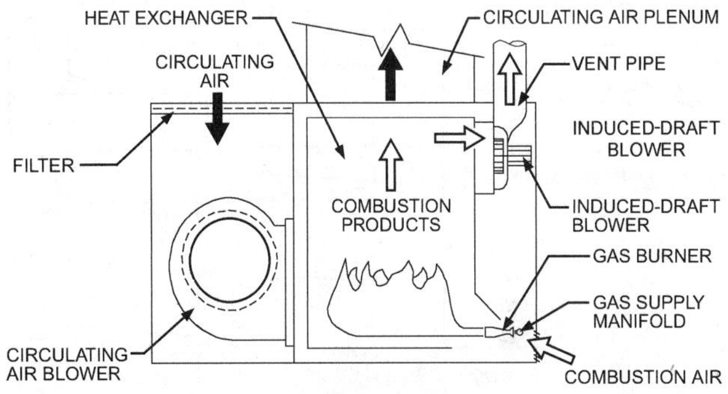

Fig. 5 Basement (Lowboy) Category I Furnace with Induced-Draft Blower

not in use, thus reducing off-cycle losses. More information about the energy-saving potential of this accessory is included in the section on Technical Data.

Airflow Variations

The components of a forced-air furnace can be arranged in a variety of configurations to suit a residential heating system. The relative positions of the components in the different types of furnaces are as follows:

- **Upflow or "highboy" furnace.** In an upflow furnace (Figure 2), the blower is located beneath the heat exchanger and discharges vertically upward. Air enters through the bottom or the side of the blower compartment and leaves at the top. This furnace may be used in closets and utility rooms on the first floor or in basements, with the return air ducted down to the blower compartment entrance.
- **Downflow furnace.** In a downflow furnace (Figure 3), the blower is located above the heat exchanger and discharges downward. Air enters at the top and is discharged vertically at the bottom. This furnace is normally used with a perimeter heating system in a house without a basement. It is also used in upstairs furnace closets and utility rooms supplying conditioned air to both levels of a two-story house.
- **Horizontal furnace.** In a horizontal furnace, the blower is located beside the heat exchanger (Figure 4). Air enters at one end, travels horizontally through the blower and over the heat exchanger, and is discharged at the opposite end. This furnace is used for locations with limited head room such as attics and crawl spaces, or is suspended under a roof or floor or placed above a suspended ceiling. These units are often designed so that the components may be rearranged to allow installation with airflow from left to right or from right to left.
- **Multiposition furnace.** A furnace that can be installed in more than one airflow configuration (e.g., upflow or horizontal; downflow or horizontal; or upflow, downflow, or horizontal) is a multiposition furnace. In some models, field conversion is necessary to accommodate an alternative installation.
- **Basement or "lowboy" furnace.** The basement furnace (Figure 5) is a variation of the upflow furnace and requires less head room. The blower is located beside the heat exchanger at the bottom. Air enters the top of the cabinet, is drawn down through the blower, is discharged over the heat exchanger, and leaves vertically at the top. This type of furnace has become less popular because of the advent of short upflow furnaces.
- **Gravity furnace.** These furnaces are no longer available, and they are not common. This furnace has larger air passages through the casing and over the heat exchanger so that the buoyancy force created by the air being warmed circulates the air through the ducts. Wall furnaces that rely on natural convection (gravity) are discussed in Chapter 33.

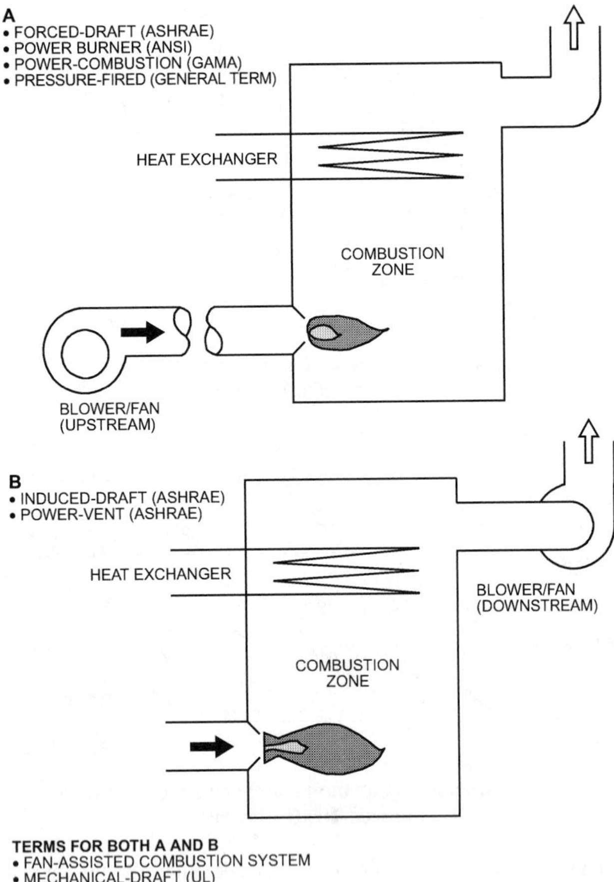

Fig. 6 Terminology Used to Describe Fan-Assisted Combustion

Combustion System Variations

Gas-fired furnaces use a natural-draft or a fan-assisted combustion system. With a natural-draft furnace, the buoyancy of hot combustion products carries these products through the heat exchanger, into the draft hood, and up the chimney.

Fan-assisted combustion furnaces have a combustion blower, which may be located either upstream or downstream from the heat exchangers (Figure 6). If the blower is located upstream, blowing the combustion air into the heat exchangers, the system is known as a forced-draft system. If the blower is downstream, the arrangement is known as an induced-draft system. Fan-assisted combustion systems have generally been used with outdoor furnaces; however, with the passage of the 1987 U.S. National Appliance Energy Conservation Act, fan-assisted combustion has become more common for indoor furnaces as well. Fan-assisted combustion furnaces do not require a draft hood, resulting in reduced off-cycle losses and improved efficiency.

Direct-vent furnaces may have either natural-draft or fan-assisted combustion. They do not have a draft hood, and they obtain combustion air from outside the structure. Mobile home furnaces must be of the direct-vent type.

Indoor/Outdoor Furnace Variations

Central system residential furnaces are designed and certified for either indoor or outdoor use. Outdoor furnaces are normally horizontal flow and convertible to downflow.

The heating-only outdoor furnace is similar to the more common indoor horizontal furnace. The primary difference is that the outdoor furnace is weatherized; the motors and controls are sealed, and the exposed components are made of corrosion-resistant materials such as galvanized or aluminized steel.

A common style of outdoor furnace is the combination package unit. This unit is a combination of an air conditioner and a gas or electric furnace built into a single casing. The design varies, but the most common combination consists of an electric air conditioner coupled with a horizontal gas or electric furnace. The advantage is that much of the interconnecting piping and wiring is included in the unit.

HEAT SOURCE TYPES

Natural Gas and Propane Furnaces

Most manufacturers have their furnaces certified for both natural gas and propane. The major difference between natural gas and propane furnaces is the pressure at which the gas is injected from the manifold into the burners. For natural gas, the manifold pressure is usually controlled at 3 to 4 in. of water; for propane, the pressure is usually 10 to 11 in. of water.

Because of the higher injection pressure and the greater heat content per volume of propane, there are certain physical differences between a natural gas furnace and a propane furnace. One difference is that the pilot and burner orifices must be smaller for propane furnaces. The gas valve regulator spring is also different. Sometimes it is necessary to change burners, but this is not normally required. Manufacturers sell conversion kits containing both the required parts and instructions to convert furnace operation from one gas to the other.

Oil Furnaces

Indoor oil furnaces come in the same configuration as gas furnaces. They are available in upflow, downflow, horizontal, and multiposition lowboy configurations for ducted systems. Oil-fired outdoor furnaces and combination units are not common.

The major differences between oil and gas furnaces are in the combustion system, heat exchanger, and barometric draft regulator used in lieu of a draft hood.

Electric Furnaces

Electric-powered furnaces come in a variety of configurations and have some similarities to gas- and oil-fired furnaces. However, when a furnace is used with an air conditioner, the cooling coil may be upstream from the blower and heaters. On gas- and oil-fired furnaces, the cooling coil is normally mounted downstream from the blower and heat exchangers so cold air leaving the cooling coil does not contact the heat exchangers, which could cause premature corrosion from condensation. If the cooling coil is upstream of the heat exchangers on a gas- or oil-fired product, the heat exchanger may require a mechanism to remove the condensed moisture.

Figure 7 shows a typical arrangement for an electric forced-air furnace. Air enters the bottom of the furnace and passes through the filter, then flows up through the cooling coil section into the blower. The electric heating elements are immediately above the blower so that the high-velocity air discharging from the blower passes directly through the heating elements.

The furnace casing, air filter, and blower are similar to equivalent gas furnace components. The heating elements are made in modular form, with 5 kW capacity being typical for each module. Electric furnace controls include electric overload protection, contactor, limit switches, and a fan control switch. The overload protection may be either fuses or circuit breakers. The contactor brings the electric heat modules on. The fan control switch and limit switch functions are similar to those of the gas furnace, but one limit switch is usually used for each heating element.

Frequently, electric furnaces are made from modular sections; for example, the coil box, blower section, and electric heat section are made separately and then assembled in the field. Regardless of

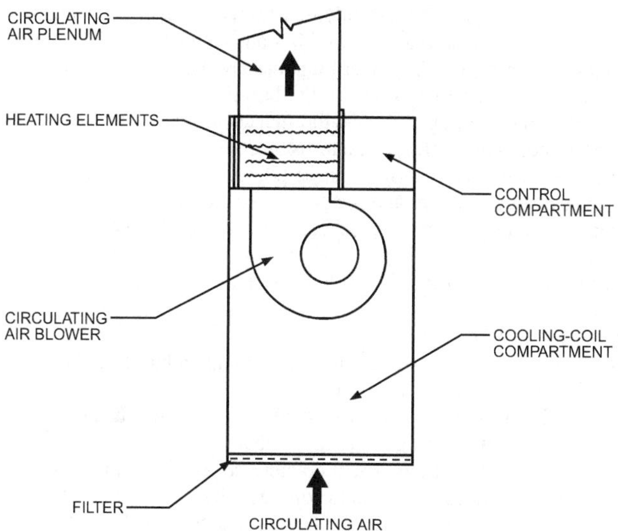

Fig. 7 Electric Forced-Air Furnace

whether the furnace is made from a single-piece casing or a modular casing, it is generally a multiposition unit. Thus, the same unit may be used for upflow, downflow, or horizontal installations.

When an electric heating appliance is sold without a cooling coil, it is known as an electric furnace. The same appliance is called a fan-coil air handler when it has an air-conditioning coil already installed. When the unit is used as the indoor section of a split heat pump, it is called a heat pump fan-coil air handler. For detailed information on heat pumps, see Chapter 48.

Electric forced-air furnaces are also used with packaged heat pumps and packaged air conditioners.

COMMERCIAL EQUIPMENT

The basic difference between residential and commercial furnaces is the size and heating capacity of the equipment. The heating capacity of a commercial furnace may range from 150,000 to over 2,000,000 Btu/h. Generally, furnaces with output capacities less than 320,000 Btu/h are classified as light commercial, and those above 320,000 Btu/h are considered large commercial equipment. In addition to the difference in capacity, commercial equipment is often constructed from material with increased structural strength and commonly has more sophisticated control systems.

Light commercial heating equipment comes in almost as many flow arrangements and design variations as residential equipment. Some are identical to residential equipment, whereas others are unique to commercial applications. Some commercial units function as a part of a ducted system, and others operate as unducted space heaters.

Ducted Equipment

Upflow Gas-Fired Commercial Furnaces. These furnaces are available up to 300,000 Btu/h and supply enough airflow to handle up to 10 tons of air conditioning. They may have high static pressure and belt-driven blowers, and frequently they consist of two standard upflow furnaces tied together in a side-by-side arrangement. They are normally incorporated into a system in conjunction with a commercial split-system air-conditioning unit and are available in either propane or natural gas. Oil-fired units may be available on a limited basis.

Horizontal Gas-Fired Duct Furnaces. Available for built-up light commercial systems, this type of furnace is not equipped with its own blower but is designed for uniform airflow across the entire furnace. Duct furnaces are normally certified for operation

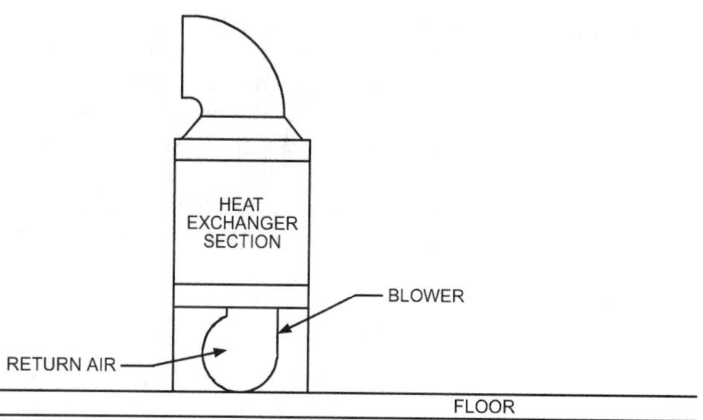

Fig. 8 Standing Floor Furnace

either upstream or downstream of an air conditioner cooling coil. If a combination blower and duct furnace is desired, a package called a blower unit heater is available. Duct furnaces and blower unit heaters are available in natural gas, propane, oil, and electric models.

Electric Duct Furnaces. These furnaces are available in a large range of sizes and are suitable for operation in upflow, downflow, or horizontal positions. These units are also used to supply auxiliary heat with the indoor section of a split heat pump.

Combination Package Units. The most common commercial furnace is the combination package unit, sometimes known as a **combination rooftop unit**. These are available as air-conditioning units with propane and natural gas furnaces, electric resistance heaters, or heat pumps. Combination oil-heat/electric-cool units are not commonly available. Combination units come in a full range of sizes covering air-conditioning ratings from 1.5, with matched furnaces supplying heat-to-cool capacity ratios of approximately 1.5 to 1.

Combination units of 15 tons and under are available as single-zone units. The entire unit must be in either heating mode or cooling mode. All air delivered by the unit is at the same temperature. Frequently, the heating function is staged so that the system operates at reduced heat output when the load is small.

Large combination units in the 15 to 50 ton range are available as single-zone units, as are small units; however, they are also available as multizone units. A multizone unit supplies conditioned air to several different zones of a building in response to individual thermostats controlling those zones. These units can supply heating to one or more zones at the same time that cooling is supplied to other zones.

Large combination units are normally available only in a curbed configuration (i.e., units are mounted on a rooftop over a curbed opening in the roof). Supply and return air enters through the bottom of the unit. Smaller units may be available for either curbed or uncurbed mounting. In either case, the unit is usually connected to ductwork in the building to distribute the conditioned air.

Unducted Heaters

Ductless furnaces, floor furnaces (Figure 8), and wall furnaces are discussed in Chapter 33. Infrared heating equipment is covered in Chapter 15.

CONTROLS AND OPERATING CHARACTERISTICS

External to Furnace

Externally, the furnace is controlled by a low-voltage room thermostat. Control can be heating-only, combination heat/cool, multistage, or night setback. Chapter 15 of the 2005 *ASHRAE Handbook—Fundamentals* discusses thermostats in more detail. A night setback thermostat can reduce the annual energy consumption, and dual setback (setting the temperature back during the night and during unoccupied periods in the day) can save even more energy. Gable and Koenig (1977) and Nelson and MacArthur (1978) estimated that energy savings of up to 30% are possible, depending on the degree and length of setback and the geographical location. The percentage of energy savings is greater in regions with mild climates; however, the total energy savings is greatest in cold regions.

Internal to Furnace

Several types of gas valves perform various functions within the furnace. The type of valve available relates closely to the type of ignition device used. **Two-stage valves**, available on some furnaces, operate at full gas input or at a reduced rate, and are controlled by either a two-stage thermostat or a software algorithm programmed in the furnace control system. They provide less heat at the reduced input and, therefore, may produce less space temperature variation and greater comfort during mild weather conditions when full heat output is not required. Two-stage control is used frequently for zoning applications. Fuel savings with two-stage firing rate systems may not be realized unless both the fuel and the combustion air are controlled.

The **fan control switch** controls the circulating air blower. This switch may be temperature-sensitive and exposed to the circulating airstream in the furnace cabinet, or it may be an electronically operated relay. Blower start-up is typically delayed about 1 min after burner start-up. This delay gives the heat exchangers time to warm up and reduces the flow of cold air when the blower comes on. Blower shutdown is also delayed several minutes after burner shutdown to remove residual heat from the heat exchangers and to improve the annual efficiency of the furnace. Constant blower operation throughout the heating season is sometimes used to improve air circulation; however, this increases fan motor energy consumption, duct conductive losses, and air distribution system air leakage losses. Electronic motors that provide continuous but variable airflow use less energy. Both strategies may be considered when air filtering performance is important.

The **limit switch** prevents overheating in the event of severe reduction in circulating airflow. This temperature-sensitive switch is exposed to the circulating airstream and shuts off the heat source (e.g., gas valve or electric element) if the temperature of air leaving the furnace is excessive. The fan control and limit switches are sometimes incorporated in the same housing and may be operated by the same thermostatic element. In the United States, the **blocked-vent shutoff switch** and **flame rollout switch** have been required on residential furnaces produced since November 1989; they shut off the gas valve if the vent is blocked or when insufficient combustion air is present.

Furnaces using fan-assisted combustion feature a **pressure switch** to verify the flow of combustion air before opening the gas valve. The ignition system has a required pilot gas shutoff feature in case the pilot ignition fails.

EQUIPMENT SELECTION

Many options are available to consumers, and careful planning is needed when selecting equipment. Some decisions can be very basic, whereas others may require research into the various kinds of equipment and features that are available. Several selection considerations are presented here.

Distribution System

A fundamental question in the selection process is whether the space to be heated uses a circulating forced-air system or a hydronic system. These two types of systems are vastly different with respect to equipment selection. For hydronic systems, refer to Chapter 35.

Forced-air systems vary widely. A forced-air distribution system typically has a central air duct, with branches feeding air to numerous supply registers. The duct branches are designed to proportion the air to the different spaces in the building, so that temperatures are best managed by a central thermostat location. Some systems are zoned, with different thermostat sensors; in these systems, dampers with electrical motors are placed at strategic branches of the distribution system and are opened or closed according to each zone's demand for heat. Choosing a zoned system affects the type of equipment selected. Chapter 9 discusses the overall design configuration and efficiency of forced-air systems.

Equipment Location

Furnaces can be installed inside or outside a building. For ideal air distribution, locate the unit in the center of the structure being heated. Furnaces are typically located in a closet, mechanical room, basement, crawlspace, garage, or outdoors.

Installation locations are characterized as either indoors, outdoors (weatherized), or isolated combustion systems (ICS). **Indoor** furnaces are installed within the heated structure, such as in a basement that is connected to the internal living space, in a utility room, or in a closet. In these locations, air that directly surrounds the furnace is in communication with the air of the heated space. Heat that is lost from the cabinet and adjacent ducts by conduction or air leakage is largely recaptured, helping to preserve furnace efficiency. Some furnaces use combustion air from inside the building. In these applications, room air is used for combustion and exhausted to the outdoors. Additionally, dilution air is also drawn from the room. Combustion/dilution makeup air is provided to the combustion appliance zone by infiltration or by a duct designed specifically to provide makeup air, as required by installation codes.

Furnaces located **outside** the conditioned space (e.g., crawlspace, attic, garage, outdoors) regain little or none of the conductive or air leakage energy losses. Furnaces typically have lower insulation levels than the ducts to which they are attached. Outdoor installations require furnaces to be qualified as weatherized. Outdoor installation locations are on rooftops, platforms, or on a pad adjacent to the heated structure. Ducts for supply and return air connections may also be exposed to the elements and should be weatherized appropriately.

Isolated combustion systems (ICS) are within the structure being heated, but the air that directly surrounds the furnace does not communicate with the heated space. Air needed for combustion and ventilation is admitted through grille openings or ducts (NFPA 54). Heat given off from the casing is not considered usable heat, and is subtracted from the furnace efficiency. Typical ICS locations include garages, attics, crawlspaces, and closets that are directly ventilated to the outdoors. These furnaces are protected from the elements (but not temperature) by the surrounding structure.

Forced-Air System Primary Use

Forced-air systems have many benefits, the most significant of which is that they can be used for both heating and cooling without needing separate duct systems and separate air-handling units. A primary function of the furnace is to circulate air through the forced-air distribution system.

Heating the air is an inherent function of a furnace. Cooling is typically a modular add-on, although some furnaces are included as a part of a packaged furnace-and-air-conditioner combination appliance. Forced-air systems also make it possible to add humidifiers and air filters or purifiers. In some cases, forced-air systems can also be connected to an additional appliance to bring clean air from the outdoors through air-to-air energy recovery heat exchangers.

Air distribution system design must take into account all the system's intended functions. The air-handling capacity must be designed to meet the demand of the highest airflow and static pressure needs. Typically, the airflow needed for heating is less than that needed for cooling. Manufacturers provide information on furnace performance capabilities, including how much airflow it can deliver at different static pressures. Furnace specification data typically describe the gas input for heating and the airflow (or cooling capacity) for which the unit is designed.

Fuel Selection

The type of fuel selected for heating is based on relative fuel cost, number of heating degree-days, and the availability of utilities in the area. The most common fuel is natural gas because of its clean burning characteristics, and because of the continuous supply of this fuel through underground distribution networks to most urban settings. Propane and oil fuels are also commonly used. These fuels require on-site storage and periodic fuel deliveries. Electric heat is also continuously available through electrical power grids and is common especially where natural gas is not provided, or where the heating demand is small relative to the cooling demand.

Furnaces are clearly marked for the type of fuel to be used. In some cases, a manufacturer-approved conversion kit may be necessary to convert a furnace from one fuel type to another. If the fuel type is changed after the original installation, the conversion must be done by a qualified service person per the manufacturer's instructions and using the manufacturer's specified conversion kit. After conversion, the unit must be properly inspected by the local code authority.

Combustion Air and Venting

All fuel-burning furnaces must be properly vented to the outdoors. Metal vents, masonry chimneys, and plastic vents are commonly used for furnaces. Manufacturers provide installation instructions for venting their furnaces, and Chapter 34 has a detailed discussion on venting.

Air for combustion enters the combustion zone through louvers in the control door or wall, or is drawn in from a different location (typically to prevent infiltration losses to the heated space and, depending on where the furnace is located, to provide clean air for combustion).

Outdoor air usually has lower levels of pollutants than are typically found in air from indoors, garages, utility rooms, and basements. If the furnace is exposed to clean air, and the heat exchanger remains dry, the heat exchanger material will have a long life and does not corrode easily. Many furnaces have a coated heat exchanger to provide extra protection against corrosion. Research by Stickford et al. (1985) indicates that chloride compounds in the condensate of condensing furnaces can cause the heat exchanger to fail unless it is made of specialty steel.

Equipment Sizing

The furnace's heating capacity (i.e., the maximum heating rate the furnace can provide) is provided on the appliance rating plate; it is also available through the Gas Appliance Manufacturers Association (GAMA 2007) [now the Air-Conditioning, Heating, and Refrigeration Institute (AHRI)] directories and manufacturers' product literature. The heating load for the intended space must be determined; variables that must be considered when calculating space load include heat gains or losses through walls, floors, and ceiling; infiltration; fenestration; ventilation; internal loads; and humidification. Chapters 29 and 30 in the 2005 *ASHRAE Handbook—Fundamentals* provide the information necessary to determine residential and nonresidential heating loads, respectively.

Other factors should be considered when determining furnace capacity. Thermostat setback recovery may require additional heating capacity. On the other hand, supplemental heat sources or off-peak storage devices may offset some of the peak demand capacity. Increasing furnace capacity may increase space temperature swing, and thus reduce comfort. Two-stage or step-modulating equipment could help by using the unit's maximum capacity to meet the setback recovery needs, and providing a lower stage of heating capacity at other times.

Finally, cooling load may need to be considered, especially in climates that have substantial cooling loads but minimal heating loads. When a large cooling load necessitates a large airflow rate, heating capacity may be greater than necessary. Two-stage or modulating heating may be a suitable alternative to reduce the potentially large temperature swings that can occur with excessively oversized furnaces.

Types of Furnaces

Fuel-burning furnaces are typically subdivided into two primary categories: noncondensing and condensing. **Condensing** furnaces have a specially designed secondary heat exchanger that extracts the heat of vaporization of water vapor in the exhaust. They typically have high efficiencies, ranging from 89 to 96%. The dew-point temperatures of flue gases of condensing furnaces are significantly above the vent temperature, so plastic or other corrosion-resistant venting material is required. Condensing furnaces must be plumbed for condensate disposal. Provisions must be taken to prevent the condensate trap and drain line from freezing if installed in a location that is likely to be below freezing at some point in the year.

Noncondensing furnaces have generally less than 85% steady-state efficiency. This type of furnace has higher flue gas temperatures and requires either metal, masonry, or a combination of the two for venting materials. Because there is no water management, noncondensing furnaces do not need freeze protection.

Consumer Considerations

Safety and Reliability. Gas furnaces sold in North America are tested and certified to the ANSI *Standard* Z21.47/CSA 2.3 standard. Oil furnaces are tested in accordance with UL *Standard* 727, and oil burners in accordance with UL *Standard* 296. These standards are intended to ensure that consumer safety and product reliability are maintained in appliance design. Because of open-flame combustion, the following safety items need to be considered: (1) the surrounding atmosphere should be free of dust or chemical concentrations; (2) a path for combustion air must be provided for both sealed and open combustion chambers; and (3) the gas piping and vent pipes must be installed according to the NFPA/AGA *National Fuel Gas Code*, local codes, and the manufacturer's instructions. For electric furnaces, safety primarily concerns proper wiring techniques. Wiring should comply with the *National Electrical Code®* (NEC) (NFPA 70) and applicable local codes.

Efficiency, Operating, and Life-Cycle Costs. Annual operating costs of furnaces must take into account both the cost of the heating fuel as well as the electrical efficiency of the blower motor. Life-cycle cost determination includes initial cost, maintenance, energy consumption, design life, and price escalation of the fuel. Procedures for establishing operating costs for use in product labeling and audits are available in the United States from the Department of Energy. Annual fuel utilization efficiency (AFUE) and energy consumption data to help calculate the annual cost for heating a building are available in the GAMA (2007; now AHRI) directory. AFUE and fuel cost are primary drivers in the operating cost. Electric furnaces are listed as nearly 100% AFUE because all of the electrical energy is converted into heat, and the only inefficiency is from cabinet conduction and air leakage losses.

Research performed on a small number of furnace installations in Florida showed cabinet air leakage that averaged 5.6% of fan flow (Cummings et al. 2002, 2003). Cabinet leakage can affect energy consumption and indoor air quality when furnaces are installed outside the conditioned living space, especially when the home and distribution system are otherwise of tight construction. Air leakage should be taken into account during system design and location. Care should also be taken to install equipment according to the manufacturer's instructions, and that gaps are not overlooked by the installer where service entries or attachments are made to the cabinet. Some manufacturers have begun improving furnace casing air leakage.

Design Life. Typically, heat exchangers made of cold-rolled steel have a design life of approximately 15 years. Special coated or alloy heat exchangers, when used for standard applications, can extend the life by several years, and are recommended for furnace applications in corrosive atmospheres.

The design life of electric furnaces depends on the durability of the contactors and heating elements. The typical design life is approximately 15 years.

Comfort. Consumer opinions of comfort vary quite a bit. Thermal comfort is affected by supply air temperature, air velocity leaving the supply registers, and proximity of the supply airflow stream to occupants. Complaints of draftiness are common when delivered supply air temperatures are low and register velocities are high. A common solution is to reduce the blower speed to get more temperature rise, while staying within the rated temperature rise range listed on the rating plate. However, reducing blower speed can lead to distribution problems. The system design, including register selection and placement, should take these issues into account to avoid comfort problems.

Large temperature swings may also cause discomfort. Factors that affect temperature swing include oversizing, thermostat cycling characteristics (related to anticipators and number of cycles per hour), and thermostatic control. Two-stage or step-modulated heating can improve comfort by reducing the wide, variable temperature swings. These control schemes reduce furnace capacity through gas and blower modulation, which reduces the amount of oversizing for the current demand.

Comfort can also be affected by the indoor air quality. Adding air filters with high minimum efficiency reporting value (MERV; see ASHRAE *Standard* 52.2) ratings reduces airborne particulate matter and allergens. Winter months typically cause drier indoor air conditions, which can be offset by adding duct-integrated humidifiers. Both filters and humidifiers affect duct resistance, which in turn affects electrical energy consumption, and therefore must be considered in system design.

Sound level can be classified as a comfort consideration. Chapter 47 of the 2007 *ASHRAE Handbook—HVAC Applications* outlines procedures for determining acceptable noise levels. In multistage systems, the lower stage may have lower sound levels.

Specialty Applications. Many options are available to increase comfort or economy. To manage comfort and cost of operation, a fuel-burning furnace can be combined with a heat pump; this takes advantage of the heat pump's relatively high efficiency during mild weather, and switches to fuel heating when outdoor temperatures drop and it becomes more difficult for heat pumps to meet the demand. To reduce peak demand for energy, off-peak storage devices may be used to decrease the required capacity of the furnace. The storage device can supply the additional capacity required during the morning recovery of a night setback cycle or reduce the daily peak loads to help in load shedding. Detailed calculations can determine the contribution of storage devices.

Selecting Furnaces for Commercial Buildings

The procedure for design and selection of a commercial furnace is similar to that for a residential furnace. First, the design capacity of the heating system must be determined, considering heat loss from the structure, recovery load, internal heat sources, humidification, off-peak storage, waste heat recovery, and back-up capacity. Because most commercial buildings use setback during weekends, evenings, or other long periods of inactivity, the recovery load is important, as are internal loads and waste heat recovery. The furnace should be sized according to the load per ACCA (2004).

Efficiency of commercial units is about the same as for noncondensing residential units. Two-stage gas valves are frequently used with commercial furnaces, but the efficiency of a two-stage system may be lower than for a single-stage system. At a reduced firing rate, the excess combustion airflow through the burners increases, decreasing the steady-state operating efficiency of the furnace. Multistage furnaces with multistage thermostats and controls may be used to more appropriately match load conditions.

The design life of commercial heating and cooling equipment is about 20 years. Most gas furnace heat exchangers are either coated steel or stainless steel. Because most commercial furnaces are made for outdoor application, the cabinets are made from corrosion-resistant coated steel (e.g., galvanized or aluminized). Blowers can be direct- or belt-driven and can deliver air at higher static pressure.

The noise level of commercial heating equipment is important in some applications, such as schools, office buildings, churches, and theaters. Unit heaters, for example, are used primarily in industrial applications where noise is less important. Duct design can greatly affect noise levels.

In many jurisdictions, safety requirements are the same for light commercial systems and residential systems. Above 400,000 Btu/h gas input, ANSI *Standard* Z21.47/CSA 2.3 requirements for gas controls are more stringent.

CALCULATIONS

Performance Criteria. Furnaces are characterized by several performance criteria, including heating capacity, annual fuel utilization efficiency (AFUE), annual fuel usage (E_f), and annual electrical energy usage (E_{ae}). Each of these are calculated using methods found in the ASHRAE *Standard* 103, and are published in the GAMA (2007; now AHRI) directories. To calculate the furnace's rated heating capacity, the steady-state efficiency must be determined. In the United States, manufacturers may publish only the AFUE as an efficiency measure. The unit capacity is proportional to the steady-state efficiency when compared to the fuel input rate. Steady-state efficiency is also calculated using the methods described in the ASHRAE *Standard* 103.

These efficiencies and other terms are generally used by the furnace industry in the following manner:

- **Steady-state efficiency (SSE).** This is the efficiency of a furnace when it is operated under equilibrium conditions based on ASHRAE *Standard* 103. It is calculated by measuring the energy input, subtracting the losses for exhaust gases and flue gas condensate (for condensing furnaces only), and then dividing by the fuel input (cabinet loss not included):

$$\text{SSE } (\%) = \frac{\text{Fuel input} - \text{Flue loss} - \text{Condensate loss}}{\text{Fuel input}} \times 100$$

- **Heating capacity.** This is the highest heating output of the furnace. It is calculated by multiplying the rated input for the furnace by the steady-state efficiency minus the heat lost from the casing through conduction (jacket loss):

$$Q_{out} = \text{Fuel input} \times (\text{SSE} - \text{Jacket loss})$$

The jacket loss is adjusted for the installation location: it is weighted more heavily for outdoor installations than for ICS, and is weighted by zero for indoor installations, where all heat lost from the casing is contained in the heated space.

- **Utilization efficiency.** This efficiency is obtained from an empirical equation developed by Kelly et al. (1978) with 100% efficiency and deducting losses for exhausted latent and sensible heat, cyclic effects, cabinet air leakage (infiltration), casing heat losses, and pilot burner effect.
- **Annual fuel utilization efficiency (AFUE).** This value is the same as utilization efficiency, except that losses from a standing

Table 1 Typical Values of Efficiency

Type of Gas Furnace	AFUE, % Indoor	ICS[a]
1. Natural-draft with standing pilot	64.5	63.9[b]
2. Natural-draft with intermittent ignition	69.0	68.5[b]
3. Natural-draft with intermittent ignition and auto vent damper	78.0	68.5[b]
4. Fan-assisted combustion with standing pilot or intermittent ignition	80.0	78.0
5. Same as 4, except with improved heat transfer	82.0	80.0
6. Direct vent, natural-draft with standing pilot, preheat	66.0	64.5[b]
7. Direct vent, fan-assisted combustion, and intermittent ignition	80.0	78.0
8. Fan-assisted combustion (induced-draft)	80.0	78.0
9. Condensing	90.0	88.0
Type of Oil Furnace	**Indoor**	**ICS[a]**
1. Standard: pre-1992	71.0	69.0[b]
2. Standard: post-1992	80.0	78.0
3. Same as 2, with improved heat transfer	81.0	79.0
4. Same as 3, with automatic vent damper	82.0	80.0
5. Condensing	91.0	89.0

[a]Isolated combustion system (estimate).
[b]Pre-1992 design (see text).

pilot during the nonheating season are deducted. This equation can also be found in Kelly et al. (1978) or ASHRAE *Standard* 103. AFUE is displayed on each furnace produced in accordance with U.S. Federal Trade Commission requirements for appliance labeling found in *Code of Federal Regulations* 16CFR305.

The AFUE is determined for residential fan-type furnaces by using ASHRAE *Standard* 103. The test procedure is also presented in *Code of Federal Regulations* Title II, 10CFR430, Appendix N, in conjunction with the amendments issued by the U.S. Department of Energy in the *Federal Register*. This version of the test method allows the rating of nonweatherized furnaces as indoor combustion systems, ICSs, or both. Weatherized furnaces are rated as outdoor.

U.S. federal law requires manufacturers of furnaces to use AFUE as determined using the isolated combustion system method to rate efficiency. Since January 1, 1992, all furnaces produced have a minimum AFUE (ICS) level of 78%. Table 1 gives efficiency values for different furnaces.

TECHNICAL DATA

Detailed technical data on furnaces are available from manufacturers, wholesalers, and dealers. The data are generally tabulated in product specification bulletins printed by the manufacturer for each furnace line. These bulletins usually include performance information, electrical data, blower and air delivery data, control system information, optional equipment information, and dimensions.

Natural Gas Furnaces

Capacity Ratings. ANSI *Standard* Z21.47/CSA 2.3 requires that the heating capacity be marked on the rating plates of commercial furnaces in the United States. The heating capacity of residential furnaces, less than 225,000 Btu/h input, is required by the Federal Trade Commission and can be found in furnace directories published semiannually by GAMA (now AHRI). Capacity is calculated by multiplying the input by the steady-state efficiency.

Residential gas furnaces with heating capacities ranging from 35,000 to 175,000 Btu/h are readily available. Some smaller furnaces are manufactured for special-purpose installations such as mobile homes. Smaller capacity furnaces are becoming common because new homes are better insulated and have lower heat loads than older homes. Larger furnaces are also available, but these are generally considered for commercial use.

Because of the overwhelming popularity of the upflow furnace, or multiposition including upflow, it is available in the greatest number of models and sizes. Downflow furnaces, dedicated horizontal furnaces, and various combinations are also available but are generally limited in model type and size.

Residential gas furnaces can be installed as heating-only systems or as part of a heat/cool system. The difference is that, in the heat/cool system, the furnace operates as the air-handling section of a split-system air conditioner. Heating-only systems typically operate with enough airflow to yield a 40 to 70°F air temperature rise through the furnace. Condensing furnaces may be designed for a lower temperature rise (as low as 35°F).

Furnaces have blowers capable of multiple speeds. When the furnace is used as the air handler for a cooling system, the blower is typically capable of delivering about 400 cfm per ton of air conditioning. Furnaces are generally available to accommodate 1.5 to 5 ton air conditioners. The blower speed for each mode of operation should be selected to provide the required airflow for both heating and cooling operation. Controls of furnaces used in a heat/cool system can be installed to operated the multispeed blower motor at the most appropriate speed for either heating or cooling when airflow requirements vary for each mode.

Efficiency Ratings. Currently, gas furnaces have steady-state efficiencies that vary from about 78 to 96%. Natural-draft and fan-assisted combustion furnaces typically range from 78 to 80% efficiency, while condensing furnaces have over 90% steady-state efficiency. Bonne et al. (1977), Gable and Koenig (1977), Hise and Holman (1977), and Koenig (1978) found that oversizing residential gas furnaces with standing pilots reduced the seasonal efficiency of heating systems in new installations with vents and ducts sized according to furnace capacity.

The AFUE of a furnace may be improved by ways other than changing the steady-state efficiency. One example is to replace a standing pilot with an intermittent ignition device. Bonne et al. (1976) and Gable and Koenig (1977) indicated that this feature can save as much as 5.6×10^6 Btu/year per furnace. Some jurisdictions require the use of intermittent ignition devices.

Another method of improving AFUE is to take all combustion air from outside the heated space (direct vent) and preheat it. A combustion air preheater incorporated into the vent system draws combustion air through an outer pipe that surrounds the flue pipe. These systems have been used on mobile home and outdoor furnaces. Annual energy consumption of a direct-vent furnace with combustion air preheat may be as much as 9% less than that of a standard furnace of the same design (Bonne et al. 1976). Direct vent without combustion air preheat is not inherently more efficient because the reduction in combustion-induced infiltration is offset by the use of colder combustion air.

An automatic vent damper (thermal or electromechanical) is another device that saves energy on indoor furnaces. This device, which is placed after the draft hood outlet, closes the vent when the furnace is not in operation. It saves energy during the *off* cycle of the furnace by (1) reducing exfiltration from the house and (2) trapping residual heat from the heat exchanger within the house rather than allowing it to flow up the chimney. These savings approach 11% under ideal conditions, where combustion air is taken from the heated space, which is under thermostat control. However, these savings are much less (estimates vary from 0 to 4%) if combustion air is taken from outside the heated space. The ICS method of determining AFUE gives no credit to vent dampers installed on indoor furnaces because it assumes the use of outdoor combustion air with the furnace installed in an unconditioned space.

The AFUE of fan-assisted combustion furnaces is higher than for standard natural-draft furnaces. Fan-assisted combustion furnaces normally have such a high internal flow resistance that combustion airflow stops when the combustion blower is off. This characteristic results in greater energy savings than those from a vent damper. Computer studies by Bonne et al. (1976), Chi (1977), and Gable and Koenig (1977) estimated annual energy savings up to 16% for fan-assisted combustion furnaces with electric ignition as compared to natural-draft furnaces with standing pilot.

Propane Furnaces

Most residential natural gas furnaces are also available in a propane version with identical ratings. The technical data for these two furnaces are identical, except for the gas control and the burner and pilot orifice sizes. Orifice sizes on propane furnaces are much smaller because propane has a higher density and may be supplied at a higher manifold pressure. The heating value and specific gravity of typical gases are listed as follows:

Gas Type	Heating Value, Btu/ft^3	Specific Gravity (Air = 1.0)
Natural	1030	0.60
Propane	2500	1.53
Butane	3175	2.00

As in natural gas furnaces, the ignition systems have a required pilot gas shutoff feature in case the pilot ignition fails. Pilot gas leakage is more critical with propane or butane gas because both are heavier than air and can accumulate to create an explosive mixture in the furnace or furnace enclosure.

Since 1978, ANSI *Standard* Z21.47/CSA 2.3 has required a gas pressure regulator as part of the propane furnace. (Previously, the pressure regulator was provided only with the propane supply system.)

Besides natural and propane, a furnace may be certified for manufactured gas, mixed gas, or propane/air mixtures; however, furnaces with these certifications are not commonly available. Mobile home furnaces are certified as convertible from natural gas to propane.

Oil Furnaces

Oil furnaces are similar to gas furnaces in size, shape, and function, but the heat exchanger, burner, and combustion control are significantly different.

Input ratings are based on the oil flow rate (gph), and the heating capacity is calculated by the same method as that for gas furnaces. The typical heating value of oil is 140,000 Btu/gal. Fewer models and sizes are available for oil than are available for gas, but residential furnaces in the range of 64,000 to 150,000 Btu/h heating capacity are common. Air delivery ratings are similar to gas furnaces.

The efficiency of an oil furnace can drop during normal operation if the burner is not maintained and kept clean. In this case, the oil does not atomize sufficiently to allow complete combustion, and energy is lost up the chimney in the form of unburned hydrocarbons. Because most oil furnaces use power burners and electric ignition, the annual efficiency is relatively high.

Oil furnaces are available in upflow, downflow, and horizontal models. The thermostat, fan control switch, and limit switch are similar to those of a gas furnace. Oil flow is controlled by a pump and burner nozzle, which sprays the oil/air mixture into a single-chamber drum-type heat exchanger. The heat exchangers are normally heavy-gage cold-rolled steel. Humidifiers, electronic air cleaners, and night setback thermostats are available as accessories.

Electric Furnaces

Residential electric resistance furnaces are available in heating capacities of 5 to 35 kW. Electric resistance furnaces are typically part of a heat/cool system and provide the appropriate airflow for both heating and cooling modes.

The only losses associated with an electric resistance furnace are the conductive and air leakage losses in the cabinet. If the furnace is located fully within the heated space, then the seasonal efficiency would be 100%.

Although the efficiency of an electric furnace is high, electricity is generally a relatively expensive form of energy. The operating cost may be reduced substantially by using an electric heat pump in place of a straight electric resistance furnace. Heat pump systems are discussed in Chapter 8.

Electric furnaces are available in upflow, downflow, or horizontal models. Internal controls include overload fuses or circuit breakers, overheat limit switches, a fan control switch, and contactors to bring on the heating elements at timed intervals.

Commercial Furnaces

Furnaces with capacities above 150,000 Btu/h are classified as commercial furnaces. The 1992 U.S. Energy Policy and Conservation Act (EPCA) prescribes minimum efficiency requirements for commercial furnaces based on ASHRAE *Standard* 90.1. Some efficiency improvement components, such as intermittent ignition devices, are common in commercial furnaces.

INSTALLATION

Installation requirements call for a forced-air heating system to meet three basic criteria: (1) the system must be safe, (2) it must provide comfort for the occupants of the conditioned space, and (3) it must be energy efficient. Location of equipment and ducts, materials selected for the distribution system, and installation practices all affect the total system efficiency. Conduction and air leakage losses can result in substantial energy and system performance degradation and deserve special attention. For maximum safety, comfort, and efficiency, proper treatment of distribution system air leakage is necessary. Two major considerations for installing furnaces are discussed here; for additional issues, see Chapter 9.

Generally, the following three categories of installation guidelines must be followed to ensure the safe operation of a heating system: (1) the equipment manufacturer's installation instructions, (2) local installation code requirements, and (3) national installation code requirements. Local code requirements may or may not be available, but the other two are always available. Depending on the type of fuel being used, one of the following national code requirements apply in the United States:

- NFPA 54-2006 (also AGA Z223.1-2006) — *National Fuel Gas Code*
- NFPA 70-2008 — *National Electrical Code®*
- NFPA 31-2006 — *Standard for the Installation of Oil-Burning Equipment*

Comparable Canadian standards are

- CAN/CSA-B149.1-05 — *Natural Gas and Propane Installation Code*
- CSA C22.1-06 — *Canadian Electrical Code*
- CAN/CSA B139-04 — *Installation Code for Oil Burning Equipment*

An additional source is the *International Fuel Gas Code* (IFGC) (ICC 2006). These regulations provide complete information about construction materials, gas line sizes, flue pipe sizes, wiring sizes, and so forth.

Proper design of the air distribution system is necessary for both comfort and safety. Chapter 35 of the 2005 *ASHRAE Handbook—Fundamentals*, Chapter 1 of the 2007 *ASHRAE Handbook—HVAC Applications*, and Chapter 9 of this volume provide information on the design of ductwork for forced-air heating systems. Forced-air furnaces provide design airflow at a static pressure as low as 0.12 in. of water for a residential unit to above 1.0 in. of water for a commercial unit. The air distribution system must handle the required volumetric flow rate within the pressure limits of the equipment. If the system is a combined heating/cooling installation, the air distribution system must meet the cooling requirement because more air is required for cooling than for heating. It is also important to include the pressure drop of the cooling coil. The Air-Conditioning and Refrigeration Institute (ARI) maximum allowable pressure drop for residential cooling coils is 0.3 in. of water.

AGENCY LISTINGS

Construction and performance of furnaces are regulated by several agencies.

The Gas Appliance Manufacturers Association [GAMA; now the Air-Conditioning, Heating, and Refrigeration Institute (AHRI)], in cooperation with its industry members, sponsors a certification program relating to gas- and oil-fired residential furnaces and boilers. This program uses an independent laboratory to verify the furnace and boiler manufacturers' certified AFUEs and heating capacities, as determined by testing in accordance with the U.S. Department of Energy's *Uniform Test Method for Measuring the Energy Consumption of Furnaces and Boilers* (Title II, 10CFR30, Subpart B, Appendix N). Gas and oil furnaces with input ratings less than 225,000 Btu/h and gas and oil boilers with input ratings less than 300,000 Btu/h are currently included in the program.

Also included in the program is the semiannual publication of the GAMA consumers' directories, which identify certified products and list the input rating, certified heating capacity, and AFUE for each furnace. Participating manufacturers are entitled to use the GAMA Certification Symbol (seal). These directories are published semiannually and distributed to the reference departments of public libraries in the United States; they are also available online.

ANSI *Standard* Z21.47/CSA 2.3 (CSA America is secretariat) gives minimum construction, safety, and performance requirements for gas furnaces. The CSA maintains laboratories to certify furnaces and operates a factory inspection service. Furnaces tested and found to be in compliance are listed in the CSA Directory and carry the Seals of Certification. Underwriters Laboratories (UL) and other approved laboratories can also test and certify equipment in accordance with ANSI *Standard* Z21.47/CSA 2.3.

Gas furnaces may be certified for standard, alcove, closet, or outdoor installation. Standard installation requires clearance between the furnace and combustible material of at least 6 in. Furnaces certified for alcove or closet installation can be installed with reduced clearance, as listed. Furnaces certified for either sidewall venting or outdoor installation must operate properly in a 31 mph wind. Construction materials must be able to withstand natural elements without degradation of performance and structure. Horizontal furnaces are normally certified for installation on combustible floors and for attic installation and are so marked, in which case they may be installed with point or line contact between the jacket and combustible constructions. Upflow and downflow furnaces are normally certified for alcove or closet installation. Gas furnaces may be listed to burn natural gas, mixed gas, manufactured gas, propane, or propane/air mixtures. A furnace must be equipped and certified for the specific gas to be used because different burners and controls, as well as orifice changes, may be required.

Sometimes oil burners and control packages are sold separately; however, they are normally sold as part of the furnace package. Pressure-type or rotary burners should bear the Underwriters Laboratory label showing compliance with ANSI/UL *Standard* 296. In addition, the complete furnace should bear markings indicating compliance with UL *Standard* 727. Vaporizing burner furnaces should also be listed under UL *Standard* 727.

UL *Standard* 1995 gives requirements for the listing and labeling of electric furnaces and heat pumps.

The following list summarizes important standards issued by Underwriters Laboratories, the Canadian Gas Association, and the Canadian Standards Association that apply to space-heating equipment:

ANSI/ASHRAE 103-1993	Method of Testing for Annual Fuel Utilization Efficiency of Residential Central Furnaces and Boilers
ANSI Z21.66-96 (R2001)/CGA 6.14-M96	Automatic Vent Damper Devices for Use with Gas-Fired Appliances
ANSI Z83.4-2003/CSA 3.7	Non-Recirculating Direct Gas-Fired Industrial Air Heaters
ANSI Z83.18-2004	Recirculating Direct Gas-Fired Industrial Air Heaters
ANSI Z83.19-2001/CSA 2.35	Gas-Fired High-Intensity Infrared Heaters
ANSI Z83.20-2001/CSA 2.34	Gas-fired Low-Intensity Infrared Heaters
ANSI Z83.8-2006/CGA 2.6	Gas Unit Heaters and Gas-Fired Duct Furnaces
ANSI Z21.47-2006/CSA 2.3	Gas-Fired Central Furnaces
ANSI/UL 296-2003	Oil Burners
ANSI/UL 307A-95	Liquid Fuel-Burning Heating Appliances for Manufactured Homes and Recreational Vehicles
ASHRAE 90.1-2007	Energy Standard for Buildings Except Low-Rise Residential Buildings
ICC	*International Fuel Gas Code*
NFPA 70-2007	*National Electrical Code*®
UL 307B-2006	Gas-Burning Heating Appliances for Manufactured Homes and Recreational Vehicles
UL 727-2006	Oil-Fired Central Furnaces
UL 1995-2005/CSA C22.2 No. 236	Heating and Cooling Equipment
CGA 3.2-1976	Industrial and Commercial Gas-Fired Package Furnaces
CSA B140.4-2004	Oil-Fired Warm Air Furnaces

REFERENCES

ASHRAE. 1993. Method of testing for annual fuel utilization efficiency of residential central furnaces and boilers. ANSI/ASHRAE *Standard* 103-1993.

ACCA. 2004. Residential equipment selection. *Manual* S. Air Conditioning Contractors of America, Arlington, VA.

Bonne, J., J.E. Janssen, A.E. Johnson, and W.T. Wood. 1976. Residential heating equipment HFLAME evaluation of target improvements. National Bureau of Standards. *Final Report*, Contract T62709. Center of Building Technology, Honeywell, Inc. Available from NIST, Gaithersburg, MD.

Bonne, U., J.E. Janssen, and R.H. Torborg. 1977. Efficiency and relative operating cost of central combustion heating systems: IV—Oil-fired residential systems. *ASHRAE Transactions* 83(1):893-904.

CFR. (Annual.) FTC appliance labeling. 16CFR305. *Code of Federal Regulations*, U.S. Government Printing Office, Washington, D.C.

CFR. (Annual.) Uniform test method for measuring the energy consumption of furnaces and boilers. Title II, 10CFR430, Subpart B, Appendix N. *Code of Federal Regulations*, U.S. Government Printing Office, Washington, D.C.

Chi, J. 1977. DEPAF—A computer model for design and performance analysis of furnaces. Presented at AICHE-ASME Heat Transfer Conference, Salt Lake City, UT (August).

Cummings, J.B., C. Withers, J. McIlvaine, J. Sonne, and M. Lombardi. 2002. Field testing and computer modeling to characterize the energy impacts of air handler leakage. *Final Report*, FSEC-CR-1357-02. Florida Solar Energy Center, Cocoa, FL.

Cummings, J.B., C. Withers, J. McIlvaine, J. Sonne, and M. Lombardi. 2003. Air handler leakage: Field testing results in residences. *ASHRAE Transactions* 109(1):496-502.

Gable, G.K. and K. Koenig. 1977. Seasonal operating performance of gas heating systems with certain energy-saving features. *ASHRAE Transactions* 83(1):850-864.

GAMA. 2007. *Consumers' directory of certified efficiency ratings for residential heating and water heating equipment*. Gas Appliance Manufacturers Association (now Air-Conditioning, Heating, and Refrigeration Institute), Arlington, VA. http://www.gamanet.org/gama/inforesources.nsf/vAllDocs/Product+Directories?OpenDocument.

Hise, E.C. and A.S. Holman. 1977. Heat balance and efficiency measurements of central, forced-air, residential gas furnaces. *ASHRAE Transactions* 83(1):865-880.

Kelly, G.E., J.G. Chi, and M. Kuklewicz. 1978. *Recommended testing and calculation procedures for determining the seasonal performance of residential central furnaces and boilers*. Available from National Technical Information Service, Springfield, VA (Order No. PB289484).

Koenig, K. 1978. Gas furnace size requirements for residential heating using thermostat night setback. *ASHRAE Transactions* 84(2):335-351.

Nelson, L.W. and W. MacArthur. 1978. Energy saving through thermostat setback. *ASHRAE Journal* (September).

NFPA/AGA. 2006. *National fuel gas code*. ANSI/NFPA 54-06. National Fire Protection Association, Quincy, MA. ANSI/AGA Z223.1-06. American Gas Association, Arlington, VA.

Stickford, G.H., B. Hindin, S.G. Talbert, A.K. Agrawal, and M.J. Murphy. 1985. Technology development for corrosion-resistant condensing heat exchangers. *Report* GRI-85-0282. Battelle Columbus Laboratories to Gas Research Institute. Available from National Technical Information Service, Springfield, VA.

BIBLIOGRAPHY

ASHRAE. 2007. Method of testing general ventilation air-cleaning devices for removal efficiency by particle size. ANSI/ASHRAE *Standard* 52.2-2007.

Benton, R. 1983. Heat pump setback: Computer prediction and field test verification of energy savings with improved control. *ASHRAE Transactions* 89(1B):716-734.

Bullock, E.C. 1978. Energy savings through thermostat setback with residential heat pumps. *ASHRAE Transactions* 84(2):352-363.

Deppish, J.R. and D.W. DeWerth. 1986. GATC studies common venting. *Gas Appliance and Space Conditioning Newsletter* 10 (September).

McQuiston, F.C. and J.D. Spitler. 1992. *Cooling and heating load calculation manual*. ASHRAE.

Schade, G.R. 1978. Saving energy by night setback of a residential heat pump system. *ASHRAE Transactions* 84(1):786-798.

CHAPTER 33

RESIDENTIAL IN-SPACE HEATING EQUIPMENT

IN-SPACE heating equipment differs from central heating in that fuel is converted to heat in the space to be heated. In-space heaters may be either permanently installed or portable and may transfer heat by a combination of radiation, natural convection, and forced convection. The energy source may be liquid, solid, gaseous, or electric.

GAS IN-SPACE HEATERS

Room Heaters

A **vented circulator room heater** is a self-contained, freestanding, nonrecessed gas-burning appliance that furnishes warm air directly to the space in which it is installed, without ducting (Figure 1). It converts the energy in the fuel gas to convected and radiant heat without mixing flue gases and circulating heated air by transferring heat from flue gases to a heat exchanger surface.

A **vented radiant circulator** is equipped with high-temperature glass panels and radiating surfaces to increase radiant heat transfer. Separation of flue gases from circulating air must be maintained. Vented radiant circulators range from 10,000 to 75,000 Btu/h.

Gravity-vented radiant circulators may also have a circulating air fan, but they perform satisfactorily with or without the fan. Fan-type vented radiant circulators are equipped with an integral circulating air fan, which is necessary for satisfactory performance.

Vented room heaters are connected to a vent, chimney, or single-wall metal pipe venting system engineered and constructed to develop a positive flow to the outside atmosphere. Room heaters should not be used in a room that has limited air exchange with adjacent spaces because combustion air is drawn from the space.

Unvented radiant or **convection heaters** range in size from 10,000 to 40,000 Btu/h and can be freestanding units or wall-mounted, nonrecessed units of either the radiant or closed-front type. Unvented room heaters require an outside air intake. The size of the fresh air opening required is marked on the heater. To ensure adequate fresh air supply, unvented gas-heating equipment must, according to voluntary standards, include a device that shuts the heater off if the oxygen in the room becomes inadequate. Unvented room heaters may not be installed in hotels, motels, or rooms of institutions such as hospitals or nursing homes.

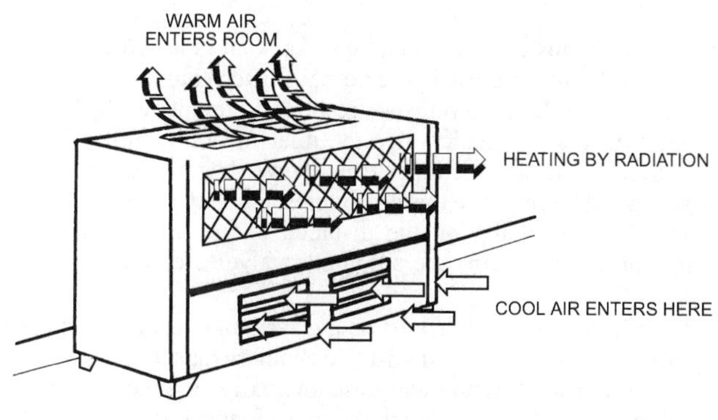

Fig. 1 Room Heater

Catalytic room heaters are fitted with fibrous material impregnated with a catalytic substance that accelerates the oxidation of a gaseous fuel to produce heat without flames. The design distributes the fuel throughout the fibrous material so that oxidation occurs on the surface area in the presence of a catalyst and room air. Catalytic heaters transfer heat by low-temperature radiation and by convection. The surface temperature is below a red heat and is generally below 1200°F at the maximum fuel input rate. The flameless combustion of catalytic heaters is an inherent safety feature not offered by conventional flame-type gas-fueled burners. Catalytic heaters have also been used in agriculture and for industrial applications in combustible atmospheres.

Unvented household catalytic heaters are used in Europe. Most of these are portable and mounted on casters in a casing that includes a cylinder of liquefied petroleum gas (LPG) so that they may be rolled from one room to another. LPG cylinders holding more than 2 lb of fuel are not permitted for indoor use in the United States. As a result, catalytic room heaters sold in the United States are generally permanently installed and fixed as wall-mounted units. Local codes and the *National Fuel Gas Code* (NFPA 54/ANSI Z223.1) should be reviewed for accepted combustion air requirements.

Wall Furnaces

A wall furnace is a self-contained vented appliance with grilles that are designed to be a permanent part of the structure of a building (Figure 2). It furnishes heated air that is circulated by natural or forced convection. A wall furnace can have boots, which may not extend more than 10 in. beyond the horizontal limits of the casing through walls of normal thickness, to provide heat to adjacent rooms. Wall furnaces range from 10,000 to 90,000 Btu/h. Wall furnaces are classified as conventional or direct vent.

Conventional vent units require approved B-1 vent pipes and are installed to comply with the *National Fuel Gas Code*. Some wall furnaces are counterflow units that use fans to reverse the natural flow of air across the heat exchanger. Air enters at the top of the furnace and discharges at or near the floor. Counterflow systems reduce heat stratification in a room. As with any vented unit, a minimum of inlet air for proper combustion must be supplied.

Vented-recessed wall furnaces are recessed into the wall, with only the decorative grillwork extending into the room. This leaves more usable area in the room being heated. Dual-wall furnaces are two units that fit between the studs of adjacent rooms, thereby using a common vent.

The preparation of this chapter is assigned to TC 6.5, Radiant and In-Space Convective Heating and Cooling.

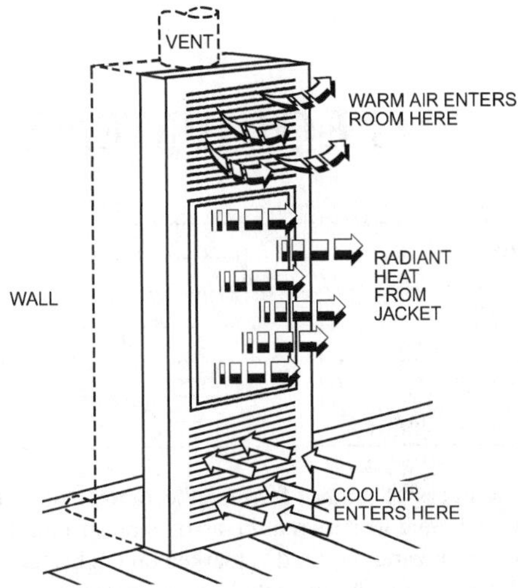

Fig. 2 Wall Furnace

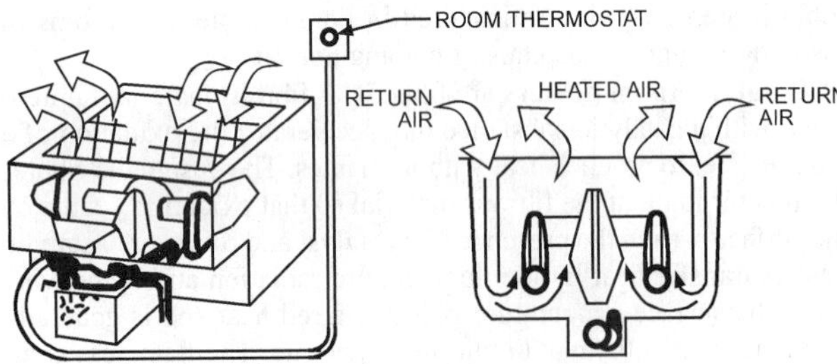

Fig. 3 Floor Furnace

Both vented-recessed and dual-wall furnaces are usually natural convection units. Cool room air enters at the bottom and is warmed as it passes over the heat exchanger, entering the room through the grillwork at the top of the heater. This process continues as long as the thermostat calls for the burners to be on. Accessory fans assist in the movement of air across the heat exchanger and help minimize air stratification.

Direct-vent wall furnaces are constructed so that combustion air comes from outside, and all flue gases discharge into the outside atmosphere. These appliances include grilles or the equivalent and are designed to be attached to the structure permanently. Direct-vent wall heaters are normally mounted on walls with outdoor exposure.

Direct-vent wall furnaces can be used in extremely tight (well-insulated) rooms because combustion air is drawn from outside the room. There are no infiltration losses for dilution or combustion air. Most direct-vent heaters are designed for natural convection, although some may be equipped with fans. Direct-vent furnaces are available with inputs of 6000 to 65,000 Btu/h.

Floor Furnaces

Floor furnaces are self-contained units suspended from the floor of the heated space (Figure 3). Combustion air is taken from outside, and flue gases are also vented outside. Cold air returns at the periphery of the floor register, and warm air comes up to the room through the center of the register.

United States Minimum Efficiency Requirements

The National Appliance Energy Conservation Act (NAECA) of 1987 mandates minimum annual fuel utilization efficiency (AFUE) requirements for gas-fired direct heating equipment (Table 1). The minimums (effective as of January 1, 1990) are measured using the U.S. Department of Energy test method (DOE 1984) and must be met by manufacturers of direct heating equipment (i.e., gas-fired room heaters, wall furnaces, and floor furnaces).

Table 1 Efficiency Requirements in the United States for Gas-Fired Direct Heating Equipment

Input, 1000 Btu/h	Minimum AFUE, %	Input, 1000 Btu/h	Minimum AFUE, %
Wall Furnace (with fan)		*Floor Furnace*	
<42	73	<37	56
>42	74	>37	57
Wall Furnace (gravity type)			
<10	59	*Room Heaters*	
10 to 12	60	<18	57
12 to 15	61	18 to 20	58
15 to 19	62	20 to 27	63
19 to 27	63	27 to 46	64
27 to 46	64	>46	65
>46	65		

CONTROLS

Valves

Gas in-space heaters are controlled by four types of valves:

The full on/off, **single-stage valve** is controlled by a wall thermostat. Models are available that are powered by a 24 V supply or from energy supplied by the heat of the pilot light on the thermocouple (self-generating).

The **two-stage control valve** (with hydraulic thermostat) fires either at full input (100% of rating) or at some reduced step, which can be as low as 20% of the heating rate. The amount of time at the reduced firing rate depends on the heating load and the relative oversizing of the heater.

The **step-modulating control valve** (with a hydraulic thermostat) steps on to a low fire and then either cycles off and on at the low fire (if the heating load is light) or gradually increases its heat output to meet any higher heating load that cannot be met with the low firing rate. This control allows an infinite number of fuel firing rates between low and high fire.

The **manual control valve** is controlled by the user rather than by a thermostat. The user adjusts the fuel flow and thus the level of fire to suit heating requirements.

Thermostats

Temperature controls for gas in-space heaters are of the following two types.

- **Wall thermostats** are available in 24 V and millivolt systems. The 24 V unit requires an external power source and a 24 V transformer. Wall thermostats respond to temperature changes and turn the automatic valve to either full-on or full-off. The millivolt unit requires no external power because the power is generated by multiple thermocouples and may be either 250 or 750 mV, depending on the distance to the thermostat. This thermostat also turns the automatic valve to either full-on or full-off.
- **Built-in hydraulic thermostats** are available in two types: (1) a snap-action unit with a liquid-filled capillary tube that responds to changes in temperature and turns the valve to either full-on or full-off; and (2) a modulating thermostat, which is similar to the snap-action unit, except that the valve comes on and shuts off at a preset minimum input. Temperature alters the input anywhere from full-on to the minimum input. When the heating requirements are satisfied, the unit shuts off.

Table 2 Gas Input Required for In-Space Supplemental Heaters

Heater Type	Average AFUE, %	Steady State Efficiency, %	Gas Consumption per Unit House Volume, Btu/h per ft³ — Outside Air Temperature, °F — Older Bungalow[a] 5	30	50	Energy-Efficient House[b] 5	30	50
Vented	54.6	73.1	6.5	3.8	1.6	2.8	1.6	0.7
Unvented	90.5	90.5	6.0	3.5	1.5	2.6	1.5	0.6
Direct vent	76.0	78.2	5.9	3.4	1.5	2.1	1.2	0.5

[a]Tested bungalow total heated volume = 6825 ft³ and $U \approx 0.3$ to 0.5 Btu/h·°F.
[b]Tested energy-efficient house total heated volume = 11,785 ft³ and $U \approx 0.2$ to 0.3 Btu/h·°F.

VENT CONNECTORS

Any vented gas-fired appliance must be installed correctly to vent combustion products. A detailed description of proper venting techniques is found in the *National Fuel Gas Code* and Chapter 34.

SIZING UNITS

The size of the unit selected depends on the size of the room, the number and direction of exposures, the amount of insulation in the ceilings and walls, and the geographical location. Heat loss requirements can be calculated from procedures described in Chapter 29 of the 2005 *ASHRAE Handbook—Fundamentals*.

DeWerth and Loria (1989) studied the use of gas-fired, in-space supplemental heaters in two test houses. They proposed a heater sizing guide, which is summarized in Table 2. The energy consumption in Table 2 is for unvented, vented, and direct vent heaters installed in (1) a bungalow built in the 1950s with average insulation, and (2) a townhouse built in 1984 with above-average insulation and tightness.

OIL AND KEROSENE IN-SPACE HEATERS

Vaporizing Oil Pot Heaters

These heaters have an oil-vaporizing bowl (or other receptacle) that admits liquid fuel and air in controllable quantities; the fuel is vaporized by the heat of combustion and mixed with the air in appropriate proportions. Combustion air may be induced by natural draft or forced into the vaporizing bowl by a fan. Indoor air is generally used for combustion and draft dilution. Window-installed units have the burner section outdoors. Both natural- and forced-convection heating units are available. A small blower is sold as an option on some models. The heat exchanger, usually cylindrical, is made of steel (Figure 4). These heaters are available as room units (both radiant and circulation), floor furnaces, and recessed wall heaters. They may also be installed in a window, depending on the cabinet construction. The heater is always vented to the outside. A 3 to 5 gal fuel tank may be attached to the heater, or a larger outside tank can be used.

Vaporizing pot burners are equipped with a single constant-level and metering valve. Fuel flows by gravity to the burner through the adjustable metering valve. Control can be manual, with an off pilot and variable settings up to maximum, or it can be thermostatically controlled, with the burner operating at a selected firing rate between pilot and high.

Powered Atomizing Heaters

Wall furnaces, floor furnaces, and freestanding room heaters are also available with a powered gun-type burner using No. 1 or No. 2 fuel oil. For more information, refer to Chapter 30.

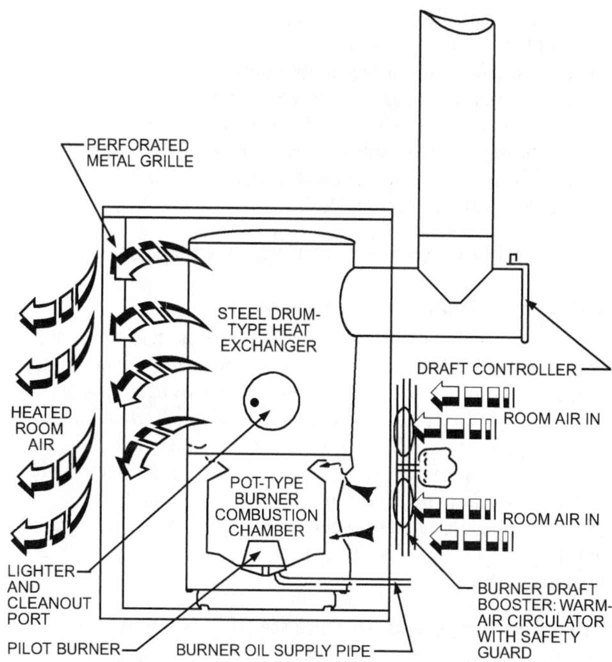

Fig. 4 Oil-Fueled Heater with Vaporizing Pot-Type Burner

Portable Kerosene Heaters

Because kerosene heaters are not normally vented, precautions must be taken to provide sufficient ventilation. Kerosene heaters are of four basic types: radiant, natural-convection, direct-fired, forced-convection, and catalytic.

The radiant kerosene heater has a reflector, while the natural convection heater is cylindrical in shape. Fuel vaporizes from the surface of a wick, which is immersed in an integral fuel tank of up to 2 gal capacity similar to that of a kerosene lamp. Fuel-burning rates range from about 5000 to 22,500 Btu/h. Radiant heaters usually have a removable fuel tank to facilitate refueling.

The direct-fired, forced-convection portable kerosene heater has a vaporizing burner and a heat-circulating fan. These heaters are available with thermostatic control and variable heat output.

The catalytic type uses a metal catalyst to oxidize the fuel. It is started by lighting kerosene at the surface; however, after a few moments, the catalyst surface heats to the point that flameless oxidation of the fuel begins.

ELECTRIC IN-SPACE HEATERS

Wall, Floor, Toe Space, and Ceiling Heaters

Heaters for recessed or surface wall mounting are made with open wire or enclosed, metal-sheathed elements. An inner liner or reflector is usually placed between the elements and the casing to promote circulation and minimize the rear casing temperature. Heat is distributed by both convection and radiation; the proportion of each depends on unit construction.

Ratings are usually 1000 to 5000 W at 120, 208, 240, or 277 V. Models with air circulation fans are available. Other types can be recessed into the floor. Electric convectors should be placed so that air moves freely across the elements.

Baseboard Heaters

These heaters consist of a metal cabinet containing one or more horizontal, enclosed, metal-sheathed elements. The cabinet is less than 6 in. in overall depth and can be installed 18 in. above the floor; the ratio of the overall length to the overall height is more than two to one.

Units are available from 2 to 12 ft in length, with ratings from 100 to 400 W/ft, and they fit together to make up any desired continuous length or rating. Electric hydronic baseboard heaters containing immersion heating elements and an antifreeze solution are made with ratings of 300 to 2000 W. The placement of any type of electric baseboard heater follows the same principles that apply to baseboard installations (see Chapter 35) because baseboard heating is primarily perimeter heating.

RADIANT HEATING SYSTEMS

Heating Panels and Heating Panel Sets

These systems have electric resistance wire or etched or graphite elements embedded between two layers of insulation. High-density thermal insulation behind the element minimizes heat loss, and the outer shell is formed steel with baked enamel finish. Heating panels provide supplementary heating by convection and radiation. They can be recessed into or surface mounted on hard surfaces or fit in standard T-bar suspended ceilings.

Units are usually rated between 250 and 1000 W in sizes varying from 24 by 24 in. to 24 by 96 in. in standard voltages of 120, 208, 240, and 277 V.

Embedded Cable and Storage Heating Systems

Ceiling and floor electric radiant heating systems that incorporate embedded cables are covered in Chapter 6. Electric storage systems, including room storage heaters and floor slab systems, are covered in Chapter 34 of the 2007 *ASHRAE Handbook—HVAC Applications*.

Cord-Connected Portable Heaters

Portable electric heaters are often used in areas that are not accessible to central heat. They are also used to maintain an occupied room at a comfortable level independent of the rest of the residence.

Portable electric heaters for connection to 120 V, 15 A outlets are available with outputs of 2050 to 5100 Btu/h (600 to 1500 W), the most common being 1320 and 1500 W. Many heaters are available with a selector switch for three wattages (e.g., 1100-1250-1500 W). Heavy-duty heaters are usually connected to 240 V, 20 A outlets with outputs up to 13,700 Btu/h (4000 W), whereas those for connection to 240 V, 30 A outlets have outputs up to 19,100 Btu/h (5600 W). All electric heaters of the same wattage produce the same amount of heat.

Portable electric heaters transfer heat by one of two predominant methods: radiation and convection. Radiant heaters provide heat for people or objects. An element in front of a reflector radiates heat outward in a direct line. Conventional radiant heaters have ribbon or wire elements. Quartz radiant heaters have coil wire elements encased in quartz tubes. The temperature of a radiant wire element usually ranges between 1200 and 1600°F.

Convection heaters warm the air in rooms or zones. Air flows directly over the hot elements and mixes with room air. Convection heaters are available with or without fans. The temperature of a convection element is usually less than 930°F.

An adjustable, built-in bimetal thermostat usually controls the power to portable electric heaters. Fan-forced heaters usually provide better temperature control because the fan, in addition to cooling the case, forces room air past the thermostat. One built-in control uses a thermistor to signal a solid logic circuit that adjusts wattage and fan speed. Most quartz heaters use an adjustable control that operates the heater for a percentage of total cycle time from 0 (off) to 100% (full-on).

Controls

Low-voltage and line-voltage thermostats with on-off operation are used to control in-space electric heaters. Low-voltage thermostats, operating at 30 V or less, control relays or contactors that carry the rated voltage and current load of the heaters. Because the control current load is small (usually less than 1 A), the small switch can be controlled by a highly responsive sensing element.

Line-voltage thermostats carry the full load of the heaters at rated voltage directly through their switch contacts. Most switches carry a listing by Underwriters Laboratories (UL) at 22 A (resistive), 277 V rating. Most electric in-space heating systems are controlled by remote wall-mounted thermostats, but many are available with integral or built-in line-voltage thermostats.

Most low-voltage and line-voltage thermostats use small internal heaters, either fixed or adjustable in heat output, that provide heat anticipation by energizing when the thermostat contacts close. The cycling rate of the thermostat is increased by the use of anticipation heaters, resulting in more accurate control of the space temperature.

Droop is an apparent shift or lowering of the control point and is associated with line-voltage thermostats. In these thermostats, switch heating caused by large currents can add materially to the amount of droop. Most line-voltage thermostats in residential use control room heaters of 3 kW (12.5 A at 240 V) or less. At this moderate load and with properly sized anticipation heaters, the droop experienced is acceptable. Cycling rates and droop characteristics have a significant effect on thermostat performance.

SOLID-FUEL IN-SPACE HEATERS

Most wood-burning and coal-burning devices, except central wood-burning furnaces and boilers, are classified as solid-fuel in-space heaters (see Table 3). An in-space heater can be either a fireplace or a stove.

FIREPLACES

Simple Fireplaces

Simple fireplaces, especially all-masonry and noncirculating metal built-in fireplaces, produce little useful heat. They lend atmosphere and a sense of coziness to a room. Freestanding fireplaces are slightly better heat producers. Simple fireplaces have an average efficiency of about 10%. In extreme cases, the chimney draws more heated air than the fire produces.

The addition of glass doors to the front of a fireplace has both a positive and a negative effect. The glass doors restrict the free flow of indoor heated air up the chimney, but at the same time they restrict the radiation of the heat from the fire into the room.

Factory-Built Fireplaces

A factory-built fireplace consists of a fire chamber, chimney, roof assembly, and other parts that are factory made and intended to be installed as a unit in the field. These fireplaces have fireboxes of refractory-lined metal rather than masonry. Factory-built fireplaces come in both radiant and heat-circulating designs. Typical configurations are open-front designs, but corner-opening, three-sided units with openings either on the front or side; four-sided units; and see-through fireplaces are also available.

Radiant Design. The radiant system transmits heat energy from the firebox opening by direct radiation to the space in front of it. These fireplaces may also incorporate such features as an outside air supply and glass doors. Radiant-design factory-built fireplaces are primarily used for aesthetic wood burning and typically have efficiencies similar to those of masonry fireplaces (0 to 10%).

Heat-Circulating Design. This unit transfers heat by circulating air around the fire chamber and releasing to the space to be heated. The air intake is generally below the firebox or low on the sides adjacent to the opening, and the heated air exits through grilles or louvers located above the firebox or high on the sides adjacent to it. In some designs, ducts direct heated air to spaces other than the area near the front of the fireplace. Some circulating units rely on natural

Table 3 Solid-Fuel In-Space Heaters

Type*	Approximate Efficiency,* %	Features	Advantages	Disadvantages
Simple fireplaces, masonry or prefabricated	−10 to +10	Open front. Radiates heat in one direction only.	Visual beauty.	Low efficiency. Heats only small areas.
High-efficiency fireplaces	25 to 45	Freestanding or built-in with glass doors, grates, ducts, and blowers.	Visual beauty. More efficient. Heats larger areas. Long service life. Maximum safety.	Medium efficiency.
Box stoves	20 to 40	Radiates heat in all directions.	Low initial cost. Heats large areas.	Fire hard to control. Short life. Wastes fuel.
Airtight stoves	40 to 55	Radiates heat in all directions. Sealed seams, effective draft control.	Good efficiency. Long burn times, high heat output. Longer service life.	Can create creosote problems.
High-efficiency catalytic wood heaters	65 to 75	Radiates heat in all directions. Sealed seams, effective draft control.	Highest efficiency. Long burn times, high heat output. Long life.	Creosote problems. High purchase price.

*Product categories are general; product efficiencies are approximate.

convection, while others have electric fans or blowers to move air. These energy-saving features typically boosts efficiency 25 to 60%.

Freestanding Fireplaces

Freestanding fireplaces are open-combustion wood-burning appliances that are not built into a wall or chase. One type of freestanding fireplace is a fire pit in which the fire is open all around; smoke rises into a hood and then into a chimney. Another type is a prefabricated metal unit that has an opening on one side. Because they radiate heat to all sides, freestanding fireplaces are typically more efficient than radiant fireplaces.

STOVES

Conventional Wood Stoves

Wood stoves are chimney-connected, solid-fuel-burning room heaters designed to be operated with the fire chamber closed. They deliver heat directly to the space in which they are located. They are not designed to accept ducts and/or pipes for heat distribution to other spaces. Wood stoves are controlled-combustion appliances. Combustion air enters the firebox through a controllable air inlet; the air supply and thus the combustion rate are controlled by the user. Conventional controlled-combustion wood stoves manufactured before the mid-1980s typically have overall efficiencies ranging from 40 to 55%.

Most **controlled-combustion** appliances are constructed of steel, cast iron, or a combination of the two metals; others are constructed of soapstone or masonry. Soapstone and masonry have lower thermal conductivities but greater specific heats (the amount of heat that can be stored in a given mass). Other materials such as special refractories and ceramics are used in low-emission appliances. Wood stoves are classified as either radiant or convection (sometimes called circulating) heaters, depending on the way they heat interior spaces.

Radiant wood stoves are generally constructed with single exterior walls, which absorb radiant heat from the fire. This appliance heats primarily by infrared radiation; it heats room air only to the extent that air passes over the hot surface of the appliance.

Convection wood stoves have double vertical walls with an air space between the walls. The double walls are open at the top and bottom of the appliance to permit room air to circulate through the air space. The more buoyant hot air rises and draws in cooler room air at the bottom of the appliance. This air is then heated as it passes over the surface of the inner radiant wall. Some radiant heat from the inner wall is absorbed by the outer wall, but the constant introduction of room temperature air at the bottom of the appliance keeps the outer wall moderately cool. This characteristic generally allows convection wood stoves to be placed closer to combustible materials than radiant wood stoves. Fans in some wood stoves augment the movement of heated air. Convection wood stoves generally provide more even heat distribution than do radiant types.

Advanced-Design Wood Stoves

Strict air pollution standards have prompted the development of new stove designs. These clean-burning wood stoves use either catalytic or noncatalytic technology to achieve very high combustion efficiency and to reduce creosote and particulate and carbon monoxide emission levels.

Catalytic combustors are currently available as an integral part of many new wood-burning appliances and are also available as add-on or retrofit units for most existing appliances. The catalyst may be platinum, palladium, rhodium, or a combination of these elements. It is bonded to a ceramic or stainless steel substrate. A catalytic combustor's function in a wood-burning appliance is to substantially lower the ignition temperatures of unburned gases, solids, and/or liquid droplets (from approximately 1000°F to 500°F). As these unburned combustibles leave the main combustion chamber and pass through the catalytic combustor, they ignite and burn rather than enter the atmosphere.

For the combustor to efficiently burn the gases, the proper amount of oxygen and a sufficient temperature to maintain ignition are required; further, the gases must have sufficient residence time in the combustor. A properly operating catalytic combustor has a temperature in the range of 1000 to 1700°F. Catalyst-equipped wood stoves have a default efficiency, as determined by the U.S. Environmental Protection Agency (EPA), of 72%, although many stoves are considerably more efficient. This EPA default efficiency is the value one standard deviation below the mean of the efficiencies from a database of stoves.

Another approach to increasing combustion efficiency and meeting emissions requirements is the use of technologically advanced internal appliance designs and materials. Generally, noncatalytic, low-emission wood-burning appliances incorporate high-temperature refractory materials and have smaller fireboxes than conventional appliances. The fire chamber is designed to increase temperature, turbulence, and residence time in the primary combustion zone. Secondary air is introduced to promote continued burning of the gases, solids, and liquid vapors in a secondary combustion zone. Many stoves add a third and fourth burn area within the firebox. The location and design of the air inlets is critical because proper air circulation patterns are the key to approaching complete combustion. Noncatalytic wood stoves have an EPA default efficiency of 63%; however, many models approach 80%.

Fireplace Inserts

Fireplace inserts are closed-combustion wood-burning room heaters that are designed to be installed in an existing masonry fireplace. They combine elements of both radiant and convection wood stove designs. They have large radiant surfaces that face the room and circulating jackets on the sides that capture heat that would otherwise go up the chimney. Inserts may use either catalytic or noncatalytic technology to achieve clean burning.

Pellet-Burning Stoves

Pellet-burning stoves burn small pellets made from wood byproducts rather than burning logs. An electric auger feeds the pellets from a hopper into the fire chamber, where air is blown through, creating very high temperatures in the firebox. The fire burns at such a high temperature that the smoke is literally burned up, resulting in a very clean burn, and no chimney is needed. Instead, the waste gases are exhausted to the outside through a vent. An air intake is operated by an electric motor; another small electric fan blows the heated air from the area around the fire chamber into the room. A microprocessor controls the operation, allowing the pellet-burning stove to be controlled by a thermostat. Pellet-burning stoves typically have the lowest emissions of all wood-burning appliances and have an EPA default efficiency of 78%. Because of the high air-fuel ratios used by pellet-burning stoves, these stoves are excluded from EPA wood stove emissions regulations.

GENERAL INSTALLATION PRACTICES

The criteria to ensure safe operation are normally covered by local codes and ordinances or, in rare instances, by state and federal requirements. Most codes, ordinances, or regulations refer to the following building codes and standards for in-space heating:

Building Codes

BOCA/National Building Code	BOCA
CABO One- and Two-Family Dwelling Code	CABO
International Building Code	ICC
National Building Code of Canada	NBCC
Standard Building Code	SBCCI
Uniform Building Code	ICBO

Mechanical Codes

National Mechanical Code	BOCA
Uniform Mechanical Code	ICBO/IAPMO
International Mechanical Code	ICC
Standard Mechanical Code	SBCCI

Electrical Codes

National Electrical Code	NFPA 70
Canadian Electrical Code	CSA C22.1

Chimneys

Chimneys, Fireplaces, Vents and Solid Fuel-Burning Appliances	NFPA 211
Chimneys, Factory-Built Residential Type and Building Heating Appliance	UL 103

Solid-Fuel Appliances

Factory-Built Fireplaces	UL 127
Room Heaters, Solid-Fuel Type	UL 1482

The chapter on Codes and Standards has further information, including the names and addresses of these agencies. Safety and performance criteria are furnished by the manufacturer.

Safety with Solid Fuels

The evacuation of combustion gases is a prime concern in the installation of solid-fuel-burning equipment. NFPA *Standard* 211, Chimneys, Fireplaces, Vents and Solid Fuel-Burning Appliances, lists requirements that should be followed. Because safety requirements for connector pipes (stovepipes) are not always readily available, these requirements are summarized as follows:

Table 4 Chimney Connector Wall Thickness*

Diameter	Gage	Minimum Thickness, in.
Less than 6 in.	26	0.019
6 to 10 in.	24	0.023
10 to 16 in.	22	0.029
16 in. or greater	16	0.056

*Do not use thinner connector pipe. Replace connectors as necessary. Leave at least 18 in. clearance between the connector and a wall or ceiling, unless the connector is listed for a smaller clearance or an approved clearance reduction system is used.

- Connector pipe is usually black (or blue) steel single-wall pipe; thicknesses are shown in Table 4. Stainless steel is a corrosion-resistant alternative that does not have to meet the thicknesses listed in Table 4.
- Connectors should be installed with the crimped (male) end of the pipe toward the stove, so that creosote and water drip back into the stove.
- The pipe should be as short as is practical, with a minimum of turns and horizontal runs. Horizontal runs should be pitched 1/4 in. per foot up toward the chimney.
- Chimney connectors should not pass through ceilings, closets, alcoves, or concealed spaces.
- When passing through a combustible interior or exterior wall, connectors must be routed through a listed wall pass-through that has been installed in accordance with the conditions of the listing, or they must follow one of the home-constructed systems recognized in NFPA *Standard* 211 or local building codes. Adequate clearance and protection of combustible materials is extremely important. In general, listed devices are easier to install and less expensive than home-constructed systems.

Creosote forms in all wood-burning systems. The rate of formation is a function of the quantity and type of fuel burned, the appliance in which it is burned, and the manner in which the appliance is operated. Thin deposits in the connector pipe and chimney do not interfere with operation, but thick deposits (greater than 1/4 in.) may ignite. Inspection and cleaning of chimneys connected to wood-burning appliances should be performed on a regular basis (at least annually).

Only the solid fuel that is listed for the appliance should be burned. Coal should be burned only in fireplaces or stoves designed specifically for coal burning. The chimney used in coal-fired applications must also be designed and approved for coal and wood.

Solid-fuel appliances should be installed in strict conformance with the clearance requirements established as part of their safety listing. When clearance reduction systems are used, stoves must remain at least 12 in. and connector pipe at least 6 in. from combustibles, unless smaller clearances are established as part of the listing.

Utility-Furnished Energy

Those systems that rely on energy furnished by a utility are usually required to comply with local utility service rules and regulations. The utility usually provides information on the installation and operation of the equipment using their energy. Bottled gas (LPG) equipment is generally listed and tested under the same standards as natural gas. LPG equipment may be identical to natural gas equipment, but it always has a different orifice and sometimes has a different burner and controls. The listings and examinations are usually the same for natural, mixed, manufactured, and liquid petroleum gas.

Products of Combustion

The combustion chamber of equipment that generates products of combustion must be connected by closed piping to the outdoors.

Gas-fired equipment may be vented through masonry stacks, chimneys, specifically designed venting, or, in some cases, venting incorporating forced- or induced-draft fans. Chapter 34 covers chimneys, gas vents, and fireplace systems in more detail.

Agency Testing

The standards of several agencies contain guidelines for the construction and performance of in-space heaters. The following list summarizes the standards that apply to residential in-space heating; they are coordinated or sponsored by ASHRAE, the American National Standards Institute (ANSI), Underwriters Laboratories (UL), the American Gas Association (AGA), and the Canadian Gas Association (CGA). Some CGA standards have a CAN1 prefix.

ANSI Z21.11.1	Gas-Fired Room Heaters, Vented
ANSI Z21.11.2	Gas-Fired Room Heaters, Unvented
ANSI Z21.44	Gas-Fired Gravity and Fan-Type Direct-Vent Wall Furnaces
ANSI Z21.48	Gas-Fired Gravity and Fan-Type Floor Furnaces
ANSI Z21.49	Gas-Fired Gravity and Fan-Type Vented Wall Furnaces
ANSI Z21.60/ CSA 2.26-M96	Decorative Gas Appliances for Installation in Solid-Fuel Burning Fireplaces
ANSI Z21.76	Gas-Fired Unvented Catalytic Room Heaters for use with Liquefied Petroleum (LP) Gases
ANSI Z21.50/ CSA 2.22-M98	Vented Gas Fireplaces
ANSI Z21.86/ CSA 2.32-M98	Vented Gas-Fired Space Heating Appliances
ANSI Z21.88/ CSA 2.33-M98	Vented Gas Fireplace Heaters
CAN1-2.1-M86	Gas-Fired Vented Room Heaters
CAN/CGA-2.5-M86	Gas-Fired Gravity and Fan Type Vented Wall Furnaces
CAN1/CGA-2.19-M81	Gas-Fired Gravity and Fan Type Direct Vent Wall Furnaces
NFPA 211	Chimneys, Fireplaces, Vents and Solid Fuel-Burning Appliances
NFPA/AGA 54	National Fuel Gas Code
ANSI/UL 127	Factory-Built Fireplaces
UL 574	Electric Oil Heaters
UL 647	Unvented Kerosene-Fired Heaters and Portable Heaters
ANSI/UL 729	Oil-Fired Floor Furnaces
ANSI/UL 730	Oil-Fired Wall Furnaces
UL 737	Fireplace Stoves
ANSI/UL 896	Oil-Burning Stoves
ANSI/UL 1042	Electric Baseboard Heating Equipment
ANSI/UL 1482	Heaters, Room Solid-Fuel Type
ASHRAE 62.1	Ventilation for Acceptable Indoor Air Quality

REFERENCES

DeWerth, D.W. and R.L. Loria. 1989. In-space heater energy use for supplemental and whole house heating. *ASHRAE Transactions* 95(1).

DOE. 1984. Uniform test method for measuring the energy consumption of vented home heating equipment. *Federal Register* 49:12, 169 (March).

BIBLIOGRAPHY

GAMA. 1995. *Directory of gas room heaters, floor furnaces and wall furnaces.* Gas Appliance Manufacturers Association, Arlington, VA.

MacKay, S., L.D. Baker, J.W. Bartok, and J.P. Lassoie. 1985. *Burning wood and coal.* Natural Resources, Agriculture, and Engineering Service, Cornell University, Ithaca, NY.

Wood Heating Education and Research Foundation. 1984. *Solid fuel safety study manual for Level I solid fuel safety technicians.* Washington, D.C.

CHAPTER 34

CHIMNEY, VENT, AND FIREPLACE SYSTEMS

A PROPERLY designed chimney or vent system provides and controls draft to convey flue gas from an appliance to the outdoors. This chapter describes the design of chimneys and vent systems that discharge flue gas from appliances and fireplace systems.

Sustainability. Good chimney and vent design is not only a safety issue, but also can enhance a building's sustainability. This chapter explains how to design vent systems to optimize and minimize the materials used to construct fuel-burning appliance vents and chimneys for low cost and long reliability, reducing the need for vent or chimney replacement, thus saving natural resources. Also, systems designed to bring outdoor air directly into the appliance space for combustion and vent gas dilution, instead of relying on air infiltration into the building, reduce heat load and conserve fuel.

TERMINOLOGY

In this chapter, **appliance** refers to any furnace, boiler, or incinerator (including the burner). Unless the context indicates otherwise, the term **chimney** includes specialized vent products such as masonry, metal, and factory-built chimneys; single-wall metal pipe; type B gas vents; special gas vents; or masonry chimney liner systems. (NFPA *Standard* 211). **Draft** is negative static pressure, measured relative to atmospheric pressure; thus, positive draft is negative static pressure. **Flue gas** is the mixture of gases discharged from the appliance and conveyed by the chimney or vent system.

Appliances can be grouped by draft conditions at the appliance flue gas outlet as follows (Stone 1971):

1. Those that require draft applied at the appliance flue gas outlet to induce air into the appliance
2. Those that operate without draft applied at the appliance flue gas outlet (e.g., a gas appliance with a draft hood in which the combustion process is isolated from chimney draft variations)
3. Those that produce positive pressure at the appliance flue gas outlet collar so that no chimney draft is needed; appliances that produce some positive outlet pressure but also need some chimney draft

In the first two configurations, hot flue gas buoyancy, induced-draft chimney fans, or a combination of both produces draft. The third configuration may not require chimney draft, but it should be considered in the design if a chimney is used. If the chimney system is undersized, draft inducers in the connector or chimney may supply draft needs. If the connector or chimney pressure requires control for proper operation, draft control devices must be used.

The preparation of this chapter is assigned to TC 6.10, Fuels and Combustion.

Vented gas-fired appliances have been grouped by draft and flue gas conditions as follows by installation codes in Canada (CSA B149.1) and in the United States (ANSI/NFPA 54/ANSI/AGA Z223.1):

1. Category I appliances operate with nonpositive vent static pressure and a vent gas temperature that avoids excessive condensate production in the vent.
2. Category II appliances operate with nonpositive vent static pressure and a vent gas temperature that may cause condensate production in the vent.
3. Category III appliances operate with positive vent static pressure and a vent gas temperature that avoids excessive condensate production in the vent.
4. Category IV appliances operate with positive vent static pressure and a vent gas temperature that may cause condensate production in the vent.

Category I venting systems are typically sized using venting tables for unobstructed vent systems, as listed in the installation codes; they are provided for fan-assisted appliances and natural-draft appliances as well as multiappliance system vent arrangements. Although these categories are intended for gas-fired appliances, they could apply to other appliances (e.g., oil- or coal-fired).

DRAFT OPERATING PRINCIPLES

Available draft D_a is the draft supplied by the vent system, available at the appliance flue gas outlet. It can be shown as

$$D_a = D_t - \Delta p - D_p + D_b \tag{1}$$

where

D_a = available draft, in. of water
D_t = theoretical draft, in. of water
Δp = flow losses, in. of water
D_p = depressurization, in. of water
D_b = boost (increase in static pressure by fan), in. of water

This equation can account for a nonneutral (nonzero) pressure difference between the space surrounding the appliance or fireplace and the atmosphere. If the surrounding space is at a lower pressure than the atmosphere (space depressurized), the pressure difference D_p should also be subtracted from D_t when calculating available draft D_a, and vice versa. This equation applies to all three appliance draft conditions at the vent system inlets; for example, in the second condition with zero draft requirement at the appliance outlet, available draft required is zero, so theoretical draft of the chimney equals the flow resistance, if no depressurization or boost is present.

Operational consequences of various values of D_a are described as follows:

- **Positive available draft (negative vent pressure); D_a is positive.** Category I fan-assisted and draft hood-equipped appliances and category II appliances can operate satisfactorily when served by venting systems having positive available draft at the appliance flue gas outlet, if the positive draft is sufficient to convey all flue gas from the appliance flue gas outlet to the outdoors and if the positive available draft does not aspirate excessive excess air to cause flame lifting or other detriments to combustion performance. Category III and IV appliances can operate satisfactorily when served by venting systems having positive available draft if the appliance flue gas discharge pressure plus positive draft is sufficient to convey all flue gas from the appliance flue gas outlet to the outdoors. If positive available draft and/or appliance flue gas discharge pressure is insufficient to convey all flue gas from the appliance flue gas outlet outdoors, incomplete combustion, flame rollout, and/or flue gas spillage can occur at the appliance.
- **Zero available draft (neutral vent pressure); $D_a = 0$.** Category I fan-assisted and draft hood-equipped appliances and Category II appliances can operate satisfactorily when served by venting systems having zero available draft (neutral draft) at the appliance flue gas outlet, if the venting system creates sufficient theoretical draft to convey all flue gas from the appliance flue gas outlet to the outdoors. If the venting system creates insufficient theoretical draft to convey all flue gas from the appliance flue gas outlet to the outdoors, incomplete combustion, flame rollout, and/or flue gas spillage can occur at the appliance. Category III and IV appliances can operate satisfactorily when served by venting systems having zero available draft, if appliance flue gas discharge pressure is great enough to overcome the vent flow pressure-drop loss.
- **Negative available draft (positive vent pressure); D_a is negative.** Category I and II appliances cannot operate satisfactorily when served by venting systems having negative available draft. Category III and IV appliances can operate satisfactorily when served by venting systems having negative draft, if appliance flue gas discharge pressure plus the theoretical draft created by the venting system is sufficient to overcome the vent negative draft at the vent inlet and the vent flow pressure-drop loss.
- If chimney height and flue gas temperatures provide **surplus available draft D_a** (excessive excess air), draft control is required.

Theoretical draft D_t is the natural draft produced by the buoyancy of hot gases in the chimney relative to cooler gases in the surrounding atmosphere. It depends on chimney height, local barometric pressure, and the **mean chimney flue gas temperature difference Δt_m**, which is the difference in temperature between the flue gas and atmospheric gases. Therefore, cooling by heat transfer through the chimney wall is a key variable in chimney design. Precise evaluation of theoretical draft is not necessary for most design calculations because of the availability of chimney design charts, computer programs, capacity tables in the references, building codes, and vent and appliance manufacturers' data sheets.

Chimney temperatures and acceptable combustible material temperatures must be known in order to determine safe clearances between the chimney and combustible materials. Safe clearances for some chimney systems, such as type B gas vents, are determined by standard tests and/or specified in building codes.

Losses from Flow Δp represent the friction losses imposed on the flue gas by flow resistance through the chimney.

Depressurization D_p is negative pressure in the space surrounding the appliance with respect to the atmosphere into which the chimney discharges. D_p can be caused by other appliances and fans operating in the building that remove or add air to the space surrounding the appliance, by building stack effect, by outdoor atmospheric effects such as wind impacting the chimney exit or the side of the building facing or leeward to the wind, and other building phenomena.

Boost D_b is the pressure boost from a mechanical-draft fan. A chimney with an forced-draft fan at the inlet of the chimney would have positive boost (increased static pressure). A chimney with an induced-draft fan at the outlet of the chimney would have negative boost (decreased static pressure).

The following sections cover the basis of chimney design for average steady-state category I and III appliance operating conditions. For other appliance operating conditions, a rigorous cyclic evaluation of the flue gas and material surface temperatures in the chimney vent system can be obtained using the VENT-II computer program (Rutz and Paul 1991). For oil-fired appliances, chimney flue gas and material surface temperature evaluations can be obtained using the OHVAP computer program (Krajewski 1996).

CHIMNEY FUNCTIONS

The proper chimney can be selected by evaluating factors such as draft, configuration, size, and operating conditions of the appliance; construction of surroundings; appliance usage classification; residential, low, medium, or high heat (NFPA *Standard* 211); and building height. The chimney designer should know the applicable codes and standards to ensure acceptable construction.

In addition to chimney draft, the following factors must be considered for safe and reliable operation: adequate air supply for combustion; building depressurization effects; draft control devices; chimney materials (corrosion and temperature resistance); flue gas temperatures, composition, and dew point; wind eddy zones; and particulate dispersion. Chimney materials must resist oxidation and condensation at both high and low fire levels at all design temperatures.

Start-Up

The equations and design charts in this chapter may be used to determine vent or chimney size for average category I and III vent system operating conditions based on steady-state operating conditions. The equations and charts do not consider modulation, cycling, or time to achieve equilibrium flow conditions from a cold start. Whereas mechanical draft systems can start gas flow, gravity systems rely on the buoyancy of hot flue gases as the sole force to displace the cold air in the chimney. Priming follows Newton's laws of motion. The time to fill a system with hot flue gases, displace the cold air, and start flow is reasonably predictable and is usually a minute or less; however, unfavorable thermal differentials, building/chimney interaction, mechanical equipment (e.g., exhaust fans), or wind forces that oppose the normal flow of vent gases can overwhelm the buoyancy force. Then, rapid priming cannot be obtained solely from correct system design. The VENT-II computer program contains detailed analysis of gas vent and chimney priming and other cold-start considerations and allows for appliance cycling and pressure differentials that affect performance (Rutz and Paul 1991). A copy of the solution methodology (Rutz 1991) for VENT-II, including equations, may be ordered from ASHRAE Customer Service.

Air Intakes

All rooms or spaces containing fuel-burning appliances must have a constant supply of combustion air from outdoors (either directly or indirectly) at adequate static pressure to ensure proper combustion. In addition, air (either directly or indirectly) is required to replace the air entering chimney systems through draft hoods and barometric draft regulators and to ventilate closely confined boiler and furnace rooms.

The U.S. *National Fuel Gas Code* (ANSI/NFPA 54/ANSI/AGA Z223.1) and the Canadian *Natural Gas and Propane Installation Code* (CSA *Standard* B149.1), along with appliance manufacturers, provide requirements for air openings. Any design must consider flow resistance of the air supply, including register-louver resistance, air duct resistance, and air inlet terminations. Compliance with these codes or the appliance manufacturers' instructions accounts for the air supply flow resistance.

Vent Size

Small residential and commercial natural-draft gas appliances need vent diameters of 3 to 12 in. U.S. and Canadian codes recommend sizes or input capacities for most acceptable gas appliance venting materials. These sizes also apply to gas appliances with integral automatic vent dampers, as well as to appliances with field-installed automatic vent dampers. Field-installed automatic vent dampers should be listed for use with a specific appliance by a recognized testing agency and installed by qualified installers.

Draft Control

Pressure, temperature, and other draft controls have replaced draft hoods in many residential furnaces and boilers to attain higher steady-state and seasonal efficiencies. Appliances that use pulse combustion or forced- or induced-draft fans, as well as those designed for sealed combustion or direct venting, do not have draft hoods but may require special venting and special vent terminals. If fan-assisted burners deliver fuel and air to the combustion chamber and also overcome the appliance flow resistance, draft hoods or other control devices may be installed, depending on the design of the appliance. Vent category II, III, and IV and some category I appliances do not use draft hoods; in such cases, the listed appliance manufacturer's vent system design requirements should be followed. The section on Vent and Chimney Accessories has information on draft hoods, barometric regulators, draft fans, and other draft control devices.

Frequently, a chimney must produce excess flow or draft. For example, dangerously high flue gas outlet temperatures from an incinerator (or normal noncondensing appliance flue gas temperatures) may be reduced by diluting the flue gas with air in the chimney (or PVC vent pipe) by applying excess draft.

Pollution Control

Where control of pollutant emissions is impossible, the chimney should be tall enough to ensure dispersion over a wide area to prevent objectionable ground-level concentrations. The chimney can also serve as a passageway to carry flue gas to pollution control equipment. This passageway must meet the building code requirements for a chimney, even at the exit of pollution control equipment, because of possible exposure to heat and corrosion. A bypass chimney should also be provided to allow continued exhaust in the event of pollution control equipment failure, repair, or maintenance.

Equipment Location

Chimney materials may allow installation of appliances at intermediate or all levels of a high-rise building without imposing weight penalties. Some gas vent systems allow individual apartment-by-apartment heating systems.

Wind Effects

Wind and eddy currents affect discharge of gases from vents and chimneys. A vent or chimney must expel flue gas beyond the cavity or eddy zone surrounding a building to prevent reentry through openings and fresh air intakes. A chimney and its termination can stabilize the effects of wind on appliances and their equipment rooms. In many locations, the equipment room air supply is not at neutral pressure under all wind conditions. Locating the chimney outlet well into the undisturbed wind stream and away from the cavity and wake zones around a building both counteracts wind effects on the air supply pressure and prevents reentry through openings and contamination of fresh air intakes.

Chimney outlets below parapet or eaves level, nearly flush with the wall or roof surface, or in known regions of stagnant air may be subjected to downdrafts and are undesirable. Caps for downdraft and rain protection must be installed according to either their listings and the cap manufacturer's instructions or the applicable building code.

Wind effects can be minimized by locating the chimney terminal and the combustion air inlet terminal close together in the same pressure zone while taking care to minimize recirculation of combustion products into the combustion air inlet.

See Chapter 16 of the 2005 *ASHRAE Handbook—Fundamentals*, for more information on wind effects.

Safety Factors

Safety factors allow for uncertainties of vent and chimney operation. For example, flue gas must not spill from a draft hood or barometric regulator, even when the chimney has very low available draft. The Table 2 design condition for natural gas vents (i.e., vent gas at 300°F rise and 5.3% CO_2 concentration) allows gas vents to operate with reasonable safety above or below the suggested temperature and CO_2 limits.

Safety factors may also be added to the system friction coefficient to account for a possible extra fitting, soot accumulation, and air supply resistance. The specific gravity of flue gas can vary depending on the fuel burned. Natural gas flue gas, for example, has a density as much as 5% less than air, whereas coke flue gas may be as much as 8% greater. However, these density changes are insignificant relative to other uncertainties, so no compensation is needed.

STEADY-STATE CHIMNEY DESIGN EQUATIONS

Chimney design balances forces that produce flow against those that retard flow (e.g., friction). **Theoretical draft** is the pressure that produces flow in gravity or natural-draft chimneys. It is defined as the static pressure resulting from the difference in densities between a stagnant column of hot flue gas and an equal column of ambient air. In the design or balancing process, theoretical draft may not equal friction loss because the appliance is frequently built to operate with some specific pressure (positive or negative) at the appliance flue gas exit. This exit pressure is added to the **available draft**, which depends on chimney conditions, appliance operating characteristics, fuel, and type of draft control.

Flow losses caused by friction may be estimated by several formulas for flow in pipes or ducts, such as the equivalent length method or the loss coefficient (velocity head) method. Chapter 34 of the 2005 *ASHRAE Handbook—Fundamentals* covers computation of flow losses. This chapter emphasizes the loss coefficient method because fittings usually cause the greater portion of system pressure drop in chimney systems, and conservative loss coefficients (which are almost independent of piping size) provide an adequate basis for design.

Rutz and Paul (1991) developed a computer program entitled VENT-II: An Interactive Personal Computer Program for Design and Analysis of Venting Systems for One or Two Gas Appliances, which dynamically predicts cyclic flows, temperatures, and pressures in venting systems. Similarly, Krajewski (1996) developed a computer program entitled OHVAP: Oil-Heat Vent Analysis Program.

For large gravity chimneys, steady-state available draft D_a may be calculated from Equation (2) or (3). Both equations use the equivalent length approach, as indicated by the symbol L_e in the flow-loss term (ASHVE 1941). These equations permit consideration of the density difference between chimney gases and ambient air and compensation for shape factors. Mean flue gas temperature T_m must be estimated separately. Starting with Equation (1), the following equations are based on zero depressurization at the appliance flue gas outlet and zero chimney draft boost.

For a cylindrical chimney,

$$D_a = 2.96HB\left(\frac{\rho_o}{T_o} - \frac{\rho_c}{T_m}\right) - \frac{0.000315w^2 T_m f L_e}{1.3 \times 10^7 d_f^5 B\rho_c} \quad (2)$$

For a rectangular chimney,

$$D_a = 2.96HB\left(\frac{\rho_o}{T_o} - \frac{\rho_c}{T_m}\right) - \frac{0.000097w^2 T_m f L_e (x+y)}{1.3 \times 10^7 (xy)^3 B\rho_c} \quad (3)$$

where

H = height of vent or chimney system above grade or system inlet, ft
B = existing or local barometric pressure, in. Hg
ρ_o = density of ambient air at design temperature and local barometric pressure conditions
ρ_c = density of chimney flue gas at average temperature and local barometric pressure, lb/ft^3
T_o = ambient air design temperature, °R
T_m = mean flue gas temperature at average conditions in system, °R
w = mass flow of flue gas, lb/h
f = Darcy friction factor
L_e = total piping length L plus equivalent length of elbows, ft
d_f = inside diameter, ft
x = length of one internal side of rectangular chimney cross section, ft
y = length of other internal side of rectangular chimney cross section, ft

In these equations, the first term determines theoretical draft, based on applicable gas and ambient density. The second term defines draft loss based on the factors for flow in a circular or rectangular duct system. To use these equations, the mass flow w for a variety of fuels and situations and the available draft needs of various types of appliances must be determined.

Equations (2) and (3) may be derived and expressed in a form that is more readily applied to the problems of chimney design, size, and capacity by considering the following factors, which are the steps used to solve the problems in the section on Chimney Capacity Calculation Examples.

1. Mass flow of combustion products in chimney
2. Chimney gas temperature and density
3. Theoretical draft
4. System pressure loss caused by flow
5. Available draft
6. Chimney gas velocity
7. System resistance coefficient
8. Final input-volume relationships

For application to system design, the chimney gas velocity step is eliminated; however, actual velocity can be found readily, if needed.

1. Mass Flow of Combustion Products in Chimneys and Vents

Mass flow in a chimney or venting system may differ from that in the appliance, depending on the type of draft control or number of appliances operating in a multiple-appliance system. Mass flow is preferred to volumetric flow because it remains constant in any continuous portion of the system, regardless of changes in temperature or pressure. For the chimney gases resulting from any combustion process, mass flow can be expressed as

$$w = IM \quad (4)$$

where

w = mass flow rate, lb/h
I = appliance heat input, Btu/h
M = ratio of mass flow to heat input, lb of combustion products per 1000 Btu of fuel burned [M depends on fuel composition and percentage excess air (or CO_2) in the chimney]

Table 1 Mass Flow Equations for Common Fuels

Fuel	Ratio M of Mass Flow to Input[a] $M = \frac{\text{lb Total Combustion Products}^b}{\text{1000 Btu Fuel Input}}$
Natural gas	$0.705\left(0.159 + \frac{10.72}{\%CO_2}\right)$
LPG (propane, butane, or mixture)	$0.706\left(0.144 + \frac{12.61}{\%CO_2}\right)$
No. 2 oil (light)	$0.72\left(0.12 + \frac{14.4}{\%CO_2}\right)$
No. 6 oil (heavy)	$0.72\left(0.12 + \frac{15.8}{\%CO_2}\right)$
Bituminous coal (soft)	$0.76\left(0.11 + \frac{18.2}{\%CO_2}\right)$
Type 0 waste or wood	$0.69\left(0.16 + \frac{19.7}{\%CO_2}\right)$

[a]Percent CO_2 is determined in combustion products with water condensed (dry basis).
[b]Total combustion products include combustion products and excess air.

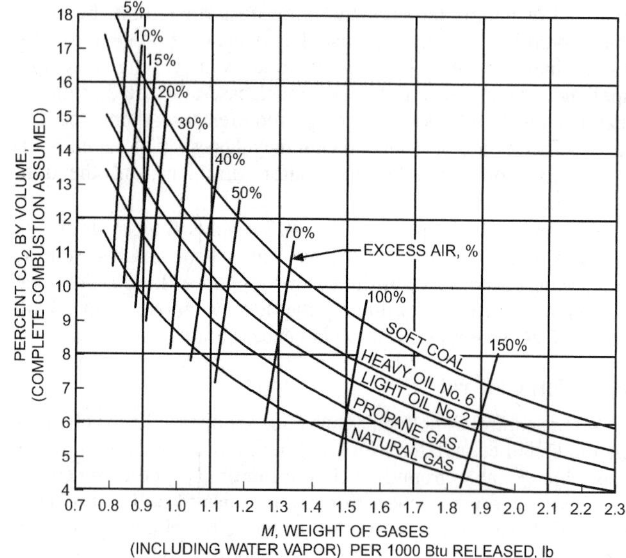

Fig. 1 Graphical Evaluation of Rate of Vent Gas Flow from Percent CO_2 and Fuel Rate

Volumetric flow rate (ft^3/min) Q can be found as follows;

$$Q = \frac{w}{60\rho_m} \quad (5)$$

where ρ_m is gas density (lb/ft^3).

In chimney system design, the composition and flow rate of the flue gas must be assumed to determine the ratio M of mass flow to input. When combustion conditions are given in terms of excess air, Figure 1 can be used to estimate CO_2. The mass flow equations in Table 1 illustrate the influence of fuel composition; however, additional guidance is needed for system design. The information provided with many heat-producing appliances is limited to whether they have been tested, certified, listed, or approved to comply with applicable standards. From this information and from the type of fuel and draft control, certain inferences can be drawn regarding the flue gas.

Table 2 suggests typical values for the vent or chimney systems for gaseous and liquid fuels when specific outlet conditions for the

Table 2 Typical Chimney and Vent Design Conditions[a]

Fuel	Appliance	% CO_2	Temperature Rise, °F	M (Mass Flow/Input), lb Total Flue Gas[b] per 1000 Btu Fuel Input	Flue Gas Density,[c] lb/ft³	Flow Rate/Unit Heat Input,[c] cfm per 1000 Btu/h at Flue Gas Temperature
Natural gas	Draft hood	5.3	300	1.54	0.0483	0.530
Propane gas	Draft hood	6.0	300	1.59	0.0483	0.547
Natural gas						
Low-efficiency	No draft hood	8.0	400	1.06	0.0431	0.409
High-efficiency	No draft hood	7.0	240	1.19	0.0522	0.381
No. 2 oil	Residential	9.0	500	1.24	0.0389	0.531
Oil	Forced-draft, over 400,000 Btu/h	13.5	300	0.85	0.0483	0.295
Waste, Type 0	Incinerator	9.0	1340	1.62	0.0213	1.267

[a]Values are for appliances with flue losses of 17% or more. For appliances with lower flue losses (high-efficiency types), see appliance installation instructions or ask manufacturer for operating data.
[b]Total flue gas includes combustion products and excess air.
[c]At sea level and 60°F ambient temperature.

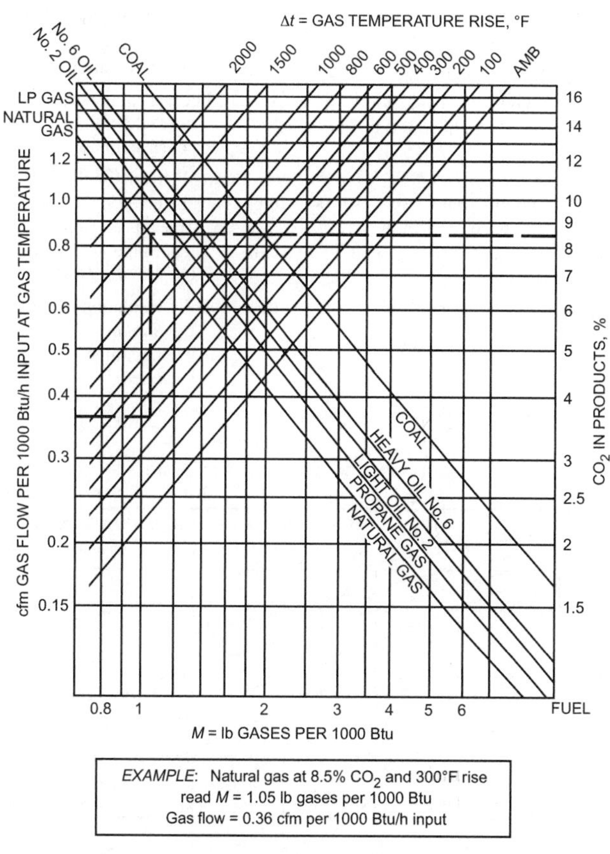

Fig. 2 Flue Gas Mass and Volumetric Flow

appliance are not known. If a gas appliance with draft hood is used, Table 2 recommends that dilution air through the draft hood reduce the CO_2 percentage to 5.3%. For appliances using draft regulators, the dilution and temperature reduction is a function of the draft regulator gate opening, which depends on excess draft. If the chimney system produces the exact draft necessary for the appliance, little dilution takes place.

For manifolded gas appliances that have draft hoods, dilution through draft hoods of inoperative appliances must be considered in precise system design. However, with forced-draft appliances having wind box or inlet air controls, dilution through inoperative appliances may be unimportant, especially if pressure at the outlet of inoperative appliances is neutral (atmospheric level).

Figure 2 can be used to estimate mass and volumetric flow. Flow conditions in the chimney connector, manifold, vent, and chimney vary with configuration and appliance design and are not necessarily the same as boiler or appliance outlet conditions.

Table 3 Mass Flow for Incinerator Chimneys

Type of Waste	Heat Value of Waste,[a] Btu/lb	Auxiliary Fuel[a] per Unit Waste, Btu/lb	Combustion Products: cfm/lb Waste at 1400°F[b]	lb per lb Waste[c]	M, lb Products per 1000 Btu
0	8500	0	10.74	13.76	1.62
1	6500	0	8.40	10.80	1.66
2	4300	0	5.94	7.68	1.79
3	2500	1500	4.92	6.25	2.50
4	1000	3000	4.14	5.33	5.33

[a]Auxiliary fuel may be used with any type of waste, depending on incinerator design.
[b]Specialized units may produce higher or lower outlet gas temperatures, which must be considered in chimney sizing, using Equation (14), Equation (23), or Figure 7.
[c]Multiply these values by pounds of waste burned per hour to establish mass flow.

Mass flow in incinerator chimneys must account for the probable heat value of the waste, its moisture content, and use of additional fuel to initiate or sustain combustion. Classifications of waste and corresponding values of M in Table 3 are based on recommendations from the Incinerator Institute of America. Combustion data given for types 0, 1, and 2 waste do not include any additional fuel. Where constant burner operation accompanies waste combustion, the additional quantity of products should be considered in the chimney design.

The system designer should obtain exact outlet conditions for the maximum rate operation of the specific appliance. This information can reduce chimney construction costs. For appliances with higher seasonal or steady-state efficiencies, however, special attention should be given to the manufacturer's venting recommendations because the flue gas may differ in composition and temperature from conventional values.

2. Mean Chimney Gas Temperature and Density

Chimney gas temperature depends on the fuel, appliance, draft control, chimney size, and configuration.

Density of gas within the chimney and theoretical draft both depend on gas temperature. Although the gases flowing in a chimney system lose heat continuously from entrance to exit, a single mean gas temperature must be used in either the design equation or the chart. Mean chimney gas density is virtually the same as air density at the same temperature. Thus, density may be found as

$$\rho_m = \rho_a\left(\frac{T_s}{T_m} \times \frac{B}{B_o}\right) = 1.325\frac{B}{T_m} \quad (6)$$

where

ρ_m = chimney gas density, lb/ft³
T_s = standard temperature = 518.67°R

Table 4 Mean Chimney Gas Temperature for Various Appliances

Appliance Type	Mean Temperature t_m* in Chimney, °F
Natural gas-fired heating appliance with draft hood (low-efficiency)	360
LP gas-fired heating appliance with draft hood (low-efficiency)	360
Gas-fired heating appliance, no draft hood	
Low-efficiency	460
High-efficiency	300
Oil-fired heating appliance (low-efficiency)	560
Conventional incinerator	1400
Controlled air incinerator	1800 to 2400
Pathological incinerator	1800 to 2800
Turbine exhaust	900 to 1400
Diesel exhaust	900 to 1400
Ceramic kiln	1800 to 2400

*Subtract 60°F ambient to obtain temperature rise for use with Figure 7.

ρ_a = air density at T_s and B_o = 0.0765 lb/ft³
B = local barometric pressure, in. Hg
T_m = mean chimney gas temperature at average system conditions, °R
B_o = standard pressure = 29.92 in. Hg

The density ρ_a in Equation (6) is a compromise value for typical humidity. The subscript m for density and temperature requires that these properties be calculated at mean gas temperature or vertical midpoint of a system (inlet conditions can be used where temperature drop is not significant).

Using a reasonably high ambient (such as 60°F) for design ensures improved operation of the chimney when ambient temperatures drop because temperature differentials and draft increase.

A design requires assuming an initial or inlet chimney gas temperature. In the absence of specific data, Table 4 provides a conservative temperature. For appliances that can operate over a range of temperatures, size should be calculated at both extremes to ensure an adequate chimney.

The drop in vent gas temperature from appliance to exit reduces capacity, particularly in sizes of 12 in. or less. In gravity type B gas vents, which may be as small as 3 in. in diameter, and in other systems used for venting gas appliances, capacity is best determined from the ANSI/NFPA 54/ANSI/AGA Z223.1 or CSA *Standard* B149.1. In these codes, the tables compensate for the particular characteristics of the chimney material involved, except for very high single-wall metal pipe. Between 12 and 18 in. diameters, the effect of heat loss diminishes greatly because there is greater gas flow relative to system surface area. For 20 in. and greater diameters, cooling has little effect on final size or capacity.

A straight vertical vent or chimney directly off the appliance requires little compensation for cooling effects, even with smaller sizes. However, a horizontal connector running from the appliance to the base of the vent or chimney has enough heat loss to diminish draft and capacity. Figure 3 is a plot of temperature correction C_u, which is a function of connector size, length, and material for either conventional single-wall metal connectors or double-wall metal connectors.

To use Figure 3, estimate connector size and length and read the temperature multiplier. For example, 16 ft of 7 in. diameter single-wall connector has a multiplier of 0.61. If inlet temperature rise Δt_e above ambient is 300°F, operating mean temperature rise Δt_m will be 0.61 × 300 = 183°F. This factor adequately corrects the temperature at the midpoint of the vertical vent for heights up to 100 ft.

The correction procedure includes the assumption that the overall heat transfer coefficient of a vertical chimney is approximately 0.6 Btu/h·ft²·°F or less (the value for double-wall metal). This procedure does not correct for cooling in very high stacks constructed entirely of single-wall metal, especially those exposed to cold ambient temperatures. For severe exposures or excessive heat loss, a trial calculation assuming a conservative operating temperature shows whether capacity problems will be encountered.

For more precise heat loss calculations, Table 5 suggests overall heat transfer coefficients for various constructions installed in typical environments at usual flue gas flow velocities (Segeler 1965). For masonry, any additional thickness beyond the single course of brick plus tile liner used in residential chimneys decreases the coefficient.

Table 5 Overall Heat Transfer Coefficients of Various Chimneys and Vents

Material	U, Btu/h·ft²·°F* Observed	U, Btu/h·ft²·°F* Design	Remarks
Industrial steel stacks	—	1.3	Under wet wind
Clay or iron sewer pipe	1.3 to 1.4	1.3	Used as single-wall material
Asbestos-cement gas vent	0.72 to 1.42	1.2	Tested per UL *Standard* 441
Black or painted steel stove pipe	—	1.2	Comparable to weathered galvanized steel
Single-wall galvanized steel	0.31 to 1.38	1.0	Depends on surface condition and exposure
Single-wall unpainted pure aluminum	—	1.0	No. 1100 or other bright-surface aluminum alloy
Brick chimney, tile-lined	0.5 to 1.0	1.0	For gas appliances in residential construction per NFPA *Standard* 211
Double-wall gas vent, 1/4 in. air space	0.37 to 1.04	0.6	Galvanized steel outer pipe, pure aluminum inner pipe; tested per UL *Standard* 441
Double-wall gas vent, 1/2 in. air space	0.34 to 0.7	0.4	
Insulated prefabricated chimney	0.34 to 0.7	0.3	Solid insulation meets UL *Standard* 103 when chimney is fully insulated

*U-factors based on inside area of chimney.

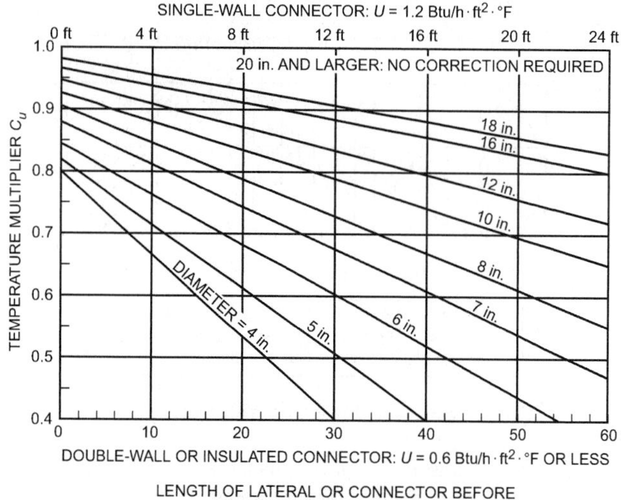

Fig. 3 Temperature Multiplier C_u for Compensation of Heat Losses in Connector

Equations (7) and (8) describe flow and heat transfer, respectively, within a venting system with $D_a = 0$.

$$q_m = 3600 T_s \sqrt{2g}\, c_p \frac{B}{B_o} \times \frac{A}{T_m}\left(\frac{H}{kT_o}\right)^{0.5} \Delta t_m^{1.5} \rho_m \quad (7)$$

$$\frac{q}{q_m} = \frac{\Delta t_e}{\Delta t_m} = \exp\left(\pi \bar{U} d_f \frac{L_m \Delta t_m}{q_m}\right) \quad (8)$$

where

A = area of passage cross section, ft²
c_p = specific heat
d_f = inside diameter, ft
$\exp x = e^x$
g = gravitational constant = 32.1740 ft/s²
H = height of chimney above grade or inlet, ft
k = fixed, dimensionless system flow resistance coefficient for pipe and fittings
L_m = length from inlet to location (in the vertical) of mean gas temperature, ft
q = heat flow rate at vent inlet, Btu/h
q_m = heat flow rate at midpoint of vent, Btu/h
$\Delta t_e = (T - T_o)$ = temperature difference entering system, °F
$\Delta t_m = (T_m - T_o)$ = chimney gas mean temperature rise, °F
T = flue gas temperature at vent inlet, °F
T_m = chimney mean flue gas temperature, °R
T_o = design ambient temperature, 520°R
T_s = 518.67°F
$\bar{U}$ = heat transfer coefficient, Btu/h·ft²·°F
ρ_m = mean flue gas density from Equation (6)

Assuming reasonable constancy of $\bar{U}$, the overall heat transfer coefficient of the venting system material, Equations (7) and (8) provide a solution for maximum vent gas capacity. They can also be used to develop cooling curves or calculate the length of pipe where internal moisture condenses. Kinkead (1962) details methods of solution and application to both individual and combined gas vents.

3. Theoretical Draft

The theoretical draft of a gravity chimney or vent is the difference in weight between a given column of warm (light) chimney gas and an equal column of cold (heavy) ambient air. Chimney gas density or temperature, chimney height, and barometric pressure determine theoretical draft; flow is not a factor. The equation for theoretical draft assumes chimney gas density is the same as that of air at the same temperature and pressure; thus,

$$D_t = 0.2554 BH\left(\frac{1}{T_o} - \frac{1}{T_m}\right) \quad (9)$$

where

B = local barometric pressure, in. Hg
D_t = theoretical draft, in. of water
H = height of chimney above grade or inlet, ft
T_m = mean flue gas temperature at average conditions in system, °R
T_o = ambient temperature, °R

Theoretical draft thus increases directly with height and with the difference in density between the hot and cold columns.

Theoretical draft D_t is always positive (unless chimney gases are colder than ambient air). Equation (9) for theoretical draft is the basis for Figure 4, which can be used up to 1000°F and 7000 ft elevation. For example, using Figure 4, ambient air temperature at 40°F, 4000 ft above sea level (25.85 in. Hg), and mean chimney gas temperature at 350°F provides D_{100} draft of 0.5 in. of water per 100 ft of height. For D_{100} draft at 25 in. Hg barometric pressure or 4900 ft above sea level, follow the intersection to the pivot line and read the new D_{100} draft of 0.48 in. of water.

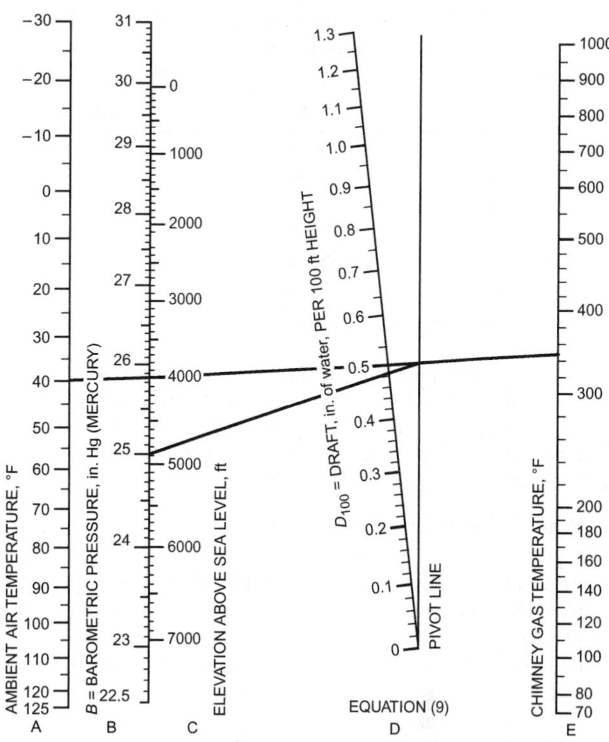

Fig. 4 Theoretical Draft Nomograph

Table 6 Approximate Theoretical Draft of Chimneys

Vent Gas Temperature Rise, °F	D_t per 100 ft, in. of water
100	0.2
150	0.3
200	0.4
300	0.5
400	0.6
500	0.7
600	0.8
800	0.9
1100	1.0
1600	1.1
2400	1.2

Notes: Ambient temperature = 60°F = 520°R
Chimney gas density = air density
Sea-level barometric pressure = 29.92 in. Hg
Equation (9) may be used to calculate exact values for D_t at any altitude.

Theoretical draft should be estimated and included in system calculations, even for appliances producing considerable positive outlet static pressure, to achieve the economy of minimum chimney size. Equation (9) may be used directly to calculate exact values for theoretical draft at any altitude. For ease of application and consistency with Figure 7, Table 6 lists approximate theoretical draft for typical gas temperature rises above 60°F ambient.

Appliances with fixed fuel beds, such as hand-fired coal stoves and furnaces, require positive available draft (negative gage pressure). Small oil heaters with pot-type burners, as well as residential furnaces with pressure-atomizing oil burners, need positive available draft, which can usually be set by following the manufacturer's instructions for setting the draft regulator. Available draft requirements for larger packaged boilers or appliances assembled from components may be negative, zero (neutral), or positive.

Table 7 Input Altitude Factor for Equation (21) Theoretical Draft

Altitude, ft	Barometric Pressure B, in. Hg	Factor*
Sea level	29.92	1.00
2,000	27.82	1.08
4,000	25.82	1.16
6,000	23.98	1.25
8,000	22.22	1.35
10,000	20.58	1.45

*Multiply operating input by factor to obtain design input.

Compensation of theoretical draft for altitude or barometric pressure is usually necessary for appliances and chimneys functioning at elevations greater than 2000 ft. Depending on the design, one of the following approaches to pressure or altitude compensation is necessary for chimney sizing.

1. Figure 7: Use sea-level theoretical draft.
2. Equation (21): Use local theoretical draft with actual energy input, or use sea level theoretical draft with energy input multiplied by ratio of sea level to local barometric pressure (Table 7 factor).
3. Equation (23): Use local theoretical draft and barometric pressure with volumetric flow at the local density.

The altitude correction multiplier for input (Table 7) is the only method of correcting to other elevations. Reducing theoretical draft imposes an incorrect compensation on the chart.

Gas appliances with draft hoods, for example, are usually derated 4% per 1000 ft of elevation above sea level when they are operated at 2000 ft altitude or above (see ANSI/NFPA 54/ANSI/AGA Z223.1 or CSA *Standard* B149.1). The altitude correction factor derates the design input so that the vent size at altitude for derated gas appliances is effectively the same as at sea level. For other appliances where burner adjustments or internal changes might be used to adjust for reduced density at altitude, the same factors produce an adequately compensated chimney size. For example, an appliance operating at 6000 ft elevation at 10×10^6 Btu/h input, but requiring the same draft as at sea level, should have a chimney selected on the basis of 1.244 times the operating input, or 12.5×10^6 Btu/h.

4. System Pressure Loss Caused by Flow

In any chimney system, flow losses, expressed as pressure drop Δp in inches of water, are the difference between theoretical and available draft:

$$\Delta p = D_t - D_a \tag{10}$$

where Δp is always positive.

In any chimney or vent system, flow losses resulting from velocity and resistance can be determined from the Bernoulli equation:

$$\Delta p = \frac{k\rho_m V^2}{5.2(2g)} \tag{11}$$

where

k = dimensionless system resistance coefficient of piping and fittings
V = system gas velocity at mean conditions, fps
g = gravitational constant = 32.1740 ft/s²
ρ_m = mean flue gas density, lb/ft³

Pressure losses are thus directly proportional to the resistance factor and to the square of the velocity.

Table 8 Pressure Equations for Δp

Required Appliance Outlet Pressure or Available Draft D_a	Δp Equation[a]: Gravity Only	Δp Equation[a]: Gravity plus Inducer[b]
Negative, needs positive draft	$\Delta p = D_t - D_a$	$\Delta p = D_t - D_a + D_b$
Zero, vent with draft hood or balanced forced draft	$\Delta p = D_t$	$\Delta p = D_t + D_b$
Positive, causes negative draft	$\Delta p = D_t + D_a$	$\Delta p = D_t + D_a + D_b$

[a]Equations use absolute pressure for D_a.
[b]D_b = static pressure boost of inducer at flue gas temperature and rated flow.

5. Available Draft

Starting with Equation (1), the difference between theoretical draft and flow losses without depressurization or boost is:

$$D_a = D_t - \Delta p$$

Available draft D_a can therefore be defined as:

$$D_a = 0.2554BH\left(\frac{1}{T_o} - \frac{1}{T_m}\right) - \frac{k\rho_m V^2}{5.2(2g)} \tag{12}$$

See Equations (2) and (3) for geometrically defined versions.

Available draft D_a can be negative, zero, or positive. The pressure difference Δp, or theoretical minus available draft, overcomes the flow losses. Table 8 lists the pressure components for three draft configurations. Table 8 applies to still air (no wind) conditions and a neutral (zero) pressure difference between the space surrounding the appliance or fireplace and the atmosphere. Columns with and without boost are provided.

The effect of a nonneutral pressure difference on capacity or draft may be included by imposing a static pressure (either positive or negative). One way to circumvent a space-to-atmosphere pressure difference is to use a sealed-combustion system or direct-vent system (i.e., all combustion air is taken directly from the outdoor atmosphere, and all flue gas is discharged to the outdoor atmosphere with no system openings such as draft hoods). The effect of wind on capacity or draft may be included by imposing a static pressure (either positive or negative) or by changing the vent terminal resistance loss. However, a properly designed and located vent terminal should cause little change in Δp at typical wind velocities.

Although small static draft pressures can be measured at the entrance and exit of gas appliance draft hoods, D_a at the appliance is effectively zero. Therefore, all theoretical draft energy produces chimney flow velocity and overcomes chimney flow resistance losses.

6. Chimney Gas Velocity

Velocity in a chimney or vent varies inversely with flue gas density ρ_m and directly with mass flow rate. The equation for flue gas velocity at mean gas temperature in the chimney is

$$V = \frac{144 \times 4w}{3600\pi\rho_m d_i^2} \tag{13}$$

where

V = flue gas velocity, fps
d_i = inside diameter, in.
ρ_m = mean flue gas density, lb/ft³
w = mass flow rate, lb/h

To express velocity as a function of input and chimney gas composition, w in Equation (13) is replaced by using Equation (4):

$$V = \frac{144 \times 4IM}{3600\pi\rho_m d_i^2} = \frac{0.0509IM}{\rho_m d_i^2} \tag{14}$$

Thus, chimney velocity depends on the product of heat input I and the ratio M of mass flow to input.

The input capacity or diameter of a chimney may usually be found without determining flow velocity. Internal or exit velocity must occasionally be known, to ensure effluent dispersal or avoid flow noise. Also, the flow velocity of incinerator chimneys, turbine exhaust systems, and other appliances with high outlet pressures or velocities is needed to estimate piping loss coefficients.

Equations (11), (13), (14), and (22) can be applied to find velocity. The velocity may also be calculated by dividing volumetric flow rate Q by chimney area A. A similar calculation may be performed when the energy input is known. For example, a 34 in. chimney serving a 10×10^6 Btu/h natural gas appliance, at 8.5% CO_2 and 300°F above ambient in the chimney, produces (from Figure 2) 0.36 cfm per 1000 Btu/h. Chimney gas flow rate is

$$Q = 10 \times 10^6\left(\frac{0.36}{1000}\right) = 3600 \text{ cfm} \quad (15)$$

Dividing by area to obtain velocity, $V = 3600/6.305 = 571$ fpm

Chimney gas velocity affects the piping friction factor k_L and also the roughness correction factor. The section on Resistance Coefficients has further information, and Example 3 illustrates how these factors are used in the velocity equations.

Chimney systems can operate over a wide range of velocities, depending on modulation characteristics of the burner system or the number of appliances in operation. Typical velocity in vents and chimneys ranges from 300 to 3000 fpm. A chimney design developed for maximum input and maximum velocity should be satisfactory at reduced input because theoretical draft is roughly proportional to flue gas temperature rise, while flow losses are proportional to the square of the velocity. Thus, as input is reduced, flow losses decrease more rapidly than system motive pressures.

Effluent dispersal may occasionally require a minimum upward chimney outlet velocity, such as 3000 fpm. A tapered exit cone can best meet this requirement. For example, to increase the outlet velocity from the 34 in. chimney ($A = 6.305$ ft^2 area) from 1600 to 3000 fpm, the cone must have a discharge area of $6.305 \times 1600/3000 = 3.36$ ft^2 and a 24.8 in. diameter. An exit cone avoids excessive flow losses because the entire system operates at the lower velocity, and a resistance factor is only added for the cone. In this case, the added resistance for a gradual taper approximates the following (see Table 9):

$$k = \left(\frac{d_{i1}}{d_{i2}}\right)^4 - 1 = \left(\frac{34}{24.8}\right)^4 - 1 = 2.53 \quad (16)$$

Noise in chimneys may be caused by turbulent flow at high velocity or by combustion-induced oscillations or resonance. Noise is seldom encountered in gas vent systems or in systems producing positive available draft, but it may be a problem with forced-draft appliances. Turbulent flow noise can be avoided by designing for lower velocity, which may entail increasing the chimney size above the minimum recommended by the appliance manufacturer. Chapter 47 of the 2007 *ASHRAE Handbook—HVAC Applications* has more information on noise control.

7. System Resistance Coefficient

The velocity head method of determining resistance losses assigns a fixed numerical coefficient (independent of velocity) or k factor to every fitting or turn in the flow circuit, as well as to piping.

The resistance coefficient k that appears in Equations (21) and (23) and Figure 7 summarizes the friction loss of the entire chimney system, including piping, fittings, and configuration or interconnection factors. Capacity of the chimney varies inversely with the square root of k, whereas diameter varies as the fourth root of k. The insensitivity of diameter and input to small variations in k simplifies design. Analyzing details such as pressure regain, increasers and reducers, and gas cooling junction effects is unnecessary if slightly high resistance coefficients are assigned to any draft diverters, elbows, tees, terminations, and, particularly, piping.

Table 9 Resistance Loss Coefficients

Component	Suggested Design Value, Dimensionless*	Estimated Span and Notes
Inlet acceleration (k_1)		
Gas vent with draft hood	1.5	1.0 to 3.0
Barometric regulator	0.5	0.0 to 0.5
Direct connection	0.0	Also dependent on blocking damper position
Round elbow (k_2)		
90°	0.75	0.5 to 1.5
45°	0.3	—
Tee or 90° connector (k_3)	1.25	1.0 to 4.0
Y connector	0.75	0.5 to 1.5
Cap, top (k_4)		
Open straight	0.0	—
Low-resistance (UL)	0.5	0.0 to 1.5
Other	—	1.5 to 4.5
Spark screen	0.5	—
Converging exit cone	$(d_{i1}/d_{i2})^4 - 1$	System designed using d_{i1}
Tapered reducer (d_{i1} to d_{i2})	$1 - (d_{i2}/d_{i1})^4$	System designed using d_{i2}
Increaser	—	See Chapter 2, 2005 *ASHRAE Handbook—Fundamentals*.
Piping (k_L)	$0.4\frac{L, \text{ft}}{d_i, \text{in.}}$	Numerical coefficient (friction factor F) varies from 0.2 to 0.5; see Figure 13, Chapter 2, 2005 *ASHRAE Handbook—Fundamentals* for size, roughness, and velocity effects.

*Initial assumption when size is unknown: $k = 5.0$ for entire system, for first trial; $k = 7.5$ for combined gas vents only

Note: For combined gravity gas vents serving two or more appliances (draft hoods), multiply total k [components + piping—see Equations (17), (18), and (19)] by 1.5 to obtain gravity system design coefficient. (This rule does not apply to forced- or induced-draft vents or chimneys.)

The flow resistance of a fitting such as a tee with flue gases entering the side and making a 90° turn is assumed to be constant at $k = 1.25$, independent of size, velocity, orientation, inlet or outlet conditions, or whether the tee is located in an individual vent or in a manifold. Conversely, if flue gases pass straight through a tee, as in a manifold, assumed resistance is zero, regardless of any area changes or flow entry from the side branch. For any chimney with fittings, the total flow resistance is a constant plus variable piping resistance—the latter being a function of centerline length divided by diameter. Table 9 suggests moderately conservative resistance coefficients for common fittings. Elbow resistance may be lowered by long-radius turns; however, corrugated 90° elbows may have resistance values at the high end of the scale. Table 9 shows resistance as a function of inlet diameter d_{i1} and outlet diameter d_{i2}.

System resistance k may be expressed as follows:

$$k = k_f + k_L \quad (17)$$

with

$$k_f = k_1 + n_2k_2 + n_3k_3 + k_4 + \cdots \quad (18)$$

and

$$k_L = \frac{FL}{d_i} \quad (19)$$

where

k_f = fixed fitting loss coefficient
k_L = piping resistance loss function (Figure 13, Chapter 2 of the 2005 *ASHRAE Handbook—Fundamentals*, adjusted for units)
k_1 = inlet acceleration coefficient
k_2 = elbow loss coefficient, n_2 = number of elbows
k_3 = tee loss coefficient, n_3 = number of tees
k_4 = cap, top, or exit cone loss coefficient
F = friction factor
L = length of all piping in chimney system, ft
d_i = inside diameter, ft

For combined gas vents using appliances with draft hoods, the summation k must be multiplied by a diversity factor of 1.5 (see Table 9 note and Example 5). This multiplier does not apply to forced- or induced-draft vents or chimneys.

When size is unknown, the following k values may be used to run a first trial estimate:

k = 5.0 for the entire system
k = 7.5 for combined gas vents only

The resistance coefficient method adapts well to systems in which the fittings cause significant losses. Even for extensive systems, an initial assumption of k = 5.0 gives a tolerably accurate vent or chimney diameter in the first trial solution. Using this diameter with the piping resistance function [Equation (19)] in a second trial normally yields the final answer.

The minimum system resistance coefficient in a gas vent with a draft hood is always 1.0 because all gases must accelerate through the draft hood from almost zero velocity to vent velocity.

For a system connected directly to the outlet of a boiler or other appliance where the capacity is stated as full-rated heat input against a positive static pressure at the chimney connection, minimum system resistance is zero, and no value is added for existing velocity head in the system.

For simplified design, a value of 0.4 for F in Equation (19) applies for all sizes of vents or chimneys and for all velocities and temperatures. As diameter increases, this function becomes increasingly conservative, which is desirable because larger chimneys are more likely to be made of rough masonry construction or other materials with higher pressure losses. The 0.4 constant also introduces an increasing factor of safety for flow losses at greater lengths and heights.

Figure 5 is a plot of friction factor F versus velocity and diameter for commercial iron and steel pipe at a flue gas temperature of 300°F above ambient (Lapple 1949). The figure shows, for example, that a 48 in. diameter chimney with a flue gas velocity of 80 fps may have a friction factor as low as 0.2. In most cases, $k_L = 0.3\ L/d_i$ gives reasonable design results for chimney sizes 18 in. and larger because systems of this size usually operate at flue gas velocities greater than 10 fps.

At 1000°F or over, the factors in Figure 5 should be multiplied by 1.2. Because Figure 5 is for commercial iron and steel pipe, an additional correction for greater or less surface roughness may be imposed. For example, the factor for a very rough 12 in. diameter pipe may be doubled at a velocity as low as 2000 fpm.

For most chimney designs, a friction factor F of 0.4 gives a conservative solution for diameter or input for all sizes, types, and operating conditions of prefabricated and metal chimneys; alternately, F = 0.3 is reasonable if the diameter is 18 in. or more. Because neither input nor diameter is particularly sensitive to the total friction factor, the overall value of k requires little correction.

Masonry chimneys, including those lined with clay flue tile, may have rough surfaces, tile shape variations that cause misalignment, and joints at frequent intervals with possible mortar protrusions. In addition, the inside cross-sectional area of liner shapes may be less than expected because of local manufacturing variations, as well as differences between claimed and actual size. To account for these characteristics, the estimate for k_L should be on the high side, regardless of chimney size or velocity.

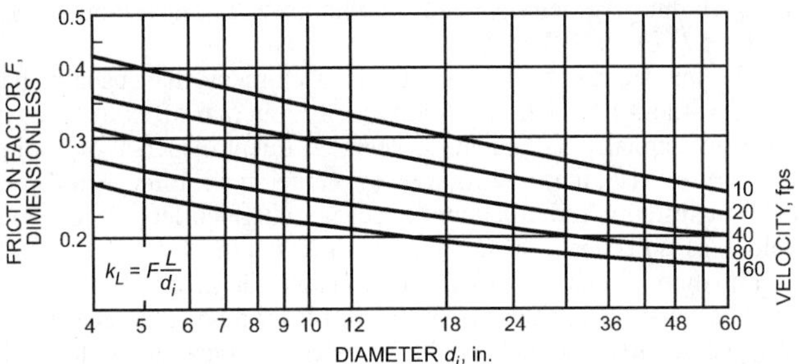

Fig. 5 Friction Factor for Commercial Iron and Steel Pipe
(Lapple 1949)

Computations should be made by assuming smooth surfaces and then adding a final size increase to compensate for shape factor and friction loss. Performance or capacity of metal and prefabricated chimneys is generally superior to that of site-constructed masonry.

Configuration and Manifolding Effects

The most common configuration is the individual vent, stack, or chimney, in which one continuous system carries products from appliance to terminus. Other configurations include the combined vent serving a pair of appliances, the manifold serving several, and branched systems with two or more lateral manifolds connected to a common vertical system. As the number of appliances served by a common vertical vent or chimney increases, the precision of design decreases because of diversity factors (variation in the number of appliances in operation) and the need to allow for maximum and minimum input operation (Stone 1957). For example, the vertical common vent for interconnected gas appliances must be larger than for a single appliance of the same input to allow for operating diversity and draft hood dilution effects. Connector rise, headroom, and configuration in the equipment room must be designed carefully to avoid draft hood spillage and related oxygen depletion problems.

For typical combined vents, the diversity effect must be introduced into Figure 7 and the equations by multiplying system resistance loss coefficient k by 1.5 (see Table 9 note and Example 5). This multiplier compensates for junction effect and part-load operation.

Manifolds for appliances with barometric draft regulators can be designed without allowing for dilution air through inoperative appliances. In this case, because draft regulators remain closed until regulation is needed, dilution under part load is negligible. In addition, airflow through any inoperative appliance is negligible because the combustion air inlet dampers are closed and the multiple-pass heat exchanger has a high internal flow resistance.

Manifold systems of oil-burning appliances, for example, have a lower flow velocity and, hence, lower losses. As a result, they produce reasonable draft at part load or with only one of several appliances in operation. Therefore, diversity of operation has little effect on chimney design. Some installers set each draft regulator at a slightly different setting to avoid oscillations or hunting possibly caused by burner or flow pulsations.

Calculation of the resistance coefficient of any portion of a manifold begins with the appliance most distant from the vertical portion. All coefficients are then summed from its outlet to the vent terminus. The resistance of a series of tee joints to flow passing horizontally straight through them (not making a turn) is the same as that of an equal length of piping (as if all other appliances were off). This assumption holds whether the manifold is tapered (to accommodate increasing input) or of a constant size large enough for the accumulated input.

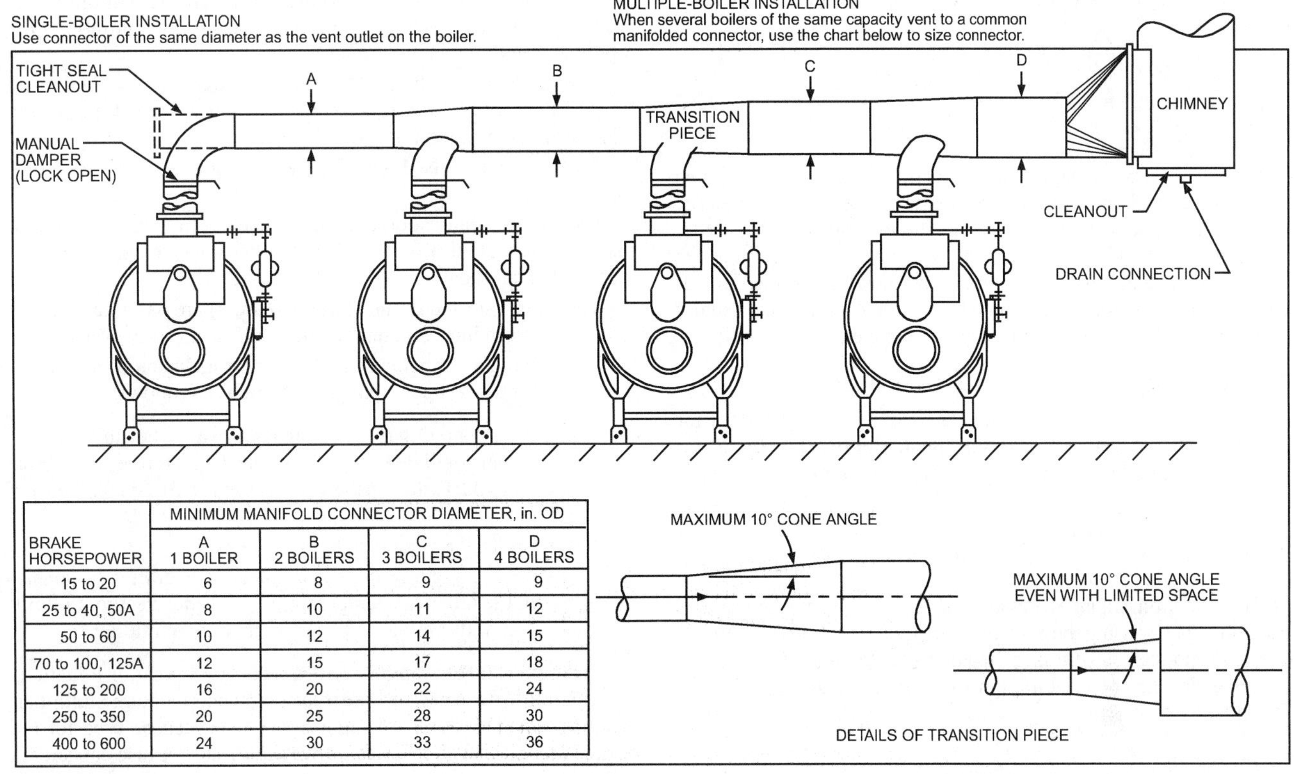

BRAKE HORSEPOWER	MINIMUM MANIFOLD CONNECTOR DIAMETER, in. OD			
	A 1 BOILER	B 2 BOILERS	C 3 BOILERS	D 4 BOILERS
15 to 20	6	8	9	9
25 to 40, 50A	8	10	11	12
50 to 60	10	12	14	15
70 to 100, 125A	12	15	17	18
125 to 200	16	20	22	24
250 to 350	20	25	28	30
400 to 600	24	30	33	36

Fig. 6 Connector Design

Coefficients are assigned only to inlet and exit conditions, to fittings causing turns, and to the piping running from the affected appliance to the chimney exit. Initially, piping shape (round, square, or rectangular) and function (for connectors, vertical piping, or both) are irrelevant.

Some high-pressure, high-velocity packaged boilers require special manifold design to avoid turbulent flow noise. In such cases, manufacturers' instructions usually recommend increaser Y fittings, as shown in Figure 6. The loss coefficients listed in Table 9 for standard tees and elbows are higher than necessary for long-radius elbows or Y entries. Occasionally, on appliances with high chimneys augmenting boiler outlet pressure, it may appear feasible to reduce the diameter of the vertical portion to below that recommended by the manufacturer. However, any reduction may cause turbulent noise, even though all normal design parameters have been considered.

The manufacturers' sizing recommendations shown in Figure 6 apply to the specific appliance and piping arrangement shown. The values are conservative for long-radius elbows or Y entries. Frequently, the boiler room layout forces use of additional elbows, requiring larger sizes to avoid excessive flow losses.

With the simplifying assumption that the maximum velocity of the flue gas (which exists in the smaller of the two portions) exists throughout the entire system, the design chart (Figure 7) and Equations (20) and (21) can be used to calculate the size of a vertical portion smaller in area than the manifold or of a chimney connector smaller than the vertical. This assumption leads to a conservative design because true losses in the larger area are lower than assumed. Further, if the size change is small, either as a contraction or expansion, the added loss coefficient for this transition fitting (see Table 9) is compensated for by reduced losses in the enlarged part of the system.

These comments on size changes apply more to individual than to combined systems because it is undesirable to reduce the vertical area of the combined type, and, more frequently, it is desirable to enlarge it. If an existing vertical chimney is slightly undersized for the connected load, the complete chart method must be applied to determine whether a pressure boost is needed, because size is no longer a variable.

Sectional gas appliances with two or more draft hoods do not pose any special problems if all sections fire simultaneously. In this case, the designer can treat them as a single appliance. The appliance installation instructions either specify the size of manifold for interconnecting all draft hoods or require a combined area equal to the sum of all attached draft hood outlet areas. Once the manifold has been designed and constructed, it can be connected to a properly sized chimney connector, vent, or chimney. If the connector and chimney size is computed as less than manifold size (as may be the case with a tall chimney), the operating resistance of the manifold will be lower than the sum of the assigned component coefficients because of reduced velocity.

The general rule for conservative system design, in which manifold, chimney connector, vent, or chimney are different sizes, can be stated as follows: Always assign full resistance coefficient values to all portions carrying combined flow, and determine system capacity from the smallest diameter carrying the combined flow. In addition, horizontal chimney connectors or vent connectors should pitch upward toward the stack at a minimum of 1/4 in. per foot of connector length.

8. Input, Diameter, and Temperature Relationships

To obtain a design equation in which all terms are readily defined, measured, or predetermined, the gas velocity and density terms must be eliminated. Using Equation (6) to replace ρ_m and Equation (14) to replace V in Equation (11) gives

$$\Delta p = \frac{k\rho_m V^2}{5.2(2g)} = \frac{k}{5.2(2g)}\left(\frac{T_m}{1.325B}\right)\left(\frac{144 \times 4\ IM}{3.6 \times 10^6 \pi d_i^2}\right)^2 \quad (20)$$

Rearranging to solve for I and including the values of π and $2g$ gives

$$I = 4.13 \times 10^5 \frac{d_i^2}{M}\left(\frac{\Delta pB}{kT_m}\right)^{1/2} \quad (21)$$

Solving for input using Equation (21) is a one-step process, given the diameter and configuration of the chimney. More frequently, however, input, available draft, and height are given and the diameter d_i must be found. Because system resistance is a function of the chimney diameter, a trial resistance value must be assumed to calculate a trial diameter. This method allows for a second (and usually accurate) solution for the final required diameter.

9. Volumetric Flow in Chimney or System

Volumetric flow Q may be calculated in a chimney system for which Equation (21) can be solved by solving Equation (11) for velocity at mean density (or temperature) conditions:

$$V = 18.3\sqrt{\frac{\Delta p}{k\rho_m}} \quad (22)$$

This equation can be expressed in terms of the same variables as Equation (21) by using the density value ρ_m of Equation (6) in Equation (22) and then substituting Equation (22) for velocity V. Area is expressed in terms of d_i. Multiplying area and velocity and adjusting for units,

$$Q = 5.2d_i^2\left(\frac{\Delta pT_m}{kB}\right)^{1/2} \quad (23)$$

where Q = volumetric flow rate, cfm. The volumetric flow obtained from Equation (23) is at mean gas temperature T_m and local barometric pressure B. Equations (21) and (23) do not account for the effects of heat transfer or cooling on flow, draft, or capacity.

Equation (23) is useful in the design of forced-draft and induced-draft systems because draft fans are usually specified in terms of volumetric flow rate at some standard ambient or selected gas temperature. An induced-draft fan is necessary for chimneys that are undersized, that are too low, or that must be operated with draft in the manifold under all conditions.

10. Graphical Solution of Chimney or Vent System

Figure 7 is a graphic solution for Equations (21) and (23) that is accurate enough for most problems. However, to use either the equations or the design chart, the details in the following sections should be understood so that proper choices can be made for mass flow, pressure loss, and heat transfer effects. The equations and Figure 7 are based on the fuel combustion products and temperatures in the chimney system. Figure 7 may also be used to determine flue gas velocity; the right-hand scale of grid E reads directly in velocity for any combination of flow and diameter. For example, at 10,000 cfm and 34 in. diameter (6.305 ft^2 area), the indicated velocity is about 1600 fpm. The velocity may also be calculated by dividing volumetric flow rate Q by chimney area A. The right-hand scale of grid E, Figure 7, may also be multiplied by the cfm per 1000 Btu/h to find velocity. For the same chimney design conditions, the scale velocity value of 1600 fpm is multiplied by 0.36 to yield a velocity of 576 fpm in the chimney.

STEADY-STATE CHIMNEY DESIGN GRAPHICAL SOLUTIONS

Design ambient temperature is 60°F (520°R), and all temperatures given are in terms of rise above this ambient. Thus, the 300°F line indicates a 360°F observed vent gas temperature.

Figure 1: Use sea-level theoretical draft. Theoretical draft may be corrected for altitude or reduced air density by multiplying the operating input by the factor in Table 7.

The resistance coefficient k in Figure 7 summarizes the friction loss of the entire chimney system, including piping, fittings, and configuration or interconnection factors.

Figure 2 can be used with Figure 7 to estimate mass and volumetric flow.

For ease of application and consistency with Figure 7, Table 6 lists approximate theoretical draft for typical gas temperature rises above 60°F ambient.

The equations and design chart (Figure 7) are based on the fuel combustion products and temperatures in the chimney system.

When using a temperature multiplier for inclusion of horizontal sections, it must be applied to grids A and C as follows:

1. In grid A, the entering temperature rise Δt_e must be multiplied by 0.61. For an appliance with an outlet temperature rise above ambient of 300°F, flow in the vent is based on $\Delta t_m = 300 \times 0.61 =$ 183°F rise.
2. This same 183°F rise must be used in grid C.
3. Determine p using a 183°F rise for theoretical draft to be consistent with the other two temperatures. (It is incorrect to multiply theoretical draft pressure by the temperature multiplier.)

The first trial solution for diameter, using Figure 7 or Equations (21) and (23), need not consider the cooling temperature multiplier, even for small sizes. A first approximate size can be used for the temperature multiplier for all subsequent trials because capacity is insensitive to small changes in temperature.

Neither Figure 7 nor the equations contain the same number or order of steps as the derivation; for example, a step disappears when theoretical and available drafts are combined into Δp. Similarly, the examples selected vary in their sequence of solution, depending on which parameters are known and on the need for differing answers, such as diameter for a given input, diameter versus height, or the amount of pressure boost D_b from a forced-draft fan. To compare the use of equations with using Figure 7 the following sample is provided. First the sample calculation is made using the provided tables and equations with variations in the original order of steps. It illustrates the direct solution for input, velocity, and volume. Secondly a calculation for input is shown that is derived from Figure 7. It differs from the first solution because of the chart arrangement.

Example 1. Find the input capacity (Btu/h) of a vertical, double-wall type B gas vent, 24 in. in diameter, 100 ft high at sea level. This vent is used with draft hood natural-gas-burning appliances.

Solution:

1. Mass flow from Table 2. M = 1.54 lb/1000 Btu for natural gas, if no other data are given.
2. Temperature from Table 4. Temperature rise = 300°F and T_m = 360 + 460 = 820°R for natural gas.
3. Theoretical draft from Table 6 or Equation (9). For 100 ft height at 300°F rise, D_t = 0.5 in. of water.
4. Available draft for draft hood appliances: D_a = 0.
5. Flow losses from Table 8. $\Delta p = D_t$ = 0.5; flow losses for a gravity gas vent equal theoretical draft at mean gas temperature.
6. Resistance coefficients from Table 9. For a vertical vent,

Draft hood	k_1 = 1.5
Vent cap	k_4 = 1.0
100 ft piping	k_L = 0.4(100/24) = 1.67
System total	k = 4.17

7. Solution for input.

Altitude: Sea level, B = 29.92 from Table 7. d_i = 24 in. These values are substituted into Equation (21) as follows:

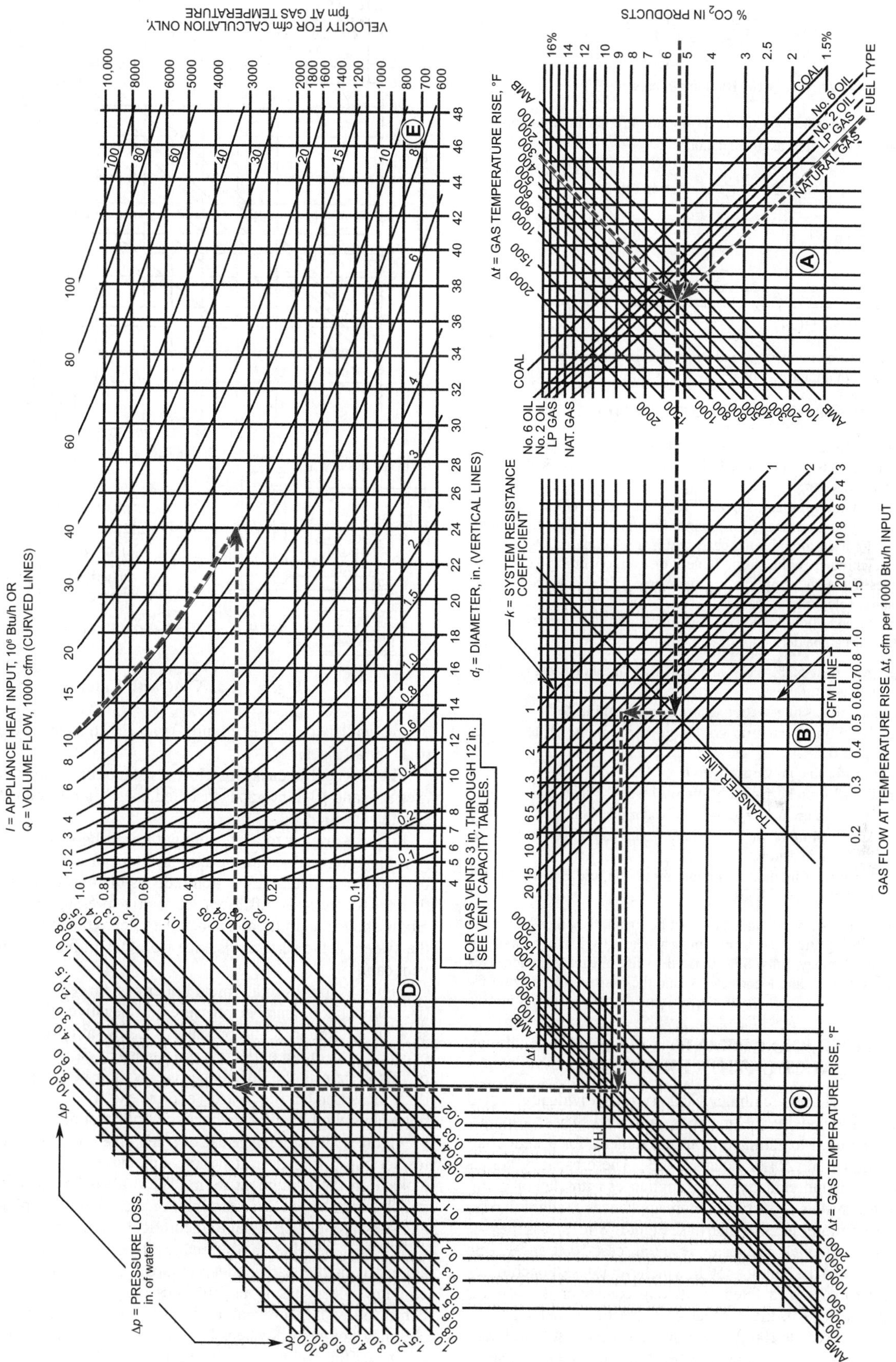

Fig. 7 Design Chart for Vents, Chimneys, and Ducts
(The dashed-line solution of Equations (21) and (23) applies to combustion products and air.)

$$I = 4.13 \times 10^5 \frac{(24)^2}{1.54}\left(\frac{(0.5)(29.92)}{(4.17)(820)}\right)^{1/2}$$

$$I = 10.2 \times 10^6 \text{ Btu/h input capacity}$$

8. A solution for velocity requires a prior solution for input to apply to Equation (14). First, using Equation (6),

$$\rho_m = 1.325\left(\frac{29.92}{820}\right) = 0.0483 \text{ lb/ft}^3$$

From Equation (14),

$$V = \frac{0.0509}{(0.0483)(24)^2} \times \frac{(10.2 \times 10^6)(1.54)}{1000} = 28.7 \text{ ft/s}$$

9. Volume flow can now be found because velocity is known. The flow area of 24 in. diameter is 3.14 ft^2, so

$$Q = (60\text{s/min})(3.14 \text{ ft}^2)(28.7 \text{ ft/s}) = 5410 \text{ cfm}$$

No heat loss correction is needed to find the new flue gas temperature because the size is greater than 20 in., and this vent is vertical with no horizontal connector.

For the same problem, Figure 7 requires a different sequence of solution. The ratio of mass flow to input for a given fuel (with parameter M) is not used directly; the chart requires selecting a CO_2 percentage in the chimney, either from Table 2 or from operating data on the appliances. Then, the temperatures are entered only as rise above ambient. Finally, calculated or estimated resistance coefficients are used to connect the conditions to the vent diameter. At this intersection, an input or flow can be derived. The solution path is as follows:

1. Enter grid A at 5.3% CO_2, and construct line horizontally to left intersecting natural gas.
2. From natural gas intersection, construct vertical line to Δt = 300°F.
3. From 300°F intersection in grid A, go horizontally left to transfer line in grid B.
4. From transfer line go vertically to k = 4.17.
5. From k = 4.17 run horizontally left to Δt = 300°F in grid C.
6. From 300°F go up to Δp = 0.54 in. of water in grid D.
7. From Δp = 0.54 in. of water in grid D, go horizontally right to intersect d_i = 24 in. in grid E.
8. Read capacity or input at 24 in. intersection in grid E as 10.2 × 10^6 Btu/h between curved lines.

If input is known and diameter must be found, the procedure is the same as with Equation (21). A preliminary k, usually 5.0 for an individual vent or chimney, must be estimated to find a trial diameter. This diameter is used to find a corrected k, and the chart is solved again for diameter.

VENT AND CHIMNEY CAPACITY CALCULATION EXAMPLES

Figures 8 to 11 show chimney capacity for individually vented appliances computed by the methods presented. These capacity curves may be used to estimate input or diameter for design chart (Figure 7), Equation (21) or Equation (23). These capacity curves apply primarily to individually vented appliances with a lateral chimney connector; systems with two or more appliances or additional fittings require a more detailed analysis. Figures 8 to 11 assume the length of the horizontal connector is (1) at least 10 ft and (2) no longer than 50% of the height or 50 ft, whichever is less. For chimney heights of 10 to 20 ft, a fixed 10 ft long connector is assumed. Between 20 and 100 ft, the connector is 50% of the height. If the chimney height exceeds 100 ft, the connector is fixed at 50 ft long.

For a chimney of similar configuration but with a shorter connector, the size indicated in the figures is slightly larger than necessary. In deriving the data for Figures 8 to 11, additional conservative assumptions were used, including the temperature correction C_u for double-wall laterals (see Figure 3) and a constant friction factor (0.4) for all sizes.

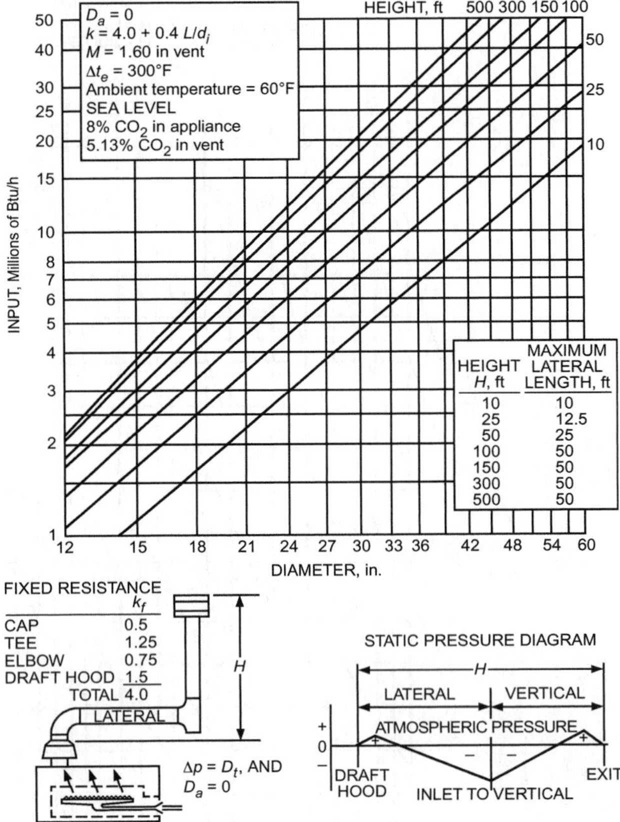

Fig. 8 Gas Vent with Lateral

The loss coefficient k_4 for a low resistance cap is included in Figures 8, 9, and 10. If no cap is installed, these figures indicate a larger size than needed.

Figure 8 applies to a gas vent with draft hood and a lateral that runs to the vertical section. Maximum static draft is developed at the base of the vertical, but friction reduces the observed value to less than the theoretical draft. Areas of positive pressure may exist at the elbow above the draft hood and at the inlet to the cap. The height of the system is the vertical distance from the draft hood outlet to the vent cap.

Figure 9 applies to a typical boiler system requiring both negative combustion chamber pressure and negative static outlet pressure. The chimney static pressure is below atmospheric pressure, except for the minor outlet reversal caused by cap resistance. Height of this system is the difference in elevation between the point of draft measurement (or control) and the exit. (Chimney draft should not be based on the height above the boiler room floor.)

Figure 10 illustrates the use of a negative static pressure connector serving a forced-draft boiler. This system minimizes flue gas leakage in the equipment room. The draft is balanced or neutral, which is similar to a gas vent, with zero draft at the appliance outlet and pressure loss Δp equal to theoretical draft.

Figure 11 applies to a forced-draft boiler capable of operating against a positive static outlet pressure of up to 0.50 in. of water. The chimney system has no negative pressure, so outlet pressure may be combined with theoretical draft to get minimum chimney size. For chimney heights or system lengths less than 100 ft, the effect of adding 0.50 in. of water positive pressure to theoretical draft causes all curves to fall into a compressed zone. An appliance that can produce 0.50 in. of water positive forced draft (negative draft) is adequate for venting any simple arrangement with up to 100 ft of

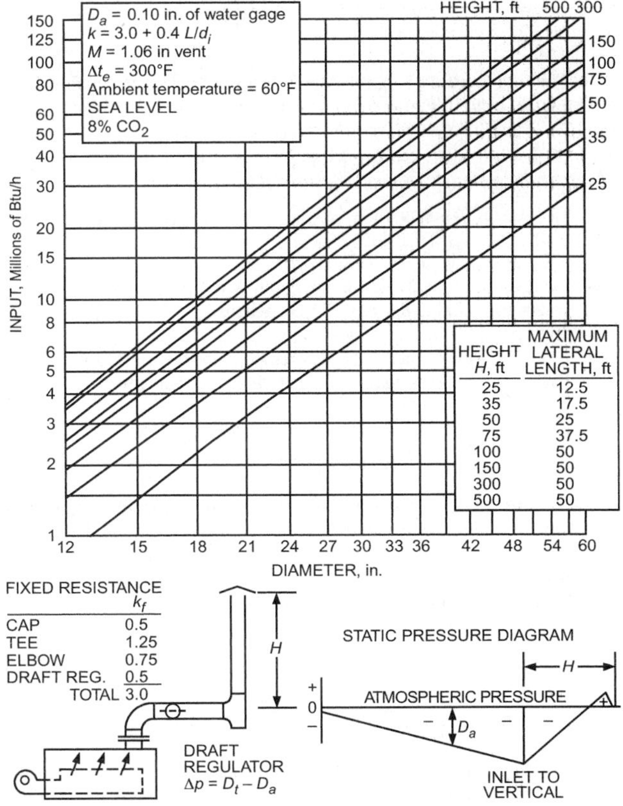

Fig. 9 Draft-Regulated Appliance with 0.10 in. of water Available Draft Required

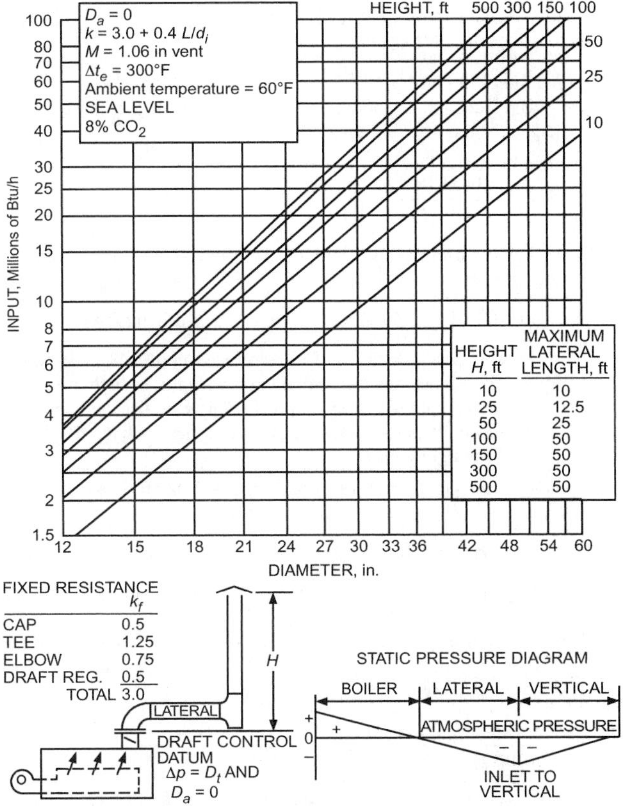

Fig. 10 Forced-Draft Appliance with Neutral (Zero) Draft (Negative Pressure Lateral)

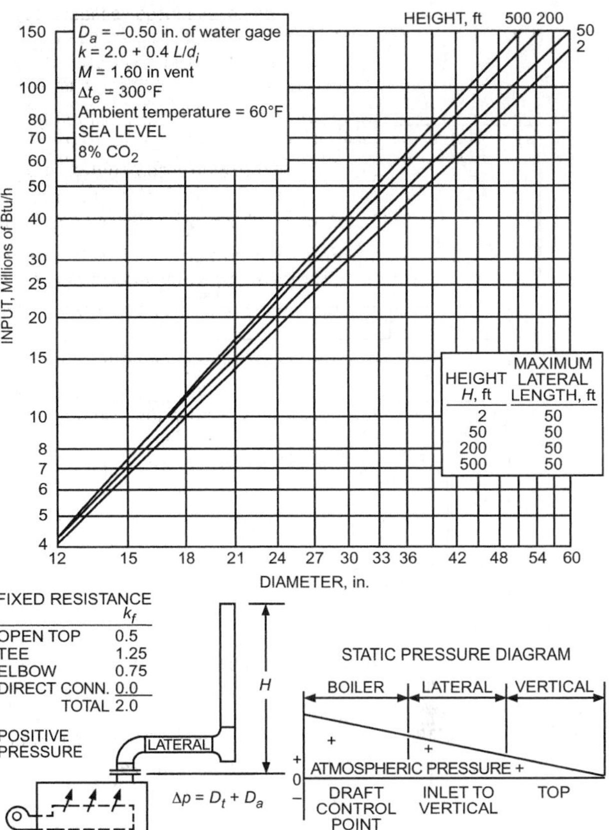

Fig. 11 Forced-Draft Appliance with Positive Outlet Pressure (Negative Draft)

flow path and no wind back pressure, for which additional forced draft is required.

The following examples illustrate the use of the design chart (Figure 7) and the corresponding equations.

Example 2. Individual gas appliance with draft hood (see Figure 12). The natural gas appliance is located at sea level and has an input of 980,000 Btu/h. The double-wall vent is 80 ft high with 40 ft lateral. Find the vent diameter.

Solution: Assume $k = 5.0$. The following factors are used successively in grids A through D of Figure 7: CO_2 = 5.3% for natural gas; flue gas temperature rise = 300°F; Transfer line: $k = 5.0$; $\Delta p = 0.537(80/100) = 0.43$. See Example 1 for the solution path.

Preliminary solution: From grid E at $I = 0.98 \times 10^6$, read diameter $d_i = 8.5$ in. Use next largest diameter, 9 in., to correct temperature rise and theoretical draft; compute new system k and Δp. From Figure 3, $C_u = 0.7$, $\Delta t = (0.7)(300) = 210$°F. This temperature rise determines the new value of theoretical draft, found by interpolation between 200 and 300°F in Table 6: D_t = (0.41 in. of water/100 ft)(80 ft) = 0.33 in. of water. The four fittings have a total fixed $k_f = 4.0$. Adding k_L, the piping component for 120 ft of 9 in. diameter, to k_f gives $k = 4.0 + 0.4(120/9) = 9.3$. System losses $\Delta p = D_t = 0.33$ in. of water.

Final solution: Returning to Figure 7, the factors are CO_2 = 5.3% for natural gas; gas temperature rise = 210°F; Transfer line: $k = 9.3$; $\Delta p = 0.33$ in. of water. At $I = 0.98 \times 10^6$, read $d_i = 10$ in., which is the correct answer. Had system resistance been found using 10 in. rather than 9 in., the final size would be less than 10 in. based on a system k of less than 9.3.

Example 3. Gravity incinerator chimney (see Figure 13).

Located at 8000 ft elevation, the appliance burns 600 lb/h of type 0 waste with 100% excess air at 1400°F outlet temperature. Ambient temperature T_o is 60°F. Outlet pressure is zero at low fire, +0.10 in. of water at high fire. The chimney will be a prefabricated medium-heat type with a 60 ft connector and a roughness factor of 1.2. The incinerator outlet is

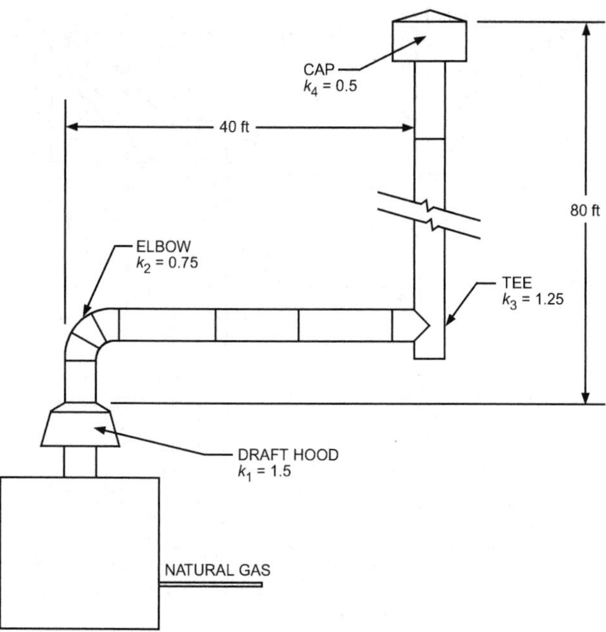

Fig. 12 Illustration for Example 2

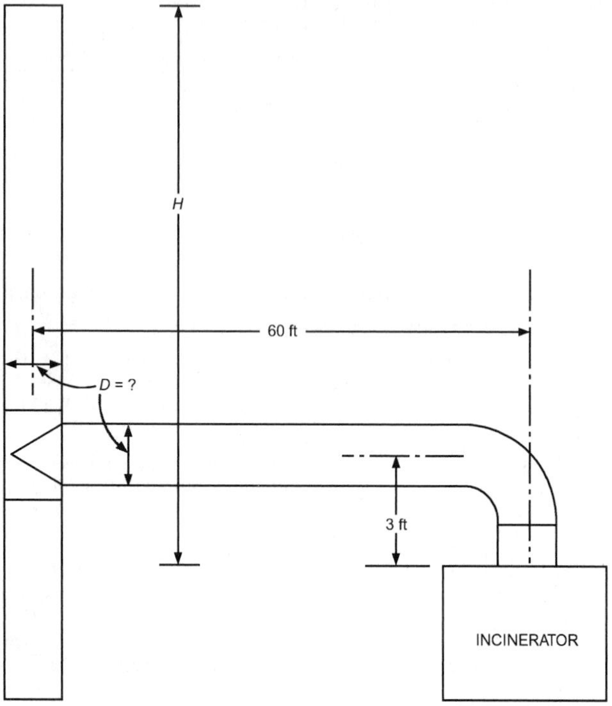

Fig. 13 Illustration for Example 3

18 in. in diameter, and it normally uses a 20 ft vertical chimney. Find the diameter of the chimney and the connector and the height required to overcome flow and fitting losses.

Solution:

1. Find mass flow from Table 3 as 13.76 lb combustion products per pound of waste, or $w = 600(13.76) = 8256$ total lb/h.
2. Find mean chimney gas temperature. Based on 60 ft length of 18 in. diameter double-wall chimney, $C_u = 0.83$ (see Figure 3). Temperature rise $\Delta t_e = 1400 - 60 = 1340$°F; thus, $\Delta t_m = 1340(0.83) = 1112$°F rise above 60°F ambient. $T_m = 1112 + 60 + 460 = 1632$°R. Use this temperature in Equation (6) to find flue gas density at 8000 ft elevation (from Table 7, $B = 22.22$ in. Hg):

$$\rho_m = 1.325\left(\frac{22.22}{1632}\right) = 0.0180 \text{ lb/ft}^3$$

3. Find the required height by finding theoretical draft per foot from Equation (9) or Figure 5 (Table 6 applies only to sea level).

$$\frac{D_t}{H} = (0.2554)(22.22)\left(\frac{1}{520} - \frac{1}{1632}\right)$$

$$= 0.0074 \text{ in. of water per ft of height}$$

4. Find allowable pressure loss Δp in the incinerator chimney for a positive-pressure appliance having an outlet pressure of +0.1 in. of water. From Table 8, $\Delta p = D_t + D_a$, where $D_t = 0.0074H$, $D_a = 0.10$, and $\Delta p = 0.0074H + 0.1$ in. of water.
5. Calculate flow velocity at mean temperature from Equation (13) to balance flow losses against diameter/height combinations:

$$V = \frac{(0.0509)(8256)}{(0.0180)(18)^2} = 72 \text{ fps}$$

This velocity exceeds the capability of a gravity chimney of moderate height and may require a draft inducer if an 18 in. chimney must be used. Verify velocity by calculating resistance and flow losses by the following steps.

6. From Table 9, resistance coefficients for fittings are

1 Tee	$k_3 = 1.25$
1 Elbow	$k_2 = 0.75$
Spark screen	$k_4 = 0.50$
Fitting total	$k_f = 2.50$

The piping resistance, adjusted for length, diameter, and a roughness factor of 1.2, must be added to the total fitting resistance. From Figure 6, find the friction factor F at 18 in. diameter and 72 fps as 0.22. Assuming 20 ft of height with a 60 ft lateral, piping friction loss is

$$k_L = \frac{1.2FL}{d_i} = \frac{(1.2)(0.22)(60 + 20)}{18} = 1.17$$

and total $k = 2.50 + 1.17 = 3.67$

Use Equation (11) to find Δp, to determine whether this chimney height and diameter are suitable.

$$\Delta p = \frac{(3.67)(0.0180)(72)^2}{(5.2)(64.4)} = 1.02 \text{ in. of water flow loss}$$

For these operating conditions, theoretical draft plus available draft yields

$$\Delta p = 0.0074(20) + 0.1 = 0.248 \text{ in. of water driving force}$$

Flow losses of 1.02 in. of water exceed the 0.248 in. of water driving force; thus, the selected diameter, height, or both are incorrect, and this chimney will not work. This can also be shown by comparing draft per foot with flow losses per foot for the 18 in. diameter configuration:

Flow losses per foot of 18 in. chimney = 1.02/80 = 0.0128 in. of water

Draft per foot of height = 0.0074 in. of water

Regardless of how high the chimney is, losses caused by a 72 fps velocity build up faster than draft.

7. A draft inducer could be selected to make up the difference between losses of 1.02 in. of water and the 0.248 in. of water driving force. Operating requirements are

$$Q = \frac{w}{(60\rho_m)} = \frac{8256}{(60 \times 0.0180)} = 7644 \text{ cfm}$$

$$\Delta p = 1.02 - 0.248 = 0.772 \text{ in. of water at 7644 cfm and 1112°F rise}$$

If the inducer selected (see Figure 27C) injects single or multiple air jets into the gas stream, it should be placed only at the chimney top or outlet. This location requires no compensation for additional air introduced by an enlargement downstream from the inducer.

Because 18 in. is too small, assume that a 24 in. diameter may work at a 20 ft height and recalculate with the new diameter.

1. As before, $w = 8256$ lb/h.
2. At 24 in. diameter, no temperature correction is needed for the 60 ft connector. Thus, $T_m = 1400 + 460 = 1860°R$ (see Table 4), and density is

$$\rho_m = 1.325\left(\frac{22.22}{1860}\right) = 0.0158 \text{ lb/ft}^3$$

3. Theoretical draft per foot of chimney height is

$$\frac{D_t}{H} = 0.2554(22.22)\left(\frac{1}{520} - \frac{1}{1860}\right) = 0.00786 \text{ in. of water per ft}$$

4. Velocity is

$$V = \frac{(0.0509)(8256)}{(0.0158)(24^2)} = 46.2 \text{ fps}$$

5. From Figure 6, the friction factor is 0.225, which, when multiplied by a roughness factor of 1.2 for the piping used, becomes 0.27. For the entire system with $k_f = 2.50$ and 80 ft of piping, find $k = 2.5 + 0.27(80/24) = 3.4$. From Equation (11),

$$\Delta p = \frac{(3.4)(0.0158)(46.2)^2}{(5.2)(64.4)} = 0.342 \text{ in. of water flow losses}$$

$D_t + D_a = (20)(0.00786) + 0.1 = 0.257$ in. of water, so driving force is less than losses. The small difference indicates that, although 20 ft of height is insufficient, additional height may solve the problem. The added height must make up for 0.085 in. of water additional draft. As a first approximation,

$$\begin{aligned} \text{Added height} &= \text{Additional draft/draft per foot} \\ &= 0.085/0.00786 = 10.8 \text{ ft} \\ \text{Total height} &= 20 + 10.8 = 30.8 \text{ ft} \end{aligned}$$

This is less than the actual height needed because resistance changes have not been included. For an exact solution for height, the driving force can be equated to flow losses as a function of H. Substituting $H = +60$ for L in Equation (17) to find k for Equation (11), the complete equations are

$$k_L = \frac{(1.2 \times 0.225 \times 60)}{24} = 0.675 \text{ connector}$$

$$k_L = \frac{(1.2 \times 0.225 \times H)}{24} = 0.01125H \text{ chimney}$$

$$k = 2.5 + 0.675 + 0.01125H = 3.175 + 0.01125H$$

$$0.10 + 0.00786H = \frac{(3.175 + 0.01125H)(0.0158)(46.2)^2}{(5.2)(64.4)}$$

$H = 32.66$ ft at 24 in. diameter.

Checking by substitution, total driving force = 0.357 in. of water and total losses, based on a system with 92.66 ft of piping, equal 0.357 in. of water.

The value of $H = 32.66$ ft is the minimum necessary for proper system operation. Because of the great variation in fuels and firing rate with incinerators, greater height should be used to ensure adequate draft and combustion control. An acceptable height would be from 40 to 50 ft.

Example 4. Two forced-draft boilers (see Figure 14).

This example shows how multiple-appliance chimneys can be separated into subsystems. Each boiler is rated 100 boiler horsepower on No. 2 oil. The manufacturer states flue gas operation at 13.5% CO_2 and 300°F temperature rise against 0.50 in. positive static pressure at the outlet. The 50 ft high chimney has a 20 ft single-wall manifold and is at sea level. Find the size of connectors, manifold, and vertical.

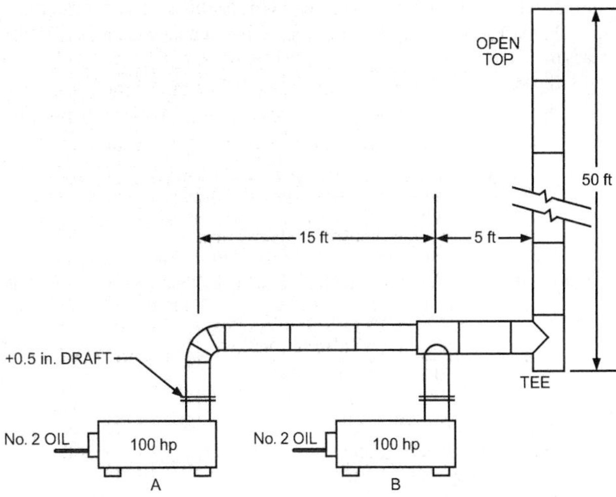

Fig. 14 Illustration for Example 4

Solution: First, find the capacity or size of the piping and fittings from boiler A to the tee over boiler B. Then, size the boiler B tee and all subsequent portions to carry the combined flow of A and B. Also, check the subsystem for boiler B; however, because its shorter length compensates for greater fitting resistance, its connector may be the same size as for boiler A.

Find the size for combined flow of A and B either by assuming $k = 5.0$ or by estimating that the size will be twice that found for boiler A operating by itself. Estimate system resistance for the combined portion by including those fittings in the B connector with those in the combined portion.

Data needed for the solution of Equation (21) for boiler A for No. 2 oil at 13.5% CO_2 include the following:

$$M = 0.72\left(0.12 + \frac{14.4}{13.5}\right) = 0.854 \text{ lb/1000 Btu} \qquad \text{(from Table 1)}$$

For a temperature rise of 300°F and an ambient of 60°F,

$$T_m = 300 + 60 + 460 = 820°R$$

From Table 6, theoretical draft = 0.5 in. of water for 100 ft of height; for 50 ft height, $D_t = (0.5)50/100 = 0.25$ in. of water.

Using $D_a = 0.5$ in. of water, $\Delta p = 0.25 + 0.5 = 0.75$ in. of water.

Assume $k = 5.0$. Assuming 80% efficiency, input is 41,800 times boiler horsepower (see the section on Conversion Factors):

$$I = 100(41{,}800) = 4.18 \times 10^6 \text{ Btu/h}$$

Substitute in Equation (21):

$$I = 4.18 \times 10^6 = 4.13 \times 10^5 \frac{(d_i)^2}{0.854}\left[\frac{(0.75)(29.92)}{(5.0)(820)}\right]^{0.5}$$

Solving, $d_i = 10.81$ in. as a first approximation. From Table 9, find correct k using next largest diameter, or 12 in.:

Inlet acceleration (direct connection)	k_1 =	0.0
90° Elbow	k_2 =	0.75
Tee	k_3 =	1.25
70 ft piping	k_L =	0.4(70/12) = 2.33
System total	k =	4.33

Note: Assume tee over boiler B has $k = 0$ in subsystem A.

Corrected temperature rise (see Figure 3 for 20 ft single-wall connector) = 0.75(300) = 225°F, and $T_m = 225 + 60 + 460 = 745°R$. This corrected temperature rise changes D_t to 0.425 per 100 ft [Table 6 or Equation (9)], or 0.21 for 50 ft. Thus, Δp becomes 0.21 + 0.50 = 0.71 in. of water.

$$4.18 \times 10^6 = 4.13 \times 10^5 \frac{(d_i)^2}{0.854}\left[\frac{(0.71)(29.92)}{(4.33)(745)}\right]^{0.5}$$

So d_i = 10.35 in., or a 12 in. diameter is adequate.

For size of manifold and vertical, starting with the tee over boiler B, assume 16 in. diameter (see also Figure 11).

System k = 3.9 for the 55 ft of piping from B to outlet:

Inlet acceleration	k_1 = 0.0
Two tees (boiler B subsystem)	k_3 = 2.5
55 ft piping	k_L = 0.4(55/16) = 1.4
System total	k = 3.9

Temperature and Δp will be as corrected (a conservative assumption) in the second step for boiler A. Having assumed a size, find input.

$$I = 4.13 \times 10^5 \frac{16^2}{0.854}\left[\frac{(0.71)(29.92)}{(3.9)(745)}\right]^{0.5} = 10.59 \times 10^6 \text{ Btu/h}$$

A 16 in. diameter manifold and vertical is more than adequate. Solving for the diameter at the combined input, d_i = 14.2 in.; thus, a 15 or 16 in. chimney must be used.

Note: Regardless of calculations, do not use connectors smaller than the appliance outlet size in any combined system. Applying the temperature correction for a single-wall connector has little effect on the result because positive forced draft is the predominant motive force for this system.

Example 5. Six gas boilers manifolded at 6000 ft elevation (see Figure 15). Each boiler is fired at 1.6 × 10^6 Btu/h, with draft hoods and an 80 ft long manifold connecting into a 400 ft high vertical. Each boiler is controlled individually. Find the size of the constant-diameter manifold, vertical, and connectors with a 2 ft rise. All are double-wall.

Solution: Simultaneous operation determines both the vertical and manifold sizes. Assume the same appliance operating conditions as in Example 1: CO_2 = 5.3%, natural gas, flue gas temperature rise = 300°F. Initially assume k = 5.0 is multiplied by 1.5 for combined vent (see note at bottom of Table 9); thus, design k = 7.5. For a gas vent at 400 ft height, $\Delta p = D_t = H \times D_t/m = 4 \times 0.5 = 2.0$ in. of water at rise 300°F in Table 6; D_t = 0.5 in. of water/ft. At 6000 ft elevation, operating input must be multiplied by an altitude correction (Table 7) of 1.25. Total design input is 1.6 × 10^6(6)(1.25) = 12 × 10^6 Btu/h. From Table 2, M = 1.54 lb/Btu at operating conditions. T_m = 300 + 60 + 460 = 820°R, and B = 29.92 in. Hg because the 1.25 input multiplier corrects back to sea level. From Equation (21),

$$12 \times 10^6 = 4.13 \times 10^5 \frac{(d_i)^2}{1.54}\left(\frac{2.0 \times 29.92}{7.5 \times 820}\right)^{0.5}$$

$$d_i = 21.3 \text{ in.}$$

Because the diameter is greater than 20 in., no temperature correction is needed.

Recompute k for 400 + 80 = 480 ft of 22 in. diameter (Table 9):

Draft hood inlet acceleration	k_1 = 1.5
Two tees (connector and base of chimney)	$2k_3$ = 2.5
Low-resistance top	k_4 = 0.5
480 ft piping	k_L = 0.4(480/22) = 8.7
System total	k = 13.2

Combined gas vent design k = multiple vent factor 1.5 × 13.2 = 19.8. Substitute again in Equation (21):

$$12 \times 10^6 = 4.13 \times 10^5 \frac{(d_i)^2}{1.54}\left(\frac{2.0 \times 29.92}{19.8 \times 820}\right)^{0.5}$$

d_i = 27.1 in.; thus use 28 in.

$$k_f = 4.5$$
$$k_L = 0.4(480/28) = 6.9$$
$$11.4$$

Design k = 11.4(1.5) = 17.1

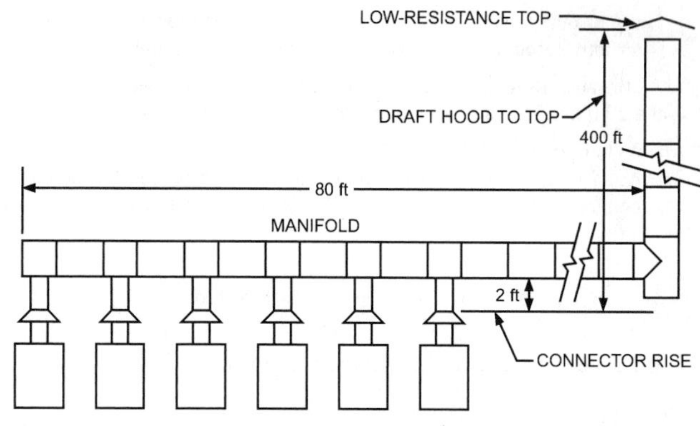

Fig. 15 Illustration for Example 5

Substitute in Equation (21) to obtain the third trial:

$$12 \times 10^6 = 4.13 \times 10^5 \frac{(d_i)^2}{1.54}\left(\frac{2.0 \times 29.92}{17.1 \times 820}\right)^{0.5}$$

$$d_i = 26.2 \text{ in.}$$

The third trial is less than the second and again shows the manifold and vertical chimney diameter to be between 26 and 28 in.

For connector size, see the *National Fuel Gas Code* for double-wall connectors of combined vents. The height limit of the table is 100 ft; do not extrapolate and read the capacity of 18 in. connector as 1,740,000 Btu/h at 2 ft rise. Use 18 in. connector or draft hood outlet, whichever is larger. No altitude correction is needed for connector size; the draft hood outlet size considers this effect.

Note: Equation (21) can also be solved at local elevation for exact operating conditions. At 6000 ft, the local barometric pressure is 23.98 in. Hg (Table 7), and assumed theoretical draft must be corrected in proportion to the reduction in pressure:

D_t = 2.0(23.98/29.92) = 1.60 in. of water. Operating input of 6(1.6 × 10^6) = 9.6 × 10^6 Btu/h is used to find d_i, again taking final k = 17.1:

$$9.6 \times 10^6 = 4.13 \times 10^5 \frac{(d_i)^2}{1.54}\left(\frac{1.60 \times 23.98}{17.1 \times 820}\right)^{0.5}$$

d_i = 26.2 (same as above)

This example illustrates the equivalence of the chart method of solution with solution by Equation (21). Equation (21) gives the correct solution using either method 1, with only the fuel input corrected back to sea level condition, or method 2, correcting Δp for local barometric pressure and using operating input at altitude. Method 1, correcting input only, is the only choice with Figure 1 because the design chart cannot correct to local barometric pressure.

Example 6. Pressure boost for undersized chimney (not illustrated).

A natural gas boiler at sea level (no draft hood) is connected to an existing 12 in. diameter chimney. Input is 4 × 10^6 Btu/h with flue gas at 10% CO_2 and 300°F temperature rise above ambient. System resistance loss coefficient k = 5.0 with 20 ft vertical chimney. The appliance operates with neutral outlet static draft, so, D_a = 0 in. of water.

a. How much draft boost is needed at operating temperature?
b. What fan rating is required at 60°F ambient temperature?
c. Where in the system should the fan be located?

Solution: Combustion data: from Figure 2, 10% CO_2 at 300°F temperature rise indicates 0.31 cfm per 1000 Btu/h.

Total flow rate Q = (0.31/1000)(4 × 10^6) = 1240 cfm at chimney gas temperature. Then, Equation (23) can be solved for the only unknown, Δp:

$$1240 = 5.2(12)^2\left[\frac{\Delta p(300 + 60 + 460)}{(5.0)(29.92)}\right]^{0.5}$$

Δp = 0.50 in. of water needed at 300°F temperature rise. For 20 ft of height at 300°F temperature rise [Table 6 or Equation (9)],

D_t = 0.5(20/100) = 0.10 in. of water

a. Pressure boost D_b supplied by the fan must equal Δp minus theoretical draft (Table 8) when available draft is zero. $D_b = \Delta p - D_t$ = 0.5 – 0.10 = 0.40 in. of water at operating temperature.

b. Draft fans are usually rated for standard ambient (60°F) air. Pressure is inversely proportional to absolute flue gas temperature. Thus, for ambient air,

$$D_b = 0.40\frac{T_m}{T_o} = 0.40\left(\frac{300+60+460}{60+460}\right) = 0.63 \text{ in. of water}$$

This pressure is needed to produce 0.40 in. of water at operating temperature. In specifying power ratings for draft fan motors, a safe policy is to select one that operates at the required flow rate at ambient temperature and pressure (see Example 7).

c. A fan can be located anywhere from boiler outlet to chimney outlet. Regardless of location, the amount of boost is the same; however, chimney pressure relative to atmosphere will change. At boiler outlet, the fan pressurizes the entire connector and chimney. Thus, the system should be gastight to avoid leaks. At the chimney outlet, the system is below atmospheric pressure; any leaks flow into the system and seldom cause problems. With an ordinary sheet metal connector attached to a tight vertical chimney, the fan may be placed close to the vertical chimney inlet. Thus, the connector leaks safely inward, while the vertical chimney is under pressure.

Example 7. Draft inducer selection (see Figure 16).

A third gas boiler must be added to a two-boiler system at sea level with an 18 in. diameter, 15 ft horizontal, and 75 ft of total height of connector and chimney system. Outlet conditions for natural gas draft hood appliances are 5.3% CO_2 at 300°F temperature rise. Boilers are controlled individually, each with 1.6 × 10^6 Btu/h, for 4.8 × 10^6 Btu/h total input. The system is currently undersized for gravity full-load operation. Find capacity, pressure, size, and power rating of a draft inducer fan installed at the outlet.

Solution: Using Equation (23) requires evaluating two operating conditions: (1) full input at 300°F rise and (2) no input with ambient air. Because the boilers are controlled individually, the system may operate at nearly ambient temperature (100°F flue gas temperature rise or less) when only one boiler operates at part load. Use the system resistance *k* for boiler 3 as the system value for simultaneous operation. It needs no compensating increased draft, as with gravity multiple venting, because a fan induces flow at all flue gas temperatures. From Table 9, the resistance summation is

Inlet acceleration (draft hoods)	k_1 = 1.50
Tee above boiler	k_3 = 1.25
Tee at base of vertical	k_3 = 1.25
90 ft of 18 in. diameter pipe	k_L = 0.4(90/18)= 2.00
System total	*k* = 6.00

At full load, T_m = 300 + 60 + 460 = 820°R; *B* = 29.92 in. Hg. Flow rate *Q* must be found for operating conditions of 1.54 lb per 1000 Btu (Table 2) at density ρ_m and full input (4.8 × 10^6)/60 Btu/min.

From Equation (6),

$$\rho_m = 1.325B/T_m = 1.325(29.92/820) = 0.0483 \text{ lb/ft}^3$$

Volumetric flow rate is

$$Q = \left(\frac{1.54}{1000}\right)\left(\frac{1}{0.0483}\right)\left(\frac{4.8\times10^6}{60}\right) = 2551 \text{ cfm}$$

For 300°F flue gas temperature rise, D_t = 0.5(75/100) = 0.375 in. of water theoretical draft in the system (Table 6). Solving Equation (23) for Δp,

$$\Delta p = \frac{(6.00)(29.92)(2551)^2}{(5.2)^2(18)^4(820)} = 0.502 \text{ in. of water}$$

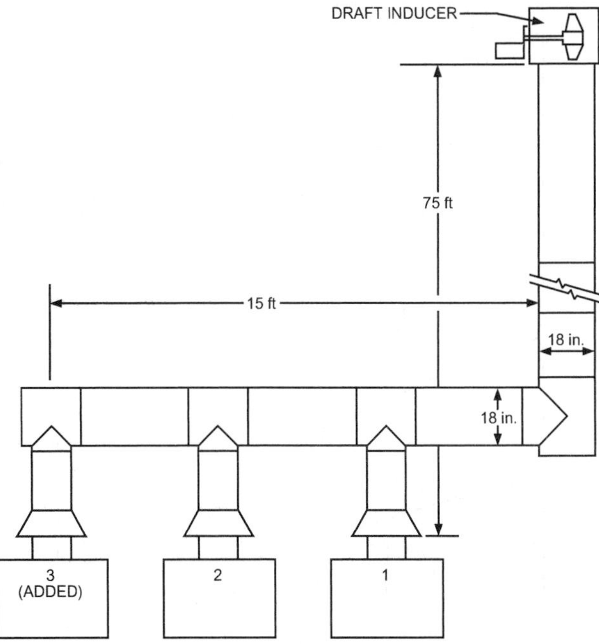

Fig. 16 Illustration for Example 7

Thus, a fan is needed because Δp exceeds D_t. Required static pressure boost (Table 8) is

D_b = 0.502 – 0.375 = 0.127 in. of water at 300°F flue gas temperature rise or a flue gas density of 0.0483 lb/ft^3

Fans are rated for ambient or standard air (60 to 70°F) conditions. Flue gas pressure is directly proportional to density or inversely proportional to absolute temperature. Moving 2551 cfm of flue gas at 300°F temperature rise against 0.127 in. of water pressure requires the static pressure boost with standard air to be

$$D_b = 0.127(820/520) = 0.20 \text{ in. of water}$$

Thus, a fan that delivers 2551 cfm of flue gas at 0.20 in. of water at 60°F is required. Figure 17 shows the operating curves of a typical fan that meets this requirement. The exact volume flow rate and pressure developed against a system *k* of 6.0 can be found for this fan by plotting airflow rate versus Δp from Equation (23) on the fan curve. The solution, at point C, occurs at 1950 cfm, where both the system Δp and fan static pressure equal 0.46 in. of water.

Although some fan manufacturers' ratings are given at standard air conditions, the motors selected will be overloaded at temperatures below 300°F air temperature rise. Figure 17 shows that power required for two conditions with ambient air is as follows:

1. 1950 cfm at 0.46 in. of water static pressure requires 0.51 hp
2. 2650 cfm at 0.25 in. of water static pressure requires 0.50 hp

Thus, the minimum size motor will be 0.5 hp and run at 1590 rpm.

Manufacturers' literature must be analyzed carefully to discover whether the sizing and selection method is consistent with appliance and chimney operating conditions. Final selection requires both a thorough analysis of fan and system interrelationships and consultation with the fan manufacturer to verify the fan and motor capacity and power ratings.

GAS APPLIANCE VENTING

In much of the United States, gas-burning appliances requiring venting of combustion products are installed and vented in accordance with the *National Fuel Gas Code* (ANSI/NFPA 54/ ANSI/AGA Z223.1), and in Canada with the *Natural Gas and Propane Installation Code* (CAN/CSA *Standard* B149.1). This

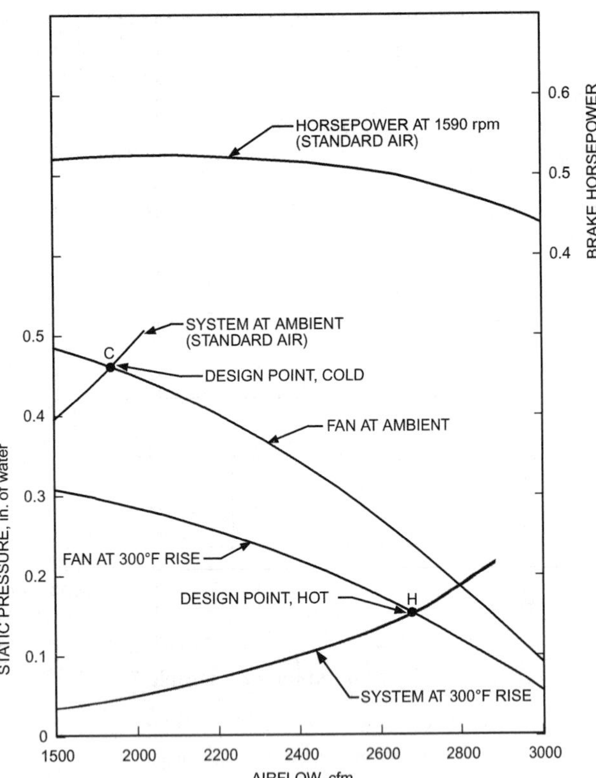

Fig. 17 Typical Fan Operating Data and System Curves

standard includes capacity data and definitions for commonly used gas vent systems.

Traditionally, gas appliances were designed with a draft hood or draft diverter and depended on natural buoyancy to vent products of combustion. The operating characteristics of many different types of appliances were similar, allowing generic venting guidelines to be applied to any gas appliance. These guidelines consisted of tables of maximum capacities, representing the largest appliance input rating that could safely be vented using a certain vent diameter, height, lateral, and material.

Small gas-fired appliance design changed significantly, however, with the increase of furnace minimum efficiency requirements to 78% annual fuel utilization efficiency (AFUE) isolated combustion system (ICS) imposed by the U.S. National Appliance Energy Conservation Act (NAECA) of 1987. Many manufacturers developed appliances with fan-assisted combustion systems (FACSs) to meet efficiency standards, which greatly changed the venting characteristics of gas appliances.

Because FACS appliances do not entrain dilution air, a FACS appliance connected to a given vent system can have a larger maximum capacity than a draft-hood-equipped appliance connected to the same vent system. The dew point of vent gases from a FACS appliance, however, remains high in the absence of dilution air, and the potential for condensation to form and remain in the vent system is much greater than with draft-hood-equipped appliances. Excessive condensate dwell time (wet time) in the vent system can cause corrosion failure of the vent system or problems with condensate runoff, either into the appliance or into the structure surrounding the vent.

New venting guidelines (*National Fuel Gas Code* and *Natural Gas and Propane Installation Code*) were developed to address the differences between draft-hood-equipped and FACS category I appliances. In addition to new maximum capacity values for FACS appliances, the guidelines include minimum capacity values for FACS appliances to ensure that vent system wet times do not reach a level that would lead to corrosion or drainage problems. Because corrosion is the principal effect of condensation in the vent connector, whereas drainage is the principal concern in the vertical portion of the vent, different wet time limits were established for the vent connector and the vertical portion of the vent.

Wet time values in the vent system were determined using the VENT-II computer program (Rutz and Paul 1991) to perform the transient analysis with the appliance(s) cycling. Gas-fired central furnace cycle times of 3.87 min on, 13.3 min off, 17.17 min total, were determined based on a design outdoor ambient of 42°F with an oversize factor of 1.7. The vent connector is required to dry before the end of the appliance *on* cycle, whereas the vertical portion of the vent is required to dry out before the end of the total cycle.

The new venting guidelines for FACS appliances severely restrict use of single-wall metal vent connectors. Also, because of excessive condensation, tile-lined masonry chimneys are not recommended in most typical installations when a FACS appliance is the only appliance connected to the venting system. FACS appliances can be used with a typical masonry chimney, however, if (1) the FACS appliance is common-vented with a draft-hood-equipped appliance or (2) a liner listed for use with gas appliances is installed in the masonry chimney.

The *National Fuel Gas Code* (NFGC) and *Natural Gas and Propane Installation Code* list both the minimum and maximum vent capacities for FACS gas appliances and the previous maximum vent capacities for gas appliances equipped with draft hoods. NFGC sections 11.2.9 and 11.3.18 of the 1999 revision, sections 13.1.9 and 13.2.20 of the 2002 revision, and sections 13.1.11 and 13.2.23 of the 2006 revision provide criteria for permitting use of masonry chimneys (either exposed or not exposed to the outdoors below the roof line) and allow alternative venting designs and installation of vents serving listed appliances in accordance with the appliance manufacturer's instructions and the terms of listing.

Vent Connectors

Vent connectors connect gas appliances to the gas vent, chimney, or single-wall metal pipe, except when the appliance flue outlet or draft hood outlet is connected directly to one of these vent systems. Materials for vent connectors for conversion burners or other appliances without draft hoods must have resistance to corrosion and heat not less than that of galvanized 0.0276 in. (24 gage) thick sheet steel. Where a draft hood is used, the connector must have resistance to corrosion and heat not less than that of galvanized 0.0187 in. (28 gage) thick sheet steel or type B vent material.

Masonry Chimneys for Gas Appliances

A masonry chimney serving a gas-burning appliance should have a tile liner and should comply with applicable building codes such as NFPA *Standard* 211. *National Fuel Gas Code*, sections 11.2.9 and 11.3.18 of the 1999 revision, sections 13.1.9 and 13.2.20 of the 2002 revision, and sections 13.1.11 and 13.2.23 of the 2006 revision have other provisions pertaining to masonry chimneys. An additional chimney liner may be needed to avoid slow priming and/or condensation, particularly for an exposed masonry chimney with high mass and low flue gas temperature. A low-temperature chimney liner may be a single-wall passage of pure aluminum or stainless steel or a double-wall type B vent.

Type B and Type L Factory-Built Venting Systems

Factory-prefabricated vents are listed by Underwriters Laboratories for use with various types of fuel-burning appliances. These vents should be installed according to the manufacturer's instructions and the appliance's listing.

Type B gas vents are listed for vented gas-burning appliances. They should not be used for incinerators, appliances readily converted to the use of solid or liquid fuel, or combination gas-oil burning appliances. They may be used in multiple gas appliance venting systems.

Type BW gas vents are listed for vented wall furnaces certified as complying with the pertinent ANSI standard.

Type L venting systems are listed by Underwriters Laboratories in 3 to 6 in. sizes and may be used for those oil- and gas-burning appliances (primarily residential) certified or listed as suitable for this type of venting. Under the terms of the listing, a single-wall connector may be used in open accessible areas between the appliance's outlet and the type L material in a manner analogous to type B. Type L piping material is recognized in the *National Fuel Gas Code* and NFPA *Standard* 211 for certain connector uses between appliances such as domestic incinerators and chimneys.

Gas Appliances Without Draft Hoods

Figure 1 or the equations may be used to calculate chimney size for nonresidential gas appliances with the draft configurations listed as 1 and 3 at the beginning of this chapter. Draft conditions 1 and 3 (see draft conditions 1, 2, and 3 under Terminology and Table 8) for residential gas appliances, such as boilers and furnaces, may require special vent systems. The appliance test and certification standards include evaluation of manufacturers' appliance installation instructions (including the vent system) and of operating and application conditions that affect venting. The instructions must be followed strictly.

Conversion to Gas

Installation of conversion burner equipment requires evaluating for proper chimney draft and capacity by the methods indicated in this chapter or by conformance to local regulations. The physical condition and suitability of an existing chimney must be checked before it is converted from a solid or liquid fuel to gas. For masonry chimneys, local experience may indicate how well the construction will withstand the lower temperature and higher moisture content of natural or liquefied petroleum gas combustion products. The section on Masonry Chimneys for Gas Appliances has more details.

The chimney should be relined, if required, with corrosion-resistant masonry or metal to prevent deterioration. The liner must extend beyond the top of the chimney. The chimney drop-leg (bottom of the chimney) must be at least 4 in. below the bottom of the connection to the chimney. The chimney should be inspected and, if needed, cleaned. The chimney should also have a cleanout at the base.

OIL-FIRED APPLIANCE VENTING

Oil-fired appliances requiring venting of combustion products must be installed and vented in accordance with ANSI/NFPA *Standard* 31. The standard offers recommendations for metal relining of masonry chimneys.

Recommendations for minimum chimney areas for oil-fired natural-draft appliances are offered in Tables 3 and 4 in the HYDI (1989).

Implementation of the U.S. National Appliance Energy Conservation Act (NAECA) of 1987 brought attention to heating appliance efficiency and the effect of NAECA on existing chimney systems. Oil-fired appliances maintained a steady growth in efficiency since the advent of the retention-head oil burner and its broad application in both new appliances and the replacement of older burners in existing appliances. Higher appliance efficiencies brought about lower flue gas temperatures. Reduced firing rates became more common as heating appliances are more closely matched to the building heating load. Burner excess air levels also dropped, which resulted in lower mass flows through the chimney and additional reductions in the flue gas temperature. However, the improvements in overall appliance efficiency have not been accompanied by upgrades in existing chimney systems, and upgraded systems are not commonly applied in new construction. An upgrade in a vent system probably involves application of corrosion-resistant materials and/or the reduction in heat loss from the vent system to maintain draft and reduce condensation on interior surfaces of the vent system.

Condensation and Corrosion

Condensation and corrosion within the vent system are of growing concern as manufacturers of oil-fired appliances strive to improve equipment efficiencies. The conditions for condensation of the corrosive components of the flue gas produced in oil-fired appliances involve a complex interaction of the water formed in the combustion process and the sulfur trioxide formed from small quantities of sulfur in the fuel oil. The sulfur is typically less than 0.5% by mass for no. 2 fuel oil. The dew point of the two-component system (sulfuric acid and water) in the flue gas resulting from the combustion of this fuel is about 225 to 240°F. This is similar to the effect on dew point of fuel gas sulfur (see Figure 4 in Chapter 18 of the 2005 *ASHRAE Handbook—Fundamentals*). In determining the proper curve for fuel oil in that figure, a value of 18 may be used. This value applies to no. 2 fuel oil with 0.5% sulfur by mass.

The effect of post-combustion air dilution on the dew point characteristics of flue gas from oil-fired appliances is highly dependent on the presence of sulfur in the fuel. Verhoff and Banchero (1974) developed an equation relating flue gas dew point to the partial pressures of both water and sulfuric acid present in the flue gas. Predictions obtained using this equation are in good agreement with experimental data. Applying this equation to the flue gas from the combustion of no. 2 fuel oil without sulfur and with varying sulfur contents reveals that a broad range of dew point temperatures is possible.

For a fuel oil with zero sulfur and 20% excess combustion air, dew point of the combustion products is calculated as approximately 114°F, typical for the presence of water formed in combustion. With the addition of post-combustion dilution air to a level equivalent to increasing the excess air in the flue gas to 100%, the apparent water dew point is reduced to 99°F. For fuel oil with 0.25% sulfur and 20% excess combustion air, the dew point is elevated to 229°F because of formation of dilute sulfuric acid in the flue gas. With the addition of post-combustion dilution air, the dew point is 237°F at 100% excess air and 234°F at 200% excess air. The implication of these calculations is that, for combustion of fuel oil, the flue gas dew-point temperature can range from a low at the apparent water dew point (calculated with no fuel sulfur) up to some elevated dew point caused by the presence of sulfur in the fuel. With no sulfur in the fuel, adding post-combustion dilution air to the flue gas has some effect on depressing the apparent water dew-point temperature. Adding post-combustion dilution air has no significant effect on the elevated flue gas dew point when sulfur is present in the fuel.

It is difficult to meet the flue-gas-side material surface temperatures required to exceed the elevated dew point at all points in the venting system of an oil-fired appliance. Even if the dew-point temperature is not reached at these surfaces, however, a surface temperature approaching 200°F allows condensation of sulfuric acid at higher concentrations and results in lower rates of corrosion.

The condensed acid concentration is critical to the applicability of plain carbon and stainless steels in vent connectors and chimney liners. The flue-gas-side surface temperatures of conventional connectors and masonry chimney tile liners are often at or below the dew point for some portion of the burner *on* period. During cooldown (burner *off* period), these surface temperatures can drop to below the apparent water dew point and, in some cases, to ambient conditions. This is of great concern because, while system surfaces are below the apparent water dew point during burner operation, the condensed sulfuric acid is formed in concentrations well below limits acceptable for steel connectors. This can be seen from an interpretation of a sulfuric acid/water phase diagram presented by Land (1977). An estimate of the condensed liquid sulfuric acid concentration at a condensing surface temperature of 120°F, for example, shows a concentration for the condensed liquid acid of about 10 to 20%.

According to Fontana and Greene (1967), the relative corrosion of plain carbon steel rises rapidly at sulfuric acid concentrations below 65%. According to Land (1977), for the condensed liquid acid concentration to rise above 65%, the condensing surface temperature must be above 200°F. However, according to Fontana and Greene, at acid concentrations above 65%, corrosion rates increase at metal surface temperatures above 175°F. This presents the designer with a restrictive operating range for steel surfaces (i.e., between 175 and 200°F). This is a compromise that does not completely satisfy either the acid concentration or the metal temperature criterion, but should minimize the corrosion rates induced by each.

Another important phenomenon is that when vent system surfaces are at or below the apparent water dew point, a large amount of water condensation occurs on these surfaces. This condensate contains, in addition to sulfuric acid, quantities of sulfurous, nitrous, carbonic, and hydrochloric acids. Under these conditions, the corrosion rate of commonly used vent materials is severe. Koebel and Elsener (1989) found that corrosion rates increase by a factor of 10 when material temperatures on the flue gas side are allowed to fall below the apparent water dew point.

The applicability of ordinary stainless steels is very limited and generally follows that of plain carbon steel. Nickel-molybdenum and nickel-molybdenum-chromium alloys show good corrosion resistance over a wide range of sulfuric acid concentrations for surface temperatures up to 220°F. This is also true for high-silicon cast iron, used in heat exchangers for oil-fired appliances, and which might find application as a liner system for masonry and metal chimneys.

Connector and Chimney Corrosion

Water and acid condensation can each result in corrosion of the connector wall and deterioration of the chimney material. Although there is little documentation of specific failures, concern in the industry is growing. The volume of anecdotal information regarding corrosive failures is significant and well supported by findings from the heating industry.

For oil-fired appliances, the rate of acid corrosion in the connector and chimney is a function of two groups of contributing factors: combustion factors and operational factors. The sulfur content of the fuel and the percent excess combustion air are the major combustion-related factors. The frequency and duration of equipment *on* and *off* periods, draft control dilution air, and rate of heat loss from the vent system are the major operational factors. In terms of combustion, Butcher and Celebi (1993) found a direct correlation of acid deposition (condensation) rate and subsequent corrosion to fuel sulfur content and excess combustion air. In general, reductions in fuel sulfur and excess combustion air reduce the amount of sulfuric acid produced in combustion and delivered to condensing surfaces in the appliance heat exchanger and carried over into the vent system. From an operational standpoint, long equipment *on* and short *off* periods and low vent system heat loss result in shorter warm-up transients and higher end-point temperatures for surfaces exposed to the flue gas. Within the limits of frictional loss, reduced vent sizes increase flue gas velocities and vent surface temperatures.

Vent Connectors

An oil-fired appliance is commonly connected to a chimney through a connector pipe. Generally, a draft control device in the form of a barometric damper is included as a component part of the connector assembly. With the advent of new power burners having high static pressure capability, draft control devices in the vent system have become less important, although many local codes still require their use. The portion of the connector assembly between the appliance flue collar and the draft control is called the flue connector; the portion between the draft control and the chimney is called the stack or chimney connector.

Chimney connectors are usually of single-wall galvanized steel. The required wall thickness for these connectors varies as a function of pipe diameter. For example, in accordance with Table 5-2.2.3 in NFPA *Standard* 211, the material thickness for galvanized steel pipe connectors between 6 and 10 in. in diameter is set at 0.023 in. (24 gage).

In accordance with the 2006 *National Fuel Gas Code*, Paragraph 12.11.2.4(2), vent, chimney, stack, and flue connectors should not be covered with insulation except listed insulated connectors installed according to the terms of their listing. Because single-wall connectors must remain uninsulated for inspection, substantial cooling of the connector wall and the flue gas can occur, especially with long connector runs through spaces with low ambient temperatures. Close examination of the connector joints, seams, and surfaces is essential whenever the heating appliance is serviced. If the connector is left unrepaired, corrosion damage can cause a complete separation failure of the connector and leakage of flue gas into the occupied space.

Where corrosion in the connector has proven to be a chronic problem, consider replacing the connector with a type L vent pipe or its listed equivalent. One product configuration consists of connector pipe with a double wall (stainless steel inner and galvanized steel outer with 0.25 in. gap). The insulated gap of this type of double-wall connector elevates inner wall surface temperature and reduces the overall connector heat loss.

Masonry Chimneys for Oil-Fired Appliances

A masonry chimney serving an oil-fired appliance should have a tile liner and should comply with applicable building codes such as NFPA *Standard* 211. An additional listed chimney liner may be needed to improve thermal response (warm-up) of the inner chimney surface, thereby reducing transient low draft during start-up and acid/water condensation during cyclic operation. This is particularly true for exposed exterior high-mass chimneys but does not exclude cold interior chimneys serving oil-fired appliances that produce relatively low flue gas temperatures. Application of insulation around tile liners within masonry chimneys is common in Europe and may be worth considering in chimney replacement or new construction.

A computational analysis by Krajewski (1996) using OHVAP (Version 3.1) to analyze a series of masonry chimney systems with various firing rates and exit temperatures revealed that current applications of modern oil-fired heating appliances may have problems with acid/water condensation during winter operation. For residential oil-fired heating appliance firing rates below 1.25 to 1.5 gph with flue-loss steady-state efficiencies of 82% or higher, exterior masonry chimneys may need special treatment to reduce condensation. For conservative design, listed chimney liners and listed type L connectors may be required for some exterior chimneys serving equipment operating under these conditions.

Replacement of Appliances

The physical condition and suitability of an existing chimney must be checked before installation of a new oil-fired appliance. The chimney should be inspected and, if necessary, cleaned. In accordance with NFPA *Standard* 211, Section 3-2.6, the chimney drop-leg (bottom of the chimney flue) must be at least 8 in. below the bottom of the appliance connection to the chimney. The liner must be continuous, properly aligned, and intact and must extend beyond the top of the chimney. The chimney should also have a properly installed, reasonably airtight clean-out at the base.

For masonry chimneys, local experience may indicate how well the construction has withstood the lower temperatures produced by a modern oil-fired appliance. Evidence of potential or existing chimney damage can be procured by visual examination. Exterior indicators such as missing or loose mortar/bricks, white deposits (efflorescence) on brickwork, a leaning chimney, or water stains on interior building walls should be investigated further. Interior chimney examination with a mirror or video camera can reveal damaged or missing liner material. Any debris collected in the chimney base, drop-leg, or connector should be removed and examined. If any doubt exists about the chimney's condition, examination by an experienced professional is recommended. Kam et al. (1993) offer specific guidance on the examination and evaluation of existing masonry chimneys in the field.

FIREPLACE CHIMNEYS

This section is condensed from an *ASHRAE Journal* article; for more details, please see Stone (1969).

Fireplaces with natural-draft chimneys follow the same gravity fluid flow law as gas vents and thermal flow ventilation systems. All thermal or buoyant energy is converted into flow, and no draft exists over the fire or at the fireplace inlet. Formulas have been developed to study a wide range of fireplace applications, but the material in this section covers general cases only.

Mass flow of hot flue gases through a vertical pipe is a function of rate of heat release and the chimney area, height, and system pressure loss coefficient k. The flow induced in a vertical pipe has a limiting value. A fireplace may be considered as a gravity duct inlet fitting with a characteristic entrance-loss coefficient and an internal heat source. A fireplace functions properly (does not smoke) when adequate intake or face velocity across those critical portions of the frontal opening nullifies external drafts and internal convection effects.

In a fireplace-chimney system, the equations assume that all potential buoyant energy is converted into flow as controlled by various losses. This system is analogous to a gas venting system with a draft hood, thereby allowing use of similar concepts as a starting point for size or capacity analysis. The amount of available draft ahead of the fireplace opening is insignificant and need not be considered. Because chimney efficiency, by one definition, equals available draft divided by theoretical draft, the numerical efficiency value approaches zero. Thus, the flow conversion basis is preferable for design over the efficiency approach.

System parameters for preventing flue gas spillage from a draft hood or similar collection fitting, can be computed with considerable certainty when heat input is constant or cannot exceed a predictable limiting value. Fireplaces, however, can be fired at extremes ranging from smoldering embers to flash fires of newspapers or dry kindling. Normal opening width and length allow greater access of combustion air to fires than a typical chimney can carry away; thus, combustion overloading occasionally leads to some smoking. At low rates of combustion, airflow velocities into the fireplace face are less than the velocity of natural convection currents induced at side walls by heat stored in the brick, allowing wisps of smoke to stray away from the combustion products' main flow path. Smoking tendencies are compounded at low firing rates by indoor/outdoor pressure differentials caused by winds, thermal forces, or fans, because the accompanying thermal force (buoyancy) of low combustion products temperature is insufficient to overcome strong wind or fan effects.

In the following analysis, note that fireplaces are primarily air-collecting hoods, diluting a small amount of combustion products with large amounts of air. Maximum mass flow of air into any given fireplace chimney not only is limited, but actually diminishes past a certain maximum. Thus, as combustion rate increases, combustion product temperatures rise to the point where masonry cracks; metals overexpand, warp, and oxidize; and steady smoking can occur because heated flue gases evolve beyond the limited capacity of the chimney.

An inoperative fireplace is completely at the mercy of indoor/outdoor pressure differences caused by winds, building stack effects, and operation of forced-air heating systems or mechanical ventilation. Thus, the complaint of smoking during start-up can have complex causes seldom related to the chimney. Increasingly in new homes and especially in high-rise multiple family construction, fireplaces of normal design cannot cope with mechanically induced reverse flow or shortages of combustion air. It is mandatory in these circumstances to treat and design a fireplace as a constantly operating mechanical exhaust system, with induced-draft blowers (mechanical-draft systems) that can overpower other mechanized air-consuming systems, and can develop sufficient flow to avoid smoking and excessive flue temperatures.

The gravity-flow capacity equation (Kinkead 1962) of a fireplace-chimney system equates mass flow with the resultant system driving forces and losses.

$$w = A_c\left(\frac{2gH}{k}\right)^{1/2}[\rho_m(\rho_o - \rho_m)]^{1/2} \tag{24}$$

where

w = flue gas flow rate, lb/s
A_c = chimney flue cross-sectional area, ft^2
g = gravitational constant, 32.1740 ft/s^2
H = height of chimney above lintel, ft
k = system equivalent resistance coefficient, dimensionless
ρ_m = flue gas density at mean temperature, lb/ft^3
ρ_o = air density of ambient temperature, lb/ft^3

From Equation (24), it is possible to develop a relationship for average frontal velocity, maximum chimney capacity, and variation of gas temperature with changes in fuel input rate.

Using resistance coefficients in these compact systems is preferable to the usual method of equivalent lengths. The summation term k in a vertical chimney is the total of four individual resistance factors:

- Acceleration k_a of ambient air to flue gas velocity, a constant value that must always be included in the total; k_a = 1.0
- Inlet loss coefficient k_i for fireplace configuration, including smoke shelf

  Cone-type fireplaces $k_i = 0.5$
  Masonry (damper throat = 2 × flue area) $k_i = 1.0$
  Masonry (damper throat = flue area) $k_i = 2.5$
- Chimney flue pipe friction k_c at a typical Reynolds number of 10,000 and roughness of 0.001, where r_h = hydraulic radius, ft

$$k_c = 0.0083\left(\frac{H}{r_h}\right)$$

- Termination coefficient k_t

  For open top pipe or tile, same size as chimney $k_t = 0$
  Disk or cone cap at $D/2$ above outlet $k_t = 0.5$
  Manufactured caps $k_t = 0$ to 4.0

For a 12 ft high open top chimney, 12 in. in diameter on a typical fireplace, the system resistance is

k_a =	1.0
k_i =	2.5
k_c = 0.0083 (12/0.25) =	0.4
k_t =	0.0
	k = 3.9 summation

Note that in a short chimney, the wall friction coefficient k_c is only 0.4 and has a minor effect on system flow. Greater or lesser chimney roughness, or a change from low to high heat loss materials will have little bearing on fireplace effectiveness in short chimneys. In some situations, it may be necessary, for completeness, to include a k term for air supply resistance.

To determine the frontal velocity V_F of ambient air, the term w is replaced using the substitution

$$w = \rho_o A_F V_F \quad (25)$$

where

A_F = fireplace frontal opening area, ft²
V_F = fireplace mean frontal velocity, ft/s

Accordingly, mean frontal velocity V_F becomes

$$V_F = \frac{A_c}{A_F}\left(\frac{2gH}{k}\right)^{1/2}\left[\frac{\rho_m(\rho_o - \rho_m)}{\rho_o}\right]^{1/2} \quad (26)$$

For present purposes, the molecular weight or specific gravity of dilute combustion products is practically the same as that of air, and both can be expressed with adequate accuracy in terms of absolute gas temperature by the same relationship:

$$\rho_o = 1.325\frac{B_o}{T_o} \quad (27)$$

$$\rho_m = 1.325\frac{B_o}{T_m} \quad (28)$$

where

B_o = existing barometric pressure, in. Hg
T_o = ambient air temperature, °R
T_m = mean combustion products temperature, °R

Substitution of the density/temperature relationships [Equations (27) and (28)], into Equation (26) allows further simplification, leading to the general frontal velocity expression:

$$V_F = \frac{A_c}{A_F}\left(\frac{2gH}{k}\right)^{1/2}\frac{[T_o(T_m - T_o)]^{1/2}}{T_m} \quad (29)$$

Here, frontal velocity is a function of the product of three terms:

- Dimensionless area ratio A_c/A_F
- Height/resistance term $(2gH/k)^{1/2}$
- Dimensionless temperature effects $[T_o(T_m - T_o)]^{1/2}/T_m$

For the 12 ft high example chimney, assume ambient temperature (for calculation purposes) is 70°F, or T_o = 530°R indoors and outdoors, with no air supply resistance. Equation (29) expresses variation in V_F with gas temperature as shown in Figure 19. Assume also A_c/A_F = 0.10, so that frontal opening is ten times chimney area.

The mean airflow velocity into a fireplace frontal opening is nearly constant from 300°F flue gas temperature rise up to any higher temperature. Local velocities vary within the opening. Depending on design, air typically enters horizontally along the hearth and then flows into the fire and upward, clinging to the back wall (see Figure 18). A recirculating eddy forms just inside the upper front half of the opening, induced by the high velocity of flow along the back. Restrictions or poor construction in the throat area between the lintel and damper also increase the eddy. Because the eddy moves smoke out of the zone of maximum velocity, the tendency of this smoke to escape must be counteracted by some minimum inward air movement over the entire front of the fireplace, particularly under the lintel.

Construction of a fireplace, its internal configuration, damper location, height and location of lintel, slope of back and sides, and so on all affect minimum frontal velocity to prevent smoking with ordinary fires. It seems desirable to maintain a smooth, gradual tapering transition between the hearth or flame region up into the damper location. A sudden transition, unless it is well above the lintel, induces velocity components that tend to increase eddying. With a shallow chamber between lintel and damper zone, there is insufficient volume for convection currents, and some flue gases may be diverted

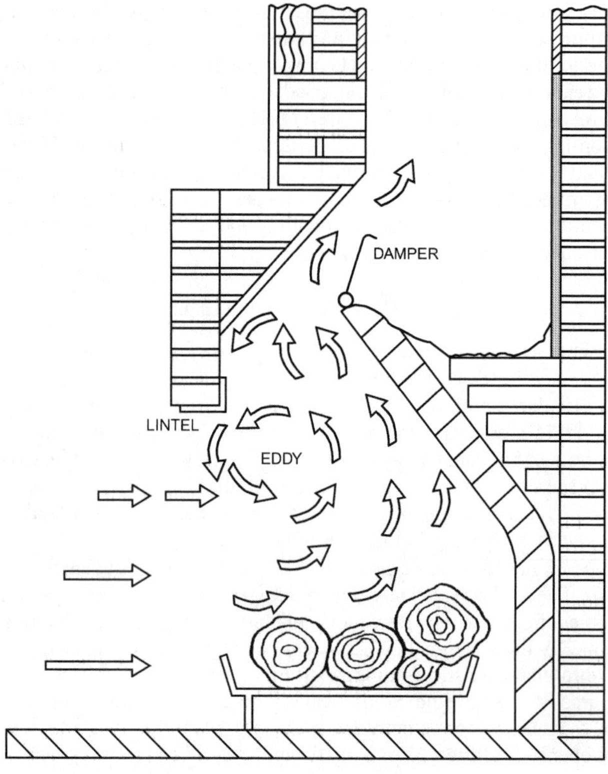

Fig. 18 Eddy Formation

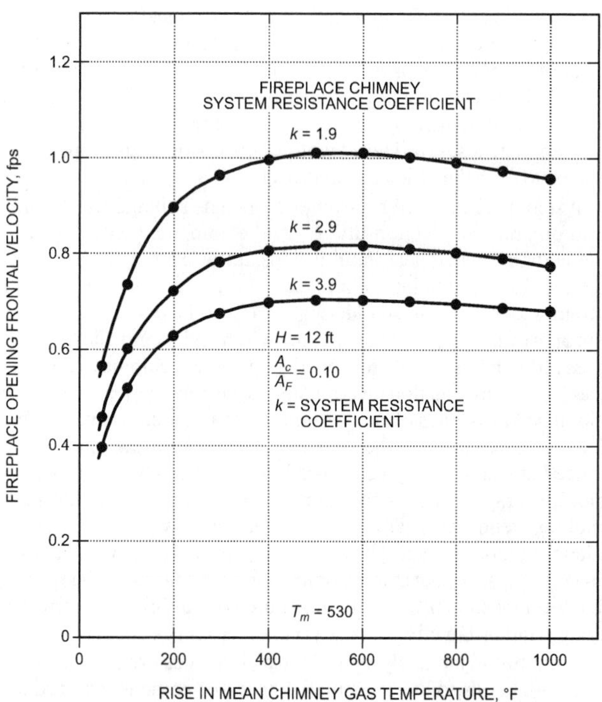

Fig. 19 Effect of Chimney Gas (Combustion Products) Temperature on Fireplace Frontal Opening Velocity

horizontally before being captured by the main flow. Masons following the dimensional parameters recommended by Ramsey and Sleeper (1956) or in damper literature can avoid these design flaws. With prefabricated fireplace systems, which often tend to be unconventional, careful testing is essential to ensure safe, smoke-free performance at minimum chimney height.

A minimum mean frontal inlet velocity of 0.8 fps, in conjunction with a chimney flue gas temperature of at least 300 to 500°F above ambient, should control smoking in a well-constructed conventional masonry fireplace. As noted in Figure 19, this velocity can be achieved even in low chimneys by system resistance of 2.9 or less, in conjunction with rates of combustion producing flue gas temperatures above a certain minimum level.

Figure 19 also illustrates why increases in flue gas temperature rise greater than 300°F have no perceptible effect on fireplace smoking, because combustion air mass flow and face velocity actually decrease at flue gas temperature rises higher than 500°F. For practical purposes, chimney flue gas temperature has little influence on fireplace performance, if flue gas temperature is at least 300°F above ambient. Fireplace performance analysis thus can be continued assuming a constant flue gas temperature of 500°F rise above ambient:

$$\frac{[T_o(T_m - T_o)]^{1/2}}{T_m} = 0.5$$

where T_m = 1030°R and T_o = 530°R.

Assuming 70°F ambient (T_o = 530°R), and factoring out 2g, Equation (29) becomes roughly

$$V_F = 4.0\frac{A_c}{A_F}\left(\frac{H}{k}\right)^{1/2} \tag{30}$$

This expression states that relative fireplace performance is purely a matter of geometry. It permits evaluation of the opening size to maintain minimum frontal velocity as a function of height, while permitting quick analysis of effect of frontal area on velocity.

Figure 20 shows that 0.8 fps face velocity requires a relatively small A_F/A_c area ratio at 8 ft chimney height, whereas the fireplace frontal opening area can be nearly twice as large with a 40 ft high chimney. To compute this curve, system resistance is assumed as

$$k = 2.5 + 0.033H/D$$

$$D = 1.0 \text{ ft diameter}$$

This expression of resistance assumes the fully open free area of the damper throat to be twice chimney flue area.

A corollary application of Equation (30) assumes fixed chimney size and height, and explores variation in frontal velocity with changes in area ratio. Figure 21 shows that a 12 in. diameter, 15 ft high chimney cannot produce adequate velocities for frontal area ratios greater than 11. These curves point out the possibility of further simplification to yield a fireplace-chimney design equation for a constant face velocity of 0.8 fps.

$$A_F = 5.0A_c\left(\frac{H}{k}\right)^{1/2} \tag{31}$$

This equation allows permissible frontal area A_F to be determined as a function of chimney area, height, and system resistance. It can also be used to determine A_c, if A_F is known. However, in this latter case, Equations (32) and (33) are preferable, because chimney size (D_c, A_c, and R_h) appears twice in the right-hand side of each equation.

For circular flues

$$A_F = 3.93D_c^2\left(\frac{H}{2.5 + 0.033\frac{H}{D_c}}\right)^{1/2} \tag{32}$$

For other flue shapes

$$A_F = 5A_c\left(\frac{H}{2.5 + 0.0083\frac{H}{R_h}}\right)^{1/2} \tag{33}$$

where D_c is chimney inside diameter, ft, R_h is inside hydraulic radius, ft, frontal velocity V_F is 0.8 fps at maximum combustion air mass flow rate, and damper opening free area is twice chimney flue area combustion air A_F.

These relationships clearly reveal the origin of rules of thumb specifying opening area as 8, 10, or 12 times chimney area. Some

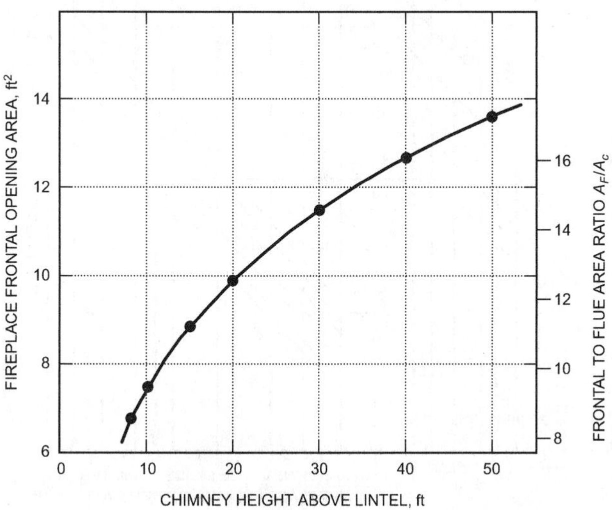

Fig. 20 Permissible Fireplace Frontal Opening Area for Design Conditions (0.8 fps mean frontal velocity with 12 in. inside diameter round flue)

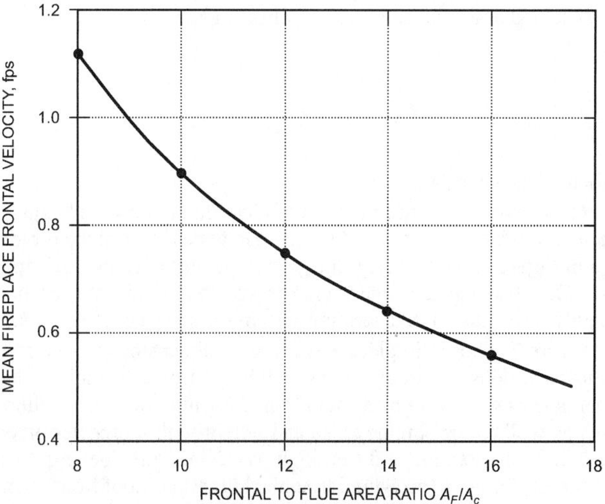

Fig. 21 Effect of Area Ratio on Frontal Velocity (for chimney height of 15 ft with 12 in. inside diameter round flue)

design guides go beyond and classify chimneys by height groups so that short ones serve smaller fireplace openings. Using a mean face velocity of 0.8 fps yields ratios and heights that fall well within the limits of such rules, as well as providing a unifying concept for computing design charts.

The previous relationships primarily apply to masonry single-face fireplaces of conventional construction, but are applicable to other types with considerable validity, if face or opening area is properly treated. Corner or double-face designs and many free-standing types embody conventional smokeshelf construction with similar resistance coefficients.

The preceding equations can be applied to illustrate the effect of excessive firing rates on chimney flue gas temperature. Masonry fireplaces are highly inefficient as heating devices, and tests show that, over a wide range of controlled fuel inputs using a drilled-port nonaerated gas burner, 75 to 80% of the gross heating value goes up the chimney. With constant flue heat loss of 80%, 70% of heat loss is sensible heat, which produces the rise in flue gas temperature. This heat input/temperature relationship may be developed by expressing the system flow relationship in terms of heat input:

$$w = \frac{q}{c_p(T_m - T_o)} \tag{34}$$

where

q = heat content of chimney flue gases, Btu/s

c_p = specific heat of chimney gases, assumed to be approximately 0.25 Btu/lb·°F

Equating Equations (34) and (24), and eliminating w,

$$\frac{q}{c_p(T_m - T_o)} = A_c\left(\frac{2gH}{k}\right)^{1/2}[\rho_m(\rho_o - \rho_m)]^{1/2} \tag{35}$$

Substituting Equations (27) and (28) for density and solving for q,

$$q = \frac{1.325 B_o A_c c_p}{(T_o)^{0.5}}\left(\frac{2gH}{k}\right)^{1/2}\frac{(T_m - T_o)^{3/2}}{T_m} \tag{36}$$

For any given system, all terms except $(T_m - T_o)^{3/2}/T_m$ may be considered fixed; therefore, gas temperature rise for a fireplace can be obtained as a concise function of heat input:

$$q = \overline{C}\frac{(T_m - T_o)^{1.5}}{T_m} \tag{37}$$

where $\overline{C}$ is a constant.

The flow-temperature function of a carefully controlled experimental fireplace [Equation (36) or (37)] in which $\overline{C}$ as fixed is plotted in Figure 22 as chimney flue gas temperature versus heat input rate. Data taken on a fireplace-chimney system in this manner may readily be evaluated to determine system resistance coefficient k.

Figure 23 shows fireplace and chimney dimensions for the specific conditions of circular flues at 0.8 fps frontal velocity. This chart solves readily for maximum frontal opening for a given chimney, as well as for chimney size and height with a predetermined opening. For example, a 30 in. high by 42 in. wide opening for a 12 ft high chimney (measured from the highest point of front opening) requires a 12.5 in. flue. Figure 23 assumes no wind or air supply difficulties. For other face velocities, A_F is found by multiplying frontal area (center scale) by velocity ratio $0.8/V_F$. To confirm the example results, calculate from Equation (32) as follows:

$$A_F = 3.93 D_c^2 \left[\frac{H}{(2.5 + 0.033H)/D_c}\right]^{1/2}$$

where H is 12 ft and D_c is 14 in.

$$A_F = 8.75\ \text{ft}^2 = \frac{30\ \text{in.} \times 42\ \text{in.}}{12\ \text{in.}/1\ \text{ft}}$$

Although derived specifically for circular flues, A_F applies with negligible sacrifice in performance to chimney flue cross sections such as squared or rounded ovals, because flue area is a much more important factor than friction caused by changes in hydraulic radius. For example, in a 20 ft high chimney, assuming a square flue section equal in area to an 8 in. circle, frontal area is reduced from 4.16 ft² with the round, to 4.09 ft² with the square, or about 2%, a difference

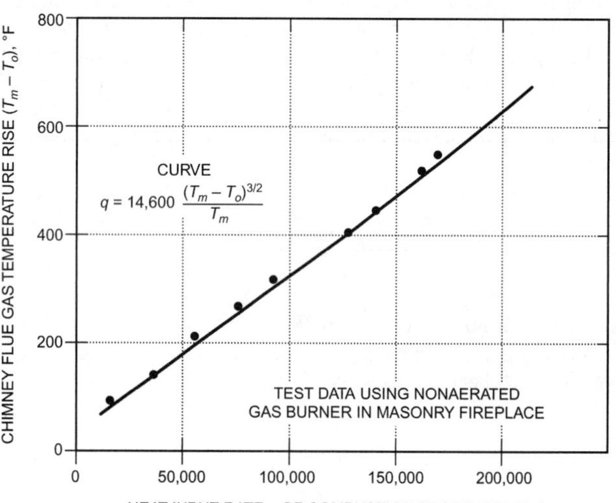

Fig. 22 Variation of Chimney Flue Gas Temperature with Heat Input Rate of Combustion Products

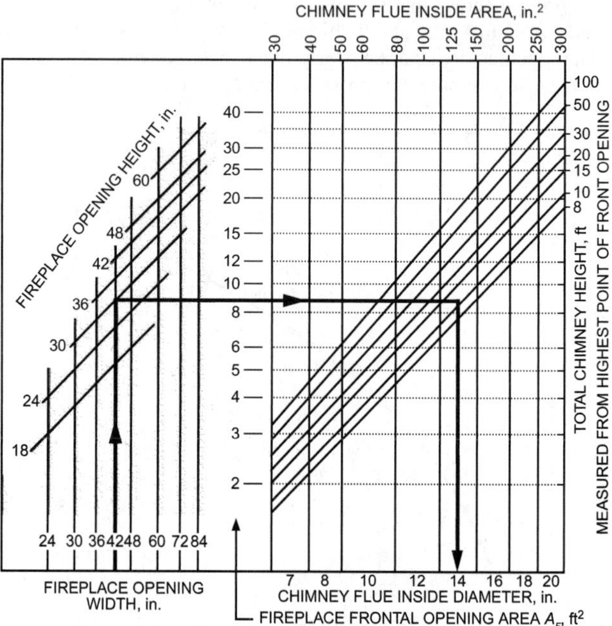

Fig. 23 Chimney Sizing Chart for Fireplaces
Mean Face Velocity = 0.8 fps
(Stone 2005)

that is hardly observable. For some typical constructions, Figure 24 suggests methods of estimating frontal area.

These relationships apply to steady-state conditions, which are obtained only after warm-up. Igniting a rapidly flammable charge in a cold system creates pulses of expanding hot combustion products, which frequently escape from the fireplace. In a typical chimney, priming time (time to accelerate from no flow to full upward velocity) is around 5 to 10 s. Thus, initial or intermittent smoking caused by momentarily excessive combustion rates occurs because of system inability to increase flue gas velocity in pace with combustion surges.

Flue or chimney material is of little relevance to fireplace-chimney operation. Materials to which these equations and charts apply include the very hazardous uninsulated single-wall metal, through conventional masonry, as well as the various constructions of lightweight, insulated, factory-built low-heat-space-appliance chimneys. (Safety standards classify fireplaces as low-heat appliances.) Because most fireplace chimneys are short and vertical, neither heat loss nor wall roughness has any important effect on flow. The governing factors in chimney selection for fireplaces are mainly safety, installation, convenience, and esthetics.

Indoor/outdoor pressure differences caused by winds, kitchen or bath exhaust fans, building stack effects, and operation of forced-air heating systems or mechanical ventilation affect the operation of a fireplace. Thus, smoking during start-up can be caused by factors unrelated to the chimney. Frequently, in new homes (especially in high-rise multiple-family construction), fireplaces of normal design cannot cope with mechanically induced reverse flow or shortages of combustion air. In these circumstances, a fireplace should include an induced-draft blower able to overpower other mechanized air-consuming systems. An inducer for this purpose is best located at the chimney outlet and should produce 0.8 to 1.0 fps fireplace face velocity of ambient air in an individual flue or 10 to 12 fps chimney velocity.

In conventional fireplaces, the greater the frontal velocity, the more freedom from smoking. The damper free area, together with

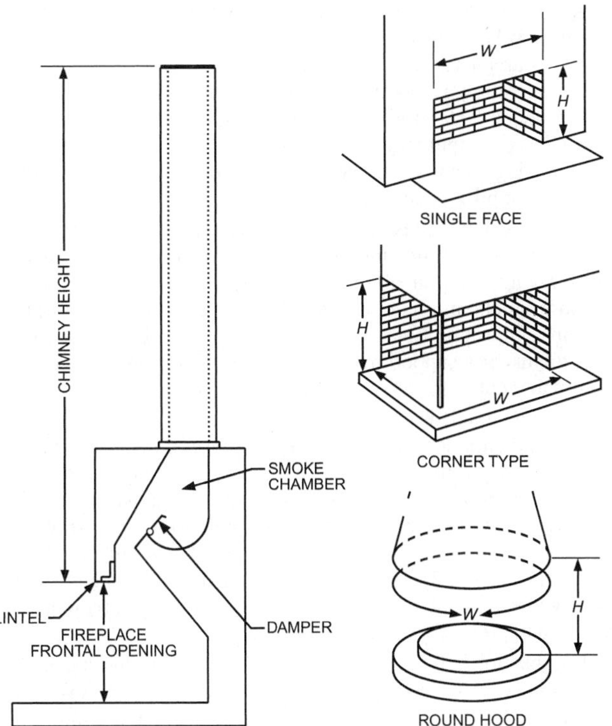

Fig. 24 Estimation of Fireplace Frontal Opening Area

its resultant resistance coefficient, are thus major factors in obtaining good masonry fireplace performance, especially with short chimneys. Tests show that damper free area need not exceed twice the required flue area, because little further resistance reduction occurs past this limit.

If the damper selected has a free area equal to or less than required area, it will be definitely restrictive, despite complete adequacy of other factors. Manufacturers' literature seldom includes damper free area or opening dimensions, and the dimensions may vary further after installation because of interferences with lintels and other parts. It is expedient to select dampers of adequate free area for best results.

Partially closing a damper during a vigorous fire illustrates this point; what is not so obvious is that greater damper openings may be needed in some cases to control smoke by achieving adequate frontal velocities.

Many free-standing fireplaces are built without the usual smokeshelf/throat damper configuration. The same parameters and relationships apply to free-standing fireplaces as to masonry fireplaces; however, in many designs it is difficult to assign a true frontal area for velocity analysis. Where freestanding fireplaces include back outlets, or require a horizontal run to reach the chimney, compensation is necessary for the flow resistance or pressure losses caused by turns. As a rough approximation, increase the system resistance coefficient k about 1.5 for two 90° elbows, or a back outlet with lateral run into a tee.

Losses caused by lateral connectors and tees generally result in a one-size increase (e.g., moving from 7 to 8) being sufficient. Collar sizes generally determine correct chimney size, and instructions are furnished to cover special situations. More sophisticated prefabricated fireplace designs are provided with matching correctly sized chimneys, and are intended for installation in conventional wood-frame residences.

The equations and design charts presented here assume no wind or air supply difficulties. Lack of replacement air, competing ventilation exhaust fans, and negative interior pressures caused by winds are all obvious causes of smoking or poor fireplace priming. Even when fully primed and hot, thermal forces in a fireplace chimney can be overpowered by a combination of adverse influences. In modern high-rise residential apartments, where an effort has been made to provide all amenities, fireplaces may have to cope simultaneously with all of these troubles. Continuous induced draft for the chimney alleviates most of these problems by maintaining a chimney prime at all times. An inducer for this purpose should be able to produce 0.8 to 1.0 fps fireplace frontal velocity of ambient air in any individual flue. Where multiple flues are installed in a chase, a single, larger inducer serving the chase can be sized for the combined fireplace frontal opening area. Even where flues are of different heights and sizes, a draft inducer selection, assuming all flues to be some compromise median height and size, produces far greater user satisfaction than reliance on gravity alone.

In single-family dwellings, fireplace problems are more frequently caused by reduced interior pressures from wind effects than by poor chimney terminal location or characteristics. Efforts to cure smoking, slow priming, or blowback of ashes usually involve one of the myriad forms of stationary or rotating caps, cowls, or chimney pots. This questionable expedient has contradictory effects. Usually, the added still-air resistance of a cap reduces fireplace frontal velocity, which limits combustion airflow rate, and thus may tend to increase smoking. On the other hand, a cap reduces air dilution of smaller fires, raises chimney temperature, and improves stability of flame, thus tending to mitigate wind impulses that cause momentary flow reduction. The usual fireplace damper can also be used to restrict flow, and thus raise temperature.

Remedies for fireplace malfunctions may be analyzed using Equation (30). For example, it is apparent that any change of parameters on the right-hand side that might decrease V_F can increase smoking tendencies. If frontal area A_F or k increases, there will be a corresponding decrease in V_F. Similarly, if chimney area A_c or chimney height H are reduced, then V_F decreases. Further, because frontal velocity varies as the square root of the term H/k, it is more effective to reduce frontal area, thereby increasing A_c/A_F, than to increase H or reduce k.

Logical expedients for increasing V_F frontal velocity, and thus improving performance of fireplaces and chimneys, include the following:

- Increase chimney height (using the same flue area) and extend the last tile 6 in. upward, or more.
- Decrease frontal opening by lowering the lintel, or raising the hearth. (Glass doors may help by increasing V_F.)
- Increase free area through damper. (Check that it opens fully without interferences.)

CSA *Standard* P.4.1-02 can be used for measuring annual efficiency in Canada.

AIR SUPPLY TO FUEL-BURNING APPLIANCES

Failure to supply outdoor air for combustion may result in erratic or even dangerous operating conditions. A correctly designed gas appliance with a draft hood can function with short vents (5 ft high) using an outdoor air supply opening as small in area as the vent outlet collar. Such an orifice, when equal to vent area, has a resistance coefficient in the range of 2 to 3. If the air supply opening is as much as twice the vent area, however, the coefficient drops to 0.5 or less.

The following rules may be used as a guide:

1. Residential heating appliances installed in unconfined spaces in buildings of conventional construction do not ordinarily require ventilation other than normal air infiltration. In any residence or building that has been built or altered to conserve energy or minimize infiltration, the heating appliance area should be considered a confined space. The air supply should be installed in accordance with ANSI/NFPA 54/ANSI/AGA Z223.1, CSA *Standard* B149.1, or the following recommendations.
2. Residential heating appliances installed in a confined space having unusually tight construction require two permanent openings to an unconfined space or to the outdoors. An unconfined space has a volume of at least 50 ft^3 per 1000 Btu/h of the total input rating of all appliances installed in that space. Free opening areas must be greater than 1 in^2 per 4000 Btu/h input with vertical ducts or 1 in^2 per 2000 Btu/h with horizontal ducts to the outdoors. The two openings communicating directly with sufficient unconfined space must be greater than 1 in^2 per 1000 Btu/h. Upper openings should be within 12 in. of the ceiling; lower openings should be within 12 in. of the floor.
3. Complete combustion of natural and propane gas or fuel oil requires approximately 1 ft^3 of air, at standard conditions, for each 100 Btu of fuel burned, but excess air is usually required for proper burner operation.
4. The size of these air openings may be modified if special engineering ensures an adequate supply of air for combustion, dilution, and ventilation or if local ordinances apply to boiler and machinery rooms.
5. In calculating free area of air inlets, consider the blocking effect of louvers, grilles, or screens protecting openings. Screens should not be smaller than 1/4 in. mesh. If the free area through a particular louver or grille is known, it should be used in calculating the size opening required to provide the free area specified. If the free area is not known, assume that wood louvers have 20 to 25% free area and metal louvers and grilles have 60 to 75% free area.
6. Mechanical ventilation systems serving the fuel-burning appliance room or adjacent spaces should not be allowed to create negative appliance room air pressure. The appliance room may require tight self-closing doors and provisions to supply air to spaces under negative pressure so fuel-burning appliances and venting operate properly.
7. Fireplaces may require special consideration. For example, a residential attic fan can be hazardous if it is inadvertently turned on while a fireplace is in use.
8. In buildings where large quantities of combustion and ventilation or process air are exhausted, a sufficient supply of fresh uncontaminated makeup air, warmed if necessary to the proper temperature, should be provided. It is good practice to provide about 5 to 10% more makeup air than the amount exhausted.

VENT AND CHIMNEY MATERIALS

Factors to be considered when selecting chimney materials include (1) the temperature of flue gases; (2) their composition and propensity for condensation of water vapor from combustion products (dew point); (3) presence of sulfur, halogens, and other fuel and air contaminants that lead to corrosion of the chimney vent system; and (4) the appliance's operating cycle (condensate dwell time).

Figure 7 covers materials for vents and chimneys in the 4 to 48 in. size range; these include single-wall metal, various multiwall air- and mass-insulated types, and precast and site-constructed masonry. Each has different characteristics, such as frequency of joints, roughness, and heat loss, but the type of materials used for systems 14 in. and larger is relatively unimportant in determining draft or capacity. This does not preclude selecting a safe product or method of construction that minimizes heat loss and fire hazard in the building.

National codes and standards classify heat-producing appliances as low, medium, and high heat, with appropriate reference to chimney and vent constructions permitted with each. These classifications are primarily based on size, process use, or combustion temperature. In many cases, the appliance classification gives little information about outlet gas temperature or venting needed. The designer should, wherever possible, obtain gas outlet temperature conditions and properties that apply to the specific appliance, rather than going by code classification only.

Where building codes permit engineered chimney systems, chimney material selection based on gas outlet temperature can save space as well as reduce structural and material costs. For example, in some jurisdictions, approved gas-burning appliances with draft hoods operating at inputs over 400,000 Btu/h may be placed in a heat-producing classification that prohibits use of type B gas vents. An increase in input may not cause an increase in outlet temperature or in venting hazards, and most building codes recommend correct matching appliance and vent.

Single-wall uninsulated steel stacks can be protected from condensation and corrosion internally with refractory firebrick liners or by spraying calcium aluminate cement over a suitable interior expanded metal mesh or other reinforcement. Another form of protection applies proprietary silica or other prepared refractory coatings to pins or a support mesh on the steel. The material must then be suitably cured for moisture and heat resistance.

Moisture condensation on interior surfaces of connectors, vents, stacks, and chimneys is a more serious cause of deterioration than heat. Chimney wall temperature and flue gas velocity, temperature, and dew point affect condensation. Contaminants such as sulfur, chlorides, and fluorides in the fuel and combustion air raise the flue gas dew point. Studies by Beaumont et al. (1970), Mueller (1968), Pray et al. (1942-53), and Yeaw and Schnidman (1943) indicate the variety of analytical methods as well as difficulties in predicting the causes and probability of actual condensation.

Combustion products from any fuel containing hydrogen condense onto cold surfaces or condense in bulk if the main flow of flue gas is cooled sufficiently. Because flue gas loses heat through walls, condensation, which first occurs on interior wall surfaces cooled to the flue gas dew point, forms successively a dew and then a liquid film and, with further cooling, causes liquid to flow down into zones where condensation would not normally occur.

Start-up of cold interior chimney surfaces is accompanied by transient dew formation, which evaporates on heating above the dew point. This phenomenon causes little corrosion when very low sulfur fuels are used. Proper selection of chimney dimensions and materials minimizes condensation and thus corrosion.

Experience shows a correlation between the sulfur content of the fuel and the deterioration of interior chimney surfaces. Figure 4 in Chapter 18 of the 2005 *ASHRAE Handbook—Fundamentals* illustrates one case, which applies to any fuel gas. The figure shows that the flue gas dew point increases at 40% excess air from 127°F with zero sulfur to 160°F with 15 grains of sulfur per 100 ft^3 of fuel having a heat value of 550 Btu/ft^3.

The figure can also be used to approximate the effect of fuel oil sulfur content on flue gas dew point. For example, fuel oil with a sulfur content of 0.5% (by mass) contains about 252 grains of sulfur per gallon or 25,200 grains of sulfur per 100 gallons. If the fuel heat value is 140,000 Btu/gal, the ratio that defines the curves in the figure gives a curve value of 18. Estimates for lower percentages of sulfur (0.25, 0.05) can be formed as factors of the value 18.

Because the corrosion mechanism is not completely understood, judicious use of resistant materials, suitably insulated or jacketed to reduce heat loss, is preferable to low-cost single-wall construction. Refractory materials and mortars should be acid-resistant, while steels should be resistant to sulfuric, hydrochloric, and hydrofluoric acids; pitting; and oxidation. Where low flue gas temperatures are expected together with low ambients, an air space jacket or mineral fiber lagging, suitably protected against water entry, helps maintain surface and flue gas temperatures above the dew point. Using low-sulfur fuel, which is required in many localities, reduces both corrosion and air pollution.

Type 1100 aluminum alloy or any other non-copper-bearing aluminum alloy of 99% purity or better provides satisfactory performance in prefabricated metal gas vent products. For chimney service, flue gas temperatures from appliances burning oil or solid fuels may exceed the melting point of aluminum; therefore, steel is required. Stainless steels such as type 430 or 304 give good service in residential construction and are referenced in UL-listed prefabricated chimneys. Where more corrosive substances (e.g., high-sulfur fuel or chlorides from solid fuel, contaminated air, or refuse) are anticipated, type AL 29-4CR® or equivalent stainless steel offers a good match of corrosion resistance and mechanical properties.

As an alternative to stainless steel, porcelain enamel offers good resistance to corrosion if two coats of acid-resistant enamel are used on all surfaces. A single coat, which always has imperfections, allows base metal corrosion, spalling, and early failure.

Prefabricated chimneys and venting products are available that use light corrosion-resistant materials, both in metal and masonry. The standardized, prefabricated, double-wall metal type B gas vent has an aluminum inner pipe and a coated steel outer casing, either galvanized or aluminized. Standard air space from 1/4 to 1/2 in. is adequate for applicable tests and a wide variety of exposures.

Air-insulated all-metal chimneys are available for low-heat use in residential construction. Thermosiphon air circulation or multiple reflective shielding with three or more walls keeps these units cool. Insulated, double-wall residential chimneys are also available. The annulus between metal inner and outer walls is filled with insulation and retained by coupler end structures for rapid assembly.

Prefabricated, air-insulated, double-wall metal chimneys for multifamily residential and larger buildings, classed as building heating appliance chimneys, are available (Figure 25). Refractory-lined prefabricated chimneys (medium-heat type) are also available for this use.

Commercial and industrial incinerators, as well as heating appliances, may be vented by prefabricated metal-jacketed cast refractory chimneys, which are listed in the medium-heat category and are

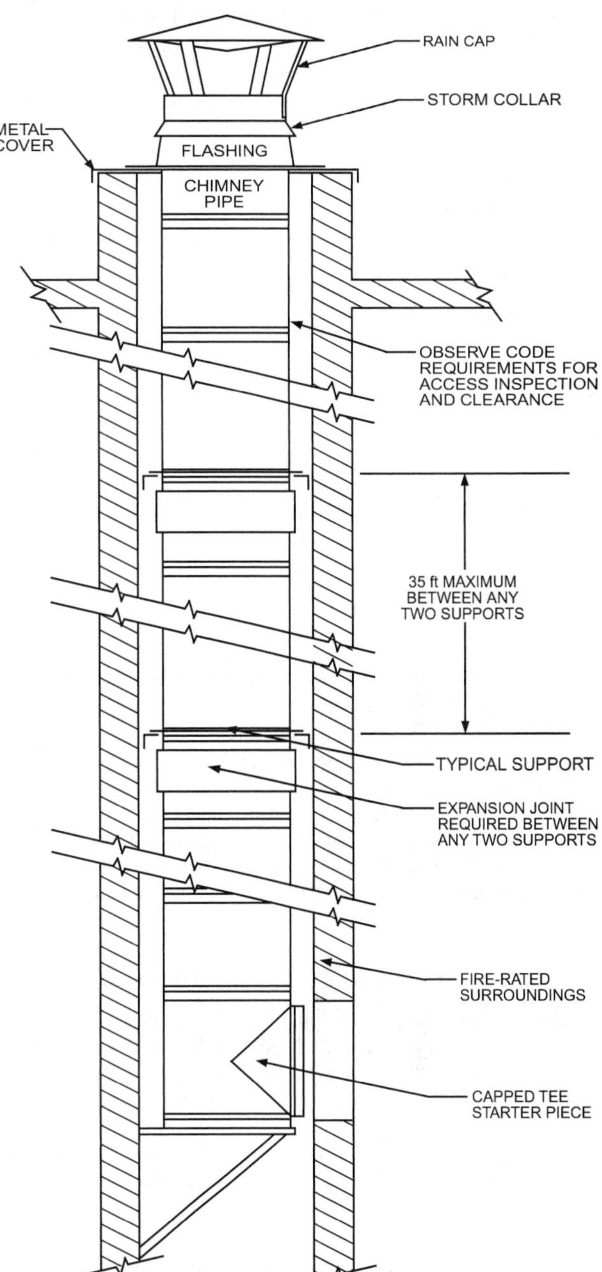

Fig. 25 Building Heating Appliance, Medium-Heat Chimney

suitable for intermittent flue gas temperatures to 2000°F. All prefabricated chimneys and vents carrying a listing by a recognized testing laboratory have been evaluated for class of service regarding temperature, strength, clearance to adjacent combustible materials, and suitability of construction in accordance with applicable national standards.

Underwriters Laboratories (UL) standards, listed in Table 10, describe the construction and temperature testing of various classes of prefabricated vent and chimney materials. Standards for some related parts and appliances are also included in Table 10 because a listed factory-built fireplace, for example, must be used with a specified type of factory-built chimney. The temperature given for the steady-state operation of chimneys is the lowest in the test sequence. Factory-built chimneys under UL *Standard* 103 are also required to demonstrate adequate safety during a 1 h test at 1330°F rise and to

Table 10 Underwriters Laboratories Test Standards

No.	Subject	Steady-State Appliance Flue Gas Temperature Rise, °F	Fuel
103	Chimneys, factory-built, residential type (includes building heating appliances)	930	All
127	Fireplaces, factory-built	930	Solid or gas
311	Roof jacks for manufactured homes and recreational vehicles	930	Oil, gas
378	Draft equipment (such as regulators and inducers)	—	All
441	Gas vents (type B, BW)	480	Gas only
641	Low-temperature venting systems (type L)	500	Oil, gas
959	Chimneys, factory-built, medium-heat	1730	All
1738	Venting systems for gas-burning appliances, categories II, III, and IV	140 to 480	Gas only

withstand a 10 min simulated soot burnout at either 1630 or 2030°F rise.

These product tests determine minimum clearance to combustible surfaces or enclosures, based on allowable temperature rise on combustibles. They also ensure that the supports, spacers, and parts of the product that contact combustible materials remain at safe temperatures during operation. Product markings and installation instructions of listed materials are required to be consistent with test results, refer to types of appliances that may be used, and explain structural and other limitations.

VENT AND CHIMNEY ACCESSORIES

Vent or chimney system design must consider the existence of or need for accessories such as draft diverters, draft regulators, induced-draft fans, blocking dampers, expansion joints, and vent or chimney terminals. Draft regulators include barometric draft regulators and furnace sequence draft controls, which monitor automatic flue dampers during operation. The design, materials, and flow losses of chimney and vent connectors are covered in previous sections.

Draft Hoods

The draft hood isolates the appliance from venting disturbances (updrafts, downdrafts, or blocked vent) and allows combustion to start without venting action. Suggested general dimensions of draft hoods are given in ANSI *Standard* Z21.12, which describes certification test methods for draft hoods. In general, the pipe size of the inlet and outlet flues of the draft hood should be the same as that of the appliance outlet connection. The vent connection at the draft hood outlet should have a cross-sectional area at least as large as that of the draft hood inlet.

Draft hood selection comes under the following two categories:

1. Draft hood supplied with a design-certified gas appliance: certification of a gas appliance design under pertinent national standards includes its draft hood (or draft diverter). Consequently, the draft hood should not be altered or replaced without consulting the manufacturer and local code authorities.
2. Draft hoods supplied separately for gas appliances: listed draft hoods for existing vent or chimney connectors should be installed by experienced installers in accordance with accepted practice standards.

Every design-certified gas appliance requiring a draft hood must be accompanied by a draft hood or provided with a draft diverter as an integral part of the appliance. The draft hood is a vent inlet fitting as well as a safety device for the appliance, and assumptions can be made regarding its interaction with a vent. First, when the hood is operating without spillage, the heat content of flue gases (enthalpy relative to dilution air temperature) leaving the draft hood is almost the same as that entering. Second, safe operation is obtained with 40 to 50% dilution air. It is unnecessary to assume 100% dilution air for gas venting conditions. Third, during certification tests, the draft hood must function without spillage, using a vent with not over 5 ft of effective height and one or two elbows. Therefore, if vent heights appreciably greater than 5 ft are used, an individual vent of the same size as the draft hood outlet may be much larger than necessary.

When vent size is reduced, as with tall vents, draft hood resistance is less than design value relative to the vent; the vent tables in the *National Fuel Gas Code* (ANSI/NFPA 54/ANSI/AGA Z223.1) give adequate guidance for such size reductions.

Despite its importance as a vent inlet fitting, the draft hood designed for a typical gas appliance primarily represents a compromise of the many design criteria and tests solely applicable to that appliance. This allows considerable variation in resistance loss; thus, catalog data on draft hood resistance loss coefficients do not exist. The span of draft hood loss coefficients, including inlet acceleration, varies from the theoretical minimum of 1.0 for certain low-loss bell or conical shapes to 3 or 4, where the draft hood relief opening is located within a hot-air discharge (as with wall furnaces) and high resistance is needed to limit sensible heat loss into the vent.

Draft hoods must not be used on appliances having draft configuration 1 or 3 (see conditions 1, 2, and 3 under Terminology and Table 8) that is operated with either power burners or forced venting, unless the appliances have fan-assisted burners that overcome some or most of the appliance flow resistance and create a pressure inversion ahead of the draft hood or barometric regulator.

Gas appliances with draft hoods must have excess chimney draft capacity to draw in adequate draft hood dilution air. Failure to provide adequate combustion air can cause oxygen depletion and spillage of flue gases and flame rollout from the combustion air inlet at the burner(s).

Draft Regulators

Appliances requiring draft at the appliance flue gas outlet generally use barometric regulators for combustion stability. A balanced hinged gate in these devices bleeds air into the chimney automatically when pressure decreases. This action simultaneously increases vent gas flow and reduces temperature. Well-designed barometric regulators provide constant flue gas static pressure over a span of impressed vent gas draft of about 0.2 in. of water, where impressed vent gas draft is that which would exist without regulation. A regulator can maintain 0.06 in. of water draft for impressed drafts from 0.06 to 0.26 in. of water. If the chimney system is very high or otherwise capable of generating available draft in excess of the pressure span capability of a single regulator, additional or oversize regulators may be used. Figure 26 shows proper locations for regulators in a chimney manifold.

Barometric regulators are available with double-acting dampers, which also swing out to relieve momentary internal pressures or divert continuing downdrafts. In the case of downdrafts, temperature safety switches actuated by hot gases escaping at the regulator sense and limit malfunctions.

Vent Dampers

Electrically, mechanically, and thermally actuated automatic vent dampers can reduce energy consumption and improve seasonal efficiency of gas- and oil-burning appliances. Vent dampers reduce loss of heated air through gas appliance draft hoods and loss of specific

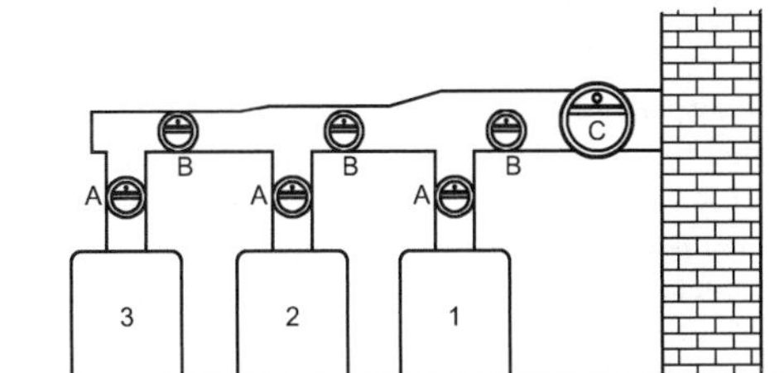
Commercial and industrial furnaces and boilers are often installed in multiples, as shown below. For best results, place a draft control between the outlet and the chimney connector at each point A. If the uptake is too short to install a control at point A, install a separate control for each boiler on the main chimney connector at each point B. If crowding or another factor prevents placing controls at points A or B, install a single large control at point C.
A
B
C
3
2
1

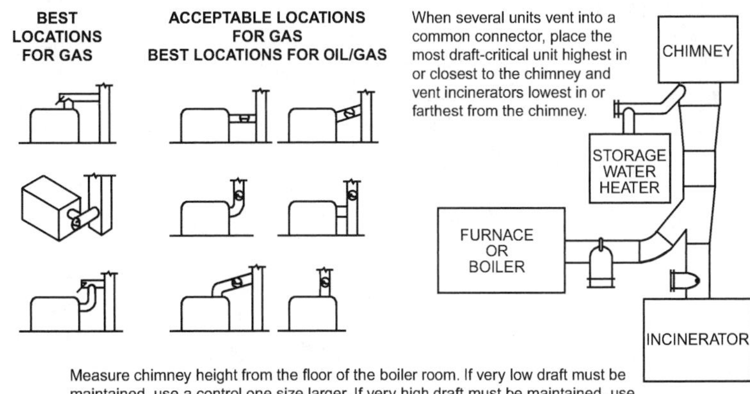
BEST LOCATIONS FOR GAS
ACCEPTABLE LOCATIONS FOR GAS
BEST LOCATIONS FOR OIL/GAS
When several units vent into a common connector, place the most draft-critical unit highest in or closest to the chimney and vent incinerators lowest in or farthest from the chimney.
CHIMNEY
STORAGE WATER HEATER
FURNACE OR BOILER
INCINERATOR
Measure chimney height from the floor of the boiler room. If very low draft must be maintained, use a control one size larger. If very high draft must be maintained, use a control one size smaller.

Fig. 26 Use of Barometric Draft Regulators

heat from the appliance after the burner stops firing. These dampers may be retrofit devices or integral components of some appliances.

Electrically and mechanically actuated dampers must open before main burner gas ignition and must not close during burner operation. These safety interlocks, which electrically interconnect with existing control circuitry, include an additional main control valve, if called for, or special gas pressure-actuated controls.

Vent dampers that are thermally actuated with bimetallic elements and have spillage-sensing interlocks with burner controls are available for draft hood-type gas appliances. These dampers open in response to gas temperature after burner ignition. Because thermally actuated dampers may exhibit some flow resistance, even at equilibrium operating conditions, carefully follow instructions regarding allowable heat input and minimum required vent or chimney height.

Special care must be taken to ensure that safety interlocks with appliance controls are installed according to instructions. Spillage-free gas venting after the damper has been installed must be verified with all damper types.

Energy savings of a vent damper can vary widely. Dampers reduce energy consumption under one or a combination of the following conditions:

- Heating appliance is oversized.
- Chimney is too high or oversized.
- Appliance is located in heated space.
- Two or more appliances are on the same chimney (a damper must be installed on each appliance connected to that chimney).
- Appliance is located in building zone at higher pressure than outdoors. This positive pressure can cause steady flow losses through the chimney.

Energy savings may not justify the cost of installing a vent damper if one or more of the following conditions exist:

- All combustion and ventilation air is supplied from outdoors to direct-vent appliances or to appliances located in an isolated, unheated room.
- Appliance is in an unheated basement that is isolated from the heated space.
- A one-story flat-roof house has a short vent, which is unlikely to carry away a significant amount of heated air.

For vents or chimneys serving two or more appliances, dampers (if used) should be installed on all attached appliances for maximum effectiveness. If only one damper is installed in such systems, loss of heated air through an open draft hood may negate a large portion of the potential energy savings.

Heat Exchangers or Flue Gas Heat Extractors

Sensible heat available in flue gas of properly adjusted furnaces burning oil or gas is about 10 to 15% of the rated input. Small accessory heat exchangers that fit in the connector between the appliance outlet and the chimney can recover some of this heat for localized use; however, they may cause some adverse effects.

All gas vent and chimney size or capacity tables assume the gas temperature or heat available to create theoretical draft is not reduced by a heat transfer device. In addition, the tables assume flow resistance for connectors, vents, and chimneys, comprising typical values for draft hoods, elbows, tees, caps, and piping with no allowance for added devices placed directly in the flue gas stream. Thus, heat exchangers or flue gas extractors should offer no flow resistance or negligible resistance coefficients when they are installed.

A heat exchanger that is reasonably efficient and offers some flow resistance may adversely affect the system by reducing both flow rate and flue gas temperature. This may cause moisture condensation in the chimney, draft hood spillage, or both. Increasing heat transfer efficiency increases the probability of the simultaneous occurrence of both effects. An accessory heat exchanger in a solid-fuel system, especially a wood stove or heater, may collect creosote or cause its formation downstream.

Retrofitting heat exchangers in gas appliance venting systems requires careful evaluation of heat recovered versus both installed cost and the potential for chimney safety and operating problems. Every heat exchanger installation should undergo the same spillage tests given a damper installation. In addition, the flue gas temperature should be checked to ensure it is high enough to avoid condensation between the exchanger outlet and the chimney outlet.

DRAFT FANS

The selection of draft fans, blowers, or inducers must consider (1) types and combinations of appliances, (2) types of venting material, (3) building and safety codes, (4) control circuits, (5) gas temperature, (6) permissible location, (7) noise, and (8) power cost. Besides specially designed fans and blowers, some conventional fans can be used if the wheel and housing materials are heat- and corrosion-resistant and if blower and motor bearings are protected from adverse effects of the flue gas stream.

Small draft inducers for residential gas appliance and unit heater use are available with direct-drive blower wheels and an integral device to sense flow (Figure 27A). The control circuit for these applications must provide adequate vent gas flow both before and while fuel flows to the main burner. Other types of small inducers are either saddle-mounted blower wheels (Figure 27B) or venturi ejectors that induce flow by jet action (Figure 27C). An essential

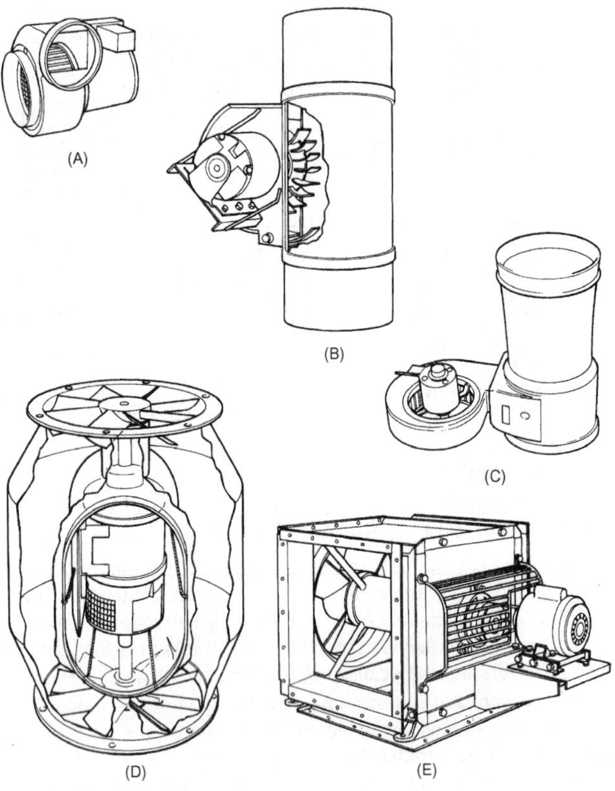

Fig. 27 Draft Inducers

safety requirement for inducers serving draft hood gas appliances does not permit appliance interconnections on the discharge or outlet side of the inducer. This requirement prevents backflow through an inoperative appliance.

With prefabricated sheet metal venting products such as type B gas vents, the vent draft inducer should be located at or downstream from the point the vent exits the building. This placement keeps the indoor system below atmospheric pressure and prevents flue gas from escaping through seams and joints. If the inducer cannot be placed on the roof or outdoor wall, metal joints must be reliably sealed in all pressurized parts of the system.

Pressure capability of residential draft inducers is usually less than 1 in. of water at rated flow. Larger inducers of the fan, blower, or ejector type have greater pressure capability and may be used to reduce system size as well as supplement available draft. Figure 27D shows a specialized axial-flow fan capable of higher pressures. This unit is structurally self-supporting and can be mounted in any position in the connector or stack because the motor is in a well, separated from the flue gas stream. A right-angle fan, as shown in Figure 27E, is supported by an external bracket and adapts to several inlet and exit combinations. The unit uses the developed draft and an insulated tube to cool the extended shaft and bearings.

Pressure, volume, and power curves should be obtained to match an inducer to the application. For example, in an individual chimney system (without draft hood) in which a directly connected inducer only handles combustion products, calculation of the power required for continuous operation need only consider volume at operating flue gas temperature. An inducer serving multiple, separately controlled draft hood gas appliances must be powered for ambient temperature operation at full flow volume in the system. At any input, the inducer for a draft hood gas appliance must handle about 50% more standard chimney gas volume than a directly connected inducer. At constant volume with a given size inducer or fan, these demands follow the Fan Laws (see Chapter 20) applicable to power venting as follows:

- Pressure difference developed is directly proportional to gas density
- Pressure difference developed is inversely proportional to absolute gas temperature
- Pressure developed diminishes in direct proportion to drop in absolute atmospheric pressure, as with altitude
- Required power is directly proportional to gas density
- Required power is inversely proportional to absolute gas temperature

Centrifugal and propeller draft inducers in vents and chimneys are applied and installed the same as in any heat-carrying duct system. Venturi ejector draft boosters involve some added consideration. An advantage of the ejector is that motor, bearings, and blower blades are outside the contaminated flue gas stream, thus eliminating a major source of deterioration. This advantage causes some loss of efficiency and can lead to reduced capacity because undersized systems, having considerable resistance downstream, may be unable to handle the added volume of the injected airstream without loss of performance. Ejectors are best suited for use at the chimney or vent exit or where there is an adequately sized chimney or vent to carry the combined discharge.

If total pressure defines outlet conditions or is used for fan selection, the relative amounts of static pressure and velocity pressure must be factored out; otherwise, the velocity head method of calculation does not apply. To factor total pressure into its two components, either the discharge velocity in an outlet of known area or the flow rate must be known. For example, if an appliance or blower produces a total pressure of 0.25 in. of water at 1500 fpm discharge velocity, the velocity pressure component can be found from Figure 7. Enter at the horizontal line marked V.H. in grid C, and move horizontally to ambient operating temperature. Read up from the appropriate temperature rise line to the 1500 fpm horizontal velocity line in grid D. Here the velocity pressure reads 0.14 in. of water (calculates to 0.143 at ambient standard conditions), so static pressure is 0.25 – 0.14 = 0.11 in. of water. Because this static pressure is part of the system driving force, it combines with theoretical draft to overcome losses in the system.

Draft inducer fans can be operated either continuously or on demand. In either case, a safety switch that senses flue gas flow or pressure is needed to interrupt burner controls if adequate draft fails. Demand operation links the thermostat with the draft control motor. Once flow starts, as sensed with a flow or pressure switch, the burner is allowed to start. A single draft inducer, operating continuously, can be installed in the common vent of a system serving several separately controlled appliances. This simplifies the circuitry because only one control is needed to sense loss of draft. However, the single fan increases appliance standby loss (especially in boilers) and heat losses via ambient air drawn through inoperative appliances.

TERMINATIONS: CAPS AND WIND EFFECTS

The vent or chimney height and method of termination is governed by a variety of considerations, including fire hazard; wind effects; entry of rain, debris, and birds; and operating considerations such as draft and capacity. For example, the 3 ft height required for residential chimneys above a roof is necessary so that small sparks will burn out before they fall on the roof shingles.

Many vent and chimney malfunctions are attributed to interactions of the chimney termination or its cap with winds acting on the roof or with adjoining buildings, trees, or mountains. Because winds fluctuate, no simple method of analysis or reduction to practice exists for this complex situation. Figures 28 to 30 show some of the complexities of wind flow contours around simple structural shapes.

Figure 28 shows three zones with differing degrees of flue gas dispersion around a rectangular building: the cavity or eddy zone, the wake zone, and the undisturbed flow zone (Clarke 1967). In addition, a fourth flow zone of intense turbulence is located downwind of the cavity. Chimney flue gases discharged into the wind at a point close to the roof surface in the cavity zone may be recirculated locally. Higher in the cavity zone, wind eddies can carry more dilute flue gas to the lee side of the building. Flue gases discharged into the wind in the wake zone do not recirculate into the immediate vicinity, but may soon descend to ground level. Above the wake zone, dispersal into the undisturbed wind flow carries and dilutes the flue gases over a wider area. The boundaries of these zones vary with building configuration and wind direction and turbulence; they are strongly influenced by surroundings.

The possibility of air pollutants reentering the cavity zone because of plume spread or of air pollution intercepting downwind cavities associated with adjacent structures or downwind buildings should be considered. Thus, the design criterion of elevating the stack discharge above the cavity is not valid for all cases. Consult a meteorologist experienced with dispersion processes near buildings for complex cases and for cases involving air contaminants.

As chimney height increases through the three zones, draft performance improves dispersion, while additional problems of gas

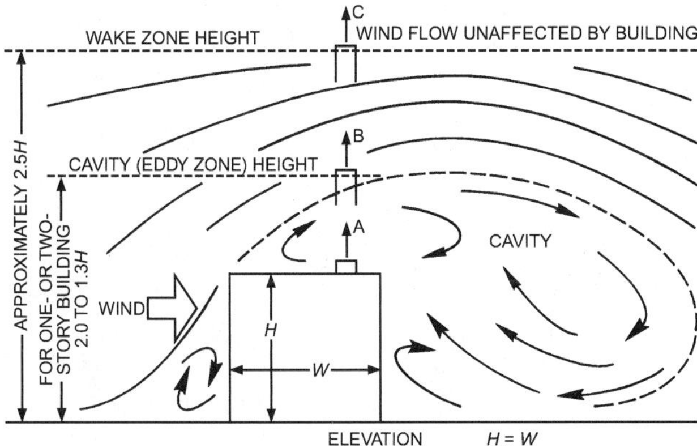

Chimney heights:

A: Discharge into cavity should be avoided because reentry will occur. Dispersion equations do not apply.

B: Discharge above cavity is good. Reentry is avoided, but dispersion may be marginal or poor from standpoint of air pollution. Dispersion equations do not apply.

C: Discharge above wake zone is best; no reentry; maximum dispersion.

Fig. 28 Wind Eddy and Wake Zones for One- or Two-Story Buildings and Their Effect on Chimney Gas Discharge

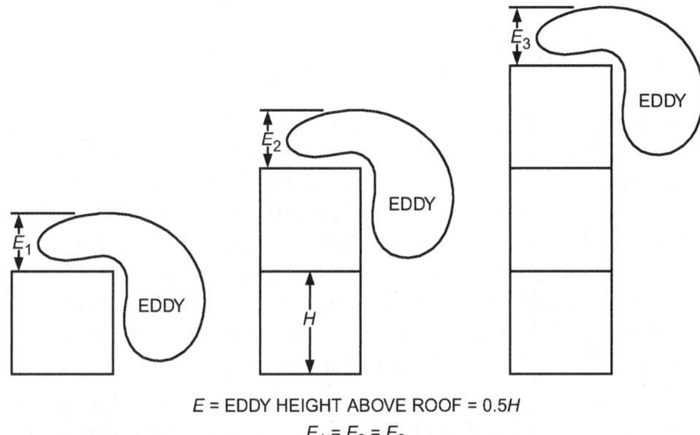

Studies found for a single cube-shaped building (length equals height) that (1) the height of the eddy above grade is 1.5 times the building height and (2) the height of unaffected air is 2.5 times height above grade. The eddy height above the roof equals 0.5*H*, and does not change as building height increases in relation to building width.

Fig. 29 Height of Eddy Currents Around Single High-Rise Buildings

cooling, condensation, and structural wind load are created. As building height increases (Figure 29), the eddy forming the cavity zone no longer descends to ground level. For a low, wide building (Figure 30), wind blowing parallel to the long roof dimension can reattach to the surface; thus, the eddy zone becomes flush with the roof surface (Evans 1957). For satisfactory dispersion with low, wide buildings, chimney height must still be determined as if $H = W$ (Figure 28).

Chien et al. (1951) and Evans (1957) studied pitched roofs in relation to wind flow and surface pressures. Because the typical residence has a pitched roof and probably uses natural gas or a low-sulfur fossil fuel, dispersion is not important because combustion products are relatively free of pollutants. For example, ANSI/NFPA 54/ANSI/AGA Z223.1 requires a minimum distance between the gas vent termination and any air intake, but it does not require penetration above the cavity zone.

Flow of wind over a chimney termination can impede or assist draft. In regions of stagnation on the windward side of a wall or a steep roof, winds create positive static pressures that impede established flow or cause backdrafts in vents and chimneys. Locating a chimney termination near the surface of a low, flat roof can aid draft because the entire roof surface is under negative static pressure. Velocity is low, however, because of the cavity formed as wind sweeps up over the building. With greater chimney height, termination above the low-velocity cavity or negative-pressure zone subjects the chimney exit to greater wind velocity, thereby increasing draft from two causes: (1) height and (2) wind aspiration over an open top. As the termination is moved from the center of the building to the sides, its exposure to winds and pressure also varies.

Terminations on pitched roofs may be exposed to either negative or positive static pressure, as well as to variation in wind velocity and direction. On the windward side, pitched roofs vary from complete to partial negative pressure as pitch increases from approximately flat to 30° (Chien et al. 1951). At a 45° pitch, the windward pitched roof surface is strongly positive; beyond this slope, pressures approach those observed on a vertical wall facing the wind. Wind pressure varies with its horizontal direction on a pitched roof, and on the lee (sheltered) side, wind velocity is very low, and static pressures are usually negative. Wind velocities and pressures vary not only with pitch, but with position between ridge and eaves. Reduction of these observed external wind effects to simple rules of termination for a wide variety of chimney and venting systems requires many compromises.

In the wake zone or any higher location exposed to full wind velocity, an open top can create strong venting updrafts. The updraft effect relative to wind dynamic pressure is related to the Reynolds number. Open tops, however, are sensitive to the wind angle as well as to rain (Clarke 1967), and many proprietary tops have been

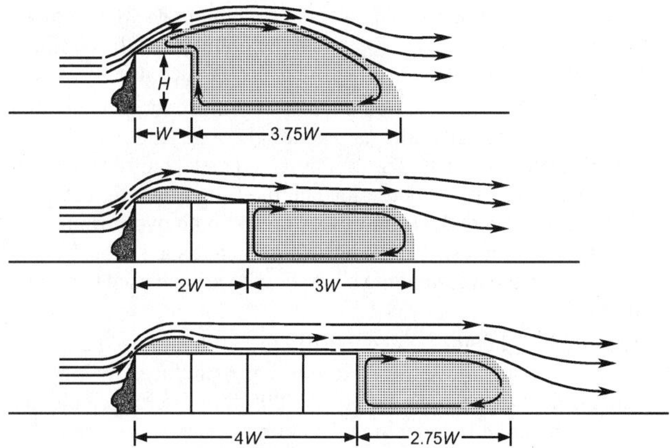

Fig. 30 Eddy and Wake Zones for Low, Wide Buildings

Table 11 List of U.S. National Standards Relating to Installation[a]

Subject	Materials Covered	NFPA[b]	ANSI[c]	CSA[f]
Oil-burning equipment	Type L listed chimneys, single-wall, masonry	31	—	—
Gas appliances and gas piping	Type B, L listed chimneys, single-wall, masonry	54	Z223.1[d]	B149.1
Chimneys, fireplaces, vents, and solid-fuel-burning appliances	All types	211	—	CAN/ULC S605-M91
Recreational vehicles	Roof jacks and vents	501C	—	—
Gas piping and gas equipment on industrial premises	All types	54	Z223.1[d]	B149.1
Gas conversion burners	Chimneys		Z21.8[e]	—
	Safe design		Z21.17[e]	—
Draft hoods	Part dimensions		Z21.12[e]	CAN1-6.2
Automatic vent dampers for use with gas-fired appliances	Construction and performance		Z21.66[e]	—

[a]These standards are subject to periodic review and revision to reflect advances in industry, as well as for consistency with legal requirements and other codes.
[b]National Fire Protection Association, Quincy, MA.
[c]American National Standards Institute, New York.
[d]Available from American Gas Association, Washington, D.C.
[e]Available from CSA America, Cleveland, OH.
[f]Canadian Standards Association, Mississauga, ON.

designed to stabilize wind effects and improve the performance. Because of the many compromises made in vent termination design, this stability is usually achieved by sacrificing some of the updraft created by the wind. Further, locating a vent cap in a cavity region frequently removes it from the zone where wind velocity could have a significant effect.

Performance optimization studies of residential vent cap design indicate that the following performance features are important: (1) still-air resistance, (2) updraft ability with no flow, and (3) discharge resistance when vent gases are carried at low velocity in a typical wind (10 fps vent velocity in a 20 mph wind). Tests in UL *Standard* 441 for proprietary gas vent caps consider these three aspects of performance to ensure adequate vent capacity.

Frequently, air supply to an appliance room is difficult to orient to eliminate wind effects. Therefore, the vent outlet must have a certain updraft capability, which can help balance a possible adverse wind. When wind flows across an inoperative vent termination, a strong updraft develops. Appliance start-up reduces this updraft, and in typical winds, the vent cap may develop greater resistance than it would have in still air. Certain vent caps can be made with very low still-air resistance, yet exhibit excessive wind resistance, which reduces capacity. Finally, because the appliance operates whether or not there is a wind, still-air resistance must be low.

Some proprietary air ventilators have excessive still-air resistance and should be avoided on vent and chimney systems unless a considerably oversized vent is specified. Vertical-slot ventilators, for example, have still-air resistance coefficients of about 4.5. To achieve low still-air resistance on vents and chimneys, the vertical-slot ventilator must be 50% larger than the diameter of the chimney or vent unless it has been specifically listed for such use.

Freestanding chimneys high enough to project above the cavity zone require structurally adequate materials or guying and bracing for prefabricated products. The prefabricated metal building heating appliance chimney places little load on the roof structure, but guying is required at 8 to 12 ft intervals to resist both overturning and oscillating wind forces. Various other expedients, such as spiral baffles on heavy-gage freestanding chimneys, have been used to reduce oscillation.

The chimney height needed to carry the effluent into the undisturbed flow stream above the wake zone can be reduced by increasing the effluent discharge velocity. A 3000 fpm discharge velocity avoids downward eddying along the chimney and expels the effluent free of the wake zone. Velocity this high can be achieved only with forced or induced draft.

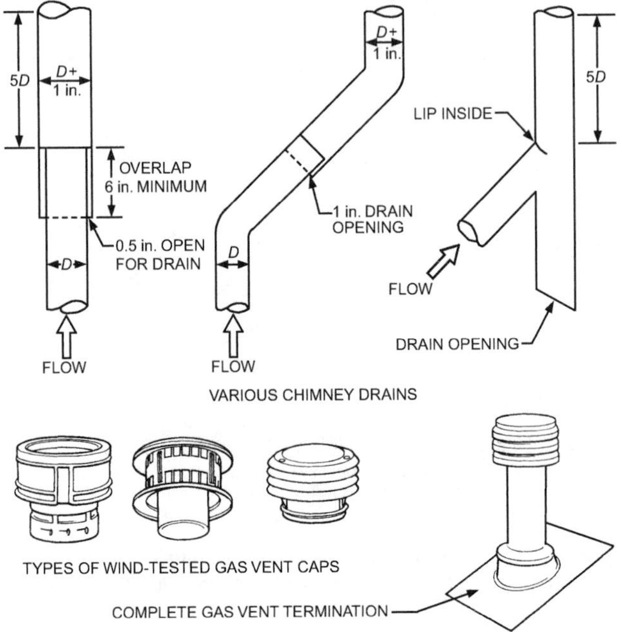

Fig. 31 Vent and Chimney Rain Protection

Rain entry is a problem for open, low-velocity, or inoperative systems. Good results have been obtained with drains that divert the water onto a roof or into a collection system leading to a sump. Figure 31 shows several configurations (Clarke 1967; Hama and Downing 1963). Runoff from stack drains contains acids, soot, and metallic corrosion products, which can cause roof staining. Therefore, these methods are not recommended for residential use. An alternative procedure is to allow all water to drain to the base of the chimney, where it is piped from a capped tee to a sump.

Rain caps prevent vertical discharge of high-velocity flue gases. However, caps are preferred for residential gas-burning equipment because it is easier to exclude rain than to risk rainwater leakage at horizontal joints or to drain it. Also, caps keep out debris and bird nests, which can block the chimney. Satisfactory vent cap performance can be achieved in the wind by using one of a variety of standard configurations, including the A cap and the wind band ventilator, or one of the proprietary designs shown in Figure 31.

Where partial rain protection without excessive flow resistance is desired, and either wind characteristics are unimportant or wind-flow is horizontal, a flat disk or cone cap 1.7 to 2.0 diameters across located 0.5 diameter above the end of the pipe has a still-air resistance loss coefficient of about 0.5.

Consult Chapter 44 of the 2007 *ASHRAE Handbook—HVAC Applications* for additional information on vent and chimney termination and wind effects.

CODES AND STANDARDS

Building and installation codes and standards prescribe the installation and safety requirements of heat-producing appliances and their vents and chimneys. Chapter 51 lists the major national building codes, one of which may be in effect in a given area. Some jurisdictions either adopt a national building code with varying degrees of revisions to suit local custom or, as in many major metropolitan areas, develop a local code that agrees in principle but shares little common text with the national codes. Familiarity with applicable building codes is essential because of the great variation in local codes and adoption of modern chimney design practice.

The national standards listed in Table 11 give greater detail on the mechanical aspects of fuel systems and chimney or vent construction. Although these standards emphasize safety aspects, especially clearances to combustibles for various venting materials, they also recognize the importance of proper flow, draft, and capacity.

CONVERSION FACTORS

The following conversion factors have been simplified for chimney design.

$$\text{Btu/h input} = 3{,}347{,}500\ \frac{\text{Boiler horsepower}}{\text{Percent efficiency}}$$

$$= 44{,}600 \times \text{Boiler horsepower (approx. 75\% eff.)}$$

$$= 41{,}800 \times \text{Boiler horsepower (approx. 80\% eff.)}$$

$$= 37{,}200 \times \text{Boiler horsepower (approx. 90\% eff.)}$$

$$= 140{,}000 \times \text{No. 1 and 2 oil (gph)}$$

$$= 150{,}000 \times \text{No. 4, 5, and 6 oil (gph)}$$

$$= 13{,}000 \times \text{Coal (lb/h)}$$

$$= 1000 \times \text{Natural gas (ft}^3\text{/h)}$$

$$= 3.412 \times \text{Watt rating}$$

$$\text{kW input} = \text{kW output/efficiency}$$

$$\text{kW} = 9.81 \times \text{Boiler horsepower}$$

SYMBOLS

A = area of passage cross section, ft^2
A_c = chimney flue cross-sectionally area, ft^2
A_F = fireplace frontal opening area, ft^2
B = existing or local barometric pressure, in. Hg
B_o = standard pressure, 29.92 in. Hg
C_u = temperature multiplier for heat loss, dimensionless
c_p = specific heat of gas at constant pressure, Btu/lb · °F
d_f = inside diameter, ft
d_i = inside diameter, in.
D_a = available draft, in. of water
D_b = boost (increase in static pressure by fan), in. of water
D_p = depressurization, in. of water
D_t = *theoretical draft, in. of water*
F = friction factor for L/d_i
f = Darcy friction factor from Moody diagram (Figure 13, Chapter 2, 2005 *ASHRAE Handbook—Fundamentals*)
g = gravitational constant, 32.174 ft/s^2
H = height of vent or chimney system above grade or system inlet, ft
I = operating heat input, Btu/h
k = system resistance loss coefficient, dimensionless
k_f = fitting friction loss coefficient, dimensionless
k_L = piping friction loss coefficient, dimensionless
L = length of all piping in chimney system from inlet to exit, linear ft
L_e = total equivalent length, L plus equivalent length of elbows, etc. [for use in Equations (2) and (3) only], ft
L_m = length of system from inlet to midpoint of vertical or to location of mean gas temperature, ft
M = ratio of mass flow to heat input, lb of combustion products per 1000 Btu of fuel burned
Δp = system flow losses or pressure drop, in. of water
q = sensible heat at a particular point in vent, Btu/h
q_m = sensible heat at average temperature in vent, Btu/h
Q = volumetric flow rate, cfm
Δt = temperature difference, °F
Δt_e = temperature difference entering system, °F
Δt_m = temperature difference at average temperature location in system, °F [$\Delta t_m = T_m - T_o$]
T = absolute temperature, °R
T_m = mean flue gas temperature at average conditions in system, °R
T_o = ambient temperature, °R
T_s = standard temperature, °R (518.67°R)
U = overall heat transfer coefficient of vent or chimney wall material, referred to inside surface area, Btu/h · ft^2 · °F
$\bar{U}$ = heat transfer coefficient for Equation (8), Btu/h · ft^2 · °F
V = velocity of gas flow in passage, fps
V_F = frontal velocity of ambient air, fps
w = mass flow of gas, lb/h
x = length of one internal side of rectangular chimney cross section, ft
y = length of other internal side of rectangular chimney cross section, ft
ρ_a = density of air at 59°F and 29.92 in. Hg, lb/ft^3 (0.0765 lb/ft^3)]
ρ_c = density of chimney gas at 0°F and 29.92 in. Hg, lb/ft^3 [0.09 lb/ft^3 in Equations (2) and (3)]
ρ_m = density of chimney gas at average temperature and local barometric pressure, lb/ft^3
ρ_o = density of air at 0°F and 29.92 in. Hg, lb/ft^3 (0.0863 lb/ft^3)

REFERENCES

ANSI. 1994. Draft hoods. *Standard* Z21.12b-1994. Canadian Standards Association, Cleveland, OH.

ASHVE. 1941. *Heating ventilating air conditioning guide*. American Society of Heating and Ventilating Engineers. Reference available from ASHRAE.

Beaumont, M., D. Fitzgerald, and D. Sewell. 1970. Comparative observations on the performance of three steel chimneys. *Institution of Heating and Ventilating Engineers* (July):85.

Butcher, T., and Y. Celebi. 1993. Fouling of oil-fired boilers and furnaces. *Proceedings of the 1993 Oil Heat Technology Conference and Workshop* (March): 140.

CSA. 2005. Natural gas and propane installation code. *Standard* B149.1-2005. Canadian Standards Association International, Mississauga, ON.

CSA. 2006. Testing method for measuring annual fireplace efficiency. *Standard* P.4.1-02 (R2006). Canadian Standards Association International, Mississauga, ON.

Chien, N., Y. Feng, H. Wong, and T. Sino. 1951. *Wind-tunnel studies of pressure on elementary building forms*. Iowa Institute of Hydraulic Research.

Clarke, J.H. 1967. Air flow around buildings. *Heating, Piping and Air Conditioning* (May):145.

Evans, B.H. 1957. Natural air flow around buildings. Texas Engineering Experiment Station *Research Report* 59.

Fontana, M.G. and N.D. Greene. 1967. *Corrosion engineering*, p. 223. McGraw-Hill, New York.

Hama, G.M. and D.A. Downing. 1963. The characteristics of weather caps. *Air Engineering* (December):34.

HYDI. 1989. *Testing and rating standard for heating boilers*. Hydronics Institute [now part of Air Conditioning, Heating and Refrigeration Institute (AHRI)], Berkeley Heights, NJ.

Kam, V.P., R.A. Borgeson, and D.W. DeWerth. 1993. Masonry chimney for category I gas appliances: Inspection and relining. *ASHRAE Transactions* 99(1):1196-1201.

Kinkead, A. 1962. Gravity flow capacity equations for designing vent and chimney systems. *Proceedings of the Pacific Coast Gas Association* 53.

Koebel, M. and M. Elsener. 1989. Corrosion of oil-fired central heating boilers. *Werkstoffe und Korrosion* 40:285-94.

Krajewski, R.F. 1996. Oil heat vent analysis program (OHVAP) users manual and engineering report. BNL *Informal Report* BNL-63668. Available from http://www.osti.gov/bridge.

Land, T. 1977. The theory of acid deposition and its application to the dew-point meter. *Journal of the Institute of Fuel* (June):68.

Lapple, C.E. 1949. Velocity head simplifies flow computation. *Chemical Engineering* (May):96.

Mueller, G.R. 1968. Charts determine gas temperature drops in metal flue stacks. *Heating, Piping and Air Conditioning* (January):138.

NFPA. 2006. Chimneys, fireplaces, vents, and solid fuel-burning appliances. ANSI/NFPA *Standard* 211-2006. National Fire Protection Association, Quincy, MA.

NFPA. 2006. Installation of oil-burning equipment. ANSI/NFPA *Standard* 31-2006. National Fire Protection Association, Quincy, MA.

NFPA/AGA. 2006. *National fuel gas code*. ANSI/NFPA 54-2006/ ANSI/AGA *Standard* Z223.1-2006. National Fire Protection Association, Quincy, MA, and American Gas Association, Washington, D.C.

Pray, H.R. et al. 1942-1953. The corrosion of metals and materials by the products of combustion of gaseous fuels. *Battelle Memorial Institute Reports* 1, 2, 3, and 4 to the American Gas Association.

Ramsey, C.G. and H.R. Sleeper. 1956. *Architectural graphic standards*. John Wiley & Sons, New York.

Rutz, A.L. 1992. VENT-II *Solution methodology*. Available from ASHRAE Handbook Editor.

Rutz, A.L. and D.D. Paul. 1991. *User's manual for VENT-II (Version 4.1) with diskettes: An interactive personal computer program for design and analysis of venting systems of one or two gas appliances*. GRI-90/0178. Gas Technology Institute, Des Plaines (formerly Gas Research Institute, Chicago), IL. Available as No. PB91-509950 from National Technical Information Services, U.S. Department of Commerce, 5285 Port Royal Road, Springfield, VA 22161.

Segeler, C.G., ed. 1965. *Gas engineers' handbook*, Section 12. Industrial Press, New York.

Stone, R.L. 1957. Design of multiple gas vents. *Air Conditioning, Heating and Ventilating* (July).

Stone, R.L. 1969. Fireplace operation depends upon good chimney design. *ASHRAE Journal* (February):63.

Stone, R.L. 1971. A practical general chimney design method. *ASHRAE Transactions* 77(1):91-100.

Stone, R.L. 2005. Letter to ASHRAE editor, revised Figure 19 (July 27, 2005).

UL. 2001. Factory-built chimneys for residential type and building heating appliances. ANSI/UL *Standard* 103-2001. Underwriters Laboratories, Northbrook, IL.

UL. 1999. Gas vents. *Standard* 441-99. Underwriters Laboratories, Northbrook, IL.

Verhoff, F.H. and J.T. Banchero. 1974. Predicting dew points of flue gases. *Chemical Engineering Progress* (August):71.

Yeaw, J.S. and L. Schnidman. 1943. Dew point of flue gases containing sulfur. *Power Plant Engineering* 47(I and II).

BIBLIOGRAPHY

ANSI. 2006. Gas-fired central furnaces. *Standard* Z21.47b-1994/CSA 2.3b-2006. Canadian Standards Association, Cleveland, OH.

Briner, C.F. 1984. Heat transfer and fluid flow analysis of sheet metal chimney systems. ASME *Paper* 84-WA/SOL-36. American Society of Mechanical Engineers, New York.

Briner, C.F. 1986. Heat transfer and fluid flow analysis of three-walled metal factory-built chimneys. *ASHRAE Transactions* 92(1B):727-38.

Hampel, T.E. 1956. Venting system priming time. *Research Bulletin* 74, Appendix E, 146. American Gas Association Laboratories, Cleveland, OH.

Jakob, M. 1955. *Heat transfer*, vol. I. John Wiley & Sons, New York.

Paul, D.D. 1992. *Venting guidelines for Category I gas appliances*. GRI-89/0016. Gas Technology Institute, Des Plaines (formerly Gas Research Institute, Chicago), IL.

Ramsey, C.G. and H.R. Sleeper. 1970. *Architectural graphic standards*, 6th ed. John Wiley & Sons, New York.

Reynolds, H.A. 1960. Selection of induced draft fans for heating boilers. *Air Conditioning, Heating and Ventilating* (December):51.

Sepsy, C.F. and D.B. Pies. 1972. *An experimental study of the pressure losses in converging flow fittings used in exhaust systems*. Ohio State University College of Engineering, Columbus (December).

CHAPTER 35

HYDRONIC HEAT-DISTRIBUTING UNITS AND RADIATORS

RADIATORS, convectors, and baseboard and finned-tube units are heat-distributing devices used in low-temperature and steam water-heating systems. They supply heat through a **combination of radiation and convection** and maintain the desired air temperature and/or mean radiant temperature in a space without fans. Figures 1 and 2 show sections of typical heat-distributing units. In low-temperature systems, radiant panels are also used. Units are inherently self-adjusting in the sense that heat output is based on temperature differentials; cold spaces receive more heat and warmer spaces receive less heat.

DESCRIPTION

Radiators

The term *radiator*, though generally confined to sectional cast-iron column, large-tube, or small-tube units, also includes flat-panel types and fabricated steel sectional types. Small-tube radiators, with a length of only 1.75 in. per section, occupy less space than column and large-tube units and are particularly suited to installation in recesses (see Table 1). Column, wall-type, and large-tube radiators are no longer manufactured, although many of these units are still in use. Refer to Tables 2, 3, and 4 in Chapter 28 of the 1988 *ASHRAE Handbook—Equipment*, Byrley (1978), or Hydronics Institute (1989) for principal dimensions and average ratings of these units.

The following are the most common types of radiators:

Sectional radiators are fabricated from welded sheet metal sections (generally 2, 3, or 4 tubes wide), and resemble freestanding cast-iron radiators.

Panel radiators consist of fabricated flat panels (generally 1, 2, or 3 deep), with or without an exposed extended fin surface attached to the rear for increased output. These radiators are most common in Europe.

Tubular steel radiators consist of supply and return headers with interconnecting parallel steel tubes in a wide variety of lengths and heights. They may be specially shaped to coincide with the building structure. Some are used to heat bathroom towel racks.

Specialty radiators are fabricated of welded steel or extruded aluminum and are designed for installation in ceiling grids or floor-mounting. An array of unconventional shapes is available.

Pipe Coils

Pipe coils have largely been replaced by finned tubes. See Table 5 in Chapter 28 of the 1988 *ASHRAE Handbook—Equipment* for the heat emission of such pipe coils.

Convectors

A convector is a heat-distributing unit that operates with gravity-circulated air (natural convection). It has a heating element with a large amount of secondary surface and contains two or more tubes with headers at both ends. The heating element is surrounded by an enclosure with an air inlet opening below and an air outlet opening above the heating element.

Convectors are made in a variety of depths, sizes, and lengths and in enclosure or cabinet types. The heating elements are available in fabricated ferrous and nonferrous metals. The air enters the enclosure below the heating element, is heated in passing through the element, and leaves the enclosure through the outlet grille located above the heating element. Factory-assembled units comprising a heating element and an enclosure have been widely used. These may be freestanding, wall-hung, or recessed and may have outlet grilles or louvers and arched inlets or inlet grilles or louvers, as desired.

Baseboard Units

Baseboard (or baseboard radiation) units are designed for installation along the bottom of walls in place of the conventional baseboard. They may be made of cast iron, with a substantial portion of the front face directly exposed to the room, or with a finned-tube element in a sheet metal enclosure. They use gravity-circulated room air.

Baseboard heat-distributing units are divided into three types: radiant, radiant convector, and finned tube. The **radiant** unit, which is made of aluminum, has no openings for air to pass over the wall side of the unit. Most of this unit's heat output is by radiation.

The **radiant-convector** baseboard is made of cast iron or steel. The units have air openings at the top and bottom to permit circulation of room air over the wall side of the unit, which has extended surface to provide increased heat output. A large portion of the heat emitted is transferred by convection.

The **finned-tube** baseboard has a finned-tube heating element concealed by a long, low sheet metal enclosure or cover. A major portion of the heat is transferred to the room by convection. The output varies over a wide range, depending on the physical dimensions and the materials used. A unit with a high relative output per unit length compared to overall heat loss (which would result in a concentration of the heating element over a relatively small area) should be avoided. Optimum comfort for room occupants is obtained when units are installed along as much of the exposed wall as possible.

Finned-Tube Units

Finned-tube (or fin-tube) units are fabricated from metallic tubing, with metallic fins bonded to the tube. They operate with gravity-circulated room air. Finned-tube elements are available in several tube sizes, in either steel or copper (1 to 2 in. nominal steel or 3/4 to 1 1/4 in. nominal copper) with various fin sizes, spacings, and materials. The resistance to the flow of steam or water is the same as that through standard distribution piping of equal size and type.

Finned-tube elements installed in occupied spaces generally have covers or enclosures in a variety of designs. When human contact is unlikely, they are sometimes installed bare or provided with an expanded metal grille for minimum protection.

The preparation of this chapter is assigned to TC 6.1, Hydronic and Steam Equipment and Systems.

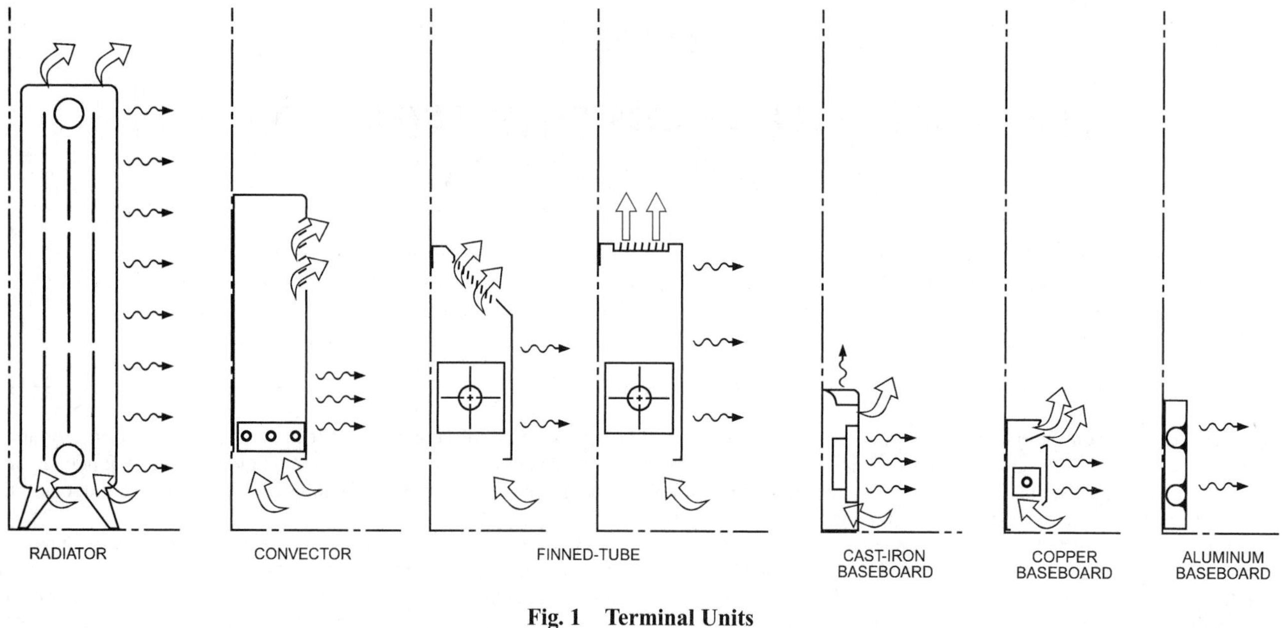

Fig. 1 Terminal Units

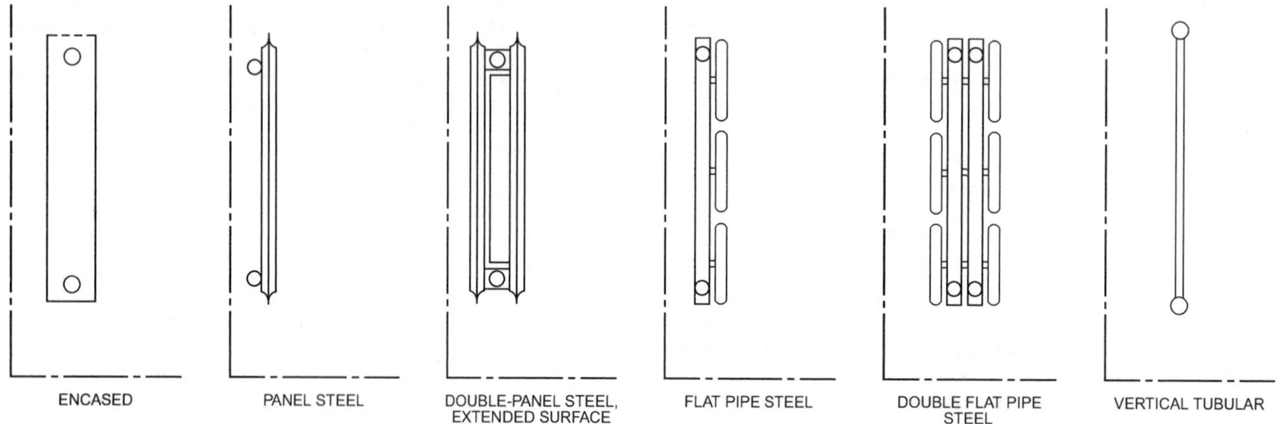

Fig. 2 Typical Radiators

A cover has a portion of the front skirt made of solid material. The cover can be mounted with clearance between the wall and the cover, and without completely enclosing the rear of the finned-tube element. A cover may have a top, front, or inclined outlet. An enclosure is a shield of solid material that completely encloses both the front and rear of the finned-tube element. An enclosure may have an integral back or may be installed tightly against the wall so that the wall forms the back, and it may have a top, front, or inclined outlet.

Heat Emission

These heat-distributing units emit heat by a combination of radiation to the surfaces and occupants in the space and convection to the air in the space.

Chapter 3 of the 2005 *ASHRAE Handbook—Fundamentals* covers the heat transfer processes and the factors that influence them. Those units with a large portion of their heated surface exposed to the space (i.e., radiator and cast-iron baseboard) emit more heat by radiation than do units with completely or partially concealed heating surfaces (i.e., convector, finned-tube, and finned-tube baseboard). Also, finned-tube elements constructed of steel emit a larger portion of heat by radiation than do finned-tube elements constructed of nonferrous materials.

The heat output ratings of heat-distributing units are expressed in Btu/h, MBh (1000 Btu/h), or in square feet equivalent direct radiation (EDR). By definition, 240 Btu/h = 1 ft^2 EDR with 1 psig steam.

RATINGS OF HEAT-DISTRIBUTING UNITS

For convectors, baseboard units, and finned-tube units, an allowance for heating effect may be added to the **test capacity** (the heat extracted from the steam or water under standard test conditions). This **heating effect** reflects the ability of the unit to direct its heat output to the occupied zone of a room. The application of a heating effect factor implies that some units use less steam or hot water than others to produce an equal comfort effect in a room.

Radiators

Current methods for rating radiators were established by the U.S. National Bureau of Standards publication, *Simplified Practices Recommendation* R174-65, Cast-Iron Radiators, which has been withdrawn (see Table 1).

Convectors

The generally accepted method of testing and rating ferrous and nonferrous convectors in the United States was given in *Commercial Standard* CS 140-47, Testing and Rating Convectors (Dept. of

Table 1 Small-Tube Cast-Iron Radiators

Number of Tubes per Section	Catalog Rating per Section,[a] ft²	Btu/h	A Height, in.[b]	B Width, in. Min.	Max.	C Spacing, in.[c]	D Leg Height, in.[b]
3	1.6	384	25	3.25	3.50	1.75	2.50
4	1.6	384	19	4.44	4.81	1.75	2.50
	1.8	432	22	4.44	4.81	1.75	2.50
	2.0	480	25	4.44	4.81	1.75	2.50
5	2.1	504	22	5.63	6.31	1.75	2.50
	2.4	576	25	5.63	6.31	1.75	2.50
6	2.3	552	19	6.81	8	1.75	2.50
	3.0	720	25	6.81	8	1.75	2.50
	3.7	888	32	6.81	8	1.75	2.50

[a]These ratings are based on steam at 215°F and air at 70°F. They apply only to installed radiators exposed in a normal manner, not to radiators installed behind enclosures, behind grilles, or under shelves. For Btu/h ratings at other temperatures, multiply table values by factors found in Table 2.

[b]Overall height and leg height, as produced by some manufacturers, are 1 in. greater than shown in Columns A and D. Radiators may be furnished without legs. Where greater than standard leg heights are required, leg height shall be 4.5 in.

[c]Length equals number of sections multiplied by 1.75 in.

Commerce 1947), but it has been withdrawn. This standard contained details covering construction and instrumentation of the test booth or room and procedures for determining steam and water ratings.

Under the provisions of *Commercial Standard* CS 140-47, the rating of a top outlet convector was established at a value not in excess of the test capacity. For convectors with other types of enclosures or cabinets, a percentage that varies up to a maximum of 15% (depending on the height and type of enclosure or cabinet) was added for heating effect (Brabbee 1926; Willard et al. 1929). The addition made for heating effect must be shown in the manufacturer's literature.

The testing and rating procedure set forth by *Commercial Standard* CS 140-47 does not apply to finned-tube or baseboard radiation.

Baseboard Units

The generally accepted method of testing and rating baseboards in the United States is covered in the *Testing and Rating Standard for Baseboard Radiation* (Hydronics Institute 1990a). This standard contains details covering construction and instrumentation of the test booth or room, procedures for determining steam and hot-water ratings, and licensing provisions for obtaining approval of these ratings. Baseboard ratings include an allowance for heating effect of 15% in addition to the test capacity. The addition made for heating effect must be shown in the manufacturer's literature.

Finned-Tube Units

The generally accepted method of testing and rating finned-tube units in the United States is covered in the *Testing and Rating Standard for Finned-Tube (Commercial) Radiation* (Hydronics Institute 1990b). This standard contains details covering construction and instrumentation of the test booth or room, procedures for determining steam and water ratings, and licensing provisions for obtaining approval of these ratings.

The rating of a finned-tube unit in an enclosure that has a top outlet is established at a value not in excess of the test capacity. For finned-tube units with other types of enclosures or covers, a percentage is added for heating effect that varies up to a maximum of 15%, depending on the height and type of enclosure or cover. The addition made for heating effect must be shown in the manufacturer's literature (Pierce 1963).

Other Heat-Distributing Units

Unique radiators and radiators from other countries generally are tested and rated for heat emission in accordance with prevailing-standards. These other testing and rating methods have basically the same procedures as the Hydronics Institute standards, which are used in the United States. See Chapter 6 for information on the design and sizing of radiant panels.

Corrections for Nonstandard Conditions

The heating capacity of a radiator, convector, baseboard, finned-tube heat-distributing unit, or radiant panel is a power function of the temperature difference between the air in the room and the heating medium in the unit, shown as

$$q = c(t_s - t_a)^n \quad (1)$$

where

q = heating capacity, Btu/h
c = constant determined by test
t_s = average temperature of heating medium, °F. For hot water, the arithmetic average of the entering and leaving water temperatures is used.
t_a = room air temperature, °F. Air temperature 60 in. above the floor is generally used for radiators, whereas entering air temperature is used for convectors, baseboard units, and finned-tube units.
n = exponent that equals 1.3 for cast-iron radiators, 1.4 for baseboard radiation, 1.5 for convectors, 1.0 for ceiling heating and floor cooling panels, and 1.1 for floor heating and ceiling cooling panels. For finned-tube units, n varies with air and heating medium temperatures. Correction factors to convert heating capacities at standard rating conditions to heating capacities at other conditions are given in Table 2.

Equation (1) may also be used to calculate heating capacity at nonstandard conditions.

DESIGN

Effect of Water Velocity

Designing for high temperature drops through the system (drops of as much as 60 to 80°F in low-temperature systems and as much as 200°F in high-temperature systems) can result in low water velocities in the finned-tube or baseboard element. Application of very short runs designed for conventional temperature drops (i.e., 20°F) can also result in low velocities.

Table 2 Correction Factors c for Various Types of Heating Units

Steam Pressure (Approx.)		Steam or Water Temp., °F	Cast-Iron Radiator Room Temp., °F					Convector Inlet Air Temp., °F					Finned-Tube Inlet Air Temp., °F					Baseboard Inlet Air Temp., °F				
			80	75	70	65	60	75	70	65	60	55	75	70	65	60	55	75	70	65	60	55
		100											0.10	0.12	0.15	0.17	0.20	0.08	0.10	0.13	0.15	0.18
		110											0.15	0.17	0.20	0.23	0.26	0.13	0.15	0.18	0.21	0.25
		120											0.20	0.23	0.26	0.29	0.33	0.18	0.21	0.25	0.28	0.31
		130											0.26	0.29	0.33	0.36	0.40	0.25	0.28	0.31	0.34	0.38
		140											0.33	0.36	0.40	0.42	0.45	0.31	0.34	0.38	0.42	0.45
in. Hg Vac.	**psia**																					
22.4	3.7	150	0.39	0.42	0.46	0.50	0.54	0.35	0.39	0.43	0.46	0.50	0.40	0.42	0.45	0.49	0.53	0.38	0.42	0.45	0.49	0.53
20.3	4.7	160	0.46	0.50	0.54	0.58	0.62	0.43	0.47	0.51	0.54	0.58	0.45	0.49	0.53	0.57	0.61	0.45	0.49	0.53	0.57	0.61
17.7	6.0	170	0.54	0.58	0.62	0.66	0.69	0.51	0.54	0.58	0.63	0.67	0.53	0.57	0.61	0.65	0.69	0.53	0.57	0.61	0.65	0.69
14.6	7.5	180	0.62	0.66	0.69	0.74	0.78	0.58	0.63	0.67	0.71	0.76	0.61	0.65	0.69	0.73	0.78	0.61	0.65	0.69	0.72	0.78
10.9	9.3	190	0.69	0.74	0.78	0.83	0.87	0.67	0.71	0.76	0.81	0.85	0.69	0.73	0.78	0.81	0.86	0.69	0.73	0.78	0.82	0.86
6.5	11.5	200	0.78	0.83	0.87	0.91	0.95	0.76	0.81	0.85	0.90	0.95	0.77	0.81	0.86	0.90	0.95	0.81	0.86	0.92	0.95	1.00
psig	**psia**																					
1	15.6	215	0.91	0.95	1.00	1.04	1.09	0.90	0.95	1.00	1.05	1.10	0.91	0.94	1.00	1.06	1.11	0.91	0.95	1.00	1.05	1.09
6	21	230	1.04	1.09	1.14	1.18	1.23	1.05	1.10	1.15	1.20	1.26	1.03	1.08	1.14	1.19	1.24	1.04	1.09	1.14	1.19	1.25
15	30	250	1.23	1.28	1.32	1.37	1.43	1.27	1.32	1.37	1.43	1.47	1.20	1.26	1.31	1.37	1.43	1.22	1.27	1.32	1.37	1.43
27	42	270	1.43	1.47	1.52	1.56	1.61	1.47	1.54	1.59	1.67	1.72	1.38	1.44	1.50	1.56	1.62	1.43	1.47	1.52	1.59	1.64
52	67	300	1.72	1.75	1.82	1.89	1.92	1.85	1.89	1.96	2.04	2.08	1.67	1.73	1.79	1.86	1.92	1.75	1.82	1.89	1.92	1.96

Note: Use these correction factors to determine output ratings for radiators, convectors, and finned-tube and baseboard units at operating conditions other than standard. Standard conditions in the United States for a radiator are 215°F heating medium temperature and 70°F room temperature (at the center of the space and at the 5 ft level).

Standard conditions for convectors and finned-tube and baseboard units are 215°F heating medium temperature and 65°F inlet air temperature at 29.92 in. Hg atmospheric pressure. Water flow is 3 fps for finned-tube units. Inlet air at 65°F for convectors and finned-tube or baseboard units represents the same room comfort conditions as 70°F room air temperature for a radiator. Standard conditions for radiant panels are 122°F heating medium temperature and 68°F for room air temperature; *c* depends on panel construction.

To determine the output of a heating unit under conditions other than standard, multiply the standard heating capacity by the appropriate factor for the actual operating heating medium and room or inlet air temperatures.

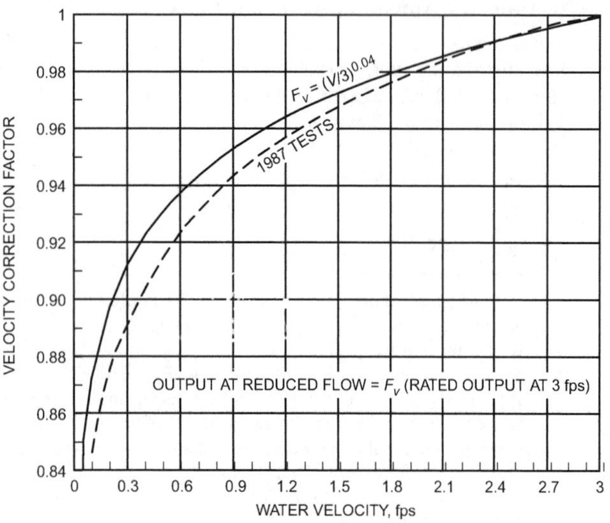

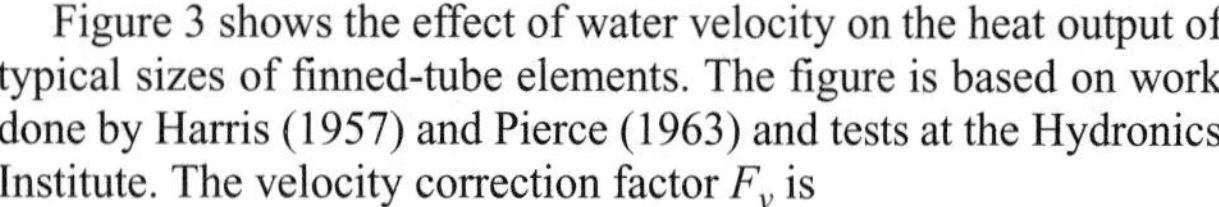

Fig. 3 Water Velocity Correction Factor for Baseboard and Finned-Tube Radiators

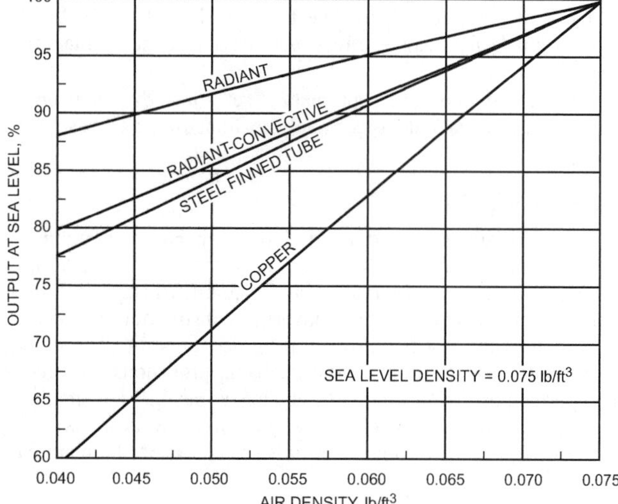

Fig. 4 Effect of Air Density on Radiator Output

Figure 3 shows the effect of water velocity on the heat output of typical sizes of finned-tube elements. The figure is based on work done by Harris (1957) and Pierce (1963) and tests at the Hydronics Institute. The velocity correction factor F_v is

$$F_v = (V/3.0)^{0.04} \tag{2}$$

where V = water velocity, fps.

Heat output varies little over the range from 0.5 to 3 fps, where F_v ranges from 0.93 to 1.00. The factor drops rapidly below 0.5 fps because the flow changes from turbulent to laminar at around 0.1 fps. Such a low velocity should be avoided because the output is difficult to predict accurately when designing a system. In addition, the curve is so steep in this region that small changes in actual flow have a significant effect on output. Not only does the heat transfer rate change, but the temperature drop and, therefore, the average water temperature change (assuming a constant inlet temperature).

The designer should check water velocity throughout the system and select finned-tube or baseboard elements on the basis of velocity as well as average temperature. Manufacturers of finned-tube and baseboard elements offer a variety of tube sizes, ranging from 0.5 in. copper tubes for small baseboard elements to 2 in. for large finned-tube units, to aid in maintenance of turbulent flow conditions over a wide range of flow.

Effect of Altitude

The effect of altitude on heat output varies depending on the material used and the portion of the unit's output that is radiant rather than convective. The reduced air density affects the convective portion. Figure 4 shows the reduction in heat output with air density (Sward and Decker 1965). The approximate correction factor F_A for determining the reduced output of typical units is

$$F_A = (p/p_o)^n \tag{3}$$

where

p = local station atmospheric pressure
p_o = standard atmospheric pressure
n = 0.9 for copper baseboard or finned tube
= 0.5 for steel finned-tube or cast-iron baseboard
= 0.2 for radiant baseboard and radiant panels

The value of p/p_o at various altitudes may be calculated as follows:

$$p/p_o = e^{-3.73 \times 10^{-5} h} \tag{4}$$

where h = altitude, ft. The following are typical values of p/p_o:

Altitude h, ft	p/p_o
2000	0.93
4000	0.86
5000	0.83
6000	0.80

Sward and Harris (1970) showed that some components of heat loss are affected in the same manner.

Effect of Mass

Mass of the terminal unit (typically cast-iron versus copper-aluminum finned element) affects the heat-up and cool-down rates of the equipment. It is important that high- and low-mass radiation not be mixed in the same zone. High-mass systems have historically been favored for best comfort and economy, but tests by Harris (1970) have shown no measurable difference. The thermostat or control can compensate by changing the cycle rate of the burner. The effect of mass is further reduced by constantly circulating modulated temperature water to the unit. The only time that mass can have a significant effect is in response to a massive shift in load. In that situation, the low-mass unit will respond faster.

Performance at Low Water Temperatures

Table 2 summarizes the performance of baseboard and finned-tube units with an average water temperature down to 100°F. Solar-heated water, industrial waste heat in a low- to medium-temperature district heating system, heat pump system cooling water, and ground-source heat pumps are typical applications in this range. To compensate for heating capacity loss, either heat-distributing equipment should be oversized, or additional heat-distributing units should be installed. Capital and operating costs should be minimized (Kilkis 1998).

Effect of Enclosure and Paint

An enclosure placed around a direct radiator restricts the airflow and diminishes the proportion of output resulting from radiation. However, enclosures of proper design may improve the heat distribution within the room compared to the heat distribution obtained with an unenclosed radiator (Allcut 1933; Willard et al. 1929).

For a radiator or cast-iron baseboard, the finish coat of paint affects the heat output. Oil paints of any color give about the same results as unpainted black or rusty surfaces, but an aluminum or a bronze paint reduces the heat emitted by radiation. The net effect may reduce the total heat output of the radiator by 10% or more (Allen 1920; Rubert 1937; Severns 1927).

APPLICATIONS

Radiators

Radiators can be used with steam or hot water. They are installed in areas of greatest heat loss: under windows, along cold walls, or at doorways. They can be installed freestanding, semirecessed, or with decorative enclosures or shields, although the enclosures or shields affect the output (Willard et al. 1929).

Unique and imported radiators are generally not suitable for steam applications, although they have been used extensively in low-temperature water systems with valves and connecting piping left exposed. Various combinations of supply and return locations are possible, which may alter the heat output. Although long lengths may be ordered for linear applications, lengths may not be reduced or increased by field modification. The small cross-sectional areas often inherent in unique radiators require careful evaluation of flow requirements, water temperature drop, and pressure drops.

Convectors

Convectors can be used with steam or hot water. Like radiators, they should be installed in areas of greatest heat loss. They are particularly applicable where wall space is limited, such as in entryways and kitchens.

Baseboard Radiation

Baseboard units are used almost exclusively with hot water. When used with one-pipe steam systems, tube sizes of 1.25 in. NPS must be used to allow drainage of condensate counterflow to the steam flow.

The basic advantage of the baseboard unit is that its normal placement is along the cold walls and under areas where the greatest heat loss occurs. Other advantages are that it (1) is inconspicuous, (2) offers minimal interference with furniture placement, and (3) distributes the heat near the floor. This last characteristic reduces the floor-to-ceiling temperature gradient to about 2 to 4°F and tends to produce uniform temperature throughout the room. It also makes baseboard heat-distributing units adaptable to homes without basements, where cold floors are common (Kratz and Harris 1945).

Heat loss calculations for baseboard heating are the same as those used for other types of heat-distributing units. The Hydronics Institute (1989) describes a procedure for designing baseboard heating systems.

Finned-Tube Radiation

The finned-tube unit can be used with either steam or hot water. It is advantageous for heat distribution along the entire outside wall, thereby preventing downdrafts along the walls in buildings such as schools, churches, hospitals, offices, airports, and factories. It may be the principal source of heat in a building or a supplementary heater to combat downdrafts along the exposed walls in conjunction with a central conditioned air system. Its placement under or next to windows or glass panels helps to prevent fogging or condensation on the glass.

Normal placement of a finned tube is along the walls where the heat loss is greatest. If necessary, the units can be installed in two or three tiers along the wall. Hot-water installations requiring two or three tiers may run a serpentine water flow if the energy loss is not excessive. A header connection with parallel flow may be used, but the design must not (1) permit water to short circuit along the path of least resistance, (2) reduce capacity because of low water velocity in each tier, or (3) cause one or more tiers to become air-bound.

Many enclosures have been developed to meet building design requirements. The wide variety of finned-tube elements (tube size and material, fin size, spacing, fin material, and multiple tier installation), along with the various heights and designs of enclosures, give great flexibility of selection for finned-tube units that meet the needs of load, space, and appearance.

In areas where zone control rather than individual room control can be applied, all finned-tube units in the zone should be in series. In such a series loop installation, however, temperature drop must be considered in selecting the element for each separate room in the loop.

Radiant Panels

Hydronic radiant heating panels are controlled-temperature surfaces on the floor, walls, or ceiling; a heated fluid circulates through a circuit embedded in or attached to the panel. More than 50% of the total heating capacity is transmitted by radiant heat transfer. Usually, 120°F mean fluid temperature delivers enough heat to indoor surfaces. With such low temperature ratings, hydronic radiant panels are suitable for low-temperature heating. See Chapter 6 for more information.

REFERENCES

Allcut, E.A. 1933. Heat output of concealed radiators. School of Engineering *Research Bulletin* 140. University of Toronto, Canada.

Allen, J.R. 1920. Heat losses from direct radiation. *ASHVE Transactions* 26:11.

Brabbee, C.W. 1926. The heating effect of radiators. *ASHVE Transactions* 32:11.

Department of Commerce. 1947. Commercial standard for testing and rating convectors. *Standard* CS 140-47. Withdrawn. Washington, D.C.

Harris, W.H. 1957. Factor affecting baseboard rating test results. Engineering Experiment Station *Bulletin* 444. University of Illinois, Urbana-Champaign.

Harris, W.H. 1970. *Operating characteristics of ferrous and non-ferrous baseboard*. IBR #8. University of Illinois, Urbana-Champaign.

Kilkis, B.I. 1998. Equipment oversizing issues with hydronic heating systems. *ASHRAE Journal* 40(1):25-31.

Kratz, A.P. and W.S. Harris. 1945. A study of radiant baseboard heating in the IBR research home. Engineering Experiment Station *Bulletin* 358. University of Illinois, Urbana-Champaign.

National Bureau of Standards (currently NIST). 1965. Cast-iron radiators. *Simplified Practices Recommendation* R 174-65. Withdrawn. Available from Hydronics Institute, Berkeley Heights, NJ.

Pierce, J.S. 1963. Application of fin tube radiation to modern hot water systems. *ASHRAE Journal* 5(2):72.

Rubert, E.A. 1937. Heat emission from radiators. Engineering Experiment Station *Bulletin* 24. Cornell University, Ithaca, NY.

Severns, W.H. 1927. Comparative tests of radiator finishes. *ASHVE Transactions* 33:41.

Sward, G.R. and A.S. Decker. 1965. Symposium on high altitude effects on performance of equipment. ASHRAE.

Sward, G.R. and W.S. Harris. 1970. Effect of air density on the heat transmission coefficients of air films and building materials. *ASHRAE Transactions* 76:227-239.

Willard, A.C., A.P. Kratz, M.K. Fahnestock, and S. Konzo. 1929. Investigation of heating rooms with direct steam radiators equipped with enclosures and shields. *ASHVE Transactions* 35:77 or Engineering Experiment Station *Bulletin* 192. University of Illinois, Urbana-Champaign.

BIBLIOGRAPHY

Byrley, R.R. 1978. *Hydronic rating handbook*. Color Art Inc., St. Louis.

Hydronics Institute. 1989. *Installation guide for residential hydronic heating systems*. IBR 200, 1st ed. Hydronics Institute, Berkeley Heights, NJ.

Hydronics Institute. 1990a. *Testing and rating standard for baseboard radiation*, 11th ed. Hydronics Institute, Berkeley Heights, NJ.

Hydronics Institute. 1990b. *Testing and rating standard for finned-tube (commercial) radiation*, 5th ed. Hydronics Institute, Berkeley Heights, NJ.

Kratz, A.P. 1931. Humidification for residences. Engineering Experiment Station *Bulletin* 230:20, University of Illinois, Urbana-Champaign.

Laschober, R.R. and G.R. Sward. 1967. Correlation of the heat output of unenclosed single- and multiple-tier finned-tube units. *ASHRAE Transactions* 73(I):V.3.1-15.

Willard, A.C, A.P. Kratz, M.K. Fahnestock, and S. Konzo. 1931. Investigation of various factors affecting the heating of rooms with direct steam radiators. Engineering Experiment Station *Bulletin* 223. University of Illinois, Urbana-Champaign.

CHAPTER 36

SOLAR ENERGY EQUIPMENT

SOLAR energy use is becoming more economical as the cost of energy continues to climb, especially with increasing government and utility incentives as well as growing interest in green and/or sustainable construction. In addition, many countries consider solar and renewable energy as a security measure to ensure the availability of power under adverse conditions. While the United States continues to grow its solar industry, China, Europe, Asia, and the Mediterranean basin are leading development of advanced manufacturing techniques and applications. However, equipment and systems are still very similar in all markets; therefore, this chapter primarily discusses the basic equipment used, with particular attention to collectors. More detailed descriptions of systems and designs can be found in Chapter 33 of the 2007 *ASHRAE Handbook—HVAC Applications.*

Commercial and industrial solar energy systems are generally classified according to the heat transfer medium used in the collector loop (i.e., air or liquid). Although both systems share basic fundamentals of conversion of solar radiant energy, the equipment used in each is entirely different. Air systems are primarily limited to forced-air space heating and industrial and agricultural drying processes. Liquid systems are suitable for a broader range of applications, such as hydronic space heating, service water heating, industrial process water heating, energizing heat-driven air conditioning, and pool heating, and as a heat source for series-coupled heat pumps. Because of this wide range in capability, liquid systems are more common than air systems in commercial and industrial applications.

Photovoltaic systems, an entirely different class of solar energy equipment, convert light from the sun directly into electricity for a wide variety of applications.

SOLAR HEATING SYSTEMS

Solar energy system design requires careful attention to detail because solar radiation is a low-intensity form of energy, and the equipment to collect and use it can be expensive. A brief overview of air and liquid systems is presented here to show how the equipment fits into each type of system. Chapter 33 of the 2007 *ASHRAE Handbook—HVAC Applications* covers solar energy use, and books on design, installation, operation, and maintenance are also available (ASHRAE 1988, 1990, 1991).

Solar energy and HVAC systems often use the same components and equipment. This chapter covers only the following elements, which are either exclusive to or have specific uses in solar energy applications:

- Collectors and collector arrays
- Thermal energy storage
- Heat exchangers
- Controls

The preparation of this chapter is assigned to TC 6.7, Solar Energy Utilization.

Thermal energy storage is also covered in Chapter 34 of the 2007 *ASHRAE Handbook—HVAC Applications*. Heat exchangers are also covered in Chapter 47 of this volume, as well as pumps in Chapter 43, and fans in Chapter 20.

AIR-HEATING SYSTEMS

Air-heating systems circulate air through ducts to and from an air heating collector (Figure 1). Air systems are effective for space heating because a heat exchanger is not required and the collector inlet temperature is low throughout the day (approximately room temperature). Air systems do not need protection from freezing, overheat, or corrosion. Furthermore, air costs nothing and does not cause disposal problems or structural damage. However, air ducts and air-handling equipment require more space than pipes and pumps, ductwork is hard to seal, and leaks are difficult to detect. Fans consume more power than the pumps of a liquid system, but if the unit is installed in a facility that uses air distribution, only a slight power cost is chargeable against the solar space-heating system. Thermal storage for hot-air systems has been problematic as well because of the difficulty in controlling humidity and mold growth in pebble beds and other such devices, particularly in humid climates.

Most air space-heating systems also preheat domestic hot water through an air-to-liquid heat exchanger. In this case, tightly fitting dampers are required to prevent reverse thermosiphoning at night, which could freeze water in the heat exchanger coil. If this system heats only water in the summer, the parasitic power consumption must be charged against the solar energy system because no space heating is involved and there are no comparable energy costs associated with conventional water heating. In some situations, solar water-heating systems could be more expensive than conventional water heaters, particularly if electrical energy costs are high. To reduce parasitic power consumption, some systems use the low speed of a two-speed fan.

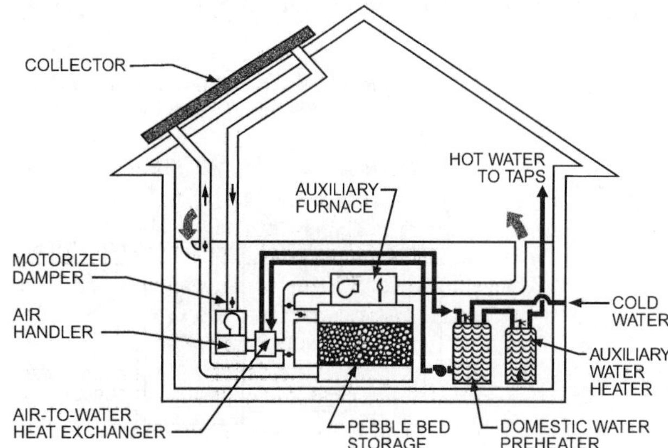

Fig. 1 Air-Heating Space and Domestic Water Heater System

LIQUID-HEATING SYSTEMS

Liquid-heating systems circulate a liquid, often a water-based fluid, through a solar collector (Figures 2 and 3). The liquid in solar collectors must be protected against freezing, which could damage the system.

Freezing is the principal cause of liquid system failure. For this reason, freeze tolerance is an important factor in selecting the heat transfer fluid and equipment in the collector loop. A solar collector radiates heat to the cold sky and freezes at air temperatures well above 32°F. Where freezing conditions are rare, small solar heating systems are often equipped with low-cost protection devices that depend on simple manual, electrical, and/or mechanical components (e.g., electronic controllers and automatic valves) for freeze protection. Because of the large investment associated with most commercial and industrial installations, solar designers and installers must consider designs providing reliable freeze protection, even in the warmest climates.

Direct and Indirect Systems

In a direct liquid system, city water circulates through the collector. In an indirect system, the collector loop is separated from the high-pressure city water supply by a heat exchanger. In areas of poor water quality, isolation protects the collectors from fouling by minerals in the water. Indirect systems also offer greater freeze protection, so they are used almost exclusively in commercial and industrial applications.

Freeze Protection

Direct systems, used where freezing is infrequent and not severe, can avoid freeze damage by (1) recirculating warm storage water through the collectors, (2) continually flushing the collectors with cold water, or (3) isolating collectors from the water and draining them. Systems that can be drained to avoid freeze damage are called **draindown** systems. Although all of these methods can be effective, none of them are generally approved by sanctioning bodies or recommended by manufacturers. Use caution when designing direct systems in freezing climates with any of these freeze protection schemes.

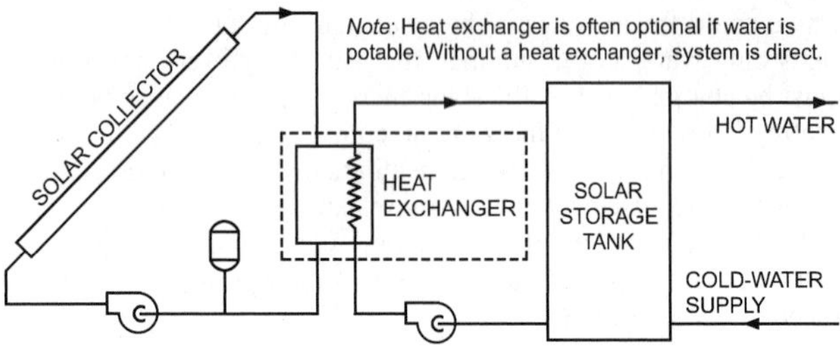

Fig. 2 Simplified Schematic of Indirect Nonfreezing System

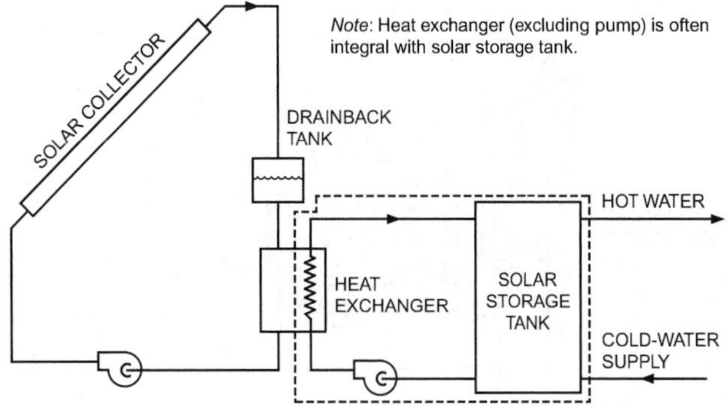

Fig. 3 Simplified Schematic of Indirect Drainback Freeze Protection System

Indirect systems use two methods of freeze protection: (1) nonfreezing fluids and (2) drainback.

Nonfreezing Fluid Freeze Protection. The most popular solar energy system for commercial use is the indirect system with a nonfreezing heat transfer fluid to transmit heat from the solar collectors to storage (Figure 2). The most common heat transfer fluid is water/propylene glycol, although other heat transfer fluids such as silicone oils, hydrocarbon oils, or refrigerants can be used. Because the collector loop is closed and sealed, the only contribution to pump pressure is friction loss; therefore, the location of solar collectors relative to the heat exchanger and storage tank is not critical. Traditional hydronic sizing methods can be used for selecting pumps, expansion tanks, heat exchangers, and air removal devices, as long as the heat transfer liquid's thermal properties are considered.

When the control system senses an increase in solar panel temperature, the pump circulates the heat transfer liquid, and energy is collected. The same control also activates a pump on the domestic water side that circulates water through the heat exchanger, where it is heated by the heat transfer fluid. This mode continues until the temperature differential between the collector and the tank is too slight for meaningful energy to be collected. At this point, the control system shuts the pumps off. At low temperatures, the nonfreezing fluid protects the solar collectors and related piping from bursting. Because the heat transfer fluid can affect system performance, reliability, and maintenance requirements, fluid selection should be carefully considered.

Because the collector loop of the nonfreezing system remains filled with fluid, it allows flexibility in routing pipes and locating components. However, a double-separation (double-wall) heat exchanger is generally required (by local building codes) to prevent contamination of domestic water in the event of a leak. The double-wall heat exchanger also protects the collectors from freeze damage if water leaks into the collector loop. However, the double-wall heat exchanger reduces efficiency by forcing the collector to operate at a higher temperature. The heat exchanger can be placed inside the tank, or an external heat exchanger can be used, as shown in Figure 2. The collector loop is closed and, therefore, requires an expansion tank and pressure-relief valve. Air purge is also necessary to expel air during filling and to remove air that has been absorbed into the heat transfer fluid.

Overtemperature protection is necessary to ensure that the system operates within safe limits and to prevent collector fluid from corroding the absorber or heat exchanger. For maximum reliability, glycol should be replaced every few years.

In some cases, systems have failed because the collector fluid in the loop thermosiphoned and froze the water in the heat exchanger. This disastrous situation must be avoided by design if the water side is exposed to the city water system because the collector loop eventually fills with water, and all freeze protection is lost.

Drainback Freeze Protection. A drainback solar water-heating system (Figure 3) uses ordinary water as the heat transport medium between the collectors and thermal energy storage. Reverse-draining (or back-siphoning) the water into a drainback tank located in a nonfreezing environment protects the system from freezing whenever the controls turn off the circulator pump or a power outage occurs.

For drainback systems with a large amount of working fluid in the collector loop, heat loss can be significantly decreased and overall efficiency increased by including a tank for storing the heat transfer fluid at night. Using a night storage tank in large systems is an appropriate strategy even for regions with favorable meteorological conditions.

The drainback tank can be a sump with a volume slightly greater than the collector loop, or it can be the thermal energy storage tank. The collector loop may or may not be vented to the atmosphere. Many designers prefer the nonvented drainback loop because makeup water is not required and the corrosive effects of air that would otherwise be ingested into the collector loop are eliminated.

The drainback system is virtually fail-safe because it automatically reverts to a safe condition whenever the circulator pump stops. Furthermore, a 20 to 30% glycol solution can be added to drainback loops for added freeze protection in case of controller or sensor failure. Because the glycol is not exposed to stagnation temperatures, it does not decompose.

A drainback system requires space for the necessary pitching of collectors and pipes for proper drainage. Also, a nearby heated area must have a room for the pumps and the drainback tank. Plumbing exposed to freezing conditions drains to the drainback tank, making the drainback design unsuitable for sites where the collector cannot be elevated above the storage tank.

Both dynamic and static pressure losses must be considered in drainback system design. Dynamic pressure loss is due to friction in the pipes, and static pressure loss is associated with the distance the water must be lifted above the level of the drainback tank to the top of the collector. There are two distinct designs of drainback systems: the oversized downcomer (or open-drop) and the siphon return. The static head requirement remains constant in the open-drop system and decreases in the siphon return.

Drainback performs better than other systems in areas with low temperatures or high irradiance. Drainback has the advantage that time and energy are not lost in reheating a fluid mass left in the collector and associated piping (as in the case of antifreeze systems). Also, water has a higher heat transfer capacity and is less viscous than other heat transfer fluids, resulting in smaller parasitic energy use and higher overall system efficiency. In closed-return (indirect) designs, there is also less parasitic energy consumption for pumping because water is the heat transfer fluid. Drainback systems can be worked on safely under stagnation conditions, but should not be restarted during peak solar conditions to avoid unnecessary thermal stress on the collector.

SOLAR THERMAL ENERGY COLLECTORS

Collector Types

Solar collectors depend on air heating, liquid heating, or liquid-vapor phase change to transfer heat. The most common type for commercial, residential, and low-temperature (<200°F) industrial applications is the flat-plate collector.

Liquid-Heating Collectors. Figure 4A shows a cross section of a flat-plate liquid collector. A flat-plate collector contains an absorber plate covered with a black coating and one or more transparent covers. The covers are transparent to incoming solar radiation and relatively opaque to outgoing (long-wave) radiation, but their principal purpose is to reduce convection heat loss. The collector box is insulated to prevent conduction heat loss from the back and edge of the absorber plate. This type of collector can supply hot water or air at temperatures up to 200°F, although efficiency diminishes rapidly above 160°F. The advantages of flat-plate collectors are simple construction, low relative cost, no moving parts, relative ease of repair, and durability. They also absorb diffuse radiation, which is a distinct advantage in cloudy climates.

Another type of collector is the integral collector storage (ICS) system. These collectors incorporate thermal storage within the collector itself. The storage tank surface serves as the absorber surface. Most ICS systems use only one tank, but some use several tanks in series. As with flat-plate collectors, insulated boxes enclose the tanks with transparent coverings on the side facing the sun. Although the simplicity of ICS systems is attractive, they are generally suitable only for applications in mild climates with small thermal storage requirements. Freeze protection by manually draining the unit is necessary in colder climates.

Still another type of collector is the evacuated tube, where the absorber mechanism is encased in a glass vacuum tube (Figure 5). Absorbers may be a simple copper fin tube on a copper sheet, a large copper cylinder in ICS applications, or a heat pipe. The latter uses a phase-change fluid to transfer heat to a common manifold where the working fluid circulates. The vacuum envelope reduces convection and conduction losses, so the tubes can operate at higher temperatures than flat-plate collectors. Like flat-plate collectors, they collect both direct and diffuse radiation. Because of its high-temperature capability, the evacuated-tube collector is favored for energizing heat-driven air-conditioning equipment, particularly when combined with concentrating reflectors behind the tubes.

Flat-plate and evacuated-tube collectors are usually mounted in a fixed position. Concentrating collectors are available that must be arranged to track the movement of the sun. These are mainly used for high-temperature industrial applications above 240°F. For more information on concentrating collectors, see Chapter 33 of the 2007 *ASHRAE Handbook—HVAC Applications*.

Finally, unglazed flat-plate absorbers are often used for low-temperature applications such as pool heating. The most popular configurations are made of plastic structured to create an almost fully wetted surface or uses very small fins on extruded tubes. Metal

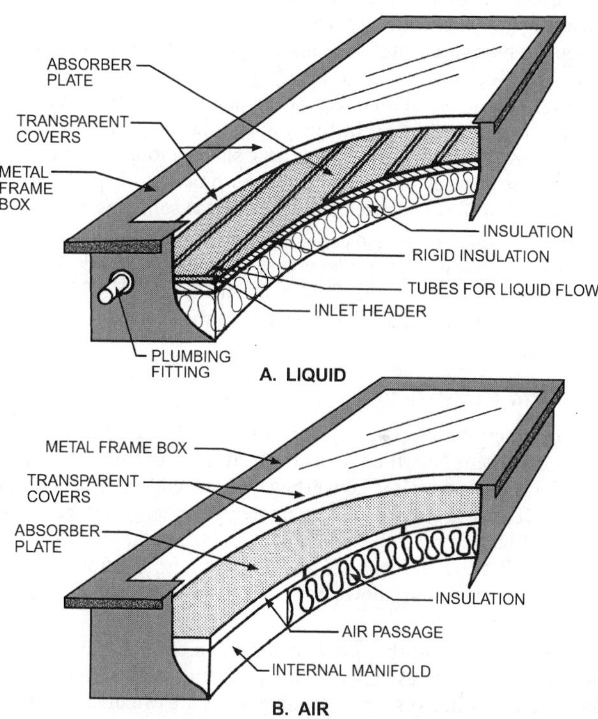

Fig. 4 Solar Flat-Plate Collectors

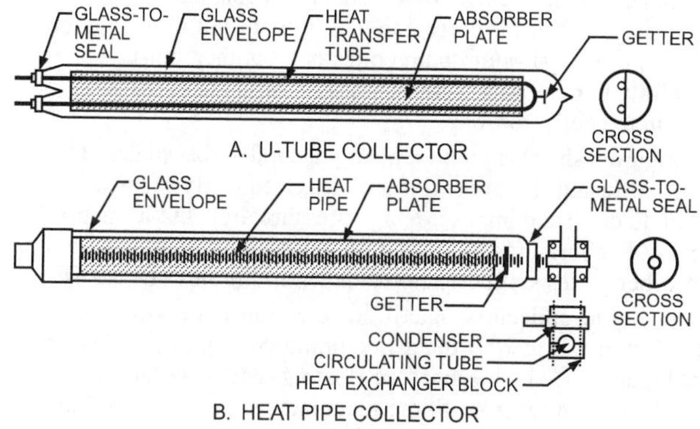

Fig. 5 Evacuated-Tube Collector

absorber plates from glazed collectors are also used, but generally with a heat exchanger to avoid contact with the pool water chemicals.

Air-Heating Collectors. Air-heating collectors are also contained in a box, covered with one or more glazings, and insulated on the sides and back. Figure 4B shows a cross section of a flat-plate air collector. The primary differences from liquid-heating collectors are in the design of the absorber plate and flow passages. Because the working fluid (air) has poor heat transfer characteristics, it flows over the entire absorber plate, and sometimes on both the front and back of the plate, to make use of a larger heat transfer surface. In spite of the larger surface area, air collectors generally have poorer overall heat transfer than liquid collectors. However, they are usually operated at a lower temperature for space heating applications because they require no intervening heat exchangers.

Transpired air collectors can be attached to the roof or side of a building to create a plenum for the preheated air. The collector is simply the siding material with many small holes through which the air is drawn, thus warming the air for space-heating or drying applications.

Liquid-Vapor Collectors. A third class of collectors uses liquid-vapor phase change to transfer heat at high efficiency. These collectors feature a heat pipe (a highly efficient thermal conductor) placed inside a vacuum sealed tube (Figure 5B). The heat pipe contains a small amount of fluid (e.g., methanol) that undergoes an evaporating/condensing cycle. In this cycle, solar heat evaporates the liquid, and the vapor travels to a heat sink region, where it condenses and releases its latent heat. This process is repeated by a return feed of the condensed fluid back to the solar absorber.

Most phase-change fluids have low freezing temperatures, so the heat pipe offers inherent protection to the tubes from freezing and overheating, but not to the fluid being heated. This self-limiting temperature control is a unique feature of the evacuated heat pipe collector.

Collector Construction

Absorber Plates. The key component of a flat-plate collector is the absorber plate. It contains the heat transfer fluid and serves as a heat exchanger by converting radiant energy into thermal energy. It must maintain structural integrity at temperatures ranging from below freezing to well above 300°F. Chapter 33 of the 2007 *ASHRAE Handbook—HVAC Applications* illustrates typical liquid collectors and shows the wide variety of absorber plate designs in use.

Materials for absorber plates and tubes are usually highly conductive metals such as copper, aluminum, and steel, although low-temperature collectors for swimming pools are usually made from extruded elastomeric material such as ethylene-propylene terpolymer (EPDM) and polypropylene. Flow passages and fins are usually copper, but aluminum fins are sometimes inductively welded to copper tubing. Occasionally, fins are mechanically attached, but there is potential for corrosion with this design. A few manufacturers produce all-aluminum collectors, but they must be checked carefully to determine whether they incorporate corrosion protection in the collector loop.

Figure 6 shows a plan view of typical absorber plates. The serpentine design (Figure 6A) is used less frequently because it is difficult to drain and imposes a high pressure drop, but it can produce higher temperatures at lower flow rates. Most manufacturers use absorber plates similar to those shown in Figures 6D and 6E.

In liquid collectors, manifold selection is important because the design can restrict the array piping configuration. The manifold must be drainable and free-floating, with generous allowance for thermal expansion. Some manufacturers provide a choice of manifold connections to give designers flexibility in designing arrays.

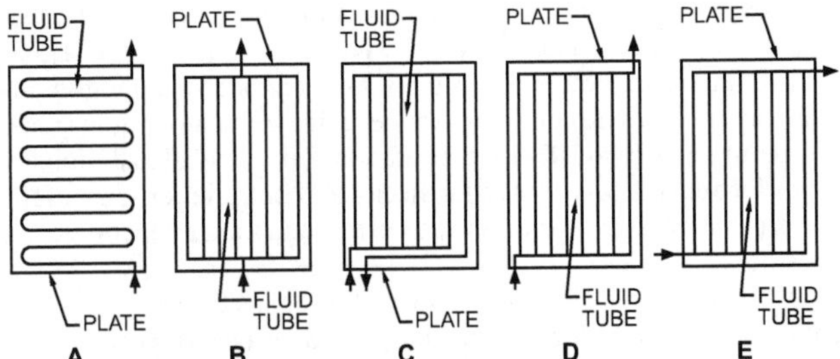

Fig. 6 Plan View of Liquid Collector Absorber Plates

In air collectors, most manufacturers increase the heat transfer area by using fins, matrices, or corrugated surfaces (Figure 7). Many of these designs increase air turbulence, which improves collector efficiency (at the expense of increased fan power).

Figure 7 also shows cross sections of typical solar water collectors. Water passages can be integrated with the absorber plate to ensure good thermal contact (Figure 7E), or they can be soldered, brazed, or otherwise fastened to the absorber plate (Figure 7F). Manufactured collectors are made in modular sizes, typically 4 by 8 ft, and these are connected to form a bank or row. Because most commercial-scale systems involve more than 1000 ft^2 of collector, there can be numerous piping connections, depending on design.

Absorber plates may be coated with spectrally selective or nonselective materials. Selective coatings are more efficient, but cost more than nonselective, or flat black, coatings. However, their higher efficiencies may reduce the overall size of the array, resulting in lower total cost. The Argonne National Laboratory has published detailed guidelines on various coating materials used in solar applications (ANL 1979a).

Housing. The collector housing is the container that provides structural integrity for the collector assembly. The housing must be structurally sound, weathertight, fire-resistant, and mechanically connectable to a substructure to form an array. Collector housing materials include the following:

- Galvanized or painted steel
- Aluminum folded sheet stock or extruded wall materials
- Various plastics, either molded or extruded
- Composite wood products
- Standard elements of the building

Extruded anodized aluminum, including extruded channels, offers durability and ease of fabrication. Grooves are sometimes included in the extruded channels to accommodate proprietary mounting fixtures. The high temperatures of a solar collector deteriorate wood housings; consequently, wood is often forbidden by fire codes.

Glazing. Solar collectors for domestic hot water are usually single-glazed to reduce absorber plate convective and radiative losses. Some collectors have double glazing to further reduce these losses; however, double glazing should be restricted to applications where the value of $(t_i - t_a)/I_t$ exceeds that of domestic hot-water applications (e.g., space heating or activating absorption refrigeration). Glazing materials are plastic, plastic film, or glass. Glass can absorb the long-wave thermal radiation emitted by the absorber coating, but it is not affected by ultraviolet (UV) radiation. Because of their impact tolerance, only tempered, low-iron glass covers should be considered. These covers have a solar transmission rating of 84 to 91%.

If the probability of vandalism is high, polycarbonate, which has high impact resistance, should be considered. Unfortunately, its transmittance is not as high as that of low-iron glass, and it is susceptible to long-term UV degradation.

Insulation. Collector enclosures must be well insulated to minimize heat losses. The insulation adjacent to the absorber plate must

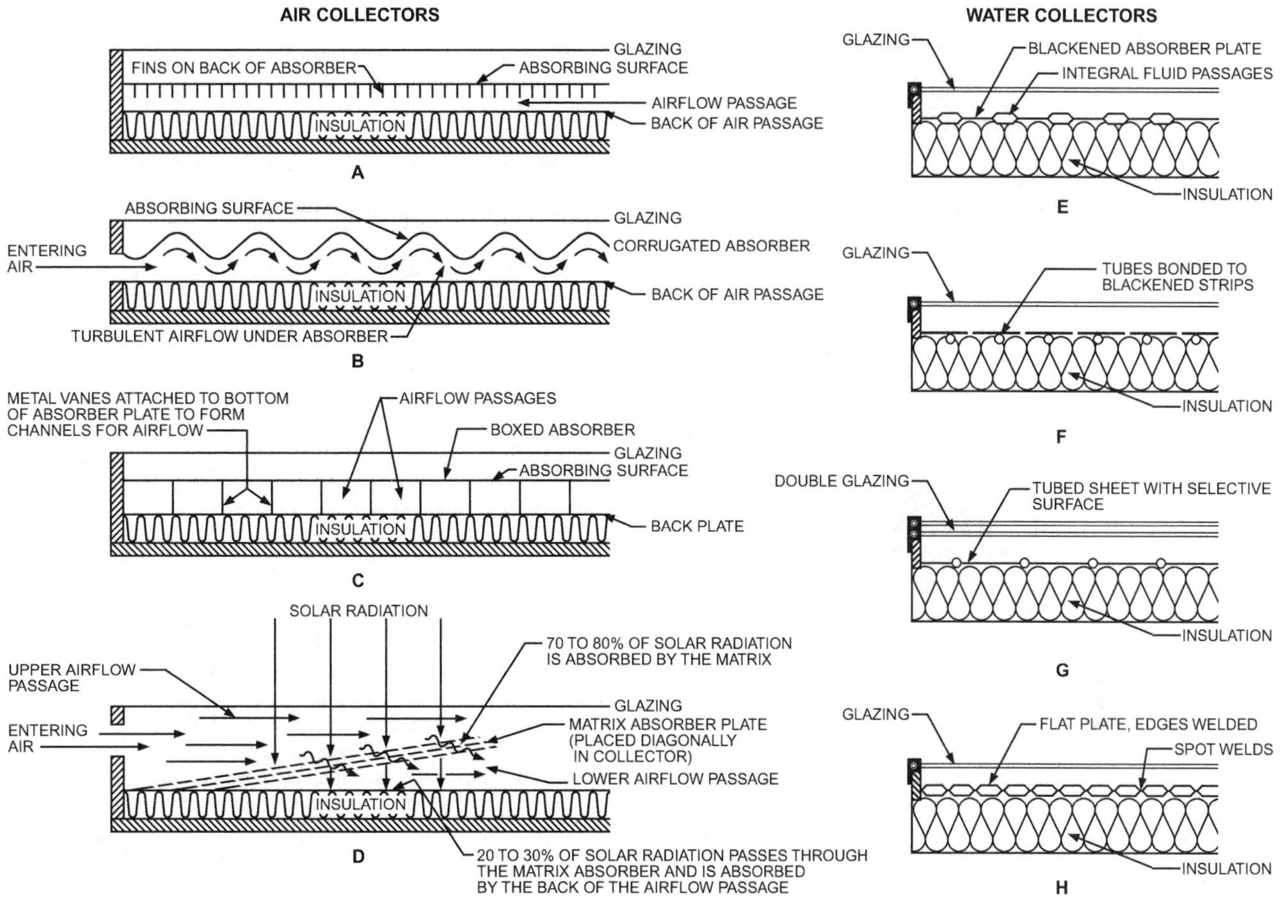

Fig. 7 Cross Sections of Various Solar Air and Water Heater

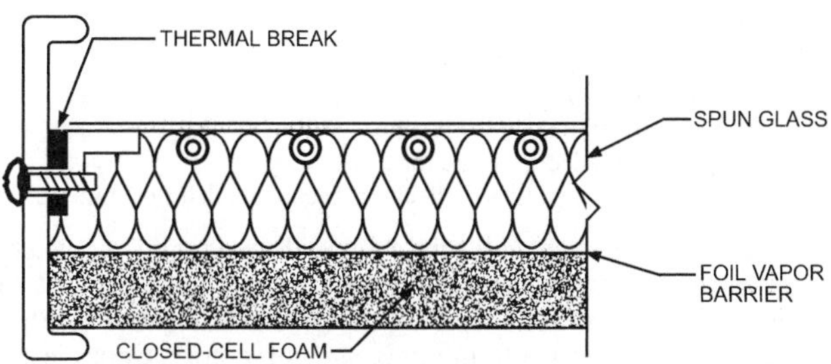

Fig. 8 Cross Section of Suggested Insulation to Reduce Heat Loss from Back Surface of Absorber

withstand temperatures up to 400°F and, most importantly, must not outgas volatile products within this range. Many insulation materials designed for construction applications are not suitable for solar collectors because the binders outgas volatiles at normal collector operating temperatures.

Solar collector insulation is typically made of mineral fiber, ceramic fiber, glass foam, plastic foam, or fiberglass. Fiberglass is the least expensive insulation and is widely used in solar collectors. For high-temperature applications, rigid fiberglass board with a minimum of binder is recommended. Also, a layer of polyisocyanurate foam in collectors is often used because of its superior R-value. Because it can outgas at high temperatures, the foam must not be allowed to contact the collector plate.

Figure 8 illustrates the preferred method of combining fiberglass and foam insulations to obtain both high efficiency and durability. Note that the absorber plate should be free-floating to avoid thermal stresses. Despite attempts to make collectors watertight, moisture is always present in the interior. This moisture can physically degrade mineral wool and reduce the R-value of fiberglass, so drainage and venting are crucial.

ROW DESIGN

Piping Configuration

Most commercial and industrial systems require a large number of collectors. Connecting the collectors with one set of manifolds makes it difficult to ensure drainability, balanced flow, and low pressure drop. An array usually includes many individual groups of collectors, called rows, to provide the necessary flow characteristics. Rows can be grouped into (1) parallel flow or (2) combined series-parallel flow. Parallel flow is the most frequently used because it is inherently balanced, has low pressure drop, and is drainable. Figure 9 illustrates various collector header designs for forming a parallel-flow row (the flow is parallel but the collectors are connected in series).

Generally, flat-plate collectors connect to the main piping in one of the methods shown in Figure 9. The **external manifold** collector has a small-diameter connection meant to carry only the flow for one collector. It must be connected individually to the manifold piping, which is not part of the collector panel, as depicted in Figure 9A.

Internal manifold collectors incorporate the manifold piping integral with each collector (Figure 9B). Headers at either end of the collector distribute flow to the risers. Several collectors with large headers can be placed side by side to form a continuous supply and return manifold. With 1 in. headers, four to six 40 ft^2 collectors can be placed side by side. Collectors with 1 in. headers can be mounted in a row without producing unbalanced flow. Most collec-

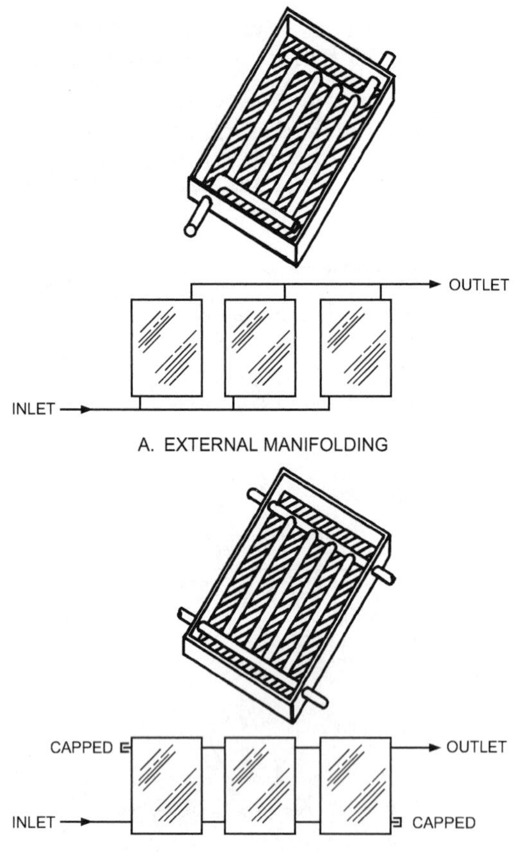

Fig. 9 Collector Manifolding Arrangements for Parallel-Flow Row

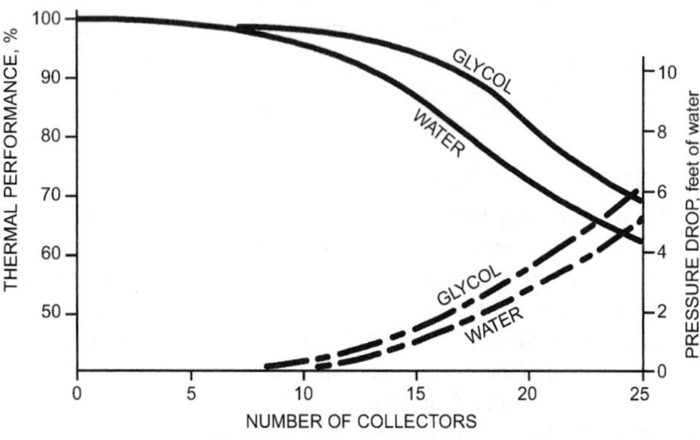

Fig. 10 Pressure Drop and Thermal Performance of Collectors with Internal Manifolds Numbers

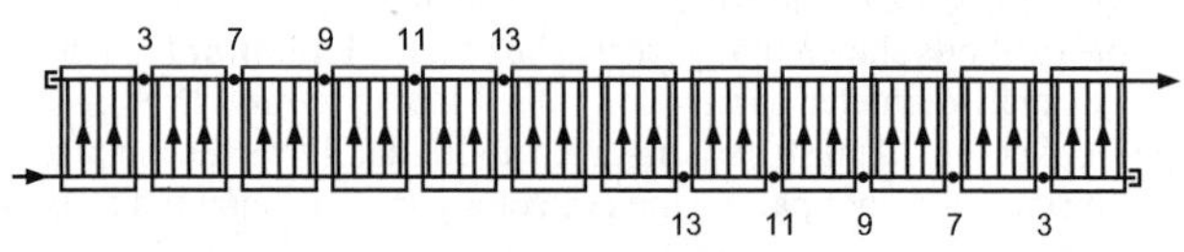

Fig. 11 Flow Pattern in Long Collector Row with Restrictions

tors have four plumbing connections, some of which may be capped if the collector is located on the end of the array. Internally manifolded collectors have the following advantages:

- Piping costs are lower because fewer fittings and less insulation are required.
- Heat loss is less because less piping is exposed.
- Installation is more attractive.

Some of their disadvantages are as follows:

- For drainback systems, the entire row must be pitched, thus complicating mounting.
- Flow may be imbalanced if too many collectors are connected in parallel.
- Removing the collector for servicing may be difficult.
- Stringent thermal expansion requirements must be met if too many collectors are combined in a row.

Velocity Limitations

Fluid velocity limits the number of internally manifolded collectors that can be contained in a row. For 1 in. headers, up to eight 40 ft^2 collectors can usually be connected for satisfactory performance. If too many are connected in parallel, the middle collectors will not receive enough flow, and performance will decrease. Connecting too many collectors also increases pressure drop. Figure 10 shows the effect of collector number on performance and pressure drop for one particular design. Newton and Gilman (1983) describe a general method to determine the number of internally manifolded collectors that can be connected.

Flow restrictors can be used to accommodate a large number of collectors in a row. The flow distribution in the 12 collectors of Figure 11 would not be satisfactory without the flow restrictors shown at the interconnections. The flow restrictors are barriers with a drilled hole of the diameter indicated. Some manufacturers calculate the required hole diameters and provide predrilled restrictors.

Chapter 35 of the 2001 *ASHRAE Handbook—Fundamentals* gives information on sizing piping. Knowles (1980) provides the following expression for the minimum acceptable header diameter:

$$D = 0.24(Q/\Delta p)^{0.45}N^{0.64} \tag{1}$$

where

D = header diameter, in.
N = number of collectors in module
Q = recommended flow rate for collector, gpm
Δp = pressure drop across collector at recommended flow rate, psi

Because pipe is available in a limited number of diameters, selection of the next larger size ensures balanced flow. Usually, the sizes of supply and return piping are graduated to maintain the same pressure drop while minimizing piping cost. Complicated configurations may require a hydraulic static regain calculation.

Thermal Expansion

Thermal expansion will affect the row shown in Figure 11. Thermal expansion (or contraction) of a module of collectors in parallel may be estimated by the following equation:

$$\Delta = 0.000335n(t_c - t_i) \tag{2}$$

where

Δ = expansion or contraction of collector array, in.
n = number of collectors in array
t_c = collector temperature, °F (generally the maximum stagnation temperature)
t_i = installation temperature of collector array, °F (generally the lowest possible air temperature)

Because absorbers are rigidly connected, the absorber must have sufficient clearance from the side frame to allow the expansion indicated in Equation (2). Information on dealing with expansion in piping may also be found in Chapter 45.

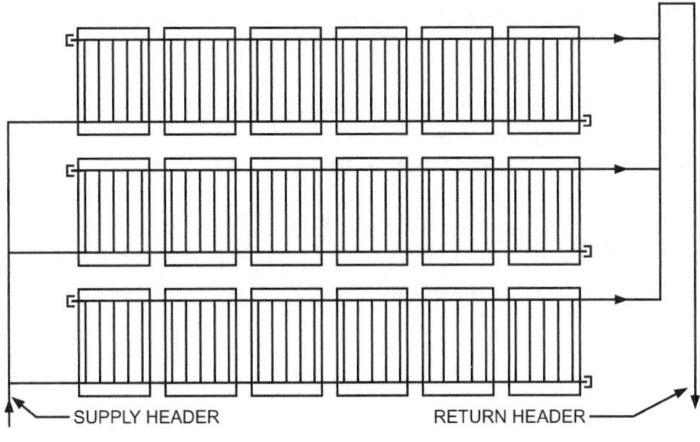

Fig. 12 Reverse-Return Array Piping

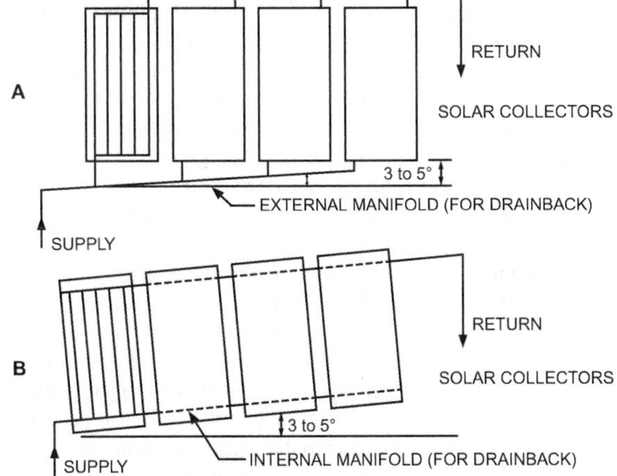

Fig. 13 Mounting for Drainback Collector Modules

ARRAY DESIGN

Piping Configuration

Liquid Systems. To maintain balanced flow, an array or field of collectors should be built from identical rows configured as described in previous sections. Whenever possible, rows must be connected in reverse-return fashion (Figure 12), which ensures that the array is self-balanced. With proper care, an array can drain, which is an essential requirement for drainback freeze protection.

Piping to and from the collectors must be sloped properly in a drainback system. Typically, piping and collectors must slope to drain at 1/4 in. per linear foot. Elevations throughout the array, especially the highest and lowest point of the piping, should be noted on the drawings.

The external manifold collector has different mounting and plumbing considerations from the internal manifold collector (Figure 13). A row of externally manifolded collectors can be mounted horizontally, as shown in Figure 13A. The lower header must be pitched as shown. The pitch of the upper header can be either horizontal or pitched toward the collectors so it can drain back through the collectors.

Arrays with internal manifolds pose a greater challenge in designing and installing the collector mounting system. For these collectors to drain, the entire bank must be tilted, as shown in Figure 13B.

Reverse return always implies an extra pipe run. Sometimes, it is more convenient to use direct return (Figure 14). In this case, balancing valves are needed to ensure uniform flow through the rows. The balancing valves *must* be connected at the row outlet to

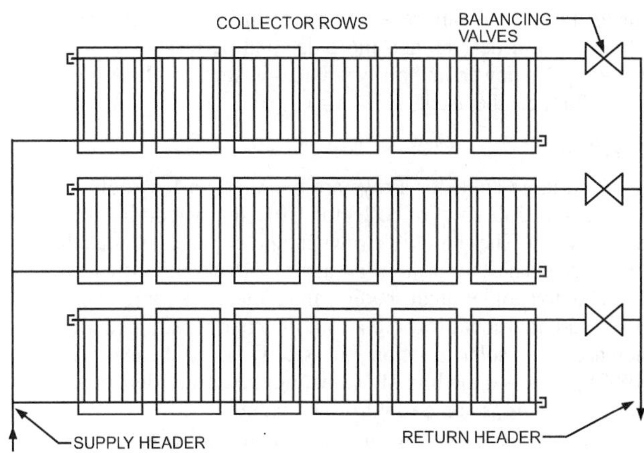

Fig. 14 Direct-Return Array Piping

provide the flow resistance necessary to ensure filling of all rows on pump start-up.

It is often impossible to configure parallel arrays because of the presence of rooftop equipment, roof penetrations, or other building-imposed constraints. Although the list is not complete, the following requirements should be considered when developing the array configuration:

- Strive for a self-balancing configuration.
- For drainback, design rows, subarrays, and arrays to be individually and collectively drainable.
- Always locate collectors or rows with high flow resistance at the outlet to improve flow balance and ensure the drainback system fills.
- Minimize flow and heat transfer losses.

In general, it is easier to configure complex array designs for nonfreezing fluid systems. Newton and Gilman (1983) provide some typical examples. However, with careful attention to these criteria, it is also possible to design successful large drainback arrays.

Air Systems. Air distribution within the collector array is the most critical feature of an air system for effective operation. Proper airflow must take into account the overall pressure drop to allow efficient fan motor sizing. Balancing, automatic, backdraft, and fire dampers are usually needed. Air leaks (both into and out of the ducts and from component to component) must be minimized.

For example, some air collector systems contain a water coil for preheating water. Despite the inclusion of automatic and backdraft dampers, leakage within the system can freeze coils. One possible solution is to position the coil near the warm end of the storage bin. Another is to circulate an antifreeze solution in the coil. Whatever the solution, the designer must remember that even the lowest-leakage dampers will leak if improperly installed or adjusted.

As with liquid systems, the main supply and return ducts should be connected in a reverse-return configuration, with balancing dampers on each supply branch to the collector modules. If reverse return is not feasible, the main ducts should include balancing dampers at strategic locations. Here too, having fewer branch ducts reduces balancing needs and costs.

Unlike liquid systems, air collectors can be built on site. Although material and cost savings can be substantial with site-built collectors, extreme care must be taken to ensure long life, low leakage, and proper air distribution. Quality control in the field can be a problem, so well-trained designers and installers are critical to the success of these systems.

Performance. The effect of array and air distribution system designs on the overall system performance must be considered. Beckman et al. (1977) give standard procedures for estimating the impact of series connection and duct thermal losses. The effect on fan operation and fan power is more difficult to determine. For unique system designs and more detailed performance estimates, including fan power, an hourly simulation such as TRNSYS (Klein et al. 2004) can be used.

Shading

When large collector arrays are mounted on flat roofs or level ground, multiple rows of collectors are usually installed in a sawtooth fashion. These multiple rows should be spaced so that they do not shade each other at low sun angles. However, it is usually not desirable to avoid mutual shading altogether. It is sometimes possible to add additional rows to a roof or other constrained area; this increases the solar fraction but sacrifices efficiency. Kutscher (1982) presents a method of estimating the energy delivered annually when there is some row-to-row shading in the array. The most common practice is to base the row spacing on the lowest sun angle during the most important time of the year. For example, water and pool heating systems that operate mostly in the summer can use the higher sun angles of summer or spring and fall. Space-heating systems may want to maximize exposure in the winter, so they should use the lowest sun angle at the winter solstice. A system that operates year round may use the spring equinox sun angle as a reasonable compromise to allow more rows.

Thermal Collector Performance

Under steady-state flow conditions, the useful heat delivered by a solar collector is equal to the energy absorbed in the heat transfer fluid minus the direct and indirect heat losses from the surface to the surroundings. This principle can be stated in the following relationship:

$$q_u = A_c[I_t\tau\alpha - U_L(\bar{t}_p - t_a)] \tag{3}$$

where

q_u = useful energy delivered by collector, Btu/h
A_c = total aperture collector area, ft²
I_t = irradiance, total (direct plus diffuse) solar energy incident on upper surface of sloping collector structure, Btu/h·ft²
τ = transmittance (fraction of incoming solar radiation that reaches absorbing surface), dimensionless
α = absorptance (fraction of solar energy reaching surface that is absorbed), dimensionless
U_L = overall heat loss coefficient, Btu/h·ft²·°F
$\bar{t}_p$ = average temperature of absorbing surface of absorber plate, °F
t_a = atmospheric temperature, °F

Except for average plate temperature $\bar{t}_p$, these terms can be readily determined. For convenience, Equation (3) can be modified by substituting inlet fluid temperature for the average plate temperature, if a suitable correction factor is included. The resulting equation is

$$q_u = F_R A_c[I_t\tau\alpha - U_L(t_i - t_a)] \tag{4}$$

where

F_R = correction factor, or collector heat removal efficiency factor, having a value less than 1.0
t_i = temperature of fluid entering collector, °F

The heat removal factor F_R can be considered the ratio of the heat actually delivered to that delivered if the collector plate were at a uniform temperature equal to that of the entering fluid. An F_R of 1.0 is theoretically possible if (1) the fluid is circulated at such a high rate that its temperature rises a negligible amount, and (2) the heat transfer coefficient and fin efficiency are so high that the temperature difference between the absorber surface and the fluid is negligible.

In Equation (4), the temperature t_i of the inlet fluid depends on the characteristics of the complete solar heating system and the building's heat demand. However, F_R is affected only by the solar collector characteristics, fluid type, and fluid flow rate through the collector.

Solar air heaters remove substantially less heat than liquid collectors. However, their lower collector inlet temperature makes their system efficiency comparable to that of liquid systems for space-heating applications.

Equation (4) may be rewritten in terms of the instantaneous efficiency of total solar radiation collection by dividing both sides of the equation by $I_t A_c$. The result is

$$\eta = F_R\tau\alpha - F_R U_L\frac{(t_i - t_a)}{I_t} \tag{5}$$

where η = collector efficiency, dimensionless.

Equation (5) plots as a straight line on a graph of efficiency versus the heat loss parameter $(t_i - t_a)/I_t$. Plots of Equation (5) for various liquid collectors are shown in Figure 15. The intercept (intersection of the line with the vertical efficiency axis) equals $F_R\tau\alpha$. The slope of the line (i.e., any efficiency difference divided by the corresponding horizontal scale difference) equals $-F_R U_L$. If experimental data on collector heat delivery at various temperatures and solar conditions are plotted, with efficiency as the vertical axis and $(t_i - t_a)/I_t$ as the horizontal axis, the best straight line through the data points correlates collector performance with solar and temperature conditions. The intersection of the line with the vertical axis is where the temperature of the fluid entering the collector equals the ambient temperature, and collector efficiency is at its maximum. At the intersection of the line with the horizontal axis, collection efficiency is zero. This condition corresponds to such a low radiation level, or to such a high temperature of the fluid into the collector, that heat losses equal solar absorption, and the collector delivers no useful heat. This condition, normally called **stagnation**, usually occurs when no coolant flows to a collector. Solving Equation (5) for t_i under the hottest conditions yields the stagnation temperature for the collectors at that site.

Equation (5) includes all important design and operational factors affecting steady-state performance except collector flow rate and solar incidence angle. Flow rate indirectly affects performance through the average plate temperature. If the heat removal rate is

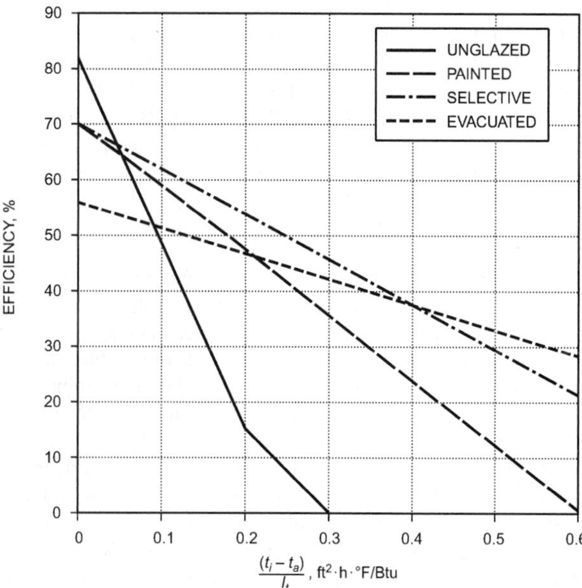

Fig. 15 Solar Collector Type Efficiencies

reduced, the average plate temperature increases, and more heat is lost. If the flow is increased, collector plate temperature and heat loss decrease.

Solar Incidence Angle. These relationships assume that the sun is perpendicular to the plane of the collector, which rarely occurs. For glass cover plates, specular reflection of radiation occurs, thereby reducing the $\tau\alpha$ product. The incident angle modifier $K_{\tau\alpha}$, defined as the ratio of $\tau\alpha$ at some incidence angle θ to $\tau\alpha$ at normal radiation $(\tau\alpha)_n$, is described by the following simple expression for specular reflection:

$$K_{\tau\alpha} = \frac{\tau\alpha}{(\tau\alpha)_n} = 1 + b_o\left[\frac{1}{\cos\theta} - 1\right] \tag{6}$$

For a single glass cover, b_o is approximately –0.10. Many flat-plate collectors, particularly evacuated tubes, have some limited focusing capability. The incident angle modifiers for these collectors are not modeled well by Equation (6), which is a linear function of $(1/\cos\theta) - 1$.

Cellular Flow Rate. Equation (5) is not convenient for air collectors when it is desirable to present data based on collector outlet temperature t_{out} rather than inlet temperature, which is the commonly measured variable for liquid systems. The relationship between the heat removal factors for these two cases is

$$F_R\tau\alpha = \frac{(F_R\tau\alpha)'}{1 + (F_R U_L)'/\dot{m}c_p} \tag{7a}$$

$$F_R U_L = \frac{(F_R U_L)'}{1 + (F_R U_L)'/\dot{m}c_p} \tag{7b}$$

where $\dot{m}c_p$ is mass flow times specific heat of air, F_RU_L and $F_R\tau\alpha$ apply to $t_i - t_a$ in Equation (5), and $(F_RU_L)'$ and $(F_R\tau\alpha)'$ apply to $t_{out} - t_a$.

Testing Methods

ASHRAE *Standard* 93 gives information on testing solar energy collectors using single-phase fluids and no significant internal storage. The data can be used to predict performance in any location and under any conditions where load, weather, and insolation are known.

The standard presents efficiency in a modified form of Equation (5). It specifies that the efficiency be reported in terms of gross collector area A_g rather than aperture collector area A_c. The reported efficiency is lower than the efficiency based on net area, but the total energy collected does not change by this simplification. Therefore, gross collector area must be used when analyzing performance of collectors based on experiments that determine $F_R\tau\alpha$ and F_RU_L according to ASHRAE *Standard* 93.

Standard 93 suggests that testing be done at 0.03 gpm per square foot of gross collector area for liquid systems, and that the test fluid be water. Although it is acceptable to use lower flow rates or a heat transfer fluid other than water, the designer must adjust the F_R for a different heat removal rate based on the product $\dot{m}c_p$ of mass flow and specific heat. The following approximate approach may be used to estimate small changes in $\dot{m}c_p$:

$$\frac{(F_R U_L)_2}{(F_R U_L)_1} = \frac{1 - \exp[-A_c(F_R U_L)'/(\dot{m}c_p)_2]}{1 - \exp[-A_c(F_R U_L)'/(\dot{m}c_p)_1]} \tag{8}$$

Air collectors are tested at a flow rate of 2 scfm per square foot, and the same relationship applies for adjusting to other flow rates. Annual compilations of collector test data that meet the criteria of ASHRAE *Standards* 93 and 96 may be obtained from Solar Rating and Certification Corporation (www.solar-rating.org).

Another source for this information is the collector manufacturer. However, a manufacturer sometimes publishes efficiency data at a much higher flow rate than the recommended design value, so collector data should be obtained from an independent laboratory qualified to conduct testing prescribed by ASHRAE *Standard* 93.

Collector Test Results and Initial Screening Methods

Final collector selection should be made only after energy analyses of the complete system, including realistic weather conditions and loads, have been determined. Also, a preliminary screening of collectors with various performance parameters should be conducted to identify those that best match the load. One way to accomplish this is to identify the expected range of heat loss parameter $(t_i - t_a)/I_t$ for the load and climate on a plot of efficiency η as a function of heat loss parameter, as indicated in Figure 15.

Ambient temperature during the swimming season may vary by 18°F above or below pool temperature. The corresponding parameter values range from 0.15 Btu/h·ft²·°F on cool overcast days (low I_t) to as low as –0.15 Btu/h·ft²·°F on hot overcast days. For most swimming pool heating, unglazed collectors offer the highest performance and are the least expensive.

The heat loss parameter for service water heating can range from 0.05 to 0.35 Btu/h·ft²·°F, depending on the climate at the site and the desired hot-water delivery temperature. Convective space heating requires an even greater collector inlet temperature than water heating, and the primary load coincides with lower ambient temperature. In many areas of the United States, space heating coincides with low radiation values, which further increases the heat loss parameter. However, many space-heating systems are accompanied by water heating, and some space-heating systems (e.g., radiant panels) are available at a lower $(t_i - t_a)/I_t$ value.

Air conditioning with solar-activated absorption equipment is economical only when the cooling season is long and solar radiation levels are high. These devices require at least 180°F water. Thus, on an 80°F day with radiation at 300 Btu/h·ft², the heat loss parameter is 0.3. Higher operating temperatures are desirable to prevent excessive derating of the air conditioner. Only the most efficient (low F_RU_L) collector is suitable. Derating may also be minimized by using convective-radiant hybrid air-conditioning systems.

Collector efficiency curves may be used as an initial screening device. However, efficiency curves illustrate only instantaneous performance of a collector. They do not include incidence angle effects (which vary throughout the year); heat exchanger effects; probabilities of occurrence of t_i, t_a, and I_t; system heat loss; or control strategies. Final selection requires determining a collector's long-term energy output as well as performing cost-effectiveness studies. Estimating a particular collector's and system's annual performance requires appropriate analysis tools such as *f*-Chart (Beckman et al. 1977) or TRNSYS (Klein et al. 2004). The Solar Rating and Certification Corporation (SRCC), International Organization for Standardization (ISO), European Committee for Standardization (CEN), and other rating agencies provide day-long collector outputs under various solar and operating conditions. These ratings can greatly simplify selection of the proper collector type, including cost considerations. Table 1 shows a comparison of generic types of liquid flat-plate collectors.

Generic Test Results

The generic types of liquid flat-plate collectors are **unglazed**, **single-glazed painted-** and **selective-surface absorbers**, and **evacuated tubes**. Table 1 shows the average output, intercept, and slope for these. The normalized day-long energy output for a clear day in category C points out the difficulties in using only the instantaneous efficiency data. Category C holds the temperature difference constant at 36°F all day, and a clear day is defined as 2000 Btu/ft²·day, which is high for most of the United States. Note that evacuated-tube collectors do not perform very well under these

Table 1 Average Performance Parameters* for Generic Types of Liquid Flat-Plate Collectors

Collector Type	Vertical Intercept	Slope, Btu/h·ft²·°F	Clear-Day Category C Output, Btu/h·ft²·°F
Unglazed	0.807	–3.289	420
Glazed	0.701	–1.15	927
Painted absorber			
Selective-surface absorber	0.699	–0.814	1007
Evacuated tube	0.554	–0.443	840

*Derived from data of Solar Rating and Certification Corporation (SRCC), www.solar-rating.org (Oct. 2006).

Table 2 Thermal Performance Ratings* for Generic Types of Liquid Flat-Plate Collectors, Btu/ft²·day

	Solar Day, Btu/ft²·day		
Category*	Clear Day 2000	Mildly Cloudy 1500	Cloudy Day 1000
Unglazed			
A	1600	1200	900
B	1000	700	400
C	400	200	—
D	—	—	—
E	—	—	—
Painted			
A	1285	971	658
B	1128	815	533
C	908	595	282
D	407	157	—
E	—	—	—
Selective surface			
A	1316	971	658
B	1191	877	564
C	1003	689	376
D	595	313	63
E	219	31	—
Evacuated tube			
A	872	655	436
B	841	623	405
C	810	592	374
D	685	467	280
E	592	374	156

*Categories	$T_i - T_a$, °F	Application
A	–9	Pool heating, warm climate
B	9	Pool heating, cool climate
C	36	Water heating, warm climate
D	90	Water heating, cool climate
E	144	Air conditioning

*Derived from data of Solar Rating and Certification Corporation (SRCC), www.solar-rating.org (Oct. 2006).

circumstances. SRCC and other rating agencies publish a matrix of collector outputs for various temperature differences and solar days. Table 2 shows the SRCC matrix for each generic collector type. Further explanation of the categories is available on the SRCC Web site, www.solar-rating.org, or in their standards.

The daily outputs in Table 2 show a clear advantage for each type of collector starting from the upper left corner and proceeding diagonally down and across. At low temperature differences, for categories A and B under clear and mildly cloudy conditions, unglazed collectors provide a clear advantage. The middle of the matrix is a bit less definitive, but painted and selective-surface collectors tend to perform best in the range. At the lower right corner, evacuated tubes perform best under high temperature differences and low solar inputs. Determining where a project falls in this rating matrix is very important in selecting the proper collector; the designer should be familiar with the location's average daily solar input and the desired operating temperature. For example, selective-surface collectors might work well for a high-temperature application in Phoenix in the summer because the daytime temperature difference is low and the solar input is high, whereas evacuated tubes are a better choice in Atlanta.

THERMAL ENERGY STORAGE

Design and selection of thermal storage equipment is one of the most neglected elements of solar energy systems. In fact, the energy storage system has an enormous influence on overall system cost, performance, and reliability. Furthermore, the storage system design is highly interactive with other system elements such as the collector loop and thermal distribution system. Thus, it should be considered within the context of the total system.

Energy can be stored in liquids, solids, or phase-change materials (PCMs). Water is the most frequently used liquid storage medium, although the collector loop may contain water, oils, or aqueous glycol as a collection fluid. For service water heating and most building space heating, water is normally contained in some type of tank. Air systems typically store heat in rocks or pebbles, but sometimes use the building's structural mass. Chapter 33 of the 2007 *ASHRAE Handbook—HVAC Applications* and ASHRAE (1991) cover this topic in more detail.

Air System Thermal Storage

The most common storage media for air collectors have been rocks or a regenerator matrix made from concrete masonry units (CMUs). Gravel was widely used as a storage medium because it is plentiful and relatively inexpensive; however, the inherent cost and likelihood of mold growth in humid climates put an end to its use. Other possible media include PCMs, water, and the inherent building mass.

In places where large interior temperature swings are tolerable, the inherent mass of the building may be sufficient for thermal storage. Designated storage may also be eliminated where the array output seldom exceeds the concurrent demand. Loads requiring no storage are usually the most cost-effective applications of air collectors, and heated air from the collectors can be distributed directly to the space.

Water can be used as a storage medium for air collectors by using a conventional heating coil to transfer heat from the air to the water in the storage tank. Advantages of water storage include compatibility with hydronic heating systems and relative compactness (roughly one-third the volume of pebble beds).

Liquid System Thermal Storage

For units large enough for commercial liquid systems, the following factors should be considered:

- Pressurized versus unpressurized storage
- External versus internal heat exchanger
- Single versus multiple tanks
- Steel versus nonmetallic tank(s)
- Type of service [i.e., service hot water (SHW), building space heating (BSH), or a combination of the two]
- Location, space, and accessibility constraints imposed by architectural limitations
- Interconnect constraints imposed by existing mechanical systems
- Limitations imposed by equipment availability

The following sections present examples of the more common configurations.

Pressurized Storage. Defined here as storage that is open to the city water supply, pressurized storage is preferred for small service water heating systems because it is convenient and provides an economical way of meeting ASME *Boiler and Pressure Vessel Code* requirements with off-the-shelf equipment. Typical storage size is about 1 to 2 gal per square foot of collector area. The largest off-the-shelf water heater is 120 gal; however, no more than three or four of these should be connected in parallel. Hence, the largest storage that can be considered with off-the-shelf water heater tanks is about 360 to 480 gal, and must be compared to a single tank with the additional labor and material to connect them, plus the amount of floor space required. For larger solar hot-water and combined systems, the following concerns are important when selecting storage:

- Higher cost per unit volume of ASME rated tanks in sizes greater than 120 gal
- Handling difficulties because of large weight
- Accessibility to locations suitable for storage
- Interfacing with existing SHW and BSH systems
- Corrosion protection for steel tanks

The choice of pressurized storage for medium-sized systems is based on the availability of suitable low-cost tanks near the site. Identifying a suitable supplier of low-cost tanks can extend the advantages of pressurized storage to larger SHW installations.

Storage pressurized at city water supply pressure is not practical for building space heating, except for small applications such as residences, apartments, and small commercial buildings.

With pressurized storage, the heat exchanger is always located on the collector side of the tank. Either the internal or external heat exchanger configuration can be used. Figure 16 illustrates the three principal types of internal heat exchanger concepts: an immersed coil, a wraparound jacket, and a tube bundle. Small tanks (less than 120 gal) are available with either of the first two heat exchangers already installed. For larger tanks, a large assortment of tube bundle heat exchangers are available that can be incorporated into the tank design by the manufacturer.

Sometimes, more than one tank is needed to meet design requirements. Additional tanks offer the following benefits:

- Added storage volume
- Increased heat exchanger surface
- Reduced pressure drop in the collection loop
- Increased stratification for larger volumes

Figure 17 illustrates the multiple-tank configuration for pressurized storage. The exchangers are connected in reverse-return fashion to minimize flow imbalance. A third tank may be added. Additional tanks have the following disadvantages compared to a single tank of the same volume:

- Higher installation costs
- Greater space requirements
- Higher heat losses (reduced performance)

An external heat exchanger provides greater flexibility because the tank and heat exchanger can be selected independently of each other (Figure 18). Flexibility is not achieved without cost, however, because an additional pump, with its parasitic energy consumption, is required.

When selecting an external heat exchanger for a system protected by a nonfreezing liquid, the following factors related to startup after at least one night in extremely cold conditions should be considered:

- Freeze-up of the water side of the heat exchanger
- Performance loss due to extraction of heat from storage

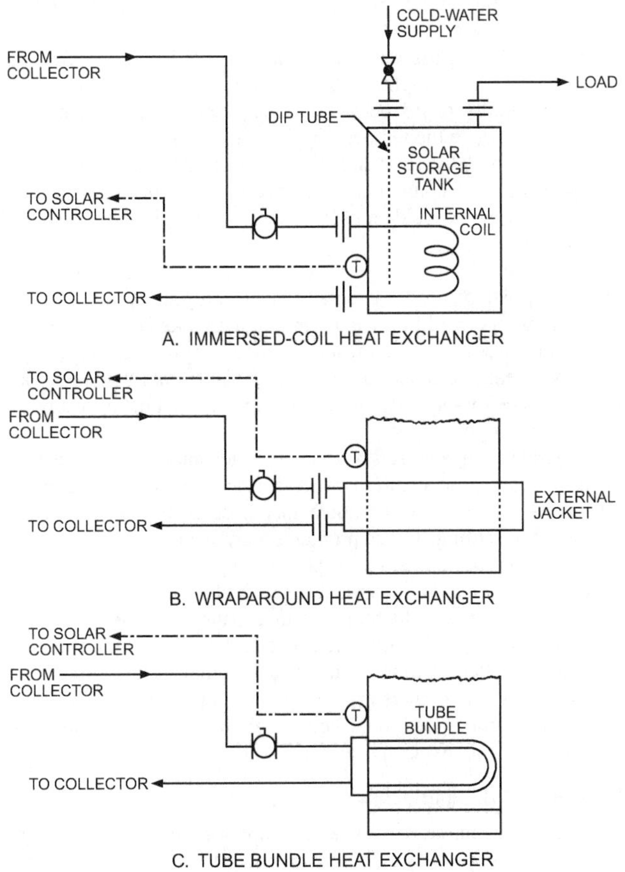

A. IMMERSED-COIL HEAT EXCHANGER

B. WRAPAROUND HEAT EXCHANGER

C. TUBE BUNDLE HEAT EXCHANGER

Fig. 16 Pressurized Storage with Internal Heat Exchanger

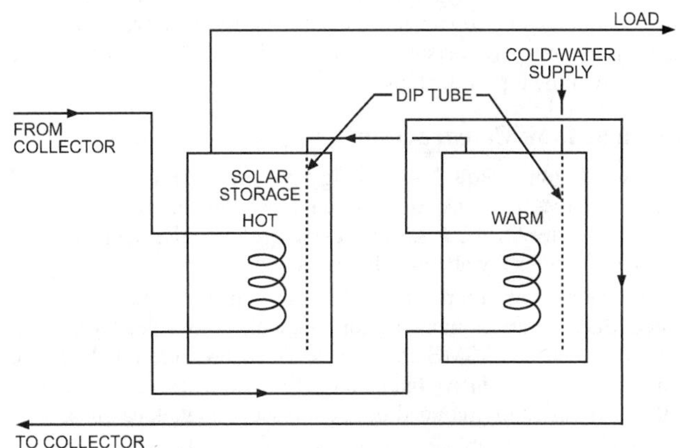

Fig. 17 Multiple Storage Tank Arrangement with Internal Heat Exchangers

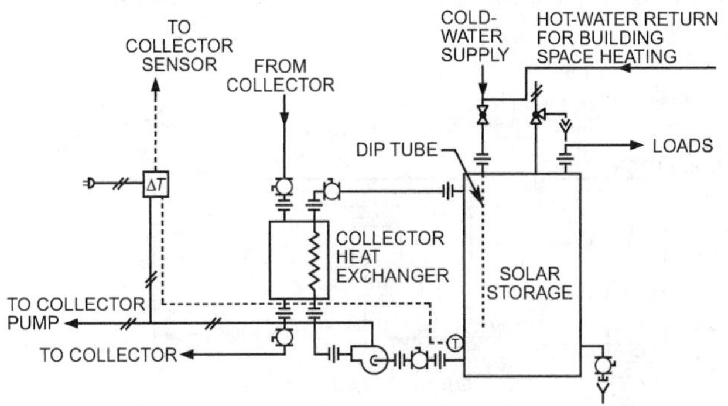

Fig. 18 Pressurized Storage System with External Heat Exchanger

For small systems, an internal heat exchanger/tank arrangement prevents the water side of the heat exchanger from freezing. However, the energy required to maintain the water above freezing must be extracted from storage, thereby decreasing overall performance. With the external heat exchanger/tank combination, a bypass can be arranged to divert cold fluid around the heat exchanger until it has been heated to an acceptable level, such as 80°F. When the heat transfer fluid has warmed to this level, it can enter the heat exchanger without causing freezing or extraction of heat from storage. If necessary, this arrangement can also be used with internal heat exchangers to improve performance.

Unpressurized Storage. For systems greater than about 1000 ft^3 (1500 gal storage volume minimum), unpressurized storage is usually more cost-effective than pressurized. As used in this chapter, the term *unpressurized* means tanks at or below the pressure expected in an unvented drainback loop.

Unpressurized storage for water and space heating implies a heat exchanger on the load side of the tank to isolate the high-pressure (potable water) loop from the low-pressure collector loop. Figure 19 illustrates unpressurized storage with an external heat exchanger. In this configuration, heat is extracted from the top of the solar storage tank, and cooled water is returned to the bottom. On the load side of the heat exchanger, water to be heated flows from the bottom of the backup storage tank, and heated water returns to the top. The heat exchanger may have a double wall to protect a potable water supply. A differential temperature controller controls the two pumps on either side of the heat exchanger. When small pumps are used, both may be controlled by the same controller without overloading it.

The external heat exchanger shown in Figure 19 provides good system flexibility and freedom in component selection. In some cases, system cost and parasitic power consumption may be reduced by an internal heat exchanger. Some field-fabricated heat exchangers use coiled soft copper tube. For larger systems, where custom fabrication is more feasible, a specified heat exchanger can be installed at the top of the tanks.

Storage Tank Construction

Steel. For most liquid solar energy systems, steel is the preferred material for storage tank construction. Steel tanks are relatively easy to fabricate to ASME *Pressure Vessel Code* requirements, readily available, and easily attached by pipes and fittings.

Steel tanks used for pressures of 30 psi and above must be ASME rated. Because water main pressure is usually above this level, open systems must use ASME code tanks. These pressure-rated storage tanks are more expensive than nonpressurized types. Costs can usually be significantly reduced if a nonpressurized tank can be used.

Steel tanks are subject to corrosion, however. Because corrosion rates increase with temperature, the designer must be particularly aware of corrosion protection methods for solar energy applications. A steel tank must be protected against (1) electrochemical corrosion, (2) oxidation (rusting), and (3) galvanic corrosion (ANL 1979b).

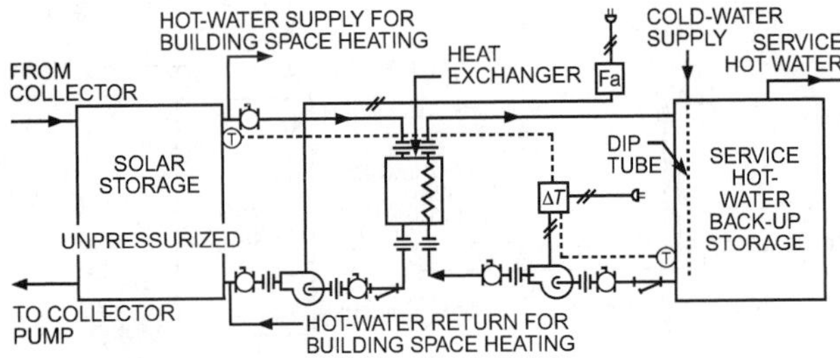

Fig. 19 Unpressurized Storage System with External Heat Exchanger

The pH of the liquid and the electric potential of the metal are the primary governing factors of electrochemical corrosion. A sacrificial anode fabricated from a metal more reactive than steel can provide protection. Magnesium is recommended for solar applications. Because protection ends when the anode has dissolved, the anode must be inspected annually.

Oxygen can enter the tank through air dissolved in the water entering the tank or through an air vent. In pressurized storage, oxygen is continually replenished by the incoming water. Besides causing rust, oxygen catalyzes other types of corrosion. Unpressurized storage systems are less susceptible to corrosion caused by oxygen because they can be designed as unvented systems, so corrosion is limited to the small amount of oxygen contained in the initial fill water.

Coatings on the tank interior protect it from oxidation. The following are some of the most commonly used coatings:

- **Phenolic epoxy** should be applied in four coats.
- **Baked-on epoxy** is preferred over painted-on epoxy.
- **Glass lining** offers more protection than the epoxies and can be used under severe water conditions.
- **Hydraulic stone** provides the best protection against corrosion and increases the tank's heat retention capabilities. Its weight may cause handling problems in some installations.

The coating should either be flexible enough to withstand extreme thermal cycling or have the same coefficient of expansion as the steel tank. In the United States, all linings used for potable water tanks should be approved by the Food and Drug Administration (FDA) for the maximum temperature expected in the tank.

Dissimilar materials in an electrolyte (water) are in electrical contact with each other and corrode by galvanic action. Copper fittings screwed into a steel tank corrode the steel, for example. Galvanic corrosion can be minimized by using dielectric bushings to connect pipes to tanks.

Plastic. Fiberglass-reinforced plastic (FRP) tanks offer the advantages of low weight, high corrosion resistance, and low cost. Premium-quality resins permit operating temperatures as high as 210°F, well above the temperature imposed by flat-plate solar collectors. Before delivery of an FRP tank is accepted, the tank should be inspected for damage that may have occurred during shipment. The gel coat on the inside must be intact, and no glass fibers should be exposed.

Concrete. Concrete vessels lined with a waterproofing membrane may also be used (in vented systems only) to contain a liquid thermal storage medium. Concrete storage tanks are inexpensive, and can be shaped to fit almost any retrofit application. Also, they have excellent resistance to the loading that occurs when they are placed in a below-grade location. Concrete tanks must have smooth corners and edges.

Concrete storage vessels have some disadvantages. Seepage often occurs unless a proper waterproofing surface is applied. Waterproofing paints are generally unsatisfactory because the concrete often cracks after settling. Other problems may occur because of poor workmanship, poor location, or poor design. Usually, tanks should stand alone and not be integrated with a building or other structure. Careful attention should be given to expansion joints and seams because they are particularly difficult to seal. Sealing at pipe taps can be a problem. Penetrations should be above liquid level, if possible.

Finally, concrete tanks are heavy and may be more difficult to support in a proper location. Their weight may make insulating the tank bottom more expensive and difficult.

Storage Tank Insulation

Heat loss from storage tanks and appurtenances is one of the major causes of poor system performance. The average R-value of storage tanks in solar applications is about half the insulation design value because of poorly insulated supports. Different standards recommend various design criteria for tank insulation. The Sheet Metal and Air Conditioning Contractors National Association (SMACNA) recommendation of a 2% loss in 12 h is generally accepted because it is more stringent than other standards. The following equation can be used to calculate the insulation R-value for this requirement:

$$\frac{1}{R} = \frac{fQ}{A\theta}\frac{1}{(t_{avg} - t_a)} \tag{9}$$

where

R = thermal resistivity of insulation, $ft^2 \cdot h \cdot °F/Btu$
f = specified fraction of stored energy that can be lost in time θ
Q = stored energy, Btu
A = exposed surface area of storage unit, ft^2
θ = given time period, h
t_{avg} = average temperature in storage unit, °F
t_a = ambient temperature surrounding storage unit during heating season, °F

The insulation factor $fQ/A\theta$ is found from Table 3 for various tank shapes (ANL 1980).

Most solar water-heating systems use large steel pressure vessels, which are usually shipped uninsulated. Materials suitable for field insulation include fiberglass, rigid foam, and flexible foam blankets. Fiberglass is easy to transport and make fire-retardant, but it requires significant labor to apply and seal.

Another widely used insulation consists of rigid sheets of polyisocyanurate foam cut and taped around the tank. Material that is 3 to 4 in. thick can provide R-20 to R-30 insulation value. Rigid foam insulation is sprayed directly onto the tank from a foaming truck or in a shop. It bonds well to the tank surface (no air space between tank and insulation) and insulates better than an equivalent thickness of fiberglass. When most foams are exposed to flames, they ignite and/or produce toxic gases. When located in or adjacent to a living space, they often must be protected by a fire barrier and/or sprinkler system.

Table 3 Insulation Factor $fQ/A\theta$ for Cylindrical Water Tanks

	Horizontal Tank Insulation Factor, $Btu/h \cdot ft^2$			
Size, gal	D × D	D × 2D	D × 4D	D × 6D
250	3.63	3.46	3.05	2.77
500	4.57	4.36	4.84	3.49
750	5.24	4.99	4.40	3.99
1000	5.76	5.49	4.84	4.39
1500	6.60	6.28	5.54	5.03
2000	7.26	6.92	6.10	5.53
3000	8.31	7.92	6.98	6.33
4000	9.15	8.71	7.68	6.97
5000	9.86	9.39	8.28	7.51

	Vertical Tank Insulation Factor, $Btu/h \cdot ft^2$			
Size, gal	D to 3D	D/2	D/3	D/4
80	2.10	1.88	2.15	1.97
120	2.39	2.15	2.46	2.26
250	3.07	2.74	2.46	2.88
500	3.87	3.46	3.96	3.63
750	4.43	3.96	4.53	4.16
1000	4.87	4.36	4.99	4.57
1500	5.58	4.99	5.71	5.24
2000	6.13	5.49	6.28	5.76
3000	7.03	6.28	7.19	6.60
4000	7.73	6.92	7.92	7.26
5000	8.33	7.45	8.53	7.82

Some tank manufacturers and suppliers offer custom tank jackets of flexible foam with zipper-like connections that provide quick installation and neat appearance. Some of the fire considerations for rigid foam apply to these foam blankets.

The tank supports are a major source of heat loss. To provide suitable load-bearing capability, the supports must be in direct contact with the tank wall or attached to it. Thermal breaks must be provided between the supports and the tank (Figure 20). If insulating the tank from the support is impractical, the external surface of the supports must be insulated, and the supports must be placed on insulative material capable of supporting the compressive load. Wood, foam glass, and closed-cell foam can be used, depending on the compressive load.

Stratification and Short Circuiting

Because hot water rises and cold water sinks in a vessel, the pipe to the collector inlet should always be connected to the bottom of the tank. The collector then operates at its best efficiency because it is always at the lowest possible temperature. The return from the collector should always run near the bottom of the tank to encourage stratification and avoid introducing cooler fluid to the top of the tank after a draw occurs. Similarly, the load should be extracted from the top of the tank, and cold makeup water should be introduced at the bottom. Because of increased static pressure, vertical storage tanks enhance stratification better than horizontal tanks.

If a system has a rapid tank turnover or if the buffer tank of the system is closely matched to the load, thermal stratification does not offer any advantages. However, in most solar energy systems, thermal stratification increases performance because temperature differences are relatively small. Thus, any enhancement of temperature differentials improves heat transfer efficiency.

Thermal stratification can be enhanced by the following:

- Using a tall vertical tank
- Situating the inlet and outlet piping of a horizontal tank to minimize vertical fluid mixing
- Sizing the inlets and outlets such that exhaust flow velocity is less than 2 fps
- Using flow diffusers (Figure 21)
- Plumbing multiple tanks in series

Multiple tanks generally yield the greatest temperature difference between cold water inlet and hot water outlet. With the piping/tank configuration shown in Figure 17, the difference can approach 30°F. In addition to the cost of the second tank, the cost of extra footings, insulation, and sensors must be considered. However, multiple tanks may save labor at the site because of their small diameter and shorter length. Smaller tanks require less demolition to install in retrofit (e.g., access through doorways, hallways, and windows).

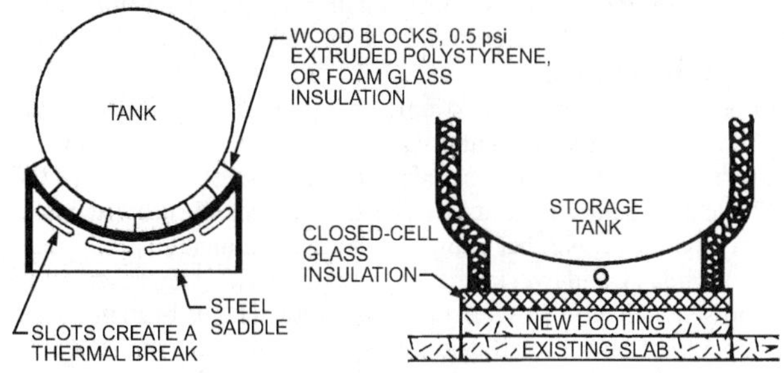

Fig. 20 Typical Tank Support Detail

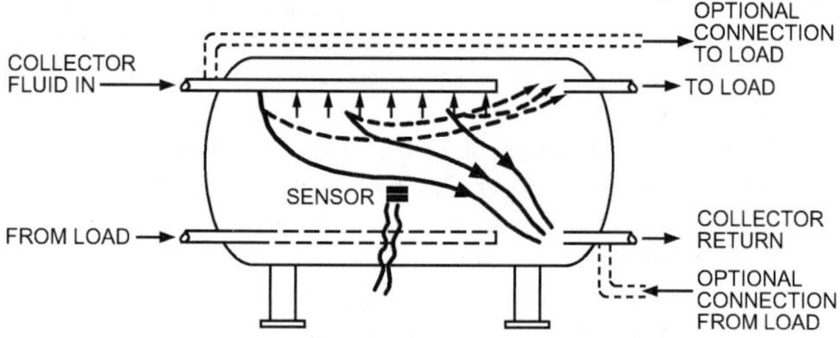

Fig. 21 Tank Plumbing Arrangements to Minimize Short Circuiting and Mixing
(Adapted from Kreider 1982)

Additional information on stratification and thermal storage may be found in Chapter 34 of the 2007 *ASHRAE Handbook—HVAC Applications*.

Storage Sizing

The following specific site factors are related to system performance or cost.

Solar Fraction. Most solar energy systems are designed to produce anywhere from 50 to 90% of the energy required to satisfy the load. If a system is sized to produce less than this, there must be greater coincidence between load and available solar energy. Therefore, smaller, lower-cost storage can be used with low-solar-fraction systems without impairing system performance.

Load Matching. The load profile of a commercial application may be such that less storage can be used without incurring performance penalties. For example, a company cafeteria or a restaurant serving large luncheon crowds has its greatest hot-water usage during, and just after, midday. Because the load coincides with the availability of solar radiation, less storage is required for carryover. For loads that peak during the morning, collectors can be rotated toward the southeast (in the northern hemisphere), sometimes improving output for a given array.

Storage Cost. If a solar storage tank is insulated adequately, performance generally improves with increased storage volume, but the savings must justify the investment. The cost of storage can be minimized by using multiple low-cost water heaters and unpressurized storage tanks made from materials such as fiberglass or concrete.

The performance and cost relationships between solar availability, collector design, storage design, and load are highly interactive. Furthermore, the designer should maximize the benefits from investment for at least one year, rather than for one or two months. For this reason, solar energy system performance models combined with economic analyses should optimize storage size for a specific site. Models that can determine system performance include *f*-Chart, SOLCOST, RETSCREEN and TRNSYS. The *f*-Chart (Beckman et al. 1977) model is readily available either in worksheet form for hand calculations or as a computer program; however, it contains a built-in daily hot-water load profile of three equal draws specifically for residential applications. Studies have shown that the daily hot-water profile may affect system performance by up to 20% more if the load is heavier in the evening and 20% less if the load occurs in the morning.

For daily profiles with extreme deviations from the residential profile, such as the company cafeteria example used previously, *f*-Chart may not provide satisfactory results. SOLCOST (DOE 1978) can accommodate user-selected daily profiles, but it is not widely available. Neither program can accommodate special days (e.g., holidays and weekends) when no service hot-water loads are present. TRNSYS (Klein et al. 2000), which has been modified for operation on personal computers, is readily available and well supported. It is time-consuming to set up and run for the first time, but after it is set up, various design parameters, such as collector area and storage size, can be easily evaluated for a given system configuration.

Many hydronic solar energy systems are used with a series-coupled heat pump or an absorption air-conditioning system. In the case of a heat pump, lower storage temperatures, which create higher collector efficiencies, are desirable. In this case, the collector size may be greater than recommended for conventional SHW or hydronic building space heating (BSH) systems. In contrast, absorption air conditioners require much higher temperatures (>175°F) than BSH or hydronic BSH systems to activate the generator. Therefore, the storage-to-collector ratio can be much less than recommended for BSH. For absorption air conditioning, thermal energy storage may act as a buffer between the collector and generator, which prevents cycling caused by frequent changes in insolation level.

HEAT EXCHANGERS

Requirements

The heat exchanger transfers heat from one fluid to another. In closed solar energy systems, it also isolates circuits operating at different pressures and separates fluids that must not be mixed. Heat exchangers are used in solar applications to separate the heat transfer fluid in the collector loop from the domestic water supply in the storage tank (pressurized storage) or the domestic water supply from the storage (unpressurized storage). Heat exchangers for solar applications may be placed either inside or outside the storage or drainback tank.

Selecting a heat exchanger involves the following considerations:

Performance. Heat exchangers always degrade the performance of a solar system; therefore, selecting an adequate size is important. When in doubt, an oversized heat exchanger should be selected.

Guaranteed separation of fluids. Many code authorities require a vented, double-wall heat exchanger to ensure fluid isolation. System protection requirements may also dictate the need for guaranteed fluid separation.

Thermal expansion. The temperature in a heat exchanger may vary from below freezing to the boiling temperature of water. The design must withstand these thermal cycles without failing.

Materials. Galvanic corrosion is always a concern in liquid solar energy systems. Consequently, the piping, collectors, and other hydronic component materials must be compatible.

Space constraints. Often, limited space is available for mounting and servicing the heat exchanger. Physical size and configuration must be considered when selection is made.

Serviceability. The water side of a heat exchanger is exposed to the scaling effects of dissolved minerals, so design must provide access for cleaning and scale removal.

Pressure loss. Energy consumed in pumping fluids reduces system performance. Pressure drop through the heat exchanger should be limited to 1 to 2 psi to minimize energy consumption.

Pressure capability. Because the heat exchanger is exposed to cold-water supply pressure, it should be rated for pressures above 75 psig.

Internal Heat Exchanger

The internal heat exchanger can be a coil inside the tank or a jacket wrapped around the pressure vessel (see Figure 16). Several manufacturers supply tanks with either type of internal heat exchanger. However, the maximum size of pressurized tanks with internal heat exchangers is usually about 120 gal. Heat exchangers may be installed with relative ease inside nonpressurized tanks that open from the top. Figures 22 and 23 illustrate methods of achieving double-wall protection with either type of internal heat exchanger.

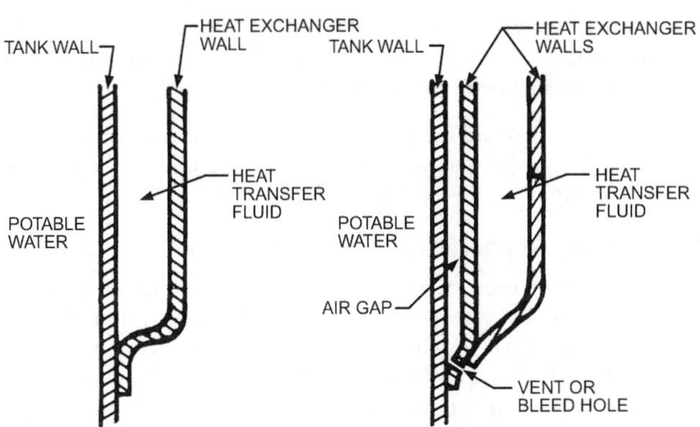

Fig. 22 Cross Section of Wraparound Shell Heat Exchangers

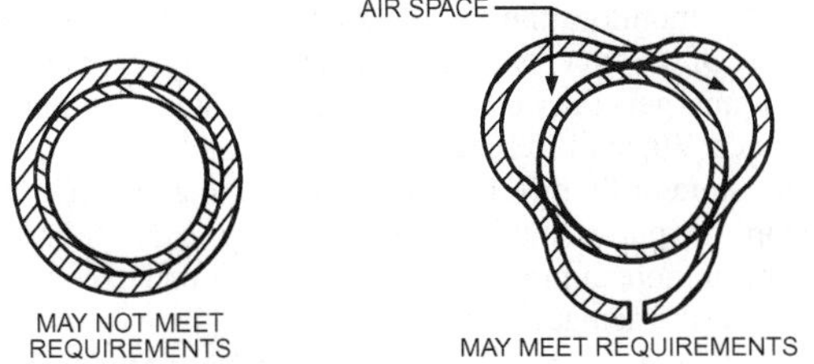

Fig. 23 Double-Wall Tubing

For installations with larger tanks or heat exchangers, a tube bundle is required. However, it is not always possible to find a heat exchanger of the desired area that will fit within the tank. Consequently, a horizontal tank with the tube bundle inserted from the tank end can be used. A second option is to place a shroud around the tube bundle and to pump fluid around it (Figure 24). Such an approach combines the performance of an external heat exchanger with the compactness of an internal heat exchanger. Unfortunately, this approach causes tank mixing and loss of stratification.

External Heat Exchanger

The external heat exchanger offers a greater degree of design flexibility than the internal heat exchanger because it is detached from the tank. For this reason, it is preferred for most commercial applications. Shell-and-tube, tube-and-tube, and plate-and-frame heat exchangers are used in solar applications. Shell-and-tube heat exchangers are found in many solar designs because they are economical, easy to obtain, and constructed with suitable material. One limitation of the shell-and-tube heat exchanger is that it normally does not have double-wall protection, which is often required for potable-water-heating applications. A number of manufacturers produce tube-and-tube heat exchangers that offer high performance and the double-wall safety required by many code authorities, but they are usually limited in size. The plate-and-frame heat exchanger is more cost-effective for large potable-water applications where positive separation of heat transfer fluids is required. These heat exchangers are compact and offer excellent heat transfer performance.

Shell-and-Tube. Shell-and-tube heat exchangers accommodate large heat exchanger areas in a compact volume. The number of shell-side and tube-side passes (i.e., the number of times the fluid changes direction from one end of the heat exchanger to the other) is a major variable in shell-and-tube heat exchanger selection. Because the exchanger must compensate for thermal expansion, flow in and out of the tube side are generally at the same end of the exchanger. Therefore, the number of tube-side passes is even. By appropriate baffling, two, four, or more tube passes may be created. However, as the number of passes increases, the path length grows,

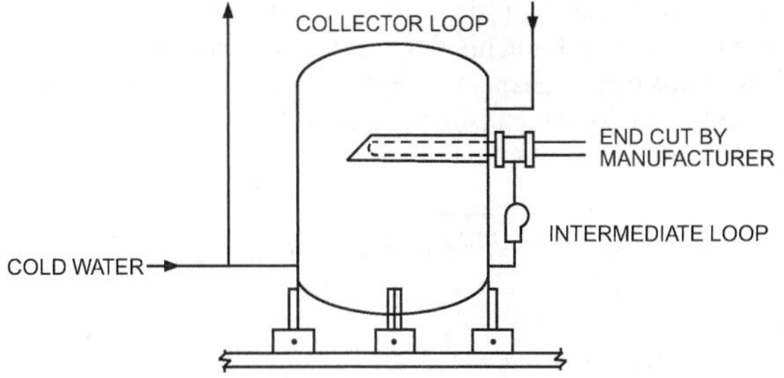

Fig. 24 Tube Bundle Heat Exchanger with Intermediate Loop

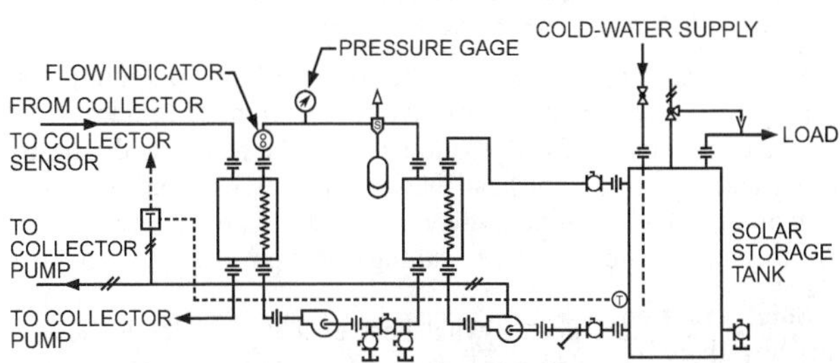

Fig. 25 Double-Wall Protection Using Two Heat Exchangers in Series

resulting in greater pressure drop of the tube-side fluid. Unfortunately, double-wall shell-and-tube heat exchangers are hard to find.

Tube-and-Tube. Fluids in the tube-and-tube heat exchanger run counterflow, which gives closer approach temperatures. The exchanger is also compact and only limited by system size. Several may be piped in parallel for higher flow, or in series to provide approach temperatures as close as 15°F. Many manufacturers offer the tube-and-tube with double-wall protection.

The counterflow configuration operates at high efficiency. For two heat exchangers in series, effectiveness may reach 0.80. For single heat exchangers or multiple ones in parallel, effectiveness may reach 0.67. Tube-and-tube heat exchangers are cost-effective for smaller residential systems.

Plate-and-Frame. These heat exchangers are suitable for pressures up to 300 psig and temperatures to 400°F. They are economically attractive when a high-quality, heavy-duty construction material is required or if a double-wall heat exchanger is needed for leak protection. Typical applications include food industries or domestic water heating when any possibility of product contamination must be eliminated. This exchanger also gives added protection to the collector loop.

Contamination is not possible when an intermediate loop is used (Figure 25) and the integrity of the plates is maintained. A colored fluid in the intermediate loop gives a visual means of detecting a leak through changes in color. The sealing mechanism of the plate-and-frame heat exchanger prevents cross-contamination of heat transfer fluids. Plate-and-frame heat exchangers cost the same as or less than equivalent shell-and-tube heat exchangers constructed of stainless steel.

Heat Exchanger Performance

Solar collectors perform less efficiently at high fluid inlet temperatures. Heat exchangers require a temperature difference between the two streams to transfer heat. The smaller the heat transfer surface area, the greater the temperature difference must be to transfer the same amount of heat and the higher the collector inlet temperature must be for a given tank temperature. As the solar collector is forced to operate at the progressively higher temperature associated with a smaller heat exchanger, its efficiency is reduced.

In addition to size and surface area, the heat exchanger's configuration is important for achieving maximum performance. The performance of a heat exchanger is characterized by its effectiveness (a type of efficiency), which is defined as follows:

$$E = \frac{q}{(\dot{m}c_p)_{min}(t_{hi} - t_{ci})} \quad (10)$$

where

E = effectiveness
q = heat transfer rate, Btu/h
$\dot{m}c_p$ = minimum mass flow rate times fluid specific heat, Btu/h·°F
t_{hi} = hot (collector-loop) stream inlet temperature, °F
t_{ci} = cold (storage) stream inlet temperature, °F

For heat exchangers located in the collector loop, minimum flow usually occurs on the collector side rather than the tank side.

The effectiveness E is the ratio between the heat actually transferred and the maximum heat that could be transferred for given flow and fluid inlet temperature conditions. E is relatively insensitive to temperature, but it is a strong function of heat exchanger design.

A designer must decide what heat exchanger effectiveness is required for the specific application. A method that incorporates heat exchanger performance into a collector efficiency equation [Equation (5)] uses storage tank temperature t_s as the collector inlet temperature with an adjusted heat removal factor F_R. Equation (11) relates F_R and heat exchanger effectiveness:

$$\frac{F_R'}{F_R} = \frac{1}{1 + (F_R U_L/\dot{m}c_p)[A\dot{m}c_p/E(\dot{m}c_p)_{min} - 1]} \quad (11)$$

The heat exchanger effectiveness must be converted into heat transfer surface area. SERI (1981) provides details of shell-and-tube heat exchangers and heat transfer fluids.

For more information on heat exchangers, see Chapter 47.

CONTROLS

The brain of an active solar energy system is the automatic temperature control. Numerous studies and reports of operational systems show that faulty controls, usually the sensors, are often the cause of poor performance. Reliable controllers are available, and with a full understanding of each system function, proper control systems can be designed. In general, control systems should be simple; additional controls are not a good solution to a problem that can be solved by better mechanical design. The following key considerations pertain to control system design:

- Collector sensor location/selection
- Storage sensor location
- Overtemperature sensor location
- On/off controller characteristics
- Selection of reliable solid-state devices, sensors, controllers, etc.
- Control panel location in heated space
- Connection of controller according to manufacturer's instructions
- Design of control system for all possible system operating modes, including heat collection, heat rejection, power outage, freeze protection, auxiliary heating, etc.
- Selection of alarm indicators for pump failure, low and high temperatures, loss of pressure, controller failure, nighttime operation, etc.

The following control categories should be considered when designing automatic controls for solar energy systems:

- Collection to storage
- Storage to load
- Auxiliary energy to load
- Alarms
- Miscellaneous (e.g., for heat rejection, freeze protection, draining, and overtemperature protection)

Types of controllers include snap switches, timers, photovoltaic (PV) modules, and differential temperature controllers. The latter is the most common, with PV popular for residential systems.

Differential Temperature Controllers

Most controls used in solar energy systems are similar to those for HVAC systems. The major exception is the differential temperature controller (DTC), which is the basis of solar energy system control. The DTC is a comparing controller with at least two temperature sensors that controls one or several devices. Typically, the sensors are located at the solar collectors and storage tank (Figure 26). On unpressurized systems, other DTCs may control the extraction of heat from the storage tank.

The DTC monitors the temperature difference, and when the temperature of the panel exceeds that of the storage by the predetermined amount (generally 8 to 20°F), the DTC switches on the actuating devices. When the temperature of the panel drops to 3 to 10°F above the storage temperature, the DTC, either directly or indirectly, stops the pump. Indirect control through a control relay may operate one or more pumps and possibly perform other control functions, such as the actuation of control valves.

The manufacturer's predetermined set point of the DTC may be adjustable or fixed. If the controller set point is a fixed temperature differential, the controller selected should correspond to the requirements of the system. An adjustable differential set point makes the controller more flexible and allows it to be adjusted to the specific system. The optimum *off* temperature differential should be the minimum possible; the minimum depends on whether there is a heat exchanger between the collectors and storage.

If the system requires a heat exchanger, the energy transferred between two fluids raises the differential temperature set point. The minimum, or *off*, temperature differential is the point at which pumping the energy costs as much as the value of the energy being pumped. For systems with heat exchangers, the *off* set point is generally between 5 and 10°F. If the system does not have a heat exchanger, a range of 3 to 6°F is acceptable for the *off* set point. The heat lost in piping and the power required to operate the pump should also be considered.

The optimum differential *on* set point is difficult to calculate because of the changing variables and conditions. Typically, the *on* set point is 10 to 15°F above the *off* set point. The optimum *on* set point is a balance between optimum energy collection and avoiding short-cycling the pump. ASHRAE's *Active Solar Heating System Design Manual* (1988) describes techniques for minimizing short cycling.

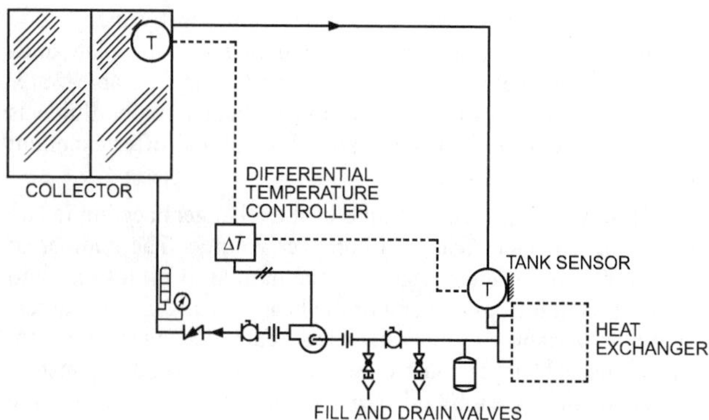

Fig. 26 Basic Nonfreezing Collector Loop for Building Service Hot Water Heating—Nonglycol Heat Transfer Fluid

Photovoltaically Powered Pumps

The ability to drive a pump in direct proportion to the available irradiance without an external electric power makes PV pumps very popular in residential applications. The lack of larger DC motors with low turning amp draws inhibits their adoption in commercial solar systems. Advantages of using PV pumps include an increase in day-long energy production and the security of having power during power outages. Disadvantages include higher first cost and the need for more frequent maintenance.

Overtemperature Protection

Overheating may occur during periods of high insolation and low load; thus, all portions of the solar energy system require protection against overheating. Liquid expansion or excessive pressure may burst piping or storage tanks, and steam or other gases within a system may restrict liquid flow, making the system inoperable. Glycols break down and become corrosive if subjected to temperatures greater than 240°F. The system can be protected from overheating by (1) stopping circulation in the collection loop until storage temperature decreases, (2) discharging overheated water from the system and replacing it with cold makeup water, or (3) using a heat exchanger coil as a means of heat rejection to ambient air.

The following questions should be answered to determine whether overtemperature protection is necessary.

- Is the load ever expected to be off, such that the solar input will be much higher than the load? The designer must determine possibilities based on the owner's needs and a computer analysis of system performance.
- Do individual components, pumps, valves, circulating fluids, piping, tanks, and liners need protection? The designer must examine all components and base the overtemperature protection set point on the component that has the lowest specified maximum operating temperature. This may be a valve or pump with a 180 to 300°F maximum operating temperature. Sometimes, this criterion may be met by selecting components capable of operating at higher temperatures.
- Is steam formation or discharging boiling water at the tap possible? If the system has no mixing valve that mixes cold water with the solar-heated water before it enters the tap, the water must be maintained below boiling temperature. Otherwise, the water will flash to steam as it exits the tap and, most likely, scald the user. Some city codes require a mixing valve to be placed in the system for safety.

Differential temperature controllers are available that sense overtemperature. Depending on the controller used, the sensor may be mounted at the bottom or the top of the storage tank. If it is mounted at the bottom, the collector-to-storage differential temperature sensor can be used to sense overtemperature. Input to a DTC mounted at the top of the tank is independent of the bottom-mounted sensor, and the sensor monitors the true high temperature.

The normal action taken when the DTC senses an overtemperature is to turn off the pump to stop heat collection. After the panels in a drainback system are drained, they attain stagnation temperatures. Although drainback is not desirable, the panels used for these systems should be designed and tested to withstand overtemperature. In addition, drainback panels should withstand the thermal shock of start-up when relatively cool water enters the panels while they are at stagnation temperatures. The temperature difference can range from 75 to 300°F. Such a difference could warp panels made with two or more materials of different thermal expansion coefficients. If the solar panels cannot withstand the thermal shock, an interlock should be incorporated into the control logic to prevent this situation. One method uses a high-temperature sensor mounted on the collector absorber that prevents the pump from operating until the collector temperature drops below the sensor set point.

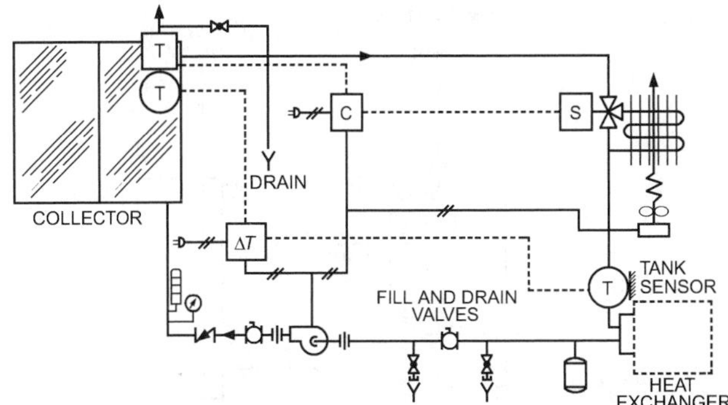

Fig. 27 Heat Rejection from Nonfreezing System Using Liquid-to-Air Heat Exchanger

If circulation stops in a closed-loop antifreeze system that has a heat exchanger, high stagnation temperatures will occur. These temperatures could break down the glycol heat transfer fluid. To prevent damage or injury caused by excessive pressure, a pressure-relief valve must be installed in the loop, and a means of rejecting heat from the collector loop must be provided. The section on Hot Water Dump describes a common way to relieve pressure. Pressure increases because of thermal expansion of any fluid; when water-based absorber fluids are used, pressure builds from boiling.

The pressure-relief valve should be set to relieve at or below the maximum operating pressure of any component in the closed-loop system. Typical settings are around 50 psig, corresponding to a temperature of approximately 300°F. However, these settings should be checked. When the pressure-relief valve does open, it discharges expensive antifreeze solution. Glycol antifreeze solutions damage many types of roof membranes. The discharge can be piped to large containers to save the antifreeze, but this design can create dangerous conditions because of the high pressures and temperatures involved.

If a collector loop containing glycol stagnates, chemical decomposition raises the liquid's fusion point, and freezing becomes possible. An alternative method continues fluid circulation but diverts flow from storage to a heat exchanger that dumps heat to the ambient air or other sink (Figure 27). This wastes energy, but protects the system. A sensor on the solar collector absorber plate that turns on the heat rejection equipment can provide control. The temperature sensor set point is usually 200 to 250°F and depends on the system components. When the sensor reaches the high-temperature set point, it turns on pumps, fans, alarms, or whatever is necessary to reject the heat and warn of the overtemperature. The dump continues to operate until the overtemperature control in the collector loop DTC senses an acceptable drop in tank temperature and is reset to its normal state.

Hot-Water Dump

If water temperatures above 200°F are allowed, the standard temperature-pressure (210°F, 150 psig) safety relief valve may operate occasionally. If these temperatures are reached, the valve opens, and some of the hot water vents out. However, these valves are designed for safety purposes, and after a few openings, they leak hot water. Thus, they should not be relied on as the only control device. An aquastat that controls a solenoid, pneumatic, or electrically actuated valve should be used instead.

Heat Exchanger Freeze Protection

The following factors should be considered when selecting an external heat exchanger for a system protected by a nonfreezing fluid that is started after an overnight, or longer, exposure to extreme cold.

- Freeze-up of the water side of the heat exchanger
- Performance loss due to extraction of heat from storage

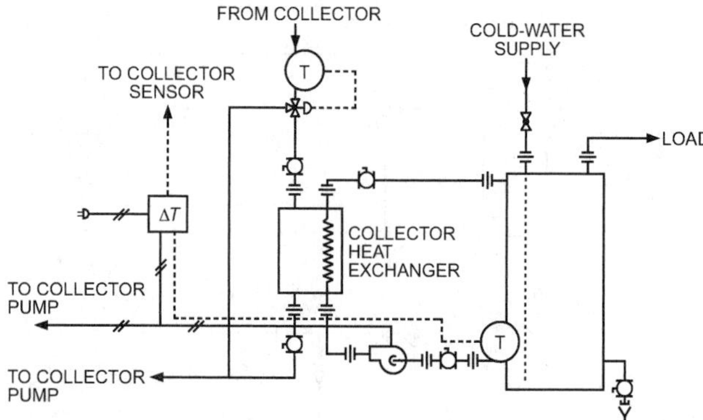

Fig. 28 Nonfreezing System with Heat Exchanger Bypass

An internal heat exchanger/tank has been placed on the water side of the heat exchanger of small systems to prevent freezing. However, the energy required to maintain the water above freezing must be extracted from storage, which decreases overall performance. With the external heat exchanger/tank combination, a bypass can be installed, as illustrated in Figure 28. The controller positions the valve to bypass the heat exchanger until the fluid in the collector loop attains a reasonable level (e.g., 80°F). When the heat transfer fluid has warmed to this level, it can enter the heat exchanger without freezing or extracting heat from storage. The arrangement in Figure 28 can also be used with an internal heat exchanger, if necessary, to improve performance.

PHOTOVOLTAIC SYSTEMS

Photovoltaic (PV) devices, or cells, convert light directly into electricity. These cells are connected in series and parallel strings and packaged into modules to produce a specific voltage and current when illuminated. PV modules can also be connected in series or in parallel into arrays to produce larger voltages or currents. PV systems rely on sunlight, have few or no moving parts, are modular to match power requirements on any scale, are reliable and long-lived, and are easily produced. Photovoltaic systems can be used independently or in conjunction with other electrical power sources.

Most early PV installations were **stand-alone** systems, (i.e., located at sites where utility power was unavailable). Some stand-alone applications powered by PV systems include communications, remote power, remote monitoring, lighting, water pumping, battery charging, and cathodic protection. These systems could include a battery bank to provide power during nighttime hours. Excess power had to be generated during the day to keep the battery bank charged through a charge controller.

More recently, **utility-interconnected** systems predominate. These systems use an inverter to produce the AC waveform to match that of the utility. If excess power is produced during the day, it is fed back to the utility. Typically, power needed at night is drawn from the utility. If utility power is lost, these systems automatically shut down to prevent backfeeding power into the lines, to protect utility personnel.

Some **hybrid** systems are utility interactive but also operate in stand-alone mode. If utility power is lost, they automatically isolate critical circuits from the utility service, and power only those circuits. They have a smaller battery bank, sized to handle just the critical loads. These systems normally keep the battery bank fully charged to use at night as back-up if utility power fails.

Fundamentals of Photovoltaics

Photovoltaic Effect. When a photon enters a PV material, the photon can be reflected, absorbed, or transmitted through. If the photon is absorbed, an electron, with its additional energy, leaves its atom. It may also jump across an established layer of junctions (within the structure of the PV cell) laid down during the manufacturing process. With many electrons jumping the junction layer, an electric charge is established between the front and back of the PV cell. Conductors placed on the front and back of the cell provide a path through which the electrons that have built up can create a current through a load from one side of the cell to the other. Without a connection to provide an electron flow, electrons build up on one side of the junction layer until their like-charge field is large enough to repel further electrons from jumping the junctions, and the number that jump over are equaled by the number that jump back and recombine with the originating atoms. This charge field represents the open-circuit voltage of the cell.

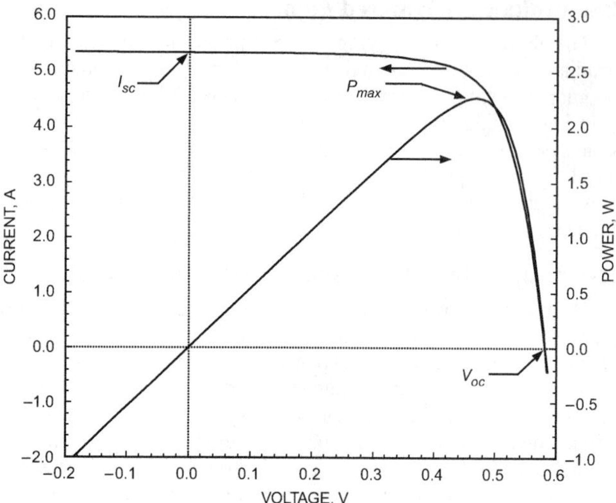

Fig. 29 Representative Current-Voltage and Power-Voltage Curves for Photovoltaic Device

Cell Design. A PV cell consists of the active PV material, metal grids, antireflection coatings, and supporting material. The complete cell is optimized to maximize both the amount of light entering the cell and the power out of the cell. The photovoltaic material can be one of a number of compounds. Metal grids enhance current collection from the front and back of the solar cell. Grid design varies among manufacturers. The antireflective coating is applied to the top of the cell to maximize the light going into the cell; typically, this coating is a single layer optimized for sunlight. As a result, PV cells range in color from black to blue. In some, the top of the cell is covered with a semitransparent conductor that functions as both the current collector and the antireflection coating. A completed PV cell is a two-terminal device with positive and negative leads, and its output is direct current.

Current-Voltage Curves. The current from a PV module depends on the external voltage applied and the amount of light on the module. When the module is short-circuited, the current is at maximum (short-circuit current I_{sc}), and the voltage across the module is zero. When the PV module circuit is open, with the leads not making a circuit, the voltage is at a maximum (open-circuit voltage V_{oc}), and the current is zero. In either case, at open or short circuit, the power (current times voltage) is zero. Between open and short circuit, the power output is greater than zero. A current-voltage curve (Figure 29) represents the range of combinations of current and voltage.

Power versus voltage can be plotted on the same graph. There exists an operating point P_{max}, I_{max}, V_{max} at which the output power is maximized. Given P_{max}, an additional parameter, fill factor FF, can be calculated such that

$$P_{max} = I_{sc} V_{oc} \mathrm{FF} \tag{12}$$

By illuminating and loading a PV module so that the voltage equals the PV cell's V_{max}, the output power is maximized. The module can be loaded using resistive loads, electronic loads, or batteries. The typical parameters of a single-crystal solar cell are current density $J_{sc} = 206$ mA/in^2; $V_{oc} = 0.58$ V; $V_{max} = 0.47$ V; FF = 0.72; and $P_{max} = 2273$ mW.

Efficiency and Power. Efficiency is another measure of PV cells. Efficiency is the maximum electrical power output divided by the incident light power. It is commonly reported for a PV cell temperature of 77 °F and incident light at an irradiance of 92.94 W/ft^2 with a spectrum of 1.5 atmospheres equivalent spectral absorption (close to that of sunlight at solar noon).

Photovoltaic Cells and Modules

Photovoltaic Cells. In general, all PV cells perform similarly. However, the choice of the PV material can have important effects on system design and performance. Both the composition of the material and its atomic structure are influential. PV materials include silicon, gallium arsenide, copper indium diselenide, cadmium telluride, indium phosphide, and many others. The atomic structure of a PV cell can be single-crystal, polycrystalline, or amorphous (i.e., having no crystalline structure). The most commonly produced PV modules are made of crystalline silicon, either single-crystal or polycrystalline. A distinction is made in the industry between cells made with junctions embedded in the crystals, and "thin film," made by laminating layers to produce the junctions.

Because of quantum effects, electrons designated to leave their atoms in a PV cell only absorb photon energy of a certain band of wavelengths; the other light energy is not absorbed. Multijunction cells (thin-film cells with two or more layers of junctions laid on top of one another) can have higher efficiency because each junction layer can be designed to respond to different wavelengths of incoming radiation.

Module Construction. A module is a collection of PV cells that protects the cells and provides a usable operating voltage. PV cells can be fragile and susceptible to corrosion by humidity or fingerprints and can have delicate wire leads. Also, the operating voltage of a single PV cell is less than 1 V, making it unusable by itself for many applications. Depending on the manufacturer and type of PV material, modules have different appearances and performance characteristics. Also, modules may be designed for specific conditions, such as hot or marine environments.

Single-crystal and polycrystalline silicon cells are produced individually. Some amorphous silicon cells, although manufactured in large rolls, are later processed as individual cells. These cells are connected in series and parallel to produce the desired operating voltage and wattage. These strings of cells are then encapsulated with a polymer, a front glass piece, and a back material.

Cells made of amorphous silicon, cadmium telluride, or copper indium diselenide are manufactured on large rolls of material that become either the front or the back of the module. A large area of PV material is divided into smaller cells by scribing or cutting the material into electrically isolated cells. Later processing steps automatically interconnect the cells in series and parallel. Depending on the manufacturer, a front glass or a back material is glued on using polymers. Modules include a robust junction box attached to the back of the module for wiring to other modules or other electrical equipment.

Module Reliability and Durability. PV modules are reliable and durable when properly made. As with any electrical equipment, the designer must verify that the product is tested or rated for the intended application. A PV module is a large-area semiconductor with no internal moving parts. Even a concentrator module has no internal moving parts, although there are external parts that keep the module aimed at the sun.

PV modules are designed for outdoor use in harsh surroundings such as marine, tropic, arctic, and desert environments. There are many international standards that test the suitability of PV modules for outdoor use in different climates. A buyer or user should check the suitability of the module for a given climate.

Modules are also tested for electrical safety. All modules have a maximum voltage rating supplied by the manufacturer or labeled on the module. This becomes important when interconnecting several modules into an array and connecting arrays to other electrical equipment. Most modules are tested for ground isolation to prevent personal injury or property damage. To meet the *National Electrical Code®*, modules must be listed by Underwriters Laboratories (UL).

Module Performance. The performance output of a module is provided by the manufacturer and shown on a label on each module. The value provided is in watts per square meter, and represents the expected peak power point output of the module in watts at standard test conditions (STC). These conditions, established by international standard, are defined as 92.94 W/ft^2 incident irradiation at a spectrum of 1.5 atmospheres equivalent spectral absorption, with a cell surface temperature of 68°F. Modules are typically marketed and sold by this peak watt value. Designers are cautioned that the actual performance of a module may be as much as 10% below the manufacturer's listed value. Performance values independently tested at STC are available for some modules from the Florida Solar Energy Center.

The performance output is provided for new modules with only brief exposure. Modules with crystalline cells are quite stable over exposure time, but thin-film cells degrade in performance over time of exposure. Some thin-film manufacturers derate their module ratings to account for this exposure degradation; others do not. Improving the stability of thin-film modules over exposure time is an active area of research in the PV industry.

Related Equipment

Batteries. Batteries are used in some PV systems to supply power at night or when the PV system cannot meet the demand. The selection of battery type and size depends primarily on load and availability requirements. In all cases, batteries must be located in an area without extreme temperatures and with some ventilation.

Battery types include lead-acid, nickel cadmium, nickel hydride, lithium, and many others. For a PV system, the main requirement is that the batteries be capable of repeated deep discharges without damage. Starting-lighting-ignition batteries (car batteries) are not designed for this and should not be used.

Deep-cycle lead-acid batteries are commonly used. These batteries can be flooded or valve-regulated (sealed) batteries and are commercially available in a variety of sizes. Flooded, or wet, batteries require greater maintenance but can last longer with proper care. Valve-regulated batteries require less maintenance. For more capacity, batteries can be arranged in parallel.

Battery Charge Controllers. Charge controllers regulate power from the PV modules to prevent the batteries from overcharging. The controller can be a shunt or series type and can also function as a low-battery-voltage disconnect to prevent battery overdischarge. The controller is chosen for the correct capacity and desired features.

Normally, controllers allow battery voltage to determine the PV system's operating voltage. However, battery voltage may not be the optimum PV operating voltage. Some stand-alone charge controllers can optimize the operating voltage of the PV modules independently of battery voltage so that the PV operates at its maximum power point. (See the Peak Power Tracking section following.)

Inverters. An inverter is used to convert direct current (dc) into alternating current (ac) electricity. The inverter's output can be single-phase or multiphase, with a voltage of 120 V, 220 V, 440 V, etc., and a frequency of 50 or 60 Hz. Inverters are rated by total power capacity, which ranges from hundreds of watts to megawatts. Some inverters have good surge capacity for starting motors; others have limited surge capacity. The designer should specify both the type and the size of the load the inverter is intended to service.

The output waveform of the inverter, though still ac, can be square, modified square (also called modified sine wave), or sine. For some small applications, modified sine wave inverters are available. Inverters should be selected with care, because equipment that contains silicon-controlled rectifiers (SCRs) or variable-speed motors, such as some vacuum cleaners and laser printers, can be damaged by square- and modified-square-wave inverter outputs. Electric utilities supply sine wave output, and most manufacturers produce inverters with sine wave outputs that avoid unexpected equipment damage. Stand-alone inverters can supply the ac waveform independently. Utility interactive inverters synchronize the waveform frequency to another ac power supply such as the electric utility or perhaps can be matched to a portable electrical generator.

Peak Power Tracking. Peak power tracking varies the operating voltage of the PV array to optimize the current for maximum power production. Typically, the PV array output wattage is sampled and the voltage is adjusted automatically many times a second. Peak power trackers can be purchased separately or specified as an option with battery-charge controllers. Off-the-shelf inverters typically have peak power tracking circuitry built in.

Balance-of-System (BOS) Components. These components include the mounting structures and wiring, and are just as important as the other major PV system components. PV systems should be designed and installed with the intent of only minimal maintenance.

Photovoltaic modules can be mounted on the ground or on a building or can be included as part of a building. A structural engineer can design the support structures in accordance with local building codes. Wind and snow loading are a major design consideration. PV modules can last 20 or more years; the support structure and building should be designed for at least as long a lifetime.

In the United States, wire sizing, insulation, and use of fuses, circuit breakers, and other protective devices are covered by the *National Electrical Code*® (NFPA *Standard* 70).

REFERENCES

ANL. 1979a. *Reliability and materials design guidelines for solar domestic hot water systems.* ANL/SDP-9. Argonne National Laboratory, Argonne, IL.

ANL. 1979b. *Design and installation manual for thermal energy storage.* ANL-79-15. Argonne National Laboratory, Argonne, IL.

ANL. 1980. *Design and installation manual for thermal storage*, 2nd ed. ANL-79-15. Argonne National Laboratory, Argonne, IL.

ASHRAE. 1988. *Active solar heating systems design manual.* In cooperation with Solar Energy Industries Association and American Consulting Engineers Council Research & Management Foundation.

ASHRAE. 1990. *Guide for preparing active solar heating systems operation and maintenance manuals.*

ASHRAE. 1991. *Active solar heating systems installation manual.*

ASHRAE. 2003. Methods of testing to determine the thermal performance of solar collectors. ANSI/ASHRAE *Standard* 93-2003.

ASME. 2001. *Boiler and pressure vessel code.* American Society of Mechanical Engineers, New York.

ASTM. 2001. Standard test method for photovoltaic modules in cyclic temperature and humidity environments. ASTM *Standard* E1171-01. ASTM International.

ASTM. 1998. Standard test method for saltwater immersion and corrosion testing of photovoltaic modules for marine environments. ASTM *Standard* E1524-98. ASTM International.

Beckman, W.A., S. Klein, and J.A. Duffie. 1977. *Solar heating design by the f-Chart method.* Wiley-Interscience, New York.

Howard, B.D. 1986. Air core systems for passive and hybrid energy-conserving buildings. *ASHRAE Transactions* 92(2B):815-831.

Huggins, J.C. and D.L. Block. 1983. Thermal performance of flat plate solar collectors by generic classification. *Proceedings of the ASME Solar Energy Division Fifth Annual Conference*, Orlando, FL.

Johnston, S.A. 1982. *Passive solar and hybrid manufactured building systems: Precast concrete applications.* American Solar Energy Society, Boulder, CO.

Klein, S.A., et al. 2004. *TRNSYS 16 reference manual.* Solar Energy Laboratory, University of Wisconsin-Madison.

Knowles, A.S. 1980. *A simple balancing technique for liquid cooled flat plate solar collector arrays.* International Solar Energy Society, Phoenix, AZ.

Kreider, J. 1982. *The solar heating design process—Active and passive systems.* McGraw-Hill, New York.

Kutscher, C.F. 1982. *Design approaches for industrial process heat systems.* SERI/TR-253-1356. Solar Energy Research Institute, Golden, CO.

Logee, T.L. and P.W. Kendall. 1984. *Component report performance of solar collector arrays and collector controllers in the National Solar Data Network.* SOLAR/0015-84/32. Vitro Corporation, Silver Spring, MD.

NFPA. 2001. *National electric code*®. *Standard* 70-98. National Fire Protection Association, Quincy, MA.

SERI. 1981. *Solar design workbook.* Solar Energy Research Institute, Golden, CO.

BIBLIOGRAPHY

Architectural Energy Corporation. 1991. *Maintenance and operation of stand-alone photovoltaic systems*, vol. 5. Naval Facilities Engineering Command, Charleston, SC.

ASHRAE. 1994. *Design guide for cool thermal storage.*

Davidson, J. 1990. *The new solar electric home—The photovoltaics how-to handbook.* Aatech Publications, Ann Arbor, MI.

DOE. 1978. *SOLCOST—Solar hot water handbook: A simplified design method for sizing and costing residential and commercial solar hot water systems*, 3rd ed. DOE/CS 0042/2. U.S. Department of Energy.

ICBO. 1997. *Uniform building code.* International Conference of Building Officials, Whittier, CA.

Marion, W. and S. Wilcox. 1994. *Solar radiation data manual for flat-plate and concentrating collectors.* NREL/TP-463-5607. National Renewable Energy Laboratory, Golden, CO.

Marion, W. and S. Wilcox. 1995. *Solar radiation data manual for buildings.* NREL/TP-463-7904. National Renewable Energy Laboratory, Golden, CO.

Risser, V. and H. Post, eds. 1995. *Stand-alone photovoltaic systems: A handbook of recommended design practices.* SAND87-7023. Photovoltaic Design Assistance Center, Sandia National Laboratories, Albuquerque, NM.

SEL. 1983. TRNSYS: A transient simulation program. Engineering Experiment Station. *Report* 38-12. Solar Energy Laboratory, University of Wisconsin-Madison.

Strong, S. 1994. *The solar electric house: Energy for the environmentally-responsive, energy-independent home*, 2nd ed. Chelsea Green, White River Junction, VT.

Wiles, J. 2003. *Photovoltaic systems and the* National Electrical Code—*Suggested practices.* Photovoltaic Design Assistance Center, Sandia National Laboratories, Albuquerque, NM.

CHAPTER 37

COMPRESSORS

A COMPRESSOR is one of the four essential components of the basic vapor compression refrigeration system; the others are the condenser, evaporator, and expansion device. The compressor circulates refrigerant through the system and increases refrigerant vapor pressure to create the pressure differential between the condenser and evaporator. This chapter describes the design features of several categories of commercially available refrigerant compressors.

There are two broad categories of compressors: positive displacement and dynamic. **Positive-displacement compressors** increase refrigerant vapor pressure by reducing the volume of the compression chamber through work applied to the compressor's mechanism. Positive-displacement compressors include many styles of compressors currently in use, such as reciprocating, rotary (rolling piston, rotary vane, single screw, twin screw), and orbital (scroll, trochoidal).

Dynamic compressors increase refrigerant vapor pressure by continuous transfer of kinetic energy from the rotating member to the vapor, followed by conversion of this energy into a pressure rise. Centrifugal compressors function based on these principles.

There are many reasons to consider each compressor style. Some compressors have physical size limitations that may limit their application to smaller equipment; some have associated noise concerns; and some have efficiency levels that make them more or less attractive. Each piece of equipment using a compressor has a certain set of design parameters (refrigerant, cost, performance, sound, capacity, etc.) that requires the designer to evaluate various compressor characteristics and choose the best compressor type for the application.

Figure 1 addresses volumetric flow rate of the compressor as a function of the differential pressure (discharge pressure minus suction pressure) against which the compressor is required to work. Three common compressor styles are represented on the chart. The positive-displacement compressors tend to maintain a relatively constant volumetric flow rate over a wide range of differential pressures. This is because a positive-displacement compressor draws a predetermined volume of vapor into its chamber and compresses it to a reduced volume mechanically, thereby increasing the pressure. This helps to keep the equipment operating near its design capacity regardless of the conditions. Centrifugal compressors dynamically compress the suction gas by converting velocity energy to pressure energy. Therefore, they do not have a fixed volumetric flow rate, and the capacity can vary over a range of pressure ratios. This tends to make centrifugal-based equipment much more application specific.

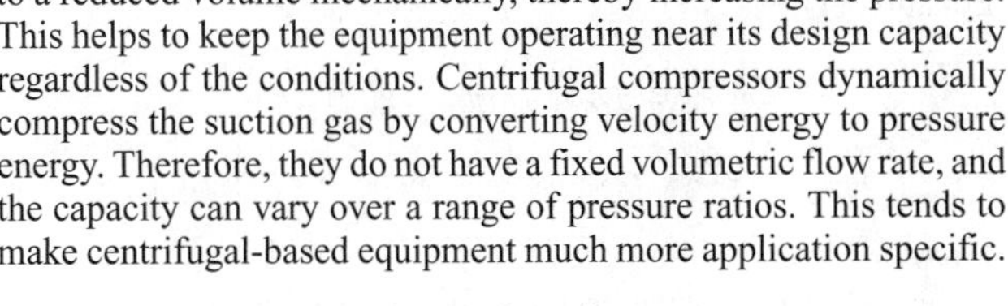

The preparation of this chapter is assigned to TC 8.1, Positive Displacement Compressors, and TC 8.2, Centrifugal Machines.

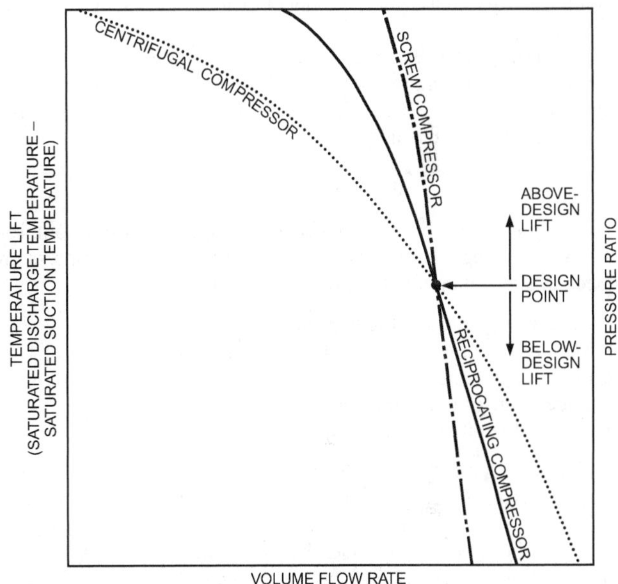

Fig. 1 Comparison of Single-Stage Centrifugal, Reciprocating, and Screw Compressor Performance

POSITIVE-DISPLACEMENT COMPRESSORS

Types of positive-displacement compressors classified by compression mechanism design are shown in Figure 2.

Compressors also can be further classified as single-stage or multi-stage, and by type of motor drive (electrical or mechanical), capacity control (single speed, variable speed, single speed with adjustable volume of the compression chamber), and drive enclosure (hermetic, semihermetic, and open). The most widely used compressors (for halocarbons) are manufactured in three types: (1) open, (2) semihermetic or bolted hermetic, and (3) welded-shell hermetic.

Ammonia compressors are manufactured only in the open design because of the incompatibility of the refrigerant and hermetic motor materials.

Open compressors are those in which the shaft or other moving part extends through a seal in the crankcase for an external drive. Ammonia compressors are manufactured only in the open design because of the incompatibility of the refrigerant and hermetic motor materials. Most automotive compressors are also open-drive type.

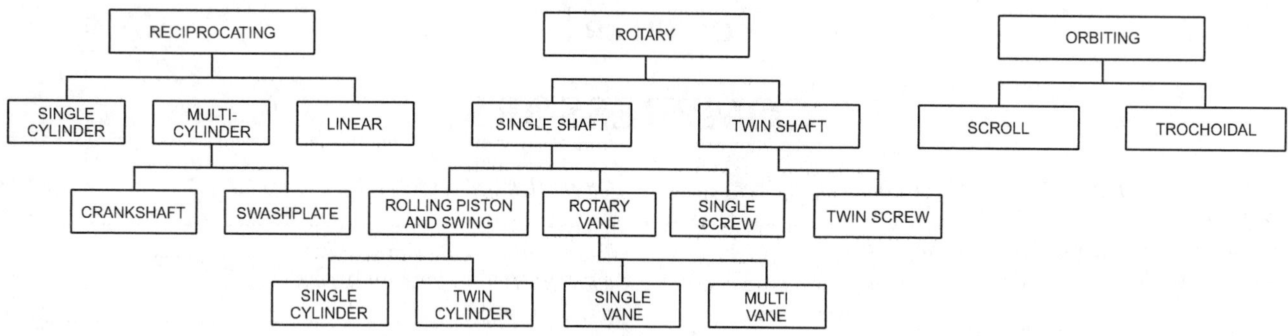

Fig. 2 Types of Positive-Displacement Compressors (Classified by Compression Mechanism Design)

Hermetic compressors contain the motor and compressor in the same gastight housing, which is permanently sealed with no access for servicing internal parts in the field, with the motor shaft integral with the compressor crankshaft and the motor in contact with the refrigerant. Hermetic compressors normally have the motor-compressor pump assembly mounted inside a steel shell, which is sealed by welding.

A **semihermetic compressor** (also called bolted, accessible, or serviceable) is a compressor of bolted construction that is sealed by gasketed joints amenable to field repair. The seal in the bolted joints is provided by O rings or gaskets.

PERFORMANCE

Compressor performance depends on an array of design compromises involving characteristics of the refrigerant, compression mechanism, and motor. The goal is to provide the following:

- Greatest trouble-free life expectancy
- Most refrigeration effect for least power input
- Lowest applied cost
- Wide range of operating conditions
- Acceptable vibration and sound level

Two useful measures of compressor performance are the coefficient of performance (COP) and the ratio of power required per unit of refrigerating capacity (power input/refrigeration output). The COP is a dimensionless number that is the ratio of the compressor's refrigerating capacity to the heat rate equivalent of the input power. The COP for a hermetic or semihermetic compressor includes the combined operating efficiencies of the motor and the compressor:

$$\text{COP (hermetic or semihermetic)} = \frac{\text{Capacity, Btu/h}}{\text{Input power to motor, Btu/h}}$$

The COP for an open compressor does not include motor efficiency:

$$\text{COP (open)} = \frac{\text{Capacity, Btu/h}}{\text{Input power to shaft, Btu/h}}$$

Because capacity and motor/shaft power vary with operating conditions, COP also varies with operating conditions.

Power input per unit of refrigerating capacity (bhp/ton) is used to compare different compressors at the same operating conditions, primarily with open-drive industrial equipment.

$$\frac{\text{bhp}}{\text{ton}} = \frac{\text{Power input to shaft, bhp}}{\text{Compressor capacity, ton}}$$

Ideal Compressor

During operation, pressure and volume in the compression chamber vary as shown in Figure 3. There are four sequential processes: first, gas is drawn into the compression chamber during the suction process (1–2); next is compression (2–3); and then higher-pressure gas is pushed out during the discharge process (3–4), followed by the next cycle.

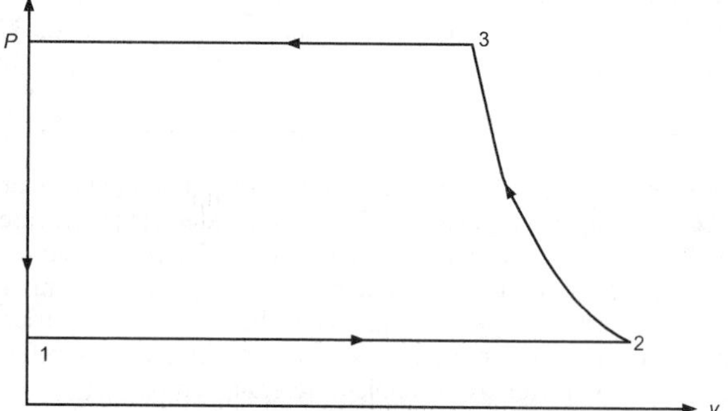

Fig. 3 Ideal Compressor Cycle

The capacity of a compressor at a given operating condition is a function of the mass of gas compressed per unit time. Ideally, mass flow is equal to the product of the compressor displacement per unit time and the gas density, as shown in Equation (1):

$$\dot{m} = \rho_s V_d \quad (1)$$

where

$\dot{m}$ = ideal mass flow of compressed gas, lb/h
ρ_s = density of gas entering compressor (at suction port), lb/ft^3
V_d = geometric displacement of compressor, ft^3/h

The ideal refrigeration cycle, discussed in detail in Chapter 1 of the 2005 *ASHRAE Handbook—Fundamentals*, consists of four processes, as shown in Figure 4:

1–2: isentropic (reversible and adiabatic) compression
2–3: desuperheating, condensing, and subcooling at constant pressure
3–4: adiabatic expansion
4–1: boiling and superheating at constant pressure

The following quantities can be determined from the pressure-enthalpy diagram in Figure 4 using m, the mass flow of gas from Equation (1),

$$Q_o = mQ_{refrigeration\ effect} = m(h_1 - h_4) \quad (2)$$

$$P_o = mQ_{work\ of\ compression} = m(h_2 - h_1) = mw_{oi} \quad (3)$$

where

w_{oi} = specific work of isentropic compression, Btu/lb
Q_o = ideal capacity, Btu/h
P_o = ideal power input, Btu/h

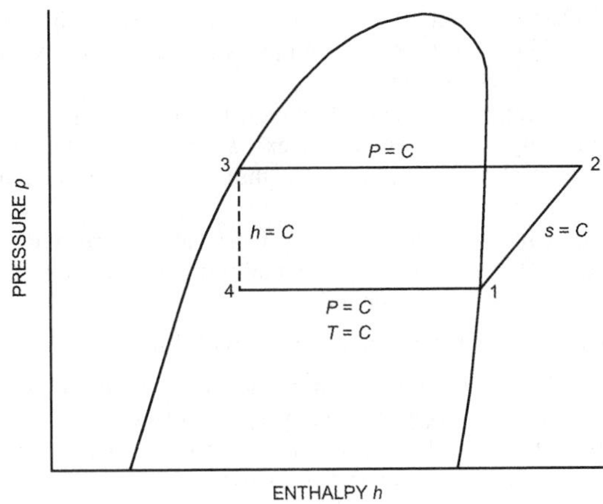

Fig. 4 Pressure-Enthalpy Diagram for Ideal Refrigeration Cycle

Actual Compressor

Ideal conditions never occur, so actual compressor performance differs from ideal performance. Various factors contribute to decreased capacity and increased power input. Depending on compressor type, some or all of the following factors can have a major effect on compressor performance.

- **Pressure drops in compressor**
  - Through shutoff valves
  - Through suction accumulator
  - Across suction strainer/filter
  - Across motor (hermetic compressor)
  - In manifolds (suction and discharge)
  - Through valves and valve ports (suction and discharge)
  - In internal muffler
  - Through internal lubricant separator
  - Across check valves
- **Heat gain by refrigerant from**
  - Cooling the hermetic motor
  - Internal heat exchange between compressor and suction gas
- **Power losses because of**
  - Friction
  - Lubricant pump power consumption
  - Motor losses
- **Valve inefficiencies** caused by imperfect mechanical action
- **Internal gas leakage**
- **Oil circulation**
- **Reexpansion** (clearance losses). The gas remaining in the compression chamber after discharge reexpands into the compression chamber during the suction cycle and limits the mass of fresh gas that can be brought into the compression chamber.
- **Over- and undercompression.** Overcompression occurs when pressure in the compression chamber reaches discharge pressure before finishing the compression process. Undercompression occurs when the compression chamber reaches the discharge pressure after finishing the compression process.
- **Deviation from isentropic compression.** In the actual compressor, the compression process deviates from isentropic compression primarily because of fluid and mechanical friction and heat transfer in the compression chamber. The actual compression process and work of compression must be determined from measurements.

Compressor Efficiency, Subcooling, and Superheating

Deviations from ideal performance are difficult to evaluate individually. They can, however, be grouped together and considered by category. Their effect on ideal compressor performance is characterized by the following efficiencies:

Volumetric efficiency (η_v) is the ratio of actual volumetric flow to ideal volumetric flow (i.e., the geometric compressor displacement).

Compression isentropic efficiency (η_{oi}) considers only what occurs within the compression volume and is a measure of the deviation of actual compression from isentropic compression. It is defined as the ratio of work required for isentropic compression of the gas (w_{io}) to work delivered to the gas within the compression volume (w_a).

$$\eta_{oi} = w_{oi}/w_a \tag{4}$$

(as obtained by measurement).

For a multicylinder or multistage compressor, this equation applies only for each individual cylinder or stage.

Mechanical efficiency (η_m) is the ratio of work delivered to the gas (measured) to work input to the compressor shaft (w_m).

$$\eta_m = w_a/w_m \tag{5}$$

Isentropic (reversible adiabatic) efficiency (η_i) is the ratio of work required for isentropic compression of the gas (w_{oi}) to work input to the compressor shaft (w_m).

$$\eta_i = w_{oi}/w_m \tag{6}$$

Motor efficiency (η_e) is the ratio of work input to the compressor shaft (w_m) to work input to the motor (w_e).

$$\eta_e = w_m/w_e \tag{7}$$

Total compressor efficiency (η_{com}) is the ratio of work required for isentropic compression (w_{oi}) to work input to the motor (w_e).

$$\eta_{com} = w_{oi}/w_e \tag{8}$$

Actual shaft compressor power is a function of the power input to the ideal compressor and the compression, mechanical, and volumetric efficiencies of the compressor, as shown in the following equation:

$$P_e = P_{ov}\,\eta_v/\eta_{com} = P_{ov}\,\eta_v/(\eta_e\eta_i) = P_{oi}\,\eta_v/(\eta_{oi}\eta_m\eta_e) \tag{9}$$

or

$$P_e = P_m/\eta_e = P_a/(\eta_m\eta_e) = P_{oi}/(\eta_{oi}\eta_m\eta_e) \tag{10}$$

where

P_e = power input to motor
P_m = power input to shaft
P_{oi} = power required for isentropic compression

Actual capacity is a function of the ideal capacity and volumetric efficiency η_v of the compressor:

$$Q = Q_o\eta_v \tag{11}$$

Total heat rejection is the sum of refrigeration effect and heat equivalent of power input to the compressor. Heat radiation or using means for additional cooling may reduce this value. The quantity of heat rejection must be known in order to size condensers.

Note that compressor capacity with a given refrigerant depends on saturation suction temperature (SST), saturation discharge temperature (SDT), superheating (SH), and subcooling (SC). **Saturation suction temperature (SST)** is the temperature of two-phase liquid/gas refrigerant at suction pressure. SST is often called evaporator temperature; however, in real systems, there is a difference because of pressure drop between evaporator and compressor. **Saturated discharge temperature (SDT)** is the temperature of two-phase liquid/gas refrigerant at discharge pressure. SDT is often called condensing temperature; however, in real systems, there is a difference because of the pressure drop between compressor and condenser.

Liquid subcooling is not accomplished by the compressor. However, the effect of liquid subcooling is included in compressor ratings by some manufacturers. *Note*: Air-Conditioning and Refrigeration Institute (ARI) *Standard* 540 and European Committee for Standardization (CEN) *European Norm* (EN) 12900 do not include subcooling.

Suction Superheat. No liquid refrigerant should be present in suction gas entering the compressor, because it causes oil dilution and gas formation in the lubrication system. If liquid carryover is severe enough to reach the cylinders, excessive wear of valves, stops, pistons, and rings can occur; liquid slugging can break valves, pistons, and connecting rods. Measuring suction superheat can be difficult, and the indication of a small superheat (<40°F) does not necessarily mean that liquid is not present. An effective suction separator may be necessary to remove all liquid.

Some compressors are specifically designed to operate without suction superheat. In this case, special design features are introduced to keep liquid from reaching suction valves and cylinders; oil viscosity must also be adjusted to anticipate its dilution with refrigerant.

High suction superheat may result in dangerously high discharge temperatures and, in hermetic compressors, high motor temperatures.

ABNORMAL OPERATING CONDITIONS, HAZARDS, AND PROTECTIVE DEVICES

To operate through the entire range of conditions for which the compressor was designed and to obtain the desired service life, it is important that mating components in the system be correctly designed and selected. Suction superheat must be controlled, lubricant must return to the compressor, and adequate protection must be provided against abnormal conditions. Chapters 1 to 4 of the 2006 *ASHRAE Handbook—Refrigeration* provide more information on protection against abnormal conditions. Chapter 6 of that volume gives details of cleanup in the event of a hermetic motor burnout.

Compressors are provided with one or more of the following devices for protection against abnormal conditions and to comply with various codes.

- **High-pressure protection** as required by Underwriters Laboratories and per ARI standards and ASHRAE *Standard* 15. This may include the following:
  - A high-pressure cutout [a pressure sensor that sends a signal to the switch to cut power to the compressor motor (drive)].
  - A high- to low-side internal relief valve, external relief valve, or rupture member to comply with ASHRAE *Standard* 15. The differential pressure setting depends on the refrigerant used and operating conditions. Care must be taken to ensure that the relief valve will not accidentally blow on a fast pulldown. Some welded hermetic compressors have an internal high- to low-pressure relief valve to limit maximum pressure in units not equipped with other high-pressure control devices.
  - A relief valve assembly on the oil separator of a screw compressor unit.
- **High-temperature control** devices to protect against overheating and oil breakdown.
  - Motor overtemperature protective devices are addressed in the section on Integral Thermal Protection in Chapter 44.
  - To protect against lubricant and refrigerant breakdown, a temperature sensor is sometimes used to stop the compressor when discharge temperature exceeds safe values. The switch may be placed internally (near the compression chamber) or externally (on the discharge line).
  - On larger compressors, lubricant temperature may be controlled by cooling with a heat exchanger or direct liquid injection, or the compressor may shut down on high lubricant temperature.
  - Where lubricant sump heaters are used to maintain a minimum lubricant sump temperature, a thermostat may be used to limit the maximum lubricant temperature.
- **Low-pressure protection** may be provided for
  - *Suction pressure.* Many compressors or systems are limited to a minimum suction pressure by a protective switch. Motor and compressor mechanism cooling, freeze-up, or pressure ratio usually determine the pressure setting.
  - *Compressor.* Forced-feed lubrication systems use lubricant-pressure, minimum-flow, or minimum-level protectors to prevent the compressor from operating with insufficient lubricant pressure.
- **Time delay or lockouts with manual resets** prevent damage to both compressor motor and contactors from repetitive rapid-starting cycles. Fixed-speed compressor motors experience a significant inrush electrical current during start-up. This current can reach the level of locked-rotor amps. If the motor restarts rapidly, without adequate cooling, overheating damage can occur. Time delays should be set for an appropriate interval to avoid this hazard.
- **Low-voltage and phase-loss or reversal protection** is used on some systems. Phase-reversal protection is used with multiphase devices to ensure proper direction of rotation.
- **Suction line strainer.** Some compressors are provided with a strainer at the suction inlet to remove any dirt that might be in suction line piping. Factory-assembled units with all parts cleaned at the time of assembly may not require the suction line strainer. A suction line strainer is normally required in all field-assembled systems.

Liquid Hazard

Liquid is essentially incompressible, so damage may occur when a compressor is handling liquid. This damage depends on the quantity of liquid, frequency with which it occurs, and type of compressor. Slugging, floodback, and flooded starts are three ways liquid can damage a compressor.

Slugging is the short-term pumping of a large quantity of liquid refrigerant and/or lubricant. It can occur just after start-up if refrigerant accumulated in the evaporator during shutdown returns to the compressor. It can also occur when system operating conditions change radically, such as during a defrost cycle. Slugging can also occur with quick changes in compressor loading.

Floodback is the continuous return of liquid refrigerant mixed with suction gas. It is a hazard to compressors that depend on maintaining a certain amount of lubricant for bearing surfaces. A properly sized suction accumulator can be used for protection.

Flooded start occurs when refrigerant is allowed to migrate to the compressor during shutdown. Compressors can be protected with crankcase heaters and automatic pumpdown cycles, where applicable.

Suction and Discharge Pulsations

Suction pulsation occurs because of the sudden flow and slight pressure drop in the suction line at the end of the reverse portion of the cycle and during suction. The frequency of this pulsation corresponds to the frequency of the compression cycle. Amplitude of the pulsation can reach 1.5 psi, especially if the compression chamber serves as a resonator. Suction pulsation may affect volumetric efficiency and a propagating compression wave may create structural problems caused by vibration of the suction line and evaporator.

Suction pulsation has a negative effect on compressor sound and vibration levels. Specially designed suction mufflers can solve the problems induced by pressure pulsation. Also, a significant volume in the suction line, such as in a suction accumulator, or enclosed by the compressor shell, can reduce pulsation level.

Discharge pulsation occurs because of the sudden flow and pressure fluctuation in the discharge line at the end of the compression portion of the cycle and during discharge. Over- or undercompression has a significant effect on these pulsations. The frequency of the discharge pulsation corresponds to the frequency of the compression cycle. Amplitude of the pulsation can reach 14.5 psi. In many cases, discharge pulsation is a major contributor to compressor sound and vibration levels. Discharge pulsation can be very destructive (structural damage, damage to sensors, etc.). Discharge mufflers can alleviate the problems induced by pressure pulsation.

Noise

An acceptable sound level is a basic requirement of good design and application. The major contributors to compressor noise are internal turbulence, impact (valves), friction, and the electric motor. Using sound shields or blankets may be feasible in certain applications. Chapter 47 of the 2007 *ASHRAE Handbook—HVAC Applications* covers design criteria in more detail.

Vibration

Compressor vibration results from gas-pressure pulses and inertia associated with moving parts. If the compressor is considered as a cylindrical body, vibration can be differentiated as axial, radial, and torsional. With increase of pressure differential between discharge and suction, vibration increases, especially axial and radial components. At fixed suction and discharge conditions, lower compressor speed usually leads to higher vibration amplitude, especially in the torsional component. Vibration problems can be handled in the following ways.

Isolation. With this common method, the compressor is resiliently mounted in the unit by springs, synthetic rubber mounts, etc. In hermetic reciprocating compressors, the internal compressor assembly is usually spring-mounted within the welded shell, and the entire unit is externally isolated. Use of flexible suction and discharge tubes may be feasible in certain applications.

Amplitude Reduction. The amount of movement can be reduced by adding mass to the compressor. Mass is added either by rigidly attaching the compressor to a base, condenser, or chiller, or by providing a solid foundation. When structural transmission is a problem, particularly with large machines, the entire assembly is then resiliently mounted.

Balancing. Proper balancing of inertial forces is important in reducing vibration. Counterweights are often used in rotary and scroll compressors.

Chapter 47 of the 2007 *ASHRAE Handbook—HVAC Applications* has further information.

Shock

In designing for shock, three types of dynamic loads are recognized:

- Suddenly applied loads of short duration
- Suddenly applied loads of long duration
- Sustained periodic varying loads

Because the forces are primarily inertial, the basic approach is to maintain low equipment mass and make the strength of the carrying structure as great as possible. The degree to which this is done depends on the amount of shock loading.

Commercial Units. The major shock loading to these units occurs during shipment or when they operate on commercial carriers. Train service provides a severe test because of low forcing frequencies and high shock load. Shock loads as high as 10*g* have been recorded; 5*g* can be expected.

Trucking service results in higher forcing frequencies, but shock loads can be equal to, or greater than, those for rail transportation. Aircraft service forcing frequencies generally range from 20 to 60 Hz with shocks to 3*g*.

Military Units. Requirements are given in detail in specifications that exceed anything expected of commercial units. In severe applications, deformation of supporting members and shock isolators may be tolerated, provided that the unit performs its function.

Basically, the compressor components must be rigid enough to avoid misalignment or deformation during shock loading. Therefore, structures with low natural frequencies should be avoided.

Testing and Operating Requirements

Compressor tests are of two types: rating (performance) and reliability.

Standard rating conditions, which are usually specified by compressor manufacturers, include the maximum possible compression ratio, maximum operating pressure differential, maximum permissible discharge pressure, and maximum inlet and discharge temperature.

Lubrication requirements, which are prescribed by compressor manufacturers, include the type, viscosity, and other characteristics of the lubricants suitable for use with the many different operating levels and the specific refrigerant being used.

Power requirements for compressor starting, pulldown, and operation vary, because unloading means differ in the many styles of compressors available. Manufacturers supply full information covering the various methods used.

Testing for ratings must be in accordance with ASHRAE *Standard* 23.

Manufacturers normally publish performance data at test conditions specified in ARI *Standard* 520, ARI *Standard* 540, or at other industry standard conditions.

Additional compressor tests might address the following characteristics:

- Compressor performance over a range of conditions (performance curves)
- Sound level
- Durability or reliability
- Operational limits (operating envelope)
- Lubrication requirements (oil type, viscosity, amount, etc.)
- Electrical power requirements (start-up current draw, running current measurements, etc.)

Operating envelope shows compressor operating range as a function of saturation suction temperature or suction pressure (SST or SSP) and saturation discharge temperature or pressure (SDT or SDP). An example of the operating envelope is shown in Figure 5.

This envelope is defined by the following extreme condition points:

High load (HL) is defined as intersection of maximum operating SDT or SDP and maximum operating SST or SSP. At this condition, the compressor experiences high power input, high average torque, and high average bearing load.

High flow (HF) is defined as intersection of minimum operating SDT or SDP and maximum operating SST or SSP. At this condition, the compressor experiences high mass flow of the refrigerant and has high cooling capacity.

High pressure differential (HPD) is defined as intersection of maximum operating SDT or SDP and the maximum discharge temperature line. At this condition, the compressor experiences high pressure differential between discharge and suction and the highest allowable discharge temperature.

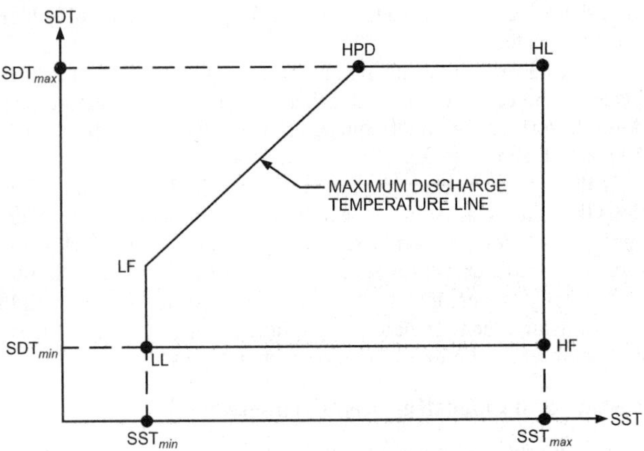

Fig. 5 Example of Compressor Operating Envelope

Low flow (LF) is defined as intersection of the maximum discharge temperature line and minimum operating SST or SSP. The combination of high pressure ratio and low refrigerant flow creates an extreme condition for compressor cooling and oil stability. Cooling capacity at this operating point is the lowest.

Low load (LL) is defined as intersection of minimum operating SDT or SDP and minimum operating SST or SSP. At this point, the compressor experiences low power input and low average torque and bearing load

The common operating range for the air-conditioning SST is 15 to 60°F. The common operating range for the air-conditioning SDT is 75 to 160°F. There are compressor design options that can allow extension of these operating ranges.

Note that compressor capacity (with given refrigerant) depends on SST, SDT, superheating, and subcooling.

MOTORS

Motors for positive-displacement compressors range from fractional to thousands of horsepower. When selecting a motor for driving a compressor, consider the following factors:

- Power and rotational speed.
- Voltage and number of phases. Smaller motors are usually 115 or 230 V and single phase. Large motors are usually three phase, 200 to 575 V.
- Starting and pull-up torques. Special attention should be paid to starting at extreme hot (highest system pressures) and cold (highest oil viscosity) temperatures. Consideration for system pressures at start-up (whether equalized or differential pressure exists across the compressor) is very important for single-phase compressors; start assist may be required with differential-pressure starting.
- Ambient and maximum temperature of the coolant. In hermetic and semihermetic compressors, the motor is usually cooled by refrigerant flow. This flow can be before suction (low-side compressors) or after discharge (high-side compressors). Note that there is no clear advantage or disadvantage in low- or high-side cooling. In low-side compressors, the motor is efficiently cooled by cold refrigerant returned from the evaporator. The cooler the motor, the higher its efficiency η_e. However, superheated refrigerant after the motor has lower density, which reduces the volumetric efficiency η_v. In high-side compressors, it is an opposite effect: higher η_v and lower η_e (motor runs hotter). In both cases, the total effect on compressor efficiency is inconclusive and must be resolved by detailed thermal analysis or testing. Sometimes, selection of motor cooling is dictated by pure reliability or minimum size requirements (e.g., motors bigger than 5 hp usually require low-side cooling to be compact).
- Cost and availability.
- Insulation. In addition to electrical requirements (dielectric properties, high potential, etc.), motor insulation should be compatible with the refrigerant/oil mixture and should withstand a temperature up to 300°F without losing effectiveness.
- Efficiency and performance. The electrical motor performance curves (torque, efficiency, and power factor versus motor speed) must be analyzed against the compressor's operating requirements. For some applications, apparent efficiency (efficiency multiplied by power factor) is important, because often the limiting factor is available kilovolt-amperes, not just power (watts).
- Locked-rotor amps. This important motor characteristic allows proper selection of motor protection devices or a current breaker. During the start of an ac motor without start assistance or a soft-start package, there is a significant inrush current. The amplitude of this current can be very close to locked-rotor amps. The duration of the inrush is also very important. It must be determined by testing, because it is a function of the individual compressor starting torque and operating speed.
- Type of protection required (fuse, internal or external current/temperature switch, etc.). See Chapter 44 for more information.
- Single-speed, multispeed, variable-speed, or linear. Single-speed motors can be one, two, or three phase. Multispeed motors contain two or three sets of windings with different number of poles. Motors with two-pole windings provide speed equivalent to the electrical current (e.g., four-pole rotates at half the speed of two-pole). The following equation calculates motor speed based on electrical frequency and number of poles:

$$\text{Motor speed (rpm)} = \frac{2 \times \text{Electrical frequency, Hz}}{60 \times \text{Number of poles}}$$

  Multispeed motors require relays for switching the windings. These relays can be conventional or solid state (to avoid the effect of vibration).

  Variable-speed motors can be ac or brushless dc (BLDC) type. They require a special electrical controller/drive to operate. Any three-phase motor can be considered as an ac variable-speed motor. Control is comparatively simple, but motor efficiency usually does not exceed 90% because of slippage and rotor losses. BLDC motors are much more efficient (up to 96%), but their control can be more involved. Permanent magnets are used in the rotors of BLDC motors. Some BLDC motors require Hall Effect sensors or resolvers to identify rotor position, or sensorless technology can be used, which complicates the control. All BLDC motors are synchronous, which means that their rotation is proportional to the frequency of the current (i.e., no slip).

  Linear motors require a special controller and drive to operate.
- Location (high-pressure versus low-pressure side). High-side motors operate at much higher temperature than those on the low side, so their efficiency is noticeably lower. (See the discussion in this section in the previous paragraph on ambient and maximum temperature of the coolant.)

Although all motors can be started across-the-line, local utilities, local codes, or specification by the end user may require that motors be started at reduced power levels. These typically include part-winding, wye-delta, double-delta, autotransformer, and solid-state starting methods, all designed to limit inrush starting current. The chosen starting method must supply enough torque to accelerate the motor and overcome the torque required for compression.

Hermetic motors can be more highly loaded than comparably sized open motors because of the refrigerant/oil mixture used for cooling.

For effective hermetic motor application, the maximum design load should be as close as possible to the breakdown torque at the lowest voltage used. This approach yields a motor design that operates better at lighter loads and higher voltage. Overtorqued designs may increase discharge gas temperature at light loads and reduce compressor efficiency.

The single-phase motor presents more design problems than the polyphase, because the relationship between main and auxiliary windings becomes critical. Sometimes it is necessary to use starting equipment in this case.

The rate of temperature rise at the locked-rotor condition must be kept low enough to prevent excessive motor temperature with the motor protection available. The maximum temperature under these conditions should be held within the limits of the materials used. With better protection (and improved materials), a higher rate of rise can be tolerated, and a less expensive motor can be used.

Some types of hermetic motors commonly selected for various applications are as follows:

Small refrigeration compressors (single-phase)

Low to medium torque—Split-phase or PSC (permanent split-capacitor)

High torque—CSCR (capacitor-start/capacitor-run) and CSIR (capacitor-start/induction-run)

Room air conditioner compressors (single-phase)

PSC or CSCR

Two-speed, pole-switching variable-speed

Central air conditioning and commercial refrigeration

Single-phase, PSC and CSCR to 4500 W

Three-phase, 2 hp and above, across-the-line start

10 hp and above; part-winding; wye-delta, double-delta, and across-the-line start

Two-speed (pole-switching) or variable-speed

Electronically commutated dc motors (ECMs)

For further information on motors and motor protection, see Chapter 44. Also see Chapter 55 of the 2007 *ASHRAE Handbook—HVAC Applications* for more on motor-starting effects.

RECIPROCATING COMPRESSORS

Table 1 lists typical design features of reciprocating compressors.

Most reciprocating compressors are single-acting, using pistons that are driven directly through a pin and connecting rod from the crankshaft. Double-acting compressors that use piston rods, crossheads, stuffing boxes, and oil injection are not used extensively and, therefore, are not covered here. Figures 6 and 7 show the basic structure and pumping cycle for a typical reciprocating compressor piston.

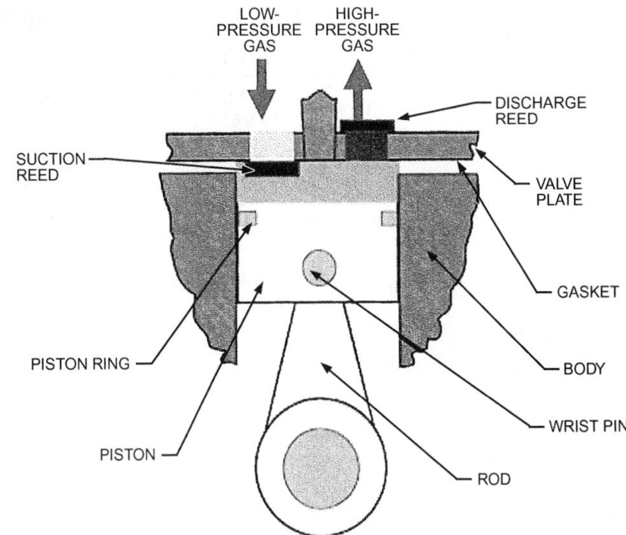

Fig. 6 Basic Reciprocating Piston with Reed Valves

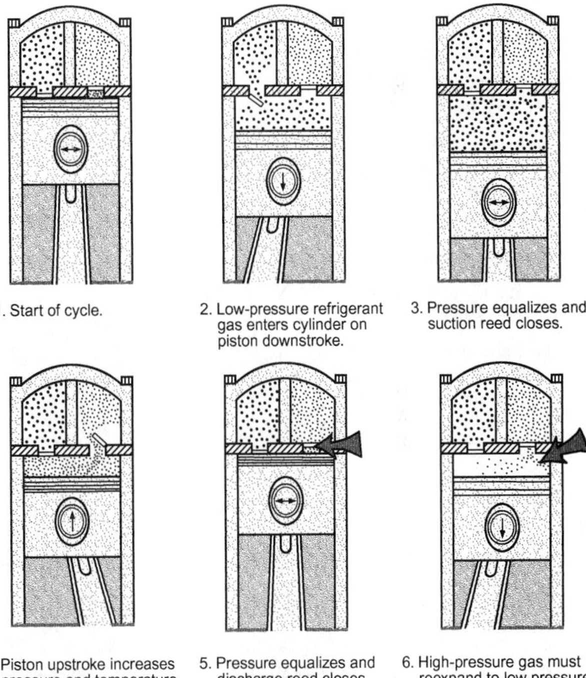

1. Start of cycle.
2. Low-pressure refrigerant gas enters cylinder on piston downstroke.
3. Pressure equalizes and suction reed closes.
4. Piston upstroke increases pressure and temperature of refrigerant in cylinder and causes discharge reed to open.
5. Pressure equalizes and discharge reed closes. Small volume of high-pressure gas remains in valve plate port and clearance between piston and valve plate.
6. High-pressure gas must reexpand to low pressure before suction reed can open. This "volumetric expansion loss" limits the compressor's efficiency.

Fig. 7 Pumping Cycle of Reciprocating Compressor

Single-stage compressors are primarily used for medium temperatures (0 to 30°F) and in air-conditioning applications, but can achieve temperatures below –30°F for refrigeration applications with suitable refrigerants. Chapters 2 and 3 of the 2006 *ASHRAE Handbook—Refrigeration* have information on other halocarbon and ammonia systems.

Booster compressors are typically used for low-temperature applications with R-22 or ammonia. Saturated suction at –85°F can be achieved by using R-22, and –65°F saturated suction is possible using ammonia.

The booster raises refrigerant pressure to a level where further compression can be achieved with a high-stage compressor, without exceeding the pressure-ratio limits of the respective machines.

Because superheat is generated as a result of compression in the booster, intercooling is normally required to reduce the refrigerant stream temperature to the level required at the inlet to the high-stage unit. Intercooling methods include controlled liquid injection into the intermediate stream, mixing-type heat exchangers, and heat exchangers where no fluid mixing occurs.

Integral two-stage compressors achieve low temperatures (–20 to –80°F) using appropriate refrigerants within the frame of a single compressor. Cylinders in the compressor are divided into respective groups so that the combination of volumetric flow and pressure ratios are balanced to achieve booster and high-stage performance effectively. Refrigerant connections between high-pressure suction and low-pressure discharge stages allow an interstage gas cooling system to be connected to remove superheat between stages. The intercooling in this case is similar to the methods used for individual high-stage and booster compressors.

Capacity reduction with reciprocating compressors may be achieved by cylinder unloading, as in the case of single-stage compressors. Special consideration must be given to maintaining the correct relationship between high- and low-pressure stages.

Table 1 Typical Design Features of Reciprocating Compressors

	Refrigerant Type			
	Halo-, Fluoro-, or Hydrocarbon			Ammonia
Item	**Open**	**Semi-hermetic**	**Welded Hermetic**	**Open**
1. Number of cylinders—one to:	16	12	6	16
2. Power range	0.17 hp and up	0.5 to 150 hp	0.17 to 25 hp	10 hp and up
3. Cylinder arrangement				
a. Vertical, V or W, radial	X	X		
b. Radial, horizontal opposed			X	
c. Horizontal, vertical V or W		X		X
4. Drive				
a. Hermetic compressors, electric motor		X	X	
b. Open compressors—direct drive, V belt chain, gear, by electric motor or engine	X			X
5. Lubrication—splash or force feed, flooded	X	X	X	X
6. Suction and discharge valves—ring plate or ring or reed flexing	X	X	X	X
7. Suction and discharge valve arrangement				
a. Suction and discharge valves in head	X	X	X	X
b. Uniflow—suction valves in top of piston, suction gas entering through cylinder walls; discharge valves in head	X			X
8. Cylinder cooling				
a. Suction-gas-cooled	X	X	X	X
b. Water jacket cylinder wall, head, or cylinder wall and head	X			X
c. Air-cooled	X	X	X	X
d. Refrigerant-cooled heads	X			X
9. Cylinder head				
a. Spring-loaded	X	X	X	X
b. Bolted	X	X	X	X
10. Bearings				
a. Sleeve, antifriction	X	X	X	X
b. Tapered roller	X			X
11. Capacity control, if provided—manual or automatic				
a. Suction valve lifting	X	X	X	X
b. Bypass-cylinder heads to suction	X	X	X	X
c. Closing inlet	X	X		X
d. Adjustable clearance	X	X		X
e. Variable-speed	X	X	X	X
12. Materials				
Motor insulations and rubber materials must be compatible with refrigerant and lubricant mixtures; otherwise, no restrictions		X	X	
No copper or brass				X
13. Lubricant return				
a. Crankcase separated from suction manifolds, oil return check valves, equalizers, spinners, foam breakers	X	X		X
b. Crankcase common with suction manifold			X	
14. Synchronous fixed speeds, rpm	250 to 3600	1500 to 3600	1500 to 3600	250 to 1500
15. Pistons				
a. Aluminum or cast iron	X	X	X	X
b. Ringless	X	X	X	X
c. Compression and oil-control rings	X	X	X	X
16. Connecting rod				
Split rod with removable cap or solid eccentric strap	X	X	X	X
17. Mounting				
Internal spring mount		X	X	
External spring mount		X	X	
Rigidly mounted on base	X	X		X

Performance Data

Figure 8 shows a typical set of capacity and power curves for a four-cylinder semihermetic compressor, 2.38 in. bore, 1.75 in. stroke, 1740 rpm, operating with R-22. Compressor curves should contain the following information:

- Compressor identification
- Degrees of subcooling and correction factors for zero or other subcool temperatures
- Degrees of superheat
- Compressor speed
- Refrigerant
- Suction gas superheat and correction factors
- Compressor ambient
- External cooling requirements (if any)
- Maximum power or maximum operating conditions
- Minimum operating conditions at fully loaded and fully unloaded operation

Motor Performance

Motor efficiency is usually a compromise between cost and size. Generally, the physically larger a motor is for a given rating, the more efficient it can be. The accepted efficiency range for ac motors is 85 to 95%. Uneven loading has a marked effect on motor efficiency. It is important that cylinders be spaced evenly. Also, the more cylinders there are, the smaller the impulses become. Greater moments of inertia of moving parts and higher speeds reduce the impulse effect. Small, evenly spaced impulses also help reduce noise and vibration.

Because many compressors start against load, it is desirable to estimate starting torque. The following equation is for a single-cylinder compressor. It neglects friction, valve losses, leakage, and the fact that tangential force at the crankpin is not always equal to normal force at the piston. This equation also assumes considerable gas leakage at the discharge valves but little or no leakage past the piston rings or suction valves. It gives only a preliminary estimate.

$$T_s = \frac{(p_2 - p_1)As}{2N_2/N_1} \tag{12}$$

where

T_s = starting torque, $lb_f \cdot in.$
p_2 = discharge pressure, psi
p_1 = pressure on other side of piston, psi
A = area of cylinder, in^2
s = stroke of compressor, in.
N_2 = motor speed, rpm
N_1 = compressor speed, rpm

Equation (12) shows that when pressures are balanced or almost equal ($p_2 = p_1$), torque requirements are considerably reduced. Thus, a pressure-balancing device on an expansion valve or a capillary tube that equalizes pressures at shutdown allows the compressor to be started without excessive effort. For multicylinder compressors, both the number of cylinders that might be on a compression stroke and the position of the rods at start must be analyzed. Because the force needed to push the piston to the top dead center is a function of how far the rod is from the cylinder centerline, the worst possible angles these might assume can be graphically determined by torque diagrams. The torques for some arrangements are shown in Table 2.

Table 2 Motor-Starting Torques

No. Cylinders	Arrangement of Crank Throws	Angle Between Cylinders	Approximate Torque from Equation (12)
1	Single		T_s
2	Single	90°	$1.025T_s$
2	180° apart	0° or 180°	T_s
3	Single	60°	$1.225T_s$
3	120° apart	120°	T_s
4	180°, 2 rods/crank	90°	$1.025T_s$
6	180°, 3 rods/crank	60°	$1.23T_s$

Pull-up torque is an important characteristic of motor starting strength because it represents the lowest torque capability of the motor and occurs between 25 and 75% of the operating speed. The motor's pull-up torque must exceed the torque requirement of the compressor or the motor will cease to accelerate and trip the safety overload protection device.

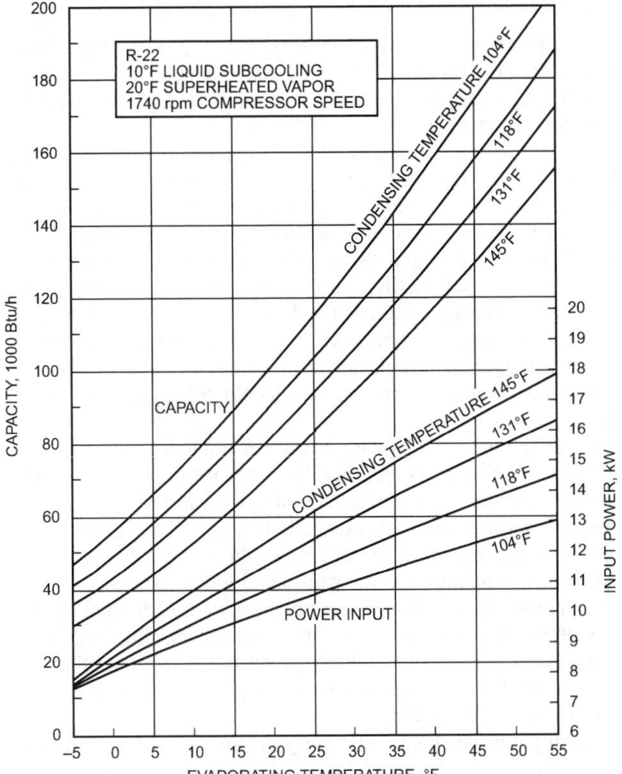

Fig. 8 Capacity and Power-Input Curves for Typical Semihermetic Reciprocating Compressor

Features

Crankcases. The crankcase or, in a welded hermetic compressor, the cylinder block is usually of cast iron. Aluminum is also used, particularly in open compressors for transportation refrigeration and welded hermetic compressors. Open and semihermetic crankcases enclose the running gear, oil sump, and, in the latter case, the hermetic motor. Access openings with removable covers are provided for assembly and service purposes. Welded hermetic cylinder blocks are often just skeletons, consisting of cylinders, main bearings, and either a barrel into which the hermetic motor stator is inserted or a surface to which the stator can be bolted.

Cylinders can be integral with the crankcase or cylinder block, in which case a material that provides a good sealing surface and resists wear must be provided. In aluminum crankcases, cast-in liners of iron or steel are usual. In large compressors, premachined cylinder sleeves inserted in the crankcase are common. With halocarbon refrigerants, excessive cylinder wear or scoring is not much of a problem and the choice of integral cylinders or inserted sleeves is often based on manufacturing considerations.

Crankshafts. Crankshafts are made of either forged steel with hardened bearing surfaces finished to 8 μin or iron castings. Grade 25 to 40 (25,000 to 40,000 psi) tensile gray iron can be used where the lower modulus of elasticity can be tolerated. Nodular iron shafts approach the stiffness, strength, and ductility of steel and should be polished in both directions of rotation to 16 μin maximum for best results. Crankshafts often include counterweights and should be dynamically and/or statically balanced.

A safe maximum stress is important in shaft design, but it is equally important to prevent excessive deflection that can edge-load bearings to failure. In hermetics, deflection can permit motor air gap to become eccentric, which affects starting, reduces efficiency, produces noise, and further increases bearing edge-loading.

Generally, the harder the bearing material used, the harder the shaft. With bronze bearings, a journal hardness of 350 Brinell is usual, whereas unhardened shafts at 200 Brinell in babbitt bearings are typical. Many combinations of materials and hardnesses have been used successfully.

Main Bearings. Both the crank and drive means may be overhung with bearings between; however, usual practice places the cylinders between main bearings. Main bearings are made of steel-backed babbitt, steel-backed or solid bronze, or aluminum. Bearings are usually integral to an aluminum crankcase. By automotive standards, unit loadings are low. The oil/refrigerant mixture frequently provides only marginal lubrication, but 8000 h/year operation in commercial refrigeration service is possible. For conventional shaft diameters and speeds, 600 psi main bearing loading based on projected area is not unusual. Running clearances average 0.001 in. per inch of diameter with steel-backed babbitt bearings and a steel or iron shaft. Bearing oil grooves placed in the unloaded area are usual. Feeding oil to the bearing is only one requirement; another is venting evolved refrigerant gas and lubricant escape from the bearing to carry away heat.

In most compressors, crankshaft thrust surfaces (with or without thrust washers) must be provided in addition to main bearings. Thrust washers may be steel-backed babbitt, bronze, aluminum, hardened steel, or polymer and are usually stationary. Oil grooves are often included in the thrust face.

Connecting Rods and Eccentric Straps. Connecting rods have the large end split and a bolted cap for assembly. Unsplit eccentric straps require the crankshaft to be passed through the big bore at assembly. Rods or straps are of steel, aluminum, bronze, nodular iron, or gray iron. Steel or iron rods often require inserts of bearing material such as steel-backed babbitt or bronze, whereas aluminum and bronze rods can bear directly on the crankpin and piston pin. Refrigerant compressor service limits unit loading to 3000 psi based on projected area with a bronze bushing in the rod small bore and a hardened steel piston pin. An aluminum rod load at the piston pin of 2000 psi has been used. Large end-unit loads are usually under 1000 psi.

The Scotch yoke type of piston-rod assembly has also been used. In small compressors, it has been fabricated by hydrogen-brazing steel components. Machined aluminum components have been used in large hermetic designs.

Piston, Piston Ring, and Piston Pin. Pistons are usually made of cast iron or aluminum. Cast-iron pistons with a running clearance of 0.0004 in. per inch of diameter in the cylinder will seal adequately without piston rings. With aluminum pistons, rings are required because a running clearance in the cylinder of 0.002 in. per inch or more of diameter may be necessary, as determined by tests at extreme conditions. A second or third compression ring may add to power consumption with little increase in capacity; however, it may help oil control, particularly if drained. Oil-scraping rings with vented grooves may also be used. Cylinder finishes are usually obtained by honing, and a 12 to 40 μin range will give good ring seating. An effective oil scraper can often be obtained with a sharp corner on the piston skirt.

Minimum piston length is determined by the side thrust and is also a function of running clearance. Where clearance is large, pistons should be longer to prevent slap. An aluminum piston (with ring) having a length equal to 0.75 times the diameter, with a running clearance of 0.002 in. per inch of diameter, and a rod-length-to-crank-arm ratio of 4.5 has been used successfully.

Piston pins are steel, case-hardened to Rockwell *C* 50 to 60 and ground to a 8 μin finish or better. Pins can be restrained against rotation in either the piston bosses or the rod small end, be free in both, or be full-floating, which is usually the case with aluminum pistons and rods. Retaining rings prevent the pin from moving endwise and abrading the cylinder wall.

There is no well-defined limit to piston speed; an average velocity of 1200 fpm, determined by multiplying twice the stroke in feet by the rpm, has been used successfully.

Suction and Discharge Valves. These valves are important components of a reciprocating compressor. Successful designs provide high volumetric efficiency and low pressure loss. Improper timing of opening/closing and excessive leakage significantly affect volumetric efficiency. Excessive pressure loss across the valve results from high gas velocities, poor mechanical action, or both.

Note that the valve flow area gradually increases during opening, which creates conditions for an overpressure pulse in discharge valves and underpressure pulse for suction valves. These pulses significantly affect compressor performance, noise, and vibration. Minimizing such pulses usually leads to better overall compressor performance.

A valve should meet the following requirements:

- Sufficient flow areas with shortest possible path
- Straight gas flow path, minimum directional changes
- Valve mass and lift should satisfy timing requirements
- Symmetry of design with minimum pressure imbalance
- Minimum clearance volume
- Durability
- Low cost
- Tight sealing at ports (low leakage)
- Minimum valve flutter

Most valves in use today fall in one of the following groups:

- **Free-floating reed valve**, with backing to limit movement, seats against a flat surface with circular or elongated ports. It is simple, and stresses can be readily determined, but it is limited to relatively small ports; therefore, multiples are often used. Totally backed with a curved stop, it can stand considerable abuse.
- **Reed, clamped at one end**, with full backstop support or a stop at the tip to limit movement, has a more complex motion than a free-floating reed; it may have multiple modes of deformation. Considerable care must be taken in design to ensure reliability.
- **Ring valve** usually has a spring return. A free-floating ring is seldom used because of its high leakage loss. Improved performance is obtained by using spring return, in the form of coil springs or flexing backup springs, with each valve. Ring-type valves are particularly adaptable to compressors using cylinder sleeves.
- **Valve formed as a ring** has part of the valve structure clamped. Generally, full rings are used with one or more sets of slots arranged in circles. By clamping the center, alignment is ensured and a force is obtained that closes the valve. To limit stresses, the valve proportions, valve stops, and supports are designed to control and limit valve motion.

Lubrication. Lubrication systems range from a simple splash system to the elaborate forced-feed systems with filters, vents, and equalizers. The type of lubrication required depends largely on bearing load and application.

For low to medium bearing loads and factory-assembled systems where cleanliness can be controlled, the splash system gives excellent service. Bearing clearances must be larger, however; otherwise, oil does not enter the bearing readily. Thus, the splashing effect of the dippers in the oil and the freer bearings cause the compressor to operate somewhat noisily. Furthermore, the splash at high speed encourages frothing (foaming) and oil pumping; this is not a problem in packaged equipment but may be in remote systems where gas lines are long. Foaming can also help reduce compressor noise, and sometimes foaming agents can be added to the oil just for this purpose.

Lubrication for a **flooded system** includes disks, screws, grooves, oil-ring gears, or other devices that lift the oil to the shaft or bearing level. These devices flood the bearing and are not much better than splash systems, except that the oil is not agitated as violently, so operation is quieter. Because little or no pressure is developed by this method, it is not considered forced-feed.

In **forced-feed lubrication**, a pump gear, vane, or plunger develops pressure, which forces oil into the bearing. Smaller bearing clearances can be used because adequate pressure feeds oil in sufficient quantity for proper bearing cooling. As a result, compressor operation may be quieter.

Gear pumps are common. Spur gears are simple but tend to promote flashing of refrigerant dissolved in the oil because of the sudden opening of the tooth volume as two teeth disengage. This disadvantage is not apparent in internal eccentric gear or vane pumps where the suction volume gradually opens. The eccentric gear pump, vane pump, or piston pump therefore give better performance than simple gear pumps when the pump is not submerged in the oil.

Oil pumps must be made with proper clearances to pump a mixture of gas and oil. The pump discharge should have provision to bleed a small quantity of oil into the crankcase. A bleed vents the pumps, prevents excess pressure, and ensures faster priming.

A strainer should be inserted in the suction line to keep foreign substances from the pump and bearings. If large quantities of fine particles are present and bearing load is high, it may be necessary to add an oil filter to the discharge side of the pump.

Oil must return from suction gas into the compressor crankcase. A flow of gas from piston leakage opposes this oil flow, so leakage gas velocity must be low to permit oil to separate from the gas. A separating chamber may be built as part of the compressor to help separate oil from the gas.

In many designs, a check valve is inserted at the bottom of the oil return port to prevent a surge of crankcase oil from entering the suction. This check valve must have a bypass, which is always open, to allow the check valve to open wide after the oil surge has passed. When a separating chamber is used, the oil surge is trapped before it can enter the suction port, thus making a check valve less essential.

Seals. Stationary and rotary seals have been used extensively on open reciprocating compressors. Older stationary seals usually used metallic bellows and a hardened shaft for a wearing surface. Their use has diminished because of high cost.

The rotary seal costs less and is more reliable. A synthetic seal tightly fitted to the shaft prevents leakage and seals against the back face of the stationary member of the seal. The front face of this carbon nose seals against a stationary cover plate. This design has been used on shafts up to 4 in. in diameter. The rotary seal should be designed so that the carbon nose is never subjected to the full thrust of the shaft, the spring should be designed for minimum cocking force, and materials should be selected to minimize swelling and shrinking.

Special Devices

Capacity Control. Capacity control may be obtained by (1) controlling suction pressure by throttling, (2) controlling discharge pressure, (3) returning discharge gas to suction, (4) adding reexpansion volume, (5) changing the stroke, (6) opening a cylinder discharge port to suction while closing the port to discharge manifold, (7) changing compressor speed, (8) closing off cylinder inlet, and (9) holding the suction valve open.

The most commonly used methods are opening the suction valves by some external force (9), gas bypassing within the compressor (6), suction shutoff, gas bypassing outside the compressor (3), and variable speed (7).

When capacity control compressors are used, system design becomes more important, and the following must be considered:

- Possible increase in compressor vibration (lower capacity) and sound level (higher capacity)
- Minimum operating conditions as limited by discharge or motor temperatures (or both) at part-load conditions
- Good oil return at lowest capacity
- Rapid cycling of unloaders
- Refrigerant metering device (TXV or capillary tube) capable of controlling at minimum capacity

Crankcase Heaters. When it is possible that refrigerant can accumulate in the compressor crankcase (cold start, gravitation, etc.), dilute the oil excessively, and result in flooded starts, a crankcase heater should be used. The heater should maintain the oil at least 20°F above the rest of the system at shutdown and well below the breakdown temperature of the oil at any time.

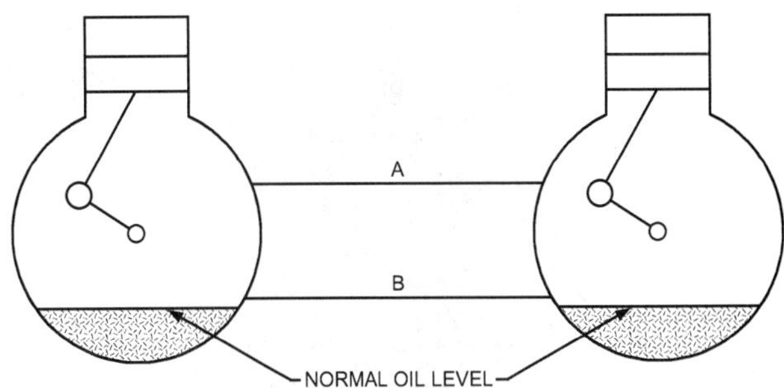

Fig. 9 Modified Oil-Equalizing System

Internal Centrifugal Separators. Some compressors are equipped with antislug devices in the gas path to the cylinders. This device centrifugally separates oil and liquid refrigerant from the flow of foam during a flooded start and thus protects the cylinders. It does not eliminate other hazards caused by liquid refrigerant in gas compressors.

Automatic Oil Separators. Oil separators are used most often to reduce the amount of oil discharged into the system by the compressor and to return oil to the crankcase. They are recommended for all field-erected systems and on packaged equipment where lubricant contamination has a negative effect on evaporator capacity and/or where lubricant return at reduced capacity is marginal.

Application

Parallel Operation. Where multiple compressors are used, the trend is toward completely independent refrigerant circuits. This has an obvious advantage in the case of hermetic motor burnout and with lubricant equalization.

Parallel operation of compressors in a single system has some operational advantage at part load. Careful attention must be given to apportioning returned oil to the multiple compressors so that each always has an adequate quantity. Figure 9 shows the method most widely used. Line A connects the tops of the crankcases and tends to equalize the pressure above the oil; line B permits oil equalization at the normal level. Lines of generous diameter must be used. Generally, line A is a large diameter, and line B is a small diameter, which limits possible blowing of oil from one crankcase to the other.

A central reservoir for returned oil may also be used with means (such as crankcase float valves) for maintaining the proper levels in the various compressors. With staged systems, the low-stage compressor oil pump can sometimes deliver a measured amount of oil to the high-stage crankcase. The high-stage oil return is then sized and located to return a slightly greater quantity of oil to the low-stage crankcase. Where compressors are at different elevations and/or staged, pumps in each oil line are necessary to maintain adequate crankcase oil level. In both cases, proper gas equalization must be provided.

Operation at Low Suction Pressure. Because reciprocating compressors do not rely on refrigerant flow to deliver oil and cool the surfaces with boundary lubrication, they are uniquely suitable for low-suction-pressure applications, such as medium- and low-temperature refrigeration. Efficiency of these compressors is lower than more advanced designs, but reliability is extremely high.

ROTARY COMPRESSORS

ROLLING-PISTON COMPRESSORS

Rolling-piston, or fixed-vane, rotary compressors are used in household refrigerators and air-conditioning units in sizes up to about 3 hp (Figure 10). This type of compressor uses a roller mounted on the eccentric of a shaft with a single vane or blade suitably positioned in the nonrotating cylindrical housing, generally

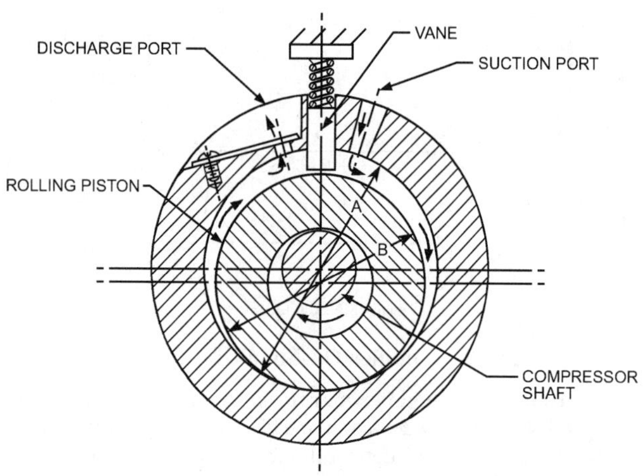

Fig. 10 Fixed-Vane, Rolling-Piston Rotary Compressor

called the cylinder block. The blade reciprocates in a slot machined in the cylinder block. This reciprocating motion is caused by the eccentrically moving roller.

Displacement for this compressor can be calculated from

$$V_d = \pi H(A^2 - B^2)/4 \tag{13}$$

where

V_d = displacement
H = cylinder block height
A = cylinder diameter
B = roller diameter

The drive motor stator and compressor are solidly mounted in the compressor housing. This design feature can lead to significant torsional vibration, and special measures should be implemented to avoid damage to suction and discharge tubes. A discharge tube attached to the compressor should be in the form of "C" coil. If a suction accumulator is used, it must be attached to the compressor shell by supporting brackets of sufficient strength. Special grommets should be used to avoid transmitting the vibration to the compressor support. Using flexible suction and discharge tubes can significantly reduce vibration transmission and noise.

Because the amplitude of the torsional vibration increases at lower compressor speed, it may be a problem to operate a single-cylinder rotary compressor below 30 revolutions per second. Twin-cylinder rotary compressors with eccentrics on the same shaft in opposite directions can produce less torsional vibration and are suitable to operate at as low as 10 revolutions per second.

Rotary compressors are usually located on the high side. Suction gas is piped directly into the suction port of the compressor, and compressed gas is discharged into the compressor housing shell. This high-side shell design is used because, in this case, the vane does not require a strong spring (after discharge pressure is built up, it creates enough force to maintain engagement between the roller and vane without the spring force) and lubrication of the vane in the slot is ensured by the pressure differential between the oil sump (high pressure) and suction chamber.

To avoid significant discharge pulsations inside the shell and the consequent high noise level, a special discharge muffler is installed to cover the discharge valve. Discharge mufflers can be made from different materials, including plastics.

Maintaining the proper oil level in rotary compressors is extremely important. The oil level should be high enough to cover the vane, providing adequate lubrication and leakage reduction, but not higher than the discharge port, because excessive oil will be pushed out of the compressor by discharge gas and may contaminate the heat exchangers. In some designs, the electrical rotor is used as an oil separator (special blades are installed on the rotor top), or a special plate is installed to shield the discharge tube entry and force separation of refrigerant and oil by inertia.

Internal leakage is controlled through hydrodynamic sealing and selection of mating parts for optimum clearance. Hydrodynamic sealing depends on clearance, surface speed, surface finish, and oil viscosity. Close tolerance and low-surface-finish machining is necessary to support hydrodynamic sealing and to reduce gas leakage.

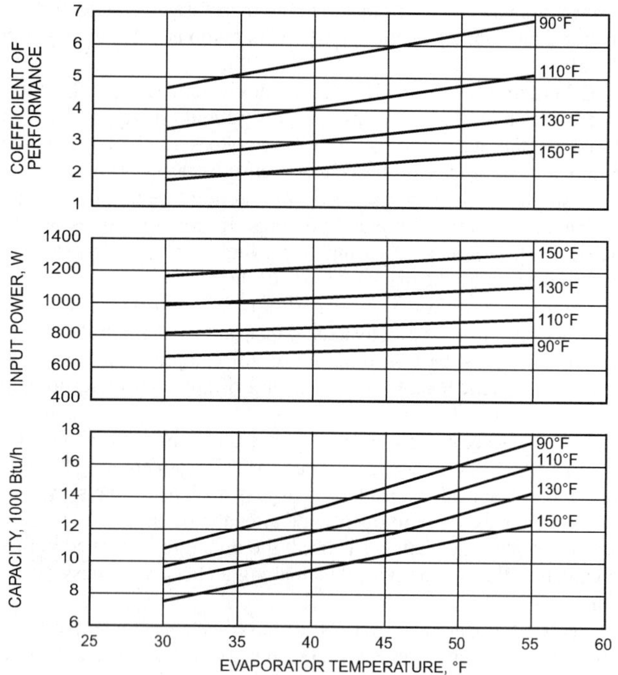

Fig. 11 Performance Curves for Typical Rolling-Piston Compressor

Table 3 Typical Rolling-Piston Compressor Performance

Compressor speed	3450 rpm
Refrigerant	R-22
Condensing temperature	130°F
Liquid refrigerant temperature	115°F
Evaporator temperature	45°F
Suction pressure	90.7 psia
Suction gas temperature	65°F
Evaporator capacity	12,000 Btu/h
Energy efficiency ratio	11.0 Btu/W·h
Coefficient of performance	3.22
Input power	1090 W

Performance

Rotary compressors have a high volumetric efficiency because of the small clearance volume and correspondingly low reexpansion losses inherent in their design. Figure 11 and Table 3 show performance of a typical rolling-piston rotary compressor, commercially

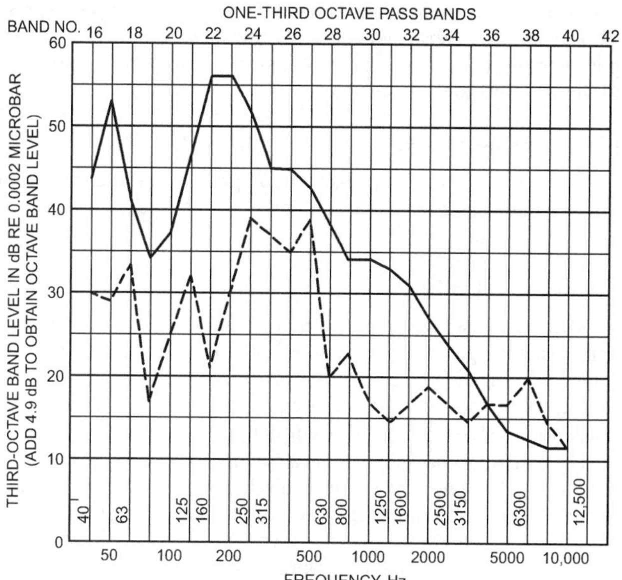

Fig. 12 Sound Level of Combination Refrigerator-Freezer with Typical Rotary Compressor

available for room air conditioning and small, packaged heat pump applications.

An acceptable sound level is important in the design of any small compressor. Figure 12 illustrates the influence of a compressor on home refrigerator noise. Because gas flow is continuous and no suction valve is required, rotary compressors can be relatively quiet. The sound spectrum (sound "signature") of the rotary compressors is quite distinctive. At 60 revolutions per second, sound level elevation between 600 and 900 Hz can be explained by gas flow and pulsations, from 2000 to 2500 Hz by valve impact, and above 3500 Hz mostly by friction.

Features

Shafts and Journals. Shaft deflection under load is caused by compression gas loading of the roller and the torsional and side-pull loading of the motor rotor. Design criteria must require minimum oil film under the maximum run and starting loads. The motor rotor should have minimal deflection, to eliminate starting problems under extreme conditions of torque; the air gap between rotor and stator should be maintained as constant as possible around the circumference.

Shafts are generally made from steel forging and nodular cast iron. Journals are ground round to high precision and polished to a finish of 10 μin or better. The lower portion of the shaft (eccentric and lower journals) is hollow inside and serves as an oil pump using centrifugal force to deliver oil to the thrust surfaces and journals through labyrinth holes. Vertical or angular grooves are made in the journals or bearings, to aid in distribution of the oil.

Bearings. The bearing must support the rotating member under all conditions. Powdered metal has been extensively used for these components, because its porous properties help in lubrication. This material can also be formed into complex bearing shapes with little machining required. Cast iron is also widely used, especially for the lower bearing.

Vanes. Vanes are designed for reliability by the choice of materials and lubrication. The vanes are hardened, ground, and polished to the best finish obtainable. Steel, powdered metal, and aluminum

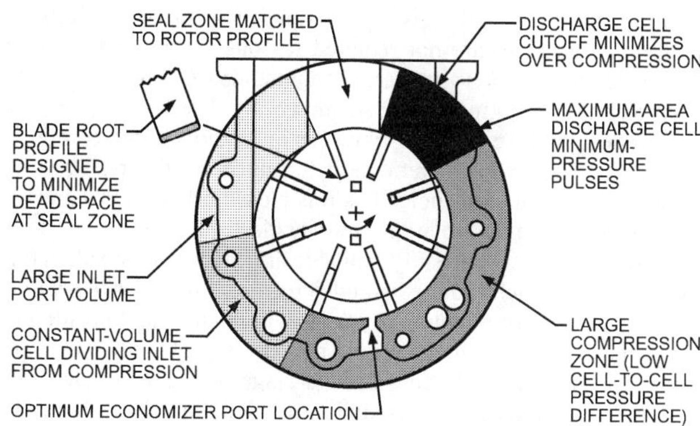

Fig. 13 Rotary-Vane Compressor

alloys have been used; however, the best results are achieved by using M2 tool steel. Special coatings and excessive hardness (above 70 Rockwell C) have not shown meaningful improvement in reliability or long-term performance. Aluminum is impractical because of its significant difference from iron in thermal expansion, which requires higher clearance between the vane and vane slot and, consequently, leads to higher leakage and inconsistent lubrication.

The vane tip radius should be selected to minimize Hertz stresses between the roller and the vane, but the sharp edge of the vane should never be in contact with the roller.

Vane Springs. Vane springs force the vane to stay in contact with the roller during start-up. The spring rate depends on the inertia of the vane. Two types of springs are widely used: C-type and helicoil.

Valves. Only discharge valves are required by rolling-piston rotary compressors. They are usually simple reed valves made of high-grade steel. Proper design of the valve stop is important for valve reliability and noise reduction.

Lubrication. A properly designed lubrication system circulates an ample supply of clean oil to all working surfaces, bearings, vanes, vane slots, and seal faces. High-side pressure in the housing shell ensures a sufficient pressure differential across the passageways that distribute oil to the bearing surfaces.

Mechanical Efficiency. High mechanical efficiency depends on minimizing friction losses. Friction losses occur in the bearings and between the vane and slot wall, vane tip, and roller wall, and roller and bearing faces. The amount and distribution of these losses vary based on the geometry of the compressor.

Motor Selection. Breakdown torque requirements depend on the displacement of the compressor, the refrigerant, and the operating range. Domestic refrigerator compressors typically require a breakdown torque of about 36 to 38 oz · ft per cubic inch of compressor displacement per revolution. Similarly, larger compressors using R-22 for window air conditioners require about 67 to 69 oz · ft breakdown torque per cubic inch of compressor displacement per revolution.

Rotary machines do not usually require complete unloading for successful starting. Pressure differentials up to 14.5 psi can be tolerated. The starting torque of standard split-phase motors is ample for small compressors. Permanent split-capacitor motors for air conditioners of various sizes provide sufficient starting and improve the power factor to the required range.

ROTARY-VANE COMPRESSORS

Rotary-vane compressors have a low weight-to-displacement ratio, which, in combination with compact size, makes them suitable for transport application. Small compressors in the 3 to 50 hp range are single-staged for a saturated suction temperature range of −13 to 59°F at saturated condensing temperatures up to 167°F. Currently, R-22, R-134a, R-404A, R-407, R-410A, R-507, and R-717 are the refrigerants used for these compressor applications.

Figure 13 is a cross-sectional view of an eight-bladed compressor. The eight discrete volumes are referred to as cells. A single shaft rotation produces eight distinct compression strokes. Although conventional valves are not required for this compressor, suction and discharge check valves are recommended to prevent reverse rotation and oil logging during shutdown.

The design of the compressor results in a fixed, built-in compression ratio. Compression ratio is determined by the relationship between the volume of the cell as it is closed off from the suction port to its volume before it opens to the discharge port.

The compressors currently available are of an oil-flooded, open-drive design, which requires an oil separator. Single-stage separators are used in close-coupled systems with high saturation suction temperature (SST), where oil return is not a problem. Two-stage separators with a coalescing second stage are used in low-SST systems, in ammonia systems, and in flooded evaporators likely to trap oil.

Rotary-vane compressors are more complicated (more parts) and, consequently, more expensive than rolling-piston or scroll compressors. Furthermore, the efficiency of rotary-vane compressors is comparatively low. Therefore, many of these compressors are being replaced with more advanced scroll or rotary types. However, the reliability of well-established rotary-vane designs is on par with other types.

SINGLE-SCREW COMPRESSORS

Screw compressors for refrigeration and air-conditioning applications are of two distinct types: single-screw and twin-screw. Both are conventionally used with fluid injection where sufficient fluid cools and seals the compressor. Screw compressors have the capability to operate at pressure ratios above 20:1 single stage.

Description

A single-screw compressor consists of a single cylindrical main rotor that works with one or a pair of gate rotors. Both the main rotor and gate rotor(s) can vary widely in terms of form and mutual geometry. Figure 14 shows the design normally encountered in refrigeration.

The main rotor has helical grooves, with a cylindrical periphery and a globoid (or hourglass) root profile. The two identical gate rotors are located on opposite sides of the main rotor. The casing enclosing the main rotor has two slots, which allow the teeth of the gate rotors to pass through them.

The compressor is driven through the main rotor shaft, and the gate rotors follow by direct meshing action with the main rotor. The geometry of the single-screw compressor is such that gas compression power is transferred directly from the main rotor to the gas. No power (other than small frictional losses) is transferred across the meshing points to the gate rotors.

Compression Process

The operation of the single-screw compressor can be divided into three distinct phases: suction, compression, and discharge. The process is shown in Figure 15.

Mechanical Features

Rotors. The screw rotor is normally made of cast or ductile iron, and the mating gate rotors are made from an engineered plastic. The inherent lubricating quality of the plastic, as well as its compliant nature, allow the single-screw compressor to achieve close clearances with conventional manufacturing practice.

The gate rotors are mounted on a metal support designed to carry the differential pressure between discharge pressure and suction pressure. The gate rotor function is equivalent to that of a piston in that it sweeps the groove and causes compression to occur. Furthermore, the gate rotor is in direct contact with the screw groove flanks

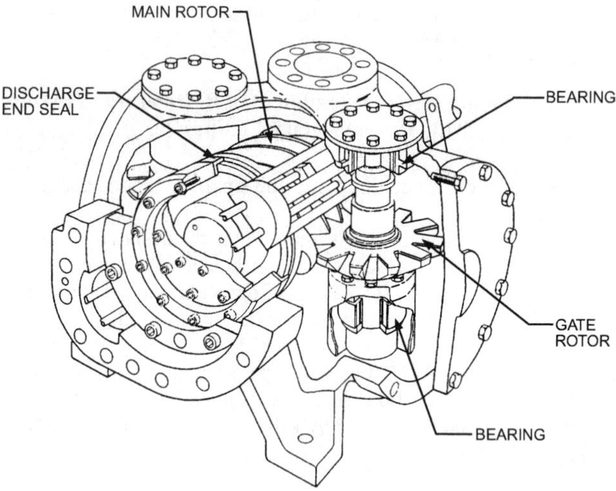

Fig. 14 Section of Single-Screw Refrigeration Compressor

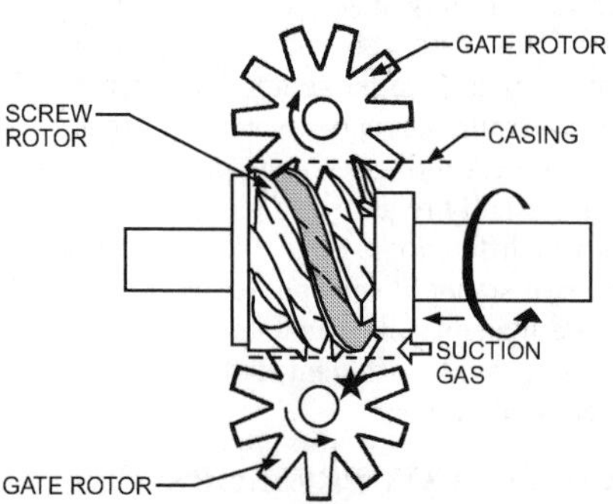

Suction. During rotation of the main rotor, a typical groove in open communication with the suction chamber gradually fills with suction gas. The tooth of the gate rotor in mesh with the groove acts as an aspirating piston.

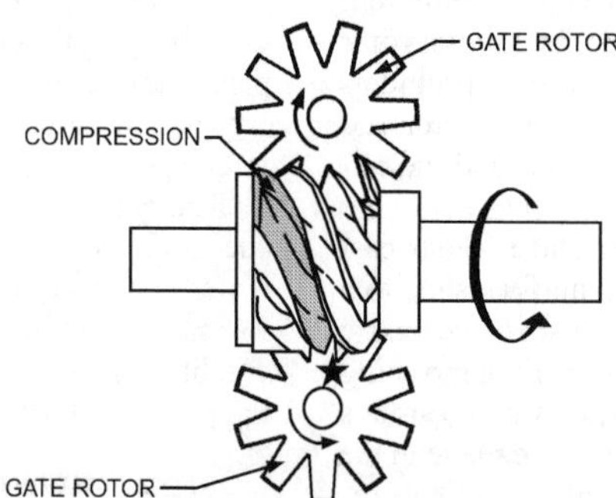

Compression. As the main rotor turns, the groove engages a tooth on the gate rotor and is covered simultaneously by the cylindrical main rotor casing. The gas is trapped in the space formed by the three sides of the groove, the casing, and the gate rotor tooth. As rotation continues, the groove volume decreases and compression occurs.

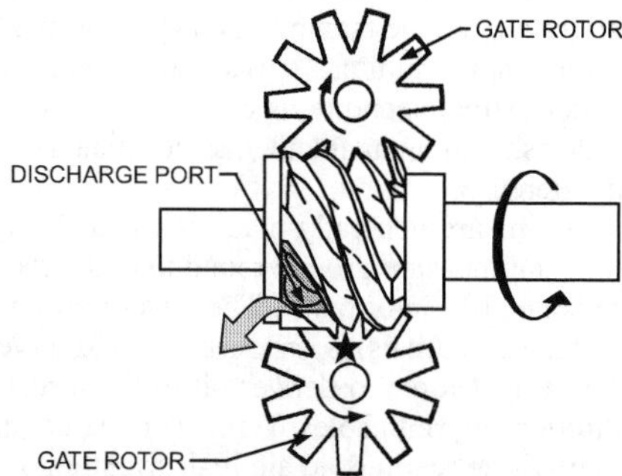

Discharge. At the geometrically fixed point where the leading edge of the groove and the edge of the discharge port coincide, compression ceases, and the gas discharges into the delivery line until the groove volume has been reduced to zero.

Fig. 15 Sequence of Compression Process in Single-Screw Compressor

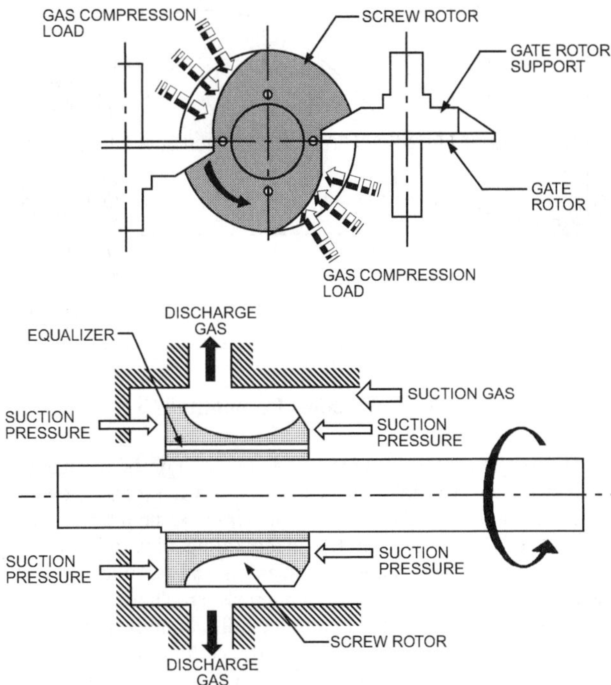

Fig. 16 Radial and Axially Balanced Main Rotor

and thus also acts as a seal. Each gate rotor is attached to its support by a simple spring and dashpot mechanism, allowing the gate rotor, with a low moment of inertia, to have an angular degree of freedom from the larger mass of the support. This method of attachment allows the gate rotor assemblies to pass a significant amount of liquid slug during transient operation without damage or wear.

Bearings. In a typical open or semihermetic single-screw compressor, the main rotor shaft contains one pair of angular contact ball bearings (an additional angular contact or roller bearing is used for some heat pump semihermetics). On the opposite side of the screw, one roller bearing is used.

Note that compression takes place simultaneously on each side of the main rotor of the single-screw compressor. This balanced gas pressure results in virtually no load on the rotor bearing during full load and while symmetrically unloaded as shown in Figure 16. Should the compressor be unloaded asymmetrically (see economizer operation below 50% capacity), the designer is not restricted by the rotor geometry and can easily add bearings with a long design life to handle the load. Axial loads are also low because the grooves terminate on the outer cylindrical surface of the rotor and suction pressure is vented to both ends of the rotor (Figure 16).

The gate rotor bearing must overcome a small moment force caused by the gas acting on the compression surface of the gate rotor. Each gate rotor shaft has at least one bearing for axial positioning (usually a single angular contact ball bearing can perform the axial positioning and carry the small radial load at one end), and one roller or needle bearing at the other end of the support shaft also carries the radial load. Because the single-screw compressor's physical geometry places no constraints on bearing size, it allows design of bearings with long lives.

Cooling, Sealing, and Bearing Lubrication. A major function of injecting a fluid into the compression area is removing heat of compression. Also, because a single-screw compressor has fixed leakage areas, the fluid helps seal leakage paths. Fluid is normally injected into a closed groove through ports in the casing or in the moving capacity control slide. Most single-screw compressors can use many different injection fluids, oil being the most common, to suit the nature of the gas being compressed.

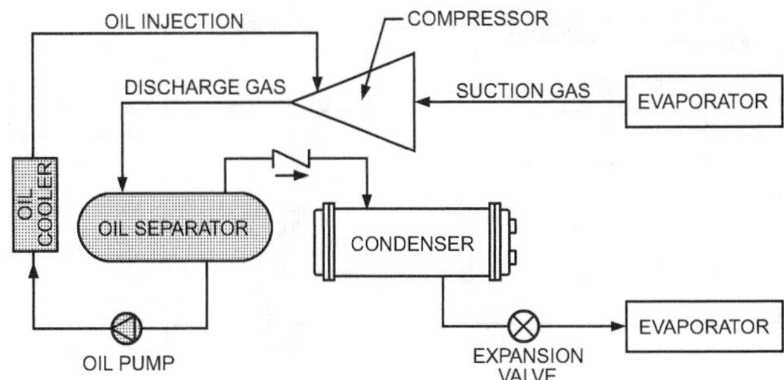

Fig. 17 Oil and Refrigerant Schematic of Oil Injection System

Oil-Injected Compressors. Oil is used in single-screw compressors to seal, cool, lubricate, and actuate capacity control. It gives a flat efficiency curve over a wide compression ratio and speed range, thus decreasing discharge temperature and noise.

Oil-injected single-screw compressors operate at high head pressures using common high-pressure refrigerants such as R-22, R-134a, and R-717. They also operate effectively at high pressure ratios because the injected oil cools the compression process.

Oil injection requires an oil separator to remove the oil from the high-pressure refrigerant (Figure 17). For applications with exacting demands for low oil carryover, separation equipment is available to leave less than 5 ppm oil in the circulated refrigerant.

With most compressors, oil can be injected automatically without a pump because of the pressure difference between the oil reservoir (discharge pressure) and the reduced pressure in a flute or bearing assembly during compression. However, this injection should be done after the compression chamber is closed (to avoid capacity loss) but before the pressure is higher than discharge (in case of overcompression), which may be very difficult to design. To avoid this limitation, a continuously running oil pump is used in some compressors to generate oil pressure 30 to 45 psi over compressor discharge pressure. This pump requires 0.3 to 1.0% of the compressor's motor power.

External oil cooling between the oil reservoir and the point of injection is possible. Various heat exchangers are available to cool the oil: (1) separate water supply, (2) chiller water on a packaged unit, (3) condenser water on a packaged unit, (4) water from an evaporative condenser sump, (5) forced air-cooled oil cooler, and (6) high-pressure liquid recirculation (thermosiphon). Heat added to the oil during compression is the amount usually removed in the oil cooler.

Oil-Injection-Free Compressors. Although single-screw compressors operate well with oil injection, they also operate with good efficiency in an oil-injection-free (OIF) mode with many common halocarbon refrigerants. This means that fluid injected into the compression chamber is the condensate of the fluid being compressed. For air conditioning and refrigeration, where pressure ratios are in the range of 2 to 8, the oil normally injected into the casing may be replaced by liquid refrigerant. Not much lubrication is required because the only power transmitted from the screw to the gate rotors is that needed to overcome small frictional losses. Thus, the refrigerant need only cool and seal the compressor. The liquid refrigerant may still contain a small amount of oil to lubricate the bearings (0.1 to 1%, or higher, depending on compressor design). A typical OIF circuit is depicted in Figure 18.

This method is used when an oil separator is not available. In this case, the liquid refrigerant carries a significant amount of oil with it. The methods of refrigerant injection include the following:

- **Direct injection of liquid refrigerant** into the compression process. Injection is controlled directly from the compressor discharge temperature, and loss of compressor capacity is minimized as injection takes place in a closed flute just before discharge occurs. This

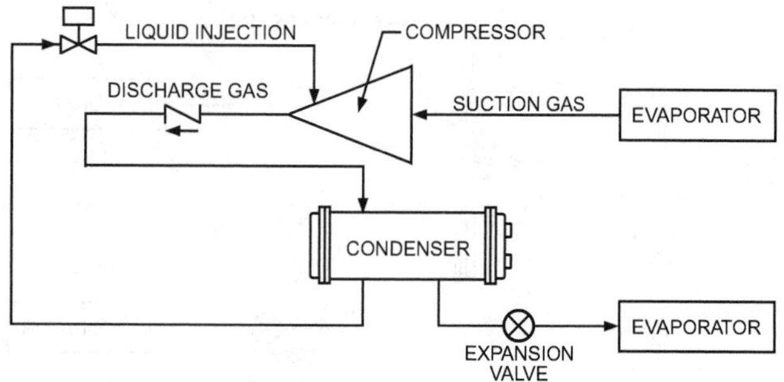

Fig. 18 Schematic of Oil-Injection-Free Circuit

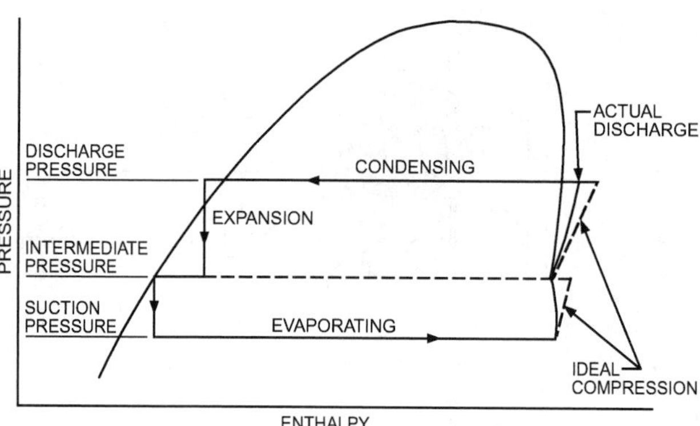

Fig. 19 Theoretical Economizer Cycle

method requires very little power (typically less than 5% of compressor power).

- A **small refrigerant pump** draws liquid from the receiver and injects it directly into the compressor discharge line. The injection rate is controlled by sensing discharge temperature and modulating the pump motor speed. The power penalty in this method is the pump power (about 1 hp for compressors up to 1000 hp), which can result in energy savings over refrigerant injected into the compression chamber.

Oil-injection-free operation has the following advantages:

- It requires no discharge oil separator, unless an oil separator is needed to reduce the oil circulation rate in the system.
- Compressors require no oil or refrigerant pumps.
- External coolers are not required.

Economizers. Screw compressors are available with a secondary suction port between the primary compressor suction and discharge port. This port, when used with an economizer, provides the means to increase compressor capacity efficiency.

In operation, gas is drawn into the rotor grooves in the normal way from the suction line. The grooves are then sealed off in sequence, and compression begins. An additional charge is added to the closed flute through a suitably placed port in the casing by an intermediate gas source at a slightly higher pressure than that reached in the compression process at that time. The original and the additional charge are then compressed together to discharge conditions. The pumping capacity of the compressor at suction conditions is not affected by this additional flow through the economizer port.

When the port is used with an economizer, the effective refrigerating capacity of the economized compressor is increased over the noneconomized compressor by the increased heat absorption capability H of the liquid entering the evaporator. Furthermore, the only additional mass flow the compressor must handle is flash gas entering a closed flute, which is above suction pressure. Thus, under most conditions, the capacity improvement also improves efficiency. Economizers become effective when the pressure ratio is 3.5 and above.

Figure 19 shows a pressure-enthalpy diagram for a flash tank economizer. In it, high-pressure liquid passes through an expansion device and enters a tank at an intermediate pressure between suction and discharge. This pressure is maintained by pressure in the compressor's closed flute (closed from suction). The gas generated from the expansion enters the compressor through the economizer port. When passed to the evaporator, the liquid (which is now saturated at the intermediate pressure) gives a larger refrigeration capacity per pound. In addition, the percentage increase in power input is lower than the percentage capacity increase.

As a screw compressor is unloaded, economizer pressure falls toward suction pressure. As a result, the additional capacity and improved efficiency of the economizer fall to zero at 70 to 80% of full-load capacity.

The single-screw compressor has two compression chambers, each with its own slide valve. Each slide valve can be operated independently, which allows economizer gas to be introduced into one side of the compressor. By operating the slide independently, the chamber without the economizer gas can be unloaded to 0% capacity (50% capacity of the compressor). The other chamber remains at full capacity and retains the full economizer effect, making the economizer effective below 50% compressor capacity.

The secondary suction port may also be used for (1) a system-side load or (2) a second evaporator that operates at a temperature above that of the primary evaporator.

Centrifugal Economizer. Some single-screw compressor designs use a patented centrifugal economizer that replaces the force of gravity in a flash economizer with centrifugal force to separate flash gas generated at an intermediate pressure from liquid refrigerant before liquid enters the evaporator. The centrifugal economizer thereby uses a much smaller pressure vessel and, in some designs, the economizer fits within the envelope of a standard motor housing without having to increase its size.

Separation is achieved by a centrifugal impeller mounted on the compressor shaft (see Figure 27); a special valve maintains a uniformly thick liquid ring around the circumference of the impeller, ensuring that no gas leaves with the liquid going to the evaporator. Flash gas is then ducted to closed grooves in the compressor screws. Some designs use flash gas with a similar liquid refrigerant to cool the motor before the gas enters the closed compression groove.

Volume (Compression) Ratio. The degree of compression in the rotor grooves is predetermined for a particular port configuration on screw compressors having fixed suction and discharge ports. A characteristic of the compressor is the volume, or compression ratio V_i, which is defined as the ratio of the volume of the groove at the start of compression to the volume of the same groove when it first begins to open to the discharge port. Hence, the volume ratio is determined by the size and shape of the discharge port.

For maximum efficiency, pressure generated within the grooves during compression should exactly equal the pressure in the discharge line at the moment when the groove opens to it. If this is not the case, either over- or undercompression occurs, both resulting in internal losses (overcompression can harm the compressor). Such losses increase power consumption and noise and reduce efficiency.

Volume ratio selection should be made according to operating conditions.

Compressors equipped with slide valves (for capacity modulation) usually locate the discharge port at the discharge end of the slide valve. Alternative port configurations yielding the required volume ratios are then designed into the capacity control components, thus providing easy interchangeability both during construction and after installation (although partial disassembly is required).

Single-screw compressors in refrigeration and process applications are equipped with a simple slide valve to vary compressor

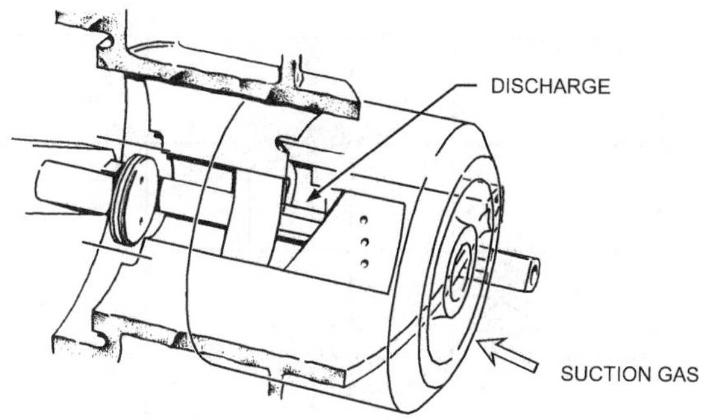

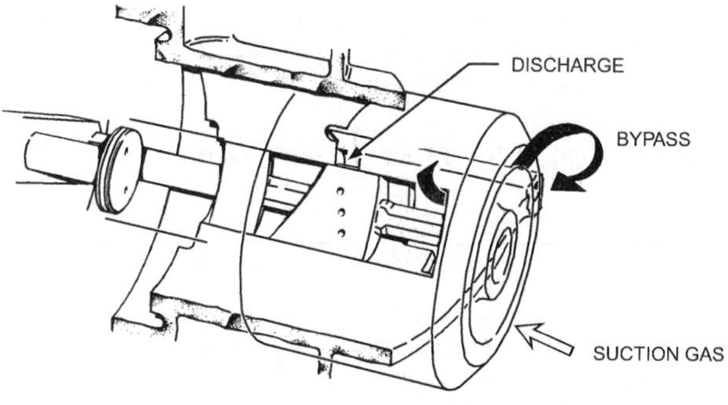

Fig. 20 Capacity-Control Slide Valve Operation

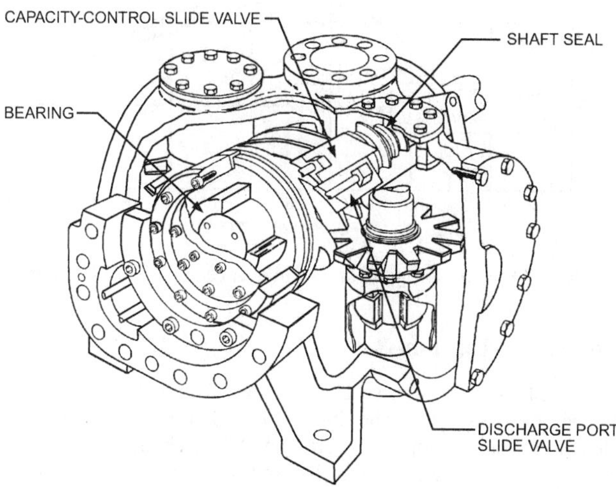

Fig. 21 Refrigeration Compressor Equipped with Variable-Capacity Slide Valve and Variable-Volume-Ratio Slide Valve

volume ratio while the compressor is running. The slide valve advances or delays the discharge port opening (Figure 20). Note that a separate slide has been designed to modulate the capacity independently of the volume ratio slide (see Figures 21 to 23). Having independent modulation of volume ratio (through discharge port control) and capacity modulation (through a completely independent slide that only varies the position where compression begins) allows the single-screw compressor to achieve efficient volume ratio control when capacity is less than full load.

Capacity Control. As with all positive-displacement compressors, both speed modulation and suction throttling can be used. Ideal capacity modulation for any compressor includes (1) continuous modulation from 100% to less than 10%, (2) good part-load efficiency, (3) unloaded starting, and (4) unchanged reliability.

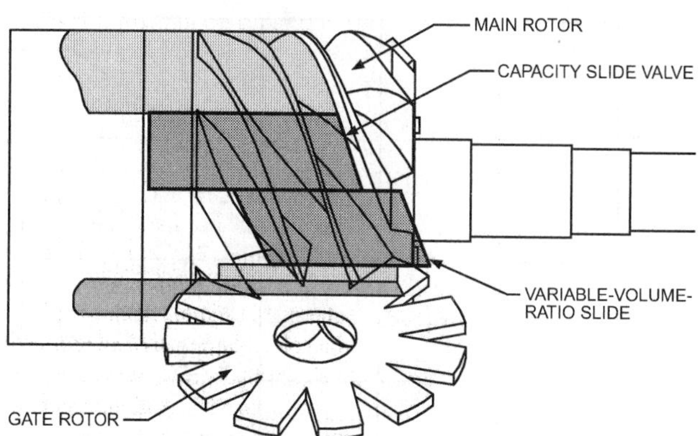

Fig. 22 Capacity Slide in Full-Load Position and Volume Ratio Slide in Intermediate Position

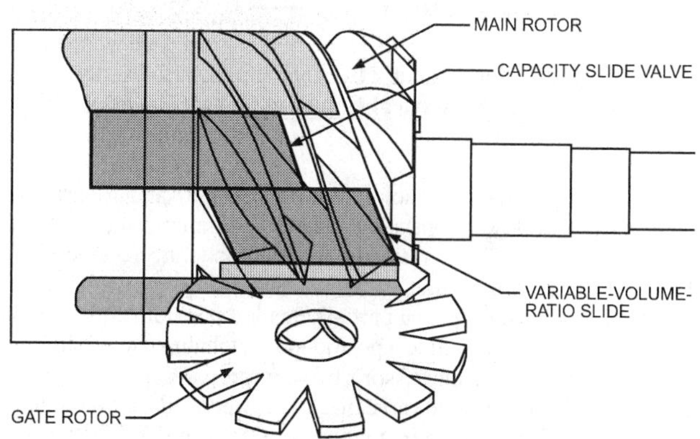

Fig. 23 Capacity Slide in Part-Load Position and Volume Ratio Slide Positioned to Maintain System Volume Ratio

Variable compressor displacement, the most common means for meeting these criteria, usually takes the form of two movable slide valves in the compressor casing (the single-screw compressor has two gate rotors forming two compression areas). At part load, each slide valve produces a slot that delays the point at which compression begins. This reduces groove volume, and hence compressor throughput. As suction volume is displaced before compression takes place, little or no thermodynamic loss occurs. However, if no other steps were taken, this mechanism would result in an undesirable drop in the effective volume ratio in undercompression and inefficient part-load operation.

This problem is avoided either by arranging that the capacity modulation valve reduces the discharge port area as the bypass slot is created (Figure 20) or having one valve control capacity only and a second valve independently modulate volume ratio (Figures 21 to 23). A full modulating mechanism is provided in most large single-screw compressors, whereas two-position slide valves are used where requirements allow. The specific part-load performance is affected by a compressor's built-in volume (compression) ratio, evaporator temperature, and condenser temperature, and whether the slide valves are symmetrically or asymmetrically controlled.

Detailed design of the valve mechanism differs between makes of compressors but usually consists of an axial sliding valve along each side of the rotor casing (Figure 20). This mechanism is usually operated by a hydraulic or gas piston and cylinder assembly in the compressor itself or by a positioning motor. The piston is actuated by oil, discharge gas, or high-pressure liquid refrigerant at discharge pressure driven in either direction according to the operation of a four-way solenoid valve.

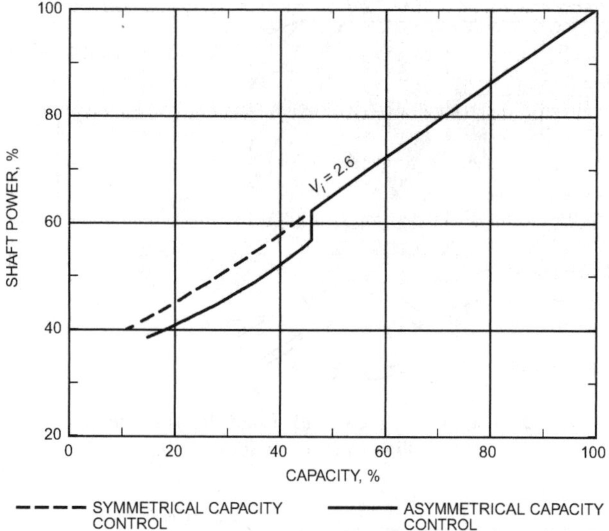

Fig. 24 Part-Load Effect of Symmetrical and Asymmetrical Capacity Control

Figure 22 shows a capacity slide valve (top) and a variable-volume-ratio slide (bottom). The capacity slide is in the full-load position, and the volume ratio slide is at a moderate volume ratio. Figure 23 depicts the same system as shown in Figure 22, except that the capacity slide is in a partially loaded position, and the volume ratio slide has moved to a position to match the new conditions.

The single-screw compressor's two compression chambers, each with its own capacity slide valve that can be operated independently, permits one slide valve to be unloaded to 0% capacity (50% compressor capacity) while the other slide valve remains at full capacity. This asymmetrical operation improves part-load efficiency below 50% capacity, and further part-load efficiency gains are realized when the economizer gas is only entered into a closed groove on the side that is unloaded second (see explanation in the section on Economizers). Figure 24 demonstrates the effect of asymmetrical capacity control of a single-screw compressor.

Performance. Figures 25 and 26 show typical efficiencies of all single-screw compressor designs. High isentropic and volumetric efficiencies result from internal compression, the absence of suction or discharge valves and their losses, and extremely small clearance volumes. The curves show the importance of selecting the correct volume ratio in fixed-volume-ratio compressors.

Manufacturers' data for operating conditions versus speed should not be extrapolated. Screw compressor performance at reduced speed is usually significantly different from that specified at the normally rated point because of the significant effect of leakage. Performance data normally include information about the degree of liquid subcooling and suction superheating assumed in data.

Applications. Single-screw compressors are widely used as refrigeration compressors, using halocarbon refrigerants, ammonia, and hydrocarbon refrigerants. A single gate rotor semihermetic version is increasingly being used in large supermarkets.

Oil-injected and oil-injection-free (OIF) semihermetic compressors are widely used for air-conditioning and heat pump service, with compressor sizes ranging from 40 to 500 tons.

Semihermetic Design. Figure 27 shows a semihermetic single-screw compressor. Figure 28 shows a semihermetic single-screw compressor using only one gate rotor. This design has been used in large supermarket rack systems. The single-gate-rotor compressor exhibits high efficiency and has been designed for long bearing life, which compensates for the unbalanced load on the screw rotor shaft with increasing bearing size.

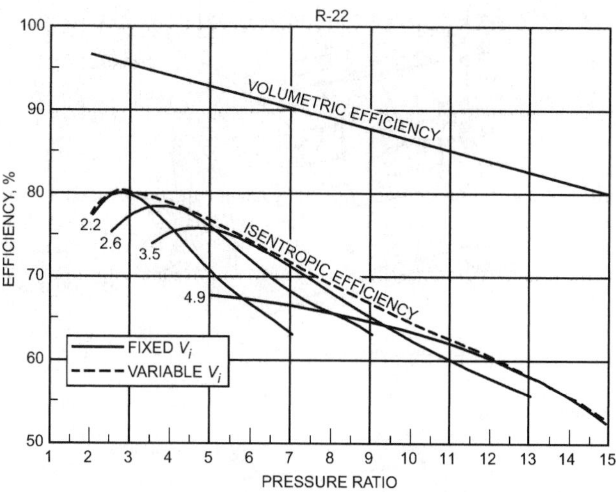

Fig. 25 Typical Open-Compressor Performance on R-22

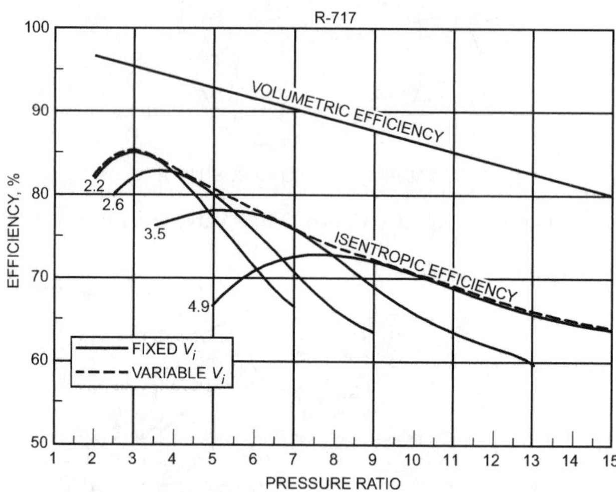

Fig. 26 Typical Compressor Performance on R-717 (Ammonia)

Noise and Vibration

The inherently low noise and vibration of single-screw compressors are due to small torque fluctuation and no valving required in the compression chamber. In particular, OIF technology eliminates the need for oil separators, which have traditionally created noise.

TWIN-SCREW COMPRESSORS

Twin screw is the common designation for double helical rotary screw compressors. A twin-screw compressor consists of two mating helically grooved rotors: male (lobes) and female (flutes or gullies) in a stationary housing with inlet and outlet gas ports (Figure 29). Gas flow in the rotors is mainly in an axial direction. Frequently used lobe combinations are 4 + 6, 5 + 6, and 5 + 7 (male + female). For instance, with a four-lobe male rotor, the driver rotates at 3600 rpm; the six-lobe female rotor follows at 2400 rpm. The female rotor can be driven through synchronized timing gears or directly driven by the male rotor on a light oil film. In some applications, it is practical to drive the female rotor, which results in a 50% speed and displacement increase over the male-driven compressor, assuming a 4 + 6 lobe combination. Geared speed increasers are also used on some applications to increase the capacity delivered by a particular compressor size.

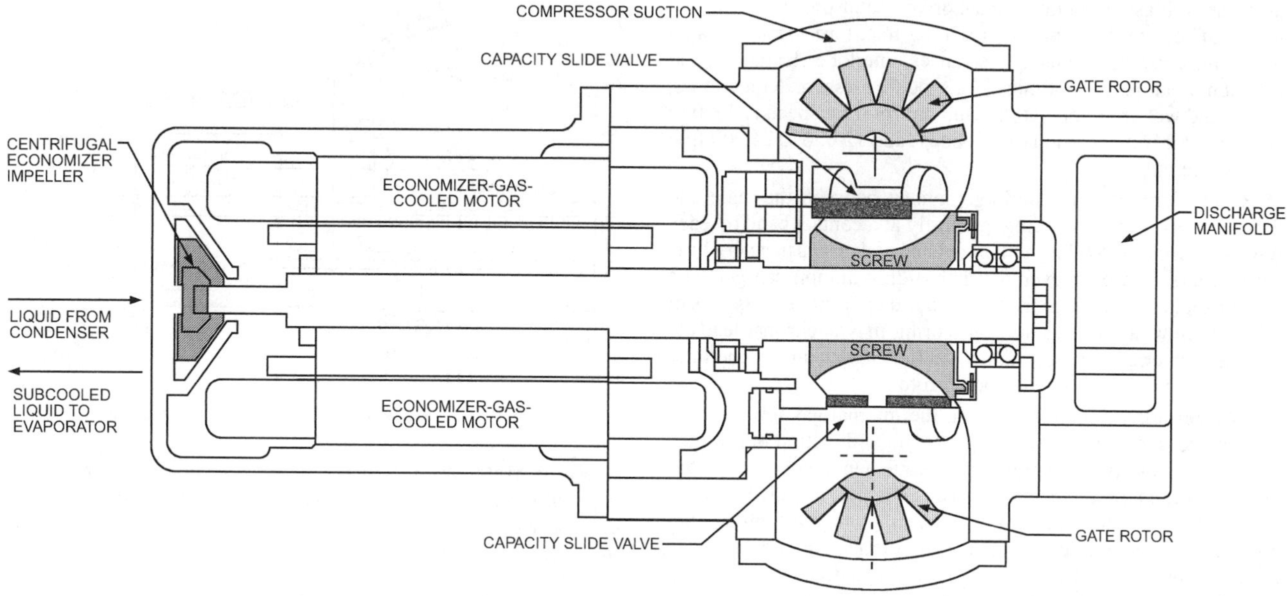

Fig. 27 Typical Semihermetic Single-Screw Compressor

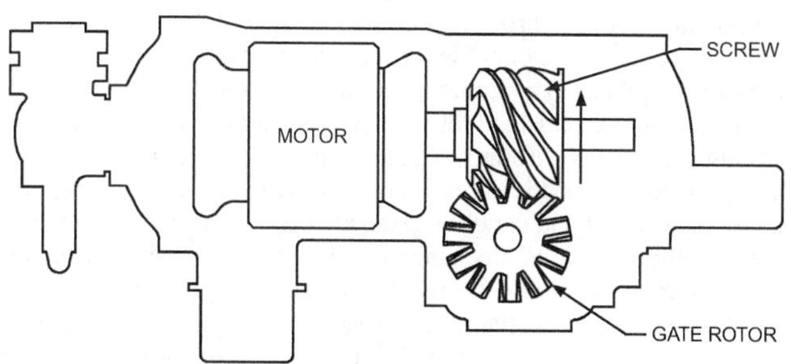

Fig. 28 Single-Gate-Rotor Semihermetic Single-Screw Compressor

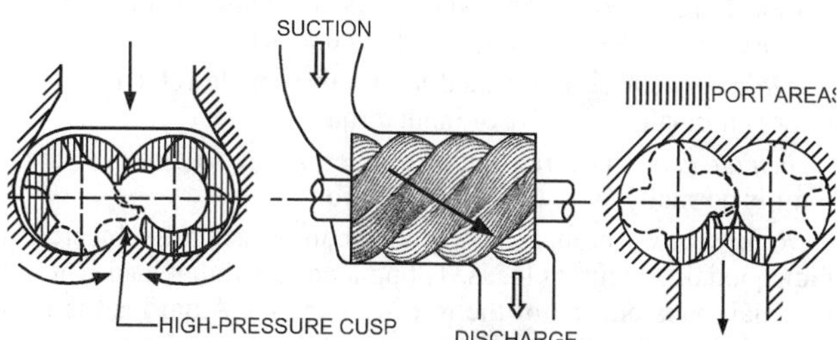

Fig. 29 Twin-Screw Compressor

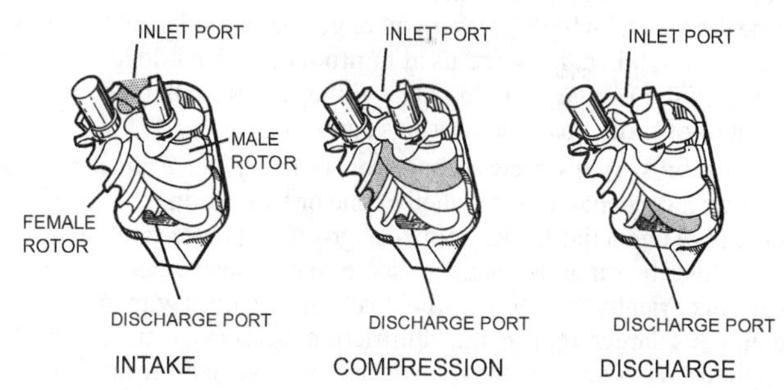

Fig. 30 Twin-Screw Compression Process

Twin helical screws find application in many air-conditioning, refrigeration, and heat pump applications, typically in the industrial and commercial market. Machines can be designed to operate at high or low pressure and are sometimes applied below 2:1 and above 20:1 compression ratios single-stage. Commercially available compressors are suitable for application on the majority of refrigerants.

Compression Process

Compression is obtained by direct volume reduction with pure rotary motion. For clarity, the following description of the three basic compression phases is limited to one male rotor lobe and one female rotor interlobe space (Figure 30).

Suction. As the rotors begin to unmesh, a void is created on both the male side (male thread) and the female side (female thread), and gas is drawn in through the inlet port. As the rotors continue to turn, the interlobe space increases, and gas flows continuously into the compressor. Just before the point at which the interlobe space leaves the inlet port, the entire length of the interlobe space is completely filled with gas.

Compression. Further rotation starts meshing another male lobe with another female interlobe space on the suction end and progressively compresses the gas in the direction of the discharge port. Thus, the occupied volume of the trapped gas within the interlobe space decreases and the gas pressure consequently increases.

Discharge. At a point determined by the designed built-in volume ratio, the discharge port is uncovered and the compressed gas is discharged by further meshing of the lobe and interlobe space.

During the remeshing period of compression and discharge, a fresh charge is drawn through the inlet on the opposite side of the meshing point. With four male lobes rotating at 3600 rpm, four interlobe volumes are filled and give 14,400 discharges per minute. Because the intake and discharge cycles overlap effectively, gas flow is smooth and continuous.

Mechanical Features

Rotor Profiles. Helical rotor design started with an asymmetrical point-generated rotor profile. This profile was only used in compressors with timing gears. The symmetrical, circular rotor profile was introduced because it was easier to manufacture than the preceding profile, and it could be used with or without timing gears.

Current rotor profiles are normally asymmetrical line-generated profiles, giving higher performance because of better rotor dynamics and decreased leakage area. This design allows female rotor drive, as well as conventional male drive. Rotor profile, blowhole, length of sealing line, quality of sealing line, torque transmission between rotors, rotor-housing clearances, interlobe clearances, and lobe combinations are optimized for specific pressure, temperature, speed, and wet or dry operation. Optimal rotor tip speed is 3000 to 8000 fpm for wet operation (oil-flooded) and 12,000 to 24,000 fpm for dry operation.

Rotor Contact and Loading. Contact between the male and female rotors is mainly rolling, primarily at a contact band on each rotor's pitch circle. Rolling at this contact band means that very little wear occurs. However, even a minor sliding motion can generate significant local heat from friction, and if there is not enough gas or lubricant flow at this area, a local scoring may occur that leads to increased internal leakage, higher friction losses, and sometimes mechanical failure because of locked rotors.

Gas Forces. On the driven rotor, the internal gas force always creates a torque in a direction opposite to the direction of rotation. This is known as positive or braking torque. On the undriven rotor, the design can be such that the torque is positive, negative, or zero, except on female-drive designs, where zero or negative torque does not occur. Negative torque occurs when internal gas force tends to drive the rotor. If the average torque on the undriven rotor is near zero, this rotor is subjected to torque reversal as it goes through its phase angles. Under certain conditions, this can cause instability. Torque transmitted between the rotors does not create problems because the rotors are mainly in rolling contact.

Male drive. The transmitted torque from male rotor to female rotor is normally 5 to 25% of input torque.

Female drive. The transmitted torque from female rotor to male rotor is normally 50 to 60% of input torque.

Rotor loads. Rotors in an operating compressor are subjected to radial, axial, and tilting loads. Tilting loads are radial loads caused by axial loads outside of the rotor center line. Axial load is normally balanced with a balancing piston for larger high-pressure machines (rotor diameter above 4 in. and discharge pressure above 160 psi). Balancing pistons are typically close-tolerance, labyrinth-type devices with high-pressure oil or gas on one side and low pressure on the other. They are used to produce a thrust load to offset some of the primary gas loading on the rotors, thus reducing the amount of thrust load the bearings support.

Bearings. Twin-screw compressors normally have either four or six bearings, depending on whether one or two bearings are used for the radial and axial loads. Some designs incorporate multiple rows of smaller bearings per shaft to share loads. Sleeve bearings were used historically to support radial loads in machines with male rotor diameters larger than 6 in.; antifriction bearings were generally applied to smaller machines. However, improvements in antifriction designs and materials have led to compressors with up to 14 in. rotor diameter with full antifriction bearing designs. Cylindrical and tapered roller bearings and various types of ball bearings are used in screw compressors for carrying radial loads. The most common thrust or axial load-carrying bearings are angular-contact ball bearings, although tapered rollers or tilting pad bearings are used in some machines. Use of moldable polymeric materials with higher temperature limits for bearing cages allows high-speed operation of the compressor at higher discharge temperatures.

General Design. Screw compressors are often designed for particular pressure ranges. Low-pressure compressors have long, high displacement rotors and adequate space to accommodate bearings to handle the relatively light loads. They are frequently designed without thrust balance pistons, because the bearings alone can handle the low thrust loads and still maintain good life.

High-pressure compressors have short, strong rotors (shallow grooves) and therefore have space for large bearings. Larger compressors are normally designed with balancing pistons for high thrust bearing life.

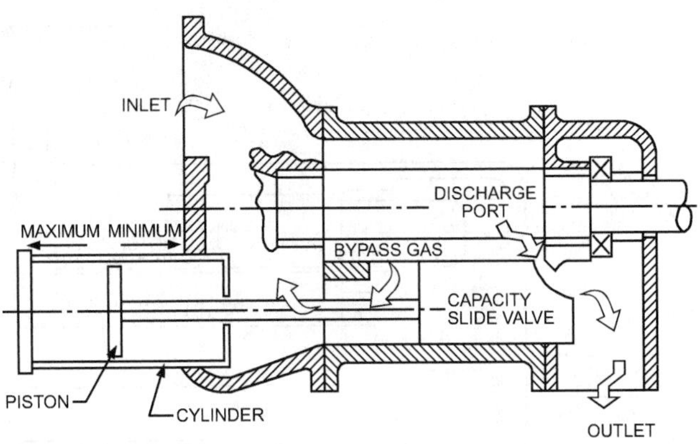

Fig. 31 Slide Valve Unloading Mechanism

Rotor Materials. Rotors are normally made of steel, but aluminum, cast iron, and nodular iron are used in some applications. Special surface treatment or coatings can be used to reduce wear and improve oil adhesion.

Capacity Control

As with all positive-displacement compressors, both speed modulation and suction throttling can reduce the volume of gas drawn into a screw compressor. Ideal capacity modulation for any compressor would be (1) continuous modulation from 100% to less than 10%, (2) good part-load efficiency, (3) unloaded starting, and (4) high reliability throughout the operational range. However, not all applications need ideal capacity modulation. Variable compressor displacement and variable speed are the best means for meeting these criteria. Various mechanisms achieve variable displacement, depending on the requirements of a particular application.

Capacity Slide Valve. A slide valve for capacity control is a valve with sliding action parallel to the rotor bores, within or close to the high-pressure region. It reduces the active length of the housing profile, thus controlling compressor displacement and capacity. There are two types of capacity slide valves:

- *Capacity slide valves regulating discharge port* are located in the high-pressure cusp region. They control capacity as well as the location of the radial discharge port at part load. The axial discharge port is designed for a volume ratio giving good part-load performance without losing full-load performance. Figure 31 shows a schematic of the most common arrangement.
- *Capacity slide valves not regulating discharge port* outside the high-pressure cusp region control only capacity.

The first type is most common. It is generally the most efficient capacity reduction method, because of its indirect correction of built-in volume (compression) ratio at part load and its ability to give large volume reductions without large movement of the slide valve.

Capacity Slot Valve. A capacity slot valve consists of a number of slots that follow the rotor helix and face one or both rotor bores. The slots are gradually opened or closed with a plunger or turn valve. These recesses in the casing wall increase the volume of compression space and also create leakage paths over the lobe tips. The result is somewhat lower full-load performance compared to a design without slots.

Capacity Lift Valve. Capacity lift valves or plug valves are movable plugs in one or both rotor bores (with radial or axial lifting action) that regulate the actual start of compression. These valves control capacity in a finite number of steps, rather than by the infinite control of a conventional slide valve (Figure 32).

Neither slot nor lift valves offer quite as good efficiency at part load as a slide valve, because they do not relocate the radial discharge

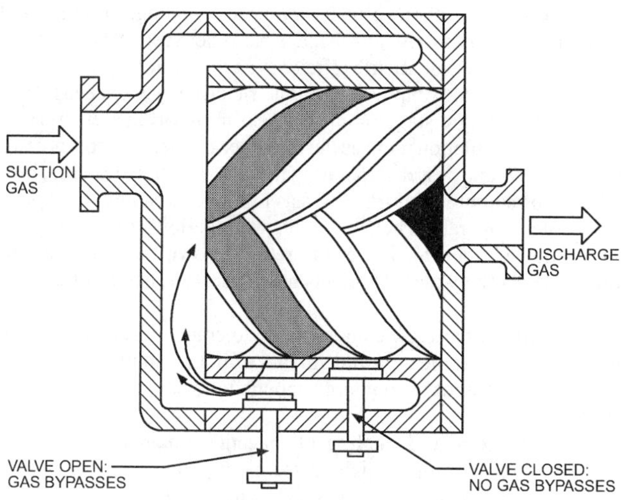

Fig. 32 Lift Valve Unloading Mechanism

port. Thus, undercompression losses at part load can be expected if machines have the correct volume (compression) ratio for full-load operation and the pressure ratio at part load does not reduce.

Volume (Compression) Ratio

In all positive-displacement rotary compressors with fixed port location, the degree of compression in the rotor thread is determined by the location of suction and discharge ports. The built-in volume (compression) ratio of screw compressors is defined as the ratio of volume of the thread at the start of compression to the volume of the same thread when it begins to open to the discharge port. The suction port must be located to trap the maximum suction charge; hence, the compression ratio is determined by the location of the discharge port.

Only suction pressure and compression volume ratio determine the internal pressure achieved before opening to discharge. However, condensing and evaporating temperatures determine discharge pressure and compression ratio in piping that leads to the compressor. Any mismatch between internal and system discharge pressures results in under- or overcompression loss and lower efficiency.

If the operating conditions of the system seldom change, it is possible to specify a fixed-volume-ratio compressor that will give good efficiency. Compressor manufacturers normally make compressors with three or four possible discharge port sizes that correspond to system conditions encountered frequently. Generally, the designer is responsible for specifying a compressor that most closely matches expected pressure conditions.

The required compression ratio for a particular application can be determined as follows:

$$\mathrm{CR} = \mathrm{PR}^{1/k} \tag{14}$$

where

CR = compression ratio
PR = pressure ratio $= p_d/p_s$
p_d = expected discharge pressure (absolute)
p_s = expected suction pressure (absolute)
k = isentropic coefficient for refrigerant used, from refrigerant tables [e.g., Lemmon et al. (2002)]

Usually, in slide-valve-equipped compressors, the radial discharge port is located in the discharge end of the slide valve. For a given ratio L/D of rotor length to rotor diameter and a given stop position, a short slide valve gives a low volume ratio, and a long slide valve gives a higher compression ratio. The difference in length basically locates the discharge port earlier or later in the compression process. Different-length slide valves allow changing the compression ratio of a given compressor, although disassembly is required.

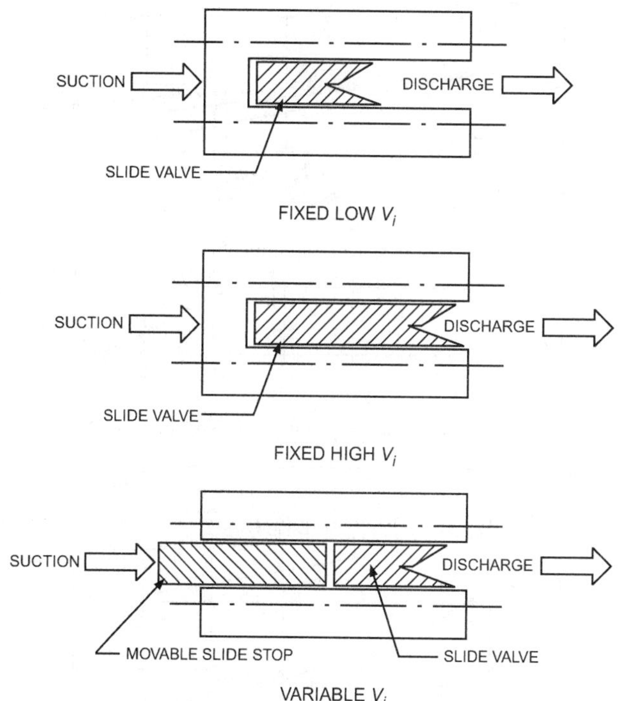

Fig. 33 View of Fixed- and Variable-Volume-Ratio (V_i) Slide Valves from Above

Variable Volume (Compression) Ratio. While operating, some twin-screw compressors adjust the compressor volume ratio to the most efficient ratio for whatever pressures are encountered.

In fixed-volume-ratio compressors, the slide valve motion toward the inlet end of the machine is stopped when it comes in contact with the rotor housing in that area. In most common variable-volume-ratio machines, this portion of the rotor housing has been replaced with a second slide, the movable slide stop, which can be actuated to different locations in the slide valve bore (Figure 33).

By moving the slides back and forth, the radial discharge port can be relocated during operation to match the compressor volume ratio to the optimum. This added flexibility allows operation at different suction and discharge pressure while maintaining maximum efficiency. Comparative efficiencies of fixed- and variable-volume-ratio screw compressors are shown in Figure 34 for full-load operation on ammonia and R-22 refrigerants. The figure shows that a variable-volume-ratio compressor efficiency curve encompasses the peak efficiencies of compressors with fixed volume ratio over a wide range of pressure ratio. Following are other secondary effects of a variable volume ratio:

- Less oil foam in oil separator (no overcompression)
- Less oil carried over into the refrigeration system (because of less oil foam in oil separator)
- Extended bearing life; minimized load on bearings
- Extended efficient operating range with economizer discharge port corrected for flash gas from economizer, as well as gas from suction
- Less noise
- Lower discharge temperatures and oil cooler heat rejection

The greater the change in either suction or condensing pressure, the more benefits are possible with a variable volume ratio. Efficiency

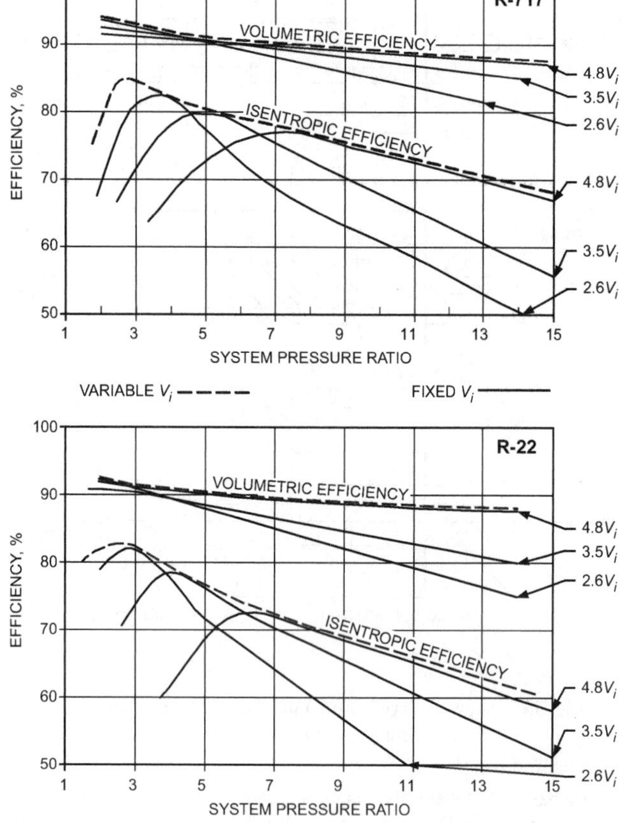

Fig. 34 Twin-Screw Compressor Efficiency Curves

improvements as high as 30% are possible, depending on the application, refrigerant, and operating range.

Oil Injection

Two primary types of compressor lubrication systems are used in twin-screw compressors: dry and oil-flooded.

Dry Operation (No Rotor Contact). Because the two rotors in twin-screw compressors are parallel, timing gears are a practical means of synchronizing the rotors so that they do not touch each other. Eliminating rotor contact eliminates the need for lubrication in the compression area; however, a small amount of oil may be needed to provide sealing. Initial screw compressor designs were based on this approach, and dry screws are still used in the gas process industry and in some transportation applications.

Synchronized twin-screw compressors once required high rotor tip speed to minimize leakage, and thus were noisy. However, with current profile technology, the synchronized compressor can run at a lower tip speed and higher pressure ratio, giving quieter operation. The added cost of timing gears and internal seals generally make the dry screw more expensive than an oil-flooded screw for normal refrigeration or air conditioning.

Oil-Flooded Operation. The oil-flooded twin-screw compressor is the most common type of screw compressor used in refrigeration and air conditioning. Oil-flooded compressors typically have oil supplied to the compression area at a volume rate of about 0.5% of the displacement volume. Part of this oil is used for lubricating bearings and shaft seal. Different oils are used, as required by the refrigerant or gas application. In the case of oil injection, oil is normally injected into a closed thread through ports in the moving slide valve and/or through stationary ports in the casing.

The oil fulfills three primary purposes: sealing, cooling, and lubrication. It tends to fill any leakage paths between and around the rotors. This keeps volumetric efficiency high, even at high compression ratios. Often, compressor volumetric efficiency exceeds 85%, even at 25:1 single stage (ammonia, 7.6 in. rotor diameter). High internal oil circulation reduces the influence of speed on compressor performance and lessens operational noise. Oil transfers much of the heat of compression from the gas to the oil, keeping the typical discharge temperature below 190°F, which allows high compression ratios without the danger of breaking down the refrigerant or oil. Lubrication by the oil protects bearings, seals, and rotor contact areas.

The ability of a screw compressor to tolerate oil also permits the compressor to handle a certain amount of liquid floodback, as long as the liquid quantity is not large enough to lock the rotors hydraulically.

Oil Separation and Cooling. Oil injection requires an oil separator to remove oil from the high-pressure refrigerant. Coalescing separation equipment routinely gives less than 5 ppm oil in the circulated refrigerant. Compressors used in direct-expansion (DX) systems and/or on packaged units have less-efficient separation capability.

Oil injection is normally achieved by one of two methods: (l) with a continuously running oil pump capable of generating an oil pressure of 30 to 45 psi over compressor discharge pressure, representing 0.3 to 1.0% of compressor motor power; or (2) with some compressors, oil can be injected automatically without a pump because of the pressure difference between the oil reservoir (discharge pressure) and the reduced pressure in an intermediate compression chamber.

Depending on the refrigerant and operating conditions, screw compressors can operate with or without oil cooling. There are performance advantages in maintaining low discharge-gas temperature by oil cooling. One cooling method is by direct injection of liquid refrigerant into the compression process. The amount of liquid injected is normally controlled by sensing the discharge temperature and injecting enough liquid to maintain a constant temperature. Some injected liquid mixes with the oil and leaks to lower-pressure threads, where it tends to raise pressure and reduce the amount of gas the compressor can draw in. Also, any liquid that has time to absorb heat and expand to vapor must be recompressed, which tends to raise absorbed power levels. Compressors are designed with liquid injection ports that inject liquid as late as possible in the compression to minimize capacity and power penalties. Typical penalties for liquid injection are in the 1 to 10% range, depending on the compression ratio. Because of this, and the danger of excessive lubricant dilution, direct liquid injection has more stringent limits than other oil-cooling methods.

Another method of oil cooling draws liquid from the receiver with a small refrigerant pump and injects it directly into the compressor discharge line. The power penalty in this method is the pump power (about 1 hp for compressors up to 1000 hp).

In the third method, oil is cooled outside the compressor between the oil reservoir and the point of injection. Various configurations of heat exchangers are available for this purpose, and oil cooler heat rejection can be accomplished by (1) separate water supply, (2) chiller water on a packaged unit, (3) condenser water on a packaged unit, (4) water from an evaporative condenser sump, (5) forced air-cooled oil cooler, (6) liquid refrigerant, and (7) high-pressure liquid recirculation (thermosiphon).

External oil coolers using water or other means from a source independent of the condenser allow condenser size to be reduced by an amount corresponding to the oil cooler capacity. Where oil is cooled within the refrigerant system by means such as (1) direct injection of liquid refrigerant into the compression process or the discharge line, (2) direct expansion of fluid in an external heat exchanger, (3) using chiller water on a packaged unit, (4) recirculating high-pressure liquid from the condenser, or (5) water from an evaporative condenser sump, the condenser must be sized for the total heat rejection (i.e., evaporator load plus shaft power for open compressors, and input power for hermetic compressors).

With an external oil cooler, the mass flow rate of oil injected into the compressor is usually determined by the desired discharge temperature rather than by the compressor sealing requirements, because oil acts predominantly as a heat transfer medium. Conversely, with direct liquid injection cooling, the oil requirement is dictated by the compressor lubrication and sealing needs.

Economizers

Twin-screw compressors are available with a secondary suction port between the primary compressor suction and discharge ports. This port can accept a second suction load at a pressure above the primary evaporator, or flash gas from a liquid subcooler vessel, known as an economizer.

In operation, gas is drawn into the rotor thread from the suction line. The thread is then sealed in sequence and compression begins. An additional charge may be added to the closed thread through a suitably placed port in the casing or sliding valve. The port is connected to an intermediate gas source at a pressure slightly higher than that reached in the compression process at that time. Both original and additional charges are then compressed to discharge conditions.

When the port is used as an economizer, some high-pressure liquid is vaporized at the side port pressure and subcools the remaining high-pressure liquid nearly to the saturation temperature at operating-side port pressure. Because this has little effect on compressor suction capacity, the effective refrigerating capacity of the compressor is increased by the increased heat absorption capacity of the liquid entering the evaporator. Furthermore, the only additional mass flow the compressor must handle is the flash gas entering a closed thread, which is above suction pressure. Thus, under most conditions, the capacity improvement is accompanied by an efficiency improvement.

Economizers become effective when the pressure ratio is about two and above (depending on volume ratio). Subcooling can be made with a direct-expansion shell-and-tube or plate heat exchanger, flash tank, or shell-and-coil intercooler.

As twin-screw compressors are unloaded, economizer pressure with a fixed port falls toward suction pressure. The additional capacity and improved efficiency of the economizer system is no longer available below a certain percentage of capacity, depending on design. Some compressors have the economizer port in the slide valve. This allows the economizer to be active down to the lowest percentage of capacity.

Hermetic and Semihermetic Compressors

Hermetic screw compressors are commercially available through 200 tons of refrigeration effect using R-22 or equivalent HFCs. Hermetic motors can operate under discharge, suction, or intermediate pressure. Motor cooling can be with gas, oil, and/or liquid refrigerant. Oil separation for these types of compressors may be accomplished with either an integrated oil separator or a separately mounted oil separator in the system. Figures 35 to 37 show three types of twin-screw compressors. For the lower capacity range (7 to 17 tons), welded-shell horizontal hermetic compressors are also available. Because of their smooth running characteristics, small size, and good capability for frequency inverter drive (1200 to 5220 rpm), they are preferable for railway and other transportation air-conditioning systems.

Performance Characteristics

Figure 34 shows the full-load efficiency of a modern twin-screw compressor. Both fixed- and variable-volume-ratio compressors without economizers are indicated. High isentropic and volumetric efficiencies result from internal compression, the absence of suction or discharge valves, and small clearance volume. The curves show that although volumetric efficiency depends little on the choice of volume ratio, isentropic efficiency depends strongly on it.

Performance data usually note the degree of liquid subcooling and suction superheating assumed. If an economizer is used, the pressure drop and the temperature of liquid approaching the economizer should be specified.

Noise

The most significant sources of noise in screw compressors are rotor contact and discharge pulsations. Adequate lubrication and discharge port design can significantly alleviate this issue. If screws are driven by gears, gear noise may become dominant.

ORBITAL COMPRESSORS

SCROLL COMPRESSORS

Description

Scroll compressors are orbital motion, positive-displacement machines that compress with two interfitting, spiral-shaped scroll members (Figure 38). They are currently used in residential and commercial air-conditioning, refrigeration, and heat pump applications as well as in automotive air conditioning. Capacities range

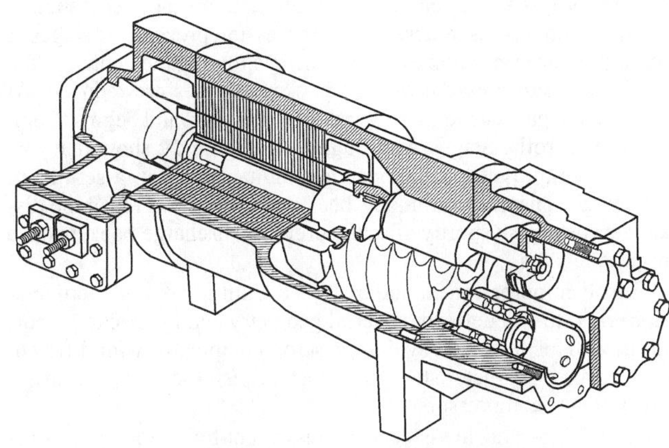

Fig. 35 Semihermetic Twin-Screw Compressor with Suction-Gas-Cooled Motor

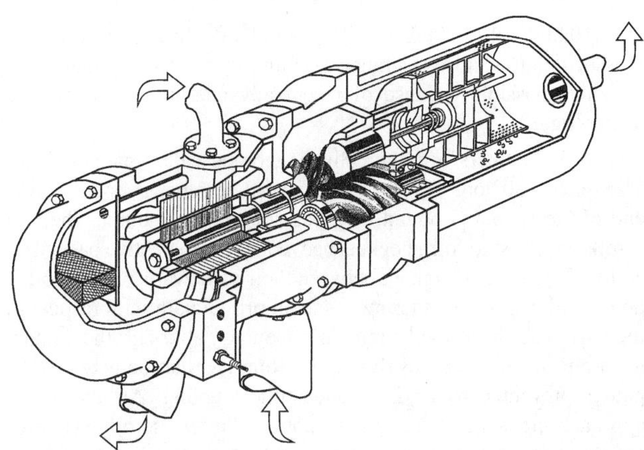

Fig. 36 Semihermetic Twin-Screw Compressor with Motor Housing Used as Economizer; Built-In Oil Separator

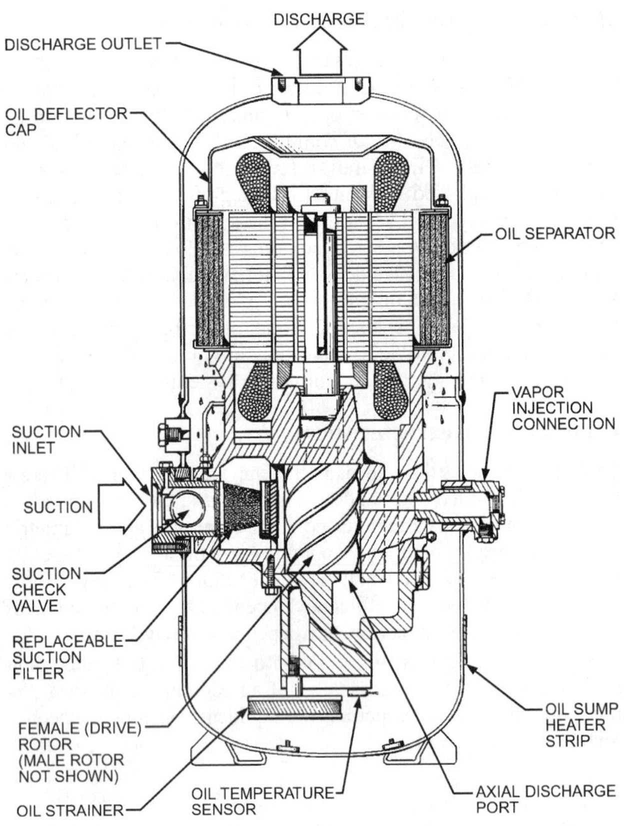

Fig. 37 Vertical, Discharge-Cooled, Hermetic Twin-Screw Compressor

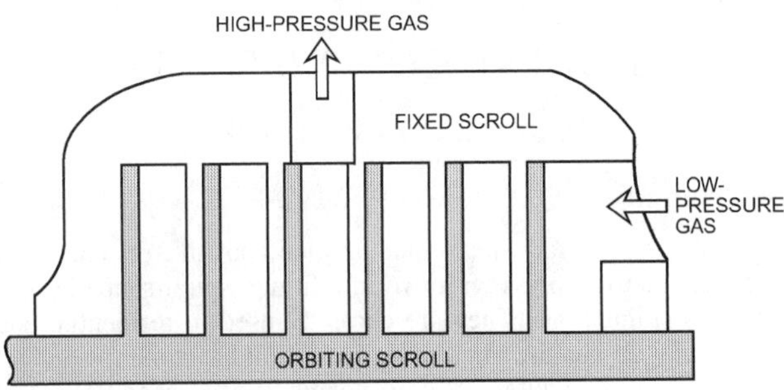

Fig. 38 Interfitted Scroll Members
(Purvis 1987)

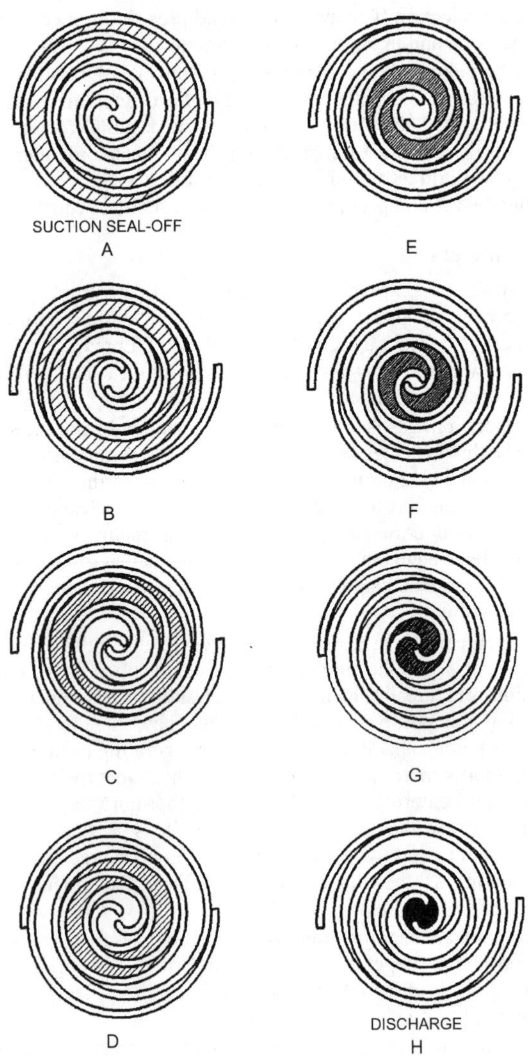

Fig. 39 Scroll Compression Process
(Purvis 1987)

from 10,000 to 170,000 Btu/h. To function effectively, a scroll compressor requires close-tolerance machining of the scroll members, which is possible because of recent advances in manufacturing technology.

Scroll members are typically a geometrically matched pair, assembled 180° out of phase. Each scroll member is open on one end of the vane and bound by a base plate on the other. The two scrolls are fitted to form pockets between their respective base plate and various lines of contact between their vane walls. One scroll is held fixed, while the other moves in an orbital path with respect to the first. The scroll flanks remain in contact, although the contact locations move progressively inward. Relative rotation between the pair is prevented by an interconnecting coupling. An alternative approach creates relative orbital motion via two scrolls synchronously rotating about noncoincident axes. As in the former case, an interconnecting coupling maintains a relative angle between the pair of scrolls (Morishita et al. 1988).

Compression is accomplished by sealing suction gas in pockets of a given volume at the outer periphery of the scrolls and progressively reducing the size of those pockets as scroll relative motion moves them inwards toward the discharge port. Figure 39 shows the sequence of suction, compression, and discharge phases. As the outermost pockets are sealed off (Figure 39A), trapped gas is at suction pressure and has just entered the compression process. At stages B through F, orbiting motion moves the gas toward the center of the scroll pair, and pressure rises as pocket volumes are reduced. At stage G, the gas reaches the central discharge port and begins to exit from the scrolls. Stages A through H in Figure 39 show that two distinct compression paths operate simultaneously in a scroll set. Discharge is nearly continuous, because new pockets reach the discharge stage very shortly after the previous discharge pockets have been evacuated.

Scroll compression embodies a fixed, built-in volume (compression) ratio that is defined by scroll geometry and by discharge port location. This feature provides the scroll compressor with different performance characteristics than those of reciprocating or conventional rotary compressors.

Both high- and low-side compressor configurations are available. In the former, the entire compressor is at discharge pressure, except for the outer areas of the scroll set. Suction gas is introduced into the suction port of the scrolls through piping, which keeps it discrete from the rest of the compressor. Discharge gas is directed into the compressor shell, which acts as a plenum. In the low-side type, most of the shell is at suction pressure, and discharge gas exiting from the scrolls is routed outside the shell, sometimes through a discrete or integral plenum.

A three-dimensional (3D) scroll compressor has recently been developed. The 3D scroll can compress refrigerant not only radially, but also axially, therefore developing higher compression ratio and large capacity.

Mechanical Features

Scroll Members. Gas sealing is critical to the performance advantage of scroll compressors. Sealing within the scroll set must be accomplished at flank contact locations and between the vane tips and bases of the intermeshed scroll pair. Tip/base sealing is generally considered more critical than flank sealing. The method used to seal the scroll members tends to separate scroll compressors into compliant and noncompliant designs.

Scrolls are usually of cast iron; however, aluminum alloys and other advanced materials show some promising results. Aluminum scrolls could potentially reduce bearing load, weight, and power loss.

Noncompliant Designs. In designs lacking compliance, the orbiting scroll takes a fixed orbital path. In the radial direction, sealing small irregularities between the vane flanks (caused by flank machining variation) can be accomplished with oil flooding. In the axial direction, the position of both scrolls remains fixed, and flexible seals fitted into machined grooves on the tips of both scrolls accomplish tip sealing. The seals are pressure-loaded to enhance uniform contact (McCullough and Shaffer 1976; Sauls 1983).

Radial Compliance. This feature enhances flank sealing and allows the orbiting scroll to follow a flexible path defined by its own contact with the fixed scroll. In one type of radial compliance, a sliding "unloader" bushing is fitted onto the crankshaft eccentric pin in such a way that it directs the radial motion of the orbiting scroll. The orbiting scroll is mounted over this bushing through a drive bearing, and the scroll may now move radially in and out to accommodate variations in orbit radius caused by machining and assembly discrepancies. This feature tends to keep the flanks constantly in contact, and reduces impact on the flanks that can result from intermittent contact. Sufficient clearance in the pin/unloader assembly allows the scroll flanks to separate fully when desired.

In some designs, the mass of the orbiting scroll is selected so that centrifugal force overcomes radial gas compression forces that would otherwise keep the flanks separated.

In other designs, the drive is designed so that the influence of centrifugal force is reduced, and drive force overcomes the radial gas compression force (McCullough 1975). Radial compliance has the added benefit of increasing resistance to slugging and contaminants, because the orbiting scroll can "unload" to some extent as it encounters obstacles or nonuniform hydraulic pressures (Bush and Elson 1988).

Axial Compliance. With this feature, an adjustable axial pressure maintains sealing contact between the scroll tips and bases while running. This pressure is released when the unit is shut down, allowing the compressor to start unloaded and to approach full operational speed before a significant load is encountered. This scheme obviates the use of tip seals, eliminating them as a potential source of wear and leakage. With the scroll tips bearing directly on the opposite base plates and with suitable lubrication, sealing tends to improve over time. Axial compliance can be implemented on either the orbiting or fixed scroll (Caillat et al. 1988; Tojo et al. 1982). Axial compliance requires auxiliary sealing of the discharge side with respect to the suction side of the compressor.

Antirotation Coupling. To ensure relative orbital motion, the orbiting scroll must not rotate in response to gas loading. This rotation is most commonly accomplished by an Oldham coupling mechanism, which physically connects the scrolls and allows all planar motion, except relative rotation, between them.

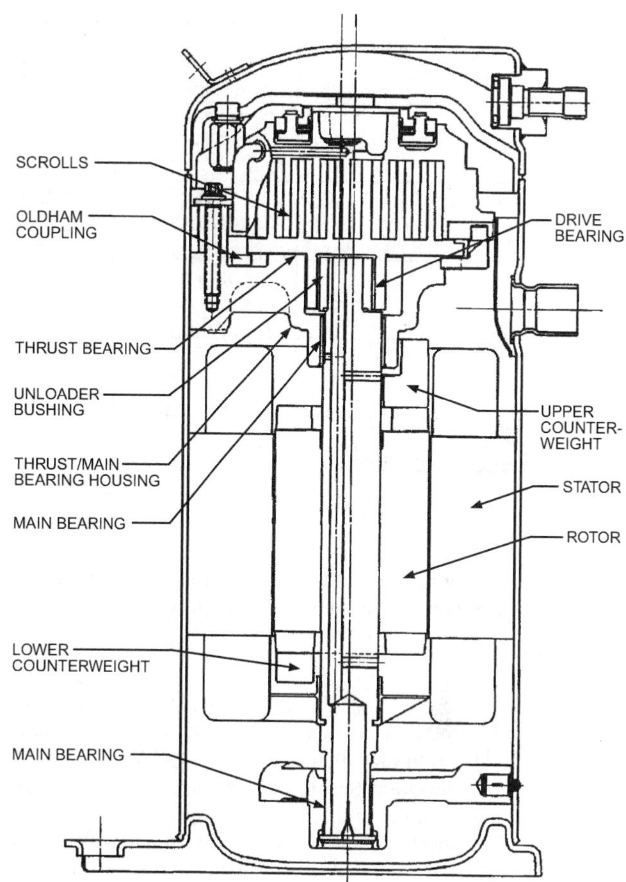

Fig. 40 Bearings and Other Components of Scroll Compressor
(Elson et al. 1990)

Bearing System. The bearing system consists of a drive bearing mounted in the orbiting scroll and generally one of two main bearings. The main bearings are either of the cantilevered type (main bearings on same side of the motor as the scrolls) or consist of a main bearing on either side of the motor (Figure 40). All bearing load vectors rotate through a full 360° because of the nature of the drive load.

The orbiting scroll is supported axially by a thrust bearing on a housing which is part of the internal frame or is mounted directly to the compressor shell.

Capacity Control

Compressor capacity control is used where applications require more precise temperature and humidity control than fixed-speed compressors can provide. Capacity controls currently in use include the following:

Variable-Speed Scroll Compressor. A variable-speed scroll compressor uses an inverter drive to convert a fixed-frequency alternating current into one with adjustable voltage and frequency, which allows variation of the motor's rotating speed. The compressor uses either an induction or a permanent magnet motor. Compressor manufacturers establish maximums and minimums of the operating frequency range based on the compressor and motor design characteristics. Capacity is nearly directly proportional to running frequency, with a slight increase at higher frequencies because of reduced leakage in proportion to refrigerant flow. Thus, virtually infinite capacity steps are possible for the system with a variable-speed compressor. Using a multiwinding motor (two or

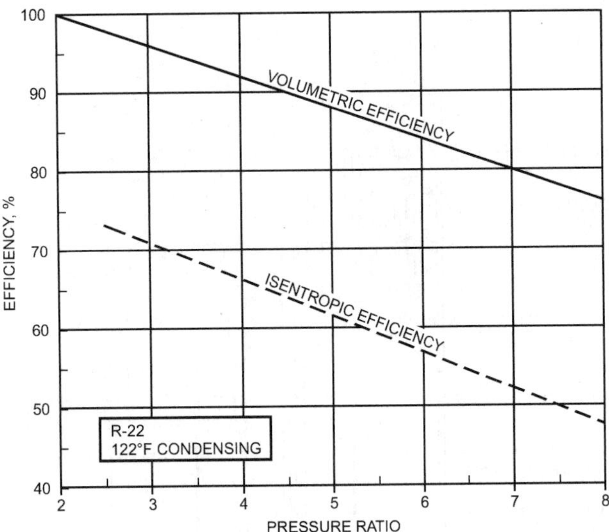

Fig. 41 Volumetric and Isentropic Efficiency Versus Pressure Ratio for Scroll Compressors
(Elson et al. 1990)

three windings) allows changing the motor speed by switching from one winding to another; for example, switching from two-pole to four-pole winding decreases compressor speed by half.

Variable-Displacement Scroll Compressor. This mechanism incorporates porting holes in the fixed scroll member. The control mechanism disconnects or connects compression chambers to the suction side by respectively closing or opening the porting holes. When all porting holes are closed, the compressor runs at full capacity; opening all porting holes to the suction side yields the smallest capacity. Thus, by opening or closing a different number of porting holes, variable cooling or heating capability is provided. The number of different capacities and extent of the capacity reduction available is governed by the locations of the ports in reference to full-capacity suction seal-off.

Pulse Width Modulation. The mechanism modulates the axial pressure that maintains sealing contact between scroll tips and bases (see the paragraph on Axial Compliance in the section on Mechanical Features). The mechanism controls capacity by cycling the loading and unloading of the fixed scroll without changing the motor speed. The mechanism receives a signal from an electronic control module that communicates with the system. When there is a call for cooling from the system, the module controls the cycle of loading and unloading of scrolls to deliver the exact capacity required to match the demand.

Performance

Scroll technology offers a performance advantage for a number of reasons. Large suction and discharge ports reduce pressure losses incurred in suction and discharge. Also, physical separation of these processes reduces heat transfer to suction gas. The absence of valves and reexpansion volumes and the continuous-flow process result in high volumetric efficiency over a wide range of operating conditions. Figure 41 illustrates this effect. The built-in volume ratio can be designed for lowest over- or undercompression at typical demand conditions (2.5 to 3.5 pressure ratio for air conditioning). Isentropic efficiency in the range of 70% is possible at such pressure ratios, and it remains quite close to the efficiency of other compressor types at high pressure ratio (Figure 41). Scroll compressors offer a flatter capacity versus outdoor ambient curve than reciprocating products, which means that they can more closely approach indoor requirements at high demand conditions. As a result, the heat pump mode requires less supplemental heating; the cooling mode is more comfortable, because cycling decreases as demand decreases (Figure 42).

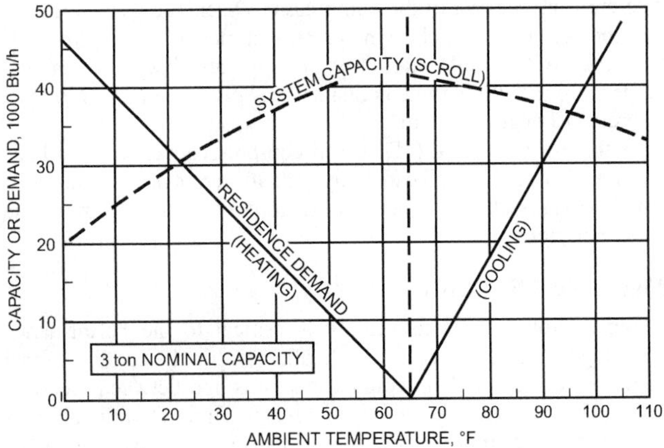

Fig. 42 Scroll Capacity Versus Residence Demand
(Purvis 1987)

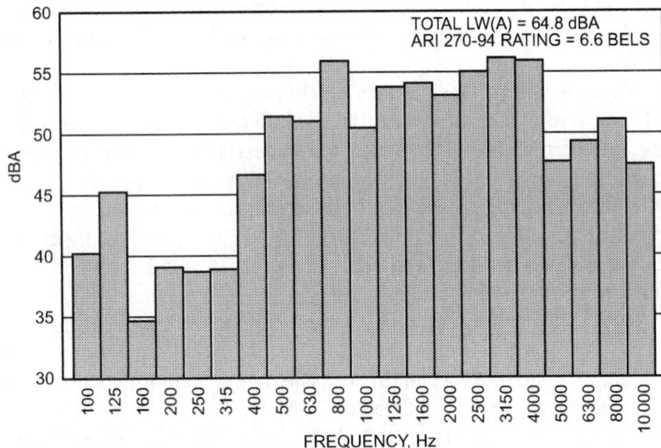

Fig. 43 Typical Scroll Sound Spectrum

Scroll compressors available for the North American market are typically specified as producing ARI operating efficiencies [coefficients of performance (COPs)] in the range of 3.10 to 3.34.

Noise and Vibration

Scroll compressors have an inherent potential for low sound and vibration. They include fewer moving parts compared to some other compressor technologies, and because scroll compression requires no valves, impact noise and vibration are completely eliminated. A continuous suction-compression-discharge process and low-gas-pressure pulsation help keep vibration low. Good dynamic balancing of the orbiting scroll with counterweights significantly reduces vibration caused by rotating parts. Also, smooth surface finish and accurate machining of the vane profiles and base plates of both scroll members (requirement for small leakage) minimizes friction impact on compressor noise. A typical sound spectrum of a scroll compressor is shown in Figure 43.

Operation and Maintenance

Most scroll compressors used today are hermetic, which require virtually no maintenance. However, the compressor manufacturer's operation and application manual should be followed.

TROCHOIDAL COMPRESSORS

Trochoidal compressors are small, rotary, positive-displacement compressors that can run at high speed up to 9000 rpm. They are

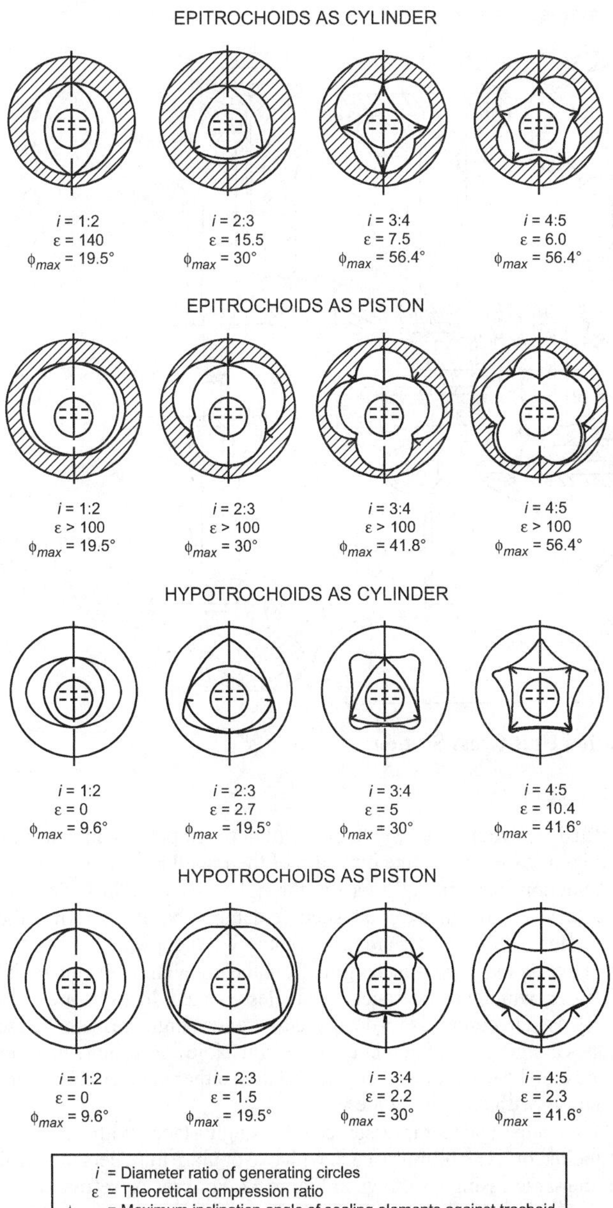

Fig. 44 Possible Versions of Epitrochoidal and Hypotrochoidal Machines

manufactured in various configurations. Trochoidal curvatures can be produced by the rolling motion of one circle outside or inside the circumference of a basic circle, producing either epitrochoids or hypotrochoids, respectively. Both types of trochoids can be used either as a cylinder or piston form, so that four types of trochoidal machines can be designed (Figure 44).

In each case, the counterpart of the trochoid member always has one apex more than the trochoid itself. In the case of a trochoidal cylinder, the apexes of the piston show slipping along the inner cylinder surface; for trochoidal piston design, the piston shows a gear-like motion. As seen in Figure 44, a built-in compression pressure ratio disqualifies many configurations as valid concepts for refrigeration compressor design. Because of additional valve ports, clearances, etc., and the resulting decrease in the built-in maximum theoretical compression ratio, only the first two types with epitrochoidal cylinders, and all candidates with epitrochoidal pistons, can be used for compressor technology. The latter, however, require

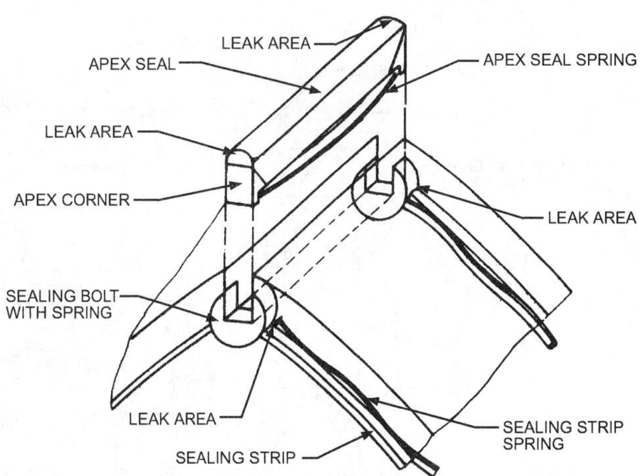

Fig. 45 Wankel Sealing System for Trochoidal Compressors

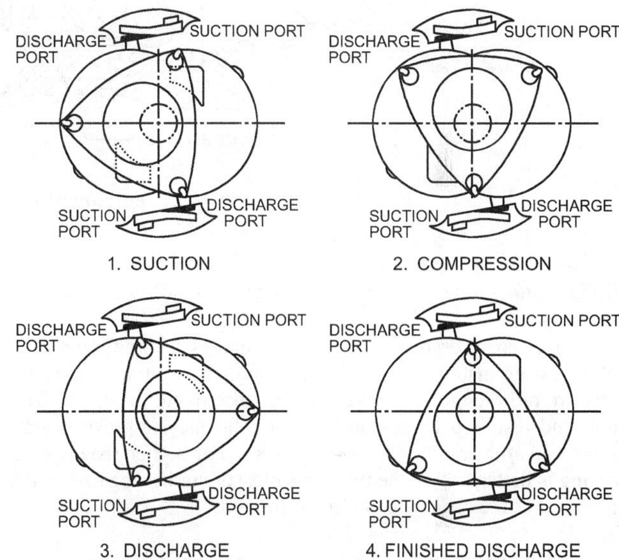

Fig. 46 Sequence of Operation of Wankel Rotary Compressor

sealing elements on the cylinder as well as on the side plates, which does not allow the design of a closed sealing borderline.

In the past, trochoidal machines were designed much like those of today. However, like other positive-displacement rotary concepts that could not tolerate high internal oil circulation, early trochoidal compressors failed because of sealing problems. The invention of a closed sealing border by Wankel changed this (Figure 45). Today, the Wankel trochoidal compressor with a three-sided epitrochoidal piston (motor) and two-envelope cylinder (casing) is built in capacities of up to 2 tons.

Description and Performance

Compared to other compressors of similar capacity, trochoidal compressors have many advantages typical of reciprocating compressors. Because of the closed sealing border of the compression space, these compressors do not require extremely small, expensive manufacturing tolerances; neither do they need oil for sealing, keeping them at low-pressure side with the advantage of low solubility and high viscosity of the oil-refrigerant mixture. Valves are usually used on a high-pressure side while suction is ported. A valveless trochoidal compressor can also be built. Figure 46 shows the operation

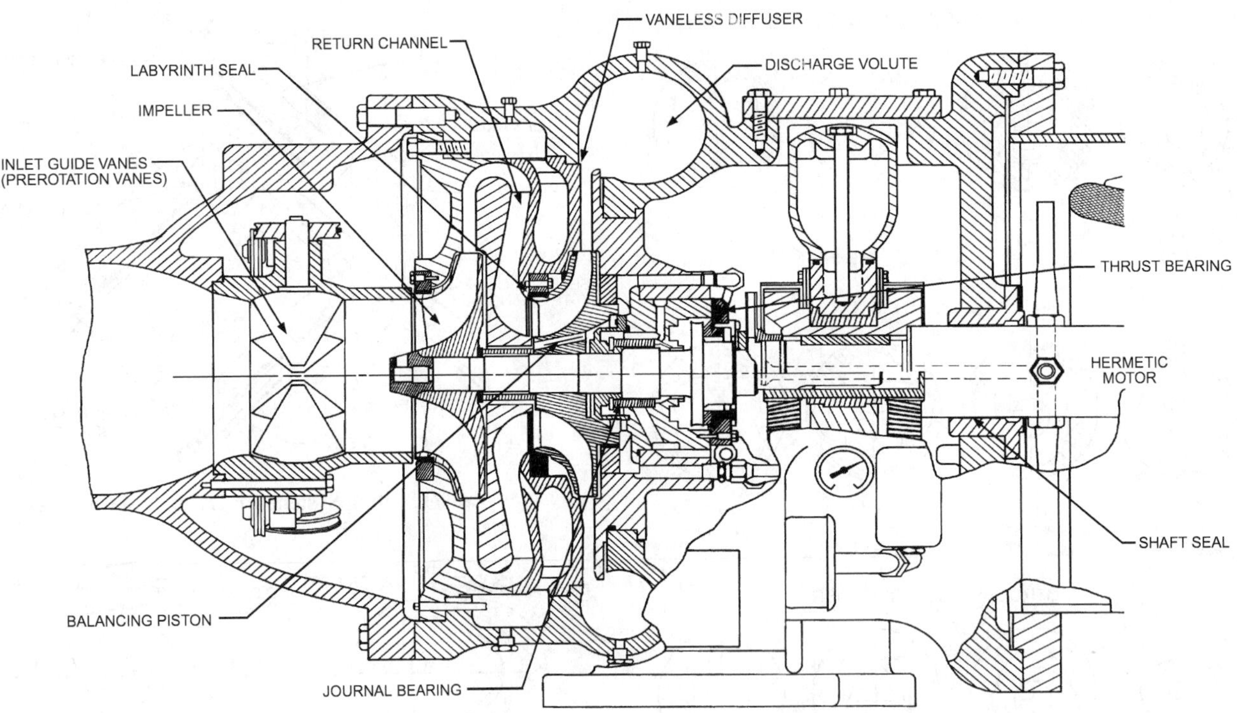

Fig. 47 Centrifugal Refrigeration Unit Cross Section

of the Wankel rotary compressor (2:3 epitrochoid) with discharge reed valves.

Wankel compressor performance compares favorably with the reciprocating piston compressors at a higher speed and moderate pressure ratio range. A smaller number of moving parts, less friction, and resulting higher mechanical efficiency improve overall isentropic efficiency. This can be observed at higher speed when sealing is better, and in the moderate pressure ratio range when the influence of the clearance volume is limited.

CENTRIFUGAL COMPRESSORS

Centrifugal compressors, sometimes called turbocompressors, belong to a family of turbomachinery that includes fans, propellers, and turbines. These are classified as "dynamic" machines because they continuously exchange angular momentum between a rotating mechanical element and a steadily flowing fluid. For effective momentum exchange, their rotating speeds must be higher, but little vibration or wear results because of the steady motion and the absence of contacting parts such as pistons or vanes. Because their flows are continuous, turbomachines have greater volumetric capacities, size for size, than do positive-displacement devices.

Centrifugal compressors are used in a variety of refrigeration and air-conditioning installations. Suction flow ranges between 60 and 30,000 cfm, with rotational speeds between 1800 and 90,000 rpm. However, the high angular velocity associated with a low volumetric flow establishes a minimum practical capacity for most centrifugal applications. The upper capacity limit is determined by physical size, a 30,000 cfm compressor having a diameter of 6 or 7 ft.

Centrifugal compressors are well-suited for air-conditioning and refrigeration applications because of their ability to produce a high pressure ratio. Suction flow enters the rotating element (impeller) axially, and is discharged radially at a higher velocity. The change in diameter through the impeller increases the gas velocity. This velocity (dynamic) pressure is then converted to static pressure through diffusion, which generally begins within the impeller and ends in a radial diffuser and volute outboard of the impeller.

Suction gas generally passes through a set of adjustable inlet guide vanes or an external suction damper before entering the impeller. These devices are used for capacity control.

High-velocity gas discharging from the impeller enters the radial diffuser, which can be vaned or vaneless. Vaned diffusers are typically used in compressors designed to produce high pressure. These vanes are generally fixed but can be adjustable. Adjustable diffuser vanes can be used for capacity modulation either in lieu of or in conjunction with the inlet guide vanes.

A centrifugal compressor can be single-stage, with only one impeller, or it can be multistage, with two or more impellers mounted in the same casing, as shown in Figure 47. For process refrigeration, a compressor can have as many as 10 stages.

In multistage compressors, gas discharged from the first stage is directed to the inlet of the second stage through a return channel. The return channel can contain a set of fixed-flow straightening vanes or an additional set of adjustable inlet guide vanes. Once the gas reaches the last stage, it is discharged from the impeller into a volute or collector chamber. From there, the high-pressure gas passes through the compressor discharge connection.

When multistage compressors are used, interstage gas flows can be introduced between stages so that one compressor performs several functions at several temperatures.

Refrigeration Cycle

Typical applications might involve a single-, two-, or three-stage halocarbon compressor or a seven-stage ammonia compressor. Figure 48 illustrates a simple vapor compression cycle in which a centrifugal compressor operates between states 1 and 2.

Figure 49 shows a more complex cycle, with two stages of compression and interstage liquid flash cooling. This cycle has a higher coefficient of performance than the simple cycle and is frequently used with two- through four-stage halocarbon and hydrocarbon compressors.

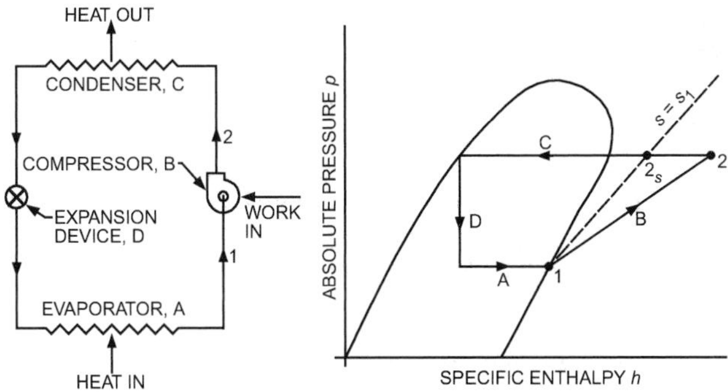

Fig. 48 Simple Vapor Compression Cycle

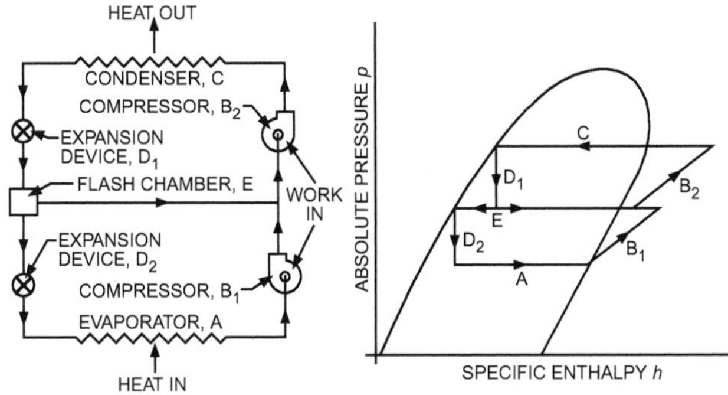

Fig. 49 Compression Cycle with Flash Cooling

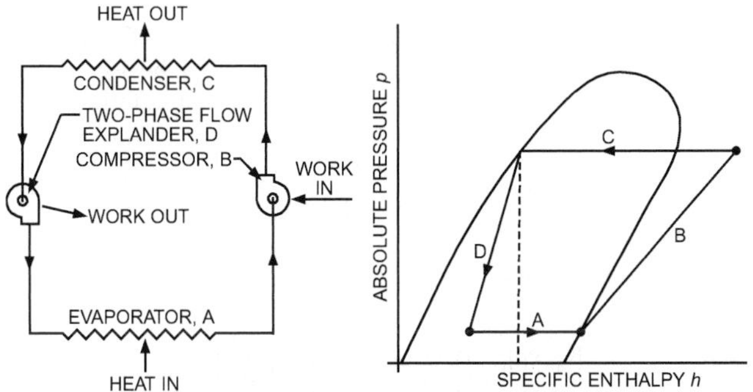

Fig. 50 Compression Cycle with Power Recovery Expander

Figure 50 shows a vapor compression cycle in which the expansion device is replaced by a power-recovering two-phase-flow turbine. The power recovered by the turbine is used to reduce the required compressor input work (Brasz 1995). Although not commonly applied on commercial centrifugals, power recovery during expansion reduces the enthalpy of the two-phase-flow mixture, thus increasing the refrigeration effect of this cycle.

More than one stage of flash cooling can be applied to compressors with more than two impellers. Liquid subcooling and interstage desuperheating can also be advantageously used. For more information on refrigeration cycles, see Chapter 1 of the 2005 *ASHRAE Handbook—Fundamentals*.

Angular Momentum

The momentum exchange, or energy transfer, between a centrifugal impeller and a flowing refrigerant is expressed by

$$W_i = u_i c_u / g_c \tag{15}$$

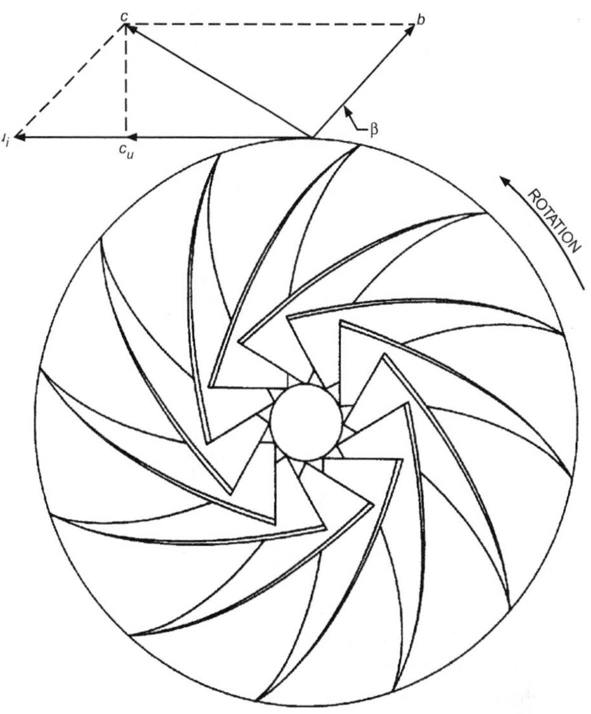

Fig. 51 Impeller Exit Velocity Diagram

where

W_i = impeller work input per unit mass of refrigerant, $\text{ft} \cdot \text{lb}_f/\text{lb}_m$
u_i = impeller blade tip speed, fps
c_u = tangential component of refrigerant velocity leaving impeller blades, fps
g_c = gravitational constant = 32.17 $\text{lb}_m \cdot \text{ft}/\text{lb}_f \cdot \text{s}^2$

These velocities are shown in Figure 51, where refrigerant flows out from between the impeller blades with relative velocity b and absolute velocity c. The relative velocity angle β is a few degrees less than the blade angle because of a phenomenon known as slip.

Equation (15) assumes that refrigerant enters the impeller without any tangential velocity component or swirl. This is generally the case at design flow conditions. If the incoming refrigerant were already swirling in the direction of rotation, the impeller's ability to impart angular momentum to the flow would be reduced. A subtractive term would then be required in the equation. Likewise, flow swirling in the direction opposite rotation would theoretically yield a positive effect on the angular momentum imparted.

Some of the work done by the impeller increases refrigerant pressure; the remainder only increases its kinetic energy. The ratio of pressure-producing work to total work is known as the impeller reaction. Because this varies from about 0.4 to about 0.7, an appreciable amount of kinetic energy leaves the impeller with magnitude $c^2/2g_c$.

To convert this kinetic energy into additional pressure, a diffuser is located after the impeller. Radial vaneless diffusers are most common, but vaned, volute, scroll, and conical diffusers are also used.

In a multistage compressor, flow leaving the first diffuser is guided to the inlet of the second impeller and so on through the machine, as shown in Figure 51. The total compression work input per unit mass of refrigerant is the sum of the individual stage inputs:

$$W = \sum W_i \tag{16}$$

provided that mass flow rate is constant throughout the compressor.

ISENTROPIC ANALYSIS

The static pressure resulting from a compressor's work input or, conversely, the amount of work required to produce a given pressure rise, depends on compressor efficiency and the thermodynamic properties of the refrigerant. For an adiabatic process, the work input required is minimal if compression is isentropic. Therefore, actual compression is often compared to an isentropic process, and performance thus evaluated is based on an isentropic analysis.

The reversible work required by an isentropic compression between states 1 and 2_s in Figure 48 is known as the **adiabatic work**, as measured by the enthalpy difference between the two points:

$$W_s = J(h_{2_s} - h_1) \tag{17}$$

where J = 778.16 ft·lb$_f$/Btu. Assuming negligible cooling occurs, the irreversible work done by the actual compressor is

$$W_s = J(h_2 - h_1) \tag{18}$$

Flash-cooled compressors cannot be analyzed by this procedure unless they are subdivided into uncooled segments with the cooling effects evaluated by other means. Compressors with side flows must also be subdivided. In Figure 49, the two compression processes must be analyzed individually.

Equation (18) also assumes a negligible difference in kinetic energies of refrigerant at states 1 and 2. If this is not the case, a kinetic energy term must be added to the equation. All thermodynamic properties discussed in this section are static properties as opposed to stagnation properties; the latter includes kinetic energy.

The ratio of isentropic work to actual work is the **adiabatic efficiency**:

$$\eta_s = \frac{h_{2_s} - h_1}{h_2 - h_1} \tag{19}$$

This varies from about 0.62 to about 0.83, depending on the application. Because of the thermodynamic properties of gases, a compressor's overall adiabatic efficiency does not completely indicate its individual stage performance. The same compressor produces different adiabatic results with different refrigerants and with the same refrigerant at different suction conditions.

In spite of its shortcomings, isentropic analysis has a definite advantage in that adiabatic work can be read directly from thermodynamic tables and charts similar to those presented in Chapter 20 of the 2005 *ASHRAE Handbook—Fundamentals*. Where these are unavailable for the particular gas or gas mixture, they can be accurately calculated and plotted using thermodynamic relationships.

POLYTROPIC ANALYSIS

Polytropic analysis is of benefit when multistage compressors are being evaluated. The computational effort is less, because the average stage efficiency can be applied to all stages.

The path equation for this reversible process is

$$\eta = v(dp/dh) \tag{20}$$

where η is the **polytropic efficiency** and v is the specific volume of the refrigerant. Reversible work done along the polytropic path is known as **polytropic work** and is given by

$$W_p = \int_{p_1}^{p_2} v\,dp \tag{21}$$

It follows from Equations (18), (20), and (21) that the polytropic efficiency is the ratio of reversible work to actual work:

$$\eta = \frac{W_p}{h_2 - h_1} \tag{22}$$

Equation (20) can be approximated by

$$\frac{p^m}{T} = \frac{p_1^m}{T_1} = \frac{p_2^m}{T_2} \tag{23}$$

$$pv^n = p_1 v_1^n = p_2 v_2^n \tag{24}$$

where

$$m = \frac{ZR}{c_p}\left(\frac{1}{\eta} + X\right) = \frac{(k - 1/k)(1/\eta + X)Y}{(1 + X)^2} \tag{25}$$

$$n = \frac{1}{Y - (ZR/c_p)(1/\eta + X)(1 + X)} = \frac{1 + X}{Y[(1/k)(1/\eta + X) - (1/\eta - 1)]} \tag{26}$$

and

$$X = \frac{T}{v}\left(\frac{\partial v}{\partial p}\right)_p - 1 \tag{27}$$

$$Y = -\frac{p}{v}\left(\frac{\partial v}{\partial p}\right)_T \tag{28}$$

$$Z = \frac{pv}{RT} \tag{29}$$

Also, R is the gas constant and k is the ratio of specific heats; all properties are at temperature T. These equations can be used to permit integration so that Equation (21) can be written as follows:

$$W_p = \frac{n}{(n-1)} p_1 v_1 \left[\left(\frac{p_2}{p_1}\right)^{(n-1)/n} - 1\right] \tag{30}$$

Further manipulation eliminates the exponent:

$$W_p = \left[\frac{p_2 v_2 - p_1 v_1}{\ln(p_2 v_2 / p_1 v_1)}\right] \ln\left(\frac{p_2}{p_1}\right) \tag{31}$$

For greater accuracy in handling gases with properties known to deviate substantially from those of a perfect gas, a more complex procedure is required. The accuracy with which Equations (23) and (24) represent Equation (20) depends on the constancy of m and n along the polytropic path. Because these exponents usually vary, mean values between states 1 and 2 should be used.

Compressibility functions X and Y have been generalized for gases in corresponding states by Schultz (1962) and their equivalents are listed by Edminster (1961). For usual conditions of refrigeration interest (i.e., for $p < 0.9p_c$, $T < 1.5T_c$, and $0.6 < Z$), these functions can be approximated by

$$X = 0.1846(8.36)^{1/Z} - 1.539 \tag{32}$$

$$Y = 0.074(6.65)^{1/Z} + 0.509 \tag{33}$$

Compressibility factor Z has been generalized by Edminster (1961) and Hougen et al. (1959), among others. Generalized corrections for specific heat at constant pressure c_p can also be found in these works.

Equations (23) and (24) make possible the integration of Equation (21):

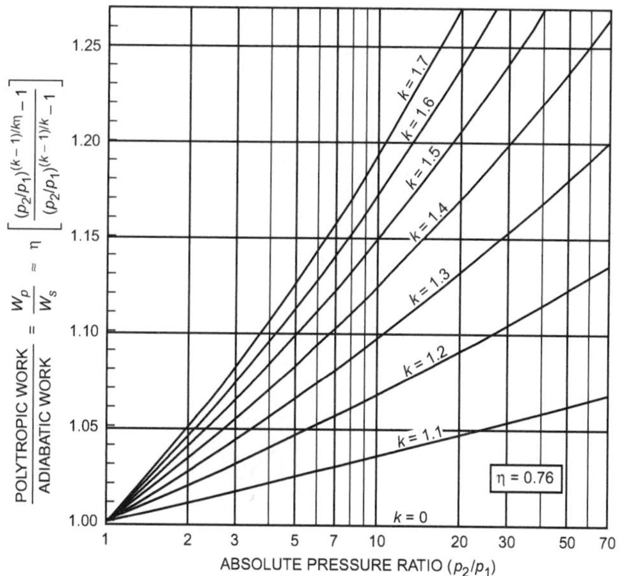

Fig. 52 Ratio of Polytropic to Adiabatic Work

$$W_p = f\left(\frac{n}{n-1}\right)p_1 v_1\left[\left(\frac{p_2}{p_1}\right)^{(n-1)/n} - 1\right] \tag{34}$$

In Equation (34), polytropic work factor f corrects for whatever error may result from the approximate nature of Equations (23) and (24). Because the value of f is between 1.00 and 1.02 in most refrigeration applications, it is generally neglected.

Once the polytropic work has been found, efficiency follows from Equation (22). Polytropic efficiencies range from about 0.70 to about 0.84, a typical value being 0.76.

The highest efficiencies are obtained with the largest compressors and densest refrigerants because of a Reynolds number effect discussed by Davis et al. (1951). A small number of stages is also advantageous because of parasitic loss associated with each stage.

Overall, polytropic work and efficiency are more consistent from one application to another because they represent an average stage aerodynamic performance.

Instead of using Equations (20) to (34), it is easier and often more desirable to determine the adiabatic work by isentropic analysis and then convert to polytropic work by

$$\frac{W_p}{W_s} \approx \eta\left[\frac{(p_2/p_1)^{(k-1)/k\eta} - 1}{(p_2/p_1)^{(k-1)/k} - 1}\right] \tag{35}$$

Equation (35) is strictly correct only for an ideal gas, but because it is a ratio involving comparable errors in both numerator and denominator, it is of more general utility. Equation (35) is plotted in Figure 52 for $\eta = 0.76$. To obtain maximum accuracy, the ratio of specific heats k must be a mean value for states 1, 2, and 2_s. If c_p is known, k can be determined by

$$k = \frac{1}{1 - (ZR/c_p)(1 + X)^2/Y} \tag{36}$$

Gas compression power is

$$P = wW \tag{37}$$

where w is mass flow. To obtain total shaft power, add the mechanical friction loss. Friction loss varies from less than 1% of gas power to more than 10%. A typical estimate is 3% for compressor friction losses.

Nondimensional Coefficients

Some nondimensional performance parameters used to describe centrifugal compressor performance are flow coefficient, polytropic work coefficient, Mach number, and specific speed.

Flow Coefficient. Desirable impeller diameters and rotational speeds are determined from blade tip velocity by a dimensionless flow coefficient Q/ND^3 in which Q is the volumetric flow rate. Practical values for this coefficient range from 0.02 to 0.35, with good performance between 0.11 and 0.21 and optimum results between 0.15 and 0.18. Impeller diameter D_i and rotational speed N follow from

$$Q/ND^3 = \pi Q_i/u_i D_i^2 = \pi^3 Q_i N^2/u_i^3 \tag{38}$$

where u_i is tip speed.

The maximum flow coefficient in multistage compressors is found in the first stage and the minimum in the last stage (unless large side loads are involved). For high-pressure ratios, special measures may be necessary to increase the last stage (Q/ND^3) to a practical level, as stated previously. Side loads are beneficial in this respect, but interstage flash cooling is not.

Polytropic Work Coefficient. Polytropic work and polytropic head can be used interchangeably. Because this chapter is concerned with polytropic work, that term will be used exclusively. Polytropic head is related to it by $H_p = W_p/g$.

Besides the power requirement, polytropic work also determines impeller blade tip speed and number of stages. For an individual stage, stage work is related to speed by

$$W_{pi} = \mu_i u_i^2 \S g_c \tag{39}$$

where μ_i is the stage work coefficient.

The overall polytropic work is the sum of the stage works:

$$W_p = \sum W_{pi} \tag{40}$$

and the overall work coefficient is

$$\mu = g_c W_p \sum u_i^2 \tag{41}$$

Values for μ (and μ_i) range from about 0.42 to about 0.74, with 0.55 representative for estimating purposes. Compressors designed for modest work coefficients have backward-curved impeller blades. These impellers tend to have greater part-load ranges and higher efficiencies than do radial-bladed designs.

Maximum tip speeds are limited by strength considerations to about 1400 fps. For cost and reliability, 980 fps is a more common limitation. On this basis, the maximum polytropic work capability of a typical stage is about 15,000 ft·lb$_f$/lb.

A greater restriction on stage work capability is often imposed by the impeller Mach number M_i. For adequate performance, M_i must be limited to about 1.8 for stages with impellers overhung from the ends of shafts and to about 1.5 for impellers with shafts passing through their inlets because flow passage geometries are moved out. For good performance, these values must be even lower. Such considerations limit maximum stage work to about $1.5a_i^2$, where a_i is the acoustic velocity at the stage inlet.

Specific Speed. This nondimensional index of optimum performance characteristic of geometrically similar stages is defined by

$$N_s = N\sqrt{Q_i}/W_{pi}^{0.75} = (1/\pi^3 \mu_i^{0.75})\sqrt{Q_i/ND_i^3} \tag{42}$$

The highest efficiencies are generally attained in stages with specific speeds between 600 and 850.

Table 4 Acoustic Velocity of Saturated Vapor, fps

Refrigerant	Evaporator Temperature, °F						
	−200	−150	−100	−50	0	50	100
11			386	409	430	446	456
12		388	413	433	446	448	437
13	390	417	434	436	417	369	
22		470	500	522	535	534	515
23	487	522	544	548	526	470	
113					362	377	388
114						383	384
123		337	361	383	402	416	424
124	329	357	382	403	417	422	416
125			405	420	421	404	361
134a		418	446	468	481	480	463
142b	391	422	451	475	492	501	498
152a		530	565	593	611	614	600
245fa				406	425	438	443
500	397	429	457	478	489	489	472
600	504	547	587	620	646	661	662
600a	509	551	588	620	642	651	643
717			1174	1242	1293	1323	1327
718						1365	1428

Source: Lemmon et al. (2002).

Mach Number

Two different Mach numbers are used. The flow Mach number M is the ratio of flow velocity c to acoustic velocity a at a particular point in the fluid stream:

$$M = c/a \tag{43}$$

where

$$a = v\sqrt{-g(\partial p/\partial v)_s} = \sqrt{n_s gpv} \tag{44}$$

Values of acoustic velocity for a number of saturated vapors at various temperatures are presented in Table 4.

The flow Mach number in a typical compressor varies from about 0.3 at the stage inlet and outlet to about 1.0 at the impeller exit. With increasing flow Mach number, the losses increase because of separation, secondary flow, and shock waves.

The impeller Mach number M_i, which is a pseudo Mach number, is the ratio of impeller tip speed to acoustic velocity a_i at the stage inlet:

$$M_i = u_i/a_i \tag{45}$$

Performance

From an applications standpoint, more useful parameters than μ and (Q/ND^3) are Ω and Θ (Sheets 1952):

$$\Omega = gW_p/a_i^2 = \mu\left(\sum u_i^2/a_i^2\right) \tag{46}$$

$$\Theta = Q_1/a_1D_1^2 = (M_1/\pi)(Q_1/ND_1{}^3) \tag{47}$$

They are as general as the customary test coefficients and produce performance maps like the one in Figure 53, with speed expressed in terms of first-stage impeller Mach number M_1.

A compressor user with a particular installation in mind may prefer more explicit curves, such as pressure ratio and power versus volumetric flow at constant rotational speed. Plots of this sort may require fixed suction conditions to be entirely accurate, especially if discharge pressure and power are plotted against mass flow or refrigeration effect.

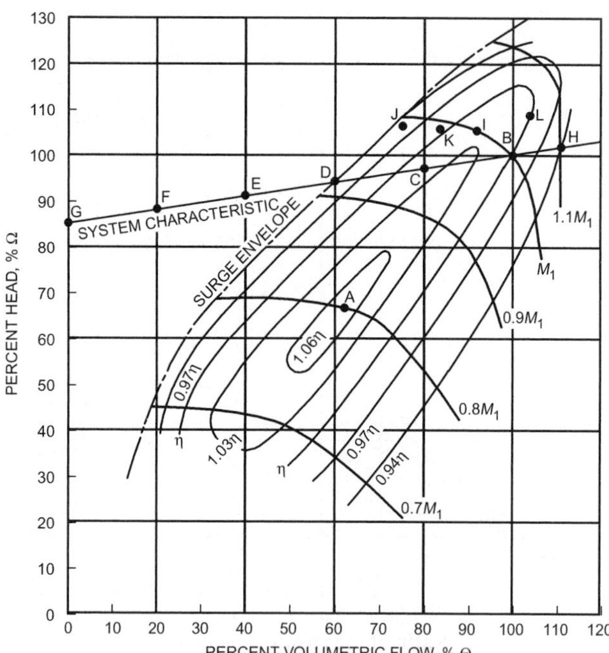

Fig. 53 Typical Compressor Performance Curves

A typical compressor performance map is shown in Figure 53, where percent of rated work is plotted with efficiency contours against percent of rated volumetric flow at various speeds. Point A is the design point at which the compressor operates with maximum efficiency. Point B is the selection or rating point at which the compressor is being applied to a particular system. From the application or user's point of view, Ω and Θ have their 100% values at point B.

To reduce first cost, refrigeration compressors are selected for pressure and capacity beyond their peak efficiency, as shown in Figure 53. The opposite selection would require a larger impeller and additional stages. Refrigerant acoustic velocity and the ability to operate at a high enough Mach number are also of concern. If the compressor shown in Figure 52 were of a multistage design, M_1 would be about 1.2; for a single-stage compressor, it would be about 1.5.

Another acoustical effect is seen on the right of the performance map, where increasing speed does not produce a corresponding increase in capacity. The maximum flow at M_1 and $1.1M_1$ approach a limit determined by the relative velocity of the refrigerant entering the first impeller. As this velocity approaches a sonic value, flow becomes choked and further increases become impossible. Another common term for this phenomenon is *stonewalling*; it represents the maximum capacity of an impeller.

Testing

When a centrifugal compressor is tested, overall μ and η versus $Q_1/ND_1{}^3$ at constant M_1 are plotted. They are useful because test results with one gas are sometimes converted to field performance with another. When side flows and cooling are involved, the overall work coefficient is found from Equations (40) and (41) by evaluating mixing and cooling effects between stages separately. The **overall efficiency** in such cases is

$$\eta = \frac{\sum w_i W_{pi}}{\sum w_i W_i} \tag{48}$$

Testing with a fluid other than the design refrigerant is a common practice known as **equivalent performance testing**. Its need arises from the impracticability of providing test facilities for the complete range of refrigerants and input power for which centrifugal compressors are designed. Equivalent testing is possible because a given compressor produces the same μ and η at the same (Q/ND^3) and M_i with any fluids whose volume ratios (v_1/v_2) and Reynolds numbers are the same.

Thermodynamic performance of a compressor can be evaluated according to either the stagnation or static properties of the refrigerant, and it is important to distinguish between these concepts. The **stagnation efficiency**, for example, may be higher than the **static efficiency**. The safest procedure is to use static properties and evaluate kinetic energy changes separately.

Surging

Part-load range is limited (on the left side of the performance map) by a **surge envelope**. Satisfactory compressor operation to the left of this line is prevented by unstable **surging** or **hunting**, in which refrigerant alternately flows backward and forward through the compressor, accompanied by increased noise, vibration, and heat. Prolonged operation under these conditions can damage the compressor.

Flow reverses during surging about once every 2 s. Small systems surge at higher frequencies and large systems at lower. Surging can be distinguished from other kinds of noise and vibration by the fact that its flow reversals alternately unload and load the driver. Motor current varies markedly during surging, and turbines alternately speed up and slow down.

Another kind of instability, **rotating stall**, may occur slightly to the right of the true surge envelope. This phenomenon forms rotating stall pockets or cells in the diffuser. It produces a roaring noise at a frequency determined by the number of cells formed and the impeller running speed. Driver load is steady during rotating stall, which is harmless to the compressor, but may vibrate components excessively.

System Balance and Capacity Control

In a centrifugal compressor, system balance is achieved through compressor capacity control. The method of capacity control selected depends on the intended refrigeration system characteristic for the application and the associated economics of the various control strategies. Capacity control methods affect compressor head capability and efficiency, which must be considered when selecting the appropriate method for an application.

Refrigeration and volumetric capacity are directly related to compressor speed, but the compressor's ability to produce pressure is a function of the square of a change in compressor speed. For example, operating at half compressor speed results in half the volumetric capacity and one-quarter the available pressure ratio. Variable-speed control requires a system characteristic of decreasing isentropic or polytropic work (pressure difference) with decreasing flow (capacity) to perform more efficiently. Reducing condensing pressure generally decreases isentropic or polytropic work. Impeller tip speed must remain constant if lift requirements do not change.

Methods of capacity control include speed variation, prerotation vanes, suction throttling, adjustable diffuser vanes, movable diffuser walls, impeller throttling sleeves, and combinations of these, such as prerotation vanes with variable speed. Each method has advantages and disadvantages in terms of performance, complexity, and cost that should be carefully considered before deciding on a capacity control strategy. The most common capacity control methods use speed variation and prerotation vanes.

Speed variation modulates capacity by adjusting the compressor drive speed to match the system characteristic. Speed variation is typically done by a variable-speed motor drive package, a turbine drive, or a generator-driven motor.

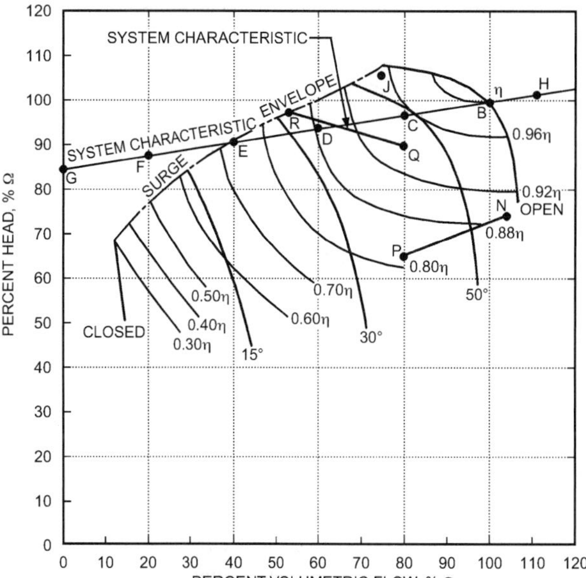

Fig. 54 Typical Compressor Performance with Various Prerotation Vane Settings

Figure 53 shows a centrifugal compressor performance map using speed variation to modulate capacity without prerotation vanes. In addition to the head and flow characteristic, it shows the speed and efficiency at which the compressor operates in that particular application. A refrigeration system characteristic for a typical brine cooling system curve has been overlaid on the map, passing through points B, C, D, E, F, G, and H. With increased speed, the compressor at point H produces more than its rated capacity; with decreased speeds at points C and D, it produces less. Because of surging, the compressor cannot be operated satisfactorily at points E, F, or G. The system can be operated at these capacities, however, by using hot-gas bypass. Volume flow at the compressor suction must be at least that for point D in Figure 53; this volume flow is reached by adding hot gas from the compressor discharge to the evaporator, or compressor suction piping. When hot-gas bypass is used, no further power reduction occurs as load decreases. The compressor is artificially loaded to stay out of the surge envelope. The increased volume caused by hot-gas recirculation performs no useful refrigeration.

Prerotation vanes (see Figure 47), also known as **inlet guide vanes**, modulate capacity by altering the direction of the fluid flow entering the impeller relative to the impeller blade leading edge. Setting the vanes to swirl flow in the direction of rotation produces a new compressor performance curve without any change in speed. Controlled positioning of the vanes can be done by pneumatically, electrically, or hydraulically.

Figure 54 shows a centrifugal compressor performance map at constant driver speed, using prerotation vanes to modulate capacity. Typical curves for five different vane positions are shown in Figure 54 for the compressor in Figure 53 at the constant speed M_1. In addition to the head and flow characteristic, it shows the prerotation vane position and efficiency at which the compressor operates in that particular application. With prerotation vanes wide open, the performance curve is identical to the M_1 curve in Figure 53. The other curves are different, as are the efficiency contours and the surge envelope. The same system characteristic has been superimposed on this performance map, as in Figure 53, to provide a comparison of these two modes of operation. In Figure 54, point E can be reached with prerotation vanes; point H cannot. Theoretically, turning the vanes against rotation would produce a performance line passing through point H, but sonic relative inlet velocities prevent

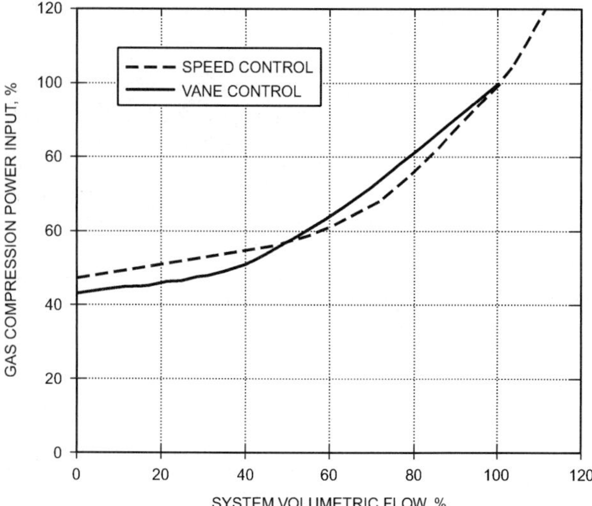

Fig. 55 Typical Part-Load Gas Compression Power Input for Speed and Vane Capacity Controls

this unless operating at low Mach numbers. Hot-gas bypass is still necessary at points F and G with prerotation vane control, but to a lesser extent than with variable speed.

Gas compression powers for both control methods are depicted in Figure 55. For the compressor and system assumed in this example, Figure 55 shows that speed control requires less gas compression power down to about 55% of rated capacity. Prerotation vane control requires less power below 55%. Complete analysis must also consider friction loss and driver efficiency. Typical losses in a variable-speed drive and motor combination increase the full-speed power consumption of the system by 2.0 to 3.5%.

In applications where pressure requirements do not vary significantly at part load, prerotation vanes alone are typically suitable. When pressure requirements vary at part load, variable speed can provide distinct operational and economic advantages. In practice, variable speed is typically used in combination with prerotation vanes. In either case, a thorough energy analysis should be performed for the specific application to ensure that the selected capacity control method and system balance most economically and efficiently meet the application requirements.

APPLICATION

Critical Speed

Centrifugal compressors are designed so that the first lateral critical speed is either well above or well below the operating speed. Operation at a speed between 0.8 and 1.1 times the first lateral speed is generally unacceptable from a reliability standpoint. The second lateral critical speed should be at least 25% above the operating speed of the machine.

Manufacturers have full responsibility for making sure critical speeds are not too close to operating speeds. Operating speed depends on the required flow of the application. Thus, the designer must ensure that the critical speed is sufficiently far away from the operating speed.

In applying open-drive machines, it is also necessary to consider torsional critical speed, which is a function of the designs of the compressor, drive turbine or motor, and coupling(s). In geared systems, gearbox design is also involved. Manufacturers of centrifugal compressors use computer programs to calculate torsional natural frequencies of the entire system, including the driver, coupling(s), and gears, if any. Responsibility for performing this calculation and ascertaining that the torsional natural frequencies are sufficiently far away from torsional exciting frequencies should be shared between the compressor manufacturer and the designer.

For engine drives, it may be desirable to use a fluid coupling to isolate the compressor (and gear set) from engine torque pulsations. Depending on compressor bearing design, there may be other speed ranges that should be avoided to prevent the nonsynchronous shaft vibration commonly called *oil whip* or *oil whirl*.

Vibration

Excessive vibration of a centrifugal compressor indicates malfunction, which may lead to failure. Periodic checking or continuous monitoring of the vibration spectrum at suitable locations is, therefore, useful in ascertaining the operational health of the machine. The relationship between internal displacements and stresses and external vibration is different for each compressor design. In a given design, this relationship also differs for the various causes of internal displacements and stresses, such as imbalance of rotating parts (either inherent or caused by deposits, erosion, corrosion, looseness, or thermal distortion), bearing instability, misalignment, distortion because of piping loads, broken motor rotor bars, or cracked impeller blades. It is, therefore, impossible to establish universal rules for the level of vibration considered excessive.

To establish meaningful criteria for a given machine or design, it is necessary to have baseline data indicative of proper operation. Significant increases of any frequency component of the vibration spectrum above the baseline then indicates a deterioration in the machine's operation; the frequency component for which this increase occurs is a good indication of the part of the machine deteriorating. Increases in the component at the fundamental running frequency, for instance, are usually because of deterioration of balance. Increases at approximately one-half the fundamental running frequency are caused by fluid-film bearing instability, and increases at twice the running frequency usually result from deterioration of alignment, particularly coupling alignment. Electrically induced vibration is typically at twice the fundamental frequency (e.g., 120 Hz for a 60 Hz line frequency).

As a general guide to establishing satisfactory vibration, a constant velocity criterion is sometimes used when the operating speed is between 600 and 60,000 rpm. Below 600 rpm, displacement is typically used as the criterion; above 60,000 rpm, acceleration is typically the criterion. In most cases, a velocity amplitude of 0.2 in/s is a reasonable criterion for vibration measured on the bearing housing.

Although measuring vibration amplitude on the bearing housing is convenient, the value of such measurements is limited because the stiffness of the bearing housing in typical centrifugal compressors is generally considerably larger than that of the oil film. Thus, vibration monitoring systems often use noncontacting sensors, which measure displacement of the shaft relative to the bearing housing, either instead of, or in addition to, monitoring bearing housing vibration (Mitchell 1977). Such sensors are also useful for monitoring axial displacement of the shaft relative to the thrust bearing.

In some applications, compressor vibration, which is perfectly acceptable from a reliability standpoint, can cause noise problems if the machine is not properly isolated from the building. Vibration tests of the installed machine under operating conditions give a base comparison for future reference.

Noise

Satisfactory application of centrifugal compressors requires careful consideration of noise control, especially if compressors are located near a noise-sensitive area of a building. The noise of centrifugal compressors is primarily of aerodynamic origin, principally gas pulsations associated with the impeller frequency and gas flow noise. Most predominant noise sources are of a sufficiently high frequency (above 1000 Hz) so that noise can be significantly reduced by carefully designed acoustical and structural isolation of the machine. Although the noise originates within the compressor proper, most is usually radiated from the discharge line and condenser shell. Equipment room noise can be reduced by up to 10 dB by covering the discharge line and condenser shell with acoustical insulation. In geared compressors, gear-mesh noise may also contribute to high-frequency noise; however, these frequencies are often above the audible range. This noise can be reduced by applying sound insulation material to the gear housing.

There are two important aspects in noise considerations for centrifugal compressors applications. In the equipment room, OSHA regulations specify employer responsibilities with regard to exposure to high sound levels. Increasing liability concerns are making designers more aware of compressor sound level considerations. Another important consideration is noise travel beyond the immediate equipment room.

Noise problems with centrifugal refrigeration equipment can occur in noise-sensitive parts of the building, such as a nearby office or conference room. The cost of controlling compressor noise transmission to such areas should be considered in building layout and weighed against cost factors for alternative locations of the equipment in the building.

If the equipment room is close to noise-sensitive building areas, it is usually cost-effective to have noise and vibration isolation designed by an experienced acoustical consultant, because small errors in design or execution can make the results unsatisfactory (Hoover 1960).

Blazier (1972) covers general information on typical noise levels near centrifugal refrigeration machines. Schaffer (2005) is another available source. Data on the noise output of a specific machine should be obtained from the manufacturer; the request should specify that measurements be in accordance with the current edition of ARI *Standard* 575.

Drivers

Centrifugal compressors are driven by almost any prime mover: a motor, turbine, or engine. Power requirements range from 33 to 12,000 hp. Sometimes the driver is coupled directly to the compressor; often, however, there is a gear set between them, usually because of low driver speed. Flexible couplings are required to accommodate the angular, axial, and lateral misalignments that may arise within a drive train. Additional information on prime movers may be found in Chapters 7 and 44.

Centrifugal refrigeration compressors are used in many special applications. These units use single-, two-, and three-stage compressors driven by open and hermetic motors. These designs have internal gears and direct drives, both of which are quieter, less costly, and more compact than external gearboxes. Internal gears are used when compressors operate at rotative speeds higher than two-pole motor synchronous speed. Chapter 42 discusses centrifugal water-chilling systems in greater detail.

A hermetic compressor absorbs motor heat because the motor is cooled by the refrigerant. An open motor is cooled by air in the equipment room; heat rejected by a hermetic motor must be considered in the refrigeration system design. Heat from an open compressor must usually be removed from the equipment room, generally by mechanical ventilation. Because they operate at a lower temperature, hermetic motors are usually smaller than open motors for a given power rating. If a motor burns out, a hermetic system will require thorough cleaning, whereas an open motor will not. When serviced or replaced, an open motor must be carefully aligned to ensure reliable performance.

Starting torque must be considered in selecting a driver, particularly a motor or single-shaft gas turbine. Compressor torque is roughly proportional both to speed squared and to refrigerant density. The latter is often much higher at start-up than at rated operating conditions. If prerotation vanes or suction throttling cannot provide sufficient torque reduction for starting, the standby pressure must be lowered by some auxiliary means.

In certain applications, a centrifugal compressor drives its prime mover backward at shutdown. The compressor is driven backward by refrigerant equalizing through the machine. The extent to which reverse rotation occurs depends on the kinetic energy of the drive train relative to the expansive energy in the system. Large installations with dense refrigerants are most susceptible to running backward, a modest amount of which is harmless if suitable provisions have been made. Reverse rotation can be minimized or eliminated by closing discharge valves, side-load valves, and prerotation vanes at shutdown and opening hot-gas bypass valves and liquid refrigerant drains.

Paralleling

Problems associated with paralleling turbine-driven centrifugal compressors at reduced load are illustrated by points I and J in Figure 53. These represent two identical compressors connected to common suction and discharge headers and driven by identical turbines. A single controller sends a common signal to both turbine governors so that both compressors should be operating at part-load point K (full load is at point L). The I machine runs 1% faster than its twin because of their respective governor adjustments, whereas the J compressor works against 1% more pressure difference because of the piping arrangement. The result is a 20% discrepancy between the two compressor loads.

One remedy is to readjust the turbine governors so that the J compressor runs 0.5% faster than the other unit. A more permanent solution, however, is to eliminate one of the common headers and to provide either separate evaporators or separate condensers. This increases the compression ratio of whichever machine has the greater capacity, decreases the compression ratio of the other, and shifts both toward point K.

The best solution is to install a flowmeter in the discharge line of each compressor and to use a master/slave control in which the original controller signals only one turbine, the master, while a second controller makes the slave unit match the master's discharge flow.

The problem of imbalance, associated with turbine-driven centrifugal compressors, is minimal in fixed-speed compressors with vane controls. A loading discrepancy comparable to this example would require a 25% difference in vane positions.

Paralleling centrifugal compressors offers advantages in redundancy and improved part-load operation. This arrangement provides the capability of efficiently unloading to a lower percentage of total load. When the unit requirement reduces to 50%, one compressor can carry the complete load and operates at a higher percent volumetric flow and efficiency than a single large compressor.

Means must be provided to prevent refrigeration flow through the idle compressor to prevent inadvertent flow of hot-gas bypass through the compressor. In addition, isolation valves should be provided on each compressor to allow removal or repair of either compressor.

Other Specialized Applications

Centrifugal compressors are used in petroleum refineries, marine refrigeration, and in the chemical industry, as covered in Chapters 31 and 37 of the 2006 *ASHRAE Handbook—Refrigeration*. Marine requirements are also detailed in ASHRAE *Standard* 26.

MECHANICAL DESIGN

Impellers

Impellers without covers, such as the one shown in Figure 51, are known as open or unshrouded designs. Those with covered blades (see Figure 47) are known as shrouded impellers. Open models must operate close to contoured stationary surfaces to avoid excessive leakage around their vanes. Shrouded designs must be fitted with labyrinth seals around their inlets for a similar purpose. Labyrinth seals behind each stage are required in multistage compressors.

Impellers must be shrunk, clamped, keyed, or bolted to their shafts to prevent loosening caused by thermal and centrifugal expansions. Generally, they are made of cast or brazed aluminum or of cast, brazed, riveted, or welded steel. Aluminum has a higher strength-weight ratio than steel, up to about 300°F, which permits higher rotating speeds with lighter rotors. Steel impellers retain their strength at higher temperatures and are more resistant to erosion. Lead-coated and stainless steels can be selected in corrosive applications.

Casings

Centrifugal compressor casings are about twice as large as their largest impellers, with suction and discharge connections sized for flow Mach numbers between 0.1 and 0.3. They are designed for the pressure requirements of ASHRAE *Standard* 15. A hydrostatic test pressure 50% greater than the maximum design working pressure is customary. If the casing is listed by a nationally recognized testing laboratory, a hydrostatic test pressure three times the working pressure is required.

Cast iron is the most common casing material, used for temperatures as low as –150°F and pressures as high as 300 psia. Nodular iron and cast or fabricated steel are also used for low temperatures, high pressures, high shock, and hazardous applications. Multistage casings are usually split horizontally, although unsplit barrel designs can also be used.

Lubrication

Like motors and gears, bearings and lubrication systems of centrifugal compressors can be internal or external, depending on whether they operate in refrigerant atmospheres. For simplicity, size, and cost, most air-conditioning and refrigeration compressors have internal bearings, as shown in Figure 47. In addition, they often have internal oil pumps, driven either by an internal motor or the compressor shaft; the latter arrangement is typically used with an auxiliary oil pump for starting and/or back-up service.

Most refrigerants are soluble in lubricating oils, the extent increasing with refrigerant pressure and decreasing with oil temperature. A compressor's oil may typically contain 20% refrigerant (by mass) during idle periods of high pressure and 5% during normal operation. Thus, refrigerant will come out of solution and foam the oil when such a compressor is started.

To prevent excessive foaming from cavitating the oil pump and starving the bearings, oil heaters minimize refrigerant solubility during idle periods. Standby oil temperatures between 130 and 150°F are required, depending on pressure. Once a compressor starts, its oil should be cooled to increase oil viscosity and maximize refrigerant retention during pulldown.

A sharp reduction in pressure before starting tends to supersaturate the oil. This produces more foaming at start-up than would the same pressure reduction after the compressor has started. Machines designed for a pressure ratio of 20 or more may reduce pressure so rapidly that excessive oil foaming cannot be avoided, except by maintaining a low standby pressure. Additional information refrigerant solubility in oil can be found in Chapters 1, 2, and 7 of the 2006 *ASHRAE Handbook—Refrigeration*.

External bearings avoid the complications of refrigerant-oil solubility at the expense of some oil recovery problems. Any nonhermetic compressor must have at least one shaft seal. Mechanical seals are commonly used in refrigeration machines because they are leak-tight during idle periods. These seals require some lubricating oil leakage when operating, however. Shaft seals leak oil out of compressors with internal bearings and into compressors with external bearings. Means for recovering seal oil leakage with minimal refrigerant loss must be provided in external bearing systems.

Bearings

Centrifugal compressors use hydrodynamic, rolling element, and magnetic bearings to support radial and thrust loads. Radial loads are a result of static weight of the rotating assembly, gear mesh separation forces (if so configured) and, to a much lesser extent, aerodynamic loads. Thrust loads are primarily the result of the pressure field behind an impeller exceeding the combined pressure and momentum forces acting on the impeller inlet. In multistage designs, each impeller adds to the total thrust, unless some are mounted in the opposite direction to oppose the thrust. In some designs, a balancing piston is used behind the last stage impeller to reduce the overall thrust loads (see Figure 47). To avoid axial rotor vibration, some net axial load must be retained on the rotating assembly. This can be achieved using preloaded bearings or careful consideration of the thrust characteristics over the machine operating range. Regardless of the bearing system chosen, the bearings' dynamic stiffness and dampening characteristics must be considered when determining compressor critical speeds, to ensure stable turbomachinery operation over the operating range.

Accessories

The minimum accessories required by a centrifugal compressor are an oil filter, oil cooler, and three safety controls. Oil filters are usually rated for 15 to 20 μm or less. They may be built into the compressor but are more often externally mounted. Dual filters can be provided for industrial applications so that one can be serviced while the other is operating.

Single or dual oil coolers usually use condenser water, chilled water, refrigerant, or air as their cooling medium. Water- and refrigerant-cooled models may be built into the compressor, and refrigerant-cooled oil coolers may be built into a system heat exchanger. Many oil coolers are mounted externally for maximum serviceability.

Safety controls, with or without anticipatory alarms, must include a low-oil-pressure cutout, a high-oil-temperature switch, and high-discharge- and low-suction-pressure (or temperature) cutouts. A high-motor-temperature device is necessary in a hermetic compressor. Other common safety controls and alarms sense discharge temperature, bearing temperature, oil filter pressure differential, oil level, low oil temperature, shaft seal pressure, balancing piston pressure, surging, vibration, and thrust bearing wear.

Pressure gages and thermometers are useful indicators of critical items monitored by the controls. Suction, discharge, and oil pressure gages are the most important, followed by suction, discharge, and oil thermometers. Suction and discharge instruments are often attached to components rather than to the compressor itself, but they should be provided. Interstage pressures and temperatures can also be helpful, either on the compressor or on the system. Electronic components may be used for all safety and operating controls. Electronic sensors and displays may be used for pressure and temperature monitoring.

OPERATION AND MAINTENANCE

Refer to the compressor manufacturer's operating and maintenance instructions for recommended procedures. A planned maintenance program, as described in Chapter 38 of the 2007 *ASHRAE Handbook—HVAC Applications*, should be established. As part of this program, operating documentation should be kept, tabulating pertinent unit temperatures, pressures, flows, fluid levels, electrical data, and refrigerant added. ASHRAE *Guideline* 4 has further information on documentation. These can be compared periodically with values recorded for the new unit. Gradual changes in data can signify the need for routine maintenance; abrupt changes indicate system or component difficulty. A successful maintenance program requires the operating engineer to recognize and identify the reason for these data trends. In addition, by knowing the component parts and their operational interaction, the designer can use these symptoms to prescribe proper maintenance procedures.

The following items deserve attention in establishing a planned compressor maintenance program:

- A tight system is important. Leaks on compressors operating at subatmospheric pressures allow noncondensables and moisture to enter the system, adversely affecting operation and component life. Leakage in higher-pressure systems allows oil and refrigerant loss. ASHRAE *Guideline* 3 can be used as a guide to ensure system tightness. Vacuum leaks can be detected by a change in operational pressures not supported by corresponding refrigerant temperature data or the frequency of purge unit operation. Pressure leaks are characterized by symptoms related to refrigerant charge loss such as low suction pressures and high suction superheat. Such leaks should be located and fixed to prevent component deterioration.
- Compliance with the manufacturer's recommended oil filter inspection and replacement schedule allows visual indication of the compressor lubrication system condition. Repetitive clogging of filters can mean system contamination. Periodic oil sample analysis can monitor acid, moisture, and particulate levels to assist in problem detection.
- Operating and safety controls should be checked periodically and calibrated to ensure reliability.
- Electrical resistance of hermetic motor windings between phases and to ground should be checked (megged) regularly, following the manufacturer's outlined procedure. This helps detect any internal electrical insulation deterioration or the formation of electrical leakage paths before a failure occurs.
- Water-cooled oil coolers should be systematically cleaned on the water side (depending on water conditions), and operation of any automatic water control valves should be checked.
- For some compressors, periodic maintenance (e.g., manual lubrication of couplings and other external components, shaft seal replacement) is required. Prime movers and their associated auxiliaries all require routine maintenance. Such items should be made part of the planned compressor maintenance schedule.
- Periodic vibration analysis can locate and identify trouble (e.g., unbalance, misalignment, bent shaft, worn or defective bearings, bad gears, mechanical looseness, electrical unbalance). Without disassembling the machine, such trouble can be found early, before machinery failure or damage can occur. Dynamic balancing can restore rotating equipment to its original efficient, quiet operating mode. Such testing can help avoid costly emergency repairs, pinpoint irregularities before major problems arise, and increase the useful life of components.
- The necessary steps for preparing the unit for prolonged shutdown (i.e., winter) and specified instruction for starting after this standby period, should both be part of the program. With compressors that have internal lubrication systems, provisions should be made to have their oil heaters energized continuously throughout this period or to have their oil charges replaced prior to putting them back into operation.

SYMBOLS

a = acoustic velocity at a particular point
a_i = acoustic velocity at impeller inlet
c = flow velocity
C_p = specific heat at constant pressure
D = impeller diameter
f = polytropic work factor
g_c = gravitation constant
h = enthalpy at a specific state point
J = mechanical equivalent of heat, 778.1 $\text{ft}\cdot\text{lb}_f/\text{Btu}$
k = ratio of specific heats, Equation (36)
m = exponent, Equation (25)
M = flow Mach number, Equation (43)
M_i = flow Mach number impeller, Equation (45)
n = exponent, Equation (26)
N = rotational speed
N_s = specific speed
P = gas compression power
p = pressure at a specific state point
p^m = pressure raised to power of m
Q = volumetric flow rate
Q/ND^3 = dimensionless flow coefficient
Q_i = volumetric flow rate in impeller
R = gas constant
T = absolute temperature at specific state point
u_i = impeller tip speed
v = specific volume
V^n = volume raised to power of n
W = total work input
w = mass flow
W_i = impeller work input
W_p = polytropic work input
W_{pi} = polytropic work by impeller
W_s = adiabatic work input
X = compressibility function, Equation (27)
Y = compressibility function, Equation (28)
Z = compressibility function, Equation (29)
η = polytropic efficiency
η_s = adiabatic efficiency
Θ = flow parameter, Equation (47)
μ = overall work coefficient
Ω = head parameter, Equation (46)

REFERENCES

ARI. 1997. Standard for positive displacement condensing units. *Standard* 520-1997. Air-Conditioning and Refrigeration Institute, Arlington, VA.

ARI. 1999. Standard for positive displacement refrigerant compressors and compressor units. ANSI/ARI *Standard* 540-1999. Air-Conditioning and Refrigeration Institute, Arlington, VA.

ARI. 1994. Method of measuring machinery sound within an equipment space. *Standard* 575-94. Air-Conditioning and Refrigeration Institute, Arlington, VA.

ASHRAE. 1996. Reducing emission of halogenated refrigerants in refrigeration and air-conditioning equipment and systems. *Guideline* 3-1996.

ASHRAE. 1993. Preparation of operating and maintenance documentation for building systems. *Guideline* 4-1993.

ASHRAE. 2001. Safety standard for mechanical refrigeration. ANSI/ASHRAE *Standard* 15-2001.

ASHRAE. 1993. Methods of testing for rating positive displacement refrigerant compressors and condensing units. ANSI/ASHRAE *Standard* 23-1993.

ASHRAE. 1996. Mechanical refrigeration and air-conditioning installations aboard ship. ANSI/ASHRAE *Standard* 26-1996.

Blazier, W.E., Jr. 1972. Chiller noise: Its impact on building design. *ASHRAE Transactions* 78(1):268.

Brasz, J.J. 1995. Improving the refrigeration cycle with turbo-expanders. *Proceedings of the 19th International Congress of Refrigeration.*

Bush, J. and J. Elson. 1988. Scroll compressor design criteria for residential air conditioning and heat pump applications. *Proceedings of the 1988 International Compressor Engineering Conference*, vol. 1, pp. 83-97. Office of Publications, Purdue University, West Lafayette, IN.

Caillat, J., R. Weatherston, and J. Bush. 1988. Scroll-type machine with axially compliant mounting. U.S. *Patent* 4,767,292.

CEN. 1999. Refrigerant compressors—Rating conditions, tolerances and presentation of manufacturer's performance data. EN *Standard* 12900:1999. European Committee for Standardization, Brussels.

Davis, H., H. Kottas, and A.M.G. Moody. 1951. The influence of Reynolds number on the performance of turbomachinery. *ASME Transactions* (July):499.

Edminster, W.C. 1961. *Applied hydrocarbon thermodynamics*, 22 and 52. Gulf Publishing, Houston.

Elson, J., G. Hundy, and K. Monnier. 1990. Scroll compressor design and application characteristics for air conditioning, heat pump, and refrigeration applications. *Proceedings of the Institute of Refrigeration* (London) 2.1-2.10 (November).

Hoover, R.M. 1960. Noise levels due to a centrifugal compressor installed in an office building penthouse. *Noise Control* (6):136.

Hougen, O.A., K.M. Watson, and R.A. Ragatz. 1959. *Chemical process principles, Part II—Thermodynamics*. John Wiley & Sons, New York.

Lemmon, E.W., M.O. McLinden, and M.L. Huber. 2002. *NIST reference fluid thermodynamic and transport properties*—REFPROP, version 7.0. Standard Reference Data Program. National Institute of Standards and Technology, Gaithersburg, MD.

McCullough, J. 1975. Positive fluid displacement apparatus. U.S. *Patent* 3,924,977.

McCullough, J. and R. Shaffer. 1976. Axial compliance means with radial sealing for scroll type apparatus. U.S. *Patent* 3,994,636.

Mitchell, J.S. 1977. Monitoring machinery health. *Power* 121(I):46; (II):87; (III):38.

Morishita, E., Y. Kitora, T. Suganami, S. Yamamoto, and M. Nishida. 1988. Rotating scroll vacuum pump. *Proceedings of the 1988 International Compressor Engineering Conference* (July). Office of Publications, Purdue University, West Lafayette, IN.

Purvis, E. 1987. Scroll compressor technology. Heat Pump Conference, New Orleans.

Sauls, J. 1983. Involute and laminated tip seal of labyrinth type for use in a scroll machine. U.S. *Patent* 4,411,605.

Schaffer, M.E. 2005. *A practical guide to noise and vibration control for HVAC systems*, 2nd ed. ASHRAE.

Schultz, J.M. 1962. The polytropic analysis of centrifugal compressors. *ASME Transactions* (January):69 and (April):222.

Sheets, H.E. 1952. Nondimensional compressor performance for a range of Mach numbers and molecular weights. *ASME Transactions* (January):93.

Tojo, K., T. Hosoda, M. Ikegawa, and M. Shiibayashi. 1982. Scroll compressor provided with means for pressing an orbiting scroll member against a stationary scroll member and self-cooling means. U.S. *Patent* 4,365,941.

CHAPTER 38

CONDENSERS

THE CONDENSER in a refrigeration system is a heat exchanger that rejects all the heat from the system. This heat consists of heat absorbed by the evaporator plus the heat from the energy input to the compressor. The compressor discharges hot, high-pressure refrigerant gas into the condenser, which rejects heat from the gas to some cooler medium. Thus, the cool refrigerant condenses back to the liquid state and drains from the condenser to continue in the refrigeration cycle.

Condensers may be classified by their cooling medium as (1) water-cooled, (2) air-cooled, (3) evaporative (air- and water-cooled), and (4) refrigerant-cooled (cascade systems). The first three types are discussed in this chapter; see Chapter 39 in the 2006 *ASHRAE Handbook—Refrigeration* for a discussion of cascade-cooled condensers.

WATER-COOLED CONDENSERS

HEAT REMOVAL

The heat rejection rate in a condenser for each unit of heat removed by the evaporator may be estimated from the graph in Figure 1. The theoretical values shown are based on Refrigerant 22 with 10°F suction superheat, 10°F liquid subcooling, and 80% compressor efficiency. Actually, the heat removed is slightly higher or lower than these values, depending on compressor efficiency. Usually, the heat rejection requirement can be accurately determined by adding the known evaporator load and the heat equivalent of the actual power required for compression (obtained from the compressor manufacturer's catalog). (Note that heat from the compressor is reduced by any independent heat rejection processes such as oil cooling, motor cooling, etc.)

The volumetric flow rate of condensing water required may be calculated as follows:

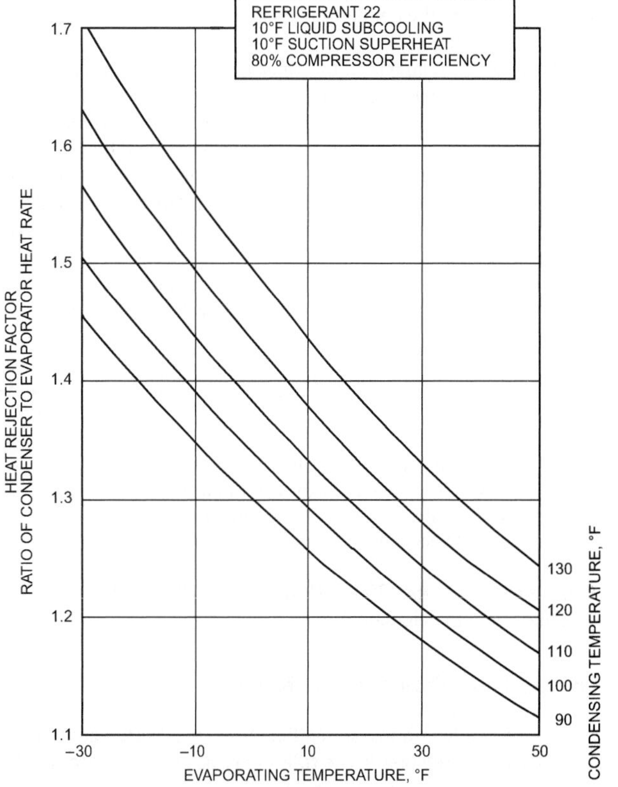

Fig. 1 Heat Removed in Condenser

$$Q = \frac{q_o}{\rho c_p (t_2 - t_1)} \quad (1)$$

where

Q = volumetric flow rate of water, ft³/h (multiply ft³/h by 0.125 to obtain gpm)

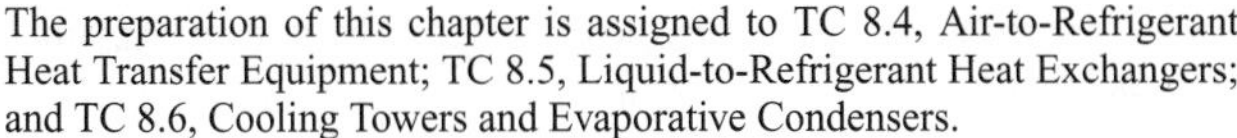

The preparation of this chapter is assigned to TC 8.4, Air-to-Refrigerant Heat Transfer Equipment; TC 8.5, Liquid-to-Refrigerant Heat Exchangers; and TC 8.6, Cooling Towers and Evaporative Condensers.

q_o = heat rejection rate, Btu/h
ρ = density of water, lb/ft^3
t_1 = temperature of water entering condenser, °F
t_2 = temperature of water leaving condenser, °F
c_p = specific heat of water at constant pressure, Btu/lb·°F

Example 1. Estimate volumetric flow rate of condensing water required for the condenser of an R-22 water-cooled unit operating at a condensing temperature of 105°F, an evaporating temperature of 40°F, 10°F liquid subcooling, and 10°F suction superheat. Water enters the condenser at 86°F and leaves at 95°F. The refrigeration load is 100 tons.

Solution: From Figure 1, the heat rejection factor for these conditions is about 1.19.

q_o = 100 × 1.19 = 119 tons
ρ = 62.1 lb/ft^3 at 90.5°F
c_p = 1.0 Btu/(lb·°F)

From Equation (1):

$$Q = \frac{1496 \times 119}{62.1 \times 1.0(95 - 86)} = 319 \text{ gpm}$$

Note: The value 1496 is a unit conversion factor.

HEAT TRANSFER

A water-cooled condenser transfers heat by sensible cooling in the gas desuperheating and condensate subcooling stages and by transfer of latent heat in the condensing stage. Condensing is by far the dominant process in normal refrigeration applications, accounting for 83% of the heat rejection in Example 1. Because the tube wall temperature is normally lower than the condensing temperature at all locations in the condenser, condensation takes place throughout the condenser.

The effect of changes in the entering gas superheat is typically insignificant because of an inverse proportional relationship between temperature difference and heat transfer coefficient. As a result, an average overall heat transfer coefficient and the mean temperature difference (calculated from the condensing temperature corresponding to the saturated condensing pressure and the entering and leaving water temperatures) give reasonably accurate predictions of performance.

Subcooling affects the average overall heat transfer coefficient when tubes are submerged in liquid. The heat rejection rate is then determined as

$$q = UA\Delta t_m \tag{2}$$

where

q = total heat transfer rate, Btu/h
U = overall heat transfer coefficient, Btu/h·ft^2·°F
A = heat transfer surface area associated with U, ft^2
Δt_m = mean temperature difference, °F

Chapter 3 of the 2005 *ASHRAE Handbook—Fundamentals* describes how to calculate Δt_m.

Overall Heat Transfer Coefficient

The overall heat transfer coefficient U_o in a **water-cooled condenser with water inside the tubes** may be computed from calculated or test-derived heat transfer coefficients of the water and refrigerant sides, from physical measurements of the condenser tubes, and from a fouling factor on the water side, using the following equation:

$$U_o = \frac{1}{\left(\frac{A_o}{A_i}\frac{1}{h_w}\right) + \left(\frac{A_o}{A_i}r_{fw}\right) + \left(\frac{A_o}{A_m}\frac{t}{k}\right) + \left(\frac{1}{h_r\phi_s}\right)} \tag{3}$$

where

U_o = overall heat transfer coefficient, based on external surface and mean temperature difference between external and internal fluids, Btu/h·ft^2·°F
A_o/A_i = ratio of external to internal surface area
h_w = internal or water-side film coefficient, Btu/h·ft^2·°F
r_{fw} = fouling resistance on water side, ft^2·h·°F/Btu
t = thickness of tube wall, ft
k = thermal conductivity of tube material, Btu/h·ft·°F
A_o/A_m = ratio of external to mean heat transfer surface areas of metal wall
h_r = external or refrigerant-side coefficient, Btu/h·ft^2·°F
ϕ_s = surface fin efficiency (100% for bare tubes)

For **tube-in-tube condensers or other condensers where refrigerant flows inside the tubes**, the equation for U_o, in terms of water-side surface, becomes

$$U_o = \frac{1}{\left(\frac{A_o}{A_i}\frac{1}{h_r}\right) + r_{fw} + \left(\frac{t}{k}\right) + \left(\frac{1}{h_w}\right)} \tag{4}$$

where

h_r = internal or refrigerant-side coefficient, Btu/h·ft^2·°F
h_w = external or water-side coefficient, Btu/h·ft^2·°F

For **brazed or plate and frame condensers** $A_0 = A_i$; therefore the equation for U_o is

$$U_o = \frac{1}{(1/h_r) + r_{fw} + (t/k) + (1/h_w)} \tag{5}$$

where t is plate thickness.

Water-Side Film Coefficient

Values of the water-side film coefficient h_w may be calculated from equations in Chapter 3 of the 2005 *ASHRAE Handbook—Fundamentals*. For turbulent flow, at Reynolds numbers exceeding 10,000 in horizontal tubes and using average water temperatures, the general equation (McAdams 1954) is

$$\frac{h_w D}{k} = 0.023\left(\frac{DG}{\mu}\right)^{0.8}\left(\frac{c_p\mu}{k}\right)^{0.4} \tag{6}$$

where

D = inside tube diameter, ft
k = thermal conductivity of water, Btu/h·ft·°F
G = mass velocity of water, lb/h·ft^2
μ = viscosity of water, lb/ft·h
c_p = specific heat of water at constant pressure, Btu/lb·°F

The constant 0.023 in Equation (6) reflects plain inner diameter (ID) tubes. Bergles (1995) and Pate et al. (1991) discuss numerous water-side enhancement methods that increase the value of this constant.

Because of its strong influence on the value of h_w, a high water velocity should generally be maintained without initiating erosion or excessive pressure drop. Typical maximum velocities from 6 to 10 fps are common with clean water. Experiments by Sturley (1975) at velocities up to approximately 26 fps showed no damage to copper tubes after long operation. Water quality is the key factor affecting erosion potential (Ayub and Jones 1987). A minimum velocity of 3 fps is good practice when water quality is such that noticeable fouling or corrosion could result. With clean water, the velocity may be lower if it must be conserved or has a low temperature. In some cases, the minimum flow may be determined by a lower Reynolds number limit.

For brazed or plate and frame condensers, the equation is similar to Equation (6). However, the diameter D is replaced by H, which is the characteristic spacing between plates.

Refrigerant-Side Film Coefficient

Factors influencing the value of the refrigerant-side film coefficient h_r are

- Type of refrigerant being condensed
- Geometry of condensing surface [plain tube outer diameter (OD); finned-tube fin spacing, height, and cross-sectional profile; and plate geometry]
- Condensing temperature
- Condensing rate in terms of mass velocity or rate of heat transfer
- Arrangement of tubes in bundle and location of inlet and outlet connections
- Vapor distribution and rate of flow
- Condensate drainage
- Liquid subcooling

Values of refrigerant-side coefficients may be estimated from correlations in Chapter 4 of the 2005 *ASHRAE Handbook—Fundamentals*. Information on the effects of refrigerant type, condensing temperature, and loading (temperature drop across the condensate film) on the condensing film coefficient is in the section on Condensing in the same chapter. Actual values of h_r for a given physical condenser design can be determined from test data using a Wilson plot (Briggs and Young 1969; McAdams 1954).

The type of condensing surface has a considerable effect on the condensing coefficient. Most halocarbon refrigerant condensers use finned tubes where the fins are integral with the tube. Water velocities normally used are large enough for the resulting high water-side film coefficient to justify using an extended external surface to balance the heat transfer resistances of the two surfaces. Pearson and Withers (1969) compared refrigerant condensing performance of integral finned tubes with different fin spacing. Some other refrigerant-side enhancements are described by Pate et al. (1991) and Webb (1984a). The effect of fin shape on the condensing coefficient is addressed by Kedzierski and Webb (1990). Ghaderi et al. (1995) reviewed in-tube condensation heat transfer correlations for smooth and augmented tubes.

In the case of brazed-plate or plate-and-frame condensers inlet nozzle size, chevron angle, pitch, and depth of the nozzles are important design parameters. For a trouble-free operation, refrigerant should flow counter to the water flow. Little specific design information is available; however, film thickness is certainly a factor in plate condenser design because of the falling film nature along the vertical surface. Kedzierski (1997) showed that placing a brazed condenser in a horizontal position improved U_o by 17 to 30% because of the shorter film distance.

Huber et al. (1994a) determined condensing coefficients for R-134a, R-12, and R-11 condensing on conventional finned tubes with a fin spacing of 26 fins per inch (fpi) and a commercially available tube specifically developed for condensing halocarbon refrigerants (Huber et al. 1994b). This tube has a sawtooth-shaped outer enhancement. The data indicated that the condensing coefficients for the sawtoothed tube were approximately three times higher than for the conventional finned tube exchanger and two times higher for R-123.

Further, Huber et al. (1994c) found that for tubes with 26 fpi the R-134a condensing coefficients are 20% larger than those for R-12 at a given heat flux. However, on the sawtoothed tube, the R-134a condensing coefficients are nearly two times larger than those for R-12 at the same heat flux. The R-123 condensing coefficients were 10 to 30% larger than the R-11 coefficients at a given heat flux, with the largest differences occurring at the lowest heat fluxes tested. The differences in magnitude between the R-123 and R-11 condensing coefficients were the same for both the 26 fpi tube and the sawtoothed tube.

Physical aspects of a given condenser design (e.g., tube spacing and orientation, shell-side baffle arrangement, orientation of multiple water-pass arrangements, refrigerant connection locations, and number of tubes high in the bundle) affect the refrigerant-side coefficient by influencing vapor distribution and flow through the tube bundle and condensate drainage from the bundle. Butterworth (1977) reviewed correlations accounting for these variables in predicting the heat transfer coefficient for shell-side condensation. These effects are also surveyed by Webb (1984b). Kistler et al. (1976) developed analytical procedures for design within these parameters.

As refrigerant condenses on the tubes, it falls on the tubes in lower rows. Because of the added resistance of this liquid film, the effective film coefficient for lower rows should be lower than that for upper rows. Therefore, the average overall refrigerant film coefficient should decrease as the number of tube rows increases. Webb and Murawski (1990) present row effect data for five tube geometries. However, the additional compensating effects of added film turbulence and direct contact condensation on the subcooled liquid film make actual row effect uncertain.

Huber et al. (1994c, 1994d) determined that the row effect on finned tubes is nearly negligible when condensing low-surface-tension refrigerants such as R-134a. However, the finned-tube film coefficient for higher-surface-tension refrigerants such as R-123 can drop by as much as 20% in lower bundle rows. The row effect for the sawtoothed condensing tube is quite large for both R-134a and R-123, as the film coefficient drops by nearly 80% from top to bottom in a 30-row bundle.

Honda et al. (1994, 1995) demonstrated that row effects caused by condensate drainage and inundation are less for staggered tube bundles than for in-line tubes. In addition, performance improvements as high as 85% were reported for optimized two-dimensional fin profiles compared to conventional fin profiles. The optimized fin profiles differed from the sawtoothed profile tested by Huber et al. (1994b) primarily in that external fins were not notched. This observation coupled, with the differences in row effects between sawtoothed and conventional fin profiles reported by Huber et al. (1994c, 1994d), suggested that there is opportunity for further development and commercialization of two-dimensional fin profiles for large shell-and-tube heat exchanger applications.

Liquid refrigerant may be subcooled by raising the condensate level to submerge a desired number of tubes. The refrigerant film coefficient associated with the submerged tubes is less than the condensing coefficient. If the refrigerant film coefficient in Equation (3) is an average based on all tubes in the condenser, its value decreases as a greater portion of the tubes is submerged.

Tube-Wall Resistance

Most refrigeration condensers, except those using ammonia, have relatively thin-walled copper tubes. Where these are used, the temperature drop or gradient across the tube wall is not significant. If the tube metal has a high thermal resistance, as does 70/30 cupronickel, a considerable temperature drop occurs or, conversely, an increase in the mean temperature difference Δt_m or the surface area is required to transfer the same amount of heat as copper. Although the tube-wall resistance t/k in Equations (3) and (4) is an approximation, as long as the wall thickness is not more than 14% of the tube diameter, the error will be less than 1%. To improve the accuracy of the wall resistance calculation for heavy tube walls or low-conductivity material, see Chapter 3 of the 2005 *ASHRAE Handbook—Fundamentals*.

Surface Efficiency

For a finned tube, a temperature gradient exists from the root of a fin to its tip because of the fin material's thermal resistance. The surface efficiency ϕ_s can be calculated from the fin efficiency, which accounts for this effect. For tubes with low-conductivity material, high fins, or high values of fin pitch, the fin efficiency becomes increasingly significant. Wolverine Tube (1984) and Young and Ward (1957) describe methods to evaluate these effects.

Fouling Factor

Manufacturers' ratings are based on clean equipment with an allowance for possible water-side fouling. The fouling factor r_{fw} is a thermal resistance referenced to the water-side area of the heat transfer surface. Thus, the temperature penalty imposed on the condenser equals the water-side heat flux multiplied by the fouling factor. Increased fouling increases overall heat transfer resistance because of the parameter $(A_o/A_i)r_{fw}$ in Equation (3). Fouling increases the Δt_m required to obtain the same capacity (with a corresponding increase in condenser pressure and system power) or lowers capacity.

Allowance for a given fouling factor has a greater influence on equipment selection than simply increasing the overall resistance (Starner 1976). Increasing the surface area lowers the water velocity. Consequently, the increase in heat transfer surface required for the same performance derives from both fouling resistance and the additional resistance that results from lower water velocity.

For a given tube surface, load, and water temperature range, the tube length can be optimized to give a desired condensing temperature and water-side pressure drop. The solid curves in Figure 2 show the effect on water velocity, tube length, and overall surface required due to increased fouling. A fouling factor of 0.00072 ft²·h·°F/Btu doubles the required surface area compared to that with no fouling allowance.

A worse case occurs when an oversized condenser must be selected to meet increased fouling requirements but without the flexibility of increasing tube length. As shown by the dashed lines in Figure 2, water velocity decreases more rapidly as the total surface increases to meet the required performance. Here, the required surface area doubles with a fouling factor of only 0.00049 ft²·h·°F/Btu. If the application can afford more pumping power, water flow may be increased to obtain a higher velocity, which increases the water film heat transfer coefficient. This factor, plus the lower leaving water temperature, reduces the condensing temperature.

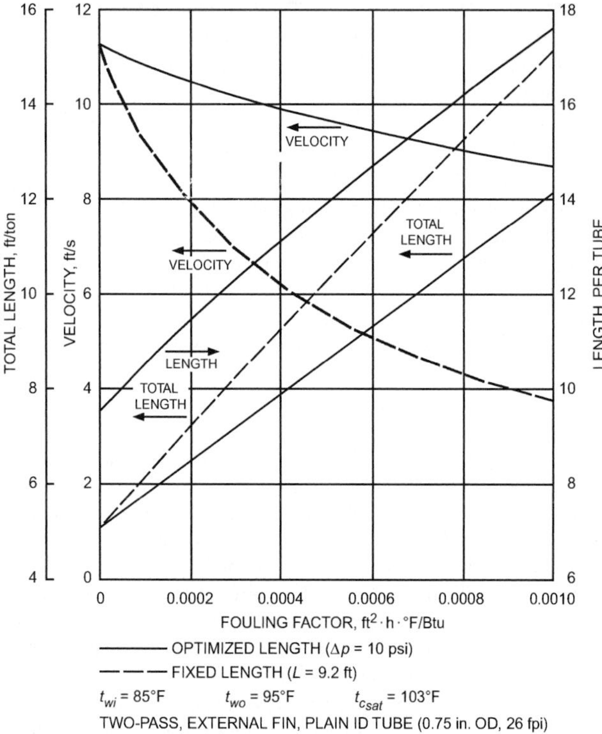

Fig. 2 Effect of Fouling on Condenser

Fouling is a major unresolved problem in heat exchanger design (Taborek et al. 1972). The major uncertainty is which fouling factor to choose for a given application or water condition to obtain expected performance from the condenser: too low a fouling factor wastes compressor power, whereas too high a factor wastes heat exchanger material.

Fouling may result from sediment, biological growth, or corrosive products. Scale results from deposition of chemicals from the cooling water on the warmer surface of the condenser tube. Chapter 48 of the 2007 *ASHRAE Handbook—HVAC Applications* discusses water chemistry and water treatment factors that are important in controlling corrosion and scale in condenser cooling water.

Tables of fouling factors are available; however, in many cases, the values are greater than necessary (TEMA 1999). Extensive research generally found that fouling resistance reaches an asymptotic value with time (Suitor et al. 1976). Much fouling research is based on surface temperatures that are considerably higher than those found in air-conditioning and refrigeration condensers. Coates and Knudsen (1980) and Lee and Knudsen (1979) found that, in the absence of suspended solids or biological fouling, long-term fouling of condenser tubes does not exceed 0.0002 ft²·h·°F/Btu, and short-term fouling does not exceed 0.0001 ft²·h·°F/Btu (ASHRAE 1982). These studies have resulted in a standard industry fouling value of 0.00025 ft²·h·°F/Btu for condenser ratings (ARI *Guideline* E-1997). Periodic cleaning of condenser tubes (mechanically or chemically) usually maintains satisfactory performance, except in severe environments. ARI *Standard* 450 for water-cooled condensers should be consulted when reviewing manufacturers' ratings. This standard describes methods to correct ratings for different values of fouling.

WATER PRESSURE DROP

Water (or other fluid) pressure drop is important for designing or selecting condensers. Where a cooling tower cools condensing water, water pressure drop through the condenser is generally limited to about 10 psi. If condenser water comes from another source, pressure drop through the condenser should be lower than the available pressure to allow for pressure fluctuations and additional flow resistance caused by fouling.

Pressure drop through horizontal condensers includes loss through the tubes, tube entrance and exit losses, and losses through the heads or return bends (or both). The effect of tube coiling must be considered in shell-and-coil condensers. Expected pressure drop through tubes can be calculated from a modified Darcy-Weisbach equation:

$$\Delta p = N_p\left(K_H + f\frac{L}{D}\right)\frac{\rho V^2}{2g} \tag{7}$$

where

Δp = pressure drop, lb_f/ft^2
N_p = number of tube passes
K_H = entrance and exit flow resistance and flow reversal coefficient, number of velocity heads ($V^2/2g$)
f = friction factor
L = length of tube, ft
D = inside tube diameter, ft
ρ = fluid density, lb/ft^3
V = fluid velocity, fps
g = gravitational constant = 32.17 $lb_m \cdot ft/(lb_f \cdot s^2)$

For tubes with smooth inside diameters, the friction factor may be determined from a Moody chart or various relations, depending on the flow regime and wall roughness (see Chapter 2 of the 2005 *ASHRAE Handbook—Fundamentals*). For tubes with internal enhancement, the friction factor should be obtained from the tube manufacturer.

The value of K_H depends on tube entry and exit conditions and the flow path between passes. A minimum recommended value is 1.5. This factor is more critical with short tubes.

Predicting pressure drop for shell-and-coil condensers is more difficult than for shell-and-tube condensers because of the curvature of the coil and flattening or kinking of the tubes as they are bent. Seban and McLaughlin (1963) discuss the effect of curvature or bending of pipe and tubes on the pressure drop.

For brazed and plate-and-frame condensers, the total pressure drop is the sum of the pressure drop in the plate region and the pressure drops associated with the entry/exit ports. The general form of equation is similar to Equation (7) except $N_p = 1$.

LIQUID SUBCOOLING

The amount of condensate subcooling provided by the condensing surface in a shell-and-tube condenser is small, generally less than 2°F. When a specific amount of subcooling is required, it may be obtained by submerging tubes in the condensate. Tubes in the lower portion of the bundle are used for this purpose. If the condenser is multipass, then the subcooling tubes should be included in the first pass to gain exposure to the coolest water.

When means are provided to submerge the subcooler tubes to a desired level in the condensate, heat is transferred principally by natural convection. Subcooling performance can be improved by enclosing tubes in a separate compartment in the condenser to obtain the benefits of forced convection over the enclosed tubes.

Segmental baffles may be provided to produce flow across the tube bundle. Kern and Kraus (1972) describe how heat transfer performance can be estimated analytically by use of longitudinal or cross-flow correlations, but it is more easily determined by test because of the large number of variables. The refrigerant pressure drop along the flow path should not exceed the pressure difference permitted by the saturation pressure of the subcooled liquid.

CIRCUITING

Varying water flow rate in a condenser can significantly affect the saturated condensing temperature, which affects performance. Figure 3 shows the change in condensing temperature for one, two, or three passes in a particular condenser. For example, at a loading of 22,000 Btu/h per tube, a two-pass condenser with a 10°F range would have a condensing temperature of 102.3°F. At the same load with a one-pass, 5°F range, this unit would have a condensing temperature of 99.3°F. The one-pass option does, however, require twice the water flow rate with an associated increase in pumping power. Three-pass design may be favorable when costs associated with water flow outweigh gains from a lower condensing temperature. Hence, different numbers of passes (if an option) and ranges should be considered against other parameters (water source, pumping power, cooling tower design, etc.) to optimize overall performance and cost.

CONDENSER TYPES

The most common types of water-cooled refrigerant condensers are (1) shell-and-tube, (2) shell-and-coil, (3) tube-in-tube, and (4) brazed-plate. The type selected depends on the cooling load, refrigerant used, quality and temperature of the available cooling water, amount of water that can be circulated, location and space allotment, required operating pressures (water and refrigerant sides), and cost and maintenance concerns.

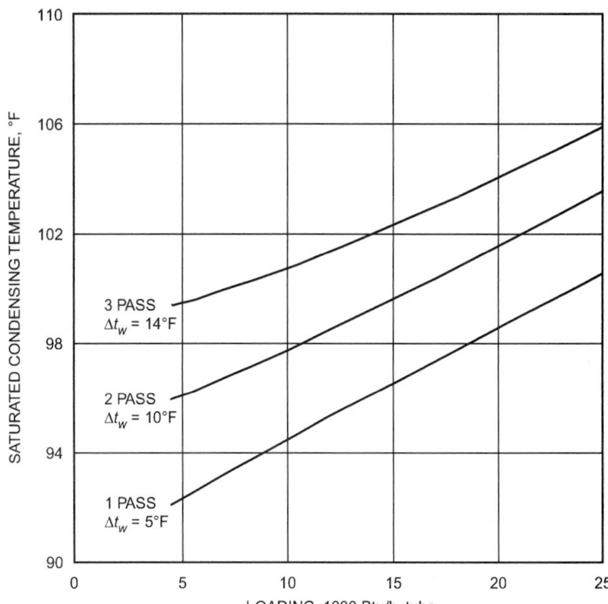

Fig. 3 Effect of Condenser Circuiting

Shell-and-Tube Condensers

Built in sizes from 1 to 10,000 tons, these condensers condense refrigerant outside the tubes and circulate cooling water through the tubes in a single or multipass circuit. Fixed-tube-sheet, straight-tube construction is usually used, although U-tubes that terminate in a single tube sheet are sometimes used. Typically, shell-and-tube condenser tubes run horizontally. Where floor installation area is limited, the tubes may be vertical, but this orientation has poor condensate draining, which reduces the refrigerant film coefficient. Vertical condensers with open water systems have been used with ammonia.

Gas inlet and liquid outlet nozzles should be located with care. The proximity of these nozzles may adversely affect condenser performance by requiring excessive amounts of liquid refrigerant to seal the outlet nozzle from inlet gas flow. This effect can be diminished by adding baffles at the inlet and/or outlet connection.

Halocarbon refrigerant condensers have been made with many materials, including all prime surface or finned, ferrous, or nonferrous tubes. Common tubes are nominal 0.75 and 1.0 in. OD copper tubes with integral fins on the outside. These tubes are often available with fin heights from 0.035 to 0.061 in. and fin spacings of 19, 26, and 40 fins/in. For ammonia condensers, prime surface steel tubes, 1.25 in. OD and 0.095 in. average wall thickness, are common.

Many tubes designed for enhanced heat transfer are available (Bergles 1995). On the inside of the tube, common enhancements include longitudinal or spiral grooves and ridges, internal fins, and other devices to promote turbulence and augment heat transfer. On the refrigerant side, condensate surface tension and drainage are important in design of the tube outer surface. Tubes are available with the outsides machined or formed specifically to enhance condensation and promote drainage. Heat transfer design equations should be obtained from the manufacturer.

The electrohydrodynamics (EHD) technique couples a high-voltage, low-current electric field with the flow field in a fluid with low electrical conductivity to achieve higher heat transfer coefficients. Ohadi et al. (1995) experimentally demonstrated enhancement factors in excess of tenfold for both boiling and condensation of refrigerants such as R-134a. For condensation, the technique responds equally well to augmentation of condensation over (or inside) both vertical and horizontal tubes. Most passive augmentation techniques perform poorly for condensation enhancement in vertical orientation. Additional details of the EHD technique are in Chapter 3 of the 2005 *ASHRAE Handbook—Fundamentals*.

Because water and refrigerant film resistances with enhanced tubes are reduced, the effect of fouling becomes relatively great. Where high levels of fouling occur, fouling may easily account for over 50% of the total resistance. In such cases, the advantages of enhancement may diminish. On the other hand, water-side augmentation, which creates turbulence, may reduce fouling. The actual value of fouling resistances depends on the particular type of enhancement and service conditions (Starner 1976; Watkinson et al. 1974).

Similarly, refrigerant-side enhancements may not show as much benefit in very large tube bundles as in smaller bundles. This is because of the row effect addressed in the section on Refrigerant-Side Film Coefficient.

The tubes are either brazed into thin copper, copper alloy, or steel tube sheets, or rolled into heavier nonferrous or steel tube sheets. Straight tubes with a maximum OD less than the tube hole diameter and rolled into tube sheets are removable. This construction facilitates field repair in the event of tube failure.

The required heat transfer area for a shell-and-tube condenser can be found by solving Equations (1), (2), and (3). The mean temperature difference is the logarithmic mean temperature difference, with entering and leaving refrigerant temperatures taken as the saturated condensing temperature. Depending on the parameters fixed, an iterative solution may be required.

Shell-and-U-tube condenser design principles are the same as those outlined for horizontal shell-and-tube units, with one exception: water pressure drop through the U-bend of the U-tube is generally less than that through the compartments in the water header where the direction of water flow is reversed. The pressure loss is a function of the inside tube diameter and the ratio of the inside tube diameter to bending centers. Pressure loss should be determined by test.

Shell-and-Coil Condensers

Shell-and-coil condensers circulate cooling water through one or more continuous or assembled coils contained within the shell. The refrigerant condenses outside the tubes. Capacities range from 0.5 to 15 tons. Because of the type of construction, the tubes are neither replaceable nor mechanically cleanable.

Again, Equations (1), (2), and (3) may be used for performance calculations, with the saturated condensing temperature used for the entering and leaving refrigerant temperatures in the logarithmic mean temperature difference. The values of h_w (the water-side film coefficient) and, especially, the pressure loss on the water side require close attention; laminar flow can exist at considerably higher Reynolds numbers in coils than in straight tubes. Because the film coefficient for turbulent flow is greater than that for laminar flow, h_w as calculated from Equation (6) will be too high if the flow is not turbulent. Once flow has become turbulent, the film coefficient will be greater than that for a straight tube (Eckert 1963). Pressure drop through helical coils can be much greater than through smooth, straight tubes for the same length of travel. The section on Water Pressure Drop outlines the variables that make accurate determination of the pressure loss difficult. Pressure loss and heat transfer rate should be determined by test because of the many variables inherent in this condenser.

Tube-in-Tube Condensers

These condensers consist of one or more assemblies of two tubes, one within the other, in which the refrigerant vapor is condensed in either the annular space or the inner tube. These units are built in sizes ranging from 0.3 to 50 tons. Both straight-tube and axial-tube (coaxial) condensers are available.

Equations (1) and (2) can be used to size a tube-in-tube condenser. Because the refrigerant may undergo a significant pressure loss through its flow path, the refrigerant temperatures used to calculate the mean temperature difference should be selected carefully. Refrigerant temperatures should be consistent with the model used for the refrigerant film coefficient. The logarithmic mean temperature difference for either counterflow or parallel flow should be used, depending on the piping connections. Equation (3) can be used to find the overall heat transfer coefficient when water flows in the tubes, and Equation (4) may be used when water flows in the annulus.

Tube-in-tube condenser design differs from those outlined previously, depending on whether the water flows through the inner tube or through the annulus. Condensing coefficients are more difficult to predict when condensation occurs within a tube or annulus, because the process differs considerably from condensation on the outside of a horizontal tube. Where water flows through the annulus, disagreement exists regarding the appropriate method that should be used to calculate the waterside film coefficient and the water pressure drop. The problem is further complicated if the tubes are spiraled.

The water side is mechanically cleanable only when the water flows in straight tubes and cleanout access is provided. Tubes are not replaceable.

Brazed-Plate and Plate-and-Frame Condensers

Brazed-plate condensers are constructed of plates brazed together to form an assembly of separate channels. Capacities range from 0.5 to 100 tons. The plate-and-frame condenser is a standard design in which plate pairs are laser-welded to form a single cassette. Refrigerant is confined to the space between the welded plates and is exposed to gaskets only at the ports. Such condensers have a higher range of capacity.

The plates, typically stainless steel, are usually configured with a wave pattern, which results in high turbulence and low susceptibility to fouling. The design has some ability to withstand freezing and, because of the compact design, requires a low refrigerant charge. The construction of brazed units does not allow mechanical cleaning, and internal leaks usually cannot be repaired. Thus, it can be beneficial to use filtration or separators on open cooling towers to keep the water clean, or to use closed-circuit cooling towers. Plate-and-frame units can be cleaned on the water side.

Performance calculations are similar to those for other condensers; however, very few correlations are available for the heat transfer coefficients.

NONCONDENSABLE GASES

When first assembled, most refrigeration systems contain gases, usually air and water vapor. These gases are detrimental to condenser performance, so it is important to evacuate the entire refrigeration system before operation.

For low-pressure refrigerants, where the operating pressure of the evaporator is less than ambient pressure, even slight leaks can be a continuing source of noncondensables. In such cases, a purge system, which automatically expels noncondensable gases, is recommended. Figure 4 shows some examples of refrigerant loss associated with using purging devices at various operating conditions. As a general rule, purging devices should emit less than one part refrigerant per part of air as rated in accordance with ARI *Standard* 580 (see ASHRAE *Guideline* 3).

When present, noncondensable gases collect on the high-pressure side of the system and raise the condensing pressure above that corresponding to the temperature at which the refrigerant is

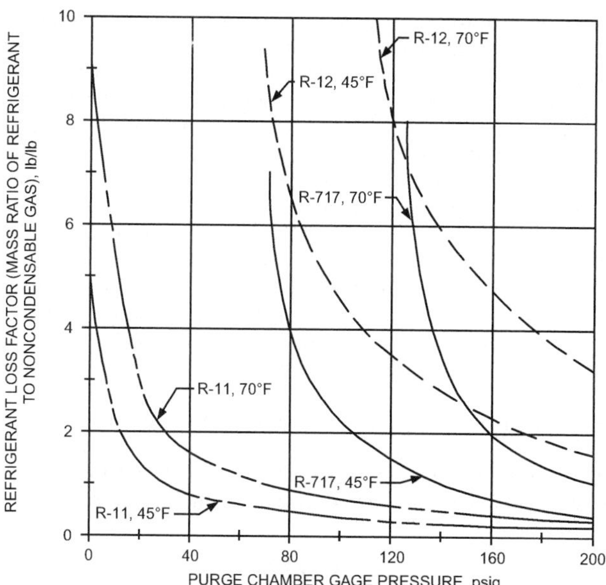

Fig. 4 Loss of Refrigerant During Purging at Various Gas Temperatures and Pressures

actually condensing. This increases power consumption and reduces capacity. Also, if oxygen is present at a point of high discharge temperature, the oil may oxidize.

The excess pressure is caused by the partial pressure of the noncondensable gas. These gases form a resistance film over some of the condensing surface, thus lowering the heat transfer coefficient. Webb et al. (1980) showed how a small percentage of noncondensables can cause major decreases in the refrigerant film coefficient in shell-and-tube condensers. (See Chapter 4 of the 2005 *ASHRAE Handbook—Fundamentals*.) The noncondensable situation of a given condenser is difficult to characterize because such gases tend to accumulate in the coldest and least agitated part of the condenser or in the receiver. Thus, a fairly high percentage of noncondensables can be tolerated if the gases are confined to areas far from the heat transfer surface. One way to account for noncondensables is to treat them as a refrigerant or gas-side fouling resistance in Equation (3). Some predictions are presented by Wanniarachchi and Webb (1982).

As an example of the effect on system performance, experiments performed on a 250 ton R-11 chiller condenser revealed that 2% noncondensables by volume caused a 15% reduction in the condensing coefficient. Also, 3% and 8% noncondensables by volume caused power increases of 2.6% and 5%, respectively.

Huber et al. (1994b) determined that noncondensable gases have a more severe effect on the condensing coefficient of tubes with saw-toothed enhancements than on conventional finned tubes. For a noncondensable gas concentration of 0.5%, the coefficient for the tube with a fin spacing of 26 fpi dropped by 15%, whereas the coefficient for the sawtoothed tube dropped by nearly 40%. At a noncondensable concentration of 5%, degradation was similar for both tubes.

The presence of noncondensable gases can be tested by shutting down the refrigeration system while allowing the condenser water to flow long enough for the refrigerant to reach the same temperature as the water. If the condenser pressure is higher than the pressure corresponding to the refrigerant temperature, noncondensable gases are present. This test may not be sensitive enough to detect the presence of small amounts of noncondensables, which can, nevertheless, decrease shell-side condensing coefficients.

CODES AND STANDARDS

Pressure vessels must be constructed and tested under the rules of appropriate codes and standards. The introduction of the current ASME *Boiler and Pressure Vessel Code*, Section VIII, gives guidance on rules and exemptions.

The more common applicable codes and standards include the following:

- ARI *Standard* 450 covers industry criteria for standard equipment, standard safety provisions, marking, fouling factors, and recommended rating points for water-cooled condensers.
- ASHRAE *Standard* 22 covers recommended testing methods.
- ARI *Standard* 580 covers methods of testing, evaluating, and rating the efficiency of noncondensate gas purge equipment.
- ASHRAE *Standard* 15 specifies design criteria, use of materials, and testing. It refers to the ASME *Boiler and Pressure Vessel Code*, Section VIII, for refrigerant-containing sides of pressure vessels, where applicable. Factory test pressures are specified, and minimum design working pressures are given by this code. This code requires pressure-limiting and pressure-relief devices on refrigerant-containing systems, as applicable, and defines the setting and capacity requirements for these devices.
- ASME *Boiler and Pressure Vessel Code,* Section VIII covers the safety aspects of design and construction. Most states require condensers to meet these requirements if they fall within the scope of the ASME code. Some of the exceptions from meeting the requirements listed in the ASME code are as follows:
  - Condenser shell ID is 6 in. or less.
  - Working pressure is 15 psig or less.
  - The fluid (water) portion of the condenser need not be built to the requirements of the ASME code if the fluid is water or an aqueous glycol solution, the design pressure does not exceed 300 psig, and the design temperature does not exceed 210°F.

  Condensers meeting the requirements of the ASME code will have an ASME stamp. The ASME stamp is a U or UM inside a four-leaf clover. The U can be used for all condensers; the UM can be used (considering local codes) for those with net refrigerant-side volume less than 1.5 ft^3 if less than 600 psig, or less than 5 ft^3 if less than 250 psig.
- UL *Standard* 207 covers specific design criteria, use of materials, testing, and initial approval by Underwriters Laboratories. A condenser with the ASME U stamp does not require UL approval.

Design Pressure

Refrigerant-side pressure should be chosen per ASHRAE *Standard* 15. Standby temperature and temperatures encountered during shipping of units with a refrigerant charge should also be considered.

Required fluid- (water-) side pressure varies, depending largely on the following conditions: static head, pump head, transients due to pump start-up, and valve closing. A common water-side design pressure is 150 psig, although with taller building construction, requirements for 300 psig are not uncommon.

OPERATION AND MAINTENANCE

When a water-cooled condenser is selected, anticipated operating conditions, including water and refrigerant temperatures, have usually been determined. Standard practice allows for a fouling factor in the selection procedure. A new condenser, therefore, operates at a condensing temperature lower than the design point because it has not yet fouled. Once a condenser starts to foul or scale, economic considerations determine how frequently it should be cleaned. As scale builds up, the condensing temperature and subsequent power increase while the unit capacity decreases. This effect can be seen in Figure 5 for a condenser with a design fouling factor of 0.00025 $ft^2 \cdot h \cdot °F/Btu$.

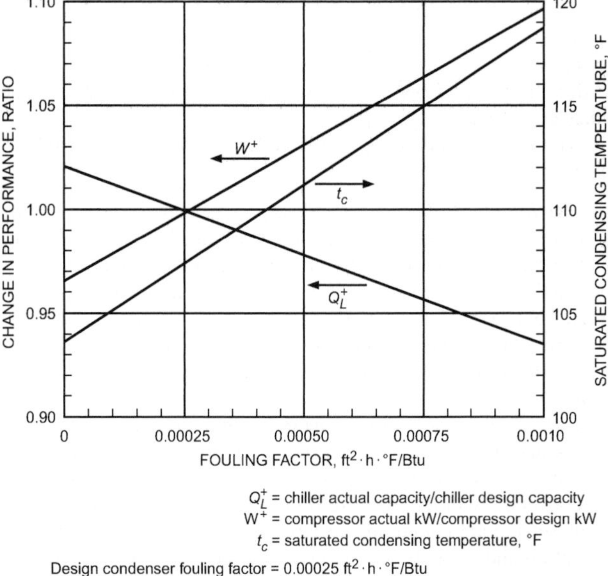

Fig. 5 Effect of Fouling on Chiller Performance

At some point, the increased cost of power can be offset by the labor cost of cleaning.

Local water conditions, as well as the effectiveness of chemical water treatment, if used, make the use of any specific maintenance schedule difficult. Cleaning can be done either mechanically with a brush or chemically with an acid solution. In applications where water-side fouling may be severe, online cleaning can be accomplished by brushes installed in cages in the water heads. By using valves, flow is reversed at set intervals, propelling the brushes through the tubes (Kragh 1975). The most effective method depends on the type of scale formed. Expert advice in selecting the particular method of tube cleaning is advisable.

Occasionally, one or more tubes may develop leaks because of corrosive impurities in the water or through improper cleaning. These leaks must be found and repaired as soon as possible; this is normally done by replacing the leaky tubes, a procedure best done through the original condenser manufacturer. In large condensers, where the contribution of a single tube is relatively insignificant, a simpler approach may be to seal the ends of the leaking tube.

If the condenser is located where water can freeze during the winter, special precautions should be taken when it is idle. Opening all vents and drains may be sufficient, but water heads should be removed and tubes blown free of water.

If refrigerant vapor is to be released from the condenser and there is water in the tubes, the pumps should be on and the water flowing. Otherwise, freezing can occur.

The condenser manufacturer's installation recommendations on orientation, piping connections, space requirements for tube cleaning or removal, and other important factors should be followed.

AIR-COOLED CONDENSERS

An air-cooled condenser uses ambient air to remove the heat of condensation from the refrigerant in a compression-type refrigeration system. Individual air-cooled condensers range in size from a few hundred Btu/h to about 100 tons, though individual units may be coupled together for larger systems. The smallest air-cooled condensers are used in residential refrigerators and may have no fans, relying only on gravity circulation of the ambient air. Larger condensers almost always use one or more motor-driven fans for air circulation. Midsized condensers are frequently provided as part of an integral package (called a condensing unit) with the compressor. All but the smallest condensers use finned coils, which may be formed or otherwise bent around a compressor/fan combination to form outdoor residential condensing units up to about 5 tons. Larger air-cooled condensers almost always use coils with a rectangular profile, either flat or with a single corner bent shape, and a minimum number of rows of finned tubing. Fans may be positioned to either draw or blow the air though the condenser coil. Fan motors are generally connected to operate when the compressor operates.

Air-cooled condensers may be located either adjacent to or remote from the compressor and may be designed for indoor or outdoor operation. For indoor use, larger condensers use centrifugal fans for ducted discharge to the outdoors. Outdoor coil orientation is generally horizontal, with propeller fans above the coil providing draw-through air circulation. For nonstandard applications, air discharge may be either vertical (top) or horizontal (side). Interconnecting piping connects the condenser coil to the compressor and to the expansion device or a liquid receiver.

TYPES

Air-cooled condensers may vary in air- and refrigerant-side design, but there are three main coil construction types: plate-and-fin, integral-fin, and microchannel.

Plate-and-Fin

Coils are commonly constructed of copper, aluminum, or steel tubes, ranging from 0.25 to 0.75 in. in diameter. For fluorocarbon and hydrocarbon refrigerants, copper is most common. Aluminum or steel tube condensers are generally used for ammonia. Fins attached to the tubes can be aluminum or copper.

Copper tubes generally have a round profile, with fins perpendicularly bound to their exterior. This arrangement normally requires no further air-side protection but may be prone to corrode when subject to some industrial or furnace flue gases, or salt-atmosphere-induced oxidation. Some inherent corrosion protection is provided by the slight galvanic effect of aluminum fins. Aluminum round-tube coil manufacturing is similar to that of round copper tube coils, except gas shield arc welding or special brazing alloys are used for joints. Corrosion protection is necessary for aluminum-to-copper joints. Steel tube condensers with steel fins are painted for indoor use or hot-dip galvanized for outdoor applications. Brazed-steel condensers are very common on small refrigeration units.

Tube diameter selection compromises between factors such as compactness, manufacturing tooling cost, header arrangement, air resistance, and refrigerant flow resistance. A smaller diameter gives more flexibility in coil circuit design and a lower refrigerant charge, but at increased cost.

Other core tubing choices include copper tube that is internally enhanced by fins or grooves known as microfin tubing. The helical microfinning is very small, on the order of 0.008 in. height in a tiny spiral with an optimum helical angle of 15 to 20° throughout the internal surface of the tube. Over 60 such grooved fins could fit in a 3/8 in. tube's cross section. This pattern is common on seamless (drawn) tubes; on welded (strip) tubes, internal grooving can be augmented with a double cross-hatch fin pattern. Although a microfin tube has a significant refrigerant-side heat transfer advantage, tube-side effect does not have dominant heat transfer conductance in the coil's overall heat transfer function. The majority of heat transfer resistance is on the air side, primarily because of an extended fin surface. Usually, an air-cooled condenser's fin-to-tube b ratio is around 25 to 1. Therefore, a 50% increase in tube-side heat transfer by microfinning may result in only a 5 to 15% overall heat transfer coefficient gain in the coil design, depending on refrigerant type, its transport velocities, and the fin surface enhancement used. Occasionally, a bare-tube condenser (i.e., without a secondary surface) is used where airborne dirt loading is expected to be excessive. For these applications, enhanced bare-tube coils could be half the size of regular bare-tube coils.

When conditions are normal, and even in a saline atmosphere, a copper tube aluminum-finned coil is generally used. The most common is the aluminum plate fin, from 0.012 to 0.006 in. thick. Most plate fins have extruded tube collars. Straight tubes are passed through these nested fin collars and expanded to fit the collars by either mechanical or, less frequently, hydraulic expansion. This results in a rigid coil assembly that has maximum tube-to-fin thermal conductance. Spiral- and spline-fin coils, tightly wound onto the individual tubes that make up the coil core, are also used. Common fin spacing for each type ranges from 8 to 20 fpi.

Stock fin coatings, as well as entire (completed) coil dip-process coatings, are readily available from specialty chemical processors. These coatings are for use in specific corrosive atmospheres, especially in a dry-surface coil application, such as aluminum-finned air-cooled condenser coils in seacoast environments

Integral-Fin

Integral-fin condensers can be made of either copper of aluminum. These condensers are made by extruding or forming fins directly from the tube material. Copper tubing can be formed using cold compression to form large fins on the outside of the tubing. Where the internal heat transfer coefficient is important, these formed tubes can be stacked to form condensers. Using aluminum, the common method of forming condensers is to rake the surface of the tubing to form protrusions. Because aluminum is a very soft material, these tubes with the “spiny” fins can be coiled to form complete condensers.

Microchannel

Small all-aluminum condensers, used in automotive and aviation applications where lightness and compactness are paramount, are made with flattened tubes having hydraulic radii from 0.01 to 0.12 in. These oval tubes are formed into serpentines with zigzag aluminum fins nested horizontally between tubing runs. New, improved versions of these heat exchangers feature individual oval tubes connected to manifold header assemblies. The entire assembly is furnace-brazed and a diffusion layer of zinc applied for corrosion protection. This type of flat tube and horizontal fin arrangement has evolved into **microchannel coils**, which are used in residential condensing units and commercial air-conditioning systems. Although an advancement, they have some limitations that are uncommon in conventional coils.

Functioning solely as a condenser, microchannel coils have a higher heat transfer efficiency than a conventional plate or spine-fin coil, both per unit volume and per unit surface area. Manufacturers' laboratory tests have shown microchannel coils, compared to a conventional 12 fpi, flat-fin, 3/8 in. copper tube core, to be 90% higher in heat transfer coefficient for similar face areas. Compared to an 18-spine fin spiral design, 3/8 in. aluminum tube, this brazed aluminum design appears to be 44% higher in heat transfer coefficient. In addition to lower air-side pressure drop, microchannel refrigerant-side pressure drop tends to be lower, depending on circuitry. However, equal refrigerant distribution in the inlet header is less favorable than with conventional coils. Another use restriction is in the heat pump reverse cycle: horizontal-tube and horizontal-fin construction gives microchannel coils a less than desirable high defrost (condensate) water retention ability. Versatility is thus restricted to a specific unit's size and design, and not one-fits-many as for conventional coil designs. The equipment designer should consider how to avoid such deficits, including field repair not being an option, when applying microchannel coils at the system level.

FANS AND AIR REQUIREMENTS

Condenser coils can be cooled by natural convection or wind or by propeller, centrifugal, and vaneaxial fans. Because efficiency increases sharply by increasing air speed across the coil, forced convection with fans predominates.

Unless unusual operating or application conditions exist, fan/motor selections are made by balancing operating and first costs with size and sound requirements. Common air quantities are 600 to 1200 cfm/ton [14,400 Btu/h at 30°F temperature difference (TD)] at 400 to 800 fpm. Fan power requirements generally range from 0.1 to 0.2 hp/ton.

The type of fan depends primarily on static pressure and unit shape requirements. Propeller fans are well suited to units with a low static-pressure drop and free air discharge. Fan blade speeds are selected in the range of 515 to 1750 rpm (1140 rpm is common). Direct-drive propeller fans up to 36 in. in diameter are used in single and multiple assemblies in virtually all sizes of condenser units. Because of the propeller fan's low starting torque requirements, the most common is a permanent split-capacitor (PSC) fractional horsepower motor, having internal inherent protection. These motors are most commonly used up to about 3/4 hp, with shaded-pole motors used in the smallest sizes. Larger motors are generally polyphase induction or synchronous for the lowest-speed fans. In outdoor upward-blow applications, special attention must be paid to motor thrust bearing selection and to weather protection. When the fan blade is positioned above a vertically installed motor, a shaft-mounted slinger disk is sometimes supplied to further protect the motor and its bearings. In industrial applications, totally enclosed weather-resistant motors are sometimes used.

For free air discharge, direct-drive propeller fans are more efficient than centrifugal fans. A belt drive arrangement on propeller and centrifugal fan(s) offers more flexibility but usually requires greater maintenance, and is used most frequently on centrifugal and propeller fans over 36 in. in diameter. Centrifugal fans are used in noise-sensitive applications and/or where significant external air resistance is expected, as with extended ductwork. Vaneaxial fans are often more efficient than centrifugal fans and are frequently used in the largest sizes, although special provision for the weight of the fans, motors, and belt drive must be made. A partially obstructed fan inlet or outlet can drastically increase the noise level. Support brackets close to the fan inlet or partially obstructing the fan outlet can noticeably effect fan noise and efficiency.

HEAT TRANSFER AND PRESSURE DROP

In condensers, heat is transferred by (1) desuperheating, (2) condensing, and (3) subcooling. Figure 6 shows the changes of state of R-134a passing through an air-cooled condenser coil and the corresponding temperature change of the cooling air as it passes through the coil. Desuperheating and subcooling zones vary from 5 to 10% of the total heat transfer, depending on the entering gas and leaving liquid temperatures, but Figure 6 is typical. Good design usually has full condensing occurring in approximately 85% of the condenser area, based on an azeotropic refrigerant. Generally this happens at a fairly constant temperature, though some drop in saturated condensing temperature through the condenser coil can occur when there is significant pressure drop in the condensing coil.

Chapter 3 of the 2005 *ASHRAE Handbook—Fundamentals* lists equations for heat transfer coefficients in the single-phase flow sections. For condensation from saturated vapors, one of the best calculation methods is the Shah correlation (Shah 1979, 1981). Also, Ghaderi et al. (1995) reviewed available correlations for condensation heat transfer in smooth and enhanced tubes.

The overall heat transfer coefficient of an air-cooled condenser can be expressed by Equation (5c) in Chapter 22. It is suggested that direct use of experimental testing for extracted coinciding thermal flow resistance data is preferable to the completely theoretical approach.

For optimum vapor desuperheating and liquid subcooling (i.e., inlet and outlet regions of the coil) a cross-counterflow arrangement of refrigerant and air produces the best performance. In the condensing portion (i.e., halfway through the coil core), the latent heat dissipates and the change of state occurs, with a concurrent refrigerant velocity drop. Because of the change of state, the velocity and volume of the refrigerant drop substantially as it becomes a liquid. In general, most air-cooled condenser designs provide an adequate heat transfer surface area and complementary pressure drop to cause the refrigerant to subcool an average of 2 to 6°F before it leaves the condenser coil. The subcooling section may be integrated into the main coil by routing the liquid outlet tubes through coil rows on the air inlet side of the coil. Some manufacturers use "tripod" circuitry, where two or more parallel tubes of he condensing portion are joined to form a single continuing circuit. Overall capacity increases about 0.5% per 1°F subcooling at the same suction and discharge pressure. To ensure proper operation, liquid quality needs to be 100% at the entrance to the evaporator's inlet flow control device.

Proper condenser design also involves calculating pressure drops for the gas and liquid flow areas in the coil. Chapter 4 of the 2005 *ASHRAE Handbook—Fundamentals* describes an estimation of two-phase pressure drop. The overall pressure drop in a typical condenser ranges from 1 to 4°F equivalent saturated temperature change.

CONDENSERS REMOTE FROM COMPRESSOR

Remote air-cooled condensers are used for refrigeration and air conditioning from 0.5 to over 500 tons. Larger systems use multiple condensers. Larger air-cooled condensers with multiple individual circuits are used when the installation has an array of independent refrigeration systems (e.g., supermarkets). The most common coil-fan arrangements are horizontal coils with upflow air (top discharge), and vertical coil with horizontal airflow (front discharge). The choice of design depends primarily on the intended application and the surroundings.

Refrigerant piping and electrical wiring are the interconnections between the compressor and remote condenser unit. Receivers are most often installed close to the compressor and not the condenser. When selecting installation points for any high-side equipment, major concerns include direct sunlight (high solar load), summer and winter conditions, prevailing wind, elevation differences, and

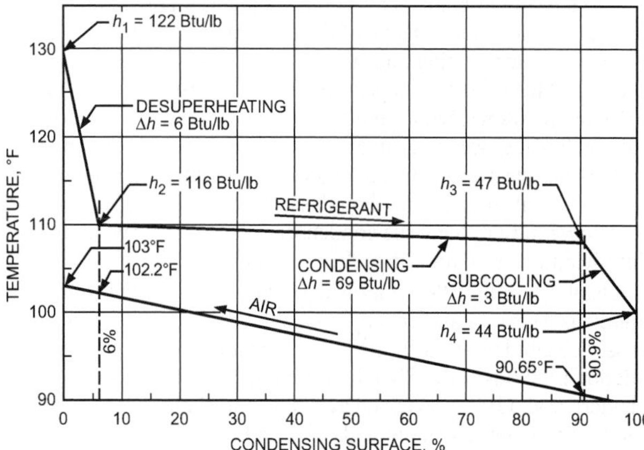

Fig. 6 Temperature and Enthalpy Changes in Air-Cooled Condenser with R-134a

length of piping runs. Recirculated condenser air is a common cause of excessive air temperatures entering the condenser and usually results from a poor choice in locating the condenser, and/or badly accommodating its airflow pattern.

CONDENSERS AS PART OF CONDENSING UNIT

When the condenser coil is included with the compressor as a packaged assembly, it is called a condensing unit. Condenser units are further categorized as indoor or outdoor units, depending on rain protection, electrical controls, and code approvals. Factory-assembled condensing units consist of one or more compressors, a condenser coil, electrical controls in a panel, a receiver, shutoff valves, switches, and sometimes other related units. Precharged interconnecting refrigerant lines to and from the evaporator may be offered as a kit with smaller units. Open, semihermetic or full-hermetic compressors are used for single or parallel (multiplexing) piping connections, in which case an accumulator, oil separator, and its crankcase oil return is likely to be included in the package. Indoor units lack weather enclosures, allowing easy access for service to all components; outdoor unit access is obtained by removing cabinet panels or opening hinged covers.

Noise is a major concern in both the design and installation of condensing units, mainly because of the compressor. The following can be done to reduce unwanted sound:

- Avoid a straight-line path from the compressor to the listener.
- Use acoustical material inside the cabinet and to envelop the compressor. Streamline air passages as much as possible. Use top-mounted, vertical air discharge.
- Cushion or suspend the compressor on a suitable base.
- Use a lighter-gage base rather than a heavier one to minimize vibration transmission to other panels.
- Where possible, ensure that natural frequencies of panels and refrigerant lines are different from basic compressor and fan frequencies.
- Avoid refrigerant lines with many bends, which are more likely to produce numerous pulsation forces and natural frequencies that cause audible sounds.
- Be aware that a condenser with a very low refrigerant pressure drop may have one or two passes resonant with the compressor discharge pulsations.
- Use wire basket fan assembly supports to help isolate noise and vibration.
- Fan selection is important; steeply pitched propeller blades and unstable centrifugal fans can be serious noise producers.
- Fan noise is more objectionable at night. Because air temperatures are generally lower at night, fan speed can usually be reduced then to reduce noise levels without negatively affecting performance.
- Design to avoid fan and compressor short-cycling.

See Chapter 47 of the 2007 *ASHRAE Handbook—HVAC Applications* for more information on sound and vibration control.

WATER-COOLED VERSUS AIR-COOLED CONDENSING

Where small units (less than 3 hp) are used and abundant low-cost water is available, both first and operating costs may be lower for water-cooled equipment. Air-cooled equipment generally requires up to a 20% larger compressor and/or longer run time, thus increasing costs. When comparing systems, operation and maintenance of the complete system as well as local conditions need to be considered. Chronic water shortages can affect selection. When cooling towers are used to produce condenser cooling water, the initial cost of water cooling may be higher. Also, the lower operating cost of water cooling may be offset by cooling tower pump, and maintenance costs. Water-cooled condensers are discussed in detail at the beginning of this chapter.

TESTING AND RATING

Condenser and condensing unit manufacturers specify, select, and test motor and fan combinations as complete assemblies for specific published capacities. Test measurements of capacity should be made in accord with ASHRAE *Standard* 20. Other common tests by manufacturers include motor winding and bearing temperature rise during highest ambient operating conditions, duty cycling at lowest ambient, and mounting positions and noise amplitude. To ensure user safety, most condenser coil designs are submitted for listing by Underwriters Laboratories.

Condensers are rated in terms of total heat rejection (THR), which is the total heat removed in desuperheating, condensing, and subcooling the refrigerant. This value is the product of the mass flow of the refrigerant and the difference in enthalpies of the refrigerant vapor entering and the liquid leaving the condenser coil.

A condenser may also be rated in terms of net refrigeration effect (NRE) or net heat rejection (NHR), which is the total heat rejection less the heat of compression added to the refrigerant in the compressor. This is the typical expression of a refrigeration system's capacity.

For open compressors, the THR is the sum of the actual power input to the compressor and the NRE. For hermetic compressors, the THR is obtained by adding the NRE to the total motor power input and subtracting any heat losses from the compressor and discharge line surfaces. Surface heat losses are generally 0 to 10% of the power consumed by the motor. All quantities must be expressed in consistent units. Table 1 recommends factors for converting condenser THR ratings to NRE for both open and hermetic reciprocating compressors.

Air-cooled condenser ratings are based on the temperature difference (TD) between the dry-bulb temperature of air entering the coil and the saturated condensing temperature (which corresponds to the refrigerant pressure at the condenser outlet). Typical TD values are 10 to 15°F for low-temperature systems at a –20 to –40°F evaporator temperature, 15 to 20°F for medium-temperature systems at a 20°F evaporator temperature, and 25 to 30°F for air-conditioning systems at a 45°F evaporator temperature. The THR capacity of the condenser is considered proportional to the TD. That is, the capacity at 30°F TD is considered to be double the capacity for the same condenser selected for 15°F TD.

The 1% design day condition is suggested for condenser selection.

The specific design dry-bulb temperature must be selected carefully, especially for refrigeration serving process cooling. An entering air temperature that is higher than expected quickly causes compressor discharge pressure and power to exceed design. This can cause an unexpected shutdown, usually at a time when it can least be tolerated. Also, congested or unusual locations may create entering air temperatures higher than general ambient conditions. Recirculated condenser air is usually the result of a poor choice in locating the installation.

The capacity of an air-cooled condenser equipped with an integral subcooling circuit varies depending on the refrigerant charge. The charge is greater when the subcooling circuit is full of liquid, which increases subcooling. When the subcooling circuit is used for condensing, the refrigerant charge is lower, condensing capacity is greater, and liquid subcooling is reduced. Laboratory testing should be in accordance with ASHRAE *Standard* 20, which requires liquid leaving the condenser coil (under test). Currently no industry certification program exists for air-cooled condensers.

CONTROL OF AIR-COOLED CONDENSERS

For a refrigeration system to function properly, the condensing pressure and temperature must be maintained within certain limits. An increase in condensing temperature causes a loss in capacity, requires extra power, and may overload the compressor motor. A condensing pressure that is too low hinders flow through conventional liquid feed devices. This hindrance starves the evaporator and causes a loss of capacity, unbalances the distribution of refrigerant in the evaporator, possibly causes strips of ice to form across the face of a freezer coil, and trips off the unit on low pressure.

Some systems use low-pressure-drop thermostatic expansion valves (balance port TXVs). These low-head (usually surge-receiving) systems require an additional means of control to ensure that liquid line subcooled refrigerant enters the expansion valve. Supplemental electronic controls, valves, or, at times, liquid pumps are included. This flow control equipment is often found on units that operate year-round to provide medium- and low-temperature food processing refrigeration, where a precise temperature must be held.

Table 1 Net Refrigeration Effect Factors for Reciprocating Compressors Used with Air-Cooled and Evaporative Condensers

Saturated Suction Temperature, °F	Condensing Temperature, °F										
	85	90	95	100	105	110	115	120	125	130	135
					Open Compressors[a,b]						
–40	0.71	0.70	0.69	0.68	0.67	0.65	0.64	0.63	0.62	0.60	—
–20	0.77	0.76	0.74	0.73	0.72	0.71	0.70	0.69	0.67	0.66	—
0	0.82	0.80	0.79	0.78	0.77	0.76	0.75	0.74	0.73	0.71	—
+20	0.86	0.85	0.84	0.83	0.82	0.81	0.79	0.78	0.77	0.76	0.75
+40	0.91	0.90	0.89	0.87	0.86	0.85	0.84	0.83	0.82	0.81	0.80
					Sealed Compressors[a]						
–40	0.55	0.54	0.53	0.52	0.51	0.50	0.49	0.47	0.46	0.44	—
–20	0.65	0.64	0.62	0.61	0.60	0.59	0.58	0.55	0.53	0.51	—
0	0.72	0.71	0.70	0.69	0.67	0.66	0.64	0.62	0.60	0.58	—
+20	0.77	0.76	0.75	0.74	0.72	0.71	0.69	0.68	0.66	0.64	0.62
+40	0.81	0.80	0.79	0.78	0.77	0.75	0.74	0.72	0.71	0.70	0.68
+50	0.83	0.82	0.81	0.80	0.79	0.78	0.76	0.75	0.74	0.73	0.72

[a] For R-22, factors are based on 15°F superheat entering the compressor. [b] For ammonia (R-717), factors are based on 10°F superheat.

Notes:

1. These factors should be used only for air-cooled and evaporative condensers connected to reciprocating compressors.
2. Condensing temperature is that temperature corresponding to saturation pressure as measured at compressor discharge.
3. Net refrigeration effect factors are an approximation only. Represent net refrigeration capacity as *approximate only* in published ratings.
4. For more accurate condenser selection and for condensers connected to centrifugal compressors, use total heat rejection of condenser and compressor manufacturer's total rating, which includes heat of compression.

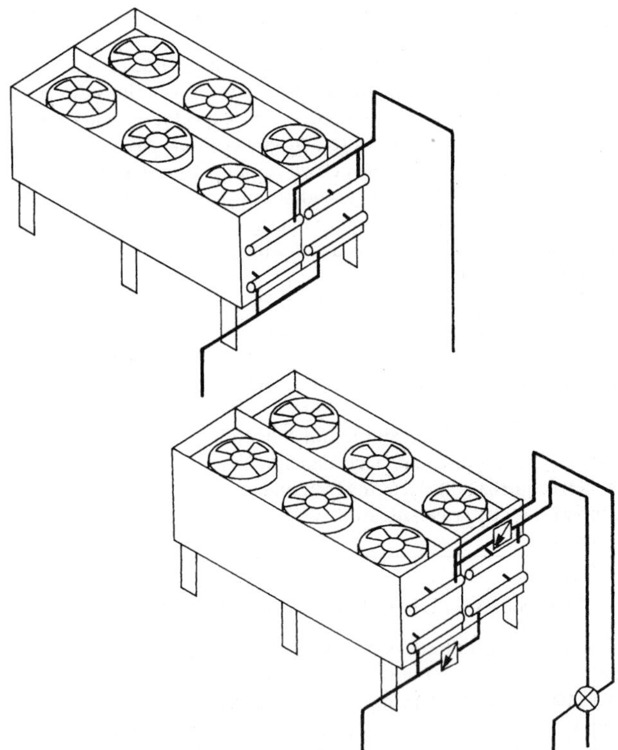

Fig. 7 Equal-Sized Condenser Sections Connected in Parallel and for Half-Condenser Operation During Winter

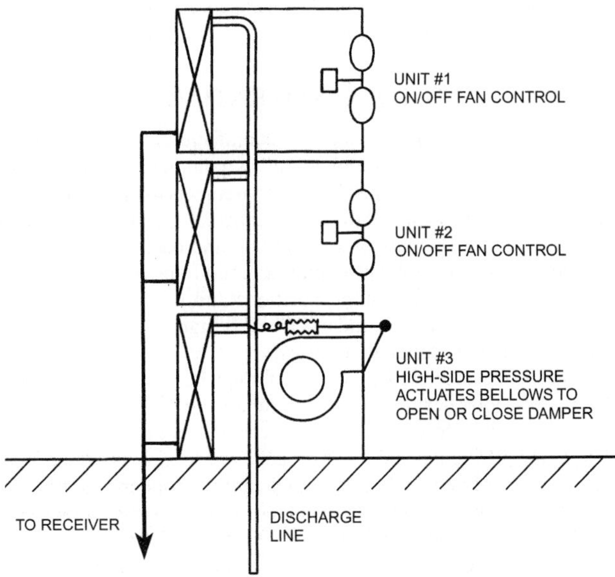

Fig. 8 Unit Condensers Installed in Parallel with Combined Fan Cycling and Damper Control

To prevent excessively low head pressure during winter operation, two basic control methods are used: (1) refrigerant-side control and (2) air-side control.

Refrigerant-Side Control. Control on the refrigerant side may be accomplished in the following ways:

- By modulating the amount of active condensing surface available for condensing by flooding the coil with liquid refrigerant. This method requires a receiver and a larger charge of refrigerant. Several valving arrangements give the required amount of flooding to meet the variable needs. Both temperature and pressure actuation are used.
- By going to one-half condenser operation. The condenser is initially designed with two equal parallel sections, each accommodating 50% of the load during normal summer operation. During winter, solenoid or three-way valves block off one condenser section (as well as its pumpdown to suction). This saves the flooding overcharge and also allows shutdown of fans on the inactive condenser side (Figure 7). Similar splits, such as one-third or two-thirds, are also possible. Anticipated load variations, along with climate conditions, usually dictate the preferred split arrangement for each specific installation.

  This split-condenser design has gained popularity, not only for its cold-weather control, but also for its ability to reduce the refrigerant charge. Part-load operating conditions also benefit when using a split condenser.

Air-Side Control. Control on the air side may be accomplished by one of three methods, or a combination of two of them: (1) fan cycling, (2) modulating dampers, and (3) fan speed control. Any airflow control must be oriented so that the prevailing wind does not cause adverse operating conditions.

Fan cycling in response to outdoor ambient temperature eliminates rapid cycling, but is limited to use with multiple fan units or is supplemental to other control methods. A common method for control of a two-fan unit is to cycle only one fan. A three-fan unit may cycle two fans. During average winter conditions, large multifan outdoor condensers have electronic controllers programmed to run the first fan (or pair of fans) continuously, at least at low fan speed. The cycle starts with the fan(s) closest to the refrigerant connecting piping as the first fan(s) on, last fan(s) off. The remaining fan(s) may cycle as required on ambient or high-side pressure. Airflow through the condenser may be further reduced by modulating airflow through the uncontrolled fan section, either by controlling the speed of some or all motors, or by dampers on either the air intake or discharge side (Figure 8).

In multiple, direct-drive motor-propeller arrangements, idle (off-cycle) fans should not be allowed to rotate backward; otherwise, air will short-circuit through them. Motor starting torque may be insufficient to overcome this reverse rotation. To eliminate this problem, designs for large units incorporate baffles to separate each fan assembly into individual compartments.

Air dampers controlled in response to either receiver pressure, ambient conditions, or liquid temperatures are also used to control compressor head pressure. These devices throttle airflow through the condenser coil from 100% to zero. Motors that drive propeller fans for such an application should have a flat power characteristic or should be fan-cooled so they do not overheat when the damper is nearly closed.

Fan speed can also be used to control compressor pressure. Because fan power increases in proportion to the cube of fan speed, energy consumption can be reduced substantially by slowing the fan at low ambient temperatures and during part-load operation. Solid-state controls can modulate frequency along with voltage to vary the speed of synchronous motors. Two-speed fan motors also produce energy savings. Chapter 44 has more information on speed control.

Parallel operation of condensers, especially with capacity control devices, requires careful design. Connecting only identical condensers in parallel reduces operational problems. (The section on Multiple-Condenser Installations has further information. Figures 13 and 14 also apply to air-cooled condensers.)

These control methods maintain sufficient condenser and liquid line pressure for proper expansion valve operation. During off cycles, when the outdoor temperature is lower than that of the indoor space to be conditioned or refrigerated, refrigerant migrates from the evaporator and receiver and condenses in the cold condenser. System pressure drops to what may correspond to the outdoor temperature. On start-up, insufficient feeding of liquid refrigerant to the evaporator, because of low compressor head pressure, causes low-pressure cut-out cycling of the compressor until the suction pressure reaches an operating level above the setting of the low pressurestat. At extremely low outdoor temperatures, system pressure may be below the cut-in point of the pressurestat and the compressor will not start. This difficulty is solved by (1) bypassing the low-pressure switch on start-up, and/or (2) using the condenser isolation method of control.

- **Low-pressure switch bypass.** On system start-up, a time-delay relay may bypass the low-pressure switch for 3 to 5 min to allow head pressure to build up and allow uninterrupted compressor start-up operation.
- **Fan cycle switch.** On system start-up, a high-side pressure-sensing switch can delay condenser fan(s) operation until a minimum preset head pressure is reached.
- **Condenser isolation.** This method uses a check valve arrangement to isolate the condenser. This prevents refrigerant from migrating to the condenser coil during the off cycle.

When the liquid temperature coming from the outdoor condenser is lower than the evaporator operating temperature, consider using a noncondensable-charged thermostatic expansion valve bulb. This type of bulb prevents erratic operation of the expansion valve by eliminating the possibility of condensation of the thermal bulb charge on the head of the expansion valve.

INSTALLATION AND MAINTENANCE

Installation and maintenance of remote condensers require little labor because of their relatively simple design. Remote condensers are located as close as possible to the compressor, either indoors or outdoors. They may be located above or below the level of the compressor, but always above the level of the receiver. During manufacturing, tubesheets or endplates are fastened to the finned-tube core at each end to complete the coil assembly. These provide a means of mounting the condenser enclosure or, in themselves, can form a part of the condenser unit enclosure. Center supports of similar design also support the condenser cabinet and, on larger units, allow fact section compartentalizing.

Installation and maintenance concerns include the following:

Indoor Condenser. When condensers are located indoors, conduct warm discharge air to the outdoors. An outdoor air intake opening near the condenser is provided and may be equipped with shutters. Indoor condensers can be used for space heating during winter and for ventilation during summer (Figure 9).

Outdoor Installations. In outdoor installations of vertical-face condensers, ensure that prevailing winds blow toward the air intake, or discharge shields should be installed to deflect opposing winds.

Piping. Piping practice with air-cooled condensers is identical to that established by experience with other remote condensers. Size discharge piping to the condenser for a total pressure drop equivalent no more than about 2°F saturated temperature drop. Liquid line drop leg to the receiver should incorporate an adequately sized check valve. Follow standard piping procedures and good piping practices. For pipe sizing, refer to Chapter 36 of the 2005 *ASHRAE Handbook—Fundamentals*.

A pressure tap valve should be installed at the highest point in the discharge piping run to facilitate removal of inadvertently trapped noncondensable gases. Purging should only be done with the compressor system off and pressures equalized. *Note*: This is only to be done by qualified personnel having the proper reclaim/recovery equipment as mandated by the appropriate agency, such as the Environmental Protection Agency in the United States.

Receiver. A condenser may have its receiver installed close by, but most often it will be separate and remotely installed. (Most water-cooled systems function without a receiver, because it is an integral part of the water-cooled condenser.) If a receiver is used and located in a comparatively warm ambient temperature, the liquid drain line from the condenser should be sized for a liquid velocity of under 100 fpm and designed for gravity drainage as well as for venting to the condenser. The receiver should be equipped with a pressure-relief valve assembly that meets applicable codes and discharge requirements, sight glass(es), and positive-action in-and-out shutoff valves. Requirements for such valving are outlined in ASHRAE *Standard* 15.

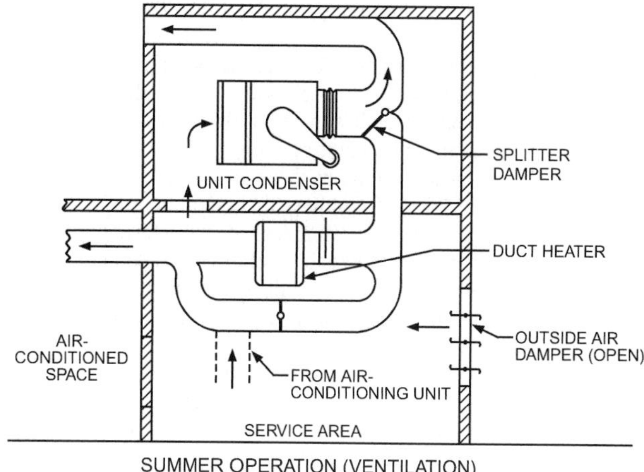

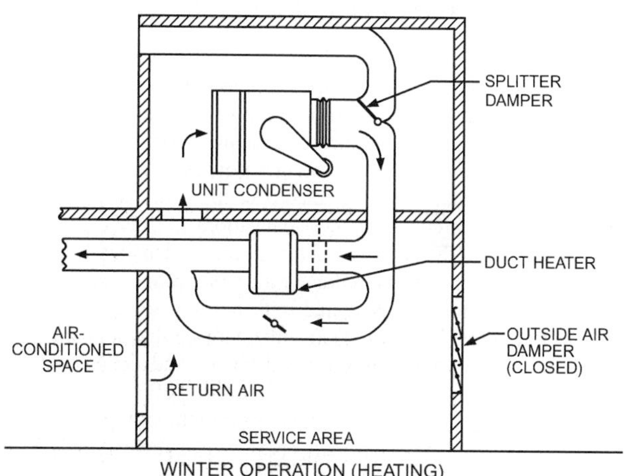

Fig. 9 Air-Cooled Unit Condenser for Winter Heating and Summer Ventilation

The receiver's manufacturer can recommend mechanical subcooling and/or reduced refrigerant charge ideas incorporating improved ambient subcooling for condenser coils.

Maintenance. Schedule periodic inspection and lubrication of fan motor and fan bearings. Lube if required, following the instructions found either on the motor nameplate or in the operation manual. Condensers also require periodic removal of lint, leaves, dirt, and other airborne materials from their inlet coil surface by brushing the air entry side of the coil with a long-bristled brush. Washing with a low-pressure water hose or blowing with compressed air is more effectively done from the side opposite the air entry face. A mild cleaning solution is required to remove restaurant grease and other oil films from the fin surface. Indoor condensers in dusty locations, such as processing plants and supermarkets, should have adequate access for unrestricted coil cleaning and fin surface inspection. In all cases, extraneous material on fins reduces equipment performance, shortens operating life, increases running time, and causes more energy use.

When a condenser coil fails (develops leaks), it is usually at the point where the tube is in contact with one of the supporting tubesheets. If not neutralized in some way, these wear points tend to leak after several years of full-time operation. Air-cooled condenser tube fractures are caused mainly by thermal stresses in the coil core, or by excessive vibration. To address this, some manufacturers of larger coils either minimize or eliminate the tube-to-metal contact wear point at tube traverse points of endplates and center supports by substantially oversizing some or all of the endplate and center support tube holes. This design substantially limits refrigerant tube fractures in the coil.

Other potential failure points are at inlet and outlet manifolds. Failure is likely to occur when the manifold or connecting pipe is not adequately supported and/or discharge pulsing is not properly controlled, or when fan unbalance causes vibration.

EVAPORATIVE CONDENSERS

As with water- and air-cooled condensers, evaporative condensers reject heat from a condensing vapor into the environment. In an evaporative condenser, hot, high-pressure vapor from the compressor discharge circulates through a condensing coil that is continually wetted on the outside by a recirculating water system. As seen in Figure 10, air is simultaneously directed over the coil, causing a small portion of the recirculated water to evaporate. This evaporation removes heat from the coil, thus cooling and condensing the vapor.

Evaporative condensers reduce water pumping and chemical treatment requirements associated with cooling tower/refrigerant condenser systems. In comparison with an air-cooled condenser, an evaporative condenser requires less coil surface and airflow to reject the same heat, or alternatively, greater operating efficiencies can be achieved by operating at a lower condensing temperature.

An evaporative condenser can operate at a lower condensing temperature than an air-cooled condenser because the condensing temperature in an air-cooled condenser is determined by the ambient dry-bulb temperature. In the evaporative condenser, heat rejection is limited by the ambient wet-bulb temperature, which is normally 14 to 25°F lower than the ambient dry bulb. Also, evaporative condensers typically provide lower condensing temperatures than the cooling tower/water-cooled condenser because the heat and mass transfer steps (between refrigerant and cooling water and between water and ambient air) are more efficiently combined in a single piece of equipment, allowing minimum sensible heating of the cooling water. Evaporative condensers are, therefore, the most compact for a given capacity.

HEAT TRANSFER

In an evaporative condenser, heat flows from the condensing refrigerant vapor inside the tubes, through the tube wall, to the water film outside the tubes, and then from the water film to the air. Figure 11 shows typical temperature trends in a counterflow evaporative condenser. The driving potential in the first step of heat transfer is the temperature difference between the condensing refrigerant and the surface of the water film, whereas the driving potential in the second step is a combination of temperature and water vapor enthalpy difference between the water surface and air. Sensible heat transfer at the water/air interface occurs because of the temperature gradient, and mass transfer (evaporation) of water vapor from the water/air interface to the airstream occurs because of the enthalpy gradient. Commonly, a single enthalpy driving force between air saturated at the temperature of the water-film surface and the enthalpy of air in contact with that surface is applied to simplify an analytical approach. Exact formulation of the heat and mass transfer process requires considering the two forces simultaneously.

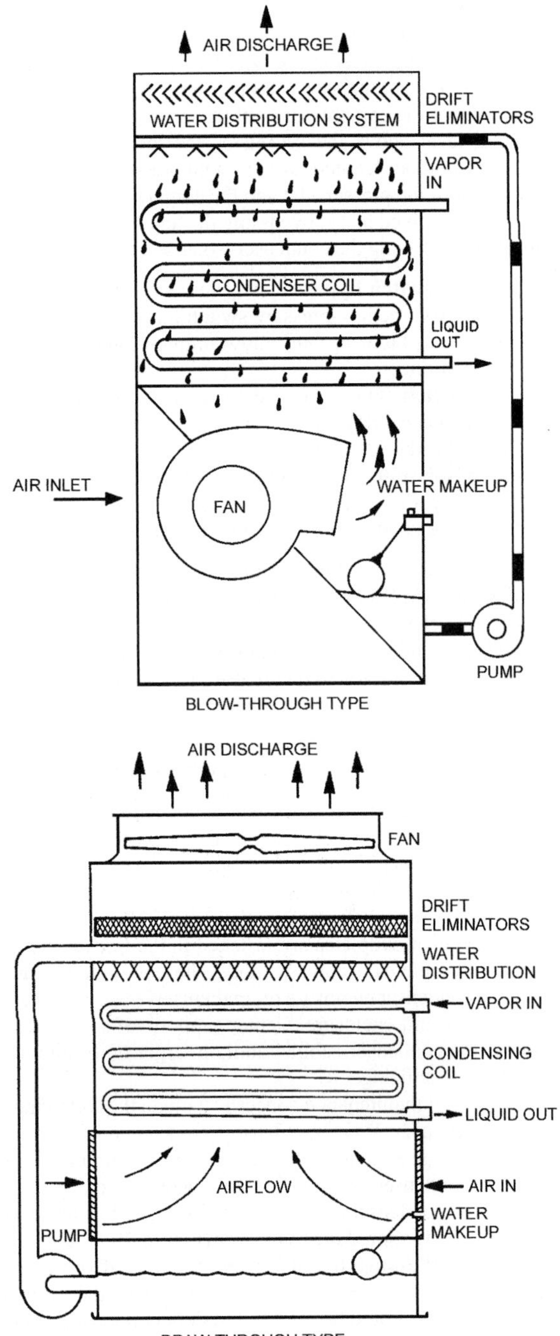

Fig. 10 Functional Views of Evaporative Condenser

Because evaporative condenser performance cannot be represented solely by a temperature difference or an enthalpy difference, simplified predictive methods can only be used for interpolation of data between test points or between tests of different-sized units, provided that air velocity, water flow, refrigerant velocity, and tube bundle configuration are comparable.

The rate of heat flow from the refrigerant through the tube wall and to the water film can be expressed as

$$q = U_s A(t_c - t_s) \tag{8}$$

where

q = rate of heat flow, Btu/h
U_s = overall heat transfer coefficient, Btu/h·ft^2·°F

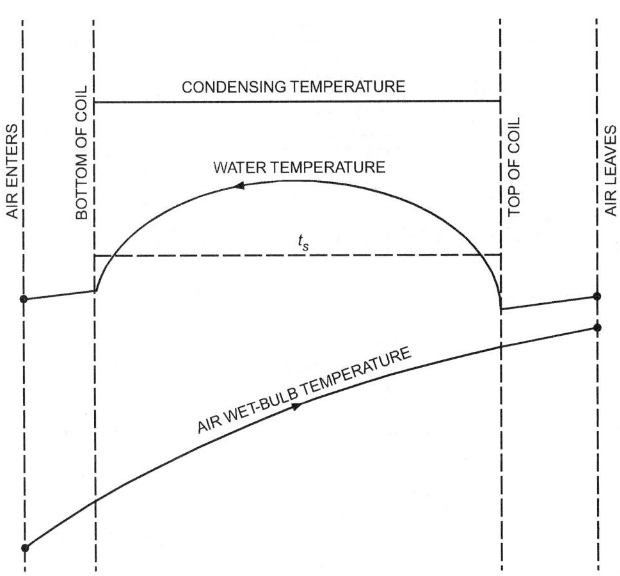

Fig. 11 Heat Transfer Diagram for Evaporative Condenser

t_c = saturation temperature at pressure of refrigerant entering condenser, °F
t_s = temperature of water film surface, °F
A = outside surface area of condenser tubes, ft²

The rate of heat flow from the water-air interface to the airstream can be expressed as

$$q = U_c A(h_s - h_e) \tag{9}$$

where

q = heat input to condenser, Btu/h
U_c = overall transfer coefficient from water-air interface to airstream, Btu/h·ft²·°F divided by enthalpy difference Δh in Btu/lb
h_s = enthalpy of air saturated at t_c, Btu/lb
h_e = enthalpy of air entering condenser, Btu/lb

Equations (8) and (9) have three unknowns: U_s, U_c, and t_s (h_s is a function of t_s). Consequently, the solution requires an iterative procedure that estimates one of the three unknowns and then solves for the remaining. This process is further complicated because of some of the more recently developed refrigerant mixtures for which t_c changes throughout the condensing process.

Korenic (1980), Leidenfrost and Korenic (1979, 1982), and Leidenfrost et al. (1980) evaluated heat transfer performance of evaporative condensers by analyzing the internal conditions in a coil. Further work on evaporative condenser performance modeling is under way in the industry.

CONDENSER CONFIGURATION

Principal components of an evaporative condenser include the condensing coil, fan(s), spray water pump, water distribution system, cold-water sump, drift eliminators, and water makeup assembly.

Coils

Generally, evaporative condensers use condensing coils made from bare pipe or tubing without fins. The high rate of energy transfer from the wetted external surface to the air eliminates the need for an extended surface. Bare coils also sustain performance better because they are less susceptible to fouling and are easier to clean. The high rate of energy transfer from the wetted external surface to the air makes finned coils uneconomical when used exclusively for wet operation. However, partially or wholly finned coils are sometimes used to reduce or eliminate plumes and/or to reduce water consumption by operating the condenser dry during off-peak conditions.

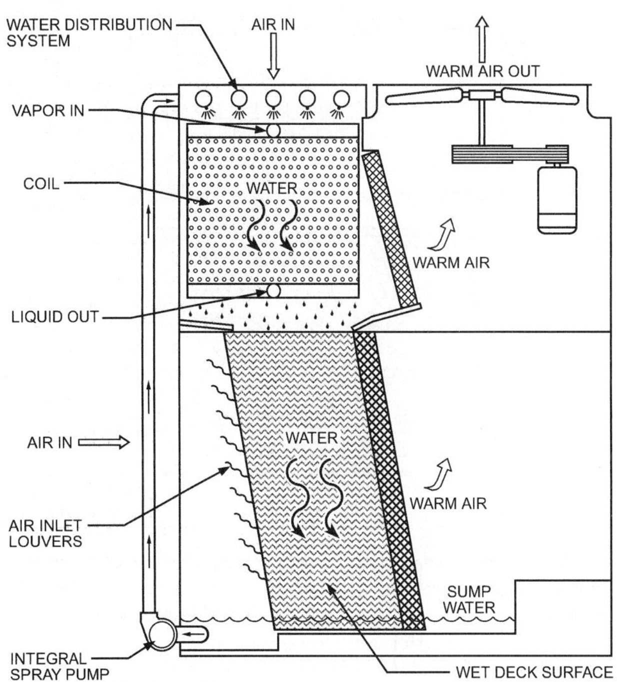

Fig. 12 Combined Coil/Fill Evaporative Condenser

Coils are usually made from steel tubing, copper tubing, iron pipe, or stainless steel tubing. Ferrous materials are generally hot-dip galvanized for exterior protection.

Method of Coil Wetting

The spray water pump circulates water from the cold water sump to the distribution system located above the coil. Water descends through the air circulated by the fan(s), over the coil surface, and eventually returns to the pan sump. The distribution system is designed to completely and continuously wet the full coil surface. Complete wetting ensures the high rate of heat transfer achieved with wet tubes and prevents excessive scaling, which is more likely to occur on intermittently or partially wetted surfaces. Such scaling is undesirable because it decreases heat transfer efficiency (which tends to raise the condensing temperature) of the unit. Water lost through evaporation, drift, and blowdown from the cold water sump is replaced through a makeup assembly that typically consists of a mechanical float valve or solenoid valve and float switch combination.

Airflow

All commercially available evaporative condensers use mechanical draft; that is, fans move a controlled flow of air through the unit. As shown in Figure 10, these fans may be on the air inlet side (forced draft) or the air discharge side (induced draft). The type of fan selected, centrifugal or axial, depends on external pressure needs, energy requirements, and permissible sound levels. In many units, recirculating water is distributed over the condensing coil and flows down to the collecting basin, counterflow to the air flowing up through the unit.

In an alternative configuration (Figure 12) a cooling tower fill augments the condenser's thermal performance. In this unit, one airstream flows down over the condensing coil, parallel to the recirculating water, and exits horizontally into the fan plenum. The recirculating water then flows down over cooling tower fill, where it is further cooled by a second airstream before it is reintroduced over the condenser coil.

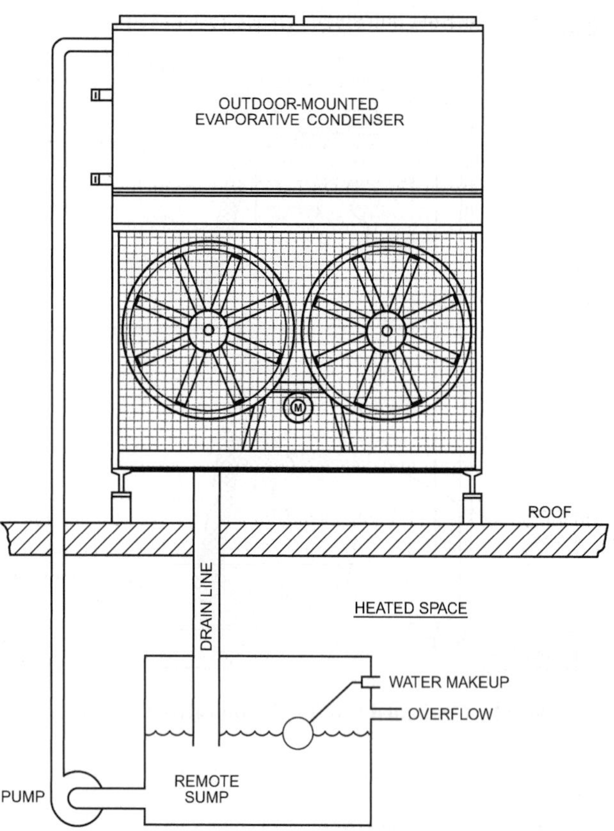

Fig. 13 Evaporative Condenser Arranged for Year-Round Operation

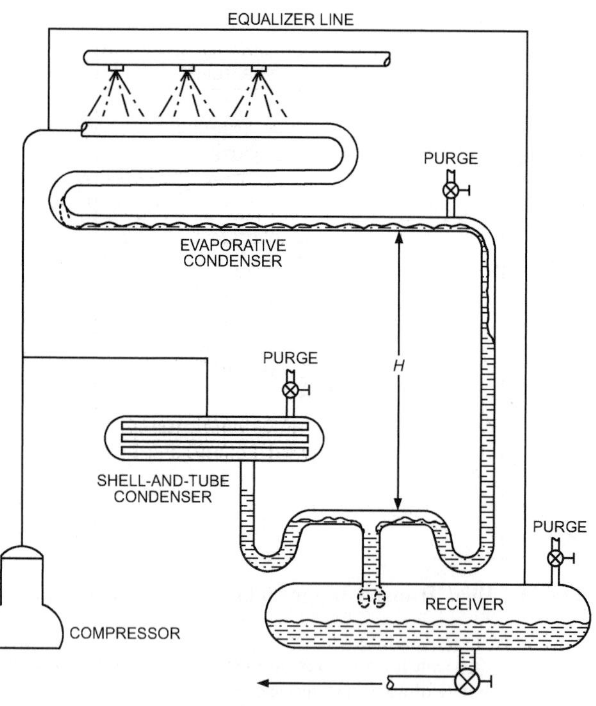

Pressure created by fluid height (H) above trap must be greater than internal resistance of condenser. Note purge connections at both condensers and receiver.

Fig. 14 Parallel Operation of Evaporative and Shell-and-Tube Condenser

Drift eliminators strip most water droplets from the discharge airstream, but some escape as drift. The rate of drift loss from an evaporative condenser is a function of unit configuration, eliminator design, airflow rate through the evaporative condenser, and water flow rate. Generally, an efficient eliminator design can reduce drift loss to a range of 0.001 to 0.2% of the water circulation rate. If the air inlet is near the sump, louvers or deflectors may be installed to prevent water from splashing from the unit.

CONDENSER LOCATION

Most evaporative condensers are located outdoors, frequently on the roofs of machine rooms. They may also be located indoors and ducted to the outdoors. Generally, centrifugal-fan models must be used for indoor applications to overcome the static resistance of the duct system.

Evaporative condensers installed outdoors can be protected from freezing in cold weather by a remote sump arrangement in which the water and pump are located in a heated space that is remote from the condensers (Figure 13). Piping is arranged so that whenever the pump stops, all the water drains back into the sump to prevent freezing. Where remote sumps are not practical, reasonable protection can be provided by sump heaters, such as electric immersion heaters, steam coils, or hot-water coils. Water pumps and lines must also be protected, for example with electric heat-tracing tape and insulation.

Where the evaporative condenser is ducted to the outdoors, moisture from the warm saturated air can condense in condenser discharge ducts, especially if the ducts pass through a cool space. Some condensation may be unavoidable even with short, insulated ducts. In such cases, the condensate must be drained. Also, in these ducted applications, the drift eliminators must be highly effective.

MULTIPLE-CONDENSER INSTALLATIONS

Large refrigeration plants may have several evaporative condensers connected in parallel with each other or with shell-and-tube condensers. In these systems, unless all condensers have the same refrigerant-side pressure drop, condensed liquid refrigerant will flood back into the condensing coils of those condensers with the highest pressure loss. Also, in periods of light load when some condenser fans are off, liquid refrigerant will flood the lower coil circuits of the active condensers.

Trapped drop legs, as illustrated in Figures 14 and 15, provide proper control of such multiple installations. The effective height H of drop legs must equal the pressure loss through the condenser at maximum loading. Particularly in cold weather, when condensing pressure is controlled by shutting down some condenser fans, active condensers may be loaded considerably above nominal rating, with higher pressure losses. The height of the drop leg should be great enough to accommodate these greater design loads.

RATINGS

Heat rejected from an evaporative condenser is generally expressed as a function of the saturated condensing temperature and entering air wet bulb. The type of refrigerant has considerable effect on ratings; this effect is handled by separate tables (or curves) for each refrigerant, or by a correction factor when the difference is small.

Evaporative condensers are commonly rated in terms of total heat rejection when condensing a particular refrigerant at a specific condensing temperature and entering air wet-bulb temperature. Many manufacturers also provide alternative ratings in terms of evaporator load for a given refrigerant at a specific combination of saturated suction temperature and condensing temperature. These

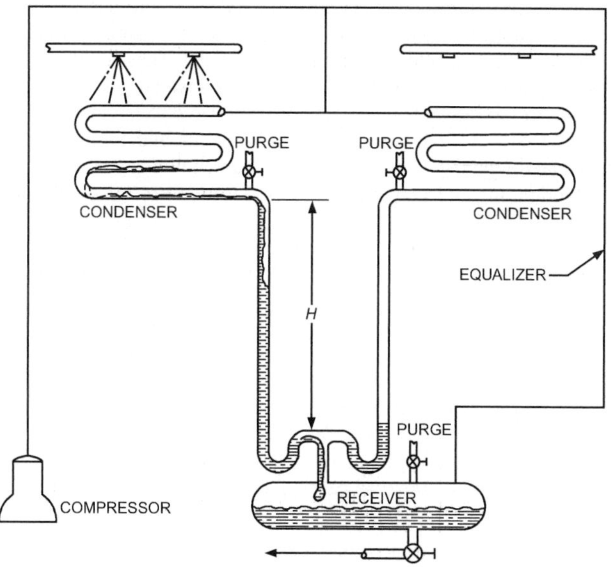

Each condenser can operate independently. Height (H) above trap, for either condenser, must be not less than internal resistance of condenser. Note purge connections at both condensers and receiver.

Fig. 15 Parallel Operation of Two Evaporative Condensers

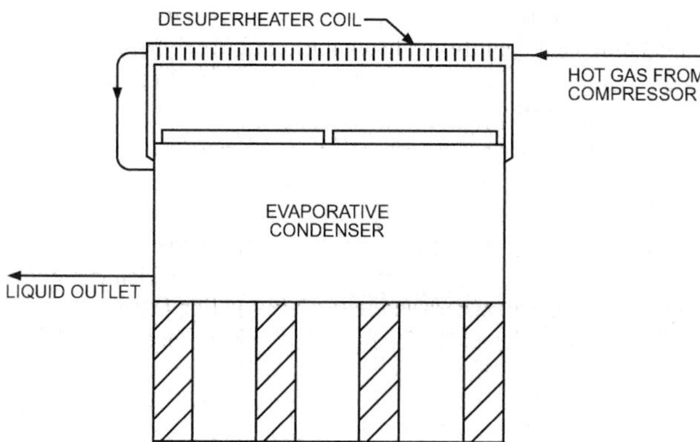

Fig. 16 Evaporative Condenser with Desuperheater Coil

ratings include an assumed value for heat of compression, typically based on an open, reciprocating compressor, and should only be used for that type of unit. Where another type of compression equipment is used, such as a gas-cooled hermetic and semihermetic compressor, the total power input to the compressor(s) should be added to the evaporator load and this value should be used to select the condenser.

Rotary screw compressors use oil to lubricate moving parts and provide a seal between the rotors and the compressor housing. This oil is subsequently cooled either in a separate heat exchanger or by refrigerant injection. In the former case, the amount of heat removed from the oil in the heat exchanger is subtracted from the sum of the refrigeration load and compressor brake power to obtain the total heat rejected by the evaporative condenser. When liquid injection is used, the total heat rejection is the sum of the refrigeration load and brake horsepower. Heat rejection rating data together with any ratings based on refrigeration capacity should be included.

DESUPERHEATING COILS

A desuperheater is an air-cooled finned coil, usually installed in the discharge airstream of an evaporative condenser (Figure 16). Its primary function is to increase condenser capacity by removing some of the superheat from the discharge vapor before the vapor enters the wetted condensing coil. The amount of superheat removed is a function of the desuperheater surface, condenser airflow, and temperature difference between the refrigerant and air. In practice, a desuperheater is limited to reciprocating compressor ammonia installations where discharge temperatures are relatively high (250 to 300°F).

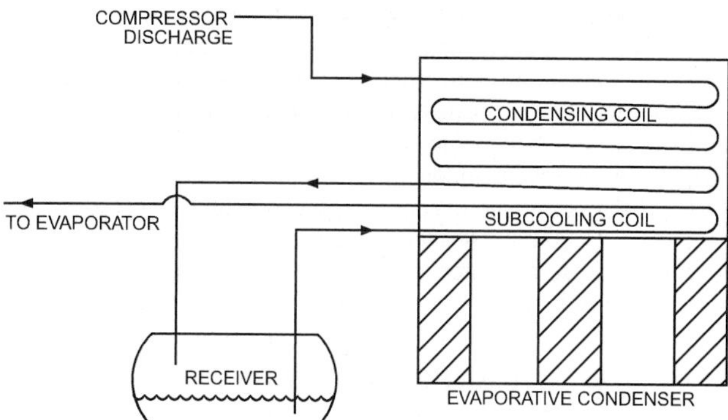

Fig. 17 Evaporative Condenser with Liquid Subcooling Coil

REFRIGERANT LIQUID SUBCOOLERS

The refrigerant pressure at the expansion device feeding the evaporator(s) is lower than that in the receiver because of the pressure drop in the liquid line. If the liquid line is long or if the evaporator is above the receiver (which further decreases refrigerant pressure at the expansion device), significant flashing can occur in the liquid line. To avoid flashing where these conditions exist, the liquid refrigerant must be subcooled after it leaves the receiver. The minimum amount of subcooling required is the temperature difference between the condensing temperature and saturation temperature corresponding to the saturation pressure at the expansion device. Subcoolers are often used with halocarbon systems but seldom with ammonia systems for the following reasons:

- Because ammonia has a relatively low liquid density, liquid line static pressure loss is small.
- Ammonia has a very high latent heat; thus, the amount of flash gas resulting from typical pressure loss in the liquid line is extremely small.
- Ammonia is seldom used in a direct-expansion feed system where subcooling is critical to proper expansion valve performance.

One method commonly used to supply subcooled liquid for halocarbon systems places a subcooling coil section in the evaporative condenser below the condensing coil (Figure 17). Depending on the design wet-bulb temperature, condensing temperature, and subcooling coil surface area, the subcooling coil section normally furnishes 10 to 15°F of liquid subcooling. As shown in Figure 17, a receiver must be installed between the condensing coil and subcooling coil to provide a liquid seal for the subcooling circuit.

MULTICIRCUIT CONDENSERS AND COOLERS

Evaporative condensers and evaporative fluid coolers, which are essentially the same, may be multicircuited to condense different refrigerants or cool different fluids simultaneously, but only if the difference between the condensing temperature and leaving fluid temperature is small. Typical multicircuits (1) condense different refrigerants found in food market units, (2) condense and cool different fluids in separate circuits (e.g., condensing refrigerant from a screw compressor in one circuit and cooling water or glycol for its oil cooler in another circuit), and (3) cool different fluids in separate circuits such as in many industrial applications where it is necessary to keep the fluids from different heat exchangers in separate circuits.

A multicircuit unit is usually controlled by sensing sump water temperature and using capacity control dampers, fan cycling, or a variable-speed drive to modulate airflow and match unit capacity to the load. Two-speed motors or separate high- and low-speed motors are also used. Sump water temperature has an averaging effect on control. Because all water in the sump is recirculated once every minute or two, its temperature is a good indicator of changes in load.

WATER TREATMENT

As recirculated water evaporates in an evaporative condenser, the dissolved solids in the makeup water continually increase as more water is added. Continued concentration of these dissolved solids can lead to scaling and/or corrosion problems. In addition, airborne impurities and biological contaminants are often introduced into recirculated water. If these impurities are not controlled, they can cause sludge or biological fouling. Simple blowdown (discharging a small portion of the recirculating water to a drain) may be adequate to control scale and corrosion on sites with good-quality makeup water, but it does not control biological contaminants such as *Legionella*. All evaporative condensers should be treated to restrict biological growth and many benefit from chemical treatment to control scale and corrosion. Chapter 48 of the 2007 *ASHRAE Handbook—HVAC Applications* covers water treatment in more detail. Also see ASHRAE *Guideline* 12 for specific recommendations on *Legionella* control. Specific recommendations on water treatment can be obtained from water treatment suppliers.

WATER CONSUMPTION

For the purpose of sizing makeup water piping, all heat rejected by an evaporative condenser is assumed to be latent heat (approximately 1050 Btu/lb of water evaporated). The heat rejected depends on operating conditions, but can range from 14,000 Btu/h per ton of air conditioning to 17,000 Btu/h per ton of freezer storage. The evaporated water ranges from about 1.6 to 2 gph per ton of refrigeration. In addition, a small amount of water can be lost in the form of drift through the eliminators. With good-quality makeup water, bleed rates may be as low as one-quarter to one-half the evaporation rate, and total water consumption would range from 2 gph/ton for air conditioning to 3 gph/ton for refrigeration.

CAPACITY MODULATION

To ensure operation of expansion valves and other refrigeration components, extremely low condensing pressures must be avoided. Capacity can be controlled in response to condensing pressure on single-circuit condensers and in response to the temperature of the spray water on multicircuited condensers. Means of controlling capacity include (1) intermittent fan operation; (2) a modulating damper in the airstream to reduce the airflow (centrifugal fan models only); and (3) fan speed control using variable-speed motors and/or drives, two-speed motors, or additional lower-power pony motors (small modular motors on the same shaft). Two-speed fan motors usually operate at 100 and 50% fan speed, which provides 100% and approximately 60% condenser capacity, respectively. Pony motors also provide some redundancy in case one motor fails. With the fans off and the water pump operating, condenser capacity is approximately 10%.

Often, a two-speed fan motor or supplementary (pony) motor that operates the fan at reduced speed can provide sufficient capacity control, because condensing pressure seldom needs to be held to a very tight tolerance other than to maintain a minimum condensing pressure to ensure refrigerant liquid feed pressure for the low-pressure side and/or sufficient pressure for hot-gas defrost requirements. For applications requiring close control of condensing pressure, virtually infinite capacity control can be provided by using frequency-modulating controls to control fan motor speed. These controls may also be justified economically because they save energy and extend the life of the fan and drive assembly compared to fan cycling with two-speed control. However, special concerns and limitations are associated with their use, which are discussed in Chapter 44.

Modulating air dampers also offer closer control on condensing pressure, but they do not offer as much fan power reduction at part load as fan speed control. Water pump cycling for capacity control is not recommended because the periodic drying of the tube surface promotes scale build-up.

The designer should research applicable codes and standards when applying evaporative condensers. Most manufacturers provide sound level information for their equipment. Centrifugal fans are inherently quieter than axial fans, but axial fans generally have a lower power requirement for a given application. Low-speed operation and multistaging usually lowers sound levels of condensers with axial fans. In some cases, factory-supplied sound attenuators may need to be installed.

PURGING

Refrigeration systems operating below atmospheric pressure and systems that are opened for service may require purging to remove air that causes a high condensing pressure. With the system operating, purging should be done from the top of the condensing coil outlet connection. On multiple-coil condensers or multiple-condenser installations, one coil at a time should be purged. Purging two or more coils at one time equalizes coil outlet pressures and can cause liquid refrigerant to back up in one or more of the coils, thus reducing operating efficiency.

Noncondensables can be removed from the condenser using a mechanical, automatic operational purge unit for halogenated refrigerants. Additional information on purge units is located in Chapter 42. Manual purging, which involves opening a manual valve and releasing noncondensables (and refrigerant) to the atmosphere, is not acceptable for halogenated refrigerants because environmental laws prohibit intentional venting of refrigerant from the system.

Purging may also be done from the high point of the evaporative condenser refrigerant feed, but it is only effective when the condenser is not operating. During normal operation, noncondensables are dispersed throughout the high-velocity vapor, and excessive refrigerant would be lost if purging were done from this location (see Figures 14 and 15). All codes and ordinances governing discharge, recovery, and recycling of refrigerants must be followed.

MAINTENANCE

Evaporative condensers are often installed in remote locations and may not receive the routine attention of operating and maintenance personnel. Programmed maintenance is essential, however, and the manufacturer's operation and maintenance guidelines (often downloadable from their Web sites) should be consulted. Table 2 shows a typical checklist for evaporative condensers. Chapter 39 also has information on maintenance that applies to evaporative condensers.

CODES AND STANDARDS

If state or local codes do not take precedence, design pressures, materials, welding, tests, and relief devices should be in accordance with the ASME *Boiler and Pressure Vessel Code*, Section VIII, Division 1. Evaporative condensers are typically exempt, however, from the ASME Code on the basis of Item (c) of the Scope of the Code, which states that if the inside diameter of the condenser shell is 6 in. or less, it is not governed by the Code. (Coils can be built in

Table 2 Typical Maintenance Checklist

Maintenance Item	Frequency
1. Check fan and motor bearings, and lubricate if necessary. Check tightness and adjustment of thrust collars on sleeve-bearing units and locking collars on ball-bearing units.	Q
2. Check belt tension.	M
3. Clean strainer. If air is extremely dirty, strainer may need frequent cleaning.	W
4. Check, clean, and flush sump, as required.	M
5. Check operating water level in sump, and adjust makeup valve, if required.	W
6. Check water distribution, and clean as necessary.	W
7. Check bleed water line to ensure it is operative and adequate as recommended by manufacturer.	M
8. Check fans and air inlet screens and remove any dirt or debris.	D
9. Inspect unit carefully for general preservation and cleanliness, and make any needed repairs immediately.	R
10. Check operation of controls such as modulating capacity control dampers.	M
11. Check operation of freeze control items such as pan heaters and their controls.	Y
12. Check the water treatment system for proper operation.	W
13. Inspect entire evaporative condenser for spot corrosion. Treat and refinish any corroded spot.	Y

D = Daily; W = Weekly; M = Monthly; Q = Quarterly;
Y = Yearly; R = As required.

compliance with ASME *Code* B31.5, however.) Other rating and testing standards include

- ASHRAE *Standard* 15, Safety Code for Mechanical Refrigeration.
- ASHRAE *Standard* 64, Methods of Testing Remote Mechanical-Draft Evaporative Refrigerant Condensers.
- ARI *Standard* 490, Remote Mechanical-Draft Evaporative Refrigerant Condensers.

REFERENCES

ARI. 1997. Fouling factors: A survey of their application in today's air conditioning and refrigeration industry. *Guideline* E-1997. Air-Conditioning and Refrigeration Institute, Arlington, VA.

ARI. 2007. Water-cooled refrigerant condensers, remote type. *Standard* 450-2007. Air-Conditioning and Refrigeration Institute, Arlington, VA.

ARI. 2001. Non-condensable gas purge equipment for use with low pressure centrifugal liquid chillers. *Standard* 580. Air-Conditioning and Refrigeration Institute, Arlington, VA.

ASHRAE. 1996. Reducing emission of halogenated refrigerants in refrigeration and air conditioning equipment and systems. *Guideline* 3-1996.

ASHRAE. 2000. Minimizing the risk of Legionellosis associated with building water systems. *Guideline* 12-2000.

ASHRAE. 1997. Methods of testing for rating remote mechanical-draft air-cooled refrigerant condensers. ANSI/ASHRAE *Standard* 20-1997.

ASHRAE. 1982. Waterside fouling resistance inside condenser tubes: Research Note 31 (RP-106). *ASHRAE Journal* 24(6):61.

ASME. 2001. *Boiler and pressure vessel code.* American Society of Mechanical Engineers, New York.

Ayub, Z.H. and S.A. Jones. 1987. Tubeside erosion/corrosion in heat exchangers. *Heating/Piping/Air Conditioning* (December):81.

Bergles, A.E. 1995. Heat transfer enhancement and energy efficiency—Recent progress and future trends. In *Advances in Enhanced Heat/Mass Transfer and Energy Efficiency*, M.M. Ohadi and J.C. Conklin, eds. HTD-vol. 320/PID-vol. 1. American Society of Mechanical Engineers, New York.

Briggs, D.E. and E.H. Young. 1969. Modified Wilson plot techniques for obtaining heat transfer correlations for shell and tube heat exchangers. *Chemical Engineering Progress Symposium Series* 65(92):35.

Butterworth, D. 1977. Developments in the design of shell-and-tube condensers. *ASME Paper No.* 77-WA/HT-24. American Society of Mechanical Engineers, New York.

Coates, K.E. and J.G. Knudsen. 1980. Calcium carbonate scaling characteristics of cooling tower water. *ASHRAE Transactions* 86(2).

Eckert, E.G. 1963. *Heat and mass transfer.* McGraw-Hill, New York.

Ghaderi, M., M. Salehi, and M.H. Saeedi. 1995. Review of in-tube condensation heat transfer correlations for smooth and enhanced tubes. In *Advances in Enhanced Heat/Mass Transfer and Energy Efficiency*, M.M. Ohadi and J.C. Conklin, eds. HTD-Vol. 320/PID-Vol. 1. American Society of Mechanical Engineers, New York.

Honda, H., H. Takamatsu, and K. Kim. 1994. Condensation of CFC-11 and HCFC-123 in in-line bundles of horizontal finned tubes: Effect of fin geometry. *Enhanced Heat Transfer* 1(2):197-209.

Honda, H., H. Takamatsu, N. Takada, O. Makishi, and H. Sejimo. 1995. Film condensation of HCFC-123 in staggered bundles of horizontal finned tubes. *Proceedings of the ASME/JSME Thermal Engineering Conference* 2:415-420.

Huber, J.B., L.E. Rewerts, and M.B. Pate. 1994a. Shell-side condensation heat transfer of R-134a. Part I: Finned tube performance. *ASHRAE Transactions* 100(2):239-247.

Huber, J.B., L.E. Rewerts, and M.B. Pate. 1994b. Shell-side condensation heat transfer of R-134a. Part II: Enhanced tube performance. *ASHRAE Transactions* 100(2):248-256.

Huber, J.B., L.E. Rewerts, and M.B. Pate. 1994c. Shell-side condensation heat transfer of R-134a. Part III: Comparison of R-134a with R-12. *ASHRAE Transactions* 100(2):257-264.

Huber, J.B., L.E. Rewerts, and M.B. Pate. 1994d. Experimental determination of shell side condenser bundle design factors for refrigerants R-123 and R-134a (RP-676). ASHRAE Research Project, *Final Report.*

Kedzierski, M.A. 1997. Effect of inclination on the performance of a compact brazed plate condenser and evaporator. *Heat Transfer Engineering* 18(3):25-38.

Kedzierski, M.A. and R.L. Webb. 1990. Practical fin shapes for surface drained condensation. *Journal of Heat Transfer* 112:479-485.

Kern, D.Q. and A.D. Kraus. 1972. *Extended surface heat transfer*, Chapter 10. McGraw-Hill, New York.

Kistler, R.S., A.E. Kassem, and J.M. Chenoweth. 1976. Rating shell-and-tube condensers by stepwise calculations. *ASME Paper No.* 76-WA/HT-5. American Society of Mechanical Engineers, New York.

Korenic, B. 1980. *Augmentation of heat transfer by evaporative coolings to reduce condensing temperatures*. Ph.D. dissertation, Purdue University, West Lafayette, IN.

Kragh, R.W. 1975. Brush cleaning of condenser tubes saves power costs. *Heating/Piping/Air Conditioning* (September).

Lee, S.H. and J.G. Knudsen. 1979. Scaling characteristics of cooling tower-water. *ASHRAE Transactions* 85(1).

Leidenfrost, W. and B. Korenic. 1979. Analysis of evaporative cooling and enhancement of condenser efficiency and of coefficient of performance. *Wärme und Stoffübertragung* 12.

Leidenfrost, W. and B. Korenic. 1982. Experimental verification of a calculation method for the performance of evaporatively cooled condensers. *Brennstoff-Wärme-Kraft* 34(1):9. VDI Association of German Engineers, Dusseldorf.

Leidenfrost, W., K.H. Lee, and B. Korenic. 1980. Conservation of energy estimated by second law analysis of a power-consuming process. *Energy* 47.

McAdams, W.H. 1954. *Heat transmission*, 3rd ed. McGraw-Hill, New York.

Ohadi, M.M., S. Dessiatoun, A. Singh, K. Cheung, and M. Salehi. 1995. EHD enhancement of boiling/condensation heat transfer of alternate refrigerants/refrigerant mixtures. *Progress Report* 9, submitted to U.S. DOE, Grant No. DOE/DE-FG02-93CE23803.A000.

Pate, M.B., Z.H. Ayub, and J. Kohler. 1991. Heat exchangers for the air-conditioning and refrigeration industry: State-of-the-art design and technology. *Heat Transfer Engineering* 12(3):56-70.

Pearson, J.F. and J.G. Withers. 1969. New finned tube configuration improves refrigerant condensing. *ASHRAE Journal* 11(6):77.

Seban, R.A. and E.F. McLaughlin. 1963. Heat transfer in tube coils with laminar and turbulent flow. *International Journal of Heat and Mass Transfer* 6:387.

Shah, M.M. 1979. A general correlation for heat transfer during film condensation in tubes. *International Journal of Heat and Mass Transfer* 22(4):547.

Shah, M.M. 1981. Heat transfer during film condensation in tubes and annuli: A review of the literature. *ASHRAE Transactions* 87(1).

Starner, K.E. 1976. Effect of fouling factors on heat exchanger design. *ASHRAE Journal* 18(May):39.

Sturley, R.A. 1975. Increasing the design velocity of water and its effect in copper tube heat exchangers. *Paper* 58. International Corrosion Forum, Toronto.

Suitor, J.W., W.J. Marner, and R.B. Ritter. 1976. The history and status of research in fouling of heat exchangers in cooling water service. *Paper* 76-CSME/CS Ch E-19. National Heat Transfer Conference, St. Louis, MO.

Taborek, J., F. Voki, R. Ritter, J. Pallen. and J. Knudsen. 1972. Fouling—The major unresolved problem in heat transfer. *Chemical Engineering Progress Symposium Series* 68, Parts I and II, Nos. 2 and 7.

TEMA. 1999 *Standards of the Tubular Exchanger Manufacturers Association*, 8th ed. Tubular Exchanger Manufacturers Association, New York.

Wanniarachchi, A.S. and R.L. Webb. 1982. Noncondensable gases in shell-side refrigerant condensers. *ASHRAE Transactions* 88(2):170-184.

Watkinson, A.P., L. Louis, and R. Brent. 1974. Scaling of enhanced heat exchanger tubes. *The Canadian Journal of Chemical Engineering* 52:558.

Webb, R.L. 1984a. Shell-side condensation in refrigerant condensers. *ASHRAE Transactions* 90(1B):5-25.

Webb, R.L. 1984b. The effects of vapor velocity and tube bundle geometry on condensation in shell-side refrigeration condensers. *ASHRAE Transactions* 90(1B):39-59.

Webb, R.L. and C.G. Murawski. 1990. Row effect for R-11 condensation on enhanced tubes. *Journal of Heat Transfer* 112(3).

Webb, R.L., A.S. Wanniarachchi, and T.M. Rudy. 1980. The effect of non-condensible gases on the performance of an R-11 centrifugal water chiller condenser. *ASHRAE Transactions* 86(2):57.

Wolverine Tube, Inc. 1984. *Engineering data book* 11. Section I, 30.

Young, E.H. and D.J. Ward. 1957. Fundamentals of finned tube heat transfer. *Refining Engineer* I (November).

BIBLIOGRAPHY

ARI. 2003. Remote mechanical-draft evaporative refrigerant coolers. *Standard* 490-2003. Air-Conditioning and Refrigeration Institute, Arlington, VA.

ARI. 2005. Remote mechanical-draft air-cooled refrigerant condensers. *Standard* 460-2005. Air-Conditioning and Refrigeration Institute, Arlington, VA.

ASHRAE. 2007. Safety standard for refrigeration systems. ANSI/ASHRAE *Standard* 15-2007.

ASHRAE. 2003. Methods of testing for rating water-cooled refrigerant condensers. *Standard* 22-2003.

ASHRAE. 2005. Methods of testing remote mechanical-draft evaporative refrigerant condensers. ANSI/ASHRAE *Standard* 64-2005.

ASME. 2006. Refrigeration piping and heat transfer components. *Code* B31.5. American Society of Mechanical Engineers, New York.

Eckels, S.J. and M.B. Pate. 1991. Evaporation and condensation of HFC-134a and CFC-12 in a smooth tube and a micro-fin tube. *ASHRAE Transactions* 97(2):71-81.

UL. 2001. Standard for refrigerant-containing components and accessories, nonelectrical. ANSI/UL *Standard* 207.

Webb, R.L. and S.H. Jung. 1992. Air-side performance of enhanced brazed aluminum heat exchangers. *ASHRAE Transactions: Symposia* 98(2):391-401.

CHAPTER 39

COOLING TOWERS

MOST air-conditioning systems and industrial processes generate heat that must be removed and dissipated. Water is commonly used as a heat transfer medium to remove heat from refrigerant condensers or industrial process heat exchangers. In the past, this was accomplished by drawing a continuous stream of water from a utility water supply or a natural body of water, heating it as it passed through the process, and then discharging the water directly to a sewer or returning it to the body of water. Water purchased from utilities for this purpose has now become prohibitively expensive because of increased water supply and disposal costs. Similarly, cooling water drawn from natural sources is relatively unavailable because the ecological disturbance caused by the increased temperature of discharge water has become unacceptable.

Air-cooled heat exchangers cool water by rejecting heat directly to the atmosphere, but the first cost and fan energy consumption of these devices are high and the plan area required is relatively large. They can economically cool water to within approximately 20°F of the ambient dry-bulb temperature—too high for the cooling water requirements of most refrigeration systems and many industrial processes.

Cooling towers overcome most of these problems and therefore are commonly used to dissipate heat from water-cooled refrigeration, air-conditioning, and industrial process systems. The water consumption rate of a cooling tower system is only about 5% of that of a once-through system, making it the least expensive system to operate with purchased water supplies. Additionally, the amount of heated water discharged (**blowdown**) is very small, so the ecological effect is greatly reduced. Lastly, cooling towers can cool water to within 4 to 5°F of the ambient wet-bulb temperature, or about 35°F lower than can air-cooled systems of reasonable size. This lower temperature improves the efficiency of the overall system, thereby reducing energy use significantly and increasing process output.

PRINCIPLE OF OPERATION

A cooling tower cools water by a combination of heat and mass transfer. Water to be cooled is distributed in the tower by spray nozzles, splash bars, or film-type fill, which exposes a very large water surface area to atmospheric air. Atmospheric air is circulated by (1) fans, (2) convective currents, (3) natural wind currents, or (4) induction effect from sprays. A portion of the water absorbs heat to change from a liquid to a vapor at constant pressure. This heat of vaporization at atmospheric pressure is transferred from the water remaining in the liquid state into the airstream.

Figure 1 shows the temperature relationship between water and air as they pass through a counterflow cooling tower. The curves indicate the drop in water temperature (A to B) and the rise in the air wet-bulb temperature (C to D) in their respective passages through the tower. The temperature difference between the water entering and leaving the cooling tower (A minus B) is the **range**. For a steady-state system, the range is the same as the water temperature rise through the load heat exchanger, provided the flow rate through the cooling tower and heat exchanger are the same. Accordingly, the range is determined by the heat load and water flow rate, not by the size or thermal capability of the cooling tower.

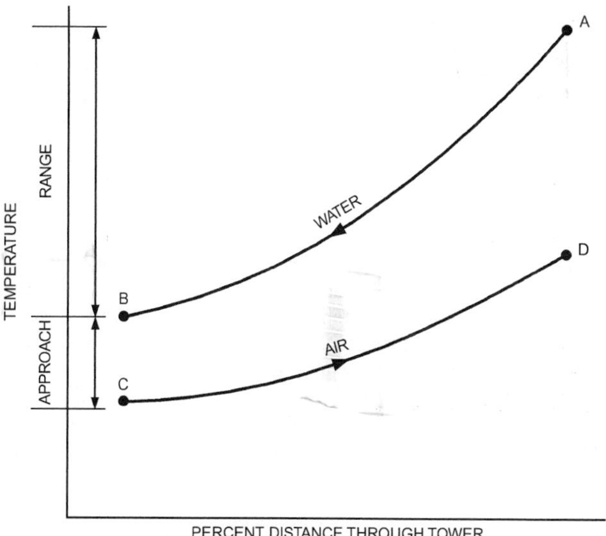

Fig. 1 Temperature Relationship Between Water and Air in Counterflow Cooling Tower

The difference between the leaving water temperature and entering air wet-bulb temperature (B minus C) in Figure 1 is the **approach to the wet bulb** or simply the **approach** of the cooling tower. The approach is a function of cooling tower capability, and a larger cooling tower produces a closer approach (colder leaving water) for a given heat load, flow rate, and entering air condition. Thus, the amount of heat transferred to the atmosphere by the cooling tower is always equal to the heat load imposed on the tower, whereas the temperature level at which the heat is transferred is determined by the thermal capability of the cooling tower and the entering air wet-bulb temperature.

Thermal performance of a cooling tower depends principally on the entering air wet-bulb temperature. The entering air dry-bulb temperature and relative humidity, taken independently, have an insignificant effect on thermal performance of mechanical-draft cooling towers, but do affect the rate of water evaporation in the cooling tower. A psychrometric analysis of the air passing through a cooling tower illustrates this effect (Figure 2). Air enters at the ambient condition Point A, absorbs heat and mass (moisture) from the water, and exits at Point B in a saturated condition (at very light loads, the discharge air may not be fully saturated). The amount of heat transferred from the water to the air is proportional to the difference in enthalpy of the air between the entering and leaving

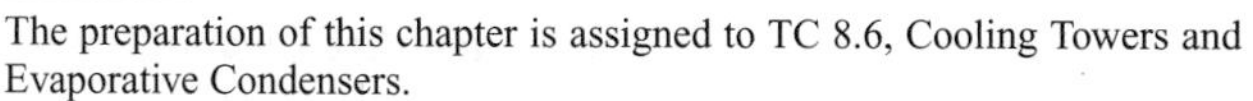

The preparation of this chapter is assigned to TC 8.6, Cooling Towers and Evaporative Condensers.

conditions ($h_B - h_A$). Because lines of constant enthalpy correspond almost exactly to lines of constant wet-bulb temperature, the change in enthalpy of the air may be determined by the change in wet-bulb temperature of the air.

Air heating (Vector AB in Figure 2) may be separated into component AC, which represents the sensible portion of the heat absorbed by the air as the water is cooled, and component CB, which represents the latent portion. If the entering air condition is changed to Point D at the same wet-bulb temperature but at a higher dry-bulb temperature, the total heat transfer (Vector DB) remains the same, but the sensible and latent components change dramatically. DE represents sensible **cooling** of air, while EB represents latent heating as water gives up heat and mass to the air. Thus, for the same water-cooling load, the ratio of latent to sensible heat transfer can vary significantly.

The ratio of latent to sensible heat is important in analyzing water usage of a cooling tower. Mass transfer (evaporation) occurs only in the latent portion of heat transfer and is proportional to the change in specific humidity. Because the entering air dry-bulb temperature or relative humidity affects the latent to sensible heat transfer ratio, it also affects the rate of evaporation. In Figure 2, the rate of evaporation in Case AB ($W_B - W_A$) is less than in Case DB ($W_B - W_D$) because the latent heat transfer (mass transfer) represents a smaller portion of the total.

The evaporation rate at typical design conditions is approximately 1% of the water flow rate for each 12.5°F of water temperature range; however, the average evaporation rate over the operating season is less than the design rate because the sensible component of total heat transfer increases as entering air temperature decreases.

In addition to water loss from evaporation, losses also occur because of liquid carryover into the discharge airstream and blowdown to maintain acceptable water quality. Both of these factors are addressed later in this chapter.

DESIGN CONDITIONS

The thermal capability of any cooling tower may be defined by the following parameters:

- Entering and leaving water temperatures
- Entering air wet-bulb or entering air wet-bulb and dry-bulb temperatures
- Water flow rate

The entering air dry-bulb temperature affects the amount of water evaporated from any evaporative cooling tower. It also affects airflow through hyperbolic towers and directly establishes thermal capability in any indirect-contact cooling tower component operating in a dry mode. Variations in tower performance associated with changes in the remaining parameters are covered in the section on Performance Curves.

The thermal capability of a cooling tower used for air conditioning is often expressed in nominal cooling tower tons. A nominal cooling tower ton is defined as cooling 3 gpm of water from 95°F to 85°F at a 78°F entering air wet-bulb temperature. At these conditions, the cooling tower rejects 15,000 Btu/h per nominal cooling tower ton. The historical derivation of this 15,000 Btu/h cooling tower ton, as compared to the 12,000 Btu/h evaporator ton, is based on the assumption that at typical air-conditioning conditions, for every 12,000 Btu/h of heat picked up in the evaporator, the cooling tower must dissipate an additional 3000 Btu/h of compressor heat. For specific applications, however, nominal tonnage ratings are not used, and the thermal performance capability of the tower is usually expressed as a water flow rate at specific operating temperature conditions (entering water temperature, leaving water temperature, entering air wet-bulb temperature).

TYPES OF COOLING TOWERS

Two basic types of evaporative cooling devices are used. The first of these, the **direct-contact** or **open cooling tower** (Figure 3), exposes water directly to the cooling atmosphere, thereby transferring the source heat load directly to the air. The second type, often called a **closed-circuit cooling tower**, involves **indirect contact** between heated fluid and atmosphere (Figure 4), essentially combining a heat exchanger and cooling tower into one relatively compact device.

Of the direct-contact devices, the most rudimentary is a **spray-filled tower** that exposes water to the air without any heat transfer medium or fill. In this device, the amount of water surface exposed to the air depends on the spray efficiency, and the time of contact depends on the elevation and pressure of the water distribution system.

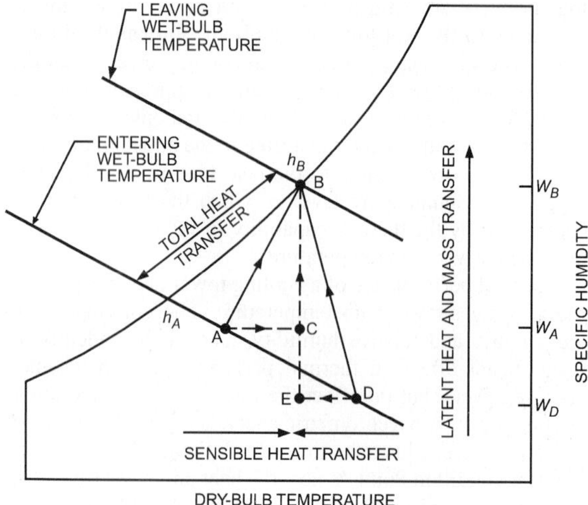

Fig. 2 Psychrometric Analysis of Air Passing Through Cooling Tower

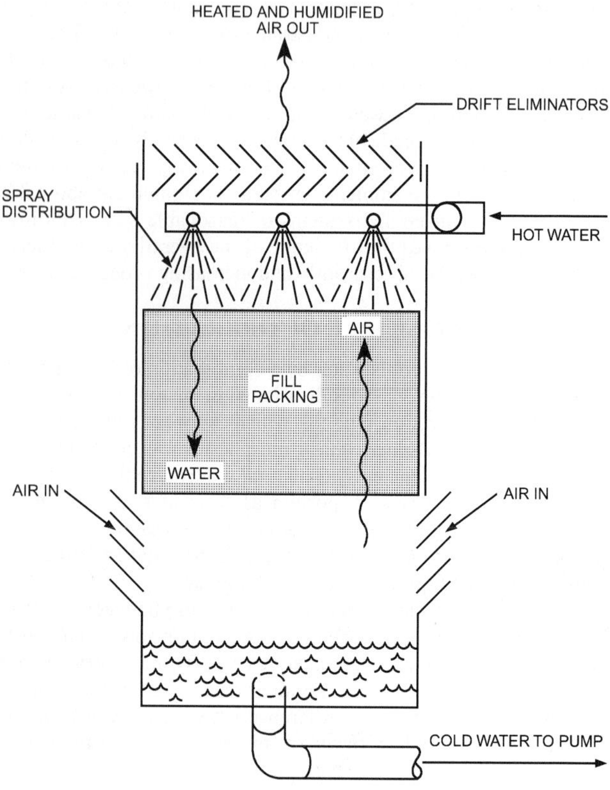

Fig. 3 Direct-Contact or Open Evaporative Cooling Tower

To increase contact surfaces as well as time of exposure, a heat transfer medium, or **fill**, is installed below the water distribution system, in the path of the air. The two types of fill in use are splash-type and film-type (Figure 5). **Splash-type fill** maximizes contact area and time by forcing the water to cascade through successive elevations of splash bars arranged in staggered rows. **Film-type fill** achieves the same effect by causing the water to flow in a thin layer over closely spaced sheets, principally polyvinyl chloride (PVC), that are arranged vertically.

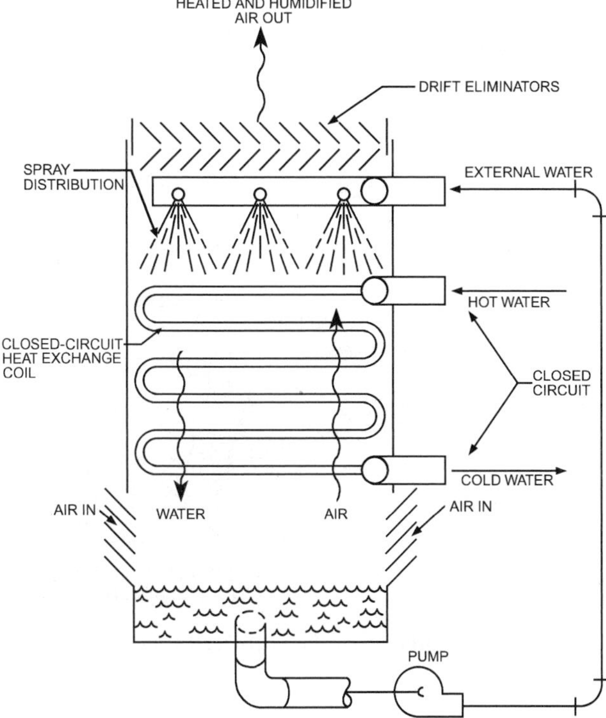

Fig. 4 Indirect-Contact or Closed-Circuit Evaporative Cooling Tower

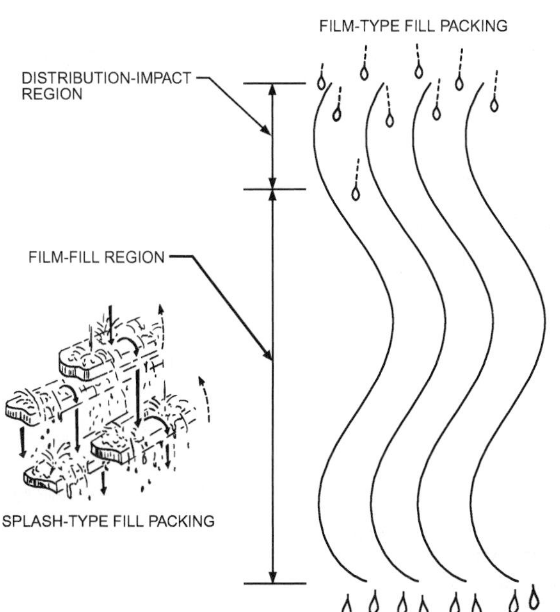

Fig. 5 Types of Fill

Either type of fill can be used in counterflow and cross-flow towers. For thermal performance levels typically encountered in air conditioning and refrigeration, a tower with film-type fill is usually more compact. However, splash-type fill is less sensitive to initial air and water distribution and, along with specially configured, more widely spaced film-type fills, is preferred for applications that may be subjected to blockage by scale, silt, or biological fouling.

Indirect-contact (closed-circuit) cooling towers contain two separate fluid circuits: (1) an external circuit, in which water is exposed to the atmosphere as it cascades over the tubes of a coil bundle, and (2) an internal circuit, in which the fluid to be cooled circulates inside the tubes of the coil bundle. In operation, heat flows from the internal fluid circuit, through the tube walls of the coil, to the external water circuit and then, by heat and mass transfer, to atmospheric air. As the internal fluid circuit never contacts the atmosphere, this unit can be used to cool fluids other than water and/or to prevent contamination of the primary cooling circuit with airborne dirt and impurities. Some closed-circuit cooling tower designs include cooling tower fill to augment heat exchange in the coil (Figure 6).

Types of Direct-Contact Cooling Towers

Non-Mechanical-Draft Towers. Aspirated by sprays or a differential in air density, these towers do not contain fill and do not use a mechanical air-moving device. The aspirating effect of the water spray, either vertical (Figure 7) or horizontal (Figure 8), induces airflow through the tower in a parallel flow pattern.

Because air velocities for the **vertical spray tower** (both entering and leaving) are relatively low, such towers are susceptible to adverse wind effects and, therefore, are normally used to satisfy a low-cost requirement when operating temperatures are not critical to the system. Some **horizontal spray towers** (Figure 8) use high-pressure sprays to induce large air quantities and improve air/water contact. Multispeed or staged pumping systems are normally recommended to reduce energy use in periods of reduced load and ambient conditions.

Chimney (hyperbolic) towers have been used primarily for large power installations, but may be of generic interest (Figure 9). The heat transfer mode may be counterflow, cross-flow, or parallel flow. Air is induced through the tower by the air density differentials that exist between the lighter, heat-humidified chimney air and the outside atmosphere. Fill can be splash or film type.

Primary justification of these high first-cost products comes through reduction in auxiliary power requirements (elimination of

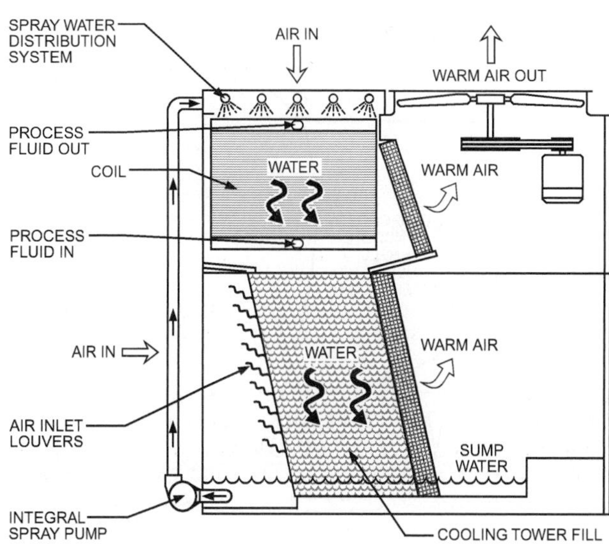

Fig. 6 Combined Flow Coil/Fill Evaporative Cooling Tower

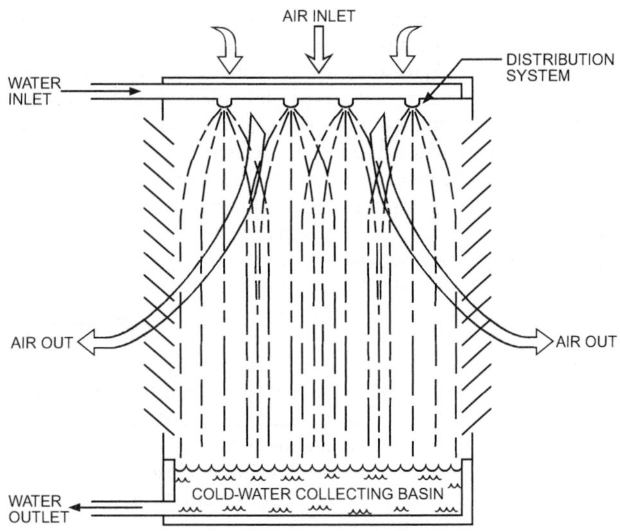

Fig. 7 Vertical Spray Tower

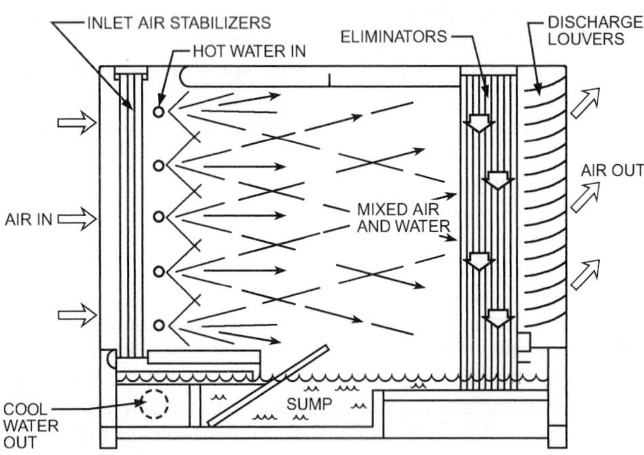

Fig. 8 Horizontal Spray Tower

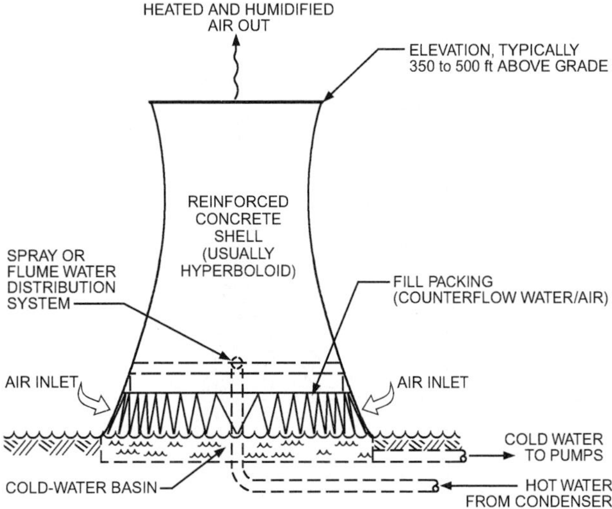

Fig. 9 Hyperbolic Tower

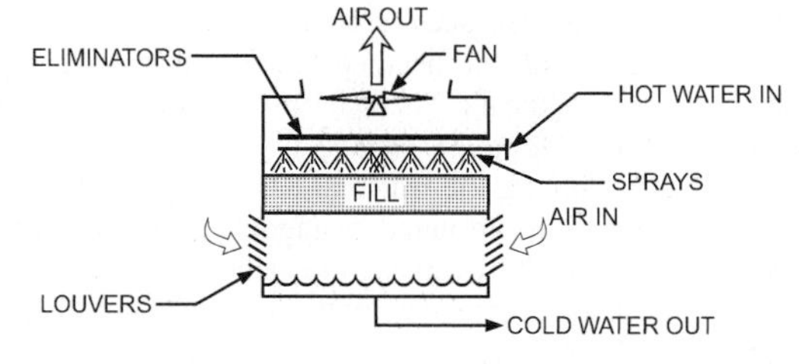

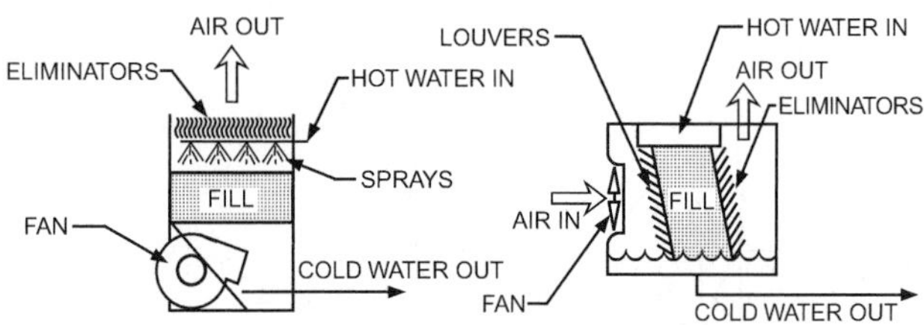

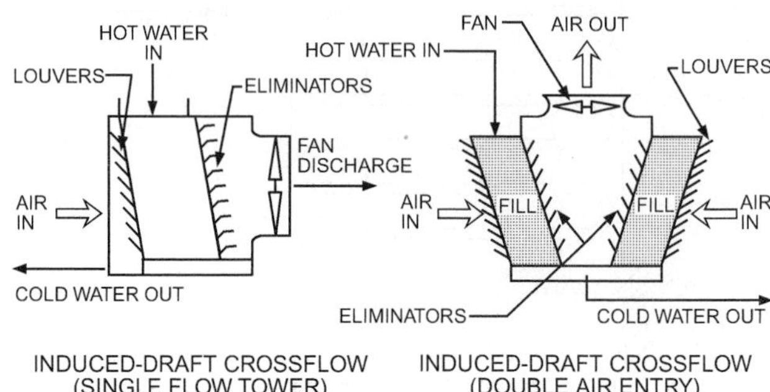

Fig. 10 Conventional Mechanical-Draft Cooling Towers

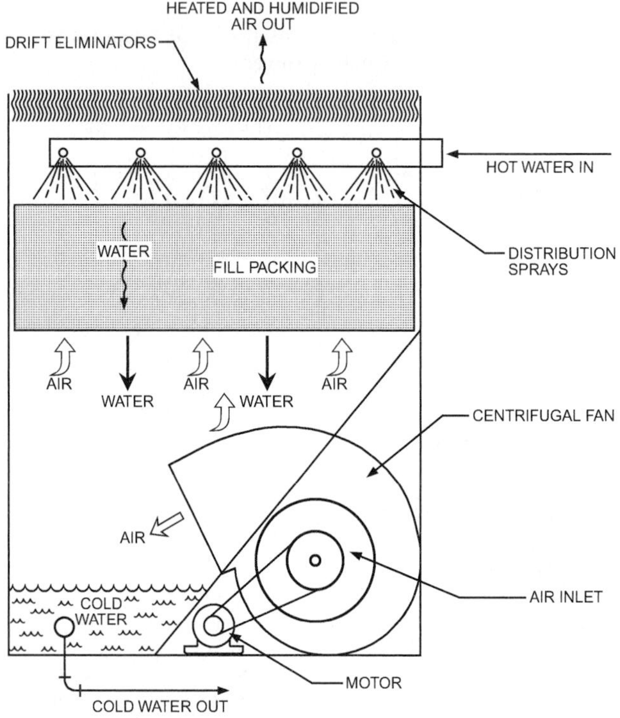

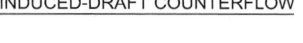

Fig. 11 Factory-Assembled Counterflow Forced-Draft Tower

fan energy), reduced property area, and elimination of recirculation and/or vapor plume interference. Materials used in chimney construction have been primarily steel-reinforced concrete; early-day timber structures had size limitations.

Mechanical-Draft Towers. Figure 10 shows five different designs for mechanical-draft (conventional) towers. Fans may be on the inlet air side (forced-draft) or the exit air side (induced-draft). The type of fan selected, either centrifugal or axial, depends on external pressure needs, permissible sound levels, and energy usage requirements. Water is downflow; the air may be upflow (counterflow heat transfer) or horizontal flow (cross-flow heat transfer). Air entry may be through one, two, three, or all four sides of the tower. All four combinations (i.e., forced-draft counterflow, induced-draft counterflow, forced-draft cross-flow, and induced-draft cross-flow) have been produced in various sizes and configurations.

Towers are typically classified as either factory-assembled (Figure 11), where the entire tower or a few large components are factory-assembled and shipped to the site for installation, or field-erected (Figure 12), where the tower is constructed completely on site.

Most factory-assembled towers are of metal construction, usually galvanized steel. Other constructions include stainless steel and fiberglass-reinforced plastic (FRP) towers and components. Field-erected towers are predominantly framed of preservative-treated Douglas fir or redwood, with FRP used for special components and casing materials. Environmental concerns about cutting timber and wood preservatives leaching into cooling tower water have led to an increased number of cooling towers having FRP structural framing. Field-erected towers may also be constructed of galvanized steel or stainless steel. Coated metals, primarily steel, are also used for complete towers or components. Concrete and ceramic materials are usually restricted to the largest towers (see the section on Materials of Construction).

Special-purpose towers containing a conventional mechanical-draft unit in combination with an air-cooled (finned-tube) heat exchanger are **wet/dry towers** (Figure 13). They are used for either vapor plume reduction or water conservation. The hot, moist plumes discharged from cooling towers are especially dense in cooler weather. On some installations, limited abatement of these plumes is required to avoid restricted visibility on roadways, on bridges, and around buildings.

A vapor plume abatement tower usually has a relatively small air-cooled component that tempers the leaving airstream to reduce the relative humidity and thereby minimize the fog-generating potential of the tower. Conversely, a water conservation tower usually requires a large air-cooled component to significantly reduce water consumption and provide plume abatement. Some designs can handle heat loads entirely by the nonevaporative air-cooled heat exchanger portion during reduced ambient temperature conditions.

A variant of the wet/dry tower is an evaporatively precooled/air-cooled heat exchanger. It uses an adiabatic saturator (air precooler/humidifier) to enhance the summer performance of an air-cooled exchanger, thus conserving water compared to conventional cooling towers (annualized) (Figure 14). Evaporative fill sections usually operate only during specified summer periods, whereas full dry operation is expected below 50 to 70°F ambient conditions. Integral water pumps return the lower basin water to the upper distribution systems of the adiabatic saturators in a manner similar to closed-circuit fluid cooler and evaporative condenser products.

Other Methods of Direct Heat Rejection

Ponds, Spray Ponds, Spray Module Ponds, and Channels. Heat dissipates from the surface of a body of water by evaporation, radiation, and convection. Captive lakes or ponds (man-made or

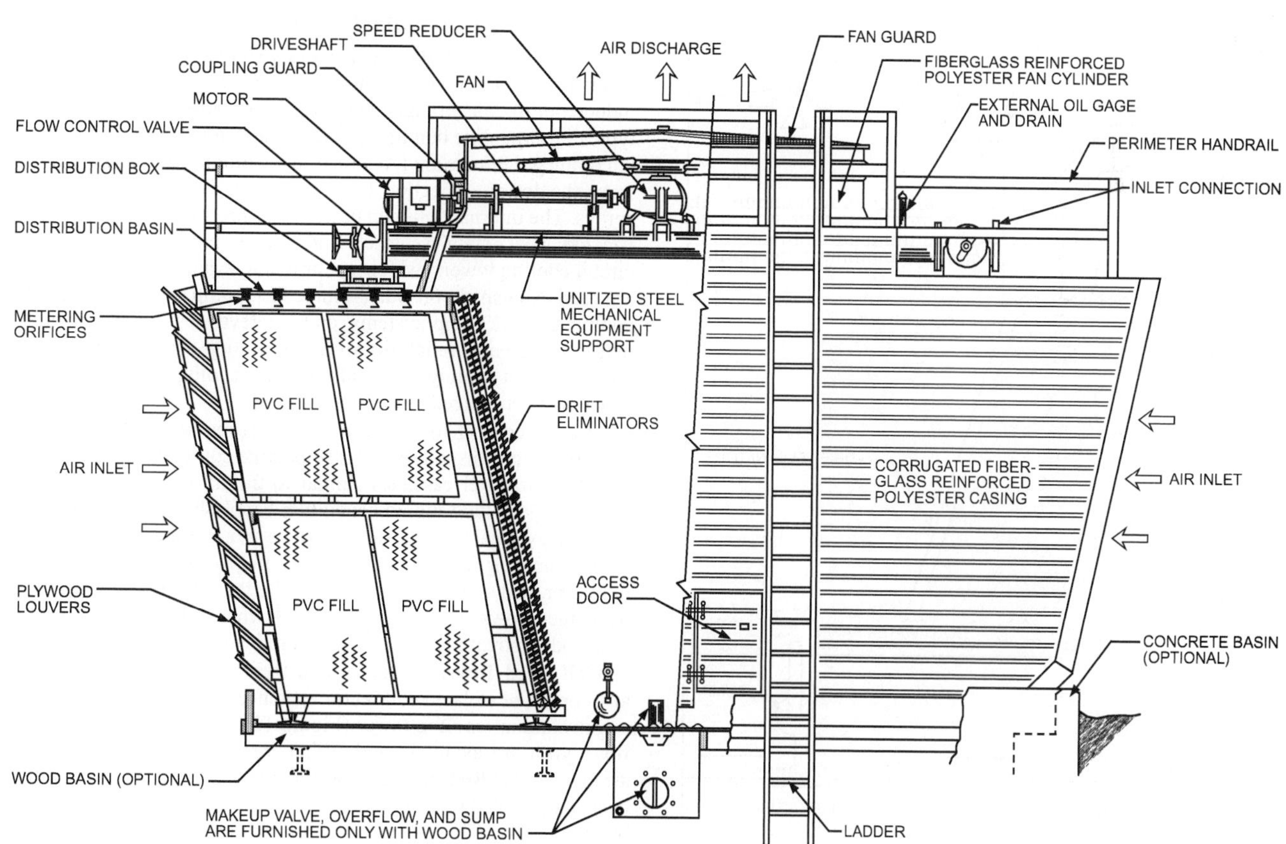

Fig. 12 Field-Erected Cross-Flow Mechanical-Draft Tower

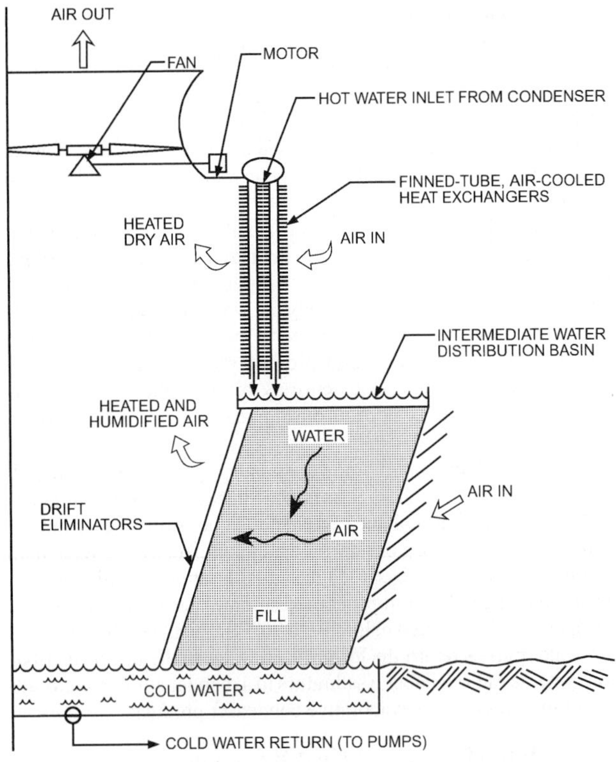

Fig. 13 Combination Wet-Dry Tower

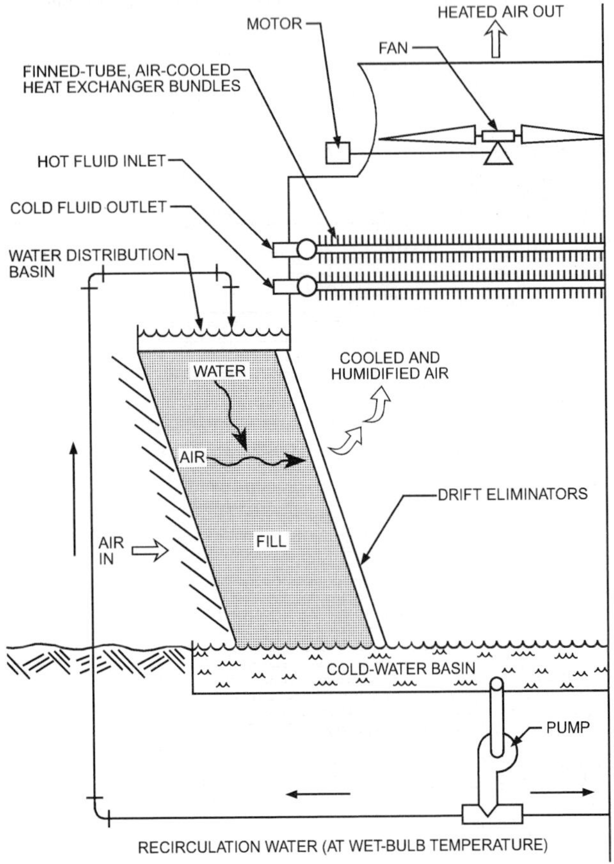

Fig. 14 Adiabatically Saturated Air-Cooled Heat Exchanger

natural) are sometimes used to dissipate heat by natural air currents and wind. This system is usually used in large plants where real estate is not limited.

A pump-spray system above the pond surface improves heat transfer by spraying water in small droplets, thereby extending the water surface and bringing it into intimate contact with the air. Heat transfer is largely the result of evaporative cooling (see the section on Cooling Tower Theory). The system is a piping arrangement using branch arms and nozzles to spray circulated water into the air. The pond acts largely as a collecting basin. Temperature control, real estate demands, limited approach to the wet-bulb temperature, and winter operational difficulties have ruled out the spray pond in favor of more compact and more controllable mechanical- or natural-draft towers.

Empirically derived relationships such as Equation (1) have been used to estimate cooling pond area. However, because of variations in wind velocity and solar radiation as well as the overall validity of the relationship itself, a substantial margin of safety should be added to the result.

$$w_p = \frac{A(95 + 0.425v)}{h_{fg}}[p_w - p_a] \tag{1}$$

where

w_p = evaporation rate of water, lb/h
A = area of pool surface, ft^2
v = air velocity over water surface, fpm
h_{fg} = latent heat required to change water to vapor at temperature of surface water, Btu/lb
p_a = saturation vapor pressure at dew-point temperature of ambient air, in. Hg
p_w = saturation vapor pressure at temperature of surface water, in. Hg

Types of Indirect-Contact Towers

Closed-Circuit Cooling Towers (Mechanical Draft). Both counterflow and cross-flow arrangements are used in forced- and induced-draft fan arrangements. The tubular heat exchangers are typically serpentine bundles, usually arranged for free gravity internal drainage. Pumps are integrated in the product to transport water from the lower collection basin to upper distribution basins or sprays. The internal coils can be of any of several materials, but galvanized steel, stainless steel, and copper predominate. Closed-circuit cooling towers, which are similar to evaporative condensers, are used extensively on water-source heat pump systems and screw compressor oil pump systems, and wherever the reduced maintenance and greater reliability of a closed-loop system are desired. Closed-circuit cooling towers also provide cooling for multiple heat loads on a centralized closed-loop system.

Indirect-contact towers (see Figure 4) require a closed-circuit heat exchanger (usually tubular serpentine coil bundles) that is exposed to air/water cascades similar to the fill of a cooling tower. Some types include supplemental film or splash fill sections to augment the external heat exchange surface area (Figure 5). In Figure 6, for instance, air flows down over the coil, parallel to the recirculating water, and exits horizontally into the fan plenum. Recirculating water then flows over cooling tower fill, where it is further cooled by a second airstream before being reintroduced over the coil.

Coil Shed Towers (Mechanical Draft). Coil shed towers usually consist of isolated coil sections (nonventilated) located beneath a conventional cooling tower (Figure 15). Counterflow and cross-flow types are available with either forced- or induced-draft fan arrangements. Redistribution water pans, located at the tower's base, feed cooled water by gravity flow to the tubular heat exchange bundles (coils). These units are similar in function to closed-circuit fluid coolers, except that supplemental fill is always required, and the airstream is directed only through the fill regions of the tower.

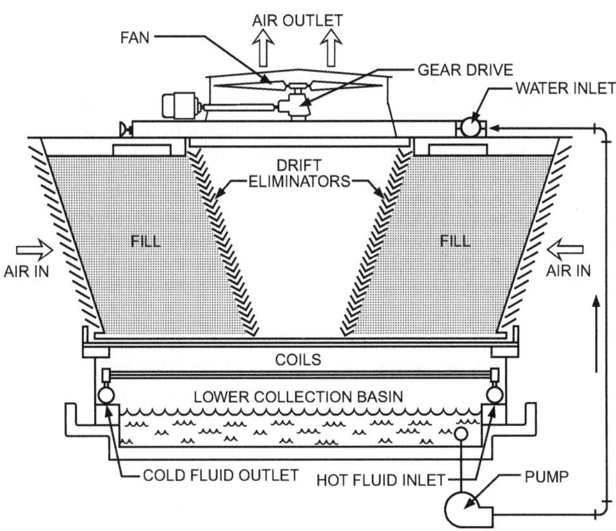

Fig. 15 Coil Shed Cooling Tower

Typically, these units are arranged as field-erected, multifan cell towers and are used primarily in industrial process cooling.

MATERIALS OF CONSTRUCTION

Materials for cooling tower construction are usually selected to meet the expected water quality and atmospheric conditions.

Wood. Wood has been used extensively for all static components except hardware. Redwood and fir predominate, usually with post-fabrication pressure treatment of waterborne preservative chemicals, typically chromated copper arsenate (CCA) or acid copper chromate (ACC). These microbicidal chemicals prevent the attack of wood-destructive organisms such as termites or fungi.

Metals. Steel with galvanized zinc is used for small and medium-sized installations. Hot-dip galvanizing after fabrication is used for larger weldments. Hot-dip galvanizing and cadmium and zinc plating are used for hardware. Brasses and bronzes are selected for special hardware, fittings, and tubing material. Stainless steels (principally 302, 304, and 316) are often used for sheet metal, drive shafts, and hardware in exceptionally corrosive atmospheres or to extend unit life. Stainless steel cold-water basins are increasingly popular. Cast iron is a common choice for base castings, fan hubs, motor or gear reduction housings, and piping valve components. Metals coated with polyurethane and PVC are used selectively for special components. Two-part epoxy compounds and epoxy-powdered coatings are also used for key components or entire cooling towers.

Plastics. Fiberglass-reinforced plastic (FRP) materials are used for components such as structure, piping, fan cylinders, fan blades, casing, louvers, and structural connecting components. Polypropylene and acrylonitrile butadiene styrene (ABS) are specified for injection-molded components, such as fill bars and flow orifices. PVC is typically used as fill, eliminator, and louver materials. Polyethylene is now used for both hot- and cold-water basins. Reinforced plastic mortar is used in larger piping systems, coupled by neoprene O-ring-gasketed bell and spigot joints.

Graphite Composites. Graphite composite drive shafts are available for use on cooling tower installations. These shafts offer a strong, corrosion-resistant alternative to steel/stainless steel shafts and are often less expensive, more forgiving of misalignment, and transmit less vibration.

Concrete, Masonry, and Tile. Concrete is typically specified for cold-water basins of field-erected cooling towers and is used in piping, casing, and structural systems of the largest towers, primarily in the power and process industries. Special tiles and masonry are used when aesthetic considerations are important.

SELECTION CONSIDERATIONS

Selecting the proper water-cooling equipment for a specific application requires consideration of cooling duty, economics, required services, environmental conditions, maintenance requirements, and aesthetics. Many of these factors are interrelated, but they should be evaluated individually.

Because a wide variety of water-cooling equipment may meet the required cooling duty, factors such as height, length, width, volume of airflow, fan and pump energy consumption, materials of construction, water quality, and availability influence final equipment selection.

The optimum choice is generally made after an economic evaluation. Chapter 36 of the 2007 *ASHRAE Handbook—HVAC Applications* describes two common methods of economic evaluation—life-cycle costing and payback analysis. Each of these procedures compares equipment on the basis of total owning, operating, and maintenance costs.

Initial-cost comparisons consider the following factors:

- Erected cost of equipment
- Costs of interface with other subsystems, which include items such as
  - Basin grillage and value of the space occupied
  - Pumps and prime movers
  - Electrical wiring to pump and fan motors
  - Electrical controls and switchgear
  - Piping to and from the tower (some designs require more inlet and discharge connections than others, thus affecting the cost of piping)
  - Tower basin, sump screens, overflow piping, and makeup lines, if not furnished by the manufacturer
  - Shutoff and control valves, if not furnished by the manufacturer
  - Walkways, ladders, etc., providing access to the tower, if not furnished by the manufacturer
  - Fire protection sprinkler system

In evaluating owning and maintenance costs, consider the following major items:

- System energy costs (fans, pumps, etc.) on the basis of operating hours per year
- Energy demand charges
- Expected equipment life
- Maintenance and repair costs
- Money costs

Other factors are (1) safety features and safety codes; (2) conformity to building codes; (3) general design and rigidity of structures; (4) relative effects of corrosion, scale, or deterioration on service life; (5) availability of spare parts; (6) experience and reliability of manufacturers; (7) independent certification of thermal ratings; and (8) operating flexibility for economical operation at varying loads or during seasonal changes. In addition, equipment vibration, sound levels, acoustical attenuation, and compatibility with the architectural design are important. The following section details many of these more important considerations.

APPLICATION

This section describes some of the major design considerations, but the cooling tower manufacturer should be consulted for more detailed recommendations.

Siting

When a cooling tower can be located in an open space with free air motion and unimpeded air supply, siting is normally not an obstacle to satisfactory installation. However, towers are often situated indoors, against walls, or in enclosures. In such cases, the following factors must be considered:

- Sufficient free and unobstructed space should be provided around the unit to ensure an adequate air supply to the fans and to allow proper servicing.
- Tower discharge air should not be deflected in any way that might promote recirculation [a portion of the warm, moist discharge air reentering the tower (Figure 16)]. Recirculation raises the entering wet-bulb temperature, causing increased hot water and cold water temperatures, and, during cold weather operation, can promote the icing of air intake areas. The possibility of air recirculation should be considered, particularly on multiple-tower installations.

Additionally, cooling towers should be located to prevent introducing the warm discharge air and any associated drift, which may contain chemical and/or biological contaminants, into the fresh air intake of the building that the tower is serving or into those of adjacent buildings.

Location of the cooling tower is usually determined by one or more of the following: (1) structural support requirements, (2) rigging limitations, (3) local codes and ordinances, (4) cost of bringing auxiliary services to the cooling tower, and (5) architectural compatibility. Sound, plume, and drift considerations are also best handled by proper site selection during the planning stage. For additional information on seismic and wind restraint, see Chapter 54 of the 2007 *ASHRAE Handbook—HVAC Applications*.

Piping

Piping should be adequately sized according to standard commercial practice. All piping should be designed to allow expansion and contraction. If the tower has more than one inlet connection, balancing valves should be installed to balance the flow to each cell properly. Positive shutoff valves should be used, if necessary, to isolate individual cells for servicing.

When two or more towers operate in parallel, an equalizer line between the tower sumps handles imbalances in the piping to and from the units and changing flow rates that arise from obstructions such as clogged orifices and strainers. All heat exchangers, and as much tower piping as possible, should be installed below the operating water level in the cooling tower to prevent overflowing of the cooling tower at shutdown and to ensure satisfactory pump operation during start-up. Tower basins must carry the proper amount of water during operation to prevent air entrainment into the water suction line. Tower basins should also have enough reserve volume between the operating and overflow levels to fill riser and water distribution lines on start-up and to fulfill the water-in-suspension requirement of the tower. Unlike open towers, closed-circuit cooling towers can be installed anywhere, even below the heat exchangers, as the fluid to be cooled is contained in a closed loop; the external spray water is self-contained within the closed-circuit cooling tower.

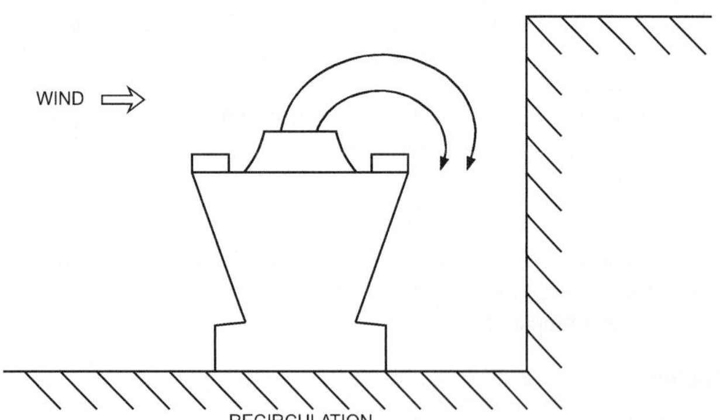

Fig. 16 Discharge Air Reentering Tower

Capacity Control

Most cooling towers encounter substantial changes in ambient wet-bulb temperature and load during the normal operating season. Accordingly, some form of capacity control may be required to maintain prescribed condensing temperatures or process conditions.

Fan cycling is the simplest method of capacity control on cooling towers and is often used on multiple-unit or multiple-cell installations. In nonfreezing climates, where close control of the exit water temperature is not essential, fan cycling is an adequate and inexpensive method of capacity control. However, motor burnout from too-frequent cycling is a concern.

Two-speed fan motors or additional lower-power pony motors, in conjunction with fan cycling, can double the number of steps of capacity control compared to fan cycling alone. This is particularly useful on single-fan motor units, which would have only one step of capacity control by fan cycling. Two-speed fan motors are commonly used on cooling towers as the primary method of capacity control, and provide the added advantage of reduced energy consumption at reduced load. Pony motors also provide some redundancy in case one motor fails.

It is more economical to operate all fans at the same speed than to operate one fan at full speed before starting the next. Figure 17 compares cooling tower fan power versus speed for single-, two-, and variable-speed fan motors.

Frequency-modulating controls for fan motor speed can provide virtually infinite capacity control and energy management. Previously, automatic, variable-pitch propeller fans were the only way to do this; however, these mechanically complex drive systems are more expensive and have higher sound levels, because they operate at full design speed only. They have largely been replaced by variable-frequency drives (VFDs) coupled with a standard fixed-pitch fan, thereby saving more fan energy and operating significantly more quietly when the tower is at less than full load.

Variable-frequency fan drives are economical and can save considerable energy as well as extend the life of the motor, fan, and drive (gearbox or V-belt) assembly compared to fan cycling with two-speed control. However, special considerations that should be discussed with the tower manufacturer and the supplier of the VFD include the following:

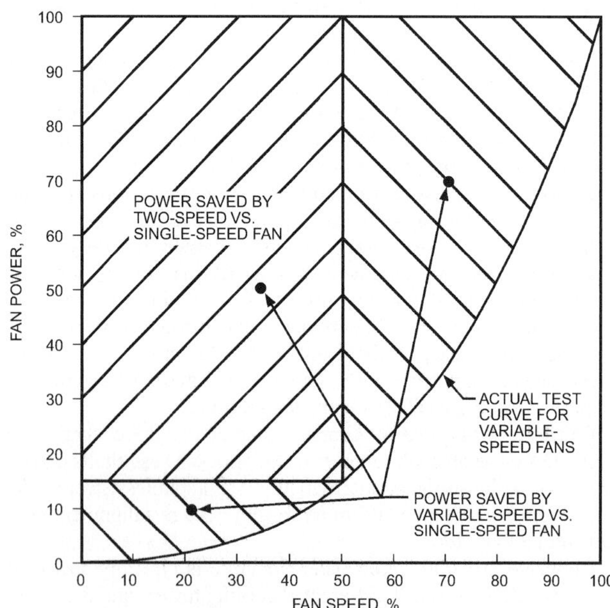

Fig. 17 Cooling Tower Fan Power Versus Speed
(White 1994)

- Care must be taken to avoid operating the fan system at a critical speed or a multiple thereof. **Critical speeds** are fan operating speeds identical to one of the natural frequencies of the fan assembly and/or supporting structure. At these speeds, fan resonance occurs, resulting in excessive vibration and possibly fan system failure, sometimes very quickly. Consult the tower manufacturer on what speeds (if any) must be avoided. Alternatively, the tower can be tested at start-up using an accelerometer to identify critical frequencies throughout the full speed range, thought this is generally not necessary with pre-engineered, factory-assembled units. Critical frequencies, identified either by the manufacturer or through actual testing, must be locked out in the VFD skip frequency program.
- Some VFDs, particularly pulse-width modulating (PWM) drives, create overvoltages at the motor that can cause motor and bearing failures. The magnitude of these overvoltages increases significantly with the length of cable between the controller and the motor, so lead lengths should be kept as short as possible. Special motors, filters, or other corrective measures may be necessary to ensure dependable operation. Consult the cooling tower manufacturer and/or the VFD supplier. Chapter 44 also has more information on variable-frequency drives.
- A VFD-compatible motor should be specified on all cooling towers with variable-frequency drives.
- Variable-frequency drives should not be operated at speeds below approximately 25% of the full design fan speed. Although most VFDs can modulate down to 10% or less of motor speed, a 25% lower limit is recommended to maintain proper air and water distribution. At very low airflow rates, the water and air can channel into separate streams, especially in counterflow units where uniform water distribution may depend on the airflow rate. Uneven water distribution can lead to wet/dry areas and scaling. In freezing climates, these separate streams could also result in localized ice formation. Additionally, the fan energy saving at 10% compared to 25% of full design speed is typically insignificant.

  This precaution is particularly true for units with gear speed reducers because low speeds can cause rotational "cogging," vibration, noise or "gear chatter," and pose a lubrication problem if the gear is either not equipped with an electric oil pump or not specifically designed to run at these low speeds. Consult the tower manufacturer for proper selection and adjustments of the VFD.

Modulating dampers in the discharge of centrifugal blower fans are also used for cooling tower capacity control, as well as for energy management. In some cases, modulating dampers may be used with two-speed motors. Note that modulating dampers have increasingly been replaced by variable-frequency drives for these purposes.

Cooling towers that inject water to induce airflow through the cooling tower have various pumping arrangements for capacity control. Multiple pumps in series or two-speed pumping provide capacity control and also reduce energy consumption.

Modulating water bypasses for capacity control should be used only after consultation with the cooling tower manufacturer. This is particularly important at low ambient conditions in which the reduced water flows can promote freezing within the tower.

Water-Side Economizer (Free Cooling)

With an appropriately equipped and piped system, using the tower for free cooling during reduced load and/or reduced ambient conditions can significantly reduce system energy consumption. Because the tower's cold-water temperature drops as the load and ambient temperature drop, the water temperature will eventually be low enough to serve the load directly, allowing the energy-intensive chiller to be shut off. Figures 18, 19, and 20 outline three methods of free cooling but do not show all of the piping, valving, and controls that may be necessary for the functioning of a specific system.

Maximum use of free-cooling operation occurs when a drop in the ambient temperature reduces the need for dehumidification. Therefore, higher temperatures in the chilled-water circuit can normally be tolerated during the free-cooling season and are beneficial to the system's heating/cooling balance. In many cases, typical 45°F chilled water temperatures are allowed to rise to 55°F or higher in free cooling. This maximizes tower usage and minimizes system energy consumption. Some applications require a constant chilled-water supply temperature, which can reduce the hours of free cooling operation, depending on ambient temperatures.

If the spray water temperature is allowed to fall too low, freezing may be a concern. Close control of spray water temperature per the manufacturer's recommendations minimizes unit icing and helps ensure trouble-free operation. Refer to the guidelines from the manufacturer and to the section on Winter Operation in this chapter.

Indirect Free Cooling. This type of cooling separates the condenser-water and chilled-water circuits and may be accomplished in the following ways:

1. A separate heat exchanger in the system (usually plate-and-frame) allows heat to transfer from the chilled-water circuit to

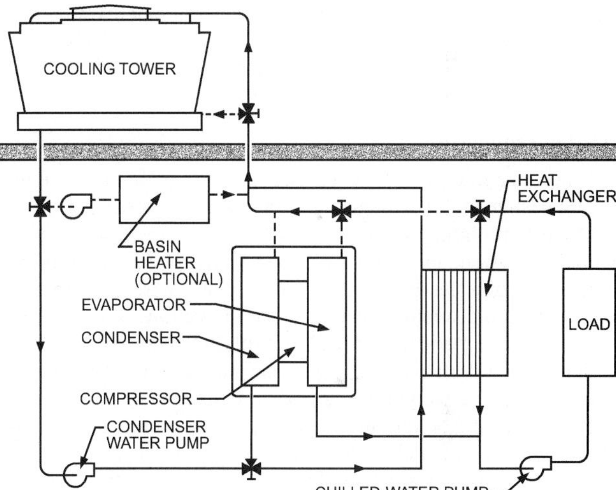

Fig. 18 Free Cooling by Use of Auxiliary Heat Exchanger

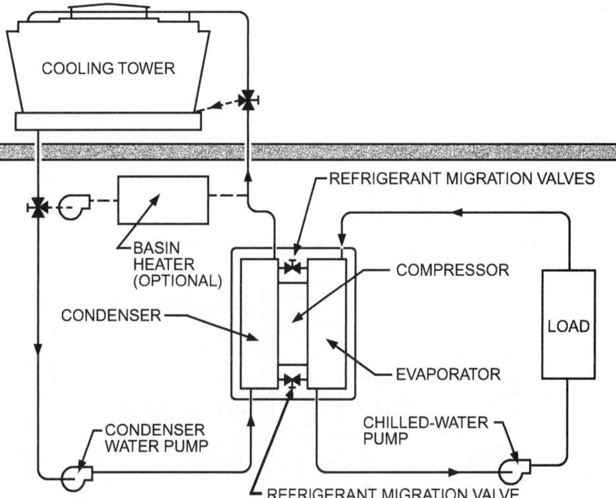

Fig. 19 Free Cooling by Use of Refrigerant Vapor Migration

the condenser-water circuit by total bypass of the chiller system (Figure 18).

2. An indirect-contact, closed-circuit evaporative cooling tower (Figures 4 and 6) also permits indirect free cooling. Its use is covered in the following section on Direct Free Cooling.
3. In vapor migration system (Figure 19), bypasses between the evaporator and condenser permit migratory flow of refrigerant vapor to the condenser; they also allow gravity flow of liquid refrigerant back to the evaporator without compressor operation. Not all chiller systems are adaptable to this arrangement, and those that are may offer limited load capability under this mode. In some cases, auxiliary pumps enhance refrigerant flow and, therefore, load capability.

Direct Free Cooling. This type of cooling involves interconnecting the condenser-water and chilled-water circuits so the cooling tower water serves the load directly (Figure 20). In this case, the chilled-water pump is normally bypassed so design water flow can be maintained to the cooling tower. The primary disadvantage of the direct free-cooling system is that it allows the relatively dirty condenser water to contaminate the clean chilled-water system. Although filtration systems (either side-stream or full-flow) minimize this contamination, many specifiers consider it to be an overriding concern. Using a closed-circuit (indirect-contact) cooling tower eliminates this contamination. During summer, water from the tower is circulated in a closed loop through the condenser. During winter, water from the tower is circulated in a closed loop directly through the chilled-water circuit.

Winter Operation

When a cooling tower is to be used in freezing climates, the following design and operating considerations are necessary.

Open Circulating Water. Direct-contact cooling towers can be winterized by a suitable method of capacity control that maintains the temperature of water leaving the tower well above freezing. In addition, during cold weather, regular visual inspections of the cooling tower should be made to ensure all controls are operating properly.

On induced-draft axial fan towers, fans may be periodically operated in reverse, usually at low speed, to deice the air intake areas. Using fan cycling or variable-frequency drives minimizes the possibility of icing by matching tower capability with the load. Forced-draft centrifugal fan towers should also be equipped with a means of capacity control, such as variable-frequency drives or capacity-control dampers for this reason.

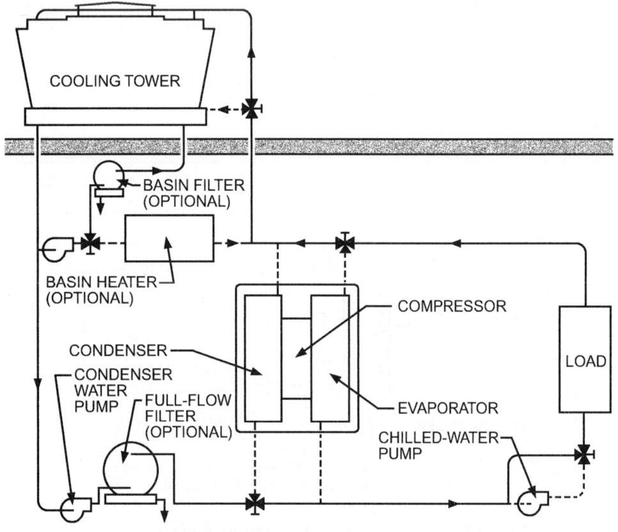

Fig. 20 Free Cooling by Interconnection of Water Circuits

Recirculation of moist discharge air on forced-draft equipment can cause ice formation on inlet air screens and fans. Installing vibration cut-out switches can minimize the risk of damage from ice formation on rotating equipment.

Closed Circulating Water. Precautions beyond those mentioned for open circulating water must be taken to protect the fluid inside the heat exchanger of a closed-circuit fluid cooler. When system design permits, the best protection is to use an antifreeze solution. When this is not possible, supplemental heat must be provided to the heat exchanger, and the manufacturer should be consulted about the amount of heat input required. Positive closure damper hoods are also available from many manufacturers to reduce heat loss from the coil section and thus reduce the amount of heat input required.

All exposed piping to and from the cooler should be insulated and heat traced. In case of a power failure during freezing weather, the heat exchanger should include an emergency draining system.

Sump Water. Freeze protection for the sump water in an idle tower or closed-circuit fluid cooler can be obtained by various means. A good method is to use an auxiliary sump tank located in a heated space. When a remote sump is impractical, auxiliary heat must be supplied to the tower sump to prevent freezing. Common sources are electric immersion heaters and steam and hot-water coils. Consult the tower manufacturer for the exact heat requirements to prevent freezing at design winter temperatures.

All exposed water lines susceptible to freezing should be protected by electric heat tape or cable and insulation. This precaution applies to all lines or portions of lines that have water in them when the tower is shut down.

Sound

Sound has become an important consideration in the selection and siting of outdoor equipment such as cooling towers and other evaporative cooling devices. Many communities have enacted legislation that limits allowable sound levels of outdoor equipment. Even if legislation does not exist, people who live and work near a tower installation may object if the sound intrudes on their environment. Because the cost of correcting a sound problem may exceed the original cost of the cooling tower, sound should be considered in the early stages of system design.

To determine the acceptability of tower sound in a given environment, the first step is to establish a noise criterion for the area of concern. This may be an existing or pending code or an estimate of sound levels that will be acceptable to those living or working in the area. The second step is to estimate the sound levels generated by the tower at the critical area, taking into account the effects of the tower installation geometry and the distance from the tower to the critical area. Often, the tower manufacturer can supply sound rating data on a specific unit that serve as the basis for this estimate. Lastly, the noise criterion is compared to the estimated tower sound levels to determine the acceptability of the installation.

In cases where the installation may present a sound problem, several potential solutions are available. It is good practice to situate the tower as far as possible from any sound-sensitive areas. Two-speed fan motors should be considered to reduce tower sound levels (by a nominal 12 dB) during light-load periods, such as at night, if these correspond to critical sound-sensitive periods. However, fan motor cycling should be held to a minimum because a fluctuating sound is usually more objectionable than a constant sound. Using variable-frequency drives can also reduce noise by matching fan speed to capacity and by minimizing sound fluctuation caused by fan cycling.

In critical situations, effective solutions may include barrier walls between the tower and the sound-sensitive area, acoustical treatment of the tower, or using low-sound fans. Attenuators specifically designed for the tower are available from most manufacturers. It may be practical to install a tower larger than would normally be required and lower the sound levels by operating the unit at reduced fan speed. For additional information on sound control, see Chapter 47 of the 2007 *ASHRAE Handbook—HVAC Applications*.

Drift

Water droplets become entrained in the airstream as it passes through the tower. Although eliminators strip most of this water from the discharge airstream, some discharges from the tower as drift. The rate of drift loss from a tower is a function of tower configuration, eliminator design, airflow rate through the tower, and water loading. Generally, an efficient eliminator design reduces drift loss to between 0.001 and 0.005% of the water circulation rate.

Because drift contains the minerals of the makeup water (which may be concentrated three to five times) and often contains water treatment chemicals, cooling towers should not be placed near parking areas, large windowed areas, or architectural surfaces sensitive to staining or scale deposits.

Fogging (Cooling Tower Plume)

Warm air discharged from a cooling tower is essentially saturated. Under certain operating conditions, the ambient air surrounding the tower cannot absorb all of the moisture in the tower discharge airstream, and the excess condenses as fog.

Fogging may be predicted by projecting a straight line on a psychrometric chart from the tower entering air conditions to a point representing the discharge conditions (Figure 21). A line crossing the saturation curve indicates fog generation; the greater the area of intersection to the left of the saturation curve, the more intense the plume. Fog persistence depends on its original intensity and on the degree of mechanical and convective mixing with ambient air that dissipates the fog.

Methods of reducing or preventing fogging have taken many forms, including heating the tower exhaust with natural gas burners or hot-water or steam coils, installing precipitators, and spraying chemicals at the tower exhaust. However, such solutions are generally costly to operate and are not always effective.

On larger, field-erected installations, combination wet-dry cooling towers, which combine the normal evaporative portion of a tower with a finned-tube dry surface heat exchanger section (in series or in parallel), afford a more practical means of plume control. In such units, the saturated discharge air leaving the evaporative section is mixed within the tower with the warm, relatively dry air off the finned-coil section to produce a subsaturated air mixture leaving the tower. In some closed-circuit hybrid cooling tower designs, the dry and wet heat exchange sections combine to abate plume. This is accomplished by (1) reducing the amount of water evaporated by the wet coil by handling some of the heat load sensibly with the dry heat exchanger, and (2) simultaneously heating the discharge air with the incoming process fluid in the dry heat exchanger. In colder weather, these units can operate completely dry, eliminating plume altogether.

Often, however, the most practical solution to tower fogging is to locate the tower where visible plumes, should they form, will not be objectionable. Accordingly, when selecting cooling tower sites, the potential for fogging and its effect on tower surroundings, such as large windowed areas or traffic arteries, should be considered.

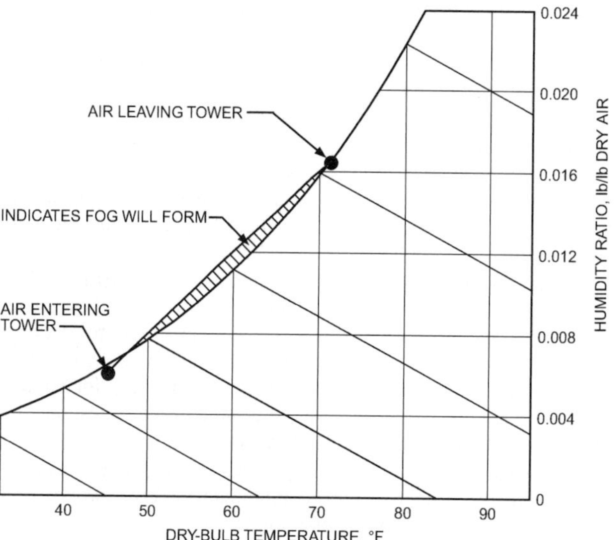

Fig. 21 Fog Prediction Using Psychrometric Chart

Maintenance

Usually, the tower manufacturer furnishes operating and maintenance manuals that include recommendations for procedures and intervals as well as parts lists for the specific unit. These recommendations should be followed when formulating the maintenance program for the cooling tower. When such instructions are unavailable, the schedule of services in Table 1 and the following discussion can guide the operator in establishing a reasonable inspection and maintenance program.

Efficient operation and thermal performance of a cooling tower depend not only on mechanical maintenance, but also on cleanliness. Accordingly, cooling tower owners should incorporate the following as a basic part of their maintenance program.

- Periodic inspection of mechanical equipment, fill, and both hot water and cold water basins to ensure that they are maintained in a good state of repair.
- Periodic draining and cleaning of wetted surfaces and areas of alternate wetting and drying to prevent the accumulation of dirt, scale, or biological organisms, such as algae and slime, in which bacteria may develop.
- Proper treatment of circulating water for biological control and corrosion, in accordance with accepted industry practice.
- Systematic documentation of operating and maintenance functions. This is extremely important because without it, no policing can be done to determine whether an individual has actually adhered to a maintenance policy.

Inspections

The following should be checked daily (no less than weekly) in an informal walk-through inspection. Areas requiring attention have been loosely grouped for clarity, although category distinctions are often hazy because the areas are interdependent.

Performance. Optimum performance and safety depend on the operation of each individual component at its designed capability. A single blocked strainer, for instance, can adversely affect the capacity and efficiency of the entire system. Operators should always be alert to any degradation in performance, as this usually is the first sign of a problem and is invaluable in pinpointing minor problems before they become major. Consult the equipment manufacturers to obtain specific information on each piece of equipment (for both maintenance and technical characteristics), and keep manuals handy for quick reference.

Check and record all water and refrigerant temperatures, pump pressures, outdoor conditions, and pressure drops (differential pressure) across condensers, heat exchangers and filtration devices. This record helps operators become familiar with the equipment as it operates under various load conditions and provides a permanent record that can be used to calculate flow rates, assess equipment efficiency, expedite diagnostic procedures, and adjust maintenance and water treatment regimens to obtain maximum performance from the system.

Table 1 Typical Inspection and Maintenance Schedule*

	Fan	Motor	Gear Reducer	Drive Shaft	V-Belt Drives	Fan Shaft Bearings	Drift Eliminators	Fill	Cold Water Basin	Distribution System	Structural Members	Casing	Float Value	Bleed Rate	Flow Control Valves	Suction Screen
1. Inspect for clogging							W	W		W						W
2. Check for unusual noise or vibration	D	D	D	D							Y					
3. Inspect keys and set screws		S	S	S	S	S										
4. Lubricate		Q				Q									S	
5. Check oil seals			S													
6. Check oil level			W													
7. Check oil for water and dirt			M													
8. Change oil (at least)			S													
9. Adjust tension					M											
10. Check water level									W	W						
11. Check flow rate														M		
12. Check for leakage									S	S			S			
13. Inspect general condition				S	M		Y	Y	Y		S	Y	Y		S	
14. Tighten loose bolts	S	S	S	S		S					Y	R				
15. Clean	R	S	R	R			R	R	S	R			R		R	W
16. Repaint	R	R	R	R					R		R	R				
17. Completely open and close															S	
18. Make sure vents are open			M													

D—daily; W—weekly; M—monthly; Q—quarterly; S—semiannually; Y—yearly; R—as required.
*More frequent inspections and maintenance may be desirable.

For those units with water-side economizers using plate heat exchangers, check temperature and pressure differentials daily for evidence of clogging or fouling.

Major Mechanical Components. During tower inspections, be alert for any unusual noise or vibration from pumps, motors, fans, and other mechanical equipment. This is often the first sign of mechanical trouble. Operators thoroughly familiar with their equipment generally have little trouble recognizing unusual conditions. Also listen for cavitation noises from pumps, which can indicate blocked strainers.

Check the tower fan and drive system assembly for loose mounting hardware, condition of fasteners, grease and oil leaks, and noticeable vibration or wobble when the fan is running. Excessive vibration can rapidly deteriorate the tower.

Observe at least one fan start and stop each week. If a fan has a serious problem, lock it out of operation and call for expert assistance. To be safe, do not take chances by running defective fans.

Fan and drive systems should be professionally checked for dynamic balance, alignment, proper fan pitch (if adjustable), and vibration whenever major repair work is performed on the fan or if unusual noises or vibrations are present. It is good practice to have these items checked at least once every third year on all but the smallest towers. Any vibration switches should be checked for proper operation at least annually.

Verify calibration of the fan thermostat periodically to prevent excessive cycling and to ensure that the most economical temperature to the chiller is maintained.

Tower Structure. Check the tower structure and casing for water and air leaks as well as deterioration. Inspect louvers, fill, and drift eliminators for clogging, excessive scale or algal growth. Clean as necessary, using high-pressure water and taking care not to damage fragile fill and eliminator components.

Watch for excessive drift (water carryover), and take corrective action as required. Drift is the primary means of *Legionella* transmittal by cooling towers and evaporative condensers (see ASHRAE *Guideline* 12 for recommendations on control of *Legionella*). Deteriorated drift eliminators should be replaced. Many older towers have drift eliminators that contain asbestos. In the United States, deteriorated asbestos-type eliminators should as a rule be designated friable material and be handled and disposed of in a manner approved by the Environment Protection Agency (EPA) and the Occupational Safety and Health Administration (OSHA).

Check the tower basin, structural members and supports, fasteners, safety rails, and ladders for corrosion or other deterioration and repair as necessary. Replace deteriorated tower components as required.

Water Distribution and Quality. Check the hot-water distribution system frequently, and clear clogged nozzles as required. Water distribution should be evenly balanced when the system is at rated flow and should be rechecked periodically. Towers with open distribution pans benefit from covers because this retards algal growth.

The sump water level should be within the manufacturer's range for normal operating level, and high enough to allow most solids to settle out, thereby improving water quality to the equipment served by the tower.

Tower water should be clear, and the surface should not have an oily film, excessive foaming, or scum. Oil inhibits heat transfer in cooling towers, condensers, and other heat exchangers and should not be present in tower water. Foam and scum can indicate excess organic material that can provide nutrients to bacteria (Rosa 1992). If such conditions are encountered, contact the water treatment specialist, who will take steps to correct the problem.

Check the cold-water basin in several places for corrosion, accumulated deposits, and excessive algae, because sediments and corrosion may not be uniformly distributed. Corrosion and microbiological activity often occur under sediments. Tower outlet strainers should be in place and free of clogging.

Do not neglect the strainers in the system. In-line strainers may be the single most neglected component in the average installation. They should be inspected and, if necessary, cleaned each time the tower is cleaned. Pay particular attention to the small, fine strainers used on auxiliary equipment such as computer cooling units and blowdown lines.

Blow down chilled water risers frequently, particularly on systems using direct free cooling. Exercise all valves in the system periodically by opening and closing them fully.

For systems with water-side economizers, maintaining good water quality is paramount to prevent fouling of the heat exchanger or chilled-water system, depending on the type of economizer used.

Check, operate, and enable winterization systems well before freezing temperatures are expected, to allow time to obtain parts and make repairs as necessary. Ensure that tower sediments do not build up around immersion heater elements, because this will cause rapid failure of the elements.

Maintain sand filters in good order, and inspect the media bed for channeling at least quarterly. If channeling is found, either replace or clean the media as soon as possible. Do not forget to carefully clean the underdrain assembly while the media is removed. If replacing the media, use only that which is specified for cooling towers. Do not use swimming pool filter sand in filters designed for cooling towers and evaporative condensers.

Centrifugal separators rarely require service, although they must not be allowed to overfill with contaminants. Verify proper flow rate, pressure drop, and purge operation.

Check bag and cartridge filters as necessary. Clean the tower if it is dirty.

Cleanliness. Towers are excellent air washers, and the water quality in a given location quickly reflects that of the ambient air (Hensley 1985). A typical 200 ton cooling tower operating 1000 hours may assimilate upwards of 600 lb of particulate matter from airborne dust and the makeup water supply (Broadbent et al. 1992). Proximity to highways and construction sites, air pollution, and operating hours are all factors in tower soil loading.

Design improvements in cooling towers that increase thermal performance also increase air scrubbing capability (Hensley 1985). Recommendations by manufacturers regarding cleaning schedules are, therefore, to be recognized as merely guidelines. The actual frequency of cleanings should be determined at each location by careful observation and system history. Do not expect sand filters, bag filters, centrifugal separators, water treatment programs, and so forth, to take the place of a physical cleaning. They are designed to improve water quality and as such should increase the time interval between cleanings. Conversely, regular cleanings should not be expected to replace water treatment.

Towers should not be allowed to become obviously fouled, but should be cleaned often enough that sedimentation and visible biological activity (algae and slime) are easily controlled by water treatment between cleanings. The tower is the only component in the condenser loop that can be viewed easily without system shutdown, so it should be considered an indicator of total system condition and cleanliness.

Water treatment should not be expected to protect surfaces it cannot reach, such as the metal or wood components under accumulated sediments. Biocides are not likely to be effective unless used in conjunction with a regular cleaning program. Poorly maintained systems create a greater demand upon the biocide because organic sediments neutralize the biocide and tend to shield bacterial cells from the chemical, thus requiring higher and more frequent doses to keep microbial populations under control (Broadbent et al. 1992; McCann 1988). High concentrations of an oxidizing biocide can contribute to corrosion. Keeping the tower clean reduces the breeding grounds and nutrients available to the microbial organisms (ASHRAE 1989; Broadbent 1989; Meitz 1986, 1988).

Proper cleaning procedures address the entire tower, including not only the cold-water basin but also the distribution system, strainers, eliminators, casing, fan and fan cylinders, and louvers. The water treatment specialist should be advised and consulted prior to and following the cleaning.

It is recommended that personnel involved wear high-efficiency particulate air (HEPA) type respirators, gloves, goggles, and other body coverings approved by the appropriate agency, such as the U.S. Department of Labor Occupational Health & Safety Administration (OSHA) or National Institute for Occupational Safety and Health (NIOSH) in the United States. This is especially true if the cleaning procedures involve the use of high-pressure water, air, and steam (ASHRAE 1989) or wet-dry vacuum equipment. If any chemicals are used, they must be handled according to their material safety data sheets (MSDSs), available from the chemical supplier.

Operation in Freezing Weather. During operation in freezing weather, the tower should be inspected more frequently, preferably daily, for ice formation on fill, louvers, fans, etc. This is especially true when the system is being operated outside the tower design parameters, such as when the main system is shut down and only supplementary units (e.g., computer cooling equipment) are operating. Ice on fan and drive systems is dangerous and can destroy the fan. Moderate icing on fill is generally not dangerous but can cause damage if allowed to build up.

Follow the manufacturer's specific recommendations both for operation in freezing temperatures and for deicing methods such as low-speed reversal of fan direction for short periods of time. Monitor the operation of winterization equipment, such as immersion heaters and heat-tracing tape on makeup lines, to ensure that they are working properly. Check for conditions that could render the freeze protection inoperable, such as tripped breakers, closed valves, and erroneous temperature settings.

Help from Manufacturers. Equipment manufacturers will provide assistance and technical publications on the efficient operation of their equipment; some even provide training. Also, manufacturers can often provide names of reputable local service companies that are experienced with their equipment. Most of these services are free or of nominal cost.

Water Treatment

The quality of water circulating through an evaporative cooling system significantly affects the overall system efficiency, degree of maintenance required, and useful life of system components. Because the water is cooled primarily by evaporation of a portion of the circulating water, the concentration of dissolved solids and other impurities in the water can increase rapidly. Also, appreciable quantities of airborne impurities, such as dust and gases, may enter during operation. Depending on the nature of the impurities, they can cause scaling, corrosion, and/or silt deposits.

Simple blowdown (discharge of a small portion of recirculating water to a drain) may be adequate to control scale and corrosion on sites with good-quality makeup water, but it will not control biological contaminants, including *Legionella pneumophila.* All cooling tower systems should be treated to restrict biological growth, and many benefit from treatment to control scale and corrosion. For a complete and detailed description of water treatment, see Chapter 48 of the 2007 *ASHRAE Handbook—HVAC Applications.* ASHRAE *Guideline* 12 should also be consulted for recommendations regarding control of *Legionella*. Specific recommendations on water treatment, including control of biological contaminants, can be obtained from any qualified water treatment supplier.

PERFORMANCE CURVES

The combination of flow rate and heat load dictates the range a cooling tower must accommodate. The entering air wet-bulb temperature and required system temperature level combine with cooling tower size to balance the heat rejected at a specified approach. The performance curves in this section are typical and may vary from project to project. Computerized selection and rating programs are also available from many manufacturers to generate performance ratings and curves for their equipment.

Cooling towers can accommodate a wide diversity of temperature levels, ranging as high as 150 to 160°F hot-water temperature in the hydrocarbon processing industry. In the air-conditioning and refrigeration industry, towers are generally used in the range of 90 to 115°F hot water temperature. A typical standard design condition for such cooling towers is 95°F hot water to 85°F cold water, and 78°F wet-bulb temperature.

A means of evaluating the typical performance of a cooling tower used for a typical air-conditioning system is shown in Figures 22 to 27. The example tower was selected for a flow rate of 3 gpm per nominal ton when cooling water from 95 to 85°F at 78°F entering wet-bulb temperature (Figure 22).

When operating at other wet bulbs or ranges, the curves may be interpolated to find the resulting temperature level (hot and cold water) of the system. When operating at other flow rates (2, 4, and 5 gpm per nominal ton), this same tower performs at the levels described by the titles of Figures 23, 24, and 25, respectively. Intermediate flow rates may be interpolated between charts to find resulting operating temperature levels.

The format of these curves is similar to the predicted performance curves supplied by manufacturers of cooling towers; the difference is that only three specific ranges (80%, 100%, and 120% of design range) and only three charts are provided, covering 90%, 100%, and 110% of design flow. The curves in Figures 22 to 25, therefore, bracket the acceptable tolerance range of test conditions and may be interpolated for any specific test condition within the scope of the curve families and chart flow rates.

The curves may also be used to identify the feasibility of varying the parameters to meet specific applications. For example, the subject tower can handle a greater heat load (flow rate) when operating in a lower ambient wet-bulb region. This may be seen by comparing the intersection of the 10°F range curve with 73°F wet bulb at 85°F cold water to show the tower is capable of rejecting 33% more heat load at this lower ambient temperature (Figure 24).

Similar comparisons and cross-plots identify relative tower capacity for a wide range of variables. The curves produce accurate comparisons within the scope of the information presented but should not be extrapolated outside the field of data given. Also, the curves are based on a typical mechanical-draft, film-filled, cross-flow, medium-sized, air-conditioning cooling tower. Other types and sizes of towers produce somewhat different balance points of temperature level. However, the curves may be used to evaluate a tower for year-round or seasonal use if they are restricted to the

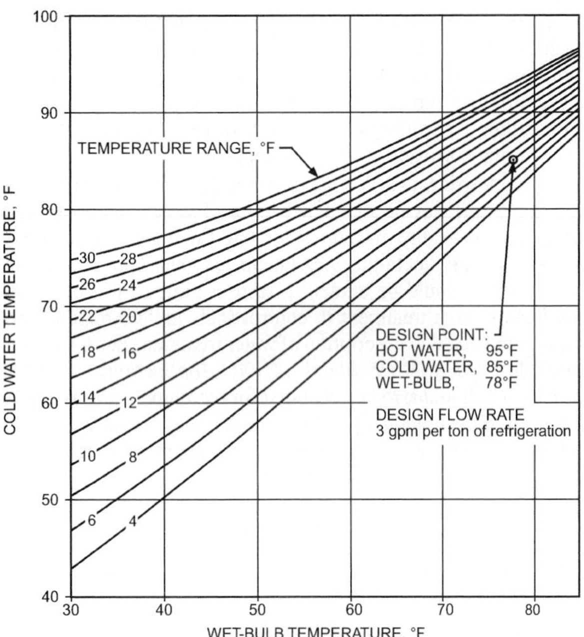

Fig. 22 Cooling Tower Performance—100% Design Flow

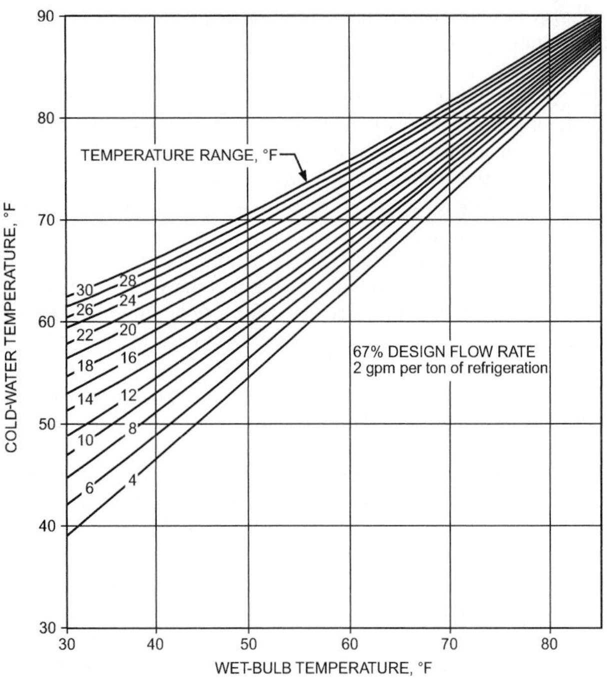

Fig. 23 Cooling Tower Performance—67% Design Flow

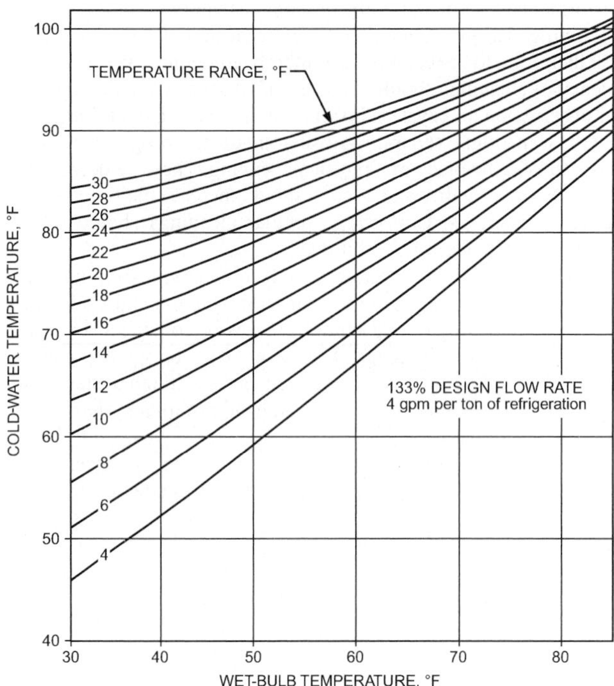

Fig. 24 Cooling Tower Performance—133% Design Flow

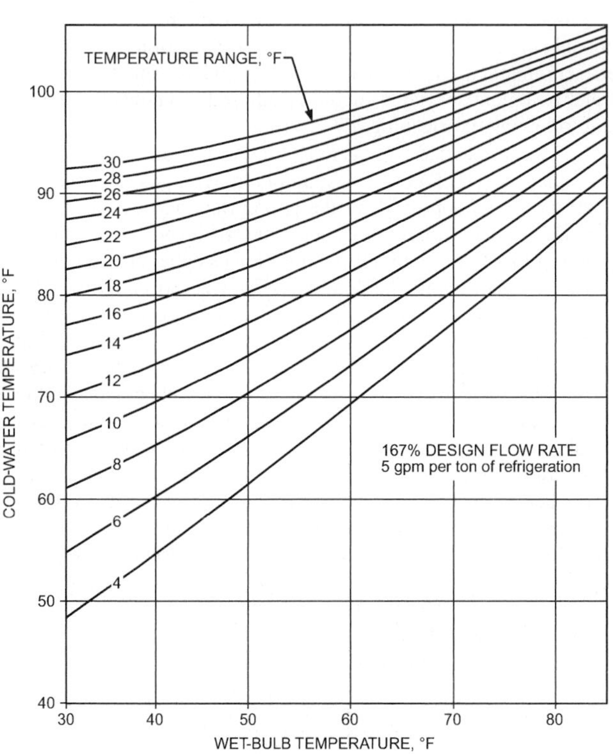

Fig. 25 Cooling Tower Performance—167% Design Flow

general operating characteristics described. (See specific manufacturer's data for maximum accuracy when planning for test or critical temperature needs.)

A cooling tower selected for a specified design condition will operate at other temperature levels when the ambient temperature is off-design or when heat load or flow rate varies from the design condition. When flow rate is held constant, range falls as heat load falls, causing temperature levels to fall to a closer approach. Hot- and cold-water temperatures fall when the ambient wet bulb falls at constant heat load, range, and flow rate. As water loading to a particular tower falls at constant ambient wet bulb and range, the tower cools the water to a lower temperature level or closer approach to the wet bulb.

COOLING TOWER THERMAL PERFORMANCE

Three basic alternatives are available to a purchaser/designer seeking assurance that a cooling tower will perform as specified: (1) certification of performance by an independent third party such as the Cooling Technology Institute (CTI), (2) an acceptance test performed at the site after the unit is installed, or (3) a performance bond. Codes and standards that pertain to performance certification and field testing of cooling towers are listed in Chapter 51.

Certification. The thermal performance of many commercially available cooling tower lines, both open- and closed-circuit, is certified by CTI in accordance with their *Standard* STD-201, which applies to mechanical-draft, open- and closed-circuit water cooling towers. It is based on entering wet-bulb temperature and certifies tower performance when operating in an open, unrestricted environment. Independent performance certification eliminates the need for field acceptance tests and performance bonds.

Field Acceptance Test. As an alternative to certification, tower performance can be verified after installation by conducting a field acceptance test in accordance with one of the two available test standards. Of the two standards, CTI *Standard* ATC-105 is more commonly used, although American Society of Mechanical Engineers (ASME) *Standard* PTC-23 is also used. These standards are similar in their requirements, and both base the performance evaluation on entering wet-bulb temperature. ASME *Standard* PTC-23, however, provides an alternative for evaluation based on ambient wet-bulb temperature as well.

With either procedure, the test consists of measuring the hot-water temperature in the inlet piping to the tower or in the hot-water distribution basin. Preferably, the cold-water temperature is measured at the discharge of the circulating pump, where there is much less chance for temperature stratification. The wet-bulb temperature is measured by an array of mechanically aspirated psychrometers. The recirculating water flow rate is measured by any of several approved methods, usually a pitot-tube traverse of the piping leading to the tower. Recently calibrated instruments should be used for all measurements, and electronic data acquisition is recommended for all but the smallest installations.

For an accurate test, the tower should be running under a steady heat load combined with a steady flow of recirculating water, both as near design as possible. Weather conditions should be reasonably stable, with prevailing winds of 10 mph or less. The tower should be clean and adjusted for proper water distribution, with all fans operating at design speed. Both the CTI and ASME standards specify maximum recommended deviations from design operating conditions of range, flow, wet-bulb temperature, heat load, and fan power.

Performance Bond. A few manufacturers offer a performance bond, which provides for seeking redress from a surety in the event the cooling tower does not meet the manufacturer's rated thermal performance.

COOLING TOWER THEORY

Baker and Shryock (1961) developed the following theory. Consider a cooling tower having one square foot of plan area; cooling volume V, containing extended water surface per unit volume a; and water mass flow rate L and air mass flow rate G. Figure 26 schematically shows the processes of mass and energy transfer. The bulk water at temperature t is surrounded by the bulk air at dry-bulb temperature t_a, having enthalpy h_a and humidity ratio W_a. The interface is assumed to be a film of saturated air with an intermediate temperature t'', enthalpy h'', and humidity ratio W''. Assuming a constant value of 1 Btu/lb·°F for the specific heat of water c_p, the total energy transfer from the water to the interface is

$$dq_w = Lc_p\,dt = K_L a(t - t'')dV \tag{2}$$

where

q_w = rate of total heat transfer, bulk water to interface, Btu/h
L = inlet water mass flow rate, lb/h
K_L = unit conductance, heat transfer, bulk water to interface, Btu/h·ft²·°F
V = cooling volume, ft³
a = area of interface, ft²/ft³

The heat transfer from interface to air is

$$dq_s = K_G a(t'' - t_a)dV \tag{3}$$

where

q_s = rate of sensible heat transfer, interface to airstream, Btu/h
K_G = overall unit conductance, sensible heat transfer, interface to main airstream, Btu/h·ft²·°F

The diffusion of water vapor from film to air is

$$dm = K'a(W'' - W_a)dV \tag{4}$$

where

m = mass transfer rate, interface to airstream, lb/h
K' = unit conductance, mass transfer, interface to main airstream, lb/h·ft²·(lb/lb)
W'' = humidity ratio of interface (film), lb/lb
W_a = humidity ratio of air, lb/lb

The heat transfer due to evaporation from film to air is

$$dq_L = r\,dm = rK'a(W'' - W_a)dV \tag{5}$$

where

q_L = rate of latent heat transfer, interface to airstream, Btu/h
r = latent heat of evaporation (constant), Btu/lb

The process reaches equilibrium when $t_a = t$, and the air becomes saturated with moisture at that temperature. Under adiabatic

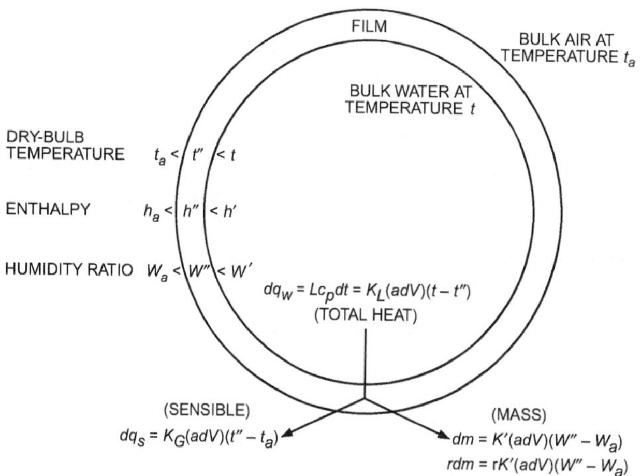

Fig. 26 Heat and Mass Transfer Relationships Between Water, Interfacial Film, and Air
(Baker and Shryock 1961)

Table 2 Counterflow Integration Calculations for Example 1

1	2	3	4	5	6	7	8	9
Water Temperature t, °F	Enthalpy of Film h', Btu/lb	Enthalpy of Air h_a, Btu/lb	Enthalpy Difference $h' - h_a$, Btu/lb	$\frac{1}{(h' - h_a)}$	Δt, °F	$NTU = \frac{c_p \Delta t}{(h' - h_a)_{avg}}$	ΣNTU	Cumulative Cooling Range, °F
85	49.4	38.6	10.8	0.0926				
					1	0.0921	0.0921	1
86	50.7	39.8	10.9	0.0917				
					1	0.0917	0.1838	2
87	51.9	41.0	10.9	0.0917				
					1	0.0913	0.2751	3
88	53.2	42.2	11.0	0.0909				
					1	0.0901	0.3652	4
89	54.6	43.4	11.2	0.0893				
					1	0.0889	0.4541	5
90	55.9	44.6	11.3	0.0885				
					2	0.1732	0.6273	7
92	58.8	47.0	11.8	0.0847				
					2	0.1653	0.7925	9
94	61.8	49.9	12.4	0.0806				
					2	0.1569	0.9493	11
96	64.9	51.8	13.1	0.0763				
					2	0.1477	1.097	13
98	68.2	54.2	14.0	0.0714				
					2	0.1376	1.2346	15
100	71.7	56.6	15.1	0.0662				

conditions, equilibrium is reached at the temperature of adiabatic saturation or at the thermodynamic wet-bulb temperature of the air. This is the lowest attainable temperature in a cooling tower. The circulating water rapidly approaches this temperature when a tower operates without heat load. The process is the same when a heat load is applied, but the air enthalpy increases as it moves through the tower so the equilibrium temperature increases progressively. The approach of the cooled water to the entering wet-bulb temperature is a function of the capability of the tower.

Merkel (1925) assumed the Lewis relationship to be equal to one in combining the transfer of mass and sensible heat into an overall coefficient based on enthalpy difference as the driving force:

$$K_G/(K'c_{pm}) = 1 \qquad (6)$$

where c_{pm} is the humid specific heat of moist air in Btu/lb·°F (dry air basis).

Equation (5) also explains why the wet-bulb thermometer closely approximates the temperature of adiabatic saturation in an air-water vapor mixture. Setting water heat loss equal to air heat gain yields

$$Lc_p\,dt = G\,dh = K'a(h'' - h_a)dV \qquad (7)$$

where G is the air mass flow rate in lb/h.

The equation considers the transfer from the interface to the airstream, but the interfacial conditions are indeterminate. If the film resistance is neglected and an overall coefficient K' is postulated, based on the driving force of enthalpy h' at the bulk water temperature t, the equation becomes

$$Lc_p\,dt = G\,dh = K'a(h' - h_a)dV \qquad (8)$$

or

$$K'aV/L = \int_{t_1}^{t_2} \frac{c_p}{h' - h_a} dt \qquad (9)$$

and

$$K'aV/G = \int_{h_1}^{h_2} \frac{dh}{h' - h_a} \qquad (10)$$

In cooling tower practice, the integrated value of Equation (8) is commonly referred to as the **number of transfer units (NTU)**. This value gives the number of times the average enthalpy potential $(h' - h_a)$ goes into the temperature change of the water (Δt) and is a measure of the difficulty of the task. Thus, one transfer unit has the definition of $c_p\Delta t/(h' - h_a)_{avg} = 1$.

The equations are not self-sufficient and are not subject to direct mathematical solution. They reflect mass and energy balance at any point in a tower and are independent of relative motion of the two fluid streams. Mechanical integration is required to apply the equations, and the procedure must account for relative motion. Integration of Equation (8) gives the NTU for a given set of conditions.

Counterflow Integration

The counterflow cooling diagram is based on the saturation curve for air-water vapor (Figure 27). As water is cooled from t_{w1} to t_{w2}, the air film enthalpy follows the saturation curve from A to B. Air entering at wet-bulb temperature t_{aw} has an enthalpy h_a corresponding to C′. The initial driving force is the vertical distance BC. Heat removed from the water is added to the air, so the enthalpy increase is proportional to water temperature. The slope of the air operating line CD equals L/G.

Counterflow calculations start at the bottom of a tower, the only point where the air and water conditions are known. The NTU is calculated for a series of incremental steps, and the summation is the integral of the process.

Example 1. Air enters the base of a counterflow cooling tower at 75°F wet-bulb temperature, water leaves at 85°F, and L/G (water-to-air ratio) is 1.2, so $dh = 1.2 \times 1 \times dt$, where 1 Btu/lb·°F is the specific heat c_p of water. Calculate the NTU for various cooling ranges.

Solution: The calculation is shown in Table 2. Water temperatures are shown in column 1 for 1°F increments from 85 to 90°F and 2°F increments from 90 to 100°F. The corresponding film enthalpies, obtained from psychrometric tables, are shown in column 2.

The upward air path is shown in column 3. The initial air enthalpy is 38.6 Btu/lb, corresponding to a 75°F wet bulb, and increases by the relationship $\Delta h = 1.2 \times 1 \times \Delta t$.

The driving force $h' - h_a$ at each increment is listed in column 4. The reciprocals $1/(h' - h_a)$ are calculated (column 5), Δt is noted (column 6), and the average for each increment is multiplied by $c_p\Delta t$ to obtain the NTU for each increment (column 7). The summation of the

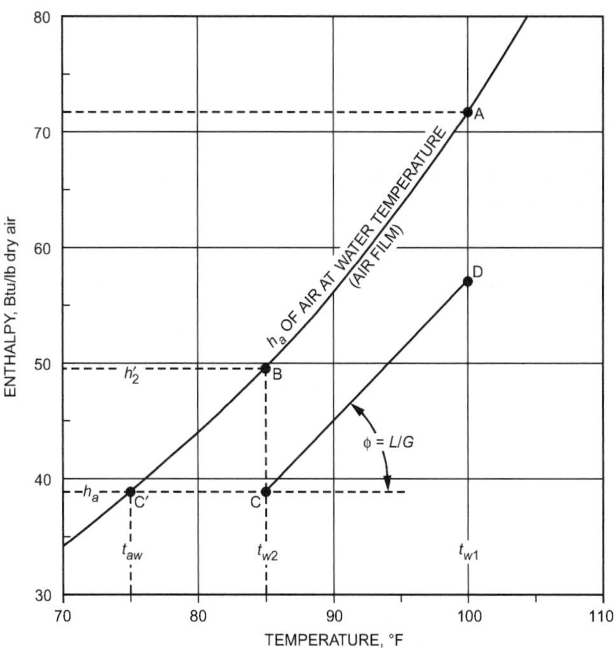

Fig. 27 Counterflow Cooling Diagram

incremental values (column 8) represents the NTU for the summation of the incremental temperature changes, which is the cooling range given in column 9.

Because of the slope and position of CD relative to the saturation curve, the potential difference increases progressively from the bottom to the top of the tower in this example. The degree of difficulty decreases as this driving force increases, reflected as a reduction in the incremental NTU proportional to a variation in incremental height. This procedure determines the temperature gradient with respect to tower height.

The procedure of Example 1 considers increments of temperature change and calculates the coincident values of NTU, which correspond to increments of height. Baker and Mart (1952) developed a unit-volume procedure that considers increments of NTU (representing increments of height) with corresponding temperature changes calculated by iteration. The unit-volume procedure is more cumbersome but is necessary in cross-flow integration because it accounts for temperature and enthalpy change, both horizontally and vertically.

Cross-Flow Integration

In a cross-flow tower, water enters at the top; the solid lines of constant water temperature in Figure 28 show its temperature distribution. Air enters from the left, and the dotted lines show constant enthalpies. The cross section is divided into unit volumes in which dV becomes $dx\,dy$ and Equation (7) becomes

$$c_pL\ dt\ dx = G\ dh\ dy = Ka(h' - h_a)dx\ dy \tag{11}$$

The overall L/G ratio applies to each unit-volume by considering $dx/dy = w/z$. The cross-sectional shape is automatically considered when an equal number of horizontal and vertical increments are used. Calculations start at the top of the air inlet and proceed down and across. Typical calculations are shown in Figure 29 for water entering at 100°F, air entering at 75°F wet-bulb temperature, and L/G = 1.0. Each unit-volume represents 0.1 NTU. Temperature change vertically in each unit is determined by iteration from

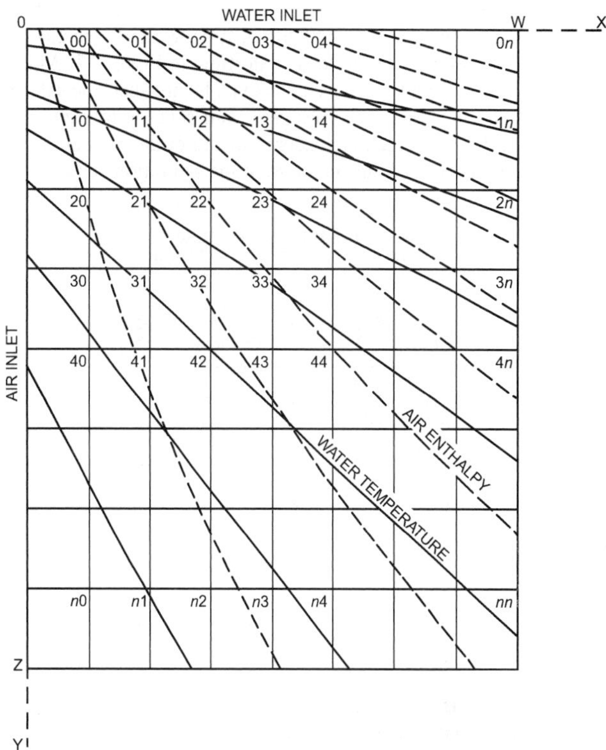

Fig. 28 Water Temperature and Air Enthalpy Variation Through Cross-Flow Cooling Tower
(Baker and Shryock 1961)

$$\Delta t = 0.1(h' - h_a)_{avg} \tag{12}$$

The expression $c_p(L/G)dt = dh$ determines the horizontal change in air enthalpy. With each step representing 0.1 NTU, two steps down and across equal 0.2 NTU, etc., for conditions corresponding to the average leaving water temperature.

Figure 28 shows that air flowing across any horizontal plane moves toward progressively hotter water, with entering hot water temperature as a limit. Water falling through any vertical section moves toward progressively colder air that has the entering wet-bulb temperature as a limit. This is shown in Figure 30, which is a plot of the data in Figure 29. Air enthalpy follows the family of curves radiating from Point A. Air moving across the top of the tower tends to coincide with OA. Air flowing across the bottom of a tower of infinite height follows a curve that coincides with the saturation curve AB.

Water temperatures follow the family of curves radiating from Point B, between the limits of BO at the air inlet and BA at the outlet of a tower of infinite width. The single operating line CD of the counterflow diagram in Figure 27 is replaced in the cross-flow diagram (Figure 30) by a zone represented by the area intersected by the two families of curves.

TOWER COEFFICIENTS

Calculations can reduce a set of conditions to a numerical value representing degree of difficulty. The NTU corresponding to a set of hypothetical conditions is called the **required coefficient** and evaluates degree of difficulty. When test results are being considered, the NTU represents the available coefficient and becomes an evaluation of the equipment tested.

The calculations consider temperatures and the L/G ratio. The minimum required coefficient for a given set of temperatures occurs at L/G = 0, corresponding to an infinite air rate. Air enthalpy does not increase, so the driving force is maximum and the degree of difficulty is minimum. Decreased air rate (increase in L/G) decreases the driving force, and the greater degree of difficulty shows as an increase in NTU. This situation is shown for counterflow in Figure 31. Maximum L/G (minimum air rate) occurs when CD intersects the saturation curve. Driving force becomes zero, and NTU is infinite. The point of zero driving force may occur at the air outlet or at an intermediate point because of the curvature of the saturation curve.

100°F HOT-WATER TEMPERATURE

75°F WET-BULB TEMPERATURE

	71.73 h'	71.73 h'	71.73 h'	71.73 h'	71.73 h'
38.59 h_a	1,1 2.92 Δt 97.08°F 66.68 h' 2.92 Δh 41.51 h	1,2 2.66 Δt 97.37°F 67.11 h' 2.66 Δh 44.17 h	1,3 2.42 Δt 97.58°F 67.52 h' 2.42 Δh 46.59 h	1,4 2.21 Δt 97.79°F 67.87 h' 2.21 Δh 48.80 h	1,5 2.02 Δt 97.98°F 68.20 h' 2.02 Δh 50.82 h
38.59 h_a	2,1 2.49 Δt 94.59°F 62.67 h' 2.49 Δh 41.08 h	2,2 2.30 Δt 95.07°F 63.42 h' 2.30 Δh 43.38 h	2,3 2.13 Δt 95.45°F 64.03 h' 2.13 Δh 45.51 h	2,4 1.97 Δt 95.82°F 64.62 h' 1.97 Δh 47.48 h	2,5 1.83 Δt 96.15°F 65.15 h' 1.83 Δh 49.32 h
38.59 h_a	3,1 2.14 Δt 92.45°F 59.43 h' 2.14 Δh 40.72 h	3,2 2.01 Δt 93.06°F 60.33 h' 2.01 Δh 42.73 h	3,3 1.89 Δt 93.56°F 61.09 h' 1.89 Δh 44.62 h	3,4 1.77 Δt 94.05°F 61.84 h' 1.77 Δh 46.39 h	3,5 1.67 Δt 94.48°F 62.50 h' 1.67 Δh 48.06 h
38.59 h_a	4,1 1.85 Δt 90.55°F 56.69 h' 1.85 Δh 40.44 h	4,2 1.77 Δt 91.29°F 57.74 h' 1.77 Δh 42.21 h	4,3 1.68 Δt 91.88°F 58.59 h' 1.68 Δh 43.89 h	4,4 1.60 Δt 92.45°F 59.43 h' 1.60 Δh 45.59 h	4,5 1.51 Δt 92.97°F 60.20 h' 1.51 Δh 47.00 h
38.59 h_a	5,1 1.62 Δt 88.93°F 54.46 h' 1.62 Δh 40.21 h	5,2 1.56 Δt 89.73°F 55.55 h' 1.56 Δh 41.77 h	5,3 1.50 Δt 90.38°F 56.45 h' 1.50 Δh 43.27 h	5,4 1.45 Δt 91.00°F 57.33 h' 1.45 Δh 44.72 h	5,5 1.38 Δt 91.59°F 58.17 h' 1.38 Δh 46.10 h

90.53°F AVERAGE WATER TEMPERATURE

L/G = 1.0 Δt (Vertically) = 0.1 $(h' - h_a)_{avg}$ Δh (Horizontally) = Δt

Fig. 29 Cross-Flow Calculations
(Baker and Shryock 1961)

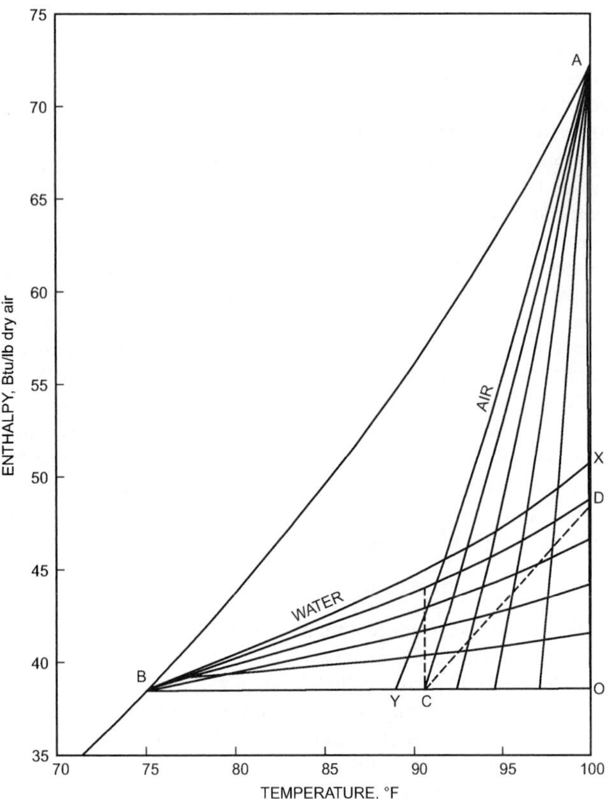

Fig. 30 Counterflow Cooling Diagram for Constant Conditions, Variable L/G Ratios

Similar variations occur in cross-flow cooling. Variations in L/G vary the shape of the operating area. At L/G = 0, the operating area becomes a horizontal line, which is identical to the counterflow diagram (Figure 31), and both coefficients are the same. An increase in L/G increases the height of the operating area and decreases the width. This continues as the areas extend to Point A as a limit. This maximum L/G always occurs when the wet-bulb temperature of the air equals the hot-water temperature and not at an intermediate point, as may occur in counterflow.

Both types of flow have the same minimum coefficient at L/G = 0, and both increase to infinity at a maximum L/G. The maximums are the same if the counterflow potential reaches zero at the air outlet, but the counterflow tower will have a lower maximum L/G when the potential reaches zero at an intermediate point, as in Figure 31. A cooling tower can be designed to operate at any point within the two

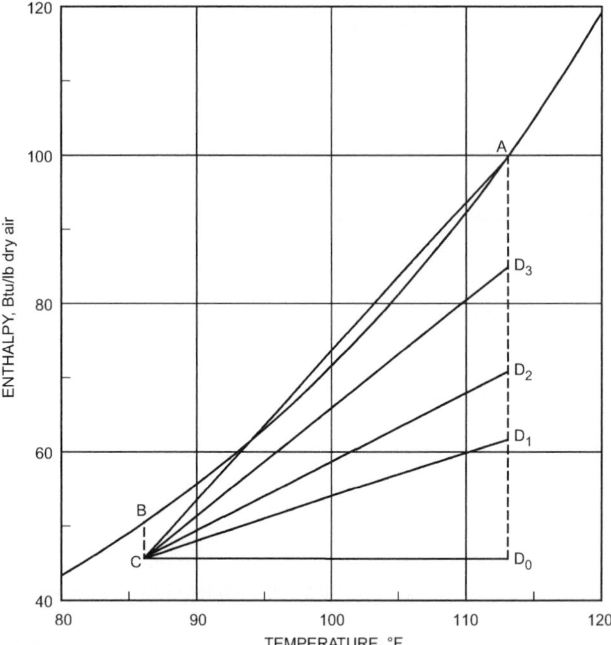

Fig. 31 Cross-Flow Cooling Diagram

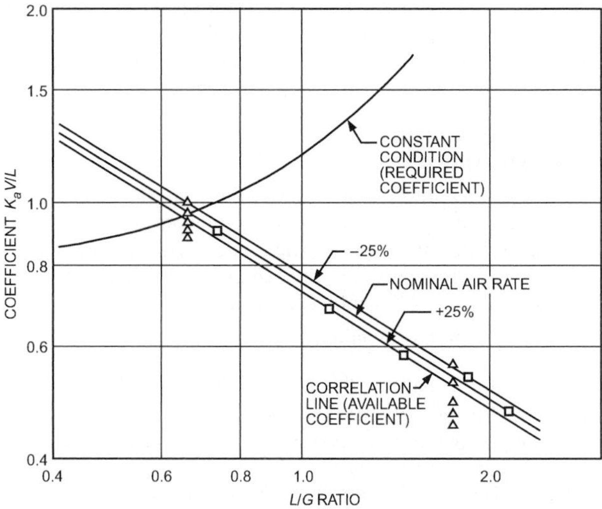

Fig. 32 Tower Characteristic, KaV/L Versus L/G
(Baker and Shryock 1961)

limits, but most applications restrict the design to much narrower limits determined by air velocity.

A low air rate requires a large tower, while a high air rate in a smaller tower requires greater fan power. Typical limits in air velocity are about 300 to 700 fpm in counterflow and 350 to 800 fpm or more in cross-flow.

Available Coefficients

A cooling tower can operate over a wide range of water rates, air rates, and heat loads, with variation in the approach of the cold water to the wet-bulb temperature. Analysis of a series of test points shows that the available coefficient is not a constant but varies with the operating conditions, as shown in Figure 32.

Figure 32 is a typical correlation of a tower characteristic showing the variation of available KaV/L with L/G for parameters of constant air velocity. Recent fill developments and more accurate test methods have shown that some of the characteristic lines are curves rather than a series of straight, parallel lines on logarithmic coordinates.

Ignoring the minor effect of air velocity, a single average curve may be considered:

$$Ka\,V/L \sim (L/G)^n \tag{13}$$

The exponent n varies over a range of about –0.35 to –1.1 but averages between –0.55 and –0.65. Within the range of testing, –0.6 has been considered sufficiently accurate.

The family of curves corresponds to the following relation:

$$Ka\,V/L \sim (L)^n (G)^m \tag{14}$$

where m varies slightly from n numerically and is a positive exponent.

The triangular points in Figure 32 show the effect of varying temperature at nominal air rate. The deviations result from simplifying assumptions and may be overcome by modifying the integration procedure. Usual practice, as shown in Equation (9), ignores evaporation and assumes that

$$G\,dh = c_p L\,dt \tag{15}$$

The exact enthalpy rise is greater than this because a portion of the heat in the water stream leaves as vapor in the airstream. The correct heat balance is as follows (Baker and Shryock 1961):

$$G\,dh = c_p L\,dt + c_p L_E (t_{w2} - 32) \tag{16}$$

where L_E is the mass flow rate of water that evaporates, in lb/h. This reduces the driving force and increases the NTU.

Evaporation causes the water rate to decrease from L at the inlet to $L - L_E$ at the outlet. The water-to-air ratio varies from L/G at the water inlet to $(L - L_E)/G$ at the outlet. This results in an increased NTU.

Basic theory considers the transfer from the interface to the airstream. As the film conditions are indeterminate, film resistance is neglected as assumed in Equation (7). The resulting coefficients show deviations closely associated with hot water temperature and may be modified by an empirical hot water correction factor (Baker and Mart 1952).

The effect of film resistance (Mickley 1949) is shown in Figure 33. Water at temperature t is assumed to be surrounded by a film of saturated air at the same temperature at enthalpy h' (Point B on the saturation curve). The film is actually at a lower temperature t'' at enthalpy h'' (Point B′). The surrounding air at enthalpy h_a corresponds to Point C. The apparent potential difference is commonly considered to be $h' - h_a$, but the true potential difference is $h'' - h_a$ (Mickley 1949). From Equations (1) and (6),

$$\frac{h'' - h_a}{t - t''} = \frac{K_L}{K'} \tag{17}$$

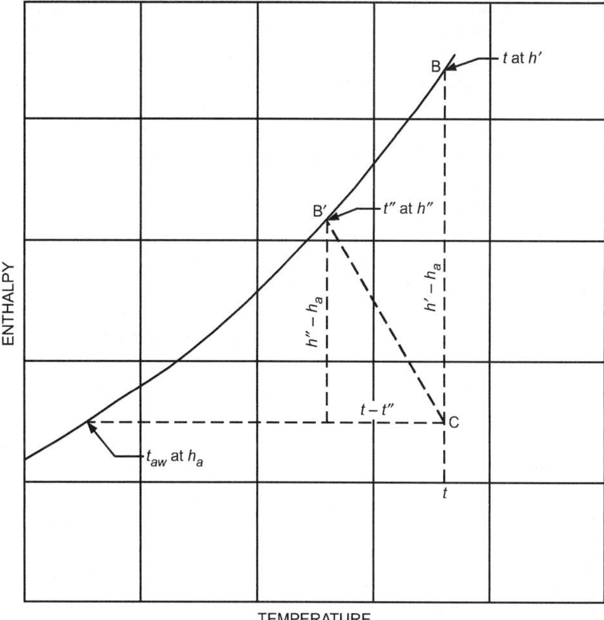

Fig. 33 True Versus Apparent Potential Difference
(Baker and Shryock 1961)

The slope of CB′ is the ratio of the two coefficients. No means to evaluate the coefficients has been proposed, but a slope of −11.1 for cross-flow towers has been reported (Baker and Shryock 1961).

Establishing Tower Characteristics

The performance characteristic of a fill pattern can vary widely because of several external factors. For a given volume of fill, the optimal thermal performance is obtained with uniform air and water distribution throughout the fill pack. Irregularities in either alter the local L/G ratios within the pack and adversely affect the overall thermal performance of the cooling tower. Accordingly, the design of the tower water distribution system, air inlets, fan plenum, and so forth is very important in ensuring that the tower will perform to its potential.

In counterflow towers, some cooling occurs in the spray chamber above the fill and in the open space below the fill. This additional performance is erroneously attributed to the fill itself, which can lead to inaccurate predictions of a tower's performance in other applications. A true performance characteristic for the total cooling tower can only be developed from full-scale tests of the actual cooling tower assembly, typically by the tower manufacturer, and not by combining performance data of individual components.

ADDITIONAL INFORMATION

The Cooling Technology Institute (CTI) offers a reference guide on CD-ROM containing information on pyschrometrics, Merkel KaV/L calculation, characteristic curve performance evaluation, and performance curve evaluation, along with general information on cooling towers.

REFERENCES

ASHRAE. 1998. *Legionellosis position document.*

ASHRAE. 2000. Minimizing the risk of Legionellosis associated with building water systems. *Guideline* 12.

ASME. 2003. Atmospheric water cooling equipment. ANSI/ASME *Standard* PTC 23-2003. American Society of Mechanical Engineers, New York.

Baker, D.R. and H.A. Shryock. 1961. A comprehensive approach to the analysis of cooling tower performance. *ASME Transactions, Journal of Heat Transfer* (August):339.

Baker, D.R. and L.T. Mart. 1952. Analyzing cooling tower performance by the unit-volume coefficient. *Chemical Engineering* (December):196.

Broadbent, C.R. 1989. Practical measures to control Legionnaire's disease hazards. *Australian Refrigeration, Air Conditioning and Heating* (July): 22-28.

Broadbent, C.R. et al. 1992. *Legionella* ecology in cooling towers. *Australian Refrigeration, Air Conditioning and Heating* (October):20-34.

CTI. 1996. Acceptance test code for closed circuit cooling towers. *Standard* ATC-105S-96. Cooling Tower Institute, Houston.

CTI. 1997. Acceptance test code for water-cooling towers. *Standard* ATC-105-97, vol. 1. Cooling Tower Institute, Houston.

CTI. 1996. Standard for the certification of water-cooling tower thermal performance. *Standard* STD-201-96. Cooling Tower Institute, Houston.

Hensley, J.C., ed. 1985. *Cooling tower fundamentals*, 2nd ed. Marley Cooling Tower Company, Kansas City.

McCann, M. 1988. Cooling towers take the heat. *Engineered Systems* 5(October):58-61.

Meitz, A. 1986. Clean cooling systems minimize *Legionella* exposure. *Heating, Piping and Air Conditioning* 58(August):99-102.

Merkel, F. 1925. Verduftungskühlung. *Forschungarbeiten* 275.

Mickley, H.S. 1949. Design of forced-draft air conditioning equipment. *Chemical Engineering Progress* 45:739.

Rosa, F. 1992. Some contributing factors in indoor air quality problems. *National Engineer* (May):14.

BIBLIOGRAPHY

Baker, D.R. 1962. Use charts to evaluate cooling towers. *Petroleum Refiner* (November).

Braun, J.E. and Diderrich, G.T. 1990. Near-optimal control of cooling towers for chilled water systems. *ASHRAE Transactions* 96(2):806-813.

CIBSE. 1991. Minimizing the risk of Legionnaire's disease. *Technical Memorandum* TM13. The Chartered Institute of Building Services Engineers, London.

CTI. 2003. *CTI ToolKit*, v. 3.0. Cooling Technology Institute, Houston.

Fliermans, C.B. et al. 1981. Measure of *Legionella pneumophila* activity in situ. *Current Microbiology* 6:89-94.

Fluor Products Company. 1958. *Evaluated weather data for cooling equipment design.*

Kohloss, F.H. 1970. Cooling tower application. *ASHRAE Journal* (August).

Landon, R.D. and J.R. Houx, Jr. 1973. *Plume abatement and water conservation with the wet-dry cooling tower.* Marley Cooling Tower Company, Mission, Kansas City.

Mallison, G.F. 1980. Legionellosis: Environmental aspects. *Annals of the New York Academy of Science* 353:67-70.

McBurney, K. 1990. Maintenance suggestions for cooling towers and accessories. *ASHRAE Journal* 32(6):16-26.

Meitz, A. 1988. Microbial life in cooling water systems. *ASHRAE Journal* 30(August):25-30.

White, T.L. 1994. Winter cooling tower operation for a central chilled water system. *ASHRAE Transactions* 100(1):811-816.

CHAPTER 40

EVAPORATIVE AIR-COOLING EQUIPMENT

THIS chapter addresses direct and indirect evaporative equipment, air washers, and their associated equipment used for air cooling, humidification, dehumidification, and air cleaning. Residential and industrial humidification equipment are covered in Chapter 21.

Principal advantages of evaporative air conditioning include

- Substantial energy and cost savings
- Reduced peak power demand
- Improved indoor air quality
- Life cycle cost effectiveness
- Easily integrated into built-up systems
- Wide variety of packages available
- Provide humidification and dehumidification when needed
- Easy to use with direct digital control (DDC)
- Reduced pollution emissions
- No chlorofluorocarbon (CFC) usage

Packaged direct evaporative air coolers, air washers, indirect evaporative air coolers, evaporative condensers, vacuum cooling apparatus, and cooling towers exchange sensible heat for latent heat. This equipment falls into two general categories: those for (1) air cooling and (2) heat rejection. This chapter addresses air-cooling equipment.

Adiabatic evaporation of water provides the cooling effect of evaporative air conditioning. In **direct evaporative cooling**, water evaporates directly into the airstream, reducing the air's dry-bulb temperature and raising its humidity. Direct evaporative equipment cools air by direct contact with the water, either by an extended wetted-surface material (e.g., packaged air coolers) or with a series of sprays (e.g., an air washer).

In **indirect evaporative cooling**, secondary air removes heat from primary air using a heat exchanger. In one indirect method, water is evaporatively cooled by a cooling tower and circulates through a heat exchanger. Supply air to the space passes over the other side of the heat exchanger. In another common method, one side of an air-to-air heat exchanger is wetted and removes heat from the conditioned supply airstream on the dry side. Even in regions with high wet-bulb temperatures, indirect evaporative cooling can be economically feasible.

Direct and indirect evaporative processes can be **combined (indirect/direct)**. The first stage (indirect) sensibly cools the air, which is then passed through the second stage (direct) and evaporatively cooled further. Combination systems use both direct and indirect evaporative principles as well as secondary heat exchangers and cooling coils. Secondary heat exchangers enhance both cooling and heat recovery (in winter), and the coils provide additional cooling/ dehumidification as needed. Used in both dual-duct and unitary systems, secondary heat exchangers can also save energy by eliminating the need for terminal reheat in some applications (in such systems, air may exit below the initial wet-bulb temperature).

Direct evaporative coolers for residences in low-wet-bulb regions typically require 70% less energy than direct-expansion air conditioners. For instance, in El Paso, Texas, the typical evaporative cooler consumes 609 kWh per cooling season, compared to 3901 kWh per season for a typical vapor-compression air conditioner with a seasonal energy-efficiency ratio (SEER) of 10. This equates to an average demand of 0.51 kW based on 1200 operating hours, compared to an average of 3.25 kW for a vapor-compression air conditioner.

Depending on climatic conditions, many buildings can use indirect/direct evaporative air conditioning to provide comfort cooling. Indirect/direct systems achieve a 40 to 50% energy savings in moderate humidity zones (Foster and Dijkstra 1996).

DIRECT EVAPORATIVE AIR COOLERS

In direct evaporative air cooling, air is drawn through porous wetted pads or a spray and its sensible heat energy evaporates some water; the heat and mass transfer between the air and water lowers the air dry-bulb temperature and increases the humidity at a constant wet-bulb temperature. The dry-bulb temperature of the nearly saturated air approaches the ambient air's wet-bulb temperature. The process is adiabatic, so no sensible cooling occurs.

Saturation effectiveness is a key factor in determining evaporative cooler performance. The extent to which the leaving air temperature from a direct evaporative cooler approaches the thermodynamic wet-bulb temperature of the entering air, or the extent to which complete saturation is approached, is expressed as the **direct saturation effectiveness**, which is defined as:

$$\varepsilon_e = 100\frac{t_1 - t_2}{t_1 - t_s'} \qquad (1)$$

where

ε_e = direct evaporative cooling or saturation effectiveness, %
t_1 = dry-bulb temperature of entering air, °F
t_2 = dry-bulb temperature of leaving air, °F
t_s' = thermodynamic wet-bulb temperature of entering air, °F

An efficient wetted pad (with a high saturation effectiveness) can reduce the air dry-bulb temperature by as much as 95% of the wet-bulb depression (ambient dry-bulb temperature less wet-bulb temperature), although an inefficient and poorly designed pad may only reduce this by 50% or less.

Although direct evaporative cooling is simple and inexpensive, its cooling effect is insufficient for indoor comfort when the ambient wet-bulb temperature is higher than about 70°F; however, cooling is still sufficient for relief cooling applications (e.g., greenhouses, industrial cooling). Direct evaporative coolers should not recirculate indoor air.

The preparation of this chapter is assigned to TC 5.7, Evaporative Cooling.

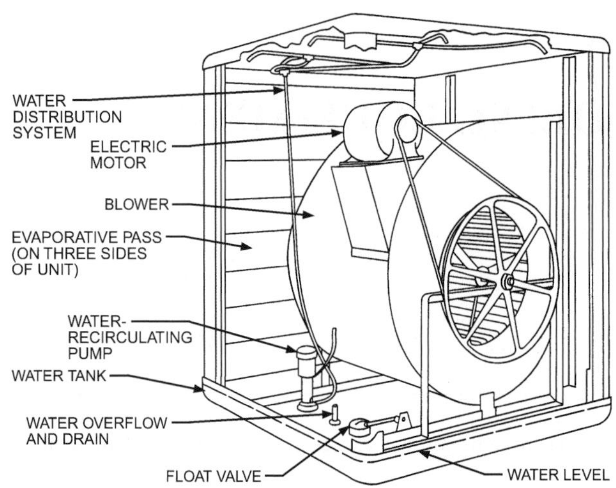

Fig. 1 Typical Random-Media Evaporative Cooler

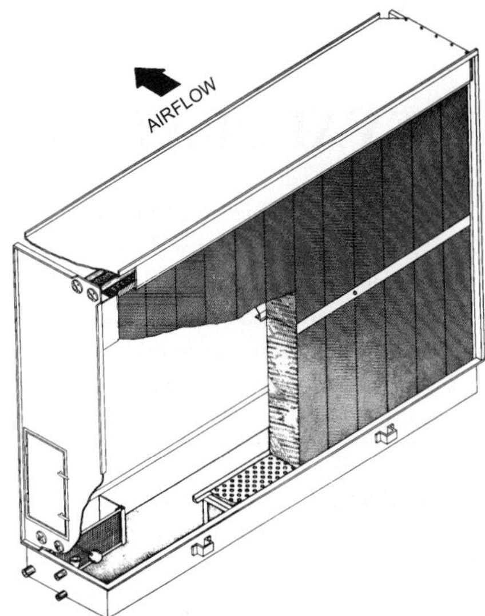

Fig. 2 Typical Rigid-Media Air Cooler

Random-Media Air Coolers

These coolers contain evaporative pads, usually of aspen wood or absorbent plastic fiber/foam (Figure 1). A water-recirculating pump lifts sump water to a distributing system, and it flows down through the pads back to the sump.

A fan in the cooler forces air through the evaporative pads and delivers it to the space to be cooled. The fan discharges either through the side of the cooler cabinet or through the sump bottom. Random-media packaged air coolers are made as small tabletop coolers (50 to 200 cfm), window units (100 to 4500 cfm), and standard duct-connected coolers (5000 to 18000 cfm). Cooler selection should be based on a capacity rating from an independent agency.

When clean and well maintained, commercial random-media air coolers operate at approximately 80% effectiveness and remove 10 μm and larger particles from the air. In some units, supplementary filters before or after the evaporative pads keep particles from entering the cooler, even when it is operated without water to circulate fresh air. Evaporative pads may be chemically treated to increase wettability. An additive may be included in the fibers to help them resist attack by bacteria, fungi, and other microorganisms.

Random-media coolers are usually designed for an evaporative pad face velocity of 100 to 250 fpm, with a pressure drop of 0.1 in. of water. Aspen fibers are packed to approximately 0.3 to 0.4 lb/ft^2 of face area based on a 2 in. thick pad. Pads are mounted in removable louvered frames, which are usually made of painted galvanized steel or molded plastic. Troughs distribute water to the pads. A centrifugal pump with a submerged inlet pumps water through tubes that provide an equal flow of water to each trough. It is important that pumps are thermally protected. The sump or water tank has a water makeup connection, float valve, overflow pipe, and drain. Provisions to bleed water to prevent the buildup of minerals, dirt, and microbial growth are typically incorporated in the design.

The fan is usually a forward-curved, centrifugal fan, complete with motor and drive. The V-belt drive may include an adjustable-pitch motor sheave to allow fan speed to increase to use the full motor capacity at higher airflow resistance. The motor enclosure may be drip-proof, totally enclosed, or a semi-open type specifically designed for evaporative coolers.

Rigid-Media Air Coolers

Blocks of corrugated material make up the wetted surface of rigid-media direct evaporative air coolers (Figure 2). Materials include cellulose, plastic, and fiberglass that have been treated to absorb water yet resist its weathering effects. The medium is cross-corrugated to maximize mixing of air and water. In the direction of airflow, the depth of medium is commonly 12 in., but it may be between 4 and 24 in. The medium has the desirable characteristics of low resistance to airflow, high saturation effectiveness, and self-cleaning by flushing the front face of the pad. The rigid medium is usually designed for a face velocity of 400 to 600 fpm.

Direct evaporative air coolers using this material are built to handle as much as 600,000 cfm with or without fans. Saturation effectiveness varies from 70 to over 95%, depending on media depth and air velocity. Air flows horizontally while the recirculating water flows vertically over the medium surfaces by gravity feed from a flooding header and water distribution chamber. The header may be connected directly to a pressurized water supply for once-through operation (e.g., gas turbines and clean rooms), or a pump may recirculate the water from a lower reservoir constructed of heavy-gage corrosion-resistant material. The reservoir is also fitted with overflow and positive flowing drain connections. The upper media enclosure is of reinforced galvanized steel or other corrosion-resistant sheet metal, or of plastic.

Flanges at the entering and leaving faces allow the unit to be connected to ductwork. In recirculating water systems, a float valve maintains proper water level in the reservoir, makes up water that has evaporated, and supplies fresh water for dilution to prevent an overconcentration of solids and minerals. Because the water recirculation rate is low and because high-pressure nozzles are not needed to saturate the medium, pumping power is low compared to spray-filled air washers with equivalent evaporative cooling effectiveness.

Remote Pad Evaporative Cooling Equipment

Greenhouses, poultry or hog buildings, and similar applications use exhaust fans installed in the wall or roof of the structure. Air is evaporatively cooled as it is drawn through pads located on the other end of the building. The pads are wetted from above by a perforated pipe, and excess water is collected for recirculation. In some cases, the pads are wetted with high-pressure fogging nozzles, which provide additional cooling. Water from fogging nozzles must never be recirculated. The pad should be sized for an air velocity of approximately 150 fpm for random-media pads, 250 fpm for 4 in. rigid media, and 425 fpm for 6 in. rigid media.

INDIRECT EVAPORATIVE AIR COOLERS

Packaged Indirect Evaporative Air Coolers

An indirect evaporative cooling (IEC) heat exchanger is illustrated in Figure 3. This cross-flow, plate-type heat exchanger uses a recirculation sump water pump to wet the inside of the heat

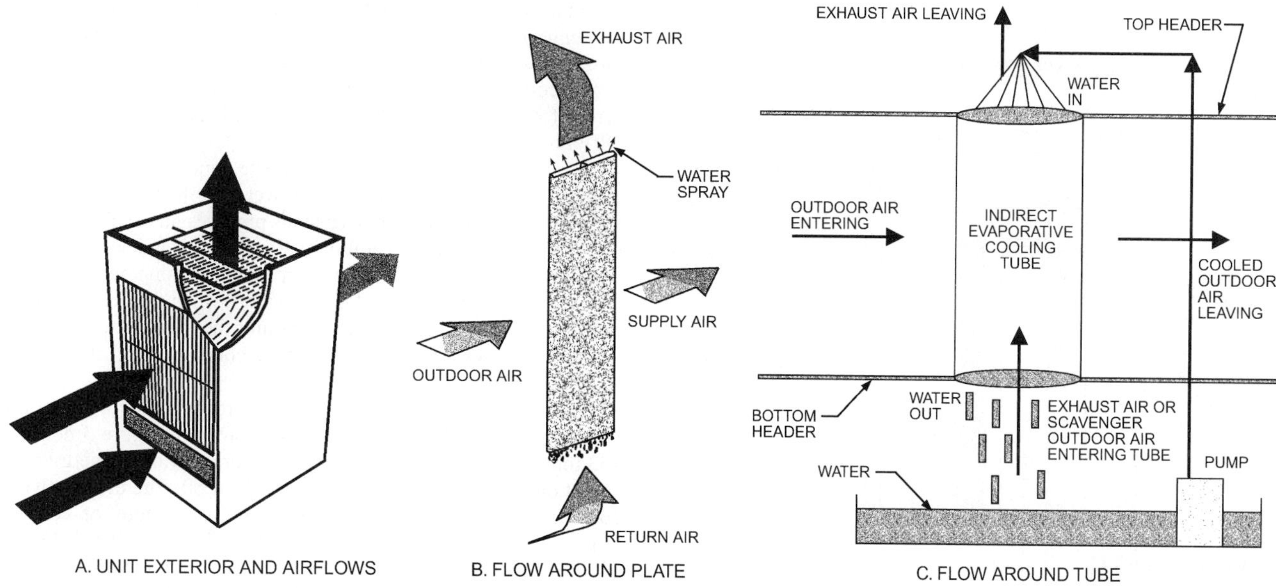

Fig. 3 Indirect Evaporative Cooling (IEC) Heat Exchanger
(Courtesy Munters/Des Champs)

exchanger tubes. Either building return or outdoor air may be drawn up through the inside of the tubes by a secondary air fan. Outdoor air entering the building is sensibly cooled by the exterior surface of the tubes, which are chilled by water evaporating off their interior surface. Latent cooling may also occur if the secondary air wet-bulb temperature is below the outdoor air dew point.

These heat exchangers are capable of a 60 to 80% approach of the ambient dry-bulb temperature to the secondary airflow entering wet-bulb temperature. This is called **wet-bulb depression efficiency (WBDE)**, and is calculated as follows:

$$WBDE = 100\frac{(t_1 - t_2)}{(t_1 - t_s')} \qquad (2)$$

where

WBDE = wet-bulb depression efficiency, %
t_1 = dry-bulb temperature of entering primary air, °F
t_2 = dry-bulb temperature of leaving primary air, °F
t_s' = wet-bulb temperature of entering secondary air, °F

Supply-air-side static pressure losses for these heat exchangers range from 0.25 to 0.75 in. of water. Wet-side air flow penalties range from 0.8 to 0.9 in. of water. Secondary airflow ratios are selected in the range of 1 to 1 down to a low of 1 cfm of outdoor air (OA) to 0.7 cfm of secondary airflow. Cooling energy efficiency ratios (EER) for this type of heat exchanger range from 40 to 80.

With DX Refrigeration. Figure 4 illustrates a package unit design that combines the plate-type indirect evaporative cooling heat exchanger with a direct-expansion (DX) refrigeration final stage of cooling. The geometry of the plate-type heat exchanger usually limits the size of this application to less than 20,000 cfm of supply air.

By placing the condenser coil in the wet-side air path off the heat exchanger, the mechanical cooling component's coefficient of performance (COP) is significantly increased over that of an air-cooled condenser system with the coil in the ambient air. When building return air is used as the secondary airflow, compressor energy inputs are often reduced from 1.1 kW per ton to 0.70 kW per ton or lower, because building return air from an air-conditioned building has wet-bulb conditions in the range of 60 to 65°F at a 75°F room dry-bulb temperature. The wet-side air leaving the heat exchanger is usually in the range of 70 to 75°F db, but at 80 to 90% rh, depending

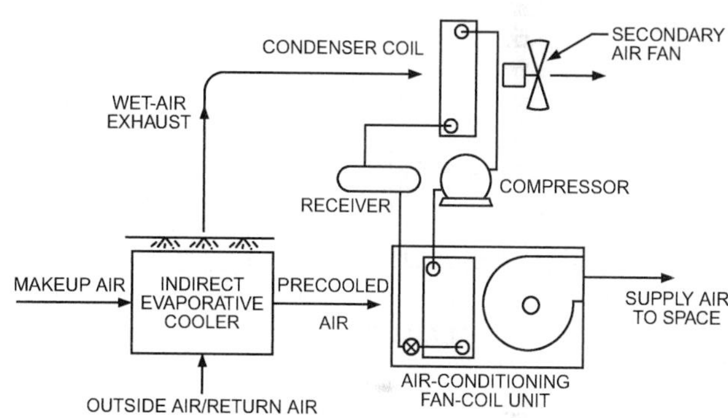

Fig. 4 Indirect Evaporative Cooler Used as Precooler

on the heat exchanger's wetting efficiency. Because refrigeration air-cooled condenser coils are unaffected by humidity, this cooler airstream may be used to reduce the refrigeration condensing temperature of the DX system, which increases compressor capacity and life by reducing vapor compression temperature lift.

Figure 5 shows how a heat-pipe, indirect evaporative cooling heat exchanger may be packaged with a DX-type refrigeration system, using building return air, to minimize cooling energy consumption for an all-outdoor-air design such as may be required for a laboratory or hospital application. The geometry of the heat pipe lends itself to the treatment of larger airflow quantities. The dimensions shown in Figure 5 are for a nominal 50,000 cfm supply air system with 220 tons of total load.

In addition, the heat pipe heat exchanger has the distinct advantage over other air-to-air heat exchangers of being able to isolate contaminated exhaust air from clean makeup air with a double-walled partition at the center bulkhead separating the two air flows. For laboratory applications, supply air fans should be positioned to blow through the heat pipe, to allow the heat pipe indirect evaporative cooler to remove some of the supply fan heat from the air before its delivery to the DX evaporator coil.

As an example, Figure 5 shows state-point conditions at each stage of the process, assuming a required 55°F db supply air temperature and an outdoor air (OA) inlet condition at summer design of 103°F db and 69.9°F wb. The indirect-cooling heat pipe reduces

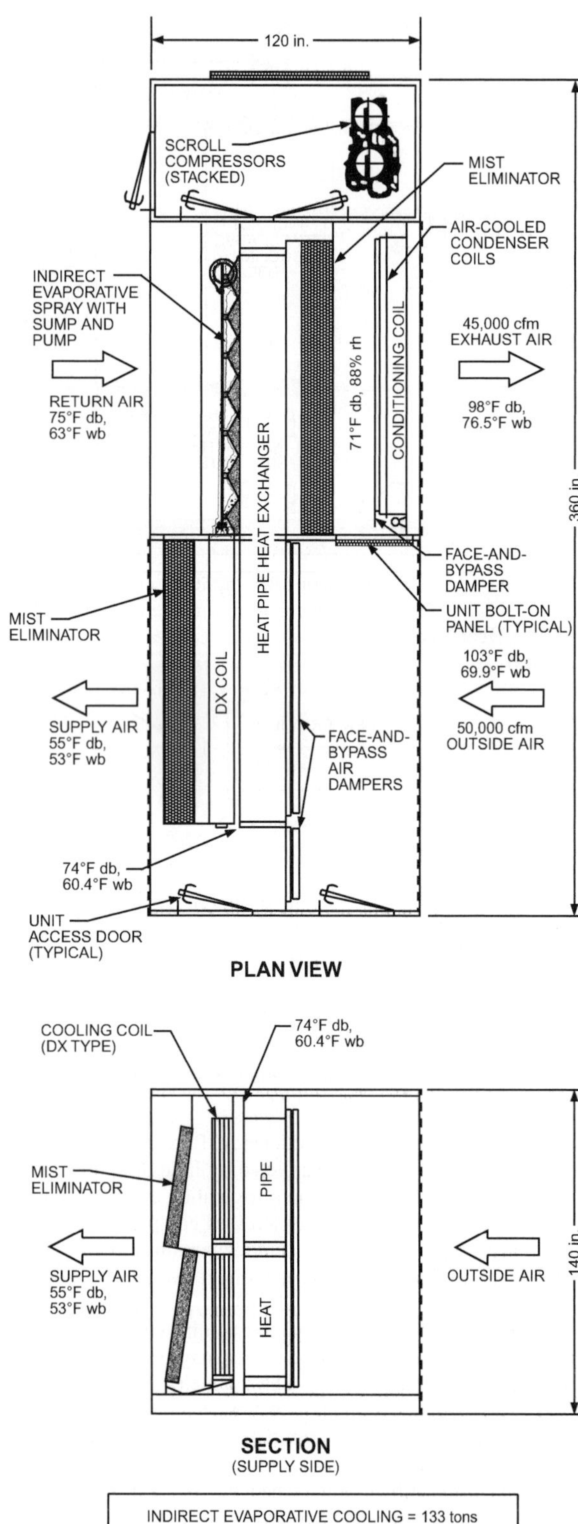

Fig. 5 Heat Pipe Indirect Evaporative Cooling (IEC) Heat Exchanger Packaged with DX System

the outdoor air to 74°F db, 60.4°F wb where it enters the direct-expansion (DX) cooling evaporator coil. The refrigeration coil sensibly cools the outdoor air to 55°F db and 53°F wb, which for 50,000 cfm would require 1,045,000 Btu/h or 87 tons. All values are for a sea-level application.

On the return air side of the heat pipe heat exchanger, the condition entering the heat pipe is 75°F db and 63°F wb. After passing through the wet side of the heat pipe, the return air enters the condenser coil at 71°F and 88% rh. The heat of compression (110 tons) is rejected to the 45,000 cfm airflow and exhausted at a condition of 98°F db, 76.5°F wb.

A mist eliminator downstream of the sprayed heat pipe keeps water droplets from carrying over to wet the refrigeration condensing coil. This cool, humid exhaust air provides an excellent source into which the condenser coil may reject heat. Condenser coil face and bypass dampers are used to control condensing head pressure within an acceptable range. During winter, when the heat pipe heat exchanger is used for heat recovery and the sprays are off, these dampers are both open to minimize the condenser coil static pressure penalty.

Many applications below 200 tons use roof-mounted, air-cooled condensers. The 50,000 cfm IEC unit shown in Figure 5 delivers 133 tons of sensible cooling to the outdoor air with an energy consumption of 0.2 kW per ton and an EER of 60. The evaporatively cooled refrigeration provides the remaining 87 tons of cooling required on the hottest day of the summer. To deliver 55°F db and 53°F wb to the building, the energy consumed for the refrigeration component is 0.7 kW per ton, with EER of 17.1. A conventional air-cooled condensing unit on the roof in 100°F ambient temperatures would typically require 1.1 kW per ton to deliver 220 tons of total load, or a total peak demand of 242 kW. By comparison, on the hottest day of the year, the heat pipe IEC and the evaporatively cooled refrigeration design would only consume a total of 87.55 kW for a combined EER of 30.2. The total peak demand reduction for an all-outdoor-air design in this example is 154.45 kW.

Because the wet side of the heat pipe has a surface temperature of 70 to 75°F when subjected to 100°F ambient air temperatures, scale and fouling of the exhaust-side surface progress very slowly. Systems of this type have been in successful service for over 25 years at various sites in North America.

For sprayed heat-pipe applications, a one-piece heat pipe is recommended. All-aluminum heat pipes are available constructed of series 3003 alloy. The fin surface is extruded directly from the heat tube wall. Corrosion-resistant coating for the wet-side surface may be necessary in some hard-water applications. Wastewater bleed rates should be field set based on the water chemistry analysis. Water consumption in the range of 1 to 1.5 gpm per 10,000 cfm of supply air is typical, for both evaporation and bleed, for an IEC system.

Chapter 51 of the 2007 *ASHRAE Handbook—HVAC Applications* includes sample evaporative cooling calculations. Manufacturers' data should be followed to select equipment for cooling performance, pressure drop, and space requirements.

Manufacturers' ratings require careful interpretation. The basis of ratings should be specified because, for the same equipment, performance is affected by changes in primary and secondary air velocities and mass flow ratios, wet-bulb temperature, altitude, and other factors.

Typically, air resistance on both primary and secondary sections ranges between 0.2 and 2.0 in. of water. The ratio of secondary air to conditioned primary air may range from less than 0.3 to greater than 1.0, and has an effect on performance (Peterson 1993). Based on manufacturers' ratings, available equipment may be selected for indirect evaporative cooling effectiveness ranging from 40 to 80%.

Heat Recovery

Indirect evaporative cooling has been used in a number of heat recovery systems, including plate heat exchangers (Scofield and DesChamps 1984), heat pipe heat exchangers (Mathur 1991; Scofield 1986), rotary regenerative heat exchangers (Woolridge et al. 1976), and two-phase thermosiphon loop heat exchangers (Mathur 1990). Indirect evaporative cooling/heat recovery can be retrofitted on existing systems, lowering operational cost and peak demand. For new installations, equipment can be downsized, lowering overall project and operational costs. Chapter 25 has more information on using indirect evaporative cooling with heat recovery.

Cooling Tower/Coil Systems

Combining a cooling tower or other evaporative water cooler with a water-to-air heat exchanger coil and water circulating pump is another type of indirect evaporative cooling. Water is pumped from the cooling tower reservoir to the coil and returns to the tower's upper distribution header. Both open-water and closed-loop systems are used. Coils in open systems should be cleanable.

Recirculated water is evaporatively cooled to within a few degrees of the wet-bulb temperature as it flows over the wetted surfaces of the cooling tower. As cooled water flows through the tubes of the coil in the conditioned airstream, it picks up heat from the conditioned air. The water temperature increases, and the primary air is cooled without adding moisture to it. The water is again cooled as it recirculates through the cooling tower. A float valve controls the fresh-water makeup, which replaces evaporated water. Bleedoff prevents excessive concentration of minerals in recirculated water.

One advantage of a cooling tower, especially for retrofits, large built-up systems, and dispersed air handlers, is that it may be remotely located from the cooling coil. Also, the tower is more accessible for maintenance. Overall WBDE ε_e may range between 55% and 75% or higher. If return air is sent to the cooling tower of an indirect cooling system before being discharged outside, the cooling tower should be specifically designed for this purpose. These coolers wet a medium that has a high ratio of wetted surface area per unit of medium volume. Performance depends on depth of the medium, air velocity over the medium surface, water flow to airflow ratio, wet-bulb temperature, and water-cooling range. Because of the close approach of the water temperature to the wet-bulb temperature, overall effectiveness may be higher than that of a conventional cooling tower.

Other Indirect Evaporative Cooling Equipment

Other combinations of evaporative coolers and heat exchangers can accomplish indirect evaporative cooling. Heat pipes and rotary heat wheels, two-phase thermosiphon coil loops, plate and pleated media, and shell-and-tube heat exchangers have all been used. If the conditioned (primary) air and the exhaust or outside (secondary) airstream are side by side, a heat pipe or heat wheel can transfer heat from the warmer air to the cooler air. Evaporative cooling of the secondary airstream by spraying water directly on the surfaces of the heat exchanger or by a direct evaporative cooler upstream of the heat exchanger may cool the primary air indirectly by transferring heat from it to the secondary air.

INDIRECT/DIRECT COMBINATIONS

In a two-stage indirect/direct evaporative cooler, a first-stage indirect evaporative cooler lowers both the dry- and wet-bulb temperature of the incoming air. After leaving the indirect stage, the supply air passes through a second-stage direct evaporative cooler; Figure 6 shows the process on a psychrometric chart. First-stage cooling follows a line of constant humidity ratio because no moisture is added to the primary airstream. The second stage follows the wet-bulb line at the condition of the air leaving the first stage.

The indirect evaporative cooler may be any of the types described previously. Figure 7 shows a cooler using a rotary heat wheel or heat pipe. The secondary air may be exhaust air from the conditioned space or outdoor air. When the secondary air passes through the direct evaporative cooler, the dry-bulb temperature is lowered by evaporative cooling. As this air passes through the heat wheel, the mass of the medium is cooled to a temperature approaching the wet-bulb temperature of the secondary air. The heat wheel rotates (note, however, that a heat pipe has no moving parts) so that its cooled mass enters the primary air and, in turn, sensibly cools the primary (supply) air. After the heat wheel or pipe, a direct evaporative cooler further reduces the dry-bulb temperature of the primary air. This method can lower the supply air dry-bulb temperature by 10°F or more below the secondary air wet-bulb temperature.

In areas where the 0.4% mean coincident wet-bulb design temperature is 66°F or lower, average annual cooling power consumption of indirect/direct systems may be as low as 0.22 kW/ton of refrigeration. When the 0.4% mean coincident wet-bulb temperature is as high as 74°F, indirect/direct cooling can have an average annual cooling power consumption as low as 0.81 kW/ton. By comparison, the typical refrigeration system with an air-cooled condenser may have an average annual power consumption greater than 1.0 kW/ton.

In dry environments, indirect/direct evaporative cooling is usually designed to supply 100% outdoor air to the conditioned spaces of a building. In these once-through applications, space latent loads and return air sensible loads are exhausted from the building rather than returned to the conditioning equipment. Consequently, the cooling capacity required from these systems may be less than from a conventional refrigerated cooling system. Design features that should be considered in systems such as the one in Figure 7 include air filters on the entering side of each heat wheel or pipe. Systems without the direct evaporative cooler on the exhaust side have been successfully used in many laboratory applications in the southwestern United States. Maintenance inside a laboratory exhaust airstream can be hazardous; therefore, the fewer components in the airstream needing maintenance, the lower the risk to staff.

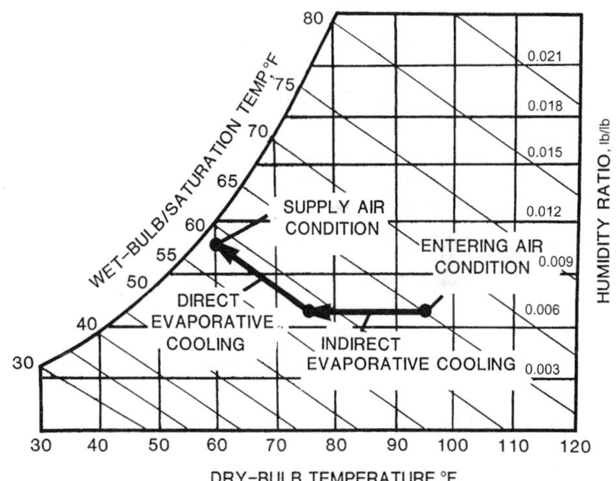

Fig. 6 Combination Indirect/Direct Evaporative Cooling Process

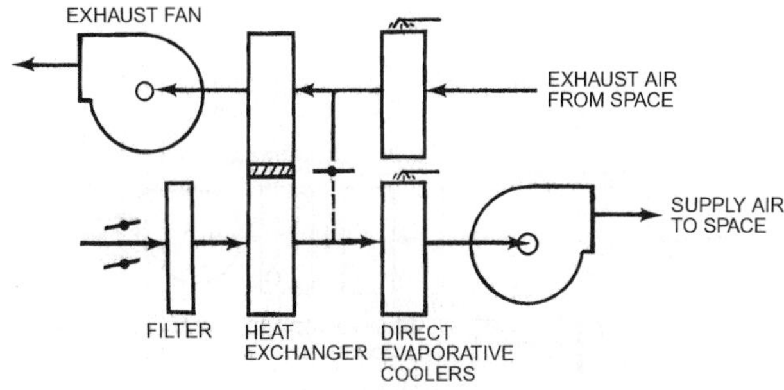

Fig. 7 Indirect/Direct Evaporative Cooler with Heat Exchanger (Rotary Heat Wheel or Heat Pipe)

In areas with a higher wet-bulb design temperature or where the design requires a supply air temperature lower than that attainable using indirect/direct evaporative cooling, a third cooling stage may be required. This stage may be a direct-expansion refrigeration unit or a chilled-water coil located either upstream or downstream from the direct evaporative cooling stage, but always downstream from the indirect evaporative stage. Refrigerated cooling is energized only when evaporative stages cannot achieve the required supply air temperature. Figure 8 shows a three-stage configuration (indirect/direct, with optional third-stage refrigerated cooling). The third-stage refrigerated cooling coil is downstream from the direct evaporative cooler. This requires careful selection and adjustment of controls to avoid excess moisture removal by the refrigerated cooling coil than can be added by the direct evaporative cooling components. Analysis of static pressure drop through all components during design is critical to maintain optimum system total pressure loss and overall system efficiency. Note the face-and-bypass damper in Figure 8 around the indirect evaporative cooler.

A single coil may be used to cool return chilled water with cooling tower water and a plate heat exchanger (a form of indirect evaporative cooling). This hybrid, three-stage configuration allows indirect cooling when the wet-bulb temperature is low and mechanical cooling when the wet bulb is high or when dehumidification is necessary.

The designer should consider using building exhaust and/or outside air as secondary air (whichever has the lower wet-bulb temperature) for indirect evaporative cooling. If possible, the indirect evaporative cooler should be designed to use both outside air and building exhaust as the secondary airstream; whichever source has the lower wet-bulb temperature would be used. Dampers and an enthalpy sensor are used to control this process. If the latent load in the space is significant, the wet-bulb temperature of the building exhaust air in cooling mode may be higher than that of the outside air. In this case, outside air may be used more effectively as secondary air to the indirect evaporative cooling stage.

Custom indirect/direct and three-stage configurations are available to allow many choices for location of the return, exhaust, and outside air; mixing of airstreams; bypass of components, or variable-volume control. Controllable elements include

- Modulating outside air and return air mixing dampers
- Secondary air fans and recirculating pumps of an indirect evaporative stage
- Recirculating pumps of a direct evaporative cooling stage
- Face-and-bypass dampers for the direct or indirect evaporative stage
- Chilled-water or refrigerant flow for a refrigerated stage
- System or individual terminal volume with variable-volume terminals, adjustable pitch fans, or variable-speed fans

For sequential control in indirect/direct evaporative cooling, the indirect evaporative cooler is energized for first-stage cooling, the direct evaporative cooler for second-stage cooling, and the refrigeration coil for third-stage cooling. In some applications, reversing the sequence of the direct and indirect evaporative coolers may reduce the first-stage power requirement. These systems are typically unfamiliar to most operations and maintenance staff, so special training may be needed.

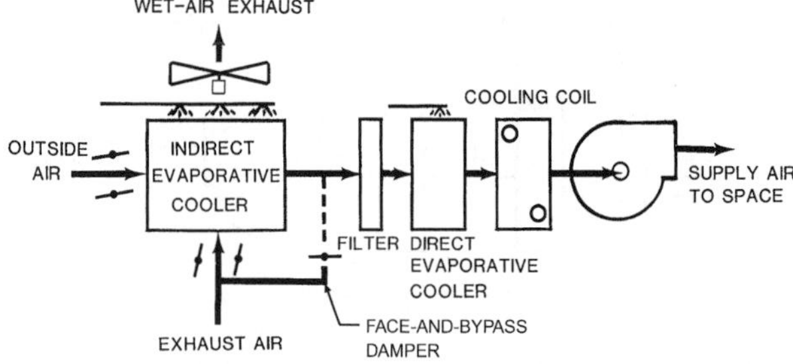

Fig. 8 Three-Stage Indirect/Direct Evaporative Cooler

Precooling and Makeup Air Pretreatment

Evaporative cooling may be used to increase capacity and reduce the electrical demand of a direct expansion air conditioner or chiller. Both the condenser and makeup air may be evaporatively cooled by direct and/or indirect means.

The condenser may be cooled by adding a direct evaporative cooler (usually without a fan) to the condenser fan inlet. The direct evaporative cooler must add very little resistance to the airflow to the condenser, and face velocities must be well below velocities that would entrain liquid and carry it to the condenser. Condenser cooler maintenance should be infrequent and easy to perform. A well-designed direct evaporative cooler can reduce electrical demand and energy consumption of refrigeration units from 10 to 30%.

Makeup air cooling with an indirect/direct evaporative unit can be applied both to standard packaged units and to large built-up systems. Either outside air or building exhaust air (whichever has the lower wet-bulb temperature) can be used as the secondary air source. Outside air is generally easier to cool, and in some cases is the only option because the building exhaust is hazardous (e.g., from a laboratory) or remote from the makeup air inlet. If building exhaust air can be used as the secondary air source, it has the potential of heat recovery during cold weather. In general, outside air cooling has higher energy savings and lower electrical demand savings than return air cooling. These systems can significantly reduce the outside air load and should be analyzed using a psychrometric process for the region and climate being considered.

AIR WASHERS

Spray Air Washers

Spray air washers consist of a chamber or casing containing spray nozzles, a tank for collecting spray water as it falls, and an eliminator section for removing entrained drops of water from the air. A pump recirculates water at a rate higher than the evaporation rate. Intimate contact between the spray water and the air causes heat and mass transfer between the air and water (Figure 9). Air washers are commonly available from 2000 to 250,000 cfm capacity, but specially constructed washers can be made in any size. No standards exist; each manufacturer publishes tables giving physical data and ratings for specific products. Therefore, air velocity, water-spray density, spray pressure, and other design factors must be considered for each application.

The simplest design has a single bank of spray nozzles with a casing that is usually 4 to 7 ft long. This type of washer is applied primarily as an evaporative cooler or humidifier. It is sometimes used as an air cleaner when the dust is wettable, although its air-cleaning efficiency is relatively low. Two or more spray banks are generally used when a very high degree of saturation is necessary and for cooling and dehumidification applications that require chilled water. Two-stage washers are used for dehumidification when the quantity of chilled water is limited or when the water temperature is above that required for the single-stage design. Arranging the two stages for water counterflow allows use of a small quantity of water with a greater water temperature rise.

The lengths of washers vary considerably. Spray banks are spaced from 2.5 to 4.5 ft apart; the first and last banks of sprays are located about 1 to 1.5 ft from the entering or leaving end of the washer. In addition, air washers may be furnished with heating or cooling coils in the washer chamber, which may affect the overall length of the washer.

Some water (even very soft water) should always be bled off (continually and/or by using a dump or purge cycle) to prevent mineral build-up and to retard microbial growth. When the unit is shut down, all water should drain from the pipes. Low spots and dead ends must be avoided. Because an air washer is a direct-contact heat exchanger, water treatment is critical for proper operation as well as good hygiene. Algae and bacteria can be controlled by a chemical or ozone treatment program and/or regularly scheduled mechanical cleaning. Make sure that any chemicals used are compatible with all components in the air washer.

The resistance to airflow through an air washer varies with the type and number of baffles, eliminators, and wetted surfaces; the number of spray banks and their direction and air velocity; the size and type of other components, such as cooling and heating coils; and other factors, such as air density. Pressure drop may be as low as 0.25 in. of water or as high as 1 in. of water. The manufacturer should be consulted regarding the resistance of any particular washer design combination.

The casing and tank may be constructed of various materials. One or more doors are commonly provided for inspection and access. An air lock must be provided if the unit is to be entered while it is running. The tank is normally at least 16 in. high with a 14 in. water level; it may extend beyond the casing on the inlet end to make the suction strainer more accessible. The tank may be partitioned by a weir (usually in the entering end) to permit recirculation of spray water for control purposes in dehumidification work. The excess then returns over the weir to the central water-chilling machine.

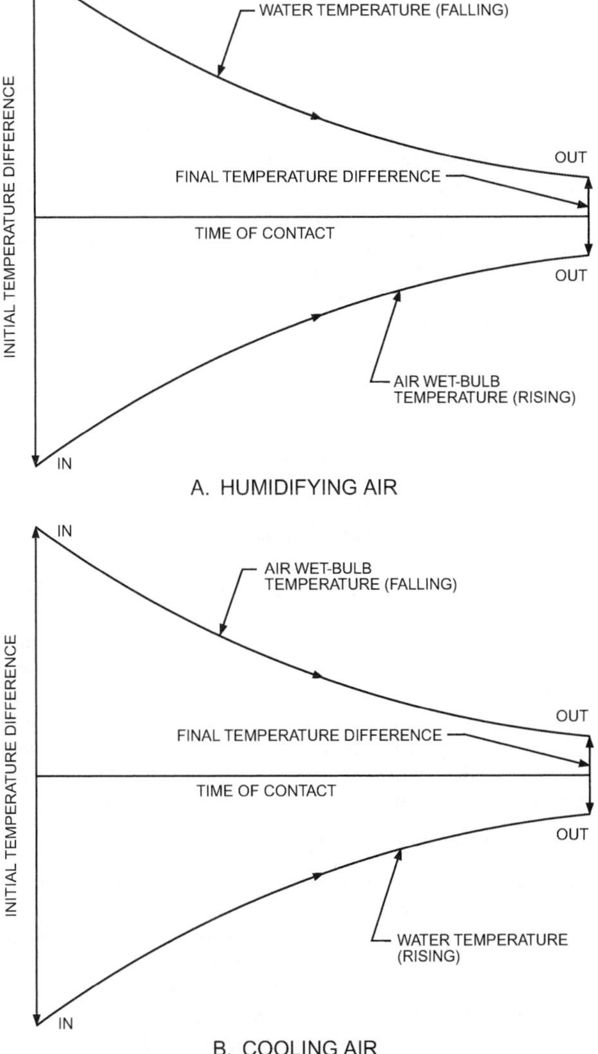

Fig. 9 Interaction of Air and Water in Air Washer Heat Exchanger

Eliminators consist of a series of vertical plates that are spaced about 0.75 to 2 in. on centers at the exit of the washer. The plates are formed with numerous bends to deflect air and obtain impingement on the wetted surfaces. Hooks on the edge of the plates improve moisture elimination. Perforated plates may be installed on the inlet end of the washer to obtain more uniform air distribution through the spray chamber. Louvers, which prevent backlash of spray water, may also be installed for this purpose.

High-Velocity Spray-Type Air Washers

High-velocity air washers generally operate at air velocities in the range of 1200 to 1800 fpm. Some have been applied as high as 2400 fpm, but 1200 to 1600 fpm is the most accepted range. The reduced cross-sectional area of high-velocity air washers allows them to be used in smaller equipment than those operating with lower air velocities. High capacities per unit of space available from high-velocity spray devices allow practical prefabrication of central station units in either completely assembled and transportable form or, for large-capacity units, easily handled modules. Manufacturers supply units with capacities of up to 150,000 cfm shipped in one piece, including spray system, eliminators, pump, fan, dampers, filters, and other functional components. Such units are self-housed, prewired, prepiped, and ready for hoisting into place.

The number and arrangement of nozzles vary with different capacities and manufacturers. Adequate values of saturation effectiveness and heat transfer effectiveness are achieved by using higher spray density.

Eliminator blades come in varying shapes, but most are a series of aerodynamically clean, sinusoidal shapes. Collected moisture flows down grooves or hooks designed into their profiles, then drains into the storage tank. Washers may be built with shallow drain pans and connected to a central storage tank. High-velocity washers are rectangular in cross section and, except for the eliminators, are similar in appearance and construction to conventional lower-velocity types. Pressure loss is in the range of 0.5 to 1.5 in. of water. These washers are available either as freestanding separate devices for incorporation into field-built central stations or in complete preassembled central station packages from the factory.

HUMIDIFICATION/DEHUMIDIFICATION

Humidification with Air Washers and Rigid Media

Air can be humidified with air washers and rigid media by (1) using recirculated water without prior heating of the air, (2) preheating the air and humidifying it with recirculated water, or (3) preheating recirculated water. Precise humidity control may be achieved by arranging rigid media in one or more banks in depth, height, or width, or by providing a controlled bypass. Each bank is activated independently of the others to achieve the desired humidity. In any evaporative humidification application, air should not be permitted to enter the process with a wet-bulb temperature of less than 39°F, or the water may freeze.

Recirculation Without Preheating. Except for the small amount of energy added by the recirculating pump and the small amount of heat leakage into or from the apparatus (including the pump and its connecting piping), the process is adiabatic. Water temperature in the collection basin closely approaches the thermodynamic wet-bulb temperature of the entering air, but it cannot be brought to complete saturation. The psychrometric state point of the leaving air is on the constant thermodynamic wet-bulb temperature line with its end state determined by the saturation effectiveness of the device. Leaving humidity conditions may be controlled using the saturation effectiveness of the process by bypassing air around the evaporative process.

Preheating Air. Preheating the air entering an evaporative humidifier increases both the dry- and wet-bulb temperatures and lowers the relative humidity, but it does not alter the humidity ratio (mass ratio of water vapor to dry air) of the air. As a result, preheating permits more water to be absorbed per unit mass of dry air passing through the process at the same saturation effectiveness. Control is achieved by varying the amount of air preheating at a constant saturation effectiveness. Control precision is a direct function of saturation effectiveness and a high degree of correlation may be achieved between leaving air and leaving dew-point temperatures when high saturation effectiveness devices are used.

Heated Recirculated Water. If heat is added to the water, the process state point of the mixture moves toward the temperature of the heated water (Figure 9A). Elevating the water temperature makes it possible to raise the air dry- and wet-bulb temperatures above the dry-bulb temperature of the entering air with the leaving air becoming fully saturated. Relative humidity of the leaving air can be controlled by (1) bypassing some of the air around the media banks and remixing the two airstreams downstream by using dampers or (2) by automatically reducing the number of operating media banks through pump staging or by operating valves in the different distribution branches.

The following table shows the saturation or humidifying effectiveness of a spray air washer for various spray arrangements. The degree of saturation depends on the extent of contact between air and water. Other conditions being equal, a low-velocity airflow is conducive to higher humidifying effectiveness.

Bank Arrangement	Length, ft	Effectiveness, %
1 downstream	4	50 to 60
	6	60 to 75
1 upstream	6	65 to 80
2 downstream	8 to 10	80 to 90
2 opposing	8 to 10	85 to 95
2 upstream	8 to 10	90 to 98

Dehumidification with Air Washers and Rigid Media

Air washers and rigid-media direct evaporative coolers may also be used to cool and dehumidify air. Compared to a typical chilled-water or direct-expansion (DX) cooling coil, direct-contact dehumidification can significantly reduce fan power requirements, static pressure losses, and energy consumption (El-Morsi et al. 2003). As shown in Figure 9B, heat and moisture removed from the air raise the water temperature. If the entering water temperature is below the entering air dew point, both the dry- and wet-bulb temperatures of the air are reduced, resulting in cooling and dehumidification. The vapor pressure difference between the entering air and water cools the air. Moisture is transferred from the air to the water, and condensation occurs. Air leaving an evaporative dehumidifier is typically saturated, usually with less than 1°F difference between leaving dry- and wet-bulb temperatures.

The difference between the leaving air and water temperatures depends on the difference between entering dry- and wet-bulb temperatures and the process effectiveness, which may be affected by factors such as length and height of the spray chamber, air velocity, quantity of water flow, and spray pattern. Final water conditions are typically 1 to 2°F below the leaving air temperature, depending on the saturation effectiveness of the device used.

The common design value for the water temperature rise is usually between 6 and 12°F for refrigerant-chilled water and normal air-conditioning applications, although higher rises are possible and have been used successfully. A smaller rise may be considered when water is chilled by mechanical refrigeration. If warmer water is used, less mechanical refrigeration is required; however, a larger quantity of chilled water is needed to do the same amount of sensible cooling. An economic analysis may be required to determine the best alternative. For humidifiers receiving water from a thermal storage or other low-temperature system, a design with a high temperature rise and minimum water flow may be desirable.

Performance Factors. An evaporative dehumidifier has a performance factor of 1.0 if it can cool and dehumidify the entering air to a wet-bulb temperature equal to the leaving water temperature. This represents a theoretical maximum value that is thermodynamically impossible to achieve. Performance is maximized when both water surface area and air/water contact is maximized. The actual performance factor F_p of any evaporative dehumidifier is less than one and is calculated by dividing the actual air enthalpy change by the theoretical maximum air enthalpy change where

$$F_p = \frac{h_1 - h_2}{h_1 - h_3} \tag{3}$$

where

h_1 = enthalpy at wet-bulb temperature of entering air, Btu/lb
h_2 = enthalpy at wet-bulb temperature of leaving air at actual condition, Btu/lb
h_3 = enthalpy of air at wet-bulb temperature leaving a dehumidifier with $F_p = 1.0$, Btu/lb

Air Cleaning

Air washers and rigid-media direct evaporative cooling equipment can remove particulate and gaseous contaminants with varying degrees of effectiveness through wet scrubbing (which is discussed in Chapter 29). Particle removal efficiencies of rigid media and air washers differ due to differences in equipment construction and principles of operation. Removal also depends largely on the size, density, wettability, and/or solubility of the contaminants to be removed. Large, wettable particles are the easiest to remove. The primary mechanism of separation is by impingement of particles on a wetted surface, which includes eliminator plates in air washers and corrugations of wetted rigid media. Spraying is relatively ineffective in removing most atmospheric dusts. Because the force of impact increases with the size of the solid, the impact (together with the adhesive quality of the wetted surface) determines the device's usefulness as a dust remover.

In practice, air-cleaning results of air washers and rigid-media direct-evaporative coolers are typical of comparable impingement filters. Air washers are of little use in removing soot particles because of the lack of adhesion to a greasy surface. They are also relatively ineffective in removing smoke because the particles are too small (less than 1 μm) to impact and be retained on the wet surfaces.

Despite their air-cleaning performance, rigid media should not be used for primary filtering. When a rigid-media cooler is placed in an unfiltered airstream, it can quickly become fouled with airborne dust and fibrous debris. When wet, debris can collect in the recirculation basin and in the media, feeding bacterial growth. Bacteria in the air can propagate in waste materials and debris and cause microbial slimes. Filtering entering air is the most effective way to keep debris from accumulating in rigid media. With high-efficiency filters upstream from the cells, most microbial agents and nutrients can be removed from the airstream. Replace rigid media if the corrugations are filled with contaminants when they are dry.

MAINTENANCE AND WATER TREATMENT

Regular inspection and maintenance of evaporative coolers, air washers, and ancillary equipment ensures proper service and efficiency. Manufacturers' recommendations for maintenance and operation should be followed to help ensure safe, efficient operation. Water lines, water distribution troughs or sumps, pumps, and pump filters must be clean and free of dirt, scale, and debris. They must be constructed so that they can be easily flushed and cleaned. Inadequate water flow causes dry areas on the evaporative media, which reduces the saturation effectiveness. Motors and bearings should be lubricated and fan drives checked periodically.

Water and air filters should be cleaned or replaced as required. The sump water level must be kept below the bottom of the pads, yet high enough to prevent air from short-circuiting below the pads. Bleeding off some water is the most practical means to minimize scale accumulation. The bleed rate should be 5 to 100% of the evaporation rate, depending on water hardness and airborne contaminant level. The water circulation pump should be used to bleed off water (suction by a draw-through fan will otherwise prevent the bleed system from operating effectively). A flush-out cycle that runs fresh water through the pad every 24 h when the fan is off may also be used. This water should run for 3 min for every foot of media height.

Regular inspections should be made to ensure that the bleed rate is adequate and is maintained. Some manufacturers provide a purge cycle in which the entire sump is purged of water and accumulated debris. This cycle helps maintain a cleaner system and may actually save water compared to a standard bleed system. Purge frequency depends on water quality as well as the amount and type of outside contaminants. Sumps should have drain couplings on the bottom rather than on the side, to drain the sump completely. Additionally, the sump bottom should slope toward the drain (approximately 0.25 in. per foot of sump length) to facilitate complete draining.

Water Treatment. An effective water treatment and biocide program for cooling towers is not necessarily good practice for evaporative coolers. Evaporative coolers and cooling towers differ significantly: evaporative coolers are directly connected with the supply airstream, whereas cooling towers only indirectly affect the supply air. The effect a biocide may have on evaporative media (both direct and indirect systems) as well as the potential for offensive and/or harmful residual off-gassing must be considered.

Pretreatment of a water supply with chemicals intended to hold dissolved material in suspension is best prescribed by a water treatment specialist. Water treated by a zeolite ion exchange softener should not be used because the zeolite exchange of calcium for sodium results in a soft, voluminous scale that may cause dust problems downstream. Any chemical agents used should not promote microbial growth or harm the cabinet, media, or heat exchanger materials. This topic is discussed in more detail in Chapter 48 of the 2007 *ASHRAE Handbook—HVAC Applications*. Consider the following factors for water treatment:

- Use caution when using very pure water from reverse osmosis or deionization in media-based evaporative coolers. This water does not wet random media well, and it can deteriorate many types of media because of its corrosive nature. The same problem can occur in a once-through water distribution system if the water is very pure.
- Periodically check for algae, slime, and bacterial growth. If required, add a biocide registered for use in evaporative coolers by an appropriate agency, such as the U.S Environmental Protection Agency (EPA).

Ozone-generation systems have been used as an alternative to standard chemical biocide water treatments. Ozone can be produced on site (eliminating chemical storage) and injected into the water circulation system. It is a fast-acting oxidizer that rapidly breaks down to nontoxic compounds. In low concentrations, ozone is benign to humans and to the materials used in evaporative coolers.

Algae can be minimized by reducing the media and sump exposure to nutrient and light sources (by using hoods, louvers and prefilters), by keeping the bottom of the media out of standing water in the sump, and by allowing the media to completely dry out every 24 h.

Scale. Units that have heat exchangers with a totally wetted surface and materials that are not harmed by chemicals can be descaled periodically with a commercial descaling agent and then flushed out. Mineral scale deposits on a wetted indirect evaporative heat exchanger are usually soft and allow wetting through to and evaporation at the surface of the heat exchanger. Excess scale thickness reduces heat transfer and should be removed.

Air Washers. The air washer spray system requires the most attention. Partially clogged nozzles are indicated by a rise in spray pressure; a fall in pressure is symptomatic of eroded orifices. Strainers can minimize this problem. Continuous operation requires either a bypass around pipeline strainers or duplex strainers. Air washer tanks should be drained and dirt deposits removed regularly. Eliminators and baffles should be periodically inspected and repainted to prevent corrosion damage.

Freeze Protection. In colder climates, evaporative coolers must be protected from freezing. This is usually done seasonally by simply draining the cooler and the water supply line with solenoid valves. Often an outside air temperature sensor initiates this action. It is important that drain solenoid valves be of zero-differential design. If a heat exchanger coil is used, the tubes must be horizontal so they will drain to the lowest part of their manifold.

Legionnaires' Disease

Legionnaires' disease is contracted by inhaling into the lower respiratory system an aerosol (1 to 5 μm in diameter) laden with sufficient *Legionella pneumophila* bacteria. Evaporative coolers do not provide suitable growth conditions for the bacteria and generally do not release an aerosol. A good maintenance program eliminates potential microbial problems and reduces the concern for disease transmittal (ASHRAE 1998, 2000; Puckorius et al. 1995). There have been no known cases of Legionnaires' disease with air washers or wetted-media evaporative coolers/humidifiers, and there is no positive association of Legionnaires' disease with indirect evaporative coolers (ASHRAE *Guideline* 12-2000).

The following precautions and maintenance procedures for water systems also improve cooler performance, reduce microbial growth and musty odors, and prolong equipment life:

- Run fans after turning off water until the media completely dries.
- Thoroughly clean and flush the entire cooling water loop regularly (minimum monthly). Disinfect before and after cleaning.
- Avoid dead-end piping, low spots, and other areas in the water distribution system where water may stagnate during shutdown.
- Obtain and maintain the best available mist elimination technology, especially when using misters and air washers.
- Do not locate the evaporative cooler inlet near a cooling tower outlet.
- Maintain system bleedoff and/or purge consistent with makeup water quality.
- Maintain system cleanliness. Deposits from calcium carbonate, minerals, and nutrients may contribute to growth of molds, slime, and other microbes annoying to building occupants.
- Develop a maintenance checklist, and follow it on a regular basis.
- Consult the equipment or media manufacturer for more detailed assistance in water system maintenance and treatment.

REFERENCES

ASHRAE. 1998. *Legionellosis: Position statement.*

ASHRAE. 2000. *Minimizing the risk of Legionnaires' disease.*

ASHRAE. 2000. Minimizing the risk of legionellosis associated with building water systems. *Guideline* 12-2000.

El-Morsi, M., S.A. Klein, and D.T. Reindl. 2003. Air washers—A new look at a vintage technology. *ASHRAE Journal* 45(10):32-36.

Foster, R.E. and E. Dijkstra. 1996. Evaporative air-conditioning fundamentals: Environmental and economic benefits worldwide. *Refrigeration Science and Technology Proceedings*. International Institute of Refrigeration, Danish Technological Institute, Danish Refrigeration Association, Aarhus, Denmark, pp. 101-110.

Mathur, G.D. 1990. Indirect evaporative cooling with two-phase thermosiphon coil loop heat exchangers. *ASHRAE Transactions* 96(1):1241-1249.

Mathur, G.D. 1991. *Indirect evaporative cooling with heat pipe heat exchangers*. ASME Book No. NE(5):79-85.

Peterson, J.L. 1993. An effectiveness model for indirect evaporative coolers. *ASHRAE Transactions* 99(2):392-399.

Puckorius, P.R., P.T. Thomas, and R.L. Augspurger. 1995. Why evaporative coolers have not caused Legionnaires' disease. *ASHRAE Journal* 37(1): 29-33.

Scofield, M. 1986. The heat pipe used for dry evaporative cooling. *ASHRAE Transactions* 92(1B):371-381.

Scofield, M. and N.H. DesChamps. 1984. Indirect evaporative cooling using plate type heat exchangers. *ASHRAE Transactions* 90(1):148-153.

Woolridge, M.J., H.L. Chapman, and D. Pescod. 1976. Indirect evaporative cooling systems. *ASHRAE Transactions* 82(1):146-155.

BIBLIOGRAPHY

Anderson, W.M. 1986. Three-stage evaporative air conditioning versus conventional mechanical refrigeration. *ASHRAE Transactions* 92(1B):358-370.

Eskra, N. 1980. Indirect/direct evaporative cooling systems. *ASHRAE Journal* 22(5):21-25.

Felver, T., M. Scofield, and K. Dunnavant. 2001. Cooling California's computer centers. *HPAC Engineering* (March):59.

Scofield, M. 1987. Unit gives 45 tons of cooling without a compressor. *Air Conditioning, Heating and Refrigeration News* (December):23.

Scofield, M. and N.H. DesChamps. 1980. EBTR compliance and comfort cooling too! *ASHRAE Journal* 22(6):61-63.

Supple, R.G. 1982. Evaporative cooling for comfort. *ASHRAE Journal* 24(8):36.

Watt, J.R. 1986. *Evaporative air conditioning handbook*. Chapman and Hall, London.

CHAPTER 41

LIQUID COOLERS

A LIQUID cooler (hereafter called a cooler) is a heat exchanger in which refrigerant is evaporated, thereby cooling a fluid (usually water or brine) circulating through the cooler. This chapter addresses the performance, design, and application of coolers.

TYPES OF LIQUID COOLERS

Various types of liquid coolers and their characteristics are listed in Table 1 and described in the following sections.

Direct-Expansion

Refrigerant evaporates inside the tubes of a direct-expansion cooler. These coolers are usually used with positive-displacement compressors, such as reciprocating, rotary, or rotary screw compressors, to cool various fluids, such as water, water/glycol mixtures, and brine. Common configurations include shell-and-tube, tube-in-tube, and brazed-plate.

Figure 1 shows a typical **shell-and-tube** cooler. A series of baffles channels the fluid throughout the shell side. The baffles create cross flow through the tube bundle and increase the velocity of the fluid, thereby increasing its heat transfer coefficient. The velocity of the fluid flowing perpendicular to the tubes should be at least 2 fps to clean the tubes and less than the velocity limit of the tube and baffle materials, to prevent erosion.

Refrigerant distribution is critical in direct-expansion coolers. If some tubes are fed more refrigerant than others, refrigerant may not fully evaporate in the overfed tubes, and liquid refrigerant may escape into the suction line. Because most direct-expansion coolers are controlled by an expansion valve that regulates suction superheat, the remaining tubes must produce a higher superheat to evaporate the liquid escaping into the suction line. This unbalance causes poor overall heat transfer. Uniform distribution is usually achieved by adding a distributor, which creates sufficient turbulence to promote a homogeneous mixture so that each tube gets the same mixture of liquid and vapor.

The number of refrigerant passes is another important item in direct-expansion cooler performance. A single-pass cooler must evaporate all the refrigerant before it reaches the end of the first pass; this requires long tubes. A multiple-pass cooler is significantly shorter than a single-pass cooler, but must be properly designed to ensure proper refrigerant distribution after the first pass. Internally and externally enhanced tubes can also be used to reduce cooler size.

A **tube-in-tube** cooler is similar to a shell-and-tube design, except that it consists of one or more pairs of coaxial tubes. The fluid usually flows inside the inner tube while the refrigerant flows in the annular space between the tubes. In this way, the fluid side can be mechanically cleaned if access to the header is provided.

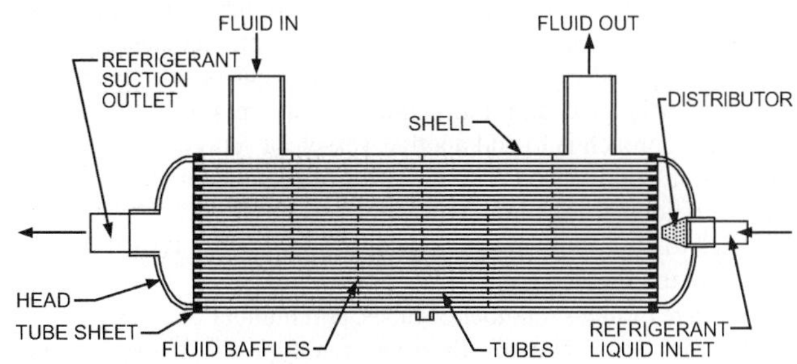

Fig. 1 Direct-Expansion Shell-and-Tube Cooler

The preparation of this chapter is assigned to TC 8.5, Liquid-to-Refrigerant Heat Exchangers.

Table 1 Types of Coolers

Type of Cooler	Subtype	Usual Refrigerant Feed Device	Usual Capacity Range, tons	Commonly Used Refrigerants
Direct-expansion	Shell-and-tube	Thermal expansion valve	2 to 1000	12, 22, 134a, 404A, 407C, 410A, 500, 502, 507A, 717
		Electronic modulation valve	2 to 1000	
	Tube-in-tube	Thermal expansion valve	5 to 25	12, 22, 134a, 717
	Brazed-plate	Thermal expansion valve	0.6 to 200	12, 22, 134a, 404A 407C, 410A, 500, 502, 507A, 508B, 717, 744
	Semiwelded plate	Thermal expansion valve	50 to 1990	12, 22, 134a, 500, 502, 507A, 717, 744
Flooded	Shell-and-tube	Low-pressure float	25 to 2000	11, 12, 22, 113, 114
		High-pressure float	25 to 6000	123, 134a, 500, 502, 507A, 717
		Fixed orifice(s)	25 to 6000	
		Weir	25 to 6000	
	Spray shell-and-tube	Low-pressure float	50 to 10,000	11, 12, 13B1, 22
		High-pressure float	50 to 10,000	113, 114, 123, 134a
	Brazed-plate	Low-pressure float	0.6 to 200	12, 22, 134a, 500, 502, 507A, 717, 744
	Semiwelded plate	Low-pressure float	50 to 1990	12, 22, 134a, 500, 502, 507A, 717, 744
Baudelot	Flooded	Low-pressure float	10 to 100	22, 717
	Direct-expansion	Thermal expansion valve	5 to 25	12, 22, 134a, 717
Shell-and-coil	—	Thermal expansion valve	2 to 10	12, 22, 134a, 717

Brazed- or **semiwelded-plate** coolers are constructed of plates brazed or laser-welded together to make an assembly of separate channels. Semiwelded designs have the refrigerant side welded and the fluid side gasketed and allow contact of the refrigerant with the fluid-side gaskets. These designs can be disassembled for inspection and mechanical cleaning of the fluid side. Brazed types do not have gaskets, cannot be disassembled, and are cleaned chemically. Internal leaks in brazed plates typically cannot be repaired. This type of evaporator is designed to work in a vertical orientation. Uniform distribution in direct-expansion operation is typically achieved by using a special plate design or distributor insert; flooded and pumped overfeed operations do not require distribution devices. Plate coolers are very compact and require minimal space.

Most tubular direct-expansion coolers are designed for horizontal mounting. If they are mounted vertically, performance may vary considerably from that predicted because two-phase flow heat transfer is a direction-sensitive phenomenon and dryout begins earlier in vertical upflow.

Flooded

In a flooded **shell-and-tube** cooler, refrigerant vaporizes on the outside of the tubes, which are submerged in liquid refrigerant in a closed shell. Fluid flows through the tubes as shown in Figure 2. Flooded coolers are usually used with rotary screw or centrifugal compressors to cool water, water/glycol mixtures, or brine.

A refrigerant liquid/vapor mixture usually feeds into the bottom of the shell through a distributor that distributes the mixture equally under the tubes. The relatively warm fluid in the tubes heats the refrigerant liquid surrounding the tubes, causing it to boil. As bubbles rise through the space between tubes, the liquid surrounding the tubes becomes increasingly bubbly (or foamy, if much oil is present).

The refrigerant vapor must be separated from the mist generated by the boiling refrigerant. The simplest separation method is provided by a dropout area between the top row of tubes and the suction connections. If this dropout area is insufficient, a coalescing filter may be required between the tubes and connections. Perry and Green (2007) give additional information on mist elimination. Another approach is to add another vessel, or "surge drum," above the suction connections. The diameter of this vessel is selected so that the velocity of the liquid droplets slows to the point where they fall back to the bottom of the surge drum. This liquid is then drained back into the flooded cooler.

The size of tubes, number of tubes, and number of passes should be determined to maintain fluid velocity typically between 3 and 10 fps for copper alloy tubing. Velocities beyond these limits may be used if the fluid is free of suspended abrasives and fouling substances (Ayub and Jones 1987; Sturley 1975) or if the tubing is manufactured from special alloys, such as titanium and stainless steel, that have better resistance to erosion. In some cases, the minimum velocity may be determined by a lower Reynolds number limit.

One variation of this cooler is the **spray shell-and-tube** cooler. In large-diameter coolers where the refrigerant's heat transfer coefficient is adversely affected by the refrigerant pressure, liquid can be sprayed to cover the tubes rather than flooding them. A mechanical pump circulates liquid from the bottom of the cooler to the spray heads.

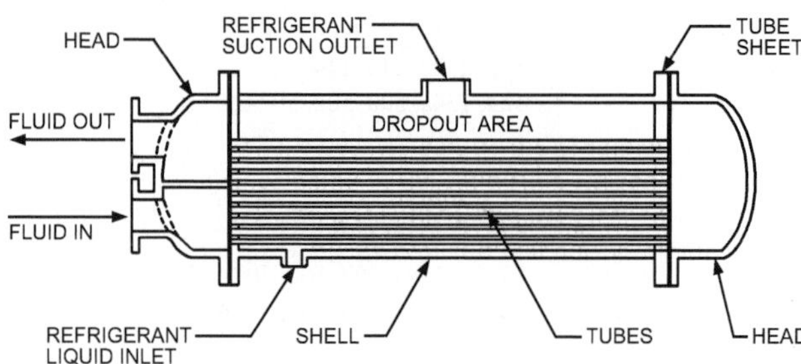

Fig. 2 Flooded Shell-and-Tube Cooler

Flooded shell-and-tube coolers are generally unsuitable for other than horizontal orientation.

In a flooded **plate** cooler (Figure 3), refrigerant vaporizes in vertical channels between corrugated plates with the liquid inlet at the bottom and the vapor outlet at the top (i.e., vertical upflow). The warm fluid flow may be either counter or parallel to the refrigerant flow. Both thermosiphon (gravity feed) and pumped overfeed operation are used. Surge drums are required for pumped overfeed operation but usually not for thermosiphon operation because the corrugated plates demist flow under most conditions.

Baudelot

Baudelot coolers (Figure 4) are used to cool a fluid to near its freezing point in industrial, food, and dairy applications. In this cooler, fluid circulates over the outside of vertical plates, which are easy to clean. The inside surface of the plates is cooled by evaporating the refrigerant. The fluid to be cooled is distributed uniformly along the top of the heat exchanger and then flows by gravity to a collection pan below. The cooler may be enclosed by insulated walls to avoid unnecessary loss of refrigeration.

R-717 (ammonia) is commonly used with flooded Baudelot coolers using conventional gravity feed with a surge drum. A low-pressure float valve maintains a suitable refrigerant liquid level in the surge drum. Baudelot coolers using other common refrigerants are generally direct-expansion, with thermostatic expansion valves.

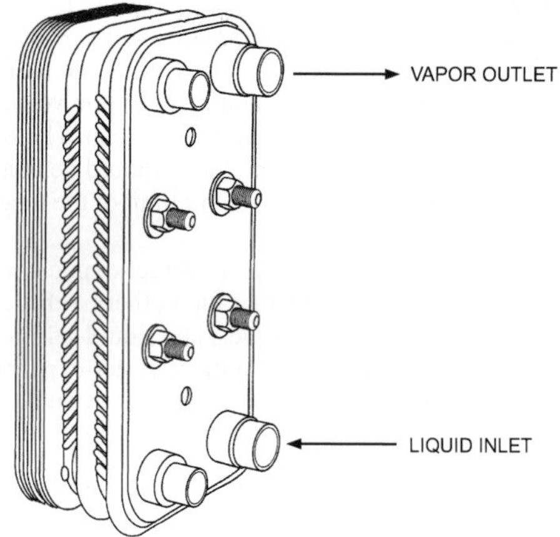

Fig. 3 Flooded Plate Cooler

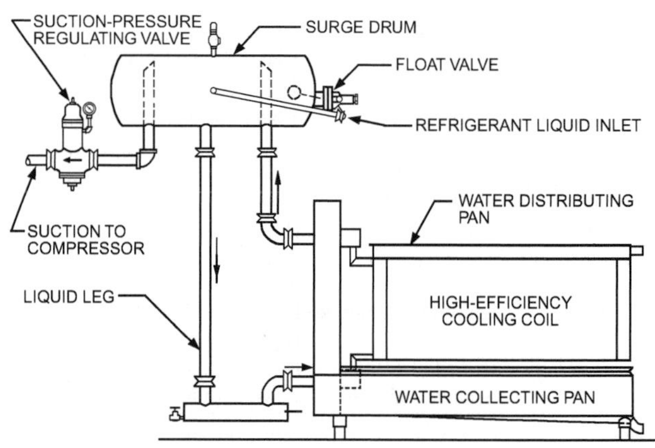

Fig. 4 Baudelot Cooler

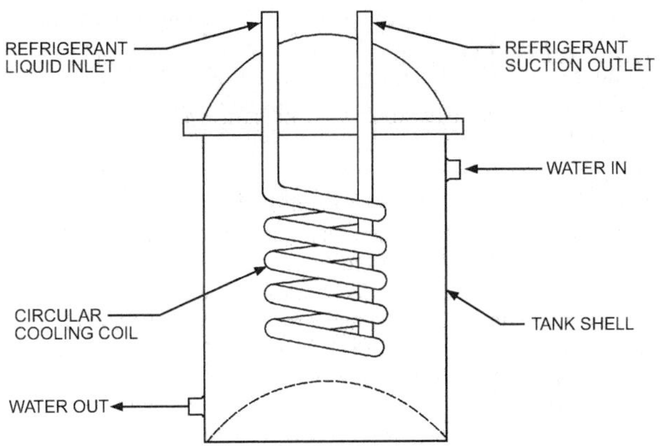

Fig. 5 Shell-and-Coil Cooler

Shell-and-Coil

A shell-and-coil cooler is a tank containing the fluid to be cooled with a simple coiled tube used to cool the fluid. This type of cooler has the advantage of cold fluid storage to offset peak loads. In some models, the tank can be opened for cleaning. Most applications are at low capacities (e.g., for bakeries, for photographic laboratories, and to cool drinking water).

The coiled tube containing the refrigerant can be either inside the tank (Figure 5) or attached to the outside of the tank in a way that allows heat transfer.

HEAT TRANSFER

Heat transfer for liquid coolers can be expressed by the following steady-state heat transfer equation:

$$q = UA\Delta t_m \tag{1}$$

where

q = total heat transfer rate, Btu/h
Δt_m = mean temperature difference, °F
A = heat transfer surface area associated with U, ft²
U = overall heat transfer coefficient, Btu/h·ft²·°F

The area A can be calculated if the geometry of the cooler is known. Chapter 3 of the 2005 *ASHRAE Handbook—Fundamentals* describes the calculation of the mean temperature difference.

This chapter discusses the components of U, but not in depth. U may be calculated by one of the following equations.

Based on inside surface area

$$U = \frac{1}{1/h_i + [A_i/(A_o h_o)] + (t/k)(A_i/A_m) + r_{fi}} \tag{2}$$

Based on outside surface area

$$U = \frac{1}{[A_o/(A_i h_i)] + 1/h_o + (t/k)(A_o/A_m) + r_{fo}} \tag{3}$$

where

h_i = inside heat transfer coefficient based on inside surface area, Btu/h·ft²·°F
h_o = outside heat transfer coefficient based on outside surface area, Btu/h·ft²·°F
A_o = outside heat transfer surface area, ft²
A_i = inside heat transfer surface area, ft²
A_m = mean heat transfer area of metal wall, ft²
k = thermal conductivity of heat transfer material, Btu/h·ft·°F
t = thickness of heat transfer surface (tube wall thickness), ft
r_{fi} = fouling factor of fluid side based on inside surface area, ft²·h·°F/Btu
r_{fo} = fouling factor of fluid side based on outside surface area, ft²·h·°F/Btu

Note: If fluid is on inside, multiply r_{fi} by A_o/A_i to find r_{fo}.
If fluid is on outside, multiply r_{fo} by A_i/A_o to find r_{fi}.

These equations can be applied to incremental sections of the heat exchanger to include local effects on the value of U, and then the increments summed to obtain a more accurate design.

Heat Transfer Coefficients

The refrigerant-side coefficient usually increases with (1) an increase in cooler load, (2) a decrease in suction superheat, (3) a decrease in oil concentration, or (4) an increase in saturated suction temperature. The amount of increase or decrease depends on the type of cooler. Schlager et al. (1989) and Zürcher et al. (1998) discuss the effects of oil in direct-expansion coolers. Flooded coolers have a relatively small change in heat transfer coefficient as a result of a change in load, whereas a direct-expansion cooler shows a significant increase with an increase in load. A Wilson plot of test data (Briggs and Young 1969; McAdams 1954) can show actual values for the refrigerant-side coefficient of a given cooler design. Collier and Thome (1994), Thome (1990, 2003), and Webb (1994) provide additional information on predicting refrigerant-side heat transfer coefficients.

The fluid-side coefficient is determined by cooler geometry, fluid flow rate, and fluid properties (viscosity, specific heat, thermal conductivity, and density) (Palen and Taborek 1969; Wolverine Tube 1984). For a given fluid, the fluid-side coefficient increases with fluid flow rate because of increased turbulence and with fluid temperature because of improvement of fluid properties as temperature increases.

The heat transfer coefficient in direct-expansion and flooded coolers increases significantly with fluid flow. The effect of flow is smaller for Baudelot and shell-and-coil coolers.

An enhanced heat transfer surface can increase the heat transfer coefficient of coolers in the following ways:

- It increases heat transfer area, thereby increasing the overall heat transfer rate and reducing the thermal resistance of fouling, even if the refrigerant-side heat transfer coefficient is unchanged.
- Where flow of fluid or refrigerant is low, it increases turbulence at the surface and mixes fluid at the surface with fluid away from the surface. For stratified internal flows of refrigerants, it may convert the flow to complete wetting of the tube perimeter.
- In flooded coolers, an enhanced refrigerant-side surface may provide more and better nucleation points to promote boiling of refrigerant.

Pais and Webb (1991) and Thome (1990) describe many enhanced surfaces used in flooded coolers. The enhanced surface geometries provide substantially higher boiling coefficients than integral finned tubes. Nucleate pool boiling data are provided by Webb and Pais (1991). The boiling process in the tube bundle of a flooded cooler may be enhanced by forced-convection effects. This is basically an additive effect, in which the local boiling coefficient is the sum of the nucleate boiling coefficient and the forced-convection effect. Webb et al. (1989) describe an empirical method to predict flooded cooler performance, recommending row-by-row calculations.

Based on this model, Webb and Apparao (1990) present the results of calculations using a computer program. The results show some performance differences of various internal and external surface geometries. As an example, Figure 6 shows the contribution of nucleate pool boiling to the overall refrigerant heat transfer coefficient for an integral finned tube and an enhanced tube as a function of the tube row. Forced convection predominates with the integral finned tube.

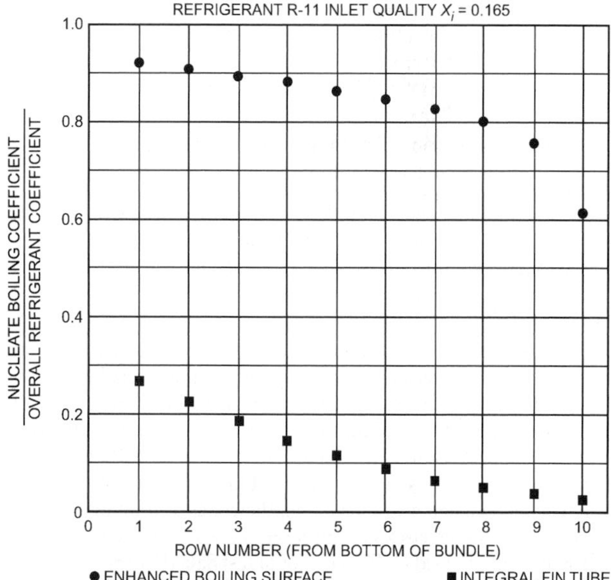

Fig. 6 Nucleate Boiling Contribution to Total Refrigerant Heat Transfer

ASHRAE research projects RP-725 (Chyu 1995) and RP-668 (Moeykens et al. 1995) studied spray evaporation performance in ammonia and halocarbon refrigerant systems. Moeykens et al. investigated shell-side heat transfer performance for commercially available enhanced surface tubes in a spray evaporation environment. The study determined that spray evaporation heat transfer can yield shell-side heat transfer coefficients equal to or greater than those found with enhanced nucleate boiling surface tubes in the flooded boiling environment. Moeykens and Pate (1996) describe an enhancement to shell-side heat transfer performance generated with small concentrations of oil (<2.5%) in spray evaporation. They attribute the improvement to foaming, which enhances heat transfer performance in the upper rows of large tube bundles operating in flooded boiling mode.

Gupte and Webb (1995a, 1995b) investigated convective vaporization in triangular enhanced tube bundles. They proposed a modified Chen superposition model that predicts the overall convective/vaporization coefficient as the sum of the single tube nucleate pool boiling coefficient and a weighted contribution of a single-phase convective coefficient.

Casciaro and Thome (2001a, 2001b) provide a recent comprehensive review of thermal design methods for flooded evaporators. For direct-expansion evaporators, Kattan et al. (1998) proposed a flow-pattern-based method that includes effects of flow stratification at low flow rates and onset of dryout at high vapor qualities. Zürcher et al. (1998) in ASHRAE research project RP-800 added a method to predict the adverse effect of oil on local flow boiling heat transfer coefficients to the Kattan-Thome-Favrat model.

Fouling Factors

Over time, most fluids foul the fluid-side heat transfer surface, reducing the cooler's overall heat transfer coefficient. If fouling is expected to be a problem, a mechanically cleanable cooler should be used, such as a flooded, Baudelot, or cleanable direct-expansion tube-in-tube cooler. Direct-expansion shell-and-tube, shell-and-coil, and brazed-plate coolers can be cleaned chemically. Flooded coolers and direct-expansion tube-in-tube coolers with enhanced fluid-side heat transfer surfaces tend to be self-cleaning because of high fluid turbulence, so a smaller fouling factor can probably be used for these coolers. Water quality in closed chilled-water loops has been studied as part of ASHRAE-sponsored research (RP-560). Haider et al. (1991) found little potential for fouling in such systems in a field survey. Experimental work with various tube geometries by Haider et al. (1992) confirmed that negligible fouling occurs in closed-loop evaporator tubes at 3 to 5 fps and 7 fps water velocities. ARI *Standard* 480 discusses fouling calculations.

The refrigerant side of the cooler is not subject to fouling, and a fouling factor need not be included for that side.

Wall Resistance

Typically, the t/k term in Equations (2) and (3) is negligible. However, with low thermal conductivity material or thick-walled tubing, it may become significant. Refer to Chapter 3 of the 2005 *ASHRAE Handbook—Fundamentals* and to Chapter 38 of this volume for further details.

PRESSURE DROP

Fluid Side

Pressure drop is usually minimal in Baudelot and shell-and-coil coolers but must be considered in direct-expansion and flooded coolers. Both direct-expansion and flooded coolers rely on turbulent fluid flow to improve heat transfer. This turbulence is obtained at the expense of pressure drop.

For air-conditioning, pressure drop is commonly limited to 10 psi to keep pump size and energy cost reasonable. For flooded coolers, see Chapter 38 for a discussion of pressure drop for flow in tubes. Pressure drop for fluid flow in shell-and-tube direct-expansion coolers depends greatly on tube and baffle geometry. The following equation projects the change in pressure drop caused by a change in flow:

$$\text{New pressure drop} = \text{Original pressure drop}\left[\frac{\text{New rate}}{\text{Original rate}}\right]^{1.8} \quad (4)$$

Refrigerant Side

The refrigerant-side pressure drop must be considered for direct-expansion, shell-and-coil, brazed-plate, and (sometimes) Baudelot coolers. When there is a pressure drop on the refrigerant side, the refrigerant inlet and outlet pressures and corresponding saturated temperature are different. This difference changes the mean temperature difference, which affects the total heat transfer rate. If pressure drop is high, expansion valve operation may be affected because of reduced pressure drop across the valve. This pressure drop varies, depending on the refrigerant used, operating temperature, and type of tubing. For flooded evaporators, Casciaro and Thome (2001b) summarize prediction methods as part of ASHRAE research project RP-1089. For direct-expansion evaporators, Ould Didi et al. (2002) describe seven prediction methods, compare them to data for five refrigerants, and provide recommendations on the best choice.

VESSEL DESIGN

Mechanical Requirements

Pressure vessels must be constructed and tested under the rules of national, state, and local codes. The ASME *Boiler and Pressure Vessel Code*, Section VIII, gives guidance on rules and exemptions.

The more common applicable codes and standards are as follows:

1. ARI *Standard* 480 covers industry criteria for standard equipment, standard safety provisions, marking, and recommended rating requirements.
2. ASHRAE *Standard* 24 covers recommended testing methods for measuring liquid cooler capacity.

3. ASHRAE *Standard* 15 involves specific design criteria, use of materials, and testing. It refers to the ASME *Boiler and Pressure Vessel Code*, Section VIII, for refrigerant-containing sides of pressure vessels, where applicable. Factory test pressures are specified, and minimum design working pressures are given. This code requires pressure-limiting and pressure-relief devices on refrigerant-containing systems, as applicable, and defines setting and capacity requirements for these devices.
4. ASME *Boiler and Pressure Vessel Code*, Section VIII, Unfired Pressure Vessels, covers safety aspects of design and construction. Most states require coolers to meet ASME requirements if they fall within the scope of the ASME code. Some of the exceptions from meeting the ASME requirements are as follows:
   - Cooler shell inner diameter (ID) is 6 in. or less.
   - Pressure is 15 psig or less.
   - The fluid portion of the cooler need not be built to the requirements of the ASME code if the fluid is water, the design pressure does not exceed 300 psig, and the design temperature does not exceed 210°F.

   Coolers meeting ASME code requirements have an ASME stamp, which is a U or UM inside a four-leaf clover. The U can be used for all coolers, and the UM can be used for small coolers.
5. Underwriters Laboratories (UL) *Standard* 207 covers specific design criteria, use of materials, testing, and initial approval by Underwriters Laboratories. A cooler with the ASME U or UM stamp does not require UL approval.

Design Pressure. On the refrigerant side, design pressure should be chosen per ASHRAE *Standard* 15. Standby temperature and the temperature encountered during shipping of chillers with a refrigerant charge should also be considered.

Required fluid-side (water-side) pressure varies, depending largely on (1) static pressure, (2) pump pressure, (3) transients caused by pump start-up, and (4) valve closing.

Chemical Requirements

The following chemical requirements are given by NACE (1985) and Perry and Green (2007).

R-717 (Ammonia). Carbon steel and cast iron are the most widely used materials for ammonia systems. Stainless steel alloys are satisfactory but more costly. Copper and high-copper alloys are avoided because they are attacked by ammonia when moisture is present. Aluminum and aluminum alloys may be used with caution.

Halocarbon Refrigerants. Almost all common metals and alloys are used satisfactorily with these refrigerants. Exceptions where water may be present include magnesium and aluminum alloys containing more than 2% magnesium. Zinc is not recommended for use with R-113; it is more chemically reactive than other common construction metals and, therefore, it is usually avoided when other halogenated hydrocarbons are used as refrigerants. Under some conditions with moisture present, halocarbon refrigerants form acids that attack steel and even nonferrous metals. This problem does not commonly occur in properly cleaned and dehydrated systems. ASHRAE *Standard* 15, paragraph 9.1.2, states that "aluminum, zinc, magnesium, or their alloys shall not be used in contact with methyl chloride," or magnesium alloys with any halogenated refrigerant.

Water. Relatively pure water can be satisfactory with both ferrous and nonferrous metals. Brackish water, seawater, and some river water are quite corrosive to iron and steel and also to copper, aluminum, and many alloys of these metals. A reputable water consultant who knows the local water condition should be contacted. Chemical treatment by pH control, inhibitor applications, or both may be required. Where this is not feasible, more noble construction material or special coatings must be used. Aluminum should not be used in the presence of other metals in water circuits.

Brines. Ferrous metal and a few nonferrous alloys are almost universally used with sodium chloride and calcium chloride brines. Copper alloys can be used if adequate quantities of sodium dichromate are added and caustic soda is used to neutralize the solution. Even with ferrous metal, brines should be treated periodically to hold pH near neutral.

Ethylene glycol and propylene glycol are stable compounds that are less corrosive than chloride brines.

Electrical Requirements

When the fluid being cooled is electrically conductive, the system must be grounded to prevent electrochemical corrosion.

APPLICATION CONSIDERATIONS

Refrigerant Flow Control

Direct-Expansion Coolers. The constant superheat thermal expansion valve is the most common control used, located directly upstream of the cooler. A thermal bulb strapped to the suction line leaving the cooler senses refrigerant temperature. The valve can be adjusted to produce a constant suction superheat during steady operation. If refrigerant pressure drop between the expansion valve and the thermal bulb is significant, an externally equalized expansion valve should be used.

The thermal expansion valve adjustment is commonly set at a suction superheat of 8 to 10°F, which is sufficient to ensure that liquid is not carried into the compressor. Direct-expansion cooler performance is affected greatly by superheat setting. Reduced superheat improves cooler performance; thus, suction superheat should be set as low as possible while avoiding liquid carryover to the compressor.

Flooded Coolers. As the name implies, flooded coolers must have good liquid refrigerant coverage of the tubes to achieve good performance. Liquid level control in a flooded cooler becomes the principal issue in flow control. Some systems are designed critically charged, so that when all the liquid refrigerant is delivered to the cooler, it is just enough for good tube coverage. In these systems, an orifice is often used as the throttling device between condenser and cooler.

Level-sensing devices are another option. Typically, a high-side float valve or level-sensing system can meter flow to the cooler at a controlled rate based on the condenser liquid level. For more direct control of liquid level in the cooler, a low-side float valve or level-sensing system can be used to control flow of entering refrigerant based on the level in the cooler itself. Because of the volatile nature of boiling inside the flooded cooler and the potential for false level readings, consideration must be given to design and installation of a low-side liquid level sensing device.

Freeze Prevention

Freeze prevention must be considered for coolers operating near the fluid's freezing point. In some coolers, freezing causes extensive damage. Two methods can be used for freeze protection: (1) hold saturated suction pressure above the fluid freezing point or (2) shut the system off if fluid temperature approaches the freezing point.

A suction-pressure regulator can hold the saturated suction pressure above the fluid's freezing point. A low-pressure cutout can shut the system off before the saturated suction pressure drops to below the freezing point of the fluid. The leaving fluid temperature can be monitored to cut the system off before a danger of freezing, usually about 4°F above the fluid freezing temperature. It is recommended that both methods be used.

Baudelot, shell-and-coil, brazed-plate, and direct-expansion shell-and-tube coolers are all somewhat resistant to damage caused by freezing and ideal for applications where freezing may be a problem.

If a cooler is installed in an unconditioned area, possible freezing caused by low ambient temperature must be considered. If the cooler is used only when ambient temperature is above freezing, the fluid should be drained from the cooler for cold weather. Alternatively, if the cooler is used year-round, the following methods can be used to prevent freezing:

- Heat tape or other heating device to keep cooler above freezing
- For water, adding an appropriate amount of ethylene glycol
- Continuous pump operation

Oil Return

Most compressors discharge a small percentage of oil in the discharge gas. This oil mixes with condensed refrigerant in the condenser and flows to the cooler. Because the oil is nonvolatile, it does not evaporate and may collect in the cooler.

In direct-expansion coolers, gas velocity in the tubes and suction gas header is usually sufficient to carry oil from the cooler into the suction line. From there, with proper piping design, it can be carried back to the compressor. At light load and low temperature, oil may gather in the superheat section of the cooler, detracting from performance. For this reason, operating refrigerant circuits at light load for long periods should be avoided, especially under low-temperature conditions. Some oil hold-up threshold measurements were presented by Zürcher et al. (1998) in ASHRAE research project RP-800.

In flooded coolers, vapor velocity above the tube bundle is usually insufficient to return oil up the suction line, and oil tends to accumulate in the cooler. With time, depending on the compressor oil loss rate, oil concentration in the cooler may become large. When concentration exceeds about 1%, heat transfer performance may be adversely affected if enhanced tubing is used.

It is common in flooded coolers to take some oil-rich liquid and return it to the compressor on a continuing basis, to establish a rate of return equal to the compressor oil loss rate.

Maintenance

Cooler maintenance centers around (1) safety and (2) cleaning the fluid side. The cooler should be inspected periodically for any weakening of its pressure boundaries. Visual inspection for corrosion, erosion, and any deformities should be included, and any pressure relief device should also be inspected. The insurer of the cooler may require regular inspection. If the fluid side is subjected to fouling, it may require periodic cleaning by either mechanical or chemical means. The manufacturer or a service organization experienced in cooler maintenance should have details for cleaning.

Insulation

A cooler operating at a saturated suction temperature lower than the ambient-air dew point should be insulated to prevent condensation. Direct-expansion coolers installed where the ambient temperature may drop below the process fluid's freezing point should also be insulated and wrapped with heat tape, to prevent the fluid from freezing and damaging the cooler during *off* periods. Chapter 23 of the 2005 *ASHRAE Handbook—Fundamentals* describes insulation in more detail.

REFERENCES

ARI. 2001. Remote type refrigerant-cooled liquid coolers. *Standard* 480-01. Air-Conditioning and Refrigeration Institute, Arlington, VA.

ASHRAE. 2000. Methods of testing for rating liquid coolers. ANSI/ASHRAE *Standard* 24-2000.

ASHRAE. 2007. Safety standards for refrigeration systems. ANSI/ASHRAE *Standard* 15-2007.

ASME. 2007. Rules for construction of pressure vessels. ANSI/ASME *Boiler and Pressure Vessel Code*, Section VIII. American Society of Mechanical Engineers, New York.

Ayub, Z.H. and S.A. Jones. 1987. Tubeside erosion/corrosion in heat exchangers. *Heating/Piping/Air Conditioning* (December):81.

Briggs, D.E. and E.H. Young. 1969. Modified Wilson plot techniques for obtaining heat transfer correlations for shell and tube heat exchangers. *Chemical Engineering Symposium Series* 65(92):35-45.

Casciaro, S. and J.R. Thome. 2001a. Thermal performance of flooded evaporators, Part 1: Review of boiling heat transfer studies. *ASHRAE Transactions* 107(1):903-918.

Casciaro, S. and J.R. Thome. 2001b. Thermal performance of flooded evaporators, Part 2: Review of void fraction, two-phase pressure drop and flow pattern studies. *ASHRAE Transactions* 107(1):919-930.

Chyu, M. 1995. Nozzle-sprayed flow rate distribution on a horizontal tube bundle. *ASHRAE Transactions* 101(2):443-453.

Collier, J.G. and J.R. Thome. 1994. *Convective boiling and condensation*, 3rd ed. Oxford University Press.

Gupte, N.S. and R.L. Webb. 1995a. Shell-side boiling in flooded refrigerant evaporators—Part I: Integral finned tubes. *International Journal of HVAC&R Research* (now *HVAC&R Research*) 1(1):35-47.

Gupte, N.S. and R.L. Webb. 1995b. Shell-side boiling in flooded refrigerant evaporators—Part II: Enhanced tubes. *International Journal of HVAC&R Research* (now *HVAC&R Research*) 1(1):48-60.

Haider, S.I., R.L. Webb, and A.K. Meitz. 1991. A survey of water quality and its effect on fouling in flooded water chiller evaporators. *ASHRAE Transactions* 97(1):55-67.

Haider, S.I., R.L. Webb, and A.K. Meitz. 1992. An experimental study of tube-side fouling resistance in water-chiller-flooded evaporators. *ASHRAE Transactions* 98(2):86-103.

Kattan, N., J.R. Thome, and D. Favrat. 1998. Flow boiling in horizontal tubes, Part 3: Development of a new heat transfer model based on flow patterns. *Journal of Heat Transfer* 120(1):156-165.

McAdams, W.H. 1954. *Heat transmission*, 3rd ed. McGraw-Hill, New York.

Moeykens, S.A., B.J. Newton, and M.B. Pate. 1995. Effects of surface enhancement, film-feed supply rate, and bundle geometry on spray evaporation heat transfer performance. *ASHRAE Transactions* 101(2):408-419.

Moeykens, S.A. and M.B. Pate. 1996. Effects of lubricant on spray evaporation heat transfer performance of R-134a and R-22 in tube bundles. *ASHRAE Transactions* 102(1):410-426.

NACE. 1985. *Corrosion data survey—Metal section*, 6th ed. D.L. Graver, ed. National Association of Corrosion Engineers, Houston.

Ould Didi, M.B., N. Kattan, and J.R. Thome. 2002. Prediction of two-phase pressure gradients of refrigerants in horizontal tubes. *International Journal of Refrigeration* 25(7):935-947.

Pais, C. and R.L. Webb. 1991. Literature survey of pool boiling on enhanced surfaces. *ASHRAE Transactions* 97(1):79-89.

Palen, J.W. and J. Taborek. 1969. Solution of shell side pressure drop and heat transfer by stream analysis method. *Chemical Engineering Progress Symposium Series* 65(92).

Perry, R.H. and D.W. Green. 2007. *Perry's chemical engineers handbook*, 8th ed. McGraw-Hill, New York.

Schlager, L.M., M.B. Pate, and A.E. Bergles. 1989. A comparison of 150 and 300 SUS oil effects on refrigerant evaporation and condensation in a smooth tube and a micro-fin tube. *ASHRAE Transactions* 95(1).

Sturley, R.A. 1975. Increasing the design velocity of water and its effect on copper tube heat exchangers. *Paper* 58, International Corrosion Forum, Toronto, Canada.

Thome, J.R. 1990. *Enhanced boiling heat transfer*. Hemisphere, New York.

Thome, J.R. 2003. Boiling. In *Heat transfer handbook*, pp. 635-717. A. Bejan and A.D. Krause, eds. Wiley Interscience, New York.

UL. 2001. Refrigerant-containing components and accessories, nonelectrical. *Standard* 207-01. Underwriters Laboratories, Northbrook, IL.

Webb, R.L. 1994. *Principles of enhanced heat transfer*. John Wiley & Sons, New York.

Webb, R.L. and T. Apparao. 1990. Performance of flooded refrigerant evaporators with enhanced tubes. *Heat Transfer Engineering* 11(2):29-43.

Webb, R.L. and C. Pais. 1991. Pool boiling data for five refrigerants on three tube geometries. *ASHRAE Transactions* 97(1):72-78.

Webb, R.L., K.-D. Choi, and T. Apparao. 1989. A theoretical model for prediction of the heat load in flooded refrigerant evaporators. *ASHRAE Transactions* 95(1):326-338.

Wolverine Tube, Inc. 2001. *Wolverine engineering data book II*. http://www.wlv.com/products/databook/databook.pdf.

Zürcher, O., J.R. Thome, and D. Favrat. 1998. Intube flow boiling of R-407C and R-407C/oil mixtures, Part II: Plain tube results and predictions (RP-800). *International Journal of HVAC&R Research* (now *HVAC&R Research*) 4(4):373-399.

CHAPTER 42

LIQUID-CHILLING SYSTEMS

LIQUID-CHILLING systems cool water, brine, or other secondary coolant for air conditioning or refrigeration. The system may be either factory-assembled and wired or shipped in sections for erection in the field. The most frequent application is water chilling for air conditioning, although brine cooling for low-temperature refrigeration and chilling fluids in industrial processes are also common.

The basic components of a vapor-compression, liquid-chilling system include a compressor, liquid cooler (evaporator), condenser, compressor drive, liquid-refrigerant expansion or flow-control device, and control center; it may also include a receiver, economizer, expansion turbine, and/or subcooler. In addition, auxiliary components may be used, such as a lubricant cooler, lubricant separator, lubricant-return device, purge unit, lubricant pump, refrigerant transfer unit, refrigerant vents, and/or additional control valves.

For information on absorption equipment, see Chapter 41 of the 2006 *ASHRAE Handbook—Refrigeration*.

GENERAL CHARACTERISTICS

PRINCIPLES OF OPERATION

Liquid (usually water) enters the cooler, where it is chilled by liquid refrigerant evaporating at a lower temperature. The refrigerant vaporizes and is drawn into the compressor, which increases the pressure and temperature of the gas so that it may be condensed at the higher temperature in the condenser. The condenser cooling medium is warmed in the process. The condensed liquid refrigerant then flows back to the evaporator through an expansion device. Some of the liquid refrigerant changes to vapor (flashes) as pressure drops between the condenser and the evaporator. Flashing cools the liquid to the saturated temperature at evaporator pressure. It produces no refrigeration in the cooler. The following modifications (sometimes combined for maximum effect) reduce flash gas and increase the net refrigeration per unit of power consumption.

Subcooling. Condensed refrigerant may be subcooled below its saturated condensing temperature in either the subcooler section of a water-cooled condenser or a separate heat exchanger. Subcooling reduces flashing and increases the refrigeration effect in the chiller.

Economizing. This process can occur either in a direct-expansion (DX), an expansion turbine, or a flash system. In a **DX system**, the main liquid refrigerant is usually cooled in the shell of a shell-and-tube heat exchanger, at condensing pressure, from the saturated condensing temperature to within several degrees of the intermediate saturated temperature. Before cooling, a small portion of the liquid flashes and evaporates in the tube side of the heat exchanger to cool the main liquid flow. Although subcooled, the liquid is still at the condensing pressure.

An **expansion turbine** extracts rotating energy as a portion of the refrigerant vaporizes. As in the DX system, the remaining liquid is supplied to the cooler at intermediate pressure.

In a **flash system**, the entire liquid flow is expanded to intermediate pressure in a vessel that supplies liquid to the cooler at saturated intermediate pressure; however, the liquid is at intermediate pressure.

Flash gas enters the compressor either at an intermediate stage of a multistage centrifugal compressor, at the intermediate stage of an integral two-stage reciprocating compressor, at an intermediate pressure port of a screw compressor, or at the inlet of a high-pressure stage on a multistage reciprocating or screw compressor.

Liquid Injection. Condensed liquid is throttled to the intermediate pressure and injected into the second-stage suction of the compressor to prevent excessively high discharge temperatures and, in the case of centrifugal machines, to reduce noise. For screw compressors, condensed liquid is injected into a port fixed at slightly below discharge pressure to provide lubricant cooling.

COMMON LIQUID-CHILLING SYSTEMS

Basic System

The refrigeration cycle of a basic system is shown in Figure 1. Chilled water enters the cooler at 54°F, for example, and leaves at 44°F. Condenser water leaves a cooling tower at 85°F, enters the condenser, and returns to the cooling tower near 95°F. Condensers may also be cooled by air or evaporation of water. This system, with a single compressor and one refrigerant circuit with a water-cooled condenser, is used extensively to chill water for air conditioning because it is relatively simple and compact.

The preparation of this chapter is assigned to TC 8.1, Positive Displacement Compressors, and TC 8.2, Centrifugal Machines.

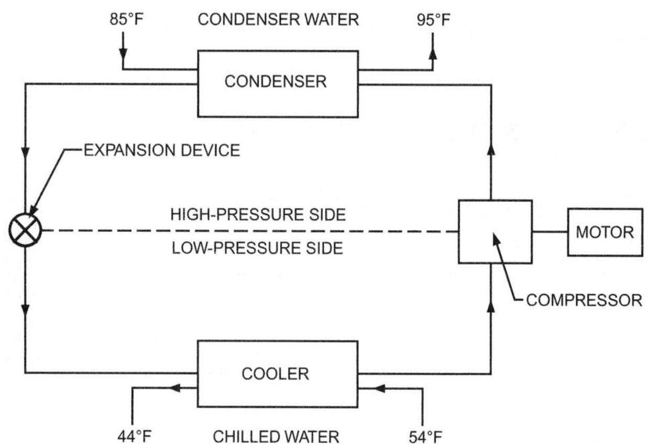

Fig. 1 Equipment Diagram for Basic Liquid Chiller

Multiple-Chiller Systems

A multiple-chiller system has two or more chillers connected by parallel or series piping to a common distribution system. Multiple chillers offer operational flexibility, standby capacity, and less disruptive maintenance. The chillers can be sized to handle a base load and increments of a variable load to allow each chiller to operate at its most efficient point.

Multiple-chiller systems offer some standby capacity if repair work must be done on one chiller. Starting in-rush current is reduced, as well as power costs at partial-load conditions. Maintenance can be scheduled for one chiller during part-load times, and sufficient cooling can still be provided by the remaining unit(s). These advantages require an increase in installed cost and space, however. Traditionally, flow was held constant through the chillers for stable control. Today, variable-flow chilled-water systems are finding favor in some applications. Both variable-flow and primary/secondary hydronic systems are discussed in further detail in Chapter 12.

When design chilled-water temperature is above about 45°F, all units should be controlled by the combined exit water temperature or by the return water temperature (RWT), because overchilling will not cause dangerously low water temperature in the operating machine(s). Chilled-water temperature can be used to cycle one unit off when it drops below a capacity that can be matched by the remaining units.

When the design chilled-water temperature is below about 45°F, each machine should be controlled by its own chilled-water temperature, both to prevent dangerously low evaporator temperatures and to avoid frequent shutdowns by low-temperature cutout. The temperature differential setting of the RWT must be adjusted carefully to prevent short-cycling caused by the step increase in chilled-water temperature when one chiller is cycled off. These control arrangements are shown in Figures 2 and 3.

In the **series arrangement**, the chilled-liquid pressure drop may be higher if shells with fewer liquid-side passes or baffles are not available. No overchilling by either unit is required, and compressor power consumption is lower than for the parallel arrangement at partial loads. Because evaporator temperature never drops below the design value (because no overchilling is necessary), the chances of evaporator freeze-up are minimized. However, the chiller should still be protected by a low-temperature safety control.

Water-cooled condensers in series are best piped in a counterflow arrangement so that the lead machine is provided with warmer condenser and chilled water and the lag machine is provided with colder entering condenser and chilled water. Refrigerant compression for each unit is nearly the same. If about 55% of design cooling capacity is assigned to the lead machine and about 45% to the lag machine,

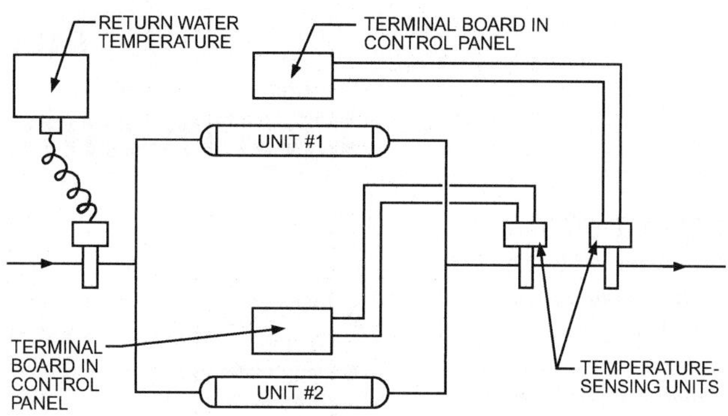

Fig. 2 Parallel-Operation High Design Water Leaving Coolers (Approximately 45°F and Above)

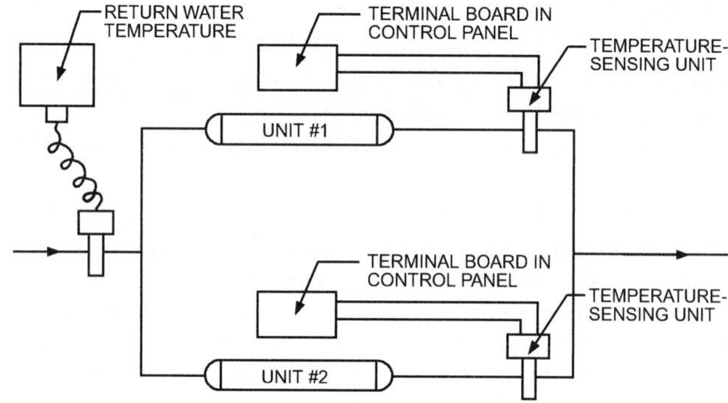

Fig. 3 Parallel-Operation Low Design Water Leaving Coolers (Below Approximately 45°F)

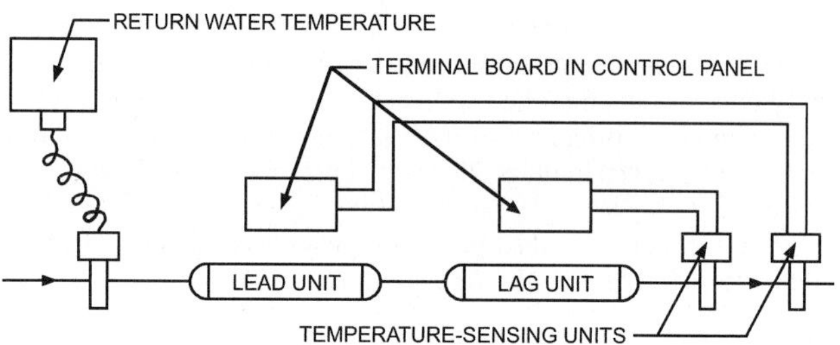

Fig. 4 Series Operation

identical units can be used. In this way, either machine can provide the same standby capacity if the other is down, and lead and lag machines may be interchanged to equalize the number of operating hours on each.

A control system for two machines in series is shown in Figure 4. (On reciprocating chillers, RWT sensing is usually used instead of leaving water sensing because it allows closer temperature control.) Both units are modulated to a certain capacity; then, one unit shuts down, leaving less than 100% load on the operating machine.

One machine should be shut down as soon as possible, with the remaining unit carrying the full load. This not only reduces the number of operating hours on a unit, but also leads to less total power consumption because the COP tends to decrease below full-load value when unit load drops much below 50%.

Heat Recovery Systems

Any building or plant requiring simultaneous operation of heat-producing and cooling equipment has the potential for a heat recovery installation.

Heat recovery systems extract heat from liquid being chilled and reject some of that heat, plus the energy of compression, to a warm-water circuit for reheat or heating. Air-conditioned spaces thus furnish heating for other spaces in the same building. During the full-cooling season, all heat must be rejected outside, usually by a cooling tower. During spring or fall, some heat is required inside, while some heat extracted from air-conditioned spaces must be rejected outside.

Heat recovery offers a low heating cost and reduces space requirements for equipment. The control system must be designed carefully, however, to take the greatest advantage of recovered heat and to maintain proper temperature and humidity in all parts of the building. Chapter 8 covers balanced heat recovery systems.

Because cooling tower water is not satisfactory for heating coils, a separate, closed warm-water circuit with another condenser bundle or auxiliary condenser, in addition to the main water chiller condenser, must be provided. In some cases, it is economically feasible to use a standard condenser and a closed-circuit water cooler.

Instead of rejecting all heat extracted from the chilled liquid to a cooling tower, a separate, closed condenser cooling water circuit is heated by the condensing refrigerant for comfort heating, preheating, or reheating. Some factory packages include an extra condenser water circuit, either a double-bundle condenser or an auxiliary condenser.

A centrifugal heat recovery package is controlled as follows:

- **Chilled-liquid temperature** is controlled by a sensor in the leaving chilled-water line signaling the capacity control device.
- **Hot-water temperature** is controlled by a sensor in the hot-water line that modulates a cooling tower bypass valve. As the heating requirement increases, hot-water temperature drops, opening the tower bypass slightly. Less heat is rejected to the tower, condensing temperature increases, and hot-water temperature is restored as more heat is rejected to the hot-water circuit.

The hot-water temperature selected has a bearing on the installed cost of the centrifugal package, as well as on the power consumption while heating. Lower hot-water temperatures of 95 to 105°F result in a less expensive machine that uses less power. Higher temperatures require greater compressor motor output, perhaps higher-pressure condenser shells, sometimes extra compression stages, or a cascade arrangement. Installed cost of the centrifugal heat recovery machine increases as a result.

Another concern in design of a central chilled-water plant with heat recovery centrifugal compressors is the relative size of cooling and heating loads. These loads should be equalized on each machine so that the compressor may operate at optimum efficiency during both full cooling and full heating seasons. When the heating requirement is considerably smaller than the cooling requirement, multiple packages lower operating costs and allow less expensive standard air-conditioning centrifugal packages to be used for the rest of the cooling requirement. In multiple packages, only one unit is designed for heat recovery and carries the full heating load.

Another consideration for heat recovery chiller systems is the potential for higher cooling energy use. A standard commercial building water chiller operates with condenser water temperatures at or below 100°F. For heat recovery to be of practical use, it may be necessary for the condenser water to operate at higher temperatures. This increases the chiller's energy consumption. The design engineer must examine the tradeoff between higher cooling energy use versus lower heating energy use; the heat recovery chiller system may not necessarily be attractive.

SELECTION

The largest factor that determines total liquid chiller owning cost is the cooling load size; therefore, the total required chiller capacity should be calculated accurately. The practice of adding 10 to 20% to load estimates is unnecessary because of the availability of accurate load estimating methods, and it proportionately increases costs of equipment purchase, installation, and the poor efficiency resulting from wasted power. Oversized equipment can also cause operational difficulties such as frequent on/off cycling or surging of centrifugal machines at low loads. The penalty for a small underestimation of cooling load, however, is not serious. On the few design-load days of the year, increased chilled-liquid temperature is often acceptable. However, for some industrial or commercial loads, a safety factor can be added to the load estimate.

The life-cycle cost as discussed in Chapter 36 of the 2007 *ASHRAE Handbook—HVAC Applications* should be used to minimize overall purchase and operating costs. Total owning cost is comprised of the following:

- **Equipment price.** Each machine type and/or manufacturer's model should include all necessary auxiliaries such as starters and vibration mounts. If these are not included, their price should be added to the base price. Associated equipment, such as condenser water pump, tower, and piping, should be included.
- **Installation cost.** Factory-packaged machines are both less expensive to install and usually considerably more compact, thus saving space. The cost of field assembly must also be evaluated.
- **Energy cost.** Using an estimated load schedule and part-load power consumption curves furnished by the manufacturer, a year's energy cost should be calculated.
- **Water cost.** With water-cooled towers, the cost of acquisition, water treatment, tower blowdown, and overflow water should be included.
- **Maintenance cost.** Each bidder may be asked to quote on a maintenance contract on a competitive basis.
- **Insurance and taxes.**

For packaged chillers that include heat recovery, system cost and performance should be compared in addition to equipment costs. For example, the heat recovery chiller installed cost should be compared with the installed cost of a chiller plus a separate heating system. The following factors should also be considered: (1) energy costs, (2) maintenance requirements, (3) life expectancy of equipment, (4) standby arrangement, (5) relationship of heating to cooling loads, (6) effect of package selection on sizing, and (7) type of peripheral equipment.

Condensers and coolers are often available with either **liquid heads**, which require water pipes to be disconnected for tube access and maintenance, or **marine-type water boxes**, which allow tube access with water piping intact. The liquid head is considerably less expensive. The cost of disconnecting piping must be greater than the additional cost of marine-type water boxes to justify using the latter. Typically, an elbow and union or flange connection can be installed immediately next to liquid heads facilitate removing heads. By making sure that all specialty piping components (valves, controls, strainers, etc.) fall outside the tube bundle boundary, the liquid heads can be removed with very minimal pipe disassembly.

Figure 5 shows types of liquid chillers and their ranges of capacities.

For air-cooled condenser duty, brine chilling, or other high-pressure applications from 80 to about 200 tons, scroll and screw liquid chillers are more frequently installed than centrifugals. Centrifugal liquid chillers (particularly multistage machines), however, may be applied quite satisfactorily at high pressures.

Advancements in technology, refrigerants, and manufacturer offerings all affect which compression technology is best suited for a given liquid chiller application. Centrifugal packages are typically available to about 3500 tons, and field-assembled machines to about 10,000 tons.

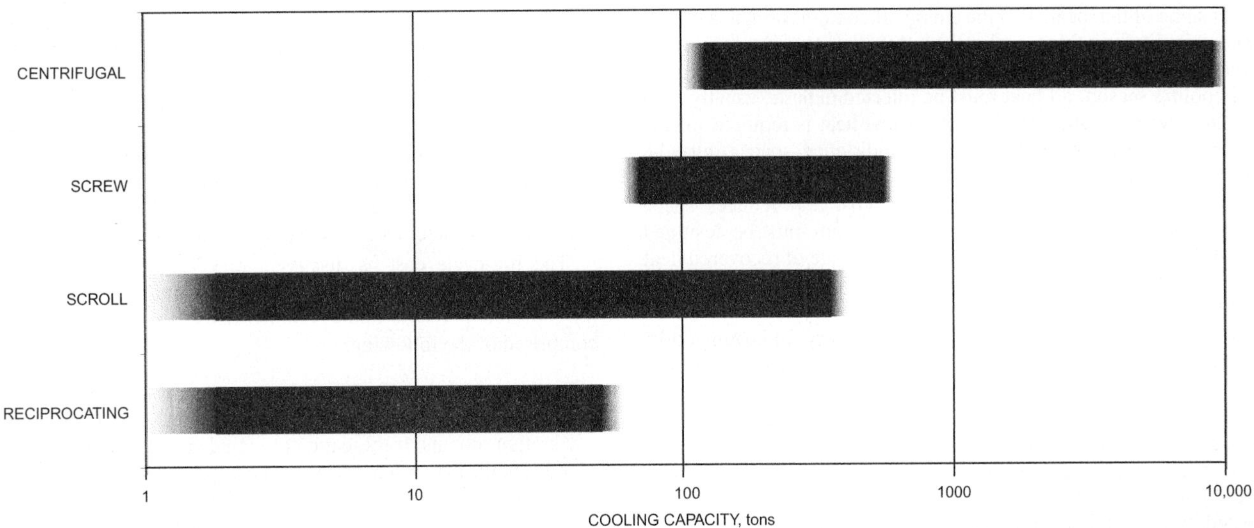

Fig. 5 Approximate Liquid Chiller Availability Range by Compressor Type

CONTROL

Liquid Chiller Controls

The **chilled-liquid temperature sensor** sends an air pressure (pneumatic control) or electrical signal (electronic control) to the control circuit, which then modulates compressor capacity in response to leaving or return chilled-liquid temperature change from its set point.

Compressor capacity is adjusted differently on the following liquid chillers:

Reciprocating chillers use combinations of cylinder unloading and on/off compressor cycling of single or multiple compressors.

Centrifugal liquid chillers, driven by electric motors, commonly use adjustable prerotation vanes, which are sometimes combined with movable diffuser walls. Turbine and engine drives and inverter-driven, variable-speed electric motors allow use of speed control in addition to prerotation vane modulation, reducing power consumption at partial loads.

Screw compressor liquid chillers include a slide valve that adjusts the length of the compression path. Inverter-driven, variable-speed electric motors and turbine and engine drives can also modulate screw compressor speed to control capacity.

In air-conditioning applications, most centrifugal and screw compressor chillers modulate from 100% to approximately 10% load. Although relatively inefficient, hot-gas bypass can be used to reduce capacity to nearly 0% with the unit in operation.

Reciprocating chillers are available with simple on/off cycling control in small capacities and with multiple steps of unloading down to 12.5% in the largest multiple-compressor units. Most intermediate sizes provide unloading to 50, 33, or 25% capacity. Hot-gas bypass can reduce capacity to nearly 0%.

The **water temperature controller** is a thermostatic device that unloads or cycles the compressor(s) when the cooling load drops below minimum unit capacity. An **antirecycle timer** is sometimes used to limit starting frequency.

On centrifugal or screw compressor chillers, a **current limiter** or **demand limiter** limits compressor capacity during periods of possible high power consumption (such as pulldown) to prevent current draw from exceeding the design value; such a limiter can be set to limit demand, as described in the section on Centrifugal Liquid Chillers.

Controls That Influence the Liquid Chiller

Condenser cooling water may need to be controlled to avoid falling below the manufacturer's recommended minimum limit, to regulate condenser pressure. Normally, the temperature of water leaving a cooling tower can be controlled by fans, dampers, or a water bypass around the tower. Tower bypass allows the water velocity through the condenser tubes to be maintained, which prevents low-velocity fouling.

A flow-regulating valve is another common means of control. The orifice of this valve modulates in response to condenser pressure. For example, reducing pressure decreases water flow, which, in turn, raises condenser pressure to the desired minimum level.

For air-cooled or evaporative condensers, compressor discharge pressure can be controlled by cycling fans, shutting off circuits, or flooding coils with liquid refrigerant to reduce heat transfer.

A reciprocating chiller usually has a thermal expansion valve, which requires a restricted range of pressure to avoid starving the evaporator (at low pressure).

An expansion valve(s) usually controls a screw compressor chiller. Cooling tower water temperature can be allowed to fall with decreasing load from the design condition to the chiller manufacturer's recommended minimum limit.

Screw compressor chillers above 150 tons may use flooded evaporators and evaporator liquid refrigerant controls similar to those used on centrifugal chillers.

A thermal expansion valve may control a centrifugal chiller at low capacities. Higher-capacity machines may use either a pilot-operated thermal control valve, an electronically controlled valve, fixed orifice(s), a high-pressure float, or even a low-side float valve to control refrigerant liquid flow to the cooler. These latter types of controls allow relatively low condenser pressures, particularly at partial loads. Also, a centrifugal machine may surge if pressure is not reduced when cooling load decreases. In addition, low pressure reduces compressor power consumption and operating noise. For these reasons, in a centrifugal installation, cooling tower water temperature should be allowed to fall naturally with decreasing load and wet-bulb temperature, except that the liquid chiller manufacturer's recommended minimum limit must be observed.

Safety Controls

Older systems often used dedicated control devices for each function of the chiller. Modern chiller systems typically use a microprocessor control center that can handle many control functions at once and can combine several control points into a single sensor. Some or all of the following safety algorithms or cutouts may be provided in a liquid-chilling package to stop compressor(s) automatically. Cutouts may be manual or automatic reset.

- **High condenser pressure.** This pressure switch opens if the compressor discharge pressure exceeds the value prescribed in ASHRAE *Standard* 15. It is usually a dedicated pressure switch that interrupts the chiller main run circuit to ensure a positive shutdown in an overpressure situation.
- **Low refrigerant pressure (or temperature).** This device opens when evaporator pressure (or temperature) reaches a minimum safe limit.
- **High lubricant temperature.** This device protects the compressor if loss of lubricant cooling occurs or if a bearing failure causes excessive heat generation.
- **High motor temperature.** If loss of motor cooling or overloading because of a failure of a control occurs, this device shuts down the machine. It may consist of direct-operating bimetallic thermostats, thermistors, or other sensors embedded in the stator windings; it may be located in the discharge gas stream of the compressor.
- **Motor overload.** Some small, reciprocating-compressor hermetic motors may use a directly operated overload in the power wiring to the motor. Some larger motors use pilot-operated overloads. Centrifugal and screw-compressor motors generally use starter overloads or current-limiting devices to protect against overcurrent.
- **Low lubricant sump temperature.** This switch is used either to protect against lubricant heater failure or to prevent starting after prolonged shutdown before lubricant heaters have had time to drive off refrigerant dissolved in the lubricant.
- **Low lubricant pressure.** To protect against clogged lubricant filters, blocked lubricant passageways, loss of lubricant, or a lubricant pump failure, a switch shuts down the compressor when lubricant pressure drops below a minimum safe value or if sufficient lubricant pressure is not developed shortly after the compressor starts.
- **Chilled-liquid flow interlock.** This device may not be furnished with the liquid-chilling package, but it is needed in external piping to protect against cooler freeze-up in case the liquid stops flowing. An electrical interlock is typically installed either in the factory or in the field. Most chiller control panels include a terminal for field-connecting a flow switch.
- **Condenser water flow interlock.** This device, similar to the chilled-liquid flow interlock, is sometimes used in external piping.
- **Low chilled-liquid temperature.** Sometimes called **freeze protection**, this cutout operates at a minimum safe value of leaving chilled-liquid temperature to prevent cooler freeze-up in the case of an operating control malfunction.
- **Relief valves.** In accordance with ASHRAE *Standard* 15, relief valves, rupture disks, or both, set to relieve at shell design working pressure, must be provided on most pressure vessels or on piping connected to the vessels. Fusible plugs may also be used in some locations. Pressure relief devices should be vented outdoors or to the low-pressure side, in accordance with regulations or the standard.

STANDARDS AND TESTING

ARI *Standard* 550/590 provides guidelines for rating and testing liquid-chilling machines. Design and construction of refrigerant pressure vessels are governed by ASME *Boiler and Pressure Vessel Code*, Section VIII, except when design working pressure is 15 psig or less (as is usually the case for R-123 liquid-chilling machines). Water-side design and construction of a condenser or evaporator are not within the scope of the ASME code unless design pressure is greater than 300 psi or design temperature is greater than 210°F.

ASHRAE *Standard* 15 applies to all liquid chillers and new refrigerants on the market. Requirements for equipment rooms are included. Methods for measuring unit sound levels are described in ARI *Standard* 575.

GENERAL MAINTENANCE

The following maintenance specifications apply to reciprocating, centrifugal, and screw chillers. Equipment should be neither overmaintained nor neglected. A preventive maintenance schedule should be established; items covered can vary with the nature of the application. The list is intended as a guide; in all cases, the manufacturer's specific recommendation should be followed.

Continual Monitoring

- Condenser water treatment: treatment is determined specifically for the condenser water used.
- Operating conditions: daily log sheets should be kept (either manually or automatically) to indicate trends and provide advance notice of deteriorating chillers.
- Brine quality for concentration and corrosion inhibitor levels.

Periodic Checks

- Leak check
- Purge operation
- System dryness
- Lubricant level
- Lubricant filter pressure drop
- Refrigerant quantity or level
- System pressures and temperatures
- Water flows
- Expansion valves operation

Regularly Scheduled Maintenance

- Condenser and lubricant cooler cleaning
- Evaporator cleaning on open systems
- Calibrating pressure, temperature, and flow controls
- Tightening wires and power connections
- Inspection of starter contacts and action
- Safety interlocks
- Dielectric checking of hermetic and open motors
- Tightness of hot gas valve
- Lubricant filter and drier change
- Analysis of lubricant and refrigerant
- Seal inspection
- Partial or complete valve or bearing inspection, as per manufacturer's recommendations
- Vibration levels

Extended Maintenance Checks

- Compressor guide vanes and linkage operation and wear
- Eddy current inspection of heat exchanger tubes
- Compressor teardown and inspection of rotating components
- Other components as recommended by manufacturer

RECIPROCATING LIQUID CHILLERS

EQUIPMENT

Components and Their Functions

The reciprocating compressor described in Chapter 37 is a positive-displacement machine that maintains fairly constant-volume flow rate over a wide range of pressure ratios. The following types of compressors are commonly used in liquid-chilling machines:

- Welded hermetic, to about 25 tons chiller capacity
- Semihermetic, to about 200 tons chiller capacity
- Direct-drive open, to about 450 tons chiller capacity

Open motor-driven liquid chillers are usually more expensive than hermetically sealed units, but can be more efficient. Hermetic motors are generally suction-gas-cooled; the rotor is mounted on the compressor crankshaft.

Condensers may be evaporative, air- or water-cooled. Water-cooled versions may be tube-in-tube, shell-and-coil, shell-and-tube, or plate heat exchangers. Most shell-and-tube condensers can be repaired; others must be replaced if a refrigerant-side leak occurs.

Air-cooled condensers are much more common than evaporative condensers. Less maintenance is needed for air-cooled heat exchangers than for the evaporative type. Remote condensers can be applied with condenserless packages. (Information on condensers can be found in Chapter 38.)

Coolers are usually direct-expansion, in which refrigerant evaporates while flowing inside tubes and liquid is cooled as it is guided several times over the outside of the tubes by shell-side baffles. Flooded coolers are sometimes used on industrial chillers. Flooded coolers maintain a level of refrigerant liquid on the shell side of the cooler, while liquid to be cooled flows through tubes inside the cooler. Tube-in-tube coolers are sometimes used with small machines; they offer low cost when repairability and installation space are not important criteria. Chapter 41 describes coolers in more detail.

The **thermal expansion** valve, capillary, or other device modulates refrigerant flow from the condenser to the cooler to maintain enough suction superheat to prevent any unevaporated refrigerant liquid from reaching the compressor. Excessively high values of superheat are avoided so that unit capacity is not reduced. (For additional information, see Chapter 44 in the 2006 *ASHRAE Handbook—Refrigeration*.)

Lubricant cooling is not usually required for air conditioning. However, if it is necessary, a refrigerant-cooled coil in the crankcase or a water-cooled cooler may be used. Lubricant coolers are often used in applications that have a low suction temperature or high pressure ratio when extra lubricant cooling is needed.

Capacities and Types Available

Available capacities range from about 2 to 450 tons. Multiple reciprocating compressor units are popular for the following reasons:

- The number of capacity increments is greater, resulting in closer liquid temperature control, lower power consumption, less current in-rush during starting, and extra standby capacity.
- Multiple refrigerant circuits are used, resulting in the potential for limited servicing or maintenance of some components while maintaining cooling.

Selection of Refrigerant

R-12 and R-22 have been the primary refrigerants used in chiller applications. CFC-12 has been replaced with HFC-134a, which has similar properties. However, R-134a requires synthetic lubricants because it is not miscible with mineral oils. R-134a is suitable for both open and hermetic compressors.

R-22 provides greater capacity than R-134a for a given compressor displacement. R-22 is used for most open and hermetic compressors, but as an HCFC, it is scheduled for phaseout in the future (see Chapter 19 of the 2005 *ASHRAE Handbook—Fundamentals* for more information on refrigerants and phaseout schedules). R-717 (ammonia) has similar capacity characteristics to R-22, but, because of odor and toxicity, R-717 use in public or populated areas is restricted. However, R-717 chillers are becoming more popular because of bans on CFC and HCFC refrigerants. R-717 units are open-drive compressors and are piped with steel because copper cannot be used in ammonia systems.

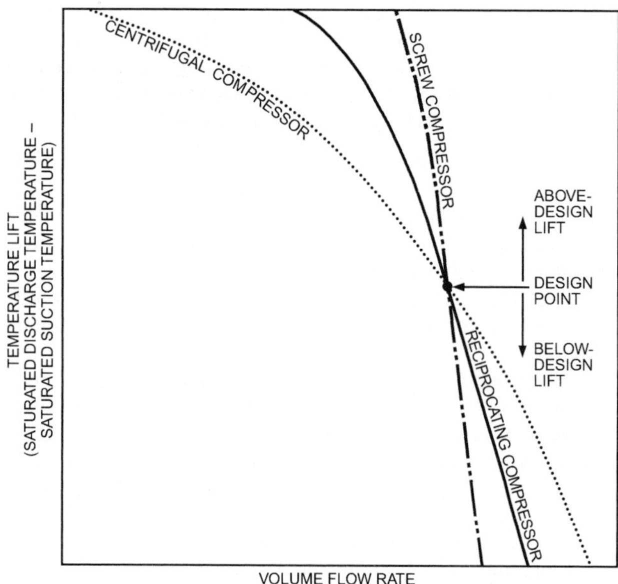

Fig. 6 Comparison of Single-Stage Centrifugal, Reciprocating, and Screw Compressor Performance

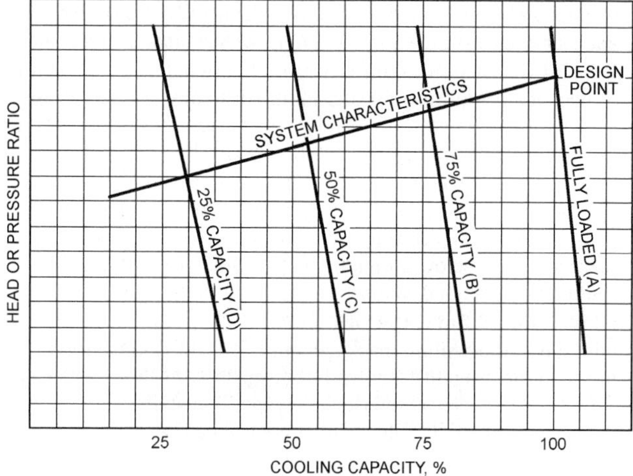

Fig. 7 Reciprocating Liquid Chiller Performance with Three Equal Steps of Unloading

PERFORMANCE CHARACTERISTICS AND OPERATING PROBLEMS

A distinguishing characteristic of the reciprocating compressor is its pressure rise versus capacity. Pressure rise has only a slight influence on the volume flow rate of the compressor, and, therefore, a reciprocating liquid chiller retains nearly full cooling capacity, even on above-design-wet-bulb days. It is well-suited for air-cooled condenser operation and low-temperature refrigeration. Typical performance is shown in Figure 6 and compared with centrifugal and screw compressors. Capacity control methods include the following:

- Unloading compressor cylinders (one at a time or in pairs)
- On/off cycling of compressors
- Hot-gas bypass
- Compressor speed control
- Combination of the previous methods

Figure 7 illustrates the relationship between system demand and performance of a compressor with three steps of unloading.

As cooling load drops to the left of fully loaded compressor line A, compressor capacity is reduced to that shown by line B, which produces the required refrigerant flow. Because cooling load varies continuously whereas machine capacity is available in fixed increments, some compressor on/off cycling or successive loading and unloading of cylinders is required to maintain fairly constant liquid temperature. In practice, a good control system minimizes load/unload or on/off cycling frequency while maintaining satisfactory temperature control.

METHOD OF SELECTION

Ratings

Two types of ratings are published. The first, for a packaged liquid chiller, lists values of capacity and power consumption for many combinations of leaving condenser water and chilled-water temperatures (ambient dry-bulb temperatures for air-cooled models). The second type of rating shows capacity and power consumption for different condensing and chilled-water temperatures. This type of rating allows selection with a remote condenser that can be evaporative, water-, or air-cooled. Sometimes the required rate of heat rejection is also listed to aid in selecting a separate condenser.

Power Consumption

With all liquid-chilling systems, power consumption increases as condensing temperature rises. Therefore, the smallest package, with the lowest ratio of input to cooling capacity, can be used when condenser water temperature is low, the remote air-cooled condenser is relatively large, or when leaving chilled-water temperature is high. The cost of the total system, however, may not be low when liquid chiller cost is minimized. Increases in cooling tower or fan-coil cost will reduce or offset the benefits of reduced compression ratio. Life-cycle costs (initial cost plus operating expenses) should be evaluated.

Fouling

A fouling allowance of 0.00025 $ft^2 \cdot °F \cdot h/Btu$ is included in manufacturers' ratings in accordance with ARI *Standard* 550/590. However, fouling factors greater than 0.00025 should be considered in the selection if water conditions are not ideal.

CONTROL CONSIDERATIONS

A reciprocating chiller is distinguished from centrifugal and screw compressor-operated chillers by its use of increments of capacity reduction rather than continuous modulation. Therefore, special arrangements must be used to establish precise chilled-liquid temperature control while maintaining stable operation free from excessive on/off cycling of compressors or unnecessary loading and unloading of cylinders.

To help provide good temperature control, return chilled-liquid temperature sensing is normally used by units with steps of capacity control. The resulting flywheel effect in the chilled-liquid circuit damps out excessive cycling. Leaving chilled-liquid temperature sensing prevents excessively low leaving chilled-liquid temperatures if chilled-liquid flow falls significantly below the design value. It may not provide stable operation, however, if rapid load changes are encountered.

An example of a basic control circuit for a single-compressor packaged reciprocating chiller with three steps of unloading is shown in Figure 8. The on/off switch controls start-up and starts the programmed timer. Assuming that the flow switch, field interlocks, and chiller safety devices are closed, pressing the momentarily closed reset button energizes control relay C1, locking in the safety circuit and the motor-starting circuit. When the timer completes its program, timer switch 1 closes and timer switch 2 opens. Timer relay TR energizes, stopping the timer motor. When timer switch 1 closes, the motor-starting circuit is completed and the motor contactor holding coil is energized, starting the compressor.

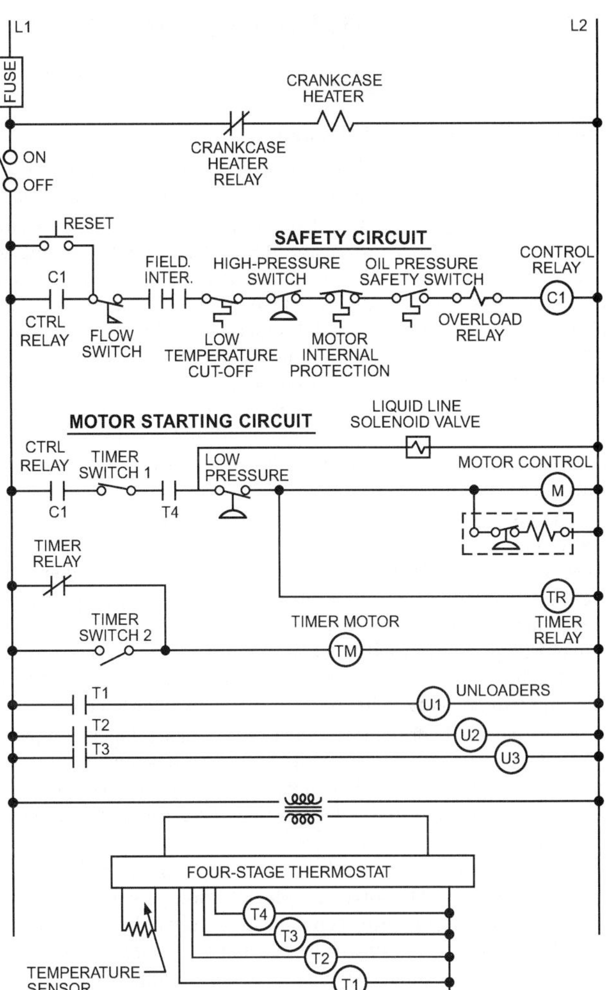

Fig. 8 Reciprocating Liquid Chiller Control System

The four-stage thermostat controls the compressor capacity in response to demand. Cylinders are loaded and unloaded by deenergizing and energizing the unloader solenoids. If load is reduced so that return water temperature drops to a predetermined setting, the unit shuts down until demand for cooling increases.

Opening a device in the safety circuit deenergizes control relay C1 and shuts down the compressor. The liquid line solenoid is also deenergized. Manual reset is required to restart. The crankcase heater is energized whenever the compressor is shut down.

If the automatic reset, low-pressure cutout opens, the compressor shuts down, but the liquid line solenoid remains energized. The timer relay TR is deenergized, causing the timer to start and complete its program before the compressor can be restarted. This prevents rapid cycling of the compressor under low-pressure conditions. A time delay low-pressure switch can also be used for this purpose with the proper circuitry.

SPECIAL APPLICATIONS

For multiple-chiller applications and a 10°F chilled-liquid temperature range, a parallel chilled-liquid arrangement is common because of the high cooler pressure drop resulting from the series arrangement. For a large (18°F) range, however, the series arrangement eliminates the need for overcooling when only one unit is operating. Special coolers with low water-pressure drop may also be used to reduce total chilled-water pressure drop in the series arrangement.

CENTRIFUGAL LIQUID CHILLERS

EQUIPMENT

Components and Their Function

Chapter 37 describes centrifugal compressors. Because they are not constant-displacement, they offer a wide range of capacities continuously modulated over a limited range of pressure ratios. By altering built-in design items (e.g., number of stages, compressor speed, impeller diameters, and choice of refrigerant), they can be used in liquid chillers having a wide range of design chilled-liquid temperatures and design cooling fluid temperatures. The ability to vary capacity continuously to match a wide range of load conditions with nearly proportional changes in power consumption makes a centrifugal compressor desirable for both close temperature control and energy conservation. Its ability to operate at greatly reduced capacity allows it to run most of the time with infrequent starting.

The hour of day for starting an electric-drive centrifugal liquid chiller can often be chosen by the building manager to minimize peak power demands. It has a minimum of bearing and other contacting surfaces that can wear; this wear is minimized by providing forced lubrication to those surfaces before start-up and during shutdown. Bearing wear usually depends more on the number of start-ups than the actual hours of operation. Thus, reducing the number of start-ups extends system life and reduces maintenance costs.

Both open and hermetic compressors are made. Open compressors may be driven by steam turbines, gas turbines or engines, or electric motors, with or without speed-changing gears. (Engine and turbine drives are covered in Chapter 7 and electric motor drives in Chapter 44.)

Packaged electric-drive chillers may be open or hermetic and use two-pole, 50 or 60 Hz polyphase electric motors, with or without speed-increasing gears. Hermetic units use only polyphase motors. Speed-increasing gears may be installed in a separate gearbox from the compressor. Several types of starters are commonly used with water-cooled chillers; starter selection depends on many variables, including cost, electrical system characteristics, voltage, and power company regulations at the installation.

For larger chillers, starters may be unit-mounted or remote-mounted from the chiller. Unit mounting saves space and reduces installation costs, and can increase the reliability of the chiller system. Unit-mounted starters are very popular on centrifugal chillers because the entire chiller's electrical requirements can be supplied with power through the starter (single-point connection). Several electrical connections are required for remote-mounted starters, and separate electrical feeds are needed for the compressor, oil pump, and unit controls. These separate wiring connections must be field-installed between the remote starter and the chiller.

Flooded coolers are commonly used, although direct-expansion coolers can also be used. The typical flooded cooler uses copper or copper alloy tubes that are mechanically expanded into the tube sheets, and, in some cases, into intermediate tube supports, as well.

Because liquid refrigerant that flows into the compressor increases power consumption and may cause internal damage, mist eliminators or baffles are often used in flooded coolers to minimize refrigerant liquid entrainment in the suction gas. (Additional information on coolers for liquid chillers is found in Chapter 37.)

The condenser is generally water-cooled, with refrigerant condensing on the outside of copper tubes. Large condensers may have refrigerant drain baffles, which direct condensate from within the tube bundle directly to liquid drains, reducing the liquid film thickness on the lower tubes.

Air-cooled condensers can be used with units that use higher-pressure refrigerants, but with considerable increase in unit energy consumption at design conditions. Operating costs should be compared with systems using cooling towers and condenser water circulating pumps.

System modifications, including subcooling and economizing (described under Principles of Operation), are often used to conserve energy by enhancing the refrigeration cycle efficiency. Some units combine the condenser, cooler, and refrigerant flow control in one vessel; a subcooler may also be incorporated. (Additional information about thermodynamic cycles is in Chapter 1 of the 2005 *ASHRAE Handbook—Fundamentals*. Chapter 38 in this volume has information on condensers and subcoolers.)

Capacities and Types Available

Centrifugal packages are available from about 80 to 4000 tons at nominal conditions of 44°F leaving chilled-water temperature and 95°F leaving condenser water temperature, but these limits are continually changing. Field-assembled machines extend to about 10,000 tons. Single- and two-stage internally geared machines and two- and three-stage direct-drive machines are commonly used in packaged units. Electric motor-driven machines constitute the majority of units sold.

Selection of Refrigerant

Information on numerous refrigerants can be found in Chapters 19 and 20 of the 2005 *ASHRAE Handbook—Fundamentals* and Chapter 5 of the 2006 *ASHRAE Handbook—Refrigeration*.Three refrigerants are widely used in centrifugal chillers for comfort air conditioning of commercial and institutional buildings: (1) R-123, which is a low-pressure refrigerant that replaced R-11 in the early 1990s; (2) R-134a, which replaced R-12; and (3) R-22, which is also commonly available but is not commonly used in new equipment. New refrigerants are also being developed as other alternatives.

All three refrigerants have advantages and disadvantages, which must be carefully considered when choosing a refrigerant and a liquid-chilling system. Legislative phaseout requirements also differ, based on environmental properties such as ozone depletion potential (ODP), direct global warming potential (DGWP), and indirect global warming potential (IGWP).

Table 1 summarizes these values for various refrigerants.

Ozone depletion potential refers to a refrigerant's potential to deplete stratospheric ozone, and is based on chlorine content and stability in the troposphere. These factors are weighted and compared relative to R-11. A lower number indicates a lower potential to deplete stratospheric ozone. HFC-134a has negligible ODP, because it contains no chlorine. HCFCs have much lower ODP compared to the CFCs that they replaced. HCFC-123 and HCFC-22 each contain chlorine, but HCFC-22 has a longer atmospheric life

Table 1 Environmental Properties of Various Refrigerants

Refrigerant	ODP	GWP	Atmospheric Life, years	COP	Operating Pressure, psia	
					Evap. (sat.) 41°F	Cond. (sat.) 95°F
CFC-11	1.000	4,750	45	6.60	7.2	21.6
CFC-12	1.000	10,890	100	6.26	52.5	122.7
HCFC-22	0.050	1,810	12	6.19	84.7	196.5
HCFC-123	0.020	77	1.3	6.54	5.9	18.9
HFC-134a	0.000	1,430	14	6.26	50.7	128.7

and thus has a higher ODP than HCFC-123. The Montreal Protocol, as well as local country requirements, has legislated a phaseout schedule for CFCs and HCFCs because of their effects on the ozone layer. The United States eliminated production of CFCs and has set national reduction benchmarks for the use of HCFCs in HVAC applications (EPA 2007). A thorough comparison of refrigerant characteristics is presented in Chapter 19 of the 2005 *ASHRAE Handbook—Fundamentals*.

The U.S. Clean air Act (EPA 1990) established the following national schedule for phasing out HCFC refrigerants in chillers:

- **R-11 and R-12:** Use in new equipment and for service was allowed until 1996. After 1996, service use was restricted to recycled, recovered, and stockpiled supplies.
- **R-123:** On January 1, 2020, there can be no production or importing of R-123 except for use in equipment manufactured before that date. From 2020 to 2030, production and importing will be restricted to servicing existing equipment. On January 1, 2030, and thereafter, no production or importing of R-123 will be allowed, although use of recycled R-123 will be allowed after 2030 for any application.
- **R-22:** On January 1, 2010, production and importing of R-22 will cease, except for use in equipment manufactured before that date, and no production or importing of new equipment that uses R-22 will be allowed. On January 1, 2020, no production or importing of R-22 will be allowed, although the use of recycled R-22 will be allowed after 2020 for any application.
- **R-134a:** This is an HFC refrigerant with negligible ozone depletion potential, and has no scheduled phaseout.

In other countries, consult with the applicable governing body.

Global warming is a major global environmental concern as well. No global-warming-based phaseouts are currently in effect for air-conditioning refrigerants in stationary applications. Refrigerants contribute to the greenhouse effect both directly (e.g., from refrigerant leakage into the atmosphere during operation, maintenance, or at end of life) and indirectly (from energy used to operate air-conditioning equipment). A less efficient chiller requires more power to be generated at the local power plant, and thus has a greater indirect contribution to global warming. R-123 has a lower direct global warming value than R-22 and R-134a. The total warming effect of a chiller should take into account the chiller's annual energy efficiency, GWP, and the refrigerant's emissive potential.

Additional refrigerant-selection methods intended to reduce ozone depletion, support early compliance with the Montreal Protocol, and minimize direct contributions to global warming are available (USGBC 2005).

Safety is also an important consideration. Regardless of the refrigerant selected, refrigerant leak detectors, alarms, and emergency ventilation are now required by code in many applications. Safety classifications of refrigerants are categorized by a code, with a letter designating toxicity levels and a numeral indicating flammability ranking. For example, R-123 has a B1 classification, and R-22 and R-134a have A1 classifications, as described in ASHRAE *Standard* 15. With proper safety procedures, R-123, R-22, and R-134a are all permitted under most North American codes.

Chiller operating pressure also affects pressure vessel requirements, emissive potential, and ancillary equipment.

- During normal operation, pressure in an R-123 evaporator is less than atmospheric, and pressure in the condenser is slightly higher than atmospheric. Therefore, a purge device is required to remove noncondensables, which may leak into the machine.
- Any chiller using R-134a or R-22 operates at positive pressure, on the order of 10 atm. Therefore, a purge device is not required, but the chiller must be constructed to a pressure vessel code.
- A chiller's emissive potential is related to the selected refrigerant's molecular weight and saturation pressure range, coupled with the machine's hermetic integrity and installation, maintenance, and service practices. In general, all chillers have the potential for extremely low emissions. ASHRAE *Standard* 147 provides methods for design, manufacturing and operational practices to achieve low leakage rates. Typical refrigerant leakage rates vary between 0.5 to 2.0% per year.

Refrigerant stability and **material compatibility** with selected refrigerants are also important considerations in chiller design; the means for controlling typical contaminants must be considered, as well. Various contaminants and their control are discussed in Chapter 6 of the 2006 *ASHRAE Handbook—Refrigeration*. Selection of elastomers and electrical insulating materials require special attention because many of these materials are affected by the refrigerants. Additional information on material selection can be found in Chapter 5 of the 2006 *ASHRAE Handbook—Refrigeration*, and information on testing methods can be found in ASHRAE *Standard* 97.

Energy efficiency is a factor when selecting a refrigerant and chiller system. Each refrigerant discussed in this section has a different theoretical or baseline energy performance, according to its thermodynamic and thermophysical properties. At temperatures and pressures commonly applied in commercial comfort air-conditioning applications, R-123, R-134a, and R-22 are listed from highest to lowest theoretical COP. From that baseline, chiller manufacturers enhance their designs to optimize refrigerant properties. Furthermore, some chillers are more efficient at peak load, whereas others perform better at off-peak conditions, so an accurate load model is necessary to make a fully informed choice. Chiller performances at peak and off-peak operating conditions are a function of specific chiller and compressor design, not refrigerant type. More thorough data on refrigerant properties are available in Chapters 19 and 20 of the 2005 *ASHRAE Handbook—Fundamentals*.

PERFORMANCE AND OPERATING CHARACTERISTICS

Figure 9 illustrates a compressor's performance at constant speed with various inlet guide vane settings. Figure 10 illustrates a compressor's performance at various speeds in combination with inlet guide vanes. Capacity is modulated at constant speed by automatic adjustment of prerotation vanes that swirl the refrigerant gas at the impeller eye. This effect matches demand by shifting the compressor performance curve downward and to the left (as shown in Figure 9). Compressor efficiency, when unloaded in this manner, is superior to

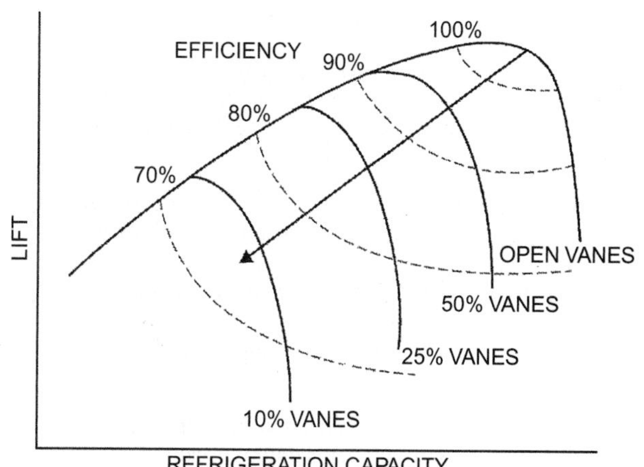

Fig. 9 Typical Centrifugal Compressor Performance at Constant Speed
(Carrier 2004)

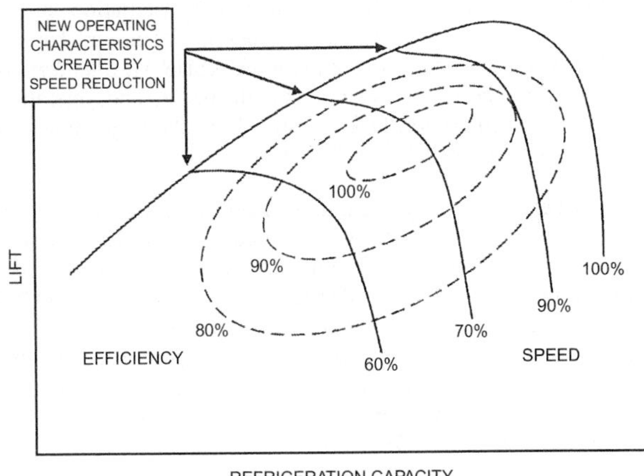

Fig. 10 Typical Variable-Speed Centrifugal Compressor Performance
(Carrier 2004)

suction throttling. Some manufacturers automatically reduce diffuser width or throttle the impeller outlet with decreasing load.

Speed control for a centrifugal compressor offers even lower power consumption. Variable-frequency drive (VFD) control continuously reduces the compressor's capacity, keeping operation in the maximum efficiency region over a much broader range of operation. Essentially, the VFD adjusts the compressor's speed to keep the inlet guide vanes (IGVs) as open as possible to meet the system lift requirements, with the lowest power consumption. Combined with the drop in condenser water temperature that occurs naturally in an air-conditioning system, the variable-speed centrifugal compressor more efficiently meets the flow and lift condition or state point required by the system.

Although capacity is directly related to a change in speed, the lift produced is proportional to the square of the change in speed.

Hot-gas bypass allows the compressor to operate down to zero load. This feature is a particular advantage for intermittent industrial applications such as cooling quenching tanks. Bypass vapor obtained by either method maintains power consumption at the same level attained just before starting bypass, regardless of load reductions. At light loads, some bypass vapor, if introduced into the cooler below the tube bundle, may increase the evaporating temperature by agitating the liquid refrigerant and thereby more thoroughly wetting the tube surfaces.

Figure 11 shows how **temperature lift** varies with load. A typical reduction in entering condenser water temperature of 10°F helps to reduce temperature lift at low load. Other factors producing lower lift at reduced loads include the following:

- Reduced condenser cooling water range (difference between entering and leaving temperatures, resulting from decreasing heat rejection)
- Decreased temperature difference between condensing refrigerant and leaving condenser water
- Similar decrease between evaporating refrigerant and leaving chilled-liquid temperature

In many cases, the actual reduction in temperature lift is even greater because the wet-bulb temperature usually drops with cooling load, producing a greater decrease in entering condenser water temperature.

Power consumption is reduced when the coldest possible condenser water is used, consistent with the chiller manufacturer's recommended minimum condenser water temperature. In cooling tower

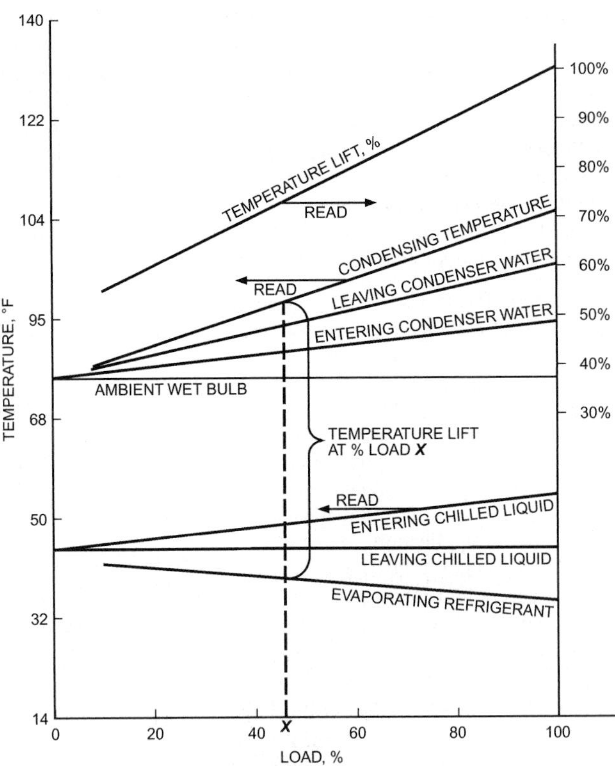

Fig. 11 Temperature Relations in a Typical Centrifugal Liquid Chiller

applications, minimum water temperatures should be controlled by a cooling tower bypass and/or by cooling tower fan control, not by reducing water flow through the condenser. Maintaining a high flow rate at lower temperatures minimizes fouling and the increase in power requirements caused by fouling.

Surging occurs when the system-specific work becomes greater than the compressor developed specific work or above the surge line indicated in Figures 9 and 10. Excessively high temperature lift and corresponding specific work commonly originate from

- Excessive condenser or evaporator water-side fouling beyond the specified allowance
- Inadequate cooling tower performance and higher-than-design condenser water temperature
- Noncondensables in the condenser, which increase condenser pressure
- Condenser flow less than design

SELECTION

Ratings

A centrifugal chiller with specified details is typically selected using a manufacturer's computer-generated selection program, many of which are ARI certified. Capacity, efficiency requirements, stability requirements, number of passes, water-side pressure drop in each of the heat exchangers, and desired electrical characteristics are input to select the chiller.

Stability is important in evaluating the part-load operating condition for a centrifugal chiller. If head pressure during part-load operation is higher than the chiller was selected for, the impeller may not be able to overcome the lift, and the chiller may begin unstable operation, causing the compressor to surge. For humid regions, typical stability is chosen at approximately 50% of full load at design entering condenser water, to guard against surge conditions.

Centrifugal chillers are typically selected for full- and/or part-load coefficient of performance (COP) targets. Then they are checked for part-load stability using software provided by the chiller manufacturer. A typical part-load stability check may involve running the chiller at part-load points at entering condenser water temperatures that follow a relief profile representative of the project geography.

Most manufacturers offer variations of evaporators, condensers, tube counts, tube types, compressor gears, impellers, etc. All of these permutations create an enormous product offering that is much too difficult to fit into a tabular format. For this reason, computer programs are the norm for chiller selections and ratings because they can analyze hundreds of combinations in a very short time.

Fouling

In accordance with ARI *Standard* 550/590, a fouling allowance of 0.00025 $ft^2 \cdot °F \cdot h/Btu$ is included in manufacturers' ratings for condenser fouling. (Chapter 38 has further information about fouling factors.) To reduce fouling, a minimum water velocity of about 3.3 ft/s is recommended in condensers. Maximum water velocities exceeding 11 ft/s are not recommended because of potential erosion problems with copper tubes.

Proper water treatment and regular tube cleaning are recommended for all liquid chillers to reduce power consumption and operating problems. Chapter 48 of the 2007 *ASHRAE Handbook—HVAC Applications* has water treatment information.

Continuous or daily monitoring of the quality of the condenser water is desirable. Checking the quality of the chilled liquid is also desirable. Intervals between checks become greater as the possibilities for fouling contamination become less (e.g., an annual check should be sufficient for closed-loop water-circulating systems for air conditioning). Corrective treatment is required, and periodic, usually annual, cleaning of the condenser tubes usually keeps fouling within the specified allowance. In applications where more frequent cleaning is desirable, an on-line cleaning system may be economical.

Noise and Vibration

The chiller manufacturer's recommendations for mounting should be followed to prevent transmission or amplification of vibration to adjacent equipment or structures. Auxiliary pumps, if not connected with flexible fittings, can induce vibration of the centrifugal unit, especially if the rotational speed of the pump is nearly the same as either the compressor prime mover or the compressor. Flexible tubing becomes less flexible when it is filled with liquid under pressure and some vibration can still be transmitted. General information on noise, measurement, and control may be found in Chapter 7 of the 2005 *ASHRAE Handbook—Fundamentals*, Chapter 47 of the 2007 *ASHRAE Handbook—HVAC Applications*, and ARI *Standard* 575.

CONTROL CONSIDERATIONS

In centrifugal systems, the **chilled-liquid temperature sensor** is usually placed in thermal contact with the leaving chilled water. In electrical control systems, the electrical signal is transmitted to an electronic control module, which controls the operation of an electric motor(s) positioning the capacity-controlling inlet guide vanes. A current limiter is usually included on machines with electric motors. An electrical signal from a current transformer in the compressor motor controller is sent to the electronic control module. The module receives indications of both leaving chilled-water temperature and compressor motor current. The part of the electronic control module responsive to motor current is called the current limiter. It overrides the demands of the temperature sensor.

Inlet guide vanes, independent of demands for cooling, do not open more than the position that results in the present setting of the current limiter. The chilled-liquid temperature sensor provides a signal. The controlling module receives both that signal and the motor current electrical signal and controls the positioning of the inlet guide vanes.

The **current limiter** on most machines can limit current draw during periods of high electrical demand charges. This control can be set from about 40 to 100% of full-load current. When power consumption is limited, cooling capacity is correspondingly reduced. If cooling load only requires 50% of the rated load, the current (or demand) limiter can be set at 50% without loss of cooling. By setting the limiter at 50% of full current draw, any subsequent high demand charges are prevented during pulldown after start-up. Even during periods of high cooling load, it may be desirable to limit electrical demand if a small increase in chiller liquid temperature is acceptable. If temperature continues to decrease after capacity control reaches its minimum position, a low-temperature control stops the compressor and restarts it when a rise in temperature indicates the need for cooling. Manual controls may also be provided to bypass temperature control. Provision is included to ensure that capacity control is at its minimum position when the compressor starts to provide an unloaded starting condition.

Additional operating controls are needed for appropriate operation of lubricant pumps, lubricant heaters, purge units, and refrigerant transfer units. An **antirecycle timer** should also be included to prevent frequent motor starts. Multiple-unit applications require additional controls for capacity modulation and proper unit sequencing. (See the section on Multiple-Chiller Systems.)

Safety controls protect the unit under abnormal conditions. Safety cutouts may be required for high condenser pressure, low evaporator refrigerant temperature or pressure, low lubricant pressure, high lubricant temperature, high motor temperature, and high discharge temperature. Auxiliary safety circuits are usually provided on packaged chillers. At installation, the circuits are field-wired to field-installed safety devices, including auxiliary contacts on the pump motor controllers and flow switches in the chilled-water and condenser water circuits. Safety controls are usually provided in a lockout circuit, which trips out the compressor motor controller and prevents automatic restart. The controls reset automatically, but the circuit cannot be completed until a manual reset switch is operated and the safety controls return to their safe positions.

AUXILIARIES

Purge units are required for centrifugal liquid-chilling machines to maintain system hermetic chemistry integrity and efficiency. ASHRAE *Standard* 147 requires purge units for liquid-chilling machines using refrigerants with working pressures below atmospheric pressure (e.g., R-11, R-113, R-123, R-245fa). If a purge unit were not used, air and moisture would accumulate in the refrigerant side. Noncondensables collect in the condenser during operation, reducing the heat-transfer coefficient and increasing condenser pressure as a result of both their insulating effect and the partial pressure of the noncondensables. Compressor power consumption increases, capacity decreases, and surging may occur.

Free moisture may build up once the refrigerant becomes saturated. Acids produced by a reaction between free moisture and the refrigerant then cause internal corrosion. A purge unit prevents accumulation of noncondensables and ensures internal cleanliness of the chiller. However, a purge unit does not reduce the need to check for leaks and repair them, which is required maintenance for any liquid chiller. Purge units may be manual or automatic, compressor-operated, or compressorless. To reduce the potential for air leaks when chillers are off, chillers may be heated externally to pressurize them to atmospheric pressure.

ASHRAE *Standard* 15 requires most purge units and rupture disks to be vented outdoors. Because of environmental concerns and the increasing cost of refrigerants, high-efficiency (air to refrigerant) purges are available that reduce refrigerant losses during normal purging.

Lubricant coolers may be water-cooled, using condenser water when the quality is satisfactory, or chilled water when a small loss in net cooling capacity is acceptable. These coolers may also be refrigerant- or air-cooled, eliminating the need for water piping to the cooler.

A **refrigerant transfer unit** may be provided for maintenance of centrifugal liquid chillers. The unit consists of a small reciprocating compressor with electric motor drive, a condenser (air- or water-cooled), a lubricant reservoir and separator, valves, and interconnecting piping. Refrigerant transfers in three steps:

1. **Gravity drain.** When the receiver is at the same level as or below the cooler, some liquid refrigerant may be transferred to the receiver by opening valves in the interconnecting piping.
2. **Pressure transfer.** By resetting valves and operating the compressor, refrigerant gas is pulled from the receiver to pressurize the cooler, forcing refrigerant liquid from the cooler to the storage receiver. If the chilled-liquid and condenser water pumps can be operated to establish a temperature difference, refrigerant migration from the warmer vessel to the colder vessel can also be used to help transfer refrigerant.
3. **Pump-out.** After the liquid refrigerant has been transferred, valve positions are changed and the compressor is operated to pump refrigerant gas from the cooler to the transfer unit condenser, which sends condensed liquid to the storage receiver. If any chilled liquid (water, brine, etc.) remains in the cooler tubes, pump-out must be stopped before cooler pressure drops below the saturation condition corresponding to the chilled liquid's freezing point.

If the saturation temperature corresponding to cooler pressure is below the chilled-liquid freezing point when recharging, refrigerant gas from the storage receiver must be introduced until cooler pressure is above this condition. The compressor can then be operated to pressurize the receiver and move refrigerant liquid into the cooler without danger of freezing.

Water-cooled transfer unit condensers provide fast refrigerant transfer. Air-cooled condensers eliminate the need for water, but they are slower and more expensive.

SPECIAL APPLICATIONS

Free Cooling

Cooling without operating the compressor of a centrifugal liquid chiller is called free cooling. When a supply of condenser water is available at a temperature below the needed chilled-water temperature, some chillers can operate as a thermal siphon. Low-temperature condenser water condenses refrigerant, which is either drained by gravity or pumped into the evaporator. Higher-temperature chilled water causes the refrigerant to evaporate, and vapor flows back to the condenser because of the pressure difference between the evaporator and the condenser. This free-cooling accessory is limited to a fraction of the chiller design capacity, and this option is not available from all manufacturers. Free-cooling capacity depends on chiller design and the temperature difference between the desired chilled-water temperature and the condenser water temperature. Free cooling is also available external to the chiller using either direct or indirect methods, as described in Chapter 39.

Air-Cooled System

Two types of air-cooled centrifugal systems are used. One consists of a water-cooled centrifugal package with a closed-loop condenser water circuit. Condenser water is cooled in a water/air heat exchanger. This arrangement results in higher condensing temperature and increased power consumption. In addition, winter operation requires using glycol in the condenser water circuit, which reduces the heat transfer coefficient of the unit.

The other type of unit is directly air-cooled, which eliminates the intermediate heat exchanger and condenser water pumps, resulting in lower power requirements. However, condenser and refrigerant piping must be leak-free.

Because a centrifugal machine will surge if it is subjected to a pressure appreciably higher than design, the air-cooled condenser must be designed to reject the required heat. In common practice, selection of a reciprocating air-cooled machine is based on an outside dry-bulb temperature that will be exceeded 5% of the time. A centrifugal chiller may be unable to operate during such times because of surging, unless the chilled-water temperature is raised proportionately. Thus, the compressor impeller(s) and/or speed should be selected for the maximum dry-bulb temperature to ensure that the desired chilled-water temperature is maintained at all times. In addition, the condenser coil must be kept clean.

An air-cooled centrifugal chiller should allow the condensing temperature to fall naturally to about 70°F during colder weather. The resulting decrease in compressor power consumption is greater than that for reciprocating systems controlled by thermal expansion valves.

During winter shutdown, precautions must be taken to prevent cooler liquid freezing caused by a free cooling effect from the air-cooled condenser. A thermostatically controlled heater in the cooler, in conjunction with a low-refrigerant-pressure switch to start the chilled-liquid pumps, will protect the system.

Other Coolants

Centrifugal liquid-chilling units are most frequently used for water-chilling applications, but they are also used with secondary coolants such as calcium chloride, methylene chloride, ethylene glycol, and propylene glycol. (Chapter 21 of the 2005 *ASHRAE Handbook—Fundamentals* describes properties of secondary coolants.) Coolant properties must be considered in calculating heat transfer performance and pressure drop. Because of the greater temperature rise, higher compressor speeds and possibly more stages may be required for cooling these coolants. Compound and/or cascade systems are required for low-temperature applications.

Vapor Condensing

Many process applications condense vapors such as ammonia, chlorine, or hydrogen fluoride. Centrifugal liquid-chilling units are used for these applications.

OPERATION AND MAINTENANCE

Proper operation and maintenance are essential for reliability, longevity, and safety. Chapter 38 of the 2007 *ASHRAE Handbook—HVAC Applications* includes general information on principles, procedures, and programs for effective maintenance. The manufacturer's operation and maintenance instructions should also be consulted for specific procedures. In the United States, Environmental Protection Agency (EPA) regulations require (1) certification of service technicians, (2) a statement of minimum pressures necessary during system evacuation, and (3) definition of when a refrigerant charge must be removed before opening a system for service. All service technicians or operators maintaining systems must be familiar with these regulations.

Normal operation conditions should be established and recorded at initial startup. Changes from these conditions can be used to signal the need for maintenance. One of the most important items is to maintain a leak-free unit.

Leaks on units operating at subatmospheric pressures allow air and moisture to enter the unit, which increases condenser pressure. Although the purge unit can remove noncondensables sufficiently to prevent an increase in condenser pressure, continuous entry of air and attendant moisture into the system promotes refrigerant and lubricant breakdown and corrosion. Leaks from units that operate above atmospheric pressure may release environmentally harmful refrigerants. Regulations require that annual leakage not exceed a percentage of the refrigerant charge. It is good practice, however, to find and repair all leaks.

Periodic analysis of the lubricant and refrigerant charge can also identify system contamination problems. High condenser pressure or frequent purge unit operation indicate leaks that should be corrected as soon as possible. With positive operating pressures, leaks result in loss of refrigerant and operating problems such as low evaporator pressure. A leak check should also be included in preparation for a long-term shutdown. (Chapter 6 in the 2006 *ASHRAE Handbook—Refrigeration* discusses the harmful effects of air and moisture.)

Normal maintenance should include periodic lubricant and refrigerant filter changes as recommended by the manufacturer. All safety controls should be checked periodically to ensure that the unit is protected properly.

Cleaning inside tube surfaces may be required at various intervals, depending on water condition. Condenser tubes may only need annual cleaning if proper water treatment is maintained. Cooler tubes need less frequent cleaning if the chilled-water circuit is a closed loop.

If the refrigerant charge must be removed and the unit opened for service, the unit should be leak-checked, dehydrated, and evacuated properly before recharging. Chapter 45 of the 2006 *ASHRAE Handbook—Refrigeration* has information on dehydrating, charging, and testing.

SCREW LIQUID CHILLERS

EQUIPMENT

Components and Their Function

Single- and twin-screw compressors are positive-displacement machines with nearly constant flow performance. Compressors for liquid chillers can be both lubricant-injected and lubricant-injection-free. (Chapter 37 describes screw compressors in detail.)

The cooler may be flooded or direct-expansion. No particular design has a cost advantage over the other. The flooded cooler is more sensitive to freezing, requires more refrigerant, and requires closer evaporator pressure control, but its performance is easier to predict and it can be cleaned. The direct-expansion cooler requires closer mass flow control, is less likely to freeze, and returns lubricant to the lubricant system rapidly. The decision to use one or the other is based on the relative importance of these factors on a given application.

Screw coolers have the following characteristics: (1) high maximum working pressure, (2) continuous lubricant scavenging, (3) no mist eliminators (flooded coolers), and (4) distributors designed for high turndown ratios (direct-expansion coolers). A suction-gas, high-pressure liquid heat exchanger is sometimes incorporated into the system to provide subcooling for increased thermal expansion valve flow and reduced power consumption. (For further information on coolers, see Chapter 41.)

Flooded coolers were once used in units with a capacity larger than about 400 tons. Direct-expansion coolers are also used in larger units up to 800 tons with a servo-operated expansion valve having an electronic controller that measures evaporating pressure, leaving secondary coolant temperature, and suction gas superheat.

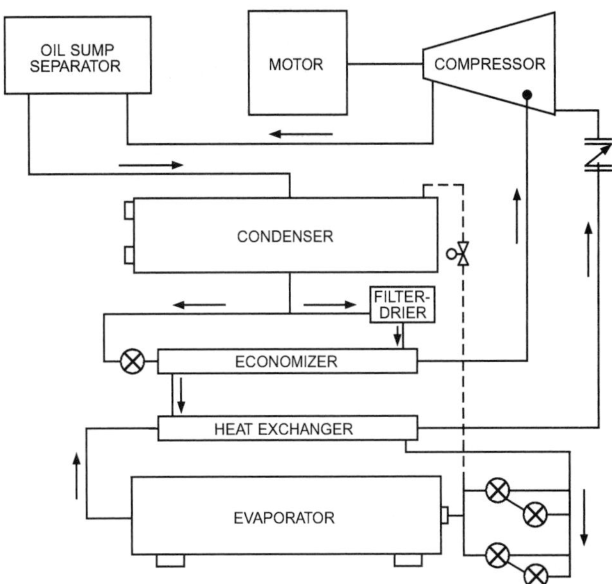

Fig. 12 Refrigeration System Schematic

The condenser may be included as part of the liquid-chilling package when water-cooled, or it may be remote. Air-cooled liquid chilling packages are also available. When remote air-cooled or evaporative condensers are applied to liquid-chilling packages, a liquid receiver generally replaces the water-cooled condenser on the package structure. Water-cooled condensers are the cleanable shell-and-tube type (see Chapter 38).

Lubricant cooler loads vary widely, depending on the refrigerant and application, but they are substantial because lubricant injected into the compressor absorbs part of the heat of compression. Lubricant is cooled by one of the following methods:

- Water-cooled using condenser water, evaporative condenser sump water, chilled water, or a separate water- or glycol-to-air cooling loop
- Air-cooled using a lubricant-to-air heat exchanger
- Refrigerant-cooled (where lubricant cooling load is low)
- Liquid injection into the compressor
- Condensed refrigerant liquid thermal recirculation (thermosiphon), where appropriate compressor head pressure is available.

The latter two methods are the most economical both in first cost and overall operating cost because cooler maintenance and special water treatment are eliminated.

Efficient lubricant separators are required. The types and efficiencies of these separators vary according to refrigerant and application. Field-built systems require better separation than complete factory-built systems. Ammonia applications are most stringent because no appreciable lubricant returns with the suction gas from the flooded coolers normally used in ammonia applications. However, separators are available for ammonia packages, which do not require the periodic addition of lubricant customary on other ammonia systems. The types of separators used are centrifugal, demister, gravity, coalescer, and combinations of these.

Hermetic compressor units may use a centrifugal separator as an integral part of the hermetic motor while cooling the motor with discharge gas and lubricant simultaneously. A schematic of a typical refrigeration system is shown in Figure 12.

Capacities and Types Available

Screw compressor liquid chillers are available as factory-packaged units from about 30 to 1250 tons. Both open and hermetic styles are manufactured. Packages without water-cooled condensers, with receivers, are made for use with air-cooled or evaporative condensers. Most factory-assembled liquid chilling packages use R-22; some use R-134a.

Additionally, compressor units, comprised of a compressor, hermetic or open motor, and lubricant separator and system, are available from 20 to 2000 tons. These are used with remote evaporators and condensers for low-, medium-, and high-evaporating-temperature applications. Condensing units, similar to compressor units in range and capacity but with water-cooled condensers, are also built. Similar open motor-drive units are available for ammonia, as are booster units.

Selection of Refrigerant

The refrigerants most commonly used with screw compressors on liquid chiller applications are R-22, R-134a, and R-717. The active use of R-12 and R-500 has been discontinued for new equipment.

PERFORMANCE AND OPERATING CHARACTERISTICS

The screw compressor operating characteristic shown in Figure 6 is compared with reciprocating and centrifugal performance. Additionally, because the screw compressor is a positive-displacement compressor, it does not surge. Because it has no clearance volume in the compression chamber, it pumps high volumetric flows at high pressure. Consequently, screw compressor chillers suffer the least capacity reduction at high condensing temperatures.

The screw compressor provides stable operation over the whole working range because it is a positive-displacement machine. The working range is wide because discharge temperature is kept low and is not a limiting factor because of lubricant injection into the compression chamber. Consequently, the compressor is able to operate single-stage at high pressure ratios.

An economizer can be installed to improve capacity and lower power consumption at full-load operation. An example is shown in Figure 12, where the main refrigerant liquid flow is subcooled in a heat exchanger connected to the intermediate pressure port in the compressor. The evaporating pressure in this heat exchanger is higher than the suction pressure of the compressor.

Lubricant separators must be sized for the compressor size, type of system (factory-assembled or field-connected), refrigerant, and type of cooler. Direct-expansion coolers have less stringent separation requirements than do flooded coolers. In a direct-expansion system, refrigerant evaporates in the tubes, which means that velocity is kept so high that lubricant rapidly returns to the compressor. In a flooded evaporator, the refrigerant is outside the tubes, and an external lubricant-return device must be used to minimize the concentration of lubricant in the cooler. Suction or discharge check valves are used to minimize backflow and lubricant loss during shutdown.

Because the lubricant system is on the high-pressure side of the unit, precautions must be taken to prevent lubricant dilution. Dilution can also be caused by excessive floodback through the suction or intermediate ports; unless properly monitored, it may go unnoticed until serious operating or mechanical problems are experienced.

SELECTION

Ratings

Screw liquid chiller ratings are generally presented similarly to those for centrifugal chiller ratings. Tabular values include capacity and power consumption at various chilled-water and condenser water temperatures. In addition, ratings are given for packages without the condenser that list capacity and power versus chilled-water temperature and condensing temperature. Ratings for compressors alone are also common, showing capacity and power consumption versus suction temperature and condensing temperature for a given refrigerant.

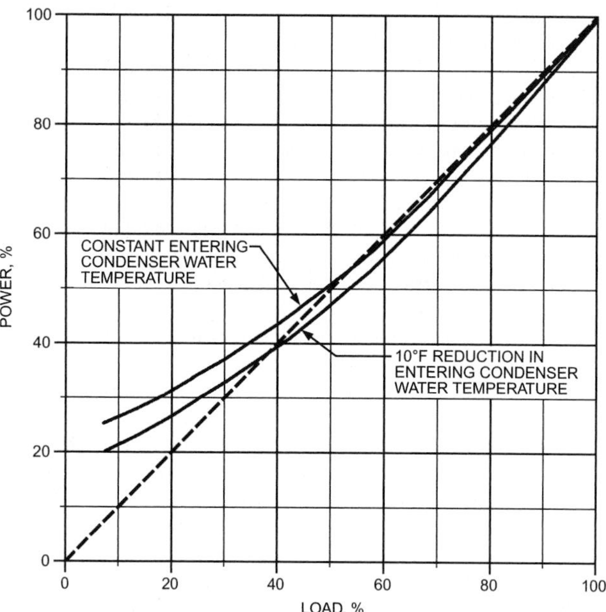

Fig. 13 Typical Screw Compressor Chiller Part-Load Power Consumption

Power Consumption

Typical part-load power consumption is shown in Figure 13. Power consumption of screw chillers benefits from reducing condensing water temperature as the load decreases, as well as operating at the lowest practical pressure at full load. However, because direct-expansion systems require a pressure differential, the power consumption saving is not as great at part load as shown.

Fouling

A fouling allowance of 0.00025 $ft^2 \cdot °F \cdot h/Btu$ is incorporated in screw compressor chiller ratings. Excessive fouling (above design value) increases power consumption and reduces capacity. Fouling water-cooled lubricant coolers results in higher than desirable lubricant temperatures.

CONTROL CONSIDERATIONS

Screw chillers provide continuous capacity modulation, from 100% capacity down to 10% or less. The leaving chilled-liquid temperature is sensed for capacity control. Safety controls commonly required are (1) lubricant failure switch, (2) high-discharge-pressure cutout, (3) low-suction-pressure switch, (4) cooler flow switch, (5) high-lubricant- and discharge-temperature cutout, (6) hermetic motor inherent protection, (7) lubricant pump and compressor motor overloads, and (8) low-lubricant-temperature (floodback/dilution protection). The compressor is unloaded automatically (slide valve driven to minimum position) before starting. Once it starts operating, the slide valve is controlled hydraulically by a temperature-load controller that energizes the load and unload solenoid valves.

The current limit relay protects against motor overload from higher than normal condensing temperatures or low voltage, and also allows a demand limit to be set. An antirecycle timer is used to prevent overly frequent recycling. Lubricant sump heaters are energized during the off cycle. A hot-gas-capacity control is optionally available and prevents automatic recycling at no-load conditions

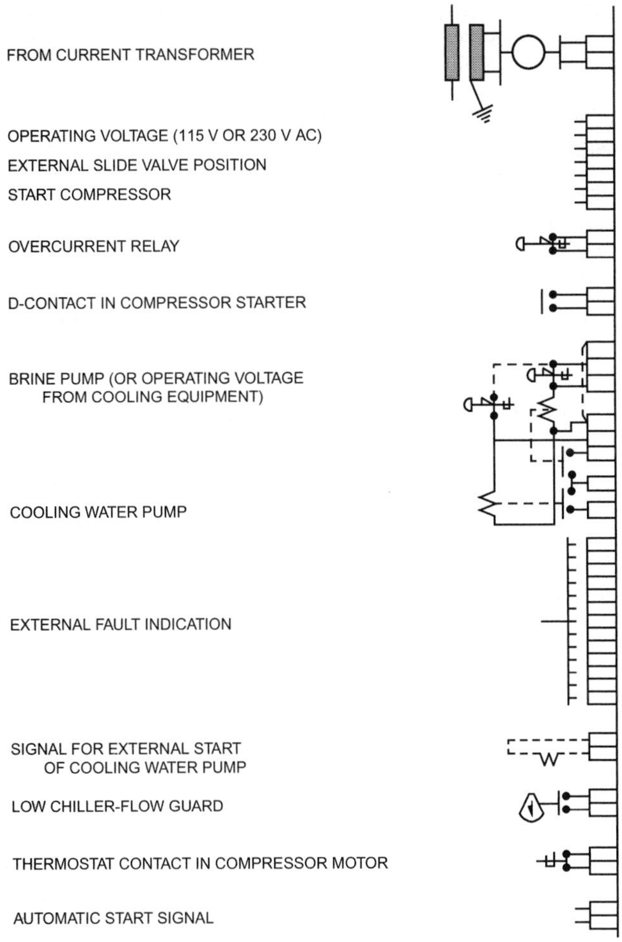

FROM CURRENT TRANSFORMER
OPERATING VOLTAGE (115 V OR 230 V AC)
EXTERNAL SLIDE VALVE POSITION
START COMPRESSOR
OVERCURRENT RELAY
D-CONTACT IN COMPRESSOR STARTER
BRINE PUMP (OR OPERATING VOLTAGE FROM COOLING EQUIPMENT)
COOLING WATER PUMP
EXTERNAL FAULT INDICATION
SIGNAL FOR EXTERNAL START OF COOLING WATER PUMP
LOW CHILLER-FLOW GUARD
THERMOSTAT CONTACT IN COMPRESSOR MOTOR
AUTOMATIC START SIGNAL

Fig. 14 Typical External Connections for Screw Compressor Chiller

such as is often required in process liquid chilling. A suction-to-discharge starting bypass sometimes aids starting and allows use of standard starting torque motors.

Some units are equipped with electronic regulators specially developed for the screw compressor characteristics. These regulators include PI (proportional-integrating) control of leaving brine temperature and functions such as automatic/manual control, capacity indication, time circuits to prevent frequent recycling and to bypass the lubricant pressure cutout during start-up, switch for unloaded starting, etc. (Typical external connections are shown in Figure 14.)

AUXILIARIES

A **refrigerant transfer unit** is similar to the unit described in the section on Auxiliaries under Centrifugal Liquid Chillers, and is designed for R-22 operating pressure. Its flexibility is increased by including a reversible liquid pump on the unit. It is available as a portable unit or mounted on a storage receiver.

A **lubricant-charging pump** is useful for adding lubricant to the pressurized lubricant sump. Two types are used: a manual pump and an electric motor-driven positive-displacement pump.

Acoustical enclosures are available for installations that require low noise levels.

SPECIAL APPLICATIONS

Because of the screw compressor's positive-displacement characteristic and lubricant-injected cooling, its use for high-pressure-differential applications is limited only by power considerations and maximum design working pressures. Therefore, it is used for many special applications because of reasonable compressor cost and no surge characteristic. Some of the fastest-growing areas include the following:

- Heat recovery installations
- Air-cooled split packages with field-installed interconnecting piping, and factory-built rooftop packages
- Low-temperature brine chillers for process cooling
- Ice rink chillers
- Power transmission line lubricant cooling

High-temperature compressor and condensing units are used increasingly for air conditioning because of the higher efficiency of direct air-to-refrigerant heat exchange resulting in higher evaporating temperatures. Many of these installations have air-cooled condensers.

MAINTENANCE

Manufacturer's maintenance instructions should be followed, especially because some items differ substantially from reciprocating or centrifugal units. Water-cooled condensers must be cleaned of scale periodically (see the section on General Maintenance). If condenser water is also used for the lubricant cooler, this should be considered in the treatment program. Lubricant coolers operate at higher temperatures and lower flows than condensers, so it is possible that the lubricant cooler may have to be serviced more often than the condenser.

Because large lubricant flows are a part of the screw compressor system, the lubricant filter pressure drop should be monitored carefully and the elements changed periodically. This is particularly important in the first month or so after start-up of any factory-built package, and is essential on field-erected systems. Because the lubricant and refrigeration systems merge at the compressor, loose dirt and fine contaminants in the system eventually find their way to the lubricant sump, where they are removed by the lubricant filter. Similarly, the filter-drier cartridges should be monitored for pressure drop and moisture during initial start and regularly thereafter. Generally, if a system reaches acceptable dryness, it stays that way unless it is opened.

It is good practice to check the lubricant for acidity periodically, using commercially available acid test kits. Lubricant does not need to be changed unless it is contaminated by water, acid, or metallic particles. Also, a refrigerant sample should be analyzed yearly to determine its condition.

Procedures that should be followed yearly or during a regularly scheduled shutdown include checking and calibrating all operation and safety controls, tightening all electrical connections, inspecting power contacts in starters, dielectric checking of hermetic and open motors, and checking the alignment of open motors.

Leak testing of the unit should be performed regularly. A water-cooled package used for summer cooling should be leak-tested annually. A flooded unit with proportionately more refrigerant in it, used for year-round cooling, should be tested every four to six months. A process air-cooled chiller designed for year-round operation 24 h per day should be checked every one to three months.

Based on 6000 operating hours per year and depending on the above considerations, a typical inspection or replacement timetable is as follows:

Shaft seals	1.5 to 4 yr	Inspect
Hydraulic cylinder seals	1.5 to 4 yr	Replace
Thrust bearings	4 to 6 yr	Check preload via shaft end play every 6 mo and replace as required
Shaft bearings	7 to 10 yr	Inspect

REFERENCES

ARI. 1998. Standard for water chilling packages using the vapor compression cycle. *Standard* 550/590. Air-Conditioning and Refrigeration Institute, Arlington, VA.

ARI. 1994. Method of measuring machinery sound within an equipment space. *Standard* 575. Air-Conditioning and Refrigeration Institute, Arlington, VA.

ASHRAE. 2007. Safety code for mechanical refrigeration. ANSI/ASHRAE *Standard* 15.

ASHRAE. 2007. Sealed glass tube method to test the chemical stability of materials for use within refrigerant systems. ANSI/ASHRAE *Standard* 97.

ASHRAE. 2002. Reducing the release of halogenated refrigerants from refrigerating and air-conditioning equipment and systems. ANSI/ASHRAE *Standard* 147.

ASME. 2001. *Boiler and pressure vessel code*, Section VIII. American Society of Mechanical Engineers, New York.

Carrier. 2004. *Water-cooled chillers*. Technical Development Program TDP-623.

EPA. 1990. *Clean air act.* U.S. Environmental Protection Agency, Washington, D.C. http://www.epa.gov/air/caa/.

EPA. 2007. *Phaseout of class II ozone-depleting substances.* U.S. Environmental Protection Agency, Washington, D.C. http://www.epa.gov/ozone/title6/phaseout/classtwo.html.

USGBC. 2005. *Green building rating system for new construction and major renovations*, v. 2.2. U.S. Green Building Council, Leadership in Energy and Environmental Design, Washington, D.C.

BIBLIOGRAPHY

ASHRAE. 1995. Method of testing liquid-chilling packages. ANSI/ASHRAE *Standard* 30.

ASHRAE. 2007. Designation and safety classification of refrigerants. ANSI/ASHRAE *Standard* 34.

ASHRAE. 2001. *Position document on ozone-depleting substances*. Available from http://www.ashrae.org/aboutus/page/335.

Calm, J.M. 2007. Resource, ozone, and global warming implications of refrigerant selection for large chillers. *Paper* ICR07-E1-535, *Proceedings of the 22nd International Congress of Refrigeration, Beijing*. Chinese Association of Refrigeration, Beijing, and International Institute of Refrigeration, Paris.

Calm, J.M. and G. C. Hourahan. 2007. Refrigerant data update. *HPAC Engineering* 79(1):50-64.

IPCC. 2005. *Safeguarding the ozone layer and the global climate system: Issues related to hydrofluorocarbons and perfluorocarbons*. Intergovernmental Panel on Climate Change.

ONLINE RESOURCE

Ozone Secretariat, U.N. Environment Programme. http://ozone.unep.org/.

CHAPTER 43

CENTRIFUGAL PUMPS

CENTRIFUGAL pumps provide the primary force to distribute and recirculate hot and chilled water in a variety of space-conditioning systems. The pump provides a predetermined flow of water to the space load terminal units or to a thermal storage chamber for release at peak loads. The effect of centrifugal pump performance on the application, control, and operation of various terminal units is discussed in Chapter 12. Other hydronic systems that use pumps include (1) condensing water circuits to cooling towers (Chapters 13 and 39), (2) water-source heat pumps (Chapter 8), (3) boiler feeds, and (4) condensate returns (Chapter 10). Boiler feed and condensate return pumps for steam boilers should be selected based on boiler manufacturer's requirements.

CONSTRUCTION FEATURES

The construction features of a typical centrifugal pump are shown in Figure 1. These features vary according to the manufacturer and the type of pump.

The preparation of this chapter is assigned to TC 6.1, Hydronic and Steam Equipment and Systems.

Materials. Centrifugal pumps are generally available in bronze-fitted or iron-fitted construction. The choice of material depends on those parts in contact with the liquid being pumped. In bronze-fitted pumps, the impeller and wear rings (if used) are bronze, the shaft sleeve is stainless steel or bronze, and the casing is cast iron. All-bronze construction is often used in domestic water applications.

Seals. The **stuffing box** is that part of the pump where the rotating shaft enters the pump casing. To seal leaks at this point, a mechanical seal or packing is used. **Mechanical seals** are used predominantly in clean hydronic applications, either as unbalanced or balanced (for higher pressures) seals. Balanced seals are used for high-pressure applications, particulate-laden liquids, or for extended seal life at lower pressures. Inside seals operate inside the stuffing box, while outside seals have the rotating element outside the box. Pressure and temperature limitations vary with the liquid pumped and the style of seal. **Packing** is used where abrasive substances (that are not detrimental to operation) could damage mechanical seals. Some leakage at the packing gland is needed to lubricate and cool the area between the packing material and shaft. Some designs use a large seal cavity instead of a stuffing box.

Shaft sleeves protect the motor or pump shaft.

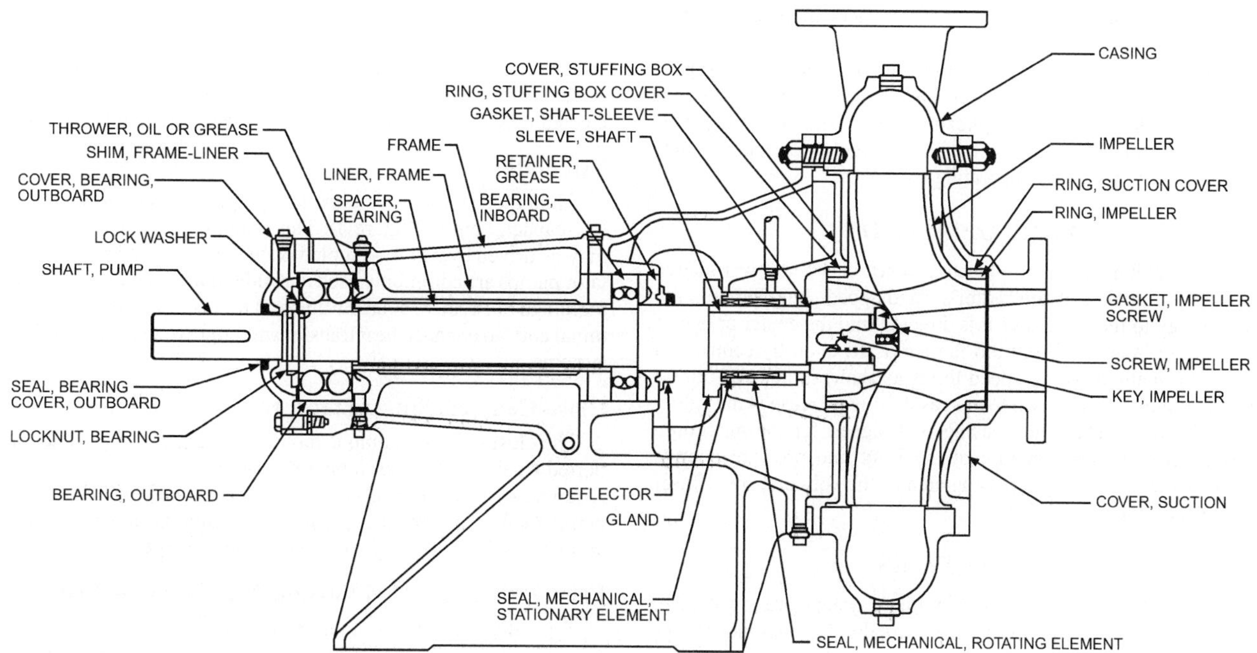

Fig. 1 Cross Section of Typical Overhung-Impeller End-Suction Pump

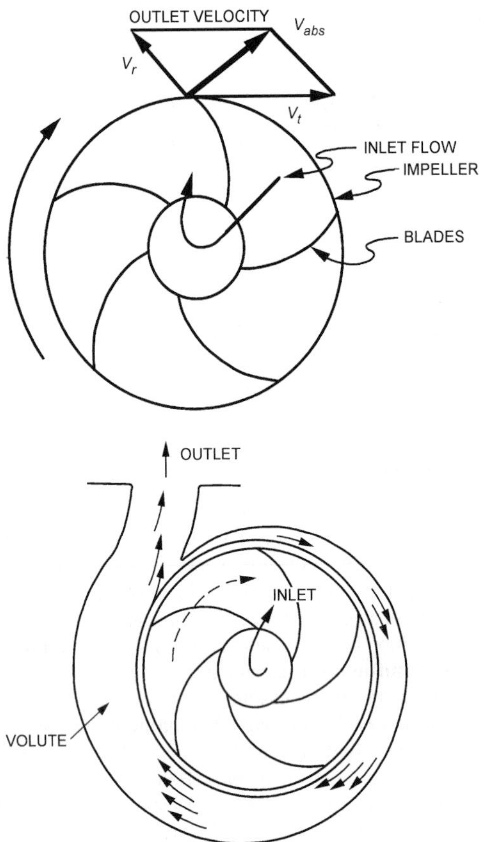

Fig. 2 Impeller and Volute Interaction

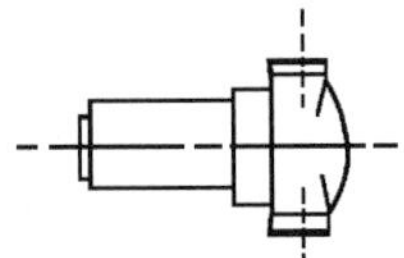

Fig. 3 Circulator Pump (Pipe-Mounted)

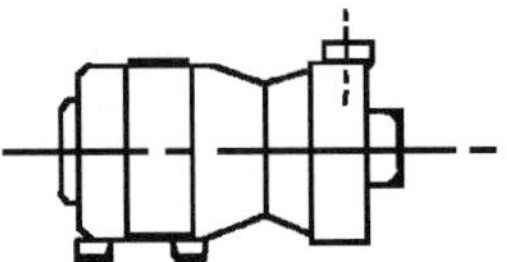

Fig. 4 Close-Coupled End-Suction Pump

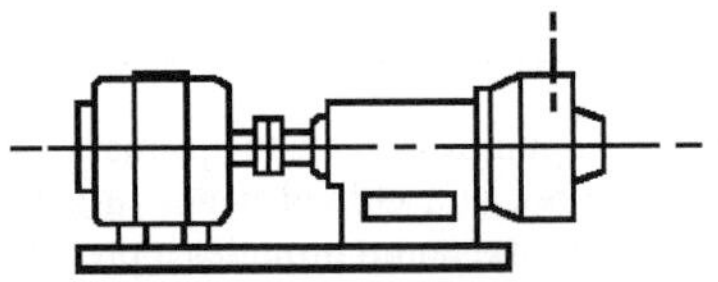

Fig. 5 Frame-Mounted End-Suction Pump on Base Plate

Wear rings prevent wear to the impeller and/or casing and are easily replaced when worn.

Ball bearings are most frequently used, except in low-pressure circulators, where motor and pump bearings are the sleeve type.

A **balance ring** placed on the back of a single-inlet enclosed impeller reduces the axial load, thereby decreasing the size of the thrust bearing and shaft. Double-inlet impellers are inherently axially balanced.

Normal operating speeds of motors may be selected from 600 to 3600 rpm. The pump manufacturer can help determine the optimum pump speed for a specific application by considering pump efficiency, the available pressure at the inlet to prevent cavitation, maintenance requirements, and operating cost.

PUMP OPERATION

In a centrifugal pump, an electric motor or other power source rotates the impeller at the motor's rated speed. Impeller rotation adds energy to the fluid after it is directed into the center or eye of the rotating impeller. The fluid is then acted upon by centrifugal force and rotational or tip speed force, as shown in the vector diagram in Figure 2. These two forces result in an increase in the velocity of the fluid. The pump casing is designed for the maximum conversion of velocity energy of the fluid into pressure energy, either by the uniformly increasing area of the volute or by diffuser guide vanes (when provided).

PUMP TYPES

Most centrifugal pumps used in hydronic systems are single-stage pumps with a single- or double-inlet impeller. Double-inlet pumps are generally used for high-flow applications, but either type is available with similar performance characteristics and efficiencies.

A centrifugal pump has either a volute or diffuser casing. Pumps with volute casings collect water from the impeller and discharge it perpendicular to the pump shaft. Casings with diffusers discharge water parallel to the pump shaft. All pumps described in this chapter have volute casings except the vertical turbine pump, which has a diffuser casing.

Pumps may be classified as close-coupled or flexible-coupled to the electric motor. The close-coupled pump has the impeller mounted on a motor shaft extension, and the flexible-coupled pump has an impeller shaft supported by a frame or bracket that is connected to the electric motor through a flexible coupling.

Pumps may also be classified by their mechanical features and installation arrangement. One-horsepower and larger pumps are available as close-coupled or base-mounted. Close-coupled pumps have an end-suction inlet for horizontal mounting or a vertical in-line inlet for direct installation in the piping. Base-mounted pumps are (1) end-suction, frame-mounted or (2) double-suction, horizontal or vertical split-case units. Double-suction pumps can also be arranged in a vertical position on a support frame with the motor vertically mounted on a bracket above the pump. Pumps are usually labeled by their mounting position as either horizontal or vertical.

Circulator Pump

Circulator is a generic term for a pipe-mounted, low-pressure, low-capacity pump (Figure 3). This pump may have a wet rotor or may be driven by a close-coupled or flexible-coupled motor. Circulator pumps are commonly used in residential and small commercial buildings to circulate source water and to recirculate the flow of terminal coils to enhance heat transfer and improve the control of large systems.

Close-Coupled, Single-Stage, End-Suction Pump

The close-coupled pump is mounted on a horizontal motor supported by the motor foot mountings (Figure 4). Mounting usually requires a solid concrete pad. The motor is close-coupled to the pump shaft. This compact pump has a single horizontal inlet and vertical discharge. It may have one or two impellers.

Frame-Mounted, End-Suction Pump on Base Plate

Typically, the motor and pump are mounted on a common, rigid base plate for horizontal mounting (Figure 5). Mounting requires a solid concrete pad. The motor is flexible-coupled to the pump shaft

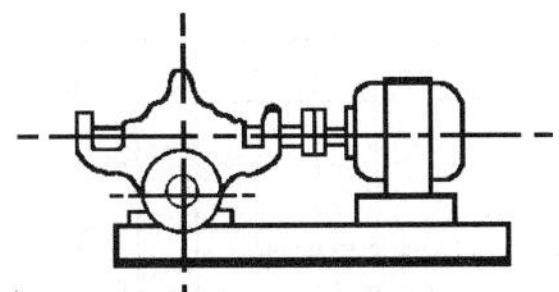

Fig. 6 Base-Mounted, Horizontal (Axial), Split-Case, Single-Stage, Double-Suction Pump

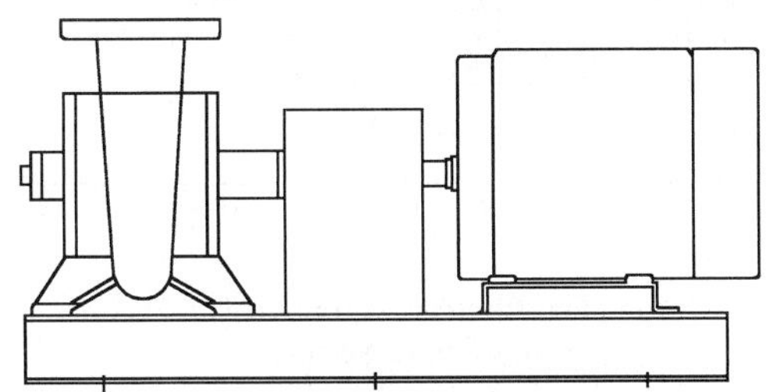

Fig. 7 Base-Mounted, Vertical, Split-Case, Single-Stage, Double-Suction Pump

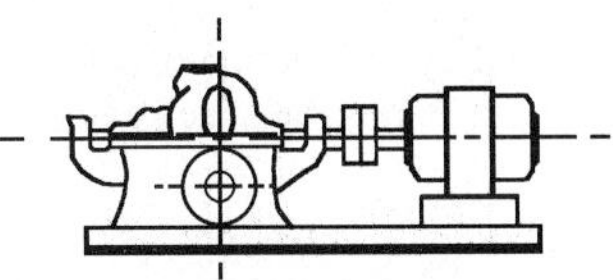

Fig. 8 Base-Mounted, Horizontal, Split-Case, Multistage Pump

and should have an OSHA-approved guard. For horizontal mounting, the piping is horizontal on the suction side and vertical on the discharge side. This pump has a single suction.

Base-Mounted, Horizontal (Axial) or Vertical, Split-Case, Single-Stage, Double-Suction Pump

The motor and pump are mounted on a common, rigid base plate for horizontal mounting (Figures 6 and 7). Sometimes axial pumps are vertically mounted with a vertical pump casing and motor mounting bracket. Mounting requires a solid concrete pad. The motor is flexible-coupled to the pump shaft, and the coupling should have an OSHA-approved guard. A split case permits complete access to the impeller for maintenance. This pump may have one or two double suction impellers.

Base-Mounted, Horizontal, Split-Case, Multistage Pump

The motor and pump are mounted on a common, rigid base plate for horizontal mounting (Figure 8). Mounting requires a solid concrete pad. The motor is flexible-coupled to the pump shaft, and the coupling should have an OSHA-approved guard. Piping is horizontal on both the suction and discharge sides. The split case permits complete access to the impellers for maintenance. This pump has a single suction and may have one or more impellers for multistage operation.

Vertical In-Line Pump

This close-coupled pump and motor are mounted on the pump casing (Figure 9). The unit is compact and depends on the connected piping for support. Mounting requires adequately spaced pipe hangers and, sometimes, a vertical casing support. The suction and discharge piping is horizontal. The pump has a single or double suction impeller.

Vertical Turbine, Single- or Multistage, Sump-Mounted Pump

Vertical turbine pumps have a motor mounted vertically on the pump discharge head for either wet-sump mounting or can-type

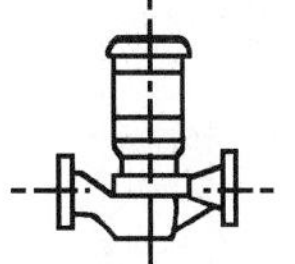

Fig. 9 Vertical In-Line Pump

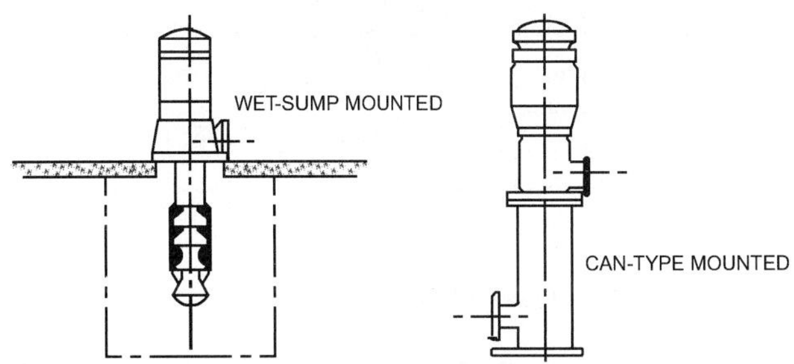

Fig. 10 Vertical Turbine Pumps

mounting (Figure 10). This single-suction pump may have one or multiple impellers for multistage operation. Mounting requires a solid concrete pad or steel sole plate above the wet pit with accessibility to the screens or trash rack on the suction side for maintenance. Can-type mounting requires a suction strainer. Piping is horizontal on the discharge side and on the suction side. The sump should be designed according to Hydraulic Institute (1994) recommendations.

PUMP PERFORMANCE CURVES

Performance of a centrifugal pump is commonly shown by a manufacturer's performance curve (Figure 11). The figure displays the pump power required for a liquid with a specific gravity of 1.0 (water) over a particular range of impeller diameters and flows. The curves are generated from a set of standard tests developed by the Hydraulic Institute (1994). The tests are performed by the manufacturer for a given pump volute or casing and several impeller diameters, normally from the maximum to the minimum allowable in that volute. The tests are conducted at a constant impeller speed for various flows.

Pump curves represent the average results from testing several pumps of identical design under the same conditions. The curve is sometimes called the head-capacity curve (H-Q) for the pump. Typically, the discharge head of the centrifugal pump, sometimes called the total dynamic head (TDH), is measured in feet of water flowing at a standard temperature and pressure. TDH represents the difference in total head between the suction side and the discharge side of the pump. This discharge head decreases as the flow increases (Figure 12). Motors are often selected to be non-overloading at a specified impeller size and maximum flow to ensure safe motor operation at all flow requirements.

The pump characteristic curve may be further described as flat or steep (Figure 13). Sometimes these curves are described as a normal rising curve (flat), a drooping curve (steep), or a steeply rising curve. The pump curve is considered flat if the pressure at shutoff is about 1.10 to 1.20 times the pressure at the best efficiency point. Flat characteristic pumps are usually installed in closed systems with modulating two-way control valves. Steep characteristic pumps are usually installed in open systems, such as cooling towers (see Chapter 13), where higher head and constant flow are usually desired.

Pump manufacturers may compile performance curves for a particular set of pump volutes in a series (Figure 14). The individual curves are shown in the form of an envelope consisting of the maximum and minimum impeller diameters and the ends of their curves. This set of curves is known as a family of curves. A family of curves

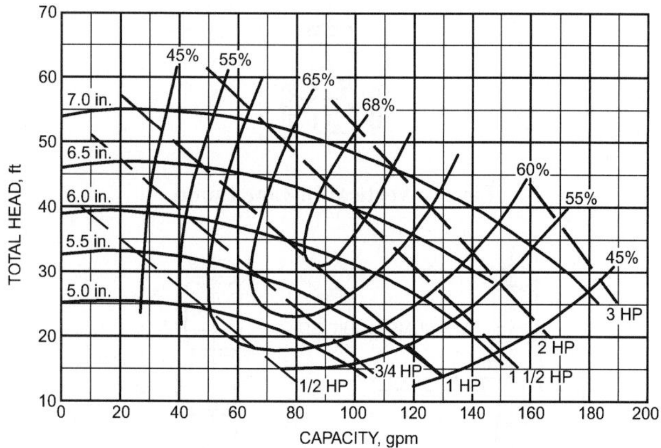

Fig. 11 Typical Pump Performance Curve

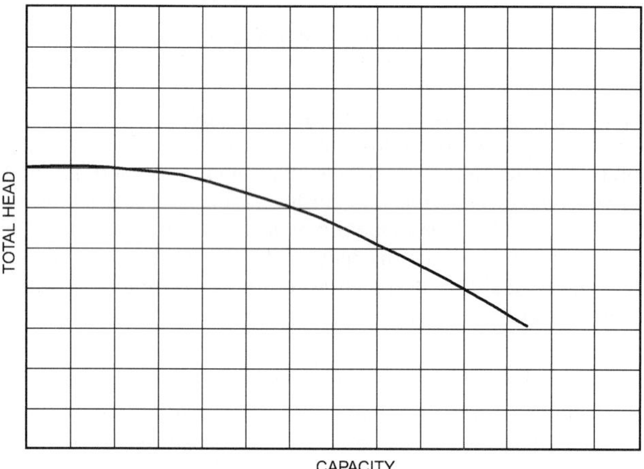

Fig. 12 Typical Pump Curve

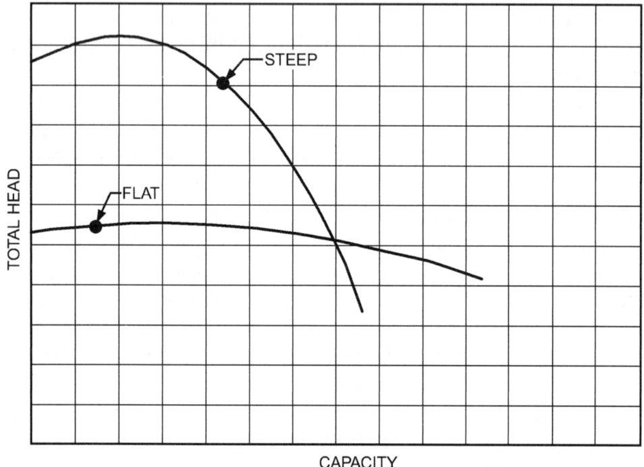

Fig. 13 Flat Versus Steep Performance Curves

is useful in determining the approximate size and model required, but the particular pump curve (Figure 11) must then be used to confirm an accurate selection.

Many pump manufacturers and HVAC software suppliers offer electronic versions for pump selection. Pump selection software typically allows the investigation of different types of pumps and operating parameters. Corrections for fluid specific gravity, temperature, and motor speeds are easily performed.

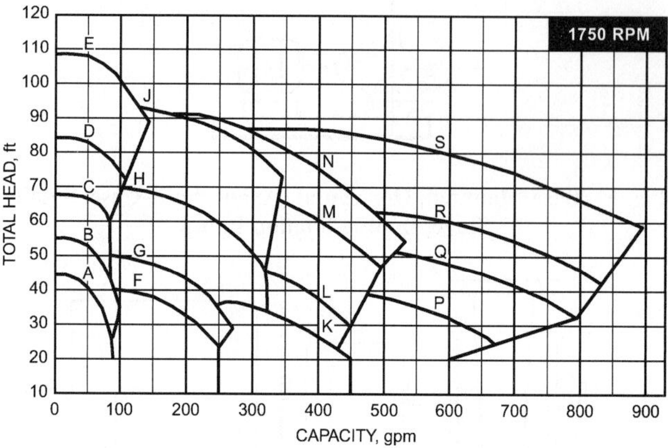

Fig. 14 Typical Pump Manufacturer's Performance Curve Series

HYDRONIC SYSTEM CURVES

Pressure drop caused by the friction of a fluid flowing in a pipe may be described by the Darcy-Weisbach equation:

$$\Delta p = f \frac{L}{D} \frac{\rho}{g} \frac{V^2}{2} \tag{1}$$

Equation (1) shows that pressure drop in a hydronic system (pipe, fittings, and equipment) is proportional to the square of the flow (V^2 or Q^2 where Q is the flow). Experiments show that pressure drop is more nearly proportional to between $V^{1.85}$ and $V^{1.9}$, or a nearly parabolic curve as shown in Figures 15 and 18. The design of the system (including the number of terminals and flows, the fittings and valves, and the length of pipe mains and branches) affects the shape of this curve.

Equation (1) may also be expressed in head or specific energy form:

$$\Delta h = \frac{\Delta p}{\rho} = f \frac{L}{D} \frac{V^2}{2g} \tag{2}$$

where

Δh = head loss through friction, ft (of fluid flowing)
Δp = pressure drop, lb/ft^2
ρ = fluid density, lb/ft^3
f = friction factor, dimensionless
L = pipe length, ft
D = inside diameter of pipe, ft
V = fluid average velocity, ft/s
g = gravitational acceleration, 32.2 ft/s^2

The system curve (Figure 15) defines the system head required to produce a given flow rate for a liquid and its characteristics in a piping system design. To produce a given flow, the system head must overcome pipe friction, inside pipe surface roughness, actual fitting losses, actual valve losses, resistance to flow due to fluid viscosity, and possible system effect losses. The general shape of this curve is parabolic since, according to the Darcy-Weisbach equation [Equation (2)], the head loss is proportional to the square of the flow.

If static pressure is present due to the height of the liquid in the system or the pressure in a compression tank, this head is sometimes referred to as **independent head** and is added to the system curve (Figure 16).

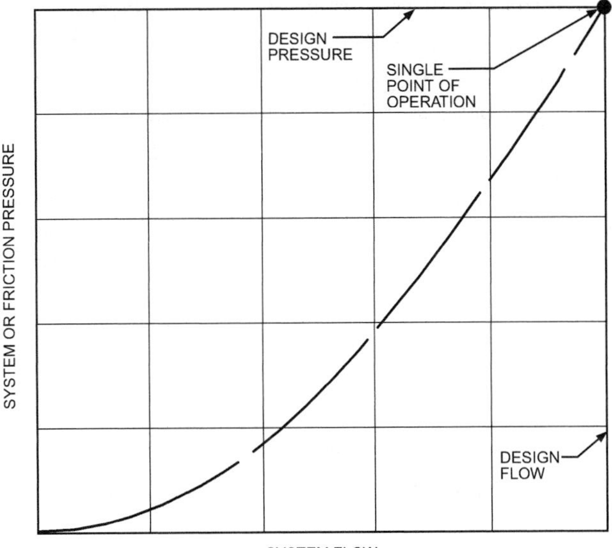

Fig. 15 Typical System Curve

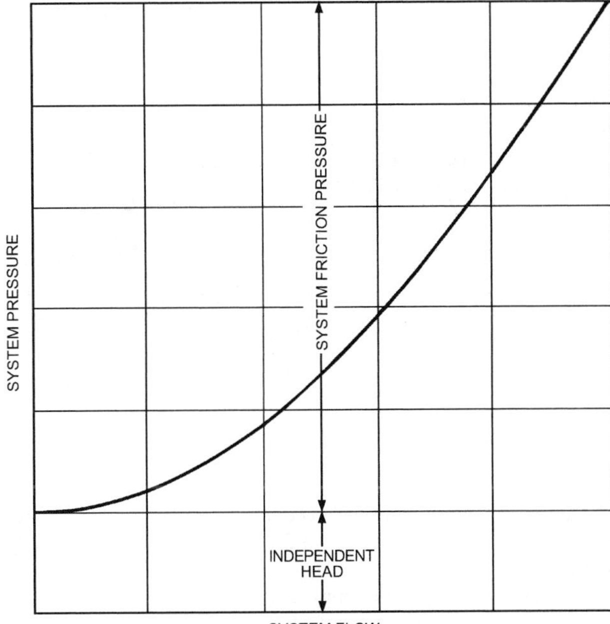

Fig. 16 Typical System Curve with Independent Head

PUMP AND HYDRONIC SYSTEM CURVES

The pump curve and the system curve can be plotted on the same graph. The intersection of the two curves (Figure 17) is the system **operating point**, where the pump's developed head matches the system's head loss.

In a typical hydronic system, a thermostat or controller varies the flow in a load terminal by positioning a two-way control valve to match the load. At full load the two-way valves are wide open, and the system follows curve A in Figure 18. As the load drops, the terminal valves begin closing to match the load (part load). This increases the friction and reduces the flow in the terminals. The system curve gradually changes to curve B.

The operating point of a pump should be considered when the system includes two-way control valves. Point 1 in Figure 19 shows the pump operating at the design flow at the calculated design head loss of the system. But typically, the actual system curve is slightly different than the design curve. As a result, the pump operates at point 2 and produces a flow rate higher than design.

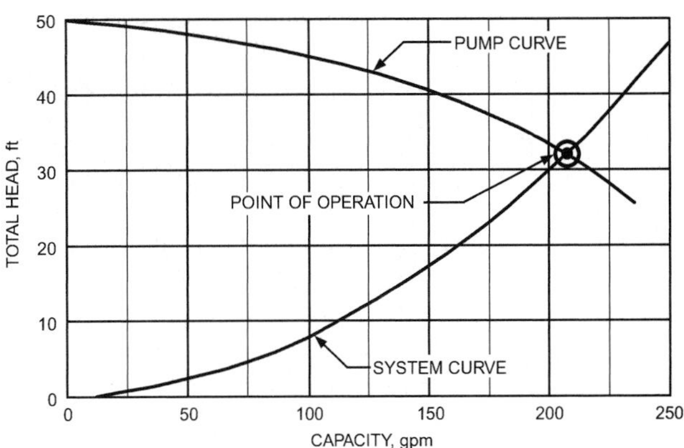

Fig. 17 System and Pump Curves

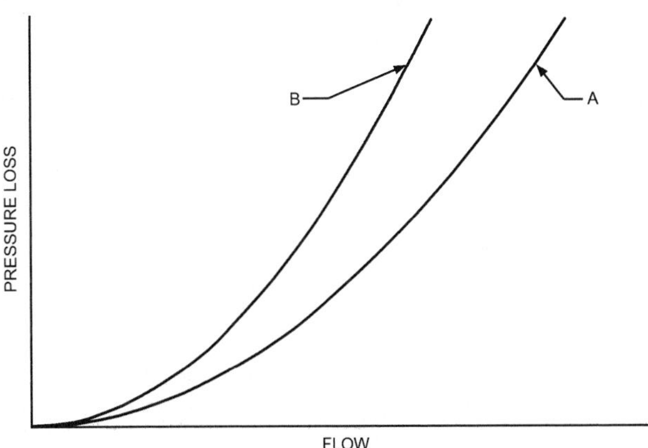

Fig. 18 System Curve Change due to Part-Load Flow

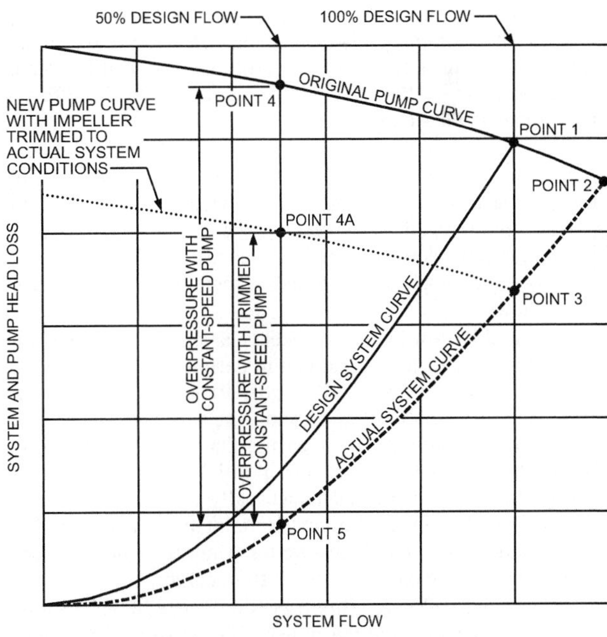

Fig. 19 Pump Operating Points

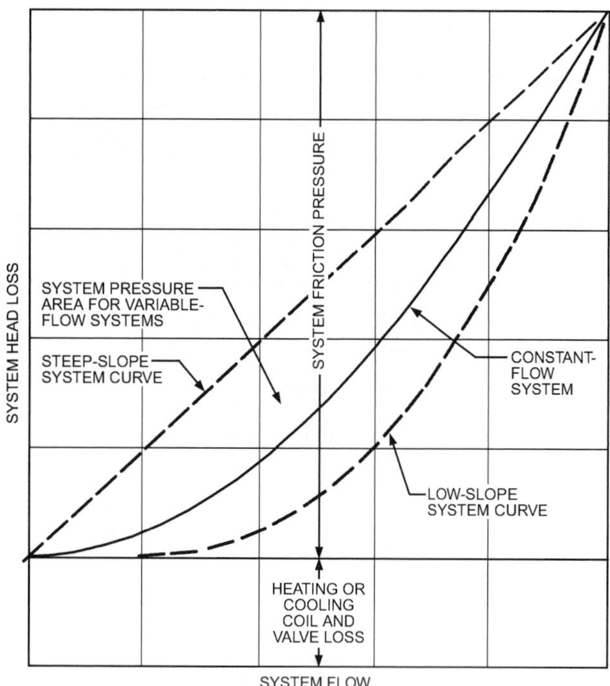

Fig. 20 System Curve, Constant and Variable Head Loss

To reduce the actual flow to the design flow at point 1, a balancing valve downstream from the pump can be adjusted while all the terminal valves are in a wide-open position. This pump discharge balancing valve imposes a pressure drop equal to the pressure difference between point 1 and point 3. The manufacturer's pump curve shows that the capacity may be reduced by substituting a new impeller with a smaller diameter or by trimming the existing pump impeller. After trimming, reopening the balancing valve in the pump discharge then eliminates the artificial drop and the pump operates at point 3. Points 3 and 4A demonstrate the effect a trimmed impeller has on reducing flow.

Figure 20 is an example of a system curve with both fixed head loss and variable head loss. Such a system might be an open piping circuit between a refrigerating plant condenser and its cooling tower. The elevation difference between the water level in the tower pan and the spray distribution pipe creates the fixed head loss. The fixed loss occurs at all flow rates and is, therefore, an independent head as shown.

Most variable-flow hydronic systems have individual two-way control valves on each terminal unit to permit full diversity (random loading from zero to full load). Regardless of the load required in most variable-flow systems, the designer establishes a minimum pressure difference to ensure that any terminal and its control valve receive the design flow at full demand. When graphing a system curve for a nonsymmetrically loaded variable-flow system (Figure 20), the Δh (minimum maintained pressure difference) is treated like a fixed pressure loss (independent head), and becomes the starting datum for the system curve.

The low slope and steep slope curves in Figure 20 represent the boundaries for operation of the system. The net vertical difference between the curves is the difference in friction loss developed by the distribution mains for the two extremes of possible loads. The area in which the system operates depends on the diverse loading or unloading imposed by the terminal units. This area represents the pumping energy that can be conserved with one-speed, two-speed, or variable-speed pumps after a review of the pump power, efficiency, and affinity relationships.

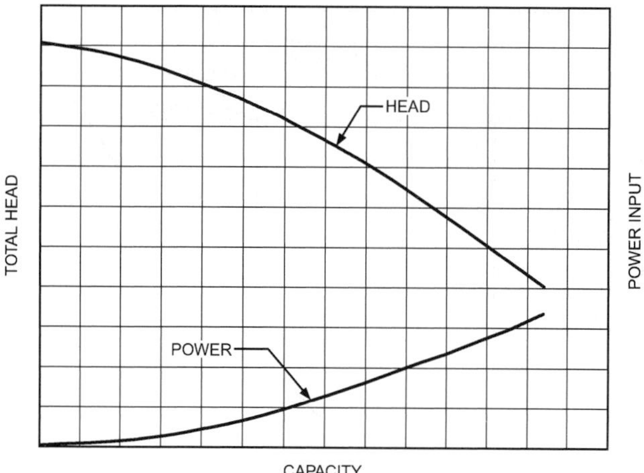

Fig. 21 Typical Pump Water Power Increase with Flow

PUMP POWER

The theoretical power to circulate water in a hydronic system is the **water horsepower** (whp) and is calculated as follows:

$$\text{whp} = \frac{\dot{m}\Delta h}{33{,}000} \tag{3}$$

where

$\dot{m}$ = mass flow of fluid, lb/min
Δh = total head, ft of fluid
33,000 = units conversion, ft·lb/min per hp

At 68°F, water has a density of 62.3 lb/ft^3, and Equation (3) becomes

$$\text{whp} = \frac{Q\Delta h}{3960} \tag{4}$$

where

Q = fluid flow rate, gpm
3960 = units conversion, ft·gpm per hp

Figure 21 shows how water power increases with flow. At other water temperatures or for other fluids, Equation (4) is corrected by multiplying by the specific gravity of the fluid.

The **brake horsepower** (bhp) required to operate the pump is determined by the manufacturer's test of an actual pump running under standard conditions to produce the required flow and head as shown in Figure 11.

PUMP EFFICIENCY

Pump efficiency is determined by comparing the output power to the input power:

$$\text{Efficiency} = \frac{\text{Output}}{\text{Input}} = \frac{\text{whp}}{\text{bhp}} \times 100\% \tag{5}$$

Figure 22 shows a typical efficiency versus flow curve.

The pump manufacturer usually plots the efficiencies for a given volute and impeller size on the pump curve to help the designer select the proper pump (Figure 23). The best efficiency point (BEP) is the optimum efficiency for this pump—operation above and below this point is less efficient. The locus of all the BEPs for each impeller size lies on a system curve that passes through the origin (Figure 24).

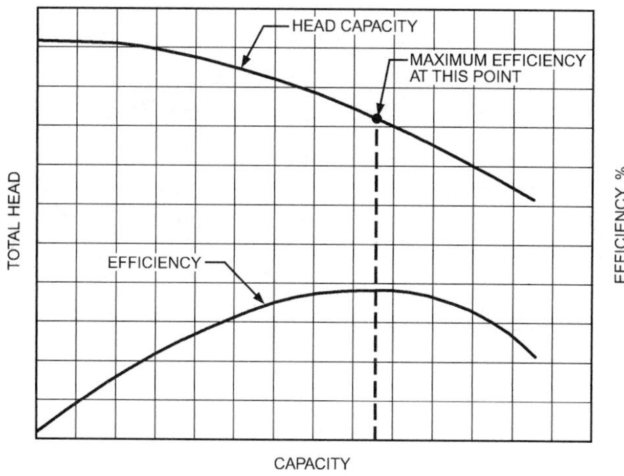

Fig. 22 Pump Efficiency Versus Flow

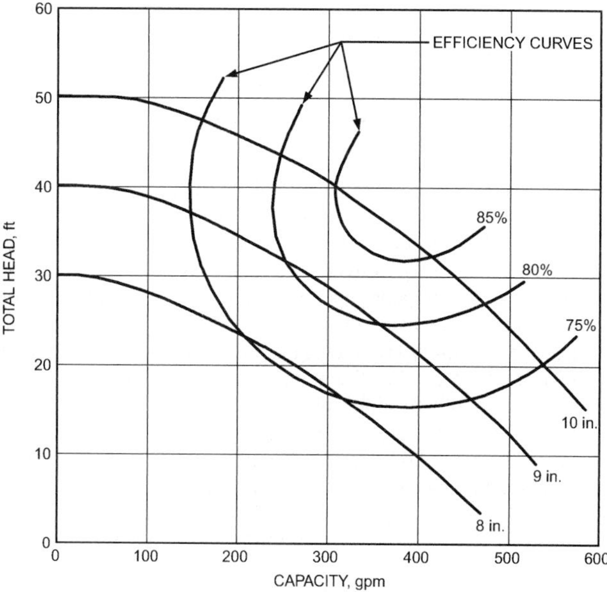

Fig. 23 Pump Efficiency Curves

AFFINITY LAWS

The centrifugal pump, which imparts a velocity to a fluid and converts the velocity energy to pressure energy, can be categorized by a set of relationships called **affinity laws** (Table 1). The laws can be described as similarity processes that follow these rules:

1. Flow (capacity) varies with rotating speed N (i.e., the peripheral velocity of the impeller).
2. Head varies as the square of the rotating speed.
3. Brake horsepower varies as the cube of the rotating speed.

The affinity laws are useful for estimating pump performance at different rotating speeds or impeller diameters D based on a pump with known characteristics. The following two variations can be analyzed by these relationships:

1. By changing speed and maintaining constant impeller diameter, pump efficiency remains the same, but head, capacity, and brake horsepower vary according to the affinity laws.
2. By changing impeller diameter and maintaining constant speed, the pump efficiency for a diffuser pump is not affected if the impeller diameter is changed by less than 5%. However, efficiency changes if the impeller size is reduced enough to affect the clearance between the casing and the periphery of the impeller.

Table 1 Pump Affinity Laws

Function	Speed Change	Impeller Diameter Change
Flow	$Q_2 = Q_1\left(\frac{N_2}{N_1}\right)$	$Q_2 = Q_1\left(\frac{D_2}{D_1}\right)$
Head	$h_2 = h_1\left(\frac{N_2}{N_1}\right)^2$	$h_2 = h_1\left(\frac{D_2}{D_1}\right)^2$
Horsepower	$\text{bhp}_2 = \text{bhp}_1\left(\frac{N_2}{N_1}\right)^3$	$\text{bhp}_2 = \text{bhp}_1\left(\frac{D_2}{D_1}\right)^3$

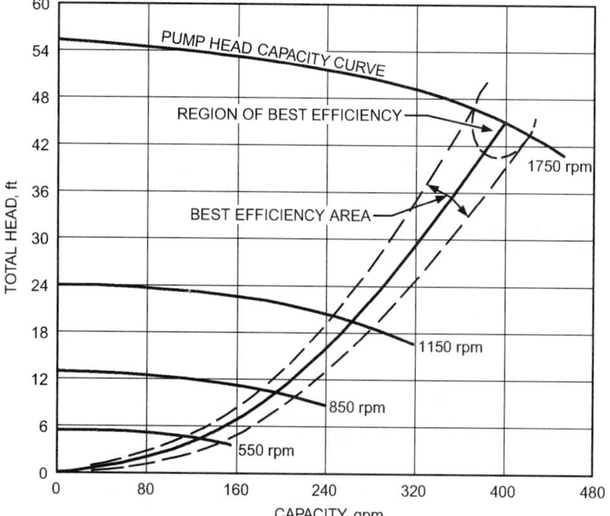

Fig. 24 Pump Best Efficiency Curves

The affinity laws assume that the system curve is known and that head varies as the square of flow. The operating point is the intersection of the total system curve and the pump curve (Figure 17). Because the affinity law is used to calculate a new condition due to a flow or head change (e.g., reduced pump speed or impeller diameter), this new condition also follows the same system curve. Figure 25 shows the relationship of flow, head, and power as expressed by the affinity laws.

The affinity laws can also be used to predict the BEP at other pump speeds. As discussed in the section on Pump Efficiency, the BEP follows a parabolic curve to zero as the pump speed is reduced (Figure 24).

Multiple-speed motors can be used to reduce system overpressure at reduced flow. Standard two-speed motors are available with speeds of 1750/1150 rpm, 1750/850 rpm, 1150/850 rpm, and 3500/1750 rpm. Figure 26 shows the performance of a system with a 1750/1150 multiple-speed pump. In the figure, curve A shows a system's response when the pump runs at 1750 rpm. When the pump runs at 1150 rpm, operation is at point 1 and not at point 2 as the affinity laws predict. If the system were designed to operate as shown in curve B, the pump would operate at shutoff and be damaged if run at 1150 rpm. This example demonstrates that the designer must analyze the system carefully to determine the pump limitation and the effect of lower speed on performance.

Variable-speed drives have a similar effect on pump curves. These drives normally have an infinitely variable speed range, so that the pump, with proper controls, follows the system curve without any overpressure. Figure 27 shows operation of the pump in Figure 19 at 100%, 80%, 64%, 48% and 32% of the speed.

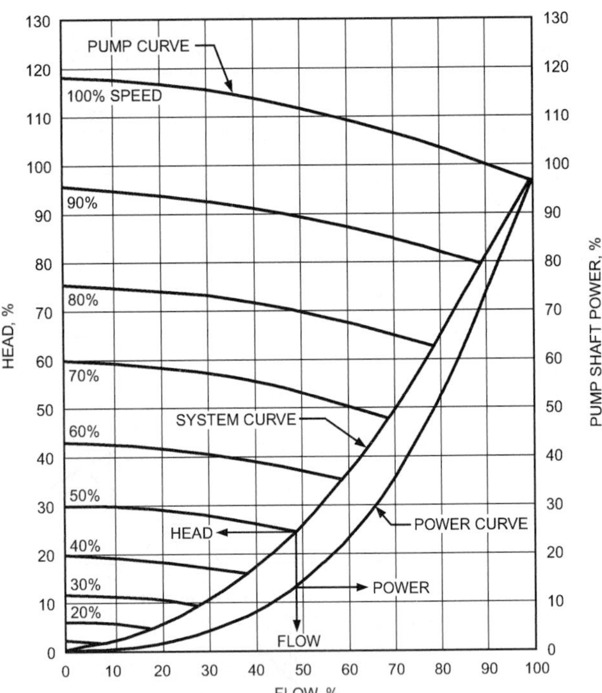

Fig. 25 Pumping Power, Head, and Flow Versus Pump Speed

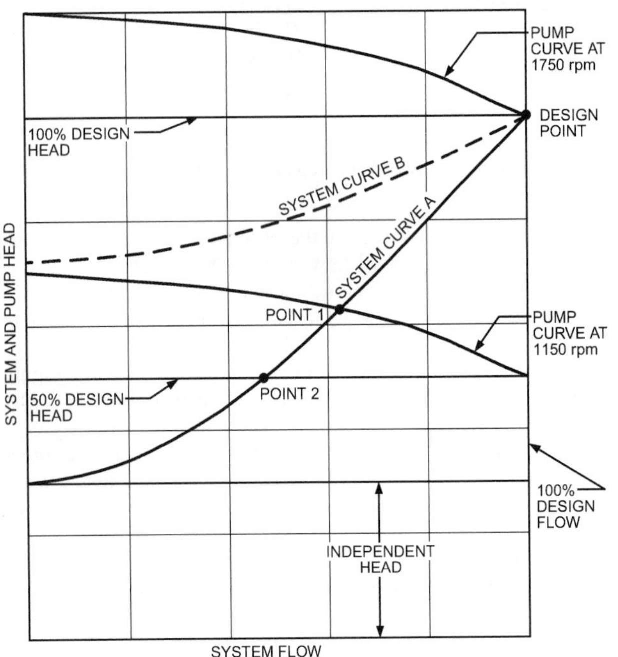

Fig. 26 Example Application of Affinity Law

Although the variable-speed motor in this example can correct overpressure conditions, about 20% of the operating range of the variable-speed motor is not used. The maximum speed required to provide design flow and pressure is 80% of full speed (1400 rpm) and the practical lower limit is 30% (525 rpm) due to the characteristics of the system curve as well as pump and motor limitations. The variable-speed drive and motor should be sized for actual balanced hydronic conditions.

The affinity laws can also be used to predict the effect of trimming the impeller to reduce overpressure. If, for example, the system shown in Figure 27 were allowed to operate at point 2, an excess flow of 15% to 25% would occur, depending on the shapes of the system and pump curves. The pump affinity laws (Table 1) show that pump capacity varies directly with impeller diameter. In Figure 27, the correct size impeller operates at point 3. A diameter ratio of 0.8 would reduce the overcapacity from 125% to 100%.

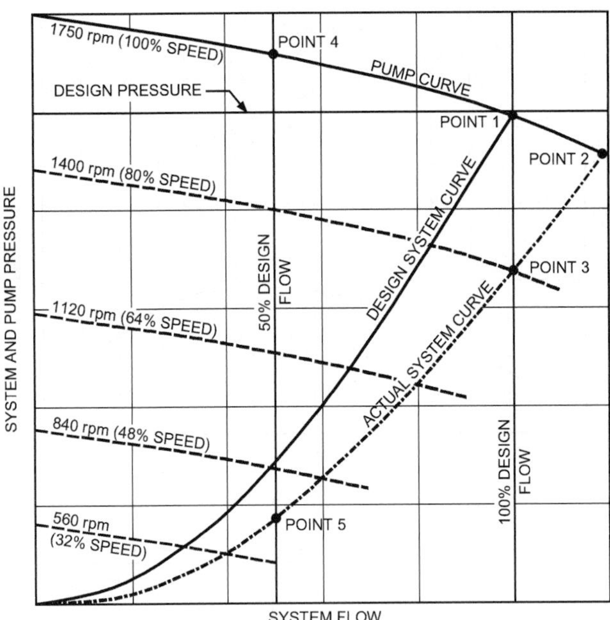

Fig. 27 Variable-Speed Pump Operating Points

The second affinity law shows that head varies as the square of the impeller diameter. For the pump in Figure 27 and an impeller diameter ratio of 0.8, the delivered head of the pump is $(0.8)^2 = 0.64$ or 64% of the design pump head.

The third affinity law states that pump power varies as the cube of the impeller diameter. For the pump in Figure 27 and an impeller diameter ratio of 0.8, the power necessary to provide the design flow is $(0.8)^3 = 0.512$ or 51.2% of the original pump's power.

RADIAL THRUST

In a single-volute centrifugal pump, uniform or near-uniform pressures act on the impeller at design capacity, which coincides with the BEP. However, at other capacities, the pressures around the impeller are not uniform and there is a resultant radial reaction.

Figure 28 shows the typical change in radial thrust with changes in the pumping rate. Specifically, radial thrust decreases from shutoff to the design capacity (if chosen at the BEP) and then increases as flow increases. The reaction at overcapacity is roughly opposite that at partial capacity and is greatest at shutoff. The radial forces at extremely low flow can cause severe impeller shaft deflection and, ultimately, shaft breakage. This danger is even greater with high-pressure pumps.

NET POSITIVE SUCTION CHARACTERISTICS

Particular attention must be given to the pressure and temperature of the water as it enters the pump, especially in condenser towers, steam condensate returns, and steam boiler feeds. If the absolute pressure at the suction nozzle approaches the vapor pressure of the liquid, vapor pockets form in the impeller passages. The collapse of the vapor pockets (**cavitation**) is noisy and can be destructive to the pump impeller.

The amount of pressure in excess of the vapor pressure required to prevent vapor pockets from forming is known as the net

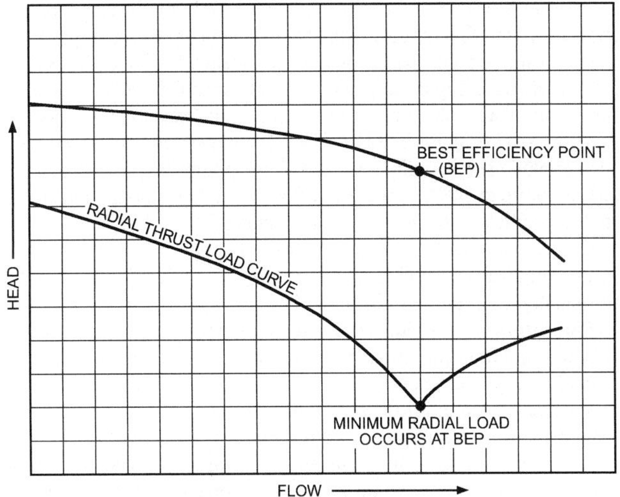

Fig. 28 Radial Thrust Versus Pumping Rate

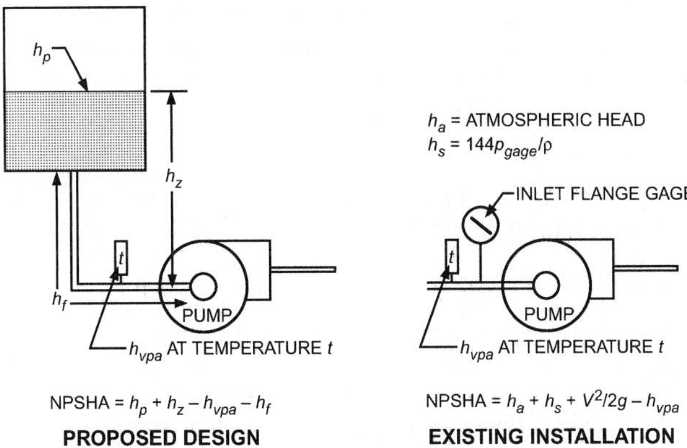

Fig. 29 Net Positive Suction Head Available

positive suction head required (NPSHR). NPSHR is a characteristic of a given pump and varies with pump speed and flow. It is determined by the manufacturer and is included on the pump performance curve.

NPSHR is particularly important when a pump is operating with hot liquids or is applied to a circuit having a suction lift. The vapor pressure increases with water temperature and reduces the net positive suction head available (NPSHA). Each pump has its NPSHR, and the installation has its NPSHA, which is the total useful energy above the vapor pressure at the pump inlet.

The following equation may be used to determine the NPSHA in a proposed design (see Figure 29):

$$\text{NPSHA} = h_p + h_z - h_{vpa} - h_f \tag{6}$$

where

h_p = absolute pressure on surface of liquid that enters pump, ft of head
h_z = static elevation of liquid above center line of pump (h_z is negative if liquid level is below pump center line), ft
h_{vpa} = absolute vapor pressure at pumping temperature, ft
h_f = friction and head losses in suction piping, ft

To determine the NPSHA in an existing installation, the following equation may be used (see Figure 29):

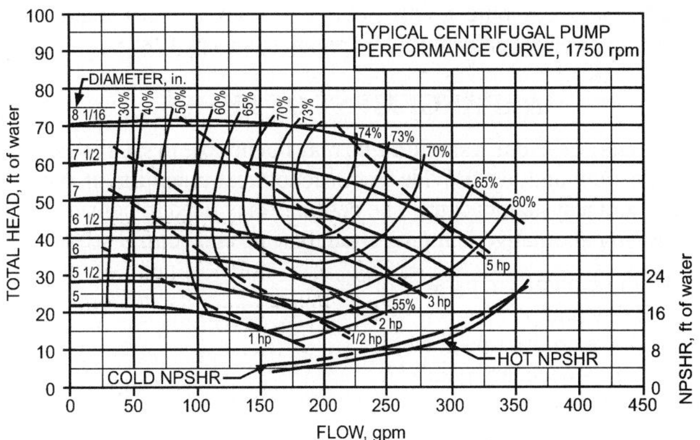

Fig. 30 Pump Performance and NPSHR Curves

$$\text{NPSHA} = h_a + h_s + \frac{V^2}{2g} - h_{vpa} \tag{7}$$

where

h_a = atmospheric head for elevation of installation, ft
h_s = head at inlet flange corrected to center line of pump (h_s is negative if below atmospheric pressure), ft
$V^2/2g$ = velocity head at point of measurement of h_s, ft

If the NPSHA is less than the pump's NPSHR, cavitation, noise, inadequate pumping, and mechanical problems will result. **For trouble-free design, the NPSHA must always be greater than the pump's NPSHR.** In closed hot and chilled water systems where sufficient system fill pressure is exerted on the pump suction, NPSHR is normally not a factor. Figure 30 shows pump curves and NPSHR curves. Cooling towers and other open systems require calculations of NPSHA.

SELECTION OF PUMPS

A substantial amount of data is required to ensure that an adequate, efficient, and reliable pump is selected for a particular system. The designer should review the following criteria:

- Design flow
- Pressure drop required for the most resistant loop
- Minimum system flow
- System pressure at maximum and minimum flows
- Type of control valve—two-way or three-way
- Continuous or variable flow
- Pump environment
- Number of pumps and standby
- Electric voltage and current
- Electric service and starting limitations
- Motor quality versus service life
- Water treatment, water conditions, and material selection

When a centrifugal pump is applied to a piping system, the operating point satisfies both the pump and system curves (Figure 17). As the load changes, control valves change the system curve and the operating point moves to a new point on the pump curve. Figure 31 shows the optimum regions to use when selecting a centrifugal pump. The areas bounded by lines AB and AC represent operating points that lie in the preferred pump selection range. But, because pumps are only manufactured in certain sizes, selection limits of 66% to 115% of flow at the BEP are suggested. The satisfactory range is that portion of a pump's performance curve where the combined effect of circulatory flow, turbulence, and friction losses are minimized. Where possible, pumps should be chosen to operate to the left of the BEP because the pressure in the actual system may be less than design due to overstated data for pipe friction and for other equipment. Otherwise, the pump operates at a higher flow and possibly in the turbulent region (Stethem 1988).

ARRANGEMENT OF PUMPS

In a large system, a single pump may not be able to satisfy the full design flow and yet provide both economical operation at partial loads and a system backup. The designer may need to consider the following alternative pumping arrangements and control scenarios:

- Multiple pumps in parallel or series
- Standby pump
- Pumps with two-speed motors
- Primary-secondary pumping
- Variable-speed pumping
- Distributed pumping

Parallel Pumping

When pumps are applied in parallel, each pump operates at the same head and provides its share of the system flow at that head (Figure 32). Generally, pumps of equal size are recommended, and the parallel pump curve is established by doubling the flow of the single pump curve.

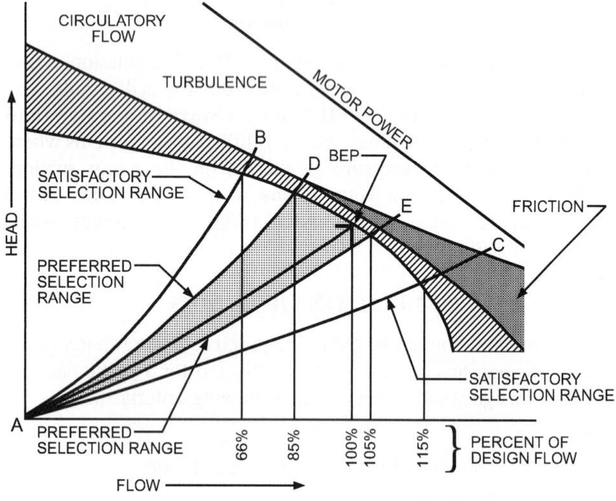

Fig. 31 Pump Selection Regions

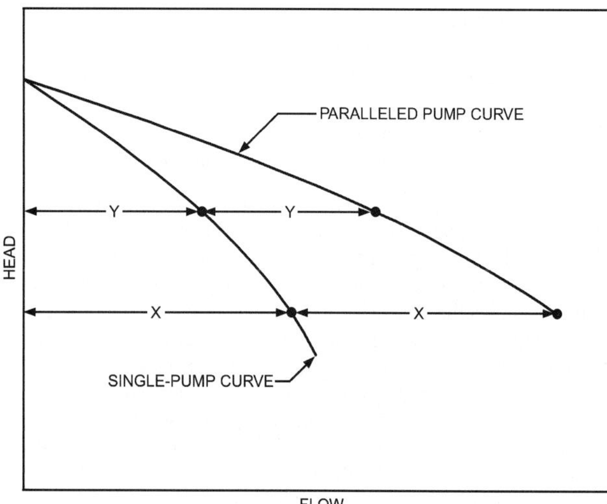

Fig. 32 Pump Curve Construction for Parallel Operation

Plotting a system curve across the parallel pump curve shows the operating points for both single and parallel pump operation (Figure 33). Note that single pump operation does not yield 50% flow. The system curve crosses the single pump curve considerably to the right of its operating point when both pumps are running. This leads to two important concerns: (1) the motor must be selected to prevent overloading during operation of a single pump and (2) a single pump can provide standby service for up to 80% of the design flow, the actual amount depending on the specific pump curve and system curve.

Construction of the composite curve for two dissimilar parallel pumps requires special care; for example, note the shoulder in the composite pump curve in Figure 34.

Operation. The piping of parallel pumps (Figure 35) should permit running either pump. A check valve is required in each pump's discharge to prevent backflow when one pump is shut down. Hand valves and a strainer allow one pump to be serviced while the other is operating. A strainer protects a pump by preventing foreign material from entering the pump. Gages or a common gage with a trumpet valve, which includes several valves as one unit, or pressure taps permits checking pump operation.

Flow can be determined (1) by measuring the pressure increase across the pump and using a factory pump curve to convert the pressure to flow, or (2) by use of a flow-measuring station or multipurpose valve. Parallel pumps are often used for hydronic heating and cooling. In this application, both pumps operate during the cooling season to provide maximum flow and pressure, but only one pump operates during the heating season.

Series Pumping

When pumps are applied in series, each pump operates at the same flow rate and provides its share of the total pressure at that

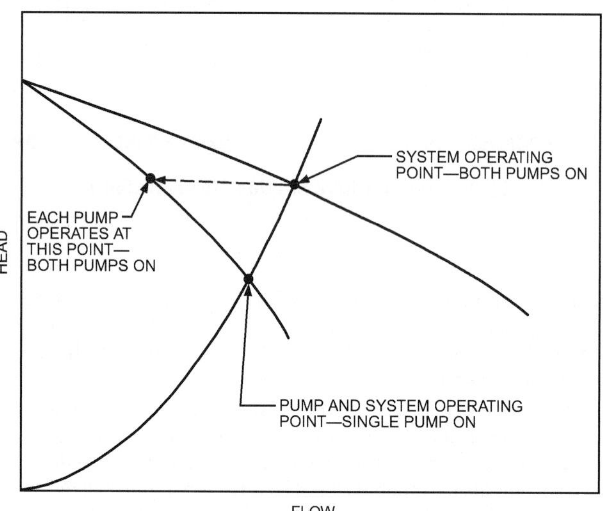

Fig. 33 Operating Conditions for Parallel Operation

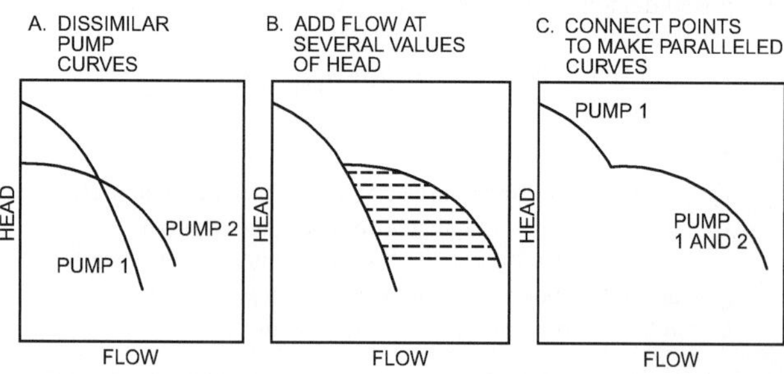

Fig. 34 Construction of Curve for Dissimilar Parallel Pumps

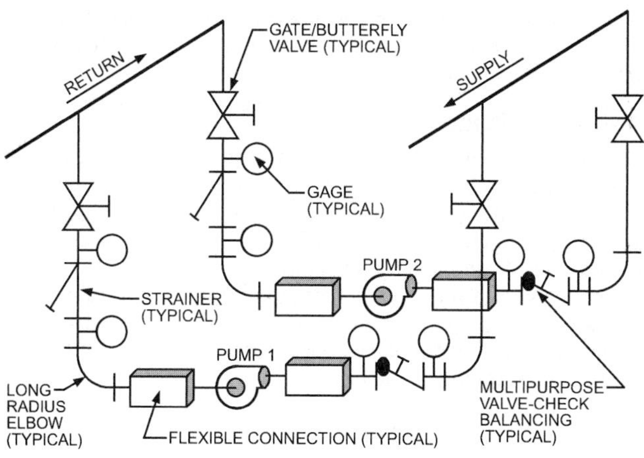

Fig. 35 Typical Piping for Parallel Pumps

flow (Figure 36). A system curve plot shows the operating points for both single and series pump operation (Figure 37). Note that the single pump can provide up to 80% flow for standby and at a lower power requirement.

As with parallel pumps, piping for series pumps should permit running either pump (Figure 38). A bypass with a hand valve permits servicing one pump while the other is in operation. Operation and flow can be checked the same way as for parallel pumps. A strainer prevents foreign material from entering the pumps.

Note that both parallel and series pump applications require that the pump operating points be used to accurately determine the actual pumping points. The manufacturer's pump test curve should be consulted. Adding too great a safety factor for pressure, using improper pressure drop charts, or incorrectly calculating pressure drops may lead to a poor selection. In designing systems with multiple pumps, operation in either parallel or series must be fully understood and considered by both designer and operator.

Standby Pump

A backup or standby pump of equal capacity and pressure installed in parallel to the main pump is recommended to operate during an emergency or to ensure continuous operation when a pump is taken out of operation for routine service. A standby pump installed in parallel with the main pump is shown in Figure 35.

Pumps with Two-Speed Motors

A pump with a two-speed motor provides a simple means of reducing capacity. As discussed in the section on Affinity Laws, pump capacity varies directly with impeller speed. At 1150 rpm, the capacity of a pump with a 1750/1150 rpm motor is 1150/1750 = 0.657 or 66% of the capacity at 1750 rpm.

Figure 39 shows an example (Stethem 1988) with two parallel two-speed pumps providing flows of 2130 gpm at 75 ft of head, 1670 gpm at 50.5 ft, 1250 gpm at 33 ft, and 985 gpm at 26.2 ft of head. Points A, B, C, and D will move left along the pump curve as loading changes.

Primary-Secondary Pumping

In a primary-secondary or compound pumping arrangement, a secondary pump is selected to provide the design flow in the load coil from the common pipe between the supply and return distribution mains (Figure 40) (Coad 1985). The pressure drop in the common pipe should not exceed 1.5 ft (Carlson 1972).

In circuit A of the figure, a two-way valve permits a variable flow in the supply mains by reducing the source flow; a secondary pump provides a constant flow in the load coil. The source pump at the chiller or boiler is selected to circulate the source and mains,

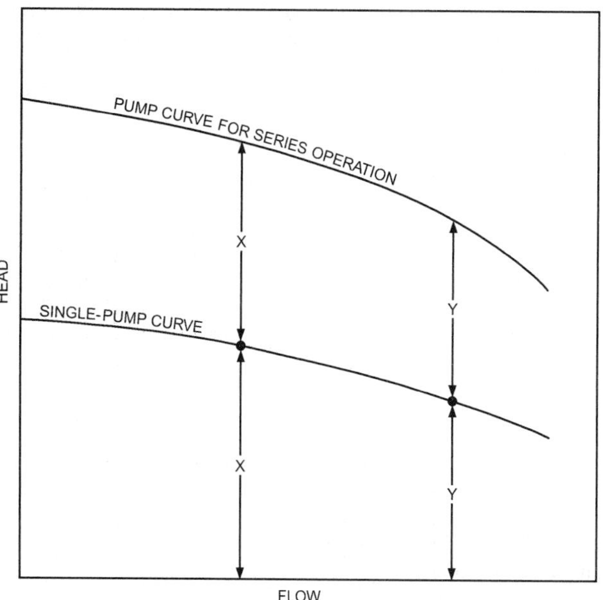

Fig. 36 Pump Curve Construction for Series Operation

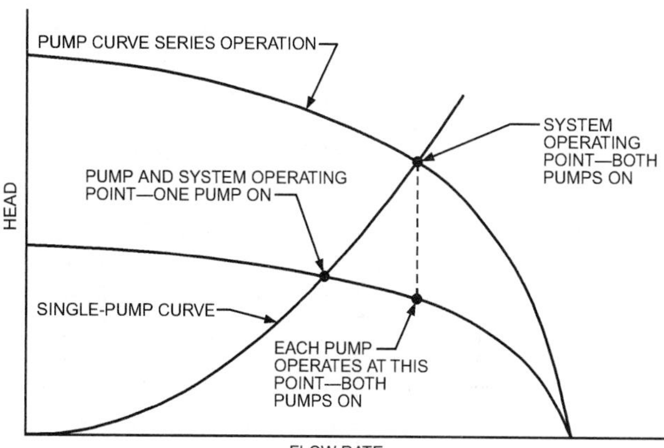

Fig. 37 Operating Conditions for Series Operation

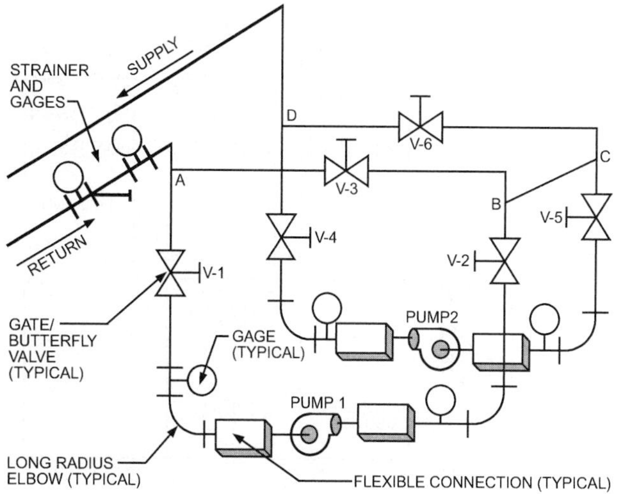

Fig. 38 Typical Piping for Series Pumps

and the secondary pump is sized for the load coil. Three-way valves in circuits B and C provide a constant source flow regardless of the load.

Variable-Speed Pumping

In a variable-speed pumping arrangement, constant flow pump(s) recirculate the chiller or boiler source in a primary source loop, and a variable-speed distribution pump located at the source plant draws flow from the source loop and distributes to the load terminals as shown in Figure 41. The speed of the distribution pump is determined by a controller measuring differential pressure across the supply-return mains or across selected critical zones. Two-way control valves are installed in the load terminal return branch to vary the flow required in the load.

Distributed Pumping

In a variable-speed distributed pumping arrangement, constant flow pump(s) recirculate the chiller or boiler source in a primary source loop, and a variable-speed zone or building pump draws flow from the source loop and distributes to the zone load terminals as shown in Figure 42. The speed of the zone or building distribution pump is determined by a controller measuring zone differential pressure across supply-return mains or across selected critical zones. Two-way control valves in the load terminal return branch vary the flow required in the load.

MOTIVE POWER

Figure 43 demonstrates the improvement in efficiency of four-pole (1800 rpm) 25 to 125 hp motors from old, standard efficiency models to current standards and available premium efficiency models.

Electric motors drive most centrifugal pumps for hydronic systems. Internal combustion engines or steam turbines power some pumps, especially in central power plants for large installations. Electric motors for centrifugal pumps can be any of the horizontal or vertical electric motors described in Chapter 44. The sizing of electric

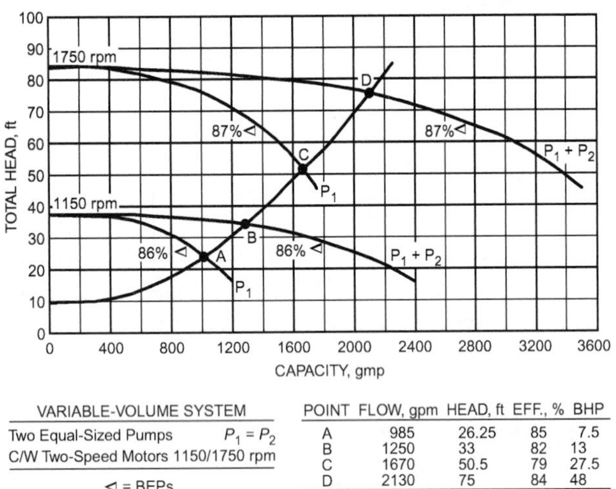

POINT	FLOW, gpm	HEAD, ft	EFF., %	BHP
A	985	26.25	85	7.5
B	1250	33	82	13
C	1670	50.5	79	27.5
D	2130	75	84	48

Fig. 39 Example of Two Parallel Pumps with Two-Speed Motors

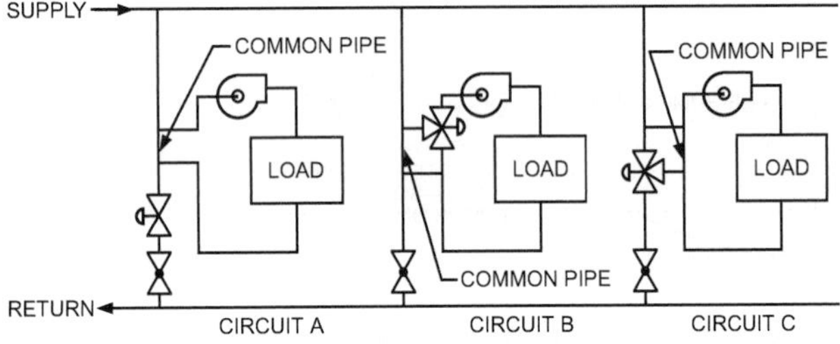

Fig. 40 Primary-Secondary Pumping

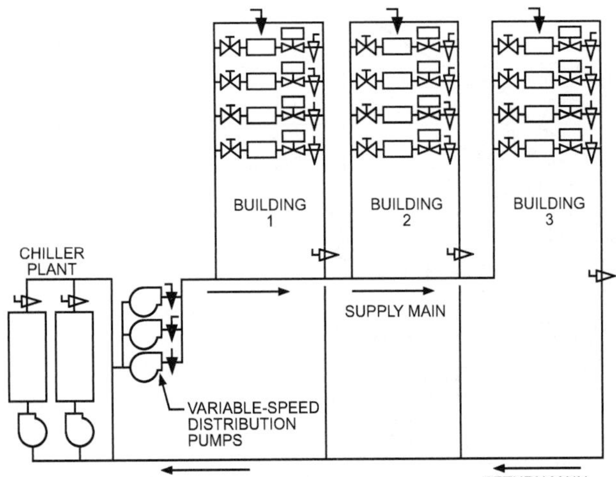

Fig. 41 Variable-Speed Source-Distributed Pumping

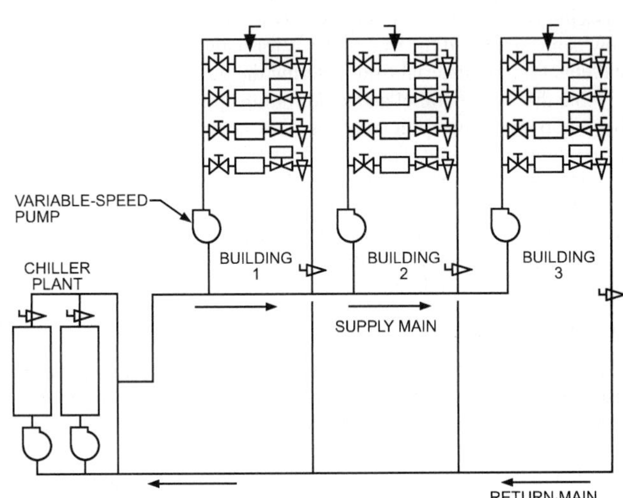

Fig. 42 Variable-Speed Distributed Pumping

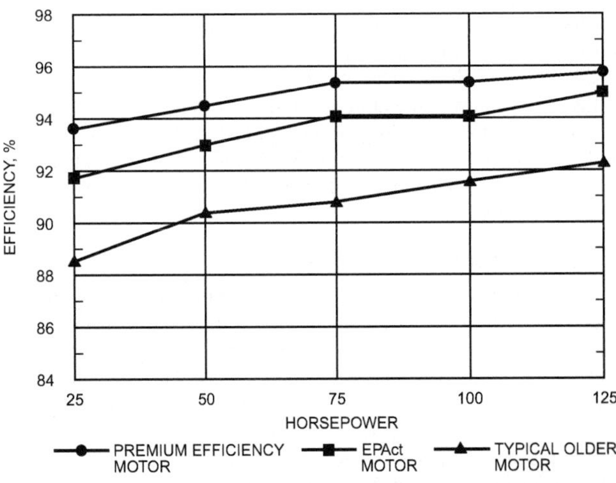

Fig. 43 Efficiency Comparison of Four-Pole Motors

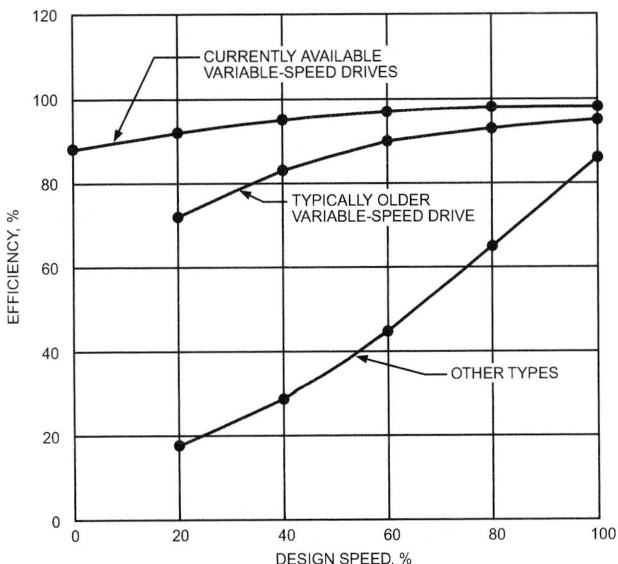

Fig. 44 Typical Efficiency Range of Variable-Speed Drives

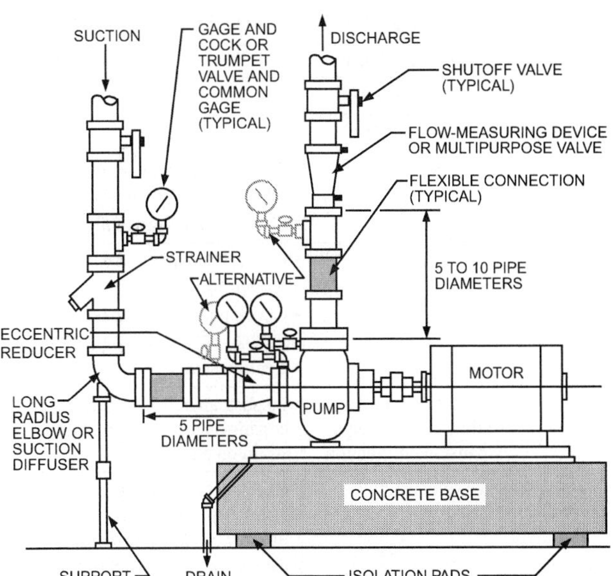

Fig. 45 Base Plate-Mounted Centrifugal Pump Installation

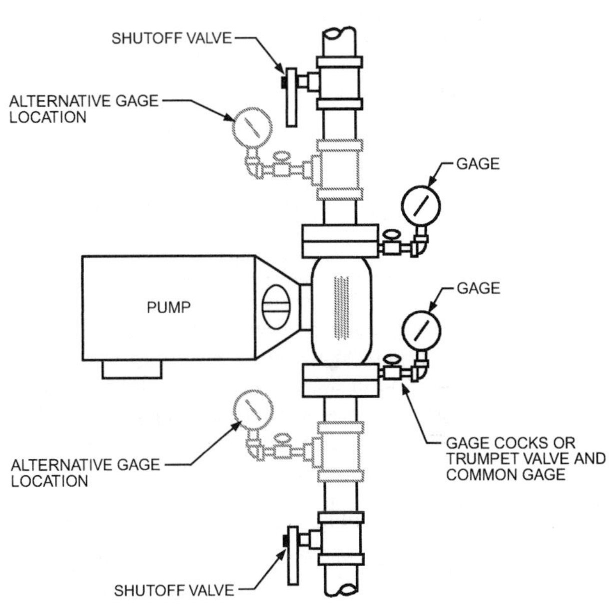

Fig. 46 In-Line Pump Installation

motors is critical because of the cost of electric power and the desire for improved efficiency. Non-overloading motors should be used; that is, the motor nameplate rating must exceed the pump brake horsepower (kW) at any point on the pump curve.

Many pumps for hydronic systems are close-coupled, with the pump impeller mounted on the motor shaft extension. Other pumps are flexible-coupled to the electric motor through a pump mounting bracket or frame. A pump on a hydronic system with variable flow has a broad range of power requirements, which results in reduced motor loading at low flow.

Many variable-speed drives (VSDs) are available for operating centrifugal pumps. Primarily, these include variable-frequency drives, and, occasionally, direct-current, wound rotor, and eddy current drives. Each drive has specific design features that should be evaluated for use with hydronic pumping systems. The efficiency range from minimum to maximum speed (as shown in Figure 44) should be investigated.

ENERGY CONSERVATION IN PUMPING

Pumps for heating and air conditioning consume appreciable amounts of energy. Economical use of energy depends on the efficiency of pumping equipment and drivers, as well as the use of the pumping energy required. Equipment efficiency (sometimes called the wire to water efficiency) shows how much energy applied to the pumping system results in useful energy distributing the water. For an electric-driven, constant speed pump, the equipment efficiency is

$$\eta_e = \eta_p \eta_m \tag{8}$$

where

η_e = equipment efficiency, 0 to 1
η_p = pump efficiency, 0 to 1
η_m = motor efficiency, 0 to 1

For a variable-speed pump, the variable-speed drive efficiency η_v (0 to 1) must be included in the equipment efficiency equation:

$$\eta_e = \eta_p \eta_m \eta_v \tag{9}$$

INSTALLATION, OPERATION, AND COMMISSIONING

1. Pumps may be base plate-mounted (Figure 45), either singly or in packaged sets, or installed in-line directly in the piping system (Figure 46). Packaged sets include multiple pumps, accessories, and electrical controls shipped to the job site on one frame. Packaged pump sets may reduce the requirements for multiple piping and field electrical connections and can be factory tested to ensure specified performance.
2. A concrete pad provides a secure mounting surface for anchoring the pump base plate and raises the pump off the floor to permit housekeeping. The minimum weight of concrete that should be used is 2.5 times the weight of the pump assembly. The pad should be at least 4 in. thick and 6 in. wider than the pump base plate on each side.
3. In applications where the pump bolts rigidly to the pad base, level the pad base, anchor it, and fill the space between pump base and the concrete with a non-shrink grout. Grout prevents the base from shifting and fills in irregularities. Pumps mounted on vibration isolation bases require special installation (see the section on Vibration Isolation and Control in Chapter 47 of the 2007 *ASHRAE Handbook—HVAC Applications*).

Table 2 Pumping System Noise Analysis Guide

Complaint	Possible Cause	Recommended Action
Pump or system noise	Shaft misalignment	• Check and realign
	Worn coupling	• Replace and realign
	Worn pump/motor bearings	• Replace, check manufacturer's lubrication recommendations • Check and realign shafts
	Improper foundation or installations	• Check foundation bolting or proper grouting • Check possible shifting caused by piping expansion/contraction. • Realign shafts
	Pipe vibration and/or strain caused by pipe expansion/contraction	• Inspect, alter, or add hangers and expansion provision to eliminate strain on pump(s)
	Water velocity	• Check actual pump performance against specified, and reduce impeller diameter as required • Check for excessive throttling by balance valves or control valves
	Pump operating close to or beyond end point of performance curve	• Check actual pump performance against specified, and reduce impeller diameter as required
	Entrained air or low suction pressure	• Check expansion tank connection to system relative to pump suction • If pumping from cooling tower sump or reservoir, check line size • Check actual ability of pump against installation requirements • Check for vortex entraining air into suction line

Table 3 Pumping System Flow Analysis Guide

Complaint	Possible Cause	Recommended Action
Inadequate or no circulation	Pump running backward (3-phase)	• Reverse any two motor leads
	Broken pump coupling	• Replace and realign
	Improper motor speed	• Check motor nameplate wiring and voltage
	Pump (or impeller diameter) too small	• Check pump selection (impeller diameter) against specified requirements
	Clogged strainer(s)	• Inspect and clean screen
	Clogged impeller	• Inspect and clean
	System not completely filled	• Check setting of PRV fill valve • Vent terminal units and piping high points
	Balance valves or isolating valves improperly set	• Check setting and adjust as required
	Air-bound system	• Vent piping and terminal units • Check location of expansion tank connection line relative to pump suction • Review provisions to eliminate air
	Air entrainment	• Check pump suction inlet conditions to determine if air is being entrained from suction tanks or sumps
	Insufficient NPSHR	• Check NPSHR of pump • Inspect strainers and check pipe sizing and water temperature

4. Support in-line pumps independently from the piping so that pump flanges are not overstressed.
5. Once the pump has been mounted to the base, check the alignment of the motor to the pump. Align the pump shaft couplings properly and shim the motor base as required. Incorrect alignment may cause rapid coupling and bearing failure.
6. Pump suction piping should be direct and as smooth as possible. Install a strainer (coarse mesh) in the suction to remove foreign particles that can damage the pump. Use a straight section of piping at least 5 to 10 diameters long at the pump inlet and long radius elbows to ensure uniform flow distribution. Suction diffusers may be installed in lieu of the straight pipe requirement where spacing is a constraint. Eccentric reducers at the pump flange reduce the potential of air pockets forming in the suction line.
7. If a flow-measuring station (venturi, orifice plate, or balancing valve) is located in the pump discharge, allow 10 diameters of straight pipe between the pump discharge and the flow station for measurement accuracy.
8. Pipe flanges should match the size of pump flanges. Mate flat-face pump flanges with flat-face piping flanges and full-face gaskets. Install tapered reducers and increasers on suction and discharge lines to match the pipe size and pump flanges.
9. If fine mesh screen is used in the strainer at initial start-up to remove residual debris, replace it with normal size screen after commissioning to protect the pump and minimize the suction pressure drop.
10. Install shutoff valves in the suction and discharge piping near the pump to permit removing and servicing the pump and strainer without draining the system. Install a check valve in the pump discharge to prevent reverse flow in a non-running pump when multiple pumps are installed.
11. Install vibration isolators in the pump suction and discharge lines to reduce the transmission of vibration noise to building spaces (Figure 45). Properly located pipe hangers and supports can reduce the transmission of piping strains to the pump.
12. Various accessories need to be studied as alternates to conventional fittings. A suction diffuser in the pump inlet is an alternate to an eccentric reducer and it contains a strainer. Separate strainers can be specified with screen size. A multipurpose valve in the pump discharge is an alternate way to combine the functions of shutoff, check, and balancing valves.
13. Each pump installation should include pressure gages and a gage cock to verify system pressures and pressure drop. As a minimum, pressure taps with an isolation valve and common gage should be available at the suction and discharge of the pump. An additional pressure tap upstream of the strainer permits checking for pressure drop.

TROUBLESHOOTING

Table 2 lists possible causes and recommended solutions for pump or system noise. Table 3 lists possible causes and recommended solutions for inadequate circulation.

REFERENCES

Carlson, G.F. 1972. Central plant chilled water systems—Pumping & flow balance, Part I. *ASHRAE Journal* (February):27-34.

Coad, W.J. 1985. Variable flow in hydronic systems for improved stability, simplicity, and energy economics. *ASHRAE Transactions* 91(1B):224-237.

Hydraulic Institute. 1994. Centrifugal pumps. *Standards* HI 1.1 to 1.5. Parsippany, NJ.

Stethem, W.C. 1988. Application of constant speed pumps to variable volume systems. *ASHRAE Transactions* 94(2):1458-66.

BIBLIOGRAPHY

ASHRAE. 1985. Hydronic systems: Variable-speed pumping and chiller optimization. *ASHRAE Technical Data Bulletin* 1(7).

ASHRAE. 1991. Variable-flow pumping systems. *ASHRAE Technical Data Bulletin* 7(2).

ASHRAE. 1998. *Fundamentals of water system design*. Self-Directed Learning Course.

Beaty, F.E., Jr. 1987. *Sourcebook of HVAC details*. McGraw-Hill, New York.

Carrier. 1965. *Carrier handbook of air conditioning system design*. McGraw-Hill, New York.

Clifford, G. 1990. *Modern heating, ventilating and air conditioning*. Prentice Hall, New York.

Crane Co. 1988. Flow of fluids through valves, fittings, and pipe. *Technical Paper* 410. Joliet, IL.

Dufour, J.W. and W.E. Nelson. 1993. *Centrifugal pump sourcebook*. McGraw-Hill, New York.

Garay, P.N. 1990. *Pump application desk book*. Fairmont Press, Lilburn, GA.

Haines, R.W. and C.L. Wilson. 1994. *HVAC systems design handbook*, 2nd ed. McGraw-Hill, New York.

Hegberg, R.A. 1991. Converting constant-speed hydronic pumping systems to variable-speed pumping. *ASHRAE Transactions* 97(1):739-745.

Karassik, I.J. 1989. Centrifugal pump clinic. Marcel Dekker Inc., New York.

Karassik, I.J., W.C. Krutzsch, W.H. Fraser, and J.P. Messina, eds. 1986. *Pump handbook*. McGraw-Hill, New York.

Levenhagen, J. and D. Spethmann. 1993. *HVAC controls and systems*. McGraw-Hill, New York.

Lobanoff, V.S. and R.R. Ross. 1986. *Centrifugal pumps design and application*. Gulf Publishing, Houston.

Matley, J., ed. 1989. *Progress in pumps*. Chemical Engineering, McGraw-Hill, New York.

McQuiston, F.C. and J.D. Parker. 1988. *Heating, ventilating and air conditioning*. John Wiley & Sons, New York.

Monger, S. 1990. *Testing and balancing HVAC air and water systems*. Fairmont Press, Lilburn, GA.

Rishel, J.B. 1994. Distributed pumping for pumping for chilled- and hot-water systems. *ASHRAE Transactions* 100(1):1521-1527.

Trane Co. 1965. *Trane manual of air conditioning*. La Crosse, WI.

CHAPTER 44

MOTORS, MOTOR CONTROLS, AND VARIABLE-SPEED DRIVES

MOTORS

MANY TYPES of alternating-current (ac) motors are available; direct-current (dc) motors are also used, but to a more limited degree. NEMA *Standard* MG 1 provides technical information on all types of ac and dc motors.

ALTERNATING-CURRENT POWER SUPPLY

Important characteristics of an ac power supply include (1) voltage, (2) number of phases, (3) frequency, (4) voltage regulation, and (5) continuity of power.

According to ARI *Standard* 110, the **nominal system voltage** is the value assigned to the circuit or system to designate its voltage class. The voltage at the connection between supplier and user is the **service voltage. Utilization voltage** is the voltage at the line terminals of the equipment. Utilization voltages are about 5% lower than their corresponding nominal voltages, to allow for distribution system impedance.

Single- and three-phase motor and control voltage ratings shown in Table 1 are adapted to the nominal voltages indicated. Motors with these ratings are considered suitable for ordinary use on their corresponding systems; for example, a 230 V motor should generally be used on a nominal 240 V system. A 230 V motor should not be installed on a nominal 208 V system because the utilization voltage is below the tolerance on the voltage rating for which the motor is designed. Such operation generally results in overheating and a serious reduction in torque. Single- and three-phase 200 V motors are designed for nominal 208 V systems. Three-phase models up to at least 100 hp are available in NEMA Premium® efficiencies.

Motors are usually guaranteed to operate satisfactorily and to deliver their full power at the rated frequency and at a voltage 10% above or below their rating, or at the rated voltage and plus or minus 5% frequency variation. Some U.S. single-phase HVAC components that are dual-voltage rated (e.g., 208/230-1-60) may carry a minus 5% voltage allowance (at rated frequency) from the lower voltage rating of 208 volts. Table 2 shows the effect of voltage and frequency variation on induction motor characteristics.

Phase voltages of three-phase motors should be balanced. If not, a small voltage imbalance can cause a large current imbalance. This leads to high motor operating temperatures that can result in nuisance overload trips or motor failures and burnouts. Motors should not be operated where the voltage imbalance is greater than 1%. If an imbalance does exist, contact the motor manufacturer for recommendations. Voltage imbalance is defined in NEMA *Standard* MG 1, Paragraph 14.34, as

$$\text{\% Voltage imbalance} = 100 \times \frac{\text{Maximum voltage deviation from average voltage}}{\text{Average voltage}}$$

Table 1 Motor and Motor Control Equipment Voltages (Alternating Current)

System Nominal Voltage	U.S. Domestic Equipment Nameplate Voltage Ratings (60 Hz): Integral Horsepower		Fractional Horsepower	
	Three-Phase	Single-Phase	Three-Phase	Single-Phase
120	—	115	—	115
208	208/230 or 200/230	208/230 or 200/230	208/230 or 200/230	208/230 or 200/230
240	208/230 or 200/230	208/230 or 200/230	208/230 or 200/230	208/230 or 200/230
277	—	265	—	265
480	460	—	460	—
600*	575	—	575	—
2,400	2,300	—	—	—
4,160	4,000	—	—	—
4,800	4,600	—	—	—
6,900	6,600	—	—	—
13,800	13,200	—	—	—

*Some control and protective equipment has maximum voltage limit of 600 V. Consult manufacturer, power supplier, or both to ensure proper application.

System Nominal Voltage	International Equipment Nameplate Voltage Ratings: 50 Hz		60 Hz	
	Three-Phase	Single-Phase	Three-Phase	Single-Phase
127	—	127	—	127
200	220/200	200	230/208 or 230/200	—
220	220/240	220/240 or 230/208	230/208 or 230/200	230/208
230	230/208	220/240 or 230/208	230/208 or 230/200	230/208
240	230/208	220/240	230/208	230/208
250	—	250	—	—
380	380/415	—	460/380	—
400	380/415	—	—	—
415	380/415	—	—	—
440	440	—	460	—
480	500	—	—	—

Note: Primary operating voltage for a dual-voltage rating is usually listed first (e.g., 220 is primary for a 220/240 volt rating).

The preparation of this chapter is assigned to TC 1.11, Electric Motors and Motor Control.

Table 2 Effect of Voltage and Frequency Variation on Induction Motor Characteristics

Voltage and Frequency Variation		Starting and Maximum Running Torque	Synchronous Speed	% Slip	Full-Load Speed	Efficiency: Full Load	Efficiency: 0.75 Load	Efficiency: 0.5 Load
Voltage variation	120% Voltage	Increase 44%	No change	Decrease 30%	Increase 1.5%	Small increase	Decrease 0.5 to 2%	Decrease 7 to 20%
	110% Voltage	Increase 21%	No change	Decrease 17%	Increase 1%	Increase 0.5 to 1%	Practically no change	Decrease 1 to 2%
	Function of voltage	Voltage^2	Constant	1/Voltage^2	Synchronous speed slip	—	—	—
	90% Voltage	Decrease 19%	No change	Increase 23%	Decrease 1.5%	Decrease 2%	Practically no change	Increase 1 to 2%
Frequency variation	105% Frequency	Decrease 10%	Increase 5%	Practically no change	Increase 5%	Slight increase	Slight increase	Slight increase
	Function of frequency	1/Frequency^2	Frequency	—	Synchronous speed slip	—	—	—
	95% Frequency	Increase 11%	Decrease 5%	Practically no change	Decrease 5%	Slight decrease	Slight decrease	Slight decrease

Voltage and Frequency Variation		Power Factor: Full Load	Power Factor: 0.75 Load	Power Factor: 0.5 Load	Full-Load Current	Starting Current	Temperature Rise, Full Load	Maximum Overload Capacity	Magnetic Noises, No Load in Particular
Voltage variation	120% Voltage	Decrease 5 to 15%	Decrease 10 to 30%	Decrease 15 to 40%	Decrease 11%	Increase 25%	Decrease 5 to 6 K	Increase 44%	Noticeable increase
	110% Voltage	Decrease 3%	Decrease 4%	Decrease 5 to 6%	Decrease 7%	Increase 10 to 12%	Decrease 3 to 4 K	Increase 21%	Increase slightly
	Function of voltage	—	—	—	—	Voltage	—	Voltage^2	—
	90% Voltage	Increase 3%	Increase 2 to 3%	Increase 4 to 5%	Increase 11%	Decrease 10 to 12%	Increase 6 to 7 K	Decrease 19%	Decrease slightly
Frequency variation	105% Frequency	Slight increase	Slight increase	Slight increase	Decrease slightly	Decrease 5 to 6%	Decrease slightly	Decrease slightly	Decrease slightly
	Function of frequency	—	—	—	—	1/Frequency	—	—	—
	95% Frequency	Slight decrease	Slight decrease	Slight decrease	Increase slightly	Increase 5 to 6%	Increase slightly	Increase slightly	Increase slightly

Note: Variations are general and differ for specific ratings.

In addition to voltage imbalance, current imbalance can be present in a system where Y-Y transformers without tertiary windings are used, even if the voltage is in balance. Again, this current imbalance is not desirable. If current imbalance exceeds either 10% or the maximum imbalance recommended by the manufacturer, corrective action should be taken (see NFPA *Standard* 70).

$$\text{\% Current imbalance} = 100 \times \frac{\text{Maximum current deviation from average current}}{\text{Average current}}$$

Another cause of current imbalance is normal winding impedance imbalance, which adds or subtracts from the current imbalance caused by voltage imbalance.

CODES AND STANDARDS

The *National Electrical Code*® (NEC) (NFPA *Standard* 70) and *Canadian Electrical Code*, Part I (CSA *Standard* C22.1) are important in the United States and Canada. The NEC contains minimum recommendations considered necessary to ensure safety of electrical installations and equipment. It is referred to in the Occupational Safety and Health Administration (OSHA 2007) electrical standards and, therefore, is part of OSHA requirements. In addition, practically all communities in the United States have adopted the NEC as a minimum electrical code.

Underwriters Laboratories (UL) promulgates standards for various types of equipment. UL standards for electrical equipment cover construction and performance for the safety of such equipment and interpret requirements to ensure compliance with the intent of the NEC. A complete list of available standards may be obtained from UL, which also publishes lists of equipment that comply with their standards. Listed products bear the UL label and are recognized by local authorities.

The *Canadian Electrical Code*, Part I, is a standard of the Canadian Standards Association (CSA). It is a voluntary code with minimum requirements for electrical installations in buildings of every kind. The *Canadian Electrical Code*, Part II, contains specifications for construction and performance of electrical equipment, in compliance with Part I. UL and CSA standards for electrical equipment are similar, so equipment designed to meet the requirements of one code may also meet the requirements of the other. However, agreement between the codes is not complete, so individual standards must be checked when designing equipment for use in both countries. The CSA examines and tests material and equipment for compliance with the *Canadian Electrical Code*.

MOTOR EFFICIENCY

Some of the many factors that affect motor efficiency include (1) sizing the motor to the load, (2) type of motor specified, (3) motor design speed, (4) number of rewinds, (5) voltage imbalance, (6) current imbalance, and (7) type of bearing specified. Oversizing a motor may reduce efficiency. As shown in the performance characteristic curves for single-phase motors in Figures 1, 2, and 3, efficiency usually falls off rapidly at loads lower than the rated full load. Three-phase motors usually reach peak efficiency around 75% load, and the efficiency curve is usually fairly flat from 50 to 100% (Figure 4). Motor performance curves (available from the motor manufacturer) can help in specifying the optimum motor

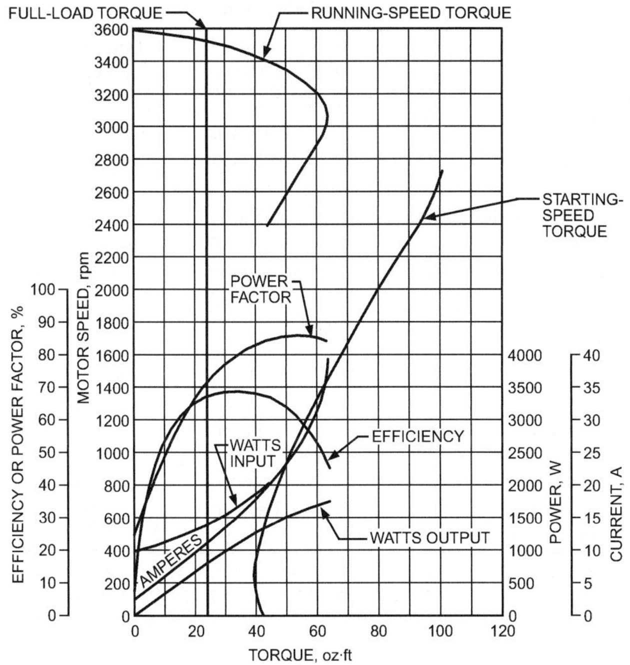

Fig. 1 Typical Performance Characteristics of Capacitor-Start/Induction-Run Two-Pole General-Purpose Motor, 1 hp

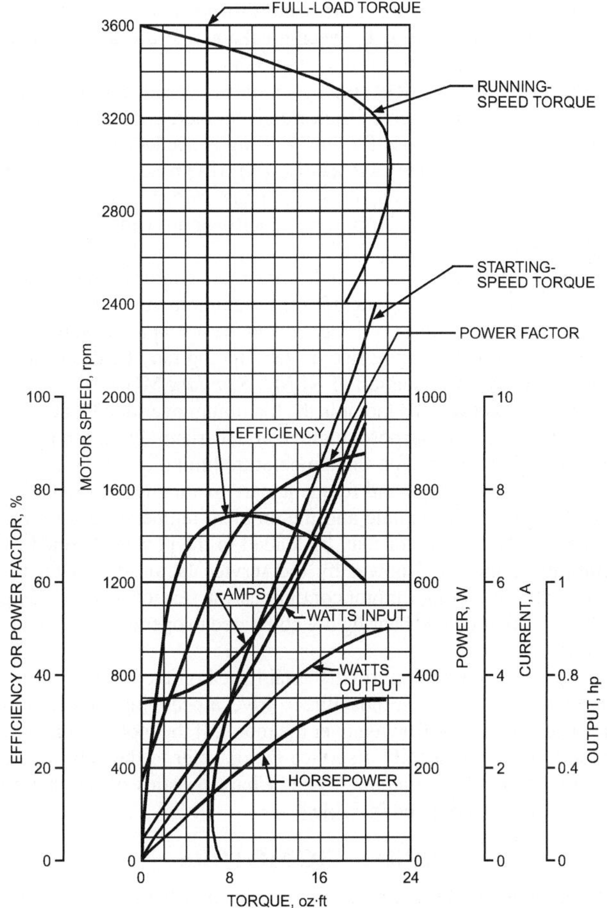

Fig. 2 Typical Performance Characteristics of Resistance-Start Split-Phase Two-Pole Hermetic Motor, 0.25 hp

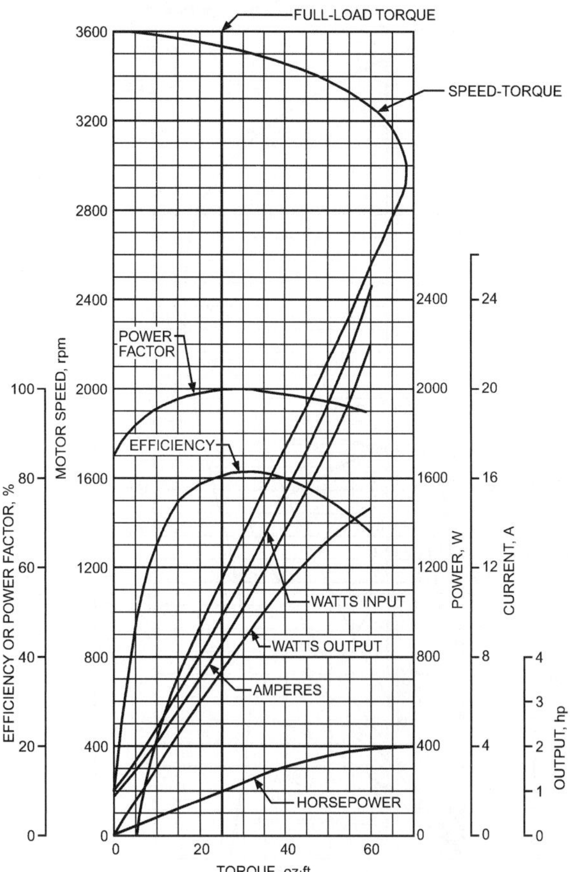

Fig. 3 Typical Performance Characteristics of Permanent Split-Capacitor Two-Pole Motor, 1 hp

for an application. The U.S. Department of Energy's (DOE) MotorMaster+ software gives part-load efficiency as well as efficiency at rated load. Larger-output motors tend to be more efficient than smaller motors at the same percentage load. Four-pole induction motors tend to have the highest range of efficiences.

It is important to understand motor types before specifying one. For example, a permanent split-capacitor motor is more efficient than a shaded-pole fan motor. A capacitor-start/capacitor-run motor is more efficient than either a capacitor-start or a split-phase motor. Three-phase motors are much more likely to have published efficiency: NEMA (National Electrical Manufacturers Association) and the DOE promulgate efficiency standards for three-phase motors between 1 and 500 hp.

Motor manufacturers offer motors over a range of efficiencies. NEMA *Standard* MG 1 describes two efficiency categories: energy-efficient and premium. These standards pertain to most three-phase induction motors between 1 and 500 hp. Note that "energy-efficient" no longer represents a remarkable level of efficiency; it was made a mandatory minimum for general-purpose induction motors from 1 to 200 hp in the United States by the Energy Policy Act of 1992. Today, it has been significantly exceeded by the NEMA premium standard.

Higher-efficiency motors are available in standard frame sizes and performance ratings. Premium-rated motors are more costly than less efficient counterparts, but the additional costs are usually recovered by energy savings very early in the motor's service life; most manufacturers also cite extra reliability features added into premium-rated motors. NEMA *Standards* MG 10 and MG 11 have more information on motor efficiency for single-phase and three-phase motors, respectively.

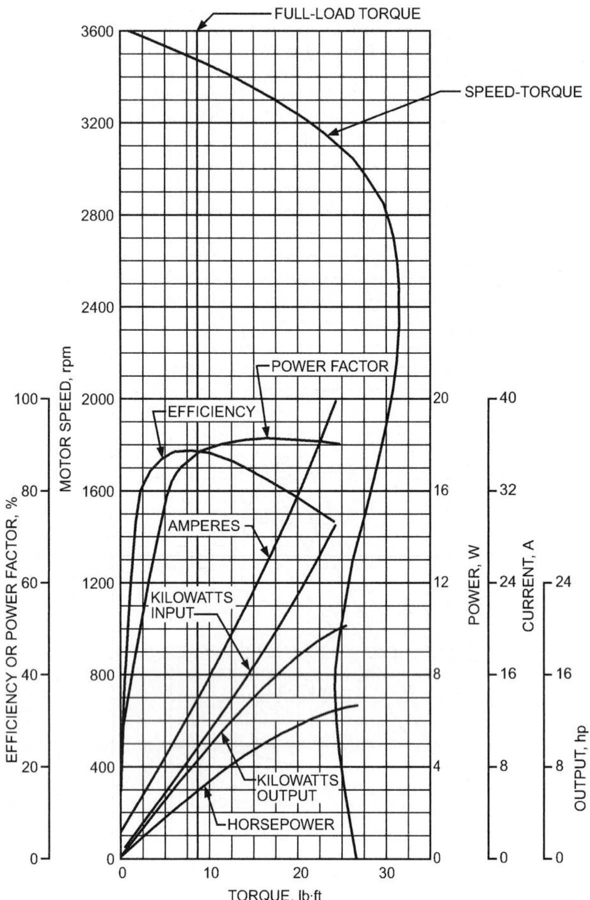

Fig. 4 Typical Performance Characteristics of Three-Phase Two-Pole Motor, 5 hp

GENERAL-PURPOSE INDUCTION MOTORS

The electrical industry classifies motors as **small kilowatt** (**fractional horsepower**) or **integral kilowatt** (**integral horsepower**). In this context, *kilowatt* refers to power output of the motor. Small kilowatt motors have ratings of less than 1 hp at 1700 to 1800 rpm for four-pole and 3500 to 3600 rpm for two-pole machines. Single-phase motors are readily available through 5 hp and are most common through 0.75 hp, because motors larger than 0.75 hp are usually three phase.

Table 3 lists motors by types indicating the normal power range and type of power supply. All motors listed are suitable for either direct or belt drive, except shaded-pole motors (limited by low starting torque).

Application

When applying an electric motor, the following characteristics are important: (1) mechanical arrangement, including position of the motor and shaft, type of bearing, portability desired, drive connection, mounting, and space limitations; (2) speed range desired; (3) power requirement; (4) torque; (5) inertia; (6) frequency of starting; and (7) ventilation requirements. Motor characteristics that are frequently applied are generally presented in curves (see Figures 1 through 4).

Torque. The torque required to operate the driven machine at all times between initial breakaway and final shutdown is important in determining the type of motor. The torque available at zero speed or standstill (**starting torque**) may be less than 100% or as high as

Table 3 Motor Types

Type	Range, hp	Type of Power Supply
Fractional Sizes		
Split-phase	0.05 to 0.5	Single phase
Capacitor-start	0.05 to 1.5	Single phase
Repulsion-start	0.13 to 1.5	Single phase
Permanent split-capacitor	0.05 to 1.5	Single phase
Shaded-pole	0.01 to 0.25	Single phase
Squirrel cage induction	0.17 to 1.5	Three phase
Direct current	0.5 to 1.5	DC
Integral Sizes		
Capacitor-start/capacitor-run	1 to 5	Single phase
Capacitor-start	1 to 5	Single phase
Squirrel cage induction (normal torque)	1 and up	Three phase
Slip-ring	1 and up	Three phase
Direct current	1 and up	DC
Permanent split-capacitor	1 to 5	Single phase

400% of full-load torque, depending on motor design. The **starting current**, or **locked-rotor current**, is usually 400 to 600% of the current at rated full load.

Full-load torque is the torque developed to produce the rated power at the rated speed. **Full-load speed** also depends on the design of the motor. For induction motors, a speed of 1750 rpm is typical for four-pole motors, and a speed of 3450 rpm is typical for two-pole motors at 60 Hz.

Motors have a **maximum or breakdown torque**, which cannot be exceeded. The relation between breakdown torque and full-load torque varies widely, depending on motor design.

Power. The power delivered by a motor is a product of its torque and speed. Because a given motor delivers increasing power up to maximum torque, a basis for power rating is needed. The National Electrical Manufacturers Association (NEMA) bases **power rating** on breakdown torque limits for single-phase motors, 10 hp and less. All others are rated at their power capacity within voltage and temperature limits as listed by NEMA.

Full-load rating is based on the maximum winding temperature. If the nameplate marking includes the maximum ambient temperature for which the motor is designed and the insulation designation, the maximum temperature rise of the winding may be determined from the appropriate section of NEMA *Standard* MG 1.

Service Factor. This factor is the maximum overload that can be applied to general-purpose motors and certain definite-purpose motors without exceeding the temperature limitation of the insulation. When the voltage and frequency are maintained at the values specified on the nameplate and the ambient temperature does not exceed 104°F, the motor may be loaded to the power obtained by multiplying the rated power by the service factor shown on the nameplate. Operating a motor continuously at service factor loading reduces insulation and bearing life compared to operation within the load rating.

The power rating is normally established on the basis of a test-run in still air. However, most direct-drive, air-moving applications are checked with air flowing over the motor. If the motor nameplate marking does not specify a service factor, refer to the appropriate section of NEMA *Standard* MG 1. Characteristics of alternating current motors are given in Table 4.

HERMETIC MOTORS

A hermetic motor is a partial motor usually consisting of a stator and a rotor without shaft, end shields, or bearings. It is for installation in hermetically sealed refrigeration compressor units. With the motor and compressor sealed in a common chamber, the winding

Table 4 Characteristics of AC Motors (Nonhermetic)

	Split-Phase	Permanent Split-Capacitor	Capacitor-Start/ Induction-Run	Capacitor-Start/ Capacitor-Run	Shaded-Pole	Three Phase
Connection Diagram						
Typical Speed Torque Curves						
Starting Method	Centrifugal switch	None	Centrifugal switch	Centrifugal switch	None	None
Ratings, hp	0.05 to 0.5	0.05 to 0.1	0.125 to 5	0.125 to 5	0.01 to 0.25	0.5 and up
Approximate Full-Load Speeds at 60 Hz (Two-Pole/Four-Pole)	3450/1725	3450/1725	3450/1725	3500/1750	3100/1550	3500/1750
Torque*						
Locked Rotor	125 to 150%	30 to 150%	250 to 350%	250%	25%	150 to 350%
Breakdown	250 to 300%	250 to 300%	250 to 300%	250%	125%	250 to 350%
Speed Classification	Constant	Constant	Constant	Constant	Constant or adjustable	Constant
Full-Load Power Factor	60%	95%	65%	95%	60%	80%
Efficiency	Medium	High	Medium	High	Low	High-Medium

*Expressed as percent of rated horsepower torque.

insulation system must be impervious to the action of the refrigerant and lubricating oil. Hermetic motors are used in both welded and accessible hermetic (semihermetic) compressors.

Application

Domestic Refrigeration. Hermetic motors up to 0.33 hp are used. They are split-phase, permanent split-capacitor, or capacitor-start motors for medium or low starting torque compressors and capacitor-start and special split-phase motors for high starting torque compressors.

Room Air Conditioners. Motors from 0.33 to 3 hp are used. They are permanent split-capacitor or capacitor-start/capacitor-run types. These designs have high power factor and efficiency and meet the need for low current draw, particularly on 115 V circuits.

Central Air Conditioning (Including Heat Pumps). Both single-phase (6 hp and below) and three-phase (1.5 hp and above) motors are used. The single-phase motors are permanent split-capacitor or capacitor-start/capacitor-run types.

Small Commercial Refrigeration. Practically all these units are below 5 hp, with single-phase being the most common. Capacitor-start/induction-run motors are normally used up to 0.75 hp because of starting torque requirements. Capacitor-start/capacitor-run motors are used for larger sizes because they provide high starting torque and high full-load efficiency and power factor.

Large Commercial Refrigeration. Most motors are three-phase and larger than 5 hp.

Power ratings of motors for hermetic compressors do not necessarily have a direct relationship to the thermodynamic output of a compressor. Designs are tailored to match the compressor characteristics and specific applications. Chapter 37 briefly discusses hermetic motor applications for various compressors.

INTEGRAL THERMAL PROTECTION

The *National Electrical Code* (NEC) and UL standards cover motor protection requirements. Separate, external protection devices include the following:

Thermal Protectors. These protective devices are an integral part of a motor or hermetic motor refrigerant compressor. They protect the motor against overheating caused by overload, failure to start, or excessive operating current. Thermal protectors are required to protect three-phase motors from overheating because of an open phase in the primary circuit of the supply transformer. Thermal protection is accomplished by either a line break device or a thermal sensing control circuit.

The protector of a hermetic motor-compressor has some unique capabilities compared to nonhermetic motor protectors. The refrigerant cools the motor and compressor, so the thermal protector may be required to prevent overheating from loss of refrigerant charge, low suction pressure and high superheat at the compressor, obstructed suction line, or malfunction of the condensing means.

Article 440 of the NEC limits the maximum continuous current on a motor-compressor to 156% of rated load current if an integral thermal protector is used. NEC Article 430 limits the maximum continuous current on a nonhermetic motor to different percentages of full-load current as a function of size. If separate overload relays and fuses are used for protection, Article 430 limits maximum continuous current to 140% and 125%, respectively, of rated load.

UL *Standard* 984 specifies that the compressor enclosure must not exceed 302°F under any conditions. The motor winding temperature limit is set by the compressor manufacturer based on individual compressor design requirements. UL *Standard* 547 sets the limit for the motor winding temperature for open motors as a function of the class of the motor insulation used.

Line-Break Protectors. Integral with a motor or motor-compressor, line-break thermal protectors that sense both current and temperature are connected electrically in series with the motor; their contacts interrupt the total motor line-current. These protectors are used in small, single-phase and three-phase motors up through 15 hp.

Protectors installed inside a motor-compressor are hermetically sealed because exposed arcing in the presence of refrigerant cannot be tolerated. They provide better protection than the external type for loss of charge, obstructed suction line, or low voltage on the stalled rotor. This is due to low current associated with these fault conditions, hence the need to sense the motor temperature increase by thermal contact. Protection inside the compressor housing must withstand pressure requirements established by UL.

Protectors mounted externally on motor-compressor shells, sensing only shell temperature and line current, are typically used on smaller compressors, such as those in household refrigerators and small room air conditioners. One benefit occurs during high-head-pressure starting conditions, which can occur if voltage is lost momentarily or if the user inadvertently turns off the compressor with the temperature control and then turns it back on immediately. Usually, these units do not start under these conditions. When this happens, the protector takes the unit off the line and resets automatically when the compressor cools and pressures have equalized to a level that allows the compressor to start.

Protectors installed in nonhermetic motors may be attached to the stator windings or may be mounted off the windings but in the motor housing. Those protectors placed on the winding are generally installed before stator varnish dip and bake, and their construction must prevent varnish from entering the contact chamber.

Because the protector carries full motor line current, its size is based on adequate contact capability to interrupt the stalled current of the motor on continuous cycling for periods specified in UL *Standards* 547 and 984.

The compressor or motor manufacturer applies and selects appropriate motor protection in cooperation with the protector manufacturer. Any change in protector rating, by other than the specifying manufacturer after the proper application has been made, may result in either overprotection and frequent nuisance tripouts or underprotection and burnout of the motor windings. Connections to protector terminals, including lead wire sizes, should not be changed, and no additional connections should be made to the terminals. Any change in connection changes the terminal conditions and affects protector performance.

Control Circuit Protectors. Protection systems approved for use with a motor or motor-compressor, either sensing both current and temperature or sensing temperature only, are used with integral horsepower single-phase and three-phase motors.

The current and temperature protector uses a bimetallic temperature sensor installed in the motor winding in conjunction with thermal overload relays. The sensors are connected in series with the control circuit of a magnetic contactor that interrupts the motor current. Thermostat sensors of this type, which depend on their size and mass, are capable of tracking motor winding temperature for running overloads. When a rotor is locked (when the rate of change in winding temperature is rapid), the temperature lag is usually too great for such sensors to provide protection when they are used alone. However, when the bimetallic sensor is used with separate thermal overload or magnetic time-delay relays that sense motor current, the combination provides excellent protection. On a locked rotor condition, the current-sensing relay protects for the initial cycle, and the combined functioning of relay and thermostat protects for subsequent cycles.

The temperature-only protector uses the resistance change of a thermistor-type sensor to provide a switching signal to an electronic circuit, whose output is in series with the control circuit of a magnetic contactor used to interrupt the motor current. The output of the electronic protection circuitry (module) may be an electromechanical relay or a power triac. The sensors may be installed directly on the stator winding end turns or buried inside the windings. Their small size and good thermal transfer allow them to track the temperature of the winding for locked rotor, as well as running overload.

Three types of sensors are available. One type uses a ceramic material with a positive temperature coefficient of resistance; the material exhibits a large, abrupt change in resistance at a particular design temperature. This change occurs at the **anomaly point**, which is inherent in the sensor. The anomaly point remains constant once the sensor is manufactured; sensors are produced with anomaly points at different temperatures to meet different requirements. However, a single module calibration can be supplied for all anomaly temperatures of a given sensor type.

Another type of sensor uses a metal wire, which has a linear increase in resistance with temperature. The sensor assumes a specified value of resistance corresponding to each desired value of response or operating temperature. It is used with an electronic protection module calibrated to a specific resistance. Modules supplied with different calibrations are used to achieve various values of operating temperatures.

A third type is a negative temperature coefficient of resistance sensor, which is integrated with electronic circuitry similar to that used with the metal wire sensor.

More than one sensor may be connected to a single electronic module in parallel or series, depending on design. However, the sensors and modules must be of the same design and intended for use with the particular number of sensors installed and the wiring method used. Electronic protection modules must be paired only with sensors specified by the manufacturer, unless specific equivalency is established and identified by the motor or compressor manufacturer.

MOTOR PROTECTION AND CONTROL

In general, four functions are accomplished by motor protection and control. Separate or integral control components are provided to (1) disconnect the motor and controller from the power supply and protect the operator; (2) start and stop the motor and, in some applications, control the speed or direction of rotation; (3) protect motor branch circuit conductors and control apparatus against short-circuiting; and (4) protect the motor itself from overloading and overheating.

Separate Motor Protection

Most air-conditioning and refrigeration motors or motor-compressors, whether open or hermetic, are equipped with integral motor protection by the equipment manufacturer. If this is not the case, separate motor-protection devices, sensing current only, must be used. These consist of thermal or magnetic relays, similar to those used in industrial control, that provide running overload and stalled-rotor protection. Because hermetic motor windings heat rapidly because of the loss of the cooling effect of refrigerant gas flow when the rotor is stalled, **quick-trip devices** must be used.

Thermostats or **thermal devices** are sometimes used to supplement current-sensing devices. Supplements are necessary (1) when automatic restarting is required after trip or (2) to protect from abnormal running conditions that do not increase motor current. These devices are discussed in the section on Integral Thermal Protection.

Protection of Control Apparatus and Branch Circuit Conductors

In addition to protection of the motor itself, Articles 430 and 440 of the *National Electrical Code* require the control apparatus and branch circuit conductors to be protected from overcurrent resulting from motor overload or failure to start. This protection can be given by some thermal protective systems that do not allow a continuous current in excess of required limits. In other cases, a current-sensing device, such as an overload relay, a fuse, or a circuit breaker, is used.

Circuit Breakers. These devices are used for disconnecting as well as circuit protection, and are available in ratings for use with small household refrigerators as well as in large commercial and industrial installations. Manual switches for disconnecting and fuses for short-circuit protection are also used. For single-phase motors up to 3 hp, 230 V, an attachment plug is an acceptable disconnecting device.

Controllers. The motor control used is determined by the size and type of motor, power supply, and degree of automation. Control may be manual, semiautomatic, or fully automatic.

Central air conditioners are generally located some distance from the controlled space environment control, such as room thermostats. Therefore, **magnetic controllers** must be used in these installations. Also, all dc and all large ac installations must be equipped with in-rush **current-limiting controllers**, which are discussed later. **Synchronous motors** are sometimes used to improve the power factor. **Multispeed motors** provide flexibility for many applications.

Manual Control. For an ac or dc motor, manual control is usually located near the motor. If so, an operator must be present to start and stop or change the motor speed by adjusting the control mechanism.

Manual control is the simplest and least expensive control method for small ac motors, both single-phase and three-phase, but it is seldom used with hermetic motors. The manual controller usually consists of a set of main line contacts, which are provided with thermal overload relays for motor protection.

Manual speed controllers can be used for large air conditioners using **slip-ring motors**; they may also provide reduced-current starting. Different speed points are used to vary the amount of cooling provided by the compressor.

Across-the-Line Magnetic Controllers. These controllers are widely used for central air conditioning. They may be applied to motors of all sizes, provided power supply and motor are suitable to this type of control. Across-the-line magnetic starters may be used with automatic control devices for starting and stopping. Where push buttons are used, they may be wired for either low-voltage release or low-voltage protection.

Three-Phase Motor-Starting and Control Methods

One advantage of three-phase induction motors is their inherently good starting torque without special coils or components. However, some applications require current reduction or additional starting torque.

Full-Voltage and Reduced-Voltage Starting. For motors, full-voltage starting is preferable because of its lower initial cost and simplicity of control. Except for dc machines, most motors are mechanically and electrically designed for full-voltage starting. The starting current, however, is limited in many cases by power company requirements made because of voltage fluctuations, which may be caused by heavy current surges. Therefore, the starting current must often be reduced below that obtained by across-the-line starting, to meet the limitations of power supply. Many methods are available to accomplish this.

Primary Resistance Starting. One of the simplest ways to make this reduction is to place resistors in the primary circuit. As the motor accelerates, the resistance is cut out by the use of timing or current relays.

Autotransformer Motor Controllers. Another method of reducing the starting current for an ac motor uses an **autotransformer** motor controller. Starting voltage is reduced, and, when the motor accelerates, it is disconnected from the transformer and connected across the line by timing or current relays. Primary resistor starters are generally smaller and less expensive than autotransformer starters for moderate size motors. However, primary resistor starters require more line current for a given starting torque than do autotransformer starters.

Solid-State Electronic Soft Starters. Soft starters are also available that can ramp the supply voltage at preprogrammed rates to reduce in-rush current and provide optimum torque for each application.

Star-Delta (Wye Delta) Motor Controllers. These controllers limit current efficiently, but they require motors configured with extra leads for this type of starting. They are particularly suited for centrifugal, rotary screw, and reciprocating compressor drives starting without load.

Part-Winding Motor Controllers (or Incremental Start Controllers). These controllers limit line disturbances by connecting only part of the motor winding to the line and connecting the second motor winding to the line after a time interval of 1 to 3 s. If the motor is not heavily loaded, it accelerates when the first part of the winding is connected to the line; if it is too heavily loaded, the motor may not start until the second winding is connected to the line. In either case, the voltage sag is less than the sag that would result if a standard squirrel-cage motor with an across-the-line starter were used. Part-winding motors may be controlled either manually or magnetically. The magnetic controller consists of two contactors and a timing device for the second contactor.

Multispeed Motor Controllers. Multispeed motors provide flexibility in many types of drives in which variation in capacity is needed. Two types of multispeed motors are used: (1) motors with one reconnectable winding and (2) motors with two separate windings. Motors with separate windings need a contactor for each winding, and only one contactor can be closed at any time. Motors with a reconnectable winding are similar to motors with two windings, but the contactors and motor circuits are different.

Slip-Ring Motor Controllers. Slip-ring ac motors provide reduced-current starting with high torque during acceleration and variable speed after acceleration. The wound rotor of these motors functions in the same manner as in the squirrel-cage motor, except that the rotor windings are connected through slip rings and brushes to external circuits with resistance to vary the motor speed. Increasing resistance in the rotor circuit reduces motor speed, and decreasing resistance increases motor speed. When resistance is shorted out, the motor operates with maximum speed, efficiency, and power factor. On some large installations, manual drum controllers are used as speed-setting devices. Complete automatic control can be provided with special control devices for selecting motor speeds. Operation at reduced speed is at reduced efficiency. These controllers have become less common with the advent of variable-frequency drives, which provide low-current, high-torque starting with good efficiency at reduced speed.

Direct-Current Motor-Starting and Control Methods

These motors have favorable speed-torque characteristics, and their speed can be precisely controlled by varying voltage in the field, armature, or both. Large dc motors are started with resistance in the armature circuit, which is reduced step by step until the motor reaches its base speed. Higher speeds are provided by weakening the motor field. These systems are becoming less common as better speed control strategies in ac motor drive systems develop.

Single-Phase Motor-Starting Methods

Motor-starting switches and relays for single-phase motors must provide a means for disconnecting the starting winding of split-phase or capacitor-start/induction-run motors or the start capacitor of capacitor-start/capacitor-run motors. Open machines usually have a centrifugal switch mounted on the motor shaft, which disconnects the starting winding at about 70% of full-load speed.

The starting methods by use of relays are as follows:

Thermally Operated Relay. When the motor is started, a contact that is normally closed applies power to the starting winding. A thermal element that controls these contacts is in series with the motor and carries line current. Current flowing through this element heats it until, after a definite time, it is warmed sufficiently to open the contacts and remove power from the starting winding. The running current then heats the element enough to keep the contacts open. Setting the time for the starting contacts to open is determined by tests on the components (i.e., the relay, motor, and compressor) and is based on a prediction of the time delay required to bring the motor up to speed.

An alternative form of a thermally operated relay is a positive temperature coefficient of resistance (PTC) starting device. This device has a ceramic element with low resistance at room temperature that increases about 1000 times when it is heated to a predetermined temperature. It is placed in series with the start winding of split-phase motors and allows current flow when power is applied. After a definite period, the self-heating of the PTC resistive element causes it to reach its high-resistance state, which reduces current flow in the start winding. The small residual current maintains the PTC element in the high-resistance state while the motor is running. A PTC starting device may also be connected in parallel with a run capacitor, and the combination may be connected in series with the starting winding. It allows the motor to start like a split-phase motor and then, when the PTC element reaches the high-resistance state, operate as a capacitor-run motor. When power is removed, the PTC element must be allowed to cool to its low resistance state before restarting the motor.

Current-Operated Relay. In this type of connection, a relay coil carries the line current going to the motor. When the motor is started, the in-rush current to the running winding passes through the relay coil, causes the normally open contacts to close, and applies power to the starting winding. As the motor comes up to speed, the current decreases until, at a definite calibrated value of current corresponding to a preselected speed, the magnetic force of the coil diminishes to a point that allows the contacts to open to remove power from the starting winding. This relay takes advantage of the **main winding current** versus **speed** characteristics of the motor. The current/speed curve varies with line voltage, so the starting relay must be selected for the voltage range likely to be encountered in service. Ratings established by the manufacturer should not be changed because this may result in undesirable starting characteristics. They are selected to disconnect the starting winding or start capacitor at approximately 70 to 90% of synchronous speed for four-pole motors.

Voltage-Operated Relay. Capacitor-start and capacitor-start/capacitor-run hermetically sealed motors above 0.5 hp are usually started with a normally closed contact voltage relay. In this method of starting, the relay coil is connected in parallel with the starting winding. When power is applied to the line, the relay does not operate because it is calibrated to operate at a higher voltage. As the motor comes up to speed, the voltage across the starting winding and relay coil increases in proportion to the motor speed. At a definite voltage corresponding to a preselected speed, the relay opens, thereby opening the starting winding circuit or disconnecting the starting capacitor. The relay keeps these contacts open because sufficient voltage is induced in the starting winding when the motor is running to hold the relay in the open position.

AIR VOLUME CONTROL

This section uses fan and air volume control as an example, but the same principles apply to centrifugal pumps and compressors.

The fan laws (Chapter 20) show that volume delivered by a fan is directly proportional to its speed, pressure is proportional to the square of the speed, and power is proportional to the cube of the speed. According to these laws, a fan operating at 50% volume requires only 12.5% of the power required at 100% volume.

Although the fan in a typical VAV system is sized to handle peak volume, the system operates at reduced volume most of the time. For example, Figure 5 shows the volume levels of a typical VAV system operating below 70% volume over 87% of the time. Thus, adjustable-speed operation of the fan for this duty cycle could provide a significant energy saving.

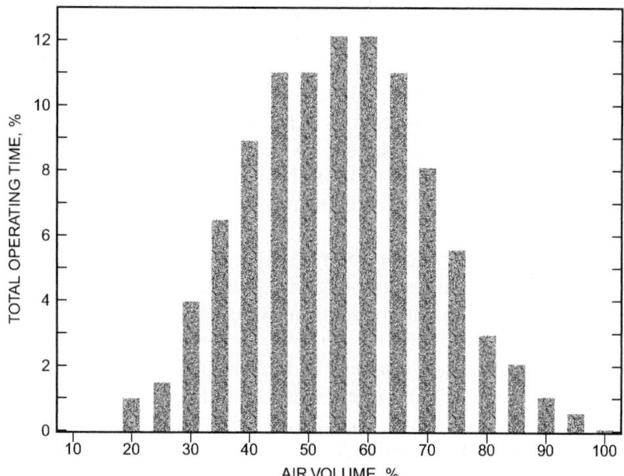

Fig. 5 Typical Fan Duty Cycle for VAV System

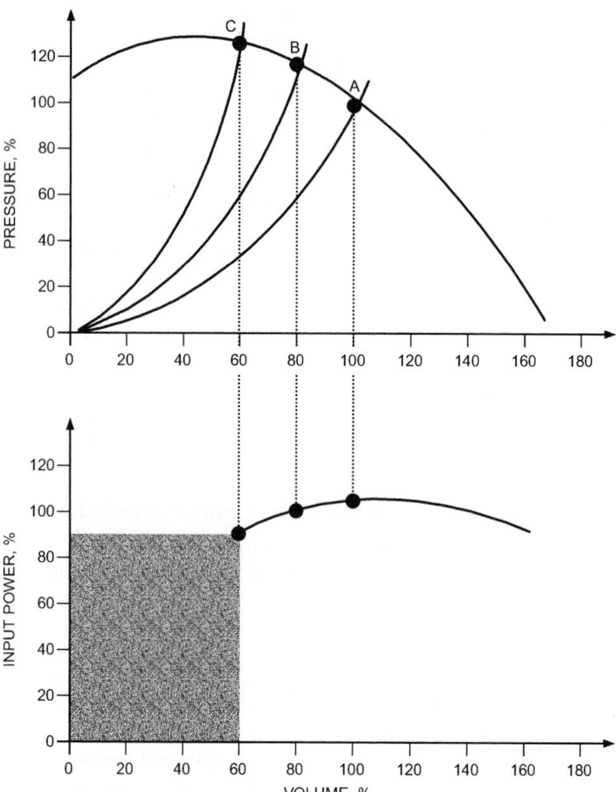

Fig. 6 Outlet Damper Control

Centrifugal fans have usually been driven by fixed-speed ac motors, and volume has been varied by outlet dampers, variable inlet guide vanes, or eddy current couplings.

Outlet dampers are mounted in the airstream on the outlet side of the fan. Closing the damper reduces the volume, but at the expense of increased pressure. Points B and C on the fan performance curve in Figure 6 show the modified system curves for two closed damper positions. The natural operating point corresponds to a wide-open damper position (point A). The input power profile is also shown for the referenced points.

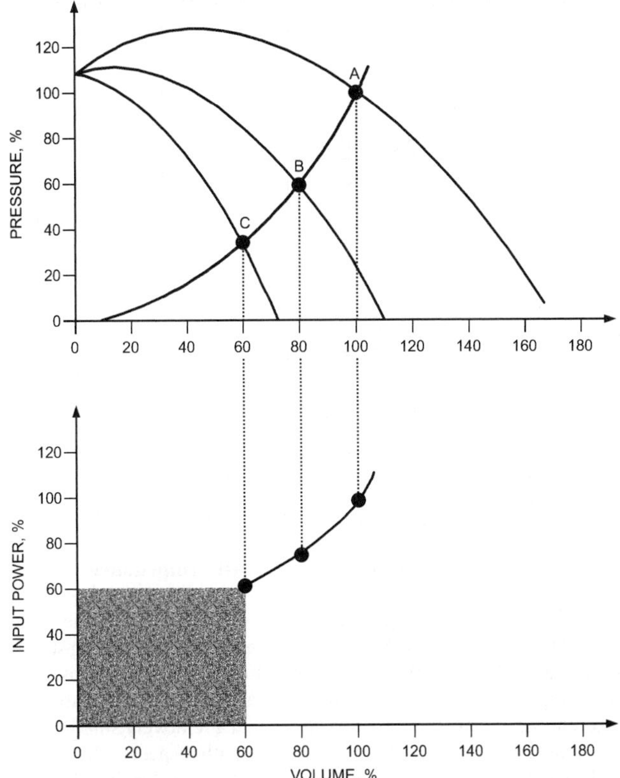

Fig. 7 Variable Inlet Vane Control

Variable inlet vanes are mounted on the fan inlet to control air volume. Altering the pitch of the vane imparts a spin to air entering the fan wheel, which results in a family of fan performance curves as shown in Figure 7. With reference to the required power at reduced flows, the inlet vane is more efficient than an outlet damper.

An **eddy current coupling** connects an ac-motor-driven fixed-speed input shaft to a variable-speed output shaft through a magnetic flux coupling. Reducing the level of flux density in the coupling increases slip between the coupling's input and output shafts and reduces speed. **Slip** is wasted energy in the form of heat that must be dissipated by fan cooling or by water cooling for large motors.

Figure 8 shows that reducing fan speed also generates a family of performance curves, but the required input power still remains relatively high because the speed of the induction motor remains relatively constant.

VARIABLE-SPEED DRIVES (VSD)

An alternative to VAV flow control methods is the variable-speed drive. [In this section, the term variable-speed drive (VSD) is considered synonymous with variable-frequency drive (VFD), pulse-width-modulated drive (PWM drive), adjustable-speed drive (ASD), and adjustable-frequency drive (AFD).] An alternating-current variable-speed drive consists of a diode bridge ac to dc converter, and a pulse-width modulation (PWM) controller with fast-rise power transistors, usually insulated-gate bipolar transistors (IGBTs). These very fast-switching power transistors generate a variable-voltage, variable-frequency waveform that changes the speed of the ac motor. As shown in Figure 9, as speed decreases, input power is reduced substantially because the power required varies as the cube of the speed (plus losses).

Comparison of Figures 6, 7, 8, and 9 shows that significant energy can be saved by using a VSD to achieve variable-air-volume control. Very high efficiencies can be achieved by using the VSD,

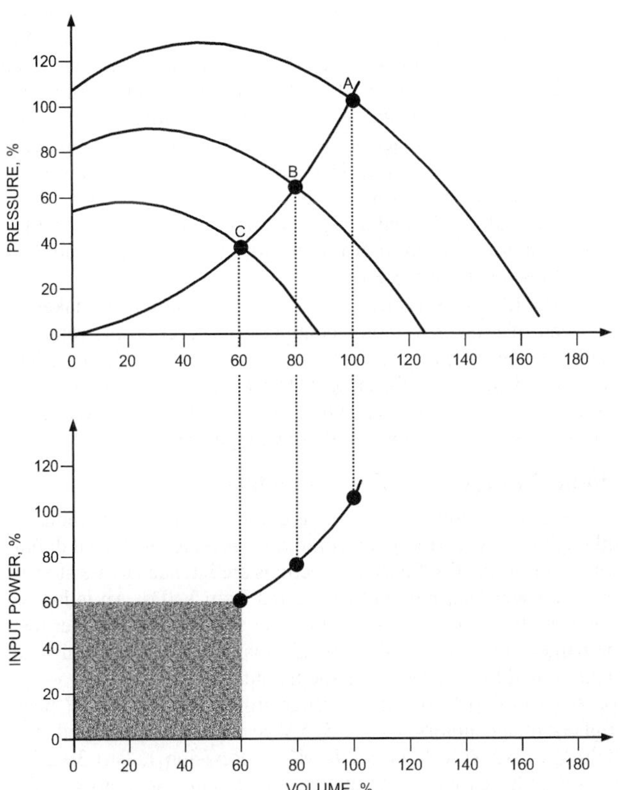

Fig. 8 Eddy Current Coupling Control

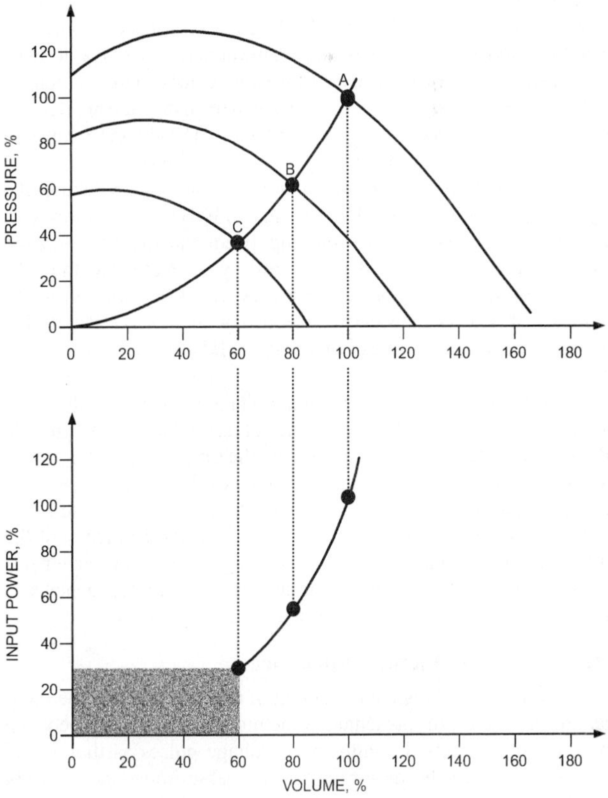

Fig. 9 AC Drive Control

Table 5 Comparison of VAV Energy Consumption with Various Volume Control Techniques

	Outlet Damper	Inlet Guide Vane	Eddy Current Coupling	ac PWM Drive
% Input Power	85	62	40	30
Annual kWh	335,000	244,000	158,000	118,000

NEMA Premium 100 hp motor producing 60% flow for 5000 h, driving fan system that requires 100 hp at unrestricted flow.

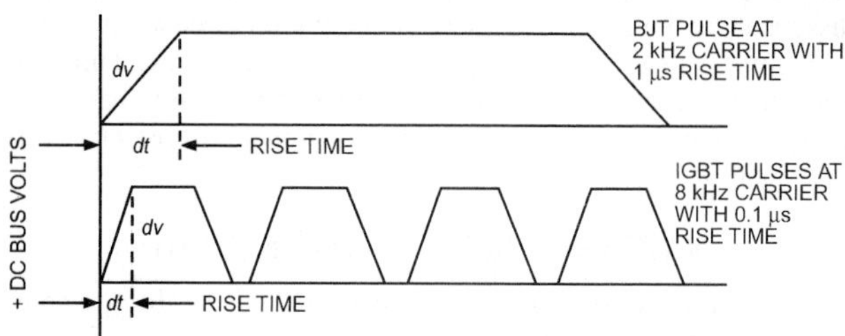

Fig. 10 Bipolar Versus IGBT PWM Switching

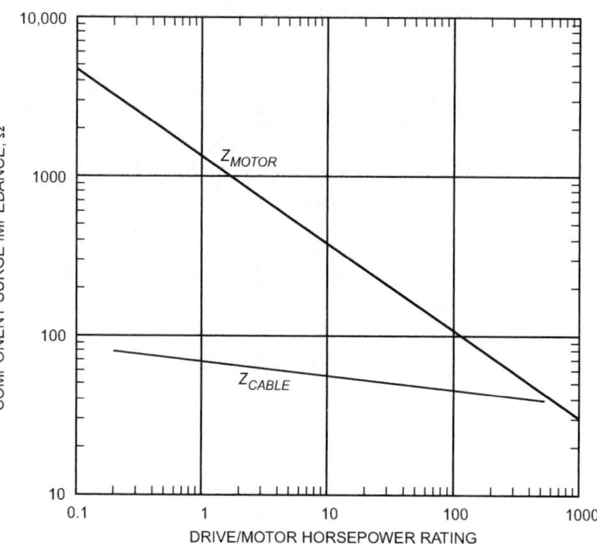

Fig. 11 Motor and Drive Relative Impedance

which is typically over 96% efficient, controlling with a NEMA premium-rated ac motor. Table 5 shows typical annual energy use for the four VAV control techniques.

Power Transistor Characteristics

The key technology used to generate the output waveform is the IGBT. This transistor changes the characteristics of waveforms applied to a motor by varying (modulating) the width of pulses applied to the motor over each cycle of drive output voltage. Pulse-width modulation has been used for many years for variable-speed drives; however, as transistor switching speeds increased, the pulse repetition rate (also known as the carrier or switching frequency) used also tended to increase, from 1 or 2 kHz to 8, 15, or as high as 20 kHz. This allowed motor drive manufacturers to provide a purer motor current waveform from the drive. With increased transistor switching speed and higher carrier frequencies came concerns over phenomena previously seen only in wave transmission devices such as antennae and broadcast signal equipment, and began to change the application variables such as drive-to-motor lead length. These factors must be considered when applying newer IGBT-based VSDs.

Switching Times and *dv/dt*. Figure 10 shows the switching of a bipolar junction transistor (BJT) versus an IGBT as an example of how increased power device switching speeds can affect turn-on and turn-off times as a ratio of the overall cycle. Note that the BJT switches at 1.0 µs at a carrier frequency of 2 kHz, and the IGBT switches at 1.0 µs at a carrier frequency of 8 kHz. The IGBT switches at a speed 10 times faster than the BJT and at a rate 4 times faster.

The rate of change of drive output voltage as the power device is switching is known as the *dv/dt* of the voltage pulse. The magnitude of the *dv/dt* is determined by measuring the time difference between 10 and 90% of the steady-state magnitude of the output pulses, and dividing this time difference into the 90%/10% steady-state pulse voltage magnitude. Note that the *dv/dt* and carrier frequency of the pulses are both a function of the drive design. Often, the carrier frequency is user-settable. The maximum design carrier frequency sets the limits on how fast a transistor must cycle on and off.

Motor and Conductor Impedance

The waveform shown at the output of the drive may not be identical to the waveform presented at the motor terminals. Impedance in ac circuits affects the high-speed voltage pulses as they travel from the drive to the motor. When the cable impedance closely matches the motor impedance, the voltage pulses received at the motor closely approximate those generated by the inverter. However, when the motor surge impedance is much larger than the cable surge impedance, the drives' pulses may be reflected, causing standing waves and very high peak motor voltages. Figure 11 shows the surge impedance of both the motor and a specific type of cable for different-sized drives and motors. Note that a relatively small motor (less than 2 hp) has a very high impedance with respect to the typical cable and can be problematic. Larger motors (greater than 100 hp) closely match cable impedance values and are generally less of a concern.

Potential for Damaging Reflected Waves. Reflected waves damage motors because transmitted and reflected pulses can add together, causing very high voltages to occur at the motor terminals and within the motor to drive wiring. Because these voltage pulses are transmitted through the conductor over specific distances, cable length and type are both variables when examining the potential for damaging voltages. Figure 12 shows the typical relationship between cable type and distance, power device switching times, and ratio of peak motor voltage at motor terminals to peak voltage generated at the drive's output. Damaging reflected waves are most likely to occur in smaller motors because of the mismatch in surge impedance values. Special design techniques are required if multiple small motors are run from a single drive because the potential for reflected waves is even higher.

Figure 13 shows typical oscilloscope measurements taken at each end of a drive-to-motor conductor to describe the reflected wave phenomena. The time scale is set to display a single pulse. The two traces demonstrate the effect of transmitted and reflected pulses adding together to form damaging voltages. The induction motor must be designed to withstand these voltage levels.

Motor Ratings and NEMA Standards

The term "inverter duty motor" is commonly used in the industry, although there is currently no commonly accepted technical definition of this term. The following sections are intended to assist engineers in specifying motors that are fed from VSDs. An induction motor is often constructed to withstand voltage levels higher than the nameplate suggests. The specific maximum voltage withstand value should be obtained from the manufacturer, but typical values for 208 and 460 V ac motors range from 1000 to 1800 V peak. Higher-voltage motors, such as 575 V ac motors, may be rated up to 2000 V peak. NEMA *Standard* MG 1, Revision 1, Part 30.2.2.8, gives established voltage limits for general-purpose motors, which are shown graphically in Figure 14. For motors rated less than 600 V, there is a peak of 1000 V and a minimum rise time of 2 µs.

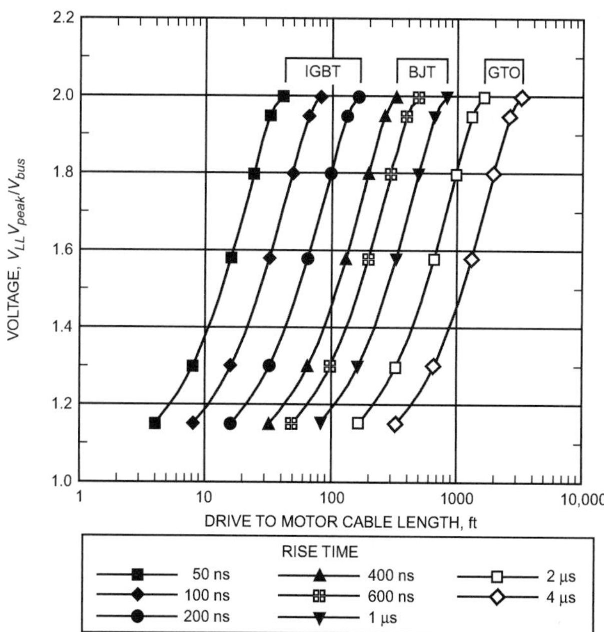

Fig. 12 Typical Switching Times, Cable Distance, and Pulse Peak Voltage

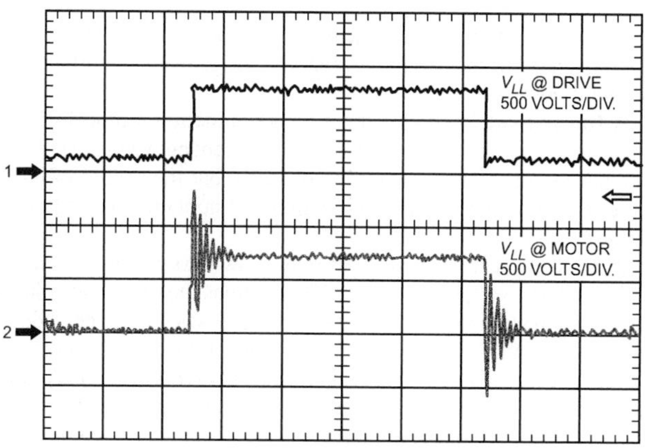

Fig. 13 Typical Reflected Wave Voltage Levels at Drive and Motor Insulation

Revision 1, Part 31, of this standard gives requirements for definite-purpose inverter-fed motors, which are required to have a somewhat higher voltage withstand value. Part 31.4.4.2 states that these motors must withstand a peak of 3.1 times rated voltage (e.g., 1426 V for a 460 V rating). The minimum rise time for these motors is 1 μs. When specifying motors for operation on variable-speed PWM drives, the voltage withstand level (based on the drive's dv/dt and the known cable type and distance) should be specified.

Motor Insulation Breakdown. If reflected waves generate voltage levels higher than the allowable peak, insulation begins to break down. This phenomenon is known as **partial discharge (PD)** or **corona**. When two phases or two turns in the motor pass next to each other, high voltage peaks can ionize the intervening air and cause localized arcing, damaging the insulation. The voltage at which this effect begins is referred to as the **corona inception voltage (CIV)** rating of the motor (Figure 15).

Insulation subjected to PD eventually erodes, causing phase-to-phase or turn-to-turn short circuits. This causes microscopic insulation breakdown, which may not be detected by the drive current sensors and may result in nuisance overcurrent drive trips.

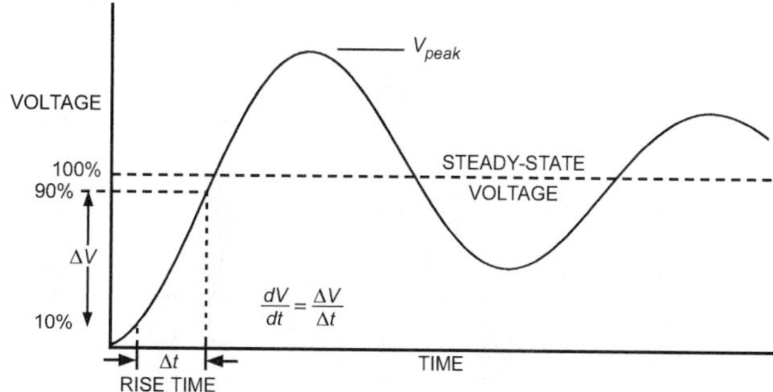

Fig. 14 Motor Voltage Peak and dv/dt Limits
(Reprinted from NEMA *Standard* MG 1, Part 30, Figure 30-5 by permission of the National Electrical Manufacturers Association)

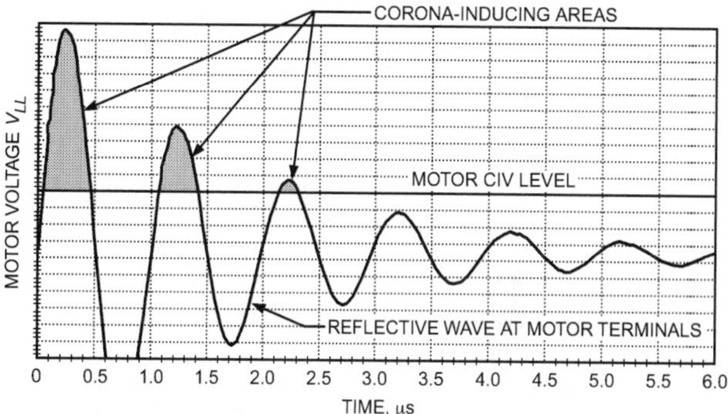

Fig. 15 Damaging Reflected Waves above Motor CIV Levels

Under this short-circuit condition, a motor may operate properly when run across the line or in bypass mode but consistently trip when run from drive power. Factory testing or special diagnostic equipment may be required to confirm this failure mode.

For short cable lengths and slower rise times, general-purpose motors may operate safely without reaching CIV. With longer cable lengths and higher rise times, even definite-purpose inverter-fed motors require mitigation. If details of the motor and cable run are specified, a VSD vendor should be able to prescribe any necessary mitigation filters to keep motor terminal peak voltage within a safe level.

Motor Noise and Drive Carrier Frequencies

Early PWM drives produced extreme motor noise at objectionable frequencies. IGBT technology allows drive designers to increase the carrier frequency to levels that minimize objectionable noise in the human hearing spectrum. Drive designs can switch up to 20 kHz, if required; however, some engineering compromises must be made to optimize the design. During the transition between turning off and on, the transistor generates heat that must be dissipated. This heat loss rises with the carrier frequency. Although higher carrier frequencies do eliminate objectionable audible noise, they also require larger heat sinks and yield lower efficiency.

Audible noise measured in the dBA-weighted scale does not increase proportionally with drive carrier frequency. Additionally, concern with noise may not be over the measured total mean pressure level but a particular frequency band that is objectionable.

Figure 16 shows typical audible noise test results measured on a 100 hp energy-efficient motor. Note that the dominant octave band is at the drive carrier frequency setting. Sine wave power is used as a reference point on the left side of the graph. When running at 2 kHz, the total sound pressure is almost 6 dBA over the sine wave power recordings. This represents 4 times the sound pressure from the motor, because the scale is logarithmic and an increase of 3 dBA

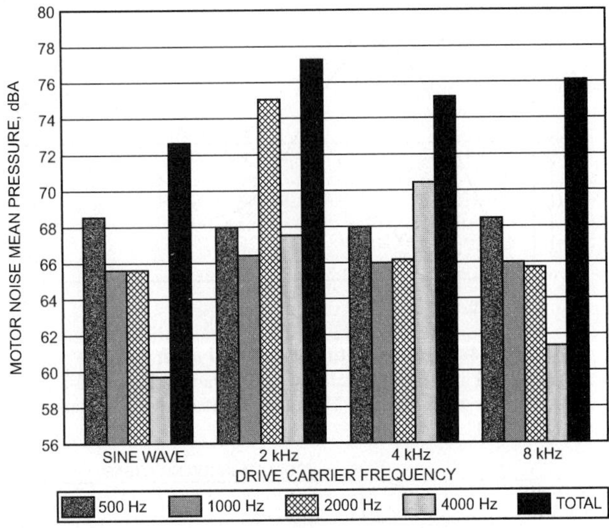

Fig. 16 Motor Audible Noise

doubles the mean pressure level. By comparison, running the drive at 4 kHz increases the mean pressure by only 3 dBA, or half the mean pressure of the 2 kHz setting. (For reference, a 10 dB rise in sound pressure is perceived by the human ear as being twice as loud.)

High Carrier Frequencies and Subharmonics. At high (above 5 kHz) carrier frequencies, harmonics can create vibration forces that match the natural mechanical resonant frequency of the stator and cause sound pressure to exceed 85 dB. The likelihood of subharmonics increases as carrier frequency approaches 20 kHz. If subharmonic vibrations appear, the carrier frequency setting should be decreased to lower the sound pressure generated from the motor.

Carrier Frequencies and Drive Ratings

In some manufacturers' drives, the carrier frequency is user-selectable. However, as carrier frequency increases, drive output ampere ratings often decrease, largely because of the additional heat that must be dissipated from the IGBT. If the rated carrier frequency of a drive is 2 kHz, setting the carrier frequency up to 8 kHz decreases the ampere output. Generally, for every 1 kHz increase in carrier frequency, the drive output current must be derated by 2%, although the specific derating should be determined by the drive manufacturer. As an example, a 10 hp, 460 V drive rated at 2 kHz may have an output of 14 A. If this drive is run at 10 kHz, or an increase of 8 kHz, it must be derated to 11.76 A, or a 16% decrease in current. If the motor nameplate full load were 14 A, this drive would not generate enough output current to obtain the full 10 hp. In effect, the drive and motor would only generate 8.4 hp continuously. This may not be enough power to drive a fan or pump at the performance specified for the application. For this reason, the specifying engineer should always state the desired audible sound level of the motor as applied to the drive to ensure proper operation.

POWER DISTRIBUTION SYSTEM EFFECTS

Variable-frequency drives draw harmonic current from the power line. It is important to distinguish these lower-order harmonics from high-frequency disturbances on the motor side of the drive caused by the PWM inverter. Line harmonics are particularly critical to ac drive users for the following reasons:

- Current harmonics cause additional heating in transformers, conductors, and switchgear. Current harmonics flowing through the impedance of the power system cause voltage harmonics in accord with Ohm's law.

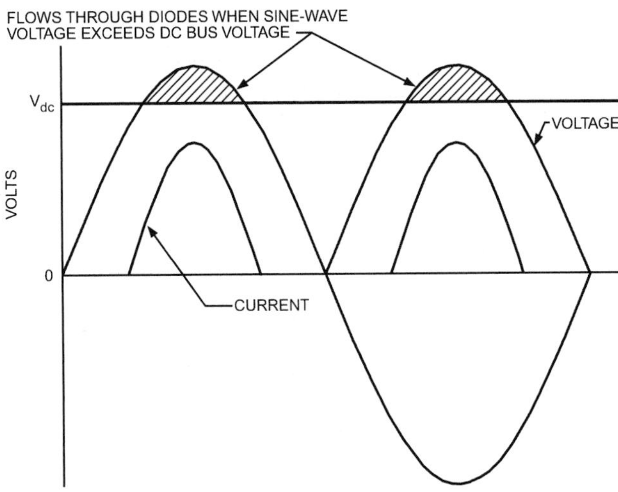

Fig. 17 Voltage Waveform Distortion by Pulse-Width-Modulated VSD

- Voltage harmonics upset the smooth, predictable voltage waveform in a normal sine wave. A power system severely distorted by voltage harmonics may damage components connected to the line or cause erratic operation of some equipment.
- High-frequency components of voltage distortion can interfere with signals transmitted on the ac line for some control systems.

However, PWM ac drives with built-in bus reactors or external reactors ahead of the drive significantly mitigate any disturbance to the input power.

A **linear load**, such as a three-phase induction motor operated across the line, may cause a phase displacement between the voltage and current waveforms (phase lag or lead), but the shapes of these waveforms are nearly pure sine waves and contain very little harmonics.

In contrast, a **nonlinear load** may draw current only from the peaks of the ac voltage sine wave. This flattens the top of the voltage waveform in single-phase circuits, and depresses it on either side of the peaks in three-phase circuits. Nonlinear loads draw currents from the power source that are rich in current harmonics. Many nonlinear loads connected to a power system can cumulatively inject harmonics. Single-phase equipment (e.g., TVs, VCRs, computers, electronic lighting) and three-phase equipment [e.g., VSDs, uninterruptible power supplies (UPSs), electric arc furnaces, electric heaters, welders] convert ac voltage to dc voltage and contain circuitry that draws current in a nonlinear fashion. Figure 17 shows how the current drawn by a PWM full wave rectification VSD may distort the voltage waveform measured at the input terminals.

A single-phase load is not necessarily too small to be of concern. With ac-to-dc converters, the demand current occurs around the peak of the voltage sine wave. A thousand 100 W fluorescent light fixtures consume 100 kW of power. If the lights are nonlinear loads, the peaks add directly and cause the voltage waveform to dip. This distortion in the single-phase voltage waveform contributes to the harmonic distortion of the three-phase power source. On single-phase harmonic distortion, these loads produce even-numbered harmonics such as 2nd, 4th, 6th, etc. Thus, if a balanced system is experiencing even-numbered harmonics, they must originate from a single-phase load and not from the drives. These loads may also use the neutral connection of the power source to conduct current; the neutral connection may overload if proper precautions are not taken to alleviate harmonic currents drawn by nonlinear loads.

VSDs and Harmonics

Figure 18 shows the basic elements of any solid-state drive. The converter section (for conversion of ac line power to dc) and the

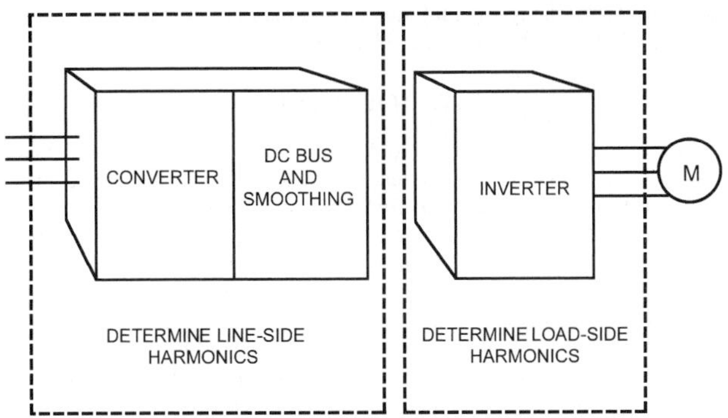

Fig. 18 Basic Elements of Solid-State Drive

inverter section (for conversion of dc to variable frequency ac) both contain nonlinear devices that cause harmonics on the input and output lines, respectively. Input-line harmonics are caused solely by the converter section and are usually referred to as **line-side harmonics**. Output-line harmonics are caused solely by the inverter section and are known as **load-side** or **motor harmonics**.

These effects are isolated from each other by a dc bus capacitor and in some designs by a dc choke so that load-side harmonics only affect equipment driven by the VSD and line-side harmonics affect the power system as a whole.

Effects of Load-Side Harmonics. Load-side voltage harmonics generated by the inverter section of a VSD are of concern for the motor. The low-order load-side voltage harmonics are minimal and only slightly decrease motor life because of the additional heating created. The much higher load-side frequencies from the PWM inverter have minimal distorting effect on the current wave form and are less an energy concern than a potential source of motor damage. However, the use of NEMA premium-rated or definite-purpose inverter-fed motors significantly compensates for any damaging effects. Additionally, hermetic refrigerant-cooled motors, as used in some variable-speed chiller designs, often experience insignificant increases in motor heat because a high degree of cooling is available. Selection and matching of both the motor and drive should account for these effects and ensure that motor performance and equipment life are not compromised when applying variable speed. Retrofit applications should be engineered to ensure that the motor and drive can provide enough power to the connected load.

As discussed in the section on Motor and Conductor Impedance, a second phenomenon associated with inverters on the load side is the effect of high voltage spikes on motor life. The fast-switching capability of the inverter combined with long power lines between the drive and motor can produce reflected waves that have high peak voltages. If these voltages are large enough, they produce potentially destructive stresses in the motor insulation.

Effects of Line-Side Harmonics. Generally, PWM ac drives that contain internal bus reactors or three-phase ac input line reactors help minimize electrical interference with other electrical equipment. But any harmonic current flowing through the source impedance causes a voltage drop that results in harmonic distortion of the supply voltage waveform. In general, the lower the drive's input current harmonics, the lower the risk of creating interference with other equipment through harmonic distortion. Ideally, the drive's input current waveform should be purely sinusoidal and contain no harmonic current distortion, similar to operation of a motor connected directly to the power source [current total harmonic distortion (THD) is ideally 0%]. IF VSDs are large or numerous, or if electrical system impedance is high, additional harmonic mitigation strategies may be necessary. A distorted supply voltage waveform can have the following undesirable effects on some equipment connected to the power line:

- Communications equipment, computers, and diagnostic equipment are "sensitive" (i.e., have a low tolerance to harmonics). Typical effects include receipt of false commands and data corruption.
- Transformers may experience trouble caused by possible additional heating in the core and windings. Many transformer manufacturers rate special transformers by K-factor, which indicates the transformer's ability to withstand degradation due to harmonics. Special cores to reduce eddy currents, specially designed windings that reduce heating, and an oversized neutral bus are some of the special design features found in some K-factor transformers. Other manufacturers simply derate their standard transformers to compensate for harmonic effects.
- Standby generators operate at frequencies that change with load. When a VSD is switched onto generator power, the frequency fluctuation could affect the VSD converter. Standby generators also have voltage regulators that are susceptible to harmonics. In addition, generators have very high impedance compared to the normal power. The harmonic currents flowing in this higher impedance can give rise to harmonic voltages three to four times the normal levels. Compounding this problem is the fact that standby generators are usually installed where sensitive equipment is prevalent (e.g., in hospitals and computer centers). Emergency power systems should be specified and tested to ensure they can serve the harmonic current load and still provide voltage clean enough for critical loads.

Any VSD application with standby generators requires careful design, and the following information should be gathered:

- Power output (kW, MW or kVA) of the generator
- Subtransient reactance
- How the generator is applied in reference to the VSD; what is the worst-case running condition of the drives (number of drives running at one time and the load on these drives)

Additional problems can be caused by resonance that can occur when **power factor correction capacitors (PFCCs)** are installed. Resonance can severely distort the voltage waveform. PFCCs may fail prematurely, or capacitor fuses may blow. Additionally, because VSDs have an inherent high displacement power factor (typically 0.96 or greater), PFCCs should never be required or used with a drive. They can even cause the drive to fail if installed on the load side of the VSD. If an older motor is retrofitted with capacitors, PFCCs should be removed because they are no longer required.

Only the fundamental current transmits power to the load. Harmonic currents increase the equipment input kVA without contributing to input power. Operating with a high harmonic content is much like operating at a low input power factor. High harmonic content means that higher total current is required to deliver a given amount of power because of equipment heat losses; thus, the true power factor (kW/kVA) can be low even if the displacement power factor (cos θ) is high or unity. All components of the power distribution system must be oversized to handle the additional current. If the utility meters are able to measure the harmonic content and/or power factor, they may assess a distortion (demand) charge or power factor penalty.

Effect of Harmonics on a System. In most applications, no harmonic problems occur with six-pulse PWM VSDs that use a series reactor in the dc bus or in the input ac line. A 3 or 5% impedance ac reactor is often offered as an option on drives. Using a dc bus or ac reactor typically reduces the current THD level between 25 and 30%. If additional harmonic reduction is desired, passive harmonic filters or higher-order multipulse inputs of 12 and 18 pulse are sometimes offered, which reduce the input current THD to 8 to 12%. Active harmonic filters are expensive but extremely effective, reducing harmonic distortion to levels of 3 to 5%.

A study can be performed of system harmonic performance to determine the expected contributions of nonlinear equipment. IEEE *Standard* 519 establishes levels for harmonic contribution by a customer's power system onto the power grid. These levels are directly related to the strength of the connected power grid. This guideline establishes the **point of common coupling (PCC)** as the primary of the transformer feeding that power system. The purpose of the study is to anticipate any potential harmonic issues, and any mitigation requirements. These studies should always be performed with a minimum 1% line-to-line voltage imbalance. As stated earlier, a small voltage imbalance can cause a large current imbalance or a large difference in harmonic contribution, and affect the performance of harmonic mitigation equipment.

With other converter loads (e.g., arc furnaces, dc drives, current source drives) and other high-reactive-current loads, harmonic problems may exist. The following problems, typically more common on single-phase systems, may indicate a harmonic condition, but they may also indicate line voltage unbalance or overloaded conditions:

- Nuisance input fuse blowing or circuit breaker tripping
- Power factor capacitor overheating, or fuse failure
- Overheating of supply transformers
- Overheating neutral conductors and connectors (normally just on single-phase systems)

Problems that are not usually harmonic problems include

- Overcurrent tripping of VSDs
- Interference with AM radio reception
- Wire failure in conduits

REFERENCES

ARI. 2002. Air-conditioning and refrigerating equipment nameplate voltages. *Standard* 110-2002. Air-Conditioning and Refrigeration Institute, Arlington, VA.

CSA. 2005. Canadian electrical code, part I. *Standard* C22.1-98. Canadian Standards Association, Etobicoke, ON.

CSA. 1996. Hermetic refrigerant motor-compressors. *Standard* C22.2 No.140.2-96, 4th ed. Canadian Standards Association, Etobicoke, ON.

IEEE. 1992. Recommended practices and requirements for harmonic control in electrical power systems. *Standard* 519-1992. Institute of Electrical and Electronics Engineers, New York.

NEMA. 2004. Motors and generators. *Standard* MG 1-2003, Rev. 1—2004. National Electrical Manufacturers Association, Rosslyn, VA.

NEMA. 2001. Energy management guide for selection and use of polyphase motors. *Standard* MG 10-2001. National Electrical Manufacturers Association, Rosslyn, VA.

NEMA. 2001. Energy management guide for selection and use of single-phase motors. *Standard* MG 11-2001. National Electrical Manufacturers Association, Rosslyn, VA.

NFPA. 2005. National electrical code®. NFPA *Standard* 70-2005. National Fire Protection Association, Quincy, MA.

OSHA. 2007. Occupational safety and health standards, subpart S—Electrical. 29CFR1910. *Code of Federal Regulations*, Occupational Safety and Health Administration, Washington, D.C.

UL. 1991. Thermal protectors for electric motors. UL *Standard* 547. Underwriters Laboratories, Northbrook, IL.

UL. 1996. Hermetic refrigerant motor-compressors. UL *Standard* 984, 7th ed. Underwriters Laboratories, Northbrook, IL.

BIBLIOGRAPHY

Ahmed, S., W. Choi, H. Toliyat, and P. Enjeti. 2002. Characterization of non-sinusoidal measurement station component requirements and errors (RP-1095). *ASHRAE Transactions* 108(1):891-896.

DOE. 2007. *MotorMaster+* and *MotorMaster+ International*. U.S. Department of Energy, Office of Energy Efficiency and Renewable Energy, Washington, D.C. Available at http://www1.eere.energy.gov/industry/bestpractices/software.html.

Evon, S., D. Kempke, L. Saunders, and G. Skibinski. 1996. IGBT drive technology demands new motor and cable considerations. IEEE Petroleum & Chemical Industry Conference. Institute of Electrical and Electronics Engineers, New York.

Kerkman, R., D. Leggate, and G. Skibinski. 1997. Cable characteristics and their influence on motor over-voltages. IEEE Applied Electronic Conference (APEC). Institute of Electrical and Electronics Engineers, New York.

Kerkman, R., D. Leggate, and G. Skibinski. 1996. Interaction of drive modulation & cable parameters on ac motor transients. IEEE Industry Application Society Conference. Institute of Electrical and Electronics Engineers, New York.

Lowery, T. 1999. Design considerations for motors and variable speed drives. *ASHRAE Journal* 41(2):28-32.

Malfait, A., R. Reekmans, and R. Belmans. 1994. Audible noise and losses in variable speed induction motor drives with IGBT inverters—Influence of the squirrel cage design and the switching frequency. *Proceedings of Industry Applications* 1:693-700. Institute of Electrical and Electronics Engineers, New York.

Mays, M. 1998. Identifying noise problems in adjustable speed drives. *ASHRAE Journal* 40(10):57-60.

NEMA. 2001. *Application guide for ac adjustable speed drive systems*. National Electrical Manufacturers Association, Rosslyn, VA.

NEMA. 1995. Electrical power systems and equipment—Voltage ratings (60 Hz). ANSI/NEMA *Standard* C84.1-1995. National Electrical Manufacturers Association, Rosslyn, VA.

Saunders, L., G. Skibinski, R. Kerkman, D. Schlegel, and D. Anderson. 1996. Modern drive application issues and solutions. IEEE PCIC Conference. Tutorial on Reflected Wave, Motor Failure, CM Electrical Noise, Motor Bearing Current. Institute of Electrical and Electronics Engineers, New York.

Sung, J. and S. Bell. 1996. Will your motor insulation survive a new adjustable frequency drive? IEEE Petroleum & Chemical Industry Conference. Institute of Electrical and Electronics Engineers, New York.

Takahashi, T., G. Wagoner, H. Tsai, and T. Lowery. 1995. Motor lead length issues for IGBT PWM drives. IEEE Pulp and Paper Conference. Institute of Electrical and Electronics Engineers, New York.

Toliyat, H., S. Ahmed, W. Choi, and P. Enjeti. 2002. Instrument selection criteria for non-sinusoidal power measurements (RP-1095). *ASHRAE Transactions* 108(1):897-903.

Wang, J., S. McInerny, and R. Stauton. 2002. Early detection of insulation breakdown in low-voltage motors, part I: Background, experimental design and preliminary results (RP-1078). *ASHRAE Transactions* 108(1): 875-882.

Wang, J., S. McInerny, and R. Stauton. 2002. Early detection of insulation breakdown in low-voltage motors, part II: Analysis and results (RP-1078). *ASHRAE Transactions* 108(1):883-890.

U.S. Department of Commerce. 2002. *Electric current abroad*. Available at http://www.ita.doc.gov/media/Publications/pdf/current2002FINAL.pdf.

CHAPTER 45

PIPES, TUBES, AND FITTINGS

THIS CHAPTER covers the selection, application, and installation of pipe, tubes, and fittings commonly used for heating, air-conditioning, and refrigeration. Pipe hangers and pipe expansion are also addressed. When selecting and applying these components, applicable local codes, state or provincial codes, and voluntary industry standards (some of which have been adopted by code jurisdictions) must be followed.

The following organizations in the United States issue codes and standards for piping systems and components:

ASME	American Society of Mechanical Engineers
ASTM	American Society for Testing and Materials
NFPA	National Fire Protection Association
BOCA	Building Officials and Code Administrators, International
MSS	Manufacturers Standardization Society of the Valve and Fittings Industry, Inc.
AWWA	American Water Works Association

Parallel federal specifications also have been developed by government agencies and are used for many public works projects. Chapter IV of ASME *Standard* B31.9 lists applicable U.S. codes and standards for HVAC piping. In addition, it gives the requirements for the safe design and construction of piping systems for building heating and air conditioning. ASME *Standard* B31.5 gives similar requirements for refrigerant piping.

PIPE

Steel Pipe

Steel pipe is manufactured by several processes. Seamless pipe, made by piercing or extruding, has no longitudinal seam. Other manufacturing methods roll a strip or sheet of steel (skelp) into a cylinder and weld a longitudinal seam. A continuous-weld (CW) furnace butt-welding process forces and joins the edges together at high temperature. An electric current welds the seam in electric-resistance-welded (ERW) pipe. ASTM *Standards* A53 and A106 specify steel pipe. Both standards specify A and B grades. The A grade has a lower tensile strength and is not widely used.

The ASME pressure piping codes require that a longitudinal joint efficiency factor E (Table 1) be applied to each type of seam when calculating the allowable stress. ASME *Standard* B36.10M specifies the dimensional standard for steel pipe. Through 12 in. diameter, nominal pipe sizes (NPS) are used, which do not match the internal or external diameters. For pipe 14 in. and larger, the size corresponds to the outside diameter.

Steel pipe is manufactured with wall thicknesses identified by schedule or weight class. Although schedule numbers and weight class designations are related, they are not constant for all pipe sizes. Standard weight (STD) and Schedule 40 pipe have the same wall thickness through NPS 10. For 12 in. and larger standard weight pipe, the wall thickness remains constant at 0.375 in., whereas Schedule 40 wall thickness increases with each size. A similar equality exists between Extra Strong (XS) and Schedule 80 pipe through 8 in.; above 8 in., XS pipe has a 0.500 in. wall, whereas Schedule 80 increases in wall thickness. Table 2 lists properties of representative steel pipe.

Joints in steel pipe are made by welding or by using threaded, flanged, or grooved fittings. Unreinforced welded-in branch connections weaken a main pipeline, and added reinforcement is necessary, unless the excess wall thickness of both mains and branches is sufficient to sustain the pressure.

ASME *Standard* B31.1 gives formulas for determining whether reinforcement is required. Such calculations are seldom needed in HVAC applications because (1) standard-weight pipe through NPS 20 at 300 psig requires no reinforcement; full-size branch connections are not recommended; and (2) fittings such as tees and reinforced outlet fittings provide inherent reinforcement.

Type F steel pipe is not allowed for ASME *Standard* B31.5 refrigerant piping.

Copper Tube

Because of their inherent resistance to corrosion and ease of installation, copper and copper alloys are often used in heating, air-conditioning, refrigeration, and water supply installations. There

The preparation of this chapter is assigned to TC 6.1, Hydronic and Steam Equipment and Systems.

Table 1 Allowable Stresses[a] for Pipe and Tube

ASTM Specification	Grade	Type	Manufacturing Process	Available Sizes, in.	Minimum Tensile Strength, psi	Basic Allowable Stress S, psi	Joint Efficiency Factor E	Allowable Stress[b] S_E, psi	Allowable Stress Range[c] S_A, psi
A53 Steel	—	F	Cont. Weld	1/2 to 4	45,000	11,250	0.6	6,800	16,900
A53 Steel	B	S	Seamless	1/2 to 26	60,000	15,000	1.0	15,000	22,500
A53 Steel	B	E	ERW	2 to 20	60,000	15,000	0.85	12,800	22,500
A106 Steel	B	S	Seamless	1/2 to 26	60,000	15,000	1.0	15,000	22,500
B88 Copper	—	—	Hard Drawn	1/4 to 12	36,000	9,000	1.0	9,000	13,500

[a]Listed stresses are for temperatures to 650°F for steel pipe (to 400°F for Type F) and to 250°F for copper tubing.
[b]To be used for internal pressure stress calculations in Equations (1) and (2).
[c]To be used only for piping flexibility calculations; see Equations (3) and (4).

are two principal classes of copper tube. ASTM *Standard* B88 includes Types K, L, M, and DWV for water and drain service. ASTM *Standard* B280 specifies air-conditioning and refrigeration (ACR) tube for refrigeration service.

Types K, L, M, and DWV designate descending wall thicknesses for copper tube. All types have the same outside diameter for corresponding sizes. Table 3 lists properties of ASTM B88 copper tube. In the plumbing industry, tube of nominal size approximates the inside diameter. The heating and refrigeration trades specify copper tube by the outside diameter (OD). ACR tubing has a different set of wall thicknesses. Types K, L, and M tube may be hard drawn or annealed (soft) temper.

Copper tubing is joined with soldered or brazed, wrought or cast copper capillary socket-end fittings. Table 4 lists pressure/temperature ratings of soldered and brazed joints. Small copper tube is also joined by flare or compression fittings.

Hard-drawn tubing has a higher allowable stress than annealed tubing, but if hard tubing is joined by soldering or brazing, the annealed allowable stress should be used.

Brass pipe and copper pipe are also made in steel pipe thicknesses for threading. High cost has eliminated these materials from the market, except for special applications.

The heating and air-conditioning industry generally uses Types L and M tubing, which have higher internal working pressure ratings than the solder joints used at fittings. Type K may be used with brazed joints for higher pressure-temperature requirements or for direct burial. Type M should be used with care where exposed to potential external damage.

Copper and brass should not be used in ammonia refrigerating systems. The section on Special Systems covers other limitations on refrigerant piping.

Ductile Iron and Cast Iron

Cast-iron soil pipe comes in XH or service weight. It is not used under pressure because the pipe is not suitable and the joints are not restrained. Cast-iron pipe and fittings typically have bell and spigot ends for lead and oakum joints or elastomer push-on joints. Cast-iron pipe and fittings are also furnished with *no-hub* ends for joining with *no-hub* clamps. Local plumbing codes specify permitted materials and joints.

Ductile iron has now replaced cast iron for pressure pipe. Ductile iron is stronger, less brittle, and similar to cast iron in corrosion resistance. It is commonly used for buried pressure water mains or in other locations where internal or external corrosion is a problem. Joints are made with flanged fittings, mechanical joint (MJ) fittings, or elastomer gaskets for bell and spigot ends. Bell and spigot and MJ joints are not self-restrained. Restrained MJ systems are available. Ductile-iron pipe is made in seven thickness classes for different service conditions. AWWA *Standard* C150/A21.50 covers the proper selection of pipe classes.

FITTINGS

The following standards give dimensions and pressure ratings for fittings, flanges, and flanged fittings. These data are also available from manufacturers' catalogs.

Applicable Standards for Fittings

Steel[a]	**ASME *Std.***
Pipe Flanges and Flanged Fittings	B16.5
Factory-Made Wrought Steel Buttwelding Fittings	B16.9
Forged Fittings, Socket-Welding and Threaded	B16.11
Wrought Steel Buttwelding Short Radius Elbows and Returns	B16.28
Cast Iron, Malleable Iron, Ductile Iron[b]	
Cast Iron Pipe Flanges and Flanged Fittings	B16.1
Malleable Iron Threaded Fittings	B16.3
Gray Iron Threaded Fittings	B16.4
Cast Iron Threaded Drainage Fittings	B16.12
Ductile Iron Pipe Flanges and Flanged Fittings, Classes 150 and 300	B16.42
Copper and Bronze[c]	
Cast Bronze Threaded Fittings, Classes 125 and 25	B16.15
Cast Copper Alloy Solder Joint Pressure Fittings	B16.18
Wrought Copper and Copper Alloy Solder Joint Pressure Fittings	B16.22
Cast Copper Alloy Solder Joint Drainage Fittings, DWV	B16.23
Cast Copper Alloy Pipe Flanges and Flanged Fittings, Classes 150, 300, 400, 600, 900, 1500, and 2500	B16.24
Cast Copper Alloy Fittings for Flared Copper Tubes	B16.26
Wrought Copper and Wrought Copper Alloy Solder Joint Drainage Fittings	B16.29
Nonmetallic[d]	**ASTM *Std.***
Threaded PVC Plastic Pipe Fittings, Schedule 80	D2464
Threaded PVC Plastic Pipe Fittings, Schedule 40	D2466
Socket-Type PVC Plastic Pipe Fittings, Schedule 80	D2467
Reinforced Epoxy Resin Gas Pressure Pipe and Fittings	D2517
Threaded CPVC Plastic Pipe Fittings, Schedule 80	F437
Socket-Type CPVC Plastic Pipe Fittings, Schedule 40	F438
Socket-Type CPVC Plastic Pipe Fittings, Schedule 80	F439
Polybutylene (PB) Plastic Hot- and Cold-Water Distribution Systems	D3309
Plastic Insert Fittings for Polybutylene Tubing	F845
Solvent Cements for PVC Plastic Piping Systems	D2564
Solvent Cements for CPVC Plastic Pipe and Fittings	F493

[a]Wrought steel butt-welding fittings are made to match steel pipe wall thicknesses and are rated at the same working pressure as seamless pipe. Flanges and flanged fittings are rated by working steam pressure classes. Forged steel fittings are rated from 2000 to 6000 psi in classes and are used for high-temperature and high-pressure service for small pipe sizes.

[b]The class numbers refer to the maximum working saturated steam gage pressure (in psi). For liquids at lower temperatures, higher pressures are allowed. Groove-end fittings of these materials are made by various manufacturers who publish their own ratings.

[c]The classes refer to maximum working steam gage pressure (in psi). At ambient temperatures, higher liquid pressures are allowed. Solder joint fittings are limited by the strength of the soldered or brazed joint (see Table 4).

[d]Ratings of plastic fittings match the pipe of corresponding schedule number.

JOINING METHODS

Threading

Threading as per ASME *Standard* B1.20.1 is the most common method for joining small-diameter steel or brass pipe. Pipe with a wall thickness less than standard weight should not be threaded. ASME *Standard* B31.5 limits the threading for various refrigerants and pipe sizes.

Soldering and Brazing

Copper tube is usually joined by soldering or brazing socket end fittings. Brazing materials melt above 1000°F and produce a stronger joint than solder. Table 4 lists soldered and brazed joint strengths. ASME *Standard* B16.22 specified wrought copper solder joint fittings and ASME *Standard* B16.18 specified cast copper solder joint fittings are pressure rated the same way as annealed Type L copper tube of the same size. Health concerns have caused many jurisdictions to ban solder containing lead or antimony for joining pipe in potable-water systems. Lead-based solder, in particular, must not be used for potable water.

Flared and Compression Joints

Flared and compression fittings can be used to join copper, steel, stainless steel, and aluminum tubing. Properly rated fittings can keep the joints as strong as the tube.

Table 2 Steel Pipe Data

Nominal Size, in.	Pipe OD, in.	Schedule Number or Weight[a]	Wall Thickness t, in.	Inside Diameter d, in.	Surface Area		Cross Section		Weight		Working Pressure[c] ASTM A53 B to 400°F		
					Outside, ft²/ft	Inside, ft²/ft	Metal Area, in²	Flow Area, in²	Pipe, lb/ft	Water, lb/ft	Mfr. Process	Joint Type[b]	psig
1/4	0.540	40 ST	0.088	0.364	0.141	0.095	0.125	0.104	0.424	0.045	CW	T	188
		80 XS	0.119	0.302	0.141	0.079	0.157	0.072	0.535	0.031	CW	T	871
3/8	0.675	40 ST	0.091	0.493	0.177	0.129	0.167	0.191	0.567	0.083	CW	T	203
		80 XS	0.126	0.423	0.177	0.111	0.217	0.141	0.738	0.061	CW	T	820
1/2	0.840	40 ST	0.109	0.622	0.220	0.163	0.250	0.304	0.850	0.131	CW	T	214
		80 XS	0.147	0.546	0.220	0.143	0.320	0.234	1.087	0.101	CW	T	753
3/4	1.050	40 ST	0.113	0.824	0.275	0.216	0.333	0.533	1.13	0.231	CW	T	217
		80 XS	0.154	0.742	0.275	0.194	0.433	0.432	1.47	0.187	CW	T	681
1	1.315	40 ST	0.133	1.049	0.344	0.275	0.494	0.864	1.68	0.374	CW	T	226
		80 XS	0.179	0.957	0.344	0.251	0.639	0.719	2.17	0.311	CW	T	642
1-1/4	1.660	40 ST	0.140	1.380	0.435	0.361	0.669	1.50	2.27	0.647	CW	T	229
		80 XS	0.191	1.278	0.435	0.335	0.881	1.28	2.99	0.555	CW	T	594
1-1/2	1.900	40 ST	0.145	1.610	0.497	0.421	0.799	2.04	2.72	0.881	CW	T	231
		80 XS	0.200	1.500	0.497	0.393	1.068	1.77	3.63	0.765	CW	T	576
2	2.375	40 ST	0.154	2.067	0.622	0.541	1.07	3.36	3.65	1.45	CW	T	230
		80 XS	0.218	1.939	0.622	0.508	1.48	2.95	5.02	1.28	CW	T	551
2-1/2	2.875	40 ST	0.203	2.469	0.753	0.646	1.70	4.79	5.79	2.07	CW	W	533
		80 XS	0.276	2.323	0.753	0.608	2.25	4.24	7.66	1.83	CW	W	835
3	3.500	40 ST	0.216	3.068	0.916	0.803	2.23	7.39	7.57	3.20	CW	W	482
		80 XS	0.300	2.900	0.916	0.759	3.02	6.60	10.25	2.86	CW	W	767
4	4.500	40 ST	0.237	4.026	1.178	1.054	3.17	12.73	10.78	5.51	CW	W	430
		80 XS	0.337	3.826	1.178	1.002	4.41	11.50	14.97	4.98	CW	W	695
6	6.625	40 ST	0.280	6.065	1.734	1.588	5.58	28.89	18.96	12.50	ERW	W	696
		80 XS	0.432	5.761	1.734	1.508	8.40	26.07	28.55	11.28	ERW	W	1209
8	8.625	30	0.277	8.071	2.258	2.113	7.26	51.16	24.68	22.14	ERW	W	526
		40 ST	0.322	7.981	2.258	2.089	8.40	50.03	28.53	21.65	ERW	W	643
		80 XS	0.500	7.625	2.258	1.996	12.76	45.66	43.35	19.76	ERW	W	1106
10	10.75	30	0.307	10.136	2.814	2.654	10.07	80.69	34.21	34.92	ERW	W	485
		40 ST	0.365	10.020	2.814	2.623	11.91	78.85	40.45	34.12	ERW	W	606
		XS	0.500	9.750	2.814	2.552	16.10	74.66	54.69	32.31	ERW	W	887
		80	0.593	9.564	2.814	2.504	18.92	71.84	64.28	31.09	ERW	W	1081
12	12.75	30	0.330	12.090	3.338	3.165	12.88	114.8	43.74	49.68	ERW	W	449
		ST	0.375	12.000	3.338	3.141	14.58	113.1	49.52	48.94	ERW	W	528
		40	0.406	11.938	3.338	3.125	15.74	111.9	53.48	48.44	ERW	W	583
		XS	0.500	11.750	3.338	3.076	19.24	108.4	65.37	46.92	ERW	W	748
		80	0.687	11.376	3.338	2.978	26.03	101.6	88.44	43.98	ERW	W	1076
14	14.00	30 ST	0.375	13.250	3.665	3.469	16.05	137.9	54.53	59.67	ERW	W	481
		40	0.437	13.126	3.665	3.436	18.62	135.3	63.25	58.56	ERW	W	580
		XS	0.500	13.000	3.665	3.403	21.21	132.7	72.04	57.44	ERW	W	681
		80	0.750	12.500	3.665	3.272	31.22	122.7	106.05	53.11	ERW	W	1081
16	16.00	30 ST	0.375	15.250	4.189	3.992	18.41	182.6	62.53	79.04	ERW	W	421
		40 XS	0.500	15.000	4.189	3.927	24.35	176.7	82.71	76.47	ERW	W	596
18	18.00	ST	0.375	17.250	4.712	4.516	20.76	233.7	70.54	101.13	ERW	W	374
		30	0.437	17.126	4.712	4.483	24.11	230.3	81.91	99.68	ERW	W	451
		XS	0.500	17.000	4.712	4.450	27.49	227.0	93.38	98.22	ERW	W	530
		40	0.562	16.876	4.712	4.418	30.79	223.7	104.59	96.80	ERW	W	607
20	20.00	20 ST	0.375	19.250	5.236	5.039	23.12	291.0	78.54	125.94	ERW	W	337
		30 XS	0.500	19.000	5.236	4.974	30.63	283.5	104.05	122.69	ERW	W	477
		40	0.593	18.814	5.236	4.925	36.15	278.0	122.82	120.30	ERW	W	581

[a]Numbers are schedule numbers per ASME *Standard* B36.10M; ST = Standard Weight; XS = Extra Strong.

[b]T = Thread; W = Weld

[c]Working pressures were calculated per ASME B31.9 using furnace butt-weld (continuous weld, CW) pipe through 4 in. and electric resistance weld (ERW) thereafter. The allowance A has been taken as

(1) 12.5% of t for mill tolerance on pipe wall thickness, *plus*

(2) An arbitrary corrosion allowance of 0.025 in. for pipe sizes through NPS 2 and 0.065 in. from NPS 2½ through 20, *plus*

(3) A thread cutting allowance for sizes through NPS 2.

Because the pipe wall thickness of threaded standard pipe is so small after deducting the allowance A, the mechanical strength of the pipe is impaired. It is good practice to limit standard weight threaded pipe pressure to 90 psig for steam and 125 psig for water.

Table 3 Copper Tube Data

Nominal Diameter, in.	Type	Wall Thickness t, in.	Diameter		Surface Area		Cross Section		Weight		Working Pressure[a,b,c] ASTM B88 to 250°F	
			Outside D, in.	Inside d, in.	Outside, ft²/ft	Inside, ft²/ft	Metal Area, in²	Flow Area, in²	Tube, lb/ft	Water, lb/ft	Annealed, psig	Drawn, psig
1/4	K	0.035	0.375	0.305	0.098	0.080	0.037	0.073	0.145	0.032	851	1596
	L	0.030	0.375	0.315	0.098	0.082	0.033	0.078	0.126	0.034	730	1368
3/8	K	0.049	0.500	0.402	0.131	0.105	0.069	0.127	0.269	0.055	894	1676
	L	0.035	0.500	0.430	0.131	0.113	0.051	0.145	0.198	0.063	638	1197
	M	0.025	0.500	0.450	0.131	0.118	0.037	0.159	0.145	0.069	456	855
1/2	K	0.049	0.625	0.527	0.164	0.138	0.089	0.218	0.344	0.094	715	1341
	L	0.040	0.625	0.545	0.164	0.143	0.074	0.233	0.285	0.101	584	1094
	M	0.028	0.625	0.569	0.164	0.149	0.053	0.254	0.203	0.110	409	766
5/8	K	0.049	0.750	0.652	0.196	0.171	0.108	0.334	0.418	0.144	596	1117
	L	0.042	0.750	0.666	0.196	0.174	0.093	0.348	0.362	0.151	511	958
3/4	K	0.065	0.875	0.745	0.229	0.195	0.165	0.436	0.641	0.189	677	1270
	L	0.045	0.875	0.785	0.229	0.206	0.117	0.484	0.455	0.209	469	879
	M	0.032	0.875	0.811	0.229	0.212	0.085	0.517	0.328	0.224	334	625
1	K	0.065	1.125	0.995	0.295	0.260	0.216	0.778	0.839	0.336	527	988
	L	0.050	1.125	1.025	0.295	0.268	0.169	0.825	0.654	0.357	405	760
	M	0.035	1.125	1.055	0.295	0.276	0.120	0.874	0.464	0.378	284	532
1-1/4	K	0.065	1.375	1.245	0.360	0.326	0.268	1.217	1.037	0.527	431	808
	L	0.055	1.375	1.265	0.360	0.331	0.228	1.257	0.884	0.544	365	684
	M	0.042	1.375	1.291	0.360	0.338	0.176	1.309	0.682	0.566	279	522
	DWV	0.040	1.375	1.295	0.360	0.339	0.168	1.317	0.650	0.570	265	497
1-1/2	K	0.072	1.625	1.481	0.425	0.388	0.351	1.723	1.361	0.745	404	758
	L	0.060	1.625	1.505	0.425	0.394	0.295	1.779	1.143	0.770	337	631
	M	0.049	1.625	1.527	0.425	0.400	0.243	1.831	0.940	0.792	275	516
	DWV	0.042	1.625	1.541	0.425	0.403	0.209	1.865	0.809	0.807	236	442
2	K	0.083	2.125	1.959	0.556	0.513	0.532	3.014	2.063	1.304	356	668
	L	0.070	2.125	1.985	0.556	0.520	0.452	3.095	1.751	1.339	300	573
	M	0.058	2.125	2.009	0.556	0.526	0.377	3.170	1.459	1.372	249	467
	DWV	0.042	2.125	2.041	0.556	0.534	0.275	3.272	1.065	1.416	180	338
2-1/2	K	0.095	2.625	2.435	0.687	0.637	0.755	4.657	2.926	2.015	330	619
	L	0.080	2.625	2.465	0.687	0.645	0.640	4.772	2.479	2.065	278	521
	M	0.065	2.625	2.495	0.687	0.653	0.523	4.889	2.026	2.116	226	423
3	K	0.109	3.125	2.907	0.818	0.761	1.033	6.637	4.002	2.872	318	596
	L	0.090	3.125	2.945	0.818	0.771	0.858	6.812	3.325	2.947	263	492
	M	0.072	3.125	2.981	0.818	0.780	0.691	6.979	2.676	3.020	210	394
	DWV	0.045	3.125	3.035	0.818	0.795	0.435	7.234	1.687	3.130	131	246
3-1/2	K	0.120	3.625	3.385	0.949	0.886	1.321	8.999	5.120	3.894	302	566
	L	0.100	3.625	3.425	0.949	0.897	1.107	9.213	4.291	3.987	252	472
	M	0.083	3.625	3.459	0.949	0.906	0.924	9.397	3.579	4.066	209	392
4	K	0.134	4.125	3.857	1.080	1.010	1.680	11.684	6.510	5.056	296	555
	L	0.110	4.125	3.905	1.080	1.022	1.387	11.977	5.377	5.182	243	456
	M	0.095	4.125	3.935	1.080	1.030	1.203	12.161	4.661	5.262	210	394
	DWV	0.058	4.125	4.009	1.080	1.050	0.741	12.623	2.872	5.462	128	240
5	K	0.160	5.125	4.805	1.342	1.258	2.496	18.133	9.671	7.846	285	534
	L	0.125	5.125	4.875	1.342	1.276	1.963	18.665	7.609	8.077	222	417
	M	0.109	5.125	4.907	1.342	1.285	1.718	18.911	6.656	8.183	194	364
	DWV	0.072	5.125	4.981	1.342	1.304	1.143	19.486	4.429	8.432	128	240
6	K	0.192	6.125	5.741	1.603	1.503	3.579	25.886	13.867	11.201	286	536
	L	0.140	6.125	5.845	1.603	1.530	2.632	26.832	10.200	11.610	208	391
	M	0.122	6.125	5.881	1.603	1.540	2.301	27.164	8.916	11.754	182	341
	DWV	0.083	6.125	5.959	1.603	1.560	1.575	27.889	6.105	12.068	124	232
8	K	0.271	8.125	7.583	2.127	1.985	6.687	45.162	25.911	19.542	304	570
	L	0.200	8.125	7.725	2.127	2.022	4.979	46.869	19.295	20.280	224	421
	M	0.170	8.125	7.785	2.127	2.038	4.249	47.600	16.463	20.597	191	358
	DWV	0.109	8.125	7.907	2.127	2.070	2.745	49.104	10.637	21.247	122	229
10	K	0.338	10.125	9.449	2.651	2.474	10.392	70.123	40.271	30.342	304	571
	L	0.250	10.125	9.625	2.651	2.520	7.756	72.760	30.054	31.483	225	422
	M	0.212	10.125	9.701	2.651	2.540	6.602	73.913	25.584	31.982	191	358
12	K	0.405	12.125	11.315	3.174	2.962	14.912	100.554	57.784	43.510	305	571
	L	0.280	12.125	11.565	3.174	3.028	10.419	105.046	40.375	45.454	211	395
	M	0.254	12.125	11.617	3.174	3.041	9.473	105.993	36.706	45.863	191	358

[a]When using soldered or brazed fittings, the joint determines the limiting pressure.

[b]Working pressures were calculated using ASME *Standard* B31.9 allowable stresses. A 5% mill tolerance has been used on the wall thickness. Higher tube ratings can be calculated using the allowable stress for lower temperatures.

[c]If soldered or brazed fittings are used on hard drawn tubing, use the annealed ratings. Full-tube allowable pressures can be used with suitably rated flare or compression-type fittings.

Table 4 Internal Working Pressure for Copper Tube Joints

Alloy Used for Joints	Service Temperature, °F	Internal Working Pressure, psi — Water and Noncorrosive Liquids and Gases[a] — Nominal Tube Size (Types K, L, M), in. 1/4 to 1	1 1/4 to 2	2 1/2 to 4	5 to 8[a]	10 to 12[a]	Sat. Steam and Condensate 1/4 to 8
50-50 Tin/lead[b] solder (ASTM B32 Gr 50A)	100	200	175	150	130	100	—
	150	150	125	100	90	70	—
	200	100	90	75	70	50	—
	250	85	75	50	45	40	15
95-5 Tin/antimony[c] solder (ASTM B32 Gr 50TA)	100	500	400	300	270	150	—
	150	400	350	275	250	150	—
	200	300	250	200	180	140	—
	250	200	175	150	135	110	15
Brazing alloys melting at or above 1000°F	100 to 200	d	d	d	d	d	—
	250	300	210	170	150	150	—
	350	270	190	150	150	150	120

Source: Based on ASME *Standard* B31.9, Building Services Piping
[a]Solder joints are not to be used for
(1) Flammable or toxic gases or liquids
(2) Gas, vapor, or compressed air in tubing over 4 in., unless max. pressure is limited to 20 psig.
[b]Lead solders must not be used in potable-water systems.
[c]Tin/antimony solder is allowed for potable-water supplies in some jurisdictions.
[d]Rated pressure for up to 200°F applies to the tube being joined.

Flanges

Flanges can be used for large pipe and all piping materials. They are commonly used to connect to equipment and valves, and wherever the joint must be opened to permit service or replacement of components. For steel pipe, flanges are available in pressure ratings to 2500 psig. High tensile strength bolts must be used for high pressure flanged joints.

For welded pipe, weld neck, slip-on, or socket weld flanges are available. Thread-on flanges are available for threaded pipe.

Flanges are generally flat faced or raised face. Flat-faced flanges with full-faced gaskets are most often used with cast iron and materials that cannot take high bending loads. Raised-face flanges with ring gaskets are preferred with steel pipe because they facilitate increasing the sealing pressure on the gasket to help prevent leaks. Other facings, such as O ring and ring joint, are available for special applications.

All flat-faced, raised-face, and lap-joint flanges require a gasket between the mating flange surfaces. Gaskets are made from rubber, synthetic elastomers, cork, fiber, plastic, teflon, metal, and combinations of these materials. The gasket must be compatible with the flowing media and the temperatures at which the system operates.

Welding

Welded-steel pipe joints offer the following advantages:

- Do not age, dry out, or deteriorate as do gasketed joints
- Can accommodate greater vibration and water hammer and higher temperatures and pressures than other joints
- For critical service, can be tested by any of several nondestructive examination (NDE) methods, such as radiography or ultrasound
- Provide maximum long-term reliability

The applicable sections of the ASME *Standard* B31series and the ASME *Boiler and Pressure Vessel Code* give rules for welding. ASTM *Standard* B31 requires that all welders and welding procedure specifications (WPS) be qualified. Separate WPS are needed for different welding methods and materials. The qualifying tests and the variables requiring separate procedure specifications are set forth in the ASME *Boiler and Pressure Vessel Code*, Section IX. The manufacturer, fabricator, or contractor is responsible for the welding procedure and welders. ASME *Standard* B31.9 requires visual examination of welds and outlines limits of acceptability.

The following welding processes are often used in the HVAC industry:

SMAW: shielded metal arc welding (stick welding). The molten weld metal is shielded by the vaporization of the electrode coating.

GMAW: gas metal arc welding, also called MIG. The electrode is a continuously fed wire, which is shielded by argon or carbon dioxide gas from the welding gun nozzle.

GTAW: gas tungsten arc welding, also called TIG or Heliarc. This process uses a nonconsumable tungsten electrode surrounded by a shielding gas. The weld material may be provided from a separate noncoated rod.

Reinforced Outlet Fittings

Reinforced outlet fittings are used to make branch and take-off connections and are designed to permit welding directly to pipe without supplemental reinforcing. Fittings are available with threaded, socket, or butt-weld outlets.

Other Joints

Grooved joints require special grooved fittings and a shallow groove cut or rolled into the pipe end. These joints can be used with steel, cast iron, ductile-iron, copper, and plastic pipes. A segmented clamp engages the grooves and a special gasket designed so that internal pressure tightens the seal. Some clamps are designed with clearance between tongue and groove to accommodate misalignment and thermal movements, and others are designed to limit movement and provide a rigid system. Manufacturers' data gives temperature and pressure limitations.

Another form of mechanical joint consists of a **sleeve** slightly larger than the outside diameter of the pipe. The pipe ends are inserted into the sleeve, and gaskets are packed into the annular space between the pipe and coupling and held in place by retainer rings. This type of joint can accept some axial misalignment, but it must be anchored or otherwise restrained to prevent axial pullout or lateral movement. Manufacturers provide pressure/temperature data.

Ductile-iron pipe is furnished with a spigot end adapted for a gasket and retainer ring. This joint is also not restrained.

Unions

Unions allow disassembly of threaded pipe systems. Unions are three-part fittings with a mating machined seat on the two parts that thread onto the pipe ends. A threaded locking ring holds the two ends tightly together. A union also allows threaded pipe to be turned at the last joint connecting two pieces of equipment. Companion flanges (a pair) for small pipe serve the same purpose.

SPECIAL SYSTEMS

Certain piping systems are governed by separate codes or standards, which are summarized below. Generally, any failure of the piping in these systems is dangerous to the public, so some local areas have adopted laws enforcing the codes.

- **Boiler piping.** ASME *Standard* B31.1 and the ASME *Boiler and Pressure Vessel Code* (Section I) specify the piping inside the code-required stop valves on boilers that operate above 15 psig with steam, or above 160 psig or 250°F with water. These codes require fabricators and contractors to be certified for such work. The field or shop work must also be inspected while it is in progress by inspectors commissioned by the National Board of Boiler and Pressure Vessel Inspectors.
- **Refrigeration piping.** ASME *Standard* B31.5 and ASHRAE *Standard* 15 cover the requirements for refrigerant piping.
- **Plumbing systems.** Local codes cover piping for plumbing.
- **Sprinkler systems.** NFPA *Standard* 13 covers this field.
- **Fuel gas.** NFPA 54/ANSI Z223.1, *National Fuel Gas Code*, prescribes fuel gas piping in buildings.

SELECTION OF MATERIALS

Each HVAC system and, under some conditions, portions of a system require a study of the conditions of operation to determine suitable materials. For example, because the static pressure of water in a high-rise building is higher in the lower levels than in the upper levels, different materials may be required along vertical zones.

The following factors should be considered when selecting material for piping:

- Code requirements
- Working fluid in the pipe
- Pressure and temperature of the fluid
- External environment of the pipe
- Installation cost

Table 5 lists materials used for heating and air-conditioning piping. The pressure and temperature rating of each component selected must be considered; the lowest rating establishes the operating limits of the system.

Table 5 Application of Pipe, Fittings, and Valves for Heating and Air Conditioning

Application	Pipe Material	Weight	Joint Type	Fitting Class	Fitting Material	System Temperature, °F	System Maximum Pressure at Temperature,[a] psig
Recirculating Water	Steel (CW)	Standard	Thread	125	Cast iron	250	125
2 in. and smaller	Copper, hard	Type L	Braze or silver solder[b]		Wrought copper	250	200
	PVC	Sch 80	Solvent	Sch 80	PVC	75	
	CPVC	Sch 80	Solvent	Sch 80	CPVC	150	
	PB	SDR-11	Heat fusion		PB	160	
			Insert crimp		Metal	160	
2.5 to 12 in.	A53 B ERW Steel	Standard	Weld	Standard	Wrought steel	250	400
			Flange	150	Wrought steel	250	250
			Flange	125	Cast iron	250	175
			Flange	250	Cast iron	250	400
			Groove		MI or ductile iron	230	300
	PB	SDR-11	Heat fusion		PB	160	
Steam and Condensate	Steel (CW)	Standard[c]	Thread	125	Cast iron		90
2 in. and smaller			Thread	150	Malleable iron		90
	A53 B ERW Steel	Standard[c]	Thread	125	Cast iron		100
			Thread	150	Malleable iron		125
	A53 B ERW Steel	XS	Thread	250	Cast iron		200
			Thread	300	Malleable iron		250
2.5 to 12 in.	Steel	Standard	Weld	Standard	Wrought steel		250
			Flange	150	Wrought steel		200
			Flange	125	Cast iron		100
	A53 B ERW Steel	XS	Weld	XS	Wrought steel		700
			Flange	300	Wrought steel		500
			Flange	250	Cast iron		200
Refrigerant	Copper, hard	Type L or K	Braze		Wrought copper		
	A53 B SML Steel	Standard	Weld		Wrought steel		
Underground Water							
Through 12 in.	Copper, hard	Type K	Braze or silver solder[b]		Wrought copper	75	350
Through 6 in.	Ductile iron	Class 50	MJ	MJ	Cast iron	75	250
	PB	SDR 9, 11	Heat fusion		PB	75	
		SDR 7, 11.5	Insert crimp		Metal	75	
Potable Water, Inside Building	Copper, hard	Type L	Braze or silver solder[b]		Wrought copper	75	350
	Steel, galvanized	Standard	Thread	125	Galv. cast iron	75	125
				150	Galv. mall. iron	75	125
	PB	SDR-11	Heat fusion		PB	75	
			Insert crimp		Metal	75	

[a]Maximum allowable working pressures have been derated in this table. Higher system pressures can be used for lower temperatures and smaller pipe sizes. Pipe, fittings, joints, and valves must all be considered.

[b]Lead- and antimony-based solders should not be used for potable-water systems. Brazing and silver solders should be employed.

[c]Extra-strong pipe is recommended for all threaded condensate piping to allow for corrosion.

Table 6 Suggested Hanger Spacing and Rod Size for Straight Horizontal Runs

NPS, in.	Hanger Spacing, ft Standard Steel Pipe* Water	Standard Steel Pipe* Steam	Copper Tube Water	Rod Size, in.
1/2	7	8	5	1/4
3/4	7	9	5	1/4
1	7	9	6	1/4
1 1/2	9	12	8	3/8
2	10	13	8	3/8
2 1/2	11	14	9	3/8
3	12	15	10	3/8
4	14	17	12	1/2
6	17	21	14	1/2
8	19	24	16	5/8
10	20	26	18	3/4
12	23	30	19	7/8
14	25	32		1
16	27	35		1
18	28	37		1 1/4
20	30	39		1 1/4

Source: Adapted from MSS *Standard* SP-69

*Spacing does not apply where span calculations are made or where concentrated loads are placed between supports such as flanges, valves, specialties, etc.

PIPE WALL THICKNESS

The primary factors determining pipe wall thickness are hoop stress due to internal pressure and longitudinal stresses due to pressure, weight, and other sustained loads. Detailed stress calculations are seldom required for HVAC applications because standard pipe has ample thickness to sustain the pressure and longitudinal stress due to weight (assuming hangers are spaced in accordance with Table 6).

STRESS CALCULATIONS

Although stress calculations are seldom required, the factors involved should be understood. The main areas of concern are (1) internal pressure stress, (2) longitudinal stress due to pressure and weight, and (3) stress due to expansion and contraction.

ASME B31 standards establish a basic allowable stress S equal to one-fourth of the minimum tensile strength of the material. This value is adjusted, as discussed in this section, because of the nature of certain stresses and manufacturing processes.

Hoop stress caused by internal pressure is the major stress on pipes. As certain forming methods form a seam that may be weaker than the base material, ASME *Standard* B31.9 specifies a joint efficiency factor E which, multiplied by the basic allowable stress, establishes a maximum allowable stress value in tension S_E. (Table A-1 in ASME B31.9 lists values of S_E for commonly used pipe materials.) The joint efficiency factor can be significant; for example, seamless pipe has a joint efficiency factor of 1, so it can be used to the full allowable stress (one-quarter of the tensile strength). In contrast, butt-welded pipe has a joint efficiency factor of 0.60, so its maximum allowable stress must be derated ($S_E = 0.6S$).

Equation (1) determines the minimum wall thickness for a given pressure. Equation (2) determines the maximum pressure allowed for a given wall thickness.

$$t_m = \frac{pD}{2S_E} + A \tag{1}$$

$$p = \frac{2S_E(t_m - A)}{D} \tag{2}$$

where

t_m = minimum required wall thickness, in.
S_E = maximum allowable stress, psi
D = outside pipe diameter, in.
A = allowance for manufacturing tolerance, threading, grooving, and corrosion, in.
p = internal pressure, psi

Both equations incorporate an allowance factor A to compensate for manufacturing tolerances, material removed in threading or grooving, and corrosion. For the seamless, butt-welded, and electric resistance welded (ERW) pipe most commonly used in HVAC work, the standards apply a manufacturing tolerance of 12.5%. Working pressure for steel pipe, as listed in Table 2, has been calculated using a manufacturing tolerance of 12.5%, standard allowance for depth of thread (where applicable), and a corrosion allowance of 0.065 in. for pipes 2 1/2 in. and larger and 0.025 in. for pipes 2 in. and smaller. Where corrosion is known to be greater or smaller, pressure rating can be recalculated using Equation (2). Higher pressure ratings than shown in Table 2 can be obtained (1) by using ERW or seamless pipe in lieu of continuous-weld (CW) pipe 4 in. and less and seamless pipe in lieu of ERW pipe 5 in. and greater (due to higher joint efficiency factors), or (2) by using heavier wall pipe.

Longitudinal stresses caused by pressure, weight, and other sustained forces are additive, and the sum of all such stresses must not exceed the basic allowable stress S at the highest temperature at which the system will operate. Longitudinal stress caused by pressure equals approximately one-half the hoop stress caused by internal pressure, which means that at least one-half the basic allowable stress is available for weight and other sustained forces. This factor is taken into account in Table 6.

Stresses caused by expansion and contraction are cyclical, and, because creep allows some stress relaxation, the ASME B31 standards permit designing to an allowable stress range S_A as calculated by Equation (3). Table 1 lists allowable stress ranges for commonly used piping materials.

$$S_A = 1.25S_c + 0.25S_h \tag{3}$$

where

S_A = allowable stress range, psi
S_c = allowable cold stress at coolest temperature the system will experience, psi
S_h = allowable hot stress at hottest temperature the system will experience, psi

PLASTIC PIPING

Nonmetallic pipe is widely used in HVAC and plumbing. Plastic is light in weight, generally inexpensive, and corrosion-resistant. Plastic also has a low "C" factor (i.e., its surface is very smooth), which results in lower pumping power and smaller pipe sizes. The disadvantages of plastic pipe include the rapid loss of strength at temperatures above ambient and the high coefficient of linear expansion. The modulus of elasticity of plastics is low, resulting in a short support span. Some jurisdictions do not allow certain plastics in buildings because of toxic products emitted under fire conditions.

Plastic piping materials fall into two main categories: thermoplastic and thermosetting. Thermoplastics melt and are formed by extruding or molding. They are usually used without reinforcing filaments. Thermosets are cured and cannot be reformed. They are normally used with glass fiber reinforcing filaments.

For the purposes of this chapter, thermoplastic piping is made of the following materials:

PVC polyvinyl chloride
CPVC chlorinated polyvinyl chloride
PB polybutylene
PE polyethylene
PP polypropylene
ABS acrylonitrile butadiene styrene
PVDF polyvinylidene fluoride

Thermosetting piping used in HVAC is called (1) reinforced thermosetting resin (RTR) and (2) fiberglass-reinforced plastic (FRP). RTR and FRP are interchangeable and refer to pipe and fittings commonly made of (1) fiberglass-reinforced epoxy resin, (2) fiberglass-reinforced vinyl ester, and (3) fiberglass-reinforced polyester.

Pipe and fittings made from epoxy resin are generally stronger and operate at a higher temperature than those made from polyester or vinyl ester resins, so they are more likely to be used in HVAC.

Table 7 lists properties of various plastic piping materials. Values for steel and copper are given for comparison.

Allowable Stress

Both thermoplastics and thermosets have an allowable stress derived from a hydrostatic design basis stress (HDBS). The HDBS is determined by a statistical analysis of both static and cyclic stress rupture test data as set forth in ASTM *Standard* D2837 for thermoplastics and ASTM *Standard* D2992 for glass-fiber-reinforced thermosetting resins.

The allowable stress, which is called the hydrostatic design stress (HDS), is obtained by multiplying the HDBS by a service factor. The HDS values recommended by some manufacturers and those allowed by the ASME B31, *Code for Pressure Piping*, are listed in Table 7.

The pressure design thickness for plastic pipe can be calculated using the code stress values and the following formula:

$$t = pD/(2S + p) \tag{4}$$

where

t = pressure design thickness, in.
p = internal design pressure, psig
D = pipe outside diameter, in.
S = design stress (HDS), psi

The minimum required wall thickness can be found by adding an allowance for mechanical strength, threading, grooving, erosion, and corrosion to the calculated pressure design thickness.

As there are many formulations of the polymers used for piping materials and different joining methods for each, manufacturers' recommendations should be observed. Most catalogs give the pressure ratings for pipe and fittings at various temperatures up to the maximum the material will withstand.

Plastic Material Selection

The selection of a plastic for a specific purpose requires attention. All are suitable for cold water. Plastic pipe should not be used for compressed gases or compressed air if the pipe is made of a material subject to brittle failure. For other liquids and chemicals, refer to charts provided by plastic pipe manufacturers and distributors. Table 7 gives properties of the various plastics discussed in this section. The last column gives the relative cost of small pipe in each category. Table 8 lists some applications pertinent to HVAC. The following are brief descriptions of common uses for the various materials.

PVC. Because polyvinyl chloride has the best overall range of properties at the lowest cost, it is the most widely used plastic. It is joined by solvent cementing, threading, or flanging. Gasketed push-on joints are also used for larger sizes.

CPVC. Chlorinated polyvinyl chloride has the same properties as PVC but can withstand a higher temperature before losing strength. It is joined by the same methods as PVC.

PB. A lightweight, flexible material, polybutylene can be used up to 210°F. It is used for both hot and cold plumbing water piping. It is joined by heat fusion or mechanical means, can be bent to a 10 diameter radius, and is provided in coils.

PE. Low-density polyethylene (LDPE) is a flexible lightweight tubing with good low-temperature properties. It is used in the food and beverage industry and for instrument tubing. It is joined by mechanical means such as compression fittings or push-on connectors and clamps.

High-density polyethylene (HDPE) is a tough, weather-resistant material used for large pipelines in the gas industry. Fabricated fittings are available. It is joined by heat fusion for large sizes; flare, compression, or insert fittings can be used on small sizes.

PP. Polypropylene is a lightweight plastic used for pressure applications and also for chemical waste lines, because it is inert to a wide range of chemicals. A broad variety of drainage fittings are available. For pressure uses, regular fittings are made. It is joined by heat fusion.

ABS. Acrylonitrile butadiene styrene is a high-strength, impact- and weather-resistant material. Some formulations can be used for compressed air. ABS is also used in the food and beverage industry. A wide range of fittings is available. It is joined by solvent cement, threading, or flanging.

PVDF. Polyvinylidene fluoride is widely used for ultrapure water systems and in the pharmaceutical industry and has a wide temperature range. This material is over 20 times more expensive than PVC. It is joined by heat fusion, and fittings are made for this purpose. For smaller sizes, mechanical joints can be used.

PIPE-SUPPORTING ELEMENTS

Pipe-supporting elements consist of (1) hangers, which support from above; (2) supports, which bear load from below; and (3) restraints, such as anchors and guides, which limit or direct movement, as well as support loads. Pipe-supporting elements withstand all static and dynamic conditions including the following:

- Weight of pipe, valves, fittings, insulation, and fluid contents, including test fluid if using a heavier-than-normal media
- Occasional loads such as ice, wind, and seismic forces
- Forces imposed by thermal expansion and contraction of pipe bends and loops
- Frictional, spring, and pressure thrust forces imposed by expansion joints in the system
- Frictional forces of guides and supports
- Other loads, such as water hammer, vibration, and reactive force of relief valves
- Test load and force

In addition, pipe-supporting elements must be evaluated in terms of stress at the point of connection to the pipe and the building structure. Stress at the point of connection to the pipe is especially important for base elbow and trunnion supports, because the limiting and controlling parameter is usually not the strength of the structural member, but the localized stress and the point of attachment to the pipe. Loads on anchors, cast-in-place inserts, and other attachments to concrete should not be more than one-fifth the ultimate strength of the attachment, as determined by manufacturers' tests. All loads on the structure should be communicated to and coordinated with the structural engineer.

The ASME B31 standards establish criteria for the design of pipe-supporting elements and the Manufacturers Standardization Society of the Valve and Fittings Industry (MSS) has established standards for the design, fabrication, selection, and installation of pipe hangers and supports based on these codes.

MSS *Standard* SP-69 and the catalogs of many manufacturers illustrate the various hangers and components and provide information on the types to use with different pipe systems. Table 6 shows suggested pipe support spacing, and Table 9 provides a maximum safe load for threaded steel rods.

The loads on most pipe-supporting elements are moderate and can be selected safely in accordance with manufacturers' catalog data and the information presented in this section; however, some loads and forces can be very high, especially in multistory buildings and for large-diameter pipe, especially where expansion

Table 7 Properties of Plastic Pipe Materials[a]

Material			Tensile Strength, psi (at 73°F)	Hydrostatic[b] Design Stress, psi (at 73°F)		Upper Temperature Limit, °F		HDS[b] Upper Limit, psi	Specific Gravity[c]	Impact Strength, ft·lb/in (at 73°F)	Modulus of Elasticity, psi (at 73°F)	Coefficient of Expansion, in/10⁶ in·°F	Thermal Conductivity, Btu·in/h·ft²·°F	Relative Pipe Cost[d]
Designation	Type and Grade	Cell No.		Mfr.	ASME B31	Mfr.	ASME B31							
Thermoplastics														
PVC 1120	T I,G1	12454-B	7,500	2,000	2,000	140	150	440	1.40	0.8	420,000	30.0	1.1	1.0
PVC 1200	T I,G2	12454-C			2,000		150				410,000	35.0		
PVC 2120	T II,G1	14333-D			2,000		150					30.0		
CPVC 4120	T IV,G1	23447-B	8,000	2,000	2,000	210	210	320	1.55	1.5	423,000	35.0	0.95	2.9
PB 2110	T II,G1		4,800	1,000	1,000	180	210	<500	0.93		38,000	72.0	1.5	2.9
PE 2306	Gr. P23				630		140				90,000	80.0		
PE 3306	Gr. P34				630		160				130,000	70.0		
PE 3406	Gr. P33				630		180				150,000	60.0		
HDPE 3408	Gr. P34	355434-C	5,000	1,600	800	140	180	800	0.96	12	110,000	120.0	2.7	1.1
PP			5,000	705		212	210		0.91	1.3	120,000	60.0	1.3	2.9
ABS	Duraplus	6-3-3	5,500			176			1.06	8.5	240,000	56.0	1.7	3.4
ABS 1210	T I,G2	5-2-2			1,000		180	640			250,000	55.0		
ABS 1316	T I,G3	3-5-5			1,600		180	1,000			340,000	40.0		
ABS 2112	T II,G1	4-4-5			1,250		180	800				40.0		
PVDF			7,000	1,275		280	275	306	1.78	3.8	125,000	79.0	0.8	28.0
Thermosetting														
Epoxy-Glass	RTRP-11AF		44,000	8,000			300	7,000			1,000,000	9 to 13	2.9	
Polyester-Glass	RTRP-12EF		44,000	9,000		200	200	5,000			1,000,000	9 to 11	1.3	
For Comparison														
Steel	A 53 B	ERW	60,000		12,800		800	9,200	7.80	30.0	27,500,000	6.31	344	1.3
Copper	Type L	Drawn	36,000		9,000		400	8,200	8.90		17,000,000	9.5		3.5

[a] The properties listed are for the specific materials listed as each plastic has other formulations. Consult the manufacturer of the system chosen. These values are for comparative purposes.
[b] The hydrostatic design stress (HDS) is equivalent to the allowable design stress.
[c] Relative to water at 62.4 lb/ft³.
[d] Based on the cost of pipe only, without factoring in fittings, joints, hangers, and labor.

Table 8 Manufacturers' Recommendations[a,b] for Plastic Materials

	PVC	CPVC	PB	HDPE	PP	ABS	PVDF	RTRP
Cold water service	R	R	R	R	R	R	R	R
Hot (140°F) water	N	R	R	R	R	R	R	R
Potable-water service	R	R	R	R	R	R	R	R
Drain, waste, and vent	R	R	N	—	R	R	—	—
Demineralized water	R	R	—	—	R	R	R	—
Deionized water	R	R	—	—	R	R	R	R
Salt water	R	R	R	R	R	R	—	R
Heating (200°F) hot water	N	N	N	N	N	N	—	R
Natural gas	N	N	N	R	N	N	—	—
Compressed air	N	N	—	R	N	R	—	—
Sunlight and weather resistance	N	N	N	R	—	R	R	R
Underground service	R	R	R	R	R	R	—	R
Food handling	R	R	—	—	R	R	R	R

R = Recommended N = Not recommended — = Insufficient information

[a]Before selecting a material, check the availability of a suitable range of sizes and fittings and of a satisfactory joining method. Also have the manufacturer verify the best material for the purpose intended.

[b]Local building codes should be consulted for compliance of the materials listed.

Table 9 Capacities of ASTM A36 Steel Threaded Rods

Rod Diameter, in.	Root Area of Coarse Thread, in^2	Maximum Load,* lb
1/4	0.027	240
3/8	0.068	610
1/2	0.126	1130
5/8	0.202	1810
3/4	0.302	2710
7/8	0.419	3770
1	0.552	4960
1-1/4	0.889	8000

*Based on an allowable stress of 12,000 psi reduced by 25% using the root area in accordance with ASME *Standard* B31.1 and MSS *Standard* SP-58.

Table 10 Thermal Expansion of Metal Pipe

			Linear Thermal Expansion, in/100 ft		
	Saturated Steam Pressure, psig	Temperature, °F	Carbon Steel	Type 304 Stainless Steel	Copper
		–30	–0.19	–0.30	–0.32
		–20	–0.12	–0.20	–0.21
		–10	–0.06	–0.10	–0.11
		0	0	0	0
		10	0.08	0.11	0.12
		20	0.15	0.22	0.24
	–14.6	32	0.24	0.36	0.37
	–14.6	40	0.30	0.45	0.45
	–14.5	50	0.38	0.56	0.57
	–14.4	60	0.46	0.67	0.68
	–14.3	70	0.53	0.78	0.79
	–14.2	80	0.61	0.90	0.90
Vacuum	–14.0	90	0.68	1.01	1.02
	–13.7	100	0.76	1.12	1.13
	–13.0	120	0.91	1.35	1.37
	–11.8	140	1.06	1.57	1.59
	–10.0	160	1.22	1.79	1.80
	–7.2	180	1.37	2.02	2.05
	–3.2	200	1.52	2.24	2.30
	0	212	1.62	2.38	2.43
	2.5	220	1.69	2.48	2.52
	10.3	240	1.85	2.71	2.76
	20.7	260	2.02	2.94	2.99
	34.6	280	2.18	3.17	3.22
	52.3	300	2.35	3.40	3.46
	75.0	320	2.53	3.64	3.70
	103.3	340	2.70	3.88	3.94
	138.3	360	2.88	4.11	4.18
	181.1	380	3.05	4.35	4.42
	232.6	400	3.23	4.59	4.87
	666.1	500	4.15	5.80	5.91
	1528	600	5.13	7.03	7.18
	3079	700	6.16	8.29	8.47
		800	7.23	9.59	9.79
		900	8.34	10.91	11.16
		1000	9.42	12.27	12.54

joints are used at a high operating pressure. Consequently, a qualified engineer should design, or review the design of, all anchors and pipe-supporting elements, especially for the following:

- Steam systems operating above 15 psig
- Hydronic systems operating above 160 psig or 250°F
- Risers over 10 stories or 100 ft
- Systems with expansion joints, especially for pipe diameters 3 in. and greater
- Pipe sizes over 12 in. diameter
- Anchor loads greater than 10,000 lb (10 kips)
- Moments on pipe or structure in excess of 1000 ft·lb

PIPE EXPANSION AND FLEXIBILITY

Temperature changes cause dimensional changes in all materials. Table 10 shows the coefficients of expansion for piping materials commonly used in HVAC. For systems operating at high temperatures, such as steam and hot water, the rate of expansion is high, and significant movements can occur in short runs of piping. Even though rates of expansion may be low for systems operating in the range of 40 to 100°F, such as chilled and condenser water, they can cause large movements in long runs of piping, which are common in distribution systems and high-rise buildings. Therefore, in addition to design requirements for pressure, weight, and other loads, piping systems must accommodate thermal and other movements to prevent the following:

- Failure of pipe and supports from overstress and fatigue
- Leakage of joints
- Detrimental forces and stresses in connected equipment

An unrestrained pipe operates at the lowest overall stress level. Anchors and restraints are needed to support pipe weight and to protect equipment connections. The anchor forces and the bowing of pipe anchored at both ends are generally too large to be acceptable, so general practice is to *never anchor a straight run of steel pipe at both ends*. Piping must be allowed to expand or contract through thermal changes. Ample flexibility can be attained by designing pipe bends and loops or by including supplemental devices, such as expansion joints.

End reactions transmitted to rotating equipment, such as pumps or turbines, may deform the equipment case and cause bearing misalignment, which may ultimately cause the component to fail. Consequently, manufacturers' recommendations on allowable forces and movements that may be placed on their equipment should be followed.

PIPE BENDS AND LOOPS

Detailed stress analysis requires involved mathematical analysis and is generally performed by computer programs. However, such involved analysis is not required for most HVAC systems because the piping arrangements and temperature ranges at which they operate are simple to analyze.

L Bends

The guided cantilever beam method of evaluating L bends can be used to design L bends, Z bends, pipe loops, branch take-off connections, and some more complicated piping configurations.

Equation (5) may be used to calculate the length of leg BC needed to accommodate thermal expansion or contraction of leg AB for a guided cantilever beam (Figure 1).

$$L = \sqrt{\frac{3\Delta DE}{(144\ \text{in}^2/\text{ft}^2)S_A}} \tag{5}$$

where

L = length of leg BC required to accommodate thermal expansion of long leg AB, ft
Δ = thermal expansion or contraction of leg AB, in.
D = actual pipe outside diameter, in.
E = modulus of elasticity, psi
S_A = allowable stress range, psi

For the commonly used A53 Grade B seamless or ERW pipe, an allowable stress S_A of 22,500 psi (see Table 1) can be used without overstressing the pipe. However, this can result in very high end reactions and anchor forces, especially with large-diameter pipe. Designing to a stress range S_A of 15,000 psi and assuming $E = 27.9 \times 10^6$ psi, Equation (5) reduces to Equation (6), which provides reasonably low end reactions without requiring too much extra pipe. In addition, Equation (6) may be used with A53 butt-welded pipe and B88 drawn copper tubing.

Thus, for A53 continuous (butt-) welded, seamless, and ERW pipe, and B88 drawn copper tubing,

$$L = 6.225\sqrt{\Delta D} \tag{6}$$

The guided cantilever method of designing *L* bends assumes no restraints; therefore, care must be taken in supporting the pipe. For horizontal L bends, it is usually necessary to place a support near point B (see Figure 1), and any supports between points A and C must provide minimal resistance to piping movement; this is done by using slide plates or hanger rods of ample length, with hanger components selected to allow for swing no greater than 4°.

For L bends containing both vertical and horizontal legs, any supports on the horizontal leg must be spring hangers designed to support the full weight of pipe at normal operating temperature with a maximum load variation of 25%.

The force developed in an L bend that must be sustained by anchors or connected equipment is determined by the following equation:

$$F = \frac{12E_c I\Delta}{(1728\ \text{in}^3/\text{ft}^3)L^3} \tag{7}$$

where

F = force, lb
E_c = modulus of elasticity, psi
I = moment of inertia, in^4
L = length of offset leg, ft
Δ = deflection of offset leg, in.

In lieu of using Equation (7), for L bends designed in accordance with Equation (6) for 1 in. or more of offset, a conservative estimation of force is 500 lb per diameter inch (e.g., a 3 in. pipe would develop 1500 lb of force).

Z Bends

Z bends, as shown in Figure 2, are very effective for accommodating pipe movements. A simple and conservative method of sizing Z bends is to design the offset leg to be 65% of the values used for an L bend in Equation (5), which results in

$$L = 4\sqrt{\Delta D} \tag{8}$$

where

L = length of offset leg, ft
Δ = anchor-to-anchor expansion, in.
D = pipe outside diameter, in.

The force developed in a Z bend can be calculated with acceptable accuracy as follows:

$$F = C_1\Delta(D/L)^2 \tag{9}$$

where

C_1 = 4000 lb/in.
F = force, lb
D = pipe outside diameter, in.
L = length of offset leg, ft
Δ = anchor-to-anchor expansion, in.

U Bends and Pipe Loops

Pipe loops or U bends are commonly used in long runs of piping. A simple method of designing pipe loops is to calculate the anchor-to-anchor expansion and, using Equation (5), determine the length *L* necessary to accommodate this movement. The pipe loop dimensions can then be determined using $W = L/5$ and $H = 2W$.

Note that guides must be spaced no closer than twice the height of the loop, and piping between guides must be supported, as described in the section on L Bends, when the length of pipe between guides exceeds the maximum allowable hanger spacing for the size pipe.

Table 11 lists pipe loop dimensions for pipe sizes 1 in. through 24 in. and anchor-to-anchor expansion (contraction) of 2 in. through 12 in.

No simple method has been developed to calculate pipe loop force; however, it is generally low. A conservative estimate is 200 lb per diameter inch (e.g., a 2 in. pipe will develop 400 lb of force and a 12 in. pipe will develop 2400 lb of force).

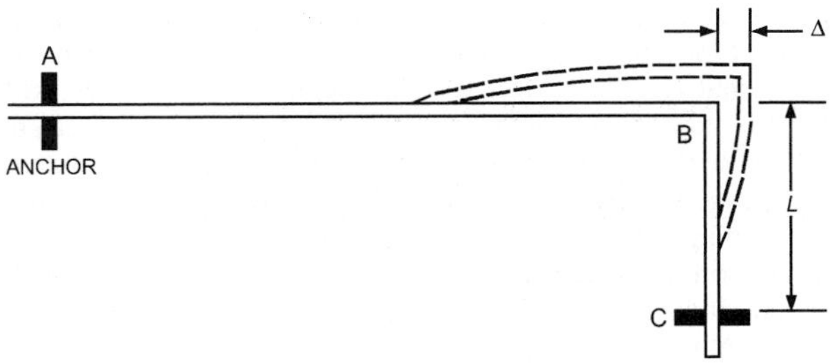

Fig. 1 Guided Cantilever Beam

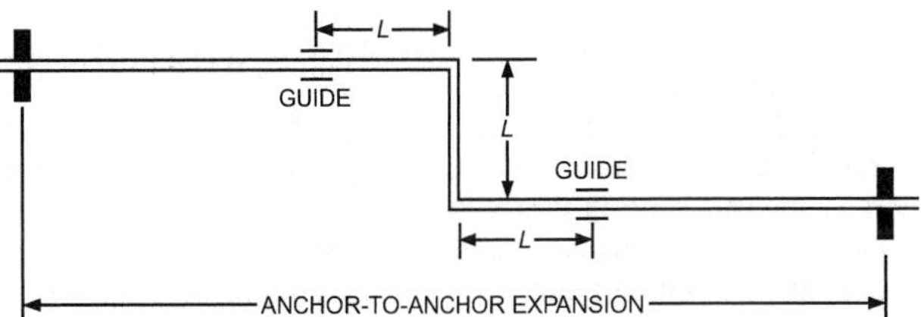

Distance from guides, if used, to offset should be equal or exceed length of offset.
Offset piping must be supported with hangers, slide plates, and spring hangers similar to those for L bends.

Fig. 2 Z Bend in Pipe

Table 11 Pipe Loop Design for A53 Grade B Carbon Steel Pipe Through 400°F

Pipe Size, in.	Anchor-to-Anchor Expansion, in. 2		4		6		8		10		12	
	W	*H*	*W*	*H*	*W*	*H*	*W*	*H*	*W*	*H*	*W*	*H*
1	2	4	3	6	3.5	7	4	8	4.5	9	5	10
2	3	6	4	8	5	10	5.5	11	6	12	7	14
3	3.5	7	5	10	6	12	6.5	13	7.5	15	8	16
4	4	8	5.5	11	6.5	13	7.5	15	8.5	17	9	18
6	5	10	6.5	13	8	16	9	18	10	20	11	22
8	5.5	11	7.5	15	9	18	10.5	21	12	24	13	26
10	6	12	8.5	17	10	20	11.5	23	13	26	14	28
12	6.5	13	9	18	11	22	12.5	25	14	28	15.5	31
14	7	14	9.5	19	11.5	23	13	26	15	30	16	32
16	7.5	15	10	20	12.5	25	14	28	16	32	17.5	35
18	8	16	11	22	13	26	15	30	17	34	18.5	37
20	8.5	17	11.5	23	14	28	16	32	18	36	19.5	39
24	9	18	12.5	25	14.5	29	17.5	35	19.5	39	21	42

Notes: *W* and *H* dimensions are feet.
L is determined from Equation (4). $W = L/5$ $H = 2W$ $2H + W = L$

Approximate force to deflect loop = 200 lb/diam. in.
For example, 8 in. pipe creates 1600 lb of force.

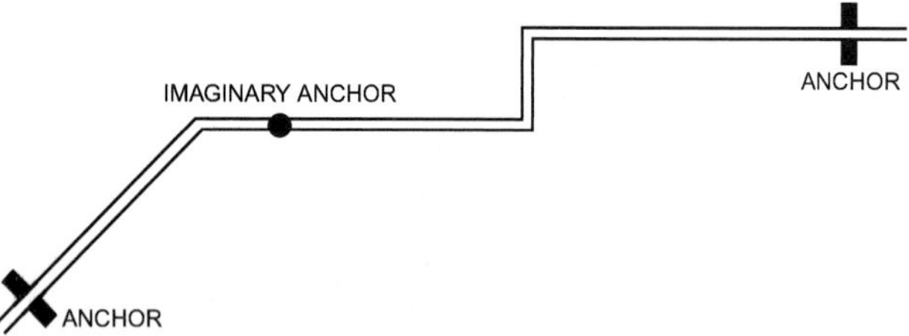

Fig. 3 Multiplane Pipe System

Cold Springing of Pipe

Cold springing or cold positioning of pipe consists of offsetting or springing the pipe in a direction opposite the expected movement. Cold springing is not recommended for most HVAC piping. Furthermore, *cold springing does not allow designing a pipe bend or loop for twice the calculated movement.* For example, if a particular L bend can accommodate 3 in. of movement from a neutral position, cold springing does not allow the L bend to accommodate 6 in. of movement.

Analyzing Existing Piping Configurations

Piping is best analyzed using a computer stress analysis program because these provide all pertinent data including stress, movements, and loads. Services can perform such analysis if programs are not available in-house. However, many situations do not require such detailed analysis. A simple, yet satisfactory method for single and multiplane systems is to divide the system with real or hypothetical anchors into a number of single-plane units, as shown in Figure 3, which can be evaluated as L and Z bends.

EXPANSION JOINTS AND EXPANSION COMPENSATING DEVICES

Although the inherent flexibility of the piping should be used to the maximum extent possible, expansion joints must be used where movements are too large to accommodate with pipe bends or loops or where insufficient room exists to construct a loop of adequate size. Typical situations are tunnel piping and risers in high-rise buildings, especially for steam and hot water pipes where large thermal movements are involved.

Packed and packless expansion joints and expansion compensating devices are used to accommodate movement, either axially or laterally.

In the **axial method** of accommodating movement, the expansion joint is installed between anchors in a straight line segment and accommodates axial motion only. This method has high anchor loads, primarily because of pressure thrust. It requires careful guiding, but expansion joints can be spaced conveniently to limit movement of branch connections. The axial method finds widest application for long runs without natural offsets, such as tunnel and underground piping and risers in tall buildings.

The **lateral** or **offset method** requires the device to be installed in a leg perpendicular to the expected movement and accommodates lateral movement only. This method generally has low anchor forces and minimal guide requirements. It finds widest application in lines with natural offsets, especially where there are few or no branch connections.

Packed expansion joints depend on slipping or sliding surfaces to accommodate the movement and require some type of seals or packing to seal the surfaces. Most such devices require some maintenance but are not subject to catastrophic failure. Further, with most packed expansion joint devices, any leaks that develop can be repacked under full line pressure without shutting down the system.

Packless expansion joints depend upon the flexing or distortion of the sealing element to accommodate movement. They generally do not require any maintenance, but maintenance or repair is not usually possible. If a leak occurs, the system must be shut off and drained, and the entire device must be replaced. Further, catastrophic failure of the sealing element can occur and, although likelihood of such failure is remote, it must be considered in certain design situations.

Packed expansion joints are preferred where long-term system reliability is of prime importance (using types that can be repacked under full line pressure) and where major leaks can be life threatening or extremely costly. Typical applications are risers, tunnels, underground pipe, and distribution piping systems. Packless expansion joints are generally used where even small leaks cannot be tolaerated (e.g., for gas and toxic chemicals), where temperature limitations preclude the use of packed expansion joints, and for very large-diameter pipe where packed expansion joints cannot be constructed or the cost would be excessive.

In all cases, expansion joints should be installed, anchored, and guided in accordance with expansion joint manufacturers' recommendations.

Packed Expansion Joints

There are two types of packed expansion joints: packed slip expansion joints and flexible ball joints.

Packed Slip Expansion Joints. These are telescoping devices designed to accommodate axial movement only. Some sort of packing seals the sliding surfaces. The original packed slip expansion joint used multiple layers of braided compression packing, similar to the stuffing box commonly used with valves and pumps; this arrangement requires shutting and draining the system for maintenance and repair. Advances in design and packing technology have eliminated these problems, and most current packed slip joints use self-lubricating semiplastic packing, which can be injected under full line pressure without shutting off the system (see Figure 4). (Many manufacturers use asbestos-based packings, unless requested otherwise. Asbestos-free packings, such as flake graphite, are available and, although more expensive, should be specified in lieu of products containing asbestos.)

Standard packed slip expansion joints are constructed of carbon steel with weld or flange ends in sizes 1.5 to 36 in. for pressures up to 300 psig and temperatures up to 800°F. Larger sizes, higher-temperature, and higher-pressure designs are available. Standard single joints are generally designed for 4, 8, or 12 in. axial traverse; double joints with an intermediate anchor base can accommodate twice these movements. Special designs for greater movements are available.

Flexible Ball Joints. These joints are used in pairs to accommodate lateral or offset movement and must be installed in a leg perpendicular to the expected movement. The original flexible ball joint design incorporated only inner and outer containment seals that could not be serviced or replaced without removing the ball joint from the system. The packing technology of the packed slip expansion joint, explained previously, has been incorporated into the flexible ball joint design; now, packed flexible ball joints have self-lubricating semiplastic packing that can be injected under full line pressure without shutting off the system (see Figure 5).

Standard flexible ball joints are available in sizes 1 1/4 through 30 in. with threaded (1 1/4 to 2 in.), weld, and flange ends for pressures to 300 psig and temperatures to 750°F. Flexible ball joints are available in larger sizes and for higher temperature and pressure ranges.

Packless Expansion Joints

Metal bellows expansion joints, rubber expansion joints, and flexible hose or pipe connectors are some of the packless expansion joints available.

Metal Bellows Expansion Joints. These expansion joints have a thin-wall convoluted section that accommodates movement by bending or flexing. The bellows material is generally Type 304, 316, or 321 stainless steel, but other materials are commonly used to satisfy service conditions. Small-diameter expansion joints in sizes 3/4 through 3 in. are generally called *expansion compensators* and are available in all-bronze or steel construction. Metal bellows expansion joints can generally be designed for the pressures and temperatures commonly encountered in HVAC systems and can also be furnished in rectangular configurations for ducts and chimney connectors.

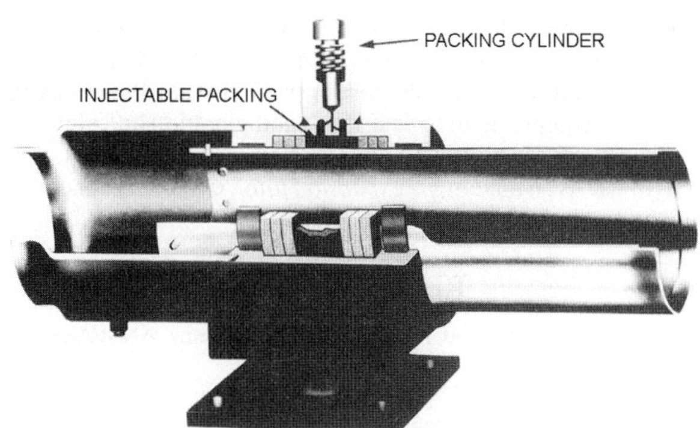

Fig. 4 Packed Slip Expansion Joint

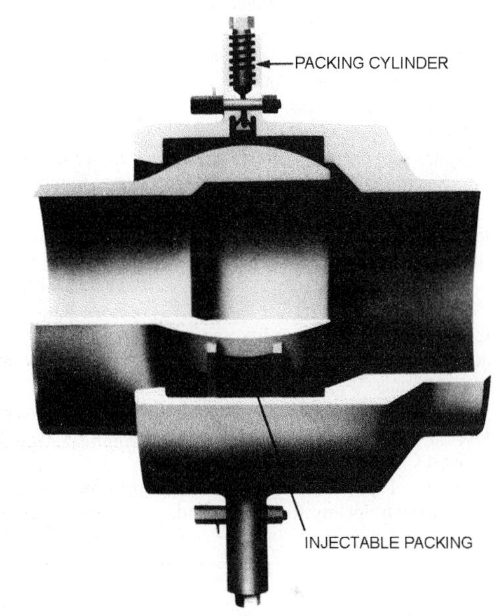

Fig. 5 Flexible Ball Joint

Overpressurization, improper guiding, and other forces can distort the bellows element. For low-pressure applications, such distortion can be controlled by the geometry of the convolution or the thickness of the bellows material. For higher pressure, internally pressurized joints require reinforcing. Externally pressurized designs are not subject to such distortion and are not generally furnished without supplemental bellows reinforcing.

Single- and double-bellows expansion joints primarily accommodate axial movement only, similar to packed slip expansion joints. Although bellows expansion joints can accommodate some lateral movement, the **universal tied bellows expansion joint** accommodates large lateral movement. This device operates much like a pair of flexible ball joints, except that bellows elements are used instead of flexible ball elements. The tie rods on this joint contain the pressure thrust, so anchor loads are much lower than with axial-type expansion joints.

Rubber Expansion Joints. Similar to single-metal bellows expansion joints, rubber expansion joints incorporate a nonmetallic elastomeric bellows sealing element and generally have more stringent temperature and pressure limitations. Although rubber expansion joints can be used to accommodate expansion and contraction of the piping, they are primarily used as flexible connectors at equipment to isolate sound and vibration and eliminate stress at equipment nozzles.

Flexible Hose. This type of hose can be constructed of elastomeric material or corrugated metal with an outer braid for reinforcing and end restraint. Flexible hose is primarily used as a flexible connector at equipment to isolate sound and vibration and eliminate stress at equipment nozzles; however, flexible metal hose is well suited for use as an *offset-type expansion joint*, especially for copper tubing and branch connections off risers.

REFERENCES

ASHRAE. 2001. Safety standard for refrigeration systems. ANSI/ASHRAE *Standard* 15-2001.

ASME. 1983. Pipe threads, general purpose (inch). *Standard* B1.20.1. American Society of Mechanical Engineers, New York.

ASME. 1998. Power piping. *Standard* B31.1. American Society of Mechanical Engineers, New York.

ASME. 1992. Refrigeration piping. *Standard* B31.5. American Society of Mechanical Engineers, New York.

ASME. 1996. Building services piping. *Standard* B31.9. American Society of Mechanical Engineers, New York.

ASME. 1996. Welded and seamless wrought steel pipe. *Standard* B36.10M. American Society of Mechanical Engineers, New York.

ASME. 1998. Rules for construction of power boilers. *Boiler and Pressure Vessel Code*. Section I. American Society of Mechanical Engineers, New York.

ASME. 1998. Qualification standard for welding and brazing procedures, welders, brazers, and welding and brazing operators. *Boiler and Pressure Vessel Code,* Section IX. American Society of Mechanical Engineers, New York.

ASTM. 1998. Standard specification for pipe, steel, black and hot-dipped, zinc-coated welded and seamless. *Standard* A53. American Society for Testing and Materials, West Conshohocken, PA.

ASTM. 1997. Standard specification for seamless carbon steel pipe for high-temperature service. *Standard* A106. American Society for Testing and Materials, West Conshohocken, PA.

ASTM. 1996. Standard specification for seamless copper water tube. *Standard* B88. American Society for Testing and Materials, West Conshohocken, PA.

ASTM. 1997. Standard specification for seamless copper tube for air conditioning and refrigeration field service. *Standard* B280. American Society for Testing and Materials, West Conshohocken, PA.

ASTM. 1998. Standard test method for obtaining hydrostatic design basis for thermoplastic pipe materials. *Standard* D2837. American Society for Testing and Materials, West Conshohocken, PA.

ASTM. 1996. Standard practice for obtaining hydrostatic or pressure design basis for "fiberglass" (glass-fiber-reinforced thermosetting-resin) pipe and fittings. *Standard* D2992. American Society for Testing and Materials, West Conshohocken, PA.

ASTM. 1996. Standard specification for polybutylene (PB) plastic hot- and cold-water distribution systems. *Standard* D3309 REV A. American Society for Testing and Materials, West Conshohocken, PA.

ASTM. 1996. Standard specification for plastic inserts fittings for polybutylene (PB) tubing. *Standard* F845. American Society for Testing and Materials, West Conshohocken, PA.

AWWA. 1996. Thickness design of ductile-iron pipe. *Standard* C150/A21.50. American Water Works Association, Denver.

MSS. 1993. Pipe hangers and supports—Materials, design and manufacture. *Standard* SP-58. Manufacturers Standardization Society of the Valve and Fittings Industry, Vienna, VA.

MSS. 1996. Pipe hangers and supports—Selection and application. *Standard* SP-69. Manufacturers Standardization Society of the Valve and Fittings Industry, Vienna, VA.

NFPA. 1996. Installation of sprinkler systems. *Standard* 13. National Fire Protection Association, Quincy, MA.

NFPA/AGA. 2002. *National fuel gas code*. ANSI/NFPA 54. National Fire Protection Association, Quincy, MA. ANSI/AGA *Standard* Z223.1-2002. American Gas Association, Arlington, VA.

BIBLIOGRAPHY

ASTM. 1996. Standard specification for poly(vinyl chloride) (PVC) plastic pipe, Schedules 40, 80, and 120. *Standard* D1785. American Society for Testing and Materials, West Conshohocken, PA.

ASTM. 2002. Standard specification for chlorinated poly(vinyl chloride) (CPVC) plastic pipe, Schedules 40 and 80. *Standard* F441/F441M. American Society for Testing and Materials, West Conshohocken, PA.

Crane Co. 1988. Flow of fluids through valves, fittings, and pipe. *Technical Paper* 410. Joliet, IL.

CHAPTER 46

VALVES

FUNDAMENTALS

VALVES are the manual or automatic fluid-controlling elements in a piping system. They are constructed to withstand a specific range of temperature, pressure, corrosion, and mechanical stress. The designer selects and specifies the proper valve for the application to give the best service for the economic requirements.

Valves have some of the following primary functions:

- Starting, stopping, and directing flow
- Regulating, controlling, or throttling flow
- Preventing backflow
- Relieving or regulating pressure

The following service conditions should be considered before specifying or selecting a valve:

1. Type of liquid, vapor, or gas
   - Is it a true fluid or does it contain solids?
   - Does it remain a liquid throughout its flow or does it vaporize?
   - Is it corrosive or erosive?
2. Pressure and temperature
   - Will these vary in the system?
   - Should worst case (maximum or minimum values) be considered in selecting correct valve materials?
3. Flow considerations
   - Is pressure drop critical?
   - Should valve design be chosen for maximum wear?
   - Is the valve to be used for simple shutoff or for throttling flow?
   - Is the valve needed to prevent backflow?
   - Is the valve to be used for directing (mixing or diverting) flow?
4. Frequency of operation
   - Will the valve be operated frequently?
   - Will valve normally be open with infrequent operation?
   - Will operation be manual or automatic?

Nomenclature for basic valve components may vary from manufacturer to manufacturer and according to the application. Figure 1 shows representative names for various valve parts.

Body Ratings

The rating of valves defines the pressure-temperature relationship within which the valve may be operated. The valve manufacturer is responsible for determining the valve rating. ASME *Standard* B16.34 should be consulted, and a valve pressure class should be identified. Inlet pressure ratings are generally expressed in terms of the ANSI/ASME class ratings and range from ANSI Class 150 through 2500, depending on the style, size, and materials of construction, including seat materials. Automatic control valves are usually either Class 125 or Class 250. Tables in the standard and in various books show pressure ratings at various operating temperatures (ASME *Standard* B16.34; Lyons 1982; Ulanski 1991).

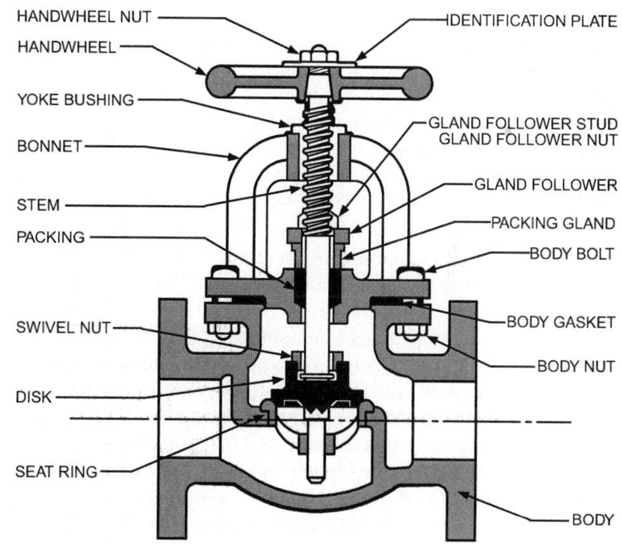

Fig. 1 Valve Components
(Courtesy Anvil Int'l.)

Materials

ASME *Standard* B16.34 addresses requirements for valves made from forgings, castings, plate, bar stock and shapes, and tubular products. This standard identifies acceptable materials from which valves can be constructed. In selecting proper valve materials, the valve body-bonnet material should be selected first and then the valve plug and seat trim.

Other factors that govern the basic materials selection include

- Pressure-temperature ratings
- Corrosion-resistance requirements
- Thermal shock
- Piping stress
- Fire hazard

Types of materials typically available include

- Carbon steel
- Ductile iron
- Cast iron
- Stainless steels
- Brass
- Bronze
- Polyvinyl chloride (PVC) plastic

Bodies. Body materials for small valves are usually brass, bronze, or forged steel and for larger valves, cast iron, cast ductile

The preparation of this chapter is assigned to TC 6.1, Hydronic and Steam Equipment and Systems.

iron, or cast steel as required for the pressure and service. Listings of typical materials are given in Lyons (1982) and Ulanski (1991).

Seats. Valve seats can be machined integrally of the body material, press-fitted, or threaded (removable). Seats of different materials can be selected to suit difficult application requirements. The valve seat and the valve plug or disk are sometimes referred to as the valve trim and are usually constructed of the same material selected to meet the service requirements. The trim, however, is usually of a different material than the valve body. Replaceable composition disks are used in conjunction with the plug in some designs in order to provide adequate close-off.

Maximum permissible leakage ratings for control valve seats are defined in Fluid Controls Institute (FCI) *Standard* 70-2.

Stems. Valve stem material should be selected to meet service conditions. Stainless steel is commonly used for most HVAC applications, and bronze is commonly used in ball valve construction.

Stem Packings and Gaskets. Valve stem packings undergo constant wear because of the movement of the valve stem, and both the packings and body gaskets are exposed to pressure and pressure variations of the control fluid. Manufacturers can supply recommendations regarding materials and lubricants for specific fluid temperatures and pressures.

Flow Coefficient and Pressure Drop

Flow through any device results in some loss of pressure. Some of the factors affecting pressure loss in valves include changes in the cross section and shape of the flow path, obstructions in the flow path, and changes in direction of the flow path. For most applications, the pressure drop varies as the square of the flow when operating in the turbulent flow range. For check valves, this relationship is true only if the flow holds the valve in the full-open position.

For convenience in selecting valves, particularly control valves, manufacturers express valve capacity as a function of a flow coefficient C_v. By definition in the United States, C_v is the flow of water in gallons per minute (at 60°F) that causes a pressure drop of 1 psi across a fully open valve. Manufacturers may also furnish valve coefficients at other pressure drops. Flow coefficients apply only to water. When selecting a valve to control other fluids, be sure to account for differences in viscosity.

Figure 2 shows a typical test arrangement to determine the C_v rating with the test valve wide open. Globe valve HV-1 allows adjusting the supply gage reading (e.g., to 10 psi); HV-2 is then adjusted (e.g., to 9 psi return gage) to allow a test run at a pressure drop of 1 psi. A gravity storage tank may be used to minimize supply pressure fluctuations. The bypass valve allows fine adjustment of the supply pressure. A series of test runs is made with the weighing tank and a stopwatch to determine the flow rate. Further capacity test detail may be found in International Society for Measurement and Control (ISA) *Standard* S75.02.

Cavitation

Cavitation occurs when the pressure of a flowing fluid drops below the vapor pressure of that fluid (Figure 3). In this two-step process, the pressure first drops to the critical point, causing cavities of vapor to form. These are carried with the flow stream until they reach an area of higher pressure. The bubbles of vapor then suddenly collapse or implode. This reduction in pressure occurs when the velocity increases as the fluid passes through a valve. After the fluid passes through the valve, the velocity decreases and the pressure increases. In many cases, cavitation manifests itself as noise. However, if the vapor bubbles are in contact with a solid surface when they collapse, the liquid rushing into the voids causes high localized pressure that can erode the surface. Premature failure of the valve and adjacent piping may occur. The noise and vibration caused by cavitation have been described as similar to those of gravel flowing through the system.

Water Hammer

Water hammer is a series of pressure pulsations of varying magnitude above and below the normal pressure of water in the pipe. The amplitude and period of the pulsation depend on the velocity of the water as well as the size, length, and material of the pipe.

Shock loading from these pulsations occurs when any moving liquid is stopped in a short time. In general, it is important to avoid quickly closing valves in an HVAC system to minimize the occurrence of water hammer.

When flow stops, the pressure increase is independent of the working pressure of the system. For example, if water is flowing at 5 fps and a valve is instantly closed, the pressure increase is the same whether the normal pressure is 100 psig or 1000 psig.

Water hammer is often accompanied by a sound resembling a pipe being struck by a hammer—hence the name. The intensity of the sound is no measure of the magnitude of the pressure. Tests indicate that even if 15% of the shock pressure is removed by absorbers or arresters, adequate relief is not necessarily obtained.

Velocity of pressure wave and maximum water hammer pressure formulas may be found in the *Hydraulic Handbook* (Fairbanks Morse 1965).

Noise

Chapter 36 of the 2005 *ASHRAE Handbook—Fundamentals* points out that limitations are imposed on pipe size to control the level of pipe and valve noise, erosion, and water hammer pressure. One recommendation places a velocity limit of 4 fps for pipe 2 in. and smaller, and a pressure drop of 4 ft water/100 ft length for piping over 2 in. in diameter. Velocity-dependent noise in piping and piping systems results from any or all of four sources: turbulence, cavitation, release of entrained air, and water hammer (see Chapter 47 of the 2007 *ASHRAE Handbook—HVAC Applications*).

Some data are available for predicting hydrodynamic noise generated by control valves. ISA *Standard* 75.01 compiled prediction correlations in an effort to develop control valves for reduced noise levels.

Body Styles

Valve bodies are available in many configurations depending on the desired service. Usual functions include stopping flow, allowing

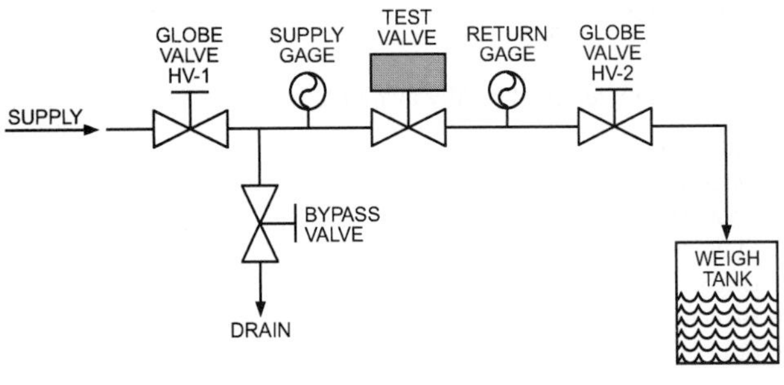

Fig. 2 Flow Coefficient Test Arrangement

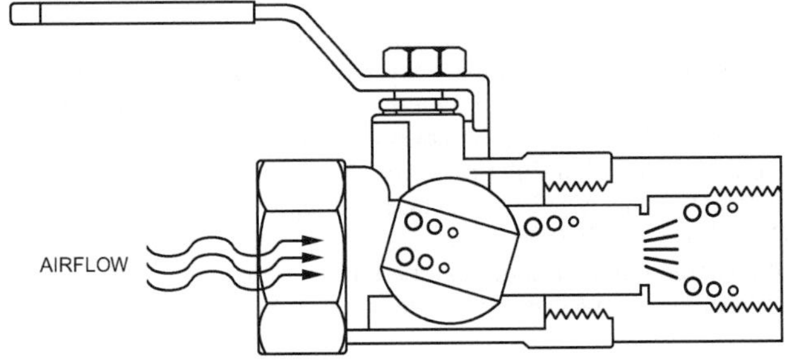

Fig. 3 Valve Cavitation at Sharp Curves

full flow, modulating flow between extremes, and directing flow. The operation of a valve can be automatic or manual.

The shape of bodies for automatic and manual valves is dictated by the intended application. For example, angle valves are commonly provided for radiator control. The principle of flow is the same for angle and straight-through valve configurations; the manufacturer provides a choice in some cases as a convenience to the installer.

The type or design of body connections is dictated primarily by the proposed conduit or piping material. Depending on material type, valves can be attached to piping in one of the following ways:

- Bolted to the pipe with companion flange.
- Screwed to the pipe, where the pipe itself has matching threads (male) and the body of the valve has threads machined into it (female).
- Welded, soldered, or sweated.
- Flared, compression, and/or various mechanical connections to the pipe where there are no threads on the pipe or the body.
- Valves of various plastic materials are fastened to the pipe if the valve body and the pipe are of compatible plastics.

MANUAL VALVES

Selection

Each valve style has advantages and disadvantages for the application. In some cases, the design documents provide inadequate information, so that selection is based on economics and local stock availability by the installer and not on what is really required. Good submittal practice and approval by the designer are required to prevent substitutions. The questions listed in the section on Fundamentals must be evaluated carefully.

Globe Valves

In a globe valve, flow is controlled by a circular disk forced against or withdrawn from an annular ring, or **seat**, that surrounds an opening through which flow occurs (Figure 4). The direction of movement of the disk is parallel to the direction of the flow through the valve opening (or seat) and normal to the axis of the pipe in which the valve is installed.

Globe valves are most frequently used in smaller diameter pipes but are available in sizes up to 12 in. They are used for throttling duty where positive shutoff is required. Globe valves for controlling service should be selected by class, and whether they are of the straight-through or angle type, composition disk, union or gasketed bonnet, threaded, and solder or grooved ends. Manually operated flow control valves are also available with fully guided V-port throttling plugs or needle point stems for precise adjustment.

Gate Valves

A gate valve controls flow by means of a wedge disk fitting against machined seating faces (Figure 5). The straight-through opening of the valve is as large as the full bore of the pipe, and the gate movement is perpendicular to the flow path.

Gate valves are intended to be fully open or completely closed. They are designed to allow or stop flow, and should not be used to regulate or control flow. Various wedges for gate valves are available for specific applications. Valves in inaccessible locations may be provided with a chain wheel or with a hammer-blow operator. More detailed information is available from valve manufacturers.

Plug Valves

A plug valve is a manual fluid flow control device (Figure 6). It operates from fully open to completely shut off within a 90° turn. The capacity of the valve depends on the ratio of the area of the orifice to the area of the pipe in which the valve is installed.

The cutaway view of a plug valve shows a valve with an orifice that is considerably smaller than the full size of the pipe. Lubricated plug valves are usually furnished in gas applications. A plug valve is selected as an on/off control device because (1) it is relatively inexpensive; (2) when adjusted, it holds its position; and (3) its position is clearly visible to the operator. The effectiveness of this valve as a flow control device is reduced if the orifice of the valve is fully ported (i.e., the same area as the pipe size).

Ball Valves

A ball valve contains a precision ball held between two circular seals or seats. Ball valves have various port sizes. A 90° turn of the handle changes operation from fully open to fully closed. Ball valves for shutoff service may be fully ported. Ball valves for throttling or controlling and/or balancing service should have a reduced

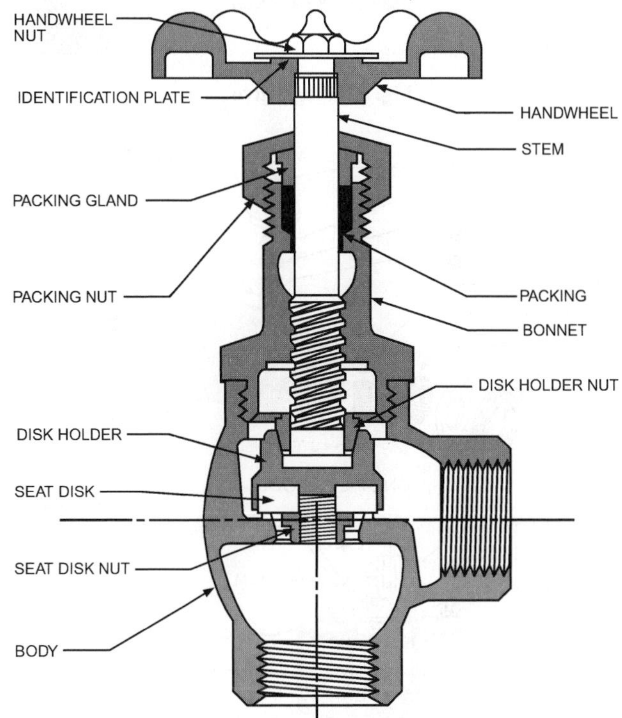

Fig. 4 Globe Valve
(Courtesy Anvil Int'l.)

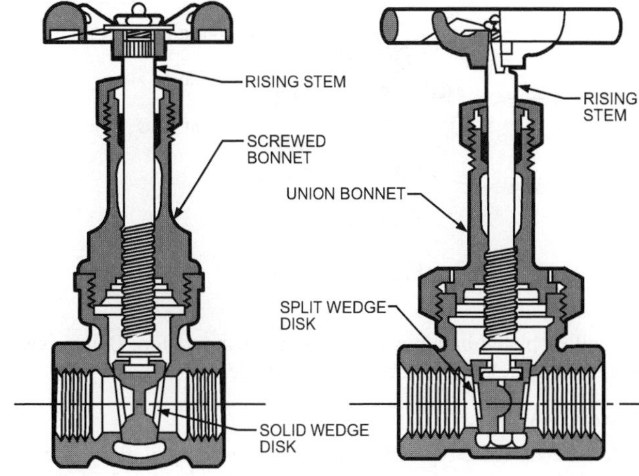

Fig. 5 Two Variations of Gate Valve

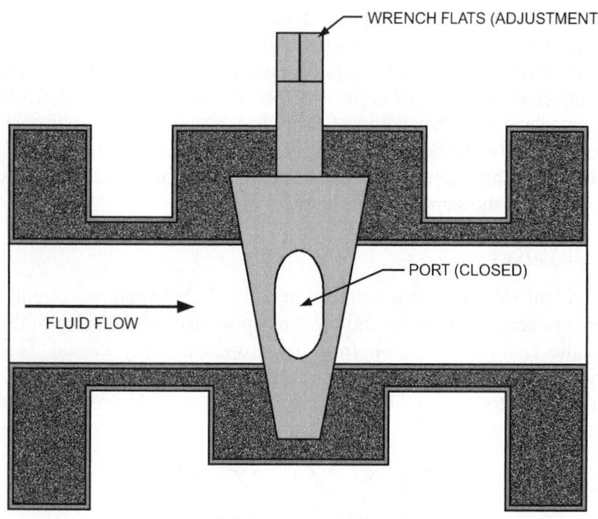

Fig. 6 Plug Valve

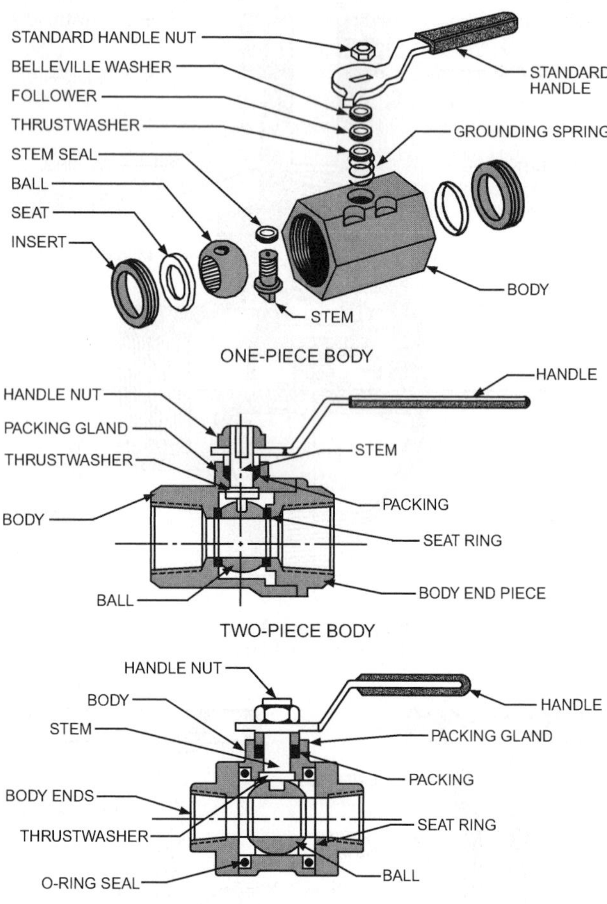

Fig. 7 Ball Valve

port with a plated ball and valve handle memory stop. Ball valves may be of one-, two-, or three-piece body design (Figure 7).

Butterfly Valves

A butterfly valve typically consists of a cylindrical, flanged-end body with an internal, rotatable disk serving as the fluid flow-regulating device (Figure 8). Butterfly valve bodies may be **wafer**

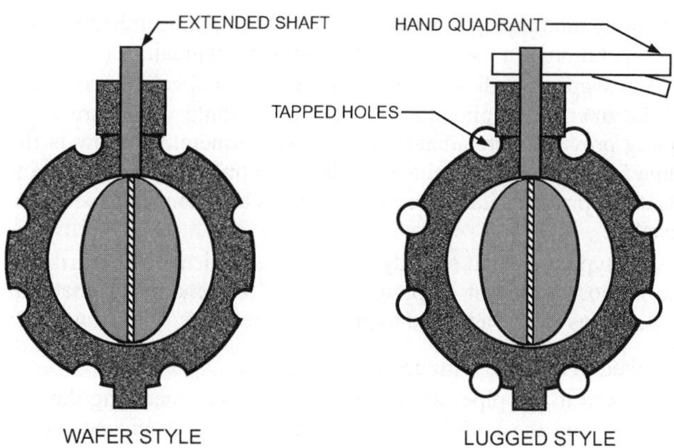

Fig. 8 Butterfly Valve

style, which is clamped between two companion flanges whose bolts carry the pipeline tensile stress and place the wafer body in compression, or **lugged style**, with tapped holes in the wafer body, which may serve as a future point of disconnection. The disk's axis of rotation is the valve stem; it is perpendicular to the flow path at the center of the valve body. Only a 90° turn of the valve disk is required to change from the full-open to the closed position. Butterfly valves may be manually operated with **hand quadrants** (levers) or provided with an extended shaft for automatic operation by an actuator. Special attention should be paid to manufacturers' recommendations for sizing an actuator to handle the torque requirements.

Simple and compact design, a low corresponding pressure drop, and fast operation characterize all butterfly valves. Quick operation makes them suitable for automated control, whereas the low pressure drop is suitable for high flow. Butterfly valve sizing for on/off applications should be limited to pipe sizing velocities given in Chapter 36 of the 2005 *ASHRAE Handbook—Fundamentals*; on the other hand, for throttling control applications, the valve coefficient sizing presented in the section on Automatic Valves must be followed.

Pinch Valves

Two styles of pinch valve bodies are normally used: the jacket pinch and the Saunders-type bodies used for slurry control in many industries. Pressure-squeezing the flexible tube jacket of a pinch valve reduces its port opening to control flow. The Saunders type uses an actuator to manually or automatically squeeze the diaphragm against a weir-type port. These valves have limited HVAC application.

AUTOMATIC VALVES

Automatic valves are commonly considered as control valves that operate in conjunction with an automatic controller or device to control the fluid flow. The "control valve" as used here actually consists of a valve body and an actuator. The valve body and actuator may be designed so that the actuator is removable and/or replaceable, or the actuator may be an integral part of the valve body. This section covers the most common types of valve actuators and control valves with the following classifications:

- Two-way bodies (single- and double-seated)
- Three-way bodies (mixing and diverting)
- Ball valves
- Butterfly arrangements (two- and three-way)

Actuators

The valve actuator converts the controller's output, such as an electric or pneumatic signal, into the rotary or linear action required by the valve (stem), which changes the control variable (flow). Actuators cover a wide range of sizes, types, output capabilities, and control modes.

Sizes. Actuators range in physical size from small solenoid or clock motor self-operated types, to large pneumatic actuators with 100 to 200 in^2 of effective area.

Types. The most common types of actuators used on automatic valve applications are solenoid, thermostatic radiator, pneumatic, electric motor, electronic, and electrohydraulic.

Output (Force) Capabilities. Although the smallest actuators, designed for unitary commercial HVAC and residential control applications, are capable of only a very small output, larger pneumatic or electrohydraulic actuators are capable of great force. The overall force ranges from a few ounces to over 0.5 ton of force.

Pneumatic Actuators

Pneumatic or diaphragm valve actuators are available with diaphragm sizes ranging from 3 to 200 in^2. The design consists of a flexible diaphragm clamped between an upper and a lower housing. On direct-acting actuators, the upper housing and diaphragm create a sealed chamber (Figure 9). A spring opposing the diaphragm force is positioned between the diaphragm and the lower housing. Increasing air pressure on the diaphragm pushes the valve stem down and overcomes the force of the load spring to close a direct-acting valve. Springs are designated by the air pressure change required to open or close the valve. A 5 lb spring requires a 5 psi control pressure change at the actuator to operate the valve. Some valves have an adjustable spring feature; others are fixed. Springs for commercial control valves usually have ±10% tolerance, so the 5 lb spring setting is 5 psi ± 0.5 psi. Two valves in a control may be sequenced simply with adjustable actuator springs.

Reverse-acting valves may use a direct-acting actuator if they have reverse-acting valve bodies; otherwise, the actuator must be reverse-acting and constructed with a sealed chamber between the lower housing and the diaphragm.

The valve close-off point shifts as the supply and/or the differential pressure increases across a single-seated valve because of the fixed areas of the actuator and the valve seat. The manufacturer's close-off rating tables need to be consulted to determine if the actuator is of an adequate size or if a larger actuator or a pneumatic positioner relay is required.

A **pneumatic positioner relay** may be added to the actuator to provide additional force to close or open an automatic control valve (Figure 9). Sometimes called positive positioners or pilot positioners, pneumatic positioners are basically high-capacity relays that add air pressure to or exhaust air pressure from the actuator in relation to the stroke position of the actuator. Their application is limited by the supply air pressure available and by the actuator's spring.

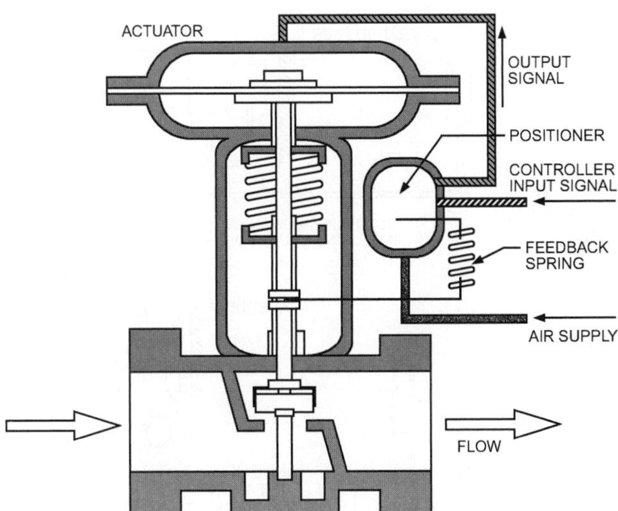

Fig. 9 Two-Way, Direct-Acting Control Valve with Pneumatic Actuator and Positioner

Electric Actuators

Electric actuators usually consist of a double-wound electric motor coupled to a gear train and an output shaft connected to the valve stem with a cam or rack-and-pinion gear linkage (Figure 10). For valve actuation, the motor shaft typically drives through 160° of rotation. The use of damper actuators with 90° full stroke rotation is rapidly increasing in valve control applications. Gear trains are coupled internally to the electric actuators to provide a timed movement of valve stroke to increase operating torque and to reduce overshooting of valve movement. Gear trains can be fitted with limit switches, auxiliary potentiometers, etc., to provide position indication and feedback for additional system control functions.

In many instances, a linkage is required to convert rotary motion to the linear motion required to operate a control valve (except ball and butterfly valves). Electric valve actuators operate with two-position, floating, proportional electric, and electronic control systems. Actuators usually operate with a 24 V (ac) low-voltage control circuit. Actuator time to rotate (or drive full stroke) ranges from 30 s to 4 min, with 60 s being most common.

Electric valve actuators may have a spring return, which returns the valve to a normal position in case of power failure, or it may be powered with an electric relay and auxiliary power source. Because the motor must constantly drive in one direction against the return spring, spring return electric valve actuators generally have only approximately one-third of the torque output of non-spring return actuators.

Electrohydraulic Actuators

Hydraulic actuators combine characteristics of electric and pneumatic actuators. In essence, hydraulic actuators consist of a sealed housing containing the hydraulic fluid, pump, and some type of metering or control apparatus to provide pressure control across a piston or piston/diaphragm. A coil controlled by a low- to medium-level dc voltage usually activates the pressure control apparatus.

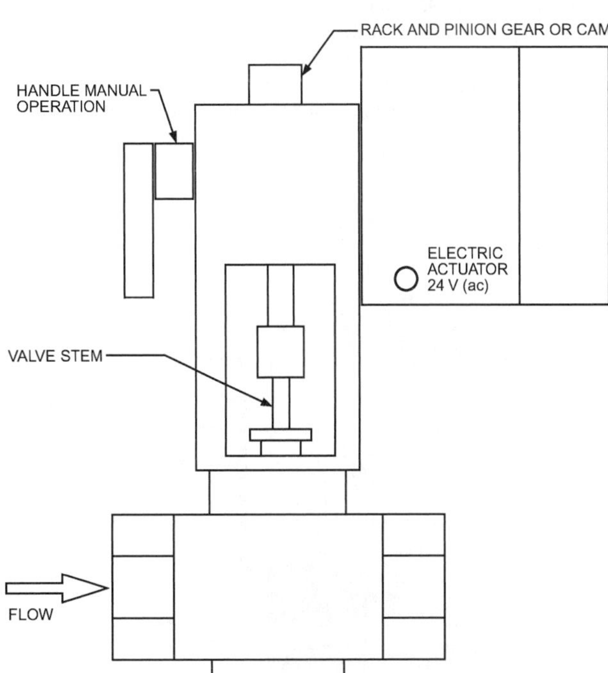

Fig. 10 Two-Way Control Valve with Electric Actuator

Solenoids

A solenoid valve is an electromechanical control element that opens or closes a valve on the energization of a solenoid coil. Solenoid valves are used to control the flow of hot or chilled water and steam and range in size from 1/8 to 2 in. pipe size. Solenoid actuators themselves are two-position control devices and are available for operation in a wide range of alternating current voltages (both 50 and 60 Hz) as well as direct current. Operation of a simple two-way, direct-acting solenoid valve in a deenergized state is illustrated in Figure 11.

Thermostatic Radiator Valves

Thermostatic radiator valves are self-operated and do not require external energy. They control room or space temperature by modulating the flow of hot water or steam through free-standing radiators, convectors, or baseboard heating units. Thermostatic radiator valves are available for a variety of installation requirements with remote-mounted sensors or integral-mounted sensor and remote or integral set point adjustment (Figure 12).

Control of Automatic Valves

Computer-based control of automatic control valves is replacing older technologies and provides many benefits, including speed, accuracy, and data communication. However, care must be used in selecting the value of control loop parameters such as loop speed and dead band (allowable set-point deviation). High loop speed coupled with zero dead band can cause the valve-actuator to seek a new control position with each control loop cycle unless the actuator itself has some type of built-in protection against this. For example, a 1 s control loop with zero dead band could result in 30,000,000 repositions (corrections) in 1 year of service.

Computer-based control systems should be tuned to provide the minimum acceptable level of response and accuracy required for the application in order to achieve maximum valve and actuator service life.

Two-Way Valves (Single- and Double-Seated)

In a two-way automatic valve, the fluid enters the inlet port and exits the outlet port either at full or reduced volume, depending on the position of the stem and the disk in the valve. Two-way valves may be single- or double-seated.

In the single-seated valve, one seat and one plug-disk close against the stream. The style of the plug-disk varies depending on the requirements of the designer and the application. For body comparison, refer to Figure 8 in Chapter 15 of the 2005 *ASHRAE Handbook—Fundamentals*.

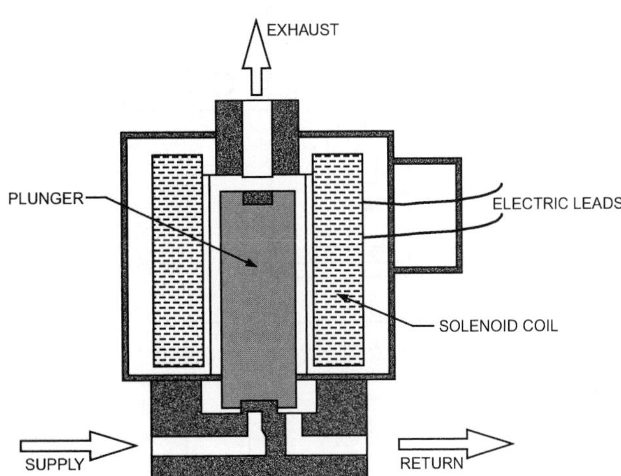

Fig. 11 Electric Solenoid Valve

The double-seated valve is a special application of the two-way valve with two seats, plugs, and disks. It is generally applied to cases where the close-off pressure is too high for the single-seated valve.

Three-Way Valves

Three-way valves either mix or divert streams of fluid. Figure 13 shows some common applications for three-way valves. Figure 9 in Chapter 15 of the 2005 *ASHRAE Handbook—Fundamentals* shows typical cross sections of three-way mixing and diverting valves.

The **three-way mixing valve** blends two streams into one common stream based on the position of the valve plug in relation to the upper and lower seats of the valve. A common use is to mix chilled or hot water. The valve controls the temperature of the single stream leaving the valve.

The **three-way diverting or bypass valve** takes one stream of fluid and splits it into two streams for temperature control. In some limited applications, such as a cooling tower control, a diverting or bypass valve must be used in place of a mixing valve. In most cases, a mixing valve can perform the same function as a diverting or bypass valve if the companion actuator has a very high spring rate. Otherwise, water hammer or noise may occur when operating near the seat.

Special-Purpose Valves

Special-purpose valve bodies may be used on occasion. One type of four-way valve is used to allow separate circulation in the boiler loop and a heated zone. Another type of four-way valve body is used as a changeover refrigeration valve in heat pump systems to reverse the evaporator to a condenser function.

Float valves are used to supply water to a tank or reservoir or serve as a boiler feed valve to maintain an operating water level at the float level location (Figure 14).

Ball Valves

Ball valves coupled with rotary actuators have seen increasing use in HVAC control applications. A reduced port should be used on

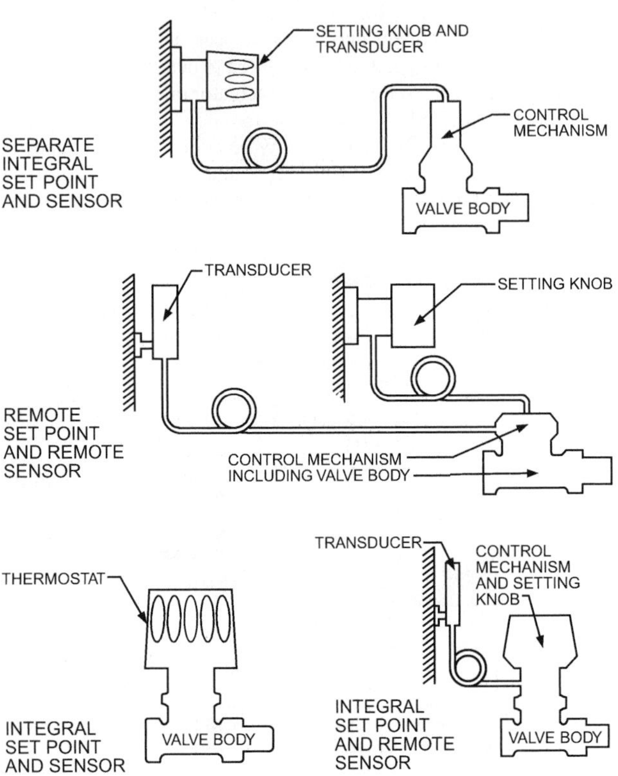

Fig. 12 Thermostatic Valves

the ball valve, or the valve should be sized smaller than the piping system to achieve adequate control (pressure drop). In some cases, the packing system of ball valves has been redesigned to accommodate the modulating control action inherent in HVAC. The control characteristic for full-ported ball valves is equal percentage, but modified seats, ball ports, or inserts are available to provide other characteristics (e.g., linear, modified linear, etc.).

Butterfly Valves

In some applications, it is not possible to use standard three-way mixing or bypass valves because of size limitations or space constraints. In these cases, two butterfly valves are mounted on a piping tee and cross-linked to operate as either three-way mixing or three-way bypass valves (Figure 15). Butterfly valves have different flow characteristics from standard seat and disk-type valves, so they may be used only where their flow characteristics suffice.

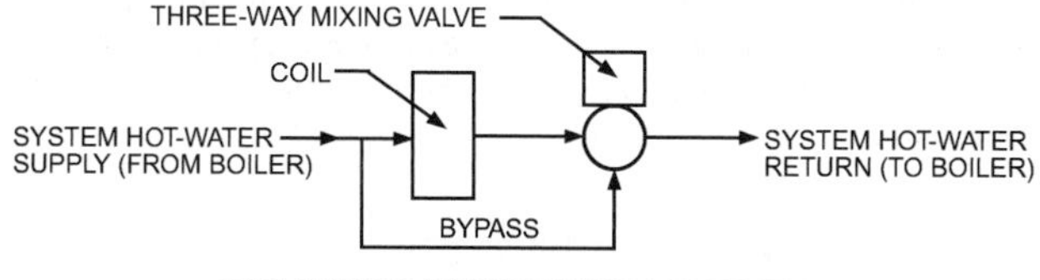

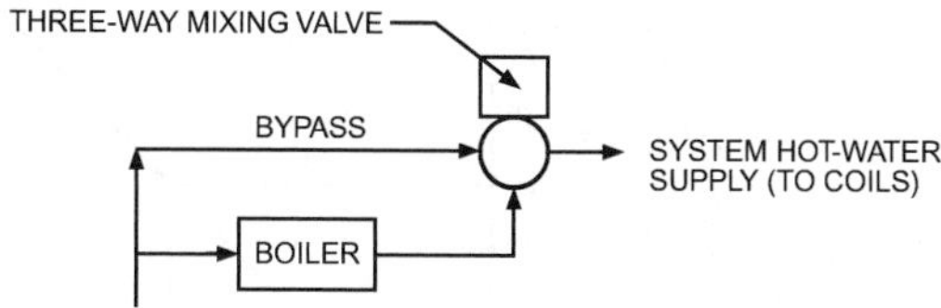

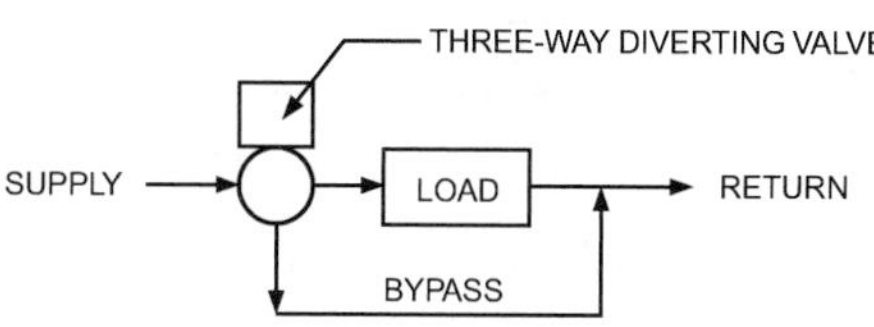

Fig. 13 Typical Three-Way Control Applications

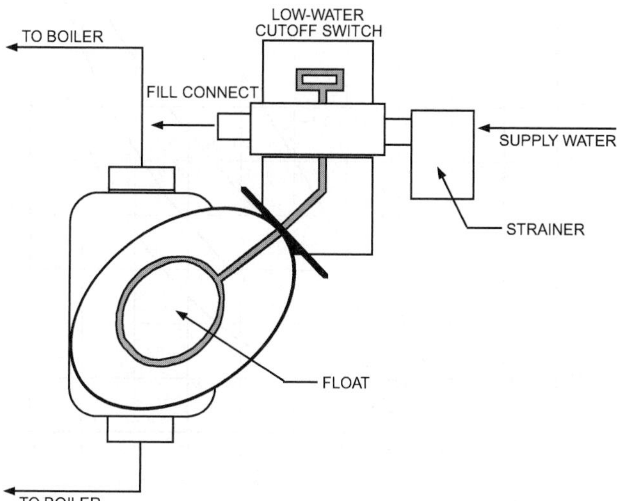

Fig. 14 Float Valve and Cutoff Steam Boiler Application

Control Valve Flow Characteristics

Generally, valves control the flow of fluids by an actuator, which moves a stem with an attached plug. The plug seats within the valve port and against the valve seat with a composition disk or metal-to-metal seating. Based on the geometry of the plug, three distinct flow conditions can be developed (Figure 16):

- **Quick Opening.** When started from the closed position, a quick-opening valve allows a considerable amount of flow to pass for small stem travel. As the stem moves toward the open position, the rate at which the flow is increased per movement of the stem is reduced in a nonlinear fashion. This characteristic is used in two-position or on/off applications.
- **Linear.** Linear valves produce equal flow increments per equal stem travel throughout the travel range of the stem. This characteristic is used on steam coil terminals and in the bypass port of three-way valves.

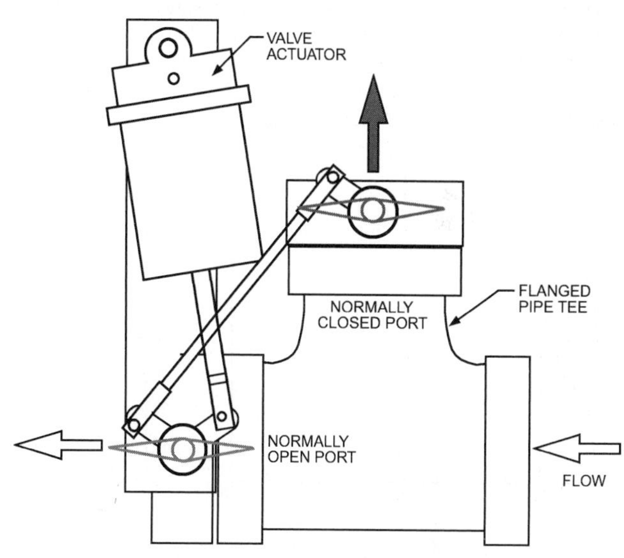

Fig. 15 Butterfly Valves, Diverting Tee Application

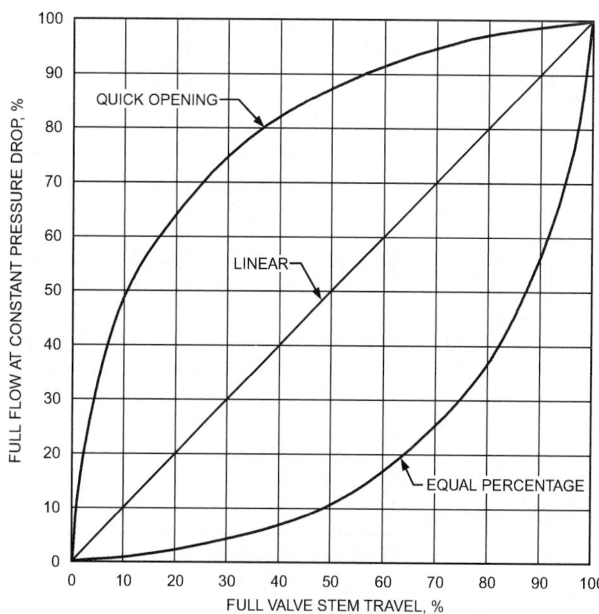

Fig. 16 Control Valve Flow Characteristics

- **Equal Percentage.** This type of valve produces an exponential flow increase as the stem moves from the closed position to the open. The term equal percentage means that for equal increments of stem travel, the flow increases by an equal percentage. For example, in Figure 16, if the valve is moved from 50 to 70% of full stroke, the percentage of full flow changes from 10 to 25%, an increase of 150%. Then, if the valve is moved from 80 to 100% of full stroke, the percentage of full flow changes from 40 to 100%, again, an increase of 150%. This characteristic is recommended for control on hot and chilled water terminals.

Control valves are commonly used in combination with a coil and another valve within a circuit to be controlled. The designer should combine the valve flow characteristics with coil performance curves (heating or cooling) because the resulting energy output profile of the circuit versus the stem travel improves (Figure 17). For a typical hydronic heating or cooling coil, the equal percentage results in the closest to a linear change and provides the most efficient control (Figure 17).

The three flow patterns are obtained by imposing a constant pressure drop across the modulating valve, but in actual conditions, the pressure drop across the valve varies between a maximum (when it is controlling) and a minimum (when the valve is near full open). The ratio of these two pressure drops is known as **authority**. Figures 18 and 19 show how linear and equal-percentage valve flow characteristic are distorted as the control valve authority is reduced because of a reduction in valve pressure drop. The quick-opening characteristic, not shown, is distorted to the point that it approaches two-position or on/off control. The selection of the control valve pressure drop directly affects the valve authority and should be at least 25 to 50% of the system loop pressure drop (i.e., the pressure drop from the pump discharge flange, supply main, supply riser, supply branch, heat transfer coil, return branch, fittings, balancing valve, and return main to the pump suction flange). The location of the control valve in the system results in unique pressure drop selections for each control valve. A higher valve pressure drop allows a smaller valve pipe size and better control.

Control Valve Sizing

Liquids. A valve creates fluid resistance in a circuit to limit the flow of the fluid at a calculated pressure drop. Each passive element in a circuit creates a pressure drop according to the following general equation:

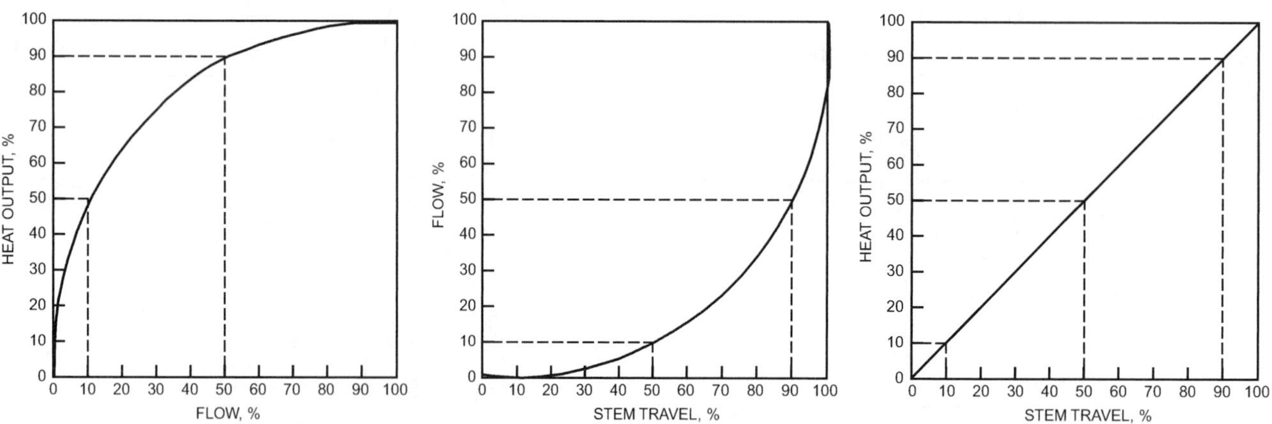

Fig. 17 Heat Output, Flow, and Stem Travel Characteristics of Equal Percentage Valve

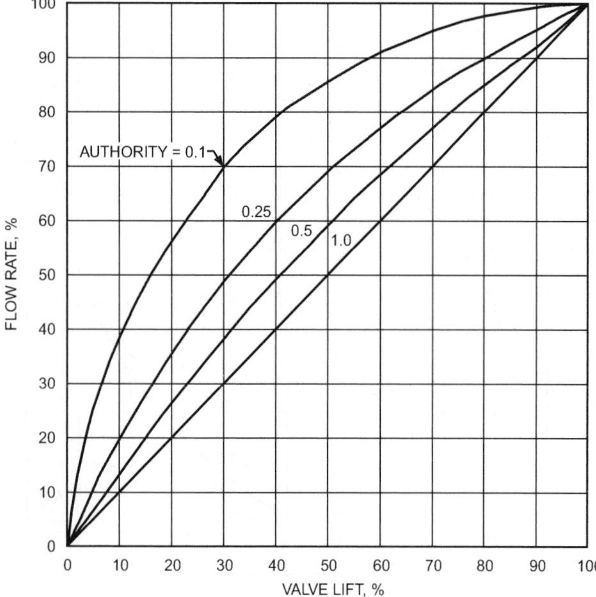

Fig. 18 Authority Distortion of Linear Flow Characteristics

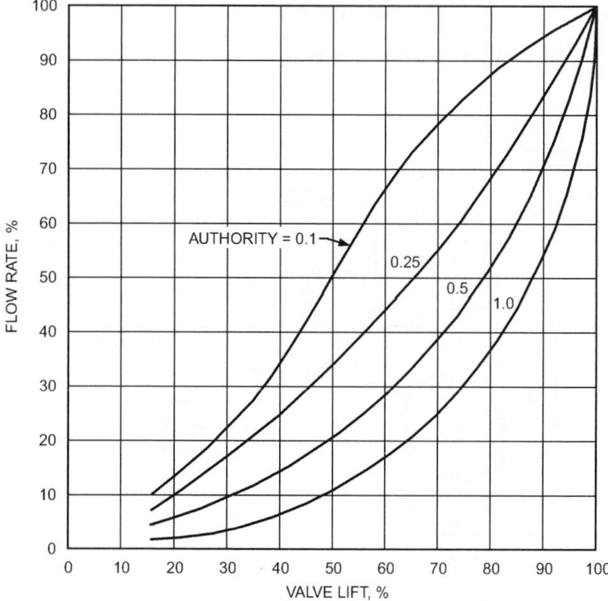

Fig. 19 Authority Distortion of Equal-Percentage Flow Characteristic

$$\Delta p = RQ^n \left(\frac{\rho}{\rho_w}\right) \quad (1)$$

where

Δp = pressure drop, psi
R = resistance
ρ = fluid density, lb/ft^3
ρ_w = density of water at 60°F, lb/ft^3
Q = volumetric flow, gpm
n = system coefficient

For turbulent flows, n is assumed to be 2, although for steel pipes $n = 1.85$.

For a valve, assuming $n = 2$, Equation (1) can be solved for flow:

$$Q = \sqrt{\left(\frac{\Delta p}{R}\right)\left(\frac{\rho_w}{\rho}\right)} \quad (2)$$

The term $\sqrt{1/R}$ can be replaced by the flow coefficient C_v, the ratio ρ/ρ_w is approximately 1 for water at temperatures below 250°F, and Equation (2) becomes

$$Q = C_v \sqrt{\Delta p} \quad (3)$$

or

$$Q = 0.67 C_v \sqrt{\Delta h} \quad (4)$$

where Δh = pressure drop, ft of water.

The control valve size should be selected by calculating the required C_v to provide the design flow at an assumed pressure drop Δp. A pressure drop of 25 to 50% of the available pressure between the supply and return riser (pump head) should be selected for the control valve. This pressure drop gives the best flow characteristic as described in the section on Control Valve Flow Characteristics.

For liquids with a viscosity correction factor V_f,

$$Q = \frac{C_v}{V_f} \sqrt{\Delta p \left(\frac{\rho_w}{\rho}\right)} \quad (5)$$

Steam. For steam flow,

$$w_s = 2.1 \frac{C_v}{K} \sqrt{\Delta p (P_1 + P_2)} \quad (6)$$

where

w_s = steam flow, lb/h
K = 1 + 0.0007 × (°F of superheat)
C_v = flow coefficient, gpm at Δp = 1 psi
P_1 = entering steam absolute pressure
P_2 = leaving steam absolute pressure
Δp = steam pressure drop across valve, $P_1 - P_2$

Note: Some manufacturers list the constant in Equation (6) as high as 3.2, but most agree on 2.1. As part of good practice, always confirm valve sizing with the manufacturer.

Steam reaches critical or sonic velocity when the downstream pressure is 58% or less of the absolute inlet pressure. If the downstream pressure is below the critical pressure, increasing the pressure drop produces no further increase in flow. As a result, when $P_2 \le 0.58 P_1$, the following critical pressure drop formula is used:

$$C_v = \frac{w_s}{1.61 P_1} \quad (7)$$

Applications

Automatically controlled valves are applied to control many different variables, the most common being temperature, humidity, flow, and pressure. However, a valve can be used directly to control only flow or pressure. When flow is controlled, a pressure drop is implied, and when pressure is controlled, some maximum flow rate is implied. These two factors must be considered in selecting control valves. For some typical valve applications, refer to Chapter 46 of the 2007 *ASHRAE Handbook—HVAC Applications*.

Although the discussion in this chapter applies to hot water, chilled water, and steam, control valves can be used with virtually any fluid. The fluid characteristics must be considered in selecting materials for the valve. The requirements are particularly strict for use with high-temperature water and high-pressure steam.

Steam is controlled in two ways:

1. When steam pressure is too high for use in a specific application, the pressure must be reduced by a **pressure-reducing valve** (PRV). This is normally a globe-type valve, because modulating control is required. The valve may be externally or internally piloted and is usually self-contained, using the steam pressure to drive the actuator. The load may vary, so it is sometimes desirable to use two or more valves in parallel, adjusted to open in sequence, for more accurate control.
2. Steam flow to a heat exchanger may be controlled in response to temperature or humidity requirements. In this case, an external control system is used with the steam valve as the controlled device. In selecting a steam valve, the maximum flow rate for the specific valve and entering steam pressure must be considered. These factors are determined from the critical pressure drop, which limits the flow.

Hot and chilled water are usually controlled in response to temperature or humidity requirements. When selecting a valve for controlling water flow, a pressure drop sufficiently large to allow the valve to control properly should be specified. The response of the heat exchanger coil to a change in flow is not linear; therefore, an equal-percentage plug should be used, and the temperature of the water supply should be as high (hot water) or as low (chilled water) as required by the load conditions.

BALANCING VALVES

Two approaches are available for balancing hydronic systems: (1) a manual valve with integral pressure taps and a calibrated port, which allows field proportional balancing to the design flow conditions; (2) or an automatic flow-limiting valve selected to limit the circuit's maximum flow to the design flow.

Manual Balancing Valves

Manual balancing valves can be provided with the following features:

- Manually adjustable stems for valve port opening or a combination of a venturi or orifice and an adjustable valve
- Stem indicator and/or scale to indicate the relative amount of valve opening
- Pressure taps to provide a readout of the pressure difference across the valve port or the venturi/orifice
- Capability to be used as a shutoff for future service of the heat transfer terminal
- Locking device for field setting the maximum opening of a valve
- Body tapped for attaching drain hose

Manual balancing valves may have rotary, rising, or nonrising stems for port adjustment (Figure 20).

Meters with various scale ranges, a field carrying case, attachment hoses, and fittings for connecting to the manual balancing valve should be used to determine its flow by reading the differential pressure. Some meters use analog measuring elements with direct-reading mechanical dual-element Bourdon tubes. Other meters are electronic differential pressure transducers with a digital data display.

Many manufacturers of balancing valves produce circular slide rules to calculate circuit flow based on pressure difference readout across the balancing valve, its stem position, and/or the valve's flow coefficient. This calculator can also be used for selecting the size and setting of the valve when the terminal design flow conditions are known.

Automatic Flow-Limiting Valves

A **differential pressure-actuated flow control valve**, also called an automatic flow-limiting valve (Figure 21), regulates the flow of fluid to a preset value when the differential pressure across it is varied. This regulation (1) helps prevent an overflow condition in the circuit where it is installed and (2) aids the overall system balance when other components are changing (modulating valves, pump staging, etc.).

Typically, the valve body contains a moving element containing an orifice, which adjusts itself based on pressure forces so that the flow passage area varies.

The area of an orifice can be changed by either (1) a piston or cup moving across a shear plate or (2) increased pressure drop to squeeze the rubber orifice in rubber grommet valves.

A typical performance curve for the valve is shown in Figure 22. The flow rate for the valve is set. The flow curve is divided into three ranges of differential pressure: the start-up range, the control range, and the above-control range.

Balancing Valve Selection

The balancing valve is a flow control device that is selected for a lower pressure drop than an automatic control valve (5 to 10% of the available system pressure). Selection of any control valve is based on the pressure drop at maximum (design) flow to ensure that the valve provides control at all flow rates. A properly selected balancing valve can proportionally balance flow to its terminal with flow to the adjacent terminal in the same distribution zone. Refer to Chapter 37 of the 2007 *ASHRAE Handbook—HVAC Applications* for balancing details.

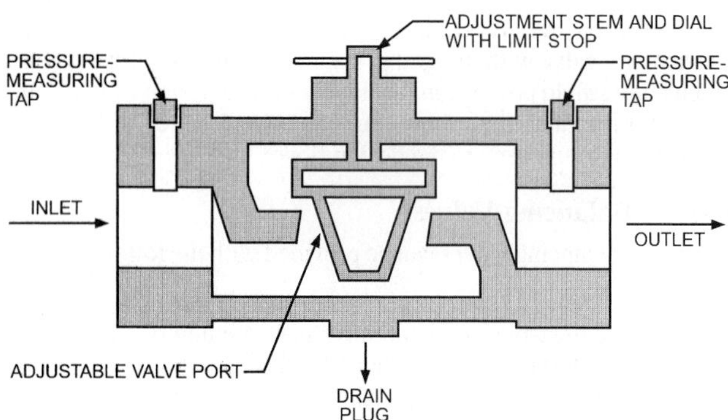

Fig. 20 Manual Balancing Valve

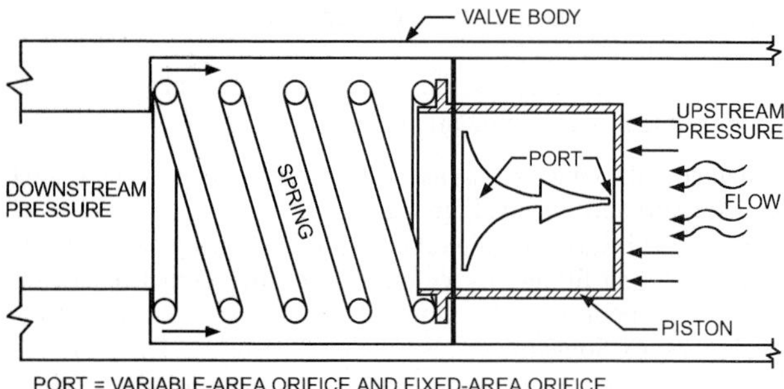

Fig. 21 Automatic Flow-Limiting Valve

MULTIPLE-PURPOSE VALVES

Multiple-purpose valves are made in straight pattern or angle pattern. The valves can provide shutoff for servicing or can be partially closed for balancing. Pressure gage connections to read the pressure drop across the valve can be used with the manufacturer's calibration chart or meter to estimate the flow. Means are provided to return the valve to its as-balanced position after shutoff for servicing. The valve also acts as a check valve to prevent backflow when parallel pumps are used and one of the pumps is cycled off.

Figure 23 shows a straight pattern multiple-purpose valve designed to be installed 5 to 10 pipe diameters from the pump discharge of a hydronic system.

Figure 24 shows an angle pattern multiple-purpose valve installed 5 to 10 pipe diameters downstream of the pump discharge with a common gage and a push button trumpet valve manifold to measure the differential pressure across the strainer, pump, or multiple-purpose valve. From this, the flow can be estimated. The differential pressure across the pump suction strainer can also be estimated to determine whether the strainer needs servicing.

SAFETY DEVICES

The terms safety valve, relief valve, and safety relief valve are sometimes used interchangeably, and although the devices generally provide a similar function (safety), they have important differences in their modes of operation and application in HVAC systems (Jordan 1998).

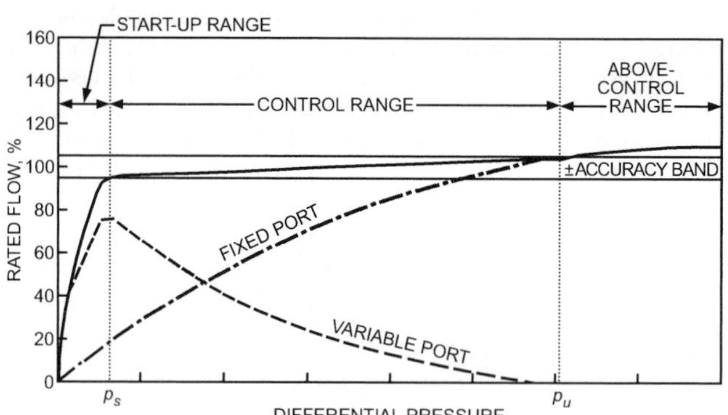

Fig. 22 Automatic Flow-Limiting Valve Curve

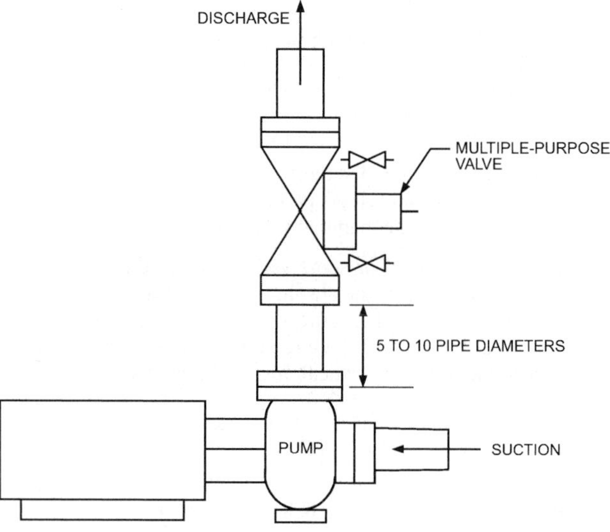

Fig. 23 Typical Multiple-Purpose Valve (Straight Pattern) on Discharge of Pump

Safety valves open rapidly (pop-action). They are used for gases and vapors (e.g., compressed air and steam).

Relief valves open or close gradually in proportion to excessive pressure. They are used for liquids (e.g., unheated water).

Safety relief valves perform a dual function: they open rapidly (pop-action) for gases and vapors and gradually for liquids. Typical HVAC application is for heating water.

Temperature-actuated pressure relief valves (or temperature and pressure safety relief valves) are activated by excessive temperature or pressure. They are commonly used for potable hot water.

Application of these safety devices must comply with building codes and the ASME *Boiler and Pressure Vessel Code*. For the remainder of this discussion, the term "safety valve" is used generically to include any or all of the four types described.

Safety valve construction, capacities, limitations, operation, and repair are covered by the ASME *Boiler and Pressure Vessel Code*. For pressures above 15 psig, refer to Section I. Section IV covers steam boilers for pressures less than 15 psig. Unfired pressure vessels (such as heat exchange process equipment or pressure-reducing valves) are covered by Section VIII.

The capacity of a safety valve is affected by the equipment on which it is installed and the applicable code. Valves are chosen based on accumulation, which is the pressure increase above the maximum allowable working pressure of the vessel during valve discharge. Section I valves are based on 3% accumulation. Accumulation may be as high as 33.3% for Section IV valves and 10% for Section VIII. To properly size a safety valve, the required capacity and set pressure must be known. On a pressure-reducing valve station, the safety valve must have sufficient capacity to prevent an unsafe pressure rise if the reducing valve fails in the open position.

The safety valve set pressure should be high enough to allow the valve to remain closed during normal operation, yet allow it to open and reseat tightly when cycling. A minimum differential of 5 psi or 10% of inlet pressure (whichever is greater) is recommended.

When installing a safety valve, consider the following:

- Install the valve vertically with the drain holes open or piped to drain.
- The seat can be distorted if the valve is overtight or the weight of the discharge piping is carried by the valve body. A drip-pan elbow on the discharge of the safety valve prevents the weight of the discharge piping from resting on the valve (Figure 25).
- Use a moderate amount of pipe thread lubricant (first 2 to 3 threads) on male threads only.
- Install clean flange connections with new gaskets, properly aligned and parallel, and bolted with even torque to prevent distortion.
- Wire cable or chain pulls attached to the test levers should allow for a vertical pull, and their weight should not be carried by the valve.

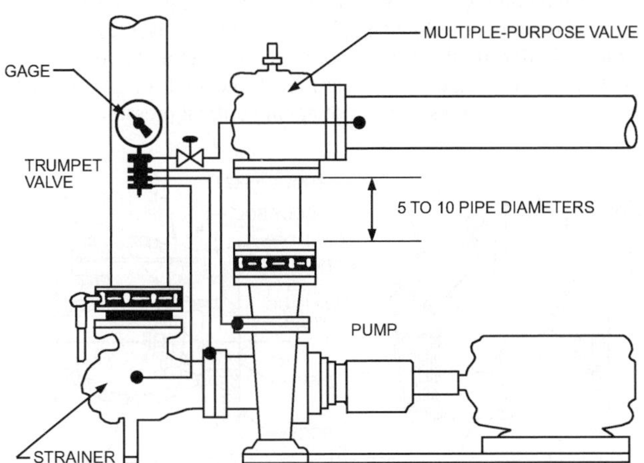

Fig. 24 Typical Multiple-Purpose Valve (Angle Pattern) on Discharge of Pump

Testing of safety valves varies between facilities depending on operating conditions. Under normal conditions, safety valves with a working pressure under 400 psig should be tested manually once per month and pressure-tested once each year. For higher pressures, the test frequency should be based on operating experience.

When steam safety valves require repair, adjustment, or set pressure change, the manufacturer or approved stations holding the ASME V, UV, and/or VR stamps must perform the work. Only the manufacturer is allowed to repair Section IV valves.

SELF-CONTAINED TEMPERATURE CONTROL VALVES

Self-contained or self-operated temperature control valves do not require an outside energy source such as compressed air or electricity (Figure 26). They depend on a temperature-sensing bulb and

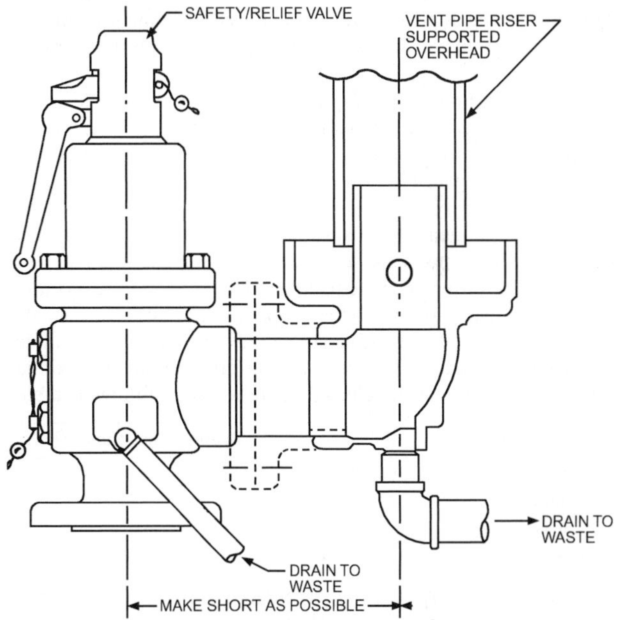

Fig. 25 Safety/Relief Valve with Drip-Pan Elbow

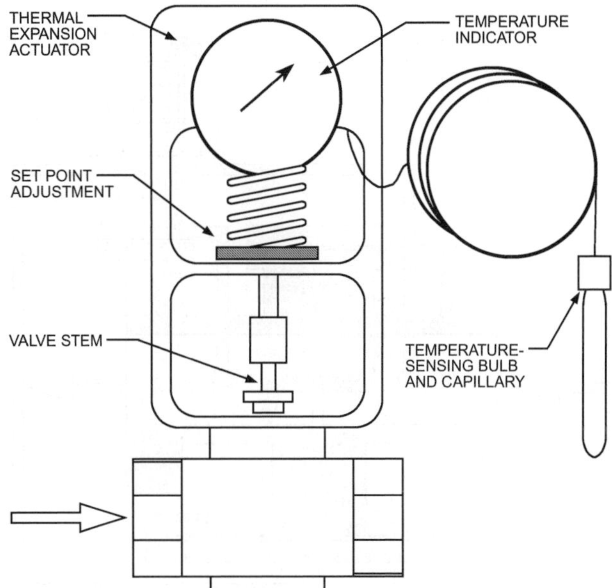

Fig. 26 Self-Operated Temperature Control Valve

capillary tube filled with either an oil or a volatile liquid. In an **oil-filled** system, the oil expands as the sensing bulb is heated. This expansion is transmitted through the capillary tube to an actuator bellows in the valve top, which causes the valve to close. The valve opens as the sensing bulb cools and the oil contracts; a spring provides a return force on the valve stem.

A volatile-liquid control system is known as a **vapor pressure** or **vapor tension** system. When the sensing bulb is warmed, some of the volatile liquid vaporizes, causing an increase in the sealed system pressure. The pressure rise is transmitted through the capillary tube to expand the bellows, which then moves the valve stem and closes the valve. Thermal systems actuate the control valve either directly or through a pilot valve.

In a **direct-actuated** design, the control directly moves the valve stem and plug to close or open the valve. These valves must compensate for the steam pressure force acting on the valve seat by generating a greater force in the bellows to close the valve. An adjustable spring adjusts the temperature set point and provides the return force to move the valve stem upward as the temperature decreases.

A **pilot-operated valve** (Figure 27) uses a much smaller intermediate pilot valve that controls the flow of steam to a large diaphragm that then acts on the valve stem. This allows the control system to work against high steam pressures caused by the smaller area of the pilot valve.

For self-contained temperature valves to operate as proportional controls, the bulb must sense a change in the temperature of the process fluid. The difference in temperature from no-load to maximum controllable load is known as the **proportional band**. Because the size of this proportional band can be varied depending on valve size, the accuracy is variable. Depending on the application, proportional bands of 2 to 18°F may be selected, as shown in the following table:

Application	Proportional Band, °F
Domestic hot-water heat exchanger	6 to 14
Central hot water	4 to 7
Space heating	2 to 5
Bulk storage	4 to 18

Although their response time, accuracy, and ease of adjustment may not be as good as those of electrically or pneumatically actuated valves, self-contained steam temperature controls are widely accepted for many applications.

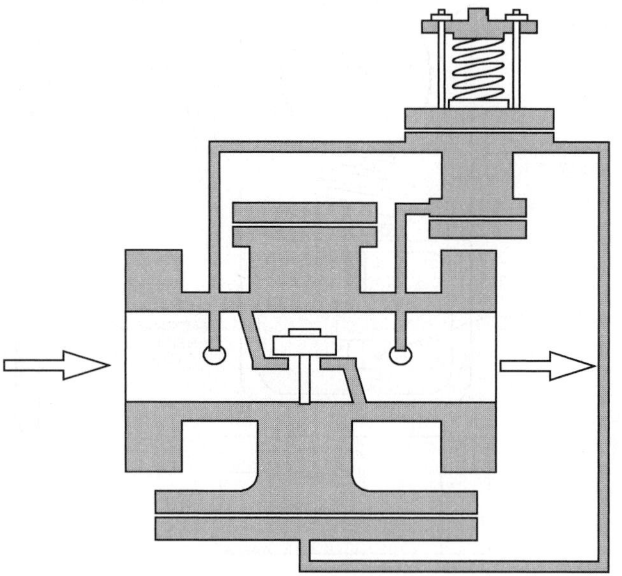

Fig. 27 Pilot-Operated Steam Valve

PRESSURE-REDUCING VALVES

Should steam pressure be too high for a specific process, a self-contained pressure-reducing valve (PRV) may be used to reduce this pressure, which will also increase the available latent heat. These valves may be direct-acting or pilot-operated (Figure 27), much like temperature control valves. To maintain set pressure, the downstream pressure must be sensed either through an internal port or an external line.

The amount of pressure drop below the set pressure that causes the valve to react to a load change is called **droop**. As a general rule, pilot-operated valves have less droop than direct-acting types. To properly size these valves, only the mass flow of steam, the inlet pressure, and the required outlet pressure must be known. Valve line size can be determined by consulting manufacturers' capacity charts.

Because of their construction, simplicity, accuracy, and ease of installation and maintenance, these valves have been specified for most steam-reducing stations.

Makeup Water Valves

A pressure-reducing valve is normally provided on a hydronic heating or cooling system to automatically fill the system with domestic or city water to maintain a minimum system pressure. This valve may be referred to as a fill valve, PRV fill valve, or automatic PRV makeup water valve, and is usually located at or near the system expansion tank. Local plumbing codes may require a backflow prevention device where the city water connects to the building domestic water system (see the section on Backflow Prevention Devices).

CHECK VALVES

Check valves prevent reversal of flow, controlling the direction of flow rather than stopping or starting flow. Some basic types include swing check, ball check, wafer check, silent check, and stop-check valves. Most check valves are available in screwed and flanged body styles.

Swing check valves have hinge-mounted disks that open and close with flow (Figure 28). The seats are generally made of metal, whereas the disks may be of metallic or nonmetallic composition materials. Nonmetallic disks are recommended for fluids containing dirt particles or where tighter shutoff is required. The Y-pattern check valve has an access opening to allow cleaning and regrinding in place. Pressure drop through swing check valves is lower than that through lift check valves because of the straight-through design. Weight- or spring-loaded lever arm check valves are available to limit objectionable slamming or chattering when pulsating flows are encountered.

Lift check valves have a body similar in design to a globe or angle valve body with a similar disk seating. The guided valve disk is forced open by the flow and closes when flow reverses. Because of the body design, the pressure drop is higher than that of a swing check valve. Lift check valves are recommended for gas or compressed air or in fluid systems not having critical pressure drops.

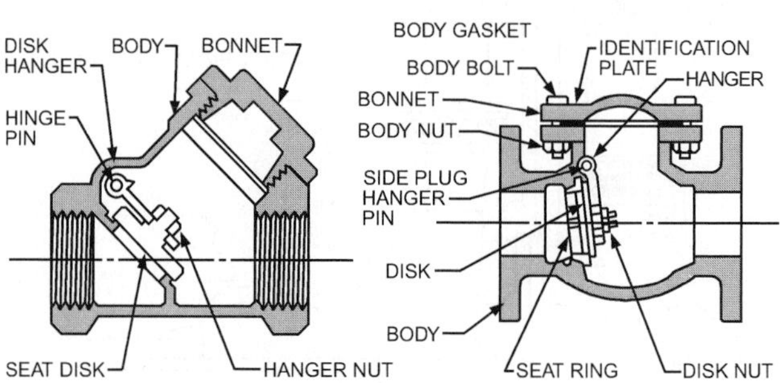

Fig. 28 Swing Check Valves
(Courtesy Anvil Int'l.)

Ball check valves are similar to lift checks, except that they use a ball rather than a disk to accomplish closure. Some ball checks are specifically designed for horizontal flow or vertical upflow installation.

Wafer check valves are designed to fit between pipe flanges similar to butterfly valves and are used in larger piping (4 in. diameter and larger). Wafer check valves have two basic designs: (1) dual spring-loaded flapper, which operates on a hinged center post, and (2) single flapper, which is similar to the swing check valve.

In **silent or spring-loaded check valves**, a spring positively and rapidly closes a guided, floating disk. This valve greatly reduces water hammer, which may occur with slow-closing check valves like the swing check. Silent check valves are recommended for use in pump discharge lines.

STOP-CHECK VALVES

Stop-check valves can operate as both a check valve and a stop valve. The valve stem does not connect to the guided seat plug, allowing the plug to operate as a conventional lift check valve when the stem is in the raised position. Screwing the stem down can limit the valve opening or close the valve. Stop-check valves are used for shutoff service on multiple steam boiler installations, in accordance with the ASME *Boiler and Pressure Vessel Code*, to prevent backflow of steam or condensate from an operating boiler to a shutdown boiler. They are mandatory in some jurisdictions. Local codes should be consulted.

BACKFLOW PREVENTION DEVICES

Backflow prevention devices prevent reverse flow of the supply in a water system. A **vacuum breaker** prevents back siphonage in a nonpressure system, while a **backflow preventer** prevents backflow in a pressurized system (Figure 29).

Selection

Vacuum breakers and backflow preventers should be selected on the basis of the local plumbing codes, the water supply impurities involved, and the type of cross-connection.

Impurities are classified as (1) contaminants (substances that could create a health hazard if introduced into potable water) and (2) pollutants (substances that could create objectionable conditions but not a health hazard).

Cross-connections are classified as nonpressure or pressure connections. In a nonpressure cross-connection, a potable-water pipe connects or extends below the overflow or rim of a receptacle at atmospheric pressure. When this type of connection is not protected by a minimum air gap, it should be protected by an appropriate vacuum breaker or an appropriate backflow preventer.

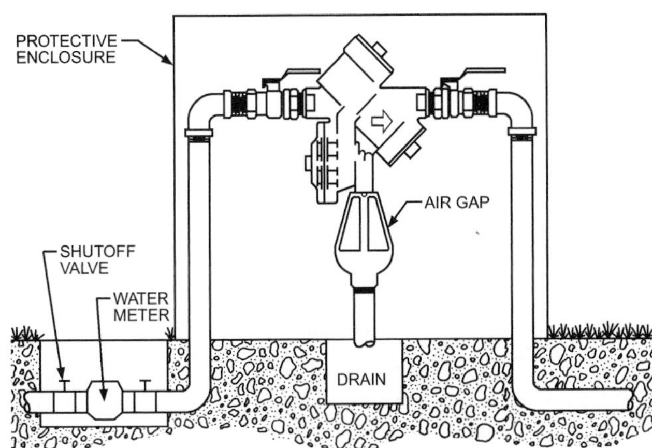

Fig. 29 Backflow Prevention Valve

In a pressure cross-connection, a potable-water pipe is connected to a closed vessel or a piping system that is above atmospheric pressure and contains a nonpotable fluid. This connection should be protected by an appropriate backflow preventer only. Note that a pressure vacuum breaker should not be used alone with a pressure cross-connection.

Vacuum breakers should be corrosion-resistant. Backflow preventers, including accessories, components, and fittings that are 2 in. and smaller, should be made of bronze with threaded connections. Those larger than 2 in. should be made of bronze, galvanized iron, or fused epoxy-coated iron inside and out, with flanged connections. All backflow prevention devices should meet applicable standards of the American National Standards Institute, the Canadian Standards Association, or the required local authorities.

Installation

Vacuum breakers and backflow preventers equipped with atmospheric vents, or with relief openings, should be installed and located to prevent any vent or relief opening from being submerged. They should be installed in the position recommended by the manufacturer.

Backflow preventers may be double check valve (DCV) or reduced pressure zone (RPZ) types. Refer to manufacturers' information for specific application recommendations and code compliance.

STEAM TRAPS

For a description and diagram of these traps, refer to Chapter 10.

REFERENCES

ASME. 1996. Valves—Flanged, threaded, and welding end. ANSI/ASME *Standard* B16.34-1988. American Society of Mechanical Engineers, New York.

ASME. 1998. *Boiler and pressure vessel code*. American Society of Mechanical Engineers, New York.

ASME. 1998. Rules for construction of power boilers. *Boiler and pressure vessel code*, Section I-98. American Society of Mechanical Engineers, New York.

ASME. 1998. Rules for construction of heating boilers. *Boiler and pressure vessel code*, Section IV-98. American Society of Mechanical Engineers, New York.

ASME. 1998. Rules for construction of pressure vessels. *Boiler and pressure vessel code*, Section VIII-98. American Society of Mechanical Engineers, New York.

Fairbanks Morse Pump Company. 1965. *Hydraulic handbook*. Catalog C and #65-26313. Beloit, WI.

FCI. 1991. Control valve seat leakage. ANSI/FCI *Standard* 70-2-91. Fluid Controls Institute, Cleveland.

ISA. 1995. Flow equations for sizing control valves. ANSI/ISA *Standard* S75.01-85 (R 1995). International Society for Measurement and Control, Research Triangle Park, NC.

ISA. 1996. Control valve capacity test procedures. ANSI/ISA *Standard* S75.02-96. International Society for Measurement and Control, Research Triangle Park, NC.

Jordan, C.H. 1998. Terminology of pressure relief devices. *Heating/Piping/Air Conditioning* 70(9):47-48.

Lyons, J.L. 1982. *Lyons' valve designer's handbook*, pp. 92-93, 209-210. Van Nostrand Reinhold, New York.

Ulanski, W. 1991. *Valve and actuator technology*. McGraw-Hill, New York.

BIBLIOGRAPHY

ASHRAE. 1988. Practices for measurement, testing, adjusting and balancing of building heating, ventilation, air-conditioning and refrigeration systems. ANSI/ASHRAE *Standard* 111-1988.

ASHRAE. 1996. The HVAC commissioning process. *Guideline* 1-1996.

ASME. 1994. Pressure relief devices. ANSI/ASME *Performance Test Code* PTC 25-94. With Special Addenda (1998). American Society of Mechanical Engineers, New York.

Baumann, H.D. 1998. *Control valve primer*, 3rd ed. International Society of Measurement and Control (ISA), Research Triangle Park, NC.

Borden, G., Jr. and P.G. Friedmann, eds. *Control valves*. International Society of Measurement and Control (ISA), Research Triangle Park, NC.

CIBSE. 1985. Automatic controls and their implications for system design. *Applications Manual* AM1. The Chartered Institution of Building Services Engineers, London.

CIBSE. 1986. *Guide B: Installation and equipment data*. The Chartered Institution of Building Services Engineers, London.

Crane. Flow of fluids through valves, fittings and pipe. *Technical Paper* 410. Crane Valve Group, Long Beach, CA.

Gupton, G. 1987. *HVAC controls: Operation and maintenance*. Van Nostrand Reinhold, New York.

Haines, R.W. 1987. *Control systems for HVAC*. Van Nostrand Reinhold, New York.

IEC. 1987. Control valve terminology and general considerations. Industrial-process control valves, Part I. IEC *Publication* 60534-1. International Electrotechnical Commission, Geneva, Switzerland.

Matley, J. and Chemical Engineering. 1989. *Valves for process control and safety*. McGraw-Hill, New York.

McQuiston, F.C. and J.D. Parker. 1993. *Heating, ventilating, and air conditioning: Analysis and design*, 4th ed. John Wiley & Sons, New York.

Merrick, R.C. 1991. *Valve selection and specification guide*. Van Nostrand Reinhold, New York.

Miller, R.W. 1983. *Flow measurement engineering handbook*. McGraw-Hill, New York.

Zappe, R.W. 1998. *Valve selection handbook*, 3rd ed. Gulf Publishing, Houston.

CHAPTER 47

HEAT EXCHANGERS

HEAT EXCHANGERS transfer heat from one fluid to another without the fluids coming in direct contact with each other. Heat transfer occurs in a heat exchanger when a fluid changes from a liquid to a vapor (evaporator), a vapor to a liquid (condenser), or when two fluids transfer heat without a phase change. The transfer of energy is caused by a temperature difference.

In most HVAC&R applications, heat exchangers are selected to transfer either sensible or latent heat. Sensible heat applications involve transfer of heat from one liquid to another. Latent heat transfer results in a phase change of one of the liquids; transferring heat to a liquid by condensing steam is a common example.

This chapter describes some of the fundamentals, types, components, applications, selection criteria, and installation of heat exchangers. Chapter 3 of the 2005 *ASHRAE Handbook—Fundamentals* covers the subject of heat transfer. Specific applications of heat exchangers are detailed in other chapters of this and other volumes of the Handbook series.

FUNDAMENTALS

When heat is exchanged between two fluids flowing through a heat exchanger, the rate of heat transferred may be calculated using

$$Q = UA\Delta t_m \tag{1}$$

where

U = overall coefficient of heat transfer from fluid to fluid
A = heat transfer area of the heat exchanger associated with U
Δt_m = log mean temperature difference (LMTD)

For a heat exchanger with a constant U, the Δt_m is calculated as

$$\Delta t_m = C_f \frac{(T_1 - t_2) - (T_2 - t_1)}{\ln(T_1 - t_2)/(T_2 - t_1)} \tag{2}$$

where the temperature distribution is as shown in Figure 1 and C_f is a correction factor (less than 1.0) that is applied to heat exchanger configurations that do not follow a true counterflow design.

Figure 1 illustrates a **temperature cross**, where the outlet temperature of the heating fluid is less than the outlet temperature of the fluid being heated ($T_2 < t_2$). A temperature cross can only be obtained with a heat exchanger that has a 100% true counterflow arrangement.

The overall coefficient U is affected by the physical arrangement of the surface area A. For a given load, not all heat exchangers with equal surface areas perform equally. For this reason, load conditions must be defined when selecting a heat exchanger for a specific application.

The load for each fluid stream can be calculated as

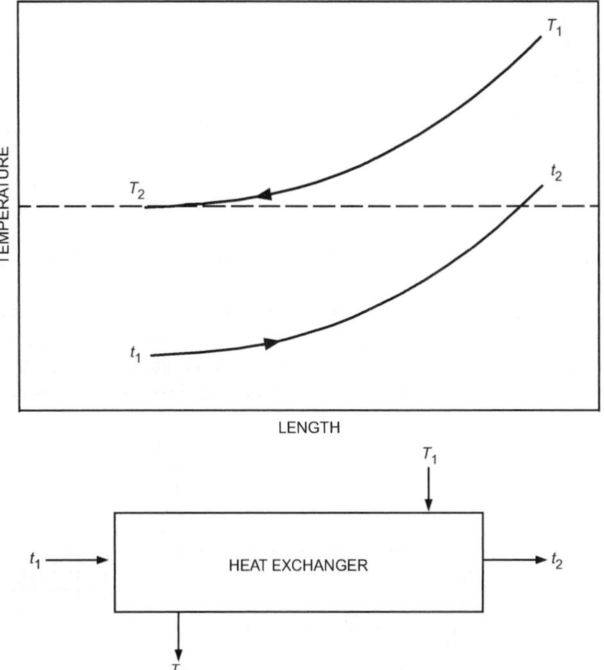

Fig. 1 Temperature Distribution in Counterflow Heat Exchanger

$$Q = mc_p(t_{in} - t_{out}) \tag{3}$$

The value of Δt_m is an important factor in heat exchanger selection. If the value Δt_m is high, a relatively small heat exchange surface area is required for a given load. The economic effect is that the heat exchanger must be designed to accommodate the forces and movements associated with large temperature differences. When the **approach temperature** (the difference between T_2 and t_1) is small, Δt_m is also small and a relatively large A is required.

Chapter 3 of the 2005 *ASHRAE Handbook—Fundamentals* describes an alternative method of evaluating heat exchanger performance that involves the exchanger heat transfer effectiveness ε and number of exchanger transfer units (NTU). This method is based on the same assumptions as the logarithmic mean temperature difference method described previously.

TYPES OF HEAT EXCHANGERS

Most heat exchangers for HVAC&R applications are counterflow shell-and-tube or plate units. While both types physically separate the fluids transferring heat, their construction is very different, and each has unique application and performance qualities.

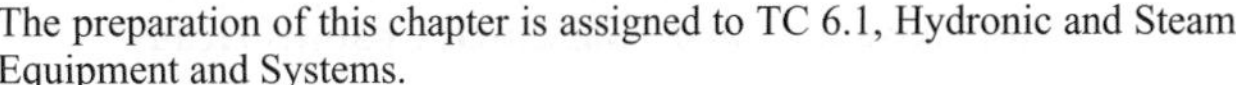

The preparation of this chapter is assigned to TC 6.1, Hydronic and Steam Equipment and Systems.

Shell-and-Tube Heat Exchangers

Figure 2 illustrates the counterflow path of a shell-and-tube heat exchanger. The fluid at temperature T_1 enters one end of the shell, flows outside the tubes and inside the shell, and exits at the other end at temperature T_2. The other fluid flows inside the tubes, entering one end at temperature t_1 and exiting at the opposite end at temperature t_2.

In a shell-and-tube heat exchanger, a tube bundle assembly is welded or bolted inside a tubular shell. The bundle is constructed of metal tubes mechanically rolled or welded at one (U-tube) or both ends (straight-tube) into tubesheet(s) that function as headers. The shell is usually a length of pipe that has inlet and outlet connections located along one or more of its longitudinal centerlines.

The shell is flanged at one or both open ends to accommodate a head assembly. The tube bundle is positioned between the shell and head assemblies so that the tube wall of the bundle mechanically separates the two flow paths.

The tube bundle is assembled with tube supports, which are held together with tie rods and spacers. Units with liquid on the shell side have baffles for tube supports that direct the flow. Condensers must have baffles that have been notched on the bottom to allow the liquid condensate to flow freely to the exit nozzle.

The head assembly directs the other fluid across the tubesheet(s) into and out of the tube bundle. Head assemblies are designed with pass partitions to isolate sections of the tube bundle such that the fluid must traverse the length of the unit one, two, four or more times before exiting.

One of two types of head assemblies is mechanically attached to the shell. Units with multiple tube-side pass construction have a head with both an inlet and outlet connection bolted at one end with a welded cap (U-tube) or bolted reversing head (straight-tube) at the opposite shell end. Single-pass units have an inlet head attached at one shell end and an outlet head attached at the other end.

Many variations of the shell-and-tube design are available, some of which are described in the following paragraphs.

U-Tube. Figure 3 illustrates a U-tube removable-bundle shell-and-tube heat exchanger. These units are commonly called **converters**. Figures 4 and 5 illustrate modifications of the U-tube design.

Tank heaters are U-tube heat exchangers with the shell replaced by a mounting collar, which is welded to a tank. A hot fluid or steam flows inside the tubes heating the fluid in the tank by natural convection. The tank heater manufacturer should be consulted about optimizing the bundle length. Although it is desirable for the bundle to significantly extend into the tank, the designer must consider the need for additional bundle support.

Tank suction heaters differ from tank heaters because they have an additional opening that allows fluid being heated to be pumped across the outside tube wall resulting in improved thermal performance.

Straight-Tube. Figures 6 and 7 illustrate two common designs of straight-tube, shell-and-tube exchangers, one with a fixed and the other with a removable tube bundle assembly.

Some straight-tube, shell-and-tube heat exchangers have a floating head bolted with a gasket to a floating tubesheet or a shell-side expansion joint. This configuration is expensive and is rarely specified in HVAC applications.

Shell-and-Coil. The tubes in this heat exchanger are coiled in a helical configuration around a small core. A spacer is placed between

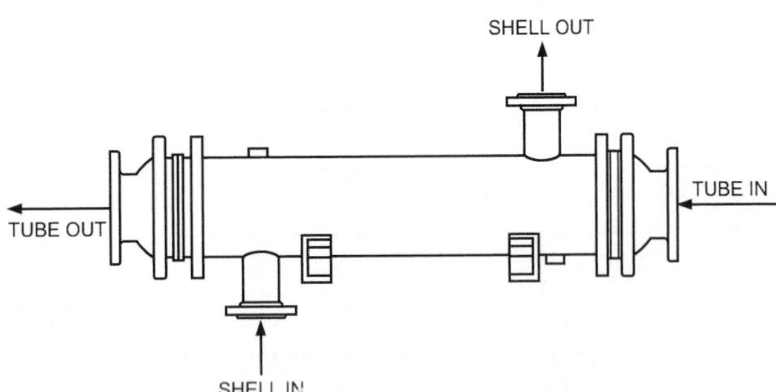

Fig. 2 Counterflow Path in Shell-and-Tube Heat Exchanger

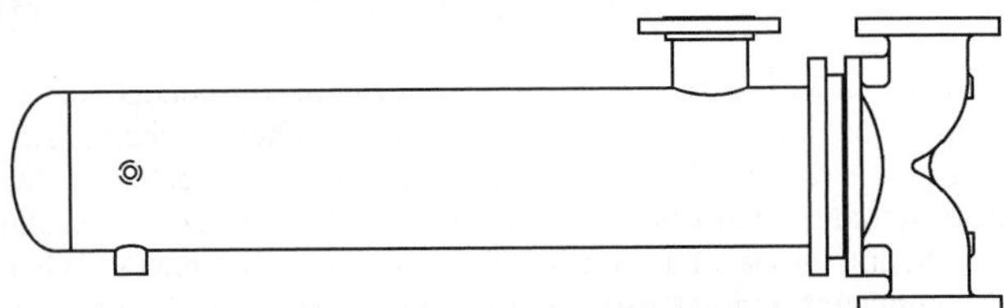

Fig. 3 U-Tube Shell-and-Tube Heat Exchanger with Removable Bundle Assembly and Cast K-Pattern Flanged Head

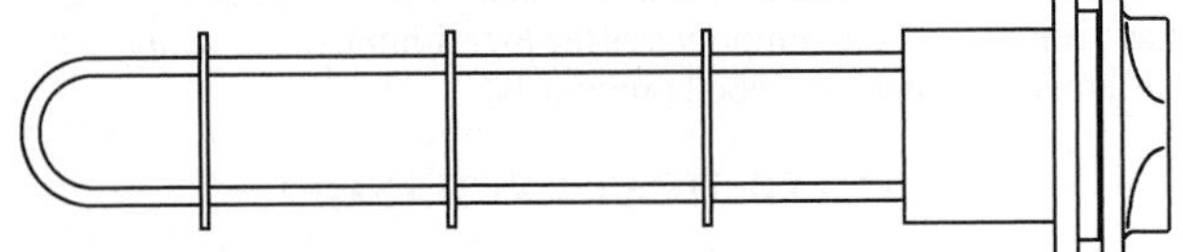

Fig. 4 U-Tube Tank Heater with Removable Bundle Assembly and Cast Bonnet Head

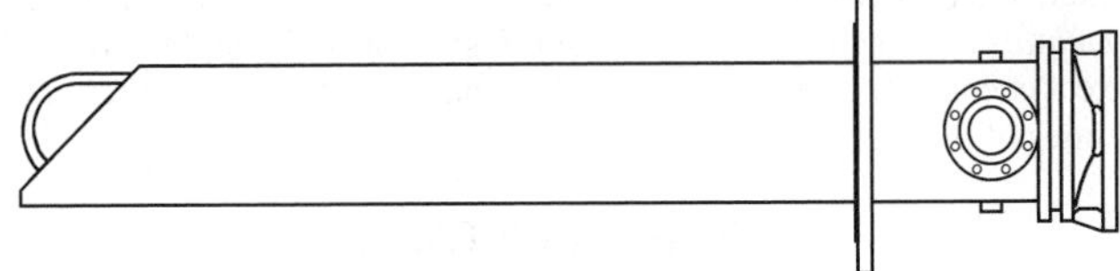

Fig. 5 U-Tube Tank Suction Heater with Removable Bundle Assembly and Cast Flanged Head

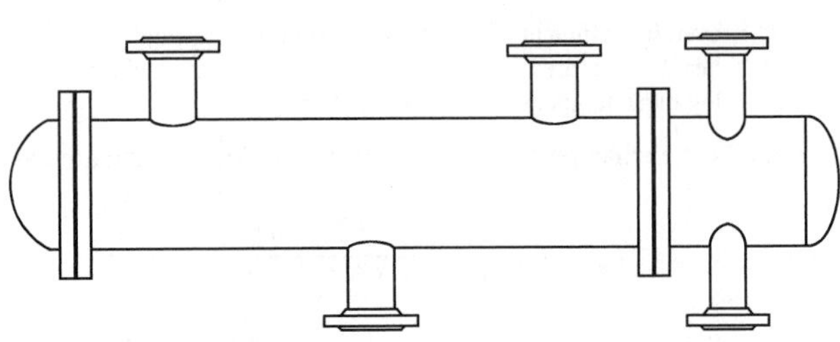

Fig. 6 Straight-Tube Fixed Tubesheet Shell-and-Tube Heat Exchanger with Fabricated Bonnet Heads and Split-Shell Flow Design

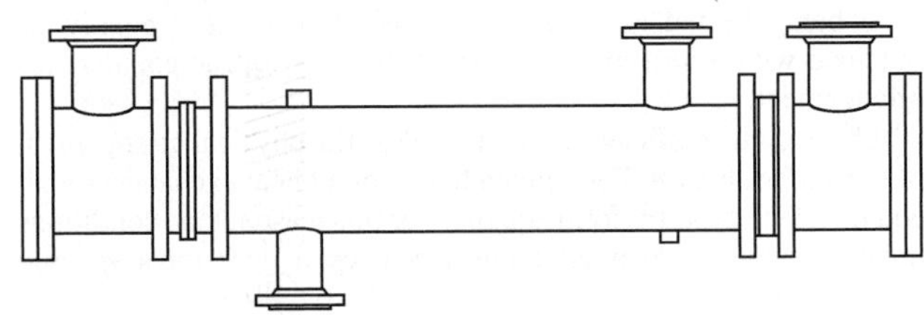

Fig. 7 Straight-Tube Floating Tubesheet Shell-and-Tube Heat Exchanger with Removable Bundle Assembly and Fabricated Channel Heads

the tube layers. In some designs the tubes have an oval cross section. These heat exchangers are very compact and have a relatively large surface area for their size. Figure 5 in Chapter 41 illustrates a shell-and-coil heat exchanger.

Plate Heat Exchangers

Plate heat exchangers consist of metal plate pairs arranged to provide separate flow paths (channels) for two fluids. Heat transfer occurs across the plate walls. The exchangers have multiple channels in series that are mounted on a frame and clamped together. The rectangular plates have an opening or port at each corner. When assembled, the plates are sealed so that the ports provide manifolds to distribute fluids through the separate flow paths. Figure 8 illustrates the flow paths.

The multiple plates, called a **plate pack**, are supported by a carrying bar and contained by pressure plates at each end. This design allows the units to be opened for maintenance or addition or removal of plate pairs. The adjoining plates are gasketed, welded, or brazed together.

Gasketed plate heat exchangers are typically limited to design pressures of 300 psig. The type of gasket material used limits the operating temperature. Brazed plate units are designed for pressures up to 450 psig and temperatures up to 500°F.

Gasketed. The most common plate heat exchanger is the gasketed plate unit. Typically, nitrile butyl rubber (NBR) gaskets are used in applications up to 230°F. Ethylene-propylene terpolymer (EPDM) gaskets are available for temperatures up to 320°F. The gaskets are glued or clipped onto the plates. The gasket pattern on each plate creates the counterflow paths illustrated in Figure 8.

Welded. Two plates can be welded together at the edges into an assembly called a **cassette**. This flow channel contains fluids when appropriate gasket material is not available such as for handling corrosive fluids. The channels containing the non-aggressive fluids are sealed with standard gaskets. Welded units can also be used for refrigeration applications. Figure 9 shows the flow path of a welded plate heat exchanger.

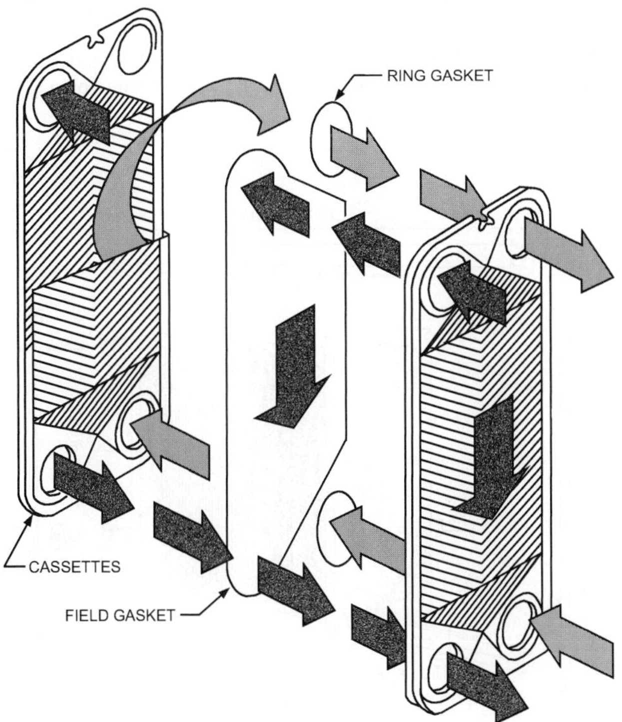

Fig. 8 Flow Path of Gasketed Plate Heat Exchanger

Brazed. Brazed-plate heat exchangers have neither gaskets nor frames (Figure 10). They consist of plates brazed together with a copper or nickel flux. This design can be very cost effective in closed-system applications where lack of maintenance is not a concern.

Double-Wall Heat Exchangers

Double-wall heat exchangers have a leakage path that warns of mechanical failure before fluids can be cross contaminated. Both shell-and-tube and plate heat exchangers are available. The overall thermal performance of a double-wall unit is less than a comparable single-wall design. Double-wall units cost significantly more than single-wall units.

A double-wall U-tube unit (Figure 11) consists of a tube-in-tube design with double tubesheets. The outer tube is rolled into the inner tubesheet. The inner tube is finned or has grooves cut in it. It is rolled into the outer tubesheet to provide a vented leak path between tubesheets to provide a visible indication of a failure of either tube.

A double-wall plate heat exchanger (Figure 12) is constructed by welding two standard channel plates together at the four port openings to form a leak path between the plates should a plate fail.

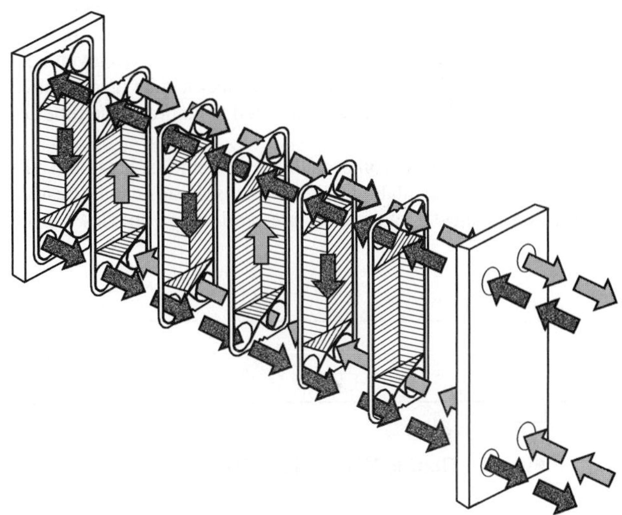

Fig. 9 Flow Path of Welded Plate Heat Exchanger

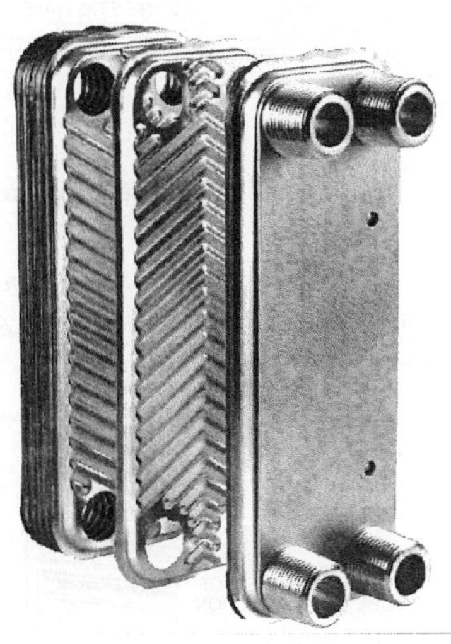

Fig. 10 Brazed-Plate Heat Exchanger

COMPONENTS

Heat exchangers for HVAC applications should be constructed and labeled according to the applicable ASME *Boiler and Pressure Vessel Code* and rated for 150 psig at 375°F. Heat exchangers operating at elevated temperatures or pressures require special construction.

Shell-and-Tube Components

Figure 13 illustrates the various components of a shell-and-tube exchanger, which include the following:

- **Shells** are usually made of steel pipe; brass and stainless steel are also used. The inlet and outlet nozzles can be made with standard flange openings in various orientations to suit piping needs. The nozzles are sized to avoid excessive fluid velocity and impingement on the tubes opposite a shell inlet connection.
- **Baffles, tube supports, tie rods, and spacers** are usually made of steel; brass and stainless steel are also available. The number and spacing of baffles controls the velocity and, therefore, a significant portion of the shell-side heat transfer coefficient and pressure drop.
- **Tubes** are usually made of copper; special grades of brass and stainless steel can be specified. The tube diameter, gage, and material affect the heat transfer coefficient and performance.
- **Tubesheets** are available in the same materials as baffles, although the materials do not have to be the same in a given heat exchanger. Tubesheets are drilled for a specific tube layout called **pitch**. The holes are sometimes serrated to improve the tube-to-tubesheet joint.
- **Heads** are usually cast iron or fabricated steel. Cast brass and cast stainless steel are available in limited sizes. Heads can be custom fabricated in most metals. The inlet and outlet nozzles can be made with standard flange openings. Figures 3, 4, and 5 illustrate three different head configurations that offer different levels of serviceability and ease of installation.

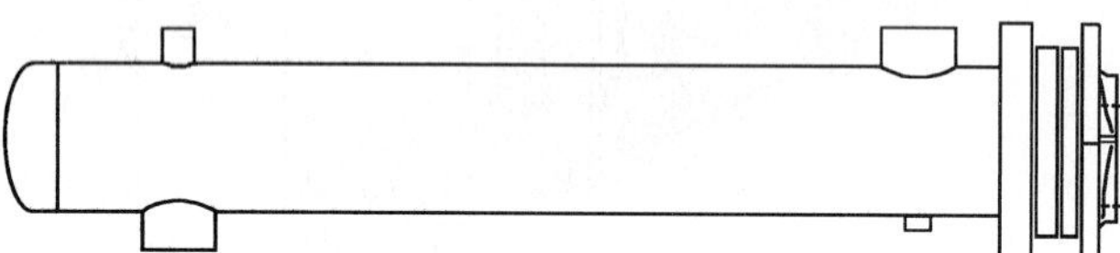

Fig. 11 Double-Wall U-Tube Heat Exchanger

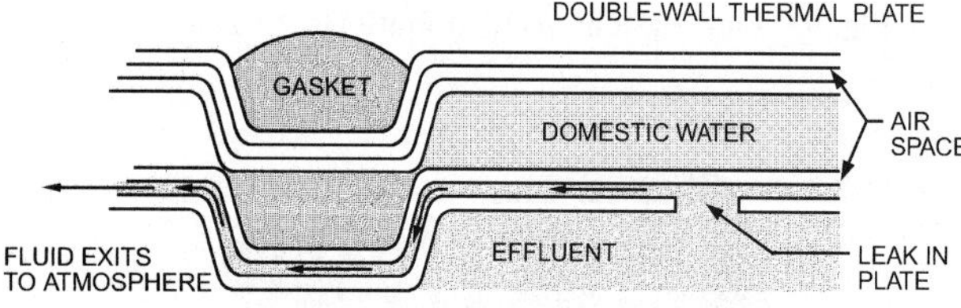

Fig. 12 Double-Wall Plate Heat Exchanger

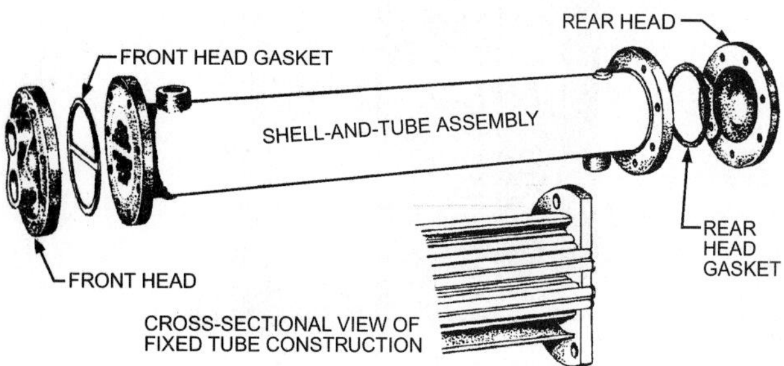

Fig. 13 Exploded View of Straight-Tube Heat Exchanger

Plate Components

Figure 14 illustrates the various components of a gasketed plate and frame heat exchanger. The materials of construction and purpose of the components are as follows:

- **Fixed frame plates** are usually made of carbon steel. Single-pass units have inlet and outlet connections for both fluids located on the fixed frame plate. Connections are usually NPT or stud port design to accommodate ANSI flanges. NPT connections are carbon steel or stainless steel. Stud port connections can be lined with metallic or rubber-type materials to protect against corrosion.
- **Movable pressure plates** can be moved along the length of the carrying bar to allow removal, replacement, or addition of plates. They are made of carbon steel. Multiple-pass units have some connections located on the movable pressure plate.
- **Plate packs** are made up of multiple heat transfer (channel) plates and gaskets. Plates are made of pressable metals, such as 316 or 304 stainless steel or titanium. They are formed with corrugations, typically in a herringbone or chevron pattern. The angle of these patterns affects the thermal performance and pressure drop of a given flow channel.
- **Compression bolts** compress the plate back between the moveable pressure and fixed frame plates. The dimension between the two is critical and is specified by the unit manufacturer for a given plate pack configuration.
- **Carrying and guide bars** support and align the channel plates. The upper bar is called a carrying bar, the lower a guide bar. They are made of stainless steel, aluminum, or carbon steel with zinc chromate finish.
- **Support columns** support the carrying and guide bars on larger plate heat exchangers.
- **Splashguards** are required in the United States by OSHA to enclose exterior channel plate and gasket surfaces. They are usually formed from aluminum.
- **Drip pans** made of stainless steel are often installed under plate heat exchangers to contain leakage on start-up or shut down, gasket failure, or condensation.

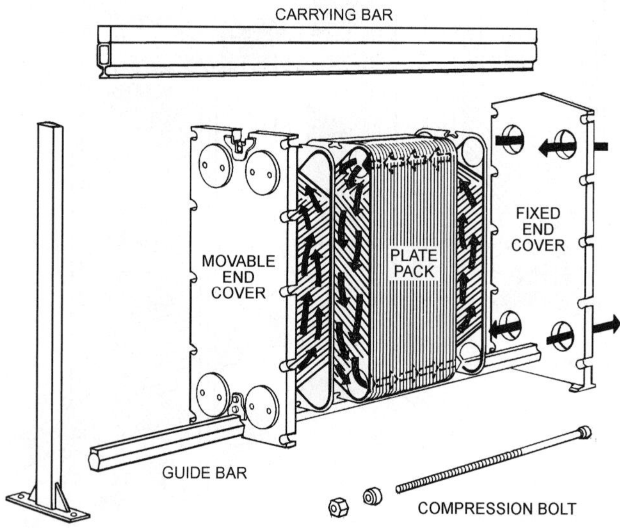

Fig. 14 Components of a Gasketed Plate Heat Exchanger

APPLICATION

Heat exchangers are used when the primary energy source is available for multiple purposes, uses a different medium, or its temperature or pressure is not in the design limits. Most of the following examples are discussed in other chapters and volumes of the *ASHRAE Handbook*. Heat exchangers are used

- To condense steam from a boiler to produce hot water for central water systems
- For service water for potable and nonpotable applications, which is often heated by a converter and hot-water or steam boilers, with or without a storage tank
- To meet special temperature requirements of parts of a system or to protect against freezing in isolated terminal units (coils) and cooling tower basins
- To isolate two systems operating at different pressures while transferring thermal energy between them
- In energy-saving applications such as condensate cooling, vent condensing, boiler blowdown, thermal storage, and chiller bypass (free cooling)
- In many refrigeration applications as evaporators, condensers, and liquid coolers

SELECTION CRITERIA

A heat exchanger is often selected by a computer program that optimizes the selection for the given design. A manufacturer should provide detailed selection guidance for both a shell-and-tube and plate exchanger for a given set of conditions.

Thermal/Mechanical Design

Shell-and-tube heat exchangers are designed first to be pressure vessels and second to transfer heat. Plate heat exchangers are designed to transfer heat efficiently within certain temperature and pressure limits.

Thermal Performance. The thermal performance of a heat exchanger is a function of the size and geometry of the heat transfer surface area. Heat transfer surface materials also affect performance; for instance, copper has a higher coefficient of heat transfer than stainless steel.

Flow rates (velocity), viscosity, and thermal conductivity of the fluids are significant factors in determining the overall heat transfer coefficient U. In addition, the fluid to be heated should be on the tube side because the overall U of a shell-and-tube unit is often reduced if the fluid to be heated is on the shell side.

Properly selected shell-and-tube heat exchangers use tube pass options and shell-side baffle spacing to maximize velocity (turbulence) without causing tube erosion. The ability to maximize velocity on each side of a heat exchanger is particularly important when the two fluids' flow rates are dissimilar. However, fluid velocity in the shell-and-tube heat exchanger is limited to avoid tube erosion. U-tube exchangers have lower tube-side velocity limits than straight-tube units due to the thinner tube wall in the U bends.

Shell-and-tube heat exchangers can be constructed for split-shell flow design (see Figure 6) to accommodate unusual conditions. These units have one shell inlet connection and two outlet connections.

Plate heat exchangers typically have U-factors three to five times higher than shell-and-tube heat exchangers. The high turbulence created by the corrugated plate design increases convection and increases the U-factor. The plate design achieves a large temperature cross at a 2°F approach because of the counterflow fluid path and high U-factor.

Thermal Stress. Heat exchangers must accommodate the thermal stresses associated with large temperature differences. U-tube units offer superior economic performance over straight-tube units with removable tube bundles under extreme conditions. Units with fixed tubesheets do not handle large temperature differences well.

Gasketed plate units have a **differential pressure/temperature limitation (DPTL)**, which is the maximum difference in operating pressure of the two fluids at a specific temperature. A unit rated for 300 psig at 260°F might have a DPTL of 220 psig at 200°F.

Pressure Drop. Fluid velocity and normal limitations on tube length tend to result in relatively low pressure drops in shell-and-tube heat exchangers. Plate units tend to have larger pressure drops unless the velocity is limited. Often a pressure drop limitation rather than a thermal performance requirement determines the surface area in a plate unit.

Fouling. Often, excess surface area is specified to allow for scale accumulation on heat transfer surfaces without a significant reduction of performance. This fouling factor or allowance is applied when sizing the unit. Fouling allowance is better specified as a percentage of excess area rather than as a resistance to heat transfer.

Shell-and-tube exchangers with properly sized tubes can handle suspended solids better than plate units with narrow flow channels. The high fluid velocity and turbulence in plate exchangers make them less susceptible to fouling.

The addition of surface area (tube length) to a shell-and-tube exchanger does not affect fluid velocity, and, therefore, has little effect on thermal performance. This characteristic makes a fouling allowance practical. This is not the case in plate units, for which the number of parallel flow channels determines velocity. This means that as plate pairs are added to meet a load (heat transfer surface area) requirement, the number of channels increases and results in decreased fluid velocity. This lower velocity reduces performance and requires additional plate pairs, which further reduces performance.

Cost

On applications with temperature crosses and close approaches, plate heat exchangers usually have the lowest initial cost. Wide temperature approaches often favor shell-and-tube units. If the application requires stainless steel, the plate unit may be more economical.

Serviceability

Shell-and-tube heat exchangers have different degrees of serviceability. The type of header used facilitates access to the inside of the tubes. The heads illustrated in Figures 3, 6, and 7 can be easily removed without special pipe arrangements. The tube bundles in all of the shell-and-tube units illustrated, except the fixed-tubesheet unit (Figure 6), can be replaced after the head is removed if they are piped with proper clearance.

The diameter and configuration of the tubes are significant in determining whether the inside of tubes of straight-tube units can be mechanically cleaned. Figure 7 shows a type of head that allows cleaning or inspection inside tubes after the channel cover is removed.

Plate heat exchangers can be serviced by sliding the movable pressure plate back along the carrying bars. Individual plates can be removed for cleaning, regasketing, or replacement. Plate pairs can be added for additional capacity. Complete replacement plate packs can be installed.

Space Requirements

Cost-effective and efficient shell-and-tube heat exchangers have small-diameter, long tubes. This configuration often challenges the designer when allocating space required for service and maintenance. For this reason, many shell-and-tube selections have large diameters and short lengths. Although this selection performs well, it often costs more than a smaller-diameter unit with equal surface area. Be careful to provide adequate maintenance clearance around heat exchangers. For shell-and-tube units, space should be left clear so the tube bundle can be removed.

Plate heat exchangers tend to provide the most compact design in terms of surface area for a given space.

Steam

Most HVAC applications using steam are designed with shell-and-tube units. Plate heat exchangers are used in specialized industrial and food processes with steam.

INSTALLATION

Control. Heat exchangers are usually controlled by a valve with a temperature sensor. The sensor is placed in the flow stream of the fluid to be heated or cooled. The valve regulates flow on the other side of the heat exchanger to achieve the sensor set-point temperature. Chapter 46 discusses control valves.

Piping. Heat exchangers should be piped such that air is easily vented. Pipes must be able to be drained and accessible for service.

Pressure Relief. Safety pressure relief valves should be installed on both sides between the heat exchanger and shutoff valves to guard against damage from thermal expansion when the unit is not in service, as well as to protect against overpressurization.

Flow Path. The intended flow path of each fluid on both sides of a heat exchanger design should be followed. Failure to connect to the correct inlet and outlet connections may reduce performance.

Condensate Removal. Heat exchangers that condense steam require special installation. Proper removal of condensate is particularly important. Inadequate drainage of condensate can result in significant loss of capacity and even in mechanical failure.

Installing a vacuum breaker aids in draining condensate, particularly when modulating steam control valves are used. Properly sized and installed steam traps are critical. Chapter 10 discusses steam traps and condensate removal.

Insulation. Heat exchangers are often insulated. Chapter 23 of the 2005 *ASHRAE Handbook—Fundamentals* has further information on insulation.

CHAPTER 48

UNITARY AIR CONDITIONERS AND HEAT PUMPS

UNITARY air conditioners are factory-made assemblies that normally include an evaporator or cooling coil and a compressor/condenser combination, and possibly provide heating as well. An **air-source unitary heat pump** normally includes an indoor conditioning coil, compressor(s), and an outdoor coil. It must provide heating and possibly cooling as well. A **water-source heat pump** rejects or extracts heat to and from a water loop instead of from ambient air. A unitary air conditioner or heat pump with more than one factory-made assembly (e.g., indoor and outdoor units) is commonly called a **split system**.

Unitary equipment is divided into three general categories: residential, light commercial, and commercial. Residential equipment is single-phase unitary equipment with a cooling capacity of 65,000 Btu/h or less and is designed specifically for residential application. Light commercial equipment is generally three-phase, with cooling capacity up to 135,000 Btu/h, and is designed for small businesses and commercial properties. Commercial unitary equipment has cooling capacity higher than 135,000 Btu/h and is designed for large commercial buildings.

GENERAL DESIGN CONSIDERATIONS

User Requirements

The user primarily needs either space conditioning for occupant comfort or a controlled environment for products or manufacturing processes. Cooling, dehumidification, filtration, and air circulation often meet those needs, although heating, humidification, and ventilation are also required in many applications.

Application Requirements

Unitary equipment is available in many configurations, such as

- **Single-zone, constant-volume**, which consists of one controlled space with one thermostat that controls to maintain a set point.
- **Multizone, constant-volume**, which has several controlled spaces served by one unit that supplies air of different temperatures to different zones as demanded (Figure 1).
- **Single-zone, variable-volume**, which consists of several controlled spaces served by one unit. Supply air from the unit is at a constant temperature, with air volume to each space varied to satisfy space demands (Figure 2).
- **Multisplit,** which consists of several controlled spaces, each served by a separate indoor unit. All indoor units are connected to an outdoor condensing unit. When each indoor unit varies its refrigerant flow in response to heating or cooling load demand, the system is called **variable refrigerant flow (VRF)**.

The preparation of this chapter is assigned to TC 8.11, Unitary and Room Air Conditioners and Heat Pumps.

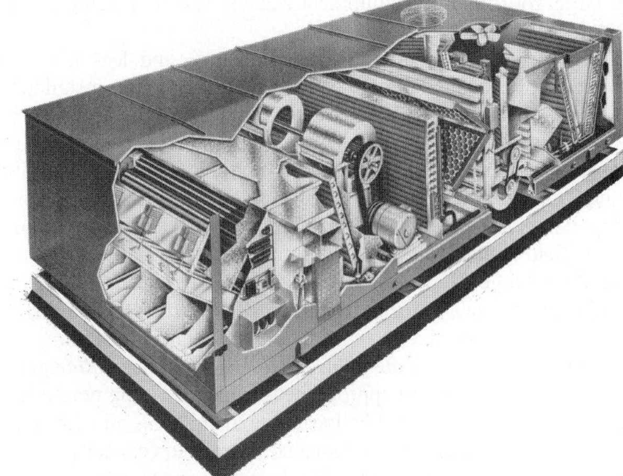

Fig. 1 Typical Rooftop Air-Cooled Single-Package Air Conditioner (Multizone)

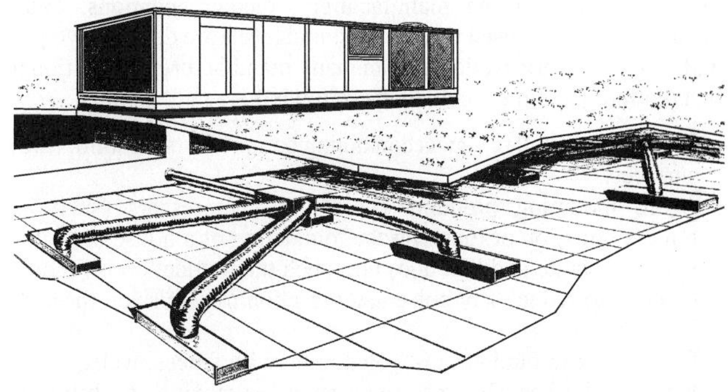

Fig. 2 Single-Package Air Equipment with Variable Air Volume

Factors such as size, shape, and use of the building; availability and cost of energy; building aesthetics (equipment located outdoors); and space available for equipment are considered to determine the type of unitary equipment best suited to a given application. In general, roof-mounted single-package unitary equipment is limited to five or six stories because duct space and available blower power become excessive in taller buildings. Split units are limited by the maximum distance allowed between the indoor and outdoor sections. Indoor, single-zone equipment is generally less expensive to maintain and service than multizone units located outdoors.

The building load and airflow requirements determine equipment capacity, whereas the availability and cost of fuels determine the energy source. Control system requirements must be established, and any unusual operating conditions must be considered early in the planning stage. In some cases, custom-designed equipment may be necessary.

Manufacturers' literature has detailed information about geometry, performance, electrical characteristics, application, and operating limits. The system designer selects suitable equipment with the capacity for the application.

Installation

Unitary equipment is designed to keep installation costs low. Equipment must be installed properly so that it functions in accordance with the manufacturer's specifications. Interconnecting diagrams for the low-voltage control system should be documented for proper future servicing. Adequate planning is important for installing large, roof-mounted equipment because special rigging equipment is frequently required.

The refrigerant circuit must be clean, dry, and leak-free. An advantage of packaged unitary equipment is that proper installation minimizes the risk of field contamination of the circuit. Care must be taken to properly install split-system interconnecting tubing (e.g., proper cleanliness, brazing, and evacuation to remove moisture and other noncondensables). Some residential split systems have precharged line sets and quick-connection couplings, which reduce the risk of field contamination of the refrigerant circuit. Split systems should be charged according to the manufacturer's instructions.

When installing split, multisplit, and VRF systems, lines must be properly routed and sized to ensure good oil return to the compressor. Chapters 2 and 3 of the 2006 *ASHRAE Handbook—Refrigeration* have more details on appropriate refrigerant piping practices.

Unitary equipment must be located to avoid noise and vibration problems. Single-package equipment of over 20 ton capacity should be mounted on concrete pads if vibration control is a concern. Large-capacity equipment should be roof-mounted only after the roof's structural adequacy has been evaluated. If they are located over occupied space, roof-mounted units with return fans that use ceiling space for the return plenum should have a lined return plenum according to the manufacturer's recommendations. Duct silencers should be used where low sound levels are desired. Weight and sound data are available from many manufacturers. Additional installation guidelines include the following:

- In general, install products containing compressors on solid, level surfaces.
- Avoid mounting products containing compressors (such as remote units) on or touching the foundation of a house or building. A separate pad that does not touch the foundation is recommended to reduce any noise and vibration transmission through the slab.
- Do not box in outdoor air-cooled units with fences, walls, overhangs, or bushes. Doing so reduces the air-moving capability of the unit, reducing efficiency.
- For a split-system remote unit, choose an installation site that is close to the indoor part of the system to minimize pressure drop in the connecting refrigerant tubing.
- For VRF units, locate the refrigerant pipes' headers so that the length of refrigerant pipes is minimized.
- Contact the unitary equipment manufacturer or consult installation instructions for further information on installation procedures.

Unitary equipment should be listed or certified by nationally recognized testing laboratories to ensure safe operation and compliance with government and utility regulations. Equipment should also be installed to comply with agency standards' rating and application requirements to ensure that it performs according to industry criteria. Larger and more specialized equipment often does not carry agency labeling. However, power and control wiring practices should comply with the *National Electrical Code*® (NFPA *Standard* 70). Local codes should be consulted before the installation is designed; local inspectors should be consulted before installation.

Service

A clear and accurate wiring diagram and well-written service manual are essential to the installer and service personnel. Easy and safe service access must be provided in the equipment for cleaning, lubrication, and periodic maintenance of filters and belts. In addition, access for replacement of major components must be provided and preserved.

Availability of replacement parts aids proper service. Most manufacturers offer warranties covering 1 year of operation after installation. Extended compressor warranties may be standard or optional.

Service personnel must be qualified to repair or replace mechanical and electrical components and to recover and properly recycle or dispose of any refrigerant removed from a system. They must also understand the importance of controlling moisture and other contaminants in the refrigerant circuit; they should know how to clean a hermetic system if it has been opened for service (see Chapter 6 of the 2006 *ASHRAE Handbook—Refrigeration*). Proper service procedures help ensure that the equipment will continue operating efficiently for its expected life.

Sustainability

Unitary equipment should be properly sized; oversizing should be avoided. Only environmentally friendly refrigerants should be used. Equipment performance should be carefully evaluated at all expected load conditions, and equipment should be selected to achieve the most efficient operation at all expected occupancy conditions.

TYPES OF UNITARY EQUIPMENT

Table 1 shows the types of unitary air conditioners available, and Table 2 shows the types of unitary heat pumps available. The following variations apply to some types and sizes of unitary equipment.

Arrangement. Major unit components for various unitary air conditioners are arranged as shown in Table 1 and for unitary heat pumps as shown in Table 2.

Heat Rejection. Unitary air conditioner condensers may be air-cooled, evaporatively cooled, or water-cooled; the letters A, E, or W follow the Air-Conditioning and Refrigeration Institute (ARI) designation.

Heat Source/Sink. Unitary heat pump outdoor coils are designated as air-source or water-source by an A or W, following ARI practice. The same coils that act as a heat sink in the cooling mode act as the heat source in the heating mode.

Unit Exterior. The unit exterior should be decorative for in-space application, functional for equipment room and ducts, and weatherproofed for outdoors.

Placement. Unitary equipment can be mounted on floors, walls, ceilings, roofs, or a pad on the ground.

Inside Air. Equipment with fans may have airflow arranged for vertical upflow or downflow, horizontal flow, 90° or 180° turns, or multizone. Indoor coils without fans are intended for forced-air furnaces or blower packages. Variable-volume blowers may be incorporated with some systems.

Location. Unitary equipment intended for indoor use may be placed in the conditioned space with plenums or furred-in ducts or concealed in closets, attics, crawlspaces, basements, garages, utility rooms, or equipment rooms. Wall-mounted equipment may be attached to or built into a wall or transom. Outdoor equipment may be mounted on roofs or concrete pads on the ground. Installations must conform with local codes.

Table 1 ARI *Standard* 210/240 Classification of Unitary Air Conditioners

System Designation	ARI Type*	Heat Rejection	Arrangement	
Single package	SP-A	Air	Fan	Comp
	SP-E	Evap Cond	Evap	Cond
	SP-W	Water		
Refrigeration chassis	RCH-A	Air		Comp
	RCH-E	Evap Cond	Evap	Cond
	RCH-W	Water		
Year-round single package	SPY-A	Air	Fan	
	SPY-E	Evap Cond	Heat	Comp
	SPY-W	Water	Evap	Cond
Remote condenser	RC-A	Air	Fan	
	RC-E	Evap Cond	Evap	Cond
	RC-W	Water	Comp	
Year-round remote condenser	RCY-A	Air	Fan	
	RCY-E	Evap Cond	Evap	Cond
	RCY-W	Water	Heat	
			Comp	
Condensing unit, coil alone	RCU-A-C	Air	Evap	Cond
	RCU-E-C	Evap Cond		Comp
	RCU-W-C	Water		
Condensing unit, coil and blower	RCU-A-CB	Air	Fan	Cond
	RCU-E-CB	Evap Cond	Evap	Comp
	RCU-W-CB	Air		
Year-round condensing unit, coil and blower	RCUY-A-CB	Air	Fan	
	RCUY-E-CB	Evap Cond	Evap	Cond
	RCUY-W-CB	Water	Heat	Comp

*Adding a suffix of "-O" to any of these classifications indicates equipment not intended for use with field-installed duct systems.

Table 2 ARI *Standard* 210/240 Classification of Air-Source Unitary Heat Pumps

	ARI Type*						
Designation	**Heating and Cooling**	**Heating Only**	**Arrangement**				
Single package	HSP-A	HOSP-A	Fan	Comp			
			Indoor Coil	Outdoor Coil			
Remote outdoor coil	HRC-A-CB	HORC-A-CB	Fan				
			Indoor Coil	Outdoor Coil			
			Comp				
Remote outdoor coil with no indoor fan	HRC-A-C	HORC-A-C	Indoor Coil	Outdoor Coil			
			Comp				
Split system	HRCU-A-CB	HORCU-A-CB	Fan	Comp			
			Indoor Coil	Outdoor Coil			
Split system, no indoor fan	HRCU-A-C	HORCU-A-C		Comp			
			Indoor Coil	Outdoor Coil			
Through-the-wall heat pump	TTW-HSP-A	TTW-HOSP-A	Fan	Comp	or	Fan	Comp
	TTW-HRCU-A-C	TTW-HORCU-A-C	Indoor Coil	Outdoor Coil		Indoor Coil	Outdoor Coil
	TTW-HRCU-A-CB	TTW-HORCU-A-CB					
Space-constrained products	SCP-HSP-A	SCP-HOSP-A	Fan	Comp	or	Fan	Comp
	SCP-HRCU-A-C	SCP-HORCU-A-C	Indoor Coil	Outdoor Coil		Indoor Coil	Outdoor Coil
	SCP-HRCU-A-CB	SCP-HORCU-A-CB					
Small-duct, high-velocity system	SDHV-HSP-A	SDHV-HOSP-A	Fan	Comp	or	Fan	Comp
	SDHV-HRCU-A-C	SDHV-HORCU-A-C	Indoor Coil	Outdoor Coil		Indoor Coil	Outdoor Coil
	SDHV-HRCU-A-CB	SDHV-HORCU-A-CB					

*A suffix of "-O" following any of these classifications indicates equipment not intended for use with field-installed duct systems.

Heat. Unitary systems may incorporate gas-fired, oil-fired, electric, hot-water coil, or steam coil heating sections. In unitary heat pumps, these heating sections supplement the heating capability.

Ventilation Air. Outside air dampers may be built into equipment to provide outside air for cooling or ventilation.

Refrigerant Flow. Unitary equipment can use either constant or variable refrigerant flow.

Desuperheaters. Desuperheaters may be applied to unitary air conditioners and heat pumps. These devices recover heat from the compressor discharge gas and use it to heat domestic hot water. The desuperheater usually consists of a pump, heat exchanger, and controls, and can produce about 5 to 6 gph of heated water per ton of air conditioning (heating water from 60 to 130°F). Because desuperheaters improve cooling performance and reduce the degrading effect of cycling during heating, they are best applied where cooling requirements are high and where a significant number of heating hours occur above the building's balance point (Counts 1985). Although properly applied desuperheaters can improve cooling efficiency, they can also reduce space-heating capacity. This causes the unit to run longer, which reduces system cycling above the balance point.

Ductwork. Unitary equipment is usually designed with fan capability for ductwork, although some units may be designed to discharge directly into the conditioned space.

Accessories. Consult the manufacturer of any unitary equipment before installing any accessories or equipment not specifically approved by the manufacturer. These installations may not only void the warranty, but also could cause the unitary equipment to function improperly or create fire or explosion hazards.

Combined Space-Conditioning/Water-Heating Systems

Unitary systems are available that provide both space conditioning and potable-water heating. These systems are typically heat pumps, but some are available for cooling only. One type of combined system includes a full-condensing water-heating heat exchanger integrated into the refrigerant circuit of the space-conditioning system. Full-condensing system heat exchangers are larger than desuperheaters; they are generally sized to take the full condensing output of the compressor. Thus, they have much greater water-heating capacity. They also have controls that allow them to heat water year-round, either independently or coincidentally with space heating or space cooling. In spring and fall, the system is typically operated only to heat water.

Another type of combined system incorporates a separate, ancillary **heat pump water heater (HPWH)**. The evaporator of this heater typically uses the return air (or liquid) stream of the space-conditioning system as a heat source. The HPWH thus cools the return stream during both space heating and cooling. In spring and fall, the space-conditioning blower (or pump) operates when water heating is needed.

As with desuperheaters, simultaneous space and water heating reduces the output for space heating. This lower output is partially compensated for by reduced cycling of the space-heating system above the balance point.

Combined systems can provide end users with significant energy savings, and electric utilities with a significant reduction in demand. The overall performance of these systems is affected by the refrigerant charge and piping, water piping, and control logic and wiring. It is important, therefore, that the manufacturer's recommendations be closely followed. One special requirement is to locate water-containing section(s) in areas not normally subjected to freezing temperatures.

Typical Unitary Equipment

Figures 1 to 6 show various types and installations of single-package equipment. Figure 7 shows a typical installation of a split-system, air-cooled condensing unit with indoor coil (the most widely used unitary cooling system). Figures 8 and 9 show split-system condensing units with coils and blower-coil units. Figure 8 also shows ductless blower-coil units, discharging air directly into the space. Ductless indoor units can be wall mounted, under ceiling mounted, or ceiling cassette. See Chapter 7 for information on engine-driven heat pumps and air conditioners.

Many special light commercial and commercial unitary installations include a single-package air conditioner for use with variable air volume systems, as shown in Figure 2. These units are often equipped with a factory-installed system for controlling air volume in response to supply duct pressure (such as dampers or variable-speed drives).

Another example of a specialized unit is the **multizone unit** shown in Figure 1. The manufacturer usually provides all controls, including zone dampers. The air path in these units is designed so that supply air may flow through a hot deck containing a means of heating or through a cold deck, which usually contains a direct-expansion evaporator coil.

To make multizone units more efficient, a control is commonly provided that locks out cooling by refrigeration when the heating unit is in operation and vice versa. Another variation to improve efficiency is the three-deck multizone. This unit has a hot deck, a

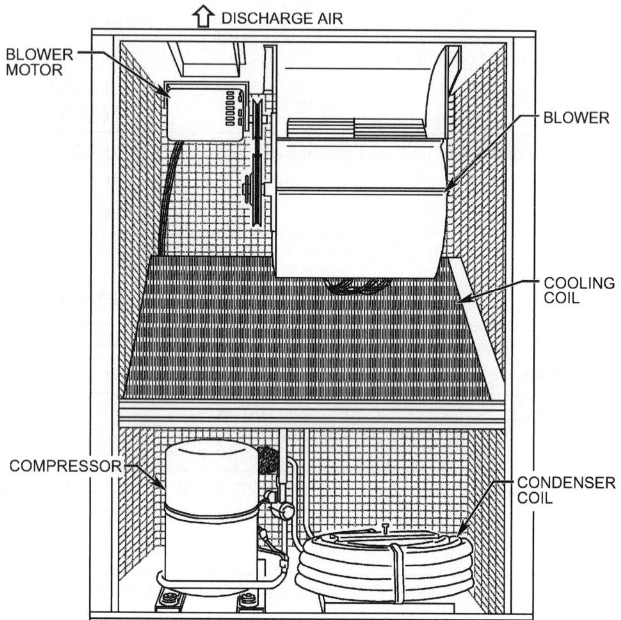

Fig. 3 Water-Cooled Single-Package Air Conditioner

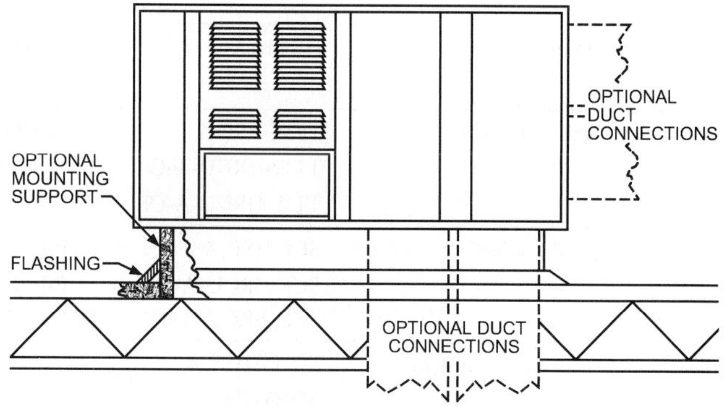

Fig. 4 Rooftop Installation of Air-Cooled Single-Package Unit

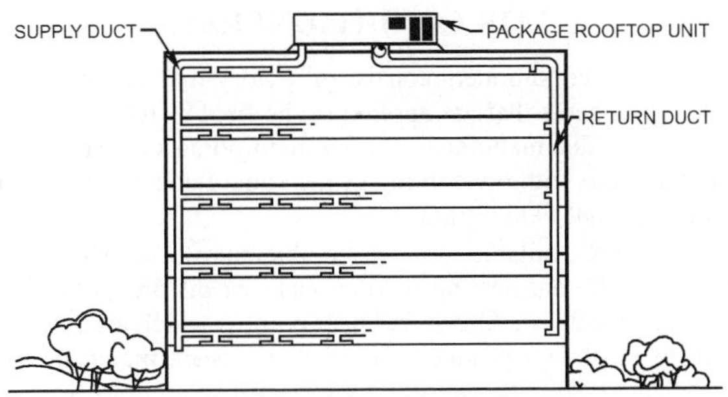

Fig. 5 Multistory Rooftop Installation of Single-Package Unit

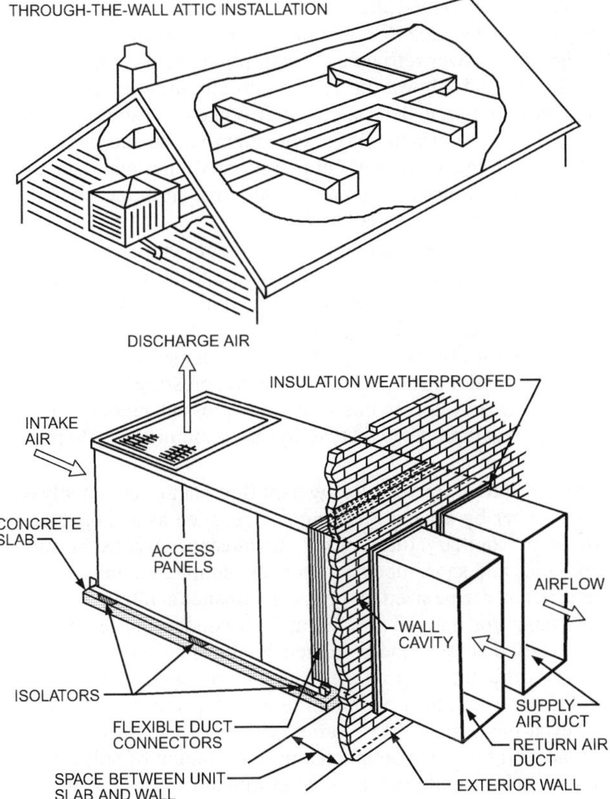

Fig. 6 Through-the-Wall Installation of Air-Cooled Single-Package Unit

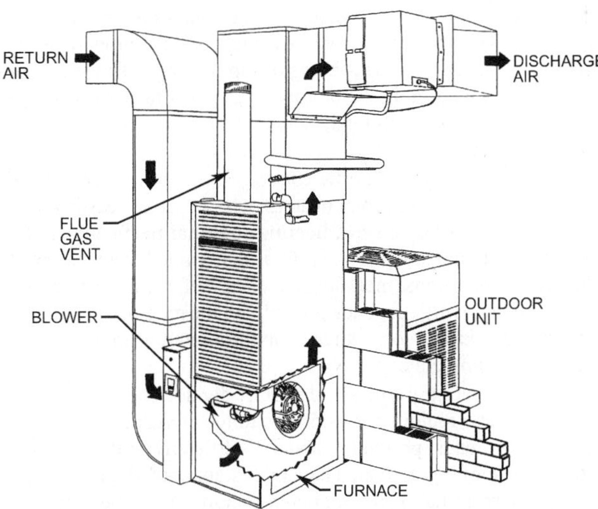

Fig. 7 Residential Installation of Split-System Air-Cooled Condensing Unit with Coil and Upflow Furnace

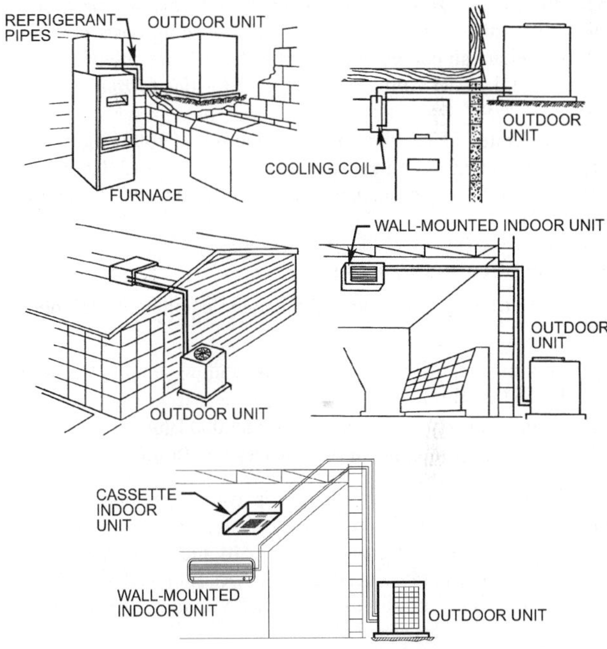

Fig. 8 Outdoor Installations of Split-System Air-Cooled Condensing Units with Coil and Upflow Furnace or with Indoor Blower-Coils

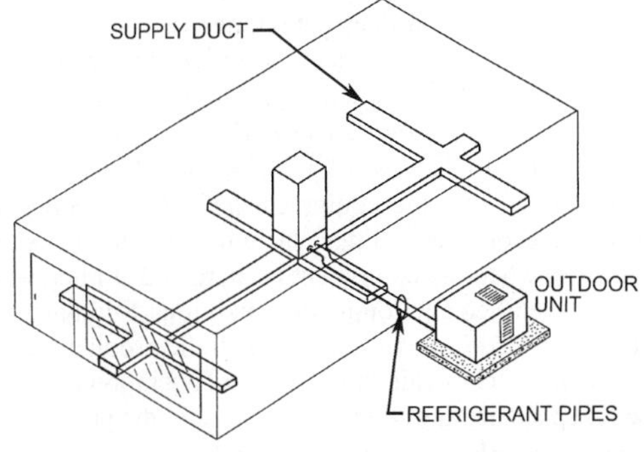

Fig. 9 Outdoor Installation of Split-System Air-Cooled Condensing Unit with Indoor Coil and Downflow Furnace

cold deck, and a neutral deck carrying return air. Hot and/or cold deck air mixes only with air in the neutral deck.

EQUIPMENT AND SYSTEM STANDARDS

Energy Conservation and Efficiency

In the United States, the Energy Policy and Conservation Act (Public Law 95-163) requires the Federal Trade Commission (FTC) to prescribe an energy label for many major appliances, including unitary air conditioners and heat pumps. The National Appliance Energy Conservation Act (NAECA) (Public Law 100-12) provides minimum efficiency standards for major appliances, including unitary air conditioners and heat pumps.

The U.S. Department of Energy (DOE) testing and rating procedure is documented in Appendix M to Subpart 430 of Section 10 of the *Code of Federal Regulations*, Uniform Test Method for Measuring the Energy Consumption of Central Air Conditioners. This testing procedure provides a seasonal measure of operating efficiency for residential unitary equipment. The **seasonal energy efficiency ratio (SEER)** is the ratio of total seasonal cooling output measured in Btu to total seasonal watt-hours of input energy. This efficiency value is developed in the laboratory by conducting tests at various indoor and outdoor conditions, including a measure of performance under cyclic operation.

Seasonal heating mode efficiencies of heat pumps are similarly expressed as the ratio of total heating output to total seasonal input energy. This is expressed as a **heating seasonal performance factor (HSPF)**. In the laboratory, HSPF is determined from test results at different conditions, including a measure of cyclic performance. The calculated HSPF depends not only on the measured equipment performance, but also on climatic conditions and heating load relative to the equipment capacity.

For HSPF rating purposes, the DOE divided the United States into six climatic regions and defined a range of maximum and minimum design loads, producing about 30 different HSPF ratings for a given piece of equipment. The DOE established Region 4 (moderate northern climate) and minimum design load as the typical climatic region and building design load to be used for comparative certified performance ratings.

SEER, HSPF, and operating costs vary appreciably with equipment design and size and from manufacturer to manufacturer. SEER and HSPF values, size ranges, and unit operating costs for DOE-covered unitary air conditioners certified by ARI are published semi-annually in the ARI *Unitary Directory of Certified Product Performance*.

In the United States, the Energy Policy Act of 1992 requires unitary equipment with cooling capacities from 65,000 to 240,000 Btu/h to meet the minimum efficiency levels prescribed by ASHRAE *Standard* 90.1.

ARI Certification Programs

ARI certification programs for unitary equipment include the following:

- Unitary air conditioners (water-, air-, and evaporatively cooled) under 65,000 Btu/h
- Air-source unitary heat pumps under 65,000 Btu/h
- Unitary large equipment at or above 65,000 Btu/h

Information on specific programs is available at http://www.ari.org/Content/CertificationPrograms_19.aspx. Smaller equipment is rated according to ARI *Standard* 210/240 and ARI *Standard* 37. Large equipment is rated according to ARI *Standard* 340/360 and 365. Sound ratings are discussed in ARI *Standard* 270, and sound application in ARI *Standard* 275.

Safety Standards and Installation Codes

Approval agencies list unitary air conditioners complying with a standard such as CSA *Standard* C22.2 No. 236/Underwriters Laboratories (UL) *Standard* 1995. Other UL standards may also apply. An evaluation of the product determines that its design complies with the construction requirements specified in the standard and that the equipment can be installed in accordance with the applicable requirements of the *National Electrical Code*®; ASHRAE *Standard* 15; NFPA *Standard* 90A; and NFPA *Standard* 90B.

Tests confirm that the equipment and all components operate within their recognized ratings, including electrical, temperature, and pressure, when the equipment is energized at rated voltage and operated at specified environmental conditions. Stipulated abnormal conditions are also imposed under which the product must perform in a safe manner. The evaluation covers all operational features (such as electric space heating) that may be used in the product.

Products complying with the applicable requirements may bear the agency listing mark. An approval agency program includes auditing continued production at the manufacturer's factory.

AIR CONDITIONERS

Unitary air conditioners consist of factory-matched refrigerant circuit components that are applied in the field to fulfill the user's requirements. The manufacturer often incorporates a heating function compatible with the cooling system and a control system that requires minimal field wiring.

Products are available to meet the objectives of nearly any system. Many different heating sections (gas- or oil-fired, electric, or condenser reheat), air filters, and heat pumps, which are a specialized form of unitary product, are available. Such matched equipment, selected with compatible accessory items, requires little field design or field installation work.

Refrigerant Circuit Design

Chapters 22, 37, and 38 describe coil, compressor, and condenser designs. Chapters 2, 5, 6, and 7 of the 2006 *ASHRAE Handbook—Refrigeration* cover refrigerant circuit piping selection, chemistry, cleanliness, and lubrication. Proper coil circuiting is essential for adequate oil return to the compressor. Crankcase heaters are usually incorporated to prevent refrigerant migration to the compressor crankcase during shutdown. Oil pressure switches and pumpdown or pumpout controls are used when additional ensurance of reliability is economical and/or required.

Safety Controls. High-pressure and high-temperature limiting devices, internal pressure bypasses, current-limiting devices, and devices that limit compressor torque prevent excessive mechanical and electrical stresses. Low-pressure or temperature cutout controllers may be used to protect against loss of charge, coil freeze-up, or loss of evaporator airflow. Because suction pressures drop momentarily during start-up, circuits with a low-pressure cutout controller may require a time-delay relay to bypass it momentarily to prevent nuisance tripping.

Flow Control Devices. Refrigerant flow is most commonly controlled either by a fixed metering device, such as a short-tube restrictor or capillary tube, or by thermostatic expansion valves. Capillaries and short-tube restrictors are simple, reliable, and economical, and can be sized for peak performance at rating conditions. The evaporator may be overfed at high condensing temperatures and underfed at low condensing temperatures because of changing pressure differential across the fixed metering device. Under these conditions, a less-than-optimum cooling capacity usually results. However, the degree of loss varies with condenser design, system volume, and total refrigerant charge. The amount of unit charge is critical, and a capillary-controlled evaporator must be matched to the specific condensing unit.

Properly sized thermostatic expansion valves provide constant superheat and good control over a range of operating conditions. Superheat is adjusted to ensure that only superheated gas returns to the compressor, usually with 7 to 14°F superheat at the compressor inlet at normal rating conditions. This superheat setting may be higher at a lower outdoor ambient temperature (cooling tower water temperature for water-cooled products) or indoor wet-bulb temperature. Compressor loading can be limited with vapor-charged thermostatic expansion valves. Low-discharge-pressure (low ambient) operation decreases pressure drop across valves and capillaries so that full flow is not maintained. Decreased capacity, low coil temperatures, and freeze-up can result unless low ambient condensing-pressure control is provided.

Properly designed unitary equipment allows only a minimum amount of liquid refrigerant to return through the suction line to the compressor during non-steady-state operation. Normally, the heat absorbed in the evaporator vaporizes all the refrigerant and adds a few degrees of superheat. However, any conditions that increase refrigerant flow beyond the heat transfer capabilities of the evaporator can cause liquid carryover into the compressor return line. This increase may be caused by a poorly positioned thermal element of an expansion valve or by an increase in the condensing pressure of a capillary system, which may be caused by fouled condenser surfaces, excessive refrigerant charge, reduced flow of condenser air or water, or the higher temperature of the condenser cooling medium. Heat transfer at the evaporator may be reduced by dirty surfaces, low-temperature air entering the evaporator, or reduced airflow caused by a blockage in the air system.

Piping. Transient flow conditions are a special concern. During *off* periods, refrigerant migrates and condenses in the coldest part of the system. In an air conditioner in a cooled space, this area is typically the evaporator. When the compressor starts, the liquid tends to return to the compressor in slugs. The severity of **slugging** is affected by temperature differences, *off* time, component positions, and traps formed in suction lines. Various methods such as suction-line accumulators, specially designed compressors, the refrigerant pumpdown cycle, nonbleed port thermostatic expansion valves, liquid-line solenoid valves, or limited refrigerant charge are used to avoid equipment problems associated with excessive liquid return. Chapter 2 of the 2006 *ASHRAE Handbook—Refrigeration* has further information on refrigerant piping.

Strainers and filter-driers minimize the risk of foreign material restricting capillary tubes and expansion valves (e.g., small quantities of solder, flux, and varnish). Overheated and oxidized oil may dissolve in warm refrigerant and deposit at lower temperatures in capillary tubes, expansion valves, and evaporators. Filter-driers are highly desirable, particularly for split-system units, to remove any moisture introduced during installation or servicing. Moisture contamination can cause oil breakdown, motor insulation failure, and freezing or other restrictions at the expansion device.

Capacity Control. Buildings with high internal heat loads require cooling even at low outdoor temperatures. The capacity of air-cooled condensers can be controlled by changing airflow or flooding tubes with refrigerant. Airflow can be changed by using dampers, adjusting fan speed, or stopping some of the fan motors in a multifan system.

In cool weather, air conditioners operate for short periods only. If the weather is also damp, high humidity levels with wide variances may occur. Properly designed capacity-controlled units operate for longer periods, which may improve humidity control and comfort. In any case, cooling equipment should not be oversized.

Units with two or more separate refrigerant circuits allow independent operation of the individual systems, which reduces capacity while better matching changing load conditions. Larger, single-compressor systems may offer capacity reduction through cylinder-unloading compressors, variable-speed or multispeed compressors, multiple compressors, or hot-gas bypass controls. At full-load operation, efficiency is unimpaired. However, reduced-capacity operation may increase or decrease system efficiency, depending on the capacity-reduction method used.

Variable-speed, multispeed, and staged multiple compressors can improve efficiency at part-load operation. Cylinder unloading can increase or decrease system performance, depending on the particular method used. Multispeed and variable-speed compressors, multiple compressors, and cylinder unloading generally produce higher comfort levels through lower cycling and better matching of capacity to load. Hot-gas bypass does not reduce capacity efficiently, although it generally provides a wider range of capacity reduction. Units with capacity-reduction compressors usually have capacity-controlled evaporators; otherwise evaporator coil temperatures may be too high to provide dehumidification. Capacity-controlled evaporators are usually split, with at least one of the expansion valves controlled by a solenoid valve. Evaporator capacity is reduced by closing the solenoid valve. Compressor capacity-reduction controls or hot-gas bypass system then provides maximum dehumidification, while the evaporator coil temperature is maintained above freezing to avoid coil frosting. Chapter 2 of the 2006 *ASHRAE Handbook—Refrigeration* has details on hot-gas bypass.

Air-Handling Systems

High airflow, low-static-pressure performance, simplicity, economics, and compact arrangement are characteristics that make propeller fans particularly suitable for nonducted air-cooled condensers. Small-diameter fans are direct-driven by four-, six-, or eight-pole motors. Low starting-torque requirements allow use of single-phase shaded pole and permanent split-capacitor (PSC) fan motors and simplify speed control for low-outdoor-temperature operation. Many larger units use multiple fans and three-phase motors. Larger-diameter fans are belt-driven at a lower rpm to maintain low tip speeds and quiet operation.

Centrifugal blowers meet the higher static-pressure requirements of ducted air-cooled condensers, forced-air furnaces, and evaporators. Indoor airflow must be adjusted to suit duct systems and plenums while providing the required airflow to the coil. Some small blowers are direct-driven with multispeed motors. In ductless indoor units, such as wall-mounted, under-ceiling-mounted, and ceiling cassette units, aluminum or fire-retardant plastic centrifugal fans are used. Large blowers are always belt-driven and may have variable-pitch motor pulleys or a variable-frequency drive (VFD) for airflow adjustment. Vibration isolation reduces the amount of noise transmitted by bearings, motors, and blowers into cabinets. (See Chapter 20 for details of fan design and Chapters 18 and 19 for information on air distribution systems.)

Disposable fiberglass filters are popular because they are available in standard sizes at low cost. Cleanable filters can offer economic advantages, especially when cabinet dimensions are not compatible with common sizes. Charcoal filters can help remove volatile organic compounds (VOCs) or microparticles. Electronic or other high-efficiency air cleaners are used when a high degree of cleaning is desired. Larger equipment frequently is provided with automatic roll filters or high-efficiency bag filters. (See Chapter 28 for additional details about filters.)

Many units have a way to introduce outside air for economizer cooling and/ or ventilation; rooftop units are particularly adaptable for receiving outside air. Air-to-air heat exchangers can be used to reduce energy losses from ventilation. Ductless units are provided with either a factory-mounted or a field-installed outside air kit. Some units have automatically controlled dampers to allow cooling by outside air, which increases system efficiency.

Electrical Design

Electrical controls for unitary equipment are selected and tested to perform their individual and interrelated functions properly and safely over the entire range of operating conditions. Internal line-break thermal protectors provide overcurrent protection for most single-phase motors, smaller three-phase motors, and hermetic compressor motors. These rapidly responding temperature sensors, embedded in motor windings, can provide precise locked rotor and running overload protection.

Branch-circuit, short-circuit, and ground-fault protection is commonly provided by fused disconnect switches. Time-delay fuses allow selection of fuse ratings closer to running currents and thus provide backup motor overload protection, as well as short-circuit and ground-fault protection. Circuit breakers may be used in lieu of fuses where allowed by codes.

Some larger compressor motors have dual windings and contactors for step starting. A brief delay when energizing contactors reduces the magnitude of inrush current.

Using 24 V (NEC Class 2) control circuitry is common for room thermostats and interconnecting wiring between split systems. It offers advantages in temperature control, safety, and ease of installation. Electronic, communicating microprocessor thermostats and control systems are common.

Motor-speed controls are used to vary evaporator airflow of direct-drive fans, air-cooled condenser airflow for low-outdoor-temperature operation, and compressor speed to match load demand. Multitap motors and autotransformers provide one or more speed steps. Solid-state speed control circuits provide a continuously variable speed range. However, motor bearings, windings, overload protection, and motor suspension must be suitable for operation over the full speed range.

In addition to speed control, solid-state circuits can provide reliable temperature control, motor protection, and expansion valve refrigerant control. Complete temperature-control systems are frequently included with the unit. Features such as automatic night setback, economizer control sequence, and zone demand control of multizone equipment contribute to improved comfort and energy savings. Chapter 46 of the 2007 *ASHRAE Handbook—HVAC Applications* has additional information on control systems.

Mechanical Design

Cabinet height is important for rooftop and ceiling-suspended units. Size limitations of truck bodies, freight cars, doorways, elevators, and various rigging practices must be considered in large-unit design. In addition, structural strength of both the unit and the crate must be adequate for handling, warehouse stacking, shipping, and rigging.

Additionally, (1) cabinet insulation must prevent excessive sweating in high-humidity ambient conditions, (2) insulated surfaces exposed to moving air should withstand air erosion, (3) air leakage around panels and at cabinet joints should be minimized, (4) cabinet and coils should be provided with a corrosion-resistant coating in highly corrosive environments, and (5) cabinet insulation must be adequate to reduce energy transfer losses from the circulating airstream.

Also, cooling-coil air velocities must be low enough to ensure that condensate is not blown off the coil. The drain pan must be sized to contain the condensate, must be protected from high-velocity air, and should be properly sloped to drain condensate to minimize standing water. Service access must be provided for installation and repair. Versatility of application, such as multiple fan discharge directions and the ability to install piping from either side of the unit, is another consideration. Weatherproofing requires careful attention and testing.

Accessories

Using standard cataloged accessories, the designer can often incorporate unitary products in special applications. Typical examples (see Figures 4, 6, 7, and 8) are plenum coil housings, return air filter grilles, and diffuser-return grilles for single-outlet units. Air duct kits offered for rooftop units (see Figure 4) allow concentric or side-by-side ducting, as well as horizontal or vertical connections. Mounting curbs are available to facilitate unit support and roof flashing. Accessories for ductless split units include wireless or wired thermostats, motorized air sweep flow louvers that automatically change airflow directions, and wind baffle kits for condensing-unit operation during high winds. Other accessories include high-static-pressure fan drives, controls for low-outdoor-temperature operation, and duct damper kits for control of outside air intakes and exhausts.

Heating

It is important to install cooling coils downstream of furnaces so that condensation does not form inside the combustion and flue passages. Upstream cooling-coil placement is permissible when the furnace has been approved for this type of application and designed to prevent corrosion. Burners, pilot flames, and controls must be protected from the condensate.

Chapter 26 describes hot-water and steam coils used in unitary equipment, as well as the prevention of coil freezing from ventilation air in cold weather. Chapters 27, 30, and 32 discuss forced-air and oil- and gas-fired furnaces commonly used with, or included as part of, year-round equipment.

AIR-SOURCE HEAT PUMPS

Capacities of unitary air-source heat pumps range from about 1.5 to 30 tons, although there is no specific limitation. This equipment is used in residential, commercial, and industrial applications. Multiunit and multisplit systems installations are particularly advantageous because they permit zoning, which allows heating or cooling in each zone on demand. Application factors unique to unitary heat pumps include the following:

- The unitary heat pump normally fulfills a dual function: heating and cooling; therefore, only a single piece of equipment is required for year-round comfort. Some regions, especially central and northern Europe and parts of North America, have little need for cooling. Some manufacturers offer heating-only heat pumps for these areas and for special applications.
- A single energy source can supply both heating and cooling requirements.
- Heat output can be as much as two to four times that of the purchased energy input.
- Vents and/or chimneys may be eliminated, thus reducing building costs.

In an air-source heat pump (Figure 10), the outdoor coil rejects heat to outside air in cooling mode and extracts heat from outside air in heating mode. Most residential applications consist of an indoor fan and coil unit, either vertical or horizontal, and an outdoor fan-coil unit. The compressor is usually in the outdoor unit. Electric

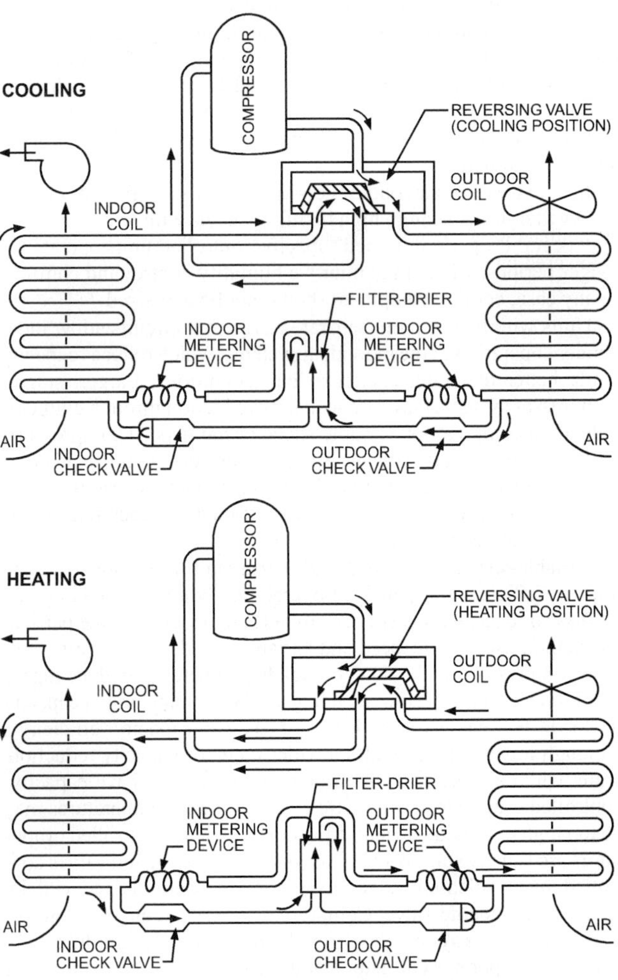

Fig. 10 Schematic Typical of Air-to-Air Heat Pump System

heaters are commonly included in the indoor unit to provide heat during defrost cycles and during periods of high heating demand that cannot be satisfied by the heat pump alone.

Add-On Heat Pumps

An air-source heat pump can be added to new or existing gas- or oil-fired furnaces. This unit, typically called an add-on, dual-fuel, or hybrid heat pump, normally operates as a conventional heat pump. During extremely cold weather, the refrigerant circuit is turned off and the furnace provides the required space heating. These add-on heat pumps share the air distribution system with the warm-air furnace. The indoor coil may be either parallel to or in series with the furnace. However, the furnace should never be upstream of the indoor coil if both systems are operated together.

Special controls are available that prevent simultaneous operation of the heat pump and furnace in this configuration. This operation raises the refrigerant condensing temperature, which could cause compressor failure. In applications where the heat pump and furnace operate simultaneously, the following conditions must be met: (1) the furnace and heat pump indoor coil must be arranged in parallel, or (2) the furnace combustion and flue passages must be designed to avoid condensation-induced corrosion during cooling operation.

Selection

Figure 11 shows performance characteristics of a single-speed, air-source heat pump, along with heating and cooling loads for a typical building. Heat pump heating capacity decreases as ambient temperature decreases. This characteristic is opposite to the trend of the building load. The outdoor temperature at which heat pump capacity equals the building load is called the **balance point**. When outdoor temperature is below the balance point, supplemental heat (usually electric resistance) must be added to make up the difference, as shown by the shaded area. The coefficient of performance (COP) shown in Figure 11 is for the refrigerant circuit only and does not include supplemental heat effects below the balance point.

In selecting the proper size heat pump, the building cooling load is calculated using standard practice. The heating balance point may be lowered by improving the structure's thermal performance or by

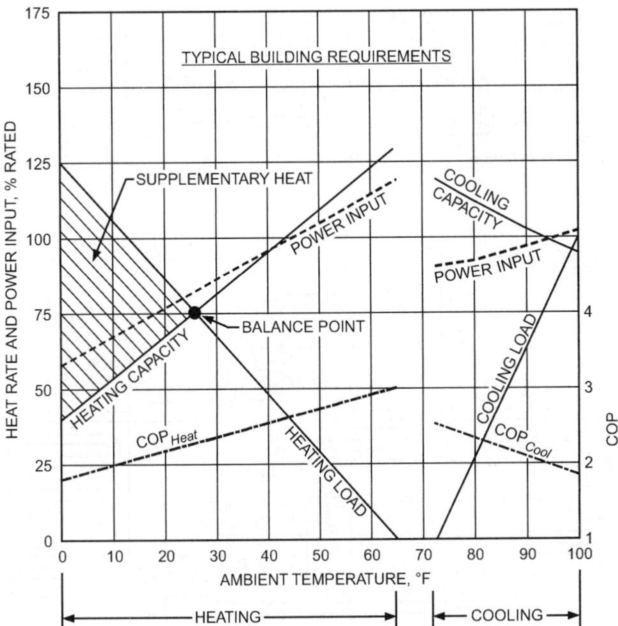

Fig. 11 Operating Characteristics of Single-Stage Unmodulated Heat Pump

choosing a heat pump larger than the cooling load requires. Excessive oversizing of cooling capacity causes excessive cycling, which results in uncomfortable temperature and humidity levels during cooling.

Using variable-speed, multispeed, or multiple compressors and variable-speed fans can improve matching of both heating and cooling loads over an extended range. This equipment can reduce cycling losses and improve comfort levels.

Building codes must be consulted before specifying indoor units that use plastic components where the proposed installation is in a return air plenum, because many codes require all materials exposed to airflow to be noncombustible or of limited combustibility, with a maximum smoke-developed index of 50.

Refrigerant Circuit and Components

In cold climates, heat pump yearly operating hours are often up to five times those of a cooling-only unit. In addition, heating extends over a greater range of operating conditions at higher-stress conditions, so the design must be thoroughly analyzed to ensure maximum reliability. Improved components and protective devices increase reliability, but the equipment designer must select components that are approved for the specific application.

For a reliable and efficient heat pump system, the following factors must be considered: (1) outdoor coil circuitry, (2) defrost and water drainage, (3) refrigerant flow controls, (4) refrigerant charge management, and (5) compressor selection.

Outdoor Coil Circuitry. When the heat pump is used for heating, the outdoor coil operates as an evaporator. The refrigerant in the coil is less dense than when the coil operates as a condenser. To avoid excessive pressure drop during heating, circuitry usually compromises between optimum performance as an evaporator and as a condenser.

Defrost and Water Drainage. During colder outdoor temperatures, usually below 40 to 50°F, and high relative humidities (above 50%), the outdoor coil operates below the frost point of the outside air. Frost that builds up on the coil surface is usually removed by **reverse-cycle defrost**. In this method, refrigerant flow in the system is reversed, and hot gas from the compressor flows through the outdoor coil, melting the frost. A typical defrost takes 4 to 10 min. The outdoor fan is normally off during defrost. Because defrost is a transient process, capacity, power, and refrigerant pressures and temperatures in different parts of the system change throughout the defrost period (Miller 1989; O'Neal et al. 1989a).

Heat pump performance during defrost can be enhanced in several ways. Defrost times and water removal can be improved by ensuring that adequate refrigerant is routed to the lower refrigerant circuits in the outdoor coil. Properly sizing the defrost expansion device is critical for reducing defrost times and energy use (O'Neal et al. 1989b). If the expansion device is too small, suction pressure can be below atmospheric, defrost times become long, and energy use is high. If the expansion device is too large, the compressor can be flooded with liquid refrigerant. During conventional reverse-cycle defrost, there is a significant pressure spike at defrost termination. Starting the outdoor fan 30 to 45 s before defrost termination can minimize the spike (Anand et al. 1989). In cold climates, the cabinet should be installed above grade to provide good drainage during defrost and to minimize snow and ice build-up around the cabinet. During prolonged periods of severe weather, it may be necessary to clear ice and snow from around the unit.

Several methods are used to determine the need to defrost. One common, simple, and reliable control method is to initiate defrost at predetermined time intervals (usually 90 min). Demand-type systems detect a need for defrosting by measuring changes in air pressure drop across the outdoor coil or changes in temperature difference between the outdoor coil and outside air. Microprocessors are used to control this function, as well as numerous other functions (Mueller and Bonne 1980). Demand defrost control is preferred because it requires less energy than other defrost methods.

Refrigerant Flow Controls. Separate refrigerant flow controls are usually used for indoor and outdoor coils. Because refrigerant flow reverses direction between heating and cooling modes, a check valve bypasses in the appropriate direction around each expansion device. Capillaries, fixed orifices, thermostatic expansion valves, or electronically controlled expansion valves may be used; however, capillaries and fixed orifices require that greater care be taken to prevent excessive flooding of refrigerant into the compressor. A check valve is not needed when an orifice expansion device or biflow expansion valve is used. The reversing valve is the critical additional component required to make a heat pump air-conditioning system.

Refrigerant Charge Management. Extra care is required to control compressor flooding and refrigerant storage in the system during both heating and cooling. The mass flow of refrigerant during cooling is greater than during heating. Consequently, the amount of refrigerant stored may be greater in heating mode than in cooling, depending on the relative internal volumes of the indoor and outdoor coils. Usually, the internal volume of indoor coils ranges from 110 to 70% of the outdoor coil volume. The relative volumes can be adjusted so the coils not only transfer heat but also manage the charge.

When capillaries or fixed orifices are used, the refrigerant may be stored in an accumulator in the suction line or in receivers that can remove the refrigerant charge from circulation when compressor floodback is imminent. Thermostatic expansion valves reduce the flooding problem, but storage may be required in the condenser. Using accumulators and/or receivers is particularly important in split systems.

To maintain performance reliability, the amount of refrigerant in the system must be checked and adjusted in accordance with the manufacturer's recommendations, particularly when charging a heat pump. Manufacturer recommendations for accumulator installation must be followed so that good oil return is ensured.

Compressor Selection. Compressors are selected on the basis of performance, reliability, and probable applications of the unit. In good design practice, equipment manufacturers often consult with compressor manufacturers during both design and application phases of the unitary equipment to verify proper application of the compressor. Compressors in a heat pump operate over a wide range of suction and discharge pressures; thus, their design parameters (e.g., refrigerant discharge temperatures, pressure ratios, clearance volume, motor-overload protection) require special consideration. In all operating conditions, compressors should be protected against loss of lubrication, liquid floodback, and high discharge temperatures.

System Control and Installation

Installation should follow the manufacturer's instructions. Because supply air from a heat pump is usually at a lower temperature (typically 90 to 100°F) than that from most heating systems, ducts and supply registers should control air velocity and throw to minimize the perception of cool drafts.

Low-voltage heating/cooling thermostats control heat pump operation. Ductless split units are usually controlled by wireless thermostats. Models that switch automatically from heating to cooling operation and manual-selection models are available. Usually, heating is controlled in two stages. The first stage controls heat pump operation, and the second stage controls supplementary heat. When the heat pump cannot satisfy the first stage's call for heat, supplementary heat is added by the second-stage control. The amount of supplementary heat is often controlled by an outdoor thermostat that allows additional stages of heat to be turned on only when required by the colder outdoor temperature.

Microprocessor technology has led to night setback modes and intelligent recovery schemes for morning warm-up on heat pump systems (see Chapter 46 of the 2007 *ASHRAE Handbook—HVAC Applications*).

WATER-SOURCE HEAT PUMPS

A water-source heat pump (WSHP) is a single-package reverse-cycle heat pump that uses water as the heat source for heating and as the heat sink for cooling. The water supply may be a recirculating closed loop, a well, a lake, or a stream. Water for closed-loop heat pumps is usually circulated at 2 to 3 gpm per ton of cooling capacity. A **groundwater heat pump (GWHP)** can operate with considerably less water flow. The main components of a WSHP refrigeration system are a compressor, refrigerant-to-water heat exchanger, refrigerant-to-air heat exchanger, refrigerant expansion devices, and refrigerant-reversing valve. Figure 12 shows a schematic of a typical WSHP system.

Designs of packaged WSHPs range from horizontal units located primarily above the ceiling or on the roof, to vertical units usually located in basements or equipment rooms, to console units located in the conditioned space. Figures 13 and 14 illustrate typical designs.

Systems

WSHPs are used in a variety of systems, such as

- Water-loop heat pump systems (Figure 15A)
- Groundwater heat pump systems (Figure 15B)
- Closed-loop surface-water heat pump systems (Figure 15C)
- Surface-water heat pump systems (Figure 15D)
- Ground-coupled heat pump systems (Figure 15E)

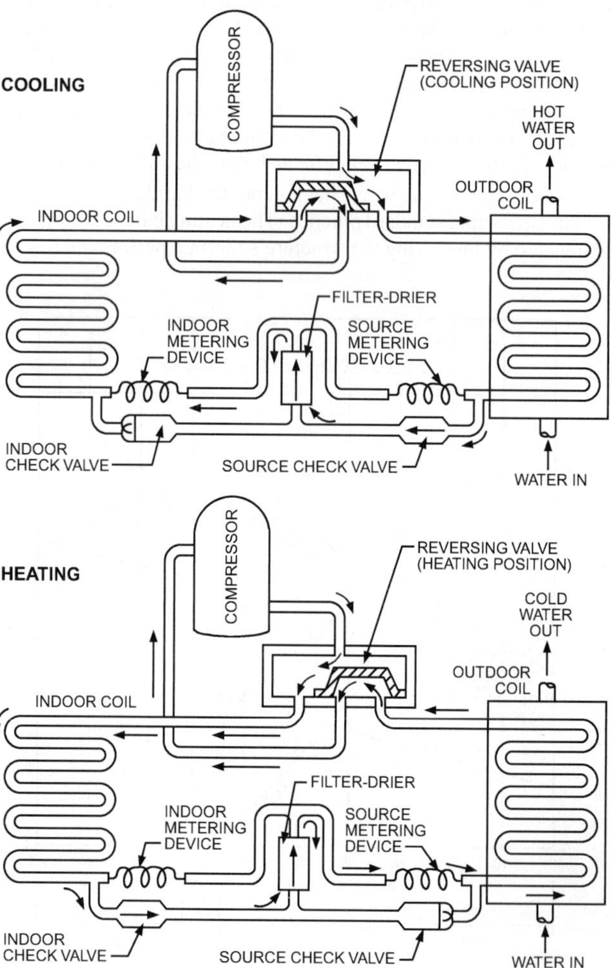

Fig. 12 Schematic of Typical Water-Source Heat Pump System

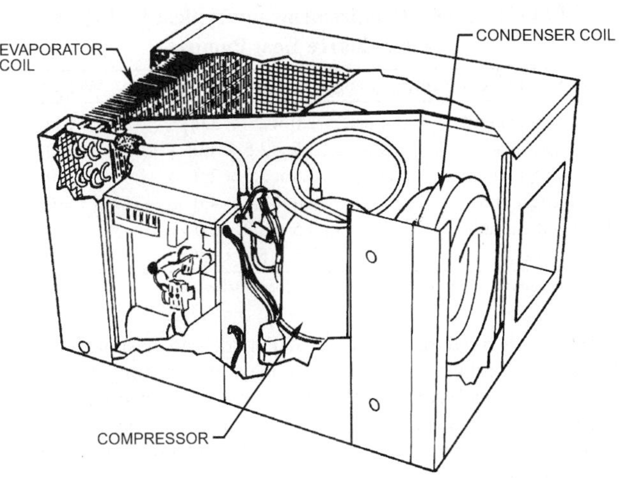

Fig. 13 Typical Horizontal Water-Source Heat Pump

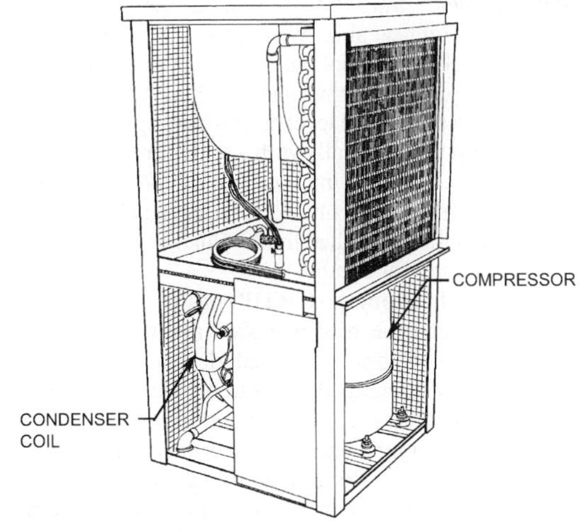

Fig. 14 Typical Vertical Water-Source Heat Pump

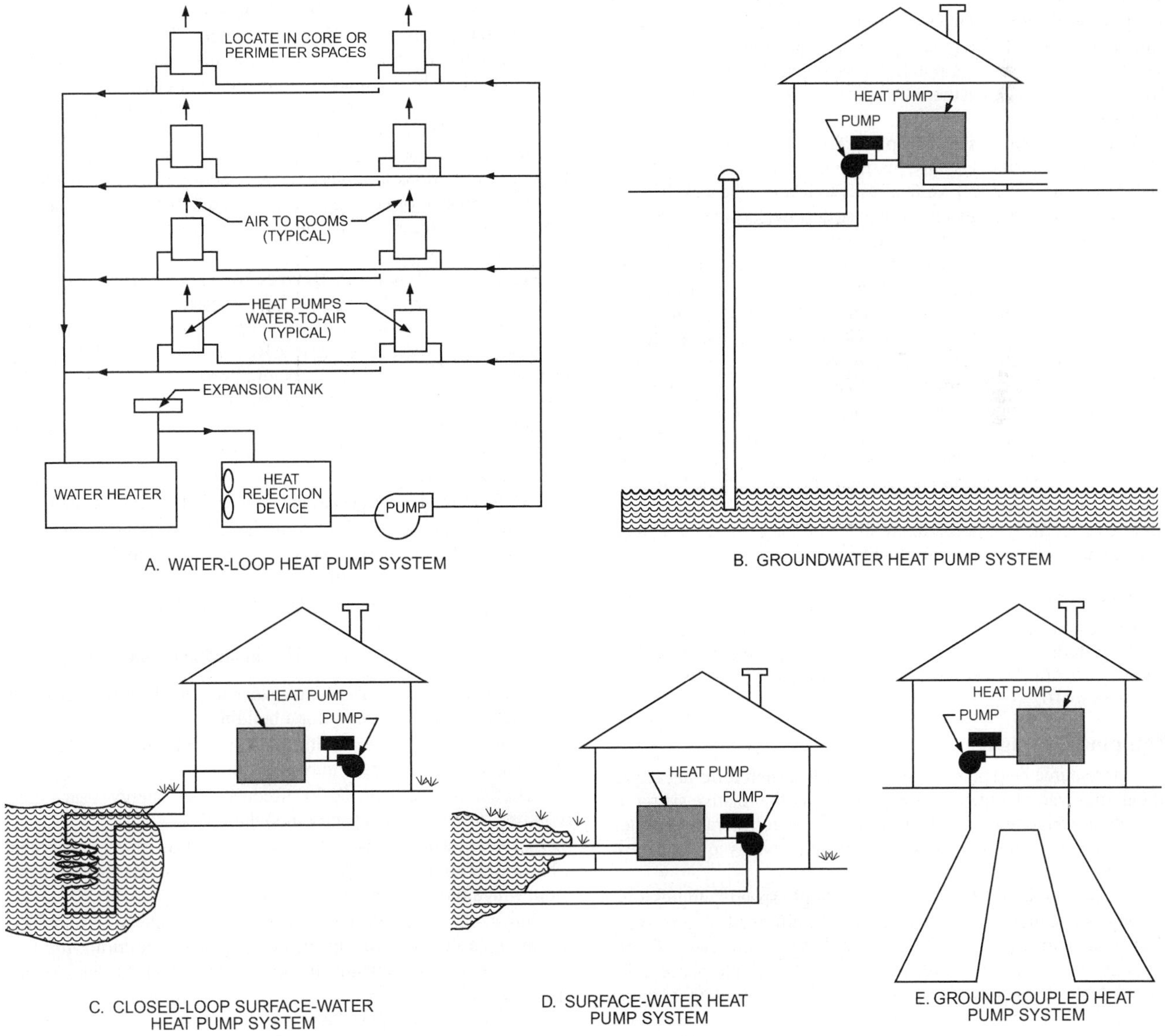

Fig. 15 Water-Source Heat Pump Systems

A **water-loop heat pump (WLHP)** uses a circulating water loop as the heat source and heat sink. When loop water temperature exceeds a certain level during cooling, a cooling tower dissipates heat from the water loop into the atmosphere. When loop water temperature drops below a prescribed level during heating, heat is added to the circulating loop water, usually with a boiler. In multiple-unit installations, some heat pumps may operate in cooling mode while others operate in heating, and controls are needed to keep loop water temperature within the prescribed limits. Chapter 8 has more information on water-loop heat pumps.

A **groundwater heat pump (GWHP)** passes groundwater from a nearby well through the heat pump's water-to-refrigerant heat exchanger, where it is warmed or cooled, depending on the operating mode. It is then discharged to a drain, stream, or lake, or is returned to the ground through a reinjection well.

Many state and local jurisdictions have ordinances about use and discharge of groundwater. Because aquifers, the water table, and groundwater availability vary from region to region, these regulations cover a wide spectrum.

A **surface-water heat pump (SWHP)** uses water from a nearby lake, stream, or canal. After passing through the heat pump heat exchanger, it is returned to the source or a drain several degrees warmer or cooler, depending on the operating mode of the heat pump. **Closed-loop** surface water heat pumps use a closed water or brine loop that includes pipes or tubing submerged in the surface water (river, lake, or large pond) that serves as the heat exchanger. The adequacy of the total thermal capacity of the body of water must be considered.

A **ground-coupled heat pump (GCHP)** system uses the earth as a heat source and sink. Usually, plastic piping is installed in either a shallow horizontal or deep vertical array to form the heat exchanger. The massive thermal capacity of the earth provides a temperature-stabilizing effect on the circulating loop water or brine. Installing this type of system requires detailed knowledge of the climate; site; soil temperature, moisture content, and thermal characteristics; and performance, design, and installation of water-to-earth heat exchangers. Additional information on GCHP systems is presented in Chapter 32 of the 2007 *ASHRAE Handbook—HVAC Applications*.

Entering Water Temperatures. These various water sources provide a wide range of entering water temperatures to WSHPs. Entering water temperatures vary not only by water source, but also by climate and time of year. Because of the wide range of entering water or brine temperatures encountered, it is not feasible to design a universal packaged product that can handle the full range of possibilities effectively. Therefore, WSHPs are rated for performance at a number of standard rating conditions.

Performance Certification Programs

ARI certifies water-to-air and brine-to-air heat pumps rated below 135,000 Btu/h. Certification is based on ISO *Standard* 13256-1. For details, go to http://www.ari.org/Content/WatertoAir andBrinetoAir HeatPumps_102.aspx.

Equipment Design

Water-source heat pumps are designed to match differing levels of entering water temperatures by optimizing the relative sizing of the indoor refrigerant-to-air heat exchanger and refrigerant-to-water heat exchangers and by matching expansion devices to refrigerant flow rates.

Compressors. WSHPs usually have single-speed compressors, although some high-efficiency models use multispeed compressors. Higher-capacity equipment may use multiple compressors. Compressors may be reciprocating, rotary, or scroll. Single-phase units are available at voltages of 115, 208, 230, and 265. All larger equipment is for three-phase power supplies with voltages of 208, 230, 460, or 575. Compressors usually have electromechanical protective devices.

Table 3 Space Requirements for Typical Packaged Water-Source Heat Pumps

Water-to-Air Heat Pump	Length × Width × Height, ft	Weight, lb
1.5 ton vertical unit	2.0 × 2.0 × 3.0	180
3 ton vertical unit	2.5 × 2.5 × 4.0	250
3 ton horizontal unit	3.5 × 2.0 × 2.0	250
5 ton vertical unit	3.0 × 2.5 × 4.0	330
11 ton vertical unit	3.5 × 3.0 × 6.0	720
26 ton vertical unit	3.5 × 5.0 × 6.0	1550

Note: See manufacturers' specification sheets for actual values.

Inside Air System. Console WSHP models are designed for free delivery of conditioned air. Other models have ducting capability. Smaller WSHPs have multispeed, direct-drive centrifugal blower wheel fan systems. Large-capacity equipment has belt-drive systems. All units have provisions for fiberglass, metal, or plastic foam air filters.

Inside Air Heat Exchanger. The indoor air heat exchanger of WSHP units is a conventional plate-fin coil of copper tubes and aluminum fins. The coil tubing is circuited so that it can function effectively as an evaporator with refrigerant flow in one direction and as a condenser when refrigerant flow is reversed.

Refrigerant-to-Water Heat Exchanger. The heat exchanger, which couples the heat pump to source/sink water, is tube-in-tube, tube-in-shell, or brazed-plate. It must function in either condensing or evaporating mode, so special attention is given to refrigerant-side circuitry. Heat exchanger construction is usually of copper and steel, and the source/sink water is exposed only to the copper portions. Cupronickel options to replace the copper are usually available for use with brackish or corrosive water. Brazed-plate heat exchangers are usually constructed of stainless steel, which reduces the need for special materials.

Refrigerant Expansion Devices. WSHPs rated in accordance with ARI *Standard* 320 operate over a narrow range of entering water temperatures, so most use simple capillaries as expansion devices. Units rated according to ARI *Standard* 325 or 330 usually use thermostatic expansion valves for improved performance over a broader range of inlet fluid temperatures.

Refrigerant-Reversing Valve. The refrigerant-reversing valves in WSHPs are identical to those used in air-source heat pumps.

Condensate Disposal. Condensate, which forms on the indoor coil when cooling, is collected and conveyed to a drain system.

Controls. Console WSHP units have built-in operating mode selector and thermostatic controls. Ducted units use low-voltage remote heat/cool thermostats.

Size. Typical space requirements and weights of WSHPs are presented in Table 3.

Special Features. Some WSHPs include the following:

Desuperheater. Uses discharge gas in a special water/refrigerant heat exchanger to heat water for a building.

Capacity modulation. Multiple compressors, multispeed compressors, or hot-gas bypass may be used.

Variable air volume (VAV). Reduces fan energy usage and requires some form of capacity modulation.

Automatic water valve. Closes off water flow through the unit when the compressor is off and permits variable water volume in the loop, which reduces pumping energy.

Outside-air economizer. Cools directly with outside air to reduce or eliminate the need for mechanical refrigeration during mild or cold weather when outside humidity levels and air quality are appropriate.

Water-side economizer. Cools with loop water to reduce or eliminate the need for mechanical refrigeration during cold weather; requires a hydronic coil in the indoor air circuit that is valved into the circulating loop when loop temperatures are relatively low and cooling is required.

Electric heaters. Used in WLHP systems that do not have a boiler as a source for loop heating.

VARIABLE-REFRIGERANT-FLOW HEAT PUMPS

A variable-refrigerant-flow (VRF) system typically consists of a condensing section housing compressor(s) and condenser heat exchanger interconnected by a single set of refrigerant piping to multiple indoor direct-expansion (DX) evaporator fan-coil units. Thirty or more DX fan coil units can be connected to a single condensing section, depending on system design, and with capacity ranging from 0.5 to 8 tons.

The DX fan coils are constant air volume, but use variable refrigerant flow through an electronic expansion valve. The electronic expansion valve reacts to several temperature-sensing devices such as return air, inlet and outlet refrigerant temperatures, or suction pressure. The electronic expansion valve modulates to maintain the desired set point.

Application

VRF systems are most commonly air-to-air, but are also available in a water-source (water-to-refrigerant) configuration. They can be configured for simultaneous heating and cooling operation (some indoor fan coil units operating in heating and some in cooling, depending on requirements of each building zone).

Indoor units are typically direct-expansion evaporators using individual electronic expansion devices and dedicated microprocessor controls for individual control. Each indoor unit can be controlled by individual thermostat. The outdoor unit may connect several indoor evaporator units with capacities 130% or more than the outdoor condensing unit capacity.

Categories

VRF equipment is divided into three general categories: residential, light commercial, and applied. Residential equipment is single-phase unitary equipment with a cooling capacity of 65,000 Btu/h or less. Light commercial equipment is generally three-phase, with cooling capacity greater than 65,000 Btu/h, and is designed for small businesses and commercial properties. Applied equipment has cooling capacity higher than 135,000 Btu/h and is designed for large commercial buildings.

Refrigerant Circuit and Components

VRF heat pump systems use a two-pipe (liquid and suction gas) system; simultaneous heat and cool systems use the same system, as well as a gas flow device that determines the proper routing of refrigerant gas to a particular indoor unit.

VRF systems use a sophisticated refrigerant circuit that monitors mass flow, oil flow, and balance to ensure optimum performance. This is accomplished in unison with variable-speed compressors and condenser fan motors. Both of these components adjust their frequency in reaction to changing mass flow conditions and refrigerant operating pressures and temperatures. A dedicated microprocessor continuously monitors and controls these key components to ensure proper refrigerant is delivered to each indoor unit in cooling or heating.

Heating and Defrost Operation

In heating mode, VRF systems typically must defrost like any mechanical heat pump, using reverse cycle valves to temporarily operate the outdoor coil in cooling mode. Oil return and balance with the refrigerant circuit is managed by the microprocessor to ensure that any oil entrained in the low side of the system is brought back to the high side by increasing the refrigerant velocity using a high-frequency operation performed automatically based on hours of operation.

REFERENCES

Anand, N.K., J.S. Schliesing, D.L. O'Neal, and K.T. Peterson. 1989. Effects of outdoor coil fan pre-start on pressure transients during the reverse cycle defrost of a heat pump. *ASHRAE Transactions* 95(2).

ARI. 2006. Performance rating of unitary air-conditioning and air-source heat pump equipment. ANSI/ARI *Standard* 210/240-2006. Air-Conditioning and Refrigeration Institute, Arlington, VA.

ARI. 1995. Sound rating of outdoor unitary equipment. *Standard* 270-95. Air-Conditioning and Refrigeration Institute, Arlington, VA.

ARI. 1997. Application of sound rating levels of outdoor unitary equipment. *Standard* 275-97. Air-Conditioning and Refrigeration Institute, Arlington, VA.

ARI. 1998. Water-source heat pumps. *Standard* 320-98. Air-Conditioning and Refrigeration Institute, Arlington, VA.

ARI. 1998. Ground water-source heat pumps. *Standard* 325-98. Air-Conditioning and Refrigeration Institute, Arlington, VA.

ARI. 1998. Ground source closed-loop heat pumps. *Standard* 330-98. Air-Conditioning and Refrigeration Institute, Arlington, VA.

ARI. 2007. Commercial and industrial unitary air-conditioning and heat pump equipment. *Standard* 340/360-2007. Air-Conditioning and Refrigeration Institute, Arlington, VA.

ARI. 2002. Commercial and industrial unitary air-conditioning condensing units. ANSI/ARI *Standard* 365-2002. Air-Conditioning and Refrigeration Institute, Arlington, VA.

ARI. Semiannual. *Unitary directory of certified product performance*. Air-Conditioning and Refrigeration Institute, Arlington, VA. Available at http:www.aridirectory.org/ari/unitary.html.

ASHRAE. 2004. Safety code for mechanical refrigeration. ANSI/ASHRAE *Standard* 15-2004.

ASHRAE. 2005. Methods of testing for rating electrically driven unitary air-conditioning and heat pump equipment. ANSI/ASHRAE *Standard* 37-2005.

ASHRAE. 2004. Energy standard for buildings except low-rise residential buildings. *Standard* 90.1-2004.

Counts, D. 1985. Performance of heat pump/desuperheater water heating systems. *ASHRAE Transactions* 91(2B):1473-1487.

CSA. 2005. Heating and cooling equipment. *Standard* C22.2 No. 236-05. Canadian Standards Association International, Toronto.

DOE. Annual. Uniform test method for measuring the energy consumption of central air conditioners. 10CFR430, Appendix M. *Code of Federal Regulations*, U.S. Department of Energy, Washington, D.C.

ISO. 1998. Water-source heat pumps—Testing and rating for performance—Part 1: Water-to-air and brine-to-air heat pumps. *Standard* 13256-1. International Organization for Standardization, Geneva.

Miller, W.A. 1989. Laboratory study of the dynamic losses of a single speed, split system air-to-air heat pump having tube and plate fin heat exchangers. ORNL/CON-253. Oak Ridge National Laboratory, Oak Ridge, TN.

Mueller, D. and U. Bonne. 1980. Heat pump controls: Microelectronic technology. *ASHRAE Journal* 22(9).

NFPA. 2008. *National electrical code®*. ANSI/NFPA *Standard* 70. National Fire Protection Association, Quincy, MA.

NFPA. 2002. Installation of air-conditioning and ventilating systems. ANSI/NFPA *Standard* 90A. National Fire Protection Association, Quincy, MA.

NFPA. 2006. Installation of warm air heating and air-conditioning systems. ANSI/NFPA *Standard* 90B. National Fire Protection Association, Quincy, MA.

O'Neal, D.L., N.K. Anand, K.T. Peterson, and S. Schliesing. 1989a. Determination of the transient response characteristics of the air-source heat pump during the reverse cycle defrost. *Final Report*, ASHRAE Research Project TRP-479.

O'Neal, D.L., N.K. Anand, K.T. Peterson, and S. Schliesing. 1989b. Refrigeration system dynamics during the reverse cycle defrost. *ASHRAE Transactions* 95(2).

UL. 2005. Heating and cooling equipment. *Standard* 1995-05. Underwriters Laboratories, Northbrook, IL.

BIBLIOGRAPHY

ANSI. 2002. Acoustics—Determination of sound power levels of noise sources using sound pressure—Precision methods for reverberation rooms. ANSI *Standard* S12.51-2002 (ISO 3741:1999 NAIS Standard). American National Standards Institute, New York.

ARI. Semiannual. *Applied directory of certified product performance*. Air-Conditioning and Refrigeration Institute, Arlington, VA. Available at http:www.aridirectory.org/ari/applied.html.

CSA. 2005. Performance of direct-expansion ground-source heat pumps. CAN/CSA *Standard* C748-94 (R2005). Canadian Standards Association International, Toronto.

CHAPTER 49

ROOM AIR CONDITIONERS AND PACKAGED TERMINAL AIR CONDITIONERS

ROOM AIR CONDITIONERS

ROOM air conditioners are encased assemblies designed primarily for mounting in a window or through a wall. They are designed to deliver cool or warm conditioned air to the room, either without ducts or with very short ducts (up to a maximum of about 48 in). Each unit includes a prime source of refrigeration and dehumidification and a means for circulating and filtering air; it may also include a means for ventilating and/or exhausting and heating.

The basic function of a room air conditioner is to provide comfort by cooling, dehumidifying, filtering or cleaning, and circulating the room air. It may also provide ventilation by introducing outdoor air into the room and/or exhausting room air to the outside. Room temperature may be controlled by an integral thermostat. The conditioner may provide heating by heat pump operation, electric resistance elements, or a combination of the two.

Figure 1 shows a typical room air conditioner in cooling mode. Warm room air passes over the cooling coil and gives up sensible and latent heat. The conditioned air is then recirculated in the room by a fan or blower.

Heat from the warm room air vaporizes the cold (low-pressure) liquid refrigerant flowing through the evaporator. The vapor then carries the heat to the compressor, which compresses the vapor and increases its temperature above that of the outdoor air. In the condenser, the hot (high-pressure) refrigerant vapor liquefies, giving up the heat from the room air to outdoor air. Next, the high-pressure liquid refrigerant passes through a restrictor, which reduces its pressure and temperature. The cold (low-pressure) liquid refrigerant then enters the evaporator to repeat the refrigeration cycle.

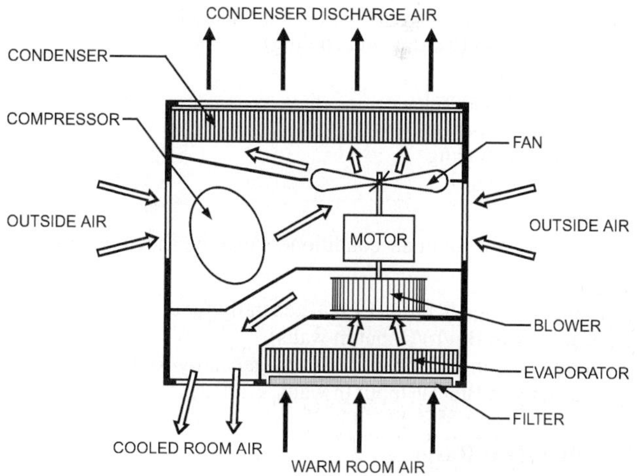

Fig. 1 Schematic View of Typical Room Air Conditioner

The preparation of this chapter is assigned to TC 8.11, Unitary and Room Air Conditioners and Heat Pumps.

SIZES AND CLASSIFICATIONS

Room air conditioners have line cords, which may be plugged into standard or special electric circuits. Most units in the United States are designed to operate at 115, 208, or 230 V; single-phase; 50 or 60 Hz power. Some units are rated at 265 V or 277 V, for which the chassis or chassis assembly must provide permanent electrical connection. The maximum amperage of 115 V units is generally 12 A, which is the maximum current permitted by the *National Electrical Code®* (NEC) for a single-outlet, 15 A circuit. Models designed for countries other than the United States are generally for 50 or 60 Hz systems, with typical design voltage ranges of 100 to 120 and 200 to 240 V, single-phase.

Popular 115 V models have capacities in the range of 5000 to 8000 Btu/h, and are typically used in single-room applications. Larger-capacity 115 V units are in the 12,000 to 15,000 Btu/h range. Capacities for 230, 208, or 230/208 V units range from 8000 to 36,000 Btu/h. These higher-voltage units are typically used in multiple-room installations.

Heat pump models are also available, usually for 208 or 230 V applications. These units are generally designed for reversed-refrigerant-cycle operation as the normal means of supplying heat, but may incorporate electrical-resistance heat either to supplement heat pump capacity or to provide the total heating capacity when outdoor temperatures drop below a set value.

Another type of heating model incorporates electrical heating elements in regular cooling units so that heating is provided entirely by electrical resistance heat.

DESIGN

Room air conditioner design is usually based on one or more of the following criteria, any one of which automatically constrains the overall system design:

- Lowest initial cost
- Lowest operating cost (highest efficiency)
- Energy-efficiency ratio (EER) or coefficient of performance (COP), as legislated by government
- Low sound level
- Chassis size

- Unusual chassis shape (e.g., minimal depth or height)
- Amperage limitation (e.g., 7.5 A, 12 A)
- Weight

The following combinations illustrate the effect of an initial design parameter on the various components:

Low Initial Cost. High airflow with minimum heat exchanger surface keeps the initial cost of a unit low. These units have a low-cost compressor, which is selected by analyzing various compressor and coil combinations and choosing the one that both achieves optimum performance and passes all tests required by Underwriters Laboratories (UL), the Association of Home Appliance Manufacturers (AHAM), and others. For example, a high-capacity compressor might be selected to meet the capacity requirement with a minimum heat transfer surface, but frost tests under maximum load may not be acceptable. These tests set the upper and lower limits of acceptability when low initial cost is the prime consideration.

Low Operating Cost. Large heat exchanger surfaces keep operating cost low. A compressor with a low compression ratio operates at low head pressure and high suction pressure, which results in a high EER.

Compressors

Room air conditioner compressors range in capacity from about 4000 to 34,000 Btu/h. Design data are available from compressor manufacturers at the following standard rating conditions:

Evaporating temperature	45°F
Compressor suction temperature	95°F
Condensing temperature	130°F
Liquid temperature	115°F
Ambient temperature	95°F

Compressor manufacturers offer complete performance curves at various evaporating and condensing temperatures to aid in selection for a given design specification.

Evaporator and Condenser Coils

These coils are generally tube-and-plate-fin, tube-and-louvered-fin, or tube-and-spine-fin. Information on the performance of such coils is available from suppliers, and original equipment manufacturers usually develop data for their own coils. Design parameters to be considered when selecting coils are (1) cooling rate per unit area of coil surface (Btu/h·ft^2), (2) dry-bulb temperature and moisture content of entering air, (3) air-side friction loss, (4) internal refrigerant pressure drop, (5) coil surface temperature, (6) airflow, and (7) air velocity.

Restrictor Application and Sizing

Three main types of restrictor devices are available to the designer: (1) a **thermostatic expansion valve**, which maintains a constant amount of superheat at a point near the outlet of the evaporator; (2) an **automatic expansion valve**, which maintains a constant suction pressure; and (3) a **restrictor tube (capillary)**. The capillary is the most popular device for room air conditioner applications because of its low cost and high reliability, even though its refrigerant control over a wide range of ambient temperatures is not optimal. A recommended procedure for optimizing charge balance, condenser subcooling, and restrictor sizing is as follows:

1. Use an adjustable restrictor (e.g., a needle valve), so that tests may be run with a flooded evaporator coil and various refrigerant charges to determine the optimum point of system operation.
2. Reset the adjustable restrictor to the optimum setting, remove it from the unit, and measure flow pressure with a flow comparator similar to that described in ASHRAE *Standard* 28.
3. Install a restrictor tube with the same flow rate as the adjustable restrictor. Usually, restrictor tubes are selected on the basis of cost, with shorter tubes generally being less expensive.

Fan Motor and Air Impeller Selection

The two types of motors generally used on room air conditioners are the (1) low-efficiency, shaded-pole type; and (2) more efficient, permanent split-capacitor type, which requires using a run capacitor. Air impellers are usually of two types: (1) forward-curved blower wheel and (2) axial- or radial-flow fan blade. In general, blower wheels are used to move small to moderate amounts of air in high-resistance systems, and fan blades move moderate to high air volumes in low-resistance applications. Blower wheels and cross-flow fans also generate lower noise levels than fan blades.

The combination of fan motor and air impellers is so important to the overall design that the designer should work closely with the manufacturers of both components. Performance curves are available for motors, blower wheels, and fans, but data are for ideal systems not usually found in practice because of physical size, motor speed, and component placement limitations.

Electronics

Microprocessors monitor and control numerous functions for room air conditioners. These microelectronic controls offer digital displays and touch panels for programming desired temperature, on-off timing, modulated fan speeds, bypass capabilities, and sensing for humidity, temperature, and airflow control.

PERFORMANCE DATA

In the United States, an industry certification program under the sponsorship of the Association of Home Appliance Manufacturers (AHAM) covers the majority of room air conditioners and certifies the cooling and heating capacities, EER, and electrical input (in amperes) of each for adherence to nameplate rating. The following tests are specified by AHAM *Standard* RAC-1:

- Cooling capacity
- Heating capacity
- Maximum operating conditions (heating and cooling)
- Enclosure sweat
- Freeze-up
- Recirculated air quantity
- Moisture removal
- Ventilating air quantity and exhaust air quantity
- Electrical input (heating and cooling)
- Power factor
- Condensate disposal
- Application heating capacity
- Outside coil deicing

Efficiency

Efficiency for room air conditioners may be shown in either of two forms:

1. Energy efficiency ratio (EER—generally for cooling)
   (Capacity in Btu/h)/(Input in watts)
2. Coefficient of performance (COP—generally for heating)
   (Capacity in Btu/h)/(Input in watts × 3.412)

Sensible Heat Ratio

The ratio of sensible heat to total heat removal is a useful performance characteristic for evaluating units for specific conditions. A low ratio indicates more dehumidification capacity, and hot, humid areas like New Orleans and arid locales like Phoenix might best be served with units having lower and higher ratios, respectively.

Energy Conservation and Efficiency

In the United States, two federal energy programs have increased the demand for higher-efficiency room air conditioners. First, the Energy Policy and Conservation Act of 2005 (Public Law 1094-58) provides a commercial building deduction for energy-efficient building improvements, and provides tax breaks for those making energy conservation improvements to their homes. Second, the National Appliance Energy Conservation Act of 1987 (NAECA) provides a single set of minimum efficiency standards for major appliances, including room air conditioners. The room air conditioner portion of NAECA originally specified minimum efficiencies for 12 classes, based on physical conformation, with minimums ranging from 8 to 9 EER and applying to all units manufactured on or after January 1, 1990.

The U.S. Department of Energy (DOE) issued increased minimum efficiency standards that became effective October 1, 2000 (*Federal Register*, September 24, 1997). (See Table 1) Four additional classes were created, two of which cover casement-type units. The minimum standards range from 8 to 9.8 EER. For the most popular classes (cooling-only units with louvered sides ranging in capacity from less than 6000 to 20,000 Btu/h) the minimum standards are either 9.7 or 9.8 EER. Table 2 shows the U.S. Environmental Protection Agency (EPA) and DOE's ENERGY STAR® requirements for room air conditioners. ENERGY STAR-qualified room air conditioners use at least 10% less energy than minimum-efficiency models. The EPA's ENERGY STAR Web site provides additional information on qualifying products (EPA 2008).

Table 1 NAECA Minimum Efficiency Standards for Room Air Conditioners

Class	Reverse Cycle	Louvered Sides	Capacity, Btu/h	Minimum EER as of January 1, 1990	Minimum EER as of October 1, 2000
1	No	Yes	< 6000	8.0	9.7
2	No	Yes	6000 to 7999	8.5	9.7
3	No	Yes	8000 to 13,999	9.0	9.8
4	No	Yes	14,000 to 19,999	8.8	9.7
5	No	Yes	≥ 20,000	8.2	8.5
6	No	No	< 6000	8.0	9.0
7	No	No	6000 to 7999	8.5	9.0
8	No	No	8000 to 13,999	8.5	8.5
9	No	No	14,000 to 19,999	8.5	8.5
10	No	No	≥ 20,000	8.2	8.5
11	Yes	Yes	< 20,000	8.5	9.0
12	Yes	No	< 14,000	8.0	8.5
13	Yes	Yes	≥ 20,000	8.5	8.5
14	Yes	No	≥ 14,000	8.0	8.0
15	Casement Only		—	*	8.7
16	Casement Slider		—	*	9.5

*Casement-only and casement-slider room air conditioners were not separate product classes under standards effective January 1, 1990. These units were subject to applicable standards in classes 1 to 14 based on unit capacity and presence or absence of louvered sides and reverse cycle.

Table 2 Room Air Conditioners ENERGY STAR Criteria

Product Class	NAECA Criteria Minimum EER	ENERGY STAR Criteria* Minimum EER
< 8000 Btu/h	9.7	10.7
8000 to 13,999 Btu/h	9.8	10.8
14,000 to 19,999 Btu/h	9.7	10.7
≥ 20,000 Btu/h	8.5	9.4

*ENERGY STAR criteria for room air conditioners are 10% above the NAECA criteria. Currently, only units *without* reverse cycle (heating function) and *with* louvered sides are considered for ENERGY STAR status. Minimum EER should be used to determine qualification for ENERGY STAR label. Values are rounded to single decimal and meet specification of 10% more efficient than new NAECA standard.

All state and local minimum efficiency standards in the United States are automatically superseded by federal standards. Many other countries have or are considering minimum efficiency standards, so such standards should be sought as part of the design process.

Whether estimating potential energy savings associated with appliance standards or estimating consumer operating costs, the annual hours H of operation of a room air conditioner are important. These figures have been compiled from various studies commissioned by DOE and AHAM for every major city and region in the United States. The national average is estimated at 750 h per year.

The estimated cost of operation is as follows:

$$C = RHW/1000$$

where

C = annual cost of operation, \$/year
R = average cost, \$/kWh
H = annual hours of operation
W = input, W

High-Efficiency Design

The EER can be affected by three design parameters. The first is **electrical efficiency**. Fan motor efficiency ranges from 25 to 65%; compressor motors range from 60 to 85%. The second parameter, **refrigerant cycle efficiency**, is increased by enhancing or enlarging the heat transfer surface to minimize the difference between the refrigerant saturation temperature and air temperature. This allows using a compressor with a smaller displacement and a high-efficiency motor. The third parameter is **air circuit efficiency**, which can be increased by minimizing pressure drop across the heat transfer surface, which reduces the load on the fan motor.

Higher EERs are not the complete answer to reducing energy costs. Energy efficiency can be increased by properly sizing the unit, keeping infiltration and leakage losses to a minimum, increasing building insulation, reducing unnecessary internal loading, providing effective maintenance, and balancing the load by using a thermostat and thermostat setback.

SPECIAL FEATURES

Some room air conditioners are designed to minimize their extension beyond the building when mounted flush with the inside wall. Low-capacity models are usually smaller and less obtrusive than higher-capacity models. Units are often installed through the wall, where they do not interfere with windows. Exterior cabinet grilles may be designed to harmonize with the architecture of various buildings.

Most units have adjustable louvers or deflectors to distribute air into the room with satisfactory throw and without drafts. Louver design should eliminate recirculation of discharge air into the air inlet. Some units use motorized deflectors for continuously changing the air direction. Discharge air speeds range from 300 to 1200 fpm, with low speeds preferred in rooms where people are at rest.

Most room air conditioners are designed to bring in outside air, exhaust room air, or both. Controls usually allow these features to function independently.

Temperature is controlled by an adjustable built-in thermostat. The thermostat and unit controls may operate in one of the following modes:

- The unit is set to the cool position, and the thermostat setting is adjusted as needed. The circulation blower runs without interruption while the thermostat cycles the compressor on and off.
- The unit uses a two- or three-stage thermostat, which reduces blower speed as room temperature approaches the set temperature, cycles the compressor off as temperature drops further, and, if temperature drops still further, finally cycles the blower off. As room temperature rises, the sequence reverses.

- The unit has, in addition to the preceding control sequence, an optional automatic fan mode in which both the blower and compressor are cycled simultaneously by the thermostat; this mode of operation requires proper thermostat sensitivity. One advantage of this arrangement is improved humidity control because moisture from the evaporator coil is not reevaporated into the room during the *off*-cycle. Another advantage is lower operating cost because the blower motor does not operate during the *off*-cycle. The effective EER may be increased an average of 10% by using the automatic fan mode (ORNL 1985).

Disadvantages of cycling fans with the compressor may be (1) varying noise level because of fan cycling and (2) deterioration in room temperature control.

Room air conditioners are simple to operate. Usually, one control selects the operating mode, and a second controls the temperature. Additional knobs or levers operate louvers, deflectors, the ventilation system, exhaust dampers, and other special features. Controls are usually arranged on the front of the unit or concealed behind a readily accessible door, but may also be arranged on the top or sides of the unit, or on an infrared remote control.

Filters on room air conditioners remove airborne dirt to provide clean air to the room and keep dirt off cooling surfaces. Filters are made of expanded metal (with or without a viscous oil coating), glass fiber, or synthetic materials; they may be either disposable or reusable. A dirty filter reduces cooling and air circulation and frequently allows frost to accumulate on the cooling coil; therefore, filter location should allow easy monitoring, cleaning, and replacement.

Some units have louvers or grilles on the outdoor side to enhance appearance and protect the condenser fins. Sometimes, these louvers separate the airstreams to and from the condenser and reduce recirculation. When provided, side louvers on the outside of the unit are an essential part of the condenser air system because they improve air movement to the condenser. Care should be taken not to obstruct air passages through these louvers.

Room air conditioners, especially those parts exposed to the weather, require a durable finish. Some manufacturers use a special grade of plastic for weather-exposed parts. If the parts are metal, good practice is to use phosphatized or zinc-coated steels with baked finishes and/or corrosion-resistant materials such as aluminum or stainless steel.

The sound level of a room air conditioner is an important factor, particularly when the unit is installed in a bedroom. A certain amount of sound can be expected because of air movement through the unit and compressor operation. However, a well-designed room air conditioner is relatively quiet, and the sound emitted is relatively free of high-pitched and metallic noise. Usually, fan motors with two or more speeds are used to provide a slower fan speed for quieter operation. To avoid rattles and vibration in the building structure, units must be installed correctly (see the section on Installation and Service).

Excessive outdoor sound (condenser fan and compressor) can be irritating to neighbors. In the United States, many local and state outdoor noise ordinances limit outdoor sound levels.

SAFETY CODES AND STANDARDS

United States. The *National Electrical Code*® (NFPA *Standard* 70), ASHRAE *Standard* 15, and UL *Standard* 484 pertain to room air conditioners. Local regulations may differ with these standards, but the basic requirements are generally accepted throughout the United States.

Canada. CSA International developed the standard for Room Air Conditioners (CSA *Standard* C22.2 No. 117), which forms part of the *Canadian Electrical Code*.

International. Two useful documents that might assist the designer are (1) International Electrotechnical Commission (IEC) *Standard* 60335-2-40 and (2) International Organization for Standardization (ISO) *Standard* 5151.

Product Standards

CSA *Standard* C22.2 No. 117 and UL *Standard* 484 are similar in content. In *Standard* 484, the construction section involves items such as the unit enclosure (including materials), ability to protect against contact with moving and uninsulated live parts, and means for installation or attachment. Attention is also given to the refrigeration system's ability to withstand operating pressures, system pressure relief in a fire, and refrigerant toxicity. Electrical considerations include supply connections, grounding, internal wiring and wiring methods, electrical spacings, motors and motor protection, uninsulated live parts, motor controllers and switching devices, air-heating components, and electrical insulating materials.

The performance section of the standard includes a rain test for determining the unit's ability to stand a beating rain without creating a shock hazard because of current leakage or insulation breakdown. Other tests include (1) leakage current limitations based on UL *Standard* C101, (2) measurement of input currents for the purpose of establishing nameplate ratings and for sizing the supply circuit for the unit, (3) temperature tests to determine whether components exceed their recognized temperature limits and/or electrical ratings (AHAM *Standard* RAC-1), and (4) pressure tests to ensure that excessive pressure does not develop in the refrigeration system.

Abnormal conditions are also considered, such as (1) failure of the condenser fan motor, which may lead to excessive pressure in the system; and (2) possible ignition of combustibles in or adjacent to the unit on air heater burnout. A static load test is also conducted on window-mounted room air conditioners to determine whether the mounting hardware can adequately support the unit. As part of normal production control, tests are conducted for refrigerant leakage, dielectric strength, and grounding continuity.

Plastic materials are receiving increased consideration in the design and fabrication of room air conditioners because of their ease in forming, inherent resistance to corrosion, and decorative qualities. When considering using plastic, the engineer should consider the tensile, flexural, and impact strength of the material; its flammability characteristics; and (concerning degradation) its resistance to water absorption and exposure to ultraviolet light, ability to operate at elevated temperatures, and thermal aging characteristics. Considering product safety, some of these factors are less important because failure of the part will not cause a hazard. However, for some parts (e.g., bulkhead, base pan, and unit enclosure) that either support components or provide structural integrity, all these factors must be considered, and a complete analysis of the material must be made to determine whether it is suitable for the application.

INSTALLATION AND SERVICE

Installation procedures vary because units can be mounted in various ways. It is important to select the mounting for each installation that best satisfies the user and complies with applicable building codes. Common mounting methods include the following:

- **Inside flush mounting.** Interior face of conditioner is approximately flush with inside wall.
- **Balance mounting.** Unit is approximately half inside and half outside window.
- **Outside flush mounting.** Outer face of unit is flush with or slightly beyond outside wall.
- **Special mounting.** Examples include casement windows, horizontal sliding windows, office windows with swinging units (or swinging windows) to allow window washing, and transoms over doorways.
- **Through-the-wall mounts or sleeves.** This mounting is used for installing window-type chassis, complete units, or consoles in walls of apartment buildings, hotels, motels, and residences.

Although very similar to window-mounted units, through-the-wall models do not have side louvers for condenser air; air comes from the outdoor end of the unit.

Room air conditioners have become more compact to minimize both loss of window light and projection inside and outside the structure. Several types of expandable mounts are now available for fast, dependable installation in single- and double-hung windows, as well as in horizontal sliding windows. Installation kits include all parts needed for structural mounting, such as gaskets, panels, and seals for weathertight assembly.

Adequate wiring and proper fuses must be provided for the service outlet. Necessary information is usually given on instruction sheets or stamped on the air conditioner near the service cord or on the serial plate. It is important to follow the manufacturer's recommendation for size and type of fuse. All units are equipped by the manufacturer with grounding plug caps on the service cord. Receptacles with grounding contacts correctly designed to fit these plug caps should be used when units are installed.

Units rated 265 or 277 V must provide for permanent electrical connection with armored cable or conduit to the chassis or chassis assembly. Manufacturers usually provide an adequate cord and plug cap in the chassis assembly to facilitate installation and service.

One type of room air conditioner is the **integral chassis** design, with the outer cabinet fastened permanently to the chassis. Most electrical components can be serviced by partially dismantling the control area without removing the unit from the installation. Another type is the **slide-out chassis** design, which allows the outer cabinet to remain in place while the chassis is removed for service.

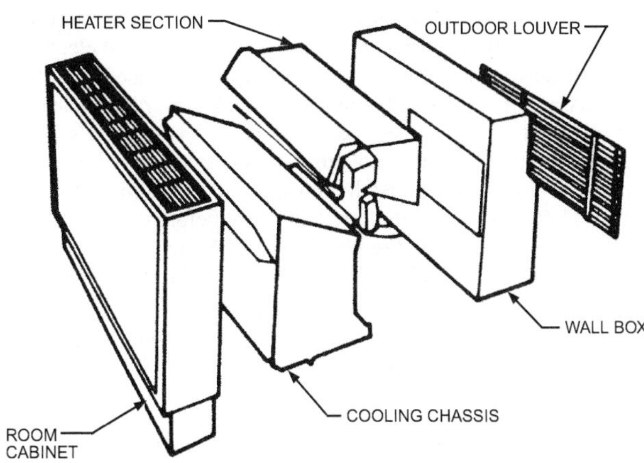

Fig. 2 Sectional Packaged Terminal Air Conditioner

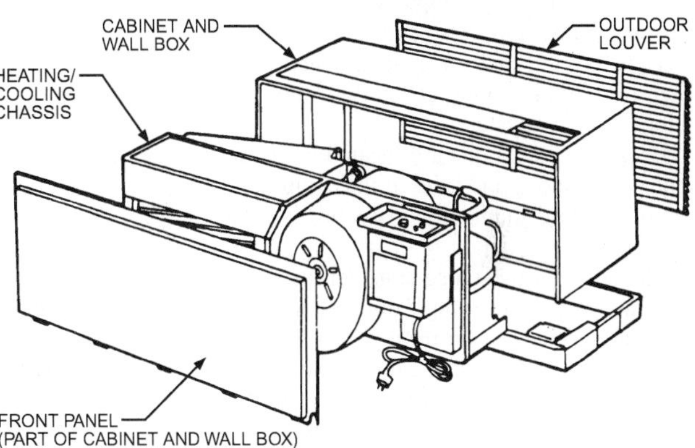

Fig. 3 Integrated Packaged Terminal Air Conditioner

PACKAGED TERMINAL AIR CONDITIONERS

The Air-Conditioning and Refrigeration Institute (ARI) defines a packaged terminal air conditioner (PTAC) as a wall sleeve and a separate unencased combination of heating and cooling assemblies intended for mounting through the wall. A PTAC includes refrigeration components, separable outdoor louvers, forced ventilation, and heating by hot water, steam, or electric resistance. PTAC units with direct-fired gas heaters are also available from some manufacturers. Cooling-only PTACs need not include heating elements. A packaged terminal heat pump (PTHP) is a heat pump version of a PTAC that provides heat with a reverse-cycle operating mode. A PTHP should provide a supplementary heat source, which can be hot water, steam, electric resistance, or another source.

PTACs are designed primarily for commercial installations to provide the total heating and cooling functions for a room or zone and are specifically for through-the-wall installation. The units are mostly used in relatively small zones on the perimeter of buildings such as hotels and motels, apartments, hospitals, nursing homes, and office buildings. In larger buildings, they may be combined with nearly any system selected for environmental control of the building core.

PTACs and PTHPs are similar in design and construction. The most apparent difference is the addition of a refrigerant-reversing valve in the PTHP. Optional components that control the heating functions of the heat pump include an outdoor thermostat to signal the need for changes in heating operating modes, and, in more complex designs, frost sensors, defrost termination devices, and base pan heaters.

SIZES AND CLASSIFICATIONS

Packaged terminal air conditioners are available in a wide range of rated cooling capacities, typically 6000 to 18,000 Btu/h, with comparable levels of heating output. Units are available as sectional types or integrated types. A sectional-type unit (Figure 2) has a separate cooling chassis; an integrated-type unit (Figure 3) has an electric or a gas heating option added to the chassis. Hot-water or steam heating options are usually part of the cabinet or wall box. Both types include the following:

- Heating elements available in hot water, steam, electric, or gas heat
- Integral or remote temperature and operating controls
- Wall sleeve or box
- Removable (or separable) outdoor louvers
- Room cabinet
- Means for controlled forced ventilation
- Means for filtering air delivered to the room
- Ductwork

PTAC assemblies are intended for use in free conditioned-air distribution, but a particular application may require minimal ductwork with a total external static resistance up to 0.1 in. of water.

GENERAL DESIGN CONSIDERATIONS

Packaged terminal air conditioners and packaged terminal heat pumps allow the HVAC designer to integrate the exposed outdoor louver or grille with the building design. Various grilles are available to blend with or accent most construction materials. Because the product becomes part of the building's facade, the architect must consider the product during the conception of the building. Wall sleeve installation is usually done by ironworkers, masons, or carpenters. All-electric units dominate the market. Recent U.S. market statistics indicate that 45% are PTHPs, 49% are PTACs with electric resistance heat, and 6% involve other forms of heating.

All the energy of all-electric versions is dispersed through the building via electrical wiring, so the electric designer and electrical contractor play a major role. Final installation is reduced to sliding in the chassis and plugging the unit into an adjacent receptacle. For these all-electric units, the traditional HVAC contractor's work involving ducting, piping, and refrigeration systems is bypassed. This results in a low-cost installation and allows installation of the PTAC/PTHP chassis to be deferred until just before occupancy.

When comparing a gas-fired PTAC to a PTAC with electric resistance heat or a PTHP, evaluate both operating and installation costs. Generally, a gas-fired PTAC is more expensive to install but less expensive to operate in heating mode. A life-cycle cost comparison is recommended (see Chapter 36 of the 2007 *ASHRAE Handbook—HVAC Applications*).

One main advantage of the PTAC/PTHP concept is that it provides excellent zoning capability. Units can be shut down or operated in a holding condition during unoccupied periods. Present equipment efficiency-rating criteria are based on full-load operation, so an efficiency comparison to other approaches may suffer.

The designer must also consider that total capacity is the sum of the peak loads of each zone rather than the peak load of the building. Therefore, total cooling capacity of the zonal system will exceed that of a central system.

Because PTAC units are located in the conditioned space, both appearance and sound level of the equipment are important considerations. Sound attenuation in ducting is not available with the free-discharge PTAC units.

The designer must also consider the added infiltration and thermal leakage load resulting from perimeter wall penetrations. These losses are accounted for during the *on*-cycle in equipment cooling ratings and PTHP heating ratings, but during the *off*-cycle or with other forms of heating, they could be significant.

Widely dispersed PTAC units also present challenges for effective condensate disposal.

Most packaged terminal equipment is designed to fit into a wall aperture approximately 42 in. wide and 16 in. high. Although unitary products can increase in size with increasing cooling capacity, PTAC/PTHP units, regardless of cooling capacity, are usually constrained to a few cabinet sizes. The exterior of the equipment must be essentially flush with the exterior wall to meet most building codes. In addition, cabinet structural requirements and the slide-in chassis reduce the available area for outdoor air inlet and relief to less than a total of 3.5 ft^2. Manufacturers' specification sheets should be consulted for more accurate and detailed information.

Outdoor air recirculation must be minimized, and attention must be given to architectural appearance. These factors may increase air-side pressure drop. With a cooling capacity range that usually spans about a 3 to 1 ratio, this makes maintaining unit efficiency at higher levels of output more difficult.

DESIGN OF PTAC/PTHP COMPONENTS

Compressor. PTAC units are designed with single-speed compressors, either reciprocating or rotary. They normally operate on a single-phase power supply and are available in 208, 230, and 265 V versions. Smaller units are available in 115 V versions. Compressors usually have electromechanical protective devices, with some of the more advanced models using electronic protective systems.

Fan Motor(s). In some PTACs, a single, double-shafted direct-drive fan motor drives both the indoor and outdoor air-moving devices. This motor usually has two speeds, which affect the equipment sound level, throw of conditioned air, cooling capacity, efficiency, sensible-to-total heat capacity ratio, and condensate disposal.

Full-featured models have two fan motors: one moves indoor air and the other brings in outdoor air. Two motors provide greater flexibility in locating components, because indoor and outdoor fans are no longer constrained to the same rotating axis. They also allow different fan speeds for the indoor and outdoor systems. In this case, the outdoor fan motor is usually single-speed, and the indoor fan motor has two or more speeds. Also, the designer has a broader selection of air-moving devices and can provide the user with a wider range of sound level and conditioned air throw options. Efficiency can be maintained at lower indoor fan speeds. When heating (other than with a heat pump), the outdoor fan motor can be switched off automatically to reduce electrical energy consumption, decrease infiltration and heat transmission losses through the PTAC unit, and prevent ice from entrapping the fan blade.

Indoor Air Mover. Airflow quantity, air-side pressure rise, available fan motor speed, and sound level requirements of the indoor air system of a PTAC indicate that a centrifugal blower wheel provides reasonable indoor air performance. In some cases, proprietary mixed-flow blowers are used. Dual-fan motor units permit the use of multiple centrifugal blower wheels or a cross-flow blower to provide more even discharge of the conditioned air.

Indoor Air Circuit. PTACs have an air filter of fiberglass, metal, or plastic foam, which removes large particles from the circulating airstream. In addition to improving indoor air quality, this filter also reduces fouling of the indoor heat exchanger. The PTAC also provides mechanical means of introducing outdoor air into the indoor airstream. This air, which may or may not be filtered, controls infiltration and pressurization of the conditioned space.

Outdoor Air Mover. Outdoor air movers may be either centrifugal blower wheels, mixed-flow blowers, or axial-flow fans.

Heat Exchangers. PTACs may use conventional plate-fin heat exchangers, which have either copper or aluminum tubes. The fins are usually aluminum, which may be coated to retard corrosion. Because PTACs are generally restricted in size, performance improvements based on increasing heat exchanger size are limited. Therefore, to improve performance or reduce costs, some manufacturers install heat exchangers with performance enhancements on the air side (lanced fins, spine fin, etc.) and/or the refrigerant side (internal finning or rifling).

Refrigerant Expansion Device. Most PTACs use a simple capillary as an expansion device. Off-rating-point performance is improved if expansion valves are used.

Condensate Disposal. Condensate forms on the indoor coil when cooling. Some PTACs require a drain to be installed to convey condensate to a disposal point. Other units spray condensate on the outdoor coil, where it is evaporated and dispersed to the outdoor ambient air. This evaporative cooling of the condenser enhances performance, but the potentially negative effects of fouling and corrosion of the outdoor heat exchanger must be considered. This problem is especially severe in a coastal installation where salt spray could mix with the condensate and, after repeated evaporation cycles, build a corrosive saltwater solution in the condensate sump.

A PTHP also produces condensate in heating mode. If the outdoor coil operates below freezing, the condensate forms frost, which is melted during defrost. This water must be disposed of in some manner. If drains are used and the heat pump operates in below-freezing weather, the drain lines must be protected from freezing. Outside drains cannot be used in this case, unless they have drain heaters. Some PTHPs introduce condensate formed during heating onto the indoor coil, which humidifies the indoor air at the expense of heating capacity. Inadequate condensate disposal can lead to overflow at the unit and potential staining of the building facade.

Controls. PTAC units usually have a built-in manual mode selector (cool, heat, fan only, and off) and a manual fan-speed selector. A thermostat adjustment is provided with set points (usually subjective, such as *high*, *normal*, and *low*). Some units offer automatic changeover from heating to cooling. Most manufacturers offer low-voltage remote heat/cool thermostats. Some units have electronic controls that provide room temperature limiting, evaporator freeze-up protection, compressor lockout in the case of actual or impending compressor malfunction, and service diagnostic aids. Advanced master controls at a central location allow an operator to override the control settings registered by the occupant. These master controls may limit operation when certain room temperature limits are exceeded, adjust thermostat set points during unoccupied periods, and turn off certain units to limit peak electrical demand.

Wall Sleeves. A wall sleeve is a required part of a PTAC unit. It becomes an integral part of the building structure and must be designed with sufficient strength to maintain its dimensional integrity after installation. It must withstand the potential corrosive effects of other building materials, such as mortar, and must endure long-term exposure to the outdoor elements.

Outdoor Grille. The outdoor grille or louvers must be compatible with the building's architecture. Most manufacturers provide options in this area. A properly designed grille prevents birds, vermin, and outdoor debris from entering; impedes entry of rain and snow; and, at the same time, provides adequate free area for the outdoor airstream to enter and exit with minimum recirculation.

Slide-In Components. The interface between the wall sleeve and slide-in component chassis allows the components to be easily inserted and later removed for service and/or replacement. In the event of serious malfunction, the slide-in component can be quickly replaced with a spare, and the repair can be made off the premises. An adequate seal at the interface is essential to exclude wind, rain, snow, and insects without jeopardizing the slide-in/slide-out feature.

Indoor Appearance. Because PTACs are located in the conditioned space, their indoor appearance must blend with the room's decor. Manufacturers provide a variety of indoor treatments, which include variations in shape, style, and materials.

HEAT PUMP OPERATION

Basic PTHP units operate in heat pump mode down to an outdoor temperature just above the point at which the outdoor heat exchanger would frost. At that point, heat pump mode is locked out, and other forms of heating are required. Some units include two-stage indoor thermostats and automatically switch from heat pump mode to an alternative heat source if space temperature drops too far below the first-stage set point. Some PTHPs use control schemes that extend heat pump operation to lower temperatures. One approach allows heat pump operation down to outdoor temperatures just above freezing. If the outdoor coil frosts, it is defrosted by shutting down the compressor and allowing the outdoor fan to continue circulating outdoor air over the coil. Another approach allows heat pump operation to even lower outdoor temperatures by using a reverse-cycle defrost sequence. In those cases, the heat pump mode is usually locked out for outdoor temperatures below 10°F.

PERFORMANCE AND SAFETY TESTING

PTACs and PTHPs may be rated in accordance with ARI *Standard* 310/380, which is equivalent to CSA International *Standard* C744.

ARI issues a *Directory of Certified Applied Air-Conditioning Products* semiannually that lists cooling capacity, efficiency, and heating capacity for each participating manufacturer's PTAC models. The listings of PTHP models also include the heating COP.

Cooling and heating capacities, as listed in the ARI directory, must be established in accordance with ASHRAE *Standard* 37 for unitary equipment, or *Standard* 16 for room air conditioners and PTACs. All standard heating ratings should be established in accordance with ASHRAE *Standard* 58.

Additionally, PTAC units should be constructed in accordance with ASHRAE *Standard* 15 and should comply with the safety requirements of UL *Standard* 484.

REFERENCES

AHAM. 2003. Room air conditioners. ANSI/AHAM *Standard* RAC-1-2003. Association of Home Appliance Manufacturers, Chicago.

ARI/CSA International. 2004. Standard for packaged terminal air-conditioners and heat pumps. ARI *Standard* 310/380-04. CSA International *Standard* C744-04. Air-Conditioning and Refrigeration Institute, Arlington, VA.

ASHRAE. 2004. Safety standard for refrigeration systems. ANSI/ASHRAE *Standard* 15-2004.

ASHRAE. 1983. Method of testing for rating room air conditioners and packaged terminal air conditioners. ANSI/ASHRAE *Standard* 16-1983 (RA 1999).

ASHRAE. 1996. Method of testing flow capacity of refrigerant capillary tubes. ANSI/ASHRAE *Standard* 28-1996 (RA 2006).

ASHRAE. 2005. Methods of testing for rating electrically driven unitary air-conditioning and heat pump equipment. ANSI/ASHRAE *Standard* 37-2005.

ASHRAE. 1986. Method of testing for rating room air conditioner and packaged terminal air conditioner heating capacity. ANSI/ASHRAE *Standard* 58-1986 (RA 1999).

CSA International. 1970. Room air conditioners. *Canadian Electrical Code* C22.2 No. 117-1970 (R2007). Canadian Standards Association, Etobicoke, ON.

EPA. 2008. *ENERGY STAR*®. http://www.energystar.gov.

IEC. 2006. Household and similar electrical appliances—Safety—Part 2-40: Particular requirements for electrical heat pumps, air conditioners, and dehumidifiers. *Standard* 60335-2-40 (2006). International Electrotechnical Commission, Geneva.

ISO. 1994. Non-ducted air conditioners and heat pumps—Testing and rating for performance. *Standard* 5151: 1994. International Organization for Standardization, Geneva.

NFPA. 2008. *National electrical code*®. NFPA *Standard* 70-2008. National Fire Protection Association, Quincy, MA.

ORNL. 1985. Room air conditioner lifetime cost considerations: Annual operating hours and efficiencies. *Report* ORNL-NSF-EP-85. Oak Ridge National Laboratory, Oak Ridge, TN.

UL. 2002. Standard for leakage current for appliances. *Standard* 101-02. Underwriters Laboratories, Northbrook, IL.

UL. 2007. Room air conditioners. ANSI/UL *Standard* 484-07. Underwriters Laboratories, Northbrook, IL.

BIBLIOGRAPHY

AHAM. Semiannually. *Directory of certified room air conditioners*. Association of Home Appliance Manufacturers, Chicago.

ARI. Semiannually. *Directory of certified applied air-conditioning products*. Air-Conditioning and Refrigeration Institute, Arlington, VA.

UL. 1998. *Electrical appliance and utilization equipment directory*. Underwriters Laboratories, Northbrook, IL.

CHAPTER 50

THERMAL STORAGE

THERMAL storage systems remove heat from or add heat to a storage medium for use at another time. Thermal energy storage for HVAC applications can involve various temperatures associated with heating or cooling. High-temperature storage is typically associated with solar energy or high-temperature heating, and cool storage with air-conditioning, refrigeration, or cryogenic-temperature processes. Energy may be charged, stored, and discharged daily, weekly, annually, or in seasonal or rapid batch process cycles. Currently, most use of thermal storage is cool storage for comfort and process cooling applications as a way to reduce the total utility bill and/or size of cooling equipment, and much of the discussion in this chapter pertains specifically to cool storage. The *Design Guide for Cool Thermal Storage* (Dorgan and Elleson 1993) covers cool storage issues and design parameters in detail.

A properly designed and installed thermal storage system can

- Reduce operating or initial costs
- Reduce size of electric service and cooling or heating equipment
- Increase operating flexibility
- Provide back-up capacity
- Extend the capacity of an existing system

Benefits are discussed further in the Benefits of Thermal Storage section.

Thermal storage may be a particularly attractive approach to meeting heating or cooling loads if one or more of the following conditions apply:

- Loads are of short duration
- Loads occur infrequently
- Loads are cyclical
- Loads are not coincident with energy source availability
- Energy costs are time-dependent (e.g., time-of-use energy rates)
- Charges for peak power demand are high
- Utility rebates, tax credits, or other economic incentives are provided for using load-shifting equipment
- Energy supply is limited, thus limiting or preventing the use of full-size nonstorage systems
- Facility expansion is planned, and the existing heating or cooling equipment is insufficient to meet the new peak load but has spare nonpeak capacity
- Interruption in cooling water cannot be tolerated by a mission-critical operation

Terminology

Charging. Storing cooling capacity by removing heat from a cool storage device, or storing heating capacity by adding heat to a heat storage device.

Chiller priority. Control strategy for partial storage systems that uses the chiller to directly meet as much of the load as possible, normally by operating at full capacity most of the time. Thermal storage is used to supplement chiller operation only when the load exceeds the chiller capacity.

Cool storage. As used in this chapter, storage of cooling capacity in a storage medium at temperatures below the nominal temperature of the space or process.

Demand limiting. A partial storage operating strategy that limits capacity of refrigeration equipment during the on-peak period. Refrigeration equipment capacity may be limited based on its cooling capacity, its electric demand, or the facility demand.

Design load profile. Calculated or measured hour-by-hour cooling loads over a complete cooling cycle that are considered to be the desired total cooling load that must be met by mechanical refrigeration and capacity from a cool thermal storage system.

Design operating profile. Equipment hour-by-hour operation, including mechanical refrigeration equipment, the thermal storage system's charge or discharge rate, and temperatures over the entire cooling system operating period.

Discharging. Using stored cooling capacity by adding thermal energy to a cool storage device or removing thermal energy from a heat storage device.

Encapsulated storage. A latent storage technology that consists of plastic containers of water or other phase-change material that are alternately frozen and melted by the influence of glycol or other secondary coolant medium in which they are immersed.

Full storage. A cool storage sizing strategy that meets the entire cooling load with discharge from the thermal storage system for some period, typically hours when loads are high and electric power is expensive.

Fully charged condition. State of a cool thermal storage system at which, according to design, no more heat is to be removed from the storage device. This state is generally reached when the control system stops the charge cycle as part of its normal control sequence, or when the maximum allowable charging period has elapsed.

Fully discharged condition. State of a cool thermal storage system at which no more usable cooling capacity can be delivered from the storage device.

Heat storage. As used in this chapter, storage of thermal energy at temperatures above the nominal temperature of the space or process.

Ice harvester. Machine that cyclically forms a layer of ice on a smooth cooling surface, using refrigerant inside the heat exchanger, then delivers it to a storage container by heating the surface of the cooling plate, normally by reversing the refrigeration process and delivering hot gases inside the heat exchanger.

Ice-on-coil (ice-on-pipe). Ice storage technology that forms and stores ice on the outside of tubes or pipes submerged in an insulated water tank.

Ice-on-coil, external melt. Ice storage technology in which tubes or pipes (coil) are immersed in water and ice is formed on the outside of the tubes or pipes by circulating colder secondary medium or refrigerant inside the tubing or pipes, and is melted externally by circulating unfrozen water outside the tubes or pipes to the load.

The preparation of this chapter is assigned to TC 6.9, Thermal Storage.

Ice-on-coil, internal melt. Ice storage technology in which tubes or pipes (coil) are immersed in water and ice is formed on the outside of the tubes or pipes by circulating colder secondary medium or refrigerant inside the tubing or pipes, and is melted internally by circulating the same secondary coolant or refrigerant to the load.

Latent energy storage (latent heat storage). A thermal storage technology in which energy is stored within a medium, normally associated with a phase change (usually between solid and liquid states), for use in cooling or heating the secondary liquid being circulated through the system.

Load leveling. A partial storage sizing strategy that minimizes storage equipment size and storage capacity. The system operates with refrigeration equipment running at full capacity for 24 h to meet the normal cooling minimum load profile and, when load is less than the chiller output, excess cooling is stored. When load exceeds chiller capacity, the additional cooling requirement is obtained from the thermal storage.

Load profile. Compilation of instantaneous thermal loads over a period of time, normally 24 h.

Mass storage. Storage of energy, in the form of sensible heat, in building materials, interior equipment, and furnishings.

Maximum usable cooling supply temperature. Maximum fluid supply temperature at which the cooling load can be met. This is generally determined by the requirements of the air-side distribution system or the process.

Maximum usable discharge temperature. Highest temperature at which beneficial cooling can be obtained from the thermal storage device.

Nominal chiller capacity. (1) Chiller capacity at standard Air-Conditioning and Refrigeration Institute (ARI) rating conditions. (2) Chiller capacity at a given operating condition selected for the purpose of quick chiller sizing selections.

Partial storage. A cool storage sizing strategy in which only a portion of the on-peak cooling load is met from thermal storage, with the rest being met by operating the chilling equipment.

Phase-change material (PCM). A substance that undergoes a change of state, normally from solid to liquid or liquid to solid, while absorbing or rejecting thermal energy at a constant temperature.

Pulldown load. Unmet cooling or heating load that accumulates during a period when a cooling or heating system has not operated, or operated in a thermostat setback mode, and which must be met on system start-up before comfort conditions can be achieved. Maximum pulldown load generally occurs on a Monday morning or following an extended shutdown.

Sensible energy storage (sensible heat storage). A heating or cooling thermal storage technology in which all energy stored is in the form of a measurable temperature difference between the hot or chilled water circulating through the system and the storage medium.

Storage cycle. A period in which a complete charge and discharge of a thermal storage device has occurred, beginning and ending at the same state.

Storage inventory. Amount of usable heating or cooling capacity remaining in a thermal storage device.

Storage priority. A control strategy that uses stored cooling to meet as much of the load as possible. Chillers operate only if the load exceeds the storage system's available cooling capacity.

Stratified chilled-water storage. A method of sensible cool thermal energy storage that achieves and maintains an acceptable separation between warm (discharged) and cool (charged) water by forming a thermocline by density differences alone, and not by mechanical separation.

Thermal storage capacity. A value indicating the maximum amount of cooling (or heating) that can be achieved by the stored medium in the thermal storage device.

Discharge capacity. The maximum rate at which cooling can be supplied from a cool storage device.

Nominal storage capacity. A theoretical capacity of the thermal storage device. In many cases, this may be greater than the usable storage capacity. This measure should not be used to compare usable capacities of alternative storage systems.

System capacity. Maximum amount of cooling that can be supplied by the entire cooling system, which may include chillers and thermal storage.

Usable storage capacity. Total amount of beneficial cooling able to be discharged from a thermal storage device. (This may be less than the nominal storage capacity because the distribution header piping may not allow discharging the entire cooling capacity of the thermal storage device.)

Thermal storage device. A container plus all its contents used for storing heating or cooling energy. The heat transfer fluid and accessories, such as heat exchangers, agitators, circulating pumps, flow-switching devices, valves, and baffles that are integral with the container, are considered a part of the thermal storage device.

Thermocline. Thermal layer of water in a chilled-water thermal storage tank, separating warmer water at the top and cooler water at the bottom. The depth of this layer depends on the effectiveness and efficiency of the upper and lower diffusers, which are designed to supply and discharge water with minimal mixing. The typical thermocline is 18 to 24 in., and rises and falls when charging and discharging the storage tank.

Total cooling load. Integrated thermal load that must be met by the cooling plant over a given period of time.

Usable cooling. Amount of energy that can actually be delivered to the system or process to meet cooling requirements. Normally, this is the total value of the energy retrieved by supplying the medium at or below the maximum beneficial cooling temperature.

Classification of Systems

Thermal storage systems can be classified according to the type of thermal storage medium, whether they store primarily sensible or latent energy, or the way the storage medium is used. Cool storage media include chilled water, aqueous or nonaqueous fluids, ice, and phase-change materials. Heat storage media include water, brick, stone, and ceramic materials. These media differ in their heat storage capacities, the temperatures at which energy is stored, and physical requirements of storing energy in the media. Types of cool storage systems include chilled-water storage, chilled-fluid storage, ice harvesting, internal- and external-melt ice-on-coil, encapsulated ice, ice slurry, and phase-change material systems. Types of heat storage systems include brick storage heaters, water storage heaters, and radiant floor heating systems, as well as solar space and water heating and thermally charged water storage tanks.

Storage Media

A wide range of materials can be used as thermal storage media. Materials used for thermal storage should be

- Commonly available
- Low cost
- Environmentally benign
- Nonflammable
- Nonexplosive
- Nontoxic
- Compatible with common HVAC equipment construction materials
- Noncorrosive
- Inert

An ideal thermal storage medium should also have

- Well-documented physical properties
- High density

- High specific heat (for sensible heat storage)
- High heat of fusion (for latent heat storage)
- Good heat transfer characteristics
- Stable properties that do not change over many thermal cycles

Common storage media for sensible energy storage include water, fluid, soil, rock, brick, ceramics, concrete, and various portions of the building structure (or process fluid) being heated or cooled. In HVAC applications, such as air conditioning, space heating, and water heating, water is often the chosen sensible storage medium because it provides many of these desirable characteristics when kept between its freezing and boiling points. For high-temperature energy storage, the storage medium is often rock, brick, or ceramic materials for residential or small commercial applications; oil, oil/rock combinations, or molten salt are often used for large industrial or solar energy power plant applications. Using the building structure itself as passive thermal storage offers advantages under some circumstances (Morris et al. 1994).

Common storage media for latent energy storage include ice, aqueous brine/ice solutions, and other PCMs such as hydrated salts and polymers. Carbon dioxide and paraffin waxes are among the alternative storage media used for latent energy storage at various temperatures. For air-conditioning applications, ice is the most common latent storage medium, because it provides many of the previously listed desirable characteristics.

Basic Thermal Storage Concepts

The fundamental characteristic of thermal storage systems is that they separate the time(s) of generation of heating or cooling from the time(s) of its use. This separation allows thermal storage systems to generate heating or cooling during periods when conditions are most favorable (e.g., the primary energy source is more available or less expensive), which can be independent of the instantaneous thermal load.

A thermal storage system can meet the same total heating or cooling load as a nonstorage system over a given period of time with smaller primary equipment. The total capacity distributed over the period is matched more closely to the total load encountered in the same period. The reduced size and cost of heating or cooling equipment can partially or completely offset the cost of the storage equipment.

Benefits of Thermal Storage

Properly designed, installed, and operated thermal storage systems offer the following benefits for heating and cooling systems.

Energy Cost Savings. The primary benefit of thermal storage is its ability to substantially reduce total operating costs, particularly for systems using electricity as the primary energy source. Thermal storage systems reduce the demand for expensive on-peak electric power, substituting less expensive nighttime power to do the same job. In addition, in many cases thermal energy storage systems actually consume less energy to deliver the same amount of cooling, primarily because of more efficient use of cooling equipment and the constant temperature of the delivered medium.

Reduced Equipment Size. Heating or cooling equipment can be sized to meet the average load rather than the peak load, and the thermal storage system can be sized to pick up the difference in the load.

Capital Cost Savings. The capital savings from downsizing cooling equipment can more than offset the added cost of the storage system (Andrepont 2005). Cool storage integrated with low-temperature air and water distribution systems can also provide an initial cost savings by using smaller chillers, pumps, piping, ducts, and fans. Smaller cooling equipment often allows designers to downsize the transformers and electrical distribution systems that supply power for cooling. This economic benefit can be significant, both in new construction, and in existing facilities where electrical systems are at or near their full capacity. In some areas, government or utility company incentives are available that can further reduce capital costs.

Energy Savings. Although thermal storage systems are generally designed primarily to shift energy use rather than to conserve energy, they can fill both roles. Cool storage systems allow chillers to operate more at night, when lower condensing temperatures improve equipment efficiency. In addition, thermal storage allows operation of equipment at full load, avoiding inefficient part-load performance. Documented examples include chilled-water storage installations that reduce annual energy consumption on a ton-hour basis for air conditioning by up to 12% (Bahnfleth and Joyce 1994; Fiorino 1994). Heat recovery from chiller condensers can also reduce, or even eliminate, the need for heating equipment and associated energy use by combining cooling thermal storage with heating thermal storage (Goss et al. 1996).

Improved HVAC Operation. Thermal storage allows the thermal load profile to be decoupled from operation of the heating or cooling equipment. This decoupling enables plants to use the optimum combination of primary equipment and storage at any given time, providing increased flexibility, reliability, and efficiency in all seasons.

Back-Up Capacity. Cool storage can substantially reduce the installed cost for back-up or emergency cooling systems by reducing the need for redundant electrical and mechanical equipment. Back-up cooling for critical areas (e.g., computer rooms, hospital operating rooms) can be immediately available from the storage reservoir.

Extending Existing Systems' Capacity. The apparent capacity of existing systems can often be increased by installing cool storage at less cost than adding conventional nonstorage equipment. A chilled-water storage tank adds cooling capacity to an existing system and may avoid the high expense of installing new chillers. With this form of thermal storage, existing chillers can now operate during normally off hours to generate the additional cooling needed. The cooling capability of existing ductwork and piping can also be increased by using cool storage. Supplying chilled water and air at lower temperatures allows existing distribution systems to meet substantially higher loads and reduces pump or fan energy consumption.

Other Benefits. Thermal storage can bring about other beneficial synergies. As already noted, cool storage can be integrated with cold-air distribution. A chilled-water storage tank can serve double duty by providing a fire protection water supply at a much lower cost than installing separate systems (Holness 1992; Hussain and Peters 1992; Meckler 1992). Some cool storage systems can be configured for charging with free cooling. Thermal storage can also be used to recover waste energy from base-loaded steam plants by storing chilled water produced by an absorption chiller for later use on demand.

Design Considerations

Thermal storage systems must be designed to meet the total integrated load as well as the peak hourly load. To properly size a thermal storage system, a designer must calculate an accurate load profile and analyze the thermal performance of the storage equipment over the entire storage cycle. Proper sizing is essential. An undersized thermal storage system is limited in its ability to recover when the load exceeds its capacity. On the other hand, excessive oversizing quickly diminishes the economic benefits of thermal storage. Load calculations and system sizing are discussed in Dorgan and Elleson (1993).

Thermal performance of a thermal storage device varies, depending on the current inventory of stored energy and rate of discharge. Therefore, the total capacity of a given thermal storage device depends on the load profile to which it is subjected. There is no standard method for rating cool storage capacity, so performance specifications must detail the required thermal performance for each hour of the storage cycle. Thermal performance specifications for cool storage systems are discussed in ASHRAE *Guideline* 4.

A thermal storage system must be controlled according to two separate time schedules. Generation of heating or cooling is controlled on a different schedule from the distribution and utilization.

Thermal storage offers the flexibility to supply heating or cooling from storage, from primary equipment, or from both. With this flexibility comes the need to define how the system will be controlled at any given time. The intended strategy for operation and control must be defined as part of the design procedure, as described in Dorgan and Elleson (1993).

SENSIBLE THERMAL STORAGE TECHNOLOGY

Sensible Energy Storage

Water is well suited for both hot and cold sensible energy storage applications and is the most common sensible storage medium, in part because it has the highest specific heat (1 Btu/lb·°F) of all common materials. ASHRAE *Standard* 94.3 covers procedures for measuring thermal performance of sensible heat storage. Tanks are available in many shapes; however, vertical cylinders are the most common. Tanks can be located above ground, partially buried, or completely buried. They can also be incorporated into the building structure. For economic reasons, they usually operate at atmospheric pressure and may have clear-span dome roofs, column-supported shallow cone roofs, or column-supported flat roofs. Sensible thermal storage vessels must separate the cooler and warmer volumes of storage medium. This section focuses on chilled-water thermal energy storage because it is the most common system type. However, similar techniques apply to hot-water sensible energy storage for heating systems and to aqueous and nonaqueous stratification fluids for cool storage.

Temperature Range and Storage Size

The cooling capacity of a chilled-water storage vessel is proportional to the volume of water stored and the temperature differential (Δt) between the stored cool water and returning warm water. There is a direct relationship between Δt and the size of sensible storage system components. For economical storage, therefore, the cooling loads served by the system should provide as large a Δt as practical. Many chilled-water systems are designed to operate with a Δt between supply and return of 12 to 16°F. Designing for an achievable 20 to 24°F Δt has the potential to decrease storage vessel size by as much as 50% relative to this typical range. The cost of the extra coil surface required to provide this larger Δt can be more than offset by cost reductions of a smaller tank, reduced pipe and insulation size, and significantly reduced pumping energy. Storage is likely to be uneconomical if the Δt is less than about 10°F because the tank must be so large (Caldwell and Bahnfleth 1997). Typical design and operating conditions result in a storage density of approximately 100 gal/ton-h. Maintaining a Δt as large as possible is very important in achieving the intended beneficial performance of the system. The total performance depends on the behavior of connected loads and needs to be addressed as a system issue.

The initial cost of chilled-water cool storage benefits from a dramatic economy of scale; that is, large installations can be much less expensive per amount of discharge than equivalent nonstorage chilled-water plants (Andrepont 1992, 2005; Andrepont and Kohlenberg 2005; Andrepont and Rice 2002).

Techniques for Thermal Separation in Sensible Storage Devices

The following methods have been applied in chilled-water storage. Thermal stratification has become the dominant method because of its simplicity, reliability, efficiency, and low cost. The other methods are described only for reference.

Thermal Stratification. In stratified cool storage, warmer, less dense return water floats on top of denser chilled water. Cool water from storage is supplied and withdrawn at low velocity, in essentially horizontal flow, so buoyancy forces dominate inertial effects. Pure water is most dense at 39.2°F; therefore, colder water introduced into the bottom of a stratified tank tends to mix to this temperature with warmer water in the tank. However, low-temperature fluids (LTFs), typically water with various admixtures, can be used to achieve lower delivery temperatures, larger temperature differences, and thus smaller storage volumes per ton-hour (Andrepont 2000; Borer and Schwartz 2005).

When the stratified storage tank is charged, chilled supply water, typically between 38 and 44°F, enters through the diffuser at the bottom of the tank (Figure 1), and the warmer return water exits to the chiller through the diffuser at the top of the tank. Typically, incoming water mixes with water in the tank to form a 1 to 2 ft thick thermocline, which is a region with sharp vertical temperature and density gradients (Figure 2). The thermocline minimizes further mixing of the water above it with that below it. The thermocline rises as charging continues and subsequently falls during discharging. It may thicken somewhat during charging and discharging because of heat conduction through the water and heat transfer to and from the walls of the tank. The storage tank may have any cross section, but the walls are usually vertical. Horizontal cylindrical tanks are generally not good candidates for stratified storage, because of the ratio of the volume of water within and outside the thermocline.

Flexible Diaphragm. A flexible barrier, normally made from a rubber composition material, installed horizontally inside the water storage tank, separates the colder and warmer water. The diaphragm must be sized to float from the top of the tank to the bottom as the tank is fully charged to fully discharged. Potential for the diaphragm to hang up during travel through the tank and lost volume at the top and bottom of its travel reduce the effectiveness of this method.

Multiple Compartments or Multiple Tanks. Multiple compartments in a single tank or a series of two or more tanks can also be used for chilled-water storage. Water may flow through compartments in series, with opposite flow directions during charging and discharging. In empty tank systems, there is one more compartment than required to contain the full stored volume. During charging and discharging, one of the full compartments is emptied while the empty compartment is filled. This is done sequentially until charging or discharging is complete. Consequently, warmer and cooler water never occupy a tank at the same time. This procedure allows using horizontal or vertical tanks of any configuration and size; drawbacks are the complexity of controls and the need to have an empty tank.

Labyrinth Tank. This storage tank has both horizontal and vertical traverses. The design commonly takes the form of successive

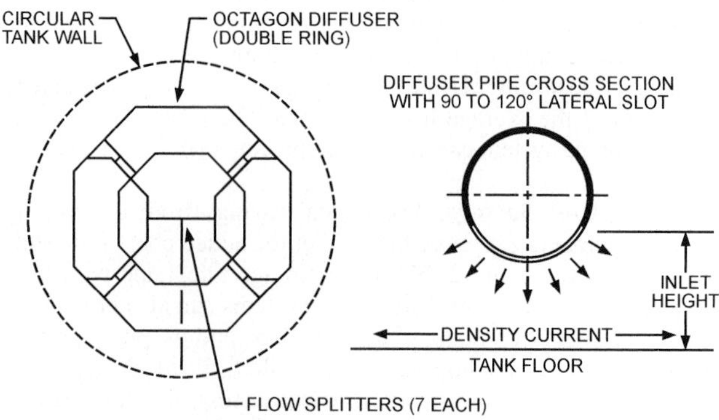

Fig. 1 Typical Two-Ring Octagonal Slotted Pipe Diffuser

cubicles with high and low ports, requiring the water to move from a high to a low port as the storage tank is being charged and the opposite as the tank is being discharged. Strings of cylindrical tanks and tanks with successive vertical weirs may also be used for this technique.

Performance of Chilled-Water Storage Systems

A perfect sensible storage device would deliver water at the same temperature at which it was initially stored. This would also require that water returning to storage neither mix with nor exchange heat with stored water in the tank or the surroundings of the tank. In practice, however, all three types of heat exchange occur.

Typical temperature profiles of water entering and leaving a storage tank are shown in Figure 3. Tran et al. (1989) tested several large chilled-water storage systems and developed the **figure of merit (FOM)**, which is used as a measure of the amount of cooling available from the storage tank.

$$\text{FOM} = \frac{\text{Area between } A \text{ and } C}{\text{Area between } A \text{ and } D} \times 100 \quad (1)$$

Well-designed storage tanks have figures of merit of 90% or higher for daily complete charge/discharge cycles.

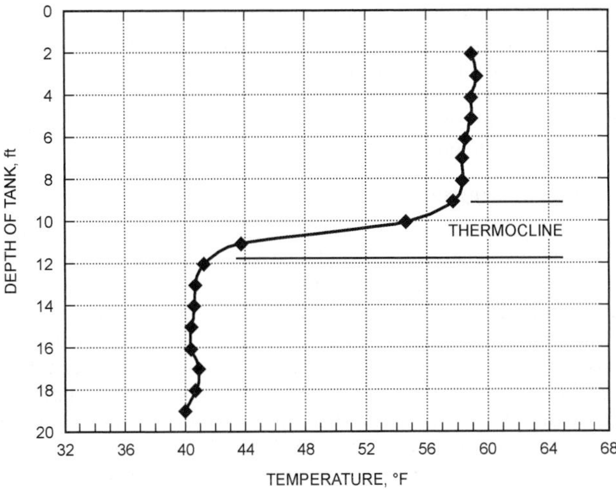

Fig. 2 Typical Temperature Stratification Profile in Storage Tank

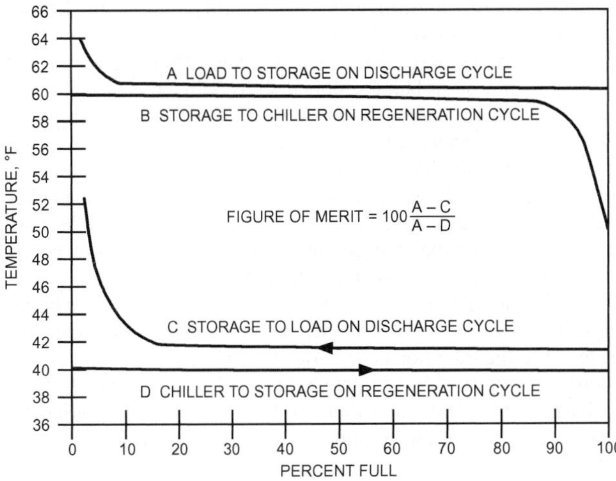

Fig. 3 Typical Chilled-Water Storage Profiles

Design of Stratification Diffusers

Diffusers must be designed and constructed to produce and maintain stratification at the maximum expected flow rate through storage. Two main styles are in widespread use today: the octagonal pipe diffuser (see Figure 1) and the radial disk diffuser (Figure 4). Inlet and outlet streams must be kept at sufficiently low velocities, so buoyancy predominates over inertia to produce a gravity current (density current) across the bottom or top of the tank.

The inlet Froude number Fr is defined as

$$\text{Fr} = \frac{Q}{\sqrt{gh^3(\Delta\rho/\rho)}} \quad (2)$$

where

Q = volume flow rate per unit length of diffuser, $\text{ft}^3/\text{s}\cdot\text{ft}$
g = gravitational acceleration, ft/s^2
h = inlet opening height, ft
ρ = inlet water density, lb/ft^3
$\Delta\rho$ = difference in density between stored water and incoming or outflowing water, lb/ft^3

The density difference $\Delta\rho$ can be obtained from Table 1. The inlet Reynolds number Re is defined as follows:

$$\text{Re} = Q/\nu \quad (3)$$

where ν = kinematic viscosity, ft^2/s.

Designers typically select values for Re, which determines the diffuser length, then select a value for the diffuser height to create an inlet Froude number of 1.0 or less. However, values up to 2.0 have been successfully applied (Yoo et al. 1986). Some experimental evidence indicates that the intensity of mixing near the inlet diffuser is influenced by the inlet Reynolds number [Equation (3)].

Wildin and Truman (1989), observing results from a 15 ft deep, 20 ft diameter vertical cylindrical tank, found that reduction of the inlet Reynolds number from 850 (using a radial disk diffuser) to 240 (using a diffuser comprised of pipes in an octagonal array) reduced mixing to negligible proportions. This is consistent with subsequent results obtained by Wildin (1991, 1996) in a 3 ft deep scale model

Table 1 Chilled-Water Density Table

°F	lb/ft³	°F	lb/ft³	°F	lb/ft³
32	62.419	44	62.424	58	62.378
34	62.424	46	62.421	60	62.368
36	62.426	48	62.417	62	62.357
38	62.427	50	62.411	64	62.344
39	62.428	52	62.404	66	62.331
40	62.427	54	62.396	68	62.316
42	62.426	56	62.387		

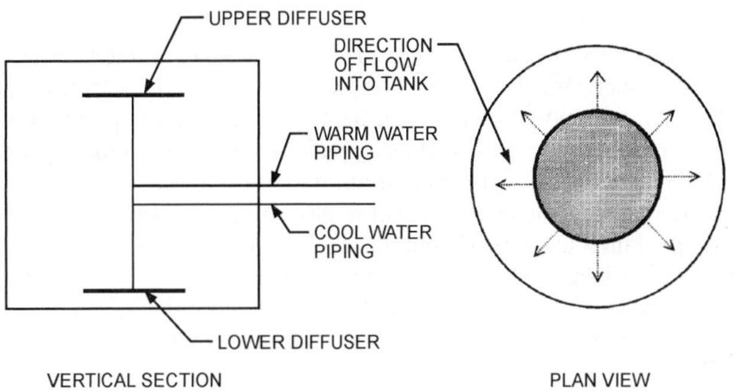

Fig. 4 Radial Disk Diffuser

tank, which indicated negligible mixing at Reynolds numbers below approximately 450, with the best performance achieved during testing at a Reynolds number of 250.

Bahnfleth and Joyce (1994), Musser and Bahnfleth (1998, 1999), and Stewart (2001) documented successful operation of tanks with water depths greater than 45 ft for design inlet Reynolds numbers as high as 10,000, although thermal performance of these systems improved at lower inlet Reynolds numbers.

A parametric study of radial diffusers by Musser and Bahnfleth (2001) found that using Froude numbers less than 1.0 significantly improved performance. Their work, based on field measurements and simulation, confirmed that the Froude number is a parameter of first-order significance for radial diffuser inlet thermal performance, but did not indicate strong effects of varying the Re. They found that parameters relating diffuser and tank dimensions (i.e., ratio of diffuser diameter to diffuser height and ratio of diffuser diameter to tank diameter) had a stronger effect on performance and could explain some of the behavior attributed to Re by earlier research. A similar study of octagonal diffusers by Bahnfleth et al. (2003a) also indicated that diffuser/tank parameters not previously used in design could be of greater significance than Re. They also concluded that diffuser/tank interaction parameters could be important factors in octagonal diffuser performance, and that radial and pipe diffusers are not well described by a single design method.

Because a chilled-water system may experience severe pressure spikes (water hammer), such as during rapid closing of control or isolation valves, the structural design of diffusers should consider this potential event. In addition, the diffusers themselves can also be subjected to buoyancy effects in the rare instance where air entrained in the chilled-water system becomes trapped in the piping and flows to the diffuser. Design considerations should be made to allow any trapped air to escape from the diffuser piping with little or no effect on diffuser performance.

Storage Tank Insulation

Exposed tank surfaces should be well insulated to help maintain the temperature differential in the tank. Insulation is especially important for smaller storage tanks because the ratio of surface area to stored volume is relatively high. Heat transfer between the stored water and tank contact surfaces (including divider walls) is a primary source of capacity loss. In addition to heat losses or gains by conduction through the floor and wall, heat flows vertically along the tank walls from the warmer to the cooler region. Exterior insulation of the tank walls does not totally inhibit this heat transfer.

The contents of chilled-water storage tanks are typically colder than the ambient dew-point temperature, so it is important that the insulation system use a high-integrity exterior vapor barrier to minimize ingress of moisture and condensation into the insulation system.

Other Factors

The cost of chemicals for water treatment may be significant, especially if the tank is filled more than once during its life. A filter system helps keep stored water clean. Exposure of stored water to the atmosphere may require the occasional addition of biocides. Although tanks should be designed to prohibit leakage, the designer should understand the potential effect of leakage on the selection of chemical water treatment. Water treatment based on requirements for a truly open system such as a condenser water system may be excessive, because water in a stratified tank is not aerated and is exposed to the atmosphere only through a small vent at the top of the storage tank. Owners should be careful to consider all factors involved when implementing a water treatment plan. The water treatment should be flexible based on the quantity of makeup water added to the system. See the Water Treatment section for more information.

The storage circulating pumps should be installed below the minimum operating water level of the lowest tank to ensure a continuously pressurized flooded suction. The required net positive suction head (NPSH) must be maintained to avoid cavitation of the pumps.

Chilled-Water Storage Tanks

Chilled-water storage systems are typically of large capacity and volume. As a result, many stratified chilled-water storage systems are located outdoors (e.g., in industrial plants or suburban campus locations). A tall tank is desirable for stratification, but a buried tank may be required for architectural, aesthetic, or zoning reasons. Tanks are typically constructed of steel or prestressed concrete to specifications used for municipal water storage tanks. Prestressed concrete tanks can be partially or completely buried below grade. For tanks that are completely buried, the tank roof can be free-standing dome type construction, or column-supported for heavy roof loads such as parking lots, tennis courts, or parks.

Low-Temperature Fluid Sensible Energy Storage

Low-temperature fluids (LTFs) may be used as a sensible cool storage medium instead of water. Using an LTF can allow sensible energy thermal storage at temperatures below 39.2°F, the temperature at which the maximum density of plain water occurs. Thus, LTFs can allow lower-temperature applications of sensible energy storage, such as for low-temperature air conditioning and some food processing applications. LTFs can be either aqueous solutions containing a chemical additive, or nonaqueous chemicals. A number of potential LTFs have been identified by Stewart (2000). One LTF has been in continuous commercial service in a stratified thermal storage tank in a very large district cooling system since 1994, and has exhibited good corrosion inhibition and microbiological control properties (Andrepont 2000, 2006).

Storage in Aquifers

Aquifers are underground, water-yielding geological formations, either unconsolidated (gravel and sand) or consolidated (rocks), that can be used to store large quantities of thermal energy. In general, the natural aquifer water temperature is slightly warmer than the local mean annual air temperature. Aquifer thermal energy storage has been used for process cooling, space cooling, space heating, and ventilation air preheating (Jenne 1992). Aquifers can be used as heat pump sinks or sources and to store energy from ambient winter air, waste heat, and renewable sources. For more information, see Chapter 32 of the 2007 *ASHRAE Handbook—HVAC Applications*.

The thickness and porosity of the aquifer determine the storage volume. Two separate wells are normally used to charge and discharge aquifer storage. A well pair may be pumped either (1) constantly in one direction or (2) alternately from one well to the other, especially when both heating and cooling are provided. Backflushing is recommended to maintain good well efficiency.

The length of storage depends on the local climate and the type of building or process being supplied with cooling or heating. Aquifer thermal energy storage may be used on a short- or long-term basis, as the sole source of energy or as partial storage, and at a temperature useful for direct application or needing augmentation by boilers or chillers. It may also be used in combination with a dehumidification system, such as desiccant cooling. The cost effectiveness of aquifer thermal energy storage is based on the avoidance of equipment capital cost and on lower operating cost.

Aquifer thermal energy storage may be incorporated into a building system in a variety of ways, depending on the other components present and the intentions of the designer (CSA 1993; Hall 1993). Control is simplified by separate hot and cold wells that operate on the basis that the last water in is the first water out. This principle ensures that the hottest or coldest water is always available when needed.

Seasonal storage has a large savings potential and began as an environmentally sensitive improvement on the large-scale mining of groundwater (Hall 1993). The current method reinjects all pumped water to attempt annual thermal balancing (Morofsky 1994; Public Works Canada 1991, 1992; Snijders 1992). Rock caverns have also been used successfully in a manner similar to aquifers to store energy. For example, Oulu, Finland, stores hot water in a cavern for district heating.

Latent Cool Storage Technology

Latent cool storage systems achieve most of their capacity from the latent heat of fusion of a phase-change material, although sensible heat contributes significantly to many designs. The high energy density of latent storage systems allows compact installations and makes factory-manufactured components and systems practical. Latent storage devices are available in a wide variety of distinct technologies that sometimes defy accurate classification. Current categories include internal-melt ice-on-coil, external-melt ice-on-coil, encapsulated, ice harvester, and ice slurry technologies.

A challenge common to all latent energy storage methods is to find an efficient and economical means of achieving the heat transfer necessary to alternately freeze and melt the storage medium. Various methods have been developed to limit or deal with the heat transfer approach temperatures associated with freezing and melting; however, leaving fluid temperatures (from storage during melting) must be higher than the freezing point, and entering fluid temperatures (to storage during freezing) must be lower than the freezing point. Ice storage can provide leaving temperatures well below those normally used for comfort and nonstorage air-conditioning applications. However, entering temperatures are also much lower than normal. Some PCM storage systems can be charged using temperatures near those for comfort cooling, but they produce warmer leaving temperatures.

Water as Phase-Change Thermal Storage Medium

An overwhelming majority of latent storage systems use plain water as the storage medium. A number of other materials have been developed over the years and are sometimes used in unique applications. However, water's combination of reliability, stability, insignificant cost, high latent heat capacity of 144 Btu/lb, high specific heat, high density, safety, and appropriate fusion temperature have proven desirable for many typical HVAC cooling applications.

Water has a very stable melting point of 32°F at sea level. Under some conditions, supercooling can depress the initial freezing point by 2 to 5°F, but impurities found in potable water, random disturbances, minor vibrations, surface interactions, or intentional agitation are generally sufficient to initiate nucleation of ice crystals. Supercooling most often occurs during an initial reduction of the liquid water temperature below 32°F. Residual ice reduces or eliminates the supercooling tendency for subsequent ice-making periods. The amount of supercooling, if experienced at all, is typically 2 or 3°F, and the phase-change temperature rapidly returns to 32°F immediately after this initial nucleation period. With relatively pure water that is isolated from any external influence, a proprietary agent may be used to induce nucleation.

Water slightly decreases in density below 39°F and expands by about 9% on freezing. This expansion is often used as an indicator of ice inventory where the ice remains completely submerged, such as with the ice-on-coil systems or ice-harvesting systems. If the ice is allowed to float in liquid water, the liquid level will remain constant as ice is frozen or melted. Ice slightly increases in density below the freezing temperature, but this is of little interest in storage systems, which all operate close to the phase-change temperature, and the variation is minor.

Many ice storage technologies use secondary coolants for heat transfer. This allows use of standard chillers, including centrifugals, for ice building and simplifies the application of the technology to typical chilled-water cooling systems. Common secondary coolants have well-documented properties and provide reliable freeze and corrosion protection when properly maintained and installed. The proper concentration for the coolant is selected to provide a freeze point well below the lowest expected coolant temperature. This temperature is determined by the heat transfer characteristics of the storage device, flow rates, and chiller capacity.

Internal Melt Ice-On-Coil

Internal-melt ice-on-coil thermal storage devices circulate a secondary coolant or refrigerant through a tubular heat exchanger that is submersed in a cylindrical or rectangular tank of water. The phase-change water remains in the tank; both charging (ice making) and discharging (ice melting) are accomplished by circulating the secondary heat transfer fluid. Ice forms on the heat exchanger tubes during charging, and melts from the inside out (hence the term internal-melt) during discharging.

Internal-melt storage devices may be provided as complete, modular, factory-assembled units, or may consist of field-erected tanks equipped with prefabricated heat exchangers. Most current designs use secondary coolants (e.g., 25% ethylene glycol/75% water) as the heat transfer fluid.

Storage devices using refrigerant as the heat transfer fluid have also been developed for applications typically served by direct expansion equipment, such as residential and small commercial installations. However, these devices have not achieved wide commercial acceptance to date.

Storage device manufacturers typically provide performance data, including at least the average and final charging temperatures as a function of charging rate and flow. The average temperature is useful for selecting chiller capacity and estimating energy consumption. The lowest anticipated temperature is needed to establish chiller protection settings, specify coolant freeze protection requirements, and verify the capabilities of chiller selections, which is particularly important for centrifugal chillers. Glycols are currently the most commonly used secondary coolants, and design practices are conventional, with no exceptional material limitations other than the exclusion of galvanized steel for surfaces in contact with the heat transfer fluid. Proper application of some secondary coolants is addressed in Chapter 4 of the 2006 *ASHRAE Handbook—Refrigeration*.

The tubular heat exchanger is typically constructed of polyethylene or polypropylene plastic, or galvanized steel. Refrigerant-based systems may use copper. The heat exchanger typically occupies about 10% of the tank volume. Tube spacing is generally closer than in external-melt heat exchangers to compensate for the indirect heat exchange and the gradual development of a liquid annulus in the discharge mode. There is no need to maintain a liquid channel between individual tubes. The entire water volume between heat exchanger tubes is frozen. Approximately 9% of the tank volume, all above the heat exchanger, is reserved for expansion of the water as it freezes. Heat exchange surface does not usually extend into this expansion volume. If water is frozen in this area of the tank, subsequent melting may result in ice remaining in the expansion area, with voids forming around the heat exchange surfaces below. Some manufacturers provide means to prevent this possibility. Because the heat exchanger is prevented from floating, structurally or by its own weight, the rise in water level is proportional to the ice inventory. Level sensors can provide inventory information to the automatic control and/or energy management systems.

Several combinations of tanks and heat exchangers are available for secondary coolant systems. Cylindrical plastic tanks provide a combination of structure, water containment, and corrosion protection, and use a polyolefin heat exchanger, typically formed into a spiral configuration to conform to the tank shape. Structural steel tanks are usually rectangular; water containment can be achieved with either a metal or flexible liner. Steel tanks are often hot-dip galvanized for corrosion protection, and can use either galvanized steel or polyolefin heat exchangers.

Some care is required in the design of external piping because these are often modular systems, sometimes with large numbers of individual tanks arrayed in a variety of patterns. Reverse-return piping is often used to simplify balancing.

An important characteristic of the internal-melt design is the relationship between the charging and discharging processes (Figure 5). During discharge, ice in contact with the heat exchanger is melted first. Initial discharge rates can be very high, and initial available temperatures approach the phase-change temperature. However, as ice melts, an annulus of water develops between the heat exchanger and ice surfaces. This results in variable performance for the internal-melt design. Manufacturers have developed various methods of incorporating the effect of ice inventory levels in product performance descriptions. The temperature of coolant leaving the storage device varies with flow rate, supply temperature, and ice inventory. A mixing or diverting valve is often included to automatically adjust flow through the storage device, controlling both the leaving temperature and discharge rate. Because the storage flow can be completely independent of the chiller flow, cooling load can be imposed on either component in any desired proportion, allowing the designer to optimize discharge of storage for a particular utility rate and building load.

When the internal-melt system returns to ice-making (charging) mode, ice first forms directly on the heat exchanger surface and gradually accumulates, perhaps to a point where it joins with ice remaining from previous charges. Chilled-coolant temperatures are always at their most efficient levels for every ice-building period, and every charge can be carried to completion without penalty. There is usually no benefit to limiting the amount of ice produced, so the storage can be fully charged every night. Because the internal storage temperature is always at the phase-change temperature, standby losses are virtually independent of ice inventory. This simplifies control logic, maximizes efficiency, and eliminates the potential liabilities of predicting the next day's required storage. The close tube spacing needed for good discharge performance provides excellent ice-making performance, because the thermal conductivity of ice is about 3.5 times greater than that of liquid water. Although coolant temperatures depend on many factors, a typical charge cycle begins with coolant temperatures of 26°F entering storage and leaving near 32°F. For most of the charge period, the coolant temperatures are fairly stable, gradually diminishing but with an average supply temperature of approximately 24°F. As the end of the charge mode nears, temperatures decrease at much faster rates to perhaps a minimum supply temperature of 22°F. The rapid temperature decrease at the end of the charge cycle is caused by the exhaustion of latent heat and the lower specific heat of ice compared to liquid water. Depending on system type, a full charge may be indicated by an inventory measurement or a final leaving temperature. The ice-making chiller or compressor is generally fully loaded throughout the ice-making period. Once the ice inventory is completely restored, ice-making chiller operation is terminated to prevent cycling the chiller in the ice-making mode.

External-Melt Ice-On-Coil

External-melt ice-on-coil thermal storage devices also consist of a tubular heat exchanger submersed in a cylindrical or rectangular tank of water. However, these devices circulate a secondary coolant through the tubes only for charging. Discharging is accomplished by circulating the water surrounding the heat exchanger. Ice melts from the outside in (Figure 6).

Traditional external-melt storage devices build ice by circulating cold liquid refrigerant through the heat exchangers. Although this approach is still commonly used in industrial applications, most external-melt systems for HVAC applications currently use glycol secondary coolants for charging. Water surrounding the heat exchanger is both the phase-change material and the heat transfer fluid for the discharge mode.

The charging heat transfer fluid is contained in a separate circuit from the water in the storage tank. Typically, a heat exchanger is installed between the two circuits to allow the same chiller to charge storage and help meet the cooling load directly. Some designs use separate chillers, or two separate evaporators served by the same compressor, to fulfill these functions. It is also possible to simply continue to circulate the heat transfer fluid through the storage heat exchanger to assist in direct cooling, but this approach requires that the refrigeration equipment continuously operate at subfreezing temperatures, incurring substantial capacity and efficiency penalties.

The individual galvanized steel heat exchangers or coils are assembled from a number of separate tubing circuits, and are usually stacked into large, site-fabricated concrete tanks. The tanks operate at atmospheric pressure, and additional measures may be needed to accommodate pressurized systems or elevation differences. The heat exchanger tube cross section can be circular or elliptical. Many different coil lengths, widths, and heights are available to conform to tank dimensions while simplifying coil connections, providing desired flow and pressure drop, and minimizing shipping costs.

Some form of agitation is needed to promote even ice building and melting, although continuous agitation may not be necessary during charging. Agitation is usually provided by distributing compressed air through plastic pipes under of each stack of heat exchangers.

The annular thickness of ice formed on the heat exchanger can vary widely, but an average of about 1.4 in. is a typical dimension used in published ratings. Proper ice-making control is particularly important for external melt systems. Ice-making temperatures are driven lower by excessive ice thickness. Because any new ice must be formed on the surface of existing ice, there is a benefit to limiting the ice thickness to what is needed for the following discharge period, while still ensuring an adequate storage inventory. Also, ice bridging must be prevented. A flow passage of liquid water must be

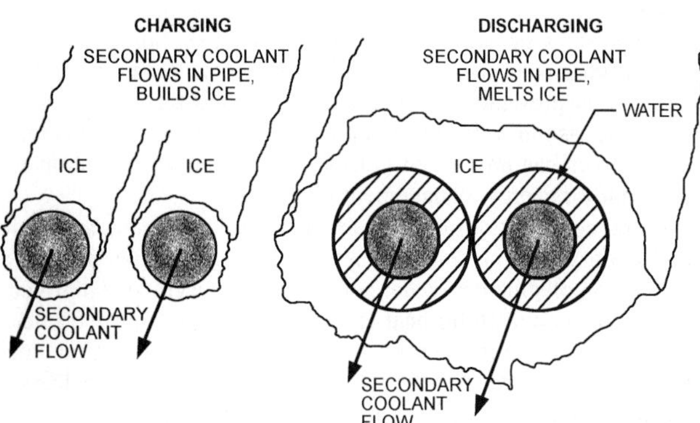

Fig. 5 Charge and Discharge of Internal-Melt Ice Storage

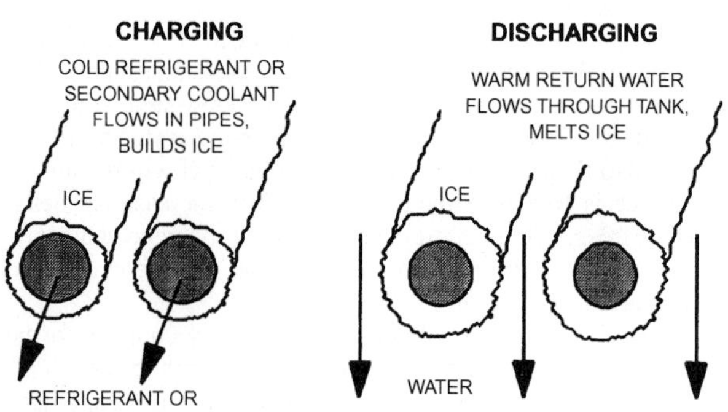

Fig. 6 Charge and Discharge of External-Melt Ice Storage

preserved between adjacent circuits of ice-building coils. Once flow is stopped or restricted, the ability to melt any ice can be significantly limited. Maintaining minimum ice inventory levels discourages ice bridging and promotes efficiency. Ice thickness controllers that sense a change in electrical conductivity are commonly used to control ice building, but other types are available. Manufacturers specify the number and location of thickness controllers. Continuous inventory measurement can be obtained by load cells or strain gages that measure the weight of the ice and steel coils, multiple thickness sensors, or by the change in water level, compensating for compressed air introduced into the tank.

Because the storage tank is at atmospheric pressure, some method may be needed to connect to a higher-pressure distribution water loop. The simplest methods are locating the tank at the highest elevation or adding a chilled-water heat exchanger. Adding pressure-sustaining valves on the tank return or regenerative turbine pumping are other possibilities.

The wider coil spacing typical of external-melt heat exchangers results in lower ice-making temperatures than in internal-melt systems. However, the direct-contact heat exchange in melting provides consistently low chilled-water temperatures of approximately 34°F and high discharge rate capacity. Separate flow circuits for the chiller and storage provide flexibility in assigning cooling load to each device in any desired proportion.

Encapsulated Ice

Encapsulated storage systems use a phase-change material, usually water, contained in relatively small polymeric vessels. A large number of these primary containers are placed in one or more tanks through which a secondary coolant circulates, alternately freezing and melting the water in the smaller containers. Standard chillers are typically used for ice production.

Currently available shapes are basically spherical and are factory-filled with water and a nucleating agent. The sealed containers are shipped to the installation site, where they are placed into the insulated tank, occupying perhaps 60% of the internal tank volume. There is usually no attempt to position the containers in their most compact configuration, instead relying on simple random packing. Some designs have additional surface features. One product uses a 4 in. diameter sphere with deformable depressions or concave hollows arrayed around the surface (Figure 7). Liquid water fills virtually the entire internal volume with the hollows in a depressed position. As water is frozen and expands, the depressions shift from concave to convex. They return to the original shape as the water melts. Besides accommodating water expansion, the variable-volume shape displaces the surrounding coolant. Displaced coolant can be used to measure ice inventory. For pressurized tanks, an additional expansion tank is included for this purpose; atmospheric tanks use the change in the coolant level for the primary tank as an inventory indicator. Other products use a more rigid shape and accommodate water expansion by leaving a void space within the container. Systems with rigid spheres use an energy balance method for inventory determination. Spheres with base diameters of 3 to 4 in. are available. The spheres are buoyant when surrounded by the secondary coolant. A third method, recently introduced, uses an internal collapsible chamber in a 5.4 in. sphere to accommodate expansion.

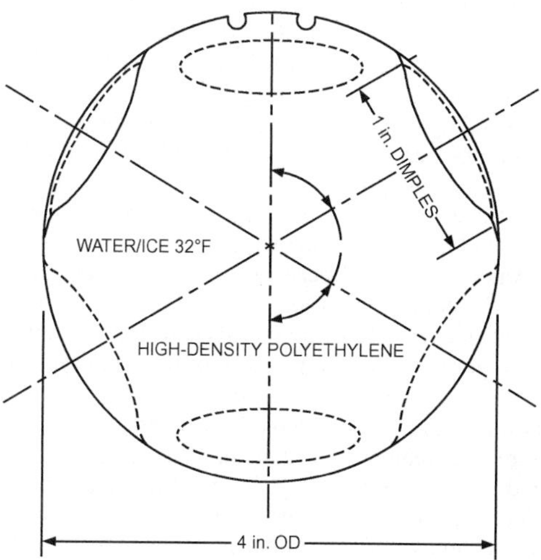

Fig. 7 Encapsulated Ice: Spherical Container
(Courtesy Cryogel)

A wide variety of vertical and horizontal tanks, both pressurized and atmospheric, can be used. Internal headers are designed to distribute flow evenly throughout the storage volume. Some tanks can be completely buried. Atmospheric tanks must have a grid near the top of the tank to prevent flotation of the buoyant water-filled spheres. Although requiring more heat transfer fluid, the large cross sectional flow area results in low pressure drops of approximately 7 to 9 feet.

Charging and discharging are similar to internal-melt ice-on-coil systems, using similar secondary coolants (e.g., 25% ethylene glycol/75% water). Flow typically proceeds through the tank from bottom to top. As with internal-melt systems, the temperature of coolant leaving the storage device varies with flow rate, supply temperature, and ice inventory. Because the shape and volume of the storage tanks varies over a wide range, manufacturers often provide performance data on a per-ton-hour basis.

Ice Harvesters

Ice-harvesting systems separate the formation of ice from its storage. Ice is generated by circulating 32°F water from the storage tank over the surfaces of plate or cylindrical evaporators arranged in vertical banks above the storage tank. Ice is typically formed to a thickness between 0.25 and 0.40 in. over a 10 to 30 min build cycle. The ice is harvested by a hot-gas defrost cycle, which melts the bond between the ice and the evaporator surface and allows the ice to drop into the storage tank below. Other types of ice harvesters separate ice from the evaporator surface mechanically.

Typically, the evaporators are grouped in sections that are defrosted individually so that the heat of rejection from active sections provides the energy for defrost. Tests done by Stovall (1991) indicate that, for a four-section plate ice harvester, the total heat of rejection is introduced into the evaporator during the defrost cycle. To maximize efficiency and ice production capacity, harvest time should be kept to a minimum.

Figure 8 shows an ice-harvesting schematic. Chilled water is pumped from the storage tank to the load and returned to the ice generator. A low-pressure recirculation pump is used to provide minimum flow for wetting the evaporator in ice-making mode. This system may be applied to load-leveling or load-shifting applications.

In load-leveling applications, ice is generated and the storage tank charged when there is little or no building load. When a building load is present, the return chilled water flows directly over the evaporator surface, and the ice generator functions either as a chiller or as both an ice generator and a chiller. The defrost cycle is energized if the return water temperature is low enough that the exit water from the evaporator is within a few degrees of freezing. When operated as a chiller only, maximum capacity and efficiency are obtained with minimum water flow and highest entering water temperature. In load-shifting applications, the compressors are turned off during the utility company's on-peak period, and ice water from the tank is circulated to the building load.

During the first part of the charge cycle, ice floats in the tank with about 10% of its volume above the water, and the water level remains constant. As ice continues to build, the ice comes to rest on

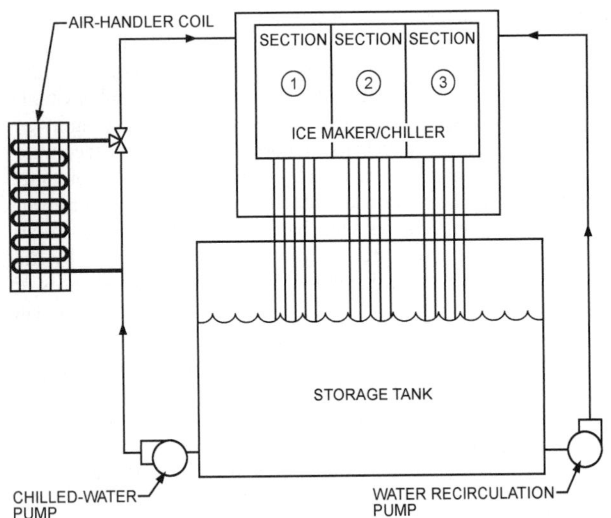

Fig. 8 Ice-Harvesting Schematic
(Courtesy Paul Mueller Company)

the bottom of the tank and builds up above the water level. The ice generation mode is terminated when the ice reaches the high ice level, as determined by a mechanical, optical, or electronic high-level sensor.

Because the ice is free to float in the tank, ice inventory cannot be determined by measuring the water level. Ice inventory may be tracked by maintaining a running energy balance calculation that monitors cooling discharged from and stored in the tank. The amount of cooling discharged can be determined by measuring chilled-water flow rate and supply and return temperatures. Cooling stored in the tank can be measured as the difference between the energy input to the compressor and the heat rejection from the condenser, or estimated based on compressor performance data and operating conditions. Ice inventory may also be tracked by measuring changes in water conductivity as the ice freezes and melts.

Although tank water level cannot be used to indicate the ice inventory during the storage cycle, the water level at the end of charging is a good indicator of system conditions. In systems with no gain or loss of water, the shutdown level should be consistent, and can be used as a back-up to determine when the tank is full for shutdown requirements. Conversely, a change in level at shutdown can indicate a water gain or loss.

Ice harvester systems usually use positive-displacement compressors with saturated suction temperatures usually between 18 and 22°F. The condensing temperature should be kept as low as possible to reduce energy consumption. The minimum allowable condensing temperature depends on the type of refrigeration used and the system's defrost characteristics. Several systems operating with evaporatively cooled condensers have operated with a compressor specific power consumption of 0.9 to 1.0 kW/ton (Knebel 1986, 1988; Knebel and Houston 1989).

Ice-harvesting systems can discharge stored cooling very quickly. Individual ice fragments are characteristically less than 6 by 6 by 0.25 in. and provide a large surface area per ton-hour of ice stored. When properly wetted, a 24 h charge of ice can be melted in less than 30 min for emergency cooling demands.

Ice harvesters typically use field-built concrete tanks or prefabricated steel tanks. The thermal storage capacity of usable ice that can be stored in a tank depends on its shape, location of the ice entrance to the tank, angle of repose of the ice (between 15 and 30°, depending on the shape of ice fragments), and water level in the tank. If the water level is high, voids occur under the water because of the ice's buoyancy. The ice harvester manufacturer may assist in tank design and piping distribution in the tank. Gute et al. (1995) and Stewart et al. (1995a, 1995b) describe models for determining the amount of ice that can be stored in rectangular tanks and the discharge characteristics of various tank configurations. Dorgan and Elleson (1993) provide further information.

The tank may be completely or partially buried or installed above ground. A minimum of 2 in. of closed-cell insulation should be applied to the external surface. Because shifting ice creates strong dynamic forces, internal insulation should not be used except on the underside of the tank cover, and only very rugged components should be placed inside the tank. Exiting water distribution headers should be constructed of stainless steel or rugged plastic suitable for the cold temperatures encountered. PVC is not an acceptable material because of its extreme brittleness at the ice-water temperature.

As with field-built concrete tanks for other storage applications, close attention to design and construction is important to prevent leakage. An engineer familiar with concrete construction requirements should monitor each pour and check all water stops and pipe seals. Unlined tanks that do not leak can be built. If liners are used, the ice equipment suppliers will provide assistance in determining a suitable type, and the liner should be installed only by a qualified installer trained by the liner manufacturer.

The sizing and location of the ice openings are critical; all framed openings must be verified against the certified drawings before the concrete is poured.

An ice harvester is generally installed by setting in place a prepackaged unit that includes the ice-making surface, refrigerant piping, refrigeration equipment, and, in some cases, heat rejection equipment and prewired controls. To ensure proper ice harvesting, the unit must be properly positioned relative to the ice drop opening. The internal piping is not normally insulated, so the drop opening should extend under the piping so condensate drops directly into the tank. A grating below the piping is desirable. To prevent air or water leakage, gasketing and caulking must be installed in accordance with the manufacturer's instructions. External piping and power and control wiring are required to complete the installation.

Ice Slurry Systems

Ice slurry is a suspension of ice crystals in liquid. An ice slurry system has the advantage of separating production of ice from its storage without the control complexity and efficiency losses associated with the ice harvester's defrost cycle. Slurries also offer the possibility of increased energy transport density by circulating the slurry itself, rather than just the circulating secondary liquid. However, a heat exchanger is usually used to separate the storage tank flow from the cooling load distribution loop. Like harvesters, slurry systems have very high discharge rates and provide coolant consistently close to the phase-change temperature.

In general, the working fluid's liquid state consists of a solvent (water) and a solute such as glycol, sodium chloride, or calcium carbonate. Depending on the specific slurry technology, the initial solute concentration varies from 2% to over 10%, by mass. The solute depresses the freezing point of the water and buffers the production of ice crystals. Slurry generation begins by lowering the temperature of the working fluid to its initial freezing point. Upon further cooling, water begins to freeze out of the solution. As freezing progresses, the concentration of solute increases and the freezing point at which ice crystals are produced decreases. Ice/water slurries have been reported to have a porosity of 0.50 and typical storage densities of 2.92 ft^3/ton·hour (Gute et al. 1995).

One design uses a 7% solution of propylene glycol that provides an initial freezing temperature of about 28°F. A final freezing point of 25°F indicates that a full charge has been reached and about 50% of the volume has been frozen. Ice is formed near the inside surfaces of multiple cylinders of the orbital rod evaporator. Rotating rods travel along the inner evaporator surfaces, creating an agitated flow that enhances heat transfer and prevents adhesion of ice crystals. The compliant method of connecting the rod prevents damage in the event of evaporator freeze-up. This arrangement reduces or eliminates the need for a defrost cycle, but adds mechanical components. The evaporators are typically arranged in vertical banks above the storage tank, and the slurry can also be pumped into an adjoining tank.

Other Phase-Change Materials

Water now dominates the market as the storage material for HVAC applications, but industrial or commercial processes sometimes require coolant temperatures that are not possible using plain water. A substantial amount of effort has been invested in the development of materials that would freeze above 32°F and still provide appropriate temperatures in the discharge mode. There are currently no commercially available systems using these materials. Phase-change materials with fusion temperatures above 32°F were usually inorganic salt hydrates that had high anhydrous salt fractions and limited latent heat capacities.

Materials with fusion temperatures below 32°F are more practical to develop and apply. These are often true salt eutectics and typically require lower percentages of the anhydrous components than materials with fusion temperatures above 32°F. Container geometry, or some chemical or mechanical means, prevents eventual stratification of components. Because these materials have salt constituents, they are more corrosive than plain water and are currently only available in encapsulated or all-plastic internal-melt ice-on-coil devices. The latent heat capacities for these materials are usually close to the volumetric heat capacities available from plain water. Supercooling is frequently associated with eutectics and hydrates and appropriate nucleating agents are often added to the material.

Unfortunately, storage at subfreezing temperatures cannot be achieved by adding a freeze depressant, such as ethylene glycol, to water. Unless the additive exists at a eutectic concentration, only the initial freezing point is affected, with the temperature gradually decreasing as the freezing process continues (see the section on Ice Slurry Systems). In most cases, the loss in volumetric heat capacity for a useful range of storage charge and discharge temperatures makes this an impractical solution.

HEAT STORAGE TECHNOLOGY

The most common current application of heat storage is in electrically charged, thermally discharged storage devices used for service water heating and space heating. Thermal storage systems for space heating can be classified as brick storage heaters (electric thermal storage or ETS heaters), water storage heaters, or radiant floor heating systems. Other applications include solar space and water heating, and thermally charged water storage tanks.

The choice of storage medium generally depends on the type of heating system. Air heaters use brick as the storage medium, whereas water heaters may use water or a PCM. Combination brick and hydronic systems use brick for the storage medium and transfer the heat to water as needed.

Solar space and water heating systems use rock beds, water tanks, or PCMs to store heat. Various applications are described in Chapter 36; also see Chapter 33 of the 2007 *ASHRAE Handbook—HVAC Applications*.

Thermally charged hot-water storage tanks are similar in design to the cool storage tanks described in the section on Storage of Heat in Cool Storage Units. Many are also used for cooling. Unlike off-peak cooling, off-peak heating seldom reduces the size of the heating plant or the utility bill (except in the case of collecting waste heat, such as applications associated with industrial processes or a turbine generator). The size of a storage tank used for both heating and cooling is based more on cooling than on heating for lowest life-cycle cost, except in some residential applications.

Sizing Heat Storage Systems

For electrically charged devices that use a solid mass as the storage medium, equipment size is typically specified by the nominal power rating (to the nearest kilowatt) of the internal heating elements. The nominal storage capacity is taken as the amount of energy supplied during an 8 h charge. For example, a 5 kW heater would have a nominal storage capacity of 40 kWh. ASHRAE *Standard* 94.2 describes methods for testing these devices. If multiple charge/off-peak periods are available during a 24 h period, an alternative method that considers not only the nominal power rating, but also fan discharge rate and storage capacity, yields a more accurate estimate of equipment size. Consult the equipment manufacturer for more information on calculating the capacity of these devices.

A rational design procedure requires an hourly simulation of the design heat load and a known discharge capacity of the storage device. The first criterion for satisfactory performance is that the discharge rate must be no less than the design heating load at any hour. During the off-peak period, energy is added to storage at the rate determined by the connected load of the resistance elements less the design heating load. The second criterion for satisfactory performance is that energy added to storage during one day's off-peak period is sufficient to meet the total load for the entire 24 h.

A simplified procedure is recommended by Hersh et al. (1982) for typical residential designs. For each zone of the building, the design heat loss is calculated in the usual manner and multiplied by the selected sizing factor. The resulting value (rounded to the next kilowatt) is the required storage heater capacity. Sizing factors in the United States range from 2.0 to 2.5 for an 8 h charge period and from 1.6 to 2.0 for a 10 h charge period. If multiple off-peak periods occur during a 24 h period, this method may not accurately reflect equipment sizing. The designer should consult the manufacturer for specific sizing information based on on-peak and off-peak hours available. Figure 9 shows a typical sizing factor selection graph for an application that has a single charging time block per day.

Service Water Heating

The tank-equipped service water heater, which is the standard residential water heater in North America, is a thermal storage device, and some electric utilities provide incentives for off-peak water heating. To meet that requirement, the heater is equipped with

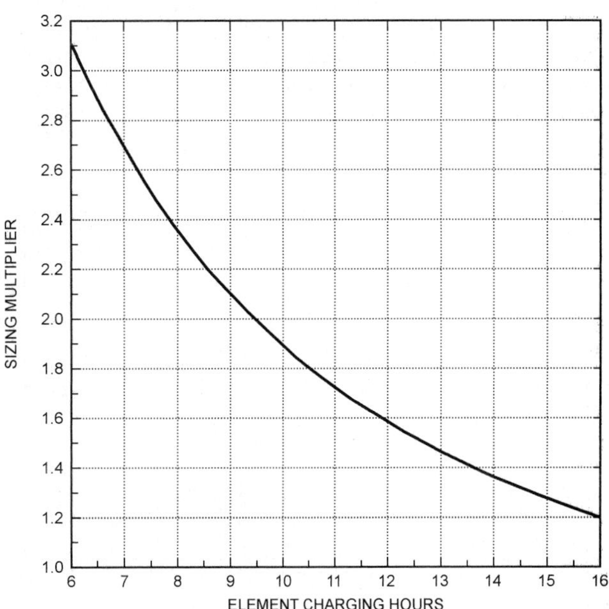

Fig. 9 Representative Sizing Factor Selection Graph for Residential Storage Heaters

a control system that is activated by a clock or by a signal directly from the electric utility to curtail power use during peak demand. An off-peak water heater generally requires a larger tank than a conventional water heater.

An alternative system of off-peak water heating consists of two tanks connected in series, with the hot-water outlet of the first tank supplying water to the second tank (ORNL 1985). This arrangement minimizes mixing of hot and cold water in the second tank. Tests performed on this configuration show that it can supply 80 to 85% of rated capacity at suitable temperatures, compared to 70% for a single-tank configuration. The wiring for the heating elements in the two tanks may have to be modified to accommodate the dual-tank configuration.

Brick Storage (ETS) Heaters

Brick storage heaters [commonly called electric thermal storage heaters (ETS)] are electrically charged and store heat during off-peak times. In this storage device, air circulates through a hot brick cavity and then discharges into the area in which heat is desired. The ceramic brick has a very dense magnetite or magnesite composition. The brick's high density and ability to store heat at a high temperature gives the heater a larger thermal storage capacity. Ceramic brick can be heated to approximately 1400°F during off-peak hours by resistance heating elements. Space requirements for brick storage heaters are usually much less than for other heat storage media.

Figure 10 shows the operating characteristics of an electrically charged room storage heater. Curve 1 represents theoretical performance with no discharge. In reality, radiation and convection from the exterior surface of the device continually supplies heat to the room during charging as well. Curve 2 shows this static discharge. When the thermostat calls for heat, the internal fan operates. The resulting faster dynamic discharge corresponds to curves 3 and 4. Because the heating elements of electrically charged room storage heaters are energized only during off-peak periods, they must store the total daily heating requirement during this period. The dynamic and static charging and discharging rates vary by equipment manufacturer.

The four types of brick storage heaters currently available are room storage heaters (room units), heat pump boosters, central furnaces, and radiant hydronic (brick to water).

Room Storage Heaters (Room Units). Room storage heaters (commonly called room units) have magnetite or magnesite brick cores encased in shallow metal cabinets (Figure 11). The core can be heated to 1400°F during off-peak hours by resistance heating elements located throughout the cabinet. Room units are generally small heaters that are placed in a particular area or room. These heaters have well-insulated storage cavities, which help retain the heat in the brick cavity. Although the brick inside the units gets very hot, the outside of the heater is relatively cool, with surface temperatures generally below 165°F. Storage heaters are discharged by natural convection, radiation, and conduction (static heaters) or, more commonly, by a fan. Air flowing through the core is mixed with room air to limit the outlet air temperature to a comfortable range.

Storage capacities range from 13.5 to 60 kWh. Inputs range from 0.8 to 10.8 kW. In the United States, 120, 208, 240, and 277 V units are commonly available. The 120 V model is useful for heating smaller areas or in geographical areas with moderate heating days. Room storage heaters are used in residential and commercial applications, including motels, hotels, apartments, churches, courthouses, and offices. Systems are used for both new construction and retrofit applications. A common commercial application is replacement of an aging boiler or hot-water heating system.

Operation is relatively simple. When a room thermostat calls for heat, fans in the lower section of the room unit discharge air through the ceramic brick core and into the room. Depending on the charge level of the brick core, a small amount of radiant heat may also be delivered from the surface of the unit. The amount of heat stored in the brick core of the unit can be regulated either manually or automatically in relation to the outside temperature or the known or estimated space requirements.

These units fully charge in about 7 h (Figure 12), and they can be fully depleted in as little as 5 h. The equipment retains heat for up to 72 h, if it has no fan discharge (Figure 13).

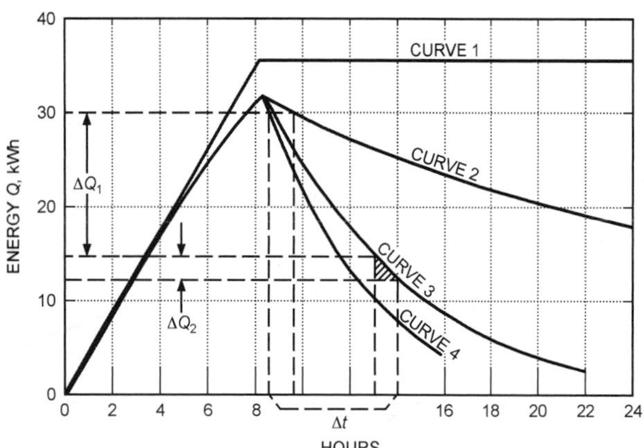

Fig. 10 Typical Storage Heater Performance Characteristics
(Hersh et al. 1982)

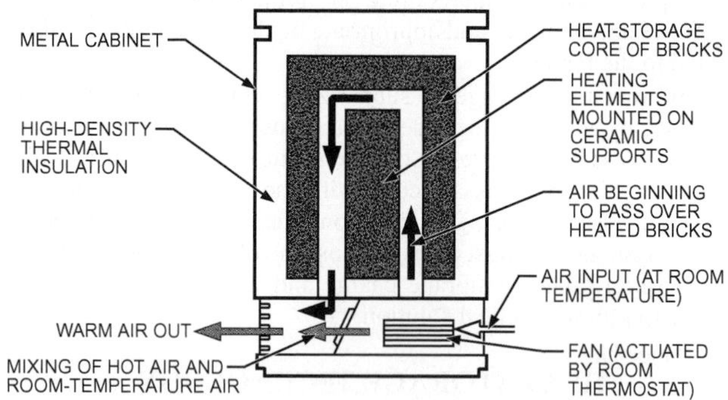

Fig. 11 Room Storage Heater
(Courtesy Steffes Corporation)

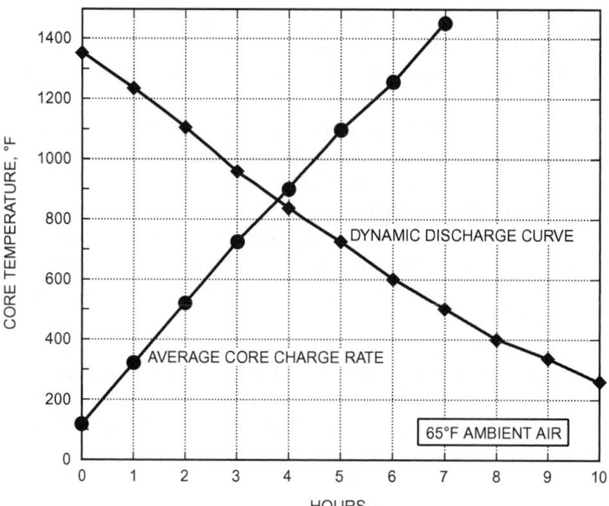

Fig. 12 Room Storage Heater Dynamic Discharge and Charge Curves

Choosing the appropriate size of room unit(s) depends on the length of the applicable on-peak and off-peak periods, local climate, and heat loss of the area or space. The designer must follow generally accepted practices for calculating design load or heat loss rate to ensure a properly functioning system that delivers satisfactory results. This equipment is generally configured according to one of the following two concepts:

Whole-House Concept. Under this strategy, room units are placed throughout the home. Units are sized according to a room-by-room heat loss calculation. This method is used in areas with long on-peak periods, generally 12 h or more.

Warm-Room Concept. Under this strategy, one or two room units are used as the primary heating source during on-peak periods. Units are placed in the area most often occupied (main area). Adjacent areas generally are kept cooler and have no operable heating source during the on-peak period; however, some heat migrates from the warmer main area.

When determining the heat loss of the main area, an additional sizing factor of approximately 25% should be added to allow for migration of heat to adjacent areas. Under the warm-room concept, sizing factors vary, depending on the rate structure of the power company and equipment performance under those conditions. Consult the equipment manufacturer for specific sizing information.

The warm-room concept is the most common method used by power companies for their load management and off-peak marketing programs. It is successful in areas that have a small number of consecutive hours of control (generally less than 12 h of on-peak time), or have a midday block of off-peak time during which the equipment can recharge. The advantage of the warm-room concept is that it requires smaller equipment and fewer heating units than the whole-house concept, thereby reducing system cost.

Heat Pump Boosters. Air-to-air heat pumps generally perform well when outside temperatures are relatively warm; however, as the outside temperature decreases, the efficiency, output capacity, and supply air temperature from a heat pump also decline. When the output of the heat pump drops below the heat loss of the structure, supplemental heat, typically from an electric resistance unit, must be added to maintain comfort. To eliminate the use of direct electric resistance heat during on-peak periods, storage heaters, such as the heat pump boosters (HPBs), can supplement the output of a heat pump. Additionally, the HPB monitors supply air temperatures and adds heat to the supply air as needed to maintain comfortable delivery air temperature at all times. The HPB can also be used as a booster for stand-alone forced-air furnaces or to back up electric or fossil fuel equipment in dual-fuel programs.

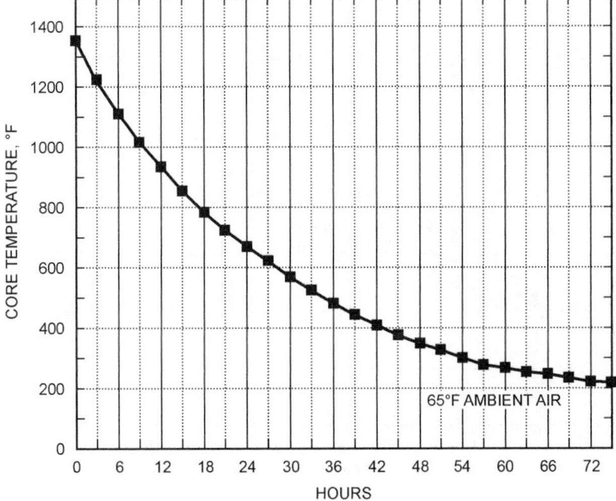

Fig. 13 Static Discharge from Room Storage Heater

Core charging of the HPB is regulated automatically based on outdoor temperature. The brick storage core is well insulated so radiant or static heat discharge is minimal. Common residential equipment input power ratings range from 14 to 46 kW. Storage capacities range from 86 to 240 kWh. Larger commercial and industrial sizes have input ratings of 53 to 160 kW and storage capacities of up to 960 kWh.

Central Furnace. A central storage furnace is a forced-air heat storage product for residential, commercial, and industrial applications. These units are available with input ratings ranging from 14 to 46 kW. Storage capacities range from 86 to 960 kWh.

Radiant Hydronic Heaters. A ceramic brick storage system is available that uses an air-to-water heat exchanger to transfer heat from the storage system to a water loop. The off-peak heated water can be used in conventional hot-water systems, including floor warming (radiant floors), hydronic baseboards, hot-water radiators, indirect domestic hot-water heating, etc. An air handler can be added to these systems to achieve hydronic and forced-air heating from the same unit. In commercial and industrial applications, these systems can be used to preheat incoming fresh air in makeup air systems. These units are available in input ratings ranging from 20 to 46 kW with storage capacities from 120 to 240 kWh.

Pressurized Water Storage Heaters

This storage device consists of an insulated cylindrical steel tank containing immersion electrical resistance elements near the bottom of the tank and a water-to-water heat exchanger near the top (Figure 14). During off-peak periods, the resistance elements are sequentially energized until the storage water reaches a maximum temperature of 280°F, corresponding to 50 psig. The *ASME Boiler Code* considers such vessels unfired pressure vessels, so they are not required to meet the provisions for fired vessels. The heaters are controlled by a pressure sensor, which minimizes problems that could be caused by unequal temperature distribution. A thermal controller gives high-limit temperature protection. Heat is withdrawn from storage by running service water through the heat exchangers and a tempering device that controls the output temperature to a predetermined level. The storage capacity of the device is the sensible heat of water between 10°F above the desired output water temperature and 280°F. The output water can be used for space heating or service hot water. The water in the storage tank is permanently treated and sealed, requires no makeup, and does not interact with the service water.

In most applications, electrically charged pressurized water storage tanks must be able to be recharged during the off-peak period to meet the full daily heating requirement, while supplying heat to

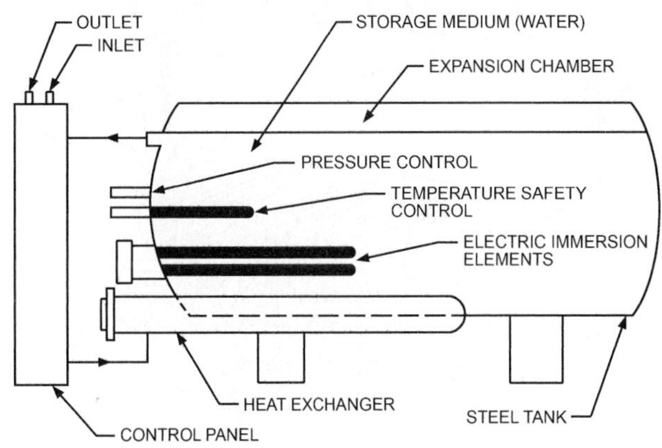

Fig. 14 Pressurized Water Heater

meet the load during the on-peak period. The heat exchanger design allows a constant discharge rate, which does not decrease near the end of the cycle.

Underfloor Heat Storage

This storage method typically uses electric resistance cables or hydronic tubing buried in a bed of sand 1 to 3 ft below the concrete floor of a building. Underfloor heat storage is suitable for single-story buildings, such as residences, churches, offices, factories, and warehouses. An underfloor storage heater acts as a flywheel. Although the unit is charged only during the nightly off-peak period, it maintains the top of the floor slab at a constant temperature slightly higher than the desired space temperature. Because the cables spread heat in all directions, they do not have to cover the entire slab area. For most buildings, a cable or tube location of 18 in. below the floor elevation is optimum. The sand bed should be insulated along its perimeter with 2 in. of rigid, closed-cell foam insulation, to a depth of 4 ft (Figure 15). Even with a well-designed and well-constructed underfloor storage, 10% or more of the input heat may be lost to the ground.

Building Mass Thermal Storage

The thermal storage capabilities inherent in the mass of a building structure can have a significant effect on the temperature in the space as well as on the performance and operation of the HVAC system. Effective use of structural mass for thermal storage reduces building energy consumption and reduces and delays peak heating and cooling loads (Braun 1990). In some cases, it improves comfort (Morris et al. 1994; Simmonds 1991). Perhaps the best-known use of thermal mass to reduce energy consumption is in buildings that include passive solar techniques (Balcomb 1983).

Cooling energy can be reduced by precooling the structure at night using ventilation air. Andresen and Brandemuehl (1992), Braun (1990), and Ruud et al. (1990) suggested that mechanical precooling of a building can reduce and delay peak cooling demand; Simmonds (1991) suggested that the correct building configuration may even eliminate the need for a cooling plant. Mechanical precooling may require more energy use; however, the reduction in electrical demand costs may give lower overall energy costs. Moreover, the installed capacity of air-conditioning equipment may also be reduced, providing lower installation costs. Braun (2003) provided an overview of related research, and found that (1) there is a tremendous opportunity for reductions in on-peak energy and peak demand, and (2) the savings potential is very sensitive to the utility rates, building and plant characteristics, weather conditions, and occupancy schedule.

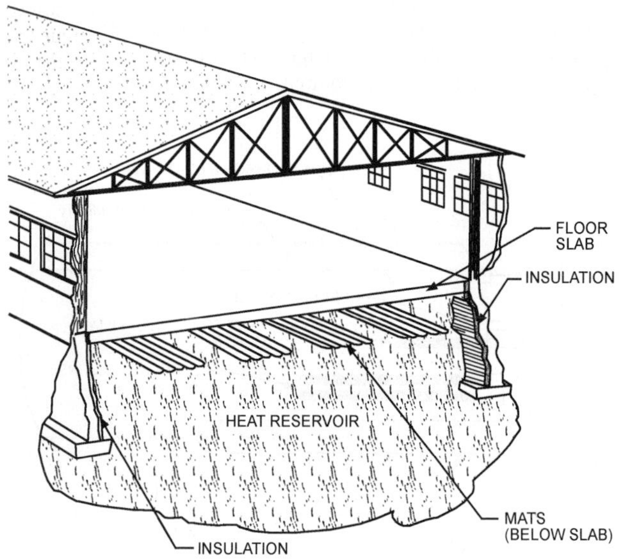

Fig. 15 Underfloor Heat Storage

The potential energy cost savings from precooling depends on the control strategies used to charge and discharge the building thermal mass. Several studies of building pre-cooling controls have been reported (Braun 1990; Braun et al. 2001; Conniff 1991; Keeney and Braun 1996; Morgan and Krarti 2006; Morris et al. 1994; Rabl and Norford 1991; and Ruud et al. 1990). Precooling's potential in reducing electric energy costs is generally well documented. Reported investigations have shown that energy costs have been reduced by 10 to 50% under dynamic rates, and on-peak demand reduced by 10 to 35%, just by precooling building thermal mass.

In particular, Morgan and Krarti (2006) performed both simulation analyses and field testing to evaluate various precooling strategies. They found that the energy cost savings associated with precooling thermal mass depend on several factors, including thermal mass level, climate, and utility rate. For time-of-use utility rates, they found that energy cost savings are affected by the ratio of on-peak to off-peak demand charges R_d, as well as ratio of on-peak to off-peak energy charges R_e. Figures 16 and 17 show the variation of the annual energy cost savings from 4 h precooling an office building with heavy mass as a function of R_d and R_e, respectively.

Henze et al. (2004, 2005) and Kintner-Meyer and Emery (1995) evaluated the combined use of both building thermal mass (passive TES) and ice or chilled-water storage tanks (active TES), a load management strategy to shift on-peak building HVAC cooling load to off-peak time. Thermal mass use can be considered incidental and not be considered in the heating or cooling design, or it may be intentional and form an integral part of the design. Effective use of building structural mass for thermal energy storage depends on factors such as (1) physical characteristics of the structure, (2) dynamic

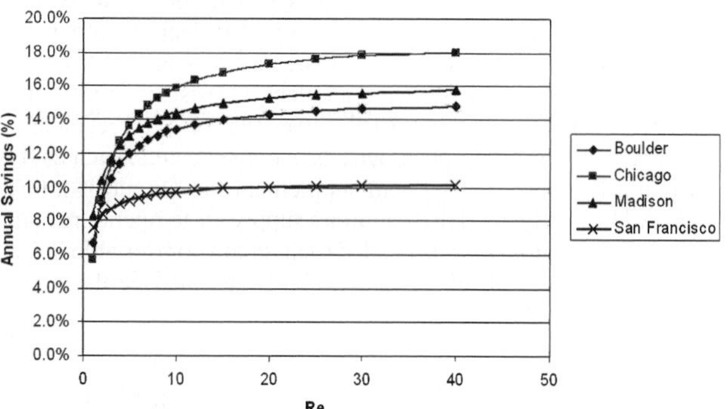

Fig. 16 Annual Energy Cost Savings from Precooling, Relative to Conventional Controls, as Function of R_e

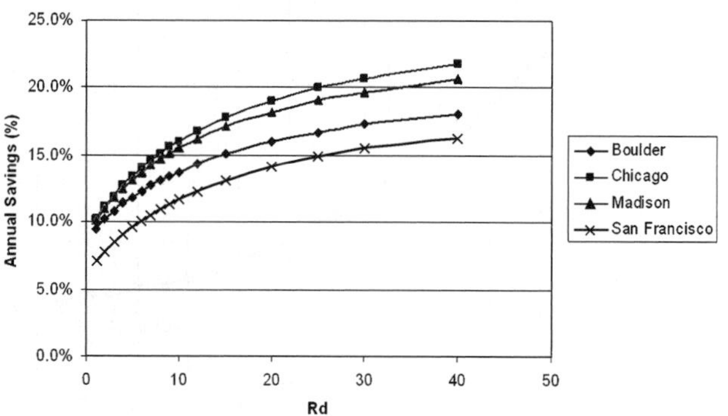

Fig. 17 Annual Energy Cost Savings from Precooling, Relative to Conventional Controls, as Function of R_d

nature of the building loads, (3) coupling between the building mass and zone air (Akbari et al. 1986), and (4) strategies for charging and discharging the stored thermal energy. Some buildings, such as frame buildings with no interior mass, are inappropriate for this type of thermal storage. Many other physical characteristics of a building or an individual zone, such as carpeting, ceiling plenums, interior partitions, and furnishings, affect the quantity of thermal storage and the coupling of the building mass with zone air.

Incidental Thermal Mass Effects. A greater amount of thermal energy is required to bring a room in a heavyweight building to a suitable condition before occupancy than for a similarly sized lightweight building. Therefore, the system must either start conditioning the spaces earlier or operate at a greater-capacity output. During the occupied period, a heavyweight building requires a smaller output, because a larger proportion of heat gains or losses is absorbed by the thermal mass.

Advantage can be taken of these effects if low-cost electrical energy is available during the night and the air-conditioning system can be operated during this period to precool the building mass. This can reduce both the peak electric demand and total energy required during the following day (Braun 1990; Morgan and Krarti 2006), but it may not always be energy efficient.

Intentional Thermal Mass Effects. To make best use of thermal mass, the building should be designed with this objective in mind. Intentional use of the thermal mass can be either passive or active. Passive solar heating is a common application that uses the thermal mass of the building to provide warmth outside the sunlit period. This effect is discussed in further detail in Chapter 33 of the 2007 *ASHRAE Handbook—HVAC Applications*. Passive cooling applies the same principles to limit the temperature rise during the day. In some climate zones, the spaces can be naturally ventilated overnight to remove surplus heat from the building mass. This technique works well in moderate climates with a wide diurnal temperature swing and low relative humidity, but it is limited by the lack of control over the cooling rate.

Active systems overcome some of the disadvantages of passive systems by using (1) mechanical power to help heat and cool the building and (2) appropriate controls to limit output during the release or discharge period.

Storage Charging and Discharging

The building mass can be charged (cooled or warmed) either indirectly or directly. Indirect charging is usually accomplished by heating or cooling either the bounded space or an adjacent void. Almost all passive and some active cooling systems are charged by cooling the space overnight (Arnold 1978). Most indirect active systems charge by ventilating the void beneath a raised floor (Crane 1991; Herman 1980). Where this is an intermediate floor, cooling can be radiated into the space below and convected from the floor void the following day. By varying the rate of ventilation through the floor void, the rate of storage discharge can be controlled. Proprietary floor slabs commonly have hollow cores (Anderson et al. 1979; Willis and Wilkins 1993). The cores are continuous, but when used for thermal storage, they are plugged at each end, and holes are drilled to provide the proper airflow. Charging is carried out by circulating cool or warm air through the hollow cores and exhausting it into the room. Fan discharge air can be controlled by a ducted switching unit that directs air through the slab or directly into the space.

A directly charged slab, used commonly for heating and occasionally for cooling, can be constructed with an embedded hydronic coil. Refer to Chapter 6 for design details. The temperature of the slab is only cycled 3 to 5°F to either side of the daily mean temperature of the slab. Consequently, the technique can use very-low-grade free cooling (approximately 66°F) (Meierhans 1993) or low-grade heat (approximately 82°F) rejected from condensers. In cooling applications the slab is used as a cool radiant ceiling, and for warming it is usually a heated floor. Little control is necessary because of the small temperature differences and the high heat storage capacity of the slab.

The amount of heat stored in a slab equals the product of mass, specific heat, and temperature rise. The amount of heat available to the space depends on the rate at which heat can be transferred between the slab and the surrounding spaces. The combined radiant/convective heat transfer coefficient is generally quite low; for example, a typical value for room surfaces is 1.4 Btu/h·ft^2·°F.

Design Considerations

Many factors must be considered when an energy source is time dependent. The minimum temperature occurs just before dawn, which may be at the end of the off-peak period, and the optimum charge period may run into the working day. Beginning the charge earlier may be less expensive but also less energy efficient. In addition, energy stored in the building mass is neither isolated nor insulated, so some energy is lost during charging, and the amount of available free energy (cooler outside air when suitable and available) varies and must be balanced against the energy cost of mechanical power. As a result, there is a tradeoff that varies with time between the amount of free energy that can be stored and the power necessary for charging.

The cooling capacity is, in effect, embedded in the building thermal mass; therefore conventional techniques of assessing the peak load cannot be used. Detailed weather records that show peaks over several 3 to 5 day periods, as well as data on either side of the peaks, should be examined to ensure that (1) the temperature at which the building fabric is assumed to be before the peak period is realistic and (2) the consequences of running with exhausted storage after the peak period are considered. This level of analysis can only be carried out effectively using a dynamic simulation program. Experience has shown that these programs should be used with a degree of caution, and the results should be compared with both experience and intuition.

Factors Favoring Thermal Storage

Thermal storage is particularly attractive in situations where one or more of the following conditions are present:

New Investment in Chiller Plant Required. A key factor favoring thermal storage is whether the owner must invest in a conventional chiller system (new or an expansion) if thermal storage is not used. This typically occurs during (1) new construction, (2) facility expansions, and (3) chiller plant expansions and rehabs.

Thermal storage systems generally require smaller refrigerating equipment than nonstorage systems. Air and water distribution equipment may also be smaller, and electrical primary power distribution requirements may be significantly smaller. In many situations, the reduced cost of this equipment can partially, completely, or more than completely offset the cost of the thermal storage equipment (Andrepont 2005; Andrepont and Rice 2002), and the operating cost benefits of thermal storage are thereby achieved at a lower cost. For an existing plant that is to be expanded, the additional loads can often be met by thermal storage at a lower cost than adding new chillers, pumps, cooling towers, and mechanical room space (Andrepont and Kohlenberg 2005).

Peak Load Higher Than Average Load. Because thermal storage systems are generally sized for the average expected load, they can provide significant equipment size reductions and cost savings for applications with high peak load spikes. Buildings such as churches, entertainment and sports facilities, convention centers, and some industrial processes can achieve particular benefits. In addition, most commercial and educational facilities also have load profiles conducive to thermal storage.

Favorable Utility Rates. Utility rates often include on-peak and off-peak schedules. On-peak demand and energy charges are higher than off-peak charges, reflecting the utility company's increased cost of meeting high daytime power demands. Some rate schedules include a ratcheted demand charge, which normally bases the demand charge for the current month on the highest demand incurred during any of the previous 12 months. High demand charges and ratcheted rates, as well as large differences between on-peak and off-peak energy charges, favor the use of cool storage to reduce operating costs.

Real-time pricing structures, which substitute time-varying energy charges for the more prevalent demand/energy structure, are becoming more common. Real-time pricing structures that feature high daily price spikes of relatively short duration also favor the use of cool storage.

Owners with cool storage systems can sometimes negotiate more favorable electric rates from their energy suppliers based on their capability to reduce their electric load during certain peak electric demand periods.

In areas where the electricity supply is limited, thermal storage is attractive because it allows a building to be conditioned with a much lower electrical peak demand.

Back-Up or Redundant Cooling Desirable. A properly designed thermal storage system can easily be maintained over long periods and discharged quickly when needed.

Thermal storage can provide cooling if the emergency back-up power supply is sized to run just the pumps. Thermal storage can also reduce the size of a full back-up power supply, because of the reduced power of the thermal storage system compared to a traditional full chiller system.

Thermal storage in combination with conventional chiller(s) can provide a larger portion of the daytime cooling required if one chiller goes down. For example, in a typical traditional system with two chillers, each is sized for 50% of the load. In a thermal storage system with two chillers, with each sized for 25% of the peak load, and thermal storage, if one chiller breaks down there will be 25% capacity from the chiller plus 50% capacity from the thermal storage, for a total of 75% of full capacity. If the chiller remains out of service for an extended period of time, the one chiller can typically recharge more than half the thermal storage so the capacity available is always above 50%. In the nonstorage system, there would only be 50% capacity available.

Cold-Air Distribution Desirable. Cold-air distribution, with supply air temperatures below the traditional 55°F, allows smaller fans and ducts to cool the spaces, and the system can maintain a lower space humidity (Kirkpatrick and Elleson 1997). Thermal storage and cold-air distribution should be considered for applications that have limited space for fans and ducts, require increased capacity from existing fans and ducts, or require (or could benefit from) lower space relative humidity.

Storage for Fire Protection Beneficial. Chilled-water thermal storage can be configured to provide a fire-protection water supply at a much lower cost than installing separate systems by equipping the storage tank with fire nozzles (Holness 1992; Hussain and Peters 1992; Meckler 1992).

Environmental Benefits Sought. Thermal storage can provide environmental benefits related to reduced source energy use, decreased refrigerant charge, and improved efficiency of the energy supply.

Thermal storage systems can operate with lower energy use than comparable nonstorage systems for several reasons. Cool storage systems benefit from operating chillers at lower condensing temperatures during nighttime operation. Thermal storage provides the opportunity to run a heating or cooling plant at its peak efficiency during 100% of its operating period, for a much higher dynamic operating efficiency. Because equipment in a nonstorage system has to follow the building load profile, the majority of its operation is at part-load conditions, which, for most systems, is much less efficient. Thermal energy storage also enables the practical incorporation of other high-efficiency technologies such as cold-air distribution and nighttime heat recovery. Several facilities have demonstrated site energy reductions with the application of thermal energy storage (Bahnfleth and Joyce 1994; Fiorino 1994; Goss et al. 1996).

Most thermal energy storage systems reduce the size of refrigerating equipment needed. Reduced refrigeration equipment size means less on-site refrigerant usage and lower probability of environmental problems caused by a leak.

In addition to the potential reduction in energy use at the building site, thermal storage systems can reduce source energy use and emissions at the generating plant. Typically, the base-load power generating equipment that generates most off-peak electricity is more efficient and less polluting than the peaking plants that are used to meet high daytime demand. Transmission and distribution losses are also lower when the supply networks are more lightly loaded and when ambient temperatures are lower. Because thermal storage systems shift the site energy use from on-peak to off-peak periods, the total source energy and power plant emissions required to deliver cooling to the facility will be lower (Gansler et al. 2001; Reindl et al. 1995). In addition, by reducing peak demand, thermal storage helps prevent or delay the need to construct additional power generation and transmission equipment.

Thermal storage also has beneficial effects on combined heat and power systems by matching the thermal and electric load profiles.

The U.S. Green Building Council's (USGBC) Leadership in Energy and Environmental Design (LEED) rating system seeks to encourage sustainable design. The system awards points for various sustainable design features, including reductions in annual energy cost compared to a standard reference building. Thermal storage is a good way for a building to reduce energy cost and comply with the LEED rating system, and has been used in many buildings that have received LEED certification.

Factors Discouraging Thermal Storage

Thermal storage may be difficult to justify economically if

- Cooling load profile is flat, with very little difference between the peak and average loads
- Available utility rates do not differ between night and day electrical costs
- Space is not available indoors or outdoors for thermal storage tank(s)

Note that even when one or more of these conditions are true, thermal storage may still be justified by the factors favoring thermal storage. In addition, thermal storage should be integrated into the initial cooling system design for it to achieve the greatest economic benefit for the owner and operator.

Typical Applications

Few applications are not suited to thermal storage. Particularly well suited to thermal storage are sports and entertainment facilities, convention centers, churches, airports, schools and universities, military bases, and office buildings, because they have widely varying loads. The average cooling load for these buildings is much lower than the peak load. In addition, using high temperature differentials for chilled-water and air distribution can significantly reduce construction costs, and decrease utility bills.

Health care facilities, data centers, and hotels having loads with only a slight variation between day and night may also be suited for thermal storage. Normally, the thermal storage should be sized to handle the variation in load, not the total load. Thermal storage can provide a steady supply temperature when rapid changes in operating or meeting room temperatures are required. High temperature differentials for chilled-water and air distribution can significantly reduce construction costs, and the lower electrical demand can also reduce the size of emergency back-up generators.

District Cooling Systems. District cooling systems benefit from thermal storage in several ways. The cost of the distribution system in district cooling applications is a large percentage of the total system cost. As a result, when higher temperature differentials available from thermal storage media are used, life-cycle costs can be lower by using smaller pipes, valves, pumps, heat exchangers, motors, starters, etc. Energy costs of chilled-water distribution are thereby reduced. Thermal storage can provide a steady supply temperature when rapid changes in load occur. The lower electrical demand may reduce the size of emergency back-up generators.

Furthermore, thermal storage may be used to increase the available cooling capacity of an existing chilled-water distribution system by providing cooler supply water. This modification has little effect on the distribution system itself (i.e., pumping and piping) during peak load, because the existing system flow rates and pressure drops do not change. During off-peak cooling load conditions, pumping energy costs can be greatly reduced, because the flow rate can be reduced proportionate to the reduced load.

Most storage tanks are open, so attention should be given to system hydraulics when connecting several plants to a common distribution network where storage tank water levels are at different elevations. Chapter 11 has additional information on district systems and related topics.

Industrial/Process Cooling. Industrial refrigeration and process cooling typically require lower temperatures than environmental air conditioning. Applications such as vegetable hydrocooling, milk cooling, carcass spray-cooling, and storage room dehumidification can use cold water from an ice storage system. Lower temperatures for freezing food, storing frozen food, and so forth can be obtained by using a low-temperature sensible energy storage fluid, a low-temperature hydrated salt PCM charged and discharged by a secondary coolant, or a nonaqueous PCM such as CO_2.

Some of the advantages of cool storage may be particularly important for industrial refrigeration and process cooling applications. Cool storage can provide a steady supply temperature regardless of large, rapid changes in cooling load or return temperature. Charging equipment and discharging pumps can be placed on separate electrical service to economically provide cooling during a power failure or to take advantage of low, interruptible electric rates. Storage tanks can also be oversized to provide water for emergency cooling or fire fighting. Cooling can be extracted very quickly from a thermal storage system to satisfy a very large load and then be recharged slowly, using a small charging system. System design may be simplified when the load profile is determined by production scheduling rather than occupancy and outdoor temperatures.

Combustion Turbine Inlet Air Cooling. Cool storage can increase the capacity and efficiency of combustion turbines by precooling the inlet air. Oil- or gas-fired combustion turbines typically generate full rated output at an inlet air temperature of 59°F, and above rated output below 59°F. At higher inlet air temperatures, the density and thus the mass flow of air is reduced, decreasing the shaft power and fuel efficiency. Capacity and efficiency can be increased by precooling inlet combustion air with chilled-water coils or direct-contact cooling. The optimum inlet air temperature is typically in the range of 40 to 50°F, depending on the turbine model.

On a 100°F summer day, a combustion turbine driving a generator may lose up to 25% of its rated output. This drop in generator output occurs at the most inopportune time, when an electrical utility or industrial user is most in need of the additional electrical output available from combined cycle or peaking turbines. This lost capacity can be regained if the inlet air is artificially cooled (Andrepont 1994; Ebeling et al. 1994; MacCracken 1994; Mackie 1994). Cool storage systems using chilled water or ice are well suited to this application because the cooling is generated and stored during off-peak periods, when demand for the generator output is lower, and the maximum net increase in generator capacity is obtained by turning off the refrigeration compressor during peak demand periods, providing the added generator capacity at the time it is needed most (Andrepont 2001; Cross et al. 1995; Liebendorfer and Andrepont 2005). Some parasitic power is required to operate water pumps to use the stored cooling energy.

Although instantaneous cooling can also be used for inlet air precooling, this refrigeration equipment often uses 25 to 50% of the increased turbine power output. It is especially taxing because the maximum need occurs during times of highest cost for energy and lowest output of the direct cooling compressor, because of the higher ambient temperature. Cool storage systems can meet all or part of the load requirements for inlet air precooling. Evaporative coolers are also a simple and well-established technology for inlet air precooling, but they cannot deliver the low air temperatures available from refrigeration systems. Additional information can be found in Chapter 17 and in the *Design Guide for Combustion Turbine Inlet Air Cooling Systems* (Stewart 1999).

Mission-Critical Operations. Thermal storage systems can be used for emergency back-up cooling during unplanned downtime of the chillers that serve mission-critical operations such as data processing centers. Such operations cannot tolerate even short periods of cooling interruptions. If a chiller serving one of these areas unexpectedly goes off-line, it will take at least 20 min to restart the chiller, during which time some kind of back-up cooling is required. Providing emergency back-up cooling with cool storage is typically more economical to install and operate than a redundant air-conditioning system.

SIZING COOL STORAGE SYSTEMS

A first step in sizing a cool storage system is to determine the amount of the design load to be met from storage. Proper sizing requires calculating a design load profile and determining the required chiller and storage capacity to meet the load profile.

Sizing Strategies

Cool storage systems can be sized for full or partial storage operation on the design day. A full storage, or load-shifting system, meets the entire design day on-peak cooling load from storage. A partial storage system meets a portion of the design day on-peak cooling load from storage, with the remainder of the load met by operating the chilling equipment. A load-leveling design is a partial storage approach that minimizes equipment size and storage capacity. The system operates with refrigeration equipment running at full capacity for 24 h to meet the design load profile. When the load is less than the chiller output, the excess cooling is stored. When the load exceeds the chiller capacity, the additional requirement is discharged from storage.

Calculating Load Profiles

An accurate calculation of the design load profile is important to the success of a cool storage design, for several reasons, such as the following:

- To determine the amount of storage capacity required, the total integrated cooling load must be known.
- To ensure that the storage capacity is available when it is needed, the timing of the loads must be known.
- To ensure that cool storage suppliers provide equipment with the appropriate performance, the load profile must be specified.

Knowledge of the cooling load is required for each hour of the storage cycle, which, for most systems, is a 24 h design day. A cool storage system based on an erroneous load calculation may be undersized and unable to meet the full load, or it may be oversized and unnecessarily expensive and inefficient. Note, however, that in some cases designers may intentionally specify excess capacity to provide for future load increases.

The shape of the load profile also affects the performance characteristics of storage systems. For example, many cool storage systems respond differently to a constant load than to a load that increases to a peak and then decreases to a lower level. Therefore, it is important to accurately calculate the load for each hour of the storage and discharge cycles to allow designing to the proper total and instantaneous capacities.

Calculating loads for storage systems is similar to that for nonstorage designs, except that all heat gains for the entire design day, on an hourly basis, must be considered.

Select design temperatures with care, based on the allowable number of hours that the load may exceed the system capacity. Note that system performance is dependent on the sequence of weather conditions as well as the peak daily condition. Colliver et al. (1998) developed design weather profiles for locations across North America, with design weather conditions for one, three, five, and seven consecutive days.

Establish accurate schedules of occupancy, lighting, and equipment use. For example, if equipment, such as computers, operates continuously, heat gains during unoccupied periods are significant. The schedules should account for any changes to occur in the foreseeable future.

Include all sources of heat in the conditioned space. Nearly all electric input to the building, as well as external heat gains from conduction and solar radiation, will eventually become a load on the cool storage system. Include reheat load when reheat is to be used and include heat gains from the surroundings to the storage tank. Gatley and Riticher (1985) present detailed lists of often-overlooked heat sources that can significantly affect the load profile.

Consider any pull-down loads, which accumulate when systems are shut down during unoccupied periods, and must be met during the first minutes or hours of system operation.

Use the expected relative humidity in the building to calculate latent heat gains from infiltration, especially if supply air temperatures are to be reduced.

Note that there are two types of design days for cool storage applications: the day with the maximum 1 h cooling load for the year, and the day with the maximum integrated load over a 24 h period. In some buildings, these two types of peaks will not occur on the same day. Systems must be sized to provide sufficient cooling capacity on the day with the highest integrated cooling load as well as the day with the highest hourly load.

Recognize that the calculated building load does not represent the chilled-water load on the central plant. To obtain the load profile for the cooling plant, the designer must consider the operation of the air and water distribution systems that are used to meet the loads. This analysis should include

- Control sequences for chilled water coils and space temperatures, including set points and throttling ranges
- Fan and duct heat gains
- Pump and piping heat gains
- Heat gains to the storage system from external conduction to the storage tank

Load calculations for cool storage systems are discussed further in the *Design Guide for Cool Thermal Storage* (Dorgan and Elleson 1993). General air-conditioning cooling load calculations are discussed in detail in Chapter 30 of the 2005 ASHRAE *Handbook—Fundamentals*.

When cool storage is to be installed in an existing facility, determining loads by direct measurement may be preferable to estimating the loads. Load data may be available from logs collected by the building automation system, from chiller logs maintained by operating personnel, or from on-site measurements made over a period of design or near-design weather conditions. Field measurements should be verified by comparison with reference instruments of known accuracy, energy balance calculations, or some other external verification. Note that, if the load on an existing system exceeds the capacity of the existing equipment, measured loads will be limited by the equipment capacity and may not reflect the actual loads.

Sizing Equipment

Quick sizing formulas are available to estimate the chiller and storage capacities necessary to meet a given load profile. However, accurate evaluation of chiller performance and storage at the appropriate conditions for each hour of the design cooling cycle is required for proper sizing of equipment.

Unlike most air-conditioning and refrigeration equipment, cool storage devices have no sustained, steady-state operating condition at which equipment performance can be characterized using standard parameters. The *usable* storage capacity of a cool storage device may vary appreciably depending on the application (ARI *Guideline* T).

Usable capacity depends on the load profile and temperature requirements that the cool storage equipment must meet. Net usable capacity is typically not equal to, and in fact is nearly always less than, the nominal capacity. The designer must carry out an hourly analysis to determine the nominal storage capacity and discharging capacity required to meet a given load profile. Hourly analysis is also necessary to specify the thermal performance that cool storage equipment must provide.

Specifications for cool storage equipment must include the operating profile that the equipment is required to meet. (Omitting the operating profile from the cool storage specification is comparable to omitting the design chilled-water and condenser water temperatures from a chiller specification.) The design team must prepare an hourly operating profile to select the appropriate chiller and storage capacities to meet the load. The operating profile defines the system operating conditions each hour of the complete storage cycle. Most cool storage systems are designed for a one-day storage cycle, and the operating profile includes data for a 24 h period.

In addition to determining capacity requirements, designers use the hourly operating profile to

- Develop control sequences
- Define performance requirements that suppliers of storage equipment must meet and against which commissioners must validate actual performance
- Determine requirements for components such as pumps and heat exchangers
- Evaluate system performance after it is installed

The operating profile must include the following information for each hour of the storage cycle:

- Ambient dry- and wet-bulb temperatures
- Total system load
- Load met by chiller(s)
- Load met by storage equipment
- Fluid temperature(s) entering chiller(s)
- Fluid temperature(s) leaving chiller(s)
- Fluid flow rate to chiller(s)
- Fluid temperature(s) entering storage
- Fluid temperature(s) leaving storage
- Fluid flow rate to storage (charging mode)
- Fluid flow rate leaving storage (discharging mode)
- Fluid pressure drop through storage, chiller and piping system
- Total capacity (ton-hours) of usable stored cooling
- Maximum cooling capacity or electric demand for chilling equipment

The hourly operating profile should be developed using manufacturers' performance data for specific chiller and storage selections. Typically, designers must iterate between chiller data and storage data to determine the balance between the components. For example, the storage performance characteristics determine the temperatures required to charge storage with a given chiller capacity. Chiller capacity varies with fluid flow rate and temperature entering the evaporator, and with condensing temperature.

The capacity and performance of existing chillers and cooling towers may not be equal to the original ratings, because of fouling or other deterioration. Performance ratings also need to be adjusted to account for the new anticipated operating conditions. Capacity of the existing chillers should be verified by test before completing the sizing of the storage system.

ARI *Guideline* T defines the minimum information to be provided by prospective users when specifying performance requirements for cool storage equipment, and the minimum information to be provided by suppliers about their equipment's thermal performance. The hourly operating profile as described previously includes user-specified data requirements from *Guideline* T, as well as additional information useful for design.

The project specification should require that submittals for cool storage equipment provide all supplier-specified information required by *Guideline* T. The project specification should also require that the equipment be guaranteed to perform as represented by the submittal data when tested according to a specified procedure. Test methods are discussed in ASHRAE *Standards* 94.2 and 94.3.

The capacity of chilling equipment is an important aspect of sizing a cool storage system. ARI *Standard* 550 allows actual capacities to deviate by approximately ±5% from the manufacturer's ratings. A 5% shortfall in chiller capacity, accumulated over one or more storage cycles, can result in a significant shortage of cooling capacity, which may make it impossible to meet the design cooling loads.

Designers may wish to include a requirement for zero negative capacity tolerance in chiller specifications. Such a requirement should also include provisions for factory testing and certification of the actual chiller's capacity. This may be a consultant- and/or customer-witnessed test. Some designers may elect to include a safety factor of 5 to 10% in the chiller capacity specification. This can be accomplished, for example, by specifying a chiller capable of cooling 105% of the design flow rate at the design temperatures (Dorgan and Elleson 1993).

Judicious use of safety factors is good engineering practice. However, the economics of cool storage are particularly sensitive to misapplication of safety factors. Safety factors should not be used as a substitute for sound engineering principles. The most successful systems are designed with modest safety factors for chiller and storage capacity.

Successful systems are also designed with flexibility to change operating strategies if operating requirements change or if loads increase. In cases with a high probability of increased future loads, designs should include provisions for adding capacity. Operating strategies should, as a minimum, be designed to ensure that excess capacity is used to full advantage, if actual loads are lower than projected. This further reduces operating costs.

APPLICATION OF THERMAL STORAGE SYSTEMS

Chilled-Water Storage Systems

Sensible storage may be applied to any cooling load within its application temperature range. Stratified chilled-water storage systems have been successfully applied in hospitals, schools, industrial facilities, campus and district cooling systems, power generation (combustion turbine inlet air cooling), emergency cooling systems, and others. Sensible storage tends to be the most competitive in large applications (above 3000 ton-hours or millions of gallons of storage medium) because of the low cost per ton-hour of large storage volumes. However, smaller systems are installed occasionally for a variety of reasons. For example, emergency cooling systems that may be required to discharge for only 10 or 20 min may be quite small relative to typical load-shifting systems. Various thermal storage project delivery methods are possible. Numerous manufacturers produce turnkey storage tanks, and they may team with engineering and construction firms in a design/build relationship. At the other extreme, many systems have been designed successfully by mechanical engineering firms using generic water storage tanks, possibly with the assistance of a specialty consultant to design diffusers and advise on interface design and, possibly, the controls. It is particularly important that the storage design be such that the storage system will be properly integrated with the existing system, so the storage system designer must be expert in chilled-water system design generally.

Sensible storage vessels can be connected to chilled-water systems in a variety of ways. One common approach is to associate the storage system with the chilled-water plant so it functions as an alternative source of capacity but has no effect on distribution of flow to end users. Sensible storage can also be located remotely from the chilled-water plant, in which case it functions like a satellite plant. This approach has been used in systems that were experiencing distribution system capacity limitations (Andrepont and Kohlenberg 2005; Borer and Schwartz 2005). Storage can be charged when loads are low, and used to backfeed the distribution system during peak periods. Examples of several Australian systems with satellite storage are described by Bahnfleth et al. (2003b).

A typical inside-the-plant connection scheme for a stratified sensible storage system serving a primary/secondary chilled-water system is shown in Figure 18. As indicated by flow arrows, the direction of flow between the storage tank and the system must reverse between the charge and discharge modes. The storage vessel is connected to the system by a transfer pumping interface, which controls the direction of flow between the tank and the system and accommodates pressure differences between the system and the tank. This is necessary in stratified tanks (typically open to the atmosphere) only when the tank water level is below the highest point in the chilled-water system. Under such conditions, an interface is needed to exchange thermal energy with the system while preventing the elevated system water from draining into the tank and overflowing.

The transfer pumping interface in Figure 18 may have several configurations, the two most common of which are shown in Figures 19 to 22. These figures are schematic. They do not show pump trim items such as check valves and strainers. One pump is shown, but multiple pumps in parallel are used in many systems. Likewise, parallel control valves may be specified in some systems to provide the required rangeability.

Figure 19 shows a direct interface, which allows water to flow between the tank and the system while maintaining tank and system pressure levels by using modulating pressure-sustaining valves. The example shown has four two-position valves and two pressure-sustaining valves. It is configured so the pump always pumps out of

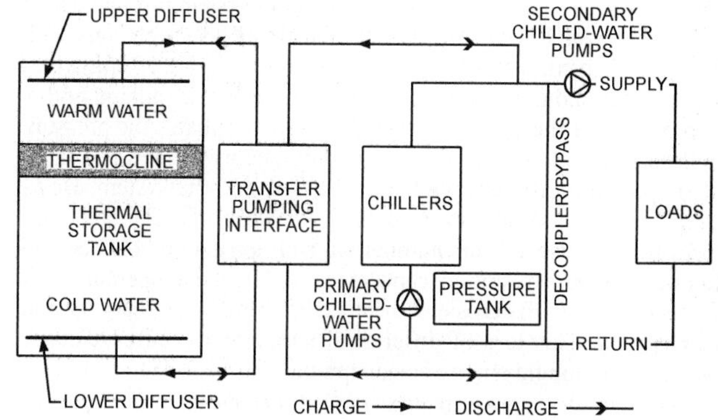

Fig. 18 Typical Sensible Storage Connection Scheme

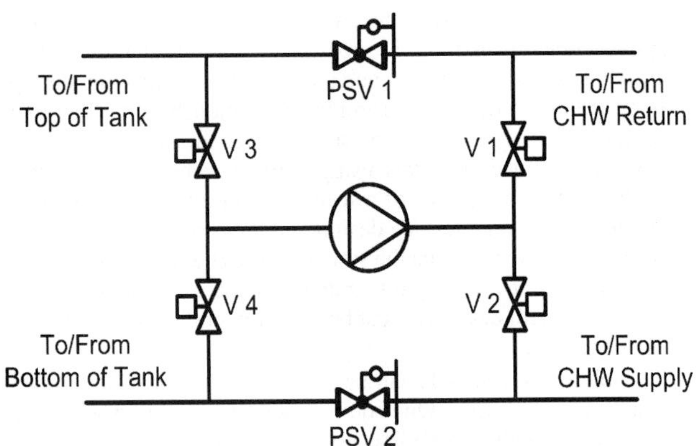

Fig. 19 Direct Transfer Pumping Interface

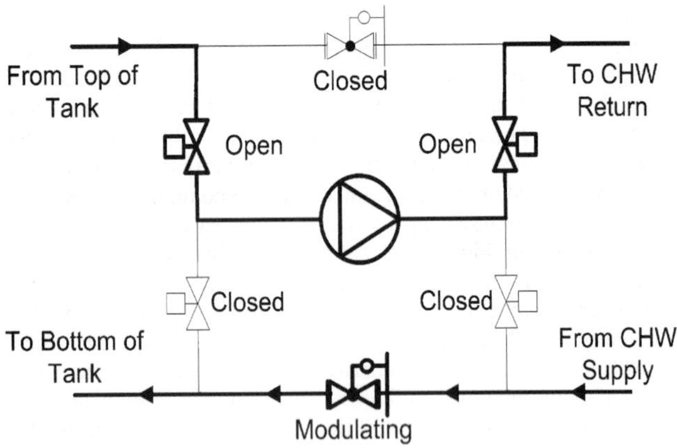

Fig. 20 Charge Mode Status of Direct Transfer Pumping Interface

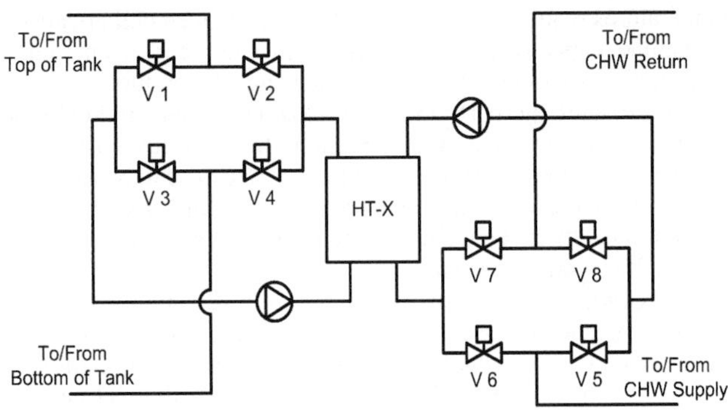

Fig. 21 Indirect Transfer Pumping Interface

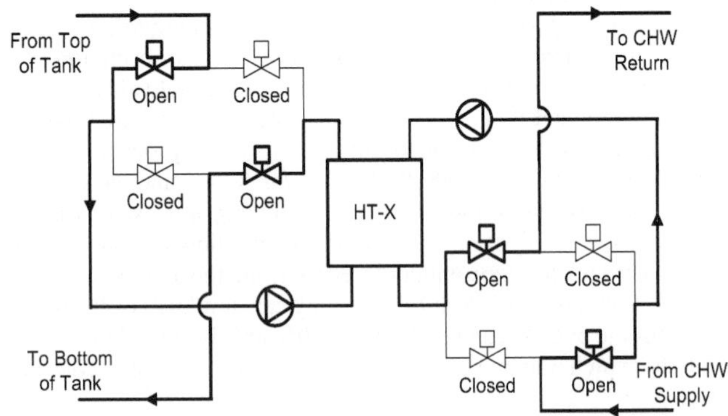

Fig. 22 Charge Mode Status of Indirect Transfer Pumping Interface

the tank into the system (i.e., from the lower-pressure zone to the higher-pressure zone). Figure 20 illustrates valve status and flow paths for charge mode. One pair of two-position valves is open and the other pair is closed, and one pressure-sustaining valve modulates and the other is closed. The status of all valves is reversed during discharge mode.

If the water level in the storage tank is at or above the water level in the tallest building or highest air handler or piping run, pressure-sustaining valves are unnecessary. In this case, the only valves required would be to change flow direction when the system is switched from the charge mode to the discharge mode and vice versa. In some cases (see Figure 23), no valves are needed to change the direction of flow.

Figure 21 is an example of an indirect interface that connects the thermal storage system and the chilled-water system through a heat exchanger, which is typically a plate-and-frame heat exchanger with a 2 to 3°F maximum approach temperature. Figure 22 illustrates valve status and flow paths for charge mode. The advantage of this approach over the direct interface is that it eliminates the pressure control problem, and may save pumping energy if there is a large static pressure difference between the tank and the system. However, the added installation and energy cost of this scheme is significant, because of the number of additional components and unavoidable loss of storage capacity caused by the temperature differences (4 to 6°F) across the heat exchanger during charging and discharging. This loss of differential temperature causes a major penalty in pumping energy consumption.

For example, suppose that the chilled-water system in Figure 18 has a supply (and charging) temperature of 40°F and a return temperature of 60°F. If the heat exchanger in an indirect interface has a 2°F approach, the temperature of water in the charged tank will be 42°F and after complete discharge it will be 58°F. Consequently, the maximum temperature difference to which the storage medium is subjected is 16°F, rather than the 20°F that could be achieved with a direct interface. To have the same capacity as a direct interface system, the tank of the indirect interface system needs to be 25% larger.

Bahnfleth and Kirchner (1999) performed a parametric study of transfer pumping interface economics for direct and indirect interfaces as well as direct interfaces with hydraulic energy recovery. Considering the increased cost of the storage tank to achieve equal storage capacity when an indirect interface is used, this study concluded that, until static pressure differential reaches several hundred feet of water, the life-cycle cost of a direct interface is much more acceptable than for an indirect interface system.

An alternative to the centralized indirect interface shown in Figures 18 and 21 is to isolate loads that are at higher elevation than the tank water level. This may result in lower system cost than a central heat exchanger, and significantly lower pumping energy consumption than a direct interface system. An analysis must be made to determine the total cost and energy consumption of the distributed pump/pressure-sustaining valve approach relative to centrally located valves at the storage tank to ensure that the proper scheme is being specified.

When it is possible to design the thermal storage tank surface to be the high water level in the system, it can function as the expansion tank and possibly provide a much simpler system design. One particularly attractive arrangement is to use the thermal storage tank as the decoupler in a primary/secondary system, as shown in Figure 23. In this system, if the flow demanded by the secondary pump exceeds the flow provided by the primary pump serving the chiller, the difference is withdrawn from the bottom of the tank and replaced by return water at the top. During charging, any surplus flow available from the chiller after the system load is met enters the bottom of the tank, displacing warm water, which flows from the top of the tank to the chiller.

Figure 23 also represents a possible configuration for a stratified emergency cooling system of the type used in some data centers and other critical loads to cover transitions to emergency systems during an equipment failure. These systems may discharge at a very high rate for a period of only a few minutes. A piping and control configuration that allows stored capacity to be available instantly, without any valve repositioning, is required.

Ice (and PCM) Storage Systems

Partial storage applications present the greatest challenge in allocating the relative contributions of chiller and storage to the cooling load. The goal is to minimize energy cost while preventing premature exhaustion of the storage capacity. Storage devices have a finite total capacity in ton-hours and an instantaneous capacity (tons) that may vary with storage inventory and coolant temperature. Ice storage systems that use secondary coolants incorporate most of the essential design elements, and the following discussion assumes this type of storage system. Note that, even on the design day, there are hours that are less than peak load. A conventional-system chiller would unload during these hours, but partial storage sizing often takes advantage of the fact that the chiller can be fully loaded throughout the design day, minimizing the investment in equipment and maximizing the efficiency of chiller operation.

On the design day, the operating logic is usually predetermined and obvious. The challenge in maximizing energy and demand savings usually occurs on days with reduced load, which, of course, comprises most of the operating hours. Control schemes can be as simple or as complex as desired, consistent with the utility rate; building load patterns; and the training, experience, and commitment of the system operators. Very effective control schemes have been as straightforward as hot day/mild day/cool day. The chiller is fully loaded on a hot day, half-loaded on a mild day, and off on a cool day. An increased level of complexity might attempt to limit demand for each billing period. There is a minimum chiller demand that can be predicted for any billing period, either by analysis of cooling loads, from experience, or established by a demand ratchet from a previous month

Because there is no avoidable demand penalty (kilowatts) for chiller operation up to this level, the cost of energy (kilowatt-hours) becomes the dominant influence when lower-load days are addressed. If there is a significant difference in on-peak and off-peak rates, further reduction in the hours of on-peak chiller operation is warranted, if practical. If the rates are approximately equal, chiller operation up to this preestablished limit carries no penalty. In either case, demand savings are maximized. Even more complex methods are available that track storage inventory, cooling load, outdoor conditions, etc., and then modulate chiller loading to maximize savings under specific utility rates. In most cases, simple control schemes are very effective and more long-lasting with the operators.

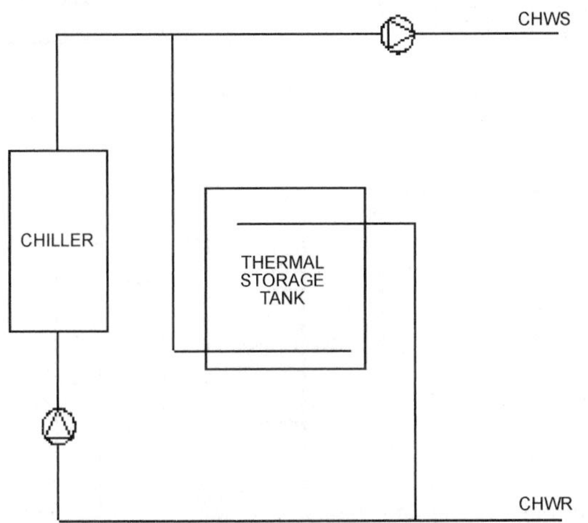

Fig. 23 Primary/Secondary Chilled-Water Plant with Stratified Storage Tank as Decoupler

The three-way storage valve (V1 in Figures 24 to 26) responds to two separate system characteristics: (1) the variation in the required contribution from storage as the building load ramps up and down and as the chiller capacity varies, and (2) the storage system's variable performance.

The temperature of coolant exiting the storage device is a function of flow, inlet temperature, and ice inventory. The temperature-modulating valve automatically compensates for all of these effects, in addition to providing isolation of storage when necessary.

Ice Making. Each system depicted in Figures 24 to 26 includes a temperature-blending valve (V1) around the storage module. This valve is simply driven to direct all flow through the storage device during the ice-making period. Termination of ice making may be indicated by an inventory measurement or by the attainment of a specific coolant temperature. The chiller is typically fully loaded for the duration of ice making. Once a full charge is reached, further ice-making operation is prevented until the next scheduled ice-making period.

It is often necessary to serve a cooling load during the charging mode. The ability to efficiently meet small night loads is a major advantage of storage systems. The temperature of coolant circulating in the primary loop during charge mode is below 32°F and considerably lower than that normally delivered to the secondary load loop. Although a separate chiller operating at a higher temperature can be used to meet night loads, an additional three-way valve (V2) and bypass (illustrated in Figures 24 to 26) is often placed in the secondary loop so warm return fluid can temper the coolant delivered to the load from the ice-making primary loop. This is critical when pure water load loops are served through a heat exchanger. If a night load is present after charging is completed, valve V1 can be positioned to bypass storage, and the chiller can be operated at standard daytime temperatures. If no night loads (i.e., during ice-making) are anticipated, this valve and bypass can be eliminated, because all daytime temperature control can be accomplished with the chiller or storage three-way valve.

Series Configuration. A common piping scheme is to place the chiller and storage in series. This arrangement provides several advantages. It does not require a change in flow path during charging when both storage and chiller must be in series. A parallel arrangement usually necessitates a change in flow path as the system cycles between charge and discharge. Simple control mechanisms can be used to vary the contribution of each component to the cooling load. All modes of operation (chiller only, ice only, or any ratio of chiller and ice) are easily implemented. Also, the manufacturer may recommend that flow through the storage equipment be in the same direction for charge and discharge. The series arrangement automatically accomplishes this.

Partial storage systems use on-peak chiller capacities that typically vary from 40 to 70% of the peak load. Because it would be difficult to direct full system flow through the smaller chiller and storage at the common 10 or 12°F temperature range, series systems often use a wider differential Δt in supply and return temperatures. Ranges of 14 to 16°F are fairly common, with up to 20°F occasionally used. Flow rates are consequently lowered to levels compatible with the equipment, and pumping energy can be reduced throughout the system. For the series flow examples, 42°F supply and 58°F return at peak load are assumed.

Series Flow, Chiller Upstream. In the upstream position (Figure 24), the chiller often operates at higher daytime evaporator temperatures than it would have in the conventional system, although there may be a negative effect on total storage capacity.

If the maximum on-peak chiller capacity is half of the on-peak load, the chiller reduces the return temperature by half the design Δt, to 50°F at peak load. However, if the chiller leaving chilled-water temperature (LCWT) is simply set to 50°F, the chiller will unload any time the load is less than peak, as the return temperature decreases. This may shift cooling load to storage that should have been served by the chiller, resulting in premature depletion of storage capacity and the inability to meet the cooling load later in the day. Alternatively, by setting the chiller LCWT at 42°F, the chiller meets all cooling loads up to its full capacity, before any load is imposed on storage. In some cases, this can be the extent of the discharge control logic. In fact, through simple adjustment of the chiller supply water temperature set point or chiller demand limits, cooling load can be shifted between chiller and storage in any desired proportion to best exploit the electric rate in response to daily or seasonal load changes. As the load fluctuates and the related chiller contribution varies, the storage three-way modulating valve automatically directs sufficient flow through the storage system to maintain 42°F coolant delivered to the load.

Series Flow, Chiller Downstream. Reversing the arrangement retains all of the control flexibility, because the storage modulating valve can be used to manage the relative contributions of storage and chiller (Figure 25). Storage capacity is maximized at the expense of some chiller efficiency, which in many cases is minimal. At peak load, the storage blending valve is set to 50°F, and the chiller will be fully loaded as it further reduces the temperature to 42°F. Maintaining this storage temperature set point keeps the chiller fully loaded throughout the day, minimizing use of storage capacity. On cooler days, the storage blended outlet temperature can be reduced. Chiller capacity and electric demand are also reduced. Alternatively, chiller demand limiting can be used and the storage blending valve temperature sensor can be located downstream of the chiller. The storage automatically meets any load in excess of the chiller limit to maintain the desired supply water temperature.

In either the chiller upstream or downstream arrangement, simple temperature control or chiller demand limiting can easily manipulate the cooling loads imposed on either storage or the chiller system.

Parallel Flow. There are times when a parallel flow configuration is preferred, perhaps to address a retrofit application with a fixed-distribution Δt. Chillers in parallel load and unload in response to cooling needs. In storage applications, it is essential that storage is not simultaneously unloaded as the chiller is unloaded.

There are many variations on the parallel flow theme. Referring to the simplified schematic of Figure 26, a two position, three-way valve (V3) at the chiller outlet is included to redirect flow for the charge and discharge modes. This example assumes that the chiller meets 60% of the daytime peak load.

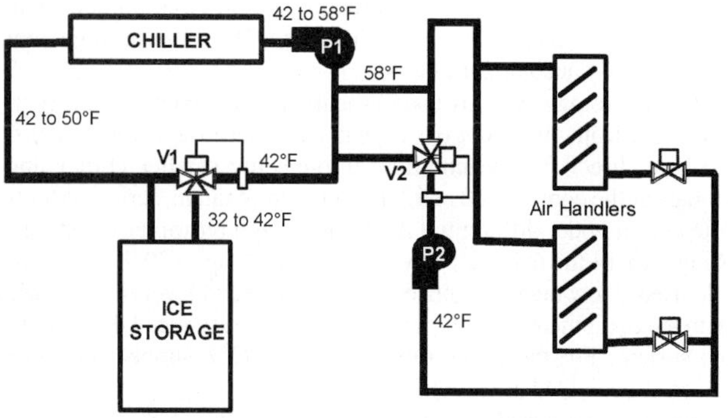

During discharge, in parallel, the same return-temperature fluid enters both the chiller and storage. With no control other than a fixed leaving temperature for both storage and chiller, the contributions of storage and chiller vary in a constant ratio as the return temperature varies. Because even on a design day, the return temperature may be reduced during much of the day, it is apparent that the chiller will unload. If the original selection assumed full capacity from the chiller during all hours, the system would then be undersized.

To avoid this problem, chiller temperature can be allowed to drop to maintain full load, and the storage blending valve sensor can be repositioned to sense the chiller/storage combined flow temperature. Variable-speed pumping of the storage loop is also possible, to maintain a fully loaded chiller at a constant chiller leaving temperature. Storage systems typically have a broad range of acceptable flow rates.

Further complications arise when chillers are demand-limited to optimize utility cost. In a constant-flow chiller loop, the chiller leaving temperature generally rises when the chiller is unloaded, necessitating a drop in storage temperature to compensate. The storage system may not have been selected to operate at the lower temperature. Variable chiller flow can provide a full system Δt at reduced capacity. In any case, it becomes more awkward to optimize load sharing between the chiller and the storage. The best approach is, regardless of system configuration, to calculate flows and temperatures resulting from the proposed control logic at a variety of cooling loads, and ensure that results are consistent with the assumptions made during equipment selection.

A versatile storage design allows the system to shift load between chiller and storage to best exploit the utility rate, thereby optimizing the total utility bill. Accompanying this versatility is the responsibility to ensure that loads are properly shared by chiller and storage under all conditions.

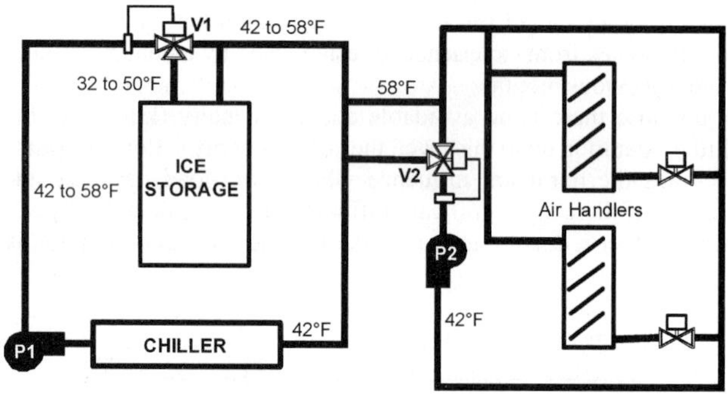

Fig. 25 Series Flow, Chiller Downstream

Fig. 24 Series Flow, Chiller Upstream

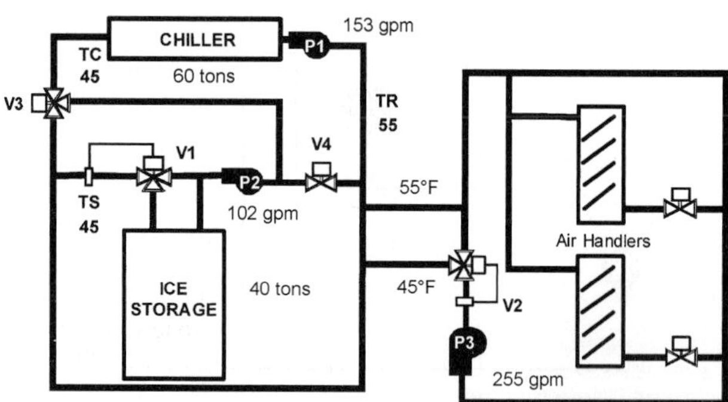

Fig. 26 Parallel Flow for Chiller and Storage

OPERATION AND CONTROL

Thermal storage operation and control is generally more schedule-dependent than that of instantaneous systems. Because thermal storage systems separate the generation of heating and cooling from its use, control of each of these functions must be considered individually. Many thermal storage systems also offer the ability to provide heating or cooling either directly or from storage. This flexibility makes it necessary to define how loads will be met at any time under various scenarios.

ASHRAE research project RP-1054 (Dorgan et al. 1999, 2001) developed a framework for describing and characterizing the many methods for controlling thermal storage systems. The methods were defined in terms of the following levels of control:

- Available **operating modes**. A thermal storage operating mode describes which of several possible functions the system is performing at a given time (e.g., charging storage, meeting load directly without storage, meeting load from discharging storage, etc.).
- **Control strategies** used to implement the operating modes. A thermal storage control strategy is the sequence of operating modes implemented under specific conditions of load, weather, season, etc. For example, different control strategies might be implemented on a design day in the summer, a cool day in the summer, and a winter day.
- The **operating strategy** that defines the overall method of control for the thermal storage system to achieve the design intent. The operating strategy provides the logic used to determine when the various operating modes and control strategies are selected. A system designed to minimize on-peak demand uses a different operating strategy than one designed to minimize use of on-peak energy or installed equipment capacity.

The design of a thermal storage system should include a detailed description of the intended operating strategy and its associated control strategies and operating modes. Dorgan et al. (1999, 2001) provides specific recommendations for documenting each of these elements of operation and control. The researchers also recommend that graphical illustrations be used to describe each operating mode and the logic used to select each mode. The documentation should address control for the complete storage cycle under full- and part-load operation, including variations related to seasonal conditions.

The operating mode, control strategy, and operating strategy framework is useful for describing the operation and control of sophisticated thermal storage systems, and for comparing the operation and control of multiple systems. Many systems use a limited number of operating modes and control strategies. For example, a simple storage system may have a single control strategy with just two modes of operation: charging and discharging (meeting load). Although the complete framework is not necessary to describe the control of such a system, it is still important for the designer to recognize and document how a system is intended to be controlled under all conditions.

Design documentation should also include the basic parameters that helped determine the operating strategy, sizing strategy (full or partial storage), the energy cost structure and the general design intent.

Operating Modes

The five most common thermal storage operating modes are described in Table 2. The operating modes used in each individual thermal storage system design vary. Some systems may use fewer modes; for example, the option to meet a building load while charging may not be available for a specific system design, when there is no cooling load during the time the system is being charged. Many systems operate under only two modes of operation: daytime and nighttime.

Table 2 Common Thermal Storage Operating Modes

Operating Mode	Heat Storage	Cool Storage
Charging storage	Operating heating equipment to add heat to storage	Operating cooling equipment to remove heat from storage
Charging storage while meeting loads	Operating heating equipment to add heat to storage *and* meet loads	Operating cooling equipment to remove heat from storage *and* meet loads
Meeting loads from storage only	Discharging (removing heat from) storage to meet loads without operating heating equipment	Discharging (adding heat to) storage to meet loads without operating cooling equipment
Meeting loads from storage and direct equipment operation	Discharging (removing heat from) storage *and* operating heating equipment to meet loads	Discharging (adding heat to) storage *and* operating cooling equipment to meet loads
Meeting loads from direct equipment operation only	Operating heating equipment to meet loads (no fluid flow to or from storage)	Operating cooling equipment to meet loads (no fluid flow to or from storage)

Many systems also include additional operating modes, such as

- Charging cool storage from free cooling (hydronic economizer)
- Charging cool storage while recovering condenser heat
- Charging heat storage with recovered condenser heat
- Discharging at distinct supply temperatures
- Discharging in conjunction with various combinations of equipment

In general, the control system selects the current operating mode based on the time of day and day of the week. Other factors that may be considered in this selection include outdoor air temperature, current system load, total facility utility demand at the billing meter, and amount of storage available compared to the storage system capacity. The logic for mode selection is defined as part of the operating strategy.

In some cases, particularly for large cool storage systems with multiple chillers and multiple loads, different operating modes or control strategies may be applied to different parts of the system at one time. For example, one chiller may operate in charging-only mode, while another chiller may operate only to meet a building load.

An operating mode is defined by its control sequence, or sequence of operation. The specified control sequence defines the specifics of what equipment is to be operated, the values of the control system set points, the function of each control valve and other control device, and any other function necessary to achieve the design intent for that operating mode. These all result in a description of individual control loops and their responses to changes in loads or other variables.

Charging Storage with No Load. Control sequences for the storage charging-only mode are generally easily defined. Typically, the generation equipment (e.g., chillers or boilers) operates at full capacity with a constant supply temperature set point and constant flow through the storage system. Charging begins at a specified time and continues until the storage is considered fully charged or the period available for charging has ended. Under this scenario, all equipment designated as being available for charging is operated for storage purposes.

Charging Storage While Meeting the Load. This mode is more complex than the charging-only mode, but is necessary in systems that must meet loads during the charging period, such as in hospitals, data centers, etc. Control sequences for this mode generally operate all designated generation equipment at its maximum capacity. Capacity that is not needed to meet the load is diverted to charging storage. Depending on the system design, the load may be piped in series or in parallel with the storage tank. Some systems may have unique requirements in this mode to allow both charging storage and distribution of capacity to meet the load. For example, for an ice storage system using a heat exchanger between a glycol/water mixture and chilled-water loop, the designer may need to address freeze protection on the chilled-water loop side of the heat exchanger.

Meeting Load from Discharging Only. A control sequence for the discharging-only mode (full storage or load shifting) is generally straightforward. The generating equipment is not operated and the entire load is met from the storage system.

Meeting Load from Discharging and Direct Equipment Operation. Control sequences for this mode, used in partial storage operation, must regulate the portion of the load at any time that will be met from storage and the proportion that will be met from direct generation. These control sequences are typically more complex, because more variables come into play in selecting what equipment to operate and at which capacity. These sequences have generally been applied to cool storage systems, although they could be equally applied to heat storage systems. The discussion here is based on cool storage applications. There are three common control sequences used for this mode: chiller-priority, storage-priority, and constant-proportion control.

Chiller-priority control includes operating the chilling equipment, up to the designated available capacity, to meet loads. Cooling loads in excess of this capacity are then met by the storage system. If a demand limit is in place, either on the facility, cooling plant, or individual chilling equipment, the available chilling equipment capacity can be less than the maximum system capacity.

Chiller-priority control can be implemented with any storage configuration; however, it is most commonly applied with the refrigeration equipment in series with and upstream of the storage components. A simple implementation method is to set both the chiller and storage discharge temperature set points equal to the desired system supply water temperature. When the load is less than that available from the cooling equipment, the equipment operates as required to maintain the set point. When the system load exceeds the available cooling equipment capacity, the supply temperature increases above its set point, which requires some flow to be diverted through the storage system to satisfy the load. It is important to note that sensing errors in downstream temperature measurement may cause the unintentional use of storage before the available equipment capacity has been fully used. When premature depletion happens, the usable storage capacity is not available when it was intended and originally scheduled, thereby resulting in a demand penalty, which will show up on the utility bill.

Storage-priority control meets the load from storage up to its available maximum discharge rate. When the load exceeds this discharge rate, the cooling equipment must operate to meet the remaining load. A storage-priority control sequence must ensure that the storage system is not depleted prematurely during the discharge cycle. Failure to properly control or limit the maximum discharge rate could result in loss of control of the cooling load, unexpected demand charges, or both. Load forecasting is required to optimize the benefits of this control method. Chapter 41 of the 2007 *ASHRAE Handbook—HVAC Applications* describes a method for forecasting diurnal energy requirements. Simpler storage-priority control sequences have also been used that include using constant system discharge rates, predetermined system discharge rate schedules, or other load predictive and control methodologies.

Constant-proportion control or **proportional control** sequences divide the load between cooling equipment and storage. The load may be divided equally or in some other specific proportion that can then be changed in response to changing conditions. Limits on the storage system discharge rate or a demand-limiting requirement on the cooling equipment, plant or facility wide, may be applied in determining these proportions.

Demand-limiting control may be applied to any of the aforementioned sequences. This control method attempts to limit the facility or cooling plant demand by either setting a maximum capacity above which the cooling equipment cannot operate or by adjusting the cooling equipment discharge temperature set point. Because chiller demand can be a significant portion of overall system demand, large demand savings are possible with this control sequence. The demand limit may be a fixed value or may change in response to predefined changes in specific conditions. Demand limiting is generally most effective when chiller capacity is controlled in response to the total demand for a facility at its billing meter. A simpler approach, which generally results in reduced demand savings, is to limit chiller capacity based on its electric demand without consideration of the facility's demand.

The storage system discharge rate may also be limited, similar to the approach taken with chiller demand limiting. Such a limit is typically used with storage priority control to ensure that sufficient capacity is available for the entire discharge cycle. The discharge rate limit may change with time based on changing conditions. The discharge rate limit can be defined as the maximum instantaneous cooling capacity supplied from the storage system. Alternatively, it may be expressed as the maximum discharge flow rate through the storage system or the minimum mixed-fluid temperature in the common discharge of the storage system and the associated bypass.

Operating strategies that continually seek to optimize system operation frequently recalculate the demand and discharge limits during the discharge period.

Nearly all partial-storage system sequences can be described by indicating (1) whether chiller-priority, storage-priority, or proportional control is to be used; and (2) by specifying the applicable cooling equipment demand limit and storage system discharge rate limit.

Applicable utility rates and system efficiency in the various operating modes ultimately determine the selection of the preferred control sequences. If on-peak energy cost is significantly higher than off-peak energy cost, use of stored energy should be maximized during on-peak period(s), and a storage-priority sequence is appropriate. If on-peak energy cost is not significantly more than off-peak energy cost, a chiller-priority sequence is appropriate. When demand charges are high, some type of demand-limiting control scheme should be implemented.

Control Strategies

A thermal storage **control strategy** is the sequence of operating modes implemented under specific conditions of load, weather, season, etc. For example, a control strategy for a summer design day might specify a discharging mode, using storage-priority control during daytime on-peak hours to minimize or eliminate on-peak chiller operation. A winter day control strategy for the same system might use a chiller-priority discharge mode to minimize the storage system operation, because the utility rates are normally close to the same value day and night, thereby negating much of the benefit of thermal storage.

Operating Strategies

It is important to distinguish between the operating strategy, which defines the higher-level logic by which the system operates, and the various control strategies, which implement the specific operating modes. The operating strategy outlines the overall method of control of the thermal storage system necessary to meet the design intent. The operating strategy determines the logic that dictates when each operating mode is to be energized, as well as the control strategy that is implemented in each mode.

Dorgan and Elleson (1993) used the term "operating strategy" to refer to full-storage and partial-storage operation. That discussion focused on design-day operation and did not discuss operation

Table 3 Recommended Accuracies of Instrumentation for Measurement of Cool Storage Capacity

	Temperature, °F	Temperature Differential, °F	Flow Rate, % of reading
Accuracy	±0.3	±0.2	±5%
Precision	±0.2	±0.15	±2%
Resolution	±0.1	±0.1	±0.1%

Notes: *Accuracy* is an instrument's ability to repeatedly indicate the true value of the measured quantity. *Precision* is closeness of agreement among repeated measurements of the same variable. *Resolution* is the smallest repeatable increment in the value of a measured variable that can be accurately measured and reported by an instrument. Values stated are maximum acceptable deviations of sensed and transmitted value from actual value, as measured by a sensor meeting industry standard for testing instrumentation and that is certified periodically to ensure its own accuracy. See Chapter 14 of the 2005 *ASHRAE Handbook—Fundamentals* for more information on instrumentation and measurement principles.

under all conditions. For example, a system designed for partial-storage operation on the design day may operate under a full-storage strategy during other times of the year.

Operating strategies that use sophisticated control routines to optimize the use of storage have been investigated (Drees 1994; Henze et al. 2005). The cost savings benefits of such optimal strategies are often small in comparison to well-designed basic strategies that rely on simpler control routines.

Instrumentation Requirements

The control system for a thermal storage system must include at least measurement points for the following fluid temperatures:

- Entering and leaving chiller(s)
- Entering and leaving storage
- To and from load

In addition, the system should include the ability to measure flow rates at the chiller(s), storage, and load, as well as electric power to the chiller(s) and auxiliaries. Some method of measuring the current storage inventory is also necessary.

Systems with control strategies that rely on measurement of load and capacity may require flow and temperature sensors of higher quality than those normally provided with HVAC control systems. The more accurate instruments are readily available, the cost above standard sensors is nominal compared to the overall cost of the total control system, and the benefits realized are extremely valuable over the life of the system. The accuracy requirements must be included in the specifications to ensure that appropriate instrumentation is provided. Field validation is recommended.

For most applications, the recommended sensor accuracies listed in Table 3 provide an uncertainty of under ±10% in cool storage capacity measurements. When accuracy requirements are critical, the design team should carry out an uncertainty analysis to determine the required accuracy of individual sensors. Note that the accuracy of the system for transmitting and reporting the measured quantities must also be considered. It is not wise to assume that all systems provide the same level of accuracy, regardless of the cost of the system or the components.

If the control system is to be used for performance monitoring, it should be selected for its ability to

- Collect and store required amount of trend log data without interfering with system control functions
- Report averages of measured quantities over a selected interval
- Calculate amount of cooling supplied to or from load, chillers, and storage
- Provide trend log reports in custom formats to fit users' needs
- Provide maximum daily cooling water temperature provided to load, and time of occurrence
- Provide maximum daily water temperature provided from storage to load, and time of occurrence

OTHER DESIGN CONSIDERATIONS

Hydronic System Design for Open Systems

For reasons of cost and ease of construction, large ice storage and chilled-water storage tanks are typically vented to the atmosphere. When, as frequently occurs, the fluid level in a storage tank is not the highest point in the system, special attention must be paid to pumping system design to ensure that the different pressure levels are maintained and the tank does not overflow. The overview of possible transfer pumping interfaces presented in this chapter identifies the main issues requiring attention in open-system design and some ways they have been successfully addressed in chilled-water storage systems installed to date. The design guide of Gatley and Mackie (1995) provides a comprehensive discussion of open systems.

Cold-Air Distribution

The low cooling supply temperatures available from ice storage and some other cool storage media allow the use of lower supply air temperatures, which permits decreased supply air volume and smaller air distribution equipment. This approach leads to the following benefits:

- **Reduced mechanical system costs**, particularly for fans and ductwork. The cost for electrical distribution to fans may also be reduced.
- **Reduction of 30 to 40% in fan electrical demand and energy consumption.**
- **Reduced space requirement for fans and ductwork.** In some cases, this can reduce the required floor-to-floor height, with significant savings in structural, envelope, and other building systems costs. In this case, if the building is being built in an area with strict height limitations, additional floor(s) may be possible, thereby greatly increasing the value of the building.
- **Improved comfort with lower space relative humidity.** Laboratory research has shown that people feel cooler and more comfortable at lower relative humidities, and air is judged to be fresher with a more acceptable air quality, as experienced with cold-air distribution systems.
- **Increased cooling capacity for existing distribution systems.** By reducing supply air temperatures, owners can avoid the expense of replacing or supplementing existing air distribution equipment.

The cost reduction enabled by cold-air distribution can make thermal storage systems competitive with nonstorage systems on an initial-cost basis (Landry and Noble 1991; Nelson 2000).

Lower supply temperatures provide greater benefits from reduced size and energy use; however, this technology involves greater departure from standard engineering practice and requires increased attention to condensation control and other critical considerations. Nominal supply air temperatures as low as 42°F can be achieved with 36°F fluid, which can be supplied from ice storage; however, 45°F supply air is a more efficient target. Supply air temperatures of 48 to 50°F can be achieved with minimal departure from standard practice and still provide significant cost reductions. The optimum supply air temperature should be determined through an analysis of initial and operating costs for the various design options.

The minimum achievable supply air temperature is determined by the chilled-fluid temperature and temperature rise between the cooling plant and terminal units. With ice storage systems, in which the cooling fluid temperature rises during discharge, the design fluid supply temperature should be based on the flow-temperature profiles of the secondary fluid and the characteristics of the selected storage device.

A heat exchanger, which is sometimes required with storage devices that operate at atmospheric pressure or between a secondary coolant and chilled-water system, adds at least 2°F to the final chilled-fluid supply temperature above the generated temperature. The temperature difference between supply fluid entering the cooling coil and air leaving the coil is generally 6 to 10°F. A closer approach (smaller temperature differential) can be achieved with more rows or a larger face area on the cooling coil, but extra heat transfer surface to provide a closer approach is often uneconomical. A 3°F temperature rise in the air between the coil discharge and supply to the space can be assumed for preliminary design analysis. This is from heat gain in the duct between the cooling coil and terminal units. With careful design and adequate insulation, this rise can be reduced to as little as 1°F.

For a given supply air volume, the desired face velocity determines the size of the coil. Coil size determines the size of the air-handling unit. A lower face velocity achieves a lower supply air temperature, whereas a higher face velocity results in smaller equipment and lower first costs. Face velocity is limited by the possibility of moisture carryover from the coil. For cold-air distribution, the face velocity should be 350 to 450 fpm, with an upper limit of 550 fpm

Cold primary air can be tempered with room air or plenum return air using fan-powered mixing boxes or induction boxes. The energy use of fan-powered mixing boxes is significant and negates much of the savings from downsizing central supply fans (Elleson 1993; Hittle and Smith 1994). Diffusers designed with a high ratio of induced room air to supply air can provide supply air directly to the space without drafts, thereby eliminating the need for fan-powered boxes.

If the supply airflow rate to occupied spaces is expected to be below 0.4 cfm/ft^2, fan-powered or induction boxes should be used to boost the air circulation rate. At supply airflow rates of 0.4 to 0.6 cfm/ft^2, a diffuser with a high ratio of induced room air to supply air should be used to ensure adequate dispersion of ventilation air throughout the space. A diffuser that relies on turbulent mixing rather than induction to temper the primary air may not be effective at this flow rate.

Cold-air distribution systems normally maintain space relative humidity between 30 and 45%, as opposed to the 50 to 60% rh generally maintained by other systems. At this lower relative humidity level, equivalent comfort conditions are provided at a higher dry-bulb temperature. The increased dry-bulb set point generally results in decreased energy consumption.

In most buildings using cold-air distribution, relative humidity is reduced throughout the building. At these low dew-point conditions, the potential for condensation is no greater than in a higher-humidity space with higher supply air temperatures. However, spaces that may be subject to infiltration of humid outside air (e.g., entryways) may need tempered supply air to avoid condensation.

Surfaces of any equipment that may be cooled below the ambient dew point, including air-handling units, ducts, and terminal boxes, should be adequately insulated. All vapor barrier penetrations should be sealed to prevent migration of moisture into the insulation. Prefabricated, insulated round ducts should be insulated externally at joints where internal insulation is not continuous. Access doors should also be well insulated.

Duct leakage is undesirable because it represents cooling capacity that is not being delivered to the conditioned space. In cold-air distribution systems, leaking air can cool nearby surfaces to the point that condensation forms, which can lead to many other problems. Designers should specify acceptable methods of sealing ducts and air-handling units, and establish allowable leakage rates and test procedures. During construction, appropriate supervision and inspection personnel should ensure that these specifications are being followed.

The same indoor air quality considerations must be applied to a cold-air distribution design as for a higher-temperature design. The required volume of outdoor ventilation air is normally the same, although the percentage is higher for cold-air distribution systems, because the total air volume is lower. As with any design, systems serving multiple spaces with different occupant densities may require special attention to ensure adequate outside air is delivered to each space.

Designs using reheat should incorporate strategies to reset the supply air temperature at low cooling loads, to avoid increased energy use associated with reheating minimum supply air quantities.

Storage of Heat in Cool Storage Units

Some cool storage installations may be used to provide storage for heating duty and/or heat reclaim. Many commercial buildings have areas that require cooling even during the winter, and the refrigeration plant must, therefore, be in operation year-round. When cool storage is charged during the off-peak period, the rejected heat may be directed to the heating system or to storage rather than to a cooling tower. Depending on the building loads, the chilled water or ice could be considered a by-product of heating. One potential use of the heat is to provide morning warm-up on the following day.

Heating, in conjunction with a cool storage system, can be achieved using heat pumps or heat reclaim equipment, such as double-bundle condensers, two condensers in series on the refrigeration side, or a heat exchanger in the condenser water circuit. Cooling is withdrawn from storage as needed; when heat is required, the cool storage is recharged and used as the heat source for the heat pump. If needed, additional heat must be obtained from another source because this type of system can supply usable heating energy only when the heat pump or heat recovery equipment is running. Possible secondary heat sources include a heat storage system, solar collectors, or waste heat recovery from exhaust air. When more cooling than heating is required, excess energy can be rejected through a cooling tower.

Individual storage units may be alternated between heat storage and cool storage service. Stored heat can be used to meet the building's heating requirements, which may include morning warm-up and/or evening heating, and stored cooling can provide midday cooling. If both the chiller and the storage are large enough, all compressor and boiler operation during the on-peak period can be avoided, and the heating or cooling requirements can be satisfied from storage. This method of operation replaces the air-side economizer cycle that uses outside air to cool the building when the enthalpy of outside air is lower than that of return air.

The extra cost of equipping a chilled-water storage facility for heat storage is small. Necessary plant additions may include a partition in the storage tank to convert a portion to warm-water storage, some additional controls, and a chiller equipped for heat reclaim. The additional cost may be offset by savings in heating energy. In fact, adding heat storage may increase the economic advantage of installing cool storage.

Care should be taken to avoid thermal shock when using cast-in-place concrete tanks, one specific type of concrete tank. Tamblyn (1985) showed that the seasonal change from heat storage to cool storage caused sizable leaks to develop if the cooldown period was fewer than five days. Raising the temperature of a cast-in-place concrete tank causes no problems because it generates compressive stresses, which concrete can sustain. Cooldown, on the other hand, causes tensile stresses, under which cast-in-place concrete has low strength and, therefore, is more prone to failure. However, precast prestressed concrete tanks are not prone to this type of failure because the walls of the tank are in constant compression from the prestressing wires that are in tension and encircle the exterior of the tank walls.

System Interface

Open Systems. Chilled-water, salt and polymeric PCMs, external-melt ice-on-coil, and ice-harvesting systems are all open (vented to the atmosphere) systems. Draindown of water from the higher-pressure system back into the storage tank(s) must be prevented by isolation valves, pressure-sustaining valves, or heat exchangers. Because of the potential for draindown, the open nature of the system, and the fact that the water being pumped may be saturated with air, the construction contractor must follow the piping details carefully to prevent pumping or piping problems.

Closed Systems. Closed systems normally circulate an aqueous secondary coolant (25 to 30% glycol solution) either directly from the storage circuit to the cooling coils or to a heat exchanger interfaced with the chilled-water system. A domestic water makeup system should not be the automatic makeup to the secondary coolant system. An automatic makeup unit that pumps a premixed solution into the system is recommended, along with an alarm signal to the building automation system to indicate makeup operation. The secondary coolant must be an industrial solution (not automotive antifreeze) with inhibitors to protect the steel and copper found in the piping. The water should be deionized; portable deionizers can be rented and the solution can be mixed on site. A calculation, backed up by metering the water as it is charged into the piping system for flushing, is needed to determine the specified concentration. Premixed coolant made with deionized water is also available, and tank truck delivery with direct pumping into the system is recommended on large systems. An accurate estimate of volume is required.

Insulation

Because the chilled-water, secondary coolant, or refrigerant temperatures are generally 10 to 20°F below those found in nonstorage systems, special care must be taken to prevent damage. Although fiberglass or other open-cell insulation is theoretically suitable when supplied with an adequate vapor barrier, experience has shown that its success is highly dependent on workmanship. Therefore, a two-layer, closed-cell material with staggered, carefully sealed joints is recommended. A thickness of 1.5 to 2 in. is normally adequate to prevent condensation in a normal room. Provisions should be made to ensure that the relative humidity in the equipment room is less than 80%, either by heating or by cooling and dehumidification.

Special attention must be paid to pump and heat exchanger insulation covers. Valve stem, gage, and thermometer penetrations and extensions should be carefully sealed and insulated to prevent condensation. PVC covering over all insulation in the mechanical room improves appearance, provides limited protection, and is easily replaced if damaged. Insulation located outdoors should be protected by an aluminum jacket.

COST CONSIDERATIONS

Cool storage system operating costs are lower in general than those of nonstorage systems. On-peak demand charges are reduced, and energy use is shifted from expensive on-peak periods to less expensive off-peak times.

Operating cost estimates can be developed to varying levels of detail and accuracy. In some cases, simple estimates based on peak demand savings may be adequate. At the other extreme, an hour-by-hour analysis, including energy consumption of chillers, pumps, and air-handler fans, may be desirable.

In many applications, most of the operating cost savings come from reductions in demand charges. Demand savings for each month of the year are calculated based on estimates or calculations of load variations throughout the year and on knowledge of system control strategies during nonpeak months. It is generally straightforward to calculate annual demand savings by repeating the design-day analysis with appropriate load profiles and rate structures for each month.

For some applications, particularly where there is a large differential between on- and off-peak energy costs, chiller energy savings are significant. For many cool storage designs, pumping energy savings are also important. Added savings in fan energy should also be considered if cold-air distribution is used.

Annual energy savings can be estimated using energy modeling programs. However, programs currently available typically do not model the unique performance characteristics of the various storage technologies. Most programs do not consider the interaction between chiller and storage performance; the effects of chiller-upstream or chiller-downstream configuration; or the performance impacts of chiller-priority, storage-priority, and demand-limiting control strategies. In some cases, these shortcomings may offset the benefits of performing a detailed computer simulation. Additional information on calculating annual energy use is given in Dorgan and Elleson (1993) and Elleson (1997).

MAINTENANCE CONSIDERATIONS

Following the manufacturer's maintenance recommendations is essential to satisfactory long-term operation. These recommendations vary, but their objective is to maintain the refrigeration equipment, refrigeration charge, coolant circulation equipment, ice builder surface, water distribution equipment, water treatment, and controls so they continue to perform at or near the same level as they did when the system was commissioned. The control systems must be kept in calibration, water and heat transfer fluids must be treated regularly, and any valves or pumps must be maintained per manufacturer's instructions. Monitoring ongoing performance against commissioned kilowatt-hours per ton-hour of cooling capacity delivered gives a continuing report of the system's dynamic performance.

The heat transfer fluid vendor should be consulted for information on treating the heat transfer fluid. Most heat transfer fluids are supplied with required additives by the manufacturer and require annual analysis.

History of previously constructed systems indicates the importance of careful observance to the following items:

- Start-up in accordance with manufacturer's directions
- Introduction of additives in accordance with manufacturer's directions
- Regular inspections (confined-space entry program should be reviewed and in place): visual periodic electronic monitors and/or divers
- Older systems: as applicable, review for upgrade modifications to (1) diffuser system and (2) control system
- Interior and exterior materials (corrosive and noncorrosive) (piping, fasteners, supports, etc.): check for container wall undercover insulation corrosion; repair damaged insulation
- Appurtenances (vents, ladders, overflows, temperature sensors): all well secured and sensors calibrated
- Electrical systems: well secured and connections secure

Water Treatment

Water treatment for thermal storage systems is not fundamentally different from water treatment for nonstorage systems, except that there is generally a greater volume of water to be treated. Although water close to freezing has very low corrosion, treatment for biological matter, scale, and corrosion must be used. Initially, many water treatment representatives felt the systems required far more chemical treatment than actually required. This was evidently because they were considering the "open" nature of the system to be a large surface area, with the associated large volume of chemical evaporation. Ahlgren (1987, 1989) provides details on water treatment for thermal storage systems. For further information, see Chapter 48 of the 2007 *ASHRAE Handbook—HVAC Applications*.

Cleaning. Starting with a clean system at start-up minimizes problems throughout the life of the system. Ahlgren (1987) provides the following list of major steps in preoperational cleaning:

1. Remove all extraneous loose debris, construction material, trash, and dirt from tanks, piping, filters, etc., before the system is filled. Removing as much dry material as possible prevents transfer to hard-to-reach portions of the system.
2. Flush the fill water pipeline separately to drain. If a new water line has been installed, be sure that rust and debris from it is not washed into the thermal storage system.
3. Fill the system with soft, clean, fresh water. Open all system valves and lines to get thorough, high-velocity recirculation.
4. Add prescribed cleaning chemicals to circulating water. Most cleaners are a blend of alkaline detergents, wetting agents, and dispersants. Be sure cleaning chemical is dissolved and distributed thoroughly so it does not settle out in one part of the system.
5. Circulate cleaning solution for manufacturer's recommended time, frequently 8 to 24 h. Check during recirculation for any plugging of filters, strainers, etc.
6. While water is recirculating at a high rate, open drain valves at the lowest points in the system and drain cleaning solution as rapidly as possible. Draining while under recirculation prevents settling of solids in remote portions of the system.
7. Open and inspect system for thoroughness of cleaning. Refill with water and start rinse recirculation. If significant amounts of contaminants are still present, repeat cleaning and draining procedure until clean.
8. When cleaning has been thoroughly accomplished, refill system with fresh water for recirculation rinse. Drain rinse water and add fresh makeup, repeating the process until all signs of cleaning chemicals have been removed.
9. System is now in a clean, unprotected state. Fill with makeup water and proceed with passivating to develop protective films on all metallic surfaces.

***Caution*:** (1) Water containing any contaminants, detergent, or disinfectant must be drained to an appropriate site. Check with local authorities before draining the system. (2) Water should be treated as soon as possible after cleaning the system. Cleaning removes any protective films from equipment surfaces, leaving them susceptible to corrosion. Especially when there is an extended period between cleaning and system start-up, corrosion and biofouling can become significant problems if water treatment is not initiated promptly.

Open Systems. Water treatment must be given close scrutiny in open systems. Although the evaporation and concentration of solids associated with cooling towers does not occur, the water may be saturated with air, so the corrosion potential is greater than in a closed system. Treatment against algae, scale, and corrosion must be provided. Water treatment must be operational immediately after cleaning. Corrosion coon assemblies should be included to monitor treatment effectiveness. Water testing and service should be performed at least once a month, preferably weekly immediately following initial start-up, by the water treatment supplier.

Systems open to the atmosphere require periodic applications of a biocide. Potable water is usually of acceptable quality for storage tanks that are filled at the installation site. Where the water is in contact with either a metallic heat exchanger or tank, additional treatment may be needed to regulate pH, chlorides, or other water quality parameters.

Closed Systems. The secondary coolant should be pretreated by the supplier. A complete analysis should be done annually. Monthly checks on the solution concentration should be made using a refractive indicator. (Automotive-type testers are not suitable.) For normal use, the solution should be good for many years without adding new inhibitors. However, provision should be made for injection of new inhibitors through a shot feeder if recommended by the manufacturer. The need for filtering, whether by including a filtering system or filtering the water or solution before it enters the system, should be carefully considered. Combination filter/feeders and corrosion coupon assemblies may be needed to monitor the effect of the solution on the copper and steel in the system.

COMMISSIONING

Commissioning is a process whereby all of the subsystems and components of the system are incorporated into a whole system that functions according to the design intent. The complete commissioning process, as outlined in ASHRAE *Guideline* 1, begins in the program phase and lasts through training of the building operators and the first year of operation of the building. Commissioning strives for improved communication among parties to the design procedure, providing for clear definition of the functional needs of the HVAC system, and documentation of design criteria, assumptions, and decisions.

Cool storage systems can especially benefit from commissioning. Some storage systems, particularly partial-storage systems, may have less reserve cooling capacity than nonstorage systems. A nonstorage system has excess capacity in every hour that is not a design hour. Storage systems benefit from the use of stored cooling to achieve a closer match of total capacity to total load. These benefits are achieved at the price of some additional complexity to control the inventory of stored cooling, and a reduced safety factor. Therefore, there is a need for increased care in design, installation, and operation, which the commissioning process helps to ensure.

Commissioning should also provide documentation of actual system capacities, which allows control strategies to be optimized to take full advantage of system capabilities.

Additional information on the commissioning process is given in Chapter 42 of the 2007 *ASHRAE Handbook—HVAC Applications*, and detailed information on commissioning and performance testing for cool storage systems is provided in ASHRAE *Guideline* 1 and *Standard* 150, and by Elleson (1997).

Statement of Design Intent

The statement of design intent defines in detail the performance requirements of the cool storage system. It is the master reference document that shapes all future work and is the benchmark by which success of the project is judged. The process of defining the design intent should begin as early as possible in the program phase, and continue throughout the design phase.

The initial statement of design intent is prepared from information developed in the owner's program and the results of the feasibility study. It describes the facility's functional needs, as well as the owner's requirements for environmental control and system performance. These items include

- Temperature and relative humidity requirements
- Cooling load parameters
- Acceptable hours (if any) that loads may exceed system capacity
- Occupancy schedules
- Reliability or redundancy requirements
- Needs for operational flexibility
- Financial criteria
- Energy performance criteria
- Demand shift

During the design phase, the following items are added:

- Narrative description of the system
- Performance goals for energy consumption and electric demand
- Hourly operating profile for design day and minimum-load day
- Schematic diagram of piping system, including cool storage system
- Description of control strategies for all possible modes and conditions

As the design develops, these items are refined and expanded, so the final statement of design intent provides a detailed description of the intended configuration, operation, and control of the system.

Commissioning Specification

The commissioning specification defines the contractual relationships by which the final system is brought into conformance with the design intent. It delineates the responsibilities of each party; specifies the requirements for construction observation, start-up, and acceptance testing; and defines the criteria by which the performance of the system is evaluated.

ASHRAE *Guideline* 1 provides additional details on and a sample of a commissioning specification.

Required Information

Certain system-specific design information must be assembled before running performance tests. These data are required to determine the test conditions and requirements for a specific system. The following information must be supplied:

- Load profile against which storage device or system is to be tested
- Tests to be performed; users may elect to perform one or more individual tests
- Boundary of system or portion of system to be tested by system capacity test
- Components with energy inputs to be included in system efficiency test
- System parameters, such as
  - Maximum usable discharge temperature
  - Maximum usable cooling supply temperature
  - Criteria for determining fully charged and fully discharged conditions
  - Maximum amount of time available for charging storage

Much of this information is normally defined in the statement of design intent.

Performance Verification

Before performance testing, verification should confirm that each component has been installed properly and that all individual components, equipment, and systems actually function in accordance with the contract documents and with the manufacturers' specifications. This verification should include items such as the following:

- Chillers maintain correct set points for each operating mode.
- Chiller capacity is within specified tolerances.
- Pumps, valves, and chillers sequence correctly in each operating mode and when changing between modes.
- Glycol concentration is as specified.
- Storage tank inlet and outlet temperatures are correct during both charging and discharging modes.
- Controllers and control valves and dampers maintain temperature set points.
- Flow switches function correctly.
- Flow rates are as reported in testing, adjusting, and balancing (TAB) report.
- Flow meters and temperature sensors report accurate measurements.
- Pressure-sustaining valve maintains correct pressure.
- Heat exchangers provide and maintain specified differential temperatures.
- Freeze protection sequence operates correctly.

Functional performance testing demonstrates that performance of the system as a whole satisfies the statement of design intent and contract documents. The functional performance test determines whether all components of the system function together, as designed, to meet the cooling load while satisfying the targets for electric demand and energy consumption. Functional performance testing is intended to answer basic questions such as the following:

- Does system function according to statement of design intent?
- Does system meet each hour's instantaneous load, at required fluid temperature?
- What is usable storage capacity of the storage tanks?
- What is available system capacity to meet the load, instantaneous and average for the day?
- What is system's efficiency in meeting the load, instantaneous and average for the day?
- What is daily tank standby loss?
- What is chiller capacity in charge mode and in direct cooling mode?

Functional performance testing also helps establish the final operating set points for control parameters such as chiller demand limit, ice-making cutout temperature, etc.

Because the performance of a cool storage system depends on the load to which it is subjected, the system must be tested with a design cooling load profile, or an equivalent profile.

ASHRAE *Standard* 150 provides detailed test methods for accurately determining a system's ability to meet a given cooling load. The standard requires some documentation before a performance test is carried out, including the design load profile, criteria for determining the fully charged and fully discharged conditions, and a schematic diagram of the system illustrating intended measuring points. This standard defines procedures for determining the charge and discharge capacities of a storage device and the total capacity and efficiency of the entire system. The starting and ending points for the performance tests are defined in terms of two distinctive states of charge: fully charged and fully discharged.

The **fully charged** condition is the state at which, according to the design, no more heat is to be removed from the tank. This state is generally reached when the control system stops the charge cycle as part of its normal control sequence, or when the maximum allowable charging period has elapsed.

The **fully discharged** condition is the state at which no more usable cooling capacity can be recovered from the tank. This condition is reached when the fluid temperature leaving the tank rises above the maximum usable chilled-water supply temperature.

Performance testing of cool storage systems tends to be more susceptible to sensor errors than testing of nonstorage systems, because data must be collected and accumulated over an extended time. Sensor accuracy should be verified in the field, particularly when the results of performance tests will have contractual or financial implications. Elleson et al. (2002) discuss verifying sensor accuracy.

Sample Commissioning Plan Outline for Chilled-Water Plants with Thermal Storage Systems

The following is an outline of a generic commissioning plan for cool storage systems.

A. Identify owner project requirements (OPR).

1. Site objectives.
2. Functional uses and operational capabilities.
3. Indoor environmental requirements.
4. Environmental and utility cost, and energy and demand savings goals.
5. Level of control desired.
6. Training expectations.
7. Expected/current capabilities of owner's O&M staff.
8. Include specific measurement performance criteria, such as discharge rate, daily cooling capacity available from storage, capacity shifted from peak to off-peak, and average daily plant performance (be explicit as to what equipment is to be included).
9. If project is a retrofit, identify current systems and their operating schedules.

B. Identify economic assumptions; perform economic feasibility analysis.
C. Identify approximate schedule of operation.
D. Define expected scope of commissioning services, systems, and equipment included, as well as commissioning team members.
E. Conduct commissioning kick-off meeting: establish schedule for future meeting; discuss how performance criteria are to be verified, monitored, and maintained.
F. Develop design intent document (OPR).
G. Conduct design review: focus on adherence to OPR, and efficacy of design parameters, including design load profile, clarity of control sequence descriptions, and maintainability of design.
H. Develop/review commissioning related specifications: include project-specific training requirements and required programming of operational block trends.
I. Conduct prebid meeting: identify commissioning requirements and expectations with potential bidders.
J. Review submittal for adherence to specification and OPR.
K. Negotiate desired exceptions to specification.
L. Prepare site specific checks and test procedures. Potential checks and tests to consider include
   1. Verify proper placement and calibration of critical sensors.
   2. Verify proper setup of valve and damper actuators.
   3. Verify proper programming of schedules, set points, key sequences of control, and graphics. Verify proper implementation of sensor calibration parameters and PID loop tuning parameters.
   4. Verify communication speeds, system response times, and maximum trending capability.
   5. Use operational trends to verify proper operation of economizers, set points, schedules, and staging.
   6. Specialized checks and tests include
      - Verify proper concentration of secondary coolant
      - Restart after power outage
      - Chiller capacity
      - Cooling capacity available from storage
      - Daily plant performance
M. Review operational block trend data. Program new trends as necessary. Conduct installation checks and functional testing. Develop summary of activities and key findings in problem log format. Discuss results with client and contractor. After contractor has made necessary corrections, review corrections and retest as necessary. Note that costs for retesting should be obligation of contractor.
N. Review as-builts and O&M documentation: be sure that clear operating plan is developed and available to operating staff.
O. Conduct/participate in/verify/document facilities staff and user training.
P. Prepare commissioning report: identify unresolved issues.
Q. Prepare systems manual.
R. Complete off-season testing and update systems manual with relevant operation and maintenance information as required.
S. Conduct one-year warranty review.
T. Conduct project wrap-up meeting. Finalize accountability requirements for maintaining performance.

GOOD PRACTICES

Elleson (1997) explored factors determining the success of cool storage systems and identified the following actions that contribute to successful systems:

- Calculating an accurate load profile
- Using an hourly operating profile to size and select equipment
- Developing a detailed description of the control strategy
- Producing a schematic diagram
- Producing a statement of design intent
- Using safety factors with care
- Planning for performance monitoring
- Producing complete design documents
- Retaining an experienced cool storage engineer to review the design before bidding the project

A number of pitfalls were also identified that can jeopardize success:

- Owner not willing to commit resources to support high-quality design process
- Owner does not identify and communicate all requirements for cooling system early in project
- Economic value of project not presented in appropriate financial terms; project goes unfunded
- Project justified on basis of incomplete or inaccurate feasibility study
- Assumptions made in feasibility study not documented and not verified in design phase
- Design intent for system not updated as design progresses, or recorded in contract documents
- System design based on estimates and assumptions rather than on engineering analysis
- Design not reviewed by qualified, independent reviewer
- Complete description of intended control strategies is not included in specifications
- Performance testing not required in contract documents, or test requirements not adequately defined
- Equipment submittals not reviewed for required conformance with specifications, or inadequate equipment substituted to reduce costs
- Operation and maintenance personnel not included in design process, or operator training is neglected
- Owner not willing to complete acceptable level of commissioning before accepting system for day-to-day use
- Owner not willing to support qualified, well-trained staff to achieve optimum performance

REFERENCES

Akbari, H., D. Samano, A. Mertol, F. Bauman, and R. Kammerud. 1986. The effect of variations in convection coefficients on thermal energy storage in buildings: Part 1—Interior partition walls. *Energy and Buildings* 9:195-211.

Ahlgren, R.M. 1987. *Water treatment technologies for thermal storage systems*. EPRI EM-5545. Electric Power Research Institute, Palo Alto, CA.

Ahlgren, R.M. 1989. Overview of water treatment practices in thermal storage systems. *ASHRAE Transactions* 95(2):963-968.

Anderson, L.O., K.G. Bernander, E. Isfalt, and A.H. Rosenfeld. 1979. Storage of heat and cooling in hollow-core concrete slabs: Swedish experience and application to large, American style building. 2nd International Conference on Energy Use Management, Los Angeles.

Andrepont, J.S. 1992. Central chilled water plant expansions and the CFC refrigerant issue—Case studies of chilled-water storage. *Proceedings of the Association of Higher Education Facilities Officers 79th Annual Meeting,* Indianapolis, IN.

Andrepont, J.S. 1994. Performance and economics of CT inlet air cooling using chilled water storage. *ASHRAE Transactions* 100(1):587.

Andrepont, J.S. 2000. Long-term performance of a low temperature fluid in thermal storage and distribution applications. *Proceedings of the 13th Annual College/University Conference of the International District Energy Association*.

Andrepont, J.S. 2001. Combustion turbine inlet air cooling (CTIAC): Benefits and technology options in district energy applications. *ASHRAE Transactions* 107(1).

Andrepont, J.S. 2005. Developments in thermal energy storage: Large applications, low temps, high efficiency, and capital savings. *Proceedings of Association of Energy Engineers (AEE) World Energy Engineering Congress*, September.

Andrepont, J.S. 2006. Stratified low-temperature fluid thermal energy storage (TES) in a major convention district—Aging gracefully, as fine wine. *ASHRAE Transactions* 112(2).

Andrepont, J.S. and M.W. Kohlenberg. 2005. A campus district cooling system expansion: Capturing millions of dollars in net present value using thermal energy storage. *Proceedings of International District Energy Association (IDEA) 18th Annual Campus Energy Conference*, March.

Andrepont, J.S. and K.C. Rice. 2002. A case study of innovation and success: A 15 MW demand-side peaking power system. *Proceedings of POWER-GEN International*, December.

Andresen, I. and M.J. Brandemuehl. 1992. Heat storage in building thermal mass: A parametric study. *ASHRAE Transactions* 98(1):910-918.

ARI. 2002. Specifying the thermal performance of cool storage equipment. *Guideline* T-2002. Air-Conditioning and Refrigeration Institute, Arlington, VA.

ARI. 2003. Performance rating of water-chilling packages using the vapor compression cycle. *Standard* 550/590-2003. Air-Conditioning and Refrigeration Institute, Arlington, VA.

Arnold, D. 1978. Comfort air conditioning and the need for refrigeration. *ASHRAE Transactions* 84(2):293-303.

ASHRAE. 1996. The HVAC commissioning process. *Guideline* 1-1996.

ASHRAE. 1993. Preparation of operating and maintenance documentation for building systems. *Guideline* 4-1993.

ASHRAE. 1981. Methods of testing thermal storage devices with electrical input and thermal output based on thermal performance. ANSI/ASHRAE *Standard* 94.2-1981 (RA 96).

ASHRAE. 1986. Methods of testing active sensible thermal energy storage devices based on thermal performance. ANSI/ASHRAE *Standard* 94.3-1986 (RA 96).

ASHRAE. 2000. Method of testing the performance of cool storage systems. ANSI/ASHRAE *Standard* 150-2000 (RA 2004).

ASME. Annual. *Boiler and pressure vessel codes.* American Society of Mechanical Engineers, New York.

Bahnfleth, W.P. and W.S. Joyce. 1994. Energy use in a district cooling system with stratified chilled-water storage. *ASHRAE Transactions* 100(1): 1767-1778.

Bahnfleth, W. and C. Kirchner. 1999. Analysis of transfer pumping interfaces for stratified chilled-water thermal storage systems—Part 2: Parametric study. *ASHRAE Transactions* 105(1):19-40.

Bahnfleth, W., J. Song, and J. Climbala. 2003a. Thermal performance of single pipe diffusers in stratified chilled-water storage tanks (RP-1185). ASHRAE Research Project, *Final Report*.

Bahnfleth, W., G. McLeod, and S. Bowins. 2003b. Chilled-water storage in western Australia. *ASHRAE Transactions* 109(1):617-625.

Balcomb, J.D. 1983. *Heat storage and distribution inside passive solar buildings*. Los Alamos National Laboratory, Los Alamos, NM.

Borer, E. and J. Schwartz. 2005. High marks for chilled-water system: Princeton upgrades and expands. *District Energy* 91(1):14-18.

Braun, J.E. 1990. Reducing energy costs and peak electrical demand through optimal control of building thermal storage. *ASHRAE Transactions* 96(2):876-888.

Braun, J.E. 2003. Load control using building thermal mass. *ASME Journal of Solar Energy Engineering* 125:292-301.

Braun, J.E., K.W. Montgomery, and N. Chaturvedi. 2001. Evaluating the performance of building thermal mass control strategies. *International Journal of HVAC&R* (now *HVAC&R Research*) 7(4):403-428.

Caldwell, J. and W. Bahnfleth. 1997. Chilled-water storage feasibility without electric rate incentives or rebates. *ASCE Journal of Architectural Engineering* 3(3):133-140.

Colliver, D.G., R.S. Gates, H. Zhang, and T. Priddy. 1998. Sequences of extreme temperature and humidity for design calculations (RP-828). *ASHRAE Transactions* 104(1):133-144.

Conniff, J.P. 1991. Strategies for reducing peak air conditioning loads by using heat storage in the building structure. *ASHRAE Transactions* 97(1):704-709.

Crane, J.M. 1991. The Consumer Research Association new office/laboratory complex, Milton Keynes: Strategy for environmental services. *CIBSE National Conference,* Canterbury, pp. 2-29.

Cross, J.K., W. Beckman, J. Mitchell, D. Reindl, and D. Knebel. 1995. Modeling of hybrid combustion turbine inlet air cooling systems. *ASHRAE Transactions* 101(2):1335-1341.

CSA. 1993. Design and construction of earth energy heat pump systems for commercial and institutional buildings. *Standard* C447-93. Canadian Standards Association, Rexdale, Ontario.

Dorgan, C.E. and J.S. Elleson. 1993. *Design guide for cool thermal storage.* ASHRAE.

Dorgan, C.B., C.E. Dorgan, and I.B.D. McIntosh. 1999. Cool storage operating and control strategies (RP-1054). ASHRAE Research Project, *Final Report.*

Dorgan, C.B., C.E. Dorgan, and I.B.D. McIntosh. 2001. A descriptive framework for cool storage operating and control strategies (RP-1045). *ASHRAE Transactions* 107(1):93-101.

Drees, K.H. 1994. *Modeling and control of area constrained ice storage systems.* M.S. thesis, Purdue University, West Lafayette, IN.

Ebeling, J.A., L. Beaty, and S.K. Blanchard. 1994. Combustion turbine inlet air cooling using ammonia-based refrigeration for capacity enhancement. thermal energy storage. *ASHRAE Transactions* 100(1):583-586.

Elleson, J.S. 1993. Energy use of fan-powered mixing boxes with cold air distribution. *ASHRAE Transactions* 99(1):1349-1358.

Elleson, J.S. 1997. *Successful cool storage projects: From planning to operation.* ASHRAE.

Elleson, J.S, J.S. Haberl, and T.A. Reddy. 2002. Field monitoring and data validation for evaluating the performance of cool storage systems. *ASHRAE Transactions* 108(1):1072-1084.

Fiorino, D.P. 1994. Energy conservation with thermally stratified chilled-water storage. *ASHRAE Transactions* 100(1):1754-1766.

Gansler, R.A., D.T. Reindl, and T.B. Jekel. 2001. Simulation of source energy utilization and emissions for HVAC systems. *ASHRAE Transactions* 107(1):39-51.

Gatley, D.P. and Mackie, E.I. 1995. *Cool storage open hydronic systems design guide.* EPRI TR 104906. Electric Power Research Institute, Palo Alto, CA.

Gatley, D.P. and J.J. Riticher. 1985. Successful thermal storage. *ASHRAE Transactions* 91(1B):843-855.

Goss, J.O., L. Hyman, and J. Corbett. 1996. Integrated heating, cooling thermal energy storage with heat pump provides economic and environmental solutions at California State University, Fullerton. *EPRI International Conference on Sustainable Thermal Energy Storage,* pp. 163-167.

Gute, G.D., W.E. Stewart, Jr., and J. Chandrasekharan. 1995. Modeling the ice-filling process of rectangular thermal energy storage tanks with multiple ice makers. *ASHRAE Transactions* 101(1):56-65.

Hall, S.H. 1993. *Feasibility studies for aquifer thermal energy storage.* PNL-8364. Pacific Northwest Laboratory, Richland, WA.

Henze, G.P., F. Clemens, and G. Knabe. 2004. Evaluation of optimal control for active and passive building thermal storage. *International Journal of Thermal Sciences* 43:173-181.

Henze, G.P., D. Kalz, S. Liu, and C. Felsmann. 2005. Experimental analysis of model based predictive optimal control for active and passive building thermal storage inventory. *International Journal of HVAC&R Research* (now *HVAC&R Research*) 11(2):189-214.

Herman, A.F.E. 1980. Underfloor, structural storage air-conditioning systems. *FRIGAIR '80 Proceedings*, Paper 7. Pretoria, South Africa.

Hersh, H., G. Mirchandani, and R. Rowe. 1982. *Evaluation and assessment of thermal energy storage for residential heating.* ANL SPG-23. Argonne National Laboratory, Argonne, IL.

Hittle, D.C. and T.R. Smith. 1994. Control strategies and energy consumption for ice storage systems using heat recovery and cold air distribution. *ASHRAE Transactions* 100(1):1221-1229.

Holness, G.V.R. 1992. Case study of combined chilled water thermal energy storage and fire protection storage. *ASHRAE Transactions* 98(1):1119-1122.

Hussain, M.A. and D.C.J. Peters. 1992. Retrofit integration of fire protection storage as chilled-water storage—A case study. *ASHRAE Transactions* 98(1):1123-1132.

Jenne, E.A. 1992. Aquifer thermal energy (heat and chill) storage. Paper presented at the 1992 Intersociety Energy Conversion Engineering Conference, PNL-8381. Pacific Northwest Laboratory, Richland, WA.

Keeney, K.R. and J.E. Braun. 1996. A simplified method for determining optimal cooling control strategies for thermal storage in building mass. *International Journal of HVAC&R Research* (now *HVAC&R Research*) 2(1):59-78.

Kintner-Meyer, M. and A.F. Emery. 1995. Optimal control of an HVAC system using cold storage and building thermal capacitance. *Energy and Buildings* 23(1):19-31.

Kirkpatrick, A.T. and J.S. Elleson. 1997. *Cold air distribution system design guide.* ASHRAE.

Knebel, D.E. 1986. Thermal storage—A showcase on cost savings. *ASHRAE Journal* 28(5):28-31.

Knebel, D.E. 1988. Economics of harvesting thermal storage systems: A case study of a merchandise distribution center. *ASHRAE Transactions* 94(1):1894-1904. Reprinted in *ASHRAE Technical Data Bulletin* 5(3): 35-39, 1989.

Knebel, D.E. and S. Houston. 1989. Case study on thermal energy storage—The Worthington Hotel. *ASHRAE Journal* 31(5):34-42.

Landry, C.M. and C.D. Noble. 1991. Case study of cost-effective low-temperature air distribution, ice thermal storage. *ASHRAE Transactions* 97(1):854-862.

Liebendorfer, K.M. and J.S. Andrepont. 2005. Cooling the hot desert wind: Turbine inlet cooling with thermal energy storage (TES) increases net power plant output 30%. *ASHRAE Transactions* 111(2):545-550.

MacCracken, C.D. 1994. An overview of the progress and the potential of thermal storage in off-peak turbine inlet cooling. *ASHRAE Transactions* 100(1):569-571.

Mackie, E.I. 1994. Inlet air cooling for a combustion turbine using thermal storage. *ASHRAE Transactions* 100(1):572-582.

Meckler, M. 1992. Design of integrated fire sprinkler piping and thermal storage systems: Benefits and challenges. *ASHRAE Transactions* 98(1): 1140-1148.

Meierhans, R.A. 1993. Slab cooling and earth coupling. *ASHRAE Transactions* 99(2):511-518.

Morgan, S. and M. Krarti. 2006. Impact of electricity rate structures on energy cost savings of pre-cooling controls for office buildings. *Energy and Buildings* 38.

Morofsky, E. 1994. *Procedures for the environmental impact assessment of aquifer thermal energy storage.* PWC/RDD/106E. Environment Canada and Public Works and Government Canada, Ottawa.

Morris, F.B., J.E. Braun, and S.J. Treado. 1994. Experimental and simulated performance of optimal control of building thermal storage. *ASHRAE Transactions* 100(1):402-414.

Musser, A. and W. Bahnfleth. 1998. Evolution of temperature distributions in a full-scale stratified chilled-water storage tank. *ASHRAE Transactions* 104(1):55-67.

Musser, A. and W. Bahnfleth. 1999. Field-measured performance of four full-scale cylindrical stratified chilled-water thermal storage tanks. *ASHRAE Transactions* 105(2):218-230.

Musser, A. and W. Bahnfleth. 2001. Parametric study of charging inlet diffuser performance in stratified chilled-water storage tanks with radial diffusers: Part 2—Dimensional analysis, parametric simulations and simplified model development. *ASHRAE Transactions* 107(2):41-58.

Nelson, K.P. 2000. Dynamic ice, low-cost alternative for new construction. *ASHRAE Transactions* 106(2):920-926.

ORNL. 1985. *Field performance of residential thermal storage systems.* EPRI EM 4041. Oak Ridge National Laboratories, Oak Ridge, TN.

Public Works Canada. 1991. Workshop on generic configurations of seasonal cold storage applications. International Energy Agency, Energy Conservation Through Energy Storage Implementing Agreement (Annex 7). PWC/RDD/89E (September). Public Works Canada, Ottawa.

Public Works Canada. 1992. *Innovative and cost-effective seasonal cold storage applications: Summary of national state-of-the-art reviews.* International Energy Agency, Energy Conservation Through Energy Storage Implementing Agreement (Annex 7). PWC/RDD/96E (June). Public Works Canada, Ottawa.

Rabl, A. and L.K. Norford. 1991. Peak load reduction by preconditioning buildings at night. *International Journal of Energy Research* 15:781-798.

Reindl, D.T., D.E. Knebel, and R.A. Gansler. 1995. Characterizing the marginal basis source energy and emissions associated with comfort cooling systems, *ASHRAE Transactions* 101(1):1353-1363.

Ruud, M.D., J.W. Mitchell, and S.A. Klein. 1990. Use of building thermal mass to offset cooling loads. *ASHRAE Transactions* 96(2):820-828.

Simmonds, P. 1991. The utilization and optimization of a building's thermal inertia in minimizing the overall energy use. *ASHRAE Transactions* 97(2):1031-1042.

Snijders, A.L. 1992. Aquifer seasonal cold storage for space conditioning: Some cost-effective applications. *ASHRAE Transactions* 98(1):1015-1022.

Stewart, W. E., G.D. Gute, J. Chandrasekharan, and C.K. Saunders. 1995a. Modeling of the melting process of ice stores in rectangular thermal energy storage tanks with multiple ice openings. *ASHRAE Transactions* 101(1):56-65.

Stewart, W.E., G.D. Gute, and C.K. Saunders. 1995b. Ice melting and melt water discharge temperature characteristics of packed ice beds for rectangular storage tanks. *ASHRAE Transactions* 101(1):79-89.

Stewart, W.E. 1999. *Design guide for combustion turbine inlet air cooling systems.* ASHRAE.

Stewart, W.E. 2000. Improved fluids for naturally stratified chilled-water storage systems. *ASHRAE Transactions* 106(1).

Stewart, W.E. 2001. Operating characteristics of five stratified chilled-water thermal storage tanks. *ASHRAE Transactions* 107(2):12-21.

Stovall, T.K. 1991. *Turbo Refrigerating Company ice storage test report.* ORNL/TM-11657. Oak Ridge National Laboratory, Oak Ridge, TN.

Tamblyn, R.T. 1985. College Park thermal storage experience. *ASHRAE Transactions* 91(1B):947-951

Tran, N., J.F. Kreider, and P. Brothers. 1989. Field measurements of chilled-water storage thermal performance. *ASHRAE Transactions.* 95(1):1106-1112.

Wildin, M.W. 1991. *Flow near the inlet and design parameters for stratified chilled-water storage.* ASME 91-HT-27. American Society of Mechanical Engineers, New York.

Wildin, M.W. 1996. Experimental results from single-pipe diffusers for stratified thermal energy storage. *ASHRAE Transactions* 102(2): 123-132.

Wildin, M.W. and C.R. Truman. 1989. Performance of stratified vertical cylindrical thermal storage tanks—Part 1: Scale model tank. *ASHRAE Transactions* 95(1):1086-1095.

Willis, S. and J. Wilkins. 1993. Mass appeal. *Building Services Journal* 15(1):25-27.

Yoo, J., M. Wildin, and C.W. Truman. 1986. Initial formation of a thermocline in stratified thermal storage tanks. *ASHRAE Transactions* 92(2A): 280-290.

BIBLIOGRAPHY

Arnold, D. 1999. Building mass cooling—Case study of alternative slab cooling strategies. *Engineering in the 21st Century—The Changing World,* CIBSE National Conference, Harrogate, pp. 136-144.

Arnold, D. 2000. Thermal storage case study: Combined building mass and cooling pond. *ASHRAE Transactions* 106(1):819-827.

ASHRAE. 2007. Specifying direct digital control systems. *Guideline* 13-2007.

ASHRAE. 2007. Safety standard for mechanical refrigeration. ANSI/ASHRAE *Standard* 15-2007.

Athienitis, A.K. and T.Y. Chen. 1993. Experimental and theoretical investigation of floor heating with thermal storage. *ASHRAE Transactions* 99(1):1049-1060.

Bellecci, C. and M. Conti. 1993. Transient behaviour analysis of a latent heat thermal storage module. *International Journal of Heat and Mass Transfer* 36(15):38-51.

Benjamin, C.S. 1994. Chilled-water storage at Washington State University: A case study. *Proceedings of the Eighty-Fifth Annual Conference of the International District Heating and Cooling Association*, pp. 330-341.

Berglund, L.G. 1991. Comfort benefits for summer air conditioning with ice storage. *ASHRAE Transactions* 97(1):843-847.

Bhansali, A. and D.C. Hittle. 1990. Estimated energy consumption and operating cost for ice storage systems with cold air distribution. *ASHRAE Transactions* 96(1):418-427.

Brady, T.W. 1994. Achieving energy conservation with ice-based thermal storage. *ASHRAE Transactions* 100(1):1735-1745.

Cai, L., W.E. Stewart, Jr., and C.W. Sohn. 1993. Turbulent buoyant flows into a two-dimensional storage tank. *International Journal of Heat and Mass Transfer* 36(17):4247-4256.

Cao, Y. and A. Faghri. 1992. A study of thermal energy storage systems with conjugate turbulent forced convection. *Journal of Heat Transfer* 114(4): 10-19.

Dorgan, C.E. and C.B. Dorgan. 1995. *Cold air distribution design guide.* TR-105604. Electric Power Research Institute, Palo Alto, CA.

Ebeling, J.A., L. Beaty, and S.K. Blanchard. 1994. Combustion turbine inlet air cooling using ammonia-based refrigeration for capacity enhancement. *ASHRAE Transactions* 100(1):583-586.

Fiorino, D.P. 1991. Case study of a large, naturally stratified, chilled-water thermal energy storage system. *ASHRAE Transactions* 97(2):1161-1169.

Fiorino, D.P. 1992. Thermal energy storage program for the 1990s. *Energy Engineering* 89(4):23-33:

Gallagher, M.W. 1991. Integrated thermal storage/life safety systems in tall buildings. *ASHRAE Transactions* 97(1):833-838.

Gatley, D.P. 1992. *Cool storage ethylene glycol design guide.* TR-100945. Electric Power Research Institute, Palo Alto, CA.

Gillespe, K.L., S.L. Blanc, and S. Parker. 1999. Performance of a hotel chilled water plant with cool storage. *ASHRAE Transactions* 99(2):SE-99-18-01.

Goncalves, L.C.C. and S.D. Probert. 1993. Thermal energy storage: Dynamic performance characteristics of cans each containing a phase-change material, assembled as a packed bed. *Applied Energy* 45(2):117.

Gretarsson, S.P., C.O. Pedersen, and R.K. Strand. 1994. Development of a fundamentally based stratified thermal storage tank model for energy analysis calculations. *ASHRAE Transactions* 100(1):1213-1220.

Guven, H. and J. Flynn. 1992. Commissioning TES systems. *Heating/Piping/Air Conditioning* (January):82-84.

Hall, S.H. 1993. *Environmental risk assessment for aquifer thermal energy storage.* PNL-8365. Pacific Northwest Laboratory, Richland, WA.

Harmon, J.J. and H.C. Yu. 1989. Design considerations for low-temperature air distribution systems. *ASHRAE Transactions* 95(1):1295-1299.

Harmon, J.J. and H.C. Yu. 1991. Centrifugal chillers and glycol ice thermal storage units. *ASHRAE Journal* 33(12):25-31.

Hensel, E.C. Jr., N.L. Robinson, J. Buntain, J.W. Glover, B.D. Birdsell, and C.W. Sohn. 1991. Chilled-water thermal storage system performance monitoring. *ASHRAE Transactions* 97(2):1151-1160.

Henze, G.V.R., M. Krarti, and M.J. Brandemuehl. 2003. Guidelines for improved performance of ice storage systems. *Energy and Buildings* 35(2):111-127.

Ihm, P., M. Krarti, and G.P. Henze. 2004. Development of a thermal energy storage model for EnergyPlus. *Energy and Buildings,* 36(8):807-808.

Jekel, T.B., J.W. Mitchell, and S.A. Klein. 1993. Modeling of ice-storage tanks. *ASHRAE Transactions* 99(1):1016-1024.

Kamel, A.A., M.V. Swami, S. Chandra, and C.W. Sohn. 1991. An experimental study of building-integrated off-peak cooling using thermal and moisture (enthalpy) storage systems. *ASHRAE Transactions* 97(2): 240-244.

Kirshenbaum, M.S. 1991. Chilled water production in ice-based thermal storage systems. *ASHRAE Transactions* 97(2):422-427.

Kleinbach, E., W. Beckman, and S. Klein. 1993. Performance study of one-dimensional models for stratified thermal storage tanks. *Solar Energy* 50(2):155.

Knebel, D.E. 1988. Optimal control of harvesting ice thermal storage systems. *AICE Proceedings*, 1990, pp. 209-214.

Knebel, D.E. 1991. *Optimal design and control of ice harvesting thermal energy storage systems.* ASME 91-HT-28. American Society of Mechanical Engineers, New York.

MacCracken, M.M. 2003. Thermal energy storage myths. *ASHRAE Journal* 45(9):36-43.

Mirth, D.R., S. Ramadhyani, and D.C. Hittle. 1993. Thermal performance of chilled water cooling coils operating at low water velocities. *ASHRAE Transactions* 99(1):43-53.

Molson, J.W., E.O. Frind, and C.D. Palmer. 1992. Thermal energy storage in an unconfined aquifer—II: Model development, validation, and application. *Water Resources Research* 28(10):28-57.

Potter, R.A., D.R Weitzel, D.J. King, and D.D. Boettner. 1995. Study of operational experience with thermal storage systems (RP-766). *ASHRAE Transactions* 101(2):549-557.

Prusa, J., G.M. Maxwell, and K.J. Timmer. 1991. A mathematical model for a phase-change, thermal energy storage system using rectangular containers. *ASHRAE Transactions* 97(2):245-261.

Rogers, E.C. and B.A. Stefl. 1993. Ethylene glycol: Its use in thermal storage and its impact on the environment. *ASHRAE Transactions* 99(1): 941-949.

Ruchti, T.L., K.H. Drees, and G.M. Decious. 1996. Near optimal ice storage system controller. *EPRI international Conference on Sustainable Thermal Energy Storage,* pp. 92-98.

Rudd, A.F. 1993. Phase-change material wallboard for distributed thermal storage in buildings. *ASHRAE Transactions* 99(2):339-346.

Rumbaugh, J., M. Blaha, W. Premerlani, F. Eddy, and W. Lorensen. 1991. *Object-oriented modeling and design.* Prentice Hall, Englewood Cliffs, NJ.

Silvetti, B. 2002. Application fundamentals of ice-based thermal storage. *ASHRAE Journal* 44(2):30-35.

Simmonds, P. 1994. A comparison of energy consumption for storage-priority and chiller-priority for ice-based thermal storage systems. *ASHRAE Transactions* 100(1):1746-1753.

Slabodkin, A.L. 1992. Integrating an off-peak cooling system with a fire suppression system in an existing high-rise office building. *ASHRAE Transactions* 98(1):1133-1139.

Snyder, M.E. and T.A. Newell. 1990. Cooling cost minimization using building mass for thermal storage. *ASHRAE Transactions* 96(2):830-838.

Sohn, C.W., D.M. Underwood, and M. Lin. 2006. An ice-ball storage cooling system for laboratory complex. *ASHRAE Transactions* 112(2).

Sohn, C.W. and J. Nixon. 2001. A long-term experience with an external-melt ice-on-coil storage cooling system. *ASHRAE Transactions* 107(1): 532-537.

Sohn, C.W., J. Fuchs, and M. Gruber. 1999. Chilled-water storage cooling system for an Army installation. *ASHRAE Transactions.* 105(2):1126-1133.

Sohn, M.S., J. Kaur, and R.L. Sawhney. 1992. Effect of storage on thermal performance of a building. *International Journal of Energy Research* 16(8):697.

Somasundaram, S., D. Brown, and K. Drost. 1992. Cost evaluation of diurnal thermal energy storage for cogeneration applications. *Energy Engineering* 89(4):8.

Sozen, M., K. Vafai, and L.A. Kennedy. 1992. Thermal charging and discharging of sensible and latent heat storage packed beds. *Journal of Thermophysics and Heat Transfer* 5(4):623.

Stovall, T.K. 1991. *Baltimore Aircoil Company (BAC) ice storage test report.* ORNL/TM-11342. Oak Ridge National Laboratory, Oak Ridge, TN.

Stovall, T.K. 1991. *CALMAC ice storage test report.* ORNL/TM-11582. Oak Ridge National Laboratory, Oak Ridge, TN.

Strand, R.K., C.O. Pedersen, and G.N. Coleman. 1994. Development of direct and indirect ice-storage models for energy analysis calculations. *ASHRAE Transactions* 100(1):1230-1244.

Tamblyn, R.T. 1985. Control concepts for thermal storage. *ASHRAE Transactions* 91(1B):5-11. Reprinted in *ASHRAE Technical Data Bulletin: Thermal Storage* 1985 (January):1-6. Also reprinted in *ASHRAE Journal* 27(5):31-34.

Tamblyn, R.T. 1990. Optimizing storage savings. *Heating/Piping/Air Conditioning* (August):43-46.

Wildin, M.W., E.I. Mackie, and W.E. Harrison. 1990. Thermal storage forum—Stratified thermal storage: A new/old technology. *ASHRAE Journal* 32(4):29-39.

Wildin, M.W. 1990. Diffuser design for naturally stratified thermal storage. *ASHRAE Transactions* 96(1):1094-1102.

CHAPTER 51

CODES AND STANDARDS

THE Codes and Standards listed here represent practices, methods, or standards published by the organizations indicated. They are useful guides for the practicing engineer in determining test methods, ratings, performance requirements, and limits of HVAC&R equipment. Copies of the standards can be obtained from most of the organizations listed in the Publisher column, from Global Engineering Documents at **global.ihs.com**, or from CSSINFO at **cssinfo.com**. Addresses of the organizations are given at the end of the chapter. A comprehensive database with over 250,000 industry, government, and international standards is at **www.nssn.org**.

Selected Codes and Standards Published by Various Societies and Associations

Subject	Title	Publisher	Reference
Air Conditioners	Commercial Application, Systems, and Equipment, 1st ed.	ACCA	ACCA Manual CS
	Residential Equipment Selection, 2nd ed.	ACCA	ANSI/ACCA Manual S
	Methods of Testing for Rating Ducted Air Terminal Units	ASHRAE	ANSI/ASHRAE 130-1996 (RA06)
	Non-Ducted Air Conditioners and Heat Pumps—Testing and Rating for Performance	ISO	ISO 5151:1994
	Ducted Air-Conditioners and Air-to-Air Heat Pumps—Testing and Rating for Performance	ISO	ISO 13253:1995
	Guidelines for Roof Mounted Outdoor Air-Conditioner Installations	SMACNA	SMACNA 1998
	Heating and Cooling Equipment (2005)	UL/CSA	ANSI/UL 1995/C22.2 No. 236-05
Central	Performance Standard for Single Package Central Air-Conditioners and Heat Pumps	CSA	CAN/CSA-C656-05
	Performance Standard for Rating Large and Single Packaged Air Conditioners and Heat Pumps	CSA	CAN/CSA-C746-06
	Performance Standard for Split-System and Single-Package Central Air Conditioners and Heat Pumps	CSA	CAN/CSA-C656-05
	Heating and Cooling Equipment (2005)	UL/CSA	ANSI/UL 1995/C22.2 No. 236-05
Gas-Fired	Gas-Fired, Heat Activated Air Conditioning and Heat Pump Appliances	CSA	ANSI Z21.40.1-1996 (R2002)/CGA 2.91-M96
	Gas-Fired Work Activated Air Conditioning and Heat Pump Appliances (Internal Combustion)	CSA	ANSI Z21.40.2-1996 (R2002)/CGA 2.92-M96
	Performance Testing and Rating of Gas-Fired Air Conditioning and Heat Pump Appliances	CSA	ANSI Z21.40.4-1996 (R2002)/CGA 2.94-M96
Packaged Terminal	Packaged Terminal Air-Conditioners and Heat Pumps	ARI/CSA	ARI 310/380-04/CSA C744-04
Room	Room Air Conditioners	AHAM	ANSI/AHAM RAC-1-2008
	Method of Testing for Rating Room Air Conditioners and Packaged Terminal Air Conditioners	ASHRAE	ANSI/ASHRAE 16-1983 (RA99)
	Method of Testing for Rating Room Air Conditioner and Packaged Terminal Air Conditioner Heating Capacity	ASHRAE	ANSI/ASHRAE 58-1986 (RA99)
	Method of Testing for Rating Fan-Coil Conditioners	ASHRAE	ANSI/ASHRAE 79-2002 (RA06)
	Performance Standard for Room Air Conditioners	CSA	CAN/CSA-C368.1-M90 (R2007)
	Room Air Conditioners	CSA	C22.2 No. 117-1970 (R2007)
	Room Air Conditioners (2007)	UL	ANSI/UL 484
Unitary	Unitary Air-Conditioning and Air-Source Heat Pump Equipment	ARI	ANSI/ARI 210/240-2006
	Sound Rating of Outdoor Unitary Equipment	ARI	ARI 270-95
	Application of Sound Rating Levels of Outdoor Unitary Equipment	ARI	ARI 275-97
	Commercial and Industrial Unitary Air-Conditioning and Heat Pump Equipment	ARI	ARI 340/360-2007
	Methods of Testing for Rating Electrically Driven Unitary Air-Conditioning and Heat Pump Equipment	ASHRAE	ANSI/ASHRAE 37-2005
	Methods of Testing for Rating Heat-Operated Unitary Air-Conditioning and Heat Pump Equipment	ASHRAE	ANSI/ASHRAE 40-2002 (RA06)
	Methods of Testing for Rating Seasonal Efficiency of Unitary Air Conditioners and Heat Pumps	ASHRAE	ANSI/ASHRAE 116-1995 (RA05)
	Method of Testing for Rating Computer and Data Processing Room Unitary Air Conditioners	ASHRAE	ANSI/ASHRAE 127-2007
	Method of Rating Unitary Spot Air Conditioners	ASHRAE	ANSI/ASHRAE 128-2001
Ships	Specification for Mechanically Refrigerated Shipboard Air Conditioner	ASTM	ASTM F1433-97 (2004)
Accessories	Flashing and Stand Combination for Air Conditioning Units (Unit Curb)	IAPMO	IAPMO PS 120-2004
Air Conditioning	Commercial Application, Systems, and Equipment, 1st ed.	ACCA	ACCA Manual CS
	Heat Pump Systems: Principles and Applications, 2nd ed.	ACCA	ACCA Manual H
	Residential Load Calculation, 8th ed.	ACCA	ANSI/ACCA Manual J
	Commercial Load Calculation, 4th ed.	ACCA	ACCA Manual N
	Comfort, Air Quality, and Efficiency by Design	ACCA	ACCA Manual RS
	Environmental Systems Technology, 2nd ed. (1999)	NEBB	NEBB

Selected Codes and Standards Published by Various Societies and Associations (*Continued*)

Subject	Title	Publisher	Reference
	Installation of Air Conditioning and Ventilating Systems	NFPA	NFPA 90A-02
	Standard of Purity for Use in Mobile Air-Conditioning Systems	SAE	SAE J1991-1999
	HVAC Systems—Applications, 1st ed.	SMACNA	SMACNA 1987
	HVAC Systems—Duct Design, 4th ed.	SMACNA	SMACNA 2006
	Heating and Cooling Equipment (2005)	UL/CSA	ANSI/UL 1995/C22.2 No. 236-05
Aircraft	Air Conditioning of Aircraft Cargo	SAE	SAE AIR806B-1997
	Aircraft Fuel Weight Penalty Due to Air Conditioning	SAE	SAE AIR1168/8-1989
	Air Conditioning Systems for Subsonic Airplanes	SAE	SAE ARP85E-1991
	Environmental Control Systems Terminology	SAE	SAE ARP147E-2001
	Testing of Airplane Installed Environmental Control Systems (ECS)	SAE	SAE ARP217D-1999
	Guide for Qualification Testing of Aircraft Air Valves	SAE	SAE ARP986C-1997
	Control of Excess Humidity in Avionics Cooling	SAE	SAE ARP987A-1997
	Engine Bleed Air Systems for Aircraft	SAE	SAE ARP1796-2007
	Aircraft Ground Air Conditioning Service Connection	SAE	SAE AS4262A-1997
	Air Cycle Air Conditioning Systems for Military Air Vehicles	SAE	SAE AS4073-2000
Automotive	Refrigerant 12 Automotive Air-Conditioning Hose	SAE	SAE J51-2004
	Design Guidelines for Air Conditioning Systems for Off-Road Operator Enclosures	SAE	SAE J169-1985
	Test Method for Measuring Power Consumption of Air Conditioning and Brake Compressors for Trucks and Buses	SAE	SAE J1340-2003
	Information Relating to Duty Cycles and Average Power Requirements of Truck and Bus Engine Accessories	SAE	SAE J1343-2000
	Rating Air-Conditioner Evaporator Air Delivery and Cooling Capacities	SAE	SAE J1487-2004
	Recovery and Recycle Equipment for Mobile Automotive Air-Conditioning Systems	SAE	SAE J1990-1999
	R134a Refrigerant Automotive Air-Conditioning Hose	SAE	SAE J2064-2005
	Service Hose for Automotive Air Conditioning	SAE	SAE J2196-1997
Ships	Mechanical Refrigeration and Air-Conditioning Installations Aboard Ship	ASHRAE	ANSI/ASHRAE 26-1996 (RA06)
	Practice for Mechanical Symbols, Shipboard Heating, Ventilation, and Air Conditioning (HVAC)	ASTM	ASTM F856-97 (2004)
Air Curtains	Laboratory Methods of Testing Air Curtains for Aerodynamic Performance	AMCA	AMCA 220-05
	Air Terminals	ARI	ARI 880-98
	Standard Methods for Laboratory Airflow Measurement	ASHRAE	ANSI/ASHRAE 41.2-1987 (RA92)
	Method of Testing the Performance of Air Outlets and Inlets	ASHRAE	ANSI/ASHRAE 70-2006
	Rating the Performance of Residential Mechanical Ventilating Equipment	CSA	CAN/CSA C260-M90 (R2007)
	Air Curtains for Entranceways in Food and Food Service Establishments	NSF	NSF/ANSI 37-2007
Air Diffusion	Air Distribution Basics for Residential and Small Commercial Buildings, 1st ed.	ACCA	ACCA Manual T
	Test Code for Grilles, Registers and Diffusers	ADC	ADC 1062:GRD-84
	Method of Testing the Performance of Air Outlets and Inlets	ASHRAE	ANSI/ASHRAE 70-2006
	Method of Testing for Room Air Diffusion	ASHRAE	ANSI/ASHRAE 113-2005
Air Filters	Comfort, Air Quality, and Efficiency by Design	ACCA	ACCA Manual RS
	Industrial Ventilation: A Manual of Recommended Practice, 26th ed. (2007)	ACGIH	ACGIH
	Air Cleaners	AHAM	ANSI/AHAM AC-1-2006
	Residential Air Filter Equipment	ARI	ARI 680-2004
	Commercial and Industrial Air Filter Equipment	ARI	ARI 850-2004
	Agricultural Cabs—Engineering Control of Environmental Air Quality—Part 1: Definitions, Test Methods, and Safety Procedures	ASABE	ANSI/ASAE S525-1.2-2003
	Part 2: Pesticide Vapor Filters—Test Procedure and Performance Criteria	ASABE	ANSI/ASAE S525-2-2003
	Gravimetric and Dust-Spot Procedures for Testing Air-Cleaning Devices Used in General Ventilation for Removing Particulate Matter	ASHRAE	ANSI/ASHRAE 52.1-1992
	Method of Testing General Ventilation Air-Cleaning Devices for Removal Efficiency by Particle Size	ASHRAE	ANSI/ASHRAE 52.2-2007
	Code on Nuclear Air and Gas Treatment	ASME	ASME AG-1-2003
	Nuclear Power Plant Air-Cleaning Units and Components	ASME	ASME N509-2002
	Testing of Nuclear Air-Treatment Systems	ASME	ASME N510-2007
	Specification for Filter Units, Air Conditioning: Viscous-Impingement and Dry Types, Replaceable	ASTM	ASTM F1040-87 (2007)
	Test Method for Air Cleaning Performance of a High-Efficiency Particulate Air Filter System	ASTM	ASTM F1471-93 (2001)
	Specification for Filters Used in Air or Nitrogen Systems	ASTM	ASTM F1791-00 (2006)
	Method for Sodium Flame Test for Air Filters	BSI	BS 3928:1969
	Particulate Air Filters for General Ventilation: Determination of Filtration Performance	BSI	BS EN 779:2002
	Electrostatic Air Cleaners (2000)	UL	ANSI/UL 867
	High-Efficiency, Particulate, Air Filter Units (1996)	UL	ANSI/UL 586
	Air Filter Units (2004)	UL	ANSI/UL 900
	Exhaust Hoods for Commercial Cooking Equipment (1995)	UL	UL 710

Selected Codes and Standards Published by Various Societies and Associations (*Continued*)

Subject	Title	Publisher	Reference
	Grease Filters for Exhaust Ducts (2000)	UL	UL 1046
Air-Handling Units	Commercial Application, Systems, and Equipment, 1st ed.	ACCA	ACCA Manual CS
	Central Station Air-Handling Units	ARI	ANSI/ARI 430-99
	Non-Recirculating Direct Gas-Fired Industrial Air Heaters	CSA	ANSI Z83.4-2003/CSA 3.7-2003
Air Leakage	Residential Duct Diagnostics and Repair (2003)	ACCA	ACCA
	Air Leakage Performance for Detached Single-Family Residential Buildings	ASHRAE	ANSI/ASHRAE 119-1988 (RA04)
	Method of Determining Air Change Rates in Detached Dwellings	ASHRAE	ANSI/ASHRAE 136-1993 (RA06)
	Test Method for Determining Air Change in a Single Zone by Means of a Tracer Gas Dilution	ASTM	ASTM E741-00 (2006)
	Test Method for Determining Air Leakage Rate by Fan Pressurization	ASTM	ASTM E779-03
	Test Method for Field Measurement of Air Leakage Through Installed Exterior Window and Doors	ASTM	ASTM E783-02
	Practices for Air Leakage Site Detection in Building Envelopes and Air Retarder Systems	ASTM	ASTM E1186-03
	Test Method for Determining the Rate of Air Leakage Through Exterior Windows, Curtain Walls, and Doors Under Specified Pressure and Temperature Differences Across the Specimen	ASTM	ASTM E1424-91 (2000)
	Test Methods for Determining Airtightness of Buildings Using an Orifice Blower Door	ASTM	ASTM E1827-96 (2007)
	Practice for Determining the Effects of Temperature Cycling on Fenestration Products	ASTM	ASTM E2264-05
	Test Method for Determining Air Flow Through the Face and Sides of Exterior Windows, Curtain Walls, and Doors Under Specified Pressure Differences Across the Specimen	ASTM	ASTM E2319-04
	Test Method for Determining Air Leakage of Air Barrier Assemblies	ASTM	ASTM E2357-05
Boilers	Packaged Boiler Engineering Manual (1999)	ABMA	ABMA 100
	Selected Codes and Standards of the Boiler Industry (2001)	ABMA	ABMA 103
	Operation and Maintenance Safety Manual (1995)	ABMA	ABMA 106
	Fluidized Bed Combustion Guidelines (1995)	ABMA	ABMA 200
	Guide to Clean and Efficient Operation of Coal Stoker-Fired Boilers (2002)	ABMA	ABMA 203
	Guideline for Performance Evaluation of Heat Recovery Steam Generating Equipment (1995)	ABMA	ABMA 300
	Guidelines for Industrial Boiler Performance Improvement (1999)	ABMA	ABMA 302
	Measurement of Sound from Steam Generators (1995)	ABMA	ABMA 304
	Guideline for Gas and Oil Emission Factors for Industrial, Commercial, and Institutional Boilers (1997)	ABMA	ABMA 305
	Combustion Control Guidelines for Single Burner Firetube and Watertube Industrial/Commercial/Institutional Boilers (1999)	ABMA	ABMA 307
	Combustion Control Guidelines for Multiple-Burner Boilers (2001)	ABMA	ABMA 308
	Boiler Water Quality Requirements and Associated Steam Quality for Industrial/ Commercial and Institutional Boilers (2005)	ABMA	ABMA 402
	Commercial Application, Systems, and Equipment, 1st ed.	ACCA	ACCA Manual CS
	Method of Testing for Annual Fuel Utilization Efficiency of Residential Central Furnaces and Boilers	ASHRAE	ANSI/ASHRAE 103-1993
	Boiler and Pressure Vessel Code—Section I: Power Boilers; Section IV: Heating Boilers	ASME	BPVC-2007
	Fired Steam Generators	ASME	ASME PTC 4-1998
	Boiler, Pressure Vessel, and Pressure Piping Code	CSA	CSA B51-2003 (R2007)
	Testing Standard for Commercial Boilers, 2nd ed. (2007)	HYDI	HYDI BTS-2007
	Rating Procedure for Heating Boilers, 6th ed. (2005)	HYDI	IBR
	Prevention of Furnace Explosions/Implosions in Multiple Burner Boilers	NFPA	ANSI /NFPA 8502-99
	Heating, Water Supply, and Power Boilers—Electric (2004)	UL	ANSI/UL 834
	Boiler and Combustion Systems Hazards Code	NFPA	NFPA 85-07
Gas or Oil	Gas-Fired Low-Pressure Steam and Hot Water Boilers	CSA	ANSI Z21.13-2004/CSA 4.9-2004
	Controls and Safety Devices for Automatically Fired Boilers	ASME	ASME CSD-1-2006
	Industrial and Commercial Gas-Fired Package Boilers	CSA	CAN 1-3.1-77 (R2006)
	Oil-Burning Equipment: Steam and Hot-Water Boilers	CSA	B140.7-2005
	Single Burner Boiler Operations	NFPA	ANSI/NFPA 8501-01
	Prevention of Furnace Explosions/Implosions in Multiple Burner Boilers	NFPA	ANSI/NFPA 8502-99
	Oil-Fired Boiler Assemblies (1995)	UL	UL 726
	Commercial-Industrial Gas Heating Equipment (2006)	UL	UL 795
	Standards and Typical Specifications for Tray Type Deaerators, 7th ed. (2003)	HEI	HEI 2954
Terminology	Ultimate Boiler Industry Lexicon: Handbook of Power Utility and Boiler Terms and Phrases, 6th ed. (2001)	ABMA	ABMA 101
Building Codes	ASTM Standards Used in Building Codes	ASTM	ASTM
	Practice for Conducting Visual Assessments for Lead Hazards in Buildings	ASTM	ASTM E2255-04
	Standard Practice for Periodic Inspection of Building Facades for Unsafe Conditions	ASTM	ASTM E2270-05
	Structural Welding Code—Steel	AWS	AWS D1.1M/D1.1:2008
	BOCA National Building Code, 14th ed. (1999)	BOCA	BNBC

Selected Codes and Standards Published by Various Societies and Associations (*Continued*)

Subject	Title	Publisher	Reference
	Uniform Building Code, vol. 1, 2, and 3 (1997)	ICBO	UBC V1, V2, V3
	International Building Code (2006)	ICC	IBC
	International Code Council Performance Code (2006)	ICC	ICC PC
	International Existing Building Code (2006)	ICC	IEBC
	International Energy Conservation Code (2006)	ICC	IECC
	International Property Maintenance Code (2006)	ICC	IPMC
	International Residential Code (2006)	ICC	IRC
	Directory of Building Codes and Regulations, State and City Volumes (annual)	NCSBCS	NCSBCS (electronic only)
	Building Construction and Safety Code	NFPA	ANSI/NFPA 5000-2006
	National Building Code of Canada (2005)	NRCC	NRCC
	Standard Building Code (1999)	SBCCI	SBC
Mechanical	Safety Code for Elevators and Escalators	ASME	ASME A17.1-2004
	Natural Gas and Propane Installation Code	CSA	CAN/CSA-B149.1-05
	Propane Storage and Handling Code	CSA	CAN/CSA-B149.2-05
	Uniform Mechanical Code (2006)	IAPMO	IAPMO
	International Mechanical Code (2006)	ICC	IMC
	International Fuel Gas Code (2006)	ICC	IFGC
	Standard Gas Code (1999)	SBCCI	SBC
Burners	Guidelines for Burner Adjustments of Commercial Oil-Fired Boilers (1996)	ABMA	ABMA 303
	Domestic Gas Conversion Burners	CSA	ANSI Z21.17-1998 (R2004)/ CSA 2.7-M98
	Installation of Domestic Gas Conversion Burners	CSA	ANSI Z21.8-1994 (R2002)
	Installation Code for Oil Burning Equipment	CSA	CAN/CSA-B139-06
	Oil-Burning Equipment: General Requirements	CSA	CAN/CSA-B140.0-03
	Vapourizing-Type Oil Burners	CSA	B140.1-1966 (R2006)
	Oil Burners: Atomizing-Type	CSA	CAN/CSA-B140.2.1-M90 (R2005)
	Pressure Atomizing Oil Burner Nozzles	CSA	B140.2.2-1971 (R2006)
	Oil Burners (2003)	UL	ANSI/UL 296
	Waste Oil-Burning Air-Heating Appliances (1995)	UL	ANSI/UL 296A
	Commercial-Industrial Gas Heating Equipment (2006)	UL	UL 795
	Commercial/Industrial Gas and/or Oil-Burning Assemblies with Emission Reduction Equipment (2006)	UL	UL 2096
Chillers	Commercial Application, Systems, and Equipment, 1st ed.	ACCA	ACCA Manual CS
	Absorption Water Chilling and Water Heating Packages	ARI	ARI 560-2000
	Water Chilling Packages Using the Vapor Compression Cycle	ARI	ARI 550/590-2003
	Method of Testing Liquid-Chilling Packages	ASHRAE	ANSI/ASHRAE 30-1995
	Performance Standard for Rating Packaged Water Chillers	CSA	CAN/CSA C743-02 (R2007)
Chimneys	Specification for Clay Flue Liners	ASTM	ASTM C315-07
	Specification for Industrial Chimney Lining Brick	ASTM	ASTM C980-88 (2007)
	Practice for Installing Clay Flue Lining	ASTM	ASTM C1283-07a
	Guide for Design and Construction of Brick Liners for Industrial Chimneys	ASTM	ASTM C1298-95 (2007)
	Guide for Design, Fabrication, and Erection of Fiberglass Reinforced Plastic Chimney Liners with Coal-Fired Units	ASTM	ASTM D5364-93 (2002)
	Chimneys, Fireplaces, Vents, and Solid Fuel-Burning Appliances	NFPA	ANSI/NFPA 211-06
	Medium Heat Appliance Factory-Built Chimneys (2001)	UL	ANSI/UL 959
	Factory-Built Chimneys for Residential Type and Building Heating Appliance (2001)	UL	ANSI/UL 103
Cleanrooms	Practice for Cleaning and Maintaining Controlled Areas and Clean Rooms	ASTM	ASTM E2042-04
	Practice for Design and Construction of Aerospace Cleanrooms and Contamination Controlled Areas	ASTM	ASTM E2217-02 (2007)
	Practice for Tests of Cleanroom Materials	ASTM	ASTM E2312-04
	Practice for Aerospace Cleanrooms and Associated Controlled Environments—Cleanroom Operations	ASTM	ASTM E2352-04
	Test Method for Sizing and Counting Airborne Particulate Contamination in Clean Rooms and Other Dust-Controlled Areas Designed for Electronic and Similar Applications	ASTM	ASTM F25-04
	Practice for Continuous Sizing and Counting of Airborne Particles in Dust-Controlled Areas and Clean Rooms Using Instruments Capable of Detecting Single Sub-Micrometre and Larger Particles	ASTM	ASTM F50-07
	Procedural Standards for Certified Testing of Cleanrooms, 2nd ed. (1996)	NEBB	NEBB
Coils	Forced-Circulation Air-Cooling and Air-Heating Coils	ARI	ARI 410-2001
	Methods of Testing Forced Circulation Air Cooling and Air Heating Coils	ASHRAE	ANSI/ASHRAE 33-2000
Comfort Conditions	Threshold Limit Values for Physical Agents (updated annually)	ACGIH	ACGIH
	Good HVAC Practices for Residential and Commercial Buildings (2003)	ACCA	ACCA
	Comfort, Air Quality, and Efficiency by Design (1997)	ACCA	ACCA Manual RS

Selected Codes and Standards Published by Various Societies and Associations (*Continued*)

Subject	Title	Publisher	Reference
	Thermal Environmental Conditions for Human Occupancy	ASHRAE	ANSI/ASHRAE 55-2004
	Classification for Serviceability of an Office Facility for Thermal Environment and Indoor Air Conditions	ASTM	ASTM E2320-04
	Hot Environments—Estimation of the Heat Stress on Working Man, Based on the WBGT Index (Wet Bulb Globe Temperature)	ISO	ISO 7243:1989
	Ergonomics of the Thermal Environment—Analytical Determination and Interpretation of Thermal Comfort Using Calculation of the PMV and PPD Indices and Local Thermal Comfort Criteria	ISO	ISO 7730:2005
	Ergonomics of the Thermal Environment—Determination of Metabolic Rate	ISO	ISO 8996:2004
	Ergonomics of the Thermal Environment—Estimation of the Thermal Insulation and Water Vapour Resistance of a Clothing Ensemble	ISO	ISO 9920:2007
Compressors	Displacement Compressors, Vacuum Pumps and Blowers	ASME	ASME PTC 9-1970 (RA97)
	Performance Test Code on Compressors and Exhausters	ASME	ASME PTC 10-1997 (RA03)
	Compressed Air and Gas Handbook, 6th ed. (2003)	CAGI	CAGI
Refrigerant	Positive Displacement Condensing Units	ARI	ARI 520-2004
	Positive Displacement Refrigerant Compressors and Compressor Units	ARI	ARI 540-2004
	Safety Standard for Refrigeration Systems	ASHRAE	ANSI/ASHRAE 15-2004
	Methods of Testing for Rating Positive Displacement Refrigerant Compressors and Condensing Units	ASHRAE	ANSI/ASHRAE 23-2005
	Testing of Refrigerant Compressors	ISO	ISO 917:1989
	Refrigerant Compressors—Presentation of Performance Data	ISO	ISO 9309:1989
	Hermetic Refrigerant Motor-Compressors (1996)	UL/CSA	UL 984/C22.2 No.140.2-96 (R2001)
Computers	Method of Testing for Rating Computer and Data Processing Room Unitary Air Conditioners	ASHRAE	ANSI/ASHRAE 127-2007
	Method of Test for the Evaluation of Building Energy Analysis Computer Programs	ASHRAE	ANSI/ASHRAE 140-2007
	Protection of Electronic Computer/Data Processing Equipment	NFPA	NFPA 75-03
Condensers	Commercial Application, Systems, and Equipment, 1st ed.	ACCA	ACCA Manual CS
	Water-Cooled Refrigerant Condensers, Remote Type	ARI	ARI 450-2007
	Remote Mechanical-Draft Air-Cooled Refrigerant Condensers	ARI	ARI 460-2005
	Remote Mechanical Draft Evaporative Refrigerant Condensers	ARI	ARI 490-2003
	Safety Standard for Refrigeration Systems	ASHRAE	ANSI/ASHRAE 15-2007
	Method of Testing for Rating Remote Mechanical-Draft Air-Cooled Refrigerant Condensers	ASHRAE	ANSI/ASHRAE 20-1997 (RA06)
	Methods of Testing for Rating Water-Cooled Refrigerant Condensers	ASHRAE	ANSI/ASHRAE 22-2007
	Methods of Laboratory Testing Remote Mechanical-Draft Evaporative Refrigerant Condensers	ASHRAE	ANSI/ASHRAE 64-2005
	Steam Surface Condensers	ASME	ASME PTC 12.2-1998
	Standards for Steam Surface Condensers, 10th ed.	HEI	HEI 2629
	Standards for Direct Contact Barometric and Low Level Condensers, 7th ed. (1995)	HEI	HEI 2634
	Refrigerant-Containing Components and Accessories, Nonelectrical (2001)	UL	ANSI/UL 207
Condensing Units	Commercial Application, Systems, and Equipment, 1st ed.	ACCA	ACCA Manual CS
	Commercial and Industrial Unitary Air-Conditioning Condensing Units	ARI	ARI 365-2002
	Methods of Testing for Rating Positive Displacement Refrigerant Compressors and Condensing Units	ASHRAE	ANSI/ASHRAE 23-2005
	Heating and Cooling Equipment (2005)	UL/CSA	ANSI/UL 1995/C22.2 No. 236-95
Containers	Series 1 Freight Containers—Classifications, Dimensions, and Ratings	ISO	ISO 668:1995
	Series 1 Freight Containers—Specifications and Testing; Part 2: Thermal Containers	ISO	ISO 1496-2:1996
	Animal Environment in Cargo Compartments	SAE	SAE AIR1600A-1997
Controls	Temperature Control Systems (2002)	AABC	National Standards, Ch. 12
	BACnet®—A Data Communication Protocol for Building Automation and Control Networks	ASHRAE	ANSI/ASHRAE 135-2004
	Method of Test for Conformance to BACnet®	ASHRAE	ANSI/ASHRAE 135.1-2007
	Temperature-Indicating and Regulating Equipment	CSA	C22.2 No. 24-93 (R2003)
	Performance Requirements for Electric Heating Line-Voltage Wall Thermostats	CSA	C273.4-M1978 (R2003)
	Performance Requirements for Thermostats Used with Individual Room Electric Space Heating Devices	CSA	CAN/CSA C828-06
	Solid-State Controls for Appliances (2003)	UL	UL 244A
	Limit Controls (1994)	UL	ANSI/UL 353
	Primary Safety Controls for Gas- and Oil-Fired Appliances (1994)	UL	ANSI/UL 372
	Temperature-Indicating and -Regulating Equipment (2007)	UL	UL 873
	Tests for Safety-Related Controls Employing Solid-State Devices (2004)	UL	UL 991
	Control Centers for Changing Message Type Electric Signals (2003)	UL	UL 1433
	Automatic Electrical Controls for Household and Similar Use; Part 1: General Requirements (2002)	UL	UL 60730-1A
	Process Control Equipment (2002)	UL	UL 61010C-1

Selected Codes and Standards Published by Various Societies and Associations (*Continued*)

Subject	Title	Publisher	Reference
Commercial and Industrial	Guidelines for Boiler Control Systems (Gas/Oil Fired Boilers) (1998)	ABMA	ABMA 301
	Guideline for the Integration of Boilers and Automated Control Systems in Heating Applications (1998)	ABMA	ABMA 306
	Industrial Control and Systems: General Requirements	NEMA	NEMA ICS 1-2000 (R2005)
	Preventive Maintenance of Industrial Control and Systems Equipment	NEMA	NEMA ICS 1.3-1986 (R2001)
	Industrial Control and Systems, Controllers, Contactors, and Overload Relays Rated Not More than 2000 Volts AC or 750 Volts DC	NEMA	NEMA ICS 2-2000 (R2004)
	Industrial Control and Systems: Instructions for the Handling, Installation, Operation and Maintenance of Motor Control Centers Rated Not More than 600 Volts	NEMA	NEMA ICS 2.3-1995 (R2002)
	Industrial Control Equipment (1999)	UL	ANSI/UL 508
Residential	Manually Operated Gas Valves for Appliances, Appliance Connector Valves and Hose End Valves	CSA	ANSI Z21.15-1997 (R03)/CGA 9.1-1997
	Gas Appliance Pressure Regulators	CSA	ANSI Z21.18-2007/CSA 6.3-2007
	Automatic Gas Ignition Systems and Components	CSA	ANSI Z21.20-2007/C22.2 No. 199-2007
	Gas Appliance Thermostats	CSA	ANSI Z21.23-2000 (R2005)
	Manually-Operated Piezo-Electric Spark Gas Ignition Systems and Components	CSA	ANSI Z21.77-2005/CGA 6.23-2005
	Manually Operated Electric Gas Ignition Systems and Components	CSA	ANSI Z21.92-2005/CSA 6.29-2005 (R2007)
	Residential Controls—Electrical Wall-Mounted Room Thermostats	NEMA	NEMA DC 3-2003
	Residential Controls—Surface Type Controls for Electric Storage Water Heaters	NEMA	NEMA DC 5-2002
	Residential Controls—Temperature Limit Controls for Electric Baseboard Heaters	NEMA	NEMA DC 10-1983 (R2003)
	Hot-Water Immersion Controls	NEMA	NEMA DC 12-1985 (R2002))
	Line-Voltage Integrally Mounted Thermostats for Electric Heaters	NEMA	NEMA DC 13-1979 (R2002)
	Residential Controls—Class 2 Transformers	NEMA	NEMA DC 20-1992 (R2003)
	Safety Guidelines for the Application, Installation, and Maintenance of Solid State Controls	NEMA	NEMA ICS 1.1-1984 (R2003)
	Electrical Quick-Connect Terminals (2003)	UL	ANSI/UL 310
Coolers	Refrigeration Equipment	CSA	CAN/CSA-C22.2 No. 120-M91 (R2004)
	Unit Coolers for Refrigeration	ARI	ARI 420-2000
	Refrigeration Unit Coolers (2004)	UL	ANSI/UL 412
Air	Methods of Testing Forced Convection and Natural Convection Air Coolers for Refrigeration	ASHRAE	ANSI/ASHRAE 25-2001 (RA06)
Drinking Water	Methods of Testing for Rating Drinking-Water Coolers with Self-Contained Mechanical Refrigeration	ASHRAE	ANSI/ASHRAE 18-2006
	Drinking-Water Coolers (1993)	UL	ANSI/UL 399
	Drinking Water System Components—Health Effects	NSF	NSF/ANSI 61-2007a
Evaporative	Method of Testing Direct Evaporative Air Coolers	ASHRAE	ANSI/ASHRAE 133-2001
	Method of Test for Rating Indirect Evaporative Coolers	ASHRAE	ANSI/ASHRAE 143-2007
Food and Beverage	Terminology for Milking Machines, Milk Cooling, and Bulk Milk Handling Equipment	ASABE	ASAE S300.3-2003
	Methods of Testing for Rating Vending Machines for Bottled, Canned, and Other Sealed Beverages	ASHRAE	ANSI/ASHRAE 32.1-2004
	Methods of Testing for Rating Pre-Mix and Post-Mix Beverage Dispensing Equipment	ASHRAE	ANSI/ASHRAE 32.2-2003 (RA07)
	Manual Food and Beverage Dispensing Equipment	NSF	NSF/ANSI 18-2005
	Commercial Bulk Milk Dispensing Equipment	NSF	NSF/ANSI 20-2007
	Refrigerated Vending Machines (1995)	UL	ANSI/UL 541
Liquid	Refrigerant-Cooled Liquid Coolers, Remote Type	ARI	ARI 480-2007
	Methods of Testing for Rating Liquid Coolers	ASHRAE	ANSI/ASHRAE 24-2000 (RA05)
	Liquid Cooling Systems	SAE	SAE AIR1811A-1997
Cooling Towers	Cooling Tower Testing (2002)	AABC	National Standards, Ch 13
	Commercial Application, Systems, and Equipment, 1st ed.	ACCA	ACCA Manual CS
	Bioaerosols: Assessment and Control (1999)	ACGIH	ACGIH
	Atmospheric Water Cooling Equipment	ASME	ASME PTC 23-2003
	Water-Cooling Towers	NFPA	NFPA 214-05
	Acceptance Test Code for Water Cooling Towers	CTI	CTI ATC-105 (00)
	Code for Measurement of Sound from Water Cooling Towers (2005)	CTI	CTI ATC-128 (05)
	Acceptance Test Code for Spray Cooling Systems (1985)	CTI	CTI ATC-133 (85)
	Nomenclature for Industrial Water Cooling Towers (1997)	CTI	CTI NCL-109 (97)
	Recommended Practice for Airflow Testing of Cooling Towers (1994)	CTI	CTI PFM-143 (94)
	Fiberglass-Reinforced Plastic Panels (2002)	CTI	CTI STD-131 (02)
	Certification of Water Cooling Tower Thermal Performance (R2004)	CTI	CTI STD-201 (04)
Crop Drying	Density, Specific Gravity, and Mass-Moisture Relationships of Grain for Storage	ASABE	ANSI/ASAE D241.4-2003
	Dielectric Properties of Grain and Seed	ASABE	ASAE D293.2-1989 (R2005)
	Thermal Properties of Grain and Grain Products	ASABE	ASAE D243.4-2003
	Moisture Relationships of Plant-Based Agricultural Products	ASABE	ASAE D245.5-19995 (R2001)
	Construction and Rating of Equipment for Drying Farm Crops	ASABE	ASAE S248.3-1976 (R2005)

Selected Codes and Standards Published by Various Societies and Associations (*Continued*)

Subject	Title	Publisher	Reference
	Cubes, Pellets, and Crumbles—Definitions and Methods for Determining Density, Durability, and Moisture Content	ASABE	ASAE S269.4-1991
	Resistance to Airflow of Grains, Seeds, Other Agricultural Products, and Perforated Metal Sheets	ASABE	ASAE D272.3-1996
	Shelled Corn Storage Time for 0.5% Dry Matter Loss	ASABE	ASAE D535-2005
	Moisture Measurement—Unground Grain and Seeds	ASABE	ASAE S352.2-2003
	Moisture Measurement—Meat and Meat Products	ASABE	ASAE S353-2003
	Moisture Measurement—Forages	ASABE	ASAE S358.2-2003
	Moisture Measurement—Peanuts	ASABE	ASAE S410.1-2003
	Energy Efficiency Test Procedure for Tobacco Curing Structures	ASABE	ASAE S416-2003
	Thin-Layer Drying of Agricultural Crops	ASABE	ANSI/ASAE S448.1-2001 (R2006)
	Moisture Measurement—Tobacco	ASABE	ASAE S487-2003
	Thin-Layer Drying of Agricultural Crops	ASABE	ASAE S488-1990 (R2005)
	Temperature Sensor Locations for Seed-Cotton Drying Systems	ASABE	ASAE 530.1-2007
Dehumidifiers	Commercial Application, Systems, and Equipment, 1st ed.	ACCA	ACCA Manual CS
	Bioaerosols: Assessment and Control (1999)	ACGIH	ACGIH
	Dehumidifiers	AHAM	ANSI/AHAM DH-1-2008
	Method of Testing for Rating Desiccant Dehumidifiers Utilizing Heat for the Regeneration Process	ASHRAE	ANSI/ASHRAE 139-2007
	Moisture Separator Reheaters	ASME	PTC 12.4-1992 (RA04)
	Dehumidifiers	CSA	C22.2 No. 92-1971 (R2004)
	Performance of Dehumidifiers	CSA	CAN/CSA C749-07
	Dehumidifiers (2004)	UL	ANSI/UL 474
Desiccants	Method of Testing Desiccants for Refrigerant Drying	ASHRAE	ANSI/ASHRAE 35-1992
Documentation	Preparation of Operating and Maintenance Documentation for Building Systems	ASHRAE	ASHRAE *Guideline* 4-1993
Driers	Liquid-Line Driers	ARI	ANSI/ARI 710-2004
	Method of Testing Liquid Line Refrigerant Driers	ASHRAE	ANSI/ASHRAE 63.1-1995 (RA01)
	Refrigerant-Containing Components and Accessories, Nonelectrical (2001)	UL	ANSI/UL 207
Ducts and Fittings	Hose, Air Duct, Flexible Nonmetallic, Aircraft	SAE	SAE AS1501C-1994
	Ducted Electric Heat Guide for Air Handling Systems, 2nd ed.	SMACNA	SMACNA 1994
	Factory-Made Air Ducts and Air Connectors (2005)	UL	ANSI/UL 181
Construction	Industrial Ventilation: A Manual of Recommended Practice, 26th ed. (2007)	ACGIH	ACGIH
	Preferred Metric Sizes for Flat, Round, Square, Rectangular, and Hexagonal Metal Products	ASME	ASME B32.100-2005
	Sheet Metal Welding Code	AWS	AWS D9.1M/D9.1:2006
	Fibrous Glass Duct Construction Standards, 5th ed.	NAIMA	NAIMA AH116
	Residential Fibrous Glass Duct Construction Standards, 3rd ed.	NAIMA	NAIMA AH119
	Thermoplastic Duct (PVC) Construction Manual, 2nd ed.	SMACNA	SMACNA 1995
	Accepted Industry Practices for Sheet Metal Lagging, 1st ed.	SMACNA	SMACNA 2002
	Fibrous Glass Duct Construction Standards, 7th ed.	SMACNA	SMACNA 2003
	HVAC Duct Construction Standards, Metal and Flexible, 3rd ed.	SMACNA	SMACNA 2005
	Rectangular Industrial Duct Construction Standards, 2nd ed.	SMACNA	SMACNA 2004
Industrial	Round Industrial Duct Construction Standards, 2nd ed.	SMACNA	SMACNA 1999
	Rectangular Industrial Duct Construction Standards, 2nd ed.	SMACNA	SMACNA 2004
Installation	Flexible Duct Performance and Installation Standards, 4th ed.	ADC	ADC-91
	Installation of Air Conditioning and Ventilating Systems	NFPA	NFPA 90A-06
	Installation of Warm Air Heating and Air-Conditioning Systems	NFPA	NFPA 90B-06
Material Specifications	Specification for General Requirements for Flat-Rolled Stainless and Heat-Resisting Steel Plate, Sheet and Strip	ASTM	ASTM A480/A480M-06b
	Specification for General Requirements for Steel, Sheet, Carbon, and High-Strength, Low-Alloy, Hot-Rolled and Cold-Rolled	ASTM	ASTM A568/A568M-07a
	Specification for Steel Sheet, Zinc-Coated (Galvanized) or Zinc-Iron Alloy-Coated (Galvannealed) by the Hot-Dipped Process	ASTM	ASTM A653/A653M-07
	Specification for General Requirements for Steel Sheet, Metallic-Coated by the Hot-Dip Process	ASTM	ASTM A924/A924M-07
	Specification for Steel, Sheet and Strip, Cold-Rolled, Carbon, Structural, High-Strength Low-Alloy and High-Strength Low-Alloy with Improved Formability	ASTM	ASTM A1008/A1008M-07a
	Specification for Steel, Sheet and Strip, Hot-Rolled, Carbon, Structural, High-Strength Low-Alloy and High-Strength Low-Alloy with Improved Formability	ASTM	ASTM A1011/A1011M-07
	Practice for Measuring Flatness Characteristics of Coated Sheet Products	ASTM	ASTM A1030/A1030M-05
System Design	Installation Techniques for Perimeter Heating and Cooling, 11th ed.	ACCA	ACCA Manual 4
	Residential Duct Systems	ACCA	ANSI/ACCA Manual D
	Commercial Low Pressure, Low Velocity Duct System Design, 1st ed.	ACCA	ACCA Manual Q
	Air Distribution Basics for Residential and Small Commercial Buildings, 1st ed.	ACCA	ACCA Manual T
	Method of Test for Determining the Design and Seasonal Efficiencies of Residential Thermal Distribution Systems	ASHRAE	ANSI/ASHRAE 152-2004
	Closure Systems for Use with Rigid Air Ducts (2005)	UL	ANSI/UL 181A

Selected Codes and Standards Published by Various Societies and Associations (*Continued*)

Subject	Title	Publisher	Reference
	Closure Systems for Use with Flexible Air Ducts and Air Connectors (2005)	UL	ANSI/UL 181B
Testing	Duct Leakage Testing (2002)	AABC	National Standards, Ch 5
	Residential Duct Diagnostics and Repair (2003)	ACCA	ACCA
	Flexible Air Duct Test Code	ADC	ADC FD-72 (R1979)
	Test Method for Measuring Acoustical and Airflow Performance of Duct Liner Materials and Prefabricated Silencers	ASTM	ASTM E477-06a
	Method of Testing to Determine Flow Resistance of HVAC Ducts and Fittings	ASHRAE	ANSI/ASHRAE 120-1999
	Method of Testing HVAC Air Ducts	ASHRAE	ANSI/ASHRAE/SMACNA 126-2000
	HVAC Air Duct Leakage Test Manual, 1st ed.	SMACNA	SMACNA 1985
	HVAC Duct Systems Inspection Guide, 3rd ed.	SMACNA	SMACNA 2005
Electrical	Electrical Power Systems and Equipment—Voltage Ratings	ANSI	ANSI C84.1-2006
	Test Method for Bond Strength of Electrical Insulating Varnishes by the Helical Coil Test	ASTM	ASTM D2519-07
	Standard Specification for Shelter, Electrical Equipment, Lightweight	ASTM	ASTM E2377-04
	Canadian Electrical Code, Part I (20th ed.)	CSA	C22.1-06
	Part II—General Requirements	CSA	CAN/CSA-C22.2 No. 0-M91 (R2006)
	ICC Electrical Code, Administrative Provisions (2006)	ICC	ICCEC
	Enclosures for Electrical Equipment (1000 Volts Maximum)	NEMA	ANSI/NEMA 250-2003
	Low Voltage Cartridge Fuses	NEMA	NEMA FU 1-2002 (R2007)
	Industrial Control and Systems: Terminal Blocks	NEMA	NEMA ICS 4-2005
	Industrial Control and Systems: Enclosures	NEMA	ANSI/NEMA ICS 6-1993 (R2006)
	Application Guide for Ground Fault Protective Devices for Equipment	NEMA	ANSI/NEMA PB 2.2-2004
	General Color Requirements for Wiring Devices	NEMA	NEMA WD 1-1999 (R2005)
	Wiring Devices—Dimensional Requirements	NEMA	ANSI/NEMA WD 6-2002
	National Electrical Code	NFPA	NFPA 70-08
	National Fire Alarm Code	NFPA	NFPA 72-07
	Compatibility of Electrical Connectors and Wiring	SAE	SAE AIR1329A-1988
	Molded-Case Circuit Breakers, Molded-Case Switches, and Circuit-Breaker Enclosures	UL	ANSI/UL489
Energy	Air-Conditioning and Refrigerating Equipment Nameplate Voltages	ARI	ARI 110-2002
	Comfort, Air Quality, and Efficiency by Design	ACCA	ACCA Manual RS
	Energy Standard for Buildings Except Low-Rise Residential Buildings	ASHRAE	ANSI/ASHRAE/IESNA 90.1-2004
	Energy-Efficient Design of Low-Rise Residential Buildings	ASHRAE	ANSI/ASHRAE/IESNA 90.2-2007
	Energy Conservation in Existing Buildings	ASHRAE	ANSI/ASHRAE/IESNA 100-2006
	Methods of Measuring, Expressing, and Comparing Building Energy Performance	ASHRAE	ANSI/ASHRAE 105-2007
	Method of Test for the Evaluation of Building Energy Analysis Computer Programs	ASHRAE	ANSI/ASHRAE 140-2007
	Method of Test for Determining the Design and Seasonal Efficiencies of Residential Thermal Distribution Systems	ASHRAE	ANSI/ASHRAE 152-2004
	Fuel Cell Power Systems Performance	ASME	PTC 50-2002
	International Energy Conservation Code (2006)	ICC	IECC
	Uniform Solar Energy Code (2000)	IAPMO	IAPMO
	Energy Management Guide for Selection and Use of Fixed Frequency Medium AC Squirrel-Cage Polyphase Induction Motors	NEMA	NEMA MG 10-2001 (R2007)
	Energy Management Guide for Selection and Use of Single-Phase Motors	NEMA	NEMA MG 11-1977 (R2007)
	HVAC Systems—Commissioning Manual, 1st ed.	SMACNA	SMACNA 1994
	Building Systems Analysis and Retrofit Manual, 1st ed.	SMACNA	SMACNA 1995
	Energy Systems Analysis and Management, 1st ed.	SMACNA	SMACNA 1997
	Energy Management Equipment (2007)	UL	UL 916
Exhaust Systems	Fan Systems: Supply/Return/Relief/Exhaust (2002)	AABC	National Standards, Ch 10
	Commercial Application, Systems, and Equipment, 1st ed.	ACCA	ACCA Manual CS
	Industrial Ventilation: A Manual of Recommended Practice, 26th ed. (2007)	ACGIH	ACGIH
	Fundamentals Governing the Design and Operation of Local Exhaust Ventilation Systems	AIHA	ANSI/AIHA Z9.2-2006
	Safety Code for Design, Construction, and Ventilation of Spray Finishing Operations	AIHA	ANSI/AIHA Z9.3-2007
	Laboratory Ventilation	AIHA	ANSI/AIHA Z9.5-2003
	Recirculation of Air from Industrial Process Exhaust Systems	AIHA	ANSI/AIHA Z9.7-2007
	Method of Testing Performance of Laboratory Fume Hoods	ASHRAE	ANSI/ASHRAE 110-1995
	Ventilation for Commercial Cooking Operations	ASHRAE	ANSI/ASHRAE 154-2003
	Performance Test Code on Compressors and Exhausters	ASME	PTC 10-1997 (RA03)
	Flue and Exhaust Gas Analyses	ASME	PTC 19.10-1981
	Mechanical Flue-Gas Exhausters	CSA	CAN B255-M81 (R2005)
	Exhaust Systems for Air Conveying of Vapors, Gases, Mists, and Noncombustible Particulate Solids	NFPA	ANSI/NFPA 91-04
	Draft Equipment (2006)	UL	UL 378

Selected Codes and Standards Published by Various Societies and Associations (*Continued*)

Subject	Title	Publisher	Reference
Expansion Valves	Thermostatic Refrigerant Expansion Valves	ARI	ANSI/ARI 750-2007
	Method of Testing Capacity of Thermostatic Refrigerant Expansion Valves	ASHRAE	ANSI/ASHRAE 17-1998 (RA03)
Fan-Coil Units	Industrial Ventilation: A Manual of Recommended Practice, 26th ed. (2007)	ACGIH	ACGIH
	Room Fan-Coils	ARI	ARI 440-2005
	Methods of Testing for Rating Fan-Coil Conditioners	ASHRAE	ANSI/ASHRAE 79-2002 (RA06)
	Heating and Cooling Equipment (2005)	UL/CSA	ANSI/UL 1995/C22.2 No. 236-95
Fans	Residential Duct Systems	ACCA	ANSI/ACCA Manual D
	Commercial Low Pressure, Low Velocity Duct System Design, 1st ed.	ACCA	ACCA Manual Q
	Industrial Ventilation: A Manual of Recommended Practice, 26th ed. (2007)	ACGIH	ACGIH
	Standards Handbook	AMCA	AMCA 99-03
	Drive Arrangements for Centrifugal Fans	AMCA	ANSIAMCA 99-2404-03
	Inlet Box Positions for Centrifugal Fans	AMCA	ANSI/AMCA 99-2405-03
	Designation for Rotation and Discharge of Centrifugal Fans	AMCA	ANSI/AMCA 99-2406-03
	Motor Positions for Belt or Chain Drive Centrifugal Fans	AMCA	ANSI/AMCA 99-2407-03
	Operating Limits for Centrifugal Fans	AMCA	AMCA 99-2408-69
	Drive Arrangements for Tubular Centrifugal Fans	AMCA	ANSI/AMCA 99-2410-03
	Impeller Diameters and Outlet Areas for Centrifugal Fans	AMCA	ANSI/AMCA 99-2412-03
	Impeller Diameters and Outlet Areas for Industrial Centrifugal Fans	AMCA	ANSI/AMCA 99-2413-03
	Impeller Diameters and Outlet Areas for Tubular Centrifugal Fans	AMCA	ANSI/AMCA 99-2414-03
	Dimensions for Axial Fans	AMCA	ANSI/AMCA 99-3001-03
	Drive Arrangements for Axial Fans	AMCA	ANSI/AMCA 99-3404-03
	Air Systems	AMCA	AMCA 200-95 (R2007)
	Fans and Systems	AMCA	AMCA 201-02 (R2007)
	Troubleshooting	AMCA	AMCA 202-98 (R2007)
	Field Performance Measurement of Fan Systems	AMCA	AMCA 203-90 (R2007)
	Balance Quality and Vibration Levels for Fans	AMCA	ANSI/AMCA 204-05
	Laboratory Methods of Testing Air Circulator Fans for Rating	AMCA	ANSI/AMCA 230-07
	Laboratory Method of Testing Positive Pressure Ventilators for Rating	AMCA	ANSI/AMCA 240-06
	Reverberant Room Method for Sound Testing of Fans	AMCA	AMCA 300-05
	Methods for Calculating Fan Sound Ratings from Laboratory Test Data	AMCA	AMCA 301-06
	Application of Sone Ratings for Non-Ducted Air Moving Devices	AMCA	AMCA 302-73 (R2008)
	Application of Sound Power Level Ratings for Fans	AMCA	AMCA 303-79 (R2008)
	Recommended Safety Practices for Users and Installers of Industrial and Commercial Fans	AMCA	AMCA 410-96
	Industrial Process/Power Generation Fans: Site Performance Test Standard	AMCA	AMCA 803-02
	Mechanical Balance of Fans and Blowers	ARI	ARI *Guideline* G-2002
	Acoustics—Measurement of Noise and Vibration of Small Air-Moving Devices—Part 1: Airborne Noise Emission	ASA	ANSI S12.11-2003/Part 1/ISO 10302:1996 (MOD)
	Part 2: Structure-Borne Vibration	ASA	ANSI S12.11-2003/Part 2
	Laboratory Methods of Testing Fans for Aerodynamic Performance Rating	ASHRAE/ AMCA	ANSI/ASHRAE 51-1999 ANSI/AMCA 210-99
	Laboratory Method of Testing to Determine the Sound Power in a Duct	ASHRAE/ AMCA	ANSI/ASHRAE 68-1997 ANSI/AMCA 330-97
	Methods of Testing Fan Vibration—Blade Vibrations and Critical Speeds	ASHRAE	ANSI/ASHRAE 87.1-1992
	Laboratory Methods of Testing Fans Used to Exhaust Smoke in Smoke Management Systems	ASHRAE	ANSI/ASHRAE 149-2000 (RA05)
	Ventilation for Commercial Cooking Operations	ASHRAE	ANSI/ASHRAE 154-2003
	Fans	ASME	ANSI/ASME PTC 11-1984 (RA03)
	Fans and Ventilators	CSA	C22.2 No. 113-M1984 (R2004)
	Rating the Performance of Residential Mechanical Ventilating Equipment	CSA	CAN/CSA C260-M90 (R2007)
	Energy Performance of Ceiling Fans	CSA	CAN/CSA C814-96 (R2007)
	Electric Fans (1999)	UL	ANSI/UL 507
	Power Ventilators (2004)	UL	ANSI/UL 705
Fenestration	Test Method for Accelerated Weathering of Sealed Insulating Glass Units	ASTM	ASTM E773-01
	Practice for Calculation of Photometric Transmittance and Reflectance of Materials to Solar Radiation	ASTM	ASTM E971-88 (2003)
	Test Method for Solar Photometric Transmittance of Sheet Materials Using Sunlight	ASTM	ASTM E972-96 (2007)
	Test Method for Solar Transmittance (Terrestrial) of Sheet Materials Using Sunlight	ASTM	ASTM E1084-86 (2003)
	Practice for Determining the Load Resistance of Glass in Buildings	ASTM	ASTM E1300-07e1
	Practice for Installation of Exterior Windows, Doors and Skylights	ASTM	ASTM E2112-07
	Test Method for Insulating Glass Unit Performance	ASTM	ASTM E2188-02
	Test Method for Testing Resistance to Fogging Insulating Glass Units	ASTM	ASTM E2189-02
	Specification for Insulating Glass Unit Performance and Evaluation	ASTM	ASTM E2190-02
	Guide for Assessing the Durability of Absorptive Electrochemical Coatings within Sealed Insulating Glass Units	ASTM	ASTM E2354-04

Selected Codes and Standards Published by Various Societies and Associations (*Continued*)

Subject	Title	Publisher	Reference
	Tables for Reference Solar Spectral Irradiance: Direct Normal and Hemispherical on 37° Tilted Surface	ASTM	ASTM G173-03e1
	Windows	CSA	A440-08
	Energy Performance of Windows and Other Fenestration Systems	CSA	A440.3-04
	Window, Door, and Skylight Installation	CSA	A440.4-98
	Energy Performance Evaluation of Swinging Doors	CSA	A453-95 (R2000)
Filter-Driers	Flow-Capacity Rating of Suction-Line Filters and Suction-Line Filter-Driers	ARI	ARI 730-2005
	Method of Testing Liquid Line Filter-Drier Filtration Capability	ASHRAE	ANSI/ASHRAE 63.2-1996 (RA06)
	Method of Testing Flow Capacity of Suction Line Filters and Filter-Driers	ASHRAE	ANSI/ASHRAE 78-1985 (RA07)
Fireplaces	Factory-Built Fireplaces (1996)	UL	ANSI/UL 127
	Fireplace Stoves (2007)	UL	ANSI/UL 737
Fire Protection	Test Method for Surface Burning Characteristics of Building Materials	ASTM/NFPA	ASTM E84-08
	Test Methods for Fire Test of Building Construction and Materials	ASTM	ASTM E119-08
	Test Method for Room Fire Test of Wall and Ceiling Materials and Assemblies	ASTM	ASTM E2257-03
	Test Method for Determining Fire Resistance of Perimeter Fire Barriers Using Intermediate-Scale Multi-Story Test Apparatus	ASTM	ASTM E2307-04e1
	Guide for Laboratory Monitors	ASTM	ASTM E2335-04
	Test Method for Fire Resistance Grease Duct Enclosure Systems	ASTM	ASTM E2336-04
	Practice for Specimen Preparation and Mounting of Paper or Vinyl Wall Coverings to Assess Surface Burning Characteristics	ASTM	ASTM E2404-07a
	BOCA National Fire Prevention Code, 11th ed. (1999)	BOCA	BNFPC
	Uniform Fire Code	IFCI	UPC 1997
	International Fire Code (2006)	ICC	IFC
	International Mechanical Code (2006)	ICC	IMC
	International Urban-Wildland Interface Code (2006)	ICC	IUWIC
	Fire-Resistance Tests—Elements of Building Construction; Part 1: Gen. Requirements	ISO	ISO 834-1:1999
	Fire-Resistance Tests—Door and Shutter Assemblies	ISO	ISO 3008:2007
	Reaction to Fire Tests—Ignitability of Building Products Using a Radiant Heat Source	ISO	ISO 5657:1997
	Fire-Resistance Tests—Ventilating Ducts	ISO	ISO 6944:1985
	Fire Service Annunciator and Interface	NEMA	NEMA SB 30-2005
	Fire Protection Handbook (2008)	NFPA	NFPA
	National Fire Codes (issued annually)	NFPA	NFPA
	Fire Protection Guide to Hazardous Materials	NFPA	NFPA HAZ-01
	Uniform Fire Code	NFPA	NFPA 1-06
	Installation of Sprinkler Systems	NFPA	NFPA 13-2007
	Flammable and Combustible Liquids Code	NFPA	NFPA 30-08
	Fire Protection for Laboratories Using Chemicals	NFPA	NFPA 45-04
	National Fire Alarm Code	NFPA	NFPA 72-07
	Fire Doors and Fire Windows	NFPA	NFPA 80-07
	Health Care Facilities	NFPA	NFPA 99-05
	Life Safety Code	NFPA	NFPA 101-06
	Methods of Fire Tests of Door Assemblies	NFPA	NFPA 252-08
	Standard Fire Code (1999)	SBCCI	SFPC
	Fire, Smoke and Radiation Damper Installation Guide for HVAC Systems, 5th ed.	SMACNA	SMACNA 2002
	Fire Tests of Door Assemblies (2008)	UL	ANSI/UL 10B
	Heat Responsive Links for Fire-Protection Service (2003)	UL	ANSI/UL 33
	Fire Tests of Building Construction and Materials (2003)	UL	ANSI/UL 263
	Fire Dampers (2006)	UL	ANSI/UL 555
	Fire Tests of Through-Penetration Firestops (2003)	UL	ANSI/UL 1479
Smoke Management	Commissioning Smoke Management Systems	ASHRAE	ASHRAE *Guideline* 5-1994 (RA01)
	Laboratory Methods of Testing Fans Used to Exhaust Smoke in Smoke Management Systems	ASHRAE	ANSI/ASHRAE 149-2000 (RA05)
	Recommended Practice for Smoke-Control Systems	NFPA	NFPA 92A-06
	Smoke Management Systems in Malls, Atria, and Large Areas	NFPA	NFPA 92B-05
	Ceiling Dampers (2006)	UL	ANSI/UL 555C
	Smoke Dampers (1999)	UL	ANSI/UL 555S
Freezers	Energy Performance and Capacity of Household Refrigerators, Refrigerator-Freezers, and Freezers	CSA	C300-00 (R2005)
	Energy Performance Standard for Food Service Refrigerators and Freezers	CSA	C827-98 (R2003)
	Refrigeration Equipment	CSA	CAN/CSA-C22.2 No. 120-M91 (R2004)
Commercial	Dispensing Freezers	NSF	NSF/ANSI 6-2007
	Commercial Refrigerators and Freezers	NSF	NSF/ANSI 7-2007
	Commercial Refrigerators and Freezers (2006)	UL	ANSI/UL 471
	Ice Makers (1995)	UL	ANSI/UL 563

Selected Codes and Standards Published by Various Societies and Associations (*Continued*)

Subject	Title	Publisher	Reference
	Ice Cream Makers (2005)	UL	ANSI/UL 621
Household	Household Refrigerators, Refrigerator-Freezers and Freezers	AHAM	ANSI/AHAM HRF-1-2007
	Household Refrigerators and Freezers (1993)	UL/CSA	ANSI/UL 250/C22.2 No. 63-93 (R1999)
Fuels	Threshold Limit Values for Chemical Substances (updated annually)	ACGIH	ACGIH
	International Gas Fuel Code (2006)	AGA/NFPA	ANSI Z223.1/NPFA 54-2006
	Reporting of Fuel Properties when Testing Diesel Engines with Alternative Fuels Derived from Biological Materials	ASABE	ASAE EP552-1996
	Coal Pulverizers	ASME	PTC 4.2 1969 (RA03)
	Classification of Coals by Rank	ASTM	ASTM D388-05
	Specification for Fuel Oils	ASTM	ASTM D396-08
	Test Method for Determination of Homogeneity and Miscibility in Automotive Engine Oils	ASTM	ASTM D922-00a (2006)
	Specification for Diesel Fuel Oils	ASTM	ASTM D975-07b
	Specification for Gas Turbine Fuel Oils	ASTM	ASTM D2880-03
	Specification for Kerosene	ASTM	ASTM D3699-07
	Practice for Receipt, Storage and Handling of Fuels	ASTM	ASTM D4418-00 (2006)
	Test Method for Determination of Yield Stress and Apparent Viscosity of Used Engine Oils at Low Temperature	ASTM	ASTM D6896-03 (2007)
	Test Method for Total Sulfur in Naphthas, Distillates, Reformulated Gasolines, Diesels, Biodiesels, and Motor Fuels by Oxidative Combustion and Electrochemical Detection	ASTM	ASTM D6920-07
	Test Method for Measurement of Hindered Phenolic and Aromatic Amine Antioxidant Content in Non-Zinc Turbine Oils by Linear Sweep Voltammetry	ASTM	ASTM D6971-04
	Practice for Enumeration of Viable Bacteria and Fungi in Liquid Fuels—Filtration and Culture Procedures	ASTM	ASTM D6974-04a
	Test Method for Evaluation of Aeration Resistance of Engine Oils in Direct-Injected Turbocharged Automotive Diesel Engine	ASTM	ASTM D6984-07a
	Specification for Middle Distillate Fuel Oil-Military Marine Applications	ASTM	ASTM D6985-04a
	Test Method for Determination of Ignition Delay and Derived Cetane Number DCN of Diesel Fuel Oils by Combustion in a Constant Volume Chamber	ASTM	ASTM D6890-07b
	Test Method for Determination of Total Sulfur in Light Hydrocarbon, Motor Fuels, and Oils by Online Gas Chromatography with Flame Photometric Detection	ASTM	ASTM D7041-04
	Test Method for Sulfur in Gasoline and Diesel Fuel by Monochromatic Wavelength Dispersive X-Ray Fluorescence Spectrometry	ASTM	ASTM D7044-04a
	New Draft Standard Test Method for Flash Point by Modified Continuously Closed Cup Flash Point Tester	ASTM	ASTM D7094-04
	Test Method for Determining the Viscosity-Temperature Relationship of Used and Soot-Containing Engine Oils at Low Temperatures	ASTM	ASTM D7110-05a
	Test Method for Determination of Trace Elements in Middle Distillate Fuels by Inductively Coupled Plasma Atomic Emission Spectrometry (ICPAES)	ASTM	ASTM D7111-05
	Test Method for Determining Stability and Compatibility of Heavy Fuel Oils and Crude Oils by Heavy Fuel Oil Stability Analyzer (Optical Detection)	ASTM	ASTM D7112-05a
	Test Method for Determination of Intrinsic Stability of Asphaltene-Containing Residues, Heavy Fuel Oils, and Crude Oils	ASTM	ASTM D7157-05
	Test Method for Hydrogen Content of Middle Distillate Petroleum Products by Low-Resolution Pulsed Nuclear Magnetic Resonance Spectroscopy	ASTM	ASTM D7171-05
	Gas-Fired Central Furnaces	CSA	ANSI Z21.47-2006/CSA 2.3-2006
	Gas Unit Heaters and Gas-Fired Duct Furnaces	CSA	ANSI Z83.8-2006/CSA-2.6-2006
	Industrial and Commercial Gas-Fired Package Furnaces	CSA	CGA 3.2-1976 (R2003)
	Uniform Mechanical Code (2006)	IAPMO	Chapter 13
	Uniform Plumbing Code (2006)	IAPMO	Chapter 12
	International Fuel Gas Code (2006)	ICC	IFGC
	Standard Gas Code (1999)	SBCCI	SGC
	Commercial-Industrial Gas Heating Equipment (2006)	UL	UL 795
Furnaces	Commercial Application, Systems, and Equipment, 1st ed.	ACCA	ACCA Manual CS
	Residential Equipment Selection, 2nd ed.	ACCA	ANSI/ACCA Manual S
	Method of Testing for Annual Fuel Utilization Efficiency of Residential Central Furnaces and Boilers	ASHRAE	ANSI/ASHRAE 103-1993
	Prevention of Furnace Explosions/Implosions in Multiple Burner Boilers	NFPA	NFPA 8502-99
	Residential Gas Detectors (2000)	UL	ANSI/UL 1484
	Heating and Cooling Equipment (2005)	UL/CSA	ANSI/UL 1995/C22.2 No. 236-95
	Single and Multiple Station Carbon Monoxide Alarms (2008)	UL	ANSI/UL 2034
Gas	International Gas Fuel Code (2006)	AGA/NFPA	ANSI Z223.1/NFPA 54-2006
	Gas-Fired Central Furnaces	CSA	ANSI Z21.47-2006/CSA 2.3-2006
	Gas Unit Heaters and Gas-Fired Duct Furnaces	CSA	ANSI Z83.8-2006/CSA-2.6-2006
	Industrial and Commercial Gas-Fired Package Furnaces	CSA	CGA 3.2-1976 (R2003)

Selected Codes and Standards Published by Various Societies and Associations (*Continued*)

Subject	Title	Publisher	Reference
	International Fuel Gas Code (2006)	ICC	IFGC
	Standard Gas Code (1999)	SBCCI	SGC
	Commercial-Industrial Gas Heating Equipment (2006)	UL	UL 795
Oil	Specification for Fuel Oils	ASTM	ASTM D396-08
	Specification for Diesel Fuel Oils	ASTM	ASTM D975-07b
	Test Method for Smoke Density in Flue Gases from Burning Distillate Fuels	ASTM	ASTM D2156-94 (2003)
	Standard Test Method for Vapor Pressure of Liquefied Petroleum Gases (LPG) (Expansion Method)	ASTM	ASTM D6897-2003a
	Oil Burning Stoves and Water Heaters	CSA	B140.3-1962 (R2006)
	Oil-Fired Warm Air Furnaces	CSA	B140.4-04
	Installation of Oil-Burning Equipment	NFPA	NFPA 31-06
	Oil-Fired Central Furnaces (2006)	UL	UL 727
	Oil-Fired Floor Furnaces (2003)	UL	ANSI/UL 729
	Oil-Fired Wall Furnaces (2003)	UL	ANSI/UL 730
Solid Fuel	Installation Code for Solid-Fuel-Burning Appliances and Equipment	CSA	B365-01 (R2006)
	Solid-Fuel-Fired Central Heating Appliances	CSA	CAN/CSA-B366.1-M91 (R2007)
	Solid-Fuel and Combination-Fuel Central and Supplementary Furnaces (2006)	UL	ANSI/UL 391
Heaters	Gas-Fired High-Intensity Infrared Heaters	CSA	ANSI Z83.19-2001/CSA 2.35-2001 (R2005)
	Gas-Fired Low-Intensity Infrared Heaters	CSA	ANSI Z83.20-2008/CSA 2.34-2008
	Threshold Limit Values for Chemical Substances (updated annually)	ACGIH	ACGIH
	Industrial Ventilation: A Manual of Recommended Practice, 26th ed. (2007)	ACGIH	ACGIH
	Thermal Performance Testing of Solar Ambient Air Heaters	ASABE	ANSI/ASAE S423-1991
	Air Heaters	ASME	ASME PTC 4.3-1968 (RA91)
	Guide for Construction of Solid Fuel Burning Masonry Heaters	ASTM	ASTM E1602-03
	Non-Recirculating Direct Gas-Fired Industrial Air Heaters	CSA	ANSI Z83.4-2003/CSA 3.7-2003
	Electric Duct Heaters	CSA	C22.2 No. 155-M1986 (R2004)
	Portable Kerosene-Fired Heaters	CSA	CAN3-B140.9.3 M86 (R2006)
	Standards for Closed Feedwater Heaters, 7th ed. (2004)	HEI	HEI 2622
	Electric Heating Appliances (2005)	UL	ANSI/UL 499
	Electric Oil Heaters (2003)	UL	ANSI/UL 574
	Oil-Fired Air Heaters and Direct-Fired Heaters (1993)	UL	UL 733
	Electric Dry Bath Heaters (2004)	UL	ANSI/UL 875
	Oil-Burning Stoves (1993)	UL	ANSI/UL 896
Engine	Electric Engine Preheaters and Battery Warmers for Diesel Engines	SAE	SAE J1310-1993
	Selection and Application Guidelines for Diesel, Gasoline, and Propane Fired Liquid Cooled Engine Pre-Heaters	SAE	SAE J1350-1988
	Fuel Warmer—Diesel Engines	SAE	SAE J1422-1996
Nonresidential	Installation of Electric Infrared Brooding Equipment	ASABE	ASAE EP258.3-2004
	Gas-Fired Construction Heaters	CSA	ANSI Z83.7-00 (R2005)/CSA 2.14-00 (R2006)
	Recirculating Direct Gas-Fired Industrial Air Heaters	CSA	ANSI Z83.18-2004
	Portable Industrial Oil-Fired Heaters	CSA	B140.8-1967 (R2006)
	Fuel-Fired Heaters—Air Heating—for Construction and Industrial Machinery	SAE	SAE J1024-1989
	Commercial-Industrial Gas Heating Equipment (2006)	UL	UL 795
	Electric Heaters for Use in Hazardous (Classified) Locations (2006)	UL	ANSI/UL 823
Pool	Methods of Testing and Rating Pool Heaters	ASHRAE	ANSI/ASHRAE 146-2006
	Gas-Fired Pool Heaters	CSA	ANSI Z21.56-2006/CSA 4.7-2006
	Oil-Fired Service Water Heaters and Swimming Pool Heaters	CSA	B140.12-03
Room	Specification for Room Heaters, Pellet Fuel Burning Type	ASTM	ASTM E1509-04
	Gas-Fired Room Heaters, Vol. II, Unvented Room Heaters	CSA	ANSI Z21.11.2-2007
	Gas-Fired Unvented Catalytic Room Heaters for Use with Liquefied Petroleum (LP) Gases	CSA	ANSI Z21.76-1994 (R2006)
	Vented Gas-Fired Space Heating Appliances	CSA	ANSI Z21.86-2004/CSA 2.32-2004
	Vented Gas Fireplace Heaters	CSA	ANSI Z21.88-2005/CSA 2.33-2005
	Unvented Kerosene-Fired Room Heaters and Portable Heaters (1993)	UL	UL 647
	Movable and Wall- or Ceiling-Hung Electric Room Heaters (2000)	UL	UL 1278
	Fixed and Location-Dedicated Electric Room Heaters (1997)	UL	UL 2021
	Solid Fuel-Type Room Heaters (1996)	UL	ANSI/UL 1482
Transport	Heater, Airplane, Engine Exhaust Gas to Air Heat Exchanger Type	SAE	SAE ARP86-1996
	Installation, Heaters, Airplane, Internal Combustion Heater Exchange Type	SAE	SAE ARP266-2001
	Heater, Aircraft, Internal Combustion Heat Exchanger Type	SAE	SAE AS8040-1996
	Motor Vehicle Heater Test Procedure	SAE	SAE J638-1998
	Heater, Aircraft, Internal Combustion Heat Exchanger Type	SAE	SAE AS8040-1996
Unit	Gas Unit Heaters and Gas-Fired Duct Furnaces	CSA	ANSI Z83.8-2006/CSA-2.6-2006

Selected Codes and Standards Published by Various Societies and Associations (*Continued*)

Subject	Title	Publisher	Reference
	Oil-Fired Unit Heaters (1995)	UL	ANSI/UL 731
Heat Exchangers	Remote Mechanical-Draft Evaporative Refrigerant Condensers	ARI	ARI 490-2003
	Method of Testing Air-to-Air Heat Exchangers	ASHRAE	ANSI/ASHRAE 84-1991
	Boiler and Pressure Vessel Code—Section VIII, Division 1: Pressure Vessels	ASME	ASME BPVC-2007
	Single Phase Heat Exchangers	ASME	ASME PTC 12.5-2000 (RA05)
	Air Cooled Heat Exchangers	ASME	ASME PTC 30-1991 (RA05)
	Standard Methods of Test for Rating the Performance of Heat-Recovery Ventilators	CSA	C439-00 (R2005)
	Standards for Power Plant Heat Exchangers, 4th ed. (2004)	HEI	HEI 2623
	Standards of Tubular Exchanger Manufacturers Association, 9th ed. (2007)	TEMA	TEMA
	Refrigerant-Containing Components and Accessories, Nonelectrical (2001)	UL	ANSI/UL 207
Heating	Commercial Application, Systems, and Equipment, 1st ed.	ACCA	ACCA Manual CS
	Comfort, Air Quality, and Efficiency by Design	ACCA	ACCA Manual RS
	Residential Equipment Selection, 2nd ed.	ACCA	ANSI/ACCA Manual S
	Heating, Ventilating and Cooling Greenhouses	ASABE	ANSI/ASAE EP406.4-2003
	Heater Elements	CSA	C22.2 No. 72-M1984 (R2004)
	Determining the Required Capacity of Residential Space Heating and Cooling Appliances	CSA	CAN/CSA-F280-M90 (R2004)
	Heat Loss Calculation Guide (2001)	HYDI	HYDI H-22
	Residential Hydronic Heating Installation Design Guide	HYDI	IBR Guide
	Radiant Floor Heating (1995)	HYDI	HYDI 004
	Advanced Installation Guide (Commercial) for Hot Water Heating Systems (2001)	HYDI	HYDI 250
	Environmental Systems Technology, 2nd ed. (1999)	NEBB	NEBB
	Pulverized Fuel Systems	NFPA	NFPA 8503-97
	Aircraft Electrical Heating Systems	SAE	SAE AIR860-2000
	Heating Value of Fuels	SAE	SAE J1498-2005
	Performance Test for Air-Conditioned, Heated, and Ventilated Off-Road Self-Propelled Work Machines	SAE	SAE J1503-2004
	HVAC Systems—Applications, 1st ed.	SMACNA	SMACNA 1987
	Electric Baseboard Heating Equipment (1994)	UL	ANSI/UL 1042
	Electric Duct Heaters (2004)	UL	ANSI/UL 1996
	Heating and Cooling Equipment (2005)	UL/CSA	ANSI/UL 1995/C22.2 No. 236-95
Heat Pumps	Commercial Application, Systems, and Equipment, 1st ed.	ACCA	ACCA Manual CS
	Geothermal Heat Pump Training Certification Program	ACCA	ACCA Training Manual
	Heat Pumps Systems, Principles and Applications, 2nd ed.	ACCA	ACCA Manual H
	Residential Equipment Selection, 2nd ed.	ACCA	ANSI/ACCA Manual S
	Industrial Ventilation: A Manual of Recommended Practice, 26th ed. (2007)	ACGIH	ACGIH
	Water-Source Heat Pumps	ARI	ARI 320-98
	Ground Water-Source Heat Pumps	ARI	ARI 325-98
	Ground Source Closed-Loop Heat Pumps	ARI	ARI 330-98
	Commercial and Industrial Unitary Air-Conditioning and Heat Pump Equipment	ARI	ARI 340/360-2007
	Methods of Testing for Rating Electrically Driven Unitary Air-Conditioning and Heat Pump Equipment	ASHRAE	ANSI/ASHRAE 37-2005
	Methods of Testing for Rating Seasonal Efficiency of Unitary Air-Conditioners and Heat Pumps	ASHRAE	ANSI/ASHRAE 116- (RA05)
	Performance Standard for Split-System and Single-Package Central Air Conditioners and Heat Pumps	CSA	CAN/CSA-C656-05
	Installation Requirements for Air-to-Air Heat Pumps	CSA	C273.5-1980 (R2002)
	Performance of Direct-Expansion (DX) Ground-Source Heat Pumps	CSA	C748-94 (R2005)
	Water-Source Heat Pumps—Testing and Rating for Performance, Part 1: Water-to-Air and Brine-to-Air Heat Pumps	CSA	CAN/CSA C13256-1-01
	Part 2: Water-to-Water and Brine-to-Water Heat Pumps	CSA	CAN/CSA C13256-2-01 (R2005)
	Heating and Cooling Equipment (2005)	UL/CSA	ANSI/UL 1995/C22.2 No. 236-95
Gas-Fired	Gas-Fired, Heat Activated Air Conditioning and Heat Pump Appliances	CSA	ANSI Z21.40.1-1996 (R2002)/CGA 2.91-M96
	Gas-Fired, Work Activated Air Conditioning and Heat Pump Appliances (Internal Combustion)	CSA	ANSI Z21.40.2-1996 (R2002)/CGA 2.92-M96
	Performance Testing and Rating of Gas-Fired Air Conditioning and Heat Pump Appliances	CSA	ANSI Z21.40.4-1996 (R2002)/CGA 2.94-M96
Heat Recovery	Gas Turbine Heat Recovery Steam Generators	ASME	ANSI/ASME PTC 4.4-1981 (RA03)
	Water Heaters, Hot Water Supply Boilers, and Heat Recovery Equipment	NSF	NSF/ANSI 5-2007
Humidifiers	Commercial Application, Systems, and Equipment, 1st ed.	ACCA	ACCA Manual CS
	Comfort, Air Quality, and Efficiency by Design	ACCA	ACCA Manual RS
	Bioaerosols: Assessment and Control (1999)	ACGIH	ACGIH
	Humidifiers	AHAM	ANSI/AHAM HU-1-2006

Selected Codes and Standards Published by Various Societies and Associations (*Continued*)

Subject	Title	Publisher	Reference
	Central System Humidifiers for Residential Applications	ARI	ARI 610-2004
	Self-Contained Humidifiers for Residential Applications	ARI	ARI 620-2004
	Commercial and Industrial Humidifiers	ARI	ANSI/ARI 640-2005
	Humidifiers (2001)	UL/CSA	ANSI/UL 998/C22.2 No. 104-93
Ice Makers	Performance Rating of Automatic Commercial Ice Makers	ARI	ARI 810-2007
	Ice Storage Bins	ARI	ARI 820-2000
	Methods of Testing Automatic Ice Makers	ASHRAE	ANSI/ASHRAE 29-1988 (RA05)
	Refrigeration Equipment	CSA	CAN/CSA-C22.2 No. 120-M91 (R2004)
	Performance of Automatic Ice-Makers and Ice Storage Bins	CSA	C742-98 (R2003)
	Automatic Ice Making Equipment	NSF	NSF/ANSI 12-2007
	Ice Makers (1995)	UL	ANSI/UL 563
Incinerators	Incinerators and Waste and Linen Handling Systems and Equipment	NFPA	NFPA 82-04
	Residential Incinerators (2006)	UL	UL 791
Indoor Air Quality	Good HVAC Practices for Residential and Commercial Buildings (2003)	ACCA	ACCA
	Comfort, Air Quality, and Efficiency by Design (Residential) (1997)	ACCA	ACCA Manual RS
	Bioaerosols: Assessment and Control (1999)	ACGIH	ACGIH
	Ventilation for Acceptable Indoor Air Quality	ASHRAE	ANSI/ASHRAE 62.1-2007
	Ventilation and Acceptable Indoor Air Quality in Low-Rise Residential Buildings	ASHRAE	ANSI/ASHRAE 62.2-2007
	Test Method for Determination of Volatile Organic Chemicals in Atmospheres (Canister Sampling Methodology)	ASTM	ASTM D5466-01 (2007)
	Guide for Using Probability Sampling Methods in Studies of Indoor Air Quality in Buildings	ASTM	ASTM D5791-95 (2006)
	Guide for Using Indoor Carbon Dioxide Concentrations to Evaluate Indoor Air Quality and Ventilation	ASTM	ASTM D6245-07
	Guide for Placement and Use of Diffusion Controlled Passive Monitors for Gaseous Pollutants in Indoor Air	ASTM	ASTM D6306-98 (2003)
	Test Method for Determination of Metals and Metalloids Airborne Particulate Matter by Inductively Coupled Plasma Atomic Emissions Spectrometry (ICP-AES)	ASTM	ASTM D7035-04
	Test Method for Metal Removal Fluid Aerosol in Workplace Atmospheres	ASTM	ASTM D7049-04
	Practice for Emission Cells for the Determination of Volatile Organic Emissions from Materials/Products	ASTM	ASTM D7143-05
	Practice for Collection of Surface Dust by Micro-Vacuum Sampling for Subsequent Metals Determination	ASTM	ASTM D7144-05a
	Test Method for Determination of Beryllium in the Workplace Using Field-Based Extraction and Fluorescence Detection	ASTM	ASTM D7202-06
	Practice for Referencing Suprathreshold Odor Intensity	ASTM	ASTM E544-99 (2004)
	Guide for Specifying and Evaluating Performance of a Single Family Attached and Detached Dwelling—Indoor Air Quality	ASTM	ASTM E2267-04
	Classification for Serviceability of an Office Facility for Thermal Environment and Indoor Air Conditions	ASTM	ASTM E2320-04
	Practice for Continuous Sizing and Counting of Airborne Particles in Dust-Controlled Areas and Clean Rooms Using Instruments Capable of Detecting Single Sub-Micrometre and Larger Particles	ASTM	ASTM F50-07
	Ambient Air—Determination of Mass Concentration of Nitrogen Dioxide—Modified Griess-Saltzman Method	ISO	ISO 6768:1998
	Air Quality—Exchange of Data	ISO	ISO 7168:1999
	Environmental Tobacco Smoke—Estimation of Its Contribution to Respirable Suspended Particles—Determination of Particulate Matter by Ultraviolet Absorptance and by Fluorescence	ISO	ISO 15593:2001
	Indoor Air—Part 3: Determination of Formaldehyde and Other Carbonyl Compounds—Active Sampling Method	ISO	ISO 16000-3:2001
	Workplace Air Quality—Sampling and Analysis of Volatile Organic Compounds by Solvent Desorption/Gas Chromatography—Part 1: Pumped Sampling Method	ISO	ISO 16200-1:2001
	Part 2: Diffusive Sampling Method	ISO	ISO 16200-2:2000
	Workplace Air Quality—Determination of Total Organic Isocyanate Groups in Air Using 1-(2-Methoxyphenyl) Piperazine and Liquid Chromatography	ISO	ISO 16702:2007
	Installation of Household Carbon Monoxide (CO) Warning Equipment	NFPA	NFPA 720-2005
	Indoor Air Quality—A Systems Approach, 3rd ed.	SMACNA	SMACNA 1998
	IAQ Guidelines for Occupied Buildings Under Construction, 1st ed.	SMACNA	SMACNA 1995
	Single and Multiple Station Carbon Monoxide Alarms (1996)	UL	ANSI/UL 2034
Aircraft	Guide for Selecting Instruments and Methods for Measuring Air Quality in Aircraft Cabins	ASTM	ASTM D6399-04
	Guide for Deriving Acceptable Levels of Airborne Chemical Contaminants in Aircraft Cabins Based on Health and Comfort Considerations	ASTM	ASTM D7034-05
Insulation	Guidelines for Use of Thermal Insulation in Agricultural Buildings	ASABE	ANSI/ASAE S401.2-2003
	Terminology Relating to Thermal Insulating Materials	ASTM	ASTM C168-05a

Selected Codes and Standards Published by Various Societies and Associations (*Continued*)

Subject	Title	Publisher	Reference
	Test Method for Steady-State Heat Flux Measurements and Thermal Transmission Properties by Means of the Guarded-Hot-Plate Apparatus	ASTM	ASTM C177-04
	Test Method for Steady-State Heat Transfer Properties of Horizontal Pipe Insulations	ASTM	ASTM C335-05ae1
	Practice for Prefabrication and Field Fabrication of Thermal Insulating Fitting Covers for NPS Piping, Vessel Lagging, and Dished Head Segments	ASTM	ASTM C450-02
	Test Method for Steady-State and Thermal Transmission Properties by Means of the Heat Flow Meter Apparatus	ASTM	ASTM C518-04
	Specification for Preformed Flexible Elastometric Cellular Thermal Insulation in Sheet and Tubular Form	ASTM	ASTM C534-07a
	Specification for Cellular Glass Thermal Insulation	ASTM	ASTM C552-07
	Specification for Rigid, Cellular Polystyrene Thermal Insulation	ASTM	ASTM C578-07
	Practice for Inner and Outer Diameters of Rigid Thermal Insulation for Nominal Sizes of Pipe and Tubing (NPS System)	ASTM	ASTM C585-90 (2004)
	Specification for Unfaced Preformed Rigid Cellular Polyisocyanurate Thermal Insulation	ASTM	ASTM C591-07
	Practice for Determination of Heat Gain or Loss and the Surface Temperature of Insulated Pipe and Equipment Systems by the Use of a Computer Program	ASTM	ASTM C680-04e4
	Specification for Adhesives for Duct Thermal Insulation	ASTM	ASTM C916-85 (2007)
	Classification of Potential Health and Safety Concerns Associated with Thermal Insulation Materials and Accessories	ASTM	ASTM C930-05
	Practice for Thermographic Inspection of Insulation Installations in Envelope Cavities of Frame Buildings	ASTM	ASTM C1060-90 (2003)
	Specification for Fibrous Glass Duct Lining Insulation (Thermal and Sound Absorbing Material)	ASTM	ASTM C1071-05
	Specification for Faced or Unfaced Rigid Cellular Phenolic Thermal Insulation	ASTM	ASTM C1126-04
	Practice for Installation and Use of Radiant Barrier Systems (RBS) in Building Construction	ASTM	ASTM C1158-05
	Test Method for Steady-State and Thermal Performance of Building Assemblies by Means of a Hot Box Apparatus	ASTM	ASTM C1363-05
	Specification for Perpendicularly Oriented Mineral Fiber Roll and Sheet Thermal Insulation for Pipes and Tanks	ASTM	ASTM C1393-00a (2006)
	Guide for Measuring and Estimating Quantities of Insulated Piping and Components	ASTM	ASTM C1409-98 (2003)
	Specification for Cellular Melamine Thermal and Sound Absorbing Insulation	ASTM	ASTM C1410-05a
	Guide for Selecting Jacketing Materials for Thermal Insulation	ASTM	ASTM C1423-98 (2003)
	Specification for Preformed Flexible Cellular Polyolefin Thermal Insulation in Sheet and Tubular Form	ASTM	ASTM C1427-07
	Specification for Polyimide Flexible Cellular Thermal and Sound Absorbing Insulation	ASTM	ASTM C1482-04
	Specification for Cellulosic Fiber Stabilized Thermal Insulation	ASTM	ASTM C1497-04
	Test Method for Characterizing the Effect of Exposure to Environmental Cycling on Thermal Performance of Insulation Products	ASTM	ASTM C1512-07
	Specification for Flexible Polymeric Foam Sheet Insulation Used as a Thermal and Sound Absorbing Liner for Duct Systems	ASTM	ASTM C1534-07
	Standard Guide for Development of Standard Data Records for Computerization of Thermal Transmission Test Data for Thermal Insulation	ASTM	ASTM C1558-03 (2007)
	Guide for Determining Blown Density of Pneumatically Applied Loose Fill Mineral Fiber Thermal Insulation	ASTM	ASTM C1574-04
	Test Method for Determining the Moisture Content of Inorganic Insulation Materials by Weight	ASTM	ASTM C1616-07e1
	Specification for Cellular Polypropylene Thermal Insulation	ASTM	ASTM C1631-05
	Classification for Rating Sound Insulation	ASTM	ASTM E413-04
	Test Method for Determining the Drainage Efficiency of Exterior Insulation and Finish Systems (EIFS) Clad Wall Assemblies	ASTM	ASTM E2273-03
	Practice for Use of Test Methods E96 for Determining the Water Vapor Transmission (WVT) of Exterior Insulation and Finish Systems	ASTM	ASTM E2321-03
	Thermal Insulation—Vocabulary	ISO	ISO 9229:2007
	National Commercial and Industrial Insulation Standards, 6th ed.	MICA	MICA
	Accepted Industry Practices for Sheet Metal Lagging, 1st ed.	SMACNA	SMACNA 2002
Louvers	Laboratory Methods of Testing Dampers for Rating	AMCA	AMCA 500-D-07
	Laboratory Methods of Testing Louvers for Rating	AMCA	AMCA 500-L-07
Lubricants	Methods of Testing the Floc Point of Refrigeration Grade Oils	ASHRAE	ANSI/ASHRAE 86-1994 (RA06)
	Test Method for Pour Point of Petroleum Products	ASTM	ASTM D97-07
	Classification of Industrial Fluid Lubricants by Viscosity System	ASTM	ASTM D2422-97 (2007)
	Test Method for Relative Molecular Weight (Relative Molecular Mass) of Hydrocarbons by Thermoelectric Measurement of Vapor Pressure	ASTM	ASTM D2503-92 (2007)
	Test Method for Determination of Moderately High Temperature Piston Deposits by Thermo-Oxidation Engine Oil Simulation Test	ASTM	ASTM D7097-06a
	Petroleum Products—Corrosiveness to Copper—Copper Strip Test	ISO	ISO 2160:1998

Selected Codes and Standards Published by Various Societies and Associations (*Continued*)

Subject	Title	Publisher	Reference
Measurement	Industrial Ventilation: A Manual of Recommended Practice, 26th ed. (2007)	ACGIH	ACGIH
	Engineering Analysis of Experimental Data	ASHRAE	ASHRAE *Guideline* 2-2005
	Standard Method for Measurement of Proportion of Lubricant in Liquid Refrigerant	ASHRAE	ANSI/ASHRAE 41.4-1996 (RA06)
	Standard Method for Measurement of Moist Air Properties	ASHRAE	ANSI/ASHRAE 41.6-1994 (RA06)
	Method of Measuring Solar-Optical Properties of Materials	ASHRAE	ANSI/ASHRAE 74-1988
	Methods of Measuring, Expressing, and Comparing Building Energy Performance	ASHRAE	ANSI/ASHRAE 105-2007
	Method for Establishing Installation Effects on Flowmeters	ASME	ASME MFC-10M-2000
	Test Uncertainty	ASME	ASME PTC 19.1-2005
	Measurement of Industrial Sound	ASME	ANSI/ASME PTC 36-2004
	Test Methods for Water Vapor Transmission of Materials	ASTM	ASTM E96/E96M-05
	Specification for Temperature-Electromotive Force (EMF) Tables for Standardized Thermocouples	ASTM	ASTM E230-03
	Practice for Continuous Sizing and Counting of Airborne Particles in Dust-Controlled Areas and Clean Rooms Using Instruments Capable of Detecting Single Sub-Micrometre and Larger Particles	ASTM	ASTM F50-07
	Use of the International System of Units (SI): The Modern Metric System	IEEE/ASTM	IEEE/ASTM-SI10-2002
	Ergonomics of the Thermal Environment—Instruments for Measuring Physical Quantities	ISO	ISO 7726:1998
	Ergonomics of the Thermal Environment—Determination of Metabolic Rate	ISO	ISO 8996:2004
	Ergonomics of the Thermal Environment—Estimation of the Thermal Insulation and Water Vapour Resistance of a Clothing Ensemble	ISO	ISO 9920:2007
Fluid Flow	Standard Methods of Measurement of Flow of Liquids in Pipes Using Orifice Flowmeters	ASHRAE	ANSI/ASHRAE 41.8-1989
	Calorimeter Test Methods for Mass Flow Measurements of Volatile Refrigerants	ASHRAE	ANSI/ASHRAE 41.9-2000 (RA06)
	Flow Measurement	ASME	ASME PTC 19.5-2004
	Glossary of Terms Used in the Measurement of Fluid Flow in Pipes	ASME	ASME MFC-1M-2003
	Measurement Uncertainty for Fluid Flow in Closed Conduits	ASME	ANSI/ASME MFC-2M-1983 (RA01)
	Measurement of Fluid Flow in Pipes Using Orifice, Nozzle, and Venturi	ASME	ASME MFC-3M-2004
	Measurement of Liquid Flow in Closed Conduits Using Transit-Time Ultrasonic Flowmeters	ASME	ASME MFC-5M-1985 (RA01)
	Measurement of Fluid Flow in Pipes Using Vortex Flowmeters	ASME	ASME MFC-6M-1998 (RA05)
	Fluid Flow in Closed Conduits: Connections for Pressure Signal Transmissions Between Primary and Secondary Devices	ASME	ASME MFC-8M-2001
	Measurement of Liquid Flow in Closed Conduits by Weighing Method	ASME	ASME MFC-9M-1988 (RA01)
	Measurement of Fluid Flow by Means of Coriolis Mass Flowmeters	ASME	ASME MFC-11M-2006
	Measurement of Fluid Flow Using Small Bore Precision Orifice Meters	ASME	ASME MFC-14M-2003
	Measurement of Fluid Flow in Closed Conduits by Means of Electromagnetic Flowmeters	ASME	ASME MFC-16M-1995 (R01)
	Measurement of Fluid Flow Using Variable Area Meters	ASME	ASME MFC-18M-2001
	Test Method for Determining the Moisture Content of Inorganic Insulation Materials by Weight	ASTM	ASTM C1616-07e1
	Test Method for Indicating Wear Characteristics of Petroleum Hydraulic Fluids in a High Pressure Constant Volume Vane Pump	ASTM	ASTM D6973-05
	Test Method for Dynamic Viscosity and Density of Liquids by Stabinger Viscometer and the Calculation of Kinematic Viscosity	ASTM	ASTM D7042-04
	Test Method for Indicating Wear Characteristics of Petroleum and Non-Petroleum Hydraulic Fluids in a Constant Volume Vane Pump	ASTM	ASTM D7043-04a
	Practice for Calculating Viscosity of a Blend of Petroleum Products	ASTM	ASTM D7152-05e1
	Test Method for Same-Different Test	ASTM	ASTM E2139-05
	Practice for Field Use of Pyranometers, Pyrheliometers, and UV Radiometers	ASTM	ASTM G183-05
Gas Flow	Standard Methods for Laboratory Airflow Measurement	ASHRAE	ANSI/ASHRAE 41.2-1987 (RA92)
	Method of Test for Measurement of Flow of Gas	ASHRAE	ANSI/ASHRAE 41.7-1984 (RA06)
	Measurement of Gas Flow by Turbine Meters	ASME	ANSI/ASME MFC-4M-1986 (RA03)
	Measurement of Gas Flow by Means of Critical Flow Venturi Nozzles	ASME	ANSI/ASME MFC-7M-1987 (RA01)
Pressure	Standard Method for Pressure Measurement	ASHRAE	ANSI/ASHRAE 41.3-1989
	Pressure Gauges and Gauge Attachments	ASME	ASME B40.100-2005
	Pressure Measurement	ASME	ANSI/ASME PTC 19.2-1987 (RA04)
Temperature	Standard Method for Temperature Measurement	ASHRAE	ANSI/ASHRAE 41.1-1986 (RA06)
	Thermometers, Direct Reading and Remote Reading	ASME	ASME B40.200-2001
	Temperature Measurement	ASME	ASME PTC 19.3-1974 (RA04)
	Total Temperature Measuring Instruments (Turbine Powered Subsonic Aircraft)	SAE	SAE AS793-2001
Thermal	Method of Testing Thermal Energy Meters for Liquid Streams in HVAC Systems	ASHRAE	ANSI/ASHRAE 125-1992 (RA06)
	Test Method for Steady-State Heat Flux Measurements and Thermal Transmission Properties by Means of the Guarded-Hot-Plate Apparatus	ASTM	ASTM C177-04
	Test Method for Steady-State Heat Flux Measurements and Thermal Transmission Properties by Means of the Heat Flow Meter Apparatus	ASTM	ASTM C518-04

Selected Codes and Standards Published by Various Societies and Associations (*Continued*)

Subject	Title	Publisher	Reference
	Practice for In-Situ Measurement of Heat Flux and Temperature on Building Envelope Components	ASTM	ASTM C1046-95 (2007)
	Practice for Determining Thermal Resistance of Building Envelope Components from In-Situ Data	ASTM	ASTM C1155-95 (2007)
	Test Method for Thermal Performance of Building Materials and Envelope Assemblies by Means of a Hot Box Apparatus	ASTM	ASTM C1363-05
Mobile Homes and Recreational Vehicles	Residential Load Calculation, 8th ed.	ACCA	ANSI/ACCA Manual J
	Recreational Vehicle Cooking Gas Appliances	CSA	ANSI Z21.57-2007
	Oil-Fired Warm Air Heating Appliances for Mobile Housing and Recreational Vehicles	CSA	B140.10-06
	Mobile Homes	CSA	CAN/CSA-Z240 MH Series-92 (R2005)
	Recreational Vehicles	CSA	CAN/CSA-Z240 RV Series-08
	Gas Supply Connectors for Manufactured Homes	IAPMO	IAPMO TS 9-2003
	Fuel Supply: Manufactured/Mobile Home Parks & Recreational Vehicle Parks	IAPMO	Chapter 13, Part II
	Manufactured Housing Construction and Safety Standards	ICC/ANSI	ICC/ANSI 2.0-1998
	Manufactured Housing	NFPA	NFPA 501-05
	Recreational Vehicles	NFPA	NFPA 1192-08
	Plumbing System Components for Recreational Vehicles	NSF	NSF/ANSI 24-2006
	Low Voltage Lighting Fixtures for Use in Recreational Vehicles (2005)	UL	ANSI/UL 234
	Liquid Fuel-Burning Heating Appliances for Manufactured Homes and Recreational Vehicles (1995)	UL	ANSI/UL 307A
	Gas-Burning Heating Appliances for Manufactured Homes and Recreational Vehicles (2006)	UL	UL 307B
	Gas-Fired Cooking Appliances for Recreational Vehicles (2006)	UL	UL 1075
Motors and Generators	Installation and Maintenance of Farm Standby Electric Power	ASABE	ANSI/ASAE EP364.3-2006
	Nuclear Power Plant Air-Cleaning Units and Components	ASME	ASME N509-2002
	Testing of Nuclear Air Treatment Systems	ASME	ASME N510-2007
	Fired Steam Generators	ASME	ASME PTC 4-1998
	Gas Turbine Heat Recovery Steam Generators	ASME	ASME PTC 4.4-1981 (RA03)
	Test Methods for Film-Insulated Magnet Wire	ASTM	ASTM D1676-03
	Test Method for Evaluation of Engine Oils in a High Speed, Single-Cylinder Diesel Engine—Caterpillar 1R Test Procedure	ASTM	ASTM D6923-05
	Test Method for Evaluation of Diesel Engine Oils in the T-11 Exhaust Gas Recirculation Diesel Engine	ASTM	ASTM D7156-07a
	Energy Efficiency Test Methods for Three-Phase Induction Motors	CSA	C390-98 (R2005)
	Motors and Generators	CSA	C22.2 No. 100-04
	Emergency Electrical Power Supply for Buildings	CSA	CSA C282-05
	Energy Efficiency Test Methods for Single- and Three-Phase Small Motors	CSA	CAN/CSA C747-94 (R2005)
	Standard Test Procedure for Polyphase Induction Motors and Generators	IEEE	IEEE 112-1996
	Motors and Generators	NEMA	NEMA MG 1-2006
	Energy Management Guide for Selection and Use of Fixed Frequency Medium AC Squirrel-Cage Polyphase Industrial Motors	NEMA	NEMA MG 10-2001 (R2007)
	Energy Management Guide for Selection and Use of Single-Phase Motors	NEMA	NEMA MG 11-1977 (R2007)
	Magnet Wire	NEMA	NEMA MW 1000-2003
	Motion/Position Control Motors, Controls, and Feedback Devices	NEMA	NEMA ICS 16-2001
	Electric Motors (1994)	UL	UL 1004
	Electric Motors and Generators for Use in Division 1 Hazardous (Classified) Locations (2003)	UL	ANSI/UL 674
	Overheating Protection for Motors (1997)	UL	ANSI/UL 2111
Pipe, Tubing, and Fittings	Scheme for the Identification of Piping Systems	ASME	ASME A13.1-2007
	Pipe Threads, General Purpose (Inch)	ASME	ANSI/ASME B1.20.1-1983 (RA01)
	Wrought Copper and Copper Alloy Braze-Joint Pressure Fittings	ASME	ASME B16.50-2001
	Power Piping	ASME	ASME B31.1-2007
	Fuel Gas Piping	ASME	ASME B31.2-1968
	Process Piping	ASME	ASME B31.3-2006
	Refrigeration Piping and Heat Transfer Components	ASME	ASME B31.5-2006
	Building Services Piping	ASME	ASME B31.9-2004
	Practice for Obtaining Hydrostatic or Pressure Design Basis for "Fiberglass" (Glass-Fiber-Reinforced Thermosetting-Resin) Pipe and Fittings	ASTM	ASTM D2992-06
	Specification for Welding of Austenitic Stainless Steel Tube and Piping Systems in Sanitary Applications	AWS	AWS D18.1:1999
	Standards of the Expansion Joint Manufacturers Association, 8th ed. (2003)	EJMA	EJMA
	Pipe Hangers and Supports—Materials, Design and Manufacture	MSS	MSS SP-58-2002
	Pipe Hangers and Supports—Selection and Application	MSS	ANSI/MSS SP-69-2003
	General Welding Guidelines (2002)	NCPWB	NCPWB

Selected Codes and Standards Published by Various Societies and Associations (*Continued*)

Subject	Title	Publisher	Reference
	International Fuel Gas Code	AGA/NFPA	ANSI Z223.1/NFPA 54-2006
	Refrigeration Tube Fittings—General Specifications	SAE	SAE J513-1999
	Seismic Restraint Manual—Guidelines for Mechanical Systems, 2nd ed.	SMACNA	ANSI/SMACNA 001-2000 (1998)
	Tube Fittings for Flammable and Combustible Fluids, Refrigeration Service, and Marine Use (1997)	UL	ANSI/UL 109
Plastic	Specification for Acrylonitrile-Butadiene-Styrene (ABS) Plastic Pipe, Schedules 40 and 80	ASTM	ASTM D1527-99 (2005)
	Specification for Poly (Vinyl Chloride) (PVC) Plastic Pipe, Schedules 40, 80, and 120	ASTM	ASTM D1785-06
	Specification for Polyethylene (PE) Plastic Pipe, Schedule 40	ASTM	ASTM D2104-03
	Test Method for Obtaining Hydrostatic or Pressure Design Basis for Thermoplastic Pipe Products	ASTM	ASTM D2837-04e1
	Specification for Polybutylene (PB) Plastic Hot- and Cold-Water Distribution Systems	ASTM	ASTM D3309-96a (2002)
	Specification for Perfluoroalkoxy (PFA)-Fluoropolymer Tubing	ASTM	ASTM D6867-03
	Specification for Polyethylene Stay in Place Form System for End Walls for Drainage Pipe	ASTM	ASTM D7082-04
	Specification for Chlorinated Poly (Vinyl Chloride) (CPVC) Plastic Pipe, Schedules 40 and 80	ASTM	ASTM F441/F441M-02
	Specification for Crosslinked Polyethylene/Aluminum/Crosslinked Polyethylene Tubing OD Controlled SDR9	ASTM	ASTM F2262-05
	Test Method for Evaluating the Oxidative Resistance of Polyethylene (PE) Pipe to Chlorinated Water	ASTM	ASTM F2263-07e1
	Specification for 12 to 60 in. Annular Corrugated Profile-Wall Polyethylene (PE) Pipe and Fittings for Gravity-Flow Storm Sewer and Subsurface Drainage Applications	ASTM	ASTM F2306/F2306M-07
	Specification for Series 10 Poly (Vinyl Chloride) (PVC) Closed Profile Gravity Pipe and Fittings Based on Controlled Inside Diameter	ASTM	ASTM F2307-03
	Standard Test Method for Evaluating the Oxidative Resistance of Multilayer Polyolefin Tubing to Hot Chlorinated Water	ASTM	ASTM F2330-04
	Test Method for Determining Chemical Compatibility of Thread Sealants with Thermoplastic Threaded Pipe and Fittings Materials	ASTM	ASTM F2331-04e1
	Test Method for Determining Thermoplastic Pipe Wall Stiffness	ASTM	ASTM F2433-05
	Specification for Steel Reinforced Polyethylene (PE) Corrugated Pipe	ASTM	ASTM F2435-07
	Electrical Polyvinyl Chloride (PVC) Tubing and Conduit	NEMA	NEMA TC 2-2003
	PVC Plastic Utilities Duct for Underground Installation	NEMA	NEMA TC 6 and 8-2003
	Smooth Wall Coilable Polyethylene Electrical Plastic Duct	NEMA	NEMA TC 7-2005
	Fittings for PVC Plastic Utilities Duct for Underground Installation	NEMA	NEMA TC 9-2004
	Electrical Nonmetallic Tubing (ENT)	NEMA	NEMA TC 13-2005
	Plastics Piping System Components and Related Materials	NSF	NSF/ANSI 14-2007
	Rubber Gasketed Fittings for Fire-Protection Service (2004)	UL	ANSI/UL 213
Metal	Welded and Seamless Wrought Steel Pipe	ASME	ASME B36.10M-2004
	Stainless Steel Pipe	ASME	ASME B36.19M-2004
	Specification for Pipe, Steel, Black and Hot-Dipped, Zinc-Coated, Welded and Seamless	ASTM	ASTM A53/53M-07
	Specification for Seamless Carbon Steel Pipe for High-Temperature Service	ASTM	ASTM A106/A106M-06a
	Specification for Pipe, Steel, Electric-Fusion Arc-Welded Sizes NPS 16 and Over	ASTM	ASTM A1034-05b
	Specification for Steel Line Pipe, Black, Furnace-Butt-Welded	ASTM	ASTM A1037/A1037M-05
	Specification for Composite Corrugated Steel Pipe for Sewers and Drains	ASTM	ASTM A1042/A1042M-04
	Specification for Seamless Copper Pipe, Standard Sizes	ASTM	ASTM B42-02e1
	Specification for Seamless Copper Tube	ASTM	ASTM B75-02
	Specification for Seamless Copper Water Tube	ASTM	ASTM B88-03
	Specification for Seamless Copper Tube for Air Conditioning and Refrigeration Field Service	ASTM	ASTM B280-03
	Specification for Hand-Drawn Copper Capillary Tube for Restrictor Applications	ASTM	ASTM B360-01
	Specification for Welded Copper Tube for Air Conditioning and Refrigeration Service	ASTM	ASTM B640-07
	Specification for Copper-Beryllium Seamless Tube UNS Nos. C17500 and C17510	ASTM	ASTM B937-04
	Test Method for Rapid Determination of Corrosiveness to Copper from Petroleum Products Using a Disposable Copper Foil Strip	ASTM	ASTM D7095-04
	Thickness Design of Ductile-Iron Pipe	AWWA	ANSI/AWWA C150/A21.50-02
	Fittings, Cast Metal Boxes, and Conduit Bodies for Conduit and Cable Assemblies	NEMA	NEMA FB 1-2007
	Polyvinyl-Chloride (PVC) Externally Coated Galvanized Rigid Steel Conduit and Intermediate Metal Conduit	NEMA	NEMA RN 1-2005
Plumbing	Backwater Valves	ASME	ASME A112.14.1-2003
	Plumbing Supply Fittings	ASME	ASME A112.18.1-2005
	Plumbing Waste Fittings	ASME	ASME A112.18.2-2005
	Performance Requirements for Backflow Protection Devices and Systems in Plumbing Fixture Fittings	ASME	ASME A112.18.3-2002
	Uniform Plumbing Code (2006) (with IAPMO Installation Standards)	IAPMO	IAPMO
	International Plumbing Code (2006)	ICC	IPC

Selected Codes and Standards Published by Various Societies and Associations (*Continued*)

Subject	Title	Publisher	Reference
	International Private Sewage Disposal Code (2006)	ICC	IPSDC
	2006 National Standard Plumbing Code (NSPC)	PHCC	NSPC 2003
	2006 National Standard Plumbing Code—Illustrated	PHCC	PHCC 2003
	Standard Plumbing Code (1997)	SBCCI	SPC
Pumps	Centrifugal Pumps	ASME	ASME PTC 8.2-1990
	Specification for Horizontal End Suction Centrifugal Pumps for Chemical Process	ASME	ASME B73.1-2001
	Specification for Vertical-in-Line Centrifugal Pumps for Chemical Process	ASME	ASME B73.2-2003
	Specification for Sealless Horizontal End Suction Metallic Centrifugal Pumps for Chemical Process	ASME	ASME B73.3-2003
	Specification for Thermoplastic and Thermoset Polymer Material Horizontal End Suction Centrifugal Pumps for Chemical Process	ASME	ASME B73.5M-1995 (RA01)
	Liquid Pumps	CSA	CAN/CSA-C22.2 No. 108-01
	Energy Efficiency Test Methods for Small Pumps	CSA	CAN/CSA C820-02 (R2007)
	Performance Standard for Liquid Ring Vacuum Pumps, 3rd ed. (2005)	HEI	HEI 2854
	Centrifugal Pumps for Nomenclature and Definitions	HI	ANSI/HI 1.1-1.2 (2000)
	Centrifugal Pumps for Design and Applications	HI	ANSI/HI 1.3 (2000)
	Centrifugal Pumps for Installation, Operation, and Maintenance	HI	ANSI/HI 1.4 (2000)
	Vertical Pumps for Nomenclature and Definitions	HI	ANSI/HI 2.1-2.2 (2000)
	Vertical Pumps for Design and Application	HI	ANSI/HI 2.3 (2000)
	Vertical Pumps for Installation, Operation, and Maintenance	HI	ANSI/HI 2.4 (2000)
	Rotary Pumps for Nomenclature, Definitions, Application, and Operation	HI	ANSI/HI 3.1-3.5 (2000)
	Sealless Rotary Pumps for Nomenclature, Definitions, Application, Operation, and Test	HI	ANSI/HI 4.1-4.6 (2000)
	Sealless Centrifugal Pumps for Nomenclature, Definitions, Application, Operation, and Test	HI	ANSI/HI 5.1-5.6 (2000)
	Reciprocating Pumps for Nomenclature, Definitions, Application, and Operation	HI	ANSI/HI 6.1-6.5 (2000)
	Direct Acting (Steam) Pumps for Nomenclature, Definitions, Application, and Operation	HI	ANSI/HI 8.1-8.5 (2000)
	Pumps—General Guidelines for Types, Definitions, Application, Sound Measurement and Decontamination	HI	ANSI/HI 9.1-9.5 (2000)
	Centrifugal and Vertical Pumps for Allowable Nozzle Loads	HI	ANSI/HI 9.6.2 (2001)
	Centrifugal and Vertical Pumps for Allowable Operating Region	HI	ANSI/HI 9.6.3 (1997)
	Centrifugal and Vertical Pumps for Vibration Measurements and Allowable Values	HI	ANSI/HI 9.6.4 (2001)
	Centrifugal and Vertical Pumps for Condition Monitoring	HI	ANSI/HI 9.6.5 (2000)
	Pump Intake Design	HI	ANSI/HI 9.8 (1998)
	Engineering Data Book, 2nd ed.	HI	HI (1990)
	Circulation System Components and Related Materials for Swimming Pools, Spas/Hot Tubs	NSF	NSF/ANSI 50-2007
	Pumps for Oil-Burning Appliances (1997)	UL	UL 343
	Motor-Operated Water Pumps (2002)	UL	ANSI/UL 778
	Swimming Pool Pumps, Filters, and Chlorinators (2008)	UL	ANSI/UL 1081
Radiators	Testing and Rating Standard for Baseboard Radiation, 8th ed. (2005)	HYDI	IBR
	Testing and Rating Standard for Finned Tube (Commercial) Radiation, 6th ed. (2005)	HYDI	IBR
Receivers	Refrigerant Liquid Receivers	ARI	ARI 495-2005
	Refrigerant-Containing Components and Accessories, Nonelectrical (2001)	UL	ANSI/UL 207
Refrigerants	Threshold Limit Values for Chemical Substances (updated annually)	ACGIH	ACGIH
	Specifications for Fluorocarbon Refrigerants	ARI	ARI 700-2006
	Refrigerant Recovery/Recycling Equipment	ARI	ARI 740-98
	Format for Information on Refrigerants	ASHRAE	ASHRAE *Guideline* 6-1996
	Method of Testing Flow Capacity of Refrigerant Capillary Tubes	ASHRAE	ANSI/ASHRAE 28-1996 (RA06)
	Designation and Safety Classification of Refrigerants	ASHRAE	ANSI/ASHRAE 34-2007
	Sealed Glass Tube Method to Test the Chemical Stability of Materials for Use Within Refrigerant Systems	ASHRAE	ANSI/ASHRAE 97-2007
	Refrigeration Oil Description	ASHRAE	ANSI/ASHRAE 99-2006
	Reducing the Release of Halogenated Refrigerants from Refrigerating and Air-Conditioning Equipment and Systems	ASHRAE	ANSI/ASHRAE 147-2002
	Test Method for Acid Number of Petroleum Products by Potentiometric Titration	ASTM	ASTM D664-07
	Test Method for Concentration Limits of Flammability of Chemical (Vapors and Gases)	ASTM	ASTM E681-04
	Refrigerant-Containing Components for Use in Electrical Equipment	CSA	C22.2 No. 140.3-M1987 (R2004)
	Refrigerants—Designation System	ISO	ISO 817:2005
	Procedure Retrofitting CFC-12 (R-12) Mobile Air-Conditioning Systems to HFC-134a (R-134a)	SAE	SAE J1661-1998
	Recommended Service Procedure for the Containment of CFC-12 (R-12)	SAE	SAE J1989-1998
	Standard of Purity for Recycled HFC-134a for Use in Mobile Air-Conditioning Systems	SAE	SAE J2099-1999
	HFC-134a (R-134a) Service Hose Fittings for Automotive Air-Conditioning Service Equipment	SAE	SAE J2197-1997
	Recommended Service Procedure for the Containment of HFC-134a	SAE	SAE J2211-1998
	HFC-134a (R-134a) Recovery/Recycling Equipment for Mobile Air-Conditioning Systems	SAE	SAE J2210-1999

Selected Codes and Standards Published by Various Societies and Associations (*Continued*)

Subject	Title	Publisher	Reference
	CFC-12 (R-12) Refrigerant Recovery Equipment for Mobile Automotive Air-Conditioning Systems	SAE	SAE J2209-1999
	Refrigerant-Containing Components and Accessories, Nonelectrical (2001)	UL	ANSI/UL 207
	Refrigerant Recovery/Recycling Equipment (2005)	UL	ANSI/UL 1963
	Field Conversion/Retrofit of Products to Change to an Alternative Refrigerant—Construction and Operation (1993)	UL	ANSI/UL 2170
	Field Conversion/Retrofit of Products to Change to an Alternative Refrigerant—Insulating Material and Refrigerant Compatibility (1993)	UL	ANSI/UL 2171
	Field Conversion/Retrofit of Products to Change to an Alternative Refrigerant—Procedures and Methods (1993)	UL	ANSI/UL 2172
	Refrigerants (2006)	UL	ANSI/UL 2182
Refrigeration	Safety Standard for Refrigeration Systems	ASHRAE	ANSI/ASHRAE 15-2007
	Mechanical Refrigeration Code	CSA	B52-05
	Refrigeration Equipment	CSA	CAN/CSA-C22.2 No. 120-M91 (R2004)
	Equipment, Design and Installation of Ammonia Mechanical Refrigerating Systems	IIAR	ANSI/IIAR 2-1999
	Refrigerated Medical Equipment (1993)	UL	ANSI/UL 416
Refrigeration Systems	Ejectors	ASME	ASME PTC 24-1976 (RA82)
	Reducing the Release of Halogenated Refrigerants from Refrigerating and Air-Conditioning Equipment and Systems	ASHRAE	ANSI/ASHRAE 147-2002
	Testing of Refrigerating Systems	ISO	ISO 916-1968
	Standards for Steam Jet Vacuum Systems, 6th ed.	HEI	HEI 2866-1
Transport	Mechanical Transport Refrigeration Units	ARI	ARI 1110-2006
	Mechanical Refrigeration and Air-Conditioning Installations Aboard Ship	ASHRAE	ANSI/ASHRAE 26-1996 (RA06)
	General Requirements for Application of Vapor Cycle Refrigeration Systems for Aircraft	SAE	SAE ARP731-2003
	Safety Standard for Motor Vehicle Refrigerant Vapor Compression Systems	SAE	SAE J639-2005
Refrigerators	Method of Testing Commercial Refrigerators and Freezers	ASHRAE	ANSI/ASHRAE 72-2005
Commercial	Energy Performance Standard for Commercial Refrigerated Display Cabinets and Merchandise	CSA	C657-04
	Energy Performance Standard for Food Service Refrigerators and Freezers	CSA	C827-98 (R2003)
	Gas Food Service Equipment	CSA	ANSI Z83.11-2006/CSA 1.8A-2006
	Mobile Food Carts	NSF	NSF/ANSI 59-2002e
	Food Equipment	NSF	NSF/ANSI 2-2007
	Commercial Refrigerators and Freezers	NSF	NSF/ANSI 7-2007
	Refrigeration Unit Coolers (2004)	UL	ANSI/UL 412
	Refrigerating Units (2006)	UL	ANSI/UL 427
	Commercial Refrigerators and Freezers (2006)	UL	ANSI/UL 471
Household	Household Refrigerators, Refrigerator-Freezers and Freezers	AHAM	ANSI/AHAM HRF-1-2007
	Refrigerators Using Gas Fuel	CSA	ANSI Z21.19-2002/CSA1.4-2002
	Energy Performance and Capacity of Household Refrigerators, Refrigerator-Freezers, and Freezers	CSA	CAN/CSA C300-00 (R2005)
	Household Refrigerators and Freezers (1993)	UL/CSA	ANSI/UL 250-1997/C22.2 No. 63-93
Retrofitting			
Building	Residential Duct Diagnostics and Repair (2003)	ACCA	ACCA
	Good HVAC Practices for Residential and Commercial Buildings (2003)	ACCA	ACCA
	Building Systems Analysis and Retrofit Manual, 1st ed.	SMACNA	SMACNA 1995
Refrigerant	Procedure for Retrofitting CFC-12 (R-12) Mobile Air Conditioning Systems to HFC-134a (R-134a)	SAE	SAE J1661-1998
	Field Conversion/Retrofit of Products to Change to an Alternative Refrigerant—Construction and Operation (1993)	UL	ANSI/UL 2170
	Field Conversion/Retrofit of Products to Change to an Alternative Refrigerant—Insulating Material and Refrigerant Compatibility (1993)	UL	ANSI/UL 2171
	Field Conversion/Retrofit of Products to Change to an Alternative Refrigerant—Procedures and Methods (1993)	UL	ANSI/UL 2172
Roof Ventilators	Commercial Low Pressure, Low Velocity Duct System Design, 1st ed.	ACCA	ACCA Manual Q
	Power Ventilators (2004)	UL	ANSI/UL 705
Solar Equipment	Thermal Performance Testing of Solar Ambient Air Heaters	ASABE	ANSI/ASAE S423-1991
	Testing and Reporting Solar Cooker Performance	ASABE	ASAE S580-2003
	Method of Measuring Solar-Optical Properties of Materials	ASHRAE	ASHRAE 74-1988
	Methods of Testing to Determine the Thermal Performance of Solar Collectors	ASHRAE	ANSI/ASHRAE 93-2003
	Methods of Testing to Determine the Thermal Performance of Solar Domestic Water Heating Systems	ASHRAE	ANSI/ASHRAE 95-1987
	Methods of Testing to Determine the Thermal Performance of Unglazed Flat-Plate Liquid-Type Solar Collectors	ASHRAE	ANSI/ASHRAE 96-1980 (RA89)

Selected Codes and Standards Published by Various Societies and Associations (*Continued*)

Subject	Title	Publisher	Reference
	Methods of Testing to Determine the Thermal Performance of Flat-Plate Solar Collectors Containing a Boiling Liquid	ASHRAE	ANSI/ASHRAE 109-1986 (RA03)
	Practice for Installation and Service of Solar Space Heating Systems for One and Two Family Dwellings	ASTM	ASTM E683-91 (2007)
	Practice for Evaluating Thermal Insulation Materials for Use in Solar Collectors	ASTM	ASTM E861-94 (2007)
	Practice for Installation and Service of Solar Domestic Water Heating Systems for One and Two Family Dwellings	ASTM	ASTM E1056-85 (2007)
	Reference Solar Spectral Irradiance at the Ground at Different Receiving Conditions—Part 1: Direct Normal and Hemispherical Solar Irradiance for Air Mass 1.5	ISO	ISO 9845-1:1992
	Solar Collectors	CSA	CAN/CSA F378-87 (R2004)
	Solar Domestic Hot Water Systems (Liquid to Liquid Heat Transfer)	CSA	CAN/CSA F379.1-88 (R2006)
	Seasonal Use Solar Domestic Hot Water Systems	CSA	CAN/CSA F379.2-M89 (R2006)
	Installation Code for Solar Domestic Hot Water Systems	CSA	CAN/CSA F383-87 (R2005)
	Solar Heating—Domestic Water Heating Systems—Part 2: Outdoor Test Methods for System Performance Characterization and Yearly Performance Prediction of Solar-Only Systems	ISO	ISO 9459-2:1995
	Test Methods for Solar Collectors—Part 1: Thermal Performance of Glazed Liquid Heating Collectors Including Pressure Drop	ISO	ISO 9806-1:1994
	Part 2: Qualification Test Procedures	ISO	ISO 9806-2:1995
	Part 3: Thermal Performance of Unglazed Liquid Heating Collectors (Sensible Heat Transfer Only) Including Pressure Drop	ISO	ISO 9806-3:1995
	Solar Water Heaters—Elastomeric Materials for Absorbers, Connecting Pipes and Fittings—Method of Assessment	ISO	ISO 9808:1990
	Solar Energy—Calibration of a Pyranometer Using a Pyrheliometer	ISO	ISO 9846:1993
Solenoid Valves	Solenoid Valves for Use with Volatile Refrigerants	ARI	ARI 760-2007
	Methods of Testing Capacity of Refrigerant Solenoid Valves	ASHRAE	ANSI/ASHRAE 158.1-2004
	Electrically Operated Valves (1999)	UL	UL 429
Sound Measurement	Threshold Limit Values for Physical Agents (updated annually)	ACGIH	ACGIH
	Specification for Sound Level Meters	ASA	ANSI S1.4-1983 (R2006)
	Specification for Octave-Band and Fractional-Octave-Band Analog and Digital Filters	ASA	ANSI S1.11-2004
	Microphones, Part 1: Specifications for Laboratory Standard Microphones	ASA	ANSI S1.15-1997/Part 1 (R2006)
	Part 2: Primary Method for Pressure Calibration of Laboratory Standard Microphones by the Reciprocity Technique	ASA	ANSI S1.15-2005/Part 2
	Specification for Acoustical Calibrators	ASA	ANSI S1.40-2006
	Measurement of Industrial Sound	ASME	ASME PTC 36-2004
	Test Method for Measuring Acoustical and Airflow Performance of Duct Liner Materials and Prefabricated Silencers	ASTM	ASTM E477-06a
	Test Method for Determination of Decay Rates for Use in Sound Insulation Test Methods	ASTM	ASTM E2235-04e1
	Sound and Vibration Design and Analysis (1994)	NEBB	NEBB
Fans	Reverberant Room Method for Sound Testing of Fans	AMCA	AMCA 300-05
	Methods for Calculating Fan Sound Ratings from Laboratory Test Data	AMCA	AMCA 301-06
	Application of Sone Ratings for Non-Ducted Air Moving Devices	AMCA	AMCA 302-73 (R2008)
	Application of Sound Power Level Ratings for Fans	AMCA	AMCA 303-79 (R2008)
	Acoustics—Measurement of Noise and Vibration of Small Air-Moving Devices—Part 1: Airborne Noise Emission	ASA	ANSI S12.11/1-2003/ISO 10302:1996 (MOD-2003)
	Part 2: Structure-Borne Vibration	ASA	ANSI S12.11/2-2003
	Laboratory Method of Testing to Determine the Sound Power in a Duct	ASHRAE/ AMCA	ANSI/ASHRAE 68-1997/ AMCA 330-97
Other Equipment	Sound Rating of Outdoor Unitary Equipment	ARI	ARI 270-95
	Application of Sound Rating Levels of Outdoor Unitary Equipment	ARI	ARI 275-97
	Sound Rating and Sound Transmission Loss of Packaged Terminal Equipment	ARI	ARI 300-2000
	Sound Rating of Non-Ducted Indoor Air-Conditioning Equipment	ARI	ARI 350-2000
	Sound Rating of Large Outdoor Refrigerating and Air-Conditioning Equipment	ARI	ARI 370-2001
	Method of Rating Sound and Vibration of Refrigerant Compressors	ARI	ARI 530-2005
	Method of Measuring Machinery Sound Within an Equipment Space	ARI	ARI 575-94
	Statistical Methods for Determining and Verifying Stated Noise Emission Values of Machinery and Equipment	ASA	ANSI S12.3-1985 (R2006)
	Sound Level Prediction for Installed Rotating Electrical Machines	NEMA	NEMA MG 3-1974 (R2006)
Techniques	Preferred Frequencies, Frequency Levels, and Band Numbers for Acoustical Measurements	ASA	ANSI S1.6-1984 (R2006)
	Reference Quantities for Acoustical Levels	ASA	ANSI S1.8-1989 (R2006)
	Measurement of Sound Pressure Levels in Air	ASA	ANSI S1.13-2005
	Procedure for the Computation of Loudness of Steady Sound	ASA	ANSI S3.4-2007
	Criteria for Evaluating Room Noise	ASA	ANSI S12.2-1995 (R1999)
	Methods for Determining the Insertion Loss of Outdoor Noise Barriers	ASA	ANSI S12.8-1998 (R2003)

Selected Codes and Standards Published by Various Societies and Associations (*Continued*)

Subject	Title	Publisher	Reference
	Engineering Method for the Determination of Sound Power Levels of Noise Sources Using Sound Intensity	ASA	ANSI S12.12-1992 (R2007)
	Procedures for Outdoor Measurement of Sound Pressure Level	ASA	ANSI S12.18-1994 (R2004)
	Methods for Measurement of Sound Emitted by Machinery and Equipment at Workstations and Other Specified Positions	ASA	ANSI S12.43-1997 (R2007)
	Methods for Calculation of Sound Emitted by Machinery and Equipment at Workstations and Other Specified Positions from Sound Power Level	ASA	ANSI S12.44-1997 (R2007)
	Acoustics—Determination of Sound Power Levels of Noise Sources Using Sound Pressure—Precision Method for Reverberation Rooms	ASA	ANSI S12.51-2002 (R2007)/ISO 3741:1999
	Acoustics—Determination of Sound Power Levels of Noise Sources—Engineering Methods for Small, Movable Sources in Reverberant Fields—Part 1: Comparison Method for Hard-Walled Test Rooms	ASA	ANSI S12.53/Part 1-1999 (R2004)/ISO 3743-1:1994
	Part 2: Methods for Special Reverberation Test Rooms	ASA	ANSI S12.53/Part 2-1999 (R2004)/ISO 3743-2:1994
	Acoustics—Determination of Sound Power Levels of Noise Sources Using Sound Pressure—Engineering Method in an Essentially Free Field over a Reflecting Plane	ASA	ANSI S12.54-1999 (R2004)/ISO 3744:1994
	Acoustics—Determination of Sound Power Levels of Noise Sources Using Sound Pressure—Survey Method Using an Enveloping Measurement Surface over a Reflecting Plane	ASA	ANSI S12.56-1999 (R2004)/ISO 3746:1995
	Test Method for Impedance and Absorption of Acoustical Materials by the Impedance Tube Method	ASTM	ASTM C384-04
	Test Method for Sound Absorption and Sound Absorption Coefficients by the Reverberation Room Method	ASTM	ASTM C423-07a
	Test Method for Measurement of Airborne Sound Insulation in Buildings	ASTM	ASTM E336-07
	Test Method for Impedance and Absorption of Acoustical Materials Using a Tube, Two Microphones and a Digital Frequency Analysis System	ASTM	ASTM E1050-08
	Test Method for Evaluating Masking Sound in Open Offices Using A-Weighted and One-Third Octave Band Sound Pressure Levels	ASTM	ASTM E1573-02
	Test Method for Measurement of Sound in Residential Spaces	ASTM	ASTM E1574-98 (2006)
	Acoustics–Measurement of Sound Insulation in Buildings and of Building Elements; Part 1: Requirements for Laboratory Test Facilities with Suppressed Flanking Transmission	ISO	ISO 140-1:1997
	Part 4: Field Measurements of Airborne Sound Insulation Between Rooms	ISO	ISO 140-4:1998
	Part 5: Field Measurements of Airborne Sound Insulation of Facade Elements and Facades	ISO	ISO 140-5:1998
	Part 6: Laboratory Measurements of Impact Sound Insulation of Floors	ISO	ISO 140-6:1998
	Part 7: Field Measurements of Impact Sound Insulation of Floors	ISO	ISO 140-7:1998
	Part 8: Laboratory Measurements of the Reduction of Transmitted Impact Noise by Floor Coverings on a Heavyweight Standard Floor	ISO	ISO 140-8:1997
	Acoustics—Method for Calculating Loudness Level	ISO	ISO 532:1975
	Acoustics—Determination of Sound Power Levels of Noise Sources Using Sound Intensity; Part 1: Measurement at Discrete Points	ISO	ISO 9614-1:1993
	Part 2: Measurement by Scanning	ISO	ISO 9614-2:1996
	Procedural Standards for Measurement and Assessment of Sound and Vibration, 2nd ed. (2006)	NEBB	NEBB
Terminology	Acoustical Terminology	ASA	ANSI S1.1-1994 (R2004)
	Terminology Relating to Environmental Acoustics	ASTM	ASTM C634-02e1
Space Heaters	Methods of Testing for Rating Combination Space-Heating and Water-Heating Appliances	ASHRAE	ANSI/ASHRAE 124-2007
	Gas-Fired Room Heaters, Vol. II, Unvented Room Heaters	CSA	ANSI Z21.11.2-2007
	Vented Gas-Fired Space Heating Appliances	CSA	ANSI Z21.86-2004/CSA 2.32-2004
	Movable and Wall- or Ceiling-Hung Electric Room Heaters (2000)	UL	UL 1278
	Fixed and Location-Dedicated Electric Room Heaters (1997)	UL	UL 2021
Symbols	Graphic Electrical/Electronic Symbols for Air-Conditioning and Refrigerating Equipment	ARI	ARI 130-88
	Graphic Symbols for Heating, Ventilating, Air-Conditioning, and Refrigerating Systems	ASHRAE	ANSI/ASHRAE 134-2005
	Graphical Symbols for Plumbing Fixtures for Diagrams Used in Architecture and Building Construction	ASME	ANSI/ASME Y32.4-1977 (RA04)
	Symbols for Mechanical and Acoustical Elements as Used in Schematic Diagrams	ASME	ANSI/ASME Y32.18-1972 (RA03)
	Practice for Mechanical Symbols, Shipboard Heating, Ventilation, and Air Conditioning (HVAC)	ASTM	ASTM F856-97 (2004)
	Standard Symbols for Welding, Brazing, and Nondestructive Examination	AWS	AWS A2.4:2007
	Standard Letter Symbols for Quantities Used in Electrical Science and Electrical Engineering	IEEE	IEEE 280-1982 (R2003)
	Graphic Symbols for Electrical and Electronics Diagrams	IEEE	ANSI 315-1975 (R1986)/IEEE 315A-1986
	Standard for Logic Circuit Diagrams	IEEE	IEEE 991-1986 (R1994)
	Use of the International System of Units (SI): The Modern Metric System	IEEE/ASTM	IEEE/ASTM-SI10-2002

Selected Codes and Standards Published by Various Societies and Associations (*Continued*)

Subject	Title	Publisher	Reference
	Abbreviations and Acronyms	ASME	ASME Y14.38-1999
	Engineering Drawing Practices	ASME	ASME Y14.100-2004
	Safety Color Code	NEMA	ANSI/NEMA Z535-2002
Terminals, Wiring	Electrical Quick-Connect Terminals (2003)	UL	ANSI/UL 310
	Wire Connectors (2003)	UL	ANSI/UL 486A-486B
	Splicing Wire Connectors (2004)	UL	ANSI/UL 486C
	Equipment Wiring Terminals for Use with Aluminum and/or Copper Conductors (1994)	UL	ANSI/UL 486E
Testing and Balancing	AABC National Standards for Total System Balance (2002)	AABC	AABC
	Industrial Process/Power Generation Fans: Site Performance Test Standard	AMCA	AMCA 803-02
	Guidelines for Measuring and Reporting Environmental Parameters for Plant Experiments in Growth Chambers	ASABE	ANSI/ASAE EP411.4-2002
	The HVAC Commissioning Process	ASHRAE	ASHRAE *Guideline* 1-1996
	Practices for Measurement, Testing, Adjusting, and Balancing of Building Heating, Ventilation, Air-Conditioning, and Refrigeration Systems	ASHRAE	ANSI/ASHRAE 111-1988
	Practices for Measuring, Testing, Adjusting, and Balancing Shipboard HVAR&R Systems	ASHRAE	ANSI/ASHRAE 151-2002
	Centrifugal Pump Tests	HI	ANSI/HI 1.6 (M104) (2000)
	Vertical Pump Tests	HI	ANSI/HI 2.6 (M108) (2000)
	Rotary Pump Tests	HI	ANSI/HI 3.6 (M110) (2000)
	Reciprocating Pump Tests	HI	ANSI/HI 6.6 (M114) (2000)
	Pumps—General Guidelines for Types, Definitions, Application, Sound Measurement and Decontamination	HI	HI 9.1-9.5 (M117) (2000)
	Submersible Pump Tests	HI	ANSI/HI 11.6 (M126) (2001)
	Procedural Standards for Certified Testing of Cleanrooms, 2nd ed. (1996)	NEBB	NEBB
	Procedural Standards for Testing, Adjusting, Balancing of Environmental Systems, 7th ed. (2005)	NEBB	NEBB
	HVAC Systems Testing, Adjusting and Balancing, 3rd ed.	SMACNA	SMACNA 2002
Thermal Storage	Thermal Energy Storage: A Guide for Commercial HVAC Contractors	ACCA	ACCA
	Method of Testing Active Latent-Heat Storage Devices Based on Thermal Performance	ASHRAE	ANSI/ASHRAE 94.1-2002 (RA06)
	Method of Testing Thermal Storage Devices with Electrical Input and Thermal Output Based on Thermal Performance	ASHRAE	ANSI/ASHRAE 94.2-1981 (RA06)
	Method of Testing Active Sensible Thermal Energy Devices Based on Thermal Performance	ASHRAE	ANSI/ASHRAE 94.3-1986 (RA06)
	Practices for Measurement, Testing, Adjusting, and Balancing of Building Heating, Ventilation, Air-Conditioning, and Refrigeration Systems	ASHRAE	ANSI/ASHRAE 111-1988
	Method of Testing the Performance of Cool Storage Systems	ASHRAE	ANSI/ASHRAE 150-2000 (RA04)
Transformers	Minimum Efficiency Values for Liquid-Filled Distribution Transformers	CSA	CAN/CSA C802.1-00 (R2005)
	Minimum Efficiency Values for Dry-Type Transformers	CSA	CAN/CSA C802.2-06
	Maximum Losses For Power Transformers	CSA	CAN/CSA C802.3-01 (R2007)
	Guide for Determining Energy Efficiency of Distribution Transformers	NEMA	NEMA TP-1-2002
Turbines	Steam Turbines	ASME	ASME PTC 6-2004
	Steam Turbines for Combined Cycle	ASME	ASME PTC 6.2-2004
	Hydraulic Turbines and Pump-Turbines	ASME	ASME PTC 18-2002
	Gas Turbines	ASME	ASME PTC 22-2005
	Wind Turbines	ASME	ASME PTC 42-1988 (RA04)
	Specification for Stainless Steel Bars for Compressor and Turbine Airfoils	ASTM	ASTM A1028-03
	Specification for Gas Turbine Fuel Oils	ASTM	ASTM D2880-03
	Land Based Steam Turbine Generator Sets, 0 to 33,000 kW	NEMA	NEMA SM 24-1991 (R2002)
	Steam Turbines for Mechanical Drive Service	NEMA	NEMA SM 23-1991 (R2002)
Valves	Face-to-Face and End-to-End Dimensions of Valves	ASME	ASME B16.10-2000 (R03)
	Valves—Flanged, Threaded, and Welding End	ASME	ASME B16.34-2004
	Manually Operated Metallic Gas Valves for Use in Aboveground Piping Systems up to 5 psi	ASME	ASME B16.44-2002
	Pressure Relief Devices	ASME	ASME PTC 25-2001
	Methods of Testing Capacity of Refrigerant Solenoid Valves	ASHRAE	ANSI/ASHRAE 158.1-2004
	Relief Valves for Hot Water Supply	CSA	ANSI Z21.22-1999 (R2003)/ CSA 4.4-M99 (R2004)
	Control Valve Capacity Test Procedures	ISA	ANSI/ISA-S75.02-1996
	Flow Equations for Sizing Control Valves	ISA	ANSI/ISA-S75.01.01-2002
	Industrial Valves—Part-Turn Actuator Attachments	ISO	ISO 5211-1:2001
	Metal Valves for Use in Flanged Pipe Systems—Face-to-Face and Centre-to-Face Dimensions	ISO	ISO 5752:1982
	Safety Valves for Protection Against Excessive Pressure, Part 1: Safety Valves	ISO	ISO 4126-1:2004
	Oxygen System Fill/Check Valve	SAE	SAE AS1225A-1997
	Valves for Anhydrous Ammonia and LP-Gas (Other Than Safety Relief) (2007)	UL	ANSI/UL 125

Selected Codes and Standards Published by Various Societies and Associations (*Continued*)

Subject	Title	Publisher	Reference
	Safety Relief Valves for Anhydrous Ammonia and LP-Gas (2007)	UL	ANSI/UL 132
	LP-Gas Regulators (1999)	UL	ANSI/UL 144
	Electrically Operated Valves (1999)	UL	UL 429
	Valves for Flammable Fluids (2007)	UL	ANSI/UL 842
Gas	Manually Operated Metallic Gas Valves for Use in Gas Piping Systems up to 125 psig (Sizes NPS 1/2 through 2)	ASME	ASME B16.33-2002
	Large Metallic Valves for Gas Distribution (Manually Operated, NPS-2 1/2 to 12, 125 psig Maximum)	ASME	ANSI/ASME B16.38-2007
	Manually Operated Thermoplastic Gas Shutoffs and Valves in Gas Distribution Systems	ASME	ASME B16.40-2002
	Manually Operated Gas Valves for Appliances, Appliance Connection Valves, and Hose End Valves	CSA	ANSI Z21.15-1997 (R2003)/CGA 9.1-M97
	Automatic Valves for Gas Appliances	CSA	ANSI Z21.21-2005/CGA 6.5-2005
	Combination Gas Controls for Gas Appliances	CSA	ANSI Z21.78-2005/CGA 6.20-2005
	Convenience Gas Outlets and Optional Enclosures	CSA	ANSI Z21.90-2001/CSA 6.24-2001 (R2005)
Refrigerant	Thermostatic Refrigerant Expansion Valves	ARI	ARI 750-2007
	Solenoid Valves for Use with Volatile Refrigerants	ARI	ARI 760-2007
	Refrigerant Pressure Regulating Valves	ARI	ARI 770-2001
Vapor Retarders	Practice for Selection of Vapor Retarders for Thermal Insulation	ASTM	ASTM C755-03
	Practice for Determining the Properties of Jacketing Materials for Thermal Insulation	ASTM	ASTM C921-03a
	Specification for Flexible, Low Permeance Vapor Retarders for Thermal Insulation	ASTM	ASTM C1136-06
	Test Method for Water Vapor Transmission Rate of Flexible Barrier Materials Using an Infrared Detection Technique	ASTM	ASTM F372-99 (2003)
Vending Machines	Methods of Testing for Rating Vending Machines for Bottled, Canned, and Other Sealed Beverages	ASHRAE	ANSI/ASHRAE 32.1-2004
	Methods of Testing for Rating Pre-Mix and Post-Mix Beverage Dispensing Equipment	ASHRAE	ANSI/ASHRAE 32.2-2003 (RA07)
	Vending Machines	CSA	C22.2 No. 128-95 (R2004)
	Energy Performance of Vending Machines	CSA	CAN/CSA C804-96 (R2007)
	Vending Machines for Food and Beverages	NSF	NSF/ANSI 25-2007
	Refrigerated Vending Machines (1995)	UL	ANSI/UL 541
	Vending Machines (1995)	UL	ANSI/UL 751
Vent Dampers	Automatic Vent Damper Devices for Use with Gas-Fired Appliances	CSA	ANSI Z21.66-1996 (R2001)/CSA 6.14-M96
	Vent or Chimney Connector Dampers for Oil-Fired Appliances (1994)	UL	ANSI/UL 17
Ventilation	Commercial Application, Systems, and Equipment, 1st ed.	ACCA	ACCA Manual CS
	Commercial Low Pressure, Low Velocity Duct System Design, 1st ed.	ACCA	ACCA Manual Q
	Comfort, Air Quality, and Efficiency by Design	ACCA	ACCA Manual RS
	Guide for Testing Ventilation Systems (1991)	ACGIH	ACGIH
	Industrial Ventilation: A Manual of Recommended Practice, 26th ed. (2007)	ACGIH	ACGIH
	Design of Ventilation Systems for Poultry and Livestock Shelters	ASABE	ASAE EP270.5-2003
	Design Values for Emergency Ventilation and Care of Livestock and Poultry	ASABE	ANSI/ASAE EP282.2-2004
	Heating, Ventilating and Cooling Greenhouses	ASABE	ANSI/ASAE EP406.4-2003
	Guidelines for Selection of Energy Efficient Agricultural Ventilation Fans	ASABE	ASAE EP566-1997
	Uniform Terminology for Livestock Production Facilities	ASABE	ASAE S501-1990
	Agricultural Ventilation Constant Speed Fan Test Standard	ASABE	ASABE S565-2005
	Ventilation for Acceptable Indoor Air Quality	ASHRAE	ANSI/ASHRAE 62.1-2007
	Ventilation and Acceptable Indoor Air Quality in Low-Rise Residential Buildings	ASHRAE	ANSI/ASHRAE 62.2-2007
	Method of Testing for Room Air Diffusion	ASHRAE	ANSI/ASHRAE 113-2005
	Measuring Air Change Effectiveness	ASHRAE	ANSI/ASHRAE 129-1997 (RA03)
	Method of Determining Air Change Rates in Detached Dwellings	ASHRAE	ANSI/ASHRAE 136-1993 (RA06)
	Ventilation for Commercial Cooking Operations	ASHRAE	ANSI/ASHRAE 154-2003
	Residential Mechanical Ventilation Systems	CSA	CAN/CSA F326-M91 (R2005)
	Parking Structures	NFPA	NFPA 88A-07
	Installation of Air Conditioning and Ventilating Systems	NFPA	NFPA 90A-02
	Ventilation Control and Fire Protection of Commercial Cooking Operations	NFPA	NFPA 96-08
	Food Equipment	NSF	NSF/ANSI 2-2007
	Class II (Laminar Flow) Biosafety Cabinetry	NSF	NSF/ANSI 49-2007
	Aerothermodynamic Systems Engineering and Design	SAE	SAE AIR1168/3-1989
	Heater, Airplane, Engine Exhaust Gas to Air Heat Exchanger Type	SAE	SAE ARP86-1996
	Test Procedure for Battery Flame Retardant Venting Systems	SAE	SAE J1495-2005
Venting	Commercial Application, Systems, and Equipment, 1st ed.	ACCA	ACCA Manual CS
	Draft Hoods	CSA	ANSI Z21.12-1990 (R2000)
	National Fuel Gas Code	AGA/NFPA	ANSI Z223.1/NFPA 54-2006

Selected Codes and Standards Published by Various Societies and Associations (*Continued*)

Subject	Title	Publisher	Reference
	Explosion Prevention Systems	NFPA	NFPA 69-08
	Smoke and Heat Venting	NFPA	NFPA 204-07
	Chimneys, Fireplaces, Vents and Solid Fuel-Burning Appliances	NFPA	NFPA 211-06
	Guide for Steel Stack Construction, 2nd ed.	SMACNA	SMACNA 1996
	Draft Equipment (2006)	UL	UL 378
	Gas Vents (1996)	UL	ANSI/UL 441
	Type L Low-Temperature Venting Systems (1995)	UL	ANSI/UL 641
Vibration	Balance Quality and Vibration Levels for Fans	AMCA	ANSI/AMCA 204-05
	Techniques of Machinery Vibration Measurement	ASA	ANSI S2.17-1980 (R2004)
	Mechanical Vibration and Shock—Resilient Mounting Systems—Part 1: Technical Information to Be Exchanged for the Application of Isolation Systems	ASA	ISO 2017-1:2005
	Evaluation of Human Exposure to Whole-Body Vibration—Part 2: Vibration in Buildings (1 Hz to 80 Hz)	ISO	ISO 2631-2:2003
	Guidelines for the Evaluation of the Response of Occupants of Fixed Structures, Especially Buildings and Off-Shore Structures, to Low-Frequency Horizontal Motion (0.063 to 1 Hz)	ISO	ISO 6897:1984
	Procedural Standards for Measurement and Assessment of Sound and Vibration, 2nd ed. (2006)	NEBB	NEBB
	Sound and Vibration Design and Analysis (1994)	NEBB	NEBB
Water Heaters	Desuperheater/Water Heaters	ARI	ARI 470-2006
	Safety for Electrically Heated Livestock Waterers	ASABE	ASAE EP342.2-1995 (R2005)
	Methods of Testing to Determine the Thermal Performance of Solar Domestic Water Heating Systems	ASHRAE	ANSI/ASHRAE 95-1987
	Method of Testing for Rating Commercial Gas, Electric, and Oil Service Water Heating Equipment	ASHRAE	ANSI/ASHRAE 118.1-2003
	Method of Testing for Rating Residential Water Heaters	ASHRAE	ANSI/ASHRAE 118.2-2006
	Methods of Testing for Rating Combination Space-Heating and Water-Heating Appliances	ASHRAE	ANSI/ASHRAE 124-1991
	Methods of Testing for Efficiency of Space-Conditioning/Water-Heating Appliances That Include a Desuperheater Water Heater	ASHRAE	ANSI/ASHRAE 137-1995 (RA04)
	Boiler and Pressure Vessel Code—Section IV: Heating Boilers	ASME	BPVC-2007
	Section VI: Recommended Rules for the Care and Operation of Heating Boilers	ASME	BPVC-2007
	Gas Water Heaters—Vol. I: Storage Water Heaters with Input Ratings of 75,000 Btu per Hour or Less	CSA	ANSI Z21.10.1-2004/CSA 4.1-2004
	Vol. III: Storage, with Input Ratings Above 75,000 Btu per Hour, Circulating and Instantaneous Water Heaters	CSA	ANSI Z21.10.3-2004/CSA 4.3-2004
	Oil Burning Stoves and Water Heaters	CSA	B140.3-1962 (R2006)
	Oil-Fired Service Water Heaters and Swimming Pool Heaters	CSA	B140.12-03
	Construction and Test of Electric Storage-Tank Water Heaters	CSA	CAN/CSA-C22.2 No. 110-94 (R2004)
	Performance of Electric Storage Tank Water Heaters for Household Service	CSA	C191-04
	Energy Efficiency of Electric Storage Tank Water Heaters and Heat Pump Water Heaters	CSA	CSA C745-03
	One Time Use Water Heater Emergency Shut-Off	IAPMO	IGC 175-2003
	Water Heaters, Hot Water Supply Boilers, and Heat Recovery Equipment	NSF	NSF/ANSI 5-2007
	Household Electric Storage Tank Water Heaters (2004)	UL	ANSI/UL 174
	Oil-Fired Storage Tank Water Heaters (1995)	UL	ANSI/UL 732
	Commercial-Industrial Gas Heating Equipment (2006)	UL	UL 795
	Electric Booster and Commercial Storage Tank Water Heaters (2004)	UL	ANSI/UL 1453
Welding and Brazing	Boiler and Pressure Vessel Code—Section IX: Welding and Brazing Qualifications	ASME	BPVC-2007
	Structural Welding Code—Steel	AWS	AWS D1.1M/D1.1:2008
	Specification for Welding of Austenitic Stainless Steel Tube and Piping Systems in Sanitary Applications	AWS	AWS D18.1:1999
Wood-Burning Appliances	Threshold Limit Values for Chemical Substances (updated annually)	ACGIH	ACGIH
	Specification for Room Heaters, Pellet Fuel Burning Type	ASTM	ASTM E1509-04
	Guide for Construction of Solid Fuel Burning Masonry Heaters	ASTM	ASTM E1602-03
	Installation Code for Solid-Fuel-Burning Appliances and Equipment	CSA	CAN/CSA-B365-01 (R2006)
	Solid-Fuel-Fired Central Heating Appliances	CSA	CAN/CSA-B366.1-M91 (R2007)
	Chimneys, Fireplaces, Vents, and Solid Fuel-Burning Appliances	NFPA	ANSI/NFPA 211-06
	Commercial Cooking, Rethermalization and Powered Hot Food Holding and Transport Equipment	NSF	NSF/ANSI 4-2007e

ORGANIZATIONS

Abbrev.	Organization	Address	Telephone	http://www.
AABC	Associated Air Balance Council	1518 K Street NW, Suite 503 Washington, D.C. 20005	(202) 737-0202	aabchq.com
ABMA	American Boiler Manufacturers Association	8221 Old Courthouse Road, Suite 207 Vienna, VA 22182	(703) 356-7171	abma.com
ACCA	Air Conditioning Contractors of America	2800 Shirlington Road, Suite 300 Arlington, VA 22206	(703) 575-4477	acca.org
ACGIH	American Conference of Governmental Industrial Hygienists	1330 Kemper Meadow Drive Cincinnati, OH 45240	(513) 742-2020	acgih.org
ADC	Air Diffusion Council	1901 North Roselle Road, Suite 800 Schaumburg, IL 60195	(847) 706-6750	flexibleduct.org
AGA	American Gas Association	400 N. Capitol Street NW, Suite 400 Washington, D.C. 20001	(202) 824-7000	aga.org
AHAM	Association of Home Appliance Manufacturers	1111 19th Street NW, Suite 402 Washington, D.C. 20036	(202) 872-5955	aham.org
AIHA	American Industrial Hygiene Association	2700 Prosperity Avenue, Suite 250 Fairfax, VA 22031	(703) 849-8888	aiha.org
AMCA	Air Movement and Control Association International	30 West University Drive Arlington Heights, IL 60004-1893	(847) 394-0150	amca.org
ANSI	American National Standards Institute	1819 L Street NW, 6th Floor Washington, D.C. 20036	(202) 293-8020	ansi.org
ARI	Air-Conditioning and Refrigeration Institute	4100 North Fairfax Drive, Suite 200 Arlington, VA 22203	(703) 524-8800	ari.org
ASA	Acoustical Society of America	2 Huntington Quadrangle, Suite 1NO1 Melville, NY 14747-4502	(516) 576-2360	asa.aip.org
ASABE	American Society of Agricultural and Biological Engineers	2950 Niles Road St. Joseph, MI 49085-9659	(269) 429-0300	asabe.org
ASHRAE	American Society of Heating, Refrigerating and Air-Conditioning Engineers	1791 Tullie Circle, NE Atlanta, GA 30329	(404) 636-8400	ashrae.org
ASME	ASME	3 Park Avenue New York, NY 10016-5990	(973) 882-1167	asme.org
ASTM	ASTM International	100 Barr Harbor Drive, P.O. Box C700 West Conshohocken, PA 19428-2959	(610) 832-9500	astm.org
AWS	American Welding Society	550 N.W. LeJeune Road Miami, FL 33126	(305) 443-9353	aws.org
AWWA	American Water Works Association	6666 W. Quincy Avenue Denver, CO 80235	(303) 794-7711	awwa.org
BOCA	Building Officials and Code Administrators International	(*see* ICC)		
BSI	British Standards Institution	389 Chiswick High Road London W4 4AL, UK	44(0)20-8996-9001	bsi-global.com
CAGI	Compressed Air and Gas Institute	1300 Sumner Avenue Cleveland, OH 44115-2851	(216) 241-7333	cagi.org
CSA	Canadian Standards Association International	5060 Spectrum Way Mississauga, ON L4W 5N6, Canada	(416) 747-4000	csa.ca
	Also available from CSA America	8501 East Pleasant Valley Road Cleveland, OH 44131-5575	(216) 524-4990	csa-america.org
CTI	Cooling Technology Institute	P.O. Box 73383 Houston, TX 77273-3383	(281) 583-4087	cti.org
EJMA	Expansion Joint Manufacturers Association	25 North Broadway Tarrytown, NY 10591	(914) 332-0040	ejma.org
HEI	Heat Exchange Institute	1300 Sumner Avenue Cleveland, OH 44115-2815	(216) 241-7333	heatexchange.org
HI	Hydraulic Institute	9 Sylvan Way Parsippany, NJ 07054-3802	(973) 267-9700	pumps.org
HYDI	Hydronics Institute Division of GAMA	2107 Wilson Boulevard, Suite 600 Arlington, VA 22201	(703) 525-7060	gamanet.org
IAPMO	International Association of Plumbing and Mechanical Officials	5001 E. Philadelphia Street Ontario, CA 91761-2816	(909) 472-4100	iapmo.org

ORGANIZATIONS (*Continued*)

Abbrev.	Organization	Address	Telephone	http://www.
ICBO	International Conference of Building Officials	(*see* ICC)		
ICC	International Code Council	500 New Jersey Ave NW, 6th Floor Washington, D.C. 20001	(888) 422-7233	iccsafe.org
IEEE	Institute of Electrical and Electronics Engineers	45 Hoes Lane Piscataway, NJ 08854-4141	(732) 981-0060	ieee.org
IESNA	Illuminating Engineering Society of North America	120 Wall Street, Floor 17 New York, NY 10005-4001	(212) 248-5000	iesna.org
IFCI	International Fire Code Institute	(*see* ICC)		
IIAR	International Institute of Ammonia Refrigeration	1110 North Glebe Road, Suite 250 Arlington, VA 22201	(703) 312-4200	iiar.org
ISA	The Instrumentation, Systems, and Automation Society	67 Alexander Drive, P.O. Box 12777 Research Triangle Park, NC 27709	(919) 549-8411	isa.org
ISO	International Organization for Standardization	1, ch. de la Voie-Creuse, Case postale 56 CH-1211 Geneva 20, Switzerland	41-22-749-01 11	iso.org
MCAA	Mechanical Contractors Association of America	1385 Piccard Drive Rockville, MD 20850	(301) 869-5800	mcaa.org
MICA	Midwest Insulation Contractors Association	16712 Elm Circle Omaha, NE 68130	(800) 747-6422	micainsulation.org
MSS	Manufacturers Standardization Society of the Valve and Fittings Industry	127 Park Street NE Vienna, VA 22180-4602	(703) 281-6613	mss-hq.com
NAIMA	North American Insulation Manufacturers Association	44 Canal Center Plaza, Suite 310 Alexandria, VA 22314	(703) 684-0084	naima.org
NCPWB	National Certified Pipe Welding Bureau	1385 Piccard Drive Rockville, MD 20850-4340	(301) 869-5800	mcaa.org/ncpwb
NCSBCS	National Conference of States on Building Codes and Standards	505 Huntmar Park Drive, Suite 210 Herndon, VA 20170	(703) 437-0100	ncsbcs.org
NEBB	National Environmental Balancing Bureau	8575 Grovemont Circle Gaithersburg, MD 20877	(301) 977-3698	nebb.org
NEMA	National Electrical Manufacturers Association	1300 North 17th Street, Suite 1752 Rosslyn, VA 22209	(703) 841-3200	nema.org
NFPA	National Fire Protection Association	1 Batterymarch Park Quincy, MA 02169-7471	(617) 770-3000	nfpa.org
NRCC	National Research Council of Canada, Institute for Research in Construction	1200 Montreal Road, Bldg M-58 Ottawa, ON K1A 0R6, Canada	(877) 672-2672	nrc-cnrc.ca
NSF	NSF International	P.O. Box 130140, 789 N. Dixboro Road Ann Arbor, MI 48113-0140	(734) 769-8010	nsf.org
PHCC	Plumbing-Heating-Cooling Contractors National Association	180 S. Washington Street, P.O. Box 6808 Falls Church, VA 22046	(703) 237-8100	phccweb.org
SAE	Society of Automotive Engineers International	400 Commonwealth Drive Warrendale, PA 15096-0001	(724) 776-4841	sae.org
SBCCI	Southern Building Code Congress International	(*see* ICC)		
SMACNA	Sheet Metal and Air Conditioning Contractors' National Association	4201 Lafayette Center Drive Chantilly, VA 20151-1209	(703) 803-2980	smacna.org
TEMA	Tubular Exchanger Manufacturers Association	25 North Broadway Tarrytown, NY 10591	(914) 332-0040	tema.org
UL	Underwriters Laboratories	333 Pfingsten Road Northbrook, IL 60062-2096	(847) 272-8800	ul.com

ASHRAE HANDBOOK

Additions and Corrections

This report includes additional information, and technical errors found between June 15, 2005, and April 1, 2008, in the inch-pound (I-P) editions of the 2005, 2006, and 2007 *ASHRAE Handbook* volumes. Occasional typographical errors and nonstandard symbol labels will be corrected in future volumes. The most current list of Handbook additions and corrections is on the ASHRAE Web site (www.ashrae.org).

The authors and editor encourage you to notify them if you find other technical errors. Please send corrections to: Handbook Editor, ASHRAE, 1791 Tullie Circle NE, Atlanta, GA 30329, or e-mail mowen@ashrae.org.

2005 Fundamentals

p. 1.17, Eq. (63). $\dot{Q}_{evap}$ should be $\dot{Q}_{cond}$.

p. 1.20, Symbols. Units for V should be ft/s.

p. 2.7, Table 2. Values for ε (right column) should be 60, 1800, 6000, and 10,200 μin.

p. 2.11, definitions for Eq. (38). In the definition for Δh, there should be parentheses around $p_1 - p_2$.

p. 3.1, definitions for Eq. (1a). Units for thermal conductivity k should be Btu/h·ft·°F.

p. 3.2, Thermal Conduction, 2nd line from bottom. Change "steady" to "steady-state."

p. 3.5, Eq. (10). The equation should be as follows:

$$c_1 = \frac{2\,\text{Bi}}{(\mu_1^2 + \text{Bi}^2)\,J_0(\mu_1)}$$

p. 3.13, 1st col. Delete first repeated paragraph after Equation (30).

p. 3.21, 2nd col., last full sentence. Change to, "Depending on frequency and amplitude of vibration, forced convection from a wire to air is enhanced by up to 300% (Nesis et al. 1994)."

p. 3.28, Eq. (44). Delete second equals sign and second fraction.

p. 7.11, definitions for Eq. (42). Add "loss" to definition for TL.

p. 7.20, Eqs. (52) and (53). Revise the equations as follows:

$$\text{TL} \approx 10\log\left(1 + \frac{Z_f}{Z_I}\right) = 10\log\left(1 + \frac{2\pi f Z_f}{k}\right) \qquad (52)$$

$$\text{TL} \approx 10\log\left(1 + \frac{d_I}{d_f}\right) \qquad (53)$$

p. 7.20, 1st col., last full paragraph. Change third sentence to read, "the resonance frequency of the system is maintained at 3.13 Hz, and the force transmitted to the structure remains at 12.5 lb_f." Change sixth sentence to read, "where a is acceleration, the maximum dynamic displacement of the mounted equipment is reduced by a factor of (M_1/M_2), where M_1 and M_2 are the masses before and after mass is added, respectively."

p. 15.3, Eq. (3) and following text. Change K_a to K_d in the equation, definitions, and following paragraph (three places total).

p. 16.11, Symbols. Add the following definitions:

h_s = exhaust stack height (typically above roof unless otherwise specified), ft (see Figure 3, and Chapter 44 in the 2003 *ASHRAE Handbook—HVAC Applications*)

S = stretched-string distance; shortest distance from exhaust to intake over obstacles and along building surface, ft [see Figure 3, and Equation (22) in the 2003 *ASHRAE Handbook—HVAC Applications*)

p. 19.8, Table 7. Replace the table with the one supplied on p. A.2.

p. 23.6, Fig. 1. Change caption from "Adsorption Isotherms" to "Typical Adsorption."

p. 26.3, Eq. (2). The correct equation is

$$R_T = 12\ln(D_3/D_2)/(2\pi k)$$

p. 27.21, Tables 4 and 5. Reverse the order of these two tables.

p. 27.14, Fig. 9. Air leakage should be at 0.2 in. of water.

p. 28.2, Table 1 (and all data tables), cols. 13a, c, and e. Because of a data processing error, the enthalpy values in these columns are systematically low by 7.687 Btu/lb. Thus, all enthalpy values in Table 1 and in all design climatic condition tables on the accompanying CD-ROM should be increased by that amount.

p. 28.6, Eq. (1), definitions. The definition for F should have a plus, not a minus, within the brackets, as follows:

$$F = -\frac{\sqrt{6}}{\pi}\left\{0.5772 + \ln\left[\ln\left(\frac{n}{n-1}\right)\right]\right\}$$

p. 29.8, Table 7. Units for OF_b should be °F.

p. 29.9, Table 9. The correct numbers for the last line of the table are as follows:

E_t	326	325	321	314	305	293	279	262	243

p. 29.9, Eq. (23). Units for CF_{slab} should be Btu/h·ft²; 0.51 is a constant with units of Btu/h·ft²; and 2.5 is a factor with units of °F.

p. 30.2, 2nd col., 6th para., last line. Change "with" to "without."

p. 30.4, bottom of 2nd col. The reference to Equation (3) should be to Equation (4).

p. 30.12, 1st col., Infiltration. The cross reference to Table 3 in Chapter 27 should be to Table 1 in Chapter 27 of the 2001 *ASHRAE Handbook—Fundamentals.*

p. 30.27, Table 19. Footnote 7 should refer to Table 3 in Chapter 39.

p. 30.32, Central Plant, Piping. The cross reference should be to Chapter 26, not Chapter 23.

p. 30.34, Part 1, Solution. In the equations, change all "1500 W" to "440 W" (1500 is correct for the solutions for q_7 to q_{18}).

p. 31.6, Table 2. In the footnote for Winter Conditions, add the following: $h_i = h_{ic} + h_{iR} = 0.30(\Delta T/L)^{0.25} + \varepsilon\Gamma(T_i^4 - T_g^4)/\Delta T$, where $\Delta T = T_i - T_g$, °R; L = glazing height, ft; T_g = glass temperature, °R.

p. 31.13, after Eq. (10). The definition of LST should be "local standard time, decimal hours."

Table 7 Comparative Refrigerant Performance per Ton of Refrigeration

(2005 Fundamentals, Ch. 19, p. 8)

Refrigerant No.	Refrigerant Chemical Name or Composition (% by mass)	Evaporator Pressure, psia	Condenser Pressure, psia	Compression Ratio	Net Refrigerating Effect, Btu/lb	Refrigerant Circulated, lb/min	Liquid Circulated, gal/min	Specific Volume of Suction Gas, ft^3/lb	Compressor Displacement, ft^3/min	Power Consumption, hp	Coefficient of Performance	Compressor Discharge Temp., °F
170	Ethane	233.2	672.8	2.88	69.5	2.85	1.22	0.541	1.54	1.72	2.70	121.73
744	Carbon dioxide	326.9	1041.4	3.19	57.3	1.79	0.36	0.269	0.48	0.91	2.69	157.73
1270	Propylene	51.9	189.1	3.64	123.0	1.62	0.39	2.081	3.37	1.04	4.50	107.33
290	Propane	41.5	155.9	3.76	119.5	1.65	0.41	2.502	4.13	1.03	4.50	96.53
502	R-22/115 (48.8/51.2)	49.7	190.3	3.83	45.6	4.40	0.44	0.814	3.58	1.08	4.38	100.13
507A	R-125/143a (50/50)	55.0	211.6	3.85	47.4	4.22	0.50	0.814	3.44	1.13	4.18	94.73
404A	R-125/143a/134a (44/52/4)	52.9	206.0	3.89	49.1	4.08	0.48	0.860	3.51	1.12	4.21	96.53
410A	R-32/125 (50/50)	69.3	271.5	3.92	72.2	2.71	0.31	0.873	2.37	1.05	4.41	123.53
125	Pentafluoroethane	58.5	226.4	3.87	36.7	5.31	0.55	0.631	3.35	1.15	3.99	87.53
22	Chlorodifluoromethane	42.8	172.2	4.02	69.9	2.85	0.29	1.248	3.56	1.01	4.66	127.13
12	Dichlorodifluoro-methane	26.3	107.5	4.09	50.3	3.94	0.37	1.479	5.83	1.00	4.70	100.13
500	R-12/152a (73.8/26.2)	31.0	127.1	4.09	60.1	3.31	0.35	1.504	4.98	1.00	4.66	105.53
407C	R-32/125/134a (23/25/52)	41.8	182.7	4.38	70.2	2.85	0.30	1.289	3.67	1.05	4.50	118.13
600a	Isobutane*	12.8	58.5	4.58	113.5	1.76	0.39	6.524	11.48	1.01	4.62	85.73
134a	Tetrafluoroethane	23.6	111.2	4.71	63.6	3.13	0.31	1.945	6.09	1.02	4.60	98.33
124	Chlorotetrafluoroethane*	12.8	64.3	5.03	50.7	3.90	0.35	2.741	10.69	1.01	4.62	85.73
717	Ammonia	34.1	168.5	4.94	474.3	0.42	0.08	8.197	3.44	1.00	4.76	209.93
600	Butane*	8.1	41.0	5.05	125.6	1.65	0.35	10.325	17.04	1.03	4.74	85.73
11	Trichlorofluoro-methane	2.9	18.1	6.25	67.0	2.95	0.24	12.317	36.34	0.93	5.02	109.13
123	Dichlorotrifluoro-ethane	2.3	15.8	6.81	61.2	3.27	0.27	14.279	46.69	0.96	4.90	91.13
113	Trichlorotrifluoroethane*	1.0	7.8	7.71	52.7	3.66	0.28	26.940	98.60	0.94	4.81	85.73

*Superheat required.

p. 31.15. All references to Figure 9 should be to Figure 8, and references to Figure 10 should be to Figure 9.

p. 31.16, 1st col., 1st full paragraph. The reference to Figure 11 should be to Figure 10.

p. 31.16, Example 5, Solution. After "Eastern daylight time of 3:00 P.M.," add "(i.e., standard time of 2:00 P.M.)."

p. 31.26, Table 13. For ID 1b, change the following values:

T	0.77	0.75	0.73	0.68	0.58	0.35	0.69
R^f	0.07	0.08	0.09	0.13	0.24	0.48	0.13
R^b	0.07	0.08	0.09	0.13	0.24	0.48	0.13

p. 31.40, Table 16, last two rows, center column. Change "0.42" to "0.16," and "0.44" to "0.10."

p. 32.5, 1st col. Cross references to the following equations in Chapter 30 should be as follows (the chapter number should stay the same; only the equation numbers should be updated):

Equation (36)	Equation (27)
Equation (35)	Equation (26)
Equation (34)	Equation (25)

p. 33.3, below Eq. (6). The definition for C_o should be $A_c C_d R_{fa}$ = effective area of stream at discharge from an open-end duct or at a contracted section, ft^2

p. 35.6, Eq. (18), definition for D_h. The reference should be to Equation (22).

p. 35.9, above Eq. (23). The reference should be to Equation (23).

p. 38.1, 1st col. The definition of an acre should be 43,560 ft^2. The conversion factor for ft of water to Pa should be 2989.

p. 40.26. The URL for CSA America should be csa-america.org.

2006 Refrigeration

p. 2.22, Table 21. The data for R-22 and R-134a were transposed; please reverse the order of the data columns, as shown on p. A.3.

p. 47.4, Testing for Leaks, 2nd paragraph. Change the first sentence to read, "ASHRAE *Standard* 147 established. . . ."

p. 47.7, References. Add the following source:

ASHRAE. 2002. Reducing the release of halogenated refrigerants from refrigerating and air-conditioning equipment and systems. ANSI/ ASHRAE *Standard* 147-2002.

2007 HVAC Applications

Contributors List. Wayne Lawton of X-nth should be listed as contributor for Chapters 29 and 30.

Ch. 14, Laboratories. Revisions were inadvertently omitted from the 2007 volume; please go to http://www.ashrae.org/publications/page/158 to download the revised chapter.

p. 17.3, Table 1, Relative humidity control range. The maximum dew point for Class 1 should be 63°F.

p. 17.16, References. Please update the following two URLs:

For LBNL 2003,
http://hightech.lbl.gov/dc-benchmarking-results.html.

For NIOSH 1986,
http://www.cdc.gov/niosh/86-113.html.

p. 32.14, 2nd col., 1st paragraph. The reference to Table 5 should be to Table 4.

p. 32.15, Table 6. Units for "Bore Fill Conductivity" should be Btu/h·ft·°F.

Table 21 Refrigerant Flow Capacity Data for Defrost Lines
(2006 Refrigeration, Ch. 2, p. 22, 1st two columns)

Pipe Size		R-22 Mass Flow Data, lb/h			R-134a Mass Flow Data, lb/h		
		Velocity, fpm			Velocity, fpm		
Copper[a]		1000	2000	3000	1000	2000	3000
1/2		110	220	330	150	300	450
5/8		170	350	520	240	480	720
3/4		260	510	770	350	710	1060
7/8		360	720	1090	500	1000	1500
1 1/8		620	1230	1850	850	1700	2550
1 3/8		940	1880	2820	1300	2590	3890
1 5/8		1330	2660	3990	1840	3670	5510
2 1/8		2310	4630	6940	3190	6390	9580
2 5/8		3570	7140	10,700	4930	9850	14,800
3 1/8		5100	10,200	15,300	7030	14,100	21,100
3 5/8		6900	13,800	20,700	9510	19,000	28,500
4 1/8		9000	17,900	26,900	12,400	24,700	37,100
5 1/8		14,000	27,900	41,900	19,300	38,500	57,800
6 1/8		20,100	40,100	60,200	27,700	55,400	83,100
8 1/8		35,100	70,100	105,200	48,400	96,700	145,100
Steel							
IPS	SCH						
3/8	80	110	210	320	150	290	440
1/2	80	180	350	530	240	480	720
3/4	80	320	650	970	450	890	1340
1	80	540	1080	1610	740	1480	2230
1 1/4	80	1120	2240	3360	1540	3090	4630
1 1/2	80	1520	3050	4570	2100	4200	6300
2	40	2510	5020	7530	3460	6930	10,400
2 1/2	40	3580	7160	10,700	4940	9870	14,800
3	40	5530	11,100	16,600	7620	15,200	22,900
4	40	9520	19,000	28,600	13,100	26,300	39,400
5	40	15,000	29,900	44,900	20,600	41,300	61,900
6	40	21,600	43,200	64,800	29,800	59,600	89,400
8	40	37,400	74,800	112,300	51,600	103,300	154,900
10	40	59,000	118,00	176,900	81,400	162,800	244,100
12	ID[b]	84,600	169,200	253,800	116,700	233,400	350,200
14	30	—	—	—	—	—	—
16	30	—	—	—	—	—	—

Note: Refrigerant flow data based on saturated condensing temperature of 70°F.[a]For brazed Type L copper tubing for defrost service, see Safety Requirements section.[b]Pipe inside diameter is same as nominal pipe size.

p. 32.15, 1st col., 1st and 2nd lines, and following Eq. (4). Change "outside pipe" to "borehole." In the definitions for Eq. (4), units should be ft²/day for α_g, days for τ, and ft for d, which is defined as borehole diameter.

p. 41.22, Example 3. The reference to Table 6 should be to Table 5. In the solution, change 0.8 to 0.85 in both equations.

p. 41.25, Eq. (26). The second term on the left side of the equation should be $\dot{Q}_{blr,i}$.

p. 41.30, Eq. (33). Change D to Δ.

p. 47.26, Eq. (26). The final number subtracted should be 0.5, not 11.

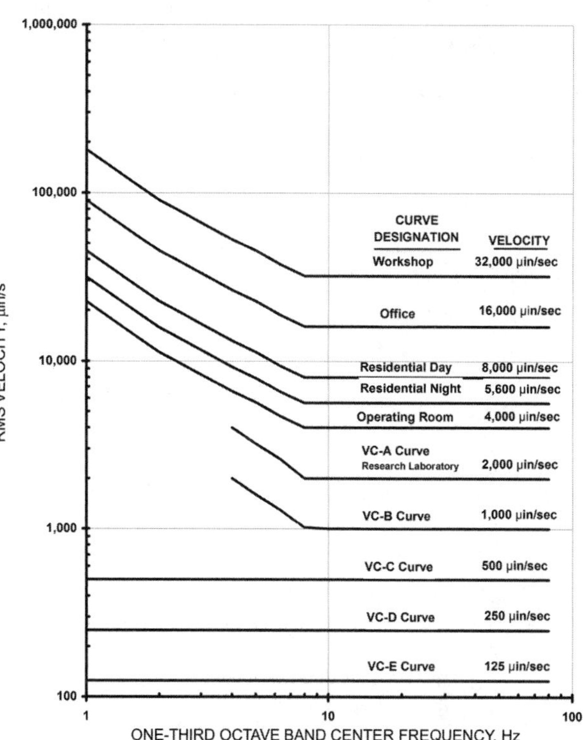

Fig. 37 Building Vibration Criteria for Vibration Measured on Building Structure
(2007 HVAC Applications, Chapter 47, p. 39)

p. 47.37, Eq. (29). The final term should be $10^{L_{w2}/10}$.

p. 47.38, Vibration Criteria, 1st paragraph. ANSI *Standard* 3.29 is the former designation; update this to ANSI *Standard* S2.71-1983 (R2006).

p. 47.39, Fig. 37. The corrected figure is as shown above.

p. 47.40, Table 48, Packaged AH, AC, H and V Units. In the Horsepower and Other column, change all ≤ to ≥ (i.e., should be ≥15, ≥4 in. SP).

p. 49.17, Fig. 21. Along the (horizontal) x axis, the labels should be 0, 12, 24, 36, and 48.

p. 49.24, Figs. 25 and 26. The current Figure 25 should be Figure 26; its caption should refer to Figure 25, not Figure 22. The current Figure 26 should be Figure 25.

p. 52.10, Example 7, 3rd paragraph. The equation for A_{bo} should be $6030(0.17 \times 10^{-3}) = 1.025$ ft². Consequently, the pressure difference Δp_{sbt} (third line from end of example) should be 0.331 in. of water, and the flow of pressurization air (last line of example) should be 5585 cfm.

p. 52.18, Symbols. The variable for floor-to-ceiling height should be H, and the variable for number of floors should be N.

p. 54.15, Example 1. For calculations using Eqs. (2) and (3), S_{DS} should be 0.623, not 0.85. The result using Eq. (2) should be 1495 lb, and the result using Eq. (3) should be 280 lb.

p. 54.17, Example 2. For calculations using Eqs. (2) and (3), S_{DS} should be 0.623, not 0.85. The result using Eq. (2) should be 1495 lb, and the result using Eq. (3) should be 280 lb.

p. 54.18, Example 3. For calculations using Eqs. (2) and (3), S_{DS} should be 0.623, not 0.85. The result using Eq. (2) should be 3738 lb, and the result using Eq. (3) should be 701 lb.

p. 54.19, Example 4. For calculations using Eqs. (2) and (3), S_{DS} should be 0.623, not 0.85. The result using Eq. (2) should be 747 lb, and the result using Eq. (3) should be 140 lb.

COMPOSITE INDEX
ASHRAE HANDBOOK SERIES

This index covers the current Handbook series published by ASHRAE. The four volumes in the series are identified as follows:

F = 2005 Fundamentals

R = 2006 Refrigeration

A = 2007 HVAC Applications

S = 2008 HVAC Systems and Equipment

Alphabetization of the index is letter by letter; for example, **Heaters** precedes **Heat exchangers**, and **Floors** precedes **Floor slabs**.

The page reference for an index entry includes the book letter and the chapter number, which may be followed by a decimal point and the beginning page in the chapter. For example, the page number S31.4 means the information may be found in the 2008 HVAC Systems and Equipment volume, Chapter 31, beginning on page 4.

Each Handbook volume is revised and updated on a four-year cycle. Because technology and the interests of ASHRAE members change, some topics are not included in the current Handbook series but may be found in the earlier Handbook editions cited in the index.

Comment Page

ASHRAE publications strive to present the most current and useful information possible. If you would like to comment on chapters in this or any volume of the *ASHRAE Handbook*, please use one of the following methods:

- Fill out the comment form on the ASHRAE Web site (www.ashrae.org)
- E-mail the editor at mowen@ashrae.org
- Cut out this page and fax it to the editor at 678-539-2187, or mail it to

Handbook Editor
ASHRAE
1791 Tullie Circle
Atlanta, GA 30329 USA

Please provide your contact information if you would like a response. (Personal identification information will not be used for any purpose beyond responding to your comments.)

Name: ____________________ **Phone:** ____________________

E-mail: ____________________ **Fax:** ____________________

Address: ____________________ **Preferred Contact Method(s):** ____________________

Comment Page

ASHRAE publications strive to present the most current and useful information possible. If you would like to comment on chapters in this or any volume of the *ASHRAE Handbook*, please use one of the following methods:

- Fill out the comment form on the ASHRAE Web site (www.ashrae.org)
- E-mail the editor at mowen@ashrae.org
- Cut out this page and fax it to the editor at 678-539-2187, or mail it to

Handbook Editor
ASHRAE
1791 Tullie Circle
Atlanta, GA 30329 USA

Please provide your contact information if you would like a response. (Personal identification information will not be used for any purpose beyond responding to your comments.)

Name: ____________________ **Phone:** ____________________

E-mail: ____________________ **Fax:** ____________________

Address: ____________________ **Preferred Contact Method(s):** ____________________

Comment Page

ASHRAE publications strive to present the most current and useful information possible. If you would like to comment on chapters in this or any volume of the *ASHRAE Handbook*, please use one of the following methods:

- Fill out the comment form on the ASHRAE Web site (www.ashrae.org)
- E-mail the editor at mowen@ashrae.org
- Cut out this page and fax it to the editor at 678-539-2187, or mail it to

Handbook Editor
ASHRAE
1791 Tullie Circle
Atlanta, GA 30329 USA

Please provide your contact information if you would like a response. (Personal identification information will not be used for any purpose beyond responding to your comments.)

Name: ____________________ **Phone:** ____________________

E-mail: ____________________ **Fax:** ____________________

Address: ____________________ **Preferred Contact Method(s):** ____________________

LICENSE AGREEMENT
2008 *ASHRAE Handbook—HVAC Systems and Equipment* CD-ROM

The 2008 *ASHRAE Handbook—HVAC Systems and Equipment* is distributed with an accompanying CD-ROM, which provides electronic access to the volume's content.

The License for this CD-ROM is for personal use only; this CD-ROM may not be used on a LAN or WAN.

Using the CD-ROM indicates your acceptance of the terms and conditions of this agreement. If you do not agree with them, you should not use this CD-ROM.

The title and all copyrights and ownership rights in the program and data are retained by ASHRAE. You assume responsibility for the selection of the program and data to achieve your intended results and for the installation, use, and results obtained from the program and data.

You may use the program and data on a single machine. You may copy the program and data into any machine-readable form for back-up purposes in support of your use of the program or data on a single machine. You may not copy or transfer the program or data except as expressly provided for in this license. To do so will result in the automatic termination of your license, and ASHRAE will consider options available to it to recover damages from unauthorized use of its intellectual property.

Specifically, you may not copy nor transfer the program or data onto a machine other than your own unless the person to whom you are copying or transferring the program or data also has a license to use them.

Distribution to third parties of ASHRAE intellectual property in print or electronic form from this CD-ROM is also prohibited except when authorized by ASHRAE.

If you wish to reprint data from this CD-ROM in print or electronic form (such as posting content on a Web site), visit www.ashrae.org/permissions and go to Handbook Reprint Permissions.

If you wish to use this CD-ROM on a network, you must purchase the network version, which is available at www.ashrae.org.